INSTRUMENT
ENGINEERS'
Handbook

PROCESS CONTROL

INSTRUMENT ENGINEERS' *Handbook*

BÉLA G. LIPTÁK
EDITOR-IN-CHIEF

KRISZTA VENCZEL
ASSOCIATE EDITOR

Revised Edition

CHILTON BOOK COMPANY   RADNOR, PENNSYLVANIA

Copyright © 1969, 1985 by Béla G. Lipták
Revised Edition   All Rights Reserved
Published in Radnor, Pennsylvania, 19089, by Chilton Book Company

Library of Congress Cataloging in Publication Data
Main entry under title:
Instrument engineers' handbook.
Includes bibliographies and index.
1. Process control—Handbooks, manuals, etc.
2. Measuring instruments—Handbooks, manuals, etc.
I. Lipták, Béla G. II. Venczel, Kriszta.
TS156.8.I56   1985   629.8   83-43297
ISBN 0-8019-7290-6

Designed by William E. Lickfield
Drawings by Adrian J. Ornik
Manufactured in the United States of America

3 4 5 6 7 8 9 0   4 3 2 1 0 9 8

After the First World War
Hungary was dismembered. Millions
of my people became persecuted
minorities without moving
from their homeland, because national
boundaries were redrawn around them.
I would like to dedicate this book
to a coming age when all divided nations
are reunited and when international
public opinion protects not only
the endangered animal species but also
the identity and cultural heritage
of national minorities.

CONTENTS

CONTRIBUTORS

JAMES B. ARANT
 BSChE, PE
 Senior Consultant–Engineering
  Department
 E.I. DuPont de Nemours & Co.
 (Sections 4.4, 4.6, 4.7, 4.9, 4.14)

ROBERT J. BAKER*
 BS
 Retired from
  Wallace & Tiernan Division, Pennwalt
   Corp.
 (Section 8.22)

HANS D. BAUMANN
 PhDME, PE
 President
  H.D. Baumann Assoc., Ltd.
 (Section 4.14)

CHESTER S. BEARD*
 BSEE
 Author, Retired from
  Bechtel Corp.
 (Sections 4.2, 4.3, 4.8, 4.12)

PAUL B. BINDER
 BSME
 Program Manager
  Leeds & Northrup Instruments,
   A Unit of General Signal
 (Sections 2.4, 4.1)

BENJAMIN BLOCK*
 BS, MSChE, MBA, PE
 Consulting Engineer
 (Sections 5.5, 8.9)

RICHARD W. BORUT
 Section Manager–Instrument Division
  M. W. Kellogg Co.
 (Section 3.2)

AUGUST BRODGESELL*
 BSEE
 President
  CRB Systems, Inc.
 (Section 5.8)

ROBERT D. BUCHANAN
 BSCE, MS, PE
 Environmental Engineer, Consultant
 Self Employed
 (Sections 4.10, 5.8†)

ANTHONY M. CALABRESE
 BSChE, BSEE, MSChE, PE
 Senior Engineering Manager
  M. W. Kellogg Co.
 (Section 8.12)

WILLIAM N. CLARE
 BSChE, MSChE
 Engineering Manager
  Automated Dynamics Corp.
 (Section 6.4)

ARMANDO B. CORRIPIO*
 BSChE, MSChE, PhDChE, PE
 Professor of Chemical Engineering
  Louisiana State University
 (Section 7.8)

---

*These authors' contributions have been updated from the previous edition of the *Instrument Engineers' Handbook*.

†These authors' contributions have been updated from the *Environmental Engineers' Handbook*, Béla G. Lipták, ed., Chilton Book Company, Radnor, Pa., 1974.

NICHOLAS O. CROMWELL*
BSChE
Principal Systems Architect
The Foxboro Co.
(Section 3.3)

LOUIS D. DiNAPOLI
BSEE, MSEE
Manager, Pressure Transmitter &
Flowmeter Development
Leeds & Northrup Instruments,
A Unit of General Signal
(Sections 2.4, 5.3)

ROBERT G. DITTMER
BSEE
Corporate Process Control Engineer
PPG Industries, Chemical Division
(Section 7.11)

LAWRENCE S. DYSART*
BSEE
Retired from
Robertshaw Controls Co.
(Section 5.12)

GEORG F. ERK
BSME, MSChE, PE
Manager, Instrumentation and Control
Systems Engineering
Sun Refining and Marketing Company
(Sections 3.8, 5.9, 5.10)

EDWARD J. FARMER
BSEE, PE
President
Ed Farmer & Associates, Inc.
(Sections 7.9, 8.21)

JEHUDA FRYDMAN
Manager, Electrical Engineering Section
Mobil Chemical Company, Plastics
Division
(Section 6.7)

CHARLES E. GAYLOR*
BSChE, PE
Manager of Engineering Services
Hooker Chemical Corp.
(Sections 4.3, 4.11)

RICHARD A. GILBERT
PhD
Assistant Professor
University of South Florida
(Sections 6.2, 6.3, 6.6, 6.8)

DAVID M. GRAY
BSChE
Engineer, Specialist
Leeds & Northrup Instruments,
A Unit of General Signal
(Section 8.15)

JOSEPH A. GUMP*
BSChE
Process Panel Manager
Control Products Corp.
(Section 3.1)

BHISHAM P. GUPTA
BS, MSME, DSc, PE
Instrument Engineer
ARAMCO
(Section 8.5)

DIANE M. HANLON
BSEE, BSME
Engineer
E.I. DuPont de Nemours & Co.
(Section 5.11)

HEROLD I. HERTANU
MSEE, PE
Manager, Corporate Computer
Engineering
Crawford & Russell, Inc.
(Section 8.18)

PER A. HOLST*
MSEE
Director of Computing Technology
The Foxboro Co.
(Sections 7.2, 7.8)

MICHAEL F. HORDESKI
BSEE, MSEE, PE
Control Systems Consultant
Siltran Digital
(Sections 1.4, 1.5, 7.7)

FRANKLIN B. HOROWITZ*
BSChE, MSChE, PE
Assistant Vice President, Manufacturing
Systems Group
Crawford & Russell, Inc.
(Sections 8.5, 8.17)

DAVID L. HOYLE
BSChE
Principal Field Application Engineer
The Foxboro Co.
(Section 8.16)

STUART P. JACKSON*
　　BSEE, MSEE, PhD, PE
　　President
　　　Engineering and Marketing Corp. of
　　　　Virginia
　　(Section 3.10)

ERIC JENETT*
　　BSChE, MSChE, PE
　　Vice-President—Support Services
　　　Brown & Root, Inc.
　　(Section 5.14)

KENNETH J. JENTZEN
　　BSEE, MSEE
　　Manager, Project Engineering and
　　　Maintenance
　　　Mobil Chemical Co.
　　(Section 8.20)

DONALD R. JONES
　　BSME, PE
　　Consultant
　　　Powell-Process Systems, Inc.
　　(Section 4.5)

VAUGHN A. KAISER*
　　BSME, MSE, PE
　　Member Technical Staff
　　　Profimatics, Inc.
　　(Sections 7.3, 7.6)

RICHARD K. KAMINSKI
　　BA
　　Instrument Engineer
　　　Dravo Engineers, Inc.
　　(Section 2.8)

LES A. KANE
　　BSEE
　　Mechanical/Electrical Editor
　　　Hydrocarbon Processing
　　(Section 8.8)

GENE T. KAPLAN
　　BSME
　　Project/Design Engineer
　　　Automated Dynamics Corp.
　　(Section 6.4)

DAVID S. KAYSER*
　　BSEE
　　Senior Instrument Engineer
　　　Texas City Refining, Inc.
　　(Sections 5.4, 5.7, 5.15)

CHANG H. KIM*
　　BSChE
　　Manager, Technical Services
　　　ARCO Chemical Company, Division
　　　　of Atlantic Richfield Co.
　　(Section 8.1)

CULLEN G. LANGFORD
　　BSME, PE
　　Systems Consultant
　　　E.I. DuPont de Nemours & Co.
　　(Sections 4.15, 4.16)

AVERLES G. LASPE
　　BS, PE
　　Senior Control Application Consultant
　　　Retired from Honeywell, Inc.
　　(Section 1.10)

DONALD W. LEPORE*
　　BSME
　　Design Engineer
　　　The Foxboro Co.
　　(Sections 3.7, 3.9)

BÉLA G. LIPTÁK
　　ME, MME, PE
　　President
　　　Lipták Associates, P.C.
　　　*(Introduction, Sections 1.2, 2.3, 2.4,
　　　2.6, 2.7, 2.10, 4.1, 4.13, 5.1,
　　　8.2, 8.4, 8.6, 8.13, 8.14, 8.17,
　　　8.18, 8.19)*

ORVAL P. LOVETT, JR.*
　　BSChE
　　Consulting Engineer, Instrumentation
　　　and Control Systems
　　　Retired from E.I. DuPont de
　　　　Nemours & Co.
　　(Section 4.2)

VICTOR J. MAGGIOLI
　　BSEE
　　Principal Consultant
　　　E.I. DuPont de Nemours & Co.
　　(Section 2.5)

CHARLES L. MAMZIC
　　BSME, CA
　　Director, Marketing and Application
　　　Engineering
　　　Moore Products, Inc.
　　(Sections 2.1, 2.6, 2.7)

ALLAN F. MARKS*
  BSChE, PE
  Engineering Specialist, Control Systems
    Bechtel Petroleum, Inc.
  (Section 2.2)

FRED D. MARTON*
  Dipl. Ing.
  Former Managing Editor
    Instruments and Control Systems
  (Sections 3.6, 4.3)

CHARLES F. MOORE*
  BSChE, MSChE, PhDChE
  Professor of Chemical Engineering
  University of Tennessee
  (Sections 1.1, 1.2)

JOHN A. MOORE
  BSChE, PE
  Senior Application Specialist
    Leeds & Northrup Instruments,
      A Unit of General Signal
  (Section 7.4)

THOMAS J. MYRON, JR.*
  BSChE
  Senior Systems Design Engineer
    The Foxboro Co.
  (Section 8.10)

C.M. JACQUES OUDAR
  Ingénieur EEMI
  President
    MTL Incorporated
  (Section 5.2)

GLENN A. PETTIT*
  Retired from
    Rosemount Engineering Co.
  (Section 8.11)

GEORGE PLATT
  BChE
  Staff Engineer
    Bechtel Power Corp.
  (Section 6.1)

HOWARD C. ROBERTS
  BAEE, PE
  Consultant, Lecturer
    University of Colorado at Denver
  (Section 6.5)

DOUGLAS D. RUSSELL*
  BSEE, MSEE
  Group Leader
    The Foxboro Co.
  (Section 3.3)

DONALD R. SADLON
  BSEE
  General Manager
    Automated Dynamics Corp.
  (Section 6.4)

CHAKRA J. SANTHANAM
  BSChE, MSChE, PE
  Management Staff
    Arthur D. Little, Inc.
  (Sections 5.6†, 8.3, 8.7)

WALTER F. SCHLEGEL*
  BE
  Assistant Chief Process Engineer
    CPI Plants, Inc.
  (Section 5.13)

PHILLIP D. SCHNELLE, JR.
  BSEE, MSEE, PE
  Technical Service Engineer
    E.I. DuPont de Nemours & Co.
  (Sections 1.3, 1.9, 1.11, 1.12)

FRANCIS G. SHINSKEY*
  BSChE
  Chief Application Engineer
    The Foxboro Co.
  (Sections 1.6, 1.7)

JOSEPH P. SHUNTA
  BSChE, MSChE, PhDChE, PE
  Senior Consultant
    E.I. DuPont de Nemours & Co.
  (Section 1.8)

JAMES E. TALBOT*
  BSEP
  Senior Account Executive
    Lewis, Gilman & Kynett
  (Section 2.3)

MAURO C. TOGNERI
  BSEE
  President
    Powell-Process Systems
  (Section 3.5)

JOHN VEUCNIL
    BSEE, MSEE
    Senior RF Engineer
      Automation Industries
      (Sections 2.9, 3.4)

MICHAEL H. WALLER
    BME, SM, ME, PE
    Associate Professor
      Miami University
      (Section 7.10)

ANDREW C. WIKTOROWICZ
    BS, PE
    President
      Automated Dynamics Corp.
      (Section 6.4)

THEODORE J. WILLIAMS
    BSChE, MSChE, MSEE, PhDChE, PE
    Professor of Engineering, Director of
      Purdue Laboratory for Applied
      Industrial Control
      Purdue University
      (Sections 7.1, 7.5)

ROBERT A. WILLIAMSON*
    BSME, BA, PE
    Supervisor, Electromechanical,
      Packaging Group
      The Foxboro Co.
      (Sections 3.7, 3.9)

INTRODUCTION

Béla G. Lipták

The *Process Measurement* volume of the INSTRUMENT ENGINEERS' HANDBOOK describes the various sensors, analyzers, and detectors that are available to the instrument engineer. In this volume, *Process Control*, the emphasis is on the other elements of the control loop, such as transmitters, telemetering systems and data highways, analog and shared digital controllers, microprocessors and computers, logic devices and PLCs, safety and alarm devices, control panel displays and CRTs, regulators and safety valves, control valves, dampers and variable speed final control elements. In addition to these hardware-related topics, specific control systems for boilers, chillers, centrifuges, compressors, cooling towers, crystallizers, distillation towers, dryers, evaporators, extruders, furnaces, heat exchangers, HVAC systems, ORP and pH controls, pumps, reactors, semiconductor manufacturing and speed control systems, turbine controls, and water treatment are described, together with control theory, dynamic analysis, multivariable constraint control, and optimization.

The two volumes of this HANDBOOK are meant to complement each other, and between these two volumes the reader should find all the information that is needed to solve problems related either to process measurement or process control.

In 1959 when I started to work as an instrument engineer, the profession was very young. Donald Eckman's classic on process control was still a new book, the Ziegler and Nichols method of tuning represented the state of the art, Greg Shinskey had not yet joined the Foxboro Company, and C. B. Moore was not yet recognized as an ISA fellow for his inventions, such as the inverse derivative relay. Universities did not offer degrees in instrumentation, one could not obtain a PE license in process control, the advocates of "bang-bang" control were still numerous and vocal, and if plants had central control rooms at all, they contained large-case pneumatic instruments.

Much has changed since those early days. In today's distributed control systems, sensors can be self-calibrating and self-diagnosing detectors or analyzers; transmission can be through data highways; control algorithms can be self-tuning, multivariable, envelope-optimized models; and the final control elements can range from throttling solenoids or digital control valves to variable speed motors. Process control theory is beginning to be applied to nonindustrial processes, such as the social sciences, economics, and medicine.

The winds of change within our profession will not diminish in the coming years, but will probably increase in speed. In fact, one can already foresee some new chapter titles in the next edition of this HANDBOOK dealing with unattended systems, robots, and the like. Yet these changes affect only the tools of our profession. A competent instrument engineer will accept and use these new tools but will rely on his or her experience in deciding on how to select and apply them. In some ways experience is more valuable than the most fashionable tools. It is for this reason that I would like to share with the readers of this HANDBOOK some conclusions drawn from my years of experience as an instrumentation engineer. These I call:

The Instrument Engineers' Commandments

- Before one can control a process, one must understand it.
- Outdated control concepts implemented in modern hardware will give outdated performance.
- Being progressive is different from being a guinea pig.
- An instrument engineer is a good professional if he tells people what they need to know and not what they want to hear.
- Even the best-trained operators and the highest-quality instruments will fail sometimes. Therefore safety is gained through backup.
- Optimization means to adapt efficiently to changing conditions. Constant is the enemy of efficiency. The goal is not to keep flows and levels constant, but to maximize efficiency and productivity. Process properties should therefore be allowed to float as they follow the load.
- All sales representatives are guilty until proven innocent. Trust your common sense, not the sales

literature. Independent performance evaluation (SIREP-WIB) should be done *before* installation, not after it. The right time for "business lunches" is *after* start-up, not before the issue of the purchase order.

- If an instrument is worth installing, it must also be worth maintaining and calibrating. No device can outperform the reference against which it was calibrated.

- All manmade sensors detect relative values, and therefore the error contribution of references and compensators must always be considered.

- Sensors with errors expressed as percent of actual reading are preferred over those with percent of full-scale errors. If the error increases as the reading moves down-scale, the loop performance will also deteriorate.

- All sensors, including analyzers, should be placed directly into the process, without sampling systems. Inline instruments should have no dead-ended cavities, obstructions, flowing junctions or moving parts, but should be solid state with self-diagnostic and self-calibrating capabilities.

- It is easier to drive within the limits of a lane than to follow a single line. Similarly, control is easier and more stable if the single setpoint is replaced by a control gap. Stability and self-regulation are gained if the control valve is left in its last position, as long as the variable is within an acceptable control gap.

- Designating a valve on a flowsheet as a TCV will not suspend the laws of nature, and this arbitrary designation will not, for example, prevent it from affecting the process pressure. Similarly, the number of available control valves in a process will not necessarily coincide with the number of process properties that need to be controlled. Multivariable herding or envelope control overcomes this limitation of the uncoordinated single loop controllers and lets us control all variables that need to be controlled, while minimizing interactions.

- Annunciators do not correct emergencies; they just throw into the laps of the operators those problems that the designers did not know how to handle. The smaller the annunciator, the better the design.

One could add to these "commandments" all of Murphy's laws, just to further emphasize the importance of common sense in all instrumentation design.

It is hoped that this revised HANDBOOK will contribute to the professional standing of the community of instrument engineers. We know that the social benefit of a competent and advanced instrument engineering profession can be substantial. We also know that our greatest national resource is our combined knowledge and professional dedication. The productivity and safety of many industries could be doubled in some cases through the exploitation of state-of-the-art instrumentation and control. Similarly, the raw material and energy cost of some processes could be cut in half merely through optimized controls. We have the tools, the knowledge, and the need to strive for these goals.

Chapter I

CONTROL THEORY

M. F. HORDESKI

C. G. LASPE

C. F. MOORE

P. D. SCHNELLE, JR.

F. G. SHINSKEY

J. P. SHUNTA

CONTENTS OF CHAPTER I

1.1 BASIC CONCEPTS, VARIABLES, AND DEGREES OF FREEDOM

Basic Concepts

The development of automatic control systems in the past 50 years has been equated in importance to the industrial revolution in the nineteenth century. In many respects the introduction of automatic control systems was a second industrial revolution of a sort. While the first was an extension of man's muscle, the second was an extension of his brain. In the nineteenth century we learned to harness and use various forms of natural energy, and in the twentieth century we learned to make devices that could make the decisions necessary to control the various forms of energy.

Principles used in automatic control cut across virtually every scientific field and in the process created a new field of its own. Today the basic principles of automatic control have a wide range of applications and interests, including process control, manufacturing control, aircraft and satellite control, traffic control, and biomedical control.

Why Automatic Control?

The need for automatic control is perhaps more obvious in the manufacturing and aerospace industries than it is in the process industries. In assembly line manufacturing facilities, the need for automation is quite apparent. A machine in many cases is more suited, both for economic and safety considerations, to performing the numerous tedious and monotonous tasks involved in production. It is also fairly clear to the casual observer that the control of a supersonic aircraft is much too complicated to be left entirely in the hands of a human pilot. However, in typical process applications the need for automatic control is perhaps a little less apparent.

Since most process equipment operates at a constant load, one might think that the best solution to the control problem is to set all the variables that affect the process to their proper positions and forget about the process. The difficulty with this reasoning is that seldom can all the inputs to the system be fixed. Most process equipment is subject to many inputs, some of which can be manipulated (or set at a fixed value) and some of which will change without regard to the operator's desires.

Changes in such variables result in disturbances in the process, unless corrective action of some sort is taken.

Consider the simple direct contact water heater shown in Figure 1.1a. The heater consists of a tank from which hot water is obtained by bubbling live steam directly into the tank which is full of water. Cool water enters at the bottom of the tank and the hot water leaves the top. A valve is available by which to regulate the flow rate of steam into the heater. In this example, if all other factors were constant the temperature of the outlet could be controlled simply by placing the steam valve at the proper setting. Note, however, that if the temperature of the inlet water changes, the outlet temperature would eventually change by the same amount unless corrective actions were taken. Other variables besides the inlet water temperature that could disturb the process are the flow rate of the water, the steam supply pressure, the steam quality, and the ambient temperature. A change in any one of these variables will cause a change in the water outlet temperature unless some correction is made.

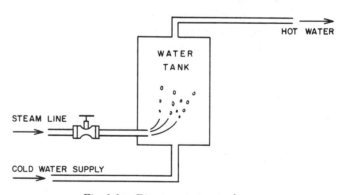

Fig. 1.1a   Direct contact water heater

Feedback Control

Two concepts provide the basis for most automatic control strategies: feedback (closed-loop) control and feedforward (open-loop) control. Feedback control is the more commonly used technique of the two and is the underlying concept on which much of today's automatic control theory is based. Feedback control is a strategy designed to achieve and maintain a desired process condition by measuring the process condition, comparing the measured condition with the desired condition, and initiating corrective action based on the difference between the desired and the actual condition.

The feedback strategy is very similar to the actions of a human operator attempting to control a process manually. Consider the procedure an individual might employ in the control of the direct contact hot water heater described earlier. The operator would read the temperature indicator in the hot water line and compare its value with the temperature he desires (Figure 1.1b). If the temperature were too high, he would reduce the steam flow, and if the temperature were too low, he would increase it. Using this strategy he will manipulate the steam valve until the error is eliminated.

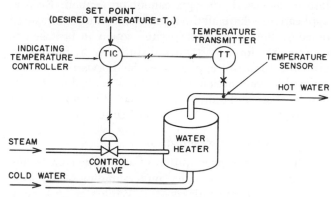

Fig. 1.1c   Automatic feedback control

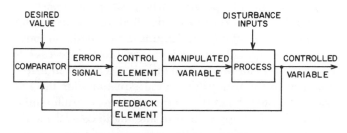

Fig. 1.1d   Basic components of a feedback control loop

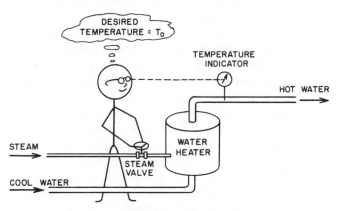

Fig. 1.1b   Manual feedback control

An automatic feedback control system would operate in much the same manner (Figure 1.1c). The temperature of the hot water is measured and a signal is fed back to a device that compares the measured temperature with the desired temperature. If an error exists, a signal is generated to change the valve position in such a manner that the error is eliminated. The only real distinction between the manual and automatic means of controlling the heater is that the automatic controller is more accurate, consistent, and not as likely to become tired or distracted. Otherwise, both systems contain the essential elements of a feedback control loop (Figures 1.1c and d).

Feedback control has definite advantages over other techniques (such as feedforward control) in relative simplicity and potentially successful operation in the face of unknown contingencies. In general, it works well as a regulator to maintain a desired operating point by compensating for various disturbances that affect the system, and it works equally well as a servo system to initiate and follow changes demanded in the operating point.

Feedforward Control

Feedforward control is another basic technique used to compensate for uncontrolled disturbances entering the

system. In this technique the control action is based on the state of a disturbance input without reference to the actual system condition. In concept, feedforward control yields much faster correction than feedback control, and in the ideal case compensation is applied in such a manner that the effect of the disturbance is never seen in the process output.

A skillful operator could use a simple feedforward strategy to compensate for changes in inlet water temperature of a direct contact water heater. Detecting a change in inlet water temperature, he would increase or decrease the steam rate to counteract the change (Figure 1.1e). The same compensation could be made automatically with an inlet temperature detector designed to initiate the appropriate corrective adjustment in the steam valve opening.

The concept of feedforward control is very powerful, but unfortunately it is difficult to implement in a pure form in most process control applications. In many cases disturbances cannot be accurately measured, and there-

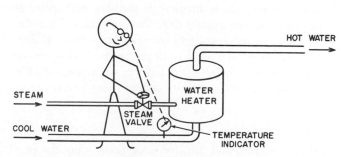

Fig. 1.1e   Concept of feedforward control

fore feedforward concepts cannot be applied. Even in applications where all the inputs can be either measured or controlled, the "appropriate" action to be taken to compensate for a particular disturbance is not always obvious. In many applications, feedforward control is utilized in conjunction with feedback control in order to handle those unknown contingencies that might otherwise disturb the pure feedforward control system.

Mathematical Representation

A fundamental prerequisite of automatic control theory application is a detailed understanding of the operation of the process under control. While standard equipment and process design requires a detailed knowledge of the equipment operation when inputs are constant, automatic control requires a detailed knowledge of the equipment operation when inputs are changing in time. This time-varying behavior is referred to as "process dynamics" and can be conveniently summarized in mathematical terms using differential equations (or the Laplace transfer function representation of differential equations). The application of automatic control theory presumes a knowledge of the entire control system mathematics, and without such knowledge automatic control theory is to a large extent useless.

In many cases, the mathematical description (mathematical model) of the various components of the control loop can be obtained analytically based entirely on the physics of the process components. In other cases, models can be obtained by experimental testing procedures in which the actual response of the process is analyzed in some manner to extract the desired dynamic information.

Variables and Degrees of Freedom

This section concerns itself with the nature of chemical processes, and in particular with those requirements and limitations that the process places on the associated automatic control systems.

The state of a process may be described by the values of its variables. This state is defined when each of the system's degrees of freedom is specified. The number of degrees of freedom of a process represents the maximum number of independently acting automatic controllers that can be placed on the process.

The transient behavior of a process reveals its dynamic characteristics. These involve its stability and speed of response, and are usually described in terms of inertia, resistance, capacitance, and transportation time. Differential equations are used to describe these characteristics, and depending on their nature, we distinguish the various process responses by calling them first order or second order.

Process Variables

Many external and internal conditions affect the performance of a process unit. These conditions may be expressed in terms of process variables such as temperature, pressure, flow, concentration, weight, level, etc. The process is usually controlled by measuring one of the variables that represent the state of the system and then by automatically adjusting one of the variables that determine the state of the system. Typically, the variable chosen to represent the state of the system is termed the *controlled variable* and the variable chosen to control the system's state is termed the *manipulated variable*.

The manipulated variable can be any process variable that causes a reasonably fast response and is fairly easy to manipulate. The controlled variable should be the variable that best represents the desired state of the system. Consider the water cooler shown in Figure 1.1f. The purpose of the cooler is to maintain a supply of water at a constant temperature. The variable that best represents this objective is the temperature of the exit water, T_{wo}, and it should be selected as the controlled variable. In other cases, direct control of the variable that best represents the desired condition is not possible. Consider the chemical reactor shown in Figure 1.1g. The variable that is directly related to the desired state of the product is the composition of the product stream; however, in this case a direct measurement of product composition is not always possible. If product composition is to be controlled, some other process variable must be used that is related to composition. A logical choice for this chemical reactor might be to hold the pressure constant and use reactor temperature as an indication of composition. Such a scheme is often used in the indirect control of composition.

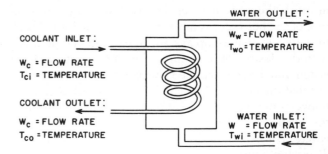

Fig. 1.1f   Process variables in a simple water cooler

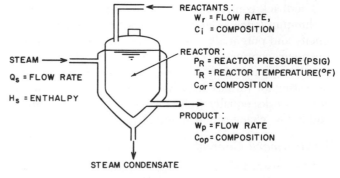

Fig. 1.1g   Process variables in a simple chemical reactor

Degrees of Freedom

Degrees of freedom is a very important concept in process control, but its importance is frequently not appreciated. Mathematically the number of degrees of freedom is defined as

$$df = v - e \qquad 1.1(1)$$

where

df = number of degrees of freedom of a system,

v = number of variables that describe the system, and

e = number of independent relationships that exist among the various variables.

A clear understanding of this principle and its application can give a keener insight into the steady state and dynamic behavior of processes and a clearer understanding of the basic framework around which a control system must be designed.

To illustrate the general principle, consider the degrees-of-freedom equation applied to three common examples: the motion of an airplane, a boat, and a train. For each vehicle the variables are the same, namely, latitude, longitude, and altitude. A value for each of the three navigational coordinates specifies the exact position of the vehicle. Therefore, in each case the value of v is 3. The number of independent equations relating the three space coordinates differs in each case. For an airplane (Figure 1.1h) on a casual flight (with no specific flight plan), there are no fixed relationships between its latitude, longitude, and altitude at any moment of time. Therefore, the number of degrees of freedom of the plane is

$$df = v - e = 3 - 0 = 3$$

This answer is intuitively correct since the plane is free to fly in virtually any direction in the three-dimensional space. The boat (Figure 1.1i) has the same three navigational variables as does the plane, but assuming the

VARIABLES:
ALTITUDE |
LATITUDE |
LONGITUDE |
                  3
EQUATIONS:
ALTITUDE = SEA LEVEL   −1
DEGREES OF FREEDOM   2

Fig. 1.1i Degrees of freedom of a boat

water is relatively calm the altitude is fixed. Therefore the following equation is established:

$$\text{altitude of boat} = \text{sea level}$$

Hence, the degrees of freedom for the boat are

$$df = v - e = 3 - 1 = 2$$

The boat is free to move in two directions but must remain confined to the water's surface.

The train (Figure 1.1j) also has the same three navigational variables but is confined to a track which has a fixed location described by two equations:

$$\text{altitude of track} =$$
$$\text{geological contour of earth}$$
$$\text{latitude of track} =$$
$$\text{specific function of the longitude}$$

The first equation fixes the altitude to be equal to the local ground surface elevation. The second equation describes the exact route of the track in terms of the latitude and longitude. Therefore, the train has the following degrees of freedom:

$$df = v - e = 3 - 2 = 1$$

The train has only one degree of freedom and can only move along the path specified by the position of the track.

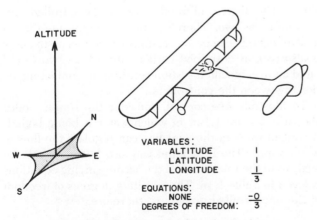

VARIABLES:
ALTITUDE |
LATITUDE |
LONGITUDE |
                  3
EQUATIONS:
NONE   −0
DEGREES OF FREEDOM: 3

Fig. 1.1h Degrees of freedom of an airplane

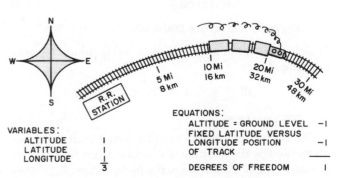

VARIABLES:
ALTITUDE |
LATITUDE |
LONGITUDE |
                  3
EQUATIONS:
ALTITUDE = GROUND LEVEL   −1
FIXED LATITUDE VERSUS
LONGITUDE POSITION   −1
OF TRACK
DEGREES OF FREEDOM   I

Fig. 1.1j Degrees of freedom of a train

The exact position of the train can be specified by the relative distance down the track.

DIRECT CONTACT WATER HEATERS

As an example of analysis for degrees of freedom inherent in a process unit, consider the direct contact hot water heater shown in Figure 1.1k.

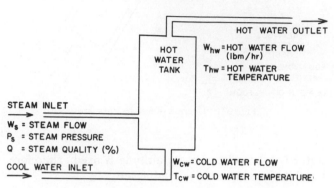

Fig. 1.1k   Degrees of freedom in a direct contact water heater

The first step in the analysis is to define the problem and list all the pertinent variables that affect the system. Table 1.1l lists such variables for this particular problem and also notes the corresponding equations relating these variables. Based on these tabulations, the degrees of freedom for the hot water heater are defined as follows:

$$df = v - e = 7 - 5 = 2$$

Note that to a large extent the number of variables included in the list is arbitrary. In this particular example we could have included the enthalpy of the three streams:

H_s, the enthalpy of steam (BTU/lbm)
H_{cw}, the enthalpy of inlet water (BTU/lbm)
H_{hw}, the enthalpy of outlet water (BTU/lbm)

However, in considering these three new variables, *three more independent equations* must also be considered, namely,

$$H_{cw} = c_p(T_{cw} - T_r) \qquad 1.1(2)$$
$$H_{hw} = c_p(T_{hw} - T_r) \qquad 1.1(3)$$
$$H_s = \text{function of } P_s \text{ and } Q \text{ (found in steam tables)} \qquad 1.1(4)$$

where T_r and c_p are constants.

Hence, the number of degrees of freedom remains unchanged:

$$df = v - e = 10 - 8 = 2$$

CONTROLLERS

The control of a direct contact water heater was implemented by the addition of a feedback control loop that measures the temperature of the outlet water (T_{hw}), compares it with some predetermined value, and adjusts the

Table 1.1l
DEGREES-OF-FREEDOM ANALYSIS OF A DIRECT
CONTACT WATER HEATER

Variables	Number of Variables
W_s = flow rate of steam	1
W_{cw} = flow rate of cool water	1
W_{hw} = flow rate of hot water	1
Q = quality of steam	1
P_s = supply pressure of steam	1
T_{cw} = temperature of cool water inlet	1
T_{hw} = temperature of hot water outlet	1
Total number of variables	7

Equations	Number of Equations
Constant terms P_s, Q, T_{cw}	3
Material balance (conservation of mass)	1
Energy balance (conservation of energy)	1
Total number of equations	5

Degrees of freedom = 2

steam flow according to the error (see Figure 1.1c). Therefore, in terms of degrees of freedom, the control loop adds another independent equation,

$$F = \text{function of } (T_{hw} - \text{setpoint}) \qquad 1.1(5)$$

The net effect on the overall degrees of freedom is a reduction by one. Therefore, this controlled process has one remaining degree of freedom.

The remaining degree of freedom could be removed by the addition of another controller on the inlet water line such that the water flow is maintained at a constant value.

OVERCONTROL

Perhaps the most beneficial result of understanding the concept of degrees of freedom is a knowledge of what can be controlled and what cannot. According to the degrees-of-freedom rule, *the number of independently acting automatic controllers on a system or on part of a system may not exceed the number of degrees of freedom.* To illustrate this rule, consider the water heater analyzed above. The degrees of freedom of the uncontrolled process unit was two. Therefore the maximum number of controllers that can be successfully employed on the tank is also two, as described earlier. More than two would yield an overspecified system resulting in conflicting actions between the controllers.

To illustrate overcontrol, consider the simple boiler shown in Figure 1.1m, in which water is being boiled. A control valve in the vent line can regulate the flow of water vapor. The heat input can also be regulated. An analysis of the variables and the accompanying equations is given in Table 1.1n. The resulting degrees of freedom of the uncontrolled process is, therefore,

$$df = v - e = 6 - 4 = 2$$

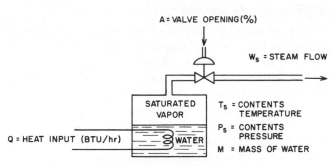

Fig. 1.1m Degrees of freedom in a simple water boiler

Table 1.1n
DEGREES-OF-FREEDOM ANALYSIS OF A SIMPLE
WATER BOILER

Variables	Number of Variables
M = mass of water in tank at any point in time	1
Q = heat input into tank	1
W_s = flow rate of vapor out of tank	1
T_s = temperature of tank contents	1
P_s = pressure in tank	1
A = valve opening	1
Total number of variables	6

Equations	Number of Equations
Material balance	1
Energy balance	1
Valve equation, $V = f(A, P_s)$	1
Vapor pressure vs temperature data (steam tables)	1
Total number of equations	4

Degrees of freedom = 2

The analysis suggests that since the process has only two degrees of freedom, only two independent controllers can be added. In this case there might be a natural tendency to attempt to control directly all of the system outputs such as temperature, pressure, and vapor flow rate; however, the number of degrees of freedom dictates that a maximum of two such control loops can be used. The possible combinations to be considered are pressure and flow rate, temperature and flow rate, or temperature and pressure. A superficial analysis indicates that any two of these combinations would be satisfactory. The individual equations involved point out another basic lesson concerning the degrees of freedom of processes. An analysis of the pressure vs. temperature data indicates that there is only one degree of freedom between temperature T_s and pressure P_s. Therefore, an attempt to control both independently would result in an overcontrolled system, in spite of the overall degrees of freedom being satisfied. The concept of degrees of freedom should be applied at all levels in the system analysis.

Process Dynamics

Controlling processes would be trivial if all systems responded instantaneously to changes in the process inputs. The difficulty in control lies in the fact that all processes to one degree or another tend to delay and retard the changes in process variables. This time-dependent characteristic of the process is termed process dynamics, and its evaluation is essential to the understanding and application of automatic control.

The dynamic characteristics of all systems, whether mechanical, chemical, thermal, or electrical, can be attributed to one or more of the following effects: (1) inertia, (2) capacitance, (3) resistance, and (4) transportation time.

INERTIA

Inertia effects pertain to Newton's second law, governing the motion of matter:

$$\Sigma F = (M \times a) \qquad 1.1(6)$$

where

ΣF = net force acting on a mass,
M = total mass, and
a = acceleration of that mass.

Inertia effects are most commonly associated with mechanical systems involving moving components, but they are also important in some flow systems in which fluids must be accelerated or decelerated.

RESISTANCE AND CAPACITANCE

Resistance and capacitance are perhaps the most important effects in industrial processes involving heat transfer, mass transfer, and fluid flow operations. Those parts of the process that have the ability to store energy or mass are termed capacities, and those parts that resist transfer of energy or mass are termed resistances. The combined effect of supplying a capacity through a resistance is a time retardation, which is very basic to most dynamic systems found in industrial processes. Consider, for example, the water heater system shown in Figure 1.1o, where the capacitance and resistance terms can be readily identified. The capacitance is the ability of the

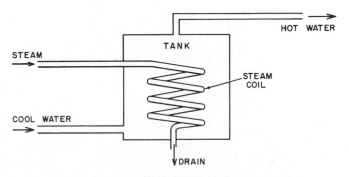

Fig. 1.1o Resistance and capacitance effects in a water heater

tank and of the water in the tank to store heat energy. A second capacitance is the ability of the steam coil and its contents to store heat energy. A resistance can be identified with the transfer of energy from the steam coil to the water due to the insulating effect of a stagnant layer of water surrounding the coil. If an instantaneous change is made in the steam flow rate, the temperature of the hot water will also change, but the change will not be instantaneous. It will be sluggish, requiring a finite period of time to reach a new equilibrium. The behavior of the system during this transition period will depend on the amount of material that must be heated in the coil and in the tank (determined by the capacitance) and on the rate at which heat can be transferred to the water (determined by the resistance).

TRANSPORTATION TIME

A contributing factor to the dynamics of many processes involving the movement of mass from one point to another is the transporation lag, or deadtime. In the simple heater problem discussed earlier, consider the effect of piping on the heated water to reach a location some distance away from the heater (Figure 1.1p). The effect of a change in steam rate on the water temperature at the end of the pipe will not only depend on the resistance and capacitance effects in the tank, but will also be influenced by the length of time necessary for the water to be transported through the pipe. All lags associated with the heater system will be seen at the end of the pipe, but they will be delayed. The length of this delay is called the transportation lag, or deadtime. The magnitude is determined as the distance over which the material is transported, divided by the velocity at which the material travels. In the heater example,

$$\theta = v/L \qquad 1.1(7)$$

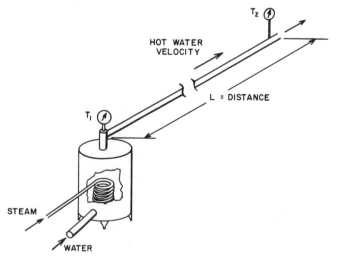

Fig. 1.1p   Transportation time effects on a water heater

Differential Equations

The quantitative effect of resistance, capacitance, transportation lag, and inertia on a process can be expressed in terms of the differential equations that describe the process. Such equations can be developed by applying Newton's law, the law of conservation of mass, and the law of conservation of energy along with specific equations related to a particular process (i.e., equations describing heat transfer, fluid flow through valves, chemical kinetics, thermodynamic equilibrium, etc.) To demonstrate the general procedure involved, several simple examples are given of methods for developing differential equations that describe process units.

CATALYST PREPARATION TANK

Consider a catalyst preparation tank shown in Figure 1.1q. The system is used to upgrade a low-concentration catalyst stream by the addition of a relatively small amount of a highly concentrated stream. The tank is assumed to be well mixed, and the flow rate of highly concentrated material is considered very small in comparison with the flow of material out of the tank.

The only equation needed to describe the system is the material balance on the catalyst. The law of conservation of mass states that:

$$\begin{pmatrix} \text{Flow of catalyst} \\ \text{into the tank} \end{pmatrix} - \begin{pmatrix} \text{Flow of catalyst} \\ \text{from the tank} \end{pmatrix}$$

$$= \begin{pmatrix} \text{Accumulation of} \\ \text{catalyst in tank} \end{pmatrix}$$

$$\left(F_h C_m + C_i F_l \right) - \left(C_0 F_l \right) = \left(\frac{d}{dt} H C_0 \right) \qquad 1.1(8)$$

Since the holdup, H, can be considered constant, the equation can be written

$$F_h C_m + C_i F_l = C_0 F_l + H \frac{d}{dt} C_0 \qquad 1.1(9)$$

where

F_h = flow rate of highly concentrated makeup stream,

F_l = flow rate of low concentrated streams,

C_m = concentration of catalyst in the makeup stream,

C_i = concentration of catalyst in the inlet stream,

C_0 = concentration of catalyst in the outlet stream, and

H = holdup of tank.

WATER HEATERS

The equations describing the direct contact hot water heater shown in Figure 1.1r are developed as follows.

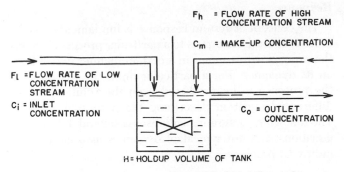

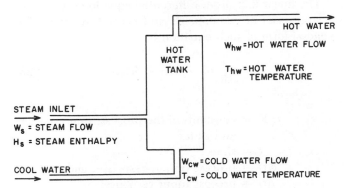

Fig. 1.1q   Catalyst preparation tank described by differential equation

Fig. 1.1r   Direct contact water heater described by differential equation

The tank is assumed to be at uniform temperature, and the quantity of water added to the system by stream condensation is assumed to be comparatively small. For this simple example an energy balance is all that is necessary to describe the entire system.

$$\left(\begin{array}{c}\text{Rate of energy}\\ \text{into heater}\end{array}\right) - \left(\begin{array}{c}\text{Rate of energy}\\ \text{out of heater}\end{array}\right) =$$

$$= \left(\begin{array}{c}\text{Rate of energy}\\ \text{accumulation}\\ \text{in the heater}\end{array}\right)$$

$$\left(H_sW_s + W_{cw}C_p(T_i - T_r)\right) -$$

$$- \left(W_{hw}C_p(T_{hw} - T_r)\right) =$$

$$= \left(\frac{d}{dt}\rho\, HC_p(T_{hw} - T_r)\right) \quad 1.1(10)$$

Or, since ρ, C_p, and T_r are constants,

$$H_sW_s + W_{cw}C_pT_i =$$

$$= W_{hw}C_pT_{hw} + H\rho C_p\frac{d}{dt}(T_{hw}) \quad 1.1(11)$$

where

$$\begin{aligned} H_s &= \text{enthalpy of steam,}\\ W_s &= \text{flow rate of steam,}\\ W_{hw} &= W_{cw} = \text{flow rate of water,}\end{aligned}$$

$$\begin{aligned} T_{cw} &= \text{inlet water temperature,}\\ T_{hw} &= \text{outlet water temperature,}\\ C_p &= \text{heat capacity of water,}\\ \rho &= \text{density of water,}\\ H &= \text{volume of tank, and}\\ T_r &= \text{reference temperature.}\end{aligned}$$

CHEMICAL REACTORS

Figure 1.1s illustrates a simple back-mix chemical reactor in which component "a" reacts to form component "b" by the following first-order reaction:

$$a \longrightarrow b$$

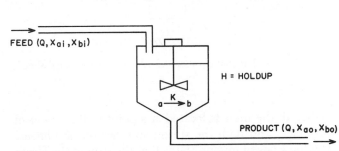

Fig. 1.1s   Back-mix reactor described by first-order differential equation

The feed is a binary mixture with composition X_{ai}, X_{bi}. The product stream is also a binary mixture containing "b" and unreacted "a". Due to the well-mixed conditions in the reactor, composition of the reactor contents and the product stream composition are identical (X_{ao}, X_{bo}). A material balance of component "a" applied to the reactor is as follows:

$$\left(\begin{array}{c}\text{Rate of "a"}\\ \text{into reactor}\end{array}\right) - \left(\begin{array}{c}\text{Rate of "a"}\\ \text{out of reactor}\end{array}\right) =$$

$$= \left(\begin{array}{c}\text{Rate of accumulation}\\ \text{of "a" in reactor}\end{array}\right)$$

$$(QX_{ai}) - (QX_{ao} + r_aH) = \left(H\frac{d}{dt}(X_{ao})\right) \quad 1.1(12)$$

where r_aH represents the amount of component "a" consumed by the reaction. No actual flow is involved, but it does represent the removal of "a" from the system and therefore must be considered. The rate of consumption of "a" is described as

$$r_a = kX_a \quad 1.1(13)$$

and can be substituted into equation 1.1(12) to yield the complete differential equation that describes the system:

$$QX_{ai} = (Q + k)X_{ao} + H\frac{d}{dt}X_{ao} \quad 1.1(14)$$

MASS-SPRING DASHPOTS

To illustrate a typical mechanical system in which inertia effects are important, consider the mass-spring dashpot of Figure 1.1t. Consider the system to be ideal

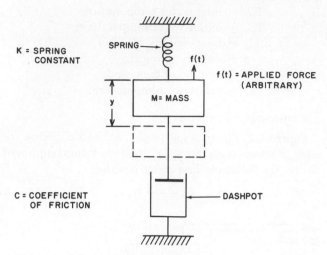

Fig. 1.1t   Differential equation describing a mass-spring dashpot

in that all the mass is located at a point in the center of the block, and both the spring and dashpot are linear. Newton's law of motion states that the sum of the forces on the mass must be equal to the product of the mass times its acceleration:

$$\Sigma F = (M)(a) \qquad 1.1(15)$$

where

$$a = \frac{d^2}{dt^2}(y) \qquad 1.1(16)$$

An analysis of the system indicates that the four forces acting on the mass are

1. Force of gravity
   (acting downward) $= (M)(g) = $ constant 1.1(17)
2. Force of spring
   (acting upward) $= (M)(g) - (K)(y) \qquad 1.1(18)$
   When the mass is in the rest position, the upward force of the spring equals the downward force due to gravity.
3. Force of dashpot $= -C\dfrac{d}{dt}(y)$
   (acting in a direction
   opposite the direction
   of movement)   1.1(19)
4. Arbitrary force
   imposed on mass $= f(t) \qquad 1.1(20)$

Substituting the force terms into Newton's equation (considering all forces acting upward to be positive) yields

$$-(M)(g) - (K)(y) + (M)(g) -$$
$$-C\frac{d}{dt}(y) + f(t) = M\frac{d^2}{dt^2}(y) \qquad 1.1(21)$$

or

$$M\frac{d^2}{dt^2}(y) + C\frac{d}{dt}(y) + (K)(y) = f(t) \qquad 1.1(22)$$

Response

The concept of system response is fundamental in automatic control. As inputs to a particular process change, the process will respond in a certain manner depending on its dynamics. For a particular input the response of the process can be predicted from the solution of the differential equation describing the process dynamics. The following paragraphs will discuss several differential equations that are typical of dynamics associated with industrial processes.

FIRST-ORDER RESPONSE

The linear first-order differential equation is typical of a large class of components and control systems. The general form of such equation is

$$\tau\frac{d}{dt}c(t) + c(t) = Kr(t) \qquad 1.1(23)$$

where

τ, K = constants of the process, time constant and gain,
t = time,
$c(t)$ = process output response, and
$r(t)$ = process input response.

Process elements of this description are common and are generally referred to as first-order lags. The response of a first-order system is characterized by two constants: a time constant τ and a gain K. The gain is related to the amplification associated with the process and has no effect on the time characteristics of the response. The time characteristics are related entirely to the time constant. The time constant is a measure of the time necessary for the component or system to adjust to an input, and it may be characterized in terms of the capacitance and resistance (or conductance) of the process:

$$\tau = \text{resistance} \times \text{capacitance} = \frac{\text{capacitance}}{\text{conductance}} \qquad 1.1(24)$$

To illustrate the nature of a first-order system, consider the response that results from an input of the following form:

$$r(t) = 0.0 \qquad t \le 0$$
$$r(t) = R_0 \qquad t > 0 \qquad 1.1(25)$$

The solution of the first-order differential equation for such an input, considering the initial value of c to be zero, is

$$c(t) = KR_0(1.0 - e^{-t/\tau}) \qquad \text{for} \qquad t > 0 \quad 1.1(26)$$

Details of the solution are given in Section 1.3.

In their responses (Figure 1.1u), two characteristics distinguish the first-order systems. (1) The maximum rate of change of the output occurs immediately following the step input. (Note also that if the initial rate were unchanged the system would reach the final value in a

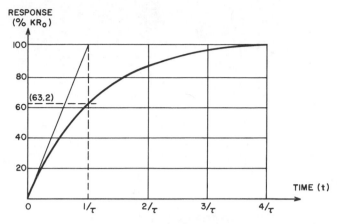

Fig. 1.1u   First-order lag step response

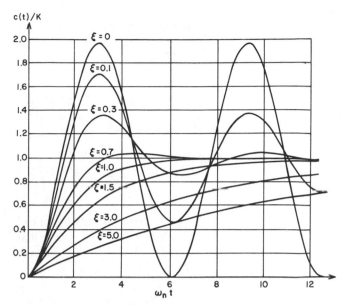

Fig. 1.1v   Second-order lag step response

period of time equal to the time constant of the system.) (2) The actual response obtained, when the time lapse is equal to the time constant of the system, is 63.2 percent of the total response. These two characteristics are common to all first-order processes.

SECOND-ORDER RESPONSE

Due to inertia effects and various interactions between first-order resistance and capacitance elements, some processes are second order in nature and are described by the following differential equation:

$$\frac{d^2}{dt^2}c(t) + 2\xi\omega_n\frac{d}{dt}c(t) + \omega_n^2c(t) = K\omega_n^2r(t) \quad 1.1(27)$$

where

ω_n = the natural frequency of the system,
ξ = the damping ratio of the system,
K = the system gain,
t = time,
r(t) = input response of system, and
c(t) = output response of system.

The solution of equation 1.1(27), for a step change in r(t) with all initial conditions zero can be any one of a family of curves shown in Figure 1.1v. (For details of the solution refer to Section 1.3.)

In the actual solution three possible cases must be considered, depending on the value of the damping ratio:

1. When $\xi < 1.0$, the system is said to be underdamped and will overshoot the final steady-state value. If $\xi < 0.707$, the system will not only overshoot but will oscillate about the final steady-state value.

2. When $\xi > 1.0$, the system is said to be "overdamped" and will not oscillate or overshoot the final steady-state position.

3. When $\xi = 1.0$, the system is said to be "critically damped" and yields the fastest response without overshoot or oscillation.

The "natural frequency" term ω_n in the second-order equation is related to the speed of the response for a particular value of ξ. The response illustrated in Figure 1.1v is plotted against a normalized time in which the actual time is divided by the natural frequency, and therefore a large frequency tends to squeeze the response and a small frequency tends to stretch the response. The natural frequency is defined in terms of the "perfect" or "frictionless" situation, where $\xi = 0.0$. In such a situation the response is a sustained sinusoid with a frequency of oscillation equal to ω_n. For the case where ξ is not zero, the actual frequency of an underdamped response is related to the natural frequency by

$$\omega = \omega_n \sqrt{1 - \xi^2} \quad 1.1(28)$$

BIBLIOGRAPHY

Buckley, P.S., "Dynamics of Pneumatic Control Systems," Paper 55-6-2, Proceedings of the 10th Annual ISA Instrument-Automation Conference and Exhibit, September, 1955, Los Angeles, CA.

———, *Techniques of Process Control*, John Wiley & Sons, New York, 1964.

Caldwell, W.I., Coon, G.A., and Zoss, L.M., *Frequency Response for Process Control*, McGraw-Hill Book Co., New York, 1959.

D'Azzo, J.J., and Houpis, C.H., *Feedback Control System Analysis and Synthesis*, McGraw-Hill Book Co., New York, 1960.

Eckman, D.P., *Automatic Process Control*, John Wiley & Sons, New York, 1958.

Harriott, P., *Process Control*, McGraw-Hill Book Co., New York, 1964.

Hougen, J.O., *Measurements and Control Applications for Practicing Engineers*, Cahners Books, Boston, 1972.

Jones, B.E., *Instrumentation, Measurement and Feedback*, McGraw-Hill Book Co., New York, 1977.

Warnock, J.D., "How Pneumatic Tubing Size Influences Controllability," *Instrumentation Technology*, February, 1967.

Wightman, E.J., *Instrumentation in Process Control*, Butterworth, 1972.

1.2 CONTROL MODES AND CONTROLLERS

In automatic control the device used to initiate control action is a special-purpose analog or digital computer that uses the difference between the desired value and the actual value of a controlled variable to eliminate the error in the actual value. In general, such controllers can be classified in two ways: in terms of the physical mechanism the controller employs (pneumatic, electronic, hydraulic, digital, etc.) or in terms of the manner in which the controller reacts to an error signal. The method by which a controller counteracts a deviation from set point is called the control mode. This section emphasizes the latter (control mode) distinction between controllers and presents the mathematical description of controllers commonly used in industry.

Modes of Control

Proportional, Integral, and Derivative Modes

The three most commonly used modes of feedback control are the proportional, integral, and derivative modes, which are defined thus:

Proportional mode:

$$m(t) = K_c[C_R - c(t)] + M_o \qquad 1.2(1)$$

Integral mode:

$$m(t) = \frac{1}{T_i} \int [C_R - c(t)] + M_o \qquad 1.2(2)$$

Derivative mode:

$$m(t) = T_d\frac{d}{dt}[C_R - c(t)] + M_o \qquad 1.2(3)$$

where

t = time,
C_R = required variable value (set point),
$m(t)$ = output of controller,
$c(t)$ = the signal fed back to the controller representing the controlled variable (measurement signal),
M_o = constant,

K_c = proportional gain, ⎫
T_i = integral time, and ⎬ adjustable controller parameters
T_d = derivative time. ⎭

Various combinations of these modes comprise most of the controllers found in industry. Some typical combinations are shown in Table 1.2a.

The proportional mode alone is the simplest of the three. It is characterized by a continuous linear relationship between the controller input and output. Several synonymous names in common usage are proportional action, correspondence control, droop control, and modulating control. The adjustable parameter of the proportional mode, K_c, is called the proportional gain, or proportional sensitivity. It is frequently expressed in terms of percent proportional band, PB, which is related to the proportional gain:

$$PB = (1/K_c)100 \qquad 1.2(4)$$

"Wide bands" (high percentage of PB) correspond to less "sensitive" controller settings, and "narrow bands" (low percentages) correspond to more "sensitive" controller settings.

The integral mode is sometimes used as a single-mode controller but is more commonly found in combination with the proportional mode. The proportional-plus-integral controller is perhaps the most widely used combination. The integral mode is synonymous with the terms reset action and floating control. The adjustable parameter associated with the integral mode is the integral time, T_i, or the reset rate, $1/T_i$.

The derivative mode of control is most commonly referred to as rate control, or preact control, because its output is based on the rate of change of the input variable. Since the output of this mode alone would be zero for a constant value of input, this mode is never used alone and is commonly found in combination with proportional mode.

The response of the individual modes along with the typical combinations of modes is shown in Table 1.2b for several inputs.

Table 1.2a
DESCRIPTIONS OF
CONVENTIONAL CONTROL MODES

Symbol	Description	Mathematic Expression
	ONE MODE	
P	Proportional	$m = K_c e$
I	Integral (reset)	$m = \dfrac{1}{T_i} \int e\, dt$
	TWO MODE	
PI	Proportional-plus-integral	$m = K_c \left[e + \dfrac{1}{T_i} \int e\, dt \right]$
PD	Proportional-plus-derivative	$m = K_c \left[e + T_d \dfrac{d}{dt} e \right]$
	THREE MODE	
PID	Proportional-plus-integral-plus-derivative	$m = K_c \left[e + \dfrac{1}{T_i} \int e\, dt + T_d \dfrac{d}{dt} e \right]$

Inverse Derivative Control Mode

A special-purpose control action used on extremely fast processes is the so-called inverse derivative mode. As the name implies, it is the exact opposite of the derivative mode. Where the output of the derivative mode is directly proportional to the rate of change in error, the output of the inverse derivative mode is inversely proportional to the rate of change in error.

Proportional-plus-derivative:

$$m = K_c \left(e + T_d \frac{d}{dt} e \right) \qquad 1.2(5)$$

Table 1.2b
RESPONSE OF PROPORTIONAL, INTEGRAL, AND
DERIVATIVE MODES

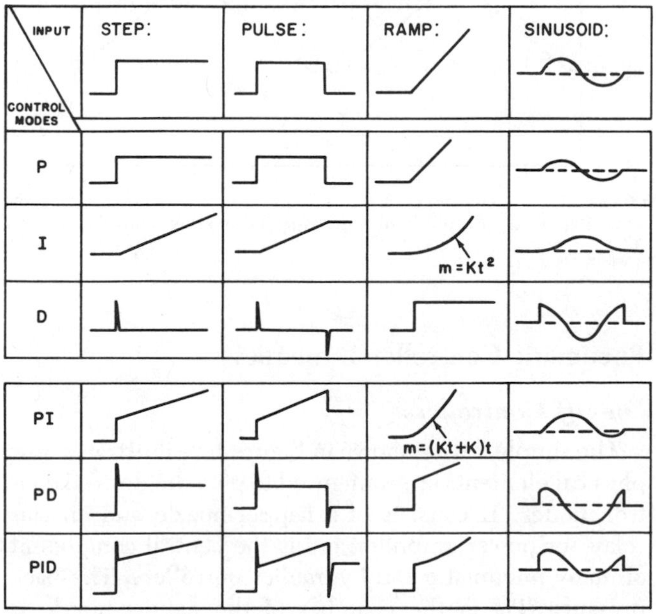

The transfer function describing most present-day analog controllers appears below:

$$\frac{m(s)}{e(s)} = \frac{K_c}{T_r s} (1 + T_r s) \left(\frac{1 + T_d s}{1 + T_d s / \alpha} \right) \qquad 1.2(6)$$

where

m = controller output,
e = error signal,
K_c = proportional gain,
s = laplace operator,
T_r and T_d = reset and derivative time constants, and
α = the maximum dynamic gain of the derivative function.

To provide conventional derivative action, α can range from 6 to 30.

Inverse derivative is used to reduce the gain of a controller at high frequencies and is therefore useful in stabilizing a flow loop. It is obtained by the same function, but with α selected as 0.5 or less. The gain of the inverse derivative controller decreases from K_c, at low frequency to a limiting value of K_c/α at high frequency. The proportional-plus-inverse-derivative controller provides high gain to minimize offset at low frequency and low gain to stabilize at high frequency.

Inverse derivative can also be added to a proportional-plus-reset controller to stabilize flow and other loops requiring very low proportional gain for stability. One should add inverse derivative only when the loop is unstable at the minimum gain setting of the proportional-plus-reset controller. Since inverse derivative is available in a separate unit, it can be added to the loop when stability problems are encountered. Interestingly, the addition of inverse derivative, properly tuned, has little effect on the natural frequency of the loop.

Two-Position (On-Off) Controllers

The two-position controller is extensively used. In its simplest form, it is the kind of control system used in domestic heating systems, refrigerators and water tanks. A perfect on-off controller is "on" when the measurement is below the set point and the manipulated variable is therefore at its maximum value. When the measured variable is above the set point, the controller is "off" and the manipulated variable is at its minimum value.

$$e > 0 \qquad m = \text{maximum value}$$
$$e < 0 \qquad m = \text{minimum value} \qquad 1.2(7)$$

In most practical applications, due to mechanical friction or arcing of electrical contacts, there is a narrow band (around zero error) that the error must pass through before a change will occur. This band is known as the differential gap, and its presence is sometimes desirable to minimize the cycling tendency of the two-position controller. Figure 1.2c shows the response of a simple on-off controller to a sinusoidal input.

Single-Speed Floating Control

For self-regulating processes with little or no capacitance, the single-speed floating controller is used. The output of this controller is either increasing or decreasing at a certain rate. Such control is commonly associated with systems in which the final control element is a single-speed reversible motor. The controller usually contains a neutral zone for which the output of the controller is zero; otherwise the manipulated variable would be changing continually in one direction or the other. The output of the reversible motor is either forward, reverse,

or off. Mathematically the single-speed floating control is expressed as follows:

$$e > +\epsilon/2 \qquad m = (T_i \times t) + M_{o1}$$
$$e < -\epsilon/2 \qquad m = -(T_i \times t) + M_{o2} \qquad 1.2(8)$$
$$-\epsilon/2 \le e \le +\epsilon/2 \qquad m = 0 + M_{o3}$$

where

$$t = \text{time,}$$
$$m = \text{controller output (manipulated variable),}$$
$$e = \text{controller input (error signal),}$$
$$\epsilon = \text{constant defining neutral zone,}$$
$$T_i = \text{controller speed constant, and}$$
$$M_{o1},\ M_{o2},\ M_{o3} = \text{constants of integration.}$$

The response of a single-speed floating controller to a sinusoidal input is shown in Figure 1.2d.

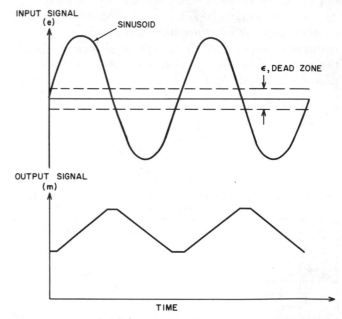

Fig. 1.2d   Response of a single-speed floating controller

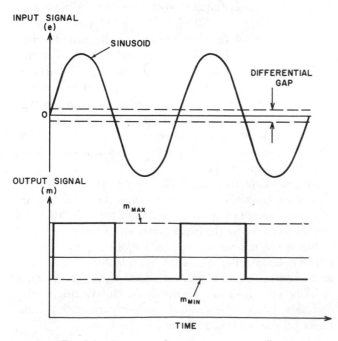

Fig. 1.2c   Response of a two-position controller

Pneumatic Controller Dynamics

On-Off Controllers

The simple device shown in Figure 1.2e illustrates how physical elements are configured to give the desired control modes. It consists of a flapper-nozzle and an "air relay" or power amplifier and is the central component of many pneumatic and hydraulic controllers. The back pressure (P_b) in the chamber of the flapper-nozzle is controlled by the position of the flapper with respect to

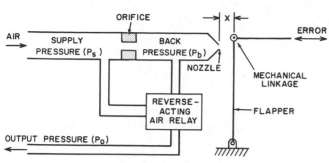

Fig. 1.2e Pneumatic two-position controller

the nozzle. If the flapper is fully closed (i.e., if $X = 0$), the back pressure will be equal to the supply pressure (P_s). If the flapper is fully open (i.e., if X is very large), the back pressure (P_b) will be approximately equal to the ambient pressure (P_a). The output of the air relay (P_o) is a direct function of the back pressure. The relay is essentially a power amplifier necessary to supply the air flow required of any pneumatic controller. The relay is termed reverse acting because for an increase in P_b there is a corresponding decrease in P_o.

In establishing the equations describing the flapper-nozzle arrangement, its operation will be assumed to be isothermal and the air will be considered as an ideal gas.

A material balance on the nozzle chamber yields

$$w_1 - w_o = \frac{d}{dt} M \qquad 1.2(9)$$

where

w_1 = deviation in weight rate of air flow into chamber, from the normal operating flow rate of W_{1i},

w_o = deviation in weight rate of air flow out of chamber from the normal operating flow rate of W_{oi}, and

M = mass of air in chamber.

The weight rate of flow into the chamber is a function of the pressure drop across the chamber, which can be considered linear over the region of interest:

$$w_1 = K_1 p_b \qquad 1.2(10)$$

The flow of air out of the nozzle chamber will be a function of the pressure P_b in the chamber and of the position X of the flapper:

$$w_o = f(P_b, X) \qquad 1.2(11)$$

In a linearized form this can be written as

$$w_o = K_2 p_b + K_3 X \qquad 1.2(12)$$

where

$$K_2 = \left. \frac{\partial w_o}{\partial P_b} \right|_i \qquad 1.2(12a)$$

and

$$K_3 = \left. \frac{\partial w_o}{\partial X} \right|_i \qquad 1.2(12b)$$

(For details of the linearization of nonlinear equations see Section 1.3.) The mass M can be defined in terms of the ideal gas law:

$$M = \frac{29 P_b V}{RT} \qquad 1.2(13)$$

or, in terms of deviations from normal operating conditions,

$$m = K_4 p_b \qquad 1.2(14)$$

where

$K_4 = 29V/RT$,

V = volume of nozzle chamber,

T = temperature in nozzle chamber,

R = universal gas constant, and

29 = molecular weight of air.

Substituting equations 1.2(10), (12) and (14) into the general material balance equation 1.2(9) yields

$$(K_1 + K_2) p_b + K_4 \frac{d}{dt} p_b = -K_3 X \qquad 1.2(15)$$

This can be written as a general first-order lag:

$$p_b + \tau \frac{d}{dt} p_b = KX \qquad 1.2(16)$$

where

$$\tau = \frac{K_4}{K_1 + K_2} \qquad 1.2(16a)$$

and

$$K = \frac{-K_3}{K_1 + K_2} \qquad 1.2(16b)$$

Considering the relatively small volume of the nozzle, the time constant τ will be negligible with respect to typical process time constants. Therefore, for all practical purposes the equation can be written as

$$-p_b = p_o = -KX \qquad 1.2(17)$$

In this equation K is typically in the neighborhood of 5,000 to 8,000 PSI/in (1.4 to 2.2 MPa/mm). This high sensitivity makes the device undesirable for proportional control, but it can be used as a simple two-position or bang-bang controller because any small positive value of X will result in the maximum value of P_o and any small negative value of X will result in the minimum value of P_o. (The terms negative and positive are used relatively to the normal flapper position.)

Proportional Controllers

The modified configuration shown in Figure 1.2f obtains proportional control from the flapper-nozzle arrangement. A feedback bellows and a spring has been

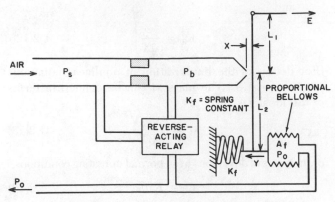

Fig. 1.2f Pneumatic proportional controller

added to the bottom of the flapper. The flapper has also been extended so that the input signal E, which positions the flapper, is mechanically linked to the flapper above the nozzle. The relative position X of the flapper to the nozzle is determined by both the position Y of the feedback bellows and the position E of the input signal.

$$X = f(E, Y) \qquad 1.2(18)$$

Using simple geometry, this can be expressed as a linear function of the deviations in X, E, and Y about normal operating conditions X_i, E_i, and Y_i.

$$X = \left(\frac{L_2}{L_1 + L_2}\right) E - \left(\frac{L_1}{L_1 + L_2}\right) Y \quad 1.2(19)$$

The position of the spring-bellows arrangement for a given pressure P_o can be determined by force balance. Since inertia effects are zero, force balance means that the force exerted by the bellows is equal to the force exerted by the spring.

$$A_f P_o = K_f Y \qquad 1.2(20)$$

In terms of deviations about a normal operating condition, equation 1.2(20) becomes:

$$A_f p_o = K_f y \qquad 1.2(21)$$

where

$$
\begin{aligned}
A_f &= \text{cross-sectional area of the bellows,} \\
K_f &= \text{spring constant,} \\
P_o \text{ and } p_o &= \text{absolute output pressure and de-} \\
&\quad \text{viation in output pressure,} \\
Y \text{ and } y &= \text{absolute position of flapper and} \\
&\quad \text{deviation in flapper position.}
\end{aligned}
$$

Substituting equations 1.2(19) and (21) into the equation for the flapper-nozzle, 1.2(17), yields:

$$p_o = -K \left(\frac{L_2}{L_1 + L_2}\right) e +$$
$$+ \left(\frac{L_1}{L_1 + L_2}\right)\left(\frac{A_f}{K_f}\right) p_o \quad 1.2(22)$$

or

$$p_o = \frac{-K\left(\dfrac{L_2}{L_1 + L_2}\right)}{1 - K\left(\dfrac{L_1}{L_1 + L_2}\right)\left(\dfrac{A_f}{K_f}\right)} e \qquad 1.2(23)$$

Since K is a very large value, the equation reduces to an expression equivalent to that of a proportional controller:

$$p_o = \frac{L_2 K_f}{L_1 A_f} e \qquad 1.2(24)$$

The addition of the internal feedback mechanism effectively reduced the sensitivity of the device to an acceptable range for proportional control. Note that the proportional gain in the above equation can be easily adjusted by the ratio L_1/L_2, which can be changed by varying the position on the flapper at which the error signal is applied.

Proportional-Plus-Derivative Controllers

With slight modifications additional modes of operation can be incorporated into the proportional controller shown in Figure 1.2f. The addition of a variable restriction in the line leading to the feedback bellows (Figure 1.2g), offers resistance to flow into the bellows and therefore creates a time lag effect.

The material balance on the bellows is

$$W_b = \frac{d}{dt} M_b \qquad 1.2(25)$$

or, in terms of deviations about a normal operating point,

$$w_b = \frac{d}{dt} m_b \qquad 1.2(26)$$

where

$$
\begin{aligned}
W_b, w_b &= \text{the weight rates of air flow into the} \\
&\quad \text{bellows and} \\
M_b, m_b &= \text{the weights of air in the bellows.}
\end{aligned}
$$

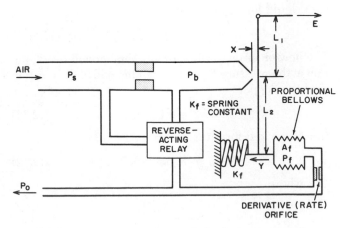

Fig. 1.2g Pneumatic proportional-plus-derivative controller

The rate of flow into the bellows will be a function of the pressure drop across the restriction in the line leading to the bellows:

$$W_b = K_5(P_o - P_f) \qquad 1.2(27)$$

Expressing the equation in terms of a linear deviation from a normal operating point yields

$$w_b = K_5(p_o - p_f) \qquad 1.2(28)$$

The mass of air can be expressed in terms of the ideal gas law:

$$M = \frac{29P_fV}{RT} \qquad 1.2(29)$$

where

V = volume of bellows and
T = temperature in the bellows.

Linearization of this equation yields

$$m = K_6p_f + K_7v \qquad 1.2(30)$$

The variation in the feedback bellows volume may be expressed in terms of the variation in bellows position

$$v = A_f y \qquad 1.2(31)$$

Equations 1.2(26), (28), (30), and (31) can be combined with the force balance on the bellows expressed by equation 1.2(32),

$$A_f p_f = K_f y \qquad 1.2(32)$$

to yield equation 1.2(33), which describes the behavior of the bellows:

$$\tau \frac{d}{dt} y + y = \frac{A_f}{K_f} p_o \qquad 1.2(33)$$

where

$$\tau = \left(\frac{K_6}{K_5} + \frac{K_7 A_f^2}{K_5 K_f}\right) \qquad 1.2(33a)$$

As before in equation 1.2(19),

$$x = \left(\frac{L_2}{L_1 + L_2}\right) e - \left(\frac{L_1}{L_1 + L_2}\right) y \qquad 1.2(34)$$

and, in accordance with equation 1.2(17),

$$p_o = -Kx \qquad 1.2(35)$$

Therefore the equation that describes the entire system is

$$p_o + \tau_s \frac{d}{dt} p_o' = K_c \left[e + T_d \frac{d}{dt} e\right] \qquad 1.2(36)$$

where

$$\tau_s = \frac{T_d}{1 - K\left(\frac{L_1}{L_1 + L_2}\right)\left(\frac{A_f}{K_f}\right)} \qquad 1.2(36a)$$

and

$$K_c = \frac{-K\left(\frac{L_2}{L_1 + L_2}\right)}{1 - K\left(\frac{L_1}{L_1 + L_2}\right)\left(\frac{A_f}{K_f}\right)} \qquad 1.2(36b)$$

But because K is very large, the constants in the equation above can be approximated as

$$\tau_s \cong 0.0 \qquad 1.2(36c)$$

$$K_c \cong \left(\frac{L_2}{L_1}\right)\left(\frac{K_f}{A_f}\right) \qquad 1.2(36d)$$

Therefore, the equation for a proportional-plus-derivative controller can be written as

$$p_o = K_c \left[e + T_d \frac{d}{dt} e\right] \qquad 1.2(37)$$

The proportional gain K_c can be adjusted in the same manner as was the proportional controller, by varying the L_2/L_1 ratio, and the derivative time T_d can be adjusted by varying the resistance of the restriction in the air line to the bellows.

Proportional-Plus-Integral Controllers

Another modification of the basic flapper-nozzle device is illustrated in Figure 1.2h, where an additional bellows included in the feedback loop is mounted to act opposite the proportional bellows.

Similar to the previous approach, a force balance and a mass balance on this bellows is utilized to determine the position of the flapper. The resulting equation 1.2(38) describes the controller dynamics:

$$y + \tau_i \frac{d}{dt} y = \left(\frac{A_f K_6}{K_f K_5}\right) \frac{d}{dt} p_o \qquad 1.2(38)$$

where τ_i is the time constant of the bellows as defined earlier by equation 1.2(33a):

$$\tau_i = \left(\frac{K_6}{K_5} + \frac{K_7 A_f^2}{K_5 K_f}\right) \qquad 1.2(33a)$$

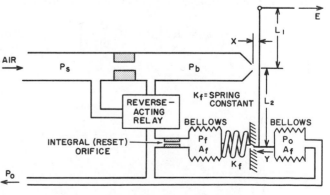

Fig. 1.2h Pneumatic proportional-plus-integral controller

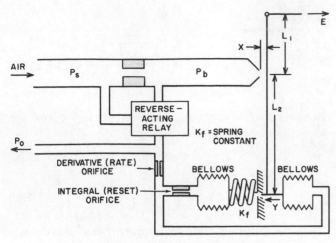

Fig. 1.2i   Pneumatic proportional-plus-integral-plus-derivative controller

All other equations remain unchanged, and the entire system equation can be written as

$$G_1 p_o + \frac{d}{dt} p_o = K_c \left(e + T_i \frac{d}{dt} e \right) \qquad 1.2(39)$$

where

$$G_1 = \frac{1}{\tau - K \left(\dfrac{L_1}{L_1 + L_2} \right) \left(\dfrac{A_f K_6}{K_f K_5} \right)} \qquad 1.2(39a)$$

$$K_c = G_2 = \frac{-K \left(\dfrac{L_2}{L_1 + L_2} \right)}{\tau - K \left(\dfrac{L_1}{L_1 + L_2} \right) \left(\dfrac{A_f K_6}{K_f K_5} \right)} \qquad 1.2(39b)$$

But because K is very large, the constants in equation 1.2(39) can be approximated as

$$G_1 \cong 0.0 \qquad 1.2(39c)$$

$$K_c = G_2 \cong \frac{L_2 K_f K_5}{L_1 A_f K_6} \qquad 1.2(39d)$$

Therefore the simplified equation can be written as

$$\frac{d}{dt} p_o = K_c \left[e + T_i \frac{d}{dt} e \right] \qquad 1.2(40)$$

Integrating both sides yields an equation in the more familiar form of a proportional-plus-integral controller:

$$p_o = K_c \left[e + \frac{1}{T_i} \int e\, dt \right] \qquad 1.2(41)$$

PID Controllers

In a similar manner, the basic pneumatic elements can be modified to yield a response approximating the proportional-plus-integral-plus-derivative action (Figure 1.2i).

The descriptive equation for such a three-mode controller is

$$p_o = K_c \left[e + \frac{1}{T_i} \int e\, dt + T_d \frac{d}{dt} e \right] \qquad 1.2(42)$$

BIBLIOGRAPHY

Andreiev, N., "A New Dimension: A Self-Turning Controller That Continually Optimizes PID Constants," *Control Engineering*, August, 1981.

Beard, M.J., "Analog Controllers Develop New Wrinkles," *Instruments & Control Systems*, November, 1977.

Buckley, P.S., "Dynamic Design of Pneumatic Control Loops: Parts I and II," *InTech*, April and June, 1975.

———, "A Modern Perspective on Controller Tuning," Texas A&M 30th Annual Symposium on Instrumentation for the Process Industries, January, 1973.

———, *Techniques of Process Control*, John Wiley & Sons, New York, 1964.

Caldwell, W.I., Coon, G.A., and Zoss, L.M., *Frequency Response for Process Control*, McGraw-Hill Book Co., New York, 1959.

Eckman, D.P., *Automatic Process Control*, John Wiley & Sons, New York, 1958.

Harriott, P., *Process Control*, McGraw-Hill Book Co., New York, 1964.

Jones, B.E., *Instrumentation, Measurement and Feedback*, McGraw-Hill Book Co., New York, 1977.

Kaminski, R.K., and Talbot, J.E., "Electronic Controllers for Process Industries," *InTech*, October, 1968.

Lipták, B.G., "How to Set Process Controllers," *Chemical Engineering*, November 23, 1964.

Luyben, W.L., "A Proportional-Lag Level Controller," *Instrumentation Technology*, December 1977.

McCauley, A.P., Jr., and Persik, S.R., "From Linear to Gain-Adaptive Control: A Case History," *Instrumentation Technology*, November, 1977.

Merritt, R., "Electronic Controller Survey," *Instrumentation Technology*, June, 1977.

Randhawd, R.S., "Understanding the Lazy Integral," *Instruments & Control Systems*, January, 1983.

Shinskey, F.G., "Control Topics," Publ. No. 413-1 to 413-8, Foxboro Company.

———, Process Control Systems, McGraw-Hill Book Co., New York, 1979.

1.3 TRANSFORM THEORY, TRANSFER FUNCTIONS, LINEARIZATION, AND STABILITY ANALYSIS

Most techniques used in the analysis of control problems are dependent on the existence of descriptive mathematical equations. In this section the reader will find some of the mathematical tools needed for the analysis of control problems.

One of the main topics of this section is therefore the subject of transforms, which allows for simple handling of difficult problems. As the use of logarithms simplifies the handling of multiplications to a requirement for addition only, so do Laplace and z transforms perform a similar function in the solution of differential and difference equations respectively.

Another useful method for expressing multiple input multiple output difference and differential equations is referred to as the state space representation. This mathematical convenience permits large systems to be handled as a first-order matrix system using linear algebra techniques.

Some of the other control system analysis tools discussed here are the block diagram for signal flow representation and linearization, which is a technique to convert nonlinear equations into linear form. There are several graphical techniques that are also extremely useful in the analysis dynamic systems. The Bode diagram, Nyquist plot, and Nyquist array methods are covered in this section.

One of the main purposes of control system analysis is to guarantee that the particular system will be stable in its operation. Stability criteria discussed in this section include the Descartes, Routh's, and Nyquist criterions.

Laplace Transforms

One very useful tool in the analysis of differential equations is the principle of Laplace transforms. The Laplace transform concept is widely used in process control and provides the basic framework upon which most automatic control theory is based.

The principle of any transform operation is to transform a difficult problem to a form more convenient to handle. Once the desired manipulations or results have been obtained from the transformed problem, an inverse transformation can be made to determine the solution of the original problem. For example, logarithms are a transform operation by which problems of multiplication and division can be transformed to problems of addition and subtraction. Laplace transforms perform a similar function in the solution of differential equations. The Laplace transform of a linear ordinary differential equation results in a linear algebraic equation. The algebraic problem is usually much simpler to solve than the corresponding differential equation. Once the Laplace domain solution has been found, the corresponding time domain solution can be determined by an inverse transformation. There also exist very powerful control analysis and design techniques available for s-domain and frequency domain models. These include Bode and Nyquist analyses to be discussed later.

The Laplace transform of a time domain function $f(t)$ will be noted by the symbol $F(s)$, defined as follows:

$$F(s) = \mathscr{L}[f(t)] = \int_0^\infty f(t)e^{-st}dt \qquad 1.3(1)$$

where $\mathscr{L}[f(t)]$ is the symbol for indicating the Laplace transformation of the function in brackets. The variable s is a complex variable ($s = a + jb$) introduced by the transformation. All time-dependent functions in the time domain become functions of s in the Laplace domain (s-domain).

For the mathematical concept of the Laplace transformation to be meaningful, certain restrictions are placed on the function $f(s)$. However, in most practical control work no such difficulties are encountered and therefore this will not be considered in this treatment. Only concepts normally used in process control will be covered.

A number of theorems exist that facilitate the use of Laplace transform techniques. The following are some of the most useful ones.

Linearity theorem:

$$\mathscr{L}[K\,f(t)] = K\mathscr{L}[f(t)] = K\,F(s) \quad (K = \text{constant}) \qquad 1.3(2)$$

$$\mathscr{L}[f_1(t) \pm f_2(t)] = F_1(s) \pm F_2(s) \qquad 1.3(3)$$

Table 1.3a
LAPLACE TRANSFORM PAIRS

Transform, $F(s)$	Function, $f(t)$	Transform, $F(s)$	Function, $f(t)$
1	$\delta(t)$	$\dfrac{s + a}{(s + a)^2 + b^2}$	$e^{-at} \cos bt$
$\dfrac{1}{s}$	$u(t)$		
$\dfrac{1}{s^n}$ $(n = 1, 2, \ldots)$	$\dfrac{t^{n-1}}{(n-1)!}$	$\dfrac{1}{(s + a)^n}$	$\dfrac{1}{(n-1)!} t^{n-1} e^{-at}$
$\dfrac{1}{s \pm a}$	$e^{\mp at}$	$\dfrac{ab}{(s + a)(s + b)}$	$\dfrac{1}{b - a}(e^{-at} - e^{-bt})$
$\dfrac{1}{s(s \pm a)}$	$\dfrac{1}{\pm a}(1 - e^{\mp at})$	$\dfrac{e^{-as}}{s}$	$u(t - a)$
$\dfrac{s}{s^2 + a^2}$	$\cos at$	$\dfrac{e^{-as}}{s^2}$	$\begin{cases} 0 & (0 < t < a) \\ t - a & (t > a) \end{cases}$
$\dfrac{a}{s^2 + a^2}$	$\sin at$	$\dfrac{1 - e^{-as}}{s}$	$\begin{cases} 1 \ (0 < t < a) \\ 0 \ (t > a) \end{cases}$
$\dfrac{s}{s^2 - a^2}$	$\cosh at$	$\log \dfrac{s - a}{s - b}$	$\dfrac{1}{t}(e^{bt} - e^{at})$
$\dfrac{a}{s^2 - a^2}$	$\sinh at$	$\tan^{-1} \dfrac{a}{s}$	$\dfrac{1}{t} \sin at$
$\dfrac{1}{(s + a)^2 + b^2}$	$\dfrac{1}{b} e^{-at} \sin bt$		

Real differentiations theorem:

First-order differential

$$\mathscr{L}\left[\frac{d}{dt} f(t)\right] = sF(s) - f(0) \qquad 1.3(4)$$

General nth-order differential

$$\mathscr{L}\left[\frac{d^n}{dt^n} f(t)\right] = s^n F(s) -$$

$$- s^{n-1} f(0) - s^{n-2} \frac{d}{dt} f(0) -$$

$$- s \frac{d^{n-2}}{dt^{n-2}} f(0) - \frac{d^{n-1}}{dt^{n-1}} f(0) \quad 1.3(5)$$

Real integration theorem:

$$\mathscr{L}\left[\int f(t)\, dt\right] = \frac{F(s)}{s} \qquad 1.3(6)$$

In general,

$$\mathscr{L}\left[\int^1 \int^2 \cdots \int^n f(t)\, dt^n\right] = \frac{1}{s^n} F(s) \quad 1.3(7)$$

Initial value theorem:

$$f(0) = \lim_{s \to \infty} s\, F(s) \qquad 1.3(8)$$

Final value theorem:

$$f(\infty) = \lim_{s \to 0} s\, F(s) \qquad 1.3(9)$$

The direct and inverse Laplace transformation of a particular function can be obtained from direct integration of equation 1.3(1) and/or the applications of the above theorems. Extensive tabulations of specific transform pairs are also available in most mathematical tables. A few of the more common transform pairs encountered in control analysis are tabulated in Table 1.3a.

First-Order Lag

The following two examples illustrate the solution of differential equations using the underlying principle of the Laplace transform technique. First consider a simple first-order lag described earlier:

$$\tau \frac{d}{dt} c(t) + c(t) = K r(t) \qquad 1.3(10)$$

where

$$c(0) = 0.0$$

and

$$r(t) = \begin{cases} 0 & t < 0 \\ 1 & t > 0 \end{cases} \qquad 1.3(11)$$

The general procedure in the solution of the above equation will be to

1. Transform the differential equation to the Laplace domain.
2. Solve the resulting algebraic equations for the system output, C(s).
3. Take inverse Laplace transformation of the expression describing C(s) to determine the corresponding time domain solution.

Since Theorem 1 indicates linearity, each term in the differential equation can be transformed individually:

$$\mathscr{L}\left[\tau \frac{d}{dt} c(t)\right] + \mathscr{L}[c(t)] = \mathscr{L}[K\ r(t)] \qquad 1.3(12)$$

The first term can be determined by the use of Theorems 1 and 2:

$$\mathscr{L}\left[\tau \frac{d}{dt} c(t)\right] = \tau\mathscr{L}\left[\frac{d}{dt} c(t)\right] = \tau(sC(s) = c(\cancel{0})^0) \qquad 1.3(13)$$

The second term by definition is

$$\mathscr{L}[c(t)] = C(s) \qquad 1.3(14)$$

The third term can be determined from the table of transform pairs in Table 1.3a.

$$\mathscr{L}[K\ r(t)] = \frac{K}{s} \qquad 1.3(15)$$

The entire Laplace domain equation therefore will read

$$\tau s\ C(s) + C(s) = \frac{K}{s} \qquad 1.3(16)$$

This can be solved for $C(s)$:

$$C(s) = \frac{K}{s(1 + \tau s)} \qquad 1.3(17)$$

For the inverse transformation we note from Table 1.3a the following transform pair:

$$\mathscr{L}\left[\frac{1}{\pm a} (1 - e^{\pm at})\right] = \frac{1}{s(s \pm a)} \qquad 1.3(18)$$

Therefore,

$$c(t) = K(1 - e^{-t/\tau}) \quad t > 0 \qquad 1.3(19)$$

Partial Fraction Expansion

Frequently the necessary inverse transformation may not be directly available in the Laplace transform tables at hand. In such cases the function must be expanded in terms of the roots of the denominator of the Laplace expression, namely,

$$F(s) = \frac{A(s)}{B(s)} = \frac{C_1}{(s + r_1)} + \frac{C_2}{(s + r_2)} +$$
$$+ \frac{C_3}{(s + r_3)} + \cdots + \frac{C_0}{(s + r_0)} \qquad 1.3(20)$$

where

$$B(s) = (s - r_1)(s - r_2)(s - r_3)$$
$$\cdots (s - r_n)$$
$$r_1, r_2, r_3, \ldots r_n = \text{roots of } B(s)$$
$$C_1, C_2, C_3, \ldots C_n = \text{constants in partial}$$
$$\text{fraction} \qquad 1.3(21)$$

The procedure for evaluating the constants in the expansion depends on the nature of the roots, which can be (1) real and distinct, (2) real and repeated, (3) complex conjugates.

When the roots are real and distinct, the expansion of $F(s)$ is

$$F(s) = \frac{A(s)}{(s - r_1)(s - r_2) \cdots (s - r_n)} \qquad 1.3(22)$$

$$F(s) = \frac{C_1}{(s - r_1)} + \frac{C_2}{(s - r_2)} +$$
$$+ \cdots + \frac{C_n}{(s - r_n)} \qquad 1.3(23)$$

and the inverse transformation is

$$f(t) = C_1 e^{r_1 t} + C_2 e^{r_2 t} + \cdots + C_n e^{r_n t} \qquad 1.3(24)$$

where

$$\left.\begin{aligned} C_1 &= \lim_{s \to r_1} [(s - r_1)\ F(s)] \\[4pt] C_2 &= \lim_{s \to r_2} [(s - r_2)\ F(s)] \\ &\quad\vdots \\ C_n &= \lim_{s \to r_n} [(s - r_n)\ F(s)] \end{aligned}\right\} \qquad 1.3(25)$$

For the case when the roots are real and repeated,

$$F(s) = \frac{A(s)}{(s - r_1) \cdots (s - r_j)^q \cdots (s - r_n)} \qquad 1.3(26)$$

(The jth root is repeated q times.)

$$F(s) = \frac{C_1}{(s - r_1)} + \cdots +$$
$$+ \left[\frac{C_q'}{(s - r_j)^q} + \frac{C_{q-1}'}{(s - r_j)^{q-1}} + \cdots + \frac{C_1'}{(s - r_j)}\right] +$$
$$+ \cdots + \frac{C_n}{(s - r_n)} \qquad 1.3(27)$$

The inverse transform is

$$f(t) = C_1 e^{r_1 t} + \cdots + [C_q' t^{q-1} + C_{q-1}' t^{q-2} +$$
$$+ \cdots + C_1'] e^{r_j t} + \cdots + C_1 e^{r_n t} \qquad 1.3(28)$$

where

$$C_q' = \lim_{s \to r_j} \{(s - r_j)^q\ F(s)\} \qquad 1.3(29)$$

$$C_{q-1}' = \lim_{s \to r_j} \left\{\frac{1}{1!} \frac{d}{ds} [(s - r_j)^q\ F(s)]\right\} \qquad 1.3(30)$$

$$C_{q-k} = \lim_{s \to r_j} \left\{\frac{1}{k!} \frac{d^k}{ds^k} [(s - r_j)^q\ F(s)]\right\} \qquad 1.3(31)$$

where $k! = 1 \cdot 2 \cdot 3 \cdot 4 \cdot 5 \cdots k$.

For the case when the roots are complex,

$F(s) =$

$$= \frac{A(s)}{(s - r_1) \cdots (s - a - jb)(s - a + jb) \cdots (s - r_n)}$$
$$1.3(32)$$

$$F(s) = \frac{C_1}{(s - r_1)} + \cdots +$$
$$+ \frac{1}{2jb} |\pi_j| \left(\frac{e^{j\alpha}}{s - a - jb} - \frac{e^{-j\alpha}}{s - a + jb} \right) +$$
$$+ \cdots + \frac{C_n}{(s - r_n)} \quad 1.3(33)$$

The inverse transformation results in

$$f(t) = C_1 e^{r_1 t} + \cdots + \frac{1}{b} |\pi_j| \, e^{at} \sin(bt + \alpha) +$$
$$+ \cdots + C_n e^{r_n t} \quad 1.3(34)$$

$$\pi_j = \lim_{s \to a + jb} [(s - a - jb)(s - a + jb) F(s)] \quad 1.3(35)$$

This results in a complex polynomial that can be expressed in terms of a magnitude, $|\pi_j|$, and an angle, α. For example,

$$F(s) = \frac{10}{s(s^2 + 0.5s + 0.4)} =$$
$$= \frac{10}{s(s + 0.25 - j0.58)(s + 0.25 + j0.58)} \quad 1.3(36)$$

$$F(s) = \frac{C_1}{s} +$$
$$+ \frac{1}{2jb} |\pi_j| \left(\frac{e^{j\alpha}}{s - a - jb} - \frac{e^{-j\alpha}}{s - a + jb} \right) \quad 1.3(37)$$

where

$$a = -0.25$$
$$b = 0.58$$

$$C_1 = \lim_{s \to 0} \left[s \left(\frac{10}{s(s^2 + 0.5s + 0.4)} \right) \right] = 25 \quad 1.3(38)$$

$$\pi_j = \lim_{s \to 0.25 - j0.58}$$
$$\left[(s^2 + 0.5s + 0.4) \left(\frac{10}{s(s^2 + 0.5s + 0.4)} \right) \right] =$$
$$= \frac{10}{-0.25 + j0.58} \quad 1.3(39)$$

$$|\pi_j| = \frac{10}{\sqrt{(0.25)^2 + (0.58)^2}} = \frac{10}{0.63} = 15.9 \quad 1.3(40)$$

$$\alpha = \text{angle of } \pi_j = 0 - \tan^{-1} \frac{0.58}{-0.25} = -113.5$$
$$1.3(41)$$

Therefore,

$$c(t) = 25 + \left(\frac{15.9}{0.58} \right) e^{-0.25t} \sin(0.58t - 113.5°)$$
$$1.3(42)$$

z Transforms

We have seen that the Laplace transformation is a powerful tool for the solution of linear *differential* equations and is particularly useful when these equations represent the dynamic behavior of continuous systems. For certain kinds of discontinuous systems, whose dynamic behavior can be defined by linear *difference* equations, there exists another kind of transformation calculus, the z-transformation. It is particularly applicable to the study and design of sampled data control. Let us suppose that we have a continuous function, f(t), such as that shown in Figure 1.3b(1) and that this signal is to be fed into a sampling device such as that shown in Figure 1.3b(2).

At fixed intervals of time, period being T, the switch closes briefly, allowing the signal f(t) to be transmitted. If the length of time the switch is closed is infinitesimally short, the switch or sampler may be treated as an impulse modulator. That is to say, the sampler puts out a train of impulses I(t), spaced from one another by the time interval T (Figure 1.3b(3)). The intermittent function at the switch output is designated f*(t).

The sampling is represented mathematically

$$I(t) = U'(t - T) + U'(t - 2T) + U'(t - 3T) +$$
$$+ \cdots + U'(t - kT) + \cdots \quad 1.3(43)$$

$$= \sum_{n=0}^{+\infty} U'(t - nT) \quad 1.3(44)$$

where U' is the unit impulse. We may now define f*(t) as the product of the input signal and the switching function:

$$f*(t) = f(t)I(t) \quad 1.3(45)$$

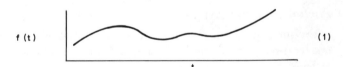
f (t)   (1)

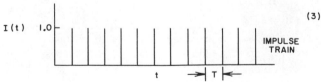
(2)

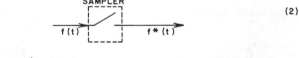
I (t) 1.0   (3)   IMPULSE TRAIN

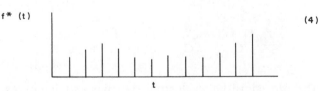
f* (t)   (4)

Fig. 1.3b   Sampling of continuous signals

Substituting equation 1.3(44) into equation 1.3(45) we obtain

$$f^*(t) = \sum_{n=0}^{\infty} f(nT)U'(t-nT) \qquad 1.3(46)$$

The Laplace transformation of both sides leads to

$$f^*(s) = \sum_{n=0}^{\infty} f(nT)e^{-nTs} \qquad 1.3(47)$$

The form of this equation is not particularly convenient for the study of sampled data control systems. Let us therefore introduce a new variable z, defined

$$z = e^{Ts} \qquad 1.3(48)$$

Then by substitution into equation 1.3(47) we obtain the z transform.

$$f^*(s) = \sum_{n=0}^{\infty} f(nT)z^{-n} \qquad 1.3(49)$$

$$= F(z) \qquad 1.3(50)$$

F(z) is seen to be an infinite series in powers of z^{-1} Three useful theorems will now be given.

1. The shifting theorem. If $z[f(t)] = F(z)$ then

$$z[f(t+T)] = z[F(Z) - f(0)] \qquad 1.3(51)$$

2. The initial value theorem

$$\lim_{t \to 0} f^*(t) = \lim_{z \to \infty} f(z) \qquad 1.3(52)$$

3. The final value theorem

$$\lim_{t \to \infty} f^*(t) = \lim_{z \to 1} \left(\frac{z-1}{z} \right) F(z) \qquad 1.3(53)$$

If the sampling frequency is at least twice as high as the highest frequency of interest in the continuous input signal, the information contained in the z transform of the sampler output is equivalent to that contained in the Laplace transform of the sampler output. To put it another way, if T is the sampling period then a complete reconstruction of the input signal can be obtained from the sampler output signal for input sinusoidal frequencies up to $\omega = (2\pi/T)1/2 = \pi/T$ radians/unit time.

As with the Laplace transform, the z transforms and inverse z transform pairs are tabulated extensively. Table 1.3c contains a partial list of these transform pairs.

First-Order Lag (with Ramped Input Function)

Solve the following first-order linear difference equation, given $w(0) = -1$:

$$w^*(t + T) + 2w^*(t) = t^* \qquad 1.3(54)$$

Table 1.3c
Z TRANSFORM PAIRS

$f(t)$	$F(z) = \mathscr{Z}[f(t)] = \sum_{n=0}^{\infty} f(nT)z^{-n}$
$\delta(t - nT)$	$\dfrac{1}{z^n}$
$U(t)$	$\dfrac{z}{z-1}$
ϵ^{at}	$\dfrac{z}{z - \epsilon^{aT}}$
$\sin \omega t$	$\dfrac{z \sin \omega T}{z^2 - 2z \cos \omega T + 1}$
$\cos \omega t$	$\dfrac{z(z - \cos \omega T)}{z^2 - 2z \cos \omega T + 1}$
$\sinh bt$	$\dfrac{z \sinh bT}{z^2 - 2z \cosh bT + 1}$
$\cosh bt$	$\dfrac{z(z - \cosh bT)}{z^2 - 2z \cosh bT + 1}$
t	$\dfrac{Tz}{(z-1)^2}$
t^2	$\dfrac{T^2 z(z+1)}{(z-1)^3}$
$\epsilon^{-at}f(t)$	$F(\epsilon^{aT}z)$
$f(t + T)$	$z[F(z) - f(0)]$
$f(t + 2T)$	$z^2[F(z) - f(0)] - zf(T)$
$f(t - nT)U(t - nT)$	$z^{-n}F(z)$

Taking the z transformation of equation 1.3(54), we have

$$z[W(z) + 1] + 2W(z) = \frac{Tz}{(z-1)^2}$$

or

$$(z + 2) W(z) = \frac{Tz}{(z-1)^2} - z$$

from which W(z) can be solved:

$$W(z) = \frac{Tz}{(z + 2)(z - 1)^2} - \frac{z}{z + 2} \qquad 1.3(55)$$

To find the inverse transform of the first term, we expand it (without the constants and the z in the numerator) in partial fractions:

$$\frac{1}{(z + 2)(z - 1)^2} =$$

$$= \frac{K_1}{z + 2} + \frac{K_{22}}{(z - 1)^2} + \frac{K_{21}}{z - 1} \qquad 1.3(56)$$

Using the method from the past sections we find

$$K_1 = \frac{1}{(z-1)^2}\bigg|_{z=-2} = \frac{1}{9},$$

$$K_{22} = \frac{1}{z+2}\bigg|_{z=1} = \frac{1}{3},$$

$$K_{21} = \left[\frac{d}{dz}\left(\frac{1}{z+2}\right)\right]_{z=1} = -\frac{1}{9}$$

Substituting these constants in equation 1.3(56), we can now write equation 1.3(55) in its partial-fraction form:

$$W(z) = \frac{T}{9}\left[\frac{z}{z+2} + \frac{3z}{(z-1)^2} - \frac{z}{z-1}\right] - \frac{z}{z+2}$$

$$1.3(57)$$

Finally, we obtain from the transform Table 1.3c

$$w^*(t) = \frac{T}{9}\left[\left(1 - \frac{9}{T}\right)(-2)^{t/T} + 3\frac{t}{T} - 1\right]^*$$

$$= \frac{T}{9}\sum_{n=0}^{\infty}\left[\left(1 - \frac{9}{T}\right)(-2)^n + 3n - 1\right]\delta(t - nT)$$

$$1.3(58)$$

The solution in equation 1.3(58) is given in the form of a train of impulses of varying strength.

Inverse z Transform

There are two basic ways of performing the inverse z transformation. The first of these is analogous to the inverse Laplace transformation (i.e., partial fractions). The z transform, $F(z)$, is expressed in partial-fraction terms, each sufficiently simple that it may be found in transform tables.

The other method involves expressing $F(z)$ as a sequence of z^{-1}, by dividing the denominator of $F(z)$ into the numerator. Then

$$F(z) = a_0 + \frac{a_1}{z} + \frac{a_2}{z^2} + \frac{a_3}{z^3} + \frac{a_4}{z^4} + \cdots \frac{a_n}{z^n} + \cdots$$

$$1.3(59)$$

$$\frac{1}{z^k} = e^{-kTs} \qquad 1.3(60)$$

$$\mathscr{L}^{-1}\left(\frac{1}{z^k}\right) = U'(t - kT) \qquad 1.3(61)$$

$$f^*(t) = a_0 U'(0) + a_1 U'(t - T) + a_2 U'(t - 2T) + \cdots$$

$$1.3(62)$$

This method therefore leads to an intermittent function, a series of impulses at the sampling instants. $f^*(t)$ may therefore be written $f^*(nT)$. Note that the sequence $f^*(t)$ or $f^*(nT)$ may or may not be convergent and that the coefficients $a_0, a_1, a_2, \ldots a_n$ are the amplitudes of $f(t)$ at $t = 0, t, 2T, \ldots, nT$.

As an example, find

$$Z^{-1}\left[\frac{Tz}{(z-1)^2}\right]$$

Before we perform the long division, it is advantageous to arrange both the numerator and the denominator in ascending powers of z^{-1}:

$$\frac{Tz}{(z-1)^2} = T\frac{z^{-1}}{(1-z^{-1})^2} = T\frac{z^{-1}}{1 - 2z^{-1} + z^{-2}} =$$
$$= T(z^{-1} + 2z^{-2} + 3z^{-3} + \cdots + kz^{-k} + \cdots) =$$
$$= \sum_{n=0}^{\infty}(nT)z^{-n} \qquad 1.3(63)$$

Hence

$$f^*(t) = Z^{-1}\left[\frac{Tz}{(z-1)^2}\right] =$$
$$= \sum_{n=0}^{\infty}(nT)\,\delta(t - nT) \qquad 1.3(64)$$

State Space Representation

Application and design of control systems is predicated on the description of the "process" by means of difference and/or differential equations. This section introduces state space notation as a convenient formulation for representing these mathematical systems. State space notation allows nth-order difference/differential equations and coupled sets of difference/differential equations to be expressed as vector-matrix equations. This then enables these dynamic systems to be manipulated, transformed, and studied by means of simple linear algebraic procedures. This section will also develop and demonstrate state space representations for a simple second-order system and for a 2×2 (two input two output) multivariable system.

Vector and Matrix Operations

This section requires an understanding of definitions and operations pertinent to the application of vector and matrix methods in modern control. The reader requiring background in this area is referred to Reference 1. In the equations below, vector quantities are underscored and matrices are designated by capital letters.

Second-Order Model

A general continuous state space model is described by a system of linear differential equations of first order.

$$\text{State equation } \underline{\dot{x}}(t) = A\underline{x}(t) + B\underline{u}(t) \qquad 1.3(65)$$

$$\text{Output equation } \underline{y}(t) = C\underline{x}(t) \qquad 1.3(66)$$

$\underline{x}(t)$ – $n \times 1$ state variables
$\underline{u}(t)$ – $m \times 1$ manipulated variables
$\underline{y}(t)$ – $j \times 1$ output variables
A – $n \times n$ state parameter matrix

B – m × m input parameter matrix

C – j × j output parameter matrix

The discrete state space formulation is completely analogous.

$$\text{State equation } \underline{x}(n + 1) = A\underline{x}(n) + B\underline{u}(n) \quad 1.3(67)$$

$$\text{Output equation } \underline{y}(n) = C\underline{x}(n) \quad 1.3(68)$$

These are very simple state space models intended for explanation. Various augmented forms are of more practical use in general application.[2]

The $\underline{u}(t)$ vector is the manipulated or disturbance input. It can be thought of as a list of variables that are used to control or disturb our dynamic process. The $\underline{y}(t)$ vector is the output or measured variable vector. It can be thought of as a list of things we can see or measure about our process. The $\underline{x}(t)$ vector is the state vector. The state vector is of dimension consistent with the order of our dynamic process. It can be thought of as the output of the integrators or delay elements in a dynamic block diagram or signal flow diagram. The conversion of a simple second-order difference equation into discrete state space representation will help illustrate this concept.

Consider the following second-order linear difference equation:

$$w(n + 2) + b_1 w(n + 1) + b_0 w(n) = e(n) \quad 1.3(69)$$

Rearranging equation 1.3(69)

$$w(n + 2) = -b_1 w(n + 1) - b_0 w(n) + e(n) \quad 1.3(70)$$

It is also true from the definition of the time delay that

$$w(n + 1) = w(n) \quad 1.3(71)$$

Define the state vector $\underline{x}(n)$ as

$$\underline{x}(n) = \begin{bmatrix} w(n) \\ w(n + 1) \end{bmatrix} \quad 1.3(72)$$

From equation 1.3(72) it is apparent that

$$\underline{x}(n + 1) = \begin{bmatrix} w(n + 1) \\ w(n + 2) \end{bmatrix} \quad 1.3(73)$$

Define the input vector $\underline{u}(n)$ as

$$\underline{u}(n) = \begin{bmatrix} 0 \\ e(n) \end{bmatrix} \quad 1.3(74)$$

Define the output vector $\underline{y}(n)$ as

$$\underline{y}(n) = \begin{bmatrix} w(n + 2) \\ 0 \end{bmatrix} \quad 1.3(75)$$

Our state system can now be represented as

$$\underline{x}(n + 1) = A \, \underline{x}(n) + \underline{b} \, \underline{u}(n) \quad 1.3(76)$$

$$\underline{y}(n) = \underline{c}\underline{x}(n) \quad 1.3(77)$$

or combining equations 1.3(70), 1.3(71), 1.3(72) and 1.3(73)

$$x_1(n + 1) = w(n + 1) = w(n) \quad 1.3(78)$$

$$x_2(n + 1) = w(n + 2) =$$
$$= -b_1 w(n + 1) - b_0 w(n) + e(n) \quad 1.3(79)$$

$$y_1(n) = x_2(n) \quad 1.3(80)$$

Rewriting yields

$$x_1(n + 1) = x_1(n) + 0 \, x_2(n) + 0 \, e(n) \quad 1.3(81)$$

$$x_2(n + 1) = -b_0 \, x_1(n) - b_1 x_2(n) + 1 \, e(n) \quad 1.3(82)$$

$$y_1(n) = 0 \, x_1(n) + 1 \, x_2(n) + 0 \, e(n) \quad 1.3(83)$$

Putting in vector-matrix notation gives the following expression:

$$\underline{x}(n + 1) = \underbrace{\begin{bmatrix} 1 & 0 \\ -b_0 & -b_1 \end{bmatrix}}_{A} \underline{x}(n) + \underbrace{\begin{bmatrix} 0 \\ 1 \end{bmatrix}}_{\underline{b}} e(n) \quad 1.3(84)$$

$$\underline{y}(n) = \underbrace{\begin{bmatrix} 1 \\ 0 \end{bmatrix}}_{\underline{c}} x(n) \quad 1.3(85)$$

The second-order difference equation 1.3(79) is now a first-order matrix difference equation of dimension two. This state space formulation is convenient for use in many control analysis methods detailed in this and later sections.

Multiple Input Multiple Output

State space notation is particularly useful in representing multivariable dynamic systems. Multivariable implies that the system under consideration has more than one input (disturbance and/or manipulated) and more than one output (measured and/or control). Multiple input multiple output systems are typically described by sets of coupled difference or differential equations.

Consider the following set of equations:

$$\frac{dL_1}{dt} = f_1(w_1) - f_2(w_2, L_1, L_2) \quad 1.3(86)$$

$$\frac{dL_2}{dt} = f_3(w_1) - f_4(L_2) \quad 1.3(87)$$

These equations are coupled because L_2 affects L_1 through f_2 of equation 1.3(86) and L_1 affects both L_1 and L_2 through f_1 and f_3.

Assume that this coupled set of equations can be linearized (see discussion later on linearization). The linearized versions of equations 1.3(86) and 1.3(87) are given below.

$$\frac{dL_1}{dt} = a_1 L_1 + a_2 L_2 + b_1 w_1 + b_2 w_2 \quad 1.3(88)$$

$$\frac{dL_2}{dt} = a_3 L_1 + a_4 L_2 + b_3 w_1 + b_4 w_2 \quad 1.3(89)$$

Define

$$\underline{x} = \begin{bmatrix} L_1 \\ L_2 \end{bmatrix}, \quad \underline{u} = \begin{bmatrix} w_1 \\ w_2 \end{bmatrix}, \quad \underline{\dot{x}} = \begin{bmatrix} \dfrac{dL_1}{dt} \\ \dfrac{dL_2}{dt} \end{bmatrix} \qquad 1.3(90)$$

Assume we can measure L_1 and L_2 so that

$$\underline{y} = \begin{bmatrix} L_1 \\ L_2 \end{bmatrix} \qquad 1.3(91)$$

The multiple input multiple output state space vector-matrix formulation of equations 1.3(88) and 1.3(89) is

$$\underline{\dot{x}} = \underbrace{\begin{bmatrix} a_1 & a_2 \\ a_3 & a_4 \end{bmatrix}}_{A} \underline{x} + \underbrace{\begin{bmatrix} b_1 & b_2 \\ b_3 & b_4 \end{bmatrix}}_{B} \underline{u} \qquad 1.3(92)$$

$$\underline{y} = \underbrace{\begin{bmatrix} 1 & 0 \\ 0 & 1 \end{bmatrix}}_{C = I} \underline{x} \qquad 1.3(93)$$

Transfer Function

A notation often used to describe the dynamics of a particular process or system is the transfer function. For a system with an input $r(t)$ and an output $c(t)$, the transfer function $G(s)$ is defined as the ratio of the Laplace transform of the output of the system $C(s)$ divided by the Laplace transform of the input to the system $R(s)$.

$$G(s) = \frac{C(s)}{R(s)} \qquad 1.3(94)$$

An inherent assumption in applying the transfer function is that the process is initially at steady state, meaning that

$$\frac{d}{dt} c(0) = \frac{d^2}{dt^2} c(0) = \cdots = \frac{d^n}{dt^n} c(0) = 0.0 \qquad 1.3(95)$$

$$\frac{d}{dt} r(0) = \frac{d^2}{dt^2} r(0) = \cdots = \frac{d^n}{dt^n} r(0) = 0.0 \qquad 1.3(96)$$

and

$$c(0) = K r(0) \qquad 1.3(97)$$

where K is the steady-state gain of the process. Under such conditions the Laplace transform of a particular differential equation can be obtained by substituting

$$s \leftrightarrow d/dt(\ ) \qquad 1.3(98)$$

$$s^2 \leftrightarrow d^2/dt^2(\ ) \qquad 1.3(99)$$

$$s^n \leftrightarrow d^n/dt^n(\ ) \qquad 1.3(100)$$

$$1/s \leftrightarrow \int(\ )dt \qquad 1.3(101)$$

$$1/s^2 \leftrightarrow \int\int(\ )dt^2 \qquad 1.3(102)$$

$$1/s^n \leftrightarrow \int\int \cdots \int^n(\ )dt^n \qquad 1.3(103)$$

$$C(s) \leftrightarrow c(t) \qquad 1.3(104)$$

$$R(s) \leftrightarrow r(t) \qquad 1.3(105)$$

The transfer function can then be determined by solving for $C(s)/R(s)$. Consider the three examples described below.

Second-Order Lag

Consider the following second-order equation:

$$\frac{d^2}{dt^2} c(t) + 2\xi\omega_n \frac{d}{dt} c(t) + \omega_n^2 c(t) = K\omega_n^2 r(t) \qquad 1.3(106)$$

After the substitutions:

$$s^2 C(s) + 2\xi\omega_n s C(s) + \omega_n^2 C(s) = K\omega_n^2 R(s) \qquad 1.3(107)$$

Therefore,

$$\frac{C(s)}{R(s)} = \frac{K\omega_n^2}{s^2 + 2\xi\omega_n s + \omega_n^2} \qquad 1.3(108)$$

PID Controllers

As shown in later chapters, the proportional reset rate controller is given by

$$m(t) = K_c \left(e(t) + \frac{1}{T_i} \int e(t)\, dt + T_d \frac{d}{dt} e(t) \right) \qquad 1.3(109)$$

After the substitutions:

$$M(s) = K_c \left(E(s) + \frac{1}{T_i s} E(s) + T_d s E(s) \right) \qquad 1.3(110)$$

Therefore,

$$\frac{M(s)}{E(s)} = K_c \left(1 + \frac{1}{T_i s} + T_d s \right) \qquad 1.3(111)$$

Multiple Input Multiple Output

Multivariable state space representations also have transfer functions. They are referred to as *transfer function matrices*. Consider the following state space system:

$$\underline{\dot{x}} = A \underline{x} + B \underline{u} \qquad 1.3(112)$$

$$\underline{y} = C \underline{x}$$

Taking the Laplace transform of both equations yields

$$s \underline{x}(s) - \underline{x}(0) = A \underline{x}(s) + B \underline{u}(s) \qquad 1.3(113)$$

$$\underline{y}(s) = C \underline{x}(s) \qquad 1.3(114)$$

Rearranging gives

$$\underline{x}(s) = (sI - A)^{-1} \underline{x}(0) + (sI - A)^{-1} B \underline{u}(s) \qquad 1.3(115)$$

And

$$\underline{y}(s) = C(sI - A)^{-1} \underline{x}(0) + C(sI - A)^{-1} B \underline{u}(s) \qquad 1.3(116)$$

The transfer function matrix is defined as

$$G(s) = C(sI - A)^{-1} B \qquad 1.3(117)$$

Consider the 2×2 state space system in equations 1.3(92) and 1.3(93).

$$(sI - A) = \begin{bmatrix} s & 0 \\ 0 & s \end{bmatrix} - \begin{bmatrix} a_1 & a_2 \\ a_3 & a_4 \end{bmatrix} =$$

$$= \begin{bmatrix} s - a_1 & -a_2 \\ -a_3 & s - a_4 \end{bmatrix} \qquad 1.3(118)$$

$$(sI - A)^{-1} = \begin{bmatrix} s - a_4 & a_2 \\ a_3 & s - a_1 \end{bmatrix} / \text{C.E.} \qquad 1.3(119)$$

$$\underbrace{S^2 - (a_1 + a_4)s + (a_1 a_4 - a_2 a_3)}_{\text{Characteristic Equation (C.E.)}}$$

$$\text{If } C = I \text{ then} \qquad 1.3(120)$$

$$G(s) = C(sI - A)^{-1} B =$$

$$= \begin{bmatrix} \dfrac{(s - a_4)b_1 - a_2 b_3}{\text{C.E.}} & \dfrac{(s - a_4)b_2 - a_2 b_4}{\text{C.E.}} \\[2mm] \dfrac{(s - a_1)b_3 - a_3 b_1}{\text{C.E.}} & \dfrac{(s - a_1)b_4 - a_3 b_2}{\text{C.E.}} \end{bmatrix} \qquad 1.3(121)$$

Consider element 1,1 of $G(s)$:

$$G_{11}(s) = \frac{b_1(s - (a_4 b_1 + a_2 b_3))}{s^2 - (a_1 + a_4)s + (a_1 a_4 - a_2 a_3)} \quad \frac{x_1(s)}{u_1(s)}$$
$$1.3(122)$$

This is the transfer function that relates $x_1(s)$ and $u_1(s)$ in the single input single output case. It has a second-order denominator (or two poles) and a first-order numerator (or one zero) as discussed later in this chapter in the section on stability.

Block Diagrams

In the analysis of a control system, using the various mathematical equations in their conventional form is not generally convenient. The so-called block diagram representation is more desirable. Such diagrams not only present an organized picture of the flow of information

and energy, but they also, in the framework of the transfer function notation, facilitate the simultaneous solution of the differential equations that describe the system.

The conventions used in the construction of block diagrams are shown in Table 1.3d. The two main symbols are a circle, which indicates summation of two signals, and a rectangle, which indicates multiplication of a signal by a constant K or by a transfer function G(s). It is important to note that both symbols indicate linear operations. In using block diagrams to analyze control problems it is important to be able to convert from one form to another. Such manipulations are termed block diagram algebra. Some examples are shown in Table 1.3e.

Table 1.3e
MANIPULATIONS OF BLOCK DIAGRAMS

In control analysis the most frequent block diagram manipulations involve feedback loops of some sort. The rules for reducing such systems into a single transfer function can be summarized as follows.

$$\text{Output} = \frac{(\text{Input})\left(\begin{array}{c}\text{Product of blocks in the forward} \\ \text{path between input signal} \\ \text{and output signal}\end{array}\right)}{1 + \left(\begin{array}{c}\text{Product of blocks in the} \\ \text{control loop}\end{array}\right)}$$
$$1.3(123)$$

For systems with more than one input, equation 1.3(123) becomes

Table 1.3d
BLOCK DIAGRAM SYMBOLS

Symbol	Interpretation
$X(s) \longrightarrow$	Input or output signal; arrow gives direction.
$X(s) \quad X(s)$ / $X(s)$	Branch point. Division of a signal to give two or more paths without modification.
$X(s) \quad \pm \quad Z(s)$ / $Y(s)$	Summing point. $Z(s) = \pm X(s) \pm Y(s)$
$X(s) \; \boxed{G(s)} \; Y(s)$	System element. $Y(s) = G(s)X(s)$

Output =

$$\sum_{i=1}^{N} \frac{(\text{Input-i})\begin{pmatrix} \text{Product of blocks in the} \\ \text{forward path between input-i} \\ \text{and the output signal} \end{pmatrix}}{1 + \begin{pmatrix} \text{Product of blocks in the} \\ \text{control loop} \end{pmatrix}} \quad 1.3(124)$$

For example, consider the feedback type control system in Figure 1.3f, where

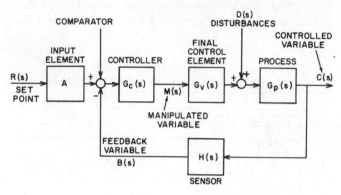

Fig. 1.3f   Block diagram of typical control loop

$$G_c(s) = K_c \text{ (proportional control)} \qquad 1.3(125)$$

$$G_p(s) = \frac{K}{1 + \tau s} \text{ (first-order process)} \qquad 1.3(126)$$

$$H(s) = \frac{K_H}{1 + \tau_H s} \text{ (first-order feedback)} \qquad 1.3(127)$$

Substituting,

$$c(s) = \frac{\dfrac{AK_cK}{1 + \tau s}}{1 + K_c\left(\dfrac{K}{1 + \tau s}\right)\left(\dfrac{K_H}{1 + \tau_H s}\right)} R(s)$$

$$+ \frac{B\left(\dfrac{K}{1 + \tau s}\right)}{1 + K_c\left(\dfrac{K}{1 + \tau s}\right)\left(\dfrac{K_H}{1 + \tau_H s}\right)} D(s) \qquad 1.3(128)$$

or

$$c(s) = \frac{AK_cK(1 + \tau_H s)}{(1 + \tau s)(1 + \tau_H s) + K_cKK_H} R(s)$$

$$+ \frac{BK(1 + \tau_H s)}{(1 + \tau s)(1 + \tau_H s) + K_cKK_H} D(s) \qquad 1.3(129)$$

Linearization

The great majority of techniques used in the analysis of control problems, including block diagrams, are dependent on the existence of mathematical equations in a linear form. By their very nature, most systems are nonlinear to one degree or another and therefore must be approximated by some linear equation. The technique used to linearize a nonlinear function can best be illustrated by considering a nonlinear equation such as

$$Y = \phi(X_1, X_2, X_3, \cdots) \qquad 1.3(130)$$

where

Y = dependent variable,

X = independent variables, and

ϕ = nonlinear function.

The equation can be expanded about a point $(X_{1i}, X_{2i}, X_{3i}, \ldots)$ using a Taylor's series expansion in which the higher order terms are ignored.

$$Y = Y_i + \frac{\delta Y}{\delta X_1}\bigg|_i (X_1 - X_{1i}) +$$

$$+ \frac{\delta Y}{\delta X_2}\bigg|_i (X_2 - X_{2i}) +$$

$$+ \frac{\delta Y}{\delta X_3}\bigg|_i (X_3 - X_{3i}) + \cdots \qquad 1.3(131)$$

or $\qquad y = K_1x_1 + K_2x_2 + K_3x_3 + \cdots \qquad 1.3(132)$

where

$$y = Y - Y_i \qquad 1.3(133)$$

$$x = X - X_i \qquad 1.3(134)$$

$$K_1 = \frac{\delta Y}{\delta X_1}\bigg|_i = \text{constant} \qquad 1.3(135)$$

$$K_2 = \frac{\delta Y}{\delta X_2}\bigg|_i = \text{constant} \qquad 1.3(136)$$

$$K_3 = \frac{\delta Y}{\delta X_3}\bigg|_i = \text{constant} \qquad 1.3(137)$$

For example, consider the development of the transfer function representation for the water tank shown in Figure 1.3g. A material balance on the tank yields

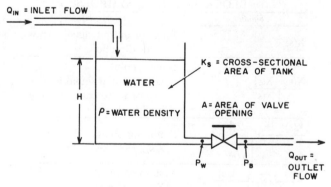

Fig. 1.3g   Nonlinear process represented by a water tank

$$\rho Q_{in} - \rho Q_{out} = \frac{d}{dt}(\rho H K_s) \qquad 1.3(138)$$

Flow out of the tank is a function of the area of the valve opening and of the pressure drop across the valve, which can be determined by the following equation:

$$Q_{out} = C_d A \sqrt{(2g_c/\rho)(P_w - P_B)} \qquad 1.3(139)$$

where

ρ = water density,
C_d = orifice discharge coefficient (constant),
P_B = back pressure (constant),
P_w = water pressure at valve, and
A = area of valve opening.

The pressure at the valve can be related to the height of water column in the tank:

$$P_w = \rho \frac{g}{g_c} H \qquad 1.3(140)$$

Therefore the valve equation is

$$Q_{out} = C_d A \sqrt{\frac{2g_c}{\rho}\left(\rho \frac{g}{g_c} H - P_B\right)} \qquad 1.3(141)$$

Substituting into the material balance yields

$$Q_{in} - C_d A \sqrt{\frac{2g_c}{\rho}\left(\rho \frac{g}{g_c} H - P_B\right)} =$$
$$= \rho K_3 \frac{d}{dt}(H) \qquad 1.3(142)$$

which is nonlinear and must be linearized before a Laplace transform can be determined.

Using the linearization technique described above, the nonlinear equation can be expressed as a linear function about a normal operating point (Q_{outi}, Q_{ini}, A_i, H_i) as follows:

$$q_{out} = K_1 a + K_2 h \qquad 1.3(143)$$

where

$$q_{out} = (Q_{out} - Q_{outi}) \qquad 1.3(144)$$

$$a = (A - A_i) \qquad 1.3(145)$$

$$h = (H - H_i) \qquad 1.3(146)$$

$$K_1 = \frac{\delta Q_{out}}{\delta A}\bigg|_i =$$
$$= C_d \sqrt{\frac{2g_c}{\rho}\left(H_i \rho \frac{g}{g_c} - P_B\right)} \qquad 1.3(147)$$

$$K_2 = \frac{\delta Q_{out}}{\delta H}\bigg|_i =$$
$$= C_d A_i \sqrt{\frac{2g_c/\rho}{H_i \rho \frac{g}{g_c} - P_B}}\left(\rho \frac{g}{g_c}\right) \qquad 1.3(148)$$

Expressing Q_{in} as a deviation from a reference operating condition, Q_{ini}, the linearized process differential equation is

$$q_{in} - K_1 a - K_2 h = K_3 \frac{dh}{dt} \qquad 1.3(149)$$

The Laplace transformation yields

$$Q_{in}(s) - K_1 A(s) - K_2 H(s) = K_3 s H(s) \qquad 1.3(150)$$

or

$$H(s) = \frac{Q_{in}(s)}{K_2 + K_3 s} - \frac{K_1 A(s)}{K_2 + K_3 s} \qquad 1.3(151)$$

Therefore the two transfer functions which describe the process are

$$\frac{H(s)}{Q_{in}(s)} = \frac{1}{K_2 + K_3 s} \qquad 1.3(152)$$

$$\frac{H(s)}{A(s)} = \frac{K_1}{K_2 + K_3 s} \qquad 1.3(153)$$

Graphical Representations of Dynamic Systems

Analytical solutions to dynamic control problems are typically very tedious. In practice these systems of equations are analyzed using computer-aided graphical techniques. The following section will briefly discuss two of these graphical methods. They are Bode plots and Nyquist plots. In the subsequent section on stability, these techniques will prove valuable for illustrating stability concepts.

Bode Plots

A very useful graphical technique for analysis of dynamic systems is referred to as the Bode plot. The Bode plot correlates certain parameters as functions of frequency. Consider the closed-loop equation based on Figure 1.3h:

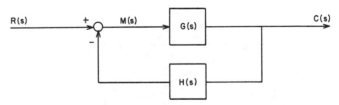

Fig. 1.3h Feedback control loop

$$\frac{C(s)}{R(s)} = \frac{G(s)}{1 + G(s)H(s)} \qquad 1.3(154)$$

If the r(t) input to this system is a sine wave, the steady-state output c(t) will be a sine wave of different amplitude and shifted in phase. If this change in amplitude and phase shift were recorded versus various different frequency r(t) inputs and plotted on rectangular coordinates, these would be the magnitude and phase Bode plots, respectively.

Equation 1.3(154) can be examined in the frequency domain by substituting $s = j\omega$.

The frequency response, magnitude versus frequency, for the control ratio

$$\frac{C(j\omega)}{R(j\omega)} = \frac{G(j\omega)}{1 + G(j\omega)H(j\omega)} \qquad 1.3(155)$$

can be determined for any given value of frequency. For each value of frequency, equation 1.3(155) yields a phasor quantity (see Nyquist Plots, below) whose magnitude is $|C(j\omega)/R(j\omega)|$ and whose phase angle α is the angle between $C(j\omega)$ and $R(j\omega)$.

For a given sinusoidal input signal, the input and steady-state output are of the following forms:

$$r(t) = R \sin \omega t \qquad 1.3(156)$$

$$c(t) = C \sin (\omega t + \alpha) \qquad 1.3(157)$$

An ideal system may be one in which

$$\alpha = 0° $$
$$R = C \qquad 1.3(158)$$

for $0 < \omega < \infty$. Curves 1 in Figure 1.3i represent the ideal system. If the above equations are analyzed, it is found that an instant transfer of energy must occur from the input to the output in zero time. This is a prerequisite for faithful reproduction of a step input signal. In reality, this generally cannot be achieved, since in any physical system there is energy dissipation and there are energy-storage elements. Curves 2 and 3 in Figure 1.3i represent actual systems.

The Bode plot provides significant information concerning the time response of the system. Two features of the frequency plot are the maximum value M_m and resonant frequency ω_m. The time response is qualitatively related to these values M_m and ω_m, which can be determined from the frequency response plots.

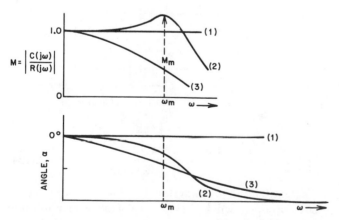

Fig. 1.3i   Bode plots (frequency response characteristic of C(jw)/R(jw) in rectangular coordinates)

The magnitude and phase versus frequency information is generated from the phasor quantity. A transfer function can be shown to equal the complex quantity.

$$G(j\omega) = \alpha + j\omega = |G(j\omega)|e^{j\theta(\omega)} \qquad 1.3(159)$$

$$|G(j\omega)| = \sqrt{\alpha^2 + \omega^2}, \text{ magnitude} \qquad 1.3(160)$$

$$\theta(\omega) = \tan^{-1} (\omega/\alpha), \text{ phase} \qquad 1.3(161)$$

The quantity typically plotted on the magnitude Bode plot is the decible, defined as

$$20 \log|G(j\omega)| \qquad 1.3(162)$$

Techniques for generating Bode plots vary from quick asymptotic approximation to relatively simple computer packages.

Nyquist Plots

Another method of representing the open-loop steady-state sinusoidal response is the polar plot or Nyquist plot. From this plot, the stability and the frequency response of the closed-loop system can be obtained (see discussion below on stability), and the closed-loop time response of the system can be predicted.

Consider the forward transfer function

$$G(j\omega) = \frac{C(j\omega)}{M(j\omega)} \qquad 1.3(163)$$

For a given frequency, $G(j\omega)$ is a phasor quantity whose magnitude is $|C(j\omega)/M(j\omega)|$ and whose phase angle is $\phi(\omega)$. Representing this quantity in polar coordinates can be done by a phasor whose length corresponds to the magnitude vector $|C(j\omega)/M(j\omega)|$ and whose angle with respect to the positive real axis corresponds to $\phi(\omega)$, as shown in Figure 1.3j(1). Thus a set of values of $G(j\omega)$ for frequencies between 0 and ∞ yields a set of phasors whose tips are connected by a smooth curve, as shown in Figure 1.3j(2). The polar plot in Figure 1.3j(2) contains all the necessary information and is of this form.

The reader has now been introduced to several forms of graphical representations employed in synthesis work. These techniques, whether computer generated or done by hand, can be extremely useful in analyzing dynamic systems.

An extension of the Nyquist plot is also very useful in analyzing interactions in multivariable systems. As defined earlier, the transfer function matrix is a matrix of single input single output transfer functions representing a multiple input multiple output system. It is possible to take this matrix of transfer functions and plot a matrix of Nyquist plots. Figure 1.3k(1) shows such a matrix for a possible dynamic system described by equation set 1.3(121). This plot alone allows the designer to study the stability and frequency response characteristics of the individual single input single output transfer functions.

Another very useful property that can be calculated and displayed graphically is the row dominance property.

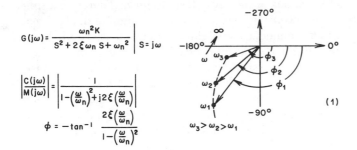

$$G(j\omega) = \frac{\omega_n^2 K}{S^2 + 2\xi\omega_n S + \omega_n^2}\bigg|_{S=j\omega}$$

$$\left|\frac{C(j\omega)}{M(j\omega)}\right| = \left|\frac{1}{1-\left(\frac{\omega}{\omega_n}\right)^2 + j2\xi\left(\frac{\omega}{\omega_n}\right)}\right|$$

$$\phi = -\tan^{-1}\frac{2\xi\left(\frac{\omega}{\omega_n}\right)}{1-\left(\frac{\omega}{\omega_n}\right)^2}$$

(1)

$$\omega_3 > \omega_2 > \omega_1$$

PROCESS PARAMETERS:

K = GAIN
ω_n = NATURAL FREQUENCY
ξ = DAMPING FACTOR
$\left|\frac{C(j\omega)}{M(j\omega)}\right|$ = FREQUENCY RESPONSE MAGNITUDE RATIO
ϕ = FREQUENCY RESPONSE PHASE ANGLE

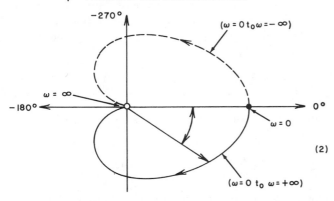

Fig. 1.3j   (1) Phasor-locus development in the complex plane. (2) the direct Nyquist plot.

Row dominance will be later used to investigate control system interaction (see the discussion of Nyquist array analysis in Section 1.11). The dominant transfer function of a given row of a transfer function matrix is the element whose time behavior would dominate (i.e., faster response within a given frequency range). Diagonal row dominance has the following definition for $Q(s)$ ($Q(s) = G(s)^{-1}$ or the inverse transfer function matrix):

$$|q_{ii}(s)| - \sum_{\substack{j=1 \\ j \neq 1}}^{m} |q_{ij}(s)| > 0 \qquad 1.3(164)$$

In words, this implies that for any s, the modulus of each element on the diagonal of $Q(s)$ exceeds the sum of the moduli of the off-diagonal elements. The row dominance can be graphically illustrated. For each Nyquist plot, at specified frequencies, the modulus of all other same-row elements is calculated. This number is used as the radius for a circle drawn on the Nyquist contours with circle center on the Nyquist contours at the frequency specified [refer to Figure 1.3k(2)]. These circles are called Gershgorin circles. When this is done for a range of frequencies, the circles sweep out a band called a Gershgorin band. If the origin does not lie within the

Gershgorin band, the element is said to be dominant on that row. For example, in Figure 1.3k(2), the g_{11} and g_{21} plots do not have Gershgorin bands that circle the origin; therefore, they are dominant on their respective rows. This matrix is then column 1 dominant, not diagonally dominant as we would like from a decoupling standpoint. Later compensation will be designed to achieve this dominance.

Multiple input multiple output Nyquist arrays bigger than 2×2 are very cumbersome to calculate by hand. Computer methods are typically necessary for these systems (see Section 1.12).

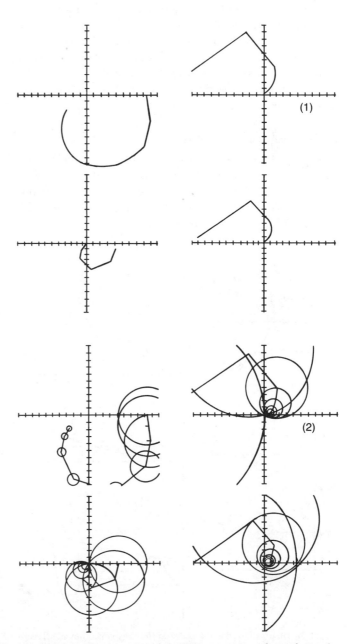

Fig. 1.3k   (1) Nyquist array plot. (2) Nyquist array plot with Gershgorin bands.

Stability

A stable linear system or element is one in which the system response is always bounded for any bounded system input. While most processes encountered in the process industries (with the exception of a few chemical reactors) are inherently stable, a feedback system employed to control the process can lead to a potentially unstable system. Mathematically, the stability of a linear system can be determined by an analysis of the roots of the "characteristic equation" from the differential equation describing the process (which corresponds to the roots of the denominator of the transfer function). Consider Figure 1.3l(1). Here the roots of the characteristic equation (for given K) are on the imaginary axis and the system is oscillating. In Figure 1.3l(2,3) the roots are to the right of the imaginary axis (i.e., the real part of the root is positive). Both systems are unstable. Their time responses are growing in amplitude. However, in many situations it may not be necessary to actually obtain the exact position of the roots in order to learn the nature of these roots.

Descartes' Rule of Signs

A simple rule-of-thumb analysis that can provide much insight into the nature of roots is Descartes' rule of signs. If the terms in the denominator are arranged in descending powers of s, Descartes' rule of signs states that the number of positive *real* roots cannot exceed the number of variations in sign from term to term. Therefore a simple *necessary* but *not sufficient* condition of stability that can be immediately assessed for any transfer function is that all the coefficients of each term in the denominator must be the same sign. If any variations in sign exist, the system is unstable. If no variations in sign exist, the system is potentially stable; however, instabilities could result from complex roots in the right-hand plane that Descartes' rule is unable to predict.

Routh's Criterion

One absolute method of determining whether complex or real roots lie in the right-hand plane is by the use of Routh's criterion. The method entails systematically generating a column of numbers that are then analyzed for sign variations. The first step is to arrange the denominator of the transfer function into descending powers of s. All terms including those that are zero should be included.

$$A(s) = a_n s^n + a_{n-1} s^{n-1} +$$
$$+ a_{n-2} s^{n-2} + \cdots + a_0 \quad 1.3(165)$$

Next, arrange the coefficients of s according to the following schedule:

$$
\begin{array}{ccccc}
a_n & a_{n-2} & a_{n-4} & a_{n-6} & \cdots \\
a_{n-1} & a_{n-3} & a_{n-5} & a_{n-7} & \cdots
\end{array} \quad 1.3(166)
$$

Now, expand according to the following manner:

$$
\begin{array}{ccccc}
a_n & a_{n-2} & a_{n-4} & a_{n-6} & \cdots \\
a_{n-1} & a_{n-3} & a_{n-5} & a_{n-7} & \cdots \\
b_1 & b_2 & b_3 & b_4 & \cdots \\
c_1 & c_2 & c_3 & & \cdots \\
d_1 & d_2 & \cdots & & \\
e_1 & e_2 & \cdots & & \\
f_1 & \cdots & & & \\
g_1 & \cdots & & &
\end{array} \quad 1.3(167)
$$

where the additional rows are calculated by

$$
\left.
\begin{aligned}
b_1 &= \frac{a_{n-1}a_{n-2} - a_n a_{n-3}}{a_{n-1}} \\[2mm]
b_2 &= \frac{a_{n-1}a_{n-4} - a_n a_{n-5}}{a_{n-1}} \\[2mm]
b_3 &= \frac{a_{n-1}a_{n-6} - a_n a_{n-7}}{a_{n-1}} \\
&\quad\vdots \\
c_1 &= \frac{b_1 a_{n-3} - a_{n-1} b_2}{b_1} \\[2mm]
c_2 &= \frac{b_1 a_{n-5} - a_{n-1} b_3}{b_1} \\
&\quad\vdots
\end{aligned}
\right\} \quad 1.3(168)
$$

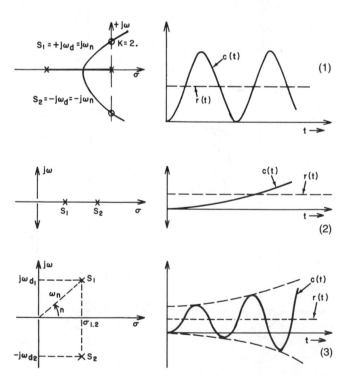

Fig. 1.3l   Roots plotted and associated time response showing (1) continuous oscillations, (2) real root unstable condition, (3) underdamped unstable pole placement

This process is continued until all new terms are zero. Routh's criterion says that the number of roots (real or complex) of the denominator of the transfer function that lie in the right-hand plane is equal to the number of changes of sign in the left-most column of the array shown in 1.3(167).

As an example, consider a system described by the following transfer function:

$$g(s) = \frac{s + 1}{s^3 + 2s^2 + 5s + 24} \qquad 1.3(169)$$

An initial analysis made using Descartes' rule of thumb indicates that, since all the signs of the denominator are the same, the system appears to be stable; however, Routh's criterion must be applied in order to determine this conclusively. The Routh array is

1	5	0
2	24	0
−7	0	
24	0	

An inspection of the left-most column indicates two sign changes (one from $+2$ to -7 and the other from -7 to $+24$); therefore, there are two roots in the right-half plane, resulting in an unstable system.

Nyquist Criterion

In the process industries the Routh criterion has the limitation of not being applicable to systems containing deadtime. Another more general stability criterion is the Nyquist criterion. This is based on the "frequency response" concept of a process and can be easily applied to deadtime or to other distributed parameter effects. To demonstrate the use of the Nyquist criterion, consider the Nyquist plot diagram of Figure 1.3j(2). From this plot the stability can be determined by an investigation of the $s = -1$ point in the complex plane. The Nyquist stability criterion is

$$N = z - P \qquad 1.3(170)$$

where

N = net number of encirclements of point s
 = -1 in a clockwise direction,

z = number of zero of $G(s)H(s)$ that lie in the right-hand plane, and

P = number of poles of $G(s)H(s)$ that lie in the right-hand plane.

The system is unstable if the contour either encircles or passes through the $(-1, 0)$ point in a clockwise direction.

REFERENCES

1. Hoffman, K., and Kunze, R., *Linear Algebra*, Prentice-Hall, Englewood Cliffs, NJ, 1961.
2. Geld, A., *Applied Optimal Estimation*, The MIT Press, Cambridge and London, 1974.

BIBLIOGRAPHY

Buckley, P.S., "Dynamics of Pneumatic Control by Pattern Recognition," *Instruments & Control Systems*, March, 1970.
———, "A Modern Perspective on Controller Tuning," Texas A&M 30th Annual Symposium on Instrumentation for the Process Industries, January, 1973.
———, *Techniques of Process Control*, John Wiley & Sons, New York, 1964.
Campbell, D.P., *Process Dynamics*, John Wiley & Sons, New York, 1958.
Chen, C.T., *Introduction to Linear System Theory*, Holt, Rinehart and Winston, New York, 1970.
Harriott, P., *Process Control*, McGraw-Hill Book Co., New York, 1964.
ISA-S26 Standard, "Dynamic Response Testing of Process Control Instrumentation," 1968.
Lipták, B.G. (ed.), *Instrument Engineer's Handbook*, vol. I, ch. I, Chilton Book Co., Philadelphia, 1969.
Looney, R., "A Thermal Sine Wave Generator for Speed of Response Studies," ASME Paper 54-5A-38, 1954.
Randhawd, R.S., "Understanding the 'Lazy' Integral," *Instruments & Control Systems*, January, 1983.
Rosenbrock, H.H., *Computer-Aided Control System Design*, Academic Press, London, 1974.

1.4 CLOSED-LOOP RESPONSE

The total performance of a control system depends on the overall response of the control loop and process acting together. Response refers to the dynamic behavior of the total system after an upset, which can be caused by a process disturbance, load change, or set-point adjustment. The response is evaluated by the speed with which the controlled variable returns to the set point and by the amount of overcorrection or overshoot that occurs that affects the stability of the system during the upset condition. Depending on the nature of the process, different control modes may be required for optimum performance. This section describes the closed-loop response using these various control modes. We will analyze the system response utilizing transfer functions to characterize the controller and the process it controls. We then develop recommendations on when to use the various control modes as a function of the type of process involved.

Control Loop Transfer Functions

In the evaluation or design of a control system, one is generally interested in the "closed-loop" response of the system to changes in the set point or to changes in the load or variables. For a particular system this response may be characterized by the differential equation or transfer function which describes the system.[1] Consider the block diagram of the general control system in Figure 1.4a. The closed-loop transfer function that describes the system is

$$C(s) = \frac{AG_c(s)G_v(s)G_p(s)}{1 + G_c(s)G_v(s)G_p(s)G_h(s)} R(s)$$

$$+ \frac{G_p(s)G_d(s)}{1 + G_c(s)G_v(s)G_p(s)G_h(s)} D(s)$$

1.4(1)

where

$C(s)$ = Laplace transform of the system output,
$R(s)$ = Laplace transform of the system set point,
$D(s)$ = Laplace transform of the disturbance input,
$G_c(s)$ = transfer function of the controller,

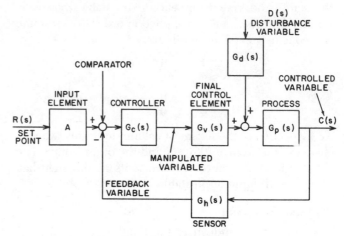

Fig. 1.4a   General control loop

$G_v(s)$ = transfer function of the final control element,
$G_p(s)$ = transfer function of the process,
$G_h(s)$ = transfer function of the feedback sensor device, and
$G_d(s)$ = transfer function of the disturbance variable.

In general the differential equation describing the behavior of a control system is a function of not only the controller but of the process dynamics and the control hardware as well. Each element in the control loop can contribute significantly to the overall performance of the system.[2]

Analysis of a Chemical Reactor

To illustrate the overall analysis of a control system consider the reactor shown in Figure 1.4b. The reactor is a simple back-mix reactor heated by a steam jacket. The reactor temperature is controlled by a feedback controller that regulates the steam flow. As indicated above, the first step in the analysis of the system is to describe each element in the system. In this case these are the reactor $G_p(s)$, the valve $G_v(s)$, the sensor $G_h(s)$ and the controller $G_c(s)$.

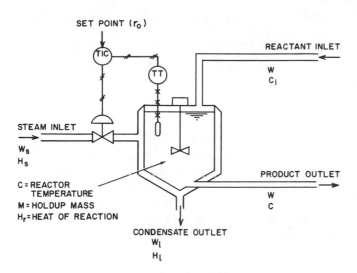

Fig. 1.4b   Temperature-controlled chemical reactor

The valve and sensor in this example will be assumed to be instantaneous.

$$G_v(s) = K_v \qquad 1.4(2)$$

$$G_h(s) = K_h \qquad 1.4(3)$$

In practice, no valve or sensor acts instantaneously, and both introduce some lag into the loop. For a highly overdamped, slow-responding process, these hardware lags may be insignificant. However, in a relatively fast process, such lags can have a considerable effect on the overall system performance and should be considered.

The description of the process is determined by an analysis of the reactor. The energy balance equations yield

$$[W_sH_s + WK(C_i - C_r)] - [WK(C - C_r) +$$
$$+ MH_r + W_1H_1] = [MK \frac{d}{dt}(C - C_r)] \quad 1.4(4)$$

or, in terms of deviations from normal operating conditions, we have

$$(H_s - H_1)w_s + (WC_p)c_i - (WC_p)c = (MC_p)\frac{d}{dt}c$$
$$\qquad\qquad 1.4(5)$$

where

W_s = flow rate of steam (variable),
W = flow rate of product (constant),
K = heat capacity of product and reactant stream (constant),
M = mass in reactor (constant),
C_i = temperature of reactant inlet stream (variable),
C = temperature of the reactor (variable),
H_s = enthalpy of steam (constant),
H_1 = enthalpy of condensate (constant),

H_r = heat of reaction (constant),
C_r = reference temperature (constant),
w_s = deviation in flow rate of steam,
c_i = deviation in temperature of inlet stream, and
c = deviation in temperature of reactor.

Assuming the system to be initially at a steady state, the Laplace transform gives

$$(H_s - H_1)W_s(s) + (WC_p)C_i(s) -$$
$$- (WC_p)C(s) = (MC_p)sC(s) \quad 1.4(6)$$

Solving for C(s):

$$C(s) = \frac{K_p}{1 + \tau_p s} W_s(s) + \frac{1.0}{1 + \tau_p s} C_i(s) \quad 1.4(7)$$

where

$$K_p = \frac{H_s - H_1}{WC_p} \text{ (process gain) and} \quad 1.4(7a)$$

$$\tau_p = M/W \text{ (process time constant)} \quad 1.4(7b)$$

Therefore, in terms of the transfer function for the general block, we have

$$G_p(s) = \frac{K_p}{1 + \tau_p s} \quad 1.4(8)$$

$$G_d(s) = \frac{1}{K_p} \quad 1.4(9)$$

The resulting block diagram for the system is shown in Figure 1.4c. The controller has not been specified. It can be any one of the types discussed in Section 1.2

$$(P): G_c(s) = K_c \quad 1.4(10)$$

$$(I): G_c(s) = \frac{1}{T_i s} \quad 1.4(11)$$

$$(PI): G_c(s) = K_c \left(1 + \frac{1}{T_i s}\right) \quad 1.4(12)$$

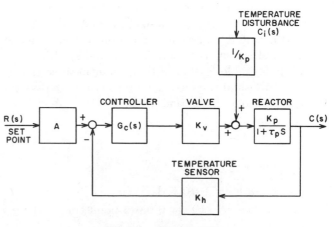

Fig. 1.4c   Reactor system

$$(PD):\ G_c(s) = K_c(1 + T_d s) \qquad 1.4(13)$$

$$(PID):\ G_c(s) = K_c\left(1 + \frac{1}{T_i s} + T_d s\right) \qquad 1.4(14)$$

The closed-loop transfer function can be determined for any of these controllers by substituting equations 1.4(2), 1.4(3), 1.4(8), and 1.4(9), as well as the appropriate controller equation into the general closed-loop transfer function of equation 1.4(1).

Proportional Control

We now consider the response of the reactor system when a proportional controller is used. Substituting equation 1.4(10) into equation 1.4(1) yields

$$C(s) = \frac{AK_c K_v \dfrac{K_p}{1 + \tau_p s}}{1 + \dfrac{K_c K_v K_p K_h}{1 + \tau_p s}} R(s) +$$

$$+ \frac{\dfrac{1}{1 + \tau_p s}}{1 + \dfrac{K_c K_v K_p K_h}{1 + \tau_p s}} C_i(s) \qquad 1.4(15)$$

or,

$$C(s) = \frac{K_{s1}}{1 + \tau_s s} R(s) + \frac{K_{s2}}{1 + \tau_s s} C_i(s) \qquad 1.4(16)$$

where

$$K_{s1} = \frac{AK_c K_v K_p}{1 + K_c K_v K_p K_h} \qquad 1.4(16a)$$

$$K_{s2} = \frac{1}{1 + K_c K_v K_p K_h} \qquad 1.4(16b)$$

$$\tau_s = \frac{\tau_p}{1 + K_c K_v K_p K_h} \qquad 1.4(16c)$$

Here the closed-loop response is described by a first-order transfer function. The rapidity of response will depend on the order of magnitude of the system time constant, τ_s. The process is also of the first order, but its time constant is larger than the time constant of the overall system; see equations 1.4(7b) and 1.4(16c).

An undesirable feature of the proportional controller is its steady-state response. The response of the system to a step change disturbance is for $t < 0$,

$$c_i(t) = 0.0$$
$$r(t) = 0.0$$

and for $t \geq 0$,

$$c_i(t) = d_0$$
$$r(t) = 0.0 \text{ (set-point constant)}$$

Therefore,

$$c(t) = K_{s1}(1 - e^{-t/\tau_s})0.0 + K_{s2}(1 + e^{-t/\tau_s})d_0 \quad 1.4(17)$$

Under steady-state conditions $(t \to \infty)$,

$$c(t) = C_{ss} = K_{s2}d_0 \qquad 1.4(18)$$

or

$$C_{ss} = \frac{1}{1 + K_c K_v K_p K_h} d_0 \qquad 1.4(19)$$

This results in a steady-state deviation from the desired set point. This error is referred to as an "offset" and is characteristic of pure proportional controllers. The offset is generally associated only with the response of the system to the disturbance variable. Errors due to set-point changes can be eliminated by calibration. The steady-state response of the system to a step change in set point is

$$C_{ss} = \underset{t \to \infty}{\text{limit}}\ K_{s1}(1 - e^{-t/\tau_s})R_0 \qquad 1.4(20)$$

$$C_{ss} = K_{s1}R_0 \qquad 1.4(21)$$

The steady-state response is equal to the set point ($C_{ss} = R_0$) if K_{s1} is calibrated at 1.0, which is achieved by setting A in equation 1.4(16a) to equal

$$A = \frac{1 + K_c K_v K_p K_h}{K_c K_v K_p} \qquad 1.4(22)$$

Integral Control

For the reactor system if an integral controller is used, equation 1.4(11) applies. The closed-loop transfer function is then

$$C(s) = \frac{A \dfrac{K_v K_p}{T_i s(1 + \tau_p s)}}{1 + \dfrac{K_v K_p K_h}{T_i s(1 + \tau_p s)}} R(s) + \frac{\dfrac{1}{(1 + \tau_p s)}}{1 + \dfrac{K_v K_p K_h}{T_i s(1 + \tau_p s)}} C_i(s)$$

$$1.4(23)$$

By rearranging, we get

$$C(s) = \frac{A \dfrac{K_v K_p}{T_i \tau_p}}{s^2 + \dfrac{1}{\tau_p} s + \dfrac{K_v K_p K_h}{T_i \tau_p}} R(s) +$$

$$+ \frac{\dfrac{s}{\tau_p}}{s^2 + \dfrac{1}{\tau_p} s + \dfrac{K_v K_p K_h}{T_i \tau_p}} C_i(s) \quad 1.4(24)$$

A major difference between the performance of the system with integral control and with proportional control is in the steady-state behavior. For a step change in both the set point and disturbance,

$$R(s) = R_0/s \qquad 1.4(25)$$

$$C_i(s) = D_0/s \qquad 1.4(26)$$

the steady-state behavior is determined using the final value theorem:

$$C_{ss} = \lim_{t \to 0} \{sC(s)\} = \frac{A}{K_h} R_0 + 0.0 D_0 \qquad 1.4(27)$$

$$= \frac{A}{K_h} R_0 \qquad 1.4(28)$$

Here, the disturbance does not ultimately affect the steady-state output of the control system. When the controller is calibrated such that $A = K_h$, the final value of the response is equal the desired value. A major drawback of the integral controller is the relatively slow response time. For the proportional control system, the response time can be extremely fast for large values of gain [equation 1.4(16c)]. The integral controller results in a second-order response that is slower and requires a longer time to reach the final steady-state value.

Proportional-Plus-Integral Control

Both proportional and integral modes can be combined to take advantage of the fast transient response of proportional control and of the offset-free steady-state behavior of integral control [equation 1.4(12)]. For the reactor example, the closed-loop transfer function is as follows:

$$C(s) = \frac{A \left(\dfrac{K_c K_v K_p}{\tau_p T_i} \right)(T_i s + 1)}{s^2 + \left(\dfrac{1 + K_h K_c K_v K_p}{\tau_p} \right) s + \dfrac{K_h K_c K_v K_p}{\tau_p T_i}} \cdot R(s)$$

$$+ \frac{\dfrac{s}{\tau_p}}{s^2 + \left(\dfrac{1 + K_h K_c K_v K_p}{\tau_p} \right) s + \dfrac{K_h K_c K_v K_p}{\tau_p T_i}} \cdot C_i(s) \qquad 1.4(29)$$

$$C_{ss} = \frac{A}{K_h} \cdot R_0 + 0.0 D_0 \qquad 1.4(30)$$

The steady-state performance is the same as the integral controller, but the transient response is much more rapid. Figures 1.4d and 1.4e compare the response of these controllers to both a change in set point and a change in disturbance. In the disturbance case, the steady-state offset of proportional control is eliminated by both the integral and proportional-plus-integral controllers. The combined modes, however, react much faster than does the integral mode alone. For the set-point change (Figure 1.4e), the best response occurs with the pure proportional control. The steady-state offset in this case can be effectively eliminated by calibration. On the other hand, integral controllers must eliminate the steady-state error by integration. Of the two controllers, the com-

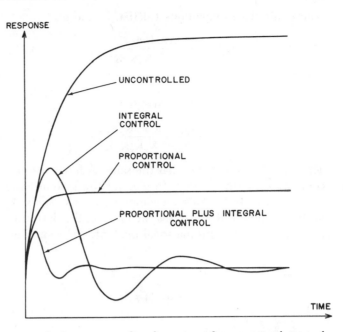

Fig. 1.4d   Response to a disturbance input for proportional, integral, and proportional-plus-integral controllers

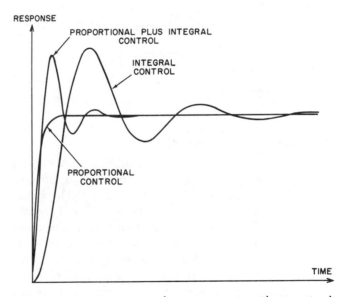

Fig. 1.4e   Response to a step change in set point with proportional, integral, and proportional-plus-integral controllers

bined controller responds faster than the pure integral one.

Derivative Control

A similar analysis of the reactor system can be made for algorithms containing derivative action. The closed-loop transfer function for a proportional-plus-derivative controller is

$$C(s) = \frac{K_{s1}(1 + T_d s)}{1 + \tau_s' s} R(s) + \frac{K_{s2}}{1 + \tau_s' s} C_i(s) \qquad 1.4(31)$$

where, similar to equations 1.4(16a, b, and c),

$$K_{s1} = \frac{AK_cK_vK_p}{1 + K_cK_vK_pK_h} \qquad 1.4(31a)$$

$$K_{s2} = \frac{1}{1 + K_cK_vK_pK_h} \qquad 1.4(31b)$$

$$\tau_s' = \frac{\tau_p + K_cK_vK_pK_hT_d}{1 + K_cK_vK_pK_h} \qquad 1.4(31c)$$

The closed-loop equation has the same order as the process, which is first order. In general the system has a fast response, but it does not compensate for steady-state offsets. Figure 1.4f illustrates the effect of the derivative mode addition to a proportional-plus-integral controller.

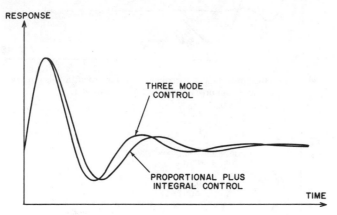

Fig. 1.4f   The effect of adding derivative mode to a two-mode controller

Selecting a Controller Mode

Selecting the proper mode of control is a critical step in the design of a control loop. There is no universal controller that is correct for all processes. The factors that the control engineer must consider are (1) control quality, (2) costs, and (3) ease of operation. For a particular controller the cost and the relative ease of operation are generally fixed, but the quality of control varies from process to process. A few general comments that might aid in the selection of the proper control mode follow.

Two-Position Control

Two-position or on-off control is the least expensive and simplest technique. It can be used with relatively slow-responding, low-order systems. It cannot be used effectively on high-order processes that do not contain a dominant time constant or on systems containing medium to large deadtimes. The areas of application for on-off controllers in the process industry are few.

Proportional Control

Pure proportional control uses the simplest regulating controller, and it has been applied with some degree of success to most processes in the process industry. It is most useful for process transfer functions containing a I/s term and having a single dominant time constant, as illustrated by the following equation:

$$G_p(s) = \frac{k}{(1 + \tau_1s)(1 + \tau_2s)(1 + \tau_3s) \cdots} \qquad 1.4(32)$$

where $\tau_1 \gg \tau_2$, $\tau_1 \gg \tau_3$. . . .

The pure proportional controller responds rapidly to both set-point and disturbance changes, but it has the undesirable characteristic of a steady-state error (offset). Tuning is relatively easy and is accomplished by the adjustment of a single parameter. Application in process control is usually limited to local regulators, where offsets can be tolerated.

Integral Control

Integral (floating) control is another relatively simple one-parameter controller. It is particularly effective for (1) very fast processes, in particular those that contain noise, (2) processes dominated by deadtime, and (3) high-order processes in which all of the time constants are roughly the same order of magnitude. Integral control does not exhibit steady-state error, but it is relatively slow responding. It decreases the stability of a system and therefore should not be used for process transfer functions which contain 1/s terms. Only a few applications exist in the process industry for this single-mode controller.

Proportional-Plus-Integral Control

The proportional-plus-integral controller is perhaps the best conventional controller in use today. It does not have the offset associated with proportional control, yet it yields a much faster dynamic response than does integral action alone. Due to the presence of the integral mode the stability of the control loop is decreased. Extreme caution should be used in applying this controller to process transfer functions containing 1/s terms.

In situations where control signal saturation may occur, saturating the final control element, the integral mode should be used with care, because of the possibility for reset windup. This is a condition in which the controller continues to integrate the error signal even though no further corrective action can be realized. The result is a confused controller that must "unwind" before effective control can be realized, resulting in a very poor transient behavior of the control system.

This two-mode controller is widely used in the process industry for controlling level, pressure, flow, and other variables that do not have very large time lags.

Proportional-Plus-Derivative Control

The proportional-plus-derivative (PD) or rate controller is effective for systems containing a number of time constants. It results in a more rapid response and less offset than is possible by pure proportional control. In general it increases the overall stability of the control loop. Care must be employed in using the derivative mode in the control of very fast processes or if the measurement signal is noisy. Applications in the process industry tend to favor the PID (proportional-plus-integral-plus-derivative) controllers rather than the PD controllers.

Inverse Derivative Control

The inverse derivative controller is a specialized two-mode controller using the derivative concept. It is used primarily to introduce a lag into the control loop when one is attempting to control very fast or noisy processes.

Proportional-Plus-Integral-Plus-Derivative Control

The proportional-plus-integral-plus-derivative (PID) controller is the most complex of the conventional control mode combinations. In theory, the PID controller can result in better control than the one- or two-mode controllers mentioned above. In practice, the control advantage can be difficult to achieve because of the difficulty of selecting the proper tuning parameters. The addition of the derivative mode to the PI controller is often specified to compensate not for process lags but for hardware lags that could be corrected at the source. The addition of the derivative mode is sometimes specified due to the incorrect notion that derivative action is helpful in overcoming process deadtimes.

PID controllers are used in the process industry to control slow variables such as temperature, pH, and other analytical variables.[3]

REFERENCES

1. Hordeski, M.F., "Fundamentals of Digital Control Loops," *Measurements and Control*, February, 1978.
2. Skrokov, M.R., (ed.), *Mini- and Microcomputer Control in Industrial Processes*. Van Nostrand Reinhold, New York, 1980.
3. Shinskey, F.G., *Process Control Systems*, McGraw-Hill Book Co., New York, 1967.

BIBLIOGRAPHY

Andreiev, N., "A New Dimension: A Self-Turning Controller that Continually Optimizes PID Constants," *Control Engineering*, August, 1981.

Buckley, P.S., "Dynamic Design of Pneumatic Control Loops: Parts I and II," *InTech*, April and June, 1975.

——, *Techniques of Process Control*. John Wiley & Sons, New York, 1964.

Caldwell, W.I., Coon, G.A. and Zoss, L.M., *Frequency Response for Process Control*, McGraw-Hill Book Co., New York, 1959.

Campbell, D.P., *Process Dynamics*, John Wiley & Sons, New York, 1958.

D'Azzo, J.J., and Houpis, C.H., *Feedback Control System Analysis and Synthesis*, McGraw-Hill Book Co., New York, 1960.

Eckman, D.P., *Automatic Process Control*, John Wiley & Sons, New York, 1958.

Harriott, P., *Process Control*, McGraw-Hill Book Co., New York, 1964.

Hordeski, M.F., "Process Controls Are Evolving Fast," *Electronic Design*, November 22, 1977.

Hougen, J.O., *Measurements and Control Applications*, ISA, Pittsburgh, 1979.

ISA-S26 Standard, "Dynamic Response Testing of Process Control Instrumentation," 1968.

Johnson, C.D., *Process Control Instrumentation Technology*. John Wiley & Sons, New York, 1977.

Jones, B.E., *Instrumentation, Measurement and Feedback*, McGraw-Hill Book Co., New York, 1977.

Kuo, B.C., *Automatic Control Systems*, Prentice-Hall, Englewood Cliffs, NJ, 1962.

Luyben, W.L., "A Proportional-Lag Level Controller," *Instrumentation Technology*, December, 1977.

McCauley, A.P., Jr., and Persik, S.R., "From Linear to Gain-Adaptive Control: A Case History," *Instrumentation Technology*, November, 1977.

Randhawd, R.S., "Understanding the 'Lazy' Integral," *Instruments & Control Systems*, January, 1983.

Shinskey, F.G., "Control Topics," Publi. No. 413-1 to 413-8, Foxboro Company.

——, *Process Control Systems*, McGraw-Hill Book Co., New York, 1979.

1.5 FEEDBACK AND FEEDFORWARD CONTROL

Sections 1.1 through 1.4 have been in a large part dedicated to the concepts of feedback control. This section discusses feedforward control theory and its application. Feedforward control complements feedback control and is generally used in conjunction with it.

Feedback Control

The principal features and limitations of feedback control formed the basis for the development of feedforward control methods. The purpose of any form of process control is to maintain the controlled quantity at a desired value, usually called the set point, in the face of disturbing forces. The control system regulates the process by opposing the disturbing forces with equivalent changes in one or more manipulated variables. (For the controlled variable to remain stationary, the controlled process must be in a balanced state. One of the means by which this balance can be achieved is feedforward control, which provides a corrective action *before* the disturbance is seen as an error in the controlled variable.)

Regulation through feedback control is achieved by acting on the change in the controlled variable that was induced by a change in load. Deviations in the controlled variable are converted into proportional changes in the manipulated variable and sent back to the process to restore the balance. Figure 1.5a shows the backward flow of information from the output of the process back to its manipulated input. The load can be divided into various components such as feed rate, feed composition, and temperatures. These may be balanced by a single manipulated variable. But there must always be at least one manipulated variable for each independently controlled variable.

Feedback Control Performance

Feedback, by its nature, is incapable of correcting a deviation in the controlled variable at the time of detection. In any process, a finite delay exists between behavior of the manipulated variable and its effect on the controlled variable. Where this delay is substantial and the process is subject to many frequent disturbances, considerable difficulty can be encountered in maintaining control. Perfect control is not even theoretically obtainable, because a deviation in the controlled variable must appear before any corrective action can begin. In addition, the value of the manipulated variable needed to balance the load must be sought by trial and error, with the feedback controller observing the effect of its output on the controlled variable.

The effectiveness of feedback control depends on the dynamic gain of the controller in relation to the frequency and magnitude of the disturbances encountered. Although a high controller gain is desirable, the dynamic gain of the closed loop at its natural period of oscillation must be less than unity if the loop is to remain stable.[1] So if the process has a high dynamic gain, the controller gain must be correspondingly low. Thus the process dictates how well it may be controlled.

The task of the feedback controller is best appreciated by defining its operation. The output m of an ideal three-mode controller is related to the deviation e between the controlled variable and its set point by

$$m = \frac{100}{PB} \left(e + \frac{1}{T_i} \int e \, dt + T_d \frac{d}{dt} e \right) \quad 1.5(1)$$

where 100/PB is equivalent to the gain K_c. The settings of the three modes are identified as the percent proportional band PB, reset time constant T_i, and the derivative time constant T_d, both in the same units as time t.

Consider the process to be in the steady state at time t_1, so that the deviation is zero. Solving equation 1.5(1) at time t_1 yields the output m_1:

$$m_1 = \frac{100}{PBT_i} \int_{t0}^{t1} e \, dt \quad 1.5(2)$$

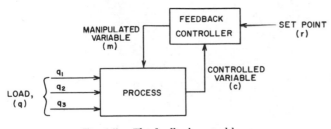

Fig. 1.5a The feedback control loop

42

The output is related to the history of the deviation e from the time control began (t_0).

Let the process encounter a sustained change in load, requiring a controller output m_2 to return the system to balance. At some time, t_2, the process will have returned to the steady state, with e again zero, so that

$$m_2 = \frac{100}{PBT_i} \int_{t0}^{t2} e \, dt \qquad 1.5(3)$$

In order to change its output from m_1 to m_2 the controller had to sustain an integrated error of

$$m_2 - m_1 = \frac{100}{PBT_i} \int_{t_1}^{t_2} e \, dt \qquad 1.5(4)$$

This integrated error appears as the area between the controlled variable and the set point in Figure 1.5b. Now let the integrated error be denoted by E and the change in output by m. Then, the integrated error caused by a load change is

$$E = \Delta m \frac{PBT_i}{100} \qquad 1.5(5)$$

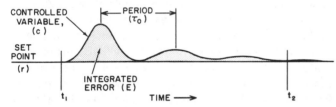

Fig. 1.5b   The integrated error resulting from a load change

Note that optimum values of PB and T_i exist for each control loop, which determine the minimum value of E developed by a given load change. Some processes may be rapid in response, such that T_i is only a few seconds. Unless the load changes are severe or frequent, feedback control is usually acceptable. Some processes are slow to respond and have a low dynamic gain, accommodating a narrow proportional band. Control of temperature in a stirred tank is an example of this type of process, and here feedback control is acceptable. Still other processes may be characterized by long delays and high sensitivity, such as the control of product composition from a distillation tower. Here a wide proportional band and long reset time can result in out-of-specification product if the load changes are frequent or severe. This type of process can benefit from feedforward control.

Feedforward Control

Feedforward provides a more direct solution to control than finding the correct value of the manipulated variable by trial and error, as occurs in feedback control. In the feedforward system, the major components of load are sensed and used to calculate the value of the manipulated

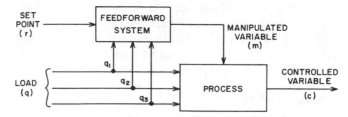

Fig. 1.5c   The feedforward control loop

variable required to maintain control at the set point. Figure 1.5c shows how information flows forward from the load to the manipulated variable input of the process. The set point is used to give the system a command. (If the controlled variable were used in the calculation instead of the incoming load, a positive feedback loop would be formed.)

A system, rather than a single controller device, is normally used for feedforward loops because it is not always convenient to provide the computing functions required by the forward loop with a single device. Instead, the feedforward system consists of several devices whose functions provide a mathematical model of the feedforward characteristic.

Load Balancing

A dynamic balance is achieved for the process by solving its material and/or energy balance equations, continuously. When a change in load is sensed, the manipulated variable is automatically adjusted to the correct value at a rate that keeps the process continually in balance. While it is theoretically possible to achieve this perfect form of control, in practice the system cannot be made to duplicate the process equations exactly. The material and energy balance equations are not usually difficult to write for a given process. Variations in non-stationary parameters like heat transfer coefficients and the efficiency of mass transfer units do not ordinarily affect the performance of a feedforward system. The load components are usually feed flow and feed composition, when the product composition is to be controlled, or feed flow and temperature where the product temperature is to be controlled. Feed flow is the primary component of load in almost every application, because it can change widely and rapidly. Feed composition and temperature are less likely to exhibit such wide excursions, and their change is always limited by upstream capacity. In some feedforward systems the secondary load components can be left out of the control loop entirely.

The output of a feedforward system should be an accurately controlled flow rate. This controlled flow rate cannot be obtained by manipulating the control valve directly, since the valve characteristics are nonlinear and changeable, and their flow is subject to external influences. Therefore, most feedforward systems depend on some measurement and feedback control of flow to obtain an accurate manipulation of the flow rate. Only when the range of the manipulated variable exceeds that which is available in flowmeters should one consider having the valves positioned directly, and in such cases care must be taken to obtain a reproducible response.

Steady-State Model

The first step in designing a feedforward control system is to form a steady-state mathematical model of the process. The model equations are solved for the manipulated variable, which is to be the output of the system. Then the set point is substituted for the controlled variable in the model.

The process will be demonstrated using the example of temperature control in a heat exchanger. A liquid flowing at rate W is to be heated from temperature T by steam at rate W_s. The energy balance, excluding the losses, is

$$WC(T_2 - T_1) = \lambda W_s \qquad 1.5(6)$$

Coefficient C is the heat capacity of the liquid and λ is the heat given up by the steam in condensing. Solving for W_s yields

$$W_s = W \frac{C}{\lambda} (T_2 - T_1) \qquad 1.5(7)$$

Now T_2 becomes the set point, and W and T_1 are components to which the control system must respond.

An implementation of equation 1.5(7) is given in Figure 1.5d. A control station introduces the set point for T_2 into the summing relay and input T_1 is subtracted. The gain of the summing relay is adjusted to obtain the constant ratio C/λ. Then, the linear liquid flow signal is multiplied by the $(C/\lambda)(T_2 - T_1)$ term to produce the required setting of steam flow. The flow signals are linear because the temperature difference signal is linear.

During start-up, if the actual value of the controlled variable does not equal the set point, an adjustment is made to the gain of the summing amplifier. Then, after making this adjustment, if the controlled variable does not return to the set point following a change in load, the presence of an error in the system is indicated. The error may be in one of the computing devices, the flowmeter, or some other factor affecting the heat balance that was not included in the system. Heat losses and variations in steam pressure are two possible sources of error in the model. Since any error can cause a proportional offset, the designer must weigh the various sources of error and compensate for the largest or most changeable components where practical. For example, if steam pressure variations were a source of error, the steam flowmeter could be pressure compensated.

Dynamic Model

Transient errors following a change in load are to be expected in feedforward control systems. A typical dynamic response of the controlled variable is shown in Figure 1.5e. Following a step increase in liquid flow, which is accompanied by a simultaneous and proportional set increase in steam flow as dictated by the feedforward system, the exit temperature will still fall temporarily. This transient error reveals a dynamic imbalance in the system, because the exchanger temporarily needs more heat than the steam flow controller is allowing. The reason for this is that the energy level of the process must be increased before an increase in heat transfer level can take place. Some of the added flow of heat from the increased steam flow is diverted in this example to increase the process energy level. Since the balance of the added heat flow is less than that calculated to maintain the exit temperature of the increased liquid flow at steady state, this exit temperature will decrease until the process energy level has been increased adequately. In order to correct the temperature error, an additional amount of energy must be added to the exchanger over what is required for the steady-state balance.

A more general explanation is provided through the use of Figure 1.5f. This figure shows the load and the manipulated variable entering the process at different points, where they encounter different dynamic ele-

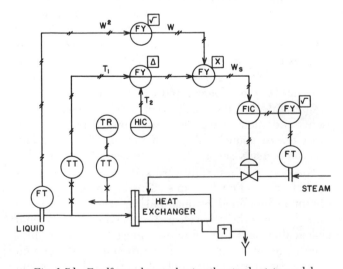

Fig. 1.5d   Feedforward control using the steady-state model

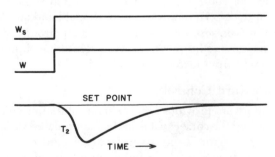

Fig. 1.5e   The dynamic response for an uncompensated feedforward system

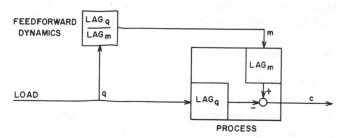

Fig. 1.5f A feedforward dynamic model for a general process

ments. In the heat exchanger example, the liquid enters the tubes while the steam enters the shell. The heat capacities of the two locations are different. As a result, the controlled variable (the liquid temperature) responds more rapidly to a change in liquid flow than to a change in steam flow. Thus, the lag characteristics of the load input is less than that of the manipulated variable.

The objective of the feedforward system is to balance the manipulated variable against the load, providing the forward loop with compensating dynamic elements. Neglecting steady-state considerations, Figure 1.5f shows what the dynamic model must contain to balance a process with two first-order lags. The lag characteristic of the load input must be duplicated in the forward loop, and the lag in the manipulated input of the process must be cancelled. Thus the forward loop should contain a lag divided by a lag. Since the inverse of a lag is a lead, the dynamic compensating function is a "lead-lag." The lead time constant should be equal to the time constant of the controlled variable in response to the manipulated variable, and the lag time constant should equal the time constant of the controlled variable in response to the load. In the case of the heat exchanger, the fact that lag m is longer than lag q causes the temperature decrease on a load increase. This is the direction in which the load change would drive the process in the absence of control.

In transfer function form, the response of a lead-lag unit is

$$G(s) = \frac{1 + \tau_1 s}{1 + \tau_2 s} \qquad 1.5(8)$$

where

τ_1 = lead time constant,
τ_2 = lag time constant, and
s = the Laplace operator.

Phase lead compensation tends to improve the rise time and overshoot of the system response, but it increases the band width. Phase lag compensation improves the steady state response or the stability margin, but it also often results in a longer rise time because of the reduced bandwidth.[2] Lead lag compensation has the advantages of both techniques and elimates some of the undesirable features of each.

The frequency response of a feedforward control is usually not critical, since forward loops cannot oscillate. A more severe test for a forward loop is a step change in load. The gain of the lead-lag unit to a step input is

$$C(t) = 1 + \frac{\tau_1 - \tau_2}{\tau_2} e^{-t/\tau_2} \qquad 1.5(9)$$

The maximum gain (at t = 0) is the lead/lag ratio τ_1/τ_2. The response curve decays exponentially from this maximum at the rate of the lag time constant τ_2. The 63 percent recovery point is reached in Figure 1.5g at t = 2.

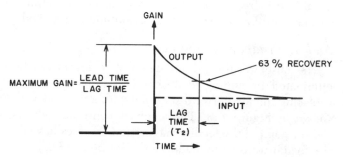

Fig. 1.5g The step response for a lead-lag unit

When it is properly adjusted, the dynamic compensation afforded by the lead-lag unit can produce the controlled variable response shown in Figure 1.5h. This is a major characteristic of a feedforward-controlled process since most processes do not consist of simple first-order lags; a first-order lead-lag unit cannot produce a perfect dynamic balance. Yet a major improvement over the uncompensated response can be attained and the error can be equally distributed on both sides of the set point.

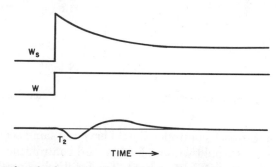

Fig. 1.5h The dynamic response for a compensated feedforward system

Adjusting the lead-lag unit requires some care. First, a load change should be introduced without dynamic compensation to observe the direction of the error. The compensation can be inhibited by setting the lead and lag time constants at equal values. If the resulting response is in the direction of the load response, the lead time should exceed the lag time; if not, the lag time should be greater. Next, measure the time required for the controlled variable to reach its maximum or minimum value. This time should be introduced into the lead-lag unit as the smaller time constant. Thus, if the lead dominates, this would be the lag setting. Set the greater time constant at twice this value and repeat the load change. If the error curve is still not equally distributed across the set point, one should increase the greater time constant and repeat the load change.

The outstanding characteristic of the response curve is that its area is equally distributed about the set point if the difference between the lead and lag settings is correct. Once this equalization is obtained, both settings should be increased or decreased with their difference constant until a minimum error amplitude is achieved.

Adding Feedback to the Loop

The offset resulting from steady-state errors can be eliminated by adding feedback. This can be done by replacing the feedforward set point with a controller, as shown in Figure 1.5i. The feedback controller adjusts the set point of the feedforward system in cascade while the feedforward system adjusts the set point of the manipulated flow controller in cascade.

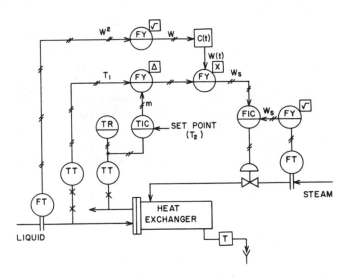

Fig. 1.5i   Feedforward-feedback control with dynamic compensation

The feedback controller should have the same control modes as it would without feedforward control, but the settings should not be as tight. The feedback controller reacts to a disturbance by creating another disturbance in the opposite direction one-half cycle later. The feedforward system then positions the manipulated variable so that the error in the controlled variable disappears. If acted upon by closely set feedback, the correct position will be altered, producing another disturbance that prolongs the settling time of the system.

The lead-lag unit is designated as C(t) in Figure 1.5i. Its settings are particular to one load input; namely the

liquid flow. Therefore it must be placed where it can modify that signal and no other. It cannot be placed anywhere in the feedback loop, as it could hinder the feedback function.

Performance

The use of feedback in a feedforward system does not detract from the performance improvement which was gained by feedforward control. Without feedforward, the control feedback loop was required to change its output to follow all changes in load. With feedforward, the feedback controller must only change its output by an amount equal to what the feedforward system fails to correct. A feedforward system applied to a heat exchanger could control the steam flow to within 2 percent of that required by the load. Thus, the feedback is only required to compensate 2 percent of a load change, rather than the full amount. This reduction of Δm in equation 1.5(5) by 50/1 results in the reduction of E by the same ratio. Reduction by 10/1 in errors resulting from load changes are relatively common, and improvements of 100/1 have been achieved in some systems.

The feedforward system is more costly and requires more engineering effort than a feedback system, so, prior to design and installation, the control improvement it brings must be determined to be worthwhile. Most feedforward systems have been applied to processes that are very sensitive to disturbances and slow to respond to corrective action and to product flows or values that are relatively high. Distillation columns of 50 trays or more have been the principal systems controlled with this technology. Boilers, multiple-effect evaporators, direct-fired heaters, waste neutralization plants, solids dryers, turbo-compressors, and other hard-to-control processes have also been controlled in this way.[3]

REFERENCES

1. Hordeski, M.F., "Fundamentals of Digital Control Loops," *Measurements and Control*, February, 1978.
2. Kuo, B.C., *Automatic Control Systems*, Prentice-Hall, Englewood Cliffs, NJ, 1962.
3. Shinskey, F.G., *Process Control Systems*, McGraw-Hill Book Co., New York, 1967.

BIBLIOGRAPHY

Anderson, N.A., "Combination Control System," *Instruments & Control Systems*, March, 1964.

Buckley, P.S., *Techniques of Process Control*, John Wiley & Sons, New York, 1964.

Catheron, A.R., "Factors in Precise Control of Liquid Flow," ISA Paper 50-8-2.

Considine, D.M., *Process Instruments & Controls Handbook*, McGraw-Hill Book Co., New York, 1957.

D'Azzo, J.J., and Houpis, C.H., *Feedback Control System Analysis and Synthesis*, McGraw-Hill Book Co., New York, 1960.

Eckman, D.P., *Automatic Process Control*, John Wiley & Sons, New York, 1958.

Gore, F., "Add Compensation for Feedforward Control," *Instruments & Control Systems,* March, 1979.

Harriott, P., *Process Control,* McGraw-Hill Book Co., New York, 1964.

Hougen, J.O., *Measurement and Control Applications,* ISA, Pittsburgh, 1979.

Johnson, C.D., *Process Control Instrumentation Technology,* John Wiley & Sons, New York, 1977.

Jones, B.E., *Instrumentation, Measurement and Feedback,* McGraw-Hill Book Co., New York, 1977.

Palmer, R., "Nonlinear Feedforward Can Reduce Servo Settling Time," *Control Engineering,* March, 1978.

Shinskey, F.G., "Controlling Unstable Processes, Part I: The Steam Jet," *Instruments & Control Systems,* December, 1974.

————, "Limit Cycles and Expanding Cycles," *Instruments & Control Systems,* November, 1971.

————, *pH and pIon Control in Process and Waste Streams,* Interscience, New York, 1973.

————, "A Self-Adjusting System for Effluent pH Control," Paper presented at the 1973 Instrument Society of America Joint Spring Conference, St. Louis, MO, April, 1973.

Skrokov, M.R., (ed), *Mini and Microcomputer Control in Industrial Processes,* Van Nostrand Reinhold, New York, 1980.

Trevathan, V.L., *pH Control in Waste Streams,* ISA Paper 72-725, 1972.

Wightman, E.J., *Instrumentation in Process Control,* Butterworth, 1972.

1.6 RATIO CONTROL

Ratio control systems maintain a relationship between two variables to provide regulation of a third variable. Ratio systems are used primarily for blending ingredients into a product or as feed controls to a chemical reactor. An example would be the addition of tetraethyl lead to a motor gasoline. A proper lead-to-gasoline ratio must be maintained to produce the desired octane number, which may or may not be measured.

Ratio systems actually portray the most elementary form of feedforward control (Section 1.5). The load input to the system (gasoline flow), if changed, would cause a variation in the controlled variable (octane number), which can be avoided by proper adjustment of the manipulated variable (additive flow). The load, or wild flow, as it is called, may be uncontrolled, controlled independently, or manipulated by another controller that responds to the variables of pressure, level, etc.

Flow Ratio Control

Ratio control is applied almost exclusively to flows. Consider maintaining a certain ratio R of ingredient B to ingredient A:

$$R = B/A \qquad 1.6(1)$$

There are two ways to accomplish this. The more common method manipulates a flow loop whose set point is calculated:

$$B = RA \qquad 1.6(2)$$

This system is shown in Figure 1.6a. The set point for the flow controller is generated by an adjustable-gain device known as a ratio station. Since the calculation is made outside of the control loop, it does not interfere with loop response.

The second method is to calculate R from the individual measurements of flows A and B, using equation 1.6(1). The calculated ratio would then be the controlled-variable input to a manually set controller. Changes in the set point would change the ratio. Such a scheme is shown in Figure 1.6b. The principal disadvantage of this system is that it places a divider inside a closed loop. If flow B responds linearly with valve B, the gain of the loop will

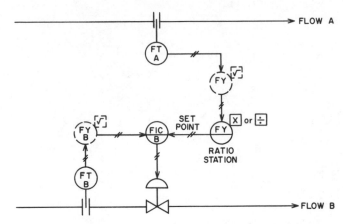

Fig. 1.6a System maintaining a constant ratio of controlled flow B to wild flow A

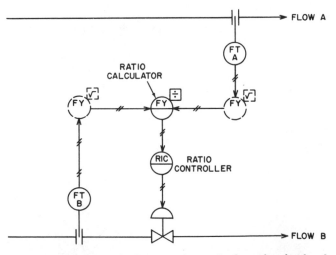

Fig. 1.6b Calculating the flow ratio places a divider within the closed loop

vary because of the divider. Differentiation of equation 1.6(1) explains why this is true:

$$\frac{dR}{dB} = \frac{1}{A} = \frac{R}{B} \qquad 1.6(3)$$

The gain varies both with the ratio and with flow B. In most instances, the ratio would not be subject to change,

48

but gain varying inversely with flow can cause instability at low rates. An equal-percentage valve must be used to overcome this danger. If the ratio were inverted,

$$R = A/B \qquad 1.6(4)$$

then

$$\frac{dR}{dB} = -\frac{A}{B^2} = -\frac{R}{B} \qquad 1.6(5)$$

and the results are essentially the same.

The square-root extractors in Figures 1.6a and b are shown in broken lines to indicate that the systems can operate without them using flow-squared signals. In this case, the controlled variable is

$$R^2 = B^2/A^2 \qquad 1.6(6)$$

The scale of the ratio controller or ratio station has to be nonlinear as a result. Differentiating,

$$\frac{d(R^2)}{dB} = \frac{2B}{A^2} = \frac{2R}{A} = \frac{2R^2}{B} \qquad 1.6(7)$$

Again, for the system in Figure 1.6b, the gain varies inversely with flow.

The principal advantage of using the ratio computing system is that the controlled variable—flow ratio—is constant and can be recorded to verify control. Using the system of Figure 1.6a, two records would have to be compared for verification.

Ratio Stations

No matter how ratio control is brought about, a computing device must be used whose scaling requires some consideration. The ratio station of Figure 1.6a normally has a gain range of about 0.3 to 3.0. Flow signal A, in percent of scale, is multiplied by the gain setting to produce a set point for flow controller B, in percent of scale. The true flow ratio must take into account the scales of the two flowmeters. The setting of the ratio station R is related to the true flow ratio by

$$R = \text{true flow ratio} \frac{\text{Flow A Scale}}{\text{Flow B Scale}} \qquad 1.6(8)$$

For example, the true flow ratio of the additive to gasoline is to be 2.0 cc/gal (0.53 cc/l). If the additive flow scale is 0–1,200 cc/min, and the gasoline flow scale is 0–500 gal/min (0–1890 l/m), then

$$R = 2.0 \text{ cc/gal} \frac{500 \text{ gal/min}}{1,200 \text{ cc/min}} = 0.833 \qquad 1.6(9)$$

When head flow signals are used, R should appear with a square root scale in order to be meaningful. Table 1.6c compares the gain of a ratio station (corresponding to linear flow ratio) with the ratio setting for head flowmeters. As shown by Table 1.6c, the available range of ratio settings is seriously limited when using squared flow signals.

Table 1.6c
RATIO SETTINGS FOR HEAD FLOWMETERS

Flow Ratio Desired	Actual Ratio (Gain) Setting Required to Achieve Desired Ratio	
	If Signals Are Linear	If Flow-squared Signals Are Used and the Scale is Linear
0.6	0.6	0.36
0.8	0.8	0.64
1.0	1.0	1.0
1.2	1.2	1.44
1.4	1.4	1.96
1.6	1.6	2.56

A device known as a ratio controller combines the ratio function and the controller in one unit. This is economical, saving not only cost, but panel space as well.

Since ratio stations are used with remote-set controllers, some means must be available for setting flow locally, for start-up, or during abnormal operation. An auto-manual station is sometimes provided for this purpose. With the ratio controller, this feature requires two scales on the set-point mechanism—one reading in ratio for remote-set operation, the other reading in flow units for local-set operation.

When using a divider as a ratio computer, with linear flow signals, the scale factor for the divider should be $\frac{1}{2}$. This places a ratio of 1.0 at midscale:

$$R = \frac{1}{2}\frac{B}{A} \qquad 1.6(10)$$

Equal flow signals will produce a 50 percent output from the divider, and the full ratio range is then 0–2.0, linear. If flow-squared signals are used, the divider should have a scale factor of $\frac{1}{3}$ to provide a full ratio range of 0–1.73, with a square root scale. This places a ratio of 1.0 (A = B) at 0.58 on the square root scale.

$$R^2 = \frac{1}{3}\frac{B^2}{A^2} \qquad 1.6(11)$$

If a scale factor of $\frac{1}{2}$ were used, the ratio range would be restricted to 0–1.41.

Setting the Ratio Remotely

When a divider is used to calculate the flow ratio, then this ratio may be set with a remote-set controller. In order to set the ratio remotely when the controlled flow set point is calculated, a multiplier must replace the ratio station. With linear flowmeters, the usual choice of scaling factor for the multiplier is 2.0:

$$B = 2RA \qquad 1.6(12)$$

In this way, a ratio of 1.0 appears at midscale of the

ratio input, i.e., where A = B, R = 50 percent of scale. If squared flow signals are used, a scaling factor of 3.0 provides a ratio range of 0–1.73:

$$B^2 = 3R^2A^2 \qquad 1.6(13)$$

One reason for using a multiplier to set the flow ratio is the availability of very narrow ratio ranges for those applications in which the need for precision, not rangeability, is paramount. Ratio ranges of 0.9–1.1, 0.8–1.2, etc. are possible. A typical application would be the accurate proportioning of ammonia and air to an oxidation reactor for the production of nitric acid. If the compositions of the individual feeds are constant, only a very fine adjustment of the ratio is required.

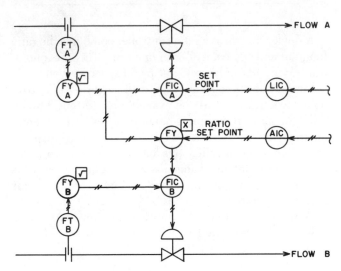

Fig. 1.6d The level controller manipulates flow rate and the composition controller manipulates flow ratio

Cascade Control of Ratio

Figure 1.6d shows two combinations of cascade and ratio control. Liquid level is affected by total flow, hence the liquid-level controller sets flow A, which in turn sets flow B proportionately. (To prevent the instability possible at low flow, when setting a head flowmeter in cascade, linear flow signals are used.)

Conversely, composition is not affected by the absolute value of either flow, but only by their ratio. Therefore, to make a change in composition, the controller must adjust the ratio set point of the multiplier. To minimize the effect of the composition controller on liquid level (through its manipulation of flow B), flow B should be the smaller of the two streams.

BIBLIOGRAPHY

Beard, M.J., "Analog Controllers Develop New Wrinkles," *Instruments & Control Systems*, November, 1977.

Buckley, P.S., "Dynamic Design of Pneumatic Control Loops: Parts I and II," *InTech*, April and June, 1975.

———, *Techniques of Process Control*, John Wiley & Sons, New York, 1964.

Eckman, D.P., *Automatic Process Control*, John Wiley & Sons, New York, 1958.

Fischer & Porter Co., "Relation and Ratio Control," Bulletin 91-10-01.

Harriott, P., *Process Control*, McGraw-Hill Book Co., New York, 1964.

Jones, B.E., *Instrumentation, Measurement and Feedback*, McGraw-Hill Book Co., New York, 1977.

Merritt, R., "Electronic Controller Survey," *Instrumentation Technology*, June, 1977.

Shinskey, F.G., "Control Topics," Publ. no. 413-1 to 413-8, Foxboro Company.

———, *Process Control Systems*, McGraw-Hill Book Co., New York, 1979.

1.7 CASCADE CONTROL

An intermediate process variable that responds both to the manipulated variable and to some disturbances can be used to achieve more effective control over the primary process variable. This technique, called cascade control, is shown in block diagram form in Figure 1.7a. Two controllers are used, but only one process variable (m) is manipulated. The primary controller maintains the primary variable c_1 at its set point by adjusting the set point r_2 of the secondary controller. The secondary controller, in turn, responds both to the output of the primary controller and to the secondary controlled variable c_2.

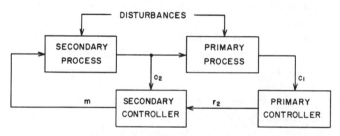

Fig. 1.7a   Process divided into two parts by the cascade control system

There are two distinct advantages gained with cascade control:

1. Disturbances affecting the secondary variable can be corrected by the secondary controller before a pronounced influence is felt by the primary variable.
2. Closing the control loop around the secondary part of the process reduces the phase lag seen by the primary controller, resulting in increased speed of response.

Primary and Secondary Loops

Figure 1.7b shows how closing the loop around the secondary part of the process can reduce its time lag. Here the response of the secondary closed loop to a step change in its set point r_2 is compared with the response of the secondary variable to an equivalent step in the manipulated variable. If one refers to Figure 1.7c, the secondary variable there is the jacket temperature and the manipulated variables are the steam and cold water flows. The secondary variable will always come to rest sooner under control, because initially the controller will demand a greater quantity of the manipulated variable than what is represented by the "equivalent" step change in the manipulated variable. This is particularly true when the secondary part of the process contains a dominant lag (as opposed to deadtime). Because the gain of the secondary controller can be high with a dominant lag, closing the loop is particularly effective. The dominant lag in the secondary loop of Figure 1.7c is the time lag associated with heat transfer.

One very important factor in designing a cascade system is the proper choice of the secondary variable. Ideally, the process should be split in half by the cascade

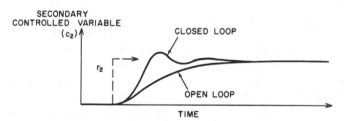

Fig. 1.7b   The secondary variable responds faster with its control loop closed

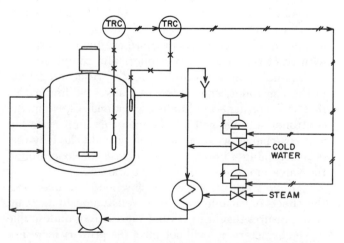

Fig. 1.7c   Cascade control of a stirred-tank reactor

loop, i.e., the secondary loop should be closed around half of the time lags in the process. To demonstrate, consider the two extremes:

1. If the secondary variable were to respond instantly to the manipulated variable, the secondary controller would accomplish nothing.
2. If the secondary loop were closed around the entire process, the primary controller would have no function.

For optimum performance, the dynamic elements in the process should also be distributed as equitably as possible between the two controllers.

An additional problem appears when most of the process dynamics are enclosed in the secondary loop. Although the response of this loop is faster than the process alone, its dynamic gain is also higher, as indicated by the damped oscillation in Figure 1.7b. This means that if stability is to be retained, the proportional band of the primary controller must be wider than it would be without a secondary loop. Proper choice of the secondary variable will allow a reduction in the proportional band of the primary controller, because the high-gain region of the secondary loop lies beyond the natural frequency of the primary loop. In essence, reducing the response time of the secondary loop moves it out of resonance with the primary loop.

Types of Secondary Loops

The most common forms of secondary loops are listed in order of their frequency of application.

1. *Valve Position:* The position assumed by the plug of a control valve is affected by forces other than the control signal, principally friction and line pressure (Section 4.2). Change in line pressure can cause a change in position and thereby upset a primary variable, and stem friction has an even more pronounced effect. Friction produces a square-loop hysteresis between the action of the control signal and its effect on the valve position. Hysteresis is a nonlinear dynamic element whose phase and gain vary with the amplitude of the control signal. Hysteresis always degrades performance, particularly where liquid level or gas pressure are being controlled with reset action in the controller. The combination of the natural integration of the process, reset integration in the controller, and hysteresis, causes a "limit cycle" that is a constant-amplitude oscillation. Adjusting the controller settings will not dampen this limit cycle but will just change its amplitude and period. The only way of overcoming a limit cycle is to close the loop around the valve motor (Section 4.15, Positioners).

2. *Flow Control:* A cascade flow loop can overcome the effects of valve hysteresis as well as a positioner can. It also insures that line pressure variations or undesirable valve characteristics will not affect the primary loop. For these reasons, in composition control systems, flow is usually set in cascade. Cascade flow loops are also used where accurate manipulation of flow is mandatory, as in the feedforward systems shown in Figures 1.5d and i.

3. *Temperature Control:* Chemical reactions are so sensitive to temperature that special consideration must be given to controlling the rate of heat transfer. The most commonly accepted configuration has the reactor temperature controlled by manipulating the coolant temperature in cascade. A typical system for a stirred tank reactor is shown in Figure 1.7c. Cascade control of coolant temperature at the exit of the jacket is much more effective than at the inlet, because the dynamics of the jacket are thereby transferred from the primary to the secondary loop. Adding cascade control to this system can lower both the proportional band and reset time of the primary controller by a factor of 2 or more. Since exothermic reactors require heating for start-up as well as cooling during the reaction, heating and cooling valves must be operated in split range. The sequencing of the valves is ordinarily done with positioners, resulting in two additional cascade sub-loops in the system.

Secondary Control Modes

Valve positioners are proportional controllers, usually with a fixed band of about 5 percent. Flow controllers invariably have both proportional and reset modes. In temperature-on-temperature systems, such as shown in Figure 1.7c, the secondary controller should not have reset. Reset is used to eliminate proportional offset, and in this situation a small amount of offset between the coolant temperature and its set point is inconsequential. Furthermore, reset adds the penalty of slowing the response of the secondary loop. The proportional band of the secondary temperature controller is usually as narrow as 10 to 15 percent. A secondary flow controller, however, with its proportional band exceeding 100 percent does definitely require reset.

Derivative action cannot be used in the secondary controller if it acts on set-point changes. Derivative action is designed to overcome some of the lag inside the control loop, and if applied to the set-point changes it results in excessive valve motion and overshoot. Some controllers are now available with derivative action on the measurement input only, and these can be used effectively in secondary loops where the measurement is sufficiently free of noise.

Instability in Cascade Loops

Adding cascade control to a system can destabilize the primary loop if most of the process dynamics are within the secondary loop. The most common example of this is the practice of using a valve positioner in a flow-control loop. Closing the loop around the valve increases its dynamic gain so much that the proportional band of the flow controller may have to be increased by a factor of 4 to maintain stability. The resulting wide proportional band means slower set-point response and deficient recovery from upsets. If a large valve motor or long pneumatic transmission lines cause problems, a volume booster should be used to load the valve motor, rather than a positioner. (See Section 4.15 on boosters and positioners).

Instability can also appear in a composition or temperature-control system where flow is set in cascade. These variables ordinarily respond linearly with flow, but the nonlinear characteristic of a head flowmeter can produce a variable gain in the primary loop. Figure 1.7d compares the manipulated flow record following a load change at 40 percent flow with a similar upset at 80 percent flow. Differential pressure h is proportional to flow squared:

$$h = kF^2 \qquad 1.7(1)$$

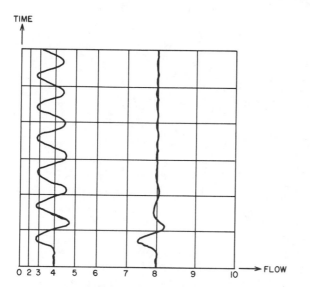

Fig. 1.7d Manipulated flow records show gain variation with load level

If the process is linear with flow, its loop gain will vary with flow when h is the manipulated variable, because flow is not linear with h:

$$\frac{dF}{dh} = \frac{1}{2kF} \qquad 1.7(2)$$

Thus, if the primary controller is adjusted for $\frac{1}{4}$-amplitude damping at 80 percent flow, the primary loop will be undamped at 40 percent flow and entirely unstable at lower rates.

Whenever a head flowmeter provides the secondary measurement in a cascade system, a square-root extractor should be used to linearize the flow signal, unless flow will always be above 50 percent of scale.

Saturation in Cascade Loops

When both the primary and secondary controllers have automatic reset, a saturation problem can develop. Should the primary controller saturate, limits can be placed on

its reset mode, or logic can be used to inhibit automatic reset as is done for controllers on batch processes. Saturation of the secondary controller poses another problem, however, because once the secondary loop is opened due to saturation, the primary controller will also saturate. A method for inhibiting reset action in the primary controller when the secondary loop is open for any reason is shown in Figure 1.7e.

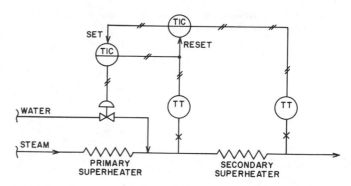

Fig. 1.7e Open secondary loop inhibits reset in the primary controller

In many pneumatic controllers, the output is fed back through a restrictor to a positive input bellows to develop automatic reset (Figure 1.2h). If the secondary controller has reset, its set point and measurement will be equal in the steady state, so the primary controller can be effectively reset by feeding back either signal. But if the secondary loop is open, so that its measurement no longer responds to its set point, the positive feedback loop to the primary controller will also open, inhibiting reset action.

Placing the dynamics of the secondary loop in the primary reset circuit is no detriment to control. In fact it tends to stabilize the primary loop by retarding the reset action. Figure 1.7e shows the first known application of this technique. The primary measurement in this example is at the outlet of a steam superheater, whereas the secondary measurement is at its inlet, with the valve delivering water. At low load, no water is necessary to keep the secondary temperature at its set point, but the controllers must be prepared to act should the load suddenly increase. In this example the proportional band of the secondary controller is generally wide enough to require reset action.

When putting a cascade system into automatic operation, the secondary controller must first be transferred to automatic. The same is true in adjusting the control modes, insofar as the secondary should always be adjusted first, with the primary in manual.

BIBLIOGRAPHY

Athans, M., and Falb, P., *Optimal Control: An Introduction to the Theory and Its Applications*, McGraw-Hill Book Co., New York, 1966.

Buckley, P.S., "Dynamic Design of Pneumatic Control Loops: Parts I and II," *InTech*, April and June, 1975.

———, *Techniques of Process Control*, John Wiley & Sons, New York, 1964.

Beveridge, G.S.G., and Schechter, R.S., *Optimization: Theory and Practice*, McGraw Hill Book Co., New York, 1970.

Eckman, D.P., *Automatic Process Control*, John Wiley & Sons, New York, 1958.

Gibson, J.E., *Nonlinear Automatic Control*, McGraw-Hill Book Co., New York, 1963.

Harriott, P., *Process Control*, McGraw-Hill Book Co., New York, 1964.

Jones, B.E., *Instrumentation, Measurement and Feedback*, McGraw-Hill Book Co., New York, 1977.

Lipták, B.G., "Process Instrumentation for Slurries and Viscous Materials," *Chemical Engineering*, January 30, 1967.

McCauley, A.P., Jr., and Persik, S.R., "From Linear to Gain-Adaptive Control: A Case History," *Instrumentation Technology*, November, 1977.

Shinskey, F.G., "Control Topics," Publ. no. 413-1 to 413-8, Foxboro Company.

———, *Process Control Systems*, McGraw-Hill Book Co., New York, 1979.

Webb, P.V., "Reducing Process Disturbances with Cascade Control," *Control Engineering*, August, 1961.

Wightman, E.J., *Instrumentation in Process Control*, Butterworth, 1972.

1.8 SELECTIVE CONTROL

For each controlled variable in a control system, there must be at least one manipulated variable. In many systems, however, the controlled variables outnumber the manipulated variables. When this happens, the system must decide how to share the manipulated variables. Switching between variables can be easily and smoothly accomplished using selective devices called signal selectors.

Signal selectors choose either the lowest, median, or highest control signal from among two or more signals. They are available in pneumatic and electronic analog hardware and in microprocessor-based digital control systems. A control loop containing this type of logic is called selective control.

Guarding Against Exceeding Constraints

Interlocks automatically shut down equipment when a hazardous temperature, pressure, composition, or level condition exists. Shutdowns can be avoided by taking corrective measures before the interlock condition is reached and keeping equipment running at a suboptimal level. Hazardous conditions that can be avoided by corrective action include

1. Flooding in a distillation column (reduce boilup or feed rate).
2. High reactor pressure or temperature (reduce heat input).
3. Low oxygen in furnace offgas (reduce fuel flow).
4. High steam header pressure (divert steam to a low pressure header or condenser).

In each of these cases, normal control is overridden by a secondary controller that has a higher priority.

Figure 1.8a shows a distillation column base level control loop where level is normally controlled by manipulating tails flow. If the tails line becomes restricted, tails flow loses its ability to control level and the level rises. The control system then selects the feed flow as a secondary manipulated variable. The reverse action proportional-only controller with a gain of 4 and the low signal selector perform the logic to override the normal feed flow controller and throttle the valve to control level.

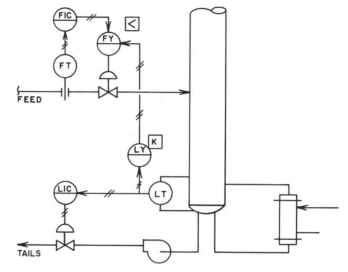

Fig. 1.8a   High base level override pinches feed valve

When the level is below 75 percent, the output of the gain 4 controller is 100 percent and is not selected. When the level rises above 75 percent, the output decreases until finally it is 0 percent when the level is 100 percent. Somewhere in between, the gain 4 controller output will be lower than the feed flow controller output and will be selected. Thus, the feed to the column will be reduced so that it just keeps up with the tails restriction, and the column does not flood.

Loops of this type are alternately called "overrides," "protective controls," or "soft constraint" controls. Signal selectors serve many purposes but, in most cases, they select the control signal of highest priority for safety or economic reasons.

Figure 1.8b illustrates a control system that protects against overpressurizing a reactor by reducing heat input. Steam is normally manipulated to control reactor temperature, but when the capacity to condense the overhead vapors decreases, causing pressure to rise, the high pressure override takes command of the steam valve and throttles it. The logic is provided by the high pressure override controller with a high set point and by a low signal selector.

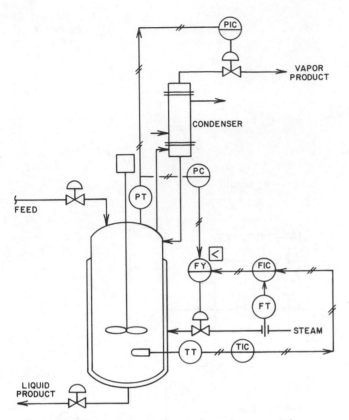

Fig. 1.8b   High pressure override pinches steam valve

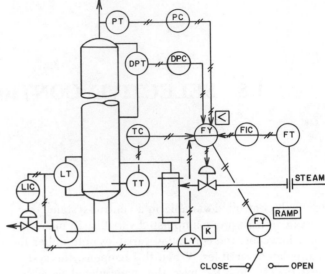

Fig. 1.8c   Automatic start-up controls

Automatic Start-up Controls

Some process start-ups are too complex to be accomplished by the operator without the aid of some automatic start-up controls. Automatic start-up controls typically include a ramping signal to open valves and override controls to prevent exceeding constraints. Signal selectors decide which control signal manipulates the valve.

Figure 1.8c illustrates automatic start-up controls on a distillation column. The low signal selector on the steam valve selects either the ramp signal, normal control signal, or one of the overrides.

Start-up proceeds like this. The operator begins to feed the column. When level is built up in the base, the steam valve ramp is energized and the valve ramps open at a predetermined rate. The normal steam flow controller is in automatic mode with the desired set point. When the valve delivers the set point flow, the flow controller output drops and takes over control of the valve. If the specified flow causes a constraint such as high column ΔP to be violated, the appropriate override controller takes over the valve. A number of override controls have been used to augment the normal controls:

1. A low-low base level (<10 percent) pinches steam.
2. A high-high pressure or ΔP pinches steam.
3. A high-high base temperature pinches steam.
4. A high-high base level pinches feed.

5. A high-high reflux tank level increases reflux or a low-low level pinches reflux if distillate flow is the normal manipulated variable.

The controls assist the operator in providing a smooth, safe start-up and reduce the level of human attention required. These benefits are particularly desirable for CRT-based distributed control systems whose window to the process is limited to only 8 to 12 loops at a time.

Protection Against Instrument Failures

Selectors are used to protect against transmitter failures by selecting a valid transmitter signal from among several. Figure 1.8d shows the feed controls to a reactor where it is critical to keep the ratio of feed streams within very tight limits. Losing a transmitter signal would cause

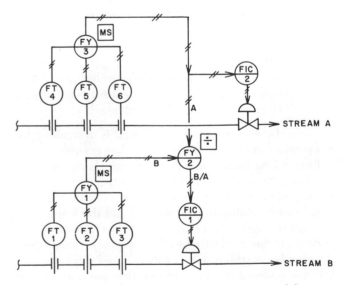

Fig. 1.8d   Redundant transmitters to protect against failures

one of the valves to open wide and exceed the ratio limit. Redundant transmitters monitor flow, and the median selector chooses the middle one for control. Two transmitters would have to fail at once for this loop to malfunction.

Redundant transmitters or sensors are commonly used in a hostile environment (high temperature or corrosive, dirty, or vibrating surroundings) where failure rates are high. High or median selectors keep the controls working when a transmitter fails, avoiding costly shutdowns.

Limitations on Manipulated Variables

In a situation of constrained control, there may be one controlled variable and a choice of manipulated variables. A typical example is the firing of a process heater with either of two fuels. The choice between the fuels is usually made on an availability basis. Fuel A may be burned to the limit of its availability and then it is supplemented with Fuel B. This limit may be set manually or by a controller responding to the storage capacity of fuel A.

A successful control system incorporating this limitation must have these two features: (1) capability of manipulating the limited variable on the allowable side of its limit and (2) smooth transition from one manipulated variable to the other without adversely affecting the controlled variable. Accommodating these features requires coordinating the manipulated variables and properly weighing their effects on the process.

A system wherein the temperature of the process is controlled by manipulating limited and supplemental fuels is shown in Figure 1.8e. Fuel A is limited by a second

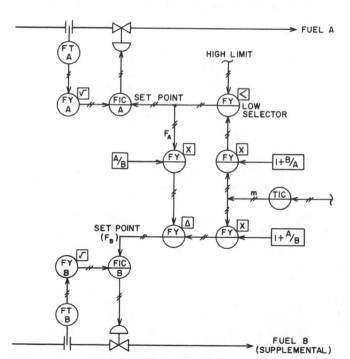

Fig. 1.8e Temperature is controlled by manipulating either limited fuel A or supplemental fuel B

(high limit) input to the low selector on its set-point signal. The limit could be set manually or could come from the output of a pressure controller on the fuel header. The objective is that fuel A, being less expensive, should be used to the limit of its availability before admitting fuel B.

The smooth transition through the limit can only be brought about by using the computing devices as shown. Any difference between the output of the temperature controller and the equivalent flow of fuel A is converted into a set-point signal for the supplemental fuel flow controller. The temperature controller output represents the total heat input to the system by the combined fuels. This is converted into the set point for fuel A by a multiplying factor of $1 + B/A$, where B/A is the ratio of full-scale heat input of fuel B to that of fuel A. This equates the heat-flow values of controller output and fuel A set point.

As long as fuel A is at or below its high limit, the two inputs to the subtracting relay will cancel one another:

$$m(1 + A/B) - m(1 + B/A)(A/B) = 0 \qquad 1.8(1)$$

When the high limit is exceeded, fuel A set point becomes constant and the fuel B set point starts increasing smoothly from zero without affecting the gain of the temperature-control loop. The system will also work with the fuel A controller in manual, as long as the heat load is above what fuel A can supply.

Selection of Extremes

Another use for signal selectors is in controlling an extreme of several similar variables. For example, the hottest temperature in a fixed-bed chemical reactor should be controlled to avoid damage, but its location can change with flow rate and time. So temperatures at several locations are measured and a high selector chooses the highest temperature for control.

Anti-Reset Windup Protection in Loops with Selectors

Reset action in a controller will cause the output to saturate if the controller is not able to adjust the manipulated variable so that the controlled variable matches the set point. This is often called "reset windup" and can occur in loops where a selector chooses between the outputs of two or more proportional-reset controllers. Only one controller at a time commands the valve and the non-selected controller will windup unless something is done to limit the reset action.

One solution to reset windup is to modify the standard proportional-reset equation by adding a term that integrates the difference between the controller output and the valve signal. When the controller is not selected, the controller output and valve signal will be different and integration occurs. The new term cancels out the normal reset action so, in effect, reset action is stopped until that controller resumes control. The valve signal must be fed back to the modified controller, hence the name "external reset feedback."

The following equations describe a standard proportional-reset controller and one with "external reset feedback."

Normal proportional-reset:

$$m(t) = K_c\, e(t) + \frac{K_c}{T_r} \int e(t)dt \qquad 1.8(2)$$

With external reset feedback:

$$m(t) = K_c\, e(t) + \frac{K_c}{T_r} \int e(t)dt +$$

$$+ \frac{1}{T_r} \int [F(t) - M(t)]dt \qquad 1.8(3)$$

Figure 1.8f shows a cascade control loop containing an override. The feedback variable (F) for the flow (slave)

controller and the override controller is the valve signal. The feedback variable for the temperature (master) controller is the flow measurement.

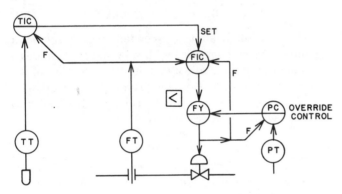

Fig. 1.8f   Override loop showing external reset feedback connections

BIBLIOGRAPHY

Anderson, L.M., "New Extraction Control System for Steam Turbines," *Hydrocarbon Processing*, February, 1976.

Buckley, P.S., "Designing Override and Feedforward Controls," *Control Engineering*, August, 1971.

———, "Protective Controls for a Chemical Reactor," *Chemical Engineering*, April 20, 1970.

———, *Techniques of Process Control*, John Wiley & Sons, New York, 1964.

Buckley, P.S., and Cox, R.K., "New Development in Overrides for Distillations," ISA Trans. 10, no. 4, 1971.

Buzzard, W.S., "Override Strategies for Analog Control," *Instrumentation Technology*, November, 1978.

Eckman, D.P., *Automatic Process Control*, John Wiley & Sons, New York, 1958.

Gaines, L.D., and Giles, R.F., "Making the Most of Changing Fuel Supplies," *Instruments & Control Systems*, February, 1978.

Giles, R.F., and Gaines, L.D., "Integral-Tracking Override is Better than Output-Tracking," *Control Engineering*, February, 1978.

Hall, C.J., "Don't Live with Reset Windup," *Instrumentation Technology*, July, 1974.

Harriott, P., *Process Control*, McGraw-Hill Book Co., New York, 1964.

Shinskey, F.G., *Distillation Control*, McGraw-Hill Book Co., New York, 1977.

———, *Energy Conservation through Control*, Academic Press, New York, 1978.

———, *Process Control Systems*, McGraw-Hill Book Co., New York, 1967.

Shunta, J.P., and Cox, R.K., "Sampled Data Controller for Loops with Overrides," Paper presented at 74th National AIChE Meeting, New Orleans, March, 1973.

Shunta, J.P., and Klein, W.F., "Microcomputer Digital Control—What It Ought To Do", ISA Trans. 18, no. 1, 1979.

Wightman, E.J., *Instrumentation in Process Control*, Butterworth, 1972.

1.9 ADAPTIVE AND NONLINEAR CONTROL

An adaptive control system is one whose parameters are automatically adjusted to meet corresponding variations in the parameters of the process being controlled in order to optimize the response of the loop. The significant point is the adjustment of *parameters*, which are normally fixed, as opposed to the outputs of a system, which are expected to vary.

The parameters set into controllers and control systems naturally reflect the characteristics of the processes they control. To maintain optimum performance, these parameters should change as the associated process characteristics change. If these changing characteristics can be directly related to the magnitude of the controller input or output, it is possible to compensate for their variation by the introduction of suitable nonlinear functions at the input or output of the controller. Examples of this type of compensation would be the introduction of a square-root converter in a differential pressure flow-control loop to linearize the measurement, or the selection of a particular valve characteristic to offset the effect of line resistance on flow. This is not considered to be adaptation, in that the controller functions remain fixed.

If the process nonlinearities are compensated by controller functions (i.e., changing controller gain, reset, or derivative tuning parameters) the controller is a nonlinear controller. An example of this would be adjusting the controller gain to compensate for the nonlinearities of a pH system titration curve. Changing controller gain as a function of the pH controller input (measured variable) is not strictly adaptive, but it is nonlinear. Nonlinear control will be considered here as a subset of adaptive control.

A system that adapts on the basis of a measurement of the disturbing factor is said to be programmed, and one which uses a measurement of its own performance is called self-adaptive. Programmed adaption of feedback control systems is widely used in the processing industry, while self-adaption techniques are slowly gaining understanding and acceptance.

Adaption of Feedback Parameters

When a feedback control loop is disturbed by a change in load or set point, a deviation from set point will appear in the controlled variable for a certain period of time. The function of the controller is to return this deviation to zero. The size of the deviation relative to the upset and the path it takes in returning to zero reveal how well the controller is doing its job. This response should be well damped, yet the time required for the controlled variable to return to the set point should be minimized. Ordinarily, the responsibility for achieving a desirable response rests with the engineer who selects and adjusts the controller. The process characteristics might change in time, and the controller loop response might deteriorate unless the engineer readjusts the controller.

It would be desirable, therefore, to have the controller settings adjusted automatically to compensate for variations in process parameters as they are detected.

Programmed Adaption

Where a measurable process variable produces a predictable effect on the gain of the control loop, compensation for its effect can be programmed into the control system. The most notable example in process plants is the variation in dynamic gain with flow in longitudinal equipment where no backmixing takes place. It is commonly seen in heat exchangers, but it causes a particularly severe problem in once-through boilers.

In a once-through boiler, feedwater enters the economizer tubing section, passes directly into the evaporative tubing, and then into the superheater tubing, and leaves as superheated steam whose temperature must be accurately controlled. No mixing takes place as in the drum boiler, so a sizable amount of deadtime exists in the temperature control loop, particularly at low flow rates. Figure 1.9a shows the response of steam temperature to changes in firing rate at two different feedwater flows. At 50 percent flow, the steady-state gain is twice as high as at 100 percent flow, because only half as much water is available to absorb the same increase in heat input. The deadtime and the dominant time constant are also twice as great at 50 percent flow.

The effect of these variable properties on the dynamic gain of the process is evidenced in Figure 1.9b. The same size load upset produces a larger excursion in temperature at 50 percent flow, indicating a higher dynamic

59

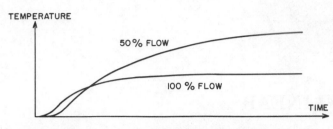

Fig. 1.9a   Step response of steam temperature to firing rate in a once-through boiler

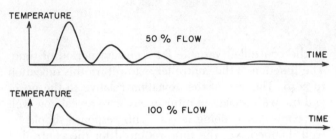

Fig. 1.9b   Response of the steam temperature loop to step changes in load without adaption

gain. The difference in damping between the two conditions also reveals the change in dynamic gain and so does the response, which is twice as fast at 100 percent flow. If oscillations were evident in the 100 percent flow response curve, their period would be much shorter than for 50 percent flow, due to the difference in deadtime. At 25 percent flow, oscillations would be still slower and damping would disappear altogether.

If the only feedback mode used in this application were proportional, the period of oscillation would vary inversely with flow, as would the dynamic gain at that period. The only adjustment that could be made would be that of the proportional band, which should vary inversely with flow. This would make the damping uniform, but nothing can change the increased sensitivity of the process to upsets at low flow rates.

Since reset and derivative modes are normally used to control temperature, some consideration should also be given to their adaption to changes in flow. The process deadtime—hence its period of oscillation under proportional control—varies inversely with flow. Therefore, the reset and derivative time constants also should vary inversely with flow for the adaption to be complete.

The equation for the flow-adapted three-mode controller is

$$m = \frac{100w}{PB}\left(e + \frac{w}{T_i}\int e\,dt + \frac{T_d}{w}\frac{d}{dt}e\right) \quad 1.9(1)$$

where w is the fraction of full-scale flow, and PB, T_i and T_d are the proportional, reset, and derivative settings at full-scale flow. Equation 1.9(1) can be rewritten to reduce the adaptive terms to two:

$$m = \frac{100}{PB}\left(we + \frac{w^2}{T_i}\int e\,dt + T_d\frac{d}{dt}e\right) \quad 1.9(2)$$

Other parameters can be substituted for flow in instances where they apply.

Gap Action

Gap action or dead band control is a frequently used form of nonlinear control. The dead band action is typically a programmed nonlinearity or adaption depending on whether dead band is set by process variable or disturbance factor. (See definition of adaptive vs. nonlinear in the beginning of this section.)

Gap action is not a stand-alone control function. Proportional, integral, and derivative modes can still be used. If this function is used, a dead band is placed around the controller error such that no control action occurs unless the error exceeds the dead band range. If the error is within the dead band, the error is set to zero. Dead band control has found application in pH control and in systems requiring two control valves, one large, one small, to overcome the rangeability problem.

For example, consider the control scheme shown in Figure 1.9c(1). Here the gap action nonlinear controller is used to drive the large valve when the output of the conventional proportional reset controller reaches a preset limit (top or bottom). The big valve makes a rough adjustment when it is active outside of its dead band or

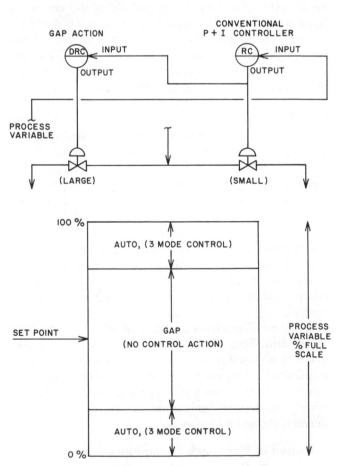

Fig. 1.9c   (1) Gap action nonlinear control application for large valve, small valve rangeability problem. (2) gap action nonlinear controller example.

gap. This causes the small valve controller to come back into its fine adjustment range, again inside the big valve controller's dead band.

Figure 1.9c(2) shows the regions where the gap action controller is effective. Depending on the relative size of the trim valve and the response desired, the gap can vary widely. Several commercial electronic panel controllers can be purchased with this gap or dead band feature. Most microprocessor and computer control packages also have this feature.

Variable Breakpoint

Another form of programmed adaptation or nonlinear control is referred to as variable breakpoint control. The proportional gain for this type of controller is changed at certain predefined values of process variable or controller error (process variable − set point). In this type of application it is usually desired to have low controller gain at or around set point, i.e., not much controller action near set point. If the process varies some programmed distance from set point, the controller gain is increased to drive the process back to set point at a faster rate. This type of control action may also be used to compensate for a very nonlinear process gain characteristic. In this case the controller gain is set high when the process gain is low and set low when the process gain is high to maintain a relatively constant closed loop gain (i.e., consistent close loop performance throughout control range).

The pH control problem is a good example of this. Consider the process gain characteristic of the titration system in Figure 1.9d. This is a typical strong acid, strong base neutralization process. The control set point is at pH = 7. The process has extremely high gain at this point (i.e., a small amount of reagent change makes a dramatic change in pH). The required controller gain between pH of 3 and 9 is small for stability. If the process were to get above 9 or below 3, this same low gain would

result in a very sluggish response because now the process gain is low. So above 9 and below 3 the required controller gain is large. The required variable breakpoint control gain for this example is shown as a dashed line on Figure 1.9d. The gain between pH of 3 and 9 is low; the gain below 3 and above 9 is high.

Again, several instrument vendors offer such electronic and microprocessor-based controllers. Computer controllers can also have this feature. It is important to point out that the implementation of this algorithm does require more than just a simple switching of gain numbers at breakpoints. Care must be taken to insure that the gain change does not result in a bump or discontinuity in control action.

Error Squared

Another useful programmed nonlinear or adaptive control algorithm is referred to as the error squared option. It again is used to modify the base proportional band and reset time constant.

A tuning parameter (RFACT) sets the severity of the nonlinearity. Proportional gain and reset time constant are multiplied by FNL and then used in the proportional reset equations discussed in Sections 1.2 and 1.4.

The equations are typically:

$$\text{Error} = \text{Set point} - \text{Measured variable}$$

$$E = \text{ABS(Error)} \text{ (normalize 0. to 1.)} \qquad 1.9(3)$$

$$\begin{aligned} \text{FNL} = {}& 1.0 + 90(\text{E/RFACT}) + \\ & + 3045(\text{E/RFACT})^2 + \\ & + 145675(\text{E/RFACT})^3 \qquad 1.9(4) \end{aligned}$$

$$\begin{aligned} 100/\text{PB} = \text{Controller gain} = {}& \text{FNL} \times \\ & \times \text{Controller gain (base)} \qquad 1.9(5) \end{aligned}$$

$$\begin{aligned} 1/T_i = \text{Controller reset} = {}& \text{FNL} \times \\ & \times \text{Controller reset (base)} \qquad 1.9(6) \end{aligned}$$

RFACT sets the control action nonlinearity by causing the factor FNL to double for every RFACT % increase in controller error (for example, RFACT = 14 means FNL = 2 when error = 14%, FNL = 4 when error = 28%, etc.). Control action becomes stronger as the error increases.

Typically the FNL factor can be applied to gain, to reset, or both. When both are chosen the algorithm is particularly suited to level control. For more detail on the control technique see Reference 1.

Computer and Microprocessor Features

Computers and microprocessors have allowed for a great deal of flexibility in the areas of programmed adaptive and nonlinear control. These methods of control typically permit the control design engineer to specify virtually any adaptive or nonlinear change in tuning parameters desired.

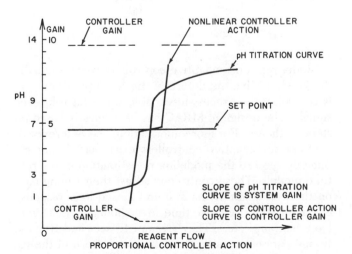

Fig. 1.9d Variable breakpoint nonlinear control action for strong acid, strong base neutralization example

With this new-found flexibility come some new risks. It is always important to consider what the effects of your programmed adaptive or nonlinear control will be at all process conditions (i.e., start-up and shutdown). It is also important to put some absolute bounds on the adaptive nature of the tuning parameter calculation. It is also a good idea never to change the tuning parameter in a discontinuous manner. This action may cause a discontinuity in control output unless smooth transitioning logic has been provided.

Self-Adaption

Where the cause of changes in control loop response is unknown or unmeasurable, strict programmed adaption cannot be used. If adaption is to be applied under these circumstances, it must be based on the response of the loop itself, i.e., it must be self-adapting. Self-adaption is a more difficult problem than programmed adaption because it requires an accurate evaluation of loop responses, ideally without knowledge of the disturbance input.

The block diagram of a self-adaptive system is given in Figure 1.9e. This system has all the problems of implementing the programmed adaption plus the problems of evaluating response and making a decision on the correct adjustment. Several self-adaptive techniques that are currently being used in the process industry are briefly discussed below. These include the self-tuning regulator, the model reference controller, and the pattern recognition adaptive controller.

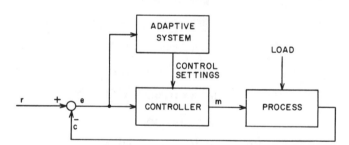

Fig. 1.9e   A self-adaptive system is a control loop around a control loop

The self-tuning regulator (STR) is a name given to a large class of self-adaptive systems. The block diagram in Figure 1.9f shows the common structure of these STR systems. The figure indicates that all STRs have an identifier section, typically consisting of a process parameter estimation algorithm. Also common to all STRs is a regulator parameter calculation section. This section calculates the new controller parameters as a function of the estimated process parameters. The methods used in these two blocks distinguish the type of STR being used. Popular varieties of STR include minimum-variance, generalized minimum-variance, detuned minimum-variance, dead beat, and generalized pole-place-

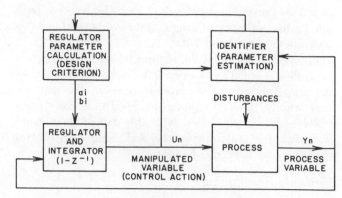

Fig. 1.9f   Self-tuning regulator

ment controllers. Reference 2 is an excellent summary paper on STR technology.

Model reference adaptive control (MRAC) offers a second possible method for the self-adaptive control problem. The controller is composed of a reference model, which specifies the desired performance, an adjustable controller whose performance should be as close as possible to that of the reference model, and an adaptation mechanism. This adaptation mechanism processes the error between the reference model and the real process in order to modify the parameter of the adjustable controller accordingly.

Figure 1.9g schematically shows how the parts of a model reference controller are organized.

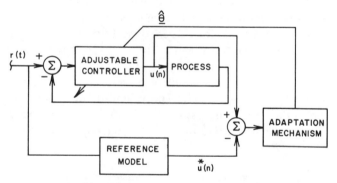

Fig. 1.9g   Model reference adaptive controller

Model reference adaptive controllers were originally designed to solve the deterministic servo problem, that is control of the process-to-variable set point reference signals. The design of MRAC has been mostly based on stability theory. For more information, see Reference 3.

Other self-adaptive controllers exist that *do not* explicitly require the modeling or estimation of discrete time models. These controllers adjust their tuning base on the evaluation of the system's closed-loop response characteristics (i.e., rise time, overshoot, settling time, loop damping, etc.). They attempt to "cut and try" the tuning parameters and recognize the pattern of the response. Thus the name "pattern recognition."

Pattern recognition controllers are commercially available from several instrument vendors. They are microprocessor based and usually heavily constrained with regard to the severity of allowable tuning parameter adjustments. They are gaining fair acceptance in operating plants.

Performance

The performance difference between programmed and self-adaptive systems is analagous to that which exists between feedforward and feedback control. A self-adaptive controller cannot make an adjustment to correct its settings until an unsatisfactory response is encountered. Two or more cycles must pass before an evaluation can be made upon which to base an adjustment. Therefore, the cycle time of the adaptive loop must be much longer than the natural period of the control loop itself. Consequently, the self-adaptive system cannot correct for the present poor response but can only prepare for the response to the next upset, assuming that the control settings presently generated will also be valid then. On the other hand, the programmed system should always have the correct settings because it responds antomatically to changes in process variables in a manner analogous to feedforward control. It does not have to "learn" the new process dynamics via adaptive loops.

REFERENCES

1. Shunta, J.P., and Fehervari, W. "Nonlinear Control of Level." *Instrumentation Technology,* January, 1976.

2. Soderstrom, T., Ljung, L., and Gustausson, I., *A Comparative Study of Recursive Identification Methods,* Division of Auto. Cont., Lund Institute of Technology, Report 7427, 1974.

3. Goodwin, G.C., *Adaptive Filtering, Prediction and Control,* Academic Press, New York, 1981.

BIBLIOGRAPHY

Borisson, U., "Self-Tuning Regulators for a Class of Multivariable Systems," *Automation,* July, 1977.

Box, G.E.P., and Jenkins, G.M., *Time Series Analysis: Forecasting and Control,* rev. ed., Holden-Day, San Francisco, 1976.

Bristol, E.H., "Adaptive Process Control by Pattern Recognition," *Instruments & Control Systems,* March, 1970.

———, "Pattern Recognition Adaptive Control," *IEEE* Trans., January, 1962.

Buckley, P.S., *Techniques of Process Control,* John Wiley & Sons, New York, 1964.

Harriott, P., *Process Control,* McGraw-Hill Book Co., New York, 1964.

Hausman, J.F., "Applying Adaptive Gain Controllers," *Instruments & Control Systems,* April, 1979.

Keviczky, L., and Hetthessy, J., "Self-Tuning Minimum Variance Control of MIMO Discrete Time Systems," *Automatic Control Theory and Applications,* May, 1977.

Lipták, B.G. (ed.), *Instrument Engineer's Handbook,* vol. I, ch. I, Chilton Book Co., Philadelphia, 1969.

McCauley, A.P., "From Linear to Gain-adaptive Control," *Instrumentation Technology,* November, 1977.

Shinskey, F.G., *Process-Control Systems,* McGraw-Hill Book Co., New York, 1967.

———, "A Self-Adjusting System for Effluent pH Control," Paper presented at the 1973 ISA Joint Spring Conference, St. Louis, April, 1973.

Shunta, J.P., "Nonlinear Control of Liquid Level," *Instrumentation Technology,* January, 1976.

Trevathan, V.L., "pH Control in Waste Streams," ISA Paper no. 72-725.

1.10 OPTIMIZING CONTROL

Optimizing control is essential in industrial processing today if an enterprise expects to retain viability and competitive vigor. The growing scarcity of raw materials aggravated by dwindling energy supplies has created a need to achieve the utmost in productivity and efficiency. "Optimization" means different things to different people. To some, it means the application of advanced regulatory control. Others have observed impressive results from applying multivariable noninteracting control. To be sure, these latter techniques have merit and may actually comprise a subset of optimization. However, in the true sense of the word, optimization implies a procedure for reaching a well-defined objective, measured by a relevant criterion, and meaningful not only for the present operation but also sensitive to time-related effects.

It is the purpose of this section to guide the practicing instrument engineer in a suitable course of action when confronted with a problem involving optimization. Topics emphasized will include:

1. Defining the problem scope
2. Choosing between feedback or feedforward optimization
3. Mathematical tools available
4. Multi-variable non-interacting control
5. Identifying and handling constraints

An extensive reference list and bibliography will be useful in amplifying the concepts and supplying details for the topics covered.

Considerations in Optimization

Optimization provides a management tool for achieving the greatest possible efficiency or profitability in the operation of any given production process. Changes in the operational environment consisting of the current constraints and values for the disturbance variables will inevitably alter the optimal position. Hence, optimizing control must be able to cope with change.

Perhaps the most difficult task in the design of an optimization control system is in the definition of the problem scope and in the subsequent choice of optimizing tactics. The need for an on-line optimizing system can only be ascertained following a reasonably in-depth feasibility study. It is the purpose of this section to provide some insights into the resolution of the above problems.

Processes which are the best candidates for optimizing control are characterized by:

1. Multiple independent control parameters
2. Frequent and sizeable changes in disturbance variables affecting plant profitability
3. A required and necessary excess in the number of degrees of freedom in the independent control parameters.

Figure 1.10a attempts to illustrate, in simplistic fashion, the interrelationships existing among the process system variables. The independent variables, $X(i)$, represent those over which control can be exercised by means of conventional regulatory systems. The disturbance variables, $Y(i)$, represent variables over which little or no control can be exercised, such as ambient conditions, feedstock quality, possible changes in market demand, etc. The intermediate variables, $W(i)$, are those that describe certain complex and calculable process conditions such as internal reflux rates, reactor conversion, and tube metal temperatures. The intermediate variables can be, and often are, constraining factors in process operation. The performance variables, $Z(i)$, represent the objective or target values for the process, such as yields, production rates, and quality of product.

Once the problem scope has been defined, the choice of either a local or global optimization strategy can be made. The local optimization problems, such as optimal

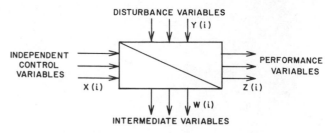

Fig. 1.10a   Process variable interrelationships

boiler fuel/air ratio control, can be implemented using simpler control algorithms and hardware. On the other hand, global optimization may be required for complex assemblages of equipment with a high degree of interaction in their operation.

The feasibility study will normally indicate whether feedforward or feedback control will be more efficacious in optimizing plant operation. As a general rule, feedback optimizing control is encountered only in the local optimization problem. Feedforward, on the other hand, is eminently suited for the larger global problems.

Feedback optimization control strategies are always applied to transient and dynamic situations. Evolutionary optimization or EVOP is a good example. Steady state optimization, on the other hand, is widely used on complex processes exhibiting long time constants and with disturbance variables that change infrequently. Hybrid strategies are also employed in situations involving both long-term and short-term dynamics. Obviously the hybrid algorithms are more complex and require custom tailoring for a truly effective implementation.

Feedback Control

Feedback control can sometimes be employed in certain situations to achieve optimal plant performance. Evolutionary optimization, or EVOP, is one such technique using feedback as the basis for its strategy. EVOP is an on-line experimentor. No extensive mathematical model is required, since small perturbations of the independent control variable are made directly upon the process itself. As in all optimizers, EVOP also requires an objective function.

EVOP does suffer certain limitations. The process must be tolerant of some small changes in the major independent variable. Secondly, it is necessary to apply EVOP or feedback control to perturb a single independent variable at a time. If a process is encountered, such that two independent variables are considered major contributors to the objective, then it may be possible to configure the controller to examine each one sequentially at alternate sampled-data periods. This latter approach is feasible only if the process dynamics are rapid when compared with the frequency of expected changes in the disturbance variables.

The author has successfully used EVOP to maximize the thermal efficiency of industrial boilers in which the fuel/air ratio was adjusted by an oxygen-trim control system. EVOP was used to perturb the oxygen set point. The objective function consisted of an on-line calculation of the current thermal efficiency.

Feedforward Optimizing Control

Multivariable processes in which there are numerous interactive effects of independent variables upon the process performance can best be optimized by the use of feedforward control. An adequate predictive mathematical model of the process is an absolute necessity. Note that the on-line control computer will evaluate the consequences of variable changes using the model rather than perturbing the process itself.

To produce a viable optimization result, the mathematical model must be an accurate representation of the process. To ensure a one-to-one correspondence with the process, the model must be updated just prior to each use. Model updating is a specialized form of feedback in which model predictions are compared with the current plant operating status. Any variances noted are then used to adjust certain key coefficients in the model to enforce the required agreement.

Figure 1.10b is a signal-flow block diagram of a computer-based feedforward optimizing control system. Process variables are measured, checked for reliability, filtered, averaged, and stored in the computer data base. A regulatory system is provided as a front line control to keep the process variables at a prescribed and desired slate of values. The conditioned set of measured variables are compared in the regulatory system with the desired set points. Errors detected are then used to generate control actions that are then transmitted to final control elements in the process. Set points for the regulatory system are derived either from operator input or from outputs of the optimization routine. Note that the optimizer operates directly upon the model in arriving at its optimal set-point slate. Also note that the model is updated by means of a special routine just prior to use by the optimizer. The feedback update feature ensures adequate mathematical process description in spite of minor instrumentation errors and, in addition, will compensate for discrepancies arising from simplifying assumptions incorporated in the model.

The mathematical bases and operating characteristics of some feedback optimization systems will be discussed more fully in the subsections to follow.

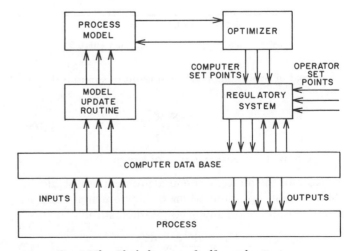

Fig. 1.10b   Block diagram—feedforward optimizer

Optimizing Tools

Calculus of variations is a classical mathematical approach that can be used to optimize the operation of dynamically changing processes. Though this technique has not found widespread use in industry, it should at least be considered in any batch chemical reactor problem in which time-temperature-concentration control must be optimized. Theoretical considerations and some application data may be found in References 1, 2, and 3.

Economic dispatch or optimal load allocation is a technique directed to the most effective use of the capabilities of parallel multiple resources to satisfy a given production requirement. As an example, the problem of steam load allocation among several boilers in an industrial utilities plant is quite frequently encountered. A typical boiler efficiency curve is shown in Figure 1.10c. For all practical purposes, the curve may be represented by a quadratic equation.

$$\eta = \eta_0 - K(S - S_0)^2 \qquad 1.10(1)$$

where

η = efficiency at given steam load, S

η_0 = maximum efficiency at S_0,

K = constant, and

S, S_0 = steam loads (any consistent units, such as tons/hour).

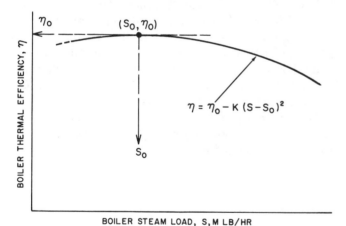

Fig. 1.10c   Typical boiler efficiency curve

The cost of producing a given amount of steam is defined by:

$$C = \frac{F \times S}{\eta} \qquad 1.10(2)$$

where F = cost per unit of steam.

Now for the multiple resource problem, use of calculus indicates an optimum when the incremental costs are equal for all boilers involved. Mathematically,

$$\frac{dC_1}{dS_1} = \frac{dC_2}{dS_2} = \frac{dC_3}{dS_3} \cdots \qquad 1.10(3)$$

Also, to satisfy the plant total steam demand, the following constraint equation must be satisfied:

$$S_1 + S_2 + S_3 + \cdots = S \qquad 1.10(4)$$

Substituting equation 1.10(1) into equation 1.10(2) and then differentiating with respect to steam load S, gives, for a given boiler,

$$IC = \frac{FK[(\eta_0 K - S_0{}^2) + S^2]}{[\eta_0 K - (S - S_0)^2]^2} \qquad 1.10(5)$$

where IC = incremental cost.

In the above equation, the parameters η_0, K, and S_0 must be evaluated from experimental test data for each boiler in the system. Figure 1.10d is a plot for a hypothetical three boiler system showing a graphical solution to the load allocation problem. The horizontal line intersecting the incremental cost curves satisfies the plant load demand expressed in equation 1.10(4).

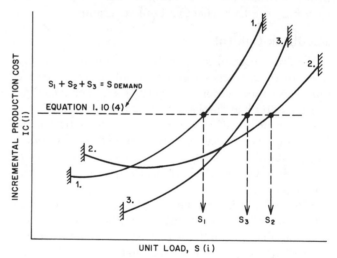

Fig. 1.10d   Graphical solution of the optimal load allocation problem

Linear programming is a mathematical technique that can be applied provided that all the equations describing the system are linear. In the field of industrial process control, the number of truly linear systems is somewhat limited, but some are encountered occasionally. Simple physical ingredient blending problems such as in cement, glass, and certain alloy manufacture are examples of these. A linear program can be used as an optimizing method if the problem can be reduced to the following set of relationships:

$$C = C_1 X_1 + C_2 X_2 + \cdots C_n X_n \qquad 1.10(6)$$

$$\begin{vmatrix} a_{11}X_1 + a_{12}X_2 + \cdots a_{1n}X_n = b_1 \\ a_{21}X_1 + a_{22}X_2 + \cdots a_{2n}X_n = b_2 \\ \cdot \qquad\qquad \cdot \qquad\qquad\quad \cdot \qquad \cdot \\ \cdot \qquad\qquad \cdot \qquad\qquad\quad \cdot \qquad \cdot \\ a_{m1}X_1 + a_{m2}X_2 + \cdots a_{mn}X_n = b_m \end{vmatrix} \qquad 1.10(7)$$

Subject to:

$$X_i \geq 0 \qquad i = 1, 2, 3, \cdots n \qquad 1.10(8)$$
$$b_i \geq 0 \qquad i = 1, 2, 3, \cdots n$$

where

x = independent variables,
c = associated cost factors,
a = linear coefficients, and
b = specified variables.

The b variables can represent specific desired concentrations, mass balance, or heat balance requirements. Note that the constants a or c can assume any value including zero.

The objective function will either be maximized or minimized depending upon whether it represents profit or costs. Many excellent textbooks outline the procedures and algorithms used in the solution of a linear program.[5,6,7,8]

Most industrial processes, especially in the chemical and petroleum industries, cannot be described by linear equations over their complete operating range. These processes therefore require some type of nonlinear optimizer to achieve maximum profitability. Two types of nonlinear optimizers—the sectionalized linear program and the gradient search—have been successfully implemented in advanced computer control schemes.

The sectionalized linear program is especially useful for those processes that exhibit slight to moderate nonlinearities in their variable relationships. By restricting the permissible range of excursion for each independent variable, the small amount of nonlinearity can be assumed to be zero. In essence, the coefficients c and a in equation 1.10(6) and equation 1.10(7) are replaced by their partial derivatives. Thus, in general, the following approximations are made:

$$a(i, j) = \left(\frac{\delta b(i)}{\delta x(j)} \right)_{x(j)} \qquad 1.10(9)$$

Note that the partial derivatives are evaluated at the current value for the independent parameter x(j). The resultant matrix or tableau of partials is used in the linear program to arrive at an interim solution. If the current or interim optimum is greater than the last value obtained, the whole procedure is repeated by "re-linearizing" the process at the newly defined operating point. Examples utilizing the above technique may be found in References 9, 10, and 11.

An alternative to the sectionalized linear program is the use of a gradient or "hill-climbing" approach. The technical literature is replete with descriptions of the mathematical bases for numerous gradient optimization methods.[12,13,14,15] Likewise, many successful industrial applications have been reviewed.[16,17,18]

In contrast to the linear program approach, which con-

siders only one variable at a time in its serial and sequential search for the optimum, the gradient methods generally perturb all the independent variables simultaneously. The magnitude of perturbation applied to each variable is directly proportional to the direction cosine of that variable. Mathematically,

$$\frac{U \cdot \dfrac{\delta c}{\delta x(i)}}{\sqrt{\displaystyle\sum_{i=1}^{i=n} \left(\frac{\delta c}{\delta x(i)} \right)^2}} \qquad 1.10(10)$$

where

U = unit step magnitude,
C = value of objective function, such as profit, and
X(i) = the ith independent control variable.

Most gradient methods, once a path of steepest ascent has been established, continue to move along that path until no further improvement in the objective function is obtained. At this juncture, another gradient is established and the entire system proceeds in this new direction until another ridge is encountered.

Constraint Handling by Gradient Methods

Optimization of a constrained nonlinear process requires a mathematical algorithm for searching along a constraint boundary. Two successful approaches that have been implemented are "hemstitching" and the use of penalty functions. Figure 1.10e is a hypothetical representation of a two-dimensional process whose permissible operating range is constrained by three variables w(1), w(2), and w(3). Parametric lines of the objective function (profit) are also plotted. Although the lines of constraint are portrayed here as linear, in most circumstances they will be markedly curved.

Beginning at the start point, the gradient search method will proceed up the hill in the direction of steepest ascent. As soon as the constraint function w(1) is violated,

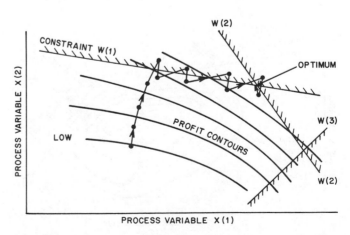

Fig. 1.10e   Gradient search using hemstitching for constraint handling

the algorithm must return the computed operating position to some point within the feasible region. Generally this is accomplished by moving perpendicularly or normal to the constraint function. The slope of the constraint is evaluated from the model by two perturbations of x(1) and x(2). Thus:

$$\text{Slope} = \left(\frac{\delta x(2)}{\delta x(1)}\right)_{w(1)} \qquad 1.10(11)$$

where the partials are evaluated at the particular constraint boundary. Now the quickest way to return to the feasible region is in the direction of the normal:

$$\text{Normal} = \frac{-1}{\text{Slope}} \qquad 1.10(12)$$

Once the constraint has been recognized, the algorithm must attempt to move along this boundary.

The method of a created-response surface suggested by C. W. Carroll[19,20] avoids many of the constraint searching problems. In this approach, the objective function is multiplied by a series of penalty terms, one for each active constraint. Thus, along each constraint boundary, the profit contour assumes a value of zero. A typical created response surface is shown in Figure 1.10f. Note that a fictitious hill is created whose summit is within the feasible operating range. Standard gradient search methods are then used to find the peak. Obviously the peak is not at the true optimum as defined by the intersection of the constraints. If the true value of the objective function is greater than the value at the starting point, then a recentering is employed. The profit contours passing through the peak are assigned a value of zero. This, along with the zero-valued constraints, permits the creation of a second hill. See Figure 1.10g. Note that the summit of the second hill is much closer to the true optimum. The foregoing recentering is repeated until the improvement in the true objective function is less than a desired and arbitrary value.

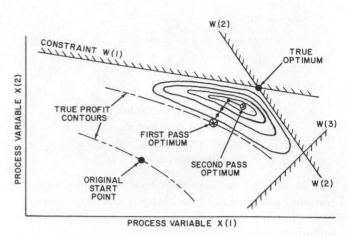

Fig. 1.10g   Gradient search using created response surface (second pass)

Multivariable Noninteracting Control

Many industrial processes exhibit considerable interaction among the control variables when attempting to regulate the values of the dependent variables. In general, a marked interaction problem will exist if the process can be described by the following set of mathematical relationships:

$$\left| \begin{array}{l} Z(i) = f_i(X_j, Y_j) \\ W(i) = g_i(X_j, Y_j) \end{array} \right| \qquad 1.10(13)$$

for $j = 1, 2, \ldots n$ and $i = 1, 2, \ldots m$.

In the general case, the functions f_i, and g_i are nonlinear in form and may be implicit in the independent variables X_j and Y_j. This situation is very complex and requires a computer iterative solution in order to decouple the interaction effects.

However, for many applications the process can be linearized around the current operating point, thus providing a much more tractable solution to the control problem. Linearization reduces the variable interrelationships to the following mathematical form:

$$\left| \begin{array}{ccc} \dfrac{\delta W_1}{\delta X_1} & \dfrac{\delta W_1}{\delta X_2} \cdot \cdot & \dfrac{\delta W_1}{\delta X_n} \\[2mm] \dfrac{\delta W_2}{\delta X_1} & \dfrac{\delta W_2}{\delta X_2} \cdot \cdot & \dfrac{\delta W_2}{\delta X_n} \\[2mm] \dfrac{\delta W_3}{\delta X_1} & \dfrac{\delta W_3}{\delta X_2} \cdot \cdot & \dfrac{\delta W_3}{\delta X_n} \end{array} \right| \times \left| \begin{array}{c} X_1 \\ X_2 \\ X_3 \\ . \\ . \\ X_n \end{array} \right| = \left| \begin{array}{c} W_1 \\ W_2 \\ W_3 \\ . \\ . \end{array} \right| \qquad 1.10(14)$$

In the above equation set, the Xs are the independent control parameters and the Ws represent the dependent controlled variables. The matrix containing the partial derivatives is known as the sensitivity matrix. The noninteracting control problem arises when it is desirable to change the value of one dependent variable without affecting the current values of the remaining dependents. Or alternatively, how should the independent X vari-

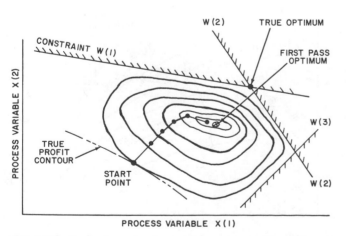

Fig. 1.10f   Gradient search using created response surface (first pass)

ables be set in order to achieve a desired slate of W values?

Using matrix algebra, the sensitivity matrix can be inverted, giving rise to the so-called control matrix:

$$\begin{vmatrix} \dfrac{\delta X_1}{\delta W_1} & \dfrac{\delta X_1}{\delta W_2} & \dfrac{\delta X_1}{\delta W_3} \\[2mm] \dfrac{\delta X_2}{\delta W_1} & \dfrac{\delta X_2}{\delta W_2} & \dfrac{\delta X_2}{\delta W_3} \\[1mm] \cdot & \cdot & \cdot \\ \cdot & \cdot & \cdot \\ \dfrac{\delta X_n}{\delta W_1} & \dfrac{\delta X_n}{\delta W_2} & \dfrac{\delta X_n}{\delta W_3} \end{vmatrix} \times \begin{vmatrix} \epsilon W_1 \\ \epsilon W_2 \\ \epsilon W_3 \end{vmatrix} = \begin{vmatrix} \Delta X_1 \\ \Delta X_2 \\ \Delta X_3 \\ \cdot \\ \Delta X_n \end{vmatrix} \quad 1.10(15)$$

The error terms of the variable W are indicated by ϵW_i, and the required change in control variable X is given by ΔX_j.

Constraint Following

Unfortunately for plant managers, but fortunately for computers, many nonlinear constrained processes can be subject to two or more constraints at the same time. As a result of the latest optimization calculation, a slate of set points is provided to achieve the stated objective, such as maximizing profit. The optimizer will at the time identify which of the control variables should be at their limiting values. A practical example of the multiconstraint situation is depicted in Figure 1.10h, which shows the profit contours bounded by six possible constraints. In this figure (for an ethylene plant pyrolysis furnace):

TMT	=	tube metal temperature,
FBT	=	fire box temperature,
COV	=	coil outlet velocity,
THR	=	total heat release,

QBET = quench boiler exit temperature, and
FMAX = maximum feed availability.

As shown in Figure 1.10h, there are four active constraints that define the feasible operating region. In addition, several additional constraints may become active if process loading or some other disturbance variable undergoes a change.

For the above example, visual inspection of Figure 1.10h indicates the optimum occurring at the intersection of the THR and COV constraints. To keep the process at optimum conditions in the interval between optimization calculations, a dual constraint-follower regulation scheme would be set in motion. The set points for these two controllers would be the maximum permitted values for THR and COV. Since there will be interaction between these two constraints, the multivariable noninteraction techniques discussed in the previous subsection would be employed.

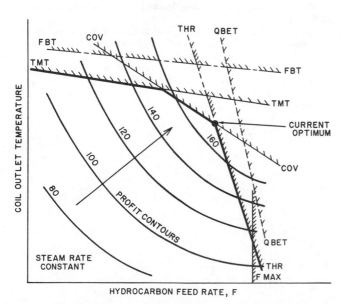

Fig. 1.10h   Typical profit surface for ethylene plant furnace cracking naphtha

REFERENCES

1. Akhiezer, N.I., *The Calculus of Variations*, Blaisdell Publishing Co., New York, 1962.
2. Elsgole, L.E., *Calculus of Variation*, Addison-Wesley Publishing Co., Reading, MA, 1962.
3. Savas, E.S., *Computer Control of Industrial Processes*, McGraw-Hill Book Co., New York, 1965.
4. Bolza, O., *Lectures on the Calculus of Variations*, Dover Publications, New York, 1961.
5. Dorfman, R., et al., *Linear Programming and Economic Analysis*, McGraw-Hill Book Co., New York, 1958.
6. Garvin, W.W., *Introduction to Linear Programming*, McGraw-Hill Book Co., New York, 1960.
7. Gass, S.I., *Linear Programming*, 2d ed., McGraw-Hill Book Co., New York, 1964.
8. Kuehn, D.R., and Porter, J., "The Application of Linear Programming Techniques in Process Control," IEEE General Meeting Paper CP63-1153, Toronto, Ontario, 1963.
9. Corbin, R.L., and Smith, F.B., "Mini-Computer System for Advanced Control of a Fluid Catalytic Cracking Unit," ISA Paper no. 73-516, presented in Houston, October, 1973.
10. Laspe, C.G., Smith, F.B., and Krall, R.A., "An Examination of Optimal Operation of an Industrial Utilities Plant," ISA Paper no. 76-524, presented at ISA 76 International Conference and Exhibit, Houston, October, 1976.
11. Horn, B.C., "On-Line Optimization of Plant Utilities," *Chemical Engineering Progress*, June, 1978, pp. 76–79.
12. Gabriel, G.A., and Ragsdell, K.M., "The Generalized Reduced Gradient Method: A Reliable Tool for Optimal Design," *Journal of Engineering for Industry*, Trans. ASME, series B, vol. 99, no. 2, May, 1977, pp. 394–400.
13. Kelley, H.J., "Method of Gradients," in Leitman, G., ed., *Optimization Techniques with Applications to Aerospace Systems*, Academic Press, New York, 1962, pp. 205–254.
14. Rosenbloom, P.C., "The Method of Steepest Descent," *Numerical Analysis*, Proceedings of the 6th Symposium on Applied Mathematics, McGraw-Hill Book Co., New York, 1956.
15. Schuldt, S.B., et al., "Application of a New Penalty Function Method to Design Optimization," *Journal of Engineering for Industry*, Trans. ASME, series B, vol. 99, no. 1, February, 1977, pp. 31–36.
16. Roberts, S.M., and Lyvers, H.I., "The Gradient Method in Process Control," *Industrial Engineering Chemistry*, November, 1961.
17. Schrage, R.W., "Optimizing a Catalytic Cracking Operation by the Method of Steepest Ascents and Linear Programming," *Operations/Research*, July–August, 1958.

18. Laspe, C.G., "Recent Experiences in On-Line Optimizing Control of Industrial Processes," Paper presented at the 5th Annual Control Conference, Purdue Laboratory, Lafayette, IN, April, 1979.
19. Carroll, C.W., "The Created Response Surface Technique for Optimizing Restrained Systems," *Operations Research*, 1961.
20. Carroll, C.W., "An Approach to Optimizing Control of a Restrained System by a Dynamic Gradient Technique," Preprint 138-LA-61, ISA Proceedings of the Instrumentation—Automation Conference, Fall, 1961.

BIBLIOGRAPHY

Balakrishnan, A.V., and Neustadt, L.W., (eds.), *Computing Methods in Optimization Problems*, Academic Press, New York, 1964.

Bellman, R., *Dynamic Programming*, Princeton University Press, Princeton, NJ, 1957.

Box, G.E.P., "Evolutionary Operation: A Method for Increasing Industrial Productivity," *Applied Statistics*, 1957.

Bryson, A.E., and Denham, W.F., "A Steepest-Ascent Method for Solving Optimum Programming Problems," Trans. ASME, *Journal of Applied Mechanics*, February, 1962.

Chatterjee, H.K., "Multivariable Process Control, Proceedings of the First IFAC Congress, Moscow, vol. 1, Butterworth.

Chestnut, H., Duersch, R.R., and Gaines, W.M., "Automatic Optimizing of a Poorly Defined Process," Proceedings of the Joint Automation Control Conference, Paper 8-1, 1962.

Chien, G.K.L., "Computer Control in Process Industries, in C.T. Leondes (ed.), *Computer Control Systems Technology*, McGraw-Hill Book Co., New York, 1961.

Douglas, J.M., and Denn, M.M., "Optimal Design and Control by Variational Methods," *Industrial Engineering Chemistry*, November, 1965.

Forsythe, G.E., and Motzkin, T.S., "Acceleration of the Optimum Gradient Method," Preliminary report (abstract) *Bulletin of the American Mathematical Society*, 1951.

Gore, F., "Add Compensation for Feedforward Control," *Instruments & Control Systems*, March, 1979.

Harriott, P., *Process Control*, McGraw-Hill Book Co., New York, 1964.

Hestenes, M.R., and Stiefel, E., "Method of Conjugate Gradients for Solving Linear Systems," *Journal of Research of the National Bureau of Standards*, 1959.

Hestenes, M.R., *Calculus of Variations and Optimal Control Theory*, John Wiley & Sons, New York, 1966.

Himmelblau, D.M., *Applied Nonlinear Programming*, McGraw-Hill Book Co., New York, 1972.

Horowitz, I.M., "Synthesis of Multivariable Feedback Control Systems," IRE Trans. Autom. Control, AC-5, 1960.

Kane, E.D., et al., "Computer Control of an FCC Unit," National Petroleum Refiners' Association Computer Conference, Technical paper 62-38, 1962.

Laspe, C.G., "Optimal Operation of Ethylene Plants," *Instrumentation Technology*, May, 1978.

Lefkowitz, I., "Computer Control," in E.M. Grabbe, S. Ramo, and D.E. Wooldridge (eds.), *Handbook of Automation, Computation, and Control*, vol. 3, John Wiley & Sons, New York, 1961.

Leitmann, G. (ed.), *Optimization Techniques*, Academic Press, New York, 1962.

Lloyd, S.G., "Basic Concepts of Multivariable Control," *Instrumentation Technology*, December, 1973.

Lipták, B.G., "Envelope Optimization for Coal Gasification," *InTech*, December, 1975.

———, "Optimizing Plant Chiller Systems," *InTech*, September 1977.

———, "Save Energy by Optimizing Boilers, Chillers and Pumps," *InTech*, March, 1981.

———, Envelope Optimization for Clean Rooms," *Instruments & Control Systems*, September, 1982.

Luntz, R., Munro, N., and McLeod, R.S., "Computer-Aided Design of Multivariable Control Systems," IEE 4th UKAC Convention, Manchester, 1971.

Luyben, W.L., "A Proportional-Lag Level Controller," *Instrumentation Technology*, December, 1977.

Mayne, D.Q., "Design of Linear Multivariable Systems," *Automatica*, vol. 9, 1973.

Merriam, C.W., III, *Optimization Theory and the Design of Feedback Control Systems*, McGraw-Hill Book Co., New York, 1964.

Munro, N., and Ibrahim, A., "Computer-Aided Design of Multivariable Sampled-Data Systems," IEE Conference on Computer Aided Control System Design, Cambridge, 1973.

Pessen, D.W., "Investigation of a Self-Adaptive Three-Mode Controller," ISA Paper, 30-361.

Roberts, S.M., and Mahoney, J.D., "Dynamic Programming Control of Batch Reaction," *Chemical Engineering Progress Symposium Series*, vol. 58, no. 37, 1962.

Shinskey, F.G., "Adaptive Nonlinear Control System," U.S. Patent 3,794,817, February 26, 1971.

———, Adaptive pH Controller Monitors Nonlinear Process," *Control Engineering*, February, 1974.

———, "Feedforward Control Applied," *ISA Journal*, November, 1963.

———, "Process Control Systems with Variable Structure," *Control Engineering*, August, 1971.

Sandgren, E., and Ragsdell, K.M., "The Utility of Nonlinear Programming Algorithms: A Comparative Study—Part 1 and 2," ASME *Journal of Mechanical Design*, July, 1980.

Wells, C.H., "Industrial Process Applications of Modern Control Theory," *Instrumentation Technology*, April, 1971.

Wilde, D.J., *Optimum Seeking Methods*, Prentice-Hall, Englewood Cliffs, NJ, 1964.

Zahradnik, R.L., Archer, D.H., and Rothfus, R.R., "Dynamic Optimization of a Distillation Column," Proceedings of the Joint Automation Control Conference, Paper 13-3, 1962.

Zellnik, H.E., Sondak, N.E., and Davis, R.S., "Gradient Search Optimization," *Chemical Engineering Progress*, August, 1962.

1.11 TUNING OF CONTROLLERS

The adjustment or tuning of single input single output (SISO) controllers is one of the least understood, poorly practiced, yet extremely important aspects of the application of automatic control theory. In the first part of this section, the objective is to present several procedures for estimating the optimum settings for a controller. Superficially, the best way to present this subject would seem to be to discuss in more detail only the best method of tuning controllers, but there is no general agreement as to which method is the best. Some methods lean heavily on experience while others rely more on mathematical considerations. Although the methods discussed in this section attempt to yield optimum settings, the only criterion stated is that the response have a decay ratio of $\frac{1}{4}$. This has been shown to be an insufficient requirement to obtain a unique combination of settings for a controller with more than one mode. When tuning a controller, one should be aware of this possibility and remember that although the response has a decay ratio of $\frac{1}{4}$, the controller may still not be at its optimum settings.

The second part of this section will address the problems of analyzing, designing, and tuning multiple input multiple output (MIMO) control systems. Control loop interaction is an increasingly common problem encountered in today's highly integrated and tightly constrained industrial processing plants. This section will discuss methods of analyzing loop interaction and designing decoupling compensation. Several of the techniques used in this MIMO control analysis section will refer to the terminology and concepts developed in Section 1.3.

Defining "Good" Control

The first problem encountered in tuning controllers is to define what is "good" control. This unfortunately differs from process to process. The adjustment of process controllers is usually based on time domain criteria. Table 1.11a gives the four most commonly used criteria. The first item, the decay ratio, has the advantage of being readily measured, as it is based on only two points on the step response. The latter three, the integral criteria, have the advantage of being more precise, that is, more than one combination of controller settings will usually

Table 1.11a
CRITERIA FOR CONTROLLER TUNING

1. Specified Decay Ratio, Usually $\frac{1}{4}$	Decay ratio = $\dfrac{\text{second peak overshoot}}{\text{first peak overshoot}}$ (see Figure 1.11b)
2. Minimum Integral of Square Error (ISE)	$\text{ISE} = \int_0^\infty [e(t)]^2 dt$ where e(t) = (set point − process output)
3. Minimum Integral of Absolute Error (IAE)	$\text{IAE} = \int_0^\infty |e(t)| dt$
4. Minimum Integral of Time and Absolute Error (ITAE)	$\text{ITAE} = \int_0^\infty |e(t)| t \, dt$

give a $\frac{1}{4}$ decay ratio, but only one combination will minimize the respective integral criteria.

The desired decay ratio is usually $\frac{1}{4}$, which is a good compromise between a rapid rise time and a short line-out time.

Although the shape of the response that minimizes the respective integral criteria differs from process to pro-

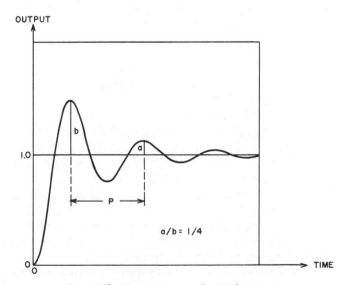

Fig. 1.11b Response curve for 1/4 decay ratio

cess, some general relative characteristics may be noted. ISE penalizes large errors whenever they occur; minimizing ISE will thus favor responses with short rise times (consequently being less damped). ITAE penalizes even small errors occurring late in time; minimizing ITAE will thus favor responses with short line-out times (highly damped). IAE is intermediate, and the corresponding response frequently has a decay ratio near $\frac{1}{4}$.

Closed-Loop Response Methods

Techniques for adjusting controllers generally fall into one of two classes. First, there are a few methods based upon parameters determined from the closed-loop response of the system, i.e., with the controller on "automatic." Second, some methods are based upon parameters determined from the open-loop response curve, commonly called the process reaction curve. To use the open-loop methods, the controller does not even have to be installed before the settings can be determined. Here we are concerned with closed-loop methods, of which the two most common are the ultimate method and the damped oscillation method.

Ultimate Method

One of the first methods proposed for tuning controllers was the ultimate method, reported by Ziegler and Nichols[1] in 1942. The term "ultimate" was attached to this method because its use requires the determination of the ultimate gain (sensitivity) and ultimate period. The ultimate sensitivity K_u is the maximum allowable value of gain (for a controller with only a proportional mode) for which the system is stable. The ultimate period is the period of the response with the gain set at its ultimate value (Figure 1.11c).

To determine the ultimate gain and the ultimate period, the gain of the controller (with all reset and derivative action turned off) is gradually adjusted until the process cycles continuously. To do this, the following steps are recommended.

1. Switch the controller on automatic.
2. Tune all reset and derivative action out of the controller, leaving only the proportional mode, i.e., set $T_d = 0$ and $T_i = \infty$.
3. With the gain at some arbitrary value, impose an upset on the process and observe the response. One easy method for imposing the upset is to move the set point for a few seconds and then return it to its original value.
4. If the resulting response curve does not damp out (as in curve A in Figure 1.11c), the gain is too high (proportional band setting too low). Therefore the proportional band setting is increased and step 3 repeated.
5. If the response curve in step 3 damps out (as in curve C in Figure 1.11c), the gain is too low

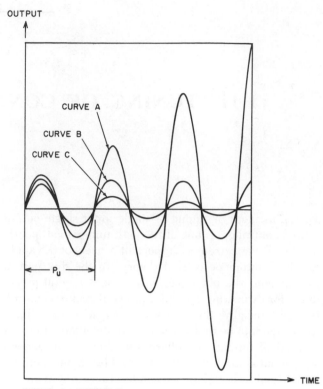

OUTPUT

CURVE A

CURVE B

CURVE C

P_u

TIME

CURVE A : UNSTABLE
CURVE B : CONTINUOUS CYCLING
CURVE C : STABLE

Fig. 1.11c Typical responses obtained when determining ultimate gain and ultimate period

(proportional band is too high). The proportional band setting is therefore decreased and step 3 repeated.

6. When a response similar to curve B in Figure 1.11c is obtained, the values of the proportional band setting and the period of the response are noted. (This must not be a limit cycle.)

There are a few exceptions to steps 4 and 5, because in some cases decreasing the gain makes the process more unstable. In these cases the "ultimate" method will not give good settings. Usually in cases of this type the system is stable at high and low values of gain but unstable at intermediate values. Thus, the ultimate gain for systems of this type has a different meaning. To use the ultimate method for these cases, the lower value of the ultimate gain is sought.

To use the ultimate gain and the ultimate period to obtain controller settings, Ziegler and Nichols correlated, in the case of the proportional controllers, the decay ratio vs. gain in the controller expressed as a fraction of the ultimate gain for several systems. From the results they concluded that a value of gain equal to one-half the ultimate gain would often give a decay ratio of $\frac{1}{4}$, i.e.,

$$K_c = 0.5K_u \quad (PB = 2PBu) \qquad 1.11(1)$$

By analogous reasoning and testing, the following equations were found to give reasonably good settings for more complex controllers:

Proportional-plus-reset:

$$K_c = 0.45K_u \qquad (PB = 2.2PBu) \qquad 1.11(2)$$

$$T_i = P_u/1.2 \qquad 1.11(3)$$

Proportional-plus-derivative:

$$K_c = 0.6K_u \qquad (PB = 1.65PBu) \qquad 1.11(4)$$

$$T_d = P_u/8.0 \qquad 1.11(5)$$

Three mode (proportional-plus-reset-plus-derivative):

$$K_c = 0.6K_u \qquad (PB = 1.65PBu) \qquad 1.11(6)$$

$$T_i = 0.5P_u \qquad 1.11(7)$$

$$T_d = P_u/8.0 \qquad 1.11(8)$$

Again it should be noted that the above equations are empirical and exceptions abound.

Damped Oscillation Method

A slight modification of the previous procedure has also been proposed by Harriott.[2] For most processes it is not feasible to allow sustained oscillations and the ultimate method cannot be used. In this modification of the ultimate method, the gain (proportional control only) is adjusted, using steps analogous to those used in the ultimate method, until a response curve with a decay ratio of $\frac{1}{4}$ is obtained. However, it is necessary to note only the period P of the response. With this value P, the reset and derivative modes are set as

$$T_d = P/6.0 \qquad 1.11(9)$$

$$T_i = P/1.5 \qquad 1.11(10)$$

After these modes are set, the sensitivity is again adjusted until a response curve with a decay ratio of $\frac{1}{4}$ is obtained. This method usually requires about the same amount of work as the ultimate method, since it is often necessary to experimentally adjust the value of the gain determined from the ultimate method to obtain a decay ratio of $\frac{1}{4}$.

In general, there are two obvious disadvantages to these methods. First, both are essentially trial and error, since several values of gain must be tested before the ultimate gain, or the gain to give a $\frac{1}{4}$ decay ratio, is determined. To make one test, especially at values near the desired gain, it is often necessary to wait for the completion of several oscillations before it can be determined if the trial value of gain is the desired one. Second, while one loop is being tested in this manner its output may affect several other loops, thus possibly upsetting an entire unit. While all tuning methods require that some changes be made in the control loop, other techniques require only one test and not several as in the closed-loop methods.

Process Reaction Curve

In contrast to the closed-loop methods, the open-loop technique necessitates only one upset to be imposed on the process. Actually, the controller is not in the loop when the process is tested. Thus, these methods seek to characterize the process and then determine controller settings from the parameters used to characterize the process. In general, it is not possible to completely characterize a process; hence, approximation techniques are employed.

Most of these techniques are based on the process reaction curve, which is the response of the process to a unit step change in the manipulated variable, i.e., the output of the controller. To determine the process reaction curve, the following steps are recommended:

1. Let the system come to steady state at the normal load level.
2. Place the controller in manual.
3. Manually set the output of the controller at the value at which it was operating in the automatic mode.
4. Allow the system to reach steady state.
5. With the controller still in manual, impose a step change in the output of the controller, which is typically the signal to the valve.
6. Record the response of the controlled variable. Although the response is usually recorded by the controller itself, it is often desirable to have a supplementary recorder or a faster chart drive for the existing controller to insure greater accuracy.
7. Return the controller output to its previous value and return the controller to automatic operation.

It is undoubtedly easier to obtain the process reaction curve than to obtain the ultimate gain.

Most open-loop methods are based on the approximating the process reaction curve by a simpler system. Several techniques are available for doing this. By far the most common approximation is that of a pure time delay plus a first-order lag. One reason for the popularity of this approximation is that a real time delay of any duration can only be represented by a pure time delay, because there is no other simple yet adequate approximation. Although it is theoretically possible to use systems higher than first order in conjunction with a pure time delay, the approximations are difficult to obtain accurately. Thus, the real system is usually approximated by a pure time delay plus a first-order lag. This approximation is easy to obtain, and it is sufficiently accurate for most purposes.

Figure 1.11d shows one popular procedure of approximating the process reaction curve by a first-order lag

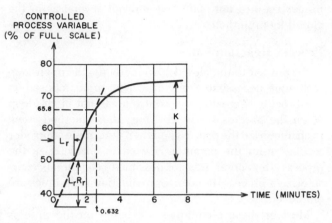

Fig. 1.11d   Process reaction curve for a step change of one unit in controller output

plus time delay. The first step is to draw a straight line tangent to the process reaction curve at its point of maximum rate of ascent (point of inflection). Although this is easy to visualize, it is quite difficult to do in practice. This is one of the main difficulties in this procedure, and a considerable amount of error can be introduced at this point. The slope of this line is termed the reaction rate R_r. The time at which this line intersects the value of the initial condition from which the process reaction curve originated is the deadtime, or time delay L_r. Figure 1.11d illustrates the determination of these values for a one-unit step change in the controller output to a process. If a different step change in controller output were used, the value of L_r would not change significantly. However, the value of R_r is essentially directly proportional to the magnitude of the change in controller output. Therefore, if a two-unit change in output were used instead of one-unit, the value of R_r would be approximately twice as large. For this reason the value of R_r used in the equations, to be presented later, must be the value that would be obtained for a one-unit change in controller output.

In addition, the value of the process gain K must be determined as follows:

$$K =$$
$$= \frac{\text{final steady-state change in controlled variable (\%)}}{\text{change in controller output (control unit)}}$$
$$\text{1.11(11)}$$

The determination of this value is also illustrated for the process reaction curve in Figure 1.11d.

There is a second method for determining the pure time delay plus first-order lag approximation. In order to distinguish between these two methods, they will be called Fit 1 (described above) and Fit 2. The only difference between the two methods is in the first-order lag time constant that is obtained. The pure time delay for both fits is the same as described before and is given by

$$\text{Fit 1 and Fit 2: } t_0 = L_r \qquad \text{1.11(12)}$$

The first-order lag time constants are given by

$$\text{Fit 1: } \tau_{F_1} = K/R_r \qquad \text{1.11(13)}$$

$$\text{Fit 2: } \tau_{F_2} = t_{0.632} - t_0 \qquad \text{1.11(14)}$$

where $t_{0.632}$ is the time necessary to reach 63.2 percent of the final value. Note that the parameters for Fit 1 are based on a single point on the response curve, which is the point of maximum rate of ascent. However, the parameters obtained with Fit 2 are based on two separate points. Studies[3] indicate that the open-loop response based on Fit 2 always provides an approximation to the actual response that is as good or better than the Fit 1 approximation. A typical curve resulting from the above procedure is shown in Figure 1.11d. From the graph, the following parameters are determined directly (response shown is from one-unit change in controller output; for different step changes, K and R_r must be adjusted accordingly):

$$L_r = 1.19 \text{ minutes}$$
$$L_rR_r = 14.9\%/\text{unit}$$
$$K = 25\%/\text{unit}$$
$$t_{0.632} = 2.58 \text{ minutes}$$

The following parameters can be calculated from those above:

From equation 1.11(12):

$$t_0 = L_r = 1.19 \text{ minutes}$$

From equation 1.11(13):

$$\tau_{F_1} = KL_r/L_rR_r = 2.00$$
$$t_0/\tau_{F_1} = \mu = L_rR_r/K = 0.595$$

From equation 1.11(14):

$$\tau_{F_2} = t_{0.632} - t_0 = 1.39$$
$$t_0/\tau_{F_2} = 0.857$$

Open-Loop Tuning

One of the earliest methods using the process reaction curve was that proposed by Ziegler and Nichols. To use their process reaction curve method, only R_r and L_r must be determined. Using these parameters, the empirical equations used to predict controller settings for a decay ratio of $\frac{1}{4}$ are given in Table 1.11e in terms of L_r and R_r and in Table 1.11f in terms of t_0 and τ.

In developing their equations, Ziegler and Nichols considered processes that were not "self-regulating." To illustrate, consider the level control of a tank with a constant rate of liquid removal. Assume that the tank is initially operating so that the level is constant. If a step change is made in the inlet liquid flow, the level in the tank will rise until it overflows. This process is not "self-regulating." On the other hand, if the outlet valve open-

Table 1.11e
EQUATIONS FOR ZIEGLER-NICHOLS AND COHEN-COON

Controller	Ziegler-Nichols	Cohen-Coon
Proportional	$K_c = \dfrac{1}{L_r R_r}$	$K_c = \dfrac{1 + \mu/3}{R_r L_r}$
Proportional +	$K_c = \dfrac{0.9}{L_r R_r}$	$K_c = 0.9 \dfrac{1 + \mu/11}{R_r L_r}$
Reset	$T_i = 3.33 \, L_r$	$T_i = 3.33 L_r \dfrac{1 + \mu/11}{1 + 11\mu/5}$
Proportional +	$K_c = \dfrac{1.2}{L_r R_r}$	$K_c = 1.35 \dfrac{1 + \mu/5}{R_r L_r}$
Reset +	$T_i = 2.0 L_r$	$T_i = 2.5 L_r \dfrac{1 + \mu/5}{1 + 3\mu/3}$
Rate	$T_d = 0.5 L_r$	$T_d = \dfrac{0.37 L_r}{1 + \mu/5}$

Table 1.11f
COMPARISON OF ZIEGLER-NICHOLS, COHEN-COON, AND 3C EQUATIONS

Controller	Ziegler-Nichols	Cohen-Coon	3C
Proportional	$KK_c = (t_0/\tau)^{-1.0}$	$KK_c = (t_0/\tau)^{-1.0} + 0.333$	$KK_c = 1.208(t_0/\tau)^{-0.956}$
Proportional + Reset	$KK_c = 0.9(t_0/\tau)^{-1.0}$	$KK_c = 0.9(t_0/\tau)^{-1.0} + 0.082$	$KK_c = 0.928(t_0/\tau)^{-0.946}$
	$\dfrac{T_i}{\tau} = 3.33(t_0/\tau)$	$\dfrac{T_i}{\tau} = \dfrac{3.33(t_0/\tau)[1 + (t_0/\tau)/11.0]}{1.0 + 2.2(t_0/\tau)}$	$\dfrac{T_i}{\tau} = 0.928(t_0/\tau)^{0.583}$
Proportional + Reset +	$KK_c = 1.2(t_0/\tau)^{-1.0}$	$KK_c = 1.35(t_0/\tau)^{-1.0} + 0.270$	$KK_c = 1.370(t_0/\tau)^{-0.950}$
	$\dfrac{T_i}{\tau} = 2.0(t_0/\tau)$	$\dfrac{T_i}{\tau} = \dfrac{2.5(t_0/\tau)[1.0 + (t_0/\tau)/5.0]}{1.0 + 0.6(t_0/\tau)}$	$\dfrac{T_i}{\tau} = 0.740(t_0/\tau)^{0.738}$
Rate	$\dfrac{T_d}{\tau} = 0.5(t_0/\tau)$	$\dfrac{T_d}{\tau} = \dfrac{0.37(t_0/\tau)}{1.0 + 0.2(t_0/\tau)}$	$\dfrac{T_d}{\tau} = 0.365(t_0/\tau)^{0.950}$

the response curve and the set point be a minimum, is a possible second constraint. This area is called the error integral or the integral of the error with respect to time.

With the proportional-plus-reset-plus-rate controller, the same problem of not having a unique solution exists even when the $\frac{1}{4}$ decay ratio and minimum error integral constraints are applied. Therefore, a third constraint must be chosen to determine a unique solution. A value 0.5 for the dimensionless group $K_c K T_d/\tau$ is one such constraint (based on the work of Cohen and Coon.)[3] The tuning relations which will result from applying these three constraints are given in Table 1.11f. This method has been referred to as the 3C method.[4,5,6]

ing and outlet back pressure are constant, the rate of liquid removal increases as the liquid level increases. Hence, in this case the level in the tank will rise to some new position but would not increase indefinitely, and the system is self-regulating. To account for self-regulation, Cohen and Coon[4] introduced an index of self-regulation μ defined as

$$\mu = R_r L_r / K \qquad 1.11(15)$$

Note that this term can also be determined from the process reaction curve. For processes originally considered by Ziegler and Nichols, μ equals zero and therefore there is no self-regulation. To account for variations in μ, Cohen and Coon suggested the equations given in Table 1.11e in terms of L_r and R_r and in Table 1.11f in terms of t_0 and τ.

For the case of proportional control, the requirement that the decay ratio be $\frac{1}{4}$ is sufficient to insure a unique solution, but for the case of proportional-plus-reset control, this restraint is not sufficient to insure a unique solution. Another constraint in addition to the $\frac{1}{4}$ decay ratio can be placed on the response to determine unique values of K_c and T_i. Requiring that the control area of the response be a minimum, meaning the area between

Integral Criteria in Tuning[7]

Tables 1.11g and h relate the controller settings that minimize the respective integral criteria to the ratio t_0/τ. The settings differ if tuning is based on load (disturbance) changes as opposed to set-point changes. Settings based on load changes will generally be much tighter than those based on set-point changes. When loops tuned to load changes are subjected to a set-point change, a more oscillatory response is observed.

The relationship between integral criteria controller settings and the ratio t_0/τ is expressed by the tuning relation given in equation 1.11(16).

$$Y = A \left(\frac{t_0}{\tau}\right)^B \qquad 1.11(16)$$

where

Y = KK_c for proportional mode, τ/T_i for reset mode, T_d/τ for rate mode,

A, B = constant for given controller and mode, and

t_0, τ = pure delay time and first-order lag time constant based on the process reaction curve $t_0 = L_r$.

Table 1.11g
TUNING RELATIONS BASED ON
INTEGRAL CRITERIA
AND LOAD DISTURBANCE[7,8,9]

Criterion	Controller	Mode	A	B
IAE	Proportional	Proportional	0.902	−0.985
ISE	Proportional	Proportional	1.411	−0.917
ITAE	Proportional	Proportional	0.490	−1.084
IAE	Proportional + Reset	Proportional Reset	0.984 0.608	−0.986 −0.707
ISE	Proportional + Reset	Proportional Reset	1.305 0.492	−0.959 −0.739
ITAE	Proportional + Reset	Proportional Reset	0.859 0.674	−0.977 −0.680
IAE	Proportional + Reset + Rate	Proportional Reset Rate	1.435 0.878 0.482	−0.921 −0.749 1.137
ISE	Proportional + Reset + Rate	Proportional Reset Rate	1.495 1.101 0.560	−0.945 −0.771 1.006
ITAE	Proportional + Reset + Rate	Proportional Reset Rate	1.357 0.842 0.381	−0.947 −0.738 0.995

Table 1.11h
TUNING RELATIONS BASED ON
INTEGRAL CRITERIA
AND SET-POINT DISTURBANCE[10]

Criterion	Controller	Mode	A	B
IAE	Proportional + Reset	Proportional Reset	0.758 1.02	−0.861 −0.323
ITAE	Proportional + Reset	Proportional Reset	0.586 1.03	−0.916 −0.165
IAE	Proportional + Reset + Rate	Proportional Reset Rate	1.086 0.740 0.348	−0.869 −0.130 0.914
ITAE	Proportional + Reset + Rate	Proportional Reset Rate	0.965 0.796 0.308	−0.855 −0.147 0.929

Digital Control Loops

Digital control loops differ from continuous control
loops in that the continuous controller is replaced by a
sampler, a discrete control algorithm calculated by the
computer, and by a hold device (usually a zero-order
hold). In such cases, Moore, et al.[5] have shown that the
open-loop tuning methods presented previously may be
used, provided the deadtime used is the sum of the true
process deadtime and one-half the sampling time, as
expressed by equation 1.11(17):

$$t'_0 = t_0 + T/2 \qquad 1.11(17)$$

where T is the sampling time and t'_0 is used in the tuning
relationships instead of t_0. (Section 1.12 deals with the
subject of controller tuning by computer.)

SISO Controller Tuning Example

For the example shown in Figure 1.11d, the model
parameters are

$L_r = 1.19$ minutes $K = 25\%$/unit

$R_r = 12.5\%$/unit $\tau = 1.39$ minute

     minute   (using Fit 2)

                      $t_0 = 1.19$ minutes

$\mu = 0.595$ $t_0/\tau = 0.857$

For a PI controller, the settings predicted by the various
tuning techniques are

Ziegler-Nichols (Table 1.11e):

$K_c = 0.9/(14.9\%/\text{unit}) = 0.0605$ unit/%

$T_i = (3.33)(1.19 \text{ min}) = 3.96$ minutes/repeat

Cohen-Coon (Table 1.11e):

$K_c = 0.9(1 + 0.595/11)/14.9\%/\text{unit} =$

$= 0.0638$ unit/%

$$T_i = \frac{3.33(1.19 \text{ min})(1 + 0.595/11)}{1 + (11)(0.595)/5} =$$

$= 1.81$ minutes/repeat

3C (Table 1.11f):

$$K_c = \frac{9.28(0.857)^{-0.946}}{25\%/\text{unit}} = 0.0430 \text{ unit/\%}$$

$T_i = (1.39 \text{ minutes})(0.928)(0.857)^{0.583} =$

$= 1.18$ minutes/repeat

IAE criterion and load disturbance (Table 1.11g):

$$K_c = \frac{(0.984)(0.857)^{-0.986}}{25\%/\text{unit}} = 0.0459 \text{ unit/\%}$$

$$T_i = \frac{1.39 \text{ minutes}}{(0.608)(0.857)^{-0.707}} = 2.05 \text{ minutes/repeat}$$

IAE criterion and set-point disturbance (Table 1.11h):

$$K_c = \frac{(0.758)(0.857)^{-0.861}}{25\%/\text{unit}} = 0.0347 \text{ unit/\%}$$

$$T_i = \frac{1.39 \text{ minutes}}{1.02 - (0.323)(0.857)} = 1.88 \text{ minutes/repeat}$$

Note that on some controllers the reset mode is calibrated in repeats/minute units. In those cases use the
$1/T_i$ values for tuning. The relationship between K_c and
proportional band is

$$PB = 100/K_c$$

Multivariable Tuning Techniques

The multiple input multiple output (MIMO) control problem has been the subject of much study in recent years. The motivation for this work stems primarily from necessity. The classic single input single output (SISO) controllers used in industrial process control frequently fall short for many interactive multivariable problems found in industry.

A commonly encountered example of this kind of process is "two-cut point" control of distillation columns. This is where the top and bottom compositions of the column are controlled to achieve minimum energy for separation. These compositions are typically controlled by reflux or distillate and heat load. Each manipulated variable has a substantial effect on both compositions. These interactions are sometimes compensated for by detuning one of the conventional PID controllers on the composition loops. Obviously, such a procedure yields poor overall control. A MIMO controller can recognize and compensate for process interaction much more efficiently than can standard SISO controllers on individual loops.

Design techniques for MIMO control have been covered rather extensively in the literature and are gaining acceptance in practical application. This section will present an overview of the field, with special emphasis on the noninteracting controller tuning with overrides and MIMO frequency response techniques.

Noninteracting and Override Control

The typical method used for dealing with MIMO control problems in industry is one of avoidance or overdesign of process equipment to bypass the problem. If a multivariable problem is encountered, the usual method employed to handle control interaction is detuning. This implies that one loop is tuned tightly (small overshoot, quick settling time) and the other loop is tuned slowly (larger overshoot, slower settling time). This is done to cause the closed-loop natural frequency of the respective loops to be different and, hopefully, noninteractive. This method frequently yields unsatisfactory control. The question becomes how to design control for a MIMO process so that critical loops can be tuned similarly. For example, how do we tune two interactive loops tightly? Before the treatment of rigorous MIMO control design methods are considered, we will first look at a common control practice of noninteractive control with overrides.

Certain MIMO control situations that appear to entail potentially troublesome, highly interactive loops can be easily handled by conventional feedforward, feedback, and selective or override control (Sections 1.5, 1.6, and 1.8, respectively). Proper, intelligent application of these techniques can go a long way to provide stable, acceptable control. There is no "modern control" technique or design method that can substitute for a good understanding of the process and control required.

Several major items must be considered when addressing the multivariable control problem using the so-called conventional approach:

1. It is important to select the proper variable pairing for interactive feedback control.
2. Application of feedforward compensation is desirable to help minimize the required effort of interactive feedback loops reacting to process disturbance.
3. The use of selective or override controls where possible will minimize the need for interactive feedback loops.
4. Using interaction compensation will frequently help provide a stabilizing effect on interactive feedback loops and allow tighter tuning.

The above items will be explained briefly in the following paragraphs.

Variable pairing is perhaps the most important aspect of multivariable control design. Faced with the choice of several variables that must be controlled and several manipulating variables that affect each of these controlled variables, the question is how we match them up. For example, do we control the distillation column top temperature with reflux, distillate, or boilup? Many methods have been devised to help us make a choice. Some of these will be covered in later sections, but it is always advisable to bring as much process understanding to this decision as possible.

One major characteristic of a two-input two-output interacting control system is the presence of significant multiple resonant frequencies (resonant frequency is defined in Section 1.3). For each loop there is a resonant frequency associated with the single-loop dynamics and a resonant frequency associated with the interactive dynamics. The interactive resonant frequency is often the lower of the two.

Choosing the best controller pairing (of the manipulated/measured variables) and tuning these controllers is closely related to the relative dominance of these resonant frequencies. The key to tuning the interactive controllers is to use a tuning philosophy that encourages the faster resonant frequencies to dominate the closed loop transient response where possible. The key to choosing the best controller pairing is to determine which configuration will have the dominant system resonant frequency once the control loops are tuned. Methods for determining dominant system resonance are discussed in the section below on Nyquist array analysis.

In many situations the variable pairing is set by process conditions and one is stuck with very interactive feedback loops. In this situation, other methods must be applied in order to achieve good response.

Feedforward techniques can be extremely valuable for

dealing with interactive multivariable loops. Interaction occurs when SISO feedback loops are attempting to satisfy themselves at the expense of the other SISO loops. If ideal feedforward control were applied, it would cause the control variables to move such that the feedback loops would not have any work to do, thus no interaction. Consider the distillation column control of Figure 1.11i. Notice that the reflux and steam are ratioed to the column feed. This is in effect feedforward. If the ratio loops adjust the reflux and steam properly, the compositions in the top and bottom of the column and pressures should remain approximately constant. This being the case, the composition feedback loops should not have much to do, thus dramatically reducing the interaction effect.

One method to avoid feedback loop interaction is to not require continuous control of certain secondary variables. If these variables require control action only to keep them within acceptable or safe ranges, then when they are within these ranges they should have no effect on control. This can be accomplished by override or selective control as discussed in Section 1.8. Consider the differential pressure control loop of Figure 1.11i. This secondary variable is used to protect the column internals. If the column differential pressure gets too high, the override "pinches" the steam flow loop back. Trying to continuously control column differential pressure would overspecify control and lead to potentially unstable interaction. Override control allows for the column differential pressure to be constrained but does not introduce additional control loop interaction for normal operation.

Control loop interaction compensation is another very important technique for improving the closed loop response of interactive feedback loops. It is still very possible to have significant loop interaction even after proper variable pairing, feedforward, and override considerations have been made. Further closed loop response improvements can be obtained using interaction compensation. Interaction compensation attempts to predict the effect of a given control action on other control variables. This predicted effect is then passed on to the other feedback controller in an attempt to minimize its disturbing effect.

Consider our distillation column example. If for some reason the base composition controller calls for additional steam flow (i.e., feed composition upset) at some later time, the reflux and make flows will have to be adjusted in order to satisfy the column material and heat balances. If this steam flow demand change is fed to the reflux and make controllers in advance, some feedback interaction could be eliminated. This is interaction compensation. The resulting benefit of this decoupling is to allow for tighter tuning of the individual feedback loops. Theoretically, if the interaction compensation is perfect, the feedback loops of an interactive process can be tuned independently (i.e., each could be tuned to achieve $\frac{1}{4}$ amplitude damping or other desired closed loop response). This total decoupling is very rarely achieved in practice; however, much improved control is possible.

The following sections will cover some analytical techniques that aid in the analysis and design of multivariable systems.

Nyquist Array Analysis

The previous section has discussed the need to analyze MIMO interaction in order to specify variable pairing, decoupling, and thus MIMO tuning philosophy. One technique that is gaining acceptance and popularity is based on frequency response analysis.

A powerful multivariable frequency response analysis technique is given in Rosenbrock.[6] We will not attempt to detail any theoretical development as that is available in the reference; rather, we will provide an approach to the study of multivariable problems. The theory is applied to the interpretation of graphically presented frequency response information, much the same way as in

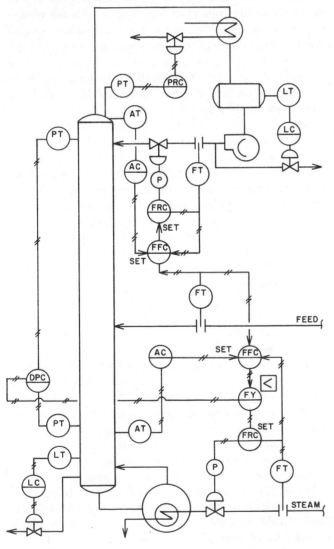

Fig. 1.11i   Multivariable distillation column control

the interpretation of single input single output frequency response plots (Bode and Nyquist plots; see Section 1.3). The discussion of graphical representations of dynamic system in Section 1.3 explains how MIMO systems can be presented as an array of Nyquist plots. This section will explain how these plots can be used to analyze and design interaction compensation.

Consider the generalized transfer function matrix in equation 1.11(18) (transfer function matrix is defined in Section 1.3).

$$G(s) = \begin{bmatrix} G_{11} & G_{12} & G_{13} & \cdots \\ G_{21} & G_{22} & G_{23} & \cdots \\ G_{31} & \cdots & & \\ \cdot & & & \\ \cdot & & & \\ \cdot & & & \end{bmatrix} \qquad 1.11(18)$$

where

$$G_{in}(s) = \frac{a_n s^n + a_{n-1} s^{n-1} + \cdots a_0}{b_{n-1} s^{n-1} + b_{n-2} s^{n-2} + \cdots b_0} \qquad 1.11(19)$$

The transfer function matrix is a dynamic representation of the MIMO process. It is typically generated from state space models or from experimental process testing (i.e., finding the process reaction curve). This matrix contains much information about the process interactions and possible control and tuning philosophy.

The property of the G(s) matrix which we are interested in is its diagonal dominance. It can be shown that the diagonal properties of the inverse of G(s), $(G(s)^{-1}K = Q(s))$ are correspondingly the same. For numerical reasons, it is advantageous to work with Q(s). The diagonal dominance of single elements of any transfer function matrix is a function of frequency and the effects that other elements have on it.

Considering row dominance, the dominant transfer function on a given row of a transfer function matrix is the transfer function element whose behavior (time behavior) would dominate (i.e., most rapid response within a given frequency range). Diagonal row dominance has the following definition for Q(s):

$$|Q_{ii}(s)| - \sum_{\substack{j=1 \\ j \neq i}}^{m} |Q_{ij}(s)| > 0 \qquad 1.11(20)$$

In words, this implies that for any s, the modulus of each element on the diagonal of Q(s) exceeds the sum of the moduli of the off diagonal elements. If a variable pairing scheme or compensator can be designed to accomplish this dominance, then the control analysis problem effectively reduces to the analysis of a system of single input single output control loops.

The dominance property of the inverse transfer function matrix can be investigated using Nyquist array analysis. The computer is used to present graphically the

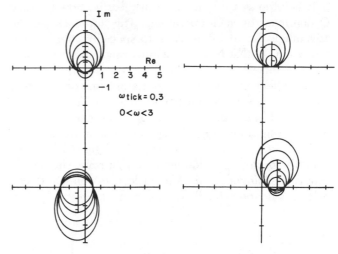

Fig. 1.11j   Inverse Nyquist array open loop

inverse transfer function matrix as a matrix of Nyquist plots (Figure 1.11j). The interaction criterion is evaluated by equation 1.11(20) in the following manner. First, each element of the transfer function matrix has its respective inverse Nyquist plot displayed over the frequency range specified. Secondly, for each Nyquist plot, at specified frequencies, the modulus of all other same-row elements is calculated. This number is used as the radius for a circle drawn on the Nyquist contours with circle center on the Nyquist contours at the frequency specified. These circles are referred to as Gershgorin circles. When this is done for a range of frequencies, the circles sweep out a band called a Gershgorin band. If the origin does not lie within the Gershgorin band, the element is said to be dominant on that row.

For example, in Figure 1.11j, the Q_{12} and Q_{22} plots do not have Gershgorin bands that circle the origin, so they are not dominant on their respective rows. This matrix is then column 2 dominant, not diagonally dominant as we would like from a decoupling standpoint.

The task is now to achieve diagonal dominance of an undiagonally dominant open loop system (i.e., Figure 1.11j). There are many ways that this can be accomplished. This design task falls in an area of considerable research from a realization or practical standpoint. In order to keep this tutorial simple, we will restrict ourselves to the design of constant precompensators.

To find a precompensator to achieve the decoupling, the following techniques can be used. First, inspect the graphical display of the element Q(s). The following operations on the elements of Q(s) correspond to elements of precompensator K. We may build up K from any sequence of these.

1. Interchange two rows (change variable pairing).
2. Add a multiple of one row (by a constant) to another row.
3. Multiply a row by a constant.

It is often possible to see, fairly easily, how to make Q dominant by such operations. This approach has the advantage that it often leads to specially simple compensators, in that it allows engineering constraints on K to be observed.

The effect of the precompensators can be analyzed using the inverse Nyquist plots and Gershgorin bands. Figure 1.11k is a plot of KQ(s) for a compensator design for the system in Figure 1.11j. Note that in the first row element kq_{11} does not have any Gershgorin bands circling the origin. It is dominant on that row. The second row is unchanged, and element kq_{22} is dominant. The matrix is now diagonally dominant as desired.

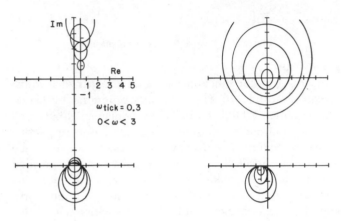

Fig. 1.11k Inverse Nyquist array compensated system

The time behavior of the uncompensated system is given in Figure 1.11l(1). The strip chart shows that as the control is tightened on each of the individual loops, they start to interact.

Figure 1.11l(2) is the exact same system (i.e., same linear system, same controller tunings) with the as-designed interaction compensator installed. The control achieved in the compensated case is better than that of the uncompensated case (i.e., the oscillation is gone).

Conclusions

Several methods have been presented for tuning automatic controllers. Although each attempts to give the optimum combination of settings, the settings vary depending on which method is used, and the question arises as to which method is best. Since fewer assumptions are involved and the basic mathematics are sounder, applying the integral criteria results in settings that are closer to the optimum combination. But *any* method will result in a controller that is more accurately and more rapidly tuned than is possible through "guessing."

The general question of nonlinearities, or "nonsymmetrical" systems, is perhaps one of the most frustrating aspects of controller tuning. It is manifest in many day-

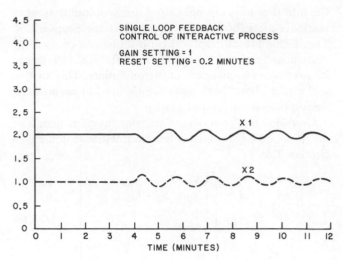

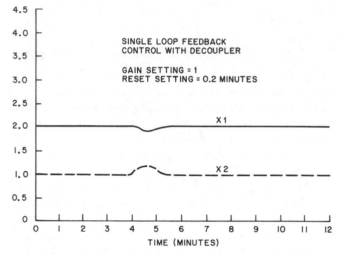

Fig. 1.11l Time behavior of uncompensated system

to-day situations in which a process is operating satisfactorily with all loops well tuned, and then a load change is experienced and all the controllers must be retuned.

When a process is nonlinear or nonsymmetrical, what is really needed is nonlinear or automatic adaptive tuning (Section 1.9). This implies that the control system will "sense its state" and automatically retune itself. There are automatic hardware systems for doing this, particularly for the controller gain, but their use is not justified in the great majority of cases. With the increased use of digital control, there is a trend toward software techniques for adaptive tuning.

In making fine adjustments to the calculated tunings, one must understand the effect of each mode upon the overall response. Thus, the following generalities about the effects of adjustments for each mode are pertinent.

Adjustment of the proportional band: Decreasing the proportional band (increasing the gain) increases the decay ratio, thus making the system less stable. However, the frequency of the response is also increased, which is usually desirable. Increasing the proportional band has an opposite effect.

Adjustment of the reset mode: When the reset time is increased, the decay ratio is decreased, thus making the system more stable. Simultaneously, the frequency increases. Decreasing the reset time has an opposite effect. One should recall that when the reset time is at its maximum value, this mode has been tuned out of the controller.

Adjustment of the derivative mode: Of all the modes, the effect of this mode is the most difficult to predict. Starting at a derivative time of zero, increasing the derivative time usually is beneficial, but not always. In almost all practical cases there is a point beyond which increasing the derivative time will prove detrimental. Thus about all that one can do is try a change in the derivative time and see what happens.

Multivariable tuning techniques require knowledge of the system loop interactions as well as an understanding of the individual feedback loop tuning. For each loop there is a resonant frequency associated with the single loop dynamics and a resonant frequency associated with the interactive dynamics. Choosing the best controller pairing and tuning is closely related to the relative dominance of the resonant frequencies. The key to tuning these interactive controllers is to use a tuning philosophy that encourages the faster resonant frequencies to dominate the closed loop transient response where possible.

The analysis of this frequency-dependent behavior can be made by studying Nyquist array plots with interaction bands. These plots reveal much about variable pairing and system loop interaction. The plot and analysis are, however, only as good as the linear representation of the MIMO process under consideration. As mentioned above, the nonlinearities and nonsymmetries that exist in the real world can give rise to problems in linear analysis. Care must be taken to investigate the extent of these problems and the effect they have on the analysis.

REFERENCES

1. Ziegler, J.G., and Nichols, N.B., "Optimum Settings for Automatic Controllers," ASME Trans., 1942, pp. 759–765.
2. Harriott, P., *Process Control*, McGraw-Hill Book Co., New York, 1964.
3. Smith, C.L., and Murrill, P.W., "An Analytic Technique for Tuning Underdamped Control Systems," *ISA Journal*, September, 1966.
4. Cohen, G.H., and Coon, G.A., "Theoretical Considerations of Retarded Control," Taylor Instrument Companies Bulletin TDS-10A102.
5. Moore, C.F., et al., "Simplification of Digital Control Dynamics for Tuning and Hardware Lag Effects," *Instrument Practice*, January, 1969, pp. 45–49.
6. Rosenbrock, H.H., *Computer-Aided Control System Design*, Academic Press, London, 1974.

BIBLIOGRAPHY

Anderson, N.A., "Step-Analysis of Finding Time Constants," *Instruments & Control Systems*, 1963.

Bjornsen, Bjorn G., "A Fluid Amplifier Pneumatic Controller," *Control Engineering*, June, 1965.
Box, G.E.P., and Jenkins, G.M., *Time Series Forecasting and Control*, Holden-Day, San Francisco, 1970.
Breakwell, J.V., et al., "Optimization and Control of Nonlinear Systems Using the Second Variation," *SIAM Journal on Control*, sec. A., vol. 1, no. 2, 1963.
Bryson, A.E., Jr., and Ho, Y.C., *Applied Optimal Control*, John Wiley & Sons, New York, 1975.
Buckley, P.S., "A Modern Perspective on Controls Tuning," Texas A&M 30th Annual Symposium on Instrumentation for the Process Industries, January, 1973.
Buckley, P.S., *Techniques of Process Control*, John Wiley & Sons, New York, 1964.
Caldwell, W.I., "Generating Control Functions Pneumatically," *Control Engineering*, September, 1954.
Caldwell, W.I., Coon, G.A., and Zoss, L.M., *Frequency Response for Process Control*, McGraw-Hill Book Co., New York, 1959.
Campbell, D.P., *Process Dynamics*, John Wiley & Sons, New York, 1958.
Chen, C.T., *Introduction to Linear System Theory*, Holt, Rinehart and Winston, New York, 1970.
Dahlin, E.B., "Designing and Tuning Digital Controllers," *Instruments & Control Systems*, June, 1968.
Fox, H.L., and Wood, O.L., "Fluid Amplifiers—The Development of Basic Devices," *Control Engineering*, September, 1964.
Gelb, A., *Applied Optimal Estimation*, The MIT Press, Cambridge & London, 1974.
Higham, J.D., "Single-Term Control of First- and Second-Order Processes with Dead Time," *Control*, February, 1968.
Hougen, J.O., "Experiences and Experiments with Process Dynamics," *Chemical Engineering Process Monograph Series*, vol. 60, no. 4, 1964.
Larson, R.E., et al., "State Estimation in Power Systems, Part I and Part II," IEEE Trans. on Power Apparatus and Systems, vol. PAS 89, no. 3, March, 1970.
Lipták, B.G., "How to Set Process Controllers," *Chemical Engineering*, November, 1964.
Lopez, A.M., et al., Controller Tuning Relationships Based on Instrumentation Technology, *Control Engineering*, November, 1967.
Mehra, R.K., "Identification of Stochastic Linear Dynamic Systems Using Kalman Filter Representation," *AIAA Journal*, January, 1971.
———, "On-Line Identification of Linear Dynamic Systems with Applications to Kalman Filtering," IEEE Trans. on Automatic Control, vol. AC 16, no. 1, February, 1971.
Miller, J.A., et al., "A Comparison of Open-Loop Techniques for Tuning Controllers," *Control Engineering*, December, 1967.
Murrill, P.W., *Automatic Control of Processes*, International Textbook Co., Scranton, PA, 1967.
Oldenbourg, R.C., and Sartorius, H., *The Dynamics of Automatic Controls*, The American Society of Mechanical Engineers, New York, 1948.
Palmer, O.J., "Which: Air or Electric Instruments?" *Hydrocarbon Processing and Petroleum Refiner*, vol. 44, no. 5, 1965.
Raven, F., *Automatic Control Engineering*, McGraw-Hill Book Co., New York, 1961.
Rovira, A.A., Ph.D. dissertation, Louisiana State University, Baton Rouge, LA, 1969.
———, Murrill, P.W., and Smith, C.L., "Tuning Controllers for Set-Point Changes," *Instruments & Control Systems*, December, 1969.
Salmon, D.M., and Kokotovic, P.V., "Design of Feedback Controllers for Nonlinear Plants," IEEE Trans. on Automatic Control, vol. AC 14, no. 3, 1969.
Shinskey, F.G., "Interaction Between Control Loops," *Instruments & Control Systems*, May and June, 1976.
Smith, C.L., "Controller Tuning That Works," *Instruments & Control Systems*, September, 1976.
———, and Murrill, P.W., "A More Precise Method for the Tuning of Controllers," *ISA Journal*, May, 1966.
———, "Controllers—Set Them Right," *Hydrocarbon Processing and Petroleum Refiner*, February, 1966.
Smith, O.J.M., "Close Control of Loops with Dead Time," *Chemical Engineering Progress*, May, 1957.

Sood, M., "Tuning Proportional-Integral Controllers for Random Load Changes," M.S. Thesis, Dept. of Chemical Engineering, University of Mississippi, 1972.

————, and Huddleston, H.T., "Optimal Control Settings for Random Disturbances," *Instrumentation Technology*, March, 1973.

Suchanti, N., "Tuning Controllers for Interacting Processes," *Instruments & Control Systems*, April, 1973.

Wells, C.H., "Application of Modern Estimation and Identification Techniques to Chemical Processes," *AIChE Journal*, 1971.

————, and Mehra, R.K., "Dynamic Modeling and Estimation of Carbon in a Basic Oxygen Furnace," Paper presented at Third International IFAC/IFIPS Conference on the Use of Digital Computers in Process Control, Helsinki, Finland, June, 1971.

Ziegler, J.G., and Nichols, N.B., "Optimal Settings for Automatic Controllers," ASME Trans., November, 1942.

————, and Nichols, N.B., "Optimal Settings for Controllers," *ISA Journal*, August, 1964.

1.12 TUNING BY COMPUTER

The design and maintenance of instrument and control systems is a task that is usually placed in the hands of instrument engineers, process engineers, and instrument technicians. An understanding of the process, control practices, and control principles is essential to the design and maintenance of workable industrial applications. With the ever-increasing complexity of process and control implementations, computers can play an important role in these functions.

In today's plant, unit operations are highly integrated to reduce capital investment and recover heat energy. Process-to-process heat exchangers, waste heat boilers, multi-effect distillation systems, and multiple reboilers are common, and these often necessitate complex multivariable control systems (see Sections 1.3 and 1.11). To design these systems and keep them in good working order, several useful approaches are detailed in the following paragraphs. They all make significant use of computer-aided techniques.

Computer programs developed for aid in process control *design* are gaining popularity and acceptance. Several good commercially available packages exist. These techniques involve nonlinear simulation tools, linearizing routines, predictor and estimation techniques, linear quadratic gaussian methods, time series and frequency response techniques. Some of these methods have functionally been demonstrated in other sections (1.3, 1.9, and 1.11). Part of this section will discuss the approach to a design problem using such methods. The theory and examples can be found in other sections.

The *maintenance* of control systems is equally important as a good workable design. Many good control schemes fall into disuse because of a lack of proper fine-tuning, failure to adapt to changing process requirements, or poor maintenance. As time progresses and the plant experiences operation changes, some controllers have to be either detuned to prevent cycling or merely switched to a manual operating mode. Providing operations with simple testing techniques and access to the computer can avoid these remedial solutions and prevent the inherent operating losses that result from having detuned or open-loop controllers. This section will also show how some of these computer-aided methods can be used to help improve utility of instrument control systems and do routine tuning maintenance.

Computer-aided control analysis tools can be thought of in two categories. First there are the techniques that are useful in control system design. These are typically used for new plant or new control system design. Such tools are most often off-line in nature as this is the safest mode for trying or investigating new control methods. The development of new methods requires good representations of the process (i.e., dynamic process models), control system design, and controller tuning techniques (i.e., Section 1.11).

The second category of computer-aided control analysis tools are those required in maintenance of existing systems. These techniques are frequently on-line methods. They involve trying to identify the process dynamics while the plant is operating using step testing techniques and/or recursive estimation method for adaptive tuning (Section 1.9). These updated process representations are then used with predefined performance requirements to calculate new control system parameters. In addition to improving control system utility by keeping controller tuned, other methods are currently being used to improve measurement utility via computer check of process measurements validity. Overall process or unit process consistency balance calculations and predictor/estimator filters (i.e., Kalman filters) are helping to provide high utility high-accuracy measurements. These better-quality measurements are contributing to higher utility control applications. Some of these techniques will be discussed in more detail in the following paragraphs.

Process Modeling

The effectiveness of any control system depends largely on how well the process under control has been mathematically modeled. Some control designs are less sensitive to model error than others. However, as control performance requirements increase, generally so does the requirement for model accuracy.

Control, instrument, and process engineers deal with many types of process models. Each type of model has

Table 1.12a
TYPES OF PROCESS MODELS USED IN CONTROL WORK

Computer Model Type	Typical Application
Large Scale Nonlinear Dynamic Simulation	Used for off-line investigation of process, process dynamics, and control application. (Large effort required to generate models.)
Frequency Response Models, Single Input Single Output (SISO) Multiple Input Multiple Output (MIMO)	Used to investigate open-loop behavior and specify closed-loop feedback control. MIMO models are used to study interaction. (Model typically generated from state space models or plant testing.)
State Space Models, MIMO Continuous and Discrete Linear Models	Used for control system design with methods such as linear quadratic optimal and pole placement control. Used for estimator/predictor models. (Models typically generated from linearization of simulation models or plant tests.)
Time Series Models, SISO Difference Equations	Used for adaptive control and state parameter estimation.
Simple Low Order Models, SISO Continuous Time	Used for quick time domain model fits to plant tests for control-loop tuning.

a different use depending on specific application. Some of the models and applications have been covered in prior sections. Table 1.12a is a partial list of these computer models and some of their typical applications.

The purpose of this section is to show how computers can be used in the design and analysis of control problems. For this task a brief overview of the modeling effort is required.

Currently the major tool employed for control system design is dynamic process simulation from basic physical laws (mass, heat, momentum, etc., time-dependent balance equations). Control is evaluated and designed via time response information generated by these dynamic simulations. This type of simulation effort is very time-consuming and requires process expertise and a facility with large-scale dynamic modeling packages. Several of these nonlinear dynamic simulation packages are available commercially. Many of these packages have several common features. First, the simulation models require input of many differential equations representing the process balance equations. The packages usually handle the numerical integration and output of the desired time variables with very little effort on the part of the user.

Once the process has been adequately modeled, control is applied to the dynamic simulation. The time behavior of the closed loop system is studied as to the adequacy of the control achieved (see Section 1.4). The control scheme can be changed in order to try new intuitively devised schemes or control designs recommended by other computer design methods such as linear quadratic optimal or frequency response methods (see Sections 1.3 and 1.11).

Frequency response models are another popular method for representing the mathematical control problem. Single input single output transfer function models allow control systems to be analyzed in the frequency domain (see Sections 1.3 and 1.11). Computer-generated graphics are used by many control analysis packages to draw root locus, Bode plot, Nichols charts, and Nyquist plots. This function removes one of the laborious tasks associated with using frequency response techniques. The individual transfer function models are typically generated from linearized differential equations, plant pulse or step testing, or transformed state space models. All of these procedures have been computerized to assist in the analysis. These techniques are very helpful in control system design and specification in the off-line mode.

State space models are also covered in Section 1.3. These models are nothing more than linearized matrix representations of multiple input multiple output or high-order differential or difference equations. The convenient matrix formulation of the state space models makes them great candidates for manipulation and study by computer. These linear models can be used for time simulation (over the range where the linearization is valid). They can be transformed to transfer function matrix representations. They can be used to formulate linear quadratic optimal controllers. They can also be used to build predictor/estimators (i.e., Kalman filters). State space is a very flexible form for control models to take and is a common starting point for many computer-aided control analysis packages. The state space model is useful for off-line design and on-line analysis and recursive study.

The above techniques are basically design tools used off-line. One popular on-line use for state space models is the estimator/predictor configuration. Here the process state space model is run along side of the real process in a computer. The measurable information about the process is fed into the computer model. The model predicts or estimates states of the process that cannot be measured. This information is then used for control of the process.

The methods used for estimating unmeasured states vary. One popular method is entitled Kalman filtering. In this method the estimated states are updated with information from the process model and measure inputs, based on statistical properties of the measurement noise. Details about Kalman filtering and other estimation/prediction techniques can be found in References 1 and 2. These filtering techniques improve measurement quality and utility, thus improving the utility of the entire control system. These on-line computer modeling techniques can also be used to check instrument measurements. Certain modeling applications can be used to indicate instrument problems by monitoring the deviation between the model predicted output and the output of the real plant. If this deviation becomes too large, possible maintenance is required on the measurement devices.

Time series models (Section 1.9) are also very useful in adaptive control applications. Discrete time models are used to represent single input single output systems of time-varying processes. The parameters of the time series models are estimated using recursive techniques. The updated model resets the on-line controller to keep the loop tuned adequately. This again is an on-line computer application that is primarily aimed at improving control loop utility.

One very popular method for generating simple models of a dynamic process is to assume a simple model representation and fit parameters. Using these simple models, classic tuning techniques can be employed, as discussed in Section 1.11. This section will show how this technique can be used in conjunction with a computer to yield a very helpful maintenance tool.

The simple low-order single input single output continuous model consists of a series of time lags associated with loop elements (valve, process, transmitter, etc.). Occasionally, these processes also involve transport delay (pure deadtime), resulting, for example, from plug flow in pipes. Unlike nuclear reactors and modern missiles, these processes are usually stable without feedback (open loop).

A general mathematical structure for these simple processes is the first-order model with deadtime given in equation 1.12(1). This model has three parameters: gain, time constant, and deadtime.

$$\frac{dc(t)}{dt} = \frac{1}{\tau}[Km(t - t_0) - c(t)] \qquad 1.12(1)$$

where

$$t = \text{variable time,}$$
$$\frac{dc(t)}{dt} = \text{derivative with respect to time of the controlled variable,}$$
$$c(t) = \text{controlled or measured process variable, process output response,}$$
$$m = \text{manipulated process variable, a function of time and deadtime,}$$
$$K = \text{process gain,}$$
$$\tau = \text{process time constant, and}$$
$$t_0 = L_r = \text{apparent process deadtime.}$$

This mathematical structure has the advantage of con-

strained dimensionality. Deadtime in the model conveniently lumps the higher-order process delays and the actual process deadtime into one parameter.

This then will be the model used to tune our feedback loop. The model is fitted to the process and the control loop tunings are computed using computer routines to achieve the desired response.

Fitting Simple Models

The conventional step response test (Figure 1.12b) typically used in industrial practice is the most popular method for explicitly fitting the three-term process model. This method works well for process loops that will readily achieve new final values (their time constant τ is small) and for processes that do not appreciably drift during the test. However, if gain K is not known from experience (it usually is not), this technique has the disadvantage that step size for a satisfactory implementation must be selected by trial and error.

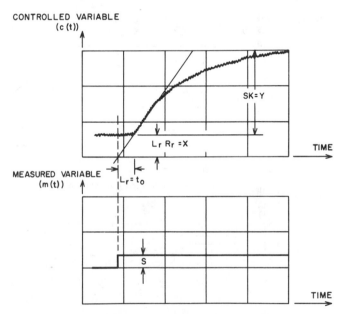

PROCESSES THAT DRIFT DURING THE CONVENTIONAL STEP RESPONSE TEST WILL INVALIDATE THE MEASURED PARAMETERS X, Y AND L_r.

Fig. 1.12b First-order response

To fit the three-parameter model using this test, the following measurements should manually or automatically be obtained: manipulated variable step size (S), apparent deadtime ($L_r = t_0$), total computed change in the controlled variable ($X = L_rR_r$) at the maximum rate of change (R_r) over the period of the apparent deadtime ($L_r = t_0$), and the final value of the controlled variable's response ($Y = K$). Then

$$R_r = \frac{X}{SL_r} \qquad 1.12(2)$$

$$K = \frac{Y}{S} \qquad 1.12(3)$$

$$\tau = \frac{K}{R_r} \qquad 1.12(4)$$

Ordinarily, the conventional step response test is not practical in loops with extremely large time constants. Furthermore, it does not provide a check on the testing accuracy itself unless it is repeated. Even if the test is repeated, errors in unit conversion are not always uncovered.

A combination of a modified step response and an ultimate sensitivity test overcomes the above limitations (Figure 1.12c). When performing the modified step response, the operator retains the step just long enough to be certain that the process response reaches the maximum rate of change. Then he returns the manipulated variable to its initial condition. Because the controlled variable does not have to achieve a new steady state,

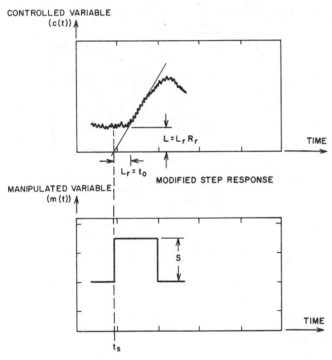

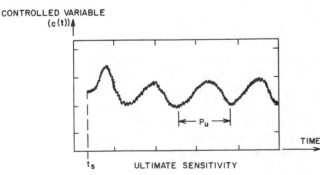

Fig. 1.12c   Modified step response test combined with ultimate sensitivity test

selection of step size S is not critical. This technique should not be confused with pulse testing, which requires analysis of the total pulse response curve. The maximum rate R_r still yields to equation 1.12(2).

The ultimate sensitivity test calls for increasing the gain of a proportional controller to K_u until the process cycles without convergence or divergence. The following equations govern the stability of the system:

$$\frac{K_u K}{\sqrt{\left(\frac{2\pi\tau}{P_u}\right)^2 + 1}} = 1 \qquad 1.12(5)$$

$$\tan^{-2}\left(\frac{2\pi\tau}{P_u}\right) + \frac{2\pi L_r}{P_u} = \pi \qquad 1.12(6)$$

where K_u is the ultimate proportional gain and P_u is the cycle period. These equations assume continuous proportional control, or sampling at such a rate that it is apparently continuous with respect to either L_r or τ. Reduction of the equations to determine the process time constant and gain gives

$$\tau = \frac{P_u}{2\pi}\left[\tan\left(\pi - \frac{2\pi L_r}{P_u}\right)\right] \qquad 1.12(7)$$

$$K = \frac{1}{K_u}\sqrt{\left(\frac{2\pi\tau}{P_u}\right)^2 + 1} \qquad 1.12(8)$$

With τ calculated from equation 1.12(7) and gain from equation 1.12(8), equation 1.12(4) gives a cross check on the accuracy of the test and on the validity of the model structure.

Another equation describing process gain may be readily obtained by discrete cycling of the loop with sample rate equal to the deadtime ($L_r = t_0$). If the gain required for discrete ultimate sensitivity is K_{ud}, then in accordance with equation 1.3(19) the conditions of stability are defined by

$$K_{ud} K[1 - e^{-t_0/\tau}] = 1 \qquad 1.12(9)$$

Process gain can be computed from this test by

$$K = \frac{1}{K_{ud}}\frac{e^{t_0/\tau}}{e^{t_0/\tau} - 1} \qquad 1.12(10)$$

The amplitude of cycling for the discrete test will be different from.that of the continuous test, giving an additional check for linearity of gain.

The degree of automatic operation possible in this sequence is strictly a function of available hardware and software.

If a valid model cannot be demonstrated, the instrument engineer should tune a controller by trial and error, or, if a more scientific approach is desired, then more general model structures and identification approaches should be tried. If gain is quite nonlinear with operating conditions, frequent adjustment or on-line adaptation should be considered.

Adjusting the Controller

For those processes that fit the first-order-plus-dead-time model, a variety of approaches are described in the literature for computing tuning adjustment coefficients. The techniques developed have been found to work well in practice with conventional controllers. A set of tables is provided for tuning controllers that relate manipulated variables M(s), with s being the Laplace operator, to process error E(s), by the following transfer functions (as derived in Sections 1.3 and 1.4):

Model(s)	Transfer Function	
P	$M(s) = K_c E(s)$	1.12(11)
PI	$M(s) = K_c(1 + 1/T_i s)E(s)$	1.12(12)
PID	$M(s) = K_c(1 + 1/T_i s + T_d s)E(s)$	1.12(13)

The tables provide tuning for a variety of tuning objectives that are summarized in Table 1.12d. This summary is an elaboration of the contents of Table 1.11a presented earlier.

Table 1.12d
CONTROLLER TUNING CRITERIA FOR PROPORTIONAL CONTROL SYSTEMS*

$$ISE - 1 = \int_0^\infty [c(t) - c(\infty)]^2 \, dt$$

$$ISE - 2 = \int_{\theta_0}^\infty [c(t) - c(\infty)]^2 \, dt$$

$$ISE - 3 = \int_0^\infty \left[\frac{c(t) - c(\infty)}{c(\infty)}\right]^2 \, dt$$

$$IAE - 1 = \int_0^\infty |c(t) - c(\infty)| \, dt$$

$$IAE - 2 = \int_{\theta_0}^\infty |c(t) - c(\infty)| \, dt$$

$$IAE - 3 = \int_0^\infty \left|\frac{c(t) - c(\infty)}{c(\infty)}\right| \, dt$$

$$ITAE - 1 = \int_0^\infty |c(t) - c(\infty)|t \, dt$$

$$ITAE - 2 = \int_{\theta_0}^\infty |c(t) - c(\infty)|t \, dt$$

$$ITAE - 3 = \int_0^\infty \left|\frac{c(t) - c(\infty)}{c(t)}\right| t \, dt$$

* Table 1.12d is provided by A. M. Lopez, J. A. Miller, C. L. Smith, and P. W. Murrill from their paper on page 57 of the November 1967 issue of *Instrument Technology*.

Tuning that will minimize the performance object we select from Table 1.12d can be obtained for any standard one- through three-mode controller by consulting (or preprogramming) the formulas given in Tables 1.12 e, f and g.

Table 1.12e
TUNING EQUATIONS FOR PROPORTIONAL CONTROL BASED ON LOAD DISTURBANCE

$$K_c = \frac{A}{K_u}\left(\frac{t_0}{\tau}\right)^B$$

	Constants	
Criterion	A	B
Ultimate	2.133	−0.877
1/4 Decay	1.235	−0.924
ISE—1	1.411	−0.917
ISE—2	0.9889	−0.993
ISE—3	0.6659	−1.027
IAE—1	0.9023	−0.985
IAE—2	0.6191	−1.067
IAE—3	0.4373	−1.098
ITAE—1	0.4897	−1.085
ITAE—2	0.4420	−1.108
ITAE—3	0.3620	−1.119

Table 1.12f
TUNING EQUATIONS FOR PI CONTROLLERS BASED ON LOAD DISTURBANCE

$$K_c = \frac{A}{K_u}\left(\frac{t_0}{\tau}\right)^B$$

$$\frac{1}{T_i} = \frac{A}{\tau}\left(\frac{t_0}{\tau}\right)^B$$

Criterion	Controller Mode	Constants	
		A	B
ISE	Proportional	1.305	−0.960
	Reset	0.492	−0.739
IAE	Proportional	0.984	−0.986
	Reset	0.608	−0.707
ITAE	Proportional	0.859	−0.977
	Reset	0.674	−0.680

Sample Problem

The previously outlined tuning method is applied here to a 1,000 lbm/hr (0.13 kg/s) pilot plant distillation unit under the direct digital control of a time-shared computer. A process operator in cooperation with a computer tunes the main temperature control loop of this unit (Figure 1.12h).

Figure 1.12i shows the modified step response and the ultimate sensitivity test data. In this example, process gain computed from equation 1.12(8) is 74 percent of that computed from equation 1.12(4), which is considered to be adequate verification.

Table 1.12g
TUNING EQUATIONS FOR PID CONTROLLERS BASED ON LOAD DISTURBANCE

$$K_c = \frac{A}{K_u} \left(\frac{t_0}{\tau} \right)^B$$

$$\frac{1}{T_i} = \frac{A}{\tau} \left(\frac{t_0}{\tau} \right)^B$$

$$T_d = \tau A \left(\frac{t_0}{\tau} \right)^B$$

Criterion	Controller Mode	Constants A	B
ISE	Proportional	1.495	−0.945
	Reset	1.101	−0.771
	Rate	0.560	1.006
IAE	Proportional	1.435	−0.921
	Reset	0.878	−0.749
	Rate	0.482	1.137
ITAE	Proportional	1.357	−0.947
	Reset	0.842	−0.738
	Rate	0.381	0.995

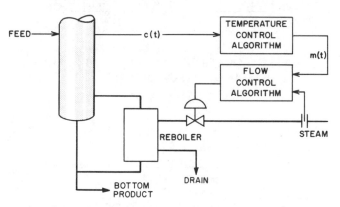

Fig. 1.12h   Operator tuning techniques demonstrated on pilot distillation unit

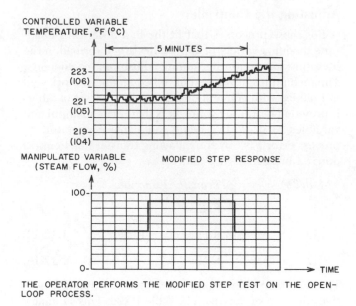

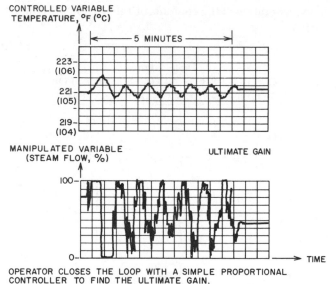

Fig. 1.12i   Controller tuning performed by process operator in cooperation with a computer

In Figure 1.12j, a PI direct digital control algorithm closes the loop, using the higher gain value for the process model. The sampling period is one second, essentially removing this parameter from the design. Controller settings, computed from Table 1.12f, are chosen to minimize the integral time multiplied by absolute error (ITAE criterion).

The operator is able to tune the loop with relative ease, since all he has to do is gather data. The computer calculates the model parameters, verifies the operator's tests, and supplies tuning adjustments for the controller.

Conclusions

With the technique described above, the operator can check model accuracy periodically or when operating conditions change. Up-to-date modeling and tuning should greatly improve process performance. The technique maintains process safety by allowing the operators to remain in the loop during the critical testing phase.

Cost reductions are realized by reducing the amount of trial-and-error tuning. The operator need not bring the process to a final value in the step response test. This saves time when dealing with slow processes. Test signal size is not critical, reducing trial-and-error selection.

Time-shared systems with ddc or supervisory control offer a convenient vehicle for implementation of operator tuning. The ddc offers a convenient vehicle for implementation of operator tuning. The ddc computer has greater scaling and sensitivity capabilities than conventional industrial instruments. Also, it can perform unit conversion and can be easily programmed to abort the tests automatically and set off alarms if necessary.

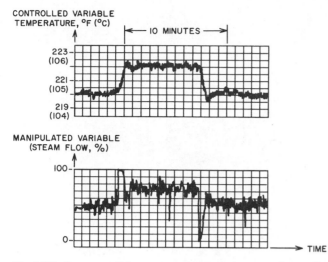

CONTROLLED VARIABLE
TEMPERATURE, °F (°C)

MANIPULATED VARIABLE
(STEAM FLOW, %)

Fig. 1.12j Response of a loop to a proportional-plus-integral controller tuned by operator and computer

In general, the computer can be a very valuable tool for control system analysis and design both on-line and off-line. Off-line techniques vary from full-blown computer simulations of process and associated control to simple linear first-order deadtime models generated from plant step testings. On-line techniques can involve complicated state space optimal control and Kalman filtering techniques or they can be as simple as the automated tuning procedure detailed in this section. The references should provide the interested reader with much more detail on this subject.

The computer has become a very important tool in recent years for the process control and instrument engineer. Off-line techniques are gaining in popularity, and the software is widely available. On-line techniques are becoming more popular as the process control computer becomes a common installation. With the need for better control of more integrated processes, these techniques will be of great practical importance.

REFERENCES

1. Gelb, A., *Applied Optimal Estimation*, The MIT Press, Cambridge & London, 1974.
2. Box, G.E.P., and Jenkins, G.M., *Time Series Analysis: Forecasting and Control*, rev. ed., Holden-Day, San Francisco, 1976.
3. Chen, C.T., *Introduction to Linear System Theory*, Holt, Rinehart and Winston, New York, 1970.
4. Bryson, A.E., Jr., Ho, Y.C., *Applied Optimal Control*, John Wiley & Sons, New York, 1975.
5. Rosenbrock, H.H., *Computer-Aided Control System Design*, Academic Press, London, 1974.
6. Lipták, B.G., (ed.), *Instrument Engineers' Handbook*, vol. I, ch. I, Chilton Book Co., Philadelphia, 1969.
7. Bakke, R.M., "Controller Tuning by Computer," *Instrument Technology*, September, 1968.

BIBLIOGRAPHY

Anderson, N.A., "Step-Analysis of Finding Time Constants," *Instruments & Control Systems*, 1963.

Bjornsen, Bjorn G., "A Fluid Amplifier Pneumatic Controller," *Control Engineering*, June, 1965.

Box, G.E.P., and Jenkins, G.M., *Time Series Forecasting and Control*, Holden-Day, San Francisco, 1970.

Breakwell, J.V., et al., "Optimization and Control of Nonlinear Systems Using the Second Variation," *SIAM Journal on Control*, sec. A., vol. 1, no. 2, 1963.

Buckley, P.S., "A Modern Perspective on Controls Tuning," Texas A&M 30th Annual Symposium on Instrumentation for the Process Industries, January, 1973.

———, *Techniques of Process Control*, John Wiley & Sons, New York, 1964.

Caldwell, W.I., "Generating Control Functions Pneumatically," *Control Engineering*, September, 1954.

Campbell, D.P., "Process Dynamics", John Wiley & Sons, New York, 1958.

Crutchley, W., "The Quest for Control," *Instruments & Control Systems*, March, 1979.

Dahlin, E.B., "Designing and Tuning Digital Controllers," *Instruments & Control Systems*, June, 1968.

Fox, H.L., and Wood, O.L., "Fluid Amplifiers—The Development of Basic Devices," *Control Engineering*, September, 1964.

Gallier, P.W., "Self-tuning Computer Adapts DDC Algorithms," *Instrumentation Technology*, February, 1968.

Higham, J.D., "Single-Term Control of First- and Second-Order Processes with Dead Time," *Control*, February, 1968.

Hougen, J.O., "Experiences and Experiments with Process Dynamics," *Chemical Engineering Progress Monograph Series*, vol. 60, No. 4, 1964.

Larson, R.E., et al., "State Estimation in Power Systems, Part I and Part II," IEEE Trans. on Power Apparatus and Systems, vol. PAS 89, no. 3, March, 1970.

Mehra, R.K., "Identification of Stochastic Linear Dynamic Systems Using Kalman Filter Representation," *AIAA Journal*, January, 1971.

———, "On-Line Identification of Linear Dynamic Systems with Applications to Kalman Filtering," IEEE Trans. on Automatic Control, vol. AC 16, no. 1, February, 1971.

Oldenbourg, R.C., and Sartorius, H., *The Dynamics of Automatic Controls*, The American Society of Mechanical Engineers, New York, 1948.

Palmer, O.J., "Which: Air or Electric Instruments?" *Hydrocarbon Processing and Petroleum Refiner*, vol. 44, no. 5, 1965.

Raven, F., *Automatic Control Engineering*, McGraw-Hill Book Co., New York, 1961.

Rothstein, M.B., "Tuning a Nearly Optimal Control by Digital Simulation," *Instrumentation Technology*, February, 1972.

Rovira, A.A., Murrill, P.W., and Smith, C.L., "Tuning Controllers for Set-Point Changes," *Instruments & Control Systems*, December, 1969.

Salmon, D.M., and Kokotovic, P.V., "Design of Feedback Controllers for Nonlinear Plants," IEEE Transactions on Automatic Control, vol. AC 14, no. 3, 1969.

Smith, O.J.M., "Close Control of Loops with Dead Time," *Chemical Engineering Progress*, May, 1957.

Sood, M., "Tuning Proportional-Integral Controllers for Random Load Changes," M.S. Thesis, Dept. of Chemical Engineering, University of Mississippi, 1972.

———, and Huddleston, H.T., "Optimal Control Settings for Random Disturbances," *Instrumentation Technology*, March, 1973.

Thayer, D., "Tuning Direct Digital Control," *Instruments & Control Systems*, October, 1967.

Wells, C.H., "Application of Modern Estimation and Identification Techniques to Chemical Processes," *AIChE Journal*, 1971.

———, and Mehra, R.K., "Dynamic Modeling and Estimation of Carbon in a Basic Oxygen Furnace," Paper presented at Third International IFAC/IFIPS Conference on the Use of Digital Computers in Process Control, Helsinki, Finland, June, 1971.

Wood, R.K., "Computerized Design and Analysis of Control Loops," *Instrumentation Technology*, 1971.

Chapter II

CONTROLLERS, TRANSMITTERS, AND TELEMETERING

P. Binder • L. D. DiNapoli

R. K. Kaminski • B. G. Lipták

V. J. Maggioli • C. L. Mamzic

A. F. Marks • J. E. Talbot

J. Venczel

CONTENTS OF CHAPTER II

Contents of Chapter II

2.1 COMPUTING RELAYS AND FUNCTION GENERATORS

LARGE-CASE, CAM ACTUATED TIME FUNCTION GENERATORS (PROGRAMMERS)

Cost: $1,200 to $1,500

Partial List of Suppliers: Bailey Controls, Div. of Babcock & Wilcox; Bristol Babcock, Inc.; Eagle Signal Industrial Controls; Foxboro Co.; Honeywell, Inc.; Leeds & Northrup Co.; Taylor Instrument Co.

LARGE-CASE, ADJUSTABLE RANGE AND HOLD PROGRAMMERS

Cost: $2,200 to $2,700

Partial List of Suppliers: Foxboro Co.; Taylor Instrument Co.

MINIATURE AND LARGE-CASE PNEUMATIC PROFILE TRACERS (PROGRAMMERS)

Inaccuracy: ±0.25% of full scale

Cost: $1,800 to $3,200

Partial List of Suppliers: Gaston County Dyeing Machine Co.; Moore Products Co.; Partlow Corp.; Pneucon, Inc.

ELECTRIC LINE AND EDGE FOLLOWER PROGRAMMERS

Inaccuracy: ±0.25% of full scale

Cost: $1,800 to $3,200

Partial List of Suppliers: Leeds & Northrup Co.; Research, Inc.

STEP PROGRAMMERS

Cost: $300 to $6,000

Partial List of Suppliers: Taylor Instrument Co.

DIGITAL TO ANALOG PROGRAMMERS

Cost: $2,700 to $7,000

Partial List of Suppliers: Barber-Coleman Co.; Honeywell Process Control Div.; Leeds & Northrup Co.

COMPUTING RELAYS

Inaccuracy: $\pm\frac{1}{2}\%$ for all types except the differentiating and integrating relays, which are ±15% if uncompensated and ±1% if compensated in specially built units

Cost: Prices shown are for pneumatic (electronic)

High and low selectors: $40 to $112 ($125 to $235)

Adding and subtracting relays: $145 to $185 ($250 to $350)

Square root extractors and function generators: $365 to $570 ($180 to $260)

Scaling and proportioning relays: $450 to $750 ($190 to $275)

Multiplying and dividing relays: $500 to $870 ($275 to $375)

Partial List of Suppliers: The letters "e" and "p" indicate whether electronic and/or pneumatic units are offered. Bailey Controls, Div. of Babcock & Wilcox (e,p); Beckman Instruments, Inc. (e,p); Bell & Howell Co., CEC Div. (e); Bristol Babcock, Inc. (e,p); Devar, Inc. (e); G. W. Dahl Co. (p); Fairchild Industrial Products Co. (p); Fischer & Porter Co. (e,p); Foxboro Co. (e,p); Leeds & Northrup Instruments, a Unit of General Signal (e); Moore Products Co. (e,p); Powell Process Systems, Inc. (e); Robertshaw Controls Co. (e,p); Rochester Instrument Systems (e); Sorteberg Controls Co. (p); Taylor Instrument Co. (e,p); Valmet IMP, Inc. (p); Westinghouse Electric Combustion Controls Div. (e)

Time Function Generators (Programmers)

The simplest and the least expensive analog time function generator is the cam-type programmer. These are assembled in large-case circular chart recorder housings (Figure 2.1a) and consist of a motor-driven cam that moves the set-point index, to which a motion transmitter is connected. The output is usually a 3–15 PSIG (0.2–1.0 bar) pneumatic set-point signal. Electric outputs are also available. The time base is a function of motor speed, and a wide selection of speeds is available. The cams can be made of plastic or metal. It is also common to incorporate an integral controller, direct-sensing element, and circular chart recorder in the same housing.

Cam programmers are usually applied to batch processes that are repeated time after time. Making up new cams and changing them is *not* a simple matter. These units are not as accurate as the profile tracer and line follower types of more recent manufacture. The cam rise is also limited for mechanical reasons to about 50-degree cam rotation for full-scale movement of the index. Curvilinear coordinates make the cams more difficult to lay out as compared with programming a device with rectilinear coordinates.

Adjustable Ramp-and-Hold Programmers

For batch processes in which the controlled variable must be made to rise at a controlled rate, then hold at some preset value and, possibly, fall at a controlled rate, programmers such as that in Figure 2.1b are often preferable to cam types, particularly if the program must be changed periodically. These, too, are usually packaged as large-case circular chart recorders. In this type programmer, the set-point index is driven by a constant-speed motor. The rate of rise is set by adjustment of an interrupter timer, which makes contact for a set percentage of the basic timer cycle time. The movement of the index is, therefore, actually in steps, but the steps are so small that the operation is, for all practical purposes, continuous. The set point rises until it coincides with the hold point index, at which point the hold timer is energized while the interrupter timer is de-energized. Controlled cooling rate requires driving the set-point index in reverse.

Usually this type of programmer comes complete with a controller element and direct sensing element.

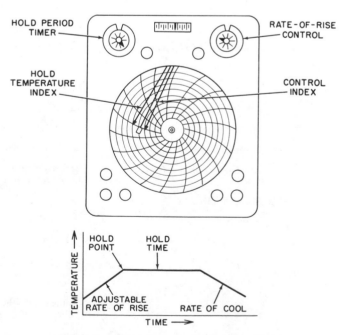

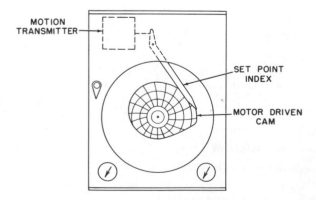

Fig. 2.1a   Cam type programmer

Fig. 2.1b   Adjustable ramp-and-hold programmer

Profile Tracer Programmers

Profile tracer programmers come in 6 × 6 in (150 × 150 mm) miniature pneumatic recorder type cases (Figure 2.1c) or in large cases. The program is stored on a laminated, endless belt plastic master. It combines an analog set-point program with up to 25 synchronized digital tracks for operation of logic circuits, auxiliary equipment, solenoid valves, lights, etc. There is no limit to the slope the programmer can follow—even slopes of 90 degrees are accommodated.

Since the master program can be quickly changed, these programmers are often used where the program does require periodic change and where accurate reproduction of the program is essential, as in textile dyeing processes. The complete program is stored on the master, thus eliminating the need for having an operator make various program settings for each change and, therefore, eliminating the chance for human error in setting the program.

These programmers are accurate to within $\frac{1}{4}$ percent of full scale, which makes them applicable when accuracy alone is the critical requirement of the operation.

The endless belt master is made up by plotting the desired analog program on the rectilinear chart and cutting the top portion away with scissors. The second layer serves as a backing and is also used to program the synchronized digital tracks. At any point of the program where a switch action is desired, a hole is punched in with a conductor's punch. The back of the analog program has a pressure-sensitive adhesive that joins the two sections. A splice finishes the make-up of the master.

In operation, a motor drives the master program. A cable-mounted tracer nozzle senses the step on the analog program profile. The back-pressure of the nozzle actuates a servo, which, through the cable drive, keeps the tracer following the profile. Operating from the same servo drive is an accurate force balance type motion detector. Sensing the back side of the digital master are a series of vertically aligned nozzles. Normally, their back pressure is high since the master baffles the nozzles. However, if a punched hole presents itself, the back pressure of that particular nozzle drops to zero, actuating the connected pressure switch.

Electric Line and Edge Follower Programmers

Electric line and edge follower programmers will perform with less than $\pm\frac{1}{4}$ percent error. In the electrostatic line follower type, Figure 2.1d, the desired program curve is etched into a conductive surface chart, dividing it into two electrically isolated surfaces. The surfaces are energized by oppositely phased AC voltages establishing a gradient across the gap. A noncontacting probe senses the electrostatic field developed by the surfaces and energizes a servo amplifier to keep the probe tracking the line, which is at zero potential. Attached to the servo drive is the wiper of a potentiometer whose output is proportional to line position. The photoelectric line follower type functions to keep the line centered between two slightly overlapping pickup heads. The detector must be manually set over the line at start-up, and slope rate is limited by the speed of the follower mechanism.

The photoelectric edge follower consists of a chart which is divided into a transparent and an opaque section at the program line. A photocell detector senses the edge and a servo system tracks it.

Up to eight digital tracks are available with the electric programmers.

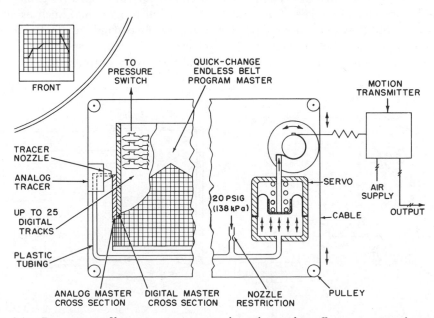

Fig. 2.1c Pneumatic profile tracer programmer with synchronized on-off sequence control switches

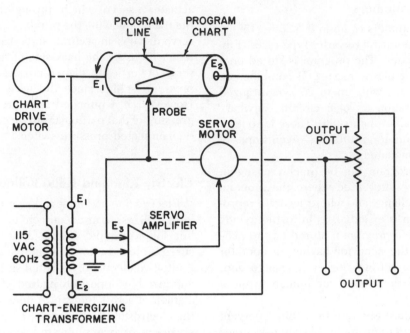

Fig. 2.1d   Electric line follower programmer

Step Programmers

Step programmers are used for on-off event sequencing. (See also Sections 6.6 and 6.8.) They do not provide an analog output. A typical type consists of a perforated drum (Figure 2.1e). Each perforation represents a step in one channel. Drums are available with from 30 to 100 steps and from 16 to 93 channels. Inserting a nylon plug into a hole results in a switch actuation on the corresponding step and channel. The stepping can be initiated by a remote sensor switch, counter, timer, or pushbutton.

These units are easily programmed and can replace complex logic and interlock circuits that are commonly implemented with electromechanical relays. They not only replace such systems, but they eliminate the need for their custom design and construction as well.

Related to the step programmer is the continuous, multi-channel cam timer in which a number of individually adjustable cams, mounted on a single drive shaft, provide event sequencing control.

Microprocessor-Based Programmer

Microprocessor-based programmers generate set points and related contact closure outputs based upon time and events. The programs are entered into the memory of the programmer through a keyboard. Once entered, the program can be copied onto a mass memory storage device such as a tape. With programs stored on tape, it becomes easy to change programs as required.

A digital programmer such as shown in Figure 2.1f can generate two independent or related 4–20 mA analog electronic set points and provide 8 contact closure or solid-state relay outputs. It can accept 16 logic inputs

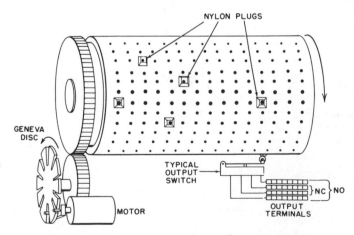

Fig. 2.1e   Drum programmer for on-off event sequencing

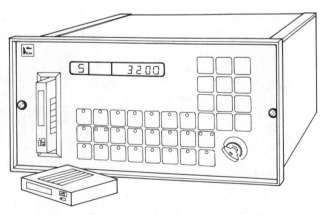

Fig. 2.1f   Microprocessor-based digital-to-analog programmer with tape memory

and accommodate 40 logic events. In addition, the programmer provides special set-point characterization for up to 7 types of thermocouples that might be used with the related temperature controller.

Computing Relays

Pneumatic Multiplying and Dividing

In the force bridge multiplier-divider shown in Figure 2.1g, input pressures act on bellows in chambers A, B, and D. The output is a feedback pressure in chamber C. The bridge consists of two weigh-beams that pivot on a common movable fulcrum, with each beam operating a separate feedback loop. Any unbalance in moments on the left-hand beam causes a movement of the fulcrum position until a moment-balance is restored. An unbalance in moments on the right-hand beam results in a change in output pressure until balance is restored. Equations which characterize the operation of the force bridge are

$$A \times a = B \times b \quad \text{and} \quad D \times a = C \times b \quad 2.1(1)$$

The equation reduces to

$$A \times C = B \times D \qquad 2.1(2)$$

or

$$C = \frac{B \times D}{A} \qquad 2.1(3)$$

Multiplication results when the two input variables are connected to chambers B and D. Division results when the dividend is connected to either chambers B or D, with the divisor connected to A. Simultaneous multiplication and division results when B, D, and A chambers are used.

The significant advantage of a cam-actuated multiplying and dividing relay is that it can operate with practically any type of nonlinear function that can be cut on a cam. This can mean operation with logarithmic functions, as in pH measurement, and computation in narrow, suppressed ranges of measurement which results in good resolution. The pure multiplier-divider in Figure 2.1g when used for temperature and pressure compensation for example, uses input signals proportional to the total absolute temperature and pressure range, starting with zero. Since the usable temperature and pressure range might be a small percentage of the total measurement range, the results might lack precision.

In Figure 2.1h, input pressure P_1 and output P_0 act on double diaphragm capsules, and the net resultant force in each is in the direction of the larger area diaphragm. Input P_1 creates force Y, which pulls the baffle, pivoted at A, away from the nozzle. Output pressure P_0 creates force X which moves the baffle closer to the nozzle. The θ input-output relationship is a function of the angle of the nozzle beam. When angle θ is 45 degrees, the relationship is 1:1. This can also be considered the multiplication factor or gain, K. At larger angles, K is greater than 1, and at smaller angles, smaller than 1. The multiplicand P_2 acts on the cam-positioning cylinder and thereby changes the nozzle beam angle in accordance with the cam characteristic. The zero adjusting springs subtract the 3 PSI (0.2 bar) zero from P_1 and set a 3 PSI (0.2 bar) zero on the output, respectively.

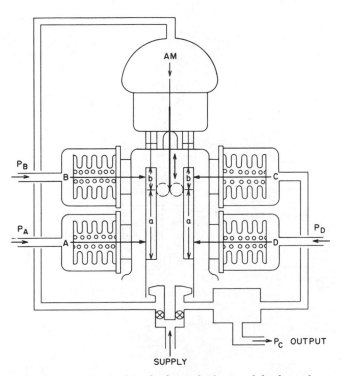

Fig. 2.1g Pneumatic force bridge multiplying and dividing relay

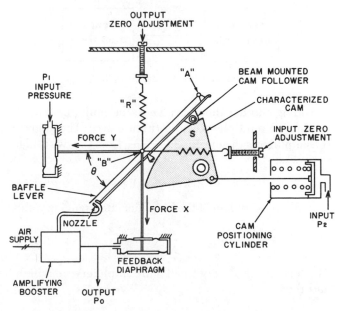

Fig. 2.1h Pneumatic cam-characterized multiplying and dividing relay

The characteristic equations are

$$(P_0 - 3) = (P_1 - 3)K \qquad 2.1(4)$$

$$\text{Cotangent } \theta = K \qquad 2.1(5)$$

$$K = f(P_2 - 3) \qquad 2.1(6)$$

Combining 2.1(4) and 2.1(6),

$$(P_0 - 3) = (P_1 - 3)f(P_2 - 3) \qquad 2.1(7)$$

$$P_0 = (P_1 - 3)f(P_2 - 3) + 3 \qquad 2.1(8)$$

Electronic Multiplying and Dividing

In Figure 2.1i, inputs e_1 and e_2 are multiplied in the diode bridge. Conduction of the diodes in the bridge is dependent upon the relative magnitude of the inputs with respect to the constant slope of the sawtooth input. The output of the diode bridge is a trapezoid, which has an area equivalent to

$$\text{Area} = e_1 e_2 \tan \theta \qquad 2.1(9)$$

The angle θ is established by the constant slope of the sawtooth, and thus

$$\text{Area} = K e_1 e_2 \qquad 2.1(10)$$

The output voltage, e_0, is amplified and filtered to a DC signal, and its voltage level will, therefore, be proportional to the area and, consequently, to the product of e_1 and e_2.

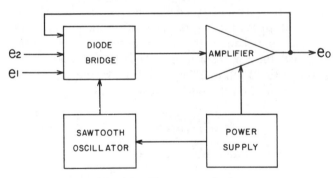

Fig. 2.1i Electronic multiplier

Adding another diode bridge to the multiplier circuit produces a multiplier/divider (Figure 2.1j). The input to the amplifier is the output difference from the two bridge networks.

$$e_0 = A(K e_1 e_2 - K e_3 e_0) \qquad 2.1(11)$$

where A = gain of the amplifier. Rearranging yields

$$e_1 e_2 = \frac{e_0}{AK} + e_3 e_0 \qquad 2.1(12)$$

The term e_0/AK is very small if the amplifier gain is high, and thus

$$e_0 = \frac{e_1 e_2}{e_3} \qquad 2.1(13)$$

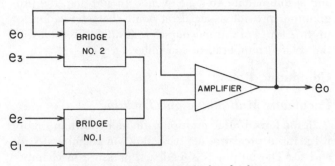

Fig. 2.1j Electronic multiplier-divider

Pneumatic Adding, Subtracting, and Inverting

In the force balance arithmetic computing relay, Figure 2.1k, a signal pressure in chamber A acts downward on a diaphragm with unit effective area. A signal in chamber B also acts downward on an annular diaphragm configuration, likewise having an effective area of unity. Signal pressures in chambers C and D similarly act upward on unit effective diaphragm areas. Any unbalance in forces moves the diaphragm assembly with its integral nozzle seat. The change in nozzle seat clearance changes the nozzle back pressure and, hence, changes the output pressure, which is fed back into chamber D until force

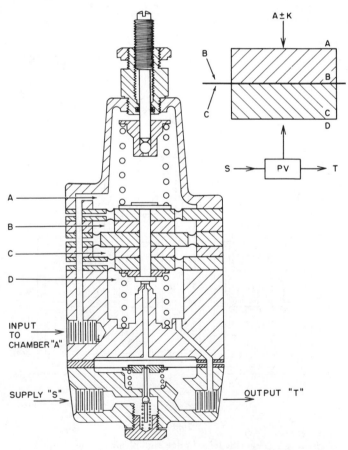

Fig. 2.1k Pneumatic adding, subtracting, inverting, and biasing relay

balance is restored. The basic equation which describes the operation of the relay is

$$T = A + B - C \pm K \qquad 2.1(14)$$

K is the spring constant. It is adjustable to give an equivalent bias of ± 18 PSI (1.24 bar).

The relay in Figure 2.1l is a modification of Figure 2.1k in that it incorporates additional input chambers and output feedback chambers. It can be used to add and/or average up to nine inputs. Figure 2.1l is an averaging relay for five inputs. The averaging feature keeps all signals in the same standard 3–15 PSIG (0.2–1.0 bar) range.

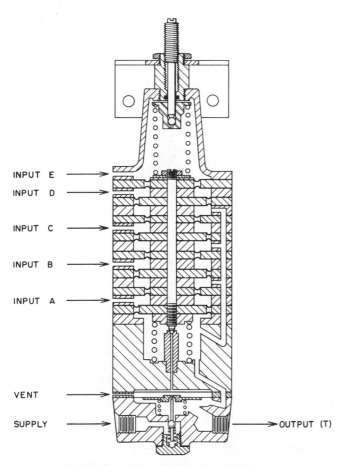

Fig. 2.1l   Pneumatic multi-input averaging relay

The characteristic equation is

$$T = \frac{A + B + C + D + E}{5} \pm K \qquad 2.1(15)$$

Figure 2.1k relay provides inverting, or reversing, action by setting the bias spring loading to a maximum and connecting the input to subtracting chamber C. If the bias is set at $+18$ PSIG (41.2 bar), then a 3–15 PSIG (0.2–1.0 bar) signal in chamber C results in a 15–3 PSIG (1.0–0.2 bar) output.

The equation describing the operation is

$$T = K - C \qquad 2.1(16)$$

Electronic Adding, Subtracting, and Inverting

In Figure 2.1m, the two input potentials e_1 and e_2 are compared in the multiple comparator, which produces a proportional output to the amplifier. The current paths of the two inputs can be the same or opposite, resulting in either an adding or subtracting circuit, respectively.

Inverting is accomplished by biasing the comparator to produce maximum output with no input. Applying a reverse input (i.e., a reverse current input with respect to bias current) causes the output to decrease with increasing input. The feedback signal is such that the amplifier acts as a unity gain network.

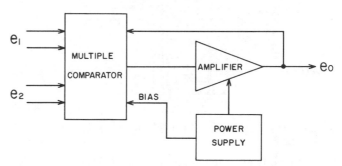

Fig. 2.1m   Electronic adder, subtracter, and inverter

Pneumatic Scaling and Proportioning

Scaling, or proportioning, involves multiplication by a constant. Several approaches are available: (1) special fixed-ratio relays, (2) pressure transmitters, (3) proportional controllers, and (4) adjustable ratio relays.

The fixed-ratio scaler is the simplest if the correct ratio is available and if adjustability and exact ratio is unnecessary. Figure 2.1n shows such a relay. The input pressure is connected to the top chamber and acts on the upper diaphragm. Output acts upward on the small bottom diaphragm. The gain is a function of the relative effective areas of the large and small diaphragm as determined by the dimensions of the diaphragm ring. The bottom spring applies a negative bias to the input and the adjustable top spring allows exact zero setting. The operating equation is

$$T = AP_1 + K \qquad 2.1(17)$$

where A is the gain constant and K is the spring setting.

Where the scaling must be exact and does not have to be adjusted periodically, pressure transmitters are an economical, reliable, and accurate choice. Where the scaling factor must be modified occasionally, conventional ratio relays, which often consist of the proportioning section of a controller, are commonly used.

Electronic Scaling and Proportioning

Simple electronic scaling or proportioning involves combining a voltage divider circuit with an amplifier. The voltage divider circuit is connected to either the

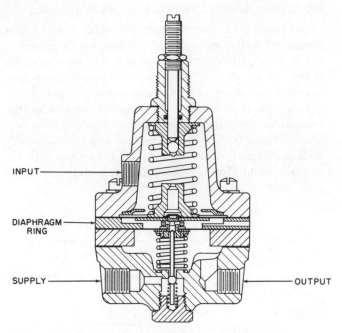

Fig. 2.1n   Pneumatic fixed-ratio amplifying relay

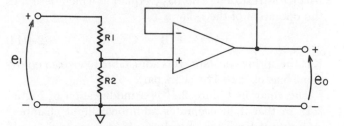

Fig. 2.1p   Electronic scaler with gain less than one

input or output side, depending upon whether the gain
is to be greater or less than unity.

In Figure 2.1o, the amplifier comes to balance when
Δe_i equals zero. Since the voltage divider is on the output, only a portion of the amplifier output is fed back to
counterbalance the input voltage. Therefore, the output
will rise above e_1, resulting in gains greater than one.
The operation can be expressed as

$$e_0 = e_1 \frac{(R_1 + R_2)}{R_1} \qquad 2.1(18)$$

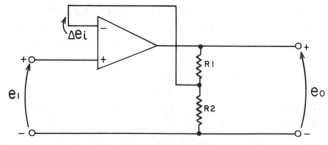

Fig. 2.1o   Electronic scaler with gain greater than one

In Figure 2.1p, the voltage divider is on the input
side, so that only a voltage equal to or less than e_1 is
impressed across the amplifier input, resulting in gains
less than one. This can be expressed as

$$e_0 = e_1 \frac{R_2}{R_1 + R_2} \qquad 2.1(19)$$

Pneumatic Differentiating

A differentiating relay produces an output proportional
to rate of change of input. Figure 2.1q shows an ideal

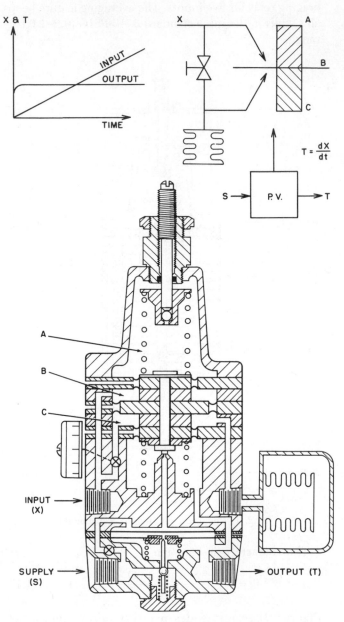

Fig. 2.1q   Pneumatic differentiating relay

pneumatic differentiating relay. The relay is basically
similar in construction to relay (Figure 2.1k), except that
the annular effective diaphragm area between chambers
B and C is more than ten times the effective area of the
small diaphragms between chambers A and B—giving a

gain of greater than ten. The input signal is transmitted unrestricted to chamber B and passes to chamber C through an adjustable restriction. When the input is steady, the forces resulting from pressures in B and C chambers cancel each other, so that the output equals the zero-spring setting (usually mid-scale, 9 PSIG [0.6 bar] if both positive and negative rates are to be measured). If the input pressure changes, a differential develops across the restriction. The relay transmits an output proportional to this differential. For accurate results, this differential must be directly proportional to the rate of change of input. Using a needle valve that produces laminar flow provides a linearly proportional volumetric flow, but the differential developed across the needle valve is a function of the mass flow, which varies with static pressure because of compressibility. This compressibility error is approximately ±15 percent. The effect can be fully compensated, however, by the addition of a variable volume to chamber C, the restricted chamber. As the static pressure increases, tending to make the differential smaller because of higher mass flow rate, the volume increases proportionately to maintain a constant differential. The needle valve setting determines the rate time constant.

The compensated relay is not a standard piece of hardware. In most cases, a noncompensated differentiating relay is satisfactory.

Electronic Differentiating

The input amplifier in Figure 2.1r is capacitor coupled so that only the rate of change of the input signal is seen by the amplifier. Two diodes in the feedback of the amplifier allow its output to go positive or negative (depending on the direction of the rate of change) by an amount equal to the forward drop across the diodes (only a few tenths of a volt). The output amplifier inverts and amplifies this signal by its open loop gain.

A small positive feedback is applied to the last amplifier to prevent output from "chattering" at the diodes' switching point.

Pneumatic Integrating

Integration, the reverse of differentiation, essentially involves measurement of accumulated pressure resulting

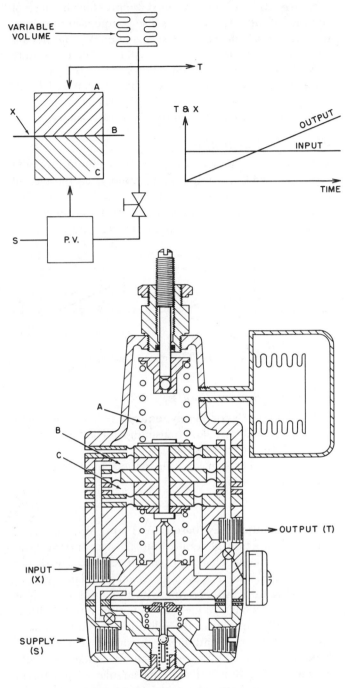

Fig. 2.1s   Pneumatic integrating relay

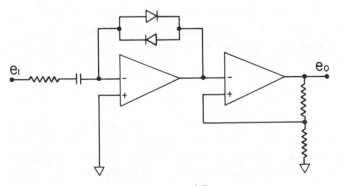

Fig. 2.1r   Electronic differentiator

from a flow that is proportional to the offset (from some chosen reference) of the input variable. Figure 2.1s shows an ideal integration relay. The input signal loads chamber B. The output, it should be noted, is the accumulated pressure in chamber A, *not* the booster pilot output. The input signal determines the pressure differential across the needle valve. As in the case of the differentiation relay, with laminar flow across the needle valve, the volumetric flow is directly related to the differential. The mass flow, however, which determines the accumulated pressure, still varies with the static pressure because of compressibility. This effect is also compensated by connecting a variable volume to chamber A. The needle valve sets the proportionality constant of the integrator.

Neither the compensated differentiating relay nor the compensated integrator is available as standard hardware. Usually the noncompensated relay (actually a proportional-speed floating controller) is satisfactory.

Electronic Integrating

The first amplifier in Figure 2.1t, a simple inverting type, performs the integration function as the charge accumulates across the capacitor of the RC network. The second amplifier is an inverting, general purpose type, which relates the output directly to the input.

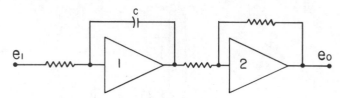

Fig. 2.1t   Electronic integrator

Pneumatic Square Root Extracting

This function is commonly required to linearize signals from differential-type flow transmitters. The force bridge, Figure 2.1g, provides square root extraction when the output is connected in common to the A and C chambers, giving the equation

$$C^2 = B \times D \qquad 2.1(20)$$

Other solutions are based on (1) use of a cam-characterized function generator and (2) a geometric relationship, namely, change in cosine compared with the change in included angle, for small angular displacements (Figure 2.1u). Starting with the input and output at 3 PSIG (0.2 bar), an increase in input causes the floating pilot link to restrict the pilot nozzle. This increases the output pressure and moves the output feedback bellows upward, until balance is restored. Since the length of the floating link is fixed, the angular displacement produced by movement of the output bellows follows the relationship

$$\cos \theta = 1 - \frac{X}{L} \qquad 2.1(21)$$

A plot of the angle θ (output displacement) versus X (input displacement) in this equation shows the relationship to be virtually an exact square root for small angular motion.

Electronic Square Root Extracting

The square root converter, Figure 2.1v, combines a DC amplifier with a negative feedback diode network.

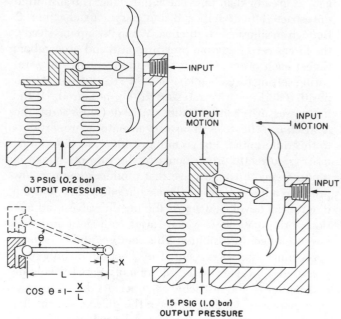

Fig. 2.1u   Pneumatic square root extractor

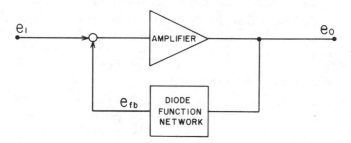

Fig. 2.1v   Electronic square root extractor

As current into the amplifier increases, the amplifier gain decreases with decreased feedback resistance in the diode network. The gain varies according to, typically, seven straight line segments that approximate a square root function. This is accomplished by having seven diode-resistance paths in the feedback network automatically parallel each other with increasing input. The output stabilizes when the diode network modified feedback counterbalances the input.

High- and Low-Pressure Selector and Limiter

Selector relays are used in override systems. The high-pressure selector relay compares two pressures and transmits the higher of the two in its full value. In Figure 2.1w, the two input pressures act against a free-floating flapper disc. The differential pressure across the flapper always results in closure of the low-pressure port.

In the low-pressure selector, Figure 2.1x, if input A is less than input B, the diaphragm assembly throttles the pilot plunger to make the output equal to input A (the conventional action of a 1:1 booster relay). If input B is less than A, the supply seat of the pilot plunger is

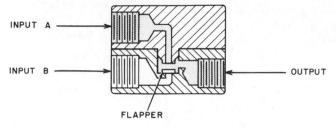

Fig. 2.1w   High-pressure selector relay

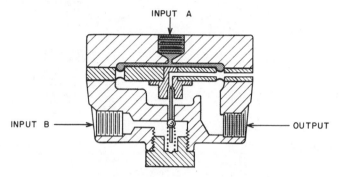

Fig. 2.1x   Low-pressure selector relay

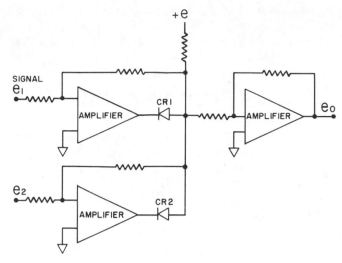

Fig. 2.1y   High-voltage selector

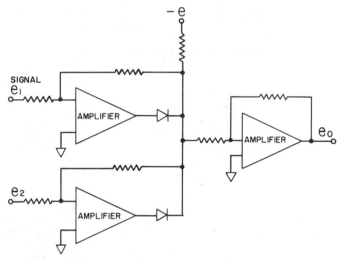

Fig. 2.1z   Low-voltage selector

wide open so that pressure B is transmitted in its full value.

By connecting a set reference pressure into one of the ports of the high-pressure selector, a low-limit relay results. Conversely, by connecting a set reference pressure into the low-pressure selector, a high-limit relay results. Limit relays are available with the reference-setting regulator built into the relay.

Electronic High- and Low-Voltage Selector and Limiter

The higher of the two positive inputs in Figure 2.1y causes a higher negative potential at the cathode of one of the diodes (CR1 or CR2). The forward bias of this diode passes the higher input and reverse biases the other diode to isolate the lower input. Thus if signal e_1 drops below signal e_2, CR2 is forward-biased to pass signal e_2 and CR1 is reverse-biased to isolate signal e_1. All the amplifiers are unity-gain inverter types.

Substituting a fixed input for one of the variables produces a low-limit relay.

To obtain a low-voltage selector (Figure 2.1z), the diodes are inverted and a negative supply (e) is used. Thus, the least positive input forward-biases one of the diodes (by the least negative potential applied to the anodes of the diodes). This automatically reverse biases the other diode and isolates the higher input from the output.

Substituting a fixed input for one of the variables produces a high-limit delay.

BIBLIOGRAPHY

Buckley, P.S., "Designing Long-Line Pneumatic Control Systems," *Instrumentation Technology*, April, 1969.

Eckman, D.P. *Automatic Process Control*, John Wiley & Sons, New York, 1958.

Farmer, E., "Pneumatics in a Digital World," *Instruments & Control Systems*, March, 1979.

Gassett, L.D., "Pneumatic or Electronic?," *Chemical Engineering*, June 2, 1969.

Mamzic, C.L., "Pneumatic Controls Interface with Computers," *Instruments & Control Systems*, October, 1975.

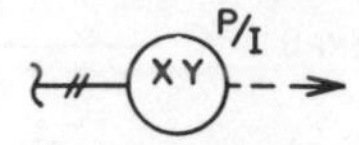

AIR TO CURRENT
CONVERTER
(WHICH BASICALLY
IS A PRESSURE
TRANSMITTER)

OTHER POSSIBLE
LETTER COMBINA-
TIONS TO DESIGNATE
CONVERTERS ARE:

A/D, D/A, D/P,
E/D, E/E, E/I, E/P,
I/D, I/E, I/I, I/P,
P/D, P/E, P/I, P/P,
R/D, R/E, R/I, R/P,
etc.,

WHERE THE LETTERS
USED REPRESENT:
A – ANALOG
D – DIGITAL
E – VOLTAGE
I – CURRENT
P – PNEUMATIC
R – RESISTANCE

2.2 CONVERTERS

Signal Ranges:	Pneumatic: 3–15, 3–27, 6–30 PSIG (0.2–1.0, 0.2–1.8, 0.4–2.0 bar)
	Voltage: 0–1 millivolts DC to 1–5 volts DC or 0–5 volts AC
	Current: 1–5, 4–20, 10–50 millivolts DC
Inaccuracy:	Generally $\frac{1}{10}$% to $\frac{1}{2}$% of full scale
Cost:	$350 to $1500
Partial List of Suppliers:	AMETEK Inc., Controls Div.; Acromag, Inc.; American Controls & Instruments, Inc.; Aro Corp.; Bailey Controls; Bellofram Corp.; Brandt Industries, Inc.; Bruce Engineering Co.; Clippard Instrument Laboratory, Inc.; Computer Instruments Corp.; Conoflow Regulators & Controls; Cutler Controls, Inc.; Dahl, G. W., Inc.; Devar, Inc.; EMX Controls, Inc.; ElectroSyn Corp.; Fairchild Industrial Products Co.; Festo Pneumatic; Fischer & Porter; Fisher Controls International, Inc.; Foxboro Co.; Foxboro/Jordan, Inc.; General Meters & Controls Co.; Honeywell, Process Control Div.; Kinetrol Ltd.; Leeds & Northrup; Mead Fluid Dynamics; Milwaukee Valve Co., Inc.; Moore Industries, Inc.; Moore Products Co.; Olympic Controls, Inc.; Pres Air. Trol Corp.; Research Inc.; Robertshaw Controls Co., Industrial Instrumentation Div.; Robinson-Halpern Co.; Rochester Instrument Systems; Rosemount Inc.; SEREG (Schlumberger Group); SMC Pneumatics Inc.; Scanivalve Corp.; Tejas Controls, Inc.; Telmar, Inc.; Thermo Electric Co., Inc.; VDO Measurement & Control, Inc.; Validyne Engineering Corp.; Valmet IMP Inc.; Ventech Controls, Inc.; Westinghouse Electric Corp., Information Center.

Pneumatic-to-Electronic Converters

The pneumatic-to-electronic transducer is used wherever pneumatic signals must be converted to electronic signals for any one of the following reasons:

1. Transmission over large distances
2. Input to an electronic logger or computer
3. Input to telemetering equipment
4. Instrument air not available at the receiver controller

In principle, any of the electronic pressure transmitters could be used, but in practice, special devices are used to improve accuracy. The air signals are at low pressure levels (3–15 PSIG or 0.2–1.0 bar), and many of the pressure detectors are not sensitive or not linear

enough at these pressures. A P/I transducer should be at least $\frac{1}{2}$ percent accurate and preferably $\frac{1}{4}$ percent to preserve the integrity of the initial signal. Since the total error is the square root of the mean squares of the individual component errors, the greater the precision of the P/I transducer, the better the signal.

Because of this need for accuracy, most P/I transducers use a bellows input and a motion balance sensor. A typical high-quality P/E converter is shown in Figure 2.2a.

Electronic-to-pneumatic converters are discussed in Sections 2.7 and 4.15.

Millivolt-to-Current Converters

Millivolt-to-current converters are widely used in the measurement of temperature, using thermocouples or

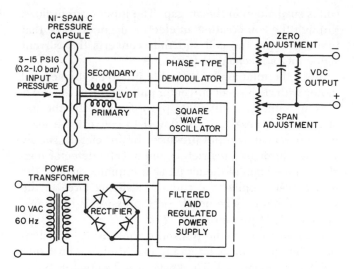

Fig. 2.2a   Pneumatic-to-electronic converter

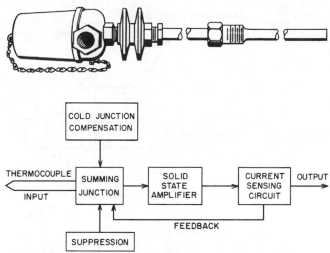

Fig. 2.2c   Locally mounted thermocouple-to-current converter

other millivolt-generating sensing elements. They are also utilized in converting the output signals of analyzers into higher-level transmission signals. A typical millivolt-to-current converter is illustrated in a block diagram form in Figure 2.2b.

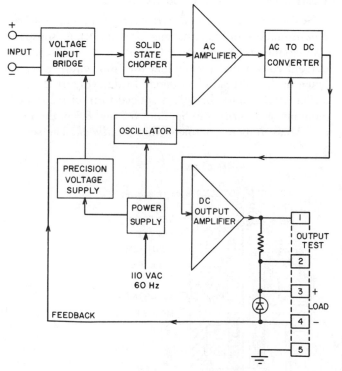

Fig. 2.2b   Millivolt-to-current converter

When these devices are used to convert thermocouple outputs, they are also referred to as temperature transmitters. In the mid-1960s several companies developed a new, miniaturized converter for local mounting in the thermocouple head directly on the thermowell (Figure 2.2c).

Voltage-to-Current Converters

Converters are also available for the conversion of higher voltages into transmission signals (Figure 2.2d). These usually consist of voltage dividers (and rectifiers if necessary) to reduce voltages to a level compatible with the receivers.

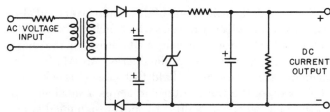

Fig. 2.2d   Solid-state voltage-to-current converter

Current-to-Current Converters

Current-to-current transducers are available to convert AC signals to DC (or DC to AC) and to amplify or reduce their levels as necessary. They are sometimes used in the power industry, but not too widely, because there are other methods of receiving current signals that are capable of handling higher power levels. Levels of alternating current can be changed, when necessary, by current transformers. A common signal level in the power industry is 0–5 amperes. Direct currents are reranged by putting a series resistor (low ohms) in the circuit and reading the voltage drop across it.

A converter of AC to DC milliamperes is shown in Figure 2.2e. This figure shows three separate devices: a current transformer, an AC to DC milliampere converter, and a current-to-air converter. The function of the transformer is to scale down the current to the range normally used for direct AC metering. The AC/DC converter makes this signal compatible with the usual DC milliampere transmission systems. DC to DC "convert-

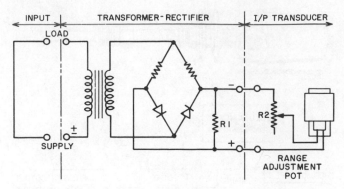

Fig. 2.2e AC-to-DC milliampere converter with integral I/P transducer

ers" are sometimes used for isolation of electrical circuits, such as with intrinsically safe systems. In this case, it is usually called a current repeater or barrier repeater.

Current-to-Air Converters

Electro-pneumatic transmitters are also called converters and transducers. They are extremely important, since they form the link between electrical measurements and pneumatic control systems. They also convert electronic controller outputs into air pressures for operation of pneumatic valves. (These devices are also discussed in Section 4.15).

Figure 2.2f illustrates one of these converters and also lists the various electric devices with which it is commonly combined. The input is usually a DC current in the range of 1–5, 4–20, or 10–50 milliamperes. An Alnico permanent magnet creates a field that passes through the steel body of the transmitter and across a small air gap to the pole piece. A multi-turn, flexure-mounted voice

coil is suspended in the air gap. The input current flows through the coil creating an electro-magnetic force that tends to repel the coil and thus converts the current signal into a mechanical force.

Since the total force obtainable in a typical voice coil motor with such small current inputs is only in the order of some ounces, a different approach, namely, the use of a reaction nozzle, is employed to convert the force into a pneumatic output pressure. In this circuit, supply air flows through a restriction and out the detector nozzle. The reaction of the air jet as it impinges against the nozzle seat supplies the counterbalancing force to the voice coil motor. The nozzle back pressure is the transmitted output pressure.

In order to make the transmitter insensitive to vibration, the voice coil is integrally mounted to a float submerged in silicone oil. The float is sized so that its buoyant force equals the weight of the assembly, leaving a zero net force.

Zero is adjusted by changing a leaf-spring force. Span is adjusted by turning the range-adjusting screw to change the gap between the screw and the magnet, thus shunting some of the magnetic field away from the pole piece.

Resistance-to-Current Converters

Resistance measurements are common in temperature measurements and in resistance or strain gauge sensors. The circuits used are similar to those of the millivolt-to-current converters, except that the front end is a resistance bridge instead of a voltage bridge (Figure 2.2g).

The strain gauge bridge, a special form of resistance element, is described in Sections 5.7 and 8.5 of the Process Measurement volume of this *Handbook*. The strain

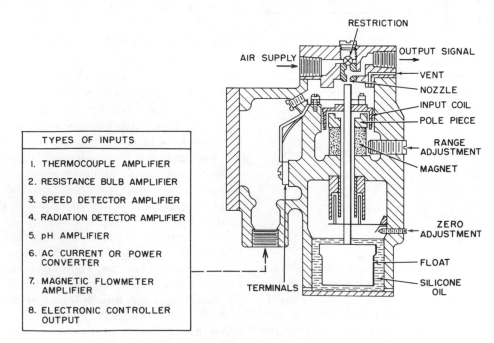

Fig. 2.2f Electric-to-pneumatic transducer-transmitter with list of typical input sources

gauge elements may take the place of two of the resistors in the resistance bridge shown in Figure 2.2g.

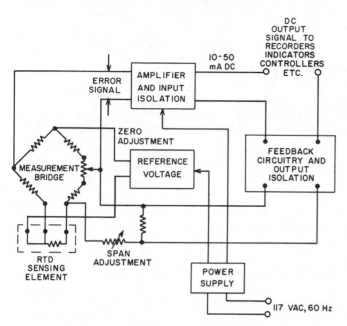

Fig. 2.2g Resistance-to-current converter

Other Converters

Analog-to-digital and digital-to-analog converters are covered in Section 7.6. Similarly, the analog-to-on-off and digital-to-on-off devices, such as monitor switches, relays, and counters, are discussed in Chapter VI.

Signal Conditioners

Pulsation Dampeners and Snubbers

Pressure systems, especially liquid-filled ones, transmit process noise very rapidly. An outstanding example is the output from a reciprocating compressor or pump. In order to obtain precise measurements in systems with pulsation problems, it is necessary to dampen the pressure pulses by the use of restricted flow passages. Some of the various types in use are illustrated in Figure 2.2h.

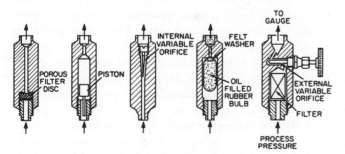

Fig. 2.2h Variations on snubber designs

One design consists of a fitting with a corrosion-resistant porous metal filter disc. By the use of such a device, the equilibrium reading on the indicator is delayed by about 10 seconds. Another snubber design depends for its damping action on a small piston in the inlet fitting, which rises and falls with pressure impulses and thereby absorbs shock and surge. Still another snubber design uses the adjustable restriction created by a microvalve in the inlet fitting to damp pulsations. This differs from a needle valve in two important ways: the filler consists of stainless steel wool to prevent plugging, and it does not quite shut off. This is to prevent "shutting-in" a false reading.

Other forms of pulsation or noise dampening can be obtained by using dashpots in motion balance devices or by restricting the flow between bellows in fluid-filled devices.

Electronic Noise Rejection

It is a general consensus that the analog electronic signal leads should be twisted pairs, and that they should be run in metallic conduit or be shielded. Less generally agreed upon is a recommendation that these conduits or shielded bundles of wiring be isolated from high-power wiring to motors, but this too is followed in many cases.

DC signals of 4–20 or 10–50 milliamperes can be run in the same conduit with telephone wires, low-voltage DC signals, and/or thermocouple leads. They should not be run in the same conduit or in the same shielded bundle with alarm signals or power wiring.

AC signals (which are much less common) should be run in twisted pairs, with each pair shielded, and then in conduit or shielded bundles. Some common mode noise rejection is built into the instruments, but this is not sufficient if computers are being used or if higher level spikes from motor loads are experienced.

The various methods of conditioning digital signals, and the subject of computer compatible wiring practices in general, is covered in Section 7.11.

BIBLIOGRAPHY

Farmer, E., "Pneumatics in a Digital World," *Instruments & Control Systems*, March, 1979.

Holben, E.F., "A Digital to Pneumatic Transducer," *Instruments & Control Systems*, January, 1970.

McNeil, E.B., "A User's View of Instrumentation," *Chemical Processing*, October, 1982.

Pinkowitz, D.C., "A Designer's Guide to Deglitching DACs," *Instruments & Control Systems*, April, 1983.

Sheridan, T.B., "Interface Needs for Coming Industrial Controls," *Instrumentation Technology*, February, 1979.

2.3 ELECTRONIC CONTROLLERS

Standard Input and Output Ranges:	4–20 mA DC
Other Input Ranges:	1–5, 10–50 mA DC; 1–5, 0–10 V DC; ±10 V DC
Other Output Ranges:	1–5, 10–50 mA DC; 1–5, ±10 V DC; ±2 mA DC
Repeatability:	0.5%
Inaccuracy:	±1%
Control Modes:	Manual, proportional, proportional-plus-integral, proportional-plus-derivative, proportional-plus-integral-plus-derivative, and integral.
Displays:	Set point, process variable, output, deviation, and balance
Cost:	$1000–$3000
Partial List of Suppliers:	Action Instruments; Bailey Controls, Div. of Babcock & Wilcox; Barber Colman Co.; Beckman Instruments Inc.; Bristol Babcock Inc.; Eurotherm Co.; Fischer & Porter Co.; Fisher Controls Co.; Foxboro Co.; Honeywell, Process Control Div.; Kent Process Control Inc.; Leeds & Northrup, Unit of General Signal; Moore Products Co.; Powell Process Systems Inc.; Robertshaw Controls Co.; Rosemount Inc.; Taylor Instrument Co.; Thermo Electric Co.; Toshiba Corp.; Tumbull Control Systems Inc.; Westinghouse Electric Corp.; Yokogava Corp. of America.

In this section the general-purpose analog electronic controllers are described. In the late 1970s, a new family of electronic controllers entered the marketplace. These are the digital, microprocessor-based units, which frequently can handle more than one loop and/or more than one variable. For this reason, they are also referred to as shared controllers. Because all microprocessor-based units are capable of communication over data highways, they are also called "distributed" controllers. The various microprocessor-based shared controller designs are described in Section 7.4; the single-loop digital controllers are also briefly described at the end of this section.

The Controller's Function

Most industrial processes require that certain variables, such as flow, temperature, level, and pressure, remain at or near some reference value, called a set point. The device that serves to maintain a process variable value at the set point is called a controller. The controller looks at a signal that represents the actual value of the process variable, compares this signal to the set point, and acts on the process to minimize any difference between these two signals.

Any simple process control loop (Figure 2.3a) contains the equivalent of a sensor, transmitter, control element (usually a valve), and a controller. The sensor measures the actual value of the process variable. The transmitter amplifies this sensed signal and transforms it into a form suitable for sending to the controller. The controller really has two inputs: this measured signal and a set-point signal. The set point, however, may be internally generated. The controller subtracts the two input signals, producing a deviation or error signal. Additional hardware in the controller shapes the deviation signal into an output according to certain control actions or modes (Section 1.2). The controller output then typically sets the position of a pneumatic control valve or other final element in a direction to decrease the error.

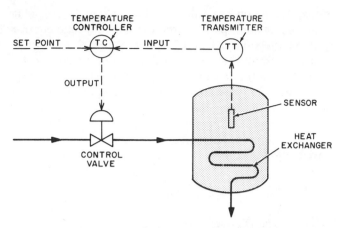

Fig. 2.3a   Standard temperature control loop

Although the function of the controller is easy to define, it may be implemented in many different ways. For example, the controller may work on pneumatic (Section 2.6), fluidic, electric, magnetic, mechanical, or electronic principles, or on combinations of these. This section, however, will confine itself primarily to a certain class of miniature electronic controllers that accept and generate standard process input and output current signals. Other kinds of electric and electronic controllers, having various forms of inputs and outputs, will be described briefly at the end of the section.

The general nature of their inputs, outputs, and control modes makes the electronic and pneumatic controllers applicable to a wide variety of process control situations. The same controller may control flow in a pipe, which responds nearly instantly to valve movements, or a temperature in a distillation column, which often takes hours to respond to disturbances. Any process variable is a candidate for control as long as it can be transduced to a standard current signal. The instrument engineer must choose and tune the control modes to fit the dynamics of the variable to be controlled (Sections 1.2 and 1.11).

Basic Parts of the Controller

The construction of electronic process controllers breaks down into six basic parts or sections: input, control, output, display, switching, and power supply (Figure 2.3b). The input section comprises the hardware for generating the set-point signal, for accepting and conditioning the process signal, and for comparing the two signals to produce a deviation signal. The heart of the control section is generally a chopper-stabilized AC amplifier that acts on the deviation signal. Associated circuitry amplifies, integrates, and differentiates the deviation signal to produce an output with the necessary proportional gain, integral (reset), or derivative (rate) control actions. (In some controllers the derivative circuit acts on the process signal rather than the deviation signal.)

The controller's output has two forms: automatic and manual. When in the manual mode, the controller's output remains at a steady level, often dependent on the shaft position of a potentiometer. Switching and balancing hardware on the front panel selects the automatic or manual output. In some cases the source of the manual output is the control amplifier (modified by switching), while in other cases the source is entirely separate, requiring only the operation of a power supply.

The display section on the front panel carries information about the set point, process variable value, deviation, and controller output. The front panel also contains mechanisms for switching between manual and automatic modes and for adjusting the value of the set-point input and the manual output. A secondary indication, balance, allows the operator to equalize the manual and automatic outputs before transferring from the automatic to the manual mode.

The power supply section serves to transform the incoming AC line voltage to the proper DC levels to operate the other controller sections. Some controller power supplies also serve to energize instrumentation external to the controller. For example, the controller may power the transmitter in the same loop. In other cases the

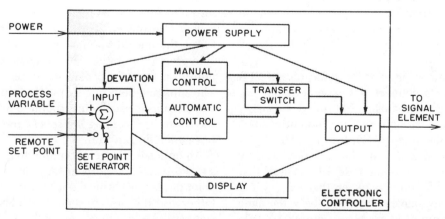

Fig. 2.3b   Basic controller components

power supply itself may be external to the controller, energizing many controllers and other instruments.

In addition to these basic parts, the controller may have various special optional or standard features. Some examples include alarms, feedforward inputs, output limits, and batching aids (described later).

Input Variations

Input and output signals for electronic controllers have more or less been standardized, through general usage, to three ranges: 1–5, 4–20, and 10–50 mA DC. Nearly all the controllers on the market will accommodate at least one of these as a standard input signal. Many offer all three as either standard or optional features. On the output side, the manufacturers generally offer only one of the signals, the most popular being 4–20 mA DC.

The use of a DC current amplifier for transmitting signals over great distances has one primary advantage over a voltage signal: small resistances that develop in the line (from switches and terminal connections and from the line itself) do not alter the signal value. Once the input signal reaches the controller, it is generally dropped across a suitable resistor to produce a voltage that matches the working voltage range of the controller's amplifier. The signal's live zero aids in troubleshooting since it differentiates between the real signal zero and a shorted or grounded conductor. Besides these three ranges, some manufacturers offer uncommon inputs such as 1–5 and 0–10 V DC, and outputs such as 1–5 V DC; centered-zero signals (± 10 V DC and ± 2 mA DC) are also available.

The controller's set-point input is either internally (local) or externally (remote) generated. Manufacturers generally provide a switch to let the user choose between the two, but some require the user to choose between two different models. In some electronic controllers the two signals can be displayed and nulled to equalize them before transferring from one to another. It is also possible, but more costly, to have the unused signal automatically track the other, so that the two signals are always equal. If the remote and local set-point signals are not equal before transfer, the controller will see a sudden set-point change. The output will therefore also change abruptly, unnecessarily disturbing or "bumping" the process.

The local set point is often merely the output of a voltage divider that is connected to the set-point adjustment and display mechanism. The divider consists of either a multi-turn variable resistor (potentiometer) or a large fixed circular slidewire with wiper. The divider's output is compared to the voltage drop produced by the process input signal, yielding a deviation signal that corresponds to the difference between the two.

The remote set-point signal must generally be the same range as the process input signal. Manufacturers provide terminals at the back of the controller to accept this

signal. Common examples in which a remote set-point signal comes into play are the cascade and ratio control systems (Sections 1.6 and 1.7, and Figure 2.3c). In the cascade system the source of the set-point signal is the output of another controller, called the primary, or master, controller. The controller that accepts this remote set point is the secondary, or slave, controller. Another common source of remote set point is a function generator or a programmer, set up to make the controlled variable follow a prescribed pattern (Section 2.1).

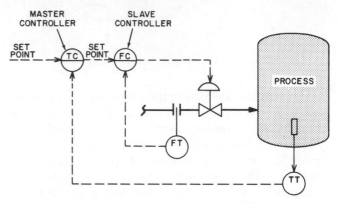

CASCADE – TEMPERATURE CONTROLLER SETS FLOW CONTROLLER

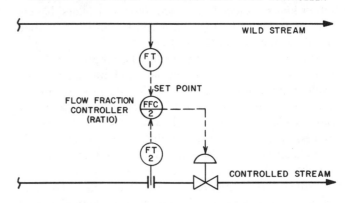

RATIO–FLOW TRANSMITTER 1 SETS FLOW CONTROLLER 2

Fig. 2.3c Cascade and ratio control systems

Blending systems are a form of ratio control. Here the set-point signal is really a second process variable from a flow transmitter. The controlled stream is forced to follow the uncontrolled (or wild) stream in some adjustable proportion (ratio). So the signal from the flow transmitter on the wild stream serves as a remote set point. The adjusting device to fix the desired ratio between the two streams is usually incorporated on the front panel of the controller, making it a special model. Calibrated ratios generally run from 0.3 to 3.0.

An optional feature pertaining to a remote set point is a servo system (motorized set point) that makes the set-point display mechanism track the remote set-point value. Otherwise the display should be blanked or marked in some way because it corresponds to the local rather than remote set point. In some controllers the set point is displayed on a galvanometer, and then the above comments do not apply. Motorized set points often find application in supervisory control systems where the remote set-point source is a computer.

Control Modes

The controller must deal with processes having wide ranges of dynamic characteristics. Adjustable control modes give it the needed flexibility. These modes act on the deviation signal, producing changes in controller output to compensate for process disturbances that have affected the controlled variable. Electronic hardware modules provide one or more of the following control actions (Sections 1.1, 1.2):

1. Proportional gain
2. Proportional gain plus integral (reset)
3. Proportional gain plus derivative (rate)
4. Proportional gain plus integral plus derivative
5. Integral (proportional speed floating)

The control action with the most widespread use is proportional gain plus integral. A few manufacturers offer proportional gain alone, and fewer yet offer pure integral.

In some cases the manufacturer provides one basic controller that has all three modes. The user then switches out derivative or integral if it is not needed. Otherwise he picks from a variety of models, depending on desired modes, fast or slow integral rates, fast or slow derivative rates, and other kinds of options. The choice depends on the dynamics of the process to be controlled.

The proportional mode varies the controller output in proportion to changes in the deviation. High gains mean that the controller output varies greatly for small changes in the deviation. Extremely high gains amount to on-off control. In processes requiring relatively low gains, the importance of the integral mode increases, because at low proportional gains the controller will be satisfied even though the controlled variable and the set point are relatively far apart (offset). The integral mode continues to drive the output until the deviation goes to zero. The derivative mode is a refinement that produces an output component proportional to the rate of change of either the deviation or the process input signal. Its effect is to anticipate changes in the process variable under control. Figure 2.3d shows a simplified circuit for implementing three-mode control.

The advantage of having the derivative mode act on the process input rather than on the deviation is that the controller does not respond directly to a set-point change. If it acts on the deviation, the derivative component of the output depends, for example, on how fast the operator turns the set-point knob. Of course he could always switch to manual before making a set-point change and then switch back to automatic. A few manufacturers can provide both kinds of derivatives as optional connections, but most offer only one or the other.

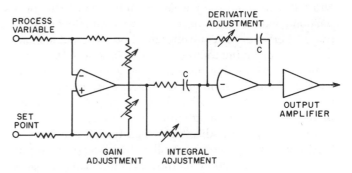

Fig. 2.3d Three-mode control circuit

Adjustments on each control-mode module let the user fit (tune) the controller to the process dynamics (Section 1.11). Continuous or multiposition knobs are located inside the case and are calibrated in terms of the degree of control-mode action. Typical units and values are:

1. Proportional gain 0.1 to 50 (proportional band 1000 to 2%)
2. Integral, 0.04 to 100 repeats per minute
3. Derivative 0.01 to 20 minutes

Higher and lower values for each mode are available, depending on the manufacturer. The term proportional band (PB) is also used; it is related inversely to proportional gain:

$$PB = \frac{100}{\text{proportional gain}}\%$$

One repeat per minute means that the integral mode will ramp the output up or down, repeating the proportional action that corresponds to a particular deviation every minute. Ramping continues until the deviation is zero. A derivative setting of one minute means that the controller output should satisfy a new set of operating conditions one minute sooner than it would without the derivative mode. Many manufacturers also have scaling switches or jumpers that multiply the effects of the integral and derivative modes by factors of two or ten. Also, the ability to switch the derivative and integral modes completely out of the control circuit is an aid for troubleshooting and calibrating.

The direction on control action depends on set point and process variable input connections. Direct control means that as the process variable increases, the controller output also increases. Reverse control means that the controller output responds in a direction opposite to that of the changing process variable. A direct-reverse switch in the controller reverses the input connections to bring about the desired controller action. If the derivative action is on the process measurement rather than the deviation, this connection must be reversed also.

Some circuit designs permit changes in the control settings while the controller is operating in automatic, without disturbing the output. Sometimes only the proportional gain changes have no immediate effect on output. Otherwise, the controller must be switched to manual before changing the settings. Usually the settings are somewhat interacting, meaning that a change in one affects the value of another. However, certain controllers incorporate features such as separate amplifiers for each mode to eliminate or minimize this interaction. These controllers have an advantage when tuning is critical.

Although the process controllers have standard output signal values, limitations on the maximum load resistance in the output circuit vary depending on the manufacturer. The load resistance is the sum of the series resistances of all the devices driven by the controller's output current. Typically the output current drives an electropneumatic converter that connects to a diaphragm-actuated control valve. These varying load requirements among manufacturers complicate matters somewhat if the user wants to buy parts of the control loop from different manufacturers.

A sampling of actual specifications for maximum load resistance includes: 75, 500, 600, 800, 1,500, and 3,000 ohms. Higher maximums, of course, give the user greater flexibility. A few manufacturers require a minimum load resistance as well. At least one vendor asks the user to adjust the load resistance in the output circuit to an exact value, which is a severe limitation.

Nonlinear Controllers

If the process tends to be sensitive (fast) at certain measurement values, while at other measurement values it is sluggish (slow to respond), then that process is nonlinear. A nonlinear process can be properly controlled only by a nonlinear controller. In order to arrive at a uniform loop gain at all measurement values, it is necessary to change the controller gain from high to low as the process measurement moves from its sluggish region to a more sensitive zone.

The best known of the nonlinear processes is acid-base neutralization, which is sensitive in the region near neutrality and therefore requires a low controller gain in that region. Figure 2.3e illustrates this nonlinear process together with the nonlinear controller response it requires. This controller has an adjustable nonlinear relationship between measurement and output, with low gain surrounding the set point and with normal gain outside of this band. The width of the "dead band" (band of low gain) is usually adjustable both manually and remotely from 0 to ±30% of measurement range, and its gain relative to the normal gain of the controller is adjustable manually from 0.02 to 0.2. If the dead band is set at zero, the result is a linear controller, illustrated by the dotted line in Figure 2.3e.

The standard control modes of proportional, reset, and derivative are superimposed on the nonlinear function. For example, if the normal gain is set at 2 and the min-

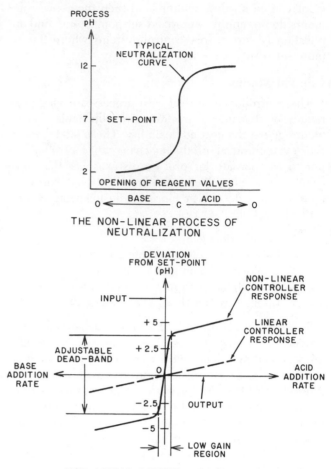

Fig. 2.3e Nonlinear acid-base neutralization process and required nonlinear controller response

imum gain within the dead zone is 0.02, then the effective gain inside that zone will be 0.04. This means that while the process measurement is inside the dead band, a 25% change in measurement will result in only a 1% change in controller output, due to the proportional mode response.

The width of the dead band can be adjusted manually or automatically in direct or reverse proportion to an external signal. For example, the control valve gain increases as that valve opens. The resulting increase in loop gain can be compensated for by widening the dead band.

Nonlinear controllers can also be used to allow the levels to float in surge tanks, while preventing the tanks from flooding or draining. The goal of such a loop is to keep the controlled flow reasonably constant while allowing it to be smoothly modified when necessary.

A nonlinear controller can also be used to filter out the noise or pulsations from flow loops, while permitting effective control over major disturbances. In order to eliminate valve cycling during normal operation, the dead band should equal the band of noise or pulsation. When used for flow control, the nonlinear controller gain inside the dead band should *not* be set at very low values because this could cause regenerative oscillation.

Special Features

Manufacturers usually offer special features, either optional or standard, to enhance the convenience and usefulness of controllers. The most common features involve alarm modules and output limits in some form. Electronic alarms can be set up on either the process measurement, the deviation, or both, depending on the supplier. Also, they can be actuated on either the high or low side of the alarm point. Some controllers have panel lights for connecting to the alarm contacts as well as terminal connections for external alarms such as lights, bells, horns, or buzzers.

Nearly all manufacturers of electronic controllers limit the high side of the output current range to the maximum standard value. Often a zener-diode arrangement serves to clamp the current at this value. Some manufacturers go a step further and place a fixed limit on the low end of the range. Still others offer adjustable output limits on both ends. Output limits serve only to constrain the range of the manipulated variable (valve stroke).

Certain limits can prevent a phenomenon called reset (integral) windup. Windup occurs if the deviation persists longer than it takes for the integral mode to drive the control amplifier to saturation. This might occur, for example, at the beginning of a batch process, under temporary loss of either controller input signal, or for a large disturbance or set-point change. Only when the deviation changes sign (process variable value crosses set point) does the integral action reverse direction. So the process variable will most likely overshoot the set point by a large margin.

If the limit acts in the feedback section of the control amplifier's integral circuit, the controller output will immediately begin to drive in the opposite direction as soon as the process signal crosses the set point. This approach is commonly referred to as *anti-reset windup*. On the other hand, if the limit acts directly on the output as discussed earlier, it essentially diverts excess current coming from the control amplifier, and therefore the output will remain at a high value for some time after the process crosses the set point. The output limit will simply divert less and less excess current. This extends the time that the output remains at a saturated value, aggravating the overshoot due to windup. Anti-reset windup circuits are usually offered only to eliminate saturated output currents at the high end of the signal range.

A few manufacturers go so far as to provide special batch controller models to minimize reset windup. Here, special circuit modifications actually begin to drive the output out of saturation *before* the process variable value crosses the set point. Reset windup can be avoided in batch applications if the operator equalizes the process variable value and set point in the manual mode before switching from manual to automatic operation.

Another controller feature gaining acceptance is a feedforward input connection. In these controllers, a feedforward input signal adds to the normal controller output signal if desired. Feedforward control (Section 1.5) assumes that the user knows exactly how much the control valve (or other final element) should change to compensate for known changes in a certain process input variable (Figure 2.3f). This relieves the controller of compensating for this particular process disturbance by normal feedback control. A few manufacturers offer gain and bias modules for conditioning the feedforward signal before it sums with the controller output signal. A direct and reverse switch is necessary to match the feedforward signal to the controller action.

Other special features offered by various manufacturers include:

Communication jack: Lets signal wires between the controller and transmitter serve also as an audio communication link; an AC carrier system does not disrupt the DC process variable signal. This eases calibration and troubleshooting of the control loop.

Trend record: Provides jacks or pins for patching a recorder to the process variable input signals available in voltage form in the controller.

Output tracking: Locks the controller output to a remote signal. This feature is useful, for example, when an electronic controller backs up a computer control installation.

Process retransmission: Converts process current signal to a low-impedance voltage signal for distribution to other parts of the control panel. Other devices can then be connected in parallel to this signal without loading down the controller input signal.

Special input modules: Extract square root, select inputs, accept resistance inputs, excite strain gages, convert pulse inputs, integrate, isolate, condition, and limit inputs. Some manufacturers offer many more options than others.

Battery backup: Automatically takes over when regular power mains fail.

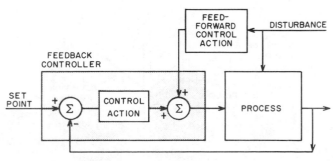

Fig. 2.3f Feedforward input connection on electronic controllers

Displays

As mentioned earlier, the controller's front panel indicates four basic signals: set point, process variable, error deviation, and controller output. The deviation meter sometimes doubles as a balance meter for equalizing the manual and automatic signals or local and remote set-point signals when the operator wants to switch from automatic to manual output. The front panel has one small meter to show the controller output. The forms of display for the other three basic variables differ among manufacturers and sometimes differ among models of the same manufacturer. Controllers without any panel displays are called blind controllers.

The majority of process controllers merge the set point, deviation, and process variable indications into one display (Figure 2.3g). This display consists essentially of a deviation meter movement combined with a long steel tape or drum scale behind the meter's pointer. The scale length ranges from 6 to 10 in. (150 to 250 mm), depending on the manufacturer, but only a portion of this scale shows; the tape or drum can be moved so that the scale value corresponding to the desired set point rests exactly at the center of the display, marked by a hairline or a transparent green band. If used, the green band masks the red pointer for small deviations. For larger deviations, the red pointer peeks glaringly from behind the band to alert the operator.

The deviation meter movement corresponds to the visible part of the scale. For example, if 50 percent of

the scale shows, the deviation movement is ±25 percent of full scale. So the pointer indicates the process variable value as long as it is on scale, the hairline at the center gives the set-point value, and the distance between the hairline and pointer indicates the deviation.

The mechanical set-point scale movement also drives components for generating the corresponding electrical set-point signal. Commonly, the scale movement drives a multi-turn potentiometer. In some controllers, a wiper attached to a movable scale drum contacts a large fixed circular slidewire connected as a voltage divider. Either design should avoid gearing between the set-point drive and electrical components, otherwise backlash would destroy their one-to-one correspondence.

The controllers having a large deviation display with a long, expanded scale offer high resolution and readability of each signal value. Of course, the deviation must be on scale to be able to read the process value at all. Also, the electrical and mechanical set-point values must be calibrated carefully or the process reading will be inaccurate.

In controllers without this kind of display, some compromise is usually made in the indication of set point, deviation, and process variable value. It is impractical and confusing to put a display for each of these signals on the controller's narrow face. The set point must be displayed, but either the process variable or deviation display can be sacrificed without great loss, since one is implied by the other once the set point is known.

The set point is either read out directly on a meter or mechanically displayed with a calibrated dial and index arrangement (Figure 2.3h). Sometimes the process variable and set-point value are shown with two pointers on a single scale. This lets the operator compare easily the

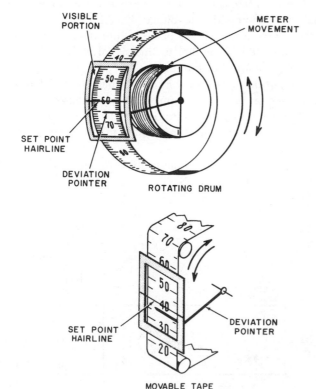

Fig. 2.3g   Expanded scale displays for showing set point, deviation, and process variable

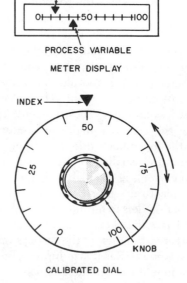

Fig. 2.3h   Standard set point displays

two for estimating the deviation. In another case it is the process value that must be estimated from markings that relate the deviation meter readings to a calibrated set point dial.

One way to circumvent the limited space on the controller's front panel is to have one meter serve to indicate more than one variable. The operator then must position a switch to choose the variable he wants to read. One manufacturer indicates five basic controller signals (including a balance indication) with one meter. To avoid confusion, a set of switches inside the case determines the direction of meter movement for different switch positions so that the meter movement always drives upscale.

The controller's output meter shows either the manual or the automatic output signal, depending on the transfer switch position. Interchangeable tags at the side of the meter remind the operator of the valve positions (open or closed) for the extreme controller outputs. In certain controllers the output meter can be physically inverted so that the pointer always drives upscale as a valve opens. Occasionally the output meter may act as a deviation meter to null the automatic and manual outputs before transfer.

The front panel may also contain alarm light displays. These are often connected to alarm modules behind the panel that monitor deviation or process variable input. Some lights may merely run to an external connection at the rear of the controller. The position and form of the lights vary greatly. One manufacturer backlights legend plates. Another backlights the top and bottom of the deviation meter to show high and low deviation. In one case, deviation alarm lights replace the deviation meter itself. A few manufacturers provide a small standard panel light to indicate whether the local or remote set point is implemented.

Balancing Methods

All electronic controllers have a switch on the front panel for transferring between automatic and manual outputs. The switch takes many forms, including lever, toggle, rotary, and pushbutton. It is often important to balance the two output signals before transferring from automatic to manual to avoid a sudden change in controller output that might disturb or "bump" the process. Controllers that feature "bumpless transfer" have some provision for balancing the two output signals just before switching.

No balancing is necessary when switching from manual to automatic output because the unused automatic controller signal takes the same value of (or tracks) the output when the controller is in the manual mode. So the operator merely switches from manual to automatic and need not concern himself with balancing procedures. However, if a deviation exists between process and set-point values at the time of transfer, the integral action in the controller (if present) will immediately begin to drive the controller in a direction to eliminate this deviation. For large deviations the integral action will often produce a controller output change fast enough to bump the process.

This balanceless, bumpless transfer from manual to automatic output depends on a modified integral circuit to make the automatic signal track the manual output. So the integral function must be present if balanceless transfer is desired.

Nearly all manufacturers require a balancing procedure for bumpless transfer when the controller is switched from automatic to manual output. The transfer switch has a balance position to accommodate this procedure. In the balance position, the controller retains its automatic output, but the functions of the display meters change. The automatic and manual output signals usually feed to either side of a deviation meter. This allows the operator to equalize the two signals by adjusting the manual output until the meter reads zero at mid-scale. This nulling method has a high sensitivity for accurate balancing. Usually the large deviation meter, if present, serves to compare the output signals, but some manufacturers connect the output meter as a deviation meter for the controller's balance state. At least one supplier places a small separate deviation meter on the panel that always shows the condition of balance.

Still other manufacturers provide a momentary pushbutton on the front panel that, when depressed, connects the manual signal to the output meter in place of the automatic signal. To balance the two signals, the operator repeatedly presses the button while adjusting the manual output. When the pointer on the output meter remains motionless, the two signals should be about equal.

A unique design features balanceless, bumpless transfer in both directions. This means that the unused output signal tracks the other signal at all times. In this case the control amplifier is switched into a simple integrator configuration when the operator transfers from automatic to manual mode. This circuit holds the latest value of the automatic output as a manual output. To change the manual value, the operator throws a momentary switch that introduces a DC input signal to the integrator circuit input. This causes the output to ramp up or down, depending on whether the input signal is positive or negative. So the operator jogs the manual output to the desired value. The control amplifier also holds the latest manual output signal at the instant the operator switches back to automatic operation, but integral action begins immediately to eliminate any existing deviation between the set point and process variable values.

Mounting

Miniature electronic controllers are designed for panel mounting in a relatively clean and safe environment such as a control room. The room should be free from corrosive vapors and excessive vibration. A few controllers have ambient temperature limitations ranging from 0 to

130°F (0 to 72°C), but others are more restricted. Some models can be mounted in areas designated by the National Electrical Code as Class I, Group C, Division 2. These areas require that equipment have no open sparking contacts, but do not require explosion proof cases. (See Section 5.2.) Here the manufacturers provide hermetically sealed on-off power switches to avoid open sparking.

The controller's small frontal size allows many controllers to be installed in a small panel space, called a high-density configuration. The controller's depth (over 20 in. or 500 mm) and weight (often over 15 lb or 6.8 kg) usually require additional support at the rear of the panel for multiple mountings. In general, the controllers can be mounted as multiple arrays in four ways (Figure 2.3i), in individual adjacent panel cutouts, side by side in a single large panel cutout, in large cases or packs that accept some multiple of controller chassis, or on shelves that accept some number of individual controller cases.

The mounting approach depends on the particular manufacturer, but some companies offer more flexibility than others. A few require individual panel cutouts for each controller. This increases panel fabrication costs. When the controllers are placed side by side in a single cutout, the manufacturer may offer outside trims to frame the installation. The company will also suggest ways to support the back ends. Packs sometimes have both vertical and horizontal capacity, while shelves are limited to horizontal mountings. Some horizontal packs may be specified for any number of controllers, but others are limited either by a maximum number or by certain multiples of controllers. Mounting in packs or shelves generally simplifies power and signal connections.

If the panel tilts too far from the vertical, the controller may not operate properly. For example, one manufacturer specifies a maximum of 60-degree panel tilt, and another stops at 75-degree tilt from the vertical. Still others say that the controller works in any position. But some companion products, like recorders, have limitations on their mounting position.

In all these electronic controllers, the chassis slides part way out of the case without interrupting the control signal. A retractable plug connects the chassis with all the necessary power, signal input, and signal output leads. This permits tuning, calibration, and realignment while the controller operates in place. In a few controllers, the panel display section can be mounted in a location different from that containing the control and output circuitry.

Although the controller is mounted in a relatively safe area (the control room), its signal must eventually go into the field to operate some final element, such as a control valve. Some control systems have been approved by either Factory Mutual or Underwriters Laboratories as intrinsically safe for hazardous field areas. (See Section 5.2.) This means that the output signal in the field cannot release sufficient electrical or thermal energy, under defined normal or abnormal conditions, to ignite a specific hazardous atmospheric mixture. Special barrier circuits and energy limiters, sometimes with zener diodes (Figure 2.3j), limit the output signal energy, even if the full voltage of the power main is somehow connected to the signal lines. Some suppliers put the barrier in the controller case; others prefer to place a barrier at the location where the signal lines leave the control room. Also, distinct mechanical separation of energy levels in the controller will add to its intrinsically safe properties.

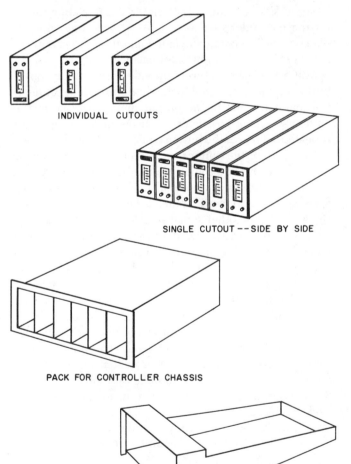

INDIVIDUAL CUTOUTS

SINGLE CUTOUT -- SIDE BY SIDE

PACK FOR CONTROLLER CHASSIS

SHELF FOR CONTROLLER CASES

Fig. 2.3i   Panel mounting arrangements

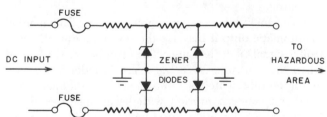

Fig. 2.3j   Barrier circuit approved in Great Britain

Servicing

The electronic controller frequently is the first to receive the blame for loop malfunctions because its panel tends to give the first indication that something is wrong. Before removing the controller, however, the user should check other loop components and the operation of main plant equipment. He should also make sure that the controller output is not shorted to ground in the field.

If the controller is at fault, the course of action depends on the manufacturer's provisions for emergency servicing. In many cases the manual section of the controller will continue to work, so the operator can set the controller output at a desired DC current value. But the instrument maintenance technician must somehow remove the automatic section so he can troubleshoot it. The following items highlight the basic approaches taken by manufacturers:

1. Manual and automatic sections are completely independent, plug-in modules. One may be removed without affecting the other. They may even have separate power supplies to ease testing of the faulty unit. Sometimes only the automatic section is removable for emergency servicing.
2. Adjustable plug-in power module takes the place of manual output when the controller is removed from its case.
3. Battery-operated or auxiliary manual station replaces controller to hold output current. In some cases the substitution can be made without interrupting the signal to the process and (by a balancing procedure) without abruptly changing (bumping) the value of the controller output signal.

A few manufacturers make no provisions for emergency servicing. The faulty unit must simply be replaced regardless of signal interruption and output bumping.

Once the controller or its automatic section is in the shop, the technician can try to isolate the trouble by hooking up suitable input signals and a dummy load to the controller, closing a mock control loop. Some suppliers offer a jig that plugs into the controller to simulate a closed loop. Terminals are provided for input signals and for monitoring the deviation and output signals. Most companies at least provide easily accessible test pins or jacks for checking and calibrating key controller signals. Instruction manuals contain guides and schematics for diagnosing the fault from the symptoms. Multimeters used for voltage signal measurements should have at least 20,000 ohms per volt sensitivity.

The best buy is the controller that requires little maintenance and can be quickly replaced and repaired when it does need servicing. All the electronic controllers will perform satisfactorily in a control loop, at least in the short run. But those that frequently break down may result in costly process downtime, a factor that tends to override any differences in initial controller purchase and installation costs. Highly modular controller designs and circuit accessibility simplifies maintenance. Rugged, high-quality, solid-state circuit designs along with dust-tight encapsulation of key parts, including meter displays, help to prevent controller breakdowns. Silicon transistors are more rugged and reliable than those made of germanium. Much depends on the quality control practices implemented by the particular supplier.

Other Electronic Controllers

The electronic controllers discussed so far were all general-purpose devices with standard inputs and outputs, analogous to the 3–15 PSIG (0.2–1.0 bar) standard signals in pneumatic control systems (Section 2.6). A wide variety of other controllers on the market are tailored to specific kinds of control by specialized input or output schemes, or both.

The most common examples by far are electronic temperature controllers for driving electric heaters (Section 5.3) and motor speed controllers (Section 5.8). The temperature controllers are designed to accept directly inputs from sensors like thermocouples, resistance bulbs, or thermistors. Their outputs may drive relays or contactors, silicon-controlled rectifiers (SCRs), or saturable reactors.

Since the sensor connects directly to the controller, the input signal lines handle very low power levels; the lines are kept short to avoid interference from other sources. The input circuit will be for a particular kind of sensor and certain range of temperature. Common thermocouple inputs include iron-constantan, copper constantan, Chromel-Alumel, and platinum-rhodium platinum pairs. The controller's input circuit will compensate for variations in the thermocouple's cold junction temperature. Resistance bulb and thermistor inputs generally feed into resistance bridge circuits.

As in all feedback controllers, the input circuit will compare the incoming signal with a reference or set-point signal. Control circuitry will act on the resulting deviation. The form of control ranges from simple two-(on-off) or three-position (high-low-off) control to sophisticated three-mode control. Some manufacturers offer three-mode control on one set-point position and on-off control on another. Proportioning, however, is generally on a time basis, called time-proportional control. Here the output is driven full on or off, but the control signal varies the percentage of on-to-off time in some duty cycle, such as 10 seconds. The *average* power level determines the load temperature. This kind of control applies to both relay and SCR outputs. However, the SCR is capable of faster switching than the relay, permitting proportioning periods of a fraction of a second and giving smoother control. Some controllers trigger the SCR to chop up each AC cycle; in others the SCR proportions the number of AC cycles that are on and off in a certain period (Figure 2.3k). Available SCRs can handle loads up to 300 kW in three line phases.

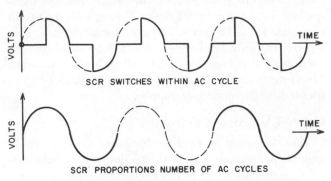

Fig. 2.3k SCR switching techniques for driving electrical heaters

A common example of an electronic controller with relay outputs is the positioner controller (Figure 2.3l). Here the controller energizes one of two relays to operate an electric motor in either direction. The motor drives some final element to a desired position. A feedback signal from the final element tells the controller when the desired position has been reached. Essentially the controller acts as a master controller in a cascade system. The relays, motor, final element, and feedback signal constitute the slave loop, and the master controller generates the slave loop's set point by conventional control action.

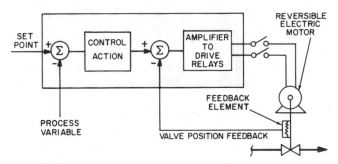

Fig. 2.3l Electronic positioner controller with relay outputs

A meter-relay combination provides an extremely simple way of implementing two- and three-position, high-gain control. Basically, the input signal, such as a thermocouple, drives a galvanometer's pointer to indicate temperature. Another pointer (or index) is manually set to the desired temperature on the same indicating scale. The idea is to operate a relay or an SCR when the two pointers are in the same position.

The system shown in Figure 2.3m optically senses the correspondence of the two pointers. The set-point arm carries a small photocell and light source. As long as light strikes the photocell, amplified current energizes the relay to supply power to a load. When the indicated temperature reaches the set-point temperature, a vane on the indicating pointer breaks the light beam, and the

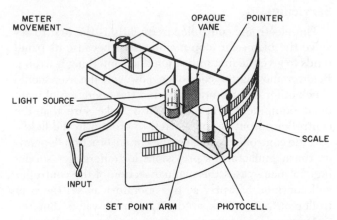

Fig. 2.3m Optical meter relay (Courtesy of West Instrument Division of Gulton Industries, Inc.)

relay drops out. Some meter relays are equipped with two set-point arms and relays to give three-position control. In a meter with an SCR output, the vane can be shaped to provide some degree of proportional control over a narrow temperature band. The optical system avoids any interaction between the process and set-point pointers, as would be present in mechanical or magnetic systems. Obviously the idea can be extended to other displays where the variable is indicated by a pointer's position. For example, it has been applied to electronic circular and strip-chart recorders.

Meters can also be purchased with built-in electronic amplifiers to boost the input signal, as in Figure 2.3n. In one meter-relay package of this type, the meter's set-point arm wipes a resistance element. This resistance element, connected as a voltage divider, provides a voltage reference signal, which is compared to the amplified input. The deviation, after further amplification, drives a transistor that operates a control relay. The meter and resistance element can be encapsulated in an inert atmosphere if the device must be installed in a relatively hostile environment.

Single-Loop Digital Controllers

The main disadvantage of the microprocessor-based shared controllers is an inherent consequence of their

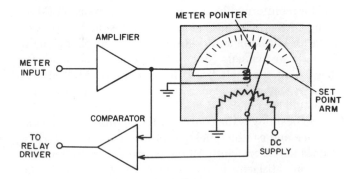

Fig. 2.3n Meter-relay package with electronic input and output circuits (usually integrated circuits)

being shared between several control loops. This results in a potential problem of reduced overall system reliability because "many eggs are placed in a single basket". The single-loop digital controllers have overcome this limitation because they can be dedicated to serve only a single output, while retaining most of the capabilities of the shared digital controllers. This gives the instrument engineer a very powerful and flexible tool for fully distributed optimal process control because all control loops can be configured and reconfigured to any desired level of sophistication while using the same controller.

Because the control functions are in software, all the control hardware can be purchased and installed before the control system itself is finalized, or an existing control system can be reconfigured without adding or revising any hardware or wiring.

The capabilities of single-loop digital controllers are very similar to the shared controllers discussed in Section 7.4 and therefore they will be discussed here only briefly, emphasizing those features that are not available in analog controllers.

A library of control algorithms is available including many of the more sophisticated functions such as lead-lag, feedforward, dead time compensation, program pattern control, sample and hold, inverse derivative, gap control, auto select or limit functions, error squared, PID, and multiple inputs and/or outputs for cascade, ratio, or batch control. Many designs are able to handle just about all the different types of input signals (including low level RTD and TC type measurements with table look-up) and are able to generate any of the various outputs.

Signals can be linearized, pressure or temperature compensated, converted to any engineering unit, totalized, time delayed, or arithmetically manipulated.

The logic interlocks associated with the loop can also be implemented because many single-loop digital controller designs have both continuous and sequential control features, utilizing all the basic logic functions. These functions are additional to the standard alarm features.

The loop configuration and programming is usually done through pushbutton operations. The problem of maintenance and spare parts storage is minimized by the complete interchangeability between units. The single-loop controllers can be integrated into larger systems through data bus communications between them, and they can also be connected to CRTs for display purposes. The data highway can also be used to provide a communication link to the plant-wide optimizing computer, which can reset the individual loops as required. Such remote signals can not only modify set points, but can also initiate auto/manual transfer, disable displays for information security, switch set points between local and remote sources, change controller actions, etc. In ad-

dition, some of the more advanced units also provide self-calibrating, self-diagnostics, and self-tuning functions.

Feature Check List

The following is a summary of the relevant features that an instrument engineer must consider when picking a controller:

> Control mode selection
> Input and output ranges (including set point)
> Output load resistance range
> Emergency service provisions
> Maintenance accessibility and convenience
> Panel readability and accuracy
> Control repeatability and accuracy
> Tuning ranges and resolution
> Electrical classification
> Power requirements; need for regulation
> Mounting flexibility and density
> Switches for local/remote set point, direct/reverse action, and manual/automatic operation
> Balancing procedures and accuracy
> Alarm modules and lights
> Output limits
> Anti-windup
> Computer compatibility

BIBLIOGRAPHY

Andreiev, N. "A New Dimension: A Self-Tuning Controller that Continually Optimizes PID Constants," *Control Engineering*, August, 1981.

————, "A Process Controller that Adapts to Signal and Process Conditions," *Control Engineering*, December, 1977.

Beard, M.J., "Analog Controllers Develop New Wrinkles," *Instruments & Control Systems*, November, 1977.

Buckley, P.S., *Techniques of Process Control*, John Wiley & Sons, New York, 1964.

Goldfeder, L.B., and Deuschle, R.C., "Analog Control Architecture: Integral vs. Split," *InTech*, April, 1977.

Harriott, P. "Process Control," *McGraw-Hill Book Co.*, New York, 1964.

Jones, B.E., "Instrumentation, Measurement, and Feedback," McGraw-Hill Book Co., New York, 1977.

Kaminski, R.K., and Talbot, T.E., "Electronic Controllers for Process Industries," *InTech*, October, 1968.

Lipták, B.G., "How to Set Process Controllers," *Chemical Engineering*, November 23, 1964.

Luyben, W.L., "A Proportional-Lag Level Controller," *Instrumentation Technology*, December, 1977.

McCauley, A.P., Jr., and Persik, S.R., "From Linear to Gain-Adaptive Control . . . A Case History," *Instrumentation Technology*, November, 1977.

Merrit, R. "Electronic Controller Survey," *Instrumentation Technology*, June, 1977.

Shinskey, F.G., "Control Topics," Publ. no. 413-1 to 413-8, Foxboro Co.

————, *Process Control Systems*, McGraw-Hill Book Co., New York, 1979.

PRESSURE TRANSMITTER

FLOW TRANSMITTER

LEVEL TRANSMITTER

INDICATING TEMPERATURE TRANSMITTER (FILLED SYSTEM)

2.4 ELECTRONIC TRANSMITTERS

Range:	Force: 15 lbf to 180,000 lbf (66.75 N to 801,000 N) Level (buoyancy): 1″ to 60′ (25mm to 18m) Motion: ⅛″ to 60″ (3.125mm to 1.5m) Pressure Differential: 0.1″ H_2O to 1,500 PSID (25 Pa to 10.35 MPa) Temperature Span (Filled): 50 to 1,000°F (28 to 555°C) within the range of absolute zero to 1,400°F (760°C)
Inaccuracy:	Generally ¼% to 1% of full scale; see Table 2.4a for details
Cost:	$500 to $1,500 (transducers in OEM quantities for automotive and similar applications are $10 to $50 each)
Partial List of Suppliers:	Action Instrument Co. Inc.; Bailey Controls, Div. of Babcock & Wilcox; Beckman Instruments Inc.; Bristol Babcock, Inc.; CEC Division of Bell and Howell; Devar, Inc.; Dwyer Instruments, Inc.; Dynisco; Fischer & Porter Co.; Fisher Controls, Inc.; Foxboro Co.; Gould, Inc.; Honeywell Process Control Div.; Hottinger Baldwin Measurements; ITT Barton; Kent Process Control, Inc.; Kulite Semiconductor Products, Inc.; Leeds & Northrup Instruments, a Unit of General Signal; MKS Instruments; Moore Products Co.; National Semiconductor; Rochester Instrument Systems, Inc.; Rosemount, Inc.; Schaevitz Engineering; SEREG; Taylor Instrument Co., Div. of Sybron Corp.; Validyne Engineering Corp.; Wallace & Tiernan Div., Pennwalt Corp.; Westinghouse Electric Corp.

A transducer is a device that receives information in one form and generates an output in response to it. A transmitter is a transducer that responds to a measurement variable and converts that input into a standardized transmission signal.

Table 2.4a summarizes the accuracies of industrial quality electronic transmitters, while Tables 2.4b and c provide performance data for electronic pressure transducers. In the coming years, the proliferation of new transducer designs and improvements, as well as the drastically lowered unit cost of these devices, are likely to have a profound impact on process control in general. This will allow the introduction of process control and optimization into processes that previously could not benefit from this technology. One such example is the

optimization of the internal combustion engine using inexpensive transducers and microprocessor-based digital controls.

An example of a simple transmitter or transducer is a thermocouple for measuring temperature. In the thermocouple, the temperature difference between the hot junction and the reference junction creates a DC voltage directly proportional to the temperature difference. This principle is explained in Chapter IV of the Process Measurement volume of this *Handbook*.

The many different kinds of transmitters will be considered in several classes—force-balance transmitters, motion-balance transmitters, physical properties transducers, chemical properties transducers, and direct electrical properties transducer/transmitters.

Table 2.4a
TYPICAL ACCURACIES OF ELECTRONIC
PRESSURE AND TEMPERATURE TRANSMITTERS

Type of Instrument	Inaccuracy	Repeat-ability	50°F (28°C) Ambient Effect	Supply Voltage Change by 1.0 Volt or by 10%
All Errors Are in Units of ±% Full Span				
Absolute PT—				
mmHg range (Pa)	1%	0.5%	1%	0.1%
PSIA range (Pa)	0.5%	0.1%	1%	0.1%
Gauge PT—				
below 2000 PSIG (14 MPa)	0.25% or 0.5%	0.1%	1%	0.1%
above 2000 PSIG (14 MPa)	0.5%	0.1%	1%	0.1%
D/P Cell—				
below 500″ H20 (125 kPa)	0.5%	0.1%	1%	0.1%
below 850″ H20 (212.5 kPa)	0.75%	0.1%	1%	0.1%
PSID range	0.5%	0.1%	1%	0.1%
Repeaters—				
up to 35 PSIA (242 kPa)	2″H20 max (0.5 kPa)	0.2″ H20 (50 Pa)	1%	N.A.
up to 100 PSIG (690 kPa)	1″H20 max (0.25 kPa)	0.3″ H20 (75 Pa)	1%	N.A.
Filled TT—				
force balance	0.5%	0.2%	0.75%	N.A.
motion balance	0.5%	0.1%	1.0%	0.1%
RTD-Based TT (platinum)	0.15%	0.05%	0.75%	0.02%
TC-Based TT	0.1% or 5 micro V	0.05%	See below	0.02%

Ambient error sources include 1) reference junction error of 40 micro Volt maximum, 2) span error of 0.5%, and 3) zero error of 0.5%.

Force-Balance Transmitters

Figure 2.4d illustrates a force-balance differential pressure transmitter, in which the measurement that produces a force tends to move the top of the force bar.[1] This tiny motion, acting through levers, moves the ferrite disc closer to the transformer, changing its output. This changes the amplitude output of the oscillator, which is rectified and then amplified to generate a DC milliampere transmitter signal. This output signal is fed back through the voice coil on the armature of the force motor, which is in series with the output terminals. When this feedback moment is equal to the moment created by the measurement force F_2, the force bar is again in its original position and the amplifier signal stablizes.

The advantage of force-balance units over motion-balance devices is that by reducing motion, one minimizes the effect of pivot friction. Further, by always returning to the same position, hysteresis is minimized and greater accuracy can be obtained. In general, a force-balance cannot be used to produce digital signals without the use of a supplementary device external to the transmitter, such as an analog-to-digital converter (ADC).

Motion-Balance Transmitters

In a motion-balance transmitter (Figure 2.4e), the process measurement produces motion against a calibration spring, resulting in a change of position corresponding to a change in the process variable. This position is detected by a transducer. The output of the transducer is amplified and an electric feedback signal is used to stabilize the amplifier. Depending upon the type of transducer and the signal level it generates, the amplifier may not be required but may be a part of the receiver.

Differential Transformer

One of the most frequently used transducer principles is known as the differential transformer.[2] Two different designs are shown in Figures 2.4f and 2.4g. Except for

Table 2.4b
ELECTRONIC PRESSURE TRANSDUCERS

Type (Operation)	Supplier(s)	Features
Strain Gauge (unbonded or bonded foil, thin film, protected by seals and capillaries)	Ametek Bell & Howell, CEC Bristol Burns Instruments Data Instruments Dynisco Gould Measuring Systems Nottinger Baldwin	5 PSI to 30,000 PSIG (173 kPa to 270 MPa) $-60°F$ to $+250°F$ ($-51°C$ to $121°C$) Error $\pm$ 0.25% FS or less Output: 3 mV/Volt excitation Excitation: 10 V DC Minimum cost in OEM quantities: $50
Solid State (diffused silicon semi-conductor, bonded semi-conductor strain gauge)	Bailey Controls Bell & Howell, CEC DJ Instruments Foxboro/ICT Motorola National Semiconductor	3″ H20 to 30,000 PSIG (0.75 kPa to 270 MPa) -60 to 250°F (650°F w/seal) Error: $\pm$ 0.1% to $\pm$ 1% Output: 3–20 mV/Volt excitation Excitation: 10–28 V DC Minimum cost: $10 (less for automotive OEM)
Resonant Wire	Foxboro	1″ H20 to 6000 PSIG (0.25 kPa to 41 MPa) Error: $\pm$ 0.2% 35 yrs life, 4–20 mA DC
Eddy Current Transducer	Kaman Instruments	To 10,000 PSIG and 1000°F (to 69 MPa and 538°C)
Piezoelectric Sensors	BBN Instruments	0.1 to 10,000 PSI and to 600°F (0.7 kPa to 69 MPa)
Optical (light emitting diode)	Dresser Industries LEL Co.	4–20 mA DC or digital output
Variable Reluctance (pressure moves armature between coils causing inductance changes)	Celesco CJ Enterprises Validyne	1″ H20 to 500 PSIG (0.25 kPa to 3.5 MPa) 4–20 mA DC
Capacitive	Kavilco MKS Setra Systems	down to 0.1″ H20 (25 Pa) OEM cost down to $50
Linear Variable Differential Transformer (LVDT) (detects position of core element inside coils)	Computer Instruments Robinson-Halpern Schaevitz Transducer Systems	4–20 mA DC 2″ H20 to 20,000 PSIG (0.5 kPa to 138 MPa)

the type of motion and the design of the transformer core and windings, these units are very similar.

The linear variable differential transformer (LVDT), illustrated in detail (A) of Figure 2.4f, consists of a transformer coil with a single primary winding and two symmetrically spaced secondary windings. The core or armature is a cylinder of magnetic material, such as ferrite, which can be moved within the "air gap" of the windings.

When AC excitation is applied to the primary winding and the armature slug is centered, or is in the "null" position, the induced AC voltage in the secondary windings is equal and is in the same or opposite direction, depending on the method of winding. If the two secondary windings are connected in series with the voltages opposed [Figure 2.4f, detail (B)], they will cancel out and give a zero or null reading when the slug is in the null position. As the slug is moved closer to coil A and further from coil B, as shown in detail (C), the output

voltage increases in the direction of the coil A output, and this increase is proportional to the displacement of the slug.

Another typical hookup is shown in detail (D), with a rectifier in the output of each secondary coil hooked up to a DC zero center meter. The meter will read zero at the null slug position and plus or minus for slug positions displaced from the center.

The DC output of such a differential transformer can be amplified and used directly as a transmitted signal (as in Figure 2.4e), or it can be used as a position detector of other devices as shown in Figure 2.4d. Another differential transformer design is shown in Figure 2.4g, and there are several others.

The excitation in a differential transformer must be supplied by an AC circuit. The AC source can be the receiving instrument using an AC transmitted signal or it can be a part of the amplifier in the transmission part of the transmitter, powered by a DC supply taken from the receiver. Alternatively, a power supply can be provided as part of the transmitter, but this requires additional wiring.

Photoelectric Transducer

Figure 2.4h shows a typical schematic of a photoelectric transducer where the position of the photocoder is proportional to the motion of a primary sensing element. Light from the source shines through perforations in the shutter to energize photoelectric cells. The output of these cells is scanned and the pulses are amplified to produce a digital signal, or they are rectified to produce a DC analog signal.

Capacitance Transducer

Figure 2.4i shows a capacitance-type pressure or differential pressure transducer, which is truly of the motion-balance type.[1] Positioned between two fixed capacitor plates is a highly prestressed thin metal diaphragm.[2] This forms the separation between two gas-tight enclosures that are connected to the process. The difference in pressure between the two chambers produces a force that causes the diaphragm to move closer to the fixed capacitor plate of the low-pressure chamber. The transducer is excited by a 10-kHz AC carrier current, and the unbalance produces a 10-kHz voltage with an amplitude proportional to the difference in pressure.

In some design, the transducer is filled with an inert fluid (usually silicone oil) to prevent contamination of the capacitor plates or transducer interior by the process fluid or gas. A metal barrier diaphragm keeps the fill fluid in the transducer and the process out of the transducer.

Potentiometric Transmitter

Figure 2.4j shows a potentiometer (resistance) driven by a Bourdon tube in a manner similar to the movement of a pressure gauge. Rotation due to a pressure change turns the shaft of a precision potentiometer, and the change in resistance is proportional to the process pressure.

Piezoelectric Transducer

Figure 2.4k shows the cross-section of a piezoelectric accelerometer transducer in which the change in the amount of strain in the piezoelectric crystal generates a minute voltage. The crystal is sandwiched between a mass and the base.[2] The entire assembly is held together and given an initial strain by the nut on the threaded center bolt. When the transducer is moved suddenly upward, the force on the crystal is increased by a force equal to the mass multiplied by the acceleration. This increased force changes the strain in the crystal and generates a voltage. Since there is no displacement other than the minute compression of the crystal, this device is a borderline case between a force-balance and a motion-balance device.

Other types of motion transducers include strain gauges and synchrotype servomechanism resolvers.

Physical Properties Transducers

Several types of transducers measure physical properties directly as the physical property changes some electrical property. The best known of these are thermocouples, resistance bulbs, and thermistors, in which a change of electrical properties occurs as a result of temperature changes. Less well known, but similar in character, are piezoelectric crystals sensitive to pressure and/or temperature, photocells sensitive to the intensity of light, capacitance devices sensing the dielectric properties of materials, density instruments, viscosity sensors, thermal conductivity detectors, mass spectrometers, and many others.

Since most of these devices perform a primary measurement function, their description can be found in the Process Measurement volume of this *Handbook*.

Chemical Properties Transducers

The pH meter exemplifies the best-known method of direct chemical measurements by electrical means. The potential produced between two electrodes is proportional to the hydrogen ion concentration. There are many other selective ion probes available and more are constantly being developed. A typical ion electrode scheme is shown in Figure 2.4l.

Electrical conductivity measurements can also be considered as chemical in nature, because the conductivity is proportional to the ion concentration. Conductivity measurement is used to detect ion concentration and is also used to sense the level of conductive liquids. Further details on conductivity type measurements are given in Sections 10.6 Process Measurement volume of this *Handbook*.

A listing of transducers commercially available in the United States and a compilation of the manufacturers' specifications has been published by the Instrument Society of America.[3]

Electronic Signal Types

There are many other types of transducers besides those that produce AC or DC output signals. Some have no output at all but are passive, exhibiting electrical properties that have to be measured, such as resistance, capacitance, reluctance, reactance, or even open and closed contacts.

Some of these electrical signals are more convenient than others, and a brief discussion of the advantages and disadvantages of each is presented in the following paragraphs.

Voltage Signals

DC voltage signals cannot be accurately measured in systems in which a current is allowed to flow, but can

Table 2.4c
ELECTRONIC TRANSDUCER PERFORMANCE CHARACTERISTICS*

| | TRANSDUCERS | | | | |
| STRAIN GAUGES | | | | | |
Unbonded	Bonded Foil	Thin Film	Diffused Semicond.	Bonded Bar Semicond.	Reluctive
0.5 psi thru 10K psi	5 psi thru 30K psi	15 psi thru 5K psi	3" H_2O thru 10K psi	3" H_2O thru 30K psi	1" H_2O thru 10K psi
−320°F to +600°F (0.005%/°F over limited compensated range)	−65°F to +250°F (0.0025%/°F over compensated range)	−320°F to +525°F (0.005%/°F over limited compensated range)	−65°F to +225°F (0.005%/°F over limited compensated range)	−65°F to +325°F (0.01%/°F over limited compensated range)	−320°F to +600°F (0.02%/°F over limited compensated range)
0.25%	0.25%	0.25%	±0.01% to ±0.35%	0.20%	0.5%
Low level 4 mV/V	Low level 3 mV/V	3 mV/V	Low to high level 3 mV/V to 20 mV/V	Low to high level 3 mV/V to 20 mV/V	40 mV/V
10 Vac-dc	10 Vac-dc	10 Vac-dc	10–28 Vdc	10 Vac to 24 Vdc	ac special
Good	Very good	Very good	Excellent	Excellent	Very good
O Hz to >2 KHz	0 Hz to >1 KHz	0 Hz to > 1 KHz	0 Hz to > 100 KHz	0 Hz to > 70 KHz	0 Hz to >1 KHz
<0.5% cal shift after 10^6 cycles	>10^6 cycles	>10^6 cycles with <0.25% cal. shift	<0.25% calibration shift after 10^6 cycles	<0.5% calibration shift after 10^6 cycles	>10^6 cycles
Accepts unidirectional and bidirectional pres meas., ac or dc excitation	Temp. effects small and linear, ac or dc excitation; rugged construction	Excellent thermal zero and sensitivity shift; ac or dc excitation	Small size, high nat. freq., steady and dynamic reliability; repeatability	Small size, high nat. freq. steady and dynamic; excellent repeatability	High output (40 mV/V excitation); rugged construction, over pressure capability
low signal level (4 mV/V)	Low signal level (3 mV/V), limited temp. range, mod natural freq.	Low natural freq., low signal level, 3 mV/V	Suscept. to handling prob.; more temp. sens. than most trans., elec. output only	More temp. sens. than most trans., requires external power source	Sensor requires ac excitation, susceptible to stray magnetic fields, ac carrier systems require balanced line for data transmission

* See Section A.1 for SI units.

			TRANSDUCERS		
Capacitive	Potentiometer	Linear Variable Diff. Trans. (LVDT)	Force Balance	Piezoelectric	Resonant Wire
.01 thru 10K psi	5 psi thru 10K psi	30–10K psi	2″ H_2O thru 5K psi	0.1 psi thru 10K psi	0 to 1″ H_2O to 750″ H_2O
0°F to 165°F Requires temperature control	−65°F to +300°F nonlinear 0.01%/F	0°F to 165°F	40°F to 375°F 1%/100°F	−450°F to +400°F 0.01%/°F	−40°F to +250°F
0.05%	1%	0.5%	0.5%	1%	±0.2% of calibrated span
Hi level (5V) Freq./bridge	Hi level	Hi level (5V) Phase Demod/bridge	Hi level (5V) with servo 4–20 mA 10–50 mA	Medium level with amp.	4–20 mA or 10–50 mA transmitter
ac-dc special	ac-dc regulated	ac special	ac line or dc power supply	dc amp and self gen. ac	Power supply dc voltage
Poor to good	Poor	Poor	Fair to good	Excellent	
0 Hz to > 100 Hz	0 Hz thru > 50 Hz	>100 Hz	0 Hz to <5 Hz	1 Hz to > 100 KHz	0.1 to 1 Hz. adjustable
>10^7 cycles with <0.05% cal. shift	<10^6 cycles low press	>10^6 cycle	>10^7 cycles with <0.5% cal. shift	Unmeas. use effects	As good as 0.1% shift over 6 months
Exc. for low press; excellent freq. response; dynamic and static meas.; output countable without A/D converter	Low cost; small size, high output without amplification	Available in rotary form; not affected by mech. overload	High accuracy; high output, stable, wide ranges	High freq. resp.; self gen. signal; small size, rugged const.	High accuracy; excellent long-term stability; ability to transmit frequency
Requires short leads from sensor; high impedance output, temp. sensitive, needs extra electronics to produce useable output	Tend. to short life due to mech. wear, noise increases with wear	Some prob. maintaining linear move. of core proport. to pressure change; will develop mech. wear	Large size; shock and vibration sensitive, low freq. resp.	Temp. sensitive, req. amplifier and spec cabling between device and amp.; slow recovery to shock and overpressure	May require compensation for temp. effect

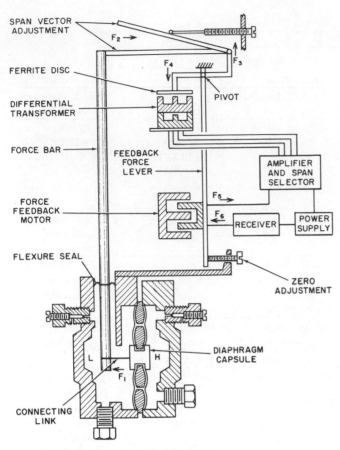

Fig. 2.4d   Force-balance differential pressure transmitter

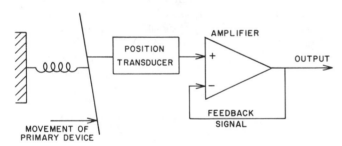

Fig. 2.4e   Motion-balance transmitter

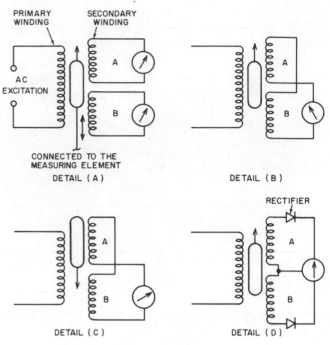

Fig. 2.4f   Linear variable differential transformer

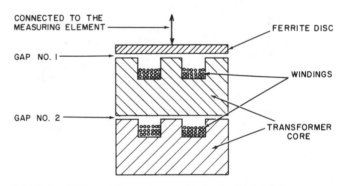

Fig. 2.4g   Differential transformer with ferrite disc. Relative spacing of gaps 1 and 2 determines the output voltage

be measured by potentiometric techniques even when the signal level is as low as 0–1 millivolts. The length and resistance of signal wires are not factors here because a potentiometer measures at zero current flow by balancing a known voltage against the unknown voltage until current flow stops. DC voltage signals must be made immune to noise by filtering techniques at the receiving instrument and by shielding the leads.

AC voltage signals, on the other hand, are not readily measured at low levels because of the difficulty in removing noise and spurious voltage peaks caused by induction onto the signal leads. Lead lengths are also limited by the inductive and capacitive effects of the wires. Twisting and shielding of the leads and the use of coaxial cables help reduce these effects.

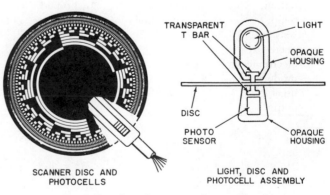

Fig. 2.4h   Photoelectric encoder transducer

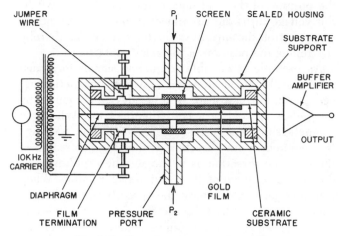

Fig. 2.4i   Capacitance pressure transmitter

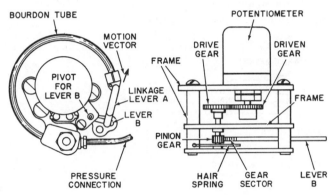

Fig. 2.4j   Potentiometric transmitter

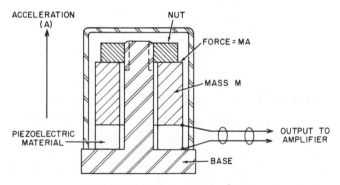

Fig. 2.4k   Piezoelectric transducer

Current Signals

DC current signals have found the widest acceptance in industrial electronic control systems in the United States, with current levels varying from 1–50 milliamperes. The most widely used current range is 4–20 mA DC because of intrinsic safety-related considerations. More detail on this subject can be found in Section 5.2 of this text. Some older systems operate on a 10–50 mA DC current signal, which is difficult to make intrinsically safe. Both of these signal levels are sufficiently high to minimize the need for special wiring, although shielding of signal cables and location away from wiring carrying heavy

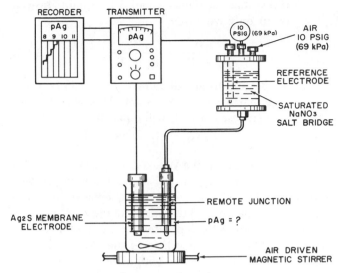

Fig. 2.4l   Selective ion electrode measurement

loads is still advisable when the signals are to be used as computer inputs. Voltage signals can easily be derived from current signals by inserting a series resistor in the wiring and measuring the voltage drop across it.

AC signals are not used for remote transmission.

Resistance Detection

Resistance is not a signal but an electrical property that can be readily measured to very high accuracies by the use of DC bridge circuits, such as the Wheatstone bridge shown in Figure 2.4m.

The classical Wheatstone bridge works as follows:

1. The battery causes a voltage difference to be exerted between points 1 and 4, and the resulting current through each branch is proportional to the resistance through that branch. In the center is a galvanometer that can measure very small currents.
2. When the galvanometer registers no current flow,

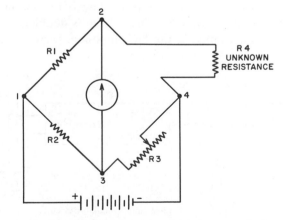

Fig. 2.4m   Resistance detection with Wheatstone bridge

the voltage at points 2 and 3 is equal. This can occur only when

$$\frac{R_3}{R_1 + R_3} = \frac{R_4}{R_2 + R_4} \qquad 2.4(1)$$

3. R_1, R_2 and R_3 are all known, and R can be made to equal R_2 and be constant. Therefore the unknown resistance R_4 will equal R_3:

$$R_4 = R_3 \qquad 2.4(2)$$

Bridge circuits such as the Wheatstone bridge have been modified by replacing the galvanometer with a differential amplifier. The battery may be replaced by a constant voltage source, and the output of the amplifier may drive a servo motor that operates potentiometer R_3. The principles, however, are still the same. Accuracies of 1 part in 10,000 are obtainable with laboratory versions of this instrument.

The strain gauge principle of measurement is also based on the change of resistances.[1] The resistor is cemented to a support or spring connected to the load so that it is distorted upon straining the material to which it is fastened.[4] Distortion of the resistor wire causes changes in its resistance. This operating principle is also described in Sections 5.7, 8.5, and 8.6 in the Process Measurement volume of this *Handbook*.

Frequency Signals

Because of the difficulties in obtaining accuracy with alternating voltage and current signals, various other methods have been developed for use in alternating current measurements. Since alternating current frequency can be accurately measured by several methods, frequency and frequency shift detection techniques can be made use of to measure variables such as capacitance, reactance, reluctance, and transformer effects. A typical block diagram for a tuned oscillator circuit is shown in Figure 2.4n.

A variable capacitance from the probe, dependent upon the material level in the tank, determines the voltage amplitude of the oscillator section in this circuit. A rising level or capacitance on the probe lowers the amplitude of the oscillator voltage. This voltage is doubled, rectified, and amplified in the DC amplifier. Depending on

the position of the "fail-safe high or low level" switch, this voltage turns the last transistor on or off. The transistor controls the current through the relay coil, thereby controlling the position of the relay contacts.

An example of variable frequency signals for transmission is the AC tachometer, in which a magnetic slug, gear tooth, key in a shaft, or other device is counted by the pulses created in a coil (Figure 2.4o). When the magnetic slug comes across the face of the coil, a voltage is generated in one direction as the slug enters and in the opposite direction as it leaves. These pulses are a direct measurement of the speed of shaft rotation. When counting gear teeth where the space between them is similar to the width of the teeth, a smooth AC without wide gaps between pulses can be generated. The design of the sensor probe has to be made with knowledge of the tooth shape, size, etc., so that good waveforms can be generated for easy measurement.

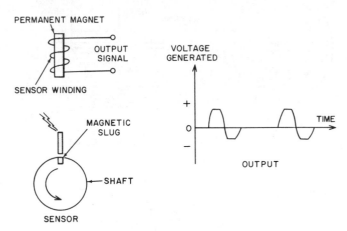

Fig. 2.4o Magnetic tachometer sensor

Phase Shift

The introduction of a reactance or a capacitance in an oscillating circuit changes the relationship between the peak voltage and the peak current flowing in the circuit. The measurement of the phase angle is a direct measurement of the variable.

Digital Signals

With the increased use of computing and logging systems, there has been a marked increase in the use of digital transmitters for high-accuracy data transmission. The early applications that led the way in digital transmission were tank farm level transmitters, using coded contact wheels, telephone dials, multiple openings of a single contact, and photo scanning. All of these were developed prior to 1960. Pulse-type tachometers were also an early development in digital transmission.

More detail on this subject can be found in Section 7.3 of this text.

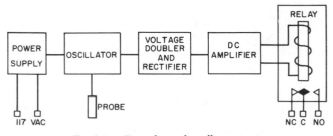

Fig. 2.4n Typical tuned oscillator circuit

On-Off Oscillating Circuits

One of the advances in capacitive and reactive transducer design and application has been the development of capacitance probes for level measurement. The principle is the same as the frequency shift principle, except that the operating frequency of the oscillator is crystal controlled so that it does not oscillate when the capacitance is changed. The output of the oscillation amplifier is sent to the relay coil. When the circuit stops oscillating, the relay opens.

Fiber Optic Transmission

Fiber optic signal transmission may be the standard in the future as EMI (electromagnetic interference) and RFI (radio frequency interference) do not affect the optical signals in the fiber optic light pipe. More information on this subject can be found in Section 2.5 of this text.

Desirable Transmitter Features

The characteristics most desirable for an electronic transmitter are high resolution, reliability, and low cost. In order to meet these requirements, a transmitter should be designed to have the following features:

1. Small size and weight for easy installation and maintenance.
2. Rugged design to withstand industrial environment.
3. Minimum dependence on environmental conditions for accuracy; this requires good temperature stability, resistance to barometric change, weatherproofing, etc.
4. Elimination of the need for adjustment due to load or line resistance variations.
5. No potential hazard to personnel or equipment in explosive atmospheres. Low-voltage operation with limited current capacity assists in eliminating these problems. By definition, intrinsically safe equipment provides all three kinds of protection. (See also Section 5.2.)
6. Convenient and accurate field calibration and maintenance. In electrical transmitters this usually means conveniently located test terminals. In explosion-proof designs, calibration should be possible without opening the housing.
7. Capacity to operate during voltage dips and power outages. Generally this is accomplished elsewhere in the control system. (Stand-by power supplies are discussed in Section 3.10.)
8. Minimum number of transmission and power wires. In most systems installed today the transmitter is powered from the receiver over the same wires that are used for transmission.
9. Output compatible with both measuring and controlling instruments.
10. Optional local indication of output signal.
11. Circuitry designed to facilitate troubleshooting and maintenance. Most systems today are solid state (no tubes) and are mounted on circuit boards or are encapsulated. Plug-in components make fast repairs easy, but encapsulated modules are considered to be throwaway items.

REFERENCES

1. Oliver, F.J., *Practical Instrumentation Transducers*, Hayden Book Co., New York, 1971.
2. Andrew, W.G., *Applied Instrumentation in the Process Industries*, Gulf Publishing Co., Houston, 1974.
3. Minar, E.J., (ed.), "ISA Transducer Compendium," latest edition.
4. Shames, I.H., *Introduction to Solid Mechanics*, Prentice-Hall, Inc., Englewood Cliffs, NJ, 1975.

BIBLIOGRAPHY

Adams, L.F., *Engineering Measurements and Instrumentation*, The English Universities Press, 1975.

Blake, H., "Transmitters and Transducers with a Purpose," Paper presented at AGA Conference on Transmission, 1968.

Combs, C.F., (ed.), *Basic Electronic Instrument Handbook*, McGraw-Hill Book Co., New York, 1972.

Demorest, W.J., "There's More to Transmitter Accuracy Than the Spec," *Instruments & Control Systems*, May, 1983.

Diefenderfer, A.J., *Principles of Electronic Instrumentation*, Saunders, Philadelphia, 1972.

Doebelin, E.O., *Measurement Systems—Application and Design*, rev. ed., McGraw-Hill Book Co., New York, 1975.

Gregory, B.A., *An Introduction to Electrical Instrumentation*, MacMillan, New York, 1973.

Hall, J. "Monitoring Pressure with Newer Technologies," *Instruments & Control Systems*, April, 1979.

Herrick, C.N., *Instrumentation and Measurement for Electronics*, McGraw-Hill Book Co., New York, 1972.

Merritt, R., "Keeping up with Pressure Sensors," *Instruments & Control Systems*, April, 1982.

National Semiconductor Corp., "Transducers, Pressure & Temperature," latest edition.

Wolf, S., *Guide to Electronic Measurements and Laboratory Practice*, Prentice-Hall, Englewood Cliffs, NJ, 1973.

2.5 FIBER OPTIC TRANSMISSION

Types:

A—Fiber optic cable
B—Fiber optic connectors
C—Fiber optic splice kits
D—Optoelectronics
E—Fiber optic kits
F—Fiber optic links
G—Fiber optic Couplers
H—Fiber optic sensors
I—Fiber optic data highways
J—System houses

Partial List of Suppliers:

Allen-Bradley Co. (H, I); American Fiber Optic Corp. (A, C); AMP, Inc. (B, E, G); Amphenol/Bunker Ramo Corp. (B); Belden Corp. (A); Berg Electronics (B); Burr Brown (F); Centronic (D); Cutler Hammer (I); Du Pont Co. (A); EOTEC (H); Galileo Electro-Optics Corp. (A, D, F, G); General Cable Corp. (A); General Electric Co. (D); General Optronics Corp. (D); Hewlett Packard Co. (D, E, F); Honeywell Corp. (E, F, G, J); Hughes Aircraft Co. (B, F, G, J); ITT (A, B, D, G); Leeds & Northrup Instruments, a Unit of General Signal (I); Laser Diode Laboratories, Inc. (D); Math Associates (J); Meret, Inc. (D, F); Motorola (D); Phalo Corp. (G); Plessey Optoelectronics & Microwave (F); RCA (D, F); Siecor (A, F); Spectronics (D, F); Texas Instruments, Inc. (D); Valtec Corp. (A, F); Vanzetti (H)

Electricity and pneumatics are commonly used to measure, control, and transmit information among process, controller, and operator. Fiber optics provide an alternate method of process measurement and information transmission. Although fiber optic (FO) transmission hardware and systems are readily available to the user, FO instruments are not yet available in the same broad spectrum as are electrical or pneumatic instruments. This is because the efforts of the FO industry have been primarily directed toward meeting the needs of their prime customers, namely AT&T (commercial FO communication) and the military (special purpose FO instrumentation). Today the industrial and commercial world lags behind the military in FO instrumentation development.

During the 80s, FO data transmission is following on the heels of microprocessor technology in modifying historical methods of instrumentation implementation. The 80s will also bring forward more off-the-shelf FO instrumentation to compete with conventional digital and analog instrumentation.

This section provides a review of fiber optics basics and its capability in the instrumentation area.

Instrumentation and Fiber Optics

Fiber optics technology provides the instrument engineer with analog and digital sensors that measure process variables, operator actuation devices, interfaces to instrument (pneumatic, electrical, electronic, mechanical) sensors (hybrid devices), and communication capability between system analog or digital controllers.

One of fiber optic's most compelling features for the instrumentation engineer is its intrinsically safe characteristics for use in hazardous environments. Moreover, the advent of digital electronics, the microprocessor, and distributed control systems has resulted in an awareness of problems associated with electronics in an industrial environment. These problems include voltage isolation between equipment, ground isolation, EMI, RFI, noise, lightning susceptibility, and EMP susceptibility. Since FO cable is a nonconductor and does not radiate energy, it can provide a built-in solution to these problems. Fiber optics are also being used today in the analog realm to transmit 0–10V and 0–20 mA information, thus minimizing use of shielded cables, lightning suppression devices, special grounding connections, dedicated raceway, etc.

The microprocessor is now implanted in all levels of the instrumentation system building block (i.e., process, controller, and operator interface). Information interchange between these blocks is implemented by various methods of communication, including 20 mA two-wire loops, serial, parallel, baseband, and broadband communication. This communication ranges from individual input/output interfaces to high-speed, high-density, long-distance communications called data highways. The ability of fiber optics to meet high information throughput requirements while maintaining its advantages in its operating and environmental capabilities is making FO an attractive alternate in communication applications.

Conventional instrumentation can be converted or expanded utilizing fiber optics because of the advances in optoelectronics. The marriage of conventional sensors and transmitters with fiber optics via optoelectronics is common and is sometimes referred to as a hybrid fiber optic link.

Fiber Optic Principles

This section provides an insight into basic fiber optic terminology and equations.

Optical fiber bandwidth (dispersion) is one of the basic parameters considered in all applications. When analyzing dispersion in an optical fiber, the *index of refraction* (n) is considered, as follows:

$$n = C/V \qquad 2.5(1)$$

where

C = speed of light in a vacuum (3×10^8 m/sec) and

V = speed of light within the optical fiber selected.

The *index profile* of an optical fiber defines how the index of refraction varies as a function of radial distance from the fiber's center. *Step index* profile refers to a fiber having an index of refraction with an abrupt change (step) at the *core* radius. The optical fiber divides into two sections, with the inner light-carrying optical conductor called the *core* and the outer jacket called the *cladding*. See Figure 2.5a for a cross section of a typical optical fiber.

Optical energy is inserted into the core and travels along the core through internal reflection at the core-cladding interface (see Figure 2.5b). Reviewing Figure 2.5a, the comparative index of refraction for core (n_1) and cladding (n_2) is shown in Figure 2.5c. Note the index of

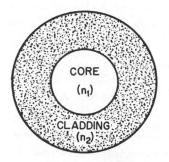

Fig. 2.5a Fiber optic cable cross section

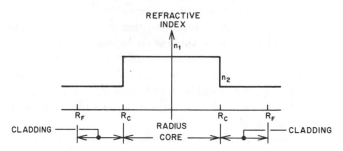

Fig. 2.5b Light energy transmission through fiber

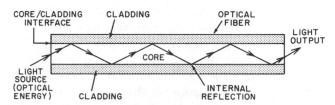

Fig. 2.5c Index of refraction comparison

refraction for the cladding (n_2) is less than that of the core (n_1).

The core to cladding index ratio is a prime factor in defining optical transmission dispersion. To illustrate this, a discussion of multimode propagation characteristics is required. In a multimode fiber, Figure 2.5b might be modified as illustrated in Figure 2.5d. Two factors must be recognized in this figure. First, not all of the light travels down the core. Some of the light is lost in the cladding (see Figure 2.5j(c)). The amount of light loss (dispersion) is a function of the angle that the light within the core hits the cladding. Snell's law provides the minimum angle that supports internal reflection, as follows:

$$\sin \theta_{min} = n_2/n_1 \qquad 2.5(2)$$

θ_{min} depends on the cable selected and the indexes of refraction of the cladding and the core. Rays striking the

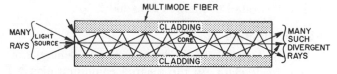

Fig. 2.5d Light transmission in a multimode fiber

core-cladding interface at angles less than θ_{min} will be lost in the cladding (see Figure 2.5e and 2.5j(c)).

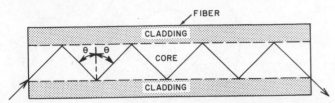

Fig. 2.5e   Minimum angle for internal reflection depends upon the indexes of refraction of cladding and core.

Second, again referring to Figure 2.5d, note that many rays are traveling down the core, varying from the axial ray to various rays striking the cladding at various acceptance angles. Because each ray travels a different distance at the same speed, their arrival times will vary. Figure 2.5d illustrates a cable that has numerous propagation modes and an abrupt change in index profile; this fiber is called a *multimode step index type fiber*. Characteristically, multimode step index fibers can cause bit smearing or intersymbol interference in a digital data system and delay distortion in an analog-modulated system.

To decrease mode volume, small glass fibers with n_1/n_2 as small as practical are constructed. If "V" is less than 2.405 m/sec, only a single-mode (axial ray) can propagate. Called a *single-mode fiber*, this design exhibits no modal dispersion at all.

Single-mode construction offers the best bandwidth features but provides design challenges in injecting light into the small-diameter core as well as fiber-to-fiber splicing. Figure 2.5f illustrates single-mode transmission.

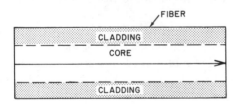

Fig. 2.5f   Single-mode (monomode) fiber

The fiber core can be constructed with an index of refraction that decreases parabolically from the center of the fiber. Light propagation now occurs through *refraction*, a continual bending of the ray toward the fiber's optical axis. This manufacturing technique provides a larger-diameter fiber than single mode and a wider bandwidth fiber than multimode step index, thus reducing the coupling problem while enhancing the distance handling capability. This product is called *graded index fiber* and is illustrated in Figures 2.5g and h.

Flux Budgeting

Flux budgeting is the budgeting of optical energy from source to detector, so the desired communication will

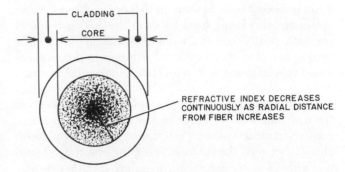

Fig. 2.5g   Graded index cable (end view)

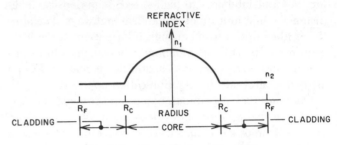

Fig. 2.5h   Graded index cable profile

be performed. Optical energy losses will occur in the source/fiber interface, the cable, fiber/detector interface, connectors, splices, couplers, etc. In addition, systems should be designed to allow for future maintenance, modification, and expansion by providing spare optical energy capacity to allow for addition of connectors, splices, taps, and so on.

The optical energy losses mentioned above can be further characterized into four groups, as follows:

1. Fiber attenuation (or cable loss).
2. Unintercepted-illumination (UI) loss resulting from an area mismatch between the source illumination spot and the fiber core. As illustrated in Figure 2.5i, the source/cable interface is a typical candidate for "UI" loss. The UI loss can be approximated in this case as follows:

$$\text{UI loss (in DB)} = 10 \log \frac{A_c}{A_s} \qquad 2.5(3)$$

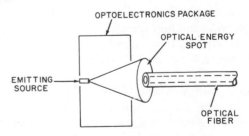

Fig. 2.5i   Source/cable interface

where

A$_c$ is fiber core area and A$_s$ is the source's spot area at the point of contact with the fiber.

3. Numerical aperture (NA) loss from light rays with angle of incidence not within the fiber's acceptance cone (see Figure 2.5j). Numerical aperture loss is calculated as follows:

$$\text{NA loss (in DB)} = 10 \log \frac{P_C}{P_T} \qquad 2.5(4)$$

where

P$_C$ = P$_T$ $[1 - (\cos \theta)^{m+1}]$ = destination power,
P$_T$ = total source power (mW), and
θ = fiber's acceptance cone half angle.

4. Reflective (R) loss is a factor in splices. Light incident on the fiber core experiences a change in index of refraction at the air/core interface; a portion of the light reflects back from the surface and is lost. Core index of refraction impacts reflected/refracted proportioning of the incident rays. Reflection coefficient "P" provides the fraction of incident light reflected from the core. Approximation can be achieved by the following:

$$P = \left(\frac{n_1 - 1}{n_1 + 1}\right)^2 \qquad 2.5(5)$$

Therefore, R (dB) = 10 log (1 − p).

Fiber Optic Components

Figure 2.5k illustrates components in a simple fiber optic system. These are identified as source, connector, cable, and detector in this basic FO circuit (link).

Source

The light source will be either an LED (light emitting diode) or ILD (injection laser diode). The major factors involved in source selection are to ensure wave length compatibility with the fiber and to minimize coupling losses. Sources should have high power outputs that are very directional and spectrally pure, and very fast response times (nanosecond range is typical).

Source selection must include an analysis of the disadvantages of LED's and ILD's. The table below represents the basic problems source manufacturers have

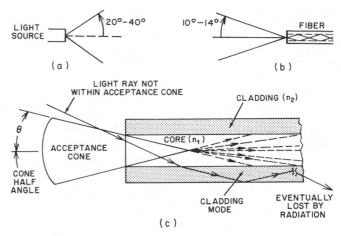

Fig. 2.5j (a) Typical source (LED) emission profile. (b) Typical acceptance cone for step-index fiber. (c) Typical source/fiber interface.

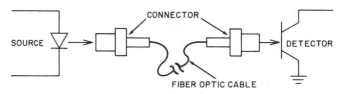

Fig. 2.5k Fiber optic link

to deal with. Each manufacturer has his own manufacturing, assembly, and circuit techniques to solve these problems. An analysis of how these are being resolved in any application is recommended.

Detector

The light detector will generally consist of a PIN diode or an avalanche photodiode (APD). Selection criteria should include the information in the table at top of page 136. The result is that APDs are generally used in optical links that are long (> 1 km) and when wide bandwidths are required; links under 1 km will generally utilize PIN diodes.

Note that "overhead" circuitry is required to supplement the sources and detector devices; design and assembly of these circuits can be included and generally obtained from the device manufacturer.

Source/detector functions can be packaged into a transceiver arrangement, commonly called a repeater. Repeaters require electrical power and are generally used in long-distance communication applications.

Source	Cost	Temperature Sensitivity	Lifetime	General Applicability	Size	Response (RISE) Time	Source to Fiber Coupling Efficiencies
ILD	Higher	Higher	Shorter	Longer Distances	Smaller	Faster	Higher
LED	Lower	Lower	Longer	≤ 1 km	Larger	20–30 nsec	Lower

Detector	Cost	Bias Circuitry	Efficiency	Temperature Range	Sensitivity	Bandwidth	Responsitivity
PIN	Lower	Simpler	Higher	Wider	Lower	$\geq$ 100M Hz	High
APD	Higher	More complex	Lower	Narrower	Higher	Higher	Higher

Connectors

Connector technology has improved dramatically in terms of material cost, installation ease, and performance. Advances in this area are being made at a rapid rate. To analyze the myriad of options available requires a continuous supply of research funds.

These development and testing funds are allocated and utilized by the military and connector manufacturers. Instrumentation companies will generally reduce their options through engineering analysis, limited testing, and use of "Beta" (test) sites. Beta sites are selected to stress the connector assemblies similar to industrial applications. However, fiber optic connectors seem to have as many "up-front" problems as do today's LSI (large-scale integrated) circuitry, and more than connector replacement is required to resolve installation problems. But connector problem analysis and solution development may result in connector redesign or replacement.

The user, selecting his own connector assemblies, should consider the following:

End surface preparation
System compatibility
Metal vs. plastic
Fiber type
Connector loss
Fiber alignment
Coefficient of thermal expansion
Bonding material

Assembly flexing
Cable dimension
Cable entry design
Installation tools
Installation complexity

Connector standardization appears to be a long way off, but several main types of connector installations can be reviewed when selecting the connectors for an application.

Cable

Single-mode, multimode step index, and multimode graded index are the three basic fiber types. Selection criteria for these fiber types would include the information in the first table below. Glass, plastic, or a combination of the two are the fiber materials used in FO cables today. Selection criteria for these fiber types include the information in the second table below.

Cable packaging can be in single plastic fibers, fiber bundles of either glass or plastic, flat-ribbon cable construction, or hybrid cable construction (includes metallic conductor or strength members with optical fiber).

Both single plastic fibers and fiber bundles serve the short-distance application and each offer good durability with a high degree of flexibility. Bundle packaging is the least popular of the two; however, its performance is not significantly degraded with breakage of a single fiber and it does offer some advantages when being coupled from LED sources.

Fiber Type	Distance	Bandwidth	Numerical Aperture	Interconnect Ease	Interconnect Loss
Single Mode	Long	Highest	Lowest	Least	Highest
Graded Index	Medium	Higher	Low	Intermediate	Intermediate
Multimode Step Index	Short < 1 km	Lowest < 10M BPS	Higher	Greatest	Lowest

Fiber Material	Cladding	Diameter	Application Distance	Raceway Routing
Glass (silica)	Glass	Smallest	Long	Suitable
Plasti-clad, Silica-core	Plastic	Intermediate	Long	Suitable
Plastic	Plastic	Largest	Short	Limited*

* May not meet code flammability requirements.

Sensors

Sensors for fiber optic transmission fall into two large classes, hybrid and fiber optic. The hybrid sensor uses existing sensor technology found in instrumentation (pneumatic, electronic, mechanical, etc.) and interfaces it to optical transmission. Optoelectronics has made these sensors available from "on–off" through 4–20 mA applications up to high-speed, high-data-rate communications.

The fiber optic sensor uses optical technology to sense process conditions and transmits data optically to a remote optoelectronic interface. Optical sensors providing "on–off" status are available today and use actuators such as pushbuttons and limit switches to interrupt the light beam. Photocells are also common sensor applications.

Replacement of the electronic transducer with a fiber optic transducer is also possible; using fiber optics to measure the infrared energy emitted by an object provides temperature measurement. Single-mode optical fibers combined with magnetostrictive jacket materials have been used as magnetic field sensors.

Generally, fiber optic sensors operate in an interferometric mode. Light from an optical source is split, with half traversing the sensing fiber and the other half routed into a length of reference fiber. Light from these two fibers is routed back together, and the sensor output generated from a comparison of the reference light source and the sensor fiber characteristics.

Splicing

Fiber optic cables can be spliced by various methods, including the following:

1. Use of male and female connector, crimped onto the fiber.
2. High-voltage fusion or heating, which melts the fiber.
3. Use of optoelectronic connector providing a light/electronic hybrid connection.
4. Use of splicing prepolymer with heatless curing (e.g., UV light).

The splicing method selected is a function of the cable used, the allowable loss per splice (see "Flux Budgeting" paragraph), and the physical and mechanical environment the splice will be subjected to.

Optical Couplers

As the sophistication of fiber optic system architecture increases, the need arises for signal splitting, such as transmitting optical energy from two or more fibers into one, or splitting energy from one into two or more channels. Couplers are utilized for these applications, providing the engineer the ability to insert taps, drop-offs, multi-access ports, and so on.

Coupler types include the T and the star configurations, with the T generally handling 4 or less interconnects and the star handling 20 or more terminals.

Coupler designs include reflective (lens and mirror beam splitting systems) and transmissive (fusing of several identical fibers into a bionical taper) schemes. Transmissive coupler ports are essentially optically isolated from each other. Transmissive star couplers with many ports are constructed from bionically tapered multimode fibers; once cut, the desired number of fibers are twisted together into a hexagonal bundle or similar close-pack formation, placed under tension, and heated. The fibers soften, the cladding fuses, and tension provides the desired bionical tapers with a central fused section.

Reflective star designs typically use a glass mixing rod with a mirror end plate that diffuses the light and reflects it back to all the other fibers; energy distribution is thus equalized among the fibers of the bus. Typically, the star coupler provides each terminal with transmit and receive capability to all other terminals.

The number of couplers allowed on a system plus the distance between couplers is determined by flux budgeting calculations. Formulas for calculating T and star coupler losses are available from coupler manufacturers.

Couplers are found in distributed networks to service many facilities. Often the mixture of source, cable, connector, and receiver loss parameters results in the need for optical-to-electronic-to-optical repeaters whose function it is to restore source energy level. The number and location of repeaters is a function of flux budgeting calculations as well as network configurations.

Links

Links are special-purpose fiber optic packages designed to do a specific function. Examples of links include a communication link to handle RS232C data transmission between microprocessor-based controllers and an input/output link to handle 0–20 mA data transmission between field instrument and controller. Links offer the advantages of being pre-engineered, pretested, application-specific assemblies. This allows their use as imbedded subsystems within equipment for sale to users, as well as direct application by users.

Kits

Fiber optic manufacturers market experimentation kits that allow potential users hands-on familiarity with fiber optic components, their assembly, and installation criteria. Kits are generally inexpensive, eliminate the component compatibility problem, and provide manageable interfaces to real-world systems through the use of optoelectronic devices.

While most kits are intended to allow users to develop learning skills, few are suitable for industrial applications; however, the industrial characteristics are available in fiber optic links.

Fiber Optic Systems

Fiber optics application designers should make use of the excellent system houses available to end users. A detailed functional specification of the fiber optic application should be included in the purchase order with the following details:

Detail description of the problem
Environmental conditions of installation
Availability requirements
Acceptance test description
Maintenance training
Flux budgeting capacity for future considerations
System life cycle (how long will it be in service)
Code requirements
Installation, checkout, and run-in instructions
Spare parts requirements
Maintenance tools and instruments
Splicing and connector technology utilized

The fiber optic functional specification provides the supplier with a detailed definition of the end user's needs and facilitates proper utilization of fiber optics technology "know-how" available in system houses.

BIBLIOGRAPHY

Amp Incorporated, "Introduction to Fiber Optics and AMP* Fiber Optic Products," HB5444.

Bliss, John, "A Designer's Guide to System Budgeting," *Electro-Optical Systems Design*, August, 1981.

Cole, James A., and Ellis, John L., "Specifying Optical Fibers and Cables," *Instruments & Control Systems*, November, 1981.

Geuchalla, M.E., "Specifying Critical Components for Fiber Optic Data Collection Systems," *InTech*, September 1983.

"Instrument Matchmaker," *Instrumentation Technology*, April, 1970.

Klaassen, M.P., "Fiber Optics in Industry", ISA, C.I. 82-864.

———, "Fiber Optics: Ready for Industry," *InTech*, May, 1983.

Kleekamp, Charles, and Metcalf, Bruce, "Designers Guide to Fiber Optics", *EDN Magazine*, December, 1977–March, 1978.

Krohn, D.A., "Fiber Optics: New Sensors for Old Problems," *InTech*, May, 1983.

———, "Inside Fiber Optic Sensors," *Instruments & Control Systems*, September, 1983.

Mirtich, Vincent L., "Designer's Guide to Fiber-Optic Data Links," *EDN Magazine*, June 20, 1980, July 20, 1980, August 20, 1980.

Montgomery, Jeff D., "Choosing Components for Industrial Fiber Optic Systems," ISA, C.I. 82-978.

Ohlhaber, Ronald, "Budgeting Optical Power for Fiber Optic Systems," *Instruments & Control Systems*, September, 1979.

Ormond, T., "Fiber Optic Components", *EDN Magazine*, March 20, 1979.

Sahm, William H., III, "The General Electric GFOD/E Series Key to Flexible, Low-Cost Fiber Optic Systems," 200.89, October, 1981.

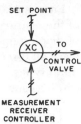

2.6 PNEUMATIC CONTROLLERS

Type:	Receiver Controllers: Indicating, recording, miniature, high-density miniature, and large case
	Direct-Connected Controllers: Blind, indicating, or recording in large or medium case for field or panel mounting
Application:	Receiver Controllers: control of any variable that can be measured and translated into an air pressure by a pneumatic transmitter; includes automatic and manual control and set-point adjustment
	Direct-Connected Controllers: Have own measuring element in contact with the process; measuring elements available include pressure, differential pressure, temperature, level, pH, thermocouple, radiation pyrometer, and humidity.
Typical Front Panel Size:	Miniature: 6″ × 6″ (150 mm × 150 mm) High-Density Miniature: 3″ × 6″ (75 mm × 150 mm) Large case: 15″ × 20″ (375 mm × 500 mm)
Minimum Response Level:	Less than 0.01% of full scale
Input Ranges:	3–15 PSIG (0.2–1.0 bar) or direct connected
Output Ranges:	3–15 PSIG (0.2–1.0 bar)
Repeatability:	±0.5%
Inaccuracy:	±1%
Displays:	Set point, process variable, output, deviation, and balance
Maximum Frequency Response:	Flat to 30 Hz
Maximum Zero Frequency Gain:	750
Control Modes:	Manual, proportional, integral (reset), derivative, floating, differential gap, two position
Costs:	Miniature Indicating Controller: $1000–$1500 Miniature Recording Controller: $1500–$2500 Miniature High-Density Indicating Controller: $1000–$2000 Large-Case Recording Controller: $1500–$2000 Large-Case Indicating Controller: $1200–$1800 Direct-Connected Indicating Field Mounted: $800–$1200
Partial List of Suppliers:	Ametek, Inc.; Bailey Controls, Div. of Babcock & Wilcox; Barton, ITT; Brandt Industries, Inc.; Bristol Babcock, Inc.; Fischer & Porter Co.; Fisher Controls Co.; Foxboro Co.; Honeywell, Process Control Div.; Kent Process Control, Inc.; Moore Products Co.; Powers Process Controls; Robertshaw Controls Co.; Taylor Instrument Co.; Westinghouse Electric Corp.

History and Development

Pneumatic controllers were first introduced at the turn of the century. They logically followed the development of diaphragm-actuated valves in the 1890s. Early types were all direct-connected, local-mounting, indicating or blind types. Large-case indicating and circular chart recording controllers appeared around 1915. All early models incorporated two-position, on-off action or proportional action. It was not until 1929 that reset action was introduced. Rate action followed around 1935.

Until the late 1930s, all controllers were direct connected and therefore had to be located close to the process. Pneumatic transmitters were not introduced until the late 1930s. To make them compatible, the large-case pressure recording and indicating controllers were easily converted into receiver controllers. This made remote mounting practicable and centralized control rooms became a reality. Because of the inherent advantages, the combination of pneumatic transmitters and receiver controllers quickly became popular. Since the recording and indicating receiver controllers were quite large, control rooms and panel boards were likewise spacious. Additionally, all control boards had a monotonous look and usually came in one color—black.

A revolution in design occurred in 1948 with the introduction of miniature instruments. Here the concept of the controller evolved into a combination of a small, approximately 6×6-in. (150×150 mm) panel front indicating and recording control station and a blind receiver controller. The station permitted the operator to monitor the measured variable, set point, and valve output; it allowed him to switch between, and operate in, either the automatic or manual control modes. Miniature controllers ushered in the era of the graphic panel, in which the instruments are inserted into graphic symbols representing the attendant process apparatus. Control rooms became more compact, control boards more meaningful and colorful, and, because operators quickly developed a "feel" for the process, training time was considerably reduced.

Nevertheless, graphic panels were also wasteful of space and presented major modification problems each time the process was changed. This led to the evolution of the semi-graphic panel, in which a graphic symbol diagram of the process appeared above the miniature instruments mounted in neatly spaced rows and columns.

In 1965, miniature, high-density mounting style stations appeared. The new lines brought with them the most advanced ideas in displays, operating safety and simplicity, packaging, installation simplicity, and servicing facility. Along with some of the standard miniature controllers, they offer computer compatibility along with some unique control capabilities that had previously been impractical.

The early 1980s saw some important new entries in the pneumatic controller market. These included pneumatic controllers using RTD and thermocouple type sensors and pneumatic controllers with microprocessor-based serial communication modules for tie-in to distributed control systems.

Pneumatic Controller Principles

A receiver type pneumatic controller is shown schematically in Figure 2.6a. A process transmitter (lower left) senses the measured variable (pressure, temperature, flow, etc.) and transmits a proportional air pressure to the measured variable (MV) bellows of the controller. The controller compares the measured variable against the set point (SP) and sends a corrective air signal to manipulate the control valve, thereby completing the feedback control loop.

The controller consists of two sets of opposed bellows of equal area, acting at opposite ends of a force beam that rotates about a movable pivot. Extending from the right end of the beam is a flapper that baffles the detector nozzle of the booster relay.

Booster Circuit

Supply air connects to the pilot valve of the booster and flows through a fixed restriction into the top housing and out the detector nozzle. The flapper is effective in changing the back pressure on the nozzle as long as the clearance is within one-fourth of the nozzle diameter. The restriction size is selected on the basis that the continuous air consumption will be reasonable and that it will be large enough not to clog with typical instrument air. The nozzle, on the other hand, must be large enough that when the flapper has a clearance of one-fourth nozzle diameter, nozzle back pressure drops practically to atmospheric. It must not be so large, however, that the seating of the flapper becomes too critical. A typical size of restriction is 0.012 in. ID (0.3 mm) while the nozzle would be 0.050 in. ID (1.3 mm). The nozzle back pressure is a function of flapper position. (For a more detailed discussion of flapper-nozzle detector circuits, refer to Section 2.2 on Pneumatic Transmitters.)

The exhaust diaphragm senses nozzle back pressure and acts on the pilot valve. If the back pressure increases, it pushes down on the valve, opening the supply port to build up the underside pressure on the diaphragm until it balances the nozzle back pressure. If the back pressure decreases, the diaphragm assembly moves upward, allowing the valve to close off the supply seat while opening an exhaust seat in the center of the diaphragm. This allows the underside pressure to exhaust through the center mesh material of the diaphragm assembly until the pressures balance.

Proportional Response

Since a pressure range of 3–15 PSIG (0.2–1.0 bar) is an almost universal standard for representing 0 to 100 percent of the range of measured variable, set point, and

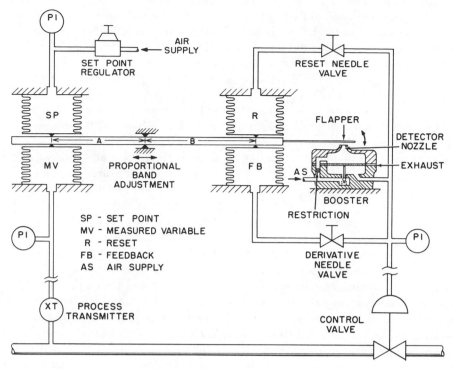

Fig. 2.6a   Moment-balance controller

output on receiver controllers, this description will assume the operation to be in this range. To understand the proportional response, first assume that the derivative needle valve is wide open and that the reset needle valve is closed with 9 PSIG (0.6 bar) mid-scale pressure, trapped in the reset bellows R (Figure 2.6a). If the set point is adjusted to 9 PSIG (0.6 bar), when the measured variable equals set point, the flapper will automatically be positioned so that the booster output, acting on the feedback bellows FB, will be equal to the reset pressure, namely 9 PSIG (0.6 bar). The reason for this is that the force beam will only come to equilibrium when all of the moments about the pivot come to balance. Any unbalance in moments causes a rotation of the beam with attendant repositioning of the flapper and change in feedback pressure until moment balance is restored.

If the pivot is positioned centrally, where moment arm A equals B, then for every 1 PSI (0.067 bar) difference between set point and measured variable there will be a 1 PSI (0.067 bar) difference between reset and controller output, or feedback. This represents a 100 percent proportional band setting, or a gain of 1. Percent proportional band is defined as the input change divided by the output change times 100. Gain equals the ratio of output change to input change. See proportional band chart, Figure 2.6b.

$$PB\% = \frac{\text{change in input}}{\text{change in output}} \times 100 \qquad 2.6(1)$$

$$\text{Gain} = \frac{\text{output change}}{\text{input change}} \qquad 2.6(2)$$

If the pivot in Figure 2.6a is shifted to the right, to the point where moment arm A is four times greater than moment arm B, then every 1 PSI (0.067 bar) change in

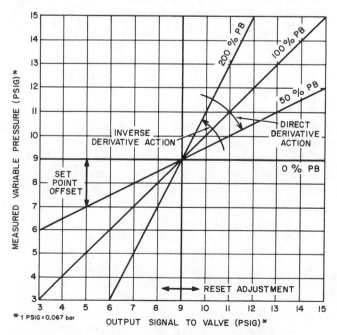

Fig. 2.6b   Graphic representation of control functions

141

measured variable results in a 4 PSI (0.27 bar) change in output. This gives a proportional band of 25 percent, or a gain of 4. If the pivot is moved to the left reversing the ratio, the proportional band would be 400 percent and the gain 0.25. If the pivot could be moved to the right to coincide with the center of the R and FB bellows, the most sensitive setting would be achieved, i.e., approaching 0 percent band or infinite gain, as the slightest difference between MV and SP would rotate the flapper to change the output to 0 PSI (0 bar) or full supply pressure, depending upon the direction of the error (on-off control).

Reset Response

If it were practical to use 1 or 2 percent proportional band on all processes, proportional action alone would be sufficient for most processes. However, most process loops become unstable at much wider bands than this. Noisy flow control loops, for example, may require more than 200 percent band for stability. Suppose the process in this case can tolerate a band no narrower than 50 percent. Then, according to the proportional band diagram, the controller can maintain the measured variable exactly at set point only when valve pressure is 9 PSIG (0.6 bar). If the valve pressure, because of conditions, had to be 5 PSIG (0.3 bar), for example, it could only be so when the measured variable deviates from set point by 2 PSI (0.14 bar) (16.7 percent of scale error). In most cases this amount of error, more correctly termed offset, is intolerable. Nevertheless, in practice, the valve pressure must change as the load changes. As an example, to hold level in a tank, the control valve on the inlet would obviously have to change opening as rate of effluent changes.

If the load were such as to require a 5 PSIG (35 KPa) valve pressure, one way of eliminating the offset would be to *manually* change the reset pressure R to 5 PSIG. The output or valve pressure would then be 5 PSIG when the measured variable equals set point. This action, which amounts to applying manual reset to the controller, was popular years ago but is rather uncommon today.

It is a simple matter to make this reset action automatic. It only requires that the reset bellows be able to communicate with the controller output pressure through some adjustable restriction such as a needle valve. The reset action must be tuned to the process in such a way as to allow the process sufficient time to respond. Too fast a reset speed, in effect, makes the controller "impatient" and results in instability. Too slow a speed results in stable operation, but the offset is permitted to persist for a longer period than necessary.

To describe the automatic reset action, assume that set point is at 9 PSIG (0.6 bar) and the reset is trapped at 9 PSIG, while valve pressure feedback, because of load, must be at 5 PSIG (0.3 bar). According to Figure 2.6b, with the proportional band at 50 percent, the mea-

sured variable will be controlled at 7 PSIG (0.46 bar). If the reset needle valve is then opened slightly, the pressure in the reset bellows will gradually decrease. As it drops from 9 to 7 PSIG (0.6 to 0.46 bar), the measured variable pressure will rise from 7 to 8 PSIG (0.46 to 0.5 bar). The reset pressure will continue to drop until it exactly equals output, i.e., 5 PSIG (0.3 bar). At this point, the measured variable will exactly equal set point, 9 PSIG (0.6 bar). With this circuit, regardless of where the valve pressure must be, the controller will ultimately provide the correct valve pressure with no offset between set point and measured variable.

Reset action can also be understood by looking at it from the opposite direction, i.e., controller configuration. The controller cannot come to equilibrium as long as there is any difference in pressure between reset and feedback, since the open communication through the reset needle valve will cause the reset to continue to change, which in turn directly reinforces the feedback pressure. Reset and feedback, in turn, cannot be equal unless set point and the measured variable are equal to each other. This fact alone assures that the controller will maintain corrective action until it makes the measured variable exactly equal to set point regardless of where feedback must be. Referring to Figure 2.6b, reset in effect shifts the proportional band lines along a horizontal axis at set-point level. It makes the center of the band coincide with the required valve pressure.

Reset time is the time required for the reset action to produce the same change in output as that resulting from proportional action as the error remains constant. For example, if an error resulting in 1 PSI (0.067 bar) output change due to proportional action is applied to a controller and the error is sustained, a 1 minute reset setting would cause the output to continue to change at the rate of an additional 1 PSI per minute in the corrective direction. The term "repeats per minute" is also used to characterize reset. This term is the reciprocal of reset time.

Derivative Response

If a needle valve is inserted between the booster output and the feedback bellows as in Figure 2.6a, it delays the rebalancing action of the feedback bellows and causes the controller to give an exaggerated response for changes in the measured variable. The degree of exaggeration is in proportion to the speed or rate at which the measured variable changes. (The term derivative action refers to the mathematical description of rate.) Derivative action is particularly effective in the slow processes, as with most temperature control loops. It compensates for lag or inertia. For a sudden change of even small magnitude, it provides an extra "kick" to the control valve because it recognizes that, with the lag that exists, even a small sudden change in measured variable signifies that a considerable exchange of energy has taken place and that the situation is likely to get worse before getting better. Conversely, as it drives the process toward set point, it begins anticipating the inertial effect and begins cutting back the valve response accordingly.

The simple method of achieving derivative action in Figure 2.6a closely resembles some of the approaches used in practice, but it has some serious limitations. In this method, the derivative action interacts with the proportional and reset responses and, in fact, follows the proportional response. It is therefore useless in preventing overshoot on start-up and on large upsets, as will be described later under special batch controllers. The derivative also interacts with set-point changes. An independent derivative unit is shown in Figure 2.6e and is described later.

Derivative action can be considered to temporarily rotate the proportional band lines in Figure 2.6b clockwise in response to rate of change, i.e., derivative temporarily narrows the band. Derivative time is the time in minutes by which the output would lead the feedback pressure during a steady ramp input change. This is described in greater detail below in the discussion on derivative relay.

Miniature Receiver Controllers

Two actual designs of force-balance receiver controllers are shown in Figures 2.6c and d. The controller in Figure 2.6c closely resembles that of Figure 2.6a, except that the bellows all act from one side against a pivoted "wobble plate." The wobble plate acts as the nozzle baffle. Rotating the pivot axis changes the proportional band. When the pivot axis coincides with the reset and feed-

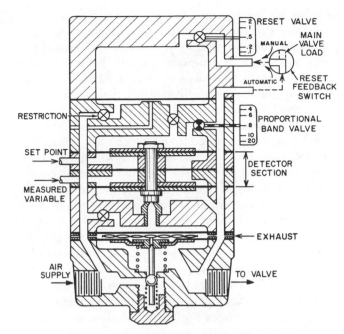

Fig. 2.6d Force-balance controller

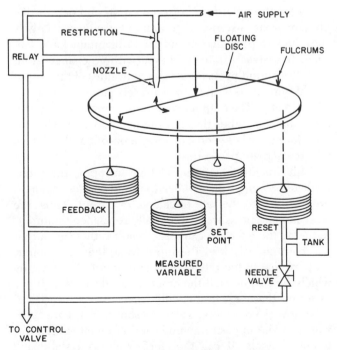

Fig. 2.6c Typical moment-balance controller

back bellows, 0 percent proportional band results; when it coincides with the measured variable and set-point bellows, infinite band results. When the axis bisects the two bellows axis, 100 percent proportional band results. Otherwise, the operation is the same as described for Figure 2.6a.

The controller in Figure 2.6d is constructed of machined aluminum rings, separated by rubber diaphragms with bolts holding the assembly together. The lower portion of the controller forms the booster section. This is quite similar to the booster in Figure 2.6a. The detector section consists of three diaphragms. The upper and lower diaphragms have equal areas while the center diaphragm has half the effective area of the other two. Reset pressure acts in the top chamber. Assume this pressure is at midscale and that the reset needle valve is closed. The reset pressure acts on the top diaphragm, which is part of a 1:1 reproducing relay. Supply air passes through a restriction and out the exhaust nozzle. The diaphragm baffles the nozzle to make the back pressure equal to the reset pressure. Assume further that the proportional band needle valve is closed. The pressure then acting on top of the detector section will be the reset pressure as reproduced by the 1:1 relay, since the two chambers are connected via a restriction. If the measured variable signal equals the set point, the detector diaphragm assembly, with its integral nozzle seat, will baffle off the nozzle so as to make the controller output, or valve pressure, equal to the reset pressure, thus balancing all of the forces acting on the detector. If the measured variable then increases by 1 PSI (0.067 bar), the increase in pressure acts downward on the lower diaphragm as well as upward on the center diaphragm. The net effect is the same as having the pressure act downward on a diaphragm of half the area of the lower diaphragm. Since the output pressure acts upward on the full area of the lower diaphragm, it need increase only $\frac{1}{2}$ PSI (0.03 bar) to bring the forces to balance. Since a 1 PSI (0.067 bar) change in input resulted in a $\frac{1}{2}$ PSI change in output, the proportional band is said to be at 200 percent.

If the proportional band needle valve were wide open, so that it provides negligible resistance, then if the measured variable pressure increases ever so slightly above set point, the full effect of the resultant change in output will be felt on top of the detector stack. This causes the output to increase further, which in turn feeds upon itself and the action continues to regenerate until the output reaches its maximum limit. Therefore, with the proportional band needle valve wide open, the narrowest proportional band is obtained.

If the proportional band needle valve is set to where its resistance equals that of the restriction separating the reproducing relay from the top of the detector section, then the following action results. If the measured variable deviates from set point by 1 PSI (0.067 bar) in an increasing direction, instantly the output will rise $\frac{1}{2}$ PSI (0.03 bar) because of the construction of the detector section. The difference in pressure between the controller output and the reproducing relay will cause a flow through the reset needle valve and the intermediate restriction. Since the resistance of the two is equal, the pressure drop will divide equally, causing a $\frac{1}{4}$ PSI (0.017 bar) increase on the top of the detector section. This $\frac{1}{4}$ PSI increase directly causes the output to increase $\frac{1}{4}$ PSI, which further causes the pressure on top of the detector to increase by $\frac{1}{8}$ PSI (0.008 bar). The action continues until equilibrium is obtained, with the output having changed a total of 1 PSI (0.067 bar) and with the pressure on top of the detector section having increased $\frac{1}{2}$ PSI (0.03 bar). Since a 1 PSI change in variable resulted in a 1 PSI change in output, this needle valve opening provides a 100 percent proportional band.

The reset action in this controller is similar to that of Figure 2.6a in that any change in reset pressure propagates down through the unit to directly affect the output. The controller will not come to equilibrium until all forces are balanced, i.e., the measured variable will have to equal set point and the reset pressure will have to equal controller output.

Derivative Relay

It was noted earlier that the type of derivative circuit used in Figure 2.6a had definite limitations. Derivative action is more effective if it is noninteracting and if it can be applied *ahead* of the proportional and reset action of the controller. Such a derivative unit can be built into the controller, or it can take the form of a separate relay as shown in Figure 2.6e. This relay employs diaphragms, but the design can be executed just as well with bellows.

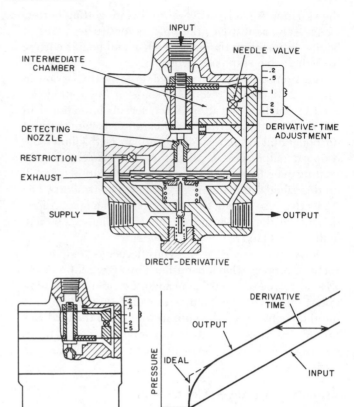

Fig. 2.6e   Direct and inverse derivative relays

The signal from some process transmitter, typically a temperature transmitter, would be connected to the input of the relay.

Due to the difference between input- and output-diaphragm areas, a step change in input produces a largely amplified step change in output to maintain force balance. In a steady-state condition (with no change in input), the output pressure acts on both sides of the output diaphragm—and output pressure rebalances input pressure directly. The output pressure is connected to the intermediate chamber through the needle valve. There is therefore a lag between a change in output and a change in intermediate pressure.

With a continuous ramp change in input, the intermediate pressure will lag the output by a constant amount, proportional to the rate of change in output. Thus, the intermediate pressure will partially rebalance the input, reducing the effective gain. The result is that the output will continuously lead the input by a definite amount, proportional to the rate of change in input. The time by which the output leads the input is the "derivative time," as set on the graduated needle valve.

An inverse derivative relay, as shown in Figure 2.6e, works in the opposite manner and attenuates high-frequency signals. It can therefore serve as a stabilizing relay in "noisy" processes.

Miniature Control Stations

Miniature control stations having a panel face of nominally 6 × 6 in. (150 × 150 mm) and inserted into individual cutouts having approximately 10 in. (250 mm) center-to-center distances are one of the common types found on central control room panels in the various process industries today. Figure 2.6f shows a typical cross section of some of the types of units available.

A typical miniature indicating control station is shown in Figure 2.6g. The measured variable is indicated on the center pointer, and the set point is indicated on the peripheral pointer of a duplex gauge. On automatic, the operator changes the set point by adjusting the set-point regulator and noting the set point on the gauge. The controller is connected to the control valve via the manual–automatic switch. The controller compares the measured variable with the set point and manipulates the control valve to bring the variable on set point. If the operator wishes to switch to manual, he notes the valve pressure by operating the upper left-hand switch on the station. He next turns the right-hand switch to "seal," which isolates the controller from the control valve. He then adjusts the regulator to match the noted valve pressure and then turns the right-hand switch to the manual position, which connects the regulator directly to the valve. In manual, the operator directly adjusts the valve while the controller reset follows whatever changes are made to the valve. In switching back to automatic, the operator goes to "seal" position and adjusts the set-point regulator to match the measured variable. If the set point and measured variable are equal at switchover, and if the reset is equal to the valve pressure, the controller output should then equal the valve pressure, and the switchover is effected without a "bump."

Four-Pipe System

Since there is a lag in the transmission of pneumatic signals, the dynamic capability of a control loop can be affected by increasing distance between the controller and the process. (The subject of the effect of transmission distance on control is covered in more detail at the end of this section.) However, the transmission lag will only be significant on the fast processes such as liquid flow control and then only as the distance exceeds 300 ft (90 m). For circuits where this lag would prove objectionable, the type of control station shown in Figure 2.6h is employed. Here the controller is mounted in the field near the measuring transmitter and control valve. Four connections are run between the control station and the field-mounted equipment. These involve the measured variable, set point, valve pressure, and relay operating pressure lines. Hence, the name "four-pipe system" is most often applied to characterize this circuit, whereas the circuit in Figure 2.6g is often referred to as a "two-pipe system." Since the lines going back to the station amount to dead-ended parallel connections, the dynam-

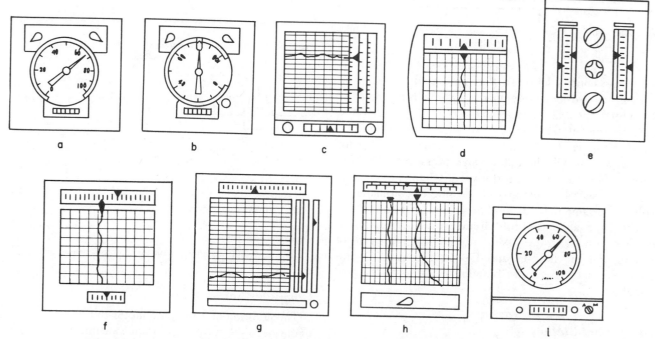

Fig. 2.6f Types of miniature pneumatic controllers. (a) Typical indicating control station. (b) Indicating control station with 12 o'clock scanning feature. (c) Recording control station with 30-day strip-chart and vertical moving pen. (d) Recording control station with horizontal moving pen and daily chart tear-off feature. (e) Indicating control station with two duplex vertical scale indicators. (f) Recording control station with no "seal" position. (g) Recording control station with servo-operated pen. (h) Recording control station with procedureless switching. (i) Indicating control station with instant procedureless switching.

S - AIR SUPPLY
SP - SET POINT
MV - MEASURED VARIABLE
CO - CONTROLLER OUTPUT
R - RESET FEEDBACK
C - CONTROLLER
------ MECHANICAL CONNECTION
⫻⫻⫻ PNEUMATIC CONNECTION
M/A - MANUAL - AUTOMATIC

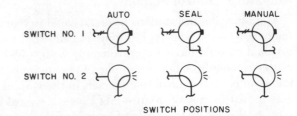

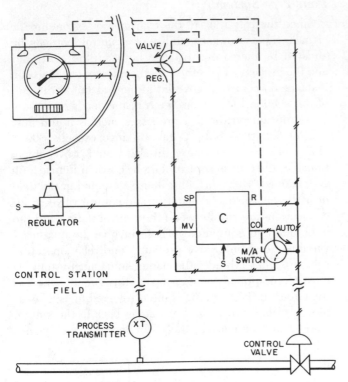

Fig. 2.6g   Miniature control station with integral-mounted controller

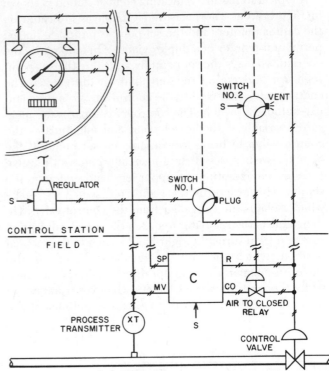

Fig. 2.6h   Miniature control station with field-mounted controller (four-pipe system)

ics of the control loop are the same as they would be with any closed-loop system. In switching from automatic to manual, the operator goes to the "seal" position, which actuates the cutoff relay to isolate the controller from the valve and permits the operator to change the set-point regulator to match the noted valve pressure. This pressure is then connected to the valve when the operator turns the switch to manual. Returning to automatic involves the same procedure as described for Figure 2.6g.

The disadvantages of this circuit are that (1) it is more costly to run four transmission tubes between the control station and the field-mounted equipment and (2) the controller settings cannot be adjusted from the control panel.

Remote-Set Station

If a modification as shown in Figure 2.6i is added to the basic station in Figure 2.6g, the control station can accommodate a remote set-point signal from sources such as a ratio or proportioning relay, primary cascade controller, or analog computer.

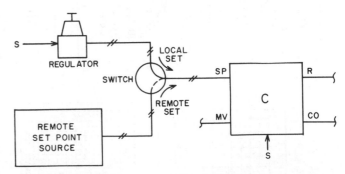

Fig. 2.6i   Control station circuit for remote set-point adjustment

Computer-Set Station

The addition of a stepping motor to the set-point regulator or to the set-point motion transmitter as in Figure 2.6j allows the control station to be set from a digital computer. The station still allows the computer to be disconnected and permits the set point to be adjusted locally and provides direct manual control. The stepper motor can be driven by time-duration signals or individ-

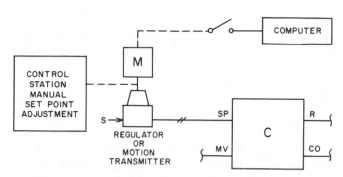

Fig. 2.6j Control station circuit for computer-adjusted set-point control

ual up-down pulses. A resolution of 1000 pulses full scale is typical.

Single-Station Cascade

Cascade control can be implemented either with two stations, such as Figure 2.6g on the primary and Figure 2.6i on the secondary, or with a single station, such as shown in Figure 2.6k. The latter scheme not only eliminates one station, but offers operating safety and convenience as well. A common problem when using two stations is that when the operator switches to manual, which he does on the secondary station, he often forgets to switch the primary station to manual as well. The primary controller then wanders around aimlessly and is not balanced for switchover back to cascade. Compounding the problem is the fact that cascade circuits are often employed on the most critical loops.

In Figure 2.6k, the regulator has three functions: (1) set point to primary controller in cascade control, (2) set point to secondary controller for independent secondary

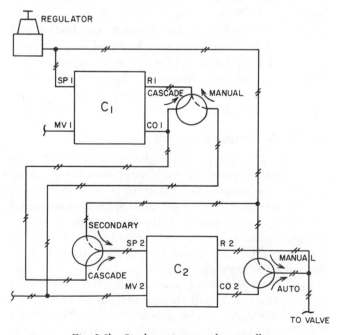

Fig. 2.6k Single-station cascade controller

control, and (3) manual valve setting. There is a seal position between each step while the regulator is set for its upcoming function. The key to this station is the concept used in making the secondary measured variable (MV2) the reset feedback of the primary controller while on manual or secondary control. Versions are also available that allow cascade, independent automatic control on the primary and manual modes of operation.

"No Seal" Station

The use of two regulators or motion transmitters in a station, Figure 2.6l, eliminates the need for a "seal" position. When the operator wishes to switch to manual, he adjusts the manual regulator to match the controller output while viewing a deviation indicator. When they are aligned, he transfers control.

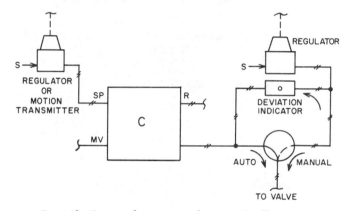

Fig. 2.6l Two-regulator station eliminates "seal" position

Procedureless Switching Station

Two methods of procedureless switching have been introduced in conventional miniature stations. With these, the operator simply turns a switch and the station automatically takes care of pressure-balancing problems. The two approaches are shown in Figure 2.6m. The mechanism on the left is a combination motion transmitter/receiver.

The motion transmitter provides set-point pressure in automatic and valve pressure in manual—just as the regulator does in Figure 2.6g. As a motion transmitter, a friction clutch holds the index lever at whatever position the operator sets it. A restriction-nozzle circuit senses the position and converts it to a proportional pneumatic output that is fed back to the rebalancing bellows. A 0 to 100 percent movement of the index gives a 3–15 PSIG (0.2–1.0 bar) output pressure. When acting as a receiver, supply pressure is cut off the restriction nozzle circuit and the pressure to be sensed is admitted to the rebalancing bellows. The friction clutch is disconnected so that the index lever can be moved by the rebalancing bellows. The unit is so designed and calibrated that a 3–15 PSIG (0.2–1.0 bar) sensed pressure produces a 0 to 100 percent index movement. Therefore, as the operator moves the

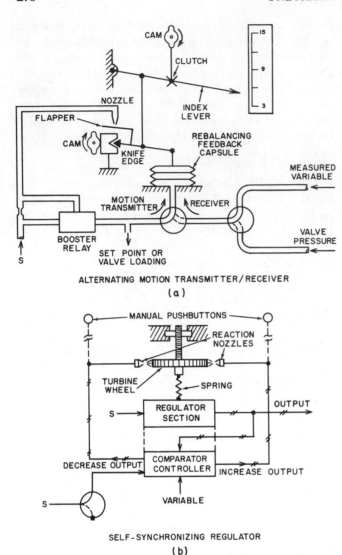

ALTERNATING MOTION TRANSMITTER/RECEIVER
(a)

SELF-SYNCHRONIZING REGULATOR
(b)

Fig. 2.6m   Two approaches to self-balancing, procedureless switching control stations

switch from automatic to manual, the following actions take place in sequence and automatically: The index is declutched; supply is cut off the restriction nozzle circuit; controller output (valve) pressure is admitted to the bellows causing the index lever to take a position proportional to the pressure; controller output is disconnected from the valve line; the clutch is engaged; supply pressure is readmitted to the restriction nozzle circuit; and the unit again acts as a motion transmitter, now providing valve pressure. Switching back to automatic involves the same sequence, in reverse. This is the same sequence as carried out in Figure 2.6g, except that here it is automatic.

A second system for procedureless switching involves the use of self-synchronizing regulators, or syncros, by which one regulator provides set point while the other is used for manual valve loading (Figure 2.6l). As the name implies, this regulator can synchronize itself to

some varying pneumatic pressure and thereby provide automatic balancing. The regulator employs a reaction nozzle circuit that results in very low spring force (approximately 1 oz or 0.28 N) to develop a 3–15 PSIG (0.2–1 bar) output. The setting spring is adjusted by rotation of a turbine wheel with an integral lead screw (Figure 2.6m). If supply is connected to the comparator controller section, air is transmitted to the increase–decrease nozzles to make the regulator section output match the variable input pressure. If supply is cut off from the comparator controller, the regulator section output remains locked in, with the memory being a function of lead screw position. The unit can then be driven manually by the operator.

On the station, when in automatic, the set-point syncro is manually adjusted by the operator, while the valve-operating syncro keeps itself matched to the controller output to allow instant transfer to manual. In manual, the operator adjusts the valve-loading syncro while the set-point syncro tracks the measured variable.

In addition to procedureless switching, these stations can be switched from a remote location or source, manually or automatically, they can be gang-switched, and they make possible the operation of stations connected in parallel on one loop as, for example, having one station in the central control room and the other in the field local to the process. Whichever station is not in active service keeps itself fully synchronized and ready to be made active at any instant.

Miniature High-Density Stations

Miniature high-density stations represent the latest developments in pneumatics. The stations have a typical panel size of 3 × 6 in. (7.5 × 150 mm), mount adjacent to each other, and allow very compact and efficient panel arrangements (Figure 2.6n). The units incorporate novel packaging features that simplify panel construction and design and facilitate servicing. Much of the design is aimed at making the job of the operator simpler, faster, and safer, in line with the present trend to consolidate control rooms, minimize the number of operators, and handle increasingly fast, complex, and critical processes.

Mid-Scale Scanning Station

Figures 2.6o, p and q show three types of high-density control stations featuring a mid-scale deviation scanning pointer. The pointer is driven either by a differential detector or differential servo that compares the measured variable against set point. If the two are equal, the red deviation pointer is positioned at mid-scale, where it is screened off by a green scan band. If there is a deviation, the red pointer stands out prominently.

In Figure 2.6o, a fixed, nominal 4-in. (100 mm) vertical scale is employed, and there are separate pointers to indicate set point and measured variable. The station uses the two-regulator approach to achieve "no seal"

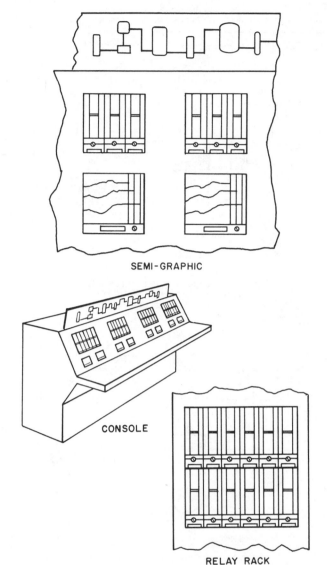

Fig. 2.6n   Typical mounting arrangements of high-density control stations

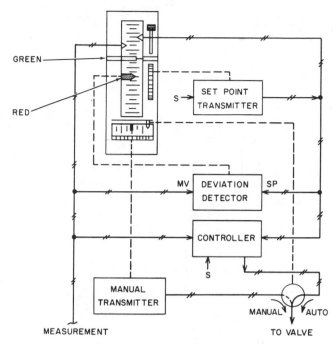

Fig. 2.6o   Functional diagram of high-density station with mid-scale scanning and individual indication of set point and measured variable

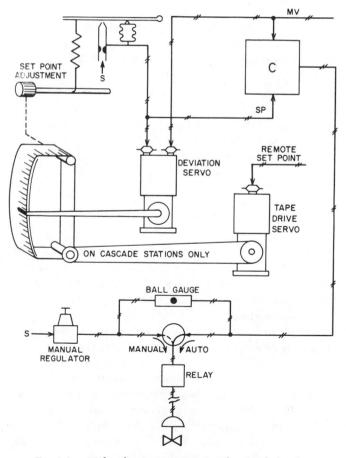

Fig. 2.6p   Mid-scale scanning station with expanded scale

switching as in Figure 2.6l. The operator does have to balance pressures before switching, however.

In Figure 2.6p, the station employs an expanded scale, which provides greater readability. The only indication on the scale is deviation, however, and this requires that the set-point transmitter scale and deviation servo stay in calibration relative to each other in order to provide an accurate reading of the variable. While the expanded scale gives greater readability, it does have to be moved to bring the reading on scale when the variable makes any excursions. This station and the one in Figure 2.6q use the two-regulator approach to eliminate the need for a seal position. In both cases the operator balances pressures prior to switchover. In Figure 2.6p, the operator notes deviation on a ball-in-tube indicator. In Figure 2.6q, the valve switch has a detent action while the indicator switch operates at the mid-throw position, so that

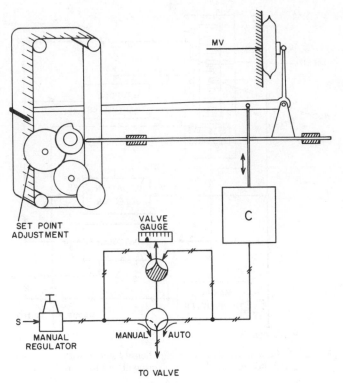

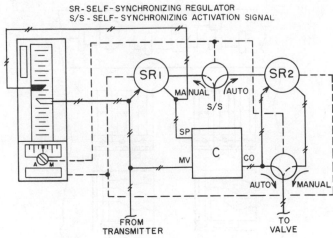

Fig. 2.6q Mid-scale scanning station, expanded scale, and deviation controller

Fig. 2.6r Self-synchronizing control station with procedureless switching

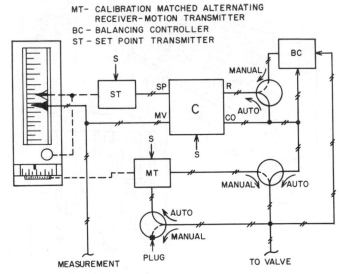

Fig. 2.6s Self-balancing control station

the operator moves the integral switch lever back and forth across center while manually matching pressures before switching. The controller in Figure 2.6q is a deviation type actuated by displacement of a deviation link in the indicator circuit. The controller acts to hold the link at its "zero" position.

While thousands of these control stations are found in control rooms, mid-scale scanning stations have evolved into a type combining mid-scale scanning with procedureless switching, using a principle similar to that shown in Figure 2.6s.

Procedureless Switching Station

Stations in Figures 2.6r and s offer procedureless switching for the operator. Both have a fixed 4-in. (100 mm) scale, separate indication of set point and measured variable, and a scanning concept in which the set-point indicator overlaps the measured variable indicator when control is normal.

The station in Figure 2.6r employs two self-synchronizing regulators or syncros (see Figure 2.6m), one for set point and the other for manual valve loading. On automatic, the operator manually adjusts the set-point syncro, while the valve-loading syncro automatically tracks controller output. On manual, the operator adjusts the valve-loading syncro while the set-point syncro tracks the measured variable. Thus the station is always balanced for instant transfer of control mode.

Like its 6 × 6-in. (150 × 150 mm) counterpart, these stations can be gang-switched, switched remotely or automatically, and operated in parallel from different locations while maintaining themselves in syncronism, and they are available in single-station cascade arrangements.

The station in Figure 2.6s employs a dual function motion transmitter/receiver for manual valve loading and valve pressure indication. This unit is similar to that described in Figure 2.6m. On automatic, the index lever is declutched and the feedback capsule, connected to the valve pressure line, moves the index accordingly. On manual, the clutch engages, and the mechanism reverts to a motion transmitter that provides manual valve loading. This allows procedureless switching to manual. Switching to automatic is also procedureless, assuming that the set point of the process does not change. If the operator wishes the controller to operate at some new value, compared to where he had the process on manual, he must obviously change the set point to that value prior to switching to automatic. However, to facilitate switching to automatic on loops where the set point remains fixed, the station incorporates a separate balancing controller that operates while the station is in manual. The balancing controller manipulates controller reset pressure to keep the controller output equal to the manual valve loading even though the measured variable may be off set point. This allows the operator to switch to automatic while off the intended set point and yet have the pressures balanced at switchover and the system return to set point without overshoot. This feature of the circuit is somewhat limited on narrow proportional band applications, which is usually the case with slow processes, as it takes little deviation from set point before the reset would have to be at either a vacuum or considerably above supply to obtain the balance between valve loading and controller output.

Large-Case Receiver Controllers

Because of their size, large-case receiver controllers are not as popular as the miniature types. They nevertheless still find considerable use in plants and industries where the 24-hour circular chart is traditional and desirable.

Most large-case controllers operate on a displacement balance principle. The set point is a mechanical index setting. The measured variable acts on a pressure spring such as a bellows, spiral, or helix that moves the recorder pen or indicator pointer. A differential linkage detects any deviation between the index and pen position and actuates the flapper-nozzle system in an effort to bring the deviation to zero.

One example of large-case controllers is shown in Figure 2.6t. If the pen moves clockwise, the differential link moves upward, and the bell-crank moves the flapper toward the nozzle. The resultant increase in nozzle back pressure is reproduced by the relay, whose output is connected to the control valve and the proportioning bellows housing. The pressure increase is transmitted through the oil to the small inner bellows, causing the two connected inner bellows to move to the right. The spring in the left inner bellows compresses while the spring in the right distends. The motion also causes the large righthand bellows to move to the right against the housed spring. As the center rod joining the inner bellows moves to the right, the flapper is moved back away from the nozzle. This negative feedback results in proportioning action. The greater this negative feedback, the greater the change that will be required in the measured variable to obtain a given change in the valve. Adjusting the linkage to change the amount of this negative feedback changes the proportional band.

Opening the adjustable restriction between the two large bellows allows oil to flow from the bellows at higher pressure to the one at lower pressure. In this example, it would flow from the left to the right bellows, causing the inner bellows to move left, moving the flapper toward the nozzle and increasing output further. This action would continue to regenerate until the pen finally returned to the index where full balance would be achieved with the oil pressure being equal in the two large bellows and with the two inner bellows centered.

Large-case controllers are also available with remote air-operated set-point adjustment as required in cascade and ratio control.

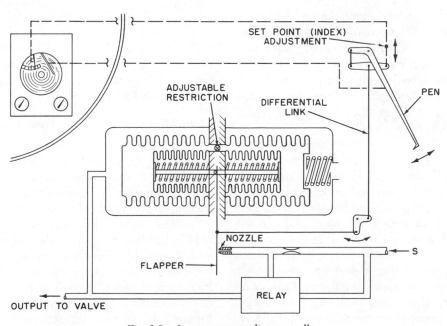

Fig. 2.6t   Large-case recording controller

Direct-Connected Large-Case Controllers

Direct-connected controllers have their own measuring elements which, as the term implies, are directly connected to the process. They therefore eliminate the need for a transmitter. However, the fact that direct-connected controllers must be located in the vicinity of the process limits their use to local rather than remote, central control rooms. Process connections must be run to the control room, which is costly, troublesome, and hazardous. These units are still used, however, usually on smaller installations and on local panels.

Direct-connected large-case controllers of the indicating and recording type predate the receiver type by 20 years. In fact, receiver controllers were simply the pressure controller version of the direct-connected controller type.

The operation of direct-connected large-case controllers is the same as that of large-case receiver controllers. The pen or pointer arm, instead of being actuated by pressure from a process transmitter, is actuated by its own built-in sensing system. A cross section of some of the measuring elements available with large-case controllers is shown in Figure 2.6u. These include elements for pressure, absolute pressure, draft, vacuum, differential pressure, liquid level, and filled systems for temperature measurement. Basic electrical measurements involved in thermocouples, resistance bulbs, radiation pyrometers, and pH probes are accommodated in the potentiometer versions of large-case pneumatic controllers.

As with the large-case receiver controller, the direct-connected type is also available with remote set-point adjustment for cascade and ratio control.

Direct-Connected, Field-Mounted Controllers

Direct-connected controllers mounted in the field, often termed pressure and temperature pilots, are smaller than the large-case instruments, come in a weatherproof case, and are available as indicating or blind types. They can be pipe-mounted, mounted directly on a valve or surface, or flush-mounted on a local panel. They include their own measuring element. Figure 2.6v shows some typical types and mounting arrangements.

These units are the least expensive pneumatic controllers, and their performance is considered less precise than those previously discussed. They find considerable use on small local installations and on many local field loops in larger plants. Every plant has such noncritical loops that do not require that the measured variable appear on the central control board, that the set point be adjusted from the board, and that it be possible to switch to manual. Combining a pressure pilot with a control valve makes a pressure regulator. Local level regulating loops are also quite common. Temperature pilots can be used to regulate temperature in preheaters, for example, where great precision is not essential.

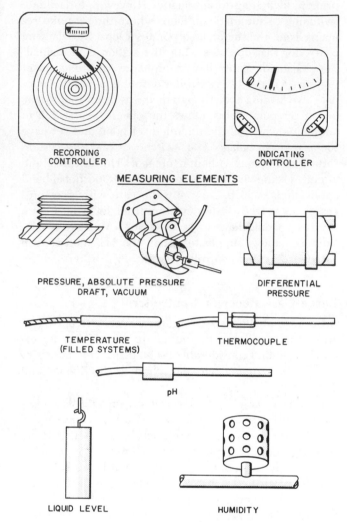

RECORDING CONTROLLER

INDICATING CONTROLLER

MEASURING ELEMENTS

PRESSURE, ABSOLUTE PRESSURE DRAFT, VACUUM

DIFFERENTIAL PRESSURE

TEMPERATURE (FILLED SYSTEMS)

THERMOCOUPLE

pH

LIQUID LEVEL

HUMIDITY

Fig. 2.6u   Direct-connected large-case recording and indicating controllers with examples of measuring elements

The pilots can be had with on-off, differential gap, proportional, reset, and derivative modes of control. The principle of operation is similar to that discussed in connection with Figures 2.6a and t.

Field-Mounted Receiver Type Controllers

In some applications, a field-mounted local controller is used, but it receives a signal from a transmitter rather than having its own measuring element. Such a controller would be applicable in cases where the signal must be transmitted to the control board for recording, alarm, or indication or where a measurement is made that is more readily handled by one of the great number of transmitters available. For these applications, there are field-mounted versions of the receiver controllers shown in Figures 2.6c and d as well as receiver versions of the direct-connected field-mounted controllers shown in Figure 2.6v. Remote adjustment of set point is also an option with these controllers.

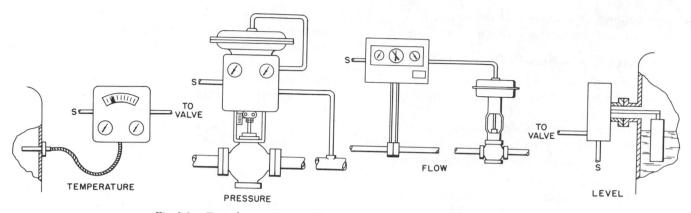

Fig. 2.6v Typical arrangements of direct-connected, locally mounted controllers

Pneumatic Controllers with Electrical Sensors

The temperature controller, illustrated in Figure 2.6w, receives a temperature measurement signal from either a thermocouple or a resistance temperature detector (RTD). These indicating pneumatic controllers provide the accuracy, convenience, and range of electrical temperature sensing, but without the need for external electrical power. A built-in generator accepts a conventional pneumatic pressure source and produces electrical power for the controller. The controller compares the process temperature signal with an operator-adjusted set point and delivers a pneumatic signal to a control element to change the process temperature toward set point. Process temperature and set point are indicated.

The controllers are available for proportional-plus-reset and proportional-plus-reset-plus-rate control, with or without a built-in bumpless and balanceless auto/manual operation capability.

Because the air supply is also used to generate the required electric power for the unit, the maximum "instantaneous" air supply requirement could reach 80 SCFH (2.3 normal m³/hr), while the maximum "steady state" air consumption is only 35 SCFH (1.0 normal m³/hr).

In this controller, a thermocouple or RTD temperature sensor senses the process temperature and produces an electrical signal that varies in proportion to the temperature. This signal is applied to the circuit board. The circuit board conditions the signal according to the settings of internal jumpers and adjustments, electrically compares the signal with the set point value, and acts on any difference with proportional and reset action to restore the process temperature to the set point value. The amount of proportional and reset action is determined by the setting of the proportional band and reset adjustments.

If the controller includes rate action, it additionally acts on changes in process variable. Rate action does not affect changes made in the set point, proportional, or

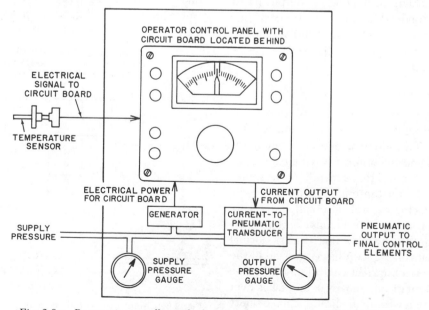

Fig. 2.6w Pneumatic controller with electronic sensors (Courtesy of Fisher Controls)

reset settings. The amount of rate action is set by the rate range selector and adjustment.

Pneumatic Controllers in Distributed Systems

With the advent of CRT-based distributed control systems, pneumatic controllers that incorporated microprocessor-based serial communications modules were introduced (Figure 2.6x). By thus making it possible to communicate over a data highway, pneumatic controls were made compatible with distributed systems.

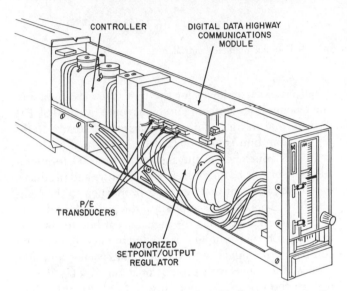

Fig. 2.6x   Pneumatic controller with digital highway communication module for use in microprocessor-based distributed control systems

The serial communications module reports, upon command, the current value of the set point, process variable, valve output, and the operating mode the station is in. The communications module also receives and executes commands to change set point or output or to change the operating mode. Miniature transducers convert the pneumatic signals to electric, from which the signals are then converted to digital. The serial link operates at 19.2 kilobaud.

Special Control Circuits

Feedforward, ratio, cascade, and other multiple loop systems are quite straightforward to implement with pneumatic hardware. Two circuits, one involving automatic selector control and the other involving batch control, receive enough attention that controllers are sometimes packaged specially for these applications.

With automatic selector control, also called override or limit control, two or more control loops are connected to a common valve in such a way that, under normal conditions, the normal control loop has command of the valve; however, if some abnormal condition arises, one of the other loops automatically moves in and takes over control to keep operation within safe limits. Unlike safety shutdown systems, normal control is only cut back as

much as necessary to stay within safe limits. When the abnormal condition abates, the normal loop resumes control. Figure 2.6y shows a system such as would be used on a booster pump station on one of the transcontinental pipelines. Normal control is on discharge pressure. If the suction pressure gets too low, however, as would be the case if the booster pump upstream failed or if a line rupture occurred, the discharge controller will open the valve wide, which will lower suction beyond the safe limit, causing cavitation, which could seriously damage the pump. To avoid this, a second loop, which senses suction pressure and whose controller set point is equivalent to the low safe limit, is coupled with the discharge pressure control loop by means of a low-pressure selector relay. Since the control valve operates air-to-open, the selector relay chooses and transmits to the valve the output of the controller that wants the valve more nearly closed. Under normal conditons, with adequate suction pressure, the discharge pressure controller will have the lower output and hence command the valve. If suction pressure drops to the set point, the suction pressure controller moves into control.

The key to correct implementation of this circuit is in making the reset feedback of both controllers common with the valve pressure. In this way the controlling unit has its reset acting normally while the stand-by controller is prevented from having its reset saturate or windup. Its reset should exactly match valve pressure at the instant it is to take over.

Figure 2.6z shows a system for batch control. The problem is that when a batch is completed and the controller is left on automatic with the manual valve closed, the reset keeps acting until it saturates, or winds up. When the system starts up again, if the controller has proportional and reset action only, or if the derivative unit is the interacting type that follows proportional and reset action, as in Figures 2.6a and c, the controller

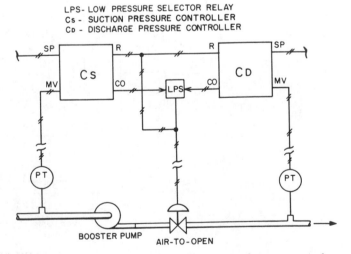

Fig. 2.6y   Automatic override or selector control circuit on pipeline booster pump

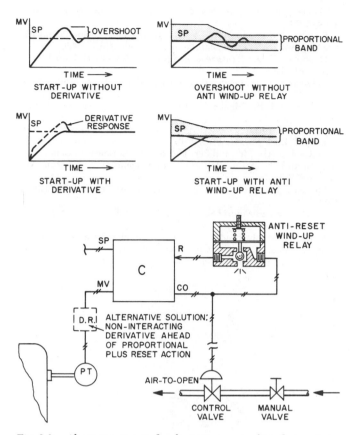

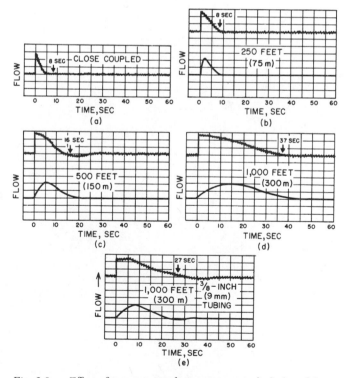

Fig. 2.6z Alternative circuits for eliminating overpeaking during start-up of batch processes

proper tuning. Too little derivative allows some overshoot; too much causes initial undershoot.

Effect of Transmission Distance on Control

Since there is a lag with pneumatic transmission, control is affected as the distance between the process and controller increases. Figure 2.6aa shows the result on recovery time as step load upsets of 10 percent are imposed on a liquid flow control system as the distance between controller and process increases. Since liquid flow control is one of the fastest processes, this amounts to a worst-case example and the effect on slower processes is proportionately less.

From the charts it can be seen that with all instruments close-coupled it took 8 seconds for the system to recover. At a transmission distance of 250 ft (75 m), the recovery time was still approximately 8 seconds. At 500 ft, (150 m) it was 16 seconds and at 1,000 feet (300 m), 37 seconds. These results were obtained using $\frac{1}{4}$-in. (6 mm)-OD tubing, which is conventional. When $\frac{3}{8}$-in. (9 mm) tubing was used, there was a significant improvement. At 1,000 ft (300 m), the recovery time was reduced from 37 to 27 seconds. An equivalent electronic control loop had an 8-second recovery time. The upper noisy record on charts (b), (c), (d), and (e) was of the flow as recorded at the transmitter. The lower smooth record was of the flow as

makes no effort whatever to move down from full scale until the error changes sign, i.e., the process crosses over set point. A considerable overshoot will obviously result.

One solution is an anti-reset windup relay. This is simply a throttling relay set to operate at 15 PSIG (1 bar), the wide-open position of the control valve. As long as the controller output is below 15 PSIG, the relay transmits the output to the reset feedback connection and reset acts normally. If the output goes above 15 PSIG, the relay begins exhausting the reset feedback line until it maintains the output at 15 PSIG. Thus it does not affect control, but when the system is shut down, it brings the reset down to whatever value it takes to limit output at 15 PSIG. This allows the proportional action to get into the act at start-up so that it can prevent overshoot. How effective it is depends upon proper tuning of the controller.

An alternative solution is to use either a separate derivative unit or a controller with a built-in derivative unit ahead of the proportional-plus-reset sections. These then act on the derivative modified signal. The derivative unit's output crosses over set point well ahead of the variable itself, and this starts the reset unwinding in time to prevent overshoot.

The effectiveness of this circuit also depends upon

Fig. 2.6aa Effect of transmission distance on control of a liquid flow control process (worst-case example) with 10 percent step upset. Upper noisy curve shows flow recorded locally; lower smooth curve shows flow recorded remotely at controller. (J. D. Warnock, "How Pneumatic Tubing Size Influences Controllability," *Instrumentation Technology,* February, 1967.)

it appeared on the recorder located remotely with the controller.

If this lag was objectionable, the four-pipe system shown in Figure 2.6h could be used, and the dynamic performance would be equivalent to that of the closed-loop system, or the installation of booster relays in the transmission lines could also be considered.

BIBLIOGRAPHY

Beard, M.J., "Analog Controllers Develop New Wrinkles," *Instruments & Control Systems*, November, 1977.

Buckley, P.S., "Dynamic Design of Pneumatic Control Loops: Parts I and II," *InTech*, April and June, 1975.

——, "A Modern Perspective on Controller Tuning," Texas A&M 30th Annual Symposium on Instrumentation for the Process Industries, January, 1973.

——, *Techniques of Process Control*, John Wiley & Sons, New York, 1964.

Buckley, P.S., and Luyben, W.L., "Designing Longline Pneumatic Control Systems," *Instrumentation Technology*, April, 1969.

Doebelin, E.D., *Measurement Systems*, McGraw-Hill Book Co., New York, 1975.

Eckman, D.P., *Automatic Process Control*, John Wiley & Sons, New York, 1958.

Goldfeder, L.B., "Analog Control Architecture: Integral vs. Split," *InTech*, April, 1977.

Jones, B.E., *Instrumentation, Measurement and Feedback*, McGraw-Hill Book Co., New York, 1977.

McCauley, A.P., Jr., and Persik, S.R., "From Linear to Gain-Adaptive Control . . . A Case History," *Instrumentation Technology*, November, 1977.

Shinskey, F.G., "Control Topics," Publ. no. 413-1 to 413-8, Foxboro Co.

——, *Process Control Systems*, McGraw-Hill Book Co., New York, 1979.

Street, C.W., Jr., and Calvin, E.L., "Fluidic Process Controller Goes On Line," *Control Engineering*, October, 1970.

Wightman, E.J., *Instrumentation in Process Control*, Butterworth, 1972.

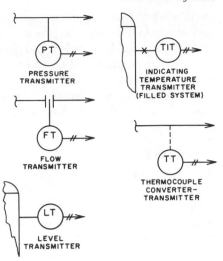

PRESSURE
TRANSMITTER

INDICATING
TEMPERATURE
TRANSMITTER
(FILLED SYSTEM)

FLOW
TRANSMITTER

THERMOCOUPLE
CONVERTER-
TRANSMITTER

LEVEL
TRANSMITTER

2.7 PNEUMATIC TRANSMITTERS

Ranges:	Force: 15 lbf to 180,000 lbf (66.75 N to 801,000 N)
	Level (buoyancy): 1″ to 60′ (25mm to 18m)
	Motion: $\frac{1}{8}$″ to 60″ (3mm to 1500mm)
	Pressure: 0.2″ H$_2$O to 80,000 PSIG (0.05 KPa to 552 Pa)
	Pressure Differential: 0.01″ H$_2$O to 1,500 PSID (2.5 Pa to 10.4 MPaD)
	Temperature Span (Filled): 50 to 1,000°F (28 to 555°C) within the range of absolute zero to 1,400°F (760°C)
Inaccuracy:	Generally $\frac{1}{4}$% to 1% of full scale
Cost:	$400 to $2000.
Partial List of Suppliers:	Ametec, Inc.; Bailey Controls Co., Babcock & Wilcox McDermott Co.; Brandt Industries, Inc.; Bristol Babcock, Inc.; Brooks Instrument, Div. of Emerson Electric Co.; Fischer & Porter Co.; Fisher Controls Co.; Foxboro Co.; Honeywell Inc., Process Control Div.; ITT Barton; Johnson Controls, Inc.; Kent Process Control, Inc.; Masoneilan Div., McGraw-Edison Co.; MKS Instruments, Inc.; Moore Products Co.; Robertshaw Controls Co.; Rochester Instrument Systems, Inc.; SEREG (Schlumberger Group), Controle Industrial; Taylor Instrument & Co.; Uniloc Div. of Rosemount, Inc.

For Current-to-Air Converters see Section 2.2.

A pneumatic transmitter is a device that senses some process variable and translates the measured value into an air pressure that is transmitted to various receiver devices for indication, recording, alarm, and control. Pneumatic controllers date back to the turn of the century. Pneumatic transmitters, however, did not make their appearance until the late 1930s—some 25 years after electric telemetering had become an established practice. Before pneumatic transmitters were introduced, controllers were all direct connected, i.e., they contained a measuring element that was connected to the process. This meant that the controllers and control boards had to be located close to the process.

Pneumatic transmitters were first developed as an alternative to expensive explosion-proof electric transmitters for use in medium-range signal transmission systems in refineries and chemical plants. It was quickly recognized that transmitters offered many advantages over the use of direct-connected controllers, recorders, and indicators, such as safety, economy, and convenience. They eliminate the need for connecting flammable, corrosive, toxic, and pressurized fluids into the control room. Furthermore, since controls can be located remotely, centralized control rooms become practical and such elements as long, gas-filled temperature-sensing bulbs with expensive armored capillary and with attendant bad ambient temperature errors and sensing lags become unnecessary. As a result, the process variable can be conveniently indicated, recorded, and controlled on relatively inexpensive standardized receiver devices. Once introduced, transmitters caught on quickly, and when miniature pneumatic controls were introduced in 1948, the concept was based on the use of pneumatic transmitters in all remotely controlled loops.

Signal Ranges

At first, each supplier settled on his own standard range of transmitter output. Generally, the span was selected to be compatible with commonly available pressure sensors and in some cases with the then commonly used operating ranges of pneumatic valve actuators. Too high a pressure would have placed extra demands on the piping and air supply system, whereas too small a span would have meant a sacrifice in resolution or accuracy. Also, the minimum pressure had to be some value above 0 PSIG (0 bar), since the detecting nozzle back pressure in a transmitter does not (theoretically) drop to atmospheric. In fact, the closer the nozzle back pressure approaches zero, the more critical the seating and the greater the change in baffle clearance per increment of output. This, in turn, results in greater error due to hysteresis and nonlinearity. Using a "live" zero provides an added benefit in that if a transmitter failed, the reading would drop below zero on the scale and give immediate evidence of failure.

Originally, ranges such as 2–14 PSIG, 3–18 PSIG, and 3–27 PSIG (0.13–0.93 bar, 0.2–1.2 bar, and 0.2–1.8 bar) were being used. The benefits from complete standardization became obvious, and by 1950 a 3–15 PSIG (0.2–1 bar) range was fast becoming the accepted standard. The 3–27 PSIG (0.2–1.8 bar) range had been used mostly in power plants and combustion-control systems, and it continued as one of the standard ranges. Formal recognition of standard ranges appeared in 1958 with the issuance of SAMA (Scientific Apparatus Makers Association) Standard RC2-11958. Three ranges were listed as standard: 3–15 PSIG, 3–27 PSIG, and 6–48 PSIG (0.2–1 bar, 0.2–1.8 bar and 0.4–3.2 bar). Today, however, 3–15 PSIG (0.2–1.0 bar) is overwhelmingly the most accepted.

Baffle-Nozzle Error Detector

The heart of practically all pneumatic transmitters and controllers is a baffle-nozzle error detector (Figure 2.7a). The circuit consists of a restriction, detecting nozzle, connecting chamber, and baffle. The baffle is effective as long as the clearance is within one-fourth of the inside diameter of the nozzle. Beyond this clearance, the annular escape area is greater than the area of the nozzle itself, and the baffle no longer provides a restrictive effect. The restriction must be small enough with respect to the nozzle so that with the baffle wide open the resultant nozzle back pressure is practically atmospheric.

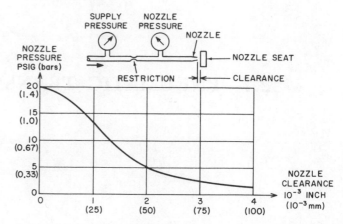

Fig. 2.7a   Relationship of nozzle seat clearance and nozzle back pressure

The restriction size also determines the continuous air consumption of the circuit, and it should be kept small for this reason. On the other hand, the restriction cannot be too small, for then it could clog easily with dirt or foreign matter that may be present in the supply air. Conversely, the nozzle should be large enough to give the proper minimum back pressure, but no so large that the clearance change for full-scale operation is so small as to require near-perfect seating. With this in mind, a typical restriction size might be 0.012 in. (0.3mm) ID and nozzle size 0.050 in. (1.3mm) ID. For such a general configuration, a plot of nozzle back pressure versus baffle clearance is given in Figure 2.7a.

Pneumatic detector circuits do take some other forms, some of which will be covered in the descriptions of individual transmitter types. The emphasis in this section is on the functioning of the pneumatic transmitter circuitry, since application and measurement details are covered in the volume on Process Measurement. It is not possible to cover all varieties of pneumatic transmitters in the space allotted, but the ones selected for discussion are typical and cover the basic types, so that, collectively, they represent a good cross section of the variation in design features.

Force-Balance Transmitters

One-to-One Repeaters

The force-balance principle is most commonly used in pneumatic transmitters. A very basic 1:1 force-balance transmitter is shown in Figure 2.7b. Other repeater designs are discussed in Sections 3.6, 3.7, and 5.5 of the Process Measurement volume of this *Handbook*. Process pressure acts downward on the flexible diaphragm, and the force resulting therefrom is counterbalanced by the force of nozzle back pressure acting upward on the diaphragm. Air, at a supply pressure slightly higher than the maximum process pressure to be measured, flows through the restriction to the underside of the diaphragm and bleeds through the nozzle to atmosphere. At equi-

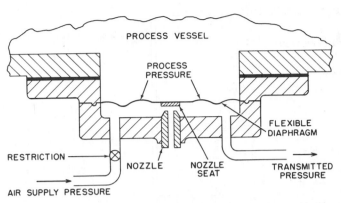

Fig. 2.7b Simplified 1:1 force-balance pressure transmitter with direct nozzle circuit

librium, the nozzle seat clearance is such that the flow of air in the nozzle is equal to the continuous flow through the restriction. If the process pressure increases, the diaphragm baffles off the nozzle, causing the back pressure to increase until a new equilibrium is achieved. If the process pressure drops, the diaphragm moves away from the nozzle, causing the back pressure to drop until equilibrium is reestablished. Since nozzle back pressure is directly related to process pressure, the signal can be used remotely for indication, recording, or control.

Though such a transmitter is accurate, it has some limitations. Namely, since all the air must flow through the restriction, the speed of response will be slow, particularly if there is much volume in the output side of the circuit. Also, a leak in the output side will cause an error or even make the unit completely inoperative. Figure 2.7c shows essentially the same repeater, but with

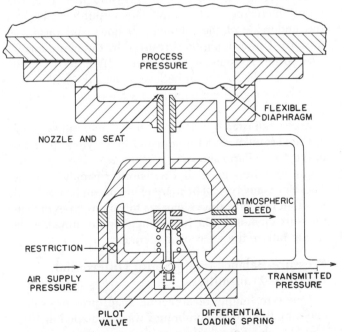

Fig. 2.7c One-to-one force-balance pressure transmitter with volume booster and constant differential nozzle circuit

refinements. This system employs a volume booster that has considerable air-handling capacity, so that it speeds up the response, minimizes transmission lag, and copes with leakage. In addition, it provides a constant pressure drop across the nozzle regardless of the level of operation. This improves accuracy with regard to linearity and hysteresis as the total diaphragm travel is lessened, and the relationship of nozzle seat position to nozzle back pressure is more linear than that of the direct nozzle circuit in Figure 2.7b.

The booster contains an exhaust diaphragm assembly consisting of two diaphragms that move integrally and that provide a bleed to atmosphere. On the underside of the diaphragm is a differential spring, which at balance exerts a force equivalent to 3 PSID (0.2 bar) acting on the diaphragm. Therefore, the nozzle back pressure that acts above the exhaust diaphragm is always nominally 3 PSI (0.2 bar) higher than the transmitted pressure. Since the nozzle bleeds into the transmitted pressure, the pressure drop across the nozzle is *always constant* regardless of output pressure. As in Figure 2.7b, the air pressure on the underside of the transmitter diaphragm counterbalances the process pressure. If the process pressure increases, the diaphragm moves the seat closer to the nozzle, increasing the nozzle back pressure. This moves the exhaust diaphragm downward, closing off the exhaust port as it contacts the pilot valve, and moves the pilot valve downward to open the supply port. The result is an increase in transmitted pressure until the transmitted pressure, which is fed back to the underside of the diaphragm, equals the process pressure. A decrease in process pressure causes the nozzle back pressure to drop, and the exhaust diaphragm moves upward so that the pilot valve closes off the supply port while the diaphragm opens the exhaust port, causing the transmitted pressure to drop until equilibrium is established. At balance, there is a continuous flow of air passing through the restriction, detection nozzle, and out to atmosphere via the exhaust diaphragm.

Pressure Transmitter (Force Balance)

The applications for pressure repeaters are limited. The more common transmitter is one which converts the measured process variable range into a standard 3–15 PSIG (0.2–1.0 bar) output pressure signal (Figure 2.7d). (Others are discussed in Chapter V of the Process Measurement volume.) It operates on a principle of force balance, or more precisely, on a principle of moment balance.

Process pressure acts on the input bellows and applies a force on the balancing beam, which rotates on the flexure pivot. A change in input pressure results in a moment that is counterbalanced by an equivalent change in moment as output pressure changes in the feedback bellows. An increase in process pressure rotates the beam counterclockwise and moves the nozzle seat closer to the

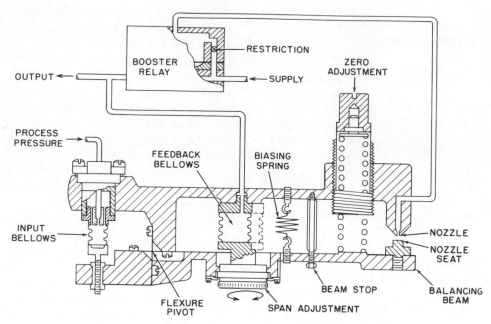

Fig. 2.7d   Force-balance pressure transmitter with variable range

nozzle, increasing the back pressure and booster output until the moments come into balance. The biasing spring assures sufficient counterclockwise moment on the balancing beam so that, even with zero process pressure, it is possible to set the output at 3 PSIG (0.2 bar) by adjusting the compression of the zero spring. Added compression of the zero spring gives an elevated zero, or range suppression.

The feedback bellows is eccentrically mounted on a rotatable seat. Rotating the assembly changes the moment arm, hence the span. The span is basically a function of the relative ratio of effective areas of the input bellows and the feedback bellows. The booster relay is an amplifying type that minimizes total nozzle clearance and improves accuracy.

Motion-Balance Pressure Transmitters

The pressure transmitter in Figure 2.7e consists of a pressure measuring element and a motion transmitter. Instead of a conventional baffle-nozzle, this transmitter employs an annular orifice with a variable restrictor called the wire pilot. Supply air passes through the fixed restriction into the follow-up bellows and out the detector nozzle. The wire pilot throttles the exhaust from the nozzle. It has a sharply tapered step so that when the large-diameter wire restricts the orifice, the back pressure rises to a maximum, and when the small diameter is effective the back pressure drops to atmospheric. At balance, therefore, the follow-up bellows moves to position the detector nozzle in line with the tapered step of the wire pilot.

Process pressure acts on the measuring diaphragm. An increase in process pressure moves the diaphragm upward, which, via the U-shaped linkage, moves the wire pilot upward. The wire pilot restricts the annular orifice and the back pressure increases. The two bellows that make up the follow-up bellows system have the same area and are connected rigidly by a center post so that nozzle back pressure has no effect on movement of the bellows assembly. The nozzle back pressure is connected to the top of the exhaust diaphragm assembly. This is an amplifying type diaphragm assembly, since the upper diaphragm has six times the effective area of the lower. Therefore, as the nozzle back pressure increases, the output increases in a 6:1 ratio. The output feeds back to the underside of the follow-up bellows and pushes it upward. Upward motion is resisted by the range spring. The spring constant is such that a 12 PSIG (0.8 bar) change on the follow-up bellows moves the bellows assembly through the nominal full-scale travel of the wire pilot.

Zero is adjusted by setting the initial operating position of the wire pilot. Span is adjusted by varying the radius of the take-off arm shown in section A-A. The total force required to operate such a transmitter is only 2 grams. Therefore, any type of primary element can be used with it—from low-force draft elements to various types of high-pressure elements. The basic range is determined by the spring rate of the measuring element.

Pressure-Balance (Solid State) Differential Pressure Transmitters

A new technique of detecting small pressure differences by the use of membranes was developed in 1972. The key element of this new device is the membrane amplifier. Membrane technology is a new pneumatic

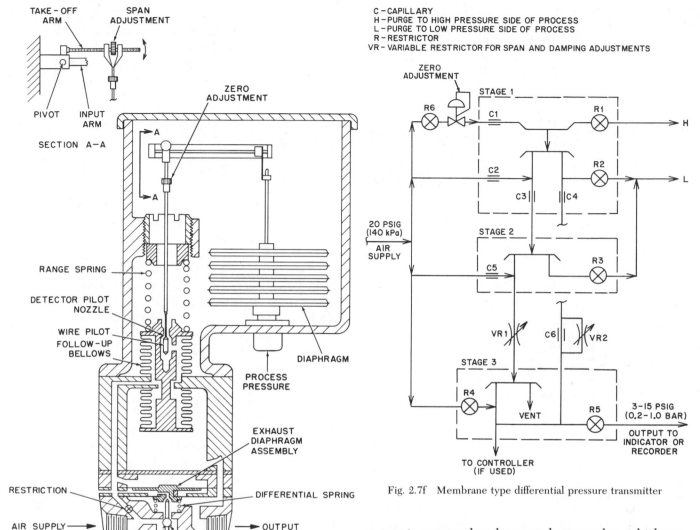

Fig. 2.7e Motion-balance type pressure transmitter

Fig. 2.7f Membrane type differential pressure transmitter

principle based upon pressure balance theory rather than force balance or diaphragm displacement. Since the membrane has virtually no movement or displacement during measurement, the valve has no moving or mechanical parts to wear or contribute to hysteresis. Thus, the membrane amplifier will accurately sense very low pressures at its input ports, yet it is insensitive to shock vibration, change in attitude, and mechanical problems. By definition, the membrane is virtually mass free and will not support a bending moment. Movement is limited to billionths of an inch (millionths of a millimeter).

A membrane type differential pressure transmitter utilizes three high-gain amplifiers, each with a specific gain according to the required span of the transmitter (Figure 2.7f). The first stage senses the input differential pressure to be measured, which may be as low as 0–0.01″ W.C. (0–2.5 Pa) full scale. It amplifies this pressure, and its

output is conveyed to the second-stage valve, which amplifies the signal further and conveys it to the third stage. This third stage amplifies the signal by a small amount, but its primary purpose is to provide "driving" power to supply a standard 3–15 PSIG (0.2–1.0 bar) transmitter output.

Transmitters Grouped by Measured Variable

Differential Pressure Transmitter

A typical force-balance differential pressure transmitter is shown in Figure 2.7g. The high and low pressures act on opposite sides of a diaphragm capsule, and the resulting differential exerts a force on the force bar. The force bar pivots on the diaphragm seal. The external end of the force bar pulls on one end of the range rod. The range rod, with its integral flapper, pivots about the range wheel. A feedback bellows acts on the opposite side of the range rod. A change in differential results in a changed flapper position, which alters the nozzle back pressure, relay output, and feedback force until all moments come to balance.

Zero is adjusted by adjusting the zero spring tension.

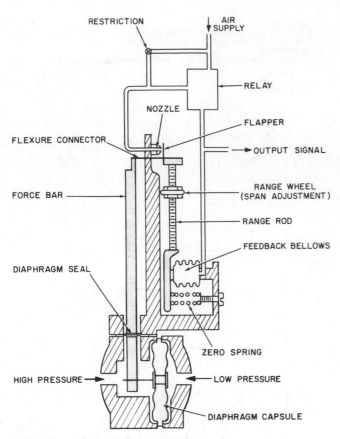

Fig. 2.7g Force-balance type differential pressure transmitter

ment of the bellows twists the torque tube, causing rotation of the torque tube shaft. It is this motion that actuates the transmitter. The bellows are filled with liquid (usually ethylene glycol), and as the bellows move, liquid is transferred from one bellows to the other via the pulsation dampener needle valve. If the normal differential pressure range is exceeded, the bellows move until an o-ring on the center shaft seals off the liquid in the bellows. The pressure can then build up to the body rating of the meter without damaging the unit.

The torque tube rotates the lever connector of the motion transmitter. A floating pilot link is socketed in the connector at one end and baffles off the detector nozzle at the other. The transmitter is calibrated so that at zero differential the output is at 3 PSIG (0.2 bar) and the pilot link is horizontal. An increase in differential rotates the connector counterclockwise. This restricts the nozzle and causes a build-up in transmitted pressure, which in turn acts on the large bellows, moving it downward until equilibrium is attained. Since the pilot link has a fixed length and nozzle clearance is for all practical purposes constant, the pilot nozzle must be moved downward according to the cosine law, i.e., at minimum input with link horizontal, there must be considerable motion of the pilot nozzle as compared with the lever connector, and therefore the gain is practically infinite. As the included angle increases, considerably less nozzle travel is required to offset lever connector motion. For small angles, the relationship is almost exactly square root.

Another transmitter having the same function employs a varying spring rate with travel. This is accomplished by picking up added leaf springs as the travel increases.

Rotameter Transmitter

When a rotameter is used as a primary flow-measuring device, a magnetic means of float position detection is utilized. The motion of the magnetic follower mechanism then is converted into a 3–15 PSIG (0.2–1.0 bar) signal. (Section 2.22 in the volume on Process Measurement contains a detailed discussion of rotameters.)

In Figure 2.7i, a permanent magnet is embedded in either the float or in an extension of the float. A magnetic steel helix is supported in an aluminum cylinder mounted between bearings. The leading edge of the helix is constantly attracted to the magnet. The vertical position of the float results in a corresponding radial position of the helix. Attached to the helix follower assembly is a cam. The profile of the cam is sensed by a pneumatic detector circuit consisting of a transmitting and a receiving nozzle. The receiving nozzle pressure serves the same function as nozzle back pressure in a conventional baffle-nozzle circuit. When the flow from the transmitting to receiving nozzle is not interrupted by the cam, the receiver pressure is a maximum. When fully interrupted, receiver pressure is 0 PSIG (Pa). At balance, the detector system is throttled by following the cam profile.

Span is adjusted by moving the range wheel, which changes the relative input/output moment arm ratio.

Since the output of the transmitter is linearly proportional to pressure differential, if the unit is used on an orifice type measurement, the flow would have to be read on a square root calibrated scale.

Square-Root-Extracting Differential Pressure Transmitter

There are differential pressure elements such as in Figure 2.7h that provide a motion output. A motion transmitter as in Figure 2.7e could be used with such a meter to give an output linearly proportional to the differential. However, the motion transmitter shown in Figure 2.7h converts linear motion of the meter into a square-root-related output so that orifice type flow can be read on a linear scale. Linear signals are preferred when flows are added, subtracted, or averaged and when other analog computing and characterizing requirements exist. They are often specified in order to give better readability and control rangeability.

The meter in Figure 2.7h consists of a high- and a low-pressure bellows joined by a common center shaft. The differential causes the bellows to move a linearly proportional amount depending upon the total spring rate of the range spring, the bellows, and torque tube. Move-

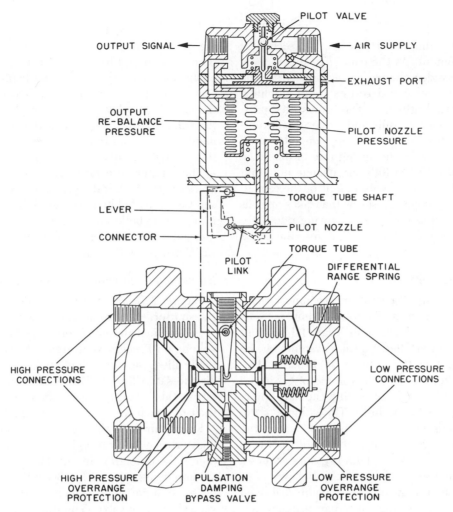

Fig. 2.7h   Linear flowmeter consisting of a combination of differential meter and square-root-extracting motion transmitter

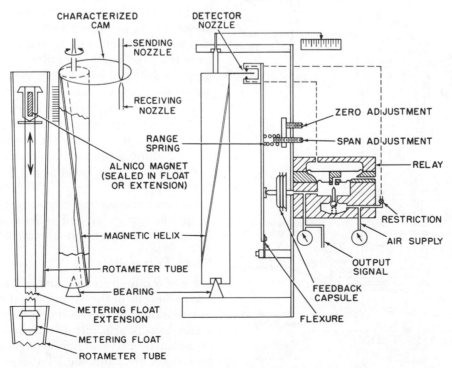

Fig. 2.7i   Rotameter transmitter with magnetic take-out and pneumatic cam follower mechanism

As the float rises, the cam rotates in a direction of decreasing displacement. As the cam edge moves away from the detector nozzles, the increasing receiver nozzle pressure acts on the relay and results in an amplified increase in transmitted pressure. The transmitted pressure acts on the feedback capsule and moves the flexure-mounted detector nozzles toward the cam edge until equilibrium is established. The spring rate of the range spring is set so that a 12 PSI (0.8 bar) change in output is required to track the full displacement of the cam.

Span is adjusted by turning a screw that takes up coils in the range spring, thus changing its spring rate. Zero is set by adjusting a second screw that sets the initial spring tension.

Temperature Transmitter (Filled)

Pneumatic temperature transmitters are almost exclusively the force-balance types. Figure 2.7j shows a transmitter with a sealed, gas-filled bulb (Class III system). (A detailed discussion of filled thermal systems can be found in Section 4.4 in the volume on Process Measurement.) Except for a negligible volume of gas in the thermal system bellows and connecting capillary, practically all the fill gas is in the bulb. The volume of the thermal system is constant, and as the bulb senses the process temperature, the pressure of the fill gas varies according to the gas laws. The fill gas pressure creates a force downward on the thermal system bellows. This force acts through the thrust rod and is counterbalanced by the force resulting from transmitted air pressure act-

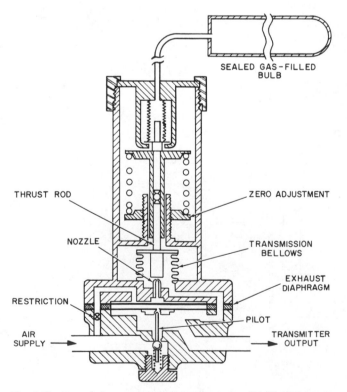

SEALED GAS-FILLED BULB

THRUST ROD

ZERO ADJUSTMENT

NOZZLE

TRANSMISSION BELLOWS

EXHAUST DIAPHRAGM

RESTRICTION

PILOT

AIR SUPPLY

TRANSMITTER OUTPUT

Fig. 2.7j   Force-balance temperature transmitter—filled bulb system

ing upward on the transmitter bellows. An increase in process temperature increases fill gas pressure, which pushes down on the thrust rod baffling off the nozzle. The consequent increase in nozzle back pressure pushes the exhaust diaphragm and pilot plunger down, closing the exhaust seat and opening the supply port until the increase in transmitted pressure acting on the transmission bellows results in force balance.

The temperature range is a function of the initial fill pressure (the higher the pressure, the narrower the span) and the ratio of effective areas of the thermal system and transmission bellows. The zero spring acts counter to the thermal system and establishes the low end of the measured temperature range.

Buoyancy Transmitter (Level or Density)

Figure 2.7k illustrates a motion balance buoyancy transmitter. (A more detailed discussion of displacement type level instruments is presented in Section 3.8 in the volume on Process Measurement.) Changes in level directly affect the net weight on the float lever as the float displaces the liquid. The float lever is connected to a torque tube. As the torque tube twists, it rotates a center shaft to which the flapper is attached.

Supply air flows through a restriction to the top of the booster relay and through a small tube inside the Bourdon to the detecting nozzle. The booster relay, which is an amplifying type with a 3:1 ratio, provides an output pressure, a portion of which is fed back to the Bourdon via a three-way valve. The three-way valve provides for span adjustment. If its plunger is moved up, closing off the exhaust seat, full transmitted pressure feeds back to the Bourdon. The result is that when the level rises and moves the flapper closer to the nozzle, the consequent increase in transmitted pressure makes the Bourdon move away from the nozzle (negative feedback) so that total flapper travel and hence measuring range is large. Adjusting the three-way valve in its other extreme position where it closes off transmitted pressure results in practically on-off action at the nozzle, representing the narrowest range of measurement. Normally the three-way valve is adjusted somewhere between these limits.

Zero is changed by rotating the Bourdon with respect to the flapper.

Buoyancy transmitters can be used to measure density or specific gravity as well as level and level interface. Force-balance designs are also available.

Force Transmitter

Force transmitters can serve as load cells in weighing applications or in the measurement of variables such as web tension. The transmitter in Figure 2.7l operates on a force-balance principle. (For a more detailed discussion of pneumatic load cells refer to Section 8.4 in the volume on Process Measurement.) Pressure from the tare regulator acts upward on the top diaphragm and is set to

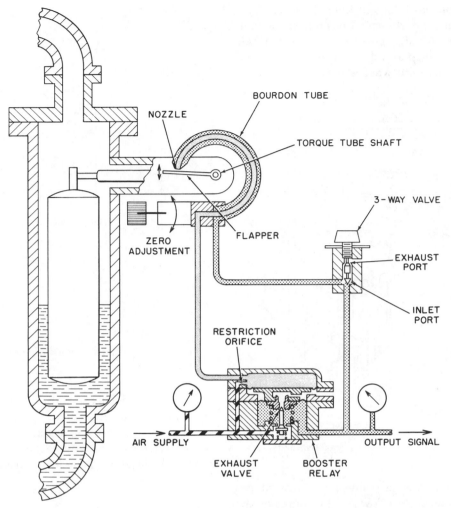

Fig. 2.7k Level transmitter with torque tube spring and pneumatic follower system

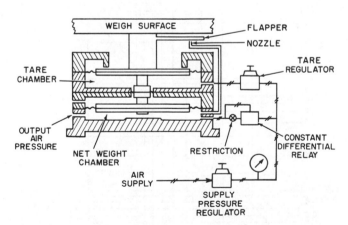

Fig. 2.7l Force-balance weight transmitter

counterbalance the fixed weight of a hopper or tank and support structure. Net weight then is counterbalanced by output air pressure acting under the bottom diaphragm. Air flows from supply through a constant differential relay, which maintains a relatively constant flow across a restriction and into the net weight chamber. The net weight chamber is connected to the detector nozzle. The nozzle baffle is attached to the supporting platform. If net weight increases, the platform and flapper move down, baffling off the nozzle and causing the back pressure to increase until force-balance is restored.

Smaller pneumatic force transmitters are also available. For stabilizing potentially noisy systems, some incorporate means of hydraulic pulsation dampening.

Motion Transmitter

A variety of approaches are used to detect motion, some of which are described in Section 9.2 in the volume on Process Measurement. A motion transmitter as shown in Figure 2.7e can be adapted to measure total motions from $\frac{1}{8}$ in. (3mm) to approximately 1 in. (25mm). This type is particularly useful where only a low force is available. Ordinary pressure regulators can be adapted as motion transmitters by substituting a sliding thrust rod for the lead screw that normally adjusts the pressure-setting spring. The most common approach, however, is to use a valve positioner and, in effect, reverse its function, i.e., instead of controlling position or motion, have it transmit a signal proportional to position or motion. Figure 2.7m shows a valve positioner connected as a motion transmitter. The pilot valve is reverse-acting, and the output feeds back into what is normally the input bellows.

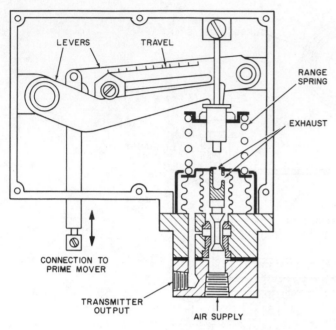

Fig. 2.7m   Valve positioner connected for motion transmitter service

If the prime mover pulls the connector downward, the parallel lever system pushes down and compresses the range spring. This opens the supply port and closes off the exhaust port of the three-way pilot, causing the output to increase. The output acts upward on the outer bellows to counteract the increase in spring force.

A roll type contact point establishes the gain of the lever system. Hence, its adjustment determines the span of measurement.

Speed Transmitter

In Figure 2.7n, a prime mover drives an input shaft that carries a multi-pole permanent magnet. The combination of magnetomotive pull and rotation of the magnet tends to turn the disc on its flexure pivot. The torque on the flexure is proportional to input shaft speed. Attached to the flexure is a radial force bar that doubles as a flapper and a rebalancing lever. Output pressure feeds back to a ball type piston, and the force derived counterbalances the input torque.

As speed increases, an increase in counterclockwise torque results. This moves the flapper toward the nozzle. The relay amplifies the nozzle back pressure change, which serves as the output and as feedback to the ball piston. The pressure increases until the feedback moment balances the input torque.

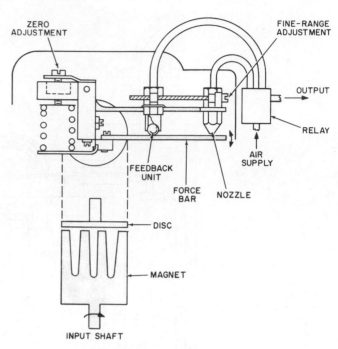

Fig. 2.7n   Speed transmitter—pneumatic force-balance type

Pneumatic speed transmitters are also made up by combining an electrical speed detector-amplifier with an electric-to-pneumatic converter. A typical detector is one that integrates the rate of magnetic pulses generated as gear teeth cut across the field of a magnetic pick-up head.

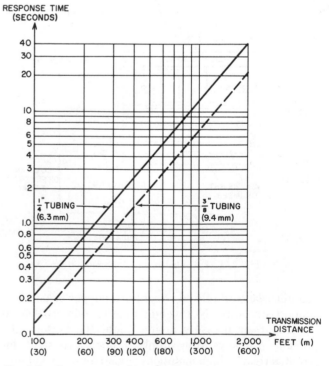

Fig. 2.7o   Response time (time for 63.2 percent complete response) versus transmission distance

(For more details refer to Section 9.9 of the volume on Process Measurement.)

Transmission Lag

With pneumatic transmitters, there is a transmission lag that increases with the length of transmission tubing. In order to get some idea of what this lag amounts to, the time for 63.2 percent complete recovery from a step input change is selected as the basis for comparison. The 63.2 percent figure defines the time constant in single-order systems, but since transmission systems do not behave as first-order systems, the figure cannot be so interpreted. Nevertheless, it does serve as an arbitrary benchmark for comparison.

A plot of response time versus transmission distance for $\frac{1}{4}$-in. and $\frac{3}{8}$-in. (6.3mm and 9.4mm) tubing is given in Figure 2.7o. Response is faster with $\frac{3}{8}$-in. tubing, but $\frac{1}{4}$-in. tubing is much more conventionally used. At 300 ft (90 m), for example, the response time for $\frac{1}{4}$-in. tubing is 1.5 seconds and for $\frac{3}{8}$-in. tubing, 0.8 seconds.

The effect of transmission distance on pneumatic control is also discussed in Sections 2.6 and 2.8.

BIBLIOGRAPHY

Adams, L.F., *Engineering Measurements and Instrumentation*, The English Universities Press, 1975.

Buckley, P.S., "Dynamic Design of Pneumatic Control Loops," *InTech*, April, 1975.

———, *Techniques of Process Control*, John Wiley & Sons, New York, 1964.

———, "A Modern Perspective on Controller Tuning," Texas A&M 30th Annual Symposium on Instrumentation for the Process Industries, January, 1973.

Buckley, P.S., and Luyben, W.L., "Designing Long-Line Pneumatic Control Systems," *Instrumentation Technology*, April, 1969.

Demarest, W.J., "There's More to Transmitter Accuracy Than the Spec," *Instruments and Control Systems*, May, 1983.

Doebelin, E.O., *Measurement Systems—Application and Design*, rev. ed., McGraw-Hill Book Co., New York, 1975.

Eckman, D.P., *Automatic Process Control*, John Wiley & Sons, New York, 1958.

Gore, F., "When to Use a Square Root Extractor," *Instruments & Control Systems*, April, 1977.

Harriott, P., *Process Control*, McGraw-Hill Book Co., New York, 1964.

Jones, B.E., *Instrumentation, Measurement and Feedback*, McGraw-Hill Book Co., New York, 1977.

Vannah, W.E., and Catheron, A.R., "Improved Flow Control with Long Lines," Paper 51-6-2, *Proceedings of the Sixth National Instrument Conference and Exhibit*, September, 1951.

Warnock, J.D., "How Pneumatic Tuning Size Influences Controlability," *Instrumentation Technology*, February, 1967.

Wightman, E.J., *Instrumentation in Process Control*, Butterworth, 1972.

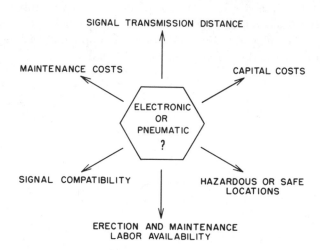

2.8 PNEUMATIC VERSUS ELECTRONIC INSTRUMENTS

Pneumatic systems have advantages over electronic instruments because of their safety in hazardous locations and the ready availability of qualified maintenance personnel. However, electronic systems are usually favored when computer compatibility, long transmission distances, or very cold ambient conditions are important factors or when long-term maintenance is to be minimized. (See Figure 2.8a.)

The old debate of when to use pneumatic and when to use electronic instruments is still with us. The pros and cons discussed in old and new articles have not changed very much, and no short cut to reaching a final decision has been developed. Consequently, this section can only review current opinion on the subject.

Perhaps "Pneumatics versus High Technology" would have been a better title for this section. If a process plant uses digital electronics, it is very difficult to justify pneumatics. In addition to analog electronics, other designs must also be considered, such as the digital controllers with or without computers, distributed control, and programmable controller hardware. In the following discussion, references are presented so the reader can study the specific topics in detail in other sources as well as other sections of this *Handbook*.

The Traditional Debate

Some authors promote electronics[1-4] and others pneumatics[5-9], and this division of opinion will probably persist for some time to come. Table 2.8b summarizes the views of nine articles that cover both pneumatic and electronic instruments.

In the area of maintenance, the literature[1-18] does not present strong enough evidence that electronics guarantees lower maintenance costs. Electronic instrumen-

Fig. 2.8a   Electronic versus pneumatic systems

tation is probably easier to maintain, or soon will be. Maintenance publications, such as those of ISA, can hasten this trend.

It is a misconception that pneumatic instruments need dry clean air—but electronic systems certainly do not, because the I/P (current-to-air) converters and electropneumatic valve positioners must have the same quality air. Deuschle,[14] in his detailed cost comparison, points out that compressor capacity for the electronic system is about 60 percent of that required for a pneumatic system. The rating approach proposed by Tompkins[17] is a quick way to assess pneumatics and electronics for a specific application.

Signals and Transmission

Putting together a pneumatic system can be relatively simple. The 3–15 PSIG (0.2–1.0 bar) signal range per-

Table 2.8b
VIEWPOINTS EXPRESSED ON ELECTRONICS VERSUS PNEUMATICS†

Features	Electronics Superior	Pneumatics Superior	About Equal	Remarks
Lowest Initial Hardware Cost		11,12,13 14,15,16 17,*		
Lowest Installation Cost	10,12,13, 14		11	Ref. 13: Cost for either varies all over the map.
Lowest Total Installed Cost			12,14,15 16,*	
Simplest System Design		13,15, 16,*		
Shortest Check-out and Start-up		11,*		
Shortest Training Period		11,*		
Highest Dependability (Reliability)			11,15,16, *	
Less Affected by Corrosive Atmospheres		13,14,17, *		Ref. 14: Air acts as a purge.
Lowest Maintenance	12,14,17	11	*	Ref. 12 cites user experience.
More Compatible with Control Valves		13,14,18, *		
Greater Accuracy	11,12,16 *			
Superior Dynamic Response	11,12,13, 14,15,16, 17,18,*			Ref. 13: Normally the fastest response is not necessary.
Better Suited for Long Transmission Distances	11,12,13, 14,15,17, 18,*			
Superior Computer Capability	11,12,13, 14,17,18, *			Ref. 13: This is the primary reason for selecting electronic instruments

† Numbers refer to the References listed at the end of this Section.
An asterisk marks the preference of this author.

mits the teeing in of different manufacturers' equipment with very few problems. Other pressures are seldom used.

Much has been written about pneumatic transmission lags.[19-23] In this regard the electronic system is superior; however, pneumatics can operate with 250–500 ft (75–150m) lines with success. Distances of 1,000 ft (300m) and more can be considered.[21,22] The longer distances can be attained by specifying such requirements as higher air capacities, use of volume boosters, $\frac{3}{8}$ in (9.38mm) OD

tubing, and, as a last resort, field-mounted controllers. According to Buckley[22], limitations on performance are caused by:

1. Control valves without positioners
2. $\frac{1}{4}$ in (6.25mm) rather than $\frac{3}{8}$ in (9.38mm) OD tubing
3. Restrictions in manual/automatic switch blocks and plug-in manifolds
4. Inadequate valve positioner air capacities

5. Single-acting (instead of double-acting) position-ers on cylinder operators
6. Inadequate field air supplies
7. Multiple process lags

For signals from electronic transmitters and controllers, 4–20 mA DC is the most common, while 10–50 mA DC is another choice. Considerable engineering must go into all-electronic systems because of impedance restrictions and polarity, power supply, shielding, and grounding requirements.

The transmission range of the electronic system can be a mile or more with no lag. However, this feature is not important for many analog control installations.

With regard to electrical noise, pneumatic instruments are of course immune. Electronic systems do experience problems if shielding and grounding is inadequate. In recent years, many manufacturers have designed instruments protected against RFI (Radio Frequency Interference).

Converters

A pneumatic system does not need additional hardware to operate an air-operated control valve, whereas an I/P converter must be installed in an electronic loop. Because many transmitter outputs are inherently electronic, some pneumatic loops also require I/P converters for applications involving temperature, flow, or analysis.

Electronic systems are superior for data loggers, computers, distributed control, and/or programmable controller applications. The pneumatic system would utilize I/Ps, P/Is, and pneumatic scanning devices for these applications.

Electrical Safety

In many process industries, hazardous locations (discussed in Section 5.2) exist because of the presence of flammable gases, combustible dusts, and ignitable fibers. Therefore, equipment in hazardous areas must be designed so that it will not cause a fire or explosion. Manufacturers and users agree that pneumatic instruments are inherently safe for installation in hazardous areas. On the other hand, analog and digital electronic instruments require special considerations if they are to operate in a hazardous environment.

Assume the area to be Class I explosion-proof, where explosive or ignitable mixtures (in air) of flammable gases or vapors constitute the hazard. The electronics must meet the requirements of either:

Division 1: Location is likely to have flammable mixtures present under normal conditions, or if frequent maintenance is necessary, or

Division 2: Location is likely to have flammable mixtures present only under abnormal conditions.

The initial and installed costs for electronic systems will be a function of the electrical classifications and the selected approaches to meet the requirements of the National Electrical Code. The choices include explosion-proof housings, purging, and intrinsically safe designs. For Division 2, nonincendive, nonarcing, and hermetically sealed equipment can also be considered. (Refer to Section 5.2 and References 24–26 for more information).

The Modern Debate

Nowadays, in selecting a control system, one must also examine both analog and digital alternatives.[27] Digital controls are compatible with computers, distributed control systems, programmable controllers, and digital controllers. Other factors are covered in Sections 2.3, 2.6, 6.4, 7.1, 7.4, 7.5 and 7.9 of this volume.

The bibliography contains additional information that will both enlighten and confuse. If anything, choosing between analog and digital systems is more difficult than choosing between pneumatic and electronic systems.

REFERENCES

1. Mathewson, C.E., "Advantage of Electronic Control," *Instruments & Automation*, February, 1955, pp. 258–269.
2. "Sun Oil Opts for Electronic Instrumentation," *Canadian Controls & Instrumentation*, February, 1968, pp. 29–31.
3. "CPI Swing to Electronic Controls," *Chemical Week*, April 13, 1968, p. 53.
4. Hileman, J.R., "Practical Aspects of Electronic Instruments," *Instrumentation Technology*, May 1968, pp. 43–47.
5. "A Plug for Pneumatics," *Chemical Week*, December 14, 1968, pp. 77 and 79.
6. Maas, M.A., "Pneumatics: It's Not Dead Yet," *Instruments & Control Systems*, November, 1974, pp. 47–52.
7. Martin, R.L., "Simple Solutions to Control Problems," *Chemical Engineering*, May 22, 1978, pp. 103–111.
8. Bailey, S., "Pneumatic Process Controls: An Eternal Case Of Borrowed Time," *Control Engineering*, October, 1978, pp. 39–42.
9. Farmer, E., "Pneumatics in a Digital World," *Instruments & Control Systems*, March, 1979, pp. 31–35.
10. Teed, J.T., "Comparative Installation Costs for Miniature Pneumatic and Electronic Instruments," Paper 1-LA61, Fall Instrument-Automation Conference, Instrument Society of America, 1961.
11. Palmer, O.J., "Instrumentation—Pneumatic Versus Electronic, Which Way To Buy?," Paper 5.4-4-64, 19th Annual Conference, Instrument Society of America, 1964.
12. Savage, H.L., "Electronic vs. Pneumatic Control Systems," unpublished paper, 1964.
13. Lynch, B., "Comparison of Electronic and Pneumatic Control Systems," Paper presented at the 6th Annual Maintenance Management Symposium, Instrument Society of America, 1971.
14. Deuschle, R., "Pneumatic? Electronic? Which to Buy?," *Instruments & Control Systems*, February, 1975, pp. 51–54.
15. Lommatsch, E.A., "Pneumatic versus Electronic Instrumentation," *Chemical Engineering*, June 21, 1976, pp. 159–162.
16. Hordeski, M., "When Should You Use Pneumatics, When Electronics?" *Instruments & Control Systems*, November, 1976, pp. 51–55.
17. Tompkins, J., "Pneumatic or Electronic Instrumentation?" *Instruments & Control Systems*, January, 1979, pp. 55–57.
18. Marcovitch, M.J., "Whither Pneumatic and Electronic Instrumentation?" *Chemical Engineering*, Deskbook Issue, October 13, 1979, pp. 45–49.
19. Harriott, P., *Process Control*, McGraw-Hill Book Co., New York, 1964, pp. 204–210.

20. Warnock, J.D., "How Pneumatic Tubing Size Influences Controllability," *Instrumentation Technology*, February, 1967, pp. 37–43.
21. Buckley, P.S., "Dynamic Design of Pneumatic Control Loops, Part I: Analysis by Segments," *Instrumentation Technology*, April, 1975, pp. 33–40.
22. ———, "Dynamic Design of Pneumatic Control Loops, Part II: Application Principles and Practices," *Instrumentation Technology*, June, 1975, pp. 39–42.
23. Shinskey, F.G., *Process Control Systems*, 2nd ed., McGraw-Hill Book Co., New York, 1979, pp. 38–40.
24. Magison, E.C., *Electrical Instruments in Hazardous Locations*, 3rd ed., Instrument Society of America, Research Triangle Park, NC, 1978.
25. Oudar, J., "Intrinsic Safety," *Measurements & Control*, October, 1981, pp. 153–150.
26. Garside, R., *Intrinsically Safe Instrumentation: A Guide*, Instrument Society of America, Research Triangle Park, NC, 1982.
27. Kaminski, R.K., "The Basic Control Choice," *Instrumentation Technology*, July, 1973, p. 41.

BIBLIOGRAPHY (Chronological)

Giles, R.G., and Bullock, J.E., "Justifying the Cost Of Computer Control," *Control Engineering*, May, 1976.

McAdams, T.H., "Digital Control at Half the Cost of Analog," *Instruments & Control Systems*, April, 1977.

Williams, T.J., "Two Decades of Change—A Review of the 20-Year History of Computer Control," *Control Engineering*, September, 1977.

Warenskjold, G.A., "Microprocessor vs. Analog Control for the Nicholasville Water Treatment Facility," Annual Conference, Instrument Society of America, 1978.

Sheridan, T.B., "Interface Needs for Coming Industrial Controls," *Instrumentation Technology*, February, 1979.

"The Configurations of Process Control: 1979," *Control Engineering*, March, 1979.

Thurston, C.W., "Experience with a Large Distributed Control System," *Control Engineering*, June, 1979.

Kompass, E.J., "The Long-Term Trends in Control Engineering," *Control Engineering*, September, 1979.

Bailey, S.J., "Process Control 1979: A Year of Drastic Change, and More To Come," *Control Engineering*, October, 1979.

Pluhar, K., "International Manufacturers Offer Many Choices in Process Control Systems," *Control Engineering*, January, 1980.

Bailey, S.J., "The Programmable Controller Is Finding Its In-Line Niche," *Control Engineering*, February, 1980.

Rapley, D.E., "How to Select Distributed Control and Data Systems," *Instrumentation Technology*, September, 1980.

Morris, H.M., "Operator Convenience Is Key as Process Controllers Evolve," *Control Engineering*, March, 1981.

Cocheo, S., "How to Evaluate Distributed Computer Control Systems," *Hydrocarbon Processing*, June, 1981.

Stanton, B.D., "Designing Distributed Computer Control Systems," *Hydrocarbon Processing*, June, 1981.

Hickey, J., "The Interconnect Dilemma," *Instrument & Control Systems*, December, 1981.

Merritt, R., "Large Control Systems: Big Price Tags but Even Bigger Returns," *Instrument & Control Systems*, December, 1981.

Hall, J., and Kuhfeld, R., "Industry Leaders Look at 1982," *Instruments & Control Systems*, January, 1982.

Smith, D.R., "Is Your Control Computer Obsolete?" *Instruments & Control Systems*, February, 1982.

Ledgerwood, B.K., "Trends in Control," *Control Engineering*, May, 1982.

"Special Report on Programmable Controllers," *Plant Services*, June, 1982.

D'Abreo, F.A., "Replacing Installed Process Control Computers," *Instrumentation Technology*, July, 1982.

Stanton, B.D., "Reduce Problems in New Control System Design," *Hydrocarbon Processing*, August, 1982.

Dartt, S.R., "Distributed Digital Control—Its Pros and Cons," *Chemical Engineering*, September 6, 1982.

McNeil, E.B., "A User's View of Instrumentation," *Chemical Processing*, October, 1982.

Neal, D., "The Datalogger Moves into Control," *Instruments & Control Systems*, November, 1982.

Miller, T.J., "New PCs Exemplify Trends to Smaller Size and Distributed Network Compatibility," *Control Engineering*, January, 1983.

Flow Sheet Symbols
Same as for other
electronic control loops

2.9 TELEMETERING SYSTEMS

Features:	Radio telemetry hardware can measure and transmit any physical phenomenon that can be transduced into an electric current, voltage, or resistance. These include acceleration, bending moment, current, proximity, resistance, shock, strain, temperature, torque, vibration, and voltage.
Signal Transmission Band:	Signal transmission is possible at any frequency allowed by FCC; however, the most frequently used bands are the 72 to 76 MHz, 88 to 108 MHz, 132 to 174 MHz, and 450 to 470 MHz.
Cost:	Transmitters: $900 to $3000
	Receivers (depending on their features and accessories): $450 to $4,000. A simple, single-channel system including transmitter, receiver, and accessories may cost between $1,500 and $3,000.
Partial List of Suppliers:	Communitronics, Ltd.; Conic Data Systems; EMR Telemetry; Leupold & Stevens, Inc.; Moore Systems, Inc.; Microdyne Corp.; Pacific Communications, Inc.; Repco, Inc.; RCI Data; RTS Systems, Inc.; Systronics, Inc.; Tele-Dynamics, Inc.; Teletronix Systems, Inc.

Introduction

In this section the principles of telemetry, its uses, advantages, and specific applications are presented. The methods of utilizing telemetry equipment in conjunction with existing transducers and indicators are also described. Following a brief presentation of what radio telemetry is, the section continues by discussing telemetry applications in chemical and other industrial processes.

Industry is turning to new techniques of measuring physical quantities at relatively inaccessible locations and recording them from a convenient distance. More precise transducers and powerful self-contained radio transmitters are used to send strain measurements back from energized high-voltage transmission lines and temperatures from inside chemical process equipment, and to warn of critical temperatures inside whirling motors. Because transmission distances are relatively short, the equipment is less elaborate and less costly than in space telemetry. But industrial environments and operation by nontechnical personnel requires most telemetry system components to be far more rugged, and to be capable of accurate measurements with only the simplest calibration techniques.

The deeper an instrumented vehicle probes into the remote reaches of outer space, the more technologically spectacular seem the achievements of telemetry. Yet the vast distances spanned by telemetry signals are less challenging technically than the stubborn problems in some industrial applications, especially the inaccessibility of the quantities being measured. Signals from a missile-launched space probe soaring toward the sun are often easier to obtain than measurements from inside a solid, earthbound motor only a foot or two away. To find the temperature of the spinning rotor, housed in a steel casing and surrounded by a strong alternating magnetic field, may require more ingenuity to transcend the operating environment than taking measurements from the most distant instrument payload speeding through the unaccommodating environment of space.

The technology that has produced missile and space telemetry is also spawning new forms of industrial radio telemetry, capitalizing on the development of new transducers, powerful miniature radio transmitters, improved self-contained power sources, and better techniques of environmental protection.

Measuring from a distance requires, first, at the remote detection point, a transducer, a device that converts the physical quantity being measured into a signal (usually an electrical one) so that it can be more conveniently transmitted. Then a connecting link is needed between the location where the measurement is being made and the point where the signal may be read or recorded. This link can be either an electrical circuit (there have been *wired* telemetry systems since long before the turn of the century), pneumatic or hydraulic lines, a beam of light, or a radio carrier for frequency or pulsed systems of measurement.

In radio telemetry, a transmitter generates the carrier, a subcarrier impresses the measurement signal onto the carrier, a radio receiver receives this carrier out of the air and reproduces the measurement signal it has borne, and a meter or recorder displays the measured quantity.

During the 1970s a new technology was developed that has a great future. This is the fiber optic communication technology described in Section 2.5. The idea is based on sending visible or infrared light impulses in a thin light conductor usually made of glass. At the transmitting end the information is transformed from an electrical signal to a light signal. This is usually done by light-emitting diodes or lasers. At the receiving end a detector transforms the light signal back to an electrical one. (See Section 2.5.)

Fiber optic systems can be used in a broad range of applications to solve many problems associated with traditional electrical hardwired or radio frequency system design. In an optical system, signals are transmitted in the form of photons (light); they have no electrical charge and therefore are not effected by electrical or magnetic noise found in high voltage environments or during a lightning discharge. Similarly, high magnetic fields from motors, machinery, transformers, and so on have no effect on optical transmission. Crosstalk is eliminated since the small flux leakage that might occur at the fiber boundary interface is retained by the opaque jacket. This factor also guarantees transmission security because the signal cannot be externally detected throughout the length of the fiber.

Telemetry Methods

Telemetry is concerned with the transmission of measured physical quantities such as temperature, displacement, velocity, humidity, blood pressure, pressure, acceleration, etc. to a convenient remote location, in a form suitable for display and analysis. The first practical telemetry applications were made by the public utilities prior to World War I. The greatest development took place, however, with the advent of high-speed aircraft, missiles, and satellites. Simultaneously, the continuing development of industrial telemetry applications proceeded at a relatively slow rate until the experience gained in military and governmental applications was applied to the industrial telemetry field. Industrial telemetry now

takes advantage of new types of transducers for measuring physical quantities to be telemetered, miniature ratio transmitters, long-life miniaturized self-contained power supplies, and better techniques of environmental protection. Industrial applications cover a broad segment of industry including utility, chemical, transportation, construction, and machinery.

Medical science is currently employing telemetry for use in experimental, clinical, and diagnostic applications. Some of the particular body characteristics telemetered include heartbeat, brainwaves, blood pressure, temperature, voice patterns, heart sounds, respiration sounds, and muscle tensions. Similar studies are being pursued in the biological and psychological fields, where greater experimental latitude permits embedding of transmitters within living animals.

The basic telemetry system consists of three building blocks. These are (1) the input transducer, (2) the transmitter, and (3) the receiving station. Transducers convert the measured physical quantity into a usable form for transmission. The conversion of the desired information into a form capable of being transmitted to the receiver is a function of the type of transducer employed. Transducers convert the physical quantities to be measured into electrical, light, pneumatic, or hydraulic energy. The type of energy conversion is determined by the type of transmission desired.

One of the most common types of transducers generates electrical signals as a function of the changing physical quantity, and one of the most common varieties of this type is the resistance wire strain gauge. In this transducer, the ability of the wire to change its dimension as it is stressed causes a corresponding change in its electrical resistance. A decrease in wire diameter generally results in greater resistance to the flow of electricity. Similarly, temperature-sensitive materials that have electrical characteristics that change with temperature make temperature detection possible.

In most transducers, the electrical output varies as a function of changes in the physical parameter. These electrical changes can be transmitted by wire direct to a control center, data display area, or a data analysis section for evaluation. However, the difficulties with use of wire in many applications have given rise to wireless telemetry.

In order to transmit the transducer information through the air, it is necessary to apply this information to a high-frequency electrical carrier as is commonly done in radio. Theoretically, any alternating current has an electromagnetic field that can be used to transmit information through open space. However, to radiate this signal effectively the transmitting antenna has to have a size around the quarter or half wavelength. The wavelength is calculated by dividing the speed of light (3×10^8 m/sec) by the frequency of the alternating current. For example, a half-wave dipole antenna for a 10 kHz radio transmitter would have to be 15,000 meter long. Although it is possible to transmit and receive electromagnetic signals with antennas much smaller than the wavelength, the efficiency is drastically reduced. As a comparison, a VHF transmitter operating at 150 MHz requires a half-wave dipole of exactly 1 meter length. It is obvious that such an antenna is much more practical.

Application of the transducer information to a high-frequency carrier is commonly called modulation. It is possible to modulate a carrier by a change in amplitude, a change in frequency, or a change in the carrier phase. The last technique is similar to the modulation employed in transmitting color by television.

In color TV the brightness signal is transmitted as amplitude modulation (AM), the sound as frequency modulation (FM), and the color as phase modulation (PM). Pulse coding is used to modulate the radio frequency carrier in either AM, FM, or PM. The various types of modulation that have been used for telemetry are shown in Figure 2.9a.

A common and extremely useful technique for increasing the information-carrying capability of a single transmitting telemetry line is called multiplexing. When it is desirable to monitor different physical parameters, such as temperature and pressure, it may be wasteful to have duplicating telemetry transmission lines. Multiplexing techniques can usually be considered to be of two types, frequency-division multiplexing and time-division multiplexing. In the frequency-division multiplexing system, different subcarrier frequencies are modulated by their respective changing physical parameter; these subcarrier frequencies are then used to modulate the carrier frequency enabling the transmission of all desired channels of information simultaneously by one carrier. At the receiver, these subcarrier frequencies must be individually removed. This is accomplished by filters that allow any one of the respective subcarrier frequencies to pass. Each subcarrier frequency is then converted back to a voltage by the discriminator. The discriminator voltages can be used to actuate recorders and/or similar devices.

Time-division telemetry systems employ either pulse modulation or pulse code modulation. In these systems the information signal is applied, in time sequence, to modulate the radio carrier. The characteristics of a pulse signal can be affected by modulating its amplitude, frequency, or phase.

Telemetry began as a wire communication technique between two widely separated stations. It now is the essential communicating link between satellites, space ships, robots, and other scientific devices.

The range of a radio link is limited by the strength of the signal radiated by the transmitter toward the receiver and by the sensitivity of that receiver. A 10-microwatt output will transmit data easily one hundred feet with a bandwidth of 100 KHz. The wider the bandwidth, the more the effect from noise, and therefore the more transmitting power required for an acceptable signal.

At the receiving station, there are usually no space restrictions in accommodating large antennas, sensitive radio tuners and recorders, and an ample power supply. But the transmitting station often must be small, possibly doughnut size but sometimes no bigger than a pea, and must be self-sufficient—carrying its own power or perhaps receiving it by radio.

On the surface, industrial radio telemetry seems to be simply a matter of hardware. And it almost is, except that the functional requirements are a lot different from those in missile and space telemetry. Distances are much shorter, a matter of a few feet to a few hundred yards; signal power can be radiated directly from the transmitter circuitry or from an antenna as simple as an inch or two of wire. Most tests are repeatable—no missile blowing up on the pad here, taking with it valuable instruments and invaluable records of the events leading up to that failure.

Quantities can be measured one or two at a time, rather than requiring an enormous amount of information to be transmitted at once. This results in relatively inefficient use of the radio link, but enables simpler circuitry at both the transmitting and receiving ends.

Surprisingly, environment plays the most critical role in industrial telemetry. It makes by far the largest difference between telemetry operations from missiles and spacecraft and those used in industrial remote measurement. While missile telemetry equipment is expected to withstand accelerations of 10 to 20 g, the rotating applications of telemetry in industry, such as the embedding of a transducer in a spinning shaft, require immunity to 10,000 or 20,000 g centrifugal accelerations.

The environmental extremes under which industrial telemeters must work are considered normal operating conditions by their users. Unlike missile telemetry equipment, which is shielded and insulated against ex-

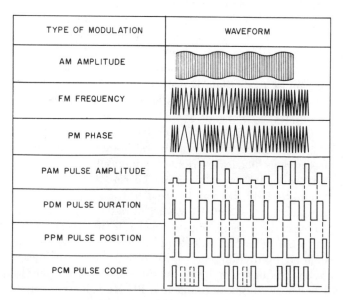

TYPE OF MODULATION	WAVEFORM
AM AMPLITUDE	
FM FREQUENCY	
PM PHASE	
PAM PULSE AMPLITUDE	
PDM PULSE DURATION	
PPM PULSE POSITION	
PCM PULSE CODE	

Fig. 2.9a   Telemetry modulations

tremes of temperature, shock, and vibrations, and which is carefully calibrated for weeks before it is used only once in an actual shot, industrial telemeters must operate repeatedly without adjustment and calibration. Used outdoors, they are often subjected to a temperature range of -40 to $+140°F$ (-40 to $+60°C$). They must operate when immersed in hot or cold fluids, and thus it is almost mandatory that they be completely encapsulated to be impervious to not only humidity and water, but to many other chemical fluids and fumes. Many lubricating oils operate at temperatures of 300 or 350°F (150 or 175°C).

We know that missile telemetry components must be small and light, yet an order of magnitude reduction in size and weight has been necessary to make telemetry suitable for high-speed rotating shafts or for biological implants. They must be so reliable that no maintenance is required, for there are no service centers set up to handle this kind of equipment, and it must work without failure in order to gain industrial acceptance.

Information theory has been used extensively to develop space telemetry for the most efficient data transmission over a maximum distance with a minimum of transmitted power. Inefficiencies, being of no real consequence in industrial telemetry, make for less elaborate, less costly equipment. Radio channels are used in a relatively inefficient manner, and the distances between transmitter and receiver are usually so short that there are few problems of weak signals. In many cases, measurement and testing via telemetry links takes place in completely shielded buildings or in metal housings.

Though telemetry is usually defined as measurement at a distance, it has gradually begun to embody the concept of control from a distance, too. In telemetry—the transmission of the value of a quantity from a remote point—it may serve merely to communicate the reading on an instrument at a distance. But the output of the instrument can also be fed into a control mechanism, such as a relay or an alarm, so that the telemetered signal can activate, stop, or otherwise regulate a process. Measurement may be taken at one location, indication provided at a second location, and the remote control function initiated at one of those two locations or at a third point.

For example, a motor might be pumping oil from one location while oil pressure is being measured at another. When the pressure reading is telemetered to a control station, a decision can be made there to reduce pump motor speed when the pressure is too high, or a valve can be opened at still another location to direct the oil to flow in another path. The decision-making controller may be an experienced pipeline dispatcher or an automatic device.

Measuring and Transmitting

Telemetry, then, really begins with measurement. A physical quantity is converted to a signal for transmission to another point. The transducer that converts this phys-ical quantity into an electrical signal could be a piezoelectric crystal, a variable resistance, or perhaps an accelerometer.

Telemetered information need be no less accurate than that obtained directly under laboratory conditions. For instance, in telemetering strain measurements, it is possible to achieve accuracies of a few microinches per inch or greater. The only limitation is usually the degree of stability in the bond of the strain gauge to the specimen, and not the strain gauge itself.

If great accuracy in temperature measurement is desired, it can be attained by choosing a transducer that provides a large variation of output signal over a small range of process property variation. The resolution that this provides may be translated into true accuracy by careful transducer calibration. Accuracy is reduced, of course, if a wider range of temperature needs to be detected. Typical single-channel analog telemetry links maintain a measurement accuracy of 1 to 5 percent. But this is not a limitation of the total system, since 1 percent of a 100°F (56°C) temperature change would only be 1°F (0.5°C), so several telemetry channels can easily share the total temperature range to be measured—say a 100°F (56°C) range divided into four 25°F (14°C) ranges, to produce an accuracy of one-fourth of a degree.

Special temperature probes have been produced for the range of 70 to 400°F (21 to 209°C) and higher to maximize the stability and accuracy of temperature telemetry. These probes, when used with the proper choice of transmitters and receivers, can provide temperature measurements to closer than 0.05°F (0.03°C).

One of the limitations to accuracy and to repeatability in telemetry is the output level of the transducer. The low electrical levels produced by thermocouples and strain gauges (millivolts) are much more difficult to telemeter than a higher-voltage level of say 5 volts. At low signal levels, extraneous electrical noises produce great degradation. This noise may be thermally generated, caused by atmospheric effects, or generated by nearby electrical equipment. When low-level transducers are used, stable amplifiers are required to raise the signal voltage to useful modulation levels.

There may be great variations in the strength of the radio signal received because of variations in distance between transmitter and receiver, or because of the interposition of metallic objects. In industrial radio telemetry transmission, these effects can be prevented from disturbing the data by resorting to frequency modulation of both the subcarrier and the carrier so that the telemetered signal is unchanged by undesirable amplitude variations. This method is called FM/FM telemetry.

If FM modulation is employed in the subcarrier of the transmitter, the transducer signal modulates the frequency of the subcarrier oscillator. This can be done by a simple resonant circuit that produces a given frequency in the audio range, say 1,000 Hz, which is varied above or below by the signal from the transducer as it responds to the variable it is measuring. If the signal were fed to a loudspeaker, a rising or falling tone could be heard. The subcarrier oscillator then modulates a radio frequency carrier, varying its frequency (FM) or its amplitude (AM) in accordance with the subcarrier signal. The radio frequency in FM industrial radio telemetry links is usually in the 88 to 108 MHz band. At the receiving end of the link, the radio receiver demodulates the signal, removing the carrier and feeding the subcarrier to a special discriminator circuit that removes the modulation and precisely reproduces the original measurement signal for calibrated indication or recording.

Multiple measurements can also be transmitted over the carrier by sampling the output of several transducers in rapid sequence, a technique called time-division multiplexing. This technique has been employed to handle as many as a million samples per second. It provides for very simple data displays and easier separation of channels for recording or analysis, and it is free of cross talk. If possible, it is advantageous to use no multiplexing at all for concurrent data taking, but to use separate radio carriers for each measurement being transmitted.

A single channel of industrial FM/FM telemetry equipment can cost between $1,500 and $3,000, depending upon the flexibility required and the measurements being made. This buys everything needed for a given remote measurement—transducer, radio link, power supply, and simple indicator.

The telemetry data received may be recorded in a number of ways, but such records must preserve the accuracy of the entire system. For example, if a 1 percent system is recorded on a graph, and $\frac{1}{64}$ in. (0.4mm) is the most that can be distinguished on the graph paper, the minimum size graph for full scale should be approximately 2 in. (50mm). Similarly, numeric data should be printed to enough decimal places to preserve the accuracy of the system.

Several examples of new industrial telemetry applications described in later paragraphs will show how systems are applied to various remote measurement problems. They will also give some idea of the specific equipment requirements for industrial environments.

High-voltage transmission lines are an excellent example of how inaccessible an object of measurement can be. These lines vibrate in the wind, and the stresses and strains require measurement under the dynamic conditions that contribute to fatigue failure. Strain tests to determine fatigue will show quickly whether the endurance limit of the line has been exceeded, and only if it is exceeded need we be concerned about fatigue failure. Therefore it is necessary to measure the number and magnitude of the strain reversals in order to predict the time of failure. Telemetry techniques permit dynamic testing under actual service conditions rather than by simulated laboratory conditions or static tests.

While the transducer that produces an electrical signal proportional to strain may have an output of 0.01 volts, the live transmission line to which it is attached may be at a potential of several hundred thousand volts. The problem is to detect this hundredth of a volt in the presence of a very large signal. In the language of the telemetry engineer, this is rejection of a common mode voltage of the order of 10^8 to 1. Then why not de-energize the line? It's a simple matter of economics—an idle line transmits no power, and the wind forces that cause the line to vibrate are neither predictable nor constant. So, weeks or months may be spent in gathering measurements for a particular set of spans. However, a ratio telemetry link makes it possible to transmit the strain signal even while power is being carried.

A self-contained FM radio transmitter is attached to the transmission line at a point adjacent to a strain gauge. All remain at the same electrical potential as the line, much like a bird sitting safely on the wire, transmitting the strain gauge output to a radio receiver and recorder located at some convenient point on the ground, where vibration analysis can be made. As a result, armor rods may be placed around the line at the vulnerable points, or vibration absorbers of the correct resonant frequency can be installed at the proper points on the line.

More down to earth, but equally inaccessible to measurement, is strain on the chain belt of an earth mover. Too light a chain will quickly fail from fatigue caused by the alternating stresses imposed by the full and empty buckets it transports. Measurements made under actual operating conditions of the earth hoist mean attaching strain gauges to a chain traveling at 500 ft per minute (150 m/min), subjecting them to violent shock and vibration. On this kind of moving equipment, slip rings and wire-link remote measurements will not work.

Here again, radio telemetry is now providing the dynamic measurements needed to test the earth-moving equipment at work. A transducer and a small, rugged transmitter are attached to points along the chain—strain varies from link to link depending on the proximity to the bucket—until the most vulnerable part of the chain is found. It is preferable to use several transducers and multiple channel telemetry equipment for such measurements to simplify correlation between load and the resulting strain at various links.

Telemetry can also determine water levels and flow rates of rivers to provide vital data for flood control or for efficient hydroelectric power generation. Data on the potential amount of water in rivers can be obtained by analyzing the water content of the snow that will eventually melt and feed them. One requirement is to measure the depth and water content of snow in the mountains, then transmit this data from remote points to a central receiving station. The snow-measuring transducer may consist of a radioactive source atop a tall pole and a radiation intensity meter on the ground beneath the snow. The gamma-ray intensity reaching the meter is a function of the height and water content of the intervening snow. Both the meter and the transmitting equipment can be powered by a storage battery and controlled by a clock timer that sets the time of transmission to a few seconds per day.

Transducers

Improvements in transducers are opening up new measurement possibilities. Typically strain gauges have consisted of a metallic element that was stretched. As its length increased, its cross section decreased (Poisson's ratio), thus increasing its electrical resistance. A metal strain gauge with a gauge factor of 2 increases resistance twice as fast as it does its length. But new semiconductor materials exhibit gauge factors as high as 100 or 200. Consequently, the output of a bridge of semiconductor materials may be used to modulate telemetry subcarrier oscillators directly without further amplification. There are some drawbacks—semiconductor materials are temperature sensitive and introduce greater drifts than metal foil. This problem is not insurmountable, for the arms of the gauge may be located at a single point for temperature compensation.

Semiconductor piezoresistivity is a property that makes possible simpler and more reliable pressure transducers than the conventional electromechanical pressure cells, which employ an elastic sensor with a deflection proportional to pressure. The ideal pressure transducer must provide a precise, repeatable measure of steady-state pressure. It must also respond linearly to large pressure variations without permitting small pressure changes to be obscured by noise or threshold effects. Furthermore, there must be a minimum of interaction between the transducer and the medium whose pressure is measured. One recent development integrates the pressure sensor and the output devices in a single silicon strip. It has a dynamic response to 6,000 Hz.

Among other promising approaches to pressure transducers is a chemical cell in which liquid displacement is used to unbalance a bridge. In this type of unit, a center electrode divides two sections of liquid that are metered through a small orifice. Electrodes at each end make up a unit that becomes half of the bridge; they are electrically connected by the liquid. The center electrode is displaced by pressure of a fluid, although it also could be moved by mechanical leverage or magnetic energy. A transducer of this type can measure absolute strain or strain rates, producing an output so high that it requires no further amplification before being fed to a telemetry transmitter. In fact, its amplification is limited only by temperature effects, which become increasingly important as the bridge configuration is physically changed to increase amplification.

Semiconductors are also appearing in light-acutated analogs of the mechanically operated potentiometer. Electromechanical potentiometers sense mechanical movement and translate it into voltage by changing the position of a sliding contact on a length of electrically resistive material. They have long been used for position measuring and as pick-offs from accelerometers, gyros, torque angle meters, etc.

In the electro-optical solid-state potentiometers, the conventional wiper arm that makes the sliding contact has been replaced with a tiny light beam. The electrical element consists of a resistance film separated from an adjacent conducting strip by a photoconductive crystal. This photoconductive strip acts as an insulator when it is dark, but where the light beam hits it, an electrical connection is made between the resistive and the conductive strips at that point. The output voltage is a linear function of the light beam displacement on the photoconductive crystal. Such a potentiometer is essentially a friction-free, noise-free device and has a marked advantage over mechanical potentiometers because it can be scanned at speeds up to 7 meters per second. Its resolution is better than 0.0005 in. (0.0125 mm), and its deviation from linearity can be made less than 0.2 percent over a range of 10 to 90 percent of full voltage. While the resistive strip is temperature sensitive, the voltage across its terminals or anywhere along the strip remains invariant. On the other hand, the temperature sensitivity of the light-detecting crystal influences the output voltage and linearity of the potentiometer. The practical temperature limits of a photopotentiometer presently range between -65 and $100°F$ (-54 and $+38°C$).

The bonded strain gauge—either wire, foil, or semiconductor—is very useful in most industrial measurements. It is applied to the actual machine part in which strain is to be measured. Strain is related to stress by means of the appropriate elasticity modulus, and many operating parameters of the equipment can be obtained. The bonded resistance strain gauge has been used in the field of stress analysis under a variety of environmental conditions from vacuum to very high pressure. Strain has been measured in the very hot environment of jet engine turbines, as well as in superheated steam. The use of these devices in cryogenic fluids at temperatures as low as $-452°F$ ($-269°C$) and under the nuclear radiation associated with reactors has also been reported. Most of the adverse environmental effects can be minimized by suitable protection of the strain gauge. The major limiting environmental variable is temperature.

Strain-Sensing Alloys

The common strain-sensing alloys undergo changes at high temperatures that alter their resistivity and temperature coefficient of resistance. The changes in the electrical properties are generally time and temperature dependent in the same way that metal physical characteristics are affected by heat treating or cold working. The change in resistivity results in a gauge resistance change with time at constant temperature, and it destroys the initial zero reference that is necessary to determine changes in specimen state of strain with time. This makes it impossible to separate stress-producing thermal or mechanical strains from the error signals produced by unwanted resistance changes due to temperature.

Fortunately, the gauge factor is not drastically changed by metallurgical property changes. Gauge factor versus temperature characteristics for the copper-nickel alloys, nickel-chrome, and platinum alloys is shown in Figure 2.9b. Copper-nickel alloys such as Advance and Cupron have a nearly uniform gauge factor change with temperature of +0.5 percent per 100°F (55.6°C) over the range of −200 to 600°F (−128 to +315°C). This alloy is the most widely used in strain gauge work because of its reasonably high resistivity, low and controllable temperature coefficient of resistance, and uniform gauge factor over an extensive strain range. The nickel-chrome alloys such as Nichrome V, Tophet A, Karma, and Evenohm commonly exhibit a reduction in gauge factor with temperature of −1.5 to −1.8 percent per 100°F (55.6°C) temperature increase, and this decrease is uniform over the temperature range from −452 to 1,200°F (−269° to +649°C). Nickel-chrome alloys are employed for cryogenic testing because of this uniformity of gauge factor and for high-temperature tests because the corrosion resistance is superior to that of the copper-nickel alloys.

The most common platinum alloy used in strain gauges is platinum-tungsten. This material has a high gauge factor (almost twice that of Advance at room temperature) and changes −2 percent per 100°F (55.6°C) temperature increase between −320 and 1,500°F (−196 and +816°C).

This alloy, while commonly used for the measurement of dynamic strain at all temperatures, is being used extensively for static strain measurements in the temperature range above 800°F (427°C) where nickel-chrome alloys become unstable. Platinum alloys are not considered suitable for use in the cryogenic range below liquid nitrogen temperature. Figure 2.9c shows the typical short-term drift rate versus temperature characteristics for all of these alloys. Above 450°F (232°C), Advance becomes electrically mobile and would be considered unsuitable if absolute stability were required. Intergranular corrosion of this copper base alloy above 600°F (316°C) drastically alters the temperature coefficient of resistance.

The temperature coefficient of nickel-chrome alloy and Nichrome V cannot be adjusted for temperature compensation by heat treatment. The material can be stabilized to produce a more drift-free condition. Note that the stable range is increased from 600 to 700°F (316 to 371°C) for material that has been stabilized.

Karma and Evenohm alloys contain small quantities of iron and aluminum in addition to the nickel and chromium. These materials are also unstable above 600°F (316°C), but can be heat treated to produce self-temperature-compensated strain gauge types. Note that 600°F (316°C) is a practical operating limit for these alloys, and that 800°F (427°C) would be considered a maximum for short-term strain measurements where zero stability is required.

The platinum alloys cannot be heat treated to produce self-temperature-compensated strain gauges, although the temperature coefficient is altered by prestabilization. Figure 2.9c indicates that reasonably stable operation can be achieved using this alloy to 800°F (427°C) and,

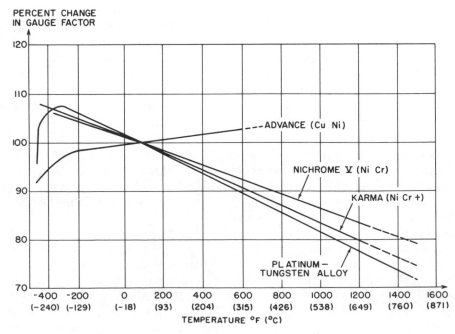

Fig. 2.9b   Typical gauge factor variation with temperature

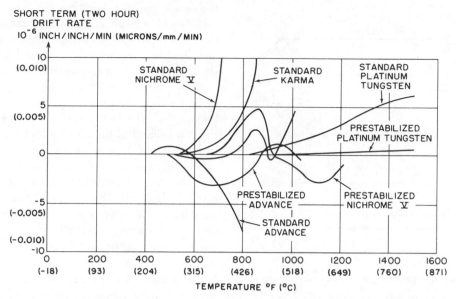

Fig. 2.9c   Typical zero drift at constant temperature

with proper treatment, an additional 400°F (204°C) can be achieved. Where the reasonably low drift of the platinum-tungsten alloys can be tolerated, 1200°F (649°C) is a practical limit for short-term measurements.

Figure 2.9d shows the typical change in slope of the apparent strain characteristic for various strain gauge alloys versus temperature. It can be seen that materials like Advance and Karma can be adjusted to produce zero slope when the strain gauges are bonded to the test

material. This is the basis for self-temperature-compensated gauge types as opposed to gauges that can be compensated by circuit techniques. It can also be seen that the platinum alloys cannot be adjusted to produce self-temperature-compensated gauges. Circuit compensation utilizing the resistance thermometer technique as shown in Figure 2.9e has been used to produce temperature-compensated gauges for specific temperature ranges on a particular test material.

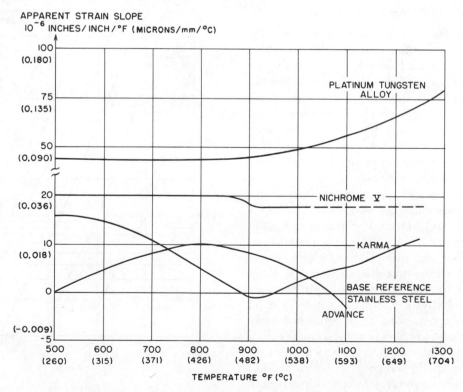

Fig. 2.9d   Typical apparent strain slope change with temperature

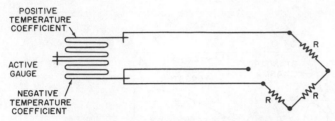

SELF TEMPERATURE COMPENSATED COMPOSITE GAUGE

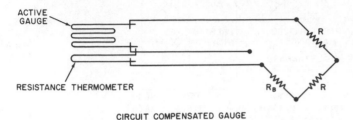

CIRCUIT COMPENSATED GAUGE

Fig. 2.9e   Temperature-compensated gauges

Karma, adjusted for a negative temperature coefficient, can be connected in series with a Nichrome V material with a positive temperature coefficient to produce a temperature-compensated gauge for cryogenic range. This composite gauge construction is shown in Figure 2.9e. The most effective temperature compensation for the cryogenic temperature range is achieved by using Nichrome V and the platinum resistance thermometry circuit compensation technique.

Bonding Cements and Carrier Materials

The strain gauge bonding cements and carrier materials must combine to faithfully transmit the outer fiber strain to the strain-sensing filament, as well as provide electrical insulation between electrically conductive materials and the strain-sensing element. The dimensional stability of these two components will contribute to the electrical stability of the strain gauge. These components should maintain their combined shear strength and electrical insulation properties over the useful operating temperature range of the alloy.

The organic backings and cements are generally limited to the temperature range between −452 and +600°F (−269 and +315°C) unless nuclear radiation and/or vacuum environments are involved. Organic materials include cellulose fiber and nitrocellulose paper carriers normally attached with nitrocellulose cements. Paper gauge installations are generally limited to the temperature range from −320 to +180°F (−196 to +82°C). There is a variety of epoxy, phenolic, and modified epoxy-phenolic carrier materials and bonding cements available. The unfilled or nonreinforced epoxy carriers are suitable for testing in the range of −300 to +200°F (−184 to +93°C). The carriers reinforced with glass cloth, or fibers bonded with filled resin cements, can extend

the operational temperature range to −452 and +600°F (−269 and +315°C). The strain range is somewhat reduced at the cryogenic temperatures, but most materials are capable of measuring at least $\frac{1}{2}$ percent strain over their entire operating temperature range.

Above 600°F (315°C), strain gauges are usually attached with ceramic cements or by the new flame-spray techniques. The limiting temperature for both of these types is related to the electrical characteristics as shown in Figure 2.9f. Foil gauges are supplied either on strippable vinyl carriers, or as "free handling" units for transfer into the ceramic cement. Transfer technique requires a great deal of skill. The commonly used ceramic cements combine metallic oxides with a phosphoric acid base binder and require a 600°F (315°C) cure temperature. These coatings are porous and hygroscopic. The useful strain range of ceramic cements is limited to $\frac{1}{2}$ percent by the brittle nature of the coatings. If a resistance to ground of one million ohms was considered a practical operating limit for electrical and dimensional stability, then foil type strain gauges bonded with ceramic cements would be limited to 1,200°F (649°C) operation. Since the resistance to ground of a strain gauge installation is a function of the distance between gauge elements and ground,

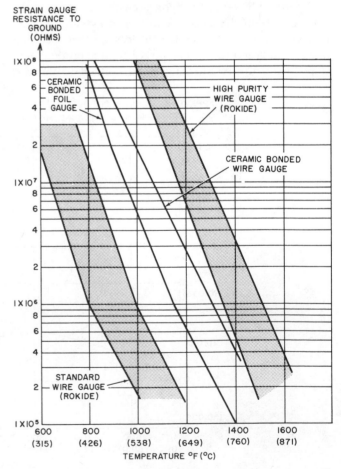

Fig. 2.9f   Typical electrical resistance to ground variation with temperature

the cross-sectional area of the conducting path, and the gauge resistance, it can be shown that wire type gauges can achieve a higher temperature limit for a given installation thickness than foil gauges.

Improved electrical properties of the aluminum oxide used in the Rokide process for flame-spray attachment of gauges has contributed greatly to the extension of high temperature limits. The flame-spray coatings exhibit the same problems with porosity and hygroscopic tendencies as ceramic cements, but are capable of 1 percent elongation instead of only $\frac{1}{2}$ percent. Gauge installations require no elevated temperature curing. This is most important in the testing of materials that will alter their basic metallurgical conditions above 600°F (315°C) cure temperature.

Lead wires also provide electrical resistance in series with the active sensing element and also have a temperature coefficient of resistance and a strain sensitivity. Selection and installation of lead wires is of extreme importance to the overall performance of the strain gauge.

Problem areas can be minimized by the use of a three-wire system as shown in Figure 2.9e. The metallic portion of the lead wire system should have a low, stable resistivity and a low temperature coefficient of resistance. For that reason copper wires clad with nickel or stainless steel, or nickel wires clad with silver have been used, depending upon the temperature range. Nickel-chrome alloy wire or lead ribbon is occasionally used at very high temperatures, but extreme care should be exercised, because this material has 40 times the resistivity of copper.

The insulation used on the lead wires is of equal importance since it must withstand the environment and provide an electrical insulation that is at least equivalent to that of the carrier and bonding cements. A variety of plastic and ceramic insulations are available for low and intermediate temperatures. Extreme care should be exercised when selecting a glass type insulation for high-temperature tests. Sodium silicate base glass insulation may have leakage problems as low as 600°F (315°C), while high-purity silica materials are capable of operation to 1,500°F (816°C). If the lead wires must be bonded to the specimen, they should be carefully installed so as not to be subjected to undue strains and thermal gradients.

In any general discussion such as this, in which state of the art technology is being considered, many special-purpose gauges designed to solve particular problems are overlooked. These range from high-elongation measurements on viscoelastic materials to interlaminate strain measurements within filament-wound rocket casings. As testing temperatures are extended, it becomes more difficult to achieve strain gauge temperature compensation, maximum stability, and repeatability. Although "active gauge–dummy gauge" or full-bridge circuit compensation techniques have not been discussed here (see Section 5.7 in the volume on Process Measurement), they represent the most accurate form of compensation for all undesirable effects. The full-bridge weldable technique is designed to take advantage of the best qualities of this type of temperature compensation and probably represents the most advanced state of the art for the measurement of static strains over wide temperature ranges.

Where weldable techniques or "active gauge–dummy gauge" compensation cannot be used, it becomes necessary to design a gauge to match the specific environmental temperature range, or other test conditions. Techniques and materials are available to solve many problems, but gauges are not, because of the tremendous number of variables to be considered for each test. Bonded resistance strain gauge types can be constructed to meet the requirements if the environmental conditions, the approximate strain range, and metallurgical characteristics of the test materials are known.

Power Sources

Power sources for the transmitter in industrial telemetry applications are seldom a problem. Batteries can be used for temporary applications and at temperatures below 200°F (93°C). Small, lightweight, rechargeable, and expendable batteries are available solidly encapsulated in epoxy resin to withdtand almost as rugged environments as the telemeter itself.

In a moving or rotating application, stationary magnets can be placed so that they generate electricity in a moving coil and are used to provide automatic power generation. If this method is not feasible, a stationary coil can be placed in the vicinity of the transmitter and fed electrical energy at a high frequency so that its field can easily couple into a moving coil in almost any environment. The stationary coil ring may be large, even encompassing a whole room; usually only one turn of wire is necessary. The stationary coil may also be made extremely small, a $\frac{1}{4}$ to $\frac{1}{2}$ in. (6 to 13 mm) in diameter, and coupled to the end of a rotating shaft. These power supplies and coil configurations are standard units and readily available.

Applications of Telemetry

In the design of machinery, one of the most difficult factors to cope with is alternating fatigue-producing stresses that occur at some parts of the machine. It has long been the custom to measure stress in equipment with bonded strain gauges to predict the failure limits before actual failure occurs. This had only been possible on those portions of the equipment that could be connected by wires. With radio telemetry, it is now possible on all members. Costly fatigue failures are now avoidable through installation of miniature telemetry components that are reliable, rugged, and accurate in heretofore inaccessible locations and environments. Industrial uses are virtually limitless; systems can be built to specifications and encapsulated to withstand the most adverse conditions. Low-cost measurement and telemetry systems have been applied to read internal vibrations and strain in rotating equipment, chains, vehicles, and projectiles—eliminating slip rings and wires. Measurement can be made under operating conditions of vibration, acceleration, strain, temperature, pressure, magnetic fields, electrical current, and voltage, under such adverse conditions as in a field of high electrical potential, in fluids, in steam, or in high-velocity gases.

Chemical Plants

In chemical processes, it is important to know the exact temperatures at various points in the process. These temperatures are obtained by using thermocouples. An interesting problem has recently occurred in a rubber treatment operation. When the temperature was measured by a fixed thermocouple, the viscosity of the material was so high that a considerable amount of heat was generated by friction with the thermocouple.

It is recognized that a stationary thermocouple will read the average temperature of the fluid passing by it; consequently, if there are hot and cold spots, these may not be detected. A telemeter and its battery can be encapsulated in a small floating ball and made to flow through the process with the fluid, reading the temperature at specific points in the moving flow without causing any stagnation heat. When this is done, it is usually necessary to provide a receiving antenna without the chemical vessel. This may be a simple insulated wire that can be stretched in the space available.

Textile Mills

Applications of telemetry in textile mills include measurement of acceleration forces and shocks on the shuttle of a loom and of tension and temperature of fibers in treating baths. As fibers are stretched and relaxed in heat-treating ovens, telemeters can measure the uniformity of this process. Telemeters also offer the ability to measure the difference between two temperatures or two strains. These quantities can be matched at the telemeter and only a difference signal transmitted. In this manner, the error involved in taking the difference between two large numbers is usually eliminated.

Conveyor Chains

Transmitters are installed in the form of links on conveyor chains (one transmitter or several per chain) to measure strain and the process speed and temperature. Receivers are located near the chain. The original purpose of telemetry for conveyor chains was to detect link strain leading to breakage. Since the points of greatest strain are at the drive pulleys, strain can be kept within acceptable limits by adjusting the drive torque of each pulley and, if necessary, increasing the number of drive points.

The instrumented link, which contains a strain gauge bridge, is placed on the side of the chain that bears against the pulleys. The lesser strains on the opposite side can be determined by laboratory measurements and extrapolation.

Conveyor chains can be several thousand feet long, with many drive pulleys and supporting blocks, and extend from one floor level to another. In automobile assembly plants the 600°F (315°C) baking oven is always at the top level, and the combination of temperature and height submits the chain and the telemetry system to the greatest strain. Link breakage is generally due to fatigue from cyclic over-stressing.

Ovens, spraying booths, and other shielded areas often have environments that are not suitable for good antenna reception. In such instances, by equipping the transmitter link with a tuned dipole antenna, the signal can be made strong enough to reach the receiver just outside the area.

The measurement system can be portable so that the measuring link can be inserted during a brief down time. The telemetry transmitter, complete with batteries and antenna, is hung on the supports reaching from the conveyor chain to the automobile bodies. Since size restrictions on the telemetry equipment are not severe, sufficient batteries can be used for the desired operating time. With rechargeable batteries, a cycle of 10 operating hours during the day, for example, and a 14-hour overnight charge can be provided.

A frequent problem with conveyor chains is lack of tension at points along the line, particularly at the lower end of descending ramps. Here the tension becomes zero and links pile up. This condition can be corrected by changing the drive torques, but the approach of a pileup cannot usually be observed. Without telemetric strain measurements, the pileups may occur regularly and become serious enough to cause the conveyor to jump the track. The presence of loose chain links in the system indicates that the other links are being overstrained. Telemetry permits the speed of the chain and drive torque at any particular point to be observed remotely, so that the operator can adjust torque and prevent pileups.

Gear Train Efficiency

To measure the efficiency of a gear train speed reducer operating between a high-speed power source, such as an electric motor, and a low-speed load, such as a ball crushing mill, radio telemetry transmitters were designed to operate from strain gauges on the input and output shafts simultaneously, and transmit torque readings to stationary indicators. A tachometer on each shaft gives the rpm of the input and output shafts. From these readings, input and output horsepower is calculated and the gear train efficiency is obtained. The telemetry transmitters can be quickly installed and removed and can be used in dirty or corrosive environments.

Shaft Horsepower

Radio telemetry is frequently the most economical method of transmitting strain and temperature signals from rotating parts. In some applications it is the only feasible method. The noise-to-signal ratio is lower than with most other methods, resulting in more accurate data. Radio telemetry represents a significant advance over previously used slip ring methods.

To operate an ocean ship efficiently, it is essential to have an accurate measurement of shaft horsepower. At the boilers and turbines, horsepower measurements are not accurate because of power losses in the speed-reducing drive. Ship speed can be slowed by hull or propeller condition, and therefore an accurate horsepower measurement of the drive shaft, correlated with engine speed measurement, is important information. Radio telemetry systems have been used to measure torque on the drive shaft of a ship. The equipment is attached to strain gauges properly mounted on the shaft, and remotely located receivers display torque and rpm readings. Horsepower is obtained by simple computation.

Telemetry equipment has also been tested and used in generators and commutators to measure shaft torques, winding temperatures, and thermal strains. Shaft torque data can be used to evaluate lubricants for use in transmissions or to measure the tension in a strip being wound into a magnetic core.

A bridge-controlled oscillator telemetry system can be used to transmit the strain readings from gauges mounted on the bars of an experimental commutator, as the temperature and speed change. The transmitters are fastened to the inside surface of the commutator. The data presented in Figure 2.9g show that a linear relationship exists between the stress in the commutator bars and the temperature of the bar. The data also show that changes in speed up to 600 rpm do not affect the measured strain. The effect of temperature on the stress was approximately 500 lbf/in.² (3.5 MPa) for every 10°F (5.6°C) rise in temperature. A maximum stress of 6,200 lbf/in.² (42.8 MPa) was measured at 144°F (62°C).

Bridge-controlled oscillator systems can also be used to transmit the torsional strain from resistance strain gauges mounted on the main shafts of the motors of hot strip mills. Motors of several thousand horsepower, which operate at 150 rpm, have shafts up to 3 ft (0.9 m) in diameter. Torque is measured and related to the electrical parameters as hot strip steel is being rolled. The transmitters are strapped to the shaft with large hose clamps. Resistance strain gauges can be used to sense the torsional strain in smaller motor shafts while they are driving a gear train in order to evaluate various dry lubricants. The bridge-controlled oscillator transmitter can be used to transmit the strain value from the motor shaft at speeds up to 2,000 rpm. Since the speed is high, the transmitter is housed inside a hollow steel cylinder, which can be keyed and bolted to the motor shaft. The strain gauges are located behind the drive pulley adjacent to the steel cylinder. The four lead wires from the strain gauge bridge pass through a hole in the pulley and steel cavity to the terminals of the transmitter. From the magnitude of the torsional strains, an evaluation of the various lubricants can be made.

Another possible application is to monitor the tension load in magnetic steel strips as the strip is wound onto a magnetic core. Resistance strain gauges are mounted on a shaft, and the shaft is calibrated for torsional strain as a function of torsional load. The shaft is then fixed to the base of the shallow steel cylinder and guided by the sleeve bearing in the spider. The mandrel for the magnetic core is fixed to the end of the shaft. As the machine rotates, the magnetic core is wound around the mandrel and the shaft is strained in torsion. From the torsional strain measurements, the tension in the magnetic strip can be calculated. A typical oscillographic trace for this application is shown in Figure 2.9h. From this record, the exact position of the core can be fixed at any time. The change in tension load in the strip, as a function of the induced friction, can also be determined. The friction or drag holding down the supply strip is produced by two pneumatic cylinders acting against a Teflon friction plate.

The voltage-controlled oscillator transmitter can be used to measure temperature in the damper windings of 100,000 horsepower size motor-generators. The motor-generators can be driven by a water turbine, and the torque in the shaft between the motor-generator and the water turbine can also be measured, using resistance strain gauges. The motor-generators, which operate at about 180 rpm, have stators about 40 ft (12 m) OD and the shaft is about 3 ft (0.9 m) in diameter. Transmitters

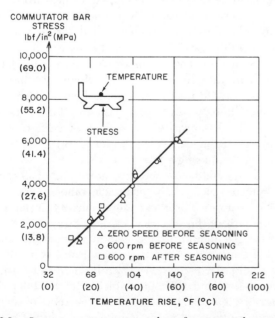

Fig. 2.9g   Stress versus temperature in bars of experimental commutator

can be fixed to the shaft and to the rotor. Test data showed that the measured shaft torque agreed well with the calculated torque, but that the damper winding temperatures were considerably lower than expected.

The biggest problems encountered in measuring strains in rotating parts involve the transmitting antenna and shielding against external electrical fields. It is found that the best configuration for the transmitting antenna is a 320 degree ring, with the center screw holding the battery and transmitter together.

A two-element rabbit-ear receiving antenna made from ¼-in. (6.25 mm) diameter brass tubing can be successfully used in turbulent air when the damper winding temperatures are measured. Two brass tubing antennas hooked in parallel will improve the reception in these measurements.

For best results, all connecting wires should be shielded and grounded. The transmitters should also be shielded to prevent electrical pickup. All wiring must be firmly fastened to prevent any motion. On shafts where torsional strains are measured, the strain gauge bridge circuit need not be wired by a continuous loop around the shaft as is normally done, but can be wired without the loop that caused electrical interference.

Transmitters used in transient temperature conditions must be calibrated under the transient conditions so that the test data can be corrected. Under transient conditions, a temperature gradient will exist over the transmitter, and the correction will depend upon the gradient.

To minimize the transient effect, the transmitter should be enclosed in a metal tube of high thermal conductivity. Although the tube distributes the temperature uniformly over the entire transmitter surface, cooling all parts uniformly, a radial gradient still exists.

The effect of thermocouple length on the accuracy of temperature measurement is negligible when it is from 4 to 12 ft (1.2 to 3.6 m). Generally, copper-constantan thermocouples with a 10-ohm resistance, which is 10 percent of the impedance of the transmitter, can be used for 1 percent accuracy. For a long copper-constantan thermocouple having a resistance greater than 10 ohms, calibration will be necessary. Thermocouples with an initial resistance less than 10 ohms have an accuracy better than 1 percent.

Power Plants

In coal-fired power plants, coal is fed to a number of hoppers by conveyor belt. A tripper on the conveyor belt diverts the coal into a particular hopper until it is full. Either an operator or a mechanical sensing device determines when the hopper is full, and a signal is transmitted to the conveyor to move to the next hopper. Before telemetering equipment was in use, costly accidents could occur if the operator were away momentarily or if the sensor failed to function. As much as six tons of coal a minute could overflow onto the power station floor.

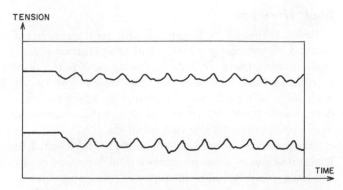

Fig. 2.9h Typical oscillograms obtained using bridge-controlled transmitters

To prevent this, pressure switches are installed in the tripper chute to activate a radio transmitter if coal backs up into the tripper. The transmitter sends its signal to a receiver located at the conveyor belt and sounds an alarm. This type of control is difficult if not impossible to achieve by wired power connections, because the tripper is moving and because the corrosive coal dust atmosphere attacks the wires. For this reason a radio transmitter equipped with long-life batteries is mounted on the tripper. The receiver at the control end is powered by AC. Subcarrier tone (frequency) coding is used to eliminate effects of interference and noise, giving positive protection at all times.

Underground Cable Tension

In some applications, strain information cannot be transmitted from the measuring transducer to the receiver and recorder by radio. When radio cannot be used, it is possible to employ a pulling cable to bring the telemetered information to a location at which it can be recorded. A telemeter developed for measuring tension in underground cables is one such application. To measure the tension in cables pulled through underground conduit, washer type load cells can be placed under the heads of each of the three pulling bolts. These load cells are then connected to the telemetry subcarrier oscillators located inside the pulling head. These oscillators include batteries to power the unit for a period of 10 hours. Each subcarrier oscillator is of a different frequency so that it can be multiplexed, with the other subcarriers, on a single line.

The center conductor and the outer braid of all cables are connected electrically, making a complete short circuit. This feature is used to conduct current through the entire length of the electrical cables. A small iron core toroid, with a high impedance winding around it, can be installed around the center conductor and beneath the outer shield. Two small wires are brought out through the insulation and the outer shield, to the telemetry subcarrier oscillators. In this manner currents of several amperes are induced in the large center conductor, circulating through the length of the line and returning through the outer shield. At the drum end of the cables, another current transformer is used to sense the flowing current and produce a voltage that is then coupled to a radio transmitter.

The radio transmitter is located on the side of the drum. It produces a low-power radio signal to permit the subcarriers, which are coupled from the conductor, to be transmitted a few feet to a receiving antenna without the use of slip rings or rotating joints. A receiving antenna is connected to an ordinary telemetry receiver and recorder. The three signals transmitted simultaneously are separated by filters in the receiver so that they can be discriminated and plotted separately.

The entire system is calibrated electrically by placing a known resistor across one of the arms of the strain gauge resistance bridge in each load cell. The shift in frequency produced by the resistor is referenced directly to the load in pounds, when the cross section (in square inches) of the load cell is known. In the very first tests of this unit, an unexpected phenomenon—large alternating stresses of "violin string effect"—were noted. These stresses varied greatly with the free length of the pulling cable and with the rate of pull.

Earth-Moving Equipment

By attempting to pick up too great a load, a crane hoist operator may damage his hoist or overturn his crane. Measurements along the crane boom or hoist bed are inaccurate at best, and connecting wires are easily broken.

Radio telemetry located in the hook (which picks up the load) transmits the magnitude of the load to the operator in his cab. A red line on a load meter indicates the danger point, and simple scale markings show the effect of the boom angle. The telemetry transmitter in the hook is equipped with rechargeable batteries, which can operate many days between chargings. Provision is made to recharge the batteries overnight from the main battery of the crane. Current requirements are negligible. Zero or tare adjustment is made by simple screwdriver control at the receiver. Recalibration or scale change is also provided as an additional receiver adjustment.

Limitations of Telemetry

The preceding paragraphs describe a number of the requirements placed upon telemetry systems by the transducers and quantities being measured. Unfortunately, the development of telemetry has not been such as to satisfy all requirements, and in many cases the telemetry system seriously limits the measurement. A compromise is therefore required between telemetry capabilities and the requirements of measurement. The shortcomings and limitations of the telemetry system place restrictions upon measurements above and beyond those encountered in the laboratory when the telemeter is not employed. In the first place, an electrical output

from the measuring device is required in order that the measurement can be placed on a radio link. Consequently, transducers that produce an electrical output in one form or another are necessary. Also, the telemetry system may not be perfectly stable down to zero frequency (DC), and transducers and methods of measurement must be chosen to minimize the effects of drift. Overmodulating the subcarrier, or the time-division multiplexer, can also affect adjacent channels, as well as produce erroneous data in its own channel. If various measuring devices are switched, the switching transients must be minimized, or the accuracy of the telemetry system could be impaired. When mechanical commutators or time multiplexers are employed, the measurement of the time occurrence of the event, such as the impact of cosmic particles or the receipt of a guidance pulse, is made more difficult, and the time ambiguity of the multiplexed system is a serious limitation.

The measurement of a large number of parameters requires extensive and bulky equipment, unless the parameters can be combined in groups of similar inputs to minimize the signal conditioning required. This fact generally dictates a relatively standard transducer rather than an optimum one for each particular measurement.

The bandwidth of the measurement, or the frequency with which the measured quantity changes, is also seriously limited by the telemeter. In the FM/FM telemeter, the permissible bandwidth varies from a relatively low value on the lower-frequency subcarriers to a reasonably high value on the high-frequency subcarriers. The bandwidth of the measurement must not exceed the subcarrier bandwidth limitations, or sidebands will be generated in adjacent channels, thereby reducing the accuracy of other measurements (if multiplexed), or interference with adjacent RF signals will be caused.

In a time-multiplexed system, the problem of "folded data" is present whenever the rate of data change is faster than one-half the sampling rate. When this occurs, it is not known whether the measured quantity has reversed itself several times between samples, or if there has been no reversal at all. It is considered desirable to limit the bandwidth of the data so that this ambiguity is not present; however, with refined techniques of analysis, this is not a rigid requirement. The form in which the data is displayed or recorded is also a limitation on measurement. In general, time-history plots of the measured quantity are desired. In this case, the speed at which the recording medium moves is often a severe limitation. If sampling is not regular, demultiplexing difficulties are magnified.

Choice of Transducers

The choice of measuring equipment is often limited by the ability to calibrate it. Calibration is usually made from a graphic plot and is applied to the data, which is then replotted in calibrated form. Calibration corrections, selected by other means, may also be applied before plotting or printing the data. Acquiring the calibration curves in the first place, however, is often a difficult procedure. The transducers are calibrated before they are installed, but the remainder of the data system must be calibrated by substitution methods. This requires accessibility for substitute transducers or signals that may be applied. Also, for this purpose, transducers that have simple simulators, such as resistors for a strain gauge, are desirable.

Within the limitations outlined above, the transducers must be chosen to match the particular telemeter employed. A variable-reluctance transducer might be quite satisfactory in an FM/FM system, but can be very difficult to use with a time-multiplexed system. Analog transducers are chosen to match levels and impedances and are often interleaved with digital transducers, such as shaft encoders and outputs from digital computers. These choices must be made to maximize the utility of the telemeter.

The accuracy of measurement may, in many cases, be limited by the telemeter rather than by the transducer. When this is the case, it is sometimes possible to use several transducers to spread the range of measurement over several telemetry channels. This is done in a manner similar to the display of a watt-hour meter, in which the reading of each dial is transmitted over a separate channel. An example of this technique is the measurement of gaseous pressure by means of a bellows gauge at high pressures, a Pirani gauge at medium vacuum pressures, and an ionization gauge at low vacuum pressures. To detect system errors, measurements of the same quantity are made by separate transducers utilizing separate telemetry channels. This form of redundant measurement has unfortunately been used very little in radio telemetry.

Transducers must also be chosen to measure the desired quantity without measuring other effects to which they are subjected. In other words, a pressure transducer should measure pressure and not be affected by temperature changes. The two principal offenders in this regard are temperature and acceleration. It is a major telemetry problem to select transducers that are free of temperature and acceleration effects, unless they are used to measure those particular quantities. Other parameters which affect transducers to a lesser degree are pressure, humidity, aging, and vibration. Tests of all these quantities can be made before the transducer is mounted, and from the results of these tests, the proper transducer can be chosen. Also, even though transducers are installed in groups with the same kind of measurements handled in the same manner, each transducer in a group may be subjected to different environmental conditions. For this reason, it is not possible to have a single "dummy" transducer to calibrate out the environmental effects on the other live transducers.

Strain and Temperature Transmission

With the development of radio telemetry, the cost of strain gauge and thermocouple transmitters to be used on rotating parts has been reduced. Various types of telemetry systems are available, including an FM/FM system that presents a DC voltage to the recording instrumentation. This means that both the subcarrier and the radio carrier are frequency modulated. The transmitters are potted in an epoxy resin of low thermal conductivity. The potted transmitter is of rugged construction for environmental extremes.

An FM/FM telemetry system is superior to one that is amplitude modulated. Electrical signal amplitudes are difficult to maintain with linearity in the presence of environmental changes, whereas frequencies are not. As a result, the FM/FM principle is widely employed, and the telemetry equipment thereby becomes insensitive to amplitude changes.

Some of the strain gauge transmitters are controlled by the impedance of a Wheatstone bridge and contain a subcarrier oscillator and a radio frequency oscillator. These transmitters can be used for both static and dynamic strain measurements. Another strain gauge transmitter is controlled by the impedance of a single resistance strain gauge and contains only a radio frequency oscillator and no subcarrier oscillator. It can be used only for dynamic strain measurement. The bridge-controlled oscillator transmitters are commonly called BCOs. The thermocouple transmitter is voltage controlled and contains a subcarrier oscillator and a radio frequency oscillator. It is called a VCO.

The FM/FM telemetry system is made up of the following elements: (1) sensing element, (2) battery, (3) transmitter, (4) receiver, (5) discriminator, (6) DC amplifier, (7) recorder, and (8) the transmitting and receiving antennas.

Sensing Elements

The sensing element is a transducer that converts the variable to be measured into a suitable electrical quantity. The sensing element for the BCO is a 120-ohm strain gauge or a resistance type temperature sensor, wired into a Wheatstone bridge type circuit with one, two, or four active gauges. When fewer than four gauges are active, the bridge must be completed with dummy gauges or resistors. The bridge resistance can range from 50 to 500 ohms, but to obtain maximum sensitivity and stability, the bridge impedance should be adjusted to 120 ohms.

The sensing element for the VCO is a copper-constantan thermocouple. The VCO has a built-in circuit for cold junction compensation. Other thermocouples can also be used, but they must be calibrated with the VCO, and the cold junction compensation has to be made manually. Other millivolt output devices can also be utilized with the VCO.

Transmitters and Batteries

The BCO transmitter can be used at temperatures from 78 to 140°F (26 to 60°C) and the VCO from −40 to 258°F (−40 to 126°C). All connections to the transmitters are made by soldered joints to pins protruding through the epoxy compound. Each transmitter has screw adjustments. One rotates a multi-turn potentiometer and adjusts the subcarrier oscillator center frequency. Another moves through the transmitter, and its position sets the transmitting radio frequency. The strain sensitivity or millivolt sensitivity can be changed by the connections to various pins protruding through the epoxy potting compound.

The transmitter is made up of two components: the subcarrier oscillator and the radio frequency oscillator. As was mentioned, the subcarrier can be bridge controlled (BCO) or voltage controlled (VCO). The subcarrier center frequency is 4,000 Hz, which frequency can be modulated ±400 Hz by the strain or voltage being measured. Using BCOs, a strain as large as 2,500 microinches per inch (2.5 microns/mm) and as small as 2 microinches per inch (0.002 microns/mm) can be measured and transmitted. The temperature measurement range of the VCO is from −200 to 4,000°F (−129 to +2209°C). With a copper-constantan thermocouple, a temperature change as small as 2°F (1.1°C) can be sensed and transmitted.

The single-resistance strain gauge transmitter does not have a subcarrier oscillator and can be used from −40 to 212°F (−40 to 100°C). It has only a radio frequency oscillator that is modulated by the sensor signal. For this reason, it is not suitable for static strain measurements and must be used for dynamic strain measurements only. It has a frequency response to 25,000 Hz or greater. A static strain signal transmitted by this device will drift. It is provided with self-contained rechargeable nickel-cadmium batteries. Pins protruding through the epoxy case are used for all electrical connections. Only one screw adjustment is provided, and this is used to set the radio frequency.

Rechargeable nickel-cadmium batteries are used with the BCO and the VCO. The BCO batteries have useful lives of 4 and 9 hours. A VCO battery has 40 hours useful life. The single-resistance strain gauge transmitter has a built-in nickel-cadmium battery with a life of 4 hours.

Receivers and Discriminators

Receivers are cabinet mounted along with the discriminators and a power supply. The receiver has a tuning range of 88 to 108 MHz, and the subcarrier discriminator is a phase-locked type with a frequency of 4,000 Hz.

When the transmitter is used in its greatest sensitivity mode, the output of the discriminator is approximately 1 volt for a 25 microinch per inch (0.025 microns/mm) strain with a single active gauge in the bridge. At the most insensitive mode one volt is obtained for approximately 500 microinches per inch (0.5 microns/mm) strain. The discriminator can withstand a 500 percent overload, which means that a five-volt signal will be obtained from a strain of 125 microinches per inch (0.125 microns/mm) at the maximum sensitivity and from 2,500 microinches per inch (2.5 microns/mm) at the minimum sensitivity.

Recorders

The recorder can be a 12-channel oscillograph using 8-in. (200 mm) recording paper. An accurate paper speed of 0.1 to 80 in. per second (2.5 mm to 2 m per second) with timing lines at 0.01, 0.1, 1.0, and 10-second intervals is a good selection. Galvanometers with a frequency response of 1,000 Hz and a coil resistance of 24 ohms are used in the oscillograph.

Antennas and Total System Operation

A nickel-cadmium battery supplies the power to the transmitter. For the BCO, the resistance change of the strain gauge changes the frequency of the subcarrier. In the case of the VCO, the millivolt output of the thermocouple changes the frequency of the subcarrier. This change modulates the radio frequency transmitted by antenna. The receiving antenna picks up the signal and conducts it by wire link to the radio receiver, which is tuned to the transmitting frequency. The radio receiver demodulates the FM carrier to reproduce the subcarrier signal. The subcarrier signal is then fed to the discriminator, which demodulates this signal to obtain a DC voltage, which is then amplified by the DC amplifier and recorded on the oscillograph. The oscillograph record, properly calibrated, is then a display of the strain in microinches per inch for the BCOs, or the temperature in degrees for the VCO. At the same time the DC signal can be read on a voltmeter and can be used as a check on the oscillograph.

The transmitter subcarrier oscillators are factory set to operate at a center frequency of about 4,000 Hz. They have a frequency range of ±400 Hz about the 4,000 Hz center frequency. The center frequency is set with a counter at the time of testing. The change of ±400 Hz is the information frequency change brought about by the change in strain or temperature measured by the sensor. It is this information frequency change that the discriminators isolate as a DC voltage change, which is proportional to the measured strain and is recorded on the oscillograph.

BIBLIOGRAPHY

Adler, A.J., "Radiotelemetry of Temperature and Strain," *Instrumentation Technology*, February, 1970.

Davis, R.E., "Large Scale Water Management Telemetry System," ISA Conference, 1976, No. 76-676.

Gruenberg, E.L., "Handbook of Telemetry and Remote Control," McGraw-Hill Book Co., New York, 1967.

Kemp, R.E., "Close Coupled Telemetry for Obtaining Large Quantities of Strain and Temperature Measurements from Rotating Components of a Gas Turbine Engine," Instrumentation in the Aerospace Industry, Vol. 26.

Proceedings of the International Telemetering Conference, San Diego, CA, October 14–18, 1980.

Strock, O.J., "Telemetry Systems, Past, Present, Future," Paper presented at the International Telemetry Conference, Los Angeles, CA, November 14–16, 1978.

2.10 THERMOSTATS AND HUMIDOSTATS

Types:	A—Conventional HVAC thermostats B—Advanced thermostats C—Conventional HVAC humidostats D—Advanced humidostats
Standard Ranges:	A—55–85°F (12–30°C) B—40–90°F (5–32°C) C and D—20–90% RH
Inaccuracies:	Sensor Error: A—±1°F (0.5°C) B—±0.5°F (0.28°C) C—±5% RH D—±2% RH Offset Error: A—±1–5°F (0.5–2.8°C) B—Zero with integral C—±2–10% RH D—Zero with integral
Air Capacities:	Relay Types (high volume): 10–30 SCFH (0.3–0.84 m³/hr) Nonrelay types (low-volume): 1 SCFH (0.028 m³/hr)
Costs:	A—$50 to $100 B—$100 and up C—$100 to $200 D—$400 and up Installed total costs vary from two to seven times the hardware costs listed above.
Partial List of Suppliers:	Barber-Colman Co.; Edwards Engineering Corp., Emerson-Chromalox Div.; The English Electric Corp.; Fenwall, Inc., Div. of Walter Kiddie & Co.; General Electric Co.; Honeywell, Inc.; ITT General Controls; Johnson Controls, Inc.; MCC Powers; Mercoid Corp.; PSG Industries, Inc.; Robertshaw Controls Co.; Staefa Control System; Westinghouse Electric Corp.

Thermostats and humidostats have been developed to serve the heating, ventilating, and airconditioning (HVAC) industry. While energy costs were low, the consideration of performance was secondary to the purchase price of these devices. When energy costs increased, new, better-quality units were developed, but the conventional HVAC devices still remained in use in existing systems and in new installations where the designers were not sensitive to the quality of performance. In this section, the conventional quality units and the more advanced designs are both described.

What Is a "Stat"?

A thermostat or humidostat is a simplified controller, having only one control mode—proportional. The pressure of the output signal from a pneumatic "stat" is a near straight-line function of the measurement, described by the following relationship:

$$0 = K_c(M - M_0) + 0_0 \qquad 2.10(1)$$

where

0 = Output signal,

K_c = Proportional sensitivity (K_c can be fixed or adjustable depending on the design),

M = Measurement (temperature or relative humidity),

M_0 = "Normal" value of measurement corresponding to the center of the throttling range, and

O_0 = "Normal" value of the output signal, corresponding to the center of the throttling range of the control valve (or damper). With a spring range of 3–15 PSIG (0.2–1.0 bar), O_0 is 9 PSIG (0.6 bar); with a 9–13 PSIG (0.6–0.9 bar) spring range, O_0 is 11 PSIG (0.75 bar); etc.

If a stat (a proportional-only controller) is controlling a valve or a damper, one needs to determine K_c, M_0, and O_0 in order to draw the straight-line relationship between measurement and output signal. Figure 2.10a illustrates the behavior of a control loop consisting of a direct acting thermostat, having a fixed K_c of 2.5 PSIG/°F (0.3 bar/°C), a "normal" measurement value of M_0 = 72°F (22°C) and a damper spring range of 8–13 PSIG (0.55–0.9 bar), resulting in O_0 = 10.5 PSIG (0.73 bar).

Does a Stat Have a Set Point?

It is a common error to call M_0 in Figure 2.10a the set point of that thermostat. Stats *do not have* set points in the sense of having a predetermined temperature value

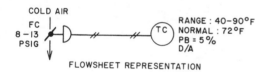

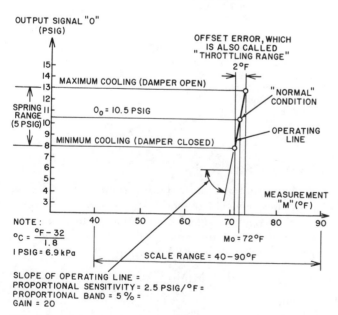

NOTE :
$$°C = \frac{°F - 32}{1.8}$$
1 PSIG = 6.9 kPa

SLOPE OF OPERATING LINE =
PROPORTIONAL SENSITIVITY: 2.5 PSIG/°F =
PROPORTIONAL BAND = 5% =
GAIN = 20

Fig. 2.10a Operating characteristics of a fixed proportional band thermostat

to which they would seek to return the condition of the controlled space. (One must add integral action in order for a controller to be able to return the measured variable to a set point after a load change.) M_0 does not represent a set point; it only identifies the space temperature that will cause the cooling damper in Figure 2.10a to be 50 percent open. This can be called a "normal" condition, because relative to this point the thermostat can both increase and decrease the cooling air flowrate as space temperature changes. If the cooling load doubles, requiring the damper to be fully open, this cannot take place until the controlled space temperature has first risen to 73°F (23°C). As long as the cooling load remains that high, the space temperature must also stay up at the 73°F (23°C) value. Similarly, the only way this thermostat can reduce the opening of the cooling damper below 50 percent is to first allow the space temperature to drift down below 72°F (22°C). Therefore stats do not have set points, but they have throttling ranges, and if a throttling range is narrow enough, this gives the appearance that the controller is keeping the variable near set point, when in fact the narrow range allows the variable to drift within limits.

Gain, Proportional Band, Proportional Sensitivity, and Throttling Range

The control response of thermostats and humidostats is described by the slope of the operating line in Figure 2.10a. This slope is described in different ways depending on the industry and manufacturer involved. People in the process control fields tend to be more familiar with terms such as "gain" and "proportional band," while individuals in the HVAC field are more used to the terms "proportional sensitivity" or "throttling range." Because all of these terms describe the same slope, each is defined below and shown in Figure 2.10a:

The *throttling range* is the gap within which the space conditions are allowed to drift as the final control element is modulated from fully closed to fully open (2°F or about 1°C in Figure 2.10a).

Proportional sensitivity is the amount of change in the output signal pressure that results from a change of one unit in the measurement (2.5 PSIG/°F or about 0.31 bar/°C in Figure 2.10a).

Gain is the ratio between the sizes of the changes in control output and measurement input. In the case of the example in Figure 2.10a, a 100 percent change in output (moving the damper from fully open to fully closed) will result from a 2°F (1°C) change in measurement, which represents only 5 percent of the thermostat span of 40 to 90°F (6.6 to 32°C). Therefore the gain of this thermostat is 20.

Proportional band is related to gain as follows: PB = 100/G = 100/20 = 5%. Therefore the proportional band of the thermostat in Figure 2.10a is 5 percent.

Having defined the "normal" conditions (M_0 and O_0)

and also the slope of the operating line, the behavior of the thermostat is now fully defined.

Stat Action

The thermostat in Figure 2.10a is said to be direct acting (D/A) because its output signal is increased as the measurement rises. If, instead of the Fail Closed (FC) damper, a Fail Open damper were used, a reversal in the thermostat action would be required, because now maximum cooling would take place when the thermostat output signal is low. Hence in that case a reverse-acting (R/A) thermostat would be used and the left-to-right slope of the operating line in Figure 2.10a would change from upward to downward.

The Effect of Spring Ranges

The operating line in Figure 2.10a varies with the spring range of the final control element. The typical spring ranges that might be used on various applications are listed below:

Fail Closed HVAC Quality Dampers
    A—3 to 7 PSIG (0.2 to 0.5 bar)
    B—5 to 10 PSIG (0.35 to 0.7 bar)
    C—8 to 13 PSIG (0.55 to 0.9 bar)

HVAC Quality Valves
    D—Fail Open, 4 to 8 PSIG (0.9 to 0.55 bar)
    E—Fail Closed, 9 to 13 PSIG (0.6 to 0.9 bar)
    F—3-way, 7 to 11 PSIG (0.5 to 0.76 bar)

Fail Closed Speed or Blade Pitch Positioners
    G—3 to 15 PSIG (0.2 to 1.0 bar)

Figure 2.10a illustrated the operating line of a thermostat with a 2.5 PSIG/°F (0.3 bar/°C) proportional sensitivity in combination with a final control element having a type C spring. If the range spring of the final control element is changed, this will also change the operating lines, as illustrated in Figure 2.10b and tabulated below:

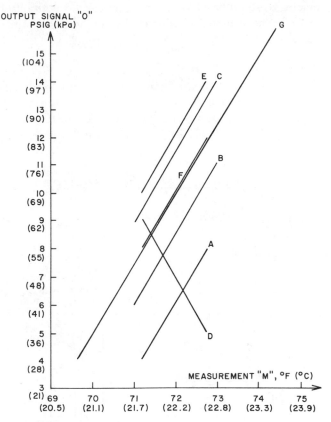

Fig. 2.10b   The effect of spring range variations on gain and throttling range

range, the larger the control error (offset or drift away from the "normal" measurement value). On the other hand, the wider the throttling range, the more stable (less cycling) the final control element is likely to be. Therefore, the tradeoff is between lower stability or higher offset error. By narrowing the spring ranges to 4 or 5 PSIG (0.3 or 0.35 bar), the conventional HVAC control manufacturers have selected lower stability as the more acceptable of the two undesirable options. The process

Range Spring Type	PB (%)	Gain	Proportional Sensitivity PSIG/°F (bar/°C)	Throttling Range °F (°C)	"Normal" Output PSIG (bar)	Spring Range PSIG (bar)
A	3.2	31	2.5 (0.3)	1.6 (0.9)	5.0 (0.35)	3–7 (0.2–0.5)
B	5.0	20	2.5 (0.3)	2.0 (1.1)	7.5 (0.52)	5–10 (0.35–0.7)
C	5.0	20	2.5 (0.3)	2.0 (1.1)	10.5 (0.72)	8–13 (0.55–0.9)
D	3.2	31	2.5 (0.3)	1.6 (0.9)	6.0 (0.4)	4–8 (0.3–0.55)
E	3.2	31	2.5 (0.3)	1.6 (0.9)	11.0 (0.76)	9–13 (0.6–0.9)
F	3.2	31	2.5 (0.3)	1.6 (0.9)	9.0 (0.6)	7–11 (0.5–0.76)
G	9.6	10	2.5 (0.3)	4.8 (2.7)	9.0 (0.6)	3–15 (0.2–1.0)

From the above, it is obvious that the range spring selection can substantially affect the throttling range and the apparent proportional band. The wider the throttling

control industry has made just the opposite choice as shown in Figure 2.10b by the 12 PSIG (0.8 bar) spring line (line G). This selection stems from the desire to maximize stability while being unconcerned with offset error because it was eliminated by the addition of the integral mode.

Sensitivity Adjustment

The simplest stats are manufactured with fixed sensitivities. Such a unit was illustrated in Figure 2.10a, having a fixed proportional sensitivity of 2.5 PSIG/°F (0.3 bar/°C). Control flexibility is improved in those designs where the proportional sensitivity is adjustable. Figure 2.10c illustrates a typical 5:1 range of adjustability applicable to standard thermostats. Lowering the proportional sensitivity tends to improve stability, but it also increases offset error due to the widening of the throttling range.

Thermostat Design Features

Figure 2.10d illustrates the design of the simplest thermostat that is manufactured. This fixed-proportional controller uses a flapper arrangement, which moves in front of a nozzle as the temperature detected by a bimetallic element changes relative to the manually adjusted "normal" value. Air at about 20 PSIG (140 kPa) is supplied through a restriction to the open nozzle, the back pressure of which is inversely proportional to the distance between nozzle and flapper. The proportional sensitivity K_c is the change in output pressure per °F measured. This is a fixed value in Figure 2.10d, because the ratio A/B is fixed, while it is adjustable in Figure 2.10e.

Due to the small restriction and nozzle openings in the design shown in Figure 2.10d, the resulting output airflow is rather small, around 1.0 SCFH (0.028 m³/hr). This type of design is also referred to as "nonrelay" or "low volume." The 1.0 SCFH air capacity is frequently insufficient to fill connecting tubing and operate final control elements. In order to increase the air capacity of a thermostat, a booster or repeater relay can be added. Such "relay type" or "high-volume" humidostats or thermostats will usually provide an output airflow of 10–30 SCFH (0.3–0.84 m³/hr), which is sufficient for most applications. In case of the relay type design (Figure 2.10e) the nozzle back pressure, instead of operating a final control element directly, is carried to the bellows chamber and acts against the area of the bellows. Because the bellows has some stiffness (spring gradient), the pilot

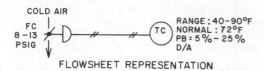

FLOWSHEET REPRESENTATION

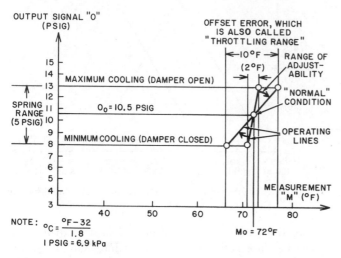

NOTE: $°C = \dfrac{°F - 32}{1.8}$

1 PSIG = 6.9 kPa

RANGE OF ADJUSTABILITY:

OFFSET ERROR OR THROTTLING RANGE: 2 TO 10°F (1.1 TO 5.6°C)
PROPORTIONAL SENSITIVITY: 2.5 TO 0.5 PSIG/°F (31 TO 6.3 kPa/°C)
PROPORTIONAL BAND: 5% TO 25%
GAIN: 20 TO 4

Fig. 2.10c   Operating characteristics of adjustable proportional band thermostats

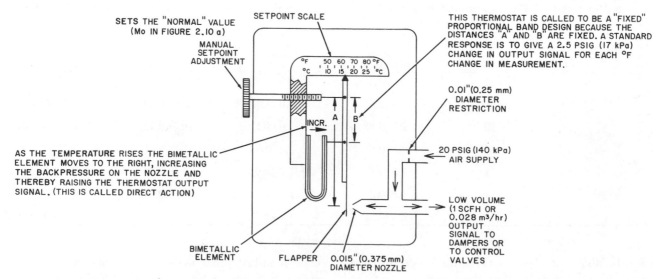

Fig. 2.10d   Direct-acting bimetallic thermostat with manual set point and fixed proportional band. The output air flow is limited because there are no relays, so this is called a "low-volume" design.

4. The proportional sensitivity determined by the A/B ratio can be fixed or adjustable
5. Feedback bellows can be added to minimize the effects of variations in supply pressure or temperature, to reduce output leakage, and to increase the range of adjustability of the proportional sensitivity setting

Accuracy of Thermostats and Humidostats

In conventional HVAC pneumatic thermostats and humidostats, the total error is the sum of the sensor error and of the offset error. Even after individual calibration, the sensor error cannot be reduced to less than $\pm 1°F$ ($\pm 0.55°C$) for thermostats or to $\pm 5\%$ RH for humidostats. Added to the sensor error is the offset error, which increases as the throttling range increases and which can be as high as $\pm 5°F$ ($\pm 2.8°C$) for thermostats and $\pm 20\%$ RH for humidostats. Therefore the total errors can approach $\pm 6°F$ ($\pm 3.3°C$) and $\pm 25\%$ RH respectively.

In conventional HVAC electrical thermostats or humidostats, the total error is the sum of three components. The first error component is the element error, which is the same as the sensor error in pneumatic designs. The second error component is the mechanical differential set in the thermostat switch, which does not modulate the final control element, but turns it on and off. The differential (or bandwidth) can vary for thermostats from 0.5 to 10°F (0.3 to 5.6°C) and for humidostats from 2 to 50% RH. The third error component is due to the fact that turning on a heat or humidity source does not instantaneously return the space to within the control differential. Instead, conditions will continue to deviate further for a while. This total error resulting from these three contributing factors can be as high as for HVAC pneumatics, although if the final control elements are sized large enough and if the control differential is set small, electrical units can outperform their pneumatic counterparts.

If space conditions are to be accurately maintained, such as in clean rooms where the temperature must be controlled within $\pm 1°F$ ($\pm 0.55°C$) and the humidity must be kept within $\pm 3\%$ RH, neither of the above designs are acceptable. In such cases it is necessary to use a resistance temperature detector type or a semiconductor transistor type temperature sensor and a proportional-plus-integral thermostat controller, which will eliminate the offset error. This can be most economically accomplished through the use of microprocessor-based shared controllers that communicate with sensors over data highways.

Electrical Thermostats and Humidostats

The movement produced by a bimetallic temperature sensor or by the expansion of a humidity-sensitive element can be used to build two-position controllers. The electric thermostats in Figure 2.10f illustrate the various

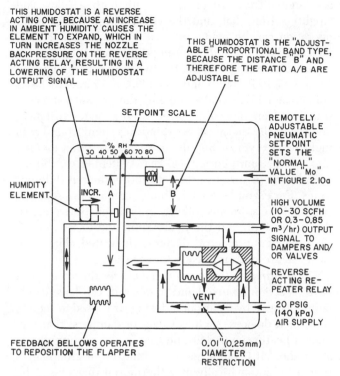

THIS HUMIDOSTAT IS A REVERSE ACTING ONE, BECAUSE AN INCREASE IN AMBIENT HUMIDITY CAUSES THE ELEMENT TO EXPAND, WHICH IN TURN INCREASES THE NOZZLE BACKPRESSURE ON THE REVERSE ACTING RELAY, RESULTING IN A LOWERING OF THE HUMIDOSTAT OUTPUT SIGNAL

THIS HUMIDOSTAT IS THE "ADJUSTABLE" PROPORTIONAL BAND TYPE, BECAUSE THE DISTANCE "B" AND THEREFORE THE RATIO A/B ARE ADJUSTABLE

SETPOINT SCALE

% RH
30 40 50 60 70 80

HUMIDITY ELEMENT

INCR. A   B

REMOTELY ADJUSTABLE PNEUMATIC SETPOINT SETS THE "NORMAL" VALUE "Mo" IN FIGURE 2.10a

HIGH VOLUME (10-30 SCFH OR 0.3-0.85 m³/hr) OUTPUT SIGNAL TO DAMPERS AND/OR VALVES

REVERSE ACTING REPEATER RELAY

VENT

20 PSIG (140 kPa) AIR SUPPLY

FEEDBACK BELLOWS OPERATES TO REPOSITION THE FLAPPER

0.01"(0.25mm) DIAMETER RESTRICTION

Fig. 2.10e   Reverse-acting humidostat with pneumatic set point and adjustable proportional band. The use of the repeater relay increases the humidostat output flow capacity and therefore this design is also referred to as "high volume."

valve plug is positioned between the inlet and outlet ports in accordance with the nozzle back pressure. A low nozzle back pressure positions the valve plug to the left, throttles the exhaust port, and causes a high output pressure. The advantages of adding the pilot are (1) the actuating signal versus output pressure relation can be made linear and (2) the capacity for air flow can be considerably increased.

Humidostat Design Features

As shown in Figure 2.10e, these units are similar in design to thermostats, except that the bimetallic measuring element is replaced by a humidity-sensitive element. A common humidity-sensitive element is cellulose acetate butyrate. This substance expands and contracts with changes in relative humidity, and the resulting movement can be used to operate the flapper of the pneumatic humidostat shown in Figure 2.10e or to open and close the contacts of electric humidostats.

The choice of features is similar to those of thermostats. Humidostats can also have the following characteristics:

1. Direct or reverse acting
2. High or low volume (relay type or nonrelay type)
3. The "normal" value can be set manually or by remotely adjustable pneumatic signal

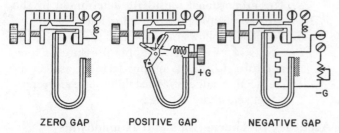

ZERO GAP   POSITIVE GAP   NEGATIVE GAP

Fig. 2.10f   Design variations of electric two-position thermostats

differential gaps in two-position control. The thermostat on the left operates load contacts directly from the bimetallic element. The set point is adjusted by setting one contact or by repositioning the whole bimetal element. The gap in this arrangement is theoretically zero, although in actual practice there is a very small gap due to contact "stickiness."

A positive differential gap is often necessary in two-position control in order to save wear on the control apparatus. The mechanism shown in Figure 2.10f employs a toggle action to produce a positive gap. The gap is adjustable by the tension of the spring because this determines the suppressive (hysteresis) force acting on the element.

A negative differential gap is often used on domestic thermostats in order to reduce an abnormally long period of oscillation. An auxiliary heater operated by the load contacts turns on a small heater within the thermostat when the contact is made. Thus the bimetal element opens the load contact before the actual room temperature attains the same point. The load contact make point is normal. This is sometimes termed "anticipation."

There are many forms of two-position thermostat or humidostat type controllers for both domestic and industrial application. These offer many arrangements of load contacts, set point, and differential gap. Generally a solenoid valve of one type or another or an electrical heating element, motor, or other power device is operated by the controller.

For floating controller action a neutral zone is usually necessary. This is obtained by a second contact position in the controller and is often obtained with two independently set on-off control mechanisms.

Single-speed floating control can be achieved by means of the same equipment as for two-position control with an electric-motor-operated valve. The difference lies only in the speed of operation of the motor or valve actuator. In two-position control the valve stroke is 120 seconds or less. In single-speed floating control the valve stroke is usually 120 seconds or more. An electrical interrupter is sometimes employed in conjunction with the motor to decrease the speed of opening and closing the valve.

Special-Purpose Thermostats

Night-day, *set-back*, or *dual room* thermostats will operate at different "normal" temperature values for day and night. They are provided with both a "day" and "night" setting dial, and the change from day to night operation can be made automatic for a group of thermostats. The *pneumatic day-night* thermostat uses a two-pressure air supply system, the two pressures often being 13 and 17 PSIG (89.6 and 117 kPa) or 15 and 20 PSIG (103.35 and 137.8 kPa). Changing the pressure at a central point from one value to the other actuates switching devices in the thermostat and indexes them from day to night or vice versa. Supply air mains are often divided into two or more circuits so that switching can be accomplished in various areas of the building at different times. For example, a school building may have separate circuits for classrooms, offices and administrative areas, the auditorium, and the gymnasium and locker rooms. In some of the electric designs, dedicated clocks and switches are built into each thermostat.

The *heating-cooling* or *summer-winter* thermostat can have its action reversed and, if desired, can have its set point changed by means of indexing. It is used to actuate controlled devices, such as valves or dampers, that regulate a heating source at one time and a cooling source at another. It is often manually indexed in groups by a switch, or automatically by a thermostat that senses the temperature of the water supply, the outdoor temperature, or another suitable variable.

In the heating-cooling design there frequently are two bimetallic elements, one being direct acting for the heating mode, the other being reverse acting for the cooling mode. The mode switching is done automatically in response to a change in the air supply pressure, similarly to the operation described for the day-night thermostats.

The *limited control range* thermostat usually limits the room temperature in the heating season to a maximum of 75°F (24°C), and in the cooling season to a minimum of also 75°F (24°C), even if the occupant of the room has set the thermostat beyond these limits. This is done internally without placing a physical stop on the setting knob.

Zero energy band (ZEB) thermostats will provide heating when the zone temperature is below 68°F (20°C), will provide cooling when it is above, say, 78°F (25.6°C) (these are adjustable settings), and in between will neither heat nor cool. This approach can reduce the yearly operating cost by about 33 percent. This method of control can also allow buildings to become self-heating by transfering interior heat to the perimeter. (For a detailed discussion of self-heating, see Section 8.14.)

A *slave* or *submaster* thermostat has its set point raised or lowered over a predetermined range, in accordance with variations in the output from a master controller. The master controller can be a thermostat, manual switch, pressure controller, or similar device. For example, a master thermostat measuring outdoor air temperature can be used to adjust a submaster thermostat controlling the water temperature in a heating system. Master-submaster combinations are sometimes designated as single-cascade action. When such action is accomplished by a single thermostat having more than one measuring element, it is known as compensated control.

Multistage thermostats are designed to operate two or more final control elements in sequence.

A *wet-bulb* thermostat is often used for humidity control with proper control of the dry-bulb temperature. A wick, or other means for keeping the bulb wet, and rapid air motion to assure a true wet-bulb measurement are essential.

A *dew-point* thermostat is a device designed to control from dew point temperatures.

A *smart thermostat* is usually a microprocessor-based unit with an RTD type or a transistorized solid state sensor. It is usually provided with its own dedicated memory and intelligence, and it can also be provided with a communication link (over a shared data bus) to a central computer. Such units can minimize building operating costs by combining time-of-day controls with intelligent comfort gap selection and with maximized self-heating.

BIBLIOGRAPHY

ASHRAE Handbook and Product Directory, 1980 Systems, American Society of Heating, Refrigerating and Air-Conditioning Engineers, Inc.

Eckman, Donald P., *Automatic Process Controls*, John Wiley & Sons, New York, 1958.

"Gas Appliance Thermostats," AGA (American Gas Association) Standard: ANSI: Z21.23-1975.

"Method of Testing for Capacity Rating of Thermostatic Refrigerant Expansion Valves," ASHRAE Standard: 17-75 (ANSI: B60.1-1975).

"Residential Controls—Integrally Mounted Thermostats for Electric Heaters," NEMA (National Electrical Manufacturers' Association) Standard: DC13-1973.

"Residential Controls—Low Voltage Room Thermostats," NEMA Standard: DC3-1972.

"Standard for Fire-Detection Thermostats," UL (Underwriters' Laboratories) Standard: UL-521 (ANSI: Z220.1-1971).

"Standard Method of Test for Flexibility of Thermostat Metals," ASTM (American Society of Testing and Materials) Standard: B106-68 (1972).

"Standard Method of Test for Thermal Deflection Rate of Spiral and Helical Coils of Thermostat Metal," ASTM Standard: B389-65 (1975).

"Thermostatic Refrigerant Expansion Valves," ARI (Air-Conditioning and Refrigeration Institute) Standard: 750-76 (1976).

Chapter III
PANELS AND DISPLAYS

R. W. Borut • N. O. Cromwell

G. F. Erk • J. A. Gump

S. P. Jackson • D. W. Lepore

F. D. Marton • D. D. Russell

M. G. Togneri • J. Venczel

R. A. Williamson

CONTENTS OF CHAPTER III

3.1 ANNUNCIATORS

Types:

AUDIOVISUAL ANNUNCIATORS: integral, remote, and semigraphic systems with audible and visual display and electromechanical (relay) or solid state (semiconductor) design.

RECORDING ANNUNCIATORS: integral, solid state, high-speed systems with recorded printout.

VOCAL ANNUNCIATORS: integral, solid state systems with audible command message.

PNEUMATIC ANNUNCIATORS: air-operated equivalent of trouble contact, logic-module, and visual indicator stages of an electrical annunciator.

Cost per Alarm Point:

integral cabinet, $100; remote system, $150; semigraphic system, $200; recording annunciator, $200; vocal annunciator, $300. Equipment cost per point decreases as system size increases. The semigraphic system cost does not include the price of the graphic display. The vocal annunciator cost does not include the price of the communication system equipment.

Partial List of Suppliers:

AUDIOVISUAL ANNUNCIATORS: Acromag, Inc.; Beta Products, Inc.; Cusco Industries, Inc.; Electro Devices, Inc.; Fisher Controls International, Inc.; Foxboro Co.; Halmar Electronics, Inc.; Kaye Instruments; Riley Panalarm Div., U.S. Riley Co.; Rochester Instrument Systems; Ronan Engineering Co.; Struthers-Dunn, Inc.; Swanson Engineering & Mfg. Co.; Technology, Inc.

PNEUMATIC ANNUNCIATORS: HEP Controls, Inc.

RECORDING ANNUNCIATORS: Hathaway Instruments, Inc.; Riley Panalarm Div., U.S. Riley Co.; Rochester Instruments Systems.

VOCAL ANNUNCIATORS: C.A. Briggs Co., Cybersonic Div.; Transmation, Inc.

History and Development

The literature contains little to document the development of current industrial annunciator systems. The term "drop" was initially applied to individual annunciator points in process applications, from which we may infer that annunciator systems developed from paging systems of the type used in hospitals and from call systems used in business establishments to summon individuals when their services were needed. These systems consisted of solenoid operated nameplates that dropped as a result of gravity when de-energized. The drops were grouped at a central location and were energized by pressing an electrical pushbutton in the location requiring service. The system also included an audible signal to sound the alert.

Similar systems were used for fire and burglar alarms. The drops were operated either by manual switches or by trouble contacts that monitored thermal and security conditions in various building locations. The use of these systems in the chemical processing industry was a logical development as the necessary monitoring switches became available.

This development, however, was preceded by explosion-proof, single-station annunciators that were designed to operate in the petroleum and organic chemical process plants constructed immediately before, during, and after World War II. They were usually installed on control panels located either outdoors at the process unit or in local control houses. These locations, being electrically hazardous, prevented the use of drop-type systems.

By the late 1940s centralized control rooms were introduced from which the plant could be remotely operated. A drop-type annunciator could be used in these general-purpose central control rooms. However, more compact, reliable, and flexible annunciators were subsequently introduced.

In the early 1950s the plug-in relay annunciator was developed. Instead of utilizing solenoid operated drops, it used electrical annunciator circuits with small telephone-type relays to operate alarm lights and to sound a horn when abnormal conditions occurred. The alarm lights installed in the front of the annunciator cabinets were either the bull's-eye type or backlighted nameplates. The annunciators were compact and reliable, and because of the hermetically sealed relay logic modules, they could be used in certain hazardous areas in addition to the general-purpose control rooms. Miniaturization of instruments and the use of graphic control panels initiated the development of remote annunciator systems consisting of a remotely mounted relay cabinet connected to alarm lights installed at appropriate points in the graphic or semigraphic diagram.

Solid state annunciator systems with semiconductor logic modules were developed in the late 1950s. These permitted additional miniaturization and lowered both the operating power requirements and the heat generated. The semigraphic annunciator was introduced in the late 1960s and fully utilized the high-density capabilities of solid state logic. It has permitted very compact and flexible semigraphic control centers. The trend toward additional miniaturization is the result of the greater availability and reliability of integrated circuit logic components.

Principles of Operation

The basic annunciator system consists of multiple individual alarm points, each connected to a trouble contact (alarm switch), a logic module, and a visual indicator (Figure 3.1a). The individual alarm points are operated from a common power supply and share a number of annunciator system components, including an audible signal generator (horn), a flasher, and acknowledge and test pushbuttons. In normal operation the annunciator system and individual alarm points are quiescent.

The trouble contact is an alarm switch that monitors a particular process variable and is actuated when the variable exceeds preset limits. In electrical annunciator systems it is normally a switch contact that closes (makes)

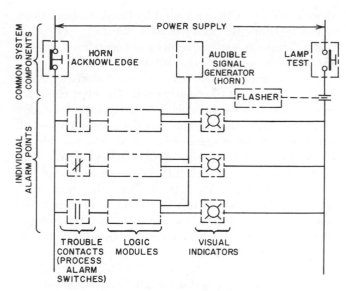

Fig. 3.1a Elements of basic annunciator system

or opens (breaks) the electrical circuit to the logic module and thereby initiates the alarm condition. In the alert stage the annunciator turns on the visual indicator for the particular alarm point and the audible signal and the flasher for the system. The visual indicator is usually a backlighted nameplate engraved with an inscription to identify the variable and the abnormal condition, but it can also be a bull's-eye light with a nameplate. The audible signal can be a horn, a buzzer, or a bell.

The flasher is common to all individual alarm points and interrupts the circuit to the visual indicator as that point goes into the alert condition. This causes the light to continue to flash intermittently until either the abnormal condition returns to normal or it is acknowledged by the operator. The horn acknowledgment pushbutton is provided with a momentary contact; when it is operated, it changes the logic module circuit to silence the audible signal, stop the flasher, and turn the visual indicator on "steady." When the abnormal condition is corrected, the trouble contact returns to normal and the visual indicator is automatically turned off. The lamp test pushbutton with its momentary contact tests for burned-out lamps in the visual indicators. When activated, the pushbutton closes a common circuit (bus) to each visual indicator in the annunciator system, turning on those lamps that are not already on as a result of an abnormal operating condition.

Operating Sequences

The operation of an individual alarm point in the normal, alert, acknowledged; and return-to-normal stages is the annunciator sequence. A wide variety of sequences can be developed from commercially available logic components; many special sequences have been designed to suit the requirements of particular process applications. The five most commonly used annunciator sequences are shown in Table 3.1b, identified by the code designation of the Instrument Society of America (ISA). (For additional details on less frequently used sequences, see the ISA recommended practice RP-18.1.)

ISA Sequence 1B, also referred to as flashing sequence A, is the one most frequently used. The alert condition of an alarm point results in a flashing visual indication and an audible signal. The visual indication turns off automatically when the monitored process variable returns to normal. ISA Sequence 1D (often referred to as a dim sequence) is identical to Sequence 1B except that ordinarily the visual indicator is dim rather than off. A dimmer unit, common to the system, is required. Because all visual indicators are always turned on—for dim (normal), flashing (alert), or steady (acknowledged)—the feature for detecting lamp failure is unnecessary. ISA Sequence 2A (commonly referred to as a ringback sequence) differs from Sequence 1B in that following acknowledgment the return-to-normal condition produces

a dim flashing and an audible signal. An additional momentary contact reset pushbutton is required for this sequence. Pushing the reset button after the monitored variable has returned to normal turns off the dim flashing light and silences the audible signal. This sequence is applied when the operator must know if normal operating conditions have been restored.

ISA Sequence 2C is like Sequence 1B except that the system must be reset manually after operation has returned to normal in order to turn off the visual indicator. This sequence is also referred to as a manual reset sequence and, like Sequence 2A, requires an additional momentary contact reset pushbutton. Sequence 2C is used when it is desirable to keep the visual indicator on (after the horn has been silenced by the acknowledgment pushbutton) even though the trouble contact has returned to normal.

ISA Sequence 4A, also known as the first out sequence, is designed to identify the first of a number of interrelated

Table 3.1b
MOST COMMONLY USED ANNUNCIATOR SEQUENCES

ISA Code for the Sequence	Annunciator Condition	Process Variable Condition (Trouble Contact)	Visual Indicator	Audible Signal	Use Frequency
1B	Normal	Normal	Off	Off	
	Alert	Abnormal	Flashing	On	
	Acknowledged	Abnormal	On	Off	55%
	Normal again	Normal	Off	Off	
	Test	Normal	On	Off	
1D	Normal	Normal	Dim	Off	
	Alert	Abnormal	Flashing	On	1%
	Acknowledged	Abnormal	On	Off	
	Normal again	Normal	Dim	Off	
2A	Normal	Normal	Off	Off	
	Alert	Abnormal	Flashing	On	
	Acknowledged	Abnormal	On	Off	
	Return to normal	Normal	Dim flashing	On	4%
	Reset	Normal	Off	Off	
	Test	Normal	On	Off	
2C	Normal	Normal	Off	Off	
	Alert	Abnormal	Flashing	On	
	Acknowledged	Abnormal	On	Off	
	Return to normal	Normal	On	Off	5%
	Reset	Normal	Off	Off	
	Test	Normal	On	Off	
4A	Normal	Normal	Off	Off	
	Alert	Abnormal			
	Initial		Flashing	On	
	Subsequent		On	Off	
	Acknowledged	Abnormal			28%
	Initial		On	Off	
	Subsequent		On	Off	
	Normal again	Normal	Off	Off	
	Test	Normal	On	Off	
All others					7%

variables that have exceeded normal operating limits. An off-normal condition in any one of a group of process variables will cause some or all of the remaining conditions in the group to become abnormal. The first alarm causes flashing, and all subsequent points in the group turn on the steady light only. This sequence monitors interrelated variables. The visual indication is turned off automatically when conditions return to normal after acknowledgment.

Optional Operating Features

Annunciator sequences may be initiated by alarm switch trouble contacts that are either open or closed during normal operations. These are referred to as normally open (NO) and normally closed (NC) sequences, respectively, and the ability to use the same logic module for either type of trouble contact is called an *NO-NC option*. It is important because some alarm switches are available with either an NO or an NC contact but not with both, and therefore without the NO-NC option in the logic module two types of logic modules would be required. The logic module is converted for use with either form of contact by a switch or wire jumper connection.

The relationship between the NO and NC sequences required in the logic module to match the various trouble contacts and analog measurement signal actions is shown in Figure 3.1c. A high alarm in a normally closed annunciator system requires a normally closed trouble contact operated by a direct-acting analog input. If an increase in the measured variable results in an increased output signal, the detector is direct-acting; if the output signal is reduced, it is a reverse-acting sensor. If the trouble contacts in all alarm switches in the plant are standardized such that normal operating conditions will cause all trouble contacts to be NC (or NO), the required annunciator sequence is also NC (or NO), and Figure 3.1c need not be consulted.

Annunciator systems are fail-safe or self-policing if they initiate an alarm when the logic module fails because of relay coil burnout. The feature is standard for most NO and NC annunciator sequences; annunciators using NC trouble contacts are also fail-safe against failures in the trouble contact circuit.

The lock-in option locks in the alert condition initiated by a momentary alarm until the horn acknowledgment button is pushed, preventing loss of a transient alarm condition until the operator can identify it. The logic module is usually changed from lock-in to non–lock-in operation by either addition of a wire jumper or operation of a switch. The lock-in feature is useful for monitoring unstable or fluctuating process variables.

The test feature in a standard annunciator serves only to test for burned-out lamps in the visual indicators. The operational test feature provides a test of the complete annunciator system, including logic modules, lamps, flasher, audible signal, and acknowledgment circuits. The operational test circuit usually requires an additional momentary contact pushbutton, which can replace the regular lamp test pushbutton. The logic module of relay-type annunciators may have spare (electrically isolated) auxiliary contacts that can operate shut-down and interlock systems when alarm conditions occur. The auxiliary contacts are wired to terminal blocks in the annunciator cabinet for connection to external circuitry.

Repeater lights may be located away from the common logic module and serve to alert operators in other areas. Annunciator cabinet terminals for connecting these repeater lights in parallel with the annunciator visual indicator are available.

It may also be desirable to actuate a horn in more than one location. The electrical load of multiple audible signals requires an interposing relay, called a horn isolating relay, operated by the logic modules. This relay has contacts of adequate capacity to operate multiple audible signals. Horn-isolating relays may be installed either in the annunciator cabinet or in a separate assembly. Annunciator systems can be used for several operational sequences without changing system wiring, and many logic modules can supply more than one operational sequence. This multiple sequence capability is sometimes useful when the sequence has not yet been determined.

Audio-Visual Annunciators

The audio-visual annunciator may be packaged as an integral, remote, or semigraphic annunciator.

Integral Annunciator

The integral annunciator, a cabinet containing a group of individual annunciator points wired to terminal blocks for connection to external trouble contacts, power sup-

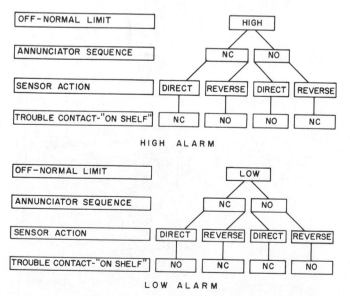

Fig. 3.1c Logic tree for NO and NC annunciator sequences

ply, horn, and acknowledge and test pushbuttons, is the most economical of the various packaging methods available in terms of cost per point. It is also the simplest and cheapest to install.

Two methods of packaging integral annunciators are illustrated in Figure 3.1d. In the nonmodular type, plug-in logic modules are installed inside the cabinet and connected to alarm windows on the cabinet door through an interconnecting wiring harness; in the modular type, individual plug-in alarm point assemblies of logic module and visual indicator are grouped together. The non-modular and modular cabinet styles are both designed for flush panel mounting with the logic modules and visual indicators accessible from the front. Electrical terminals for the external circuitry are located in the rear of the cabinet and are accessible from the back.

Integral annunciators are used on nongraphic and on semigraphic control panels in which physical association of the visual indicators with a specific location in the graphic process flow diagram is not required. Integral annunciator cabinets occupy more front but less rear panel space than the equivalent remote designs. The electrical terminals are in a general-purpose enclosure at the rear of the cabinet, and trouble contacts can be wired directly to them, thus eliminating the need for and resultant costs of intermediate terminal blocks for trouble contact wiring.

An advantage of the modular-type cabinet is that it can be expanded by enlargement of the panel cutout and addition of modular alarm point assemblies. Nonmodular cabinets cannot be expanded, and new cabinets must be installed to house additional alarm points. Consequently, one should include more spare points when specifying the cabinet size for a nonmodular system. The modular cabinet is also more compact, takes up less panel space, and has a greater visual display area per point than has the nonmodular design. Figure 3.1e illustrates various configurations of visual indicators that can be supplied with integral annunciator cabinets. Many of these groupings are also available in single-unit assemblies for remote annunciator systems.

Remote Annunciator

The remote annunciator differs from the integral annunciator in that the visual indicators are remote from the cabinet or chassis containing the logic modules. Remote annunciators were developed to allow the visual indicators to be placed in their actual process location in the graphic flow diagram. They are used with full and semigraphic control panels and in nongraphic applications in which an integral annunciator cabinet may require too much front panel space. Figure 3.1f illustrates a remote annunciator chassis with optional cabinet enclosure. The chassis contains spare positions for plug-in logic modules and a system flasher. Auxiliary system modules, such as horn isolating relays, may also be plugged into the logic module chassis positions. The chassis and

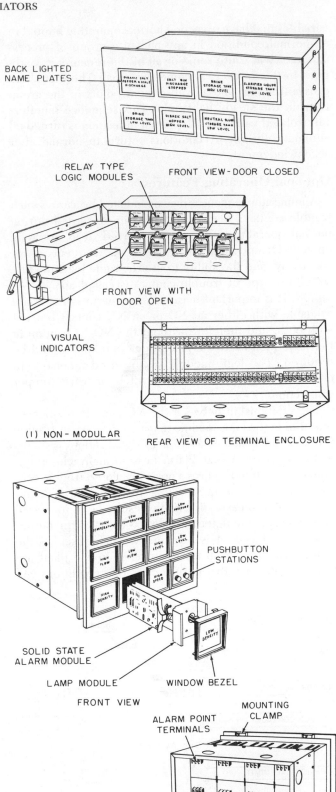

(1) NON-MODULAR

(2) MODULAR TYPE

Fig. 3.1d Integral annunciator cabinets, modular and nonmodular

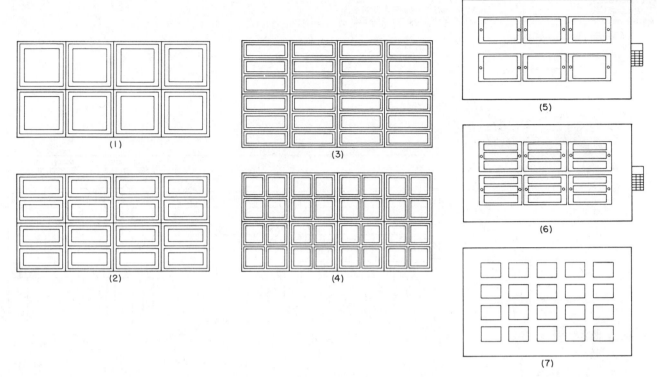

Fig. 3.1e   Integral annunciator window configurations. *1*. Modular single point annunciator. *2*. Modular double point annunciator. *3*. Modular triple point annunciator. *4*. Modular quadruple point annunciator. *5*. Nonmodular single point. *6*. Nonmodular triple point. *7*. Nonmodular single point with small nameplate.

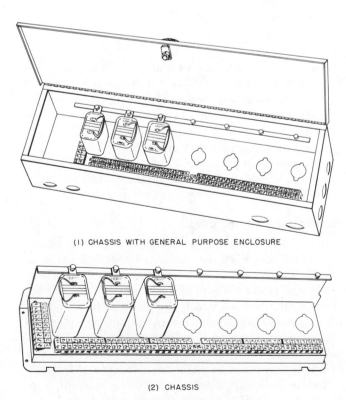

(1) CHASSIS WITH GENERAL PURPOSE ENCLOSURE

(2) CHASSIS

Fig. 3.1f   Remote annunciator cabinets

cabinet enclosure are designed for wall or surface mounting behind the control panel. Each chassis position has terminal points for connecting the visual indicator and trouble contact. In addition, the chassis has a system terminal block for connecting electrical power, horn, flasher, and acknowledge and test pushbuttons.

Remote annunciator disadvantages include higher equipment and installation costs and increased back panel space. In addition, the wiring connections from field trouble contacts must be made to intermediate terminal blocks rather than directly to the cabinet terminals, as with the integral annunciator. These terminal blocks, the terminal enclosure, and the required wiring result in higher installation costs and extra space requirements. Finally, the remote annunciator is difficult to change, and modification costs of remote systems are substantially higher than those of the integral type, partially because spare visual indicators cannot be installed initially.

Semigraphic Annunciator

The semigraphic annunciator developed in the late 1960s combines some of the advantages of the integral annunciator with the flexibility to locate visual indicators at appropriate points in a graphic flow diagram. Figure 3.1g illustrates a semigraphic annunciator. It consists of a cabinet containing annunciator logic modules wired to visual indicators placed in a ¾-inch (18.75 mm) lamp insertion matrix grid forming the cabinet front. The semigraphic display is placed between the lamp grid and a transparent protective cover plate, and the visual in-

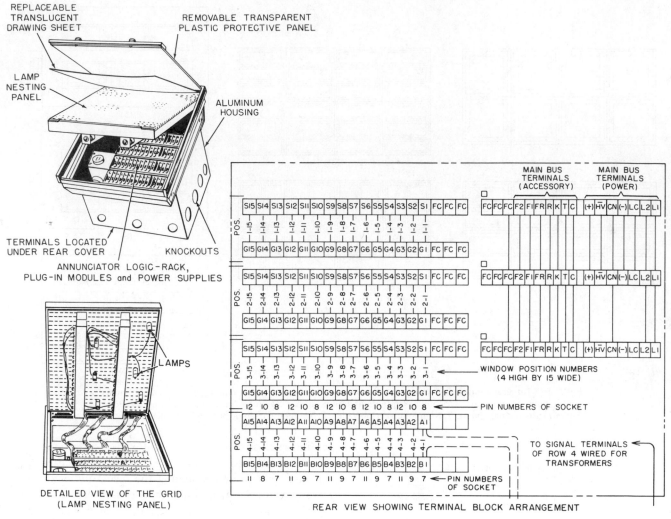

Fig. 3.1g   Typical semigraphic annunciator

dicators are positioned to backlight alarm nameplates located in the graphic display. The protective cover and lamp grid are either hinged or removable so as to provide access to the logic module and lamp assemblies. The lamp assemblies are connected to intermediate terminals located behind the lamp grid, and the terminals in turn are connected to the logic modules. Terminal points for trouble contact wiring are in the back of the cabinet.

The semigraphic annunciator is flexible, and changes in the annunciator system, graphic display, and related panel modifications can be made easily and cheaply. It is practical to prepare the graphic displays in the drafting room or model shop, thus protecting proprietary process information of a confidential nature. The graphic display has little or no effect on completing either the annunciator or the control panel because it can be installed on site or at any time. The semigraphic annunciator has a high density of 40 alarm points per linear foot (0.3 m) and a solid state rather than relay-type logic design. Power supplies are self-contained in the semigraphic annunciator cabinet.

Front panel layouts illustrating integral, remote, and semigraphic annunciators are shown in Figure 3.1h. Integral systems similar to the one shown at the left in the figure are normally specified on nongraphic control panels. The graphic panel in the center contains a remote annunciator with backlighted nameplates (shaded rectangles) and pilot lights (shaded circles) for visual indication. The remote system may also be used with miniature lamps in a semigraphic display similar to the one shown at the right.

Recording Annunciators

In recent years, solid state, high-speed recording annunciators have been developed, substituting a printed record of abnormal events for visual indication. These systems print out a record of the events and identify the variable, the time at which the alarm occurs, and the time at which the system returns to normal. They can also discriminate among a number of almost simultaneous events and print them out in the time sequence in which they occurred. A number of optional features, including secondary printers at remote locations, supplementary visual indication, and computer interfacing,

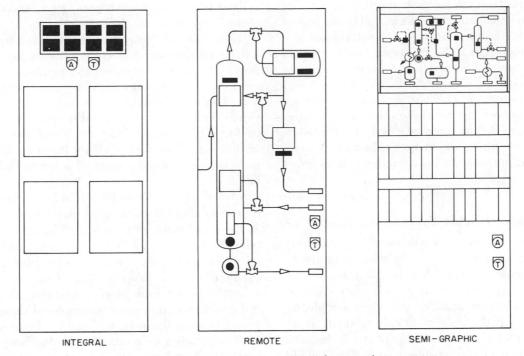

INTEGRAL   REMOTE   SEMI–GRAPHIC

Fig. 3.1h   Control panels with integral, remote, and semigraphic annunciators

are also available. The typical unit consists of logic, control, and printer sections.

The input status is continuously scanned, and if a change in the trouble contact has occurred since the preceding scan cycle, the central control operates to place the exact time, the alarm point identification, and the new status of the trouble contact (normal or abnormal) into the memory and initiates the operation of the output control unit (Figure 3.1i). The output control unit accepts the stored information and transfers it to the printer, which logs the event. Following this, the memory is automatically cleared of the data and is ready to accept new information. In addition to or in place of the printer (if a permanent record is not required), a CRT display can serve as the event readout. Trouble contacts are connected to terminal points in the logic cabinet, and a cable connects the cabinet and the printer.

A recording annunciator can perform more sophisticated monitoring than an audio-visual annunciator and is correspondingly more expensive on a per point basis. System cost per point decreases as the system size increases. Higher equipment cost, however, is offset in part by savings in control panel space and in installation costs. Recording annunciators are frequently used by the electrical power generating industry but may be applied to advantage in any industrial process that must monitor large numbers of operating variables and analyze abnormal events efficiently.

Vocal Annunciators

Vocal annunciators are unique in the type of abnormal audible message they produce. The audible output is a

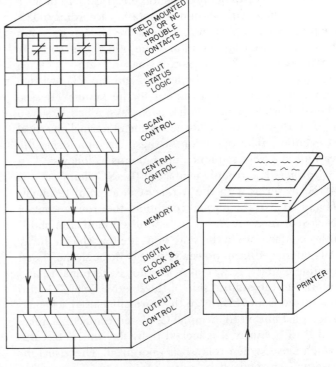

Fig. 3.1i   Functional block diagram of recording annunciator

verbal message identifying and describing the abnormal condition when it occurs and repeating the message until the operator acknowledges the difficulty. The system continuously scans the trouble contacts, and when an abnormality is found, it turns on a flashing visual indicator and selects the optional proper verbal message for broadcast. The visual indicator is turned off by the system when the point returns to normal. The control unit also arranges the messages to be broadcast in the order in which the difficulties occur. In the event of multiple alarms, the second message is played only after the first has been acknowledged. The flashing visual indicator for each point, however, turns on when the point becomes abnormal. The verbal message may be broadcast simultaneously in the control room and related operating areas, thus permitting personnel at the operating unit to correct the problem immediately.

Electromechanical Relay-Type Annunciators

The basic element of this annunciator is an electrical relay wired to provide the logic functions required to operate a particular sequence (see Section 6.5). At least two relays are required per logic module for most sequences. The relays are installed and wired in a plug-in assembly, which is the logic module for a single alarm point. The plug-in module assembly is usually hermetically sealed in an inert atmosphere to prolong the life of the relay contact. The sealing also makes the logic module acceptable in certain hazardous electrical areas.

Figure 3.1j is a semischematic electrical circuit for a remote system with sequence operation according to ISA Sequence 1B. Two logic modules are shown; one is in normally closed operation and the other is in normally open operation. The remote visual indicators for each alarm point, the horn, the flasher, and the acknowledge and test pushbuttons common to the system are also shown. Each logic module has two relays, A and B, shown in their de-energized state according to normal electrical convention. The operation of these circuits during the various stages of sequence operation is as follows:

Normal. The trouble contact of the NC alarm point is wired in series with the A relay coil at point terminals H and NC. In the normal condition, the trouble contact is closed, relays A and B are energized, and all A and B relay contacts are in the state opposite that shown. Relay A is energized from power source H through the closed trouble contact, relay coil A, resistor R, terminal K, and jumper to the neutral side of the line N. Relay B is also energized from H through the normally closed acknowledge pushbutton to terminal C, closed contact A2, relay coil B to N. Relay B is locked in by its own contact B2, which closes when relay B is energized. The visual indicator is turned off by open contact A5. The audible signal and the flasher motor are turned off by open contacts A3 and B4 in the same circuit.

Alert. The trouble contact opens, de-energizing relay A and returning all A relay contacts to the state shown in Figure 3.1j. The visual indicator is turned on, flashing through circuit H, lamp filament, terminal L, closed contacts A5 and B6, bus F, flasher contact F1 to N. The flasher motor is driven through circuit C, closed contacts

A3 and B3, R bus, flasher motor to N, and the audible signal is turned on by the same circuit.

Acknowledged. Relay B is de-energized by operating (opening) the momentary contact horn acknowledgment pushbutton and is locked out by open contact A2. All B relay contacts are returned to the state shown in Figure 3.1j. The visual indicator is turned on steady through closed contact B5 to N and is disconnected from flasher contact F1 by open contact B6. The flasher motor and audible signal are turned off by the horn acknowledgment pushbutton and remain off as a result of open contact B3.

Normal Again. When the variable condition returns to normal, the trouble contact closes to energize relay A, the visual indicator is turned off by open contact A5, relay B is energized by closed contact A2, and all circuits are again in the state described under normal.

Lamp Test. The lamp test circuit operates the visual indicators of only those alarm points that are in the normal condition. The circuit is completed through power source H, lamp filament, terminal L, closed contact A6, bus T, normally open momentary contact lamp test pushbutton to N. Closing the lamp test pushbutton to N completes the circuit and lights the visual indicators. Alarm points that are in the off-normal condition (either alert or acknowledged) do not operate because their A relays are de-energized and the A6 contact is open. The visual indicators of these abnormal alarm points are already turned on (either flashing or steady) through the operation of the alarm sequence.

Lock-in. The lock-in feature operates to prevent an annunciator alert condition (caused by a momentary alarm) from returning to normal until the horn acknowledgment button is pushed. Point terminals H and SL are jumpered to provide the lock-in feature (see Figure 3.1j). When the trouble contact opens, the A relay is de-energized, and power source H is applied to the N side of the relay through closed contacts A1 and B1. The power is dissipated through resistor R, terminal K, and jumper to N. If the trouble contact returns to normal, relay A will remain de-energized because potential H is on both sides of the coil. If the acknowledgment button is pushed before the trouble contact closes again, relay B will be de-energized, opening contact B1 and the lock-in circuit, thus permitting the system to return automatically to normal when the trouble contact closes. If the acknowledgment button is pushed after the trouble contact has reclosed, contact B1 opens momentarily, allowing the A relay to re-energize. Contact A1 opens, and the circuits are re-established in their normal operating state.

Operational Test. Full operational test is incorporated in the annunciator sequence shown by replacement of the jumper connection between main bus terminals K and N with a normally closed momentary contact pushbutton, which when pushed opens all annunciator circuits, thus initiating the alert condition of all alarm points in the normal condition.

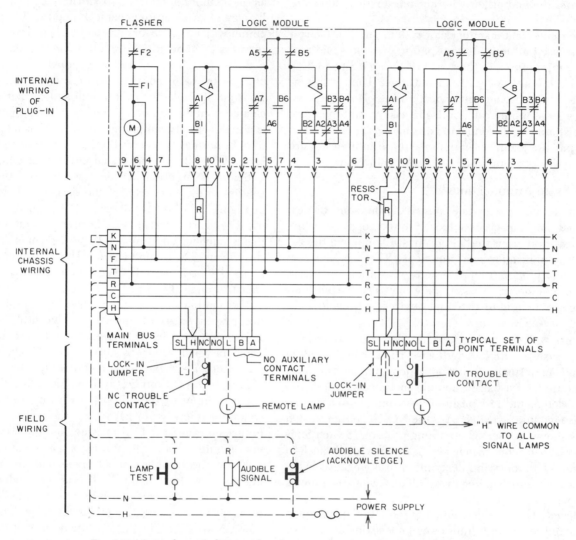

OPERATIONAL SEQUENCE

CONDITION	TROUBLE CONTACT	SIGNAL LAMPS	AUDIBLE SIGNAL	"A" RELAY	"B" RELAY
NORMAL	NORMAL	OFF	OFF	ENERGIZED	ENERGIZED
ALERT	ABNORMAL	FLASHING	ON	DEENERGIZED	ENERGIZED
AUDIBLE SILENCED (ACKNOWLEDGED)	ABNORMAL	STEADY – ON	OFF	DEENERGIZED	DEENERGIZED
NORMAL AGAIN	NORMAL	OFF	OFF	ENERGIZED	ENERGIZED
LAMP TEST	NORMAL	STEADY – ON	OFF	ENERGIZED	ENERGIZED

Fig. 3.1j   Semischematic diagram of a relay-type annunciator for ISA sequence 1B

Auxiliary Contacts. Normally closed contact A7 connected to point terminals A and B is available for auxiliary control functions.

Relay Fail-Safe Feature. Two parallel circuits, one consisting of closed contact A3 and open contact B3 and the other of open contact A4 and closed contact B4, operate an alert signal when there is a failure of either the A or the B relay coil. A failure of the former initiates a normal alert in the same way as the trouble contact. A failure of the B relay turns on the audible signal through closed contacts A4 and B4.

Normally Open Trouble Contacts. The annunciator sequence and features described for NC trouble contacts operate in essentially the same way when NO contacts are used. In the NO system, however, the trouble contact is wired in parallel with the A relay coil to point terminals H and NO, and a wire jumper is installed between point terminals H and NC. Normally, the trouble contact is open and the A relay is energized from terminals H and jumper to terminal NC. In the alert condition, the trouble contact closes to de-energize the A relay by applying power source H to the N side of the relay.

Electromechanical relays are available for use with a variety of AC and DC voltages, but 120 AC, 50 to 60 Hz, and 125 DC are the most popular. Power consumption of the logic modules is normally less than 10 volt-amperes (AC) and 10 watts (DC). Special low drain and no drain logic modules are available; these consume no power during normal operation. Visual indicators consume different amounts of power, depending on the type. Small bull's-eye lights and backlighted nameplates use approximately 3 watts, whereas large units require 6 to 12 watts, depending on whether one or two lamps are used.

Electromechanical annunciator systems are reliable and may be used at normal atmospheric pressures and ambient temperatures in the 0 to 110°F (− 17.8 to 43.3°C) range. They are not position-sensitive. They will generate a substantial amount of heat during plant shutdown when a large number of points are askew, and therefore power should be disconnected during these periods. The principal disadvantages of the relay-type annunciator are size, power consumption, and heat generation.

Solid State Annunciators

A solid state logic module consists of transistors, diodes, resistors, and capacitors soldered to the copper conductor network of a printed circuit board supplying the required annunciator logic functions. The modules terminate in a plug-in printed circuit connector for insertion into an annunciator chassis; they may also contain mechanical switching or patching devices to provide lock-in and NO-NC options.

Figure 3.1k is a semischematic electrical circuit for a remote system with ISA Sequence 1B. The logic module shown is in normally closed operation. Remote lamps for two points and a flasher-audible module, speaker, and acknowledge and test pushbuttons common to a system are also included. Switch S1 is the NO-NC option switch and is shown in the NC operating position. Switch S2 is the lock-in option switch and is shown in the lock-in position. The following description uses negative logic, i.e., a high equals a negative voltage, whereas a low is approximately zero volts.

Normal. The trouble contact of the NC alarm point is connected to an input filter circuit consisting of resistors R13 and R50 and capacitor C1. This provides transient signal suppression as well as voltage dropping. The slide switch S1 connects the trouble contact and filter network to resistor R14. In this state transistor T1 is conducting, causing the full negative voltage to be dropped across resistor R17, resulting in a low voltage at the bottom end of resistor R20. Transistors T2 and T3 are the active elements of the input memory and are roughly equivalent to the A relay of an electromechanical module (see Figure 3.1j). The base of T2 has four inputs, including resistor R20, either directly from the trouble contact in NO operation or from the collector of T1 in NC operation; resistor R19 with a locking signal from the alarm memory transistor T5; resistor R28 and capacitor C2, which form a regenerative feedback from T3; and resistor R15 from

the test circuit. The base of T3 has one input, resistor R29 from the collector of T2. In normal operation, all four inputs to the base of T2 are low, T2 is not conducting, and T3 is conducting. Conversely, when a high signal is present at any one of the four inputs to the base of T2, T2 conducts and T3 turns off.

Transistors T4 and T5 are the active elements of the alarm memory and are approximately equivalent to the B relay of an electromechanical module. T4 and T5 together with bias resistors R30 and R33 and cross-coupling resistors R31 and R32 form a bistable (flip-flop). In normal operation T4 is off and T5 conducts. The upper end of capacitor C4 is connected to the collector T2. When T2 is off, its collector is at a high and capacitor C4 will change from top to bottom, minus to plus. Transistor T7 is a high-capacity lamp amplifier and T6 is its preamplifier. In normal operation the base of T6 is high and T6 is on, T7 is off, and the visual indicator is off.

Before completion of the description of the normal condition, the operation of the flasher-audible module in Figure 3.1k will be described. The module has two oscillators. The first is a 3-Hz unit generating a signal that is amplified and supplied to the logic modules through the F1 bus. The second is a 700-Hz oscillator generating a signal that is amplified and supplied to the audible signal through the R bus.

The audible signal is a permanent magnet-type transducer (speaker), that converts the electrical energy into sound. Initiated by an audio oscillator, the active elements of which are transistors T1 and T2, these transistors together with passive components (capacitors C1 and C2 and diodes D3 and D4) form an unstable multivibrator when an input is present on the FR bus. In the normal conditions there is no FR signal, the voltage necessary to turn on transistor T1 is missing, and the oscillator will not operate. This is the normal or quiescent condition.

Alert. The trouble contact opens and the base of T1 becomes low, turning off T1. This action produces a high on the base of T2 through R20, which turns on T2 and turns off T3. The negative end of C4 is clamped to common through T2, causing a positive pulse at the base of T5 through diode D12, turning T5 off and T4 on. With T2 conducting, the base of T6 becomes low through resistor R6, and with T5 off the clamp on the flasher signal is removed through diode D4 at the junction of R1 and R2. The flasher source provides an alternating high and low voltage at the F1 bus, which turns T6 on and off, which in turn turns T7 and the light off and on. The flasher signal is generated by transistors T8 and T9, which are the active elements of an unstable multivibrator used as an on-off signal to the output driver stage. Resistor R24 and capacitor C6 decouple the oscillator from the power lines so that its frequency is not affected by that of the other oscillators. The output driver stage consists of transistors T10 and T11, a switching inverter, and an emitter follower stage, which produces an alternating

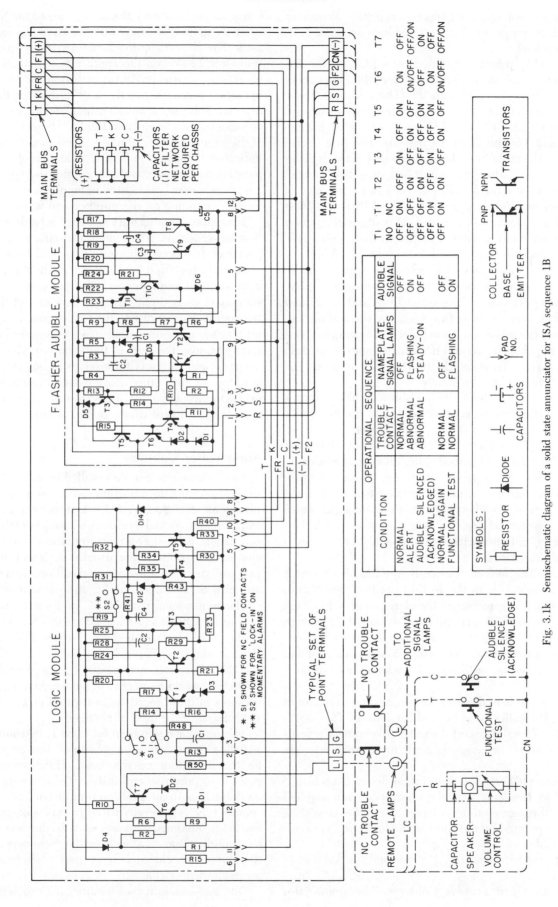

Fig. 3.1k Semischematic diagram of a solid state annunciator for ISA sequence 1B

high and low voltage at F1 bus through R23. Transistor T11 is a high-current transistor capable of driving a multiple lamp load.

The audible signal is initiated by a high on the FR bus, which turns on an audio oscillator, the elements of which are transistors T1 and T2. The audio oscillator output is amplified in an audio amplifier stage composed of transistors T3, T4, T5, and T6 connected in two pairs—one T3 and T5; the other T4 and T6. The input components to the stage from the audio oscillator are opposite each other, i.e., whenever one is high (negative voltage) the other is low (near zero), causing only one pair of transistors to conduct at a time. When T5 is off, T6 is on and the capacitor is discharged. This alternating action causes an alternating current to flow in the speaker coil, giving an audible signal.

Acknowledged. A negative voltage is applied to the base of T5 through resistor R40 by closing the acknowledge pushbutton. This turns T5 on and T4 off, and the FR bus becomes low, thus silencing the audible signal. When the point is acknowledged, T5 conducts; this restores the clamp at R1 and R2, which removes the flash source voltage. T6 is off all the time, T7 is on all the time, and the light is on steady.

Normal Again. When the variable condition returns to normal, the trouble contact closes and the base of T1 becomes high, turning T1 on. This produces a low on the base of T2, which turns T2 off and T3 on. All circuits are again in the state described under the normal condition.

Lamp Test. No separate lamp test is normally provided. One initiates a full operational test by pushing the test button, which applies an alternate input signal through resistor R15 to the base of transistor T2. This turns T2 on, initiating a full operational test of the system as already described.

Lock-in. The lock-in feature is provided by a switch S2. If the switch is in the lock-in position (see Figure 3.1k), when T5 turns off (on the alarm condition) a high at the collector of T5 is coupled to the base of T2 through R19, which keeps T2 turned on even if the trouble contact returns to normal. Transistor T5 will remain off and keep T2 turned on until the acknowledgment button is pushed.

If the switch is in the non–lock-in position, the circuit between the collector of T5 and the base of T2 is open; therefore, T2 will turn off if the trouble contact returns to normal before acknowledgment and return all circuits to normal.

Operational Test. See description under lamp test.

Auxiliary Contacts. Auxiliary contacts are not supplied as part of the logic module. Adapter assemblies consisting of relays operated by the semiconductor logic, however, are available.

Relay Fail-Safe Feature. Not available in solid state circuits.

Normally Open Trouble Contacts. The annunciator sequence and features already explained for NC trouble contacts operate in essentially the same way for NO contacts. The NO-NC option switch bypasses inverter stage transistor T1. When the contact closes on an abnormal condition, it turns on T2 through R20 and the sequence operation proceeds exactly as described for the NC operation. Solid state annunciators are for use with DC voltages ranging from 12 to 125 DC. Power consumption of the logic modules ordinarily is less than 5 watts. Visual indicators consume different amounts of power, depending on the type. Bull's-eye lights use approximately 1 watt, whereas backlighted nameplates use from 1 to 6 watts, depending on the number and wattage of lamps.

Solid state annunciators are very reliable and are not position-sensitive. They offer the advantages of compactness, low power consumption, and little heat generation, factors that make them particularly useful in large integral annunciators. The per point cost of solid state systems is slightly higher than that of their relay-type equivalent, owing to the cost of power supplies and interfacing accessories that may be required with solid state systems. The cost of the logic modules, visual indicators, and cabinets themselves is not excessive. Integrated circuit components using recently developed microcircuits will most likely reduce size, power consumption, and heat dissipation of annunciator systems.

Annunciator Cabinets

Annunciator systems are installed in areas ranging from general-purpose to hazardous. (For more on area classifications, see Section 5.2.) Annunciator cabinets are installed indoors and outdoors in a variety of dusty, moisture-laden, and other adverse environments. Industrial annunciator cabinets are usually designed for general-purpose, dry indoor use. Special cabinets and enclosures are used in hazardous, moist, and outdoor locations.

The requirements of class 1, division 2 hazardous locations as defined in Article 500 of the National Electric Code (NEC) are satisfied by the visual indicators and logic modules (either relay or solid state) of most annunciator systems. A manually operated or door-interlocked power disconnect switch is used with annunciator cabinets in those locations to turn off power when logic modules are relamped or changed.

Annunciator equipment for class 1, division 1 areas is installed in cast steel or aluminum housings approved for the hazardous environment. The housings are expensive to purchase and install. These annunciators are available in both integral and remote configurations. The remote type is generally wired to explosion-proof bull's-eye lights. Annunciator power must be disconnected either manually or automatically before the enclosures are opened to prevent an accidental arc or spark when logic modules are relamped or changed.

One can weatherproof annunciator cabinets installed

in either general-purpose or hazardous areas (class 1, division 2) either by housing them in a suitable enclosure or by covering the exposed cabinet front with a weatherproof door. Housings that comply with class 1, division 1 requirements are also weatherproof. Figure 3.1l illustrates several weatherproof and hazardous area enclosures.

FRONT VIEW

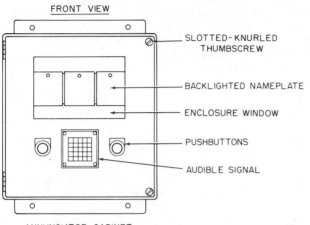

SLOTTED-KNURLED THUMBSCREW

BACKLIGHTED NAMEPLATE

ENCLOSURE WINDOW

PUSHBUTTONS

AUDIBLE SIGNAL

ANNUNCIATOR CABINET
IN WEATHERPROOF ENCLOSURE

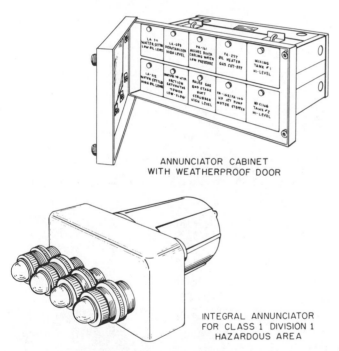

ANNUNCIATOR CABINET
WITH WEATHERPROOF DOOR

INTEGRAL ANNUNCIATOR
FOR CLASS 1 DIVISION 1
HAZARDOUS AREA

Fig. 3.1l   Weatherproof and hazardous area enclosures

Annunciators are classified as intrinsically safe if they are designed to keep the energy level at the trouble contact below that necessary to generate a hot arc or spark. Care must also be taken in installing the system to place wiring so as to prevent a high-energy arc or spark at the trouble contact caused by accidental short circuit or mechanical damage. Thus, general-purpose trouble contacts may be used with intrinsically safe annunciator systems even though they are installed in a

hazardous area. The annunciator logic modules and visual indicators, however, must conform to the electrical classification of the area in which they are installed. (For more on intrinsically safe designs, see Section 5.2.)

Pneumatic Annunciators

Pneumatic annunciators consist of air-operated equivalents of the trouble contact, logic module, and visual indicator stages of an electrical annunciator system. A single-point system furnishing high tank level monitoring is shown in Figure 3.1m. Power supply to the system is instrument air at 80 to 100 PSIG (0.6 to 0.69 MPa), which is reduced to the required operating pressure by pressure regulator (1). The operating pressure is indicated on pressure gauge (2). A 3 to 15 PSIG (0.2 to 1.0 bar) analog input signal from a direct-acting level transmitter (LT-9) enters high-pressure limit relay (4), which is normally closed and set to open when the high level limit is exceeded. When this happens (alert condition), an input at supply pressure from (4) turns on a pneumatic visual indicator (3), and a normally open high-pressure limit relay (6) allows supply air flow to air horn (7), turning it on. Simultaneously, the air output from (4) enters normally closed high-pressure limit relay (5) and momentary contact pushbutton (8), which is a normally open acknowledgment pushbutton for the system. In the alert condition, the pneumatic indicator and horn are both on.

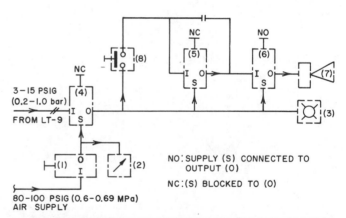

CONDITION	RELAY (4)	RELAY (5)	RELAY (6)	HORN (7)	INDICATOR (3)
NORMAL	CLOSED	CLOSED	OPEN	OFF	OFF
ALERT	OPEN	CLOSED	OPEN	ON	ON
ACKNOWLEDGE	OPEN	OPEN	CLOSED	OFF	ON
NORMAL AGAIN	CLOSED	CLOSED	OPEN	OFF	OFF

Fig. 3.1m   Pneumatic annunciator circuit

One acknowledges the alert condition by pushing button (2), closing it, and thereby opening high pressure limit relay (5). Supply air pressure from (5) closes high-pressure limit relay (6), which cuts off the operating air to the horn, thereby turning it off. Simultaneously, operating air pressure from (5) is fed back to the inlet of (5). The feedback pressure locks up (5) so that it will not close when the acknowledgment pushbutton (8) is released. In the acknowledged condition, the pneumatic indicator is on and the horn is off. The system returns to normal when the 3 to 15 PSIG (0.2 to 1.0 bar) analog input falls below the set point. This closes high-pressure limit relay (4) which turns off the pneumatic indicator (3). It also closes high pressure limit relay (5) by venting the lock-in circuit through relay (4).

Pneumatic annunciators are used when one or two alarm points are needed but electrical power is not readily available and in hazardous electrical areas where an electrical annunciator might not be practical. Pneumatic annunciators require a substantial amount of installation space and are expensive to manufacture.

BIBLIOGRAPHY

Davis, R.C., "Vocal Alarms," *InTech*, December 1969.

Jutila, J.M., "Guide to Selecting Alarms and Annunciators," *InTech*, March 1981.

National Electrical Code, Articles 725 and 760, National Fire Protection Association, 1981.

DISTRIBUTED CONTROL
OPERATE FROM CRT

MOUNTED ON MAIN
CONTROL PANEL

DISTRIBUTED CONTROL
AUXILIARY INTERFACE

MOUTED ON LOCAL
CONTROL BOARDS

COMPUTER CONTROL
OPERATE FROM CRT

MOUNTED ON REAR
OF MAIN PANEL

DISTRIBUTED CONTROL
BINARY LOGIC
OPERATE FROM CRT

MOUNTED ON REAR
OF LOCAL BOARDS

3.2 CONTROL CENTERS

Cost:	$1000 to $3500 per linear foot ($3300 to $12,000 per meter)
*Partial List of Suppliers:**	AETRON, Div. of Aerojet-General Corp.; Custom Engineering Co.; Customline Control Products Inc.; Honeywell Inc., Special Systems Div.; Lake Erie Electric Mfg. Co.; Mercury Company Inc.; Monitor Panel Co.; Swanson Engineering Co.; Sycamore Engineering & Mfg. Co.; Custom Controls Inc.

*Because many process units are constructed by members of building trade unions, prefabricated control boards will not be permitted on the site unless they have labels certifying that they were constructed by members of the required unions. The required labels are: United Association of Journeymen and Apprentices of the Plumbing and Pipe Fitting Industry of the United States and Canada (UA) and International Brotherhood of Electrical Workers (IBEW).

Early control panels, from which simple or batch-type processes were operated, required very little engineering and design for field fabrication. Most control boards contained several panels of large-case recorders or controllers, a few alarm units, and a potentiometric temperature indicator. The instrument lead lengths were short and the control philosophy uncomplicated.

Today's integrated units, with their huge through-puts and minimum holding times, require centralized control panels to keep them operating at the optimum level. The development of sophisticated pneumatic and electronic instrumentation has kept pace with the growth of the Gargantuas they control. This sophistication has engendered the growth of a number of control panel manufacturers specializing in fabricating, piping, and wiring complete control centers. The control centers must be designed for the different types of systems.

Analog: This is the conventional type of recorder, controller, and indicator, both pneumatic and electronic, which is mounted on the control panel or console in its own housing.

Analog-to-digital shared display (A/D): This system, although controlled by an individual analog controller, is displayed and manipulated via a cathode ray tube (CRT) on the control center console.

Direct digital control (DDC): This system is also a hybrid of analog and digital signals. However, the analog signals are limited to the input and output of the digitally transmitted signals to either the central processing unit (CPU) of the system itself or some other CPU. Obviously, only the simplest and least sophisticated center could possibly be field-fabricated.

The design of a control center is dependent upon many considerations, such as control room layout, panel profile, instrument type, complexity of process, sophistication of control philosophy, and aesthetics.

Control Rooms

The control room must be designed so that only those operations necessary for the direct control of the plant are performed there. The operators must not be distracted by unassociated functions. The room should have limited access and should not act as a passageway. Equipment must be arranged either in such a way that unauthorized personnel cannot tamper with the instruments or with the auxiliaries mounted close by. Figure 3.2a shows a typical control room layout for analog instruments.

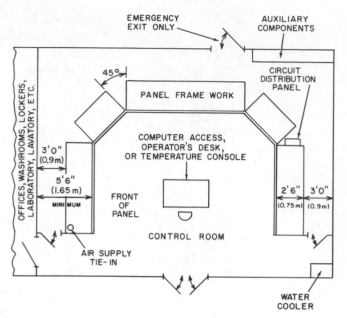

Fig. 3.2a   Control room layout

Air conditioning and room pressurization must be provided. Aside from ensuring operator comfort, a constant ambient temperature at the instruments will minimize the possibilities of signal drift. Room pressurization is normally used where the plant atmosphere is explosive or flammable and is achieved by forcing fresh air through ducts from a safe area. This permits the reduction of area classification from either "hazardous" or "semihazardous" to unclassified with commensurate savings in instrument and installation costs. (Section 10.11 in the volume on Process Measurement gives details on area classification.)

Analog Control Rooms

The illumination in the control room must be of a level consistent with close work. The panel should average 75 foot-candles (807 lx) across its face. The back of the panel area should be lighted to 30 foot-candles (322.8 lx).

The lighting system should be designed to minimize reflections on instrument cases. Point sources of light should be avoided. Continuous fluorescent lighting behind egg–crate type ceiling fixtures will give adequate light and minimize annoying highlights.

The most advantageous ratio of panel length to control room area is obtained by bending the panel to a "U" shape. Right-angled bends of the panel, as opposed to 45-degree bends, should be avoided. The slightly increased panel length that could be gained in this manner

is negated by the interferences to opening instrument doors or withdrawing the chassis. An operator can monitor a greater length of panel if it bends around him.

Panel Types

There are three basic types of control panel shapes—straight, breakfront, and console. Each type has an accompanying family of variations. The dimensions shown in the various illustrations are only typical, and instrument heights and spacing must be adjusted to suit the particular manufacturer and application.

FLAT PANELS

The type of panel shown in Figure 3.2b is the least expensive, the easiest to construct, and the simplest to design of all types. The straight, vertical plane of the panel allows an orderly layout of tubing, electrical duct work, and miscellaneous equipment. Instruments and auxiliary components can be arranged so that all are accessible for maintenance and calibration. The lower row of instruments, approximately 3 feet 3 inches (0.975 m) from the floor, should be used for recording or indicating instruments only, because this elevation is rather inconvenient for operation. Since the maximum of four horizontal rows of miniature recorders and controllers can be used, this type of panel will require more control room space and even higher instrument density layouts than some others.

BREAKFRONT PANELS

Breakfront panels allow greater use of the front plane of the board, because the instruments located in the lower rows are swung upward to a convenient height (Figure 3.2c). The top portion of the panel is swung downward to an angle normal to the line of sight, allowing better visibility. The additional rows of instrumentation obtained with this layout cut the overall panel length requirement. The higher instrument density, however, significantly reduces the space for maintenance and for mounting of auxiliary components in the back of the panel.

CONSOLES

Consoles appear to be aesthetically very pleasing to most people. They are often used with high-density instrumentation in control rooms that are limited in size. Normally, their length is set by the operator unit responsibility limits, i.e., one operator per console and one console per operator. The lengths vary from 4 to 12 feet (1.2 to 3.6 m).

Auxiliary equipment, such as transducers and pressure switches, can be installed inside the console cabinet, but the arrangement of the flush-mounted instruments as shown in Figure 3.2d severely limits the available free space. Maintenance of instruments within the console can therefore become a real problem.

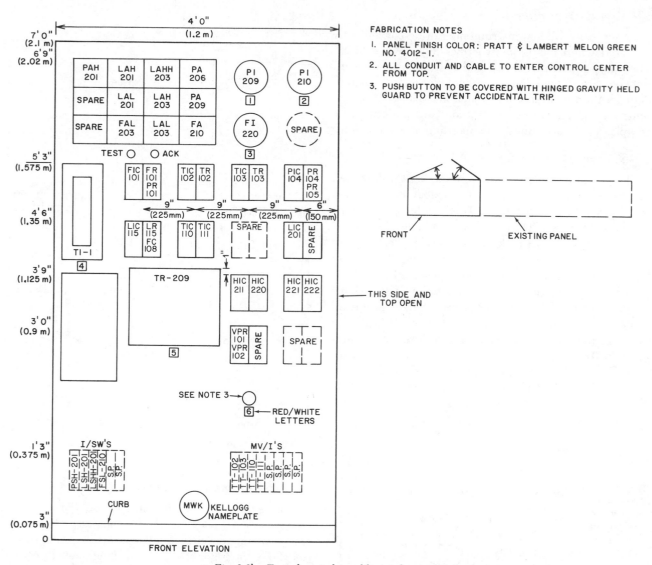

FABRICATION NOTES

1. PANEL FINISH COLOR: PRATT & LAMBERT MELON GREEN NO. 4012-1.

2. ALL CONDUIT AND CABLE TO ENTER CONTROL CENTER FROM TOP.

3. PUSH BUTTON TO BE COVERED WITH HINGED GRAVITY HELD GUARD TO PREVENT ACCIDENTAL TRIP.

Fig. 3.2b Typical control panel layout drawing

Consoles are often coupled with a conventional flat "back-up" panel, which contains the larger-sized instruments and auxiliary components. Because the average operator is approximately 5 feet 9 inches (1.7 m) tall, the "see over" console should not rise above 5 feet 0 inches (1.5 m). This will allow the operator to see the back-up panel over the top of the console.

D/A or DDC Control Rooms

Although these control rooms use significantly less space, they do require auxiliary equipment rooms adjacent to the operating area. A computer-type floor will be the most convenient for intercable routing. A significant amount of planning must be used for this type of system. Allowance must be made for the various peripheral equipment cabinets, input/output (I/O) cabinets, and the CPU and its supporting gear.

Figure 3.2e shows a possible DDC control room with an auxiliary equipment room. Obviously, the number and size of the cabinets are dependent upon the instrument vendor. Where the possibility of plant enlargement is high, adequate spare cabinet space and display area should be allotted.

Panel Layout

Currently available are three basic instrument sizes, which may be fitted to almost any panel configuration and type. Available sizes and shapes range from the large-cased conventional type (nominally 18 inches wide by 24 inches high [450 mm wide by 600 mm high]) down through the miniature type (nominally 6 inches square [150 mm]) to the high-density type (2 inches wide by 6 inches high [50 mm wide by 150 mm high]).

The same factors that determined the panel type very often also determine the instrument type.

LARGE-CASED INSTRUMENTS

Large, conventional instruments can be conveniently operated when mounted two rows high (Figure 3.2f). They do not permit a great deal of flexibility in layout, but they are rugged and are suitable for almost any outdoor location as local control stations. The control systems for most of these applications are not apt to be complex or to have numerous instrument components.

MINIATURE INSTRUMENTS

The miniature instruments were the most widely used process control devices. They allow a moderate reduction in panel length (Figure 3.2g). This is sufficient for most indoor installations. These devices can be used in almost all configurations of breakfront or console designs, but there are some mounting angle limitations that should be kept in mind when the panel is designed.

For estimating the length of control panels, a horizontal spacing of 9 inches (225 mm) between vertical center lines of miniature instruments may be used. Ten to 12 inches may be used between horizontal center lines. Instrument manufacturers have not fully standardized overall dimensions, cutouts, connection locations, and so forth. The manufacturer's spacing and installation recommendations for each instrument must therefore be checked.

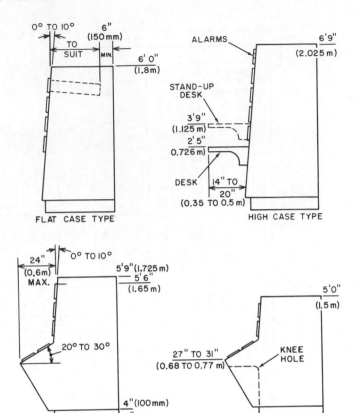

Fig. 3.2d   Variations of console shapes

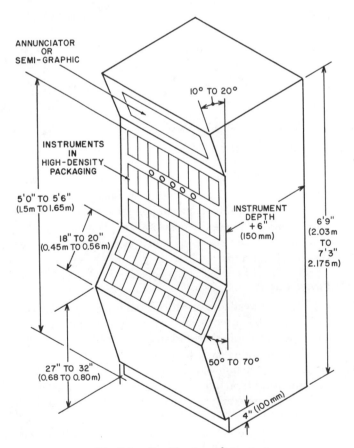

Fig. 3.2c   Breakfront panel structure

HIGH-DENSITY INSTRUMENTS

There are several high-density–type miniature instrument designs available on the market. They may be mounted in groupings to suit a particular processing unit, as shown in Figure 3.2h, or in long rows to condense panel length, as illustrated in Figure 3.2i.

This type of instrument layout requires additional space on the rear of the panel for auxiliary equipment. Additional materials are sometimes mounted on the wall behind the panel or in a peripheral equipment room remote from the panel. High-density instruments tend to have longer chassis, requiring about six additional inches in panel depth. Some of these instruments require amplifiers or converters, which must be mounted in proximity to the primary instrument.

Some high-density instrument designs do not include a similarly sized recorder. In this instance a standard two- or three-pen, 4-inch (100 mm) strip chart recorder from the miniature (6 by 6 inches or 150 by 150 mm) line may be utilized (Figure 3.2g). The use of trend recorders rather than those permanently wired will considerably reduce the overall recorder requirements.

Graphic Panels

A graphic control panel depicts a simplified flow diagram of the processing unit and its control philosophy.

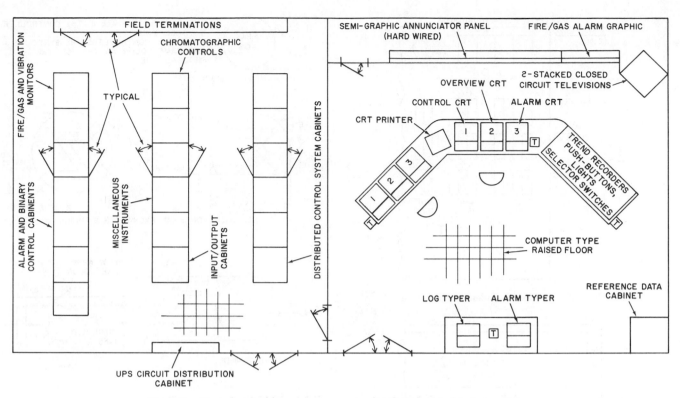

Fig. 3.2e   Typical control house arrangement—distributed control instrumentation

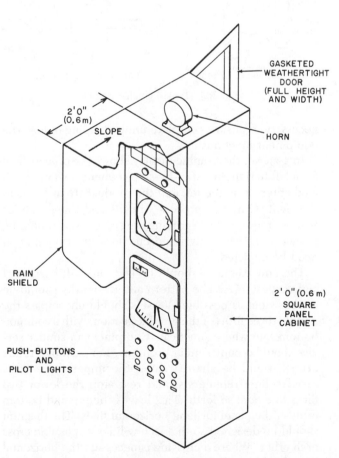

Fig. 3.2f   Local panel cabinet with large case instruments

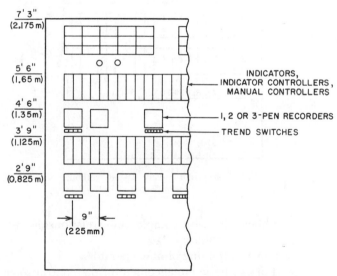

Fig. 3.2g   Panel layout using miniature instruments

The most common material used for this depiction is colored plastic or melamine. Backpainted translucent vinyl or other plastics are also used to some extent. The lines and symbols may be affixed to a removable steel, aluminum, or plastic plate.

The extra expense of a graphic panel may be justified for some of the following reasons:

1. To enable the panel operator to visualize a complex process flow pattern.
2. To make understandable a sophisticated control

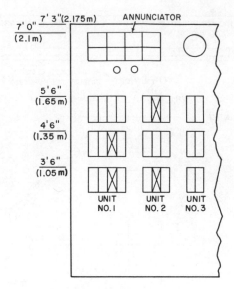

Fig. 3.2h   Grouped high-density panel layout

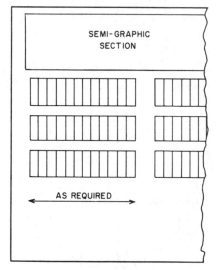

Fig. 3.2i   High-density layout without grouping

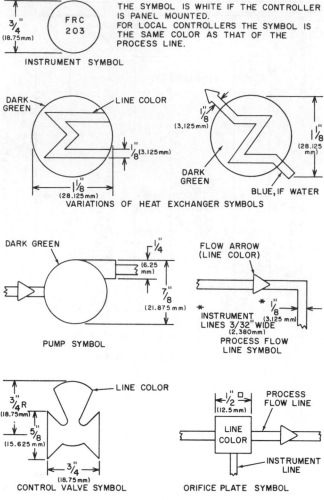

Fig. 3.2j   Typical graphic symbols

philosophy with complex interrelationships between variables.

3. For the training of new operators.
4. For aesthetic enhancement of the control room (a motivation seldom admitted but frequently present).

Although the industry has not yet standardized shapes or dimensions for equipment symbols, there is some agreement in the use and adoption of some common symbols for graphic layout purposes, such as the ones shown in Figure 3.2j. The dimensions shown here are typical and can be adjusted to suit individual requirements, as long as they are consistent throughout the panel. Symbols representing equipment, such as furnaces, vessels, drums, pumps, and compressors, are usually shown in silhouette. Internals are shown only when

necessary to ensure complete understanding of how the equipment functions.

In general, the graphic display shows the process flow from left to right in a step-by-step sequence, where fresh feed enters from the left and the product stream exits to the right. Only the process streams and those utilities that are required for clarifying unit operations should be shown. Similarly, only important local instrumentation need be depicted.

The flow diagrams should be laid out with horizontal and vertical lines. The pattern and its density should be as consistent as possible. Vessels should encompass the middle three fifths of the graphic section, with a common bottom line where practical. Equipment in similar service should occupy similar positions. For example, reflux drums would be aligned along the upper three fifths division line; reboilers would rest atop the lower two fifths line, and so forth. This leaves the top and bottom fifths of the panel for long horizontal lines. The diagram should be designed so that as few lines as possible cross each other. Where a crossing is necessary the horizontal line should break.

SEMIGRAPHIC PANELS

Semigraphic panels are either flat or of the breakfront type. The top portion of the panel is occupied by a flow diagram of the process (Figures 3.2i and 3.2k). The location of each instrument should be so selected that it is installed directly underneath its corresponding symbol in the graphic diagram (or as close to this point as is feasible). The instrument item number is clearly noted in both locations. In some instances, alarm or running lights are also installed within the graphic section in the appropriate position. The instrument density is reduced with this type of panel because of the desirability of locating the instruments in relative proximity to their graphic symbols.

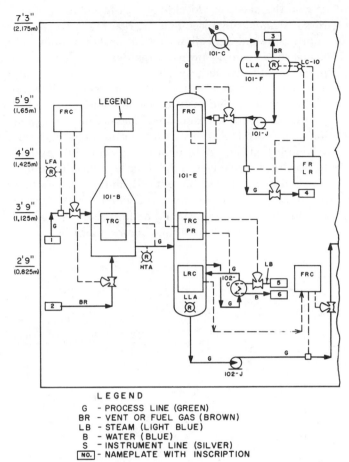

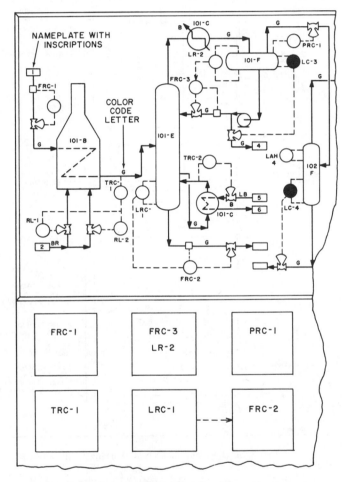

Fig. 3.2k   Semigraphic panel with miniature instruments

FULL GRAPHIC PANELS

Full graphics cover the complete front face of the control panel with the process diagram, as shown in Figure 3.2l. Instruments are positioned in the panel at the point that corresponds to their measurement or control location. As in all other panels, the instruments may be aligned in horizontal and vertical rows for ease in conduit, duct, and tube layout. The instrument density for this type of panel is extremely low and can vary from one to

LEGEND

G - PROCESS LINE (GREEN)
BR - VENT OR FUEL GAS (BROWN)
LB - STEAM (LIGHT BLUE)
B - WATER (BLUE)
S - INSTRUMENT LINE (SILVER)
NO. - NAMEPLATE WITH INSCRIPTION

Fig. 3.2l   Full graphic panel layout with miniature instruments

three instruments per linear foot of board length. This significantly extends the overall length of panel.

VARIATIONS OF GRAPHIC PANEL CONSTRUCTIONS

In addition to the glued-on plastic-type graphic panels, there are other designs available. The degree of difficulty in making changes to the graphic diagram should be considered if there is a likelihood of process revision in the future. The aforementioned plastic strips are relatively easy to change. If the lines and symbols are painted on the panel, changes can be made easily when the panel is new, but as the paint ages and fades, colors become difficult to match.

A flow diagram can also be drawn full scale on reproducible paper and placed in a semigraphic frame. It may be held in place and protected with a glass or clear rigid plastic cover. This method of graphic presentation is simple to change by revising the drawing master and acquiring a new print. This type of panel is not as aesthetically pleasing as other types because it is usually black and white rather than color-coded.

Another method of graphic process display is by the use of automatic slide projectors. This system is used with a rear-projection–type, translucent screen. In order to gain the necessary focal length distance, a set of mirrors is strategically placed on the rear of the panel. This approach to graphics allows a great deal of flexibility. As many slides can be prepared as desired, showing the process or instrument systems or subsystems in color and in as much detail as is deemed necessary. Slides can be prepared to give instructions for emergency procedures, unit operating parameters, and optimum set points for various product choices. One serious drawback to the system is that the bright light in the control room tends to "wash out" the image and reduce legibility. Therefore, the lighting level must be reduced, and the fixtures must be so located as to minimize reflections on the screen. In addition, the screen should be furnished with a black shield to reduce side lighting (Figure 3.2m).

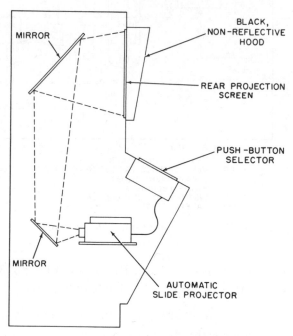

Fig. 3.2m Panel console with rear projection system for graphic presentation

Yet another type of graphic panel is fabricated by back-engraving a sheet of clear plastic. The plastic lines are then filled with the selected colors. This type of panel is easy to maintain because only the smooth surface is exposed. Changes can be made by re-engraving and by refilling. This is extremely difficult, and even when it is expertly done, telltale signs of the change remain.

The front-engraved and enamel-filled graphic line work allows vivid coloring and sharp lines, making this panel one of the most impressive and pleasing to the eye. It is not feasible to make changes on this panel, and maintenance is somewhat more difficult, since the engraved lines tend to fill with dust.

Back-of-Panel Layout

In order for the overall panel design to be well executed, sufficient attention must be given to the back-of-

panel arrangement. It must be verified that there is sufficient room to run the conduits, ductworks, air headers, tube leads, and so forth. There must also be enough room to mount the various switches, relays, converters, amplifiers, and other auxiliary components. A back-of-panel profile will sometimes assist in this (Figure 3.2n). Auxiliary equipment must be located such that its connecting wiring or tubing does not obstruct the maintenance or calibration of, and accessibility to, the other instruments.

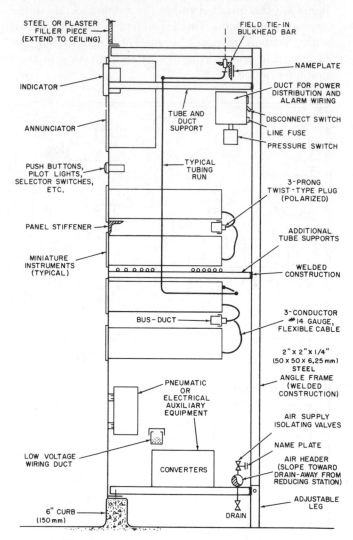

Fig. 3.2n Back of panel—typical arrangement

Human Engineering

In the past few years a significant amount of research has been done on the subject of human engineering, or ergonomics. This has proved invaluable in the design of all of the panels and consoles we have described in the preceding paragraphs. We can be assured that 95 percent of the people in the United States can easily see and conveniently operate the controls. Median heights and reaches of people from other nations should be used if the control center is to be used in another country (Figure 3.2o).

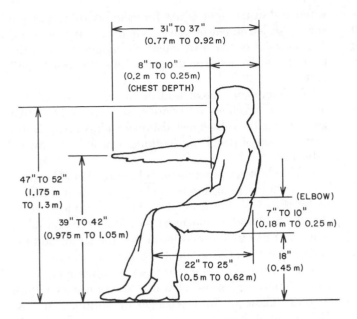

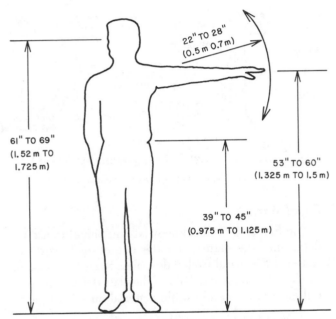

Fig. 3.2o Anthropometric figures. Mid-range of these dimensions should cover the 95 to 5 percentiles of American males.

Studies performed in nuclear energy power plant control rooms have shown that psychological as well as physical considerations are important. Many of the rules listed here will be obvious, but it is surprising how often they are overlooked.

1. Indicating lights and pushbuttons should be colored so that their relative importance is immediately recognized; i.e., green = go or safe, red = danger or stop, yellow = caution or slow.

2. Indicating scales, selector switches, and other instruments should push in or up, rotate clockwise, or slide to the right for the "increase" or

"on" setting. The opposite position should represent the "decrease" or "off" setting.

3. The style of emergency operating devices should be reviewed. Emergency trips must be designed so that they are easy and fast to operate, but they should also be located and designed to prevent an accidental trip.

Other less obvious ideas include the following:

1. *Annunciator horns:* These can be set to increase in pitch or to go on and off with increasing frequency as the emergency increases in severity.

2. *Annunciator lights:* These can be made to flash at a brighter and higher frequency.

3. *Instrument arrangement:* Instruments should be located in a panel or CRT display as though they were in a graphic panel, that is, following the process from left to right. They should be grouped in logical sets. For example, a distillation column instrument arrangement would have the column feed to the left, the overhead pressure instrument high, the reflex drum instruments high and to the right, and a reboiler instrument low and to the right. It is not possible to follow these recommendations exactly in all cases, but the devices should be located in these relative positions. Additionally, these instruments should be separated from the next column or entity by a space or a line.

The operator must be allowed to concentrate on running the plant. Distractions such as a difficult, complicated, or confusing instrument procedure should not take the operator's mind from his primary responsibility. Every facet of control center design must have the operator and the plant operation as its primary concern. The effect, if any, of every design feature on the plant operation must be considered.

Panel Piping and Tubing

Most panel manufacturers stock an acceptable line of industrial-grade tubing, pipe, and fittings. One of the most commonly used tubing materials behind the panel is copper. It is relatively resistant to corrosion, readily available, and rigid enough to require only a minimum of support, yet it is sufficiently ductile to bend to precise measurements. It is available with a tightly extruded PVC sheath for corrosive atmospheres, and with cadmium or tin plating for damp locations. Commonly used sizes are $\frac{1}{4}$ inch (6.25 mm) OD, 0.030 inch (0.75 mm) wall thickness, ASTM B68.

Another commonly used tubing material is aluminum ($\frac{1}{4}$ inch [6.25 mm] OD, 0.032 inch [0.8 mm] wall thickness, federal specification WWT-700/4c). Aluminum tubing offers good resistance to the attack of many types of chloride or ammonia atmospheres. It is somewhat softer to work with than copper and requires significantly more support. It has a tendency to harden, which makes it particularly susceptible to vibration failures. This tubing is also available with a PVC sheath.

The most frequently used plastic tubing material is polyethylene. This tubing should be of a material that has undergone environmental stress testing in accordance with ASTM D1693. Although polyethylene is markedly less expensive than the metallic tubes, additional costs are incurred in the use of an extensive support network, which consists of plastic, slotted, or sheet metal ducting. The ducting should extend to within 1 or 2 inches of the connected instrument. Unsupported lengths are to be avoided because the plastic tubing has a tendency to kink and because it is almost impossible to run this soft tubing neatly. Tubing may be color coded to conform to the ISA recommended practice ISA RP-7.2. Common sizes are ¼ inch (6.25 mm) OD, 0.040 inch (1 mm) wall thickness. Other tubing materials are available, such as stainless steel, nylon, polyvinyl, rubber, and glass, but they are not used as frequently as the aforementioned three substances.

Fittings are available for all tubing materials mentioned. The industry has generally recommended standard compression-type tube fittings for panel work rather than flared, soldered, or other specialty types. Pipe material should be seamless red brass, threaded in accordance with USAS B.2.1 and soldered or brazed. The fittings should be rated to at least 125 PSIG (865 kPa). Air supply isolating valves should be either the straight-line needle type or the packless diaphragm two-way type.

The panel air supply should consist of two sets of parallel-piped reducing valves and filters. Each should be sized and manifolded so that either set can supply the panel requirements. In general, 0.5 SCFM (0.014 m³/m) of air flow per air user is sufficient. The main air header must be sized so that the pressure drop to the farthest instrument is under 1 PSI (6.9 kPa).

The use of a plugged tee and isolating valve or a three-way switch allows for calibration of instruments or components without disconnecting them or losing the measurement signal to other loop components (Figure 3.2p). The isolating valve need be used only when the measurement signal is branching to another instrument or

when the instrument is not furnished with an integral spring-loaded air cut-off valve, as in plug-in type devices. The location of the instrument air supply tie-in to the board is usually noted on the panel drawings, as in Figure 3.2b.

The pneumatic signal lead tie-in location usually requires only a general description, since there is enough flexibility in the tubing installation for adjustment. Notes such as "top of panel" or "bottom of panel" are usually sufficient.

Each tubing termination for field leads should be identified with the item number of the instrument it serves and its function. For example, FRC-10-V could indicate that the field lead terminates at the FRC-10 control valve, whereas FRC-10-T would mean that the air signal tube is received from the so-identified flow transmitter.

There are several different types of panel-field, interface connections, some of which are shown in Figure 3.2q. To allow for future changes or additions, at least 10 percent additional tie-in points should be provided.

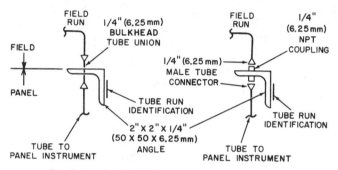

Fig. 3.2q　Typical field tie-ins to pneumatic panels

Panel Wiring

The first step in deciding what equipment to use is to determine the nature and the degree of hazard. The National Electrical Code* describes hazardous locations by class, group, and division. The class defines the physical form of the combustible material mixed with air:

Class I:　Combustible material in the form of a gas or vapor.

Class II:　Combustible material in the form of a dust.

Class III:　Combustible material in the form of a fiber, such as textile flyings.

The group subdivides the class:

Group A:　Atmospheres containing acetylene.

Group B:　Atmospheres containing hydrogen, gases, or vapors of equivalent hazard, such as manufactured gas.

Group C:　Atmospheres containing ethyl ether vapors, ethylene, or cyclopropane.

* NFPA 70-1981, ANSI/NFPA-70 National Electrical Code, 1981. Canadian equivalent: CSA Standard C22.1 Canadian Electrical Code Part 1, 1978.

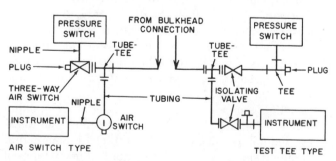

Fig. 3.2p　Typical test or calibration connections

Group D: Atmospheres containing gasoline, hexane, naphta, benzine, butane, propane, alcohol, acetone, benzol, lacquer, solvent vapors, or natural gas.

Group E: Atmospheres containing metal dust, including aluminum, magnesium, and their commercial alloys or other metals with similarly hazardous characteristics.

Group F: Atmospheres containing carbon black, coal, or coke dust.

Group G: Atmospheres containing flour starch or grain dusts.

The division defines the probability that an explosive mixture is present. For instance, a hazardous mixture is normally present in a Division 1 area but will be only accidentally present in a Division 2 area.

In addition to knowing the area classifications, one should also be aware of the National Electrical Manufacturers Association (NEMA) terminology for classifying equipment enclosures:

NEMA 1   General-purpose
NEMA 2   Drip-tight
NEMA 3   Weatherproof
NEMA 4   Watertight
NEMA 5   Dust-tight
NEMA 6   Submersible
NEMA 7   Hazardous (Class I, Groups A, B, C, or D)
NEMA 8   Hazardous (Class I, Groups A, B, C, or D)—oil-immersed
NEMA 9   Hazardous (Class II, Groups E, F, or G)
NEMA 10   Explosionproof—Bureau of Mines
NEMA 11   Acid- and fume-resistant, oil-immersed
NEMA 12   Industrial

For additional details on this subject, refer to Section 5.2.

Most panels are enclosed by or parallel to a wall, with a door on either or both ends to limit unauthorized access. Under such conditions and when the area is general-purpose and nonhazardous (see Figure 3.2a), it is permitted to reduce the mechanical protection requirements for the wiring. Electronic transmission, power, and signal wiring need not be enclosed in conduit or in thin-walled metallic tubing.

All of the wiring may be run in a sheet metal or slotted plastic duct. The insulated wire may be run exposed, from the duct to the instrument, an inch or two (25 to 50 mm) without the necessity for a conduit nipple. Bare or exposed terminals, however, are not permitted.

Panels installed in hazardous or semihazardous areas must be installed in strict adherence to the National Electrical Code requirements. Because the code does not allow much flexibility in these cases, this discussion

will be limited to the nonhazardous applications, in which the designer has some flexibility. For requirements in hazardous locations refer to Section 5.2.

POWER DISTRIBUTION

Instrument power supplies should be taken from a reliable source, with automatic switchover capability to an alternate power supply to be used upon failure of the main source. The two typical standby power supplies are a separate supply bus (fed from batteries or from a different source) and a steam- or gasoline-powered generator. A detailed coverage of back-up power supply systems is given in Section 3.10.

To reduce the cross-sectional area of the power feeders, it is often expedient to mount a three-phase (440, 208, and 120 volts transformer directly in the control house. Then only a 440-volt power supply is provided from the switch gear to the transformer (Figure 3.2r).

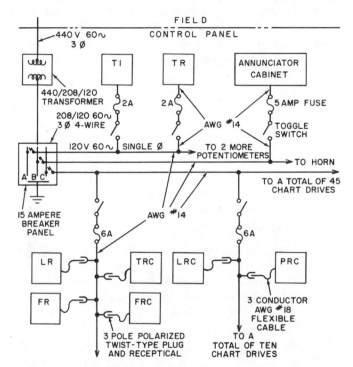

Fig. 3.2r   Typical power distribution system

A conventional lighting-type circuit breaker panel may be installed on the back of the panel to provide the necessary circuit distribution. This permits the panel manufacturer to install the complete system and significantly reduces field tie-in time. Breakers (sized to trip above 15 amperes) and AWG #14 wire gauge are generally used. Three-wire power circuits having hot, grounded neutral, and ground leads are frequently used.

To avoid the possibility of overloading, circuits should be lightly loaded to approximately half their rated capacity or to a maximum of 850 volt-amperes.

The following groups of instruments will keep the load on 15-ampere circuits within acceptable limits:

4 Potentiometer-type temperature instruments
1 Annunciator cabinet with horn
2 Analyzer circuits (900 volt-amperes maximum)
45 Miniature pneumatic recorder chart drives
5 Miscellaneous auxiliary components, at 800 volt-
  amperes maximum
10 Electronic instrument loops (500 volt-amperes
  maximum)
2 Emergency trip circuits

Secondary subcircuits must be used so that a short or ground at one instrument does not trip out the 15-ampere circuit breaker but only the associated fuse and so that each instrument (or group) can be isolated for maintenance or replacement. This isolation is accomplished with a fused disconnect device. The fuses must be coordinated so that it is not possible for the 15-ampere breaker to trip before a fuse blows. The fuse must also go before a significant voltage dip occurs. An exception to the individual fusing isolation rule is chart drive power to pneumatic instrumentation. Here, ten drives (or some similar number) may be connected to one common fuse. A three-pronged, polarized, twist-type plug can serve as a disconnect.

When instruments are internally fused, that fuse may be utilized, but the instrument circuits must be checked to verify that the fuse protects the complete chassis and not just a single critical component.

BATTERY BACK-UP

There are several factors that must be considered in deciding whether a standby power supply is required. These are as follows:

1. The power source that is used has a history of failures with up to one half hour's duration.
2. The unit uses normally energized solenoids or relays that must be manually reset.
3. There is a flame safety system.
4. The process is extremely fast-acting with electronic controllers in critical services, in which a dip in power supply could send the unit into uncontrollable cycling.
5. The process control system is computerized.

A typical uninterruptable power supply (UPS) is the battery back-up system shown in Figure 3.2s. This system consists of a battery charger, a bank of batteries, and an inverter. One of the main advantages of this system is that the AC power input phase does not have to be synchronized with the output phase. The instruments are normally powered directly through this system. Upon failure of the mains, the batteries (which have been floating on the charging current) start feeding the inverter.

The battery ampere-hour capacity should be sized using one or more of the following considerations:

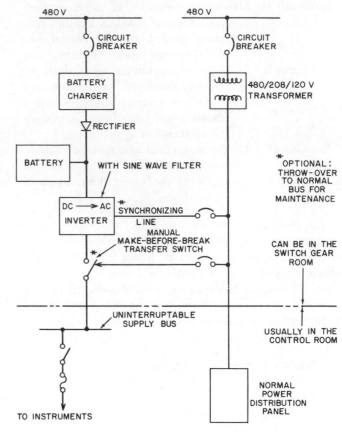

Fig. 3.2s Standby power supply system—battery back-up type

1. Length of time of average power outage × 1½.
2. Length of time plant will remain operable after power mains fail × 1½.
3. Length of time to switch to alternative power supply × 2.
4. Length of time instruments will be required to bring about an orderly shutdown.

The charger must be sized so that it can simultaneously operate the unit and recharge the battery system after a discharge. A recharge time of eight hours is reasonable. Circuit breakers and fuses downstream of the inverter must be sized and coordinated so that the available current will trip them out before a significant voltage disturbance could occur.

Isolation Transformers

Digital equipment is usually very susceptible to spikes in its power supply. The UPS system will filter out spikes from the main power source. If equipment other than the digital system is connected to the UPS bus, an isolating transformer should be used. A transformer that will filter a 50-volt peak and is furnished with a Faraday shield will often be adequate. The actual instrument manufacturer's recommendations must govern the final choice of specifications.

Grounding

Each instrument case, control panel, auxiliary cabinet, and instrument system must be safely wired to the control house ground. Low-level instrumentation and DDC equipment require their own separate grounding net. A 1-ohm resistance to ground is usually a safe number. Here again, the individual equipment manufacturer's recommendations must be followed.

WIRING AND TERMINAL IDENTIFICATIONS

Most electronic control loops are of the "two-wire" type. This means that the locally mounted transmitter or control element does not require a separate power supply but takes its actuating force from the signal wires. Therefore, the installation requires only a two-conductor cable for each transmitter or final control element. To simplify field wiring and minimize overall field installation costs, auxiliary components are usually mounted on the back of the panel. In this way the complex interconnecting wiring is installed by the panel manufacturer.

For flexibility in making loop changes and additions, one method of installation is to use a centralized terminal block for each complete instrument loop on the back of the panel. In this system, each transmission and control loop is assigned a set number of terminals in the field tie-in junction box. Each group of terminals is identically marked, and the terminal marking strip carries the instrument loop number. Each component of the loop is then wired to this terminal, as shown in Figure 3.2t. Spare terminals should also be included for future instruments and components.

Terminal strips are available with white plastic or painted marking surfaces, suitable for pencilling in identification. At all electronic instrument field tie-ins, the junction blocks should be identified with the instrument loop number, its function, and its polarity in the following manner: FRC-10-T(+) FRC-10-T(−). All other terminals are identified similarly, except that they are marked with the instrument terminal designation instead of the polarity. For equipment such as relays, switches, and other components without terminal designations, the terminal should duplicate the identification shown in the wiring diagrams, such as SV-10-VS (SV-1) or annunciator 'A'-(S1-1).

Wire identification data should duplicate the information shown on the terminal marking strips: instrument item number, function, and polarity. Function and polarity identification may be replaced with color-coded wire insulation. To be effective, the color code must be simple and consistent. A typical color code is given in Table 3.2u.

Table 3.2u
CONTROL PANEL WIRING COLOR CODE

Color of Wire Insulation	AC Service (120-volt, 60-cycle supply using AWG #14 wires)	DC Service (low voltage using 2-conductor AWG #18 wires)
Black	"A" phase—hot	Positive power, transmission or control signal
Red	"B" phase—hot	Negative power or control
Blue	"C" phase—hot	—
White	Neutral	Negative transmission signal
Brown	Annunciator common (H)	Annunciator signal
Orange	Annunciator signal	Annunciator signal
Green	Ground	Ground
Gray	Miscellaneous interconnections and jumpers	

In addition to the color coding, the wire may be identified with the instrument item number by means of a preprinted marker tape or by a plastic sleeve. The sleeves either are sized to a snug fit over the wire insulation or are of the shrink type.

Wire terminations should be made with crush-type wire lugs. Special lugs are available for solid wire. A good lug to use is a flanged spade lug. This type combines the ease of installation of a spade lug with flanges that hold the lug in place if the terminal screw becomes loose.

Where wire is run within a duct, it need not be laced. Although it gives a neater appearance, lacing is time-consuming and is a nuisance when wires must be frequently added or rerouted.

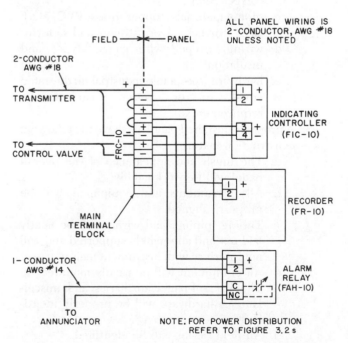

Fig. 3.2t Typical wiring diagram for an electronic instrument loop

Panel Construction Materials

The materials used most frequently for the construction of panel boards are steel and various plastics. The advantages of steel are as follows:

1. *Strength*. At instrument cutouts, where the panel is weakened, steel requires less stiffening than do other materials. Steel stands up better to the cantilever effect of the deep bezel-supported, flush-mounted instruments.

2. *Ease of construction*. Holes can be drilled or flame-cut. Auxiliary equipment supports can be welded at any point. Bend-backs on straight panels and the breakfront shapes add significantly to panel stiffness.

3. *Safety*. When grounded, the panels offer an excellent path to ground if an instrument chassis or case becomes energized as a result of a short circuit or mistermination.

4. *Attractive finish*. Long lengths can be finished in any color or hue desired, without visible seams.

5. *Ready availability*. Sheet steel of good quality and with a minimum of surface pitting is readily available.

Disadvantages of steel panels include the following:

1. *High susceptibility to corrosion*. In corrosive atmospheres, the finish is only as durable as the paint.

2. *Difficulty in adding cutouts*. Once the panel has been constructed and instruments are installed, cutting the steel for further instruments scatters steel particles that may interfere with the mechanical or electrical operation of the surrounding instruments.

The plastics available for panel boards are all melamines: Formica, Peonite, Micarta, Textolite, and so forth. They are used either as sheets $\frac{1}{2}$ inch (12.5 mm) in thickness or as laminates. The cores of the laminates are materials such as aluminum, flakeboard, steel, and plywood.

Plastic panels, like those of steel, have both good and unfavorable aspects:

1. *Great durability:* The finish is very resistant to scratching and heat.

2. *Availability in colors:* Sheets come in many colors, but if the desired color is not in stock, there may be a long wait for delivery. A color may vary slightly from one run at the factory to the next.

3. *Seam required every 4 feet (1.2 m):* This is true regardless of panel length, because sheets are usually 4 by 8 feet (1.2 by 2.4 m).

4. *Steel frame required:* The plastic panel must be bolted to a skeletal steel frame for support.

5. *Routing of holes may be necessary:* Many switches, pilot lights and other flush-mounted instruments are designed for a $\frac{3}{8}$ inch (9.36 mm) maximum panel thickness. Solid Formica and some laminates will require routing in the back of the panel, around the cutout for bezel locking rings or locknuts, because of their thickness.

Less frequently used materials used for panel construction are stainless steel, aluminum, and fiberglass.

Control Center Inspection

A complete panel inspection at the manufacturer's plant will usually pay dividends in ease of installation, field tie-ins, loop checkouts, and in a smoother plant start-up with fewer field man-hours expended.

The control center designer, familiar with the overall instrumentation and operating philosophy of the unit, can visually inspect and functionally check a control panel most expeditiously. Together with a pipefitter and an electrician, he can locate piping errors or wiring misterminations in a fraction of the time that it would take in the field.

Each panel has its own peculiarities. The following examples may be used as a guide for formulating a check list prior to inspection:

1. Panel construction dimensions should be evaluated, for example:
   a. overall dimensions,
   b. thickness of panel, and
   c. size of framing.

2. Construction materials to be assessed include:
   a. panel and framing material;
   b. panel finish (e.g., smooth, unblemished, correct color);
   c. piping materials (copper, brass, PVC, etc.): sizes, correct valves, fittings, and so forth;
   d. wiring, proper wire gauge, type, and insulation;
   e. hardware (acceptable industrial grade, rated equal to or better than service requirements).

3. A properly designed panel will have the following construction features:
   a. The finish and appearance of the overall panel will be workmanlike.
   b. All instruments and equipment will be properly aligned.
   c. Tubing, piping, and wiring will be neatly laid out and adequately supported and will not interfere with instrument maintenance.
   d. All equipment will be rigidly mounted.
   e. The back-of-panel auxiliaries and miscellaneous hardware will be properly identified by item numbers.
   f. All field tie-ins will be identified.

4. The following rules regarding instrumentation should also be followed:

a. All instruments should be installed in their proper location on the panel.

b. Correct instruments should be furnished and installed, including the charts, scales, model or type numbers, and instrument nameplate inscriptions.

5. Preliminary checks are performed as follows:

a. Power distribution is checked using these steps:

i. Verify that no one is working on the panel.

ii. Check that the panel is securely grounded.

iii. Put all disconnect switches and circuit breakers in the off position.

iv. Pull all polarized plugs.

v. With a high resistance (light or other) across the input terminals, energize the panel.

vi. If the light dims, find and remove the ground or short circuit.

vii. Energize each circuit and check each power supply subcircuit sequentially. De-energize the circuit after checking and before energizing the next one.

viii. Check for proper voltage.

b. Air supply is checked using the following steps:

i. Close all instrument air supply isolating valves.

ii. Close reducing station gate valves.

iii. Connect clean, dry air supply to the panel at the specified pressure.

iv. Open reducing station blocks and check downstream air pressure of each reducing station (set at 20 PSIG, or 138 kPa).

v. Blow down the filters and the header drain valves.

vi. Increase the header pressure until the relief valve pops.

vii. Individual air supplies may be checked as each loop is operated.

viii. Bubble-test the main air header connections for leakage.

6. Functional tests may include the following examinations, which should be performed in the order listed:

a. Assessment of pneumatic instruments:

i. Simulate the input signal at the bulkhead fitting with a 3- to 15-PSIG (21 to 104 kPa) regulator.

ii. Attach a 0- to 30-PSIG (0 to 207 kPa) gauge to controlled output at the bulkhead.

iii. Turn on the air supply.

iv. Verify that the bulkhead and air supply nameplates are correctly inscribed.

v. Vary the input signal and watch the output gauge for proper response.

b. Evaluation of alarm and 120-volt control circuits:

i. "Jumper" the input terminals one at a time to verify that the correct annunciator light flashes. The horn may be disconnected after the first alarm checkout.

ii. Energize the relay circuits on the panel by simulating the input signals.

iii. Connect a pilot light to the output terminals of outgoing signals actuating remote solenoid valve or relay. Be sure to verify output voltage so that pilot lamp will match.

iv. As each item is checked, verify the tagging of equipment and of field tie-ins.

v. Energize all chart drives, place mark on roller, and check after one hour for movement.

c. Assessment of electronic instruments:

i. Energize the loop for checking and de-energize when checked.

ii. Simulate the input signals at field tie-in points. Check input signal type, level, and voltage. This is particularly important for special instruments.

iii. Put proper resistance across output terminals.

iv. As each instrument is checked, verify the identification of equipment and of field tie-in terminals.

DDC or D/A instrumentation is checked as follows:

1. Simulate the proper field input signal at the I/O cabinet.

2. Add resistance across output terminals at the I/O cabinet.

3. Check the CRT for proper response.

4. Verify that all incoming square roots have been converted to linear signals within the distributed system.

5. Verify that all control algorithms and PID settings are correct.

6. Check the groupings to ensure that each loop appears in the proper place.

7. As each loop is checked, verify that the I/O terminals are correct.

If the inspector is unable to stay at the shop and verify that all misterminations and errors have been corrected prior to panel shipment, then a "punch list" is prepared. One copy is left with the panel manufacturer, another is kept as a record, and a third copy is sent to the field, so that the panel can be checked upon arrival at the job site.

After a proper panel inspection there should be no difficulty in hooking up the field tie-ins. Any problems in the loop checkout and calibration will most likely be external to the panel. This will significantly ease the trouble shooting in the field.

Panel Shipment

A panel should be handled as little as possible, because the chance of damage is much higher during loading and unloading and when the device is in motion on the carrier.

When the panel is to be shipped by truck and installed immediately upon arrival at the job site, only skids with a light framework holding a tarpulin are necessary. To save time and handling, the panel should travel via an air ride van. The van should be "exclusive," that is, reserved for transportation of the panel. This will ensure a direct route to the plant, without stopovers at trucking terminals, and will reduce handling. The van should also be furnished with a removable top so that the panel can be lifted out. If the panel cannot be installed immediately and must be stored at the plant site, heavier crating is required and a thicker plastic sheeting should be used. Because time is not critical and the panel is better protected, an exclusive van need not be used in this case. Shipment by train, although less expensive for long distances, requires additional handling and moving. Some trains are also severely jostled during make-up and routing.

When shipped by boat, the panel should be sent as below-decks cargo. The panel crating must be especially heavy and must be cushioned within the case. The wrapping should effectively seal out the salt air. Prior to sealing, the voids inside the wrapping should be liberally loaded with a dessicant, such as silica gel. Heavy, impregnated, water-resistant paper or 5-mil-thick polyethylene can be used for wrapping. All seams should be covered with waterproof tape. When possible, smaller shipping units should be used to ease handling. All panel equipment must be securely braced.

Air freight does not require any particular crating or wrapping other than that required for a nonexclusive van. The particular airline must be contacted and questioned regarding weight and overall size limitations for each panel and crate. The plant site airport may also be checked to verify that it is capable of receiving the type of airplane required for the shipment.

Panel Specifications

In addition to the drawings and diagrams described earlier, a written specification covering other important aspects of panel manufacturing must be developed. This specification is the document that precisely instructs the panel manufacturer as to design options and material to be used to fulfill this particular contract.

A panel specification should include a delineation of at least the following requirements:

1. *General:* Definition of the design drawing specification and codes furnished by the purchaser that the panel manufacturer is to follow.
2. *Engineering:* Description of the extent and type of engineering drawings to be developed by the panel manufacturer including whether "as-built" drawings are required. It also includes the number of prints and reproducibles required and the approval or review requirements for preliminary designs.
3. *Construction:* Description of the type of panels and their fabrication. This includes NEC area classification, ambient conditions, and similar requirements.
4. *Design:* Specification of methods of installing wiring and piping systems. This includes a listing of materials of construction for wire, pipe, tubing, ducts, nameplate inscriptions, and so forth.
5. *Materials:* Complete description of all materials to be used. A generic description is usually sufficient.
6. *Cost:* The specification should direct the bidder to delineate various costs, so that additions and deletions to the contract can be negotiated more easily. Class O Board Quotation form (such as is shown in Figure 3.2v), completed by the control center manufacturer and included with his bid, will assist these negotiations.
7. *Inspection:* Delineation of the number and types of inspections planned, which may include preliminary inspections during specific stages of construction. This section of the specification should also describe the extent of inspection required, such as visual, point-to-point checks or functional testing.
8. *Shipping:* Specification of the type of conveyance used to ship the panel to the plant site, type of crating, and protection requirements.
9. *Guarantees:* Conditions under which a panel or equipment may be rejected and the length of time during which the panel is covered by the manufacturer's warranty.

Conclusions

Control center designs need not be limited to the basic examples discussed in this section. There is no limit to the number of design variations. Each center may be formulated of new and different shapes specifically conceived and adapted to its own unique application.

JOB		
CLASS: OBQ		
DATE:		
PAGE	OF	

□ INDOOR   □ OUTDOOR

N.E.C.: □ GENERAL PURPOSE
CLASS 1 GROUP "C" OR "D" "B"
DIV □ 1   DIV □ 2

CLASS "O"
BOARD QUOTATIONS
BIDDER SHALL COMPLETE THIS FORM AND RETURN WITH QUOTE

REQUISITION SHEET NO.

DESCRIPTION OF ITEMS ADDED OR DELETED	UNIT PRICES	
	ADD	DEDUCT
MINIATURE INSTRUMENTS □ ELECTRONIC □ PNEUMATIC		
RECORDER		
RECORDER CONTROLLER		
SECOND PEN		
RATIO OR INDEX SET		
INDICATOR		
INDICATING CONTROLLER		
MANUAL CONTROLLER		
COVERPLATES		
ALARM AND BINARY CONTROL COMPONENTS		
RELAY CABINET — SELF SUPPORTED — SINGLE DOOR — ENCLOSED — 19" RELAY RACK		
ANNUNCIATOR ALARM (ASSUMING NO CABINET CHANGE) WITH PLUG-IN MODULE AND NAMEPLATES		
* PILOT LIGHT		
* PUSH BUTTON SWITCH (DPDT)		
* SELECTOR SWITCH (SPDT)		
* DISCONNECT DEVICE (FUSED)		
* DISCONNECT DEVICE (UNFUSED)		
* 120 VOLT RELAY COIL WITH DPDT CONTACTS		
* ADDITIONAL POLES FOR ALL OF ABOVE DEVICES		
* ADDITIONAL WIRE TERMINATIONS (PER WIRE)		
* SOLENOID VALVE		
AUXILIARY BACK OF PANEL MOUNTED INSTRUMENTS OR AUXILIARY BACK MOUNTED INSTRUMENTS		
* ALARM SWITCH (PNEUMATIC)		
* ALARM SWITCH (ELECTRONIC)		
EMF CONVERTER		
INTEGRATOR		
MINIATURE INSTRUMENT CUT-OUT (ON PANEL ASS'Y FLOOR)		
* NAMEPLATES - FRONT OF PANEL (WITH INSCRIPTION)		
* NAMEPLATES - BACK OF PANEL (ITEM NO. ONLY)		
AIR SWITCHES 3-WAY (INCLUDING NAMEPLATE)		
AIR SWITCHES 4-WAY (INCLUDING NAMEPLATE)		
ADDITIONAL BLANK PANEL PER LIN. FT. (PAINTED NO CUTOUTS)		

1. UNIT INSTALLATION PRICE FOR ALL ITEMS INCLUDES THE COMPLETE INSTALLATION, I.E.: PANEL CUT-OUT MOUNTING, PIPING, WIRING AND/OR POWER SUPPLY, ENGINEERING, ETC., AND WHERE NOTED THUS: (*) INCLUDE THE COST OF THE INDICATED EQUIPMENT.

2. PRICES SET FORTH HEREIN, WILL REMAIN FIRM FOR THE DURATION OF THIS PROJECT ONLY.

3. BIDDER SHALL COMPLETE ONE SHEET FOR EACH TYPE OF CONTROL PANEL SPCEIFIED, I.E. OUTDOOR-INDOOR, GENERAL PURPOSE — DIVISION 2.

4. EFFECT ON DELIVERY PROMISE TO BE ESTABLISHED AT THE TIME THE CHANGE IS MADE AND PURCHASER IS NOTIFIED.

5. NO DEVIATIONS ARE PERMITTED FROM THIS FORM WITHOUT THE PERMISSION OF M. W. KELLOGG. HOWEVER, A MULTIPLIER MAY BE ADDED TO ADJUST PRICES FOR VARIOUS STAGES OF CONSTRUCTION.

Fig. 3.2v   Class "O" Board Quotations

The multiplicity of design parameters is such that drawings and specifications cannot cover every particular feature. The only reasonable way to ensure the development of the exact control panel desired is by close cooperation of the panel manufacturer and the panel user.

BIBLIOGRAPHY

Callisen, F.I., "Control Rooms of the Future," *Chemical Engineering*, June 2, 1969.

"Control Center Design," *ISA Publications RP 60.1 through 60.11*, Instrument Society of America, Research Triangle Park, NC.

Farmer, E., "Design Tips for Modern Control Rooms," *Instruments and Control Systems*, March 1980.

"Flow Diagram Graphic Symbols for Distributed Control/Shared Display Instrumentation Logic and Computer Systems," *ISA Draft Standard dS5.3*, Instrument Society of America, Research Triangle Park, NC, 1982.

"Human Factors Affecting the Reliability and Safety of LNG Facilities," Volumes 1 and 2, Final Report, Gas Research Institute, January–September 1982.

"Human Factors Evaluation of Control Room Design and Operator Performance at Three Mile Island–2," *U.S. Department of Commerce NUREG/CR-2107*, Volumes 1, 2, and 3, January 1980.

"Human Factors Review of Nuclear Power Plant Control Room Design," *ERPI NP-309*, Palo Alto, CA, March 1977.

Military Standard, "Human Engineering Design Criteria for Military Systems, Equipment and Facilities," *MIL-STD 14728*, U.S. Department of Defense, 1974 and Later Addenda.

"Pneumatic Control Circuit Pressure Test," *ISA Publication RP 7.1*, Instrument Society of America, Pittsburgh, Pennsylvania, 1956.

Thompson, B.J., "Preparing for Computer Control," *Instruments and Control Systems*, February 1980.

3.3 CRT DISPLAYS

Types:	Storage tube, refreshed raster (T.V.) scan, refreshed X-Y positioned.
Screen Size:	From 6 × 8 inches (150 × 200 mm) to 23 × 30 inches (575 × 750 mm)
Refresh Rate:	40 to 60 Hz
Character Capability:	500 to 4800 characters
Characters per Line:	64 to 128 characters
Number of Character Lines:	12 to 74 lines
Character Set:	64 to 96 characters
Vector Modes:	Relative or absolute, or both
Vector Capability:	500 to 5000 per frame
Cost:	$10,000 to $250,000
Partial List of Suppliers:	Applied Automation Inc.; Argent Industries Inc.; Aydin Controls; Bailey Controls Co.; Babcock & Wilcox McDermott Co.; Fisher Controls Co.; Foxboro Co.; GTE Products Corp.; Instrumentation Graphics Corp.; Kent Process Control Inc.; Metra Instruments Inc.;* Process Automation Co.; Raytheon Co., Microwave & Power Tube Div.; Tektronix Inc.; Thomas Electronics Inc.; Watkins-Johnson Co.; Westinghouse Electric Corp.

*This unit displays as many as 40 analog inputs in a bar graph format.

As larger and more complex plants are built with central control rooms containing greater numbers of instruments, the man-process communication problems grow. When digital computer control was first applied in the process industry, alphanumeric control panels were used in addition to the conventional analog instrumentation displays. With these systems the user was able to display and manipulate the control loop parameters (usually, however, only singly).

Plant Communications

More recently, the man-process communication requirement has been expanded to include not only the needs of the process operator but also all communication between the process and plant personnel. Table 3.3a defines seven levels of communication in a process plant involving process operating, engineering, programming, and management personnel. A process operator, for example, would use information from levels 1 and 3, an instrument engineer would require information from level 2, and a system engineer would require information from level 4. A manager needs information from level 7.

The Total System

Although there are multi-channel bargraph instruments utilizing cathode ray tubes (CRTs) for displaying analog inputs, we will assume that the CRTs are used in a computer controlled plant and are operated by digital logic devices. A block diagram of a typical digital computer control system is shown in Figure 3.3b. The CRT hardware is contained in the two consoles and consists of a CRT display, a keyboard (containing alphanumeric, functional, and cursor control keys), alarm light switches, a refresh memory, an alphanumeric character generator

Table 3.3a
COMMUNICATION LEVELS IN A PROCESS PLANT

Level 1:   *Emergency Indicators and Alarms*
Includes both indicators and safety alarms that warn of impending difficulty; assists operator either in moving process to a safe operating point or in shutting it down.

Level 2:   *Component Diagnosis and Maintenance*
Includes information required to diagnose plant and computer system component failure; makes maintenance checks and assists maintenance personnel in correcting faults.

Level 3:   *Process Operation Information*
Includes all information required by operating personnel to keep plant running safely and as close to economic optimum as possible.

Level 4:   *Process Evaluation and Diagnosis*
Includes information needed to determine how well process is operating, to investigate potential technical or efficiency problems, and to diagnose rapidly complex process failures when they occur.

Level 5:   *Process Supervision Information*
Includes plant parameters that affect overall economy and efficiency: e.g., information on current schedules, feed stock availability and quality, utility usage, and product qualities and costs required to make day-by-day or minute-by-minute adjustments of operating conditions to achieve optimum plant operation.

Level 6:   *System Maintenance and Improvement Information*
Includes program information needed to derive the most from computer system and to make on-line system modifications as better operating methods are developed by plant personnel.

Level 7:   *Process Accounting and Scheduling*
Includes information on quantities of production, feed stock supplies, shipping, and labor to assist in establishing production schedules.

and format control, and associated control logic. A vector generator can be supplied (optional) for graphic displays. Figure 3.3c illustrates the CRT hardware.

The digital computer memory stores the operating system and data lists. An auxiliary bulk storage device (drum or disk) is sometimes used to store additional programs and data files, and the computer uses a priority interrupt scheme and two bi-directional information channels for communication with other devices. One of these channels, commonly referred to as the programmed input-output (PIO) channel, transfers control information and single data words to and from a specified register (usually an accumulator) in the computer. The second channel transfers multiple words or blocks of data to and from the computer memory and is usually referred to as the direct memory access (DMA) channel, or simply the channel input-output (CIO).

The process control program requires a list of control tasks to be performed at specified intervals. These tasks include acquiring process data through analog and digital inputs and computation of appropriate control or alarm actions, or both, based on the input data. The task list contains the necessary data for the process control programs to carry out the desired control operations. These data include input and output addresses, constants, point names, digital status information, and current input-output values. This task list is called the *process data base* and is the prime source of process data for display.

Servicing console data transmission requests from programs operating in the central processing unit and handling console keyboard interrupt requests are functions of the real-time executive system. Programs performing tasks requiring data to or from a process display console pass their requests for service to the real-time executive function programs. The calling programs receive acknowledgment of successful completion of the requested operation or indicators describing an aberrant condition.

Communication

Data Display

Cathode ray tubes display large amounts of information and are selective in displaying only relevant parameters or complex relationships between parameters. By fully exploiting the alphanumeric and graphic capabilities, the CRT is more efficient and economical than other methods of data display. Several choices of CRT implementation include a storage tube display, a raster (T.V.) scan display, and random X-Y positioned, refreshed display. What follows is a description of the random X-Y positioned, refreshed display.

The size of the usable display area, and hence the size of the CRT, is determined primarily by the size of the character, the number of characters per line, and the number of lines required. A secondary consideration is the required amount of graphic display. A typical display of a process plant unit is shown in Figure 3.3d. In order that this display be legible from a distance of 5 to 10 feet, the character height should be between $\frac{3}{8}$ and $\frac{1}{4}$ inch (9 and 6.3 mm) (see Table 3.4a). To display the information shown in Figure 3.3d, a character format of 80 to 96 characters per line and 20 to 30 lines are required. These characteristics dictate a diagonal measurement on the CRT of at least 19 or 21 inches (475 or 525 mm). The mixed display of alphanumeric characters and graphics shown in Figure 3.3e would also fit comfortably on a 21-inch (525 mm) CRT.

In a block diagram of a typical CRT display unit (Figure 3.3f), electromagnetic deflection and low-voltage electrostatic focus maintains display quality at all locations on the CRT screen. P-31 phosphor (green) is usually preferred over P-1 phosphor (white) because it is more durable. The block of input data to the display unit shown in Figure 3.3f (X and Y position data and blanking) is

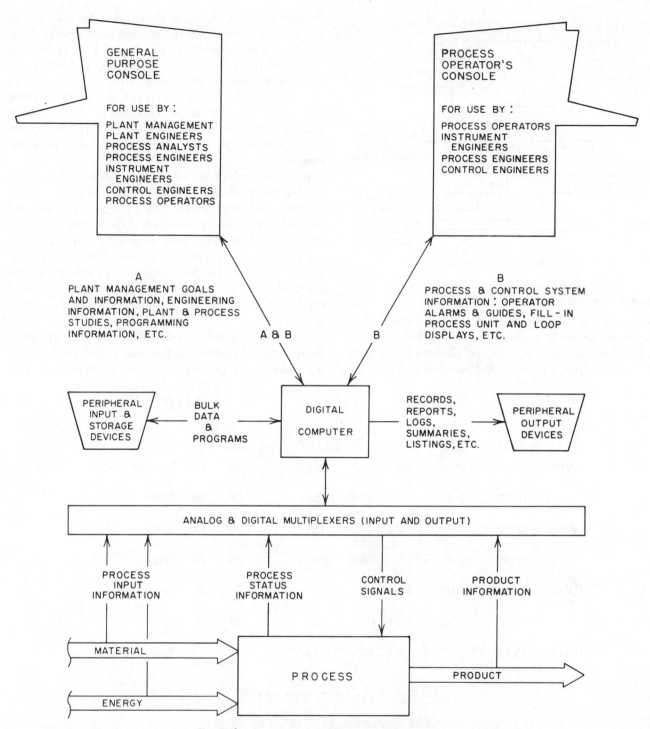

Fig. 3.3b   Computer control system with CRT display

supplied by a character generator, format control, or vector generator. The data are digital in nature. The body of X and Y position data is loaded into output registers connected to high-speed, digital-to-analog (D/A) converters, the output of which drives a linearity corrector and deflection amplifier. The linearity corrector compensates for geometrical distortion in the CRT, and the deflection amplifier must be capable of furnishing as much

as 5 amperes of current to the deflection yoke. The blanking amplifier provides a signal to turn the electron beam in the CRT either on or off.

The information supplied to the CRT display unit must be continually repeated or refreshed. So that flickering or a "swimming" effect does not occur on the display screen, the refresh rate should be synchronous with the power line frequency—ordinarily 60 (or 50) Hz.

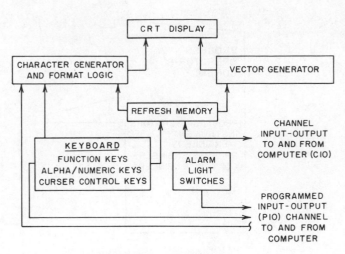

Fig. 3.3c   CRT hardware

beneath the line, value, or character to be selected for the next operation. Cursor control keys (Figure 3.3g) include the four arrow keys ←, →, ↑, ↓ for movement in one of the four primary directions, a FAST key to increase the rate of movement, a HOME key to return the cursor to the upper left corner of the CRT screen, and a JUMP key, which will be subsequently explained together with the "protect" feature. The position of the cursor can also be controlled or questioned from the computer.

It is often undesirable to enable a user to modify values or characters on the CRT screen. A "protect" feature protects characters specified by a program in the digital computer from being modified. This feature might be implemented by a bit associated with each character, such that when it is set to a "1" state, the character can be modified only by the computer, not directly by the user. This feature also enables the computer to read selectively only unprotected information in the refresh memory.

The cursor control JUMP key allows the cursor to move from a current position to the next unprotected character following a protected one, thus bypassing (pro-

Cursor Control and Data Protection

When a CRT display is used in process control, a pointer, or cursor, is required to indicate the parameter upon which the action is to occur. Cursors (see Figure 3.3d) are manipulated from a keyboard so that they are

PROCESS DISPLAY 2   FEED SPLITTER   DATE 3-1-71   TIME 1515

Loop	Block	Input	Meas	Units	Set Point	Output	Scan	CNT	Mode	Alarms ABS	Dev
FSM100 → FSC100		FSP100	340 5	PSIG	300.0	....	ON	ON	Auto	HI	
	FSC101	FSF101	5.8	TCFT/H	5.6	....	ON	ON	Comp		
FSM300 → FSC300		FSL300	12.8	FT	13.0	....	ON	ON	Auto		
	FSC301	FSF301	120	TGPH	124	....	ON	ON	Bkup		
FSM400 → FSC400		FST400	550.7	DEGF	555.0	....	ON	ON	Auto		
	FSC401	FSF401	128.5	TP/H	125.0	....	ON	ON	Comp		
FSM450	FSC450	FSF450	132.7	TGPH	134.0	....	ON	ON	Comp		

→↑ : Cursors

Fig. 3.3d   Typical process display on a CRT

PROCESS DISPLAY 3 SP TEMP LOOP FSM400

BLOCK	TYP	INPUT	MEAS	UNITS	SET PT	SCAN	CNT	MODE	ALM	HIGH	LOW	DEV
FSC400	PID	FST400	550.7	DEGF	555.0	ON	ON	AUTO		580.0	530.0	10.0
FCS401	PID	FST401	128.5	TP/H	125.0	ON	ON	COMP		N/A	N/A	10.0

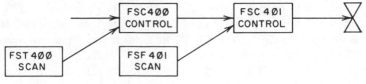

Fig. 3.3e   Mixed alphanumeric and graphic display on a CRT

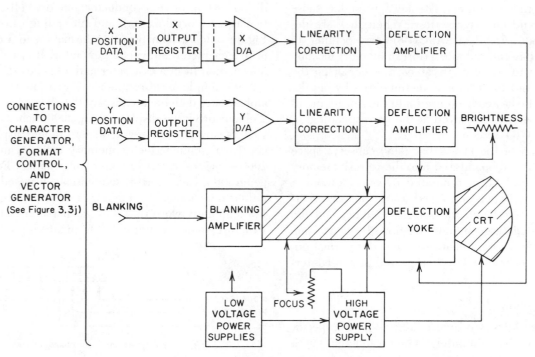

Fig. 3.3f   CRT display components

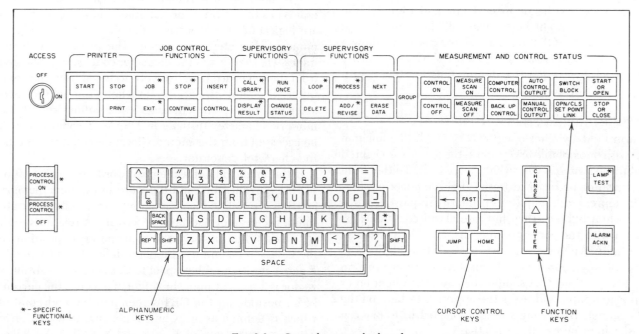

Fig. 3.3g   General-purpose keyboard

tected) characters that cannot be changed by the user—a very useful feature in a fill-in-the-blanks operation.

A "blink" feature permits individual characters displayed on the CRT to be blinked on and off several times per minute; this is useful for special conditions, such as alarm indication. This too is controlled by a bit associated with each character in the refresh memory. Supplying solid, dashed, and dotted lines is useful for graphic displays. For example, a solid line and a dashed line might

differentiate between a measurement and a set point when trend information is displayed.

Alphanumeric Keyboard

Alphanumeric keys (see Figure 3.3g) modify or make additions to the display on the CRT screen. Entries can be made only into unprotected locations and are themselves unprotected. The operations are performed by the hardware associated with the CRT and require no response from the computer. The keys resemble those commonly found on a typewriter. When they are depressed, a code (usually USASCII-8) corresponding to the key legend is entered into a refresh memory location corresponding to one directly above the cursor on the CRT screen, and the cursor is incremented by one location. With the key code entered into the refresh memory, the character is displayed at the corresponding location on the CRT screen.

A depression of the SPACE bar (key) causes a space (blank) character to be entered into the refresh memory and the cursor to be incremented by one location. A BACK SPACE key, when depressed, causes a space character to be entered into the refresh memory and the cursor to be decreased by one location. By depression of the repeat key and a character key, the normal operation of the character key is repeated at a predetermined rate.

Function Keyboard

The function keys (see Figure 3.3g) request a specific action of the digital computer. When a function key is depressed, a priority interrupt signal is sent to the computer. The computer reads a code on the PIO channel corresponding to the depressed key and executes the request, which might be to place all the control loops displayed on the CRT on manual control or to show a directory of the display library on the CRT. In other words, each key requests a unique function that is programmed in the computer.

Alarm Light Switches

Alarm light switches operate very much like function keys, with one notable exception—the former are lighted pushbutton switches whose light is controlled either from the computer or by an external (field) contact closure. When depressed, the buttons primarily request new displays; when lighted, they indicate alarm conditions associated with the corresponding display. Depressing an alarm light switch causes a hardware action identical to a function key depression. When the computer program detects conditions that should turn an alarm light on or off, the computer addresses the appropriate light on the PIO channel. By setting a unique bit to 1 or 0, one can turn the light on or off, respectively.

Data Display Methods

Refresh Memory

The refresh memory stores information (in coded form) displayed on the CRT screen. Since the duration of the CRT phosphor is several hundred microseconds, the displayed information must be regenerated and displayed at a nominal rate of 60 times per second. The refresh memory may consist of magnetic or acoustic delay lines, semiconductor shift registers, magnetic cores, magnetic disk or drum, or semiconductor memory cells. The particular size, organization, and bit coding can vary (see Figure 3.3c). It can furnish information to a computer or to a character and vector generator. It can also accept information from a computer and a keyboard.

For example, a refresh memory associated with a 2000 alphanumeric character display (80 characters per line, 25 lines) or with a display having 3000 inches (7.5 m) of vectors (straight line segments for graphic displays) may consist of semiconductor memory cells organized into 2000 words, each word 12 bits of information. For display generation, each word is sequentially accessed and sent either to a character generator or to a vector generator. For a memory word that stores a character code, the bit structure shown in Figure 3.3h might be used.

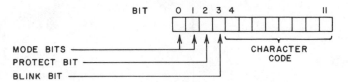

Fig. 3.3h   Bit structure for character code

The mode bits differentiate among characters and several types of vectors. For example, when the mode bits are logical 00, the word is defined as containing alphanumeric character information. The protect bit determines whether or not the character code can be changed from the keyboard or selectively accessed from the computer. The blink bit determines whether or not the character will blink. The character code defines the alphanumeric character (usually in USASCII code) that will be accessed from this memory location and displayed by the character generator.

If the mode bits of a memory word are logical 10 or 11, the current word and the word in the next memory location are defined as containing either relative or absolute vector information, respectively. A relative vector is a straight line, its origin the current beam position on the CRT screen and its end point defined as a change in X and Y position with respect to this origin. An absolute vector is a straight line, the origin of which is the current beam position on the CRT screen and the end point of which is defined as an X and Y position in a fixed grid with the grid origin of $X = 0$ and $Y = 0$ at the lower left corner of the CRT screen. The bit structure in Figure 3.3i might be used to define a vector.

If the mode bits (in word 1) are logical 10, the two words define a relative vector, and therefore the X displacement contains a ΔX value and the Y displacement contains a ΔY value. If the mode bits are logical 11, the two words define an absolute vector, and therefore the X displacement contains an X value and the Y displacement contains a Y value. The line type determines whether the vector to be generated will be a solid, dashed, dotted, or invisible line (blanked movement).

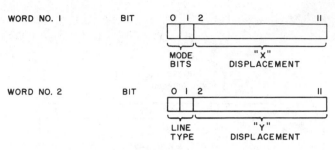

Fig. 3.3i   Bit structure for vector code

Character Generation and Format Control

Alphanumeric characters may be generated by means of several techniques. Analog stroke, a "race-track," character mask scanning, and read-only memory character generation are a few examples. The following example is based on read-only memories.

Figure 3.3j illustrates a character generator and format control logic. Since there are 2000 memory locations containing character codes for each of 2000 character positions on the CRT screen, the value contained in the refresh memory address register (Figure 3.3j) must be unique for each character position. The refresh memory is accessed at the location specified by the contents of the address register, and data are loaded into the data register from this location. The format control accepts the contents of the address register as an input and generates absolute values of X and Y data, which positions the CRT beam to the starting position of the appropriate character location on the CRT screen.

The contents of the data register are then used as an address for the read-only memory, and the body of X and Y data is accessed and loaded into the appropriate shift registers (see Figure 3.3j). As the information is

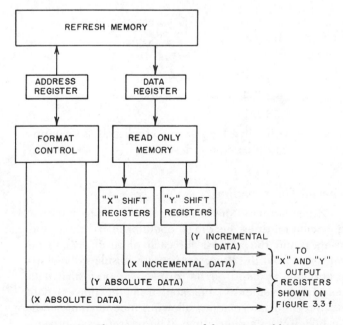

Fig. 3.3j   Character generator and format control logic

shifted out of these registers bit by bit, it is decoded and sent to the X and Y output registers (see Figure 3.3f). This mass of decoded data causes the output registers to increment (or count) up or down, which in turn causes the appropriate character (specified by the contents of the data register) to be written on the CRT screen.

Vector Generator

Vectors are straight line segments used to construct graphic displays. Typical methods of vector generation use analog ramp generators, binary rate multipliers, or digital arithmetic units. The example to be described uses the digital arithmetic units.

The vector generator receives data from the refresh memory (see Figures 3.3i and 3.3j) and based on this body of data provides incremental data to the X and Y output registers (see Figure 3.3f). When the vector generator receives the X and Y displacement information from the refresh memory, if the vector was specified as a relative vector, the information will consist of a ΔX and a ΔY value. The vector generator operates directly on this body of data. If the information received from the refresh memory has been specified as an absolute vector, an auxiliary operation of computing ΔX and ΔY will occur, by subtraction of the current beam coordinates (X and Y values) from those obtained from the refresh memory.

The operation (algorithm) of the vector generator is such that an incremental step (or movement) is made to minimize the value of ΔX and ΔY. A new value of ΔX and ΔY is then computed, and another incremental step is made to minimize the value of ΔX and ΔY. The process is repeated until the computed values of ΔX and ΔY are both zero. The result is a best fit to the desired straight line segment displayed on the CRT screen. The solid, dashed, or dotted lines are generated by turning the CRT beam on or off (blanking) at desired intervals.

Display Initiation

Process display initiation comprises a chain of events that begins with an operator key action and ends when the selected display has been transmitted to the console refresh memory and real-time update has commenced. From the operator's point of view, one or more key actions are required to fetch a display. From the point of view of program or software, these key actions identify what the operator wants to see. The operator must have a method of observing both the process variables and the response to actions taken by the control programs in the computer. The operator also needs to be notified of alarm conditions and requires a method of communicating directives to the process control program.

To accomplish these objectives the process display programs must allow the operator to (1) initiate process data display requests, which will be updated to reflect process variable changes; (2) manipulate reference values and states (on-off or automatic-manual) of block or loop records in the process data base; (3) terminate a process display; (4) request other relevant programs, such as directories or plant efficiency calculations; and (5) respond directly to an alarm.

Keys and Key Sequencing

Inherent in each of the operations just mentioned is the process display console keyboard. Considerations of key sequence include (1) how much data must be entered from memory; (2) how many key strokes are required to achieve a display response; (3) how many key strokes are required to recover from a data entry error; and (4) how many operator decisions (choices) are required to proceed through a desired sequence.

The function keys (see Figure 3.3g) may be divided and arranged in groups as shown. From the point of view of software, the keys are also arranged by purpose. Keys supplying a constant response can be grouped by key code. All other keys are conditioned response keys. It is useful to think of these two groups as specific (constant-response) keys and conditional (sequence-dependent) keys. Alarm key lights are specific keys. Conditional keys manipulate process reference values, control states, and select data from a recipe. Specific functional keys are indicated by an asterisk in Figure 3.3g; all other keys are conditional. Although in general it is desirable to minimize key operations that serve to initiate process displays, it does not always follow that every action that an operator might take should have minimum key activity. On one hand, operations such as modification of numerical values may require visual verification before entry into the process data base is attempted. On the other hand, state changes of process data blocks or loops should require minimum key actions. Thus, the design of operator key activity and key sequencing must be related to the display tasks and to the keyboard design.

The key sequence for any operation may be constructed as follows: First a specific key is used. This produces a fixed (by key code) visual response. Operator data are entered by the alphanumeric keys followed by a conditional key. The alphanumeric keys transmit data only to the display memory rather than to the central processing unit (CPU). The cursor is manipulated by the operator to enter data at appropriate display locations. Key sequence diagrams are useful in planning process display–process operator interaction. Figures 3.3k and 3.3l show two typical sequences. The alarm key sequence (Figure 3.3k) is used when an alarm key light comes on owing to a process upset. The operator presses the key, initiating a process unit display (see Figure 3.3d). The process loop display sequence (Figure 3.3l) requires entry of data before the loop to be displayed can be selected. The sequence begins with a specific key operation (loop key) and continues through entry of data (Figure 3.3m) and initiation of loop display (see Figure 3.3e). These examples are initiating sequences. Operator in-

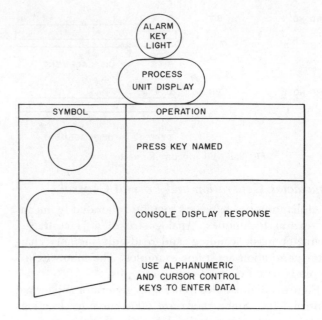

Fig. 3.3k   Alarm key sequence

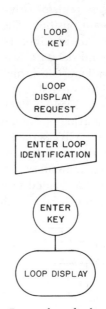

Fig. 3.3l   Process loop display sequence

teraction with a live loop or process display is a continuation of the techniques described, using conditional keys.

Format Construction

Display formats consist of the fixed or static information (column titles, headings, operator instructions, and recipes) and the address for each piece of data to be retrieved from the process data base, displayed and appropriately updated on the display. Static format data may be conveniently separated from the process data base–related information, allowing independent modification of titles and headings. The references to process

```
LOOP OR BLOCK DISPLAY REQUEST

LOOP OR BLOCK ID...............        DISPLAY TYPE...............
                                       (LEAVE BLANK FOR LOOP DISPLAYS
COMPLETE FOR TYPE 5 DISPLAY:            OR BASIC FORMAT BLOCK DISPLAYS)
TREND PEN NO...............
MEASUREMENT SCALE: MIN ........... PCT MAX........... PCT (LEAVE BLANK FOR
                                                     0 TO 100 PCT)

        TYPE NO.    DISPLAY DESCRIPTION
           1           BASIC FORMAT
           2           MEASUREMENT FORMAT
           3           BAM PROCESS OPERATORS DISPLAY
           4           BAM INITIATING DISPLAY
           5           TREND RECORDER
           6           TREND DISPLAY (CRT)
           7
           8
           9
          10
```

Fig. 3.3m   Loop display request

data base information should be symbolic. Usually, the process data base is referenced by block or loop name, which points to a complete set of measurement and control data about one process control input or output, or both. Within this set of data the references to particular information, such as a set point or measurement, should be symbolic.

Symbolic references to data items in each block record simplify the display-initiating program. The references are passed as arguments to a subroutine set that locates the appropriate item and performs internal-to-external format conversions.

Process Data Retrieval and Display

The CRT-based process display allows considerable flexibility about what process data are to be shown. Various special-purpose displays may be constructed to suit individual processes and operating policies. Individual blocks or a single piece of information for several loops (Figure 3.3n), groups of connected blocks (a control loop—

```
PROCESS DISPLAY: 8   PLANT FLOW MONITOR   DATE: 3-1   TIME: 1515
```

Point ID	Description	Status	Value	Alarm
FSF101[10]	Plant 2 feed flow	On	5.8 TCFT/H	
FSF301	Splitter flow to heater	On	120.8 TGPH	
FSF401	Splitter steam flow	On	128.5 TP/H	
FSF450	Acc flow to splitter	On	132.7 TGPH	
HRF502	Heater fuel flow	On	2500 CFT/H	
PFF600	Product A flow	On	60.7 TCFT/H	
PFF550	Rct. feed flow to fractionator	On	106 TGPH	
PFF701	Fractionator steam flow	On	107.4 TP/H	
PFF750	Product B flow	On	75.5 TGPH	

Fig. 3.3n   Plant flow monitoring display

see Figure 3.3e) or sets of data related to process unit performance (see Figure 3.3d) may be displayed. If each of these displays is considered a standard type, one may then construct displays using different sets of block or loop names for each process unit display, the format of which will remain constant—the block or set of loop data, or both, is changed to reflect the set of names chosen. Each set of blocks or loops is referred to by an identity code.

Process unit displays may be automatically shown in response to an alarm key action by construction of a list of alarm key codes and association of a "process unit display" identity code with each. The identity code retrieves the list of block or loop names to be displayed. At the same time that the name list is retrieved, the identity of the format (both the fixed part and the data base–related part) is also retrieved. Thus, *a process unit display identity is used to point to a predefined display format and a set of process data*.

The body of data necessary to initiate a display of the type shown in Figure 3.3d includes:

1. The names of the blocks or loops, or both, to be displayed;
2. The format of the display;
3. For each block in the display:
   a. the symbolic references to values in the block record, e.g., MEAS (measurement), ABS (absolute);
   b. for each symbolic block reference value, its relative or absolute display address;
   c. whether or not the value is to be updated in real time, whether or not the operator is to be allowed to modify the value, and how many characters are to be displayed.

Display Propagation and Termination

The display-initiating program retrieves the appropriate data for building the display and supplies real-time display control and functional key service data through files or lists to the respective programs. The updating program is responsible for maintaining the displayed measurements and other values in a current or real-time state. The functional key service program is responsible for all operator-requested modifications of the data display. Typically, these changes are of two types: data entry or value manipulation, and state changes (e.g., on-off or automatic-manual).

In Figures 3.3d, 3.3e, and 3.3n, unprotected underscores define the appropriate areas for data entry for the operator. All other data displayed are protected and cannot be modified or changed by the operator at the console. The display initiating program also must set a flag or bit in each block record requiring update, and this bit or flag is referred to as the display capture bit.

Data Capture, Conversion, and Routing

In display propagation the values shown are changed to reflect the variations in the controlled process. Measurements, alarm states, internally modified set points, and reference values are examples of data requiring continuous updating. Update frequency may be other than the normal processing interval, which is inconvenient and requires additional program logic. If the update frequency is the same as the processing (scan) interval, the process control program may be constructed to examine the data capture bit in each block record. When the bit is on, the block record data are set aside in a temporary display file or list. When the process control program has completed its tasks for the current interval, it calls on the display update program for execution. It should be noted that display update is called on only when data have been captured for update.

The display update program finds the block record in the display file or list and with the display control information assembled by the display initiator converts the appropriate mass of data in the captured block record to external format and transmits it to the display. Typically, for each block record* the display update program includes (1) block name; (2) display console number if more than one; (3) symbolic data names for all items to be updated; and (4) display memory address (where the data is to be displayed).

Operator Interaction, Data Entry, and State Changes

Operator interaction with a process display consists of manipulating the blocks displayed, e.g., changing a block from ON to OFF, and of entering new numerical data, e.g., changing a set point.

Figure 3.3d shows a process display. In the column labeled "set point" there is (for each block) a numerical value, and directly under the value a line of underscores that shows the operator where a new value for the set point is to be entered. The cursor is moved to the underscore field and the new value is entered, using the numerical keys. The underscores are unprotected characters and can be overwritten from the keyboard. The operator may then either move the cursor to other underscores and enter more data or press the ENTER key, which causes an interrupt to occur in the CPU. In response to the interrupt a program is called for execution that will service the operator's console. The program will in this case read the unprotected characters in the display memory and attempt to modify the appropriate values in the process data base block records.

The functional key service program uses data supplied by the display initiator to determine the set of data that

---

* Note that the use of symbolic block data item reference is carried in the display update function as well as in the display initiator.

goes with the blocks on display. Appropriate visual feedback to the operator is obtained by overwriting the existing value (the one above the entered value) with the new value and restoring the data entry area to unprotected underscores. Should the new value be unacceptable, the underscores are not restored and the offending value may be set to blink. An error diagnostic message may also be displayed.

The operator may accomplish block state changes by pointing the cursor to the block name and pressing one of the measurement and control status function keys. These keys have been appropriately labeled control on, control off, and so forth (see Figure 3.3g). The same interrupt response takes place as already described. The program responding to operator requests for service must be able to activate and deactivate function keys. Also, the interrupt-causing keys must be identified. Activating and identifying keys is performed by a key mask table for each console keyboard. It contains the code for each key that is currently active. An active key is one that has been put in the table by the program servicing the console. Thus, the servicing program can at any time determine the function keys that the operator may legitimately use. The key mask table is used by the console-interrupt-handler segment of the real-time executive to determine if the servicing program has to be called.

If many consoles are operating concurrently, the servicing program attends to each as requests occur. There need be only one servicing program. The information that it requires for each console that has a currently active process display includes (1) the total number of unprotected characters on display; (2) a sequential list of the block name and symbolic value name for each data entry field; (3) the length of each data entry field and the display address of its related protected value; and (4) access to the same data as the display update program. This set of data is needed to ensure that visual feedback for every requested state change is available. If visual feedback is not possible, the requested state change is erroneous, and an appropriate diagnostic measure is displayed.

Terminating the Display

After observation and manipulation, the operator indicates that the operation is complete by requesting another display by a specific function key or alarm key light, or both. The console-interrupt-handler segment of the real-time executive determines if process display termination is necessary by keeping track of real-time update operations on a console-by-console basis. If the current display on a console is not being updated in real time, termination is unnecessary; the requested program is responsible for clearing the display.

The process display termination program determines which blocks in the process data base were being "captured" for display on the console and resets or stops their ensuing capture and display by resetting the display capture bit in the block record. The program also purges the data files supplied by the display initiation to the update program. When termination is complete, the operator-requested function is allowed to proceed. If many process displays on many consoles are being updated in real time, the termination must take care not to terminate capture of blocks that are being displayed on other consoles.

BIBLIOGRAPHY

Aronson, R.L., "CRT Terminals Make Versatile Computer Interface," *Control Engineering*, April 1970.

Davis, R.C., "Vocal Alarms," *Instrumentation Technology*, December 1964.

Ellis, R.K., "Color Graphics CRT's Provide Window into Factory Operations," *Instruments and Control Systems*, February 1983.

Jutila, J.M., "Guide to Selecting Alarms and Annunciators," *Instrumentation Technology*, March 1981.

McCready, A.K., "Man-Machine Interfacing for the Process Industries," *InTech*, March 1982.

3.4 DIGITAL READOUTS

Types:

(a) Mechanical and electrical counters, (b) gaseous discharge displays, (c) cathode ray tubes (CRTs), (d) rear projection displays, (e) light emitting diode (LED) displays, (f) liquid crystal displays (LCD), (g) vacuum fluorescent displays (VFD).

NOTE: In this feature summary the letters (a) through (g) refer to the listed alphanumeric readout types.

Display Application:

(a) numeric; (b,e,g) alphanumeric; (c,d,f) alphanumeric and symbols.

Character Sizes:

(a) 0.1 inch to 0.5 inch (2.5 to 12.5 mm); (e,g) 0.1 inch to 0.8 inch (2.5 to 20 mm); (b,c) 0.3 inch to 2 inches (7.5 to 50 mm); (d) 0.12 inch to 3.38 inches (3 to 84.5 mm); (f) 0.3 inch to 6 inches (7.5 to 150 mm) and larger.

Maximum Viewing Distance:

(a) 30 feet (9.12 m); (e,g) 50 feet (15 m); (b,c) 100 feet (30 m); (d) 150 feet (45 m); (f) 200 feet (60 m) or more.

Viewing Angle:

(a,c,e,f) 90°; (d,g) 100°; (b) 120°.

Brightness in Foot-lamberts:

(a,f) a function of ambient lighting; (d) 75(255 cd/m^2); (c) 100(340 cd/m^2); (b) 150(510 cd/m^2); (e) 200(680 cd/m^2); (g) 300(1020 cd/m^2).

Life:

(d) 10,000 hours; (c) 20,000 hours; (a,b) 100,000 hours; (g) 300,000 hours; (e,f,) 500,000 hours.

Operating Voltage:

(e) 3 to 5 Vdc: (d,f) 5 to 28 Vdc: (a) 0 to 115 Vac or Vdc; (b) 170 to 300 Vdc; (g) 15 to 40 Vdc; (c) 3000 to 15,000 Vdc.

Relative Cost:

(a,e,f,g) low ($5 to $25); (b) medium ($15 to $30); (d) high ($25 to $50); (c) highest ($50 to $150).

Parial List of Suppliers:

Amperex Electronics Corp. (b,c); Beckman (b); Burroughs Corp. (b,e); Cherry Electronic Products (b); Dialight Corp. (e); Fairchild (f); Futaba Corp. (g); Hewlett-Packard Co. (e); Industrial Electronics Engineers, Inc. (c,d,f,g); Monsanto Co. (e); Rockwell International (g); UCE, Inc. (f).

Introduction

The development of digital technology in the 1970s was followed by the increased use of digital or alphanumeric (a/n) displays. As pocket calculators became commonplace, the light emitting diode and liquid crystal displays were produced in very large quantities. This led to increased research and development efforts and a greatly reduced price. The evolution of cheaper monolithic analog-to-digital (A/D) converters paved the way for the widespread use of digital voltmeters. This again opened up new markets for alphanumeric displays. The use of microprocessors in the last few years only added to the popularity of the digital readouts, and so the digital era has led to the obsolescence of analog art. Today the majority of instrument panels and annunciators use alphanumeric readouts.

As their name implies, a/n readouts furnish alphabetical (letters or legends) and numerical (digits or numbers) information. Common examples are the television cathode ray tube and the automobile odometer. These readouts provide accurate, easily understood displays and require no operator interpretation because they present clear, concise legends or exact numerical values.

Versatility is the primary advantage of a/n readouts because they have the capability to display many different types of information. Some readouts display only numbers, whereas others display numbers, letters, and symbols. The contents of the display can be tailored to the type of variable being displayed. For example, the readout of a digital voltmeter or electronic calculator may accommodate the maximum value and the decimal point location. In a process control display the readout may include the loop name and tag number with the value of the variable being measured.

Human Factors

To transmit information efficiently, readouts must have sharp character resolution, which is a measure of the character image sharpness or clarity and relates directly to readability. Readability is the quality that allows an observer to perceive information with speed and accuracy and is a function of character style, proportions, height, contrast, color, and refresh rate.

Characters that are pleasing to the eye are generated by simple continuous lines. To aid in character recognition, critical details should be prominent, and special features such as openings or breaks should be readily apparent. Many a/n readouts use a matrix of dots or bar segments for character generation. Closely spaced dots are more legible and natural-looking, whereas bar segments usually form box-like characters containing noticeable intersegmental spaces. Character proportions should be predicated on a height-to-width ratio of roughly 3 to 2. Line width should be approximately one seventh of character height. Minimum spacing between characters should be two line widths with approximately six line widths between words.

Size and Contrast Considerations

A guideline for determining character height based on viewing distance is given in Table 3.4a. The heights can be modified slightly for high ambient illumination or high brightness displays. A display mockup is recommended if a particular viewing distance is critical.

Contrast is the ratio between character image brightness and its background brightness when measured under normal ambient illumination. Acceptable ratios for a/n readouts are about 5 or 10 to 1. Assuming the background to be a typical control console surface with a brightness of 20 to 50 foot-lamberts (ft-L) or 68 to 170 candela per square meter (cd/m^2), a nominal brightness range for readouts should be 100 to 500 ft-L (340 to 1700 cd/m^2). Contrast depends on ambient illumination. Consequently, locations near an outside window or directly under a lighting fixture may require a filter to prevent image washout or to reduce objectionable reflections. An

Table 3.4a
CHARACTER HEIGHTS BASED ON VIEWING DISTANCE

Required Viewing Distance in Feet (m)	Nominal Character Height in Inches (mm)
2.3 (28 in or 0.7 m)	0.12 (3)
5 (1.5)	0.18 (4.5)
10 (3)	0.25 (6.25)
15 (4.5)	0.31 (7.75)
20 (6)	0.38 (9.5)
30 (9)	0.50 (12.5)
40 (12)	0.68 (17)
50 (15)	0.75 (18.75)
65 (19.5)	1.38 (34.5)
100 (30)	2.00 (50)

antireflective filter reduces reflections and improves contrast by passing proportionately more self-generated image light than reflected image background light.

The optimum readout colors are green and yellow, since the eye is most sensitive to them (Figure 3.4b). Amber, red, and orange are the next best choices. A filter can be used to obtain a desired color. A red filter, for example, will pass only the red light of an incandescent readout while it absorbs all other colors. Several types of a/n readouts generate character images by addressing time-displaced current pulses to dots or segments. The number of times (per second) that the image is generated is called the refresh rate. Low refresh rates can result in lowered brightness and occasionally cause flicker.

Two additional human factors involve viewing angle

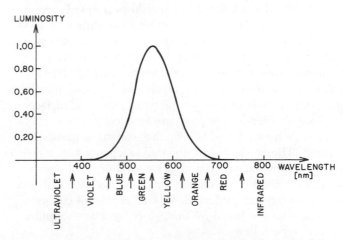

Fig. 3.4b Luminosity as a function of wavelength. The National Bureau of Standards developed the above spectral-luminosity curve based on the findings of 52 experienced observers. The curve indicates that the eye responds to light between 380 and 760 nanometers, reaching maximum sensitivity at 555 nm in the yellow-green region.

Table 3.4c
ALPHANUMERIC READOUT CHARACTERISTICS

Type/Feature	Method of Operation	Operating Characteristics	Typical Applications
Mechanical and Electrical Counters	Numbered wheels are rotated behind a viewing window.	Contain inherent memory; power required only to change display; usually illuminated by ambient light.	Digital clock; water meter; gasoline pump.
Gas Discharge Displays	Shaped electrodes, in form of characters, ionize surrounding neon gas.	Require external memory and operate on high voltages (170–300 DC).	Electronic calculators; electrical meters.
Cathode Ray Tubes	Shaped electron beam is projected on phosphor screen.	Require character generation and memory circuitry; operate on very high voltages (3000–15,000 VdC).	Computer-controlled displays.
Rear Projection Displays	Miniature optical projectors containing incandescent lamps display images on viewing screen.	Require external memory; models are available for wide range of operating voltages.	Control-console displays.
Light Emitting Diode Displays	Semiconductor junction emits light when DC current is passed through it. Individual diodes make up dot or bar matrix.	External memory and drivers are required. Low-voltage, high-current operation (2V, 15 mA per segment).	Electronic calculators, clocks, panel meters.
Liquid Crystal Displays	LCD does not generate light. It can be made to pass or block light by application of AC voltage.	External memory and AC drivers are required. Low-voltage, low-current operation (5 V, 1 $\mu A/cm^2$).	Digital watches, portable instrument displays.
Vacuum Fluorescent Displays	A triode vacuum tube with phosphor-coated anodes. Variously shaped segments form matrix.	Medium-voltage (20–40 V), low-power device. External memory and drivers are required.	Clocks, calculators, and alphanumeric displays on instruments.

and change or update rate. If the readout must be seen from several operator positions, it should have a viewing angle of approximately 120 degrees. Recessed characters are satisfactory only for direct viewing. The maximum readout update rate is twice per second if the operator is expected to read consecutive values. Numerical readouts are not recommended for determining rate of change, direction, or tracking.

Application Notes

For critical readouts there should be a test function or switch to locate burned-out lamps or tubes or lost character portions. The loss of the horizontal center section of some matrix readouts, for example, can cause an operator to misread an integer as zero when eight is really intended. The readout bezel should be aesthetically pleasing and in contrast with the panel mounting surface.

Most a/n readouts require an electrical power supply to provide memory and to keep the readout continuously lighted. The readouts are updated or changed by a decoder driver that accepts binary coded decimal inputs. Readout modules containing standard decoder drivers with memory are preferred. Characteristics of common a/n readouts are listed in Table 3.4c, and a description of each type follows.

Mechanical and Electrical Counters

Mechanical and electrical counters (Figures 3.4d and 3.4e) are relatively simple and inexpensive devices that

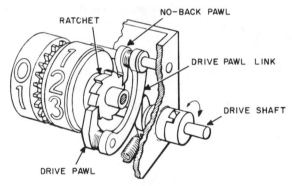

Fig. 3.4d   Mechanical counter

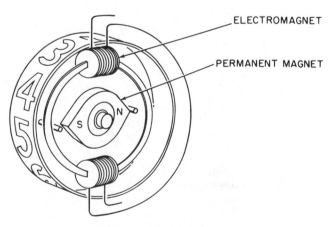

Fig. 3.4e   Electrical counter

provide a cumulative count of sequential events. Usual configurations consist of a series of numbered wheels mounted on a common input shaft rotated by mechanical or electrical pulses. The numerical readout is visible through a window placed over the foremost row of digits. Digits should be as close as possible to the window so as to provide the maximum viewing angle and to eliminate shadows from ambient lighting. Internal lamps should be supplied if these counters are to be used in darkened areas.

The driver output and the counter input should be compatible. Mechanical counters are driven by mechanical rotation or oscillation of the input shaft. Rotation changes the readout through internal gearing, and oscillations use ratchet-pawl mechanisms. Electrical counters work similarly, except that a solenoid or an electromagnet actuates the input shaft. Counters operating with mechanically rotating inputs usually provide one or ten counts per complete rotation of the input shaft. Inputs can be either clockwise or counterclockwise; consequently, the count adds in one direction and subtracts in the other. Oscillating inputs provide one count per oscillation, and counting is in one direction only.

Electrical counters ordinarily supply one count per pulse and are available for AC or DC operation with a wide selection of voltage coils. Counters with two coils will add one count for a pulse through one coil and will subtract one count for a pulse through the other coil. Zero reset counters should be used for applications requiring a cumulative count between random times or events. A single knob depression for mechanical counters or pulse for electrical counters will set all wheels to zero, and a new count can be established.

Gas Discharge Displays

The forerunners of this group of displays were the so-called Nixie tubes (Figure 3.4f). The operation of these tubes is similar to that of neon lamps. A sealed glass tube contains a common anode and a stack of ten independent cathodes shaped like numbers. When a large negative voltage is applied to a selected cathode, the gas around it glows from being ionized and emits a yellow-orange light. The gas in the tube is usually a mixture of neon and argon. A small amount of radioactive krypton and liquid mercury is added to the mixture to improve starting and to reduce cathode sputtering. These tubes were used up to the 1970s, when the planar-type matrix displays took over. These displays involve the same physical phenomenon as the Nixie tubes, except that the characters are made up of selectively lighted segments. The most common version is the seven-segment display, in which by using variable combinations of seven bar-shaped segments one can write all ten numbers and some other characters as well (Figure 3.4g).

The characters may be straight or slanted. The segments can be rectangular, mitered, or a combination of

Fig. 3.4f Gaseous discharge tube

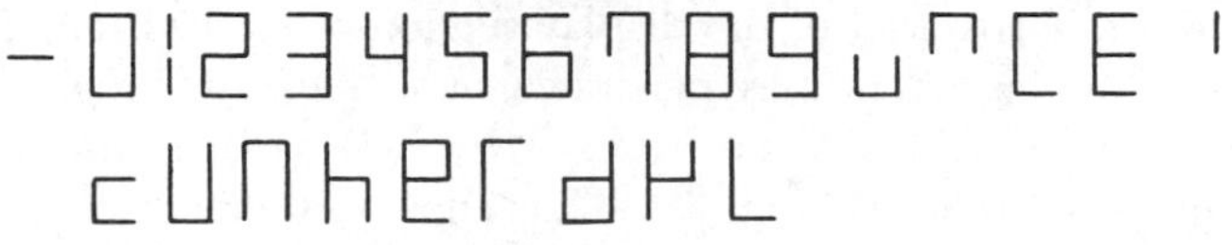

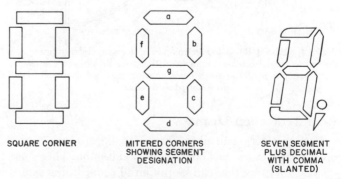

Fig. 3.4g Some symbols available from the seven-segment character

both with rounded corners (Figure 3.4h). For construction details, refer to Figure 3.4i. The whole assembly is placed in a flat, sealed glass container. The character segments are the cathodes and are located on the bottom of the glass enclosure. The anode is a fine metal mesh or a transparent tin oxide or indium coating on the inside surface of the cover glass. There are generally two or three characters to an envelope, with the connecting pins coming out on the rear side. A seal-off tube is located on the rear side for pumping the air out and inserting the gas mixture.

These devices are produced by several manufacturers. They provide high brightness, good viewing angle, and 10,000 to 30,000 hours of operation and are moderately

SQUARE CORNER

MITERED CORNERS
SHOWING SEGMENT
DESIGNATION

SEVEN SEGMENT
PLUS DECIMAL
WITH COMMA
(SLANTED)

Fig. 3.4h Typical character shapes of seven-segment displays

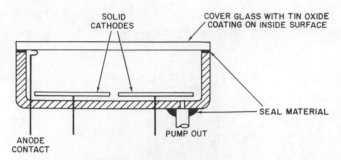

Fig. 3.4i Cross-sectional view of a popular construction technique for the gas discharge display

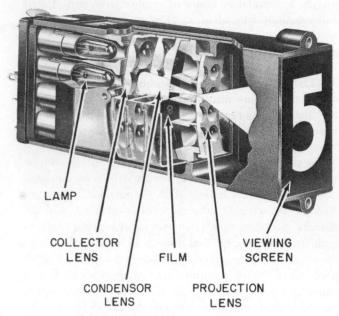

Fig. 3.4k Rear projection readout

priced ($3 to $5 per digit). Among the disadvantages are the need for a high-voltage power supply (~ 180 Vdc) and a high voltage driving circuit.

Cathode Ray Tubes

Cathode ray tubes, commonly called CRTs, are versatile alphanumeric readouts. From the outside they look like ordinary television sets. In effect, the operation is very similar to that of the display section of a television set. The characters may be drawn by a line raster that blanks and unblanks the beam or by X-Y writing, also known as stroke writing. A CRT is shown schematically in Figure 3.4j. This type of display requires the most complicated support circuits. It requires a high-voltage power supply (3 to 15 kV), horizontal and vertical deflection circuits for raster generation, and a character generator. The raster generally consists of 525 horizontal lines or 1029 lines in high-resolution displays. Digital character generators can produce any types of characters previously defined and stored in semiconductor memories. CRT displays are complicated but can display a large amount of information. Typical uses are in computer terminals and airport departure-arrival displays.

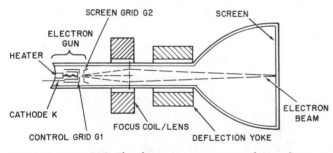

Fig. 3.4j Basic CRT. The electron gun projects an electron beam on the screen that becomes luminous at the point of impact. The electromagnetic focus lens causes beam convergence, reducing spot size to usable dimensions.

Rear Projection Displays

Rear projection displays (Figure 3.4k) are miniature optical projectors stacked in cordwood fashion. They can display anything that can be put on film, including symbols, words, and colors. Twelve incandescent lamps at

the rear of the display ordinarily illuminate the corresponding filmed messages, which are focused through a lens system and projected onto a single-plane, non-glare viewing screen. Rear projection displays exhibit excellent character resolution and readability and produce a very natural a/n readout appearance. They can be operated directly from electrical relays or from self-contained decoder drivers.

Light Emitting Diode Displays

The light emitting diode (LED) displays have a wide variety of uses. They were developed in the 1970s, and by the end of that decade they could be found in all types of products, from watches and calculators to instruments, appliances, and point-of-sale adertising. Among their advantages are that they are inexpensive and easy to drive and have reasonable power requirements. The disadvantages include a reduced viewing angle for displays with magnifiers and possible nonuniform segments for larger displays. They can be used as solid state lamps, bar-segment displays, and dot-array displays. There are three colors available for LED displays: red, yellow, and green. The red displays have the highest brightness and are the most widely used. The yellow is acceptable, and the green is the least legible. The sizes range from 0.1 inch to 0.8 inch (2.5 to 20 mm). As an example, Figure 3.4l shows a seven-segment digital LED display manufactured by Hewlett-Packard.

Liquid Crystal Displays

The liquid crystal display (LCD) is an electronic readout that uses the unique characteristics of a class of materials called liquid crystals. This display technology is unique in that electrical energy is not converted to visible

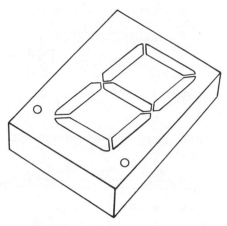

Fig. 3.4l The HDSP–3400 Series are very large 20.32mm (0.8 in.) GaAsP LED seven-segment displays. Designed for viewing distances up to 10 meters (33 feet), these single-digit displays provide excellent readability in bright ambients.

light, as with LEDs, gas discharge, and vacuum fluorescents; rather, the display modifies light. To create substantial contrast, the other display technologies require either considerable power (milliwatts per display element) or high voltage.

By taking advantage of unique characteristics of the liquid crystal materials, one can design LCD's to pass or block light. Only the orientation of the liquid crystal molecule must be altered in order for this effect to be created. This takes only microwatts of power per display element. To understand the operation of an LCD display, refer to Figures 3.4m and 3.4n. The molecules in a liquid crystal display form a helix when viewed from the front to the back of the display cell. That is, the cigar-shaped molecules at the front of the cell may be oriented vertically and those at the rear horizontally. Those in the center are somewhere in between, forming a very orderly 90-degree helix. This 90-degree angle can be controlled exactly during construction.

The key to the operation of a liquid crystal display is polarized light. When scattered light is polarized (see Figure 3.4m) in the vertical plane and passed through the liquid crystal helix described previously, light is twisted, following the helical arrangement of the molecules. The light leaving the back of the display will be polarized in the horizontal direction. If a relfector is placed behind the display, the polarized light will return along its entry path, being twisted as it passes through the helix in the opposite direction.

If the helix of molecules is placed in an electric field, the molecules will align with the electric field, destroying the molecular helix. The polarized light entering the front of the liquid crystal cell will not be twisted but will emerge from the rear of the display polarized in the vertical plane. When vertically polarized light encounters the horizontally oriented rear polarizer, the light will be attenuated (see Figure 3.4n). If a reflector is present, no light will be reflected; the display would appear dark.

The construction of the LCD involves simple implementation of the operating principles (Figure 3.4o). The liquid crystal material is encapsulated in a glass "sandwich." The glass is coated with a transparent conductive material, usually indium-tin oxide, which is etched back to the preferred pattern.

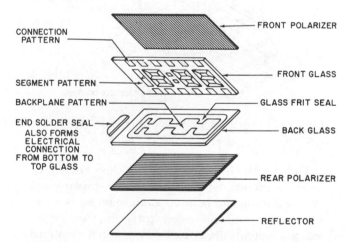

Fig. 3.4o Liquid crystal cell structure (courtesy of Fairchild Corp.)

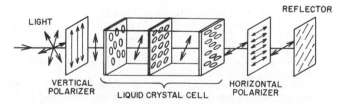

Fig. 3.4m Field effect display unenergized between segments (courtesy of Fairchild Corp.)

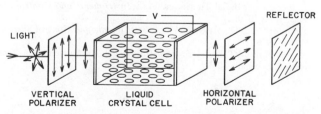

Fig. 3.4n Field effect display energized between segments and backplane (courtesy of Fairchild Corp.)

Once the electrode has been formed, the inside surface must be treated to give the molecules a 90-degree twist. After this, the package is sealed with a small gap left between the plates—generally 10 microns or less for the liquid crystal material.

Liquid crystal displays are available in several configurations; reflective, transmissive, and transreflective. The choice of configuration depends on the lighting to be used. The reflective option includes a small reflector laminated on the rear of the display. This option gives the best performance and appearance in bright light. In a dark environment, however, the display has to be lighted from the front.

In summary, LCDs are low-power, low-voltage electronic display devices. Their size ranges from fractions of an inch to a foot or larger. They do not generate light; therefore, external lighting is required in a dark environment. Their major disadvantage is the small viewing angle, generally less than 90 degrees.

Vacuum Fluorescent Displays

The vacuum fluorescent (VF) display is based on the electronic tuning eye used in radio receivers up to the 1960s. The VF display consists of a vacuum tube triode with phosphor-coated anodes arranged in a planar configuration. It is widely used in digital alarm clocks and radios and as a digital display in automobiles. The typical color is a bright bluish green. The construction of the device is shown in Figures 3.4p and 3.4q.

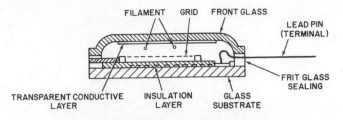

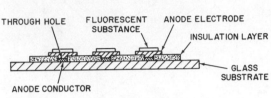

Fig. 3.4q   Cross section of a VF display (courtesy of Futaba Corp.)

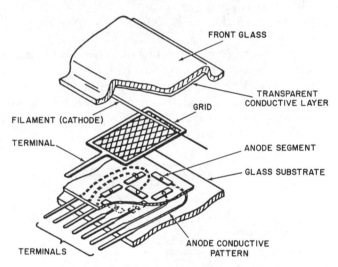

Fig. 3.4p   Construction of a vacuum fluorescent display (courtesy of Futaba Corp.)

The display is sealed in a flat glass enclosure. The anode segments, which represent the display elements, are located on the bottom. Each segment is connected to outside terminals sealed in the glass envelope. The anode segments are coated with phosphor material that emits visible light under electron bombardment. The grid is a fine mesh located between the filament (cathode) and the anode segments. The device requires a separate filament voltage, which may be AC or DC and ranges from 1.5 V to 5 V. The anode voltage is typically between 15 and 40 V. The grid is used for turning on or blanking a whole digit for multiplexing purposes, whereas the combination of anode segments defines the character.

One glass envelope generally contains several characters. The number of characters ranges from 4 to 40 or more.

There are various numbers given in the literature for life expectancy, or mean time before failure (MTBF), of a VF display. These numbers also depend on the definition of failure. Filament breakage is not a serious problem, since it rarely happens. The phosphor coating will lose its effectiveness in time, and this results in lower brightness. Futaba Corp. reports that its VF devices have a life span of 300,000 hours.

In summary, the VF displays are adequately illuminated, have a wide viewing angle, and involve a medium-voltage, low-power operation. Their disadvantage is the low contrast at normal ambient lighting. This can be improved with green or blue filters, but these can be the source of distracting glare. These devices can be quite inexpensive, especially when purchased in large quantities.

BIBLIOGRAPHY

Bylander, E. G., *Electronic Displays*, Texas Instruments Electronics Series, McGraw-Hill Book Co., New York, 1979.

Condon, E. U. and Odishaw, H., *Handbook of Physics*, McGraw-Hill Book Co., New York, 1975.

Dallimonti, R., "New Desgns for Process Control Consoles," *Instrumentation Technology*, November 1973.

Kelleway, J. L. and McMorriss, A. H., "Are Control Rooms Obsolete?" *Canadian Controls and Instrumentation*, September 1971.

Merritt, R., "CRT Terminals—an Update," *Instruments and Control Systems*, May 1979.

"Review of Digital Pressure Systems," *Instruments and Control Systems*, 1971 Buyers Guide Issue.

"Review of Digital Thermometers," *Instruments and Control Systems*, June 1970.

Sclater, N., "Displays Go Flat," *Instruments and Control Systems*, March 1975.

"Selecting CRT Based Process Interfaces," *Instrumentation Technology*, February 1979.

3.5 HUMAN ENGINEERING

Human engineering, also known as engineering psychology, human factors engineering, and ergonomics (in England), is probably the most basic form of engineering, because the fashioning of any tool to suit the hand is part of human engineering. The discipline began as a cut-and-dried effort in which modification was conveniently left to future planning instead of being made part of the original design. This approach sufficed when technological progress was sufficiently slow to allow several generations of improvement. The efficiency and cost of tools were such that improper original design was not severely punishing.

Current technology has supplied man with highly efficient, complex and, therefore, expensive tools, the ultimate success of which rests with his ability to use them. Today's tools are required not only to fit the human hand but also to reinforce many physiological and psychological characteristics, such as hearing, color discrimination, and signal acceptance rates. The complexity and cost of tools make the trial-and-error method unacceptable.

World War II introduced our first attempt at technical innovation and our first serious effort to treat human engineering as a discipline. Postwar economic pressures have been almost as grudging of critical mistakes as war has been; thus, the role of human engineering has increased with technological progress. A few highlights of human engineering described in this section will aid the instrument engineer in evaluating the impact of this discipline on process plants.

Applicability extends from instruments as sophisticated as those used in the Apollo moon flight to those as prosaic as the kitchen stove.

Man-Machine System

The man-machine system is a combination of men and equipment interacting to produce a desired result from given inputs. The majority of industrial control systems are indeed sophisticated man-machine types.

Because the purposes of an instrument and control system are limited, satisfactory definition is difficult. Subdividing the purpose into its constituent goals is an important first step in clarification. Many operators can-

not supply much more information than what is currently available and regarded as insufficient. Learning from mistakes, although necessary, is not a desirable engineering approach.

The backbone of a man-machine system is a flow chart of man-related activities. This approach concentrates design efforts on improving operator information and control capability—not just on improving instruments. Proper design criteria are obtained by asking what should be done, not what has been done. The machine should be adapted to the man. Two constraints exist: (1) the level of ability of the average operator and (2) the amount of information and control required by the process. Sight, hearing, and touch help the operator to control information storage and processing, decision making, and process control. Information is processed by man and machine in a similar manner (Figure 3.5a).

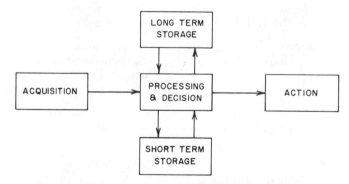

Fig. 3.5a Information processing: man or machine

Although men and machines process information similarly, their abilities to execute diverse functions vary greatly. Man is better suited than machines to process information qualitatively, an aptitude needing experience or judgment. His ability to store enormous quantities of data (10^{15} bits) and to use these facts subconsciously in decision making allows him to reach decisions with a high probability of being right, even if all necessary information is not available. Machines are suited for quantitative processing and storage of relatively small quantities of exact data (10^8 bits) with rapid access. Machine decisions must be based on complete and exact data.

In a process control system the differences are best applied by training the operator in emergency procedures and letting the machine store data on limits that, when exceeded, trigger the emergency. Table 3.5b summarizes the characteristics of men and machines. Since information is the primary quantity processed by a man-machine system, a brief account of informational theory is necessary.

Information Theory

The relatively new discipline of information theory defines the quantitative unit of evaluation as a binary digit, or "bit." For messages containing a number of equal possibilities (n), the amount of information carried (H) is defined as

$$H = \log_2 n, \qquad 3.5(1)$$

and the contact of a pressure switch (ps) carries one bit of information—

$$H_{ps} = \log_2^2 \text{ (alternatives)} = 1 \qquad 3.5(2)$$

The nature of most of the binary information in process control is such that probabilities of alternatives are equal. If this is not so, as with a pair of weighted dice, the total information carried is reduced. In an extreme case (if the switch were shorted), probability of a closed contact is 100%, probability of an open contact is 0% and total information is zero.

Characteristics of Man

Much has been learned about man as a system component since organized research in human factors was initiated during World War II. Anthropometric (body measurement) and psychometric (psychological) factors are now available to the designer of control systems, as are many references to information concerning, for instance, ideal distances for perception of color, sound, touch, and shape. The engineer should keep in mind that generally the source of data of this sort is a selected group of subjects (students, soldiers, physicians, and so on) who may or may not be representative of the operator population available for a given project.

Body Dimensions

The amount of work space and the location and size of controls or indicators are significant parameters that affect operator comfort and output and require knowledge of body size, structure, and motions (anthropometry). Static anthropometry deals with the dimensions of the body when it is motionless.

Dynamic anthropometry deals with dimensions of reach and movement of the body in motion. Of importance to the instrument engineer is the ability of the operator to see and reach panel locations while he is standing or seated. Figure 3.5c and Table 3.5d illustrate dimensions normally associated with instrumentation (for additional information, see reference 1). *The dimensions given are for the American male population. When operator populations other than American males are involved, the dimensions must be adjusted accordingly.*

Table 3.5b
CHARACTERISTIC COMPARISON:
MAN VS. MACHINE

Man Excels in	Machines Excel in
Detection of certain forms of very low energy levels	Monitoring (of both men and machines)
Sensitivity to an extremely wide variety of stimuli	Performing routine, repetitive, or very precise operations
Perceiving patterns and making generalizations about them	Responding very quickly to control signals
Detecting signals in high noise levels	Exerting great force, smoothly and with precision
Ability to store large amounts of information for long periods—and recalling relevant facts at appropriate moments	Storing and recalling large amounts of information in short periods
Ability to exercise judgment when events cannot be completely defined	Performing complex and rapid computation with high accuracy
Improvising and adopting flexible procedures	Sensitivity to stimuli beyond the range of human sensitivity (infrared, radio waves, and so forth)
Ability to react to unexpected low-probability events	Doing many different things at one time
Applying originality in solving problems, e.g., alternative solutions	Deductive processes
Ability to profit from experience and alter course of action	Insensitivity to extraneous factors
Ability to perform fine manipulation, especially when misalignment appears unexpectedly	Ability to repeat operations very rapidly, continuously, and precisely the same way over a long period
Ability to continue to perform even when overloaded	Operating in environments that are hostile to man or beyond human tolerance
Ability to reason inductively	

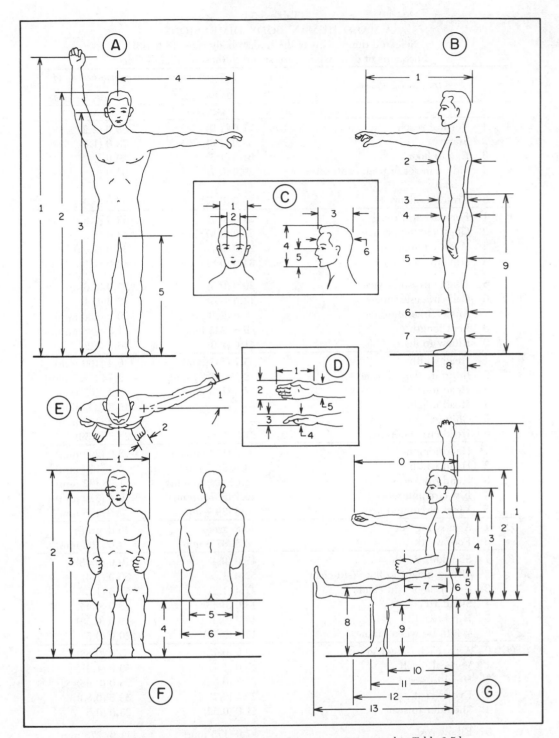

Fig. 3.5c Body dimension chart. Dimensions are summarized in Table 3.5d

Information Capability

Human information processing involves sensory media (hearing, sight, speech, and touch) and memory. The sensory channels are the means by which man recognizes events (stimuli). The range of brightness carried by a visual stimulus, for example, may be considered in degrees and expressed in bits, and discrimination between stimuli levels is either absolute or relative. In relative discrimination, the individual compares two or more stimuli; in absolute discrimination, a single stimulus is evaluated.

Man performs much better in relative than in absolute discrimination. Typically, a subject can separate 100,000 colors when comparing them with each other; the number is reduced to 15 colors when each must be considered individually. Data in Table 3.5e deal with absolute dis-

Table 3.5d
MALE HUMAN BODY DIMENSIONS
Selected dimensions of the human body (ages 18 to 45).
Locations of dimensions correspond to those in Figure 3.5c.

		Dimensional Element	Dimension in inches (m) except where noted	
			5th Percentile	95th Percentile
		Weight in pounds (kg)	132 (59.4)	201 (90.5)
A	1	Vertical reach	77.0 (1.9)	89.0 (2.2)
	2	Stature	65.0 (1.6)	73.0 (1.8)
	3	Eye to floor	61.0 (1.5)	69.0 (1.7)
	4	Side arm reach from center line of body	29.0 (0.7)	34.0 (0.8)
	5	Crotch to floor	30.0 (0.75)	36.0 (0.9)
B	1	Forward arm reach	28.0 (0.7)	33.0 (0.8)
	2	Chest circumference	35.0 (0.87)	43.0 (1.1)
	3	Waist circumference	28.0 (0.7)	38.0 (0.95)
	4	Hip circumference	34.0 (0.8)	42.0 (1.1)
	5	Thigh circumference	20.0 (0.5)	25.0 (0.6)
	6	Calf circumference	13.0 (0.3)	16.0 (0.4)
	7	Ankle circumference	8.0 (200 mm)	10.0 (250 mm)
	8	Foot length	9.8 (245 mm)	11.3 (283 mm)
	9	Elbow to floor	41.0 (1.0)	46.0 (1.2)
C	1	Head width	5.7 (143 mm)	6.4 (160 mm)
	2	Interpupillary distance	2.27 (56.75 mm)	2.74 (68.5 mm)
	3	Head length	7.3 (183 mm)	8.2 (205 mm)
	4	Head height	—	10.2 (255 mm)
	5	Chin to eye	—	5.0 (125 mm)
	6	Head circumference	21.5 (0.54)	23.5 (0.59)
D	1	Hand length	6.9 (173 mm)	8.0 (200 mm)
	2	Hand width	3.7 (92.5 mm)	4.4 (110 mm)
	3	Hand thickness	1.05 (26.25 mm)	1.28 (32 mm)
	4	Fist circumference	10.7 (267.5 mm)	12.4 (310 mm)
	5	Wrist circumference	6.3 (39.7 mm)	7.5 (186 mm)
E	1	Arm swing, aft	40 degrees	40 degrees
	2	Foot width	3.5 (87.5 mm)	4.0 (100 mm)
F	1	Shoulder width	17.0 (0.4)	19.0 (0.48)
	2	Sitting height to floor (std chair)	52.0 (1.3)	56.0 (1.4)
	3	Eye to floor (std chair)	47.4 (1.2)	51.5 (1.3)
	4	Standard chair	18.0 (0.45)	18.0 (0.45)
	5	Hip breadth	13.0 (0.3)	15.0 (0.38)
	6	Width between elbows	15.0 (0.38)	20.0 (0.5)
G	0	Arm reach (finger grasp)	30.0 (0.75)	35.0 (0.88)
	1	Vertical reach	45.0 (1.1)	53.0 (1.3)
	2	Head to seat	33.8 (0.84)	38.0 (0.95)
	3	Eye to seat	29.4 (0.7)	33.5 (0.83)
	4	Shoulder to seat	21.0 (0.52)	25.0 (0.6)
	5	Elbow rest	7.0 (175 mm)	11.0 (275 mm)
	6	Thigh clearance	4.8 (120 mm)	6.5 (162 mm)
	7	Forearm length	13.6 (340 mm)	16.2 (405 mm)
	8	Knee clearance to floor	20.0 (0.5)	23.0 (0.58)
	9	Lower leg height	15.7 (393 mm)	18.2 (455 mm)
	10	Seat length	14.8 (370 mm)	21.5 (0.54)
	11	Buttock-knee length	21.9 (0.55)	36.7 (0.92)
	12	Buttock-toe clearance	32.0 (0.8)	37.0 (0.93)
	13	Buttock-foot length	39.0 (0.98)	46.0 (1.2)

Note: All except critical dimensions have been rounded off to the nearest inch (mm).

Table 3.5e
ABSOLUTE DISCRIMINATION CAPABILITY
(Maximum Rates of Information Transfer)

Modality	Dimension	Maximum Rate (Bits/Stimulus)
Visual	Linear extent	3.25
	Area	2.7
	Direction of line	3.3
	Curvature of line	2.2
	Hue	3.1
	Brightness	3.3
Auditory	Loudness	2.3
	Pitch	2.5
Taste	Saltiness	1.9
Tactile	Intensity	2.0
	Duration	2.3
	Location on the chest	2.8
Smell	Intensity	1.53
(Multidimensional Measurements)		
Visual	Dot in a square	4.4
	Size, brightness, and hue (all correlated)	4.1
Auditory	Pitch and loudness	3.1
	Pitch, loudness, rate of interruption, on-time fraction, duration, spatial location	7.2
Taste	Saltiness and sweetness	2.3

crimination encountered in levels of temperature, sound, and brightness.

In addition to the levels of discrimination, sensory channels are limited in the acceptance rates of stimuli. The limits differ for the various channels and are affected by physiological and psychological factors. A study by Pierce and Karling suggests a maximum level of 43 bits per second.

Several methods of improving information transmission to man include coding (language), use of multiple channels (visual and auditory), and organization of information (location of lamps; sequence of events). Simultaneous stimuli, similar signals with different meanings, noise, excessive intensity, and the resulting fatigue of the sensors—stimuli that approach discrimination thresholds—are all detrimental to the human information process.

The limit of the operator's memory affects information processing and is considered here in terms of long and short span. Work by H. G. Shulmann[2] indicates that short-term memory uses a phonemic (word sounds) and long-term memory uses a semantic (linguistic) similarity. These approaches tend to make codes that are phonemically dissimilar less confusing for short-term recall and make codes that are semantically different less confusing for long-term recall. Training, for example, uses long-term memory.

Performance (Efficiency and Fatigue)

Motivation, annoyance, physical condition, and habituation influence the performance of primarily non-physical tasks. These influences make the subject of efficiency and of fatigue controversial. Both the number and frequency of occurrence of stimuli to which the operator responds affect efficiency and fatigue.

Performance increases when the number of tasks increases from a level of low involvement to one of high involvement. When rates of information processing are raised beyond the limits of the various senses, performance rapidly deteriorates.

Application of Human Engineering

System Definition

The seemingly commonplace task of defining the man-machine system is more difficult to formalize than expected because (1) operators themselves are too closely involved for objectivity; (2) engineers tend to make assumptions based on their technical background; and (3) equipment representatives focus their attention mainly on the hardware.

The definition can be more highly organized around the purpose, components, and functions of the system. The purpose of the system is its primary objective; it significantly affects design. In a system for continuous control of a process, operator fatigue and attention span are of first importance. Attention arousal and presentation of relevant data are important when the objective is the safety of personnel and equipment, and the operator must reach the right decision as quickly as possible.

Components of the system are divided into those that directly confront the operator and those that do not. The distinction is used to locate equipment. For example, a computer process console should be located for operator convenience, but maintenance consoles should be out of the way so as to reduce interference with operation.

System functions are primarily the operator's responsibility, and operator function should be maximized for tasks that involve judgment and experience and minimized for those that are repetitive and rapid. Table 3.5b is an efficient guideline for task assignment.

Statistics of Operator Population

As was mentioned in the introduction to this section, the statistical nature of human engineering data influences individual applications. A specific problem demands full evaluation of the similarity among operators. For example, if color blindness is not a criterion for rejecting potential operators, color coding must be backed by redundant means (shape) to ensure recognition, because one of four subjects has some degree of color blindness.

Variations and exceptions are to be expected and recognized, since men are significantly different in physical size, native aptitude, and technical acculturation. Control systems and instruments should be designed for the user. A frequent mistake made by the instrument engineer is to assume that the operator will have a background similar to his.

Some operators may have never ridden a bicycle, much less driven a car or taken a course in classical physics. In arctic regions, massive parkas and bulky gloves change all the figures given in body dimension tables. It is only in Western culture that left-to-right motion of controls and indicators is synonymous with increasing values.

Setting Priorities

The diverse functions of the system confronting the operator vary in importance. The most accessible and visible areas should be assigned to important and frequently used items. Vertical placement is critical and dependent on the normal operating mode (standing, sitting, or mixed). Determination of optimum viewing zones is centered on the average eye level for standing (approximately 66 inches, or 1.65 m) and sitting (approximately 48 inches, or 1.2 m). Above these areas, visibility is still good, but accessibility falls off (Figures 3.5f and 3.5g). At the lower segment, even grocers know that objects are virtually hidden. Canting the panel helps to make subordinate areas more usable. (For more on control panel design, see Section 3.2).

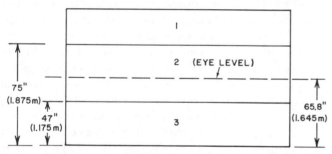

VARIABLES ARE GROUPED BY FUNCTION
- INDICATION AND CONTROLS IN AREA 2.
- SELDOM USED CONTROLS OR PUSH BUTTONS IN AREA 3.
- INDICATORS (ANALOG OR DIGITAL) ONLY IN AREA 1.
- ALARM AND ON/OFF INDICATION IN AREA 1.

Fig. 3.5f   Panel for standing operator. Variables are grouped by function, with indication and controls in area 2, seldom used controls or pushbuttons in area 3, indicators (analog or digital) only in area 1, and alarm and on-off indicators in area 1.

Instruments should be arranged from left to right (normal eye scan direction) in (1) spatial order, to represent material movement (storage and distribution); (2) functional order, to represent processing of material (distillation, reaction, and so on); and (3) sequential order, to modify the functional order approach to aid the operator in following a predetermined sequence even under stress.

Wrap-around consoles (now in common use with the advent of distributed control systems) with multifunction displays (such as cathode ray tubes) do not require operator movement; however, the primary (most used or critical) displays and controls should be centered in front of the sitting operator with auxiliary or redundant instruments on either side. Reach considerations are much more important when the operator is normally sitting (Figure 3.5h).

Locating similar functions in relatively the same place gives additional spatial cues to the operator by associating location with function. On this account it is advisable to divide alarms into groups related to meaning, including (1) alarms requiring immediate operator action (for example, high bearing temperatures), which must be read when they register; (2) alarms for abnormal conditions that require no immediate action (standby pump started on failure of primary), which must be read when they register; and (3) status annunciator for conditions that may or may not be abnormal. These are read on an as-needed basis.

Alarms can be separated from other annunciators, have a different physical appearance, and give different audible cues. For example, group 1 alarms can have flashing lamp windows measuring 2 inches × 3 inches (50 × 75 mm) with a pulsating tone. Group 2 alarms can have flashing lamp windows measuring 1 inch × 2 inches (25 × 50 mm) with a continuous tone, and group 3 can have lamp windows measuring 1 inch × 1 inch (25 × 25 mm) without visual or auditory cues.

Alarms are meaningless if they occur during shutdown or as a natural result of a primary failure. Whether usable or not, they capture a portion of the operator's attention. Several alarms providing no useful information can be replaced by a single shutdown alarm. Later, the cause of the shutdown will be of keen interest to maintenance personnel; to provide detailed information about the cause, local or back-of-panel annunciators can be used. CRT displays are generally multifunctional, and rapid operator association (especially in emergencies) with a particular situation or function can be visually facilitated by unique combinations of shapes, colors, or sounds.

Hardware Characteristics

Digital indicators are best when rapid reading or obtaining accurate values is of utmost concern. However, they give only a useless blurr at a high rate of change. Analog indications allow rapid evaluation of changing values. The reading can be related to zero and full-scale for intepretation of rate of change (estimated time to reach minimum or maximum levels). Correspondingly, analog information is difficult to read when exact values are needed.

Display patterns should be natural to the eye movement—in most cases, left-to-right. An operator scanning only for abnormal conditions will be aided by this arrangement. The horizontal scan line approach proved successful in several major lines of control instruments and is now widely used by "Overview" displays in distributed control systems (Figure 3.5i).

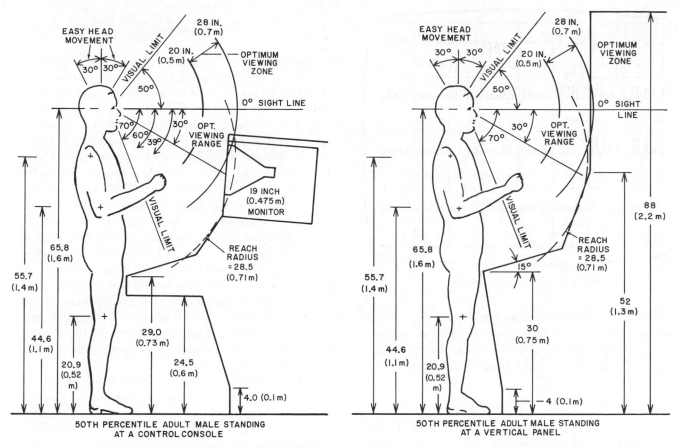

Fig. 3.5g *Left*, 50th percentile adult male standing at a control console. *Right*, 50th percentile adult male standing at a vertical panel.

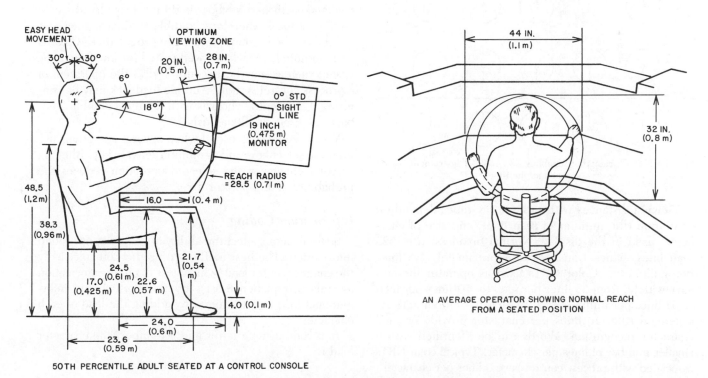

Fig. 3.5h *Left*, 50th percentile adult male seated at a control console. *Right*, normal reach from a seated position.

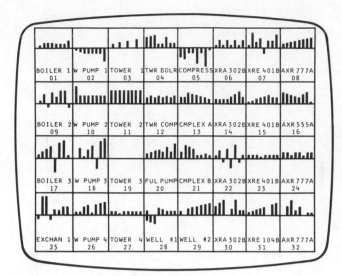

Fig. 3.5i CRT overview display

An alternative system is shown in Figure 3.5j, in which two rows are combined to give the effect of one line containing twice the number of readings. In our culture, horizontal scanning is eight times faster than vertical scanning.

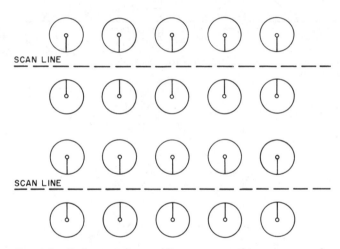

Fig. 3.5j Dial orientation enabling two rows of instruments to be scanned at one time

Viewing distances with cathode ray tubes depend on resolution (the number of scan lines) and size of characters used in the display. Normal broadcast uses 525 scan lines, which translates to approximately 1.8 lines per millimeter. Color CRTs used as operator displays vary widely, from 38 lines per inch to 254 lines per inch (1.52 lines per millimeter to 10 lines per millimeter). As a general rule, 10 lines per character provide optimal character recognition. Words can be identified with a smaller number of lines per character. Typical color CRTs associated with process control have 8 lines per character.

Viewing distance as it relates to character height was addressed by Fletcher (1982) who used the following formula:

Optimum viewing distance (inches) =

$$= \frac{\text{Character height (inches)}}{0.00291} \quad \text{3.5(3)}$$

The interactive effect of color, contrast, and lines per character adversely influences the viewing distance.

Black-and-white screens are visible at nearly any light level, provided that reflected light is eliminated by a hood or a circularly polarized filter. Flicker resulting from slow refresh rate is a potential fatigue factor. Rates above 25 flickers per second are reasonably stable. Color CRTs require a lighting level above 0.01 lumen to ensure that color vision (retinal rods) is activated and color discrimination is possible. Display of information through cathode ray tubes can be graphic (pictorial or schematic representation of the physical process) or tabular (logical arrangement of related variables by function). The graphic format is useful when the controlled process changes in operating sequence or structure (batch reactors, conveyor systems, and so forth) or if the operator is not familiar with the process. The tabular display format conveys large quantities of information rapidly to an experienced operator in continuous processes (distillation columns, boilers, and so forth).

Operator response in modern control systems occurs predominantly through specialized keyboards. Key activation can be sensed by the operator through tactile, audible, or visual feedback. Tactile feedback is usually found in discrete or conventional pushbutton keys and is well suited for rapid multiple operations, such an entry of large quantity of data. Membrane-, laminate-, or capacitance-activated keyboards do not provide adequate tactile feedback; therefore, audible feedback, such as a "beep" or "bell" sound, is usually provided. Keyboards with audible feedback are well suited for low-frequency usage, such as may be encountered in the operation of a process control system. Audible feedback may be confusing if adjacent keyboards with identical audible feedback are used simultaneously.

Visual feedback is useful when several operations use multiple select keys and a common activate key. Blinking or color change of the selected operation reduces the probability of an operator error.

Information Coding

Several factors affect the ability of people to use coded information. The best method to predict difficulty is by the channel center load, which uses the basic information formula of equation 3.5(1). By this method the informational load is determined by adding the load of each character.

In a code using a letter and a number, the character load is:

letter (alphabet) $L_h = \text{Log}_2\ 26 = 4.70$
number (0–9) $N_h = \text{Log}_2\ 10 = 3.32$
code information $(C_h) = 4.70 + 3.32 = 8.02$

This method applies when the code has no secondary meaning, such as when A 3 is used for the first planet from the sun, Mercury. When a code is meaningful in terms of the subject, such as O 1 for Mercury (the first planet), the information load is less than that defined by the formula.

Typical laboratory data indicate that when a code carries an information load of more than 20, man has difficulty in handling it. Tips for code design include: (1) using no special symbols that have no meaning in relation to the data; (2) grouping similar characters (HW5, not H5W); (3) using symbols that have a link to the data (MF for male-female, not 1–2); and (4) using dissimilar characters (avoiding zero and the letter O and I and the numeral 1).

The operator environment can be improved by attending to details such as scale factor, which is the multiplier relating the meter reading to engineering units. The offset from mid-scale for the normal value of the variable and the desired accuracy of the reading for the useful range are constraints. Within these constraints, the selected scale factor should be a one-digit number, if possible. If two significant digits are necessary, they should be divisible by (in order of preference) 5, 4, or 2. Thus, rating the numbers between 15 and 20 as to their desirability as scale factors would give 20, 15, 16, 18 (17, 19).

Operator Effectiveness

The operator is the most important system component. His functions are all ultimately related to those decisions that machines cannot make. He brings into the system many intangible parameters (common sense, intuition, judgment, and experience)—all of which relate to the ability to extrapolate from stored data, which sometimes is not even consciously available. This ability allows the operator to reach a decision that has a high probability of being correct even without all the necessary information for a quantitative decision.

The human engineer tries to create the best possible environment for decision making by selecting the best methods of information display and reducing the information through elimination of noise (information type), redundancy (even though some redundancy is necessary), and environmental control.

Operator Load

The operator's load is at best difficult to evaluate, since it is a subjective factor and because plant operation varies. During the past 20 years, automation and miniaturization have exposed the operator to a higher density of information, apparently increasing his efficiency in the process. Since increases in load increase attention and efficiency, peak loads will eventually be reached that are beyond the operator's ability to handle.

Techniques for reducing operator load include the simultaneous use of two or more sensory channels.

Vision is the most commonly used sensory channel in instrumentation. Sound is next in usage. By attracting the operator's attention only as required, it frees him from continuously having to face the display and its directional message. The best directionality is given by impure tones in the frequency range of 500 to 700 Hz.

Blinking lights, pulsating sounds, or combination of both effectively attract an operator's attention. The relative importance of these stimuli can be greatly decreased if the use is indiscriminate. (Blinking, as an example, should not be used in normal operator displays on a CRT.)

Increased frequency of a blinking light or a pulsating sound is normally associated with increased urgency; however, certain frequencies may have undesirable effects on the operator. Lights changing in intensity or flickering (commonly at frequencies of 10 to 15 cycles) can induce epileptic seizures, especially in stressful situations.

Information quantity can be decreased at peak load. In a utility failure, for example, more than half the alarms could be triggered when many of the control instruments require attention. It would be difficult for the operator to follow emergency procedures during this shutdown. A solution gaining acceptance in the industry involves selectively disabling nuisance alarms during various emergencies. Some modern high-density control rooms have more than 600 alarms without distinction as to type or importance.

Environment

The responsibilities of human engineering are broad and include spatial relations—distances that the operator

Table 3.5k
GENERAL ILLUMINATION LEVELS

Task Condition	Level (foot candles [lux])	Type of Illumination
Small detail, low contrast, prolonged periods, high speed, extreme accuracy	100 (1076)	Supplementary type of lighting; special fixture such as desk lamp
Small detail, fair contrast, close work, speed not essential	50–100 (538–1076)	Supplementary type of lighting
Normal desk and office-type work	20–50 (215–538)	Local lighting; ceiling fixture directly overhead
Recreational tasks that are not prolonged	10–20 (108–215)	General lighting; random room light, either natural or artificial
Seeing not confined, contrast good, object fairly large	5–10 (54–108)	General lighting
Visibility for moving about, handling large objects	2–5 (22–54)	General or supplementary lighting

Table 3.5l
SPECIFIC RECOMMENDATIONS, ILLUMINATION LEVELS

Location	Level (foot candles [lux])	Location	Level (foot candles [lux])
HOME		SCHOOL	
Reading	40 (430)	Blackboards	50 (538)
Writing	40 (430)	Desks	30 (323)
Sewing	75–100 (807–1076)	Drawing (art)	50 (538)
		Gyms	20 (215)
Kitchen	50 (538)	Auditorium	10 (108)
Mirror (shaving)	50 (538)	THEATRE	
		Lobby	20 (215)
Laundry	40 (430)	During intermission	5 (54)
Games	40 (430)		
Workbench	50 (538)	During movie	0.1 (1.08)
General	10 (108) or more	PASSENGER TRAIN	
		Reading, writing	20–40 (215–430)
OFFICE		Dining	15 (161)
Bookkeeping	50 (538)	Steps, vestibules	10 (105)
Typing	50 (538)		
Transcribing	40 (430)	DOCTOR'S OFFICE	
General correspondence	30 (323)	Examination room	100 (1076)
Filing	30 (323)	Dental-surgical	200 (2152)
Reception	20 (215)	Operating table	1800 (19, 368)

has to span in order to reach his equipment and his fellow operators. Controlling temperature and humidity to maintain operator and equipment efficiency is also an important responsibility of the human engineer.

Light must be provided at the level necessary without glare or superfluous eye strain. Daylight interference caused by the sun at a low angle can be an unforeseen problem. Last-minute changes in cabinet layout can create undesirable shadows and glare from glass and highly

Table 3.5n
PERMISSIBLE NOISE EXPOSURES

When the daily noise exposure is composed of two or more periods of noise exposure of different levels, their combined effect should be considered, rather than the individual effect of each. If the sum of the following fractions: $C1/T1 + C2/T2 \ldots Cn/Tn$ exceeds unity, then the mixed exposure should be considered to exceed the limit value. Cn indicates the total time of exposure at a specified noise level, and Tn indicates the total time of exposure permitted at that level.

Exposure to impulsive or impact noise should not exceed 140 dbA peak sound pressure level.

Duration per Day (hours)	Sound Level (dbA)	Duration per Day (hours)	Sound Level (dbA)
8	90	1½	102
6	92	1	105
4	95	½	110
3	97	¼ or less	115
2	100		

reflective surfaces. Flat, brushed, or textured finishes reduce glare. Poor illumination can be reduced by light colors, low light fixtures, and special fixtures for special situations. Fluorescent lighting, because of its 60-Hz flicker rate, should be supplemented by incandescent lamps to reduce eye fatigue, particularly important in the presence of rotating equipment, in order to eliminate strobing effect. Tables 3.5k and 3.5l give detailed information on typical lighting applications.

Glare is the most harmful effect of illumination (Figure 3.5m). There is a direct glare zone that can be eliminated, or at least mitigated, by proper placement of luminaires

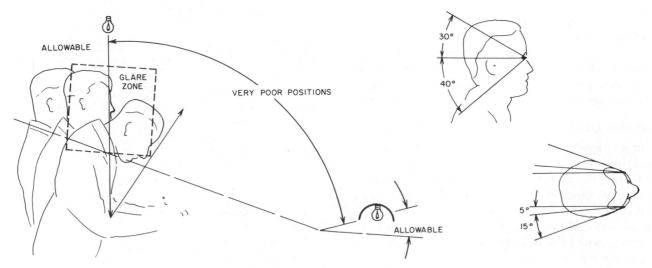

Fig. 3.5m   Typical lighting chart

and shielding, or, if luminaires are fixed, by rearrangement of desks, tables, and chairs. Overhead illumination should be shielded to approximately 45 degrees to prevent direct glare. Reflected glare from the work surface interferes with most efficient vision at a desk or table and requires special placement of luminaires.

Eyeglasses cause disturbing reflections unless the light source is 30 degrees or more above the line of sight, 40 degrees or more below, or outside the two 15-degree zones as shown in Figure 3.5m.

Noise and vibration affect performance by producing annoyance, interfering with communication, and causing permanent physical damage. Prolonged exposure to high sound levels causes hearing loss, both temporary and permanent. The noise exposure values in Table 3.5n are those stated in the Walsh Healy Public Contracts Act as being permissible. They are derived from the curves in Figure 3.5o and reflect the variation in sensitivity with frequency.

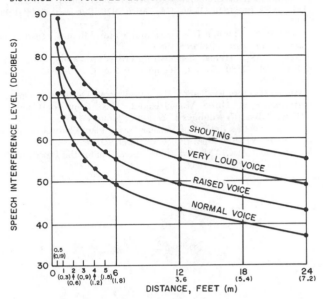

THE SPEECH INTERFERENCE LEVEL SHOULD BE LESS THAN THAT GIVEN IN ORDER TO HAVE RELIABLE CONVERSATION AT THE DISTANCE AND VOICE LEVELS SHOWN.

Fig. 3.5p   Speech interference levels

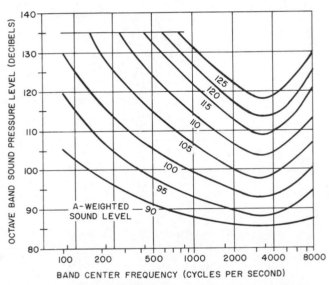

OCTAVE BAND SOUND PRESSURE LEVELS MAY BE CONVERTED TO THE EQUIVALENT A-WEIGHTED SOUND LEVEL BY PLOTTING THEM ON THIS GRAPH AND NOTING THE A-WEIGHTED SOUND LEVEL CORRESPONDING TO THE POINT OF HIGHEST PENETRATION INTO THE SOUND LEVEL CONTOURS. THIS EQUIVALENT A-WEIGHTED SOUND LEVEL, WHICH MAY DIFFER FROM THE ACTUAL A-WEIGHTED SOUND LEVEL OF THE NOISE, IS USED TO DETERMINE EXPOSURE LIMITS, FROM TABLE 3.5n.

Fig. 3.5o   Equivalent sound level contours

Annoyance and irritability levels are not easily determined, because they are subjective factors, and habituation significantly affects susceptibility. A quantitative tolerance limit has not yet been established. One aspect of background noise is deterioration of speech communication.

Noise reduction may take the form of reducing noise emission at the source, adding absorbent material to the noise path, or treating the noise receiver (having operators wear protective equipment, like ear plugs and ear muffs). The last precaution reduces both speech and noise,

but the ear is afforded a more nearly normal range of sound intensity and thus can better recognize speech. Figure 3.5p illustrates typical speech interference levels.

Summary

This section has described the tools needed, but the solutions that will fulfill the engineer's mandate from the industrial sector must come out of his own creativity. He must develop an original approach to each new assignment and introduce creative innovations consistent with sound engineering practices. Decisions must not be based solely on data from current projects reflecting only what has not worked, and human engineering must not be left to the instrument manufacturer who, however well he fabricates, still does not know how the pieces must fit together in a specific application. Compilations of data on mental and physical characteristics must be approached cautiously, because they may reflect a group of subjects not necessarily like the operators in a specific plant.

It is the engineer's task to ensure that the limitations of men and machines do not become liabilities and to seek professional assistance when means of eliminating the liabilities are not evident.

REFERENCES

1. Woodson, V.J. and Conover, H.N., *Human Engineering Guide for Equipment Designers*, 2nd ed., University of California Press, Berkeley, 1964.
2. Shulmann, H.C., Psychological Bulletin, Volume 75, 6, June 1971.
3. Chapanis, A., *Man-Machine Engineeering*, Wadsworth Publishing Company, Inc., Belmont, CA, 1965.
4. McCormick, E.J., *Human Factors Engineering*, (2nd ed., McGraw-Hill, New York, 1964.

BIBLIOGRAPHY

"Application of Ergonomics to Occupational Safety," Report No. 712F, WSA Inc. (San Diego, CA), April 1978, Retrieval #USGPB82-154253.

Farmer, E., "Design Tips for Modern Control Rooms," *Instruments and Control Systems,* March 1980.

Geer, C.W., "Human Engineering Procedure Guide," Report No. AD-A108643 D180-25471-1, Boeing Aerospace Co., Seattle, WA, September 1981.

Human Factors Evaluation of Control Room Design and Operator Performance at Three Mile Island-2, U.S. Dept. of Commerce NUREG/CR-2107, Volumes 1, 2, and 3, January 1980.

Human Factors Review of Nuclear Power Plant Control Room Design, EPRI NP-309, Palo Alto, CA, March 1977.

McCready, A.K., "Man-Machine Interfacing for the Process Industries," *InTech,* March 1982.

Military Standard—Human Engineering Design Criteria for Military Systems, Equipment and Facilities, MIL-STD 14728 and Later Addenda.

Nixon, C.W., "Hearing Protection Standards," Report No. AFAMRL-TR-80-60, Air Force Aerospace Medical Research Lab (Wright-Patterson Air Force Base Ohio), 1982; Retrieval #USGAD-A110 388/6.

Rehm, S, et al., "Third International Congress on Noise Effects as a Public Health Problem, Biological Disturbances, Effects on Behavior," Report No. BMFT-FB-HA-81-013, Mainz University (Germany), October, 1981, Retrieval #USGN82-15597/9.

Thompson, B.J., "Preparing for Computer Control," *Instruments and Control Systems,* February 1980.

Treurniet, W. C., "Review of Health and Safety Aspects of Video Display Terminals," Report No. CRC-TN-712-E, Communications Research Centre (Ottawa, Ontario, Canada), February 1982.

MULTIPOINT PANEL
INDICATOR

DIGITAL
INDICATOR

MULTIPOINT SCANNING
INDICATOR

3.6   INDICATORS

Special Features: Analog, digital, movable pointer, movable scale, parametric, scanning, multipoint, and so forth.

Partial List of Suppliers: Action Instruments; Ametek Inc. Controls Div.; Babcock & Wilcox McDermott Co.; Bailey Controls Co.; Beckman Instruments Inc.; Bristol Babcock Inc.; Fischer & Porter Co.; Fisher Controls Co.; Foxboro Co.; General Electric Co., Instrument Products Operation; Honeywell, Process Control Div.; ITT Barton; Kent Process Control Inc.; Leeds & Northrup, Unit of General Signal; Moore Products Co.; Robertshaw Controls Co.; Taylor Instrument Co. Div. of Sybron Corp.; Thermo Electric Co., Inc.; West Instruments, Div. of Gulton Industries; Westinghouse Electric Corp.; Yokogawa Corp. of America.

Terminology of Analog Indication

Indication implies a representation to the eye from which the mind infers either an individually distinct state or, in most instances, a quantity in which it is interested.

Indications can be conveyed to human perception in a variety of ways. In instrumentation, we are primarily concerned with the measurement of a quantity to which we attach a numerical value. This magnitude of measurement can be conveyed to the eye either individually by a digit, by a combination of digits, or on a graduated scale on which digits are shown in a logical sequence. In the last case, a movable reference, such as a pointer, is required to indicate the digit of interest on the scale.

Visual observation also includes an indication of the existence of a variable without necessarily attaching a quantitative value to it.

The indicating apparatus most frequently used for showing the variable is some form of a meter. When a meter also traces a record, this fact is brought out by the adjective "recording," such as in "a recording voltmeter."

The term gauge (or gage) is always used for an instrument that only indicates.

The suffix -scope refers to an instrument used for viewing only. An oscilloscope is an indicator; an oscillograph is a recorder. The oscillogram is the recording itself.

All of the aforementioned indicators are analog instruments. They use one physical phenomenon to indicate another one by analogy. One of the most widely used analog indicators is the liquid-in-glass thermometer, which uses the property of metallic media (such as mercury, mercury-thallium, or gallium) or of various organic liquids to expand under the influence of heat to indicate the temperature of another medium.

The term "analog" has become popular with the advent of the analog computer. Here an "analog" electrical quantity is used to represent any variable, since the electrical quantity lends itself to easy computation.

Other analog measuring methods include the following examples:

1. One type of viscosimeter uses the time required for a ball to sink from the surface of a liquid medium in a cylinder to the bottom as an analog measure of the viscosity of the liquid.

2. The vapor-pressure thermometer is actuated by pressure to indicate temperature.

3. The frequency of an alternating current is measured by the vibration of a reed in an instrument containing several reeds, each tuned to a different frequency. The particular reed whose resonant frequency is closest to the current frequency swings with the greatest amplitude and indicates the frequency on an adjacent scale.

Indication of Measurements

Indications on a graduated scale require the relative motion of two elements. One of the two must be a fixed reference. In most indicators, the scale is stationary and the indicating reference moves. The indicating element may be a liquid column, a float, a pointer, or a beam of light. The scale may be laid out on a straight line or on a circular arc. Figures 3.6a, 3.6b, and 3.6c illustrate some of the available design variations. The required simplicity of the movements in indicating instruments usually renders any other possibility for a different shape prohibitive. The larger the radius, the better the readability. A scale of more than 180 degrees laid out on a circular disk is usually referred to as a dial. A scale is called uniform if the graduations on it are equidistant. The increments may also be different, reflecting some other than a linear relationship. Flowmeters usually have a square root scale; vapor-pressure thermometers have an uneven scale that becomes progressively wider, calibrated according to the individual filling medium.

Fixed-Scale Indicators

A liquid column moving in a transparent, usually cylindrical tube with a stationary scale on the tube or placed inside forms the simplest fixed-scale indicator. Liquid-in-glass thermometers and manometers are examples of this. Manometer and thermometer scales are often ad-

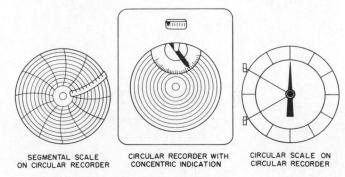

SEGMENTAL SCALE ON CIRCULAR RECORDER — CIRCULAR RECORDER WITH CONCENTRIC INDICATION — CIRCULAR SCALE ON CIRCULAR RECORDER

Fig. 3.6c   Indication methods used on recorders

justable in height for recalibration, and the tubes are usually vertical. One manometer type uses two tubes connected in the shape of a "u." However, one leg of the "u" may be bent on an angle; the device is then called an inclined-tube manometer. For very low pressures it offers the advantage of a longer, more accurately readable scale, stretched out by the reciprocal of the sine of the angle that the inclined leg forms with a horizontal line (Figure 3.6d). In some designs the inclined leg is provided with a logarithmic scale to stretch readings in the low range comparatively farther out.

Fig. 3.6d   Inclined manometer-type indicator

Readings in liquid columns are taken at the crest of the meniscus formed as a result of capillary attraction. Mercury forms a convex meniscus, most other liquids a concave meniscus.

A plummet (or plumb bob) floating in a calibrated tapered tube is used in the variable-area–type flowmeter (of which the rotameter is one version) to indicate by its position the rate of flow through the tube. The position of a plummet in a liquid inside a graduated transparent tube is an indication of fluid density.

The most commonly used moving reference for indication is a pointer (in a timepiece called a "hand"). The color of the pointer usually contrasts with the background for better readability. Pointers are frequently covered with luminescent materials to make readings in darkness possible. Transparent plastics are used in cases in which the pointer would obscure other indicators under it, e.g., when a totalizing counter is on the same dial. Incandescent, fluorescent, or neon-edge lighting is used to facilitate observation. To minimize the parallax or apparent displacement effects caused by the lens, knife-edge pointers are used. Also, mirrors are used in the same

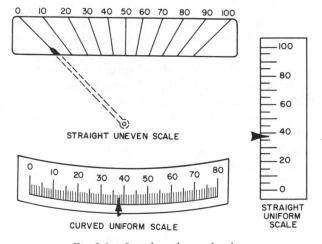

STRAIGHT UNEVEN SCALE

CURVED UNIFORM SCALE

STRAIGHT UNIFORM SCALE

Fig. 3.6a   Straight and curved scales

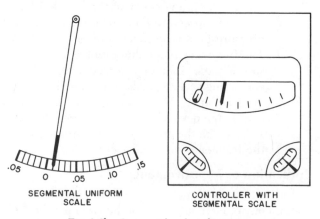

SEGMENTAL UNIFORM SCALE

CONTROLLER WITH SEGMENTAL SCALE

Fig. 3.6b   Segmental scale indicators

plane as the dial, and the readings are taken at an angle of observation where pointer and mirror image coincide. Sometimes the dial is raised to pointer level. In another meter, either a part of the rotatable scale or the pointer is optically projected onto a coated window, whereby parallax is completely eliminated.

Indicating intruments usually read from left to right or clockwise. In cases in which negative values have to be read, as in vacuum indications, graduations progress from right to left. Compound gauges indicating pressure and vacuum have the zero point near the middle of the scale. When readings near zero are of lesser importance, suppressed-zero ranges are used with narrower graduations at the beginning of the scale followed by wider increments. There are scales with extended sections and scales with condensed sections.

A multiple scale is illustrated in Figure 3.6e. Others are used for thermometers that indicate temperature in two units, such as Fahrenheit and Celsius degrees. Pressure gauges usually have a circular scale of 270 angular degrees (Figure 3.6f). Precision gauges are made with 350-degree scales and the possibility of extending the pointer rotation to two turns that, covering a total of 660 degrees, results in a scale of 80 inches (2 m) in a gauge 16 inches (400 mm) in diameter.

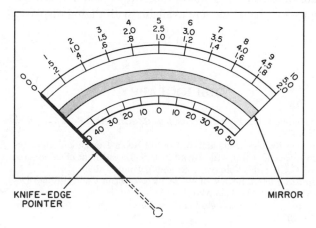

Fig. 3.6e   Multiple-scale indicator

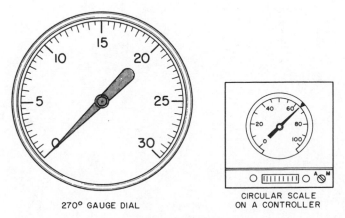

270° GAUGE DIAL

CIRCULAR SCALE ON A CONTROLLER

Fig. 3.6f   Dials

A light beam can be used as the movable reference, throwing a light spot on a scale. Readings are taken at the center of the light spot.

Multiple-scale meters show different ranges of the same variable on concentric scales. A selector switch changes the range in such electrical indicators and shows the range of the scale on which the reading is to be taken.

Movable-Scale Indicators

The alternative to the moving pointer is the moving scale and fixed reference or index (Figure 3.6g). A dipstick with a scale engraved on it used for immersion in a vessel containing a liquid is the simplest indicator in this category. A large circular disk with graduations on its entire periphery, read at an index in 12 o'clock position, affords great precision in reading fine subdivisions (Figure 3.6h). The hydrometer is also in this category. The floating vertical scale projecting through the surface of a liquid shows the specific gravity or density of the liquid as indicated by the surface of the liquid on the scale.

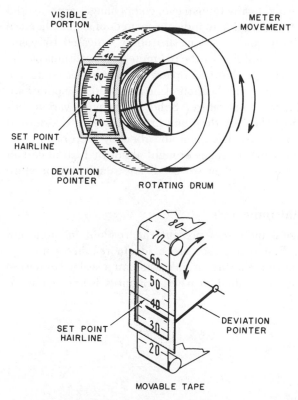

Fig. 3.6g   Movable-scale indicators

Parametric Indication

A parametric indication tells only that a state exists, e.g., whether a fluid is flowing or static. The precise level of water in a boiler may be less important than the fact that it contains an adequate quantity. The sight flowmeter equipped with a vane or a paddlewheel moving in a glass enclosure and the bubbler in which an air

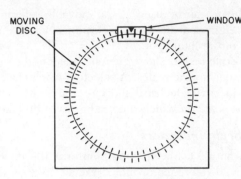

Fig. 3.6h   Rotating disc-type
movable-scale indicator

stream passing through a narrow glass tube monitors air flow do not measure but do indicate, just as the full water column is a nonquantitative indicator of liquid level. In an electrical system, an indicator light is often used to show that an electric potential exists at a certain point. A Geiger counter indicates radiation by clicking or blinking.

Various other nonmetering analog indicators are in use in industry. Because color is a phenomenon easily distinguished by the human eye, color changes are also used in instrumentation to indicate changes of state. Chemical conditions and reactions are often indicated by color. This is exemplified by the pH meter and by litmus paper.

The change that takes place in the physical state of a compound by application of heat is used to monitor approximate temperature. Solid waxes with various admixtures show by drooping in an oven that the desired amount of heat has been absorbed by the object to be heated. Color changes brought about by heat in certain materials are likewise used for rough temperature indications.

Digital Indicators

Digital indicators present the readout in numerical form. Fewer mistakes are made in reading a number than in deciphering an indication on a scale. Numerical readouts are called for when a quantity is to be counted.

Thus, counters usually present the result in figures, even though an analog indication of a variable measured at an instant offers other advantages.

A speedometer is an analog indicator for an instantaneous rate of speed, whereas mileage traveled is digitally shown on a counter.

A flowmeter customarily uses analog indication for the rate of flow (i.e., the quantity flowing at a given instant per unit time). The total quantity having passed a given point is usually indicated numerically on a separate counter, even though a somewhat complicated conversion by an integrator is required. A positive displacement meter indicates total flow by adding liquid quantities flowing through a pipe. It is generally provided with a digital readout.

Digital clocks and thermometers are easier to read than dial clocks and scale thermometers. The necessary analog-to-digital conversion is even simpler in electrical measurements.

A digital computer requires that digital data be fed to it. Many quantities easily obtained in analog form thus require conversion. Digital meters are now made for the readout of almost any variable. Voltmeters (including panel meters with gaseous numerical display devices); thermocouple thermometers; indicators of pressure, load, strain or torque; pH meters; oscilloscopes; and stroboscopes—all have become available with digital displays.

BIBLIOGRAPHY

Bylander, F.G., *Electronic Displays*, McGraw-Hill Book Co., New York, 1979.

Human Factors Evaluation of Control Room Design and Operator Performance at Three Mile Island-2, U.S. Department of Commerce NUREG/CR-2107, Volumes 1, 2, and 3, January 1980.

Human Factors Review of Nuclear Power Plant Control Room Design, ERPI NP-309, Palo Alto, CA, March 1977.

Sclater, N., "Displays Go Flat," *Instruments and Control Systems*, March 1975.

3.7 LIGHTS

Types:	(a) Incandescent, (b) neon, (c) solid state
	NOTE: In the feature summary the letters(a) to (c) refer to the listed indicator light types.
Operating Power Ranges:	(a) 1 to 120 V AC, V DC; 8 mA to 10 amps; (b) 105 to 250 V AC, 90 to 135 V DC; 0.3 to 12 mA; (c) 1 to 5 V DC; 10 to 100 mA.
Color of Unfiltered Light:	(a) White, (b orange, (c) red.
Relative Brightness:	(a) 1.0, 1.0; (b) 1.0, 0.5; (c) 2.0, 2.0.
Average Useful Life:	(a) 10 to 50,000 hours, (b) 5 to 25,000 hours, (c) 50 to 100,000 hours.
Application Limitations:	(a) Shock and vibration can cause early failures and generate considerable heat; (b) require high voltages and current-limiting resistors; have relatively low light output; (c) are expensive; brightness is high but total light output is low.
Partial List of Suppliers:	AMP Special Industries, Div. of AMP Products Corp.; Clare-Pendar Co.; Cutler-Hammer Products, Eaton Corp.; Dialight Corp.; Drake Mfg. Co.; Gemco Electric Co.; General Electric Co., Miniature Lamp Products Dept.; Hewlett-Packard Co.; Industrial Devices Inc.; Master Specialties Co.; North American Philips Lighting Corp.; Oak Technology Inc., Switch Div.; Ronan Engineering Co.; Square-D Co.; Westinghouse Electric Corp.

Lighted indicators convey several types of information to the operator, including binary information, in which an on-off or open-closed condition can be displayed by a lighted (on) or unlighted (off) indicator; and status information, in which normal, abnormal, or alarm conditions are expressed by legends, colors, and flashing or nonflashing indicators.

The amount of information that can be displayed is directly proportional to equipment or system size and complexity. Redundant indicators should be omitted, because the attention value of all indicators is reduced and confusion results if large numbers of them are used.

Characteristics of Light Sources

The most common types of lighted indicators are the incandescent, neon, and solid state lamps; their spectral response curves are shown in Figure 3.7a. The relative response ordinate at 100 gives the peak sensitivity of the human eye and the peak wavelengths emitted by the lamps. Standard vision, for example, is most sensitive at approximately 560 millimicrons, which is in the yellow and yellow-green band. The peak output wavelength of the gallium arsenide phosphide (GaAsP) solid state lamp is 650 millimicrons, which is in the red and red-orange band. The curves reflect the relative efficiencies of the light sources; the more efficient ones will have an output that matches and falls within the standard vision curve. This output is approximated both by the neon and by the solid state lamps. They convert most of their input power into light and emit little heat.

Selection and Application

Important human factors in the selection and use of lighted indicators are visibility and arrangement. To transmit information, the indicator must be visible to the operator. Variables that affect visibility are location,

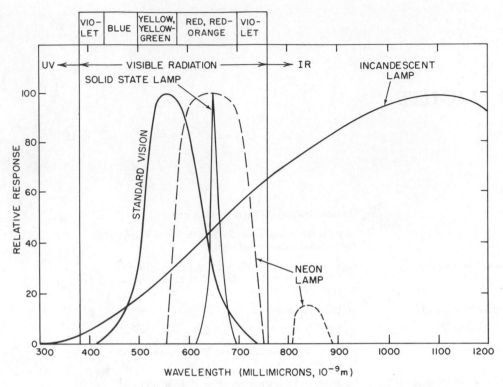

Fig. 3.7a Spectral response curves of the human eye and common lamps

brightness, contrast, color, size, and whether the indicator is flashing or nonflashing. Critical indicators should be located within 30 degrees of the line of sight and should be at least twice as bright as the surface of the mounting panel. Dark panels are recommended because they furnish strong contrast with the indicator and reflect little light to the operator. For high ambient light levels, alarm legends should have dark characters imprinted on a light background; routine messages should use the reverse combination.

Colors can be a powerful tool when properly used for lighted indicators. To avoid confusion, however, only a few colors should be used to code different types of information. General information should be lighted in white; normal conditions may be green; and for cautions or abnormal conditions amber (yellow with a reddish tint) is excellent because it affords maximum visibility. Red should be used only for critical alarms that require immediate response by the operator. The use of blue or green lenses should be kept to a minimum. All lamps emit most strongly in the red and red-orange band. Consequently, much light is lost if it is filtered so as to appear blue or green. For important indicators, one should use the largest size that is compatible with the panel scale.

Flashing greatly improves visibility, but its use should be limited to critical alarms. The rate should be 3 to 10 flashes per second with "on" time approximately equal to "off" time. Light indicators should be arranged according to a functional format. Indicators associated with a manual control (pushbutton or switch) should be placed closely above the control device. It is best to arrange related indicators into separate subpanel areas. Displays requiring sequential operator actions should be arranged in the normal reading pattern—from left to right or from top to bottom. Critical indicators sould have dual lamp assemblies for additional reliability. A lamp check switch should be supplied to test for and locate burned-out lamps. Another important design target is easy lamp replacement.

The selected lenses should be diffusive and should eliminate glare and lamp hot spots. The lens should also provide a wide angle of view (120 degrees minimum), and if side visibility is required, it should protrude over the mounting surface. The lens must be large enough to accommodate the required legends. Legends are commonly produced by hot stamping, engraving, or photographic reproduction of transparencies. Ordinarily, hot stamping is the most economical, whereas phototransparencies furnish the sharpest characters and are the most versatile.

Environmental parameters also affect the operation of the indicator lights. Special designs are available for shock, vibration, or high temperature applications. Rapid dissipation of heat is important if the indicator generates heat. Drip-proof or watertight designs should be selected if indicators are to operate in high humidity or corrosive atmospheres or if panel washdown resistance is a requirement.

Components of Indicator Lights

Major components of most indicator light assemblies (Figure 3.7b) include the lampholder, the lamp, and the lens. Panel light types are usually secured to a panel by a nut and a lockwasher. Cartridge models can be held by a speednut friction clip. Snap-in lights are retained by expandable latching fingers. Power can be supplied through wire leads and solder, screw, or quick-connect terminals.

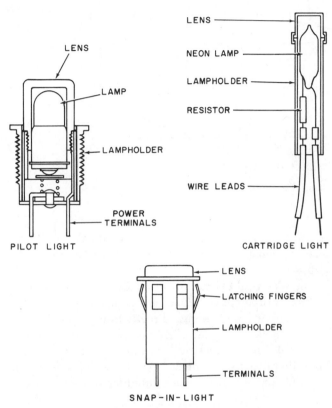

Fig. 3.7b   Indicator light assemblies

The heart of all lighted indicators is the lamp or light source. The three types (Figure 3.7c) in common use are incandescent, neon, and solid state. Major parts of a lamp are the bulb (containing the light emitter) and the base. Lamps are also classified according to bulb shape, size, and type of base. Bulb shape and size are designated by a letter describing shape and a number that gives the

nominal diameter in eighths of an inch. For example, a T-1 lamp has a tubular-shaped bulb that is one eighth of an inch in diameter. Common bases are bayonet, screw, flanged, grooved, and bi-pin. Some lamps have no base and are supplied with wire terminals.

Incandescent Lamps

The incandescent lamp (see Figure 3.7c) consists of a coiled tungsten filament mounted on two support wires in an evacuated glass envelope. When current is passed through the filament, its resistance causes it to glow and to emit both light and heat. Note from Figure 3.7a that only about one third of the radiation emitted from this lamp is in the visible band (white light); the rest is in the infrared band (heat). This means that approximately two thirds of the input power is emitted as heat. If large quantities of incandescent lamps are to be operated continuously and mounted closely together, special allowance for adequate heat dissipation is necessary.

If the lamp is to operate under shock and vibration, a low-voltage, high-current design should be used because it has strong filaments. Lamps of 6 volts or less usually have short, thick filaments, whereas lamps of more than 6 volts generally have longer and thinner filaments. In all cases, however, the lamp should be tested under simulated operating conditions.

Incandescent lamps can operate from 1 to 120 volts AC or DC. Current drain will be from 10 milliamperes to 10 amperes. Figure 3.7d shows the relationship of the applied voltage, lamp life, current, and light output. Variations in applied voltage have a drastic effect on lamp life. It is common practice to improve life and sacrifice some light by operating the lamp at slightly below rated voltage. For example, a 6-volt lamp operated at 5 volts will have roughly 8 times the normal life and will still provide 60 percent of normal light. Overvoltage results in nearly the opposite effect: A 6-volt lamp operated at 7 volts will have only about one sixth of its normal life and will provide one and one half times the normal light. Therefore, controlling applied voltage is very important.

Space permitting, a large lamp is preferred to a small one because its cost will be lower and its life and reliability higher. The large lamp will also emit less heat than the small one of equal light output, because the former has more surface area and its filament operates at a lower temperature.

Neon Lamps

The neon lamp (see Figure 3.7c) consists of two closely spaced electrodes mounted in a glass envelope filled with neon gas. When sufficient voltage is applied across the electrodes, the gas ionizes, conducts a current, and emits light and heat. All neon lamps require a current-limiting resistor in series with the lamp to guarantee the designed life and light characteristics, an example of which is the cartridge light in Figure 3.7b. The orange light emitted

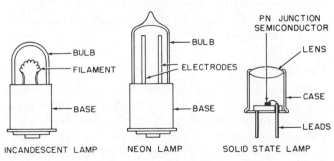

Fig. 3.7c   Common lamps

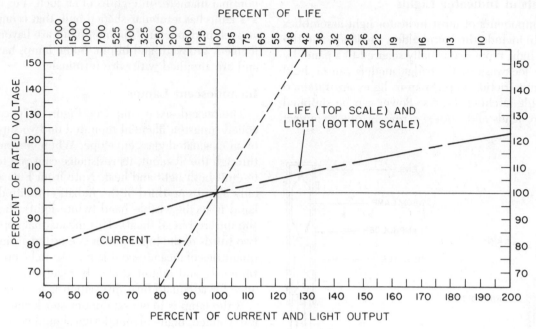

Fig. 3.7d   Incandescent lamp characteristics

by this lamp is easily seen because a large portion of it falls within the standard vision curve (see Figure 3.7a). Since most of its emitted radiation is in the visible band, little heat is emitted. An important consideration, however, is that although the neon lamp is an efficient light source, its total light output is low. A clear or lightly diffusing lens should be used with it so that only a small portion of the light is absorbed.

Neon lamps are very satisfactory for use under conditions of severe shock and vibration. The rugged mechanical construction avoids the use of the fragile filament of the incandescent lamp. Neon lamps will operate only on high voltages and commonly run directly from standard 120- or 240-volt AC line voltages. Because of the high voltage, they require little current and power.

Solid State Lamps

The solid state lamp (see Figure 3.7c) is commonly called a light emitting diode (LED), and it is a valuable by-product of semiconductor technology (see Section 6.6). It is basically a P-N junction diode mounted in a hermetically sealed case with a lens opening at one end. Light is produced at the junction of the P and N materials by two steps. A low voltage DC source increases the energy level of electrons on one side of the junction. In order to maintain equilibrium, the electrons must return to their original state; they cross the junction and give off their excess energy as light and heat. The light output of the popular gallium arsenide phosphide LED (see Figure 3.7a) has a very narrow bandwidth centered in

the red band. These lamps are very efficient and have an exceptionally long life but are small and have low light output.

The electrical characteristics of the LED are similar to those of the silicon diode, are compatible with integrated circuits, and operate directly from low-level logic circuits. Solid state lamps, like neon lamps, perform satisfactorily under shock and vibration. Currently, they are rather expensive, but the cost can be expected to fall as additional applications are found for these devices.

Checklist

1. Determine operating voltage.
2. Select lamp type and size.
3. Select lens for type, color, size, and shape. The last two features should be large enough to hold the necessary legends.
4. Select a lamp holder that is compatible with both lamp and lens. Also consider the allowable panel space and the methods of mounting and providing electrical connections.
5. Test the indicator under simulated operating conditions.

Conclusions

The possible uses of indicator lights to display information are limited only by the designer's imagination. There is usually one particular combination of lamp, lens, and lampholder that is best suited for an application. Incandescent lamps are preferred for most applications because they are available in the widest range of light output, sizes, and voltages. The low light output of both the neon and the solid state lamps limits their use to on-off indicators. Amber lenses should be widely used because they absorb little lamp light and they are the most visible of all colors. Snap-in lampholders should be used wherever possible because they require no mounting hardware and take little assembly time.

BIBLIOGRAPHY

Bylander, F.G., "Electronic Displays," McGraw Hill Book Co., New York, 1979.

Jutila, J.M., "Guide to Selecting Alarms and Annunciators," *InTech*, March 1981.

McCready, A.U., "Man-Machine Interfacing for the Process Industries," *InTech*, March 1982.

National Electrical Code, Article 410, National Fire Protection Association, 1981.

3.8   RECORDERS

Features: Circular, strip, X-Y, multiple, and operation/event recorders; chart recorders; digital, magnetic, printing, and video tape types.

Cost: $300 to $10,000, depending on type, application, accuracy, and other parameters.

Partial List of Suppliers: Bacharach Instruments; Barber-Colman Company; Beckman Instruments, Inc.; Beta Products, Inc.; Bristol Babcock, Inc.; Chessell Corporation; Dranetz Engineering Laboratories, Inc.; Fischer & Porter Company; The Foxboro Company; General Electric Company; Gould Inc., Instruments Division; Gulton Industries, Inc.; Hewlett Packard San Diego Division; Honeywell Process Control Division; International Products & Technologies, Inc.; Kaye Instruments; Kent Process Control, Inc.; Leeds & Northrup Company (a unit of General Signal); Newport Electronics, Inc.; Omega Engineering Incorporated; Rochester Instrument Systems; Soltec Corporation; Texas Instruments; Transitron Electronic Corp., Transitron Control Div.; Westinghouse Electric Corp., Combustion Control Div.; Yokogawa Corp. of America.

Charts and Coordinates

The process variable actuates a recording mechanism, such as a pen, which moves across a chart. The chart moves constantly with time. These two motions produce an analog record of variable versus time. Any point on the continuous plot obtained in this manner can be identified by two values, called coordinates. Many coordinate systems are in use, with the Cartesian coordinates most widely used in industry. If the reference lines are straight and cross each other at right angles, they are called rectilinear coordinates. If at least one of the reference lines is an arc of a circle, the coordinates are curvilinear.

The shape of the chart provides a primary means of classification into (1) circular charts and (2) rectangular charts, in sheet or strip form. Strip charts can be torn off and can be stored in rolls or folded in Z folds. Strip chart lengths vary from 100 to 250 feet (30 to 75 m).

Circular Chart Recorders

Circular chart recorders are still used because they are simple and, consequently, low-priced. Chart speeds are a uniform number of revolutions per unit time. Chart drives making one revolution in 24 hours are standard, but instruments are also equipped with drives of 15 minutes; 1, 4, 8, 12, or 48 hours; and 7, 8, or 28 days per revolution.

The best application for portable circular chart recorders may be in areas where temperature or humidity recording is needed infrequently, perhaps to set up a heating or ventilating system.

The circular chart recorder shown in Figure 3.8a produces concentric circles crossed by circular arcs whose center, when the pen crosses the arc, is the same as that of the recording arm. The tangents at the intersection points of the two curves should be as near to a right angle as possible for best readability. The concentric circles form the scale on which the variable is read. The "time arcs" divide the full circles into appropriate uniform intervals of the total period.

Chart diameters vary from as low as 3 inches (75 mm) to up to 12 inches (300 mm). Circular charts offer the advantage of a flat surface. Special features, such as automatic chart changers, allow collection of daily records in a continual manner. Contacts can also be mounted in

Fig. 3.8a Circular chart recorder

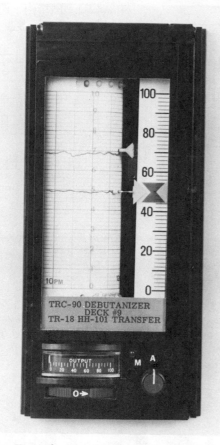

Fig. 3.8b Horizontal strip chart recorder

the pen mechanism for alarms on preset high or low signal conditions.

Strip Chart Recorders

Strip chart recorders are characterized by the uniform linear motion of the paper either horizontally (Figure 3.8b) or vertically (Figure 3.8c). The measurement lines can be straight or curved, furnishing either rectilinear (Figure 3.8c) or curvilinear (Figure 3.8b) recordings. The curvilinear method uses very simple linkage geometry and offers better readability than is obtained with circular charts. Figure 3.8d shows a two-pen recorder (second from left) and a recorder controller. Whereas the two-pen recorder records two independent variables, the recorder controller accepts one sensor signal, compares that signal with a manually operated set point, and provides a signal output for closing a control loop by actuating a final control element.

X-Y Recorders

X-Y recorders plot two variables simultaneously, such as stress versus strain or temperature versus pressure. Either the chart is stationary and the scriber is moved along both the abscissa and the ordinate by the two signals or the chart is moved in one direction while the stylus slides on an arm in the other direction.

The signals entering the function plotter can be analog or digital. Digital signals require transducers to obtain an analog plot. Likewise, digital records can be provided with analog-to-digital conversion and conventional digital printout.

There are also combination function plotters, called X-Y-Z_1 recorders. Some allow the pen to be driven along either axis at a constant speed, thus making recordings of X versus Y, Y versus T, and X versus T possible. Recorders with three independent servo systems allow the recording of two variables against a third.

X-Y recorders usually have flat beds with measurements ranging from $8\frac{1}{2}$ inches × 11 inches (212.5 × 275 mm) to 45 inches × 60 inches (1125 × 1500 mm). Some record on drums. Recorders that print from the back or on a glass screen do not obstruct visual observation.

Multiple Recorders

When several variables are to be recorded on the same chart, such as several temperatures from thermocouples in various locations, multiple recorders are used. Circular chart recorders may handle only up to four variables, whereas strip chart recorders handle as many as 24 to 36 measurements. To identify each variable, symbol or numerical coding or color printing is used, as well as full digital alphanumeric printouts on chart margin.

With the advent of microprocessor technology, multivariable recorders have become available. These allow recording of a multitude of variables, such as flow and an associated temperature and pressure. Likewise, for comparison of several variables on the same time scale

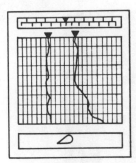
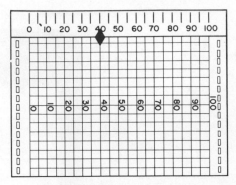
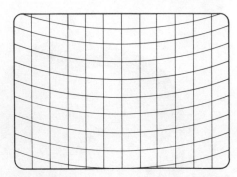

Fig. 3.8c   Strip chart recorders with vertical movement

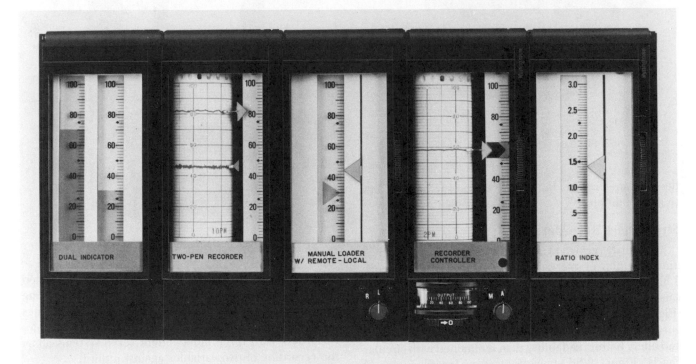

Fig. 3.8d   Two-pen recorder and recorder controller

without line crossing or overlapping, multi-channel recorders are available.

Event Recorders

Operations or event recorders mark the occurrence, duration, or type of event. They record multiple incidents, such as on-time, down-time, speed, load, and overload, on the chart. The records that are produced are usually in the form of a bar, with interruptions in a continuous line indicating a change. Microprocessor technology allows scores of points to be scanned every millisecond, with high speed printouts made for the events that occur.

Digital Recorders (Printers)

Digital recorders print the output of electronic equipment on paper in digital form. High-speed recording is achieved by combining electronic readouts with electrostatic printing. As many as 1 million characters per minute can be recorded in this way. The high-speed digital recorder may be best suited for electric power generation.

Magnetic Recording

Magnetic recording on tape is used frequently to store information. Material can be stored either in analog or in digital form.

Video Tape Recording

Video tape recording is similar to magnetic tape recording with the additional feature of reproducing both picture and sound. Industrial applications are found in closed-circuit television.

BIBLIOGRAPHY

Chiranky, L. "A Case for Linear Array Recording," *Instruments and Control Systems*, July 1982.

Cuvelier, A., "A French 'First'—A Six Color, Six Channel Recorder Synchronized by a Microprocessor," *Electronic and Applied Industry* (France), No. 249, March 1978.

Dobrowolski, M., "Guide to Selecting Strip and Circular Chart Recorders," *InTech*, vol. 28, no. 10, October 1981.

Howes, P.A., "Data Display Recording Devices," *Advances in Test Measurement*, vol. 16, Instrument Society of America, Research Triangle Park, N.C., 1979, pp. 199–210.

Krigman, A., "Multipen and Multipoint Recorders: Making the Technology Work for You," In Tech, August 1984, pp. 39–48.

McDowell, W.P., "Microprocessor Controlled Multipoint Recorder with Graphical Video Output for Reactor Core Temperatures," *IEEE Transactions of Nuclear Science*, Vol. NS-27, No. 1, February 1980.

Nylen, P., "Electronic Device for Compensating for Pen Displacement in a Multiple Channel Strip-Chart Recorder," *Review of Scientific Instruments*, Vol. 51, No. 3, March 1980.

Olabri, J.P., "Chart Recorders," *Instruments and Control Systems*, July 1982.

Quinn, G.C., "Recording Instruments," *Power*, Vol. 121, No. 12, December 1977.

3.9 SWITCHES AND PUSHBUTTONS

Types:	(a) Pushbutton, (b) toggle, (c) rotary, (d) thumbwheel.
	NOTE: In this feature summary the letters (a) to (d) refer to the listed electrical switch types.
Features:	(a) Keyboard, panel, industrial oil-tight; (b) lever, rocker, thumbwheel selectors; (c) selector, adjustor; (d) selector.
Actuation:	(a) Push in, (b) pivot, (c) twist, (d) push up and down.
Sizes:	(a) Standard, miniature; (b) standard, miniature, subminiature; (c) standard, miniature; (d) standard.
Costs:	(a) $10 to $100, (b) $5 to $40, (c) $25 to $250, (d) $10 to $20.
Partial List of Suppliers:	Allen-Bradley Co.; Automatic Switch Co.; Cherry Electrical Products Corp.; Cutler-Hammer Products; Eagle Signal Industrial Controls; Electro Switch Corp.; General Electric Co.; GTE Sylvania, Electrical Components Group; Licon Div. of Illinois Tool Works Inc.; Marcon Switches, Inc.; Master Specialties Co.; Micro Switch, Div. of Honeywell; Square-D Co.

Proper selection and use of manual controls are complex tasks. Usually a best choice exists for each application among several that would do the job.

Principles of Operation

An electric switch-type manual control (Figure 3.9a) consists of the switching contacts, an actuator to bring them into proximity, terminals to connect the contacts to the conductor of an electric circuit, the insulating mounting provisions, and the enclosure, either separate from or integral to the mounting for contacts and terminals. The gap between contacts is filled by air or an inert gas. The control is operated by moving the contacts together until the applied voltage causes an electrostatic failure of the gap. The ensuing electron flow heats the cathode and cools the anode contact until molecular welding occurs, creating a continuous metallic path and completing the circuit.

Contact resistance is a measure of the degree to which the insulating gas remains between the contacts, preventing welding over their entire bearing area. It is directly proportional to contact pressure.

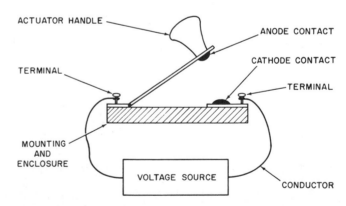

Fig. 3.9a   Operating elements of a manual control

Switching Action

Switching action refers to the positions assumed by the contact in response to actuator motion. The two basic types (Figure 3.9b) include momentary action, which provides one contact closure and subsequent reopening with one actuation; and maintained action, which requires separate actuations to transfer the contacts from one extreme position to another, and back.

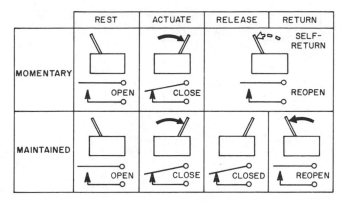

	REST	ACTUATE	RELEASE	RETURN
MOMENTARY				SELF-RETURN
	OPEN	CLOSE		REOPEN
MAINTAINED				
	OPEN	CLOSE	CLOSED	REOPEN

Fig. 3.9b Basic switching actions

	DIAGRAM	OPERATION	FEATURES
SPST NO	FORM A	MAKE (PULSE)	PULSES WHEN ACTUATOR IS DEPRESSED, BUT NOT WHEN RELEASED.
SPST NC	FORM B	BREAK (PULSE)	SAME AS FORM A EXCEPT OPENS CIRCUIT INSTEAD OF CLOSING IT
SPDT NO/NC	FORM C	BREAK-MAKE (TRANSFER)	PROVIDES COMPLETE CIRCUITS AT BOTH CONTACT POSITIONS
SPDT NO/NC	FORM D	MAKE-BEFORE-BREAK (CONTINUITY TRANSFER)	MOMENTARY OVERLAP OF CONTACTS PROVIDES CIRCUIT CONTINUITY
SPDT NO	FORM K	CENTER OFF (PULSE)	PROVIDES PULSES IN TWO CONDUCTORS FROM ONE SWITCH. USEFUL FOR "UP, DOWN" JOGGING.
DPST NO	FORM AA	DOUBLE MAKE (DOUBLE PULSE)	INTERRUPTS CURRENT IN TWO INDEPENDENT CONDUCTORS OF A CIRCUIT. REPLACES TWO FORM A SWITCHES.
DPDT NO/NC	FORM CC	DOUBLE BREAK-DOUBLE MAKE (DOUBLE TRANSFER)	SIMULTANEOUSLY OPERATES TWO CIRCUITS. SEVERAL WIRING POSSIBILITIES. REPLACES TWO FORM C SWITCHES.

Fig. 3.9c Switching contact configurations

Other special types include

1. Mechanical bail, which allows operation of only one control at a time, physically inhibiting depression of others. Automobile radio selectors are a typical example.
2. Electrical bail, which allows delayed or remote actuation, including sequencing of several controls. Solenoid operators frequently sequence rotary selectors or ganged pushbuttons.
3. Sequential action, which opens one contact set in a specified order in relation to other contacts in the same control. This type may involve simultaneous operation of several contact sets and repeated opening and closing of one contact set during one actuation.

Contact Arrangements

Switching capacity refers to the number, type, and arrangement of contacts and circuits within the control. Figure 3.9c contains the related terminology. (For more on contact configuration standards, see Figure 6.5b.)

Normally open contacts (NO). The circuit is open (current does not flow) when the contacts are not actuated. Moving them to the other extreme position completes (closes) the circuit.

Normally closed contacts (NC). The circuit is complete when the switch is not operated. Moving the contacts opens the circuit, interrupting current flow.

Number of poles (P). The number of conductors in which current will be simultaneously interrupted by one operation of the switch.

Number of throws (T). The number of positions at which the contacts will provide complete circuits. Since contacts usually have only two extreme positions, switches are either single-throw or double-throw.

Switching Elements and Circuits

Switching circuits are classified as low-energy or high-energy. Low-energy circuits do not develop sufficient volt-amperes to break down insulating contaminant film (that may have built up on the contacts) by contact heating. Consequently, infrequent use degrades the controls in these circuits. Accordingly, reliable low-energy switching elements must maintain low contact resistance. Solutions include (1) wiping action under high contact pressures; (2) contact materials with high oxidation resistance; and (3) hermetically sealed enclosures, either evacuated or filled with an inert atmosphere.

High-energy circuits generate sufficient power to sustain an arc between contacts. Therefore, a reliable high-energy switching element must perform under very high contact temperatures. With these circuits, unlike low-energy circuits, frequent operation degrades reliability. Large contact areas, thick plating, and thermally conductive alloys combat high temperatures. Special features divert, vent, or extinguish the arc.

Switching elements are summarized in Figure 3.9d. Mechanical contactors either wipe the contacts over each other or strike them together. Wiping contacts move in a series of jerks, during which the contacts alternately weld and break loose. Striking contacts are either elastically or plastically deformed on impact.

	NAME	DIAGRAM	OPERATION	OPERATING CHARACTERISTICS	ADVANTAGES AND DISADVANTAGES	APPLICATIONS
MECHANICAL CONTACTORS	MAGNETIC REED	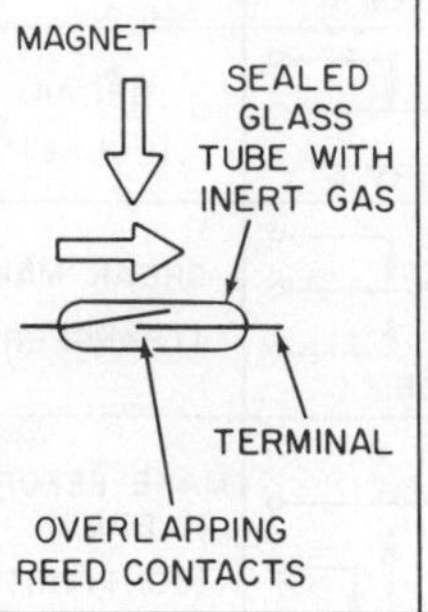 MAGNET / SEALED GLASS TUBE WITH INERT GAS / TERMINAL / OVERLAPPING REED CONTACTS	MAGNETIC INDUCTION BENDS REEDS. STRIKING CONTACTS.	ACTUATION FORCE INDEPENDENT OF CONTACT PRESSURE. HIGH CONTACT PRESSURE. FAST, PRECISE TIMING. EXCELLENT FOR LOW-ENERGY CIRCUITS.	HERMETICALLY SEALED CONTACTS. BOUNCE IS PROBLEM, REQUIRES EXTERNAL COMPENSATING CIRCUITRY. CAN BE ACTUATED BY VIBRATION. REQUIRES SHIELDING FROM EXTERNAL MAGNETS. LONG LIFE, INEXPENSIVE.	KEYBOARD PUSHBUTTON, PANEL PUSHBUTTON, (INTEGRAL).
	SNAP-ACTION	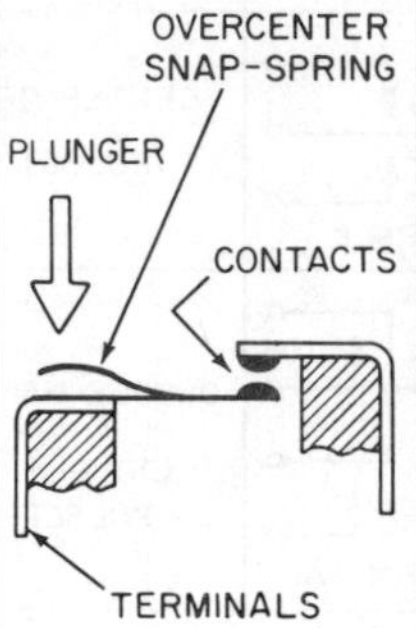 OVERCENTER SNAP-SPRING / PLUNGER / CONTACTS / TERMINALS	LEAF SPRING SNAPPED. OVER-CENTER. STRIKING CONTACTS.	IRREVERSIBLE ACTION NO "TEASING". FAST, PRECISE, SHORT, REPEATABLE TIMING AND TRAVEL. BEST FOR SIMULTANEOUS SWITCHING OF MANY CIRCUITS WITH ONE ACTUATOR. HIGH CONTACT PRESSURE, HIGH ACTUATION FORCE.	ENCLOSED CONTACTS. SUITABLE FOR LOW OR HIGH-ENERGY CIRCUITS. RELATIVELY INEXPENSIVE. VERY SMALL SIZES AVAILABLE.	TOGGLE, PANEL PUSHBUTTON (BUILT-UP), INDUSTRIAL PUSHBUTTON (MEDIUM AND HEAVY DUTY).
	WIPING BLADE	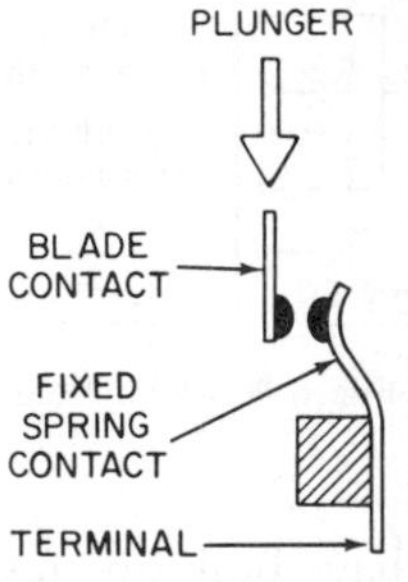 PLUNGER / BLADE CONTACT / FIXED SPRING CONTACT / TERMINAL	CANTILEVER BLADE CONTACT WIPES PAST FIXED SPRING CONTACT.	LOW CONTACT PRESSURE. WIPING CLEANS CONTACTS, BUT WEARS AWAY PLATING. TIMING AND TRAVEL NOT PRECISELY REPEATABLE.	HARD SILVER PLATING USED TO RESIST ABRASION, THUS NOT GENERALLY SUITED TO LOW ENERGY USE. ROTARY IS EXCEPTION SINCE INFREQUENT USE ALLOWS GOLD PLATING. RELATIVELY INEXPENSIVE.	INTEGRAL PANEL PUSHBUTTON, ROTARY, INDUSTRIAL PUSHBUTTON (LIGHT DUTY).
CONTACTLESS	INTEGRATED CIRCUIT	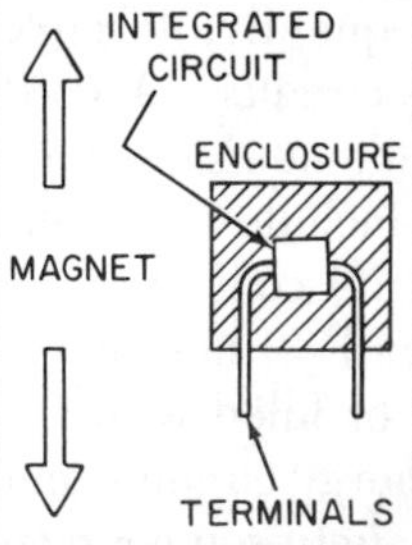 INTEGRATED CIRCUIT / ENCLOSURE / MAGNET / TERMINALS	(HALL EFFECT) VOLTAGE IS DEVELOPED ACROSS EDGES OF CURRENT-CARRYING CONDUCTOR BY MAGNETIC INDUCTION.	ACTUATOR FORCE INDEPENDENT OF SWITCHING ACTION. VERY FAST SWITCHING, PRECISE TIMING.	NO CONTACT BOUNCE. ALWAYS-ON BIAS VOLTAGE SOMETIMES REQUIRED. EXCELLENT FOR REPETITIVE, LONG-LIFE USE IN LOW-ENERGY CIRCUITS. EXPENSIVE. SEALED.	KEYBOARD PUSHBUTTON.

Fig. 3.9d   Basic switching elements

Magnetic reed contacts are overlapping ferromagnetic beams, cantilevered from a glass tube filled with a dry inert gas. A small gap separates the overlapping ends, and magnetic induction in the gap eventually overcomes their stiffness, bending them until they touch. The circuit is reopened by removal of the magnetic influence. Reed contacts provide (1) high contact pressure with low actuator force, since the actuator does not directly move the contacts; (2) very fast response; and (3) long life. Since the reeds are cantilevered springs, a major problem is contact bounce.

Snap-action switches are very rapidly struck together by an overcenter spring mechanism. They provide (1) irreversible switching (since the operator can do no more than start the switching action; (2) precise timing, including virtually simultaneous switching of many circuits with one actuator; (3) precise, consistently repeatable travel because there is only one moving part—the spring;

and (4) high load capacity and long life, since arcing is limited by the very fast contact transfer time.

Wiping blade contacts employ a stiff, blade-shaped contactor that passes over spring-mounted stationary contacts under high pressure. Alternately, the moving blade is the spring and the fixed contacts are stiff. The contacts are sometimes knurled to assist the wiping action. Hard silver plating alloys are required to resist wear, which generally restricts their use to high-energy circuits.

Contactless switches are useful in low-signal-level electronic circuits. The absence of contact bounce eliminates the need for external filtering circuits. Other advantages are virtually unlimited life, very fast switching, and freedom from the effects of contamination. Most contactless switches employ semiconductors, which require an always-on bias voltage. This can be a disadvantage when compared with a mechanical switch.

Types and Grades

Table 3.9e summarizes the features of four types of manual control devices, including pushbutton, toggle, rotary, and thumbwheel, all of which are available in several grades and types of construction.

Appliance-grade switches are not suitable for process control; they are low-cost devices designed for light, non-abusive environments in which precise timing and current control are not required, performance degradation is not critical during operating life, and exacting operator feedback is not necessary.

Table 3.9e
CHARACTERISTICS OF ELECTRIC SWITCHES

	Pushbutton	Toggle	Rotary	Thumbwheel
Typical applications	Keyboard, motor, machine tool, instrument; best for pulsing; action dynamics match digital display	Computer data register; power on-off, test; "center off" type suitable for jog, adjustment about reference	Computer control panels, ammeter adjustment (potentiometer); action dynamics match fine analog adjustment	Encoded data entry to digital computing circuits; instrument settings; exact value setting
Advantages and disadvantages	Easiest, most comfortable operation; self-contained label flexibility	Least expensive 2,3 position selector; simple nonprecision operation, reflex action; best for actuation of multiple switches simultaneously	Most capacity in one control; selector available for 4+ positions in high-energy circuits; self-contained encoding optional; also mounting for components	Most capacity in least panel area; self-contained encoding, dial readout of value setting; single alphanumeric character-position or complete messages
Handles and feedback features	Integral handle, lighting very useful; no feedback from handle position	Integral, special "decorator" style available; handle position positive feedback (indicator) of status; no light or label except rocker (marginal)	Separate, many shapes and color options; label on skirts; no lighting	Integral; small tabs adjacent to indicator dial; same number as number of positions; lighting optional
Environmental protection	Sealed switch elements standard; sealed actuators optional; standard industrial grade	Enclosed switch elements standard; sealed actuator handle optional	Enclosed switch elements optional; sealed actuator shaft optional	Enclosed switch elements and actuator handle optional
Number of positions	1,2	2,3	2–24 (potentiometer)	8–16
Comparable cost, position	2.5	1.0	5.0	—
Ratings	0.5–60A @ 125 VAC or 30 VDC; 1A @ 600 VAC; 0.5A @ 600 VDC	20A @ 125 VAC 12A @ 30 VDC	0.3A @ 125 VAC 1A @ 30 VDC	0.13A @ 125 VAC or @ 125 VDC
Life	10,000,000 operations	1,000,000 operations	100,000 360° cycles	100,000 360° cycles
Mounting	Front or back of panel; printed wiring board	Back of panel; printed wiring board	Back of panel; printed wiring board	Front or back of panel; printed wiring board

Commercial-grade switches are suitable for the control room. They provide consistent actuation and switching performance at specified reliability in average environments. Some degree of contact sealing is present, and corrosion-resistant contact materials are used. Actuation and switching mechanisms afford precise control of light to medium electrical loads.

Industrial-grade switches are specialized for local installation in harsh environments. Construction is rugged in order to accommodate abusive actuation and unusually heavy electrical loads. Actuators and contacts are sealed against liquid and solid contaminants, especially oil.

Military-grade switches are suitable for process control use but are usually overspecified in some areas and underspecified in others and carry a price premium.

Built-up construction uses separate housings for the actuator and the switching element. Circuit flexibility is the principal advantage, since switching modules can be ganged at will, although standard capacities are available. Disadvantages are unsealed interface between actuator and switch modules, back-of-panel volume, and price.

Integral construction combines the actuator and the switching element in a common housing. Advantages include satisfactory sealing and smaller overall dimensions. The major disadvantage is fixed switching capacity.

Pushbuttons

Pushbuttons provide the simplest, most naturally comfortable actuation motion. They alone accommodate extensive labeling without the use of separate nameplates. Disadvantages include a switching capacity that is lower than that found in other types because there are only two contact positions and lack of inherent indication of switching status (such as by handle position) other than lighting. Two methods of producing colored light are by transmission and by projection. Color selection has been discussed in Section 3.7.

Transmitted colors are produced by transmission of white light through colored button-lenses; they should be used whenever possible. Colors are intense, saturated, and uniformly distributed over the button; do not degrade from lamp heat; and can be read both from great distances and through wide viewing angles.

Projected colors are produced by projection of colored light on a white button-screen. The colored light is produced by transmission of white light through colored filters that slip over the lamps. Projected colors are weak, dilute, and easily overpowered by normal room ambient light; have restricted viewing distances and angles; and degrade as the colored filter is progressively destroyed by lamp heat.

PANEL PUSHBUTTONS

A variety of integral and built-up configurations is available (Figure 3.9f), incorporating molded or metal housings, various force and travel options, and a wide range of lighted and unlighted buttons. Square and rec-

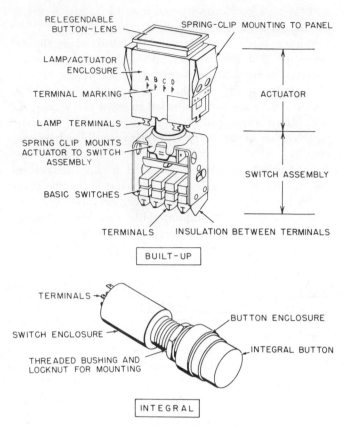

Fig. 3.9f   Panel pushbuttons

tangular buttons are most easily mounted and provide the most labeling area. Labeling is accomplished by hot-stamping, engraving, two-shot molding, silk-screening, or by use of photographic film inserts. Panel pushbuttons are usually mounted in individual openings; built-up configurations mount from the front of the panel, using spring clips or barriers. Integral configurations mount from the back of the panel, using threaded bushings and locknuts. Buttons and lamps in each configuration are accessible from the front of panel.

Each mounting method has definite panel thickness limits, which can, however, be circumvented by subassembly to a bracket that in turn mounts on the back of the panel. The buttons are available for both low-energy and high-energy circuits.

INDUSTRIAL PUSHBUTTONS

These buttons are ruggedly made to withstand abusive environments and to resist abusive operation (Figure 3.9g). Compact and standard sizes are available, and both kinds use a built-up construction to accommodate custom switching for specific application. Actuators and contacts are both liquid-tight and dust-tight, and contacts are available for three grades of duty: (1) electronic (wiping action, gold plating); (2) standard (wiping action, silver plating); and (3) heavy-duty (striking action, silver, silver alloy, or cadmium alloy plating).

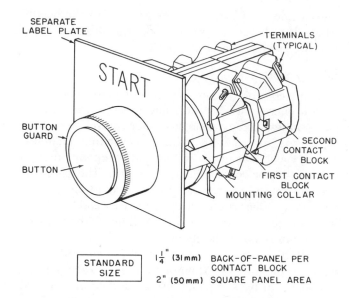

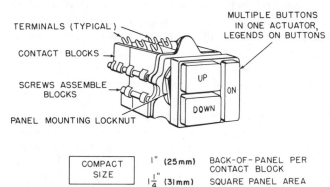

Fig. 3.9g   Industrial pushbuttons

KEYBOARD PUSHBUTTONS

These buttons are integral pushbuttons used in low-energy electronic circuits (Figure 3.9h). Force-travel characteristics are carefully designed for precise tactile feedback during repetitive, high-speed touch typing. Simplified, durable, low-cost actuator mechanisms and switching elements furnish a high degree of reliability for many operations. Construction is as functional as possible for mass production, group mounting, and group connecting. Techniques include molded housings, two-shot molded buttons and labels, and printed wiring board mounting and connecting. Both assembled keyboards and a variety of standard and custom arrangements, encoding, operating forces, and buttons are available from most manufacturers of keyboard pushbuttons.

Toggle Switches

Toggle switches are the cheapest two-position or three-position selectors for a large number of circuits of any level of complexity or energy, and integral and built-up configurations are available. A major advantage is positive indication of switching status from handle position. The three basic handle configurations include (1) lever, (2) rocker, and (3) thumbwheel.

The lever handle configuration (Figure 3.9i) is the most common and gives the most positive handle position indication. Careful consideration of length, shape, protrusion above the panel, and panel graphics design is required before the operator can capitalize on this advantage. A wide selection of shapes, sizes, colors, and trim hardware is available to create various front-of-panel appear-

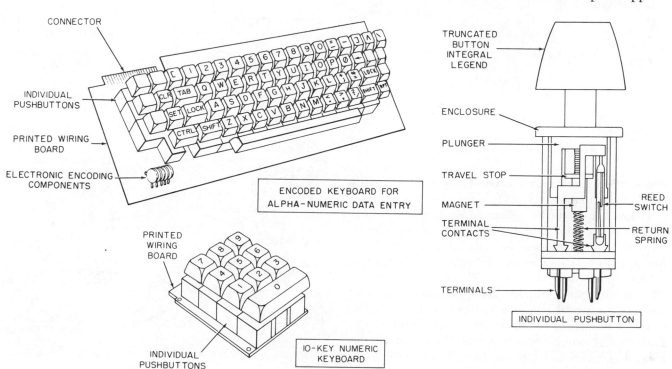

Fig. 3.9h   Keyboard pushbuttons

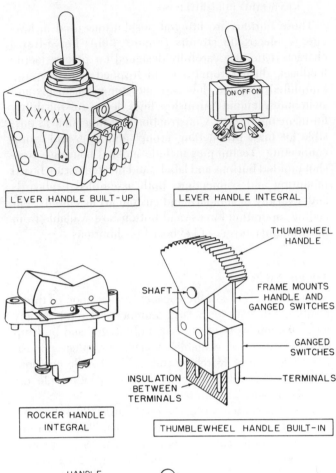

LEVER HANDLE BUILT-UP

LEVER HANDLE INTEGRAL

ON OFF ON

ROCKER HANDLE INTEGRAL

THUMBWHEEL HANDLE

SHAFT

FRAME MOUNTS HANDLE AND GANGED SWITCHES

GANGED SWITCHES

TERMINALS

INSULATION BETWEEN TERMINALS

THUMBLEWHEEL HANDLE BUILT-IN

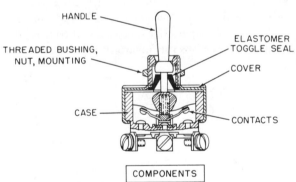

HANDLE

THREADED BUSHING, NUT, MOUNTING

ELASTOMER TOGGLE SEAL

COVER

CASE

CONTACTS

COMPONENTS

Fig. 3.9i   Toggle switches

ances, accentuate handle positions, and safeguard against accidental actuation.

The rocker handle (see Figure 3.9i) combines positive handle indication with the operating simplicity of pushbuttons. Limited labeling and lighting are also available, but both features are marginal owing to space limitations.

The thumbwheel handle (see Figure 3.9i) is occasionally used to reduce front-of-panel protrusion and to accommodate special actuation motion. The disadvantages of the thumbwheel handle are that there is no positive indication of switch position by handle position and that control of travel and force is difficult, since the finger moves in and out at the same time that it moves from

side to side. The latter characteristic makes the handle especially awkward for use as an adjustment control about a spring-loaded "center off" position.

Rotary Switches

Rotary controls (Figure 3.9j) are selectors or adjustors for low-energy circuits and afford greater switching capacity in less volume than any other control. Switching capacity is expanded by addition of switching modules, and a major advantage is that entire operational sequences can be contained on one control. The disadvantages of rotary controls are that operating force is increased with increased switching capacity and that there is no direct access to individual positions without actuation of intermediate positions.

Builtup construction is almost universal, either enclosed or open-frame. Mounting is either to printed wiring boards or back-of-panel, using standoffs or threaded bushings and locknuts. Switch modules consist of fixed contacts mounted on round printed wiring board decks. Each deck has a movable contactor attached to a common shaft. The printed wiring affords flexibility in contact arrangement, contact spacing, total rotational travel, switching action, and mounting of electronic components for self-contained encoding.

Users commonly abuse the inherent flexibility of these controls by packing too much switching capacity into one control, based on the rationale of saving the cost of another rotary switch. Angular spacing of contacts becomes so small, and operating force so high, that actuation is awkward and imprecise. Very high contact pressures result and can mechanically damage contacts and the ac-

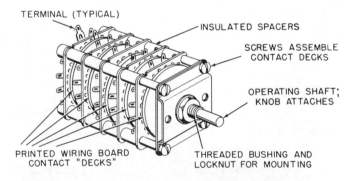

TERMINAL (TYPICAL)

INSULATED SPACERS

SCREWS ASSEMBLE CONTACT DECKS

OPERATING SHAFT; KNOB ATTACHES

PRINTED WIRING BOARD CONTACT "DECKS"

THREADED BUSHING AND LOCKNUT FOR MOUNTING

OPEN FRAME

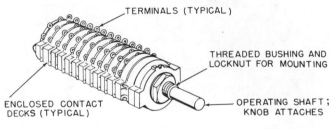

TERMINALS (TYPICAL)

THREADED BUSHING AND LOCKNUT FOR MOUNTING

ENCLOSED CONTACT DECKS (TYPICAL)

OPERATING SHAFT; KNOB ATTACHES

ENCLOSED FRAME

Fig. 3.9j   Rotary selectors

tuator mechanism. Considerable panel area is required to label a large number of positions. Large-diameter handles, although they decrease force, also consume panel area. The other extreme of excessively wide contact spacing causes awkward wrist and arm movements.

The guidelines generally accepted include:

1. Minimum angluar spacing = 15°
2. Maximum angular spacing = 90°
3. Minimum positions = 4 in 360°
4. Better minimum = 4 in 240°
5. Maximum positions = 24 in 360°
6. Better maximum = 20 in 300°

Thumbwheel Switches

Thumbwheel switches (Figure 3.9k) are a form of rotary selector for encoded data entry to low-energy electronic circuits. Encoding is accomplished by internally mounted electronic components. Thumbwheels take up very little front-of-panel area owing to parallel arrangement of contact disk, moving contactor and handle, and self-contained indicator dials. Contact disks can be stacked horizontally to create controls of any size. Thus, thumb-

wheel switches can do the job of several rotaries, pushbuttons, or toggles and use considerably less panel space. A major disadvantage of their use is that considerable concentration and dexterity are required for efficient operation. Also, (1) actuator motion is usually opposite dial motion, (2) handles are very small and must be poked at with finger tips; (3) incremental spacing on dials is small; and (4) horizontal spacing between handles is small.

Application and Selection Considerations

The goal of the selection (Figure 3.9l) is to provide the most direct link between operator and process by considering (1) what is being controlled—the process; (2) how it is being controlled—the control system; (3) how the control is used—human factors; and (4) the service conditions—electrical, mechanical, and environmental. Process response to control action determines the operator's subsequent action. Important variables include time response, amplitude response, linearity, and damping.

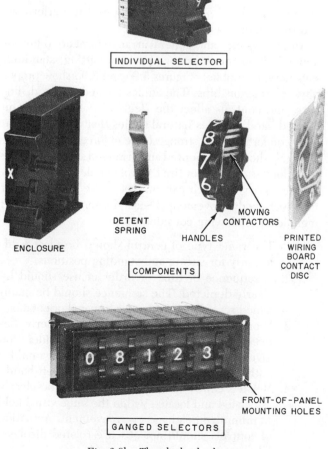

Fig. 3.9k Thumbwheel selector

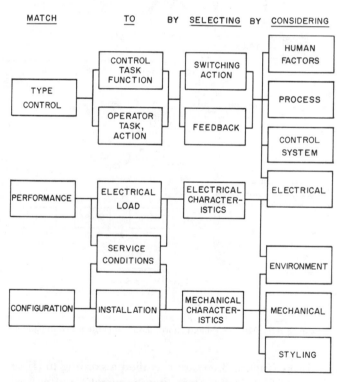

Fig. 3.9l Selection procedure

Control system translation separates the process from the operator's control action and the operator from the process response. Important factors affecting selection of manual controls involve (1) type of control—analog, digital; (2) resolution—proportional, derivative, integral, and combinations; (3) accuracy—physical losses through transmitters; (4) data processing errors—chopoff and roundoff; (5) visual feedback—display resolution, speed, accuracy; and (6) safety provisions.

Human Factors

Basic to the human factors is mind-eye-hand coordination. Figure 3.9m depicts the role of this neural aptitude in manipulating controls, and its efficiency depends on how well the controls are matched to (1) the control task and actions required to perform it, (2) the work station design, and (3) the displays associated with the controls.

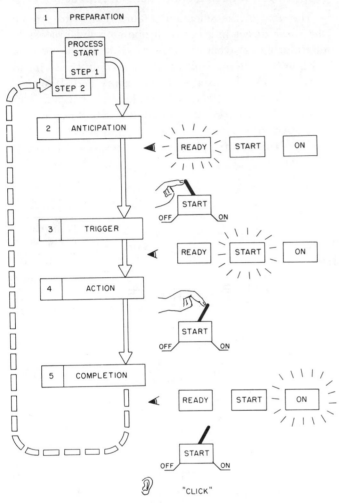

Fig. 3.9m   Mind-eye-hand coordination in control manipulation

Tasks (Figure 3.9n) are classified according to (1) selection of alternates: start, stop, automatic and manual; (2) adjustment: gross, fine, fast and slow; (3) tracking: adjustment about a moving reference; and (4) data entry: setting values, loop address, and loading computer registers.

Actions (see Figure 3.9n) required to perform these tasks are (1) sequential or simultaneous; (2) continuous or intermittent; (3) exact or approximate; (4) repetitive or different each time; (5) frequent or infrequent; (6) single-pulse or hold-down; and (7) normal or emergency.

The work station design objective is to match the physical characteristics of controls and displays to the nature

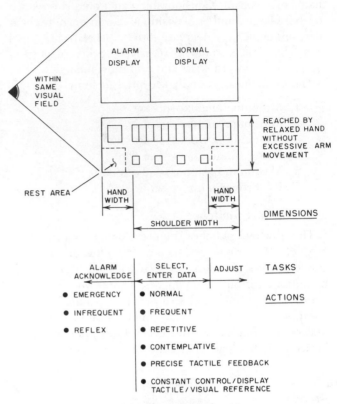

Fig. 3.9n   Example work station

of the control task, the actions required to perform it, and the operator's capabilities.

Anthropometry and general arrangement are different depending on the operator's position (sitting, standing, stationary, or mobile). Figures 3.9o and 3.9p show proper physical relationships. The angles formed by hand, fingers, and controls affect the "feel" of the control. The relaxed hand assumes natural angles that determine the most comfortable and strongest line of force (Figure 3.9q). Controls should be oriented and arranged so as to achieve these angles over the entire control panel. Compromises are necessary to reach extremities of the panel without inordinate body movement. Detailed arrangement should include the following considerations:

1. The same type of control should be placed differently for sitting and standing positions.
2. A sequence of time or order of use should be clearly depicted. The sequence should be maintained regardless of operator or panel orientation.
3. Workload should be distributed properly between left and right hands with consideration given to the fact that most people are nimbler with their right hand than with their left hand.
4. Related controls and displays should be clearly associated and located within the same visual field to minimize head and eye movement. Actuation of controls should not obscure related displays, labeling, or lighting.

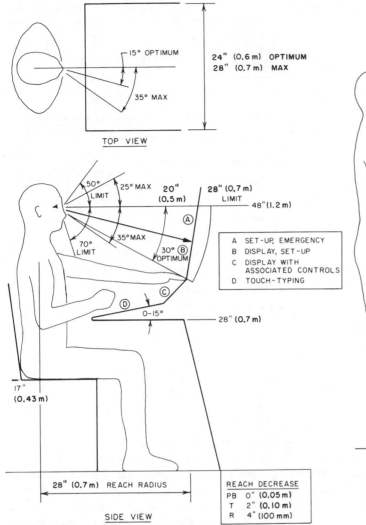

Fig. 3.9o   Sitting operating position

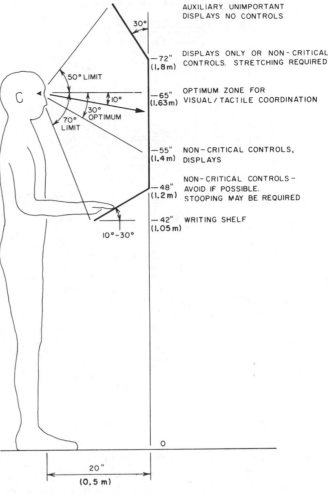

Fig. 3.9p   Standing operating position

5. Controls should be located below or to the left of the display for left-hand operation and below or to the right for right-hand operation.
6. Similar controls performing similar functions should operate in similar directions; similar controls performing dissimilar functions should operate in distinctly different directions.
7. Critical controls may require safeguards against accidental actuation by clothing, general body contact, and falling objects.

Display Movement

Control and display movement should be coordinated for direction, rate, accuracy, resolution, ratio, and linearity. Certain natural relationships are required by human habit patterns and reflexes (Figure 3.9r). Small, precise display adjustments require long-travel, low-force control motion. Short travel and low force should be used for fast displays. Pulse controls are well suited to incremental displays. Analog displays are tracked or adjusted

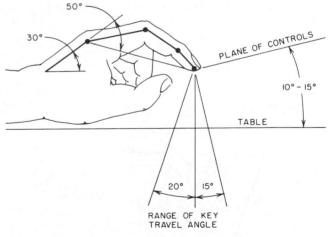

Fig. 3.9q   Natural hand and finger angles

most efficiently by a rotary potentiometer or by a control with maintained "on" positions on either side of a spring-return center "off." Efficient identification of the control, its function, and its use is a keystone of efficient operation. Considerable care is required to achieve this seem-

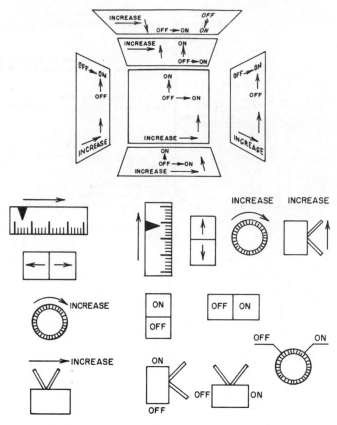

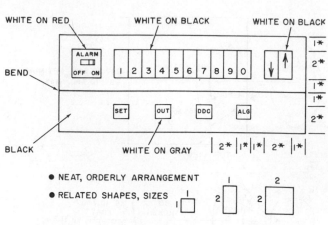

Fig. 3.9t Styling

- NEAT, ORDERLY ARRANGEMENT
- RELATED SHAPES, SIZES
- MODULAR SPACING *
- RELATED COLORS – BLACK, GRAY
- KEY GROUPING BY SPACING, COLOR, SHAPE
- LABELS CONCISE, UNAMBIGUOUS, CONSISTENT
- LABEL COLORS UNIFORM, HIGH CONTRAST
- SPECIAL COLOR CODING FOR HABIT – PATTERN
  REFLEX – RED = ALARM

Fig. 3.9r Natural direction of motion relationships

ingly commonplace goal. It is important to recognize that most identification techniques involve double duty and that redundancy of controls can both confuse the operator and lend a "busy" appearance to the panel (Figure 3.9s). Control design should blend with the overall styling theme of the work station. Handles and mounting hardware are

available in a variety of shapes, sizes, textures, materials, labeling, and lighting (Figure 3.9t).

Error Prevention

Error prevention techniques are valuable in order to forestall simultaneous and missed actuations. Both activities are especially prevalent in repetitive, high-speed keyboard operation and can be traced to speed limits imposed by switching mechanisms, actuator mechanisms, and scanning speed of external monitoring equipment.

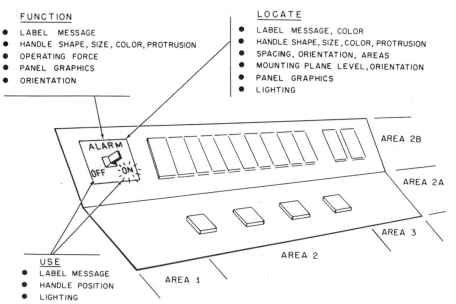

FUNCTION
- LABEL MESSAGE
- HANDLE SHAPE, SIZE, COLOR, PROTRUSION
- OPERATING FORCE
- PANEL GRAPHICS
- ORIENTATION

LOCATE
- LABEL MESSAGE, COLOR
- HANDLE SHAPE, SIZE, COLOR, PROTRUSION
- SPACING, ORIENTATION, AREAS
- MOUNTING PLANE LEVEL, ORIENTATION
- PANEL GRAPHICS
- LIGHTING

USE
- LABEL MESSAGE
- HANDLE POSITION
- LIGHTING

LABELING GUIDELINES
- TOO MUCH INFORMATION IS AS USELESS AS TOO LITTLE
- BE SPECIFIC, CONCISE, CLEAR, BUT NOT ELABORATE
- AVOID "ENGINEERING LANGUAGE," IDENTIFY WHAT IS BEING CONTROLLED NOT THE NAME OF THE CONTROL
- LABELS SHOULD BE ON CONTROLS WHENEVER POSSIBLE. WHEN ON PANEL, ASSOCIATION WITH IDENTIFIED CONTROL SHOULD BE OBVIOUS.
- ADJACENT LABELS SHOULD NOT CREATE ADDITIONAL "FALSE" LABEL
- PLACE LABELS CONSISTENTLY ABOVE OR BELOW IDENTIFIED CONTROL
- ORIENT LABELS HORIZONTAL TO READING LINE-OF-SIGHT, AVOID VERTICAL, SLANTED, OR CURVED ORIENTATIONS
- USE CAPITAL LETTERS, CORRELATE SIZE WITH VIEWING DISTANCE (REFER TO SECTION 3.4)
- COLORS OF LABELS SHOULD PROVIDE MAXIMUM CONTRAST WITH LABEL BACKGROUND

Fig. 3.9s Identification

Mechanical interlock inhibits two types of simultaneous actuations so that (1) two keys cannot be simultaneously depressed to the switching point and (2) a depressed key must be released before another can be depressed.

Electronic interlocks artificially increase the actuation speeds by delaying scanning of consecutive key outputs. "Two-key roll over" generates codes during depression of one key and release of another by blocking the second until the first is released. "One-character memory" stores a character code until another is generated, erasing the first.

Tactile feedback in the form of a properly designed force-travel relationship is the best guarantee against missed actuations. Skilled operators develop a rhythm of depressing keys. They keep several steps ahead of their actual manipulation, counting on a change in their rhythm as a signal to stop. This change is usually the absence of a completion cue for an intended actuation. Completion is signified by a sudden increase in operating force near or at the end of travel. Supplementary audible or visual cues also occur at this point. The force-travel relationship shown in Figure 3.9u is generally regarded as desirable, is available in solenoid-assisted electric keyboards, and is closely approximated by some manual keyboard controls. However, the curve shown in Figure 3.9v is the most economical (a simple linear spring) and therefore the most available. Tests have shown that it is a primary source of missed keystrokes in high-speed data entry.

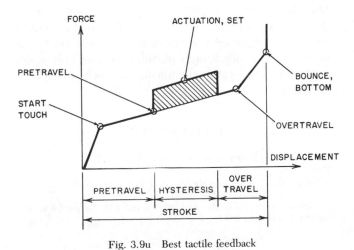

Fig. 3.9u Best tactile feedback

Environment variables affecting manipulation of controls include temperature, relative humidity, ambient illumination, and noise. Sweating may limit the amount of force the operator can apply, degrade dexterity, and cause fingers to slip off handles. Excessive light can mask lighted buttons, causing errors and loss of time and reducing safety. Noise detracts from concentration and may preclude audible feedback.

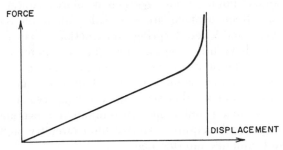

Fig. 3.9v Force-travel relationships with linear spring feedback

Electrical Rating and Performance

Electrical performance of the control is determined by current and voltage ratings, contact bounce, switching sequence, and electrical interference from other equipment. Current rating measures resistance to welding of contacts. Underrated contacts either fail to open or do not open the required number of times during a specified interval. Current ratings are determined by contact configuration, plating materials, and transfer time. Selection should account for potential overloads, which can exceed normal currents by as much as 30 times. Common sources of overload are (1) inrush starting currents of motors and tungsten-filament lamps; (2) relays and solenoid; (3) equipment malfunctions, such as short circuits and unstable power supplies; and (4) lightning.

Voltage rating reflects resistance to arcing between contacts or terminals owing to voltage surges. Ratings are determined by air gap spacing between contacts and between terminals and the insulation resistance of their mounting materials. Terminal spacing can cause a problem on subminiature switches, requiring supplementary insulating strips between terminals. Low insulation resistance can cause voltage leakage to ground or excessive dissipation within the dielectric itself. High-impedance analog computing circuits are especially sensitive to these phenomena. Certain insulating materials support formation of conductive surface films by repeated arcing, leading to eventual flashover between contacts or terminals.

Alternating current resistive loads are easily interrupted because the cyclical current reversals prevent arcing caused by continuous current buildup across the contacts. However, high AC frequencies (400 Hz) can approach DC in this respect. Direct current ratings are usually lower than AC ratings, and DC arcs tend to sustain themselves, since the applied voltage does not reverse as it does with AC. Special arc suppression techniques are sometimes required and are incorporated either within (1) the switch using surge cavities, vents, or magnets whose field opposes that generated by the arcing current, or within (2) external circuitry using resistors, capacitors, and rectifiers to reduce or surmount the arcing current.

Contact bounce occurs partially in all mechanical contactors. Reed contacts are especially vulnerable owing to their cantilevered spring construction. Snap-action contacts have little bounce, since the overcenter spring provides damping. Contactless switches, of course, are free from bounce. Bounce is equivalent to a sequence of contact openings and closings, the consequences of which are (1) reduced contact life; (2) generation of extraneous signals, which requires external filter circuitry; and (3) radio frequency interference.

The switching sequence should be chosen so as to minimize the circuit complexities that can occur during contact transfer. Make-before-break sequences furnish momentary contact overlaps, useful maneuvers when attempting simultaneous switching of several fast, low-inductance circuits with a multi-pole switch.

Mechanical Features

Mechanical features of electrical switches and of parent equipment that affect work station design include (1) type of mounting (front-of-panel, back-of-panel, printed wiring board, individual, in groups, and vibration isolation); (2) front-of-panel dimensions (spacing and size of openings); (3) back-of-panel dimensions (spacing and access); (4) access requirements (repair, replacement, maintenance and rearrangement of lamps, handles, and wiring); and (5) wiring (terminal configuration and location). Special precautions are occasionally required to preclude undesirable interaction between controls and parent equipment. Communications equipment operating at radio frequencies requires shielding from electrical noise generated by contact bounce. Reed switches must be shielded from heavy magnetic ferric materials (steel) and from magnetic fields stronger than their actuating magnet.

Environmental Considerations

Environmental factors pertinent to process control application include (1) temperature; (2) relative humidity; (3) contamination from gases, liquids, and solids; (4) barometric pressure; and (5) vibration and shock. Most commercial and industrial electrical switches function within specification in environments that the operator can also withstand. However, infrequently attended controls may be exposed to much more severe operating conditions. Electrical codes sometimes require low-energy circuits in hazardous locations, a circumstance that may necessitate intermediary isolation circuitry between control and controlled equipment, or sufficient air gap spacing to ensure potential differences below spark-generation levels.

The contact temperature is the sum of the room temperature, the rise within the parent equipment, and the rise within the switch. High temperatures accelerate contact corrosion. Unequal expansion of components can cause cracking and binding. Forced cooling may be required for densely spaced lighted pushbuttons that operate continuously.

Relative humidity and temperature cycling produce condensation. Moisture accelerates contact corrosion (decreasing current rating) and decreases insulation resistance (decreasing voltage rating). Contamination from moisture, salts, oils, corrosive gases, and solid matter degrades contacts and actuator mechanisms. Contacts are protected by glass-to-metal or plastic seals and inert atmospheres. Sealing can be done at the control panel, between actuator and contact mechanism, or within the contact enclosure. Controls should remain in their original packing materials until just before installation. Special materials, such as low-sulfur-content papers, retard contact corrosion during shipping and storage.

Vibration and shock can cause contact chatter, arcing, and outright structural failure of the contact mechanism and can affect the choice of operating force, handle shape, weight, and mounting method. Vibration damping is afforded by snap-in spring mounting and the parent equipment structure. Orientation on the control panel should place handle travel at right angles to the vibration force. Barometric pressure affects the ability of air to extinguish arcs and dissipate heat. Current and voltage ratings must be reduced for use at high altitudes; hermetically sealed enclosures are occasionally needed for such atmospheric applications.

BIBLIOGRAPHY

Bylander, F.G., "Electronic Displays," McGraw Hill Book Co., New York, 1979.

McCready, A.K., "Man-Machine Interfacing for the Process Industries," *InTech*, March 1982.

National Electrical Code, Articles 380 and 430, National Fire Protection Association, 1981.

3.10 UNINTERRUPTIBLE POWER SUPPLIES (UPS)

Cost:	For the cost of batteries, alternators, chargers, and inverters, refer to Figures 3.10a, 3.10b, 3.10c, and 3.10d.
Main Features:	For classification, refer to Tables 3.10e and 3.10f.
Partial List of Suppliers:	Cyberex Inc.; Elgar Corp.; Gould Inc., Electric Power Conversion Div.; Kohler Co., Generator Div.; NIFE Inc.; Ratelco Inc.; Teledyne Inet; Topaz Inc.; Varo Inc.; Westinghouse Electric Corp., Computer and Instrumentation Div.

Electric power lines are commonly assumed to be a perfectly reliable source of constant voltage. This assumption is valid when complete source reliability, particularly on a short-term basis, is not important. Control and instrument engineers frequently make this assumption for systems that do not satisfy this criterion. Obvious long-time power outages are fairly rare in modern power systems. Notable exceptions, however, have called attention to the complete spectrum of possibilities of power failure. Voltage dips too short to be noticed by human senses can occur frequently during the lightning season on exposed suburban and rural power systems. These systems serve many industrial plants. In addition to the obvious power failures caused by weather, there are many transient variances resulting from electrical, mechanical, and human occurrences. A system designer who assumes a perfectly reliable power source is responsible for any

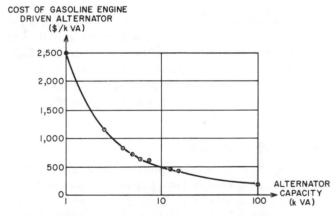

Fig. 3.10b Cost of gasoline engine–driven alternator as a function of capacity

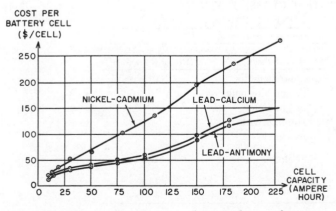

Fig. 3.10a Battery cell unit costs as a function of type and capacity

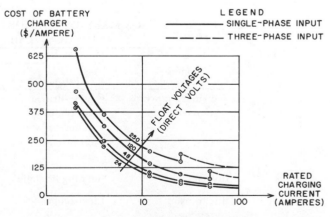

Fig. 3.10c Battery charger cost as a function of current and voltage requirements

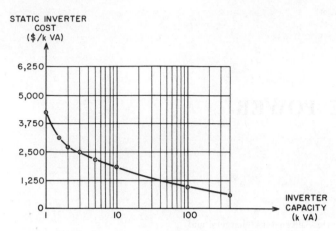

Fig. 3.10d   Static inverter cost as a function of its capacity

Table 3.10f

STANDBY SYSTEM COMPONENT REDUNDANCY

Levels of Redundancy	Input Side	Output Side
One source	Battery and battery charger	Inverter
One source with some equipment redundancy	Battery, battery charger, and rectifier	Inverter with transfer to line
Two sources	Battery, battery charger, and engine-driven generator	Inverter with transfer to alternate inverter
Two sources with some equipment redundancy	Battery, battery charger, rectifier, and engine-driven generator	Inverter with transfer to alternate inverter and then to line

loss of production and damage to machines and plant facilities resulting from power irregularities.

The trend in control system design is to use faster components operating on smaller signals. This results in increasing system sensitivity to line voltage variation and in particular to short transients. As control instrumentation technology advances, the need for standby power sources increases.

Frequently, the cause of the transients that result in control or instrument circuit complications is not the lack of voltage alone. Phase shift, a change in frequency, inadequate transient response, and noise can be equally damaging. Therefore, an adequate standby power supply system must consider all types of power failure.

Summary of Features

Standby systems can be quite complex. They involve the use of a number of components that may be of the static type, the electromechanical type, or some combination of both. The selection of the standby system and the components for it should be based upon the degree of integrity required of the application.

It is frequently possible to improve standby system reliability *and* to decrease cost. An example is that of the redundant input circuit shown in Figure 3.10g. By separating system functions, it is possible to purchase the component that does the precise job required. Overspecification of components frequently results in more expensive equipment that has less reliability.

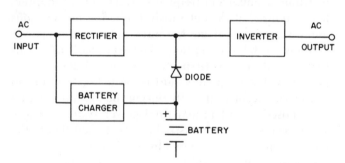

Fig. 3.10g   Standby system with battery charger redundancy

Table 3.10e

**SUMMARY OF STANDBY SYSTEM CLASSIFICATIONS
AND "MOST USED" COMPONENT PARTS**

System Component	Standby System Classification		
	Multi-Cycle	Sub-Cycle	"No-Break" (Minimum Transient)
Secondary power source	Engine-driven alternator or generator *starting* on primary source failure	Engine or motor driven alternator or generator *running* with flywheel	Battery
Inverter	Rotating or static	Rotating or static	Static
Bus transfer switches	Electromechanical	Static	Static

Much attention has been focused on the problem of failure of the incoming power line. Although this is an important consideration, it is not the only consideration. Attention should also be given to the equipment constituting the standby power supply system. Should critical pieces of the standby equipment be less reliable than the incoming power line, failure will be more frequent and little improvement will have been accomplished. Also of importance is the load circuit. If a number of load branches are connected to the output of a standby power supply system, the failure of any one load may result in the failure of the remaining loads. Proper design of the load system will minimize or eliminate this possibility.

The importance of system redundancy cannot be overemphasized. Once designed, the system must be evaluated to determine its weakest link. The cost of redundant equipment may then be assessed in light of the importance of maintaining the load.

Table 3.10e summarizes the standby system classifications and lists the most commonly used components. Obviously, special arrangements may be found that extend some of the components into an additional classification. This table is intended as a means of outlining the discussion that follows.

Various degrees of *redundancy* may be designed into the standby system. This redundancy may occur on the input side of the inverter, on the output side, or on both sides. A tabulation of components in a single- and a double-source redundant system is provided in Table 3.10f. Both input redundancy and output redundancy are included.

Power Failure Classifications

Standby power systems can be characterized according to the time it takes to achieve full output from the standby power source after failure of the primary source. This characterization by time implies that various standby systems can be discussed in terms of transfer time. This transfer time might be as long as many cycles—for large electromechanical switching devices or for engine-driven alternators or generators that start up on failure of the primary source—or as short as fractions of a cycle—for some of the solid-state switching devices or for motor-alternator or generator sets with flywheels. Additionally, no-break systems are available. Most early standby systems necessitated an interval during which there was no voltage to the load on transfer from the primary source to the standby source or on retransfer from the standby source to the primary source. As control and instrumentation circuits have become more critical, standby systems have been developed that include new techniques for transferring power sources such that essentially no transfer time occurs. For the most part, these no-break systems cost little more than those requiring a significant amount of time to transfer. In general, this character-

ization of the standby power system by transfer time is disappearing, since a large percentage of present standby systems are of the no-break variety.

Another means of characterizing standby systems is by type of failure. One's first thought is to protect the critical loads from failure of the commercial power line. A careful scrutiny into the system suggests that there are other points worthy of consideration. Among these is the failure of standby power supply components and the failure of the load.

Source Failure

A very simple standby power system is shown in Figure 3.10h. It consists of an AC power line feeding a battery charger. The battery charger, in turn, floats a battery that provides power to the inverter. The inverter provides an AC output through a distribution panel to a number of loads. Should the AC line fail, the battery charger will cease to provide the current to the inverter. The current will then be provided by the battery that is floating on the system. In this fashion, the inverter supplies the loads until such time as the AC line is re-energized, at which point the battery charger again provides the power for the inverter and for the loads and at the same time provides recharge current to the battery. Thus, the simple standby system of Figure 3.10h protects against a line failure, since there is no cessation of power to the loads when the AC line fails.

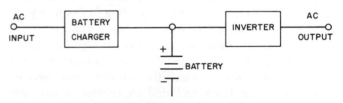

Fig. 3.10h   Basic AC standby system

A suggested improvement in this basic system is shown in Figure 3.10g. Because the functions of the battery charger are separated into (1) supplying steady-state running current to the inverter and (2) supplying recharge current to the battery, two rectifiers can be used. An unregulated rectifier is adequate to supply load current by means of the inverter. The battery charger rectifier must be regulated to ensure long battery life. Since the unregulated supply is less likely to fail, some additional reliability is gained. This can be seen when one considers the results of failure of the battery charger. In the circuit of Figure 3.10h, should the battery charger fail, the system is no longer operable after the energy stored in the battery is consumed by the load. In the case of Figure 3.10g, however, should the battery charger fail, the system continues to function as long as the AC input is available. Should the AC input source fail, the system will continue to operate until the energy stored in the battery is consumed. If, prior to this time, the AC input source is restored, the system continues to function properly but the battery charger is not capable of recharging the battery. While the system continues to operate properly, the battery charger may be repaired if it is possible to do so between the time that the failure of the battery charger is noticed and the next failure of the AC input source. Indeed, if nothing else can be done, it is generally possible to bypass the diode with some available resistance (even a light bulb) that will restore some energy to the battery. Even a small amount of battery capacity is ample for a number of short, transient outages.

In addition to the increased reliability of the two sources noted in Figure 3.10g, a lower cost for this system frequently results. This lower cost is attained because the unregulated rectifier in most cases is providing a larger current than is the battery charger. Thus, the rectifier capacity is greater than that of the battery charger. Since it is less expensive to buy unregulated power than to buy regulated power, it is possible, under many circumstances, to achieve a lower cost. This combination of lower cost and increased reliability is the optimum objective of the system designer.

The system of Figure 3.10g can be extended to include more than one AC source, as illustrated by Figure 3.10i. Alternative sources may include other AC lines and the output from engine-driven alternators, or, indeed, from any alternator regardless of the number of phases, the voltage, the frequency, or the variation in frequency. It may be desirable to "stagger" the input voltage ranges of the sources to favor one or another source. Since this "staggering" of sources results in an increase in input voltage variation over which the inverter must operate, a thyristor has been included to provide a dynamic switching of sources, minimizing the input voltage variation. The peak value of the alternative sources when rectified must be greater than the battery potential in order to "turn off" the thyristor.

The ultimate in input redundancy occurs with more than one complete power source. Shown in Figure 3.10j is a system with source, battery charger, and battery redundancy. It is also wise to separate the power feeds to inverter. This system can be extended to any desired degree.

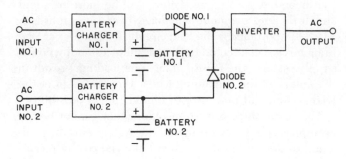

Fig. 3.10j Standby system with multiple-input redundancy

The systems of Figures 3.10g, 3.10h, 3.10i, and 3.10j offer a continuous source of power to a load without regard to the state of the AC input source as long as the standby source has sufficient energy to supply the load. The options noted provide any degree of redundancy desired for the standby power source. These figures show that this redundancy is adequate to ensure the most critical loads. The diagrams also make it clear that, should a failure occur in the inverter, the load source is no longer protected. Thus, our next concern must be the failure of equipment within the inverter block.

Equipment Failure (Inverter)

In many applications of standby power, the integrity of the line must be maintained in spite of equipment failure. Failure to preserve this integrity can result in loss of output, scrap material, plant damage, or loss of life. The degree of the protection necessary depends on the damage that can result. Process control computers are particularly important in plant operations because failure of the computer system results in an uncontrolled process. Loss may be sufficiently high to justify greater system redundancy.

The simplest form of output redundancy is illustrated by Figure 3.10k. A bypass switch is provided from the output of the inverter to the AC input line. In this dia-

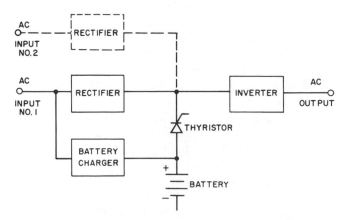

Fig. 3.10i Standby system with multiple AC inputs

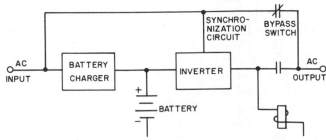

Fig. 3.10k Output redundancy to AC input

gram, and in many to follow, the symbol for an electromechanical switch (relay) will be used. This symbol should be construed to include both static and electromechanical devices.

Two items are essential in the operation of the circuit of Figure 3.10k: a synchronization circuit and a means of sensing source failure. A synchronization circuit has been added to ensure that both the AC input and the inverter are in phase in order to minimize the switching transient. The switching device or devices have also been added together with appropriate sensing circuitry.

At this point, it is easy to gloss over an essential discussion masked by the obviousness of the preceding remark. Consider for a moment the fact that both voltage waveforms must be of the same frequency and in phase. On failure of the input AC line, no transfer occurs, since the output of the inverter is not impaired. In the event of a loss of output from the inverter, transfer to the AC source takes place. Retransfer to the inverter can occur when the inverter output is re-established and when synchronization of the outputs has been restored. The retransfer may occasion an output voltage transient, as well as the transfer, since stored energy in the filter or the inertia of the rotating alternator used in the inverter requires some time to bring up to full output current.

Note that the addition of the line synchronization capability has increased the number of components in the inverter, thereby decreasing its reliability. The synchronization circuit must be carefully designed to eliminate any AC line transients that may cause a failure in inverter output.

The point of detection of inverter output failure is important. Sensing as early in the circuit as possible provides a better transfer, since energy stored in the filter may be used to reduce the transient. Early detection of thyristor failure or of abnormal vibration is preferable to the simple detection of reduced output voltage.

Difficulties can occur in providing a sensing circuit that operates on an adjustable reduced output voltage level. If no delay is built into the sensing circuit, transfer can occur on line transients, causing frequent operation. If the normal delay is included in the retransfer circuit, no system redundancy occurs in the interval defined by the time of transfer and the delay before the retransfer.

In order to minimize transfers on simple line transients, which may be caused by sudden load demands, an integrating circuit may be inserted in the voltage level sensor. Although this eliminates the transfer resulting from short transients, it causes a greater output transient when a transfer is made because of equipment failure.

The difficulties noted strongly suggest that consideration be given to early failure detection. The importance of this portion of the standby power supply system cannot be overemphasized. In many cases it is wise to have a double sensor system that will provide a delay either on failure of a component such as a thyristor or on failure

of a bearing resulting in abnormal vibration as well as a sensor circuit using an integrating circuit and detecting the reduction of the output voltage.

Input and output redundancy is shown in Figure 3.10l. Any of the input redundant schemes could be used.

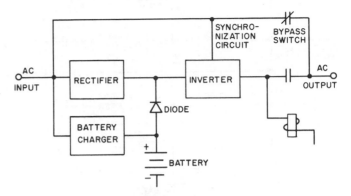

Fig. 3.10l   Input and output redundancy

With AC input redundancy, the integrity of the AC line must be considered. Naturally, if frequent line disturbances occur, little is achieved by such an arrangement. The only gain is the possibility of not having a line failure until something can be done to re-establish source redundancy.

The degree of redundancy can be improved if a back-up inverter standby system is provided. Figure 3.10m illustrates this emergency power supply backing up an emergency power supply but with dual loads. Here each inverter is capable of handling the full output capacity (both loads). On failure of either, the remaining unit assumes the full load. Momentary paralleling is possible to use the stored energy in the filter of the unit going off. As noted previously, this energy can help to minimize the transfer transient.

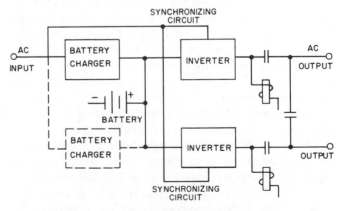

Fig. 3.10m   Redundancy with dual loads

Complete input and output redundancy are diagrammed in Figure 3.10n. Battery, battery charger, rectifier, and inverters are repeated. *Each* is sized to handle the full load. Further redundancy is provided by the AC

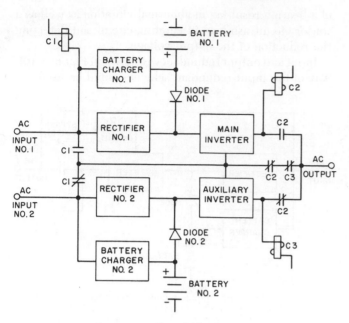

Fig. 3.10n   Three-fold redundancy

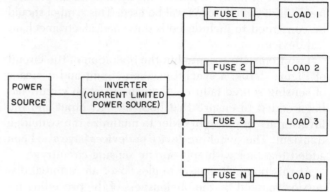

Fig. 3.10o   Inverter system supplying N fused load branches

line. If the main set fails, the auxiliary set supplies the load. If it, in turn, fails before the main is repaired, the load is supplied by the line.

Common Bus Branch (Load) Failure

In addition to providing a means of transfer from the output of an inverter that has failed to an alternate power source, the transfer switch provides a means of clearing branch circuit fuses sufficiently fast to protect other loads from a faulted load. The inverter is frequently current-limited to provide a finite overload capability. Thus, on failure of one load, the current limit provides a known amount of current for opening the fuse in the faulted branch. By referring to available fuse characteristics one derives the information that a number of branches are required with even the fastest of available fuses in order to provide load clearing within one half cycle. By using a transfer switch, which allows the transfer of the load from the inverter output to the AC line, one can, in effect, increase the short circuit current capability, thus providing a means of rapidly clearing the fuse in the branch circuit. After clearing the faulted branch, one can retransfer back to the inverter standby systems.

Ideally, a short circuit in one fused branch of an n-branch load, such as in Figure 3.10o, would have no effect upon the power supplied to the remaining branches. In reality, however, a fuse requires a finite amount of time to clear. If the power source is current-limited, the supply voltage will drop to a value near zero until the fault is cleared. Unless rapid opening of the fuse occurs, other loads will become inoperable.

Assuming, for any given supply output capacity, that all branches of an n-branch load consume nearly equal parts of the supply power, it follows that the larger the

number of load branches, the smaller the average branch fuse rating. A 10-KVA power source, for example, may have five equally loaded (2 KVA) branches. Another 10-KVA supply may have ten equally loaded (1 KVA) branches. In the latter case, the average branch fuse has approximately one half the current rating of those in the former case.

In general, the larger the current capacity of a power source, the shorter the time required to clear a fuse in a short-circuited branch. Looking at the same relationship in a slightly different way, we can see that the smaller the fuse rating of a short-circuited branch, the less time required to clear it—assuming the short-circuited supply capacity remains constant.

The preceding two paragraphs lead to the generalization that the smaller the fraction of total supply capacity carried by a fused load branch, the smaller the time required to clear a branch fuse and the less serious the load power disturbance. Conversely, the larger the fraction of total capacity carried by each branch, the larger the fuses in each and the longer the time required to clear the fault.

Figure 3.10p demonstrates this generalization. Assuming that (1) all branches consume equal fractions of load

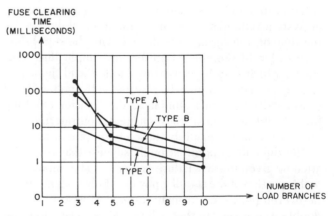

Fig. 3.10p   Fuse clearing time as a function of the number of load branches

power, (2) the source always current-limits at 15 amperes, and (3) the short circuit load impedance is zero, fuse clearing time is shown as a function of the number of load branches. The graph illustrates these data for three different fuse "speeds."

A well-designed standby power supply system requires that consideration be given to all types of failure. The most often neglected area of consideration is the load bus system. Selection of the type of branch circuit protector must be coordinated with the short circuit characteristic of the inverter as well as the requirements of the loads.

System Components

It is frequently true that the characterization of the system itself is dependent on the components available for use in the system. For each system function there are a number of components from which to choose. Arbitrary selection of any single component without regard to the others can result in an unworkable or, at best, an inefficient system. Thus, the designer's problem is to choose a compatible set of components that will satisfy his requirements. Specifically, the designer seeks the optimum compatible set of components. In order to select the appropriate component for a given function, it is necessary to understand the characteristics of the components from which one must select.

Rotating Equipment

Rotating equipment may be subdivided into two general classes. The first class includes all devices operating from a source of electric power. This set includes motors and, because of their intimate relationship, alternators and generators. The other general category of rotating equipment includes those devices driven by engines that have the ability to operate from liquid or gaseous fuels.

The preceding discussion has not covered the nature of the equipment in the blocks. As an example, the battery charger could be a motor generator set with appropriate controls. The inverter could, of course, be a DC motor driving an alternator. The selection of these components is determined by economic considerations. The economics involve not only the initial cost, operating cost, and maintenance cost of the equipment itself but also an evaluation of the need for reliability based on the importance of the load. If load failure results in a vacant lot characterized by a hole or by the need to repipe a plant because of the solidification of material that under normal circumstances would have been a usable product, the greatest possible reliability should be built into the standby power source. On the other hand, if the loss of power results in some annoyance but not in the loss of plant capacity or deterioration of product quality, then the ultimate in redundancy and reliability is not warranted.

Particular note should be taken of standby power supply systems using engines. The engine has been proved to be a reliable device in many situations and in very adverse environments. Unfortunately, the engine itself is not the most frequent cause of failure. Most complaints of poor reliability of engine-driven sets can be traced to unreliable but necessary peripheral equipment, such as fuel pumps, cooling systems, and so forth. Care should therefore be exercised in the specification of all the engine system components.

Two typical standby power supply systems involving motors, generators, and alternators are shown in Figures 3.10q and 3.10r. In Figure 3.10q, the AC line provides power until it fails. On failure, the battery supplies power to the alternator, which, in turn, supplies the output voltage. In Figure 3.10r, the AC motor has been replaced by a rectifier of the static variety. Now referring to Figure 3.10r and to Figure 3.10h, it is easy to see that the two are identical, with the motor-alternator set replacing the block marked "inverter." A system involving an engine is shown in Figure 3.10s.

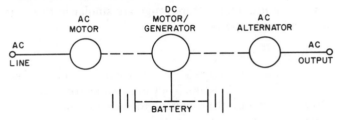

Fig. 3.10q Standby power supply system with motors, generators, and alternators

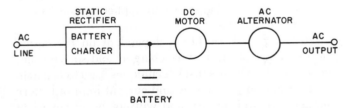

Fig. 3.10r Standby power supply system with static rectifier

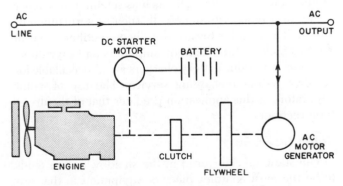

Fig. 3.10s Standby power supply system with engine

Batteries

There are three types of batteries in general use for standby systems. They are the lead antimony, lead calcium, and nickel cadmium. The lead antimony and lead calcium are lead acid batteries deriving their name from the hardening material in the lead alloy. Both have approximately the same ampere-hour characteristics on discharge. The lead antimony construction costs less, requires more maintenance because of its higher internal losses, and evolves more hydrogen than the lead calcium. The life expectancies of 14 to 30 years are frequently quoted. The life depends on the construction of the plate and the plate thickness. It also depends in large measure on the care given the batteries in service. Both types can operate over a temperature range of $-10°$ to $110°F$ ($-23°$ to $43°C$). The lead calcium cell may be floated at 2.25 volts per cell without the necessity for equalizing charge or, at worst, with long periods between equalizations.

Nickel cadmium batteries are the alkaline type. They differ from lead acid batteries in that they have a larger short-time current capability, higher cost, and lower volts per cell. Little hydrogen is generated by this cell, and frequent overcharge is recommended. Life expectancy and operating temperature range are similar to those of the lead acid types.

Battery Chargers

The battery chargers generally used for standby systems are "float" chargers. They are characterized by relatively constant output voltage to recharge the battery, because their output current varies from almost zero to rated value. Beyond their rated current, output voltage drops rapidly with increased load current. This current limit protects the charger when applied to the battery in the discharge state.

Satisfactory battery life is dependent on the design and operation of the battery charger, and so are maintenance costs. The feedback techniques used to maintain the constant output voltage and current limit of battery chargers are well known and will not be discussed in detail. The rectifiers themselves may take the form of either a polycrystalline cell, such as selenium or copper oxide, or a monocrystalline cell, such as germanium or silicon. The trend is toward the silicon rectifier. Control devices include the magnetic amplifiers and thyristors.

Motor-generator battery chargers are also available for standby system recharging service. The use of motor generators in this application predates that of the drive-type rectifiers.

Static Inverters

The static inverter used in the standby system tends to be the most complex piece of equipment in that system. Since static inverters are relatively new, a very brief discussion will be provided to show how they operate.

Transistor inverters are the least expensive of the static inverters. Their principal area of usefulness is the low input voltage (24 direct volts and less) and low output capacity (500 volt-amperes and less) range. This type of inverter can operate at high frequency and can cease operation under dangerously high output current overloads.

Figure 3.10t shows a typical circuit for the center-tap transistor inverter; N_1 and N_2 are feedback windings, R_1 is the feedback resistor, and R_2 is the starting resistor. The operating cycle can be traced by assuming Q_1 closed and Q_2 open. Substantially all the supply voltage E appears across N_3, causing a change in flux level by Faraday's law:

$$\Delta\phi = \frac{10^8 Et}{N_3} \qquad 3.10(1)$$

where

$\Delta\phi$ = change in flux level,
N = turns in N_3 and N_4 coil (identical),
E = supply voltage (appearing across N_3), and
t = time.

Eventually the core saturates, requiring an increase in exciting current. To supply this increased collector current, the transistor must have an increased base current. This cannot be supplied because of the decreased coupling between N_2 and N_3 resulting from core saturation. Q_1 begins to open, reducing exciting current. At this point, the change of flux reverses, reversing the coil voltage polarity. Q_2 is then turned on by N_1, and Q_1 is turned off by N_2. The half cycle begun in this fashion is similar to that described until a reversal occurs again, completing the cycle.

Starting resistor R_2 provides enough base bias current to allow exciting current to flow. Natural circuit imbalance ensures that only one transistor closes, thus starting the oscillator.

Should the load be short-circuited, no feedback is provided by N_1 or N_2, and oscillations cease. Other modes of operation are also possible. Care must be taken to provide the correct base current (which is dependent on transistor current gain) so that loading does not exces-

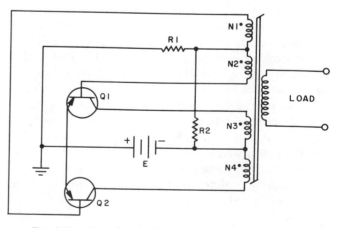

Fig. 3.10t   Typical circuit for center-tap transistor inverter

sively shift the frequency. The amount of this shift is also dependent on the "rounding" of the B-H loop.

In the center-tap inverter circuit, each transistor must withstand a voltage equal to or greater than twice the supply potential. Transistors having a sufficiently high rating to withstand a 48-volt source are more expensive. Usually, above 24 volts it is less expensive to use the bridge circuit shown in Figure 3.10u.

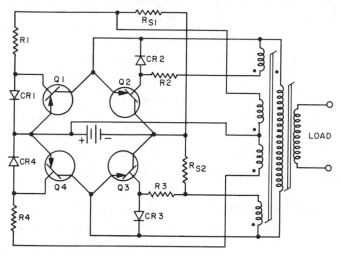

Fig. 3.10u   Bridge transistor inverter

The bridge circuit operates in a manner similar to that described for the center-tap circuit. Two starting resistors, R_{s1} and R_{s2}, are necessary. The diodes CR 1, 2, 3, 4 provide transient voltage suppression for unsymmetrically wound transformers.

A typical means of stabilizing frequency is shown in Figure 3.10v. Since the saturation flux density and turns are relatively constant, frequency is controlled by the supply voltage E. Use of a "zener" diode stabilizes this voltage and, therefore, frequency.

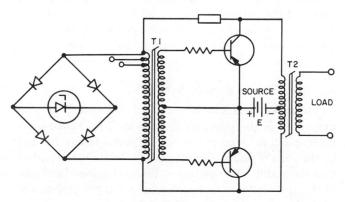

Fig. 3.10v   Typical means of stabilizing inverter frequency

Silicon-controlled rectifier-inverters are the "work horse" of static inverters. They operate efficiently and reliably at high input voltages (130 to 600 direct volts) and high output capacities (500 volt-amperes and larger).

Proper specification of the equipment is essential to obtaining reliable operation.

The operation of a static inverter may be simulated by switches, as shown in Figure 3.10w. Switches 1 and 1′ are operated in unison, as are switches 2 and 2′. When 1 and 1′ are closed and 2 and 2′ are open, load current flows in a direction shown by the arrow in Figure 3.10w. With 2 and 2′ closed and 1 and 1′ open, load current flows in the reverse direction. Thus, whereas the source current i_s flows in the same direction when either set of switches is closed, the load current i_1 reverses polarity as each set is alternately closed and opened. An inversion of current has been performed by the circuit.

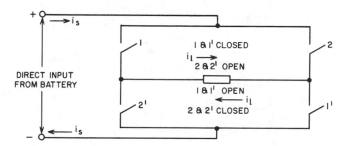

Fig. 3.10w   Simulated static inverter using mechanical switches

The switches used in Figure 3.10w are not static, since a switch contains moving parts. In Figure 3.10x, the mechanical switches have been replaced with electrical switches (silicon-controlled rectifiers). Also shown are the commutating inductors L and commutating capacitor C. These components are necessary to turn the controlled rectifiers off. Turn-on is accomplished by the application of voltage to the gate leads of the controlled rectifiers by the oscillator.

Rectifier diodes RT_1, RT_1^1, RT_2, and RT_2^1 are *not* a part of the basic inverter switching circuit. They serve to clamp the amplitude of the load voltage to a value approximately equal to the magnitude of source voltage.

Figure 3.10x shows the diagram of a bridge-connected

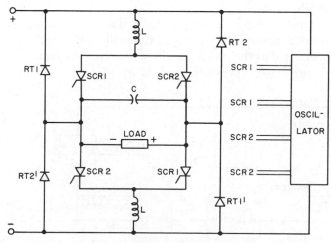

Fig. 3.10x   Bridge SCR static inverter

static inverter. This arrangement is frequently used for source voltages of 130, 260, and 600 volts. For source potentials of 12, 24, and 48 volts, the circuit of Figure 3.10y is frequently used. Its operation is seen to be similar to that of the bridge circuit. It differs in that half the number of controlled rectifiers are used, and each must hold off a voltage approximately equal to twice the supply voltage.

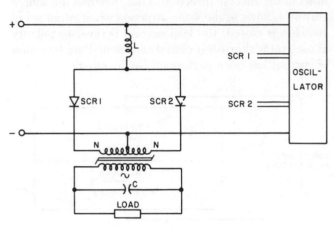

Fig. 3.10y   Center-tap SCR static inverter

Various types of output waveforms may be obtained from the square wave, which is the basic output waveform of the static inverter. Sinusoidal waveforms are most common, but triangular, sawtooth, and many rectangular combinations are also possible. Voltage stabilization may be a welcome bonus provided by the output wave-shaping circuitry. Current limiting is also possible.

Bus Transfer Switches

Bus transfer switches have historically been of the electromechanical type. Various techniques have been used to speed the transfer from one source to the other. These techniques have included a pulsing arrangement on the coil of the electromechanical switch in order to overcome the inherent inertia. "Make-before-break" sequences have also been used.

A newer development has been the use of the static switch for more rapid transfers. Generally, thyristors have been used as the power-handling switching component. Figure 3.10z represents a hybrid transfer switch on the

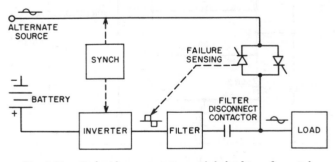

Fig. 3.10z   No-break power system with hybrid transfer switch

output side of the inverter. Momentary paralleling of sources is achieved by the "drop out" time of the electromechanical contactor. Fast turn-on is achieved by the thyristors. Sensing is performed at the output of the inverter switch prior to the filter in order to anticipate output failure.

A static set that serves the same function is shown in Figure 3.10aa. In this case, momentary paralleling is achieved by the logic circuit that supplies the gating pulses.

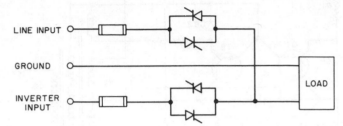

Fig. 3.10aa   Static no-break power system

Protective Components

It is essential to the proper operation of the emergency power supply system that adequate thought be given to the protective devices. Available as protective devices are fuses of various speeds and circuit breakers. The application of fuses to the load circuit was discussed earlier in this section. Although it is not possible to provide a clear definition of the components to be used for the specific applications within a system, it is necessary to urge that one pay careful attention to the selection. Inrush current on start-up, transient variation of the input voltage, and transient load variation each present particular problems. A frequent experience is that the protective system may be the least reliable of the system components. This is not intended to mean that the component itself fails, but rather that the system fails through action of the protective device under normal operating conditions. Whenever the standby power supply system fails to provide output, whether the components are damaged or not, the resulting damage is the same.

Standby Power Supply Systems

A number of standby power supply systems have been presented in the previous discussion. These have been categorized in terms of the type of power failure but not by the class of system they represent. It was noted earlier that the systems themselves could be classified by virtue of the time it takes to transfer from the primary to the secondary power source. This grouping is more nearly akin to the thinking of the purchaser than is the classification by power failure noted previously.

Multicycle Transfer System

In general, multicycle transfer systems involve either electromechanical transfer switches or engine- or motor-driven equipment that must start up. Figures 3.10bb and 3.10cc show two composite systems using both rotating and static equipment. In Figure 3.10bb both a battery input and an engine-driven generating input are provided to the inverter. *If* the engine-driven generator is started prior to the time the energy contained in the battery is completely used by the inverter and the load, *no* interrupt time occurs. Note that the commercial AC power input provides the load normally. On failure of the power lines, the contactor K_1 operates to insert the inverter. Since the inverter is started on the fallout of the contactor K_1, some start-up transient must occur in the inverter, requiring some time between the failure of the commercial AC input source and the inverter source. Note that the use of a static switch in position K_1 would increase the speed of operation of the electromechanical

contactor but would *not* materially affect the start-up of the inverter.

Note that a rearrangement of the components of Figure 3.10bb results in the system of Figure 3.10cc. This system may not have a transfer time, since the battery would normally be chosen to have a capacity sufficient to cover the energy required by the inverter and load during the time it takes to start the engine-driven alternator. Should that interval be long, multicycle start-up will result. Note, however, that in this system there is no protection for inverter failure. The system of Figure 3.10cc does have a bypass to the AC commercial line, providing some protection in the event of component failure in the system.

Subcycle Transfer System

The subcycle transfer systems are generated by the use of static switches on the output side of the inverter. Thus, the circuits of Figures 3.10k, 3.10l, 3.10m, 3.10n, 3.10aa, and 3.10bb may be subcycle transfer systems depending entirely on the arrangement of the switch itself and, in the case of Figure 3.10bb, the start-up time of the inverter.

No-Break Transfer System

The so-called no-break transfer system may be no-break in the sense that if the AC commercial source fails, no cessation in output power results. No-break may also be applied to those systems having a redundant source on the output side of the inverter. Figures 3.10g, 3.10h, 3.10i, 3.10j, and 3.10cc are examples of no-break systems with redundant input sources so that no break occurs in the output power should the AC commercial source fail. The switching systems shown in Figures 3.10z and 3.10aa can be used in many of the previously defined circuits to provide no-break switching under the right sequence of operations. Figures 3.10k, 3.10l, 3.10m, and 3.10n are of this type.

System Redundancy

The subject of system redundancy has been frequently mentioned in the previous discussions. Two basic classifications of system redundancy are made on the basis of input and output. Redundancy in the input circuit to the inverter results from the use of multiple battery sources, multiple input lines coming from separate power feeders, or various types of rotating alternators and generators with either motor or engine drives. Figures 3.10g, 3.10i, and 3.10j illustrate the various types of input redundancy.

Redundancy in the output circuit of the inverter is obtained by use of a switch that will provide a path to an alternate source. Output redundancy may involve switching from the inverter output to the power line. More complex schemes involve switching from inverter standby system to inverter standby system. Figures 3.10k,

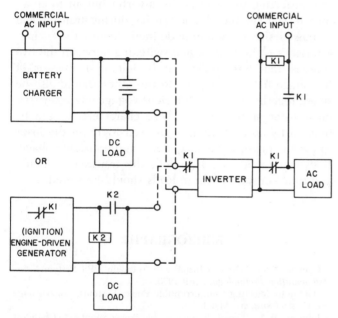

Fig. 3.10bb   Composite multi-cycle transfer system using both batteries and engine-driven generator

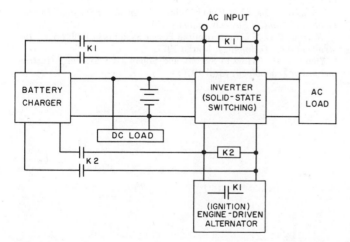

Fig. 3.10cc   Alternative arrangement of a multi-cycle transfer system

3.10l, 3.10m, and 3.10n provide examples of output redundancy.

The previous examples have shown the wide number of choices available for component redundancy in standby power supply systems. In order to select the best system for a given application, it is necessary first to evaluate the degree of integrity required for the application. It is then essential that the standby system be evaluated to determine which component is most likely to fail. A decision can then be made regarding the cost of redundancy in that area versus the needs of the application.

Specifications

Considerable activity is under way to attempt to provide specifications for standby power supply systems and system components. It is essential that the user provide in his specifications certain types of information that are important in ensuring system reliability. Some of these items will be noted.

Little need be said about the specification of the power-handling capability of the standby power supply system. Nevertheless, it is suggested that consideration be given not only to the immediate load requirements but also to future load requirements. It is generally less expensive to purchase additional capacity when the first system is acquired than to add capacity to that system at a later date.

In many cases, the characteristics of the loads are most important. Power factor as well as transient characteristics should be clearly defined. These definitions are particularly important if static equipment is involved.

Transient data on the input sources are important in proper design of the system. If a battery source is used, it is desirable to know the transients that can exist on the battery bus. These transients should be specified in terms of their maximum voltage as well as their energy content. If an existing battery installation is used, any loads that are switched will generally institute a transient voltage because of the inductance of the lines themselves. A knowledge of this transient voltage is particularly important in the proper design of static equipment.

Although it is difficult to obtain any meaningful data regarding the number and duration of outages of the utility power lines, it is necessary to provide information concerning the length of time the power supply must produce power for the loads without having the primary input source available. This evaluation can best be performed on the time required to "shut down" the load system rather than with reference to the input failure.

Particular note should be taken of the characteristics of static inverters. Three overload ratings are important. In order that sufficient commutating capacity can be designed into the inverter, it is necessary to know the maximum instantaneous current. It is also necessary to know the overload current for a one- to two-minute interval in order that ample cooling be provided to the semiconductor devices. The final overload rating of importance is the one- or two-hour overload necessary in order to provide ample thermal capacity in the magnetic components. This third overload rating is also of importance in defining rotating equipment.

In the event that the standby system includes a means of transferring the load of the inverter output to an alternate source, the characteristics during transfer and retransfer should be amply defined. Among those characteristics is the time, when switching between the two sources, that zero voltage can be tolerated. Phase shift in voltage from one source to the other should also be stated together with the transient voltage characteristics on transfer or retransfer. These characteristics can be defined by an evaluation of the sensitivity of the loads to varying phase angle, frequency, and transient voltage. If dynamic loads are included, the transient response and time constant of these loads should be stated.

BIBLIOGRAPHY

Conger, N.L., "How to Install and Maintain UPS Systems," *Instrumentation Technology*, April 1970.

"Guide to Selecting Uninterruptible Power Systems," *Instruments and Control Systems*, March 1973.

Johnson, W.J., "Use of Uninterruptible Power Supplies in Chemical Plants," ISA Conference Paper 76-835, Instrument Society of America, Houston, TX, October 1976.

"Key Items for UPS Selection," *Instruments and Control Systems*, October 1977.

Teets, R., "Don't Let Power Failures Plague Your System," *Instruments and Control Systems*, January 1977.

Warren, E.G., "Uninterruptible Power Systems in Refineries," *Instrumentation Technology*, March 1977.

Yuen, M.H., "How to Specify and Select UPS Systems," *Instrumentation Technology*, April 1970.

Chapter IV

CONTROL AND ON-OFF VALVES

J. B. Arant • H. D. Baumann

C. S. Beard • R. D. Buchanan

L. D. DiNapoli • C. E. Gayler

D. R. A. Jones • C. G. Langford

B. G. Lipták • O. P. Lovett, Jr.

F. D. Marton

CONTENTS OF CHAPTER IV

4.1 ELECTRIC, HYDRAULIC, DIGITAL, AND SOLENOID ACTUATORS

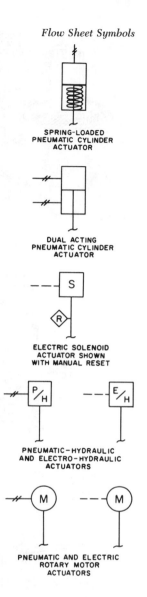

Actuator Energy Sources: Electric or electro-hydraulic

Speed Reduction Schemes: Worm gear, spur gear, or gearless

Rotary Actuator Torques: 20 in.-lb to 75,000 ft-lb (0.23 to 10,368 kg-m)

Linear Actuator Thrusts: 100 to 10,000 lb (45 to 4500 kg)

Costs: A typical rotary actuator with 300 ft-lb (41.5 kg-m) torque rating is about $1500.

Partial List of Suppliers: Automatic Switch Co.; Bailey Controls; Barber-Colman; H. Beck & Sons; G. W. Dahl Co.; Fisher Controls; Foxboro/Jordan; Honeywell; ITT General Controls; Leeds & Northrup Co.; Limitorque Corp.; Raco Intl.; Raymond Controls; Skinner Valve Div.; Staefa Control Systems; Target Rock Corp.

This section covers the broad field of electric actuators, with emphasis on the mode of speed reduction and type of output motion. Feedback elements are required so the actuator can function in a control loop. Solenoid valve operators and electro-hydraulic actuators are looked at in some detail. Digital valve operators, which are being more widely used in connection with microprocessors, are also discussed.

Electric Actuators—Rotary Output

Worm Gear Reduction

The actuator shown in Figure 4.1b is an example of the use of a double worm gear reduction to obtain output speeds of around 1 rpm with an input motor speed of 1800 rpm. The worm gear is self-locking, so it prevents the load from moving downward by backdriving the motor. However, the worm gear is less than 50 percent efficient, so more power is used compared to spur gears. Also shown is a handwheel for manual operation during power loss.

Spur Gear Reduction

The spur gear actuator in Figure 4.1c has very low power loss through the relatively efficient spur gears. Ordinarily the load could backdrive this system, but a friction member at the motor end of the gear train minimizes this undesirable action. Of course, if the load has a frictional characteristic, it will not impose a backdriving torque on the actuator.

Actuators for larger outputs use externally mounted

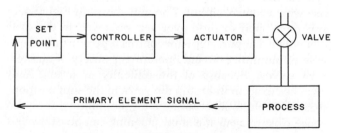

Fig. 4.1a   Typical automatic control loop, with the actuator as the end element.

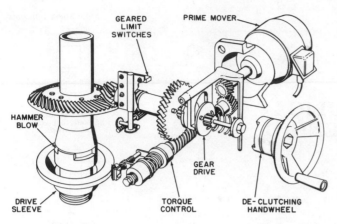

Fig. 4.1b   Electric actuator with worm gear reduction

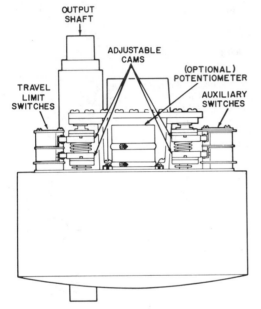

Fig. 4.1c   Spur gear reduction

motors. Two motors operating a single gear train have been used to obtain 45,000 in.-lbf (5085 N·m) of torque through 90-degree rotation in 75 sec. An actuator has been designed for accurate rotary positioning that develops 5,000 ft-lbf (6800 N·m) at stall and will rotate 90 degrees in 10 sec. SCRs energize the motor as commanded by a servo-trigger assembly housed separately.

Electrical gear may include adjustable cams to operate limit and auxiliary switches. A potentiometric feedback calibrated to the rotation of the actuator is required for use with a control circuit. The unit shown in Figure 4.1c was developed to slide over and be keyed to the shaft of a boiler damper. Actuators of this type must be adaptable to mounting on and operating a variety of quarter-turn valves. Because of the difficulty in setting limit switches to accurately stop the valve in the shut position, it is advisable to incorporate a torque-limiting device to sense closure against a stop. Opening can be controlled by a limit switch. One compact unit contains the features

noted with an output of 9,000 in.-lbf (1017 N·m) of torque. The unit is powered by a motor with a high-torque capacitor and includes a mechanical brake, feedback potentiometer, limit switches, and a declutchable handwheel.

Flex-Spline Reduction

Figure 4.1d is an example of a unique single-stage, high-reduction system. Instant breakaway and efficient transfer of prime mover power is obtained with a modified concentric planetary system consisting of a semi-flexible gear within a rigid gear. A three-lobed bearing assembly transmits power to the gears by creating a deflection wave transmission that causes a three-point mesh of the gearing teeth on 30 percent or more of the external gearing surface.

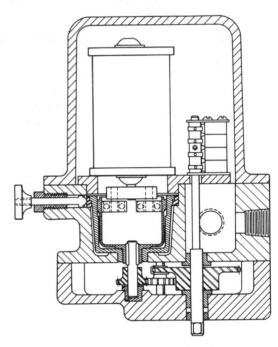

Fig. 4.1d   Flex-spline reduction

The semiflexible geared spline has fewer teeth than the nonrotating internal gear it meshes with. The spline slowly rotates as it is pressed into the larger gear by the bearings on the motor shaft.

Electric Actuators—Linear Output Solenoid

Solenoids (consisting of a soft iron core that can move within the field set up by a surrounding coil) are used extensively for moving valve stems. Although the force output of solenoids may not have many electrical or mechanical limitations, their use as valve actuators has economic and core (or stem) travel limitations, and they are expensive.

A solenoid valve consists of the valve body, a magnetic core attached to the stem and disc, and a solenoid coil (Figure 4.1e). The magnetic core moves in a tube that is closed at the top and sealed at the bottom, allowing

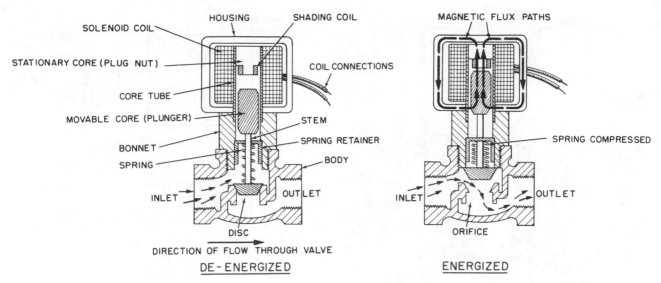

Fig. 4.1e   Direct-acting solenoid valve

the valve to be packless. A small spring assists the release and initial closing of the valve. The valve is electrically energized to open. Stronger springs are used to overcome the friction of packing when it is required (Figure 4.1f). Reversing the valve plug causes reverse action (open when de-energized). Increased stroking force is obtained by using the mechanical advantage of a lever with a strong solenoid (Figure 4.1g).

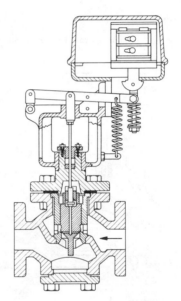

Fig. 4.1g   Solenoid valve with lever type actuator force amplifier

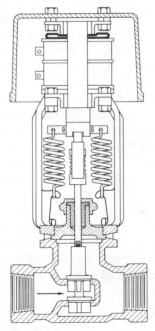

Fig. 4.1f   Solenoid valve with strong return springs

Using a solenoid to open a small pilot valve (Figure 4.1h) increases the port size and allowable pressure drop of solenoid-operated valves. Small solenoid pilot valves are widely employed to supply pressure to diaphragms or pistons for a wide range of output forces. Pilot operation applies pressure to a diaphragm or piston or may release pressure, allowing the higher upstream pressure to open the valve. A good example is the in-line valve (Figure 4.1i). Most solenoid valves are designed to be continually energized, particularly for emergency shutdown service. Thus the power output is limited to the current whose I^2R-developed heat can be readily dissipated.

Using a high source voltage and a latch-in plunger overcomes the need for continuous current. The single-pulse valve-closing solenoid is disconnected from the voltage source by a single-pulse de-latch solenoid and hence does not heat up after it is closed. A pulse to the

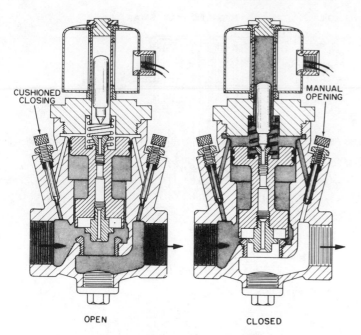

Fig. 4.1h   Pilot-operated solenoid valve

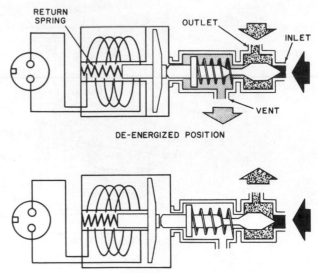

Fig. 4.1j   Normally closed three-way solenoid valve

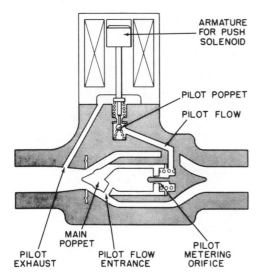

Fig. 4.1i   Pilot-operated in-line solenoid valve

de-latch solenoid permits the valve to be opened by a spring.

Three-way solenoid valves with three pipe connections and two ports are used to load or unload cylinders or diaphragm actuators (Figure 4.1j). Four-way solenoid pilot valves are used principally for controlling double-acting cylinders.

Modulating Solenoid Valves

Modulating magnetic valves (Figure 4.1k) utilize spring-loaded low-power solenoids to provide throttling action. The only moving part in this design is the valve stem, which has the valve plug attached to one end and an iron core attached to the other. The valve opening is thereby

a function of the voltage applied across the solenoid. Such throttling solenoids are frequently used in the HVAC industry and in other applications where the valve actuators do not need to be very powerful. The actuator thrust requirements are lowered by balancing the inner valve through the use of pressure equalization bellows or floating pistons. The positive failure position of these valves is provided by spring action.

Throttling solenoids are typically available in $\frac{1}{2}$ in. to 4 in. (12mm to 100mm) sizes and are also limited by the pressure difference against which they can close. They are available in two or three-way designs.

The use of throttling solenoids can completely eliminate the need for instrument air in the control loop. These valves also tend to offer a higher speed and better rangeability than their pneumatic counterparts. Some manufacturers claim a rangeability of 500:1 and a stroking time of 1 second. The design illustrated in Figure 4.1k can be driven directly from microprocessor-based building automation systems.

The design shown in Figure 4.1k also includes a separate positioner which accepts a 4-20 mA DC input from a controller and delivers a DC output signal to the throttling solenoid. Valve position feedback is obtained through the use of a linear variable differential transformer mounted directly on the valve.

Motor and Rack

Figure 4.1l uses a worm and a rack and pinion to translate horizontal shaft motor output to vertical linear motion. Maximum force output is approximately 1,500 lbf (6675 N) at about 0.1 in./min (2.5 mm/min). A continuously connected handwheel, which must rotate the rotor of the motor, can be used when there is short stem travel and relatively low force output.

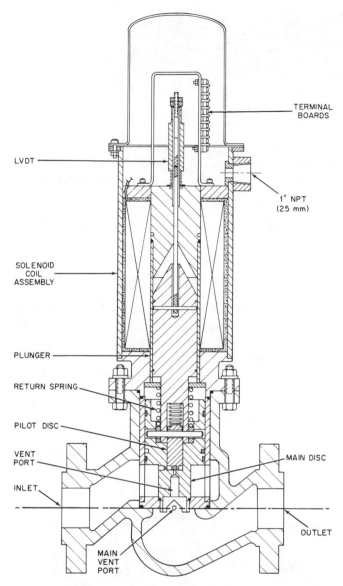

Fig. 4.1k Throttling solenoid valve with LVDT type position feedback
(*Courtesy of Target Rock Corporation*)

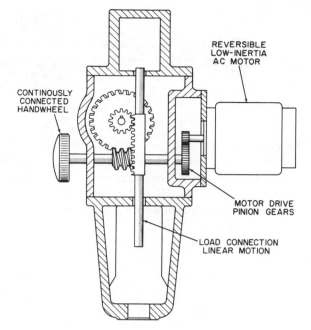

Fig. 4.1l Rack and pinion

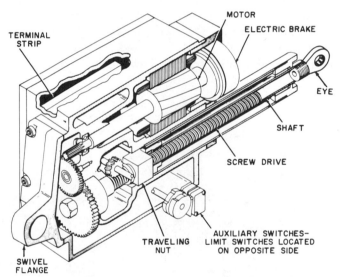

Fig. 4.1m Electric quarter-turn actuator with linear output

The actuator is designed with a conventional globe valve bonnet for ease of mounting. Units operate on 110 volts and have been adapted to proportional use with a 135-ohm Wheatstone bridge or any of the standard electronic controller outputs.

Motor and Traveling Nut

A linear unit consists of the motor, gears, and a lead screw that moves the drive shaft (Figure 4.1m). A secondary gear system rotates cams to operate limit and auxiliary switches. The unit may have a brake motor for accurate positioning and a manual handwheel. The bracket on the rear end allows the actuator to rotate on the pin of a saddle mount, so that the drive shaft can be pinned directly to the lever arm of a valve or a lever arm fixture that fits the shaft configuration of a plug cock or ball

valve. Maximum output is 1,600 lbf (7120 N) at 5 in./min (125 mm/min).

Rotating Armature

An internally threaded drive sleeve in the armature of the motor is used to obtain a linear thrust up to 6,600 lbf (29,370 N) at a rate of 10 in./min (250 mm/min). Bearings in the end cap support the drive assembly (Figure 4.1n). The drive stem is threaded to match the drive sleeve and is kept from rotating by a guide key. Thrust-limit switch assemblies are mounted in each end of the housing to locate the hollow shaft in mid-position. When the linear movement of the drive stem is restricted in

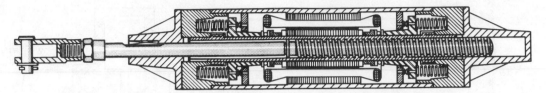

Fig. 4.1n   Electric actuator with rotating armature

either direction, the limit switch involved will operate to shut down the unit. Thermal cut-outs in the motor windings offer additional overload protection. Strokes are available from 2 to 48 in. (50 to 1200 mm). The unit has been adapted for proportional control by use of an external sensing position for feedback. For use as a valve actuator, it must be mounted so that the drive stem can be attached to the valve stem, or a suitably threaded valve stem must be supplied.

Rotary to Linear Motion

An electric proportional actuator (Figure 4.1o) is designed for continuous rotation of a drive sleeve on a ball-screw stem thread. 3000 lb of thrust (13,350 N) is obtained at a stem speed of 1 in./min (25 mm/min).

One or two DC signals are used separately or numerically added or subtracted. Triacs operate on position error to control a DC permanent magnet motor that positions a stem within an adjustable dead band. Degree of error and rate of return are sensed by a lead network to determine the direction and time that the motor must run.

Stem reversals are almost instantaneous. The back emf of the motor is used as a velocity sensor and is fed into a circuit that allows adjustment of the speed of drive sleeve rotation. Gain can be adjusted to control oscillation of the stem. Stem position feedback is by an LVDT. Use of the DC motor allows for torque control through sensing of motor current. A manual handwheel is furnished that can only be used when the unit is de-energized by a manual/automatic switch.

Application of the linear requirement of a valve necessitates a linkage for translation from rotary to linear motion. Use of a linkage (Figure 4.1p) provides a thrust for operating the valve.

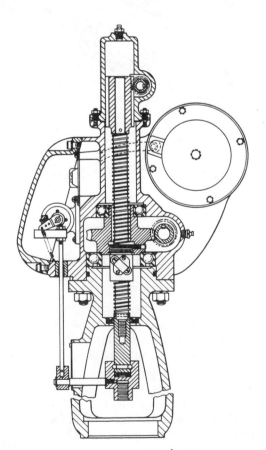

Fig. 4.1o   Rotary to linear

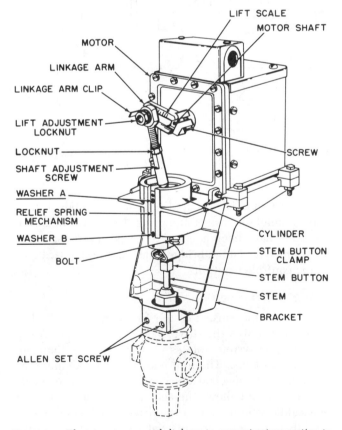

Fig. 4.1p   Electric actuator with linkage to convert rotary motion to linear

Electro-Hydraulic Actuators—External Hydraulic Source

The term "electro-hydraulic" has been applied to actuator systems in which the hydraulic pressure to one or more actuators is supplied by a hydraulic mule. The hydraulic power is supplied to the actuator by electrical control means. In the broad sense, the use of two three-way solenoids or one four-way solenoid externally mounted to the actuator constitutes an electro-hydraulic system.

More extensively, "electro-hydraulic" applies to a proportionally positioned cylinder actuator. This requires a servo-system, which is a closed loop within itself. A servo-system requires one of the standard command signals, which is usually electrical but can be pneumatic. This small signal, which often requires amplification, controls a torque motor or voice coil to position a flapper or other form of variable nozzle. This positions a spool valve or comparable device to control the hydraulic positioning of a high-pressure second-stage valve. The second-stage valve directs operating pressure to the cylinder for very accurate positioning. Closing the loop requires mechanical (Figure 4.1q) or electrical feedback to compare the piston position with the controller output signal.

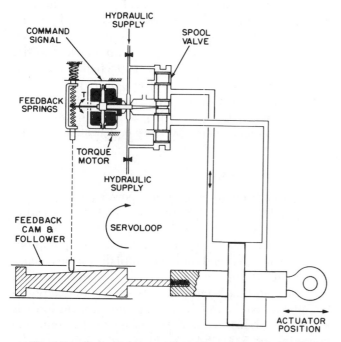

Fig. 4.1q   Two-stage servo valve with mechanical feedback

Hermetically Sealed Power Pack

A much more useful and compact type of electro-hydraulic actuator combines the electro-hydraulic power pack with the cylinder in one package. Many of these actuators are designed as a truly integral unit. An electric motor pump supplies high-pressure oil through internal ports to move the piston connected to the stem (Figure 4.1r). The small magnetic relief valve is held closed dur-

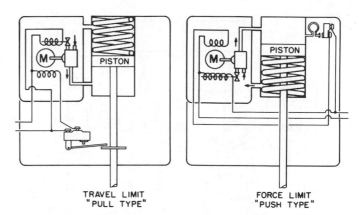

Fig. 4.1r   Hermetically sealed electro-hydraulic power pack

ing the power stroke until de-energized by an external control or emergency circuit to allow the spring to cause a "down" stroke. The same unit can be used to cause spring return to up position using a Bourdon switch to produce force limit.

Motor and Pump

REVERSIBLE MOTOR AND PUMP

A reversible motor can be used to drive a gear pump in a system to remove oil from one side of the piston and deliver it to the other side (Figure 4.1s). The check valves allow the pump to withdraw oil from the reservoir and position the directional control valve in order to pressurize the cylinder. Reversing the motor (and pump) reverses the direction. When the motor is de-energized, the system is "locked up." For proportional control, a feedback is necessary from stem position to obtain a balance with the control signal.

JET PIPE SYSTEMS

A very old control system for a cylinder, the jet pipe, is employed in an electro-hydraulic actuator. An electro-mechanical moving coil in the field of a permanent magnet is used to position a jet that can direct oil to one end or the other of the cylinder actuator (Figure 4.1t). A force balance feedback from stem position creates the balance with the controller signal.

HYDRAULIC CONTROL OF JACKETED PINCH VALVES

Controlled hydraulic positioning of a sleeve valve is obtained with a moving coil and magnet to position the pilot (Figure 4.1u), which controls pressure to the annular space of the valve. Feedback is in the form of a Bourdon tube, which senses the pressure supplied to the valve and moves the pilot valve to lock in that pressure.

MULTIPLE PUMP

The multiple pump system consists of three pumps running on the shaft of one prime mover (Figure 4.1v).

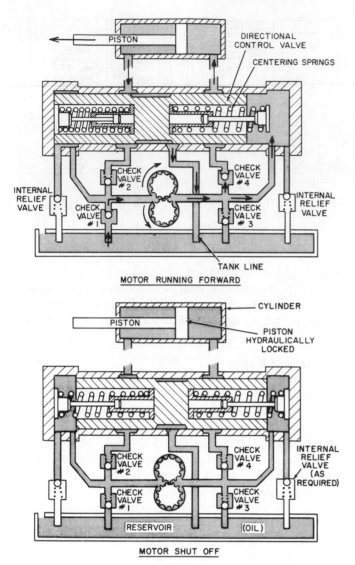

MOTOR RUNNING FORWARD

MOTOR SHUT OFF

Fig. 4.1s Pump with reversible motor and closed center

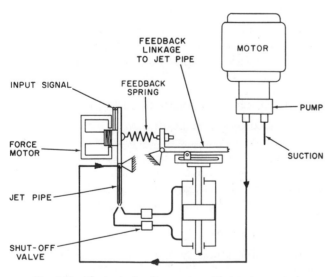

Fig. 4.1t Electro-hydraulic actuator with jet pipe control

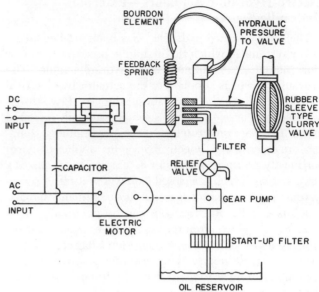

Fig. 4.1u Electro-hydraulic control of jacketed pinch valve

There is one pump for each side of the piston and one for the control circuit. The force motor tilts a flapper to expose or cover one of two control nozzles. The flow through a restricting nozzle allows pressure to be transmitted to one side of the piston or the other. Force balance feedback is created by a ramp attached to the piston shaft, which positions a cam attached to the feedback spring. Upon loss of electric power, the cylinder shutoff valves close to lock up the pressure in the cylinder and assume a status quo. A bypass valve between the cylinder chambers allows pressure equalization to make use of the manual handwheel.

VIBRATING PUMP

A vibratory pump is used in a power pack mounted on a cylinder actuator (Figure 4.1w). A 60-Hz alternating source causes a plunger to move toward the core and, upon de-energizing, a spring returns it. The pump operates on this cycle and continues until the piston reaches the end of its stroke, when a pressure switch shuts it off. The solenoid that retains the pressure is de-energized by an external control circuit. Maximum stem force is 2,500 lbf (11,125 N) at a rate of about 1-in./min (25 mm/min), or 5,000 lbf (22,250 N) at 0.3-in./min (7.5 mm/min).

TWO-CYLINDER PUMP

A two-cylinder pump, driven by a unidirectional motor, injects pressure into one end of a cylinder or the other, depending upon the positions of two solenoid relief valves (Figure 4.1x). The solenoid on the left is closed to move the piston to the right, with hydraulic pressure relieved through the other relief valve. Motion continues until the valve it is operating is seated. The buildup of cylinder pressure operates the pressure switch at a pre-

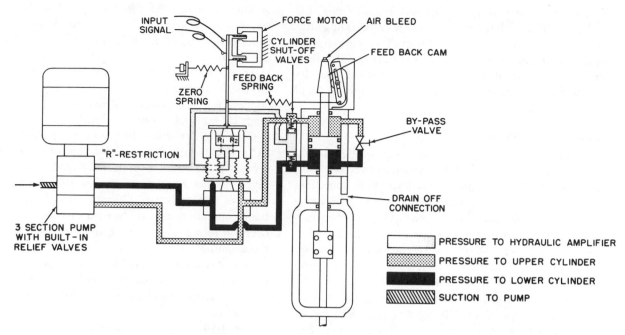

Fig. 4.1v   Multiple pump hydro-electric valve actuator

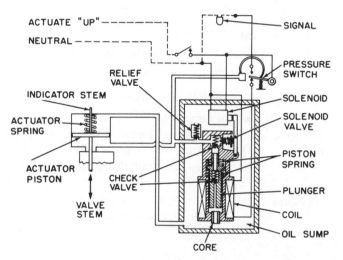

Fig. 4.1w   Electro-hydraulic actuator with vibratory pump

determined setting to de-energize the motor and both solenoid relief valves. This locks the hydraulic pressure in the cylinder. Switching of the three-way switch will start the motor and reverse the sequence of the relief valves to move the piston to the left. At full travel (which is the up position of a valve stem), a limit switch shuts down the unit. Open-centering the three-way switch at any piston position will lock the piston (and valve stem) at that point.

Remote control is accomplished by manipulating the open-center switch. Automatic control is acknowledged by including this open-center function, which may be solid state, in the control circuit. A potentiometer or LVDT that senses the stem position is required for feedback.

Stem output is 6,000 lbf (26,700 N) at a rate of 3 sec/in.

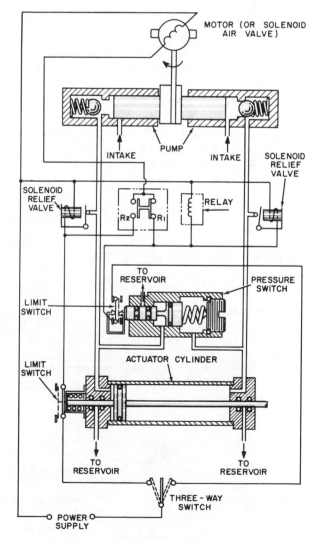

Fig. 4.1x   Two-cylinder pump type electro-hydraulic actuator

(0.12 sec/mm). Stem travels up to 7 in. (175 mm) are available, although longer travels are feasible. The entire system is designed in a very compact explosion-proof package that can be mounted on a variety of valve bonnets. The speed of response to energizing and de-energizing the control circuit makes it feasible to adapt the unit to digital impulses.

Digital Valve Actuators

Digital actuators can accept the output of digital computers directly without digital-to-analog converters. Only simple on-off elements are needed for their operation. The number of output positions that can be achieved is equal to 2^n, where n is the number of inputs. Accuracy of any position is a function of the manufacturing tolerances. Resolution is established by the number of inputs and by the operating code selected for a given requirement. The smallest move achievable is called a 1-bit move. The code may be binary, complementary binary, pulse, or special purpose. A 3-input piston adder assembly produces 8 discrete bit positions. The adders in Figure 4.1y are shown in the 6-bit extended position. The interlocking pistons and sleeves will move when vented or filled through their selector valves. This same adder can be used to position a four-way spool valve, with a mechanical bias to sense position. The spool valve controls the position of a large-diameter piston actuator or force amplifier.

Use of a DC motor featuring a disc-armature with low moment of inertia has created another valve actuator particularly adaptable to a digital input. Brushes contacting the flat armature conduct current to the armature segments. Incremental movement is caused by half-waves

at line frequency for rotation in either direction. The rotation of the armature is converted to linear stem motion by use of a hollow shaft internally threaded to match the valve stem. Actuator output is 5,000 lbf (22,250 N) maximum for noncontinuous service at a rate of 0.4 in./sec (10 mm/s) through a valve stroke of 3 in. (75 mm). The actuator is de-energized at stroke limits or at power overloads by thermal overload relays.

Operation of the actuator requires application of a thyristor (SCR) unit designed for this purpose. This unit accepts pulses from a computer or pulse generator. The thyristor unit consists of two SCRs with transformer, triggers with pulse shift circuit, facility for manual actuator operation, and the previously mentioned thermal overload relays. The output consists of half-waves to pulse the armature of the actuator.

Modules are also available for process control with the necessary stem position feedback, slow pulsing for accurate manual positioning, and full-speed emergency operation.

The digital to pneumatic transducer shown in Figure 4.1z is used to convert a controller output to a pressure signal for operating a pneumatic valve.

Selection and Application

The following are characteristics to consider in the application and selection of all types of actuators.

Speed and Torque Output Ranges

Speed requirements vary from less than 1 rpm to about 160 rpm. The upper limit of available torque seems to be about 60,000 ft-lbf (81,600 N·m). Stem thrusts and rotational drive torques are limited only by the size of

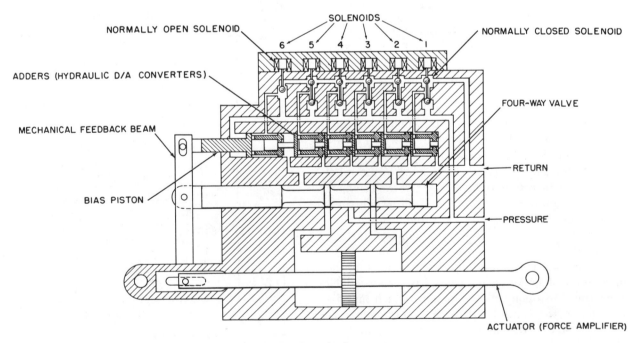

Fig. 4.1y   Digital valve actuator

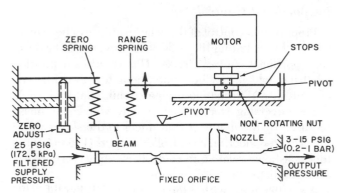

ZERO
SPRING

RANGE
SPRING

MOTOR

STOPS

PIVOT

ZERO
ADJUST
25 PSIG
(172.5 kPa)
FILTERED
SUPPLY
PRESSURE

PIVOT

BEAM

NON-ROTATING NUT

NOZZLE

3-15 PSIG
(0.2-1 BAR)

FIXED ORIFICE

OUTPUT
PRESSURE

Fig. 4.1z Digital-to-pneumatic transducer

the motor used and the ability of the gear, bearings, shafts, etc. to carry the load. Speed of operation depends on the gear ratios, adequate prime mover power, and means of overcoming the inertia of the moving system for rapid stopping. This is most important for proportional control uses. Some actuators have a limited selection of drive speeds while others are furnished in gear ratios in discrete steps of 8 to 20 percent between speeds.

Manual Operation

Manual operation is sometimes necessary for normal operational procedures, such as start-up, or under emergency conditions. Only units that are rotating very slowly or those with low output should have continuously connected handwheels. Most units have the handwheel on the actuator with a clutch for demobilizing the handwheel during powered operation. Clutches are manual engage and manual disengage, or manual engage and automatic disengage upon release of the handwheel. Others are manual engage when the handwheel is rotated, with the motor reengaging when the handwheel is not being rotated; or power reengage, which takes the drive away from the operator upon energization, leaving the handwheel freewheeling.

Electrical Equipment

Most of these actuators include much of the electric gear within the housings of the unit. Components such as limit, auxiliary, and torque switches and position or feedback potentiometers are run by gearing to stem rotation, so they must be housed on the unit. Installation is optional concerning push-buttons, reversing starters, lights, control circuit transformers, or line-disconnect devices in an integral housing on the unit. Any or all of these components may be located externally, such as in a transformer, switch, or control house. The enclosures must be designed to satisfy NEMA requirements for the area. (See Section 5.2.)

High Breakaway Force

Resistance to opening requires a method of allowing the motor and gear system to develop speed to impart

a "hammer blow" which starts motion of the valve gate or plug. Selection of motors with high starting torque is not always sufficient. The dogs of a dog-clutch rotate before picking up the load, or a pin on the drive may move within a slot before picking up the load at the end of the slot. Systems are used that delay contact for a preselected time or until the tachometer indicates the desired speed of rotation.

Torque Control for Shutdown

Torque control for shutdown at closure or due to an obstruction in the valve body is accomplished in numerous ways, but each one uses a reaction spring to set the torque. When rotation of the drive sleeve is impeded, the spring will collapse, moving sufficiently to operate a shut-off switch.

Position Indication

An indicator can be geared to the stem rotating gear, but it becomes a problem when actuator rotation varies from 90 to as much as 240 revolutions. Gearing of a cam shaft operating the position indicator or auxiliary switches must be calculated to obtain a fairly uniform angle. Upon correct gearing, an indicating arrow or transmitting potentiometer can be rotated.

Maintenance of Status Quo Position

This is no problem when the actuator includes a worm gear or an Acme stem thread. Use of spur gears can cause instability when positioning a butterfly or ball valve. Status quo is obtained by use of a motor brake or insertion of a worm gear into the system.

Protection Against Stem Expansion

The status quo ability of an Acme thread or worm gear is detrimental when the valve itself is subjected to temperatures high enough to expand the stem. This expansion, when restrained, can damage the seat or plug, bend the stem, or damage the actuator thrust bearings. One of the original patents for this type of actuator included Belleville springs to allow the drive sleeve to move with the thermal expansion and relieve the linear force.

Mounting Methods

Industry has dictated a set of dimensions for the mounting flanges and bolt holes for newly manufactured valves. Retrofit mounting requires adaption to existing valves. Mounting requires a plate to match the existing valve, which is either screwed into the yoke upon removal of the manual drive sleeve, welded or brazed to the yoke, or, for a split yoke, bolted to the yoke.

Adaptability to Control Circuits

This feature includes adaptability to many voltages and to single or polyphase supplies. Polyphase motors of 240 to 480 volts and 60 Hz predominate. Single-phase motors

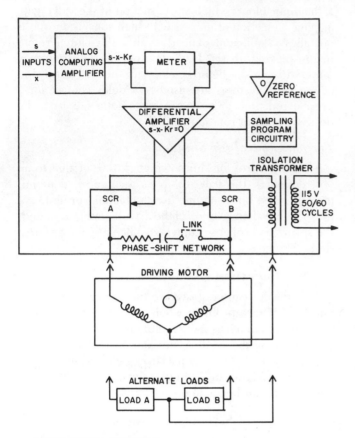

s = INSTANTANEOUS SET POINT
x = INSTANTANEOUS TRANSMITTER SIGNAL
K = ADJUSTABLE TRANSMITTER COMPENSATION
r = INSTANTANEOUS RATE OF CHANGE OF TRANSMITTER SIGNAL

Fig. 4.1aa   Proportional motor control circuit with position feedback

Proportional Control

Proportional control of these large units can be accomplished by including the coils of the reversing starter in a proportional control circuit. This requires a position feedback, which may be a potentiometer. For this type of control, a Wheatstone bridge circuit would be used.

The reversing starter controls any voltage or phase required by the motor. A transducer to transform any of the accepted electronic controller outputs (1–5 mA, etc.) to a resistance relative to controller output permits use of the actuator in these systems.

A smaller unit, with a force output of 1,460 lbf (6497 N) at stall and 500 lbf (2225 N) at a rate of 0.23-in./sec (5.75 mm/s), uses solid state control of the motor. At the same time it eliminates stem position feedback into the controller. A DC signal "x" (Figure 4.1aa) from a process transmitter provides the loop responses, such as temperature, as well as high-response systems, such as flow. A differential amplifier responds to the magnitude and the polarity of an internally modified error signal which triggers SCRs to obtain bidirectional drive. The synchronous motor can be driven in either direction, depending upon the relation to the setpoint.

BIBLIOGRAPHY

Barnes, P.L., "Protect Valves with Fire-Tested Actuators," *Instruments & Control Systems*, October, 1982.

Baumann, H.D., "Trends in Control Valves and Actuators," *Instrument & Control Systems*, November, 1975.

Carey, J.A., "Control Valve Update," *Instruments & Control Systems*, January, 1981.

Colaneri, M.R., "Solenoid Valve Basics," *Instruments & Control Systems*, August, 1979.

Fernbaugh, A., "Control Valves: A Decade of Change," *Instruments & Control Systems*, January, 1980.

Hammitt, D., "How to Select Valve Actuators," *Instruments & Control Systems*, February, 1977.

Howleski, M.F., "Adapting Electric Actuators to Digital Control," *Instrumentation Technology*, March, 1977.

Koechner, Q.V., "Characterized Valve Actuators," *Instrumentation Technology*, March, 1977.

Lipták, B.G., "Control Valves in Optimized Systems," *Chemical Engineering*, September 5, 1983.

Shinskey, F.G., "Dynamic Response of Valve Motors," *Instruments & Control Systems*, July, 1974.

Usry, J.D., "Electric Valve Actuators Can Fail Safe," *Instruments & Control Systems*, May, 1980.

————, "Stepping Motors for Valve Actuators," *Instrumentation Technology*, March, 1977.

up to about 2 horsepower (1500 w) are used. Reversing starters with mechanical interlocks are used for both proportional and on-off service. The coils that open and close the contacts are energized by an open-center double-pole switch, which can be incorporated in the automatic control circuit. For manual control, the starter may be the type that maintains contact, requiring the open or close button to be held in position. Some units are also wired for momentary depression so that the actuator runs until it reaches the limit switch, or until a stop button is depressed.

DIAPHRAGM
ACTUATOR

HAND
ACTUATOR

PRESSURE
BALANCED
DIAPHRAGM
ACTUATOR

SPRING-LOADED
PNEUMATIC CYLINDER
ACTUATOR

SINGLE-ACTING,
CUSHION LOADED
PISTON ACTUATOR

PNEUMATIC
ROTARY MOTOR
ACTUATOR

DOUBLE-ACTING
PISTON ACTUATOR

4.2 PNEUMATIC ACTUATORS

Types of Pneumatic Actuators:

A—Linear
A1—Diaphragm
A2—Piston
B—Rotary
B1—Cylinder
B2—Spline or helix
B3—Vane
B4—Pneumo-hydraulic
B5—Rotary air motor
B6—Electro-pneumatic

Partial List of Suppliers:

Andale Co. (B); Bellofram Corp. (A); Bestobell Canada Ltd. (A); Bettis, GH (B); Conoflow Regulators & Controls Div. of ITT (A); Duriron Co. (A); Fisher Controls Co. (A); Foxboro Co. (A); Hills-McCanna Co. (A); Honeywell Process Control Div. (A); Jamesburg Corp. (A); Kamyr Valves, Inc. (A); Kent Process Control, Inc. (A); Ledeen Flow Control Systems, Inc. (B); Limitorque Corp. (B); Posi-Seal International, Inc. (A); Shafer Valve Co. (B); Valtek, Inc. (A); WKM Div. of ACF Industries (A); Worchester Controls Corp. (A).

Pneumatic valve actuators respond to an air signal by moving the valve trim into a corresponding throttling position. This section covers the two basic designs most frequently applied to globe valves: the diaphragm and the piston actuators. In connection with the performance of these actuators, an analysis is presented of the various forces positioning the plug, including diaphragm, spring, and dynamic forces generated by the process fluid. An understanding of the interrelationships among these forces will allow the reader to properly size these actuators and make the correct spring selection. The failure safety of valve actuators and the relative merits of diaphragm and piston actuators are also discussed.

The discussion of diaphragm and piston type actuators is followed by the treatment of pneumatic-rotary and pneumatic-hydraulic actuators.

Definitions

An actuator is that portion of a valve that responds to the applied signal and causes the motion resulting in modification of fluid flow. Thus an actuator is any device that causes the valve stem to move. It may be a manually positioned device, such as a handwheel or lever. The manual actuator may be open-closed, or it may be manually positioned at any position between fully open and fully closed. Other actuators are operated by compressed air, hydraulics, and electricity.

The actuators discussed here are those capable of moving the valve to any position from fully closed to fully open and those using compressed air for power. Of such there are two general types: the spring and diaphragm actuator and the piston actuator.

In a spring and diaphragm actuator, variable air pressure is applied to a flexible diaphragm to oppose a spring. The combination of diaphragm and spring forces acts to balance the fluid forces on the valve.

In a piston actuator, a combination of fixed and variable air pressures is applied to a piston in a cylinder to balance the fluid forces on the valve. Sometimes springs are used, usually to assist valve closure. Excluding springs, there are two variations of piston actuators: cushion loaded and double acting. In the cushion-loaded type, a fixed air pressure, known as the cushion pressure, is opposed by a variable air pressure and is used to balance the fluid forces on the valve. In the double-acting type, two opposing variable air pressures are used to balance the fluid forces on the valve.

317

An actuator can be said to have two basic functions: (1) to respond to an external signal directed to it and cause an inner valve to move accordingly (with the proper selection and assembly of components, other functions can be obtained, such as a desired fail-safe action) and (2) to provide a convenient support for certain valve accessory items, including positioners, limit switches, solenoid valves, and local controllers.

Actuator Performance

This discussion is restricted to pneumatic actuators. The external signal, therefore, is an air signal of varying pressure. The air signal range from a pneumatic controller is commonly 0–18 PSIG (0–124 kPa). Signal or actuator input pressure starts at 0 PSIG, not 3 PSIG (21 kPa). A common mistake is to confuse the 3–15 PSIG (21–104 kPa) range of transmitter output pressure with the signal to a valve. The higher value of 18 PSIG (124 kPa) is fixed only by the air supply to the controller (or positioner), and it can easily be set to 20 PSIG (138 kPa) or higher. A variety of other input pressures are sometimes used, such as 0–30 or 0–60 PSIG (0–207 or 0–414 kPa).

Both the spring and diaphragm and the piston actuator produce linear motion to move the valve. These actuators are ideal for use on valves requiring linear travel, such as globe valves. A linkage or other form of linear-to-rotary motion conversion is required to adapt these actuators to rotary valves, such as the butterfly type.

The Steady-State Equation

In spring and diaphragm actuators stem positioning is achieved by a balance of forces acting on the stem. These forces are due to pressure on the diaphragm, spring travel, and fluid forces on the valve plug (Figure 4.2a).

Equation 4.2(1) can be derived from a summation of forces on the valve plug adopting the positive direction downward.

$$PA - KX - P_vA_v = 0 \qquad 4.2(1)$$

where A is the effective diaphragm area, A_v is the effective inner valve area, K is the spring rate, P is the diaphragm pressure, P_v is the valve pressure drop, and X is stem travel. Equation 4.2(1) applies to a push-down-to-close actuator and valve combination with flow under the plug. This type of actuator is commonly referred to as direct acting.

Another popular actuator configuration is one causing the stem to rise on an increase of air pressure. It is commonly called a reverse actuator (Figure 4.2b). By using the same sign convention, the equation for this valve configuration is seen to be

$$-PA + KX - P_vA_v = 0 \qquad 4.2(2)$$

If the flow direction is reversed in Figure 4.2a, the equation is

$$PA - KX + P_vA_v = 0 \qquad 4.2(3)$$

Likewise, reversing flow direction in Figure 4.2b results in equation 4.2(4):

$$-PA + KX + P_vA_v = 0 \qquad 4.2(4)$$

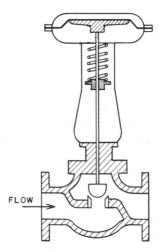

Fig. 4.2b   Reverse-acting actuator

These equations are simplified because they do not consider friction and inertia. Friction occurs in the valve stem packing, in the actuator stem guide, and in the valve plug guide or guides. Usually, for static valve actuator sizing problems, negligible error is introduced by ignoring the friction terms.

If equation 4.2(1) is plotted as signal pressure vs stem travel and if the case of no fluid forces on the plug (bench test) is assumed, then the curve shown in Figure 4.2c is obtained.

Next, consider the case of plug forces due to fluid flow, assuming that the term P_v is constant for all travel po-

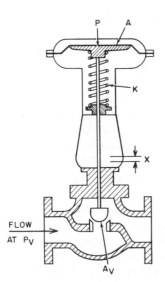

Fig. 4.2a   Forces on a spring-
and-diaphragm valve

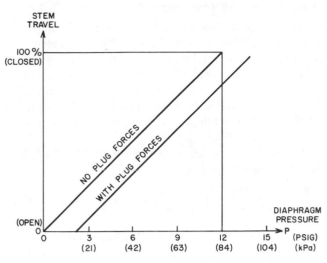

Fig. 4.2c   Diaphragm pressure vs stem travel

$$P(46) = (885)(0) + (100)(\pi/4)$$

$$P = 1.7 \text{ PSIG (18 kPa)}$$

This means that the diaphragm pressure must increase to 1.7 PSIG (18 kPa) before stem travel begins. Figure 4.2c shows this with the line labeled "With plug forces."

Actuator Nonlinearities

In practice we encounter many nonlinearities, and the ideal curves in Figure 4.2c are not obtained. These non-linearities are due to several factors, such as the variable effective diaphragm areas. The effective diaphragm area varies with travel and with the pressure level on the diaphragm. Figure 4.2d illustrates this for three different sizes of diaphragms.

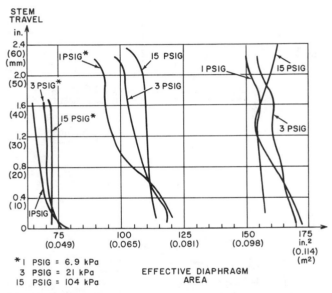

Fig. 4.2d   Nonlinearities in effective diaphragm areas

sitions. This has the effect of shifting the straight line to the right to some position depending on the magnitude of P_v. Curves similar to those in Figure 4.2c can readily be drawn for the other valve configurations represented by equations 4.2(2), 4.2(3), and 4.2(4).

Actuator Sizing Example

A typical problem uses values applicable to a 1-in. (25-mm) single-ported valve:

$$A = 46 \text{ in.}^2 (0.03 \text{ m}^2)$$

$$X = \tfrac{5}{8} \text{ in. (15.6 mm) full travel, and}$$

$$K = 885 \text{ lbf (155 N/mm)}$$

If no plug forces exist, equation 4.2(1) reduces to PA = KX. We solve for the pressure change required for full travel as follows:

$$P = \frac{KX}{A}$$

$$= \frac{(885)(\tfrac{5}{8})}{46} \qquad\qquad 4.2(4)$$

$$= 12.03 \text{ PSI (82.6 kPa)}$$

which is reasonably close to the 12 PSI (82.8 kPa) desired operating span. Practical considerations of variations in spring constants and in actuator-effective areas usually prevent such a close approach to the desired span, and frequently a ±10 percent leeway is permitted.

When there are plug forces, it is seen from Equation 4.2(1) that an additional actuator force is required to maintain balance. The actuator pressure required to begin stem motion can be calculated for the case of a 1-in.-diameter (25 mm) plug and 100 PSID (690 kPa) pressure drop. Equation 4.2(1) can be used to solve for P as follows (stem travel is zero, thus there are no spring forces):

Another nonlinearity is in valve plug forces $P_v A_v$. Figure 4.2e illustrates the variations in plug forces for two 4-in. (100 mm) valves, single ported and double ported. It also shows the effects of flow over and under the plugs of a single-ported valve.

Springs are also nonlinear in that spring rates vary with travel. By judicious selection of springs, considering their spring rate and travel, the effects of nonlinearity on the valve assembly can be minimized.

When all of these nonlinearities are considered, a typical plot of actuator travel vs diaphragm pressure would not be a straight line as shown in Figure 4..2c but *might* be a curve such as in Figure 4.2f.

A nonlinear curve, such as the one labeled "actual" in Figure 4.2f, is not necessarily objectionable. When used in an automatic control loop, the static nonlinearities are compensated for by the controller. This curve is actually a part of the gain term in the valve's transfer function, and the other part is the flow characteristic. When a

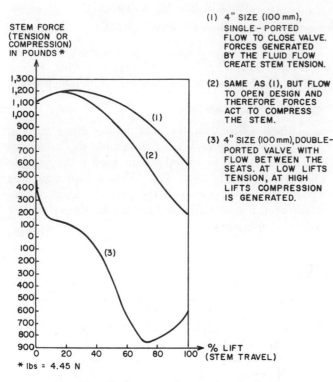

STEM FORCE
(TENSION OR
COMPRESSION)
IN POUNDS *

(1) 4" SIZE (100 mm), SINGLE-PORTED FLOW TO CLOSE VALVE. FORCES GENERATED BY THE FLUID FLOW CREATE STEM TENSION.

(2) SAME AS (1), BUT FLOW TO OPEN DESIGN AND THEREFORE FORCES ACT TO COMPRESS THE STEM.

(3) 4" SIZE (100 mm), DOUBLE-PORTED VALVE WITH FLOW BETWEEN THE SEATS. AT LOW LIFTS TENSION, AT HIGH LIFTS COMPRESSION IS GENERATED.

* lbs = 4.45 N

Fig. 4.2e   Nonlinearities due to valve plug forces

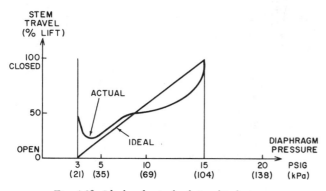

Fig. 4.2f   Ideal and actual relationship between
diaphragm pressure and stem travel

valve positioner is used, the positioner overcomes these nonlinearities, and the result is the ideal curve.

Dynamic Performance of Actuators

Several control valve subsystems must be analyzed in order to thoroughly evaluate their dynamic performance. The separate systems include:

1. The spring-mass system of the valve's moving parts.
2. The pneumatic system from controller output to valve diaphragm chamber. If a valve positioner is used, there are two separate pneumatic systems: one from the controller output to the positioner and another from the positioner output to the diaphragm chamber. The interconnecting tubing is a consideration in all of the pneumatic systems.

Spring-Mass System Dynamics

Analysis of the spring and mass system is only valid for linear systems. It is necessary either to neglect consideration of the nonlinear elements or have a system wherein the nonlinear effects are minor. In the case of control valves with sufficient power in the actuator, the latter case is approached. With such an understanding of the nonlinear effects, we proceed as though valve actuators were linear devices.

The spring-mass system is represented by the following differential equation:

$$M \frac{d^2X}{dt^2} + b \frac{dX}{dt} + KX = PA - P_v A_v \qquad 4.2(6)$$

where b is the viscous friction force and M is mass.

The static, time-independent terms of equation 4.2(6) are identical with equation 4.2(1). The transfer function of the valve actuator is the LaPlace transform of differential equation 4.2(6):

$$\frac{X(s)}{P(s)} = \frac{A/K}{(w/gK)s^2 + (b/K)s + 1} \qquad 4.2(7)$$

where g is the gravitation constant, s is the LaPlace operator, and w is the weight of moving parts. This can be written in terminology more useful to instrument engineers using the time constant τ (tau) and damping factor ζ (zeta).

$$\frac{x(s)}{P(s)} = \frac{1}{\tau^2 s^2 + 2\tau\zeta s + 1} \qquad 4.2(8)$$

The coefficient of the s^2 term in equation 4.2(7) is the square of the reciprocal of the undamped natural frequency of the spring-mass system. It is a useful number in understanding the relative importance of a control valve's dynamic components.

Table 4.2g is a list of the natural frequencies for a variety of valve sizes, considering the "average" design. The real significance of Table 4.2g is in the values listed. The largest of the sizes listed has a natural frequency of nearly 10 Hz, which is ten times faster than the typical pneumatic performance of a control valve.

Table 4.2g
NATURAL FREQUENCY OF CONTROL VALVES

Valve Size (in.)	(mm)	Undamped Natural Frequency (Hz)
1	25	32
1½	37.5	27
2	50	22
3	75	16
4	100	14
6	150	10
8	200	9

Pneumatic System Dynamics

The preceding discussion considered the transfer functions of the spring and diaphragm actuator from the actuator pressure to the resulting stem travel. Next we will consider the pneumatic transfer function of the pressure from the controller to the diaphragm pressure (Figure 4.2h).

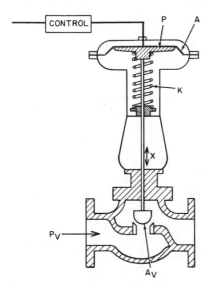

Fig. 4.2h   Forces on a spring-and-diaphragm valve

The analysis of transfer tubing dynamics involves distributed parameter systems and linear control theory analysis. Some systems, especially with short transmission lines, behave nearly like linear systems. A short tube behaves like pure resistance, and the air volume above the diaphragm is a capacitance resulting in a resistance-capacitance time constant. Some values obtained from tests with very short tubing are given in Table 4.2i. The values given are from actual tests in which the air supply was not limiting. These figures show that the valves are capable of fast response. Performance is usually limited by the controller's or positioner's ability to supply the required air fast enough. Section 2.6 contains a more detailed discussion of transmission lags and methods of boosting.

Table 4.2i
TIME CONSTANTS FOR SHORT TUBE SECTIONS

Valve Size (in.)	(mm)	Time Constant (sec)
1	25	0.03
2	50	0.05
4	100	0.8

Piston Actuators

Piston actuators are either single or double acting. The single-acting actuator, Figure 4.2j, utilizes a fixed air pressure, known as the cushion, to oppose the controller

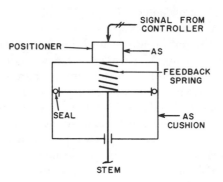

Fig. 4.2j   Single-acting piston actuator

signal. This valve does not have spring or diaphragm area nonlinearities, but it is of course subject to the same plug force nonlinearities (Figure 4.2e) as the spring and diaphragm actuator.

In order to use such an actuator for throttling purposes, it is necessary to have a positioner. The positioner senses the actuator motion and causes the valve to move accordingly. It cannot be used as a proportioning travel device without the positioner; consequently, its performance is that of the "ideal" curve in Figure 4.2f.

A double-acting piston actuator is one that eliminates the cushion regulator and uses a positioner with a built-in reversing relay. Thus the positioner has two air pressure outputs, one connected above the piston and the other below. The positioner receives its signal and senses travel in the same manner as a single-acting positioner. The difference is in the outputs; one pressure increases and the other decreases to cause piston travel.

Safe Failure

The valve application engineer must choose between the two readily available fail-safe schemes for control valves, either fail open or fail closed. The choice will be based upon process safety considerations in the event of control valve air failure. Complete plant air failure, controller signal failure, and local air supply failure must all be considered. Local failure is significant when a valve positioner is being used and when piston actuators with cushion loading are used.

The choice must be based on detailed knowledge of the valve application in the overall process or system. Two generalizations are that in a heating application, the valve should fail closed, and in a cooling application it should fail open. There are certainly applications where either failure mode is equally safe; then considerations of standardization may be used.

Fail-safe involves the selection of actions of actuator and inner valve. Both actuator and inner valve usually offer a choice of increasing air to push the stem down or up, and pushing the stem down may open or close the inner valve. The proper choice of combinations may be made by fail-safe considerations. The process application of the valve must be investigated to determine whether, on instrument air failure, it would be better to have the valve go fully open or fully closed.

There may not be much flexibility in the inner valve action. For example, a single-seated top-guided valve must have a push-down-to-close plug. There is freedom of choice, however, in either single- or double-seated top- and bottom-guided valves. Other valve bodies, such as the Saunders and pinch valve styles, must be of the push-down-to-close type. Rotary types, such as butterfly and ball valves, may be arranged either way.

The inner-valve flexibility leads to two cases: one in which either inner-valve action is permissible and one in which the inner-valve must be push-down-to-close.

When there is a choice of inner-valve action, overall valve action may be obtained by selecting the suitable inner-valve action and always using increasing air to push down the actuator. This is known as a *direct actuator*. A direct actuator is preferred because of economy reasons in spring and diaphragm actuators. The savings may be in purchase cost. It is also realized in maintenance costs, because there is no stem seal to cause possible leakage and maintenance costs. The piston type actuator is equally suitable for either direct or reverse action; if it is the actuator to be used, the application engineer has complete freedom of choice of actions.

When the inner valve must be push-down-to-close, it is necessary to use both direct and reverse actuators to accomplish the desired fail-safe actions. Figure 4.2k summarizes the possibilities available.

Valve Failure (Overall)	Fail Open		Fail Closed	
Actuator	Direct	Reverse	Reverse	Direct
Inner Valve	Direct	Reverse	Direct	Reverse

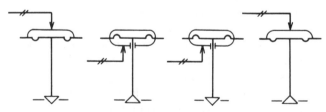

Fig. 4.2k   Overall valve failure positions achieved by various actuator and inner valve combinations

Relative Merits of Diaphragm and Piston Actuators

When choosing between piston actuators and the spring-diaphragm type, the fail-safe consideration may be the reason for the final selection. If properly designed, the spring is the best way of achieving fail-closed action. Fail-open action is less critical.

Piston actuators depend upon air lock systems to force the valve closed on air failure. Such systems may work well initially, but there are many possibilities for leaks to develop in the interconnecting tubes, fittings, and

check valves, and such piston actuator systems are not considered reliable. Air lock systems also add to the actuator's cost. Piston actuators may also be specified with closure springs to provide positive failure positions.

Valve installation in the line is also a factor to consider. Flow over the plug assists in maintaining valve closure after air failure, but the considerations involving dynamic stability are more important, and therefore the use of "flow-to-open" valves is recommended.

Other Types of Pneumatic Actuators

Cylinder Type

Increased use of ball and butterfly valves or plug cocks for control has bred a variety of actuators and applications of existing actuators for powering these designs.

Positioning a quarter-turn valve with a linear output actuator using a lever arm on the valve resolves itself into a problem of mounting and linkages. The actuator can be stationary, with a bushing to restrain lateral movement of the stem. This requires a joint between the stem and a link pinned to the lever arm. The actuator can be mounted on a gimbal mechanism to allow required movement. The actuator can be hinged to allow free rotation to allow for the arc of the lever arm.

Various Scotch yoke designs, such as the one shown in Figure 4.2l, can be used with one, two, or four cylinders. Use of rollers in the slot of the lever arm utilizes the length of the lever arm of the valve opening or closing points.

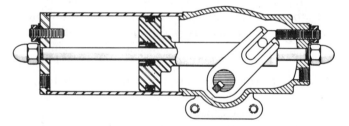

Fig. 4.2l   Pneumatic cylinder actuator with Scotch yoke

A rack and pinion can be housed with the pinion on the valve shaft and the rack positioned by almost any linear valve actuator. The rack (Figure 4.2m) can be carried by a double-ended piston or by two separate pistons (Figure 4.2n) in the same cylinder, where they move toward each other for counterclockwise rotation and away from each other for clockwise rotation. Similar action is obtained (Figure 4.2o) by two parallel pistons in separate cylinder bores. Dual cylinders are used in high-pressure actuators used to rotate ball valves as large as 16 in. (400 mm) in less than 0.5 sec. An actuator similar to the one shown in Figure 4.2p can be spring-loaded for emergency or positioning operation.

On-off operation of cylinders for quarter-turn valves requires solenoid or pneumatic pilots to inject pneumatic

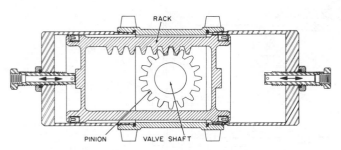

Fig. 4.2m   Rack and pinion actuator with dual-acting cylinder

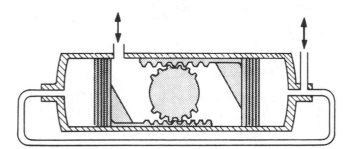

Fig. 4.2n   Rack and pinion actuator operated by two separate pistons

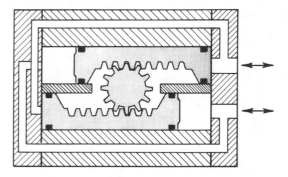

Fig. 4.2o   Rack and pinion actuator operated by
two parallel pistons

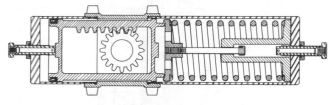

Fig. 4.2p   Rack and pinion actuator with
spring-loaded (fail-safe) cylinder

or hydraulic pressure into the cylinders. Open or closed position must be set by stops that limit shaft rotation or piston travel. Thereby the valve rotation is stopped and held in position until reverse action is initiated. Positioning action requires a calibrated spring in the piston or diaphragm actuator, a valve positioner, or a positioning valve system that loads and unloads each end of the cylinder. Rotation of the valve must be translated to the positioner by gears, direct connection, cam, or linkage. The valve positioner must be the type that includes the four-way valve. The positioning valve system can be a

four-way valve with a positioner for use with a pneumatic controller. The piston is sometimes positioned by a servo-system consisting of a servo-valve that accepts an electronic signal, a four-way valve to amplify and control the pressure to the cylinder, and a feedback signal from a potentiometer or LVDT.

An electro-hydraulic power pack or pneumatic pressure source may be used to furnish pressure to a pair of cylinders, one for each direction of rotation, as shown in Figure 4.2q.

Rotation by Spline or Helix

A multiple helical spline rotates through 90 degrees as pressure below the piston (Figure 4.2r) moves the assembly upward. A straight spline on the inside of this piston extension sleeve rotates the valve closure member through a mating spline. The cylinder is rated at 1,500 PSIG (10 MPa). The actuator will rotate 1-in. (25 mm) and $1\frac{1}{2}$-in. (37.5 mm) valve stems. The mounting configuration is designed to adapt to many quarter-turn valves.

A nonrotating cylinder (Figure 4.2s), with an internal helix to mate with a helix on a rotatable shaft, creates a form of rotating actuator. Hydraulic or pneumatic pressure in the drive end port (left) causes counterclockwise rotation; clockwise rotation is caused by pressure on the opposite side of the piston. There is a patented seal between the internal and the external bores of the cylinder and the external surface of the shaft. The unit is totally enclosed by seals to protect it from contaminated atmospheres. A hydraulic pump, reservoir, and necessary controls can be mounted integrally.

Vane Type

Injection of pressure on one side of a vane to obtain quarter-turn actuation is straightforward and obtainable with a minimum number of parts. A single vane (Figure 4.2t) can be used for 400 to 20,000 in.-lb (45.4 to 1,666.7 N·m). Units can be mounted together for double output, or a double vane design (Figure 4.2u) can also be used. The success of the vane actuator as a control device is dependent upon the control systems. By use of an auxiliary pneumatic pressure source, all types of fail-safe actions are possible although not as positively as with spring loading. Use of line pressure to create hydraulic pressure on the vane is piloted by both manual and automatic methods. Use of a rotary potentiometer to sense position and complete a bridge circuit is necessary for proportional control.

Pneumo-Hydraulic Type

An actuator with two double-acting cylinders uses an integrally designed pneumatic or electric powerpack (Figure 4.2v). This has the advantage of furnishing a constant hydraulic pressure to the cylinders regardless of the power source to the prime mover. Few cylinder sizes are needed to cover a wide range of torque outputs.

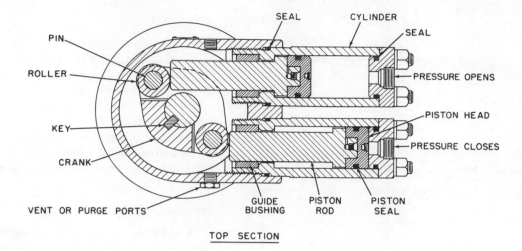

PIN

ROLLER

KEY

CRANK

VENT OR PURGE PORTS

SEAL

CYLINDER

SEAL

PRESSURE OPENS

PISTON HEAD

PRESSURE CLOSES

GUIDE BUSHING

PISTON ROD

PISTON SEAL

TOP SECTION

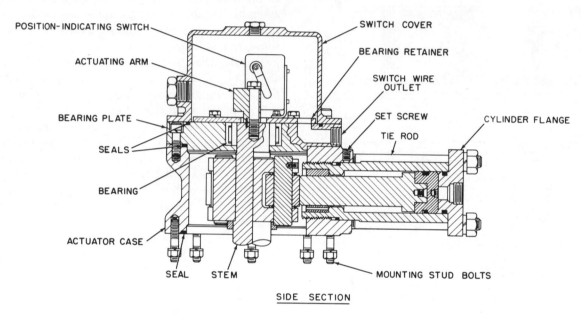

POSITION-INDICATING SWITCH

ACTUATING ARM

BEARING PLATE

SEALS

BEARING

ACTUATOR CASE

SEAL   STEM

SWITCH COVER

BEARING RETAINER

SWITCH WIRE OUTLET

SET SCREW

TIE ROD

CYLINDER FLANGE

MOUNTING STUD BOLTS

SIDE SECTION

Fig. 4.2q   Crank and roller actuator operated by a pair of cylinders

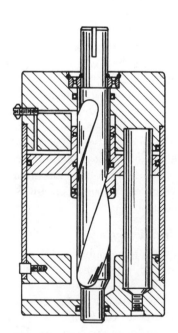

Fig. 4.2r   Helical spline actuator

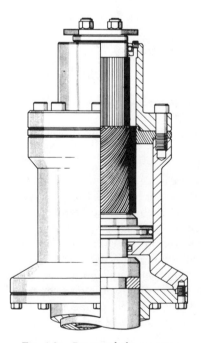

Fig. 4.2s   Rotating helix actuator

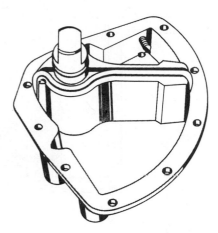

Fig. 4.2t   Single-vane quarter-turn actuator

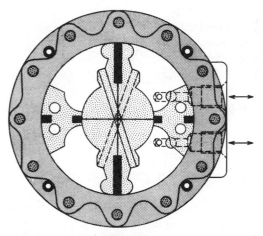

Fig. 4.2u   Double-vane quarter-turn actuator

The prime movers are sized and selected to obtain the actuator speeds desired with the pneumatic pressures or electric voltages available. Multiple auxiliary switches, position transmitter, and positioning devices are adapted to the unit.

Gas pressure is used to create hydraulic pressure using two bottles (Figure 4.2w). The stability of a hydraulically operated cylinder is utilized in this manner, using line gas pressure and the bottle size for amplification. The manual control valve can be replaced by a variety of electric or pneumatic pilot valves for automatic control. A hand pump is furnished that can take over the hydraulic operation in the absence of gas pressure or malfunction of the pilot controls. This self-sustaining approach to cylinder operation finds wide application for line break shutoff and for the various diverting and bypass operations of a compressor station.

A hydrostatic system consisting of a pneumatic prime mover (Figure 4.2x) on the shaft of a hydraulic pump to run a hydraulic motor has interesting features. This actuator incorporates many of the features of other high-force geared actuators that rotate a drive sleeve. Torque control consists of a relief valve in the hydraulic line to the motor. This eliminates the reactive force of spring-loaded torque controls. Starting torque occurs because hydraulic slippage of the pump allows the motor to reach maximum speed. Direction and deactivating control is attained with a four-way valve. As many as 16 auxiliary switches, settable at any position, are housed in the unit. Limit switches can be pneumatic or electric. A wide range of torque outputs and speeds is obtained by selection of prime mover, pump, and motor combinations. Initial success of the unit was partially due to its adaptability for retrofit to existing valves. The unit can be manually operated if required.

Rotary Pneumatic Type

Pneumatic pressure is used to power a rotary motor to drive any of the large gear actuators. Control is by a four-way valve. The motor shown in Figure 4.2y is running in one position. This will continue until the valve is repositioned or until a cam operates a shutoff valve at one end of the stroke. Reversal of the four-way valve causes reverse operation. An intermediate position causes

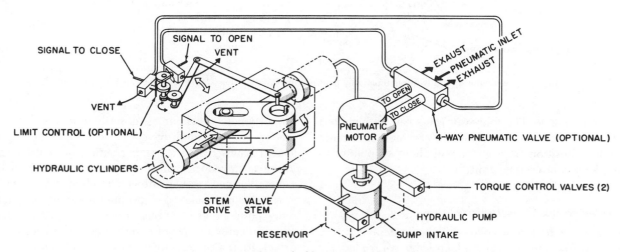

Fig. 4.2v   Actuator with two double-acting cylinders and power pack

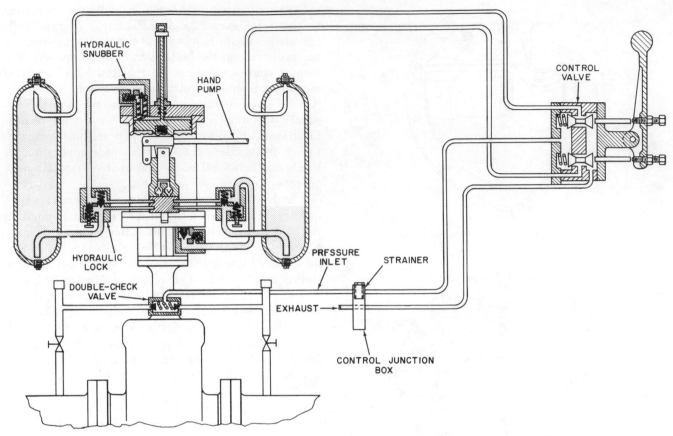

Fig. 4.2w   Pneumo-hydraulic actuator powered by line pressure (*Courtesy of Shafer Valve Co.*)

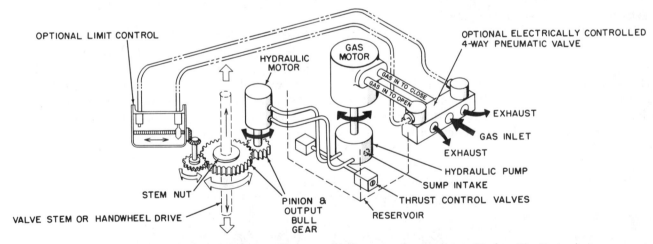

Fig. 4.2x   Actuator system with pneumatically powered hydrostatic valve (*Courtesy of Ledeen Div. Textron*)

the motor to stop. The four-way valve can be operated by pneumatic or electric actuators for remote automatic control. A position transmitter will allow adaptation to closed-loop proportional control.

Electro-Pneumatic Actuators

An actuator that defies classification, except that it is pneumatically powered and electrically controlled for proportional application, is described at this point. Operation of a threaded drive sleeve occurs when spring-loaded pawls create a jogging action on a drive gear. Pressure introduced through one of the external lines selects the pawl to become active when the rocker arm is repetitively rocked by the pneumatic motor. A lead screw positions a sliding block to operate control switches, potentiometer and position indicators. The lead screw is driven from a small spur gear and bevel gears.

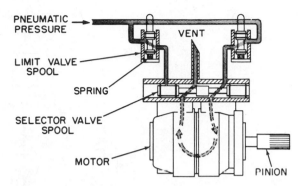

Fig. 4.2y Rotary air motor actuator

Air is supplied from 60 to 140 PSIG (414 to 966 kPa) to give torque outputs up to 360 or 720 ft-lbf (504 or 1008 J) using two actuators. Air consumption is from 0.75 SCF to 1.70 SCF (0.021 to 0.048 m) per revolution at 140 PSIG (966 kPa). Maximum valve stem diameter that can be rotated is 2 in. (50 mm). Numerous motivating combinations are used for control, including electro-pneumatic, electric, and fully pneumatic. A wide variety of components can be used to build up these systems, as shown in Figure 4.2z.

BIBLIOGRAPHY

Barnes, P.L., "Protect Valves with Fire-Tested Actuators," *Instruments & Control Systems*, October, 1982.

Baumann, H.D., "Trends in Control Valves and Actuators," *Instruments & Control Systems*, November, 1975.

Carey, J.A., "Control Valve Update," *Instruments & Control Systems*, January, 1981.

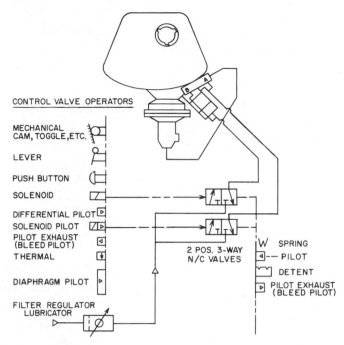

Fig. 4.2z Electro-pneumatic valve actuator (*Courtesy of OIC Corp.*)

Fernbaugh, A., "Control Valves: A Decade of Change," *Instruments & Control Systems*, January, 1980.

Hammitt, D., "How to Select Valve Actuators," *Instruments & Control Systems*, February, 1977.

Hordeski, M.F., "Adapting Electric Actuators to Digital Control," *Instrumentation Technology*, March, 1977.

Koecher, Q.V., "Characterized Valve Actuators," *Instrumentation Technology*, March, 1977.

Shinskey, F.G., "Dynamic Response of Valve Motors," *Instruments & Control Systems*, July, 1974.

Usry, J.D., "Stepping Motors for Valve Actuators," *Instrumentation Technology*, March, 1977.

STANDARD BALL
CONTROL VALVE

4.3 BALL VALVES

THREE-WAY
BALL CONTROL
VALVE

FULL-PORTED
BALL CONTROL

CONVENTIONAL

CHARACTERIZED

CAGE

Types of Ball Valves:	A—Conventional B—Characterized C—Cage
Size and Design Pressure:	A, B—$\frac{1}{2}''$ to 36″ (12.5 mm to 900 mm) in ANSI Class 150; to 12″ (300 mm) in ANSI Class 2500. Segmented ball—1″ to 24″ (50 mm to 600 mm) in ANSI Class 150; to 16″ (400 mm) in ANSI Class 300; to 2″ (50 mm) in ANSI Class 600 C—$\frac{1}{4}''$ to 14″ (6.3 mm to 350 mm)
Design Pressure:	C—Up to 2,500 PSIG (17 MPa)
Design Temperature:	A,B—Varies with size and materials C—From $-425°F$ to $1,800°F$ ($-254°C$ to $982°C$)
Rangeability:	Over 50:1
Capacity:	A,B—Standard ball: $C_v = 30\ d^2$ to $C_v = 35\ d^2$; segmented ball: $C_v = 24\ d^2$ to $C_v = 30\ d^2$; full bore ball: $C_v = 35\ d^2$ to $C_v = 40\ d^2$ C—$C_v = 20\ d^2$ (noncritical flow)
Materials of Construction:	A,B—Most metals, plastic, and glass. Also TFE faced. C—Stainless steel
Cost:	See Figure 4.3a. Cost data are based upon ANSI Class 150 flanged bodies with single-acting, spring-return actuator for on-off service. Other operators can be furnished and positioners added for throttling service. Ball and stem materials are 316 SST for both carbon and stainless steel bodies with TFE seat seals. Other alloys are available at higher cost.
Special Features:	A—Full-ported, three-way, split body, two-directional B—Depending on "contour edge" of ball, the flow characteristics vary slightly between suppliers. Slurry design provides for continuous purging of low-activity zone of valve to prevent build-up of solids, dewatering, or entrapment. C—Good resistance to cavitation and vibration
Partial List of Suppliers:	American Chain Div. of ACCO (A); Asahi America (A); Bestobell Canada Ltd. (A); Cameron Iron Works, Inc. (A); Crane Co. (A); Devar Inc. Control Products Div. (C); DeZurik, A Unit of General Signal (B); Fischer & Porter Co. (A); Fisher Controls Co. (A,B); Flodyne Controls, Inc. (A); Foxboro Co. (A); Hills-McCanna Co. (A); Honeywell Process Control Div. (A); ITT-Grinnell Valves (A); Jamesbury Corp. (A); Kamyr Valves, Inc. (A); Kieley & Mueller, Inc., Sub. of Johnson Controls, Inc. (B); KTM Industries Inc. (A); Lunkenheimer Co., Div. of Conval Corp. (A); Masoneilan Div. of Mc-Graw Edison Co. (B); Otis Engineering Corp. (C); Powers Regulator Co., Transitube Div. (C); Rockwell International Corp. (A); Rockwood Systems Corp. (A); WKM Div. of ACF Industries, Inc. (A); Worchester Controls Corp. (A).

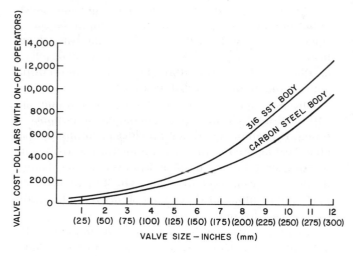

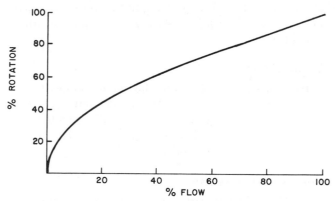

Fig. 4.3a   Cost data and inherent characteristics of conventional carbon steel ball valves

The ball valve contains a spherical plug that controls the flow of fluid through the valve body. The three basic types of ball valves manufactured are discussed below: (1) the conventional or quarter-turn pierced ball type, (2) the characterized, and (3) the cage type.

Conventional Ball Valves

The quarter turn required (90 degrees) to fully uncover or cover an opening in the valve body can be imparted to the ball either manually by turning a handle, or mechanically by use of an automatic valve actuator. Actuators used for ball valves may be the same as those used to control other valve types—pneumatic, electric (or electronic), hydraulic, or combination. The latter types include electro-pneumatic, electro-hydraulic, electro-thermal, or pneumo-hydraulic actuations, as treated in Sections 4.1 and 4.2, but of special design for rotating motion. Most of the ball valves on the market are available with built-in, or integral, valve actuators. They are designed so that they can be applied with or without an actuator or so that they can be fitted with other manufacturers' actuators. This section describes ball valves as special control valves.

Features

The spherical plug lends itself not only to precise control of the flow through the valve body but also to tight shutoff. Thus the ball valve may assume the double role of control and block valve (Figure 4.3b). Special materials used for valve seats help achieve these functions.

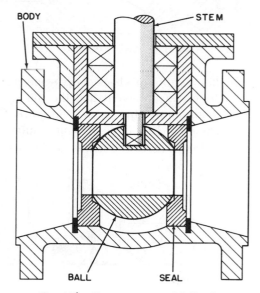

Fig. 4.3b   Top-entry pierced ball valve

Body

The body of a ball valve is predominantly made as a two-way globe valve type, but three-way and split-body valves are also manufactured. Figure 4.3c illustrates some porting arrangements of multi-port valves. Body materials are listed in the feature summary at the beginning of this section.

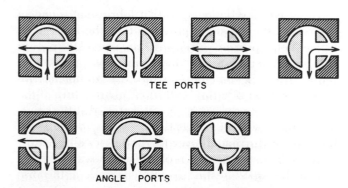

Fig. 4.3c   Porting arrangements of multi-port ball valves

Trim

The ball is cradled by seats on the inlet and the outlet side. The seats are usually made of plastic and are identical on both sides, especially in double-acting valves. Tetrafluoroethylene materials are preferred for their good resilience and low-friction properties (Figure 4.3d). Some

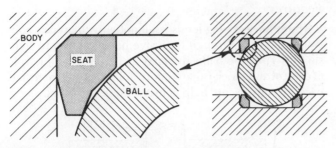

Fig. 4.3d   Ball valve seats

valves have their plastic seats backed up by metallic seats to insure tightness in the event that the soft seat gets damaged by high temperature such as in a fire (Figure 4.3e). Such precautions are imperative on shipboard, for nuclear installations, and in cryogenic applications.

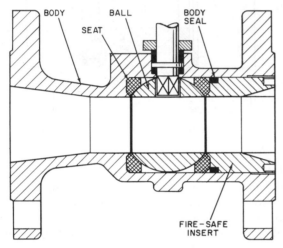

Fig. 4.3e   Fire-safe ball valve

The valves are designed so that lubrication is unnecessary and the torque required to turn the ball is negligible. Both upstream and downstream seats of the pierced ball are sometimes made freely rotatable in order to reduce wear. In some designs the seats are forcibly rotated a fraction of a turn with each quarter turn of the ball. Thus seat wear, which is concentrated at the points where the flow begins or ends on opening or closing of the valve is distributed over the periphery of the seat.

To facilitate cleaning or replacing worn seats, in some designs the whole seating assembly is made in the form of a tapered cartridge. If the valve has top-entry design, the cartridge can be removed without disturbing the valve arrangement. O-rings usually close off stem and seats, and thrust washers made of tetrafluoroethylene compensate for axial stem thrust due to line pressure and reduce stem friction to a minimum.

Some seats are preloaded by springs or are made tapered for wear compensation and leak-tight closure. Where fluids of high temperature are handled, graphite seats are recommended. They hold tight up to 1,000°F (538°C).

Ball and stem are often machined from one piece. Other designs use square ends on the stem to engage in square recesses of the ball. In this case, the ball is made floating in fixed seats, while other designs provide a fixed location of ball and stem through the application of top and bottom guiding and ball bearings.

Balls are subject to wear by friction. Where long life and dead-tight closure are of paramount importance, a design is recommended that provides for lifting the ball off its seat before it is turned. This measure also prevents freezing or galling. Lift-off is achieved by mechanical means such as an eccentric cam. This design also facilitates the handling of slurries and abrasive fluids, and it can be used for high pressures.

The proper materials for body and trim depend on the application. For handling chemicals or corrosive fluids, all wetted parts will possibly require stainless steel, plastics, or glass (borosilicate glass is preferred for impact strength).

Connections

Ball valves are made with the same connections as used in all other valve types. Where screwed connections must be used, valves manufactured with ends that take the place of unions should be preferred. Threads are either NPT or AND standard.

Flow Characteristics

The flow characteristics of a ball valve approximate those of an equal-percentage plug (Figure 4.3a). Both curves compare favorably with those of other rotary-stem valves. The flow path through a ball valve includes two orifice restriction locations (Figure 4.3f). Balls characterized either by a notch or by a noncircular bore give somewhat better characteristics. Critical flow in ball valves is encountered at $\Delta P = 0.15P_1$, far below the usual figure of 50 percent of absolute inlet pressure.

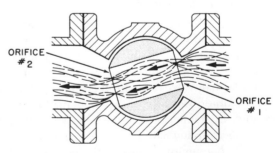

Fig. 4.3f   Ball valve throttling

Sizing

Sizing of ball valves proceeds along the lines described in Section 4.16, with the possible exception that due to the essentially straight-through flow feature, a ball valve can be chosen whose size is equal to the nominal pipe size, which might be an advantage on slurry service. The low-pressure loss and high recovery of a ball valve must be considered in the calculations.

Conclusions

The application of ball valves for control is a comparatively new field. Not many useful hints are published in manufacturers' literature, and it is recommended to proceed with caution, relying partly on one's experimentation. Where much noise is encountered, such as with natural gas flowing at high speed, it may be necessary to bury the valve.

The use of ball valves to control liquid oxygen in the experimental X-15 air-space vehicle as early as 1961, and later in the Atlas rocket, attests to the precise controllability of the ball valve. A special valve with dual-ball design for mixing liquid oxygen with liquid ammonia in precise proportions has also been used in the Atlas.

Characterized Ball Valves

Operation

The characterized ball valve category includes the V-notched ball valve and the parabolic ball valve. These valves were introduced in 1962, partially in an effort to solve the problem of valve clogging and dewatering in paper stock applications. Since then these valves have come into more widespread use as a result of increased valve rangeability and the shearing action at the sharp edges of the valve as it closes.

In essentially all characterized ball valves the "ball" has been modified so that only a portion of it is used (Figure 4.3h). The edge of the partial ball can be contoured or shaped to obtain the desired valve characteristics. The V-notching as used by one manufacturer serves this purpose as well as its shearing purpose.

This shape or contour of the valve's leading edge is the main difference between the various manufacturers' products. The ball is usually closed as it is rotated from top to bottom, although this action can be reversed.

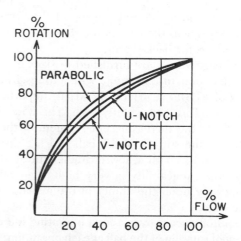

Fig. 4.3g   Inherent flow characteristics of steel body characterized ball valves

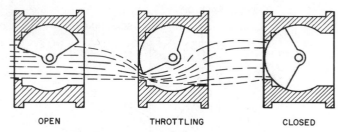

OPEN   THROTTLING   CLOSED

Fig. 4.3h   Positions of the characterized ball valve

Construction

Mechanically the characterized ball valves are very similar to their ancestor, the ball valve. However, because of the assymmetrical design, the characterized ball valve has some design problems that are not significant with the conventional ball valves. A typical characterized ball valve is shown in Figure 4.3i, in its end and side views. The main parts of a characterized ball valve are described below.

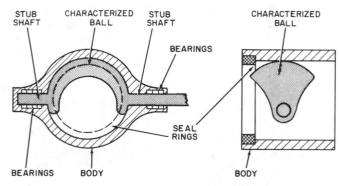

Fig. 4.3i   Characterized ball valve parts

The controlling edge of the ball can be notched or contoured to produce the desired flow characteristics. They are presently available as a V-notched and as a parabolic curve. Mechanically this part can create problems by bending under pressure and thus introducing movement into the shaft seals.

The stub shafts can be distorted by the bending of the partial ball under operating loads.

Valve bodies are still not designed for high pressure or for installations other than insertion between flanges.

The seal ring and seal-retaining ring are usually held in place by companion flanges. Damage due to overtightening of flange bolts sometimes occurs. Figure 4.3j illustrates a special sealing arrangement useful in slurry applications, due to the purging effect created by the flow into the otherwise low-activity zone, through the indent in the ball plug.

Characteristics

The flow characteristics are dependent upon the shape of the edge of the partial ball and on the method of

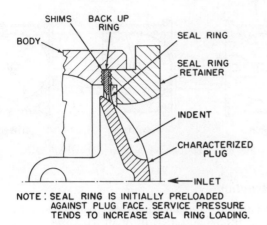

SHIMS BACK UP RING SEAL RING BODY SEAL RING RETAINER INDENT CHARACTERIZED PLUG INLET

NOTE: SEAL RING IS INITIALLY PRELOADED AGAINST PLUG FACE. SERVICE PRESSURE TENDS TO INCREASE SEAL RING LOADING.

Fig. 4.3j   Special seal ring arrangement

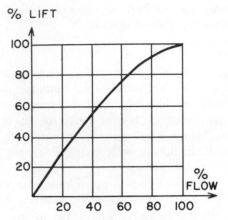

Fig. 4.3l   Inherent characteristics of ball and cage valves with 1,500 PSIG (10 MPa) grating and in stainless steel materials

installation. The shape of the V-notch at the edge of the valve varies from concave for small openings to convex for large openings. Figure 4.3k shows this together with the corresponding shapes for the parabolic ball valves.

The flow characteristics for parabolic, U- and V-notched valves are given in Figure 4.3g. These curves are based on water flow. If the characteristics were evaluated using compressible fluids at critical velocities, these curves would be flatter, closer to linear.

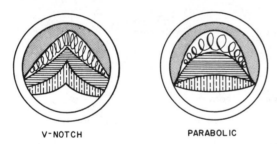

V-NOTCH   PARABOLIC

Fig. 4.3k   Shapes of throttling areas of some characterized ball valves

Ball and Cage Valves

Cage Positioned Ball Design

Positioning of a ball by a cage, in relation to a seat ring and discharge port, is used for control (Figure 4.3m). The valve consists of a venturi-ported body, two seat rings, a ball that causes closure, a cage that positions the ball, and a stem that positions the cage. Seat rings are installed in both inlet and discharge but only the discharge ring is active. The body can be reversed for utilization of the spare ring.

The cage rolls the ball out of the seat as it is lifted by the stem, positions it firmly during throttling, and lifts it out of the flow stream for full opening (Figure 4.3n). The cage is contoured for unobstructed flow in the open position. Cage design includes four inclined control surfaces. The two surfaces next to the downstream seat lift the ball out of the seat and roll it over the top edge of

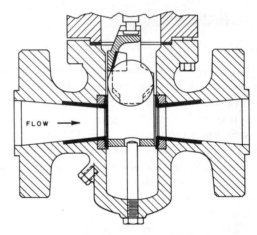

Fig. 4.3m   Cage positioned ball valve

the seat ring as the valve is opened. As the valve opens further, the ball rolls down the first two inclined surfaces to the center of the cage to rest on all four inclined surfaces. The Bernoulli effect of the flowing stream holds it cradled in this position throughout the rest of the stroke. A nonrotating slip stem is guided by a bushing at the bottom and a gland at the top of the bonnet. A machined bevel near the base of the stem acts as a travel limit and allows for back-seating.

Ball and cage valves are furnished in sizes from $\frac{1}{4}$ in. to 14-in. (6.3 mm to 350 mm), with ratings from 150 PSIG to 2,500 PSIG (1 MPa to 17 MPa). Reported flow coefficients (C_v) are consistently high. The flow characteristic reflects the increasing enlargement of the crescent between the surface of the ball and the discharge port (Figure 4.3l). With a flow characteristic starting at zero flow, the rangeability is very high, over 50 to 1, depending only upon the ability of the actuator to position the cage.

Tight shutoff occurs over a long operating life due to the continual rotation of the ball at each operation, which offers a new seating surface each time it is closed. Closure

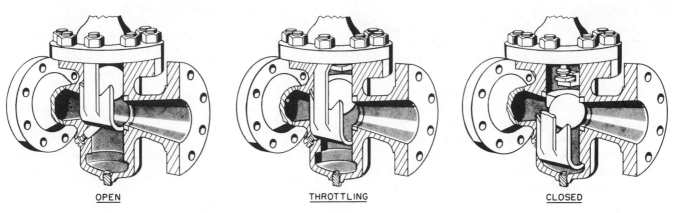

Fig. 4.3n   Throttling with cage positioned ball valve (*Courtesy of DeVar-Kinetics.*)

is positive due to the wedging of the cage in addition to line pressure. Although tightly closed, the stem force for opening is approximately 25 percent of a single-seated valve due to the manner in which the inclined surfaces of the cage roll the ball away from the seat. Opening and closing force factors have been determined for all sizes of the valve. The low opening force requirement (4.76 × 2,000, or 9,520 lb for a 4-in. valve at 2,000 PSIG or 42,364 N for a 100 mm valve at 13.8 MPa) is beneficial in selecting the actuator. The design is conducive to minimizing cavitation effects because flow tends to follow the curve of the ball, thus reducing turbulence. Cavitation tends to occur in the venturi passage, not at the seat, allowing use of hardened or replaceable throats. The expanding venturi discharge assists in handling flashing liquids.

The bonnet design readily lends itself to the adaptation of a variety of linear or rotary actuators. Because the ball must be moved completely out of the flow stream, stem travels are at least as much as the diameter of the valve throat.

Ball Unseated by Stem

The ball cage has also been used for regulators. The ball is cradled in the cage with the valve installed in the vertical position (Figure 4.3o). The stem of the regulator, coming from below the ball, forces the ball away from the seat. Flow is around the ball through the annular space, similar to the flow in a single-seated valve.

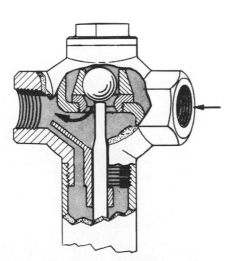

Fig. 4.3o   Ball unseated by stem
(*Courtesy of Powers Regulator Co.*)

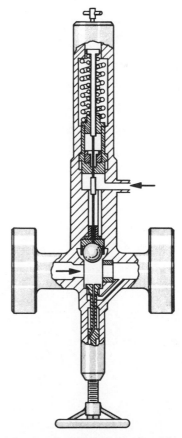

Fig. 4.3p   Ball and cage valve for emergency closure

Ball Gripped by Cage

A variation of the ball and cage design is used for emergency closure (Figure 4.3p). Separate springs and ejection pistons allow high and low limit settings. Pressure above the high setting pushes the piston down to eject the ball from the holder into the seat. Low pressure allows the low-pressure spring to push the piston down. The ball is held firmly on the seat by the differential pressure. An internal bypass is opened to equalize the system pressures. Rotation of the bypass handwheel moves the ball back into the holder, the reset rod is retracted, and the bypass valve closed for normal operation.

BIBLIOGRAPHY

Baumann, H.D., "Trends in Control Valves and Actuators," *Instruments & Control Systems*, November, 1975.

Carey, J.A., "Control Valve Update," *Instruments & Control Systems*, January, 1981.

Dobrowolski, M., "Guide to Selecting Rotary Control Valves," *Instrumentation Technology*, December, 1981.

Fernbaugh, A., "Control Valves: A Decade of Change," *Instruments & Control Systems*, January, 1980.

Hammitt, D., "Rotary Valves for Throttling," *Instruments & Control Systems*, July, 1977.

Holton, A.D., "Control Valve Update," *Instruments & Control Systems*, January, 1982.

Toyama, A., "On Extra-Large High Pressure Ball Valves," *CEER*, April, 1970.

Ytzen, G.R., "Ball Valves for Throttling Control," *19th ISA Conference*, Preprint No. 11.6-2-64.

4.4   BUTTERFLY VALVES

Types of Designs:	A—Aligned B—Offset
Sizes:	A—$\frac{3}{4}''$ to 200″ (18.8 mm to 5000 mm) B—2″ to 80″ (50 mm to 2000 mm)
Design Pressure:	A—To 6000 PSIG (40 MPa) B—To 1480 PSIG (10.2 MPa)
Design Temperature:	A—Cryogenic to 2200°F (1204°C) B—Cryogenic to 1700°F (927°C)
Relative Capacity:	A—Full open: To $C_v = 45\ d^2$; throttling: $C_v = 17\ d^2$ to $C_v = 20\ d^2$ B—$C_v = 20\ d^2$ to $C_v = 38\ d^2$
Materials of Construction:	A—Any metal, elastomer lined, TFE encapsuled B—Steel, stainless steel, alloy steel, Monel, Hastelloy C; seals of TFE or metal.
Special Features:	Reduced torque disc designs, fire-tested seals, reduced noise disc, special disc seal designs.
Cost:	See Figure 4.4a. Cost data are based upon standard high-performance, eccentric disc, soft-seat valves with double-acting piston operator and positioner. Carbon steel bodies up through 8 in. (200 mm) size have 316 SST disc and 17-4PH shaft. Above 8 in. size, discs are chrome-plated carbon steel. Stainless steel bodies have 316 SST disc with 17-4PH shaft in all sizes. Other operators can be furnished. Other alloys are available at higher cost.
Partial List of Suppliers:	Allis-Chalmers; AMRI, Inc.; Clow Corp.; Contromatics; Demag Valves; DeZurik, a Unit of General Signal; The Duriron Co., Inc.; Eur-Control; Fabri-Valve; FACE; Fisher Controls; Flowseal, Div. Mark Controls Corp.; Frisch; Jamesbury Corp.; Keystone International, Inc.; Masoneilan International, Inc.; Norris, Div. Dover Corp.; Posi-Seal International, Inc.; Shan-Rod, Inc.; Valtek, Inc.; Vanessa

The butterfly valve is one of the oldest types of valves still in use. The dictionary defines the butterfly valve as a "damper or throttle valve in a pipe consisting of a disc turning on a diametral axis" (Figure 4.4b). Butterfly valves are not only used in industry, but variations are found in consumer products such as furnace dampers, automobile carburetors, and shower heads.

Wide use of butterfly valves dates back only to the 1920s when improved designs resulted in recognition and acceptance by engineers of public waterworks. The valve was particularly applicable to the low-pressure on-off service usually encountered in waterworks applications. Today's modern butterfly valve designs are suitable for a wide variety of fluid applications including those with high pressure drop, tight shutoff, and corrosive characteristics. The straight-through design has high capacity and has advantages when erosion is considered. Especially since the development of the eccen-

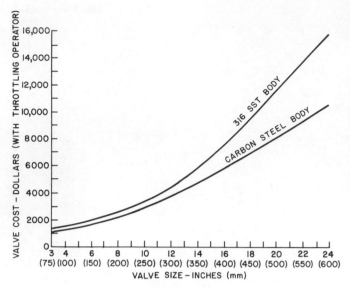

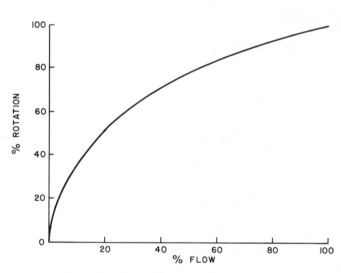

Fig. 4.4a   Cost and inherent characteristic of 60° opening standard swing-through butterfly

Fig. 4.4b   Vane positions of butterfly valve

tric shaft and disc designs known as the high-performance butterfly valve (HPBV), it is one of the fastest growing segments of the valve industry.

Operation

Butterfly valve operation is basically simple, since it involves only rotating the vane, disc, louver, or flapper by means of the shaft to which it is fastened. This may be done manually by a lever handle on small sizes or by a handwheel and rotary gear box on larger sizes. Automatic operation may be accomplished by pneumatic, hydraulic, or electrical motor drives attached to the shaft by various methods. As the disc moves through a 90-degree rotation, the valve moves from fully closed to fully open (Figure 4.4b). The free area developed by the disc as it moves toward the full open position provides the throttling operation. A plot of the free area vs percent vane rotation results in a curve very similar to the butterfly inherent characteristic curve of Figure 4.4a. However, modern valve operator positioners are often available with cams to give other characteristic curves.

Construction

Mechanically, butterfly valves vary widely in their construction features. However, common to all is the valve body, the disc and shaft, shaft support bushings and/or bearings, shaft packing, and a means of attaching an operator to the shaft. Butterfly valves also fall into two basic categories, swing-through and shutoff designs. Most swing-through designs (Figure 4.4c) have a symmetrical disc and shaft design with a certain clearance required between disc and body. The body is usually the solid ring type, which is mounted between pipe flanges. It can be either the wafer type or the single flange lug pattern where the flange bolting also goes through the valve body. Discs are cast in one piece. The thickness of the disc and hub along with the diameter of the shaft is a function of the maximum pressure drop and torque required. Careful alignment of the body, bushings, shaft, and disc eliminates binding. Hard facing materials can be applied to the disc edge and body bore where erosive fluids such as steam are involved. Refractory type linings are also available for the body.

Variations of swing-through butterfly valves are available that offer special features such as special disc shapes to reduce torque and increase throttling angle capability

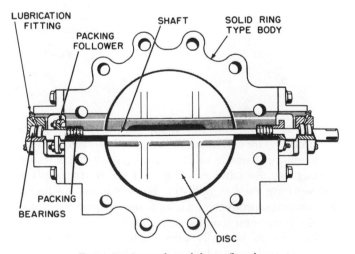

Fig. 4.4c   Swing-through butterfly valve
(*Courtesy of Fisher Controls Company*)

(Figure 4.4d). Special shutoff seals such as piston rings for high temperatures to 1500°F (816°C) and T-ring seals for tight shutoff are also available (Figure 4.4e). However, these seals are not as popular now since the development of the HPBV valves and other designs with improved high-temperature seals.

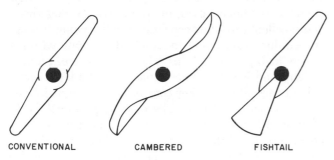

Fig. 4.4d   Disc shapes

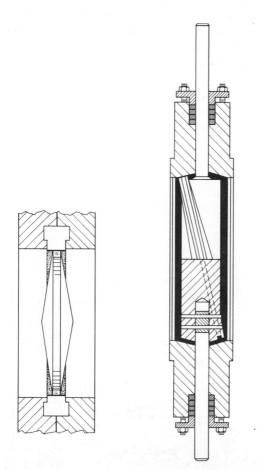

Fig. 4.4e   Special butterfly seals
(*Courtesy of Fisher Controls Company*)

Swing-through butterfly valves are normally limited to about 70 degrees open for the standard patterns and 60 degrees open for the heavy patterns with larger diameter shafts (Figure 4.4f). This is because the disc profile projection tends to "disappear" into the shaft area as the valve opens.

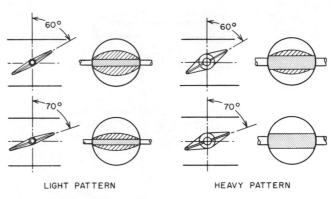

Fig. 4.4f   Effect of design pattern on flow area in butterfly valves

Shutoff Designs

Butterfly valves designed for tight shutoff generally fall into two categories. One is the valve with an elastomer or plastic liner, with the disc also encapsulated in some cases (Figure 4.4g). The other is the HPBV with the cammed disc and a separate seal ring clamped into the body (Figure 4.4h). In addition, there are some special designs with laminated seal rings located on the disc edge that wedge into a conical seat in the valve body (Figure 4.4i). These laminated seal designs are especially suitable for high pressure and temperature shutoff.

Lined butterfly valves were once the only butterfly valves specifically designed for shutoff. They use various elasatomer materials, usually with a more rigid backup ring, which completely line the bore of the valve and the valve gasket face area. Some liners are bonded to the body and others are removable. In addition, liners of plastics, such as Teflon® are available for corrosive fluids. In some cases the disc is also encapsulated in the elastomer or Teflon. Sealing in these valves is usually accomplished by a wedging action of the disc edge into the elastomeric or plastic seat. The discs may be symmetrical on the shaft (similar to swing-through), offset

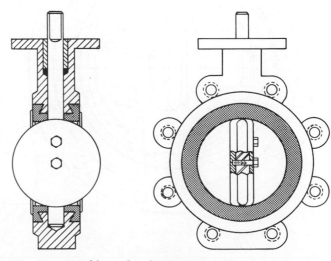

Fig. 4.4g   Lined butterfly valve (*Courtesy of Keystone Valve-USA*)

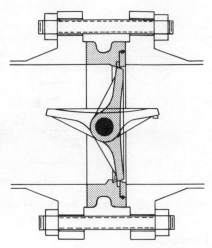

Fig. 4.4h   High-performance butterfly valve (HPBF)
with cammed disc (*Courtesy of Valtek, Inc.*)

these problems if the process fluid seeps past liner shaft seals and attacks the backing material or the body metal. By their very nature, these materials are also temperature limiting and can seldom exceed 350°F (177°C).

The most significant design advance in butterfly valves is the development of the high-performance butterfly valve (HPBV). This design concept combines the tight shutoff of the lined valves, the reduced operating torque and excellent throttling capabilities of the swing-through special disc shapes, and the ability to operate with relatively high pressure drops. The compact size, reduced weight, and lower cost have made the HPBV a formidable competitor to other control valve designs in sizes 3 in. (75 mm) and larger. There are many excellent designs available from many manufacturers. The two main characteristics of this design are the separable seat ring contained in the body and the eccentric cammed disc (Figure 4.4j and k). This camming action enables the disc to back out of and into the seat before and after the disc rotation when throttling. This is accomplished by having the shaft offset from both the centerlines of the disc and the valve body.

The seat retainer rings are fastened to the body by various means. The most common method is by countersunk screws or bolts. Usually these fasteners are within the gasket area and it is advisable to evaluate this inter-

from the shaft, or canted on the shaft. The objective of the latter two designs is to give a 360-degree seal contact on the disc edge.

Caution must be exercised in the selection of elastomers since they may be subject to attack by the process fluid. This attack may result in softening, swelling, cracking, or other effects. Plastic liners are not immune from

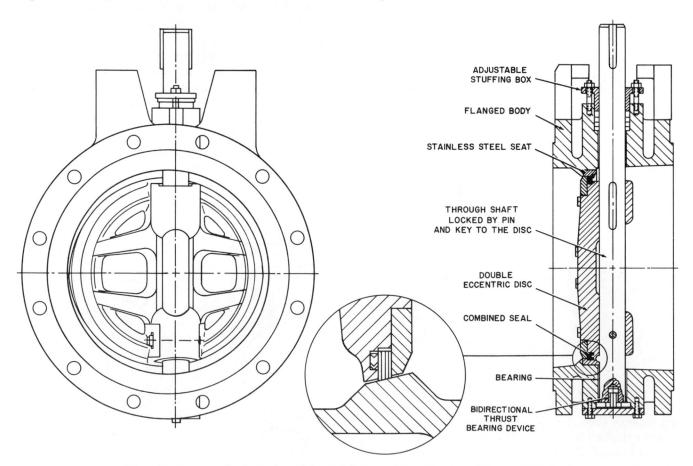

ADJUSTABLE
STUFFING BOX

FLANGED BODY

STAINLESS STEEL SEAT

THROUGH SHAFT
LOCKED BY PIN
AND KEY TO THE DISC

DOUBLE
ECCENTRIC DISC

COMBINED SEAL

BEARING

BIDIRECTIONAL
THRUST
BEARING DEVICE

Fig. 4.4i   Eccentric disc butterfly with laminated disc seal ring (*Courtesy of Vanessa USA, Inc.*)

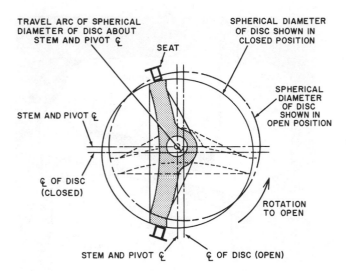

Fig. 4.4j Eccentric shaft design (*Courtesy of Valtek, Inc.*)

ference in the gasket area for the intended service. Normally this does not pose a problem, but it may require special gaskets or a change in gasket type to effect a proper seal. Other methods, such as a snap ring, friction fit, or retaining pins, do not intrude into the gasket area and may offer a better choice in some applications.

Seats are commonly made of plastic materials such as Teflon or various elastomers. Each manufacturer has a specific idea of how this seat seal should be designed and configured. These designs range from very simple to very complex shapes, and in some cases may incorporate elastomers behind the Teflon to effect a pressure-energized seal. Basically, these various seat designs are classified as ANSI Class VI, which is commonly construed to be "bubble-tight." Where temperatures are too high for the

soft seat materials, metal seats are available which can also provide excellent shutoff equivalent to ANSI Class IV. Some metal seat designs can even approach ANSI Class V.

A special design variation known as a "fire-safe" seat is also available from many manufacturers. This combines the soft seat discussed above with a backup metal seat such that fire destruction of the soft seat will enable the metal seat to minimize leakage (Figure 4.4l). These designs also incorporate special fire-resistant gaskets and stem packing to minimize external leakage during and after a fire.

Torque Characteristics

Operating torque requirements of butterfly valves require more careful consideration than any other type. The disc acts much like an airfoil or the wing of an aircraft. However, the special disc shapes and the HPBV designs already discussed have much lower torque requirements due to the shape effects, much like "spoilers" on an airfoil. The conventional symmetrical disc behavior as an airfoil results in different pressure distributions around the face of the disc, producing a torque which tends to close the valve. Only at 0 and 90 degrees are the pressures equal on both sides of the disc. Between 0 and 90 degrees the thrust load of the disc wing turned toward the upstream side is larger than that on the downstream side. This is called the unbalanced or hydraulic torque, and its magnitude is a function of the pressure drop and disc diameter.

The valve operator must have enough power to overcome the unbalanced torque and, in addition, overcome the friction of the bearings and packing on the valve shaft. The total is known as combined torque. The combined

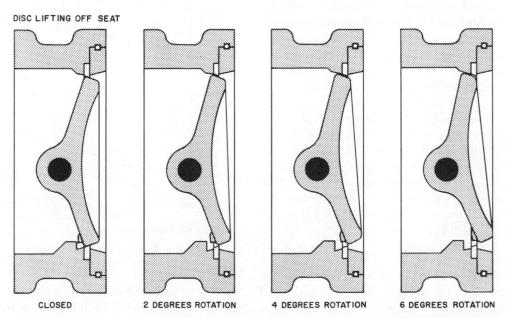

Fig. 4.4k High-performance butterfly valve cam action disc operation (*Courtesy of Valtek, Inc.*)

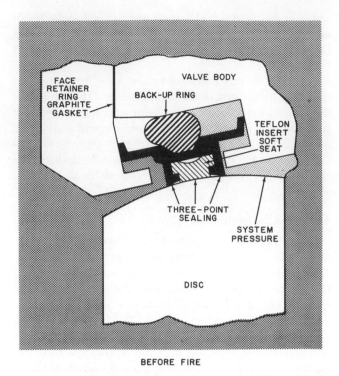

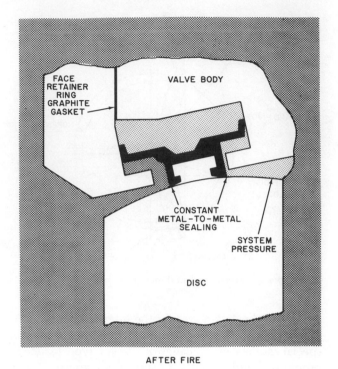

Fig. 4.4l   One type of "fire-safe" disc seal seat design (*Courtesy of Posi-Seal, Inc.*)

torque required to open the valve is larger than that required to close the valve because the unbalanced torque helps to close the valve. This difference is known as torque hysteresis. In fact, when the symmetrical disc closes, the torque may become negative somewhere around 30 to 35 degrees from the closed position, and the valve will tend to close itself. The torque characteristics also indicate that at about a 75 to 80-degree opening, the torques for both opening and closing maximize. Above this rotation angle, the torque falls quickly to zero. Thus, the torque characteristics are highly nonlinear. They pose a considerable burden on the valve automatic operator since it must cope with sharp increases and decreases in torque as well as positive and negative forces.

Typical torque curves for conventional symmetrical, special shape, and fluted discs are shown in Figure 4.4m. It should be noted that while HPBV disc designs can be considered a reduced torque disc, the amount of torque and its characteristic behavior curve is a function of whether the valve is installed with the shaft upstream or downstream. Flow tends to open the valve with the shaft downstream and tends to close the valve with the shaft upstream. However, with the shaft downstream, dynamic torques with the disc open are much lower. On gas service, the shaft can be either direction, but on liquid service the shaft should be downstream only because of the effects of liquid inertial forces.

Lined butterfly valves are not only subject to the above torque considerations but also encounter the additional torque required to seat and unseat the disc into the liner. Since the torque requirements for any butterfly valve design are subject to so many variables, they are normally

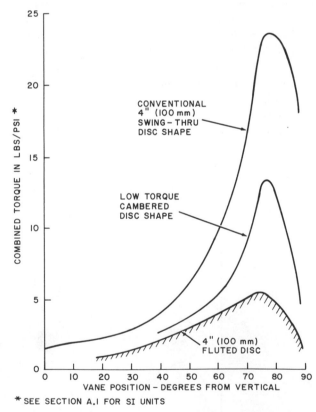

* SEE SECTION A.I FOR SI UNITS

Fig. 4.4m   Typical butterfly valve disc shape torque curves
(*Courtesy of H. D. Baumann Co.*)

rated by the manufacturer, and the manufacturer's recommendations should be followed for the operator size required to operate the valve over the application range. It is best to be conservative when sizing butterfly valve operators since operating and seating torques often are greater than predicted.

Noise Suppression

Butterfly valves will generate noise as will any other valve design when throttled under the flow rate and pressure drop conditions are conducive to noise generation (See Section 4.14). The noise characteristics of the special disc designs (such as cambered and fishtail shown in Figure 4.4d) are generally improved as compared to the conventional swing-through disc shape. However, with higher mass flows and pressure drops the noise can still be substantial.

A newer design development for the butterfly is the fluted disc shown in Figure 4.4n. For compressible fluid applications, this special disc design is capable of noise reduction of up to 10 decibels on the A-weighted scale (dBA) as shown on Figure 4.4o. In addition, the airfoil "spoiler" effect of the flutes enables this disc to have the lowest operating torque requirements of any disc design. It can be provided for valve sizes 2 in. (25 mm) through

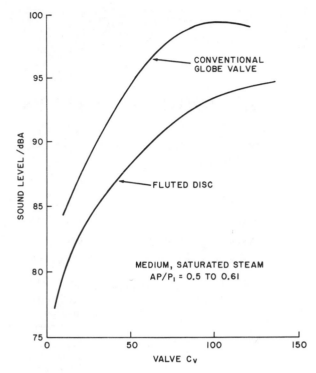

Fig. 4.4o   Typical noise reduction with a fluted disc

16 in. (400 mm). It is a more expensive design than others, but is a very useful alternate where needed.

BIBLIOGRAPHY

Baumann, H.D., "The Case for Butterfly Valves in Throttling Applications," *Instruments & Control Systems*, May, 1979.

———, "Trend in Control Valves and Actuators," *Instruments & Control Systems*, November, 1975.

Boger, H.W., "Low-Torque Butterfly Valve Design," *Instrumentation Technology*, September, 1970.

Carey, J.A., "Control Valve Update," *Instruments & Control Systems*, January, 1981.

Daneher, J.R., "Sizing Butterfly Valves," *Water & Wastes Engineering*, July and September, 1970.

Dobrowolski, M., "Guide to Selecting Rotary Control Valves," *Instrumentation Technology*, December, 1981.

Fernbaugh, A., "Control Valves: A Decade of Change," *Instruments & Control Systems*, January, 1980.

Hammitt, D., "Rotary Valves for Throttling," *Instruments & Control Systems*, July, 1977.

Hutchison, J.W., ed. *ISA Handbook of Control Valves*, 2nd ed., 1976.

Passage, D., "Butterfly Control Valves," *Instruments & Control Systems*, March, 1964.

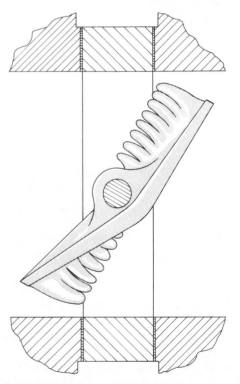

Fig. 4.4n   Fluted butterfly disc useful for valve noise reduction

4.5 DIGITAL VALVES

Size:	$\frac{3}{4}''$ to 10″ (19 mm to 250 mm) in-line and angle pattern
Design Temperature Limits:	Cryogenic to 950°F (510°C)
Design Pressure Limits:	10,000 PSIG (68,688 kPa)
Capacity:	$C_v = 13\ d^2$
Materials of Construction:	Body—Aluminum, carbon steel, stainless steel, titanium Seals—Buna, rubber, Viton, or TFE
Cost:	Typical prices for ANSI Class 600 carbon steel body, 12 bit valves are $4500 for $\frac{3}{4}$ in. (20 mm), $11,000 for 1, 1$\frac{1}{2}$, and 2 in. (25, 32, 40, and 50 mm), $21,500 for 3 in. (75 mm), $34,500 for 4 in. (100 mm), $38,500 for 6 in. (150 mm), and $51,000 for 8 in. (200 mm). For 316 SST bodies, add $1000 for $\frac{3}{4}$ in., $2,000 for 1, 1$\frac{1}{2}$, and 2 in., $3,000 for 3 in., $5,000 for 4 in., $7,000 for 6 in., and $14,000 for 8 in.
Suppliers:	Digital Valve Co.

Operation

A digital valve is a group of valve elements assembled into a common manifold. The elements have a binary relationship to each other, i.e., starting with the smallest, each increasing size element is twice as large as its next smallest neighbor. Each element is controlled by an individual electric or electronic signal. Thus, an 8-bit digital valve requires 8 parallel, on-off electric (or electronic) signals. A 12-bit digital valve will require 12 parallel signals, and a 16-bit digital valve will require 16 parallel signals.

An 8-bit digital valve system controlling a corresponding number of flow elements in a valve body (Figure 4.5a) constitutes a digital control valve with a rangeability of 255:1. This body, an in-line, top entry design, includes an air reservoir to assure the fail-safe operation of the cylinder actuators. The design includes adequate manifold area to assure consistent performance from the individual elements. The manifolds are large enough to minimize the possibility of cavitation and resulting errosion.

Each element in the array is on-off. Flow throttling is accomplished by opening enough ports to provide the exact flow area called for by the electric signals. There is a 1:1 relationship between the binary weighted signal and the binary weighted flow area. Figure 4.5b illustrates schematically the size relationship between binary elements in a digital valve.

Since a binary array of 12 bits will provide resolution of 1 in more than 4000 (the area of the least flow element in a 12-bit digital valve is less than 0.025 percent of the total flow area), it is essential that the digital valve be leak tight. Leakage rates that are acceptable in standard globe valves would make such resolution impractical. Digital valves are commonly manufactured to extremely high leak-tight standards.

In order to smooth the transfer between 49+ and 50 percent open, it is usual to use two 25-percent elements instead of one 50-percent element for the largest bit. Thus an 8-bit digital valve would have nine elements—three 25 percent, one 12.5 percent, one 6.25 percent, one 3.125 percent, one 1.56 percent, one 0.78 percent, and one 0.39 percent element, for a total of nine.

In higher resolution digital valves, multiple elements are common, i.e., a 12-bit digital valve could have 16 elements, and a 14-bit digital valve could have 18 elements. The largest element then would be 12.5 percent. Each such valve would have seven 12.5 percent elements.

The 6 in. (150 mm) cast steel body digital valve (Figure 4.5c) uses 12 elements to provide 8-bit performance. The binary series of control element capacities is 1, 2, 4, 8, 16, 32–32, 32–32, 32–32. In this way no individual element handles more than 12.5 percent of the total flow. The largest bit (50 percent) is handled by 4 elements spread out around the body. This arrangement improves redundancy in operation and uniformity in individual element sizing. Noise generation is reduced by breaking the flow up into many flow paths. The 8-bit computer word remains the same, but operates 12 control elements.

This method of breaking the flow into many streams has a number of advantages. Each element actuator (electric, hydraulic, or pneumatic) is small and very quick. The multiple element arrangement provides redundancy, usually permitting operations to continue even if one or more elements are disabled. Flow calibration requirements are affected since the calibration unit need be only $\frac{1}{8}$ as large to qualify 100-percent capacity.

A valve assembly often incorporates two or three sized elements. Flow differences between individual elements are determined by the control orifice or the nozzle size. Each element consists of a plunger and a seat. The plunger is operated by a solenoid or solenoid-piloted cylinder (pneumatic or hydraulic). Elements are usually dynamically balanced for low actuating force and are designed to assure tight shutoff.

Because of the necessary leak-tight construction, digital valves of very small C_v are practical. Standard digital valve body sizes range from $\frac{3}{4}$ in. NPT (19 mm) through 10 in. (254 mm) pipe size. Pressure ratings up to 10,000 PSIG (68,788 kPa) have been supplied. Digital valves

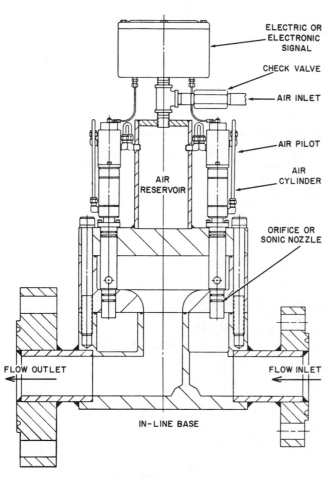

ELECTRIC OR ELECTRONIC SIGNAL

CHECK VALVE

AIR INLET

AIR PILOT

AIR CYLINDER

AIR RESERVOIR

ORIFICE OR SONIC NOZZLE

FLOW OUTLET

FLOW INLET

IN-LINE BASE

Fig. 4.5a   3" 3000 ANSI carbon steel 8-bit, 12-element digital valve
(*Courtesy of Powell Process Systems, Inc.*)

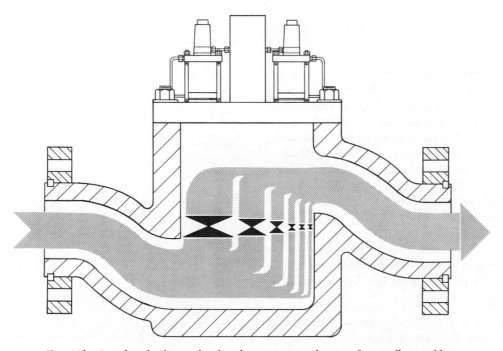

Fig. 4.5b   In a digital valve, each valve element is twice the size of its smaller neighbor.

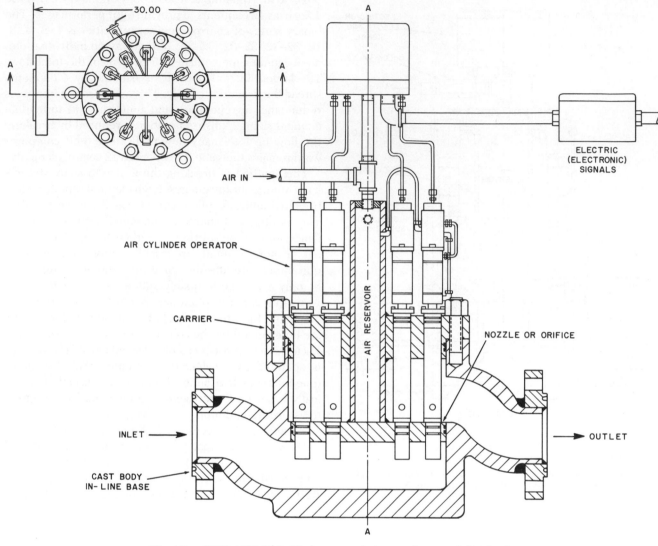

Fig. 4.5c   6″ 900 ANSI 8-bit, 12-element explosion-proof cast steel digital valve
(*Courtesy of Powell Process Systems, Inc.*)

have been used effectively in cryogenic service, and special high-temperature models are under development (to 1250°F or 677°C). Fluid temperatures above 450° to 500°F (232° to 260°C) require special seals and seal design.

Digital valves provide exactly repeatable performance, since each digital command causes the opening of a precisely defined port area. Nominal transfer time from any one position to any other position is usually under 100 milliseconds, and the transfer time is uniform from one position to another.

The binary relationship between elements provides a linear increase of port area with a linearly increasing digital signal command. Any desired valve characteristic can be provided by programming the digital command.

Flow Metering

A valve-flowmeter is a special version of the digital valve equipped with flow sensing nozzles or orifices in each element (flow port).

Gas Flow

When the sensing element installed in each port is a "sonic venturi" (Figure 4.5d), and the valve inlet pressure is high enough to induce sonic flow at the vena contracta (throat), the rate of gas flow is dependent only on the inlet conditions (i.e., absolute temperature and

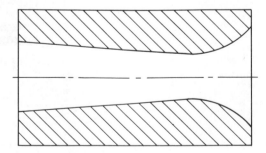

Fig. 4.5d   Sonic venturi element

absolute pressure in a given fluid). This condition of "choked flow" is independent of downstream pressure, and in most designs it can be maintained with a pressure drop across the venturi of 15 percent of the inlet absolute pressure or less (as long as the downstream pressure is less than the critical at the vena contracta). The divergent portion of the sonic venturi recovers part of the velocity head. Thus variations in the downstream pressure do not affect the flow rate.

The flow rate is so identifiable under these conditions, that digital valves so equipped are being used as transfer standards for calibrating other flow meters. An 8″ (200 mm) digital valve-flowmeter equipped with sonic venturis has been installed in a power plant fuel line to act as a combination pressure regulator and fuel flow meter.

Liquid Flow

An orifice can be installed in each element and the differential pressure across the digital valve can be used as a measure of the fluid flow through the open port area. Digital valves can be used as wide-range flowmeters by controlling the valve opening to maintain a constant pressure drop across the valve. Thus a low pressure drop measurement can have a wide flow range (up to 16,000:1 or more).

In summation, digital valves can provide high resolution, very fast response, exact repeatability, and very wide range, and they can be equipped with flow elements to provide double service. Recommended applications are flow blending, compressor surge control, gas meter-regulators, transfer-standard flow provers, and precise liquid flow rate measurement and control. In fact they are appropriate any place where speed, accuracy, and/or resolution have useful application.

BIBLIOGRAPHY

Britton, Charles L., "Sonic Nozzles," Appalachian Gas Measurement Short Course, August, 1978.

Corey, J.A., "Control Valve Update," *Instrument & Control Systems*, January, 1981.

Clark, Herbert L., "Turbine Meter Testing," AGA Transmission Measurement Conference, May, 1979.

Fernbaugh, A., "Control Valves: A Decade of Change," *Instrument & Control Systems*, January, 1980.

Jones, Donald R.A., "Digital Valves," AGA Transmission Measurement Conference, May, 1977.

Morris, Warren, "Digital Valve as a Transfer Standard," *Gas Magazine*, September, 1980.

Langill, A.W., "New Control Valve Accepts Digital Signals," *Control Engineering*, August, 1969.

4.6 GLOBE BONNET AND TRIM DESIGNS

Definitions

The valve bonnet is the top closure assembly for the globe valve, as well as for several other valve body design types. In addition to closing the valve body, the bonnet also provides the means for mounting the actuator assembly to the valve body and sealing the valve stem against process fluid leakage. The various bonnet designs will be discussed along with the subject of stem sealing utilizing packing materials, lubricants, and special seal designs.

The valve trim consists of the internal parts contained within the body and wetted by the process fluid. The main components are the plug and stem and the seat ring(s). Some globe valve body designs will also incorporate other parts such as seat retainers, spacers, guide bushings, and special elements. Most of the pressure loss dissipated in the valve is absorbed by the main trim parts. The trim design also serves to determine the inherent flow characteristics of the valve. The various aspects of trim design, construction, and selection will be discussed.

Bonnet Designs

The most common valve bonnet design is the bolted bonnet. It is usually fastened to the valve body by high-strength stud bolts and heavy nuts. Removal of the bonnet gives complete access to the valve trim for maintenance purposes. Since the bonnet is a pressure-carrying part of the overall fluid containment system, the design is calculated in accordance with the applicable ASME Code requirements. This includes flange size and thickness, wall thickness, and bolting size.

The seal between the valve bonnet and body can take several forms depending upon the valve design and application range, including containment pressure and fluid temperature. Fluid corrosion is also a factor that must be considered. The most common seal is a contained gasket, either flat or the spiral wound design (Figure 4.6a). Other designs that may be found are the API ring joint (oval or octagonal cross section), lens type, delta gasket, and Bridgeman gasket (Figure 4.6b). These latter designs are metal type gaskets. The flat gasket is usually

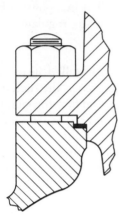

Fig. 4.6a Valve bonnet joint showing retained gasket design

either asbestos or Teflon plastic. The spiral wound is usually asbestos or Teflon in combination with an austenitic (304 or 316 SST) metal ring, although other metal alloys are obtainable for special service needs such as high temperature or severe corrosion. There is a move away from the use of asbestos due to environmental considerations, but tests indicate that substitutes do not perform as well. Proper handling and disposal should obviate any asbestos hazards as a strong factor for consideration.

Other than considerations of ASME Code design, manufacturers design their bonnets to provide features that predominate in their particular philosophy of valve design. The depth of the stuffing box and surface finish, provision of guides, method of operator attachment, packing design flexibility, packing follower design, etc. are all details that vary from manufacturer to manufacturer or even among various body designs offered by one manufacturer. Some low-pressure valves, especially in sizes below 2 in. (50 mm) can be provided with a threaded bonnet. This reduces the weight and is more economical in first cost than the flanged and bolted design. In many cases this is false economy since it is difficult to remove after extended service in high temperature, may promote crevice corrosion with corrosive fluids, and generally is a more costly valve to maintain. In rare cases the bonnet

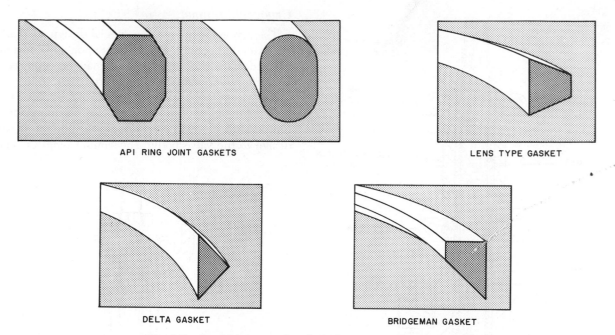

API RING JOINT GASKETS   LENS TYPE GASKET

DELTA GASKET   BRIDGEMAN GASKET

Fig. 4.6b Special bonnet gaskets for high pressure and temperature

to body joint may be seal welded to minimize leakage of extremely toxic or radioactive fluids.

Bonnets fall into basically three classifications. These are standard, extended for hot or cold service, and special designs for cryogenic service, bellows seals, etc. These classifications along with the stem seal systems, guides, and bushings will be discussed in more detail later in this section.

Standard Bonnet

The standard or plain bonnet (Figure 4.6c) is the normal bonnet design furnished on most valves. It covers the range of pressures and temperatures compatible with standard seal gaskets and stem packing materials. In general, this includes valves designed for ANSI Classes 150 through 2500 pressure rating and temperatures from −20° to 600°F (−30° to 315°C). Above 450°F (230°C) alternate packings or extended bonnets must be used. Probably over 90 percent of all control valve applications can be handled by the plain bonnet design. It may or may not incorporate stem guides or bushings and may have very broad or very limited packing configurations available, depending upon the specific manufacturer and valve design.

Extended Bonnet

The extended bonnet (Figure 4.6d) is usually required when the fluid temperature is outside the plain bonnet temperature limitations. Even when normal process temperatures are within the plain bonnet limits, it may be best to go to the extended bonnet design to protect against temperature excursions that often occur with upsets in process operation.

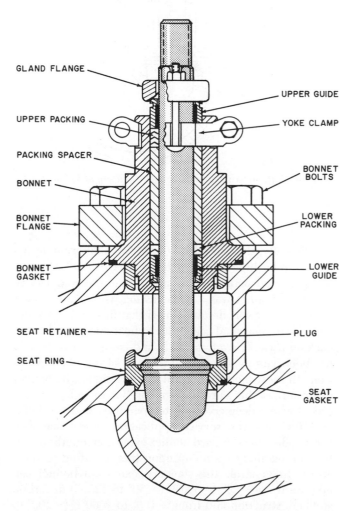

GLAND FLANGE

UPPER GUIDE

UPPER PACKING

YOKE CLAMP

PACKING SPACER

BONNET
BOLTS

BONNET

BONNET
FLANGE

LOWER
PACKING

BONNET
GASKET

LOWER
GUIDE

SEAT RETAINER

PLUG

SEAT RING

SEAT
GASKET

Fig. 4.6c Plain bonnet design using seperable instead of integral bonnet flange (*Courtesy of Valtek, Inc.*)

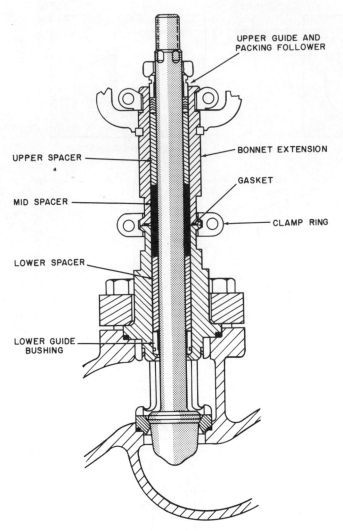

Fig. 4.6d   Standard extension bonnet (*Courtesy of Valtek, Inc.*)

Cryogenic Bonnet

The cryogenic bonnet is a special design adaptation of the extended bonnet. Depending upon a manufacturer's valve design, this style bonnet may be required with operating temperatures ranging from −150° to −300°F (−100° to −185°C) down to −425°F (−255°C) using a bolted bonnet. To go down to the maximum extreme temperature of −454°F (−270°C), the bonnet is a welded design to the valve body. The bonnet length is tailored

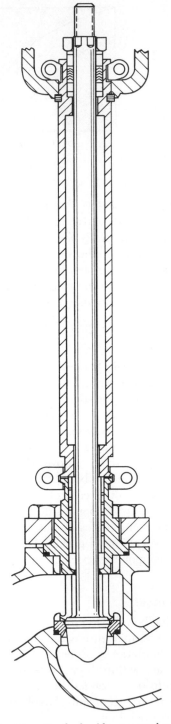

Fig. 4.6e   Standard cold extension bonnet
(*Courtesy of Valtek, Inc.*)

Originally, the extended bonnet was a different design for hot and cold service. The hot service extended bonnet was provided with "cooling fins" while the cold service extended bonnet was a plain casting without fins. Over many years it was demonstrated that fins on the bonnet added only marginal capability to heat dissipation in the packing area. It made a costly and complex casting and has been largely abandoned in favor of the plain extension. In most modern control valve designs, the bonnet extension is the same design for hot or cold service, except where deep cryogenic temperatures under −150°F (−100°C) are the service application. Some manufacturers offer two standard bonnet extension lengths (other than cryogenic) depending upon the operating temperature. In general, the standard extension bonnet can operate from −20° to 800°F (−30° to 425°C) in carbon steel construction and from −150° to 1500°F (−100° to 815°C) in austenitic stainless steel (304 or 316) construction, although some designs will go down to −300°F (−185°C) before a cryogenic extension design is needed.

to the application to suit valve body size, piping requirements, and operating temperature needs, and it will generally range from 12 in. (300 mm) to 36 in. (900 mm).

The standard and "hard" cryogenic designs are distinctly different. The standard design is usually similar to the standard extension bonnet, except much longer (Figure 4.6e). It can be cast in either one or more pieces and assembled by welding and bolted clamps or can be a completely fabricated weldment. The "hard" cryogenic extension design is a specially built-up system utilizing thin-walled stainless steel tubing (to reduce cooldown weight) that is welded directly to the body casting (Figure 4.6f). The top is welded to a flange assembly that will accept an essentially normal plain bonnet containing the stem packing system. There is a stem seal system at the plug end of the bonnet to keep the cryogenic liquid out of the bonnet and packing area. This seal may be a vented or unvented design (Figure 4.6g), but in any case allows build-up of gas pressure in the warmer bonnet area to relieve back into the valve body. The two designs are purely a result of philosophy differences between companies who design and operate cryogenic plants. However, the most widely used version is the unvented design. (See Section 4.7 for further discussion.)

In some cases, where additional insulation is needed to reduce outside heat flow, the bonnet can be fitted with a vacuum jacket. A cryogenic control valve should be installed such that the bonnet is either vertical or no nearer than 20 degrees from horizontal to insure liquid does not get up to the packing area. In general, these bonnets are limited to a maximum ANSI Class 600 pressure design. Since the extreme cold requires good Charpy impact resistance, materials of construction are limited to the austenitic stainless steels (304 or 316) and bronze.

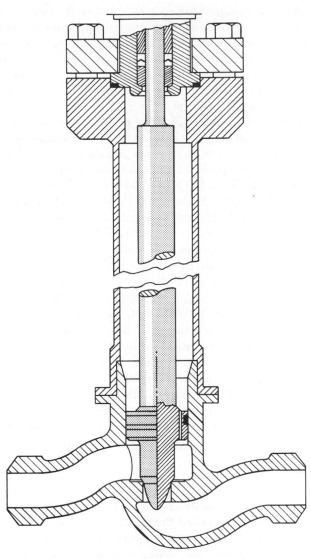

Fig. 4.6f   Typical cryogenic cold extension bonnet (*Courtesy of Valtek, Inc.*)

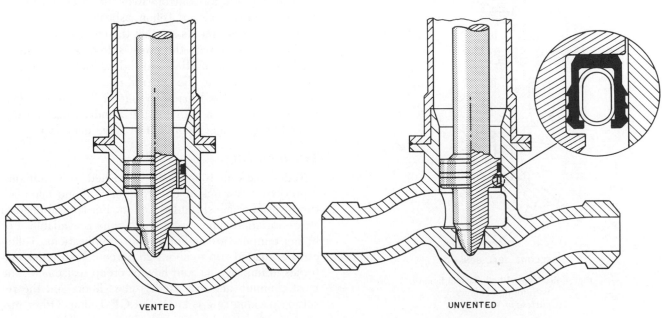

VENTED   UNVENTED

Fig. 4.6g   Plug and bonnet stem seal system for cryogenic valve (*Courtesy of Valtek, Inc.*)

Bellows Seal Bonnet

Available from many control valve manufacturers are extended bonnet designs that incorporate a bellows seal around the stem (Figure 4.6h). The bellows is usually made of stainless steel or other corrosion resistant alloys such as Hastelloy and Inconel and can be hydraulically formed (as shown) or welded in individual segments known as the nested type. The bellows is attached to one end of the stem by welding and to the other end by welding to a clamped-in fitting with an antirotation device. The antirotation device prevents the bellows from being twisted during assembly and disassembly or if the valve plug tends to want to turn due to fluid reaction forces.

Bellows seals are justified in applications involving toxic or radioactive fluids where leakage to the outside would pose personnel safety hazards. As such, these seals are usually helium leak tested using a mass spectrometer to leakage rates below 1×10^{-6} cc/sec, from atmospheric pressure to vacuum. Experience with bellows seals indicates that they may provide a false sense of security due to unpredictable lives. They should always be backed up with a full stem packing gland and a leakage monitoring port between the bellows and the packing. These seals also have pressure and temperature rating limitations, usually about 150 PSIG (1030 kPa) and 100°F (40°C) or 90 PSIG (620 kPa) and 600°F (315°C). The average full stroke cycle life will vary from 50,000 cycles for 1 in. (25 mm) and smaller valves to 8,000 cycles for 3 to 6 in. valves (75 to 150 mm). In some cases, cycle life can be improved by reducing operating pressures or by using special short stroke valve plugs. Operating pressures can be increased to as much as 2900 PSIG (20,000 kPa) and temperatures up to 1100°F (590°C) by multi-ply or heavy-wall bellows and selection of metal alloy. However, this sacrifices cycle life considerably. As a result of these various factors, metallic bellows type seals are seldom used today and have been replaced by special packing systems that will be covered in the discussion on bonnet packing. This trend away from bellows seals has been accelerated by the extremely high purchase, maintenance, and replacement costs associated with these seals.

Bonnet Packing

In order to seal the valve stem against leakage of process fluid to the atmosphere, the upper part of the bonnet contains a section called the stuffing box. This assembly consists of a packing flange, packing follower, a packing spacer or lantern ring, lower packing retainer, and a number of packing rings. Various valve packing materials are available, but for control valves the three most commonly used materials are Teflon, asbestos, and graphite. The major requirements are that the control valve packing be compatible with the process fluid, seal the stem, produce minimum starting and sliding friction, and give long service life. The one material that meets all of these conditions over the broadest range of fluid applications is Teflon. As a result, this is the standard packing furnished by the majority of control valve manufacturers.

Teflon Packing

Teflon packing is normally furnished as virgin (not reprocessed) V-rings or chevron rings. Some manufacturers also furnish shredded Teflon. For special needs a shape variation known as cup and cone is available. For higher temperature and/or pressure applications, Teflon can be "loaded" up to about 25 percent with other materials to improve its cold-flow or creep resistance. The most common material used is glass fiber, and the resulting packing ring is known as GF Teflon. Other ma-

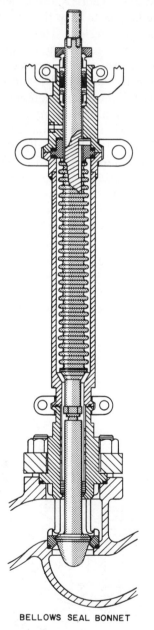

BELLOWS SEAL BONNET

Fig. 4.6h   Typical bellows seal bonnet
with formed bellows
(*Courtesy of Valtek, Inc.*)

terials that have been used are silica and carbon. As noted earlier, Teflon is limited to a maximum high temperature of about 450°F (230°C) in a plain bonnet, but can be used successfully in the GF version in an extended bonnet to about 850°F (455°C). The low temperature limit is about −300°F (−185°C).

Contrary to some popular claims, Teflon packing does not require a constant loading spring to be effective as a stem seal. Normal packing follower loading and adjustment is all that is required, especially with the V-ring or chevron shape (Figure 4.6i). The cup and cone style requires a higher loading to effectively energize the seal. The surface finish of the stem should be very fine, in the order of 4 to 8 AA, but this is true regardless of the packing material. Internal finish of the stuffing box should be at least 16 AA, but it is possible to obtain valves with a burnished finish down to 8 AA for maximum packing performance. The packing rings and stuffing box dimensions should be very accurate since V-ring packing is in effect a lip-seal labyrinth. Dimension tolerances are even more critical for the cup and cone design because of the high degree of stiffness and packing loading required. Teflon packing does not require lubrication due to its extremely low friction characteristics.

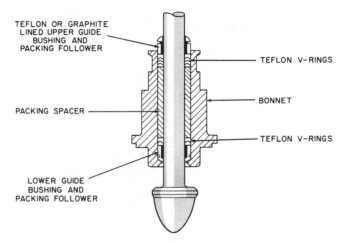

Fig. 4.6i   Typical Teflon V-ring arrangement

TEFLON OR GRAPHITE LINED UPPER GUIDE BUSHING AND PACKING FOLLOWER

TEFLON V-RINGS

PACKING SPACER

BONNET

TEFLON V-RINGS

LOWER GUIDE BUSHING AND PACKING FOLLOWER

Asbestos Packing

The oldest type of material used for valve stem packing is some form of asbestos rings. This packing is still widely available and used but it has largely been superseded by Teflon and to some degree by graphite rings at higher temperatures. Asbestos rings often incorporate binders and plasticizers to help form the rings. For high temperature services they usually include Monel or Inconel metal wire reinforcement. Rings are either compressed fiber construction or a braided construction.

Asbestos packing rings must be lubricated and this is done by various means. The most common methods of lubrication involve impregnating the asbestos fibers and the formed ring with Teflon suspensoid or graphite. The

oldest method, which is still used to some extent, involved grease compound lubricants of various types injected into the lantern ring area by a packing lubricator assembly (Figure 4.6j). A loading bolt is turned to force the grease sticks into the packing. An isolating valve should be used for safety. External lubrication works reasonably well, although it is cumbersome and requires constant maintenance for checking and reloading of the lubricator. Also, it is sometimes difficult to find lubricants compatible with the process fluid. The Teflon impregnated version has the same temperature constraints as regular Teflon packing, while the graphite impregnated and other types are prone to age hardening, especially at elevated temperatures. This type of packing deterioration results in increasing stem leakage and often scores the stem as well.

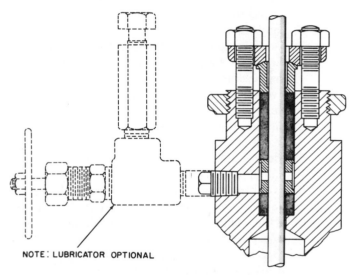

NOTE : LUBRICATOR OPTIONAL

Fig. 4.6j   Stuffing box assembly with external lubricator

Graphite Packing

The most recent addition to valve stem packing materials is graphite packing. This packing material has some virtues but also has several drawbacks. The main virtues are high temperature capability up to 1000°F (540°C) and chemical inertness, except for strong oxidizing fluids. These fluids limit the packing temperature to about 700°F (370°C). Special designs using laminated graphite rings in an extension bonnet can extend the upper temperature to 1200°F (650°C) on both oxidizing and nonoxidizing service. The low temperature limit is 0°F (−18°C).

The drawbacks of graphite packing have forced control valve manufacturers to undertake considerable developmental work in an attempt to overcome or moderate these shortcomings. Some success has been achieved, but it still remains a difficult packing to use and should be considered only if there is no viable alternate. Among these shortcomings are the following:

Relatively high stem friction

Difficulty in "energizing" the packing to give an effective stem seal

Low cycle life without leakage

Pitting of stainless stems in conductive or high temperature services

Shortened packing life due to graphite plateout on the stem

Some of the things that will help improve graphite packing performance as a result of manufacturers' development efforts or user experience are the following:

1. Use combination packing assemblies consisting of laminated graphite and braided graphite fiber rings. The braided graphite rings help as antiextrusion and graphite plateout scrubbing or wiping rings.
2. Use sacrificial zinc washers where possible or hard chrome plating or Inconel 600 stems to combat pitting.
3. Carefully torque the packing flange nuts to the minimum torque recommended by the valve manufacturer. Over-torque will result in excessive valve stem friction and may even lock the stem.
4. Remove the packing while the valve is in storage or out of service for extended time periods.
5. Clean graphite plating from the stem before installing new packing.
6. Avoid trapping air between the rings during installation. Leave the ring level with the chamfer of the stuffing box cavity and install the next ring on top.
7. "Breaking in" new packing by cycling the valve at least 50 times may help.
8. On hot services of 700°F (370°C) or higher, do not cycle the valve until the body assembly is up to temperature.

Packing Arrangements

There are a number of packing arrangement systems that have been developed to suit various types of packing and fluid containment problems (Figure 4.6k). In general, the arrangements are equivalent for either Teflon V-rings or square rings (asbestos and graphite), although the number of rings will vary for each packing type. In the standard packing arrangement, the lower set of rings serves to minimize process fluid penetration up into the stuffing box and acts as stem wipers. The upper set consists of a larger number of rings and acts as a labyrinth seal to effectively seal off any fluid from the atmosphere. Note that in the V-ring configuration, certain rings are modified to have a flat bearing surface for the packing follower, spacer or lantern ring, and lower guide bushing if one is provided.

For harder-to-hold fluids or in lieu of the use of a

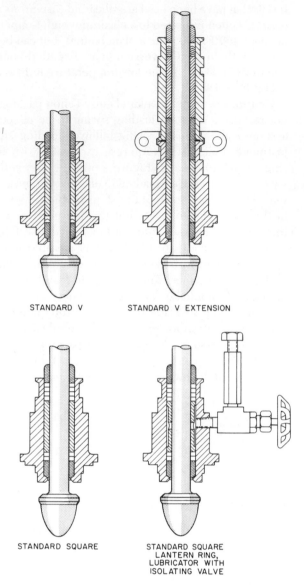

STANDARD V   STANDARD V EXTENSION

STANDARD SQUARE   STANDARD SQUARE
LANTERN RING,
LUBRICATOR WITH
ISOLATING VALVE

Fig. 4.6k   Typical standard packing arrangements
(*Courtesy of Valtek, Inc.*)

bellows seal, a packing arrangement known as the twin-seal is used (Figure 4.6l). This arrangement is also good for vacuum service. In this configuration, there are two full sets of packing installed. This system requires a control valve design with an extra deep stuffing box since space is needed for spacers or lantern rings and a lower guide bushing as well as the packing. The bottom set of packing is intended to function as the primary stem seal. In vacuum service, this set is inverted or "backward" since it is on the low pressure or secondary end of the seal system. The top set of packing is the secondary backup to eliminate fluid leakage to the atmosphere. Again, in vacuum service, the top packing is the primary seal since the pressure drive is from atmosphere to the inside of the valve.

For toxic or radioactive fluid services, the bonnet is

used, but also of the valve metallurgy and the valve and bonnet physical relationship. Heat is transmitted to the packing area by conduction through the metal, by convection via the process fluid, and by the relative heat radiation balance to the atmosphere. Stainless steel, for example, has a much lower heat conductivity coefficient than carbon steel and thus is about 20 to 30 percent less "efficient" in conducting heat into the packing area. This does not mean that the packing temperature rating can be increased, but it may serve to reduce some heat load and increase packing life. (See Section 4.13 for further discussion and illustration of packing temperature determination).

In some cases, the valve can be installed upside down so that the bonnet is below the valve body. In liquid service, this eliminates convection, and heat is transferred to the bonnet by conduction only. In vapor service (with or without superheat), it may be possible to condense sufficient vapor in a plain or extended bonnet to effect a condensate seal and lower the packing temperature to a suitable level. This approach must be evaluated carefully because alternate wetting and drying or a vapor-liquid interface may result in metallurgical problems with certain fluids.

Other Packing Materials

O-rings are used for stem sealing in some globe valve designs, usually for less demanding services such as water. General application use is not recommended, since proper selection of an elastomer for the process fluid can be a problem. The O-ring installation design will require a fully retained groove and probably retainer rings to minimize extrusion or blowout. In gas service or with liquids containing gases, many elastomers will absorb some gas. Depressuring the valve will result in sudden expansion of this gas, which will generate splits in the o-ring or, in some cases, even massive destruction. Installation requires great care to avoid cutting, twisting, or other damage. Lubrication may be needed to overcome sliding friction, and gradual loss of this lubrication in service will cause excessive o-ring wear or sticking. A standard stem seal packing ring system is usually preferable in all situations.

Trim Designs

Definition of the components that make up valve trim is given at the beginning of this section, and additional discussion of some of the more common globe valve trims will be found in Section 4.7. Globe valve trims are available in many design variations, depending upon manufacturer and the intended application of a particular valve design. A complete discussion of every trim design variation is impossible, but those with the most general application use will be covered, including special trims for severe services such as noise, cavitation, and erosion (See

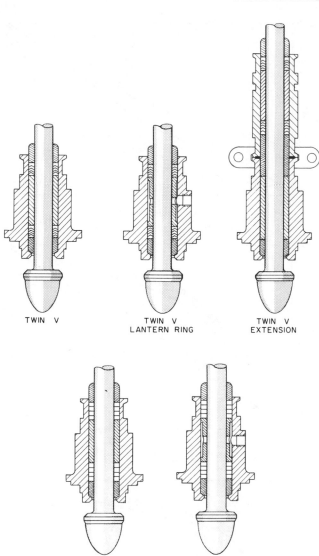

TWIN V   TWIN V   TWIN V
                LANTERN RING   EXTENSION

TWIN SQUARE   TWIN SQUARE
                LANTERN RING

Fig. 4.6l Typical twin or double packing arrangements
(*Courtesy of Valtek, Inc.*)

tapped at the lantern ring area, and this connection can be used in three different ways, depending upon specific design requirements. The tap can be used for a leakage monitor connection using a pressure gauge or switch or as a sample. More commonly, the tap is used as a leak-off connection whereby any process fluid leakage is piped to a vent disposal header or nonpressurized waste container. In some cases, the tap can be used as a purge connection. Here, an inert gas or liquid (depending upon the process) is introduced at a pressure well above the highest expected process pressure. This purge material provides an additional pressure seal, and any packing leakage will be in the direction of purge into the process.

Packing Temperature Considerations

The relationship between the process and the packing temperature is a function not only of the type of bonnet

Section 4.14 for a complete discussion of noise reduction control valves).

Trim Flow Characteristics

All control valves are essentially pressure-reducing devices or, in other words, they have to throttle the flowing fluid in order to achieve control. The most widely used form of throttling is with a single-stage orifice and plug assembly. Multiple-stage orifice elements are usually found in trim designs for combating noise, erosion, and cavitation. In all cases, the valve trim is the heart of the valve and operates to give a specific relationship between flow capacity and valve plug lift. This relationship is known as the valve flow characteristic and is achieved by different valve plug shape designs (Figure 4.6m).

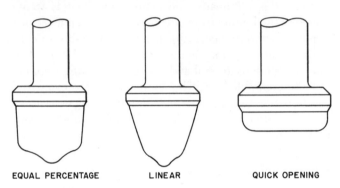

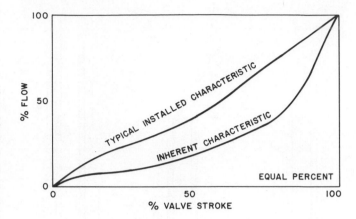

EQUAL PERCENT

Fig. 4.6m   Valve plug shapes to produce the three common flow characteristics

The term "flow characteristic" usually refers to the "inherent" characteristic, which is a function of a number of valve design and manufacturing parameters. The inherent characteristic is determined by testing the valve flow vs valve lift using a constant differential pressure drop across the valve throughout the test. Therefore, the manufacturer's trim characteristic curves should not be confused with the installed flow characteristic in the actual process fluid flow loop. In actual service, the differential pressure across the valve varies throughout the valve lift and flow range as a function of the system characteristics. This variation is due to such factors as pump head changes with flow, piping friction losses, and the hydrostatic resistance of pipe fittings, block valves, and flow measurement devices.

Control valve manufacturers currently furnish three types of inherent characteristic valve trims along with some minor variations (Figure 4.6n). These are idealized curves and do not accurately reflect the true inherent characteristic as determined by actual test. Examination of actual test data will show deviations in lift vs flow of 10 percent or more, slope variations, and other distortions from the ideal curve. This is due to a number of factors, among them the trim type and design, valve body geometry effects, manufacturing tolerances, and reproducibility and quality control. For practical purposes,

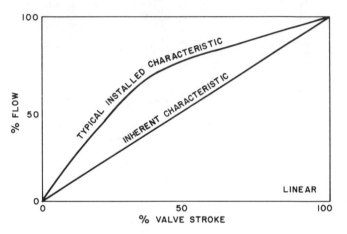

LINEAR

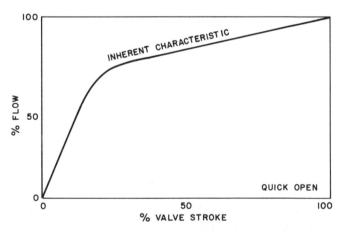

QUICK OPEN

Fig. 4.6n   Typical inherent and installed characteristics for the three common valve trims (*Courtesy of Valtek, Inc.*)

these distortions, if kept within reasonable limits, do not materially affect the valve in actual service. The typical inherent characteristic or, preferably, the typical flow vs lift test data can be used for determining the valve characteristic behavior in the installed system. This can be used to select the best valve trim to suit the process control loop as it affects overall system gain. However, where necessary, the installed valve flow characteristic can be modified by a compensator. Cams in the valve positioner are used in some cases but it is preferable to install a signal characterizer relay between the controller output and the valve positioner input (See Section 4.15). A traditional rule-of-thumb is to use a linear trim in a roughly constant control valve pressure drop situation (such as in pure pressure reducing). Where there is significant system and valve pressure drop variation with flow changes, use the equal-percentage trim.

Trim Flow Rangeability

The traditional definition of valve flow rangeability is the ratio between maximum and minimum controllable flow, where the rate of flow change follows the desired inherent flow characteristic. Generally, the minimum controllable flow is considered to be that flow just past the minimum clearance flow as the plug lifts off the seat. Figures generally quoted for single-seated turned plug control valves range from 30:1 for equal-percentage trim to 50:1 for linear trim. Actually, except for trim type comparison, these numbers have little practical use since they are dependent upon a number of variable factors. The most important factor is the true installed characteristic of the control valve in service. On a real world basis, the flow rangeability is more likely to be in the order of 7:1 to 15:1. This is more than ample since many processes do not operate much over 5:1 turndown. However, extreme conditions in some cases can require flow rangeabilities beyond the capability of one valve and will require parallel valves with split-ranging (See Section 4.15).

While linear and equal-percentage trims are designed to throttle essentially over the full valve travel, the quick-opening characteristic is designed to act more like a "bath stopper" plug. The flow characteristic is such that about 80 percent of its C_v capacity is developed in a roughly linear manner over the initial 20 to 30 percent of valve lift. The remaining capacity is "dumped" over the balance of the lift. This plug can be used for on-off service, short-stroke valves such as self-contained pressure regulators, or in some process applications where this characteristic can be useful.

Standard Trim Configurations

The valve plug configuration on most modern control valves is either contoured plug, ported plug, and piston, poppet, or curtain plug (Figure 4.6o). The turned or contoured plug is probably the most common, followed by the piston and ported plugs. The contoured plug has inherent manufacturing advantages since it can easily be turned out of barstock material, is accurately and uniformly reproduced by modern machine tools, and can be hard-faced more easily for erosive services. The contoured plug is usually used with single-seat valves, but it is also available in double-seat designs.

The ported plug must be cast or forged and it is difficult

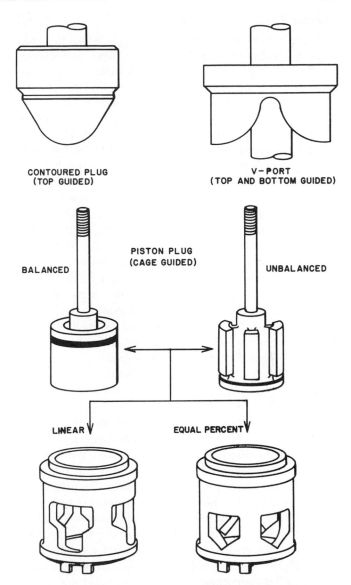

Fig. 4.6o Types of valve plug configurations for various valve designs (*Courtesy of ITT-Hammel Dahl Conoflow*)

to hard-face. This plug is most common in double-seat valves, but it is also available in single-seat designs. Both the contoured and ported plugs have the desired flow characteristic as part of the plug design as it moves in and out of the seat ring.

The piston plug is relatively easy to fabricate and is usually made from a hardenable type of stainless steel such as 410, 440, or 17-4PH. These materials are not as corrosion resistant as the austenitic stainless steels such as 316. If the service requires materials such as 316, hard-facing must be done since these plugs are normally guided in a cage assembly. The flow characteristic for this trim design is incorporated in the cage, although some poppet trims work in a contoured seat. More complete discussion of globe valves and their trim designs will be found in Section 4.7.

Special Trim Configurations

There are certain severe service applications that may require special trim configurations. These applications usually involve noise, cavitation, erosion, or combinations of these problems. Interestingly, the special trim designs for all of these are often very similar in concept although they will differ in design detail. Since noise is an especially difficult and complex problem to deal with, it will be covered in depth in Section 4.14. The following discussion will touch on noise reduction trim, but will primarily cover cavitation and erosion.

Liquid cavitation is a fluid dynamics problem that is not completely understood. However, it is fairly easy to explain the basic mechanism and understand its impact upon the valve trim and sometimes the valve body. The phenomenon of cavitation is related to Bernoulli's theorem, which explains what takes place when a fluid flowing through a pipe passes through a narrower passage, restriction, or orifice (Figure 4.6p). Simply stated, some of the pressure head is converted into velocity head or fluid acceleration. This is a transfer of energy to push the same mass flow through the smaller passage. The fluid accelerates to its maximum velocity, which is also the point of minimum pressure (vena contracta), and then gradually slows down as it again expands out to the full pipe area. The head pressure also recovers but at a lower level, since part of it is lost into internal energy by fluid friction. If the pressure head decreases to a point where it is less than the liquid vapor pressure (P_v) at that temperature, then vapor bubbles will form downstream of the restriction. If the head pressure recovers to a point greater than the vapor pressure, then the vapor bubbles will collapse back into liquid. Otherwise, the vapor remains and is known as fluid flashing and two-phase flow. The collapse of the bubbles is high-energy implosions and is called cavitation.

These implosions generate noise and fluid shock cells and jets that impinge upon the trim metal parts. It is thought that this phenomenon generates tremendous and concentrated impact forces that overstress the metal and

fracture out tiny metal particles. Indeed, cavitation damage gives a very distinctive appearance which is easily recognized, once seen. No known material will withstand continuous cavitation without damage and eventual failure. The length of time it will take is a function of the fluid, metal type, and severity level of the cavitation. Without special trims, the only mitigating actions possible were using extremely hard trim materials or overlays, maintaining sufficient downstream back pressure to reduce the bubble formation, or limiting the pressure drop by installing control valves in series to distribute the drop and reduce the vena contracta pressure in each valve. Hardened materials are still useful in combating cavitation damage, but the modern solution is to use these materials as necessary in conjunction with unique trim designs that reduce the total pressure drop by stages or by high-energy dissipation.

Some of the special trim designs for combating cavitation are the Self-Drag® (Figure 4.6q), Flash-Flo® (Figure 4.6r), Cavitrol® (Figure 4.6s), Turbo-Cascade® (Figure 4.6t), VRT® (Figure 4.6u), Hush® (Figure 4.6v), Channel Stream® (Figure 4.6w), and various staged or step type plugs and orifices (Figures 4.6x and 4.6y). The Flash-Flo and Cavitrol trims are designed to break the flow into multiple fluid jets and force the jets to impinge upon themselves with high turbulence. This high degree of turbulence changes part of the upstream energy into heat energy, disrupts vena contracta formation, and minimizes bubble formation. Those bubbles that do form tend to implode within the turbulent fluid core and this greatly reduces trim damage. However, these trims are mainly suitable for moderate cavitation. Highly cavitating services will require more sophisticated designs.

The balance of the trims listed can be considered variations of staged trims. These trims reduce the total valve pressure drop in multiple steps such that the vapor pressure of the fluid is not reached in one stage. In some cases, the fluid is also forced to undergo multiple changes in direction to promote additional head loss and energy conversion. Many of these staged trims can be likened to adiabatic flow with friction, much like pressure drop in a long pipeline. These designs vary in degree of efficient control of cavitation, but all of them have vastly improved our capability in dealing with the problem. It should be pointed out that virtually all anticavitation trims are designed using water as the test fluid. Therefore, actual performance with a different fluid cannot be absolutely extrapolated. However, since water is one of the most destructive fluids to handle under high pressure drop conditions, it is likely that a particular valve trim will be reasonably effective with any fluid. In any given process application, actual field experience is necessary to demonstrate the adequacy of any particular trim design and the need for any further changes.

Erosion of valve trim can also be caused by high-velocity liquid impingement, abrasive particles, and ero-

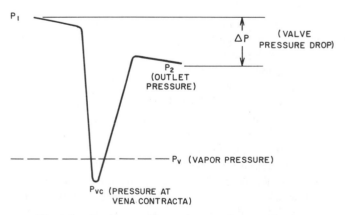

Fig. 4.6p   Pressure profile: single-seat valve experiencing cavitation

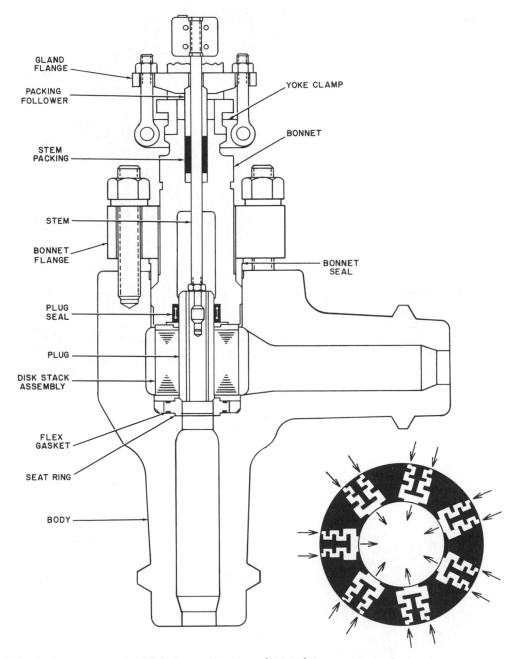

Fig. 4.6q Self-drag® valve with example of disk element (*Courtesy of Control Components, Inc.*)

sive-corrosive combination action. Erosion damage, in general, will increase as the square of the flow velocity. As in cavitation, one major key to the solution is to reduce the velocity through the trim. It is no surprise that valve and trim designs discussed above are useful on any erosion problem, although there may be some variations of design and materials of construction needed to better cope with a particular problem. Compared to the past, there are far more valve designs available today to help alleviate almost any problem. So much so, that a properly chosen and specified control valve and trim type can be one of the most reliable pieces of equipment in process service. Indeed, many companies no longer use control

valve bypasses in their plant design, except in some unique situations. However, it is an absolute requirement that there be close consultation with the control valve manufacturer in specification of special trims. High-velocity liquid impingment erosion is usually associated with high pressure drop coupled with valve geometry. High-velocity fluid jets developed by the liquid moving through the seat area will often result in erratic flow patterns and turns that allow the liquid to impinge directly on the valve trim and body. The damage is often confined to specific areas in the valve. Liquid droplets in a vapor stream can also cause impingement erosion, but it is generally spread over a greater area. Impingement dam-

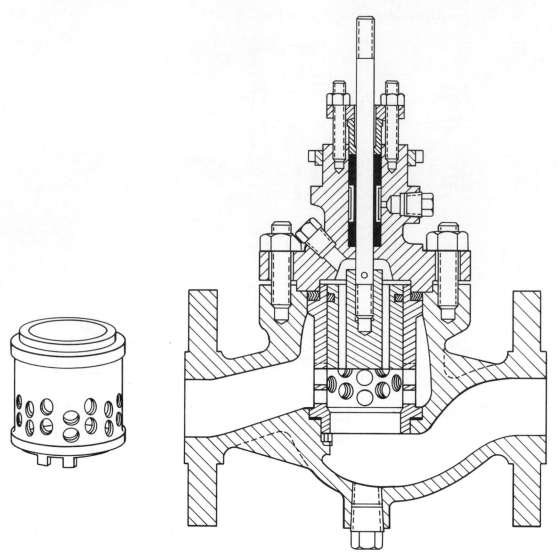

Fig. 4.6r   Flash-Flo® trim element and valve installation (*Courtesy of Hammel Dahl Conoflow*)

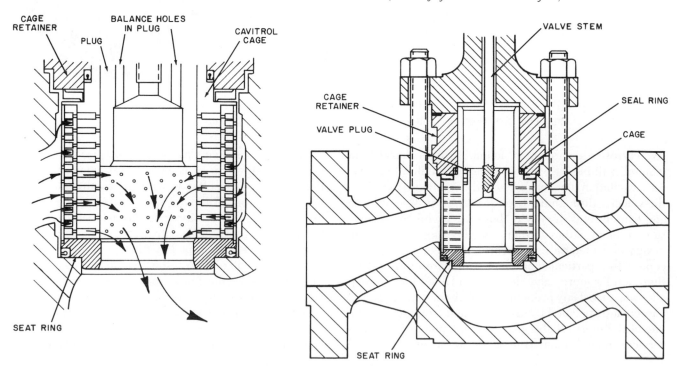

Fig. 4.6s   Cavitrol® trim element and valve installation (*Courtesy of Fisher Controls*)

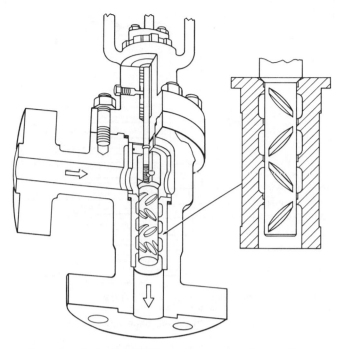

Fig. 4.6t Turbo-Cascade® trim element and valve installation
(*Courtesy of Yarway Corp.*)

age is characterized by relatively smooth grooves and
pockets worn into the metal.

Erosive-corrosive combination action is peculiar to
corrosive fluid services subject to moderate to high pressure drops, but not as high as those encountered with
impingement erosion. Most metals used in corrosive services are actually protected from continued corrosion

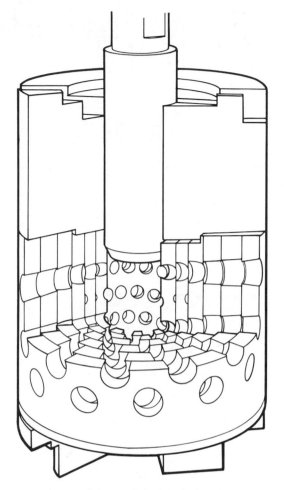

Fig. 4.6v Hush® trim element and plug; flow is into
plug bore and out (*Courtesy of Copes Vulcan, Inc.*)

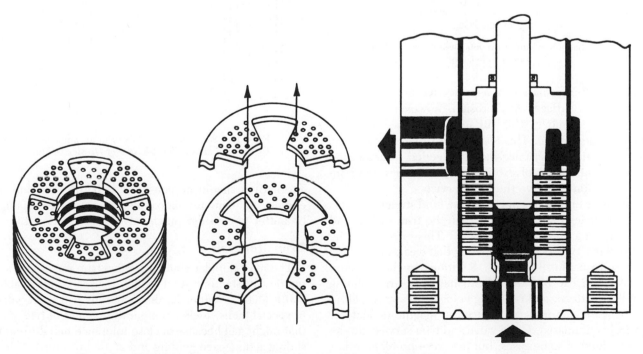

Fig. 4.6u VRT® (Variable Resistance Trim) element, plates, and assembly in a valve (*Courtesy of
Masoneilan Division of Dresser Industries*)

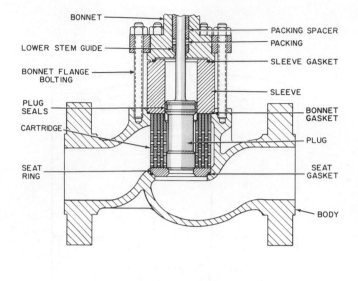

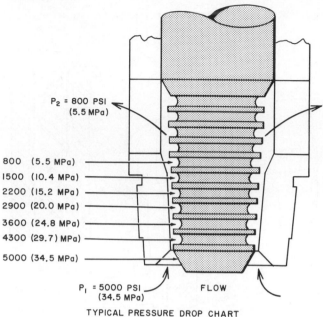

Fig. 4.6x   Cascade-Trim® step type plug for high pressure breakdown
(*Courtesy of Copes Vulcan, Inc.*)

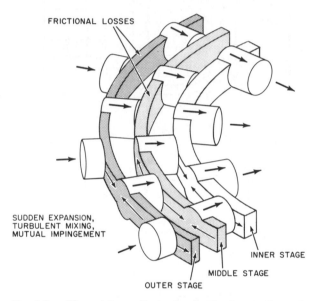

Fig. 4.6w   Channel Stream® trim detail and valve installation (*Courtesy of Valtek, Inc.*)

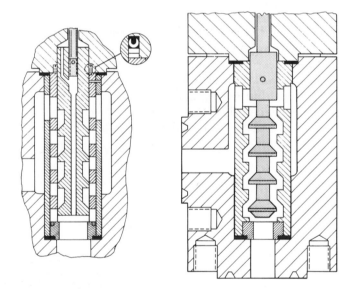

Fig. 4.6y   Step plug and orifice trim for liquid service
(*Courtesy of Masoneilan Division of Dresser Industries*)

attack by the formation of a protective or "passivated" film on the surface. Any erosion, however mild, that continually removes this surface film allows corrosion attack to accelerate. This is a complex phenomenon and actual service experience may be needed to determine which combination of exotic metals or alloys and trim configuration will give the best service.

Abrasive erosion occurs when the fluid stream contains solid particles that are harder than the trim surface and are traveling at sufficient velocity. This erosion can be likened to a type of scouring action that wears away metal similar to a file or grinder. Solutions to the problem involve the use of harder trim materials, streamlining the flow pattern, and reducing velocity. However, abrasive erosion can only be reduced in magnitude and not entirely eliminated. Good valve and trim service life can be obtained in some cases, but in severe problems other

alternatives should be considered. If the fluid and operating conditions are compatible with elastomers, it might be better to consider pinch valves (See Section 4.9).

Trim Materials

The most popular general service trim material is austenitic 316 stainless steel, which is useable up to about 750°F (400°C). Other harder materials will be required in special trims, higher temperatures, and in trim parts that might gall because of close tolerance metal-to-metal sliding action (cage guiding and guide bushings). Among these materials are 17-4PH, 410, 416, and 440-C stainless steels, Stellite, Colmonoy, tungsten carbide, and ceramics. For very corrosive services, more noble or high alloy metals are used to advantage. Among these are nickel, titanium, tantalum, Monel, Inconel, K-Monel, Hastelloy-B, Hastelloy-C, and Alloy-20. It is difficult to generalize on recommended materials or material combinations for valve trims because of the wide range of valve designs and process application requirements. The specifying engineer should utilize not only his own knowledge, but also enlist the experience and expertise of the manufacturer and material consultants (metallurgists) when needed. Fortunately, many applications are reasonably straightforward and standard trim material combinations set forth by the manufacturer can be used.

Some general guidelines can be given for certain areas of trim specification, as follows:

1. When specifying hard-faced plugs and seats, the plug and seat rings should be considered separately using the following logic (Figure 4.6z). The plug can be supplied with hard-facing alloy on the seat surface only. This may be sufficient if the valve is subject to high pressure drop primarily during shutoff. However, for continuous high pressure drop throttling, the plug should be completely full-contour covered with hard-face alloy. In addition, the lower guide area of the stem is often hard-faced, and this is a requirement if the fluid temperature is above 750°F (400°C). As with the plug, the seat ring can be hard-faced only on the seating surface. This may be ample protection on moderately high pressure drop services because the point of maximum velocity on the seat ring is protected. As the pressure drop and severity of the service increases, full-bore hard-facing of the complete inner surface of the seat ring is recommended. A

number of cobalt and boron base alloys are available for hard-facing. In some cases, coating techniques such as flame spraying will be used. The valve manufacturer's recommendations are valuable guidance.

2. Commonly used hardenable alloys for trim material are 410 and 416 stainless steels hardened up to about 38 Rockwell C, 440-C stainless steel ranging up to 60 Rockwell-C, and 17-4PH stainless steel, a precipitation-hardened material combining reasonable corrosion resistance and a hardness to about 40 Rockwell-C. The 400 series stainless steels have limited corrosion resistance and are generally used for simple shapes that can be made from barstock or forgings. Type 17-4PH is available more readily in cast form and is a useable general purpose alloy up to 750°F (400°C). It does not have as good a chemical corrosion resistance as austenitic 316 stainless steel. Where 316 or other soft alloys are indicated as the proper alloy for a cage-guided trim, hard-facing will be required to minimize metal-to-metal galling problems.

3. When tight shutoff (ANSI Class VI) is required, soft seating can be provided in the trim design. The soft seat insert may be incorporated either in the plug or in the seat ring (Figure 4.6aa). The most common soft seat insert material is Teflon, but Kel-F and various elastomers can be used in valve designs for particular services. It must be recognized that operating service temperatures,

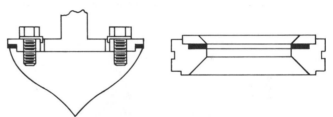

Fig. 4.6aa   Typical soft-seat insert designs (*Courtesy of Masoneilan Division of Dresser Industries*)

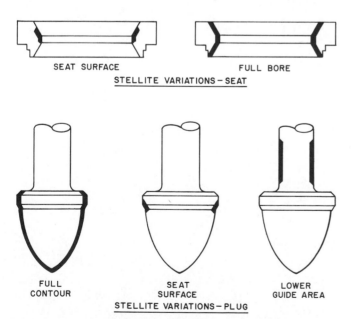

SEAT SURFACE   FULL BORE
STELLITE VARIATIONS – SEAT

FULL   SEAT   LOWER
CONTOUR   SURFACE   GUIDE AREA
STELLITE VARIATIONS – PLUG

Fig. 4.6z   Hard-facing of plug and seat rings typically used in erosive services (*Courtesy of Valtek, Inc.*)

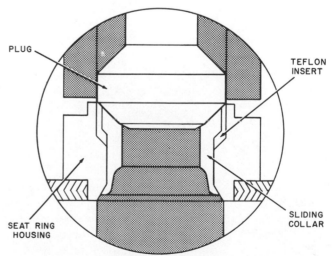

PLUG

TEFLON INSERT

SEAT RING HOUSING

SLIDING COLLAR

Fig. 4.6bb   Protected soft seat insert for high pressure drop and tight shutoff (*Courtesy of Masoneilan Division of Dresser Industries*)

high or low, will limit the use of these materials. The pressure drop across the valve is another limiting factor, although there are some "protected insert" designs (Figure 4.6bb) that will operate at very high pressure drops.

Valve Plug Stems

The valve stem connected to the plug must be heavy and stiff enough to carry the load from the actuator to plug and bear the seat loading forces without bowing. Yet it cannot be so big so as to generate excessive packing friction. Fortunately, the almost universal use of Teflon packing allows the manufacturer to design heavy-duty stems that will withstand all of the various forces to which they are subjected. Heavy-duty stem guiding along with a heavy stem resists even minor bowing and fluid dynamic vibration forces. This results in greatly increased packing seal life.

Valve stems are normally the same material as the valve plug. In some unusual corrosive services, the stem may be of a different material to better cope with packing leakage that might react with moisture in the atmosphere. Depending upon the type of valve design, the valve stem may be integral (one piece) with the plug or it can be threaded into the plug and then pinned to prevent unscrewing. Since the manufacturer does a number of things to insure proper alignment and a solid, vibration-resistant threaded connection, good maintenance practice may dictate replacement as a unit rather than separate pieces.

Stem guides in one form or another are an integral part of the trim assembly. This is treated in more depth in Section 4.7. Stem guides must be harder than the stem material to minimize metal-to-metal galling or be of some material that does not gall with metal. All metal guides may be of such materials as 17-4PH, 440-C, Stellite, or hard-chrome plated 316 stainless steel. Some metal guides may be lined with Teflon or graphite inserts. Guides of GF-Teflon are also used. The lined metal and GF-Teflon guides are usually very close tolerance with the stem and serve as stem wipers and to minimize penetration of process fluids up into the packing box area.

BIBLIOGRAPHY

Beard, Chester S., *Final Control Elements*, Philadelphia, PA: Chilton Co., 1969.

Control Components Division of IMI, Valve Catalog, Irvine, CA.

Copes-Vulcan Division of White Industries, Catalog of Valves and Desuperheaters, Lake City, PA.

Fisher Controls Co., Process Industries Catalog 61, Marshalltown, IA.

Hutchison, J.W., ed., *ISA Handbook of Control Valves*, 2nd ed., Research Triangle Park, NC, 1976.

Lyons, J.L., *Lyons' Valve Designers Handbook*, New York: Van Nostrand Reinhold, 1982.

Masoneilan Division of Dresser Industries, General Valve Catalog, Norwood, MA.

Valtek, Inc., Control Valves, Springville, UT.

4.7 GLOBE VALVES

FC THREE-WAY

FC FAIL CLOSED GLOBE VALVE

NOTE: THE LETTER "S" IF MARKED INSIDE THE VALVE SYMBOL, REFERS TO "SPLIT BODY" AND THE LETTER "C" TO CAGE DESIGN.

FO FAIL OPEN ANGLE VALVE

Sizes:	$\frac{1}{4}''$ to 16″ (6.25 mm to 400 mm); angle type to 42″ (1050 mm)
Design Pressure:	1250 PSIG (8.6 MPa) to 60,000 PSIG (414 MPa) in small sizes
Design Temperature:	Cryogenic to 1200°F (649°C)
Relative Capacity:	$C_v = 10\ d^2$ to $C_v = 14\ d^2$
Materials of Construction:	Any forged, castable, and machinable metal for the body and trim; plastic, TFE in small sizes.
Special Features:	Hard facings on trims; special trim designs for noise, cavitation, and erosion; balanced plugs; special seal designs; Teflon lining
Cost:	See Figure 4.7a. Cost data are based upon ANSI Class 600 flanged bodies with double-acting piston operator and positioner. Other operators such as single-acting piston, spring and diaphragm, and electric motor drive can be furnished. In addition, bodies are available in numerous metal alloys but at substantially higher prices.
Partial List of Suppliers:	Badger Meter, Inc.; H. D. Baumann, Inc.; Copes-Vulcan, Inc.; G. W. Dahl Co., Inc.; Fisher Controls Co.; Honeywell, Inc.; Hammel Dahl; Jamesbury, Inc.; Kent-Introl; Kieley & Mueller, Inc.; Leslie Co.; Masoneilan International; Norriseal Uniflow; Research Control Valves; Valtek, Inc.

Body Forms

The actual pressure containment and fluid conduit portion of a control valve is called the valve body assembly. This assembly consists of the body, a bonnet or top closure, sometimes a bottom flange closure, and the internal elements known as the trim. The body can have flanged, screwed, or weld-end (butt weld or socket weld) connections for installation into the piping. The trim consists of such elements as the plug and stem, guide bushings, seat rings, seat retainers, and stuffing box lantern ring.

The body configuration can be in-line, angle, offset inline, Y-shape, and three-way. Some of these will be discussed in more detail. The shape and style of the valve body assembly is usually determined by the type of trim elements it contains, piping requirements, and the application function of the valve in the process system. There are a large number of body designs on the market, including a number of special purpose designs. The end result is a device that can be fitted with a power operator and used to modulate the flow of process fluid to regulate such things as pressure, flow, temperature, liquid level, or any other variable in a process system. This is done by pressure reduction of the fluid, and a control valve is always a pressure loss device. Examples of some of the more widely used body configurations are shown in Figure 4.7b.

Control valves have varying degrees of shutoff capability, depending upon the valve and internal trim design. While tight shutoff is not always a requirement, most users prefer valve trims that will provide reasonably tight shutoff. Fortunately, there are good single-seated valve trim designs available that can give ANSI Class IV or V shutoff in nonbalanced plug design and ANSI Class III or IV shutoff in balanced plug design. Table 4.7c is a simplified seat leakage class tabulation. For valve sizes larger than 8 in. (200 mm), the maximum leakage may

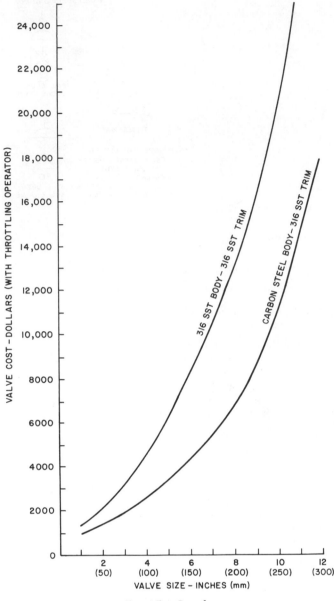

Fig. 4.7a   Cost data

be calculated using the equation $L = 0.11 \, (d^2)$, mL/min, where d is valve nominal body size in inches. It is possible to get ANSI Class VI shutoff with soft seat inserts such as Teflon or special lapped-in metal-to-metal seats. However, it should be recognized that after some period of process operation lapped-in trim will likely not survive for this level of shutoff.

Double-Seated Valves

While the double-seated valve (Figure 4.7d) is still available today from some manufacturers, it is a very old design that has generally lost favor with users. It offers little, if any, advantage that cannot be better served by the single-seated valve, butterfly valve, or other designs such as the rotary plug. Size for size, it is much larger and heavier. Shutoff is poor since it is not mechanically

possible to have both plugs contact the seat at the same time. Some special seat designs have been developed to help overcome this, but application is limited.

The double-seated valve is considered semibalanced, i.e., the hydrostatic forces acting on the upper plug tend to cancel out the forces acting on the lower plug. The result is less actuator force requirement and a smaller actuator can be used. However, there is always an unbalanced force due to the upper and lower plug diameter difference required for assembly. In addition, unbalance forces are generated by the effect of dynamic hydrostatic forces acting on the respective throttling area of each plug. Such forces can be quite high, particularly with the smooth contoured or turned plugs. These can reach as much as 40 percent of such forces of the equivalent single-seated valve plug. Characterized V-port plugs are preferred for higher pressure drops, especially as the valve goes toward the larger sizes. Double-seated valves have been built as large as 24 in. (600 mm), although most manufacturers now limit themselves to 12 in. (300 mm) as a maximum.

Reference to the illustration shows that the valve can be converted from the push-down-to-close configuration shown to a push-down-to-open type. This is done by removing the bottom closure flange, bonnet, and stem, inverting the entire assembly and reinstalling stem, flange, and bonnet. This, coupled with the use of direct (air pushes stem down) and reverse (air pushes stem up) acting operators gives full flexibility to control and air failure actions.

Single-Seated Valves

Single-seated valves are the most widely used of the available globe body patterns. There are good reasons for this. They are available in a wide variety of configurations, including special purpose trims, have good flow shutoff capability, are less subject to vibration due to reduced plug mass, and are generally easy to maintain. Single-seated valve plugs are guided in one of three ways, depending upon the particular manufacturer's design. These are stem guided, top guided, or top- and-bottom guided. The most popular modern design globe valves are stem- (Figure 4.7b) or top-guided (Figure 4.7e) types. These designs require only the one body opening for the bonnet and have one less closure subject to leakage as compared to the top-and-bottom guided configuration shown in Figure 4.7f. In addition, the valve plug mass is reduced, which increases the natural undamped frequency of the trim as compared to the top-and-bottom guided valve. Thus, it is much less susceptible to vibration. The illustrations show that the stem- and top-guided body fluid passage is more streamlined and less subject to fouling and it also eliminates one guide bushing (lower). Like its double-seated counterpart, the top-and-bottom guided valve has lost favor with users.

It should be pointed out that the stem-guided valve

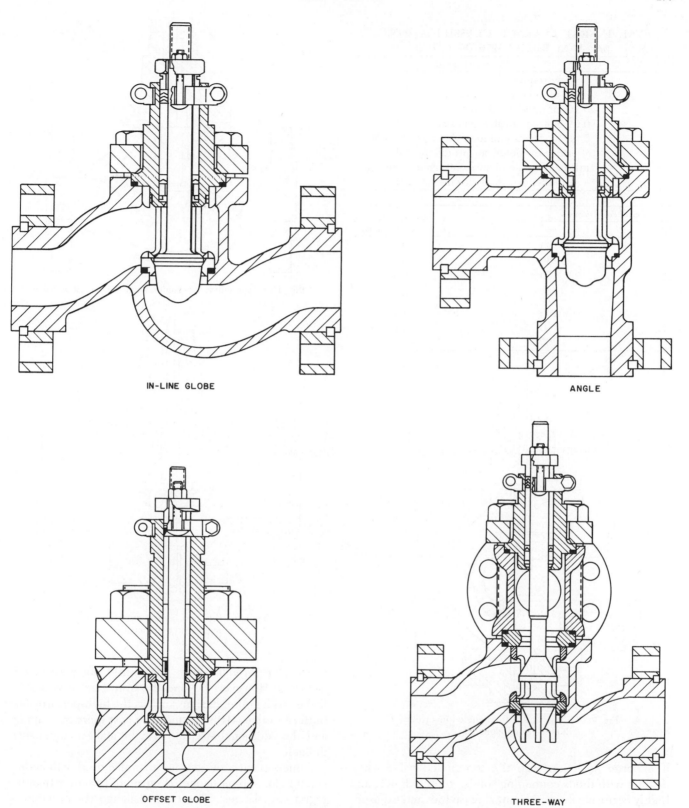

IN-LINE GLOBE

ANGLE

OFFSET GLOBE

THREE-WAY

Fig. 4.7b Typical globe body configurations (*Courtesy of Valtek, Inc.*)

Table 4.7c
VALVE SEAT LEAKAGE CLASSIFICATIONS
per ANSI B16.104-1976 (FCI 70-2)

Class	Maximum Leakage
I	No test required
II	0.5% of rated valve capacity
III	0.1% of rated valve capacity
IV	0.01% of rated valve capacity
V	5×10^{-4} mL/minute of water per inch of orifice diameter per psi differential
VI	mL per minute of air or nitrogen versus port diameter per the following tabulation

Nominal Port Diameter		Maximum Seat
In.	*mm*	Leakage, *mL/minute*
1	25	0.15
1½	38	0.30
2	50	0.45
3	75	0.90
4	100	1.70
6	150	4.00
8	200	6.75

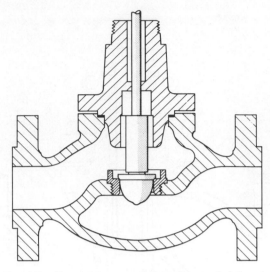

Fig. 4.7e   Top-entry, top-guided single-seated globe valve

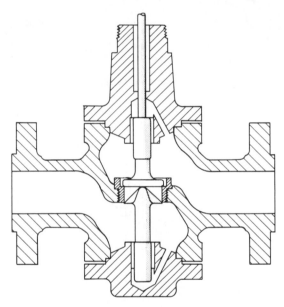

Fig. 4.7f   Top- and bottom-guided invertible single-seated globe valve

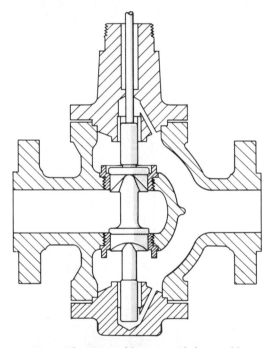

Fig. 4.7d   Top- and bottom-guided invertible double-seated globe valve

is a better selection than the top-guided valve when dealing with fluids containing solids, gummy fluids, and highly corrosive fluids. Minimizing cavities and exposure of parts in these services will help maximize the duration of trouble-free operation, since it eliminates buildup, crystallization, etc. on the top plug guide.

The current state-of-the-art top entry, single-seat globe valve offers many application and maintenance advantages for general purpose use. It is most commonly used in sizes from 1 in. (25 mm) through 6 in. (150 mm), but

is available through 12 in. (300 mm) from most manufacturers. However, beyond the 6 in. size, it is best to first consider other designs such as the high-performance butterfly valve (HPBV), since the cost increases rapidly and the additional size and weight make maintenance difficult.

Some top entry designs are manufactured with bodies suitable for slip-on flanges (Figure 4.7b) rather than integral cast flanges. This type of flange construction is discussed in more detail below under Split-Body Valves and Valve Connections. There are a number of advantages to this design.

Cage Valves

The cage valve is a variant of the single-seated valve and is the most popular design used in the process industries. The top entry bonnet and trim design makes it extremely easy to change trim or do maintenance work. There are no threaded joints such as the seat ring in conventional globe designs to corrode and make removal difficult. The design is very flexible in that it allows a variety of trim types to be installed in the body. This includes such variations as reduced trim, anticavitation, and reduced-noise elements (see Section 4.6). In addition, the overall design is very rugged. With proper specification of trim materials and type, they will provide relatively trouble-free service for extended time periods. These designs have largely eliminated the need for block and bypass valves in most cases, since their service life can be equivalent to or better than most components in the process that require periodic plant shutdown for maintenance.

There are two basic design concepts available for cage valves. One type uses the cage solely to clamp the seat ring into the valve body (Figure 4.7g). This design is usually stem or top guided and the valve plug is of conventional characterized design and does not contact the cage. The other type uses the cage to guide the plug as well as clamp the seat ring into the body (Figure 4.7h). Here, there are two design variations. In one, the cage openings are specially shaped ports to provide the desired flow characteristic, and the valve plug is a poppet or "curtain" to expose the ports. In the other, the cage openings are large-area rectangular ports to provide the needed flow area and the plug is the conventional characterized design. In this latter case, the plug has a heavy cylindrical section above the characterized portion to provide the guiding area in the cage, or else there is some form of top guiding on the stem, similar to Figure 4.7h.

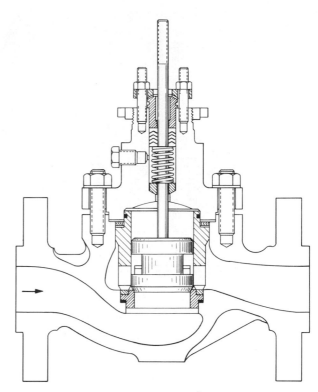

Fig. 4.7h   Cage valve with unbalanced plug and flow characterized ports (*Courtesy of Fisher Control*)

Both of the above cage designs can be provided with unbalanced plugs for best shutoff, or balanced plugs for reduced operator size or for better handling of high pressure drops and unbalance forces (Figures 4.7i and 4.7j). The balanced plug designs can be provided with a variety of balance seals and materials to meet service conditions.

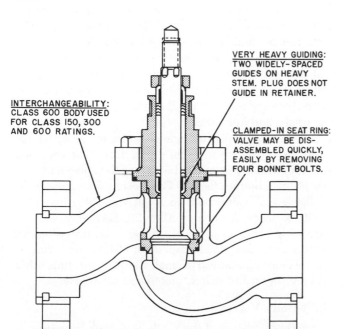

INTERCHANGEABILITY: CLASS 600 BODY USED FOR CLASS 150, 300 AND 600 RATINGS.

VERY HEAVY GUIDING: TWO WIDELY-SPACED GUIDES ON HEAVY STEM. PLUG DOES NOT GUIDE IN RETAINER.

CLAMPED-IN SEAT RING: VALVE MAY BE DISASSEMBLED QUICKLY, EASILY BY REMOVING FOUR BONNET BOLTS.

Fig. 4.7g   Cage valve with clamped-in seat ring and characterized plug (*Courtesy of Valtek, Inc.*)

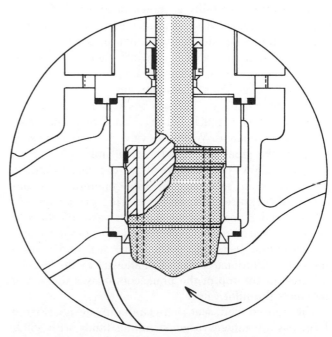

Fig. 4.7i   Cage valve with balanced plug and conventional plug characterization (*Courtesy of Valtek, Inc.*)

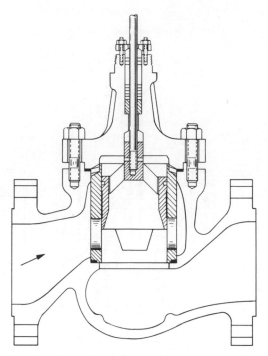

Fig. 4.7j Cage valve with balanced plug and port flow characterization (*Courtesy of Masoneilan International*)

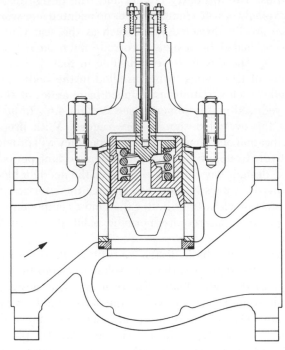

Fig. 4.7k Cage valve with pilot balanced construction and characterized ports (*Courtesy of Masoneilan International*)

These may be plastics such as Teflon, various elastomer O-rings, and metal piston rings such as Ni-resist. Generally, the use of balanced plugs will degrade the valve shutoff capability to some degree. For example, a valve rated to ANSI Class V shutoff will drop to Class IV and a Class IV rated valve drops to Class III. This is due to leakage past the balance seal, either initially or by wear with continuous operation.

One variation of the balanced seal plug that has improved shutoff capability is shown in Figure 4.7k. This is called the pilot balanced plug. This design is particularly helpful when dealing with high pressure drop situations. When the valve is shut off, there is provision for a leakage path from the upstream pressure to the area above the plug. Thus, the upstream pressure working against the plug area imposes very high seat loading forces to effect a much better shutoff than the valve operator force alone. When the valve is going from shut to open, the stem first lifts the small pilot plug off its seat. This allows the pressure above the plug to vent to the downstream side, and the valve operates as a conventional pressure balanced valve. The spring shown in this particular design keeps the pilot valve off the seat during normal operation. When the valve goes to shutoff, the cage first seats off on the valve seat and a small amount of additional stem travel, allowed by the flex clip retainer on the top of the plug, compresses the spring and seats the pilot valve.

The stem-guided, seat-ring-clamp-only design (Figure 4.7g) has advantages when handling fluids with solids such as liquid slurries, sticky fluids, other fluids subject to plate-out, and corrosive fluids. This is because of large clearances between the plug and cage and trim availability in austenitic stainless steel (304 or 316) or other alloys. Close tolerances are required in the cage-guided version, and this means that metal surfaces must be hard enough to eliminate metal-to-metal galling. In most designs, this is done by making the trim parts of hardenable high-chrome stainless steels of the 400 series, which have limited corrosion resistance, or 17-4PH, which is better but not equivalent to 304 and 316 stainless steel. The trim can be built of the austenitic stainless steels for corrosive services, with hard overlays such as Stellite, but this is a very expensive option. It is better to use a different valve design. In addition, close tolerances between plug and cage are less "forgiving" of slurries, gummy materials, and line "trash."

Another variation of a single-seated, cage-retained seat ring valve is shown in Figure 4.7l. Here, the cage and seat are installed from the bottom of the valve and supported by a bottom flange. This design was intended as a "flangeless" valve built in small sizes of exotic alloy barstock materials such as Alloy 20, Hastelloy, etc, for extremely corrosive services. It allows for provision of a bottom drain and quick inspection of the valve trim. It can be more economical in small alloy services than other globe designs, but it has limited use otherwise.

Split-Body Valves

Another variation of the single-seated globe pattern is the streamlined flow, split-body valve (Figure 4.7m).

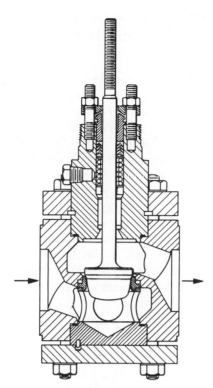

Fig. 4.7l   Quick-change trim valve
designed for bottom trim removal
(*Courtesy of Fisher Controls*)

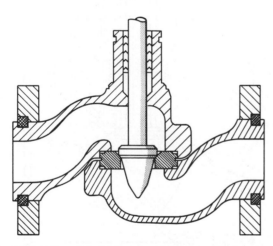

Fig. 4.7m   Split body valve with removable flange
(*Courtesy of Masoneilan International*)

This design was the original stem-guided chemical service valve intended for hard-to-handle services involving slurries, gummy fluids, and corrosive services. The seat ring is clamped between the body halves and the body is easily disassembled for maintenance. Another feature is the adaptability for building the body to use slip-on flanges. This results in cost savings when corrosion-resistant alloy castings are required. Since the flanges are not normally subject to wetting by the process fluid, they can be of cast steel or lower alloys than the body. Also, the manufacturer usually casts the body to a stan-

dard ANSI Class 600 rating and can install 150, 300, or 600 ANSI flanges as needed. This vastly reduces both manufacturing and user inventory needs for bodies and different flange service ratings. Of limited value, it is possible to rotate the lower body 90 degrees to the line axis and eliminate a pipe elbow.

While still popular for some applications, this design is not as widely used now because of limitations on installing special trim modifications, as compared to the top- or stem-guided top entry cage valve. The body flange is prone to leakage on large magnitude thermal cycling and it cannot be welded into the piping because maintenance requires body separation. Above 4 in. (100 mm) to the maximum available 10-in. (250 mm) size, integral rather than separable flanges are provided. ANSI Class 900 and 1500 ratings are available through the 10-in. size but are only available through 6 in. (150 mm) size for ANSI Class 2500 rating.

Angle Valves

Angle valves are one of the many configurations of body assemblies. Their original application was in a flow-to-close direction for high pressure drop service. This is favorable to the valve body and trim but places a high velocity burden on the downstream piping, which may result in erosion problems. This application has been reversed in favor of other special designs for handling high pressure drops within the valve and limiting exit pipe velocities. Proper applications of angle valves today are to accommodate special piping arrangements to aid drainage, for erosive service with solids impingement problems, and for other special applications such as coking hydrocarbons. Figure 4.7n shows a coking valve design with streamlined passages and a replaceable sleeve venturi outlet. The use of this valve type on general high pressure drop applications is not recommended because the streamlined flow path results in exceptionally high pressure recovery. This in turn reduces the installed flow capacity under choked flow conditions and makes the valve highly susceptible to cavitation on liquid service, even with moderate pressure drops.

Y-Style Valves

The Y-pattern valve has application in several special areas. Among these applications are those where good drainage of the body passages and/or high flow capacity is required, such as molten metals or polymer materials, cryogenic fluids, and liquid slurries. The valve can be installed in horizontal, vertical, or angled piping to suit the application requirements. Because of the very simple body and bonnet design, they are very easy to fit with thermal or vacuum jackets. Figure 4.7o shows a valve fitted with a vacuum jacket, which provides maximum thermal insulation for such cryogenic applications as liquid hydrogen. The compact design permits minimum and uniform wall thickness, which in turn enables rapid

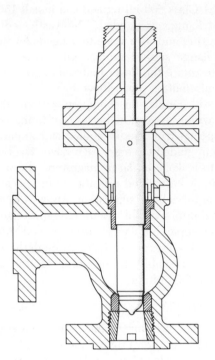

Fig. 4.7n Streamlined angle valve with lined venturi outlet

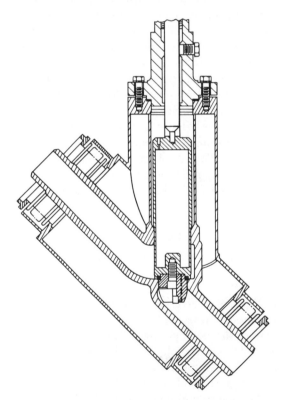

Fig. 4.7o Y-valve fitted with vacuum jacket

cool-down rates. The single-seated design allows good shutoff and can be provided with soft inserts for exceptionally tight shutoff. Note that the vacuum jacket is designed with a metal bellows to allow for thermal changes

or mechanical tolerances and has a provision for welding to the similar jacket around the adjacent piping.

Jacketed Valves

Many standard globe pattern valves can be fitted with a jacket to allow heating or cooling as required (Figure 4.7p). Normally, these jackets are for services requiring steam, Dowtherm, or similar heating fluids to prevent solidification or crystallization of certain fluids. Often the manufacturer can provide these jackets, but where this is not available, there are firms that specialize in designing and installing such jackets on valves and other equipment. These special jackets can either be designed to weld to the valve as a permanent fixture (Figure 4.7q) or as separate devices bolted or clamped to the valve body. In the latter case, it may be necessary to use a heat transfer paste between jacket and valve body to give efficient transfer by eliminating the air gap.

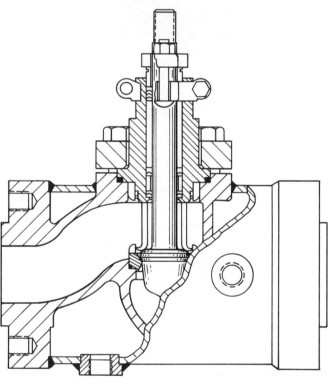

Fig. 4.7p Steam-jacketed valve (*Courtesy of Valtek, Inc.*)

Three-Way Valves

Three-way valves are another form of specialized valve body configuration. There are two basic types. One is for mixing service, that is, the combination of two fluid streams passing to a common outlet port (Figure 4.7r). The other is for diverting service, that is, taking a common stream and splitting it into two outlet ports (Figure 4.7s).

A typical mixing valve application would be the blending of two different fluids to produce a specific outlet

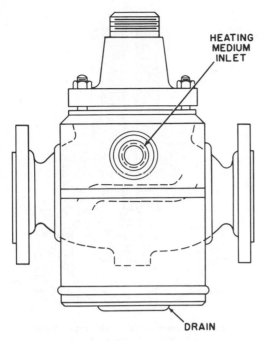

Fig. 4.7q   Special steam jacket for retrofit installation on valve

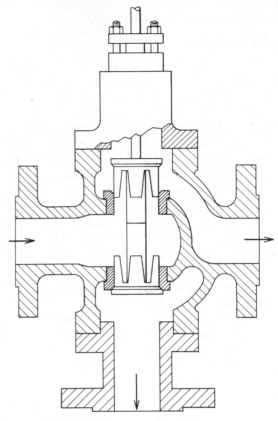

Fig. 4.7s   Three-way valve for diverting service

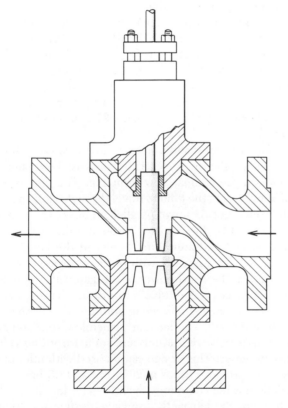

Fig. 4.7r   Three-way valve for mixing service

end product. A diverting valve might be used for switching a stream from one vessel to another vessel or for temperature control on a heat exchanger. In the latter, one portion of the fluid could go through the exchanger

and the balance bypass the exchanger. The relative split would provide the heat balance needed for good temperature control.

The forces acting on the double-seated, three-way plug do not balance. This is because different pressure levels exist in each of the three flow channels. Also, there are different flow dynamic forces acting on each plug. These valves should not be subjected to high pressure drop service. The valve plugs are usually port and top or stem guided to maintain stability in the face of the varying forces. These is no lower stem guide in order to permit unrestricted flow capacity through the lower outlet flange connection. Because of the larger size and piping complications, some users prefer to use two opposite-acting, standard two-way globe valves operating from one controller to do the same job as a three-way valve. However, the application needs to be carefully evaluated to decide which system to use.

Another version of the three-way valve, which has a universal design for mixing or diverting, is shown in Figure 4.7b. This is an excellent example of the adaptability of the modern top entry globe valve since the conversion from two-way to three-way is accomplished in the standard body by means of an insert adapter and the special three-way trim. Inspection of this design will quickly show that proper choice of ports for fluid inlet and outlet will allow it to be used for either service. Use of this design eliminates the need for different trims as shown in Figures 4.7r and 4.7s.

Small-Flow Valves

The field of small-flow control valves is a highly specialized area unlike any other application. The mechanical design and fluid flow constraints encountered essentially make these valves custom applied for the service. Small-flow valves are used in laboratories, process pilot plants, and in some areas of full scale plants. In the latter, they are normally applied to small flows such as catalysts, additives, pH control reagents, and similar applications.

The design and building of these valves test the ingenuity of any manufacturer. There are several design types available, and great care must be taken to match the application with the valve design. While the manufacturers publish C_v ratings for their valves and trims, these should be treated with extreme caution and used for reference purposes only (below 0.01 C_v). In many cases, the flow may shift from laminar to turbulent with flowing conditions and valve stroke changes. It is quite common for a laminar flow pattern to predominate, particularly with viscous fluids or in low-pressure applications. Laminar flow means the flowing quantity will vary directly with pressure drop instead of the square root of pressure drop. It is wise to devise a test procedure simulating the service application to evaluate performance of a specific valve before using it in actual service. It is not uncommon for two "identical" small valves to exhibit somewhat different C_v capacities and flow curves under the same test conditions.

One of the most common designs looks like a miniature version of the standard globe control valve (Figure 4.7t). The trim consists of a precision honed and close-fitting plug fitted into an orifice made of a hard alloy. The control area consists of a fine taper slot milled into the outer surface of the piston-shaped plug or a long shallow taper plug. In smaller C_v trims, this slot may be a calibrated scratch in the surface. It is not uncommon to find up to 30 trims available in a given body to cover the range of 0.000002 to 0.1 C_v. The rangeability, i.e., the ratio between maximum and minimum controllable flow, can be limited for this type of valve due to the inherent leakage flow between piston and orifice. The smaller the trim size, the lower the rangeability.

In contrast to the relatively long-stroke piston and orifice valve discussed, there is another valve design that has a very short and variable adjusted stroke (Figure 4.7u). Here, a synthetic sapphire ball is allowed to lift off a metal orifice and throttle the flow. The particular advantage of this valve is that the diaphragm stroke can be adjusted to produce various stem lifts with a standard 3–15 PSIG (0.2–1.0 bar) signal. Two versions are available. One can be adjusted to cover 0.07 to 0.00007 C_v and the other can be adjusted from 1 to 0.001 C_v. These can be used for high-pressure, high-drop applications

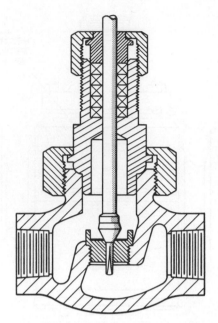

Fig. 4.7t   Low-flow valve with close clearance grooved piston plug

ranging from 3,000 to 30,000 PSIG (20.7 to 207 MPa), depending upon the model.

Another short stroke valve especially suitable for high-pressure service (up to 50,000 PSIG or 345 MPa) is shown in Figure 4.7v. Variation in C_v rating of identical plugs and seats is achieved by mechanical adjustment of a toggle arrangement. The toggle can change the valve stroke between 0.010 and 0.150 in. (0.25 to 3.75 mm). With different trim inserts, a C_v range of 1 to 0.000001 can be covered.

Finally, the latest addition to the low-flow valve design is shown in Figure 4.7w. This valve is designated to operate on the laminar flow principle. The flow is controlled by forcing the fluid to travel through a long narrow path formed between two parallel surfaces. The laminar gap is varied by the actuator through a diaphragm and o-ring, sealed hydraulic ram. Since, in the laminar regime, flow varies linearly with pressure drop across the valve and to the third power of the gap width between the two surfaces (valve travel), this design has extremely high control rangeability along with a wide C_v range of 0.02 to 0.000001. The laminar principle also eliminates cavitation effects with liquids and sonic throttling velocities with gases. This design can be used with inlet pressures and pressure drops of 2675 PSIG (18.5 bar).

While it is not commonly thought of for low-flow valve application, the labyrinth disc valve design (see Figure 4.6q) can be manufactured for this purpose. Its most common application is for reducing high pressure fluid samples for analyzers, but obviously it can be used in other services. One common factor for this design, as well as for all of the others discussed, is the need for the fluid to be extremely clean. These valves are not tolerant

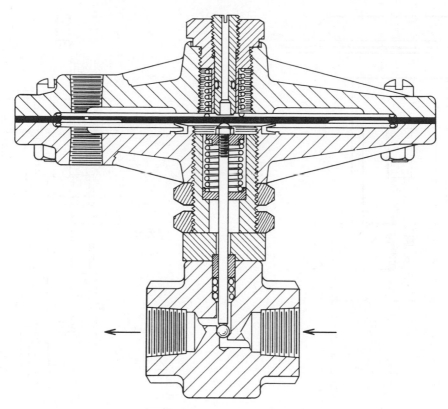

Fig. 4.7u   Low-flow valve with adjustable short-stroke actuator (*Courtesy of A. W. Cash Co.*)

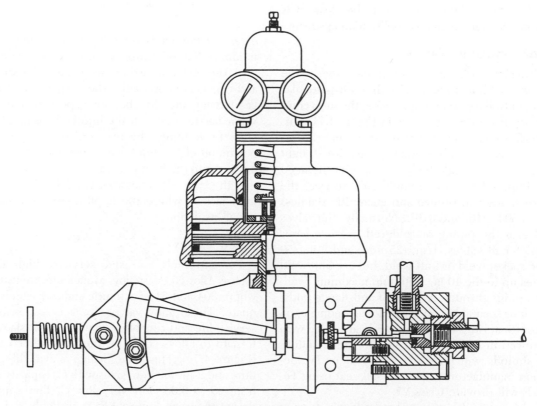

Fig. 4.7v   Low-flow valve with replaceable trim and ratio adjustable stroke linkage for high pressure
service (*Courtesy of Masoneilan International*)

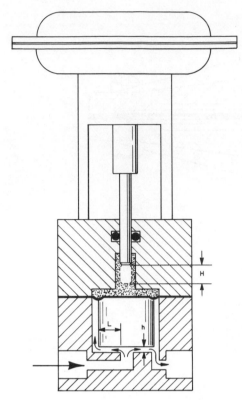

Fig. 4.7w   Low-flow valve using laminar flow element (*Courtesy of H . D . Baumann Co.*)

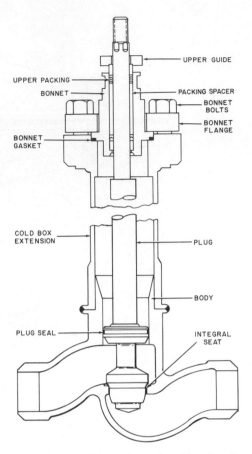

Fig. 4.7x   Cold box valve with weld ends and welded bonnet (*Courtesy of Valtek, Inc.*)

of dirt or sediment due to the small passages and close clearances. Unless the fluid is known to be clean, it is necessary to provide for a high level of filtration upstream.

Cold Box or Cryogenic Valves

A special design variation on the globe valve is the cryogenic valve. While Figure 4.7o shows a particular Y-valve design fitted for cryogenic service, the most common design for this service is shown in Figure 4.7x. This design is specific to cryogenic (down to $-454°F$ or $-270°C$) service and no other. Body configurations are straight through, as shown, or angle body. Because of the need for Charpy impact for the extremely cold service, the materials are limited to bronze and austenitic stainless steels such as 304, 316, and 316L. Normally, the valves are welded into the piping or soldered in some cases with bronze. Small valves, 1 in. (25 mm) and 2 in. (50 mm), can be socket weld or butt weld, but are butt weld in larger sizes up to the 10 in. (250 mm) maximum available. Body ratings through ANSI 600 and flange ends, either integral or separable, are available depending upon manufacturer. Seat rings may be integral hard-faced with Stellite, screwed-in metal, or soft seat for tight shutoff. The metal shutoffs will be ANSI Class III or IV, depending upon manufacturer, and the soft seat using Teflon or Kel-F will provide Class VI.

Cold box valves are designed to have a low body mass for fast cool-down and reduced heat transfer. The long

extended bonnet is provided with a plug stem seal to minimize liquid "refluxing" into the bonnet and packing area, thereby minimizing the heat loss due to conduction and convection. Actually, the small amount of liquified gas passing into the bonnet vaporizes and provides a vapor barrier between the liquified gas and the packing area. In addition, the pressure resulting from the vaporization of the liquid prevents additional liquid from passing into the bonnet area. Excess pressure vents back into the body. It is possible to fit these valves with vacuum jackets where the application requires this additional insulation.

Lined Valves

For extremely corrosive services, high alloy metals such as Alloy 20, Hastelloy, Monel, etc. are usually needed where 304 and 316 austenitic stainless steels are inadequate. These alloy valves are very expensive and the service life and reliability may be unsatisfactory. An alternate to alloys could be the Teflon-lined globe valve (Figure 4.7y). This valve design is available in 1-in. (25 mm) and 2-in. (50 mm) sizes with C_v capacity coefficients from 45 to 0.33. For very low C_v flow capacities, the design shown in Figure 4.7z is available in 1-in. size (25 mm) with trim selections from 1.0 to 0.001 C_v. These

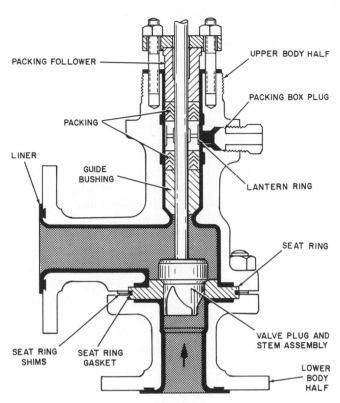

Fig. 4.7y   Standard Teflon-lined valve with Teflon encapsulated trim (*Courtesy of Fisher Controls Co.*)

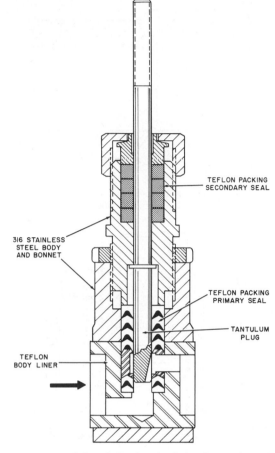

Fig. 4.7z   Small-flow Teflon-lined valve with tantulum plug (*Courtesy of H. D. Baumann Co.*)

valve designs are limited to operating temperatures in the range of 300° to 400°F (150° to 200°C) and pressure drops in the range of 125 to 250 PSI (8.5 to 17 bar). To some degree, their application overlaps with that of the Teflon tube pinch valve (see Section 4.9).

Valve Connections

The most commonly used valve body end connection for piping closure is the flange. In the United States, flange design and ratings are standardized under ANSI Standard B16.5. Other countries often use this standard, but some have their own standards such as DIN in Germany. These rating systems are different and not interchangeable. For example, ANSI B16.5 rates the flange pressure level at an elevated temperature (750°F or 399°C), whereas DIN defines the rated pressure at room temperature (100°F or 38°C maximum). Thus, an ANSI Class 150 flange can be used for higher pressures than a DIN 10 flange since the ANSI flange can be used to 320 PSIG (2.2 MPa) at room temperature.

Flanges may be flat face, raised face, ring type joint (RTJ), tongue and groove, male and female, or other configuration to suit the application needs. Generally, flanges are available from 150 through 2500 PSIG ANSI (1.0 through 17.3 MPa). Cast iron, ductile iron, and bronze are usually flat face; carbon steel or alloys are usually raised face; and above ANSI 600, the RTJ is fairly common. Gaskets may be sheet stock, such as asbestos or

Teflon, or a widely used type known as the spiral wound asbestos or Teflon contained within metal rings. RTJ ring gaskets may be oval or octagonal in cross-section and can be of any suitable metal softer than the flange.

As already discussed, some valves can be supplied with slip-on flanges with attendant benefits. Referring to Figure 4.7g, it can be seen that the valve body ends have a machined groove. Two circular half-rings are slipped into the grooves to retain the flange when it is bolted into the piping.

Flangeless bodies (clamped-in) have been used in the small barstock bodies. While this permits economies where expensive alloys are involved, it does pose some bolting and piping alignment problems. The tie-rod bolting must be high tensile strength material and the valve must be carefully centered to permit proper gasket sealing and loading. Such bolting may be subject to thermal longitudinal expansion problems with the valve body and piping and may not be compatible with the process fluid to resist corrosion in the event of a leak.

Welding ends are not common in the process industries, but are recommended where unusually high line stresses or thermal shock conditions exist, such as in many power plant applications. Socket weld ends are

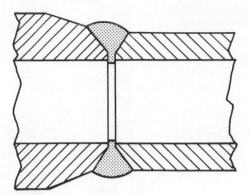

Fig. 4.7aa   Butt weld valve joint showing full penetration weld

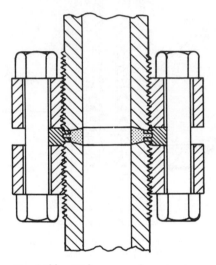

Fig. 4.7bb   High pressure lens ring type joint

easy to align and may be used in small sizes of 2 in. (25 mm) or under. However, most valves are installed with butt-weld ends where maximum joint integrity by full penetration of the weld is needed (Figure 4.7aa). Butt-weld joints are commonly checked by radiographic inspection (X-ray) to confirm full penetration and lack of meaningful defects such as cracks, voids, slag, etc. The valve body material should be selected that is suitable for welding to the adjoining piping.

Some chemical processes require pressure ratings above 5,000 PSIG (345 MPa). Some operating companies have their own proprietary fitting designs or adaptations of available commercial designs. However, one widely used high-pressure connection design is the lens-ring type fitting shown in Figure 4.7bb. This is a self-energizing type of seal, i.e., the lens-ring deforms to give a tighter seal with increase in line pressure. (Further details are given in Section 4.13). In some cases, a proprietary fitting called the Grayloc fitting is used on valves instead of flanges, where the flange rating is 900 to 2500 PSIG ANSI (6.2 to 22.5 MPa). This fitting greatly reduces the bulk of the valve as compared to the large high-pressure flanges. The Grayloc fitting utilizes a lens-ring type seal, as shown in Figure 4.7cc, but the special hub design is fastened together with a bolted clamp type fitting instead of threaded-on flanges and regular bolting. These fittings are also available in 10,000 and 15,000 PSIG (69 and 104 MPa) designs.

Screwed connections are not normally used with process valves since they can leak, are difficult to make up in the piping, and are subject to corrosion and sticking. However, some special small valve designs are furnished

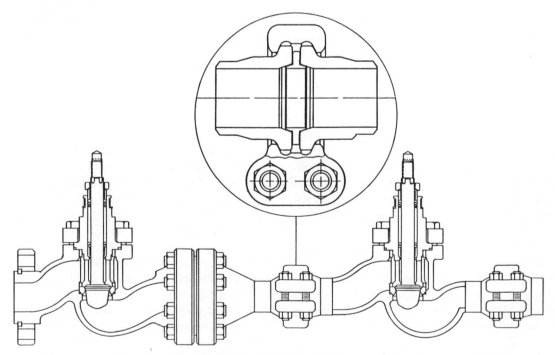

Fig. 4.7cc   Grayloc end connection detail and comparison with high-pressure flange connection (*Courtesy of Gray Tool Company*)

Table 4.7dd
COMMON VALVE BODY MATERIALS

Material	ASTM Specification	Maximum Pressure Rating ANSI	Low Temperature Limit °F (°C)	High Temperature Limit °F (°C)	Service
Cast iron	A126A	400 psig (2.8 MPa)	−20 (−28)	400 (204)	Noncorrosive, water, low-pressure steam
Cast carbon steel	A216-WCB	2500	−20 (−28)	800 (427)	Noncorrosive, general use
Cast chromemoly steel	A217-WC9	2500	−20 (−28)	1050 (566)	High temperature, flashing water
Cast stainless steel (304)	A351-CF8	2500	−450 (−268)	1500 (816)	Corrosive, low temperature
Cast stainless steel (316)	A351-CF8M	2500	−425 (−254)	1500 (816)	Corrosive, high temperature
Cast bronze	B62	300 psig (2.1 MPa)	−350 (−212)	400 (204)	Low temperature, oxygen, marine
Cast monel	A296-M35	2500	−325 (−199)	900 (482)	Corrosive, brine, chlorine
Cast Alloy-20	A296-CN-7M	2500	−50 (−45)	300 (149)	Corrosive, sulfuric acid
Cast Hastelloy-C	A494	2500	−325 (−199)	1000 (538)	Corrosive, general use

with threaded end connections, as well as various economy valves for services such as low-pressure steam, water, and gas. Where it is necessary, screwed valves can be converted to flanges with a flange welded to a pipe nipple, then screwed into the valve and seal welded. For some services, welding may not be necessary and the whole assembly can be put together with threaded fittings and a suitable pipe dope on the threads.

Materials of Construction

A valve body assembly is a pressure containment vessel. As such, material selection must follow guidelines set forth by the ASME Unfired Pressure Vessel Code. For example, a valve body material should not be used for low temperature service if its Charpy impact level is below 15 PSIG (1 bar) or for high temperature service if it is subject to carburization. Where in-line welding is required, particular attention must be paid to the weldability of body materials. For example, to avoid intergranular corrosion due to carbide precipitation in stainless steel valves, specify a low carbon austenitic stainless steel such as 304L or 316L, or in more severe corrosion situations a stabilized stainless such as 347 may be needed. Cast carbon and austenitic stainless steels will cover over 90 percent of the valve service applications.

Other special trade-name alloys are available for special corrosive applications. Valve bodies and trim materials are available from many manufacturers in such alloys as Monel, Hastelloy, nickel, Inconel, Alloy 20,

tantalum, nickel-bronze, chrome-moly steel, etc. Table 4.7dd is a very simplified guide for selection of body materials. However, when dealing with application services, one should consult with the valve manufacturer or a qualified materials specialist when necessary. Alloy selection is not always straightforward due to variations in fluid concentration, temperature, impurities, or other factors. Often, compromises must be made in materials selection to give a reasonable balance between service life and cost.

BIBLIOGRAPHY

Adams, M., "Control Valve Dynamics," *InTech*, July, 1977.

Baumann, H.D., "How to Assign Pressure Drop Across Control Valves for Liquid Pumping Services," *Proceedings* of the 29th Symposium on Instrumentation for the Process Industries, Texas A&M Univ., 1974.

Buckley, P.S., "A Control Engineer Looks at Control Valves," *Proceedings* of the 1st ISA Final Control Elements Symposium, Wilmington, DE, 1970.

Carey, J.A., "Control Valve Update," *Instruments & Control Systems*, January, 1981.

Cunningham, E.R., "Solutions to Valve Operating Problems," *Plant Engineering*, Sept. 4, 1980.

Fernbaugh, A., "Control Valves: A Decade of Change," *Instruments & Control Systems*, January, 1980.

Hutchison, J.W., ed., *ISA Handbook of Control Valves*, Instrument Society of America.

Shinskey, F.G., "Control Valves and Motors," Foxboro Publication No. 413-8.

Wolter, D.G., "Control Valve Selection," *InTech*, October, 1977.

4.8 MISCELLANEOUS VALVES

Flow Sheet Symbols

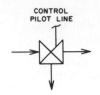

EXPANSIBLE TUBE OR
DIAPHRAGM VALVES

CONTROL
PILOT LINE

FLUID INTER ACTION VALVES

Valve Designs:	A—Expansible tube or diaphragm valve B—Fluid interaction valves C—Specialized in-line valves
Sizes:	A—1″ to 12″ (25 to 300 mm) B—½″ to 4″ (12.5 to 100 mm)
Design Pressure:	A—Up to 1,500 PSIG (10.4 MPa) B—100 PSIG (0.7 MPa) or greater (no theoretical limit) C—Up to 2,500 PSIG (17.3 MPa)
Design Temperature:	A—150°F (66°C) B—Can handle molten metals
Rangeability:	A—20:1 B—10:1
Capacity:	A—$C_v = 12d^2$ (noncritical flow) B—$C_v = 14d^2$ (noncritical flow)
Partial List of Suppliers:	American Meter Div. of Singer Co., (C); Belfab Co., (C); Bailey Controls Co., (C); G.W. Dahl, Inc., (C); Eisenwerk Heirich Schilling (C); Fisher Controls Co. (A,C); Grove Valve & Regulator Co., (A); Moore Products Co. (B)

Expansible Tube Type

Control of flow is obtained (Fig 4.8a) by use of an expansible tube that is slipped over a cylindrical metal core containing a series of longitudinal slots at each end and a separating barrier in between. A cylindrical, in-line jacket surrounds the tube so that the fluid line pressure can be introduced between the jacket and the sleeve to cause the sleeve to envelop the slots. With pressure shut off to the annular space and bled to the downstream line, the line pressure in the valve body will cause the valve to open fully (Figure 4.8b). Control of the pressure on the sleeve creates a throttled flow condition by first uncovering the inlet slots and then progressively opening the outlet slots. A continuous dynamic balance between fluid pressure on each side of the sleeve makes it possible to obtain wide rangeability from a no-flow to a fully open condition.

The basic operation of the valve can be accomplished with a manually positioned, three-way valve, or by a three-way pilot valve positioned from a remote location. A variety of automatic pilots give versatility to the basic

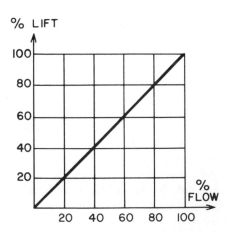

Fig. 4.8a Inherent characteristics of expansible tube valves

valve. For reduced pressure control, a pilot is used to modulate the jacket pressure in response to the sensed pressure in the downstream pipeline. As downstream pressure tends to fall below the set point, the double-acting pilot positions itself to reduce the jacket pressure.

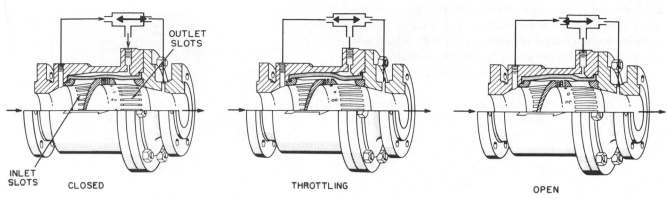

Fig. 4.8b Expansible tube valve

This allows the valve to open to a throttling position. Therefore downstream pressure increases to the set pressure, with attendant change in flow rate which holds the set pressure. The static sensing line is separate from the pilot discharge line, precluding a pressure drop effect in the sensing line.

Another form simulates a conventional regulator, in that system gas is bled into the jacket annular space through a fixed orifice and bled off through the pilot regulator. In this form, the static sensing line and pilot output are common. Double-acting pilot systems use seven control ranges from 2 to 1,200 PSIG (0.02 to 8.3 MPa) with corresponding inlet pressures up to 1,500 PSIG (10.4 MPa). The fixed orifice design is available for control from 2 oz to 600 PSIG (0.86 kPa to 4 MPa).

Back-pressure control and pressure relief are obtained in the same manner, with pilot regulators that sense upstream pressure. By using a separate sensing and bleed port, a build-up from cracking to fully open can be varied from 3 percent to 14 percent of the set pressure. Return to normal operation causes the valve to create absolute shutoff. Emergency shutoff service may use an external pressure source piloted to obtain immediate shutoff upon abnormal conditions.

A diaphragm-operated, three-way slide valve may also control jacket pressure by proportioning the inlet and outlet pressures. Application of the controller output pressure to the three-way slide valve diaphragm actuator causes the sleeve valve to open proportionally, in a manner somewhat similar to that of a conventional diaphragm control valve. In this manner, the valve becomes a control valve (Figure 4.8c).

Differential, or flow control, is accomplished by using a pilot valve in which the diaphragm is acted upon by both upstream and downstream pressures. The differential static lines may be taken at the inlet and outlet of the line valve, or, for better precision, across an orifice plate in the line.

An unlimited variety of control systems is possible with the pilots available. One of its important applications in gas distribution is pressure boosting. The line pressure

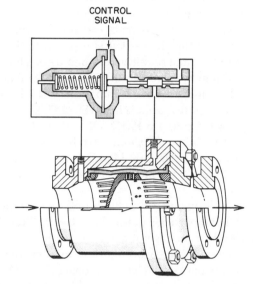

Fig. 4.8c Expansible tube valve with throttling control pilot

losses due to increased consumption can be counteracted by automatically increasing the set pressure of the distribution control valve. Normal pressure is controlled by pilot #1 in Figure 4.8d. An increased flow is sensed at

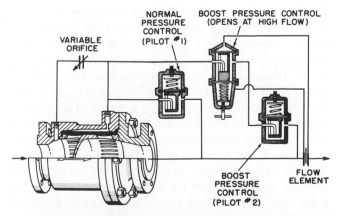

Fig. 4.8d Expansible tube valve applied to gas distribution pressure boosting service

a differential pressure-producing device, orifice, or flow tube, which opens a high-differential pilot to cut in a boost-pressure-control regulator (pilot #2), which is set at a higher control pressure. Return to normal flow cuts out the boost pressure control and reinstates the normal control pilot.

The expansible tube valve is made in sizes from 1 through 12 in. (25 through 300 mm) with pressure ratings from 200 PSIG (1.4 MPa) in iron to 1,500 PSIG (8.3 MPa) in steel construction. It is made flanged or flangeless for insertion between line flanges. The flangeless body is cradled in the studs between the line flanges. Removal of the body is made easier by expanding the flanges about $\frac{1}{8}$ in. (3 mm) using nuts on the studs inside the flanges. Tight shutoff or throttling requires a differential between the line pressure or external source used for closure and the downstream pressure sensed on the inside surface of the sleeve that is exposed to this pressure. This differential requirement for a special low-pressure 2-in. (50 mm) valve is 3.6 PSID (24.8 kPa) and 1.6 PSID (11 kPa) for a 4-in. (100 mm) valve. The low-pressure series requires from 21 PSID (145 kPa) for the 1-in. (25 mm) size to 4.6 PSID (32 kPa) for the 10- and 12-in. sizes (250 and 300 mm). High-pressure models require from 58 PSID (400 kPa) for the 1-in. (25 mm) size to 11 PSID (76 kPa) for the 10- and 12-in. sizes (250 and 300 mm).

The body design allows tight shutoff, even with comparatively large particles in the flow stream. Freezing of the pilot by hydrates is not common because the intermittent and small bleed occurs only to open the valve and as such is not conducive to freezing. The pilot may be heated or housed, or even located in a protected area close to the warm line. With its only moving part a flexible sleeve, this type of valve has no vibration to contribute to noise. The flow pattern also helps make this valve from 5 to 30 db more silent than most regulators. Flow capacities are comparable to those of single-seated as well as many double-seated regulators.

Expansible Diaphragm Design

An expansible element (Figure 4.8e) is stretched down over a dome-shaped grid causing shutoff of the valve when pressure above this resilient member overcomes the line pressure under the element. Line pressure is evenly directed over the expansible area by a series of pressure channels. Selection of the correct action on a pilot that supplies line or external pressure to the exterior of the expansible element causes the valve to control as a back pressure or reducing control valve. For back pressure control, or pressure relief, the static line is taken upstream of the valve. Increase in line pressure will increase the bleed from the annular space between the expansible element and the metal housing. Reducing regulation is accomplished by restricting the bleed upon increase in downstream pressure and increasing it upon decrease in downstream pressure.

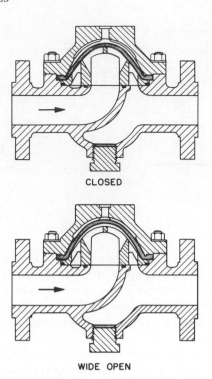

CLOSED

WIDE OPEN

Fig. 4.8e   Expansible diaphragm valve

The valve is available in iron or steel with ratings to 600 PSIG (4 MPa). Models are available from $-10°F$ to $150°F$ ($-23°C$ to $66°C$), using a molded, buna-N diaphragm. Relief valve pressures are from 30 to 300 PSIG (0.21 to 2.1 MPa), while reducing service varies from 5 to 150 PSIG (0.03 to 1 MPa). Capacity factors vary from $C_v = 11d^2$ in smaller sizes to $C_v = 14d^2$ in larger valves.

Fluid Interaction Type

The Coanda effect, the basis of fluidics, is used in diverting valves from $\frac{1}{2}$-in. through 4-in. sizes (12.5 through 100 mm). The Coanda effect means the attachment of a fluid stream to a nearby sidewall of a flow passage. This effect can be used in a so-called flip-flop valve for diverting a stream from one discharge port to another. Figure 4.8f shows the flow through the right-hand port due to both control ports being closed. Opening the right-hand port, to allow air or liquid to enter, will shift the flow. The industrial valve has rectangular diverting tubes, but the end connections may be circular. Control is maintained by opening or closing the control port or by injecting low-pressure air or liquid through a solenoid or other pilot valve.

In this valve, with the stream flowing in one diversion tube, the flow at the inlet contains some potential (pressure) energy and some kinetic (flowing) energy. Much of the potential energy is converted to kinetic energy at the nozzle. Up to 70 percent recovery of potential energy occurs in a diffuser section. Fifty percent recovery is guaranteed for commercial valves, and somewhat less for gases above critical flow and for viscous fluids ($\Delta P \approx P_1/2$).

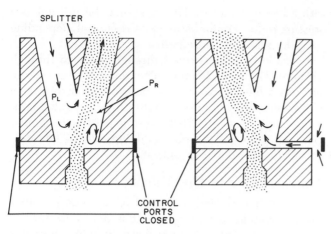

Fig. 4.8f Operation of fluid interaction valves

This design is proposed as the basis for a 100-in.^{2} (0.06 m^2) valve to control the exhaust of a high-performance turbojet engine. The fluidic valves can be used for diversion of engine exhaust gases from tailpipe propulsion nozzles to wing-mounted lift fans with ambient control flow. A four-ported valve has been used for direction control of a missile. By manipulation, a flow stream emits from each port. Upon proper injection of control streams the flow is directed to the ports required to create change in direction. A sophisticated digital control system maintains flow volume control in a given direction by controlling the accumulated time that the flow is directed to the desired port. It is evident that the ultimate use of this valve design has not been reached.

Specialized In-Line Valves

In addition to the previously discussed valves, there are some more special designs: The expandable element in-line valve, which is used in the gas pipeline field, the positioned plug in-line valve, which is most adaptable to toxic fluid services and is used in the aerospace industry, and the diaphragm-operated cylinder in-line valve, which is most suitable to high-pressure gas service where a reduced noise level is desired.

Expansible Element In-Line Valves

Streamlined flow of gas occurs in a valve in which a solid rubber cylinder is expanded or contracted to change the area of an annular space (Figure 4.8h). A stationary inlet nose and discharge bullet allow hydraulic pressure to force a slave cylinder against the rubber cylinder to vary its expansion. Control is from a diaphragm actuator, with the diaphragm plate carrying a piston. The piston

Installation must allow the recovered pressure to create the desired flow against friction effects of piping or fittings. An uninhibited flow will create some aspirating effects in the open outlet; restriction causes blocking, while excessive restriction will cause a leak or diversion to the open port.

Industrial valves, with their ability to divert in less than 100 msec, fill a wide variety of uses. The primary one is for level control in which the effluent not required for filling may be returned to storage. It is necessary only to use a dip tube set at the control point, as shown in Figure 4.8g. Lack of moving parts or of detrimental effects due to fast diversion action allows the system to provide very close control.

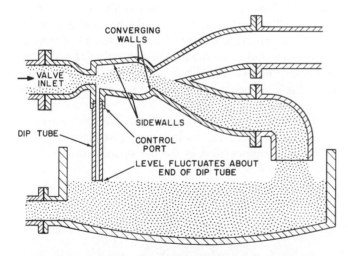

Fig. 4.8g Level control with fluid interaction diverting valve

Numerous uses of diversion valves exist, such as tank filling, which is accomplished by using an external signal. Diversion of a process stream upon contamination, sensed by a pH or other analyzer, is important in paper mills and chemical plants. The ability to divert rapidly makes this valve applicable to oscillating flows. The valve may be used if total bypassing of a heating or cooling medium is adequate for temperature control.

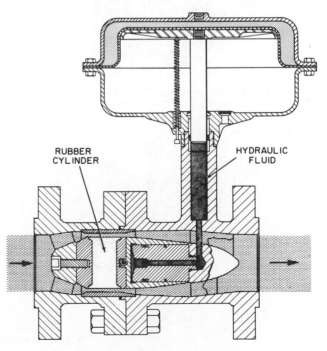

Fig. 4.8h Expandable element in-line valve

acts as a pump to supply hydraulic pressure to the slave cylinder.

The rubber cylinder offers the seating ability of a soft seat valve. It has the capability of closing over foreign matter, and the design allows for the use of a restricted throat for reduced capacity. With this design pressure-drops as high as 1,200 PSID (8.3 MPa) have been handled with a low noise level. The valve may be utilized as a pressure reducer or for back-pressure control, depending upon the system requirements.

Available sizes are from 1 through 6-in. (25 through 150 mm). The 1-in. (25 mm) valve can have screwed connections, while all sizes can be flanged. The body is steel with flange ratings to 600 PSIG (4 MPa). A valve positioner can be used by calibrating stem position to annular space reduction and thereby obtaining accurate flow control relative to controller output.

Positioned Plug In-Line Valves

The positioned plug in-line valve, excluding control units, resembles a pipe spool. It is only necessary to inject pressure into its ports for positioning the valve plug. This simple design (Figure 4.8i) requires only three pressure seals. The plug is carried on a cylinder which also includes the piston. Pressure in one port causes closing, while the opposite port is used for opening. The valve has only one moving part. The valve is available in sizes from 2 in. to 8 in. (50 to 200 mm) for use to 350 PSIG at 400°F (2.4 MPa at 204°C). Control quality is dependent upon the pilot valves and auxiliary units employed.

Another in-line valve available in small sizes (Figure 4.8j) carries the valve plug on a bridge in the operating cylinder, with the seat as part of a split body.

A spring-loaded version of this design uses the beveled end of the moving cylinder to seat on a replaceable soft seat, retained in a dam, held in position by struts from the inside wall of the valve body. The spring loading may cause fail-close or fail-open actions, as illustrated on Figure 4.8k. The unit can be powered with line fluid or by an external pressure source. A double bleed feature can be incorporated to eliminate the possibility of actuation and line fluids combining if one of the dynamic seals should fail. All seats and seals are replaceable by separation at the body flange.

The unit is particularly adaptable to fluids that are toxic or difficult to contain, such as nitrogen tetroxide, hydrogen and others used in the aerospace industry. The

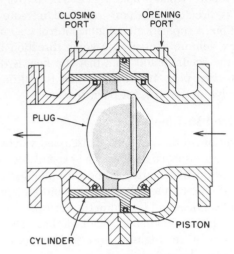

Fig. 4.8i   Positioned plug valve in fully open position. (*Courtesy of Eisenwerk Heinrich Schilling.*)

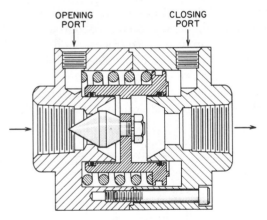

Fig. 4.8j   Positioned plug valve in closed position. (*Courtesy of G.W. Dahl Co.*)

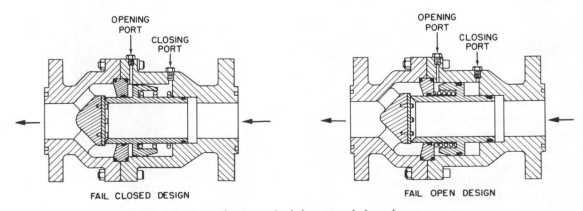

FAIL CLOSED DESIGN   FAIL OPEN DESIGN

Fig. 4.8k   Spring-loaded positioned plug valve

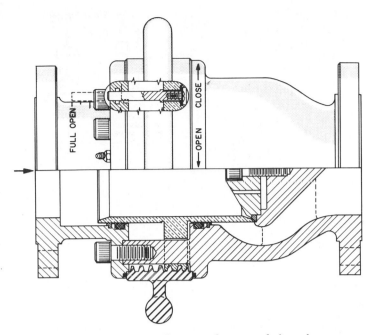

Fig. 4.8l   Manually operated positioned plug valve

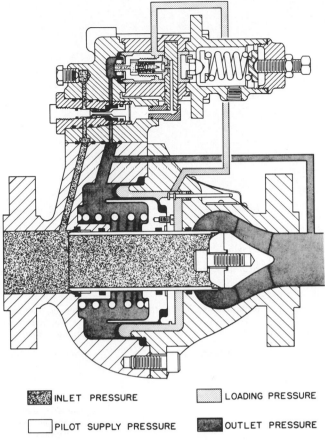

INLET PRESSURE		LOADING PRESSURE
PILOT SUPPLY PRESSURE		OUTLET PRESSURE

Fig. 4.8m   Diaphragm-operated cylinder type in-line valve

unit is furnished in sizes from 1½ in. to 18 in. (37.5 to 450 mm), with ratings to 2,500 PSIG (17.3 MPa). All types of end connections are available, and control is dependent upon the auxiliary control components selected for the application. An explosion-proof limit switch can be furnished for position indication.

A similar valve (Figure 4.8l) has been adapted to manual operation. Rotation of the external handwheel carrying internal threads moves a cylinder with external threads. The cylinder's leading edge closes on the soft seat.

Diaphragm-Operated Cylinder In-Line Valves

An in-line valve using a low convolution diaphragm for positive sealing and long travel (Figure 4.8m) is designed particularly for gas regulation. The low level of vibration, turbulence, and noise of this in-line design makes it suitable for high-pressure gas service. Inlet pressures to 1,400 PSIG (9.7 MPa) and outlet pressures to 600 PSIG (4 MPa) are possible in the 2-in. (50 mm) size. It is a high-capacity valve, as expressed by $C_v = 23d^2$. As a gas regulator the unit is supplied with a two-stage pilot to accept full line pressure. This pilot resists

freeze-up and serves as a differential limiting valve. All portions of the pilot and line valve will withstand a full body rating of 600 PSIG (4 MPa).

BIBLIOGRAPHY

Carey, J.A., "Control Valve Update," *Instruments & Control Systems*, January, 1981.

Colaveri, M.R., "Solenoid Valve Basics," *Instruments & Control Systems*, August, 1979.

Dobrowolski, M., "Guide to Selecting Rotary Control Valves," *Instrumentation Technology*, December, 1981.

Fernbaugh, A., "Control Valves: A Decade of Change," *Instruments & Control Systems*, January, 1980.

O'Connor, J., "The Turbine Control Valve," *Instrumentation Technology*, December, 1973.

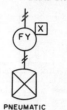

PNEUMATIC
PINCH VALVE

MECHANICAL
PINCH VALVE

4.9 PINCH VALVES

Sizes:	$\frac{1}{8}''$ to 72″ (3 mm to 1800 mm)
Design Pressure:	Up to 300 PSIG (2050 kPa or 20.5 bar)
Design Temperature:	−20° to 400°F (−29° to 204°C)
Relative Capacity:	$C_v = 60d^2$ (not useful for sizing)
Rangeability:	10:1 maximum, 5:1 average
Materials of Construction:	Sleeves: pure natural rubber, various elastomers, and Teflon; bodies: cast iron, aluminum, cast steel, and stainless steel
Special Features:	Pneumatic, hydraulic, or electric operation; reduced port designs; fire-shields; special corrosion resistant coatings on metal parts
Cost:	See Figure 4.9a. Cost data are based upon standard flanged cast or ductile iron housing, pure gum rubber sleeve, and on-off operation. Power operator on pinch clamp design is a rolling diaphragm piston. Numerous other elastomers are available for the sleeve, and throttling piston actuators are available plus electric motor drives. Other body materials are available
Partial List of Suppliers:	Badger Meter, Inc.; Clarkson Co.; Fisher Controls; Flexible Valve Corp.; Galigher Co.; Legris, Inc.; Precision Products Div.; Red Valve Co., Inc.; Resistoflex Corporation; RKL Controls, Div. of Robbins & Myers, Inc.; Valtek.

These valves are called either pinch or clamp valves depending upon the configuration of the flexible tube and the means used for tube compression. The compression can be done by mechanical clamping mechanisms of various designs or by external pneumatic or hydraulic power within a metal jacket enclosure. Pinch valves have enjoyed a greater application than in the past due to improvements in elastomers and reinforcing fabrics as well as the introduction of successful plastic tubes. Tubes can be fabricated from pure gum rubber or from a variety of rubber-like elastomers such as Buna-N, butyl, Neoprene, Nordel, Hypalon, Viton, silicone, polyurethane, polypropylene, white butyl, and odorless and tasteless white Neoprene. The latter two materials are usually found in the food and allied industries. Reinforcing fabrics may include many of the same materials used in automobile tire fabrication, such as cotton duck, rayon, nylon, fiberglass and Kevlar®, which is a new arimid

polymer material that is as strong as steel at one-sixth the weight. Another recent design uses a specially shaped Teflon plastic tube, which makes it uniquely capable of handling a wide range of highly corrosive or sticky fluids.

Two types of tubes are generally available: full-round opening and prepinched. Each type has a preferred end use. The full-round should be considered for full flow on-off service or where large solids are handled that might become lodged in the valve if full pipe opening is not available. The prepinched is better for throttling control and has a better flow characteristic. Figure 4.9a shows the typical throttling characteristics for some various tube type designs. However, it should be recognized that other flow characteristics are possible depending upon the tube design and means of actuation. Specific information should be obtained from the manufacturer.

A general statement can be made that the first 50 percent of the pinch action for standard full-round tubes

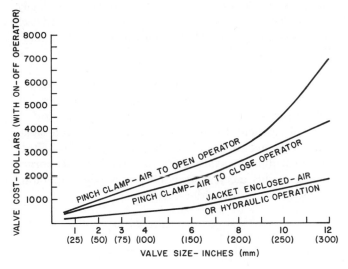

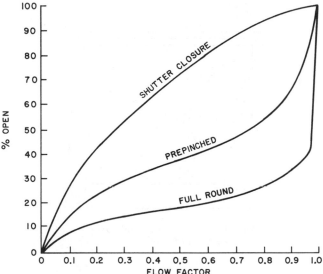

Fig. 4.9a   Cost and typical throttling characteristics for various pinch tube designs

resistance, high corrosion resistance, packless construction, high corrosion resistance, packless construction, reasonable flow control rangeability, smooth flow with no pockets or obstructions, low replacement costs, and long life with lower costs than metal valves on abrasive or corrosive services. This is not to say that the pinch valve does not have limitations. There are definite pressure and temperature restraints due to the nature of the elastomer or plastic materials involved. It must be recognized that the application of elastomers to fluid mixtures is an imperfect science involving compromises on resistance to the fluids, temperature, abrasion, corrosion, and external weathering.

Cycle life is another factor that must be considered in pinch valve application and service life. Generally, pinch valves will operate over cycle lives ranging from the tens to the hundreds of thousands or even in the millions in some cases. A reasonable minimum expectation is 50,000 open-close cycles with no evidence of cracking or incipient failure. The actual cycle life is dependent upon all the above-mentioned factors. It is not uncommon for these valves to operate 10 to 15 years in some services without failure. In the areas where abrasion and corrosion are present in combination, the pinch valve is almost unequaled and will generally outlast an alloy metal valve 5:1 or better. Many factors must be considered, and the service experience of the manufacturer is paramount to success.

Traditionally, pinch valves are used throughout the mining and ore processing industries for handling all types of slurries. These range from rather large rock gravel pieces to relatively fine mineral particles. The fluid carrier ranges from water to fairly corrosive solutions. Usually, these tubes can be closed over lumps of materials and either break up the lumps or "encapsulate" them and give very tight shutoff. The pinch valve is a natural for this industry and has given excellent service for many years.

In the water and waste treatment industry, the pinch valve handles many difficult fluids such as lime slurry, raw sewage, recycle sludge, grit and garbage particles, grease, slime, and other equally obnoxious materials. The flexing of the valve during operation enables it to shed most coatings, and the smooth, unobstructed flow passage enables it to handle anything that will move through the piping. This same capability also makes the pinch valve very useful in areas of the pulp and paper industry, where everything from wood knots and chips to wood fibers must be handled.

In the chemical industry, pinch valves are applicable to many slurries, erosive or corrosive or both. Also, the pinch valve is a good valve to consider in the handling of pneumatic-conveyed solids such as plastic pellets, powders, and similar materials. The valve will readily give tight shutoff over the pellets or powder and will not damage the product as might rotary air lock or plug valves. Further, it can be fabricated with a light, open

has little effect on the flow through the valve. The advantage of the prepinched tube is that this "dead" part of the valve control is essentially eliminated. However, very little fluid capacity is sacrificed since the shape of a full open prepinched tube has approximately 98 percent of the area of a full open standard round tube. Neither tube style generates any great turbulence at normal operating velocities in the center section of the curve when throttling. However, at the lower closed section of the curve, high turbulence can develop downstream from the pinch point at high pipe velocities.

Applications

Pinch valves can cover a broad application range in such industries as mining, water treatment, sewage and waste disposal, chemical, food, cosmetics and pharmaceuticals, and others. With proper selection of tube material and design, the pinch valve can offer high abrasion

aluminum frame and power operating cylinder and be self-supporting in the extremely lightweight pipe typical of these systems.

Valve Designs

The pinch valve falls into two basic actuation designs. The first is the mechanical clamp design, which also includes the variation known as the "accurate closure." The second is the enclosed metal housing using pneumatic or hydraulic pinching power, which includes the variation known as the "shutter closure." Each of these designs has preferable areas of application; in many areas, either is satisfactory. Careful consideration of all aspects of the application along with consultation with manufacturers should lead to the preferable choice in each case. There are few, if any, absolutes in pinch valves contrary to some claims.

Mechanical Pinch Design

Each manufacturer has designed a method of compressing the tube liner, depending upon how the tube is designed, shaped, and supported in the housing. Among the variations are screw clamps, wedges, rollers, and bars. Some ingenious mechanical designs have been developed to compress the tube effectively and at the same time minimize the effects of flexing stress for longer service life. This is important since it is readily apparent that the tube is always flexed at a fixed location. Thus a combination of compressor design and tube and fabric reinforcement design is a key to operation and cycle life.

Basically, the mechanical pinch design is either a single compressor type or a double compressor type. The single compressor design (Figure 4.9b) uses a single movable compressor device to compress the tube against a fixed restraint bar or lower tube support. The double compressor (Figure 4.9c) generally utilizes a fork-jack design that moves two compressor elements toward each other simultaneously. They usually meet at or just below the center line of the tube. In order to prevent over-pinching of the tube, which can result in damage or shortened life, there is an adjustable mechanical stop as part of the compressor design. Many compressor and tube designs also incorporate some type of lifting fork or other attachment (Figure 4.9d) to help pull the tube apart when opening. It usually takes a 10 to 15 PSIG (70 to 100 kPa) fluid pressure to restore the tube to normal shape, so this mechanism is necessary when operating at low pressures or under vacuum.

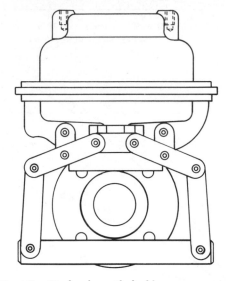

Fig. 4.9c   Pinch valve with double compressor bars
(*Courtesy of Flexible Valve Corp.*)

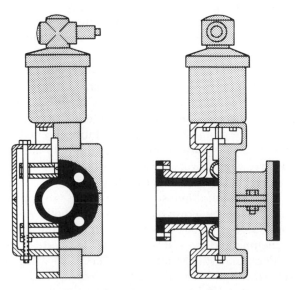

Fig. 4.9b   Single compressor pinch valve design
(*Courtesy of RKL Controls, Inc.*)

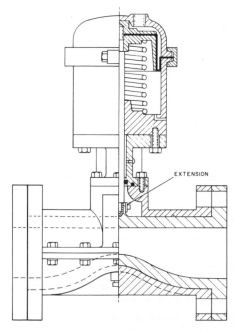

Fig. 4.9d   Mechanical pinch valve with
tube lifting mechanism for positive opening

The sleeve or tube can be fabricated in various forms depending upon the application and means of actuation. End connections can be enlarged to slip over the ends of the pipe in sizes $\frac{1}{4}$ in. through 8 in. (6–200 mm) and be retained by hose clamps (Figure 4.9e) or can be flanged to bolt into conventional piping systems (Figure 4.9d). One special flanged design for use with buried or below grade piping uses what is known as a mechanical joint end seal (Figure 4.9f). Flanged-end valves are available in the size range of $\frac{1}{4}$ in. through 18 in. (6–450 mm). The pinch valve can also be obtained in a three-way configuration for diverting service (Figure 4.9g). Here the double compressors are operated by a linkage system such that flow is always maintained in either the straight or

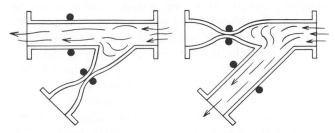

Fig. 4.9g   Three-way pinch valve (*Courtesy of RKL Controls, Inc.*)

branch leg as needed. During switching, the flow is through both legs proportional to the openings as the compressors open and close the two legs.

Although similar in principle, a more special design had to be developed for utilizing plastic tubes such as Teflon. This is because plastics, while flexible, are stiffer than elastomers and more prone to cracking with repeated flexing or pinching. Therefore, to be successful, the design had to eliminate creasing, kinks, and pinching to minimize overstress on the tube. A second Teflon tube is shrunk over the main tube for added reinforcement. Technically, this design is a clamp valve rather than a pinch valve. The normal clamping system is the double compressor type in which the top clamp travels down a stem with rotation and the bottom clamp on a yoke travels up the stem at the same time (Figure 4.9h). The valve can be operated with a rotary manual handwheel or power operator utilizing either a motor drive or an air motor drive. While normally used for on-off or fixed position service, modern electronic controls would also allow

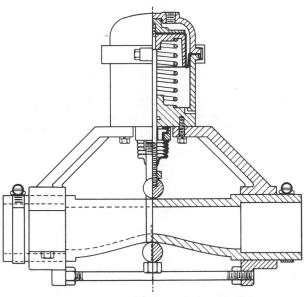

Fig. 4.9e   Pinch valve with slip-on tube
(*Courtesy of RKL Controls, Inc.*)

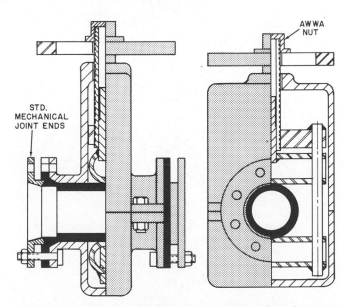

Fig. 4.9f   Mechanical joint end seal pinch valve for buried piping
(*Courtesy of RKL Controls, Inc.*)

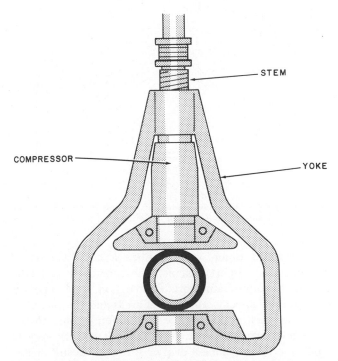

Fig. 4.9h   Teflon clamp valve double compressor clamping mechanism
(*Courtesy of Resistoflex Corp.*)

throttling control. However, a design variation specifically for throttling control utilizes a pneumatic piston linear power actuator and a fixed bottom compressor.

The heart of the Teflon tube design is the special shape (Figure 4.9i). Here the Teflon tube is wrapped around teardrop-shaped Teflon inserts and held solidly in place by external radius clamps. Thus, all flexing takes place on the centerline of the valve between the teardrops. This insures long life and minimum tube stressing. The valve is available in sizes 1 in. through 8 in. (25–200 mm). The inherent flow characteristic is essentially linear, especially between 20 to 80 percent of rated travel (Figure 4.9j). The tube is completely contained within an epoxy-coated ductile iron body, over which the ends of the tube are flared, and internal metal parts are corrosion-resistant alloys.

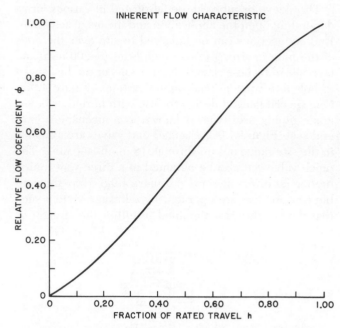

INHERENT FLOW CHARACTERISTIC

THIS CURVE SHOWS THE RELATION BETWEEN THE RELATIVE FLOW COEFFICIENT $\phi *$ AND THE FRACTION OF RATED TRAVEL $h **$. ANALYSIS OF THIS CURVE REVEALS THAT THE TFE CLAMP VALVE HAS A LINEAR FLOW CHARACTERISTIC BETWEEN $0.20 h$ AND $0.80 h$ OBEYING THE FORMULA $\phi = 1.16 h - 0.09$.

$$*\phi = \frac{\text{FLOW COEFFICIENT AT } h}{\text{RATED FLOW COEFFICIENT}}$$

$**h$ = FRACTION OF RATED TRAVEL

Fig. 4.9j   Typical throttling characteristic of the Teflon tube clamp valve (*Courtesy of Resistoflex Corp.*)

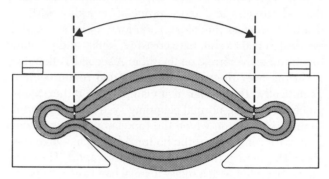

FLEXING ON CENTER LINE

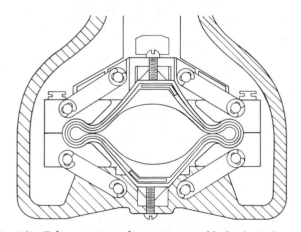

Fig. 4.9i   Tube retaining and operating assembly for the Teflon tube clamp valve (*Courtesy of Resistoflex Corp.*)

Mechanical pinch valves utilize a large variety of housing and support designs. These range from totally enclosed to open framework construction. Each type was designed for specific application areas and the manufacturer can best make service recommendations. In the open framework type (Figure 4.9k), it is very easy to spot tube failure since a drip or spray will result. Thus, a general statement is that these designs may not be appropriate where corrosive or harmful fluids are han-

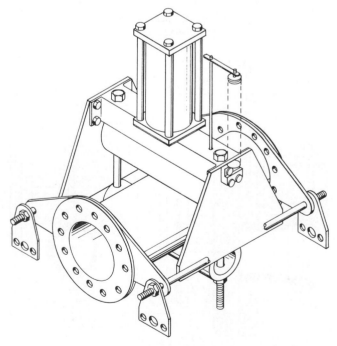

Fig. 4.9k   Open framework pinch valve used for services such as plastic pellets and powders (*Courtesy of Red Valve Co.*)

dled. The totally enclosed types are usually designed to contain the leak. Stems are provided with packing or O-ring seals, and the body cavity has one or more available drilled and tapped openings to use as vents or leakage "tell-tales." Frequent inspection is needed since the housings will probably corrode through with some types of fluid leaks.

By its nature, the pinch valve is a fail-open device since the tube will tend to assume its normal shape unless restrained. However, proper selection of the power actuator will allow it to either fail open or fail closed. In the latter case, the operator will be provided with either auxiliary spring or pneumatic pressure to hold the tube closed against line pressure. For proportioning or throttling control, the best operation is obtained by using a "double-acting" pneumatic piston actuator with a valve positioner. This type of operator uses the positioner to put air pressure on both sides of the piston and is considered a "stiff" operator with high power capability. Thus, it is able to easily overcome clamping mechanism friction, tube stiffness, and fluid pressure changes during operation. Some modern designs are available with positive failure action springs to enable proper failure action to be obtained without any problem.

In general, the mechanical pinch valve can operate with fluid pressures from vacuum to 300 PSIG (2050 kPa) depending upon design, size, tube material, and power operator. Operating temperatures can range from −20°F (−29°C) to +400°F (204°C) depending upon tube material, size, and fluid pressure. Any liquid that does not attack the tube sleeve can be handled in a completely enclosed, packless manner. Care must be taken to select a suitable tube material compatible with the service conditions even with the availability of Teflon tubes. Again, the manufacturer should be supplied with complete and specific details of the service fluid and application. It is difficult for the casual user to build up the extensive experience of the manufacturer over the broad range of service applications for which these valves can be used.

Pneumatic Pinch Design

By enclosing the tube completely within a metal housing designed for pressure containment, this design (Figure 4.9l) is operated by imposing pneumatic or hydraulic fluid pressure in the annular space. This pressure squeezes the tube completely around the annular space, thus opening, closing, or throttling the resilient sleeve. Up to 4 in. (100 mm) in size, the sleeve shape behaves similar to that shown in Figure 4.9m. Valves 6 in. (150 mm) and larger close in a three-way convolute pattern due to the larger area. These valves are particularly effective in surrounding lumps or stones and in breaking up chunks of lime, clay, or sewage solids while still preserving leak-tight operation.

Screwed bodies are available from ⅛-in. through 3-in. sizes (3–75 mm) and flanged bodies can be obtained

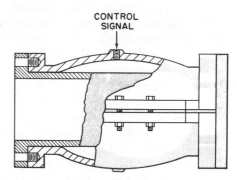

Fig. 4.9l   Pneumatic pinch valve
(*Courtesy of Red Valve Co.*)

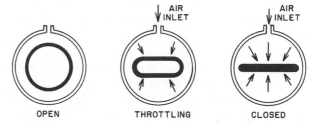

Fig. 4.9m   Pneumatic pinch valve tube shape behavior
(*Courtesy of Red Valve Co.*)

through 72 in. (1800 mm). The metal housing is commonly furnished in cast iron or aluminum for actuating pressures up to 125 PSIG (860 kPa) and in carbon steel or stainless steel for actuating pressures up to 400 PSIG (2750 kPa). Flanging can be ANSI Class 125 Cast Iron or ANSI Class 150 or 300 depending upon housing size and material. Sleeves are available in a variety of elastomers and pure gum rubber and in full round or reduced port configurations (Figure 4.9n). In general, the same considerations on tube material selection apply as previously discussed. Manufacturers have a wealth of service application experience and should always be consulted for recommendations. Corrosion or other factors may dictate selection of carbon steel or stainless steel housings rather than actuating pressure aspects.

The pneumatic pinch design can be operated directly from a 3 to 15 PSIG (20–100 kPa) pneumatic controller output if the fluid pressure is less than 1 PSIG (7 kPa).

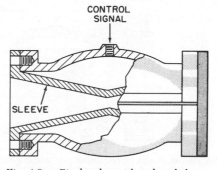

Fig. 4.9n   Pinch valve with reduced sleeve

However, it is more positive to operate through a booster relay (Figure 4.9o) and most applications are for higher line pressures. The booster relay allows the 3 to 15 PSIG (0.2 to 1.0 bar) control signal to operate a higher air pressure source in a proportional manner. Booster relays are available in ratios of 1:2, 1:3, 1:4, 1:5, and 1:6. Selection of the proper booster ratio is dependent upon line pressure and operating air supply pressure availability. For throttling only, the maximum air pressure needed is about 30 PSI (205 kPa) above line pressure; for throttling plus tight shutoff, the maximum air pressure needed is about 45 PSI (310 kPa) above line pressure. For example, if the line pressure is 15 PSIG (100 kPa), then a 60 PSIG (415 kPa) air supply is required to insure tight shutoff. A 1:4 ratio booster relay would be selected since the 3 to 15 PSIG control signal will output 12 to 60 PSIG (83–415 kPa). The valve will begin to close at about 16 PSIG (110 kPa) air pressure (4 PSIG or 28 kPa control signal) and will be completely closed at 60 PSIG air pressure (15 PSIG control signal) (415 kPa air pressure, 100 kPa control signal). A similar analysis can be performed for other line pressure levels. Since the maximum air supply pressure availability seldom exceeds 100 PSIG (690 kPa), pneumatic operated pinch valves are limited to line pressures of about 50 PSIG (345 kPa) where shutoff is required. However, hydraulic fluid power operation will allow operation up to much higher pressures subject to housing containment pressure limitations. Hydraulic control systems are more costly and complex, so there must be a definite need to justify their use.

The pneumatic pinch valve design does have some

pressure drop limitations that can produce flutter or instability. In general, the pressure drop when throttling should not exceed 25 percent of the inlet pressure in PSIG. The fluttering effect is position dependent and is generated by unstable nozzle angles since these valves close in a venturi type pattern. This is generally around the 40 and 90 percent closure positions, if the pressure drop exceeds 25 percent of the inlet pressure. In some cases, flutter is induced with back pressures under 2 PSIG (14 kPa) and can be eliminated by increasing the downstream head by some means suitable to the installation. Special designs such as the reduced port can usually eliminate flutter. However, it should be recognized that manufacturers usually do their testing with water, and the valve behavior in actual service, especially slurries, may be different. This may require some modifications to the valve or installation after being put into service. One solution can be the use of hydraulic instead of pneumatic power since it is "stiffer" and serves to contain or dampen any tendency to flutter or instability. Cavitation is definitely to be avoided with any pinch valve design since it can lead to premature failures. Conventional cavitation calculation parameters can be applied to pinch valves and cavitation index information can be provided by the manufacturer.

Shutter Closure Design

Shutter closure or round-hole configuration, rather than the slotted type closure obtained in conventional pneumatic pinch valve designs, utilizes an auxiliary rubber "muscle" as shown in Figure 4.9p. This "muscle" exerts pressure around the periphery of the sleeve. Actuation can be either pneumatic or hydraulic, depending upon system requirements. Round-hole configuration is maintained through about 50 percent reduction in diameter, which is equivalent to reducing the open condition area to 25 percent. Beyond this point, the opening tends toward a slot closure. The advantage of this design is that it will pass larger particles than the normal slotted configuration. The metal housing is designed for bolting between ANSI Class 150 flanges. The size range is from 1 through 6 in. (25–150 mm). Smaller sleeve areas are available in a full size housing where provision is required for future capacity expansion.

Accurate Closure Design

Another mechanical pinch valve design is known as the "accurate closure" type (Figure 4.9q). A rounded-end plunger, which may be compared to a powerful "thumb," compresses the sleeve until the top wall meets the bottom wall. The shape of the closure forms a crescent-shape flow area similar to going from a full moon (open) to a total eclipse (closed). The sleeve is very thick to withstand the forces and stress involved, and considerable power is required to operate the "thumb" and compress this thick wall. The flanged metal housing is

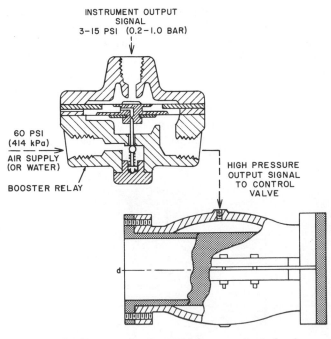

Fig. 4.9o   Booster relay operation of pneumatic pinch valve
(*Courtesy of Red Valve Co.*)

INSTRUMENT OUTPUT
SIGNAL
3–15 PSI (0.2–1.0 BAR)

60 PSI
(414 kPa)

AIR SUPPLY
(OR WATER)

BOOSTER RELAY

HIGH PRESSURE
OUTPUT SIGNAL
TO CONTROL
VALVE

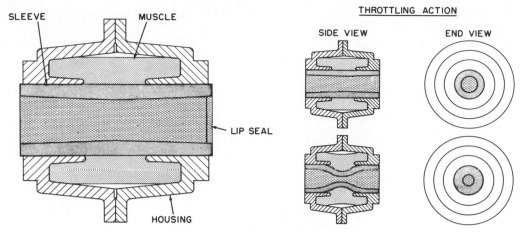

Fig. 4.9p   Pinch valve with shutter closure. (*Courtesy of Clarkson Co.*)

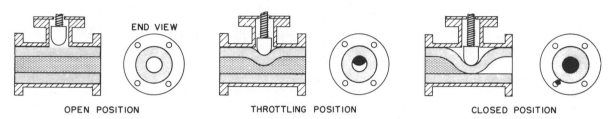

Fig. 4.9q ´ Pinch valve with "accurate closure"

available in sizes from 1 to 10 in. (25–250 mm) with ANSI Class 150 flanges. Shutoff is limited to about 50 PSIG (345 kPa) line pressure.

Sizing

Sizing of pinch valves is handled somewhat differently than with globe, butterfly, or more conventional control valves. Since a wide-open pinch valve behaves essentially like a short piece of pipe, it will pass an extremely high flow or C_v at 1 PSI (7 kPa) pressure drop. However, calculation will show fluid velocities are far in excess (around 2 to 3 times) of those considered good piping practice. Normal piping practice dictates pipe velocities in the range of 5 to 12 fps (1.5 to 3.6 m/s). Some corrosive fluids such as sulfuric acid may dictate a maximum velocity of 3 fps (0.9 m/s). Various pinch valve manufacturers have somewhat different methods of valve sizing, but all involve aspects of velocity as a limitation. There

are empirical "bias factors" for slurries that depend upon the solids percentage content. Preferable operating opening limits (usually between 10 and 70 percent) and an ideal opening for normal operation may be stipulated by the manufacturer. As in all things pertaining to pinch valves, sizing selection is best performed with full assistance from the manufacturer. Careful consideration of all operating application factors will lead to a valve that is sized for optimum operation.

BIBLIOGRAPHY

Carey, J.A., "Control Valve Update," *Instruments & Control Systems*, January, 1981.

Dobrowolski, M., "Guide to Selecting Rotary Control Valves," *Instrumentation Technology*, December, 1981.

Fernbaugh, A., "Control Valves: A Decade of Change," *Instruments & Control Systems*, January, 1980.

4.10 PLUG VALVES

Types of Design:	A—Cylindrical B—Eccentric spherical
Sizes:	A—$\frac{1}{2}''$ to 36″ (12.5 mm to 900 mm) depending on design B—1″ to 12″ (25 mm to 300 mm)
Design Pressure:	A—Varies with size, ANSI Class 125 to 15,000 PSIG (103.5 MPa) B—To 1400 PSIG (9.7 MPa)
Design Temperature:	A—Varies with material B—From −320° to 750°F (−196° to 400°C) with the eccentric rotating plug type
Rangeability:	Can approach 100:1
Capacity:	A—Approx. $C_v = 20\ d^2$ B—$C_v = 13\ d^2$
Materials of Construction:	Limited only by the application. The V-ported plug valve, for example, can be obtained in Ni-resist, Alloy-20, Monel, nickel, Hastelloy, rubber, and plastic lining, in addition to the standard materials
Costs:	Vary widely depending on use. See Figure 4.10a
Partial List of Suppliers:	Allis-Chalmers; Andale Co.; DeZurik Unit, General Signal; Duriron Co. Valve Div.; Foxboro; Hydril Co.; Masoneilan Div., Dresser Industries; Orbit Valve Co.; Wm. Powell Co.; WKM Div. ACF Ind.

The plug valve is a type of quarter-turn valve that is among the oldest designs known in engineering. Wooden plug valves were used in the water distribution system of ancient Rome and probably predate the butterfly valve. Although no longer as popular as ball or butterfly valves for general use, they lend themselves to special designs that work very well in specific control applications.

Function

Plug valves afford quick opening or closing with tight, leak-proof closures under conditions ranging from vacuum to pressures as high as 10,000 PSIG (69 MPa). Some, such as the V-ported or diamond design, can be used for throttling, while others, like the multiport, are used to direct flows to various process lines or to bypass processes altogether.

These valves are used for gases, liquids, and non-abrasive slurries. However, lubricated plug valves can be

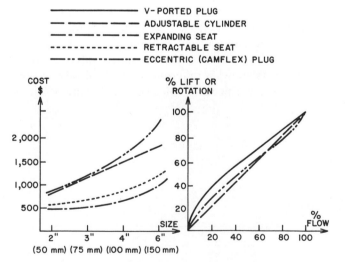

Fig. 4.10a   Cost and inherent flow characteristics of various plug control valves in standard materials

used for abrasive slurries, and eccentric and lift plugs are used for sticky fluids. Plug valves are useful also for contamination-free handling of foods and pharmaceuticals. They are good for applications requiring:

High flows at low pressure drops
Low flow control
Flow diversion
High- or low-temperature applications (when made from the proper materials)
Vibration-free operation

They are less efficient than some other types of valves for:

Flow modulation or continuous, exact flow throttling (except for some eccentric and V-ported plug and other special designs)
Maintenance-free operation (occasional lubrication is required for some types, and plugs may wear)

However, plug valves are usually lower cost and lighter weight than comparable gate or globe valves and may be the best valve to specify in many difficult service applications.

Design Characteristics

Plug valves began as a tapered or straight vertical cylinder containing a horizontal opening or flow-way inserted into the cavity of the valve body (Figure 4.10b). They have developed through time into numerous shapes and patterns, depending on the application, but almost all are adaptations of the cylindrical or tapered plug. Within that plug, however, the ports may be round, oval, rectangular, V, or diamond shaped, and flow-through or multiport. These make up the special designs described in subsequent paragraphs.

Further, plug valves can be categorized as lubricated or nonlubricated. In the lubricated type, the thin film of lubricant serves not only to reduce friction between the plug and the body, but also to form an incompressible seal to prevent gas or liquid leakage. Because the seating surfaces are not exposed in the open position, gritty slurries may be handled. The lubricant hydraulically lifts the plug against the resilient packing to prevent sticking. A special lubricant must be injected periodically while the valve is either fully open or closed.

The plugs of nonlubricated plug valves are treated with coatings such as Teflon or are specially heat-hardened and polished to prevent sticking. Often they are constructed so the tapered plug may be lifted mechanically from the seat for easier operation.

V-Ported Design

The V-ported plug valve (Figure 4.10c) is used for both on-off and throttling control of slurries and fluids containing solids in suspensions greater than 2 percent. These applications occur principally in the chemical and pulp and paper industries. A diamond-shaped opening is created by matching a V-shaped plug with a V-notched body. Straight-through flow occurs on 90-degree rotation, when the plug is swung out of the flow stream. Shearing action and a pocketless body make the valve applicable to fibrous or viscous materials. The opening develops a modified linear flow characteristic with C_v capacities approximating $17d^2$. Valves are flanged from 3 to 16 in. (75 to 400 mm) in bronze, corrosion-resistant bronze, or stainless steel. The body may be rubber-lined with a rubber-coated plug. A cylinder actuator and valve positioner are used for proportional control.

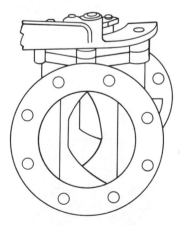

Fig. 4.10c V-ported plug valve

A variation is the true V-port opening (Figure 4.10d). It is obtained by a rotating segment closing against a straight edge. The valve can be smoothly throttled on thick stock flows without the stock packing or interfering. The valve is available from 4 through 20 in. (100 to 500 mm) with C_v stated as more than $20d^2$. The valve is

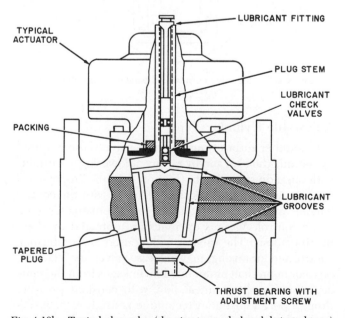

Fig. 4.10b Typical plug valve (showing tapered plug, lubricated type)

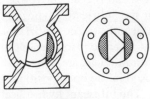
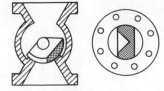
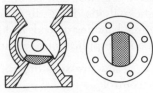

OPEN
THERE ARE NO SHARP CORNERS
OR NARROW OPENINGS TO PACK
WITH STOCK. LARGE PORT AREA
AND CLEAN INTERIOR DESIGN
ASSURE HIGH FLOW CAPACITY.

THROTTLING
CLOSE PLUG-TO-SEAT CLEARANCE
REMAINS CONSTANT. THE V-ORIFICE
RETAINS ITS SHAPE THROUGHOUT
THE CYCLE.

CLOSED
FLOW CONTINUES UNTIL THE
V-ORIFICE IN THE LEADING EDGE
OF THE PLUG ROTATES PAST THE
SEAT. THROTTLING IS SMOOTH
DOWN TO THE SHUTOFF POSITION.

Fig. 4.10d   Throttling with a V-ported plug valve

available in many materials and trims for use in the chemical and pulp and paper industries. A cylinder-operated rack and pinion is used for on-off service with the addition of a valve positioner for throttling services.

A variety of flow characteristics (Figures 4.10e, f, and g are examples) can be obtained with shaped throttling plates. Rotation of the plug is within a TFE sleeve locked into the body with recessed areas minimized. Although rangeability is stated as 20:1, this is made usable for the capacity of the valve by designing the port to handle full flow at open position. The valve is available in $\frac{1}{2}$-in. to 12-in. (12.5 to 300 mm) sizes and to 600 PSIG ANSI (4.1 MPa) rating for use to 400°F (204°C).

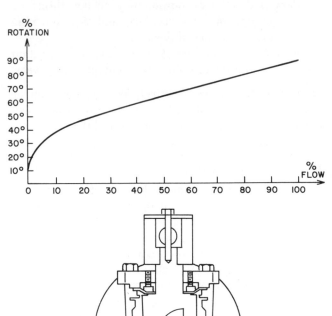

Fig. 4.10f   Plug valve for modified parabolic characteristics
at 50 percent flow

Adjustable Cylinder Type

In another form of quarter-turn valve, flow is varied by rotating the core and by raising or lowering a "curtain" with an adjusting knob (Figure 4.10h). Proportional opening at any curtain position is made with the control handle, which may be attached to an actuator for automatic control. Various openings are obtained by these manipulations. The valve is widely used for combustion control and for mixing, in which case valves are "stacked" on a common shaft or operated by linkages from the same actuator. To obtain linear flow with constant pressure drop, a port adjustment technique is used. After installation, the curtain is closed until the pressure drop across

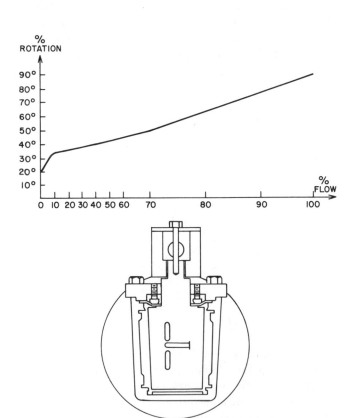

Fig. 4.10e   Plug valve with modified linear characteristics

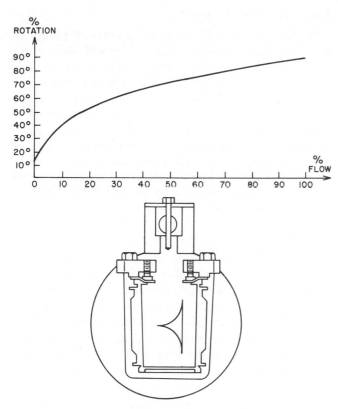

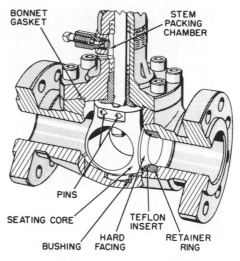

Fig. 4.10i Plug valve with semipherical plug for tight closure (*Courtesy of Orbit Valve Co.*)

Fig. 4.10g Plug valve with equal percentage characteristics for reduced flow

the seating surface. As the stem is rotated for opening, the closure surfaces separate and the pin moves into a vertical slot so that rotation occurs. A nonlubricated seal is possible with a primary Teflon seal enclosed in a body seat retainer ring. The valve is adapted for automatic operation by connecting the stem to a diaphragm actuator.

Expanding Seat Plate Design

Another means of seating uses metal-to-metal or resilient seats carried in two seating segments (Figure 4.10j). These segments are carried on a rail that is tapered so that downward stem movement forces the plates against the inlet and outlet ports. On opening, the first few turns of an actuator cause retraction of the plates and then plug rotation proceeds. These plates can be removed by merely removing the bottom plate of the valve V-body.

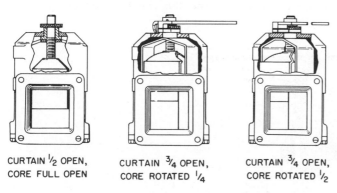

Fig. 4.10h Adjustable cylinder plug valve

the valve is one-sixth of the total pressure drop of the system using the control handle in wide open position. This provides a flow characteristic approximating linear without decreasing sensitivity by limiting valve stroke. The percentage of flow is equalized by manipulation of the linkage to the actuator.

Semispherical Plugs for Tight Closure

Various designs have been developed to obtain tight closure without the continuous friction of seals during rotation, as with most ball valves. The valve design illustrated in Figure 4.10i uses an eccentric ball. In the closed position, the rectangular end of the stem protrudes into the ball and the closure face is wedged toward

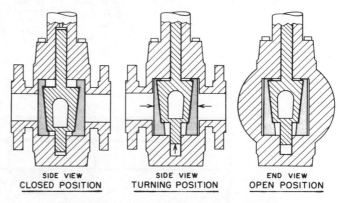

Fig. 4.10j Plug valve with expanding seat plate

Eccentric Shaft Design

Double seating has been accomplished in a spherical valve by employing an eccentric shaft (Figure 4.10k). On initial rotation of the valve, the relative offset of the seats from the shaft center provides pullout action to separate

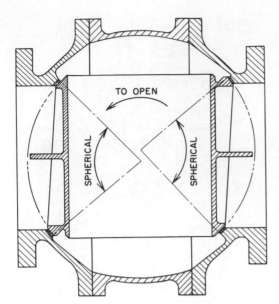

Fig. 4.10k   Plug valve with eccentric shaft design
(*Courtesy of Darling Valve Co.*)

Retractable Seat Type

Positioning a movable seal after a spherical plug is in the closed position creates tight closure with sliding friction. A trunion-mounted partial sphere is operated by spur or worm gears (Figure 4.10l). The gear system rotates the plug until it is in a closed position, at which point additional rotation of the drive creates a camming action to compress the packing ring. In the opening operation, the packing ring is released before rotation of the plug occurs. Although applicable to low pressure ranges, this valve is particularly useful up to API 5,000 PSIG (34.5 MPa) rating for 10,000 PSIG (69 MPa) service. It is used mostly in oil fields.

Overtravel Seating Design

By fabricating a cylindrical flow passage within a tapered cylindrical plug, a plug valve can be built at least through 60 in. without undue weight and attendant inertia to rotation. Rotation is caused by a rod operated by a piston pushing down on a rotator (Figure 4.10m). Continuous movement of the rod closes the plug, but leaves a small crescent (Figure 4.10n), because the plug is slightly raised from seating. In Figure 4.10o, the rod has contacted a seating adjustment to force the plug into the tapered seating surface for tight closure. The design makes possible a rapid restriction of flow area, causing about 65 percent closure with 30 percent rod travel. Complete rotation of the plug occurs at about 65 percent

the plug seat from the stationary body. This design permits building very large valves without excess weight. Standard sizes range from 4 to 48 in. (100 mm to 1.2 m) for pressures up to 300 PSIG (2.1 MPa) and temperatures to 450°F (232°C). The eccentric center allows use of trunions to carry the plug.

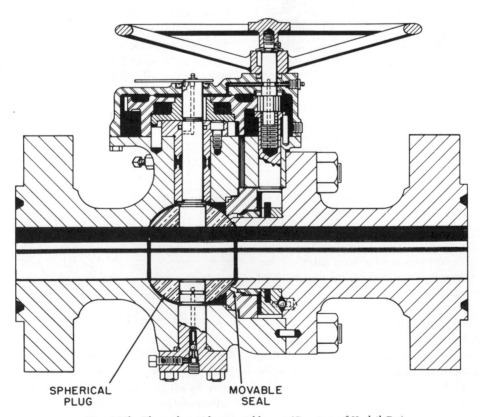

Fig. 4.10l   Plug valve with retractable seat (*Courtesy of Hydril Co.*)

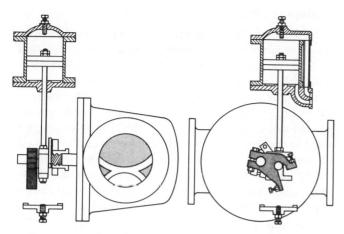

Fig. 4.10m   Plug valve of the overtravel seating design in its throttling position (*Courtesy of Allis-Chalmers*)

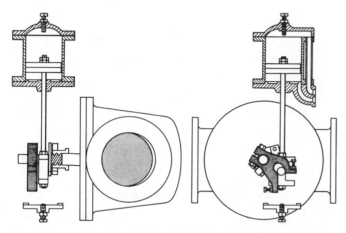

Fig. 4.10n   Overtravel seating plug valve in its closed position (*Courtesy of Allis-Chalmers*)

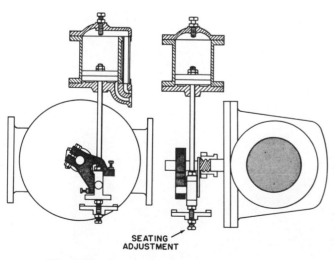

SEATING ADJUSTMENT

Fig. 4.10o   Overtravel seating plug tightly closed

stroke, with the additional stroke utilized for seating. Inasmuch as rapid closure to about 20 percent area does not create serious surge pressures, this valve can be used for emergency closure without undue consideration of

the piston speed. Rangeability well exceeding 50:1 is claimed for proportional control. Free rotation and relatively low weight of plug contribute to lower power requirements. Materials are selected by the manufacturer to suit many services.

Eccentric Rotating Spherical Segment Type

Although designed for uses comparable to those of globe or butterfly valves, the eccentric rotating spherical segment valve has the advantage of low torque requirements (Figure 4.10p).

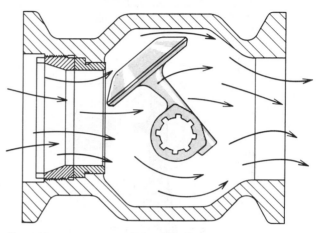

Fig. 4.10p   Eccentric rotating spherical segment type plug valve (*Courtesy of Masoneilan Division of Dresser Industries*)

This control valve makes exaggerated use of the offset center, as used in butterfly valve designs, to obtain contact at closure without rubbing. The seat portion of the plug has the form of a spherical segment that is rotated 50 degrees for maximum opening. The support arms for the plug flex upon closure to cause tighter contact with the seat upon increase in actuator force.

Flow characteristic approaches linear (Figure 4.10a). Change in characteristic is accomplished with a cam in the positioner. Capacity is between a double-seated valve and a butterfly valve. High-flow capacity is achieved with only moderate pressure increase in the body, so the critical flow factor is much higher than that of a butterfly valve throughout its throttling range.

A low-convolution cylinder actuator meets torque requirements that follow a uniform decrease from closed to open without an objectionable peak. Since the drive shaft is normally horizontal, an extended bonnet may be furnished for use at elevated temperatures.

The valve is made in sizes from 1 to 12 in. (25 to 300 mm) with a 600 PSIG (4.1 MPa) body for clamping between flanges rating from 150 PSIG to 600 PSIG (1 to 4.1 MPa). Use of a positioner allows maximum pressure drops of 600 PSIG (4.1 MPa) in up to 3-in. (75 mm) sizes or 400 PSIG (2.8 MPa) in up to 6-in. (150 mm) sizes with a high-pressure positioner. A rangeability of over 50:1 is

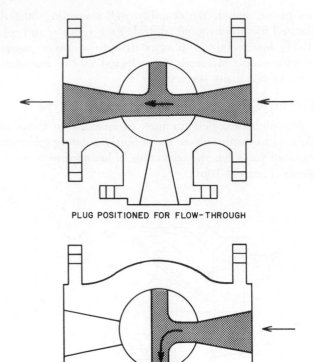

PLUG POSITIONED FOR FLOW-THROUGH

PLUG POSITIONED FOR DISCHARGE AT 90 DEGREES

Fig. 4.10q   Three-way plug valve

Multiport Design

By designing the plug with an extra port at 90 degrees from the inlet, the flow-through characteristics are modified so that flow can be directed in either of two directions (Figure 4.10q). A multitude of directions can be achieved by nesting combinations of the simple multiport valves or by using more complex designs. These include a multi-storied arrangement with the plug extending upward to connect to a series of tiered outlets. For this the plug has a long, vertical passageway connecting the horizontal ports. Another method of increasing the directions of flow is to design the plug with a sufficiently larger diameter than the ports so that intermediate ports can be placed at 45 degrees or even 30 and 60 degrees. The actuators can be programmed to serve a variety of process applications.

BIBLIOGRAPHY

Anderson, N., *Instrumentation for Process Measurement and Control*, Philadelphia: Chilton, 1980.

Baumann, H.D., "Trends in Control Valves and Actuators," *Instruments & Control Systems*, November, 1975.

Dobrowolski, M., "Guide to Selecting Rotary Control Valves," *Instrumentation Technology*, December, 1981.

Gas Engineering Handbook, New York: Industrial Press, 1966.

Hammitt, D., "Rotary Valves for Throttling," *Instruments & Control Systems*, July, 1977.

Kawamura, H., "Selecting Valves for the Hydrocarbon Processing Industry," *Hydrocarbon Processing*, August, 1980.

Lyons, J., *Lyons' Valve Designer's Handbook*, New York: Van Nostrand Reinhold, 1982.

Rase, H., *Piping Design for Process Plants*, New York: Wiley, 1963.

Tyner, M., and May, F., *Process Engineering Control*, New York: Ronald Press, 1968.

Zappe, R., *Valve Selection Handbook*, Houston, TX: Gulf Publishing, 1981.

claimed. The entire valve positioner unit is much lighter than the normal globe valve actuator and positioner.

4.11 SAUNDERS DIAPHRAGM VALVES

Sizes:	$\frac{1}{4}''$ to 20'' (6.3 mm to 500 mm)
Design Pressure:	Up to 200 PSIG (1380 kPa) (Figure 4.11e)
Design Temperature:	From $-30°$ to 350°F ($-34°$ to 177°C) if operating pressure is below 50 PSIG (345 kPa) (Figure 4.11e)
Rangeability:	15:1
Capacity:	$C_v = 22d^2$ (noncritical flow)
Materials of Construction:	Bodies: most castable or machinable materials including graphite, with a variety of linings; diaphragm: several rubber and plastic materials; lining materials include titanium
Cost:	See Figure 4.11a. Cost data are based upon standard flanged cast or ductile iron bodies with either standard elastomer or fluorocarbon lining and diaphragm assembly. These costs do not include an operator since a variety of pneumatic and electric units can be used for on-off and throttling services
Special Features:	Dual-range, full-bore, and straight-through design
Partial List of Suppliers:	Conoflow Regulators & Controls Div. of ITT Grinnell Valve Co.; Fisher Controls Co.; Foxboro Co.; Hills-McCanna Co.; Masoneilan Div., Dresser Industries; Norris Div., Dover Corp.; Taylor Instrument Co., Div. of Sybron Corp.

Operation

The Saunders valve is also referred to as a diaphragm valve and occasionally as a weir valve. Conventional Saunders valves utilize both the diaphragm and the weir for control of the flow (Figure 4.11b).

The Saunders valve is opened and closed by moving a flexible or elastic diaphragm toward or away from a weir. The elastic diaphragm is moved toward the weir by the pressure of a compressor on the diaphragm. The compressor is attached to the valve stem for this purpose. The diaphragm, which is attached to the compressor at the center, is pulled away from the weir when the compressor is withdrawn.

For high-vacuum service it is often desirable to evacuate the bonnet in order to reduce the force pulling the diaphragm away from the compressor. This is especially desirable for large valves, where otherwise the flow vacuum might be sufficient to tear the diaphragm from the compressor.

A Saunders valve can be considered as a half pinch valve. The pinch valve contains two diaphragms that move toward or away from each other, whereas the Saunders valve has only one diaphragm and a fixed weir. Because of their design similarity, their flow characteristics are also similar, as illustrated by Figures 4.11a and 4.9a. Figure 4.11c shows the three basic positions of a Saunders valve.

Construction

The body of a conventional Saunders valve (Figure 4.11d), because of its simple and smooth interior, lends itself well to lining with plastics, glass, titanium, zirconium, tantalum, and other corrosion-resistant materials. Valve bodies are available in iron, stainless and cast steels, alloys, and plastics. Iron bodies are lined with plastic, glass, special metals, and ceramics. The flow capacity of lined Saunders valves in the smaller sizes (under 2 in.

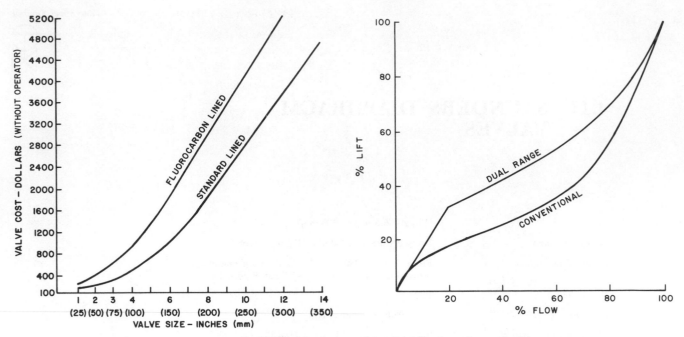

Fig. 4.11a   Cost and inherent characteristics of Saunders (diaphragm) control valves
with glass-lined iron body

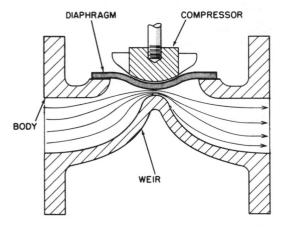

Fig. 4.11b   Main components of a Saunders valve

or 50 mm) is about 25 percent less than that of unlined ones.

The diaphragm for the conventional Saunders valve is available in a wide choice of materials. These include polyethylene, tygon, white nail rubber, gum rubber, hycar, natural rubber, neoprene, hypalon, black butyl, KEL-F, Teflon, etc., with various backings, including silicone. Some contain reinforcement fibers.

Maintenance requirements of the Saunders valve depend on diaphragm life, which is determined by the diaphragm's resistance to the flow material (which may be corrosive or erosive), pressure, and temperature (Figure 4.11e).

The compressor is designed to clear the finger plate,

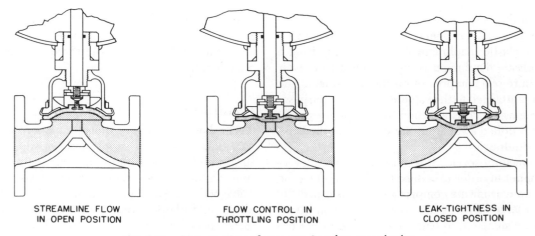

STREAMLINE FLOW
IN OPEN POSITION

FLOW CONTROL IN
THROTTLING POSITION

LEAK-TIGHTNESS IN
CLOSED POSITION

Fig. 4.11c   Main positions of weir type Saunders control valve

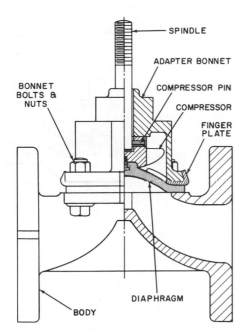

Fig. 4.11d   Main parts of a weir type Saunders valve

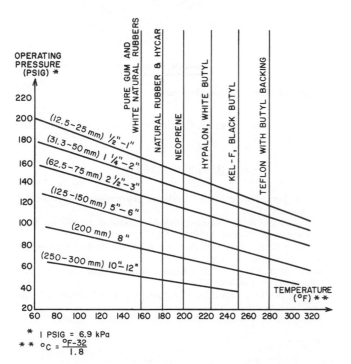

* 1 PSIG = 6.9 kPa

** °C = $\frac{°F-32}{1.8}$

Fig. 4.11e   Pressure-temperature limitations of various diaphragm
materials as a function of valve size

or diaphragm support plate, and to contour the diaphragm so that it matches the weir (Figure 4.11d). The purpose of the finger plate is to support the diaphragm when the compressor has been withdrawn. The finger plate is utilized for valve sizes 1 in. (25 mm) and larger. For valves larger than 2 in. (50 mm), the finger plate is built as part of the bonnet.

Straight-Through Design

The valve seat of the straight-through diaphragm valve is not the conventional weir but is contoured into the walls of the body itself (Figure 4.11f). The longer stem stroke of the straight-through vale necessitates a very flexible diaphragm. The increased flexure requirement tends to shorten the life of the diaphragm, but the valve's smooth, self-draining, straight-through flow pattern makes it applicable to hard-to-handle materials, such as to slurry.

The flow characteristics of the straight-through design are more nearly linear than those of the conventional Saunders valve.

Full-Bore Valve

The full-bore Saunders valve has a body design modified by special forming of the weir. As a result, the internal flow path is fully rounded at all points, permitting ball brush cleaning (Figure 4.11g). This is an important feature in the food industry, which requires valve interiors with a smooth, easy-to-clean surface.

Dual-Range Design

The rangeability and flow characteristics of a conventional Saunders valve are rather poor, and so it is not suitable for critical control applications. The dual-range design, with its improved flow characteristics, represents an improvement in this regard (Figure 4.11a). The dual-range valve contains two compressors, which provide independent control over two areas of the diaphragm

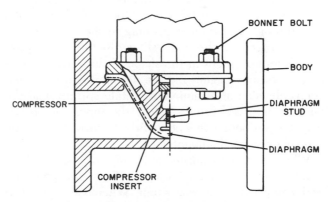

Fig. 4.11f   Straight-through Saunders valve

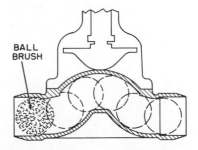

Fig. 4.11g   Full-bore Saunders valve

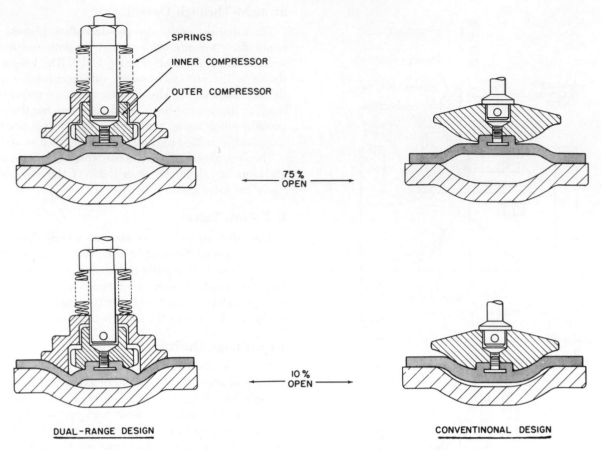

Fig. 4.11h Stroking characteristics of dual-range and conventional Saunders valves

(Figure 4.11h). The first increments of stem travel raise only the inner compressor from the weir. This allows flow through a contoured opening in the center of the valve, rather than through a slit across the entire weir.

This improvement in the shape of the valve opening helps prevent clogging and the dewatering of stock and keeps abrasion at a minimum. While springs hold the outer compressor firmly seated, the inner compressor may be positioned independently to provide accurate control over small amounts of flow.

When the inner compressor is opened to its limit, the outer compressor begins to open. From this point on, both compressors move as a unit. When wide open, this valve provides the same flow capacity as its conventional counterpart.

BIBLIOGRAPHY

Carey, J.A., "Control Valve Update," *Instruments & Control Systems*, January, 1981.

Cunningham, E.R., "Solutions to Valve Operating Problems," *Plant Engineering*, September 4, 1980.

Fernbaugh, A., "Control Valves: A Decade of Change," *Instruments & Control Systems*, January, 1980.

Hutchinson, J.W., ed., *ISA Handbook of Control Valves*, Instrument Society of America, 1971.

Sanderson, R.C., "Elastomer Coatings: Hope for Cavitation Resistance," *InTech*, April, 1983.

Walter, D.G., "Control Valve Selection," *InTech*, October, 1977.

4.12   SLIDING GATE VALVES

Sizes:	On-Off—2″ to 120″ (50 mm to 3000 mm) Throttling—$\frac{1}{2}$″ to 24″ (12.5 mm to 600 mm) Multi-orifice throttling—$\frac{1}{2}$″ to 6″ (12.5 mm to 150 mm)
Design Pressure:	On-Off—to 3600 PSIG (25 MPa) Multi-orifice—to 1440 (10 MPa)
Design Temperature:	Cryogenic—to 500°F (260°C) Multi-orifice— −20° to 1125°F (−29° to 607°C)
Rangeability:	50:1, lower with V-insert type
Capacity:	$C_v = 65 \ d^2$ (not useful for sizing) Multi-orifice—$C_v = 6 \ d^2$ to $C_v = 14 \ d^2$
Materials of Construction:	Standard metals
Cost:	See Figure 4.12a. Cost data are based upon construction with flanged carbon steel body, 304 SST liner, stem, and gate. Valve is equipped with pneumatic double-acting on-off cylinder operator. Other operators, such as pneumatic throttling cylinder with positioner and electric motor drive, can be furnished. The valves can be furnished in all stainless steel construction and some alloys
Partial List of Suppliers:	Anchor/Darling Valve Co.; Borg-Warner Energy Equipment; Clarkson Co.; DeZurik, A Unit of General Signal; Everest Valve Co.; Grove Valve & Regulator Co.; Hammond Valve Corp.; Jordan Valve Div. of Richards Industries Inc.; McCartney Mfg. Co.; Target Rock Corp.; Willis Oil Tool Corp.

Knife Gate Valves

Changing the flow rate of a valve by sliding a hole or plate past a stationary hole is a very basic approach to throttling flows. The most common valve, the gate valve, is just this. Although occasionally used for automatic control, the design is not considered to be a control valve. A form of "guillotine" gate valve (Figure 4.12b) is much used in the pulp and paper industry due to its shearing ability and nonplugging body design.

Use of a "slab-type" valve with a round hole (Figure 4.12c) gives flow characteristics created by converging circles, which is similar to equal percentage for 70 percent of its flow, and then it becomes linear. Seventy percent of the flow occurs during the last 30 percent of opening.

A V-shaped insert (Figure 4.12d) in the valve opening creates a parabolic flow characteristic somewhat similar to the V-ported globe valve. The success of these valves in proportional control is entirely dependent upon the ability of the actuator to achieve accurate positioning.

Positioned-Disc Valves

Rotation of a movable disc with two holes to progressively cover two holes in the stationary disc successfully throttles flow (Figure 4.12e). This variable choke was designed to control flow of high-pressure oil wells. Use of ceramic or tungsten carbide discs allows handling of pressures up to 10,000 PSIG (69 MPa). Such valves are presently furnished in 1-in. and 2-in. (25 and 50 mm) sizes with areas from 0.05 in.² with 0.25-in. hole (32 mm² with 6.3 mm hole) to 1.56 in.² with two 1-in. holes (1006 mm² with two 25-mm holes).

Angle design is used for proportioning control, with an actuator capable of controlling the discharge flow at

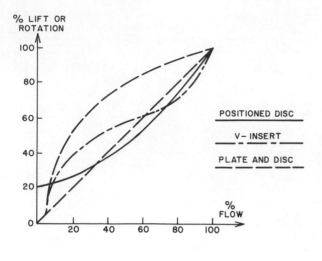

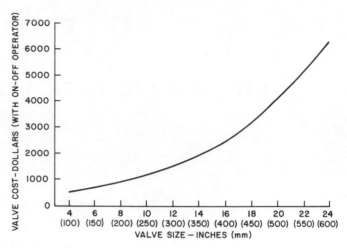

Fig. 4.12a   Cost and inherent flow characteristics of cast iron sliding gate valves

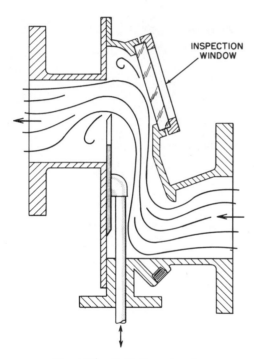

Fig. 4.12b   Guillotine gate valve

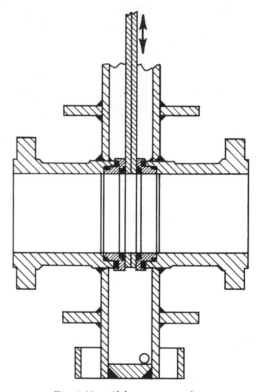

Fig. 4.12c   Slab type gate valve

quarter-turn movement. Both linear and rotary type output actuators can be used. The relation between the discs remains in the status-quo mode without power being required from the positioner. A stepping actuator (Figure 4.12f), positions in 1-degree increments with a pneumatic input to a double-acting, spring-centered piston. Rotation occurs through a rack and pinion. Limit switches are provided, and a stepping switch caan be used for position transmission and, conceivably, for feedback in control systems.

Plate-and-Disc Valves

A wide variety of flow characteristics is available using a stationary plate in the valve body and a disc made

movable by the valve stem. The plate (Figure 4.12g) is readily replaceable by removing a flanged portion of the body which retains the plate with a pressure ring. Areas of the plate are undercut to reduce friction. A circumferential groove provides flexibility and allows the plate to remain flat in spite of differential pressures, expansion or contraction of the body. The stem contacts the disc by a pin through a slot in the plate.

The disc is held in contact with the plate by upstream pressure and by retaining guides. The contacting surfaces of the disc and plate are lapped to light band flatness.

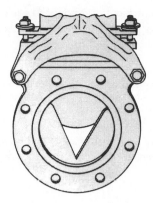

Fig. 4.12d Sliding gate valve
with V-insert

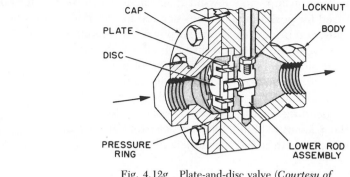

Fig. 4.12g Plate-and-disc valve (*Courtesy of
Jordan Division of Richard Industries*)

 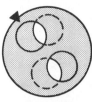 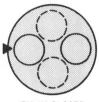

FULLY OPEN   THROTTLING   FULLY CLOSED

Fig. 4.12e Positioned-disc sliding gate valve

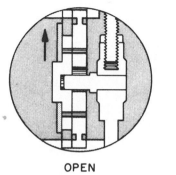 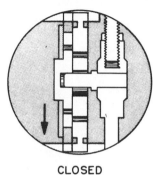

OPEN   CLOSED

Fig. 4.12h Throttling with a plate-and-disc valve
(*Courtesy of Jordan Division of Richards Industries*)

The chrome-plated surface of the stainless steel plate has a hardness comparable to 740 Brinell to resist galling and corrosion and obtain smooth movement of the disc. The material, with the registered name of Jordanite, is reported to have an extremely low coefficient of friction, is applicable to high pressure drops and has great resistance to heat and corrosion.

Flow occurs through mating slots in the disc and the plate. Positive shutoff occurs (Figure 4.12h), when the slots are separated. Flow increases on an approximately linear relationship until the slots are lined up for maximum flow. Capacities are about $C_v = 6.5d^2$ through the 2-in. (50 mm) size and about $C_v = 12d^2$ through the 6-in. (150 mm) size.

Stem travels to obtain full flow are very short due to

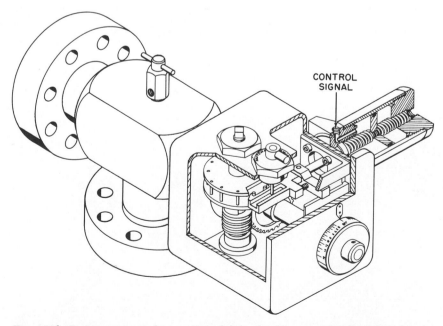

Fig. 4.12f Stepping actuator for a positioned-disc valve (*Courtesy of Willis Oil Tool Co.*)

the slot relationship, and low-lift-diaphragm actuators can be used for positioning. Forces needed for positioning are low, requiring only sufficient power to overcome friction between the plate and disc, which is right-angle motion and not opposed to the direction of flow. Valve bodies are offered in sizes between $\frac{1}{4}$ in. and 6 in. (6.3 and 150 mm) and with ratings through 300 PSIG (2.1 MPa), depending on the material, with a selection of trims and packings. Many styles of actuators are used, including one with a thermal unit and cam actuation.

This body design has been adapted for extensive use as a self-contained pressure or temperature regulator.

BIBLIOGRAPHY

Arant, J.B., "Fire-Safe Valves," *InTech*, December, 1981.

Carey, J.A., "Control Valve Update," *Instruments & Control Systems*, January, 1981.

Fernbaugh, A., "Control Valves: A Decade of Change," *Instruments & Control Systems*, January, 1980.

Hutchinson, J.W., ed., *ISA Handbook of Control Valves*, Instrument Society of America, 1971.

4.13 VALVE APPLICATION

The previous sections dealt with the design features and operating characteristics of the various types of control valves. The application considerations encountered by the instrument engineer in selecting and applying these devices are covered in this section.

Before proceeding through the steps of selecting a control valve, one should evaluate if a control valve is truly needed in the first place, or if a simpler and more elegant system will result through some other means. For example, an overflow weir can suffice to keep levels below maximum limits, and choke or restriction fittings can serve the function of pressure let down at constant loads. In other locations it might be possible to reduce the investment by using regulators instead of control valves. The advantages of regulators include their high speed (high gain) and their self-contained nature, which eliminates the need for power supplies or utilities. If remote set point adjustment is needed, regulators can be provided with air loaded pilots to accommodate that requirement. While all regulators (being proportional only controllers) will display some offset as the load changes, the amount of offset can be minimized by maximizing the regulator gain.

In still other applications it is prudent to replace whole flow control loops with positive displacement metering pumps or to replace the control valve with variable speed centrifugal pumps. The cost-effectiveness of this approach is usually found to be in lowered pumping costs because the pumping energy that was "burned up" in the form of pressure drop through the control valve is not being introduced and therefore it is saved. (Section 8.17 covers this topic in more detail.)

Collecting Process Data

In order to select the right control valve, one must fully understand the process that the valve controls. Fully understanding the process means not only understanding normal operating conditions, but also the requirements that the valve must live up to during start-up, shut-down, and emergency conditions. Therefore all anticipated values of flow rates, pressures, vapor pressures, densities, temperatures, and viscosities must be identified in the process of collecting the data for sizing. In addition, it is desirable to identify the sources and natures of potential disturbances and process upsets. One should also determine the control quality requirements, so as to identify the tolerances that are acceptable in controlling the particular variable. The process data should also state if the valve needs to give tight shut-off, if the valve noise needs to be limited, or any other factors that might not be known to the instrument engineer. These can include subjective factors, such as user preferences, or objective ones, such as spare parts availability, delivery, life expectancy, or maintenance history.

The process data should also state the position that the valve should take when the actuator power fails. The flowsheet abbreviations can be FC (fail closed), FI (fail indetermined), FL (fail in last position), and FO (fail open). Spring loaded actuators are the most convenient means of providing FC or FO action, while two-directional air or electric motors will naturally tend to fail in their last positions.

In addition to anticipating the consequences of actuator power failure, one should also consider the results of other component failures, such as the spring, diaphragm, piston, etc. When such failures occur, the ultimate valve position will *not* be a function of the actuator design, but of the process fluid forces acting upon the valve itself. The choices are FTO (flow to open), FTC (flow to close), FB (friction bound—tends to stay in last position). FTO action is available with globe valves. FTC action can be obtained from butterfly, globe, and conventional ball valves. Rotary plug, floating ball, and segmented ball valves tend to be friction bound, with the flow direction possibly affecting the torque required to open the valve.

Control Valve Performance

Good control valve performance usually means that the valve is stable across the full operating range of the process, it is not operating near to one of its extreme positions, it is fast enough to correct for process upsets or disturbances, and it will not be necessary to retune the controller every time the process load changes. In order to meet the above goals, one must consider such factors as valve characteristics, rangeability, installed gain, and actuator response. These topics will be separately addressed in the paragraphs that follow.

Theoretical Control Valve Characteristics

The inherent characteristics of a control valve describes the relationship between the controller output signal received by the valve actuator and the flow through that valve, assuming that:

a. The actuator is linear (valve travel is proportional with controller output).
b. The pressure difference across the valve is constant.
c. The process fluid is not flashing, cavitating, or approaching sonic velocity (choked flow).

Such inherent lift to flow relationships is illustrated in Figure 4.13a. In a linear valve, travel is linearly pro-

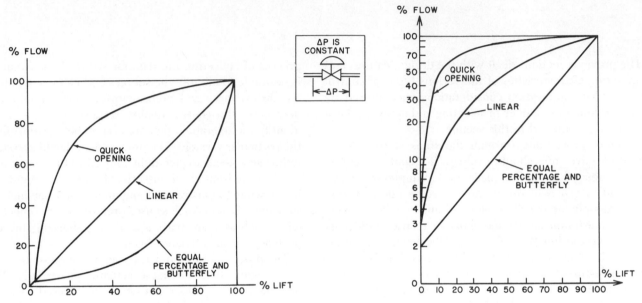

Fig. 4.13a   Inherent flow characteristics: quick-opening, linear, and equal percentage

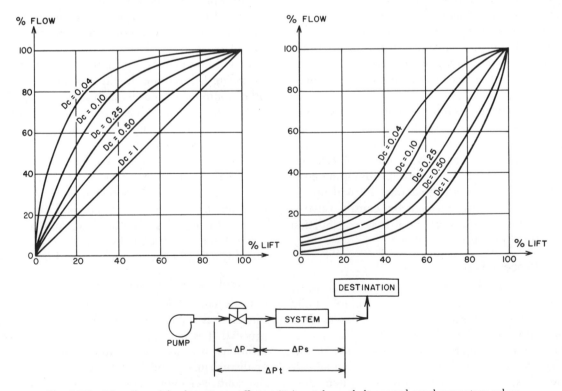

Fig. 4.13b   The effect of the distortion coefficient (D_c) on inherently linear and equal percentage valves, according to Boger. $D_c = \dfrac{(\Delta P_t)_{min}\,(\Delta P)}{(\Delta P_t)_{max}\,(\Delta P_s)}$

portional to capacity and therefore the theoretical gain is constant at all loads. (The actual gain is shown in Figure 4.13c.) In equal percentage valves, a unit change in lift will result in a change in flow, which is a fixed percentage of the flowrate at that lift. For example, in Figure 4.13d, each percent increase in lift will increase the previous flowrate by about 3 percent. Therefore the theoretical gain of equal percentage valves is directly proportional with flow (actual gain shown in Figure 4.13c) and increases as the flow increases. In quick-opening valves, the gain decreases with increasing flows. Figure 4.13a shows the quick-opening valve characteristics with the same total lift as for the other plug types. If the travel of the quick-opening plug is restricted so that the distance of 100 percent lift travel corresponds to only ¼ of

the seat diameter, then the valve characteristics will approach linear with the gain being nearly constant.

Valve and Process Characteristics

Control loops are usually tuned at normal load levels (at normal flow rates through the control valve), and it is assumed that the total loop gain will not vary with process load. This assumption is seldom completely valid, and the process gain does usually change with load. Because one cannot afford to retune the controller for each new load, it is desirable to select control valves that will compensate for these effects.

For example, when one controls a liquid-to-liquid heat exchanger, the process gain and dead time (transportation lag) will both decrease as the load increases. Therefore one should attempt to compensate for this inverse load-to-gain relationship by using a valve with a direct load-to-gain relationship, such as the equal percentage valves. With such a valve, an increase in load will decrease the process gain while increasing the valve gain, thereby reducing the total change in the loop gain. Equal percentage valves are not recommended if high turndown is required or if there are solids in the throttled process fluid.

An opposite example is a control loop whose sensor has an expanding scale, such as an orifice plate or a vapor pressure thermometer. With such sensors the process gain is increasing with load and therefore the gain of the selected valve should decrease with load. Therefore a linear or quick-opening control valve should be used.

In a fairly large number of cases, the choice of valve characteristics is of no serious consequence. Just about any characteristics will be acceptable for:

a. Processes with short time constant, such as flow control, most pressure control loops, and temperature controls through mixing.

b. Control loops operated by narrow proportional band (high gain) controllers, such as most regulators.

c. Processes with load variations of less than 2:1.

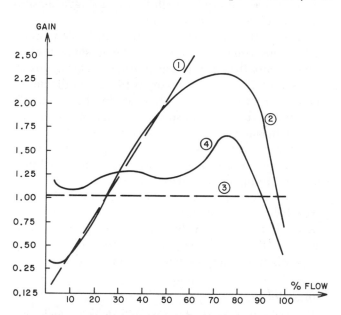

① THEORETICAL GAIN CHARACTERISTICS OF EQUAL % VALVE

② ACTUAL, INHERENT GAIN CHARACTERISTICS OF EQUAL % VALVE

③ THEORETICAL GAIN CHARACTERISTICS OF LINEAR VALVE

④ ACTUAL, INHERENT GAIN CHARACTERISTICS OF LINEAR VALVE

Fig. 4.13c   Theoretical vs actual characteristics of 2-in. (50 mm) cage-guided globe valves, according to Driskell

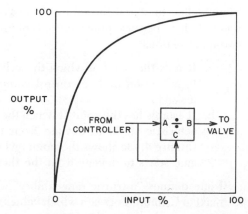

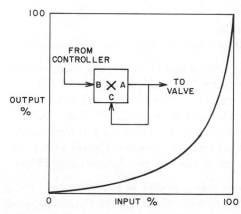

Fig. 4.13d   Modifying valve characteristics by the insertion of divider or multiplier relays into the controller output signal

According to Driskell, one can avoid a detailed dynamic analysis by using the following rules of thumb in selecting the valve characteristics for the more common loops:

Service	Valve ΔP Under 2:1	Valve ΔP Over 2:1 but Under 5:1
Orifice type flow	Quick-opening	Linear
Linear flow	Linear	Equal %
Level	Linear	Equal %
Gas pressure	Linear	Equal %
Liquid pressure	Equal %	Equal %

Installed Valve Characteristics

When the control valve is installed as part of a process plant, its flow characteristics are no longer independent of the rest of the system. The fluid flow through the valve is subject to frictional resistances in series with that of the valve. The consequence is the type of distortion illustrated in Figure 4.13b.

From these curves, one can conclude that the particular installation involved can have a very substantial affect on both flow characteristics and rangeability. Clearance flow alone can increase as much as tenfold, and equal percentage characteristics can be distorted toward linear or even quick opening under conditions of excessive distortion.

It should be emphasized that Figure 4.13b assumes the use of a constant speed pump. In variable speed pumping systems, one usually would adjust the pump speed so as to keep the valve ΔP constant, and therefore in such control systems the installed and theoretical valve characteristics are the same and no distortion is allowed to occur. This is one of the advantages of variable speed pumping systems.

The predictability of installed valve behavior is further reduced by such factors as:

a. The inherent valve characteristics deviate substantially (Figure 4.13c) from their theoretically prescribed character.
b. Actuators without positioners will introduce nonlinearities.
c. Pump curves will also introduce nonlinearities.

It should also be recognized that in order to learn the true requirements for valve characteristics, a full dynamic analysis is required. Even if one took the trouble of performing such analysis, it would probably yield a valve characteristic requirement that is not commercially available in conventional air-operated control valves. In this regard one might distinguish the characteristics which:

a. Are an intrinsic property of the valve construction, such as a butterfly or a beveled (quick-opening) disc.
b. Valves that are characterized by design, such as the linear or the equal percentage trims.
c. Digital control valves that can be characterized by software.
d. Characteristics that are superimposed through auxiliary hardware, such as function generators, characterized positioners, cams, etc.

Correcting a Wrong Valve Characteristic

The linear valve has a constant gain at all flow rates, while the gain of the equal percentage valve is directly proportional to flow. If the control loop tends to oscillate at low flow but is sluggish at high flow, one should switch to an equal-percentage valve. If, on the other hand, oscillation is encountered at high flow and sluggishness at low flow, a linear valve is needed.

Changing the valve characteristics can be done more easily by inserting accessories into the air signal leading to the actuator than by replacing the valve. One approach proposed by Fehérvári/Shinskey is to insert a divider or a multiplier into this line, as illustrated in Figure 4.13d. By adjusting the zero and span of the bellows at port C, a complete family of curves can be obtained. The divider is used to convert an air-to-open equal-percentage valve to linear, or an air-to-close linear valve to equal-percentage. Multiplier is used to convert an air-to-open linear valve to equal-percentage, or an air-to-close equal-percentage valve to linear.

According to Shinskey, both devices are perfectly standard, sensitive, stable, easy to calibrate, and "real life-savers when one needs a linear butterfly valve."

Rangeability

Rangeability as applied to control valves is usually defined in very vague terms, such as being the ratio of maximum to minimum controllable flows, where the word controllable implies that within this range the deviation from the specified inherent flow characteristic will not exceed some stated limits. The value of such a definition would be rather limited even if the limits of the deviation were internationally agreed upon. Therefore, the best thing to do is to review the subject from a common sense point of view, because rangeability is of interest for only two main reasons:

a. It tells the point at which the valve is expected to act on-off or lose control completely due to leakage.
b. It establishes the point at which the flow-lift characteristic starts to deviate from the expected. (Figure 4.13c shows the points where the actual gain starts to deviate from the theoretical.)

If one defines "intrinsic rangeability" as that ratio of $C_v(max)$ to $C_v(min)$, between which values the valve gain does not vary by more than 50 percent from the theoretical than according to Figure 4.13c, the rangeability of a linear valve is greater than that of an equal percentage. Actually, if one uses this definition, the rangeability of equal percentage valves is seldom more than 10:1, while the rangeability of some of the rotary valves can be quite high because their clearance flow tends to be less near the closed position than that of other valves, and their body losses near the wide open position also tend to be lower than that of other valve designs. In addition to the effects of leakage flow, valve rangeability can also be limited by positioning sensitivity.

Control Valve Sequencing

When the rangeability requirements of the process exceed the capabilities of a single valve, control valve sequencing loops must be designed that will keep the loop gain constant while switching valves. This requires careful thought.

Assuming that the task is to sequence two linear valves, with sizes of 1 in. and 3 in. (25 and 75 mm) having C_vs of 10 and 100 respectively, Shinskey's recommendation is the following: If the large valve was to operate from 9–15 PSIG (0.6–1.0 bar) and the small one from 3–9 PSIG (0.2–0.6 bar), the loop gain would change by 10 when passing through 9 PSIG (0.6 bar). The only way to keep loop gain constant in this example would be to operate the small valve from 0–10 percent and the large valve from 10–100 percent of controller output. This just is not a practical solution to the problem because this would result in a 3–4.2 PSIG (0.2–0.28 bar) range for the 1 in. (25 mm) and a 4.2–15 PSIG (0.28–1.0 bar) range for the 3 in. (75 mm) valve. For this reason linear valves are not recommended for this task.

Sequencing equal-percentage valves is much more reasonable. If the small valve had a rangeability of 50:1, its minimum C_v would be 10/50 = 0.200. A line drawn on semilogarithmic coordinates connecting C_v 100 and 0.20 appears in Figure 4.13e. Observe that the C_v of the small valve (10) falls slightly above the midscale of the controller output (over 9 PSIG or 0.6 bar) providing a much more favorable span for positioner calibration. In order to have the two valves act as one without disturbing the smooth equal percentage characteristics at transition, only one valve must open at any one time and the large valve must be prevented from operating in its nearly closed position because, as was shown in Figure 4.13c, its characteristics are not equal percentage in that region. Both of these problems can be circumvented, however, by opening only one valve at a time.

In the scheme shown in Figure 4.13e, the small valve is manipulated alone until the controller output reaches the value corresponding to its full opening. At this point, the pressure switch energizes both three-way solenoid valves, venting the small valve and opening the large to the same flow that the small had been delivering. Switching takes place in one second or less, adequate for all

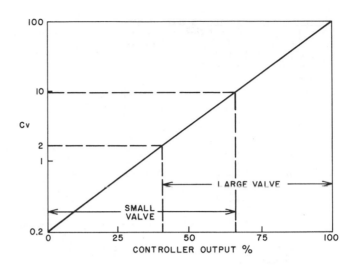

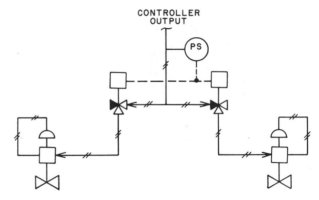

Fig. 4.13e   Sequencing two equal percentage valves, while avoiding transition problems.

but the fastest control loops. When the controller output falls to the point of minimum flow from the larger valve, the solenoids return to their original position. Thus the switch has a differential gap adjusted to equal the overlap between valve positioners. The range of the positioner for the large valve is found by locating its minimum C_v on Figure 4.13e. A rangeability of 50 would give a minimum C_v of 2.

This same approach can be used to sequence three or more valves. If linear characteristics are required, one should insert one of the relays described in Figure 4.13d into the controller output.

Some instrument engineers feel that the switching scheme in Figure 4.13e could become a maintenance problem. They use other methods of valve sequencing illustrated in Figure 4.13f. Some of these schemes include: The "split ranging" loop shown containing gain-plus-bias relays to provide the added rangeability. Even better response can be provided according to Langford by the use of "floating" valve position control where the valve controller slowly moves the larger valve to drive the smaller valve towards 50 percent open, while the small valve retains its sensitivity and response to changes within its capacity. The large valve is a sort of automatic

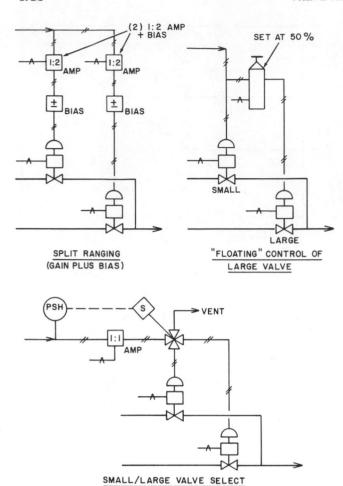

Fig. 4.13f   *Alternate methods of obtaining high turndown through the use of multiple control valves*

bypass to set capacity, with the frequency response of the larger valve being limited. The "large/small valve selection" scheme shown in Figure 4.13f differs from the one in Figure 4.13e because it uses only one solenoid pilot. It is claimed that the 1:1 amplifier eliminates the bounce when the solenoid switches.

Control Valve Sizing

Control valve sizing is addressed in depth in Section 4.16 and therefore only a few general suggestions are given here.

One should first determine both the minimum and maximum C_v requirements for the valve, considering not only normal but also start-up, and emergency conditions. The selected valve should perform adequately over a range of 0.8 [C_v(min)] to 1.2 [C_v(max)]. If this results in a high rangeability requirement, use two or more valves.

Control valves should never operate below their minimum throttling point or at an opening that exceeds their maximum gain point shown in Figure 4.13c. Driskell properly points to the fact that all "fat" settles in the control valve. In constant speed pumping systems, each

design engineer will add his own safety margin in calculating pressure drops through pipes, exchangers, and finally in selecting the pump. Therefore, the control valve will end up with all these safety margins as added pressure drops, resulting in a much over-sized valve. A highly oversized valve will operate in a nearly closed state, which is the most unstable and least desirable operating condition.

For the above reasons, the question of how much pressure drop should be assigned for a control valve can usually be answered simply: None. This is because the various safety margins by themselves will usually contribute more than what is necessary to satisfy the requirements described in Section 4.16. In variable speed pumping systems this approach does not apply because there the pump speed is adjusted to keep the control valve differential constant, and therefore the effect of accumulated safety margins is eliminated.

Process Fluid Characteristics

In selecting control valves, the properties of the process fluid must be fully considered. The process data should be carefully and accurately determined because even small variations in temperature or pressure can cause flashing or cavitation. These include such obvious variables as pressure, temperature, viscosity, slurry, or corrosive nature, or the less obvious factors of flashing, cavitation, erosion, leakage, etc. These are discussed in the paragraphs below.

High Pressure or High Differential Pressure Services

The design pressures for each valve type are listed in the corresponding feature summaries in the front of their sections. Design of high-pressure valves usually necessitates at least three considerations:

     a. Increased physical strength
     b. Selection of erosion-resistant material
     c. Use of special seals

Valve bodies can usually withstand higher pressures than the piping. Valve bodies for high-pressure service are usually forged to provide homogeneous materials free of voids and with good mechanical properties. The loads and stresses on the valve stem are also high. For this reason, higher-strength materials are used with increased diameter. As shown in Figure 4.13g, they are usually kept short and are well guided to prevent column action.

High operating pressure frequently involves high pressure drops. This usually means erosion, abrasion, or cavitation at the trim. Cavitation and erosion resistance are usually not properties of the same metal. Materials resistant to erosion and abrasion include 440C stainless steel, flame-sprayed aluminum oxide coatings (Al_2O_3), and tungsten carbide. On the stem, where the unit pres-

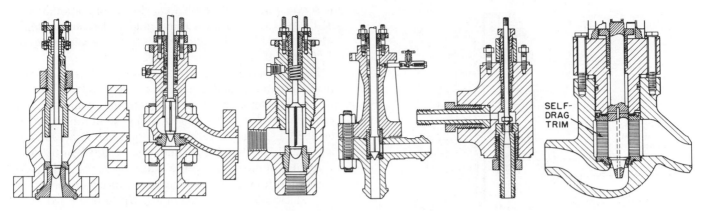

Fig. 4.13g High pressure valve designs

sure between it and the packing is high, it is usually sufficient to chrome plate the stem surface to prevent galling. Special "self-energizing" seals are used with higher pressure valves (above 10,000 PSIG or 69 MPa service) so that the seal becomes tighter as pressure rises. Popular body seal designs for such service include the delta ring closure and the Bingham closure (Figure 4.13g).

The self-energizing seals are used in connecting the high-pressure valves into the pipeline. These designs depend on the elastic or plastic deformation of the seal ring at high pressures for self-energization.

Special packing designs and materials are also required in high-pressure service, because conventional packings would be extruded through the clearances. To prevent this, the clearance between stem and packing box bore is minimized, and extrusion-resistant material, such as glass impregnated Teflon, is used for packing.

Some of the likely causes of valve failure in these services include:

a. Elastomer elements in Saunders or pinch valves (particularly if they fail open) can be ruptured.
b. The stem thrust can be excessive for globe valves. If globe valves are flow to close, high ΔP can damage the seat or prevent the actuator from opening the valve. If they are flow to open they might open against the actuator.
c. High ΔP can exceed the capabilities of plug or floating ball valves and it can bend the shafts of butterflys, damage the bearings of trunion type ball valves, or damage the seat of floating ball valves.
d. Pressure cycling generated by positive displacement pumps can cause bolt fatigue if the number of cycles per year are excessive.

High-pressure services also will increase the probability of noise, vibration, and cavitation.

Vacuum Service

Low pressures can prevent some pressure-energized seals from properly operating or they can cause leakage.

In some processes, the in-leakage from the atmosphere results in overloading the vacuum source; in others it represents a contamination that cannot be tolerated. Potential leakage sources include all gasketed areas and, to an even greater extent, the locations where packing boxes are used to isolate the process from the surroundings.

For vacuum service, valves that do not depend on stuffing boxes to seal the valve stem generally give superior performance. Such designs include Saunders valves and pinch valves. These designs unfortunately are limited in their application by their susceptibility to corrosion and their temperature and control characteristics. Their applicability to vacuum service is further limited by their design. The jacketed pinch valve versions, for example, require a vacuum source on the jacket side for proper operation, and the mechanically operated pinch and Saunders designs are limited in their capability to open the larger size units against high vacuum on the process side. The vacuum process tends to keep the valve closed, and this can result in the diaphragm breaking off the stem and rendering the valve inoperative.

For services requiring high temperatures and corrosion-resistant materials, in addition to good flow characteristics and vacuum compatibility, conventional globe valves can be considered, with special attention given to the type of seal used.

One approach to consider is the use of double packing, as shown in Figure 4.13h. The space between the two sets of packings is evacuated so that air leakage across the upper packing is eliminated. The vacuum pressures on the two sides of the lower packing are approximately equal, and therefore there is no pressure differential to cause leakage across it. Usually the space between the two packings is exposed to a slightly higher vacuum than the process, so that no in-leakage is possible.

Double packing provides reasonable protection against in-leakage under vacuum, but it does not relieve the problems associated with corrosion and high temperatures. When all three conditions exist (vacuum, corrosive flow, and high temperature), the use of bellows seals can be considered. The bellows are usually made of 316 stain-

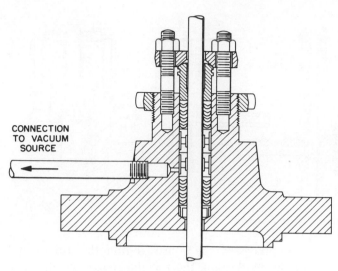

CONNECTION
TO VACUUM
SOURCE

Fig. 4.13h Double packing to seal process under vacuum

less steel and are mass spectrometer tested for leakage. They not only prevent air infiltration but also can protect some parts of the bonnet and topworks from high temperature and corrosion. Like all metallic bellows, these too have a finite life, and therefore it is recommended that a secondary stuffing box and a safety chamber be added after the bellows seal. A pressure gauge or switch can be connected to this chamber between the bellows and the packing to indicate or warn when the bellows seal begins to leak and replacement is necessary.

Considerations similar to those noted for high-vacuum service would also apply when the process fluid is toxic, explosive, or flammable.

High-Temperature Service

The temperature limitations of each valve design are listed in their feature summaries.

All process conditions involving operating temperatures in excess of 450°F (232°C) are considered high temperature. The maximum temperatures at which control valves have been successfully installed are up to 2,500°F (1371°C).

High operating temperatures necessitate the review of at least three aspects of valve design:

   a. Temperature limitations of metallic parts
   b. Packing temperature limitations
   c. Use of jacketed valves

Limitations of Metallic Parts

High temperatures can cause galling, can affect clearances, and can soften hardened trims. Temperature cycling can cause thermal ratchetting and stress resulting in body or bolting rupture if the rate or frequency of temperature cycles is high.

The high operating temperatures are considered in selecting materials for both the valve body and trim. For the body, it is suggested that bronze and iron be limited

to services under 400°F (204°C), steel to operation below 850°F (454°C), and the various grades of stainless steel, Monel, nickel or Hastelloy alloys to temperatures up to 1,200°F (649°C).

For the valve trim, 316 stainless steel is the most popular material, and it can be used up to 750°F (399°C). For higher temperatures, the following trim materials can be considered: 17-4 PH stainless steel (up to 900°F or 482°C), tungsten carbide (up to 1,200°F or 982°C), and Stellite or aluminum oxide (up to 1,800°F or 649°C).

At high temperatures, the guide bushings and guide posts tend to wear excessively, and this can be offset by the selection of proper materials. Up to 600°F (316°C), 316 stainless steel guide posts in combination with 17-4 PH stainless steel guide bushings give acceptable performance. If the guide posts are surfaced with Stellite, the above combination can be extended up to 750°F (399°C) service. At operation over 750°F (399°C), both the posts and the bushings require Stellite.

Packing Limitations

Packings and bonnets in general are discussed later in this section. Here only their suitability for high temperature service is reviewed.

The packing temperature limitation for most nonmetallic materials is in the range of 400°F to 550°F (204 to 288°C), the maximum temperature for metallic packings is around 900°F (482°C), and Teflon should not be exposed to temperatures above 450°F (232°C). Pure graphite (Graphoil®) can be used from −400° to 750°F (−240° to 399°C) in oxidizing service and up to 1200°F (649°C) in nonoxidizing service, with an ultimate potential of 3000°F (1649°C).

Bonnets can be screwed, welded, or flanged. Screwed bonnets are not recommended for high-temperature service. Finned bonnet extensions were used in the past on high-temperature services, when packing material capabilities were more limited. These finned designs were not effective and therefore with the introduction of Graphoil, their use was largely discontinued on rotary valves. For sliding stem valves, Teflon V-rings within extension bonnets are frequently selected and used up to 850°F (454°C).

On high-temperature services, it can be effective to mount the bonnet below the valve. In liquid service, with the bonnet above the valve, the packing is exposed to the full process temperature due to the natural convection of heat in the bonnet cavity. If the bonnet is mounted below the valve, no convection occurs and the heat from the process fluid is transferred by conduction in the bonnet wall only. Therefore, by this method of mounting, the allowable process temperature can be substantially increased in some processes. This is not the case for all applications, because some processes do not generate effective condensate seals, or in other services the liquid-vapor interface line can cause metallurgy problems.

Figure 4.13i provides a method for determining pack-

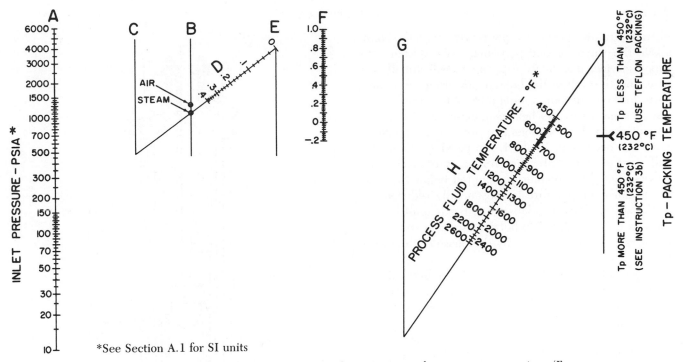

Fig. 4.13i Nomograph for packing temperature determination on hot gas or vapor services. (*From "Determination of Proper Bonnet and Packing for High-Temperature Processes," R. F. Lytle, Fisher Controls*)

*See Section A.1 for SI units

BONNET CHARACTERISTICS*

Valve Size	Bonnet Material**	Bonnet Factors	Standard Bonnet	Extension Bonnet	Radiation Fin Bonnet
1"	CS	D	0.07	0.29	
		F	.83	.25	
	SS	D	.13	.33	
		F	.65	.15	
1½"	CS	D	.07	.24	0.25
		F	.83	.39	.35
	SS	D	.12	.30	.31
		F	.69	.22	.21
2"	CS	D	.11	.29	
		F	.72	.25	
	SS	D	.17	.33	
		F	.55	.15	
3"	CS	D	.10	.24	.31
		F	.74	.39	.20
	SS	D	.17	.31	.34
		F	.57	.21	.11
4"	CS	D	.07	.25	
		F	.83	.36	
	SS	D	.14	.31	
		F	.65	.20	
6"	CS	D	.04	.23	
		F	.90	.40	
	SS	D	.08	.30	
		F	.80	.22	
8"	CS	D	.06	.23	.33
		F	.86	.40	.14
	SS	D	.11	.30	.35
		F	.72	.22	.10

* "Determination of proper bonnet and packing for high-temperature processes," R. F. Lytle, Fisher Controls.

** CS: carbon steel, SS: stainless steel.

INSTRUCTIONS

1. DETERMINE CONSTANTS FROM TABLE AT LEFT CORRESPONDING TO BONNET SELECTION TO LOCATE POINTS ON SCALES "D" and "F".
2. SOLVE NOMOGRAPH   KEY:

Line up straight edge on	Locate intersection on
A TO B	C
C TO D	E
E TO F	G
G TO H	J

3a IF Tp ≤ 450°F (232°C), USE TEFLON PACKING.
 b IF Tp > 450°F (232°C), EITHER USE HIGH-TEMPERATURE PACKING OR SELECT ANOTHER BONNET WITH SMALLER "F" VALUE AND RECHECK PACKING TEMPERATURE.

415

ing temperature, in gas or vapor service, with the bonnet above the valve for one particular valve design. With vapor service, it is likely that vapors will initially condense on the wall of the bonnet, lowering the temperature to the saturation temperature of the process fluid, but in some cases the heat conducted by the metallic bonnet wall will be sufficient to prevent this condensation from occurring. In short, the packing temperature will be at or above saturation temperature (T_s) in vapor service, but if the bonnet is mounted below the valve, the packing temperature is substantially reduced, due to the accumulated condensate. If the bonnet is below the valve, the relationship between process and packing temperature is not affected by the phase of the process fluid.

In case of ball or plug valves with double-sealing, it is important to vent the space between the seals to the line, so that damage will not be caused by thermal expansion.

Jacketed Valves

A number of control valve designs are available with heat transfer jackets. Others can be traced or jacketed by the user. Jacketed valves can be installed for either cooling or heating. When a cooling medium is circulated in the jackets, this is usually done to lower the operating temperature of the heat-sensitive working parts. Such jacketing is particularly concentrated on the bonnet, so that the packing temperature is reduced relative to the process. For certain operations at very high temperatures, intermittent valve operation is recommended, such that when the valve is closed it is cooled by the jacket, and when it is opened, it is kept open only long enough to prevent temperature equalization between the valve and the process.

Heating jackets with steam or hot oil circulation are used to prevent the formation of cold spots in the more stagnant areas of the valve or where the process fluid otherwise would be exposed to relatively large masses of cold metal. Figure 4.13j shows one of these valves, designed to prevent localized freezing or decomposition of the process fluid due to cold spots.

Low-Temperature Service

Cryogenic service is usually defined as temperatures below −150°F (−101°C). Properties of some cryogenic fluids are listed in Table 4.13k. Valve materials for operation at temperatures down to −450°F (−268°C) include copper, brass, bronze, aluminum, 300 series stainless steel alloys, nickel, Monel, Durimet, and Hastelloy. The limitation on the various steels falls between 0° and −150°F (−17° and −101°C), with cast carbon steel representing 0°F and 3½ percent nickel steel being applicable to −150°F (−101°C). Iron should not be used below 0°F (−17°C).

Conventional valve designs can be used for cryogenic

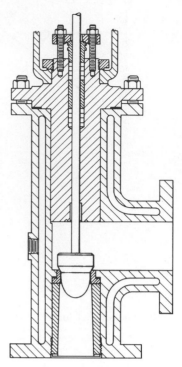

Fig. 4.13j   Jacketed control valve for high-temperature service

service with the proper selection of construction materials and with an extension bonnet to protect the packing from becoming too cold. The extension bonnet is usually installed vertically so that the boiled-off vapors are trapped in the upper part of the extension, which provides additional heat insulation between the process and the packing. If the valve is installed in a horizontal plane, a seal must be provided to prevent the cryogenic liquid from entering the extension cavity. When the valve and associated piping are installed in a large box filled with insulation ("cold box"), this requires an unusually long extension in order to keep the packing box in a warm area.

Although conventional valves with proper materials can be used on cryogenic service, there are valves designed specifically for this application. The unit shown on Figure 4.13l has some characteristics that make it superior to conventional valves on cryogenic service. Most important is its small body mass, which assures a small heat capacity and therefore a short cool-down period. In addition, the inner parts of the valve can be removed without removing the body from the pipeline, and if the valve is installed in a cold box, no leakage can occur inside this box because there are no gasketed parts.

The most effective method of preventing heat transfer from the environment into the process is by vacuum jacketing the valve and piping (Figure 4.13m). The potential leakage problems are eliminated by the fact that there are no gasketed areas inside the jacket.

For cryogenic services where tight shutoff is required,

Table 4.13k
PROPERTIES OF CRYOGENIC FLUIDS*

	Methane	Oxygen	Fluorine	Nitrogen	Hydrogen	Helium
Boiling Point (°F)	−259	−297	−307	−320	−423	−452
Critical Temperature (°F)	−117	−181	−200	−233	−400	−450
Critical Pressure (PSIA)	673	737	808	492	188	33
Heat of Vaporization at Boiling Point (BTU/lbm)	219	92	74	85	193	9
Density (lbm/ft³)						
Gas at Ambient Conditions	0.042	0.083	0.098	0.072	0.005	0.010
Vapor at Boiling Point	0.111	0.296	—	0.288	0.084	1.06
Liquid at Boiling Point	26.5	71.3	94.2	50.4	4.4	7.8

* See Section A.1 for SI units.

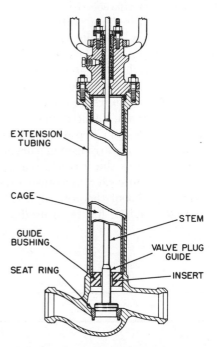

Fig. 4.13l Control valve for
cryogenic service

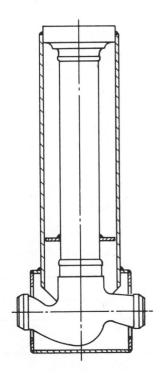

Fig. 4.13m Vacuum jacketing of
cryogenic valve

Kel-F has been found satisfactory as a soft seat material because cold can cause many other elastomer materials to harden, set, or shrink. If used as seals, this can cause leaking.

Cavitation

The flow velocity at the vena contracta of the valve tends to increase by obtaining its energy to accelerate from the potential energy of the stream, causing a localized pressure reduction, which if it drops below the fluid's vapor pressure, results in temporary vaporization. (Fluids form cavities when exposed to tensions equal to their vapor pressure). Cavitation damage always occurs downstream of the vena contracta when pressure recovery in the valve causes the temporary voids to collapse. Destruction is due to the implosions, which generate the extremely high-pressure shock waves in the substantially noncompressible stream. When these waves strike the solid metal surface of the valve or downstream piping, the damage gives a cinder-like appearance. Cavitation is usually coupled with vibration and a sound like rock fragments or gravel flowing through the valve.

Cavitation damage always occurs downstream of the vena contracta at the point where the temporarily formed voids implode. In case of flow-to-open valves, the destruction is almost always to the plug and seldom to the seat.

Methods to Eliminate Cavitation

Because no known material can remain indefinitely undamaged by severe cavitation, the best and the only sure solution is to eliminate it completely. Even mild cavitation over an extended time will attack the metal parts upon which the bubbles impinge. Hard materials survive longer than soft, but they are not an economical solution except for services with mild intermittent cavitation. Cavitation damage also varies greatly with the type of liquid flowing. The greatest damage is caused by a dense pure liquid with high surface tension (e.g. water or mercury). Density governs the mass of the microjet stream, illustrated in Figure 4.13n, and surface tension governs the more important jet velocity. Mixtures are least damaging, since the bubble cannot collapse as suddenly. As the pressure increases, partial condensation in the bubble changes the vapor composition, leaving some vapor to slow the collapse. Some applications of cavitating mixed hydrocarbons show no mechanical damage or high noise level. Cavitation can be reduced or eliminated by several methods, listed in the following paragraphs.

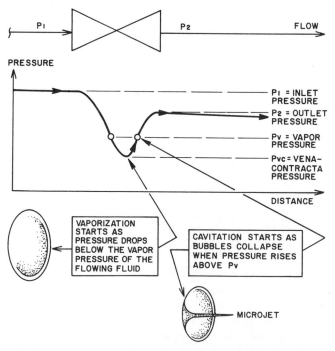

Fig. 4.13n   Cavitation: Downstream of the vena contracta the pressure rises. When vapor pressure is reached, the vapor bubbles implode, releasing microjets that will damage any metallic surface in the area.

REVISED PROCESS CONDITIONS

A slight reduction of operating temperature can usually be tolerated from a process point of view, and this might be sufficient to lower the vapor pressure sufficiently to eliminate cavitation. Similarly, increased upstream and downstream pressures, with ΔP unaffected, can also relieve cavitation. Such control valves should be installed at the lowest possible elevation in the piping system. Moving the valve closer to the pump will also serve to elevate both the up and downstream pressures.

If cavitating conditions are unavoidable from the process conditions point of view, then it is actually preferred to have not only cavitation, but also some permanent vaporization (flashing) through the valve. This can usually be accomplished by a slight increase in operating temperature or by decreasing the outlet pressure. Flashing eliminates cavitation by eliminating pressure recovery.

REVISED VALVE

Where the operating conditions cannot be changed, it is logical to review the type of valve selected in terms of its pressure recovery characteristics. The more treacherous the flow path through a particular valve design, the less likelihood exists for cavitation. Therefore, if cavitation is anticipated, one should select valves with low recovery and therefore high K_c and F_l coefficients. Different valve designs react differently to the effects of cavitation, depending upon where the bubbles collapse. If the focus is in midstream, materials may be unaffected. For example, in the "Swiss Cheese" type design, small holes in the skirt or cage are arranged in pairs on opposite sides of the centerline of the valve. Streams from opposing holes impinge on each other causing the cavities to collapse in the liquid pool (theoretically). This method, illustrated in Figure 4.13o, has been used successfully

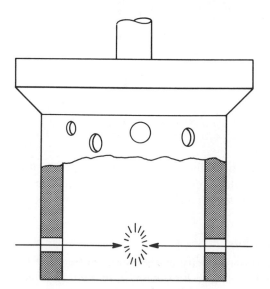

Fig. 4.13o   The "Swiss cheese" design can withstand mild cavitation

for mild cavitation. Labyrinth type valves avoid cavitation by a very large series of right-angle turns with negligible pressure recovery at each turn, but the narrow channels are subject to plugging if particulate matter is in the stream (Figure 4.13p).

Multistep valves avoid cavitation by dividing pressure drop over a series of several orifices, but flow conditions must not stray too far from design range. As shown by condition A in Figure 4.13p, this design can eliminate cavitation due to its relatively low recovery. On the other hand, condition B shows that for precisely the same reason it is possible for a multistep valve to cavitate where only flashing would occur in the conventional valve.

Another valve design variation that can alleviate cavitation is to introduce noncondensible gases or air into the region where cavitation is anticipated. The presence of this compressible gas prevents the sudden collapse of the vapor bubbles as the pressure recovers to values exceeding the vapor pressure, and, instead of implosions, a more gradual condensation process occurs. As shown

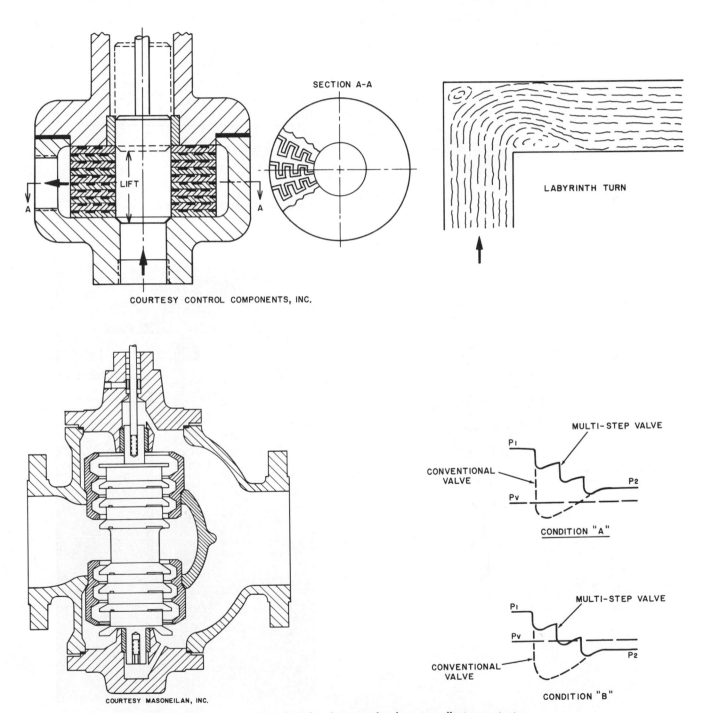

Fig. 4.13p Labyrinth and multi-step valve designs to alleviate cavitation

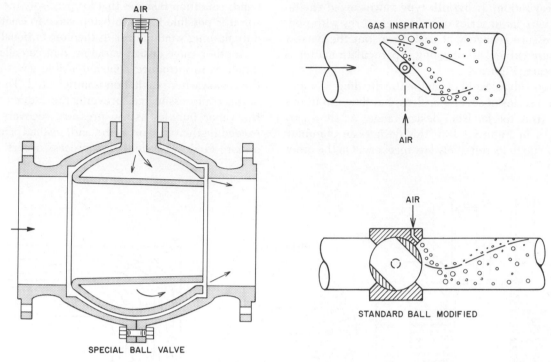

Fig. 4.13q   Valve design variations to alleviate cavitation through the admission of air into the flowing stream

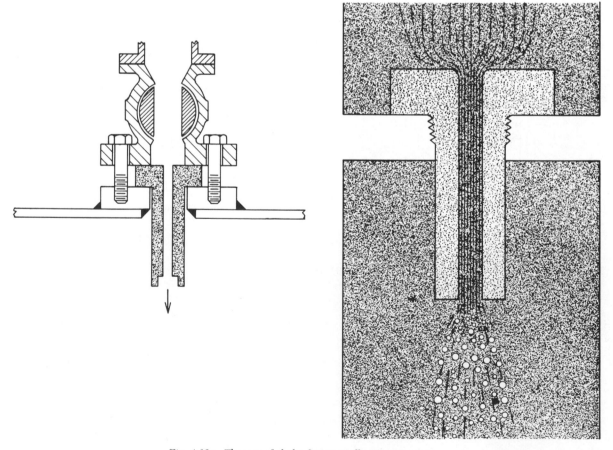

Fig. 4.13r   The use of choke fitting in alleviating cavitation

in Figure 4.13q, the gas may be admitted through the valve shaft or through downstream taps on either side of the pipe, in line with the shaft and as close to the valve as possible. Since the fluid vapor pressure is usually less than atmospheric, the air or gas need not be under pressure.

REVISED INSTALLATION

In order to eliminate cavitation, it is possible to install two or more control valves in series. Cavitation problems can also be alleviated by absorbing some of the pressure drop in breakdown orifices, chokes, or in partially open block valves upstream to the valve. The amount of cavitation damage is related to the sixth power of flow velocity or to the third power of pressure drop. This is the reason why reducing ΔP by a factor of 2, for example, will result in an 8-fold reduction in cavitation destruction.

In some high-pressure let-down stations, it might not be possible to eliminate cavitation accompanied by erosion and/or corrosion. In such installations one might consider the use of inexpensive choke fittings (shown in Figure 4.13r) instead of control valves. The fixed chokes may be of different capacities and isolated by full bore on-off valves, providing a means of matching the process flow with the opening of the required number of chokes. If n chokes are installed, this will allow for operating at $2^n - 1$ flowrates. If, for example, the chokes discharge into the vapor space of a tank, this will minimize cavitation damage because the bubbles will not be collapsing near to any metallic surfaces.

Material Selection for Cavitation

While no material known to man will stand up to cavitation, some will last longer than others. Table 4.13s shows that the best overall selection for cavitation resistance is Stellite 6B (28 percent chromium, 4 percent tungsten, 1 percent carbon, 67 percent cobalt). This is a wrought material and can be welded to form valve trims in sizes up to 3 in. (75 mm). Stellite 6 is used for hardfacing of trims and has the same chemical composition but less impact resistance. Correspondingly, its cost is lower.

In summary, the applications engineer should first review the potential methods of eliminating cavitation. These would include adjustment of process conditions, revision of valve type, or change of installation layout. Only if none of these techniques can guarantee the complete elimination of cavitating conditions should the design engineer consider accepting the presence of this phenomenon and install chokes or special valves, that can last for some reasonable period, while cavitation is occurring.

Flashing and Erosion

Cavitation occurs when (in Figure 4.13n) $P_2 > P_v$, while flashing takes place when $P_2 < P_v$. When a liquid

Table 4.13s
RELATIVE RESISTANCE OF VARIOUS MATERIALS TO CAVITATION

Trim or Valve Body Material	Relative Cavitation Resistance Index	Approximate Rockwell C Hardness Values	Corrosion Resistance	Cost
Aluminum	1	0	Fair	Low
Synthetic sapphire	5	Very high	Excellent	High
Brass	12	2	Poor	Low
Carbon steel, AISI C1213	28	30	Fair	Low
Carbon steel, WCB	60	40	Fair	Low
Nodular iron	70	3	Fair	Low
Cast iron	120	25	Poor	Low
Tungsten carbide	140	72	Good	High
Stellite #1	150	54	Good	Medium
Stainless steel, type 316	160	35	Excellent	Medium
Stainless steel, type 410	200	40	Good	Medium
Aluminum oxide	200	72	Fair	High
K-Monel	300	32	Excellent	High
Stainless steel, type 17-4 PH	340	44	Excellent	Medium
Stellite #12	350	47	Excellent	Medium
Stainless steel, type 440C	400	55	Fair	High
Stainless steel, type 329, annealed	1,000	45	Excellent	Medium
Stellite #6	3,500	44	Excellent	Medium
Stellite #6B	3,500	44	Excellent	High

flashes into vapor, there is a large increase in volume. In this circumstance the piping downstream of a valve needs to be much larger than the inlet piping in order to keep the velocity of the two-phase stream low enough to prevent erosion. Also the piping must be designed so that it is not damaged by slug flow. The impingement of liquid droplets can be erosive if the velocity is great enough (>200 fps or 60 m/s across the orifice), such as pressure let-down of gas or vapor with suspended droplets. Some metals do not corrode because of a self-regenerating protective surface film. If this film is removed by erosion faster than it is formed, the metal corrodes rapidly. Either nobler metals or ceramics should be considered in such situations. The preferred arrangement for flashing service is to use a reduced port angle valve discharging directly into a vessel or flash tank.

Corrosion

Some data on corrosion is contained in Table 4.13s. A detailed tabulation of the chemical resistance of materials is given in Appendix A.3 in the *Process Measurement* volume of this Handbook. In evaluating a particular application, one should also consider the fact that most process fluids are not pure, and that the corrosion rate is much influenced by flow velocity and the presence of dissolved oxygen.

On corrosive services, one can also consider the use of lined valves (tantalum, glass, plastics, elastomers) but one should also consider the consequences of lining failure. Damage can be caused by accidentally exposing the lining to high concentrations of inhibitors or line cleaning fluids. Gas absorption in elastomer linings can also cause blistering.

Viscous and Slurry Service

When the process stream is highly viscous or when it contains solids in suspension, the control valve is selected to provide an unobstructed streamline flow path.

The chief difficulty encountered with heavy slurry streams is plugging. Conditions that can contribute to this include a difficult flow path through the valve, shoulders, pockets, or dead-ended cavities in contact with the process stream. Valves with these characteristics must be avoided because they represent potential areas in which the slurry can accumulate, settle out, and gel, freeze, solidify, decompose or, as most frequently occurs, plug the valve completely.

The ideal slurry valve is one with the following features:

> Provides full pipeline opening in its open position.
> Provides for unobstructed and streamlined flow in its throttling position.
> Has high pressure and temperature ratings.
> Is available in corrosion-resistant materials.
> Is self-draining and has a smooth contoured flow path.
> Will fail safe.
> Has acceptable characteristics and rangeability.
> Its topworks are positively sealed from the process.
> Is not an equal percentage valve.

Unfortunately, no one valve meets all of these requirements, and the instrument engineer has to judge which features are essential and which can be compromised.

If, for example, it is essential to provide a full pipe opening when the valve is open, there are several valves that can satisfy this requirement, including the various pinch valves (A in Figure 4.13t), the full opening angle valves (B in Figure 4.13t), some of the Saunders valve designs (C in Figure 4.13t), and the full-ported ball valves. While these units all satisfy the requirement for a fully open pipeline when open, they differ in their limitations.

The pinch valves, for example, are limited in their materials of construction, pressure, temperature ratings, flow characteristic, speed of response, and rangeability, but they do provide self-cleaning streamlined flow, which in some designs resembles the characteristics of a variable venturi.

The Saunders valves and pinch valves have similar features, including the important consideration that the sealing of the process fluid does not depend on stuffing boxes. They are superior in the availability of corrosion-resistant materials, but they are inferior if completely unobstructed streamline flow is desired. Pinch valves are suitable for very low pressure drop services only, while Saunders and wedge plug valves can operate at slightly higher pressures. Lined butterfly valves are a good choice if the process pressure is high, while the valve drop is low.

The angle valve with a scooped-out plug satisfies most requirements except that its flow characteristics are not the best, and it is necessary to purge above the plug in order to prevent solids from migrating into that area.

Full-ported ball valves in their open position are as good as an open pipe section, but in their throttling positions both their flow paths and their pressure recovery characteristics are less desirable.

Valves that do not open to the full pipe diameter but still merit consideration in slurry service include the characterized ball valves (D in Figure 4.13t), various self-draining valve types (E in Figure 4.13t), the eccentric rotating plug designs (F in Figure 4.13t), and the sweep angle valves (G in Figure 4.13t).

Each of these have some features that represent an improvement over some other design. The characterized ball valve, for example, exhibits an improved and more flexible flow characteristic in comparison with the full-ported ball type. It is well-suited for process fluids containing fibers or larger particles. The self-draining valve allows slurries to be flushed out of the system periodically. Complete drainage is guaranteed by the fact that all surfaces are sloping downstream.

The sweep angle valve, with its wide-radius inlet bend and its venturi outlet, is in many ways like the angle slurry valve. Its streamlined nonclogging inner contour minimizes erosion and reduces turbulence. In order to prevent the process fluid from entering the stuffing box, a scraper can be furnished, which if necessary can also be flushed with some purge fluid. The orifice located at the very outlet is built like a choke fitting. Both orifice and plug may be made of abrasion-resistant ceramic or hard metals.

For slurries with large solid particles, the ideal orifice shape is a circle, such as that of an iris valve or a jacketed pinch (A in Figure 4.13t).

Orifice size, particle size, and rangeability are interrelated. For any particle size and orifice shape there is a minimum opening below which plugging can be expected. To get good rangeability (control at low flow rates) the valve ΔP should be made small. One way to accomplish this is by use of a "head box" (Figure 4.13u). The selection of the valve style and the piping configuration around the valve inlet must be guided by the intractability of the particular slurry.

In food processing applications, in addition to the above,

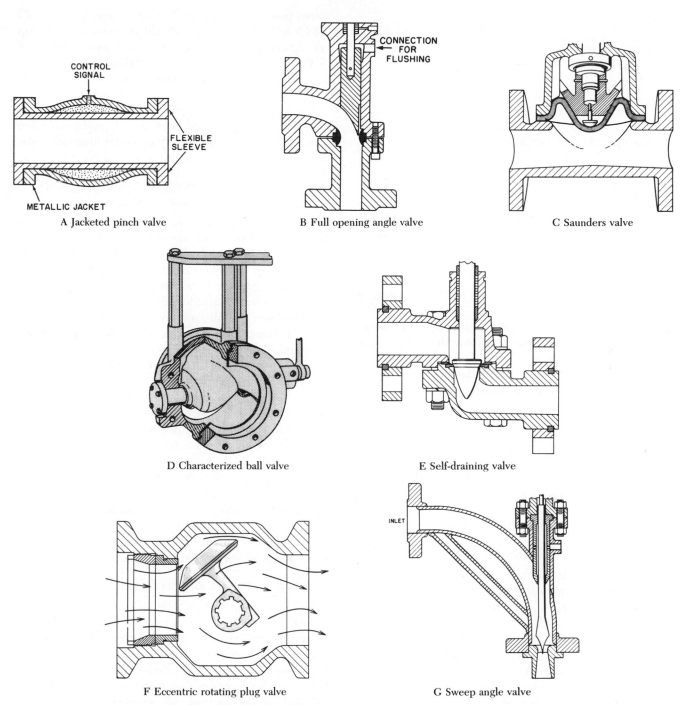

A Jacketed pinch valve

B Full opening angle valve

C Saunders valve

D Characterized ball valve

E Self-draining valve

F Eccentric rotating plug valve

G Sweep angle valve

Fig. 4.13t   Valves for viscous and slurry services

valves must not contain pockets where process material can be retained, and they should be constructed so that they can be easily sterilized and disassembled for cleaning. Materials of construction should not contain compounds which are prohibited by the FDA, such as some of the elastomer compounding materials.

Leakage

Any flow through a fully closed control valve when exposed to the operating pressure differentials and tem-peratures is referred to as leakage. It is expressed as a cumulative quantity over a specified time period for tight shutoff designs and as a percentage of full capacity for conventional control valves.

According to ANSI B16.104, valves are categorized according to their allowable leakage into six classes. These leakage limits are applicable to unused valves only:

Class I valves are neither tested nor guaranteed for leakage.

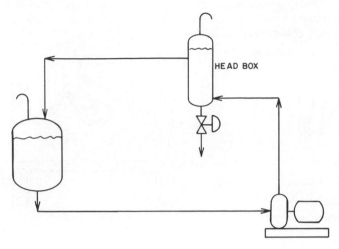

Fig. 4.13u The use of a head box on slurry service

Class II valves are rated to have less than 0.5 percent leakages.

Class III valves are allowed up to 0.1 percent leakage.

Class IV valves must not leak more than 0.01 percent of their capacity.

Class V valves are specified to have a leakage of 5×10^{-4} ml/min water flow per inch of seat diameter.

Class VI is for soft-seated valves and leakage is expressed as volumetric air flow at rated ΔP up to 50 PSI (345 kPa).

Generally the functions of tight shutoff and those of control should not be assigned to the same valve. The best shutoff valves are rotary on-off valves, which are not necessarily the best choices for control.

Factors Affecting Leakage

Some valve manufacturers list in their catalogs the valve coefficients applicable to the fully closed valve. For example, a butterfly valve supplier might list a C_v of 13.2 for a fully closed, metal-to-metal seated 24-in. (600 mm) valve. It should be realized that such figures apply only to new, clean valves operating at ambient conditions. After a few years of service, valve leakage can vary drastically from installation to installation as affected by some of the factors to be discussed.

It should also be noted that some fluids are more difficult to hold than others. Low viscosity fluids such as dowtherm, refrigerants, or hydrogen are examples of such fluids.

SOFT SEATS

One of the most widely applied techniques for providing tight shutoff over reasonable periods of time is the use of soft seats. Standard materials used for such services include Teflon and Buna-N. Teflon is superior in its corrosion resistance and in its compatibility to high-temperature services up to 450°F (232°C). Buna-N is

softer than Teflon but is limited to services at 200°F (93°C) or below. Neither should be considered for operating conditions such as static pressures of 500 PSIG (3.45 MPa) or greater, for use with fluid containing abrasive particles, or if critical flow is expected at the valve seat.

The leakage of double-ported valves is much greater than that of single-ported ones, and it can be as high as 2 to 3 percent of full capacity in metal-to-metal seated designs.

TEMPERATURE EFFECTS AND PIPELINE FORCES

It is frequently the case that either the valve body is at a different temperature than the trim or that the thermal expansion factor for the valve plug is different from the coefficient for the body material. It is usual practice in some valve designs (such as the butterfly) to provide additional clearance to accommodate the expansion of the trim when designing for hot fluid service. The leakage will therefore be substantially greater if such a valve is used at temperatures below those for which it was designed.

Temperature gradients across the valve can also generate strains that promote leakage. Such gradients are particularly likely to exist in three-way valves when they are in combining service and when the two fluids involved are at different temperatures. This is not to imply that three-way valves are inferior from a leakage point of view. Actually their shutoff tightness is comparable to that of single-seated globe valves.

Pipe strains on a control valve will also promote leakage. For this reason, it is important not to expose the valve to excessive bolting strains when placing it in the pipeline and to isolate it from external pipe forces by providing sufficient supports for the piping.

SEATING FORCES AND MATERIALS

The higher the seating force in a globe valve the less leakage is likely to occur. An average valve is 50 lbf per linear inch (8750 N/m) of seat circumference. Where necessary, a much increased seating force will create better surface contact by actually yielding the seat material. Seating forces of this magnitude (about ten times the normal) are practical only when the port is small.

Seating materials are selected for compatibility with service conditions, and Stellite or hardened stainless steel is an appropriate choice for nonlubricating, abrasive, high-temperature, and high pressure drop services. These hard surface materials also reduce the probability of nicks or cuts occurring in the seating surface, which might necessitate maintenance or replacement.

Small Flow Valves

Valves with small flow rates are found in laboratory and pilot plant applications. Even in industrial installations, the injection of small quantities of neutralizers, catalysts, inhibitors, or coloring agents can involve flows in the cubic centimeter per minute range.

Valves are usually considered miniature if their C_v is less than 1. This generally means a $\frac{1}{4}$-in. (6.25 mm) body connection and a $\frac{1}{4}$-in. (6.25 mm) or smaller trim. The topworks are selected to protect against oversizing, which could damage the precise plug.

There are at least three approaches to the design of miniature control valves: (1) the use of smooth-surfaced needle plugs, (2) the use of cylindrical plugs with a flute or flutes milled on it, and (3) positioning the plug by rotating the stem.

Needle plugs give more dependable results than the ones with grooves, scratches, or notches because the flow is distributed around the entire periphery of the profile. This results in even wear of the seating surfaces and eliminates side thrusts against the seat. The trim is machined for very small clearances, and hard materials or

facings are recommended to minimize wear and erosion. Needle plugs are available with equal percentage (down to $C_v = 0.05$), linear, and quick-opening characteristics (Figure 4.13v). Some manufacturers claim the availability of valves with coefficients of $C_v = 0.0001$ or less. At these extremely small sizes it is very difficult to characterize the plugs (equal percentage is not available), and rangeability also suffers.

It is easier to manufacture the smaller cylindrical plugs with one or more grooves (Figure 4.13v) and obtain the desired flow characteristic by varying the milling depth. Both the needle and the flute plugs are economical, but it is difficult to reproduce their characteristics and capacity accurately.

More reproducible control is provided by the valve illustrated in Figure 4.13w. Here the lateral motion of the plug is achieved by rotating the stem through a lead screw. The linear diaphragm motion is transferred into rotation by the use of a slip ball joint. Valve capacity is a function of orifice diameter (down to 0.02 in. or 0.5 mm), number of threads per inch (25 mm) in the lead screw (from 11 to 32), amount of stem rotation (from 15 to 60 degrees), and the resulting total lift, which generally varies from 0.005 to 0.02 in. (0.125 to 0.5 mm).

The extremely short distance of valve travel makes accurate positioning of the plug essential, and this necessitates a positioner. The combination of a long stem and short plug travel makes this valve sensitive to stem load and temperature effects. Because the differential thermal expansion can cause substantial errors in plug position, this valve is limited to operating temperatures below 300°F (149°C).

Sound and Noise

Sound is the sensation produced when the human ear is stimulated by a series of pressure fluctuations transmitted through the air or other medium. Sound is described by specifying the magnitude and frequency of these fluctuations. The magnitude is expressed as the level of sound pressure having the units of decibels (dB). This is a logarithmic function related to the ratio between an existing sound pressure and a reference sound pressure. The reference sound pressure has been selected

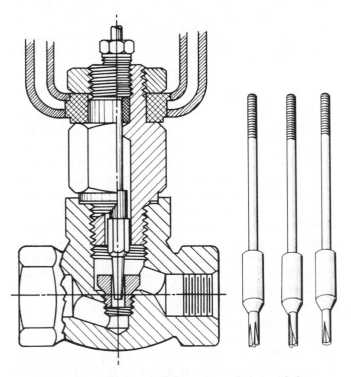

Fig. 4.13v Needle type small-flow valve and plugs with flutes milled on cylindrical surface

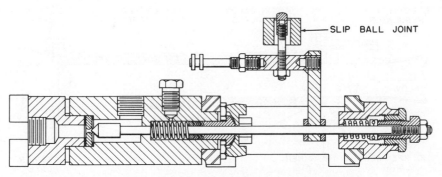

Fig. 4.13w Small-flow plug positioned by stem rotation

as 0.0002 microbars, which in more frequently used engineering units is equivalent to 0.29×10^{-8} PSI.

$$\text{decibels} = 20 \log \frac{\text{existing sound pressure}}{0.0002 \text{ microbars}}$$

OSHA exposure time limits (as of this date) are as follows:

Hours per Day	dBA
8	90
4	95
2	100
1	105
$\frac{1}{2}$	110
$\frac{1}{4}$	115 (max.)

Apparent loudness to the human ear is dependent on both the sound pressure level and its frequency. Sounds of equal sound pressure levels will appear to be louder as the frequency approaches 2,000 Hz. Apparent loudness in phons is defined as the sound pressure in decibels of a pure tone having a frequency of 1,000 Hz. For pure tones, the equal loudness contour in Figure 4.13x shows the pressure level required for the tone to sound as loud as the corresponding reference tone of 1,000 Hz.

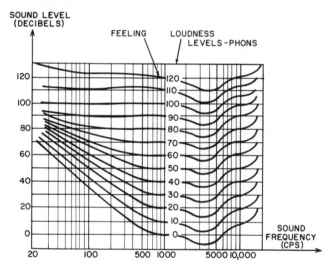

Fig. 4.13x   Apparent loudness contours

Table 4.13y relates the various ways of describing the frequency of a sound and also notes the approximate frequency levels usually generated by the various noise sources in a control valve.

Table 4.13y
SOUND FREQUENCIES AND SOURCES IN VALVES

Frequency (Hz)	Octave Band Number	Sound Description	Typical Noise Source in Valves
20–75	1	Rumble	Vertical plug oscillation
75–150	2		Cavitation*
150–300	3	Rattle	
300–600	4	Howl	Horizontal plug vibration
600–1,200	5		
1,200–2,400	6	Hiss	Flowing gas
2,400–4,800	7	Whistle	
4,800–7,000	8	Squeal	Natural frequency vibration
20,000 and up		Ultrasonic	

* The noise accompanying cavitation varies and can be much higher in frequency than noted here.

Controlling Noise

The transmission of a noise requires a source of sound, a medium through which this sound is transmitted, and a receiver. Each of these can be changed to reduce the noise level. In some cases, such as when the noise is caused by vibrating control valve components, the vibrations must be eliminated or they might result in valve failure. In other cases, such as when the source of noise is the hiss of a gas-reducing station, the accoustical treatment of the noise medium is sufficient. Figure 4.13z illustrates a silencer design that can be installed downstream from a gas-regulating valve; due to the resulting accoustical attenuation, it reduces the sound pressure by a factor of five (from 96 to 82 dB).

The higher the frequency of vibration, the more effective are the commercially available sound absorption materials. Figure 4.13aa gives an example of accoustical treatment for the outside of a pipe. It is also possible to cover the inside walls of the building with sound-

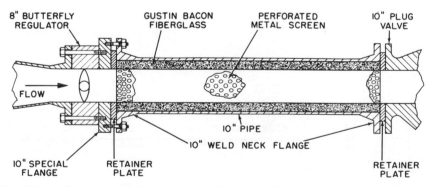

Fig. 4.13z   Silencer for gas-regulating stations (*From "Control Valve Noise," C. F. King, Fisher Controls*)

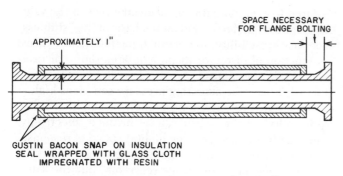

APPROXIMATELY 1"

SPACE NECESSARY
FOR FLANGE BOLTING

GUSTIN BACON SNAP ON INSULATION
SEAL WRAPPED WITH GLASS CLOTH
IMPREGNATED WITH RESIN

Fig. 4.13aa Acoustical treatment of pipe walls

absorbing materials to prevent the reflection and radiation of the sound waves. It is important to seal all openings: a 12-in. (300 mm) thick concrete wall will lose 95 percent of its effectiveness as a sound block if a 1-in. (25 mm) diameter hole is bored through it.

Personnel can be restricted from noisy areas or required to wear ear-protective devices when in them (this would be considered controlling the receiver). The most effective and desirable technique is the treatment of the noise source, and this subject is covered in detail in Section 4.14.

Piping and Installation Considerations

When the valve is larger than 4-in. (100 mm) or sometimes when it is more than one size smaller than the pipe, it is advisable to use pipe anchors to minimize force concentrations at the reducers and more frequently to relieve flange stress loading due to valve weight. The end connections on the valve should match the pipe specifications. If welded valves are specified, the nipples should be factory welded and the welds should be stress relieved. If lined valves are specified, their inside diameters should match that of the pipe to avoid extrusion. On flangeless valves, the bolting and the tightness of the gaskets can be a problem if the valve body is long.

If valves are slow to close or are fast closing (or fail) in long liquid lines, water hammer can result in the upstream pipe or vacuum can develop in the downstream line. Fast-opening steam valves can thermally shock the downstream piping. Steam traps should be provided at all low points in a steam piping network. Anchors should be provided in all locations where sudden valve repositioning can cause reaction forces to develop.

Flow-to-close single-seated valves should not be used because if operated close to the seat, hydraulic hammer can occur. If the damping effect of the actuator alone will not overcome the vertical plug oscillation, than hydraulic snubbers should be installed between the yoke and the diaphragm casing.

Climate and Atmospheric Corrosion

In humid environments such as the tropics, moisture will collect in all enclosures and therefore drains should be provided. Electrical parts should all be encapsulated where possible or be provided with suitable moisture-proof coating.

Vent openings should be provided with storage plugs and insect screens. Even with such precautions, the vents and seals will require preventive maintenance and anti-fungus treatment in some extreme cases.

In cold climates one should be prepared to experience high breakaway torques of elastomers and, in general, metals and plastics will become more brittle. Electro-hydraulic actuators will require heating because oils and greases can become very viscous.

In high-temperature environments, the weak link is usually the actuator, but liners, plastic parts, and electric components are also vulnerable. The damage is not only a function of the temperatures but also of the lengths of time periods of exposure. Diaphragm temperature limits are a function of their materials: Neoprene–200°F (93.3°C), Nordel–300°F (148.7°C), Viton–450°F (232.2°C), silicone glass–500°F (260°C). For higher temperatures, one can replace the diaphragms with pistons (or with metallic bellows in some extreme cases) or add heat shields.

In power plant applications, valves frequently must be designed to withstand anticipated seismic forces. If the atmosphere contains corrosive gases or dusts, it is desirable to enclose, purge, or otherwise protect the more sensitive parts. The stem, for example, can be protected by a boot.

In hazardous areas, all electrical devices should either be replaced by pneumatic ones or be made intrinsically safe or explosion proof.

Packing Designs

The ideal packing provides a tight seal while contributing little friction resistance to stem movement. With TFE packing, which is industry standard, the required stem finish is between 6 and 8 micro-inches RMS (0.15 and 0.2 micron).

A common packing design might consist of Teflon V-rings, as are discussed in more detail in Section 4.6. Double packing with leak-off connection in between can be used on toxic or vacuum service (Figure 4.14h). On toxic services an added seal can be provided by the injection of high viscosity silicone or plastic packings, but this is rarely done.

Solid rings or ribbons of pure graphite (Graphoil), while more expensive than Teflon, are also popular because they are suited for higher temperatures. On the other hand, they require more loading to energize the packing than does Teflon, and the resulting friction can cause stem lockup. On less demanding services, o-rings are also used, but not frequently because under pressure these elastomers will absorb gases, which can destroy the o-ring when depressurized rapidly.

Metallic bellows type seals are seldom used because of their pressure limitations and unpredictable lives. On toxic services they should be provided with automatically monitored guard packings for security.

The bonnets are usually flanged and are extended on hot or cold services, so as to bring the operating temperature of the packing closer to the ambient. Screwed bonnets are not recommended for severe duty, and welded bonnets are not used at all, except as an extreme precaution on hazardous services. The sliding stems can sometimes drag atmospheric contaminants or process materials into the packing, but this can be overcome by close tolerance guide bushings and/or wiper rings. Packing contamination is less likely with rotary valves. In case of LPG, the packing should be isolated from outboard roller bearings and the intervening space vented to protect the lubricant.

Valve Motors and Actuators

Most valve actuators display some dead band and/or hysteresis band due to packing friction. This can cause instability if a small-amplitude control signal falls within the hysteresis band width.

Most valve actuators are also velocity limited because they cannot move faster than their maximum design speed. This is true of both electric and pneumatic motors or actuators. In case of the latter, maximum speed is set by the maximum rate at which air can be supplied or vented. If the full stroking (100 percent) of a valve takes 4 seconds, then its velocity limit is 25 percent per second. Valve signal changes usually occur in small steps, and therefore the velocity limit does not represent a serious limitation because, for example, the time required to respond to a 5 percent change is only 0.2 seconds. This is fast enough for most loops. Figure 4.13bb illustrates the response of velocity-limited actuators to various types of control signals.

Actuator speeds can be further increased by enlarging the air flow ports and by installing larger pilots. On on-off valves, the addition of a quick-dump valve will dramatically increase the venting rate. The dynamic performance of the actuator can also be affected by modifying the tare volume, pressure range, or dead band. In order to reduce the dead band, one usually needs to modify piston seals, linkages, or rack and pinion connections.

The actuator is sized on the basis of the power or thrust required to overcome the unbalanced forces in the valve body, the seating force, and on the basis of the "stiffness" necessary for stability. In pneumatic actuators, the thrust is a function of piston or diaphragm area times air pressure. While the control signal is usually 3 to 15 PSIG (0.2 to 1.0 bar), the actuating pressure can be as high as the air supply pressure, if positioners or amplifier relays are installed. If their costs can be justified, electro-hydraulic actuators will give the highest speed of response.

In pneumatic actuators, the fail-safe action can come from a spring or from an air reservoir, but the latter represents added expense, complexity, and space requirements.

Electric actuators are seldom cost effective, except for such designs as the spring-loaded or modulating solenoid valves (Figure 4.13cc). Electric actuators are therefore limited to locations where air is not available, where the thrust required is less than 1000 pounds (450 kg), where it is acceptable to have the valve fail in its last position, and where slow response is not a drawback.

Positioners as Cascade Slaves

A positioner is a high gain (narrow band) position controller. Whenever a positioner is placed on a control valve, the result is a cascade loop with the position controller being the slave. As is the case with all cascade systems, this arrangement with two controllers in series will be stable if the time constant of the slave is not similar to that of the master. Therefore the addition of positioners to valves on fast loops can degrade the loop performance or result in a very low gain requirement (wide proportional band) for the controller that is not available on conventional controllers. Therefore the use of positioners on fast processes (flow, liquid pressure, or small-volume gas pressure) is unnecessary, and if the controller is not available with the required wide proportional band, the positioner can cause oscillation and cycling. The response speed of positioners can be increased by the use of high-pressure piston actuators and booster relays.

It is important to provide sufficient air supplies so that the actuator volume is not too large for the positioner and to solidly mount the positioner making sure that the linkage is not loose.

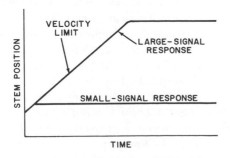
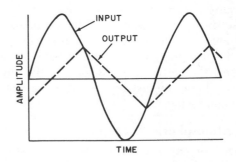

Fig. 4.13bb   Response of velocity-limited actuators to a step change and to a large amplitude sine wave in the control signal, according to Shinskey

Fig. 4.13cc   Modulating solenoid valve
(*Courtesy of Staefa Control Systems, Inc.*)

ing the pressure and/or volume of the controlling
air signal and thereby overcoming such effects
as long control signal lines, etc.

d. To modify the inherent valve characteristics
through the use of external cams or other types
of function generators.

Of the above reasons, a. and b. are legitimate justifications for considering the use of positioners, while c.
and d. are not.

Positioners Will Eliminate Dead Band

All valves and dampers will display some dead band
because of friction in their packings unless positioners
are used. Whenever the direction of the control signal
is reversed, the stem remains in its last position until
the dead band is exceeded, as shown in Figure 4.13dd.
A sine wave driving the valve motor produces stem motion that is distorted and shifted in phase. This phase
shift, when combined with the integrating characteristic
of certain processes and reset action of a controller, causes
a limit cycle to develop. According to Shinskey, widening
the proportional band will not dampen the oscillation,
but only make it slower.

The limit cycle will not appear if a proportional-only
controller is used, or if the process has no integrating
element. Processes that are prone to limit cycling in this
way are liquid-level, volume (as in digital blending), weight
(not weight-rate), and gas pressure—all of which are in-
tegrals of flow. Whenever one intends to control such a
process with a PI controller, the use of positioners should
be considered. In case of level control one can accomplish
the same goal by using a plain proportional controller
and a booster or amplifier instead of a positioner.

Positioners in general will eliminate the limit cycle by
closing a loop around the valve actuator. Positioners will
also improve the performance of valves on slow processes, such as pH or temperature. On the other hand,
dead band caused by stem friction should not be corrected by the use of positioners on fast loops, such as
flow or "fast" pressure. The positioner being a cascade
slave, as was explained above, can cause oscillation and
cycling on fast loops if the controller cannot be suffi-

Common reasons for using positioners include:

a. To protect the controlled process from upsets
caused by valve performance variations due to
stem sticking (a result of dirt build-up), changes
in valve plug positions (a result of process pressure changes) or any other cause of valve hysteresis or dead band.

b. To allow for split-range operation, where the same
control signal is sent in parallel to several valves.

c. To increase actuator speed or thrust by increas-

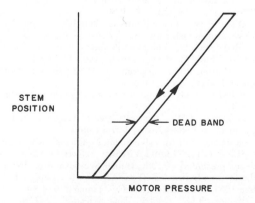

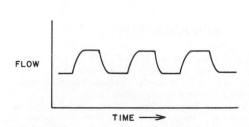

Fig. 4.13dd   Dead band in the valve (left) can result in limit cycling (right) when the loop is closed.

ciently detuned (minimum gain not low enough). Similarly, negative force reactions on the plug require an increase in actuator stiffness and not the addition of a positioner.

Split-Range Operation

The use of positioners for split-range applications is usually accepted regardless of the speed of the process. This is not entirely logical, because on fast loops the control performance can be degraded by the use of positioners. Some instrument engineers do discourage the use of positioners to implement split ranging. Instead they recommend gain-plus-bias relays, so that the positioner (the less accurate device) will operate over its full range (Figure 4.13f). This also eliminates the need for a special calibration.

One can also consider accomplishing the split-range operation through the use of different spring ranges in the valve actuators. In addition to the standard 3-15 PSIG (21-104 kPa) range spring, valves can also be obtained with the following spring ranges: 3-7 PSIG (21-49 kPa), 4-8 PSIG (28-56 kPa), 5-10 PSIG (35-70 kPa), 7-11 PSIG (49-77 kPa), 8-13 PSIG (55 to 90 kPa), 9-13 PSIG (63 to 90 kPa).

Lastly, if split-range positioners are installed on fast processes, the resulting degradation of control quality can be limited by adding a restrictor or an inverse derivative relay in the control signal to it and thereby dampening its effect by artifically making the process appear to be slower than it really is. This technique is not highly recommended (except to reduce wear and tear on the valve in noisy loops) because restrictors are prone to plugging or maladjustments.

Accessories

If the need is to increase the speed or the thrust of the actuator, it is sufficient to install an air volume booster or a pressure amplifier relay, instead of using a positioner. Boosters will give superior performance relative to positioners on fast processes, such as flow, liquid pressure, or small-volume gas pressure control, and they will not be detrimental (nor will offer advantages) if used on slow processes.

If the reason for adding a positioner is to alter or modify the control valve characteristics, this is not a valid justification on fast processes, because this aim can be satisfied by the use of external relays (Figure 4.13d), which will not degrade the quality of control.

BIBLIOGRAPHY

Adams, M. "Control Valve Dynamics," *InTech*, July, 1977.
Arant, J.B. "Fire-Safe Valves," *InTech*, December, 1981.
Baumann, H.D., "A Case for Butterfly Valves in Throttling Applications," *Instruments & Control Systems*, May, 1979.
———, "How to Assign Pressure Drop Across Control Valves for Liquid Pumping Services," Proceedings of the 29th Symposium on Instrumentation for the Process Industries, Texas A&M Univ., 1974.

———, "Effect of Pipe Reducers on Valve Capacity," *Instruments & Control Systems*, December, 1968.
———, "The Introduction of a Critical Flow Factor for Valve Sizing," *ISA Transactions*, vol. 2, no. 2, April, 1963.
Beard, C.S., *Final Control Elements*, Philadelphia: Chilton 1969.
Boger, H.W., "Recent Trends in Sizing Control Valves," 23rd Annual Symposium on Instrumentation, Texas A&M Univ., 1968.
———, "Flow Characteristics for Control Valve Installations," *ISA Journal*, November, 1966.
Buckley, P.S., "Design of Pneumatic Flow Controls," Proceedings of the 31st Annual Symposium on Instrumentation for the Process Industries, Texas A&M Univ., 1976.
———, "A Control Engineer Looks at Control Valves," Proceedings of 1st ISA Final Control Elements Symposium, Wilmington, 1970.
Buresh, J.F., and C.B. Schuder, "The Development of a Universal Gas Sizing Equation for Control Valves," *ISA Transactions*, vol. 3, no. 4, October, 1964.
Coughlin, J.L., "Control Valves and Pumps: Partners in Control," *Instruments & Control Systems*, January, 1983.
Cunningham, E.R., "Solutions to Valve Operating Problems," *Plant Engineering*, Sept. 4, 1980.
Dobrowolski, M., "Guide to Selecting Rotary Control Valves," *InTech*, December, 1981.
Driskell, L.R., "Control Valve Selection and Application," Lecture notes used in the course "Instrument Selection and Application," offered by the Center for Professional Advancement.
———, "Sizing Control Valves," *ISA Handbook of Control Valves*, 2nd ed., Pittsburgh, PA: ISA, 1976.
———, "Practical Guide to Control Valve Sizing," *Instrumentation Technology*, June, 1967.
Hammitt, D., "Key Points About Rotary Valves for Throttling Control," *Instruments & Control Systems*, July, 1977.
———, "How to Select a Valve Actuator," *Instruments & Control Systems*, February, 1977.
Hanssen, A.J., "Accurate Valve Sizing for Flashing Liquids," *Control Engineering*, February, 1961.
Hanson, C.L., and J.C. Clark, "Fast-Closing Vacuum Valve for High-Current Particle Accelerators," *Review of Scientific Instruments*, January, 1981.
Hutchison, J.W., ed., *ISA Handbook of Control Valves*, 2nd. ed., Instrument Society of America, 1976.
Jury, F.D., "Positioners and Boosters," *Instruments & Control Systems*, October, 1977.
Lipták, B.G., "Control Valves in Optimized Systems," *Chemical Engineering*, September 5, 1983.
———, *Instrument Engineers' Handbook*, vol. II, Philadelphia, PA: Chilton Co., 1970.
———, "Control Valves for Slurry and Viscous Services," *Chemical Engineering*, April 13, 1964.
———, "Control Valves in Optimized Systems," *Chemical Engineering*, Sept. 5, 1983.
———, "How to Size Control Valves for High Viscosities," *Chemical Engineering*, December 24, 1962.
———, "Valve Sizing for Flashing Liquids," *ISA Journal*, January, 1963.
Moore, R.L., "Flow Characteristics of Valves," *ISA Handbook of Control Valves*, 2nd ed., Pittsburgh, PA: ISA, 1976.
Morgenroth, J., "Quarter-Turn Plug, Ball, and Butterfly Valves," *Plant Engineering*, July 24, 1980.
O'Keefe, W., "Learn Fluid-Handling Lessons from Nuclear Isolation Valves and Actuator Systems," *Power*, January, 1981.
"Recommended Voluntary Standard Formulas for Sizing Control Valves," Fluid Controls Institute, Inc., FCI 62-1, May, 1962.
Sanderson, R.C., "Elastomer Coatings: Hope for Cavitation Resistance," *InTech*, April, 1983.
Shinskey, F.G., "Control Valves and Motors," Foxboro Publication No. 413-8.
Singleton, E.W., "Control Valve Sizing for Liquid Viscous Flow," Engineering Report No. 7, Introl Limited.
"Standard Control Valve Sizing Equations," ANSI/ISA-S75.01-1977.
Stiles, G.F., "Liquid Viscosity Effects on Control Valve Sizing," Paper presented at 19th Annual Symposium on Instrumentation for the Process Industries, Texas A&M Univ., 1964. Lab Report 3, Problem 1263, Fisher-Governor Co., January 19, 1964.
Wolter, D.G., "Control Valve Selection," *InTech*, October, 1977.

4.14 VALVE NOISE, CALCULATION, AND REDUCTION

Sizes:	1″ to 24″ (25 mm to 600 mm) in standard bodies; sizes above 24″ in special weldment fabrications
Design Pressure:	Up to 2500 ANSI (170 bar) standard; above 2500 ANSI in special designs
Materials of Construction:	Any forged, castable, and machinable metal for body and trim
Special Features:	Balanced plugs, special seal designs, hard facings, piloted inner valve, characterized flow, dual (high/low) operating conditions
Cost:	Highly variable depending upon type of design, size, metallurgy, special features. Range may be from 2 to 10 times equivalent standard valve
Partial List of Suppliers:	Control Components, Inc.; Copes-Vulcan, Inc.; Fisher Controls Co.; H.D. Baumann, Inc.; Honeywell, Inc.; Kent-Introl Div. of Brown-Boveri; Masoneilan Div. of Dresser Industries; NUOVO Pignone; Sulner; Valtek, Inc.

Noise

A weed has been defined as an unwanted plant or flower. As an analogy, noise is a sound that people do not want to hear. Sound is what we call the ear stimulation produced by some device generating pressure fluctuations normally transmitted through the air. These fluctuations vary in frequency and magnitude and produce a sensation in the brain via the mechanism of the ear. The magnitude is expressed in units of sound pressure called the decibel (dB). Frequency is expressed in cycles per second (cps) or in modern terminology, Hertz (hz), where cps and Hz are equivalent units.

Sound magnitude in decibels is a dimensionless ratio of an actual sound pressure to a reference sound pressure. The reference sound pressure has been defined internationally as 2×10^{-5} Newtons/m², 2×10^{-4} microbar, or their equivalent. The decibel is also a logarithmic function, and for every increase of 10 dB, there is a 10-fold increase in sound intensity. Thus a 100 dB sound is 10 times as intense as 90 dB and 100 times as intense as 80 dB. The human ear perceives each 10 dB increase as an approximate doubling of loudness. The term "decibel" is also used to represent two very different sound level values from a source. These two sound levels are sound power (L_w) and sound pressure (L_p).

Sound power is the total acoustic energy created by the noise source. Sound pressure is the level of sound that the receiver (human ear) actually perceives. Sound power is energy expressed in watts, and the international reference level is 10^{-12} watt.

The human ear can potentially perceive sound frequencies between 20 and 18,000 Hz. The human ear is a curious mechanism that does not give equal weight (loudness perception) to the same sound pressure level across the frequency spectrum. Studies of apparent loudness or perception of sound by many human subjects over the frequency spectrum when compared to a pure tone of 1000 Hz frequency has resulted in mapping the ear response. The resulting weighting correction numbers representing human ear response is called the A weighting network, and the corresponding decibel level is called a dBA. Apparent loudness is expressed in phons and is defined as the sound pressure in decibels of a pure tone having a frequency of 1000 Hz. For pure tones, the equal loudness contour of pressure levels required for the tone to sound as loud as the reference tone of 1000 Hz is shown in Figure 4.14a. For example, a sound at 100 Hz and 60 dB is identical in loudness level to a sound at 1000 Hz and 40 dB.

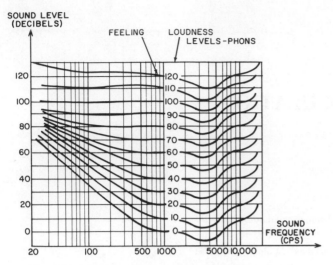

Fig. 4.14a   Apparent loudness contours for human hearing

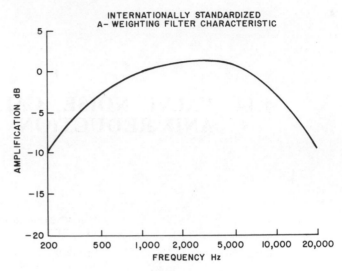

Fig. 4.14c   The A-weighting filter characteristic approximates the human ear's response to different sound frequencies

Valve Noise

While there are many noise sources in industrial and process plants, one of the main contributors can be control valves operating under conditions of high pressure drop. It is one of the few and sometimes the only source of sound levels over 100 dBA found in plants. To gain some perspective of how loud 100 dBA actually is, refer to Figure 4.14b for a comparison of common environmental sounds. As was noted earlier, the human ear response is sensitive not only to sound level but also to the frequency. We can tolerate much louder sounds at low frequencies and at very high frequencies than we can in the middle range of the spectrum. This is shown by the A weighting curve (Figure 4.14c). Note that in the 500- to 7000-Hz range, the human ear is most responsive, and this is the area where high noise level exposure can do the most damage. For this reason, the

U.S. government act known as OSHA limits noise exposure to a weighted 90 dBA maximum level exposure over an 8-hour period. Figure 4.14d shows a typical frequency octave band noise level contour that will meet this limit. Note that if the predominate noise frequency exposure is in the critical middle frequency range of 1000

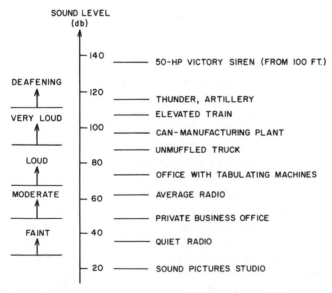

Fig. 4.14b   Common environmental sound comparisons

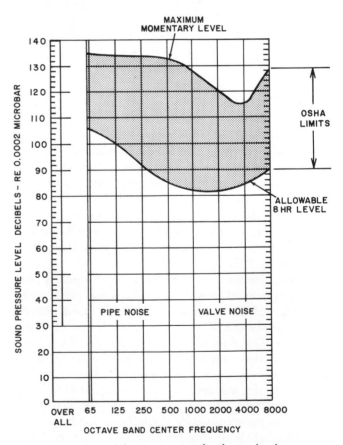

Fig. 4.14d   Typical frequency octave band noise level contours resulting in weighted average exposure to meet OSHA limits

to 5000 Hz, the allowable weighted noise level over 8 hours would be considerably less than 90 dBA.

Valve Noise Sources

The five major sources of noise in control valves are as follows:

Mechanical vibration
Resonant vibration
Inner-valve instability
Hydrodynamic noise
Aerodynamic noise

Mechanical vibration of valve internal parts is caused by unsteady flow and turbulence within the valve. It is usually unpredictable and is really a design problem for the manufacturer. Noise levels are low, usually well under 90 dBA, and in the 50- and 1500-Hz frequency range. The problem is not noise, but progressively worse vibration as guides and parts wear. The solution is a better valve design with heavy duty stems and guides.

Resonant vibration is a variation of mechanical vibration that is characterized by resonant excitation of the valve internals by the fluid flow. This is a narrow band noise with an unusually annoying single pitch tone at some frequency generally in the range of 2000 to 8000 Hz. The noise level may be in the 90- to 100-dBA range, and mechanical valve failure or failure by localized metal overheating is very likely. This can be eliminated by a change in stem diameter, change in the plug mass, or sometimes by reversal of flow direction through the valve. These changes will shift the natural frequency of the plug and stem out of the excitation range. In rare cases where this does not work, then changes in type of guiding, number of parts, or plug fabrication will be necessary. Again, this is a manufacturer's valve design problem.

Inner-valve instability is usually due to mass flow turbulence impingement on the valve plug. The relationship between velocity and static pressure forces across the plug face and the operator force balance in the vertical direction is time varying. This will produce vertical stem oscillations and a low level rattle or rumble noise under a 100-Hz frequency. This instability is detrimental to control. Correction requires changing the damping characteristics of the valve and actuator combination. This is done by providing a stiffer valve actuator. If the actuator is a spring and diaphragm type, then one can increase the nominal spring rate from 3 to 15 PSIG (20 to 100 kPa) to 6 to 30 PSIG (40 to 200 kPa). For single-acting piston actuators, the cushion air loading can be increased. If these changes do not solve the problem, then either operator can be replaced with a double-acting air piston operator. This actuator is extremely stiff and powerful and seldom bothered by this problem. In extreme cases, it may be necessary to provide a hydraulic snubber on the stem or go to an all hydraulic actuator.

Hydrodynamic noise is associated with cavitation or flashing (discussed in Section 4.6), which is avoided by use of a suitable trim or valve type (high F_L). In most cases the noise is not troublesome, although severe cavitation can produce noise in the range of 90 to 100 dBA or higher. Cavitation noise is caused by imploding vapor bubbles in the liquid fluid stream, and the noise can vary from a low frequency rumble or rattling to a high frequency squeal. This latter condition is due to resonant frequency vibration generated by the cavitating fluid. In any case, the problem is not so much noise as it is destruction of the valve trim. Reducing or eliminating the cavitation and its damage also eliminates the noise.

Flashing is rarely a significant source of valve noise, although it can cause valve trim erosion damage in some cases. Flashing results in increasing valve exit velocity and downstream piping velocity as a result of the higher specific volume of the two-phase flow. Expanded outlet valves and larger downstream piping will be required under situations where a large percentage of the liquid undergoes flashing.

Aerodynamic noise is the real control valve noise problem and the only one capable of generating noise levels of 120 dBA or greater. Noise produced by fluid turbulence in liquids is almost negligible as compared to the noise generated by the turbulence and shock cells due to the high velocity of gases and vapors passing through the valve orifice. The mechanisms of noise generation and transmission through pipe walls are highly complex and still not completely understood. As a result, prediction of valve noise levels outside the pipe or at the exit of atmospheric exhaust vents is an inexact science. However, much research is still being performed by universities, manufacturers, and interested technical societies. As more is learned, prediction methods will be refined and noise prediction accuracy will improve. Noise generation, in general, is a function of both mass flow rate and the pressure drop ratio across the valve as shown in Figure 4.14e.

Fluid sonic speed in the valve vena contracta is a function of the valve design type and its F_L value combined with the pressure ratio of upstream to downstream absolute pressure (P_1/P_2). For F_L values of 0.5 to 0.95, the pressure ratio required to generate valve sonic flow will vary from 1.15 to 1.80, respectively. Such valves are called choked valves since their capacity does not increase with increasing pressure ratio, providing the upstream pressure remains the same. Generally, choked valves are the source of the highest noise levels. Valves with pressure ratios less than that indicated for a given F_L value are called subsonic flow valves. For a given mass flow, they are less noisy than choked valves, but the noise level will dramatically increase as the pressure ratio increases to the sonic level. Expressed another way, noise is generated to a significant degree starting at pipe velocities of about mach 0.4 to mach 1.0 (sonic) as shown in Figure 4.14f. Noisy gas or vapor control valves can

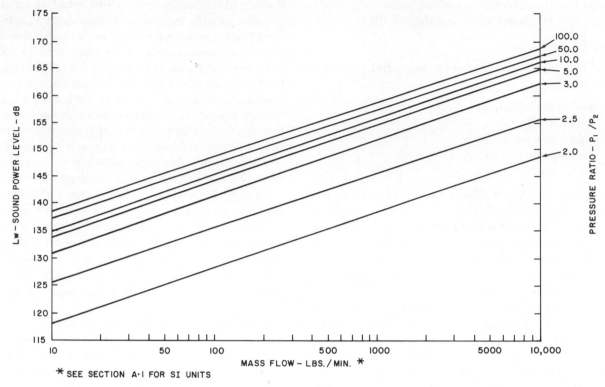

Fig. 4.14e   Relationship between mass flow and pressure ratio in generating sound power.

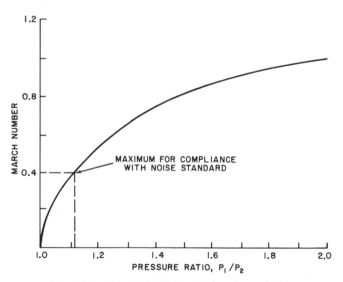

Fig. 4.14f   Relationship between pressure ratio and
gas velocity mach number

have high-velocity-induced acoustic and flow vibration damage, trim erosion, and valve instabilities. Noise with high-vibration-induced damage can drastically reduce valve service life, and in some cases it can lead to complete inner valve destruction in a matter of minutes to hours.

Valve Noise Treatment

Depending upon the magnitude of the aerodynamic noise and assuming that massive valve damage is not a factor, valve noise treatment can be accomplished either by path treatment or source treatment. Valve damage considerations are amenable only to source treatment that minimizes or eliminates the damage mechanism. There is no absolute rule that will enable one to choose between path or source treatment. However, in general, if the noise is under 100 dBA, then either a path or source treatment is a possible solution. Noise above 100 dBA almost always requires source treatment to successfully solve the noise problem. Proper choice of noise treatment method is not always easy, and the problem is compounded by the inexact science of noise level prediction and frequency spectrum. As a result, arriving at a satisfactory solution is in part good engineering expertise and in part an art based upon experience. A good solution should tend toward a conservative approach, since rework or retrofit of an inadequate system is often very expensive.

Path Treatment

Path treatment, as its name implies, does nothing to change the noise source. The intent of path treatment is to attenuate the noise transmission from the source to the receiver (ear). There are several methods of reducing noise by this method, such as the use of heavy wall pipe; installation of diffusers, mufflers, or silencers; and application of acoustical insulation. Path treatment is not necessarily a more economical solution than source treatment, and economics must be evaluated for each individual application. For existing installations, path treatment may be used, not because it is the best solution, but because it may be the only feasible one.

Heavy wall pipe reduces noise by increasing the transmission loss through the pipe wall. The amount of attenuation is a function of the stiffness and mass of the pipe. The mechanisms are complex and beyond the scope of discussion in this text. However, as a simple rule of thumb for rough evaluation, each doubling of pipe wall thickness will result in approximately 6 dBA attenuation depending upon pipe size (attenuation increases with pipe size).

Diffusers (Figure 4.14g) can be helpful in both original installation and retrofit situations. These devices can aid in reducing turbulence and/or shock cell flow exit the

valve and thus help attenuate the noise mechanism. Diffusers can also be designed to serve as a flow restrictor or pressure drop device. This will serve to reduce the total pressure drop across the control valve and thus reduce its noise generation capability. The diffuser works best in this situation when the flow rate is substantially constant or at least does not vary over a wide range. As a restrictor, its effectiveness in generating pressure drop decreases substantially as the flow reduces. This is because it follows the square law similar to an orifice plate. However, to some degree this shift of pressure drop back to the control valve does not necessarily lead to a large increase in noise because the mass flow is reducing and this reduces noise.

Mufflers or silencers (Figure 4.14h) can be used for in-line path treatment or for atmospheric vents in some cases. These are usually expensive devices, with the cost escalating dramatically with size. Dissipative or dissipative/reactive silencers are most commonly used, but a comprehensive discussion of these devices is beyond the scope of this text. However, some rough guidelines for application are discussed. Inlet velocity must be subsonic, and the silencer cannot be sized to serve as a pressure reducer. The outer shell should be heavy wall or else the silencer can act as a loud speaker if the shell wall is too light and resonates. Materials of construction to meet process conditions must be evaluated. An inlet diffuser (as shown) can be helpful, since it will break up turbulence or shock cell oscillations that generally characterize downstream sound fields and reduce the effectiveness of the unit.

Another method of increasing transmission loss at the pipe wall is the use of acoustic insulation. Even thermal insulation can add 3 to 5 dBA attenuation. Proper se-

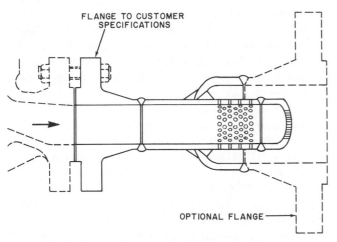

Fig. 4.14g   One type of acoustical diffuser for reducing valve exit turbulence (*Courtesy of Fisher Controls*)

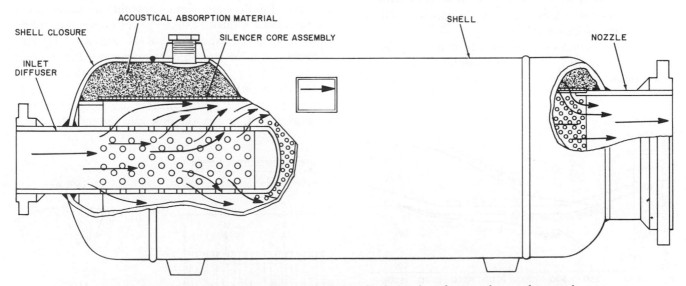

Fig. 4.14h   Silencers are devices that are installed in the flow path to dissipate the sound energy by absorbing it in an acoustical pack. They are designed to take less than 1 PSI pressure drop. In-line silencers are often the most economical approach to noise control where high mass flow rates and low pressure drops exist. These units are normally used immediately downstream of control valves, but in some cases they may also be required upstream of the valves. (*Courtesy of Fisher Controls*)

lection and application of 1 to 2 in. (25 to 50 mm) of a good acoustic insulation can reduce the noise level by roughly 10 dBA. Certain types of insulation are most effective at specific frequency bands, so this information is important for proper selection. It should be recognized that sound travels down the pipeline with very little attenuation over long distances. Therefore, increasing the pipewall thickness and/or applying acoustical insulation can be a very expensive solution. This approach is most useful when downstream piping runs are short.

Locating the valve at a distance from normal personnel working areas may be helpful. If the valve can be located far enough away, such as on top of a structure or pipe bridge, the distance attenuation factor may minimize the noise treatment at the valve source. For example, instead of a control valve with a noise specification of 85 dBA, it may be possible to relax this to 90 or 95 dBA. This can have a considerable impact on the cost of the valve or noise treatment system.

Source Treatment

Source treatment treats the noise problem by reducing its magnitude at the source. In most cases treatment consists of a special control valve and trim design or a special control valve combined with one or more special diffuser or static resistor elements. While they may differ in concept, design, and manufacturing specifics, the basic intent of these special systems is to reduce pressure drop in stages, limit fluid velocity to low subsonic levels, and reduce or eliminate the formation of high turbulence and shock cells generation. Any given design will utilize one or more of these mechanisms to reduce noise generation by the control valve. Depending upon the particular design, noise can be reduced by 7 to 10 dBA with simple element designs to as much as 30 to 40 dBA with the more sophisticated valve element designs or multi-element systems (Figure 4.14i).

Specification of source treatment valves or systems is not a simple matter. There are a number of design considerations, such as:

Application—in-line or vent
Noise reduction actually required—dBA
Valve absolute pressure ratio—P_1/P_2 or $\Delta P/P_1$
Fluid to be handled and physical properties
Temperature operating level and range
Mass flow rate and turndown
Metallurgy and mechanical design considerations
Other potential velocity-induced problems
Valve shutoff requirements, especially vents
Plant operation utility needs (valve service life)
Valve location, physical arrangement, piping arrangement, valve support and access
Economics, including purchase, installation, and maintenance costs of solutions as well as valve cost

Experience indicates that control valve vendors cannot be relied upon to propose or produce completely successful solutions in every case. Vendor proposals run the gamut from wishful thinking to serious and fairly detailed technical proposals. Thus it is up to the user to carefully weigh and evaluate all proposals. The weight attached to each factor is a matter of judgment and experience, understanding all aspects of the application and plant needs. While cost is one factor, this is not the only consideration; it may be that the more expensive equipment is the most economical solution in the long run. Plant downtime and retrofit costs for deficient valve noise solutions is usually very expensive. With these caveats, there is much good equipment available, and vendor expertise and experience can be very valuable to the user with limited experience in controlling valve noise.

Basically, source treatment control valve designs fall into three categories: multi-path, multi-stage, and combination of multi-path/multi-stage. These designs are listed in order of sophistication and capability for noise and/or valve damage reduction under severe operating conditions. As might be expected, cost also increases. The cost of these special designs tends to range from 2 to 20 times the cost of a standard valve with the same flow range.

The multi-path valve (Figure 4.14j) provides multiple orifices in parallel. These orifices are uncovered by a cylindrical plug to vary the flow rate through the valve.

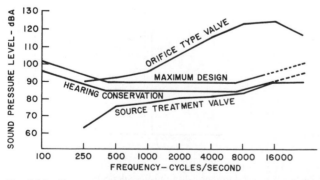

Fig. 4.14i   Test on air 300 PSIG (2.1 MPa) to atmosphere measured 3 ft (0.9 m) away. Comparison of noise levels between standard valve and source treatment valve

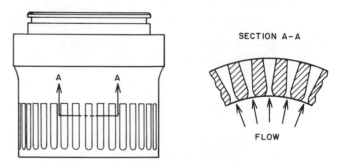

Fig. 4.14j   One type of multi-path valve design for moderate noise reduction (*Courtesy of Fisher Controls*)

Although the path shape may vary by manufacturer, the principle consists of splitting the single-path flow into a large number of small paths. The noise generated by the mass flow through each small path is much less than that generated by the total flow through the single path orifice. The combined or additive noise of all these small individual paths is still several dBA less than the noise generated by the single-orifice valve. Typical attenuation levels are 7 to 10 dBA but may reach 12 to 15 dBA in some applications. Variations of the multi-path design are used for both hydrodynamic and aerodynamic valve noise or damage problems of low to moderate severity. Typically, compressible fluid pressure ratios (P_1/P_2) of 2 to 2.5 and valve exit velocities of mach 0.33 maximum with predicted noise levels of 90 to near 100 dBA are good candidates for this design.

Pure multi-stage valves are available but they are not as widely used as the other designs. A typical example of this design is shown in Figure 4.14k. This valve provides multiple orifices in series that serve to divide the total valve system pressure drop over several stages (typically 3 to 9). Thus the reduced pressure drop per stage or "multiple velocity head-loss elements" in series results in greater friction loss, reduced velocities, and reduced

noise. The shape of the trim element is dictated by the fact that the effective flow area has to increase between the inlet and the outlet to compensate for the change in gas density and increase in specific volume. Thus the outlet flange size is larger to limit the exit velocity to a level that will not regenerate excessive noise. Typically, these valves can provide noise attenuations up to 20 to 25 dBA provided that the P_1/P_2 pressure ratio is limited to 3 or 4 maximum. Above these ratios, the exit velocity will exceed mach 0.3 and require additional path treatment or require holding the pressure ratio with an in-line restrictive device.

In addition to diffusers, as discussed earlier, there are other special design multiple orifice restrictors available (Figure 4.14l). These devices are built in a wafer design for installing between flanges and can be used as single or multi-stage elements (Figure 4.14m). The plates are designed to work with the control valve as shared pressure drop restrictors that are sized to keep the average velocity low. In addition, the design of most of these devices serves to force the fluid to seek flow passages that involve multiple changes in direction. Thus they behave somewhat like friction plates with resulting noise attenuation capability of several dBA. These multiple orifice restrictors are very useful in valve noise control work, especially for multi-element systems often required for high pressure drop and high noise attenuation applications. Like the simple diffuser, they lose effectiveness with flow turndown.

Combination multi-path and multi-stage valves are usually required for the more severe and high noise producing services. These are the workhorse designs for the really tough services, especially those that can cause extensive valve trim damage and/or noise levels in excess of 100 dBA. Various manufacturers have taken different approaches to the design of this type of valve. Most are based upon multiple orifices in series and parallel with

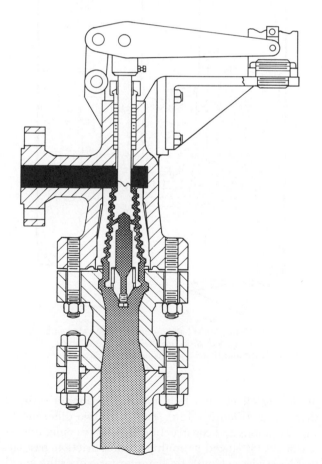

Fig. 4.14k Multi-stage step trim valve for use on compressible fluids. Outlet is expanded to compensate for volume change
(*Courtesy of Masoneilan Division of Dresser Industries*)

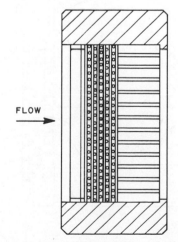

Fig. 4.14l Special resistor element used for valve back pressure and noise reduction
(*Courtesy of Masoneilan Division of Dresser Industries*)

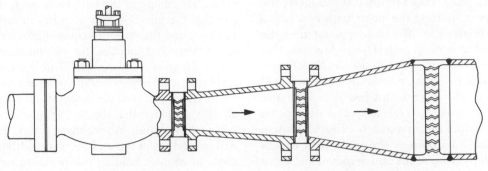

Fig. 4.14m   Use of special resistor elements in series for noise reduction
(*Courtesy of Masoneilan Division of Dresser Industries*)

the flow controlled by the number of orifices uncovered in the first stage by a close-fitting cylindrical plug (Figures 4.14n and 4.14o). One variation of this is a design that also incorporates a secondary diffuser element built into the valve (Figure 4.14p). Finally, one manufacturer has designed a valve using a standard valve trim assembly that eliminates the close-fitting cylindrical plug. Here the flow rate is modulated the same as any standard globe control valve but can be provided with a noise attenuator element ranging from 1 to 7 stages (Figures 4.14q and 4.14r). Depending upon the manufacturer and design, these multi-orifice in series valves will have 2 to 7 stage capability. A typical noise attenuation curve for 1 and 2 stage elements is shown in Figure 4.14s and for multi-stage elements is shown in Figure 4.14t.

One special multi-path/multi-stage valve design is unusual in several respects. This design with multiple orifices in series and in parallel utilizes the pressure loss

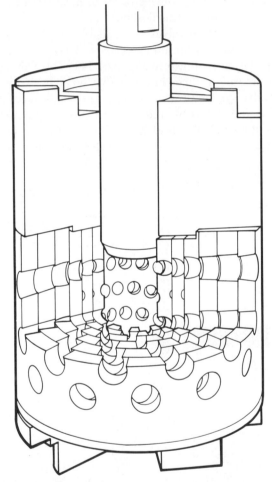

Fig. 4.14o   Multi-path and multi-stage valve with
six stages of series/parallel orifices, suitable
for both liquids and gases (*Courtesy of Copes-Vulcan*)

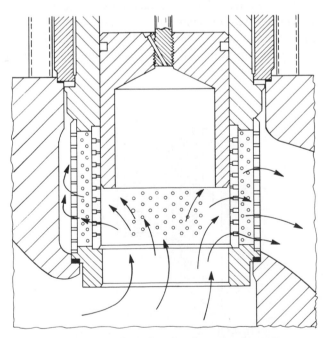

Fig. 4.14n   Multi-path and multi-stage valve with
shaped first stage holes (*Courtesy of Fisher Controls*)

producing effects of a fluid passing through a series of sharp turns (Figure 4.14u). There are two different design variations and construction. In either configuration, it can be visualized as multiple short friction passages that have a large length-to-diameter ratio over the length of the passage. Each "plate" and each passage design and configuration can be developed to suit the specific fluid

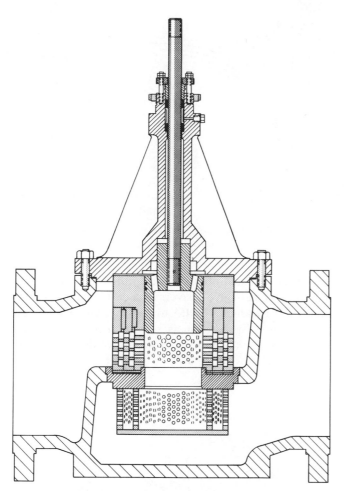

Fig. 4.14p   Multi-path and multi-stage valve with integral secondary diffuser element (*Courtesy of Kent Introl Division of Brown Boveri Corp.*)

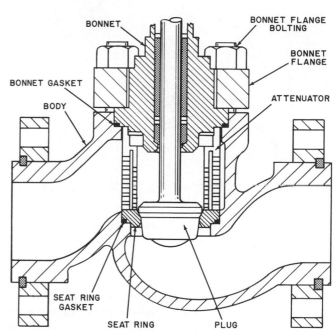

Fig. 4.14q   Two-stage noise attenuator in valve designed for standard inner valve trim assembly (*Courtesy of Valtek, Inc.*)

application. As such, the throttling velocity into, through and out of the plate, is very close to pipeline velocity. The overall trim element is sized to handle the required flow by stacking the required number of plates and using a piston type cylindrical plug to expose the plates to inlet flow (Figure 4.14v). Plate design and capacity per plate can be varied to produce a variety of flow characteristics. Plate designs are available for either liquid or compressible fluids.

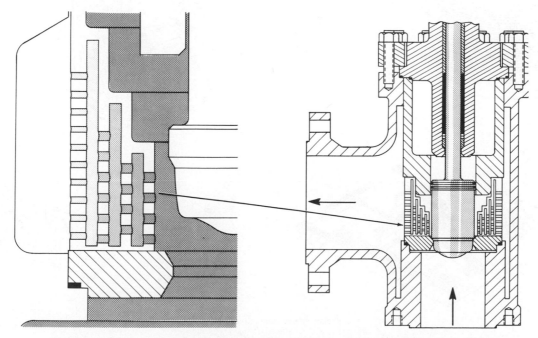

Fig. 4.14r   Multi-stage noise attenuator and detail in valve designed for standard trim assembly with pressure balance (*Courtesy of Valtek, Inc.*)

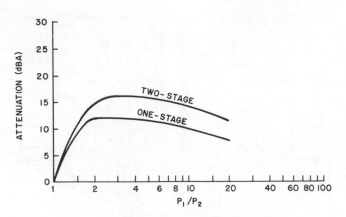

Fig. 4.14s Typical noise attenuation curves for one- and two-stage elements (*Courtesy of Valtek, Inc.*)

Fig. 4.14t Typical noise attenuation curves for multi-stage elements (*Courtesy of Valtek, Inc.*)

Valve Noise Prediction

As mentioned earlier, valve noise prediction is an inexact science because of the complex nature of noise generation by the control valve and the transmission of this noise through the pipe wall. So it is not surprising that a number of different prediction methods are used by manufacturers and others. The problem is that the various methods can give answers for a particular application that differ by around 5 dBA to 20 dBA. The whole topic of valve noise prediction is still subject to continuing research and evaluation. So what do we do? Each manufacturer claims to be able to predict the valve noise

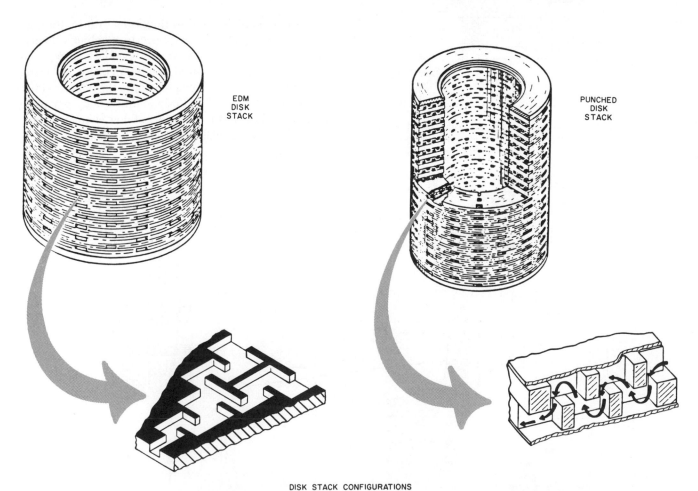

DISK STACK CONFIGURATIONS

Fig. 4.14u Special noise element design using labyrinth passages incorporated on plates (*Courtesy of Control Components Division of IMI*)

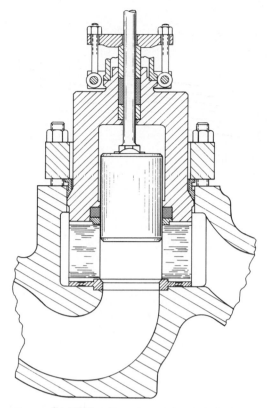

Fig. 4.14v Labyrinth passage disks are stacked into an element for valve noise attenuator. Disks are exposed to flow by cylindrical plug (*Courtesy of Control Components Division of IMI*)

problem and provide a valve design solution. Some do better in this than others. It finally falls upon the user to provide the best possible data, carefully evaluate all proposals, ask questions and resolve marked differences, and finally use good engineering judgment and experience in selecting the vendor for each application. This may mean that more than one vendor will be chosen to cover a multi-valve requirement for a project. Normally, this would number only two or three at the most. It is wise to err on the conservative side when making a final selection since the cost of mistakes and the required retrofit may far outweigh valve cost differentials. Unfortunately, there are no absolute answers or guidelines on this problem, but experience indicates that proper attention to the analysis detail outlined will result in a high percentage of success.

Valve Noise Calculation

Although there are a number of valve noise prediction methods available, the following simplified method is one that represents a reasonable state-of-the-art approach. This calculation gives adequate answers over a broad range of pressure ratios and other operating situations. It should enable the user to classify a valve noise problem and permit a more intelligent appraisal of pos-

sible methods of treatment. It will also enable the user to better assess solutions offered by various manufacturers.

The following mathematical method of estimating the aerodynamic noise level of any style of standard control valve is based upon the research conducted by M. J. Lighthill in the early 1950s. It yields reasonably accurate results considering the complexity of flow patterns in different styles of control valves. The equation covers noise in conventional single-stage valves and does not apply to the special multi-stage or multi-path valves used for noise reduction as discussed earlier in this section.

The sound pressure level for gas or vapor passing through a valve and measured at a distance of 1 m from the adjacent downstream piping is expressed by equation 4.14(1):

$$SL = 145.5 + N_M + \\ + 10 \log (C_v F_L P_1 P_2) - P_L + G \quad 4.14(1)$$

where

SL = sound level (dBA)
N_M = acoustic efficiency factor (Figure 4.14w)
C_v = valve capacity factor at operating conditions
F_L = pressure recovery factor (Table 4.14x) (Exact values of F_L can be obtained from the valve manufacturer.)
P_1 = upstream pressure (PSIA)
P_2 = downstream pressure (PSIA)
P_L = pipe transmission loss (dBA) (See below.)
G = gas property factor adjustment (dBA) (Table 4.14y)

The sound transmission loss through steel pipes surrounded by air can be calculated by equation 4.14(2):

$$P_L = 10 \log \left[\frac{10^{10} \times t^3 \times (39 + D/2)}{35.3 D^3} \right] + \\ + 10 \log \left[\frac{(4D^2) n_o}{C_v F_L} \right] \quad 4.14(2)$$

where

P_L = transmission loss, dBA
D = actual outside pipe diameter (inches)
t = pipe wall thickness (inches)
n_o = number of apparent sound producing orifices (Table 4.14z)

Table 4.14aa lists typical transmission losses which are, in effect, based upon the first part of equation 4.14(2). The specific basis for the table is sound measurement at 1 m distance from the pipe wall, air as the fluid, valve capacity of $C_v F_L = 4D^2$ and $n_o = 1$. Thus the calculation of P_L can be simplified to a degree. For typical full-ported valves ($C_v F_L > 4D^2$) and having values of n_o other than 1, add $10 \log (n_o)$ to the appropriate number from Table 4.14aa. For typical reduced-port or reduced-travel valves

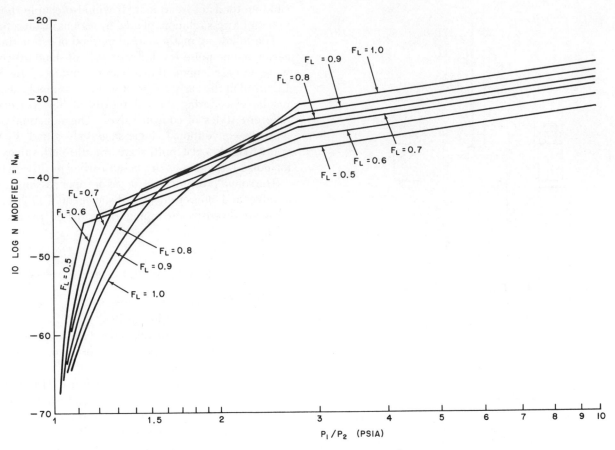

Fig. 4.14w   Acoustic efficiency factor as a function of pressure ratio

$(C_vF_L \leq 4D^2)$, add 10 log $(4D^2/C_vF_L)$ to the value from Table 4.14aa.

The following examples of estimating valve noise will illustrate the use of the equations and tables. In addition, they will illustrate how a different choice of valve may help a noise problem and how distance from the source (valve) can be a factor. The techniques utilized for valve noise estimation assume valve outlet pipe velocities are below one-third of sonic velocity (mach 0.33). If this velocity is exceeded, it is possible for the noise generated by pipe velocity to be greater than the estimated valve throttling noise.

Example 1: Determine the expected sound level for the following valve installation.

Valve type: Streamlined, reduced port, flow to close,
  angle, $n_o = 1$
Valve body size: 2″ (50 mm)
  C_v = 2.5
  F_L = 0.5 (Table 4.14x)
Pipe: 2″ (50 mm), schedule 80
Gas: Natural gas
  P_1 = 3800 PSIA (262 bar)
  P_2 = 1050 PSIA (72.4 bar)
  P_1/P_2 = 3.6
  N_M = −37 (Figure 4.14w)

P_L = 72 (Table 4.14aa) +
  $+ 10 \log \left[\dfrac{(4D^2)\, n_o}{C_vF_L} \right]$
  $= 72 + 10 \log \left[\dfrac{4 \times 2^2 \times 1}{2.5 \times 0.5} \right]$
  = 83 dBA
G = 0.5 (Table 4.14y)

From equation 4.14(1):

$SL = 145.5 - 37 + 10\log(2.5 \times 0.5 \times 3800 \times$
  $\times\ 1050) - 83 + 0.5$
  $= 145.5 - 37 + 10\log(4.98 \times 10^6) - 83 +$
  $+\ 0.5$
  $= 93$ dBA

Example 2: Determine the expected noise level for a valve installation with the following conditions.

Valve type: Cage type, full port, flow to open,
  $n_o = 4$
Valve body size: 4″ (100 mm)
  C_v = 105
  F_L = 0.85 (Table 4.14x)
Pipe: 4″ (100 mm), schedule 40
Gas: Air
  P_1 = 150 PSIA (10.35 bar)
  P_2 = 100 PSIA (6.9 bar)

Table 4.14x
TYPICAL VALVE PRESSURE RECOVERY FACTORS
(F_L)

Valve Type	F_L Fully Open Condition
Globe, single seated, flow to open, full port	0.90
Globe, single seated, flow to open, 50% reduced port	0.90
Globe, single seated, flow to open, full port cage trim	0.85
Globe, single seated, flow to open, 50% reduced cage trim	0.80
Split body, single seated, flow to open, full port	0.80
Split body, single seated, flow to open, 50% reduced port	0.90
Globe, double seated, contoured plug, full port	0.90
Globe, double seated, contoured plug, 50% reduced port	0.80
Globe, double seated, % V-port, full port	1.00
Globe, double seated, % V-port, 50% reduced port	0.95
Globe, angle, single seated, flow to close, full port	0.80
Globe, angle, single seated, flow to close, 50% reduced port	0.80
Globe, angle, single seated, flow to open, full port	0.90
Globe, angle, single seated, flow to open, 50% reduced port	0.95
Globe, streamlined angle, flow to close, reduced port	0.50
Eccentric rotating plug valve	0.85
Characterized ball valve, with D/d = 1.5	0.63
Noncharacterized throttling ball, with D/d = 1.5	0.55
Butterfly, line size, 60° open	0.70
Butterfly, line size, 90° open	0.56
Butterfly, D/d = 2, 60° open	0.62
Butterfly, D/d = 2, 90° open	0.50

Table 4.14y
GAS PROPERTY FACTOR G

Gas	G(db)	Gas	G(db)
Saturated steam	−2	Carbon dioxide	−3
Superheated steam	−3	Carbon monoxide	0
Natural gas	0.5	Helium	−9
Hydrogen	−9	Methane	2
Oxygen	−0.5	Nitrogen	0
Ammonia	1.5	Propane	−4.5
Air	0	Ethylene	−1.5
Acetylene	−0.5	Ethane	−2

Table 4.14z
NUMBER OF APPARENT SEPARATE SOUND PRODUCING VALVE ORIFICES IN PARALLEL (n_o)

	n_o	$10 \log(n_o)$
Full bore ball valve	1	0
Single-seat angle valve, flow to close	1	0
Eccentric rotary plug valve, flow to close	1	0
Segmented ball valve	1	0
Cage valve, flow to close ($P_1/P_2 > 2$)	1	0
Butterfly valve (nonfluted), 60° travel	2	3
Butterfly valve (fluted)	10 to 30	10 to 15
Single-seat globe valve, flow to open	2	3
Angle valve, flow to open	2	3
Eccentric rotary plug valve, flow to open	2	3
Double-seated globe valve (parabolic)	4	6

Cage valves, flow to open, n_o = number of windows constantly open throughout travel.

Table 4.14aa
TYPICAL PIPE TRANSMISSION LOSSES (∂BA)

Pipe Size (in.)	(mm)	Pipe Schedule (40)	(80)
1	25	71	75
1½	37.5	67	71
2	50	65	70
3	75	64	69
4	100	63	67
6	150	59	64
8	200	58	63
10	250	57	62
12	300	57	62
20	500	55	62

$P_1/P_2 = 1.5$
$N_M = -42$ (Figure 4.14w)
$P_L = 64 + 10 \log(n_O)$
$\quad = 64 + 10 \log(4)$
$\quad = 70$ dBA
$G = 0$ (Table 4.14y)

From equation 4.14(1):

$$SL = 145.5 - 42 + 10\log(105 \times 0.85 \times 150 \times \times 100) - 70 + 0$$
$$= 94.8 \text{ dBA}$$

Since this sound level exceeds the normal acceptable limit of 90 dBA, an alternate valve design can be considered. This might be a fluted disc butterfly valve (See Section 4.4), which is a low noise design with an F_L

similar to that above ($F_L = 0.8$). Since all flow conditions remain the same, we only have to consider the effect on the transmission loss (F_L) by the increase in the n_o factor from 4 in the cage valve to 16 in the fluted disc. Therefore:

$$P_L = 64 + 10 \log (16)$$
$$= 76 \text{ dBA (instead of 70)}$$

Then
$$SL = 145.5 - 42 + 10 \log (1.33 \times 10^6) -$$
$$- 76 + 0$$
$$= 88.8 \text{ dBA}$$

Thus we have gained an additional 6 dBA attenuation, which may be enough for the application.

Example 3: If the valve in Example 1 is located 100 feet (30 m) away from the normal operating location, what will the noise be at the operating location?

$$SL \text{ (at any distance)} = SL - 10 \log \left(\frac{\text{Distance in ft.}}{3} \right)$$

This equation is only valid for enclosed piping. If the noise source is an atmospheric vent, then substitute 20 log for the 10 log in the equation.

$$SL_{100} = 95 - 10 \log \left(\frac{100}{3} \right)$$
$$= 79.8 \text{ dBA}$$

BIBLIOGRAPHY

Arant, J.B., "Special Control Valves Reduce Noise and Vibration," *Chemical Engineering*, March 6, 1972.

——, "How to Cope with Control Valve Noise," *Instrument Technology*, March, 1973.

Baumann, H.D., "On the Prediction of Aerodynamically Created Sound Pressure Level of Control Valves," ASME Paper 70 WA/FE-28, December, 1970.

——, "How to Estimate Aerodynamic Valve Throttling Noise: A Fresh Look," ISA Paper 82-902, 1982.

Boyle, S.J., "Acoustical Design Considerations for Low-Noise Control Valves," *Proceedings of Noise-Con 83*, Institute of Noise Control Engineering, March, 1983.

Catalog 10 and Noise Data, Fisher Controls Co., Marshalltown, IA 50158.

Chow, G.C., and Reethof, G., "A Study of Valve Noise Generation Processes for Compressible Fluids," ASME Paper 80 WA/NC-15, 1980.

"Engineering Data 3: Control Valve Noise Prediction," Valtek, Springville, UT 84663.

Fagerlund, A.C., and Chow, D.C., "Sound Transmission Through a Cylindrical Pipe," *ASME Journal of Engineering for Industry*, vol. 103, no. 4, November, 1981.

Hutchison, J.W., ed., *ISA Handbook of Control Valves*, 2nd ed., Instrument Society of America, 1976.

Hynes, K.M., "The Development of a Low Noise Constant Area Throttling Device," ISA Paper 839-70, October, 1970.

Lighthill, M.J., "On Sound Generated Aerodynamically II, Turbulence as a Source of Sound," *Proceedings of The Royal Society of London*, vol. 222, series A, 1954.

Ng, Kam W., "Control Valve Aerodynamic Noise Generation and Prediction," *Proceedings of the National Noise and Vibration Control Conference*, INCE, Chicago, IL, 1980.

"Noise Control Manual," Bulletin 0Z3000E, 2nd ed., Masoneilan Div., McGraw-Edison Co., 1981.

"Noise Control Manual," Control Components, Inc.

Reed, C.M., "Optimizing Valve Jet Size and Spacing Reduces Valve Noise," *Control Engineering*, September, 1976.

Reethof, G., "Some Recent Developments in Valve Noise Prediction Methods," ASME Transactions of Winter Annual Meeting, November, 1982.

Saitta, G.N., "A Realistic Look at Reducing Control Valve Noise Problems," *Oil and Gas Journal*, June 18, 1979.

Sawley, R.J., and White, P.H., "The Influence of Pressure Recovery on the Development of Valve Noise Description," ISA Paper 74-834, October, 1974.

Seebold, J.G., "Reduce Noise in Process Piping," *Hydrocarbon Processing*, October, 1982.

Shea, Allan K., "A Comparative Study of Sound Level Prediction Methods for Control Valves," *Proceedings of Noise-Con 83*, Institute of Noise Control Engineering, March 1983.

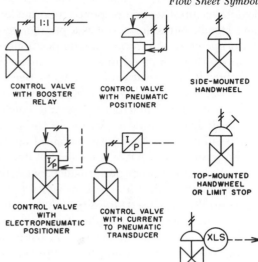

CONTROL VALVE
WITH BOOSTER
RELAY

CONTROL VALVE
WITH PNEUMATIC
POSITIONER

SIDE-MOUNTED
HANDWHEEL

CONTROL VALVE
WITH
ELECTROPNEUMATIC
POSITIONER

CONTROL VALVE
WITH CURRENT
TO PNEUMATIC
TRANSDUCER

TOP-MOUNTED
HANDWHEEL
OR LIMIT STOP

VALVE WITH
LIMIT SWITCH

FROM
SAFETY
INTERLOCKS

VALVE WITH
SOLENOID PILOT

4.15 VALVE POSITIONERS AND ACCESSORIES

Materials of Construction:	Bodies and cases: die cast zinc, "white metal," aluminum. Bellows: copper alloys. Diaphragms: elastomer-coated fabric, thin metal. Misc. parts: steel, brass, aluminum
Supply Pressure (Gauge):	Varies from 20 to 25 PSIG (140 to 170 kPa) for signal relays to 60 to 100 PSIG (400 to 700 kPa) for positioners or boosters, some to 150 PSIG (1000 kPa)
Inaccuracy:	Positioners: repeatable to ±.1 to 1% and accurate to ±.5 to 2% of span. Transducers: ±.5 to 1% of span. Boosters: ±.1 to 1% of span
Cost:	Pneumatic positioner: $200 to $300. Electropneumatic positioner: $300 to $650. Relays: $30 to $300. I/P transducer: $150 to $300
Signal Ranges (Gauge Pressure):	Pneumatic: 3–9, 3–15 (preferred), 6–30, and 9–15 PSIG (20–60, 20–100, 40–200, and 60–100 kPa). Electronic: 1–5, 4–20 (preferred), and 10–50 mA DC. Others are used such as 0–10 and 1–10 V DC
Partial List of Suppliers:	Fairchild Industrial Products (boosters, digital E/P, signal relays); Fisher Controls (positioners, signal relays, dead band booster, lock-up relay); Foxboro Co. (positioners, valve stem position transmitter, signal relays, I/P transducer); Honeywell (positioners, transducers); ITT-Conoflow (positioners, boosters, air set regulators; Masoneilan Div., Dresser Industries (positioners, transducers, lock-up relay); Moore Products Co. (positioners, boosters, transducers, signal relays); Target Rock Corp. Sub. of Curtiss-Wright Corp. (electric solenoid motor actuator/positioner); Taylor Instruments Co., Div. of Sybron Corp. (positioners); Valtek (positioners); WABCO (boosters, quick release valves).

Valve Positioners

The valve positioner is a servo-amplifier used with the valve actuator to assure that the control valve stem accurately takes the position that the input signal commands. Without a positioner, the stem position is affected by varying fluid pressures on the plug and by unpredictable friction forces. Positioners are not normally required for on-off service. Most positioners use air both as the operating fluid and as the source of power.

Hydraulic actuator/positioners can be used where large valves and high differential pressures require actuator pressure above the normally available 60–100 PSIG (0.4–0.7 MPa) instrument air supply and the largest air actuator is inadequate. Hydraulic actuator-positioners are considerably more expensive, may be noisy, and require specialized maintenance.

Electric motor operators are also available. They are used where air is not available and where their slow operating speed is acceptable. The electro-pneumatic positioner input section responds to a standard milliampere signal, but otherwise it is a pneumatic positioner. Performance of electro-pneumatic positioners has been disappointing compared to good pneumatic positioners, and they have a relatively high cost. Because of this, many users continue to use the pneumatic positioner with a separate I/P (current to pressure converter), usually located near the valve. An all-electric solenoid motor type positioner/actuator valve is available from one vendor.

The positioner provides a substantial improvement in valve and control loop performance, with the greatest improvement realized in the slow control loops and the low controller gain used typically for level or temperature control. Improvement is less for fast loops, such as flow or fast pressure control. Positioners have typical open loop gain (change in output pressure per change in input signal, with the valve stem locked in position) of 10:1 to 200:1. Dead band (the minimum input change for a detectable output change) is claimed as .1 to .5 percent of span. Vendors claim positioning accuracy of .002 to .005 in. (0.05 to 0.13 mm), under bench test conditions.

The problem in designing a positioner is to have high enough gain for high accuracy but to avoid overshoot on a large signal change and oscillations caused when gain is too high. The worst problems are caused by sticky valve packing or sloppy positioner linkage. Independent tests indicate that vendor claims for positioner performance are most often optimistic. Catalogs rarely list valve actuator/positioner frequency response data, and a special request for this information is usually required. If fast valve response (less than 10 to 20 seconds full stroke or frequency response of faster than 1 Hz) is required because of process time constants, the response requirement must be included in the purchase specification. The information required includes both the frequency response plot and the time for full stem stroke in each direction. The frequency response shows response to small signal changes. Full stroke time is set by the supply and venting capacity of the system. The error detecting mechanism and the amplifier in a modern positioner saturates (reaches its amplifying limit) at a relatively small error. This is why the response to small changes differs from the response to large changes.

Positioners must have an air supply of $\frac{3}{8}$ in. (9.4 mm) tubing as a minimum, and each one should have its own supply filter-regulator. Control valves use substantial amounts of air, and severe control interaction can occur if a changing signal to a control valve causes a common air supply pressure source to decrease, which can result in other devices changing their signal or response.

Stability

A few experts advise against using positioners in very fast control loops and recommend booster relays instead. Situations where this is called for are rare and could exist only for a large valve and slow-acting positioner or perhaps an inadequate air supply. As a general rule, valve positioners are essential for good control loop performance. They can safely be omitted only for the smallest control valves (less than 1″ or 25 mm) or for valves 2″ (50 mm) and smaller where much of the loop gain is not at the valve.

Stability problems with a positioner are caused by:

> Loose linkage
> Flexible mounting
> Inadequate supply capacity
> Inadequate positioner flow capacity
> Incorrect calibration
> Leaky air lines or fittings
> Positioner/actuator time constant equal to process time constant

In all the above, the solution is straightforward. It is usually possible to observe the positioner gauges and the valve action to find out where the problem lies. As in any control situation, instability is caused when phase (time) lag causes the control loop to react too late. A valve actuator is like a capacitance, that is, the valve does not change position when airflow starts to enter but moves only when enough air accumulates to build up pressure and cause motion. A positioner has a small input capacitance, but because it causes a large flow amplification, it reduces the effective actuator capacitance. If fast action is required for large actuators, add volume boosters between the positioner and the actuator to provide the required flow capacity. For the last problem in the above list, control becomes difficult and it will be necessary to make a change. Depending on the quality of control required and the importance of the loop, two choices are open. Either the positioner/actuator response can be speeded up by adding a booster or the positioner can be removed and a booster substituted. The booster must have considerable capacity or the performance will not be improved.

Typically, spring and diaphragm actuators are the slowest type, a piston actuator with return spring is usually faster, a piston with fixed air pressure for return is faster yet, and the double-acting piston, where the positioner supplies and exhausts air from both sides of the piston, is the fastest. Higher air supply pressure, up to the safe actuator operating pressure, improves response. An undersized or marginal actuator will exhibit slow response because it will require a greater change in pressure to cause motion, and as actuator pressure approaches supply pressure, the flow rate will decrease.

Delaying or Slowing Valve Action

There are a number of situations where it may be desirable to slow down valve action. For example, excessive pressure surge can be caused when fluid flow in a very long pipeline is stopped quickly; this is called water hammer. It could also be desirable to start or stop some process flow slowly for control or safety reasons. A very noisy process signal can lead to excessive valve wear. The "inverse derivative" relay, which passes a fraction of a sudden input change immediately and the balance of the change later, has been used to provide a low pass filtering effect. A needle valve, or restrictor, and a volume chamber will provide a resistor/capacitor type filter. Probably the best scheme is to use a reset-only controller to get a predictable, linear delay. Even if the service is on-off, an equal percentage type trim will help some in smoothing out the change in flow.

Split Range

At one time, it was common to use positioners to accomplish "split ranging." That is, with two valves controlled by one controller, each positioner moved its valve for only a part of the controller output range. One typical application is for temperature control with the cooling valve full open at 0 percent controller signal and closed at 50 percent signal, and the heating valve beginning to open at 50 percent controller signal and fully open at 100 percent. Current preferred practice is to use a fixed-gain-plus-adjustable-bias relay to convert the controller output signal. With this approach, maintenance is simplified because standard calibrations are used and full positioner accuracy is retained. See Figure 4.15a and Tables 4.15b and 4.15c for examples of this. Table 4.15c

shows the overlap of valve signals needed to reduce the discontinuity at the valve crossover value.

A number of different control design requirements can be accomplished with positioners. A reverse-acting positioner makes an air-to-open, spring-to-close valve into an air-to-close, spring-to-close valve. Combinations like this are used for interlock systems or for some batch operations.

The force balance positioner (Figure 4.15d) has an element that compares the force generated by the input signal with the force generated by the feedback spring connected to the valve stem. Figure 4.15e shows the electro-pneumatic force balance positioner. The position balance positioner (Figure 4.15f) compares the motion of an input bellows or diaphragm with linkage attached to the valve stem. Either can be very accurate. Bellows type input elements are generally thought to be more accurate than simple diaphragms, and although more likely to fail in fatigue, both types are used successfully.

Options

A number of options are available for most positioner designs such as gages showing supply pressure, signal

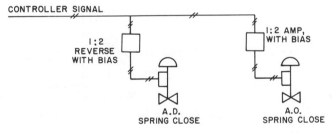

Fig. 4.15a   Split ranging two control valves with signal relays

Table 4.15b
SIGNALS IN %

Controller Output	Output Relay 1	Output Relay 2	Control Valve 1	Control Valve 2
0	100	0	100	0
25	50	0	50	0
50	0	0	0	0
75	0	50	0	50
100	0	100	0	100

Table 4.15c
SIGNALS IN PSIG (bar)

Output Controller	Output Relay 1	Output Relay 2	Control Valve 1	Control Valve 2
3 (0.2)	15 (1.0)	3 (0.2)	100%	0%
9 (0.6)	3.2 (0.22)	3.2 (0.22)	10	10
15 (1.0)	3 (0.2)	15 (1.0)	0	100

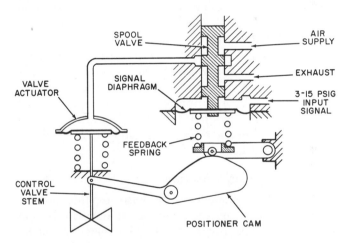

Fig. 4.15d   Force balance positioner

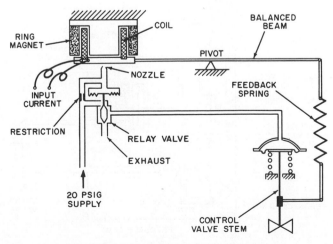

Fig. 4.15e   Electro-pneumatic force balance positioner

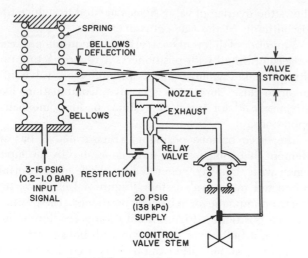

Fig. 4.15f Motion balance positioner

pressure, and output pressure. These cost little but can be a great help during checkout or for maintenance. Bypass air switches to send the controller output directly to the actuator were once popular when 15 PSIG (104 kPa) was enough pressure for the spring-diaphragm actuator and the positioner was direct acting (increase input-increase output). The bypass is rarely specified now because this feature permits only limited maintenance on the positioner, and most modern high performance positioners do not offer this option. In some corporations the only control valves now purchased without positioners are small (1″ or 25 mm and less), low pressure differential (0–20 PSI or 0–138 kPa), less critical service (comfort heating and ventilation), and where absolutely minimum cost is required.

Digital/Pneumatic Positioners

The digital positioner uses a pulse stepping motor to rotate a shaft for moving a follower to develop an input signal force. See Figure 4.15g. The remainder of the positioner is pneumatic. Application would be for ddc

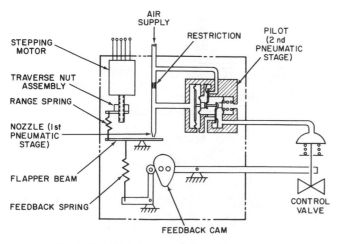

Fig. 4.15g Digital-to-pneumatic valve positioner

(direct digital control) systems. Because the motor will not turn without signals, it functions as a memory device as well as a digital-analog converter. Stroke speed may be limited by stepping motor response.

Transducers

Electro-pneumatic Transducer

The electro-pneumatic transducer (I/P) is used to convert electrical signals (usually 4–20 mA DC) into a pneumatic signal, usually 3–15 PSIG (0.2–1.0 bar). The most common application is between an electronic controller output and pneumatic control valves. Is is also used between a digital computer and a control valve. A few designs use signal feedback to improve accuracy (Figure 4.15h). Most designs are of the "motion balance" type (Figure 4.15i), where the small force developed by the milliampere current through a coil in a magnetic field causes motion of a nozzle-baffle assembly, resulting in a changing pneumatic pressure. The nozzle-baffle system is an amazingly accurate mechanism for measuring small motion.

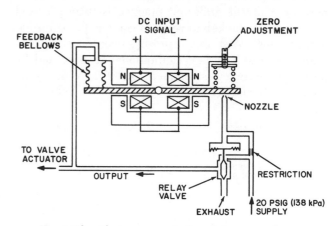

Fig. 4.15h Electro-pneumatic force balance transducer

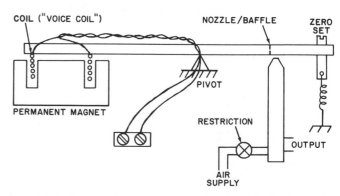

Fig. 4.15i Coil is of similar construction to voice coil of loudspeaker. Leads are arranged to go through pivot point to reduce effect on output

When an I/P is used within a control loop between the controller and the control valve, its error is combined with the valve error which is detected by the loop controller and driven towards zero error. Repeatability and reasonable linearity are required, and most I/Ps have advertised accuracies of .35 to 1.0 percent of full scale. Most I/Ps have relatively low air capacity, and a booster relay is needed to drive a pneumatic actuator unless a positioner is used.

Digital Electric to Pneumatic Transducers

A variety of devices are used to convert digital signals in addition to the electronic digital to analog converter. One device uses a stepper motor, as in the digital positioner mentioned above. A second device has a number of wires connected to it and receives data on all these wires simultaneously. This "parallel" data signal is converted to an analog current and this controls the output pressure. Another device responds to a string of pulses ("serial" data) to set the output pressure.

Booster Relay

A "booster" relay (Figure 4.15j) is a device that amplifies pneumatic signals in volume (capacity) or in pressure, or both. Most booster designs were derived from a pressure regulator with the input signal providing the loading force in place of the usual spring. Internal pressure feedback provides accuracy. The downstream-facing pitot tube compensates for flow-related pressure drop. For pressure amplification, the feedback diaphragm has a smaller area than the input diaphragm so that greater pressure is required to achieve force balance. Volume boosters with a pressure ratio of 1:1 (gain of 1) are sometimes used to speed up a valve actuator response if a positioner is not used. The booster cannot overcome inaccuracies due to friction or forces on the valve plug, but it will reduce their effects because of faster response.

With large control valves and big actuators, boosters are used between the positioner and the actuator when fast response speed is required. One special "dead band booster" (Figure 4.15k) does not respond until a 1 PSI (6.9 kPa) difference between input and output is exceeded. A built-in needle valve allows a limited air flow to bypass the booster gain portion and provides adjustable damping. The two springs provide the 1 PSI dead band with this relay. A very large volume amplification occurs for fast input signal change, but it has no gain with a slow change. One common use is with centrifugal compressor anti-surge controls where the control valve must open very quickly (1–3 seconds), but then is required to throttle smoothly. The needle valve is adjusted with the complete control system assembled and operating. Careful tuning is vital to proper operation. With the booster needle valve fully closed, the positioner-valve system may be unstable. Best operation occurs when the needle valve is open only enough to dampen the oscillations.

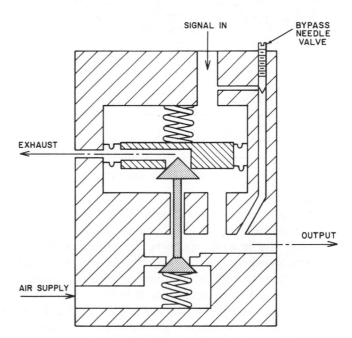

Fig. 4.15k   Dead band booster: The springs provide the 1 PSIG (6.9 kPa) dead band

Note always that no air device will operate properly without an adequate air supply. With their large air handling capacities, valve booster relays require $\frac{1}{2}''$ to $\frac{3}{4}''$ (12.5 to 18.75 mm) air supply and output tubing to the valve operator. Large filters are used to minimize any air supply restriction.

Other Relays

Besides the booster function, pneumatic relays are used in other control valve functions. The fixed-gain-plus-adjustable-bias relay for split ranging was mentioned above. Another example is the 1:2 reversing relay

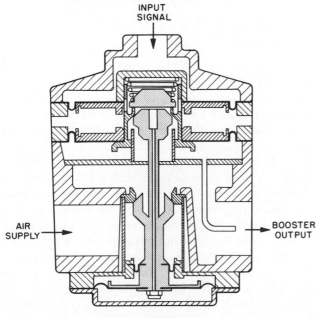

Fig. 4.15j   Volume booster relay

(Figure 4.15l) where the gain is two (0 percent input results in 100 percent output, 50 percent input results in 0 percent output), will reverse the operation of a valve, provide the gain and bias needed for split ranging, and retain the original action on loss of supply air. The relay is a force balance device. The output pressure generates a force on a diaphragm with one-half the area of the input diaphragm so that the output pressure must be twice the input pressure.

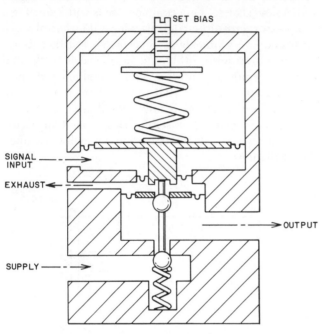

Fig. 4.15l   2:1 reversing relay

The quick release valve (Figure 4.15m) or dump valve will open a large-capacity vent when signal pressure goes to zero. This provides much faster venting of air from an actuator where this venting capacity is needed, such as in interlocking. It is not unusual for a positioner to bleed excess air for some seconds before a valve starts to move, because the actuator is completely filled with supply pressure air even if the valve does not require that pressure.

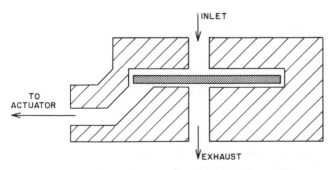

Fig. 4.15m   Quick exhaust valve: Elastomeric flapper allows air to flow to actuator while sealing exhaust, then allows actuator to exhaust when inlet pressure is low

A lock-up relay or "fail in last position" relay (Figure 4.15n) is available to seal in the existing actuator pressures if the air supply is lost, and is used to hold the valve in the last controlled position. If the valve is partly open at the time of air failure, this "frozen" position is not absolutely predictable. This is because the servo-action is lost, variable plug forces may move the stem, and air may leak out in an unpredictable manner.

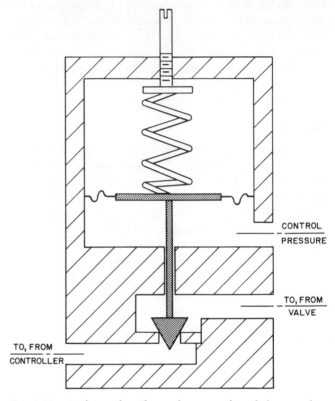

Fig. 4.15n   Lock-up relay: If control pressure drops below set, then relay valve stops flow to or from valve

Handwheels

Handwheels are available as an accessory to provide for partial or complete control of the valve to override the pneumatic actuator. Some, mounted on top of the actuator (Figure 4.15o) can only push on the valve stem to close (or open with inverted trim). Another type, side mounted, has an engagement clutch to allow the handwheel to fully stroke the valve open and closed (Figure 4.15p). A third approach has sufficient lost motion or clearance or play in the linkage that, with the handwheel in a central, neutral position, the valve stem can move over its full range. Turning the wheel one way limits the amount open, the other limits the amount closed. Where manual throttling control is intended, it is necessary to consider how the human operator will know how to set the valve and if the process can be safely controlled manually. Often this is not the case, so handwheels have limited application in modern continuous process plants.

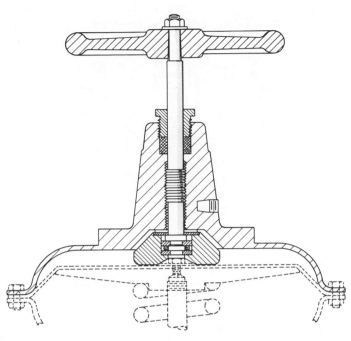

Fig. 4.15o Top-mounted handwheel for spring-opposed pneumatic actuator

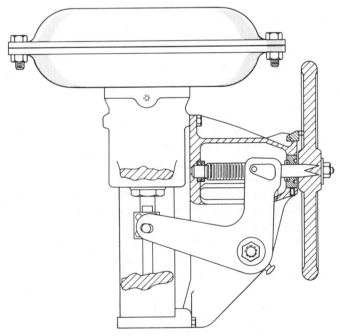

Fig. 4.15p Side-mounted handwheel for spring-opposed pneumatic actuator

Limit Switches

Switches are installed on electric motor driven valves to open the circuit and stop driving the motor when the valve is at its limit (fully open or closed) or on motor over-torque. The name "limit switch" is also used to describe switches installed to signal when a valve is at or beyond a predetermined position. These switches are used for operator information, interlock inputs, or com-

puter feedback. It is necessary to consider mounting problems, electrical classification of area, electrical characteristics of circuit, over-travel of actuating arm, and corrosive nature of the area. Usually, it is easiest to purchase the valve complete with any required switches installed. Because of environment problems, some users have been using sealed magnetically actuated or proximity switches. Note that it is difficult to adjust limit switches closer than ±5 to 10 percent and that dead bands of 2 to 5 percent are typical.

Solenoid Valves

The solenoid valve as a control valve accessory is used (1) to

operate an on/off pneumatic actuator or (2) to interrupt the action of a modulating valve by switching air or hydraulic pressures. It is common practice to use a solenoid valve as the pilot for a pneumatically operated on/off valve because of the wide choice of features and capabilities available in the pneumatic valve.

Solenoid valves are primarily used as a part of a start-up or shut-down system, interlock, or a batch system to cause the control valve to take some predetermined action under certain conditions. Two philosophies are in common use. In the first, the 3–15 PSI (0.2–1.0 bar) signal to the positioner is blocked and the downstream tubing either vented or connected to some other preset pressure (Figure 4.15q). This is reliable because the solenoid valve is lightly stressed, the positioner and valve have been in continuous use, and any failure or poor operation should have been detected during normal operation. With this scheme, the substituted signal can be any value over the operating range, the valve will go to the desired opening, and the advantages of the positioner are retained. If the actual positioner input is fed back to the controller external feedback connection, then a smooth return to normal control may be expected when the control loop is returned to normal operation.

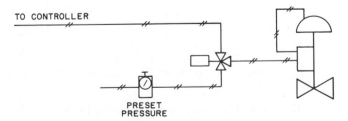

Fig. 4.15q Interlock solenoid valve in signal tubing

In the second philosophy, the solenoid valve is installed in the tubing between the positioner and the actuator, and the positioner is not used to operate the valve (Figure 4.15r). Solenoid valves with adequate pressure rating and flow capacity are required. Only three

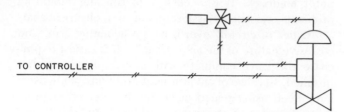

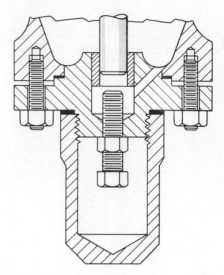

Fig. 4.15r   Interlock solenoid valve controlling actuator directly

control valve actions are possible: fully closed, fully open, or lock-up existing pressures.

For interlock safety applications, it is desirable for the solenoid valve to be continuously energized for normal operation so that deenergization, failure, loss of power, or a broken wire will cause shutdown. As a control valve accessory, usually a three-way (three connections) solenoid valve is required. Some designs require that pressure always be applied to one certain port and that another certain port always be the vent. This does not always suit the required logic, but valves can be found rated for "universal" operation. Also, note that clean, dry, oil-free instrument air provides no lubrication, and some types of solenoid valves (spool type) will have a short life without lubrication.

Air Accumulators

Some piston type actuators have no spring, and others have a spring too weak to guarantee valve action on loss of air supply. For these, an air tank (typically $\frac{1}{2}$ cubic foot or 0.014 m^3) with a check valve on the inlet may be supplied to allow limited operation, or orderly shutdown (Figure 4.15s). Because some leakage is likely, the air tank cannot be relied upon for any extended time and the process pressure must tend to hold the valve plug in the desired direction.

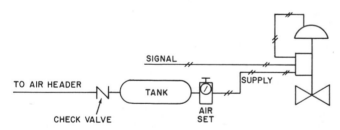

Fig. 4.15s   Air supply accumulator tank

In the case of very large valves with high capacity boosters, a large (10 cubic foot or 0.28 m^3) tank could be installed to provide local energy storage and to reduce peak demand effects on the air supply system.

Limit Stops

It is possible to get fixed limit stops to limit valve stem motion to either assure a minimum opening or to limit maximum opening (Figure 4.15t). These are usually purchased with the valve.

Fig. 4.15t   Externally adjustable limit stop in valve body subassembly

Stem Position Indicators

Nearly all valves have some sort of pointer and scale mounted on the valve or actuator to show approximate stem position. For remote indication, a valve stem position transmitter is available. Some are LVDTs, some use variable resistors, and others are valve positioners adapted to transmit a proportional pneumatic signal. It is possible to use these devices as a signal source in place of limit switches.

Air Set

The "air set" is the air regulator with filter and drip pot used to supply air to the positioner or other instrument. It is often purchased with the valve, mounted, and piped. The regulator must have an appropriate pressure range for the positioner and actuator, and it should have a built-in overpressure vent. Capacity is a problem only with very large valves. Some newer actuators and positioners will operate with up to 150 PSIG (1 MPa) supply, and there is a trend towards eliminating the regulator but retaining the filter.

BIBLIOGRAPHY

Baumann, H.D., "The Case for Butterfly Valves in Throttling Applications," *Instruments & Control Systems,* May, 1979.

Carey, J.A., "Control Valve Update," *Instruments & Control Systems,* January, 1981.

Fernbaugh, A., "Control Valves: A Decade of Change," *Instruments & Control Systems,* January, 1980.

Jury, F.D., "Positioners and Boosters," *Instruments & Control Systems,* October, 1977.

Lipták, B.G., "Control Valves in Optimized Systems," *Chemical Engineering,* September 5, 1983.

Lloyd, S.G., "Guidelines for the Use of Positioners and Boosters," *Instrumentation Technology,* December, 1968.

Schuder, C.B., "Fluid Forces in Control Valves," *Instrumentation Technology,* May, 1971.

Shinskey, F.G., "Dynamic Response of Valve Motors," *Instruments & Control Systems,* July, 1974.

4.16 VALVE SIZING

See also Sections A.3 and A.4.

Nomenclature

B	Flow per unit C_v
C_d	Relative valve capacity (C_v/d^2) (Table 4.16v)
C_f	Same as F_L, a manufacturers' term
C_v	Valve capacity parameter
C_{vn}	Valve C_v, fully open
C_{vr}	Required or calculated C_v
d	Valve nominal body size, inches
D	Fluid density, lbm/ft^3
D_1	Upstream pipe nominal size, inches
D_2	Downstream pipe nominal size, inches
F_d	Valve type coefficient
F_F	Liquid critical pressure ratio
F_k	Ratio of specific heats factor
F_L	Pressure recovery factor (Table 4.16v)
F_{LP}	Liquid combined factor
F_P	Piping geometry factor
F_R	Reynolds Number factor
F_Y	Liquid choked flow factor
g	Gravitational acceleration, 32.17 (lb-ft)/(lb-sec^2)
k	Ratio of specific heats, gases, C_p/C_v
K_{B1}	Inlet fitting Bernoulli effect coefficient
K_{B2}	Outlet fitting Bernoulli effect coefficient
K_i	Inlet combination factor
K_m	Same as F_L, a manufacturers' term
K_1	Inlet fitting friction coefficient
K_2	Outlet fitting friction coefficient
M	Mass, lbm
m	Molecular weight
P	Pressure, PSIA
P_1	Upstream pressure, PSIA
P_2	Downstream pressure, PSIA
P_c	Critical pressure, PSIA
P_{vc}	Pressure at vena contracta, PSIA
P_{vp}	Vapor pressure, flowing conditions, PSIA
P_r	Reduced pressure, actual pressure/reduced pressure (used with compressibility curves)
Q	Flow, gpm
R	Gas constant, 19.316 lb-ft^3/in^2-mol °K
R_c	Same as F_F, a manufacturers' term
R_1	Part of low Reynolds number flow calculation
Re_v	Valve equivalent Reynolds number

T	Temperature, °K = °C + 273.16
T_c	Critical temperature, °K
V	Volume, ft^3
W	Flow, lbm/h
x	Liquid portion of a steam-water mixture
X	Pressure drop ratio (P_1-P_2)/P_1
X_T	Gas factor for valve only (Table 4.16v)
X_m	X minimum, the smaller of (F_k)(X_{TP}) or (P_1-P_2)/P_1
X_{TP}	Gas combined factor
Y	Expansion factor
Z	Compressibility (ratio of actual density to that predicted by perfect gas lower)
ΔP	Pressure difference, PSI
Δpt	Pressure difference at start of choking
μ	Viscosity, centipoise

The following pages provide a method for calculating the flow through a control valve under specified conditions. This is not the equivalent to sizing the valve. When sizing a valve, other factors must be considered: Uncertainty of control valve parameters (typically ±5 percent), uncertainty of pump curve data (±10 percent), and uncertainties of other given data. For most control applications, a certain amount of excess or "catch up" capacity is needed. However, an oversized valve may cause control problems. This is not only because of the greater flow change per unit input signal, or because of excessive flow at small openings, but because the effects of any imperfections such as loose linkage, sticking, and so on will be magnified. Note that an oversize but otherwise perfect valve might well control any flow within its range; however, no valve is perfect. The various required C_v (C_{vr}) should be calculated for the full range of operating conditions before a valve body and trim are selected. If such calculations show that valve stem positions are predicted to be beyond the 5 to 95 percent range, then special wide-range arrangements such as dual valves may be required.

The study of control valve sizing is the study of complex fluid mechanics. The information in this section reflects our current knowledge. It has been developed from the ISA (Instrument Society of America) standards SP75.01 (1977), "Control Valve Sizing Equations" and

SP75.02 (1981) "Control Valve Capacity Test Procedure."[1,2] The user is cautioned that the calculations are not perfectly accurate for all conditions. Accuracy is best for fluids similar to air, water, and steam, for the simpler conventional valve designs, and for straightforward piping configurations. For fluids that behave much differently, for very large or for very small valves, and for extreme Reynolds numbers (a measure of turbulence), the calculations are less dependable. In all cases, reliable calculations require reliable information about the fluid, the flow conditions, and certain valve characteristics.

To keep the following discussion as clear as possible, one set of units will be used, as indicated under "Nomenclature" at the beginning of this section. Once the concepts are clear, other units can be substituted. Many instrument practitioners standardize on one set of units for their calculations for simplicity and to reduce errors. The Process Measurement volume of this *Handbook* has an extensive list of conversion factors.

The notation used throughout this section is almost identical to that used in the ISA Standards. In addition, the following symbols are used to help clarify certain concepts: C_{vn}, C_{vr}, X_m, C_d, B, and R1.

General Discussion

The earliest control valves were mechanized manual globe valves. Sizing was easy: a 4″ (100 mm) valve "belonged" in a 4″ (100 mm) line. It became obvious that, depending on pressure drop, this valve sometimes had too much capacity for good control. The next typical practice was to install a valve one size smaller than line size.

The basic valve capacity parameter C_v is defined as:

$$C_V = Q/(\Delta P)^{\frac{1}{2}} \qquad 4.16(1)$$

This parameter is still the standard one despite the units, which are often inconvenient. However, a serious problem limits the applicability of this equation since under certain conditions actual flow is less than predicted flow. The reasons for this are described below.

At temperatures where the vapor pressure of a liquid is less than the downstream pressure but higher than the lowest pressure existing at some point inside the valve (see Figure 4.16a), a part of the liquid is converted into vapor bubbles. Downstream where pressure is recovered, these bubbles collapse very suddenly and each one results in a tiny high velocity liquid jet. The velocity is high enough to cut even hardened metal. This process, called cavitation, usually generates a noise like the sound of rocks flowing through the valve. No known material will long survive cavitation, especially if water is the liquid. Special valve trim designs can be used where cavitation cannot be avoided. The amount of damage is proportional to the amount of energy involved.

If the vapor pressure exceeds the downstream pressure, the bubbles do not collapse but continue to grow,

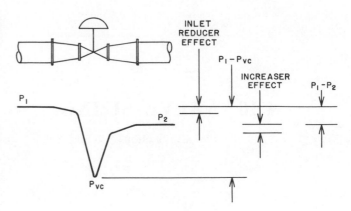

Fig. 4.16a   Pressure profile through a valve

flashing into vapor. With larger bubbles, and therefore lower average density, fluid velocity increases and the system flow capacity is limited or "choked." Very high velocity can cause excessive wear. Because the vaporization process requires heat energy and this comes from the liquid, the liquid temperature will decrease and the process may be self-limiting. The amount of flashing is calculated from an energy balance (see example 9). If the amount of liquid flashed exceeds 3 percent (by weight), this fact should be considered in valve sizing and selection and noted on the specification tab sheet. If flash exceeds 15 percent, a special valve design should be used, usually with an oversized outlet, a replaceable throat, or special trim design.

For compressible fluids, the density will change as pressure decreases through the valve. At a certain critical pressure drop, capacity becomes limited (choked) as the maximum velocity in the valve becomes equal to the speed of sound in the fluid. Sonic velocity is also associated with a sudden increase in valve noise.

Flow through the complex path of a typical control valve cannot be uniform, and the flow-limiting phenomenon must occur in one part of this path earlier than the others. That is, the change between not choked and choked is not abrupt but is gradual as flow conditions change. The working equations do not account for this, and caution is recommended in predicting capacity at near-choked conditions.

A number of new valve styles were introduced in the early 1960s. These have a higher C_v for their body size than the traditional globe style, and a lower cost. In many applications these new valves, usually rotary style, work well. In other applications flow capacity is well below predicted, and/or cavitation will occur. These valves developed a reputation for being noisy and unpredictable.

Relative valve capacity is expressed as:

$$C_d = C_v/d^2 \qquad 4.16(2)$$

(See Table 4.16v.) These new valves have a higher C_d and a higher maximum fluid velocity. Thus they have a lower minimum internal pressure, and this lower pressure causes problems. P_{vc} (pressure at the vena contracta) is the term used for this minimum pressure.

Valve Parameters

A new parameter F_L, pressure recovery factor, is defined as:

$$F_L = [(P_1-P_2)/(P_1-P_{vc})]^{\frac{1}{2}} \qquad 4.16(3)$$

Experimental measurement of P_{vc} is almost impossible. The term F_F, the liquid critical pressure ratio, is defined (see Figure 4.16b):

$$F_F = 0.96 - 0.28 (P_{vp}/P_c)^{\frac{1}{2}} \qquad 4.16(4)$$

If

$$(.7)(P_1) < P_{vp}, \text{ then } F_F = 0.96 \qquad 4.16(5)$$

The limiting value, that is the lowest P_{vc} value before choking takes place, is:

$$P_{vc} = (P_{vp})(F_F) \qquad 4.16(6)$$

Values for F_L and for P_{vc} are characteristic of a valve design and are evaluated by manufacturers flow testing. The pressure differential where choking begins (Δpt) is measured and used to define the liquid pressure recovery factor F_L:

$$F_L = \Delta pt/(P_1 - P_{vc}) \qquad 4.16(7)$$

These tests are run, per the standard (ISA, SP75.02), with cold water. For F_L values of various valve designs, see Table 4.14x.

For compressible fluids the parameter X_T is used to predict choking. The ratio of pressure drop to absolute upstream pressure is called X.

$$X = (P_1 - P_2)/P_1 \qquad 4.16(8)$$

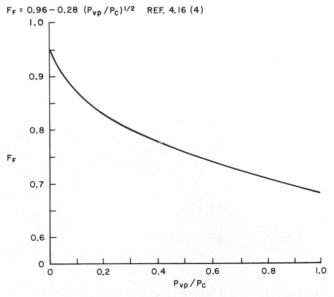

$F_F = 0.96 - 0.28 (P_{vp}/P_c)^{1/2}$ REF. 4.16 (4)

Fig. 4.16b Liquid critical pressure ratio factor as a function of the P_{vp}/P_c ratio

During the valve flow test, if P_1 is held constant and P_2 reduced, flow will increase with the pressure drop until sonic velocity is developed at some point in the valve. The value of X at this point is called X_T, the terminal or ultimate value of X. X_T is also called the pressure drop ratio factor and is a characteristic of the valve design.

Note that both F_L and X_T will vary as the valve stem position changes. This relation may be very complex. Some vendors' catalogs show curves or have tables of X_T and F_C. For globe-style valves, X_T is approximately 80 percent of $(F_L)^2$.

Reducer Effects

By convention, valve tests and calculations have included that portion of piping adjacent to the valve. P_1 is measured upstream of the pipe reducer, if used, and P_2 is downstream of the increaser. Some vendors supply tables of C_v for valves with standard pipe reducers. Where this data is based on actual test data it should be used in preference to the approximations shown below. Elbows and block valves installed near the valve will upset the flow profiles and probably cause some reduction in capacity. We have no procedure to predict these losses. It is suggested that if block and bypass valves are installed, that the path through the control valve be the simplest and straightest to reduce losses.

The standard reducer and increaser fittings have an angle of taper high enough to be classified as an abrupt change in size. For any change in cross section flow area in a filled pipe the following two things occur. First, there is an irreversible energy loss due to turbulence called friction loss. The second is the conversion between pressure energy and velocity energy. For the common situation of a control valve smaller than line size, the pressure at the valve inlet is reduced by the friction loss and is also reduced because of the smaller flow area and higher velocity.

If the downstream piping is the same size as the inlet piping, then the velocity-energy/pressure-energy exchange is reversed and it cancels out. Friction losses always add together. A number of references[3,4] show estimating methods for friction loss. For a reducer on the valve inlet, the ISA standard uses:

$$K_1 = 0.5(1 - (d/D_1)^2)^2 \qquad 4.16(9)$$

For an increaser on the valve outlet the ISA standard uses:

$$K_2 = (1 - (d/D_2)^2)^2 \qquad 4.16(10)$$

Although not explicity allowed by ISA SP75.01, an analysis of the references shows that for the unusual increaser at the valve inlet

$$K_1 = (1 - (D_1/d)^2)^2 \qquad 4.16(11)$$

and for a reducer on the outlet:

$$K_2 = 0.5(1 - (D_2/d)^2)^2 \qquad 4.16(12)$$

For the velocity-pressure exchange (Bernoulli effect) for the usual case:

$$\text{Reducer on inlet } K_{B1} = 1 - (d/D_1)^4 \qquad 4.16(13)$$

$$\text{Increaser on outlet } K_{B2} = (d/D_2)^4 - 1 \qquad 4.16(14)$$

It can also be shown from the same basic fluid mechanics for the unusual case:

$$\text{Increaser on inlet } K_{B1} = (D_1/d)^4 - 1 \qquad 4.16(15)$$

$$\text{Reducer on outlet } K_{B2} = 1 - (D_2/d)^4 \qquad 4.16(16)$$

See the curves in Figure 4.16(c).

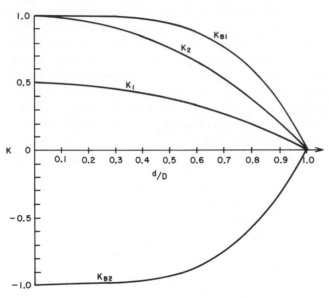

Fig. 4.16c   Reducer and increaser coefficient values as a function of the d/D ratio

Certain combinations of these coefficients are used in calculations:

$$K_i = K_1 + K_{B1} \text{ (inlet combination)} \qquad 4.16(17)$$

$$\Sigma K = K_i + K_2 + K_{B2} \qquad 4.16(18)$$

A piping geometry factor is defined as:

$$F_P = [1 + (\Sigma K)(C_d{}^2)/890]^{-\frac{1}{2}} \qquad 4.16(19)$$

Valve flow capacity is directly proportional to this factor. See Figure 4.16d. Note that capacity is reduced for high values of ΣK combined with high C_d. Because the inlet fittings affect the valve inlet pressure, a combination of F_L and F_P is used in the process of predicting choking or cavitation (see Figure 4.16e):

$$F_{LP} = [1/F_L{}^2 + (K_i)(C_d{}^2)/890]^{-\frac{1}{2}} \qquad 4.16(20)$$

Again note the large impact of reducers on valves with a high C_d. This equation is an estimate and values determined by test are preferred.

For liquids, the effect of F_{LP} on flow capacity is expressed as the liquid choked flow factor (see Figure 4.16f):

$$F_Y = (F_{LP}/F_P)((P_1 - P_{vc})/(P_1 - P_2))^{\frac{1}{2}} \qquad 4.16(21)$$

By definition, F_Y maximum is 1.

Choking starts at a pressure drop of

$$(P_1 - P_{vc})(F_{LP}/F_P)^2 = \Delta P \text{ critical} \qquad 4.16(22)$$

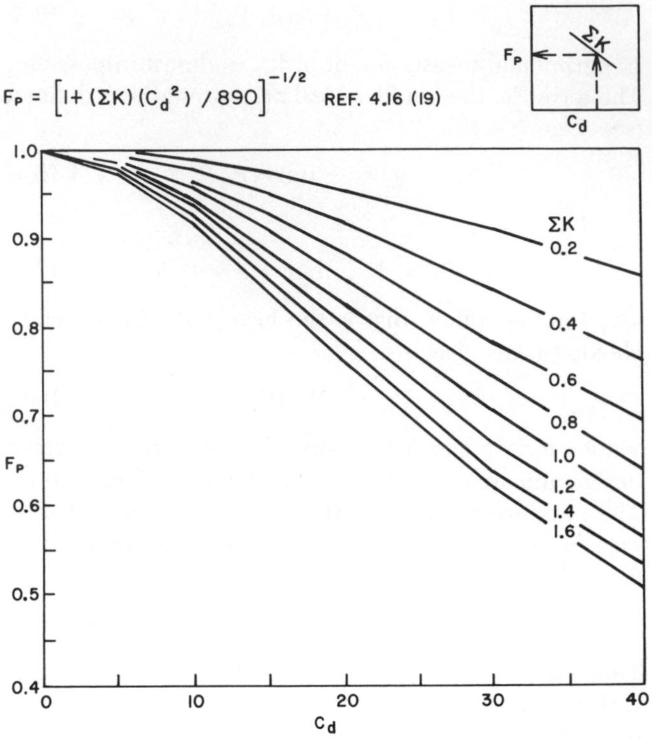

$$F_P = \left[1 + (\Sigma K)(C_d{}^2)/890\right]^{-1/2} \qquad \text{REF. 4.16 (19)}$$

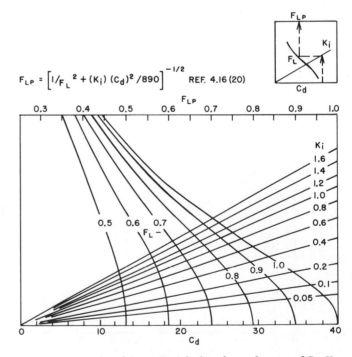

Fig. 4.16d   Piping geometry factor as a function of C_d and ΣK

$$F_{LP} = \left[1/F_L{}^2 + (K_i)(C_d)^2/890\right]^{-1/2} \qquad \text{REF. 4.16 (20)}$$

Fig. 4.16e   Combined factor (F_{LP}) for liquids as a function of C_d, K_i, and F_L.

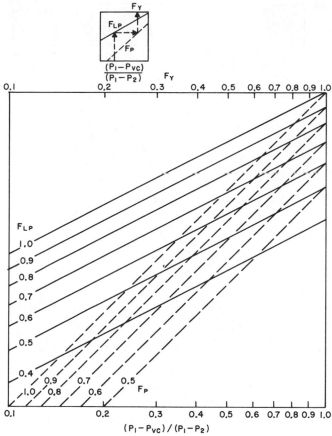

$$F_Y = (F_{LP}/F_P) [(P_1 - P_{VC})/(P_1 - P_2)]^{1/2} \quad \text{REF. 4.16 (21)}$$

Fig. 4.16f   Graph for determining the liquid choked flow factor

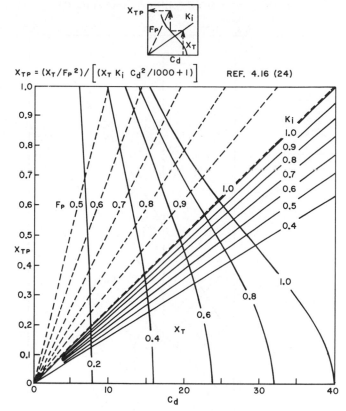

$$X_{TP} = (X_T/F_P^2)/[(X_T K_i C_d^2/1000 + 1)] \quad \text{REF. 4.16 (24)}$$

Fig. 4.16g   Graph for determining the gas combined factor

If $P_2 \leq P_{vp}$, then the liquid is flashing. If choked and not flashing, then it is cavitating.

From the above, flow capacity is defined as:

$$W = 63.3 \, F_P C_v F_Y F_R [D(P_1 - P_2)]^{\frac{1}{2}} \quad 4.16(23)$$

Compressible Fluids

For compressible fluids where the valve is not line sized, the effective X_T becomes the combined factor X_{TP} (see Figure 4.16g):

$$X_{TP} = (X_T/F_P^2)/[(X_T)(K_i)(C_d^2/1000 + 1)] \quad 4.16(24)$$

To correct for the difference in thermodynamic properties of different compressibles, the ratio of specific heats factor F_k is used:

$$F_k = k/1.4 \quad 4.16(25)$$

F_k affects capacity and is used in the calculations to determine if choking exists. For very high operating pressures (above 1000 PSIG) k may not express the thermodynamic gas properties accurately and the polytropic expansion factor n should be substituted. See Reference 5.

Gases increase in volume per unit weight as pressure decreases. This increases velocity and therefore losses.

To account for this, the expansion factor Y is defined:

$$Y = 1 - X_m/(3)(F_k)(X_{TP}) \quad 4.16(26)$$

where X_m is the smaller of F_k, X_{TP}, or X.

The minimum value of Y is $\frac{2}{3}$. See Figure 4.16h.

$$\text{If } X > (F_k)(X_{TP}), \text{ then flow is choked.} \quad 4.16(27)$$

Flow capacity is

$$W = 63.3 \, C_v \, F_P \, Y \, (P_1 X_m D)^{\frac{1}{2}} \quad 4.16(28)$$

Low Flow

High flow rates and high Reynolds numbers are common in control valves. It is, however, quite possible to have low flow rates, small valves, low pressure differential, and high viscosity. With the low Reynolds number resulting, laminar flow and severely diminished capacity can occur. The valve equivalent Reynolds number, Re_v, is defined as:

$$Re_v = \frac{34.55 \, W F_d}{\mu \, (C_v)^{\frac{1}{4}}} \left(\frac{(F_L C_d)^2}{890} + 1 \right)^{\frac{1}{4}} \quad 4.16(29)$$

For single port globe or V-port ball, $F_d = 1$; for double port globe, ball, butterfly, and other valves, $F_d = 0.7$. The "bracketed" part of this equation (R1) ranges in value only between 1.00 and 1.3 for combinations of C_d be-

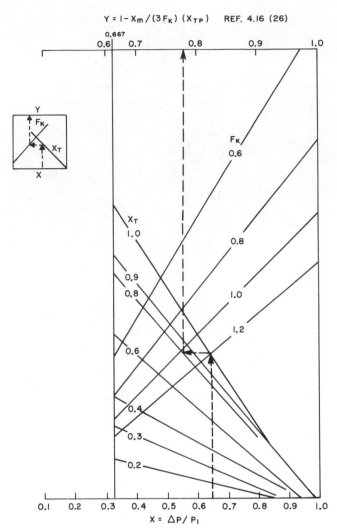

$Y = 1 - X_m / (3 F_K)(X_{TP})$ REF. 4.16 (26)

Fig. 4.16h   Graph for determining the expansion factor

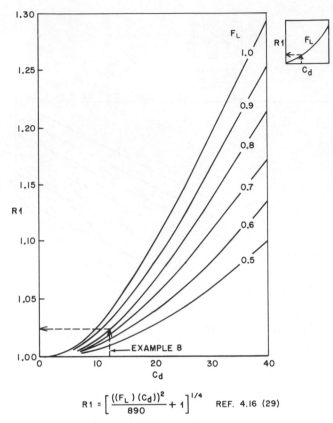

$$R1 = \left[\frac{((F_L)(C_d))^2}{890} + 1 \right]^{1/4} \quad \text{REF. 4.16 (29)}$$

Fig. 4.16i   Graph for determining the R1 factor

tween 5 and 40 and F_L between .4 and 1.0, and thus it seldom needs to be evaluated. See Figure 4.16i. For an approximate solution of the balance of the equation see Figure 4.16j. Comparatively little data on valves at low Reynolds numbers is available, and the results of these calculations should be used cautiously. Uncertainties increase as Reynolds number decreases. The relationship between Re_v (the top scale on the graph) and F_R is from *ISA Handbook of Control Valves*, 2nd ed., p. 188, Figure 3.

This analysis assumes Newtonion fluids that have constant viscosity despite fluid shear rate. Non-Newtonion fluids include most polymers and many other fluids. If no experimental flow data exists, the prudent engineer may do well to specify a line size valve and purchase extra sets of plug and seat trim with reduced C_{vn}, with the final choice of trim to be determined by test after start up.

Two-Phase Flow

Two-phase flow, where both liquid and compressible fluids exist at the valve inlet, creates valve sizing problems. Two general types of such flow can be identified. The first type consists of a liquid plus a compressible that is *not* the vapor phase of the liquid involved. In this case, it is reasonably safe to assume that the required C_v is the sum of C_v required for each phase. Note that it is quite possible for either or both fluids to be in the choked condition. In the second type, the vapor is the vapor of the liquid involved, e.g., steam and water or liquid and vapor methanol. The problem here is the severe interaction between the phases as pressures, temperatures, and the fraction that is vapor change in flowing through the valve and piping. The relative volumes of the phases are very sensitive to pressure and temperature variations. With this situation, it is difficult to accurately determine the required valve size. Some information exists that indicates that a small amount of vapor actually increases flow through a valve. See Fisher Catalog 10, ISA *Control Valve Handbook*, 2nd ed., p. 191, and the flow section in *Chemical Engineers Handbook*.[6,7,4]

Working Equations—U.S. Units

The complex nature of the ISA equations discourages manual computations. The equations listed below have

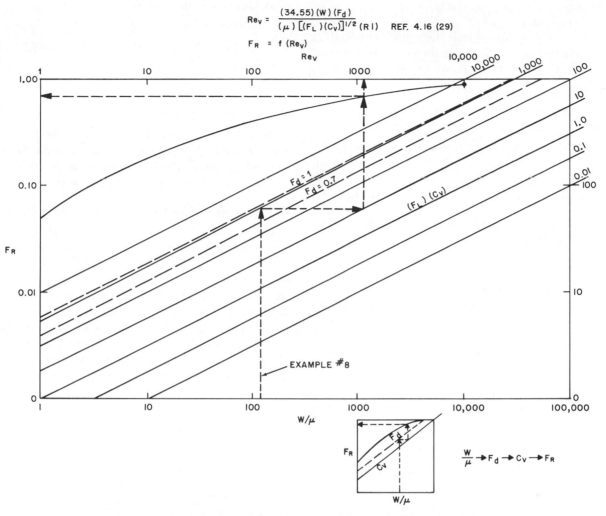

$$Re_v = \frac{(34.55)(W)(F_d)}{(\mu)\left[(F_L)(C_v)\right]^{1/2}} (RI) \quad REF. \ 4.16 \ (29)$$

$$F_R = f(Re_v)$$

Fig. 4.16j   Graph for determining the valve equivalent Reynold's number (R_{ev}).

been rearranged and rewritten somewhat to provide a single and direct path suitable for mechanization. Flow capacity limitations caused by the type of flow regime are accounted for by using limit values instead of alternate equations.

Piping Configuration

These equations apply to both liquid and compressible fluids.

Inlet: $d < D_1$?

Yes:	$K_1 = 0.5 \ (1 - (d/D_1)^2)^2$	4.16(9)
	$K_{B1} = 1 - (d/D_1)^4$	4.16(13)
No:	$K_1 = (1 - (D_1/d)^2$	4.16(11)
	$K_{B1} = (D1/d)^4 - 1$	4.16(15)

Outlet: $d < D_2$?

Yes:	$K_2 = (1 - (d/D_2)^2)^2$	4.16(10)
	$K_{B2} = (d/D_2)^4 - 1$	4.16(14)
No:	$K_2 = 0.5 \ (1 - (D_2/d)^2)^2$	4.16(12)
	$K_{B2} = (1 - D_2/d)^4$	4.16(16)

$$K_i = K_1 + K_{B1} \qquad 4.16(17)$$
$$\Sigma K = K_i + K_2 + K_{B2} \qquad 4.16(18)$$
$$C_d = C_v/d^2 \qquad 4.16(2)$$
$$F_P = \left[1 + (\Sigma K)(C_d^2)/890\right]^{-\frac{1}{2}} \qquad 4.16(19)$$
$$F_{LP} = \left[1/F_L^2 + (K_i)(C_d^2)/890\right]^{-\frac{1}{2}}$$
$$4.16(20)$$

Note: Equations 4.16(11), 4.16(12), 4.16(15) and 4.16(16) are not included in ISA SP75.01.

Liquids

A. Factors and Regime

$$F_i = (0.8)(F_L^2) \qquad 4.16(30)$$

(default value for coefficient of incipient cavitation)

If $P_2 \leq P_{vp}$, then flashing occurs. If $(P_1 - P_2) > (F_i)(P_1 - P_{vp})$, then incipient cavitation occurs.

$$F_F = 0.96 - 0.28 \ (P_{vp}/P_c)^{\frac{1}{2}} \qquad 4.16(4)$$

unless $(.7)(P_1) > P_{vp}$, then:

$$F_F = 0.96 \qquad 4.16(5)$$
$$P_{vc} = P_{vp}F_F \qquad 4.16(6)$$

At start of choking, Δp critical =
$$(P_1 - P_{vc})(F_{LP}/F_P)^2 \qquad 4.16(31)$$
If $(P_1 - P_2) > \Delta p$ critical, then flow is choked. If choked and not flashing, then cavitation occurs.
$$F_Y^2 = (F_{LP}/F_P)^2 \, (P_1 - P_{vc})/(P_1 - P_2) \qquad 4.16(21)$$

But F_Y maximum = 1.0
  F_R: See Figures 4.16i and j.

B. Capacity
$$B = W/C_v = 63.3 \, F_P F_Y F_R \, [D(P_1 - P_2)]^{\frac{1}{2}}$$
$$4.16(23)$$
$$W = C_v B \qquad 4.16(32)$$
$$CVR = W/B \qquad 4.16(33)$$

Note: Equations 4.16(32) and 4.16(33) are not included in ISA SP75.01

Compressibles

A. Factors and Regime
$$X_T = (0.8)(F_L)^2 \text{ (default value for } X_T, \text{ valid for globe style valve)} \qquad 4.16(34)$$
$$X_{TP} = (X_T/F_P^2)/[(X_T K_i C_d^2/1000) + 1] \qquad 4.16(24)$$
$$X = (P_1 - P_2)/P_1 \qquad 4.16(35)$$
$$F_k = k/1.4 \qquad 4.16(25)$$

If $F_k X_{TP} < X$, then flow is choked. $X_m = F_k X_{TP}$ if choked; otherwise $X_m = X$ (not included in ISA SP75.01). $\qquad 4.16(27)$

$$Y = 1 - X_m/3F_k X_{TP}. \text{ Y minimum} = \tfrac{2}{3} \qquad 4.16(26)$$

At start of choking ΔP critical $= P_1 X_{TP} F_k \qquad 4.16(36)$

B. Capacity
$$B = W/C_v = 63.3 \, F_P Y \, (P_1 X_m D)^{\frac{1}{2}} \qquad 4.16(37)$$
$$W = (C_v)(B) \qquad 4.16(38)$$
$$C_{vr} = W/B \qquad 4.16(39)$$

Note: Equations 4.16(37), 4.16(38) and 4.16(39) are not included in ISA SP75.01.

Working Equations—Other Units

All equations remain the same as in the previous paragraphs, with these changes:

$$C_d = C_v/N_1^2 d^2 \qquad 4.16(40)$$
$$B = W/C_v = N_2 F_P F_Y F_R \, [D(P_1 - P_2)]^{\frac{1}{2}} \text{ (liquids)}$$
$$4.16(41)$$
$$B = W/C_v = N_2 F_P Y (DP_1 X_m)^{\frac{1}{2}} \text{ (compressibles)}$$
$$4.16(42)$$
$$Re_v = N_3 W F_d/[\mu(F_P F_L C_v)^{\frac{1}{2}}] \qquad 4.16(43)$$

μ	W	D	P	d	N_1	N_2	N_3
cp	lbm/h	lbm/ft³	PSIA	inch	1	63.3	34.55
cp	kg/s	kg/m³	KPa	mm	25.4	1/1318	274,000
cp	kg/h	kg/m³	bar	mm	25.4	27.32	76.17
Pa/s	kg/s	kg/m³	KPa	mm	25.4	1/3318	274.0

(Note that using C_d in place of C_v/d^2 throughout greatly simplifies using other units.)

Gas Density Calculation

A. Perfect Gas

$$PV = MRT/m \quad D = M/V = mP/RT \quad 4.16(44)$$
$$D_2 = D_1 P_2/P_1 T_2 \qquad 4.16(45)$$

B. Real Gas

$$D = mP/ZRT \qquad 4.16(46)$$

Tables of T_c, P_c, and charts of Z are in the ISA *Control Valve Handbook* and other literature. See also *Flow Measurement Engineering Handbook* by R. W. Miller for an extensive discussion of physical properties.[8]

Assignment of Control Valve Pressure Drop

One common design problem is the selection of a pump, pipe size, and control valve for a pumped system like that in Figure 4.16k. Let us look at the various parts of this system, one at a time. A typical pump curve is shown as Figure 4.16l. The vertical axis shows pressure developed in feet of head of the fluid. The horizontal axis shows flow in gpm. Simplifying somewhat, the head at zero flow is a function of impeller diameter squared, fluid density, and rotation speed squared. The drop off in pressure with increasing flow is approximately proportional to flow squared. The effect of fluid density on head explains why pressure is shown in feet of head. The

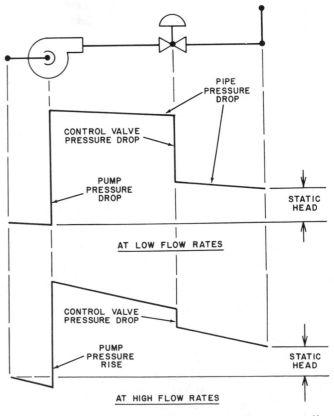

Fig. 4.16k   Pump, pipe, and control valve system and pressure profiles

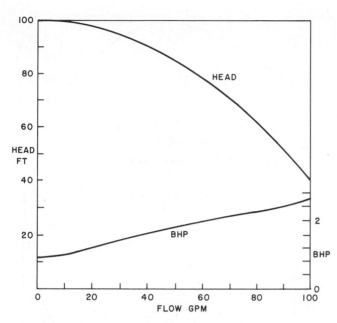

Fig. 4.16l   Typical pump curve

relation between pressure loss and volumetric flow results in expressing flow in volumetric units.

One problem to be faced often in instrument calculations is the frequent need to convert measurement units. To convert between head and pressure:

pressure, PSI = (head,ft.)(0.4335)(specific gravity)

head, ft. = (pressure,PSI)/(0.4335)(specific gravity)

To convert between gpm and mass flow:

lb/h = (gpm)(500)(specific gravity)

gpm = (lb/r)/(500)(specific gravity)

The energy put into the fluid by the pump, and removed where pressure drop occurs, is the hydraulic horsepower:

HP = (head,ft.)(specific gravity)(gpm)/3960

or

HP = (ΔP,PSI)(lb/h)/(density, lb/ft³)(13,750)

Pipe friction loss is an approximate function of flow squared, as is the pressure loss of a control valve with constant stem position. To determine the pumping requirements, it is necessary to decide (assign) just how much pressure drop will occur in the piping, the process equipment (if involved), and the control valve. In service, the control system will cause the valve to take whatever actual pressure drop is required to control flow. The pressure drops assigned during design are used to size the valve, but the final operating system will very likely differ from the system designed. The pressure drop assigned is important because it will affect:

Purchase costs (pump, pump base, motor starter, etc.)

Installation costs

Operating costs

Maintenance costs

Quality of control

Assigning pressure drop is made difficult because:

The required data is often lacking or inaccurate.

Pumps come in standard sizes (pump impellers can be machined to intermediate sizes but rarely are).

Pipe comes in standard sizes.

Valves come in standard sizes.

Undersized pumps and valves cause obvious problems.

Oversized pumps and valves also cause problems but they are often less obvious.

Figure 4.16m shows a pump curve with static head, piping friction losses, and the sum of static and friction pressure drops. Figure 4.16n shows a curve derived from Figure 4.16m showing the control valve pressure drop. It is simply the difference between the pump pressure and the sum of all the other pressure drops. Note that the pressure drop available to push fluid through the valve is a nonlinear function of flow. Also note that the highest pressure drop in this example is available for the lowest flow and that the lowest drop is available for the highest flow. For this reason, when C_{vr} (C_v required) is calculated, it becomes obvious that the ratio of maximum C_{vr} to minimum C_{vr} can be far greater than the ratio of the flows involved.

Control valves are made to conform, roughly, to several standard relationships between C_v and stem position (see Figure 4.16o). These are called the "inherent characteristics." They describe flow versus stem position only where the pressure drop is constant. In the high friction pumped system used as an example, the "equal percentage" trim, where C_v increases more rapidly as the valve opens, may partly offset the effects of system fric-

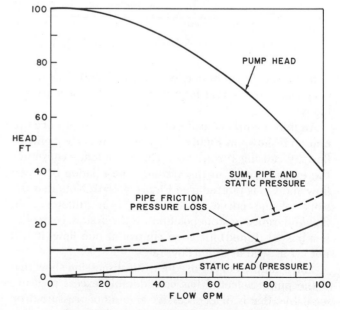

Fig. 4.16m   Pump curve plus system curve

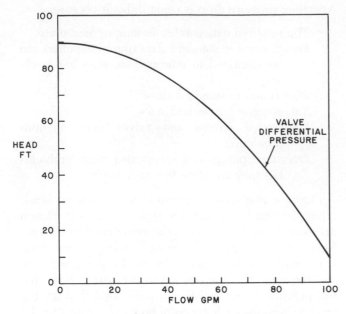

Fig. 4.16n Available valve pressure differential

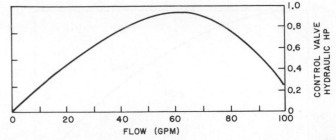

Fig. 4.16p Hydraulic power loss in control valves

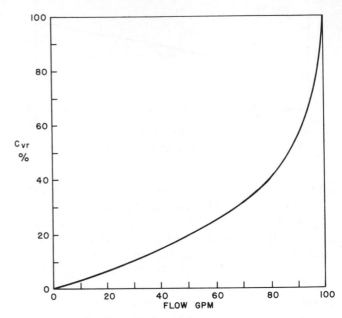

Fig. 4.16q Relationship between required C_v and system flow rate (load)

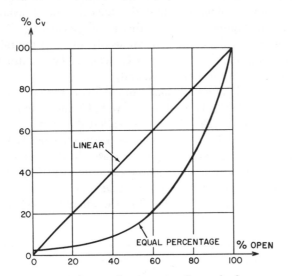

Fig. 4.16o Inherent characteristics of control valves

tion. The quality of fit between actual C_v and C_{vr} is rarely very close, but a real improvement over linear trim is likely.

An infrequently considered aspect of control valve operation is shown in Figure 4.16p. A control valve controls flow by causing turbulence. This turbulence consumes the excess energy in the system. The relation between flow and C_{vr} is plotted (see Figure 4.16q). Note that the slope of the curve is the valve gain (difference in flow/difference in stem position). If the system controller is adjusted (tuned) for good control at one flow, it will not act as well at other flow rates.

The example curves used for this discussion show that if the pump selected has considerable excess pressure available, that is, if an excessive amount of pressure drop is assigned to the valve, then the relative drop off in

valve pressure at higher flows will be less, the amount of nonlinearity will be less, and control will be easier. But a valve with excessive pressure drop is more liable to cavitation, generates more noise, requires a larger actuator, may suffer excessive wear and consumes more energy.

A pump is most efficient when it operates near its design conditions of pressure and flow. An oversize pump will operate less efficiently. A pump operated at low flow may be subject to low-flow cavitation caused by upsets in impeller flow patterns. A certain minimum flow is also required to carry away the heat caused by turbulence. Two schemes are available to reduce the effects of nonlinearity. Figure 4.16r shows how valve action can be linearized if a linear flow meter and a controller are added to replace the valve with a cascade slave loop. The flow meter need not be very accurate, and the controller can be a simple one to achieve a substantial improvement. The other approach involves modifying the controller output signal to compensate for the nonlinearities. The newer computer-based systems can do this or signal relays can be used.[9]

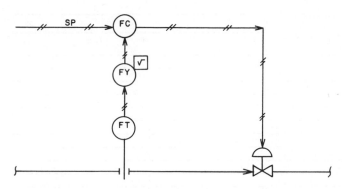

Fig. 4.16r Valves can be linearized by replacing them with a complete slave control loop

Much of the above discussion also applies to dampers and fans; the only differences are in fluid density and pressures. An amazing amount of energy can be lost in fan systems, due to incorrect sizing.

High Pressure Drop

With high flow rates and high pressure drop, a large amount of energy is dissipated in turbulence. A fraction of this energy is radiated as noise (see Section 4.14). For most (but not all) gases and conditions, one result of high pressure drop is very low outlet temperatures. See example 10 for an approximate calculation. It is not always proper to use the gas laws to predict these temperatures,[10] and instead, actual thermodynamic properties should be used. The very low temperature will cause some valve materials to become brittle, and careful selection of alloys and other materials is always required. Depending on the gas involved and other materials in the flowing fluid, hydrates or other solids may form in the valve. Liquid droplets may develop and cause erosion. It is necessary to investigate for any pecularities of the flowing fluid. A high velocity jet leaving the valve can erode downstream piping. Very high forces are developed on the valve body and internal parts and can cause valve instability. With the change in magnitude and, often, direction of the fluid, substantial reaction forces are developed. See example 11. Serious damage may result if the valve and piping are not properly restrained.

Control Valve Specification Sheet

Assembling the information to specify a control valve is best done with the aid of a specification tabulation sheet. An example, with space to show three valves is included as Figure 4.16s.

Flow Calculation Examples

Example 1a—Hot Condensate (Water)

P_1 = 100 PSIA
P_2 = 80 PSIA

T = 260° F, P_{vp} = 35 PSIA, P_c = 3206 PSIA
F_L = 0.8, D = 1/0.0171 (water data from Table A.5 in the Process Measurement volume of this *Handbook*)
C_{vn} = 12, d = D_1 = D_2 = 1 in.
C_d = 12/(1)^2 = 12   Ref. 4.16(2)
K_i = 0   Ref. 4.16(17)
ΣK = 0 because of no reducers. Ref. 4.16(18)
Therefore F_P = 1.00.   Ref. 4.16(19)

$$F_{LP} = \left(\frac{1}{(0.8)^2} + 0\right)^{-\frac{1}{2}} = 0.8000 \quad \text{Ref. 4.16(20)}$$

F_F = 0.96 − 0.28 (35/3206)$^{\frac{1}{2}}$= 0.9307
                                    Ref. 4.16(4)
P_{vc} = 35 × 0.9307 = 32.57 PSIA   Ref. 4.16(6)
ΔP critical = (100 − 32.57) [0.8000/2]^2 = 43.15
                                    Ref. 4.16(31)
Actual ΔP = 20, therefore flow is not choked and
    F_Y = 1.0.

(F_R is safely ignored until high viscosity or very low flows are involved, therefore use F_R = 1.)
B = (63.3)(1)(1)(1) [20/0.0171]$^{\frac{1}{2}}$ = 2165 lb/h/C_v
                                    Ref. 4.16(23)

If required flow is 20,000 lb/h, C_{vr} = 20,000/2165
    = 9.24
    or valve will be at (9.24/12)100 = 77% of capacity.   Ref. 4.16(33)

Or flow at 100% open:
    (12)(2165) = 25,978 lb/h (+5% tolerance allowed)   Ref. 4.16(32)
or
$$\left(\frac{0.0171 \text{ ft}^3}{\text{lb}}\right)\left(\frac{1728 \text{ in.}^3}{\text{in.}^3}\right) = \left(\frac{\text{gal}}{231 \text{ in.}^3}\right)\left(\frac{\text{h}}{60 \text{ m}}\right) =$$
    = 55 gpm

Example 1b—Same as Example 1a Except P_2 is Lower

P_2 = 32 PSIA
Then ΔP actual = 100 − 32 = 68, which exceeds
    ΔP critical = 53.94. Therefore flow is choked.
$P_{vp} > P_2$; 35 > 32. Therefore flow is not flashing
    but it *is* cavitating (cavitation destroys valves).
F_Y^2 = (0.8000/1)^2 (100 − 32.57)/68 = 0.6346 and
    F_Y = 0.7966   Ref. 4.16(21)
B = (63.3)(1)(1)(0.7966) [68/0.0171]$^{\frac{1}{2}}$ = 3180
                                    Ref. 4.16(23)

Flow at full open = (3555)(12) = 38,160 lb/h
                                    Ref. 4.16(32)

Example 2—Same as Example 1a Except 1-in. Valve in 2-in. Pipeline

P_1 = 100 PSIA, P_2 = 80 PSIA
T_1 = 260°F, P_{vp} = 35 PSIA,
           P_c = 3206 PSIA

TABULATION SHEET
CONTROL VALVE

Plant _____ Spec No. _____

Project _____ Date _____ Rev _____ Page _____

	Instrument Item No.							
	Quantity							
	Design Features							
• 1	Valve Style							
• 2	Body Size, In./Rating, PSIG							
• 3	Flange Rating, PSIG							
• 4	Trim Size/Characteristic							
5	Seat Leakage Class							
6	Actuator, Air Open/Close							
• 7	Min Supply Pressure							
• 8	Flow To Open/Close							
• 9	Face-To-Face Dimension							
• 10	Overall Height From ₵							
11	Line Size/Schedule							
	Positioner							
20	Yes-No/STD Or Hi Performance							
• 21	Supply Pressure							
	Materials							
• 30	Body							
• 31	Flanges							
• 32	Stem							
• 33	Plug							
• 34	Seat Rings							
• 35	Guides							
• 36	Gaskets							
• 37	Packing							
	Application							
40	Fluid							
41	Viscosity, Units							
42	Density, Units							
43	Norm Op Temp, Units							
44	Inlet Pressure, Units							
45	Normal Flow, Units							
46	Normal $\triangle$ P, Units							
47	Max Flow/$\triangle$ P							
48	Min Flow/$\triangle$ P							
49	Max $\triangle$ P At Shutoff							
50	P_v, Units/P_c, Units							
• 51	F_1/X_T							
52	Flash / % Solids							
• 53	C_v, Normal Flow/Full Open							
60								
61								
62								
63								
64								
65								
66								
67								
70	Materials Test Req'd							
• 71	Sound Pressure Level							
• 72	MFR Model Number							
• 73	Price							
74	Drawing Number							
	Design Area Number							

•To be filled in by vendor if not specified

Fig. 4.16s   Control valve specification sheet

$$F_L = 0.8, D = 1/0.0171 \text{ lb/ft}^3$$
$$C_{vn} = 12, d = 1 \text{ in.}, D_1 = D_2 = 2 \text{ in.}$$

$$C_d = 12/1^2 = 12 \qquad \text{Ref. } 4.16(2)$$
$$K_1 = 0.5 \left[1 - \left(\tfrac{1}{2}\right)^2\right]^2 = 0.2813$$
$$\text{Ref. } 4.16(9)$$
$$K_{B1} = 1 - \left(\tfrac{1}{2}\right)^4 = 0.9375 \qquad \text{Ref. } 4.16(13)$$
$$K_2 = \left[1 - \left(\tfrac{1}{2}\right)^2\right]^2 = 0.5625 \qquad \text{Ref. } 4.16(10)$$
$$K_{B2} = \left(\tfrac{1}{2}\right)^4 - 1 = -0.9375 \qquad \text{Ref. } 4.16(14)$$
$$K_i = 1.2188 \qquad \text{Ref. } 4.16(17)$$
$$\Sigma K = 0.8438 \qquad \text{Ref. } 4.16(18)$$
$$F_P = \left[1 + (0.8438)(12)^2/890\right]^{-\frac{1}{2}} = 0.9380$$
$$\text{Ref. } 4.16(19)$$
$$F_{LP} = (1/0.8)^2 + \left[1.2188)(12^2)/890\right]^{-\frac{1}{2}} =$$
$$= 0.7538 \qquad \text{Ref. } 4.16(20)$$
$$(F_{LP}/F_P) = 0.8038, (F_{LP}/F_P)^2 = 0.6458$$
$$F_F = 0.96 - 0.28 (35/3206)^{\frac{1}{2}} = 0.9307$$
$$\text{Ref. } 4.16(4)$$
$$P_{vc} = 32.57 \text{ PSIA} \qquad \text{Ref. } 4.16(6)$$
$$\Delta P \text{ critical} = (100 - 32.57)(F_{LP}/F_P)^2 = 43.55$$
$$\text{Ref. } 4.16(31)$$

ΔP actual $= 20$ PSIA, therefore flow is not choked and $F_Y = 1.0$.

$$B = (63.3)(0.9380)(1)(1) \left[\frac{100 - 80}{.017}\right]^{\frac{1}{2}} =$$
$$= 2030 \qquad \text{Ref. } 4.16(23)$$

Note that a capacity loss of $(2165 - 2030)/(2165) = 6.6\%$ is caused by the reducer and increaser. Use of reducers is not necessarily detrimental. However the reduced pressure drop in a 2-in. pipe might well be significant.

Flow at 100% open: $(12)(2030) = 24,360$ lb/h
$$\text{Ref. } 4.16(32)$$

Example 3—Nonsymmetrical Piping Reducers

$$P_1 = 100 \text{ PSIA}, P_2 = 80 \text{ PSIA}$$
$$T_1 = 260°F, P_{vp} = 35 \text{ PSIA},$$
$$P_c = 3206 \text{ PSIA}$$
$$F_L = 0.8, D = 1/0.0171 \text{ lb/ft}^3$$
$$C_{vn} = 12, d = 1 \text{ in.}, D_1 = 2 \text{ in.},$$
$$D_2 = 3 \text{ in.}$$
$$C_d = 12/(1)^2 = 12 \qquad \text{Ref. } 4.16(2)$$
$$K_1 = 0.5 \left[1 - \left(\tfrac{1}{2}\right)^2\right]^2 = 0.2813$$
$$\text{Ref. } 4.16(9)$$
$$K_{B1} = 1 - \left(\tfrac{1}{2}\right)^4 = 0.9375 \qquad \text{Ref. } 4.16(13)$$
$$K_2 = \left[1 - (1/3)^2\right]^2 = 0.7901 \quad \text{Ref. } 4.16(10)$$
$$K_{B2} = (1/3)^4 - 1 = -0.9877$$
$$\text{Ref. } 4.16(14)$$
$$K_i = 1.2188 \qquad \text{Ref. } 4.16(17)$$
$$\Sigma K = 1.0212 \qquad \text{Ref. } 4.16(18)$$
$$F_P = \left[1 + (1.0212)(12)^2/8901\right]^{-\frac{1}{2}} = 0.9264$$
$$\text{Ref. } 4.16(19)$$
$$F_F = 0.96 - 0.28 (35/3206)^{\frac{1}{2}} = 0.9307$$
$$\text{Ref. } 4.16(4)$$
$$P_{vc} = 32.57 \qquad \text{Ref. } 4.16(6)$$
$$F_{LP} = \left[1/(0.8)^2 + (1.2188)(12)^2/8901\right]^{-\frac{1}{2}} =$$
$$0.7538 \qquad \text{Ref. } 4.16(20)$$

$$F_{LP}/F_P = 0.8137, (F_{LP}/F_P)^2 = 0.6622$$
$$\Delta P \text{ critical} = (100 - 32.57)(F_{LP}/F_P)^2 = 44.65$$
$$\text{Ref. } 4.16(31)$$
$$B = W/C_v = (63.3)(.9264)(1)(1) \left[20/.0171\right]^{\frac{1}{2}} = 2005$$
$$\text{Ref. } 4.16(23)$$

Flow at 100% open $= 24,066$ lb/h.

Note that the outlet pipe size change to 3 in. further reduced capacity by 0.8% and ΔP critical increased slightly.

Example 4—Effect of F_L

Data same as example 1a except $F_L = 0.5$.
$$F_P = 1.0, F_{LP} = \left[1/(0.5)^2 + 0\right]^{-\frac{1}{2}} = 0.5$$
$$\text{Ref. } 4.16(20)$$
$$\Delta P \text{ critical} = (100 - 32.57)(0.5/1)^2 = 16.85 \text{ PSI}$$
$$\text{Ref. } 4.16(31)$$

Therefore flow is choked at only 20 PSIA.
$$F_Y^2 = (0.5/1)^2 (67.43)/(20) = 0.8429 \quad \text{Ref. } 4.16(21)$$
$$B = (63.3)(1)(1)(0.9181) \left[20/0.0171\right]^{\frac{1}{4}} = 1988 \text{ lb/h/C}_v$$
$$\text{Ref. } 4.16(23)$$

For the same C_v, this valve will pass 8.2% less flow because of change in F_L from 0.8 to 0.5. Further, since flow is choked and not flashing, cavitation is indicated. Therefore this valve is not acceptable for this service. Addition of reducers simply makes the situation worse. See Figure 4.16t.

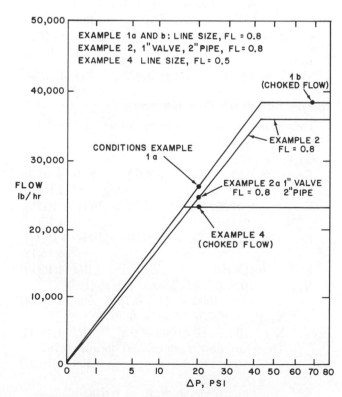

Fig. 4.16t   Flow vs pressure drop for control valve examples 1, 2, and 4.

Example 5a—Airflow

P_1 = 100 PSIA, P_2 = 80 PSIA, C_{vn} = 12

T = 100°F = 38°C = 311°K, X_T = 0.8

D = (28.95)(100)/(19.316)(311) = 0.4818 lb/ft³

Ref. 4.16(44)

k = 1.4, D = D_1 = D_2 = 1 in.

C_d = 12/1² = 12, F_k = 1.4/1.4 = 1.0

Refs. 4.16(2), 4.16(25)

F_P = 1.0, K_i = 0, ΣK = 0 Ref. 4.16(19), 4.16(17), 4.16(18)

X_{TP} = (0.8/1)/(0 + 1) = 0.8 Ref. 4.16(24)

X = (100 − 80)/(100) = 0.2 Ref. 4.16(35)

F_k (X_{TP}) = (1.0) (0.8) = 0.8

Choking will start at X = 0.8 or Δp = 80 PSI.

X_m = 0.2

Y = 1 − 0.2/(3)(0.8)(0.8) = 0.9167

Ref. 4.16(26)

B = (63.3)(1)(0.9167) [(100)(0.2)(0.4818)]^¼ =
= 180.12 lb/h/C_v Ref. 4.16(37)

At C_{vn} = 12, flow = 2161 lb/h Ref. 4.16(38)

Example 5b—Lower Outlet Pressure

Data same as in Example 5a, except P_2 = 10 PSIA.

F_P = 1, K_i = 0, ΣK = 0 Refs. 4.16(19), 4.16(17), 4.16(18)

X = (100 − 10)/(100) = 0.9, X_{TP} = 0.8

Refs. 4.16(35), 4.16(24)

Therefore flow is choked because 0.9 > 0.8. X_m = 0.8.

Y = 1 − 0.8/(3)(1)(0.8) = 1 − 1/3 = 2/3

Ref. 4.16(26)

B = (63.3)(1)(2/3) [(100)(0.8)(0.4818)]^¼ = 262

Ref. 4.16(37)

At C_{vn} = 12, flow = 3144 lb/h Ref. 4.16(38)

Example 6a—With Pipe Reducers

Data same as in example 5a, except D = 1 in., D_1 = D_2 = 2 in.

C_d = 12

K_1 = 0.2813, K_{B1} = 0.9375 Refs. 4.16(9), 4.16(13)

K_2 = 0.5625, K_{B2} = −0.9375 Refs. 4.16(10), 4.16(14)

K_i = 1.2188, ΣK = 0.8438 Refs. 4.16(17), 4.16(18)

F_P = 0.9380 Ref. 4.16(19)

X_{TP} = [0.8/(0.9380)²]/[(0.8)(1.2188)(12²)/ 1000 + 1] = 0.7973 Ref. 4.16(24)

$X_{TP}F_k$ = (0.7973)(1.0) = 0.7973

X = (100 − 80)/(100) = 0.2 Ref. 4.16(35)

Therefore flow is not choked because 0.7973 > 0.2. Ref. 4.16(27)

X_m = X = 0.2

Y = 1 − 0.2/(3)(0.7973) = 0.9164

Ref. 4.16(26)

B = (63.3)(0.9380)(0.9164) × × [(100)(0.2)(0.4818)]^¼ = 169

Ref. 4.16(37)

At C_{vn} = 12, flow = 2028 lb/h. Ref. 4.16(38)

Example 6b—Reducers and High ΔP

Data same as in example 6a except P_2 = 10.

X = (100 − 10)/100 = 0.9 Ref. 4.16(35)

Therefore flow is choked because 0.7973 < 0.9.

Ref. 4.16(27)

X_m = 0.7973, therefore Y = 2/3 Ref. 4.16(26)

B = (63.3)(0.9380)(2/3) [(100)(0.7973)(0.4818)]^¼ =
= 245 Ref. 4.16(37)

At C_{vn} = 12, flow = 2940 lb/h. Ref. 4.16(38)

Example 7a—Effect of X_T

Data same as in example 5a, except X_T = 0.25.

P_1 = 100 PSIA, P_2 = 80 PSIA, C_{vn} = 12

T = 100°F = 311°K

k = 1.4, D = 0.4818 lb/ft³

F_P = 1.0 Ref. 4.16(19)

X_{TP} = (0.25/1)/(0 + 1) = 0.25 Ref. 4.16(24)

X = (100 − 80)/(100) = 0.2 Ref. 4.16(35)

F_kX_{TP} = 0.25

Flow is not choked because 0.25 > 0.2.

Ref. 4.16(27)

Choking starts at X = 0.25

X_m = 0.2

ΔP critical = (100)(0.25)(1) = 25 PSI

Ref. 4.16(36)

Y = 1 − 0.2/(3)(0.25)(1) = 0.7333

Ref. 4.16(26)

B = (63.3)(1.0)(0.7333) [(100)(0.2) × × (0.4818)]^¼ = 144 Ref. 4.16(37)

At C_{vn} = 12, flow = 1729 lb/h Ref. 4.16(38)

Example 7b—Increased ΔP

Data is same as in example 7a except P_2 = 10 PSIA.

X = (100 − 10)/(100) = 0.9 Ref. 4.16(35)

Flow is choked, because 0.25 < 0.9.

Ref. 4.16(27)

X_m = 0.25

Y = 2/3 Ref. 4.16(26)

B = (63.3)(1)(2/3) [(100)(0.25)(0.4818)]^¼ = 146

Ref. 4.16(37)

At C_{vn} = 12, flow = 1758 lb/h. Ref. 4.16(38)

See Figure 4.16u showing Examples 5, 6, and 7 plotted in detail. The effect of reducers and increasers can be seen. For example 7, the impact of X_T is shown. This valve acts like the X_T = 0.8 valve until ΔP critical. Additional points were calculated to complete the plot.

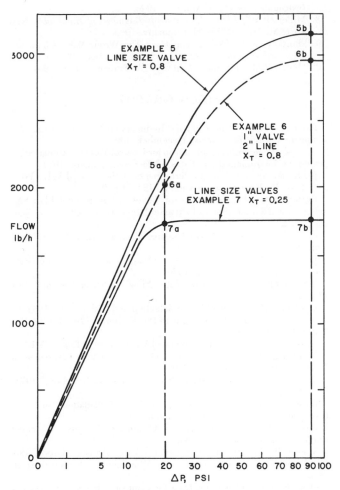

Fig. 4.16u   Capacity vs ΔP for control valve examples 5, 6, and 7.

Table 4.16v
REPRESENTATIVE VALVE FACTORS

Type	F_L	X_T	C_d
Single-seat globe	0.8–0.9	0.6–0.7	12 (0.5 to 16)
Ball	0.5–0.6	0.25	25
Butterfly	0.6–0.7	0.38	17

Example 9—Percent Flashing

Problem: Water at $P_1 = 100$ PSIA and 250°F passes through a control valve, $P_2 = 20$ PSIA. What percent of the water is converted to steam (flashes)?

Basic principle: Flow through a control valve occurs quickly. Because of the high velocity, very little heat is transferred, per pound of flowing fluid, through the valve. The heat used to vaporize the liquid must come from cooling the flowing liquid.

Let x = fraction of water remaining. Obtaining thermodynamic data from the Steam Tables in the Process Measurement volume of this *Handbook*, at 250°F water has 218.59 btu/lb, at 20 PSIA water has 196.27 btu/lb, saturated steam has 1156.3 btu/lb, and the saturation temperature is 227.96°F. Then:

Heat lost from water $= x(T_1 - T_2)$
$$\text{(specific heat)}$$
$$= x(250.0 - 227.96) \times$$
$$\times (1.0)$$

Heat to vaporize water to steam $=$
$$(1.0 - x)(1156.3 - 196.27)$$

Setting these two heat quantities equal gives:
$$x(250.0 - 227.96)(1.0)$$
$$= (1.0 - x)(1156.3 - 196.27)$$
$$x = 0.9775$$
$$1 - x = 0.0224 \text{ or } 2.24\%$$
$$\text{(weight percent of flash)}$$

The calculation has been slightly simplified, but given the usual quality of our data it is more accurate than required.

Example 10—Gas Temperature Change

Problem: Gas (molecular weight = 18, k = 1.3) at 10° C and 500 PSIG expands through a control valve to 300 PSIG. What is the (approximate) downstream temperature?

Assuming isentropic expansion:
$$(T_2/T_1) = (P_2/P_1)^{(k-1)/(k)} \qquad \text{Ref. 4.16(47)}$$
$$T_2 = (10+273)(315/515)^{(1.3-1)/(1.3)}$$
$$= 253°K \ (-20°C \text{ or } -5°F)$$

See any thermodynamics text for background.

Example 8—Fuel Oil

P_1 = 30 PSIA, P_2 = 10 PSIA
C_{vn} = 12, d = D_1 = D_2 = 1 in., F_L = 0.8
W = 24,000 lb/h, viscosity = 200 cp
F_P = 1, F_{LP} = $1/0.8^2$ = 1.5625
$$\text{Ref. 4.16(19 \& 20)}$$

C_d = 12 $\qquad$ Ref. 4.16(2)
From Figure 4.16(i), R1 = 1.03
$(R1)(W)/\mu$ = 2400/200 = 124
$C_{vn}F_L$ = 9.6

Assuming F_d of about 0.8, from Figure 4.16(j) F_R = 0.7 (approximately). Therefore when required C_v is calculated, a correction factor of $1/0.7$ will be required. There is little reason to calculate any more closely.

Note that this 30% loss in flow capacity could have been easily overlooked, resulting in an undersized or marginal value.

Example 11—Reaction Forces

Problem: 100,000 lb/h of air (mol wt = 28.9) at 1000 PSIG and 10° C is vented to atmosphere through a 4-in. globe valve, the air does not change direction. Assume outlet temperature of −30° C. Note that momentum forces are vector quantities and their direction must be considered. What reaction force is generated?

From a fluid mechanics text:

$$F = W (V_2 - V_1)/g \qquad \text{Ref. 4.16(48)}$$

where

$$V = Q/A = W/DA \qquad \text{Ref. 4.16(49)}$$
$$F = W^2(1/D_2 - 1/D_1)/gA \quad \text{Ref. 4.16(50)}$$
$$g = \text{conversion constant} = 32.17 \text{ lb-ft/lb-sec}^2$$
$$W = 27.7 \text{ lb/sec}$$
$$D_2 = mP/RT = (28.9)(14.7)/(19.316)(243)$$
$$= 0.0905 \text{ lb/ft}^3 \qquad \text{Ref. 4.16(42)}$$
$$D_1 = (28.9)(1014.7)/(19.316) \times$$
$$\times (283) = 5.3645 \text{ lb/ft}^3$$
$$A = (2/12)^2\pi = 0.0872 \text{ ft}^2$$
$$F = 2985.5 \text{ lb}$$

Note that the upstream velocity has little effect (51 lb) on the reaction forces. Mass flow appears as a squared function and the downstream temperature and pressure determine fluid density and velocity. This high reaction force requires proper restraint and support to avoid failure and damage.

1. "Standard Control Valve Sizing Equations," ANSI/ISA-S75.01-1977.
2. "Standard Control Valve Sizing Equations," ANSI/ISA-S75.02-1982.
3. Technical Paper No. 410, Crane Company.
4. Perry & Chilton, *Chemical Engineer's Handbook*, 5th ed., New York: McGraw-Hill Book Company.
5. Fagerlund, A.C., and Winkler, R.J. "The Effects of Non-Ideal Gases on Valve Sizing," ISA CI 82-903.
6. "Catalog 10, Technical Information," Fisher Controls Company.
7. J.W. Hutchison, ed., *ISA Handbook of Control Valves*, 2nd ed., Instrument Society of America, 1976.
8. Miller, R.W., *Flow Measurement Engineering Handbook*, New York: McGraw-Hill Book Company, 1983.
9. Shinskey, F.G., "When You Have The Wrong Valve Characteristic," *Instruments & Control Systems*, October, 1971.
10. Morse, P.M., *Thermal Physics*, W.A. Benjamin, Inc., 1964.

BIBLIOGRAPHY

Baumann, H.D., "Effect of Pipe Reducers on Valve Capacity," *Instruments & Control Systems*, December, 1968.

——, "Selecting Pressure Drop for Control Valves," Annual Symposium on Instrumentation, Texas A&M Univ., January, 1974.

Buchwald, W.G., "Control Performance with Reduced Valve Pressure Drop for Energy Conservation," ISA 74-610.

Buster, A.A., "Matching Control Valves to Square-Root Flow Signals," Annual Symposium on Instrumentation, Texas A&M Univ., January, 1982.

Carey, J.A., "Control Valve Update," *Instruments & Control Systems*, January, 1981.

Daneker, J.R., "Sizing Butterfly Valves," *Water and Waste Engineering*, July and September, 1970.

De Filippis, Luigi, "Control Valve Flow Theory and Sizing," ISA FCE 736454.

Driskell, L.R., "Control Valve Selection and Sizing," ISA 1983.

——, "Control Valve Sizing with ISA Formulas," *Instrumentation Technology*, July, 1974.

——, "Practical Guide to Control Valve Sizing," ISA CVS 1.

Fernbaugh, A., "Control Valves: A Decade of Change," *Instruments & Control Systems*, January, 1980.

Fosburg, H.G., "Computer Aided Control Valve Selection," 1973, ISA FCE 736469.

Hynes, K., "Solve Valve Sizing and Noise Calculations on Site," *Instruments & Control Systems*, March, 1976.

Lipták, B.G., "Control Valves in Optimized Systems," *Chemical Engineering*, September 5, 1983.

Moore, R.W., "Allocation of Control Valve Pressure Drop," ISA Final Control Elements Symposium, Wilmington, Delaware, April, 1970.

Quance, R.A., "Collecting Data for Control Valve Sizing," *Instrumentation Technology*, November, 1979.

Sanderson, R.C., "Elastomer Coatings: Hope for Cavitation Resistance," *InTech*, April, 1983.

Scheider, C.B., "Fluid Forces in Control Valves," *Instrumentation Technology*, May, 1971.

——, "Inherent Valve Rangeability," *Instruments & Control Systems*, February, 1976.

——, "Installed Valve Rangeability," *Instruments & Control Systems*, March, 1976.

Self, R.E., "Why Velocity Control," Los Alamitos, CA: Control Components, Inc.

Stiles, G.F., "Liquid Viscosity Effects on Control Valve Sizing," 19th Annual Symposium on Instrumentation, Texas A&M Univ., January 1964.

Chapter V

REGULATORS, SAFETY, AND MISCELLANEOUS CONTROL ELEMENTS

B. BLOCK • A. BRODGESELL

R. D. BUCHANAN • L. D. DINAPOLI

L. S. DYSAKT • D. M. HANLON

E. JENETT • D. S. KAYSER

B. G. LIPTÁK • C. M. J. OUDAR

C. J. SANTHANAM • W. F. SCHLEGEL

CONTENTS OF CHAPTER V

5.1 DAMPERS AS CONTROL ELEMENTS

Types of Designs:	A—Guillotine or slide-gate B—Multiblade or louvre C—Butterfly or wafer including multiple discs D—Radial vanes E—Iris dampers
Design Pressure:	Usually up to 10″ H_2O (2.5 kPa) shut-off differential; iris dampers up to 15 PSID (103 kPa)
Materials of Construction:	Steel, galvanized steel, aluminum, stainless steel
Size Range:	Up to 6 ft × 8 ft (1.8 m × 2.4 m) for HVAC industry; even larger for process applications
Leakage:	At 3″ H_2O (0.75 kPa) shut-off pressure the leakage through each ft² (0.092 m²) of damper area is: Standard design (B): 50 SCFM (250 l/s/m²) Low leakage (B): 5 SCFM (25 l/s/m²) Positive seal (C): 0.5 SCFM (2.5 l/s/m²)
Cost:	$50 to $150 per ft² of damper area ($550 to $1700 per m²), depending on size, design features, and accessories (type B)
Partial List of Suppliers:	Air Clean Damper Co.; American Foundry & Furnace Co.; American Warming & Ventilating, Inc.; Arrow Louver & Damper Corp.; Babcock & Wilcox Co.; Damper Design, Inc.; Dowco Corp.; Energy Vent, Inc.; Envirotech Corp.; FMC Corp., Material Handling Equipment Div.; Forney Engineering Co.; Honeywell, Inc.; Johnson Controls; Lundy Electronics & Systems, Inc.; Mitco Corp.; Quality Air Design; Robertshaw Controls Co.; Ruskin Mfg. Co.; Vent Products Co., Inc.; Young Regulator Co.

Dampers and louvres are used to control the flow of gases and vapors. These streams usually occur in larger-size ducts at relatively low static pressures. There are both "commercial" and "process control" quality dampers on the market. Commercial-quality units are used for the less-demanding applications, such as HVAC, while the process-control-quality units can handle higher pressures, higher temperatures, corrosive vapors, and they provide superior leakage and control characteristics. Dampers are also used to control the flow of solids or to throttle the capacity of fans and compressors.

Butterfly and Guillotine Dampers

No clear-cut criteria distinguish the family of butterfly valves from butterfly dampers or the slide-gate valves from guillotine dampers. The design features of these dampers are quite similar to their control valve counterparts, discussed in Chapter 4. Dampers in general tend to be larger in size, and they therefore tend to be limited to lower operating and shut-off pressures. The diameters of some of these dampers can exceed 20 ft (6m).

A unique variation of the butterfly design is the multiple rotating disc type damper. In this design several disc elements are distributed over the area of the damper face. One advantage of this approach is improved flow control characteristics, because each disc can have its own unique spring range and failure position. Another major advantage is the substantial reduction in leakage compared to the parallel-blade designs. At a static pressure of 3″ H_2O (0.75 kPa) the leakage can be estimated as 0.01 percent of full damper capacity, which corresponds to about 0.5 SCFM (2.5 l/s/m²) leakage per ft² (0.092 m²) of damper area.

Parallel-Blade Dampers, Louvres

The multiblade damper consists of two or more rectangular vanes mounted on shafts one above the other and interconnected so as to rotate together. The vanes are operated by an external lever, which can be positioned manually, pneumatically, or electrically. In the unirotational louvre the vanes remain parallel at all positions. In a counterrotational louvre, alternate vanes rotate in opposite directions. Both designs are illustrated in Figure 5.1a.

Flow guides are sometimes installed between adjacent vanes in order to improve the effectiveness of throttling. The flow-angle characteristic is shown in Figure 5.1b for a 90-degree unirotational louvre. It may be seen that the sensitivity is very high at midflow and that the last 30 degrees of rotation is relatively ineffective. As shown in the upper portion of the illustration, the flow characteristics of a butterfly is somewhat superior to a louvre. The lower portion of Figure 5.1b shows how an opposed blade damper with equal percentage linkage will shift its flow characteristics as less and less of the total system pressure drop is assigned to the damper.

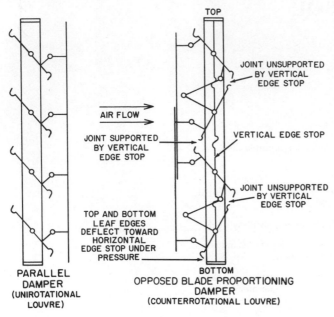

Fig. 5.1a   Unirotational and counterrotational louvre designs

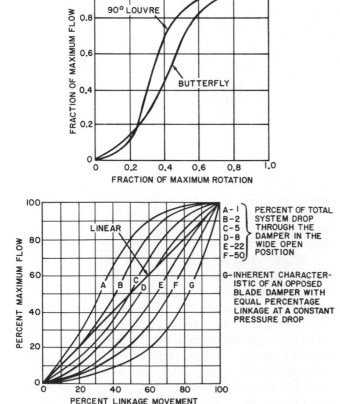

Fig. 5.1b   Damper flow characteristics and the effect of variations in system pressure drop

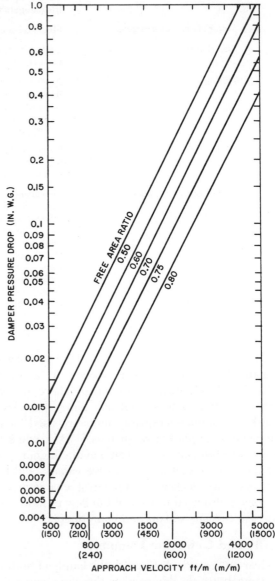

Fig. 5.1c   Pressure drop through wide open dampers. The free area ratio is the total open area of the damper between blades, divided by the nominal area.

WIDTH (W) IN./mm

Nominal H (IN/mm)	Actual H	12 / 304 (11-15/16 / 303)	15 / 381 (14-59/64 / 370)	18 / 456 (17-29/32 / 453)	21 / 533 (20-57/64 / 530)	24 / 609 (23-7/8 / 606)	27 / 686 (26-55/64 / 682)	30 / 760 (29-27/32 / 756)	33 / 838 (32-53/64 / 832)	36 / 914 (35-13/16 / 909)	42 / 1067 (41-25/32 / 1061)	48 / 1219 (47-3/4 / 1213)	60 / 1524 (59-11/16 / 1516)
6 / 152	5-31/32 / 151	.5 / .046	.62 / .058	.75 / .070	.87 / .081	1 / .093	1.12 / .104	1.25 / .116	1.37 / .127	1.5 / .139	1.75 / .163	2 / .186	—
12 / 304	11-15/16 / 303	1 / .093	1.25 / .116	1.5 / .139	1.75 / .163	2 / .186	2.25 / .209	2.5 / .232	2.75 / .255	3 / .279	3.5 / .325	4 / .372	5 / .465
18 / 456	17-29/32 / 453	1.5 / .139	1.87 / .174	2.25 / .209	2.63 / .244	3 / .279	3.37 / .313	3.75 / .348	4.12 / .383	4.5 / .418	5.25 / .488	6 / .557	7.5 / .697
24 / 609	23-7/8 / 606	2 / .186	2.5 / .232	3 / .249	3.5 / .325	4 / .372	4.5 / .418	5 / .465	5.5 / .511	6 / .557	7 / .650	8 / .743	10 / .929
30 / 760	29-27/32 / 756	2.5 / .232	3.12 / .290	3.75 / .348	4.37 / .406	5 / .465	5.62 / .522	6.25 / .581	6.87 / .638	7.5 / .697	8.75 / .813	10 / .929	12.5 / 1.16
36 / 914	35-13/16 / 907	3 / .279	3.75 / .348	4.5 / .418	5.25 / .488	6 / .557	6.75 / .627	7.5 / .697	8.25 / .766	9 / .836	10.5 / .975	12 / 1.11	15 / 1.39
42 / 1066	41-25/32 / 1061	3.5 / .325	4.37 / .406	5.25 / .486	6.12 / .569	7 / .650	7.87 / .731	8.75 / .813	9.62 / .894	10.5 / .975	12.25 / 1.14	14 / 1.30	17.5 / 1.63
48 / 1219	47-3/4 / 1213	4 / .372	5 / .465	6 / .557	7 / .650	8 / .743	9 / .836	10 / .929	11 / 1.02	12 / 1.11	14 / 1.30	16 / 1.49	20 / 1.86
60 / 1524	59-11/16 / 1516	—	—	—	—	10 / .929	11.25 / 1.05	12.5 / 1.16	13.75 / 1.28	15 / 1.39	17.5 / 1.63	20 / 1.86	25 / 2.32
72 / 1829	71-5/8 / 1820	—	—	—	—	12 / 1.11	13.5 / 1.25	15 / 1.39	16.5 / 1.53	18 / 1.67	21 / 1.95	24 / 2.23	30 / 2.79
84 / 2133	83-9/16 / 2122	—	—	—	—	14 / 1.30	15.75 / 1.46	17.5 / 1.63	19.25 / 1.79	21 / 1.95	24.5 / 2.28	28 / 2.60	35 / 3.25
96 / 2438	95-1/2 / 2425	—	—	—	—	16 / 1.49	18 / 1.67	20 / 1.86	22 / 2.04	24 / 2.23	28 / 2.60	32 / 2.97	—

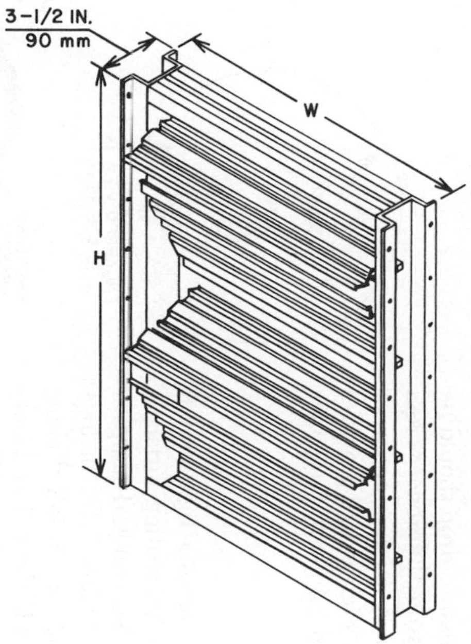

HEIGHT (H) IN./mm

Fig. 5.1d Standard single frame sizes and areas, $\frac{ft^2}{m^2}$ (Courtesy of Johnson Controls)

475

Figure 5.1c gives the pressure drop across a wide open damper. Ideally the wide open pressure drop should be between 4 and 8 percent of the pressure difference across the closed damper. If the damper is so sized, its apparent characteristic will be nearly linear. Figure 5.1d provides some dimensional data for standard commercial damper frames, including their areas.

Low-Leakage Designs

The parallel-blade damper cannot provide tight shutoff because of the long length of unsealed seating surfaces. The leakage characteristics of unsealed standard dampers are given in the lower portion of Figure 5.1e. In low-leakage damper designs, blade seals are installed along the blade seating surfaces, resulting in the reduced leakage characteristic shown in the upper portion of Figure 5.1e.

There are a number of variations in the blade edge seal designs. Some of these are illustrated in Figure 5.1f.

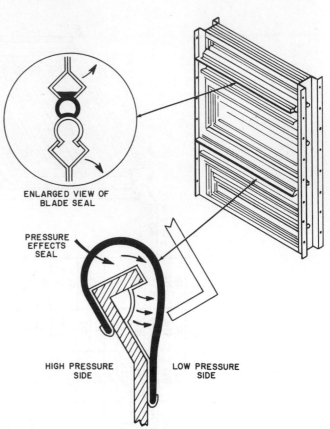

Fig. 5.1f　Variations on blade edge seal designs (Courtesy of Honeywell, Inc. and Johnson Controls)

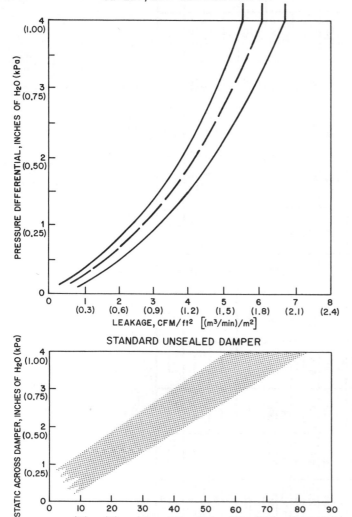

Fig. 5.1e　Leakage through sealed and unsealed dampers (Courtesy of Honeywell, Inc. and Johnson Controls)

Actuators and Accessories

Damper actuators can be manual, electric, hydraulic, or pneumatic. Standard pneumatic actuators vary their effective diaphragm area from 2 to 24 in.^{2} (13 to 155 cm^2), while their stroke length ranges from 2 to 6 in. (51 to 152 mm). The amount of force they produce ranges from about 10 to 300 lb (4.5 to 136 kg). The standard spring ranges include the spans of 3 to 7, 5 to 10, and 8 to 13 PSIG (0.2 to 0.48, 0.34 to 0.68, and 0.54 to 0.88 bars.).

For more accurate throttling, the actuators can also be provided with positioners. If remote indication of damper status is desired, limit switches can be installed to detect the blade angle. These can be pneumatic sensors of nozzle back pressure or mechanically actuated position sensors. The damper position switch can be furnished with an adjustable mounting flange; this allows the unit to be mounted through a duct wall with the trip lever positioned so that it is actuated by the damper blade itself.

Fan Suction Dampers

On blowers and fans, where throughputs must be controlled, radial vane dampers can be utilized. The damper consists of a number of radial vanes arranged to rotate about their radial axis (Figure 5.1g).

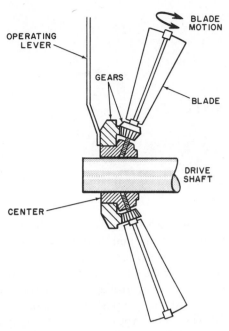

Fig. 5.1g   Fan suction damper, radial vane type

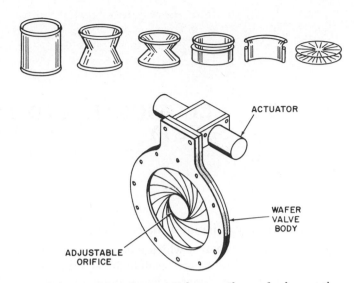

Fig. 5.1h   Variable-orifice or iris damper. Sleeve of nylon or other material with built in retaining ring at each end. With upper end fixed, lower end is rotated, gradually reducing orifice. At 180° of rotation the orifice is completely closed. If the sleeve is first turned back on itself, partly "inside out," the effect of rotation is exactly the same but in the much more compact form of a pleated duplex diaphragm.

Control quality is not very good, and leakage rates are fairly high in the closed position. Control applications can also involve such secondary services as furnace draft control.

Usually a positioner is furnished, providing a linear relationship between the control signal and the blade pitch angle. In certain packages the positioner is factory set in the reverse acting mode, meaning that an increasing control air signal will reduce the air flow by decreasing the blade pitch. In such packages, one has to install a reversing relay between the positioner and the actuator, if direct action is desired.

Variable-Orifice Damper Valves

Variable-orifice valves use the same principle as the iris diaphragm of a camera to achieve control action. The closure element moves within an annular ring in the valve body and produces a circular flow orifice of variable diameter (Figure 5.1h). The flow characteristics are similar to those of a linear valve. However, tight shutoff is not possible, and leakage rates are comparable to or greater than those of a butterfly valve of equal size. Maximum pressure differential is limited to approximately 15 PSID (104 kPa). Dual valve units are available with a common discharge port for blending two streams.

On solids service, the variable-orifice valve can be used for throttling. However, the valve must be installed in a vertical line. The shutter mechanism of the valve forms a dam, making the valve unsuitable for solids service in horizontal lines. Standard sizes range from 4 to 12 in. (100 to 300 mm).

Conclusions

Dampers are suitable for control of large flows at low pressures where high control accuracy is not a requirement. Typical applications of these units include air conditioning systems and furnace draft control. Variable-orifice or iris dampers, although smaller than other dampers, offer better control quality and can be used to control vertical solids flow.

BIBLIOGRAPHY

Brown, E.J., "Air Diffusing Equipment," Chapter 2 In 1979 Equipment Volume of ASHRAE Handbook.

Dickey, P.S., "A Study of Damper Characteristics," Reprint No. A-8, Bailey Meter Co.

Daryanani, S., et al, "Variable Air Volume Systems," *Air Conditioning, Heating and Ventilating*, March, 1966.

Lipták, B.G., "Reducing the Operating Costs of Buildings by Use of Computers," ASHRAE Transactions 1977, Vol. 83, Part 1.

———"Optimization of Semiconductor Manufacturing Plants," *Instruments & Controls Systems*, October, 1982.

5.2 ELECTRICAL AND INTRINSIC SAFETY

Electrical Safety, Fire and Shock Hazards

There are three ways in which electricity can kill or injure. Two are indirect—the result of fire and the result of explosion. One is direct—electrocution by electric shock.

As an academic matter, a fire and an explosion are basically the same, explosion being simply a very fast-spreading form of fire. They are considered separately for two reasons: one, because the results are so different, and two, perhaps more important, because the precautionary and preventive measures are quite different.

The danger of fire from electrical causes is generally confined to the supply side of instrumentation—from the point where power enters the system up to and including the power transformer found in most process control instrumentation. Transformer secondary circuits within the instrument can present a potential fire hazard, but they can be controlled with proper instrument construction. Instrument field wiring, employing the popular signals of 10–50 mA, 4–20 mA DC, and lower, operates at energy levels at which fire hazard is remote. The precautions required for personnel protection and reliability automatically result in fire safety.

Power supply to the point of entry into an instrument is subject to detailed rules and regulations. In most of the United States, the National Electrical Code has been adopted and has the effect of law. Many states and municipalities have their own codes, which differ slightly from the NEC. These are often obsolete editions of the National Electrical Code. It is always wise to check local rules and interpretations when planning an installation.

Safety depends primarily on three fundamental factors:

1. Enclosure of live parts, both to avoid personnel contact and to avoid accidental short circuiting.
2. Fuses or circuit breakers to open in case of overload.
3. Grounding of all exposed metal.

Enclosures—NEMA Types

For ordinary locations, an enclosure need only be tight enough to prevent entrance of the human finger far enough to contact live parts. Unless ventilation is required for cooling, it should also be tight enough to prevent entrance of foreign material and to prevent escape of sparks or hot material in case of internal short circuit or fire. It is particularly important to prevent the escape of flaming drops from any burning insulation or plastics. The enclosure itself must not support combustion. This does not rule out plastics but does require selectivity in their use.

Because instruments differ in use, the usual rules for electrical enclosures need modification when applied to them. For ordinary electrical equipment, it is assumed that only a qualified electrician has access to the interior and, therefore, only unusual interior hazards need be guarded. However, instruments frequently have doors for access by other than qualified technicians for purposes such as recorder chart changing, inking of pens, and controller adjustments. No live parts operating at voltage levels dangerous to personnel should be accessible during operational maintenance.

For special environmental conditions, further requirements are imposed that usually follow the terminology established by the National Electrical Manufacturers Association (NEMA) for motor starters and similar equipment. The following is excerpted from NEMA ICS-1970, "Industrial Control."

Type 1—General Purpose—A general-purpose enclosure is intended primarily to prevent accidental contact with the enclosed apparatus. It is suitable for general-purpose applications indoors where it is not exposed to unusual service conditions. A Type 1 enclosure serves as a protection against dust, light, and indirect splashing, but is not dusttight.

Type 2—Driptight—A driptight enclosure is intended to prevent accidental contact with the enclosed apparatus and, in addition, is so constructed as to exclude falling moisture or dirt. A Type 2 enclosure is suitable for application where condensation may be severe, such as is encountered in cooling rooms and laundries.

Type 3—Weather-Resistant (Weatherproof)—A weather-resistant enclosure is intended to provide suitable protection against specified weather hazards. It is suitable for use outdoors.

Type 4—Watertight—A watertight enclosure is designed to meet the hose test described in the following note. A Type 4 enclosure is suitable for application outdoors on ship docks and in dairies, breweries, etc.

Note: Enclosures shall be tested by subjection to a stream of water. A hose with a one-inch (25 mm) nozzle shall be used and shall deliver at least 65 gallons per minute (246 l/min.). The water shall be directed on the enclosure from a distance of not less than 10 feet (3m)and for a period of 5 minutes. During this period it may be directed in any one or more directions as desired. There shall be no leakage of water into the enclosure under these conditions.

Type 5—Dusttight—A dusttight enclosure is provided with gaskets or their equivalent to exclude dust. A Type 5 enclosure is suitable for application in steel mills, cement mills and other locations where it is desirable to exclude dust.

Type 6—Submersible—A Type 6 enclosure is suitable for application where the equipment may be subject to submersion, as in quarries, mines, and manholes. The design of the enclosure will depend upon the specified conditions of pressure and time.

Type 7 (A, B, C, or D) Hazardous Locations—Class I—These enclosures are designed to meet the application requirements of the National Electrical Code for Class I hazardous locations that may be in effect from time to time.

Type 8 (A, B, C, or D) Hazardous Locations—Class I Oil Immersed—These enclosures are designed to meet the application requirements of the National Electrical Code for Class I hazardous locations that may be in effect from time to time. The apparatus is immersed in oil.

Type 9 (E, F, or G) Hazardous Locations—Class II—These enclosures are designed to meet the application requirements of the National Electrical Code for Class II hazardous locations that may be in effect from time to time.

Type 10—Bureau of Mines—Explosion-Proof—A Type 10 enclosure is designed to meet the explosion-proof requirements of the U.S. Bureau of Mines that may be in effect from time to time. It is suitable for use in gassy coal mines.

Type 11—Acid- and Fume-Resistant—Oil Immersed—This enclosure provides for the immersion of the apparatus in oil such that it is suitable for application where the equipment is subject to acid or other corrosive fumes.

Type 12—Industrial Use—A Type 12 enclosure is designed for use in those industries where it is desired to exclude such materials as dust, lint, fibers and flyings, oil seepage or coolant seepage.

Type 13—Oiltight and Dusttight—Indoor—A Type 13 enclosure is intended for use indoors to protect against lint, dust, seepage, external condensation, and spraying of water, oil, or coolant.

Fuses and Circuit Breakers

The conventional 15 or 20 ampere fuse or breaker in the supply wiring to an instrument is designed to protect the wiring, not the instrument. Component failures or circuit faults within the instrument may result in total destruction of the instrument. To minimize damage and possible fire, a much smaller fuse, usually $\frac{1}{4}$ to 3 amperes, is used in the instrument.

Grounding

A low-resistance, noncurrent-carrying metallic connection to ground should be established and maintained from every exposed metallic surface that can possibly become connected to an electrical circuit. Electrical connection could occur because of a fault, such as a loose wire making electrical contact, or as a result of leakage current through insulation. Grounding is usually accomplished by bonding all elements together in a system terminated at the ground connection where power enters the premises. It may be a bare or green insulated wire. More often it is the conduit enclosing the wires. It must be securely joined, electrically and mechanically, to each piece of equipment. It is connected at the service entrance to the grounded circuit conductor (white wire) and to ground. Instead of connection to a ground at the entrance connection, other suitable earth ground connections are acceptable. Equipment mounted directly on the structural steel of a building is considered effectively grounded. Water pipes have been considered effective grounds. However, because of the increasing use of plastic pipe for water connection, this is no longer an unquestionable ground.

Grounding serves two distinct purposes, both relating to safety. First, since the ordinary power circuit has one side grounded, a fault that results in electrical contact to the grounded enclosure will pass enough current to blow a fuse. Second, possibility of shock hazard is minimized since the low-resistance path of a properly bonded and grounded system will maintain all exposed surfaces at substantially ground potential.

Grounding is effective against hazard from leakage currents. All electrical insulation is subject to some electrical leakage. This may rise to a significant level as insulation deteriorates with age, as layers of conductive dust accumulate in the presence of high humidity. A proper grounding system with low electrical resistance will conduct leakage currents to ground without developing significant potential on exposed surfaces.

Grounding of exposed metal surfaces is distinct from the grounded conductor of the ordinary power wiring. The latter is a current-carrying ground, capable of developing significant potential, particularly on long lines, and with surge or even short-circuit currents. Grounding systems are substantially noncurrent-carrying, except for possible leakage currents. Potential can build up only during the time required for a fuse to blow as a result of a specific fault that results in direct contact between power and grounding systems. Grounding is customarily not required for signal circuits where either maximum voltage is 30 volts, or maximum current under any circumstance cannot exceed 5 mA.

Personnel Safety

The electrical energy necessary to kill or injure a person varies widely with conditions of exposure, especially with contact conditions; i.e., wet or dry skin, contact area, and the path the current takes through the body. A few millivolts applied directly to the heart can cause fibrillation and death, yet it is a common, though not approved, practice among electricians to ascertain if a 120- or even a 240-volt circuit is energized by putting two fingers of the same hand in contact with the two conductors. If the person is insulated from the ground, for instance, by standing on a dry wood floor, the current path is through the fingers and the hazard is nil. If the person is standing in water or has a firm grasp on a water pipe, the current path would be through the central nervous system, and it might well be fatal. Much depends also on the nature of the contact. An electrical shock causes muscles to contract, thus a fingertip contact can be broken, but if the live part is gripped, it might be impossible to let go.

In addition to guarding against possibly lethal shocks, the surprise factor should not be overlooked. An unexpected, but harmless, shock may induce a sudden reaction and expose a worker to an entirely different hazard, such as falling off a ladder.

Energy Levels

This discussion of energy levels hazardous to humans is limited to those circumstances where the whole body is involved. For medical electronics, where electrodes may be implanted within the body, a much lower set of numbers apply. Consideration of such equipment is outside the scope of this volume.

The significant factor is current through the body. A current of less than 1 mA is imperceptible to a normal man. Above 3 mA, it becomes unpleasant. Above 10 mA, the victim is unable to let go. Above 30 mA, asphyxiation may result. Still higher levels lead to heart stoppage and death. These values are for sustained contact. Much higher levels can be tolerated for a fraction of a second.

To relate this information to circuit voltage requires knowledge of body resistance. Internal body resistance can be as low as 100 ohms, but the resistance of the whole body is primarily in the skin and in skin contact. Dry fingers grasping a wire or small terminal will have a resistance in the order of 100,000 ohms. Wetting the fingers would lower this. To lower total body resistance to 1000 ohms would require immersion of a hand or foot in water and a solid grip on a large object. As a practical working figure, the National Electrical Safety Code re-

quires guarding above 50 volts. Thirty volts is considered safe for general use, even in children's toys, though surely not for swimming pools.

The National Electrical Code permits circuits up to 150 volts if they are incapable of delivering more than 5 ma without special requirements for wire insulation.

If all line voltage circuits are enclosed in properly fused and grounded enclosures, and signal circuits not meeting either the 30 volt or the 5 ma criteria are guarded, personnel safety will be assured.

Explosion Hazards and Intrinsic Safety

Areas in which combustible gas, vapor, or dust may be present in explosive proportions are called "hazardous locations." Special precautions must be taken with electrical equipment in hazardous locations to eliminate a source of ignition that could touch off an explosion. The specific precautions vary with the nature of the combustible material and the probability of its presence.

Definitions for Hazardous Locations

The first step in deciding what equipment to use is to determine the nature and the degree of hazard. The National Electrical Code[1] describes hazardous locations by class, group, and division. The class defines the physical form of the combustible material mixed with air:

> *Class I*—Combustible material in the form of a gas or vapor.
> *Class II*—Combustible material in the form of a dust.
> *Class III*—Combustible material in the form of a fiber, such as textile flyings.

The group subdivides the class:

> *Group A*—Atmospheres containing acetylene.
> *Group B*—Atmospheres containing hydrogen, gases or vapors of equivalent hazard, such as manufactured gas.
> *Group C*—Atmospheres containing ethyl ether vapors, ethylene, or cyclopropane.
> *Group D*—Atmospheres containing gasoline, hexane, naphtha, benzine, butane, propane, alcohol, acetone, benzol, lacquer, solvent vapors, or natural gas.
> *Group E*—Atmospheres containing metal dust, including aluminum, magnesium and their commercial alloys, and other metals of similarly hazardous characteristics.
> *Group F*—Atmospheres containing carbon black, coal or coke dust.
> *Group G*—Atmospheres containing flour starch or grain dusts.

The division defines the probability of an explosive mixture being present. Only the breakdown for Class I is given because it is the most often encountered. Classes II and III are similarly subdivided.

Class I, Division 1—Locations (1) in which hazardous concentrations of flammable gases or vapors exist continuously, intermittently, or periodically under normal operating conditions, (2) in which ignitible concentrations of such gases or vapors may exist frequently because of repair or maintenance operations or because of leakage, or (3) in which breakdown or faulty operation of equipment, or processes which might release ignitible concentrations of flammable gases or vapors, might also cause simultaneous failure of electrical equipment.

Class I, Division 2—Locations (1) in which volatile flammable liquids or flammable gases are handled, processed, or used, but in which the hazardous liquids, vapors, or gases will normally be confined within closed containers or closed systems from which they can escape only in case of accidental rupture or breakdown of such containers or systems, or in case of abnormal operation of equipment, (2) in which ignitible concentrations of gases or vapors are normally prevented by positive mechanical ventilation, but which might become hazardous through failure or abnormal operation of the ventilating equipment, or (3) that are adjacent to Class I, Division 1 locations, and to which ignitible concentrations of gases or vapors might occasionally be communicated unless such communication is prevented by adequate positive-pressure ventilation from a source of clean air, and effective safeguards against ventilation failure are provided.

As a rule of thumb, any atmosphere tolerable for a person to breathe is not within the explosive range. Except for methane and hydrogen, gases and vapors become toxic or irritating well below their lower explosive limit. An explosive mixture of dust limits visibility to a few feet.

Economic reasons (i.e., cost of lost product) tend to limit Division 1 locations to an area within a few feet of probable leaks such as pump glands and valve stem packing. The American Petroleum Institute has developed a series of detailed criteria for classifying areas at such distances in typical situations.[2]

Explosions

An explosion is dependent upon the simultaneous presence of three conditions (Figure 5.2a). An oxidizer in the form of the oxygen in air is ordinarily present. Fuel, a gas, vapor, or finely divided solid is normally kept confined for economic reasons, if not for safety.

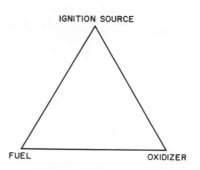

Fig. 5.2a Prerequisites for an explosion

However, by definition, a hazardous location is a place where fuel and oxidizer are present in combustible proportions, at least at times. Ignition of this dangerous combination must not be permitted. Electrical equipment must be built and operated in a manner to prevent its becoming a source of ignition. It could ignite this hazardous atmosphere in either of two ways: by surface temperatures in excess of the ignition temperature or by sparks. Some sparks are incidental to normal operation, as in the operation of switches; some are accidental, as in faulty connections. Both must be guarded against.

Protection Methods

There are several approaches to safety:

1. Confine explosions so they do no damage (explosion-proof)
2. Keep atmosphere away from ignition source (pressurization or ventilation, oil immersion, sealing, or potting)
3. Limit energy to levels incapable of ignition (intrinsic safety)
4. Miscellaneous (sand filling, increased safety, dust ignition-proof, nonincendive)

Explosion-Proof ("Flameproof" in Britain)—All equipment is contained within enclosures strong enough to withstand internal explosions without damage, and tight enough to confine the resulting hot gases so that they will not ignite the external atmosphere. This is the traditional method and is applicable to all sizes and types of equipment.

Purging, Pressurization, or Ventilation—This depends upon the maintenance of a slight positive pressure of air or inert gas within an enclosure so that the hazardous atmosphere cannot enter. Relatively recent in general application, it is applicable to any size or type of equipment.

Oil Immersion—Equipment is submerged in oil to a depth sufficient to quench any sparks that may be produced. This technique is commonly used for switchgears but it is not utilized in connection with instruments.

Sealing—The atmosphere is excluded from potential sources of ignition by sealing them in airtight containers.

Table 5.2b
REFERENCE FOR PROTECTION METHODS

Protection Method	Class I Gas/Vapor Group A, B, C, D		Class II Dust Group E, F, G		Class III Flyings and Fibers	
	Div. 1	Div. 2	Div. 1	Div. 2	Div. 1	Div. 2
Explosion-proof housings	OK	Required only for sparking or hot devices	Not applicable unless also dust ignition-proof		OK if tightly enclosed and no overheating when covered with flyings	
Dust ignition-proof	Not applicable		Dust proof and no overheating when dust covered		Not applicable	
Intrinsic safety	OK	OK	OK	OK	OK	OK
Purging	OK	OK	Subject of work by NFPA		?	?
Potting	?		OK if no overheating			
Hermetic sealing	?		OK if no overheating			
Oil immersion	Acceptable but not convenient to use for instruments				?	?
Nonincendive	Not applicable	OK	Not applicable unless dust ignition-proof	OK	OK if tightly enclosed and no overheating when covered with flyings	

This method is used for components such as relays, not for complete instruments.

Potting—Potting compound completely surrounding all live parts and thereby excluding the hazardous atmosphere has been proposed as a method of protection. There is no known usage except in combination with other means.

Intrinsic Safety—Available energy is limited under all conditions to levels too low to ignite the hazardous atmosphere. This method is useful only for low-power equipment such as instrumentation, communication, and remote control circuits.

Sand Filling—All potential sources of ignition are buried in a granular solid, such as sand. The sand acts, in part, to keep the hazardous atmosphere away from the source of ignition and, in part, as an arc quencher and flame arrester. It is used in Europe for heavy equipment. It is not used in instruments.

Increased Safety—Equipment is so built that the chance of spark or of dangerous overheating is nil. In practice, this means rugged construction, wide spacings between parts of opposite polarity, extra insulation, and good mechanical protection. Widely used in Europe for heavy equipment such as large motors. It is also recognized for instruments particularly in West Germany. It is not recognized in the United States.

Dust Ignition-Proof—Enclosed in a manner to exclude ignitable amounts of dust or amounts that might affect performance. Enclosed so that arcs, sparks, or heat otherwise generated or liberated inside of the enclosure will not cause ignition of exterior accumulations or atmospheric suspensions of dust.

Nonincendive—Equipment which in normal operations does not constitute a source of ignition; i.e., surface temperature shall not exceed ignition temperature of the specified gas to which it may be exposed, and no sliding or make-and-break contacts operating at energy levels capable of causing ignition. Used for all types of equipment in Division 2 locations. Relies on the improbability of an ignition-capable fault condition occurring simultaneously with an escape of hazardous gas.

Advantages and Disadvantages of Various Protection Methods

Table 5.2c is an attempt to rate the various protection means as used for instrumentation. Methods are rated from A to C.

Safety—All methods are safe if the equipment is properly installed and maintained. Intrinsic safety is rated A because recognized standards are most conservative and it is less dependent on day-to-day usage. Ordinary carelessness does not make intrinsically safe equipment un-

Table 5.2c
RATING THE PROTECTION METHODS

	Safety	Cost of Instrument	Cost of Installation	Maintenance	Flexibility
Intrinsic safety	A	C	A	A	C
Explosion-proof	B	B	B	B	B
Purging	C	A	C	C	A

safe. Explosion-proof equipment is worthless if the cover is left off and purging is dubious under the same circumstances. Purging is also dependent on reliability of purge air supply.

Cost of Instrument—Purging is usually lowest in cost because it requires no special construction for hazardous location use, except for an air inlet. Since an intrinsically safe instrument does not require the special housing of an explosion-proof instrument, there is a possible saving. The cost difference between an explosion-proof housing and a rugged weatherproof housing is too small to justify two separate designs for small devices, such as field-mounted transmitters. Cost of review and listing by a testing agency, such as Underwriters' Laboratories, Inc. or Factory Mutual, add to the cost of intrinsic safety.

Cost of Installation—Intrinsic safety is at least potentially lowest in cost. The National Electrical Code permits wiring for approved intrinsically safe equipment in hazardous locations to be the same as in ordinary locations, such as multi-wire cables without special protection. For explosion-proof or purged equipment, all wiring must be in rigid conduit, all fittings must be explosion-proof, and conduits must be sealed. Purged equipment also requires an air supply system with purge failure alarms and in some cases automatic shutdown.

Maintenance—Intrinsic safety is rated A because the equipment is accessible for routine calibration checks and adjustments. Explosion-proof equipment must be de-energized before being opened or maintenance must be deferred until the area is known to be safe. Purging is similarly limited and alarms and interlocks must also be maintained.

Flexibility—Purging is rated A because essentially any standard or special instrument with reasonably tight housing can be readily adapted to purging. Explosion-proofing is limited to instruments available in that construction or that can be fitted into a standard, explosion-proof box. The need for external adjustments and indication can make this very expensive. Intrinsic safety must be evaluated as a system. Therefore, one of the serious problems which has faced both manufacturers and users in applying the intrinsic safety concept has been the inability to interconnect apparatus of different manufacturers and be assured that the combination is still intrinsically safe. The marking scheme described in paragraph A-4-2 of the NFPA Standard 493, provides a convenient way to assess the compatibility of apparatus of different manufacturers with respect to intrinsic safety. This concept facilitates the connection between two-terminal devices such as a two-wire transmitter and barrier.[3]

Purging, Pressurization, or Ventilation

Any reasonably tight enclosure housing electrical equipment can be made safe by providing a continuous flow of air or inert gas. The enclosure can be of any size from a small instrument case or fractional horsepower motor to an entire building, such as a control house. The essentials are:

1. A source of clean air.
2. Sufficient initial flow to sweep out gas that may have been present.
3. Sufficient pressure buildup in the enclosure to prevent entrance of combustible atmosphere.
4. Suitable alarms and interlocks.

AIR SUPPLY

For a single instrument or other small device, the instrument air system is the best source of air for pressurization since it is clean and dry. Where a large number of units are involved or an entire control house, the large volumes required make this impractical. The best solution varies with individual plant conditions. Finding a safe place for an air intake requires careful study. An intake 25 feet (7.5 m) off the ground and not within a 45 degree shadow cone of any potential source of vapor is generally considered adequate for a refinery handling vapors heavier than air. For gases nearly as dense or lighter than air, there is no definitive answer except distance and an upwind location. In any event, the suction line must be of substantial construction and free of leaks where it passes through a potentially hazardous area.

INITIAL PURGING

Pressurized apparatus, when put in service after having been open, may contain a combustible mixture. Before power may be turned on, sufficient air must pass through it to sweep out the combustible gas or at least reduce it to a harmless concentration. This is usually achieved by requiring a time interval to elapse after closing up the apparatus and starting purge flow before circuits are energized. A flow of four times the internal volume of the case is adequate for the usual instrument housing. Large or compartmentalized enclosures require special consideration.

If natural leakage does not provide the necessary purge volume in a reasonable length of time, auxiliary vents may be provided to accelerate the operation.

PRESSURE

During operation the enclosure must be maintained under a pressure of at least 0.1″ of water column (25Pa) to prevent influx of combustible mixture. This figure is equivalent to the wind pressure at 15 mph (24 km/h). While higher wind velocities might force outside air into the housing, the potential hazard is considered negligible since at these velocities any combustible vapors would be very rapidly dispersed.

While some warning of pressurization failure is needed, the specific requirement depends on the nature of the enclosure's contents and the degree of hazard outside. ISA[4] and the NFPA[5] classify purging as follows:

Type X purging reduces the classification within an enclosure from Division 1 to nonhazardous.

Type Y purging reduces the classification within an enclosure from Division 1 to Division 2.

Type Z purging reduces the classification within an enclosure from Division 2 to nonhazardous.

Type X purging would permit an arcing switch in a general purpose housing located in a truly hazardous, Division 1 location. Since failure of the purge air supply would soon lead to disaster, immediate automatic shutdown by pressure switch or flow detector is required.

Type Y purged equipment within an enclosure does not normally constitute a source of ignition, hence purge failure presents no immediate hazard. Only a visible or audible indication is required.

Type Z purged equipment, used in Division 2, where the atmosphere is not normally hazardous, again presents no immediate hazard and visible or audible indication is sufficient.

Temperature of all parts exposed to the atmosphere, in hazardous locations, must not exceed 80 percent of the ignition temperature of the combustible material.

Explosion-Proof Components

Article 100 of the National Electrical Code defines explosion-proof apparatus as "apparatus enclosed in a case which is capable of withstanding an explosion of a specified gas or vapor which may occur within it and of preventing the ignition of a specified gas or vapor surrounding the enclosure by sparks, flashes, or explosion of the gas or vapor within, and which operates at such an external temperature that a surrounding flammable atmosphere will not be ignited thereby."

Explosion-proof apparatus is not intended to be gas tight. It is assumed that no enclosure that may have to be opened from time to time for inspection or maintenance can practically be maintained gas tight. Hence, if the surrounding atmosphere is hazardous, the atmosphere within will also become hazardous, and an internal explosion may result. If the box holds together and the only openings are long, narrow and preferably crooked paths, the escaping gases will not be hot enough to ignite the external atmosphere. The British term, "flameproof," is perhaps more descriptive of the function than the American term, "explosion-proof."

There are two broad approaches to construction: a relatively tight box with broad flanges and tightly fitted, threaded (5 thread minimum) or rabbet joints, or a rel-

atively loose box with many small passages designed to minimize pressure buildup. The first requires a very strong box to withstand full explosion pressure, up to 175 PSIG (1208kPa) for Group C or D, 1000 PSIG (6900kPa) or greater for Group B or A. This is usually cast iron or cast aluminum. Plastic is used in Europe for smaller boxes. For the second approach, the pressure rise can be kept down, to 20 PSIG (138kPa) or less, allowing lightweight construction. Though officially recognized, the latter approach has been little used because of the possibility of vent passages becoming plugged by dirt or by injudicious use of a paint brush.

Intrinsic Safety

The National Electrical Code defines intrinsic safety in Section 500-1 as follows:

Intrinsically safe equipment and wiring shall not be capable of releasing sufficient electrical or thermal energy under normal or abnormal conditions to cause ignition of a specific flammable or combustible atmospheric mixture in its most easily ignitible concentration.

Abnormal conditions shall include accidental damage to any field-installed wiring, failure of electrical components, application of overvoltage, adjustment and maintenance operations, and other similar conditions.

A quantity of a combustible mixture must be heated to its ignition temperature for an explosion to occur. A weak spark heats so little mixture that heat loss exceeds heat supply and the incipient explosion dies out. A large spark heats enough mixture for combustion to become self-sustaining and an explosion propagates. If energy is kept at a low level, ignition will not occur. This is the basis of intrinsic safety. It is not sufficient that energy be low in normal operation. It must also be low under any conceivable abnormal operation or fault condition.

Safe energy levels cannot be defined in any simple form. Ignition depends on specific gas, gas concentration, voltage, current, energy storage elements, contact material, contact size, and speed of opening or closing of contacts. Ignition of hydrogen (one of the most easily ignited gases) has been achieved under laboratory conditions (high voltage) with energy as low as 20 microjoules. For common hydrocarbons and the voltages actually encountered in instrumentation, energies circa 0.2–0.3 millijoules are required. Curves such as Figure 5.2d show limiting circuit parameters that provide ignition energy for a particular gas. They can be used safely only after careful examination of the specific equipment

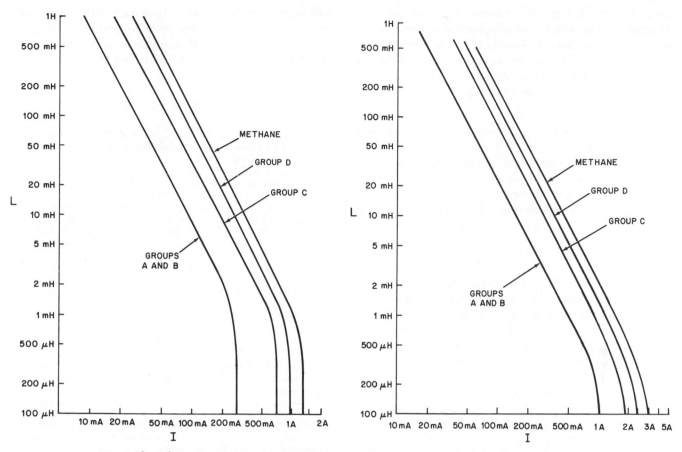

Fig. 5.2d Inductance circuits (Reprinted with permission from NFPA 493-1978, Standard for Intrinsically Safe Apparatus and Associated Apparatus for Use in Class I, II, and III, Division 1 Hazardous Locations, Copyright © 1979, National Fire Protection Association, Quincy, MA 02269. This reprinted material is not the complete and official position of the NFPA on the referenced subject, which is represented only by the standard in its entirety.) Left: Inductance circuits. (L > 1 mH). Minimum igniting currents at 24 volts. Applicable to all circuits containing cadmium, zinc or magnesium. Right: Inductance circuits. (L > 1 mH). Minimum igniting currents at 24 volts. Applicable only to circuits where cadmium, zinc or magnesium can be excluded.

by one skilled in the art and by application of an adequate safety factor. Actual ignition testing of the specific apparatus is the preferred practice.

LOOP CONCEPT

In evaluating equipment for intrinsic safety, it is always necessary to look at all elements of the complete loop. For example, an ordinary thermocouple, by itself, is unquestionably safe. Add a simple millivolt indicator and it is still safe. Connect the couple to a recorder, powered from a 120 V line, and the question of safety arises. What could happen in the recorder that might put dangerous energy on the thermocouple leads? The problem is easily solved by good design and construction in the recorder, but it cannot be overlooked.

Thus, there is no such thing as an intrinsically safe instrument unless it is one with a self-contained, low-energy power source, such as a small battery or solar cell that is isolated from all other power. All others must be

viewed as components of a loop. This, in general, precludes mixing instruments of different manufacture in the same loop. If it is to be done, someone must evaluate the combination and assume responsibility for safety.

DETERMINING INTRINSIC SAFETY

The practical answer for the instrument buyer and user is to look for certification by a qualified testing agency. In the United States these are the Underwriters' Laboratories, Inc. or Factory Mutual Research Corporation. It is the Canadian Standards Association in Canada, Physikalisch-Technische Bundesanstalt in West Germany, and BASEEFA (British Approvals Service for Electrical Equipment in Flammable Atmospheres) in Britain. Many other countries have their own agencies—the above are simply the best known.

While each testing agency has its own detailed rules and traditions, NFPA 493-1978, Intrinsically Safe Apparatus for Use in Division 1 Hazardous Locations, outlines a reasonably typical evaluation. As it says, however, "it is not an instruction manual for untrained persons but is intended to promote uniformity of practice among those skilled in the art." The following outline of the procedure will not make the reader an expert but will indicate the conservative nature of the approach.

Three steps are involved:

1. Circuit analysis—to determine worst possible fault conditions.
2. Evaluation—to insure a margin of safety under the conditions found above.
3. Construction review—to insure that critical components are reliable and that circuit will function as planned.

Circuit analysis is a review of the circuit, component by component, considering possible mode of failure of each and its effect on energy levels available in the hazardous area. It starts where line power enters the system and includes all parts of the interconnected instrument system wherever located—in the control house or in the field. The object is to pinpoint the fault condition or combination of conditions allowing highest energy in the field circuits.

Evaluation is the next step in the procedure. The purpose is to ascertain whether or not each fault condition constitutes a possibility of ignition. With the use of the actual circuit, the faults are produced by short- or open-circuiting components with the field leads connected to a test apparatus. The test apparatus consists of a pair of contacts operating in a chamber filled with the most readily ignited mixture of a suitable combustible gas with air. The contacts simulate a broken wire, a wire dragging on a surface, or the short-circuiting of a pair of leads.

An internationally accepted form of test apparatus is of West German origin. In July 1967, this test apparatus was tentatively accepted by the International Electrotechnical Commission meeting in Prague and was later adopted.

It is also possible to use measured voltages, currents, inductances, capacitances, etc., and compare them with the results of previous tests with similar circuit parameters. Suitable curves exist[3,4] (Figure 5.2d is an example) but they must be interpreted with care and a suitable factor of safety.

Construction review is the final step in the procedure. A circuit component must be of a reliable form of construction if it is to be depended upon for safety. A transformer that will withstand a high potential (1480 volts applied primary to secondary) immediately after being deliberately burned out is considered reliable. This presupposes that the transformer is so constructed that consistent performance of this nature can be anticipated. A resistor must withstand gross overloading without significantly changing in value. Live parts such as terminals must be so separated that an accidental short circuit is essentially impossible. If an instrument system survives this kind of examination its safety is assured.

REFERENCES

1. NFPA 70-1981 (ANSI C1-1981), "National Electrical Code 1981," Canadian Equivalent, CSA Standard C22.1-1982, "Canadian Electrical Code Part I."
2. American Petroleum Institute 500A, 3rd ed. April 1966 (Reaffirmed 1973), "Recommended Practice for Classification of Areas for Electrical Installations in Petroleum Refineries."
3. NFPA 493-1978, "Intrinsically Safe Apparatus and Associated Apparatus for Use in Class I, II and III, Division Hazardous Locations."
4. ISA-S12.4 Standard Practice, Instrument Purging for Reduction of Hazardous Area Classification, 1970. (Covers only small enclosures, such as instruments.)
5. NFPA 496, "Purged Enclosures for Electrical Equipment," 1982. (Similar to ISA S12.4, but expanded to cover large switchgear and complete control houses.)

BIBLIOGRAPHY

ANSI/ISA-RP12.6 (Approved May 26, 1977), "Recommended Practice, Installation of Intrinsically Safe Instrument Systems in Class I Hazardous Locations."

Burklin, C.R., "Safety Standards, Codes and Practices," *Chemical Engineering*, November 13, 1972.

Calder, W., "Electronic Instruments," *InTech*, March, 1982.

Comins, C., "International Practice for Electronic Monitoring Equipment," *InTech*, October, 1980.

———"Problems and Solutions for Intrinsically Safe Systems," *Instrumentation Technology*, August, 1978.

Cotugo, L., Rinard, G.A., and Zborovszky, Z., "Instrinsic Safety Computer Program," 1976 ISA Conference, Houston, Texas, Paper No. 76-597.

Hickes, W.F., "Use of Intrinsically Safe Refinery Equipment," *Oil and Gas International*, November, 1967.

IEC Standard 79-11:1976, "Electrical Apparatus for Explosive Gas Atmospheres, Part II: Construction and Test of Intrinsically Safe and Associated Apparatus."

ISA-S124-1970, "Instrument Purging."

ISA-RP12.1-1960, "Electrical Instruments in Hazardous Locations."

Kletz, T.A., "Hazards Associated with Instrumentation," Paper 51C, 64th AIChE Conference, San Francisco, December, 1971.

Krigman, A., "Instruments and Safety," *InTech*, February, 1983.

Magison, E.C., "Analyzing Intrinsic Safety Designs," *Instrumentation Technology*, July, 1970.

———"Engineering Instruments for Safety," *Instrumentation Technology*, February, 1969.

National Fire Protection Association, "Intrinsically Safe Process Control," NFPA-493-T.

Oudar, J., "Intrinsic Safety," *Journal of the Southern California Meter Association*, October, 1981.

Redding, R.J., "Barrier Methods of Safety," *Instruments & Control Systems*, July, 1969.

———"Intrinsic Safety and Its Significance," *Instrumentation Technology*, October, 1972.

Spadaro, B., "Process Monitoring in Hazardous Environments," *InTech*, February, 1983.

Widginton, D.W., "Some Aspects of the Design of Intrinsically Safe Circuits," SMRE Report 256, *Safety in Micro Research Establishment*, Sheffield, England, 1968.

Williams, J.R., "Intrinsic Safety," 1974 ISA Conference, New York, Paper No. 74-611.

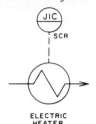

5.3 ELECTRICAL ENERGY MODULATION

Range:	Up to a few megawatts with standard units
Accuracy or Linearity:	Better than ±1% of full scale
Cost:	Proportional Control: $525 for 7 kW load; $2500 for 330 kW load for SCR power controllers
Partial List of Suppliers:	SCRs: Halmar Electronics, Inc.; Magnetics; Research Inc.; Robicon Corp., a Barber-Colman Co.; Vectrol
	Saturable Reactors: Kirkhof Co.; Niagara Transformer Corp.; Quality Transformer
	Contactors: Allan Bradley; Potter and Brumfield Div. of AMF, Inc.; Ward Leonard; Westinghouse

Electrical Energy Control Devices

Devices for modulating the flow of electrical energy or power can be grouped in two classifications: on-off devices and proportional control devices. These devices are usually used as the final control element in conjunction with a temperature controller as part of a temperature control loop, as shown in Figure 5.3b. (Characteristics of various devices are summarized in Table 5.3a.)

Modern examples of on-off final control elements include solid state relays (SSR) and electromechanical power relays or contactors. Modern proportional final control elements are almost all power controllers, with some saturable core reactors, power amplifiers, and a few ignitrons being used. On-off control always produces a cyclic response, while properly tuned proportional control produces close control with little or no cycling. For a more detailed discussion of this subject, refer to Section 1.1 on Control Theory.

Power Controllers

An assembly containing solid state switching devices, heat sinks, protection circuits, and the electromics to control the solid state switching devices is commonly called a power controller. The power controller accepts standard industrial control inputs such as 0–5mA, 1–5mA, 4–20mA, or 0–5 Volts, 0–10V, etc. and controls the power or voltage delivered to an electrical load in proportion to the input signal.[1] Power controllers are divided into two general types—phase angle fired and zero voltage

Table 5.3a
SERVICE APPLICABILITY OF VARIOUS CONTROL DEVICES

Service/Control Element	SCR	Ignitron	Saturable Core Reactor	Contactor
DC output required	✔	✔		
Operating voltage above 600V		✔	✔	✔
Small size	✔			✔
Severe overload or transients		✔	✔	
Overall efficiency	✔			
Low cost per watt	✔			✔

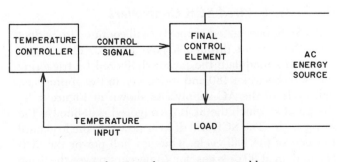

Fig. 5.3b   Typical temperature control loop

fired. Each has its advantages and disadvantages, which will be discussed in this section.

The power switching element in most power controllers is a silicon-controlled rectifier (SCR)—thus the general term "SCR power controller." The SCR is a four-layer, solid-state silicon structure consisting of alternating layers of negatively and positively doped material. It is a three-terminal device consisting of an anode, a cathode, and a gate, encapsulated and bonded to a thermally conductive base (Figure 5.3c). In order for the SCR to conduct, the anode must be positive with respect to the cathode and a trigger signal at the gate must initiate conduction. Once the gate has been pulsed by a DC voltage, conduction is self-sustained until the current flow through the SCR tries to reverse. When current attempts to reverse, the SCR "commutates" and returns to the blocking state and will remain "off" until triggered again, provided the anode is positive with respect to the cathode. In this manner an SCR can efficiently control loads from a few hundred watts to megawatts of electrical energy.

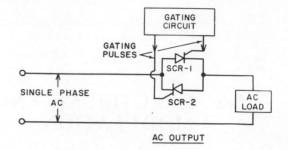

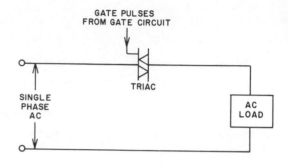

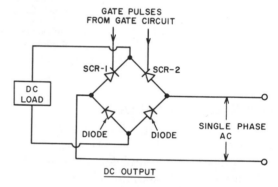

Fig. 5.3d   Power controller configurations—AC load (SCR and triac) and DC load

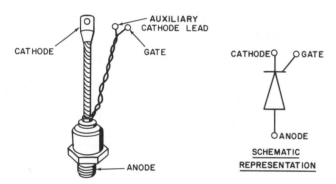

Fig. 5.3c   Silicon controlled rectifier

For DC application, two SCRs are combined with two diodes to provide a full wave rectified, modulated output. For AC loads, two SCRs are connected in an antiparallel fashion to provide modulated AC output. Below a kilowatt, some AC power controllers use a triac as the power switching element. A triac is a four-layer device with three terminals that behave as a pair of SCRs connected in an antiparallel fashion. Power controller configurations for DC loads, as well as SCR and triac AC loads are shown in Figure 5.3d.[2]

Phase Angle Fired SCR Controllers

An SCR, once gated properly, will continue to conduct until the current through it goes to zero. Phase angle fired units modulate the power delivered by triggering the SCR between 180 and 0 degrees in the appropriate half cycle of the AC supply. As shown in Figure 5.3e, the point at which the SCR is triggered is defined as the firing angle. The SCR will deliver power to the load until the end of the half cycle. Thus for half power the SCR is gated on at 90 degrees; for 75 percent power the firing

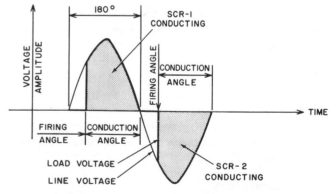

Fig. 5.3e   Definition of firing and conduction angles

angle would be 60 degrees, and the full-power firing angle would be 0 degrees. The same action occurs for the negative half cycle, thus resulting in the waveforms shown. Thus, by varying the point in each half cycle where the SCRs are turned on, the average power delivered to the load can be varied from zero to full power, in porportion to the control signal, with infinite resolution.

With the addition of a current transformer and some control circuitry, an adjustable current limit feature can be used with phase angle fired power controllers for loads that exhibit high inrush currents. These types of loads are heaters whose cold resistance is very low compared to the resistance at normal operating temperatures. Current limiting works by always gating the SCRs on at a 180-degree firing angle, then moving the firing angle back toward that desired by the control signal. As the firing angle is backed away from 180 degrees, the current is monitored and will be kept below the preset current limit before the firing angle is permitted to back away toward higher current output firing angles. This feature is also useful for transformer-coupled loads to ensure that the transformer is not driven into saturation and drawing excessive current.

SOFT START FEATURE

With phase angle fired power controller units, a soft start feature will start the firing angle at 180 degrees and ramp up slowly (typically $\frac{1}{2}$ to 1 sec) to the firing angle desired by the control signal. This feature can be incorporated with little or no additonal control circuitry and is usually standard on most phase angle fired power controllers. It helps to prevent high inrush currents experienced with transormer-coupled loads and nonlinear heating elements.

RADIO FREQUENCY INTERFERENCE (RFI)

Because a phase angle fired SCR power controller switches from the off state to the on state very quickly (less than 1 microsecond) at substantial voltage levels, harmonics in the radio frequency area of the spectrum are created. This radio frequency interference (RFI) can interfere with proper control loop performance. In severe cases or sensitive applications, filtering of the RFI by means of line chokes and bypass capacitors may be necessary.

Zero Voltage Fired Power Controllers

In zero voltage fired power controllers the SCRs are gated on at the beginning of each half cycle. Thus, each SCR will conduct for a full half cycle. Generally, power modulation is accomplished by varying the number of full cycles on versus the number of full cycles off over some time base (usually $\frac{1}{2}$ to 1 second). If we assume a one second time base, then at 50 percent power the load would receive 30 full cycles on and 30 full cycles off. At 20 percent, power would be 12 full cycles on and 48 full cycles off. Sometimes zero voltage fired units are called burst fired units because in this type of unit the load receives a burst of energy followed by some off time.

Some units proportion energy without a fixed time base by varying both the on time and off time. For this type of unit, 50 percent would be full cycle on and one full cycle off, 20 percent would be one full cycle on and four full cycles off. Typical waveforms of both methods are shown in Figure 5.3f.

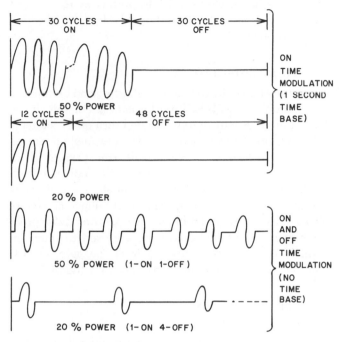

Fig. 5.3f Typical zero voltage firing output wave forms

Since a zero fired SCR controller gated the SCR on at the beginning of the cycle, current limiting cannot be done with this type of unit. For this reason zero fired units are usually applied to resistive loads and not transformer-coupled loads.[3]

Since the SCR is gated at or near zero voltage, almost no RFI is generated by this type of SCR power controller. This is the chief advantage of this type of SCR unit over the phase angle fired type.

General Considerations for SCR Power Controllers

Regardless of the type of power controller chosen, some common limitations exist that require protective devices or circuitry. These are usually provided by the manufacturer and are covered here to inform the user of the limitation and how the protective action is accomplished.

TRANSIENT VOLTAGE PROTECTION

SCRs can be damaged by the line voltage transients that exceed the voltage capability of the device. Most power controllers offer metal oxide varistor (MOV) protection circuitry across the SCR. The MOV will short circuit the voltage spike around the SCR and prevent a false turn on or permanent damage to the device.

CURRENT LIMIT FUSES

SCRs will fail due to load shorts just as mechanical contactors will. SCRs have fault current ratings that are specified by the terms I^2T and I surge. These ratings allow proper selection of the special fuses that clear in less than a half cycle in order to protect SCRs. The I^2T and I surge value of the fuse should be less than the SCR it is to protect.

DV/DT PROTECTION

A large change in voltage in a short period of time can cause problems with SCR performance. High dv/dt can trigger an SCR into an unscheduled turn on by inducing internal gate trigger currents via interjunction capacitance. Transformer-coupled loads can present dv/dt problems due to a lagging current in the inductive load. In this case the SCR will not commutate (turn off) when the gate trigger is stopped. Transient voltage spikes are also a form of dv/dt problems. To prevent dv/dt problems, gate to cathode capacitors are used to shunt internally coupled gate currents from the gate, or R-C snubber circuits are used in parallel with the SCR.

DI/DT PROTECTION

A high anode current flow during the turn on of the SCR can stress the junctions of the device. If a load requires near rated current before the entire junction area of the SCR is turned on, then the current density in the partially conducting junction is too high. This can cause localized heating and eventaully lead to failure of the device. Usually, gate drive circuitry ensures good junction turn on with a large fast-rising trigger pulse. If a load is purely resistive, then for phase angle fired units some small inductance should be put in a series with the load to prevent di/dt problems around a 90-degree firing angle. Usually zero fired power controllers do not have di/dt problems, since the voltage is near zero when the SCR is turned on.

COOLING AND HEAT SINKING

The temperature of an SCR will increase during normal operation due to the power dissipated in it. This power loss stems from the fact that there is a voltage drop of approximately one volt which the device is conducting. The current that an SCR can safely carry is dependent on the heat sinking provided. Operating at excessive temperatures will cause failures. Most power controllers up to 20 kW are convection cooled. Fan cooling is usually required for units between 30 kW and 500 kW. Above 500 kW water cooling is employed to maintain safe operating temperatures for the SCR. Suppliers can also supply thermostats mounted to the heat sinks to signal if the temperature of the SCR is getting too high.

Other Proportional-Type Final Control Elements

SCR power controllers offer such a great advantage in terms of size, cost, efficiency, and ease of maintenance that they almost always are the best choice for proportional final control element applications. Some of the other types are described in this section for completeness.

Ignitron Tube

Ignitron tubes are similar to the SCR in function, but quite different in construction. As shown in Figure 5.3g, the Ignitron is a mercury arc rectifier consisting of a graphite anode, a cathode containing a pool of mercury, and an electrode called the ignitor. Ignitron operation is achieved by applying a DC voltage pulse to the ignitor. The arc produces a large number of mercury ions that permit current to flow from anode to cathode, provided the proper potential exists. The ignitron can be subjected to 200 percent overloads for up to a minute without damage.

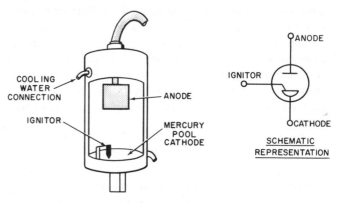

Fig. 5.3g   Ignitron tube

Because of the lower efficiency of the ignitron, the amount of cooling required is much greater than for an SCR. Normally, forced air or water cooling must be employed with the ignitron. In general, the ignitron is inferior to the SCR in terms of size, efficiency, life expectancy, and vibration resistance. It is for these reasons that the ignitron is rarely applied to new processes and is being replaced by SCRs, where possible, on existing installations.

Saturable Core Reactor

A saturable core reactor is a device consisting of an iron core, an AC or gate winding, and a control winding. The impedance of the gate winding can be changed by means of the magnetic flux produced by the DC current in the control winding. A DC control signal of a few watts can therefore control hundreds of kilowatts of AC power. Since the control signal requires watts of power, a magnetic amplifier is always used to boost the control signal from the process controller up to the power level required by the control winding of the saturable core reactor. Current limiting can be employed in a saturable core by means of a current transformer and an appropriate input to the magnetic amplifier. A diagram of a typical saturable core reactor setup is shown in Figure 5.3h.

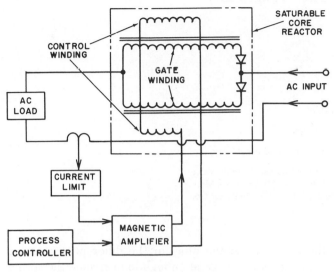

Fig. 5.3h Typical saturable core reactor control loop with current limiting

Saturable core reactors are relatively inefficient (90 percent), large in size, and require water or air cooling, depending upon power rating. Their main advantage is their immunity to modest transients and overloads. SCRs with proper design can offer similar reliability at greater efficiency and reduced size. Saturable core reactors cannot be used when the load voltage must be applied to the load. Typically, the load voltage is 5 percent of line voltage at zero input to the control winding, and it is 90 percent of line voltage at full control excitation.

Power Amplifier

A power amplifier is a combination of active devices, transistors, operational amplifiers or tubes, and passive elements, resistors, capacitors, and inductors to yield power amplification with certain characteristics such as linearity, frequency response, gain, etc. Such amplifiers are not usually used as final control elements. Reproduction of an input waveform, while important in communications, is not necessary in process control where only modulation of power is desired. For this reason the previously discussed devices offer more cost effective and efficient solutions.

On-Off Final Control Elements

As discussed earlier in this section, on-off control always results in a cyclic process response. Thus they should only be used on processes where overshoot and close control are not required. Power relays and contactors can be used for on-off final control elements. Additional coverage of relays and contactors is provided in Section 6.5.

REFERENCES

1. Donnal, C., *SCR Power Controller Manual*, Halmar, Inc.
2. *Machine Design*, Electronics Reference Issue, Chapter 4, May 13, 1982.
3. *Thyristors Manual*, General Electric Co., 1982.

BIBLIOGRAPHY

Hall, J. "Motor Drives Keep Pace with Changing Technology," *Instruments & Control Systems*, September, 1982.
Janki, C., "What's New in Motors and Motor Controls?" *Instruments & Control Systems*, November, 1979.
Kinligh, S.S., "ABC's of SCR Controllers," *Instrumentation Technology*, February, 1976.
Pendarvis, M.T. "SCR Control of Saturable Resetors," *Instruments & Control Systems*, August, 1970.

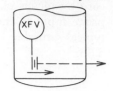

5.4   EXCESS FLOW CHECK VALVES

Materials of Construction:	Brass, steel, stainless steel
Sizes:	$\frac{1}{2}''$ to $10''$ (12.5 mm to 250 mm)
Cost:	$25 to $1200 depending on size and construction
Partial List of Suppliers:	Beckett Co.; Fisher Governor Company; Jordan Valve Div. of Richards Industries, Inc.; Metal Goods Manufacturing Co.; Trinity Steel Co.; Weatherhead Co.

Excess flow valves are in-line safety devices that act to limit flow of liquids or gases out of a pressurized system. While they will pass normal rates of flow, they will close against excess outward flow rates in the event that the pressurized system is opened to atmosphere due to pipe or hose breakage or because of system misoperation.

The excess flow valve consists of a plug, a seat, and a spring all housed or supported in a cylindrical tube. The valve may be threaded so that it can be screwed into the pipe line or can be flanged for tank nozzle mounting. See Figure 5.4a.

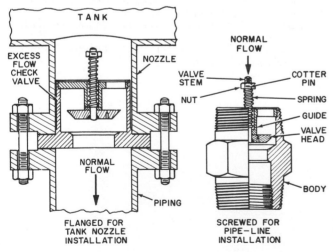

Fig. 5.4a   Excess flow check valves

Under normal operating conditions, the force generated by the flowing process fluid is directed against and tends to close the valve head. The spring is arranged to work against this force and keep the valve open. As flow increases above the normal level to an excessive rate,

the force against the plug, or the differential pressure across it, becomes great enough to overcome the spring force and the valve closes. There are one or more bleed ports around the plug so that after valve closure the bleed will allow pressure equalization across the plug and the valve may reopen. However, if a pipe break occurs, the differential across the valve will be the same as the pressure difference between the pressurized system and atmosphere, and the valve will not reopen until the pipe is repaired. Because of the equalizing feature of these valves, it should be remembered that they will not give tight shutoff.

One major application for excess flow valves is, in accordance with Factory Mutual recommended practice, on large pressurized storage tanks containing liquefied petroleum gas or other dangerous or expensive materials. L-P gas, for example, must be stored under relatively high pressures if it is to be kept in the liquid state. Propane, for one, has a vapor pressure of 192 PSIG at 100°F (1.3 MPa at 37°C). If a line to or from a propane tank opens when it is at some high storage pressure, a very large amount of propane will quickly escape to create an extremely hazardous condition. For this reason, it is good practice to install excess flow valves on every piping connection to the storage tank except for the fill and relief lines. The fill line should have a check valve (See Figure 5.4b) and the relief line should be unobstructed. Special check valves for the fill line are available from manufacturers listed at the front of this section. The check valves allow an unlimited flow in one direction and very little or none in the other. Some excess flow valves used for pressurized storage tanks are constructed so that they may be mounted internally to the tank and they are thus protected from mechanical damage. See Figure 5.4c.

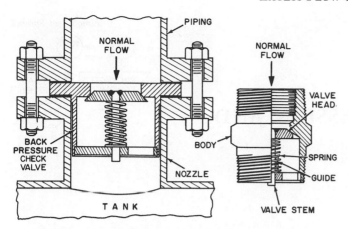

Fig. 5.4b Flanged and screwed check valves used in fill lines

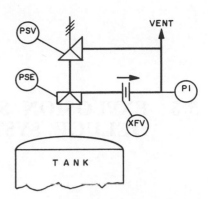

Fig. 5.4d Excess flow check valve in overpressure protection system

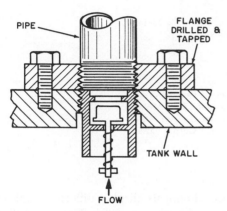

Fig. 5.4c Excess flow valve installed in tank

A second application for excess flow valves is to bleed off pressures above rupture discs. See Figure 5.4d. As noted in Section 5.14, rupture discs are actually differential pressure devices in that the set or burst pressure must appear across the disc before it will rupture. Any time the downstream side of the rupture disc is sealed away from atmospheric pressure, provision must be made to vent any pressure build-up on the downstream side to atmosphere. Examples of this are where two rupture discs are installed in series or where a rupture disc relief valve combination is used. The best way to vent this pressure is by means of an excess flow valve. The excess flow valve will allow small amounts of pressure release that are caused by rupture disc leakage or thermal breathing but will not pass the large flows that would accompany disc rupture. This installation is recommended by the ASME code for boilers and pressure vessels.

Since excess flow valves are safety devices, it is important that they be sized, selected, and installed properly. As a general rule for sizing, excess flow valves should be rated to close at about 150 to 200 percent of normal flow. The 150 percent figure should be used when the "normal" flow is well defined, or on installations involving the larger valve sizes. In the 150 to 200 percent sizing

range, the valve will be insensitive to surges during start-up and normal operation and will not chatter or restrict the flow. However, it will be sensitive to close against excess flows caused by pipe breakage. For valve selection, it is important to specify mounting orientation, flow direction, and flowing material since valve design is dependent on those operating conditions.

On the matter of installation, it is important to guarantee that the excess flow valve offers greater resistance to flow than any other pipe line item in the system.

Therefore, the downstream piping should not contain many bends, elbows, and tees, and should not be reduced in size below the line size of the excess flow valve. Excess flow valves will not necessarily respond to pipe breakage if it occurs on the discharge side of a downstream pump. This is because the pump will offer considerable resistance to flow even while running.

One way to check if an excess flow valve is sized, installed, and functioning properly is to simulate a pipe break downstream of the valve. This is done by opening a valve to atmosphere at the furthest point in the piping away from the excess flow valve. As the valve is opened, product should start to flow out of the system, but flow should then stop due to the action of the excess flow valve. This maintenance test should be conducted before start-up and then on a regular maintenance schedule.

BIBLIOGRAPHY

American Insurance Association, "Fire Prevention Code," latest edition.

American Petroleum Institute, *Manual in Installation of Refinery Instruments*, API Recommended Practice 550, latest edition.

Factory Mutual System, *Handbook of Industrial Loss Prevention*, McGraw Hill Book Co., New York, latest edition.

National Board of Fire Underwriters, "Liquefied Petroleum Gases," NBFU No. 58, latest edition.

National Fire Protection Association, "Explosion Venting," NFPA No. 68, latest edition.

National Fire Protection Association, "Flammable and Combustible Liquids Code," NFPA No. 30, latest edition.

Vervalin, C.H., *Fire Protection Manual*, Gulf Publishing Co., latest edition.

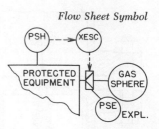

5.5 EXPLOSION SUPPRESSION AND DELUGE SYSTEMS

EXPLOSION SUPPRESSION SYSTEMS

Cost:	$10,000 to $60,000, depending upon size and complexity of system
Partial List of Suppliers:	Fenwal Electronics Div. of Kiddie, Inc.; Fike Metal Products; Graviner Mfg. Co. (England)

ULTRA-HIGH-SPEED DELUGE SYSTEMS

Cost:	$7500 to $50,000
Partial List of Suppliers:	Automatic Sprinkler Corp. of America; Conax Corp.; Fenwal Electronics Div. of Kiddie, Inc.; Fike Metal Products; Grinnell Fire Protection Systems Co.; MSA Instrument Div., Rockwood Systems Corp.; Varec Div., Emerson Electric Co.

Explosion suppression and ultra-high-speed deluge systems present a new approach in combating the hazards of explosion and fire. Traditionally, safety design has stressed two areas of concentration: prevention of explosion when possible or, if ignition does occur, application of proper measures to reduce the spread of damage. Explosion suppression and ultra-high-speed deluge systems act within milliseconds to extinguish an explosion or fire almost at its inception.

As similar as they may be in their speed of operation, the two techniques are quite different in their application. Each is discussed separately below.

Explosion Suppression Systems

Explosion suppression systems are designed to achieve a threefold purpose:

1. To confine and inhibit a primary explosion.
2. To prevent a secondary and more serious deflagration or a detonation.
3. To keep equipment damage at a minimum.

Build-up of pressure is usually kept to within 3-5 PSIG (21–104 kPa) of normal levels. Under these conditions some damage could be caused to light-walled vessels, but the danger of large-scale damage or fire is minimized.

Explosion suppression systems were developed in England shortly after the Second World War. Their first commercial application began in the mid-fifties. Subsequent installations in the United States date from 1958.

Because chemicals display different explosive characteristics and processes differ in physical dimensions, an explosion suppression system is usually a design package. In many instances, approval for insurance must be obtained from fire underwriters with evidence of design capability demonstrated in a test.

Explosions

A flame can be described in terms of its propagation from the source of ignition. There are three categories of flame behavior:

1. Burning—The flame does not spread or diffuse, but remains at an interface where fuel and oxidant are supplied in proper proportions.
2. Deflagration or explosion—The flame front advances through a gaseous mixture at subsonic speeds.
3. Detonation—Advancement of the flame front occurs at supersonic speeds.

The first task in the development of an explosion suppression system is to establish the propagation characteristics of the material in question. This is done by producing the required atmosphere in a cylindrical or spherical vessel. Oxidation is initiated by the application of energy, usually in the form of a spark. The test data are recorded through a pressure-time relationship generated by a pressure cell coupled to a high-speed oscil-

494

lograph. A typical dust explosion chart is shown in Figure 5.5a.

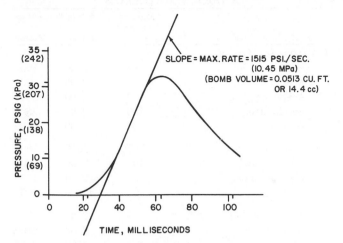

Fig. 5.5a   Typical explosion bomb test

Pressure within a spherical vessel after ignition of a quiescent fuel-air mixture can be predicted by the equation,

$$p = KS_r^3t^3P/V \qquad 5.5(1)$$

where
K is a characteristic of the system
$\quad$ S_r = radial flame speed
$\quad$ t = time
$\quad$ P = maximum pressure that would be reached within a closed container (also a function of the system but not dependent on the volume of the container).
$\quad$ V = volume of the vessel.
S_r in this equation is the radial flame speed and not the normal combustion velocity. They are related by the equation

$$S_r = S_n(T_f/T_i)(\overline{M_i}/\overline{M_f}) \qquad 5.5(2)$$

where S_n is the normal combustion velocity and the multipliers are the ratios of the initial and final (before and after combustion) absolute temperatures and average molecular weights. The difference in velocity is quite significant, since S_r will normally be on the order of 10 times S_n. Radial flame speeds for some materials are given in the table below:

RADIAL FLAME VELOCITIES OF EXPLOSIVE MIXTURES

Fuel	Oxidant	Typical Material	Radial Flame Vel. S_r ft/sec (m/s)
Organic dust	Air	Flour, starch	2–5 (0.6–1.5)
Organic vapor	Air	Propane, hexane	9–12 (2.7–3.6)
Hydrogen	Air		30 (9)
Organic vapor	Oxygen		80 (24)

For consideration of explosion suppression, it is more convenient to rearrange equation 5.5(1):

$$t = S_r(pV/KP)^{1/3} \qquad 5.5(3)$$

To be effective, maximum pressure, p, must be held to 2–4 PSIG (13.8–27.6 kPa). For a given substance, S_r, K, and P can also be considered constant, which leads to the simplified form,

$$t = CV^{1/3} \qquad 5.5(4)$$

Design of actual systems is based on producing explosions within test chambers to determine the parameters characteristic of the system. The information is then adjusted to the size of the real equipment by means of equation 5.5(4). The time required for the explosion to reach the limiting maximum pressure must be of sufficient duration to permit effective corrective action. For example, the explosion test illustrated in Figure 5.5a was performed in a small bomb with a volume of 0.0513 ft³ (11.4cc). The data from the early part of this test were then used to predict the normal curve for a vessel of 3.38 ft³ (946cc) by using equation 5.5(4):

$$t_1 = t_s(3.38/0.0513)^{1/3}$$
$$= 1.875 \, t_s \text{ (l is large, s small)}$$

The translated data were then made the basis for the test illustrated in Figure 5.5b.

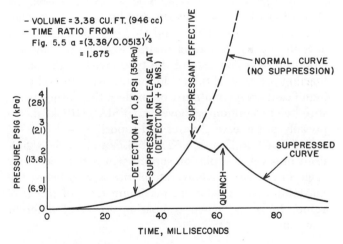

Fig. 5.5b   Explosion suppression sequence

How Suppression Works

The operation of an explosion suppression system is a race with time. On one hand, there is the physically determined build-up in pressure due to the explosion. The counterplay is detection of the explosion, application of suppressants to extinguish the deflagration, and corrective action to limit the extent of damage. Operation of a typical system is illustrated in Figure 5.5b. The basic relationships that make a process like this practical are:

1. The explosion can be detected early in the process. The pressure front advances at the speed

of sound (on the order of 1100 ft/sec or 330 m/s) while the flame front propagates at about 10 ft/sec (3 m/s).

2. The impulse received at a detecting device can be transmitted to the suppressant container at basically the speed of an electrical impulse.

3. Release of the suppressant is promoted by the explosive opening of a suppressant bottle or a high-speed hydraulically balanced system. The time period required for the triggering explosion to take effect is designed to be much less than the one in the vessel. The fill volume of the corrective explosive is kept very small.

4. The suppressant, ejected from several sources, is propelled into the explosive zone at a velocity of 200–300 ft/sec (60–90 m/s).

5. The course of events from initiation of the explosion to its complete extinction can be of very short duration. The specific time depends upon the characteristics of the material and the geometry of the system. Quench time of the explosion illustrated in Figure 5.5b was 60 milliseconds.

Considerable work has been done in measuring explosion characteristics. Typically reported values are average and maximum rate of pressure rise, and maximum pressure produced by the explosion. Some of this information is presented in Table 5.5c. A very extensive list of materials has been investigated by the U.S. Bureau of Mines. Reports RI 5753 and 5971 provide a tabulation of their results. A note of caution must be injected relating to explosion data of the type given in the table. Starting pressure of the explosion test was atmospheric. Significant correction must be made if the normal pressure before ignition is above 14.7 PSIA., (101 kPa), especially in the case of gases and vapors.

Reference to Figure 5.5b will disclose that the significant interval in explosion suppression is the very early stage of the curve where rate of rise is well below the maximum. Information on maximum rates of rise is of value, however, when comparing action of different materials.

Suppressant Chemicals

Effective explosion suppression requires getting sufficient amounts of chemical to the trouble area in a very short time, adapting required dispersing equipment to withstand the environment, and immunizing the system to outside influences, such as temperature of the vessel or its surroundings. The suppressant must also be compatible with the other chemicals in the system.

In general, an explosion is considered to be an oxidation reaction. Water and carbon dioxide, two popular materials for extinguishing fires in normal usage, are not generally utilized for explosions. Aside from a possible

Table 5.5c
EXPLOSION CHARACTERISTICS OF VARIOUS MATERIALS

Material	Maximum Pressure PSIG*	Rate of Rise, PSI/sec Maximum	Average
I. VAPORS AND GASES			
Acetaldehyde	94	2100	1900
Acetone	83	2000	1200
Acetylene	150	12000	8800
Acrylonitrile	109	2800	2600
Butane	97	2300	1700
Benzene	97	2300	1600
Butyl alcohol	104	2700	1600
Ethyl alcohol	99	2300	1550
Hydrogen	101	11000	10000
Methyl alcohol	90	3030	1500
Cyclohexane	104	2200	2000
Ethane	98	2500	2100
Ethylene	119	8500	6600
Hexane	92	2500	1500
Propane	96	2500	1700
Toluene	92	2400	920
II. AGRICULTURAL DUSTS			
Alfalfa meal	61	800	350
Cloverseed	76	1000	450
Coffee, instant spray dried	68	500	200
Corn, dust	95	6000	1700
Cornstarch, fine	145	9500	2900
Soy flour	104	1500	800
Sugar, powdered	91	5000	1700
Wheat flour	97	2800	900
III. PLASTIC DUSTS			
Cellulose acetate	108	6500	2200
Methyl methacrylate	101	1800	450
Nylon	95	3600	2200
Phenol furfural	88	8500	2000
Phenol formaldehyde	83	3600	2600
Polycarbonate	78	4700	1600
Polyethylene	82	2300	1100
Polypropylene	76	5000	1500
Polystyrene	77	5000	1500
Polyurethane	88	3700	1400
Rayon	88	1700	800
Urea formaldehyde	89	3600	1300

*1 PSIG = 6.9kPa

reactivity with the chemicals in question, relatively large quantities of water would be necessary to limit reactions. Carbon dioxide has a low effectiveness-weight ratio and would require large storage units. Materials have also been known to re-ignite after CO_2 extinguishment.

Halogenated compounds, mostly methane derivatives,

are more popular suppressants. The table below lists the properties of some of these agents.

COMPOUNDS USED AS EXPLOSION SUPPRESSANTS

Agent	Chemical Formula	Relative Effectiveness % by wt. ($CCl_4 = 100$)	UL Relative Toxicity 1 = highest 6 = lowest
Chlorobromomethane	CH_2BrCl	180	3
Bromodifluoromethane	$CHBrF_2$	161	
Bromotrifluoromethane	$CBrF_3$	195	6
Dibromodifluoromethane	CBr_2F_2	201	4
Carbon dioxide	CO_2	95	5
Water	H_2O	72	
Carbon tetrachloride	CCl_4	100	3

Chlorobromomethane and bromotrifluoromethane are most commonly used. While water owes its effectiveness to a cooling action and CO_2 relies upon its ability to exclude oxygen from the fire, the halogenated compounds seem to have a chemically inhibitive effect on the combustion reaction. Therefore, a chemical such as bromotrifluoromethane can be effective in extinguishing fires where oxidizing agents are present. Certain of the halogenated chemicals are also very low in residue so that subsequent interruptions for cleaning can sometimes be held to a minimum.

Explosion Suppression Hardware

The hardware for explosion suppression falls into three categories: detectors to discern the initiation of the explosion, control units to initiate corrective action in one or several directions, and actuated devices to blanket the protected area with suppressant. Adjacent areas are vented or isolated as required.

DETECTORS

Any physical characteristic that will give evidence of an explosion in its early stages can be used for detection. Absolute value or rate of change of pressure and temperature, and infrared and ultraviolet radiation levels have all been used in specific instances. Basic operational characteristics of these alternates are given below:

Temperature—Measurement is by means of a thermocouple with an exposed hot junction of very low mass. Even so, temperature is a slowly changing physical characteristic and would only be suitable under unusual circumstances or when other detection methods cannot be used.

Infrared Radiation—Detection by infrared radiation is extremely fast and sensitive. On the other hand, there are some factors that must be allowed for when designing a system with these detectors. Since they are a line-of-sight system, they must be placed where they "see" all of the locations where an explosion might develop. Usually multiple detectors can be used with overlapping coverage. In dusty atmospheres, precautions must be taken to assure that the lense opening is kept clean. The circuit design must incorporate a screening device to guard against false actuation by spurious infrared sources. To improve discrimination, a system of filters is often utilized and an adjustable threshold sensitivity is included.

Ultraviolet Radiation—Ultraviolet detectors are used alternately to infrared units. They are alike in their extreme speed of detection and in the design requirements typical of line-of-sight units.

Pressure—The most universally applied detectors are those relying on pressure. Activation in these devices is by means of a diaphragm and switch combination that is fast acting and has low inertial mass. The generally preferred form is by absolute pressure. Detectors for equipment that is normally at atmospheric pressure can be activated at 0.5 PSIG (3.5 kPa) or less. Where pressure fluctuations might be expected to be the normal rule, or where usual operating pressure is above atmospheric, a pressure rate-of-rise unit is required. Activation in this case is initiated by pressure drop across an orifice in excess of a preset minimum. Although pressure-activated devices do not respond as quickly as radiation detectors, they are suitable for a broader range of atmospheres.

CONTROL UNITS

The basic function of the control unit is to convert the weak signal generated by the detector into a form of energy sufficient to operate extinguishing and alarm devices. In the course of assuring system reliability, more than just the one duty is provided. The central control unit must:

1. Operate devices based on the actuating signal from a detector.
2. Monitor the suppression system for ground faults that might interfere with proper operation.
3. Contain internal and automatic battery back-up units that are activated in case of power failure.
4. Monitor the shut-down of specific pieces of equipment and to give local and remote alarm.
5. Provide a test circuit so that operation of the system components can be checked nondestructively.
6. Continually supervise integrity of all external circuitry.

ACTUATED DEVICES

Actuated devices produce a condition that limits the damage caused by the explosion. The most important of these are the suppressors and extinguishers. Additionally, there may be preaction vents, isolation valves, and other corrective measures initiated by detection of an explosion.

Suppressors and Extinguishers—The distinction between suppressors and extinguishers is basically in the method of mounting and the mechanism of release of the suppressant. Suppressors are mounted internally to the equipment being protected. They contain a relatively small volume (to 5000 cc.) and are actuated by detonation of an explosive charge within the container. These units are mounted close to possible sources of ignition and provide a fast source of extinguishing chemical. Figure 5.5d illustrates a hemispherical suppressor unit.

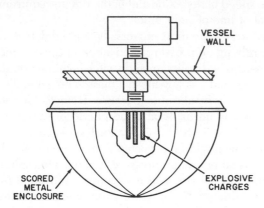

Fig. 5.5d   Hemispherical suppressor unit

Extinguishers are much larger in volume, up to 30 liters (about 7½ gallons), and are mounted outside of the equipment on a boss or flange. They are usually pressurized with nitrogen to 300 PSIG (2070 kPa) and are fitted with a diaphragm which is opened by an explosive charge. These units are used where more suppressant is required than would be available from the small suppressor unit, such as in large ducts and bag filters. They have an additional advantage in that they can be fitted with a new closure and refilled for reuse. See Figure 5.5e.

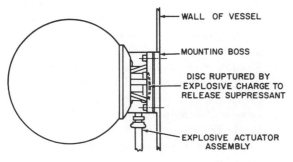

Fig. 5.5e   Spherical extinguisher

Explosive-Actuated Rupture Discs—If an explosion occurs in a pressure vessel fitted with a rupture disc, pressure in the vessel will build up to a point where it causes the disc to rupture, after which some relief of

pressure will result. Due to the short interval needed for the pressure to build compared to that required for stretching and rupturing of the disc, the actual maximum pressure reached in the vessel will be well above the rating of the disc. It is sometimes advantageous in the case of pressure vessels to include an explosive-actuated rupture disc as part of the system. On detection of an explosion, the closure is ruptured by a self-contained charge. In this way, the vent is completely open before the pressure wave reaches it and the maximum possible pressure build up is reduced. See Figure 5.5f.

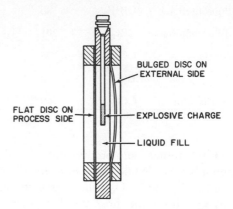

Fig. 5.5f   Explosive-actuated rupture disc

Other Auxiliary Units—Additional devices can be actuated by the system. Isolation of components can be affected by fast-acting valves and dampers, and auxiliary sprinkler systems can be activated by opening deluge valves. Pumps, blowers, agitators, and other process equipment can be interlocked into the system.

Applications

Explosion suppression is used for the protection of extremely hazardous systems in industry. The technique is primarily applied to bins, hoppers, reactors, air conveying systems, bag filters, and other closed arrangements. A particularly well-suited application is the protection of hammer mills and other grinding equipment where the elements of severe explosion are present in the form of well-mixed dust, air, and tramp metal.

There are cases where explosion suppression will not work. Decomposition usually cannot be halted because suppressant chemicals will not stop the reaction. Explosions that develop very high radial flame speeds (such as hydrogen-oxygen) are too fast for existing equipment. Many detonations (ultrasonic) also develop from an initial deflagration. It is possible to arrest the flame if detection and extinguishment can be effected before the detonation develops, but there is no means of dealing with detonation once it has developed.

The keyword in system design and application is reliability. Having a unit that is certain to work when it is needed justifies thorough investigation of the physical aspects of each case and the chemical nature of the ingredients. Reliability is assured by using devices that are known to be trouble-free, and by duplicating them. A given installation may have two or more detectors and several suppressors. Frequently, different types are installed in parallel.

Ultra-High-Speed Deluge Systems

Although ultra-high-speed deluge (UHSD) bears a good deal of similarity to explosion suppression, the unique characteristics of this system require separate study. The two methods resemble each other in the use of certain devices and in the time period in which they must function. But they differ in where they are applied and how they work.

UHSD was developed for extinguishing fires at their inception. Its point of application is generally an open area or room instead of a vessel or container. Where the room is a space capsule or hypobaric chamber, this distinction narrows. Fire in a solid rocket fuel processing plant can lead to an explosion unless, with the application of a UHSD system, it is extinguished promptly.

Suppressant for UHSD is almost always water.

Detectors

Since UHSD is applied in open areas and detection must take place within a very short time interval, detection devices that depend on pressure or temperature change are of little value. For UHSD, the speed and sensitivity advantages of ultraviolet and infrared detectors have been used successfully. (See description of these units under Explosion Suppression, above).

Control Units

The function of the control unit is basically similar to those described under Explosion Suppression. In some cases a cycle timer is also included as a part of the package. After a set time of operation the water is turned off and the detector is reinterrogated. If the alarm condition still exists, the deluge system reactivates for the set period. This feature is desirable to prevent flooding by the large quantities of water released.

Actuated Devices

Deluge systems must apply a lot of water on the source of ignition within a very short time. Where density requirements for normal high-hazard applications (Class I, Group D) may run 0.3 gpm/ft² (2×10^{-4} m³ps/m²), in the case of these special hazards the requirement is frequently 7.5 gpm/ft² (5×10^{-3} m³ps/m²). A plentiful source of water at sufficient pressure is required with the lines sized for low pressure drop. Available head is a significant factor in the speed of response since the water delivery time is proportional to the square root of the supply pressure.

There are two basic deluge system designs: the high-speed deluge valve and the pressure-balanced nozzle. Both of them depend upon a completely air-free, primed piping system to insure fast action. Tests by one firm have shown that an air pocket amounting to 5 percent of the total volume will double the operating time.

High-Speed Deluge Valve System—An explosive-actuated deluge valve is used to initiate flow. In order to prime the system, a bypass is provided around the valve and the nozzles are sealed with a protective cap. The cap is forced off by pressure in the nozzle when the system is activated. Figure 5.5g illustrates typical piping, and operation of the deluge valve is shown in Figure 5.5h.

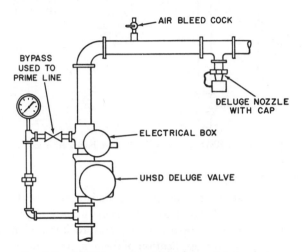

Fig. 5.5g   UHSD system with high-speed deluge valve

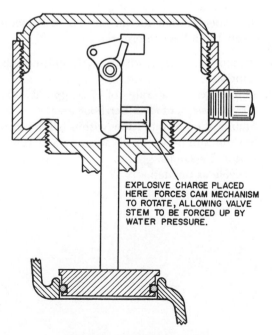

Fig. 5.5h   UHSD deluge valve

Pressure-Balanced Nozzle System—The pressures of the main water line and a pilot line are balanced at the nozzle to keep it closed. The pilot line takes off from the main riser through an orifice. Bleed cocks are provided to prime both lines. On actuation of the system, one or more solenoid valves vent the pilot line. Pressure in the main opens the nozzle to cause flow. Figures 5.5i and 5.5j show a typical system and internal construction of the nozzle.

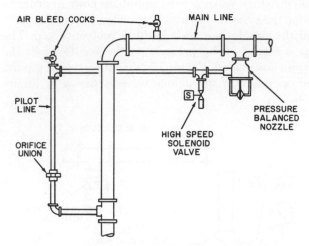

Fig. 5.5i   UHSD system with pressure-balanced nozzle

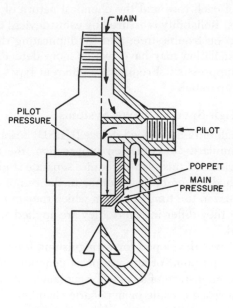

Fig. 5.5j   Pressure-balanced nozzle

Applications

UHSD systems are used in special hazard locations such as hypo- and hyperbaric chambers, munitions plants, munition stores on board ships, and rocket fuel processing plants. One application of these systems involves protection of a lathe operation where solid rocket propellant is machined.

A typical specification for an oxygen-rich operating chamber sets the following criteria:

1. System activation within 200 milliseconds of ignition.
2. Discharge at a rate of 7.5 gpm/ft² ($5 \times 10^{-3} m^3 ps/m^2$) of chamber floor area.
3. Stabilization of waterflow within half a second.
4. System shutdown in 20 seconds and resetting within 5 seconds.
5. Recycle as needed.

Operating time for UHSD depends greatly upon the system size and configuration. Water is generally applied within 15 to 200 milliseconds, with 90 milliseconds being an average for most applications.

BIBLIOGRAPHY

American Insurance Association, "Fire Prevention Code," latest edition.

American Petroleum Institute, "Manual on Installation of Refinery Instruments," API Recommended Practice 550, latest edition.

Factory Mutual System, *Handbook of Industrial Loss Prevention,* McGraw Hill Book Co., New York, latest edition.

National Board of Fire Underwriters, "Liquefied Petroleum Gases," NBFU No. 58, latest edition.

National Fire Protection Association, "Fire Facts," 1982.

National Fire Protection Association, "Explosion Venting," NFPA No. 68, latest edition.

National Fire Protection Association "Flammable and Combustible Liquids Code," NFPA No. 30, latest edition.

Vervalin, C. H., *Fire Protection Manual,* Gulf Publishing Co., latest edition.

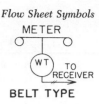
5.6 FEEDERS AS CONTROL ELEMENTS

Types of Feeders:	A-Dry feeders
	  1-shaker
	  2-vibratory
	  3-revolving plate
	  4-rotary
	  5-belt
	  6-screw
	  7-gravimetric
	B-Wet feeders
	  1-metering pumps
	  2-lime slakers
	C-Gas feeders

Types of Chemicals:	See Table 5.6a
Feeder Capacities:	In the range of a few hundred pounds per hour for types A and B
Typical Slurry Concentrations:	5 to 10%
Lime Slurry Slaking Time:	5 to 60 minutes
Feeder Costs:	$2,000 to $3,500 for 1 and 20 lb (0.45 and 9 kg) per hour capacities (C); $2,500 to $3,750 (A1,4); $2,500 to $5,000 (A2); $3,250 to $6,250 (B1); $5,000 (A6); $5,000 to $10,000 (A3); $6,250 and up (A5); $22,500 and up (B2)
Partial List of Suppliers:	BIF, a unit of General Signal Corp. (A,B,C); Capital Controls Co., Inc. (C); Carter-Day Co. (A); Clow Corp. (A); Crane Co. (B); Culligan, Inc. (C); Dravo Corp. (A); Fischer and Porter Co. (C); Mec-O-Matic Co. (B,C); Merrick Scale Mfg. Co. (A,B); Milton Roy Co. (B); Vibra Screw, Inc. (A,C); Wallace & Tiernan, Div. of Pennwalt Corp. (B,C)

Chemical feeders are subdivided into dry, wet, and gas types. The equipment can be proportioning or of the constant-rate type. Proportioning types are designed to feed chemical dosages in proportion to the influent wastewater flowrates. Constant-rate types are designed to feed at a fixed rate only. These devices stop feeding altogether when the influent water flow stops. The subject of solids flow metering is covered in Section 2.17 in the Process Measurement volume of this *Handbook*.

Dry Chemical Feeders

For a given throughput, dry chemical feeders are more compact than are solution feeders. For hydrated lime, slurried lime is often fed as a 5 percent solution. Hence, one cubic foot (0.028m³) of lime slurry carries about $3\frac{1}{2}$ pounds (1.6 kg) of calcium hydroxide, while one cubic foot (0.028m³) of hydrated lime powder contains about 35 to 40 pounds (15.6 to 18 kg) of calcium hydroxide. On the other hand, there usually are more handling problems with solids than with slurries.

Dry chemical feeders are employed for larger flows. Dry chemical feeders can be either volumetric or gravity feeders.

Since the material flow into the feeder is often uncontrolled, surge capacity ahead of the feeder is required. A feed hopper with sloping bottom is usually employed. For solids that tend to cake, vibrating bottom hoppers are used. The operating capacity range of dry feeders decreases with increasing solids particle size, and both the flow characteristics and uniformity of particle size affect the accuracy of feeding.

Table 5.6a
CHARACTERISTICS OF CHEMICALS USED IN WASTEWATER TREATMENT

Name and Formula	Molecular or Formula Weight	Grade or Size Available	Method of Shipping	Bulk Density (lb/ft³)*	Solubility in Water (g per 100g of water at 70°F or 21°C)	Remarks
Alum, Ammonium $Al_2(SO_4)_3 \cdot (NH_4)_2SO_4 \cdot 24H_2O$	906	Lump or crystal	Bags or cartons	65–70	15.1	—
Alum, Potassium $Al_2(SO_4)_3 \cdot K_2SO_4 \cdot 24H_2O$	949	Lump or crystal	Bags or cartons	65–70	11.4	—
Aluminum Sulfate (Alum) $Al_2(SO_4)_3 \cdot 18H_2O$	666	Lump, ground, or powdered solutions	Bags or in bulk; solutions in tank cars	Lump 55–65; powder 35–45	71.0	—
Ammonia NH_3	17	Liquefied gas	Cylinders or tank cars	41	34	—
Calcium Carbonate (Limestone, Calcite) $CaCO_3$	100	Pebbles, granules, or powder	Bags, drums, or bulk	Granules 100–115; powder 35–60	0.001	—
Calcium Chloride $CaCl_2$	111	Solids, flakes, or solutions	Bags or drums	25–45	74.5	—
Calcium Hydroxide (Hydrated Lime) $Ca(OH)_2$	74.1	Powder	Bags, barrels, or bulk	35–50	0.16	Used as 5 to 10% slurry
Calcium Hypochlorite $Ca(OCl)_2$	179	Powder	Cartons and drums	Granules 70–80; powder 30–50	Very soluble	Also used as basic hypochlorite
Calcium Oxide (Quicklime) CaO	56	Pebble, crushed lump, ground, or pulverized	Bags, barrels, or bulk	55–75	0.16	Forms $Ca(OH)_2$
Carbon, Activated C	12	Granules 8 × 30, 12 × 40, 14 × 40, 20 × 50, U.S. sieve mesh sizes	Bags and bulk	27–35	Insoluble	—
Chlorine Cl_2	71	Liquefied gas	Cylinders, multicar units, and bulk	97	0.72	—
Dolomitic Lime $CaO + MgO$	Varies	Pebble, lump, or powder	Bags, barrels, and bulk	Pebble 60–65 powder 37–60	Varies	Slakes to form $Ca(OH)_2$ + MgO; latter slakes very slowly.
Dolomitic Hydrated Lime $Ca(OH)_2 + Mg(OH)_2$	Varies	Powder	Bags, barrels, and bulk	27–43	Varies	Used as slurry

*1 lb/ft³ = 16.01 kg/m³.

Table 5.6a (*continued*)

Name and Formula	Molecular or Formula Weight	Grade or Size Available	Method of Shipping	Bulk Density (lb/ft³)*	Solubility in Water (g per 100g of water at 70°F or 21°C)	Remarks
			Characteristics			
Ferric Sulfate $Fe_2(SO_4)_3$	400	Granules	Bags, barrels, and bulk	60–70	Very soluble	Hydrolyzed by water. Stock solutions on 17 g per 100 cc.
Ferric Chloride $FeCl_3 \cdot 6H_2O$	270	Crystals, or solution	Bags, carboys, and bulk	Varies	Very soluble	—
Ferrous Sulfate (Copperas, Green Vitriol) $FeSO_4 \cdot 7H_2O$	278	Crystals or granules	Bags, barrels, and bulk	60–66	48	—
Hydrochloric Acid (Muriatic Acid) HCl	36.5	Anhydrous and 21 to 37% solutions	Carboys and bulk	80; solutions; 71–75	Very soluble	—
Phosphoric Acid H_3PO_4	98	50% P_2O_5 and 75% P_2O_5 solutions	Carboys and bulk	50% acid = 83; 75% acid = 98	Very miscible	—
Polyelectrolytes (Cationic, Anionic, and Nonionic)	Varies 10^6 to 20×10^6	Powder	Cartons	20–50	Varies	Used as 0.1 to 1% stock solutions
Potassium Permanganate $KMnO_4$	158	Crystals	Drums or bulk	90–100	6.3	—
Sodium Carbonate (Soda Ash) $Na_2CO_3 \cdot 10H_2O$	286	Lump and crystals	Bags, barrels, and bulk	65–70	58	—
Sodium Chloride (Common Table Salt) $NaCl$	58.5	Rock salt and granules	Bags, barrels, and bulk	50–70	36	—
Sodium Hydroxide (Caustic Soda) $NaOH$	40	Flakes, pellets and solutions	Drums and bulk	50–60	109	—
Sodium Phosphates Mono = $NaH_2PO_4 \cdot H_2O$ Di = $Na_2HPO_4 \cdot 12H_2O$ Tri = $Na_3PO_4 \cdot 12H_2O$	138 358 380	Solids and crystals	Bags, barrels, and bulk	Mono = 55–80; Di- = 46–50; Tri- = 55–60	Mono- = 98.0; Di- = 19.3; Tri- = 25.3	—
Sodium Silicate (Water Glass) $Na_2O \cdot 2$ to $5 SiO_2$	182 to 362	40°Be′ solution	Drums and bulk	Average 86	Very miscible	—
Sulfur dioxide SO_2	64.1	Liquefied gas	Cylinders, multicar units, and bulk	85	11.3	—
Sulfuric Acid H_2SO_4	98.1	98% 66°Be′ 60°Be′	Carboys, drums, and bulk	60°Be′ = 106.3; 98% = 114.5	Very soluble	—

Volumetric Feeders

The types of volumetric feeders used to handle water and wastewater treatment chemicals include shaker feeders, vibratory feeders, revolving plate feeders, rotary feeders, belt feeders, and screw feeders.

The *shaker feeder* (Figure 5.6b) consists of a shaker pan mounted beneath a hopper. The back end of the shaker is mounted on hanger rods, while the front end is carried on wheels, and it can be moved by a crank. As the pan oscillates, the material is moved forward and dropped into the feed chute.

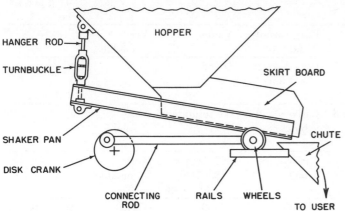

Fig. 5.6b   Shaker feeder

In most units the number of strokes of the feeder is kept constant, while the length of the stroke is varied. The angle of inclination of the pan is chosen for the particular material to be fed and varies from about 8 degrees for freely flowing materials to about 20 degrees for sticky materials. If arching is expected in the hopper, special agitator plates are installed in the hopper to break up the arches. The shaker feeder is rugged and self-containing and it can handle most types of material regardless of particle size or condition.

The *vibratory feeder* (Figure 5.6c) consists of a feed chute (either an open pan or closed tube) that is moved back and forth by the oscillating armature of an electromagnetic driver. Material transfer rate to the treatment process can be adjusted by adjusting the current input to the electromagnetic driver, which controls the pull of the electromagnet and the length of the stroke. This feature also permits flowrate adjustment or control from a distant control panel.

The feed chute can be jacketed for heating or cooling, and the tubular chutes can be made dusttight by flexible connections at both ends. The vibratory feeders can resist flooding (liquid-like flow) and are available for a wide capacity range (from ounces to tons per hour).

Revolving plate feeders (Figure 5.6d) consist of a rotating disk or table located beneath the hopper outlet. This (usually horizontal) table is driven by gears from above or below. As the table rotates, material is drawn from the hopper and is scraped off by the skirt board. Feedrate is controlled by adjusting the height of the adjustable gate of the position of the skirt board.

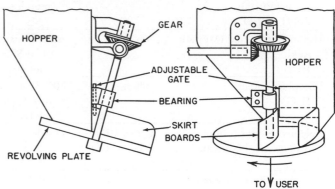

Fig. 5.6d   Revolving plate feeder

Revolving plate feeders handle both coarse and fine materials. Sticky materials are also handled satisfactorily because the skirt board is able to push them into the chute. This type of unit cannot handle materials that tend to flood. A variation of the revolving plate feeder utilizes rotating fingers to draw feed material from the bin. Revolving plate feeders can be equipped with arch breaker agitators in the conical throat section of the hopper.

Rotary feeders consist of a rotating shaft with a number of vanes that form individual pockets (Figure 5.6e). As the shaft rotates, each pocket becomes filled with ma-

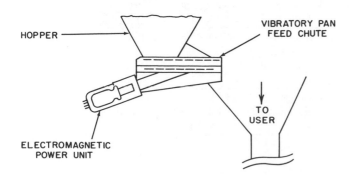

Fig. 5.6c   Vibratory feeder

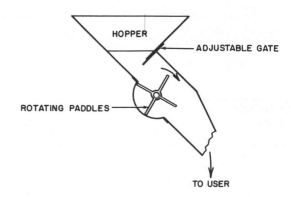

Fig. 5.6e   Rotary feeder

terial from the hopper above and transfers its contents to the chute below. The rate of flow is changed by changing the speed of rotation. This feeder provides a relatively even feed flowrate but can handle only free-flowing materials.

A variation of the rotary feeder is the rotary air-lock feeder, which is a rotary vane feeder in an airtight enclosure where the feeder serves as an air-lock gate. This type of feeder is desirable when handling extremely fine materials or materials capable of being aerated.

Belt feeders are short belt conveyors with full skirt boards (retaining walls on both sides) and flat idlers. A variation is the apron type, which utilizes a number of overlapping pans to prevent leakage of material. Belt conveyors provide a uniform feedrate, which can be adjusted by varying the belt speed. These are medium- to high-capacity machines (hundreds of pounds to hundreds of tons per hour) and do not find much application as chemical feeders in wastewater treatment.

Screw feeders can handle free-flowing materials and deliver uniform flowrates (Figure 5.6f). Rate of flow is determined by the speed of the screw shaft, which can be varied. A variable pitch screw is often employed to permit a uniform draw of material (no localized packing) throughout the length of the screw. Agitating arms in the hopper can be incorporated for materials that tend to bridge. A major advantage of screw conveyors is their resistance to flooding caused by the liquid-like flow of very fluid powders.

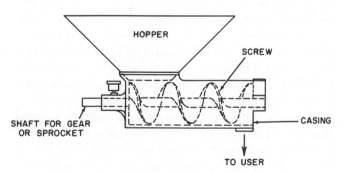

Fig. 5.6f   Screw feeder

Gravimetric Feeders

Gravimetric feeders are a combination of a volume feeder and a weighing mechanism guaranteeing mass flowrate control of the charged dry solids. They are available in capacities ranging from a few pounds per hour to several tons per hour, and their charging accuracy is fairly high, with the error being around 1 percent.

In the gravimetric *belt feeder* (Figure 5.6g), the feed material is transferred from the feed hopper by a volumetric feeder to the weigh belt. The belt can be driven by a constant speed drive (no belt slippage) and is usually mounted on load cells. The "belt loading" weight signal generated by the load cells is continuously compared by

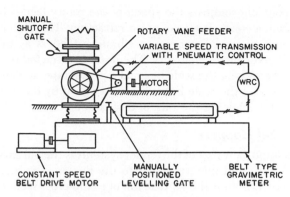

Fig. 5.6g   Gravimetric feeding system using a rotary valve volumetric feeder controlled by a belt-type gravimetric meter

the weight controller (WRC) to the set point representing the desired feed mass flowrate. This differential controls the varidrive on the volumetric screw feeder.

Another type of gravimetric feeder is the *self-powered feeder* (Figure 5.6h). These are not as accurate as the belt types, but neither do they require a power source. The dry solid material flows through the inlet spout and strikes the impact pan. If its force is greater than the preset poise weight, the impact pan and gate will rotate clockwise, thereby restricting the flow of material. If the impact force is less, the gate opens further to permit greater flow of the charged material.

Another feeder type is the continuous *loss-in-weight feeder* (Figure 5.6i). The weight of material in the hopper is counterbalanced by a retracting poise on a scale beam.

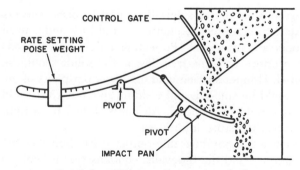

Fig. 5.6h   Self-powered gravimetric feeder

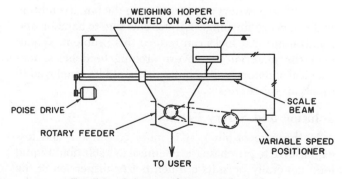

Fig. 5.6i   Continuous loss-in-weight feeder

The controller adjusts the speed of the rotary feeder to maintain a constant rate of poise retraction. This system can employ rotary, screw, or vibratory feeders as the modulated control elements. The poise is retracted by a lead screw along the scale beam at a preset rate, and any unbalance of the scale beam is counteracted by the controller by changing the rate of discharge.

Bins and Hoppers

An important part of the feeder systems is the bin from which the solid material flows to the feeder. The most common design is the vertical bin of round cross section, constructed of metal.

Hoppers are used to obtain complete discharge of bin contents. A deaerating surge hopper is shown in Figure 5.6j. Hoppers can be designed to have multiple outlets to feed more than one feeder.

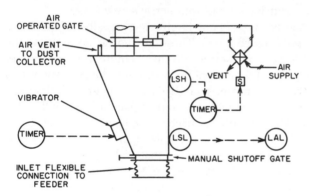

Fig. 5.6j   Deaerating surge hopper

In the handling of solids that tend to pack, corners facilitate arching; consequently, round bins and conical hoppers are preferred. Projecting ledges and tie-rods that interfere with the free flow of the solids should be avoided. Hopper surfaces and the lower periphery of the bin should be smooth, and welding splatter, rivets, and valleys should be avoided. All joints should be airtight to prevent moisture infiltration.

When a temperature differential exists between the bin contents and the bin walls, condensation is likely to occur, causing particles to agglomerate. Proper insulation and bin ventilation can alleviate this problem. Aeration of fine powders can occur inside the bin, a condition that can cause flushing. (Erratic flow due to aeration and fluidization). Air locks can reduce this problem. Whenever the material has severe arching tendencies, mechanical agitators or live bottoms should be employed to break up the arches[1,2].

Solution and Slurry Feeders

In the case of alum, it is more convenient and less expenseive to purchase the chemical as a solution. Liquid feed generally permits a much better dispersion of the chemical into the wastewater stream, regardless of the chemical additive. The feeder system for solutions and slurries generally consists of a dissolver and a feeder.

In dissolvers, some chemicals, such as alum, form a clear solution while others, such as lime, usually form thin slurries. Polyelectrolytes are often semicolloidal suspensions.

Soluble chemicals are dissolved in an agitated vertical tank; the residence time requirement varies with the ease of dissolution of the chemical and ranges from 10 to 50 minutes or more. In smaller tanks, agitators can be mounted at an angle, thus avoiding the need for baffling. Propeller-type agitators are the must common. Power requirements are usually in the range of 1 to $1\frac{1}{2}$ hp (746 to 1120W) per 1000 gallon (3780 l) of usable tank volume. For large particles or with slow dissolving materials, heating coils or jackets are employed to hasten dissolution.

In circumstances in which heat absorption or evolution accompanying the dissolving step is substantial, automatic temperature control is necessary.

Slurries, Lime Slaking

Slurry mixers or lime slakers employ proportioning devices to feed the chemical at predetermined ratios to the diluting water and are usually provided with temperature controls to better dissipate the heat. In case of substantial heat evolution, the vapor removal method of cooling can also be provided. Slaking tests should be made before starting the detailed design.[3]

In a typical lime slaker (Figure 5.6k), quicklime or hydrated lime is fed into the slaker by a volumemetric or gravimetric feeder. For quicklime, gravimetric feeders are usually employed. Lime-to-water ratio is set at about $3\frac{1}{2}$ pounds (1.58kg) of water to 1 pound (0.45kg) of lime. Water flow is ratio controlled in proportion to the lime feed. The slaker is divided into three compartments. Lime slaking takes place in the first compartment, where the lime and water are vigorously agitated by a horizontal agitator. Slaking is substantially complete in this chamber where the temperature may reach up to 185°F (85°C). Temperature is controlled by automatic introduction of more water as needed into the slurry. Suction created by the centrifugal dust blower is relieved by a vent connection at the lime feed end of the chute.

The slurry in the first chamber is fluid enough to flow under agitation and is moved across the unit by the static head caused by the introduction of new material. The slurry is next subjected to a soaking period that is considerably longer than the slaking time.

The lime slurry is finally discharged from the slaker compartments to the slurry surge tank. A grit screen retains all grit. The slurry at this point contains about 25 percent solids. Additional dilution to the desired level is done in the slurry surge tank. The lime slurry discharge rate is determined by the requirements of the wastewater treatment process. The process demand signal can set the charging rate of the lime feeder directly, or it can

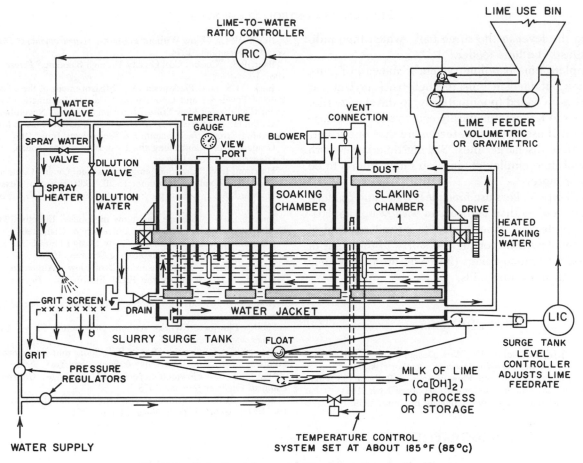

Fig. 5.6k Detention slaker used in the preparation of milk of lime

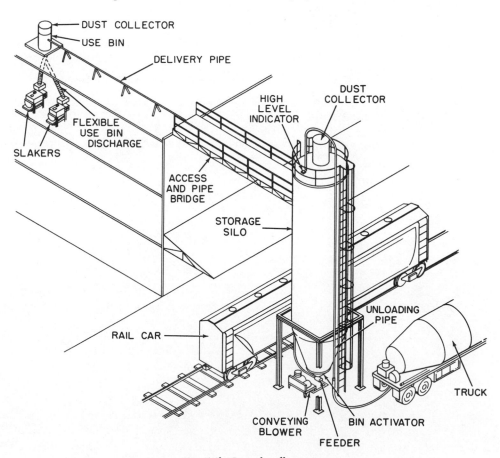

Fig. 5.6l Lime-handling system

influence the level in the surge tank, which then indirectly adjusts the lime feeder.

A complete lime handling system is shown in Figure 5.6l. Lime obtained in bulk through truck or tank car transport is unloaded (pneumatically) to the storage silo. From the silo it is pneumatically conveyed to bins close to the point of use. A bin may feed more than one slaker, using flexible discharge lines. Slaking time varies with the type of lime employed, and can range from 5 to 60 minutes or more.

Pug mill type slakers use much less water and produce lime pastes rather than slurries. In wastewater applications the paste has to be diluted to permit removal of grit and to allow transporting the slurry to the point of application. Detention slakers (Fig. 5.6k) are more widely employed than are pug mills.

REFERENCES

1. Johanson, J.R., and Colijn, H., "New Design Criteria for Hoppers and Bins," *Iron and Steel Engineering*, October, 1964.
2. Jeniche, A.W., "Storage and Flow of Solids," Bulletin 123, Utah Engineering Experiment Station, University of Utah, 1964.
3. "Chlorine Manual," New York: Chlorine Institute, 1959.

BIBLIOGRAPHY

AWWA Standard for Quicklime and Hydrated Lime, American Water Works Association, New York, 1965.

Baker, W.C., "Flow Without Fouling," *Measurements & Data*, November-December, 1974.

Baur, P.S., "Control Coal Quality Through Blending," *Power*, March, 1981.

Beck, M.S., and Plaskowski, A., "Measurement of the Mass Flow Rate of Powdered and Granular Materials in Pneumatic Conveyors Using the Inherent Flow Noise," *Instrument Review*, November, 1967.

Colijn, H., and Chase, P.W., "How to Install Belt Scales to Minimize Weighing Errors," *Instrumentation Technology*, June, 1967.

Grader, J.E., "Controlling the Flow Rate of Dry Solids," *Control Engineering*, March, 1968.

Harrison, J.W., "Coal Sampling Systems and Quality Control," *Coal Technology 80*, vol. 4, November 18-20, 1980, Houston, Texas.

Hayward, A.J., "Choose the Flowmeter Right of the Job," *Processing Journal*, 1980.

Jenicke, A.W., "Storage and Flow of Solids," Bulletin 123, Utah Engineering Experiment Station, University of Utah, 1964.

Johanson, J.R. and Colijn, H., "New Design Criteria for Hoppers and Bins," *Iron and Steel Engineering* October, 1964.

Laskaris, E.K., "The Measurement of Flow," *Automation*, 1980.

Lomas, D.J., "Selecting the Right Flowmeter," *Instrumentation Technology*, 1977.

Nolte, C.B., "Solids Flow Meter," *Instruments & Control Systems*, May 1970.

"Solids Flowmeter Works Without Obstructing Flow," *Chemical Engineering*, September 18, 1972.

Stepanoff, A.J., *Gravity Flow of Bulk Solids and Transportation of Solids in Suspension*, John Wiley & Sons, New York, 1969.

Watson, G.A., "Flowmeter Types and Their Usage," *Chartered Mechanical Engineer Journal*, 1978.

Zanetti, R.R, "Continuous Proportioning for the Food Industry,"*Instrumentation Technology*, March, 1971.

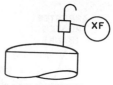

FLAME ARRESTER

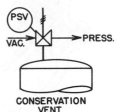

CONSERVATION VENT

EMERGENCY VENT

5.7 FLAME ARRESTERS AND CONSERVATION VENTS

FLAME ARRESTERS

Materials:	Aluminum, cast iron, and stainless steel; also available with special coatings such as heresite
Sizes:	1″ to 8″ (25mm to 200mm) with screwed or flanged connections
Cost:	$75 to $1000 (aluminum materials for the above size range)
Partial List of Suppliers:	Black, Sivalls & Bryson, Inc.; GPE Controls; Groth Equipment Corp.; Industrial Controls Div., The Johnston & Jennings Co.; The Protectoseal Co.; Varec Div. of Emerson Electric Co.

CONSERVATION VENTS

Materials:	Aluminum, cast iron, and stainless steel; plastic and rubber gaskets and coatings are also available
Sizes:	Conservation Vents: 2″ to 10″ (50mm to 250mm) Emergency Vents: 16″ to 24″ (400mm to 600mm)
Rating:	150# ASA or API Flanges
Cost:	Conservation Vents (aluminum): $150 to $750; Emergency Vents (aluminum): $350 to $750
Partial List of Suppliers:	Black, Sivalls & Bryson, Inc.; GPE Controls; Groth Equipment Corp.; Industrial Controls Div., The Johnston & Jennings Co., The Protectoseal Co.; Varec Div. of Emerson Electric Co.

Flame Arresters

Flame arresters are in-line or end-of-line venting devices provided with an internal flame-arresting grid. They are designed to prevent an external fire from entering a tank containing a flammable product. See Figures 5.7 a and b. The internal grid or bank of plates is of sufficient size and the plates are spread so that an external flame that flashes into the arrester will be cooled and extinguished before it can pass through into the storage tank. Flame arresters are available in combination flame arrester/conservation vent units. See Figure 5.7c.

There are several requirements for the sizing an application of flame arresters. They must be large enough to vent a quantity of air equal to the volume displaced by product transfer plus the volume generated by thermal expansion. The data given for conservation vents (see below) applies equally for flame arresters. It should be noted that a flame arrester of a given size has a lower capacity than the same size conservation vent. Therefore, the flame arrester will determine line and vent nozzle size if the two units are installed in series.

Factory Mutual requires that flame arresters be used under the following conditions:

1. On tanks that contain liquids with flash points below 110°F, (43°C).
2. On tanks that contain liquids with flash points above 110°F (43°C), but where the tank may be exposed to combustibles or other tanks containing liquids with flash points below 110°F (43°C).
3. On tanks where the contents can be heated to the flash point under normal operation.

The National Fire Protection Association requires that flame arresters or conservation vents be used on tanks

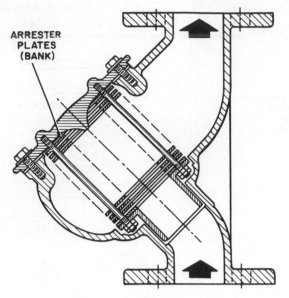

Fig. 5.7a In-line flame arrester

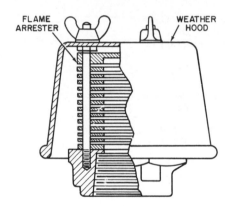

Fig. 5.7b End-of-line flame arrester

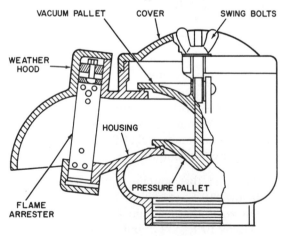

Fig. 5.7c Combination flame arrester/conservation vent

that contain liquids with flash points below 73°F (23°C). and boiling points above 100°F (38°C) and on tanks that contain liquids with flash points in the 73 to 100°F (23 to 38°C) range.

Flame arresters cannot always be relied upon to pro-

vide tank venting. Whenever the stored material in the tank can foam, plug, or freeze, it is possible that the space between the flame arrester plates will clog and restrict or stop venting. When this condition is anticipated, the flame arrester can be steam-traced to prevent product freezing. Under these conditions it is advisable to provide a secondary pressure-relieving device that will protect the tank in case of flame arrester plugging.

Conservation Vents

All large storage tanks are exposed to small changes in internal pressure resulting from (1) vacuums due to liquid thermal contraction or pump-out and (2) pressures due to liquid thermal expansion or pump-in. These changes are usually small (on the order of inches of water column), but because they act on the large internal area of the tank, their effect can be great. For example, a positive pressure of slightly over 3 in. of water column is enough to rupture the roof to side weld seam on many vertical tanks with cone roofs. Similarly, a small negative pressure over the large wall area may be enough to cause the walls to collapse. It is therefore necessary to provide some mechanism to keep the internal tank pressure within certain tolerance around the atmospheric pressure during thermal expansion and liquid transfer.

In many cases, a simple vent pipe is not enough. One of these is where the material has a low flash point and would cause a hazard if large amounts were allowed to vent continuously to atmosphere. Another is where the vapor is corrosive, toxic, or expensive and continuous venting to air is not desirable. Several tank constructions have been developed to eliminate or reduce the requirement for reliance on breather or vent equipment. Among these are various designs of floating or lifting roofs, and diaphragms that have sufficient expansion capacity to take up volume changes due to liquid transfer and thermal expansion. It should be noted that some of these tank constructions are safer and better able to conserve vapors than the standard cone roof tank. This is true even when the cone roof tank is properly vented through breather valves, emergency vents, and flame arresters.

However, there are many applications for the cone roof design and this section deals with vent sizing and application for those tanks. As noted, tank breathing capacity should be large enough to offset volume changes due to filling, emptying, and thermal expansion.

Determination of Capacity Requirements

There are several guides to tank venting requirements, but one of the most widely used is the Guide for Tank Venting of the American Petroleum Institute. Table 5.7d gives capacity requirements for thermal venting and Table 5.7e gives capacities for fire emergency venting of above-ground tanks. In addition, the following recommendations are made to cover venting required due to liquid transfer.

Table 5.7d
THERMAL VENTING CAPACITY REQUIREMENTS
(Cubic Feet per Hour of Air)*

Tank Capacity		Inbreathing (Vacuum) All Stocks	Outbreathing (Pressure)		
Barrels	Gallons		Flash Point 100°F (37.8°C) or Above	Flash Point Below 100°F (37.8°C)	
60	2,500	60	40	60
100	4,200	100	60	100
500	21,000	500	300	500
1,000	42,000	1,000	600	1,000
2,000	84,000	2,000	1,200	2,000
3,000	126,000	3,000	1,800	3,000
4,000	168,000	4,000	2,400	4,000
5,000	210,000	5,000	3,000	5,000
10,000	420,000	10,000	6,000	10,000
15,000	630,000	15,000	9,000	15,000
20,000	840,000	20,000	12,000	20,000
25,000	1,050,000	24,000	15,000	24,000
30,000	1,260,000	28,000	17,000	28,000
35,000	1,470,000	31,000	19,000	31,000
40,000	1,680,000	34,000	21,000	34,000
45,000	1,890,000	37,000	23,000	37,000
50,000	2,100,000	40,000	24,000	40,000
60,000	2,520,000	44,000	27,000	44,000
70,000	2,940,000	48,000	29,000	48,000
80,000	3,360,000	52,000	31,000	52,000
90,000	3,780,000	56,000	34,000	56,000
100,000	4,200,000	60,000	36,000	60,000
120,000	5,049,000	68,000	41,000	68,000
140,000	5,880,000	75,000	45,000	75,000
160,000	6,720,000	82,000	50,000	82,000
180,000	7,560,000	90,000	54,000	90,000

* See Section A.1 for SI units.

Table 5.7e
REQUIRED CAPACITIES FOR EMERGENCY RELIEF OF EXCESSIVE INTERNAL PRESSURE IN ABOVE-GROUND TANKS*

Tank Capacity		Total Pressure-Relief Capacity SCFH of air
Gallons	Barrels	
1,000	23.8	25,300
4,000	95.2	69,500
18,000	428	139,000
25,000	595	166,000
56,000	1,330	253,000
100,000	2,380	363,000
155,000	3,690	458,000
222,000	5,290	522,000
475,000	11,300	624,000
735,000	17,500	648,000
Unlimited		648,000

* See Section A.1 for SI units.

For Liquids with Flash Points Below 100°F (37°C)— Provide 17 scfh (0.13m³/h) of vent capacity for each gpm (l/m) of filling rate and 8 scfh (0.06m³/h) of inbreathing capacity for each gpm (l/m) of emptying rate plus sufficient capacity to meet Table 5.7d requirements.

For Liquids with Flash Points Above 100°F (37°C— Provide 8.5 scfh (0.063m³/h) of vent capacity for each gpm (l/m) of filling rate and 8 scfh (0.06m³/h) of inbreathing capacity for each gpm (l/m) of emptying rate plus sufficient capacity to meet Table 5.7d requirements.

Some other vent line requirements should be noted. In general, the vent line must be at least as large in diameter as the largest filling or emptying line to the tank. For small tanks the vent line sizes required for U.L. labeled tanks may be used. See Table 5.7f.

Table 5.7f
VENTS FOR UNDERWRITER LABORATORY LABELED TANKS*

Tank Capacity (Gallons)	Vent-Pipe Size (Inches)	
	Buried Tanks	Above-Ground Tanks
To 500	1	$1\frac{1}{4}$
501–1,000	$1\frac{1}{4}$	$1\frac{1}{2}$
1,001–3,000	$1\frac{1}{2}$	2
3,001–6,000	$1\frac{1}{2}$	$2\frac{1}{2}$
6,001–12,000	2	3
12,001–30,000	3	4

* See Section A.1 for SI units.

Types of Vents

Three kinds of vents are utilized to meet the various venting requirements. The first is a simple free vent breather. This is an open pipe with its outlet screened and pointing down, which prevents rain and debris from getting into the tank. The second type of vent used is the conservation vent or breather valve. See Figure 5.7 g. This valve is a low pressure relief and vacuum breaker valve enclosed in one housing. Operation of the conservation vent is simple and reliable. As pressure builds up in the tank there is a differential created across the pressure relief pallet. When this differential becomes great enough to lift the pallet, the pallet unseats and allows vapor flow out of the tank. Similarly, when a vacuum is developed in the tank the differential across the vacuum pallet eventually becomes great enough to lift the vacuum pallet and allows air flow in the tank.

There are several advantages to the conservation vent over the free vent. Conservation vents reduce the loss of vapors to the air and thus save product. The American Petroleum Institute did a study of the effectiveness of conservation vents in saving product. It was found, for example, that in a 30-foot-diameter (9m) tank having the capacity of 210,000 gallons (793,800l) an annual savings

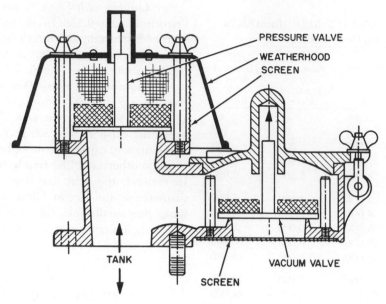

Fig. 5.7g   Conservation vent

of 1400 gallons (5292l) of material could be realized by use of a conservation vent. The study was based on a liquid with 5 PSIA (34.5kPa) vapor pressure at 60°F (16°C) and a vent pressure setting of $\frac{3}{4}$ in. H_2O (0.19kPa). By reducing product loss, the conservation vent reduces potential fire hazard and air pollution around the storage tank. Of course, the conservation vent becomes less effective as the vapor pressure of the stored liquid increases. If no liquid vaporized at ambient temperature, a free vent would be sufficient.

Whether a conservation vent is required, recommended, or not needed depends on the applicable code and judgment. API requires conservation vents on all tanks containing liquids whose flash point is below 100°F (37°C). The National Fire Protection Association requires a conservation vent on all tanks that contain a liquid with a flash point below 73°F (23°C) and a boiling point below 100°F (37°C). NFPA requires a conservation vent or flame arrester on tanks that contain liquids with flash points below 73°F (23°C) and boiling points above 100°F (37°C) and on tanks that contain liquids with flash points in the 73 to 100°F (23 to 37°C) range. (Flash point for a liquid is the lowest temperature at which sufficient vapors are given off to form a flammable vapor-air mixture above the liquid.) Conservation vents are generally recommended for liquids having flash points up to 140°F (60°C). Conservation vents should be considered for tanks where the stored liquid is heated to or near to the flash point. Stored liquids are heated to prevent freezing and to reduce viscosity.

Another application of conservation vents is on tanks that require an inert gas blanket. Inert gas blankets are used to prevent product contamination by air or by the moisture in air and to hold down the vapor layer of low flash point materials. Free vents are of no use where a gas blanket is required. One method of venting a blanketed tank is shown in Figure 5.7h.

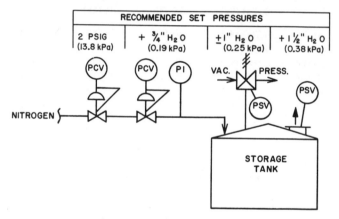

Fig. 5.7h   Venting of blanketed storage tanks

Emergency Vents

Emergency vents are a third type of vent and are used as a secondary relief device or where the storage tank may be subjected to external heating such as fire. The emergency vent is a large area cover that is opened when excessive pressure develops in the tank. See Figure 5.7i. Some constructions of emergency relief devices can be bolted to the tank manhole and double as manhole covers. NFPA requires that "Every above-ground storage tank shall have some form of construction or device that will relieve excessive internal pressure caused by external fire." This requirement may be met by use of the emergency vent or by the use of one of several special

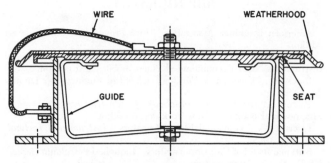

Fig. 5.7i   Emergency manhole cover

tank constructions. These special constructions are float-ing roofs or weak-seam constructions.

Desiccating Vents

The desiccant vent air dryer is a device that is some-times used on storage tanks in conjunction with vents and that may be installed in the vent piping. The des-iccant dryer may be required to prevent the contami-nation of stored liquid due to moisture entering through the tank venting system. See Figure 5.7j. It is important to note that the breather valve also has a vacuum relief setting. This is required to prevent vacuum build-up in the tank if the water in the desiccant freezes and thus plugs the dryer. The vacuum setting on the breather valve should be about 1 in. H_2O (0.25 kPa) below the setting on the vacuum valve.

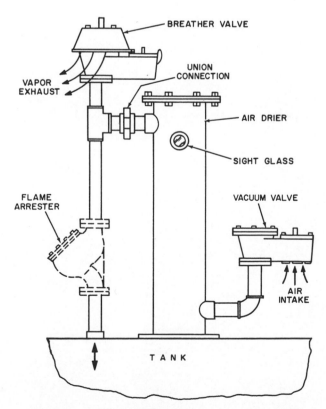

Fig. 5.7j   Breather valve installation with desiccant dryer

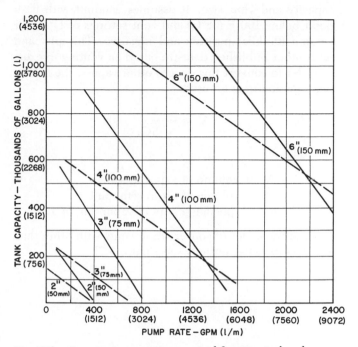

Fig. 5.7k   Conservation vent size required for given tank and pump sizes. Capacity based on SCFH (m^3/h) of air and pressure and vacuum setting at $1\frac{1}{2}''$ H_2O (0.38 kPa). Solid lines for pressure, dotted lines for vacuum relief.

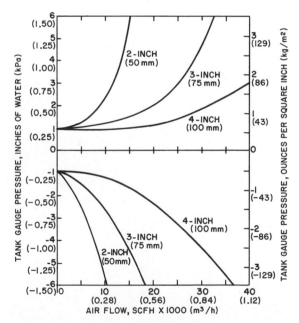

Fig. 5.7l   Approximate capacities for conservation vents with various settings and sizes.

Sizing

Most sizing data on vent devices are empirical. There-fore, once venting requirements are determined the user should refer to published capacity charts to select vent sizes. As a guide to conservation, vent sizing Figure 5.7k is presented. This is a plot of tank size versus pumping capacity and vent size. It assumes a liquid with flash point below 100°F (38°C) and vent settings of $1\frac{1}{2}''$ water column (0.4kPa) positive and negative. This plot and most other published data are based on venting free air. In order to convert this data to other vapors, the following equation may be used.

$$V = \frac{A}{\sqrt{(SG)}} \qquad 5.7(1)$$

where

$\qquad$ V = volume of vapor in scfh,

$\qquad$ A = volume of air in scfh, and

$\qquad$ SG = specific gravity of vapor (air = 1.0).

For the capacities of conservation vents with other than $1\frac{1}{2}''$ H$_2$O (0.4kPa) settings see Figure 5.7l.

BIBLIOGRAPHY

American Insurance Association, "Fire Preventsion Code," latest edition.

American Petroleum Institute, "Manual on Installation of Refinery Instruments," API Recommended Practice 550, latest edition.

Block, B., "Emergency Venting," *Chemical Engineering*, January 22, 1962.

Factory Mutual System, *Handbook of Industrial Loss Prevention*, McGraw Hill Book Co., New York, latest edition.

Lisciani, C., "Vents and Flame Arrestors," *Instrumental Technology*, July, 1968.

National Board of Fire Underwriters, "Liquefied Petroleum Gases," NBFU No. 58, latest edition.

National Fire Protection Association, "Fire Facts," 1982.

National Fire Protection Association, "Explosion Venting," NFPA No. 68, latest edition.

National Fire Protection Association, "Flammable and Combustible Liquids Code," NFPA No. 30, latest edition.

Vervalin, C. H., *Fire Protection Manual*, Gulf Publishing Co., latest edition.

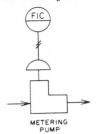

METERING
PUMP

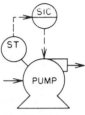

ELECTRIC SPEED
CONTROL

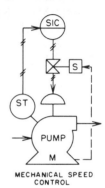

MECHANICAL SPEED
CONTROL

5.8 PUMPS AS CONTROL ELEMENTS

Types of Pumps:	A—Radial-flow centrifugals B—Axial-flow and mixed-flow centrifugals C—Reciprocating pistons or plungers D—Diaphragm pumps E—Rotary screws F—Pneumatic ejectors G—Air lifts H—Metering pumps I—Variable-speed drives
Rangeability:	Variable speed drives: From 4:1 to 40:1 (electric types), from 4:1 to 10:1 (mechanical designs) up to 40:1 (hydraulic types) Metering pumps: Can exceed 100:1
Efficiencies:	Pump efficiencies range from 85% for large capacity centrifugals (types A and B) to below 50% for many of the smaller units. For type C and H, the efficiency ranges from 30% up, depending on horsepower and number of cylinders. For D the efficiency is almost 30%, and for E, F, and G it is below 25%. For I, see Fig. 9.12m in the Process Measurement volume.
Materials of Construction:	For water using type A or B pumps: normally bronze impellers, bronze or steel bearings, stainless or carbon steel shafts, cast iron housing. For domestic wastes using type A, B, or C pumps: similar except that they are often cast iron impellers. For industrial wastes and chemical feeders using type A or C pumps: a variety of materials depending on corrosiveness. Type D similar except that diaphragm is usually rubber. Types E, F, G and H: steel. Type H: cast iron, steel, stainless steel, carpenter 20, Hastelloy, plastics, and glass.
Costs:	Vary with size and horsepower. For example, type A pumps cost from $500 to more than $25,000. Typically, a type A 10 HP (74–57W) sewage pump might cost $2,500 for a horizontal and $2,000 for a vertical model. A 10 HP (7457W) type F ejector costs about $12,500, but to do the same work as the type A pump, a 30 HP (22.3kW) $25,000 ejector is needed. A 10 HP (7457W) type C, D, or E sludge pump costs about $6,000. Prefabricated stations (including pump) range from $10,000 to $75,000. $2,500 for type H metering pump with positioner. Type I-$2,500 for 5 HP (3729W) mechanical drive, $3,000 for 5 HP (3729W) motor and SCR controller, and $50,000 for 600 HP (447kW) magnetic coupling drive.
Partial List of Suppliers:	Allis-Chalmers (A,B,); American Standard (I): Aurora Pump Unit, Gen. Signal Corp. (A,B,); Avtek Systems (I); Beckman Instruments Inc. (H); Can-Tex Industries (A,B); Chicago Pump, FMC Corp. (A,B,E,); Deming Div., Crane Co. (A,B,C,D,E); Eaton Corp. (I); Fairbanks Weighing Div., Colt Industries (A,B); Fischer & Porter Co. (H); Flygt Corp. (A); Fuller

Co., a GATX Co. (G); Gardner Denver Co. (A,C,D); General Electric Co. (I); Gerbing Manufacturing Co. (I); Gorman Rupp Industries (A,C,D); Hydr-O-Matic Pump Co. (A,D,); Ideal Electronics Corp. (I); Komline Sanderson Engr. Corp. (C,D,F); Louis Allis Div., Litton Industrial Products, Inc. (I); Marlow Div., ITT (A,C,D,); Mec-O-Matic Co. (H); Milton Ray Co. (H); Moyns Pump Div., Robbins & Meyers (E); Pacific Pumping Co. (A,B,D); Peerless Pump (I); Ramsey Controls, Inc. (I); Reliance Electric Co. (I); Robicon Corp. (I); Seiscor, Inc. (H); U. S. Motors Corp. (I); Vickers Division of Sperry Rand Corp. (I); Wallace & Tiernon Div., Pennwalt Corp. (C,D,H); Weil Pump Co. (A,F); United Weneco Div., Envirotech Corp. (A,E); Yeomans Brothers Div., Clow Corp. (A,C,F,G).

Centrifugal Pumps

Types A and B of the feature summary (Table 5.8a) fall within this classification, which is the most common type of pump. In the form of tall, slender, deep well submersibles, they pump clear water from depths greater than 2,000 feet (600 m). Horizontal centrifugals with volutes almost the size of a man can pump 9,000 gpm (0.57 m³/s) of raw sewage through municipal treatment plants. Few applications are beyond their range, including flow rates of 1 to 100,000 gpm (6.3×10^{-5} to 6.3 m³/s) and process fluids from clear water to all but the densest sludge.

Radial-Flow Centrifugals

Radial-flow pumps are designed to throw the liquid entering the center of the impeller or diffuser out into a spiral volute or bowl.[1,2] The impellers may be closed, semi-open, or open, depending on the application (Figure 5.8b). Closed impellers have higher efficiencies and are more popular than the other two types. They can readily be designed with nonclogging features. In addition, by using more than one impeller the lift characteristics can be increased. These pumps may be of horizontal or vertical design.

Table 5.8a
PUMP FEATURE SUMMARY

Type Designation	Type of Pump	For Liquid Pumped					For Capacity	For Feet of Head
		Clear Liquids—Low Viscosity	Clear Liquids—High Viscosity	Thin Slurries or Suspensions	Raw or Partially Treated Sewage and Heavy Suspensions	Viscous or Thick Slurries and Sludges	GPM*	PSIG* / Feet of Head (H₂O)*
A	Radial-flow centrifugals	✔		✔	① ✔	① ✔		
B	Axial-flow and mixed-flow centrifugal	✔		✔	✔			
C	Reciprocating pistons and plungers	② ✔ ②	②	✔	② ✔	✔		
D	Diaphragm pumps	✔	✔ ②	✔	✔ ②	✔		
E	Rotary screws		✔		✔	✔		
F	Pneumatic ejectors				✔			
G	Air lift pumps			✔	✔			

✔ Suitable for normal use. ① See text for limitations. ② Not used for this purpose in environmental engineering (with some exceptions).
Note. If not checked, either not suitable or not normally used for this purpose.
* See Section A.1 for SI units

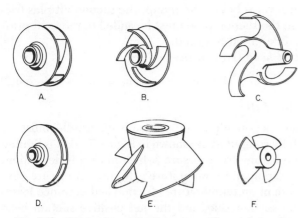

Fig. 5.8b Types of centrifugal pump impellers: (A) closed impeller, (B) semi-open impeller, (C) open impeller, (D) diffuser, (E) mixed flow impeller, (F) axial flow impeller

Axial- and Mixed-Flow Centrifugals

Axial-flow propeller pumps, although classed as centrifugals, do not truly belong in this category since the propeller thrusts rather than throws the liquid upward. Impeller vanes for mixed-flow centrifugals are shaped so as to provide partial throw, partial push of the liquid outward and upward. Axial-flow and mixed-flow designs can handle huge capacities but only at the expense of a reduction in discharge heads.[3] They are constructed vertically.

Applications

Most water and wastes can be pumped with centrifugal pumps. It is easier to list the applications for which they are not suited than the ones for which they are. They should not be used for pumping (1) very viscous industrial liquids or sludges (the efficiencies of centrifugal pumps drop to zero and therefore various positive displacement pumps are used, (2) low flows against very high heads (except for deep well applications, the large number of impellers needed put the centrifugal design at a competitive disadvantage), and (3) low to moderate flows of liquids with high solids contents (except for the recessed impeller type, rags and large particles will clog the smaller centrifugals).

Positive Displacement Pumps

Reciprocating Piston, Plunger, and Diaphragm Pumps

Almost all reciprocating pumps used in environmental engineering are either metering or power pumps. The steam-driven pump is historically interesting but rarely used in water or wastewater processing. Most frequently a piston or plunger is utilized in a cylinder, which is driven forward and backward by a crankshaft connected to an outside drive.[1,2] Metering pump flows can readily be adjusted by changing the length and number of strokes of the piston.[4] Type D is similar to C but instead of a

piston, it contains a flexible diaphragm that oscillates as the crankshaft rotates.[4]

Applications

Plunger and diaphragm pumps feed metered amounts of chemicals (acids or caustics for pH adjustment) to a water or waste stream and also pump sludge and slurries in waste treatment plants.

Rotary Screw Pumps

In this type (E), a vaned screw or rubber stator on a shaft is rotated by a motor to lift or feed sludge or solid waste material to a higher level or to the inlet of another pump.

Air Pumps

Pneumatic Ejectors

This method of pumping (type F) employs a receiver pot into which the wastes flow and an air pressure system that then blows the liquid out to a treatment process at a higher elevation. A controller is usually included, which keeps the tank vented while it is in the process of filling, and when it is full the level controller energizes a three-way solenoid valve to close the vent port to vent and open the air supply to pressurize the tank.

The air system may use plant air (or steam), a pneumatic pressure tank, or an air compressor directly. With large compressors, a capacity of 600 gpm (0.038 m³/s) with lifts of 50 feet (15 m) may be obtained. The advantage of this system is that it has no moving parts in contact with the waste and thus no impellers to clog. Ejectors are normally more maintenance free and longer lived than pumps.

Air Lifts

Air lifts consist of an updraft tube, an air line, and an air compressor or blower. Air is blown into the bottom of the submerged updraft tube, and as the air bubbles travel upward, they expand (reducing density and pressure within the tube), inducing the surrounding liquid to enter. Flows as great as 1,500 gpm (0.095 m³/s) may be lifted short distances in this way. Air lifts are of great value in waste treatment to transfer mixed liquors or slurries from one process to another.

Design of Pumping Systems

In order to choose the proper pump, the conditions that must be known include capacity, head requirements, and liquid characteristics. Liquid characteristics have already been discussed; therefore only capacity and head requirements will be considered here.

Capacity

To compute capacity, one should first determine the average flowrate for the system and then decide if adjustments are necessary. For example, when pumping wastes from a community sewage system, the pump must handle peak flows that are roughly two to five times the average flow, depending on the size of the community. Summer and winter flows and future needs may also dictate capacity, and the population trends and past flowrates should be considered in this evaluation.[5]

Head Requirements

Head describes pressure in terms of feet of lift. It is calculated by the expression:

$$\text{Head in feet} = \frac{\text{Pressure (psi)} \times 2.31}{\text{Specific Gravity}} \qquad 5.8(1)$$

The discharge head on a pump is a summation of several contributing factors[6]:

Static head (h_d) is the vertical distance through which the liquid must be lifted (Figure 5.8c).

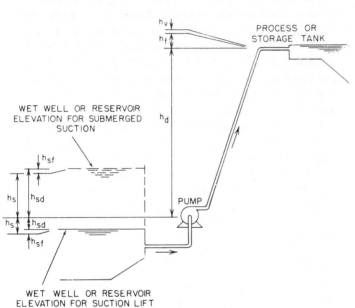

Fig. 5.8c Determination of pump discharge head requirements. Legend: h_v = velocity head; h_f = friction head; h_d = static head; h_s = suction head; h_{sd} = suction side static head; h_{sf} = suction side friction head

Friction head (h_f) is the resistance to flow caused by the friction in pipes. Entrance and transition losses may also be included. Since the nature of the fluid (density, viscosity, and temperature) and the nature of the pipe (roughness or straightness) affect the friction losses, a careful analysis is needed for most pumping systems although for smaller systems, tables can be used.[7,8]

Velocity head (h_v) is the head required to impart energy into a fluid to induce velocity. Normally this is quite small and may be ignored unless the total head is low.

Suction head (h_s), if there is a positive head on the suction side (a submerged impeller), will reduce the pressure differential that the pump has to develop. If the

water level is below the pump, the suction lift plus friction in the suction pipe must be added to the total pressure differential required.

Total head (H) is expressed by

$$H = h_d + h_f + h_v \pm h_s \qquad 5.8(2)$$

Suction Lift

The suction lift that is possible to handle must be carefully computed. As shown in Figure 5.8d, it is limited by the barometric pressure (which in turn is dependent on elevation and temperature); the vapor pressure (also dependent on temperature); friction and entrance losses on the suction side; and the net positive suction head (NPSH)—a factor that depends on the shape of the impeller and is obtained from the pump manufacturer.[1]

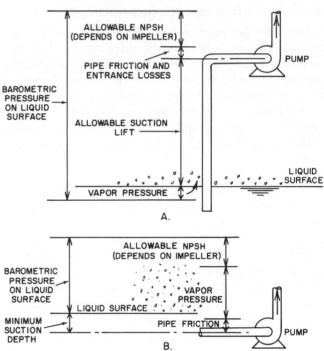

Fig. 5.8d Role played by NPSH in determining allowable suction lift. (A) Pump with suction lift, (B) Pump with submerged suction but high vapor pressure (possibly hot water).

Specific Speed

The rotational speed of the impeller affects capacity, efficiency, and the extent of cavitation. Even if the suction lift is within permissible limits, cavitation may be a problem and should be checked. The specific speed of the pump[1,2,5] is found by equation 5.8(3).

$$\text{Specific Speed, } N_s =$$
$$= \frac{\text{RPM} \times \sqrt{\text{Capacity (gpm)}}}{H^{\frac{3}{4}}} \qquad 5.8(3)$$

Charts are available showing the upper limits of specific speed for various suction lifts.[1]

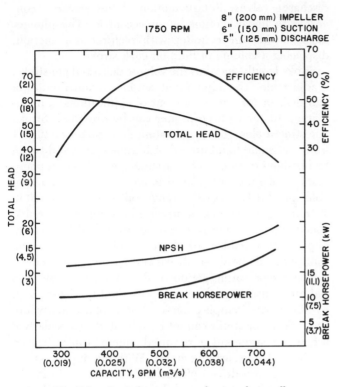

Fig. 5.8e Typical pump curve for a single impeller

Horsepower

The horsepower required to drive the pump is called brake horsepower. It is found by solving equation 5.8(4).

$$BHP = \frac{Capacity \ (gpm) \ \times \ H \ (ft) \ \times \ Sp. \ Gr.}{3960 \ \times \ Pump \ efficiency} \qquad 5.8(4)$$

Pump Curves

The essential features of a given pump are described by its performance curves. A typical centrifugal pump curve is shown in Figure 5.8e. Charts or tables that summarize pump curve data are also available.

Pumping Station Design

The typical designs are shown in Figure 5.8f. In selecting the best design for a particular application, the following factors should be considered:

1. Many gases are formed by domestic wastes, including some that are flammable. When pumps or other equipment are located in rooms below grade, the possibilities of explosion or the build-up of these gases exist and ventilation is extremely important.

2. When pumping at high velocities or through long lines, water hammer can be a problem. Valves and piping

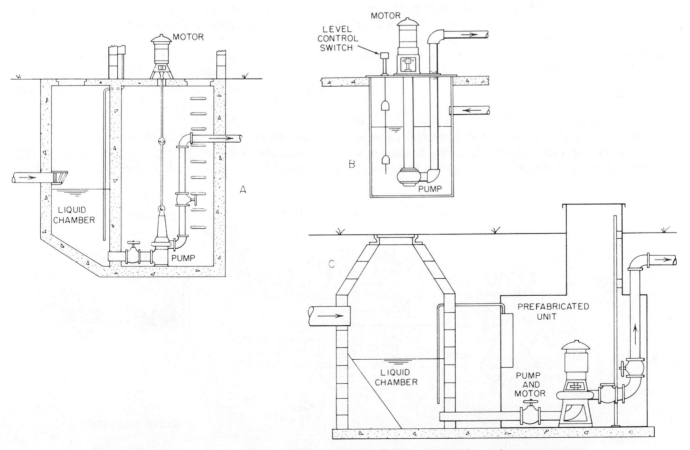

Fig. 5.8f Pumping stations: (A) dry well design, (B) wet well design, (C) prefabricated pumping station

should be designed to withstand these pressure waves. Even for pumps that discharge to atmosphere, check valves should be chosen so as to cushion the surge.[9]

3. Bar screens and comminutors are undesirable because they require maintenance, but they may be necessary for small centrifugal pump stations where the flow might get clogged.

4. Pump level controls are not fully reliable because rags can short electrodes and hang on floats. Purged air systems (air bubblers) require less maintenance but need an air compressor that must operate continuously. Therefore it is important to provide maintenance-free instrumentation.[10]

5. Charts and formulas are available for sizing wet wells, but infiltration and runoff must also be taken into account.

6. Sump pumps, humidity control, a second pump with alternator, and a pump hoisting mechanism are desirable.

7. Most states prefer the dry well designs for ease of maintenance.

Metering Pumps

Flow control of liquids and solids can be accomplished by means of pumps and feeders that incorporate the measurement and control element in a single unit. Metering pumps and feeders are designed to provide measurement and control of the process. For a measurement-oriented discussion of these pumps, refer to Section 2.11 in the volume on Process Measurement. However, it is also possible to utilize standard centrifugal pumps with variable-speed control to give the desired control action.

Plunger Pumps

Plunger pumps are suitable for use on clean liquids at high pressures and low flow rates. A typical plunger pump is shown in Figure 5.8g. The pump consists of a plunger, cylinder, stuffing box, packing, and suction and

discharge valves. Rotary motion of the driver is converted to linear motion by an eccentric. The plunger moves inside the cylinder with reciprocating motion, displacing a volume of fluid on each stroke.

Stroke length, and thus the volume delivered per stroke, is adjustable. The adjustment can be a manual indicator and dial, or for automatic control applications, a pneumatic actuator with positioner can be provided. Stroke adjustment alone offers operating flow ranges of 10 to 1 from maximum to minimum. Additional rangeability can be obtained by means of a variable-speed drive. A pneumatic stroke positioner used in conjunction with a variable-speed drive provides rangeability of at least 100 to 1. In the case of automatic stroke adjustment and variable speed, the pumping rate can be controlled by two independent variables, or the controller output can be "split-ranged" between stroke and speed adjustment.

The reciprocating action of the plunger results in a pulsating discharge flow as represented in Figure 5.8h by the dotted simplex curve. For applications where these flow pulsations cannot be tolerated, particularly if a flow measurement is required, pumps can be run in duplex or triplex arrangements. With the duplex pump, two pumps are driven off the same motor, and the discharge strokes are phased 180 degrees apart. With a triplex arrangement, three pumps are driven by one motor and the discharge strokes are 120 degrees apart. Both the duplex and triplex pumps provide a smoother flow than the single pump, as shown by the dashed and solid curves of Figure 5.8h.

For blending two or more streams, several pumps can be ganged to one motor. Stroke length adjustment can be used to control the blend ratio, and drive speed can control total flow. However, in this case rangeability is sacrificed for ration control.

Pumping efficiency is affected by leakage at the suction and discharge valves. These pumps are therefore not recommended for fluids such as slurries, which will in-

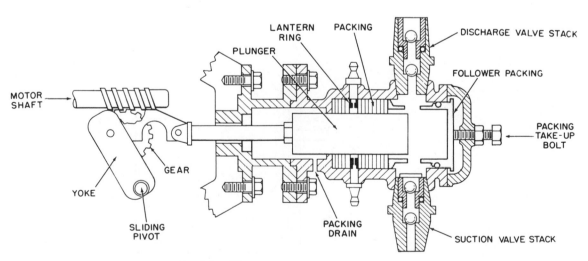

Fig. 5.8g Plunger or piston type metering pump

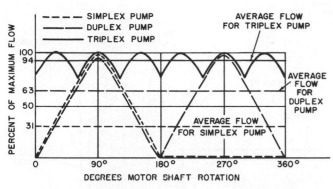

Fig. 5.8h Flow characteristics of simplex and multiple plunger pumps

terfere with proper valve seating or settle out in pump cavities.

Diaphragm Pumps

Diaphragm pumps use a flexible diaphragm to achieve pumping action. The input shaft drives an eccentric through a worm and gear. Rotation of the eccentric moves the diaphragm on the discharge stroke by means of a push rod. A spring returns the push rod and diaphragm during the suction stroke. A typical pump is shown in Figure 5.8i.

Operation of the diaphragm pump is similar to that of the plunger pump; however, discharge pressures are much lower due to the strength limitation of the diaphragm. Their principle advantage over plunger pumps is lower cost. Designs with two pumps driven by one motor can be used to advantage for increased capacity or to smooth out flow pulsations. By combining automatic stroke length adjustment with a variable-speed drive, operating ranges

can be as wide as 20 to 1. These pumps can be used only on relatively clean fluids, because solids will interfere with proper suction and discharge valve seating or may settle out in the pump cavities.

The weakness of the diaphragm pump design is in the diaphragm, which is operated directly by the push rod. The diaphragm has to be flexible for pumping and yet strong enough to deliver the pressure. The strength requirement can be reduced by using a hydraulic fluid to move the diaphragm, thereby eliminating the high differential pressures across it. This design consists basically of a plunger pump to provide hydraulic fluid pressure for diaphragm operation and the diaphragm pumping head (Figure 5.8j). The forces on the diaphragm are balanced, and discharge pressures comparable to plunger pumps are possible. The volume pumped per stroke is equal to the hydraulic fluid displaced by the plunger, and this volume is controlled by the stroke length adjustment as in the plunger pump.

A pump design using a flexible tube to achieve pumping action is shown in Figure 5.8k. Motion of the plunger displaces the diaphragm, which in turn causes the flexible tube to constrict, forcing fluid in the tube to discharge (similar to the operation of a peristaltic pump). This design is better suited for use on viscous and slurry liquids than the previously discussed types because the flow path is straight with few obstructions and no cavities, but seating of the valves can still be a problem.

General Considerations for Displacement Pumps

In order to insure a properly working installation, a number of factors associated with the physical installation and with the properties of the fluid must be considered.

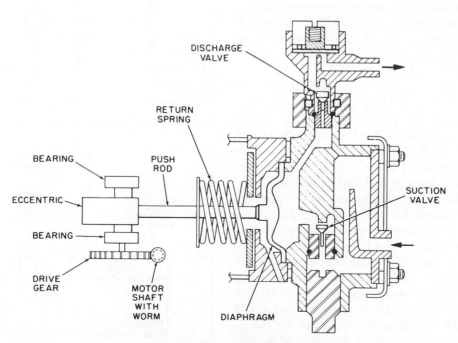

Fig. 5.8i Diaphragm type metering pump

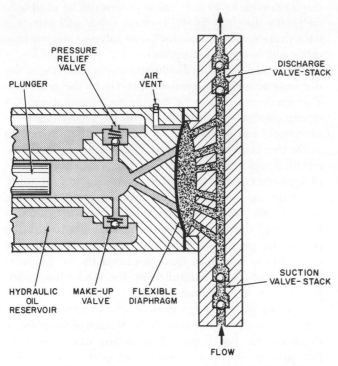

Fig. 5.8j   Diaphragm pump operated with hydraulic fluid

Some factors which can contribute to a poor installation include:

1. Long inlet and outlet piping with many fittings and valves.
2. Inlet pressure higher than outlet pressure.
3. Pocketing of suction or discharge lines.
4. Low suction head or suction lift.

A tortuous flow path in the pump suction or discharge can be troublesome when the fluid handled contains solids, is of high viscosity, or has a high vapor pressure, or if the suction head available is low. Generally, valves that offer a full flow path (such as ball valves) are preferred. Needle valves should be avoided. If the inlet pressure is higher than discharge, the fluid may flow unrestricted through the pump. Spring-loaded check valves at the pump are undesirable because the ball check should be free to rotate and find a new seating surface for increased valve life. For such applications the installations shown in Figures 5.8l and m offer solutions. In Figure 5.8l the piping arrangement will supply the head to prevent through flow and syphoning. Dimension "A" is a variable depending on pump capacity and fluid velocity. This dimension varies between 2 and 10 ft (0.6 and 3 m) and increases with capacity and velocity. Figure 5.8m illustrates the use of a spring-loaded back-pressure valve, to overcome the suction pressure. For this installation a volume chamber to dampen pulsations should be placed between the pump discharge and the valve.

Dissolved or entrained gases in the fluid can destroy metering accuracy and, if they are of sufficient volume, they can stop all pumping action. Figure 5.8n illustrates

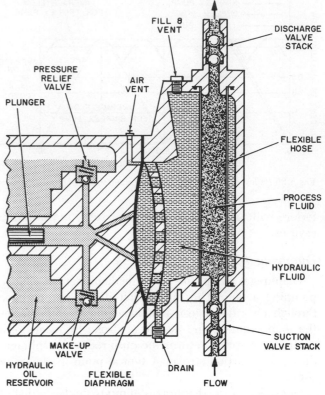

Fig. 5.8k   Diaphragm pump operated with hydraulic fluid and flexible hose element

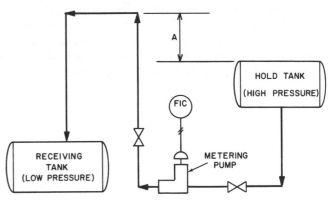

Fig. 5.8l   Piping arrangement to prevent through flow and syphoning

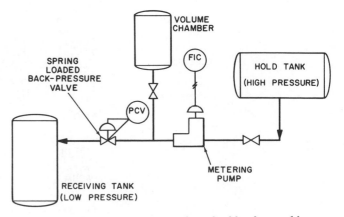

Fig. 5.8m   Metering pump with artificial head created by back-pressure valve

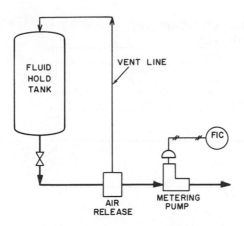

Fig. 5.8n   Elimination of entrained gases in
metering pump installations

an installation designed to vent entrained gases back to
the fluid hold tank.

It is always desirable to locate the pump below and
near the fluid hold tank. Under these conditions the fluid
will flow by gravity into the pump suction and loss of
prime is unlikely. If the pump cannot be located below
the hold tank, other measures must be taken to prevent
loss of prime.

In order for the pump to operate properly, the net
positive suction head must be above the minimum practical suction pressure of approximately 10 PSIA (69 kPa).
Net positive suction head (NPSH) is given by equation
5.8(5):

$$NPSH = P - P_v \pm P_h -$$
$$- \sqrt{\left(\frac{lvGN}{525}\right)^2 + \left(\frac{lvC}{980Gd^2}\right)^2} \quad 5.8(5)$$

where

P = feed tank pressure (PSIA),*
P_v = liquid vapor pressure at pump inlet temperature (PSIA),
P_h = head of liquid above or below the pump
centerline (PSID),
l = actual length of suction pipe (ft),
v = liquid velocity (ft/sec),
G = liquid specific gravity,
N = number of pump strokes per minute,
C = viscosity (centipoise), and
d = inside diameter of pipe (in.).

For liquids below approximately 50 centipoise (0.05 Pa·s),
viscosity effects can be neglected, and equation 5.8(5)
reduces to

$$NPSH = P - P_v \pm P_h - \frac{(lvGN)}{(525)} \quad 5.8(6)$$

The calculated value of NPSH must be above the minimum suction pressure required by the pump design.

* For SI units, see Section A.1.

In addition to multiple pumping heads, a pulsation
dampener can be used on the pump discharge to smooth
the flow pulsations. The pulsation dampener is a pneumatically charged diaphragm chamber that stores energy
on the pump discharge stroke and delivers energy on
the suction stroke, thus helping to smooth the flow pulses.
In order to be effective, however, the dampener volume
must be equal to at least five times the volume displaced
per stroke.

Opposed Centrifugal Pumps

The opposed centrifugal pump is not a device specifically designed as a control element but represents an
adaptation of a centrifugal pump to flow control. This
method of control is particularly suitable for coarse, rapidly settling slurries at low flow rates. In such instances
the conflicting requirements of control at low flow, and
large free area to pass the solids, may make it impossible
to find a suitable control valve. A system that requires
a small quantity of slurry to be fed to a receiving vessel
under controlled conditions is depicted in Figure 5.8o.
Pump P_1 continuously circulates the slurry from the feed
tank at high velocity. A branch line from the discharge
of P_1 is run to the opposed centrifugal pump P_2. Pump
P_2 is connected in opposition to the direction of slurry
flow, and pressure drop to throttle flow is obtained by
means of the mechanical energy supplied to the pump.
A variable-speed driver on pump P_2 permits changing
the pump pressure drop. At full speed the pressure difference across P_2 is sufficient to stop the branch line
slurry flow completely. A magnetic flowmeter or some
other suitable device can be used to measure the slurry
flow. An SCR speed control can be used to vary pump
speed in response to the flow controller output signal.

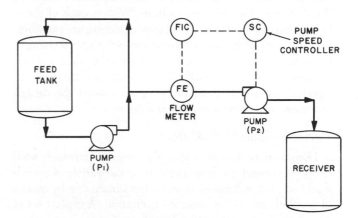

Fig. 5.8o   Opposed centrifugal pump as final control element

Variable-Speed Drives

In addition to reading the treatment below, the reader
will also want to review Section 9.12 in the Process Measurement volume of this Handbook, which deals with
variable-speed pumping.

There are a great number of methods for speed ad-

Table 5.8p
FEATURE SUMMARY AND SERVICE APPLICABILITY OF VARIOUS ROTARY DRIVES

Service and Feature	Electric Motor	Eddy Current or Magnetic Couplings	Mechanical Stepped-Speed Transmissions	Continuously Variable Mechanical Drives	Hydraulic Drives
Very wide range of speeds	✔				
Few speed steps with remote control	✔				
Few speed steps with local and manual control	✔		✔		
Very high output power at variable speed		✔			
Vibration at load or driver		✔			
Small or moderate load with narrow speed range				✔	
Accurate speed control	✔	✔			✔
Shock loads, frequent overloads					✔
Speed reversal required	✔		✔		✔

justment of rotating machines. These methods fall into two categories: speed adjustment of the prime mover and speed adjustment through a transmission connecting the driver to the driven machine. Within each of these two groups several degrees of sophistication are possible, ranging from manually actuated step-wise speed changes to continuously variable automatic speed changers. Each method of speed control offers certain advantages and disadvantages that must be weighed against the design criterion to permit a proper selection.

Variable-Speed DC Motors

Direct current motors lend themselves extremely well to speed control, as demonstrated by the variety of speed-load and load-voltage characteristics obtainable by means of parallel, series, or separate excitation. A typical set of speed-load characteristics for shunt, series, and compound motors is shown in Figure 5.8q. The inherent speed regulation, or constancy of speed under varying load conditions, of the shunt motor is shown graphically in the diagram. The speed change of the motor is only 5 percent for a load change from zero to full load. Although the methods of speed control are applicable to all three types of DC motors, the superior speed regu-

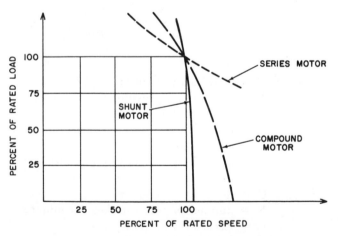

Fig. 5.8q   Speed–load characteristics of DC motors

lation of the shunt motor accounts for its wider use on control applications.

The speed of a DC motor is a function of armature voltage, current, and resistance, the physical construction of the motor, and the magnetic flux produced by the field winding. The equation relating these variables to speed is

$$S = \frac{V - IR}{K\Phi} \qquad 5.8(7)$$

where

S = motor speed,
V = armature terminal voltage,
I = armature current,
R = armature resistance,
Φ = magnetic field flux, and
K = a constant for each motor depending on physical design.

Three methods of speed control are suggested by equation 5.8(7): adjustment of field flux Φ, adjustment of armature voltage V, and adjustment of armature resistance R.

The first of these, adjustment of field flux, involves varying the field current. This can be accomplished by means of a field rheostat, or the current can be controlled electronically in response to a set point signal through an energy-throttling device such as an SCR. Open-loop control should be utilized only when the load is constant and precise speed control is not essential. Closed-loop control is generally used in conjunction with electronic devices and offers better control than open loop, but at higher cost (Figure 5.8r). Either type of control requires a measurement of the controlled speed; in closed-loop control the measured speed is compared with the set point and the field current is adjusted automatically to bring the difference to zero. For a discussion of speed sensors refer to Section 9.9 of the Process Measurement volume of this Handbook. In open-loop control a speed readout is required to permit set adjustments by the operator for the desired speed.

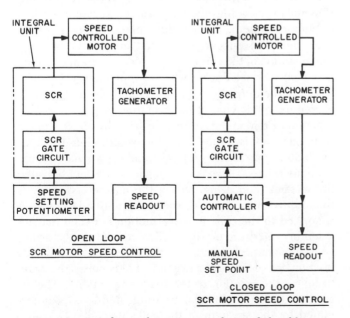

OPEN LOOP
SCR MOTOR SPEED CONTROL

CLOSED LOOP
SCR MOTOR SPEED CONTROL

Fig. 5.8r Speed control systems, open loop and closed loop

Adjustment of field flux yields a motor with constant horsepower. The allowable armature current is approximately limited to the motor rating in order to prevent overheating. The effects of changing flux and changing speed effectively cancel each other so that the allowable output horsepower—the product of armature current and induced armature voltage—is constant. Torque, however, varies directly with field flux, and therefore this type of speed control is suited for applications involving increased torque at reduced speeds. The speed range possible with field flux adjustment is approximately 4 to 1.

Armature voltage control can also be used to control motor speed. With armature voltage control, the change in speed from zero load to full load is almost entirely due to the full-load armature resistance drop, and this speed change is independent of the no-load speed. Consider two motors with identical speed changes from zero to fully loaded, but one operating at 100 rpm and the other at 1,000 rpm at zero load. In terms of percent speed change, a 10 rpm variation may be unacceptable for the lower-speed motor, but may be insignificant for the higher-speed motor.

One method used for armature voltage control is the use of a motor-generator set to supply a controlled voltage to the motor whose speed is to be regulated. An AC motor drives a DC generator whose output voltage is variable and supplies the variable-speed motor armature. An obvious disadvantage of this method is the initial investment in three full-size machines; this method is so versatile, however, that it is oftn used. Armature voltage, of course, can also be controlled electronically by one of the devices discussed in Section 5.3. These are more commonly used for speed control due to the lower investment and greater efficiency than the motor-generator set. In the motor with controlled armature voltage, both the allowable armature current and field flux remain constant. The driver therefore has a constant torque output, as opposed to the constant horsepower output of the field-controlled motor.

The armature voltage method of speed control yields speed ranges in the order of 10 to 1. By combining field flux control and armature voltage control, speed ranges of 40 to 1 are obtainable. The base speed of the motor is set at full armature voltage and full field flux; speeds above base are obtained by field flux control, speeds below base by armature voltage control. Speed ranges greater than 40 to 1 are obtainable through the addition of special motor windings and SCR controls.

Adjustment of armature resistance is another method of speed control suggested by equation 5.8(7), and it can be used to obtain reduced speeds. An external, variable resistance is inserted into the armature circuit (Figure 5.8s). Speed regulation with this method and its variants is very poor, however, and this type of speed control is not commonly used. An added disadvantage of the ar-

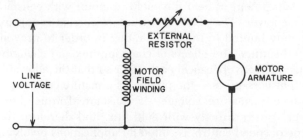

Fig. 5.8s Variable armature resistance to control shunt motor

mature resistance method is the decrease in efficiency due to the power consumption in the resistor.

STARTING CIRCUITS

DC motors must be protected from high in-rush currents during starting. The starting current through the armature is limited by resistors R1, R2, and R3 in Figure 5.8t. When the start button is depressed momentarily, relay M and time delay relays TD1, TD2, and TD3 are energized. The two M contacts close instantaneously providing power to the motor. After a time delay, contact TD1-1 closes, shunting resistor R1. Contacts TD2-1 and TD3-1 close successively until the armature is directly on the line. The number of resistors is set by the torque and current limitations of the motor and the desired smoothness of startup.

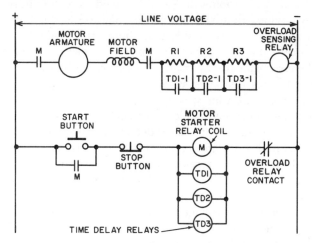

Fig. 5.8t Starting circuit for nonreversing DC motor

DYNAMIC BRAKING

The motor will also act as a generator, converting mechanical to electrical energy. This feature can be utilized to advantage where rapid stopping of the motor is required. When power is disconnected from the motor, a resistor is automatically connected across the armature and the mechanical energy of the rotating member is disspated as heat in the resistor.

REVERSE ROTATION

DC motors can be run in reverse rotation by changing the polarity of the armature voltage. This can be accom-plished by means of two contactors—one forward, one reverse. These must be either mechanically or electrically interlocked to inhibit closing both contactors simutaneously, and to prevent applying reverse line voltage to the motor prematurely.

CONTROL DEVICES

Normally, DC power for operation of the motors is not readily available. Rectifier tubes and solid state devices such as SCRs are used extensively to convert the incoming power to DC and simultaneously to throttle the current or voltage delivered to the motor in response to an external control signal. A detailed discussion of these devices is given in Section 5.3.

Speed Changes on AC Motors

AC motors, while not readily usable for continuous speed control, can be made reversing or furnished with several fixed speeds. The squirrel cage induction motor is best suited from the standpoint of reliability, and it can be furnished with up to four fixed motor speeds. Speed changes are accomplished through motor starter contacts that reconnect the windings to yield a different number of poles. In the resulting pole motor, the sections of the stator winding are interconnected through the motor starter to provide different motor speeds. On separate winding motors, a different winding with the required number of poles is energized for each motor speed. The starter contacts, however, must be interlocked to prevent two or more contactors from closing simultaneously.

Magnetic Coupling

EDDY CURRENT COUPLING

In most electrical machinery, eddy currents are detrimental to operating efficiency, and great pains are taken to eliminate them. In the eddy current coupling, however, these currents are harnessed and are the basis for the operation of the coupling. The eddy current coupling is a nonfrictional device where input energy is transferred to the output through a magnetic field. The eddy current coupling consists of a rotating magnet assembly separated from a rotating ring or drum by an air gap. In addition, a coil is wound onto the magnet assembly (Figure 5.8u), or on larger units, the coil is stationary on the coupling frame. There is no mechanical contact between the magnets and drum. When the magnet assembly is rotated, the drum remains stationary until a DC current is applied to the coil. Relative motion between magnet assembly and drum produces eddy currents in the drum, whose magnetic field attracts the magnet assembly. Attraction between the two magnetic fields causes the drum to follow the rotation of the magnet assembly. The attraction between the two rotating members is determined by the strength of the coil's magnetic field and by the difference in speed between the two members.

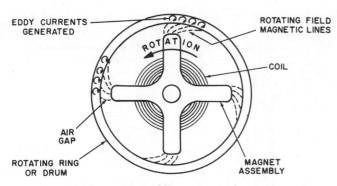

Fig. 5.8u Eddy current coupling

Thus by controlling the coil excitation, the amount of slip and, hence, the output speed can be controlled.

Being a slip device, the eddy current coupling must of necessity develop slip and reject the slip power in the form of heat. The amount of slip loss can be determined from equation 5.8(8):

$$P_s = P_L \frac{S_s}{S_0} \qquad 5.8(8)$$

where

P_s = slip loss power,
P_L = load power,
S_s = slip speed (rpm) and
S_0 = output speed (rpm).

Slip devices always generate heat, so eddy current couplings are either air or water cooled. Small air-cooled units below 5 horsepower (3.7kW) can be designed to dissipate all of the rated power. Air-cooled units above 300 horsepower (224kW) in size can dissipate only about 25 percent of the rated power capacity. For this reason, air-cooled units are not recommended where cooling capacity would be greater than about 20 percent of rated power. Water-cooled units are designed to dissipate the rated horsepower continuously. In addition, water-cooled units show a small efficiency advantage over air-cooled ones, particularly on "water-in-the-gap" types, where the water contributes slightly to the torque capability. Generally, water-cooled units are preferred to air-cooled types except when lack of coolant precludes their use or where very low slip losses are encountered.

Eddy current couplings require only a low percentage of transmitted power for excitation. Typically a 3-horsepower (2.2kw) unit will require 50 watts of excitation, while a 12,000-horsepower (8,947kW) unit will require 20 kilowatts. Eddy current couplings are readily adaptable to SCR or magnetic amplifier control, providing speed control within 1 percent accuracy, over 10 to 100 percent load change. The speed-control devices mentioned above are discussed in Section 5.3.

Efficiency of the eddy current coupling is very good, particularly in the large sizes and at full speed torque.

At lower speeds, efficiency drops considerably, but efficiencies as high as 95 percent are possible at full excitation and torque. Since there is no contact between the input and output shafts of the coupling, the unit will not transmit vibrations. On prime movers that exhibit some torsional vibration, the use of an eddy current coupling can be of advantage because it will suppress or at least attenuate these vibrations.

Integral combinations of motor, coupling, and excitation are available in small sizes of 1 horsepower (746W) or less. Air-cooled couplings range in size up to 900 horsepower (671kW) but the larger sizes are practical only where a relatively low cooling capacity is required. Liquid-cooled units can be as large as 18,000 horsepower (13,428kW) capacity. Speed-control units range from simple open-loop control to precise, automatic closed-loop control with tachometer-generator speed feedback.

MAGNETIC PARTICLE COUPLING

The magnetic particle coupling offers another solution to adjustable speed drives. Basically the coupling consists of two concentric cylinders separated by an air gap and a stationary excitation coil surrounding the cylinders. Ferromagnetic particles fill the gap between the concentric cylinders. When a controlled amount of current is used to energize the coil, the particles form chains along the magnetic lines of flux connecting the cylinder surfaces. The shear resistance of the magnetic particles is proportional to the coil excitation and provides the basis for power transmission from input to output.

The output torque of this unit is always equal to input torque regardless of speed. This fact allows the output torque to be set at standstill by controlling the coil excitation. Whenever the torque capacity of the coupling is exceeded, slip will result with accompanying heat liberation. Whenever the coupling is used for speed control, selection of a coupling with adequate cooling capacity is of great importance. Air-cooled units are available with large heat dissipation capacity; however, water cooling is more effective and therefore preferred.

Magnetic particle couplings can be used effectively for speed control, since they can provide a constant torque output independent of speed. However, a closed-looped control system is required for effective speed control in order to cancel the effect of changing torque on slip speed.

MAGNETIC FLUID CLUTCHES

The magnetic fluid clutch is similar in operating principle to the magnetic particle coupling, but either a magnetic dry powder or magnetic powder suspended in a lubricant is utilized. A typical disc type magnetic fluid clutch is shown in Figure 5.8v.

While the operating principle is similar to that of magnetic particle coupling, the clutch can be used only for low-power applications in the order of a few horsepower.

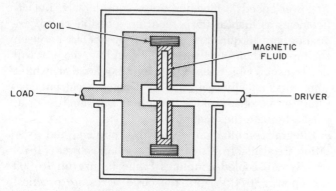

Fig. 5.8v   Disc type magnetic fluid clutch

Recharging of the clutch due to fluid deterioration is a definite disadvantage; however, recharging can be fairly easily accomplished during routine maintenance.

Mechanical Variable-Speed Drives

STEPPED SPEED CONTROL

Mechanical methods of speed adjustment offer a number of gear and pulley devices for both stepped and continuously variable-speed control. Stepped speed control methods provide setting a number of speeds very accurately. However, they are not readily adaptable to automatic process control. The stepped pulley system shown in Figure 5.8w is one of the earliest methods of speed adjustment. Its advantages are low cost and simplicity, but belt slippage contributes to inefficiency, high maintenance, and reduced speed control. Two factors to be considered in the design of a stepped pulley system are the proper ratio of pulley diameters to obtain the desired speeds and proper pulley dimensions to maintain belt tension for all positions. Pulley dimensions for a system such as shown in Figure 5.8w can be determined from the relationships in equations 5.8(9) 5.8(10).

$$\frac{\pi}{2}(R_1 + r_1) + \frac{(R_1 - r_1)^2}{4d} =$$

$$= \frac{\pi}{2}(R_2 + r_2) + \qquad 5.8(9)$$

$$\frac{(R_2 - r_2)^2}{4d}\frac{S_2}{S_1} = \frac{R_2}{r_2} \qquad 5.8(10)$$

The variables are defined in Figure 5.8w.

Gear transmissions offer high efficiency of power transmission at precise stepped speed control. For speed changes of gear train drives, either clutches or brakes are required or the change must be made at rest. Epicyclic gears, such as planetary gears, offer the most compact unit, operating quality, and high efficiency. However, auxiliary clutches and brakes are required and the cost is therefore higher than that of other gear drives. The complexity of control arrangements increases rapidly with the number of speeds required, making the planetary gear drive impractical above four or five speeds.

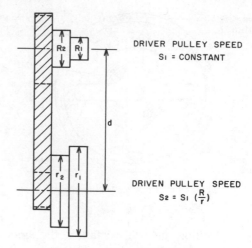

Fig. 5.8w   Stepped pulley system

CONTINUOUSLY VARIBLE-SPEED DRIVES

Cone pulley systems of the type shown in Figure 5.8x are a natural evolution of the stepped pulley system shown in Figure 5.8w. Cone pulleys are designed similarly to stepped pulleys. A series of diameters is calculated equidistant on the pulley axis. The diameter end points are joined to form the pulley contour. The cone pulley is inherently inefficient, since contact surface speed varies across the belt, causing slippage. Belts must be kept narrow to reduce slippage and wear, and thus the capacity for power transmission is reduced. Belt guides at the pulleys are required to hold the belt in position. These guides must move simultaneously for speed changes.

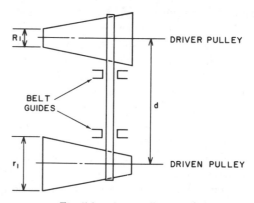

Fig. 5.8x   Cone pulley system

VARIABLE-PITCH PULLEY SYSTEMS

Cone pulleys were the forerunners of the more sophisticated variable pitch pulley systems, which permit continuous, automatic speed adjustments over wide ranges. Speed adjustment in all of these systems is obtained by means of sliding cone face pulleys or sheaves, whose effective diameter can be changed.

A simple variable-pitch sheave is shown in Figure 5.8y. Two flanges are mounted on a threaded hub. The flanges are set to the desired spacing and locked in place with

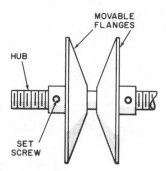

Fig. 5.8y Variable-pitch sheave

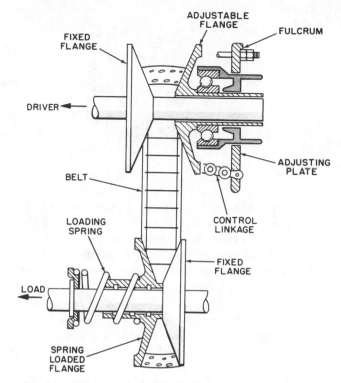

Fig. 5.8z Continuously variable mechanical speed transmission

a setscrew. Speed adjustments are accomplished by changing the flange spacing, producing in effect a pulley of different diameter. This method is very economical, but it requires stationary speed adjustment and special adjustable motor bases to maintain belt tension.

In place of the adjustable motor base, a spring-loaded flat-face idler pulley can be used to maintain belt tension when center-to-center distance must be held constant. Alignment of driving and driven sheaves is also critical, because belt wear will be severe with poor alignment. Also, speed adjustment with this drive is limited by the fact that the setscrew must engage a flat surface.

A number of designs are available with in-motion speed adjustment. The designs vary from manual speed adjustment, through a crank or handwheel, to automatic actuators. Generally these drives use one sheave with adjustable pitch and one spring-loaded sheave that automatically adjusts itself to maintain belt tension (Figure 5.8z. Speed adjustment is obtained by means of a mechanical linkage that moves one of the flanges of the driving sheave. The opposite flange on the driven sheave is spring loaded and moves to maintain belt tension and alignment. Radial motion of the belt along the flange faces in response to speed changes is facilitated by the rotary motion of the sheaves. When speed adjustments are made at rest, the belt cannot move radially and the adjusting linkage may be damaged. On automatic actuators, an interlock should be provided to vent air off the diaphragm whenever the drive is stopped and inhibit speed changes at rest.

Varible-pitch sheave drives are available as package units including motor. Speed adjustment ranges of 4 to 1 are common, but higher ranges to 10 to 1 are possible. Horsepower ranges to 100 horsepower (74.6kW) are also possible. For very narrow speed ranges, drives to 300 (224kW) horsepower are available.

Hydraulic Variable-Speed Drives

Hydraulic variable-speed drives utilize a pump with fixed or variable displacement driving a hydraulic motor

with fixed or variable displacement. The pump is driven by an electric motor at fixed speed. Output speed is controlled by changing pump or motor displacement.

Pumps and motors are virtually identical, differing only in the location of the power input and output. Several designs of variable-displacement units are possible, including axial piston, radial piston, and vane types. An axial piston unit is shown in Figure 5.8aa.

Of course, combinations of pumps and motors with variable or fixed displacement are possible for speed control. Pumps with variable displacement combined with motors with fixed displacement yield a variable-horsepower, fixed-torque drive. Motor speed reversal is possible by reversing the pump stroke.

Combinations of fixed-displacement pump with variable-displacement motors produce drives with fixed horsepower and variable torque. Speed reversal in this combination, however, requires the use of a valve to reverse the supply and return connections at the motor. A variable-displacement pump in conjunction with a variable-displacement motor has characteristics intermediate between the two previously discussed combinations. Range of speed control, however, is widest for this combination.

Pumps and motors with fixed displacement can be used for speed control by controlling the amount of fluid delivered to the motor. This can be accomplished by means of a pressure relief valve on the pump discharge and throttling valve in the oil line to the motor. Motor

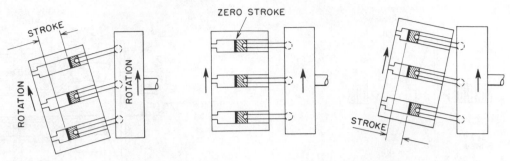

Fig. 5.8aa   Axial piston pump with variable displacement

reversal can also be accomplished by means of a four-way valve, reversing the oil supply and return lines at the motor. These methods, however, are comparatively inefficient and are not used with automatic control.

Hydraulic drives are available as pump-motor packages, or the motor can be mounted remotely and connected to the pump by hydraulic tubing. Displacement is adjustable through push-pull rods or handwheels, and both methods are adaptable for operation by hydraulic and pneumatic actuators or by electric motors for automatic speed control. Speed ranges are adjustable up to about 40 to 1 at horsepower ratings up to 4,000 (2,984kW).

Several attributes of hydraulic variable-speed drives are clear advantages over the other types of drives discussed. Hydraulic drives offer the fastest response in acceleration, deceleration, and speed reversal. They are generally better suited than other types of drives for high shock loads, frequent speed-step changes, and reversals. A disadvantage associated with hydraulic drives is frequent fluid leaks. Besides requiring maintenance, leakage of hydraulic fluid creates safety hazards that can preclude their use.

REFERENCES

1. Hydraulic Institute Standards. Hydraulic Institute, 1969.
2. Hicks, T.G., *Pump Selection and Application*, McGraw-Hill Book Co., New York, 1957.
3. *Sewage Treatment Plant Design*, American Society of Civil Engineers and Water Pollution Control Federation, 1959.
4. Lipták, B.G., (ed.), *Instrument Engineers' Handbook: Process Measurement*, Chilton Book Co., Radnor, PA, 1981.
5. Fair,G.M., and Geyer, J.C., *Water Supply and Waste-Water Disposal*, John Wiley and Sons, New York, 1954.
6. Rouse, H., *Fluid Mechanics for Hydraulic Engineers*, Dover, New York, 1961.
7. Rase, H.F., *Piping Design for Process Plants*, John Wiley and Sons, New York, 1963.
8. Williams, G.S., and Hazen, A., *Hydraulic Tables*, John Wiley and Sons, New York, 1905.
9. Parmakian, J., *Waterhammer Analysis*, Dover, New York, 1963.
10. Lipták, B.G., (ed.), *Instrument Engineers' Handbook: Process Measurement*, Chapter I, Chilton Book Co., Radnor, PA, 1981.

BIBLIOGRAPHY

Adams, R. M., "Don't Overlook Simple Regulators for Complex Fluid Control Jobs," *In Tech*, April, 1983.

Baumann, H. D., "A Case for Butterfly Valves in Throttling Applications," *Instrument & Control Systems*, May, 1979.

——,"Control Valves vs. Speed-Controlled Pumps," Texas A&M Symposium, 1981.

——,"How to Assign Pressure Drop Across Liquid Control Valves," *Proceedings* of the 29th Annual Symposium on Instrument Engineering for the Process Industry, Texas A & M University, January, 1974.

Bower,J.R., "The Economics of Variable-Speed Pumping with Speed-changing Devices," *Pump World*.

Coughlin, J.L., "Control Valves and Pumps: Partners in Control," *Instruments & Control Systems*, January, 1983.

Fischer, A.K., "Using Pumps for Flow Control," *Instruments & Control Systems*, March, 1983.

Gottliebson, M., "Explore the Use of Variable Speed Water Booster Pumps," *Water and Wastes Engineering*, May, 1978.

Hall, J., "Motor Drives Keep Pace with Changing Technology," *Instruments & Control Systems*, September, 1982.

——,"Pump Primer," *Instruments & Control Systems*, April, 1977.

Holland, F.A., "Centrifugal Pumps," *Chemical Engineering*, July 4, 1966.

Holloway, F.M., "What to Look for in Nutering Pumps," *Instruments & Control Systems*, April, 1976.

Janki C., "What's New in Motors and Motor Controls?" *Instruments & Control Systems*, November, 1979.

Lipták, B.G., "Costs and Benefits of Boiler, Chiller, and Pump Optimization," *Instrumentation in the Chemical and Petroleum Industries*, vol. 16, Instrument Society of America, Research Triangle Park, NC, 1980.

——, "Save Energy by Optimizing Boilers, Chillers, Pumps," *In-Tech*, March, 1981.

Liu, T., "Controlling Pipeline Pumps for Energy Efficiency," *In-Tech*, June, 1979.

Merritt, R., "Energy Saving Devices for AC Motors," *Instruments & Control Systems*, March, 1980.

——, "What's Happening with Pumps, *Instruments & Control Systems*, September, 1980.

Papez, J.S., and Allis, Louis, "Consideration in the Application of Variable Frequency Drive for Pipelines," ASME Paper 80-PET-78.

Rishel, J.B., "The Case for Variable-Speed Pumping Systems," *Plant Engineering*, November, 1974.

——, "Matching Pumps to System Requirements," *Plant Engineering*, October, 1975.

——, "Water System Head Analysis," *Plant Engineering*, October, 1977.

Schroeder, E.G., "Choose Variable Speed Devices for Pump and Fan Efficiency," *In Tech*, September, 1980.

Stewart, R.F., "Applying Adustable Speed AC Drives," *Instruments & Control Systems*, July, 1981.

Thurlow, C., "Pumps and the Chemical Plant," *Chemical Engineering*, May 24, 1965.

5.9   FLOW REGULATORS

Design Pressure:	Up to 3000 PSIG (20.7 MPa)
Design Temperature:	Up to 700°F (371°C)
Materials of Construction:	Aluminum, carbon steel, stainless steel
Design Flowrates:	0.02 to 600 gpm (1.2×10^{-6} to 0.038 m³/s) of unit water flow
Line Sizes:	⅛″ to 4″ (3.125 mm to 100 mm)
Cost:	$100 to $8,000, depending on type, size, and material. A 2″ (50 mm) size carbon steel self-contained flowrate controller, for example, costs about $2000.
Partial List of Suppliers:	Ametec, Inc., Schutte & Koerting Division; Brooks Instrument Div. of Emerson Electric Co.; Fischer & Porter Co.; The W. A. Kates Co.

Self-contained flowrate controllers combine an adjustable orifice and an automatic internal regulating valve in one single body. Pressure differential across the adjustable orifice is automatically maintained constant, regardless of the orifice area. Total flowrate is, therefore, directly proportional to orifice area. Orifice and automatic differential pressure regulator are enclosed in a single housing with a calibrated flowrate indicating dial.

A flowrate controller of this type is shown in Figure 5.9a. The metering orifice is formed by an arcuate slot in a cylinder (orifice sleeve) and a second slot in an inner (orifice) cylinder around which the first (orifice sleeve) can be rotated. When the two slots match or coincide, the orifice has its maximum open area; when the outer sleeve is rotated, the open length of slot and the open area decrease in proportion to the angle of rotation of the sleeve. As the slots extend 160 degrees, rotation through this angle reduces orifice area to zero.

The automatic differential pressure regulator includes the impeller, spring, valve sleeve, and valve tube. The impeller disc reciprocates in the upper part of the orifice sleeve, driven by the force balance between orifice differential pressure across it and the spring force. The impeller is rigidly connected to the valve sleeve. It closes or opens the valve ports as dictated by the force balance, maintaining orifice differential pressure and holding orifice flow at the valve determined by the differential pressure and the orifice area.

Advantages of the self-contained flowrate controllers are:

1. Accurate flowrate control by means of a self-operated single unit.
2. No requirements for straight pipe runs upstream or downstream of the unit.
3. No utility requirements to operate the unit (no electric power, pneumatic air, or hydraulic fluids).
4. Since control occurs at the point where the signal shows need for correction, maximum speed of response is assured.
5. Flow scale is linear and, therefore, the unit has a higher rangeability than an instrument based on a square root function.
6. Simple design and few parts with little calibrations and adjustments required.
7. First cost and maintenance cost are low.

A disadvantage is that there is a minimum pressure drop required for the valve to work properly.

A typical application of these devices is for flow control of sealant fluids to the rotating seals of centrifugal pumps, compressors, turbines, and other rotating machinery. Another application is in pressure filtration, where a self-

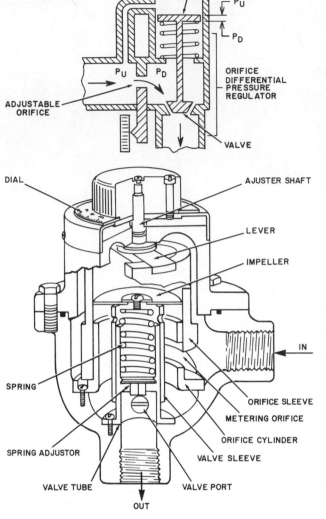

Fig. 5.9a   Self-contained flowrate controller

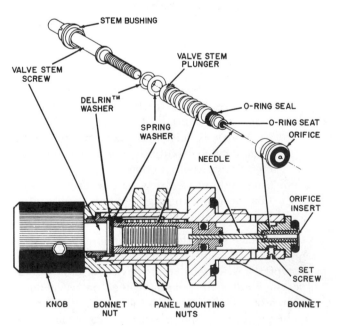

Fig. 5.9b   Nonrising stem type flow controller

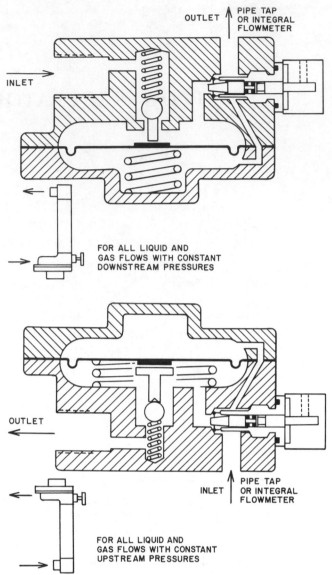

FOR ALL LIQUID AND
GAS FLOWS WITH CONSTANT
DOWNSTREAM PRESSURES

FOR ALL LIQUID AND
GAS FLOWS WITH CONSTANT
UPSTREAM PRESSURES

Fig. 5.9c   Differential pressure type flow regulator used in combination with variable area flowmeters

contained flowrate controller will keep the filter effluent constant at a constant rate as the pressure drop across the filter increases with increasing filter cake build-up. Other applications include proportionate blending, batch measurement or control, and hydraulic oil system control, where a constant flowrate to various components must always be maintained.

For accurate flowrate control in gas flow applications, the upstream gas pressure and temperature variations should be minimized. To determine the capacities for a given device, manufacturers provide gas flow equivalent graphs. These graphs show values to 1 gpm (0.00063 m^3/s) (i.e., ft^3 at stp) of water. All capacity data are based on water for purposes of standardization and convenience. Also, water is normally used as the test fluid for factory calibrations. Graphs are readily available for air, hydrogen, natural gas, nitrogen, and oxygen. For other gases, a factor (C) determined from its specific gravity is applied to the capacity data for air. Hence at a given pressure and temperature, flow (gas) = flow (air) × C where

$$C = \frac{1}{\sqrt{S.G. \text{ of gas}}}$$

In some applications it is desirable to position the orifice size remotely. Pneumatic and electric actuators are readily available for this and they can be directly mounted on the device. This allows for manual operation as well as use of an automatic controller output for flow-rate adjustments.

Another type of self-contained flow regulator is shown in Figure 5.9b. It is designed to accurately adjust and maintain small gas and liquid flows. This type utilizes a nonrising stem and features a positive, direct mechanical means of adjusting a sliding tapered needle that virtually prevents sticking caused by foreign matter in the fluid stream. These valves are particularly suitable for the precise control requirements in chromatography.

Also, variable area flowmeters (rotameters) are frequently used for flow control in conjunction with an integrally mounted diaphragm actuated control valve. See Figure 5.9c and Section 2.2 in the volume on Process Measurement.

BIBLIOGRAPHY

Adams, R.M., DeChiara, W.A., "Don't overlook Simple Regulators for Complex Fluid Control Jobs," *InTech*, April, 1983.

Hayward, A.J., "Choose the Flowmeter Right for the Job," *Processing Journal*, 1980.

Kates, W.A., "Direct-Acting Flow Rate Regulators," Paper #86-NY60, presented at ISA Conference, New York, 1960.

———, "Self-Actuated Flow Rate Controls," Paper #69-536, presented at the ISA Conference, Houston, Texas, 1969.

Laskaris, E.K., "The Measurement of Flow," *Automation*, 1980.

Lomas, D.J., "Selecting the Right Flowmeter," *Instrumentation Technology*, 1977.

Watson, G.A., "Flowmeter Types and Their Usage," *Chartered Mechanical Engineer Journal*, 1978.

Weh, T.W., and Dougherty, W.S., "Self-Contained Flow Regulator for Accurate Low-Cost Slurry Control," *Chemical Processing*, November, 1958.

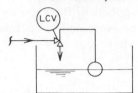

5.10 LEVEL REGULATORS

Pressure in Vessel:	Up to 500 PSIG (3.5 MPa) with float type; atmospheric for diverter and altitude valves
Maximum Pressure Drop Through Valve:	100 dPSI (0.7 dMPa) for float type; 50 dPSI (0.35 dMPa) for diverter or altitude valves
Design Temperature:	Up to 450°F (232°C) with float type; ambient for altitude valves; up to material selected on diverter valves
Materials of Construction:	Iron, steel, stainless steel for float type; unlimited for diverter valves; cast iron for altitude valves
Range:	Normally less than 12″ (300 mm) for float type; diverter valves essentially on-off; altitude valves practically on-off (some regulate)
Valve Sizes:	½″ to 42″ (12.5 mm to 1050 mm) for float and diverter type; 2″ to 42″ (50 mm to 1050 mm) for altitude valves
Cost:	$100 to $50,000 depending on valve size and materials of construction. A 30″ (750 mm) altitude valve with multiple control functions costs approximately $25,000 while a ¾″ (18.8 mm) float valve runs $100
Partial List of Suppliers:	CLA-VAL Co.; Fisher Controls International, Inc.; Kieley & Mueller Inc., Sub. of Johnson Controls Inc.; Masoneilan Div., McGraw-Edison Co.; Moore Products Co.; Ross Valve Mfg. Co., Inc.

Self-contained level regulators contain components to detect the level of the process fluid, compare that information with a desired value of level (set point), and generate a valve output to bring the measured signal to the desired level. They are similar to the self-contained flow regulators described in Section 5.9, where the conventional components are also combined into a single device. It should be noted, however, that several companies have discontinued manufacturing such self-contained devices and the "conventional" way of measuring, controlling, and adjusting a final control element has become much more prevalent in the process industry.

In the various designs described here, the energy source for operation is gained from either the static and kinetic energy of the flowing stream or from the buoyant force generated by the process fluid in the vessel. Below, we will describe three basic level regulator designs, namely the float level control valves, solid state diverter valves, and altitude valves.

Float Level Control Valves

In float level control valves the movement of the level float is transmitted through mechanical linkages to the final control element (control valve) to regulate the flow from or to the vessel. In order to fully recognize the advantages and disadvantages we need to consider the following:

1. Offset due to load changes, hysteresis, dead band, and span.
2. Power limitations from the float.
3. Stuffing box arrangements on pressurized vessels.
4. Installation requirements.

Offset, Hysteresis, Dead Band, and Span

When process load changes occur, a plain proportional controller is not capable of keeping the process properly on the set point, and a permanent offset between set point and measurement will occur. See Figure 5.10a.

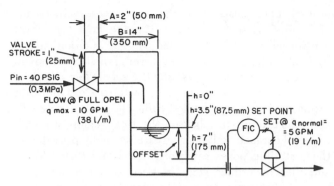

Fig. 5.10a Offset in a float level control loop

The proportional band in the control loop shown in Figure 5.10a can be defined as the level change in the tank that will result in full stroking of the valve. The valve stroke from open to close is 1 in. (25 mm) and the corresponding level change is B/A, or 7 in. (175 mm). Proportional sensitivity (K_c) can be defined as the change in flowrate through the valve that corresponds to a unit change of liquid level in the tank. Since 7 in. (175 mm) of level change will move the valve from fully closed (no flow) to fully open (maximum flow of 10 gpm or 0.00063 m^3/s), the proportional sensitivity is:

$$K_c = \frac{q}{h} = \frac{10}{7} = 1.4 \text{ gpm/in.} =$$

$$= \frac{0.00063}{175} = 3.5 \times 10^{-6} m^3/s/mm \qquad 5.10(1)$$

The flow controller on the discharge from the vessel maintains a flowrate of 5 gpm (3.2×10^{-4} m^3/s). To maintain steady state conditions, the flow into the vessel has to match that flowrate. The level set point for this process load condition is:

$$h = \frac{q \text{ normal}}{K_c} = \frac{5}{1.4} = 3.5 \text{ in.} =$$

$$= \frac{.00032}{3.5 \times 10^{-6}} = 87.5 \text{ mm} \qquad 5.10(2)$$

Load changes on the set point could result from an operator-induced change in the effluent flow rate or a change in the header pressure to the vessel. We will demonstrate the effect of load changes on the set point of the level controller caused by a change in header pressure. Assuming that the header pressure is reduced from 40 PSIG (276 kPa) to 10 PSIG (69 kPa), and the maximum flow through the fully open control valve drops from 10 gpm (0.00063 m^3/s) to 5 gpm (0.00032 m^3/s), the new proportional sensitivity is:

$$K_c = \frac{q}{h} = \frac{5}{7} = 0.7 \text{ gpm/in.} =$$

$$= \frac{0.00032}{175} = 1.7 \times 10^{-6} \ m^3/s/mm$$

The new level at which steady state conditions will be reached is:

$$h = \frac{q \text{ normal}}{R_c} = \frac{5}{0.7} = 7 \text{ in.} =$$

$$= \frac{0.00032}{1.7 \times 10^{-6}} = 175 \text{ mm}$$

Therefore, the previously maintained level cannot be kept and a permanent offset of 3½ (87.5 mm) inches results because of that load change. The amount of offset can be reduced by narrowing the proportional band, but it cannot be fully eliminated. If the proportional band was reduced to 1 in. (A–B in Figure 5.10a), the proportional sensitivity would be increased by a factor of seven and the permanent offset would be reduced to ½ in. (12.5 mm). On the other hand, with a wider proportional band (a wider range of the float controller), the permanent offset would be greater.

As with any other instrument having stuffing boxes, pivots, and mechanical linkages, the quality in terms of hysteresis, dead band, and span needs to be considered. Before a motion takes place a "dead band" or "break away torque" has to be overcome. Secondly, the hysteresis of such mechanical systems, in terms of force-to-motion relationship as affected by the direction of motion, is pronounced. Both hysteresis and dead band are substantial in float type level control valves. The span of these blind, self-contained level control devices is adjustable, but the control range is normally less than 12 in. (300 mm).

Power Generated by the Float

The buoyant force generated by the float is a function of float diameter and the process fluid density. The net power available to close the valve against the static pressure in the header is substantially less because the float weight has to be subtracted from the gross buoyant force. All the friction losses in stuffing boxes, pivots, and mechanical linkages also reduce the net power available. In addition, in some applications the total float volume cannot be used for sizing because the float is supposed to be only partially submerged when the valve is closed. For example, an 8-in. (200-mm) diameter float with 150 PSIG (1.04 MPa) pressure rating has a gross buoyancy of 9.5 lbf (42.3 N) when totally submerged. After subtracting the float weight the net buoyancy is 6 lbf (26.7 N). On a 500 PSIG (3.45 MPa) design pressure float the net buoyancy is reduced to 4.5 lbf.(20 N). This net force is then amplified through the mechanical levers, while at the same time some of it is used to overcome the system friction.

Figure 5.10b shows the relationship between size of the float valve and the maximum pressure drop against which the valve will be able to close. It shows that there is substantial difference between rotary and sliding stem

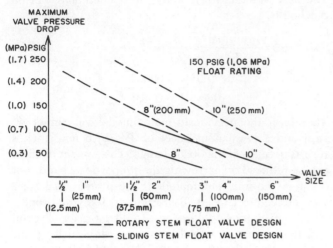

Fig. 5.10b   Valve pressure drop limitations as a function of
valve size and design

units that are isolated from the process vessels by means of frictionless seals. The seals on the float type level control valves are not frictionless. The rotation caused by float motion is transmitted through a stuffing box to the mechanical levers operating the valve. The stem friction in the stuffing box is reduced by the use of a lubricant fitting, which can be refilled while the unit is in service and which tends to lubricate the total region of metal-to-metal contact. Ball bearings are normally provided in these stuffing boxes to absorb radial and thrust loads, and thereby reduce stem friction. When repacking is needed, it can be done by breaking the packing gland union on the stuffing box and replacing the package without removing the unit from the vessel.

With operating temperatures below 30°F (-1.1°C) or above 450°F (232°C), radiation fins are provided between the vessel and the packing box to keep it closer to the ambient temperature.

design. A larger float is capable of closing the same size valve against a higher pressure. The rotary stem design shown in Figure 5.10c is more sensitive than the sliding stem design in Figure 5.10d and can handle higher pressure drops through the valve because of the relatively low friction in the lubrication stuffing box.

Stuffing Boxes

Displacer and float type level transmitters and controllers were described in Sections 3.8 and 3.9 of the Process Measurement volume. These sections covered

Installation Considerations

The installation of float control valves on open vessels is fairly simple as shown in Figure 5.10d. When the vessel is pressurized, the installation becomes more difficult because of the instrument configuration involved. Figures 5.10e, f, and g illustrate some of the standard methods of installation. On Figure 5.10e, a flange-mounted unit is shown. The float is inserted into the vessel through a nozzle. Consequently, the vessel has to be empty before the assembly can be removed for repairs or maintenance. The stuffing box is perpendicular to the float

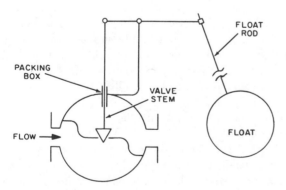

Fig. 5.10c   Rotary stem type float valve

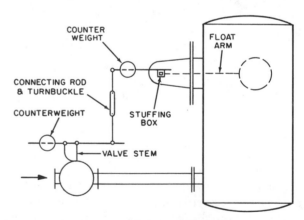

Fig. 5.10e   Flange-mounted float

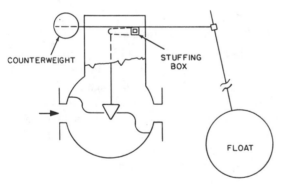

Fig. 5.10d   Sliding stem type float valve

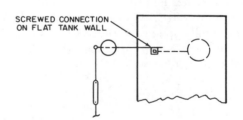

Fig. 5.10f   Direct mounting of float level control
valve on pressurized tank coupling

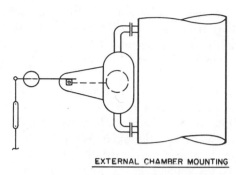

Fig. 5.10g External chamber mounting of float
level control valve on pressurized tank

arm. The proper balance of linkage, turnbuckle, float, and inner valve weights is achieved by adjusting of the counterweights. The connecting rod between the float assembly and the valve should be of minimum length and needs to be carefully aligned to avoid binding during float travel.

In Figure 5.10f, the stuffing box is screwed directly into the flat vessel wall. The float has to be installed from the inside, and the vessel must be drained for servicing.

The external cage utilized in the installation in Figure 5.10g can be provided with isolationg valves so that the instrument can be serviced without disruption of the process.

The maximum recommended lengths for the connecting rod with the turnbuckle is about 8 ft. (2.4 m). If the vertical distance between float assembly and valve is greater than that, a mechanical lever system cannot be used and a pneumatic control loop has to be installed. The float assembly with the rotating stem stuffing box is available with a pneumatic pilot. This has the same control characteristic as the mechanical unit, except for a simpler sensitivity adjustment. With a pneumatic relay, the power limitation of the float is no longer critical since air pressure is used to operate the valve. The float assembly can also be used for level alarming. A switch can be actuated by the rotating stem instead of moving the mechanical levers.

Conclusions

Float type level control can be used where the quality of the control is not critical and the narrow span, set point, offset, dead band, and hysteresis can be tolerated. Because of the large number of moving parts, pivots, levers, and the stuffing box, the device requires regular maintenance and should be used only on clean fluids where material build-up will not interfere with the operation of the moving parts. The use of these instruments is limited on pressurized vessels, in as much as it is a blind controller, the seal is not frictionless, the device is quite complex, and the physical relationship between valve and float is restricted. The economy of this design must also be balanced against the limitations of float

power and the nature of the materials in contact with the fluid. Industrial applications of float type level controllers are therefore quite limited. On open water tanks, such as cooling tower basins, the lack of readily available or economical external power sources is another limitation.

Diverter Valves

A completely solid state level regulator with no moving parts has been devised that utilizes fluidics principles. This valve works on the principle of the Coanda effect— that is, a fluid stream will attach itself to a nearby sidewall. Figure 5.10h shows this on-off level control valve. One sidewall is closer to the incoming fluid stream than the opposite sidewall. A level sensing dip pipe is connected to the control port, upstream of the sidewall and closer to the incoming liquid.

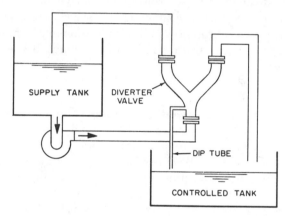

Fig. 5.10h Solid state diverter valve installed for
on-off level control

When the level in the vessel is below the dip tube opening, air from the atmosphere enters the dip tube. This air is entrained into the control port, causing the inlet jet to detach itself from the adjacent sidewall and to attach itself to the opposite sidewall, diverting the full flow to the controlled vessel. As the level in the vessel rises, it will eventually cover the end of the dip tube cutting off the air flow to the control port. As soon as this occurs, the flow stream will attach itself to the "favored" sidewall, diverting the full flow back to recirculation. Having no moving parts, the on-off cycling of the valve has no harmful effects.

The shape of the flow passage makes the valve function. With no plugs or packing as in conventional valves, the unit is essentially maintenance free. The valve is not affected by vibration or temperature (operating or ambient) and requires no external power source for operation. The flow stream is diverted in one tenth of a second, and the valve can be manufactured out of metal, plastic, or ceramic material. The design is streamlined without shoulders or pockets and so is recommended by the supplier for use not only on clean liquids, but also on fluidized solids and slurries. It should be noted, however, that for proper operation the dip tube and the control port must be open; any material build-up in them will interfere with the functioning of the device.

When the solid state diverter valve is considered for level or any other control application, the effect of back pressure must be carefully considered. Outlet pressures other than atmospheric should be evaluated as to their effect on tight shut-off and safe failure position.

The solid state diverter valves can also be used for throttling instead of on-off control. In such a design, there are two ports for each leg of the valve, and the quantity of air entering the ports determines the ratio between the streams leaving through the two legs. For throttling control, the back pressure in the controlled leg cannot be more than half of the inlet pressure to the valve in order not to spill into the unused leg. The solid state diverter valve is a reliable on-off control device for atmospheric vessels containing hot or cold, corrosive or noncorrosive clean liquids. Its application in pressurized vessels or with hard-to-handle process fluids may not be desirable.

Altitude Valves

Altitude valves are installed in supply lines to elevated basins, tanks, or reservoirs for the purpose of holding constant heads and to prevent overflow. No external energy source is required because the power for operation is gained from the pressure of the process fluid.

One of the simplest versions of this design is shown in Figure 5.10i. The purpose of this valve is to admit water into the elevated tank until a preset level is reached and to close at that point to prevent overflow. The hydraulic head in the elevated tank is balanced by the setting spring of the pilot. When the tank water level exceeds the setting of the spring, the pilot diaphragm is moved down, closing the drain port and opening the main line pressure to apply its force to the top piston. Because the top area of the piston is greater than the bottom, the main valve closes. If the head in the tank decreases, the setting spring moves the diaphragm up,

closing the main line pressure connection and opening the drain port on the pilot. This results in removing the pressure from the large end of the piston, thereby opening the main valve to supply additional water to the head tank.

Where this type of altitude valve is used, discharge from the tank is handled by a swing check valve in a bypass line around the altitude valve or by a separate line from the tank to the plant. With the check valve as shown in the figure, water will flow from the head tank to the plant whenever the demand of water used exceeds the supply available. When the plant demand falls back below the capacity of the supply, the check valve closes and the water is made up in the tank under the control of the altitude valve.

Altitude valves are multipurpose devices, and with the addition of pilots, they can satisfy a number of requirements, in addition to the simple on-off level control described above.

All altitude valves close on high tank level to prevent overflow, but they can have other features as well:

1. The valve can have a delayed opening so that it opens only when the tank level has dropped below the control point by a set amount.
2. Instead of on-off action, the valve can be closed gradually to eliminate pressure shock (water hammer).
3. Instead of opening on low level, the valve can stay closed until the inlet pressure drops below the head pressure of the tank. With this unit, there is no need for the check valve in the bypass as shown in Figure 5.10i because the valve is designed for two-way flow allowing the water to return from the tank when the distribution or supply pressure drops below that of the tank head.
4. The valve can be modified to delay its opening until the supply or distribution pressure has dropped to an adjustable preset point.
5. By using another combination of pilots, the valve will open on low level only to the extent needed to maintain constant inlet pressure to the valve (which is the distribution pressure to the plant).
6. The unit can be provided with an additional check valve feature. In this arrangement, the valve operates as described, except that it closes if the distribution pressure drops to a predetermined low point, due to line breakage or other reasons.
7. When used on aerator basins with high supply pressures, the altitude valve can also act as a pressure reducer. In this case, the valve will open on low level only to the extent needed to provide a permissible outlet pressure for the aerator nozzles.

As far as accessories are concerned, all altitude valves can be provided with solenoid pilots to close the valve on a remote electrical signal. Limit switches can also be

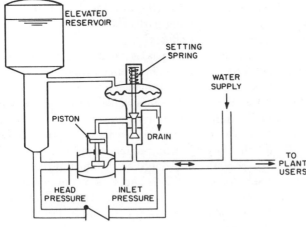

Fig. 5.10i   Typical altitude valve installation

furnished to give remote indication of valve position (whether open or closed).

The use of altitude valves is recommended for locations where only hydraulic power is available to operate large water valves controlling the level in storage reservoirs. They are not applicable to level control on pressurized vessels. The materials of construction of these units restrict their main application to water service. Due to the substantial number of components and moving parts, they require periodic cleaning and maintenance. Protection against freezing of stationary sensing lines is also required. In line sizes greater than 12 in. (300 mm), there is some economic advantage in using butterfly valves with hydraulic pilots (see Section 4.4).

Conclusions

Float-type level control can be used where the quality of the control is not critical and the narrow span, set point, "off-set," dead band and hysteresis can be tolerated. Because of the large amount of moving parts, pivots, levers, and stuffing boxes, the device requires regular maintenance and should be used only on clean fluids where material build-up will not interfere with the operation of the moving parts. The use of these instruments is limited on pressurized vessels, because it is a blind controller, the seal is not frictionless, the device is quite complex, and the physical relationship between valve and float is restricted. When the limitations of float power and the construction material of all parts in contact with the fluid are considered, the economics of this design must be balanced against all these factors. Industrial installations of these float-type level controllers are therefore quite limited on open water tanks (such as cooling tower basins) where, due to the remoteness of the installation, external power sources are not readily available or are extremely expensive to be provided.

BIBLIOGRAPHY

Akeley, L.T., "Eight Ways to Measure Liquid Level," *Control Engineering*, July, 1967.

API Guide for Inspection of Refinery Equipment, Chapter XV, "Instruments and Control Equipment."

Belsterling, C.A., "A Look at Level Measurement Methods," *Instruments & Control Systems*, April, 1981.

Cheung, T.F., and Luyben, W.L., "Liquid-Level Control in Single Tanks and Cascades of Tanks with Proportional-Only and Proportional-Integral Feedback Controllers," *Ind. Eng. Chem. Fundam.*, February, 1979, p. 15.

Considine, D.M., "Process Instrumentations; Liquid Level Measurement Systems; Their Evaluation and Selection." *Chemical Engineering*, February 12, 1968.

Hall, J., "Level Monitoring; Simple or Complex," *Instruments & Control Systems*, October, 1979.

———, "Measuring Interface Levels," *Instruments & Control Systems*, October, 1981.

Lawford, V.N., "How to Select Liquid-Level Instruments," *Chemical Engineering*, October 15, 1973.

Morris H.M., "Level Instrumentation from Soup to Nuts," *Control Engineering*, March, 1978.

Sastry, V.A., "Self-Tuning Control of Kamyr Digester Chip Level," International Symposium on Process Control, May 2-4, 1977.

Taylor, D., "Economical Method of Glass Level Control Utilizing a Laser Beam Generator and a Solid State Detector," ISA, 1977.

Yearous, H.B., "Sensors for Liquid/Solid Level Controls," *Plant Engineering*, March 8, 1979, pp. 109–112.

"Manual on Installation of Refinery Instruments and Control Systems," API Recommended Practice 550, Part I, Section 2 Level, Fourth Edition, February, 1980.

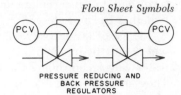

PRESSURE REDUCING AND
BACK PRESSURE
REGULATORS

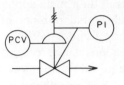

AIRSET, REGULATOR WITH
OUTLET PRESSURE RELIEF
AND INDICATOR

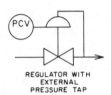

REGULATOR WITH
EXTERNAL
PRESSURE TAP

PRESSURE DIFFERENTIAL
REGULATOR WITH
INTERNAL AND EXTERNAL
PRESSURE TAPS

5.11 PRESSURE REGULATORS

Regulator Types:	A—Weight loaded B—Spring loaded C—Piloted ¼″ (6.25 mm) air regulator D—Internally piloted E—Externally piloted
Sizes:	A—½″ to 6″ (12.5 to 150 mm) B—¼″ to 4″ (6.25 to 100 mm) C—¼″ (6.25 mm) D—½″ to 6″ (12.5 to 150 mm) E—⅜″ to 12″ (9.38 to 300 mm)
Design Inlet Pressure:	B—Up to 6,000 PSIG (41.4 MPa) D and E—Up to 1,500 PSIG (10.35 MPa) A and C—Up to 500 PSIG (3.45 MPa)
Minimum Regulated Outlet Pressure:	B and E—Down to 2 PSIG (13.8 kPa) A and D—Down to 0.5 PSIG (3.45 kPa) C—Down to 0.1 PSIG (0.69 kPa)
Droop or Offset:	B—5% to 80% E—2% to 10% A and D—1% to 2% C—½%
Cost:	C—$80 B—$30 to $800 A and D—$120 to $1,500 E—$100 to $2,000
Partial List of Suppliers:	A.W. Cash Valve Mfg. Co.; Cashco, Inc.; Conoflow Regulators & Controls; Fairchild Industrial Products Co.; Fisher Controls International, Inc.; Foxboro Co.; Jordan Valve Div., Richard Industries; ITT General Controls Div.; Kieley Mueller Inc., Sales of Johnson Controls Inc.; Leslie Co.; Masoneilan Div., McGraw Edison Co.; Modern Engineering Co.; Moore Products Co.; Rockwell International, Municipal & Utility Div.; Spence Engineering Co.

Regulators vs. Control Valves

A regulator incorporates a sensor, a controller, and a control valve. Regulators are readily available to control level, vacuum, flow, mass flow, pressure, and temperature. Many applications in pressure, temperature, and flow control can be handled with either a regulator or a control valve. How does one choose between them? Which type of valve should be selected? The advantages and disadvantages are discussed below and summarized in Table 5.11a.

Advantages of Regulators

Regulators usually cost less than the combination of control valves, transmitters, and controllers. They are less expensive to buy, install, and maintain. But when the application requires a larger valve, the economics begin to change in favor of control valves.

Regulators have a built-in controller and do not require an air supply. This results in savings on purchase and installation costs. Regulators are not subject to air supply failure because the power to operate them is contained

Table 5.11a
REGULATORS VS. CONTROL VALVES

Regulators	Control Valves
Lower cost (in small sizes and ordinary materials of construction	Higher cost
Smaller size	Larger and heavier
Lower installation cost	Higher installation cost due to size and weight
No air supply required	Requires air connections
Built-in controller	External controller
Limited materials of construction	Practically no limit on materials of construction
No remote control; set point adjustment is at regulator location	Local or remote control; remote/manual control possible
Single mode (proportional only) with associated droop	Varied control modes possible
Most fail open; relief valve is recommended	No limitation in fail-safe
Limited accessories	Wide variety of accessories
Limited interchangeability	Possibility of interchangeability among services
Fewer applications	More (and more complex) applications

in the fluid. In critical fail-safe applications or at locations remote from a source of compressed air, this is an important consideration. However, diaphragm failure in a regulator usually causes valve opening, which can be unsafe.

Advantages of Control Valves

Control valves are used with an external controller. This controller has the flexibility of one-, two- or three-mode control with remote/manual operation. They are compatible with any measuring system. Materials of construction of the valve include any castable metal, and the valve may be lined. Fail-safe may be in any desired form. Valve accessories include positioners, limit switches, manual handwheels, solenoids, and local controllers. These can be specified to be interchangeable with valves in other services, resulting in reduced spare parts requirements.

Disadvantages of Regulators

The set point of a regulator is provided integrally, and remote control is not possible. The controller is single mode (proportional only), and therefore it suffers from having a set-point droop curve varying with throughput.

Materials of construction and interchangeability with other services is limited. Accessories cannot be applied to regulators. Regulators are limited to pressure, temperature, flow, and level applications and cannot be used with some of the more complicated measurements, such as analysis, or with electrical signals, such as thermocouples. Even level regulators will not suffice if a large level range is required.

Disadvantages of Control Valves

Control valves with controllers and transmitters are more expensive to purchase, install, and maintain than are regulators. They are generally larger, requiring more space for installation and making handling more difficult. They require air supplies, which results in increased cost and increased maintenance due to the possibility of leakage.

Common Terms

Some terms commonly used in connection with regulators are defined below:

> *Droop*—The amount by which the controlled variable (pressure, temperature, or liquid level) deviates from the set value at minimum controllable flow when the flow through the regulator is gradually increased from the minimum controllable flow to the rated capacity.
> *Disturbance Variable*—An undesired variable applied to a system that tends to adversely affect the value of a controlled variable.
> *Open Loop Gain*—The ratio of the change in the feedback variable to the change in the actuating signal.

The Pressure Regulator

The self-contained pressure regulator, which was developed late in the 19th century, has the advantages of being simple, dependable, rugged, and inexpensive.[1] Energy from the flowing fluid operates the regulator.

Figure 5.11b shows that the diaphragm is the "brain" of the regulator. It compares the set point, which is converted into a spring force, to the regulated pressure,

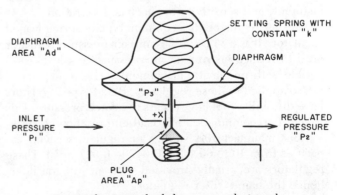

Fig. 5.11b   Spring-loaded pressure reducing valve

which is converted into a force by the diaphragm itself. It adjusts the valve opening to reduce the error between the two. The diaphragm is a feedback device, an error detecting mechanism, and an actuator. Figure 5.11b shows a spring-loaded pressure reducing valve. Regulators are also available to control back pressure, differential pressure, and vacuum. They can be applied to gasses, vapors, and liquids. Springs, weights, and gas pressures can actuate regulators. Thousands of different regulator types and designs are available. Descriptions of the most common regulator designs are given below.

Weight-Loaded Regulators

The force in a spring changes as the spring is compressed. In a weight-loaded regulator, however, a weight provides a constant actuating force on the diaphragm, thus minimizing droop. Set point changes are accomplished by adding weight or by changing the weight position (Figure 5.11c). Weight-actuated regulators are widely used to regulate gas pressures at 1 PSIG (6.9 kPa) or below when load changes are slow. It is not recommended for service where mechanical shock or vibration are present or for regulation of incompressible fluids.

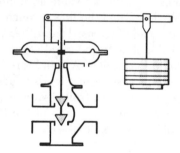

Fig. 5.11c   Weight-loaded pressure regulating valve with external pressure tap

Spring-Loaded Regulators

Spring-actuated regulators are used more than any other type because of their economy and simplicity.[2,3] They also have an extremely fast dynamic response. Disadvantages include high droop, awkward set point adjustment, and sensitivity to shock and vibration.[3]

The "air regulator" is a common type of spring-loaded regulator. It is a ¼ in. (6.25 mm) air pressure regulator used to reduce instrument air pressure to a level compatible with pneumatic instruments. The term "airset" is also applied to these regulators. Bleed-type regulators have the ability to relieve excess regulated pressure. This design is recommended for dead-end (no-flow) service. Various manufacturers supply air regulators having capacities from 10 to 60 scfm (0.047 to 0.028 m³/s). These regulators are usually provided with an integral filter, hence the name "filter regulator."

Figure 5.11d shows one design of a spring-actuated

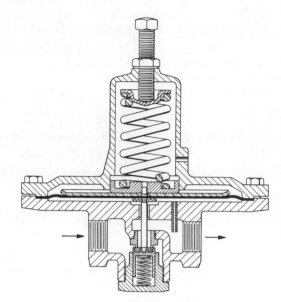

Fig. 5.11d   Spring-loaded pressure regulator

regulator. The valve is normally closed. Turning the handwheel or setscrew compresses the spring and opens the valve. Increasing downstream pressure acts beneath the diaphragm, raising it and closing the valve. Spring compression is adjusted to provide the desired downstream pressure at the demand flow.

Figure 5.11e shows a spring-actuated vacuum pressure regulator.

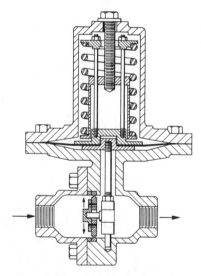

Fig. 5.11e   Spring-loaded vacuum regulator. (Courtesy of Jordan Valve Div., Richards Industries)

Pilot-Operated Regulators

The pilot-loaded regulator is a two-stage device. The first stage contains a spring-actuated pilot regulator that controls pressure on the diaphragm of the main regulating valve. The second stage contains a spring-actuated regulator that controls the main valve stem. One advantage is that the actuating fluid operating the first stage (pilot) regulator is at the upstream pressure. This provides a higher level of force to the actuating mechanism. Travel is very short, which reduces droop. More accurate regulation is possible with this design.

Figure 5.11f shows that the pilot regulator can be external or internal. Here it controls pressure to a diaphragm operating the main valve. In some designs, this diaphragm is replaced by a piston. Pilot operated regulators use the difference between upstream and downstream pressures for actuation. Therefore, a minimum differential is required for proper operation.

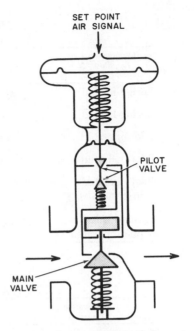

Fig. 5.11g   Pneumatically loaded remote set-point pilot regulator

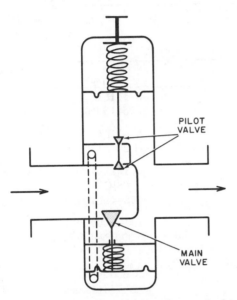

Fig. 5.11f   Pressure regulator with integral pilot

Pilot-operated regulators provide accurate regulation for a wide range of pressures and capacities. They are more expensive to purchase and install than spring-loaded regulators on the basis of valve size, but not necessarily on the basis of capacity.[1] Because the small passages and ports can become plugged, these regulators can only be used with clean fluids. Their dynamic response is slower than for spring-loaded designs.

Externally piloted regulators are available in a wide range of sizes. One advantage of this design is that the pilot regulator can be mounted at a distance from the main valve, in an accessible location away from possible pipeline shock and vibration. The external pilot also simplifies maintenance. Figure 5.11g illustrates an internally piloted regulator with a pneumatic set point loading. This design offers the convenience of remote set point adjustment at distances up to 500 ft (150 m), but the requirement for compressed air makes it more expensive to operate. Table 5.11h provides data to help in deciding between spring-loaded and pilot-loaded regulators.

Regulators for the Gas Industry

In addition to regulators available for general service, the gas utility industry utilizes an extensive family of regulators designed specifically for gas service.[4,5] The types of regulators for these services are tabulated in Table 5.11i.

Regulator Features

The self-contained regulator has no adjustments, either in terms of capacity (reduced valve trim) or stability (proportional band and reset adjustments). Service conditions must be well defined in order to specify a regulator resulting in a satisfactory installation. The following features should be considered.

Seating and Sensitivity

The balance of forces on the valve stem (Figure 5.11b) is:

$$kx = A_p(p_1 - p_2) - A_d p_3 \qquad 5.11(1)$$

where

k = spring constant (1bf/in. or N/m),
x = valve stem movement (in. or mm),
A_P = port area (in.^{2} or mm^2),
p_1 = inlet pressure (PSIG or Pa),
p_2 = outlet pressure or regulated pressure (PSIG or Pa),
p_3 = diaphragm back pressure (PSIG or Pa) and
A_d = diaphragm area (in.^{2} or mm^2).

Table 5.11h
REGULATOR SELECTION FOR VARIOUS SERVICE CONDITIONS[3]

Recommended Regulator Type	Magnitude of Pressure Reduction	Supply Pressure Variations	Load Variations
Spring-loaded	Moderate	Small	Moderate
Spring-loaded	Moderate	Large	Moderate
Pilot-operated	Moderate	Large	Large
Pilot-operated	Large (in one stage)	Moderate	Large
Pilot-operated first stage	Large (in two stages)	Large	Large (in first stage)
Spring-loaded or pilot-operated	Large (in two stages)	Large	Moderate (in second stage)

Table 5.11i
REGULATORS USED IN THE GAS INDUSTRY

Application Pressures	Inlet Pressures	Outlet
Appliance	8″ H_2O (2 kPa)	3–4″ H_2O (.75–1 kPa)
Household	2–100 PSIG (13.8–690 kPa)	5–8″ H_2O (1.25–2 kPa)
Community of users, 60 PSIG (414 kPa)	Up to 200 PSIG (1380kPa)	3″ H_2O to 60 PSIG (.75–414 kPa)
Transmission line take-off, PSIG in two stages	Up to 1,200 PSIG (8.28 MPa)	100–1,000 PSIG (690 kPa–6.9 MPa)
Transmission line take-off to individual user	Up to 1,500 PSIG (10.35 MPa)	3–600 PSIG (20.7–4140 kPa)

Inserting typical values into equation 5.8(1) can result in a relatively large product $A_p(p_1 - p_2)$ as compared to A_dp_3. The actuating force level is comparatively low, and a balanced plug becomes desirable because it eliminates the term $A_p(p_1 - p_2)$. Figure 5.11c shows an essentially balanced double-seated plug, but it is limited to continuous service, since it cannot be shut off tightly. Figure 5.11e shows a sliding gate mechanism that is also balanced, but it cannot be used with high pressure drops. Analysis of the regulator as a feedback mechanism[6] shows that the balanced plug reduces sensitivity to variations in supply pressure.

Intermittent service requiring tight shutoff necessitates using a single-seated valve. Balancing pistons, diaphragms, and ingenious seating configurations can be used to provide a balanced plug in a single-seated design, but these devices are more complex and costly. In valves with port diameters of 1 in. (25 mm) or less, the ratio of diaphragm area to port area can be made large enough so that balancing is not required for moderate pressure drops.

Droop or Offset

The regulator is a complete, self-contained feedback control loop with only proportional control action.[6] Thus the regulated pressure will be offset by changes in the disturbance variables (upstream pressure and flow demand in the case of a pressure-reducing valve). The offset in regulated pressure with changing flow is called droop. It can be expressed as follows:

$$\triangle p = \frac{(kx/q)(p_1{}^2 - p_2{}^2)}{kxp_2 + A_d(p_1{}^2 - p_2{}^2)} \triangle q \qquad 5.11(2)$$

where

q = throughput and the other factors are from equation 5.11(1)

Equation 5.11(2) shows droop to be a function of the pressures and of the regulator design parameters such as spring rate k, valve lift x, and diaphragm area A_d. Size and economic considerations limit the length of the spring (for low spring rate) and also limit the diaphragm area. Low lift, x, is necessary to reduce diaphragm fatigue as well as minimize droop. Figure 5.11j illustrates the phenomenon of droop.[7]

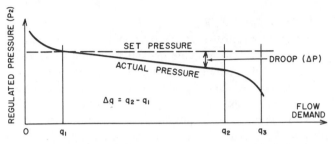

Fig. 5.11j Regulated pressure as a function of flow demand

Equation 5.11(2) applies to the flow range of q_1 to q_2 in Figure 5.11j. This is the operating span of the regulator. While ideally linear, the operating span takes on many shapes in practice because of valve plug flow characteristics and varying effective diaphragm areas. The valve is fully open at q_2 and acts as a fixed orifice from q_2 to q_3. The span from 0 to q_1 is dominated by flow-generated pressure forces on the plug[8], and q_1 is considered to be the minimum controllable flow rate. The minimum flow rate is a function of plug design and is typically 5 to 10 percent of maximum capacity, q_2.

Figure 5.11j emphasizes that maximum regulator capacity is not at full valve opening (q2), but at maximum acceptable droop. Information on droop versus flow is therefore essential for satisfactory regulator performance.

Regulator designs that minimize droop are available. Many of them place the feedback sensing line at a point of high velocity, either in the throat of a slight restriction or in the middle of the flowing fluid (Figure 5.11d). The latter design uses the aspirating effect. Droop compensation can also be provided by a moving valve seat, called a pressure-compensating orifice, which moves with upstream pressure. The roll-out diaphragm, which reduces its effective area with spring compression, also acts to reduce droop. These devices allow the specification of the droop curve for limited flow spans.

Noise

High velocities of compressible fluids are primarily responsible for noisy regulator installations, a continuing source of complaint. Aerodynamic noise caused by high-velocity gas flow in the valve body has been identified as the principal source of irritating "screaming." The volume of a gas increases in proportion to the ratio of inlet and outlet pressures. Gas velocity increases in proportion to the volume increase of the cross-sectional area if the regulator outlet port does not also increase. Increased velocity raises pressure drop until sonic velocity is reached. A tapered pipe expander downstream from the valve has been shown to increase the sonic velocity (mach 1) to supersonic velocities of mach 2 to mach 3. Velocities above sonic result in (1) noise, (2) a major portion of the energy conversion being downstream from the valve plug, (3) a static pressure build-up in the downstream piping that can far exceed its pressure rating, and (4) a "choking" (reduced capability) of the regulator due to the downstream static pressure.[9]

Since sonic velocity is known to occur when downstream pressure is less than 50 percent of upstream pressure, it does not occur at pressure reductions of less than 2 to 1. Because changes in flow path direction also contribute to noise, a maximum gas velocity of 200 fps (60 m/s) (500 fps or 150 m/s below ground) has been recommended. Gas velocity in regulators is difficult to calculate because of interacting velocity, pressure drop, and cross-sectional area. Many manufacturers provide tables of maximum capacities for quiet operation.

Maintaining gas flows to less than sonic velocities at the regulator outlet is recommended to avoid noisy installations. Since velocity is difficult to calculate, sonic velocity is avoided by limiting pressure reduction to less than the critical ratio of 2 to 1. High to low reductions are made with regulators in series or stages. Two to three stages are common to reduce noise and improve regulation. Slightly less than critical reduction is made in the second of the two stages, or in the second and third of three stages, with the remainder of the reduction across the first stage.

Noise is further reduced by eliminating changes in flow direction. As much straight pipe as practical on both sides of the regulator is recommended. Mufflers are also available for further noise reduction.

Noise in regulators on liquid service that is caused by the valve being opened or closed too quickly is called water hammer. When a valve closes quickly, a pressure wave is generated as the moving water column hits the closed plug. This travels upstream at sonic speed and is reflected by the first solid surface it comes in contact with. If this valve is closed when the reflected pressure arrives, vibration results. Another noise source is cavitation. Cavitation is caused by a localized vena contracta pressure drop and can result in vibration damage and valve metal erosion. Maximum velocities of 145 fps (43.5 m/s) are recommended to avoid these difficulties.

Sizing and Rangeability

Oversizing is the most common error in regulator selection. The droop characteristic makes a larger valve attractive, because a greater capacity is obtainable for the same droop. The larger valve also reduces noise because of its larger passages. These apparent advantages are offset by higher cost, severe seat wear, and poor regulation.

The limitation on sizing is rangeability. Rangeability varies from 4 to 1 for a steam regulator, which cannot be operated close to its seat because of wire drawing, to over 50 to 1 for an air regulator. Figure 5.11j illustrates rangeability. Minimum flow, q_1, is 5 to 10 percent of q_2, depending on the seat configuration. Maximum flow is not necessarily q_2 but is determined by maximum acceptable droop. A typical rangeability is 10 to 1. An oversized regulator may not accommodate the minimum flow requirement of certain installations.

Regulators are sized on the basis of tabulated data or valve coefficients (C_v) provided by the manufacturer. Size is chosen to accommodate the maximum flow at minimum pressure drop. The valve should ideally operate at some 50 to 60 percent open under normal conditions. Catalog information must be used judiciously, since capacity and droop may be specified at different points. Rated capacity might result in too high a velocity, or the (external) pressure feedback tap might have been located at a different point during the testing than in the application.

Rangeability is increased by using two regulators in parallel. The pressure set point of one regulator is set 10 percent higher than the other. So one regulator will be wide open while the other modulates under normal conditions. As demand is reduced, the second regulator will close and the first will modulate. Leakage in the second regulator must be a minor portion of the capacity of the first regulator.

Stability

A regulator chosen in accordance with the preceding information may perform very poorly because the regulated pressure does not stabilize. Stability of the regulator installation depends on the open loop gain.[6] For the regulator shown in Figure 5.11b, open loop gain is defined as

$$K_o = \frac{A_d(p_1{}^2 - p_2{}^2)}{kxp_2} \qquad 5.11(3)$$

where K_o is the open loop gain.

These factors are the same as those upon which the droop depends, as shown in equation 5.11(2). Comparison of the two equations indicates that parameter changes that decrease droop would increase the open loop gain and decrease stability, as expected in a feedback mechanism. Since the regulator has no controller mode adjustments, the manufacturer must choose design parameters that will provide an adequate compromise between droop and stability. It can thus be expected that the open loop gain of some applications will be too high, resulting in a noisy and cycling pressure regulator.

Guidelines for a stable installation are few. Difficulties are generally found after installation. Because the regulator has no adjustments, it is costly to alleviate problem conditions. Generally, field revisions to stabilize an installation include (1) relocation of the pressure sensing tap, (2) redesign of the downstream piping to provide more volume, and (3) restricting the pressure feedback line, either by a needle valve in an external line or by filling an internal line and redrilling a smaller hole.

Safety

Diaphragm rupture is the most common failure source in a regulator. Most regulators fail full open upon diaphragm failure, which is an unsafe condition. A relief valve on the regulated pressure is recommended if an increase in regulated pressure to the level of the supply pressure is considered unsafe. In installations where it is imperative that the user continue to be supplied even on a regulator failure, two regulators can be placed in series. The second regulator has a set point adjusted about 10 percent higher than the first, and it will remain wide open under normal circumstances. If the first regulator fails, the second will take over pressure regulation.

Installation

Regulator installation is generally easier than regulator selection. The following installation suggestions will help ensure satisfactory regulator performance:

1. Steam regulators should be preceded by a separator and a trap.
2. All regulators should be preceded by a filter or strainer.
3. A valve bypass is recommended where it is necessary to service the regulator while continuing to supply users.
4. Use straight lengths of pipe upstream and downstream to reduce noise.
5. External feedback lines should be ¼ in. (6.25 mm) pipe or tubing.
6. Locate the pressure feedback tap at a point where it will not be affected by turbulence, line losses, or sudden changes in velocity. A distance of 10 ft (3 m) from the regulator has been recommended.

REFERENCES

1. Watts, G.R., "Pressure Regulators or Control Valves?" *Instrumentation Technology*, June, 1980, p. 55–6.
2. Kubitz, R., "Valves in Gas Regulators," *Instruments & Control Systems*, April, 1966, p. 139.
3. O'Connor, J.P., "12 Points to Consider When Selecting Pressure Regulators," *Plant Engineering*, March, 1958, p. 109.
4. Beard, C.S., and Marton, F.D., *Regulators and Relief Valves*, Chilton Book Co., Philadelphia, PA, p. 79.
5. Perrine, E.B., "Pressure Regulators and Their Application," *Instruments & Control Systems*, September, 1965, p. 167.
6. Moore, R.L., "The Use and Misuse of Pressure Regulators," *Instrumentation Technology*, March, 1969, p. 52.

7. Yedidiah, S., "Effect of Seat Configuration on Pressure Operated Control Valves," *Design News*, October 11, 1967, p. 114.

8. Quentzel, D., "A Balance Factor for Pressure Reducing Valves," *Control Engineering*, May, 1965, p. 97.

9. Baumann, H.D., "Why Limit Velocities in Reducing Valves?" *Instruments & Control Systems*, September, 1965, p. 135.

BIBLIOGRAPHY

Comber, J., and Hockman, P., "Pressure Monitoring: What's Happening?" *Instruments & Control Systems*, April, 1980.

Dickinson, W., "Don't Overlook Pressure Regulators," *Instruments & Control Systems*, May, 1983.

Fleming, A., "Regulator Station Design," *Instruments & Control Systems*, April, 1961.

Fluid Controls Institute, "Definitions of Regulator Capacities," FCI-58-1, April, 1966.

———, "Guide to Material Selection for Industrial Regulators," FCI-65-1.

Hall, J., "Monitoring Pressure with Newer Technologies," *Instruments & Control Systems*, April, 1979.

Herceg, E.E., *Handbook of Measurement and Control*, Schaevitz Engineering, Pennsauken, NJ, 1972.

Ledeen, H., "A New Look at Valve Regulation," *Instruments & Control Systems*, August, 1965.

Moore, R.L., "Basic Instrumentation Lecture Notes and Study Guide—Measurement Fundamentals," ISA, 1976.

Soisson, H.E., *Instrumentation in Industry*, John Wiley & Sons, New York, 1975.

5.12   TEMPERATURE REGULATORS

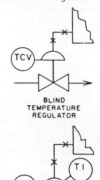

BLIND
TEMPERATURE
REGULATOR

TEMPERATURE
REGULATOR WITH
DIRECT INDICATOR

Size:	From ¼″ to 6″ (6.25 mm to 150 mm)
Features:	Two- or three-way designs with direct or reverse action, only proportional mode of control and the only adjustment capability being the changing of set point. Bulb pressure may exceed 1,000 PSIG (6.9 MPa)
Range:	− 25 to 480°F (−31.7 to 248.9°C)
Spans:	Standard from 30 to 60°F (16.7 to 33.3°C); special up to 100°F (55.6°C).
Materials of Construction:	Body in iron, bronze, steel, or stainless steel; bulb in copper, steel, stainless steel, or coated with PVC, Teflon, etc.; tube in copper or stainless steel
Tubing Length:	Usually 10 ft (3 m) with 50 ft (15m) being the maximum.
Base Cost:	$300 for standard unit in ½″ (12.5 mm) size; $650 for standard unit in 2″ (50 mm) size
Partial List of Suppliers:	Dresser Industries, Inc., Valve Division; Jordan Valve Div., Richards Industries; Leslie Co.; Powers Process Controls Unit; Robertshaw Controls Co., Fulton Sylphon Division; Sarco Co.; Spence Engineering Co.; Trerice Co.

The inherent advantages of temperature regulators and the relatively wide range of designs available recommend them for many temperature control jobs that do not require the high performance level or sophisticated features of temperature controllers and pneumatic or electronic systems.

A temperature regulator is a control device containing a primary detecting element (bulb), measuring element (thermal actuator), a reference input (adjustment), and a final control element (valve) (Figure 5.12a). It might properly be referred to as an "automatic temperature control system," but the term "regulator" is preferred.

Since they require no external power (electricity, air, etc.), regulators are described as "self-actuated." They actually "borrow" energy from the controlled medium for the required forces.

Types and Styles

Two types of regulators are available—direct-actuated and pilot-actuated. They are distinguished by the way in which the valve (the final control element) is actuated.

In the direct-actuated type, the power unit (bellows,

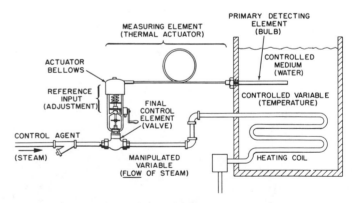

Fig. 5.12a   Self-actuated temperature regulator

diaphragm, etc.) of the thermal actuator is directly connected to the valve plug and develops the force and travel necessary to fully open and close the valve (Figure 5.12b). In the pilot-actuated type, the thermal actuator moves a pilot valve, internal or external (Figure 5.12c). This pilot controls the amount of pressure energy from the control agent (fluid through valve) to a piston or diaphragm, which in turn develops power and thrust to

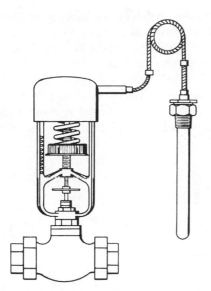

Fig. 5.12b   Direct-actuated temperature regulator

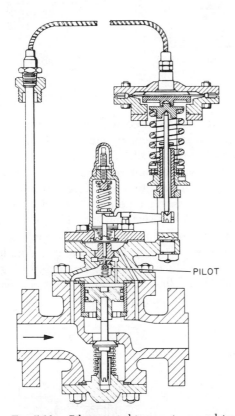

Fig. 5.12c   Pilot-actuated temperature regulator

also handle interrelated functions through use of multiple pilots, such as temperature plus pressure plus electric interlocks, etc.

There are two styles of temperature regulators—self-contained and remote-sensing. Which one is appropriate depends on the location and structure of the measuring element (thermal actuator).

Self-contained regulators contain the entire thermal actuator within the valve body, and the actuator serves as the primary detecting element (Figure 5.12d). Thus the self-contained style can sense only the temperature of the fluid flowing through the valve, and that fluid might be said to be both the controlling agent and the controlled medium. The device regulates the temperature of the fluid by regulating the fluid's flow. Self-contained regulators are generally provided with liquid expansion or fusion type thermal elements.

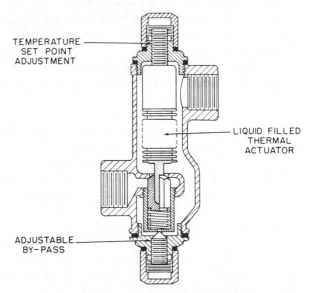

TEMPERATURE
SET POINT
ADJUSTMENT

LIQUID FILLED
THERMAL
ACTUATOR

ADJUSTABLE
BY-PASS

Fig. 5.12d   Self-contained temperature regulator

In remote-sensing regulators, the bulb (the primary detecting element) is separate from the power element (bellows, etc.) of the thermal actuator, and is usually connected to it by flexible capillary tubing (Figures 5.12a and b). This style is able to sense and regulate the temperature of a fluid (the controlled medium) distinct from that of the fluid flowing through the valve (the control agent).

The self-contained style is simpler, frequently lower in cost, and "packless" (has no stem sliding through the valve body "envelope"), but it is limited in application to such uses as regulating the temperature of coolant (water, etc.) that is leaving the engine, compressor, exothermic process, etc.

The Regulator Valve Body

Further discussion here is limited to some special forms of valves and to factors relating to selection and use due

position the main valve plug. The pilot may be internal or external. When external, independently acting multiple pilots are also available.

Direct-actuated regulators are generally simpler, lower in cost, and more truly proportional in action (with somewhat better stability), whereas pilot-actuated regulators have smaller bulbs, faster response, and shorter control (proportional) band, and they can handle higher pressures through the valve. Pilot-actuated regulators can

to force limitation of the thermal actuator. See also the discussion on globe valves in Section 4.7, which is also applicable here.

Valve action refers to the relationship of stem motion to plug and seat position. A direct-acting valve closes as the stem moves down into the valve. A reverse-acting valve opens as the stem moves down. A three-way valve combines these, one port (or set of ports) opening and the other port(s) closing on downward movement.

Valve action is produced by movement of the actuator resulting from a temperature change. Heating control is accomplished by a direct-acting regulator, using a direct-acting valve to reduce the flow of heating medium on temperature rise.

Cooling control is most frequently achieved by a reverse-acting regulator with a reverse-acting valve to increase the flow of coolant on rising temperature. However, on systems using a secondary coolant with steady flow through the process, a direct-acting regulator with its valve in a bypass line is occasionally used to vary the flow through the heat exchanger. A three-way valve provides positive control of such a cooling system by giving a full range of flows through the two legs of the cooling circuit (exchanger and bypass).

Mixing of two media at different supply temperatures to control the mixed temperature is also accomplished with three-way valves.

Single-seated valves are desirable for minimum seat leakage in the closed position. The pilot-acuated regulators always use single-seated valves (Figure 5.12e). On direct-actuated types, the use of single-seated valves is limited to a 2-in. (50 mm) maximum by two factors: the required full-open lift (approximately $\frac{1}{4}$ of valve size) and the closing force required (full port area times maximum pressure drop across the valve). A single-seated valve with piston balanced plug (Figure 5.12e) eliminates the closing force problem but requires slightly higher full-open lift.

Double-seated valves are most common on direct-actuated regulators in sizes up to 4 in. (100 mm). The reasons for this include their low full-open lift (approximately $\frac{1}{8}$ of valve size) and minimum closing force requirement. The unbalanced area of double-seated valves is the difference between the two port areas, which is approximately 6 percent of the larger of the two—hence, the term "semi-balanced" for double-seated valves (Figure 5.12f).

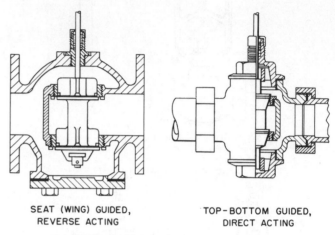

SEAT (WING) GUIDED,
REVERSE ACTING

TOP-BOTTOM GUIDED,
DIRECT ACTING

Fig. 5.12f Double-seated, two-way regulator valves

Three-way valves used with temperature regulators are of three basic designs (Figure 5.12g). Smaller valves (to 2 in. or 50 mm) are of the "unbalanced" single-seated construction with a common plug moving between two seats. In sizes from 2 to 6 in. (50 to 150 mm) there are two designs. A sleeve type plug (requiring a body seal) moves between two seats and provides near balance of closing forces, but it requires typical single-seated lift. A semi-balanced design, essentially two double-seated valves for each flow path with connected plugs, gives both low lift and low seating force. This construction is somewhat more expensive, however.

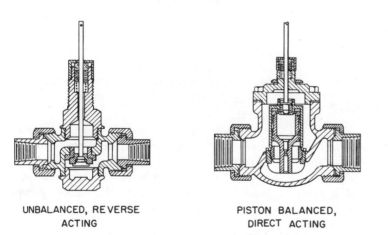

UNBALANCED, REVERSE
ACTING

PISTON BALANCED,
DIRECT ACTING

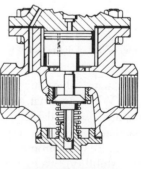

PISTON ACTUATED,
INTERNAL PILOT
CONTROLLED

Fig. 5.12e Single-seated, two-way regulator valves

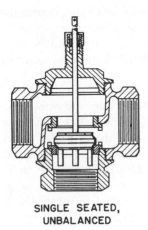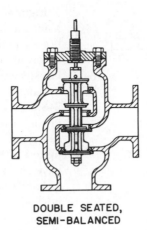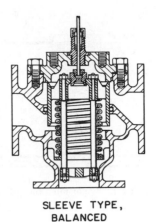

SINGLE SEATED,   DOUBLE SEATED,   SLEEVE TYPE,
UNBALANCED   SEMI-BALANCED   BALANCED

Fig. 5.12g   Three-way regulator valves

Plug guiding is considered to be a less important factor with regulators than with control valves because the applications involve relatively low pressures and velocities. Therefore, many direct-actuated single-seated valves are only stem guided. Most double-seated valves are seat ("wing") guided, but some top- and bottom-guided designs are also in use. Pilot-actuated valves are generally well guided, as are the piston balanced valves.

Flow Characteristics—Most regulator valves are of the quick opening type in order to use the shortest practical lifts. On special basis, low-lift plugs are available in some small single-seated valves and also in a few double-seated valves.

Materials of Construction—Standard materials used in body construction are bronze in sizes below 2 in. (50 mm) and cast iron in larger sizes. Steel and stainless steel are considered to be special body materials. The standard bronze valve trim is used with pressure drops below 50 PSID (345 kPa), while stainless steel is recommended for greater drops. On low pressure and temperature services, composition discs can be used for tight shutoff. Stellite or hardened alloy trim for extreme services is special, though some pilot-actuated designs offer it as standard. Monel and other special alloys are available for specialized services, such as seawater.

Pressure Rating—The pressure rating of a temperature regulator is frequently determined more by limitations in the thermal actuator power (seating force) than by the design and materials of the valve body and trim. Therefore, while catalogs will usually state the maximum temperature and pressure rating for the valve, they should also give a lower recommended maximum pressure limit for a direct-actuated regulator. This will usually decrease as valve size increases, and it should be given full cognizance. With pilot-actuated valves and piston-balanced valves (direct actuated) the actuator limitation is not a factor. However, regulators are seldom available or recommended for pressures above 250 PSIG (1.7 MPa).

End Connections—To lessen the chance of installation damage, union ends are usual for $\frac{1}{2}$ to $1\frac{1}{2}$ in. (12.5 to 37.5 mm) valves (in some cases 2 in. or 50 mm). Flanged ends are standard in sizes 2 in. (50 mm) and larger.

Valve Capacity—The valve capacity data published by manufacturers is reasonably accurate, having been determined by recognized test procedures. Most manufacturers publish capacity tables for common fluids, from which the full-open capacity figures can be read directly for various supply pressures and pressure drops. In a few cases, the information takes nomographic form. Some list the valve capacity factor C_v for each valve size and type.

Seat Leakage—Regulators should not be considered as positive shutoff devices. However, in "as new" condition, leakage across seats in the closed position for metal seated valves may be generally expected to be at the level of 0.02 percent of full capacity for single-seated and 0.5 percent for double-seated regulators. Single-seated valves with composition disc are almost "dead tight" or "bubble tight."

Thermal Systems

There are four classes of "filled system" thermal actuators used in temperature regulators. These are distinguished by the principle of power development. All develop power and movement proportional to the measured temperature, and hence temperature regulators are basically proportional in action. For a detailed discussion of filled thermal systems, refer to Section 4.4 in the Process Measurement volume of this *Handbook*. Gas-filled systems (SAMA Class III) and bimetal or "differential expansion" thermal elements are not presently used in the family of devices covered here.

Vapor-Filled System

The operation of the vapor pressure class (SAMA Class II) is illustrated in Figure 5.12h. The thermal actuator, having been evacuated to eliminate the contaminating effect of air and other gases and vapors, is partially filled

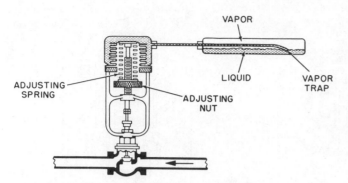

Fig. 5.12h Vapor-filled thermal actuator and element

with a volatile liquid. Liquids used are chemically stable at temperatures well above the range used.

The bellows and tubing usually contain liquid only, although with ambient conditions *above* the set point of the regulator they may contain vapor only—*never both* under conditions of normal operation. The bulb contains the liquid-vapor interface, and so the pressure within the actuator is the vapor pressure of the charge or fill liquid at the temperature sensed by the bulb. This vapor pressure, increasing with rising temperature, acts on the bellows against the force of the adjustment to produce the power required to reposition the valve plug.

Figure 5.12i shows various charging liquids used, each having a different vapor pressure curve and atmospheric boiling point. Different ranges are achieved by using the different liquids whose curves are shown, not by changing springs.

Ranges of adjustment from -25 to $+480°F$ (-31.7 to $248.9°C$) are commonly available, and common spans are in the order of 40 to $60°F$ (22 to $33°C$). In some special cases, involving small short-stroke valves, longer spans up to $100°F$ ($55.6°C$) are furnished. Other special designs have shorter ranges of 25 to $30°F$ (13.9 to $16.7°C$).

Adjustment of vapor pressure regulators is accomplished by changing the initial thrust of the adjusting spring (Figure 5.12h). Turning the adjusting nut toward the bellows to compress the spring increases the spring thrust. This requires an increased pressure to start valve movement and thus effects a higher temperature setting. Lowering the adjustment to decompress the spring permits motion of the stem with lower actuator pressure, thereby lowering the set temperature.

The rate of the adjusting spring (force required per inch of spring compression) is an important factor in determining the length of adjustment span as well as the proportional band of a regulator. The stiffer the spring, the longer the span or range, but also the longer the proportional band.

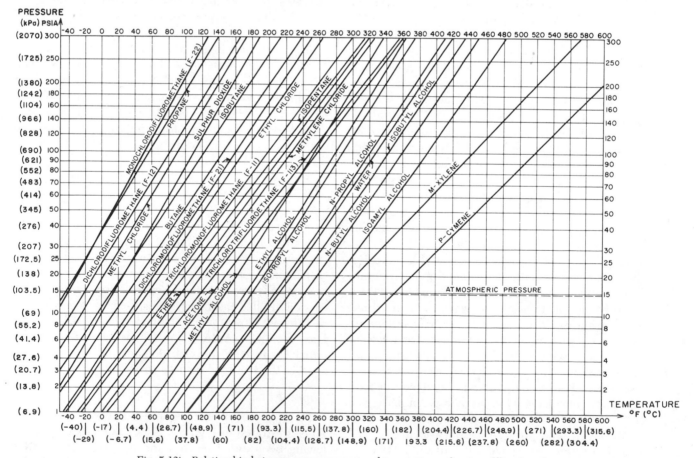

Fig. 5.12i Relationship between vapor pressure and temperature of various filling liquids

Weight-and-lever adjustment (Figure 5.12j), once very common, is seldom used today because of problems of position, location, and general vulnerability of the moving lever. Here the force produced by the weight acting on an adjustable length lever or "movement" arm opposes the thrust produced by the thermal actuator. Increasing the length of the lever raises the setting. Since the force of the weight-and-lever system has no "rate" (being almost constant for a given setting), the proportional band of a weight-adjusted regulator is narrower than for a spring-adjusted design.

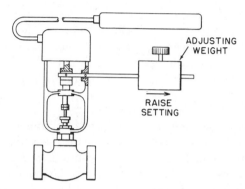

Fig. 5.12j   Weight-adjusted vapor pressure regulator

Over-temperature protection is standard in most regulators of the vapor pressure class. It consists of a special spring-loaded and compressible "overrun" section of the actuator stem. When increasing temperature produces a force beyond that for normal control action, this overrun compresses. The added movement of the actuating bellows drains the remaining liquid from the bulb, and pressure then follows the gas law instead of the vapor pressure curve. This provides protection up to 100°F (55.6°C) above the top end of the range.

Temperature indication is available as a special feature (Figure 5.12k). Generally, this is simply a compound pressure gauge, sealed in the thermal system and responding to the changing pressure of the fill. The dial is

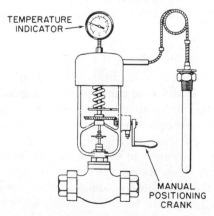

Fig. 5.12k   Temperature regulator with indicator and handwheel

calibrated to correspond to the temperature-pressure curve of the particular fill (charge) and therefore "reads" the bulb temperature. In some designs the thermometer has its own sensing element, with a separate bulb contained in or attached to the regulator bulb. Improved characteristics are claimed for this design. In either case, the "thermometer" is in the order of approximately ±2 percent inaccuracy. Thermometers are furnished with both Fahrenheit and Celsius calibration.

Actuator bellows design (diameter, area, length, etc.) is the major factor in determining the characteristics of the vapor pressure regulator. A small bellows, having limited power and stroke, results in a compact actuator, successful only for small valves and on moderate to low pressures. Increasing bellows size makes direct actuation of larger valves possible at higher pressure levels. However, since the bellows movement requires the transfer of some liquid from the bulb to the bellows (equal to the volumetric displacement), larger bellows require larger bulbs. In general, larger bellows result in shorter adjustable ranges but also narrower proportional bands. Some manufacturers offer a choice of designs based on bellows size in order to offer different characteristics.

Bulb size of the vapor pressure regulators is governed by the volumetric displacement of the bellows (area × valve travel) and varies with actuator design and with valve design and size. It is also affected by the possible need to operate under "cross-ambient" conditions; i.e., with ambient temperature either above or below the regulator's set point. This requirement produces a bulb substantially larger than that used where the ambient temperature will always be either above or below the range of the regulator. The use of vapor-filled thermal elements on cross-ambient applications is generally not recommended. Bulb sizes of this class range from $\frac{1}{2}$ in. (12.5 mm) OD by 6 in. (150 mm) long in some very small units to $1\frac{1}{2}$ in. (37.5 mm) OD by 36 in. (900 mm) long in large valves with cross-ambient fill.

Bulb construction for vapor pressure regulators includes a "vapor trap" (Figure 5.12h). This serves to give the best possible response by insuring liquid transfer of pressure changes from bulb to bellows. Since the end of the vapor trap tube must always be in liquid, it requires that the installation of the bulb be made in accordance with markings on the bulb. For installations where the tube connection end of the bulb is below the horizontal, the extended vapor trap must be omitted.

Advantages of vapor pressure regulators are (1) ambient temperature has no effect on calibration, and (2) there is a wide choice of designs, characteristics, ranges and special features available.

Liquid-Filled System

The operation of the liquid expansion class (SAMA IB) is illustrated in Figure 5.12l. The thermal actuator is completely filled with a chemically stable liquid. Chang-

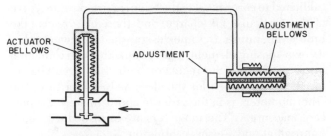

UNIFORM LIQUID EXPANSION PER DEGREE TEMPERATURE RISE
AT SENSING BULB COMPRESSES VALVE ACTUATING BELLOWS
TO CONTROL FLOW OF HEATING OR COOLING MEDIUM

Fig. 5.12l   Liquid-filled thermal actuator and element

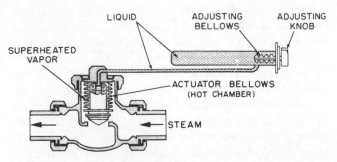

Fig. 5.12m   Hot chamber type thermal actuator and element

ing temperature at the bulb produces a volumetric change, which causes the small-area bellows to move in a corresponding direction, thereby moving the valve. The force and action are positive and linear, limited only by the components of the element, chiefly the actuator and adjustment bellows.

Ranges are from 0 to 300°F (−17.8 to 148.9°C) with spans generally from 30 to 50°F (16.7 to 27.8°C). The special 100°F (55.6°C) spans are sometimes furnished with very small, short-stroke valves. The different ranges (not field changeable) are achieved by filling procedure and sealing temperature. Hydrocarbon fills are generally used. Mercury is not used as a filling fluid for regulators because of cost and hazards.

Adjustment is achieved by moving the actuator bellows relative to the valve plug (Figure 5.12a) or by using a separate adjusting bellows (Figure 5.12l). This adjustment bellows transfers liquid into the actuator bellows to lower the set point, or vice versa. The adjustment may be located at the bulb or at the valve, or it may be separate from both.

The proportional band is generally wider than with other classes of actuators, and it is a direct function of bulb volume to actuator bellows area.

Over-temperature protection, which permits over-travel of the bellows, allows an uncontrolled temperature to overrun the set point by up to 50°F (27.8°C). Ambient temperature effects on the liquid in the tubing, bellows, and adjustment are minimized by design, but wide ambient swings do introduce errors.

Packless construction (no moving stem through the body envelope) is a feature of this class of regulator. However, use on corrosive fluids must be avoided.

Temperature indication is not generally available.

Advantages of the liquid expansion class regulators include a positive actuating force, linear movement, and packless design.

Hot Chamber System

The hot chamber class of regulator has no corresponding SAMA designation. This thermal actuator is partially filled with a volatile fluid (Figure 5.12m) and has some characteristics common to the liquid expansion class. Ris-

ing temperature at the bulb forces liquid to the actuator bellows or "hot chamber." There the heat of the control agent, always at a temperature substantially above the regulator range, flashes the liquid into a superheated vapor. The pressure increase in the bellows causes the valve to move against a return spring.

Hot chamber regulators are furnished from ½ to 1½ in. (12.5 to 37.5 mm). Because of practical considerations, only direct-acting designs are available (close on rising temperature) and they are limited to use on steam at pressures up to 75 PSIG (518kPa). Theoretically, other designs are also possible.

Ranges are from 30 to 170°F (−1.1 to 76.7°C) (to 200°F or 93°C in special cases) with spans from 30 to 60 °F (16.7 to 33.3°C).

Adjustment is achieved by the use of adjusting bellows that are usually located at the bulb but may be separate and remote.

Ambient temperature effects are minimized by design, but wide ambient swings are nevertheless detrimental. Effects of steam supply pressure variations are partially self-compensating.

Bulbs (thermostats) take several forms. For steam space heating systems the wall-mounting types or duct-mounting types are common, while for liquid heating the cylinder types with pressure-tight bushings are utilized.

Advantages of the hot chamber class include calibrated knob and dial adjustment, good response and good proportional characteristics, packless design, and compact construction.

Fusion Type System (Wax Filled)

The fusion class of regulator does not have a SAMA equivalent. In this least common class, the compact element is filled with a special wax containing a substantial amount of copper powder to improve its rather poor heat transfer characteristics (Figure 5.12n). A considerable and positive volumetric increase occurs in the wax as temperature rises through its fusion or melting range. This volumetric change produces a force used for valve actuation through the ingenious arrangement of a sealing diaphragm that acts to compress a rubber plug which "squeezes out" the highly polished stem. This element has more hysteresis and dead band than the other classes.

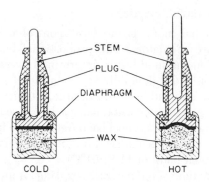

Fig. 5.12n   Wax-filled (fusion type) thermal element

Different waxes, such as natural waxes, hydrocarbons, silicones, and mixes of these are available that provide a variety of operating (control) spans between 100 and 230°F (55.6 and 127.8°C).

No adjustment of set point is provided, since the fusion (melting) span of the wax is narrow, from 6 to 15°F (3.3 to 8.3°C). A heavy return spring, which is always present in the regulator, forces the rubber components of the element to follow the wax fill on falling temperature.

Remote bulb designs have not been offered since they would provide no advantages over the other three classes. Thus fusion class regulators are avaiable only in "self-contained" styles (Figure 5.12o).

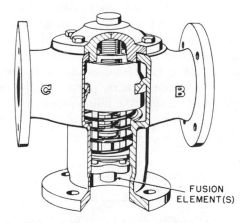

Fig. 5.12o   Self-contained temperature regulator with wax-filled (fusion type) element

Over-temperature protection up to 50°F (27.8°C) above the operating span is provided by the flat temperature expansion curve of the wax above its melting range.

Valve sizes are from 1 to 6 in. (25 to 150 mm) with the majority being of the three-way valve design in sizes from 2 to 6 in. (50 to 150 mm).

Advantages of fusion class regulators lie in their compact size and low cost of the thermal element, even where multiple elements are required to stroke larger valves.

Thermal Bulbs and Fittings

The primary sensing element of temperature regulators is usually called the bulb in remote sensing designs. The following forms, sizes, fittings and materials are available.

Forms—The cylindrical bulb form is usually standard and is suitable for most liquid control requirements (Figure 5.12p). For faster response in air or gas, a cylindrical form with metal fins is utilized. Other increased surface forms are the precoiled bulb (for liquid expansion and hot chamber classes) and the tubular bulb. The tubular bulb is made from a substantial length of flexible tubing, to be formed and supported in the field, as, for example, in and across a duct or wrapped around piping. Other forms include those with dead (inactive) extensions, flexible extensions, etc.

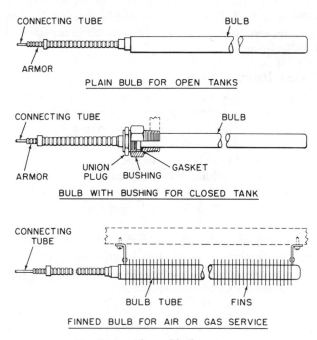

Fig. 5.12p   Thermal bulb variations

Size—Because of widely varying regulator designs and bulb volume requirements, no bulb dimension standards have been developed. Bulb sizes vary between the extremes of $\frac{1}{2}$ in. (12.5 mm) diameter by 9 in. (225 mm) long and $1\frac{1}{2}$ in. (37.5 mm) diameter by 36 in (900 mm) long. Most bulbs range from $\frac{3}{4} \times 9$ in. to $1\frac{1}{4} \times 24$ in. (18.7 × 225 mm to 31.25 × 600 mm) long. Some manufacturers will vary the diameter and length combination to fit special needs.

Fittings—For open tank use, no fittings are needed. The union type screwed bushing for pressure vessel or pipeline installation is most common (Figure 5.12q). These are generally ground joint unions but are sometimes gasketed. Wells, or sockets, are also available for higher pressure ratings, corrosion protection, and to maintain

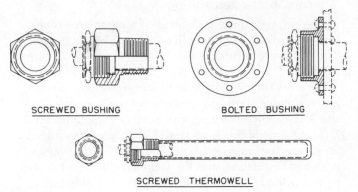

SCREWED BUSHING   BOLTED BUSHING

SCREWED THERMOWELL

Fig. 5.12q   Thermal bulb fittings

the integrity of the pressure vessel or system during service. Sanitary fittings are also available.

Materials and Pressure Ratings—Bulbs are usually built-up assemblies and are available in copper or in other materials at extra cost. Table 5.12r gives some of their features.

Transmission Tubing

The tubing that transmits the pressure from the bulb to the actuator bellows is usually covered with armor (spiral or braided) for mechanical strength. A tube with $\frac{3}{16}$ in. (4.69 mm) OD and approximately 0.050 in. (1.25 mm) wall thickness is common.

The standard tubing material is copper with brass armor, but stainless steel (tube or armor), lead-coated copper, and PVC- or Teflon-coated copper or stainless steel are also available.

Standard tubing length is 10 ft (3 m). Longer or shorter lengths are also provided, but lengths above 50 ft (15 mm) are not recommended because of increased mechanical hazards and slow response.

Control Characteristics

Proportional mode is the basic control mode of all temperature regulators. The proportional band is not adjustable in the field but is determined by various design factors, and it generally varies in direct proportion to the valve size and lift. For example, a typical 1-in. (25 mm) vapor pressure regulator, using a double-seated valve, has a mid-range proportional band (full open to full closed) of 7°F (3.9°C), and a similar 2-in. (50 mm) regulator has a proportional band of 11°F (61°C).

Response time of temperature regulators is much longer than that of corresponding pneumatic or electronic control systems. Time constants vary widely with design factors such as actuator bellows size (area), valve size (lift), transmission tube length, bulb size, and material (heat conductivity). Typical regulators have time constants in the range of 30 to 90 sec, although it might be reduced to under 10 sec under favorable conditions.

Special Features and Designs

Temperature indication is available with most vapor pressure regulators (Figure 5.12k). It usually consists of a pressure gauge sealed in the thermal system and calibrated to correspond to the particular range (charging liquid) used. Inaccuracy is about ±2 percent of indicated span.

Fail-safe design is available as a special version of the vapor pressure regulator. Here failure (loss of fill) of the thermal actuator produces downward movement of the valve stem, i.e., the valve goes to the "safe" position, producing shutoff of the heating medium or full flow of the cooling medium, depending on valve action. Adjustable ranges are short (approximately 30°F or 16.7°C), and valve actuating force is rather low, since the actuator works on the subatmospheric portion of the nonlinear vapor pressure curve of the filling medium.

Table 5.12r
CONSTRUCTION MATERIALS AND FEATURES OF THERMAL BULBS

			Rating	
Bulb Material	*Assembly*	*Fittings*[5]	*Pressure*[1], *PSIG (MPa)*	*Temperature, °F (°C)*
Copper, brass	Brazed	Brass	700 (4.8)	300 (148.9)
Stainless steel[2]	Welded	Stainless steel	1,100 (7.6)	600 (315.6)
Steel	Welded	Steel, stainless steel	1,000 (6.9)	600 (315.6)
Lead[3]	Fusion	Lead	50 (0.35)	250 (121.1)
Monel	Welded	Monel	1,000 (6.9)	600 (315.6)
PVC coated[4]	Fusion	None	—	160 (71)
Teflon coated	Shrinking	None	—	300 (148.9)

[1] Average, based on 1-in. (25 mm) diameter bulb. Larger diameters have lower ratings.
[2] Type #316 is available.
[3] Some lead bulbs are sheathed over copper.
[4] Coatings are usually over copper, but may be over other materials.
[5] Wells for sockets are available in all metals. The use of copper bulbs with other well materials is usual.

Manual handwheels are available in a few regulator versions (Figure 5.12k). These are useful to manually override the element, to provide a minimum flow valve position, or for emergency operation.

Combination temperature/pressure control is available in the pilot-actuated regulator. This is accomplished in two ways. In the external pilot version, two separately adjusted pilot units are piped in series in the pilot pressure line. The temperature pilot modulates the flow of heating medium (control agent) to regulate the process temperature, while the pressure pilot limits the discharge pressure from the main valve to its setting under maximum load conditions. In the internal pilot version, the two pilot elements are cascaded; i.e., the temperature element mechanically changes the setting of the pressure element from zero at no load to the maximum pressure for which the pressure pilot is set at full load.

Conclusions

Since the characteristics of a regulator cannot be altered in the field, greater care must be taken in proper process evaluation and regulator selection than would be the case with a pneumatic or electronic controller having an adjustable control mode or modes.

In general, temperature regulators are most successful on control applications having high "capacitance" (and, therefore, slow response requirements), such as heated or cooled storage tanks, process ovens and dryers, and/or on applications with minimum load changes, such as metal treating (cleaning and plating) tanks.

Proper installation includes good bulb (primary element) location where process temperature changes can be sensed quickly, but away from potential damage by moving equipment. Regulators in sizes 2 in. (50 mm) and greater are installed upright in horizontal lines, and regulators of all sizes should be protected by strainers.

Regulator maintenance is minimal. Direct-actuated regulators with packed valve stems may require occasional lubrication and infrequent replacement of packing and cleaning or polishing of stem. Pilot-actuated regulators with piston-actuated valves (Figure 5.12c) may require programmed cleaning to remove "glaze" and deposits that would cause the piston to stick. Depending on the service, occasional examination and servicing of valve seats may be required to maintain needed tightness.

Relative to pneumatic control loops, the advantages and limitations of temperature regulators can be summarized as follows:

The advantages of regulators include their low installed cost, ruggedness, simplicity, low maintenance, and lack of need for an external power source.

Its limitations include the fixed proportional control capability, the local set point, the limited and narrow ranges, the slow response, the size and operating pressure limitations, and the large bulb size requirements.

BIBLIOGRAPHY

Dřenberg, C.H., "Automatic Valve Control," *Chemical Processing*, August, 1980.

Farr, W.E., "Save Dollars on Plant Heaters," *Chemical Processing*, March 10, 1980.

———, "Temperature Control Valves Replace Traps," *Fond Processing*, July, 1980.

Hormuth, G.A., "Ways to Measure Temperature," *Control Engineering*, Reprint No. 948, 1971.

Jutila, J.M., "Temperature Instrumentation," *Instrumentation Technology*, February, 1980.

Plumb, H.H., "Temperature, Its Measure and Control in Science and Industry," 5th Symposium on Temperature, 1971, NBS, API, and ISA.

Schooley, J.F., "State of the Art of Instrumentation for High Temperature Thermometry," *Argonne National Laboratories Publication No. ANL-78-7, 1977 Symposium.*

Suczkstorf, D., "Temperature-Activated Valves Save Steam," *Plant Services*, July, 1980.

5.13 RELIEF CAPACITY DETERMINATION

In the analysis of process relief devices, the first concern is the purpose for which they are to be utilized. The obvious immediate answer is to protect against over-pressure. The various causes of over-pressure fall into two broad categories: fire conditions and process conditions. In this section, the various aspects of these conditions are discussed and means of determining required relief capacity are presented.

Fire Conditions

The ASME Unfired Pressure Vessel Code requires that pressure vessels covered by the code be adequately relieved. For fire conditions, the code requires that relief devices be sized such that, at maximum relieving conditions, the vessel relieving pressure does not exceed the vessel design pressure by more than 20 percent. This is referred to as 20 percent accumulation. Relief capacity under fire conditions is a function of tank area exposed to fire, of the heat flux per unit area, and of the latent heat of the process fluid.

Heat Flux

In order to define the required relief capacity, it is necessary to define heat flux. By heat flux, we mean the rate at which heat is transferred into the vessel or process equipment. A number of heat flux determination methods can be considered. The simplest technique employs a fixed heat flux regardless of the type or size of vessel. In such case, a heat flux of 20,000 BTU per hour per square foot (63,000 W/m^2) is commonly employed. Other approaches relate the magnitude of the heat flux to the size of the vessel, reasoning that the larger the tank, the more unlikely it is to be completely immersed in flame.

Bulletin API RP 520, "API Recommended Practice for the Design and Installation of Pressure-Relieving Systems in Refineries," published by the American Petroleum Institute, presents one commonly used approach for determining heat flux under fire conditions. This bulletin gives a recommendation for heat flux in BTU per hour per square foot as a function of the total wetted surface of a vessel exposed to fire, expressed in square feet. The recommendation is based on the formula:

$$q = 21,000F(A)^{-0.18} \qquad 5.13(1)$$

where q is the average heat absorption per square foot of wetted surface exposed to fire, A is the wetted surface of the vessel expressed in square feet, and F is an environmental factor relating to the type of tank installation, (see Table 5.13a). The reader should consult the API Standard for additional details.

Another standard that is commonly employed in the determination of heat flux under fire conditions is presented in the recommendations of the National Fire Protection Association (NFPA). This is an organization of insurance companies and regulatory organizations. Their recommendations should meet the requirements of most insurance companies. Values recommended in the NFPA Flammable and Combustible Liquids Code, Bulletin No. 30, are generally more conservative than the corresponding values recommended by API.

The curve expressing heat flux in BTU per hour per square foot of exposed surface area as a function of exposed surface area may be expressed on log-log paper as indicated in Table 5.13b.

Table 5.13a
ENVIRONMENTAL FACTORS FOR
TANK INSTALLATIONS

Type of Installation	Factor F
1. Bare vessel	1.0
2. Insulated vessels (listed below are three conductance values expressed in BTU/hr/ft²/°F) Conductance of 4.0 Conductance of 2.0 Conductance of 1.0	0.3 0.15 0.075
3. Water application facilities provided on bare vessels	1.0
4. Depressurizing and emptying facilities provided	1.0
5. Underground storage	0.0
6. Earth covered storage above grade	0.03

Table 5.13b
HEAT FLUX AS A FUNCTION OF
SURFACE AREA

Wetted Area, ft² (m²)	q, BTU/hr/ft²*
20–200 (1.86–18.6)	20,000
200–1000 (18.6–93)	199,300 $A^{-0.434}$
1000–2800 (93–260)	963,400 $A^{-0.662}$

* 1 BTU/hr/ft² = 3.15 W/m².

For areas exceeding 2,800 square feet (260 m²), it has been concluded that complete fire involvement is unlikely. A maximum heat input of 14,090,000 BTU (4,128,370W) per hour is recommended. For a comparison between API and NFPA heat flux recommendations, see Figure 5.13c.

Low-Pressure Tanks

For above-ground tanks and storage vessels designed to operate from atmospheric pressure to 15 PSIG (104 kPa) and used for the storage of flammable liquids, still another method of determining heat flux under fire conditions is presented. Developed by the American Petroleum Institute and presented in their bulletin API-RP-2000, the method is also referred to in the National Fire Codes of the National Fire Protection Association.

This procedure relates the required relieving rate, expressed as cubic feet of free air per hour, to the wetted area.

Wetted Area

According to NFPA, the wetted area of the tank is calculated on the basis of 55 percent of total exposed area of a sphere or spheroid, 75 percent of total exposed area of a horizontal tank, and the first 30 feet (9m) above grade of a vertical tank.

API recommendations for calculations of wetted area are similar to the above, except that for a sphere or spheroid, the total exposed surface up to the maximum horizontal diameter or to a height of 25 ft (7.5 m) is used, whichever is greater. In case of distillation towers, normally only the heights of liquid layers on the trays are considered in determining the wetted area.

Relieving rates for low-pressure tanks are tabulated as shown in Table 5.13d.

Table 5.13d is based on the physical properties of hexane, and utilizes the API recommendations for heat flux due to fire conditions. The total emergency relief capacity for any specific liquid may be calculated from Table 5.13d by the use of the following formula:

$$\text{Design cubic feet of free air per hour} = V\frac{1337}{L\sqrt{M_w}}$$

5.13(2)

where

V = tabular cubic feet of free air per hour,
L = latent heat of vaporization of the specific liquid in BTU per pound, and
M_w = molecular weight of the specific liquid.

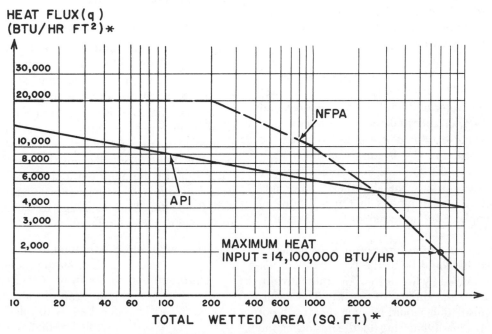

Fig. 5.13c   Heat flux for fire conditions. For SI units, see Section A.1.

Table 5.13d
WETTED AREA VS. CUBIC FEET
OF FREE AIR PER HOUR
(14.7 PSIA and 60°F or 101.4 kPa and
15.6°C)

Sq. Ft.*	SCFH**	Sq. Ft.*	SCFH**
20	21,100	350	288,000
30	31,600	400	312,000
40	42,100	500	354,000
50	52,700	600	392,000
60	63,200	700	428,000
70	73,700	800	462,000
80	84,200	900	493,000
90	94,800	1000	524,000
100	105,000	1200	557,000
120	126,000	1400	587,000
140	147,000	1600	614,000
160	168,000	1800	639,000
180	190,000	2000	662,000
200	211,000	2400	704,000
250	239,000	2800	742,000
300	265,000	& over	742,000

* ft^2 = 0.09 m^2 **SCFH = 0.028 m^3/s

For calculation convenience, a table of physical property constants presented in the National Fire Codes of the National Fire Protection Association is reproduced in Table 5.13e.

Environmental Factors

As it has been noted in equation 5.13(1) and Table 5.13a in connection with the API method, the environmental factors may be applied to the heat flux to reduce the required relieving capacity. These are suggested values, and whenever the situation is not exactly as described, the reader must exercise sound engineering judgment when applying these factors. The list of factors in Table 5.13f are those which appear in the NFPA National Fire Codes.

As noted in Table 5.13a, API allows for additional factors. A factor of 0.03 is used for earth-covered storage above grade, and a factor of 0.0 is used for underground storage. In addition, credit is taken for the thickness of insulation, and a correlation is presented relating the insulation conductance to the factor recommended. It should be noted that API does not take credit for water application facilities on bare metal surfaces. They feel that, because of the uncertainties in the reliability of effective water application, no reduction in environmental factor should be used.

In order to take insulation into account when determining heat flux, it should be of the type that will not be damaged or removed by fire or by a fire water stream, exposing the bare metal surface to the fire. The means of fastening the insulation should be such that it will not fall off under fire conditions due to the fusion of the banding material.

The steps involved in determining the relief capacity under fire conditions are summarized as follows:

1. Establish wet surface area (A) considering size, shape, and location of tank.
2. Based on this area, determine the heat flux (q) to be used. (See Figure 5.13c).
3. Based on the type of installation, select the applicable environmental factor F.
4. Relief capacity = (F)(q)(A)/L.   5.13(3)

Fluids at Critical Point

Occasionally, it is found that under relieving conditions, the liquid in a pressure vessel is above the critical point. Liquids in such state must be handled as a gas. Once the design heat flux is determined, calculations are made to determine the rate of thermal expansion of the vessel contents. In order to calculate this properly, the gas specific head and the gas compressibility factor must be known. It is inadequate to assume the vapor to be an ideal gas because just above the critical point the compressibility factor of the vapor changes very rapidly. The maximum required valve capacity is often not determined at the critical point, but at some point above critical point, where the change in vapor compressibility is less drastic. It should be noted that under these conditions, the metal wall of the vessel will rapidly approach the flame temperature and will fail prematurely because it is not being cooled by the latent heat of vaporization of the boiling liquid contents of the tank. Under these conditions it is advisable to protect the metal surface of the tank with a cooling water spray to prevent premature failure.

Process Considerations

Low-Pressure Storage Tanks

One of the most common process relief requirements involves the "breathing" of atmospheric storage tanks. In this, there are two considerations. One is the need to vent the displaced air when liquid is pumped into the vessel, and the other is the need to admit air to the tank when liquid is pumped out of the vessel. This "breathing" is needed to prevent overpressuring the vessel while pumping liquid in or collapsing the vessel due to vacuum when liquid is pumped out and the corresponding volume is not replaced with air. Referring once again to the API Guide for Tank Venting, RP-2000, the following recommendtions are offered:

Inbreathing—8 scfh (0.06 m^3/hr) air for each gpm (l/m) of maximum emptying rate.
Outbreathing—8.5 scfh (0.064 m^3/hr) air for each gpm (l/m) of maximum filling rate for fluids with a flash point of 100°F (37.8°C) or higher and 17 scfh (0.13 m^3/hr) air for each gpm (l/m) of maximum fill rate for fluids with a flash point below 100°F (37.8°C).

Table 5.13e
TABLE OF CONSTANTS*

Chemical	$L\sqrt{M_w}$	Molecular Weight	Heat of Vaporization, BTU per lbm at Boiling Point
Acetaldehyde	1673	44.05	252
Acetic acid	1350	60.05	174
Acetic anhydride	1792	102.09	177
Acetone	1708	58.08	224
Acetonitrile	2000	41.05	312
Acrylonitrile	1930	53.05	265
n-Amyl alcohol	2025	88.15	216
iso-Amyl alcohol	1990	88.15	212
Aniline	1795	93.12	186
Benzene	1493	78.11	169
n-Butyl acetate	1432	116.16	133
n-Butyl alcohol	2185	74.12	254
iso-Butyl alcohol	2135	74.12	248
Carbon disulfide	1310	76.13	150
Chlorobenzene	1422	112.56	134
Cyclohexane	1414	84.16	154
Cyclohexanol	1953	100.16	195
Cyclohexanone	1625	98.14	164
o-Dichlorobenzene	1455	147.01	120
cis-Dichloroethylene	1350	96.95	137
Diethyl amine	1403	73.14	164
Dimethyl acetamide	1997	87.12	214
Dimethyl amine	1676	45.08	250
Dimethyl formamide	2120	73.09	248
Dioxane (diethylene ether)	1665	88.10	177
Ethyl acetate	1477	88.10	157
Ethyl alcohol	2500	46.07	368
Ethyl chloride	1340	64.52	167
Ethylene dichloride	1363	98.97	137
Ethyl ether	1310	74.12	152
Furan	1362	68.07	165
Furfural	1962	96.08	200
Gasoline	1370–1470	96.0	140–150
n-Heptane	1383	100.20	138
n-Hexane	1337	86.17	144
Hydrogen cyanide	2290	27.03	430
Methyl alcohol	2680	32.04	474
Methyl ethyl ketone	1623	72.10	191
Methyl methacrylate	1432	100.14	143
n-Octane	1412	114.22	132
n-Pentane	1300	72.15	153
n-Propyl acetate	1468	102.13	145
n-Propyl alcohol	2295	60.09	296
iso-Propyl alcohol	2225	60.09	287
Tetrahydro furan	1428	72.10	168
Toluene	1500	92.13	156
Vinyl acetate	1532	86.09	165
o-Xylene	1538	106.16	149

* For SI units, see Section A.1.

Table 5.13f
ENVIRONMENTAL FACTORS (NFPA)

Installation	Factor
Drainage in accordance with NFPA No. 30 for tanks with over 200 ft² (18 m²) of wetted area	0.5
Approved water spray	0.3
Approved insulation	0.3
Approved water spray with approved insulation	0.15

In addition to this, provision must be made to accommodate the thermal venting requirements of the vessel. This is defined as the expansion or contraction of the tank vapors due to changes in the tank ambient temperature conditions. For example, at the beginning of a rain storm the vapors in the tank above the liquid would cool and contract. In order to avoid creating a vacuum in the atmospheric tank, additional air must be admitted into the vapor space of the tank.

The API recommendations are based on an inbreathing capacity of 2 cubic feet of air per hour per square foot (0.63 m³/s/m²) of total shell and roof area for very large tanks (a capacity of more than 20,000 barrels or 3180 m³), and an inbreathing requirement of 1 cubic foot of air (0.18 m³/s) per hour for each barrel (m³) of tank capacity for tanks with a capacity of less than 20,000 barrels (3180 m³). This is roughly equivalent to 1 cubic foot of air per hour (0.028 m³/s) for each 42 gallons (159l) of tank capacity. This capacity is based on a rate of change of vapor space temperature of 100°F (37.8°C) per hour. This maximum rate of temperature change is assumed to occur during a condition such as a sudden cold rain. It is further assumed that the tank roof and shell temperatures cannot rise as rapidly under any condition as they can drop, and for liquids with a flash point at 100°F (37.8°C) or above, the thermal outbreathing requirement has been assumed to be 60 percent of the inbreathing capacity requirement. For materials with a flash point below 100°F (37.8°C), the thermal outbreathing requirement has been assumed to be equal to the inbreathing requirement in order to allow for vaporization of the liquid and the fact that the specific gravity of the tank vapors with volatile hydrocarbons is greater than air. The thermal venting capacity requirements as recommended by API are given in Table 5.7d.

The total inbreathing or outbreathing requirement of a tank is to be the sum of the thermal venting capacity requirements and the requirements set by the rate of pumpage in or out of the tank.

Heat Exchangers

Heat exchangers are a class of process equipment requiring special relief considerations. Heat exchangers frequently have valves located on both the inlet and outlet piping of the exchanger. When these valves are all closed, the exchanger is "blocked in." If it is normally used in the process, relief protection for fire conditions is frequently not installed. This is done on the assumption that the exchanger will not be blocked in at the time of a fire, and vapors generated under fire conditions will be relieved through the piping systems and ultimately through a vessel vent or relief device. Relief devices are, however, frequently installed to provide protection against thermal expansion of liquids in the exchanger when the unit is blocked in. This is always done for the cold side of an exchanger where the liquid can be heated by the hot fluid on the other side, or can be heated by ambient temperature, while sitting with the inlet and outlet valves closed.

In the case of liquid refrigerants, a relief device should always be provided for the protection of the refrigerant side, which may be blocked on the inlet and outlet, whenever the vapor pressure of the refrigerant material raised to the temperature on the hot side exceeds the design pressure of the refrigerant side of the exchanger. This is also done whenever the vapor pressure of the material flowing at 100°F (37.8°C) is greater than the design pressure of the exchanger.

No relief device is necessary for the protection of either side of an exchanger that cannot be blocked in. In such installations it is assumed that the relief of the unit is taken care of by the relief device on the related tank or equipment.

Direct gas-fired tubular heaters are always protected by relief valves on the tube side. The valve is normally sized for the design heat transfer rating of the heater, and must initially handle a fluid rate corresponding to the rate of thermal expansion in the tubes when they are blocked in.

When working with exceptionally high pressures, consideration is given to relief protection of low-pressure equipment in the event an exchanger tube should rupture.

Pumps and Compressors

Another category of devices requiring overpressure protection due to process considerations are fluid flow devices. In the case of positive displacement pumps, a relief device is required to relieve the pumped liquid when the discharge line is blocked in. This relief device is sometimes provided as an integral part of the pump. In sizing the valve, the type of pumping equipment must be taken into consideration. In the case of rotary pumps with a fairly uniform instantaneous flowrate, a relief device is sized for the rated pump capacity. In case of reciprocating pumps, consideration must be given to the fact that the rated flowrate is the average of the total stroke of the piston. It is suggested that for a single piston pump, four times the average flowrate be used for relief sizing. With a duplex or triplex pump, there is some flow averaging, and the engineer must exercise his own judgment.

The pressure setting of the relief device is determined by the design pressure of the weakest part of the system. This may be the design pressure of the pump casing, the design pressure of some valving in the line, etc. Normally, a set pressure is chosen below this limit, but high enough so that the valve will not open under any normal operating conditions.

In case of turbine pumps, relief devices are generally provided to protect the pump, the associated piping, and the equipment that may be blocked in.

In case of centrifugal type pumps, it is uncommon that the maximum pump shutoff head would exceed the design pressure of any system components, but the design engineer must remain aware of the overpressure possibilities in this case.

Reciprocating compressors are also protected against overpressure on the discharge side in case the discharge piping is valved off.

Care must be exercised in the routing of the discharge side of the relief devices. In many cases, directing the relieved fluid back to the pump or compressor suction may result in dangerous overheating of the fluid because of the work input by the pump or compressor. This may result in unit overheating, fluid vaporization, seal failure, etc.

Distillation Towers

A class of vessels where concern is given for overpressure due to process conditions are factionating columns. Here, there is a normal heat input to the unit from the column reboiler. The vapors generated are normally condensed in an overhead condenser. In the event of failure of cooling water or cooling medium to the overhead condenser (or a failure of the fan drive unit for an air cooled condenser), there is a dangerous overpressure situation that may develop as a result of the continued generation of vapors in the reboiler. A relief device must be added to relieve the vapors thus generated, and it is generally rated for the normal heat input from the reboiler.

Overpressure can also develop in a fractionating column when the source of heat is continuous and the overhead vapor line from the column is blocked in. In this case, a relief device must be placed on the fractionating column to relieve vapors as they are generated by the column reboiler.

Reflux failure to a column where reflux acts as a coolant may also cause an overpressure condition.

Pipe Headers

The design engineer must always be alert to problems that may develop from instrument failure, accidentally subjecting a piece of equipment to pressures exceeding the equipment design pressure. Such an overpressure condition may exist at a steam pressure reducing station. To illustrate this, high pressure steam (150 PSIG or 1035 kPa) is commonly reduced to a lower pressure (30 PSIG or 207 kPa) for use in low-pressure equipment. Should this control station fail due to the control valve sticking open, all equipment on the low-pressure steam system could be subjected to high pressure. In such case, a relief device is normally placed on the low-pressure header near the control valve and is rated for the maximum capacity of the control valve. As another example, liquid from a high-pressure source may be admitted to a vessel on either level or flow control. Should the control valve fail, the equipment downstream of the control valve may be subjected to pressures in excess of its design pressure. Here again, relief devices capable of handling the maximum flow through the control valves are provided.

Reactors

A special case of process relief requirements, better considered an art than a science, is the relief of reactors under runaway exothermic reaction conditions. An example of such case is a polymerization reactor with a cooling water failure or an agitator failure. The problems associated with the sizing of the relief device for a polymerization reactor fall into two categories. First, it is difficult and sometimes impossible to determine the actual rate of heat evolution under runaway conditions. Data is frequently not available for such conditions at high relief temperatures. Second, the relief device seldom relieves vapor alone. When the relief device opens up, it commonly passes mixtures of liquids, gases, and solids as in the case of polymerization reactors. One approach to this problem is presented in Paper 32B, submitted at the 61st AIChE National Meeting at Houston, Texas by W. J. Boyle, Jr., of the Monsanto Company, entitled 'Sizing Relief Area for Polymerization Reactors." Generally, the approach recommends determination of the rate of heat evolution, conversion of this heat to equivalent vapor generation, and the assumption that the relief device will vent this equivalent vapor volume as 100 percent liquid. Experimental results confirmed that this approach was conservative.

Sometimes in these cases, relief devices are sized based on past practice. Table 5.13g presents typical relief sizes for PVC reactors as an example.

In cases where relief requirements cannot be quantitatively determined, the reader must be careful to use simplifying assumptions that will lead to conservative results.

Conclusions

The engineer must always exercise sound judgment when determining the basis for selecting relief capacity for sizing. There are so many combinations of installations and circumstances that it is difficult to do much more here than present generalized considerations. In many cases it is necessary to calculate relief capacity requirements based on several considerations, such as fire, cooling water failure, run away reactions, etc., and

Table 5.13g
RELIEF VALVE SIZES FOR PVC REACTORS

Nominal Reactor Volume, gallons (liters)	Relief Valve	Orifice Area, in.² (mm²)	In.²/gal (mm²/l)
1,000 (3780)	2J3	1.287 (830)	0.0013 (0.2218)
2,500 (9450)	3L4	2.85 (1838)	0.0012 (0.2048)
3,750 (14,175)	4P6	6.38 (4115)	0.0017 (0.2901)
5,200 (19,656)	4P6	6.38 (4115)	0.0012 (0.2048)

determine which will require the largest relief device. There can also be installations where the relief capacity requirements based on fire and those based on some other consideration should be added because both conditions are likely to exist simultaneously.

BIBLIOGRAPHY

ANSI B16.5, "Steel Pipe Flanges and Flanged Fittings," American National Standards Institute, New York.

API-PR-520, "Design and Installation of Pressure-Relieving Systems," American Petroleum Institute, New York, reaffirmed 1973.

API-RP-521, "Guide for Pressure Relief and Depressuring Systems," American Petroleum Institute, New York.

ASME Boiler and Pressure Vessel Code, Section I (Power Boilers), Section IV (Low Pressure Heating Boilers), Section VIII (Pressure Vessels—Division 1), American Society of Mechanical Engineers, New York.

Block, B., "Emergency Venting,"*Chemical Engineering*, January 22, 1962.

Bouilloud, P., "Calculation of Maximum Flow Rate Through Safety Valves," *British Chemical Engineering*, November, 1970.

Guide for Inspection of Refinery Equipment, Chapter XVI, "Pressure Relieving Devices," America Petroleum Institute, New York.

Isaacs, M., "Pressure-Relief Systems,"*Chemical Engineering*, February 22, 1971.

Jenett, E., "Components of Pressure Relieving Systems," Parts 1 and 2, *Chemical Engineering*, July and August, 1963.

National Board Inspection Code, "A Manual for Boiler and Pressure Vessel Inspectors," National Board of Boiler and Pressure Vessel Inspectors, Columbus, OH.

Rearick, J.S., "How to Design Pressure Relief Systems," Parts 1 and 2, *Hydrocarbon Processing*, August and September, 1969.

Weber, C.G., "How to Protect Your Pressure Vessels," *Chemical Engineering*, October, 1955.

———, "Installation and Performance of Safety Valves," 16th Southeastern ISA Conference, April 17, 1970.

———, "Set Safety-Relief Valves to Cut Leakage," *Oil and Gas Equipment*, September, 1961.

Willardson, D.F., "How to Protect from Overpressure," *Instruments & Control Systems*, August, 1977.

5.14 RELIEF VALVE FEATURES AND SIZING

Design Pressure: Screwed up to 10,000 PSIG (69 MPa), higher as special. Flanged 150# ASA through 2500# ASA, 125# and 250# in cast iron depending on size.

Design Temperature: −450 to 1000°F (−268 to 538°C) with suitable material selections for pressure parts, trim, and springs; breaks at 450°F (232°C) and 800°F (427°C).

Sizes: 0.5″ to 6″ (12.5 to 150 mm)—some mfgs. to 12″ (300 mm) for special services. Pilot-operated 1″ to 8″ (25 to 200 mm) with double outlets starting at 2″ (50 mm).

Materials of Construction: Pressure Parts: Cast iron, bronze, cast steel, 300 series stainless, nickel steels, monel, Hastelloy, high temperature carbon steel alloys. Trim: any machinable alloy.

Cost: See Figure 5.14a.

Inaccuracy: ±3% of set pressure allowed by ASME Code.

Partial List of Suppliers: *United States*: Anderson-Greenwood Co.; Crosby Valve Div., Geosouvre Inc.; Dresser Industrial Inc., Valve & Inst. Div.; Fisher Controls International, Inc.; GPE Controls; ITT General Controls; Kunkle Valve Co.; J. E. Lonergan Co.; Lunkenheimer Co.; C. A. Norguen Co.; Singer Co., American Meter Div.; Teledyne Farris Engineering; Varec Div., Emerson Electric Co.. Most U.S. manufacturers have foreign manufacturing facilities or licensees. *Belgium*: Contigea S.A. *Germany*: Bopp & Reuther, V.A.G., K.S.B. *France*: S.A.J. Robinetterie, Titan Robinetterie. *Holland*: Dikkers & Co. *Italy*: A.S.T., Pignone Sud, Sella, Spa.

Purpose

Safety relief valves, as the name implies, are commonly installed for one or more of the following reasons:

1. The safety of operating personnel with respect to overpressure.
2. The prevention of destruction of capital investment due to overpressure.
3. The avoidance of civil suits from property or personal damage external to the plant occasioned by results of overpressure.
4. The obtaining of favorable insurance treatment for the capital investment of the plant.
5. The necessity for compliance with local, national, state, and other court-enforceable regulations.
6. The conservation of material loss during and after an upset producing temporary overpressure.
7. The minimizing of unit down-time due to the results of overpressure.
8. The prevention of damage to downstream equipment due to overpressure being transmitted through the connecting equipment and piping.
9. The prevention of pollution (primarily air pollution) due to the discharge of vapors upon overpressure with subsequent rupture or damage.

It should be noted that in every event the need for relief is occasioned by the entry of energy or mass into a system over and above that which the system as designed, installed, and operating is capable of handling, and which would result in a rise in system pressure above design if not checked in some way. The design pressure of the equipment protected is usually the set pressure for the relief device. For more details on set pressure determination, see the last paragraph of this section. The usual justifications for the installation of these relieving devices are code regulation, insurance requirements, or company policy to use this device as the last, final, backup device for pressure relief. The relative simplicity of the valve and the fact that is is self-contained and self-actuating make it probably the most reliable device short of the rupture disc (which does not, however, reclose). It is important to remember that a pressure relief valve is installed only to limit pressure; it will not control, reduce, or depressure the system unless special provisions are incorporated.

Nomenclature

There is a good deal of confusion, ambiguity, and duplication in nomenclature on this subject, and in order to permit further discussion, the pertinent nomenclature is given below:

> *Pressure Relieving Device*—The broadest category in the pressure relief system, includes rupture discs and pressure relief valves of the simple spring-loaded type and certain pilot-operated types.
>
> *Pressure Relief Valve*—A generic term that includes safety valves, relief valves, and safety relief valves.
>
> *Safety Valve*—An automatic pressure relieving device actuated by the static pressure upstream of the valve and characterized by rapid full opening or pop action. It is used for steam, gas, or vapor service.
>
> *Relief Valve*—An automatic pressure relieving device actuated by the static pressure upstream of the valve, which opens in proportion to the increase in pressure over the opening pressure. It is used primarily for liquid service.
>
> *Safety Relief Valve*—An automatic pressure-actuated relieving device suitable for use as either a safety or relief valve, depending on application.
>
> *Pressure*—There are various significant pressures that differ in importance according to the use the valve is assigned to, the applicable code, the hazardous nature of the fluid, and the design of the valve. In general, it is desirable that a clear statement of these items be made as a part of the agreements to be reached.
>
>> *Start-to-Leak Pressure*—The pressure at the valve inlet at which the relieved fluid is first detected on the downstream side of the seat before normal relieving action takes place.
>>
>> *Set Pressure (Opening Pressure)*—The pressure, measured at the valve inlet, at which there is a measurable lift, or at which discharge becomes continuous as determined by seeing, feeling, or hearing. In the pop-type safety valve, it is the pressure at which

the valve moves further in the opening direction compared to corresponding movements at higher or lower pressures. A safety valve or a safety relief valve is not considered to be open when it is simmering at a pressure just below the popping point, even though the simmering may be audible.

> *Overpressure*—Pressure increase over the set pressure of the primary relieving device is overpressure. It is the same as accumulation when the relieving device is set at the maximum allowable working pressure of the vessel. Note: From this definition it will be observed that, when the set pressure of the first (primary) safety or relief valve is less than the maximum allowable working pressure of the vessel, the overpressure may be greater than 10 percent of set pressure.
>
> *Accumulation*—Pressure increase over the maximum allowable working pressure of the vessel during discharge through the pressure relief valve, expressed as a percent of that pressure or in pounds per square inch.
>
> *Relieving Pressure (Opening Pressure plus Overpressure)*—The pressure, measured at the valve inlet, at which the relieving capacity is determined.
>
> *Blowdown (Blowback)*—The difference between the set pressure and the reseating pressure of a pressure relief valve, expressed in percent of the set pressure or in pounds per square inch.
>
> *Closing Pressure (Reseat Pressure)*—The pressure, measured at the valve inlet, at which the valve closes, flow is substantially shut off, and there is no measurable lift.
>
> *Seal-off Pressure*—The pressure, measured at the valve inlet after closing, at which no further liquid, steam, or gas is detected at the downstream side of the seat.
>
> *Reopening Pressure*—The opening pressure when the pressure is raised as soon as practicable after the valve has reseated or closed from a previous discharge.
>
> *Operating Pressure*—The operating pressure of a vessel is the pressure, in pounds per square inch gauge, to which the vessel is usually subjected in service. A processing vessel is usually designed for a maximum allowable working pressure, in pounds per square inch gauge, that will provide a suitable margin above the operating pressure in order to prevent any undesirable operation of the relief device. It is suggested that this margin be approximately 10 percent, or 25 PSI (173 kPa), whichever is greater. Adequate margin will prevent undesirable open-

ing and operation of the pressure relief valve caused by minor fluctuations in the operating pressure.

Maximum Allowable Working Pressure—As defined in the construction codes for unfired pressure vessels, the maximum allowable working pressure depends on the type of material, its thickness, and the service conditions set as the basis for design. The vessel may not be operated above this pressure or its equivalent at any metal temperature other than that used in its design; consequently for that metal temperature, it is the highest pressure at which the primary pressure relief valve can be set to open.

Back Pressure—Pressure on the discharge side of a pressure relief valve.

Constant Back Pressure—Back pressure that does not change under any condition of operation whether or not the pressure relief valve is closed or open.

Variable Back Pressure—Back pressure that varies due to changes in operation of one or more pressure relief valves connected into a common discharge header.

Built-up Back Pressure—Variable back pressure which develops as a result of flow after the pressure relief valve opens.

Superimposed Back Pressure—Variable back pressure that is present before the pressure relief valve starts to open.

Lift—The rise of the disc in a pressure relief valve.

Conventional Safety Relief Valve—A safety relief valve with the bonnet vented either to atmosphere or internally to the discharge side of the valve. The performance characteristics (set pressure, blowdown, and capacity) are directly affected by changes of the back pressure on the valve.

Balanced Safety Relief Valve—A safety relief valve with the bonnet vented to atmosphere. The effect of back pressure on the performance characteristics of the valve (set pressure, blowdown, and capacity) is much less than on the conventional valve. The balanced safety relief valve is made in three designs: (1) with a balancing piston, (2) with a balancing bellows, and (3) with a balancing bellows and an auxiliary balancing piston.

Flutter—Rapid, abnormal reciprocating variations in lift during which the disc does not contact the seat.

Chatter—Rapid, abnormal reciprocating variations in lift during which the disc contacts the seat.

Simmer (Warn)—The condition just prior to opening at which a spring-loaded relief valve is at the point of having zero or negative forces hold-

ing the valve closed. As soon as the valve disc attempts to rise, the spring constant develops enough force to close the valve again.

Special Considerations

Since one of the reasons for installing a pressure relief valve is compliance with local codes and regulations, the system design must be acceptable to the codes and authorities involved. The designer may also have to consider the insurance underwriting groups and the requirements they impose. Since the object is to prevent destruction of capital investment and produce personnel safety with a pressure relieving system, such considerations as toxicity, polymerization, corrosion, and damage to other equipment in the plant must be considered when deciding where the valve will discharge.

Since unit down-time and loss of material are to be minimized, important considerations for the valve would be the requirement of tight shutoff against maximum pressure and the tolerances required for actuation. Personnel safety and the minimization of property damage lead to requirements for reliability of operation both in terms of accuracy and repeatability. A pressure relieving valve in chemical plants is rather unique in that it hopefully will never need to operate. Furthermore, it is rare that any such system be actuated for a check in its entirety although certain components can be tested periodically—generally at the cost of a more expensive installation due to valving and bypasses. Even when performed, this inspection usually occurs at rather long intervals. A pressure relief valve is not tested weekly for its condition as one might test an emergency generator or a fire pump. In spite of the extended periods of dead storage while exposed to process fluids, temperatures, and pressures, plus the full range of ambient conditions, we still expect the valve to work if called upon. Therefore it is important to consider component quality in a different light from quality in most other process plant components. This requirement that the valve work when called upon, coupled with the desire to provide safety for personnel and equipment, introduces the idea of redundancy as exemplified by duplicate or multiple relief valves, rupture discs plus safety valves, flare valves backed up by pressure relief valves, etc.

A further consideration for pressure relief valves is company practice and standards. This factor may introduce the following considerations:

1. The use of extra safety factors in sizing or rating over and above those established in codes, regulations, and recommended practices.
2. The provision of other protective facilities that may result in credits under the codes or regulations.
3. The provision of credit for redundancy or other protective facilities even though such credit may not be clearly established in codes or regulations.

4. Preferences, particularly in installation practices, based on operating or maintenance practices.

5. The establishment of minimum design pressures for various types of equipment in various services (this would affect pressure relieving systems in that operating pressure may not always bear the same relationship to design pressure) and smaller relieving valves set at the minimum design pressure may be suitable.

6. The relationship between operating pressure and relief valve set pressure as affected by the upset to operating conditions that is acceptable before relief occurs.

7. The standardization of relieving device sizes and types and of relieving device mounting nozzle sizes and locations for various types of equipment in different services.

8. The inspection and test procedures established by the company for its pressure relieving systems or the components of those systems.

9. The accounting needs and practice of the company as they might affect flow detection and metering in the pressure relieving system or at the valve.

Principles

The basic principle of the pressure relief valve is to provide a pressure-actuated relief for material flow out of the system in danger of being subjected to overpressure. The most commonly accepted means of providing this is to sense the approach to the desired limiting pressure with some sort of force-balance mechanism, which at the set point, activates to open the desired or required relieving area for fluid flow. Usually this force balance has, on one side the system, pressure acting upon a given area and, on the other side, either springs or weights. The magnitude of the forces being balanced can be reduced, and thus their control improved, through the use of secondary devices such as pilot valves, solenoids, etc. Weight-loaded valves have largely disappeared from process plant services except in the extremely low pressure range where their improved accuracy and absence of the "spring constant" effect due to valve travel make them advantageous.

The basic requisite for a direct spring-loaded pressure relieving valve is a suitable valve body, usually an angle type, having an inlet connection that is suitable for the inlet pressure/temperature requirements. The body and outlet connection as well as the bonnet is usually designed for a lower pressure than the line or the inlet connection.

The inlet incorporates a valve seat with a disc for full closure of the inlet port. The disc is usually direct spring-loaded and the spring force is applied directly on the disc by means of a stem. The disc may be either disc-guided or top-guided. A bottom (disc) guided valve has vanes or feathers for guiding in the valve bore (inlet port). Process valves are usually top-guided. Boiler valves and liquid relief valves are often disc-guided.

The bonnet is sized to accommodate the spring for the maximum pressure rating of the valve. The bonnet is used when the discharging medium must be confined within the valve body and discharge piping. Flanged valves for steam boilers usually have an open spring with a yoke in place of a closed bonnet. The spring is exposed on the steam valve, whereas the spring is totally enclosed on the bonnet type valve. Marine boiler valves are of the yoke type except that additional cover plates must be added and sealed to prevent tampering with the spring. The covers are vented (not pressure tight). All bonnet type valves have either caps over the adjusting bolt or lifting levers, either plain or packed.

The set or opening pressure is governed by the selection of the proper spring and by adjusting the bolt that compresses the spring to the correct opening pressure. Springs are classed in different spring ranges so that the spring is never overstressed and that proper clearance between the coils allows for full lift. The spring force at open position must not exceed the lifting force of the flowing medium when the valve is fully open. The spring setting may be raised or lowered 10 percent of the factory setting without jeopardizing the valve operation. Generally the outlet connection is larger than the inlet. This is necessary for expanding mediums such as gases or vapors. A safety or relief valve must always relieve to a lower pressure (usually atmospheric).

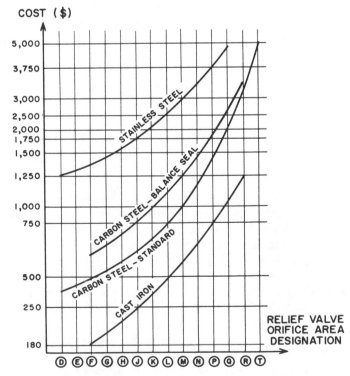

Fig. 5.14a   Approximate relative cost of relief valves

As mentioned earlier, the valve has an inlet (port) seat. This may be described as a bushing, semi-nozzle, or full nozzle. The bushing is used on bottom-guided valves. The semi-nozzle is usually found in cast iron valves. Both are screwed into the body. The full nozzle is utilized on steel valves and is the line pressure part of the valve. The seat, orifice, and flange facing is one piece. The flange of the valve is used for bolting force, which is to ASA standards. The nozzle flange construction is similar to a Van Stone type flange. The discharge area or orifice of a nozzle type valve is smaller than the nominal inlet. By using a converging nozzle, high velocity flow results in the high kinetic energy required to obtain high lift.

Since these devices are concerned with safety, their capacity must be calculated with some accuracy and their action be reliable and foolproof over the long-run. As indicated above, reliability is achieved by the maximum use of simple mechanisms, linkages, etc. Accuracy in capacity is achieved by conservatism, by treating the relief device as an orifice, and by attempting to size and intall it so that it can always *control* or establish the fluid (and thus energy) release.

A typical cross-section of this type of device is shown in Figure 5.14b. Note that there is an adjustable ring around the nozzle. Furthermore, the disc has either a fixed or adjustable deflecting lip. The purpose of this lip is to provide the pop action used to distinguish safety valves. The inlet nozzle is used to generate velocity head—and thus capacity—efficiently, while the lip and the ring on the nozzle form a secondary orifice and a device for

reconversion of this velocity energy to static pressure to provide the pop action once sufficient vapor flow develops.

The control of the so-called safety and relief valves is generally accomplished through spring loading and force balance. Being spring-loaded, they require some increase in force as movement of the spring takes place on opening the valve. The amount of this increase is determined by the spring constant and the amount of valve lift required to achieve capacity dimensions. Most safety valves reach their full capacity dimensions at about 3 percent above set pressure, with any additional pressure rise serving to increase capacity only because of nozzle flow response to higher inlet pressure; i.e., at 3 percent overpressure the valve has fully lifted so that the curtain or cylinder area between nozzle and disc is greater than the cross-sectional area of the nozzle. The so-called low-lift valves gain more capacity at higher pressure because they have not reached limiting dimensions for the curtain area at the low overpressures. In such cases it is not permitted under the ASME Code to calculate down-ratings for valves at lower pressures based on performance at a higher test pressure, although up-rating above test pressure is permitted because it will always tend to be quite conservative. In contrast to safety valves, relief valves generally do not reach their full capacity dimensions until their full nominal rating overpressure of 25 percent is reached. Down-rating to lower pressures is permitted and provided by all manufacturers.

Effect of Pressure Conditions

It is possible for back pressure on the outlet of these devices—regardless of the source—to affect their set pressure. This is illustrated in Figure 5.14c, which considers several valve designs and means for remedying this effect. Note that venting or not venting the bonnet can change the direction of the back pressure effect. If this back pressure builds up as the valve opens, the valve may chatter and close. If it is constant and present when the valve starts to open, it should be possible to compensate for it by raising or lowering the spring setting. Figure 5.14d shows that the dimensions and design of the particular valve will establish the change needed and that this change is a function of the back pressure. Therefore, it is *not* suggested to compensate for back pressure with the spring setting without consulting the manufacturer of the particular valve.

Two of the most common means of overcoming this back pressure effect on set pressure are by the use of a balancing piston or a bellows seal. Figure 5.14e shows these devices as actually installed. Note that the devices may or may not provide a true balance depending on how well they match the nozzle area. The choice of type is made by the manufacturer in view of other consideration such as capacity or full-lift dimensions at low overpressure.

Back pressure can also affect the capacity and thus the

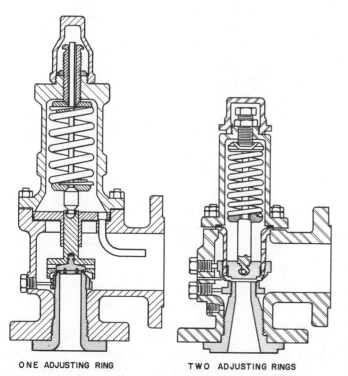

ONE ADJUSTING RING   TWO ADJUSTING RINGS

Fig. 5.14b Conventional type safety relief valves

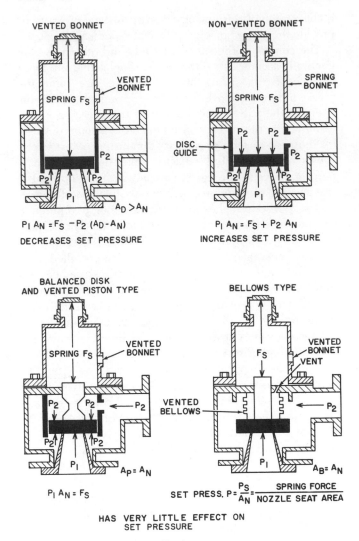

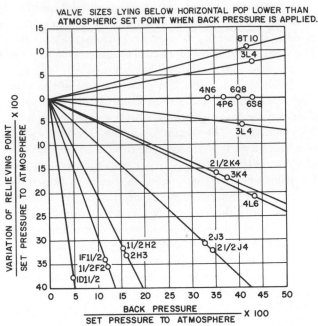

Fig. 5.14d   Manufacturer's tests on effect of back pressure on standard safety relief valves (Data courtesy Farris Engineering Corp.)

Fig. 5.14c   Effect of back pressure on set pressure

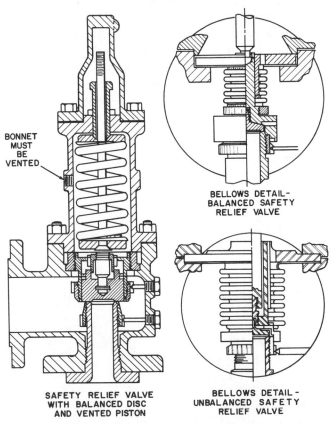

Fig. 5.14e   Balancing devices

sizing of relieving devices. This area of sizing is often a source of confusion. Definitions become extremely important. With discharge piping involved, the safety relief valve is in series with the pressure relieving system. In this sense, it is similar to an open control valve. However, instead of being a *controlling* element, it is *controlled by* the pressure conditions at the inlet and outlet. Most safety valves will handle flow very similarly to a theoretical nozzle as long as they are at full lift, so that the nozzle establishes the controlling flow area. It is known that the capacity of a theoretical nozzle is not affected by the downstream pressure so long as it does not exceed the critical pressure needed to maintain sonic flow. At higher downstream pressures reduction in flow can be calculated. A safety valve is kept open by pressure balance. Thus when back pressure is present or develops, the opening forces are affected. The effect of this force balance on a conventional safety valve with unvented bonnet is shown in Figure 5.14f. Note that the two sources of back pressure produce different effects. Because of this capacity reduction due to lift action, conventional

valves should never be used where back pressure variation can exceed 10 percent of set pressure.

In conventional pressure relief valves, constant back pressures up to the critical ratio can be compensated by spring setting without affecting the capacity. Above that

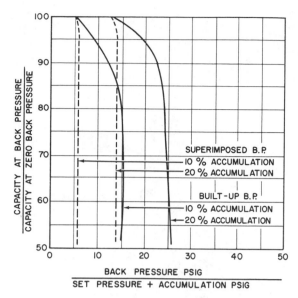

Fig. 5.14f Effect of back pressure on capacity of conventional safety relief valves (not vented to atmosphere)

do not require viscosity correction. Since corrections are needed because of flow considerations, the point at which a correction is required will be a function of orifice size, flowrate, and flowing viscosity. There is reasonably good agreement among manufacturers as to the viscosity correction, and there is general agreement that viscosities below 50-100 SSU do not warrant correction.

One area of sizing not covered by any code, regulation, or industry practice is that for flashing fluids. While the actual sizing is discussed below, certain features of the valve discharge system can affect both valve sizing and the cost of the installation. Since the amount of flash is influenced by the discharge line size through the actual back pressure developed, the ability of discharge line sizing to suppress vapor generation should be considered. Based on limited experience and some calculation, it appears that the most economical valve size results at about 30 percent back pressure. Balanced safety valves are usually used in flashing flow to minimize the effects of the flashed vapors and of the resultant backpressure on valve operation and capacity.

setting compensation can be used, but an appropriate capacity reduction factor must be employed in sizing. All manufacturers provide this reduction factor, and there is virtually unanimous agreement on the factor at any given ratio of constant back pressure to actual relieving pressure for both vapors and liquids.

When it comes to balanced or bellow valves there is a good deal less agreement as shown for vapors in Figure 5.14g. These are valves that can be used with superimposed as well as built-up back pressures with no change in spring setting needed. These variations are a function of the specific valve design, and the capacity factor for the particular valve must be used. For liquids, it is common practice to use the minimum pressure drop in the standard formula with an additional factor as function of back pressure. Here also, there is variation in the factor among vendors. Viscous materials adversely affect the capacity of both safety and relief valves. Generally, gases

Pilot-Operated Valves

Another type of safety valve that should be mentioned as it is getting increasing attention is the so-called pilot-operated pressure relieving valve. Cross sections of two typical valves are shown in Figure 5.14h. The valve on the right utilizes an integrally mounted conventional safety valve. It senses the process pressure and vents or admits pressure to the top of the main piston or disc. The valve on the left is a newly developed design that permits adjustment of blowdown without entering the valve proper and can use an external sensing line to the pilot. There are also types available in which the blowdown and relief settings are independently adjusted through pilot valves mounted on top of the main piston or disc chamber. Most of these designs have not had wide acceptance because of the large number of static and moving seals that must function and because of the small clearances existing in the pilot mechanisms. Most process streams never seem to be sufficiently clean.

Recent improvements in design, together with the increased attention to leakage because of pollution considerations, larger capacity requirements, and the tendency to operate units closer to design, have markedly improved the acceptance of this type of valve. Figure 5.14i shows graphically some of the advantages of the pilot-operated design. In addition, nozzles are not needed to generate the velocity to provide the pop action required for full opening by spring-loaded valves in vapor service. This means that larger capacities can be handled in the same valve body. The increase in possible capacity over the largest nozzle orifice in a given body size ranges from about 150 percent in the smaller sizes to about 120 percent in the larger sizes.

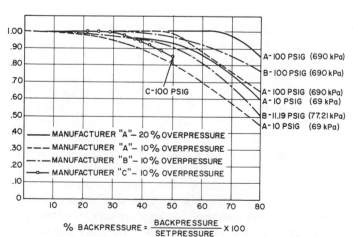

Fig. 5.14g Back pressure capacity reduction factor for balanced bellows valves

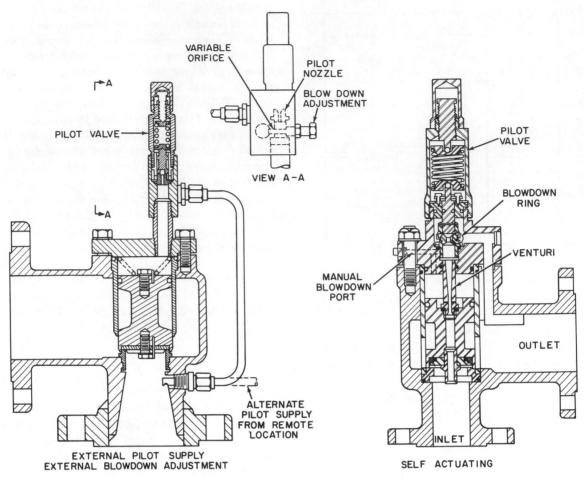

Fig. 5.14h   Pilot-operated safety relief valves

Blowdown and Chatter

The spring control of safety and relief valves creates the need for blowdown. The relieving system, when in operation, is a kinetic system with the valve at a point of high kinetic energy in contrast to the equipment it is protecting. The valve spring balance is made against a pressure that equals the equipment pressure, less the kinetic effects. The closed relief system is, of course, a static one with no kinetic effects, and pressure at the point of spring balance is equal to that in the equipment. This discrepancy between relieving and static conditions necessitates the allowance of blowdown. This is the amount the equipment pressure has dropped below start-of-relieving or set pressure when the valve reseats. It is needed to insure that the valve finds its force balance satisfied once it closes. Normal blowdown on a safety valve is 5 percent of set pressure. The control of blowdown is accomplished by the adjustable ring on the nozzle whose position establishes a secondary orifice area as the valve opens and closes. Blowdown is normally set by bringing this ring up to the disc—the maximum blowdown position—and then backing off the number of turns recommended by the vendor. This is necessary because

most test facilities have limited supplies of gas and the set pressure is usually tested at zero blowdown. The percent blowdown per turn is usually based on methane for process valves; a 5 percent blowdown setting on methane can represent about 11 percent when the valve is handling butane. Very few, if any, manufacturers have published data on this condition. If blowdown is a serious consideration, field tests must usually be made after installation. Even then it is nearly impossible to simulate actual desired or design relieving conditions. The pilot-operated valve is not troubled so greatly with these considerations, or at least can usually be set for smaller blowdowns; 3 percent is fairly standard and 1 percent can be achieved.

Energy losses in the inlet piping between the relieving valve and the source of pressure as well as an oversized valve can lead to a condition known as chatter, wherein the valve repeatedly cycles between open and closed. The use of conventional valves with variations in back pressure greater than about 10 percent while relieving will also produce an unstable force balance condition and chatter. The new Section II of API RP-520 endorses the widely used limit of 3 percent of set pressure as the

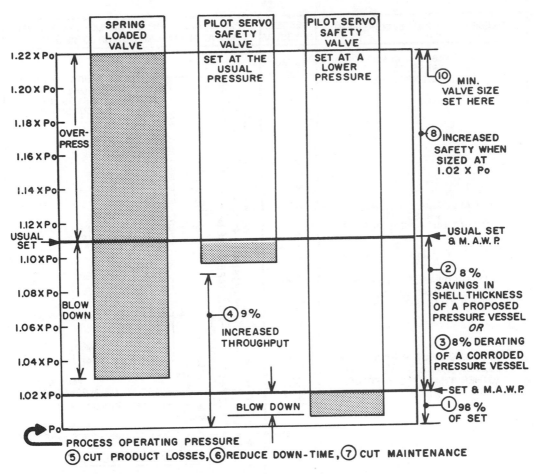

Fig. 5.14i Comparative operating ranges

maximum value for inlet piping loss. Losses much higher than this cause the pressure at the valve during relief flow to reach the blowdown reseating value. When this happens, the valve closes and immediately is subjected to a pressure rise because the kinetic effects no longer exist. This rise is often enough to cause valve opening and the cycle repeats. Oversized valves can cause this same condition because a spring-loaded valve needs about 20–30 percent of maximum flow to establish a stable force relationship and maintain disc position.

Chatter has at least two detrimental effects. Obviously the valve's seating areas will beat themselves into a condition that invites, even insures, leakage. Second, bellows failure can be expected if the disc chatters or flutters. Pilot-operated valves are not subject to chatter to the same extent. Basically they are open or closed and cannot "throttle down" to 20–30 percent flow. Thus, if they are oversized, repeated openings and closings may be needed to hold system pressure, and the net result is an action approaching chatter.

Valve Tightness and Leakage

Valve tightness in spring-loaded valves can be markedly improved if fire load is the only source of over-

pressure by taking advantage of the 20 percent that the ASME Code allows for this condition. If one valve is set for 109 percent of maximum allowable working pressure and 10 percent accumulation is allowed, this results in the same size valve as one set at 100 percent and allowing 20 percent accumulation. However, if the working pressure is at or near 90 percent of maximum allowable, the first valve will have about twice the force keeping the seat tight at working pressure as that available in the second valve. Similarly, if there is a large fire load but relatively small operating load, the installation of a small valve set at 100 percent and sized at 10 percent accumulation for the operating load, which accumulates to 20 percent under fire conditions, plus the installation of a valve for the additional fire load set at 109 percent and sized at 10 percent overpressure, will generally reduce leakage quantities. This becomes even more attractive if company policy requires two valves. Also, at low operating pressures, the API recommendation of 3 percent maximum inlet loss makes the use of two valves on tall towers attractive. These considerations are shown in Figure 5.14j (1) and (2).

Greatly influencing valve tightness are the factors of seating closure design, cleanliness of fluid handled, and

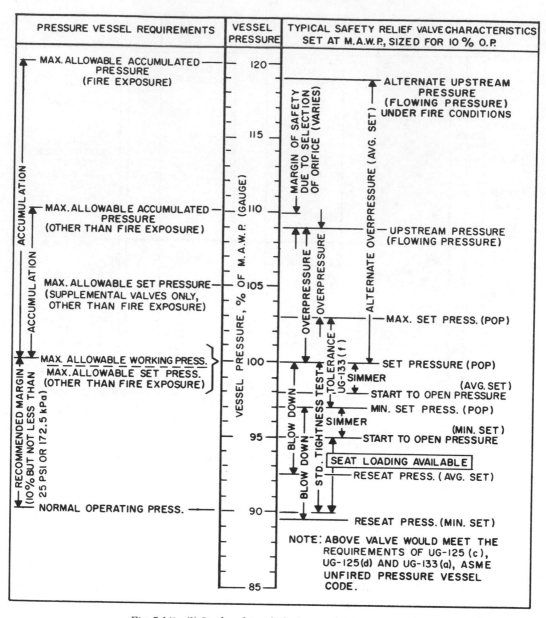

Fig. 5.14j   (1) Single safety relief valve used to protect vessel

installation practices. The areas of seating closure design, such as mechanical containment of nozzle and disc and of guiding moving parts, have provided a fertile field for manufacturers' claims. Some are very real and some are not, so that a careful engineering analysis of these areas for the designs being considered can pay off handsomely in trouble-free service. The so-called soft seat design and the o-ring seat will do much to prevent leakage as long as the valve does not relieve. However, it is well to check the design and question the vendor as to what the leakage will be after the valve reseats. Some designs actually expect the o-ring to be shredded or destroyed during relief and then provide a metal-to-metal secondary seat to minimize leakage until maintenance can replace the ring.

Temperature provides probably the most serious limitation on soft seat or o-ring materials. In addition to soft seat sealing, the deformable line contact type of seat has been well received in some services outside the temperature limits of synthetic seating material. The knife edge seat is perhaps the most desirable for services where icing or other like deposits can form on the seat during relief. This type has also been used for liquids containing fine solids in suspension.

The methods of holding the nozzle in the pressure-containing body to minimize the effects of bolting forces and piping reactions that would tend to deform the nozzle, of applying the spring load to the disc, and of disc design to compensate for differential expansion effects are all part of the refinements manufacturers have de-

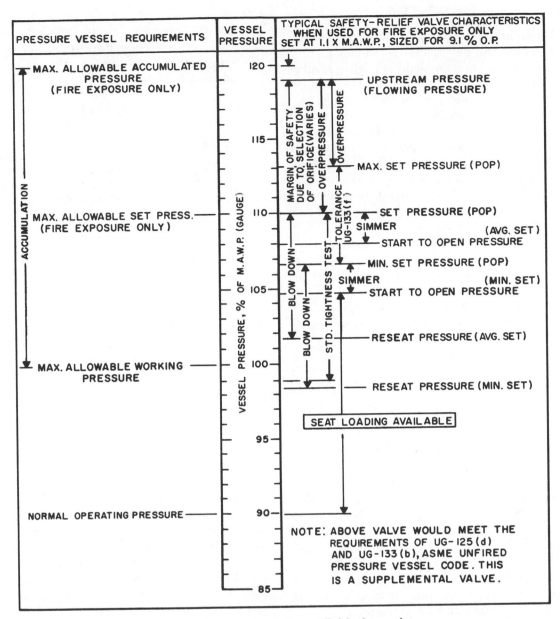

Fig. 5.14j (2) Supplemental valve installed for five conditions

veloped as users demanded tighter, longer-lasting valves while service conditions grew progressively more severe. It may sound ridiculous to talk about nozzle deformation by bolting stresses, but most process service valves as tested and shipped by the manufacturer have a seat finish of less than 5 micro inches with a flatness deviation of less than 5 to 15 millionths of an inch (0.13 to 0.38 microns).

The influence of fluid cleanliness on valve performance is illustrated by the contrast between refinery or chemical service and missile service. In a number of services in refineries and chemical plants, experience has shown that the valve will leak once it has opened and reseated; you might just as well take it down, clean and lap it, reset and reinstall it. Contrast this with the shop inspection requirements on some missile service valves

where 200 successive cyclings on clean, high-pressure nitrogen must still leave the valve virtually bubble-tight. This requirement is regularly met on a production basis.

Considerations in Selecting a Particular Valve

In selecting a particular pressure relief valve for a specific service, other aspects need consideration in addition to those discussed above. Rather than catalog a number of specific problems and solutions, a check list of the more frequent problem areas is offered. To offset what may appear to be casual treatment of mechanical features, the reader is advised to check the mechanical design of competitive offerings very carefully before selecting a particular valve for his service.

Check List for Pressure Relief Valves

1. High temperature effects on disc warpage, spring set point, and disc guiding area binding.
2. Galling or seizing problems in guiding or other close clearance areas due to materials selection, foreign matter entry, or deposit build-up.
3. Corrosion of discharge portion of valve bonnet and portions of mechanisms due to corrosion from materials possibly present in atmosphere or discharge header system though not present in service on which valve is installed. Check particularly the effect on high-stressed elements such as springs. Designs such as diaphragm seals and bellows seals are available.
4. Possibility of polymer or other material build-up in throats of valve or operating mechanism after or during relief so as to impair valve action. Again, various seals are available.
5. Resistance of valve assembly to vibration.
6. Provisions of valve design to minimize chatter on pulsating service.
7. Probable condition of valve and seating after exposure to external fire.
8. Need for steam jacketing to prevent tendencies toward solidification or crystallization within valve. Check particularly the extent of steam jacketing required.
9. Provision for accurately guiding the disc.
10. Need for valve position indicating devices.
11. Availability and desirability of augmenting normal disc seating forces, or keeping them constant until valve set point is reached so as to minimize leakage.
12. Seat design in light of ability to reseat tightly after relieving on specific service.
13. Method of countering effects of all types of back pressure and the resulting variation in set point and blowdown with back pressure.
14. Consequences of and ability to detect bellows rupture.
15. Open spring versus bonneted spring design.
16. Degree of blowdown needed. Does blowdown have to be adjusted? Can pop action be destroyed in adjusting blowdown?
17. Need for special service valves such as chlorine service, toxic material, LPG storage tank, ICC approved, Coast Guard approved, etc.
18. Need for various auxiliary features such as test gags, lifting levers open or packed, screwed vs. bolted adjusting nut caps, etc.

Sizing the Pressure Relief Valve

Vapors, Gases

The basis for almost all valve sizing for process industry service is found in the ASME Unfired Pressure Vessel Code Section VIII. The basic equation results from Par. UA 230 (1962) and Par. 10-100 (1968). This states that capacity is converted from the test medium on which the valve was officially rated to any other *vapor or gas* by the following formula:

$$W = K_b CKAP \sqrt{\frac{M}{TZ}} \qquad 5.14(1)$$

where

W = relieving flow of gas or vapor, lbm/hr,[*]

C = constant for gas or vapor which is a function of the ratio of specific heats "k" as defined below:

$$C = 520 \sqrt{k\left(\frac{2}{k+1}\right)^{k+1/k-1}} \qquad 5.14(2)$$

K = coefficient of discharge as determined from tests,

A = required nozzle area of the valve, sq. inches,

P = (set pressure × 1.10) plus atmospheric pressure, (PSIA),

M = molecular weight,

T = absolute temperature, °Rankine (°F + 460),

Z = vapor compressibility at inlet conditions, and

K_b = correction factor for constant back pressure.

This basic equation can be manipulated almost any number of ways and each manufacturer and designer has his own favorite sizing procedure. A commonly accepted and easily used graphical approach is given in Figures 5.14 k and l. For those who prefer calculations, API RP520 Part I and the manufacturers' catalogs both tabulate values of C (or k) for many of the more common gases and vapors.

This equation assumes sonic flow across the valve nozzle and does account for the effect of back pressure in reducing valve capacity. Figure 5.14m represents the correction curve for the effects of constant back pressure in reducing the flow across the nozzle of conventional, nonbalanced valves, when the spring setting has been compensated for any constant back pressure. As pointed out previously, the effect of back pressure, whether fixed or variable on bellows sealed valves, is a function of the specific valve design and size. There is general acceptance of the fact that a back pressure up to 30 percent of inlet pressure will not need correction. Above that value the specific manufacturer should be contacted, though Figure 5.14n (taken from API RP520) can also be utilized as a conservative basis.

* For SI units, see Section A.1.

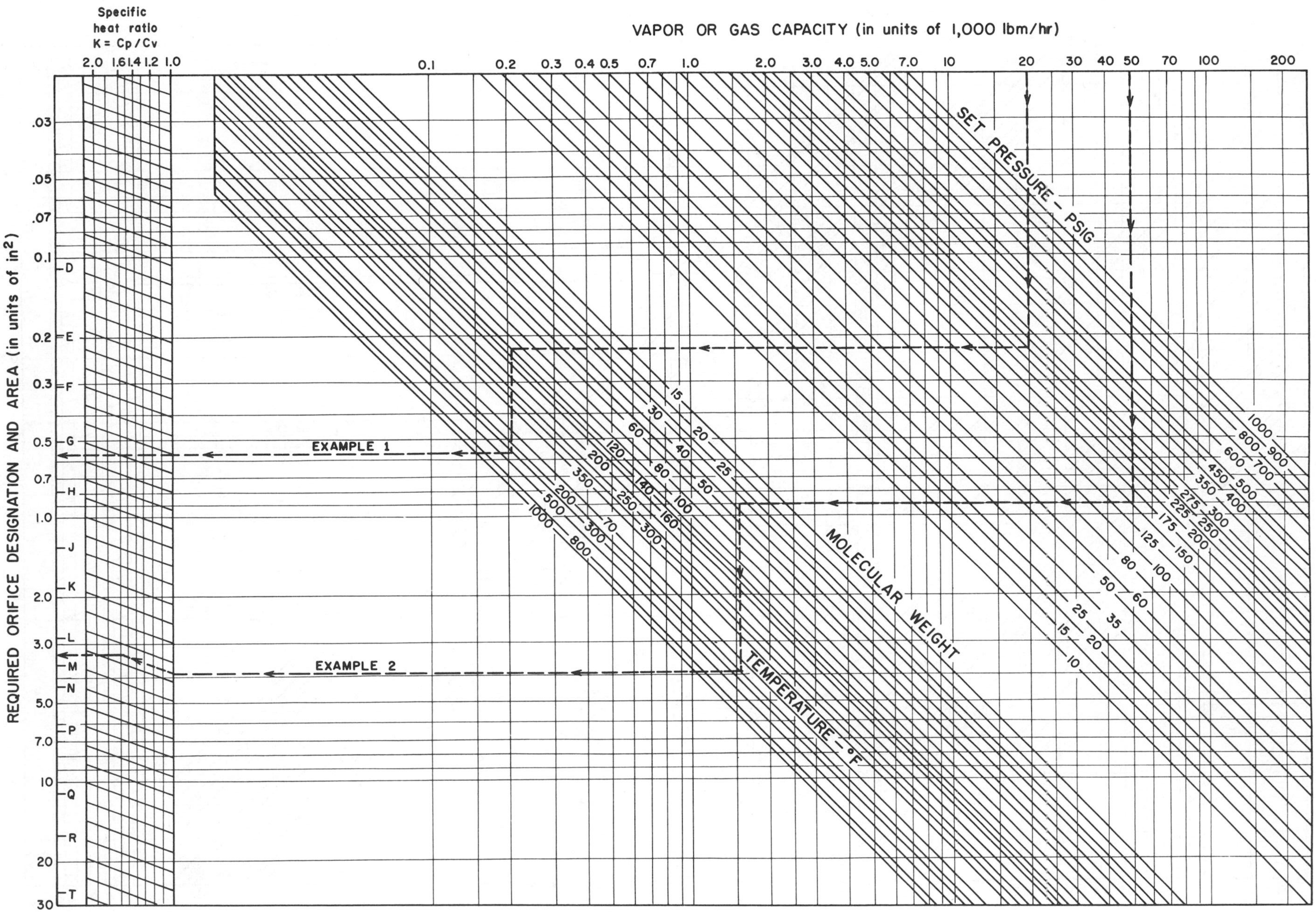

Fig. 5.14k   Safety valve sizing chart for vapors or gases (For SI units see Section A.1.)

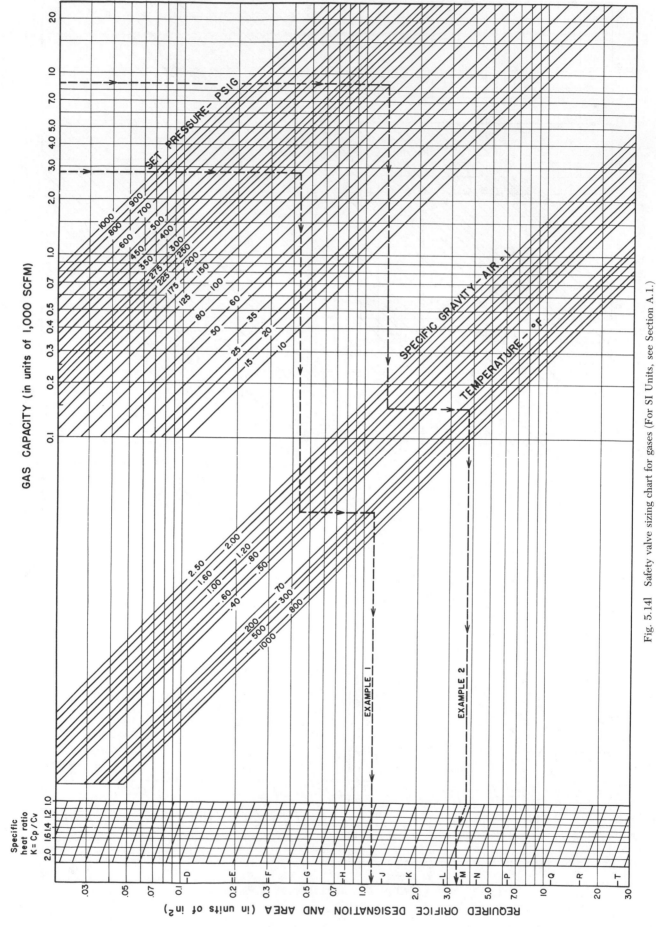

Fig. 5.14l   Safety valve sizing chart for gases (For SI Units, see Section A.1.)

578

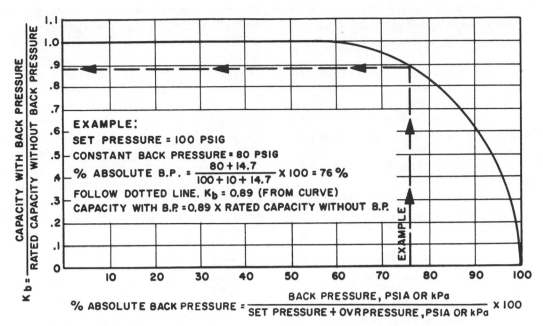

Fig. 5.14m   Constant back pressure sizing factor—conventional valves (vapors and gases)

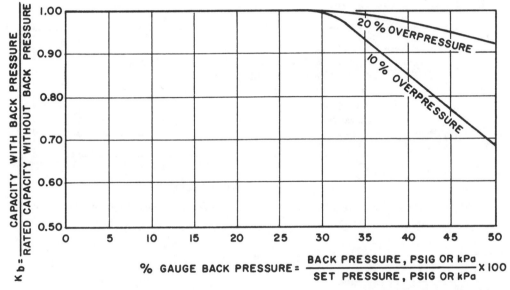

Fig. 5.14n   Variable or constant back pressure sizing factor, K_b, for balanced bellows safety relief valves (vapors and gases). The curves represent a compromise of the values recommended by a number of relief valve manufacturers and may be used when the make of a valve or the actual critical flow pressure point for the vapor or gas in unknown. When the make is known, the manufacturer should be consulted for the correction factor. These curves are for set pressures of 50 PSIG (345 kPa) and above; for set pressures lower than 50 PSIG (345 kPa) the manufacturer should be consulted for the values of K_b.

Liquids

The ASME Code does not cover liquid relief, but there has been general acceptance by all manufacturers of the following formula, which is derived from the basic orifice equation:

$$Q_g = 27.2A \sqrt{\frac{P_d}{G}} K_p K_u K_w \qquad 5.14(3)$$

where

Q_g = relieving flow of liquid, gpm,*

A = actual nozzle area of the valve, sq. inches,

P_d = inlet (relieving) pressure less any constant back pressure, PSID,

G = specific gravity of liquid at flowing conditions relative to water at 60°F,

K_p = overpressure correction factor for liquid,

* For SI units, see Section A.1.

K_w = variable or constant back pressure factor for bellows sealed valves only, and

K_u = viscosity correction factor.

Manufacturers commonly tabulate liquid capacity values for their valves at overpressures of 25 percent based on water. For any other fluid these tables can be used so long as the *equivalent water volumetric rate* in gpm is used to enter the tables. If the overpressure is other than 25 percent, the correction for a conventional or balanced valve can be taken from Figure 5.14o. If the back pressure is variable, the additional correction K_w for a bellows valve is taken from Figure 5.14p. Note the discrepancy between manufacturers. For both conven-

tional and balanced valves, it is common to use the minimum differential pressure as P_d.

The factor K_u accounts for the fact that the resistance to flow encountered when handling viscous liquids above 50–100 SSU may reduce the velocity and thus capacity enough to require a larger orifice size than the usual liquid capacity formula would indicate. Since the correction factor is a function of the flow conditions and these in turn depend on orifice diameter, a trial orifice is usually calculated based on no correction. The next larger standard orifice is then used together with the graph for liquid viscosity correction given in Figure 5.14q to find the correction factor. This procedure is then repeated, using the corrected value as a trial value as often

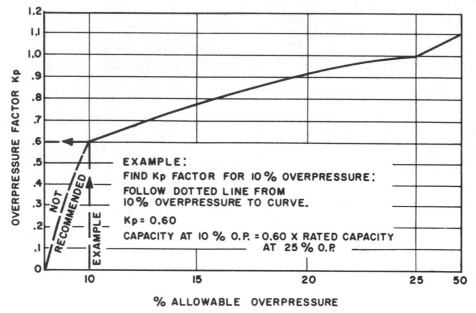

Fig. 5.14o   Overpressure sizing factor, other than 25% overpressure—conventional and balanced valves (liquids only)

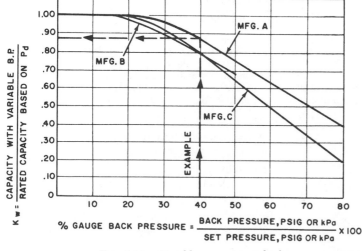

Fig. 5.14p   Variable or constant back pressure sizing factor, 25% overpressure—balanced valves (liquids only)

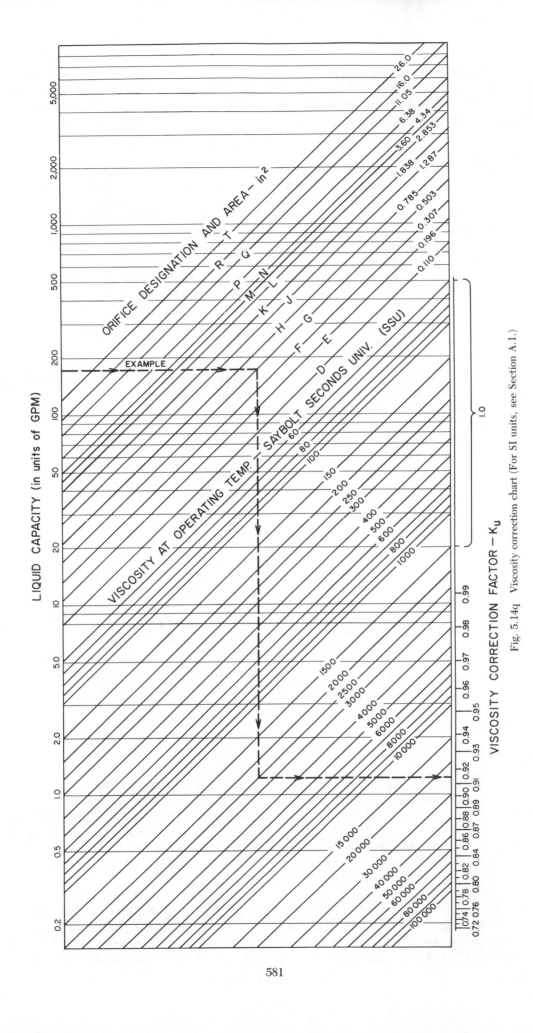

Fig. 5.14q Viscosity correction chart (For SI units, see Section A.1.)

as needed, until the corrected *required* orifice area is **less** than that of the standard orifice which was used to establish the K_u factor. The nomographs for liquid sizing are given in Figure 5.14r.

Flashing Fluids

Flashing fluids are a controversial subject in sizing relief valves. In the absence of experience in the specific service for instructions to the contrary, most manufacturers today use an enthalpy balance (isenthalpic flash) across the valve to calculate the respective volumetric rates of liquid and vapor downstream of the valve. Areas are then established independently for these two rates treated as all liquid and all vapor respectively, and a valve is selected whose discharge area is at least equal to the sum of these two calculated areas. While perhaps not theoretically refined, there have been no reported incidents where this procedure led to an undersized valve. Conceivably, it could lead to an oversized valve and possibly valve chatter. However, it is not considered good practice to install a discharge line smaller than the relief valve outlet and it is probable that the valve plus its discharge piping will come to a stable condition of balance between flash and the downstream pressure drop.

One word of caution in using this method: Liquid capacities are commonly tabulated and calculated at 25 percent overpressure while vapor capacities are commonly calculated and tabulated at 10 percent overpressure. Obviously, both sizing calculations must be made at the same overpressure with appropriate capacity correction factors applied as needed.

Steam Flow

While steam relief sizing is not very often encountered in the design of process plants, the sizing is straightforward once the required load and the relieving conditions are established. The various codes having jurisdiction need to be consulted in order to establish set pressures and allowable overpressure, valve location, etc., since these aspects are defined in considerably more detail than for process services. The basic steam sizing equation is a modification of Napiers

$$W_s = 51.5 A K P K_b K_{sh} \qquad 5.14(4)$$

where

W_s = required steam capacity, lbm/hr (kg/hr),

A = required nozzle area of the valve, sq. inches (mm²),

K = coefficient of discharge determined from test procedure (depends upon the Code and Section having jurisdiction over the valve's service),

P = relieving pressure = [set pressure × (1 + accumulation allowed by code having jurisdiction)] + atmospheric pressure (PSIA or Pa),

K_b = vapor gas correction for constant back pressure above critical pressure, and

K_{sh} = super heat correction factor (see manufacturer's tables).

Special Cases

There are a number of special cases (such as relieving polymerizing materials, handling high temperature hot water, Dowtherm systems, super pressure steam systems, cryogenic fluids, toxic materials relief, etc.) where manufacturers have developed special designs of valves and sizing experience or special sizing methods. It is recommended that the reader avail himself of as much of such advice and assistance as he can obtain. Frequently the recommendations of two manufacturers will not agree. The reader is thus alerted that he has a problem case and that further study may be warranted. Also refer to Section 4.16 on valve sizing.

Excerpts from ASME Unfired Pressure Vessel Code

UG-125(c)—All unfired pressure vessels other than unfired steam boilers shall be protected by pressure-relieving devices that will prevent the pressure from rising more than 10% above the maximum allowable working pressure, except when the excess pressure is caused by exposure to fire or other unexpected source of heat.

UG-125(d)—Where an additional hazard can be created by exposure of a pressure vessel to fire or other unexpected sources of external heat (for example, vessels used to store liquefied flammable gases), supplemental pressure-relieving devices shall be installed to protect against excessive pressure. Such supplemental pressure-relieving devices shall be capable of preventing the pressure from rising more than 20% above the maximum allowable working pressure of the vessel. A single pressure-relieving device may be used to satisfy the requirements of this paragraph and (c), provided it meets the requirements of both paragraphs.

UG-133(a)—When safety or relief valves are provided, they shall be set to blow at a pressure not exceeding the maximum allowable working pressure of the vessel at the operating temperature, except as permitted in (b). If the capacity is supplied in more than one safety or relief valve, only one valve need be set to open at a pressure not exceeding the maximum allowable working pressure of the vessel; the additional valves may be set to open at a higher pressure, but not to exceed 105% of the maximum allowable working pressure of the vessel. See Paragraph UG-125(c).

UG-133(b)—Protective devices permitted in Paragraph UG-125(d) as protection against excessive pressure caused by exposure to fire or other sources of external heat shall be set to operate at a pressure not in excess of 110% of the maximum allowable working pressure of

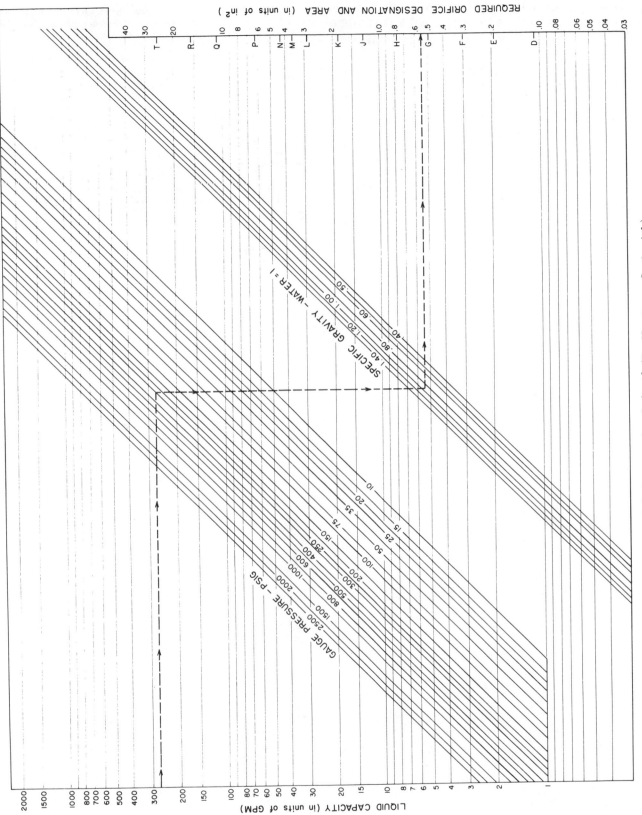

Fig. 5.14r   Relief valve sizing chart for liquids (For SI units, see Section A.1.)

583

the vessel. If such a device is used to meet the requirements of both Paragraphs UG-125(c) and UG-125(d), it shall be set to operate at not over the maximum allowable working pressure.

UG-133(f)—The set pressure tolerances, plus or minus, of safety or relief valves, shall not exceed 2 PSI (13.8 kPA) for pressures up to and including 70 PSIG (483 kPa), and 3% for pressures above 70 PSIG (483 kPa).

BIBLIOGRAPHY

API-PR-520, "Design and Installation of Pressure-Relieving Systems," American Petroleum Institute, New York, reaffirmed 1973.

API-RP-521, "Guide for Pressure Relief and Depressuring Systems," American Petroleum Institute, New York.

API Standard 526, "Flanged Steel Safety Relief Valves," American Petroleum Institute, New York.

API Standard 527, "Commercial Seat Tightness of Safety Relief Valves with Metal-to-Metal Seats," American Petroleum Institute, New York.

Bloch, B., "Emergency Venting," *Chemical Engineering*, January 22, 1962.

Bouilloud, P., "Calculation of Maximum Flow Rate Through Safety Valves," *British Chemical Engineering*, November, 1970.

Conison, J., "Factors in Sizing a Safe and Economical Vapor Relief System," API Paper, May 15, 1963.

Guide for Inspection of Refinery Equipment, Chapter XVI "Pressure-Relieving Devices," American Petroleum Institute, New York.

Isaacs, M., "Pressure-Relief Systems," *Chemical Engineering*, February 22, 1971.

Jenett, E., "Components of Pressure Relieving Systems," Parts 1 and 2, *Chemical Engineering*, July and August, 1963.

Rearich, J.S., "How to Design Pressure Relief Systems," Parts 1 and 2, *Hydrocarbon Processing*, August and September, 1969.

Weber, C.G., "Installation and Performance of Safety Valves," 16th Southeastern ISA Conference, April 17, 1970.

———, "Safety Relief Valve Sizing," ISA Paper, April 20, 1961.

Willardson, D.F., "How to Protect from Overpressure," *Instruments & Control Systems*, August, 1977.

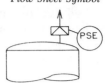
5.15 RUPTURE DISCS

Design Pressure:	2 to 100,000 PSIG (13.8 kPa to 69MPa) burst ratings available
Design Temperature:	To 1000°F (538°C) depending on materials
Sizes:	¼″ to 44″ (26.5 mm to 1.1 m)
Materials:	Most commercially available metals, either bare or plastic coated
Cost:	Approximately $100 per inch of diameter ($4 per mm of diameter) for smaller sizes and common materials of construction
Inaccuracy:	±5% of rated burst pressure
Partial List of Suppliers:	Metal: Autoclave Engineers, Inc.; BS&B Safety Systems, Inc.; Continental Disc Corp.; Crosby Valve Div., Geosource, Inc.; Engelhard Industries, Inc.; Fike Metal Products Corp.; Groth Equipment Corp.; Hydril Co.; Lamot Corp.; Super Pressure, Inc. Graphite: Carbone Corp.; Kearney Industries.

As discussed in Sections 5.13 and 5.14, any time there is a closed process system, there is the danger that system pressure will build up to an excessive level. In many cases overpressures can develop that are greater than the design or maximum allowable working pressure of one or more of the system components. Overpressure may occur due to thermal expansion, equipment or control failure, mis-operation, external fire, runaway reaction, or a combination of these. The rupture disc has been recognized for many years as a suitable device for relieving overpressure.

The rupture disc in its simplest form is a metallic or graphite membrane that is held between flanges and that is designed and manufactured to burst at some predetermined pressure. See Figure 5.15a.

Rupture discs, which are generally installed on or directly above a pressurized vessel nozzle, can be looked upon as the weak link in the pressure system. They will burst and relieve pressure before other system components fail.

Code Requirements

Several codes govern the use and installation of pressure safety devices, such as rupture discs. The most widely accepted code for pressurized vessels is the ASME Code

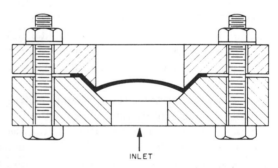

Fig. 5.15a   Metal disc and flanges

for Boilers and Pressure Vessels. This code can generally be assumed to apply throughout the United States except in areas where more stringent local codes exist. The ASME code recognizes rupture discs as means of relieving overpressure in pressurized vessels, and it states the requirements for applying these devices in Sections UG-125 through UG-134 and in Appendix M. Familiarity with this code is highly recommended. Two of the most important paragraphs in these sections are UG-125c and UG-125d. UG-125c reads, "All pressure vessels other than unfired steam boilers shall be protected by pressure relieving devices that will prevent the pressure from rising more than 10% above the maximum allowable working pressure, except when the excess pressure is caused by exposure to fire or other unexpected source of heat." To comply with this requirement, the relief device must be sized to prevent internal pressure accumulation over the maximum working pressure from exceeding 10 percent.

Paragraph UG-125d deals with vessels that may be exposed to fire. It states, "Where an additional hazard can be created by exposure of a pressure vessel to fire or other unexpected sources of external heat (for example, vessels used to store liquefied flammable gases), supplemental pressure-relieving devices shall be installed to protect against excessive pressure. Such supplemental pressure-relieving devices shall be capable of preventing the pressure from rising more than 20% above the maximum allowable working pressure of the vessel. A single pressure-relieving device may be used to satisfy the requirements of this paragraph and UG-125c, provided it meets the requirements of both paragraphs." To comply with this paragraph, the relief device must be sized to prevent internal pressures from rising above 20 percent over the maximum allowable working pressure when the vessel is exposed to fire. One relief device may be used for both requirements if it is sized for the more rigorous requirement.

Two additional ASME code paragraphs, UG-133a and b, deal with the set pressures of relief devices. These may be summarized as follows: primary or sole relief devices to meet the requirements of paragraph UG-125c or d must be set to open at the maximum allowable working pressure of the vessel. Supplemental devices to meet paragraph UG-125c may be set to open at 105 percent of maximum working pressure. Supplemental devices used to meet paragraph UG-125d requirements may be set to open at 110 percent of maximum allowable working pressure.

In other cases, relief devices are recommended, not on the basis of code requirement, but on the basis of good practice. Most important of these is on reactors where a runaway reaction is possible. Some runaway reactions occur as explosions in a fraction of a second and generate tremendous pressures and quantities of materials to be relieved. Primary relief devices may not be sized to handle the relief requirement of explosion-like runaways. For these cases, secondary relief devices may be specified and are set to burst at a higher pressure than the setting of the primary relief device.

Other good practice applications for relief devices are for overpressure relief due to thermal expansion. Thermal relief should be provided any time a liquid may be blocked in a system and the system subject to external heating. Examples of this are long piping runs that are exposed to the sun and the cooling side of heat exchangers or vessel jackets. For more details on this subject see Section 5.13.

Rupture Discs vs. Relief Valves

Rupture discs and relief valves can be used singly, in series, or in parallel. The greatest difference between relief valves and rupture discs is that relief valves are reclosing devices and rupture discs are not. As the process pressure starts to build, a relief valve will gradually open and then reseal the process once the pressure drops. The rupture disc will remain sealed tight until its burst pressure is reached. Then it will open to about 98 percent of its total area and stay open. In order to reclose the process, the disc must be replaced. It follows from this that the rupture disc can be used as a tight seal whereas the relief valve cannot. The only time a rupture disc will leak is if it develops pinhole corrosion, or if stress hairline cracks develop due to metal fatigue caused by pressure cycling.

A second area of difference is in the construction of the two relief devices. Because rupture discs have a relatively simple construction, they are commercially available in an extremely wide variety of materials at reasonable cost and on reasonable delivery schedules. This cannot be said for relief valves, which are generally limited to copper alloy, steel, and stainless steel constructions, and become very expensive, long-delivery items if special materials are required. Also, because of simple construction, the upstream side of a rupture disc can be protected against material accumulation, plugging, or freezing as illustrated on Figure 5.15b. Relief valves can plug or freeze if they are exposed to certain processes.

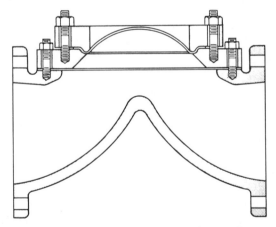

Fig. 5.15b   Non-plug, in-line rupture disc installation

There is one other small but important point. Both valves and discs are differential pressure devices in that they require a differential pressure between inlet and outlet in order to function. A rupture disc selected to burst at 100 PSIA (690 kPa) will actually burst when a differential pressure of $100 - 14.7 = 85.3$ PSI (588.57 kPa) appears across it. Anytime an unexpected back pressure can appear on the downstream side of a relief device, the ASME code requires that provision be made to overcome it or vent it to atmosphere. In the case of relief valves, a special seal may be obtained that enables the valve to open regardless of back pressure. However, for rupture discs, a separate venting valve must be installed on the downstream side. See Section 5.4 for a discussion on excess flow check valves.

When to Use a Rupture Disc

With the code requirements and the comparison between discs and valves in mind, the following rules can be used on where and when to specify rupture discs.

As a primary or sole relieving device (see Figure 5.15c)—The rupture disc may be used to relieve an inexpensive and inert material to air if process pressure loss can be tolerated. At the other extreme, rupture discs may be used to vent highly toxic, poisonous, or corrosive materials into a vent surge or flare header system. The advantage of the rupture disc for this application is that it will not allow any leakage under normal conditions.

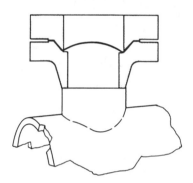

Fig. 5.15c   Rupture disc as primary relief device

As a supplemental relieving device (relief valve is primary)—Rupture discs are set at the higher relieving pressures as permitted by ASME code paragraphs UG-133a and b. The relief valve will open on mild overpressures, relieve a small amount of material, and reclose the process. The rupture disc will not function unless a more extreme condition arises. See Figure 5.15d. When such condition arises, the process must commonly be shut down. Therefore, complete system pressure loss due to disc rupture is not a serious drawback. Also, the causes of more extreme overpressure, such as external fire, generally require a larger venting area. For a given venting area, the rupture disc is less expensive than the relief valve.

As explosion protectors—Rupture discs are almost always the best selection. They open to about 98 percent of their total area in a fraction of a second and the same money will purchase more relief area if the rupture disc instead of a relief valve is used.

Upstream of a relief valve   This is a very useful application for rupture discs. See Figure 5.15e. Under normal conditions the rupture disc is sealed tight and protects the relief valve against corrosive, plugging, or freezing processes. If the maximum allowable working pressure is exceeded, the disc will break and the relief valve will start to relieve the pressure. As the pressure drops, the valve will shut and reclose the process. Thus, the best characteristics of both devices are utilized. Another important advantage in this type of installation is that the rupture disc may be used as a piping specification break. By this is meant that materials of construction for the rupture disc and the inlet flange of the disc must be compatible with the process. However, the rupture disc downstream flange, the relief valve, and all downstream piping may be garden-variety materials such as carbon steel. This is permissible because the rupture disc can be relied on to seal the process away from the downstream items. The resultant savings from using lower-grade downstream materials will in many cases more than pay for the rupture disc. Figures 5.15e and 5.15f illustrate these types of installation.

Graphite discs—Figure 5.15e shows the installation for a metal disc and Figure 5.15f shows the installation for a graphite disc. As can be noted, the graphite disc requires some extra piping to guarantee its successful

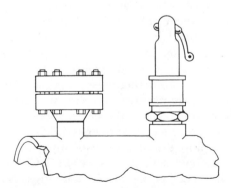

Fig. 5.15d   Rupture disc as a supplemental relief device

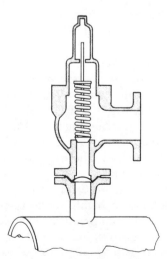

Fig. 5.15e   Metal rupture disc installed upstream of relief valve

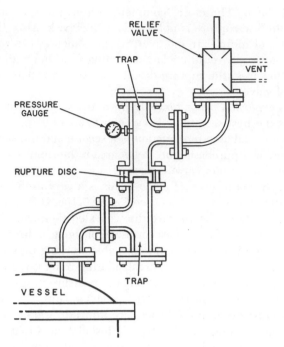

Fig. 5.15f Graphite rupture disc installed upstream of relief valve

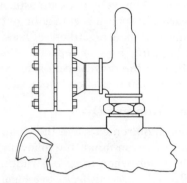

Fig. 5.15g Rupture disc downstream prevents relief valve leakage

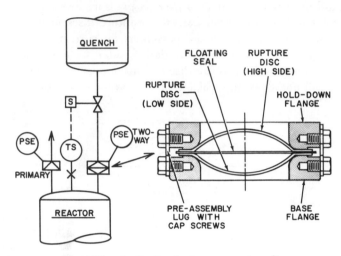

Fig. 5.15h Application for two-way rupture disc

operation. First of all, it is not good practice to allow a graphite disc to rupture directly into a relief valve. The reason is that a graphite disc will fragment rather than tear as a metal disc will. Since the graphite fragments can jam or restrict the operation of a relief valve, it is necessary to provide some kind of downstream pocketing to catch the fragments. A further difficulty arises once the relief valve recloses. If the graphite fragments are pocketed directly over the vessel nozzle, they will drop back into the tank. They can then get into the piping and damage pumps and instruments. For this reason, the entire relief assembly is shown offset from the vessel nozzle and an additional pocket is shown upstream of the graphite disc flanges. As is discussed later in this section, graphite discs are best suited for certain applications and the extra care required for their use is justified in those cases.

Downstream of a relief valve—This is a rare installation requirement that is necessary only when the relief valve discharge is piped into a vent header that might contain corrosives or where downstream relief valve leakage is to be prevented. See Figure 5.15g.

Special applications—Rupture discs may be used where two-way relief is required. There are commercially available discs for this purpose. One common installation sketch is shown in Figure 5.15h. The system requirements are as follows. A highly corrosive process is unstable above a certain temperature. Therefore, it is necessary to quench the reaction if the temperature rises above this point. Further, due to its corrosive nature, it is decided to vent the process into a surge system in case of process overpressure (rupture disc as primary relief).

This application calls for a rupture disc to vent overpressure and for a two-way rupture disc to isolate the process away from the quench valve and to prevent quench leakage into the process. The solution is to specify a two-way rupture disc that will burst into the tank if overtemperature occurs but isolate the valve under other circumstances.

Margin Between Operating and Set Pressures

There are several different constructions of rupture discs on the market to meet the applications just discussed. All constructions consist of the metal or graphite disc or with some extra features that make the particular construction suitable for a given service. One of the first problems that arises in rupture disc application is the allowable margin between normal rupture disc working pressure and rated bursting pressure. For a plain metal disc, under steady positive pressure, this margin is in the order of 66 percent of the bursting pressure. As an example, if a process is to be operated at 200 PSIA (1.38 MPa), the rupture disc and consequently all system components must be designed for 300 PSIA (2.07 MPa). This margin is necessary to prevent inelastic stress of the disc and premature failure. Obviously, it is expensive to build this kind of margin into process equipment, so considerable effort has been made to design a modified disc that will allow a closer working to rupture pressure ratio. One construction that improves the ratio is the graphite disc. Since graphite does not fatigue as metal does, most graphite disc suppliers recommend that the graphite disc be operated at 75 percent or even 80 percent of its rated bursting pressure. It might be well to note at this point that all critical or borderline rupture disc applications should be discussed with the manufacturer prior to making the installation, because published information may not be applicable to the particular service the user has in mind.

A second commercially available disc construction that may be used in the 75 to 80 percent range is as shown in Figure 5.15i. This disc consists of a relatively weak disc (seal) plus a heavier metallic back-up plate that has been scored or punched so that it will open like orange peels when burst pressure is reached. This construction is less subject to deterioration due to metal fatigue than the plain disc is.

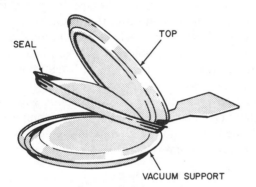

Fig. 5.15i   Composite rupture disc assembly reduces both set pressure and margin to operating pressure

A relatively new, patented rupture disc construction may be operated to 90 percent of the rated bursting pressure. See Figures 5.15j and k. This disc is a reverse buckling design. Rupture discs that are pre-bulged are normally installed with the bulge toward the outside. The reverse buckling disc is installed with the bulge facing into the process. Because of this, the disc is not as sensitive to high pressures and pressure cycling. Once the bursting pressure is reached, the reverse buckling

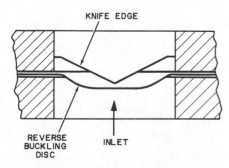

Fig. 5.15j   Reverse buckling, knife-edge design

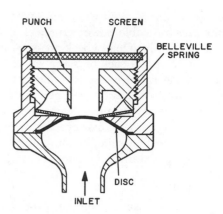

Fig. 5.15k   Belleville spring provides accurate back-loading for shear disc

disc will snap its pre-bulge to the downstream side. As the bulge passes from the up to downstream side it is cut into segments by sharp knife edges placed against the disc on the downstream side, thus allowing the process to vent. All discs mentioned have a tolerance on the bursting pressure. Tolerance of the reverse buckling disc is ±3 percent of the stamped bursting pressure and for conventional discs it is ±5 percent.

There are other disc constructions and methods of installation that allow higher operating to burst ratios. One method is to use two identical discs in series. The idea here is to assume that the first disc will fatigue and fail or leak prematurely because of high normal pressure operation, but the second one will then take over to operate satisfactorily for a period of time. The first disc must, of course, be replaced at the earliest opportunity. A pressure gauge with maximum pointer and an excess flow valve is installed in the pocket between the two discs. See Figure 5.15l.

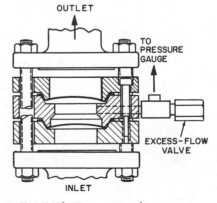

Fig. 5.15l   Two rupture discs in series

Another technique is to use explosive-actuated rupture discs that do not rupture due to the pressure forces alone. The explosion is ignited by a pressure switch that senses process overpressure and closes an electric circuit on the explosive charge. This is not a self-contained unit and its successful operation is dependent on several outside components plus a reliable power source.

Finally, it is possible to load a back pressure artificially on the rupture disc. The back pressure reduces the differential across the rupture disc and allows the normal operating pressure to be any desired percent of the rupture pressure. Back pressure is unloaded by an external venting system when process overpressure occurs. Here again, this is not a self-contained system; it relies on the proper operation of external components.

Vacuum Supports

Process vacuum pressure is another process property that must be considered when selecting a rupture disc. The reason for this is that if a vacuum appears on the upstream side of a rupture disc, the disc might collapse or implode. This is possible regardless of what the venting pressure rating is. Therefore, anytime a rupture disc is exposed to upstream vacuum, a vacuum support should be provided. A vacuum support is an assembly that supports the disc on the upstream face against collapse while adding no strength in the bursting direction. It is recommended that all rupture disc applications be checked to see if vacuum can be developed on the upstream side due to mis-operation or under normal conditions.

Pulsating or Low Pressures

Another process property to check is the frequency and amplitude of pressure cycling. As previously mentioned, disc life can be shortened by operating the disc too close to the rated rupture pressure. Pressure cycling can also shorten disc life because of metal fatigue. Disc life is a function of the severity of pressure frequency and amplitude cycling. When cycling is expected, the best selection is to use the reverse buckling knife edge design.

Sometimes it is not possible to obtain a specific rupture disc construction that will burst at low-enough pressures. For example, a 2 in. (50 mm) stainless steel disc has a minimum bursting pressure of approximately 160 PSIG (1.1 MPa) (see Table 5.15m). If the corrosion resistance of stainless steel is required, but the set pressure must be lower, it is necessary to specify some kind of laminated construction such as a teflon-faced aluminum disc or utilize the sealed low-pressure design shown on Figure 5.15i or, alternately, specify a larger diameter stainless steel disc.

Temperature and Corrosion Considerations

There are two other process characteristics that influence disc selection. One is temperature and the other corrosives. On temperature, the rupture disc is selected and sized to match the design temperature of the system. Since metals lose strength as they are heated, the disc material should be selected to withstand the highest expected process temperature with some to spare (Table 5.15m). High-temperature applications will also accelerate metal fatigue and creep. Thus temperature becomes a factor where high normal pressures or pressure cycling is encountered.

Table 5.15m
MINIMUM RUPTURE PRESSURES* AS A FUNCTION OF SIZE,
BASED ON 70°F (21°C) OPERATING TEMPERATURE AND ON
STANDARD DISC DESIGN

Size (in.)	Aluminum	Aluminum Lead Lined	Copper	Copper Lead Lined	Silver	Platinum	Nickel	Monel	Inconel	321 or 347 Stainless
$\frac{1}{4}$	310	405	500	650	485	500	950	1085	1550	1600
$\frac{1}{2}$	100	160	250	330	250	250	450	530	775	820
1	55	84	120	175	125	140	230	265	410	435
$1\frac{1}{2}$	40	60	85	120	85	120	150	180	260	280
2	33	44	50	65	50	65	95	105	150	160
3	23	31	35	50	35	45	63	74	105	115
4	15	21	28	40	28	35	51	58	82	90
6	12	17	25	25	24	26	37	43	61	70
8	9	19	35	35	27	—	30	34	48	55
10	7	16	42	42	—	—	47	28	—	45
12	6	10	55	55	—	—	—	360	—	45
16	5	8	55	55	—	—	—	270	—	33
20	3	8	70	70	—	—	—	215	—	27
24	3	8	60	60	—	—	—	178	—	65
Maximum Recommended Temperature, °F (°C)	250 (121)	250 (121)	250 (121)	250 (121)	250 (121)	600 (315)	750 (399)	800 (427)	900 (482)	600 (315)

* See Section A.1 for SI units.

Rupture discs can be manufactured of almost any metal or graphite. Graphite is one good answer to corrosion problems because it resists almost all chemicals. It is also possible to get pure or plated discs made of such materials as Hastelloy "B," tantalum, platinum, and inconel, to name a few. The more common disc materials may be lead, teflon, Kel-F, or polyethylene coated for corrosion resistance. In general, obtaining suitable corrosion resistant rupture discs is not a serious problem.

Disc Flanges and Accessories

In order for rupture discs to perform as specified, they must be installed between special flanges. The exception to this rule is the graphite disc, which is held by a shop-fabricated collar, and the collar in turn held by standard pipe flanges. For metal discs, the design of the flange throat and shoulders is important and, therefore, standard piping flanges are not sufficient. Metal rupture disc flanges come in a wide variety of materials and constructions. They can be made up into screwed, flanged, or welded pipe, and the inlet flange connection may be specified different from the outlet flange connection. For added versatility and cost reduction, there is a rupture disc flange assembly that may be installed inside the bolt hole circle of the standard piping flanges. See Figure 5.15n. In specifying rupture disc flanges, it should be remembered that the downstream flange can be of lower-grade material, such as carbon steel, since it will not ordinarily be in contact with the process.

Two accessories are available with rupture disc flanges that may be specified as required. Jack screws are used to jack the rupture disc joint apart so that the disc can be replaced and they should be considered on all installations of 2 in. (50 mm) and above. The second is the downstream flange tap that can be used for the excess flow valve and pressure gauge installation.

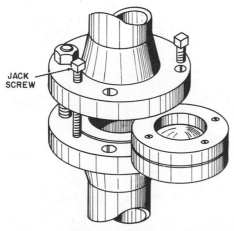

Fig. 5.15n   Insert type rupture disc flanges fit inside the bolt circle of the pipe flanges

Sizing

Rupture discs may be sized in accordance with the following formulas:

For vapor:

$$d = \sqrt{\frac{W}{146\,P}}\ \sqrt[4]{\frac{I}{M_w}} \qquad 5.15(1)$$

For steam:

Dry and saturated:

$$d = 0.205\ \sqrt{\frac{W}{P}} \qquad 5.15(2)$$

Superheated:

$$d = 0.205\ \sqrt{\frac{W(1\ +\ 0.00065Ts)}{P}} \qquad 5.15(3)$$

Wet:

$$d = 0.205\ \sqrt{\frac{W(1\ -\ 0.012y)}{P}} \qquad 5.15(4)$$

For liquids:

$$d = 0.236\sqrt{Q}\ \sqrt[4]{\frac{SG}{P_1}} \qquad 5.15(5)$$

where

d = minimum rupture disc diameter, inches,*

M_w = molecular weight,

P = relieving pressure, PSIA, including allowable accumulation (10% in normal conditions, 20% in fire conditions),

P_1 = relieving pressure, PSIG, including allowable accumulation,

Q = relieving rate, gpm, ———,

SG = liquid specific gravity where water = 1.0,

T = relieving temperature, °R (460 + °F),

Ts = degrees of superheat, °F,

W = relieving rate, lbm/hr, and

y = percent moisture (100 minus steam quality).

Note: Where rupture discs are installed upstream of a relief valve the rupture disc is normally the same size as the relief valve inlet nozzle.

BIBLIOGRAPHY

API RP 521, "Guide for Pressure Relief and Depressuring Systems," American Petroleum Institute, New York.
Block, B., "Emergency Venting," *Chemical Engineering,* January 22, 1962.

* For SI units, see Section A.1.

Diss, E., "Practical Way to Size Safety Discs," *Chemical Engineering*, September 18, 1961.

Guide for Inspection of Refinery Equipment, Chapter XVI, "Pressure-Relieving Devices," American Petroleum Institute, New York.

Jenett, E., "Components of Pressure Relieving Systems," Parts 1 and 2, *Chemical Engineering*, July and August, 1963.

Lipták, B.G., "Rupture Disk Systems," *Chemical Engineering*, July 5, 1965.

Myers, J.F., "Enhancing Accuracy of Rupture Disks," *Chemical Engineering*, November 8, 1965.

"Pushbutton Changer for Ruptured Pressure Disc," *Chemical Processing*, January, 1976.

Resnick, J.S., "How to Design Pressure Relief Systems," Parts 1 and 2, *Hydrocarbon Processing*, August and September, 1969.

Solter, R.L., "Selection and Sizing of Rupture Units," *Instruments & Control Systems*, September, 1964.

Chapter VI

LOGIC DEVICES AND PLCS

W. N. Clare • J. Frydman

R. A. Gilbert • G. T. Kaplan

G. Platt • H. C. Roberts

D. R. Sadlon • A. C. Wiktorowicz

CONTENTS OF CHAPTER VI

6.1 ISA LOGIC SYMBOLS

Introduction

This section lists the symbols used to denote binary (on-off) process operations and illustrates a typical application of the symbols to a plant process. Logic symbol diagramming is applicable to any process control system that uses switching devices to initiate normal or emergency operations. The method is primarily process-based rather than hardware-based. It describes operations in terms of the essential process functions that can be carried out by any class of hardware, whether electric, pneumatic, hydraulic, or other. The method is directed to the needs of an engineer who may have only a rudimentary knowledge of hardware circuit design but who knows what the process-sensing instruments are and how the process is supposed to operate. The hardware and circuit subfunctions needed to perform the process functions then can be detailed by the circuit designer as necessary to satisfy the instrument engineer's intentions.

Logic symbol diagrams are appropriate whenever the operating requirements of the process have to be described to operating personnel, maintenance workers, designers, or others and are particularly useful for group discussions. Knowledge of how to read relatively complex and specialized circuit diagrams is not required. However, in cases in which it is necessary to trace the actions of a circuit in detail, there is usually no substitute for an exacting circuit diagram.

Use of Logic Symbols

A logic diagram may be more or less detailed, depending on its intended use. The amount of detail in a logic diagram depends on the degree of refinement of the logic and on whether auxiliary, essentially non-logic, information is included. For example, a logic system may have two opposing inputs, a command to open and a command to close, which do not normally exist simultaneously. The logic diagram may or may not go so far as to specify the outcome if both the commands were to

Appreciation is expressed to Instrument Society of America for permission to abstract from their publication S5.2, entitled "Binary Logic Diagrams for Process Operations," Instrument Society of America © 1973.

exist at the same time. In addition, explanatory notes may be added to the diagram to record the logic rationale. Non-logic information (reference document identification, tag numbers, terminal markings, etc.) may also be added, if desired.

The existence of a logic signal may correspond physically to either the existence or the nonexistence of an instrument signal, depending on the particular type of hardware system and the circuit design philosophy that are selected. For example, a designer may choose a high-flow alarm to be actuated by an electric switch whose contacts open on high flow; on the other hand, the high-flow alarm may be designed to be actuated by an electric switch whose contacts close on high flow. Thus, the high-flow condition may be represented physically by the absence of an electric signal or by the presence of the electric signal. The logic diagram does not attempt to relate the logic signal to an instrument signal of any specific kind.

The flow of information is represented by lines that interconnect logic statements. The normal direction of flow is from left to right, or from top to bottom. Arrowheads may be added to the flow lines wherever needed for clarity and must be added to lines whose flow is not in a normal direction.

A summary of the status of an operating system may be put in the diagram wherever it is deemed useful as a reference point in the logic sequence.

A specified logic condition may be misunderstood when it involves a device that may have more than two specific alternative states. For example, if it is stated that a valve is not closed, this could mean either (a) that the valve is open fully or (b) that the valve is simply not closed: It may be in any position from almost closed to wide open. To aid accurate communication between writer and reader of the logic diagram, the diagram should be interpreted literally. Therefore, possibility (b) is the correct one.

If a valve is an open-close valve, it is necessary to do one of the following to avoid misunderstanding.

1. Develop the logic diagram in such a way that it says exactly what is intended. If the valve is intended to be open, then this should be so stated;

596

the valve should not be described as being not closed.

2. Have a separate note specifying that the valve always assumes either the fully closed or the fully open position.

By contrast, a device such as a motor-driven pump is either operating or stopped, barring some special situations. To say that the pump is not operating usually clearly denotes that it has stopped.

The following definitions apply to devices that have open, closed, or intermediate positions. The positions stated are nominal to the extent that there are differential-gap and dead band in the instrument that senses the position of the device.

Open position: a position that is 100 percent open.

Not-open position: a position that is less than 100-percent open. A dvice that is not open may or may not be closed.

Closed position: a position that is 0-percent open.

Not-closed position: a position that is more than 0-percent open. A device that is not closed may or may not be open.

Intermediate position: a specified position that is greater than 0 percent and less than 100 percent open.

Not-at-intermediate position: a position that is either above or below the *specified* intermediate position.

For a logic system having an input statement that is derived inferentially or indirectly, a condition may arise that will lead to an erroneous conclusion. For example, an assumption that flow exists because a pump motor is energized may be false because of a closed valve, a broken shaft, or another mishap. Statements based on positive measurements confirming that a certain condition actually exists or does not exist are generally more reliable.

A process operation may be affected by loss of the power supply—electric, pneumatic, or other—to memories and to other logic elements. In order to take such possibilities into account, it may be necessary to consider the effect of power loss to any logic component or to the entire logic system. In such cases, power supply or loss of power supply should be entered as logic inputs to a system or to individual logic elements. For memories, the power supply may be entered as a logic input or as shown in the diagrams. The effect of power supply restoration also might need to be shown. Logic diagrams do not necessarily have to cover the effect of logic power supplies on process systems but may do so for thoroughness.

It is recommended for clarity that a single time-function symbol be used to represent each time function in its entirety. Though not incorrect, the representation of a complex or uncommon time function by the use of one time-function symbol in immediate sequence with a second time-function symbol or with a NOT symbol should be avoided.

Definitions

Table 6.1a illustrates and defines the logic symbols and some typical uses of them. The symbols shown with three inputs, A, B, and C, are typical for the logic functions having any number of two or more inputs. In the several truth tables, 0 denotes the nonexistence of the logic input or output signal or state given at the head of the column. 1 denotes the existence of the logic input signal or state. D denotes the existence of the logic output signal or state as a result of appropriate logic inputs.

Application to a Process

The process must have high vacuum to proceed properly. Vacuum is normally maintained by an air ejector, but in case of failure or overload of the air ejector the system pressure rises. The rise is sensed by a pressure switch (PSH), which automatically starts a vacuum pump, provided that a hand-actuated control switch (HS) for the pump motor is in the AUTOMATIC position. This switch also can be used to start and stop the pump manually. However, the pump is not permitted to start or run if the discharge temperature, as sensed by a temperature switch (TSH), is high or if the motor is overloaded and its circuit breaker is not manually reset. If high pressure is maintained for ten minutes, a high-pressure alarm (PAH) is actuated. High temperature is signaled by another alarm (TAH). Pump motor overload is signaled by the alarm (IAH). If the pump control logic circuit loses power, the pump shall stop automatically but shall not be able to be restarted until the system is reset manually.

Whenever the pump is required to operate, cooling water is automatically turned on. The water flow is controlled by an air-actuated control valve (UV), which is operated by a solenoid valve (UY) that, in turn, is operated by auxiliary contacts of the pump motor circuit breaker. The water is automatically turned off when the pump is stopped. The following instruments are on the instrument board:

HS	Manual control switch for pump operation. The switch has three momentary-contact pushbuttons for START, AUTOMATIC, and STOP.
PAH	Alarm actuated upon rise of pressure to abnormal value. However, this alarm is blocked for ten minutes after a pump start is required.
TAH	Alarm that is actuated if pump discharge temperature rises to abnormal value.
XL-A	Green pilot light denoting that the pump motor circuit breaker is not closed, i.e., that pump is not operating.

Table 6.1a
LOGIC SYMBOLS

Function	Symbol	Definition and Truth Table	Example
INPUT	(Statement of input) ⊢ Statement may be preceded by instrument balloon with tag number	Logic sequence input	Start chemical injection by pushbutton: Start injection manually ⊢
OUTPUT	(Statement of output) ⊢ Statement may be followed by instrument balloon with tag number	Logic sequence output	The logic sequence causes drawoff to cease: Stop drawoff ⊢
AND	A ⊢ [A] ⊢ D B ⊢ C ⊢	Output D exists only if and while inputs A, B, and C exist (truth table) C: A B 0 1 0 0 0 0 0 1 0 0 1 0 0 0 1 1 0 D	Operate pump if feed tank level is high and provided that discharge valve is open: FIELD TANK LEVEL HIGH ⊢ [A] ⊢ OPERATE PUMP DISCHARGE VALVE OPEN ⊢
OR	A ⊢ [OR] ⊢ D B ⊢ C ⊢	Output D exists only if and while one or more of inputs A, B, and C exist (truth table) C: A B 0 1 0 0 0 D 0 1 D D 1 0 D D 1 1 D D	Stop compressor if cooling water pressure is low or bearing temperature is high: WATER PRESSURE LOW ⊢ [OR] ⊢ STOP COMPRESSOR BEARING TEMPERATURE HIGH ⊢
QUALI-FIED OR	A ⊢ [*] ⊢ D B ⊢ C ⊢ *Insert numerical quantity (see "Definition" column).	Output D exists only if and while a specified number of inputs A, B, and C exist. Mathematical symbols shall be used, as appropriate, in specifying the number of inputs, e.g., ≥ 3, to denote 3 or more.	Operate feeder if and while two and only two mills are in service: MILL N ⊢ MILL E ⊢ [=2] ⊢ OPERATE FEEDER MILL S ⊢ MILL W ⊢
NOT	A ⊢—◯—⊢ B	Output B exists only if and while input A does not exist.	Close valve if and while pressure is not high: PRESSURE HIGH ⊢—◯—⊢ CLOSE VALVE
FLIP-FLOP MEMORY	A ⊢ [S] ⊢ C B ⊢ [R] ⊢ D* *If output D is not used, it shall not be shown.	S denotes set memory and R denotes reset memory. Output C exists as soon as A exists. C continues to exist, regardless of the subsequent state of A, until the memory is reset, i.e., terminated by B existing. C remains terminated, regardless of the subsequent state of B, until A causes the memory to be set.	If standby pump operation is initiated, the pump shall operate, even on loss of the logic power supply, until the process sequence is terminated. The pump shall operate if START and STOP commands exist simultaneously.

Table 6.1a *Continued*
LOGIC SYMBOLS

Function	Symbol	Definition and Truth Table	Example

Output *D*, if used, exists when *C* does not exist, and *D* does not exist when *C* exists.

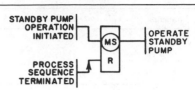

Input-Override Option

If inputs *A* and *B* exist simultaneously, and if *A* is then required to override *B*, then *S* should be encircled, i.e., $\textcircled{S}$; if *B* is to override *A*, then *R* should be encircled.

Loss-Of-Power-Supply Option

Required action of the memory on loss of power shall be symbolized by modifying the *set* letter, *S*, as follows:

Modified Symbol	Required Memory Action On Loss Of Power
S (un-changed)	Was not considered by logic designer
LS	Memory lost
MS	Memory maintained
NS	Is not significant, no preference

The use of a logic feedback to symbolize a memory that is lost in the event of loss of power is deprecated. Thus, the following shall not be used:

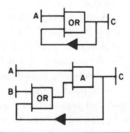

TIME ELEMENT

*Insert symbolism for specific functions and time interval (see "Definition" column).

Basic Method

This uses the following specific symbols:

Symbol	Meaning
DI	*Delay Initiation* of output. The continuous existence of input *A* for a specified time causes output *B* to exist when the time expires. *B* terminates when *A* terminates.
DT	*Delay Termination* of output. The existence of input *A* causes output *B* to exist immediately. *B* terminates when *A* has terminated and has not again existed for a specified time.
PO	*Pulse Output.* The existence of input *A* causes output *B* to exist

If vessel purge fails even momentarily, operate evacuation pump for 3 minutes and then stop the pump.

Table 6.1a *Continued*
LOGIC SYMBOLS

Function	Symbol	Definition and Truth Table	Example

immediately. *B* exists for a specified time, regardless of the state of *A*, and then terminates.

Generalized Method
This is suitable for all time functions. The following illustrations are typical but not all-inclusive.

Input logic state exists.
Input logic state does not exist.

Output logic state exists.
Output logic state does not exist.

Steam is turned on for 15 minutes beginning 6 minutes after agitator has stopped except that the steam shall be turned off if the agitator restarts.

The time at which the logic input *A* is initiated is represented by the left-hand edge of the box. Passage of time is from left to right and is usually shown unscaled.

The logic output *B* always begins and ends in the same state within the time element box.

More than one output may be shown, if required.

The timing of logic may be applied to either the existence state or the nonexistence state, as applicable.

*Show action of output *B* as required.

The action of output *B* depends on how long input *A* is in continuous existence, up to the line break for *A*. Beyond the break in *A*, the state of *A* is not significant to the completion of the *B* sequence. If a *B* time segment is required to go to completion only if *A* exists continuously, then *A* must be drawn beyond that segment. If *A* is drawn past the beginning but not beyond the end of a time segment, then the segment will be initiated and go to completion regardless of whether *A* exists only momentarily or longer.

If pH is low continuously for ½ minute, add caustic for 3 minutes.

SPECIAL

*Insert statement of special logic requirements.

Output *B* exists with a logic relationship to input *A* as specified in the statement of special requirements.

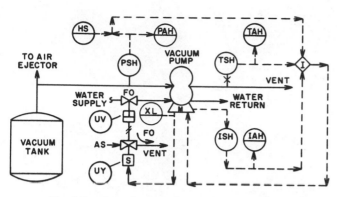

Fig. 6.1b   Control system for standby vacuum pump

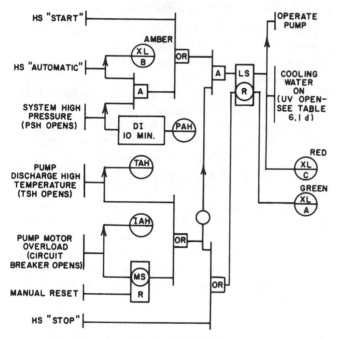

Fig. 6.1c   Logic diagram for standby vacuum pump

XL-B   Amber pilot light denoting that the pump is ready for an automatic start.

XL-C   Red pilot light denoting that the pump motor circuit breaker is closed, i.e., that pump is operating.

IAH   Alarm that is actuated upon current overload of pump motor.

The required process operations are diagrammed in Figure 6.1b, and Figure 6.1c and Table 6.1d describe logic functions.

Table 6.1d
CONTROL VALVE OPERATING SEQUENCE

Vacuum Pump	Motor Circuit Breaker Auxiliary Contacts	Solenoid Valve (UY) Coil	Control Valve (UV) Actuator	Port	Cooling Water
Off	Closed	Energized	Pressurized	Closed	Off
On	Open	De-energized	Vented	Open	On

BIBLIOGRAPHY

American National Standard Institute, "Graphic Symbols for Logic Diagrams," ANSI–Y32.14, 1973.

"Binary Logic Diagrams for Process Operations." Instrument Society of America, ANSI/ISA–S5.2, 1976 (R 1981).

Ellermeyer, W., "Manipulate Logic Without Boolean Algebra," *Control Engineering*, December 1969.

Platt, G., "Describing Interlock Sequences," *ISA Journal*, January 1966.

Steve, E. H., "Logic Diagram Boosts Process-Control Efficiency," *Chemical Engineering*, March 8, 1971.

6.2 LADDER DIAGRAMS

Ladder diagrams are one traditional method for describing control circuits. There are a few basic ladder diagram symbols that are used to express the meaning and purpose of a control circuit. This section presents some of the common symbols used in the ladder diagrams. It also presents examples that illustrate the connection between the process and the ladder diagram and indicate the procedure for analyzing a ladder diagram to determine its control functions.

Ladder Diagram Symbols

Figure 6.2a presents some of the common symbols that are used in ladder diagrams and will be studied in this section. The first symbol represents a normally open contact that is automatically operated. This symbol may signify the contacts on a starter, a limit switch, a relay switch, or any other automatic pilot device. The next symbol represents the normally closed contact switch, which is also automatically operated. The third symbol signifies normally open manually operated switches. This symbol, along with its normally closed version (shown as symbol 4), is generally used to indicate pushbutton-type switches. Symbol 5 represents a starter coil, a relay coil, or a solenoid coil (see Section 6.5). This symbol often includes a descriptive letter to indicate which type of coil is involved. The heating element of an overload device is represented by symbol 6, and symbol 7 signifies the contacts of a time delay relay (see Section 6.7). Although there are other ladder diagram symbols, the set presented here is sufficient to enable the reader to acquire an introduction to ladder diagrams and to understand the drawings presented in this section.

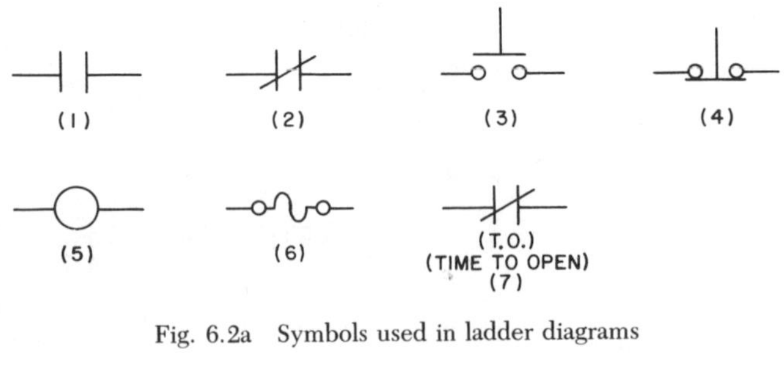

Ladder Diagram Development

One can appreciate the utility of ladder diagrams by reviewing an example and connecting it to a process step. Figure 6.2b illustrates a water tank portion of a process and the ladder diagram for its automatic control circuit. The figure shows a water storage tank, a pressure tank, a pump, and an assortment of pilot devices to provide control for the system. The ladder diagram for the system indicates that the control circuit has an automatic as well as a manual mode. The manual control in this instance is simply a start-and-stop pushbutton (P.B.) together with safety overloads for the pump. These elements are shown in the lowest horizontal line of the ladder diagram and have been arranged so that the electrical connection from L_1 to L_2 requires the stop P.B. to be inactive, the start P.B. to be pressed, and the mode selection switch to be in the A position. The normally open contacts that are activated when the pump relay is energized allow the

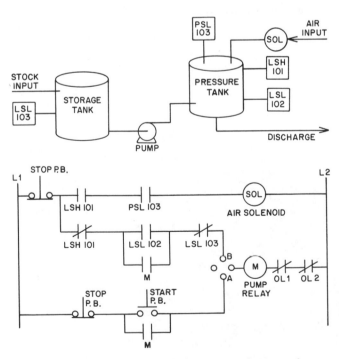

Fig. 6.2a Symbols used in ladder diagrams

Fig. 6.2b Automatic control of a pressurized water tank

602

pump to remain engaged after the start P.B. has been released.

The requirements for automatic control of this water system are more complex. The drawing for the system presented in Figure 6.2b indicates that there is high- and low-level monitoring as well as pressure control for the pressure tank and low-level monitoring for the storage tank. To obtain a ladder diagram for the control scheme, one must arrange the symbols for each of the control elements on the drawing in the proper electrical sequence. The water pump is energized only when each of the control constraints has been satisfied and the connection from L_1 through the pump starter relay to L_2 has been completed.

The automatic portion of this water control circuit is presented in the top half of the ladder diagram in Figure 6.2b. In the development of this diagram, the electrical connection between L_1 and L_2 was created with the control element symbols placed in between. The stop P.B. is usually located near L_1 before any branches are added to the drawing. The high-level sensor for the pressure tank (LSH 101) has two sets of contacts: a normally open set to control the operation of the air solenoid and a normally closed set to provide a condition for pump operation. The operation of the pump is also governed by the low-level sensors on the pressure and storage tanks, i.e., LSL 102 and LSL 103. The symbols for these elements must be placed on the drawing to indicate that the pump starter coil is energized when the water is low in the pressure tank but deactivated when the water is low in the storage tank. This requirement, which prevents pump damage because of low water levels in the storage tank and guarantees pressure stability in the pressure tank, is met by placement of LSL 103 and LSL 102 in series with the starter and the normally closed LSH 101 contacts. To provide continual pump operation after LSL 102 is momentarily activated, an interlock with a set of pump relay contacts must be provided.

These level and pump control element symbols are presented in the middle portion of the water system ladder diagram. Note that the mode selection switch is placed near the starter relay and the normally closed contacts of LSL 103 and LSH 101 are placed in series with the normally open LSL 102 contacts. As with the start P.B. in the manual portion of the drawing, a set of normally open starter relay contacts circumvent LSL 102 to ensure that the starter relay remains active when the liquid level rises above the minimum but remains below the maximum set point for the pressure tank. LSH 101 also controls the relay state and disengages the relay when the pressure tank water level reaches the high set point. This high water mark activates the normally open LSH 101 contacts, which allows the air solenoid to be energized (provided that PSL 103 indicates the tank pressure is below the desired setting).

In summary, the method of developing a ladder dia-

gram for a control scheme is to review the control functions required in the circuit, to select the control elements that accomplish this task, and then to arrange the symbols for concurrent control constraints in a sequential fashion between the voltage lines on the drawing. When necessary, interlock contacts should be provided around start pushbuttons and other momentary contacts. Stop pushbuttons and other safety interlocks should be arranged so that they are electrically close to the source voltage line. Each component of the drawing should be labeled so that it is easily associated with the actual device in the process. Particular attention should be paid to identifying all of the contact set for a specific relay coil. Finally, the normally open or closed condition of all automatic contacts should always be indicated when they are in their idle or inactive state.

Ladder Diagram Analysis

The basic procedure for the analysis of a control circuit ladder diagram is to consider the diagram one component at a time and to decide what occurs if a pushbutton is engaged or a contact switch is made or broken. If the diagram is reviewed from this point of view, with the realization that such contact activity usually closes or opens complete circuits from one voltage line through a coil to the other, then a particular coil will be energized or de-energized, depending on the continuity of the circuit. When the circuit is complete to any particular coil, that contactor, relay, or starter is energized and its contacts, wherever they may be, are now opposite their normal position. If they are normally closed contacts, they are now open, and if they are normally open, they are now closed. When a time delay relay is used in the circuit, its contacts are opened or closed after some delay, depending on their normal position and function in the circuit. If relays are used in the circuit, it is important to consider every contact that is operated by that relay whenever the coil is energized. Failure to consider all contact sets of a relay will result in a misconception of the function of the circuit. Finally, when analyzing a circuit, one should be certain that every component is considered in its normal and energized position so that the whole operation of the complete circuit can be comprehended.

Figure 6.2c contains the ladder diagram for a dynamic braking circuit of a motor.[1] This diagram can be investigated by use of the basic ladder diagram analysis procedures presented in this section. The example also provides insight into the concept of dynamic braking. Dynamic braking can be applied to any equipment when a smooth, fast stop is required or when it is desired to have the motor shaft free from manual rotation when the power is disconnected. A dynamic braking system provides a stop without any tendency to reverse and produces less shock to motor drive components than does plug stop braking.

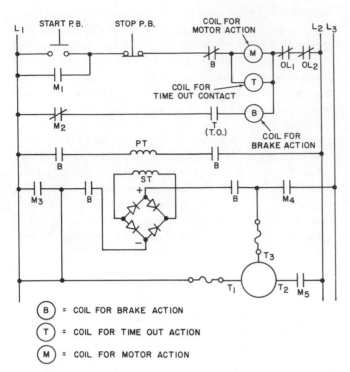

(B) = COIL FOR BRAKE ACTION

(T) = COIL FOR TIME OUT ACTION

(M) = COIL FOR MOTOR ACTION

Fig. 6.2c   Ladder of a motor diagram for a dynamic braking circuit

In a dynamic braking system a DC voltage is applied to the rotating motor to provide a smooth but positive braking action and to bring the motor to a rapid stop. The DC signal is removed when the motor is almost stopped to prevent any motor winding damage caused by overheating as a result of excessive current flow in the low-resistance windings. The diode rectifier element that produces the DC voltage is represented in the lower portion of the ladder diagram in Figure 6.2a. This section of the diagram also shows that the rectifier is isolated from motor terminals T_1 and T_3 by two sets of normally open brake coil contacts, B. There are also two sets of normally open brake contacts that isolate the primary side of the transformer from power lines L_1 and L_2. The fifth contact controlled by the brake coil is a normally closed set that is in series with the motor relay coil and the start pushbutton.

If the motor start P.B. is depressed (provided that the safety overload contacts are closed) and the stop P.B. is not also closed, then motor relay coil is engaged and the five sets of motor relay contacts are activated. The first of these motor relay contacts, M_1, is illustrated in the top horizontal line of the ladder diagram. This normally open contact closes when the coil is energized to guarantee that the coil remains hot even after the start P.B. is released. The second motor contact set, M_2, which is located on the drawing just below the first, is normally

closed but is opened on motor coil activation to prevent the brake coil from receiving the L_1 signal. It should be noted that the normally open time out contact, T, is closed because the time out relay is also energized when the motor coil is active. The last three normally open motor contact sets are connected to the motor terminals and, when closed, provide a connection from the power lines to the three motor terminals.

The braking portion of the ladder diagram includes symbols for the brake relay, the time out relay, the rectifier, the primary and secondary transformer elements, and the five brake contact sets. Figure 6.2c indicates that motor terminals T_1 and T_3 are connected to the DC terminals of the rectifier when the stop P.B. is pressed. When the stop P.B. is engaged, the motor relay is deactivated, the motor is isolated from the AC lines, the time out relay is dropped from the circuit, and the brake contacts are all activated. The brake relay is energized because M_2 has returned to its idle state and the time out contact, T, remains closed for the time out period. Thus, the activated brake coil keeps the four normally open B contacts associated with the transformer and the rectifier closed until the time delay relay has timed out to force T to its normally open position. The actual time out period is determined as a function of the time needed to bring the motor to a near stop.

In summary, the basic procedure for ladder diagram analysis involves the consideration of the circuit one component at a time to decide what happens to that component when a contact is closed, the determination of the function of each component in its normal and energized positions and, finally, the role of each component in relation to the other elements of the diagram. It is important to perform a complete analysis of the diagram without jumping to conclusions halfway through the analysis. A hurried analysis is usually disastrous, because the consideration of just one additional contact set can easily change the basic nature of the circuit.

REFERENCE

1. Liptak, B. G., ed, *Instrument Engineers' Handbook*, Vol. II, Chilton, Philadelphia, 1969, Chap. 2.

BIBLIOGRAPHY

Ellermeyer, W., "Manipulate Logic Without Boolean Algebra," *Control Engineering*, December 1969.

McIntyre, R. L., *Electric Motor Control Fundamentals*, McGraw-Hill, New York, 1960.

Platt, G., "Describing Interlock Sequences," *ISA Journal*, January 1966.

6.3 OPTIMIZATION OF LOGIC CIRCUITS

An important function of logic design is to establish the interface between the equipment, which defines a process, and the engineer, who establishes the operational parameters for that process. Classically, the design and optimization of logic circuits have centered on the synthesis of switching element (static or relay-type) interconnections that will satisfy the switching functions required by the process. However, the 1980s have introduced the use of logic circuits as the interface between the process and the process computer. The greater demand for computer control of industrial processes has increased the need for the instrument and control engineer who can create logic interface circuits that are optimally designed for reliable service.

This section outlines the fundamental concepts required to develop logic functions for logic circuits.[1] The various alphabetic and graphic logic symbols, their numerical assignments, their truth tables, and other tools that help the engineer create logic circuits are presented. The role of logic maps and Boolean algebra in removing logic redundancies and simplifying logic circuits is also outlined in this section. Since logic simplification always means less hardware, circuits that are less expensive and more reliable, and facilitation of trouble shooting, the new circuit designer is encouraged to expand his knowledge in these two areas beyond the introductory material presented here.

Logic Elements and Truth Tables

The basic logic components and operations are straightforward and have their origins in ancient Greek philosophy. Modern engineering uses of logic expand the application of the AND, NAND, OR, and NOR operations to decision making—e.g., whether equipment is to be turned on or off, a voltage level is to be high or low, or a valve is to be opened or closed.

Logic signals or logic variables, such as the ones into or out of logic switching elements, can be represented by any combinations of the symbols A,B,C through X,Y,Z. The values for these signals are often assigned as "1" or "0." Thus, the statement, "Liquid is pumped into the tank only when its level is below the high- and low-level set points of the tank," can be expressed as

$$Z = 1 \text{ only if LSL} = 1 \text{ and LSH} = 0 \qquad 6.3(1)$$

where Z represents the pump and LSL and LSH stand for level sense low and level sense high and symbolize the status of the low-level and high-level set points of the tank. Thus, LSL = 1 reveals that the liquid has dropped below the low level set point, LSH = 0 indicates that the liquid level has not exceeded the high-level set point, and Z = 1 means that the pump is operating.

The creation of a logic scheme to express the operation of the pump discussed previously is accomplished by development of a logic table or a truth table. The procedure is to list all of the possible combinations of inputs and desired outputs systematically using "1" and "0." The following truth table shows the results for equation 6.3(1):

LSL	LSH	Z	
1	1	0	
0	1	0	6.3(2)
0	0	0	
1	0	1	

The first three rows of the table summarize the possible conditions that do not result in the operation of the pump, whereas the last row of the table is the only set of conditions that is defined by equation 6.3(1). It should be noted that this example does not allow the pump to function when the liquid level is below the high-level set point but above the low-level set point. If operation of the pump under these conditions is desired, then equation 6.3(1) is not satisfactory; an alternative equation with this logic function will be presented later in this section.

A review of 6.3(2) quickly confirms that equation 6.3(1) does describe the initially stated operating conditions for the pumps and suggests the combination of basic logic functions required to create the logic function for this truth table. Once this logic function has been established, the logic circuit can be created to perform the

605

desired function and to control the pump in the prescribed manner.

The five basic logic functions used to create other logic functions can be expressed in terms of logic values and can be defined by manipulation of the input variables of a horizontal form of tabulation 6.3(2), as follows:

LSL	1	0	0	1	
LSH	1	1	0	0	
AND	1	0	0	0	6.3(3)
NAND	0	1	1	1	6.3(4)
OR	1	1	0	1	6.3(5)
NOR	0	0	1	0	6.3(6)

The NOT operation is merely the inverse of the current value of the input variable; i.e., if LSL = 0, then LSL = 1 after the NOT operation. Figure 6.3a summarized the graphic symbols for these basic logic operations.

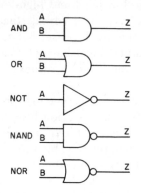

Fig. 6.3a   Logic symbol summary

Logic functions are created by combinations of these elementary logic operations. Table 6.3b presents a few examples of combined logic operations and illustrates that logic quantities and operations upon these quantities can be concisely expressed by symbols. For example,

$$Z = \overline{A \cdot B + C \cdot D} \qquad 6.3(7)$$

is an extremely succinct method of stating that two outputs from two AND operations based on variables (A, B

and C, D; respectively) are the inputs to an OR operation; the results of this are inverted to provide a value for variable Z.

Synthesis with AND/OR Logic Operations

The synthesis of a complex logic function using AND/OR logic operations is carried out with the use of two special functions, the minterm and the maxterm. As an example of AND/OR synthesis, consider the creation of the function that generates the conditions expressed in equation 6.3(1). The truth table for this system, tabulation 6.3(2), is presented again for convenience.

LSL	LSH	Z	
1	1	0	
0	1	0	6.3(2)
0	0	0	
1	0	1	

The minterm for this system is the row in the table in which there is a "1" in the output column. A minterm is implemented through an AND operation based on the row where the single "1" appeared. All of the inputs in this row that have "0" indicate the variables that have a NOT operation applied to them. For this example, the last row of the truth table is the only minterm; it is defined by equation 6.3(8).

$$Z = \overline{LSH} \cdot LSL \qquad 6.3(8)$$

Note that the NOT operation is applied to variable LSH and that this variable has a value of "0" in the last row of the table.

The synthesis of complex functions using the maxterm is a process that is the inverse of that for the minterm. An OR operation is used, and those inputs with a "1" for the row where the "0" output appears have the NOT operation applied to them. When more than one maxterm exists, the logic function is synthesized through the AND-ing together of these maxterms. (If multiple minterms were present, a logic function would be created by OR-ing the minterm contributions.) Thus, equation 6.3(1) can also be expressed by the following logic function:

$$Z = (\overline{LSL} + \overline{LSH}) \cdot (LSL + \overline{LSH}) \cdot (LSL + LSH) \qquad 6.3(9)$$

There are three separate maxterms in this equation, and they correspond to the first three rows of the truth table shown in tabulation 6.3(2).

Negation and DeMorgan's Theorem

Before synthesis through NOR and NAND operations is considered, some basic principles for inverting functions will be established.

The first principle is that two cascaded inversions cancel, since this amounts to a double negation. Stated in symbolic form:

Table 6.3b
OPERATION ELEMENTS IN LOGIC DESIGN

Word Designation	Symbolic Representation
Not A	$\overline{A}$
A and B and C	$A \cdot B \cdot C$
A or B or C	$A + B + C$
A nor B nor C	$\overline{A + B + C}$
A nand B nand C	$\overline{A \cdot B \cdot C}$

$$\overline{\overline{A}} = A \qquad\qquad 6.3(10)$$

where it is indicated that the logic signal A has been inverted twice, returning it to its original value.

The second principle is that the inversion of logic based on AND/OR operations can be obtained by inversion of every input signal and simultaneous interchanging of AND and OR operations (DeMorgan's theorem). For example, the inversion of

$$Z = A \cdot B + C \cdot D \qquad\qquad 6.3(11)$$

is obtained as:

$$\overline{Z} = (\overline{A} + \overline{B}) \cdot (\overline{C} + \overline{D}) \qquad\qquad 6.3(12)$$

These two principles can be used to obtain equivalent forms of both the NOR and the NAND operations. For the NOR operation:

$$\overline{A + B} = \overline{A} \cdot \overline{B} \qquad\qquad 6.3(13)$$

For the NAND operation:

$$\overline{A \cdot B} = \overline{\overline{\overline{A} + \overline{B}}} = \overline{A} + \overline{B} \qquad\qquad 6.3(14)$$

These two expressions are called DeMorgan equivalents. Figure 6.3c presents other logic functions with their DeMorgan equivalents.

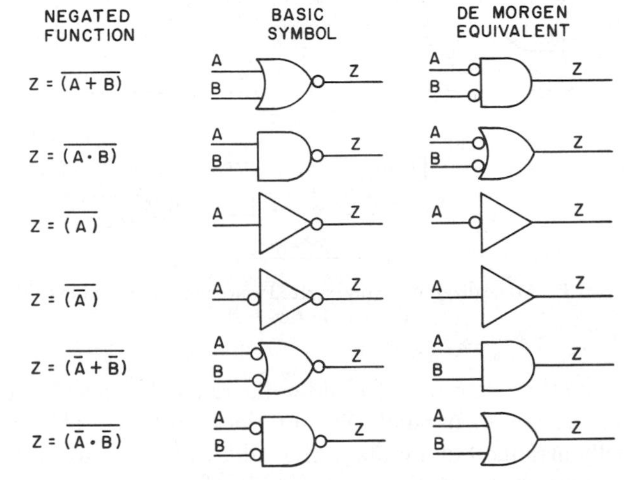

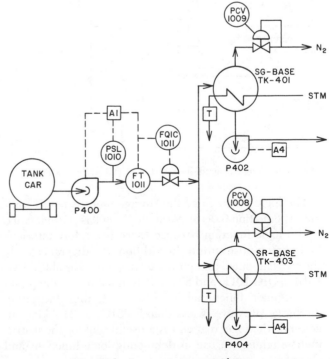

Fig. 6.3c Summary of negative logic functions and graphic symbols

Synthesis with NAND/NOR Operations

The synthesis of a logic function by NAND operations can be based on the minterm procedure explained earlier. The procedure is to construct the minterm function as before and then to replace both AND and OR operations by NAND operations. For logic function synthesis with NOR operations, the maxterms are used; then the NOR operation replaces both the OR and the AND operations.

As an example of function synthesis with NAND operations, consider the modification of equation 6.3(1) so that the pump can also function when the liquid is above the low-level set point but below the high-level set point. This modification is summarized as follows:

LSL	LSH	A	
1	1	0	
0	1	0	6.3(15)
0	0	1	
1	0	1	

where the third row of the table now reflects the modification in the desired operation of the pump system; i.e., the pump now also runs when LSL = 0 and LSH = 0. The logic function for this new system is

$$Z = (\overline{LSL} \cdot \overline{LSH}) + (LSL \cdot \overline{LSH}) \qquad 6.3(16)$$

where the first AND term represents the new minterm in the system. Now, following the procedures stated previously, the AND operations $(\overline{LSL} \cdot \overline{LSH})$ and $(LSL \cdot \overline{LSH})$ are replaced by NAND operations. The OR operation is then also replaced by a NAND operation, to yield the final NAND synthesized function:

$$Z = \overline{(\overline{LSL} \cdot \overline{LSH})} \cdot \overline{(LSL \cdot \overline{LSH})} \qquad 6.3(17)$$

Graphic Logic Functions

Logic functions can be represented graphically through logic diagrams. These diagrams are very helpful in projecting the characteristics of the logic function of the user and are created from the elementary logic drawing presented in Figure 6.3b.

The quickest method of appreciating the utility of logic diagrams is to review an example. Figure 6.3d illustrates

Fig. 6.3d A reagent storage tank system

the tank section for a chemical process. The drawing shows three pumps (equipment 400, 402, and 404), two steam lines, a low-pressure sensor (PSL 1010), two pressure control valves, (PCV 1008 and 1009), and a flow quantity indicator and controller (FQIC 1011). For simplicity, consider the filling operation for this system because it involves only one pump and two control elements, PSL 1010 and FQIC 1011. The logic function for this part of the operation is

$$Eq\ 400 = ((\overline{FQIC\ 1011} \cdot Eq\ 400 \\ \cdot \overline{PSL\ 1010}) + START\ P.B.) \cdot \overline{STOP\ P.B.} \quad 6.3(18)$$

where START P.B. and STOP P.B. are the variables assigned to the start and stop pushbuttons in the system and Eq 400 is the variable associated with the operation of the pump. Variable Eq 400 equals 1 when the pump is on and has a value of 0 when the pump is off. Although equation 6.3(18) is accurate, it is difficult to comprehend, particularly since the desired variable, Eq 400, appears on both sides of the equation. Such implicit logic functions are common and are best interpreted by use of a logic diagram.

Figure 6.3e provides the logic diagram for the storage tank system presented in Figure 6.3d. The drawing uses a second graphic symbol convention common to industrial applications. A 3 input AND, a 2 input AND, an OR, and three inversion operations are illustrated in the figure. The inversions are represented by the three circles; the OR symbol is the short vertical line in the center of the drawing.

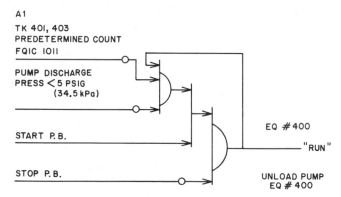

Fig. 6.3e   Logic diagram for the storage tank system in Fig. 6.3d

The logic diagram for the storage tanks is not as complex as equation 6.3(18). Variable Eq 400 is present twice in the logic function because there is a safety interlock for the stop pushbutton. In addition, the diagram clearly shows that once the pump is running, variables PSL 1010, FQIC 1011, and STOP P.B. must be 0 if the pump is to remain functional. (If PSL = 1, then the pump discharge pressure is less than 5 PSIG, or 34.5 kPa). In general, the logic diagram is a useful tool for the instrument engineer to use in debugging logic functions and circuits.

Logic Diagram Construction from Ladder Diagrams

The development of inexpensive reliable solid state logic devices (see Section 6.6) has resulted in the use of logic diagrams rather than other methods of describing logic circuits. Existing switching circuits are being updated with solid state circuits, and the associated ladder diagrams must be translated to logic diagrams. Figure 6.3f presents a logic diagram that was developed from the ladder diagram for the storage tank system illustrated in Figure 6.2b. A review of these two drawings illuminates the procedure for interchanging ladder diagram elements with logic diagram elements.

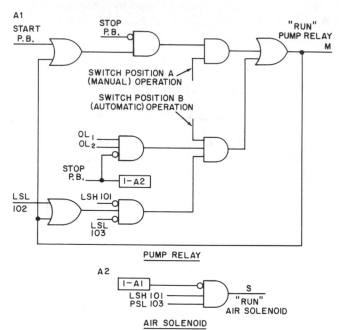

Fig. 6.3f   Logic diagram for pressurized water tank system from ladder diagram in Fig. 6.2b

The three features of a ladder diagram to be considered are the serial, parallel, and interlock portions. The serially arranged elements in the ladder diagram are used as inputs to an AND operation in the logic diagram. The parallel components of ladder diagrams become inputs to OR operations in logic diagrams. Thus, for the present example, the serially arranged elements (LSH 101, LSL 102, and LSL 103 in Figure 6.2b) become the three inputs to the AND operation shown in the bottom of plate A1 of Figure 6.3f. The elements in the solenoid branch of the ladder diagram (i.e., the solenoid relay, PSL 103, and the normally closed contacts for LSH 101), which are parallel to the elements of the pump relay portion of the ladder diagram, (i.e., the pump relay, LSL 103, LSL 102, and the normally open contacts of LSH 101) are arranged as two separate inputs to the OR operation shown on the far right side of plate A1 in the figure. Finally, it is important to note the way in which the two ladder diagram pump relay contact set interlocks, M, are represented in the logic diagram. The final output in the logic diagram is returned to the other inputs of the LSL 102 and the start pushbutton or operations.

Logic Simplification Through Logic Maps

The synthesis of logic functions based on minterms or maxterms often produces functions that are unnecessarily complex. The comparison of the two equivalent expressions, equation 6.3(8) and equation 6.3(9), dramatically demonstrates this point. In this example, the minterm synthesis was more efficient, but in many cases the designer does not intuitively select the better synthesis route. Another problem common to equation 6.3(9) and other complex functions is the presence of extra logic elements.

The existence of unnecessary inputs in a logic expression is called redundancy. Some cases of logic redundancy are easily seen by inspection of the logic function or logic table, as in the example just given. However, many cases of redundancy are difficult to recognize. The most frequently used tool for aiding in finding redundant terms is a graphical method known as the logic map, or the Karnough map.

The logic map consists of an array of squares, with each square corresponding to a column of the logic table. The best way to show the construction of the map is to number the columns of the logic table, starting from left to right. This is illustrated for three-inputs in Figure 6.3g, which shows the logic map array for three-inputs with numbers to indicate the relationship to the logic table. (This is only one of several forms of constructing a map.) Figure 6.3h then shows how the values for A, B, and C would be entered onto the map.

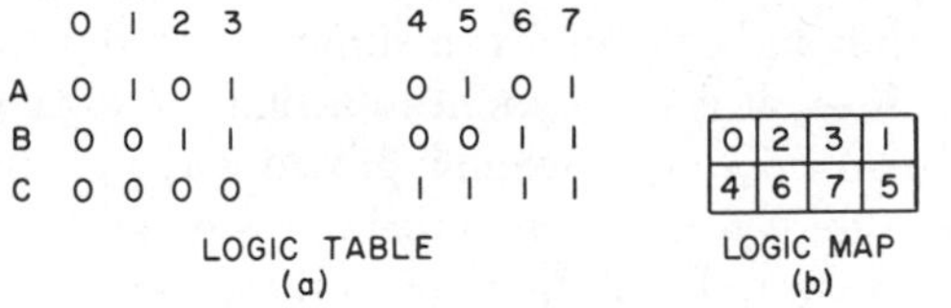

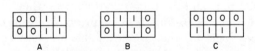

Fig. 6.3g Three-input logic map and order of entry from the logic table. *a.* A three-input table. *b.* Correspondingly numbered squares of the map.

Fig. 6.3h Three-input maps resulting when each input row (A, B, C) of the table is entered onto the map

The maps of Figure 6.3h are fixed, since they correspond to the input portion of the logic table. The true implementation of the map occurs when the output portion of the table (the "Z" row) is mapped. For example, the following tabulation of logic redundancy results in the map of Figure 6.3i:

A	0101	0101	
B	0011	0011	
C	0000	1111	6.3(19)
Z	0001	0001	

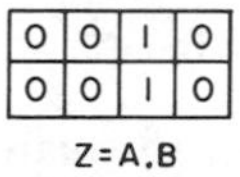

Fig. 6.3i Map resulting from the entry of Z = 0001 0001

The rule for the use of the map of Figure 6.3i is that any two adjacent squares with 1s represent an input that can be eliminated. This can be recognized on inspection of the corresponding area in the input maps in Figure 6.3h. A and B have the constant value of 1 in this area, but C has two possible values, 0 and 1. It also should be noted that squares on the sides of the map are considered contiguous, as shown in Figure 6.3j. Finally, any four areas with 1s that are symmetrical, i.e., ones that form either a square or a rectangle, eliminate two inputs.

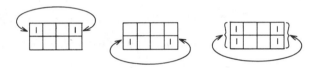

Fig. 6.3j Illustration of contiguous squares based on "wraparound"

Logic Simplification with Boolean Algebra

Boolean algebra is the class of mathematics used to manipulate logic expressions. A few of the theorems of this system are given as examples to show how Boolean algebra can be used to simplify logic and to cast it in different forms. These particular theorems are easy to remember, since they parallel familiar arithmetic and conventional algebraic operations.

The following are some examples of Boolean algebra equivalents:

$$1 \cdot A = A \qquad 6.3(20)$$

$$0 \cdot A = 0 \qquad 6.3(21)$$

$$1 + A = 1 \qquad 6.3(22)$$

$$0 + A = A \qquad 6.3(23)$$

$$\overline{A} + A = 1 \qquad 6.3(24)$$

$$A + A = A \qquad 6.3(25)$$

$$A \cdot B + A \cdot C = A \cdot (B + C) \qquad 6.3(26)$$

To show the use of these relationships for logic simplification, consider the expression defined by tabulation 6.3(19):

$$
\begin{array}{lll}
A & 0101 & 0101 \\
B & 0011 & 0011 \\
C & 0000 & 1111 \\
\hline
Z & 0001 & 0001
\end{array}
\qquad 6.3(19)
$$

The minterm expression is

$$ Z = A \cdot B \cdot \overline{C} + A \cdot B \cdot C \qquad 6.3(27) $$

Using equation 6.3(26), this can be "factored":

$$ Z = A \cdot B (C + \overline{C}) \qquad 6.3(28) $$

Equation 6.3(28) then becomes

$$ Z = A \cdot B \cdot 1 \qquad 6.3(29) $$

Consider another expression:

$$ Z = A \cdot B \cdot C + A \cdot \overline{B} \cdot C + A \cdot \overline{B} \cdot \overline{C} \qquad 6.3(30) $$

This is simplified by repetition of the middle term, which can be done without changing the value of the logic, based on equation 6.3(25):

$$
\begin{aligned}
Z = A \cdot B \cdot C &+ A \cdot \overline{B} \cdot C \\
&+ A \cdot \overline{B} \cdot C + A \cdot \overline{B} \cdot \overline{C}
\end{aligned}
\qquad 6.3(31)
$$

Then, using equation 6.3(26):

$$ Z = A \cdot C(B + \overline{B}) + A \cdot \overline{B}(\overline{C} + C) \qquad 6.3(32) $$

Equation 6.3(32) becomes, by equations 6.3(20) and 6.3(24),

$$ Z = A \cdot C + A \cdot \overline{B} \qquad 6.3(33) $$

This can be placed in still another form by an additional use of equation 6.3(26):

$$ Z = A \cdot (C + \overline{B}) \qquad 6.3(34) $$

Finally, it should be noted that the repetition of the term $A \cdot \overline{B} \cdot C$ in the original expression is the equivalent of the overlapping of areas on the logic map.

Negative Versus Positive Logic Usage

The graphic presentation of complex logic functions that describe the control of an industrial process is an essential aspect of a designer's responsibilities. Large process logic control schemes cannot be understood or maintained by the engineering support group unless a graphic version of the logic circuit is supplied. Once the designer has developed the logic circuit and has removed the redundancies, a decision must be made concerning whether to present the circuit graphically with positive or negative logic. The choice of logic is based on the purpose of the circuit and the general desire to explain the circuit graphically in a clear, concise fashion.

A positive logic circuit is one that illustrates logic elements that output 1s as a result of the input of 1s into the element. An AND is an example of a positive logic element. The only way an AND operation produces a 1 is when all of its input variables are 1. Logic functions described with ANDs and ORs are considered positive logic functions. Conversely, negative logic circuits illustrate a logic function by using logic elements that require 0s at inputs to produce 0s at their outputs.

The distinction between positive and negative logic may be initially difficult to understand. Figure 6.3k presents graphic representations of positive and negative logic elements to help explain the difference. Part a of Figure 6.3k illustrates the AND operation; it should be clear from previous discussions that C has the unique value of 1 only if A and B also have a value of 1. (All other combinations of values for A and B result in a value of 0.) Part b of the figure does not seem as simple to understand when viewed from a positive logic point of view. The addition of the three inversion circles complicates our thought process when we attempt to determine the output logic of the drawing. However, the function of part b is easily interpreted from a negative logic perspective, i.e., output variable C has a value of 0 when either input A or input B has a value of 0.

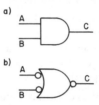

Fig. 6.3k An example of positive and negative logic operations. a. An AND presented in positive logic. b. An AND presented in negative logic.

The merit to negative logic is that it facilitates the creation of control circuits for real systems. For example, many pressure and level sensors provide zero volt signals when they indicate an alarm situation. Under these circumstances, it is convenient to assign the logic value of 0 to the sensor output when it provides a zero volt signal. If these detectors are monitoring a storage tank whose contents are to be pumped from the tank under normal level and pressure conditions, then the drawing in part a of Figure 6.3k shows the necessary logic for this transfer operation, provided the pump interface circuit is connected to C and the pump does not run when output C has a value of 0. Part b of Figure 6.3k also shows the proper logic for the circuit. If the pressure and level sensors are connected to the A and B inputs and the pump control relay is connected to C, then the pump cannot operate when either the pressure or the level sensor goes into alarm, because C is forced to 0 and this active output state stops the pump. We can match the physical characteristics and requirements of the sensors and the pump interface circuit to the negative logic drawing more easily than to the positive logic drawing even though both drawings describe identical means of control of the storage tank operation by the circuit.

REFERENCE

1. Lipták, B.G., ed., *Instrument Engineers' Handbook*, Vol. II, Chilton, Philadelphia, 1969, Chap. 4.

BIBLIOGRAPHY

Ellermeyer, W., "Manipulate Logic Without Boolean Algebra," *Control Engineering*, December 1969.

Graca, J.D., "Graphic Batch Processing Language Simplified," *Instruments and Control Systems*, December 1975.

Harrison, H.L., "Optimize Switching Circuits Using Karnough Maps," *Control Engineering*, August 1969.

Mergler, H., "The Key to Optimum Logic Design," *Control Engineering*, January 1967.

Sandige, R.S., *Digital Concepts Using Standard Integrated Circuits*, McGraw-Hill, New York, 1978.

Stoffels, R.E., "Shift Registers and Stepping Switches for Sequence Control," *Control Engineering*, January 1967.

Williams, G.E., "Digital Technology," SRA, Chicago, 1982.

6.4 PROGRAMMABLE CONTROLLERS (PCs)

Types of Input/Output (I/O):	Discrete I/O—120 V AC, 0–5 V DC, 220 V AC; Analog—4–20 mA, 1–5 V DC, thermocouple, RTD, etc.; transistor/transistor logic (TTL); stepper motor (pulse); contact output; high-speed counter; motion control.

Scan Time per 1000 Words (1 K) of Logic: 1 msec to 50 msec, depending on manufacturer and enhanced software features.

Word Size: 4 bit, 8 bit, or 16 bit (typical).

Amount of I/O: Small PC—20 to 256 I/O; medium PC—256 to 1024 I/O; large PC—1024 and greater I/O.

Size of Memory: Small PC—256 to 2 K words (K = 1024 bits, bytes, or words of digital data); medium PC—2 K to 12 K words; large PC—12 K and larger.

Type of Memory: CMOS (Complementary Metal Oxide Semiconductor); RAM (Random Access Memory); EPROM (Erasable Programmable Read Only Memory); CORE (Ferrite Cores).

Environmental Conditions: 0 to 60°C (32–140°F), relative humidity 0–95% noncondensing, 115 V AC ± 15% and 230 V AC ± 15%. |
Small PC Hardware System Cost:	Discrete I/O size 20—$2000; 32—$2250; 64—$3000; 96—$3800; 128—$4600; 256—$8000.
Medium PC Hardware System Cost:	Discrete I/O size 256—$7000; 512—$19,000; 1024—$28,000.
Large PC Hardware System Cost:	$25,000 to $80,000.
I/O Cost (Price/Pt.):	Discrete in—$20; discrete out—$25; analog in—$150; analog out—$180.
I/O Cost:	50 to 85% total PC hardware cost.
Software Costs:	Relay ladder logic—$600 to $2500 per K of memory.
Software Cost:	50 to 100% hardware costs.
Systems Engineering Cost:	25 to 50% hardware costs.
Support Equipment Cost:	Hand-held programmer—$500 to $3500; CRT programmer—$1500 to $10,000.
Documentation Cost:	25 to 35% software costs.
Partial List of Suppliers:	Allen-Bradley; August Systems; Automation Systems; Barber-Coleman Co., Industrial Instruments Div.; Cincinnati Milacron, Inc., Electric Systems Div.; Divelbiss Corp.; Dynage, Inc.; Eagle Signal Controls; Eaton Corp.; Encoder Products Co.; General Electric; Giddings & Lewis, Inc.; Gould; Iconix; Industrial Solid State Controls; Klockner-Moeller Corp.; McGill Mfg. Co.; Mitsubishi Electric; Omron Electronics, Inc.; Reliance Electric; Satt; Siemens; Square-

D Co.; Struthers-Dunn, Inc.; Telemechanique; Tenor Co., Inc.; Texas Instruments, Inc.; Westinghouse Electric Corp.

For a glossary of terms used in connection with PLC's, please consult Section 12.4 of the Process Measurement volume of this handbook.

Introduction

A programmable controller (PC) is a device that performs discrete or continuous control logic in process plant or factory environments. Programmable controllers have become popular since their inception (1969) in a wide variety of applications, from boiler control to robot control. Recent surveys by several journals have identified over 26 American and foreign manufacturers of programmable controllers;[1-3] some manufacturers offer several (three to five) different models of PCs. This section will identify the common capabilities of programmable controllers and will point out some important benefits of this emerging technology.

History

The PC was originally invented to replace electro-mechanical relays and hard-wired discrete logic devices (Table 6.4a). The automotive industry fostered the development of the PC primarily because of the massive rewiring that had to be done every time an automotive model change occurred. Solid state logic proved to be much easier to change than relay panels, and for this reason the programmable controller had significant cost advantages over traditional relay systems (Table 6.4b).

Cost considerations alone do not explain the success of programmable controllers. Compared with electro-mechanical relay systems, PCs offer the following advantages:

- Ease of programming and reprogramming in the plant (familiar programming language)
- High reliability and maintainability
- Small physical size
- Ability to communicate with computers
- Economy
- Rugged construction
- Modular design

As acceptance of the PC grew outside of the automotive industry, the features of the device were enhanced. Today's PC has the capacity to operate as more than a relay replacement. Many functions, including counting, timing, complex mathematical applications, and PID and feedforward controls, can be performed by PCs. The PC has brought solid state logic out to the plant floor and has ushered in some important new control capabilities, such as providing the powerful combination of sequential control and PID control in the same device.

Table 6.4a
PC HISTORY

• 1968	DESIGN OF PCs DEVELOPED FOR GENERAL MOTORS CORPORATION TO ELIMINATE COSTLY SCRAPPING OF ASSEMBLY-LINE RELAYS DURING MODEL CHANGEOVERS.
• 1969	FIRST PCs MANUFACTURED FOR AUTOMOTIVE INDUSTRY AS ELECTRONIC EQUIVALENTS OF RELAYS.
• 1971	FIRST APPLICATION OF PCs OUTSIDE THE AUTOMOTIVE INDUSTRY.
• 1973	INTRODUCTION OF 'SMART' PCs FOR ARITHMETIC OPERATIONS, PRINTER CONTROL, DATA MOVE, MATRIX OPERATIONS, CRT INTERFACE, ETC.
• 1975	INTRODUCTION OF ANALOG PID (PROPORTIONAL, INTEGRAL, DERIVATIVE) CONTROL, WHICH MADE POSSIBLE THE ACCESSING OF THERMOCOUPLES, PRESSURE SENSORS, ETC.
• 1976	FIRST USE OF PCs IN HIERARCHICAL CONFIGURATIONS AS PART OF AN INTEGRATED MANUFACTURING SYSTEM.
• 1977	INTRODUCTION OF VERY SMALL PCs BASED ON MICROPROCESSOR TECHNOLOGY.
• 1978	PCs GAIN WIDE ACCEPTANCE. SALES APPROACH $80 MILLION.
• 1979	INTEGRATION OF PLANT OPERATION THROUGH A PC COMMUNICATION SYSTEM.
• 1980	INTRODUCTION OF INTELLIGENT INPUT AND OUTPUT MODULES TO PROVIDE HIGH-SPEED, ACCURATE CONTROL IN POSITIONING APPLICATIONS.
• 1981	DATA HIGHWAYS ENABLE USERS TO INTERCONNECT MANY PCs UP TO 15,000 FEET FROM EACH OTHER. MORE 16 BIT PCs BECOME AVAILABLE. COLOR GRAPHIC CRTs ARE AVAILABLE FROM SEVERAL SUPPLIERS.
• 1982	LARGER PCs WITH UP TO 8192 I/O BECOME AVAILABLE.
• 1983	'THIRD PARTY' PERIPHERALS, INCLUDING GRAPHIC CRTs, OPERATORS' INTERFACES, 'SMART' I/O NETWORKS, PANEL DISPLAYS, AND DOCUMENTATION PACKAGES, BECOME AVAILABLE FROM MANY SOURCES.

Table 6.4b
COST ADVANTAGES OVER RELAY
(Source Gould)

	Relays	Solid State Controls	Micro Processor	Mini-Computer	PCs
Hardware Cost	Low	Equal	Low	High	High to low Depending on Number of Controls
Versatility	Low	Low	Yes	Yes	Yes
Usability	Yes	Yes	No	No	Yes
Maintain-Ability Troubleshoot	Yes	No	No	No	Yes
Computer-Compatible	No	No	Yes	Yes	Yes
Arithmetic Capability	No	No	Yes	Yes	Yes
Information Gathering	No	No	Yes	Yes	Yes
Industrial Environment	Yes	No	No	No	Yes
Programming Cost	(Wiring) High	(Wiring) High	Very High	High	Low
Reusable	No	No	Yes	Yes	Yes
Space Required	Largest	Large	Small	OK	Small

Functional Description

The programmable controller is difficult to characterize because of its current evolutionary nature; however, in 1978 the National Electrical Manufacturers Association (NEMA) published a standard definition (NEMA Standard ICS3–1978 part ICS3–304), stating that:

> The programmable controller is a digitally operating electronic apparatus which uses a programmable memory for the internal storage of instructions for implementing specific functions, such as logic, sequencing, timing, counting, and arithmetic, to control through digital or analog input/output modules various types of machines or processes. A digital computer which is used to perform the functions of a programmable controller is considered to be within this scope. Excluded are drum and similar mechanical type sequencing controllers.

This definition took four years to develop and is currently well accepted. Regardless of size, complexity, or evolutionary state, all PCs show certain functional similarities (Figure 6.4c).

POWER SUPPLY

The power supply may be integral or separately mounted and provides the isolation necessary to protect solid state

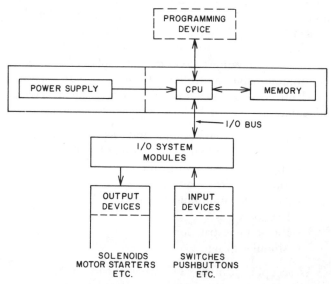

Fig. 6.4c   PC architecture

components from high-voltage line spikes. The power supply converts power line voltages to those required by the solid state components. Another important function provided by the power supply is the heat dissipation required for plant floor operation. This capability to dissipate is responsible for the high ambient temperature specifications of most PCs and represents one of the important differences between PCs and minicomputers.

The power supply drives the I/O logic signals, the central processor unit, the memory unit, and some peripheral devices. As I/O is expanded some PCs may require additional power supplies in order to maintain proper power levels. The additional power supplies may also be separate or part of the I/O structure.

INPUT/OUTPUT SYSTEM

Inputs are real-world signals giving the controller real-time status of variables. These variables can be analog, register, or discrete: maintained or momentary. Typical analog inputs can be from thermocouples (T/C), RTDs, flow, pressure, temperature transmitters, or strain gauges (direct or from transducers); some analog inputs may require 4–20 mA signals or may be direct from the T/C or RTD. Typical discrete inputs can be pushbuttons, limit switches, or even electromechanical relay contacts. Register inputs can be thumbwheels and BCD (Binary Coded Decimal) transmitters. These inputs are all converted to digital data and transmitted over the I/O bus to the central processor unit.

Instructions as to the desired status of outputs are transmitted over the I/O bus to modules, which convert these signals to desired real-world outputs. There are three categories of outputs: discrete, register, and analog. Discrete outputs can be pilot lights, solenoid valves, or annunciator windows (lamp box). Register outputs can drive panel meters or displays; analog outputs can drive signals to variable speed drives or to I/P (current to air) converters and thus to control valves.

Most I/O systems are modular in nature; that is, a system can be arranged by use of modules that contain multiples of I/O points. These modules can be composed of 1, 4, 8, or 16 points and plug into the existing bus structure. The bus structure is really a high-speed multiplexer that carries information back and forth between the I/O modules and the central processor unit.

One of the most important functions of the I/O is its ability to isolate real-world signals (0–120 VaC, 0–24 VdC, 4–20 mA, 0–10 V, and thermocouples) from the low signal levels (typically 0–5 VdC MAX) in the I/O bus. This is accomplished by use of optical isolators, which trigger a process switch to transfer data in (input module) or out (output module) to a solenoid valve without violating bus integrity. Typical discrete I/O schematics are shown in Figures 6.4d, 6.4e, and 6.4f.

CENTRAL PROCESSOR UNIT (REAL-TIME)

The central processor unit (CPU), or central control unit (CCU), performs the tasks necessary to fulfill the PC function. Among these tasks are scanning, I/O bus traffic control, program execution, peripheral and external device communications, special function or data handling execution (enhancements), and self-diagnostics.

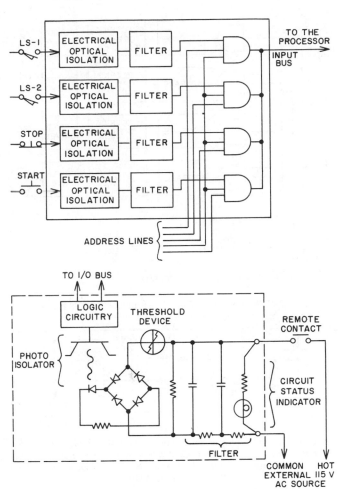

Fig. 6.4d Typical input area; typical AC input unit. (From Londerville, S., Programmable Controllers in Boiler Control Applications," Instrument Society of America, Northern California Section, ISA Days, June 2, 1982.

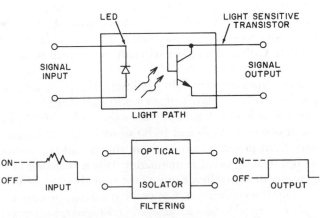

Fig. 6.4e Electrical optical isolator. (From Londerville, S., "Programmable Controllers in Boiler Control Applications," Instrument Society of America, Northern California Section, ISA Days, June 2, 1982.)

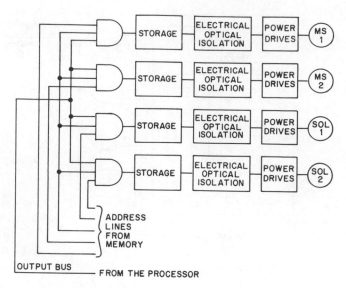

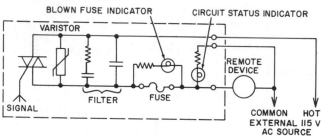

Fig. 6.4f Typical output area; typical AC output unit. (From Londerville, S., "Programmable Controllers in Boiler Control Applications," Instrument Society of America, Northern California Section, ISA Days, June 2, 1982.)

Central processing units can use TTL (Transistor-Transistor Logic), CORE (ferrite cores), or CMOS (Complementary Metal Oxide Silicon) technology or can be microprocessor-based (VLSI). Although TTL is faster (faster scan times), CORE requires no battery backup, CMOS is more compact and requires lower power levels, and microprocessor-based systems are both more powerful and more flexible. Generally, trade-offs must be made between speed and special features. It should be noted that the CPU and memory units are considered separate functions.

Scan times can vary from 1 millisecond (1/1000 second) per 1 K (1024) words of logic to more than 50 milliseconds per 1 K of logic. Word size varies from 4 bits to 16 bits (most common are 8- and 16-bit words). Special features vary from preprogrammed drum timers to PID control or full floating point mathematics. When selecting programmable controller features, the user should take into account the application. Generally speaking, process applications need to take advantage of microprocessor power, whereas machine control applications should be concerned with program execution speed. Obviously, many exceptions to this rule exist.

MEMORY UNIT

The memory unit of the PC is the library where the program is stored. The basic memory unit is the word. Words are made up of bits. A bit is a single piece of data and contains information on only two states (ONE/ZERO or YES/NO). Historically, PC manufacturers have called a word (16 bit), a byte (8 bit), or a nibble (4 bit) a "word." Combinations of bits yield coded messages, commonly referred to as words. Words can be from 4 to 16 bits long. Obviously, the longer the word, the more information that can be contained within. Combinations of words that produce control logic are called programs. Memory sizes vary from 256 words to 192 K words.

Memory can be volatile or nonvolatile. Volatile memory is erased if power is removed. Obviously, this is undesirable, and most units with volatile memory provide battery backup to ensure that there will be no loss of program in the event of a power outage. Nonvolatile memory does not change state on loss of power and is used in cases in which extended power outages or long transportation times to job site (after program entry) are anticipated.

There are three basic types of memory: Read Only Memory (ROM), Read/Write Memory (R/W), and Random Access Memory (RAM).

Read Only refers to memory that must be physically altered to change its contents and is nonvolatile. One type of ROM is UV-PROM. This technique requires erasing of PROM chips ("Programmable" Read Only Memory) by exposing the top of the chip to ultraviolet light. The chip can then be programmed, or "burned," with the revised instructions. Modern programmable controllers come with PROM burners that are already installed ("on board").

Another type of memory is the electrically alterable EAPROM, EPROM, or EROM. These varieties also allow "on-board" PROM burning.

Read/Write Memory can be changed without physical alteration and can be volatile or nonvolatile. A common R/W memory is magnetic core. Core memory is easily altered and is nonvolatile but is the most expensive type of memory. Core memories are generally used in larger PC systems.

RAM uses integrated circuit technology. This is also read/write memory, but it is volatile. CMOS (Complementary Metal Oxide Silicone) is an extremely compact, low-cost memory system. However, some form of battery backup is required.

PROGRAMMER UNITS

The programmer allows interface between the PC and the user during program development, startup, and trouble shooting. The instructions to be performed during each scan are coded and inserted into memory with this device.

Programmers vary from small, hand-held units the size of a portable calculator to large, intelligent CRT-based units with documentation, reproduction, I/O status, and on-line and off-line programming ability.

Programming units are the liaison between what the PC understands (words) and what the engineer desires to occur during the control sequence. Some programmers have the ability to store programs on other media, including cassette tapes and floppy disks. Another desirable feature is automatic documentation of the existing program. This is accomplished by a printer attached to the programmer. With off-line programming, the user can write his control program on the programming unit, then take the unit to the PC in the field and load the memory with the new program, all without removing the PC. Selection of these features depends on user requirements and budget. On-line programming allows cautious modification of the program while the PC is controlling the process or the machine.

PERIPHERAL DEVICES

As this is written, many new PC peripheral devices are being added.[5] These can be grouped in several categories.

Programming aids provide documentation and program recording capabilities. Some devices can program many models of different manufacturers' PCs. *Operational aids* include a variety of resources, from color graphics CRTs to equipment that can give the operator point access of data (timers, counters, loops, etc.) but not access to the program itself. Some PCs can issue information to dumb terminals, such as printers, in a report format. Some devices have the ability to set up an entire panel and plug into the PC through external RS232C ports, thereby saving enormous panel and wiring costs. *I/O enhancements* is the largest PC peripheral category; everything from dry contact modules to intelligent I/O to remote I/O capabilities can be found here. Also included in this category are I/O simulators used for program development and debugging. *Computer interface devices* allow peer-to-peer communications (i.e., one PC connected directly to another) and offer networking capabilities, long-distance data, and supervisory control transmission.

R. Whitehouse, in a 1978 paper,[6] discussed standardization within the fast-evolving PC industry. Currently, as then, the only "standardization" seems to be common industry trends, such as dense I/O packaging with the CPU mounted to the right or the left (within integral power supply); a wide selection of I/O types; networking with PCs; redundancy; peer-to-peer PC communications; PCs that can be easily upgraded to larger memory size; and one I/O system used for all PC models within a PC vendor's product lines. The manufacturers' objective seems to be to introduce as many new features as possible: I/O type, combinational programmer and operator interface functions (used together), 24-VdC power supply, high-level language programming, and so forth. Other PC vendors then introduce similar products. This is called "evolutionary standardization."

Differences Between Computers and PCs

PCs are often thought of as computers. To a certain extent this is true; however, there are four important differences between PCs and computers.

REAL-TIME OPERATION/ORIENTATION

The PC is designed to operate in a real-time control environment. Most PCs have internal clocks and "watchdog timers" built into their operations to ensure that some functional operation does not send the central processor into the "weeds." The first priority of the CPU is to scan the I/O for status, to make sequential control decisions (as defined by the program), to implement those decisions, and to repeat this procedure all within the allotted scan time.

ENVIRONMENTAL CONSIDERATIONS

PCs are designed to operate near the equipment they are meant to control. This means that they function in hot, humid, dirty, noisy, and dusty industrial environments. Typical PCs can operate in temperatures as high as 140°F (60°C) and as low as 32°F (0°C), with tolerable relative humidities ranging from 0% to 95% noncondensing. In addition, they have electrical noise immunities comparable with those required in military specifications. Manufacturers' experience shows mean time between failures (MTBF) ranging from 20,000 to 50,000 hours.

PROGRAMMING LANGUAGES AND TECHNIQUES

PC languages are designed to emulate the popular relay ladder diagram format. This format is read and understood worldwide by maintenance technicians as well as by engineers. Unlike computer programming, PC programming does not require extensive special training. Applications know-how is much more important. Although certain special techniques are important to programming efficiency, they are easily learned. The major goal is the control program performance. Another difference between computers and PCs is the sequential operation of the PC. Program operations are performed by the PC in the order they were programmed (Figure 6.4g). This is an extremely useful feature that allows easy programming of shift registers, ring counters, drum timers, and other useful indexing techniques for real-time control applications.

MAINTENANCE AND TROUBLE SHOOTING

As a plant floor controller, the PC must be maintainable by the plant electrician or the instrument technician. It would be highly impractical to require computer-

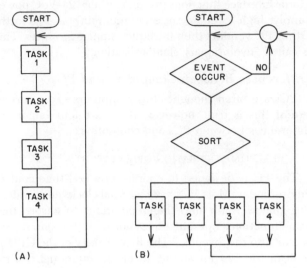

(A)   (B)

Fig. 6.4g   PC versus computer. Program structure for a programmable controller (A) requires sequential execution with a scan, starting with task 1 and proceeding through task 4. The program structure for a general-purpose computer (B) permits task execution in any order.

type maintenance service. To this end, PC manufacturers build in self-diagnostics to allow for easy trouble shooting and repair of problems. Most PC components are modular and simple to isolate; remove-and-replace (system modules) diagnostic techniques are usually implemented.

Justification for the Use of PCs

Today's programmable controllers represent a product that is maturing rapidly. Capability exists for control of everything from relay ladder logic to control loops to distributed controls capabilities. Why should one choose a PC rather than a conventional system? PCs are usually selected because of their cost, versatility, flexibility, and expandability.

Cost

One of the basic reasons for the development of PCs was to reduce the installed cost of relay systems. Not only are today's PCs inexpensive, but also they can provide a much lower per loop cost for loop control purposes as well as low per point costs for analog and digital multiplexing functions. Thus, the PC can act as a sequencer, a loop controller, a data logger and, in some cases, a multiplexer. This multifunction capability, if imaginatively applied, allows the PC user to save on hardware. PC purchasers can have 8 full PID loops, 20 analog points, 100 discrete I/O points, 2 K words of sequencer (interlock) logic, and data output to a printer for less than the price of a conventional data logger! Of course, this discussion is limited to hardware costs, but it gives an idea of the decrease in the price of an overall control system with a programmable controller.

Versatility

The multifunction capability of a PC allows control logic decision making, a versatility rarely possible with other systems. The ability to combine discrete and analog logic is a powerful tool for the controls engineer. This is especially evident in the control of batch processes. Entire startup and shutdown sequences can be performed by the sequencer logic, and analog logic can be brought in during the batch run. Control of critical startup parameters, such as temperature and pressure, can be precisely preprogrammed for each startup step. Temperature stepping is easily programmed, as are the feedforward calculations that are used in some polymer reactors. All of these types of PC applications are currently in use today and are well documented.[8-11]

Flexibility

As a process goes online and is refined, the control equipment should be easily reconfigured to accommodate such modifications. The multifunction use of the PC has already been discussed. In addition, digital blending applications, boiler control of either carbon monoxide or excess oxygen, and other forms of optimizational control are all within the capabilities of PCs. Because one common device performs multiple functions in a plant, fewer spare parts are needed, and the programming language is technician-friendly. In addition, the digital nature and self-diagnostic capabilities are strong additional justification for the PC.

Expandability

As a process matures, it is inevitable that enhancements will be added. These usually require more inputs and outputs. For hand-wired relay systems this usually necessitates extensive panel changes, which generally are problematic. A PC easily accommodates the additional I/O without requiring changes in the existing wiring: The new points are merely placed in the system. If a PID loop or two is being added, no panel rework is necessary; only the wiring of the new points and some reprogramming to incorporate them is required. Of course, if the initially selected PC is "tight," additional I/O bases might be necessary. For this reason, most manufacturers recommend sizing the system to allow for 10 to 20 percent expansion.

Another advantage of the PC is that it allows piecemeal implementation of projects. Systems can be brought up on line quickly and can be gradually converted over to the PC while on line. The ability of the PC to be reprogrammed while operating permits automation of processes that are too expensive to shut down. This technique is valuable to new as well as to retrofit projects (revamps).[12,13]

Project Execution

The PC project must take into account the important considerations of schedule and budget as is true of any major undertaking. The PC can facilitate the transition, however, by simultaneously pursuing several activities, thereby condensing the overall project schedule. A review of each major activity is presented in the following paragraphs.

Systems Analysis

The control system should be analyzed as a whole to determine plant control requirements. The PC plays an integral part in these analyses, and its capabilities should be thoroughly understood by the controls engineer. Vital to systems analysis are the process and instrument diagram (P&ID), the descriptive operational sequence, and a logic diagram or electrical schematic. Part of this evaluation will be system sizing and selection. Once the appropriate PC is selected and purchase orders are placed, two activities should begin immediately: engineering design and software development.

Engineering Design

The first step is development of the I/O list (Table 6.4h). This detailed document will be used extensively and should be developed with great care. (Once I/O numbers are assigned, it becomes very difficult to change all references to these numbers.) The I/O list is followed by the configuration drawing.

The configuration drawing (Figure 6.4i) shows the arrangement of the I/O and support hardware. The point-to-point wiring diagram (Figure 6.4j) will be used by the panel shop and the installation contractor to make the I/O device interconnect.

Panel, or enclosure, design should now be coordinated with the addition of panel instrumentation, such as light switches, meters, and recorders. Once these steps are completed, panel fabrication and assembly can begin.

Software (Program) Development

The I/O list mentioned previously will be used to begin the program development. Basic control philosophy decisions need to be made at this point. Should valves fail open or closed? What fail-safe provisions are necessary for analog control? These philosophical decisions should be documented and included with the process operational descriptions.[15] Quite often this document will be referred to as the software functional specification. Its purpose is to define, as precisely as possible, the operation of the controls.[16] It also has several other functions:

Table 6.4h
EXAMPLE OF A DETAILED I/O LIST FOR A MILLING MACHINE

Input	Definition			Used in Rung(s)*					
X0		MOA	Automatic	1 −46	40 −47	40 48	43	44	45
X1	Rt. Shot	Pin In	L.S. 1	3	−3	9			
X2	Left Shot	Pin In	L.S. 2	3	−3	9			
X3	Machine	Slide Adv	L.S. 3	7	−22				
X4	Machine	Slide Ret	L.S. 4	3	8	−39			
X5	Shot Pins	Out	L.S. 5 & 6	6	−52				
X6	Pump	#1	3M	1					
X7	Rt. Bore	Slide Ret	L.S. 7	11	46				
X8	Left Bore	Slide Ret	L.S. 8	12	47				
X9	Rt. Bore	Slide Adv	L.S. 9	3	15	−19			
X10	Left Bore	Slide Adv	L.S. 10	3	15	−21			
X11	Swing	Clamp On	P.S. 1	6					
X12	Pull	Clamp On	P.S. 2	5					
X13	Swing	Clamp Off	P.S. 3	16					
X14		Cycle	Start	2 −45	3	−17	−36	−36	−37
X15		Cycle	Stop	1	2				
X16		Auto Lube	On/Off	40	48				
X17	Lube	Complete	Pin	−40	−48				
X18		Hi Lube	Level	37	−40				
X19		Lo Lube	Level	−40	−48				
X20		Manual	Light	−1	49	50	51	52	
X21	Pull	Clamp	On/Off	49	−52				
X22		Shot Pins	In/Out	50	−52				
X23	Swing	Clamp	On/Off	51	−52				

Note: * Rung is a grouping of PC instructions which controls one output or storage bit. This is represented as one section of a logic ladder diagram.

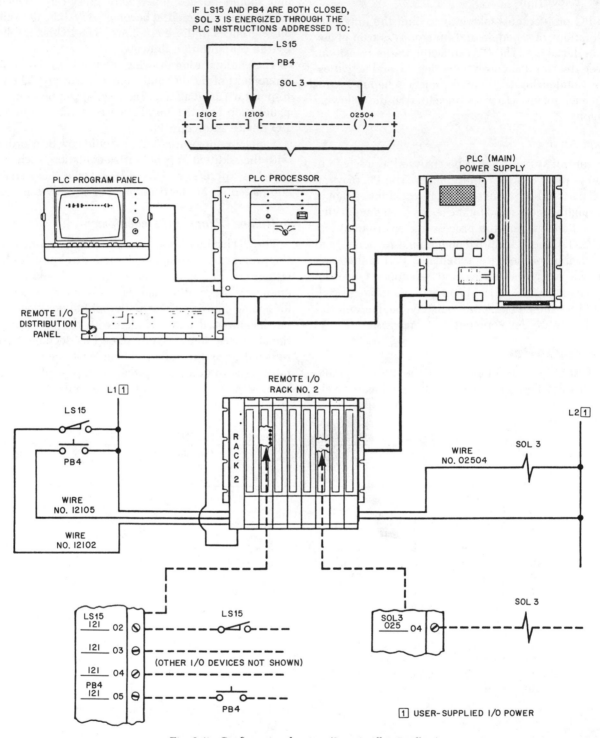

Fig. 6.4i Configuration drawing. (Source: Allen-Bradley.)

1. It communicates the functional requirements of the control system to those writing the PC code.

2. It records the thought process (regarding control) of the system designer to be used in the event of a personnel change. Such information will be invaluable.[17]

3. It provides a review document for personnel working in other capacities (mechanical, process, and project management) to ensure that they understand the operation of the controls.

4. It provides a guide for developing the operational description for the operator's manual.

After the functional specification has been reviewed and approved, a detailed operational sequence chart,

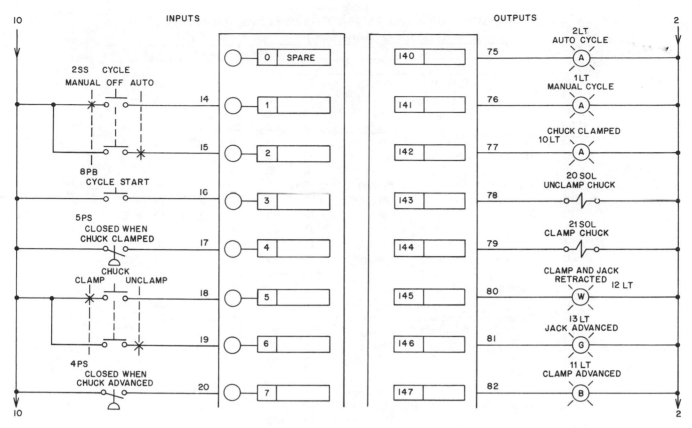

Fig. 6.4j   Point-to-point wiring diagram. (From Conrad, W., "Document Your PC Design," 9th Annual Programmable Controllers Conference Proceedings, ESD, March 1980.)

timing diagram, logic diagram, flow chart, or electrical schematic is developed from it. This schematic is translated or coded into the appropriate PC language, cross-referencing I/O with PC designator tags. The piping and instrument diagram is also cross-referenced with PC designators. In this way, future cross-referencing of system drawings and PC codes is facilitated.

As the code is entered, a memory map or register index is kept by the programmer (Table 6.4k). This map is useful for organizing program data in logical arrangements and will prove invaluable during startup, when the programmer may need to locate available blocks of memory quickly for program revisions.

Once the program is entered, a simulation is recommended, and the program check-out process is begun "on the bench." This process uses the functional specification to prove that the software is acceptable. A large percentage of the program can be proved in this manner. Program debugging can be completed before field installation. Field corrections will therefore be minimized, and high-salaried electrical and installation personnel will not be standing around waiting. The savings that can be realized are quite large.[18] There is no substitute for a bench-simulated program check. The software simulation proves the program and allows acceptance by the

customer. The program should be reproduced and documented after it has been checked.

Software/Hardware Integration

Using the PC and programming aids, the panel wiring can be "rung out" (that is, checked point by point) in through the PC. Each I/O point should be activated separately to the terminal block or from the panel controls (buttons, lights, switches). In this manner, the electrical integrity of the panel from the terminal blocks inward is ensured. If any continuity problems exist hereafter, they will be located in the field wiring.

Some organizations prefer to perform a simulated operation checkout at this time. This is a highly useful approach and can be implemented if the simulators and the I/O point arrangement are organized to simulate the process outside the panel. Occasionally, operations personnel are brought in to demonstrate this process. This is an effective way for the operations personnel to become familiar with the new system, especially in new automation projects, in which there may be some "fear of the unknown" to overcome. Occasionally, operations personnel find operational flaws or make suggestions for improvements that can be easily implemented in the panel shop but would be difficult or impossible in the

Table 6.4k
EXAMPLE OF MEMORY MAP FOR MILLING MACHINE
(Source Xcel)

Coil	Definition			Used in Rung(s):					
CR0		R.H. Head	Retract	13	15	19	−20	−33	−46
CR1		L.H. Head	Retract	14	15	21	−33	−38	−47
CR2		Clamp	Part	4	5	−34			
CR3	Cycle	Part	Unclamp	2	2	25	25		
CR4	Shot	Pins	Out	5	5	6	27	−28	
CR5	Machine	Slide Adv	Milling	6	6	7	22	24	29
				30	32	35			
CR6	Machine	Slide	Retract	7	7	8	−22	39	
CR7	Milling	Spindles	Off	9	9	10	−24	−24	−32
CR8	Start	Boring	Spindles	10	10	11	12	20	23
				31	33	38			
CR9		R.H. Bore	Complete	11	11	13	13	46	
CR11	Unclamp	Swing	Clamps	15	15	16	−23	28	−29
				−30	−31	−31			
CR12	Unclamp	Pull	Clamps	−3	16	16	17	26	
CR13		Hold	Light On	3	3	4	4	−26	34
CR15	Milling	Complete	S.Pins In	8	8	9	−27	−39	
CR17		Pump #1	Pressure	25	26	28	52		
CR18	Unclamp	Pull	Clamp	26	49				
CR19		Retract	Shot Pins	27	50				
CR20	Unclamp	Swing	Clamps	28	51				
CR29		L.H. Bore	Complete	12	12	14	14	47	
CR31	Press. Off	Pump Time	Start	17	17	18	18		
CR32	Pressure	Off	Pump #1	18	−52				
CR101		Auto Lube	Cycle	36	36	40	40	41	41
				−43	48				
CR102			Lube Off	41	42	42	−43	−44	45
				−48					
CR103			Lube On	−41	−41	42			
CR104	60 Min.	Lube	Restart	43	44				
CR105		Start/	Disable	−40	45	45			
CR106	Contin.	Lube	Restart	40	44				

field. At the conclusion of this exercise the panel is accepted by the customer and is shipped to the job site.

System Checkout and Startup

After electrical interconnections are made and point-to-point wiring is completed (mechanical completion), the system is ready for startup. The ability of the PC to operate step by step through the startup becomes very useful at this stage.

Experienced PC personnel may provide temporary STOP, CONTINUE, and STEP switches in the back of the panel in order to facilitate the startup procedures. These switches can be key-locked, software-locked, or disconnected for normal operation. They are also very useful as future maintenance and trouble-shooting tools to diagnose future problems as either hardware- or software-based.

Unanticipated circumstances always are a factor during startup. For this reason it is not uncommon to have programming personnel available at this time for imple-

mentation of any program changes that might be necessary. Quite often, these changes can be accomplished over long distance telephone lines using modems. These changes are not easily implemented without adequate documentation. After successful startup, the plant is signed off.

After Startup

Once again, it is imperative for future successful plant operation that complete, current documentation be available. This documentation should include the items discussed previously. Especially useful for future changes or additions are the startup notes and notes pertaining to future modifications.

Training of operations, maintenance, and engineering personnel should be timely and "hands-on." It is useful to videotape these training sessions for future reference (e.g., for training of new personnel). Suggested programs for PC training can be obtained from the PC vendor. This is an important function provided locally at the job site, at a nearby metropolitan area, or at the PC vendor factory.

Programmer

The programmable controller relies heavily on relay ladder diagrams as a programming language. Therefore, a working knowledge of electrical schematics would be helpful to the reader. Please refer to Sections 6.1 and 6.2 of this volume for a detailed discussion of electrical schematics and ladder diagrams.

Relationship of Electrical Schematics, Ladder Logic, and Boolean Algebra

Today's programmable controllers can be programmed in either free format or bounded format. Free format means that an unlimited number of contacts can be programmed either in series or parallel. Although this system is highly flexible, it would be impractical for documentation purposes to program in this manner. Bounded formats vary and usually limit a "network" to what can be displayed on a CRT programming screen (typically 8″ × 11″ elements). One programming technique involves defining the sequential logic in electrical schematic format, using actual tag numbers, and then translating this diagram to the appropriate programming language. Figure 6.4l shows the translation of some examples of typical circuits to ladder diagrams, Boolean algebra, and mnemonics. Because this translation is relatively simple, maintenance and engineering personnel have accepted programmable controllers, although they have not accepted computers as readily. The unknown has been replaced with the familiar.

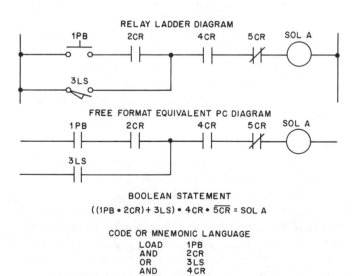

Fig. 6.4l　Ladder translation. Here is a comparison of programming languages that are used with various programmable controllers. The most popular is still the relay ladder diagram because plant personnel are more familiar with it. (From Programmable Controller Course. *Instruments and Central Systems*, Chilton Co., Radnor, PA, 1981.)

Timing and Counting

Figure 6.4m is a schematic representation of a timer and a counter. Although their formats differ, the principles are the same. The legs of the timer represent start/stop and reset. Timers can be on-delay or off-delay and can be cascaded (that is, linked together in series). Counters can be up or down and have a count leg (in which the number of switch closures is the count) and an up/down leg (in which the position of the switch determines up count or down count). A PC with arithmetic capability can use a combination timer and counter as an integrator. Figure 6.4n shows a turbine meter pulse counter turned into a low-cost integrator within the PC. Obviously, the scan rate of the PC (scans per second) must be twice the pulse rate of the turbine (pulses per second).

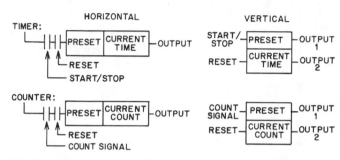

Fig. 6.4m　Timer/counter schematic. A key part of any PC programming is its capability to do timer/counter functions. The instructions are entered either horizontally or vertically, depending on the make. Horizontal programming, however, is more commonly used. (From Programmable Controller Course. *Instruments and Control Systems* Chilton Co., Radnor, PA, 1981.)

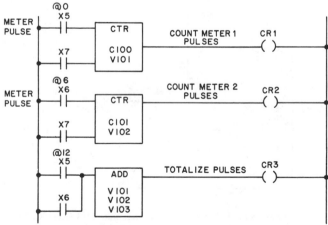

Fig. 6.4n　Pulse counter totalizer

Arithmetic Capabilities

Figure 6.4o shows an arithmetic program that permits the rapid addition of pulse counts from two counters hooked to two electric meters. The resulting sum is displayed through a panel meter. This logical addition is performed using integer mathematics (that is, no decimal calculations can be performed). Most PCs use an ap-

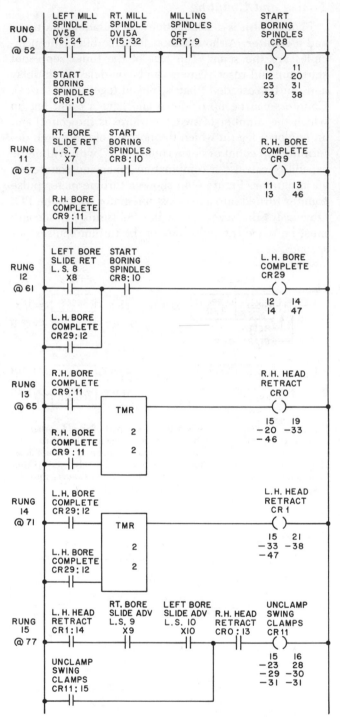

Fig. 6.4o   Sample ladder diagram—annotated. (Source: Xcel.)

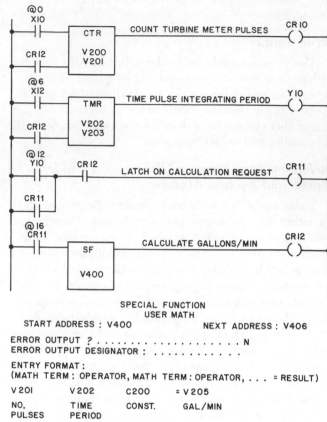

SPECIAL FUNCTION
USER MATH

START ADDRESS : V400   NEXT ADDRESS : V406

ERROR OUTPUT ? . . . . N
ERROR OUTPUT DESIGNATOR : . . . .

ENTRY FORMAT :
(MATH TERM : OPERATOR, MATH TERM : OPERATOR, . . . = RESULT)

V201	V202	C200	= V205
NO. PULSES	TIME PERIOD	CONST.	GAL/MIN

Fig. 6.4p   Floating point integrator

proximation technique called "double-precision integer mathematics" to do calculations of more complexity (such as PID). Some PCs have true floating point mathematics capability.

Floating point mathematics is a powerful tool for process applications. For example, in the integrator example in Figure 6.4p the division of pulses by elapsed time can be expressed as a decimal number rather than as a truncated integer. Feedforward calculations and PID can be performed in double-precision integer mathematics but are more memory-intense than in floating point mathematics. Floating point mathematics may require the use of a separate microprocessor within the CPU and usually involves two adjacent memory locations to store the mantissa and the abscissa in a form of scientific notation. The programmer automatically translates when the memory location is followed by a period (.), indicating a floating point number.

Advanced Capabilities

PC manufacturers have developed an ingenious method for calling up the advanced capabilities inherent within some of the more process-oriented PCs.

The use of preprogrammed special functions, or data handling features, gives the operator the ability to call up drum timers, stepper switches, shift registers, and PID loops without programming the entire exercise. The only requirement is definition of the memory locations or I/O locations that will be used. The data handling features are usually programmed within the ladder. Arithmetic logic has given many unique advantages to the programmable controller in process applications.

Programming Documentation

The following technique can be used for PC programming applications. We have found many advantages to this approach.

1. Develop detailed I/O lists. Table 6.4h shows an I/O cross-reference relating tag numbers to I/O points. This list should be used extensively; starting without it will cause confusion and errors resulting from inevitable changes.
2. Develop a detailed descriptive operational sequence of events. Figure 6.4q shows a sample sequence using a process batch application.
3. Develop electrical schematics or ladder diagrams for sequential control.
4. Develop piping and instrumentation drawings (or a logic diagram) for process control. Figure 6.4r shows a diagram that will allow maintenance personnel to find important activities quickly. Note the I/O matrix index, showing what happens inside the PC software, and the cross-referenced I/O memory locations and tag numbers.
5. Translate the drawings in steps 3 and 4 to the programmable controller language.
6. Enter the program code using a memory map (see Table 6.4k).
7. Debug the program at the programmer's facility. Use a simulator to debug the program. Run through the operational sequence defined in step 2. If it has changed, be sure to ascertain how that change has affected other parts of the program. Rewrite the sequence description to reflect current operations, if necessary.
8. Save and document the program. Reproduce the program on transportable media, such as cassettes or floppy disks. (Do this daily.) Document programming changes using peripherals, such as CRT programmers or tape loaders. These devices can be purchased or rented.
9. Enter and debug the program in the field. It is essential to note all changes made on the documentation. One of the biggest problems with relay systems is undocumented field modifications.
10. Redocument and reproduce the final program.[14,21]

The final documentation package should include the following: I/O list and cross-reference, descriptive operational sequence, electrical

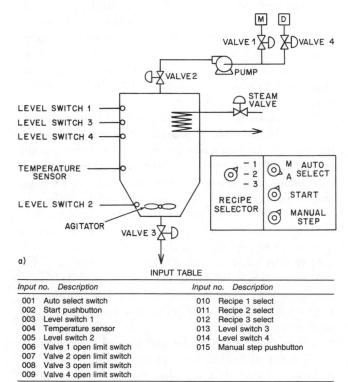

OUTPUT TABLE									
			Outputs						
Step	Description	Input conditions to advance	Valve 1 300	Valve 2 301	Valve 3 302	Pump 303	Steam 304	Agitate 305	Valve 4 306
Home	Start	Auto select and start							
1	Fill	Tank full (Level switch 1)	X	X		X			X
2	Heat	Temperature setpoint or timer		X			X	X	
3	Cool	Timer							
4	Agitate	Timer						X	
5	Empty	Tank empty (Level switch 2)				X			
6	End	Automatic reset							
		Conditional interlocks:		Valve 2 open and Valve 1 or 4 open			Temperature not at setpoint or tank not empty	Tank not empty (Level switch 2 not set)	

X = Output energized

B)

INPUT TABLE

Input no.	Description	Input no.	Description
001	Auto select switch	010	Recipe 1 select
002	Start pushbutton	011	Recipe 2 select
003	Level switch 1	012	Recipe 3 select
004	Temperature sensor	013	Level switch 3
005	Level switch 2	014	Level switch 4
006	Valve 1 open limit switch	015	Manual step pushbutton
007	Valve 2 open limit switch		
008	Valve 3 open limit switch		
009	Valve 4 open limit switch		

C)

Fig. 6.4q Batch sequence. *A*. Graphic representation of a simple batch process reactor vessel, with inputs and outputs. *B*. Outputs for each step in the process are described in the Output Table. The advance conditions for each step are shown along with input interlocks (*bottom*), which affect certain outputs. *C*. The Input Table correlates the connection terminal number with the English description of each input. (From Dietz, R.O., "Programmable Controllers in Batch Processing," *InTech*, May 1975.)

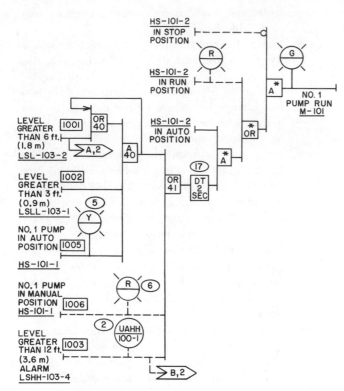

SYMBOLS DEFINED FOR PROGRAMMABLE CONTROLLER LOGIC DIAGRAM

| 1004 |, | 1008 | ETC. INDICATES PC INPUT

(5), (7), (17), (41) ETC. INDICATES PC OUTPUT

INDICATES FIELD WIRED "OR" GATE

INDICATES "AND" GATE IN PC LINE 40

THIS LOGIC DIAGRAM IS FOR SUMP PUMP

Fig. 6.4r   Process and logic diagrams. This logic diagram is for a sump pump. (From Klostermeyer, W.H., and Thurston, L.W., "PCs Prove Reliable in Chemicals and Plastics," *Control Engineering*, April 1978.)

schematics, process schematics, program listing (see Figure 6.4o for annotated final document), memory maps (showing the memory areas that have been used and those that are available), and notes for future program changes or additions.

One important word about the documentation package: A major advantage of the programmable controller is its ability to be reprogrammed as plant requirements change. Without proper documentation, previous programming efforts will have to be reproduced in order to make the changes that are required. Poor documentation results in wasteful efforts at reconstruction. The programmable controller, like all engineering tools, requires good control systems engineering practices in order for its full potential to be realized. Good documentation is an essential element of any PC project.[22]

Hardware and System Sizing and Selection

Introduction

Despite the variety of available PC models, system sizing is relatively simple. Hardware and system size can be determined by an analysis of the following system characteristics:

1. I/O quantity and type,
2. I/O remoting requirements,
3. memory quantity and type,
4. programming requirements,
5. programmers, and
6. peripheral requirements.

Although sizing is generally straightforward, selection of the right PC requires considerable judgment regarding tradeoffs between future requirements and present cost.

I/O Quantity and Type

In most programmable controllers, plug-in modules are used to convert the I/O signal level to one that is compatible with the bus architecture. These modules can be composed of 1, 4, 8, or 16 points, depending on the manufacturer's standard design. For small projects (20 to 256 I/O), I/O requirements are usually easy to define and group. A systematic approach is required for medium sized projects (256 to 1024 I/O), however, in order to avoid confusion of I/O allocation. Obviously, the organization of I/O for large systems (1024 I/O and above) requires careful planning.

The I/O base (rack or housing) is used to hold the I/O module into place and to provide a termination point for the wiring. The bases may be mounted anywhere in the control enclosure; however, there are cable length requirements that must be met. The majority of bases mount horizontally to allow proper module cooling. A terminal strip is built into the mounting base for field connections so that no wiring need be disturbed in order to remove or replace a module. These bases typically hold various quantities of I/O—anywhere from 1 to 128 I/O points. Whereas in most systems the module has the intelligence to communicate with the CPU, some systems require the use of serial interface modules. In any case, some provision is made to accept register input data from the input modules and to send these data (on or off status of field device) in serial format to the PC processor. Serial data are also converted into register data to be sent to the output module.

Input modules are typically transistor-triggered and have built-in time delays to protect against contact bounce. The input signal from a field device (limit switch) has to be energized for some amount of time in order for the module to notify the processor of a true "on" condition.

The discrete output module uses a solid state switch

(triac) to power a field device, such as a motor starter, a valve, or lights. Outputs are available for voltage ranges of 5 to 240 volts at currents up to 5 amps, with typical 120-volt outputs operating at 2 amps maximum. Solid state drivers of this type are not intended to drive large loads directly (e.g., a large motor starter). Highly inductive loads or those with a high surge current may also require an interposing dry contact relay in order to power the field device.

Manufacturers rate their equipment output current at different temperatures. All current ratings for solid state outputs will vary with ambient temperatures. Therefore, one should be sure to check the PC output rating for each application and manufacturer.

Discrete I/O modules come equipped with an LED indicator light to indicate the status of the module (on or off) for trouble shooting. The input module LED indicates field side status of the pushbutton, and the output module LED indicates logic side status.

Built-in fusing on output modules is becoming standard in the industry and provides good protection for overload conditions. The type of fuse depends on the module, and the way in which fuses are accessed varies from one PC manufacturer to another.

Field connection is made to the I/O base by way of a built-in terminal block. Some PC vendors provide wireway as part of the rack structure to "clean up" the panel design. Each new generation of I/O becomes more dense, so these newer systems become progressively harder to wire. Interposing terminal blocks between the field connections and the rack may become standard in the future. Standard cabling is provided for connection from the base to the PC processor. These cables are generally multi-bit data I/O connector/cable assemblies to simplify installation.

ANALOG I/O SYSTEMS

The analog I/O system is designed to interface with analog field devices, such as flow meters, pressure transmitters, and value positioners. This system accepts inputs of 0 to 5 V DC or 4 to 20 mA. The output analog module can output a signal of 4 to 20 mA current loop up to 750 ohms as well as 0 to 10 V DC. Some of the newer I/O modules include direct thermocouple wiring and RTDs. On-board cold junction compensation is often included with thermocouple input modules, and many different types of thermocouples can be accommodated.

These modules are sometimes referred to as A/D (analog to digital) and D/A (digital to analog) modules. They provide optical isolation for electrical noise protection and are typically arranged in a quad module or an eight-point module.

REGISTER I/O SYSTEMS

The register I/O system provides direct interface to multi-bit data field devices, such as thumbwheel switches, position encoders, and digital readouts to the PC. These devices are typically TTL level, which allows interface to other types of electronic hardware as well. Intelligent I/O and other special-purpose I/O requirements are becoming increasingly common.

The user should arrange any special I/O types as well as the commonly available modules according to an I/O matrix by logical area, as shown in Table 6.4s. In this table, I/O types are listed across the first row and plant areas are listed down the first column. In this way, one can accurately reconstruct the decision-making process concerning I/O quantity and type. It is important to include at least 10 to 20 percent spare rack space in all I/O considerations.

Table 6.4s
I/O MATRIX

Plant Area	Analog	In	Model No Qty	Analog Out	Model No. Qty	Discrete In Voltage	Model No. Qty	Discrete Out Voltage	Model No. Qty
Process area	24	4-20 mA		0		230		136	
			2			24 VDC	16	24 VDC	28
Tank farm #1	0			0		62		36	
						120 VAC	9	120 VAC	5
Tank farm #2	0			0		52		34	
						120 VAC	8	120 VAC	5
Loading station	0			0		68		24	
						120 VAC	10	120 VAC	4
								98	
								240 VAC	14

Notes: (1) Discrete in 16 pts/card
Discrete out 16 pts/card
Analog in 16 pts/card
Analog out
(2) Add model numbers
after award of contract.

For example, consider a typical process application. Assume a total I/O count of 764, broken down into 436 inputs and 328 outputs. This application falls into the *medium* PC category. Since the majority of the field devices are located a good distance from the CPU, a PC with remote I/O is desirable. The I/O requirements by locations are as follows:

- **Process Area**—390 I/O
  254 inputs; 230 at 24 Vdc, 24 at 4 to 20 mA analog
  136 outputs; all 24 Vdc
- **Tank Farm #1**—98 I/O
  62 inputs; all 120 Vac
  36 outputs; all 120 Vac
- **Tank Farm #2**—86 I/O
  52 inputs; all 120 Vac
  34 outputs; all 120 Vac
- **Loading Station**—190 I/O
  68 inputs; all 120 Vac
  122 outputs; 24 at 120 Vac, 98 at 240 Vac

Let us assume that the PC system being considered has the following features: (1) no constraints on input and output mixture; (2) the I/O modules are available in two formats, 16 points per module and 8 points per module; and (3) 10% spare I/O is required. For the sake of illustration, we will use the 16 point per module I/O structure in the process area and the 8 point per module structure in the tank farms and loading station. The I/O distribution (including spares) per location would now be as follows:

- **Process Area**—390 I/O
  24 Vdc inputs—$(230 + 10\%) \div$
    16 pts/module = 15.8   16
  4—20 mA analog inputs—$(24 +$
    $10\%) \div$ 16 pts/module = 1.6   2
  24 Vdc outputs—$(136 + 10\%) \div$
    16 pts/module = 9.4 $\dfrac{10}{28}$ 16-pt modules
- **Tank Farm #1**—98 I/O
  120 Vac inputs—$(62 + 10\%) \div$
    8 pts/module = 8.5   9
  120 Vac outputs—$(36 + 10\%) \div$
    8 pts/module = 4.9 $\dfrac{5}{14}$ 8 pt modules
- **Tank Farm #2**—86 I/O
  120 Vac inputs—$(52 + 10\%) \div$
    8 pts/module = 7.2   8
  120 Vac outputs—$(34 + 10\%) \div$
    8 pts/module = 4.7 $\dfrac{5}{13}$ 8 pt. modules
- **Loading Station**
  120 Vac inputs—$(68 + 10\%) \div$
    8 pts/module = 9.4   10
  120 Vac outputs—$(24 + 10\%) \div$
    8 pts/module = 3.3   4

240 Vac outputs—$(98 + 10\%) \div$
  8 pts/module = 13.5 $\dfrac{14}{28}$ 8 pt modules

If we assume that one remote communication channel can service up to 128 I/O in groups of sixteen 8-point modules or eight 16-point modules, the system becomes:

- **Process Area**—390 I/O; 28 16-point modules; 4 remote channels
- **Tank Farm #1**—98 I/O; 14 8-point modules; 1 remote channel
- **Tank Farm #2**—86 I/O; 12 8-point modules; 1 remote channel
- **Loading Station**—190 I/O; 28 8-point modules; 2 remote channels

I/O Remoting Requirements

A unique feature of the PC is the multiplexed nature of the I/O bus. This can be used to great advantage to reduce overall wiring cost. If I/O racks are centralized in logical clusters, plant wiring requirements can be greatly reduced. Wiring between racks and the CPU can be reduced to a few twisted pairs of wires or a single cable. The tremendous cost savings that result can be realized without a compromise of control accuracy or capability.

A system configuration diagram (such as that shown in Figure 6.4i), when used in conjunction with the I/O matrix in Figure 6.4s, aids in keeping track of the overall system configuration.

Remote I/O is broken down into two distinct types: the integral type, which allows a limited transmission distance (up to 15,000 feet, or 4500 meters); and the transmitter/receiver type, which allows virtually unlimited transmission capability. Most PC manufacturers and third party peripherals manufacturers can provide some form of either type. Technology is advancing greatly in this area as systems change from fiber optics to microwave and radio transmission.

It is important to remember the major weakness of remote I/O systems. If the bus is cut or interrupted, the effects of I/O failure will be relatively unpredictable. One must consider the effect of a possible system failure on each step in the sequence. For this reason, duplication of smaller CPUs at each remote location is often considered preferable to a large central CPU. This is actually an extension of distributed control within the network of the PC itself. This approach can be very cost-effective, since requirements for the central unit size can be reduced. Serious consideration should be given to distributed versus centralized architecture in remote I/O systems in which control system integrity is important.

Memory Quantity and Type

As discussed previously, memory types vary; however, three basic types can be identified. These are:

1. ROM—usually nonvolatile, mid-price range.
2. RAM—usually volatile, requiring battery backup (a read/write [R/W] memory).
3. CORE—nonvolatile, most expensive (another form of R/W memory).

PC manufacturers offer variations of these memory types in sizes ranging from 256 words to 192 K words. The significant point to remember is that memory size can be deceptive. The word size (from 4 bits to 16 bits) determines to a large extent the programming efficiency of a particular machine. In addition, there is a difference between executive memory and user memory.

Total memory, as stated in manufacturers' literature, does not necessarily mean that the entire content is available to the user. Some manufacturers reserve large blocks for the PC executive. A system with 4 K of 16-bit words of user memory may comfortably accommodate a program, whereas another system with 8 K of 8-bit words may have too small a memory for the same program.

Special programming language features are an important aspect of memory sizing, especially in process control. The PID algorithm is a perfect example: One manufacturer requires 33 words of user-available memory, whereas another may need in excess of 1000 words. Obviously, the memory sizing for a loop control program would vary in these two systems. Another example is the use of special functions, such as shift registers. An alternative way of developing a shift register in ladder logic is to use a special function shift register or handling data to require less user memory. Word (or register) moves are also powerful in terms of memory efficiency. Programming languages, which can be binary- or octal-based or alphanumeric Boolean, affect memory use. The closer the language is to machine code (binary-based), the more user memory is required to perform the more complex functions. The closer the language to alphanumeric Boolean, the less memory will be required for complex functions.

The best way to determine program memory prerequisites is to write a representative sample program reflecting some actual project requirements and to request information about user memory size from the various manufacturers. If the manufacturer's suggestions are followed, the user can be reasonably assured that the memory will not be undersized.

The final area of caution about memory size concerns the consideration of data storage. Data tables, scratch pads, and historical data retrieval requirements can inflate the size of the PC memory. It should be remembered that the primary task of a PC is control of the process. If data requirements are large, connection to auxiliary devices, such as mini- and microcomputers, should be given serious consideration. Many of these devices are currently available and are of an industrial grade; furthermore, the price of these systems is coming down rapidly. It is not good engineering practice to degrade control capabilities by burdening the PC with excessive data acquisition functions. As a plant goes on line, operational requirements for data generally increase astronomically. These will be easily accommodated by a mini- or microcomputer but not by the PC memory.

Programming Requirements

The user's programming requirements are an important aspect of PC selection. As already discussed, the use of a binary-based language makes implementation, debugging, and maintenance of a PID loop most difficult. However, if the extent of the user's requirements is simple ladder logic, the extra cost of sophisticated programming capabilities is not justified.

Generally, the aforementioned sample program written to size memory will "shake out" the best-matched programming capabilities to project requirements. This can be especially enlightening if the user can actually get hands-on experience with the sample program on a demonstrator unit. A few hours spent in this fashion can save much difficulty during the programming phase of the project.

Programmers

Two basic programming devices provided by the PC vendor can be used to enter a program into a PC: a hand-held programmer and a CRT programmer.

The hand-held programmer enables the operator to enter a program one contact at a time. These units are widely used because they are rugged, portable, and easy to operate. They are very cost-effective and give an engineer the capability to enter a program and to diagnose trouble in logic and field devices.

The CRT programmer provides the engineer with a visual picture of the program in the PC. Ladder diagrams are drawn on the screen, just as they would be drawn on paper. Design and trouble-shooting time is reduced with the use of the CRT. With menu-driven software, programmer training time is decreased. The CRT is designed for desktop or factory floor orientation. The exterior is made of impact-resistant foam or metal with a spill-proof tactile keyboard. The screen size varies from 4 to 9 inches (100 to 225 mm), with keylock protection to prevent unauthorized program alteration.

These units can be ordered with memory storage capabilities. Information is stored on either floppy disks or magnetic tape. With floppy disk memory, programs can be copied from one disk to another and then verified without need for loading the PC's memory. Stand-alone programming is a feature of disk memory—an engineer can develop his program on the disk and then load into the PC.

Some CRT programmers provide complete documentation capability, including ladder diagrams, cross-reference listing, and I/O listing. More sophisticated CRT programmers supply "user"-defined contacts and coil names along with commentary above each network or rung. The CRT programmer also includes an external RS232C port for connection to a printer.

The CRT screen typically shows 8 rungs of ladder logic by 11 contacts across. The ladder diagrams can be placed into the "real-time" mode, which allows visual contact status. A whole screen of contacts and coils can be updated in as fast as 40 msec. These programmers are designed for portability but do weigh 45 to 60 lbs (20 to 27 kg). With a modem connection these CRTs can be used at remote locations for programming and trouble shooting.

Peripherals

Project requirements can make this category very important. The advanced capabilities of the PC have resulted in an increase in the number of peripherals offered by PC manufacturers and third party vendors. Up-to-date manufacturer information is the best source for a description of PC capabilities at this time.

PC Installation

Installation of programmable controller systems is not a difficult or mysterious procedure, but the following general rules will save time and trouble for the systems designer or installer. The basic principles of PC installation are the same as those for installation of relay or other control systems. Safety rules and practices governing proper use of electrical control equipment in general should be observed. These include correct grounding techniques, placement of disconnect devices, proper selection of wire gauge, fusing, and logical layout of the device. PCs can often be retrofitted into existing hardwired relay enclosures because they are designed to withstand the typical plant environment. PC vendors provide installation manuals upon request.

Safety Considerations

Perhaps the most important safety feature, which is often neglected in PC system design, is emergency stop and master control relays. This feature must be included whenever a hardwired device is used in order to ensure operator protection against the unwanted application of power. Emergency stop functions should be completely hardwired (Figure 6.4t). In no way should any software functions be relied upon to shut off the process or the machine. Disconnect switches and master control relays should be hardwired to cut off power to the output supply of the PC. This is necessary because most PC manufacturers use triacs for their output switching devices, and triacs are just as likely to fail on as off.[15,20,23,24]

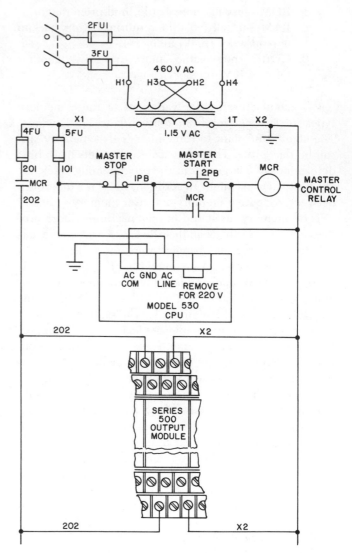

Fig. 6.4t   Master stop switch installation. (Source: Texas Instruments.)

Implementation

Planning ahead is every bit as important in designing a complete PC system as in laying out a relay logic panel. Care in counting I/O points in the beginning—and leaving a safety factor—will save headache in the panel fabrication stage. Panels should always have plenty of expansion room left over, since I/O is invariably added as the job progresses and the operators see the advantages of PCs. The designer should refer to the layout considerations provided by the manufacturer. Extra space should be left to provide access to the boards and connectors of the PC. The diagnostic and status indicators should all be visible. The designer should leave room between I/O racks for wireways and large hands.

One good technique for ensuring efficient panel layouts is to involve maintenance personnel in the design procedure. This not only optimizes the layout but also introduces the staff to the hardware (Figure 6.4u).

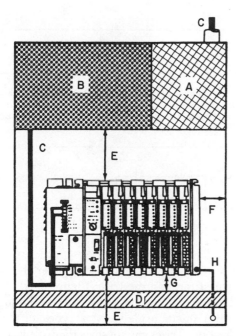

Fig. 6.4u   Typical enclosure layout. *A*. This 9″ × 12″ (225 × 300 mm) area reserved for disconnecting means. *B*. Area available for fuses, control relay, transformers, or other user devices. *C*. AC input line. Routed separately from I/O wiring. *D*. Wiring duct for I/O wiring within enclosure. *E*. Unobstructed minimum vertical space required. Above the controller: Typical = 6″ (150 mm), Worst Case = 12″ (300 mm). Below the controller: Typical = 6″ (150 mm), Worst Case = 10″ (250 mm). Between two controllers: Typical = 6″ (150 mm), Worst Case = 12″ (300 mm). *F*. Unobstructed horizontal space required: 4″ (100 mm) minimum. *G*. Minimum 2″ (50 mm) between I/O wiring duct or terminal strip and I/O chassis. *H*. 8-gauge wire or 1″ (25 mm) braid for bonding purposes. 6″ = 150 mm; 36″ = 900 mm; 25″ = 625 mm.
(Source: Allen-Bradley.)

In general, the best defense against creating a tangled mess when designing a PC system is to follow proper documentation techniques. By spending a little more time documenting panel layout, I/O counts, and wiring diagrams, one will spend a lot less time starting up the system. PCs can handle large amounts of I/O points with varying electrical characteristics, so things can get pretty confusing in a hurry. Cable requirements between hardware boxes vary from one type of PC to another, so this is an important consideration in panel layout.

Enclosure

Enclosures should nearly always be provided for the PCs themselves. This protects the electronics from moisture, oil, dust particles, and unwanted tampering. Most manufacturers recommend a NEMA 12 enclosure for the standard industrial environment. This type of enclosure is readily available in a variety of sizes and, in fact, may be already included with a new system.

Programmable controllers are designed to be located close to the machine or the process under control. This keeps the wiring runs short and aids in the troubleshooting procedure. At times, however, mounting the PC directly on the machine or too close to the process is not advisable, such as in cases of vibration inherent in the machine, electrical noise interference, or excessive heat problems. In these situations, the PC must be either moved away or successfully protected against these environmental conditions.

Temperature Considerations

Installing any solid state device requires paying attention to ambient temperatures, radiant heat bombardment, and the heat generated by the device itself. PCs are typically designed for operation over a broad range of temperatures, usually from 0 to 60°C. When analyzing the proposed PC environment, however, one should remember that enclosure temperatures usually run a few degrees higher than ambient temperatures. Radiant heat on an enclosure from surrounding tanks can raise the internal temperature beyond that specified by the manufacturer.

Heat generated by the PC is a key issue when the device is placed in ambient temperatures close to the extreme mentioned in the specifications. The temperature rise caused by the power consumption of the PC itself is not hard to estimate. In addition, most manufacturers will provide a notation of the power consumption of the triacs driving field loads. When designing the hardware layout within the panel, one should adhere to the manufacturer's suggestions regarding ways to minimize heating problems. Most PCs use convection over fins to take heat away from particular areas within the hardware. Care must be taken to ensure that no obstruction to air flow over these fins is introduced by placement of the PC in the enclosure. Wireways are typically provided with holes to allow air to pass through. Generally, one can avoid problems with PC enclosures by simply leaving plenty of air space around the heat producers.

Should all of these factors combine to cause a temperature problem, the panel can be vented, air conditioned, or moved to another location. Usually, simply blowing filtered air through the enclosure will resolve minor difficulties. If air conditioning is required, small units that are designed for cooling electronic enclosures are readily available.

Noise

Noise or unwanted electrical signals can generate problems for all solid state circuits, particularly microprocessors. Each PC manufacturer suggests methods for designing a noise-immune system. These guidelines should be strictly followed in the design and installation phases, since noise problems can be very difficult to isolate after the system is up and running. I/O systems are isolated from the field, but voltage spikes can still appear within the low-voltage environment of the PC if proper grounding practices are not followed.

A well-grounded enclosure can provide a barrier to noise bombardment from outside. Metal-to-metal contact between the PC and the panel is a must, as is a good connection from the panel to the ground. Noise producers within the panel should be noted during the panel design phase, and the PC must not be located too close to these devices. Wiring within the panel should also be diverted around noise producers so as not to pick up any stray signals. Often, it is necessary to keep AC and DC wiring bundles apart, particularly when high-voltage AC is used at the same time that low-level analog signals are present.

Line voltage variations can cause hard-to-trace problems in the operation of any computer-based system. PCs are no exception, even though they are designed to operate over a much larger variation in supply voltage. Large spikes or brownout conditions can cause errors in program execution. Most manufacturers protect against this, enabling the controller to come up running after a brownout, but these measures may not be acceptable in all applications. The designer may wish to add an isolation transformer to his PC system, sized for twice the anticipated load. This is cheap insurance, and PC manufacturers will help determine the required load.

Hookup

PC panels can be very neat and orderly if all the terminals are arranged in a logical fashion. The actual result is a direct function of the time spent during the design process. Interposing terminal blocks between the PC I/O structure and the field is suggested, since the terminations provided by PC manufacturers are shrinking in the race to provide higher-density I/O. This also gives the panel designer the ability to place the field termination points where they are easily accessed. Wiring ducts keep the panel neat and protect the wire from mishap.

Many noise problems can be averted by following good wiring practices. Low-voltage signal wiring should be kept away from noise sources. Analog signals should be shielded, with the shield terminated at an isolated ground in the panel only (to prevent shield ground loops). Again, these analog signals should be separated from power wiring.

Triac outputs require some special attention that will be new to relay users. Triacs used for AC loads typically leak a small amount of current. In the case of triac outputs from a PC, this leakage may be enough to keep panel lamps glowing or small relays energized. When a triac is used to switch the input on a PC, the leakage may be enough to make the PC "think" the input is on. A dummy load (shown in Figure 6.4v) can be used to drain this leakage when the input should be off. Whenever a mechanical contact is used in series with a load energized by a triac (as shown in Figure 6.4w), a resistance-capacitance (RC) network should be used as shown to protect the triac from inductive kickback. A varistor should be

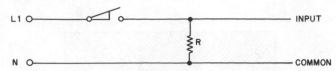

Fig. 6.4v   Dummy load for leakage. If a "leaky" triac, such as a proximity switch, is used to switch the hot side of the line, then a pulldown resistory (R) is required to counteract the leakage.

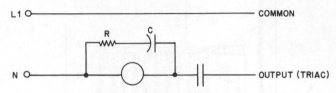

Fig. 6.4w   Mechanical contact in series, RC suppressor

provided in parallel with a load whenever the load can be "hot-wired" around the triac (Figure 6.4x). The user should check with the PC manufacturer for the suggested RC and MOV (metal oxide varistor) types for the particular application. Triacs cannot directly drive large motor starters and similar devices. PC manufacturers provide surge specifications for the various I/O cards. Sometimes an interposing relay or dry contacts will be required for large loads.

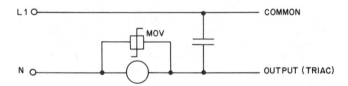

Fig. 6.4x   Mechanical contact in parallel, MOV suppressor

Programmable controllers are basically similar to most other electrical control systems. To be sure, solid state devices, microprocessors, and triacs require some special considerations during the design, installation, and start-up phases of a project, but these concepts are not too unreasonable or difficult to assimilate. As always, good design habits in the beginning will ensure a safe and reliable control system.

Applications

Applications for PCs can fill a whole book. Many sources documenting various PC applications are listed in the bibliography of this section.

PC vendors can supply numerous application notes for the products that they offer. Most major PC vendors also publish detailed articles about applications in technical journals and prepare papers for engineering societies and industrial symposia on control, automation, and so forth. Each manufacturer's software package usually has its own application programming techniques. Vendors also are a valuable source of "how-to" information, providing training courses in their local office or at the factory and actual "hands-on" experience to help users gain familiarity with the PC. Most vendors offer an applications or programming manual that gives insight on how to use available programming features. Of course, familiarity with one brand of PC will help the engineer learn to use another brand quickly.

PC Peripherals

The PC peripherals marketplace is expanding rapidly at the time of this writing; however, all current products can still fit into three major categories: operator stations, I/O enhancements, and programming and documentation tools. Operator stations are devices designed to allow operator interaction with the PC during normal operation (not necessarily only during programming or trouble shooting). I/O enhancements include all those capabilities not ordinarily supplied with PCs or those that a particular manufacturer may not carry. Programming and documentation tools include those provided both by the manufacturer and by aftermarket vendors.[5,25]

Operator Stations

Operator stations include those provided by manufacturers and intended to be used with their particular PC and those offered by third parties for use with either a particular brand or anyone's PC. These stations may include devices such as Timer/Counter Access Modules (TCAMs), Loop Access Modules (LAMs), data terminals, color graphics consoles, computers, printers, and manual back-up stations.

Most PC manufacturers provide an Operator Interface Unit (OIU) designed specifically for their PC. These are either part of the standard system or offered as an option. They are usually mounted directly on the PC but may be designed to be panel-mounted and cabled back to the controller. Functions include access to read/write register data, simple programming, and diagnostics. Some specialized devices, such as TCAMs, LAMs, and OIUs, provide operator interaction with PC internal registers and loop tables. This gives the systems designer the ability to provide real-time changing of variables, loop tuning and inspection, manual control of analog outputs, and the ability to provide batch- or menu-type information at low cost. Communications with the PC are multi-dropped over an RS422 or a similar differential line. Unauthorized data entry is prevented with software locks, keylock protection, or both.

Some PCs can support communications directly with dumb data terminals. Operators enter the data by issuing special control characters to the PC communications port. Data terminals can be provided in industrial versions intended for the plant floor or in office machines for entry of data by a supervisor. This operator interface approach is not very user-friendly and can be intimidating.

Color graphics consoles offer process graphics and communications facilities to many brands of PCs simultaneously. These systems range from those that can simply be purchased and put on line with a minimum of engineering effort to those that require some programming. The basic differences are in flexibility. Those that do not require programming may not be able to provide the custom menus and graphics that are required. The ease of communications with different types of PCs also varies according to manufacturer. Finally, the method of generating the graphics pages differs greatly. Most color graphics consoles offer multiple graphics pages that are animated by reading data tables in the PCs. Operators enter the data by means of standard keyboards, user-configurable industrial keyboards, light pens, touch screens, and the like. Different graphics pages may be selected with preformatted menus or custom menus programmed by the user or the systems house. Development stations are often required to give the final user the ability to change graphics menus or key commands after the initial project is completed.

Computer systems can be made to perform man-machine interface functions. Indeed, the color graphics consoles described in the previous paragraph are simply computers with standard graphics and communications software packages. Most PC manufacturers provide board-level additions or modules that give the PC the ability to converse via the RS232 protocol to most any computer. Of course, both the communications software and the particular applications software must be generated to provide an interface. Many vendors and systems houses are providing communications packages for various PCs to run on microcomputers and personal computers. These small systems offer low-cost operator interfaces to PCs, providing data handling capabilities and the ability to be networked into a true distributed architecture. In this way, PC purchasers can be assured that their investment will be protected from factory automation. Microcomputers that have the ability to multi-task, access large amounts of both RAM and nonvolatile memory, have proper software support, and are able to be networked will provide a good investment in terms of operator interface functions as well as total system capability.

Printers have always been an important part of the PC system both as a development tool and for handling some of the operator interface functions. Many PCs are able to provide communications directly to dumb printers. A stand-alone PC system then can often provide performance reports, alarm logging, and the like without ever involving a computer. This feature is usually somewhat limited, since PCs were designed primarily to control the process machine. Large amounts of data, sophisticated print logs, and multiple alarms are not really within the realm of a stand-alone PC system. This type of data manipulation is too cumbersome and requires too much memory for most PCs.

Manual control stations are important as backups in case of failure of the PC controlling PID loops. Loop access modules provide manual control capabilities but still rely on the integrity of the PC, so they are not truly manual in the hardwired sense. A manual control station is an important part of the distributed control system because it gives true manual control of the loops locally or in the control room, even when the local controllers are down.

I/O Enhancements

PC manufacturers are providing more and more types of input and output capabilities for their products. There are, however, many third party peripherals that aid the PC in interfacing to the field devices. New I/O capabilities that are being offered include faster response, new analog capabilities, intelligence, high-speed pulse counters, dry contact, and specialty modules.

Fast-response I/O is currently offered in both discrete and analog versions. Discrete rapid response modules are facilitated by the PC logic, but the output does not rely on ladder logic scan times to get updated. Fast analog modules provide quicker analog-to-digital (A/D) and digital-to-analog (D/A) conversions. This gives PCs the ability to control faster PID loops and to make analog measurements of assembly line parts (weight, for example).

Analog I/O capabilities for PCs are being expanded from the conventional 4–20 mA, 0–5 V, 0–10 V versions to include direct thermocouple and RTD inputs. These modules typically accept eight to ten points each, and different types of T/Cs and RTDs are accommodated.

Intelligent I/O modules include all modules that are able to perform processing functions. Because the tasks performed by the PC are further distributed, greater speed and reliability for the overall system can be realized. Intelligent I/O modules give the PC multiple additional capabilities, which may include memory storage and retrieval, computing tasks, and communications. Memory modules provide additional room to store data points, alarm messages, lookup tables, and the like. This approach leaves the main operating memory free for the control tasks. Computing modules give PCs the ability to perform true computer functions using a language like BASIC. Again, the real-time tasks are left in the main memory, but tasks such as set point calculation, formation of data, and some operator interface tasks may be placed in the computer module.

Communications modules can provide the PC with a range of capabilities, from simple ASCII output strings to communication networking. The storage of ASCII messages for a printer or display can be contained outside the main memory of the PC, and the data can be output when required. Full-system communication networking capabilities are provided with network modules, giving the designer the ability to multi-drop PCs off a single operator interface device or a supervisory computer.

High-speed pulse counter modules provide the ability to interface with turbine meters, stepper motors, and optical encoders. High-speed pulses cannot normally be interfaced to PC inputs because of the scan time of the ladder logic. These modules provide an interface that does not rely on the scan time, so that the PC is able to monitor pulses that indicate position or flow.

Dry contact modules are offered by both manufacturers and third party vendors. These modules solve the problems normally associated with triacs, low power, and uncertainty of failure state.

Specialty modules are designed to solve a single interface problem. X/Y positioner modules can be included in this category, as well as servo axis controllers, stepper motor outputs, and even maintenance access modules. These modules are a further extension of the distributed technology. Clock modules that fit into the I/O bus may be considered to be part of this group. These modules provide real-time and day/date functions upon interrogation from the PC. Most are backed up by a battery to ensure timekeeping during power outages.

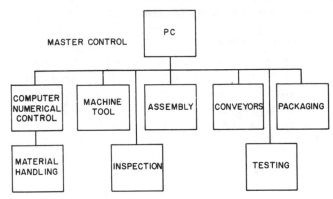

Fig. 6.4y   With master control, one PC controls a number of related machines and processes. This system is simple but may require long runs of multiple-wire cables, and the entire system is vulnerable to the failure of the one PC. (Source: Allen-Bradley.)

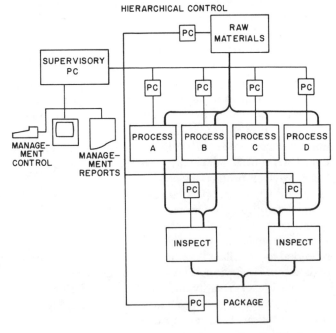

Fig. 6.4z   Hierarchical control. (Source: Allen-Bradley.)

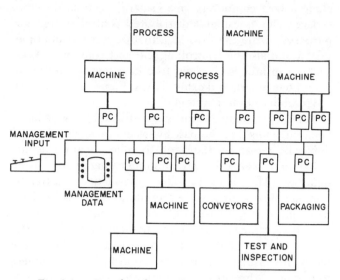

Fig. 6.4aa   Distributed control. (Source: Allen-Bradley.)

Programming and Documentation Tools

Both PC and aftermarket parties offer programming and documentation tools for the system designer or user. These tools include programmers, CRT documentors, and complete microcomputer-based systems.

Programmers are typically provided by the PC manufacturer and are designed to program a specific machine or family of machines. Some third parties are offering universal programmers and documentors. These microcomputer devices vary greatly in price and capabilities but offer on- and off-line programming to many different types of PCs, real-time status, and some very sophisticated annotation. Communications to different PCs are usually supported with different software packages. Each vendor's product offers different types and amounts of ladder and contact comments. Again, many types of cross-references are available to be printed out. Often other PC design documentation problems may be solved, such as the generation of panel configuration drawings, point-to-point wiring diagrams, and I/O layout. One system even prints out the wire labels! Some of these devices offer still other computer services—word processing, BASIC programming, and even computer aided design (CAD) facilities.

Communications Networking

Because many plants have multiple processing areas with many brands of PCs, a growing number of com-

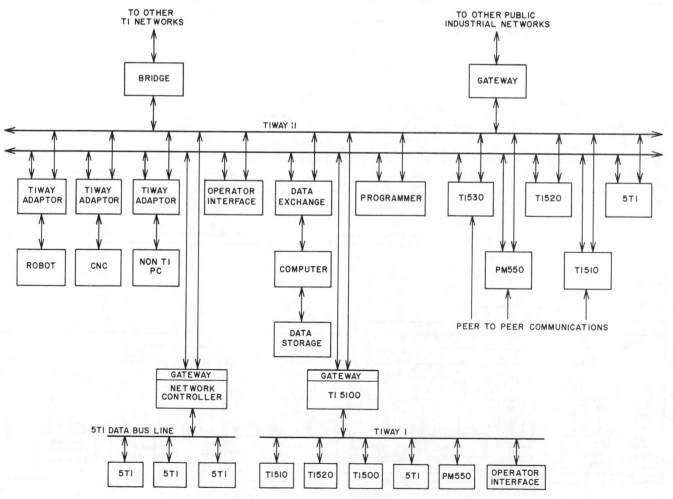

Fig. 6.4bb   Peer-to-peer communications. (Source: Texas Instruments.)

panies are moving to offer communications networking capabilities to these customers.[26-29]

History

Initially, many PC vendors offered a large PC controlling many field machines or processes. This was called master control (Figure 6.4y). As more processes or machines were added to the plant, the one large PC became too overburdened to handle all the control functions. In this case, the main PC needed to offload some of its computing power, and the result was placement of many smaller PCs for controlling the new processes.

In this way, the supervisory PC controlling a network of other subsidiary PCs evolved. The supervisory PC is called a "master," and the subsidiary PCs are called "slaves." The name for this type of system is a hierachical system (Figure 6.4z).

The largest communications networks typically include a host computer, or a "smart" man-machine interface. This is called a distributed control system for purposes of the "PC world." In this system, multiple PCs would also be controlling the host computer. Typically, a printer is involved to obtain management and operations reports and to log alarms. Figure 6.4aa depicts a typical distributed system.

Communications networking and data highways have been introduced in the past five years by many PC vendors who recognize the need to collect many existing PCs into a "network", all communicating over a "data highway." The idea is to collect information in a timely fashion within a plant-wide communication system. This field, although quite well defined in the instrumentation industries, is a fast emerging technology in the PC world as the management, engineering, and operations departments of many companies are growing accustomed to obtaining the latest information regarding plant performance from their control PCs.

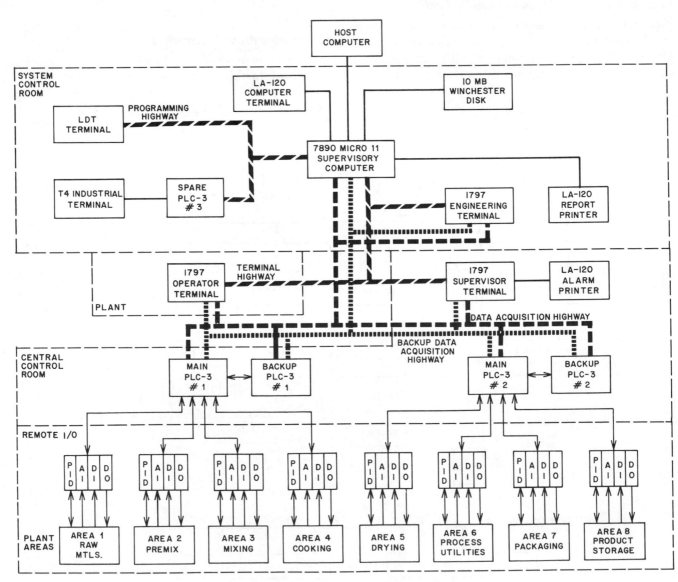

Fig. 6.4cc   Hot backup PC networking. (Source: Allen-Bradley.)

Most major PC vendors are far advanced in their ability to network their own products, and many are evaluating the possibility of incorporating multiple brands of PCs on their own network. Various network protocols exist, and the field performance of all of these systems has been proved. Standardization within PC networking is being addressed by the industry, but currently each PC vendor has his own variety.

Universal Communications Networking

Some PC manufacturers and third parties are offering universal communications networking. This is sure to be the way of the future, because the need for networking different brands of PCs together will increase.

Several exciting new developments in the networking arena are showing up. Included are peer-to-peer communications, "hot" redundancy, on-line engineering, and cost-effective networking of intelligent man-machine interfaces.

Peer-to-Peer Communications

Many of the PC manufacturers have already addressed the need for peer-to-peer communication among the PCs on a distributed network. Without this feature, every time one PC needs to know the status of another part of the machine or plant, it must interrupt the activities of the supervisory computer in order to get the information (Figure 6.4bb). The ability of one PC to "talk" to another along the data highway greatly speeds the control activities of each machine and allows the supervisory computer to "concentrate" on its tasks.

Hot Backup PCs

The new breed of data highways can provide a process with a "hot" backup PC to take over in the event of a failure. With the supervisory computer involved, sufficient intelligence exists to determine whether or not a particular PC is performing properly. Figure 6.4cc depicts a distributed system with redundant PCs.

On-Line Engineering

On-line engineering of PC systems can now be offered from a remote location with the combination of the programming/documentation tools and the distributed network. Systematic startup and debugging of processes are available with this technique. Figure 6.4dd depicts on-line engineering as part of a distributed system.

Intelligent Man-Machine Interfaces

Intelligent man-machine interfaces, or enhanced computer-based operator interfaces, can be networked into a total distributed system to give a redundant or local interface to the system (Figure 6.4ee). This technique can be used to provide the control room interface while the supervisory computer supports its own interfaces. Another powerful technique is to allow the networking of a few PCs together solely for the purpose of providing

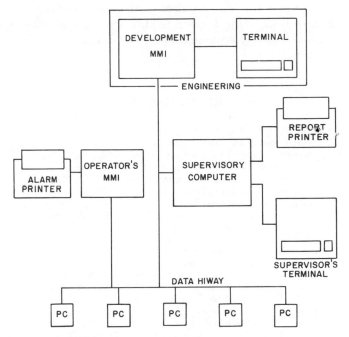

Fig. 6.4dd   Networking with on-line engineering

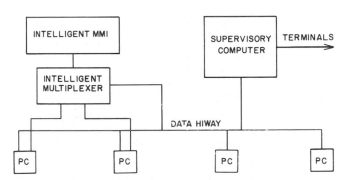

Fig. 6.4ee   Local or redundant MMI in network

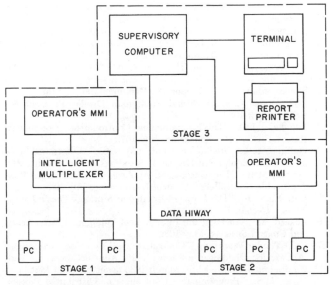

Fig. 6.4ff   Entry-level networking grows with user's needs

a single man-machine interface to all parts of the system (Figure 6.4ff). This solution is ideal for small PC users because it is very economical yet allows expansion as the plant grows.

Control Systems," International Conference on Power Transmissions, Houston, June 1982 (sponsored by *Power Magazine*).

29. Jannotta, K.L., "Factory Communications: How to Talk to the Machines," International Conference on Power Transmissions, Houston, June 1982 (sponsored by *Power Magazine*).

REFERENCES

1. "The PC User 1983 Buyer's Guide," *PC User*, December 1982.
2. "Fast Growing PC Market Encourages Wide Range of Product Offerings," *Control Engineering*, January 1983.
3. "PC Users Guide," *Instruments and Control Systems*, May 1983.
4. Londerville, S., "Programmable Controllers in Boiler Control Applications," Instrument Society of America, Northern California Section, ISA Days, June 2, 1982.
5. Laduzinsky, A.J., "Peripherals Enhance PC Performance," *Control Engineering*, January 1983.
6. Whitehouse, R.A., "The Future of Programmable Controllers," ISA 1978 Annual Conference, Advances in Instrumentation, Vol. 33, Part I, Instrument Society of America, October 15–19, 1978.
7. Korarek, L.A., "Sizing Criteria for the Evaluation of Programmable Controllers," Instruments and Control Systems Seminar/Workshop (available from Allen-Bradley).
8. Bennett, L.E. and Lockert, C.F., "Evaluating Controls for Batch Processing," *InTech*, July 1979.
9. Dietz, R.O., "Programmable Controllers in Batch Processing," *InTech*, May 1975.
10. Larsen, G.R., "A Distributed Programmable Controller System for Batch Control," *InTech*, March 1983.
11. Melville, H.J., "PC Applications in Batch Digester Control," ISA 1980 Annual Conference, Advances in Instrumentation, Vol. 35, Part II, Instrument Society of America, October 20–23, 1980.
12. Hawkinson, G.M., "Selecting a Programmable Controller for a Retrofit," 9th Annual Programmable Controllers Conference Proceedings, ESD, March 1980.
13. King, J. and Ptak, J., "Toothpaste Processing Using a PC/Mini/Microcomputer Network," 12th Annual Conference and Equipment Display Proceedings, ESD, March 1983.
14. Conrad, W., "Document Your PC Design," 9th Annual Programmable Controllers Conference Proceedings, ESD, March 1980.
15. Bryant, J.A., "Is Your Plant's Control System Safe?," *Power*, August 1979.
16. Penz, D.A., "Organizing PC Software Development," *Instruments and Control Systems*, February 1982 and March 1982.
17. Heider, R.L., "The Programmable Controller as a Process Link to a Distributed Computer Network," ISA 1980 Annual Conference, Advances in Instrumentation, Vol. 35, Part II, Instrument Society of America, October 20–23, 1980.
18. Kraemer, W.P., "Testing and Startup of Programmable Controller Systems," Paper PC 179–14, IEEE Transactions on Industry Applications, Vol. IA–16, No. 5, September/October 1980.
19. Programmable Controller Course. *Instruments and Control Systems*, Chilton Co., Radnor, PA, 1981.
20. Klostermeyer, W.H. and Thurston, C.W., "PC's Prove Reliable in Chemicals and Plastics," *Control Engineering*, April 1978.
21. Stockweather, J.W., "Proper PC Documentation for the User," 11th Annual Conference and Equipment Display Proceedings, ESD, March 1982.
22. Langhans, J.D., "The Programmable Controller in a Continuous Process," *Control Engineering*, July 1972.
23. Hoffman, E.L., "Redundant Control Systems," 10th Annual Conference and Equipment Display Proceedings, ESD, March 1981.
24. Sykora, M.R., "The Design and Application of Redundant Programmable Controllers," *Control Engineering*, July 1982.
25. Lutz-Nagy, R., "The Dawn of Intelligent Motion," *Power Transmission Design*, February 1980.
26. Eckard, M., "Tackling the Interconnect Dilemma," *Instruments and Control Systems*, May 1982.
27. ESD Vendor Workshop, "Control Networks vs. Data Acquisition (A Case for Dividing the Tasks)," 12th Annual Conference and Equipment Display, ESD, March 1983 (available from ISSC).
28. Hoag, D.S., "Programmable Controllers in Distributed Process

BIBLIOGRAPHY

Baille, N.B., "The Application of Programmable Controllers to Compressor Systems for Offshore Applications," *Measurement and Control: Journal of the Institute of Measurement and Control*, (England), Vol. 14, No. 9, September 1981.

Baker, A.T. and Rahoi, R.J., "Programmable Controller Applications for Power Plant Waste Water Treatment Facility," ISA Power Instrumentation Symposium, Chicago, May 14–16, 1980.

Barrett, L.E. and Lemmon, R.H., "Use of Programmable Controllers to Control Series Slurry Pumping Systems," 10th Annual Programmable Controller Conference and Equipment Display Proceedings, ESD, March 10–12, 1981.

Bartlett, P.G., Jr., "Installing a PC," *Instruments and Control Systems*, August 1980.

Bauman, D.E. and Mellott, M.T., "Automate Small Refinery Blending Operations," *Hydrocarbon Processing*, November 1981.

Booty, D.J., "So You're Buying a Programmable Controller," *InTech*, September 1980.

Burr, J., "Smart Programmable Controller: Key to Pipeline Pump Station Control," (available from Gould Modicon).

Carlisle, B.H., "More Power for Programmable Controllers," *Machine Design*, March 6, 1980.

Cherba, D.M., "Redundancy in Programmable Control Systems," *Instruments and Control Systems*, April 1983.

Clare, W. and Wilson, G., "Programmable Controllers Application to Water Treating Plants: What, Why, When, Where, How," 7th Annual Joint Instrumentation Conference and Workshop (ISA and AWWA), Santa Ana College, CA, August 20, 1980.

Cleaveland, P., "Programmable Controllers: A Technology Update," *Instruments and Control Systems*, May 1983.

Coelho, W.A., "Putting Programmable Controllers to Work on Material Handling," *Instrumentation Technology*, March 1974.

Cooke, E.F., Halajko, J.S., et. al., "Supervisory Programmable and Setpoint Control for Automating a Batch Plant," *InTech*, July 1980.

Daiker, J., "Operator Interface: Link-up of a Color-Graphic CRT and PLC," 9th Annual Conference and Equipment Display Proceedings, ESD, March 1980.

Dechont, G.R. and Ware, W.E., "Performing PID with a Programmable Controller on an Extrusion Line," 11th Annual Conference and Equipment Display Proceedings, ESD, March 1982.

Deltano, D., "Programming Your PC," *Instruments and Control Systems*, July 1980.

Ellis, R., "Color Graphics CRT's Provide "Window" into Factory Operation," *Instruments and Control Systems*, February 1983.

Farrar, L., "Discrete Control Products for Reliable Automation," *Instruments and Control Systems*, April 1982.

———, "Programmable Controllers in the Factory," *Instruments and Control Systems*, July 1983.

Fischer, G.I., "Use of PC's for Energy Management," 9th Annual Programmable Controllers Conference Proceedings, ESD, March 1980.

Franck, D.T., "Programmable Controllers Today and Tomorrow," *Electrical Consultant*, 1974.

Frost, H.C., "New Processes Improve Food Quality, Yields, Help Energy Efficiency, Labor Utilization," *Food Product Development*, April 1980.

Harold, R.A., "Designing Automated Systems for High Reliability and Maintainability," 12th Annual Conference and Equipment Display Proceedings, ESD, March 1983.

Hasenfuss, F.C., "Programmable Controllers Control Natural Gas Underground Storage," 10th Annual Programmable Controller Conference and Equipment Display Proceedings, ESD, March 10–12, 1981.

Henry, D., "PC Input/Output Considerations," *Instruments and Control Systems*, June 1980.

Herndon, W., "Advantages of Digital Control in Small Process Loop

Applications," ISA 1980 Annual Conference, Advances in Instrumentation, Vol. 35, Part I, Instrument Society of America, October 20–23, 1980.

Hickey, J., "Programmable Controller Update," *Instruments and Control Systems*, July 1979.

———, "Spec Correctly and You Can Design Your Own PC," *Instruments and Control Systems*, March 1976.

Hill, C., "Microprocessors Simplify an Adaptive Gain Problem," ISA 1980 Annual Conference, Advances in Instrumentation, Vol. 35, Part I, October 20–23, 1980.

Hollopeter, W.C., "Man/Machine Interfaces: Yesterday, Today, and Tomorrow," (ISA Regional Show 1979, Los Angeles), Gould Modicon, Andover, MA.

Huber, R.F., "Programmable Controls: Where the Action Is," *Production*, September 1971.

Hyde, J.F., "Sophisticated PC's: The Right Direction But Don't Stop Now," 11th Annual Conference and Equipment Display Proceedings, ESD, March 1982.

"I & CS Product Applications: Microprocessor-Controlled Reverse Osmosis Plant is a First," *Instruments and Control Systems*, June 1981.

Jannotta, K.L., "What is a PC?" *Instruments and Control Systems*, February 1980.

Juda, C., "A Guide to PC Programming Languages," *Power Transmission Design Magazine*, 1981.

———, "Sizing and Selecting a PC System," *Instruments and Control Systems*, May 1980.

Jurgen, R.K., "Industry's Workhorse Gets Smarter," *IEEE Spectrum*, February 1982.

Kajen, P., "Roasting Coffee: PCs Do it Better," *InTech*, October 1982.

Kelley, R.J. and Martin, W.E., "Programmable Controller Application in Plastic Production," International Conference on Power Transmission, Houston, June 1982 (sponsored by *Power Magazine*).

Kumontis, S.D., "PC Features," *Instruments and Control Systems*, February, March, and April, 1980.

Lapidus, G., "Programmable Logic Controllers—Painless Programming to Replace the Relay Bank," *Control Engineering*, April 1971.

Lutz-Nagy, R.C., "The Race Towards Programmable Controllers," *Automation*, December 1976.

Maggioli, V.J., "How to Apply Programmable Controllers," *Hydrocarbon Processing*, December 1978.

Martin, W.G., "PC's Provide Complete Plant Control for Drum-Mix Asphalt Facility," 12th Annual Conference and Equipment Display Proceedings, ESD, March 1983.

Martin, J.A. and Price, J.F., "Control Technologies for Small Plants: Cost, Reliability, and Versatility," *InTech*, July 1979.

Merritt, R., "Some PC Application Ideas," *Instruments and Control Systems*, January 1982.

Miller, A.R., "Do You Have a Control Problem?" 12th Annual Conference and Equipment Display Proceedings, ESD, March 1983.

Morris, H.M., "Implementing PID in PCs," *Control Engineering*, October 1981.

Nelson, J.E. and Menefee, W.F., "Application of Programmable Controllers in the Loop Supervisory Control and Monitoring System" (detailed paper available from Gould Modicon).

Obert, P., "The 12 Most Often Asked Questions About PLC's," *Instruments and Control Systems*, August 1976.

Owens, D.R., "A Case for Custom Software," *Instruments and Control Systems*, November 1981.

Palko, E., ed., "The Solid State Industrial Programmable Controller," *Plant Engineering*, April 1973.

Parkin, T.O., "How Cost-Effective Are PC's?" *Instruments and Control Systems*, April 1980.

Persson, N.C., "Programmable Controllers," *Design News*, April 1978.

Poisson, N.A. and Anderson, R., "Batch Process Control with Programmable Controllers," (available from Gould Modicon).

Programmable Controller Handbook, Cincinnati Milacron Inc., Electronics Systems Division, Lebanon, OH, 1972.

Saitta, G.N., "PC Manages Vapor Recovery System at Alaska Pipeline Terminal," *InTech*, May 1978.

Sampson, W.H., "Controlling Electric Demand with a Programmable Controller," 11th Annual Conference and Equipment Display Proceedings, ESD, March 1982.

Savelyev, M.K., "How to Use PCs for Energy Management Systems," *Control Engineering*, February 1979.

Stockweather, J.L., "Suppliers to Users: What is Their Role?" *PC User*, March 1982.

Standard ICS3–1978, Industrial Systems, National Electrical Manufacturers Association, 1978.

Taebel, T.A., "An Introduction to Programmable Controllers," presented at the ESD PC Conference, May 1977.

Thumann, A., "For Greater Control Sequence Flexibility, Use the Programmable Controller," *Electrical Consultant*, 1974.

Urmie, M., "Programmable Controllers for Batch Processing," *Measurements and Control*, December 1982.

Waite, R., "PC Study Reveals User Wants and Needs," *Instruments and Control Systems*, February 1983.

Walker, F.H., "Use of Programmable Controllers for Industrial Power Distribution System Monitoring, Control and Protection," IEEE Proceedings Conference Record 79CH1423–3 1A, Paper Number PCI–79–25, 1979.

Whitehouse, R.A., "Programmable Controllers: Cost and Maintenance," 1979 Plant Engineering and Maintenance Conference, Chicago, May 8, 1979.

Wiktorowicz, A.C. and Jeffreys, J., "Boiler Combustion Control with Programmable Controllers," International Conference on Power Transmission, Houston, June 1982 (sponsored by *Power Magazine*).

Williams, R.I., "Fundamentals of Pipeline Instrumentation, Automation and Supervisory Control," ISA 1981 Annual Conference, Advances in Instrumentation, Vol. 36, Part I, Instrument Society of America, March 23–26, 1981.

6.5 RELAYS

Available Capacities:	Relays can handle microvolts to kilovolts, microamperes to kiloamperes, with response speeds from milliseconds to any longer period.
Power and Frequency:	Relays can be actuated by microwatts but usually use milliwatts. They will operate from DC to RF.
Environmental Limits:	Some relays can operate from $-50°$ to $400°F$ ($-45°$ to $204°C$), from vacuum to 400 PSIG (2.8 MPa), and to 100 G shock or vibration.
Sizes:	From 0.03 in^3 (0.47 cc) up.
Cost:	Depending on size and capability, from about $10 to $100.
Partial List of Suppliers:	Allen-Bradley Corp., Automatic Timing and Control Co., Durakool, Inc., Furnas Electric Co., Honeywell, Inc., Leach Corp., Magnecraft Electric Co., Sigma Instruments, Inc., Wabash Relay and Electronics. Westinghouse Electric Corp., and many others.

Definition and Functions of Relays

An electrical relay is a device that initiates action in a circuit in response to some change in conditions in that circuit or in some other circuit. In addition to electrical relays, there are pneumatic, fluidic, and other varieties. Most of the relays described here are electromechanical; solid state and vacuum tube devices will not be discussed.

The function of a relay is to open or close an electrical contact or a group of contacts in consequence of a change in some electrical condition. Such a change is often called a "signal." These contact closures are used in the associated circuitry to select other circuits or functions, to turn on or off various operations, and so forth.

More generally, a relay may be considered as an amplifier and a controller; it has a power gain that is defined as the ratio of output power handled to input power required. Thus, a relay may require a coil current of 0.005 amperes at 50 volts but can control 2500 watts of power—a gain of 10,000. Among the many forms of relays many other specific functions can be identified.

Characteristics

The relays described here are either electromechanical or electrothermal. They appear in a number of forms,

but the characteristics described here are common to all of them.

Relays control the flow of electric current by means of contacts, which may be open or closed. These contacts display extremely high resistance when open (megohms) and low resistances when closed (milliohms). They may have multiple contacts (as many as eight DPDT contact assemblies are readily obtainable on stock relays), with each contact assembly electrically isolated from all others. The contacts are actuated in some definite and positive sequence.

The actuating coil can be (and usually is) completely isolated from the controlled circuit. It may be actuated by electric energy of entirely different character from the controlled circuit; e.g., milliamperes of DC may control kilowatts of RF.

Each of the various mechanical structures has certain advantages and certain limitations. Some respond rapidly—in less than 1 millisecond—but cannot safely handle large amounts of power; some handle large amounts of power but at a lower speed, and so forth. Nearly all forms can be obtained with open contacts, enclosed in dust cover, or hermetically sealed. Some are vacuum-type for handling extremely high voltages. Some have contact assemblies suited for handling radio-frequency voltages and for avoiding capacitive cross-coupling.

There are also "solid state relays," which are not electromechanical but instead use transistors, SCR's, triacs, and so forth as current-control elements. Usually in these devices the controlled circuit is isolated from the controlling circuit by a transformer (DC only), by an optical element, or perhaps by an electromechanical relay. These devices will not be described in detail in this section.

More specific information concerning the characteristics of the different relay forms will be given in the discussion of relay structures and contact materials.

Relay Types

The wide variety of available relays includes DC relays with contact capacities from microamperes to kiloamperes and coil energies from a few microwatts to several watts. With meter relays and amplifier relays the actuating energy may be a small fraction of a microwatt. There are also AC relays handling from a few watts of power to many kilowatts. Relays can control both AC and DC potentials in the thousands of volts and frequencies from DC to RF. They can be made to respond to specific frequencies only or indiscriminately to all frequencies.

Many of these varieties represent special forms for special applications. This discussion will be limited to the forms most used in typical instrumentation work—principally small DC and AC relays, sensitive relays, miniature and sealed relays, and some small power relays. In the accompanying references information may be found describing a wide variety of relay applications.

For many applications, it is convenient to consider first the actuating energy needed for a relay. A rough classification, associating coil requirement with power-handling ability, is given in Table 6.5a. The values in this table are only approximate.

Relays can also be classified by operating function. Available energy often dictates relay choice. Examples are given in the following paragraphs.

Meter relays and ultrasensitive relays, actuated by very small energies (perhaps as small as a fraction of a microwatt), are used in cases in which only a minimum signal is available, such as at the output of a transducer or bridge.

General-purpose, or small control, relays are used in cases in which no great amount of power need be handled but flexibility in application and reliability in operation are essential. Depending on the number of contacts, they usually require coil power of 200 to 800 milliwatts, which may be either AC or DC; typically they will control perhaps 5 amperes at 120 volts per contact.

Small power relays are used in cases in which more power must be controlled. These usually have coils accepting 120 or 240 volts AC; this energy may in turn be controlled by sensitive relays. These small power relays

Table 6.5a
THE RANGE OF RELAY CHARACTERISTICS

Type of Relay	Coil Power Required (Approximate)	Contact Capacity (Typical)
Meter relay	As low as 1 microwatt	Low energy only
Ultra-sensitive relay	Less than 10 milliwatts	0.5 to 1.0 amperes, noninductive
Sensitive relay	From 10 to 60 milliwatts	1.0 ampere noninductive typical
Typical crystal-can relay	100 milliwatts per form C contact	0.5 to 1.0 amperes, noninductive
Transistor-can (TO-5) or miniature relay	150 milliwatts per form C contact	0.5 ampere, noninductive
Reed relay	200 milliwatts per form C contact	0.5 to 1.0 amperes, noninductive
General-purpose relay	200 milliwatts per form C contact, up to 3 volt-amperes AC	10 amperes, 120 volts AC
Small power relay	Usually AC coils, from 1 to 10 volt-amperes	30 amperes, 240 volts AC

have contact capacities up to 30 amperes, at voltages up to 600. Much larger relays are of course available (often called magnetic contactors or controllers).

Other criteria for relay selection may be used; for example, mechanical size. Subminiature relays may be as small as 0.13 cubic inch (2.03 cc), and TO-5 size relays are the size of a TO-5 transistor case. In contrast, miniature plug-in relays measure approximately 1 cubic inch (16 cc), and general-purpose plug-in relays are about 2 cubic inches (31 cc). The type of mounting may be important (screw-mount, plug-in, or printed-circuit mounts); or the number of poles available in a single unit (as many as 48) may be the determining factor. Alternatively, extremely low (or perhaps extremely constant) contact resistance may be needed—mercury relays, or mercury-wetted contacts, have this characteristic. Reed relays and other subminiature units are suitable when many relays must be mounted on a single circuit board.

Relays may have appliance-grade, general-purpose, aerospace, or military-standard applications. There may be little difference in their reliability when they are used for their prescribed purposes, but the more expensive types will operate more satisfactorily under adverse conditions. Relays also vary in terms of mode of construction: There are clapper-type relays, telephone relays, solenoid-actuated types, reed relays, and many other forms.

Relay Contact Configurations

Electromechanical relays are produced with a wide range of contact arrangements, in various combinations. A number of these have been established as standard forms; those used most frequently are shown in Figure 6.5b with their identifying code letters. The same relay contact assembly can be described as single-pole, double-throw (SPDT); as make, break (continuity transfer); or as having form C contacts—all these terms mean the same thing. Four form C contacts means 4PDT, and so on.

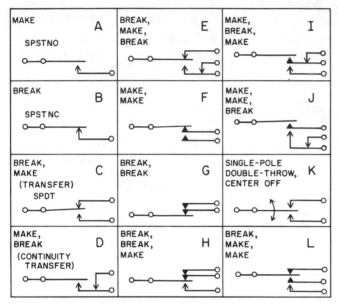

Fig. 6.5b   Standard relay contact configurations

Certain mechanical structures tolerate some contact pile-ups (the term for an assembly of leaf contacts) better than others. Small clapper-type relays seldom carry more than three form C contacts (although a modified clapper structure is used for high-density relays carrying as many as 48 contacts). Telephone-type relays may handle as many as eight form Cs, or perhaps four form Cs and one or two form Es or form Fs. These contact assemblies permit the actuation of some circuits only after others have been actuated. This ensures the proper sequence of operation; there can be no "race of contacts." Solid state relays seldom offer any wide selection of contact configurations; usually only form A or form B contacts are provided, unless the solid state relay serves as a pilot for an electromechanical relay.

Electrical insulation between contacts can be very high, so that there is little leakage even at high voltages. Contact spacing can be wide, to control high voltage without arc-over. Capacitive coupling can be kept low.

Relay Structures

Electromagnetic relays are actuated by magnetic forces that are produced by electric current flowing through wire coils. In most relays of this type, the magnetic force moves an iron armature; in a few (especially in meter relays) a coil is moved in a magnetic field.

The most widely used mechanical structures are sketched in Figure 6.5c. The elements of a clapper-type relay are shown in part A of the figure, and the elements of a telephone-type relay are illustrated in part B. When no current flows in the coil, the relay armature is held away from the core by a spring. When current flows through the coil, the magnetic field that is produced pulls the armature toward the core, decreasing the air gap. As the air gap decreases, the pull usually increases, providing appreciably more contact pressure when the relay is energized than when the contact force is only that of the return spring.

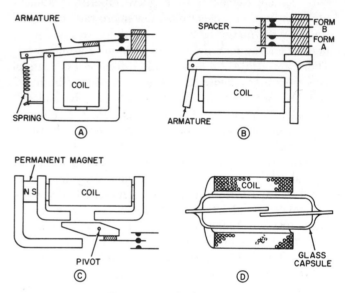

Fig. 6.5c   Typical relay structures

Both the type depicted in part A and the type shown in part B can be used with either direct or alternating current; when AC is used, a shading coil is often added to smooth out the magnetic pull and to eliminate chattering. When AC is used, the smaller air gap when the relay is actuated increases the impedance of the coil circuit and reduces the current flowing in that condition. Telephone-type relays are often fitted with shading coils (called "slugs") when used with DC; this controls the speed of response of the relay.

Two other configurations now in wide use are shown in Figure 6.5c, in parts C and D. The "balanced-force" mechanism is illustrated in part C. This structure has two mechanically stable positions. One of them is controlled by the permanent magnet, and the other is controlled by the magnetic force from the coil, which must be strong enough to overcome the pull of the permanent magnet. These relays can be made quite small in size yet can be positive in action and affected only slightly by vibration.

The reed relay is shown in the partial sectional drawing of Figure 6.5c, in part D. Two reeds of magnetic metal are mounted in a glass capsule, which is itself installed within a coil. Current flowing through the coil produces a magnetic field, magnetizing the reeds and causing them to attract each other, to bend toward each other, and to make contact. The contacting surfaces are usually plated with precious metal contact alloy. The required spring action is provided by the reeds themselves. Reed relays are among the fastest electromechanical units available; some operate in less than 500 microseconds. They are available in several contact configurations: They can be polarized, or they can be made into latching relays by the addition of small, permanent magnetic elements. Reed relays are available either with dry or with mercury-wetted contacts. More than one reed assembly can be used in a single capsule, or several capsules can be operated by a single coil assembly. Reed relays are widely used in transistorized driver systems and in printed-circuit work in general because of their small size, high reliability, and long life—100,000,000 operations are not unusual.

Specialized Relay Structures

The elementary magnetic structures described earlier are combined and elaborated to produce special-purpose relays.

If small permanent magnets are added to the relay, a relay armature may be forced in one direction for a signal of one polarity and in the opposite direction for a signal of the opposite polarity. The elements of such a polarized relay are shown in Figure 6.5d, part A.

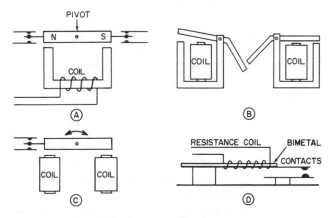

Fig. 6.5d   Some specific relay structures

Mechanical or magnetic latching devices can be applied to produce a relay. This relay can permit either one of two circuits to be actuated, but not both at the same time. A simple form of this is shown in Figure 6.5d, part B.

For certain applications, it is useful to be able to actuate one of two different circuits, depending on which of two signals is the larger—with little regard to the

absolute magnitude of either signal. Two coils, with a single tilting armature, provide this ability (Figure 6.5d, part C). Usually, a weak spring is added to ensure that when no signal at all is present the relay will go to its neutral position.

Multi-position rotary switches can be driven by magnetic ratcheting devices, permitting the selection of any desired contact position. Such stepping switches (or stepping relays, or "steppers") are available in a wide range of positions and number of circuits; they can also be secured with electrical zero-reset or for continuous rotary operation. Stepping switches are used in many communications systems, in data-handling and materials-handling equipment, and in a large number of miscellaneous applications.

Simple thermal time delay relays are useful for causing an action to be delayed for a brief period after another action, in cases in which accuracy of timing is not critical. One form consists of a thermal bimetal strip wound with a resistance coil. The coil heats on the application of current, causing the bimetal strip to bend, which closes the electrical contacts (Figure 6.5d, part D). In another popular form, the electric current flows through a stretched resistance wire, which expands and causes movement. These devices are somewhat affected by ambient temperature, and they cannot be recycled instantly. Another popular, low-cost time delay relay uses a small dashpot to delay the armature movement. Still another form (often called "slugged" relays) uses a heavy shading coil around a portion of the magnetic structure. This can produce a delay of a fraction of a second on opening or closing or on both actions; it is especially useful for avoiding any "race-of-contacts" situation. None of these low-cost, low-accuracy time delay devices should be confused with the high-accuracy timers, counters, and programmers described elsewhere.

The lowest-current relays are the moving-coil meter-type relays. These involve a d'Arsonval meter movement carrying delicate contacts. Sometimes the meter pointer is retained, and sometimes contact force is supplemented by magnetic contacts or by auxiliary coils. Such relays are susceptible to vibration and shock and to overloading. They are, however, capable of close adjustment for over-voltage or under-voltage applications and the like.

Contact Materials and Shapes

A variety of contact materials are available, with characteristics suiting them to various applications.

For very low-current, low-voltage applications (dry circuit) it is essential that one select contact materials that do not oxidize, develop insulating coatings, or erode mechanically. Some precious metals (such as gold and platinum) and some proprietary alloys satisfy these requirements. Such contacts are used in choppers and in meter relays, in which contact-sticking can be a serious problem.

Silver and silver-cadmium contacts withstand fairly high currents without overheating, but they tend to form coatings (oxide and sulphide) which, while conductive, do have appreciable resistance. Tungsten-alloy contacts usually resist pitting and erosion when used at high voltages.

Mercury-wetted contacts can be expected to have higher current ratings and lower contact resistances than do dry metal contacts of the same size; like mercury-pool contacts, they are usually less noisy and display far less bounce. Dry-metal contacts may fail by welding together; mercury contacts seldom do this. Relays with mercury contacts must usually be mounted in a nearly vertical position, and often they are vibration-sensitive.

The shapes of contacts depend on their use. Heavy-current contacts are usually dome-shaped. Low-resistance, small-current contacts are often crossed cylinders, placed so that they wipe against each other. In wire-spring relays the round wires themselves, plated locally with contact material, also form the contacts. They thus are long-lived and inexpensive. In reed relays, the flat strips, which are the reeds, also form the contacts. They may be shaped at the ends and usually are plated for good contact.

Contact mounting is an important part of relay design. In multi-contact relays it is essential that all contacts be able to bear properly on their mating contacts, without interference. In low-voltage applications it is usually desirable that a wiping contact be provided; higher voltages can break through thin surface films. In nearly all relays it is desirable that contact bounce and chatter be minimized. Some forms of reed relays, in particular, are very fast yet do not bounce.

Selection of Relays

Among the many factors affecting the selection of relays are cost, physical size, speed, and required energy. In addition, more restrictive parameters, such as mounting limitations and open or sealed contacts, are sometimes required for safety and sometimes for protection against unfavorable ambient conditions. The catalogs of the many relay manufacturers list dozens of types, forms, and sensitivities. Although relays have been in use for more than a century, new forms are still appearing. Tables 6.5e and 6.5f list a few representative types of open-contact and sealed-contact relays, respectively.

Application and Circuitry

In order for relays to be applied satisfactorily, the relay functions must be clearly understood, relay characteristics must be established, the relay must be selected to fit the need, and the circuitry must be designed to couple the relay with the rest of the system properly. Thus, it is usual to begin by determining how much energy must be controlled and how much is available as signal. In doing this one must consider the number of contacts needed. It may be necessary to use two cascaded relays if signal energy is too small or two paralleled relays to provide enough contacts. Ambient conditions must also be considered. The following questions must be addressed: Are sealed relays needed? Is there a space problem? Is there a vibration, shock, or temperature problem?

If a suitable relay has been selected and is available for the projected use, the circuit problem must be considered. In general, the same criteria may be used in designing relay circuits as in designing other circuits, yet because of some basic relay characteristics, relay circuitry involves some special problems of its own. Among the most important of these are the problems of transients across relay coils and the problem of protecting relay contacts from sparking, arcing, and welding.

In low-current, low-voltage relay circuit applications, as in most communication or logic circuits, serious problems with contacts usually do not arise, and often the most difficult problem is simply to keep the contact surfaces clean under low-current (dry-circuit) conditions. On the other hand, when larger currents must be handled (especially if the load is inductive or if it experiences an appreciable inrush for any reason), steps should be taken to protect the contacts from the effects of arcs, sparking, or welding. Under certain conditions, arcs tend to develop between contact and case or between contact and mounting. This can be prevented by proper circuit design.

One can minimize welding of contacts from high inrush currents by using sufficiently large contacts that are made of suitable materials. Occasionally, more drastic means—such as parallel contacts—must be considered. Lamp loads and the starting of single-phase motors—both drawing high inrush currents—are examples of troublesome conditions.

Sparking at contacts or arcing resulting from interruption of an inductive load can be minimized by use of spark suppressors (surge protectors, contact protectors, and so forth). These often are simple RC circuits but may be special devices—discharge tubes, diodes, or other solid state instruments. Mercury-pool relay contacts are often used for heavy currents because the circulation of mercury provides a clean contact surface for each operation. Large power relays may use double-break contacts, blowout devices, or other means that are not within the scope of this discussion.

Most relay coils possess enough inductance to produce large transients when their currents are interrupted; a 28-volt DC coil can produce a 1000-volt transient, which can endanger insulation. Such transients can be reduced by use of semiconductor devices, neon lamps, or (on DC relays) a short-circuited winding as an absorption device. The problem can be a serious one and should not be ignored.

In many relay applications, it is advantageous to apply overvoltage momentarily to ensure fast and positive action and then to reduce the coil current to a lower value to avoid overheating. There are many methods for doing

Table 6.5e
TYPICAL OPEN-CONTACT RELAYS

Relay Type	Coil Description	Contact Description	Mounting
Sensitive DC relay	1000–10,000 ohms, 20 milliwatts per form C	Up to 4 form C, $\frac{1}{2}$ ampere	Screw
Plate-circuit relay	2000–20,000 ohms, 50 milliwatts per form C	Up to 2 form C, 2 amperes	Screw
Small utility relay	AC or DC, 6–120 volts, about 250 milliwatts	Up to 2 form C, 2 amperes, 120 volts	Screw
General-purpose relay	AC or DC, about 200 milliwatts per form C	Up to 3 form C, 10 amperes, 120 volts	Screw or plug-in
Small power relay	Usually 120 volt AC, 2–10 volt-amperes	Up to 3 form C, 30 amperes, 600 volts	Screw

this; two are shown in Figure 6.5g. Shown in part A is a circuit useful only for DC relays. A 24-volt relay is fed from a 50-volt source; the capacitor (C) is initially charged to 50 volts, so that when the controlling contacts are closed, the inrush current is nearly double the normal coil current. After the capacitor has discharged to its normal value, the resistor (R) limits the coil current; if R is equal to the coil resistance, only the normal coil current flows. In this diagram, a resistor-capacitor transient suppression circuit is connected in parallel with the

Table 6.5f
TYPICAL HERMETICALLY SEALED RELAYS

Relay Type	Contact Description	Mechanical Size (inches)	Speed of Action (milliseconds)
Midget plug-in	1 form C, 0.5 to 1.0 amperes	$\frac{3}{4} \times 2$	2
Mercury-wetted plug-in	1 form C, 2 amperes	$1\frac{1}{2} \times 3\frac{1}{2}$	5
Mercury-wetted plug-in	4 form C, 2 amperes	$1\frac{3}{4} \times 3\frac{1}{2}$	5
Balanced-armature DC, sensitive	2 form C, 2 amperes	$1 \times 1 \times 2$	3 (shock-resistant)
Crystal-can relay	2 form C, 1 ampere, noninductive	$0.8 \times 0.8 \times 0.4$	1.5 (shock-resistant)
Reed relay, dry contact	1 form C, 12 volt-amperes	$0.5 \times 0.5 \times 3$	1
Transistor-can relay	2 form C, 1 ampere, noninductive	$0.3 \times 0.3 \times 0.3$	1.5
Mercury-pool power relay	1 form A or B	$2 \times 2 \times 6$	Slow. Vertical mount, affected by vibration

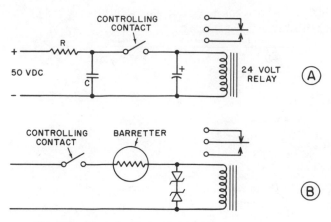

Fig. 6.5g Typical coil circuits with transient suppressors

coil. Other types of suppressors might be used with equally good results.

Circuit *B* in Figure 6.5g can be used with either AC or DC coil voltage. It involves a positive-coefficient resistor (often a small tungsten lamp, or a barretter) in series with the relay coil. When the barretter is cold, its resistance is rather low, but after current has passed through it for a few seconds, its temperature (and, hence, its resistance) increases, and the coil current is reduced to its normal value. This diagram shows two Zener diodes connected across the coil to serve as transient suppressors.

Sometimes an unused relay contact is connected to insert more resistance after a relay has been energized. It should be remembered that the inductance of an AC relay coil usually increases when the relay is closed, and this will reduce the coil current to some extent.

Relay users should not hesitate to consult the relay manufacturer for assistance in selecting and applying any of these devices. No catalog can possibly contain all the available data, and most manufacturers are happy to supply details on relay use.

Relative Costs of Relays

Small general-purpose relays can usually be purchased for less than the cost of competing devices—$6 to $10 each, with dust covers and simple contact configurations. More complicated contact assemblies, with hermetic seals, will cost a bit more. If plug-in relays are used—and many general-purpose industrial relays are plug in—the socket will add a few dollars to the cost; transient protection will cost a little more. High-sensitivity relays are yet more costly. Installation charges are difficult to estimate; they vary tremendously with locality.

Reliability

Relay reliability depends upon quality of relays but even more on correct application. Relays carrying only small currents may be expected to operate millions of times before failure; relays that control currents in the order of amperes necessarily wear faster. With proper contact protection, however, relay lives of 100,000 to 1,000,000 operations are common.

Reliability does depend upon selection of the proper relay; contact protection and transient suppression; the use of dust covers or hermetic seals when needed; proper circuit design; and observance of the usual rules of good practice as regards voltages, ambient conditions, and the like.

Relays Versus Other Logic Elements

In the field of instrumentation, relays compete primarily with various kinds of solid state devices; in particular, silicon-controlled rectifiers, silicon switches, and transistors (see Section 6.6).

Electromechanical relays out-perform solid state devices in some respects. Electromechanical relays offer both extremely low resistances (and thus low voltage drop) when closed and extremely high resistances when open. They provide essentially complete isolation of the controlled circuit from the controlling circuit without auxiliary links, and they can provide essentially simultaneous actuation of several circuits or actuation of several circuits at short intervals, without added cost. Some relays can handle extremely high voltages at a range of frequencies; some can handle high currents in more compact form than can most solid state devices. Relays provide positive circuit closures and can tolerate much abuse with only some shortening of life. Modern relays withstand extremes in ambient conditions well.

Some relays (in particular, high-density relays) are as small as or smaller than the competing solid state device when installed. In most cases, they are cheaper to procure and install than the competing devices, and some—such as latching relays or steppers—can provide other advantages, such as a functional memory that is not destroyed by a power interruption.

For extremely high-speed operation, solid state devices often are superior to relays, because only a few relay forms will operate in less than a millisecond. Solid state devices are often more resistant to vibration than are relays but may be less resistant to some other environmental conditions. For the ultimate in compactness if cost is not a factor, and for logic rather than power-handling applications, integrated solid state devices will usually be preferred. Many relays operate on ordinary line power, in contrast with solid state devices, which often require some other form—although this is frequently hidden in a packaged assembly. If only a few relays of different kinds and capacities are needed, electromechanical units may be more convenient for this reason alone.

Special Relays

A multitude of special relays are available; many of them are little known outside of their special fields of application. Some relays that might be of interest to the instrument engineer include sequence counters with transmitting contacts, opto-electronic relays (in which coupling between actuator and circuit closure is a light beam, permitting elaborate interlocking of functions; these are often part of solid state relays), meter relays with microwatt sensitivity, ultrasensitive polarized DC relays (20-microwatt sensitivity), sensitive meter relays with two or more actuating values, resonant-reed relays for remote-control switching, coaxial RF relays, vacuum relays for extremely high voltages, multipole relays that can actuate 50 or more circuits simultaneously, polarized telegraphic and pulse-forming relays, voltage-sensitive relays for power system protection, phase-monitoring relays for protection in polyphase systems, impulse-actuated relays, instrumentation relays with negligible capacity between circuits, and crossbar switches that can select any one of 100 or more circuits. Solid state components with many of the same features have become available in recent years.

BIBLIOGRAPHY

Blake, B.M., "Host of Relays for Control Jobs," *Control Engineering*, January 1967.

Deeg, W.L., "Reed Switches for Fast Relay Logic," *Control Engineering*, 1967.

Fink, D.G., *Standard Handbook for Electrical Engineers*, 10th ed., McGraw-Hill, New York, 1968.

Gaylord, M.E., *Modern Relay Techniques*. Transatlantic, New York, 1975.

Mason, C.R., *Art and Practice of Protective Relaying*. Wiley-Interscience, New York, 1956.

National Association of Relay Manufacturers, *Engineers' Relay Handbook*. Hayden Book Co., New York, 1966.

Oliver, F.J., *Practical Relay Circuits*. Hayden Book Co., New York, 1971.

Roberts, H.C., "Microwatt Relays," *ISA Journal*, May 1966.

Stoffels, R.E., "Shift Registers and Stepping Switches for Sequence Control," *Control Engineering*, January 1967.

Warrington, A.R., *Protective Relays: Their Theory and Practice*, Vols. I and II. London, Methuen, Inc., 1968 and 1978.

Yudewitz, N., "Back to Relay Basics," *Instruments and Control Systems*, February 1979.

6.6 SOLID STATE LOGIC ELEMENTS

Partial List of Suppliers: Fairchild, National Semiconductor, Square-D Co., Texas Instruments.

Solid state logic elements are electronic devices that perform control switching operations without moving parts.[1] Their advantages are: (1) high switching speeds, (2) life independent of the number of operations, (3) operation unaffected by dirt and certain corrosive atmospheres, (4) small size, and (5) low power consumption. Their disadvantages are: (1) sensitivity to electrical interference, (2) lack of input-output isolation, and (3) low power output driving capabilities.

The choice of static logic elements over mechanical relay elements is a matter of economics and requirements. In general, solid state elements are indicated when the following factors are important:

1. The switching speed requirements are high (above 100 per second). At such speeds, the life of an ordinary relay is limited to a few hundred hours.
2. The system requirements are complex, involving a large number of elements for implementation.
3. The system receives signals from a digital computer.
4. Space is limited, such as in portable equipment.
5. Process down time is costly and must be minimized.

This section will briefly discuss semiconductor materials, describe the common switching devices, define the basic switching actions, and describe circuits for carrying out these switching actions.

Semiconductors

Switching, the changing of a circuit from zero conductance to full conductance and vice versa, is accomplished with relays by opening and closing of contacts. With solid state switching devices, this is accomplished by use of the properties of a class of solids known as semiconductors. The two most significant semiconductor materials for these switching devices are germanium and silicon. Of these, silicon has been used most frequently, because devices constructed of silicon can operate at higher temperatures than can those of germanium.

The semiconductor material is modified to give switching properties by introducing some other material through a process called doping. Although the concentration of this material is very low (typically one part in 10 million), the electrical properties of the semiconductor are significantly changed.

There are two kinds of doping. The first is called N-type and results in an N-type semiconductor. It is also called donor doping, because the effect of the doping is to make available conductance electrons in the material. Materials used for donor doping are phosphorus, arsenic, and antimony. The second type of doping is termed P-type, or acceptor, doping, because the added material "accepts" electrons from the silicon, creating conductance holes. Commonly used acceptor doping materials are boron, aluminum, and gallium.

Diodes

A diode is formed by the joining together of a small amount of P-material and a similarly small amount of N-material, which results in what is termed a P-N junction. Such a junction provides a comparatively low resistance for one direction of current flow and very high resistance for the flow of current in the reverse direction. The symbols used for the diode in circuit representations are illustrated in Figure 6.6a. Shown are the voltage polarities across the diode that result in conduction and nonconduction and are called forward voltage and reverse voltage, respectively. Figure 6.6b depicts a plot of the voltage current characteristics for a typical diode. As seen in the conducting region, the current increases rapidly after a small threshold voltage is exceeded. In the opposite direction, the current is small and increases slowly with voltage. The diode is said to be reverse biased with a reverse voltage.

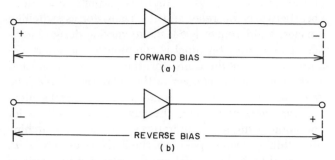

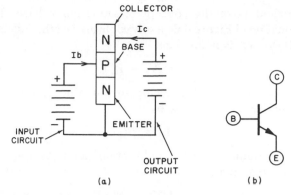

Fig. 6.6a Symbols used for diodes in circuits. *a*. Voltage polarity for conducting. *b*. Voltage polarity for nonconducting.

Fig. 6.6d Combination of P-N junctions into a transistor. *a*. Arrangement for the NPN transistor. *b*. Commonly used circuit symbol for NPN transistor.

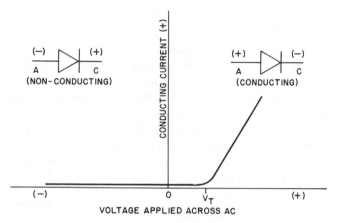

Fig. 6.6b Voltage-current characteristics of a diode

Transistors

The transistor is a semiconductor device that is formed by the combination of two P-N junctions and contains two circuits: emitter-to-base and emitter-to-collector. The emitter-to-base circuit is generally referred to as the base circuit, and the emitter-to-collector circuit is known as the power, or output, circuit. Figures 6.6c and 6.6d show the drawings for a PNP and NPN transistor circuit. The voltages applied and the resultant current flow are shown for the common-emitter connection (the emitter is com-

mon to the input and output circuits), which is the usual arrangement for switching circuits.

The internal action of the transistor is usually of little interest to the instrument engineer, but two general characteristics (which can be observed by external measurements) are significant for switching action:

1. The emitter-to-base circuit controlling I_b in Figure 6.6c closely resembles a diode.
2. The emitter-to-collector circuit may be considered an open switch whenever the voltage difference between the emitter and the base is zero or whenever the emitter-to-base circuit is reversed-biased.

Figure 6.6e illustrates a PNP transistor circuit as an interface to a pressure-sensitive switch. Because of various design considerations it is usually desirable for the transistor to be conducting when no signal is applied and to stop conducting when a signal is applied. Resistors R_2 and R_3 together with the normally open contact PSL 101 provide this function for the circuit. If properly selected, R_2 and R_3 supply the ideal negative potential value at the base in order to yield full conduction in the emitter-to-collector circuit. The closing of contact PSL 101 connects the transistor base lead to the positive line and causes the transistor to cut off, thus reducing the output to zero volts.

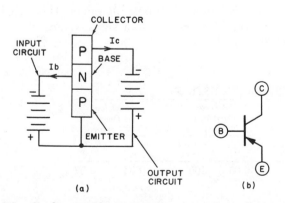

Fig. 6.6c Combination of P-N junctions into a transistor. *a*. Arrangement for the PNP transistor. *b*. Commonly used circuit symbol for PNP transistor.

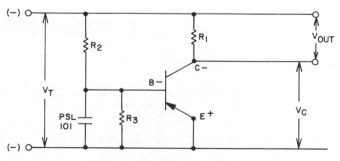

Fig. 6.6e Basic PNP transistor switch

The proper selection of R_2 and R_3 depends on an understanding of the concept of saturation voltage. Examination of Figure 6.6e easily confirms that the collector voltage, V_c, is defined as

$$V_c = V_t - I_C R_1 \qquad 6.6(1)$$

where I_C is the PNP collector current. Since I_C is related to the base current, I_B, by

$$I_C = \beta I_B \qquad 6.6(2)$$

with β defined as the amplification factor, V_C can now be expressed as:

$$V_C = V_T - \beta I_B R_1 \qquad 6.6(3)$$

The base current changes as a function of the voltage divide network that contains R_2 and R_3. As the potential at the transistor base is made more negative, I_B increases and V_C decreases to a minimum value. This minimum value, called the saturation voltage, is less than 0.2 volt for typical switching transistors, and the effective resistance between collector and emitter at saturation is accordingly small (on the order of a few ohms). On the other hand, when $I_b = 0$, the collector-to-emitter resistance is that of a reverse-biased or nonconducting diode and is in the region of millions of ohms. The collector-to-emitter circuit thus appears to open and shut under the control of I_b, and in fact an analogy can be made to a relay as shown in Figure 6.6f, in which the coil of the relay is analogous to the base-emitter input circuit and the contact of the relay is analogous to the collector-

emitter circuit. However, there is one inherent difference between the relay and the transistor as switching devices, a difference that leads to varying design methods. The emitter terminal is common to the input and output circuits of the transistor; these cannot be isolated as the coil and the contact of the relay can. Table 6.6g gives the characteristics of some commonly used switching transistors and shows the superiority of silicon for operation at higher temperatures. The lower switching times (higher switch speeds) of the 2N2894 and 2N2369A compared with the 2N404 are the result of the method of construction and reflect an improvement in fabrication techniques in the last few years.

Logic Devices

Before consideration of switching circuits formed from diodes and transistors, the nature of the switching operations themselves will be defined. The general requirement for logic devices is that they generate on-off output control signals from on-off input signals in a way described by a logic function. Section 6.3 shows how these logic functions are obtained by interconnecting groups of elementary logic operations. The five elementary and basic logic operations are:

1. NOT   The output is on if the input is off and vice versa.
2. AND   The output is on if all inputs are on.
3. OR   The output is on if one or more inputs are on.
4. NOR   The output is off if one or more inputs are on.
5. NAND The output is off if all inputs are on.

Diode Switching Circuits

Figure 6.6h shows how two diodes (and resistors) can be connected to give the AND-logic operation. The input signals range between two values, 0 and $+V_i(V_a$ or $V_b)$, and there is a supply voltage, $+ V$, which is greater than $+V_i$. Let us now consider how V_o will change as the input signals assume their possible values. There are four possible combinations of input signal values; these can be listed as follows:

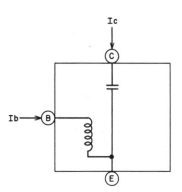

Fig. 6.6f   NPN transistor switching circuit analogy with relay coil and contacts

Table 6.6g
SWITCHING TRANSISTOR CHARACTERISTICS

"JEDEC" Transistor Type Numbers	Material	Beta	Saturation Voltage, Vs	Reverse Current, I_b (10^{-6} amperes)		Switching Times (10^{-6} seconds)	
				77°F (25°C)	185°F (85°C)	on	off
2N404	Germanium PNP	30	0.15	0.05	100	0.34	0.57
2N2894	Silicon PNP	30	0.15	0.08	5	0.06	0.09
2N2369A	Silicon NPN	40	0.2	0.01	0.1	0.01	0.03

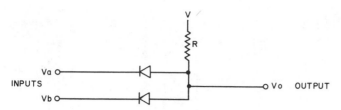

Fig. 6.6h   Circuit diagram for diode-type AND logic generator

$$V_a = 0 \quad +V_i \quad 0 \quad +V_i$$
$$V_b = 0 \quad 0 \quad +V_i \quad +V_i \qquad 6.6(4)$$

where V_a and V_b are the two inputs and 0 and $+V_i$ are the two signal values of each input.

For the first case, in which V_a and V_b both equal zero, both diodes are conducting. (The voltage across the diodes is in a direction to give conduction.) The output voltage V_o differs from the input only by the small voltage across the diodes required for conduction. This will also be true for the last case, in which V_a and V_b both equal $+V_i$, because V_i was assumed to be negative with respect to $+V$. Therefore, for the first and last columns of this tabulation, V_o is equal to 0 and $+V_i$, respectively, if the diode-conducting voltage is neglected.

For the second and third columns of tabulation 6.6(4), one of the inputs has the value 0 and the other the value $+V_i$. The significant action here is that the diode with the input of 0 will be conducting but the diode with the value of $+V_i$ will be nonconducting and will have, in effect, a reverse voltage applied. This is seen by inspection of Figure 6.6i. The conducting diode will maintain V_o as near zero, and the application of $+V_i$ on the other diode is the equivalent of applying a reverse voltage. The result is that V_o will be equal to 0 (neglecting the diode-conducting voltage) for both cases in which one input is equal to 0 and the other is equal to $+V_i$.

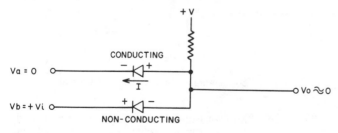

Fig. 6.6i   The circuit of Fig. 6.6h with one input at 0 and the other at V_i

The action for the four cases can be summarized by adding an "output" line to the 6.6(4) tabulation:

$$V_a = 0 \quad +V_i \quad 0 \quad +V_i$$
$$V_b = 0 \quad 0 \quad +V_i \quad +V_i \qquad 6.6(5)$$
$$V_o = 0 \quad 0 \quad 0 \quad +V_i$$

where V_o represents the output. The AND operation is thus generated if $+V_i$ is associated with the idea of being

"on." The output has the value of $+V_i$ only if both inputs are at $+V_i$.

The OR-logic action is obtained by reversing the diodes and having a negative supply voltage as shown in Figure 6.6j. The action can now be summarized a follows:

$$V_a = 0 \quad +V_i \quad 0 \quad +V_i$$
$$V_b = 0 \quad 0 \quad +V_i \quad +V_i \qquad 6.6(6)$$
$$V_o = 0 \quad +V_i \quad +V_i \quad +V_i$$

or the output is $+V_i$ if either or both of the inputs have the value V_i.

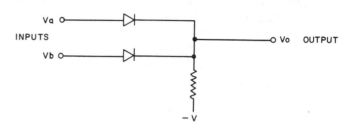

Fig. 6.6j   Diode circuit to generate OR logic

The two circuits can be cascaded to give more complex logic operations, as illustrated in Figure 6.6k, in which two AND operations involving V_a, V_b and V_c, V_d are the inputs to an OR operation.

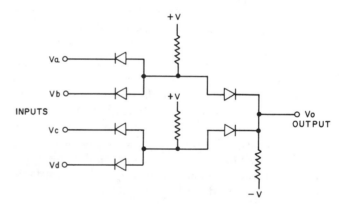

Fig. 6.6k   Cascading of diode logic circuits. Shown are two AND circuits whose outputs are to an OR circuit.

However, there are two limitations to diodes as switching elements. The "inverting" logic operations, such as the NOT function, cannot be generated. In addition, there is inherent signal deterioration because of the forward voltages necessary for diode conduction, which limits the amount of cascading and the complexity of the logic function that can be formed.

Transistor Switching Circuits

Since there are PNP and NPN transistors it is only natural that there be two types of transistor switching circuits. Figure 6.6e is an example of a PNP transistor circuit functioning as a NOT operation. When contact PSL 101 is closed, there is an input to the transistor switch, but for reasons discussed previously there is no output. When the contact is not closed, there is no input to the switch, but there is an output. Thus, the output response is opposite, or inverted from, the input. An additional PNP transistor at the output portion of this switching circuit produces a noninverted output. Figure 6.6l illustrates this point by showing how a PNP OR circuit is constructed. The figure shows the extra transistor (T_2) with its output resistor (R_4) as well as the three contacts (A, B, and C) connected in parallel to the base of T_1. The base of T_1 becomes positive when any one input or combination of these inputs is closed. If these three contacts had been connected in series to the base of T_1, then the circuit would be an AND. Finally, if the second transistor had not been used in Figure 6.6l, the remaining circuit would be a NOR.

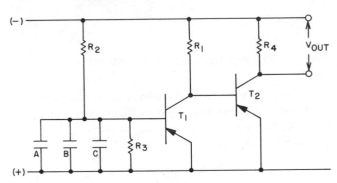

Fig. 6.6l   PNP transistor switch functioning as an OR

The basic transistor switching circuit for an NPN transistor is shown in Figure 6.6m. Referring to the previous discussion on transistor action, the output voltage V_o will be equal to $+V$ when $V_a = 0$, since both I_b and $I_c = 0$. When the input voltage is equal to $+V_i$ (assuming that this causes sufficient I_b to flow to give saturation), V_o will be close to 0, differing from it only by the small value of saturation voltage. If this is neglected, a table can be prepared to describe this action:

$$\frac{V_a = 0 \qquad +V_i}{V_o = +V \qquad 0} \qquad 6.6(7)$$

If both $+V$ and $+V_i$ are taken as "on" values, it is seen that the NOT logic operation is generated.

If a second resistor is added to the circuit of Figure 6.6n, the resultant action is easily determined: $+V_i$ ap-

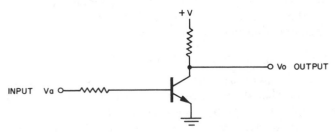

Fig. 6.6m   Basic NPN transistor circuit

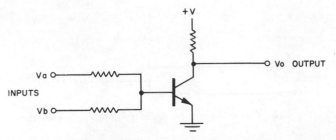

Fig. 6.6n   NPN transistor as a two-input NOR (resistor-coupled)

plied to either of the inputs will cause transistor saturation. The switching table then is

$$
\begin{array}{llcccc}
V_a = & 0 & +V_i & 0 & +V_i & \\
V_b = & 0 & 0 & +V_i & +V_i & 6.6(8) \\
\hline
V_o = & +V & 0 & 0 & 0 &
\end{array}
$$

This is recognized as the NOR-logic operation as previously defined. The output is off (output at 0) if any of the inputs are on (input at $+V_i$).

NOR circuits are readily cascaded, as shown in Figure 6.6o. This circuit has a significant advantage compared with the circuit of Figure 6.6k, in that there is no signal deterioration, since each transistor switching operation completely regenerates the signal.

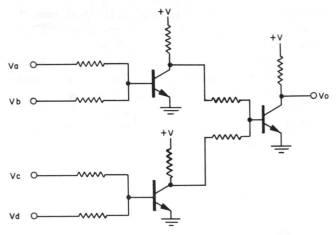

Fig. 6.6o   Example of a cascading NOR circuit

The diode circuit of Figure 6.6h can be combined with the transistor circuit of Figure 6.6m to give the NAND-logic operation as shown in Figure 6.6p. The action is that of the AND operation generated by the diode circuits, followed by a NOT operation generated by the transistor. The action can be tabulated as follows:

$$
\begin{array}{llcccc}
V_a = & 0 & +V_i & 0 & +V_i & \\
V_b = & 0 & 0 & +V_i & +V_i & 6.6(9) \\
\hline
V_o = & +V & +V & +V & 0 &
\end{array}
$$

The circuits of Figures 6.6n and 6.6p are called resistance-coupled and diode-coupled transistor switching

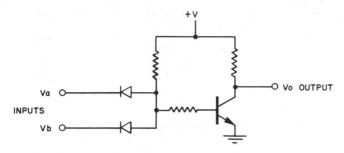

Fig. 6.6p   Circuit for obtaining NAND logic based on combining diode and transistor circuits (diode coupled)

Integrated Switching Circuits

There are four main families of integrated circuit switching in common use for control:

1. RTL (resistance-transistor logic),
2. DTL (diode-transistor logic),
3. TTL (transistor-transistor logic), and
4. HLL (high-level logic).

RTL was the first type of switching logic produced in integrated form. A typical circuit is shown in Figure 6.6q.

A modified form of the circuit of Figure 6.6p is the basis of the DTL circuit, shown in Figure 6.6r. The circuit, it will be recalled, generates the NAND operation.

circuits, respectively. Figure 6.6q shows a NOR circuit in which a transistor is incorporated for each input.

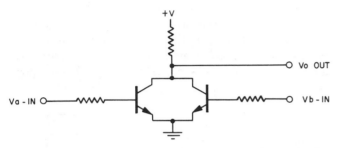

Fig. 6.6q   RTL (resistance-transistor logic) integrated circuit

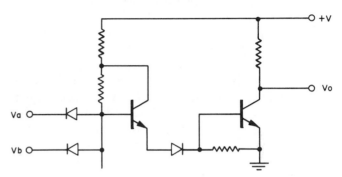

Fig. 6.6r   DTL (diode-transistor logic) integrated logic circuit

Integrated Circuits

The term "integrated circuits" refers to a way of fabricating the previously discussed circuits in such a way that the technique for fabricating a single transistor is applied to the entire circuit. Transistors, diodes, and resistors are formed and interconnected with a process that is analogous to casting a metal part, as opposed to assembling the part from subcomponents by welding or other techniques.

The integrated circuit process begins with a silicon wafer approximately 1 inch (25 mm) in diameter and 6 to 8 mils in thickness. The basis of the integrated circuit technology is the application of a set of "masked diffusions," in which a succession of photographic masks are applied to the wafer and selected areas of the wafer are doped (see discussion on semiconductors) to give a pattern of P-material and N-material joined in a way to form transistors and diodes.

The components thus formed are then interconnected by depositing a metallic pattern (again controlled by a photographic mask).

The significant nature of the process is that this is carried out for up to several hundred circuits simultaneously, since this is the number of circuits (called dice) that a wafer can hold. The result is a very small circuit (typically 0.050 in², or 31.25 mm²). After the completion of wafer processing, the dice are separated by scribing and are assembled into packages.

A more complicated circuit is the TTL circuit shown in Figure 6.6s. This circuit is characterized by a high speed of switching and generates the NAND operation.

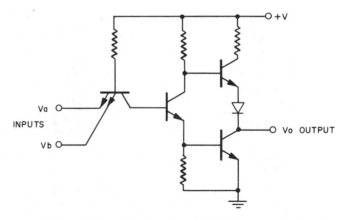

Fig. 6.6s   TTL (transistor-transistor logic) integrated logic circuit

Table 6.6t
INTEGRATED CIRCUIT CHARACTERISTICS

Type	Resistance to Noise	Switching Speed	Cost
RTL	Poor	Fair	Low
DTL	Fair	Good	Medium
TTL	Good	High	Medium
HLL	Excellent	Low	High

A special kind of circuit (HLL) is one designed to discriminate against electrical interference of the type encountered in many industrial situations. The circuit is a modification of DTL in which elements have been introduced to give large voltage swings and, in effect, to make the switching signals large compared with the interfering signals. A summary of integrated circuit characteristics is given in Table 6.6t.

The influence of the semiconductor integrated circuit devices on modern technology is reflected in the world-wide recognition of these small, rectangular plastic–packaged devices and the multi–billion-dollar industry associated with their production. Modern integrated circuit devices are reliable and convenient to use and are found in all phases of technology.

Figure 6.6u illustrates four AND devices in a common package arrangement. Each of the ANDs operates independently of the others and is manufactured using TTL and other technologies. The operation of the ANDs in this package is straightforward. The power and ground connections are provided, and when input signals that

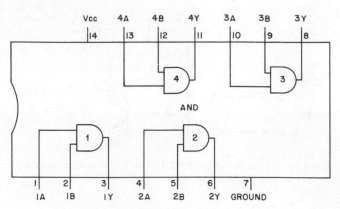

Fig. 6.6u   Illustration for AND device package

represent the 1s and 0s are applied to the appropriate connections, the device responds with the logically correct output signal. Usually, this output signal is not capable of directly driving the desired equipment but requires modification or amplification. Thus, particular attention to interface circuits between the process equip-

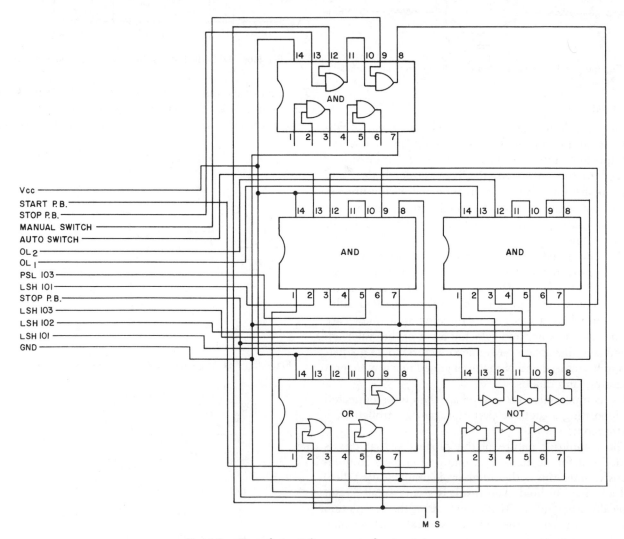

Fig. 6.6v   Control circuit for pressurized water system

ment and the logic devices is important to ensure that the correct voltage requirements for the equipment starter relays and other parts match the current and voltage output capabilities of the integrated circuit device.

To illustrate the utility of integrated circuit devices, Figure 6.6v provides a drawing of a control circuit of the pressurized water system shown in Figure 6.2b. This control circuit is a literal translation of the logic diagram presented in Figure 6.3f and is presented here for educational reasons only. Even though none of the optimization techniques discussed in Section 6.3 have been applied to this circuit, only five integrated circuit packages are required to complete the circuit, and there are several unused devices in the circuit. The circuit is constructed on a plastic card, called a printed circuit board (PCB). The conducting connections shown between pins on the device packages are etched on the PCB with the input and output leads brought to the edge of the board.

The logic packages are attached to the appropriate places on the board, and the entire circuit board is plugged into a protective chassis. Circuit repair usually encompasses replacement of the entire board.

REFERENCE

1. Lipták, B.G., ed. *Instrument Engineers' Handbook*, Vol. II, Chilton, Philadelphia, 1969, Chap. 4.

BIBLIOGRAPHY

Bishop, T., "The Logic of SSR's," *Instruments and Control Systems*, December 1977.

Gilbert, R.A. and Llewellyn, J.A., "Principles and Applications of Digital Devices," Instrument Society of America, Research Triangle Park, NC, 1982.

Reilley, C.M., *Transistor Engineering and Introduction to Integrated Semiconductor Circuits*, McGraw-Hill, New York, 1961.

6.7   TIME DELAY RELAYS

Type of Delay:	On-delay, off-delay, interval, single-shot and repeat cycle.
Range of Delay:	1 msec to 1000 hrs.
Timing Accuracy:	±0.1 to 35%.
Load Switching:	Relay contacts, snap-action switches, snap-action valves, SCRs, transistors, triacs.
Environmental Limits:	−20°C to 45°C; 5% to 95% humidity.
Mounting Configurations:	Plug-in, surface mount, panel mount, in-line module.
Cost:	$15 to $250.
List of Suppliers:	Agastat Controls Products Div., Air-O-Tronics, Allen-Bradley, AMF Inc., Amperite, Applied Electro Technology, Arrow-Hart, Artisan, Bristol Saybrook, Cornell-Dubilier, Cutler-Hammer, Deltrol Controls, Diversified Electronics, Durakool, Eagle Signal, Electromatic Components Ltd., Essex Group, General Time Corp. Industrial Controls Div., Guardian/California, Heinemann, HI G Co. Inc., Instrumentation & Control Systems, Jettron, Leach, Line Electric, Magnecraft, Midtex, Panasonic, Potter & Brumfield Div., Square-D, SSAC, Struthers-Dunn, Syracuse Electronics Corp., Tempo Instrument, Valcon Engineering, Vanguard, Xanadu Controls Div.

General Characteristics

Time delay relays are logic components that are generally considereed as special-purpose relays and have some characteristics of both relays and timers. Time is one of the variables of any process, and often it must be monitored or controlled. Some operating functions of a process may need to coincide logically, or alternatively they may need to be separated by a specific interval of time. Time delay relays are used to introduce the required timing corrections.

The two major elements of time delay relays are the circuit mechanism that produces a time interval and the load switching contacts that are actuated during or after the time interval.

For a time interval to be produced, the following elements must be present:

1. a power source (if needed in addition to signal power),
2. signal power,
3. a change of state of the time-determining device, and
4. an indication that the time-determining device has changed to the desired state.

The power or signal power can be thermal, pneumatic, or AC or DC electric current. The change of state of the time-determining device can be mechanical (e.g., rotating of a motor, motion of a plunger in a restricting fluid, or bending of a bimetal strip due to presence of heat), electrical (e.g., accumulating charge of a capacitor or count of oscillations or pulses), or software-based (e. g., involving the number of program scans or interrupt-driven). The load switching can occur via snap-action switches, snap-action valves, relay contacts (e.g., SPDT, DPDT, reed [hermetically sealed], or mercury [hermetically sealed]), or solid state devices (e.g., transistors, SCRs, or triacs).

There are several modes of timing that relate the time of application of the signal power to the time of load switching. This relationship may also depend on the continuous presence of the power source. The four most prevalent modes are on delay, off delay, interval, and single shot. A special combination of the on delay and the off delay is called a repeat cycle.

In addition to the aforementioned combinations, one must consider the following characteristics when choosing time delay relays among the vast number available in the marketplace:

1. Timing range
2. Fixed versus adjustable timing
3. Accuracy
    a. Dial setting accuracy
    b. Tolerance
    c. Repeat accuracy
    d. Time between operations
4. Environmental factors
    a. Temperature range
    b. Vibration, shock
5. Load switching
    a. Duty cycle
    b. Type of load: resistive, inductive, lamp
    c. Life expectancy with load
6. Mounting considerations
    a. Size limitations
    b. Mounting style: surface mount, panel mount, plug-in, in-line module
    c. Terminals: plug-in, quick connects, solder lugs, screw terminals
7. Housing
    a. Open chassis
    b. Dustproof
    c. Weatherproof
    d. Explosionproof
    e. Totally encapsulated

Dashpot or Pneumatic Time Delay

The armature of a normal relay can be augmented with a piston that travels in an oil or fluid in a dashpot. The piston contains several holes through which the fluid passes. The current of the coil must be applied continuously to exert a pull on the armature. The armature travels slowly because of the resistance of the fluid. The fluid has the required viscosity to produce the necessary delay. The size of the aperture holes may be modified to vary the time delay. Once the armature travels the required distance, the moving contacts engage the stationary contacts.

In a pneumatic time delay version, a diaphragm and cap encase the head in which the air that is used for timing is recirculated, making the time chamber dustproof. A dashpot time delay is shown in Figure 6.7a.

Thermostatic Time Delay Relay

In this version, one of the contacts is a bimetallic element that is wrapped with an insulated heating coil. The

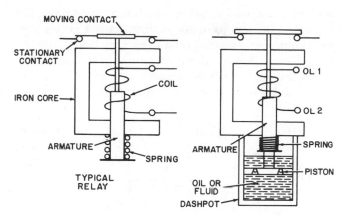

Fig. 6.7a   Dashpot time delay

signal power is applied to the heating coil. The time delay is the time required to raise the temperature of the contact high enough to produce warping, which finally causes the contact to transfer. Thermostatic time delay relays are not accurate or repeatable but are of low cost. A thermostatic time delay is depicted in Figure 6.7b.

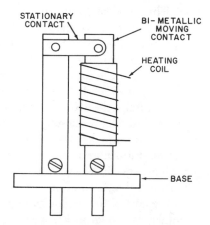

Fig. 6.7b   Thermostatic time delay

Motor-Driven Time Delay Relay

This type of relay is powered by a synchronous motor. A tripping arm is driven through a simple train of machine-cut spur gears. The arm actuates a snap-action switch, and the motor speed and the gear reduction determine the time delay. Resetting is instantaneous by means of an electromagnetic coil that clutches and declutches the trip locking gear as required. The clutch coil is often equipped with a set of auxiliary contacts. The power source is normally applied to the motor circuit; the signal power is normally applied to the clutch coil. In this manner, there are instantaneous contacts actuated by the clutch coil and time-delayed contacts actuated by the tripping arm. When the power source to the motor is interrupted while the signal power to the clutch is maintained, the time delay is increased by the delay in power interruption. Viewed otherwise, the relay accumulates all of the time that the power source is applied to the motor until it equals the preset time. The use of a synchronous motor ensures great accuracy and, with appropriate gearing, allows for very long delays, up to and including hundreds of hours. Sometimes these relays are provided with a progress pointer. A motor-driven delay is illustrated in Figure 6.7c.

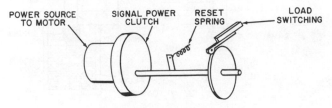

Fig. 6.7c   Motor-driven delay

Analog Solid State Time Delay

Analog time delays are based on the RC circuit, in which a capacitor is charged via a resistor until it reaches a certain voltage. When the predetermined voltage is reached, the load switching electromechanical relay or a solid state switching device turns on or off, depending on the function. This type of timing is much simpler, more reliable, and less expensive than the mechanically complex synchronous motor-driven delay relay. However, there are some shortcomings that must be considered: The time-constant curve for the capacitor charging is exponential, making it difficult to set a potentiometer (the variable R) accurately; the RC circuit is sensitive to variations in supply voltage; settability and repeat accuracy are affected by variations in the resistance because of variations in temperature; and the capacitor shifts its value as it ages. Measures can be taken to offset or lessen the adverse effects of voltage variations or temperature, but these add to the cost of the analog time delay relay. An analog time delay is shown in Figure 6.7d.

Digital Solid State Time Delay

The digital method for creating a time delay uses a frequency counting or dividing circuit. This approach, although more expensive than the analog RC circuit, provides better accuracy and repeatability and lends itself to medium- or large-scale (chip) integration. In this type of circuit, pulses from an oscillator or from the 60-Hz power line are applied to a counter or a divider chain, which is preset to output a pulse at a specific count. This output pulse will turn on or off the load switching electromechanical relay or solid state switching device.

The 60-Hz line is used because of its frequency stability. Using a higher-frequency free-running oscillator provides finer time increments. For ultimate accuracy, a crystal-controlled oscillator is used, providing absolute accuracy and repeatability, settable to the fourth significant figure. This type of delay relay would most often be supplied with a thumbwheel switch or dip switches to program the dividing counter. A digital time delay is shown in Figure 6.7e.

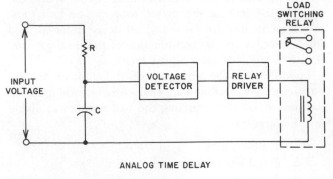

ANALOG TIME DELAY

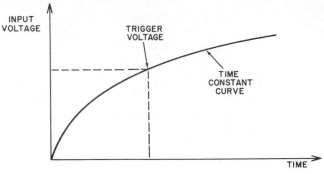

Fig. 6.7d   Analog time delay

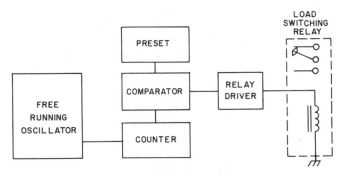

Fig. 6.7e   Digital time delay

Software-Based Delay Relay

Modern control systems make increasing use of programmable controllers or microprocessors for all the logic, sequence, and timing functions. All these functions are emulated in software by a program stored in memory. The program is merely a series of binary instructions written by the applications designer, upon which the central processing unit interprets and acts. The software-based delays use the digital method of generating delays. Internal clock pulses are accumulated in a memory register or in a counter, which subsequently transfers its count to a memory register. The program regularly compares the accumulated count with a preset value that is stored in another memory register. When the accumulated count equals or exceeds the preset value, the program branches to another set of instructions that require the delay for further logic sequence.

An error may be present in the aforementioned method, affecting accuracy and repeatability. Since the software program takes varying amounts of time for acting on a set of instructions, successive scans through the whole program involve differing amounts of time. Therefore, when the software program returns each scan to compare the actual count with the preset count, the actual may exceed the preset by an unacceptable amount for the accuracy desired. Some microprocessor-based devices have a modified digital method that corrects this problem. The actual count takes place in an integrated circuit, which creates an interrupt when it reaches the preset count. This interrupts the main software program immediately, and load switching can be executed without incurring additional time delay. A software-based delay is depicted in Figure 6.7f.

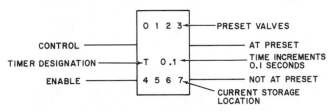

Fig. 6.7f Software-based time delay. Timer block general form as viewed on a CRT of a typical programming panel of a PC.

Time Delays Glossary and Charts

On Delay

An on delay is alternatively called delay on make, delay on operate, delay on pull-in, delay on energize, slow-acting, or slow-operating. The load switching occurs a certain time after the application of the signal power. If a power source is normally required and the power is interrupted before load switching takes place, the timing cycle needs to be repeated from zero to effect a delayed load switching (except in certain synchronous motor-driven delay relays).

Off Delay

An off delay is alternatively called delay on break, delay on de-energize, delay on drop-out, delay on release, slow release, drop-out delay, or delayed drop-out. The load switching occurs a certain time after the removal of the signal power. If the power source is normally required and the power is interrupted before the minimum amount of on time or after removal of the signal power prior to load switching, the amount of time delay may be inaccurate.

Interval

After application of the signal power and while the signal power is maintained, the load switching occurs for a certain time only, and then the load switching is de-energized. If the signal power is interrupted or the power source (if normally required) is removed before the completion of the time delay, the load switching is de-energized instantaneously.

Single Shot

A single shot is alternatively called latched interval, momentary actuation, or one shot. Load switching occurs for a certain time only, and then the load switching de-energizes after a momentary application of the signal power. The signal power may be applied longer without altering the load switching interval. If the power source is normally required and it is interrupted while the load switching is energized, it de-energizes instantaneously.

Repeat Cycle

A repeat cycle is alternatively called dual delay; combination delays; on delay, off delay; or slow acting–slow release. The load switching occurs a certain time after the application of the signal power, and remains energized for a certain other time delay after removal of the signal power. If the power source is normally required and it is interrupted while the load switching is energized, it may be de-energized prematurely.

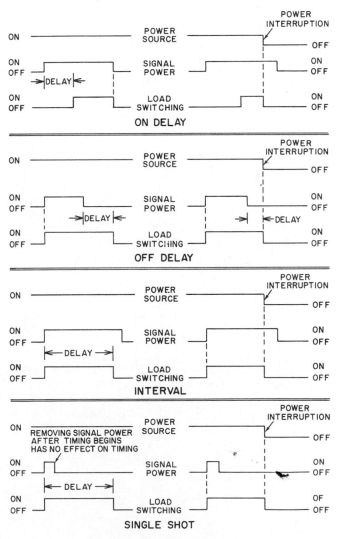

Fig. 6.7g Time delays timing charts

Recovery Time

Recovery time is the minimum amount of time between removal of the signal power and its reapplication that is necessary so that the subsequent operation will have the desired repeatability. The analog or digital solid state timing circuits have the shortest recovery time.

BIBLIOGRAPHY

Auger, R.W., *The Relay Guide*, Reinhold Publishing Co., New York, 1960.

Benoit, C.N., "Specifying Solid-State Timing Devices," *Automation*, November 1972.

Collin, M., "TDR's are Alive and Well," *Instruments and Control Systems*, December 1977.

Kosow, I.L., *Control of Electric Machines*, Prentice-Hall, Inc. Englewood Cliffs, NJ, 1974.

Lipták, B.G., *Instrument Engineers' Handbook*, Chilton, Philadelphia, 1969.

Margolin, R., "Time Delay Relays," *Electronics Products Magazine*, May 1979.

Schwartz, J.C., "Understanding Timers for Better Control," *Instruments and Control Systems*, December 1975.

Slater, N., "Solid-State Timing Shows Gains," *Design Engineering*, July 1980.

"Selecting and Specifying Time Delay Relays," *Instruments and Control Systems*, March 1979.

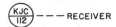
6.8 TIMERS AND PROGRAMMING TIMERS

Partial List of Suppliers: Fairchild, National Semiconductor, Square-D Co., Texas Instruments.

Timers and programming timers provide a method of controlling a process that has a prescribed set of states with defined time periods in each state. This section reviews the properties of mechanical and electronic-based timing devices. Cam-operated timers, band-programming timers, and punched-card programmers are types of mechanical sequencers that will be discussed in this section. The properties of an electronic sequencer will also be reviewed to illustrate an example of an electronic programming timer. This discussion includes the role of the astable oscillator, the decade counter, and the decoder in the timer circuit.

Mechanical Sequencers

Programming timers are capable of controlling a number of circuits on a continuing basis in cases in which many different actions are programmed at many different times during the entire operating cycle of the programmer.

Most widely used are the cam-operated programming multiple-circuit time switches, drum programmers, and band programming timers or punched-card programmers—all quite alike in principle but varying in construction.[1]

Cam-operated timers, or timeswitches, involve mechanical switches arranged with an equal number of adjustable cams on a shaft (Figure 6.8a). A clock or a synchronous motor turns the shaft at the desired speed—one revolution for the desired cycle duration. During the cycle, each of the switches is opened and closed once (occasionally two or three times is permissible) for du-

rations established by the cam settings. Such a timer can control as many as 60 circuits, with a cycle as short as a few seconds or as long as a few weeks. Usually about 95 percent of the cycle can be programmed either "on" or "off" for each circuit; the contacts can generally handle currents as great as 10 amperes at 120 volts. These timers are available in a variety of sizes and speeds.

Cam-operated timers are often used in conjunction with (either in cascade or in parallel) interval timers, which permits much greater flexibility of action. They are moderately priced, very reliable, and their timing accuracy is high-limited by facility in adjustment.

A limitation of cam-operated timers is that each switch can be actuated only a few times during each timing cycle. This limitation is entirely overcome in drum programmers, band-programming timers, and punched-card programmers.

Band-programming timers consist of a belt, or band, of flexible plastic material, several inches wide and as long as is needed to provide the required cycle length (Figure 6.8b). An array of switches is provided, each with an actuating finger resting on the band. Where the band is intact the switch is held open, but where it is perforated the switch can close. As many as 80 such switches can be used; the band can be of any reasonable length. Many hours of operation can be controlled with many distinct on-off cycles for each switch during the total

Fig. 6.8a Elements of a cam-operated timer

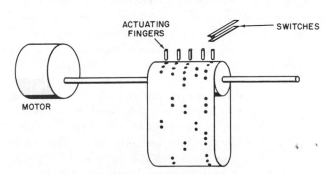

Fig. 6.8b Elements of a band programmer

cycle. Quite complicated switching patterns can be provided.

Punched-card programmers operate in a analogous manner. The cards are usually handled singly instead of in a continuous band. Such programmers are in fact quite similar to the Jacquard loom principle. Drum programmers are somewhat comparable in that a drum in which many movable studs are placed rotates in a time cycle under an array of switches.

In a band or drum programmer, the patterned band is often made to move against the switch array by a pulsed stepping motor rather than a simple synchronous motor. The stepping motor is driven by timed pulses from a separate source, perhaps varying in rate according to a subsidiary pattern. Added flexibility is thus afforded.

Electronic Sequencers

The digital sequencer is a relatively simple digital circuit that can provide time control for a sequential process. Such circuits usually contain an astable oscillator, a decade counter, and a decoder. A detailed view of the components of such a circuit is presented in Figures 6.8c, 6.8d, and 6.8e. A generalized view is depicted in Figure 6.8f.

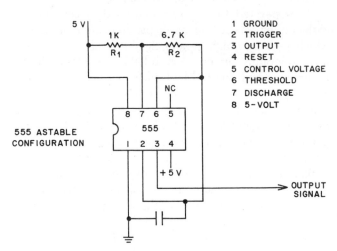

Fig. 6.8c   An astable multivibrator circuit using the 555

Figure 6.8c is an example of an astable multivibrator in a circuit configuration that generates a continuous train of clock pulses. An astable multivibrator is a device that does not have any stable output state but constantly changes output states, e.g., 5 volts, then 0 volts, then 5 volts, and so forth. The duration of time in one state is determined by the value of the resistance and capacitance in the circuit.

Figure 6.8c illustrates the pin identification as well as a pulse generation circuit of a 555. Since the device is very versatile and a component of many digital circuits, a quick review of its pin functions is in order. The supply voltage, which can range from 5 volts to 15 volts, is attached to pin 8, and the ground connection is attached

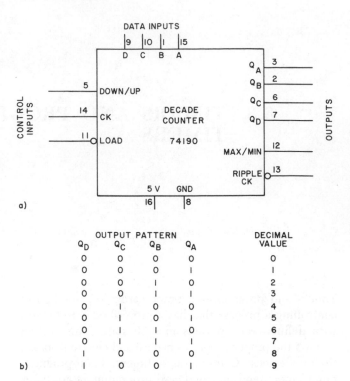

Fig. 6.8d   An example of a decade counter with its output code

	OUTPUT PATTERN			DECIMAL VALUE
Q_D	Q_C	Q_B	Q_A	
0	0	0	0	0
0	0	0	1	1
0	0	1	0	2
0	0	1	1	3
0	1	0	0	4
0	1	0	1	5
0	1	1	0	6
0	1	1	1	7
1	0	0	0	8
1	0	0	1	9

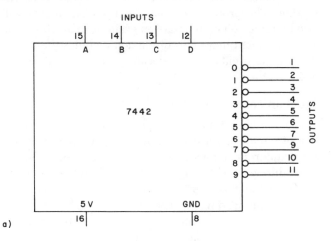

ELECTRICAL PINOUT FOR 7442
1 OF 10 DECODER

BCD INPUTS				CHANNEL OUTPUT (DECIMAL)									
D	C	B	A	0	1	2	3	4	5	6	7	8	9
0	0	0	0	L	H	H	H	H	H	H	H	H	H
0	0	0	1	H	L	H	H	H	H	H	H	H	H
0	0	1	0	H	H	L	H	H	H	H	H	H	H
0	0	1	1	H	H	H	L	H	H	H	H	H	H
0	1	0	0	H	H	H	H	L	H	H	H	H	H
0	1	0	1	H	H	H	H	H	L	H	H	H	H
0	1	1	0	H	H	H	H	H	H	L	H	H	H
0	1	1	1	H	H	H	H	H	H	H	L	H	H
1	0	0	0	H	H	H	H	H	H	H	H	L	H
1	0	0	1	H	H	H	H	H	H	H	H	H	L
HEXIDECIMAL VALUES FOR A THROUGH F				H	H	H	H	H	H	H	H	H	H

FUNCTION TABLE FOR THE 7442

Fig. 6.8e   An example of a decoder with its function table

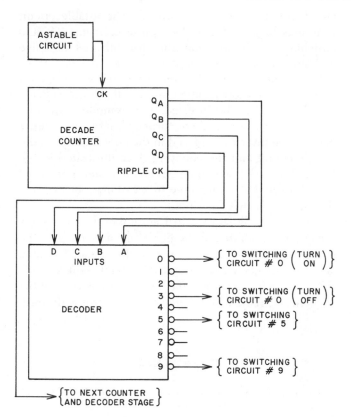

Fig. 6.8f Elements of a simple digital sequencer circuit to control switching circuits

to pin 1. The control voltage pin, pin 5, is an input that controls the threshold input voltage level. A change in the control voltage alters the voltage level needed to enable the threshold input pin. If the control voltage is 0, i.e., if pin 5 is tied to ground, then two thirds of the supply voltage activates pin 6, the threshold input.

Once activated, the threshold input brings the output line to 0 volts. This output line, pin 3, is capable of sinking up to 200 mA when it is at 0 volts or supplying up to 200 mA when it is at 5 volts. The reset, pin 4, is another input that also controls the output line. The reset is an active low input that sets the output pin and pin 7, the discharge pin, to 0 volts. The discharge pin is an active low output that is at 0 volts when on and is essentially an open circuit when it is not active.

A review of some of the circuit connections emphasizes the function of the 555. When the circuit is first turned on, the capacitor connected between pin 2 and ground discharges and pin 2 is activated as the voltage drops below 1.4 volts. Once activated, the trigger sets the output line high and brings the discharge terminal to an open circuit state. Since the connection from the 555 to the junction of R_1 and R_2 is now essentially open, the capacitor recharges through the two resistors until the threshold voltages reach approximately 3.6 volts. Once this voltage level has been reached, pin 6 is activated, pin 2 is deactivated, and the output and discharge pins move to 0 volts. The zero output on pin 7 provides a

path for the capacitor to discharge through R_2. When the capacitor discharge drops the voltage to one third the supply voltage, the trigger input is reactivated, and the process starts over. It should be noted that in the 555 astable configuration illustrated in Figure 6.8c, pin 5 is shown with no connection. Often this pin is connected to ground by means of a 0.01 mF capacitor. Pin 5 is the control voltage input for the threshold voltage and is not used in this example. It is a good practice to use the capacitor connection to ground because the capacitor and R_1 form a decoupling element that prevents power supply noise from altering the threshold-input voltage.

The clock frequency for a 555 as an astable multivibrator can be calculated according to the formula:

$$\nu = 1.44/(R_1 + 2R_2)\, C_z \qquad 6.8(1)$$

The duty cycle, which in this case is the ratio of the time spent in the logic zero state to the sum of the times spent in the logic zero and one state, has a calculation formula:

$$D = R_2/(R_1 + 2R_2) \qquad 6.8(2)$$

Thus, a duty cycle of 0.5 produces a square wave and is available when $R_2 \gg R_1$.

Figure 6.8d illustrates another important element of a digital sequencer circuit. The 74190 is a binary-coded decimal (BCD) synchronous down/up counter. The device can be used in electronic sequencer circuits because its outputs are triggered on a low-to-high level transition on the clock input pin. Thus, the output of the astable circuit shown in Figure 6.8c can be used to produce sequentially the 74190 output patterns shown in Figure 6.8d, part *b*. These patterns provide a convenient method of keeping track of the number of clock pulses generated by the astable device and can also be used to drive other digital devices in the sequencer circuit.

There are several other interesting pins on the counter that provide additional control options for the sequencer circuit. The enable G, load, and ripple clock pins can be used to start the count sequence, to enter a predetermined count into the 74190, and to activate any other counters that may be cascaded together to extend the time domain of the sequencer circuit. All three of these pins are low active. The ripple clock pin goes low on nine in the up count mode and drops at zero in the down count mode. The direction of the count is determined by the logic on the down/up pin. A high starts a down count, and a low initiates an up count. The width of the low ripple count output is equal to the low-level width of the clock input. The ripple clock pin is used to cascade the 74190/74191 chips. The maximum/minimum count output goes high just before the terminal count. This output pin is used for look-ahead operations when the count is beyond ten events.

The final element of the digital sequencer circuit is the decoder. The 7442 is a BCD-to-decimal converter;

its electrical pin arrangement and function table are shown in Figure 6.8e. The device is easy to understand and functions well if the 5-volt input, which is pin 16, and the ground contact, which is pin 8, are properly connected. The BCD code is input on pins 15 through 12. (These pins are also labeled A, B, C, and D, respectively, with pin 12 representing the most significant bit of the four-bit input code.) Once the code is presented to the input pins, the appropriate output pin moves to logic zero. There are 10 output pins, which are labeled one through seven and nine through eleven. These pins represent digits zero through six and seven through nine, respectively. The small circles in the output pins of the electrical pinout inform the user that when an output pin is activated, it goes low (i.e., the pin moves to ground or logic zero). All other times the pin is high or at logic one. The circle is a very common pinout symbol and represents inversion from expected positive logic (see Section 6.3). In the case of the 7442, if 1001 is presented to the input pins, the logic of output pin 11 moves from high to low while the remaining output pins remain high. A summary table for the operation of the 7442 is also presented in Figure 6.8e.

A functional view of the electronic sequencer is presented in Figure 6.8f. This illustration shows one stage of the sequencer. The astable circuit is set at the desired clock rate (i.e., 1 Hz, 1.667×10^{-2} Hz, or 2.778×10^{-4} Hz), and the switching circuit connections are made to the appropriate decoder output pins. For the general example shown in Figure 6.8f, if the astable circuit is set at one clock pulse per minute, then the counter will change output states every minute and the decoder output channels will sequentially go low. Thus, switching circuit number zero is turned on when the sequencer is engaged and remains on for three minutes. At that time, decoder channel number three is activated and switching

circuit number zero is turned off. The astable circuit continues to generate a clock pulse each minute, and switching circuits five and nine are activated after five and nine minutes have elapsed.

Figure 6.8f shows only one state in the sequencer. If time is needed before or after this stage, then extra counters and decoders can be inserted. For example, switching circuit number zero could be turned on at the ten-minute mark if a counter were placed in the circuit between the astable circuit and the counting stage illustrated in the figure. Similarly, switching circuits five and nine could be deactivated at a later time by an addition of stages at the end of the present circuit.

Finally, the illustration in Figure 6.8f does not show the necessary additional circuitry needed to interface the output signal from the decoder to the actual switching circuit elements These interfacing circuit requirements differ for each application and are an extra portion of the sequencer circuit when digital timer control is desired.

REFERENCE

1. Lipták, B.G., ed., *Instrument Engineers' Handbook*, Vol. II, Philadelphia, Chilton, 1969, Chap. 4.

BIBLIOGRAPHY

Gilbert, R.A. and Llewellyn, J.A., *Principles and Applications of Digital Devices*, Instrument Society of America, Research Triangle Park, NC, 1982.

Madden, D.G., "Drum Programme," *Instrumentation Technology*, August 1970.

Schwartz, J.C., "Understanding Timers for Better Control," *Instruments and Control Systems*, December 1975.

"Timers," *Instruments and Control Systems*, August 1960.

Chapter VII

COMPUTER AND DISTRIBUTED CONTROL

A. B. Corripio • R. G. Dittmer

E. J. Farmer • P. A. Holst

M. F. Hordeski • V. A. Kaiser

J. A. Moore • M. H. Waller

T. J. Williams

CONTENTS OF CHAPTER VII

Contents of Chapter VII

7.1 PROGRAM LANGUAGES AND ALGORITHMS

Process Control Programming

The purpose of computer programming is to individualize the functions carried out by a particular computer system. Making the computer do what is wanted in a particular context—that is, writing the programs that run the hardware—has become the leading cost factor in developing computer systems today, whether for scientific, business, or process control applications.

Programming methods for process control computers have tended to lag somewhat behind the state-of-the-art programming techniques developed for scientific and business applications. This is because of the smaller number of applications for any one control computer model, the correspondingly smaller size of the vendor company, and the company's resulting inability to provide the latest in programming techniques for that particular computer system. Nevertheless, the high cost of developing application programs for particular functions from scratch has led to many developments in computer programming that are designed to streamline the programming effort and minimize the individuality of any one program. New programs tend to be developed from old programs used previously for similar functions. There is also a major trend to make all programs as general as possible in order to enhance their use for related applications, to minimize their individuality, and thus to reduce the cost of developing the next system. In fact, a major subset of the computer industry has evolved solely for the purpose of producing these general programs.

Types of Programs

The desire for minimum individuality in process control computer programs has resulted in a very common program organization, consisting of three major parts: (1) the *operating system* or *executive program*, (2) the *application program*, and (3) the *system support* software.

The executive program consists of all the programs that supervise the overall operations of the computer system. The functions it performs are as follows:

1. Scheduling and actually starting the execution of the application programs.

2. Operating all hardware of the system (e.g., allocating main memory to specific functions, loading programs into main memory from external bulk memory, etc.).
3. Supervising input/output operations.
4. Servicing the priority interrupt system.
5. Loading analog and digital inputs into memory.
6. Controlling outputs to the plant actuators.

Applications programs handle all of the specialized functions required for that particular installation. These functions may include:

1. Conversion of plant input data to engineering units.
2. Scheduling, optimization, and control correction computations.
3. Operator's console, logging, and other operator and management presentations.

Even though applications programs are individualized for each installation, every effort is made to use pre-written, general programs whenever possible and to minimize the required contribution of the user.

The system support software items are those programs that help the user to prepare the application programs. They include the following:

1. Assemblers and compilers that convert programs written in a specific assembly language or a higher-level language such as FORTRAN into the machine language of the computer.
2. Editors, linking-loaders, and similar packages that allow segments of programs written separately to be incorporated into one program in the computer.
3. Programs that help debug applications programs. These programs may not be carried in the computer's memory at all times but may be entered when needed from external storage (disk or tape).

One rapidly developing programming technique is that of *host compiling* or *host programming*. Here a large computer contains in its programming system an *emulator* of the smaller computer for which software is being developed. The emulator lets the large computer behave

668

just like the smaller computer. Since the larger computer is faster, has a larger memory, and is capable of utilizing more complex and capable system support software, it can develop better programs for the smaller computer than the latter could have produced by itself. The machine language program developed by the large computer is then read into (down-loaded to) the small computer as its object program. Thus the small computer need never have any system support software and the user does not have to buy the memory to carry it.

Features of Process Control Programs

Computer systems used for process control differ from those used for general scientific and business purposes in that they contain special features that allow operation in a real-time environment. That is, process control systems perform time-relative operations that are governed by a real-time clock, and they respond to other externally generated occurrences through an external or priority interrupt system. They must also be able to read the values of external variables and transmit signals to external devices, including human interpretable systems such as an operator's console.

Obviously, the process control program used in the system must allow for these functions. Accommodating these functions comprises the major difference between programming in the process control field and programming in general scientific and business fields. Because the priority interrupt capability vitally affects the overall management of the computer system, this function is usually included within the executive program of the process control computer. The other time-based and externally stimulated functions are included in the application program of the system.

A second major difference between process control programs and typical scientific or business programs is the great dependence upon *multiprogramming* or *parallel execution* of functions. The time-based operation of the control computer, the large amount of time required to complete many functions (such as printing reports and reading process variables), and the necessity to check the status of many tasks mean that the computer must shift its attention between many different functions while completing only small parts of any one function at any one time. To keep track of the on-going status of each of the several tasks under way at any one time requires a very sophisticated executive program. All successful process control executive programs must have this capability. It is not called for in any but the most sophisticated scientific and business systems run on computers the size normally used for process control.

Figure 7.1a illustrates the operations that are executed by programming in the process computer system. The diagram shows the overall system as carried out by a single computer containing all functions. The modular system shown in the diagram allows any particular mod-

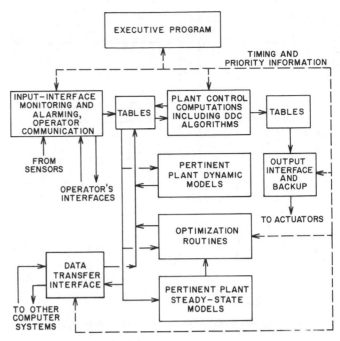

Fig. 7.1a   Modular process control programming system

ule to be modified without affecting any of the other modules. This greatly simplifies both the initial programming effort and any later required program modifications. This is made possible by the use of the data tables indicated in the diagram. A further advantage of such a program system is that programs developed by others for any of the modules can be readily integrated into the overall program. The chance of finding a suitable existing module is obviously much greater than that of finding an existing overall program for any particular application.

The Executive Program

The executive program apportions computer time to tasks that must be carried out by the computer according to their relative urgency.[1] Response-critical events occurring on-line must be taken care of at once, resulting in priority task execution. When no such urgency exists, the executive program assigns time to background operations. The executive software function is basically a timesharing one. It resolves process situation conflicts, switching computing facilities back and forth between critical and noncritical demand. Figure 7.1b is a schematic representation of this timesharing function, in which available computer time is allotted among events of varying priority. Pre-established priority rules are used to determine which situation is attended to at any point in time. As process situations are resolved, computer time is returned to noncritical operations.

The principal programs contained in the executive software are the interrupt processor, task scheduler, and error control. Figure 7.1c shows these functions, as well

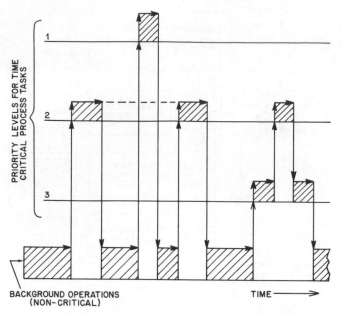

PRIORITY LEVELS FOR TIME
CRITICAL PROCESS TASKS

1

2

3

BACKGROUND OPERATIONS
(NON-CRITICAL)

TIME ⟶

Fig. 7.1b Operation of the executive program's timesharing function.

rupts as well as the periodic monitoring of live input values to determine if present limits have been violated or if control corrections are necessary. Thus the executive software must control the use of hardware timer interrupts (the real-time clock) and provide a periodic response to the user program. Aside from controlling periodic monitoring, task scheduling is usually carried out in response to random demand for program execution. Simple algorithms are often used to assign priorities to these demands. But more sophisticated applications require that the user vary priorities. In some cases user software may dynamically assign priorities that take into account complex combinations of process events.

Error control is probably the most difficult function of the executive software. It is responsible for the proper handling of errors originating in hardware or created by software. Some errors can be handled with no response from the user; others require user intervention to minimize the effect on the process being controlled. In extreme cases, error handling requires computer control back-up to avoid catastrophy. In general, error handling routines provide action that is unique to each application.

The principal purpose of the *input-output (I/O) software* is to optimize use of peripheral devices, resolve conflicts arising from concurrent requests for their use, and provide a simple interface with peripheral and process equipment for programming ease. I/O devices may be peripheral to the computer, such as printers and tape readers, or they may be functional units in the process.

The best *hardware diagnostics* can often be built into the user program. For example, input A/D can be verified for reasonableness prior to being applied to the actuation of an output device. Again, based on availability of output member feedback, the program can require

as other sources of data or routines with which the executive program regularly communicates.

The *interrupt processor* plays a central role in the control system. Events in the outside world that require response usually appear as external interrupts. The interrupt processor must be capable of sensing the degree of urgency and assigning computer time in proper order of priority. At the same time, the internal registers and the status of an interrupted program must be "saved" to properly restore service.

The *task scheduler* must accommodate random inter-

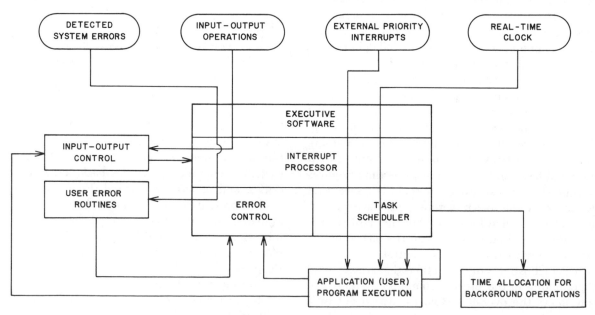

Fig. 7.1c Principal executive software functions and lines of communication with external events or other software

verification that the output motion has actually taken place.

In many process control applications, down-time for maintenance is not practical. This is true, for example, of I/O devices that are on call for random service. But devices such as typewriters, card readers, punches, and disks can be diagnosed on-line if there is a back-up capability. On-line diagnosis is also possible if the application is designed to be carried out under degraded conditions—an important consideration in control programming.

Programming Languages for Process Control

Because they are easier to use than assembly languages, the so-called higher-level languages are becoming ever more popular for all types of computer programming, including process control programming. Admittedly, assembler languages permit the knowledgeable programmer to take advantage of programming tricks and the design details of the particular computer to optimize the operating speed of the system while minimizing the need for memory. In other words, assembler languages permit more efficient use of the computer. However, the ever-increasing speed and capacity and the ever-shrinking cost of each new model of computer are reducing the need for such sophisticated programming techniques and the extensive experience required to apply them. Thus, we are sure to witness an ever-decreasing use of the assembler, except perhaps for operating systems and for a few very large microcomputer applications where any possible savings resulting from optimum use of the computer's capacity are critical. Thus our emphasis in this section will be on higher-level programming languages.

Higher-level languages for process control can be readily divided into two major branches: general-purpose languages and problem-oriented languages. Figure 7.1d shows these two branches and their further subdivision

into several different categories. Higher-level languages have two major advantages over an assembler or any other so-called lower-level language. First, the same program can be run on any type of computer for which the corresponding translator program has been written. That is, the higher-level languages are not usually machine architecture dependent as are assemblers. The translators, called compilers or interpreters, are discussed below. The second advantage is that their symbology is usually much more readily interpreted by the average user than are the assembler or machine languages they replace; that is, they are more like normal English or mathematical statements.

A compiler takes the code (program) as written by the programmer and converts it completely to the machine language of the particular computer. This translation may include converting the program into an assembler language as an intermediate step.

An interpreter, on the other hand, stores the program in the computer in approximately the form developed by the programmer. The computer then executes the program a line at a time by interpreting, or converting, each line to machine language and carrying out the instructions included therein before going on to the next line.

It can be readily understood that programs to be used with an interpreter must be somewhat simpler than programs to be used with a compiler. When compiling a program, the computer can sense the complexity of the entire program and incorporate some means to handle it. This "look-ahead" capability is, by definition, not available with the interpreter.

Compiler languages differ among themselves in terms of their capabilities. Those that permit complex programs to be readily written in them are called "systems programming languages." Some of these can be used in the preparation of very difficult programs, such as large computer operating systems or executive programs. Otherwise these programs must be written in the assembler language of the particular computer.

Problem-oriented languages are those that have been especially developed for a particular type of application. Process control is a particularly important application type in this regard. In fact, many problem-oriented languages have been especially developed for relatively narrow types of applications within process control, such as the control of batch reactors.

As indicated in Figure 7.1d, problem-oriented languages can be further divided into compiler type languages and "fill-in-the-blanks" languages. One of the major types of "fill-in-the-blanks" programs originated with the General Process Control Programming system (GPCP), developed by Humble Oil and Refining Company.[2] The system was quickly picked up by the IBM Company as PROSPRO[3] and by General Electric as BICEPS.

Compiler type languages are usually modified or ex-

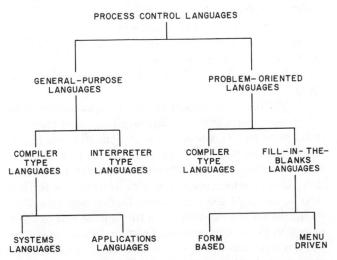

Fig. 7.1d Organization of higher-level languages for process control

tended versions of the general-purpose compiler languages. FORTRAN and BASIC have been especially popular as the basis for such efforts. A well-known example of this kind of program is AUTRAN (Automatic Utility TRANslator), which was developed originally by the Merck, Sharp and Dohme Automation Department and perfected and marketed by Control Data Corporation.[4]

The fill-in-the-blanks system is a largely prewritten, interpreter type program that offers a large number of choices in its execution in addition to a well-developed skeleton data base. By means of responding to a series of questions or a "menu" displayed on a CRT screen, the system developer, or programmer, is able to make a proper choice of algorithms for each of the control loops from the available options. The programmer can also supply all of the necessary system parameters to the computer's data base. The result is a completely operative control program for the process, developed from the CRT-displayed options. Program development may also be handled by means of a set of preprinted forms, one for each control loop, one for each analog or digital input, one for each process function, etc. These forms are then converted to a deck of punched cards that are read into the computer's memory.

The prewritten fill-in-the-blanks program must be interpretively executed since it is developed with no knowledge of how the final programmer will organize the execution of the final program or how many and what type of functions will be involved.

Correcting the Deficiencies of General-Purpose Compiler Programs

FORTRAN and BASIC have become very popular scientific computation languages because programs written in these languages strongly resemble normal mathematical expressions. As a result they are widely used as a basis for university teaching, and a great number of people are familiar with their use.

This same mathematical formula-like representation makes these languages viable candidates for process control programming. However, because they were both originally developed only for batch scientific programs, they are incapable of readily expressing many of the operations vital to process control. These operations are primarily related to the process computer's need to interface with its surroundings and to carry out logic functions. These operations can be listed as follows:

1. Bit string manipulation (i.e., all the logic functions, bit counts, bit testing, and bit movements within a computer word).
2. Input/output data handling (analog and digital input and output).
3. Task scheduling and priority interrupt functions (these are mainly executive program functions).
4. Some aspects of file handling (i.e., those not pro-

vided by FORTRAN 77, the most recent version of the FORTRAN standard).

Because of the popularity of FORTRAN and its obvious potential as a basis for a process control language, nearly every process control computer vendor provides a FORTRAN compiler as part of their programming systems. They have circumvented the deficiencies of FORTRAN for process control (as listed above) by providing nonstandard extensions to the language that mechanize these missing functions in a FORTRAN-like manner.

The first of the extended FORTRAN languages was developed by the Thompson-Ramo-Wooldridge Company in 1964 for the TRW 340 computer. It was quickly followed by others. One of the best of these extended compiler systems was that developed in 1968 by the Control Data Corporation (CDC) for the CDC 1700 process control computer system.

Since 1969 there has been a concerted effort to combine and standardize the extensions necessary to use FORTRAN for process control.[5,6,7,8] The result, as described below, has been widely accepted and is the basis for most versions of FORTRAN used in process control today. The necessary extensions have been accomplished by writing subroutine "calls" in the assembly language of the computer in question and using these as one would use FORTRAN functions. So-called function extensions were also used in some cases. Essentially the same deficiencies have been corrected, with similar results, in the extension and standardization effort on the BASIC language.[12]

Ada—The DOD-HOL Project

Software development costs have become a very serious economic and manpower problem, not only for industry, but also for the military, even with its seemingly unlimited financial resources. A large portion of these costs in the military segment are related to embedded computer systems. Embedded computer systems are those dedicated to the operation and control of missiles, aircraft, ships, tanks, etc.—in other words, they are process control systems. Because these systems are quite similar to industrial and real-time systems, it is natural that similar solutions to development costs were sought.

In 1974 the U.S. Department of Defense initiated an effort to develop a common language for embedded computer systems.[9] This project rapidly gained momentum, and in February 1978 four preliminary language designs were submitted to the technical community for evaluation. Two language proposals were selected for further development. At the same time, the functional requirements document was revised in the light of new insights gained in the design process. The two detailed language designs were submitted in February 1979, and after extensive evaluation and testing, the one by CII-Honeywell Bull was selected for final development. The resulting language, christened Ada by the Department of Defense Higher Order Language (DOD-HOL) Working Group, was completed in 1980.[10,11] Since that time, work has proceeded on the development of compilers and other programming aids to accompany it.

The name Ada (note the lack of capital letters since it is not an acronym) honors Ada Byron, the Countess Lovelace of England, daughter of the poet Lord Byron, who is credited as the first computer programmer because of her work with Charles Babbage on his mechanical computer, the "analytical engine."

To increase the efficiency and the impact of the project, several measures have been taken beyond the development of the language itself. The most important of these measures has been the definition of a software environment for the use of the new language—compilers, debugging tools, editors, operating systems, etc. A set of requirements for such software tools is under development.

The Department of Defense has required that there be no extensions or subsets of the Ada language, and it has developed a means of verifying this through a process of compiler validation. Needless to say, it has the economic clout to enforce this requirement. Thus the possibility finally exists of a truly portable set of software for process control computers if commercial users insist that vendors use DOD-certified compilers for their systems.

Standardization of Process Control Languages

Almost since the appearance of the first high-level languages and language extensions for real-time applications, there has been a demand—especially from users—to standardize the languages in the same way as languages for "conventional" applications have been standardized. There are several good reasons for this demand, but the most important one is the need for portability of user programs. Portability of user programs has been made possible in principle by higher-level languages, but it has been thwarted repeatedly by the existence of numerous high-level languages and by the appearance of "language-dialects" of otherwise standard languages.

Standardization should therefore serve two purposes: (1) it should keep the number of different programming languages for one problem area as small as possible, and (2) it should discourage the development of dialects. The benefits of standardization to users are obvious, but vendors, too, will find it profitable because they will have to provide fewer different compilers for any one specific computer.

Much work has been done over the years to develop a standard programming language for process control. Fortunately, this work has now come to fruition and has achieved wide acceptance. The first of these efforts was directed toward the possible standardization of the FORTRAN language for process control applications. This

work was carried out by the International Purdue Workshop on Industrial Computer Systems, its predecessor, and associated organizations throughout the world.

Since standardization of the syntax of the FORTRAN language is the province of the X3J3 Committee of the American National Standards Institute and since there was no process control interest in that committee, it was necessary to carry out this work through the standardization of subroutine calls and related artifacts. This work was carried out by the Industrial Real-Time FORTRAN Committee of the Purdue Workshop (TC-1), also organized as Standards Committee SP 61 of the Instrument Society of America. This committee has produced three standards documents.[5,6,7] The European branch of the Workshop, the European Workshop on Industrial Computer Systems, has collected these documents into one and submitted them for international standardization.[8]

Concurrently, the Industrial Real-Time BASIC Committee of the Workshop (TC-2) has carried out a similar development for the BASIC language, working closely with the International Standards Organization, the American National Standards Institute, and the European Computer Manufacturers Association (ECMA). This has also been submitted for international standardization.[12]

The major part of future programming by user groups will be carried out using problem-oriented languages (POLs). These can be developed specifically for individual projects. Standardization would be achieved by having their own compilers written in the overall standard language (FORTRAN or Ada). Thus even though the application of the language is specific, its compilation could be handled on any computer. Thus transportability would still be achieved.

Algorithms

An algorithm, by definition, is the description of the mechanization of a particular equation by the digital computer. It has by wide usage also come to mean an equation that has become popular for the implementation of a particular function on the computer. We will use the term in the second sense since we wish to describe several equations that have become particularly important in the process control field, especially for the implementation of direct digital control.

In direct digital control, the digital computer computes the actual movement of the valve or other final actuator. In supervisory control, on the other hand, the computer merely computes the required change in the variable's set point and the final control actuation is left to an analog electronic or pneumatic controller.

There are two major approaches for digitally representing the action of conventional three-mode pneumatic or electronic controllers: position and velocity algorithms.[13] They differ principally in the place where integration occurs. For the position algorithm, the computer output is the corrected valve (or other final element) position; integration is done in the computer. For the velocity algorithm, the computer output is the change that the valve should undergo between sampling periods; therefore integration must be done by the final element via a stepping motor, integrating amplifier, or other similar device.

The Position Algorithm

The three-mode controller can be represented by:

$$P_n = K_P e + K_D \frac{\Delta e}{\Delta t} + K_I \sum_0^n e\Delta t + P_M \qquad 7.1(1)$$

$$e = S - V \qquad\qquad\qquad\qquad\qquad 7.1(2)$$

where

$$
\begin{aligned}
P_n &= \text{valve position at time n,} \\
P_M &= \text{a median valve position,} \\
K_P &= \text{proportional-mode gain,} \\
K_I &= \text{integral-mode gain,} \\
K_D &= \text{derivative-mode gain,} \\
e &= \text{error,} \\
S &= \text{set point, and} \\
V &= \text{variable.}
\end{aligned}
$$

In the usual representation of this controller by analog hardware, there is usually some interaction between the three modes as follows:

$$P_n = K_p \left(e + T_D \frac{\Delta e}{\Delta t} + \frac{1}{T_R} \sum_0^n e\Delta t \right) + P_M \qquad 7.1(3)$$

Thus $K_D = K_P T_D$, and $K_I = K_P/T_R$. Individual mode gains cannot be changed independently. In the process industry, the terms are:

$$K_P = \frac{100}{PB}, \text{ where PB is proportional band (\%),}$$

$$T_D = \text{rate time (minutes),}$$

$$T_R = \text{reset time (minutes), and}$$

$$\frac{1}{T_R} = \text{reciprocal time (repeats per minute).}$$

Equations 7.1(1) and 7.1(2) are convenient to use with digital systems. However, reference must frequently be made to equation 7.1(3) because of the nomenclature built around it.

Because of the vagaries of digital data and particularly the sampling of noisy data, the derivative expression in equations 7.1(1) and 7.1(3) must be chosen carefully. A four-point difference technique, chosen because of convergence and accuracy considerations, is defined as follows:

$$\text{Let } V^* = \frac{V_n + V_{n-1} + V_{n-2} + V_{n-3}}{4} \qquad 7.1(4)$$

where V is a variable and $n-1$, $n-2$, $n-3$ denote times previous to time n. Then

$$\frac{\Delta V}{\Delta t} =$$

$$\frac{\dfrac{V_n - V^*}{1.5\Delta t} + \dfrac{V_{n-1} - V^*}{0.5\Delta t} + \dfrac{V^* - V_{n-2}}{0.5\Delta t} + \dfrac{V^* - V_{n-3}}{1.5\Delta t}}{4}$$

$$\frac{\Delta V}{\Delta t} = \frac{\Delta e}{\Delta t} \text{ (if the set point is constant)} \qquad 7.1(5)$$

$$\frac{\Delta e}{\Delta t} = \frac{1}{6\Delta t} (V_n - V_{n-3} + 3V_{n-1} - 3V_{n-2}) \qquad 7.1(6)$$

$$\qquad\quad = \frac{1}{6\Delta t} (e_n - e_{n-3} + 3e_{n-1} - 3e_{n-2}) \qquad 7.1(7)$$

The position algorithm requires that the computer recalculate the full value of the valve setting at each time increment. In addition, this value must be transmitted to the valve positioner as an analog signal, requiring a digital-to-analog converter or the equivalent for each output.

The Velocity Algorithm

If the computer output goes to a stepping motor or integrating amplifier, the computer simply calculates the required change of valve position. The output is a digital pulse train. Where a stepping motor is used, the stepping motor in turn drives a slide wire and the slide wire output is proportional to the correct valve position. Therefore, the combination of stepping motor (digital correction) and slide wire (analog signal to the final control element) acts as a digital-to-analog converter. No special digital-to-analog converter is required and computation word length in the computer may possibly be reduced. An integrating amplifier will perform a similar action in the electronic sense.

The velocity algorithm output (ψ_n) is found by subtracting the outputs of two successive position algorithm calculations:

$$\psi_n = P_n - P_{n-1} \qquad\qquad\qquad 7.1(8)$$

$$\psi_n = K_p(e_n - e_{n-1}) + K_I e_n \Delta t +$$

$$\qquad\quad + \frac{K_D}{6\Delta t} (e_n + 2e_{n-1} - 6e_{n-2} + \qquad 7.1(9)$$

$$\qquad\qquad + 2e_{n-3} + e_{n-4})$$

The stepping motor, or its equivalent, serves to sum or integrate the position value, and

$$P_n = P_0 + \sum_0^n \psi_t = P_{n-1} + \psi_n \qquad 7.1(10)$$

The derivative mode is optional, but some small amount of integration is always required for the velocity algorithm. The algorithm with derivative included can be written as:

$$\psi_n = K_p(V_n - V_{n-1}) + K_I(S - V_n)\,\Delta t$$

$$+ \frac{K_D}{6t}(V_n + 2V_{n-1} - 6V_{n-2} +$$

$$+ 2V_{n-3} + V_{n-4}) \qquad 7.1(11)$$

Since the set point appears only in the integral term, severe controller drift could occur unless this latter term is always included.[15]

In further discussion, equation 7.1(9) will be used as the basic control equation for the system. It can be represented:

$$\psi_{n_i} = F_i(e_i) \qquad 7.1(12)$$

where the subscript i is the index to the control loop in question. In general, where the i and j loops are considered in a cascade or a feedforward configuration, the higher index refers to the inner or downstream loop.

The velocity algorithm has a unique property. At any given sample time, the proportional and integral terms do not necessarily have the same signs as do the corresponding terms in the position algorithm. Examine the proportional and integral terms of the velocity algorithm:

$$\psi_n = K_P(e_n - e_{n-1}) + K_I e_n\,\Delta t \qquad 7.1(13)$$

These terms have the same sign when the control variable is moving away from the set point, but have opposite signs when the control variable moves toward the set point (except for the first step of the variable in the direction of the set point). See Figure 7.1e for a sketch analyzing this action. As can be seen from the figure, the proportional terms can cause an oscillation as follows:

$$\begin{aligned} \text{Output 1} &= K_P(e_1 - e_0) \\ &= -|\Delta e_1|K_P \\ \text{Output 2} &= K_P(e_2 - e_1) \\ &= +|\Delta e_2|K_P \end{aligned} \qquad 7.1(14)$$

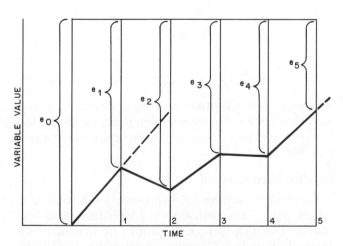

Fig. 7.1e   Action of the velocity algorithm without sign correction of the proportional term

Thus, unless strongly overbalanced by the integral term, an uncorrected velocity algorithm can give the oscillatory response illustrated in the figure. An obvious remedy to this situation is to define the proportional term as follows:

$$\begin{aligned} \text{Output P} &= K_P(e_n - e_{n-1}) \quad \text{(Modified)} \\ &= K_P|\Delta e_n| \qquad \text{(Sign of the integral term)} \quad 7.1(15) \end{aligned}$$

This property may also be used to help prevent integral overshoot and subsequent integral term oscillation. As the set point is approached, the opposing proportional term is desired for damping the integral action. When the control variable is some distance from the set point, all proportional terms are modified to have the same sign as the integral term just described.

This limited proportional action can be achieved by placing a band around the set point. Within the band all proportional contributions are accepted as calculated; outside the band all proportional terms have the same sign as the integral term. The control action can be further improved by having high proportional gain constants outside the band and much lower gain constants within the band. This allows a faster recovery from a process upset than the conventional algorithm.

This modified velocity algorithm makes possible better control action than the position algorithm. Analog values are scaled so that values from the analog-to-digital converters (ADC values) occupy the full computer variable range. This makes it possible to define the set point band width (S.B.) as some fraction of full scale. The following definitions and expressions can be used to modify the velocity algorithm in the described manner:

$$e_n = ADC_{S_n} \text{ (set point)} - ADC_n$$
$$\Delta e = e_n - e_{n-1}$$

S.B. ~ 0.07 full scale
$|e_{n-1}| \le$ S.B., add proportional term to
velocity algorithm ψ.
$|e_{n-1}| >$ S.B., do not add proportional term
to ψ (if Δe has sign different from e_n).

(*Note:* ADC as used here refers to the raw count obtained by the analog-to-digital converter without correction to engineering units.)

The set point band of 7 percent full scale gave good results on simulation tests of systems with first and second order time constants. Better control or finer tuning could be obtained if the set point band could be adjusted for each loop since high inertia systems need wider bands to get more damping of integral action. Figure 7.1f illustrates equation 7.1(16).

Relative Advantages and Disadvantages

Since both the position and velocity algorithms described above are widely used, each must have some

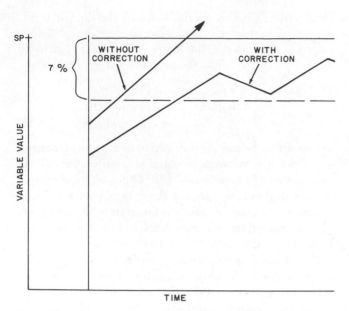

Fig. 7.1f   Effect of set-point band adjustment on velocity algorithm
response

specific advantages relative to the other to foster its selection by a particular vendor.

The major benefit of the position algorithm is that it maintains its own reference value in the term P_M of equation 7.1(3). Thus it need not depend upon an external device such as a stepping motor or integrating amplifier for a reference value. The position algorithm has two major drawbacks, however, which are thus advantages for the velocity algorithm. These are its susceptibility to reset windup and its lack of bumpless transfer. Reset windup is the saturation of the integral mode of the position algorithm if the input is connected and the output is disconnected (such as in the test mode). Then a repeatedly computed correction term without a corresponding process reaction to correct the error will eventually cause the integral term to increase to its limit (saturate or "windup"). This will result in a major upset when the output is finally connected. Since it has no internal integration, the velocity algorithm does not exhibit this problem. This problem can, of course, be corrected for the position algorithm by activating the integral term only when the output is connected. This latter modification does impair the testing capabilities of the system, however.

Whenever a major change is made suddenly in the external set point of the process, a large error is detected by the algorithm. Unless special precautions are taken, this will cause a correspondingly large correction to be computed, with a resulting "bump" to the process. This can occur when the loop is initially closed if care is not taken to adjust the set point to be close to the output value of the loop. The velocity algorithm is not susceptible to this problem since its output is a pulse train that

can be readily limited to some value within the capabilities of the process to respond.

Unlike the position algorithm, the velocity algorithm does not have an internal reference. A second major drawback to the velocity algorithm is the oscillatory response to large inputs (discussed in the previous section).

An important point to keep in mind when using these or any other control algorithm is the effect of process sampling rate on systems gain, i.e., the effective value of K_p, K_I, or K_D, for example. This can be easily understood if one considers the following example:

Suppose that one has effectively tuned a control loop at some specified sampling rate. Suppose then that the sampling rate were doubled. Since the process response rate has not changed, the amount of error detected by the controller has doubled (twice as many readings of the same average values). If the same computations are made (i.e., same controller gains), the amount of correction sent to the valves will also be doubled. Therefore we can say that overall control loop gain is a direct function of process sensor sampling rate.

Cascade and Ratio Control

Often in process control the value of one plant variable adjusts the set point of another, usually subsidiary, variable. In conventional analog-controller practice, one controller computes the set point correction of the second in essentially the same manner as if the resulting correction were being made to any final element (see Figure 7.1g). This is called cascade control. In terms of our developed nomenclature it may be expressed as follows:

$$S_{n_j} = S_{n-1_j} + \psi_{n_i} \qquad 7.1(17)$$

Another related method of conventional plant control is called ratio control, expressed as:

$$S_{n_j} = RV_{n_i} \qquad 7.1(18)$$

where R is a constant ratio. However, this is also equivalent to

$$S_{n_j} = K_P V_{n_i}$$

or

$$S_{n_j} = S_{n-1_j} + K_P(V_{n_i} - V_{n-1_i}) \qquad 7.1(19)$$

for the position algorithm or velocity algorithm. Therefore ratio control can be considered as cascade control where the set point value S_i is numerically equal to zero (see Figure 7.1h).

Feedforward Control

Feedforward control has been widely investigated in recent years. This method uses information about load changes to anticipate and, if possible, to compensate for their effects on process output. Better control results by inserting a function into the control loop that closely duplicates the process response between the load change

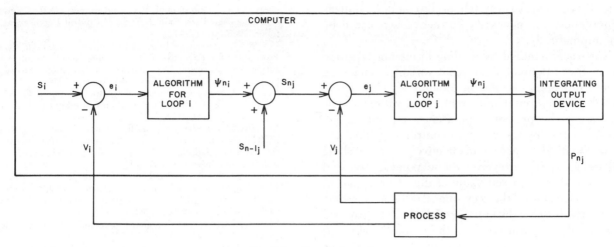

Fig. 7.1g   Cascade control

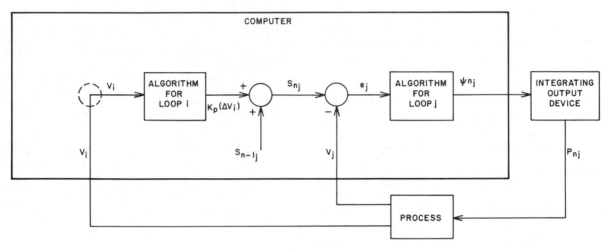

Fig. 7.1h   Ratio control

and final control element.[14,15] Corrections applied too early will anticipate the system response too much, resulting in over control and oscillation.

Unfortunately, process models must be expressed in the form of transfer functions or differential equations because of the expressed dynamic functions. Exact duplication on the control computer would require integration, imposing a severe load on computer time. This makes it necessary to devise an approximation method that will adequately describe the process, but will not heavily load the computer. The method chosen here approximates the open-loop step response of the process by a time delay followed by a ramp.

The response of chemical process systems to a step input can be approximated by one of the following transfer functions:[14,16]

$$\frac{Ke^{-T_{DT}s}}{(\tau_1 s + 1)(\tau_2 s + 1)} \qquad 7.1(20)$$

$$\frac{Ke^{-T_{DT}s}}{(\tau s + 1)^n} \qquad 7.1(21)$$

where K is a gain constant, T_{DT} is dead time, and τ, τ_1, τ_2 are time constants.

Some Other Forms of DDC Algorithms

While the algorithms described above are the basic ones applied to direct digital control, there are many others used by one or more vendors in implementing their microprocessor-based systems. Some of these are described below.

Dead Band Control. Often the establishment of a dead band about the set point is desirable to reduce noisy computer inputs. For example, consider level control of a reboiler on a distillation column. Surface turbulence caused by boiling makes the level signal quite erratic within a band of the set point. Dead band control would be implemented as follows:

If $|e| = |(S_j - V_j)| < D$ (deadband), then $\psi_{n_j} = 0$
If $|e| > D$, then use equations 7.1(3) or 7.1(9)

Emergency Response. In many plant emergency situations the correct valve action can be specified ahead of time. If a process alarm system or the computer detects an emergency, the necessary corrective action can be taken automatically.

This is easily handled by holding in memory a group of alternate set points for the plant variables, including zero and full open. When the emergency situation is detected, the computer immediately substitutes a new set point for the previous operation set point, taking corrective action by the standard control algorithm.

Error Squared. A method of effectively accomplishing the higher gain when away from the set point as discussed earlier is to use a squared value of the error as part of either the calculation of the proportional or integral modes. This effectively achieves equation 7.1(15) since squared values are always positive, as is the absolute value. At the same time, the higher value effectively acts as a higher gain on the original error value.

The technical literature available from vendors of microprocessor-based digital control systems is a useful source of additional information on available digital control algorithms.

REFERENCES

1. Weaver, S., "Importance of Manufacturer Software," *Control Engineering*, January, 1968, pp. 60–64.
2. Ewing, R.W., et al., "Generalized Process Control Programming Systems," *Chemical Engineering Progress*, January, 1967, pp. 104–110.
3. Markham, G.W., "Fill-in-the-Form Programming," *Control Engineering*, May, 1968, pp. 87–91.
4. Gaspar, T.G., Dobrohotoff, V.V., and Burgess, D.R., "New Process Language Uses English Terms," *Control Engineering*, October, 1968, pp. 118–121.
5. ISA S61.1, Standard: "Industrial Computer System FORTRAN Procedures for Executive Functions and Process Input-Output, Bit Manipulation and Data and Time Information," Instrument Society of America, 1972. Revised February, 1976.
6. ISA S61.2, Standard: "Industrial Computer System FORTRAN Procedures for Handling Random Unformatted Files," Instrument Society of America, 1978.
7. ISA S61.3, Proposed Standard: "Industrial Computer System FORTRAN Procedures for Task Management," Instrument Society of America and Purdue Laboratory for Applied Industrial Control, Purdue University, 1979.
8. "Industrial Real-Time FORTRAN," Draft Standard, European Workshop on Industrial Computer Systems, Submitted to the International Standard Organization (ISO) by the Netherlands Standards Organization (NNI), 1981.
9. Fisher, D.A., "DoD's Common Programming Language Effort," *Computer*, March, 1978, pp. 24–33.
10. Reference Manual for the Ada Programming Language, Draft Revised MIL-STD 1815, Washington, D.C.: United States Department of Defense, July, 1982.
11. Barnes, J.G.P., "An Overview of Ada," in *Software Practice and Experience*, New York: John Wiley & Sons, 1980, pp. 851–887.
12. "Industrial Real-Time BASIC," Draft Standard, European Workshop on Industrial Computer Systems, Budapest, Hungary, Central Research Institute for Physics, September, 1981.
13. Cox, J.B., et al., "A Practical Spectrum of DDC Chemical Process Control Algorithms," *ISA Journal*, October, 1966, pp. 65–71.
14. Lupfer, D.E., and Parsons, J.R., "A Predictive Control System for Distillation Columns," *Chemical Engineering Progress*, September, 1962, pp. 37–42.
15. Bollinger, R.E., and Lamb, D.E., "The Design of a Combined Feedforward-Feedback Control System," Paper presented at the Joint Automatic Control Conference, University of Minnesota, June 19–21, 1963.
16. Moczek, J.S., Otto, R.E., and Williams, T.J., "Approximation Models for the Dynamic Response of Large Distillation Columns," *Process Control and Applied Mathematics—Chemical Engineering Progress Symposium Series*, vol. 61, 1965, pp. 136–146.

7.2 ANALOG AND HYBRID COMPUTERS

Feature Summary:	For a tabulation of electronic computer component characteristics and symbols, refer to the feature summary at the beginning of Section 2.1.
Size:	Small—under 50 amplifiers Medium—from 50 to 150 amplifiers Large—over 150 amplifiers
Components:	Integrators, adders, multipliers, function generators, resolvers, analog switches, signal relays, digital attenuators, logic gates, flip-flops, counters, monostables, differentiators, pulse generators, and coefficient potentiometers
Interface Components:	Analog-to-digital and digital-to-analog converters, multiplexers, track-store amplifiers, and logic signal registers
Cost:	Computers with 20 to 100 amplifiers—$5,000 to $25,000. Computers with more than 100 amplifiers—$15,000 to $120,000. For hybrid interface to cover 10 to 50 channels—$12,000 to $60,000.
Inaccuracy:	For static, linear analog units, $\pm 0.005\%$ to $\pm 0.1\%$. For static, nonlinear analog units, $\pm 0.01\%$ to $\pm 2\%$.
Frequency:	Clock cycle frequency, 100 kHz to 2 MHz. Conversion rate, 10 kHz to 1 MHz. Full power signal bandwidth, from DC to 25 to 500 kHz.
Partial List of Suppliers:	Applied Dynamics International; Devar Inc.; Electronic Associates Inc.; The Foxboro Co.; GAP Instrument Corp.; Hitachi Ltd.; Hybrid Systems Corp.; Leads & Northrup Instruments, A Unit of General Signal; Systron Downer Corp.; Taylor Instrument Co.; Telefunken AEG; Westinghouse Electric Corp.; Zeltex, Inc.

Electronic Analog Computers

Principles

The basic building block of an analog computer is the operational amplifier. All the fundamental functions of an analog computer (summation, integration, multiplication, and function generation) can be carried out with operational amplifiers. The operational amplifier is single ended (all signals are referenced to a common ground), DC coupled wide bandwidth, and has a very high static gain (often 10^5 to 10^8). Special precautions are taken to reduce or eliminate offsets, drift with temperature changes or time, and electronic noise in the circuitry. The computer is designed to minimize cross talk (electromagnetic and static coupling) between different computing units, and other disturbing or impairing electronic effects due to the presence of a large number of components and signal sources within one system or console.

Figure 7.2a shows the primary analog computing units or functions obtained with an operational amplifier. Note that the amplifier, in addition to the high gain, also inverts the polarity of the resultant input signal combination. In case of the integrator, the actual input to the operational amplifier is switched between two points: the initial condition (e_1) input network, and the integrand (e_2) input network. In the initial condition position shown, the capacitor C charges up to a signal value e_0 (at time $t = 0$) with an electronic time constant determined by

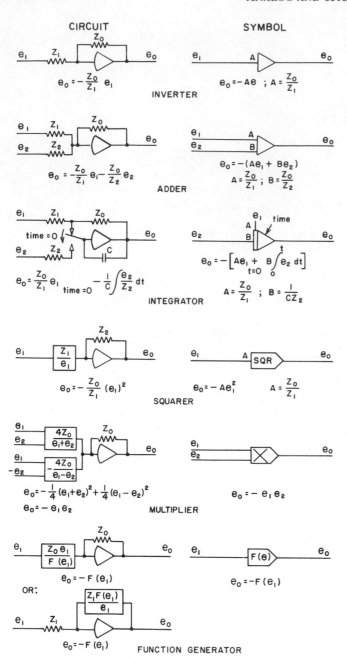

Fig. 7.2a   Primary analog computing units

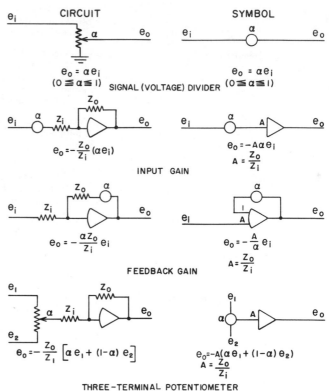

Fig. 7.2b   Potentiometer circuits

impedance according to a specified function, the program. This input network program may be either fixed, as in the case of mathematical functions, such as $\sin(x)$. $\cos(x)$, $\log_e(x)$, or variable, to allow empirical and other functions to be developed.

Coefficient and parameter values are introduced in the analog computation and simulation circuits using potentiometers, which may be adjusted manually or by a servo motor. The potentiometer serves as a signal (voltage) divider (Figure 7.2b) in different circuits, enabling a wide variety of input or feedback gains to be established. Note that in all these circuits the signal-division ratio α is defined by the potentiometer setting and the actual input network load effects due to the finite input impedances used.

Z_0C. When the switch is in the lower (operate) position, the output signal e_0 will correspond to the time integral of the integrand input e_2 with a speed or time scale determined by $1/Z_2C$.

The squarer and the multiplier are based on nonlinear input networks that cause the output signal to follow a square law relationship to the input. In the multiplier this is utilized in the quarter-share principle, which states that

$$XY = \tfrac{1}{4}[(X + Y)^2 - (X - Y)^2] \qquad 7.2(1)$$

Similarly, the function-generating unit employs an input (or feedback) network that is programmed to change

Other analog computing units (often called hybrid units) are the comparator, the analog switch, and the signal relay, shown in Figure 7.2c. These units tie the analog signal variables to logic signals (discrete, binary variables) to detect relative signal values and to introduce changes in the analog circuits.

Operational Amplifier

Operational amplifiers are used for three basic purposes in the analog computer: (1) to generate the necessary computing functions, (2) to amplify in signal level and power the analog variables, and (3) to provide iso-

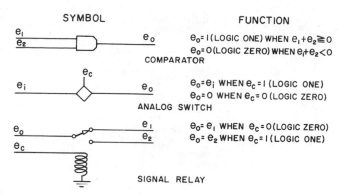

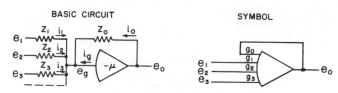

Fig. 7.2d   Operational amplifier as adder

Fig. 7.2c   Hybrid computing units

lation and unloading between the different input and output signals within the computing units.

In many analog computers all three purposes are served by the use of one amplifier type. This gives commonality in the design of the computing units, allows easy exchange of components and parts, and simplifies service and maintenance. In larger and more specialized analog computers, however, it is generally better to utilize a number of amplifier types, each with some desired characteristics for particular purposes. Examples of this are *chopper-stabilized amplifiers* for low-drift integrators, *wide-bandwidth amplifiers* in inverters and output stages, and *high-gain amplifiers* in adders.

The general features of an operational analog computer amplifier include:

Single-ended input and output
Very high gain, especially at DC
Linear, inverting output-input relationship
Low electronic noise offsets and drift characteristics
Stable and rugged in use, allowing extensive cable capacitances and long wiring in the console, with little performance deterioration
Resistance to overrange and short-circuit abuse without being impaired

Figure 7.2d shows the use of an operational amplifier as an adder. Since the amplification μ is normally in the order of 10^6 or more, the electrical potential e_g at the summing junction is virtually zero ($e_g = -e_0/\mu$). Disregarding the input current i_g, the balanced condition of the amplifier gives

$$i_0 = -(i_1 + i_2 + i_3 + \cdots) \qquad 7.2(2)$$

or

$$e_0 = -\left(\frac{z_0}{z_1} e_1 + \frac{z_0}{z_2} e_2 + \frac{z_0}{z_3} e_3 + \cdots\right) \qquad 7.2(3)$$

Introducing relative input gains $g_i = z^*/z_i$ where z^* is a conveniently selected reference value, the equation states

$$e_0 = -\frac{1}{g_0}(g_1 e_1 + g_2 e_2 + g_3 e_3 + \cdots) \qquad 7.2(4)$$

This corresponds to a weighted summation of the input variables, where the weighing factors are the input and feedback gains. The circuit shown in Figure 7.2e thus corresponds to the summation

$$e_0 = -\frac{1}{10 \times 0.250}(2e_1 - 0.1) \qquad 7.2(5)$$

or

$$e_0 = -0.8e_1 + 0.04 \qquad 7.2(6)$$

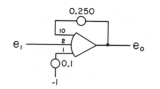

Fig. 7.2e   Example of adder

If the amplification μ is low, the summing junction potential e_g must be considered in equation 7.2(3), resulting in a correcting factor f_μ

$$e_0 = -\frac{f_\mu}{g_0}(g_1 e_1 + g_2 e_2 + g_3 e_3 + \cdots) \qquad 7.2(7)$$

where

$$f_\mu = \frac{\mu}{\mu + \dfrac{1 + g_1 + g_2 + g_3 + \cdots}{g_0}} \qquad 7.2(8)$$

Equation 7.2(8) expresses the dependence of the accurate summation of the input variables on the amplification and the input gains used in the actual analog computer circuit. Since the amplification (the open-loop gain) of the operational amplifier falls off with increasing signal frequencies, the analog computations must be carried out within the corresponding frequency limits.

An example of open-loop gain characteristic is shown in Figure 7.2(f), for an analog computer amplifier. If this amplifier is used as a summer with input gains of 10, 10, 1, and 1 (total of 22), for a summation accuracy of 0.01 percent, the open-loop gain must be at least

$$\mu = \frac{0.9999(1 + 22)}{0.0001} = 230{,}000 \qquad 7.2(9)$$

or corresponding to signal frequencies below 120 Hz. If the same amplifier is used in a unity gain inverter configuration, the necessary open-loop gain is only 20,000

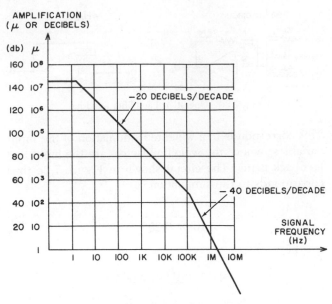

Fig. 7.2f   Typical gain characteristics of open loop

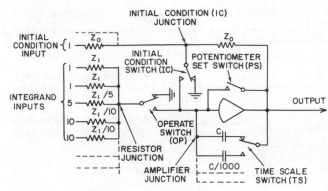

Fig. 7.2g   Integrator circuit

and the signal frequency limit extends to 1.6 kHz. Practical considerations, such as amplifier input current, accuracy of input, and feedback networks, further reduce these limits.

Computing Units

Typical analog computing units are listed in the feature summary at the beginning of this section and Section 2.1. Outmoded or specialized units, such as servomultipliers, tapped potentiometers, and CRT function generators, fall outside the scope of this section.

The key analog computing unit is the integrator, which distinguishes the analog computer from an assembly of operational amplifiers. The integrator carries out integration with respect to time, which is the only independent variable in the analog computation or simulation. All other variables appear as functions of time. The integrator is controlled by *mode commands* that specify the computational (operational) mode, such as initial condition (IC), operate (OP), or hold (HD). Other modes, such as potentiometer set (PS) and static testing (ST), are also used. Mode switching is accomplished with electronic or electromechanical switches, as shown in Figure 7.2g. The closure of the switches IC, OP, and PS gives the basic operational modes, as listed in Table 7.2h. Also included is the time scale (TS) switch, which selects one of two or more time scales to be used in the integration.

The PS mode is the neutral or passive mode that effectively cancels out any inputs to the integrator. This mode should be used whenever the integrator (or the computer) is inactive. In other modes, notably in OP and HD, the integrators may drift off scale, causing overloads.

The mode switches and the time scale switch may be commanded either by the common computer control

Table 7.2h
INTEGRATOR MODE SWITCHING

Switch Condition*				Operating Mode Required	Time Scale Required
(IC)	(OP)	(PS)	(TS)		
0	0	0	0	Hold (HD)	Normal ($\times 1$)
0	0	0	1		Fast ($\times 1,000$)
0	0	1	0	Potentiometer	
1	0	1	0	set (PS)	
0	1	1	0		
0	0	1	1		
0	1	0	0	Operate	Normal ($\times 1$)
0	1	0	1	(OP)	Fast ($\times 1,000$)
1	0	0	0	Initial condition	Normal ($\times 1$)
1	0	0	1	(IC)	Fast ($\times 1,000$)

*"0" means that switch is in condition noted in Figure 7.2g. "1" means contact has been transferred.

system (for all the integrators that operate in synchronism), or individually, by manual or logic signal, or by commands originating in a hybrid computer. Particular computer designs offer other modes, such as extended hold for long time constants, priority of IC over OP modes, and reversed mode control functions. Usually only the IC and the OP modes are critical for the proper use of the integrator.

The circuit in Figure 7.2g shows that both the initial condition junction and the resistor junction are connected to ground (reference zero) when not in the active modes. This ensures proper loading of the input networks under all conditions. There is no resistor in the circuit that shorts the PS from the amplifier output to the input (amplifier junction). If the shorting is carried out with a non-zero impedance, improper and error-causing output signals may result in the PS mode.

In Figure 7.2i are given four alternate applications of the integrator, employing the mode switches for functional and operational signal switching. Of these applications, the track-store circuit and the analog switch are the most useful. For practical reasons, only one mode switch is used in each circuit. The remaining modifications are done by regular program (patching) connections.

The circuit for the adder, or inverter, is shown in

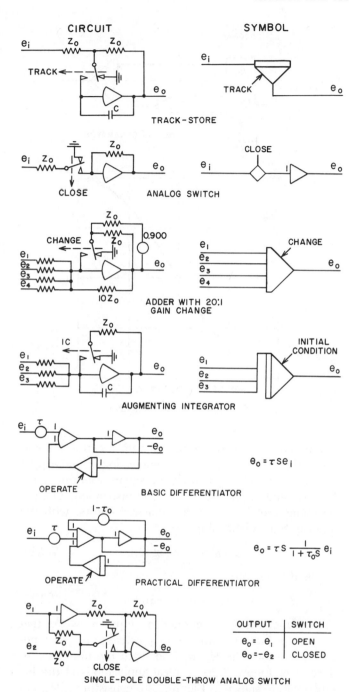

Fig. 7.2i   Integrator applications

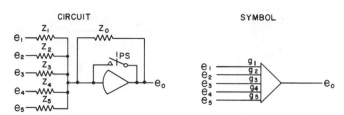

Fig. 7.2j   Adder-inverter circuit

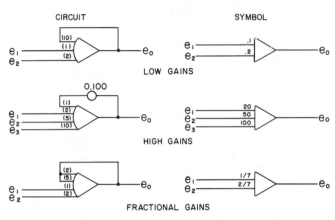

Fig. 7.2k   Input gain arrangements in adder amplifiers

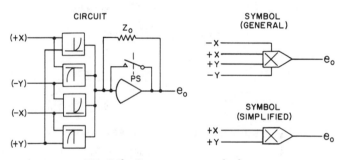

Fig. 7.2l   Quarter square multiplier

Figure 7.2j. The PS switch is used to clamp the output to zero during the potentiometer set mode. Various input gain configurations are given in Figure 7.2k for the most common adder amplifier circuits. The normal (standard) input gains are given in parentheses.

The electronic multiplier is outlined in Figure 7.2l. Four square-law networks require the inputs $+X$, $-X$, $+Y$, and $-Y$ for four quadrant operations. Some analog computer multipliers require only $+X$ and $+Y$ inputs when the necessary inverters are built in. Other multiplier types, such as those based on the transconduct-

ance principle, logarithmic functions, or time division, may require alternate inputs. In the following application circuits, only the $+X$ and the $+Y$ inputs are shown when no ambiguity exists.

Figure 7.2m shows some of the most common applications of the multiplier in analog computations and simulations. In order to preserve stability, the denominator input (e_2) must be of one polarity only and equal to or greater in value than the numerator (e_1). Similarly, the square rooter argument (e_1) must be of only one polarity.

Of special interest in process control and instrumentation applications are the sign-preserving squaring and square-root circuits shown in Figure 7.2m. In some types of computing units these functions can be obtained directly by rearranging the connections to the square-law networks shown in Figure 7.2l.

The function generator is outlined in Figure 7.2n. Two fixed or variable program networks are associated with an output (operational) amplifier circuit. The function generator unit comprises 10 to 20 segments or elements

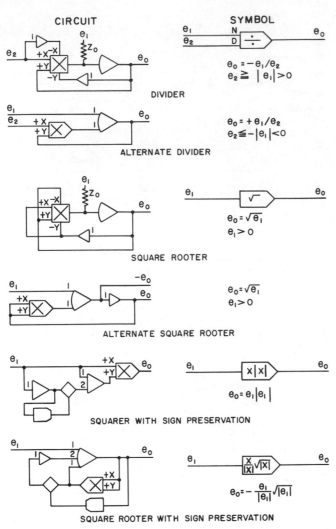

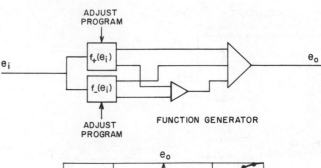

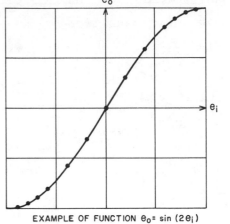

EXAMPLE OF FUNCTION $e_0 = \sin(2e_i)$

Fig. 7.2n　Function generator

Fig. 7.2m　Multiplier applications

arranged to additively contribute to the output (e_0), depending on the value of the input variable (e_1) and the program. In this way a broad range of functions can be approximated by straight line segments of appropriate lengths and slopes (gain effects). The programming thus consists of selecting the necessary number of segments and defining and adjusting the intercept points between adjoining segments over the required functional range. The example in Figure 7.2n shows the approximation of the function $e_0 = \sin(2e_1)$ with a total of 14 segments, 7 for each polarity of the input.

A variable function generator is programmed by manual adjustments of the segment intercept coordinates, using some program setup units or control systems within the analog or hybrid computer. Automated programming systems are also available by which the adjustments can be made according to information read in from punched cards, tapes, or from the hybrid computer, using either incrementally defined function values, or servo-assisted, continuously adjusted program points.

Commonly available fixed program function generators include sin(x) and cos(x) functions, $\log_e(x)$ and $\exp_e(x)$ functions, as well as squaring and square-root functions. Widely used in process control and instrumentation applications are the $\log_e(x)$ function generators, with the input-output relationship

$$e_0 = -\tfrac{1}{10}\log_e(e_1) \qquad 7.2(10)$$

and the inverse,

$$e_0 = \exp_e(-10e_1) \qquad 7.2(11)$$

These functions are generated by input networks that will accept input signals of only positive or negative polarity, as shown in Figure 7.2o, with the functions indicated in Figure 7.2p. Typical applications of the log functions are shown in Figure 7.2q, consisting of log and exp circuits. In general the circuits can be simplified and considerably improved by combining the function-generating networks in single-amplifier circuits, as is the case in the four-quadrant multiplication illustration.

Coefficient Units

The main units for establishing parametric values in the analog computing circuits are the coefficient potentiometers and the digital attenuators. The potentiometers are generally wire-wound, multi-turn, linear voltage dividers, as shown in Figure 7.2r, with the low terminal connected to analog signal ground. Ungrounded potentiometers with three terminals available for signal con-

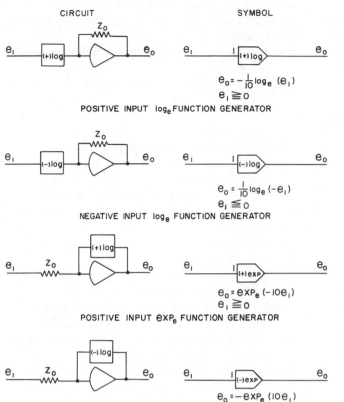

POSITIVE INPUT $\log_e$ FUNCTION GENERATOR

$$e_0 = -\frac{1}{10}\log_e(e_1)$$
$$e_1 \geqq 0$$

NEGATIVE INPUT $\log_e$ FUNCTION GENERATOR

$$e_0 = \frac{1}{10}\log_e(-e_1)$$
$$e_1 \leqq 0$$

POSITIVE INPUT $\exp_e$ FUNCTION GENERATOR

$$e_0 = \exp_e(-10e_1)$$
$$e_1 \geqq 0$$

NEGATIVE INPUT $\exp_e$ FUNCTION GENERATOR

$$e_0 = -\exp_e(10e_1)$$
$$e_1 \leqq 0$$

Fig. 7.2o $\log_e$ and $\exp_e$ function generators

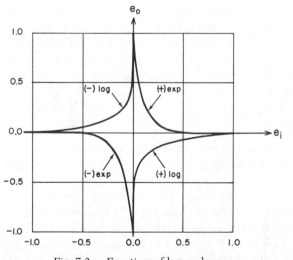

Fig. 7.2p Functions of $\log_e$ and $\exp_e$

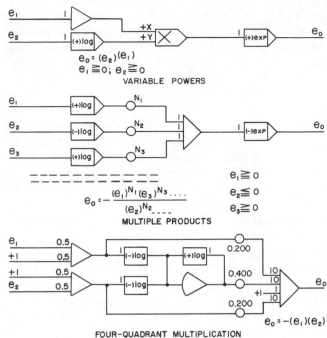

VARIABLE POWERS
$$e_0 = (e_2)^{(e_1)}$$
$$e_1 \geqq 0; \quad e_2 \geqq 0$$

MULTIPLE PRODUCTS
$$e_0 = -\frac{(e_1)^{N_1}(e_3)^{N_3}\cdots}{(e_2)^{N_2}\cdots}$$
$$e_1 \geqq 0$$
$$e_2 \leqq 0$$
$$e_3 \geqq 0$$

FOUR-QUADRANT MULTIPLICATION
$$e_0 = -(e_1)(e_2)$$

Fig. 7.2q Applications of the $\log_e$ function generator

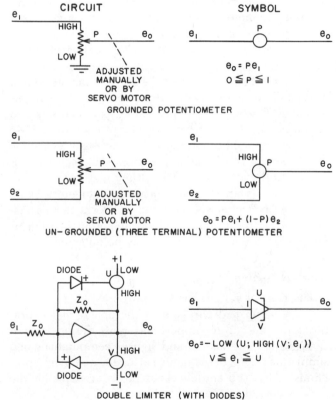

GROUNDED POTENTIOMETER
$$e_0 = Pe_1$$
$$0 \leqq P \leqq 1$$

UN-GROUNDED (THREE TERMINAL) POTENTIOMETER
$$e_0 = Pe_1 + (1-P)e_2$$

DOUBLE LIMITER (WITH DIODES)
$$e_0 = -\text{LOW}(U; \text{HIGH}(V; e_1))$$
$$V \leqq e_1 \leqq U$$

Fig. 7.2r Coefficient potentiometer

nections are often used in voltage division circuits where the reference is non-zero, as indicated by the limiter circuit in Figure 7.2r. The potentiometers may have one or more taps on the winding to provide terminals for compensation networks, which decrease the signal phase shift effects introduced by the winding construction.

Since the potentiometers are passive units with no power or signal amplification, it is important to adjust them with the desired load (input terminals) connected to the wiper and to make sure the load does not change during actual signal attenuation applications. This means the load impedance must remain constant during signal variations. Nonconstant input impedances are usually associated with multipliers, function generators, and signal switches or relays actuated during the computations. The adjustments are made by applying a fixed voltage to the high terminal of the potentiometer, with the low terminal connected to reference ground. The output voltage (e_0) is the ratio of the applied fixed voltage (e_1), measuring the actual signal voltage division (e_0/e_1) as the potentiometer setting. The control system of the analog computer provides automatic signal and ground connections to the potentiometers in certain modes, such as potentiometer set or potentiometer read. Normally the computer reference voltage will be applied as the fixed voltage, and the potentiometer setting read as a decimal fraction $0 \leq p \leq 1$. The limits 0 and 1 may be difficult to achieve in practice, and the potentiometer should not be used near these values. Best results are usually obtained with the potentiometer set between the limits of 0.100 and 0.950, or further restricted to 0.500 to 0.900. Some analog computers do not automatically provide one or both of the required connections for potentiometer setting of the three-terminal type potentiometers. Also, in some designs all the potentiometer terminals are switched at the same time in the PS mode, which means two potentiometers in cascade (series connected) cannot be properly adjusted when one loads the other.

Digital attenuators take the place of coefficient potentiometers in many hybrid computers, primarily for functional and efficiency reasons. These attenuators consist of digitally actuated switch networks that alter the attenuation (or signal conductance) in fixed and standard amounts. The digital attenuators also provide signal level and power amplification, at least in terms of unloading the effects of the input impedances that may be connected to the attenuator. Thus, they may be used with less restrictions than conventional potentiometers. Some manufacturers offer special input networks as a permanent part of the attenuators, by which input signal gains ranging from 0 to 1.6 or more may be obtained.

Control System

The control system has two purposes: (1) to provide for easy setup and checkout of the analog and hybrid programs, and (2) to support and assist the operator in carrying out the analog and hybrid computations and simulations. Due to the nature of analog and hybrid programs, the setup and checkout phase accounts for the greater part of the control system uses.

Typical applications of the control system functions include:

Inserting the analog and hybrid program signal connections

Establishing the appropriate modes of operation of the computing units

Adjusting the coefficient values of the program

Selecting the initial states of bistable units

Setting up the values of programmable function generations

Initializing the program control units within the control system

Verifying the signal connections by measuring initial computer variables

Comparing integrand values with precalculated reference values

Checking interface channel functioning

Stepping through logic unit sequences and iterative routines (slow motion)

Connecting the desired display and recording devices and setting the necessary scales and modes of operations

Energizing the appropriate operational modes of the computing units

Controlling and commanding the execution of particular computational sequences

Monitoring and reacting to the overload and underranging of the computational variables

Introducing variations in coefficient values in the analog and hybrid program

Reading and printing out the values associated with the analog and hybrid program variables, coefficients, interface channels, and program conditions

Analog computer control systems vary greatly in the degree to which these functions are supported by automated and power assisted routines within the system. Present computers will permit the operator to carry out most of these functions by centralized pushbutton and switch manipulations, with relatively few advanced and extended auxiliary systems provided. Few of the functions are operated (activated or controlled) through the hybrid interface.

Design

Most analog and hybrid computers are designed in the form of a console, arranged modularly with computing units, control systems, display, and recording devices as plug-in components. The computer consoles are constructed of rigid steel frames with removable side panels for easy access. The front sides are usually furnished with screw-in standard-size panels that mount the monitoring and control elements in an assembled and convenient arrangement. Components and modules are connected by extensive and bulky electrical cables, often furnished as elaborate wire harnesses, with individually color-coded signal and power conductors.

The computing units are plug-in individual or dual/quadruple trays, which are inserted into assigned positions or slots in the computer. All the slots are normally prewired, allowing easy and flexible expansion and modification of computing units. A disadvantage is the high initial price of an incomplete computer. Often field conversions or adaptations turn out to be impractical and expensive.

Critical to precise computations are stable and accurate power supply units in the computer, with good voltage regulation characteristics, ample power reserves, and excellent isolation from line voltage and frequency transients. Also of vital importance are signal and power ground systems, which serve as reference for all analog signal measurements and drain all electrical operating and energizing currents.

Ground systems are divided into four categories:

1. Reference ground (no current drain allowed)
2. Analog signal return (grounding of potentiometers)
3. Electrical power return
4. Operating and energizing signal returns (relays, switches)

For proper operation, a computer must be installed with adequate electrical power and ground connections to permit as noise-free and stable operations as possible and to take full advantage of the computer capabilities and signal ranges. Similarly, to ensure reliable, high-precision operation, the environment of the computer may require air conditioning or closed system circulation of air. Temperature-insensitized or compensated solid-state components have greatly alleviated some of the heat problems and have eliminated the need for special temperature-controlled chambers (ovens) within the consoles. Electrical contact contaminations, however, are still a major concern in the maintenance of electronic computers, since the designs usually include numerous multi-pin connectors for low-level signals, and large numbers of electromechanical relay contacts activate computational modes, address computing units, and implement signal connections.

Hybrid Computers

Concept

A hybrid computer is a combination in hardware and software of one or more analog and digital computers. Such a combination aims at providing faster, more efficient, and more economical computational power than is available with computers of either type alone. The results depend to a large extent on the exchange of information between the analog and the digital computers, and on the compatibility in operations and mutual interactions between the two parts.

A hybrid computer provides for the rapid exchange of information between the parallel and simultaneous computations and simulations within the analog computer and the serial and sequential operations of the digital computer. This information exchange links the two computational domains and offers the combined advantages of the fast and flexible analog computer with the precise and logic-controllable digital computer.

The extent of the information exchange between the two parts and the sophistication of the control structures

and instruction repertoires determine the capability and the capacity of the hybrid computer. Best results are obtained when both computers are designed and developed with hybrid applications as the major purpose. If a hybrid computer is made up of general-purpose analog and digital computers, with an interface tailored to these, the resulting hybrid computer often poses severe limitations in equipment complement and operational features.

Hardware

The distinguishing feature of a hybrid computer is the interface that connects the analog and digital computers. The interface consists of data communication channels in which information is passed between the two computational parts. The interface does not carry out computations, but it may contain equipment that incorporates computational units, such as multiplying digital-to-analog converters.

The interface contains a number of conversion channels in which information is converted between an analog signal (voltage) and an encoded numerical (discrete) digital computer representation, according to programmed instructions and hardware executions. The number of conversion channels states the total parallel capacity of the interface, for conversions in both directions. In practice, the number of A/D (analog-to-digital) channels may differ from the number of D/A channels, depending on applications and implementations. Since the conversion channels link parallel and concurrent analog computer variables with serial and sequential program steps in the digital computer, the interface must provide storage facilities, by which information (variables) can be stored while all channels are being prepared (loaded), so all conversions can take place simultaneously, in terms of analog variables, and sequentially, in terms of digital variables.

If the converted information is not buffered, it reflects computational variables or conditions that are not concurrent or co-referenced in time. Such a skewness is indicated in Figure 7.2s, showing the effect of a sequential conversion capability (among several time-shared units). To a degree, and at the cost of computer time, the skewness effects may be reduced by a common hold mode for all the analog computing units, during which the information conversion may take place. Fast, accurate mode control facilities that allow rapid computer interruption and resumption of the analog program are necessary for this.

Figure 7.2t shows an example of an A/D conversion channel in which the analog signal is tracked and stored in analog form. The conversion channel utilizes a multiplexer and converter that is shared among a number of channels, typically 24 to 36, and it reads the converted information into a programmed (controlled) memory location in the digital computer. From this memory location the digital program can then obtain the converted information.

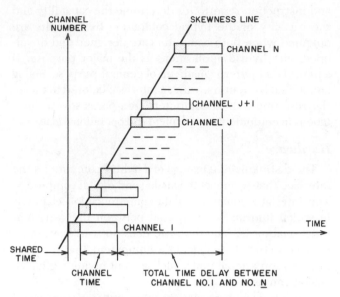

Fig. 7.2s   Skewness of converted data

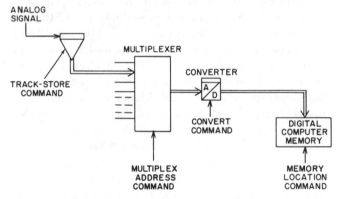

Fig. 7.2t   Components in an A/D channel

An example of a D/A conversion channel is shown in Figure 7.2u. It consists of a buffer register for the digital information and a converter unit. The buffer register holds the digital information until the moment of conversion, when it is loaded into the converter with other D/A channels. For single-channel or continuous D/A conversion, the buffer register is sometimes bypassed, and the digital information read directly (jammed) into the converter. The analog conversion output is a direct signal output or, as in a multiplying D/A, the product of an analog signal input and the converted D/A information. The conversion moment is only triggered by digital computer instructions, or indirectly, by interrupts from

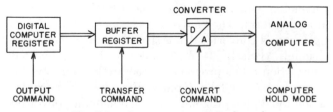

Fig. 7.2u   Components in a D/A channel

the analog computer through the digital computer program.

A number of status channels in the interface provide binary information exchange between the analog and the digital computers. This information relates to the status of particular computational variables or computing units and to the condition and progress of the programmed functions. Status channels are used to ensure proper relationships in terms of operation sequences, timing of program events, steps, and coordination of the two hybrid computer parts.

There are three main types of status channels, depending on the importance or immediacy of the reported conditions:

1. Status indicating, which at all times corresponds in status to the condition of its input variable.
2. Status retaining, which will retain the predetermined status of the input variable from the first time it is set or activated until the status channel is interrogated by the receiving computer and the channel is reset or deactivated.
3. Status responding, which will respond immediately to the condition of the input variable when this is set or activated, and will cause the receiving computer to interrupt its tasks or change its modes or functions according to the programmed actions.

Status channels are generally furnished in the interface as one-way signal lines that must be assigned to the particular conditions or computing units that are to be reported during the programming.

The interaction and control of operations between the two computers are handled by a number of binary command channels. The command channels represent direct, fixed interactive links between the control systems or command sources of the two computers. They control the executions of programs, such as start and stop computations and iterations, and the initialization or reset of programmed functions and routines.

Operation

The operational efficiency of a hybrid computer depends on the command and control (instruction) structures of the two computer parts. Analog and digital computers are sufficiently different in organization and operation to present profound problems in terms of command characteristics and functional orientation. In a hybrid computer, the analog and digital computers are linked together on the fundamental control level, providing facilities for mutual interruption and interaction of tasks.

In terms of analog computer control, the hybrid computer permits the digital computer program to carry out the control system functions previously outlined. These functions span the setup and checkout of analog and hybrid programs, the initialization and presetting of conditions for the computations and simulations, and the measurement, recording, and monitoring of the analog computer variables and functions during productive runs. Of prime importance in a hybrid computer is the ability of the computer programs to govern the progress of the computations and to take the appropriate control actions depending on the obtained results and responses.

The digital computer instruction repertoire normally includes many bit-handling instructions designed to facilitate the exchange of information through the status channels, command channels, and the direct mode control and command channels. These instructions permit multiple level priority assignments and convenient program handling of the exchanged information, either in converted data format or in single bit format. The fast access to information in the digital computer memory through direct access or cycle stealing is important for high-speed conversion channel utilization and efficient hybrid computations.

Software

Proper and complete software is required to attain the optimum operational efficiency and utilization of the hybrid computer. This is especially important in hybrid computer applications, since the complexity and extent of many hybrid programs and the sophistication of the instruction repertoire and control routines in the computers otherwise limit the usefulness and understanding of the computer capabilities.

The software is designed with practical problem-solving objectives in mind, such as handling the systems functions and presenting conditions and variables in concise and efficient formats. For best results the software is written for a particular hybrid computer and is defined and specified according to the characteristics of the hardware configuration. The software consists of three types of system programs:

1. Batch-oriented hybrid computer operation programs—organized and utilized for large, complex hybrid problems, with extensive program setup and checkout demands.
2. Conversational computer operation programs—designed for experimental and development-oriented type hybrid computations and simulations.
3. Utility routines—for efficient and convenient setup, checkout, and documentation of intermediate or limited hybrid computer programs, such as test programs, experimental circuit evaluations, and hybrid program developments.

The software enables the operator to carry out a hybrid computation or simulation of the specific problem with a minimum of effort. In a hybrid application this includes determining the signal connections and tie-lines, calculating the scale factors and coefficients of analog variables, adjusting coefficient units to the appropriate values prior to or during computations, and selecting the states or modes of the analog computing units. An important aspect of the software is the readout and documentation capabilities in terms of obtaining the values within the hybrid program, decoding or interpreting these in the language of the problem (such as in engineering units or mathematical terms), and making this information available to the operator, either by typewriter output, graphic displays, punched cards, paper tape, or other forms.

BIBLIOGRAPHY

Ashley, J.R., *Introduction to Analog Computation*, New York: John Wiley & Sons, 1963.

Blum, J.J., *Introduction to Analog Computation*, New York: Harcourt, Brace & World, 1969.

Computer Basics, Vol. I, II, and V (U.S. Navy Course), New York: H.W. Sams and Bobbs-Merrill, 1961.

Fifer, S., *Analogue Computation, I-IV*, New York: McGraw-Hill Book Co., 1961.

Giloi, W., and Lauber, R., *Analogrechnen*, Berlin/Gottingen/Heidelberg: Springer-Verlag, 1963.

Howe, R.M., *Design Fundamentals of Analog Computer Components*, New York: Van Nostrand, 1961.

Huskey, H.D., and Korn, G.A., *Computer Handbook*, New York: McGraw-Hill Book Co., 1962.

Jackson, A.S., *Analog Computation*, New York: McGraw-Hill Book Co., 1960.

James, M.L., Smith, G.M., and Wolford, J.C., *Analog Computer Simulation of Engineering Systems*, Scranton, PA: International Textbook Co., 1966.

Jenness, R.R., *Analog Computation and Simulation: Laboratory Approach*, Boston: Allyn and Bacon, 1965.

Johnson, C.L., *Analog Computer Techniques*, New York: McGraw-Hill Book Co., 1956.

Karplus, W.J., *Analog Simulation (Solution of Field Problems)*, New York: McGraw-Hill Book Co., 1958.

Korn, G.A., and Korn, T.M., *Electronic Analog Computers*, New York: McGraw-Hill Book Co., 1952.

———, *Electronic Analog and Hybrid Computers*, New York: McGraw-Hill Book Co., 1964.

———, *Random Process Simulation and Measurement*, New York: McGraw-Hill Book Co., 1966.

Lupfer, D.E., "Analog Computer Cuts Distillation Costs," *ISA Journal*, August, 1966.

MacKay, D.M., and Fisher, M.E., *Analog Computing at Ultra-High Speed*, New York: John Wiley & Sons, 1962.

McLeod, J., ed., *Simulation—The Dynamic Modeling of Ideas and Systems With Computers*, La Jolla, CA: Simulation Councils, Inc., 1968.

A Palimpsest on the Electronic Analog Art, G.A. Philbrick Researches, Inc., Boston, 1955.

Peterson, G.R., *Basic Analog Computation*, New York: Macmillan, 1967.

Rogers, A.E., and Connolly, T.W., *Analog Computation in Engineering Design*, New York: McGraw-Hill Book Co., 1960.

Scott, N.J., *Analog and Digital Computer Technology*, New York: McGraw-Hill Book Co., 1960.

Scrivers, D.B., "Special Purpose Analog Computers," *ISA Journal*, July, 1966.

Smith, G.W., and Wood, R.C., *Principles of Analog Computation*, New York: McGraw-Hill Book Co., 1959.

Soroka, W.W., *Analog Methods in Computations and Simulation*, New York: McGraw-Hill Book Co., 1954.

Tomovic, R.J., and Karplus, W.J., *High Speed Analog Computers*, New York: John Wiley & Sons, 1962.

Truitt, T.D., and Rogers, A.E., *Basics of Analog Computers*, New York: J.F. Rider, 1960.

Warfield, J.N., *Introduction to Electronic Analog Computers*, New York: Prentice-Hall, 1956.

Wass, C.A.A., *Introduction to Electronic Analogue Computers*, London: Pergamon Press, 1955.

7.3 DATA HIGHWAYS AND MULTIPLEXERS

Multiplexing Systems

Types of Systems:

A—Process control computer
B—Data acquisition-logging
C—Satellite multiplexing
D—General purpose digital communication

Stand-Alone Capabilities:

E—Slave operation only
F—Individual point addressing
G—Scan sequence control
H—Display or recording, or both
I—Data storage

Multiplexer Types:

J—Electromechanical
K—Solid state
L—Analog
M—Digital
N—Signal

Control and Data Transmission:

O—Analog
P—Digital-parallel
Q—Digital-serial
R—With format
S—Without format
T—Simplex or half-duplex
U—Duplex

Cabling and Communication:

V—Baseband
W—Amplitude shift keying (ASK)
X—Frequency shift keying (FSK)
Y—Phase shift keying (PSK)
Z—Open or twisted pairs
AA—Coaxial cable
AB—Modem, voice grade line

Cost:

Installed wiring cost on a per foot per conductor pair basis: $0.5 to $3. Analog multiplexing system cost on a per point basis: $100. Higher if simultaneous operation of several multiplexers is involved.

Partial List of Suppliers:

American Multiplex Systems, Inc. (B,C,F,G,H,I,J,K,L,M,N,Q,R,T,V,Z); Applied Systems Corp. (A); CompuDyne Controls, Inc. (C,E,K,L,M,Q,R,T,X,Z); Computer Products, Inc. (C,E,K,L,O,Z); Control Data Corp. (A); Datatron, Inc. (B); Digital Equipment Corp. (A); Electronic Modules Corp. (B); Fisher Controls, Inc. (A); The Foxboro Co. (A); General Automation, Inc. (A); General Electric Co. (A); Honeywell, Inc. (A); Houston Engineering Research Corp. (B); IBM Corp. (A); I/C Engineering Corp. (C,F,G,H,K,L,M,N,Q,R,T,X,Y,AA); Larse Corp. (D,F,G,K,M,Q,R,T,V,W,X,Y,Z,AA,AB); Leeds & Northrup Instruments, A Unit of General Signal (A); Motorola Instrumentation and Control, Inc. (B,F,G,H,I,J,L,M,O,P,T,V,Z); Process Automation Co. (A); Redcor, Inc. (A); Scanivalve, Corp. (C,E,J,L,N,O,P,V,Z); Systems Engineering Laboratories, Inc. (A); Vidar (B); Xerox Data Systems (A). *Note:* Suppliers who furnish process control computer systems (A) generally also supply data acquisition-logging systems (B); this same equipment can ordinarily be used as satellite multiplexing systems (C). Likewise, data acquisition systems (B) can also usually be supplied as satellite multiplexing systems (C).

Party Line Systems

Major Types:

A—Process control systems
  Complete systems
B—Computer systems, including interface
C—Hardware scan sequence control capability with optional computer interface
D—Multiplexing subsystems designed for operation by computer
E—Stand-alone analog systems with optional computer interface
F—Digital communication systems

Maximum Analog Scan Speeds:

G—1,000 to 5,000 points per second
H—100 to 1,000 points per second
I—1 to 100 points per second

Communication and Wiring:

J—Twisted pairs
K—Coaxial cables
L—Modem

Monitoring and Control Capabilities:

M—Analog input
N—Analog output
O—Digital Input
P—Digital output

Maximum Resolution of Analog-to-Digital Conversion (Bits):

Q—10 or less
R—11 to 13
S—14 or more

Partial List of Suppliers:

American Multiplex Systems, Inc. (A,C,H,J,M,N,O,P,Q); Bristol Babcock, Inc. (A,C,I,L,M,P,Q); CompuDyne Controls, Inc. (F,J,M,N,O,P); Control Data Corp. (A,D,G,L,M,N,O,P,R); Electronic Modules Corp. (A,E,I,J,M,O,R); General Electric Co. (A,C,I,L,M,N,O,P,R); Houston Engineering Research Corp. (A,E,I,J,M,R); I/C Engineering Corp. (A,C,G,K,M,N,O,P,R); IBM Corp. (A,D,I,J,M,N,O,P,R); Leeds & Northrup Instruments, A Unit of General Signal (A,B,C,I,J,L,M,N,O,P,S); Motorola Instrumentation & Control, Inc. (A,E,I,J,M,O,P).

Multiplexing systems and party line systems share a common characteristic—both enable multiple signals to timeshare common equipment, which may include signal conditioners, communication channels, and data processing equipment. The sharing is usually for economic reasons, with the tradeoff being a reduction in maximum speed of operations because of the necessity of performing functions sequentially rather than simultaneously. Multiplexers (Figure 7.3a) consist of terminals, to which two or more signals are connected, a switching mechanism or circuit that can route any one of the signals connected at the input terminals to common output terminals, and a driver mechanism or circuitry that operates the switching mechanism. Additions to the basic multiplexer may include multiplexer terminal address decoding and shared signal conditioning equipment at the

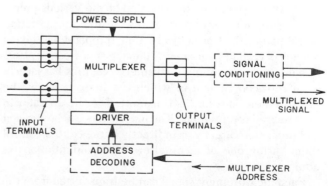

Fig. 7.3a Signal multiplexer

output terminals. A multiplexing system consists of the multiplexer itself and the enhancements provided in the package. The signals switched can be either analog (continuous) or digital (discrete) signals.

The term "party line" ordinarily refers to a method of communication rather than to a piece of equipment. A party line system is one in which a common communication channel is shared by equipment distributed along the length of the communication channel (Figure 7.3b). At each station the equipment sends and/or receives information (coded electrical signals) from other stations along the party line. Typically, a station contains a local multiplexing system, allowing multiple signals to share the station and the party line. Multiplexers are therefore an integral part of party line systems.

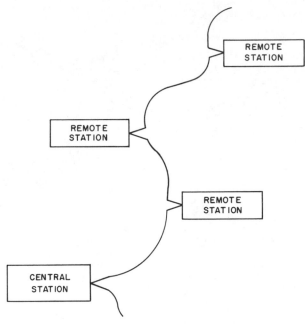

Fig. 7.3b   Party line arrangement

Multiplexing and Scanning

Operating information can be gathered in a variety of ways. Consider measurements of process parameters (temperatures, pressures, and flows) taken throughout a plant with instrumentation located at the various points of measurement. An operator responsible for operation of the plant might periodically tour the plant and take a reading of each measurement. In modern processing plants this method of scanning instrument readings has largely been replaced by one in which the information at the measurement locations is transmitted to the operator in a centralized control room. The information can be scanned and processed directly or indirectly in the central control room, using one of many alternative configurations of equipment.

Each measurement signal can be transmitted from the sensing element in the field to the control room on in-

dividual channels, typically pneumatic tubing or electrical conductors. The information can be displayed on individual indicators or recorders. Certain signals may share a display device, such as a multipoint indicator or recorder. These signals are sequentially switched into the common display, making this design financially attractive when the display costs are high relative to the costs of the equipment required to perform signal switching or multiplexing.

The multipoint thermocouple indicator is a simple example of multiplexing. Millivolt signals from the thermocouple junctions are switched into circuitry that performs reference junction temperature compensation and linearization and generates the driving signal for the indicator. The equipment that selects and switches the individual thermocouple leads is less costly than the equipment required for separate indications of each thermocouple reading. Temperature readings are scanned either manually or automatically.

In manual operation of a multiplexer, the operator addresses—usually with a pushbutton or rotary switch—one of the thermocouple signals (channels) for temperature indication. Automatic multipoint recorders are typically cyclic; at a given speed (scan rate), the signals and the print character are selected in turn and the temperature level is recorded on the chart. Multiplexer channels can therefore be either randomly or sequentially addressed and can be operated either manually or automatically.

In the data logger or process computer, the individual signals are usually multiplexed in order to timeshare amplifiers, analog-to-digital converters, and hardware or software for additional processing, which may include conversion to appropriate engineering units, checking for alarm conditions, logging operating data, and calculating performance and control data. The distinction between this equipment and the multipoint equipment discussed earlier is that the signals are converted to digital signals prior to presentation to the process operator.

Multiplexer Designs

Important characteristics to consider when selecting multiplexer equipment for a particular application include (1) accuracy, or the degree of signal alteration through the multiplexer; (2) size, or number of channels switched; (3) input signal requirements such as voltage range and polarity, source impedance, and common mode rejection; (4) channel cross-talk; (5) driving signal feed-through; (6) maximum sampling rates; (7) physical and power requirements; and (8) whether one (single-ended), two (differential), and three (including shield) lines are switched for each multiplexer point.

As discussed in Section 7.6, electromechanical switches with motor-driven commutators, crossbar switches, or relays are more accurate and less expensive, particularly for low-level (millivolt) signals than are solid state switches.

In general, electromechanical switches exhibit open-to-closed resistance ratios as high as 1×10^{14} and little feedthrough of the driving signal. Although solid state switches are faster, smaller, and consume less power, they are more likely to exhibit channel cross-talk and feed-through of the driving signal. Optical methods (e.g., light-emitting diodes optically coupled to transistors) and transformer coupling for isolation are becoming increasingly attractive as circuit costs decrease.

Packaged multiplexer systems may include address decoding capability, analog-to-digital conversion, sample-and-hold capability, scan sequence control, and power supplies. The multiplexer itself is only one component in the system. A remote multiplexer station can therefore range from a multiplexer-only operated by a separate central system to a system with stand-alone (scanning and alarming) capability and computer interface, to a complete, dedicated computer system.

Alternative System Configurations

The simplest system is probably one in which all process signals are brought to a common location and connected to terminals on a multiplexer. The multiplexer can be manually operated—as when an operator randomly selects a channel for a digital reading—or can be driven by other equipment. The multiplexer can be part of the "front-end" equipment of a data logging system or a process computer system, in which case the signals may be randomly addressed or sequentially scanned. Internal storage of the digital data obtained may or may not be a capability of the system.

This configuration (Figure 7.3c) has the advantage that all major equipment items are in one location, which facilitates environmental protection and maintenance and simplifies signal transmission. A disadvantage is that each signal must be individually transmitted to the multiplexer terminal. If the distances between the signal sources and the multiplexer are great, signal transmission costs may become prohibitive.

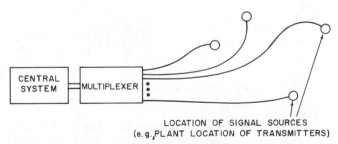

Fig. 7.3c Direct wired system

If the signals are pneumatic, they must be converted into electrical signals external to the data logger or computer. The conversion can be performed either with individual pressure-to-electrical signal converters (see Section 2.2) or with pneumatic signal multiplexers (Section

7.6). If the signals are electrical, they may be either analog or digital. Following sound wiring practices (Section 7.11), wiring costs range from about 50 cents per foot ($1.67 per meter) per conductor pair to higher than $3.00 per foot-pair ($10 per meter) including conduit or trays and installation. The cost of transmitting, say, 1,000 electrical signals an average distance of 1,000 feet (300 m) can easily exceed the cost of the remainder of the multiplexing and data logging or computer equipment associated with the installation.

If the signal sources are concentrated at a location remote from the data logging or computer equipment, significant savings in wiring costs may be obtained by locating the multiplexer near the signals (Figure 7.3d). The signals sequentially selected by the multiplexer can share a single, high-quality communication channel to the data logger or computer. The remote multiplexer will typically be a slave of the primary system, which may either address points individually or have start-stop control on sequential or cyclic scans. The remote equipment may merely multiplex the signal to the common line as received, or additional conditioning of the signal may be performed at the remote location prior to transmission.

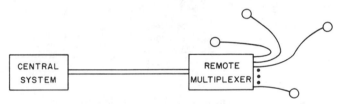

Fig. 7.3d Remote multiplexing

If the signal sources are concentrated in more than one location, it may be preferable to install more than one remote multiplexer. Each of these remote stations must communicate with the centrally located data logger or computer system. One method is to have the communication to each remote station independent of the other stations, which leads to the "radial" arrangement illustrated in Figure 7.3e. Another method, in which the communication with all stations is done through a common cable, is the party line arrangement illustrated in Figure 7.3b. A third arrangement—a combination of the first two—is illustrated in Figure 7.3f.

The radial arrangement has the advantage that the communication links between the central station and the various remote stations are independent. If a link to one remote station is broken, communication with the others is unaffected. As the number of remote stations and their location relative to the central station increase, wiring costs tend to favor the party line arrangement, or at least a combination of radial and party line communication links. Depending on the design, a break in the party line communication link may or may not interrupt commu-

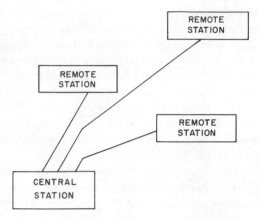

Fig. 7.3e   Radial arrangement

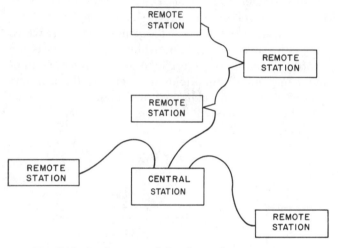

Fig. 7.3f   Combination radial and party line arrangement

nications with all stations on the party line. In one design, for example, the party line is actually a loop, and communication can be in either direction around the loop; in this case a single break interrupts no communications. The choice of configurations depends on wiring costs, security requirements, and the required number, capability, and distribution of the stations.

Remote Stations

Remote multiplexing systems are characterized primarily by (1) the amount of self-contained control logic capability, and (2) the extent of the signal conditioning performed. A rotary electromechanical thermocouple multiplexer, responding only to "start" and "stop" commands, which switches the millivolt signals directly on the common lines to the central station, is an example of minimum remote station capability. Adding the capability to identify and respond to specific and random addressing of individual multiplexer terminals, rather than responding to only a step-to-next-position command, makes a "smarter" remote station. Adding signal conditioning (for example, of the amplifier and other circuitry to effect an emf-to-current conversion of the

thermocouple millivolt signals so that current can transmit the signal to the central station) is also an extension of the basic remote station capability.

These basic configurations are illustrated in Figure 7.3a for analog multiplexing systems with optional signal conditioning and address decoding. Costs for these systems are about $100 per point for sizes of 100 to 500 points and for basic speeds of 10 to 50 points per second. Simultaneous operation of multiple small multiplexers can increase effective sampling rates to about 100 points per second. Signal conditioning and address decoding capabilities can increase prices from $200 to $250 per point range.

An additional extension of remote station capability is the transfer of the analog-to-digital conversion function from the central station to the remote station. Digital rather than analog signals are then transferred to the central station. The digital communication may be on parallel paths (Figure 7.3g), or the digital data from the analog-to-digital converter may be serialized for transmission along a single path (Figure 7.3h). Typically, terminals on remote multiplexing stations of this type are randomly addressable through serial or parallel transmission of addresses and return to the central station a message containing the requested data, the address selected, some code-checking, (e.g., parity) signals, and timing and control signals if required.

Fig. 7.3g   Parallel digital transmission

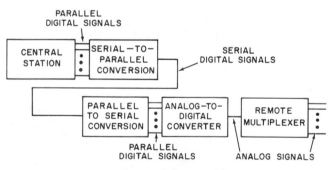

Fig. 7.3h   Serial data transmission

The functions of the remote station can be enlarged to include stand-alone capabilities, examples of which include (1) local control of timing and addressing so that periodic sequential scans can be made; (2) local (e.g., core) storage of scan data; (3) local thermocouple reference junction compensation and linearization; (4) local testing for high-low limit alarms and equipment problems, e.g., open thermocouple detection; (5) manual addressing and display capability; (6) logging capability; and

(7) computational capability. A very "smart" remote station would include a digital computer with virtually the same basic scanning, computational, and control capability as would be expected of the central computer. Two or more levels of computers result in a hierarchical arrangement.

Thus a very wide range of capabilities is available in remote multiplexing stations from stripped, basic multiplexers with no stand-alone capability to complete, full-scale computer control systems. The efficient and economic allocation of overall system functions among remote stations and between these stations and the central station is the primary task of the system designer.

Supercommutation and Subcommutation

Hardware for random addressing of multiplexer channels is an extra expense and complexity that is not always a requirement of the application. Scanning systems that select channels in a fixed sequential or cyclic order are less expensive and can be used for periodic or continuous scans of all channels.

A scanning system which is incapable of direct addressing of individual channels does not necessarily mean that all signals must be sampled at the same rate. With supercommutation and subcommutation, certain selected signals can be sampled at periods that are integer multiples or divisions of the basic scan cycle. As an example, consider a 100-point multiplexer operating with a basic scan cycle of 2 seconds, i.e., 50 points per second. If one signal were to be connected to every tenth multiplexer terminal, it (supercommutated channel) would be sampled at a rate ten times the basic scan period, or every 0.2 second. The penalty paid for this higher sampling rate is, of course, the use of more than one multiplexer terminal per channel. In the example cited, the supercommutated channel occupies 10 multiplexer terminals, leaving only 90 for the remaining signals.

Subcommutated channels are channels that share by means of another level of multiplexing a single terminal of the primary multiplexer. The subcommutated channels are sampled at a lower rate than the channels connected directly to the multiplexer. If, for example, ten subcommutated channels occupied one primary multiplexer channel, each of the ten would be sampled one-tenth as often as the basic scan cycle. Each of the subcommutated channels can in turn be further subcommutated to obtain sub-subcommutated channels.

Supercommutation allows selected points to be sampled at a higher rate than that of the basic scan cycle at the expense of multiplexer channel capacity. Subcommutation allows the expansion of channel capacity at the expense of reduced sampling rates and additional submultiplexing equipment. The alternatives and the economics associated with the implementation must be considered for each application.

Digital Signal Transmission

If analog signals are converted to digital signals at the remote locations, all communication with the central unit will be digital. There is a variety of ways in which digital information can be transmitted between the remote stations and the central unit. Data loggers and computers generally communicate internally, using binary or two-state code, with each state represented by a logic-level voltage. Internal communication and communication with local peripherals are usually done a word or byte at a time, with each of the bits that make up the word or byte transmitted simultaneously on parallel paths. The electrical signal on each path is a voltage varying with time between the two possible states. The word transfer rate on the communication channel and the bit transfer rate on each path are equal.

Parallel data transmission over pairs of twisted wires is a common practice when distances are short and required transmission rates are not high. Twisted wire can generally handle a maximum of about 2,500 bits per second reliably. Higher transfer rates are possible with more expensive coaxial cable—500,000 bits per second for distances up to 200 feet (60 m) and 200,000 bits per second for distances up to 400 feet (120 m) have been reported for coaxial cable. Beyond 400 feet (120 m), signal loss for coaxial cable becomes significant, and the cost of cable becomes a major consideration. Serial rather than parallel transmission must be considered for distances beyond 400 feet (120 m).

In serial transmission, the bits that make up a word or byte are transmitted sequentially along the same path rather than simultaneously over parallel paths. Word rates are therefore only a fraction of the bit rates. Because data logging and computing systems are generally parallel machines internally, parallel-to-serial and serial-to-parallel interface equipment is required if serial data transmission is used. The interface costs must be related to the cable costs at the distances involved and considered together with the reduced data transfer rates of serial transmission. Serial data can be transmitted with the two level or on-off (baseband) waveform, or the signals can be translated into an alternating current form through a modulation process. If the communication link is to be a modulated alternating current type, modulators and demodulators (modems) are required at each end of the communication channel.

Three basic modulation alternatives are available as shown in Figure 7.3i. The baseband signal can modulate the amplitude of a sinusoidal carrier; this is amplitude modulation or, in common data transmission parlance, amplitude shift keying (ASK). The two-level signal can control the frequency of the alternating current signal; this is frequency modulation or frequency shift keying (FSK). The third technique is control of the phase of the sinusoidal signal with the two-level signal for phase modulation or phase shift keying (PSK).

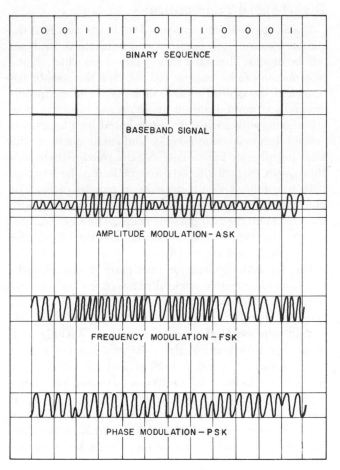

Fig. 7.3i　Modulation methods

between subchannels; frequency separation is achieved by band-pass filters. Filters of nearly exact characteristics and highly linear amplifiers are required for FDM. In time division multiplexing, bits or bytes are interleaved in time. The multiplexing can be either synchronous—in which timing is controlled by a master system clock—or isochronous (time-independent synchronous), in which storage buffers instead of a master clock maintain order.

Data concentrators perform functions similar to data multiplexers, with one major difference. Concentrators format (or reformat) the data prior to transmission. Concentrators are generally minicomputers or programmable multiplexers that take character streams and strip away the unnecessary bits and pack the remainder to transmit complete messages with formats. If, for example, the data transmitted consisted of a series of "frames," each containing a number of bits defining a character (or number), plus additional timing, control, and parity bits, the concentrator may retransmit the block of character bits (message) with timing, control, and parity bits only for the entire block.

The digital information transferred in party line systems may consist of addresses for station and point identification; data defining the status of field contacts, control commands, or the value of a converted analog signal; self-test and security coding; and system, timing, and control bits. In some cases, addresses and data use separate lines; in others, the message frame includes both address and data. Nearly all systems have some sort of security provisions—usually a cyclic (e.g., the Bose-Chandhuri) code or parity bits included in the messages transmitted. Sometimes both horizontal and vertical parity checks are made.

FSK is generally preferred for low-speed applications, i.e., when data rates are not greater than about 1,200 bits per second. ASK and PSK are most popular for higher-speed data transmission. Most low-speed data communication systems use binary coding, but more than two states (amplitudes, frequencies, and phases) can be used with any modulation technique if the level of noise and other signal distortions are not excessive. With more than two states, each time-frame carries more than one bit of information. The term baud is used for pulses (two or more states) per second; with binary transmission, bits and bauds are equivalent. If R states are used, each baud contains the equivalent information of $\log_2 R$ bits, and the bit rate is R times the baud rate.

Communication channels can be multiplexed or shared, so that more than one "message" can be transmitted over the same physical link. Two alternatives are frequency division multiplexing (FDM) and time division multiplexing. In frequency division multiplexing, the entire channel bandwidth is divided into a number of smaller bandwidth channels (subchannels), and signals are sent in parallel across the subchannels. The carrier frequencies for each channel are spaced across the entire available channel bandwidth so that no interference occurs

BIBLIOGRAPHY

Allen, D.P., "How to Choose Data Acquisition Systems," *Control Engineering*, November, 1969.

Arnett, W., "Metallic Contacts vs. Solid State Switches," *Control Engineering*, November, 1969.

Aronson, R.L., "Line-Sharing Systems for Plant Monitoring and Control," *Control Engineering*, January 1971.

Ball, J., "Tying Computers Together," *Control Engineering*, September, 1966.

Beaston, T., "Choosing a Bus for Control," *Instruments & Control Systems*, March, 1983.

Black, T., and Gorin, J., "What is a Bus-Based System?" *Instruments & Control Systems*, February, 1983.

Blasdell, J.H., Jr., "Getting the Most out of Data Links Through Multiplexing," *Computer Decisions*, January, 1971.

Buckley, J.E., "Telephone Carrier Systems," *Computer Design*, March, 1971.

Capel, A.C., "The Evolving Standard for Process Control Data Highways," *InTech*, September, 1983.

Cech, C., "Hardware for Data Bases," *Instruments & Control Systems*, April, 1983.

Cofer, J.W., "Saving Money on Data Transmission as Signals Take Turns on Party Line, *Electronics*, April 15, 1968.

"Digitized Thermocouple Compensation Yields Direct Reading for Data Logger," *Electronics*, February 2, 1970.

Eckard, M., "Tackling the Interconnect Problem," *Instruments & Control Systems*, May, 1982.

Feldman, R., "Selecting Input Hardware for DDC," *Control Engineering*, August, 1967.

Harper, W.L., "The Remote World of Digital Switching," *Datamation*, March 15, 1971.

Hersch, P., "Data Communications," *IEEE Spectrum*, February, 1971.

Hoeschele, D.F., Jr., *Analog-to-Digital/Digital-to-Analog Conversion Techniques*, New York: John Wiley & Sons, 1968.

Insose, F., Takasugi, K., and Hiroshima, M., "A Digital Data Highway System for Process Control," ISA Paper 70-510. Presented at the 1970 ISA Annual Conference, Philadelphia, October 26–29, 1970.

Kaiser, V.A., "Changing Patterns of Computer Control," *Instrumentation Technology*, February, 1975.

———, "New Configurations in Computer Control," *Instrumentation Technology*, October, 1968.

Klosky, R.A., and Green, P.M., "Computer Architecture for Process Control," ISA Paper 69–512. Presented at the 1969 ISA Annual Conference, Houston, October 27–30, 1969.

Pierce, J.R., "Some Practical Aspects of Digital Transmission," *IEEE Spectrum*, November, 1968.

Prickett, M.J., "Microprocessors of Control and Instrumentation," *InTech*, March, 1983.

Simon, H., "Multiplex Systems Save Multibucks in Refinery and Chemical Plants," ISA Paper 70–564. Presented at the 1979 ISA Annual Conference, Philadelphia, October 26–29, 1970.

Williams, T.J., "Hierarchical Distributes Control," *InTech*, March, 1983.

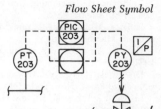

7.4 DISTRIBUTED SYSTEMS, SHARED CONTROLLERS

Analog Inputs:	Single ended, referenced to signal common
	Input ranges: 1–5 volts, 0–4 volts, thermocouple, RTD, pulse
	Input resolution: 10 bits (1 part in 1000) to 13 bits (1 part in 8000)
Analog Outputs:	4–20 mA into a 0 to 650 ohm load; raise/lower pulses. Resolution, 10 bits
Digital Inputs and Outputs:	24 volts DC; 120 volts AC; optically isolated
Number of Loops of Control:	8 to 2000
Highway Baud Rate:	100,000 to 1,000,000
CRT Characteristics:	19-inch (475 mm) diagonal measure, raster scan, RGB dot triad, shadow masked tube, multicolor, dot-matrix character generation
Normal Operating Limits:	Temperature: 0 to 50°C Relative humidity: 5 to 90 Power supply: 105 to 125 volts AC
Cost:	$1500–2500 per loop, depending on the number of controlled loops
Partial List of Suppliers:	Bailey Controls Div. of Babcock & Wilcox; Beckman Instruments, Inc.; Bristol Babcock, Inc.; Fischer and Porter Company; Fisher Controls; The Foxboro Company; Honeywell PMSD; Leeds & Northrup Instruments, A Unit of General Signal; Moore Products Company; Powell-Process Systems, Inc.; Rosemount, Inc.; Taylor Instrument Company Div. of Sybron Corp.; Westinghouse Electric Corp.; Xomox Corp., Systems Div.

Distributed Process Control

The distributed control system is a form of instrumentation used for industrial process control.[1,2] Equipment making up a distributed control system is separated, by function, into two different work areas of a processing installation. Equipment that the operator of the process uses to monitor process conditions and to manipulate the controls for the process operation is located in a central control room. From this location the operator can (1) view information transmitted from the processing area and displayed on a cathode ray tube (CRT) and (2) change control and process conditions from a keyboard. The controlling portions of the system, which are distributed at various locations throughout the pro-

cess area, perform two functions at each location: the measurement of analog variable signals and discrete inputs and the generation of output signals to actuators that can change process conditions. Input and output signals can be both analog and discrete. By means of electrical transmission, information is communicated between the central location and the remotely located controller locations. The communication path is either a cable from each remote location to the central station, or a single cable data highway interfacing all the remote stations. The parts are always sold as a system; no supplier sells only the remote portions or the centrally located portion by itself. Since the parts function together as a system, they must be discussed on the same basis.

Limitations of Other Control Forms

Since its introduction in the early 1970s, distributed control has become an accepted alternate to the other two forms of electronic control equipment used for process control, namely analog controllers and digital computers. Why does distributed control qualify as an alternate, and what are the reasons for considering its use in new projects as well as in retrofitting existing installations?

Analog Control

Surveys of control room practice have shown that instrument panel design, particularly for multiloop systems using electronic controllers, does not provide the operator the most effective interface with the process he is responsible for operating.[3] Instrument case size has shrunk along with the size of circuit components, and the instrument population of a control room has increased as electrical transmission made central panels accessible to measurements taken a mile or more away. Panels have become larger to accommodate the increase in the number of instruments, and the density of cases in a given panel area has increased as the case mounting size has decreased. Thus there is a great deal of information available to the operator, in a relatively small space.[4]

It has been found that operators cannot make full use of the amount of information available to them. Much information arrives concurrently, but the operator responds by performing a sequence of discrete actions. When standing some distance from the panel, everything associated with a group of instruments is visible, but indicated values may not be readable. When the operator moves in to make specific readings and adjustments, associated information shown on other instruments may be outside his field of vision. Since the operator must be aware of a number of things happening at once, particularly when handling a plant upset or emergency, this type of panel arrangement is unsatisfactory.

Other criticisms of instrument panel design are that expensive space may be taken up by recording instruments whose charts will be filed in equally expensive storage areas, that instrument arrangements on the panel are fixed, and that the panels do not easily permit relocating associated devices for grouping so that they can be observed as a unit. The original panel designer may have been unaware of the interrelation among devices. The requirement to run electrical wiring from every transmitter and final control device to the control room is a large contributor to the cost of a project. Other costs include the cost of real estate for the control room, the cost of the instrument panel, and the cost of utilities necessary to provide a clean and comfortable environment for the instruments.[5]

Digital Computer Control

Direct digital control (DDC) and supervisory control are two other control options. In direct digital control, a digital computer develops control signals that operate control devices directly. In supervisory control, a digital computer generates signals used as reference (set point) values for conventional analog controllers. Both options have been offered by large mainframe digital computers, with varying degrees of success.[6,7] Prospective users have at least two concerns. One is reliability: if the computer has a failure, the entire production operation is threatened. The other is programming cost. The computer must be told how to control the process, but it must also be told how to function as a process controller. It must be able to scan, move information, and store parameters so that they can be used when and where they are required. In the early years of computer application, failure to satisfy this need extended costs and delivery time unacceptably. It is still a problem to be reckoned with, because each installation is different and requires a separate programming effort.

Distributed Control Hierarchies

Before explaining how distributed control provides suitable solutions to the problems raised above, it is necessary to describe the physical relationship of the parts of distributed control systems. The point-to-point formation shown in Figure 7.4a is the simplest arrangement. It consists of a CRT display and keyboard in the central operating area, a modular control unit in the processing area, and a cable connection between them. The parts may be separated by 1,000 or more feet (300 or more meters). The remote area electronics can accommodate a combination of analog and digital input and output signals; eight control outputs (usually with a range of 4 to 20 mA) is common in many systems. The number of digital input and output signals may be anywhere from 8 to 256, depending on the supplier.

Next in complexity is the system with several remote controller files, each connected by a cable to the centrally located display unit. As many as eight remote units can be so connected, radiating from the central station in a star formation. This arrangement is also illustrated in Figure 7.4a. The operator interfaces with many control loops.

Instead of eight separate cables, each run to an individual remote controller module, it is possible to communicate over a single pair of wires running from the operating station to all the remote locations, one after the other. This is called a data highway. Many remote units can be connected to a highway (one supplier advertises 254 drops), the highway can be a mile or more in length, and several highways can be connected to one

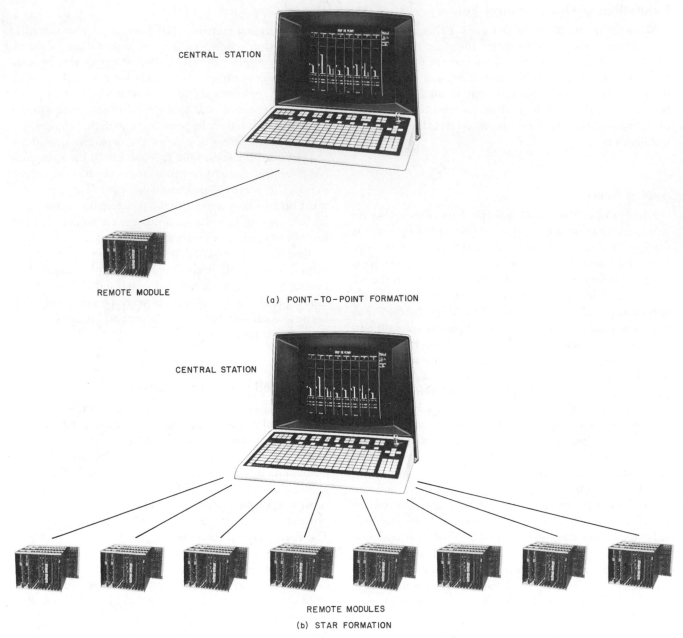

Fig. 7.4a   Nonhighway distributed control

operator station. In fact, several operator stations can be connected to one highway.

Some suppliers can interface programmable logic controllers and data acquisition units to the highway, transferring information between these devices and the operator station. Almost all suppliers provide highway access to a digital minicomputer. Theoretically, a port of a multipurpose minicomputer can be connected into the network of a large mainframe computer, making the distributed control system a satellite of the plant master computer or even of the corporate headquarters computing system. Such a system is shown in Figure 7.4b. To date, practical operating systems use a minicomputer as a satellite of the distributed control system to optimize,

schedule, and perform production logging. The technology exists for the more elaborate systems, however, and their advent is delayed only by expense and programming requirements.

Displays

The operator station at the central control location functions somewhat like the cockpit of an airplane, where the pilot can sit and observe all the information relevant to the plane's performance. The operator of a distributed control system depends on the displays on the video screen of his CRT for plant information. There are three principal types of display common to all suppliers' systems, and a number of special displays. The three prin-

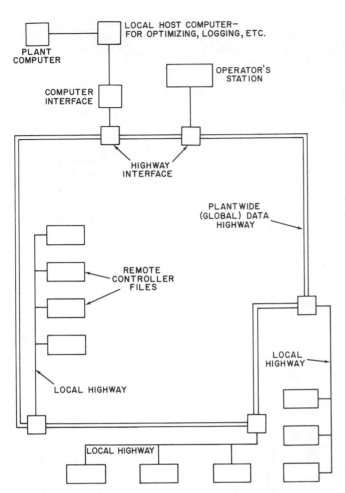

Fig. 7.4b Distributed control hierarchy: remote controllers to operator's central station control to supervisory computer control

cipal types are the group display, the overview display, and the detail display. While there are superficial differences between the pictorial representations, almost all suppliers use the same format for these three displays.

The group display shows the operating parameters of eight, twelve, or sixteen control loops, arranged in rows so that they look like the faces of instruments on an instrument panel. Figure 7.4c shows an eight-unit group display. Each of the control loops is represented by a rectangle with bar graphs to indicate the values of the process variable and the output signal. A moving index positioned beside the process variable bar shows the set point value. Engineering values for process variable, set point, and output percent are printed in or below the rectangular area. The process variable value range may be printed on the left or right side of the bar graph. Several lines of text permit a tag number and a service description to be shown. The rectangle may change color (usually to red) if an alarm condition occurs.

The array of rectangles resembles the row of instruments on a panel board of analog controllers. It can be used in the same way: from the keyboard an operator can select a loop; select an operating mode, choosing between automatic, manual, or cascade; change set point value; and change output values. These are all operations that would be available on analog instruments.

Most helpful is the capability to change the format of the loop arrangement, or even to change the content of the display. This can be done from the keyboard by a procedure called configuration. This permits any combination of loops, arranged by process association, alarm priority, sequential operation, or any other common parameter, to be included in the same group display. In fact, a loop may be included in several displays if there is a reason to do so.

If the system is one of the smaller ones with the individually cabled star formation, the display may have to correspond to the loops associated with the controller file at the end of a cable. Highway systems allow much greater flexibility, permitting loops from one file to be combined with loops from other files. It is possible to have many pages of group displays (if there are a number of displays of one type that can be consecutively brought to the screen by repeated pressing of a keyboard button, each display is called a page).

The overview display shows the bare essentials of a number of groups, each group in a separate rectangle. A typical overview display (see Figure 7.4d) will show twelve groups, although some suppliers show more, some less. The set point is shown as a straight line reference. Deviation of process variable from set point appears as a vertical bar. If a group display has eight loops in a row, the reference line will have eight segments. If the process variable for a particular loop is greater than set point, a vertical line will rise up out of the reference line from the segment corresponding to that particular loop. If the deviation is in the other direction, a vertical line will drop down from the segment. If the process variable is at set point value, there will be no vertical line. The operator, looking at an overview display, can see at a glance the condition of all the loops in a number of operating areas and can quickly spot a loop that is out of control. The overview display takes advantage of the operator's natural tendency to use pattern recognition and provides a lot of information by means of a skeleton picture of the process conditions.

Digital conditions can also be displayed on an overview display. Discrete conditions (an open or closed switch, for example) can be shown as the presence or absence of a bar rising from the reference line. Sequential events can be displayed by displaying messages that change as the sequence advances.

If the overview display indicates an alarm condition in a particular group, and the group parameters must be examined more closely, the operator can frequently call up the group display in which the alarm has occurred with a single keyboard stroke. If he wants still more detailed information, another keystroke can often be used

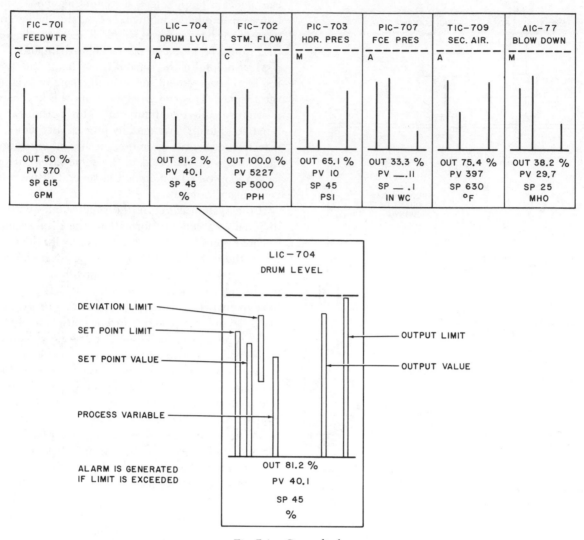

Fig. 7.4c   Group display

to call up a detailed display of the loop in the group that has generated the alarm.

The detail display is specific to a single loop or control function. Figure 7.4e shows a typical detail display. This display has the same bar graph representation that is part of the group display, but it includes additional information defining constants, limits, and other characteristics of the function. The sources of the signals coming to it are listed on the screen. Many suppliers make use of this display to enter the information that defines the function, using the keyboard like a typewriter to "fill in the blanks." The configuration of a function is done by moving a cursor to locations on the screen where values are to be entered and typing in the values on the keyboard.

Each supplier has his own complement of special lists, menus, and libraries (of units of measure and of messages, for example) that are displayed on the CRT, but all have two additional types of display that are useful to the operator: the graphic display and the trend display. While the overview, group, and detail displays are always provided, these two are optional.

Graphic display capability allows a picture to be drawn on the screen so that the operator can look at a portion of his process more realistically than by watching a row of bar graphs. Figure 7.4f is a graphic display representation of a fractionating column. Process and control information is included in the picture, and it can be interactive, dynamically changing as real time information changes. A pipeline, for example, can become filled with color when a valve is opened, the symbol of the valve can change color, and its condition can be identified by a label that indicates "on" or "off." Graphics are valuable training tools and help the operator relate to plant conditions when a number of variables are changing at one time.

Trend displays are the distributed control system equivalent of chart records. They are a profile of values of a process variable showing changes that have taken place over a period of time. Some detail displays (see Figure 7.4e) include a real-time trend graph of the process variable values during a selectable period (for example, 90 seconds, 1 hour, or 24 hours). In some dis-

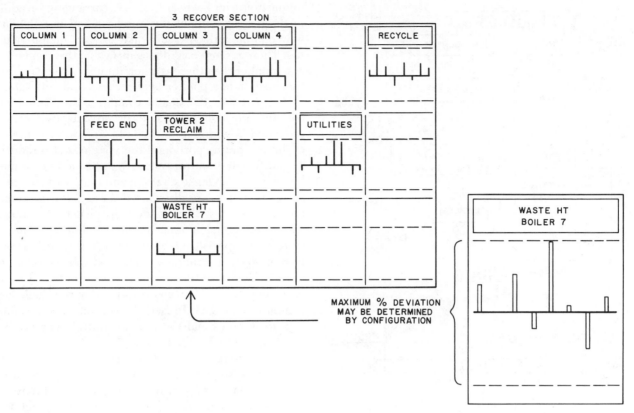

Fig. 7.4d   Overview display

CONFIGURATION WORD
ALGORITHM TYPE
HIGHWAY AND STATION ADDRESS

HWY 1   STA 33   PID   CONF 0100210003

FIC−22A
FUEL GAS

▷

OUT 0 %
PV 810
SP 750
GMP

                                                                              1000

120 MIN   90   60   30   500
            PROCESS VARIABLE TREND

	INP 1	+ 10102		K1 GAIN 0.7		L1 PV HI ALM	550
INPUT	INP 2	+ 00202		K2 RESET 2.0		L2 PV LO ALM	600
ADDRESSES	INP 3	− 15202		K3 RATE		L3 DEV HI ALM	50
	INP 4		CONSTANTS	K4 SP HI 900	LIMITS	L4 DEV LO ALM	−50
				K5 SP LO 700		L5 OUT HI LIM	90 %
				K6 SCALE 1000		L6 OUT LO LIM	10 %
						L7 OUT RATE	100 %

% DEVIATION - FULL SCALE ON OVERVIEW DISPLAY ——— OVW INDEX   50 %
DEFAULT VALUE FOR EMERGENCY ——————————————— EM DOWN   0 %

Fig. 7.4e   Detail display

plays, several trend graphs can be displayed at once, allowing comparison of the history of several variables. One supplier accommodates four trends graphs in a single display.

Figure 7.4g shows the display of two trended variables.

This information is valuable to a foreman coming on shift, to observe the recent pattern of operating history. It is valuable to an operator after an upset has occurred, allowing him to determine which of several interrelated variables was the first to be affected by the changing

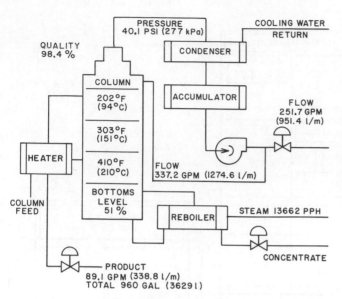

Fig. 7.4f   Graphic display

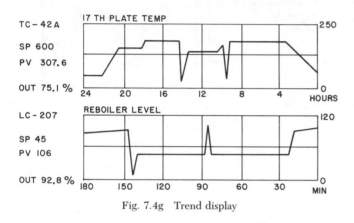

Fig. 7.4g   Trend display

ration from the keyboard allows rearranging or adding to the display without the purchase and installation of new equipment. Operators like to read digital values, too. There is more confidence in a value that can be read as a number than there is when it must be inferred from the position of a pointer between two scale divisions.

Compared to the digital computer installation, distributed control has the same cost advantages due to the reduction in wiring from the field. In addition, the cost of the computing portion of the system is also reduced. The distributed control system can be called a computer, but it is no longer a single mainframe. Instead, many small computers are distributed throughout the remote area installations, sharing the work that the mainframe had to do alone. A minicomputer connected into the system can specialize in one or two tasks, such as optimizing, logging, generating graphic displays. It does not have to handle all the information transmittal, data manipulation, and system coordination that the single large computer managed. In fact, the central station facilities can break down and the remote control operation will continue without interruption.

The programming effort required for the mainframe computer system is also eliminated. Configuration will be discussed subsequently, but it can be said here that the system programming that directs the operation of the remote controller electronics, the central station, and the communication links is furnished as part of the distributed control package. Furthermore, the programming required to tailor the system to the needs of the individual process to which it is applied can be done without knowing a high-level programming language.

Reliability

Digital computers are more reliable today than when they were first introduced, but the possibility of failure of a single piece of electronic equipment causing the shutdown of an entire production facility still raises concerns that cannot be ignored. How does distributed control satisfy the requirement for continuous production?

We have already mentioned task partitioning, the breaking up of the duties of the mainframe computer into many smaller tasks distributed throughout the system. In the individual electronic assemblies, there is further distribution of responsibility, because in each one a number of microprocessors share the functions of communication. Some suppliers include a separate microprocessor for each proportional control function, so that there is essentially single loop integrity. One loop may fail, but others emanating from the same module will remain operational.

Most suppliers subject their equipment to extensive periods of cycling at temperatures exceeding the extremes listed in equipment specifications. This weeds out the components most likely to fail. Failures of marginally operational parts will usually show up in the first

conditions. Trends over longer periods (up to one week) can be saved on floppy disk memory and displayed on command.

Attributes of Distributed Control

Distributed control has a number of advantages over electric analog or mainframe digital computer systems and satisfies several requirements that they cannot. First of all, distributed control can reduce installation cost. Less wiring is required when information is transmitted serially across the two wires of a data highway, rather than in parallel over many pairs of wires. Panel space is reduced, and so is the room size required to house it. Distributed control costs approximately $2,000 per loop. This cost is greater on a hardware basis than is the equivalent analog controller, but the overall cost including installation is lower.

From the point of view of the operator, the interface with the process is improved. The group display provides a means of viewing a combination of control loops that has meaning in terms of process associations. Configu-

few weeks of operation. Once past that period, electronic equipment seems to operate indefinitely, so long as limits of temperature and atmospheric cleanliness are observed. The advent of large-scale integration (LSI) has fostered this reliability. As size has been reduced, so has heat generation.

Nevertheless, failures will inevitably occur. It has been estimated that a reasonable life expectancy, after initial operation has been completed, for the remote location electronics is one year; and for the central station video display, about 100 days.[8]

Consequently, suppliers provide redundancy in their design, as well as backup. Some suppliers simply build two of everything, and supply it as standard. Others offer redundancy on an optional basis. Power supplies, data highways, traffic directors, and remote controller electronics are important links in the communications chain and should be considered as candidates for redundancy. The operator station itself, with its video terminal, will not shut the system down if it fails, but it will leave the operator blind to the condition of the process, and so it is another candidate for redundancy.

There should be automatic transfer between redundant parts, so that if one fails the other takes over, with no disturbance of the operation or output. At the same time, there must be some sort of alarm to alert the operator to the fact that a failure has occurred. How much redundancy is provided must depend on how much loss of production will cost. If continuous production is absolutely necessary, no expense cutting is justified.

Another form of redundancy, available from most suppliers, is controller backup. A complete modular file is mounted in the same remote location as others that are considered important enough to require backup protection. Some suppliers back up one complete set of electronics with another complete set, updating the data base of the backup unit as that of the primary is changed. The two are cabled together, and on failure of the primary, the secondary takes over automatically. Another supplier backs up eight controller files with a single backup file, and if any one of the eight fails, its place will be taken by the backup unit. Still another supplier, in addition to supplying one-for-one backup, allows portions of the backup file to back up portions of the primary controller file, freeing the unused portions of the backup file for additional control tasks.

Another type of backup is available from most suppliers in the form of manual stations. These are specific to a single control output and allow a local operator to take over from the automatically generated signal at any time. The manual station commonly has two dials, one showing process variable value and one showing percentage output value, and a manual adjustment for opening or closing the final control device. Some include set point adjustment, in addition to the final control element adjustment. If a separate power supply is available to the manual station, this type of device can be used to continue operation even if the entire distributed control system fails or, at least, to shut down the process safely.

Some systems are designed so that the output signals generated by the distributed control system can be backed up by conventional analog controllers. This permits continuous automatic control, even if there is a single microprocessor failure in an output circuit of the distributed controller. Other systems have trip circuits that actuate relays in the event of failure of an electronic controller file. The relay contacts can be used to connect actuating devices, either manual or automatic, to permit a local operator to maintain operation of the process.

It has been pointed out that availability is as important as reliability.[9] Defining availability as the ratio of mean time between failures (MTBF) to mean time between failures plus mean time to repair (MTBF + MTTR), it is clear that a system will be most available when it is very reliable (high MTBF) and can be quickly repaired (low MTTR). Since distributed control equipment is highly modular and contains many printed circuit cards, time to repair can be very short if sufficient spare parts are available. Most systems make good use of diagnostics; internal failures are reported on the CRT screen, indicating where in the system a failure has been detected and providing some clue to the cause. Substitution of printed circuit cards can often restore operation, and this can be done quickly by a serviceman who knows where to look for the trouble and has spare parts for making the substitution. The failed card can then be returned to the supplier for replacement.

Hardware at the Local Processing Area

The hardware at the process area location, that is, the remote controller, has three subdivisions: power supply, input/output handling, and computation. This equipment is usually relay rack mounted, most often in general purpose cabinets, although sometimes NEMA 12 construction is used.

Power Supplies

Well-regulated DC power supplies are included as part of the controller electronics. Frequently used values are 5 volts DC and 12 volts DC. The system power supply is commonly 24 volt DC, in 16 ampere or 20 ampere modules. Ferromagnetic transformer type supplies are sometimes used to accommodate voltage changes in the alternating current supply. For good system operation, power supplies must be properly grounded, and attention should be paid to protection from transient voltages and heavy demands from other loads on the supply line.[10]

Input/Output

Input and output terminations are made at terminals that are either part of the electronic mounting frames or on separate terminal boards. In the latter case there will usually be a cable connection between the terminal board and the electronic controller file. Connections are usually made from the front of the cabinet. An alternate method is to use a separate termination cabinet, filled with rows of terminal strips. This requires extra wiring from the termination cabinet over to the terminals in the remote controller cabinet, but it has the advantage that field wiring can be completed before the distributed control housings are delivered and installed.

Analog input and output signals will usually be carried on shielded, twisted pairs of copper wire. Digital inputs and outputs, either 120 volt AC or 24 volt DC, can be carried on twisted pairs, which do not, however, have to be shielded. Analog signals should never be run in proximity to alternating current wiring. The controller files operate almost universally on 1 to 5 volt signals, so the most common input is a 4 to 20 mA current signal, developing a 1 to 5 volt input across a 250 ohm resistor mounted on the input terminal board. Some distributed control systems can accept low level signals from RTDs and thermocouples, performing the signal amplification in their input electronic circuitry. A few systems can accept pulse inputs with frequencies sufficiently high to allow signals from turbine flow meters to be used directly.

Most suppliers offer some conditioning of signals. Taking the square root, linearizing signals derived from thermocouples and resistance thermometers, and dampening noisy inputs can be selected by configuration. Some input/output boards provide terminals with fused 24 volt DC power that can be used to supply a positive voltage to two-wire transmitters.

Separate terminal boards may also be supplied for digital input and output signals. Usually, optical isolation is provided. A DC input signal (or a rectified AC input signal) causes a light emitting diode (LED) in the isolating relay to be energized. A photoelectric device energized from the LED actuates a transistor in transistor-transistor logic (TTL) input circuitry to signal a digital input. A digital output signal is similarly isolated to actuate a transistor driver circuit for DC outputs or a triac for AC outputs. The solid state relay from which the output is generated functions like a dry contact, and the output must be powered from a separate power source.

Controller File

The solid state electronics portion of the locally mounted equipment is a special purpose minicomputer, programmed to perform the control functions for which the distributed control system is being used. It has an input section and an output section, but the major portion of the circuitry is made up of the central processing unit (CPU), read only memory (ROM), random access memory (RAM), and registers and buses that are characteristic of microprocessor-based devices. The CPU operates with a clock to repeat a sequence of programmed subroutines over a time period, typically half a second. This period is divided into a number of segments, or time slots, in each of which other subroutines specific to a control operation are carried out. The subroutines are called algorithms; they are a program of instructions designed to perform a specific function. Depending on the configuration practices of a particular supplier, time slots may be of equal length, with one algorithm per time slot, or they may have variable lengths, with a series of algorithms used to accomplish the control action.

The electronic circuitry associated with a module controlling a number of loops is contained on several complex printed circuit cards, frequently with three or four layers of foil on one card. The cards are usually plugged into a back plane containing buses that provide communication between them. As an example, we will consider one particular supplier's design. No two suppliers have exactly the same circuitry and hardware, but this description can be considered to be generic, since the end result is the same for all.

Each time slot of this controller file has a $\frac{1}{32}$ of a second period, and there are 16 time slots. Accordingly, all operations are repeated every half second. At the beginning of the half second period, all inputs (there can be 30 analog inputs and 256 digital inputs or output signals) are scanned by a multiplexing circuit, and the analog signals are changed to digital values, which are then stored in registers as binary numbers. The transition from analog to digital is performed by a 13-bit analog-to-digital converter circuit. The signals are linearized, if that is required, before being stored.

All of the information about each of the 16 time slots, as well as about each input and output value, is stored in a data base. Functional subroutines (algorithms) are stored in ROM on another card, and the operating program for communications control and information transfer is in ROM on still other cards. When a $\frac{1}{32}$ of a second period begins, the information for that particular time slot is brought from the data base into a section of RAM. The addresses of inputs are checked and the information in the addressed storage registers is fetched. The configuration is checked to ascertain which algorithms are to be used, and these subroutines are called. An output value is computed and sent to an output register. The updated function information is returned to the data base, and the time period is completed, to be followed by the next $\frac{1}{32}$ of a second operation.

The output register in this case is one of eight (the controller file services eight control loops). The binary value of each output is changed in a separate conversion circuit to a frequency, which in turn is converted to a 4 to 20 mA output signal, updated every $\frac{1}{2}$ second. Other suppliers have as many as 30 outputs developed in one controller file, while many restrict outputs to one or two. One supplier is able to develop pneumatic outputs, through a servo-operated 3 to 15 PSI (0.2 to 1.0 bar) analog controller. Another provides position type analog control, developing raise and lower signals to an electric reversing motor type drive unit. This diversity is a result of each supplier designing his equipment to satisfy the requirements of his traditional markets. The perceptive purchaser must investigate the design and specifications of a prospective supplier thoroughly, in order to evaluate the product in light of the specific application for which it will be used.

Hardware at the Central Processing Area

The principal components of the central operator station are a video screen and a keyboard. These may be supplemented with memory storage, devices for readout out and record keeping, and interfaces to equipment for data acquisition, programmable controllers, and digital computers.[11]

Video Display

The process monitor, sometimes referred to as the operator's window to the process, is usually a 19″ (475 mm) diagonal, multicolor television tube, using dot matrix character generation.[12] Some distributed control systems use smaller, monocolor screens. A few offer 25″ (625 mm) diagonal screens for wall-mounted display, observable from the control room working area. The more common 19″ (475 mm) variety is suitable for mounting on a table top or in a console.

Most suppliers depend on visible cursor control to define screen areas; some use light pens. The newest feature is the touch-sensitive screen, which gives the operator the ability to call up, for example, a group display from an overview display by touching the portion of the screen where the group picture is located. One type of touch-sensitive screen uses a grid of conductive material that changes circuit capacitance when a crossing pair is touched. The material is imbedded in a sandwich of transparent plastic. Another uses a grid of infrared light beams. Touching the screen breaks two crossing beams and triggers an appropriate response.

Flat tubes are not yet used, because of cost and resolution limitations. Neither is there yet a distributed control system that accepts voice commands and talks back to the operator. The technology exists for both, however, and at the rate of present-day technological change, these features will probably appear soon.

Keyboard

A keyboard may be incorporated in the monitor housing, or it may be separate, connected by a cable to the CRT.[13] The keys may be movable pushbuttons, or they may be printed squares on a flexible membrane.[14] The membrane type keyboard is familiar to users of microwave ovens and pocket calculators. It has a cost advantage, besides permitting the design of imaginative keyboard layouts. The switches operated by the keys may be sealed reed switches, Hall-effect switches that are operated by actuating a magnetically energized semiconductor, or capacitative switches that are operated by the motion of a plate on the end of a plunger that increases the capacitive coupling between two other plates. The membrane type switch has a flexible, hermetically sealed covering; when the pattern of a key printed on the membrane is pressed, a conductive elastomer sheet is pushed through an opening under the key picture, making contact with another conductive sheet beneath it.

Some keyboards use special function keys, programmable by the user. Others use a conventional typewriter keyboard combined with blocks of special function keys whose purpose is predefined. A list of the functions commonly performed by keys is included in Table 7.4h.

Special keyboards may be used to construct graphic display, in order to accommodate the requirement for color selection and to facilitate the construction of special symbols. Several suppliers offer devices on which the graphic picture can be traced, using a special pen. This makes the construction of the graphic display almost as easy as drawing a picture on a sketch pad.

Keyboards are usually provided with a keyswitch, to provide protection against tampering with the system configuration. When the switch is in the "Edit" position, any detail display can be called up and the information in it can be changed. When the switch is not in the "Edit" position there is no cursor, and no changes can be made. The operator can always, however, change set point values and switch from manual to automatic mode, as he would be able to do with a panel-mounted analog instrument.

Peripheral Devices

Distributed control systems require bulk memory storage that exceeds the capacity of solid state memory on the printed circuit cards making up the electronics portion of the circuitry. One supplier makes extensive use of bubble memory for this purpose, but the device most commonly used is the disk drive. Eight-inch floppy disks are used extensively for storing operating programs and algorithm configurations. To accommodate the large amount of memory required for archival trending and for many pages of graphic display use is made of the Winchester disk drive. This utilizes a hard disk, permanently sealed in an airswept housing. It is more expensive than the floppy disk drive, but it will retain many times the amount of information the floppy can hold, and accesses it much more quickly.

Another bulk memory device sometimes used is the tape cassette, but this is usually used only for recipes, configurations that are used repeatedly in batch control applications. Use of cassettes for storing programs is decreasing because of the amount of time required to access information on the tape.

Printers are another common accessory. They are used

Table 7.4h
KEY FUNCTIONS USED WITH
DISTRIBUTED CONTROL SYSTEM KEYBOARDS

Standard typewriter keyboard

Numberpad: 0–9, + and −

Cursor control: up, down, left, right, home

Clusters
  Alarm
    Acknowledge
    Defeat
    Restore
  Trending
    Trend time span
      Increase
      Decrease
      Range limit adjust
        Increase
        Decrease
  Mode select
    Automatic
    Manual
    Cascade
    Computer
  Direct input keys (input number keyed from keypad)
    Set point
    Output
    Bias
    Ratio
  Set point trim—slow and fast entry
    Raise
    Lower
  Output trim—slow and fast entry
    Raise
    Lower
  Logic
    On, Raise, Start, Reset
    Off, Lower, Stop, Reverse

Group select

Overview select

Detail select

Channel select

Special function keys

to make permanent records of alarm conditions, changes in control parameter values, and production logs.[15] The thermal head printer is the most common, with printouts on 80-column paper. Printouts with 132 columns are also available. A dot-matrix type color printer is also available as a printout device for reproducing graphic displays.

Another device used for snapshot records of screen displays is the video copier, an electronic device that prints a photographic record from a miniature, very high resolution video screen that duplicates the display on the operator station monitor. Video copiers are available in monocolor and multicolor, and some models can reproduce the screen display on $8\frac{1}{2} \times$ 11-inch paper or on 35 mm transparencies.

Unlike the solid state electronics making up the distributed control system circuitry, printers and disk drives are mechanical devices with moving parts. They are more sensitive to dust, dirt, temperature, and contaminants in the atmosphere than are the printed circuit cards, and they should be kept in the central control room area.

Communications

The data highway is the communication device that allows a distributed control system to live up to its name, permitting distribution of the controlling function throughout a large plant area.[16] Data highways vary in length as a function of traffic capability and speed of transmission. Highway lengths can be in excess of 5 miles. The most popular medium is coaxial cable. Several suppliers supply highways made of Twinax cable, a twisted and shielded coax. Shorter highways may be constructed of a pair of twisted and shielded copper wires. The most recent addition is the fiber glass cable, making use of fiber optics.[17] This is used most commonly for point-to-point connection, but at least one supplier offers a hybrid network that uses a fiber glass loop up to 20,000 feet (6,000 m) in length with electrical highways interfacing to it. Fiber optics is attractive for use as a data highway because it eliminates problems of electromagnetic and radio frequency interference, ground loops, and common mode voltages; it is safe in explosive or flammable environments; it can carry more information than copper conductors; it is inert to most chemicals; and it is lighter and easier to handle than coaxial cable.

Drops to an electrical cable can be made by splicing wires, but this is not possible when fiber glass is used. An optical electrical interface (OEI) unit is required as a node on the fiber optics highway. This uses an LED to change electrical pulses coming from a controller file or from an electrical highway into light pulses. The light pulses are changed back into electrical signals by using a PIN diode, a light-sensitive device that produces an electrical output when exposed to light. OEIs are also used as relay stations if the nodes for a 100-micron-diameter fiber glass highway are greater than 6,600 feet (1980 m) apart, because attenuation of signal takes place in fiber glass, as it does in all conductors.

Devices producing and receiving signals must be compatible with the highway transmitting the information. The information originates in the electronic devices in the form of binary coded numbers. The ones and zeros that make up the digital information are held in registers, from which they are taken for transmission. Each one and each zero is called a bit; a binary number made up of eight bits is called a byte. Transmission rate is expressed as baud rate, a measure of the number of changes of energy level per second permissible on a transmission line during transmission. This is synonymous with bits per second in most cases. It is not a measure of the amount of information that is transmitted, because some of the bits that are sent in a message act as formatting agents and are not part of the information. Teletype transmission operates at 9,600 baud. Some distributed control system highways operate at more than 1,000,000 baud.

A device is said to be compatible with a data highway at the physical level if it can be connected to the highway and can reproduce the correct sequence of binary ones and zeros from the signal appearing on the highway in the receive mode, and conversely produce the appropriate highway signal in the transmit mode. To accomplish this, devices are commonly supplied with ports and modems. A port changes the parallel form of the binary information in registers in the terminal equipment (operating stations and remote controller files) into serial form and sends it through a message formatting section at a rate specific to the highway capability. A port at the receiving end performs the same operations in reverse. A modem is a device that changes the energy level of the ones and zeros making up the binary information into a code characterized by frequency, amplitude, or phase change, for transmission over the highway; it performs the reverse operation at the receiving end.[18] A one, for example, may appear as a 0.5 MHz pulse, and a logic zero as a 1.0 MHz pulse. Ports for relatively slow devices like printers use a design conforming to the Electronic Industries Association (EIA) standard RS-232C. This is limited to transmission lengths up to 50 feet (15 m), at relatively slow rates. Most distributed control systems operating at high baud rates use ports designed to the EIA standard RS-449.[19]

In order for the information to be understood by both sender and receiver, it must be encoded, using an accepted protocol. Protocol is a kind of grammar that establishes the rules for moving data across a communications link. There are many different protocols. The one used most commonly for distributed control system highways is called High-level Data Link Control protocol (HDLC). Other protocols used in distributed control systems include the IBM Binary Synchronous Communications (BISYNC), Digital Data Communications Message Protocol (DDCMP), and Synchronous Data Link Control (SDLC). HDLC and SDLC are block transfer protocols. The format of SDLC is illustrated in Figure 7.4i.

Once the ability to send and receive messages is established by providing physical and message format compatibility, it is necessary to establish which device can transmit. Access must be guaranteed within a specified time for loop dynamics to function properly and for alarms to be sensed and acted upon. There are at least four communication methods used for distributed control highways:

1. A traffic director may be placed on the highway to grant transmitting privileges based upon polling the various stations on the highway, or following a priority controlled sequence.

2. A token is passed from station to station, granting it mastership. During the time it has the token, the station can communicate with any or all stations. When it has completed its transmitting tasks, which may not exceed a maximum time (typically a few milliseconds), it passes the token to the next station determined by a predetermined sequence. Some devices, such as an operating station, require more time than other stations, and may be in a high-priority loop, allowing them to transmit when the low-priority stations have nothing to say.

3. Carrier Sense Multiple Access is used extensively in office management communication, but is also used for some distributed control systems. As in the other methods, all stations listen all the time. Any station that needs to transmit can do so, providing no other station is already transmitting. If two stations start simultaneously, both will stop and the higher-priority station begins after waiting a period of time.

4. The fourth method broadcasts a shared memory from each station, making it available to every other station. The shared memory serves as an interface between the input circuitry of the station device and a highway controller device. Process input data is scanned and stored in shared memory. The data is taken from memory once a second and broadcast, accessible to all stations. If any other station needs information, its data highway controller listens for the data and places what it needs in its shared memory section. The station input section then takes the data from the shared memory when it needs to execute a control or output display information.

Configuration

Like any computer, distributed control equipment must be told what to do. Programming the instructions is called

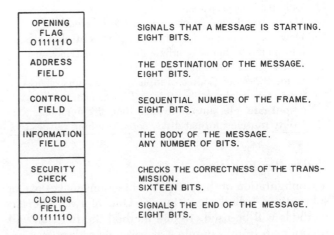

OPENING FLAG 01111110	SIGNALS THAT A MESSAGE IS STARTING. EIGHT BITS.
ADDRESS FIELD	THE DESTINATION OF THE MESSAGE. EIGHT BITS.
CONTROL FIELD	SEQUENTIAL NUMBER OF THE FRAME. EIGHT BITS.
INFORMATION FIELD	THE BODY OF THE MESSAGE. ANY NUMBER OF BITS.
SECURITY CHECK	CHECKS THE CORRECTNESS OF THE TRANSMISSION. SIXTEEN BITS.
CLOSING FIELD 01111110	SIGNALS THE END OF THE MESSAGE. EIGHT BITS.

Fig. 7.4i   SDLC (Synchronous Data Link Control) frame structure

configuring. There are two phases to configuration. First, the operating system must be configured to define the composition of groups and overviews, to define trending periods, to assign highway station priorities, to establish message tables, and so on. Second, the individual control functions must be programmed to accomplish the control strategy for which the system is to be used.

Operating System

If the system does not have a data highway, the operating system programming will be completed by the supplier. Highway systems are configured from the operator station, sometimes using a special keyboard. Files are set up for the various categories of information, and a master program opens and closes the files to use information as it is required. Configuring is usually done off-line (i.e., the control system operates autonymously, and displays are not updated), and the recipe is saved on a disk, from which it can be loaded into the operating system and the controller file data bases. The information configured for a typical system is listed in Table 7.4j.

Table 7.4j
DISTRIBUTED CONTROL FUNCTIONS ENTERED BY CONFIGURATION

1. Highway definition—Assign station numbers to the various highway stations and define the type of station (controller file, operator station, etc.).
2. Define the priority of the station.
3. Define overview index scale—This is the percent of deviation that will show on the overview display as full scale.
4. Define points to be recorded.
5. Make a table of units of measure.
6. Make a table of messages.
7. Define tag names and service information.
8. Assign points to group displays and groups to overview displays.
9. Define titles for group displays and for overview displays.
10. Define alarm points to be grouped and summarized.
11. Define alarm priorities.
12. Define trend time spans for points to be trended.
13. Download configured information from controller files into memory. Reload configuration information from memory into controller files.
14. Enter addresses (source of inputs), parameters (gain, reset, sensitivity, ratio, etc.), and limits (alarm limits, output rate, etc.) into individual time slot functions.
15. Utility routines: format disks, copy, combine recipes.

Controller Function

Configuration of the controller functions makes use of subroutines called algorithms.[20] One of two different methods will be used. In one method the function will be built up from a number of basic algorithms (for example, input module, alarm limit, error detection), se-

quencing them in much the same way that the steps of a programmable pocket calculator are sequenced. The length of the time slot may be variable. In the other method, each time slot will utilize one highly structured algorithm, with many smaller subroutines already included. The first type is somewhat analogous to the SAMA instrumentation symbology, where error signal, set point, and limit functions are shown separately. The second is similar to ISA symbology, where a circle encompasses the entire controller function. The type built up of many small subroutines is very flexible; the more highly structured type simplifies configuration. Examples of each are shown in Figure 7.4k.

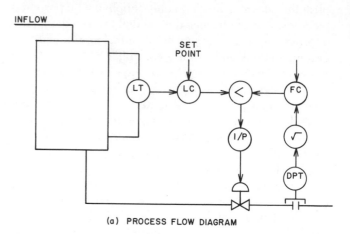

(a) PROCESS FLOW DIAGRAM

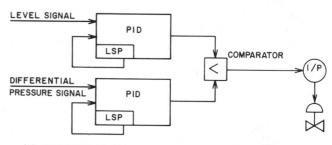

(b) ALGORITHM DIAGRAM- USING HIGHLY STRUCTURED ALGORITHMS

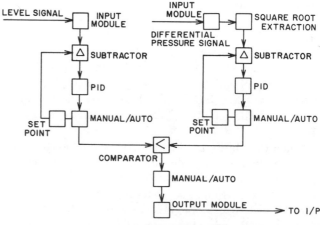

(c) ALGORITHM DIAGRAM- THE SAME FUNCTION CONFIGURED BY COMBINING ALGORITHMS

Fig. 7.4k   Algorithm configuration

Algorithms are used to satisfy three distinct process needs: analog control, digital control, and sequential control. Table 7.4l lists algorithms in a typical distributed control library.

Table 7.4l
TYPICAL ALGORITHM LIBRARY

PID Algorithms	PID Functions
PID	Output limiting
PID—ratio	Output tracking
PID—gap	Feedforward
Auto/Manual	Set point ramping
Supervisory control	Anti-reset windup
Adaptive tuning	Set point clamping
PV vs. set point error	
Ramp generation	Logical Functions
	And, or, exclusive
Computation	*or, not*
Summer	Switch—latched,
Divider	unlatched
Mass flow	Timer
Function generators	Counter
High and low selectors	Sequencer
Dead time	Flip-flop
Comparator	Motor control
Median	
Lead/Lag	
Square root	
Exponential function	
Log function	
Input/Output	
Analog input	
Analog output	
Digital input	
Digital output	

Algorithms are connected together by software addresses to form a combination of functions. This is sometimes called soft wiring. Since all information resides in memory in registers, soft wiring actually provides pointers toward specific bits or bytes in the register at a specified address. Input addresses can correspond to the output registers of other algorithms, to the input variable storage addresses, or to the status registers of algorithms, allowing much internal information to be made part of the configuration. Configuration information is entered from the operator station keyboard, usually in a "fill in the blanks" manner, to complete a listing or picture displayed on the CRT screen. A configured algorithm of the highly structured type is shown in Figure 7.4e.

The Distributed Control System

Suppliers sell the central control room portion, the local processing area portions, and the data highway as a distributed control system. Other systems, such as programmable logic controllers, data acquisition equipment, and digital computers, can be added. Everything trans-

mitting on the highway must be able to communicate compatibly, and there are many protocols that have to be interfaced.[21] In each case a subroutine must be written so that a device knows how to request information from the device to which it interfaces, and also knows how to translate it into the language of the requesting device. Some suppliers supply complete systems, using specific models of equipment they do not produce themselves. Few suppliers will supply a system that combines anything the user may desire. Even fewer will write high-level language programs to perform optimizing, production logging, or other tasks outside the normal routine of the distributed control equipment. However, some small systems houses recognize the need for this service and are serving to expand the definition of system engineering to include electronic design as well as instrument packaging.

REFERENCES

1. Baur, Paul S., "Microprocessors, Data Highways: Building Tomorrow's Control Systems," *Power*, March, 1982.
2. "Microprocessors Shape Control Technology," *Electrical World*, April, 1982.
3. Dallimonti, R., "Human Factors in Control Center Design," *Instrumentation Technology*, May, 1976.
4. ———, "Future Operator Consoles for Improved Decision-making, and Safety," *Instrumentation Technology*, August, 1972.
5. Calder, W., "Electronic Instruments: Hostile Environments Seek Vendor/User Solutions," *InTech*, March, 1982.
6. Karl, J. Lawrence, and Jones, Donald J., "Microprocessors: DDC for Large Electric Utility Boilers," *InTech*, July, 1980.
7. Skrokov, M.R., *Mini- and Microcomputer Control in Process Industries*, New York: Van Nostrand Reinhold, 1980, pp. 1–5.
8. Lynott, Frank, "Control System Dependability Considerations Encompass Availability, Integrity, Security," *Control Engineering*, January, 1982.
9. Bothne, R.E., "Distributed Control Offers System Reliability and Low System Investment," *Control Engineering*, May, 1977.
10. Teets, Rex M., "Protecting Minicomputers from Power Line Perturbations," *Computer Design*, June, 1976.
11. Andriev, N., "Data Highways Lead Microcomputer Peripherals to Industrial Control," *Control Engineering*, January, 1981.
12. Ruth, Smed, "CRT Terminals—An Overview," *Instruments & Control Systems*, May, 1981.
13. McDermott, J., "Keyboards for Microprocessor-based Systems Grow in Versatility, Performance," *Electronic Design*, December 6, 1978.
14. Abler, Robert A., "Touch Pad Control for the Harsh Environment," *Computer Design*, November, 1981.
15. Runyon, S., "Focus on Printers," *Electronic Design*, August 2, 1978.
16. Hickey, J., "Wire and Cable—What's Happening?," *Instruments & Control Systems*, August, 1980.
17. Andriev, N., "New Fiber Optic Data Highway Is a Masterless Token-passer," *Control Engineering*, May, 1982.
18. Weissberger, Alan J., "Modems: The Key to Interfacing Digital Data to Analog Telecomm Lines," *Electronic Design*, May 10, 1979.
19. Eckard, Mark, "Tackling the Interconnect Dilemma," *Instruments & Control Systems*, May, 1982.
20. King, K.L., Forney, D.E., and Scheib, T.J., "Software Structure for a General-Purpose Digital Controller," *Proceedings* of the ISA Conference and Exhibit, St. Louis, Missouri, March 23–26, 1981.
21. Moon, J., and Hedrick, J., "Standards for Distributed Control Interfaces: Are They Practical," *InTech*, June, 1981.

BIBLIOGRAPHY

Allen, B.S., "Data Highway Links Control Equipment of Any Number of Different Manufacturers," *Control Engineering*, July, 1981.

Allen, R., "Local Networks," *Electronic Design*, April 16, 1981.

Archibald, W., "Remote Multiplexing," Burr-Brown Application Note AN-80, January, 1976.

Aronson, R.L., "Line-Sharing Systems for Plant Monitoring and Control," *Control Engineering*, January, 1971.

Balph, T., Schreyer, A., and Tonn, J., "The Impact of Semiconductors on Industrial and Process Control," *Instruments & Control Systems*, September, 1981.

Barnes, G.F., "Single Loop Microprocessor Controllers," *Instrumentation Technology*, December, 1977.

Bibbero, R.J., *Microprocessors in Instruments and Control*, New York: John Wiley & Sons, 1977.

Brook, R.C., "Use of Microprocessors for Process Control," *Food Technology*, October, 1981.

Buckley, P.S., "Distillation Column Design Using Multivariable Control," *Instrumentation Technology*, September & October, 1978.

Carlo-Stella, G., "Distributed Control for Batch Systems," *InTech*, March, 1982.

Crutchley, W., "Software—A Guide Through the Maze," *Instruments & Control Systems*, February, 1979.

Davis, H., "A Sampling of Pascal Programming," *Instruments & Control Systems*, July, 1979.

———, "Introduction: Pascal," *Instruments & Control Systems*, June, 1979.

Dobrowolski, M., "Guide to Selecting Distributed Control Systems," *InTech*, June, 1981.

Fadum, O., "Plant-wide Automation," *InTech*, September, 1983.

Garrett, L.T., and McHenry, J.M., "Analyzing Costs of Digital and Analog Control Systems," *Hydrocarbon Processing*, December, 1981.

Gasperini, R., "A Guide to Digital Troubleshooting Aids," *Instruments & Control Systems*, February, 1978.

Gooze, M., and Nelson, G., "What Do You Have to Know About Serial Data Communication," *InTech*, June, 1981.

Griem, P.D., "Security Functions in Distributed Control," *InTech*, March, 1983.

Groves, B., "Microprocessors," *Instruments & Control Systems*, March, 1975.

"Guide to Selecting Temperature Multiplexers," *Instrumentation Technology*, February, 1980.

Hackmeister, D., "Focus on Printers: The Application Determines the Type and Technology to Choose," *Electronic Design*, June 21, 1979.

Higham, J., Kendall, B., and Gerdts, M., "The Data Freeway Network: A Coaxial Multi-drop Management Control Bus," *Control Engineering*, September, 1981.

Hollister, A.L., "A Primer on Microprocessor Software," *Instrumentation Technology*, October, 1977.

Hu, S.C., "Microprocessors—Characteristics and Role in Process Control," 1976 ISA Conference, Preprint No. 76-552.

Kaiser, V.A., "Changing Patterns of Computer Control," *Instrumentation Technology*, February, 1975.

Kovalcik, E.J., "Understanding Small Computers," *Instruments & Control Systems*, January, 1976.

Lagana, T., "Digital Control System Interfaces: The Same But Different," *InTech*, November, 1981.

Larsen, G.R., "A Distributed Programmable Controller for Batch Control," *InTech*, March, 1983.

Laspe, C.G., "Personal Microcomputers," *InTech*, November, 1982.

Lynn, R.L., "Guidelines for Specifying and Procuring Distributed Control Systems," *InTech*, September, 1980.

Merritt, R., "Large Control Systems," *Instruments & Control Systems*, December, 1981.

———, "Distributed Controls: A Technology Update," *Instruments & Control Systems*, September, 1983.

Myron, T.J., "Digital Technology in Process Control," *Computer Design*, November, 1981.

Ogdin, C.A., "The Highs and Lows of Microcomputer Programming Languages—Part I," *Instruments & Control Systems*, May, 1978.

———, "The Highs and Lows of Microcomputer Programming Languages—Part II," *Instruments & Control Systems*, June, 1978.

Pracht, C.P., "A Distributed Microprocessor-Based Control System," 1976 ISA Conference, Preprint No. 76-515.

Prickett, M.J., "Microprocessors for Control and Instrumentation," *InTech*, March, 1983.

Shaw, W.T., "Using Distributed Control Systems for Process Simulation," *Instruments & Control Systems*, December, 1981.

Sheridan, T.B., "Interface Needs for Coming Industrial Controls," *InTech*, February, 1979.

Shepherd, B., "Watch Those Other Factors in Your Bus System Design," *Instruments & Control Systems*, June, 1983.

Tebbett, G., "Putting the System Together," *Instruments & Control Systems*, May, 1983.

Thaler, R.M., "Things to Consider When Selecting a Minicomputer System," *Instruments & Control Systems*, January, 1981.

Troutman, P., "Distributed Digital Process Control at the Control Valve," 1978 ISA Conference, Preprint No. 78-543A.

Uyetani, A., "Multiloop Process Controller," 1976 ISA Conference, Preprint No. 76-823.

Warren, C., "Disk Drives," *Engineering Design News*, August 19, 1981.

Washburn, J., "Communications Interface Primer," *Instruments & Control Systems*, Part I, March, 1978 and Part II, April, 1978.

Williams, T.J., "Hierarchical Distributed Control," *InTech*, March, 1983.

Zielinski, M., "Microprocessor-Based Controller Troubleshooting," *Instrumentation Technology*, June, 1977.

7.5 HIERARCHY CONTROL

Hierarchy control is the use of several levels of computer systems in an extended master-slave relationship to carry out all or a large part of the possible automatable functions in the control of an industrial plant or similar enterprise. This section describes the historical development of hierarchical control and a typical form of such a control system, including the distribution of the potential duties among the several levels of computers involved.

Early Computer Control Systems

Prior to the introduction of the use of computers for industrial control applications, the standard industrial control system consisted of a large number of single-loop analog controllers, either pneumatic or electronic, as diagramed in Figure 7.5a. Note in the figure that Level 1 refers to those instruments that have no signal read-out readily available to the operator (e.g., some field-mounted controllers, ratio instruments, etc.). Level 2 devices have a regular indication to the operator, such as a strip chart, pointer, etc. These latter are the standard PID (proportional, integral, and derivative control mode) controllers.

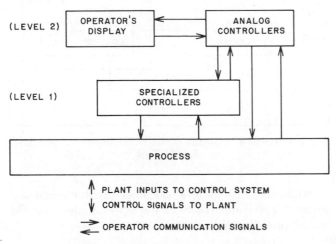

Fig. 7.5a The basic system—analog control

Computers were first used in industrial plants not as controllers but as monitors or data loggers (see Figure 7.5b). However, these were quickly followed by the first controllers (early 1960s), which used the computer in a

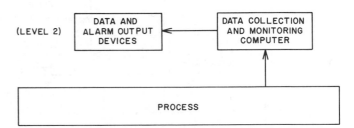

Fig. 7.5b Data collection and monitoring system

secondary role to change the set points of the analog controllers (see Figure 7.5c).

It was the dream of every control engineer connected with early computer control systems to bypass the analog controllers of Figure 7.5c and have the computer itself move the valves, as illustrated in Figures 7.5d and 7.5e. Hence the term direct digital control (DDC). Figure 7.5e includes specialized dedicated digital controllers to illustrate the relationship to Figure 7.5a. While common today, these devices were not present in the very early DDC systems. All work was carried out then in the digital computer of Level 2.

Although direct digital control systems have always had the potential for assuming an unlimited variety of complex automatic control functions in every control loop,

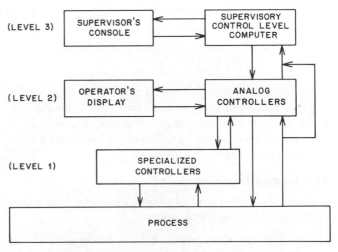

Fig. 7.5c Secondary or supervisory digital control only

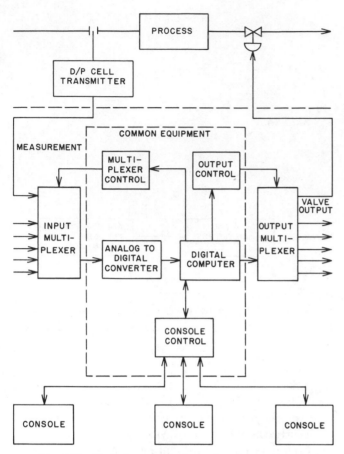

Fig. 7.5d   Block diagram of a direct digital control system

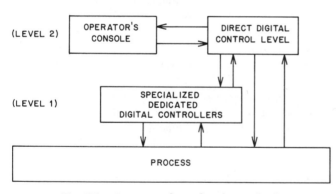

Fig. 7.5e   Primary or direct digital control only

most DDC systems have been implemented as digital approximations of conventional three-mode analog controllers. Thus the computer system of Figure 7.5e is considered most often as a direct replacement for the analog control system of Figure 7.5a.

The Centralized Computer

The early digital control systems suffered from many drawbacks. Among these were:

1. The computers were all drum machines, i.e., the memory was located on a magnetic drum. As a

result, they were very slow. A typical addition time for such machines was 4 milliseconds.
2. The memories were very small, usually 4,000 to 8,000 words of 12 to 16 bits each.
3. All programming had to be done in machine language.
4. Neither vendors nor users had any experience in computer applications. Thus it was very difficult to size the project properly within computer capabilities; most projects had to be reduced in size to fit the available machines.
5. Many of the early computer systems were very unreliable, particularly if they used temperature-sensitive germanium rather than silicon circuitry, as many did. In addition, these computers depended on unreliable mechanical devices, air conditioners, for their successful operation.

The response of the computer manufacturers to these drawbacks was to design a much larger computer with a magnetic core memory, wired-in arithmetic functions, etc., which made it much faster. But the high cost of core memories and the additional electronic curcuitry made the system much more expensive. To help justify the cost, vendors advocated the incorporation of all types of computer functions, including both supervisory control and DDC, in one computer box or mainframe at a central control room location, as shown in Figures 7.5f and 7.5g.

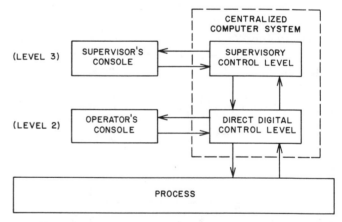

Fig. 7.5f   Complete secondary digital control: supervisory plus direct digital control

While these computers had much greater speed and memory size compared to earlier systems, their use led to still further problems:

1. Most were marketed before their designs were thoroughly proved or their programming aids (compilers, higher-level languages, etc.) were fully developed. Thus there were many frustrating delays in getting them installed and running.
2. The vast communication system required to bring

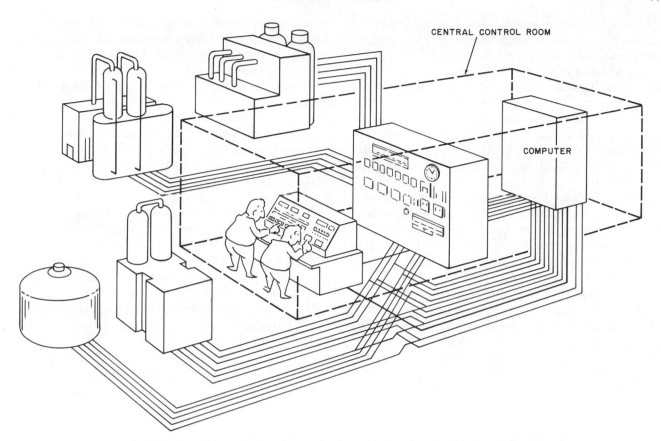

Fig. 7.5g   Typical centralized computer control system with analog panel board and backup
(circa 1965–1975)

the plant signals to the centralized computer location and return control signals to the field was very expensive, and unless it was carefully designed and installed, it was prone to electrical noise problems.

3. Because all of the control functions were located in one computer, the possibility that the computer might fail, resulted in demands for a complete analog backup system. The resulting system, illustrated in Figure 7.5h, was a combination of the systems shown in Figures 7.5f and 7.5c, which greatly increased the cost.

4. To compensate for these high costs, user and vendor alike attempted to squeeze as large as possible a project into the computer system, thus drastically complicating the programming and aggravating the difficulties described in Item 1 above.

As a result of these difficulties, management in many companies reacted sharply against computer control, and there was a hiatus in computer system installations until about the mid-1970s.

Distributed Control

Because of the problems experienced with the centralized computer systems of the late 1960s, most of the

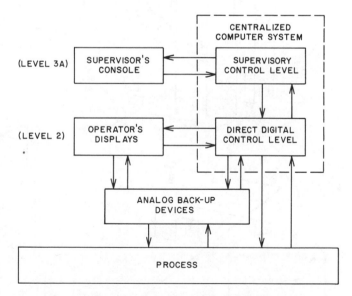

Fig. 7.5h   Complete secondary digital control: supervisory plus direct
digital control with associated analog control backup

new computer projects of the early 1970s were relatively small, specialized projects that took advantage of the capabilities of the newly arrived minicomputer. Many such projects flourished. They generally followed the lines of Figure 7.5c or 7.5h and differed from earlier undertakings mainly in their smaller size.

However, at this time, two other developments were underway that would change forever the prospects for digital computer based control and indeed for all process control. The first of these was the rapid development of the integrated circuit and the resulting production of the all-important microprocessor or microcomputer. The second development was the courageous effort begun by the Honeywell Company in 1969 to design an alternative to the unwieldy and unreliable centralized computer control system. This alternative was the distributed computer system.

The idea behind the distributed computer system was to have a set of small, widely distributed computer "boxes" containing one or more microprocessors. Each of the boxes controlled one or a very few loops. All of them were connected by a single high-speed data link that permitted communication between each of the microprocessor-based "boxes" and with a centralized operator station or console. This became the TDC 2000 system, the principles of which were widely followed by other process control system vendors. These systems solved the problems of reliability in two ways. First, all units controlled only one or a few process loops; thus

any single failure involved only those few loops. Second, a digital backup capability was developed, that is, backup computer systems were included to take over the duties of any failed components.

Figure 7.5i illustrates this concept. Comparing this figure with Figure 7.5g dramatically points up what was accomplished with this new concept. Almost universally, the distributed computer systems offer the following features and capabilities, which greatly fostered their acceptance over electronic analog or centralized computer-based control systems.

1. A modular system development capability that is easy to use, particularly with the configuration aids available from the vendor.
2. A color CRT-based, largely preprogrammed operator interface system that is easy to adapt to the individual plant situation.
3. A preprogrammed menu type of instruction system for the microcomputers of the controller box. This permits the final programming (configuration) of the total system to be done by pushing a few buttons on the keyboard.

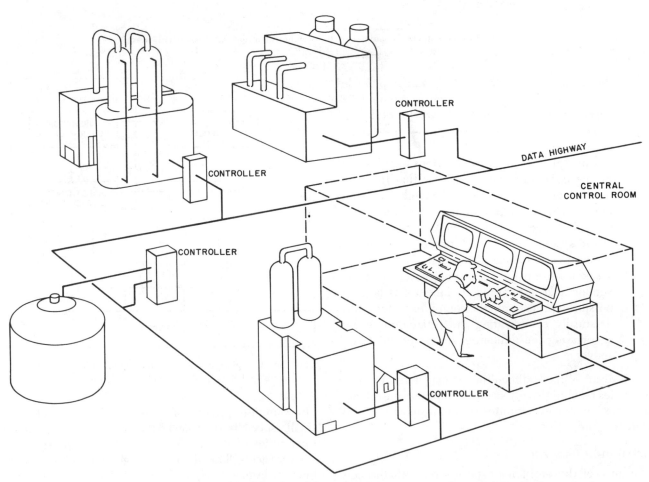

Fig. 7.5i Microprocessor-based distributed direct digital computer control system (circa 1975)

4. A very wide selection of control algorithms or computational schemes within the preprogrammed menu, which permits easy selection and testing of alternate control schemes for a process.

5. Data highway data transmission and communications capabilities between separate units of the system. Data highways provide very wide band communications and the possibility of redundancy for extra safety.

6. Relatively easy communications with mainframe computer systems for supervisory control or other higher-level process control or hierarchy control functions. However, these new control systems themselves are generally restricted to supplying the needs of the first- and second-level or dynamic control. Supervisory control is externally supplied.

7. Extensive diagnostic schemes and devices for easy and rapid maintenance through replacement of entire circuit boards.

8. Redundancy and other fail-safe techniques to help promote high system reliability. These features are often standard, but may be optional.

What has been achieved with these new systems is amply illustrated by Figures 7.5g and 7.5i, as noted above. These particular sketches are adapted from Honeywell drawings, but they are just as applicable to almost any other vendor's process control system.[3]

Hierarchical Control

The development of the distributed digital control system greatly simplified the connection of the computer to the process. (Compare Figure 7.5j to Figure 7.5c.) Also, since redundancy or other backup devices can be incorporated into the digital system directly, analog

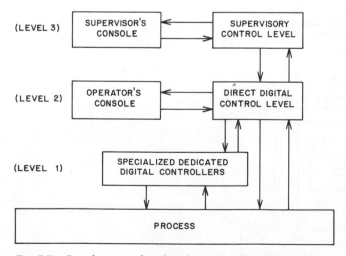

Fig. 7.5j   Complete secondary digital control: supervisory plus direct digital control

backups are no longer needed for plant safety in case of computer failure.

As illustrated in Figure 7.5j, the three levels of control devices, each with their distinct duties, form a hierarchical computer system in which upper-level computers depend on lower-level devices for process data, and the lower-level systems depend upon the higher-level systems for more sophisticated control functions such as overall plant optimization. On this basis, it is possible to envision the ultimate computer control system—one that combines the company's production scheduling and management information functions with the process control functions to form a total plant hierarchical control system, as illustrated in Figure 7.5k. This Figure outlines all the probable functions of such a hierarchical computer control system, but the magnitude of the tasks to be accomplished in the upper levels of the system are better indicated by the expanded version shown in Figure 7.5l. Such a system has not yet been developed at the time of this writing (1984), but it represents the ultimate dream of many process control system engineers. While what has already been achieved in the field of computer control fits easily within the framework shown in this diagram, this framework is only one of several possible structures for such a system and it is not necessarily the optimal one for the task. Nevertheless, it is an excellent vehicle for our purpose here since it allows us to treat all the possible benefits of the computer system with one example.

It should be noted that the several levels shown in Figure 7.5l are operational levels and do not necessarily represent separate and distinct computer or hardware levels. In large systems each level would be handled by a separate computer, but in small systems, two or more operational levels might be collapsed into one computer level. The dedicated digital controllers at Level 1 require no human intervention since their functional tasks are completely fixed by systems design and are not to be altered on-line by the operator. All other levels have human interfaces as indicated. It should also be noted that each element of the hierarchy can exist as an individual element. Indeed, all of the earlier forms of industrial digital computer systems (Figures 7.5b, 7.5c, 7.5e, and 7.5h) still exist and will no doubt continue to be applied where their particular capabilities appear to best fit the application at hand.

The paragraphs below will explain the basis for the organization of the six-level hierarchical system depicted in Figure 7.5l.

Overall Tasks of the Digital Control System

Automatic control of any modern industrial plant, whether achieved by a computer-based system or by conventional means, involves an extensive system for the automatic monitoring of a large number of variables op-

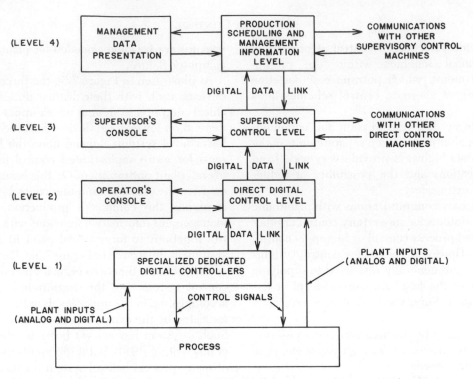

Fig. 7.5k   Hierarchical organization for a complete process computer control system

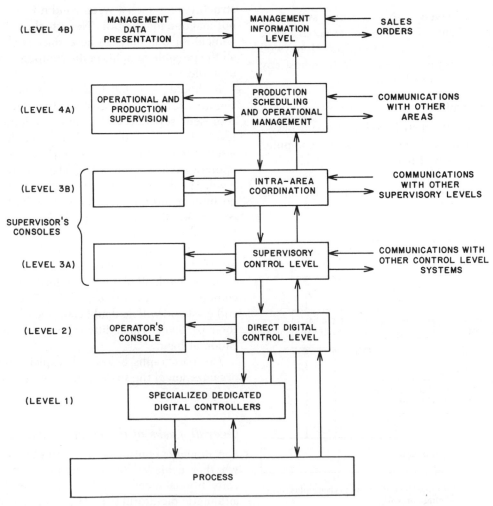

Fig. 7.5l   Hierarchical computer system structure for a large manufacturing complex

erating under a wide range of process dynamics. It requires the development of a large number of functions, some of which might be quite complex, for the translation of the plant variable values into the required control correction commands. Finally, these control corrections must be transmitted to a large number of widely scattered actuation mechanisms of various types. Because of the nature of the manufacturing processes involved, these control corrections often require the direction of the expenditure of very large amounts of material and energy. Also, both operators and managers must be kept aware of the current status of the plant and of each of its processes.

In addition, an industrial plant is faced with the continual problem of adjusting its production schedule to match its customer's needs, as expressed by the order stream being continually received, while maintaining high plant productivity and the lowest practical production costs. Production scheduling today is usually handled through a manual, computer-aided production control system in conjunction with an in-process and finished goods inventory judged adequate by plant personnel.

Another role of the digital computer control systems in industrial plants is as a "control systems enforcer." In this mode, the main task of the lower-level computers is continually to assure that the control system equipment is actually carrying out the job that it was designed to do in keeping the units of the plant production system operating at some optimal level—that is, to be sure that the controllers have not been set on manual, that the optimal set points are being maintained, etc.

Often the tasks carried out by these control systems are ones that a skilled and attentive operator could readily do himself. But the automatic control systems offer a much greater degree of attentiveness over much longer periods of time.

All of these requirements must be factored into the design and operation of the control system that will operate the plant, including the requirements for maximum productivity and minimum energy use. As the overall requirements become more complex, more sophisticated and capable control systems are necessary. Thus we enter the realm of the digital computer-based control system.

To obtain the kind of control responses mentioned above, an overall system is needed that offers the following capabilities:

1. Tight control of each operating unit of the plant to assure that it is using raw materials and energy at maximum efficiency and/or is operating at the most efficient production level, based upon the production level set by the scheduling and supervisory functions listed below. This control reacts directly to any emergencies that occur in its own unit. (This control is executed by Levels 1 and 2 of Figure 7.5l.)

2. A supervisory and coordinating system that determines and sets the local production level of all units working together between inventory locations and optimizes their operation. This system assures that no unit exceeds the general area level and thus assures that no unit uses excess energy or raw materials. This system responds to emergencies or upsets in any of the units under its control to shut down or systematically reduce the output in these and related units. (This control is executed by Levels 3A and 3B of Figure 7.5l.)

3. An overall production control system capable of carrying out the scheduling function for the plant based on customer orders or management decisions so as to produce the required products at the optimum combination of time, energy, and raw materials, suitably expressed as cost functions. (This is the specific task of Level 4A of the hierarchy).

4. A method of assuring the overall reliability and availability of the total control system through fault detection, fault tolerance, redundancy, and other applicable techniques built into the system's specification and operation. (This task is performed at all levels of the control system.)

Because of their ever-widening scope of authority, control tasks 1, 2, and 3 above can effectively become the distinct and separate levels of a hierarchical control structure. Also, in view of the amount of information that must be passed back and forth among these three control tasks, it appears that some sort of distributed computational capability, organized in a hierarchical fashion, could be a logical structure for the required control system.

In Figure 7.5l, we can see that the right-hand elements of Levels 3A to 4A are all handled by computers of increasing capability as one goes up the hierarchy. Level 4B also contains a computer, but it is used mainly for communications and data base management tasks. These elements are best handled by data processing or scientific type computers since the non-control tasks at these levels far outnumber process control tasks. These non-control computational tasks will thus determine the design and cost of these computers and how many are manufactured. As long as they are satisfactory for the purposes of process control, these established computer models should be used so that the economy gained from large-scale production can be capitalized on. This leaves Levels 1 and 2 as candidates for the application of specialized process control hardware—the distributed, microprocessor based systems discussed earlier.

Detailed Task Listings

In the context of a large industrial plant, the tasks carried out at each level of the hierarchy are as described in Tables 7.5m to 7.5r. Note that in each table the tasks are subdivided into tasks related to production scheduling, control enforcement, system coordination and re-

Table 7.5m
DUTIES OF CONTROL
COMPUTER SYSTEMS

I.   Production Scheduling

II. Control Enforcement

III. System Coordination, Reporting and
     Management Information

IV. Reliability Assurance

Table 7.5n
DUTIES OF THE CONTROL LEVELS
(LEVELS 1 AND 2)

II. Control Enforcement
  1. Maintain direct control of the plant units under
     their cognizance.
  2. Detect and respond to any emergency condition
     in these plant units.

III. System Coordination and Reporting
  3. Collect information on unit production and raw
     material and energy use and transmit to higher
     levels.
  4. Service the operator's man/machine interface.

IV. Reliability Assurance
  5. Perform diagnostics on themselves.
  6. Update any standby systems.

Table 7.5o
DUTIES OF THE SUPERVISORY LEVEL
(LEVEL 3A)

II. Control Enforcement
  1. Respond to any emergency condition in its
     region of plant cognizance.
  2. Locally optimize the operation of units under its
     control within limits of established production
     schedule; carry out all established process
     operational schemes or operating practices in
     connection with these processes.

III. Plant Coordination and Operational Data Reporting
  3. Collect and maintain data queues of production,
     inventory, and raw material and energy usage for
     the units under its control.
  4. Maintain communications with higher and lower
     levels.
  5. Service the man/machine interfaces for the units
     involved.

IV. System Reliability Assurance
  6. Perform diagnostics on itself and lower-level
     machines.
  7. Update all standby systems.

Table 7.5p
DUTIES OF THE AREA LEVEL
(LEVEL 3B)

I.   Production Scheduling
  1. Establish the immediate production schedule
     for its own area, including transportation
     needs.
  2. Locally optimize the costs for its individual
     production area as a basis for modifying the
     production schedule established by the
     production control computer system (Level 4A)
     (e.g., minimize energy usage or maximize
     production).

III. Plant Coordination and Operational Data Reporting
  3. Make area production reports.
  4. Use and maintain area practice files.
  5. Collect and maintain area data queues for pro-
     duction, inventory, raw materials usage, and en-
     ergy usage.
  6. Maintain communications with higher and lower
     levels of the hierarchy.
  7. Operations data collections and off-line analysis
     as required by engineering functions.
  8. Service the man/machine interface for the area.
  9. Carry out needed personnel functions (such as
     vacation schedule, work force schedules, and union
     line of progression).

IV. System Reliability Assurance
  10. Diagnostics of self and lower-level functions.

porting, and reliability assurance, somewhat as described in Items 1 through 4 in the previous section. Note also that the duties listed in Table 7.5n for Levels 1 and 2 begin with Item II, Control Enforcement, since the lower-level machines do not do any production scheduling. Likewise, the upper-level machines do no control enforcement since they have no direct connection to the process actuators. Finally, Level 4B does neither since its main task is management and staff function communications with a production data file maintained by Level 4A, Production Scheduling. These tables outline the tasks that must be carried out in any industrial plant, particularly at the upper levels of the hierarchy. Details of how the operations are actually carried out will vary drastically, particularly at the lowest levels, because of the nature of the actual processes being controlled. This does not change the basic definition of these tasks.

The general duties of the different levels in the hierarchical computer system are summarized in Figure 7.5s.

Tasks of the Lower-Level Computers

In the hierarchy shown in Figure 7.5l, all contact with the process being controlled is maintained through the computers of Levels 1 and 2. Likewise, the smallest basic digital control systems sold today, the distributed, mi-

Table 7.5q
DUTIES OF THE PRODUCTION SCHEDULING
AND OPERATIONAL MANAGEMENT
LEVEL (LEVEL 4A)

I. Production Scheduling
  1. Establish basic production schedule.
  2. Modify the production schedule for all units per order stream received, energy constraints, and power demand levels.
  3. Determine the optimum inventory level of goods in process at each storage point. The criteria to be used will be the trade-off between customer service (e.g., short delivery time) versus the capital cost of the inventory itself, as well as the trade-offs in operating costs versus costs of carrying the inventory level. (This is an off-line function.)
  4. Modify production schedule as necessary whenever major production interruptions occur in downstream units, where such interruptions will affect prior or succeeding units.

III. Plant Coordination and Operational Data Reporting
  5. Collect and maintain raw material use and availability inventory and provide data for purchasing for raw material order entry.
  6. Collect and maintain overall energy use data for transfer to accounting.
  7. Collect and maintain overall goods in process and production inventory files.
  8. Collect and maintain the quality control file.
  9. Maintain interfaces with management interface level function and with area level systems.

IV. System Reliability Assurance
  10. Run self check and diagnostic routines on self and lower-level machines.

Table 7.5r
REQUIRED TASKS OF THE INTRACOMPANY
COMMUNICATIONS CONTROL SYSTEM
(LEVEL 4B)

III. Plant Coordination and Operational Data Reporting
  1. Maintain interfaces with plant and company management, sales personnel, accounting and purchasing departments, and the production scheduling level (Level 4A).
  2. Supply production and status information as needed to plant and company management, sales personnel, and the accounting and purchasing departments. This information will be supplied in the form of regular production and status reports and in response to on-line inquiries.
  3. Supply order status information as needed to sales personnel.

IV. System Reliability Assurance
  4. Perform self check and diagnostic checks on itself.

croprocessor-based systems, are all effectively stand-alone Level 1 and 2 systems.

According to Table 7.5n, the tasks of these systems are to maintain direct control of the process; to detect and respond to timing signals, emergencies, and other events in the process; to collect process data for the plant operators or for higher-level functions; and to assure reliability by monitoring their own operation.

To do this, the system must monitor each important plant or process variable on a regular basis. That is, it must read the current value of the variable and compare it with, first, a set of alarm limits to detect the presence of any emergency situation and, second, with the current operating set point to determine whether any correction to the current control level is necessary. Close check is also kept of the passage of time on the system real-time clock since most systems are time coordinated. That is, the monitoring program is reinitiated at fixed time intervals based upon the required sampling interval for the process variables.

As just indicated, plant variables are normally monitored either on a fixed time schedule so that every variable is tested each second or fraction of a second (common in the microprocessor-based distributed systems), or on a variable schedule depending upon the type of variable being sensed. Table 7.5t presents a table of popular time intervals for sensing on a variable schedule. This second system has been popular for minicomputer-based digital control systems that commonly do not have the speed capability of the distributed systems. It is based on the dynamics or speed of response of the process being monitored and controlled.

The timing of the process variable sampling is normally based upon a real-time clock included in the computer system.

If an emergency is detected, the computer's program contains routines for correcting the emergency or calling the operator's attention to it. The routines are selected according to a priority interrupt scheme. Control correction computations are carried out by means of a set of control algorithms or computational routines also stored in the computer's memory.

In addition to being used for emergency detection and control corrections, the values of the process variables are stored in the computer's memory in a process data base. These data are then used for operator's console read-out functions, for data logging for historical records, for process efficiency calculations, and for read-out to higher level computers for optimization calculations, inventory monitoring, overall plant production status and historical file updates, and other necessary calculations. These calculations thus fulfill the systems coordination and reporting functions listed in Table 7.5n.

The computer uses any spare computational time to run test computations to check the validity of its own operation and of any companion computers in the sys-

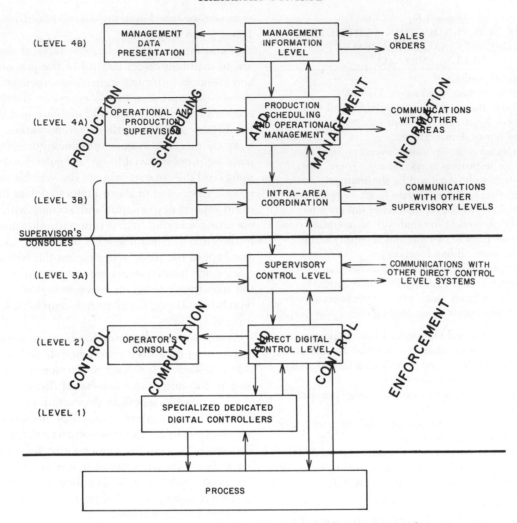

Fig. 7.5s　Summary of the real tasks of the hierarchical computer control system

Table 7.5t
SAMPLING FREQUENCY FOR
PROCESS VARIABLES

Variable	Frequency
1. Flow	Once each second
2. Level and pressure	Once each 5 seconds
3. Temperature	Once each 10 seconds
4. Queries from the operator's console	Once each second

Note: In some systems, process variables are sampled much more frequently than this. These are minimum rate values established from plant experience.

tem. These and other related tests fulfill the need for the reliability assurance functions listed in Table 7.5n.

Tasks of the Higher-Level Computers

As noted in the previous section, all contact with the process being controlled is maintained through the process input/output interfaces of the computer systems of Levels 1 and 2. These interfaces may be part of a microprocessor-based, distributed control system or of a minicomputer-based, direct digital controller. Any other layout would require a reversion to the schemes of Figures 7.5c, 7.5f, or 7.5h, all of which have distinct drawbacks compared to Figures 7.5j or 7.5l.

The upper-level machines are connected to the lower-level systems and to each other through the communications system. Their major tasks, as outlined in Tables 7.5o to 7.5r, are to carry out the extensive computations involved in (1) the optimization of productivity and of raw material and energy use in each process; (2) the development of the best production schedule for the overall plant; and (3) the minimization of the plant's in-process inventory.

An equally important task of all of these computers is the processing of the plant's production data as collected by the lower-level machines in order to supply the proper information to plant supervisory and management personnel and in order to maintain the plant's production data base for the company's production, financial, and personnel reports.

BIBLIOGRAPHY

Belick, F.M., "Computer Control," *Water & Wastes Engineering*, March, 1975.

Bennett, C.E., "Evaluating Controls for Batch Processing," *Instrumentation Technology*, July, 1977.

Chattaway, A.T., "Control System Integration," *InTech*, September, 1983.

Griem, P.D., "Security Functions in Distributed Control," *InTech*, March, 1983.

Howard, J.R., "Experience in DDC Turbine Start-up," *ISA Journal*, July, 1966.

Kaiser, V.A., "Changing Patterns of Computer Control," *Instrumentation Technology*, February, 1975.

Kovalcik, E.J., "Understanding Small Computers," *Instruments & Control Systems*, January, 1976.

Larsen, G.R., "A Distributed Programmable Controller for Batch Control," *InTech*, March, 1983.

Long, M.V., and Holzmann, E.G., "Approaching the Control Problem of the Automatic Chemical Plant," *Transactions* of ASME, October, 1953.

Luyben, W.L., "Batch Reactor Control," *Instrumentation Technology*, April, 1975.

Maurin, L.V., "Computing Control at Little Gypsy," *ISA Journal*, May, 1966.

Merritt, R., "Personal Computers Move Into Control," *Instruments & Control Systems*, June, 1983.

Miller, W., "Designing Reliability into Computer Based Systems," *Instruments & Control Systems*, May, 1982.

Miyazaki, S., "Development Project for DDC System," *CEER*, May, 1970.

Murphy, J.A., "Computer Control of an Aluminum Plant," *Instrumentation Technology*, April, 1972.

Prickett, M.J., "Microprocessors for Control and Instrumentation," *InTech*, March, 1983.

Purdue Laboratory for Applied Industrial Control, *Tasks and Functional Specifications of the Steel Plant Hierarchy Control System* (Expanded Version), Report no. 98, Purdue University, June, 1982.

Rispoli, L.M., "Hierarchical Computer Control Systems," *Instruments & Control Systems*, October, 1970.

Sayles, J.H., "Computer Control Maximizes Hydrocracker Throughput," *Instrumentation Technology*, May, 1973.

Stout, T.M., "Justifying Process Control Computers," *Chemical Engineering*, September 11, 1972.

Van Horn, L.D., "Crude Distillation Computer Control Experience," 1976 ISA Conference, Preprint No. 76-533.

Williams, D.L., and McHugh, Anne, "TDC-1000—An Overview." Minutes of the Eighth Annual Meeting, International Purdue Workshop on Industrial Computer Systems, Purdue University, September 22–25, 1980.

Williams, T.J., "Hierarchical Distributed Control," *InTech*, March, 1983.

———, and Ryan, F.M., eds., *Progress in Direct Digital Control*, ISA, 1969.

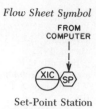
7.6 I/O HARDWARE AND SET-POINT STATIONS

Available Designs:	A—Electronic output B—Pneumatic output C—Stepping motor D—Synchronous motor E—Integrating amplifier F—Digital to pneumatic converter
Output Signal Ranges:	3–15 PSIG (0.2–1.0 bar), 1–5 mA DC, 4–20 mA DC, and 10–50 mA DC.
Speed of Full-Scale Travel:	C—9 to 33 seconds D—5 to 1,000 seconds E—under 1 second
Resolution (minimum step in set point change):	C—0.05% to 0.1% D—0.1 to 0.5%
Other Features:	Limit switches are available (C and D), available separately (C), available combined with controller (C, D and E), available combined with recorder controller (BC and BD).
Sizes:	3″ × 6″ (75 × 150 mm) (AC and DC), 2″ × 6″ (50 × 150 mm) (E, contains two units), 2″ × 6″ or 6″ × 6″ (50 × 150 mm or 150 × 150 mm) (BC and BD).
Cost:	$750 to $1300
Partial List of Suppliers:	Fairchild Industrial Products Co. (F); Foxboro Co. (AC); General Electric Co. (AC and AD); Moore Products Co. (BC and BD); Motorola Inc. (AE); Taylor Instrument Companies (AC and AD)

I/O Hardware Stations

For a computer system with digital process control to be in direct communication with process instrumentation and equipment, connections between the two are required. Types of signals that can be sent and received by the computer and the process will usually be different, and conversions of signals to the appropriate type between the sending and receiving devices are required. The I/O (input-output) interface hardware handles the necessary signal conversions so that direct communication can take place.

In a process control computer system (Figure 7.6a), information concerning process operating conditions is made available through process instrumentation, either as continuous (analog) electrical or pneumatic signals or as discrete (digital) on-off type signals. To adjust process

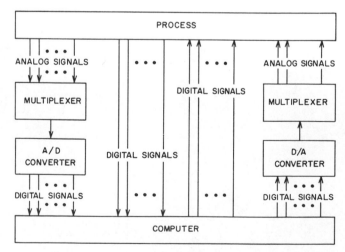

Fig. 7.6a   Schematic of computer–process interface

operating conditions, the process instrumentation may require either type of signal. The digital computer itself, on the other hand, can only accept or transmit digital signals having certain characteristics. Techniques and equipment for getting digital information to (digital input) and from (digital output) the computer, and for getting analog information to (analog input) and from (analog output) the computer are described in this section.

Digital Input

Digital information is generally electrical in nature and appears in the process as the status (open or closed) of relay contacts, as alternative voltage levels, or as a pulse or sequence of pulses. These digital signals are used to represent the status of valves (open or closed), switches (on or off), alarms (high or low), and the output of digital sensors such as turbine flowmeters and digital tachometers. The information can be scanned by the computer to determine the status of the process or to interrupt the computer in case of specific changes in status.

The most common form of digital input is the contact closure type, where an isolated set of contacts operates together with a process-related switch. The computer determines the status of these contacts by applying an electrical signal on one line and detecting with circuitry on the other line, whether an open or closed circuit exists. Typically, a current of 10 mA flows when the contacts are closed, and transistor switching at the computer interface indicates "1" for closed circuit and "0" for open circuit. Restrictions are placed on total loop resistance (maximum for closed circuit, minimum for open circuit), and maximum levels of shunt capacitance and series inductance. For voltage level inputs a similar scheme is used, except that the sensing voltage is that generated in the field, and an additional voltage supply at the computer interface is unnecessary (Figure 7.6b).

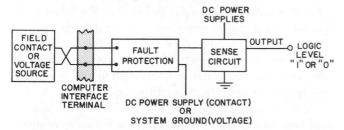

Fig. 7.6b Digital input interface for sensing contact or voltage signals

Digital inputs are usually treated at the computer end in groups corresponding to the word length of the computer. A 16-bit machine, for example, would read digital input data on 16 lines simultaneously. The data, as a series of ones and zeros, would be stored internally as one word. The correspondence between bit position in the word and field function is maintained by software. If the number of digital inputs exceeds the length of one computer word, multiplexing can be used to share one group of computer input lines among the several groups of digital input lines from the field.

Maximum scan rates depend on such factors as line and filter capacitances as well as on computer speed and interface design. Computer instructions can be used to address one word, and subroutines can be written to address, and store in memory tables, a group of words. Alternatively, hard-wired logic can be used to scan and store a group of words concurrently with the execution of other computer instructions. With short signal lines, no filtering, and direct-to-memory input channels, burst-mode scan rates on the order of 500,000 words per second are possible with modern process control computer systems.

Pulse inputs are handled in much the same way as logic voltage level inputs, except that the output of the sensing circuit is normally connected to a hardware counter rather than to a specific bit position in the input register. Each successive pulse increments or decrements the previous count by one, and computer instructions can be used to read and reset the counter register. Pulse rates are established by comparing contents of the counter register with the elapsed time between successive readings. Specifications on pulse inputs generally include voltage level ranges for both states and information on minimum pulse widths. Maximum pulse rates are limited by counter circuit requirements. If, for example, the electronics requires pulse widths of at least 100 microseconds (200 microseconds between successive pulses), the maximum pulse rate would be 5,000 pulses/second.

Change-of-state information, used to interrupt the computer, is sensed in the same way as other digital inputs. The output of the sense circuitry is, however, connected directly to the central processor control unit for interrupt signals. The occurrence of an external interrupt suspends the normal sequence of instruction execution and allows special interrupt response software routines to be executed.

Digital Outputs

Digital outputs enable the computer to provide on-off control to external devices such as relays, indicator lamps, and small DC motors to generate pulse trains for stepping motor devices and to transmit data to other computers. Digital outputs can also be used to generate analog outputs through digital-to-analog converters. In general, the computer interface provides solid-state or relay switching for a group of signal pairs that are connected to external devices. As with digital inputs, the size of a group depends on the computer word length, and several groups of field lines can share one group of computer lines.

Upon computer instruction, a word contained in an internal register or in a memory location is transferred to a digital output register.

For latching outputs, the "ones" in the word typically cause the switch or relay associated with the bit position to latch, while the "zeros" cause the switch to unlatch. In turn, external devices assigned to switches (bit positions) can be turned on or off, depending on whether the external contacts are normally open or normally closed.

In pulse outputs, the presence of a one usually causes a pulse (momentary closure) to be generated on the line corresponding to the bit position, while a zero causes no pulse. In so-called register outputs, the computer supplies power to lines corresponding to bit positions containing ones, and equipment similar to the digital input equipment at the other end of the lines reads the data to effect the transfer of information (Figure 7.6c).

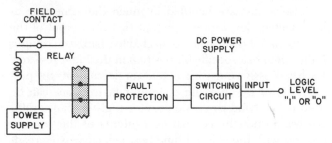

Fig. 7.6c   Digital output control

Because the switching of external power is performed at the computer interface, restrictions are placed on the voltage and current corresponding to the rating of the switching device. Maximum voltages are normally in the range of 40 to 50 volts DC, and maximum currents are on the order of 0.4 to 0.5 amperes for solid-state switches. Ratings on the order of 100 to 200 VA (volt-amperes) for relays are common.

Analog Outputs

Continuous pneumatic or electrical signals are generated from digital data using digital-to-analog (D/A) converters. If a large number of analog outputs are required, it is usually economical to share a common D/A converter among several analog channels by multiplexing the D/A output. Computer interface hardware thus consists of the logic circuitry necessary to obtain a binary digital value in a register ready for conversion, the D/A converter itself, and an analog multiplexer.

Digital to pneumatic (D/P) transducers are also available with up to 15 scfm (25.5 m³/hr) booster capacity.

The circuitry for analog output logic accepts a digital word from an internal register or computer memory, either in serial or parallel form. For signed outputs, one bit in the word indicates whether the analog output should be positive or negative. The logic circuitry recognizes this "sign bit" and switches the polarity on the reference supply accordingly. The remaining magnitude bits can be in natural binary form or in a number of other binary codes (binary coded decimal, excess three, etc.). (See

Section 7.1 for details.) Unless a digital code with multiple-valued bit weighing (e.g., the Gray code) is used, no code conversion is normally required prior to D/A conversion. For example, code conversion is not normally performed when magnitude bits are in complementary form for negative numbers, because the D/A converter can be designed to accept this single-valued digital code.

D/A converters either supply a voltage output proportional to the magnitude represented by the digital value or a current output corresponding to the digital value. For voltage outputs, the voltage drop between the reference voltage and the output is adjusted by switching resistors in and out of the circuit, with each magnitude bit either opening or closing one switch depending on whether the bit is "1" or "0." For current outputs, either summing of constant-current source outputs or switching in resistors to cause current branching can be used.

The resolution of the D/A conversion is dependent on the number of magnitude bits employed in the digital representation. For example, 10 magnitude bits can represent $2^{10} = 1,024$ possible states, or 1,023 possible changes in magnitude, between full scale and zero. For a full-scale output of 5 volts and a zero output of 0 volts, the smallest change that could be made in the analog output signal would be 5/1,023 volt, or 4.887 millivolts. Higher resolutions require more magnitude bits, and each additional bit approximately halves the smallest possible change in analog output signal. It is approximate only because of the $2^n - 1$ possible steps, where n is the number of bits converted. The maximum decoding error due to the converter resolution is $\pm\frac{1}{2}$ of the value of the least significant bit, or, for the 10-bit, 5-volt full-scale converter, ± 2.444 millivolts.

D/A voltage conversion can be performed with resistor ladder networks made up of either single- or weighted-value resistors (Figures 7.6d and e). For the former, all resistors used can be identical, making it fairly easy to obtain matched temperature coefficients for accuracy considerations. The weighed-resistor network can provide some savings in power consumption and reference supply costs, but for high-resolution conversion, the resistor values can become very high and therefore difficult to obtain. In addition, temperature coefficients are more difficult to match if resistance values differ over a large range.

Errors associated with converting digital values to analog signals accumulate from several sources. In D/A converters that use relays for the analog signal switching, most errors are due to imperfectly matched resistors. Because outputs are based on ratios of resistances, individual resistors must be matched as they age and as temperature changes. Metal film and wire-wound resistors can be manufactured with tighter and more easily matched tolerances than carbon composition or depos-

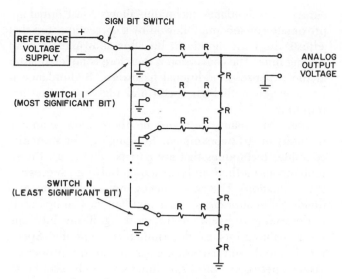

Fig. 7.6d   Single-valued resistor network for D/A conversion

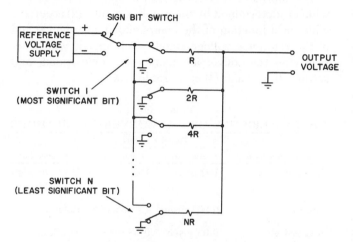

Fig. 7.6e   Weighted resistor network for D/A conversion

ited carbon resistors, and they are favored for medium- and high-accuracy converters.

Since a relay is a nearly perfect analog switch, the use of relays for analog switching in the D/A converter is preferred for high-accuracy requirements. (For details on relays, refer to Section 6.5.) Relays are, however, larger, require more operating power, have more limited switching speeds, and are less reliable for high-speed operation than solid-state switches. (For a detailed discussion of solid-state switches see Section 6.6.) Most solid-state switches are not perfect in that they exhibit a finite resistance when closed and leakage currents when open. In addition, they allow some feedthrough of the switch-driving signal. For most industrial applications, where size and power consumption are not particularly important and where reliable operation at moderate conversion speeds is important, relay-resistor D/A converters are preferred.

The accuracy and stability of the reference voltage is also of prime importance in overall D/A converter ac-

curacy. The most commonly used source for a reference voltage is the silicon Zener diode, reverse-biased at some current value where the dynamic resistance and the voltage-temperature relationship are known. If the operating temperature range is large, temperature-compensated Zener reference diodes are favored. Changes in the voltage applied to the Zener circuit can also affect overall D/A conversion accuracy, but these errors can be reduced with additional circuitry.

Analog Inputs

Continuous electrical signals are converted to digital values using analog-to-digital (A/D) converters. Multiplexing of analog signals into shared A/D converters is common because of the relative costs of multiplexers and converters. Depending on the multiplexer-converter system used, some signal conditioning may be required for amplification or impedance matching. Computer interface hardware therefore includes the A/D converter and the logic circuitry necessary to control the multiplexer and to supply the digital representation to computer memory or to an internal register.

The heart of the A/D converter is the comparator, which is an electronic circuit that compares unknown voltages and makes a binary decision as to which is larger. The comparator can be used in a number of ways in A/D conversion.

The simultaneous method of conversion uses $2^n - 1$ comparators to convert the analog signal to n bits of data (Figure 7.6f). It is fast, since voltage comparisons for each bit value are done simultaneously, but it is expensive for high-resolution conversion because of the large number of comparators required.

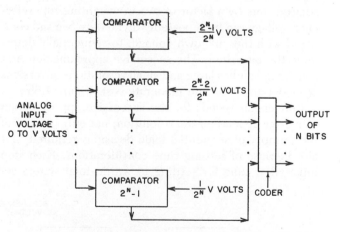

Fig. 7.6f   Simultaneous method of A/D conversion

Slower, but less expensive for high-resolution conversion, are the A/D converters that use one comparator circuit and a digital-to-analog (D/A) converter. As illustrated in Figure 7.6g, the converter operates with pulses gated to a counter, while, with each new count, the D/A unit converts the counter contents to an analog voltage

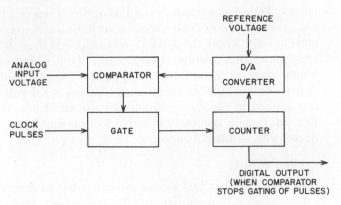

Fig. 7.6g   Counter method of A/D conversion

that is compared with the unknown voltage. When the comparator indicates agreement (within the one-bit resolution), counting ceases and the counter contains the digital value of the unknown analog signal. Conversion time depends on the size of the counter, which also determines resolution, but the converter cost is not particularly sensitive to resolution. The counter method is therefore favored when high resolution and low speed are required.

A variation on the counter method of A/D conversion is the ramp method, in which a ramp generator (e.g., an integrating operational amplifier) rather than the D/A converter supplies the signal to the comparator circuit. Because the ramp does not have to be generated by the D/A converter logic, this technique is somewhat faster than the counter method.

The most popular A/D converter design is based on the so-called successive approximation method. Since each bit position in a natural binary code differs from its adjacent bits by a factor of two in weighting, the reference voltage can be repeatedly divided by two and compared with the unknown voltage, to sequentially determine the bit values. The successive approximation A/D converter handles large and small resolution conversions at moderate speed and moderate cost (Figure 7.6h).

Conversion speeds for the successive approximation converter depend on the resolution, not only because of the number of sequential logic decisions required, but also because of settling time considerations. Each step must allow time for settling to within total system ac-

curacy. A redundancy technique allows a fast initial approximate conversion, followed by a correction step that adjusts the least significant bit after allowing sufficient settling time. The conversion is therefore completed faster at the expense of additional hardware. Redundancy is useful when both high speed and high resolution are required.

There are many variations on these more common methods of A/D conversion. Subranging, for example, combines techniques that are part parallel (e.g., the simultaneous method) and part sequential (e.g., successive approximation). The entire analog input voltage range is divided into subranges using simultaneous comparators to determine the value of the most significant bits, and the remaining bits are determined sequentially. Speed requirements and hardware costs dictate the proper division between the simultaneous and sequential operations.

The merits of A/D converters include (1) resolution, which is determined by the number of bits; (2) accuracy, which is a function of the comparator performance and of the reference voltage stability; (3) speed, which depends on the conversion technique, logic speed, and settling times; and (4) cost (Table 7.6i).

Table 7.6i
CHARACTERISTICS OF A/D CONVERTER METHODS

| Conversion Method | Time for 10 Bits | | Relative Cost |
	Aperture	Conversion	
Simultaneous	100 nsec	100 nsec	High for high resolution
Counter	0.003 msec*	1.8 msec**	Low
Successive approximation	0.03 msec	0.03 msec	Medium
Ramp	0.001 msec*	0.5 msec**	Low
Successive approximation with redundancy	0.003 msec	0.03 msec	High

*Not constant.
**Average.

Aperture time, or the time that the converter must "see" the analog voltage in order to complete a conversion, may also be an important consideration if the analog voltage can change rapidly. Aperture time can be decreased by using a sample-and-hold circuit which, in effect, charges a capacitor while sampling and holds the charge during conversion. (Sample-and-hold circuits are also used in multiplexed D/A conversion to "hold" the analog voltages on channels which share a common D/A converter.) In the successive approximation converter, aperture time can be decreased using the redundancy technique described above, where the final correction step compensates for changes in analog input voltage that occur during the initial approximate conversion.

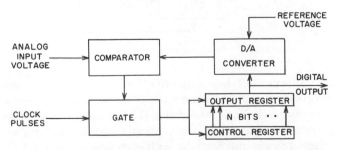

Fig. 7.6h   Successive approximation method of A/D conversion

Multiplexing of Analog Signals

Where many analog signals must be converted to or from digital values, the costs of using separate D/A or A/D converters for each analog signal are usually prohibitive. Analog multiplexers are used to time-share D/A and A/D equipment among many analog signals. The multiplexer selects and switches one analog signal at a time. The selection may be performed either sequentially or randomly, and the switching can be done with either electromechanical or solid-state devices (Figure 7.6j).

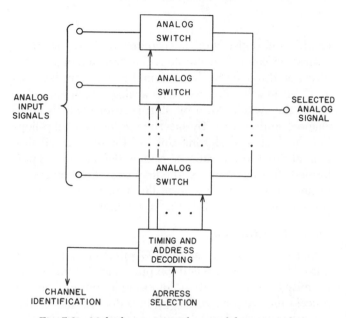

Fig. 7.6j   Multiplexer, using relay or solid-state switching

Electromechanical switches use metallic contacts that are closed when the switch is on and open when the switch is off. Because of the very low resistance of a metal-to-metal contact (typically less than one ohm), and the very high (for most practical purposes, infinite) resistance when the contacts are separated, electromechanical switches permit very accurate multiplexing.

Solid-state switches, on the other hand, can exhibit higher resistances when closed and, when open, leakage and driving signal feedthrough, and therefore permit less accurate multiplexing. (For a discussion of the relative merits of mechanical and solid-state switches, see also Sections 6.5 and 6.6). Solid-state switches, besides being smaller and consuming less power, can be operated at much higher speeds than electromechanical devices. When these attributes are important and only moderate accuracy is required, solid-state multiplexers are used.

One of the earlier electromechanical switches developed is the so-called crossbar switch, which is a three-dimensional array of contacts. One of several x-dimension and one of several y-dimension actuators are selected to close a vertical column of contacts. Further electromechanical selection chooses one of the several vertical contacts. Crossbar switches can operate at speeds of nominally 100 points/second and have small contact resistances and cross talk (Figure 7.6k).

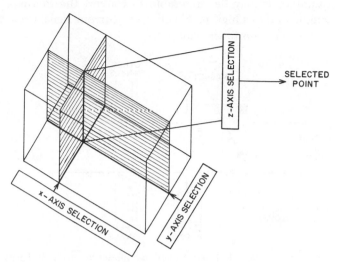

Fig. 7.6k   Multiplexer with crossbar switch

Motor-driven rotary multiplexers are used for both pneumatic and electrical signals. In pneumatic applications, the multiplexer connects one pressure signal at a time to a pressure transducer that supplies a single electrical signal to the computer for A/D conversion (Figure 7.6l). Typically, 64 pressure signals share one high-quality transducer and one input line on an electrical signal multiplexer at the computer interface. Savings in transducer and multiplexer costs are obtained at the expense of lower scanning rates, which are in the order of 5 to 10 ports per second.

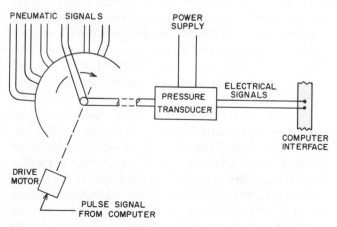

Fig. 7.6l   Electromechanical multiplexing of pneumatic signals

Electrical signal switching with motor-driven rotary multiplexers can be done with brushes or wipers, or with rotating magnets that cause reed switches to close. Analog input signals are wired to contacts circularly located on a round plate. The other side of the contacts is common and can be wired directly to the computer input terminals or to intermediate signal conditioning equipment (Figure 7.6m). With thermocouple signals that are multiplexed at some distance from the computer, for example, it may be preferable to convert the thermocouple emf to a high-level voltage or current signal prior to transmission to the computer.

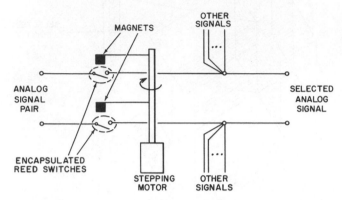

Fig. 7.6m   Rotary electromechanical multiplexer for electrical signals

Except for the fact that relay switches involve mechanical motion and solid-state analog switches do not, the basic principles of these two methods of switching are the same: each analog input signal is connected to a separate analog switch. The opposite side of the switch is common to all analog switches and carries the analog signal for which the switch has been closed. Timing and address decoding logic is used to close one analog switch at a time, and either random or sequential operation can be used.

As described previously, analog switching with relays is usually preferred for accurate but moderate-speed applications when size and power requirements are not stringent. Solid-state switching circuits include those using silicon junction diodes, bipolar transistors, and semiconductor junction and metalized-oxide semiconductor (MOS) field effect transistors (FET). MOSFET is, because of relative accuracy and circuit simplicity, often favored for multiplexing low-current analog voltages.

Random selection of multiplexer channels is performed by decoding a digital word containing the "address" of the desired channel and closing the analog switch for the selected channel. The D/A converter output, or the A/D converter input, is then connected to the desired analog signal line. An address n bits in length can contain one of 2^n different binary addresses and therefore can be used to select one of 2^n multiplexer channels (Figure 7.6n).

Sequential scanning can be performed with either software or hardware. A computer program that continually increments or decrements the multiplexer address can effect a sequential D/A or A/D converter scan. Faster scan rates are possible using separate hardware counters, which count pulses at the desired channel selection rate,

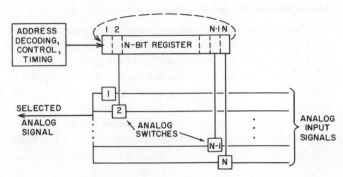

Fig. 7.6n   Multiplexer channel selection

or with shift registers or ring counters. Shift registers are a series of connected flip-flops that transfer bits in one direction through the chain. In multiplexing applications a "1" bit in a flip-flop closes an analog switch and a "0" opens the switch. An n-bit shift register controls an n-channel multiplexer. By inserting zeros in all flip-flops but the first, and shifting this "1" bit through all flip-flops at the desired scan rate, sequential scanning is performed. In ring counters, the final-stage flip-flop is connected back to the first-stage flip-flop, thus effecting a continuous scan controlled by a digital clock.

Signal Conditioning

Signal conditioning at the I/O interface may be required to make process and computer signals compatible. In pulse counter operation, for example, pulses from the process may require reshaping in order for the counter to work properly. Digital inputs and outputs may require signal level changes so that contact ratings are not exceeded and on-off logical voltage levels are sufficient to cause switching. Analog signals may require circuitry for impedance matching, amplification of low-level signals or attenuation of high-level signals, and filtering to reject noise.

In process control computer installations, signal conditioning requirements are most often associated with analog inputs. Multiplexer-converters that use solid-state switching may require high-level analog signals. For low-level signals (e.g., from thermocouples), amplification is required. With a relay multiplexer and a solid-state A/D converter, amplification is usually performed on low-level signals *after* multiplexing, since low-level signals can be accurately switched with relays and the amplifier can therefore be shared along with the A/D converter. In some cases, gain-selectable amplifiers are used and the computer program can control the gain as necessary while selecting multiplexer channels for conversion.

Some solid-state multiplexers require amplification of low-level signals prior to switching. For thermocouples, emf-to-current conversion, with voltage output across a resistor in the current loop, is used to supply the high-level voltage signal to the computer interface. If conditions warrant, it may be attractive to use remote electromechanical multiplexing of the low-level signals to share a common emf-to-current converter and one channel on the solid-state multiplexer at the I/O interface. In an application where most analog input signals are from pneumatic instrumentation and thermocouples, and high scan rates are not required, separate electromechanical multiplexing is often favored over installing individual pressure-to-current and emf-to-current transducers for each signal.

Noise considerations also influence the selection of I/O interface hardware and signal conditioning equipment. Electrical analog input signals are received from the process as a current loop or as a potential (voltage) difference on a pair of signal lines. As described previously, typical high-level current signals can be easily converted into high-level voltage signals of the proper range by the insertion of a resistor in the current loop. In some cases, all electrical signals can share a common ground potential, and single-ended multiplexing can then be used. If it is necessary to use both signal lines (e.g., thermocouples), then both lines must be switched together by the multiplexer, and differential amplifiers are required to obtain the proper voltage range for conversion.

Common-mode noise and normal-mode noise are important considerations in both system design and cabling practice (see Section 7.11). Common-mode noise does not affect the system performance unless it is converted to normal mode. Normal-mode noise, principally from common-mode conversion resulting from unbalanced line and leakage impedances, and from inductive coupling between adjacent conductors, can seriously affect performance. The use of twisted signal pairs with shielding is usually recommended to reduce electromagnetic interference. Methods for the reduction of common-mode interference include single-point grounding and isolation. The use of a differential amplifier guarded with a floating input and with a common ground for transducer (or, if floating, transducer shield) and cable and amplifier guard shields helps isolate the computer system from the effects of common-mode interference. However, since the shield must be continuous through the input section of the amplifier, expensive three-wire multiplexing is required. Because of the expense, three-wire multiplexing is avoided where possible, and attention is concentrated on cable selection, routing, and good grounding practices. Some two- and three-wire multiplexers are illustrated in Figure 7.6o.

The higher-frequency alternating current components of the normal-mode noise that do appear at the computer I/O interface can usually be inexpensively filtered out. The most common 60-Hz noise component is normally attenuated with simple, first-order, RC circuits (Figure 7.11k) in process control applications. If the process changes at frequencies that are affected by the RC noise filter, more expensive LRC filters can be used to get

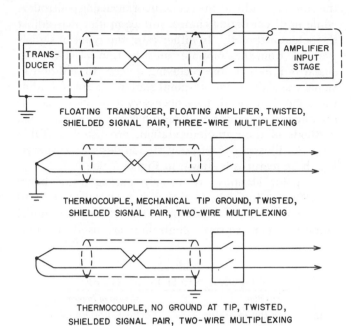

FLOATING TRANSDUCER, FLOATING AMPLIFIER, TWISTED, SHIELDED SIGNAL PAIR, THREE-WIRE MULTIPLEXING

THERMOCOUPLE, MECHANICAL TIP GROUND, TWISTED, SHIELDED SIGNAL PAIR, TWO-WIRE MULTIPLEXING

THERMOCOUPLE, NO GROUND AT TIP, TWISTED, SHIELDED SIGNAL PAIR, TWO-WIRE MULTIPLEXING

Fig. 7.6o   Shielding and grounding practices

more selective filtering. A more detailed discussion of this subject can be found in Section 7.11.

Summary

Connecting computers to external devices can involve many considerations and trade-offs in hardware costs, effective operating speeds, flexibility, resulting signal fidelity, and reliability. The many parameters in I/O interface design are, however, much reduced with the selection of a particular control computer system. In general, most control computer manufacturers and suppliers offer a configuration of computer and interface equipment design that is mutually compatible and that will interface with the normal complement of conventional process instrumentation and control devices. With the selection of a computer system, many of the design decisions (for example, an A/D conversion method) are already made, and the major attention can be directed toward communication with devices having unique requirements. Most suppliers of computer systems publish manuals describing the characteristics of the I/O interface system in detail and give advice on connecting process signals to the interface. These manuals should be consulted prior to connecting any signals from external devices to the computer system.

Set-Point Stations

Features

Set-point stations are used to interface the digital control computer with plant controllers and actuators, and they enable the computer to make changes in plant operating conditions. In supervisory control applications, the computer adjusts the set point of an analog controller, while in direct digital control, the computer may adjust the actuator directly. In either case, the types of signals available from the computer must be translated into the signal required by the receiving device. In addition to signal translation, the set-point station has several other important features that are necessary to provide efficient operator communication. These features, and the usual methods of their implementation, are listed in Table 7.6p. An illustrative set-point station design incorporating these features is shown in Figure 7.6q.

Set-point stations are usually packaged for panel mounting, using the common 2 × 6, 3 × 6, or 6 × 6-in. (50 × 150, 75 × 150 or 150 × 150 mm) nominal cutouts. Where separate controllers are used in super-

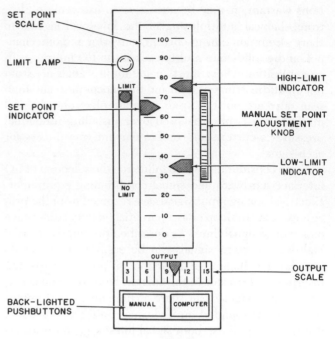

Fig. 7.6q   Computer set-point station

Table 7.6p
SET-POINT STATION FEATURES

Function	Method of Implementation
A means for the operator to observe the value of the computer-generated set point	Vertical, horizontal, or circular scales, usually with 0 to 100% range and with pointer indicating set-point position
A means for the operator to switch from computer to manual control	Selector switch or back lighted pushbutton
A means for the operator to detect at a glance if the station is on computer control	Switch position or indicating lamp, which is on when the computer controls set point
A means for the operator to manually adjust, or to over-ride, the computer's set point	Knob or push-bar for manual adjusting. Pointer on scale indicates set-point value.
A means to limit the range over which the computer can adjust the set point	Mechanical stops, with separate lamp to indicate when set point engages upper or lower limit
A means to remove the limits when the set point is under manual control	Automatic, with switching from computer to manual or a separate limit-no-limit switch
A means for the operator to observe the set-point station output signal	A scale, in output units with pointer indicating output value
A means for the station to accept electrical signals from the computer, indicating change in set point	Stepping or synchronous motors or integrating amplifiers
A means for the computer to determine status of set-point station	Contacts, indicating position of computer/manual and limit/no-limit switches and information on whether set point is at limit

visory applications, the set-point station is preferably mounted adjacent to the associated controller for ease of operator association. In new installations, it is sometimes more economical to use packages that combine the set-point station with an indicating or recording controller and require less panel space than the separately mounted units. In direct digital control applications, all set-point stations may be mounted on a separate operator's console to create a "cockpit" type plant control station.

Conventional controllers typically obtain their set-point signals from manually adjusted knobs, in which case the mechanical motion must be translated into an electrical or pneumatic signal internally, or, in case of the cascade arrangement, the electrical or pneumatic output of other controllers has to be translated. Set-point stations can be designed to interface with the electronic or pneumatic controller through either input. Several methods have been used, but the most commonly accepted one is to use the set-point station to generate, from computer output signals, a set point that can be directly used in the controller, through, for example, a remote or cascade input connection. The set-point station must therefore supply a standard electronic or pneumatic controller signal.

One earlier method of interfacing the computer with conventional controllers, though no longer acceptable, illustrates some of the functional design requirements for this interface. For electronic controllers, a voltage output was generated by the computer (via a digital-to-analog converter) proportional to the desired set point. A switch on the controller allowed the operator to select the effective set point as either the computer output signal, for computer control, or the signal from the operator-adjusted set-point potentiometer.

Fail-Safe

This computer-manual switch installation was a relatively simple modification to an existing electronic controller, and additional panel space for a separate set-point station was not required. However, this method has serious drawbacks and is not normally recommended. This controller modification scheme is not "fail-safe," in the sense that a computer system failure and loss of the computer-supplied set-point voltage could cause a "zero" set point to the controller. This sudden change on one or more controllers could cause dangerous plant conditions or initiate plant shutdown procedures. Means can and should be provided to generate automatic switching to the manually adjusted set-point potentiometer in the event of computer failure. In this case the operator is required to continually adjust the manual set point to keep it up to date with the computer's, so that bumpless transfer is guaranteed. In addition, when the transition from operator control to computer control is made, the computer must supply a signal identical to that supplied manually, in order to achieve bumpless transfer. The computer therefore should read the operator's set-point voltage through the analog-to-digital converter and then generate an identical voltage, using the digital-to-analog converter in order to "line up" with the existing set point. This procedure is difficult to perform without a resulting offset and set-point "bump."

Modern set-point stations are designed to provide fail-safe operation in case of computer failure and to furnish bumpless transfer during computer-to-manual and manual-to-computer changes in set-point source. These features are provided by designing the set-point station so that the computer must supply a live signal only while making changes in the set point, and by providing means to automatically maintain the manual set point at the computer-set value while under computer control.

Stepping and Synchronous Motors

The most common fail-safe technique used is that of operating a stepping or synchronous motor with computer digital output signals. A stepping motor receives a pulse train from the computer, with the number of pulses proportional to the change in set point to be made. Failure of the computer results in no more pulses and leaves the set point in its last position. A synchronous motor receives a timed contact closure signal from the computer, and the change made in the set point is proportional to the duration of the contact closure signal. Unless the computer system fails while changing the set point in such a way that the motor is driven all the way up or down scale, fail-safe operation is achieved. To guard against this latter type of failure, operator-adjustable mechanical stops to keep the computer set point within preselected limits are recommended, particularly with synchronous motor type set-point generators.

For electronic controllers, the motor in the set-point station drives a potentiometer to supply the appropriate DC milliampere signal to the controller set point. In pneumatic set-point stations, the motor adjusts a pressure regulator to provide the required 3–15 PSIG (0.2–1.0 bar) pressure signal. In either case, the mechanical movement provided by the motor is identical to the mechanical adjustment made when the operator manually adjusts the set point on the set-point station. Bumpless transfer is therefore inherent in this design. (In this arrangement, manual set-point adjustments are made at the set-point station rather than at the conventional controller.)

Integrating Amplifier

Another method of generating continuous set-point signals from discontinuous computer signals, indicating desired changes in set points, is to use an integrating electronic amplifier. Computer voltage pulses are integrated, and the integral voltage, representing the sum of all previous changes in set point, is continuously supplied to the controller set point. In this scheme, the amplifier drift rate may be an important consideration when the computer is not adjusting the set point often. This, of course, is not a consideration when stepping or synchronous motors are used.

Speed, Resolution, and Feedback

Maximum operating speed and resolution of set-point position can be major considerations in the selection of set-point stations. For a stepping motor station, resolution is defined by the number of steps required to drive the set point from zero to full scale. A full-scale travel of 1,000 steps, for example, gives a resolution of 0.10 percent. The maximum pulse rate is defined by the sum of the minimum "on" pulse duration and the minimum allowable time between pulses. If, for example, the sum is 30 milliseconds, the maximum pulse rate is 33 pulses/second, and assuming 0.1 percent resolution, 30 seconds would be required for a full-scale change (1,000 pulses) in set point. Depending on the application, faster changes in set points may be required.

In general, higher operating speeds and lower resolutions are characteristic of synchronous motor-operated set-point positioners. Full-scale travel times of 5 seconds are available with 0.5 percent resolution and with correspondingly higher resolutions for slower speeds (e.g., 10 seconds, 0.25 percent resolution), depending on motor speeds and gearing. Anti-backlash gearing usually provided for stepping motors is not available with synchronous motors, and gear train backlash of nominally 0.2 percent is typical.

In supervisory control, the computer program must be initialized prior to taking over control of the set points from the operator, and therefore the computer must be able to read the value of manually set set points. Also, during operation, in either supervisory or direct digital control mode, the computer should check the results of every set-point adjustment to insure that the requested change was effected accurately. Set-point feedback signals are therefore required from the set-point stations to the computer. These feedback signals can be obtained from a retransmitting slidewire to insure proper motor positioning in electronic set-point stations. For pneumatic set-point stations, the motor position provides only the input to the pressure regulator; therefore, to further check that the pressure regulator responded properly, the pneumatic output signal, through appropriate transducers, should be fed back to the computer. Computer set-point station interface signals are shown in Figure 7.6r.

Summary

Major features of some available set-point stations are listed at the beginning of this section. Indicating elec-

tronic controllers, with integral motor-driven potentiometers, are available from a number of suppliers, and the selection of a particular model will be influenced by the type of plant instrumentation currently in use and the particular features of the set-point station and controller required for the application. Separate set-point stations, particularly for pneumatic instrumentation, and set-point stations combined with recording controllers, are supplied by fewer instrument manufacturers, sometimes leaving little choice for selection. Costs for set-point stations generally range from about $750 for stripped-down separate stations, to about $2,500 for set-point stations combined with strip-chart recording controllers and with a full complement of switches, indicators, etc. for complete operator and computer communication.

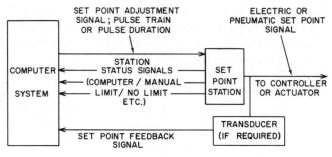

Fig. 7.6r   Set-point station interface signals

BIBLIOGRAPHY

Archibald, W., "Remote Multiplexing," Burr-Brown Application Note AN-80, January, 1976.

Aronson, R.L., "Line-Sharing Systems for Plant Monitoring and Control," *Control Engineering*, January, 1971.

Gore, F., "Computer-Process Interface Checklist," *Instruments & Control Systems*, February, 1977.

"Guide to Selecting Temperature Multiplexers," *InTech*, February, 1980.

Mamzic, C.L., "Pneumatic Controls Interface with Computers," *Instruments & Control Systems*, October, 1975.

Merritt, R., "Personal Computers Move Into Control," *Instruments & Control Systems*, June, 1983.

Soderquist, D., "Digital to Analog Converter," *Instruments & Control Systems*, March, 1977.

Sylvan, J., "Putting Intelligent I/O on the Factory Floor," *Instruments & Control Systems*, November, 1982.

Zuch, E.L., "Linking the Analog World to Digital Computers," *Instruments & Control Systems*, September, 1977.

7.7 MEMORY UNITS

Available Types:	Semiconductor: MOS, bipolar, CCD Bubble Magnetic Core Magnetic Disk: Fixed, fixed/removable, removable, $5\frac{1}{4}$, 8, and 14-in. (131.25, 200, and 350 mm) Winchester, $5\frac{1}{4}$ and 8-in. (131.25 and 200 mm) floppy Magnetic Tape: Reel-to-reel, cartridge, cassette
Capacity (8-bit Bytes/Unit Cost):	Semiconductor: MOS—1×10^6/\$19,000; bipolar—32×10^3/\$849; CCD—0.5×10^6/\$2,750 Bubble: 0.5×10^6/\$6,250 Magnetic Core: 32×10^3/\$1,350 Magnetic Disk: Fixed—1–$1,300 \times 10^6$/\$10,890–\$59,900; fixed/removable—5–656×10^6/\$6,000–\$63,000; removable—1.5–600×10^6/\$4,670–\$75,260; Winchester—$5\frac{1}{4}$ in.—0.5–13×10^6/\$2,000–\$3,800, 8-in.—0.5–160×10^6/\$8,400–\$14,600, 14-in.—9–$2,500 \times 10^6$/\$5,600–\$97,680; floppy—$5\frac{1}{4}$-in.—0.1–10×10^6/\$1,500–\$6,000, 8-in.—0.25–14×10^6/\$3,600–\$9,845
Speed (Inches/Second Unit Cost):	Magnetic Tape: reel-to-reel—25–125/\$4,500–\$29,100; cartridge—18–60/\$1,600–\$2,800; cassette—2–80/\$60–\$2,040

Memory Organization

The memory unit of a computer is the component that stores the programs that are to be executed. A small computer with a limited application may be able to fulfil its intended task without additional (external) storage capacity for programs and data. However, most computers run more efficiently if they are supplied with additional storage beyond the capacity of the main memory. There is usually not enough space in one memory unit to accommodate all the programs needed for a typical application. Also, most computer installations accumulate large amounts of information that does not need to be available to the processor all the time. Thus it is more economical to use lower-cost storage devices, rather than the computer's main memory, for storing information that is not currently needed. The memory unit that communicates directly with the computer is called the main memory. The devices that provide the addi-tional storage are the auxiliary memory. Examples of auxiliary memory devices are magnetic disks and tapes. Only the programs and data currently needed by the processor reside in main memory. The other information is stored in auxiliary memory and is transferred to main memory on demand.

The total memory capacity of a computer may be considered as a hierarchy of components. The memory hierarchy system consists of all the storage devices required by the computer system, from the slow, high-capacity auxiliary memory devices to the faster main memory, and sometimes to an even smaller and faster buffer memory.

Figure 7.7a illustrates a typical memory hierarchy. On top are the relatively slow magnetic tapes used to store removable files. Below this are the magnetic disks used as auxiliary storage. The main memory communicates directly with auxiliary devices and the processor. When

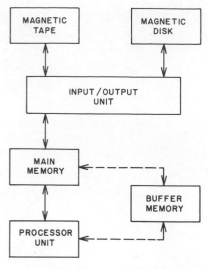

Fig. 7.7a   The memory hierarchy

programs not residing in main memory are needed, they are brought in from auxiliary memory. Programs not currently needed in main memory are transferred into auxiliary memory.

A buffer memory is sometimes used in large computer systems to compensate for the speed differential between the main memory access time and the processor logic. The processor logic is usually faster than the main memory access time, with the result that processing speed is limited primarily by the speed of main memory access. To overcome this limitation, a fast, small memory with an access speed close to that of the processor logic can be inserted between the processor and main memory. This type of memory is called a buffer or a cache memory.

In many systems the need arises for running partial programs, for varying the amount of main memory in use by a given program, and for moving programs around in the memory hierarchy. Application programs may be too long to be accommodated in the total space available in the main memory. A computer system uses not only the application programs but many system programs as well, and all of these programs cannot reside in main memory at all times. A program with its data will normally reside in auxiliary memory. When the program or a segment of the program is to be executed, it is transferred to main memory. The auxiliary memory can contain the total information stored in the computer system, and it is the task of the operating system to maintain in main memory the portion of this information that is currently active. The part of the operating system program that supervises the flow of information between the storage devices is the memory management system.

In a timesharing mode, many users communicate with the computer via remote terminals. Because of the slow human response compared to the computer speeds, the computer can respond to multiple users at, seemingly, the same time. This is accomplished by having many

programs reside in memory while the system allocates a time-slice to each program for execution.

The cost per bit of storage is roughly proportional to the memory's level in the hierarchy. It would be too expensive to maintain all programs and data in main memory, especially during the times that they are not needed by the processor. The memory management system distributes the programs and data to the various levels in the memory hierarchy according to their expected usage.

Memory Unit Characteristics

A memory unit is specified by the number of words it contains and the number of bits in each word. The selection of a specific word in the memory may be done by inserting its binary address value into selection lines. Thus, k address bits can select any one of $2^k = m$ words. Computer memories may range from 1,024 words, requiring an address of 10 bits, to 1,048,576 (2^{20}) words, requiring 20 address bits. One usually refers to the number of words in a memory unit with the unit K. K refers to $1,024 = 2^{10}$ words; thus 4K = 4,096 words and 64K = 2^{16} words.

Access time is defined as the time from when a memory unit receives a read signal to the time when the information read from memory is available in the outputs. In a destructive read memory, such as a magnetic core, the information read out is physically destroyed by the reading process. It is automatically restored, but this requires some additional time. The sum of the access time and restoration time is called the cycle time. In a nondestructive read memory, the cycle time is equal to the access time since no restoration is necessary. Typical cycle times of main memory units range from about 100 nsec to about 1 μsec.

The mode of access of a memory unit is a function of the type of components used. In a semiconductor random access memory, memory registers can be thought of as being separated in space, with each register occupying a particular location. In a sequential access memory, the information stored is not immediately accessible; it is available only at certain times. A shift register memory is an example of this type.

Semiconductor memories can be divided into three broad categories, as shown in Figure 7.7b. Most of these

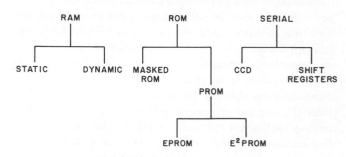

Fig. 7.7b   Types of semiconductor memory

generic categories can be implemented with either of the two major technologies MOS (metalized oxide semiconductor) and bipolar. The exceptions are the CCDs (charge coupled devices) and EPROM (erasable programmable read only memories), which are uniquely implemented with MOS technology.[1]

Charge coupled devices use small squares of aluminum that are deposited on a silicon substrate to transfer charges in a serial fashion. Because of the simple and repetitive structure, these MOS memories are characterized by a very high density. The access time is from 50 to 100 microseconds.

Random Access Memories (RAM)

A memory unit may be a collection of storage registers, together with the associated circuits needed to transfer the information in and out of the registers. The memory registers can be accessed for information transfer as required in a semiconductor random access memory (RAM).[2]

The communication between the memory unit and the processor environment is achieved through control lines, address selection lines, and data input and output lines. The control signals specify the direction of transfer required, that is, if a word is to be stored in a register or if a word previously stored is to be transferred out of a register. The address lines specify the particular word chosen out of all those available. The input lines provide the information to be stored in the memory, and the output lines supply the information coming out of the memory.

The two control signals are called read and write. The write signal specifies a transfer-in operation; the read signal specifies a transfer-out operation. The internal control circuits inside the memory unit provide the desired function. When the memory unit receives a write control signal, internal control transfers the data input bits into the word specified by the address lines. With a read control signal, the word selected by the address lines appears at the data output lines.

Some integrated circuit memories use a single line for the read/write control. One binary state specifies a read operation, and the other binary state specifies a write operation.

The internal construction of a random access memory of m words with n bits per word consists of m × n binary storage cells and the logic to select a word for writing or reading. The binary storage cell is the basic building block for these memory units. A binary cell that stores one bit of information might include five gates and a two-transistor flip-flop.[3] Cells that lose their charge and must be periodically refreshed are used in dynamic RAMs. Cells that hold their charge without refreshing are used in static RAMs. Several manufacturers offer an automatic refresh feature that includes formerly external functions in a quasi-static RAM.

Alpha radiation errors in dynamic RAMs have sparked an interest in error correction and detection. Simple parity checks can find single-bit errors, and more elaborate schemes, such as the Hamming code, may correct single-bit errors and detect double-bit errors.

The low-cost market is led by dynamic RAMs. Ultra-low power is the domain of CMOS, and high-speed MOS and ECL static RAMs head the speed market.[4] The large middle ground of slow statics is giving way to quasi-statics and high-speed CMOS RAMs (see Figure 7.7c).

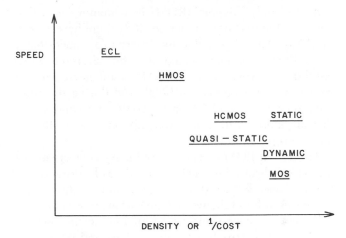

Fig. 7.7c Semiconductor RAM characteristics

IC memories retain the binary information when a word is read from memory. This type of memory has a nondestructive read property since the contents of the word in memory are not destroyed during the reading process. Another component that has been used as a binary storage cell in memory units is the magnetic core, as shown in Figure 7.7d. A magnetic core is a small toroid made of ferromagnetic material. Its magnetic property makes the core usuable for binary storage. One direction of magnetization represents a 0 and the other represents a 1. Reading out the binary information stored in a core forces the direction of magnetization to the 0 state. Therefore, a magnetic core memory with its destructive read property loses the previously stored information after the reading process. Because of this destructive read property, a magnetic core memory must provide additional control functions to restore the original con-

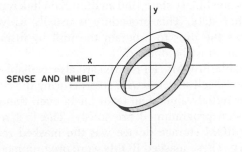

Fig. 7.7d Ferrite core

tents of the word. The cores are also more expensive to produce than integrated circuits.

The reduction in core size is limited by core manufacturing and testing problems and by the difficulty of threading wires through arrays of the individual cores in order to assemble a working core memory. For these reasons semiconductor RAM has replaced most core applications.

Read Only Memories (ROM)

A read only memory (ROM) is a memory unit that performs the read operation only; it does not have a write capability. This implies that the binary information stored in the ROM is permanent and cannot be altered by writing different words into it. A RAM is a general-purpose device whose contents can be altered during the computational process; a ROM is used only for reading words that are stored within the unit, although some ROMS may be reprogrammed.[5]

An m × n ROM is an array of binary cells organized into m words of n bits each. A ROM has k address lines to select one of 2^k words of memory, and n output lines, one for each bit of the word. The various types of ROMs include the capacitor matrix, mutual inductance matrix, transformer matrix, ferrite core matrix, and permanent magnet-twistor matrix. Most of these systems are not fast enough to operate with the presently available integrated circuits. The limits of capacity of these memories are also set by physical size and electrical delay, but the main limitation to using these ROMs is cost.

An IC ROM is fabricated with the outputs being all 1 or all 0. The particular pattern of 1s and 0s is then obtained by providing a mask in the last fabrication step. Each cell in the ROM incorporates a link that can be fused. A broken link in a cell defines one binary state, and an unbroken link represents the other state. The customer fills out a truth table. The manufacturer then makes the mask for the links to produce the 1s and 0s of each word desired. This is called custom or mask programming. Note that it is a hardware procedure even though the word "programming" is used.

For smaller quantities, it is more convenient to use a programmable ROM, or PROM. Each cell in a PROM incorporates a link that can be fused by application of a high current pulse (Figure 7.7e). A broken link in a cell defines one binary state, and an unbroken link represents the other state. This procedure is usually irreversible. It allows the user to program the unit by fusing those links that must be opened.[6]

Erasable PROMs (EPROMs) are also available. These use special cell structures for restructuring the cells back to their initial value of all 0s or all 1s even though they have been programmed previously. The first semiconductor ROM storage device was the masked read only memory. These masked ROMs were programmed by the manufacturer with instructions supplied by the buyer. Once programmed, they cannot be altered. Each pro-

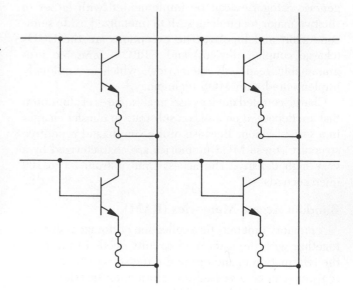

Fig. 7.7e   A section of a fusible link bipolar PROM

gram change, therefore, requires the purchase of a new ROM device, which could take several months. ROM devices are inexpensive in large quantities, but they require a large initial investment and a large quantity commitment.

The programmable ROM, or PROM, can be "burned" by the user, but they can be programmed only once. They are costlier than ROMs on a per-unit basis, but they eliminate the dependence on the manufacturer for programming.

The erasable PROMs add considerable flexibility. Like the PROMs, EPROMs can be programmed by the user, but they can be programmed many times. This eliminates the waste of PROM devices when program changes are required. EPROMs are used as a development tool for designers, who need to change programs frequently while prototyping and debugging systems. They are also shipped in production equipment due to their value to the user, who may wish to make a program change. EPROMs have become a popular program storage memory.

A drawback to EPROM devices is that they must be removed from the equipment to be reprogrammed. EPROMs are erased optically, through exposure to ultraviolet light, then rewritten electrically.

Electrically erasable PROMs (E^2PROMs) provide the benefits of the EPROM without the drawback of removal for reprogramming. The E^2PROM can be electrically reprogrammed by the user, but without removal from the system, and without the use of exterior programming devices such as PROM programmers.[7]

Applications of ROMs

The applications of read only memories are widespread. They can be programmed to perform a sequence of logic events or a mix of combinational and sequential events. The delay that occurs in some large combinational nets is avoided. Read only memories can perform as code converters, trigonometric function generators, and as storage for small programs.

Semiconductor MOS read only memories are now an integral part of logic functions in control applications because they are electrically and physically compatible with logic ICs and have self-contained decoding and sense circuitry. Earlier ROMs were generally treated as a subsystem because they required special control and sense circuits.

ROM allows more efficient use of computing hardware, enables faster overall operation of the computer, and allows flexibility in basic computer design. A general-purpose computer with ROM can be made into a special-purpose device by programming the read only memory to best suit the particular application.

Auxiliary Memory

The most common auxiliary memory devices used in computer systems are magnetic disks and magnetic tapes. Other devices are magnetic drums, core, plated wire, bubble, and laser memories. The physical mechanisms of auxiliary memory devices use a combination of magnetics, electronics, and electromechanical systems. Although the physical design of these storage devices can be quite complex, their properties can be characterized and compared by a few parameters. The important characteristics are access mode, access time, transfer rate, capacity, and cost.

The average time required to reach a storage location in memory for reading or writing is called the access time. In electromechanical devices with moving parts such as disks and tapes, the access time consists of the seek time required to position the read-write head and the time required to transfer data to or from the device. Since the seek time is usually much longer than the transfer time, auxiliary storage is organized in records or blocks. A record is a specified number of characters or words. Reading or writing is done on entire records. The transfer rate is the number of characters or words that the device can transfer per second, after it has been positioned at the beginning of the record.

Magnetic drums and disks are similar in operation. Both consist of high-speed rotating surfaces (a cylinder or a round, flat plate) coated with a magnetic recording medium. The recording surface rotates at a uniform speed and is not started or stopped during access operations. The bits are recorded as magnetic spots on the surface as it passes a stationary write head. The stored bits are detected by a change in magnetic field produced by the recorded spots on the surface as it passes by a read head. The amount of surface available for recording on a disk is greater than on a drum of equal physical size. More information can be stored on a disk than on a drum of comparable size. For this reason, disks have replaced drums.

Magnetic Disk Memories

A disk unit is an electromechanical assembly containing a flat disk coated with magnetic material. Often both sides of the disk are used, and several disks may be stacked on one spindle with read-write heads available in each surface (Figure 7.7f). The disks rotate together at high speed. The bits are stored on the magnetized surface in spots along concentric circles called tracks. The tracks are usually divided into sections called sectors. In most systems, the minimum quantity of information that may be transferred is a sector. The subdivision of one disk surface into tracks and sectors is shown in Figure 7.7g.

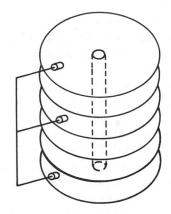

Fig. 7.7f A moving head disk memory system

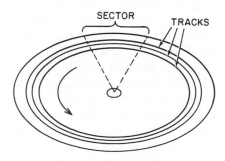

Fig. 7.7g Recording surface of disk

Permanent timing tracks are used in disks, as in drums, to synchronize the bits and recognize the sectors. A disk system is addressed by address bits that specify the disk number, the disk surface, the sector number, and the track within the sector. After the read-write heads are positioned in the specified track, the system has to wait until the rotating disk reaches the specified sector under the read-write head. Information transfer is fast once the beginning of a sector has been reached. Disks with multiple heads allow the simultaneous transfer of bits from several tracks at the same time.

A track in a given sector near the circumference is

longer than a track near the center of the disk. If the bits are recorded with equal density, some tracks will then contain more recorded bits than others. In order to make all the records in a sector of equal length, some disks use a variable recording density with a higher density on tracks near the center than on tracks near the circumference. This equalizes the number of bits on all of the tracks of a given sector. Some disks are attached to the unit assembly and cannot be removed easily.

Some units use a single read-write head for each disk surface. In these fixed units, the track address bits are used by a mechanical assembly to move the head into the specified track position before reading or writing. In other disk systems, separate read-write heads are provided for each track on each surface. The address bits then select a particular track electronically through a decoder circuit. Another type of unit, called the disk pack, allows the disk to be removed easily.

The Winchester drive uses a disk pack that also contains the recording heads. The disks themselves are aluminum, 0.075 inches (1.875 mm) thick, with a lubricating oxide surface.

A more recent type of disk storage uses a flexible or floppy disk in place of the rigid risk. The flexible disk is made of plastic and is coated with a magnetic recording medium. It is approximately the size and shape of a 45-rpm phonograph record. The flexible disk can be inserted and removed about as easily as a tape cartridge (Figure 7.7h). A typical floppy disk drive system is shown in Figure 7.7i. The drive motor rotates the spindle at 360 rpm. The average access time is about 80 milliseconds. A single 5¼-inch (131.25 mm) floppy disk has a capacity of 2,050,048 bits. There are 77 tracks with 26 sectors on each track and 128 8-bit words on each sector. The main advantage of the floppy disk is that it is inexpensive to manufacture. The drives also do not need the critical tolerances required for rigid disks in order to minimize crashing (when the head damages the recording surface). Thus the drives are inexpensive also. The main advantages of rigid disk systems are the faster access times (to 50 microseconds), larger capacity, and greater reliability for mechanical failure modes.

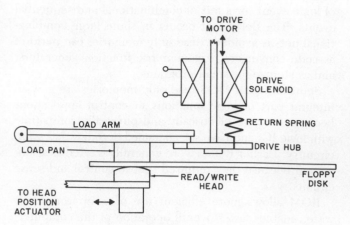

Fig. 7.7i Typical floppy disk drive system

One or both sides of a disk may be used for recording. Rigid disks that are stacked and driven together with a common drive are called disk files. Removable and replaceable disk files are called disk packs. These removable groups of disks that share a common spindle have access times between those of fixed-head disks and magnetic tape (Table 7.7j).

Table 7.7j
DISK PACKS

Capacity range	12 to 5,000 million bits
Number of disks	1 to 11
Number of recording surfaces	2 to 20
Number of tracks per surface	200/100
Weight	4 to 16 pounds (1.8 to 7.2 kg)

One type of pack consists of six 14-in. (350 mm) diameter disks mounted on a common spindle, 0.35 in. (8.75 mm) apart. The ten inside disk surfaces are used for recording, and the outer two surfaces are not used. The data is recorded on 200 tracks, at bit densities varying from 765 to 1,105 bits/in., (30.6 to 44.2 bits/mm) on a 2-in. (50 mm)-wide band on each surface. Table 7.7k lists the major features that differentiate magnetic disk units. Trade-offs between performance (primarily the access times) and costs determine the choice for an application.

Table 7.7k
MAGNETIC DISK UNIT FEATURES

Access time

Tracks/disk

Data transfer rate

Storage/pack

Drives/subsystem

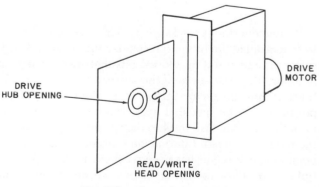

Fig. 7.7h Floppy disk and drive

Magnetic Tape Memories

A magnetic tape transport system consists of the electrical, mechanical, and electronic components that provide the control for a magnetic tape unit. The tape is a strip of plastic that is coated with a magnetic recording medium. The bits are recorded as magnetic spots on the tape along several tracks. Usually 7 or 9 bits are recorded simultaneously to form a character along with a parity bit. The read-write heads are mounted one for each track so the data can be recorded and read as a sequence of characters.

Magnetic tape units can be stopped, started to move in forward or reverse, or rewound. They cannot be started or stopped fast enough between individual characters, so the information is recorded in blocks known as records. Gaps of unrecorded tape are inserted between the records where the tape can be stopped. The tape starts moving while in a gap and attains a constant speed by the time it reaches the next record. Each record on tape has an identification bit pattern at the beginning and end. Reading the bit pattern at the beginning, the tape control identifies the record number; reading the bit pattern at the end of the record, the control recognizes the beginning of a gap.

A standard $10\frac{1}{2}$-in. (262.5 mm) reel of tape contains 2,400 ft (720 m) of $\frac{1}{2}$-in. (12.5 mm) wide tape. A typical tape stores 800 bits of information on each of the 7 or 9 channels (800 characters). One reel can therefore store some 160 to 200 million bits. Most tapes use a base, or substrate, of polyethylene terphthalate polyester film 1.42 mils (36 microns) thick and a magnetic coating of gamma ferric oxide of about 400 μin (10 microns) thick.

A tape transport mechanism is required to move the tape past the recording heads. The transport must be able to start and stop quickly to allow the tape to pass the recording heads at high speed. Supply and take-up reels, which have a high inertia, are generally isolated from the mechanism that drives the tape past the heads so that the tape can be quickly accelerated or decelerated. The tape is usually either laced around sets of tension arms (Figure 7.7l) or passed through a vacuum column (Figure 7.7m) between the reels and the continuously rotating capstans. Pressure rolls press the tape against the capstans when activated. Mechanisms with start/stop times of the order of 1 to 5 milliseconds, and tape speeds of up to 125 in./second (3.125 m/s) are available.

Mechanical buffers, tape reels, and multiple drives are not needed in another type of drive design (Figure 7.7n). The Newell drive uses a single motor to drive a large capstan against which the tape supply and take-up rolls are forced. The center of each tape roll moves toward or away from the center capstan as the tape is removed or added to the roll. Tape accelerations on the order of 1,000 in./sec² (25 m/s²) can be obtained.

A variety of other systems for handling magnetic tape are available. Most use some form of cartridge. The tape

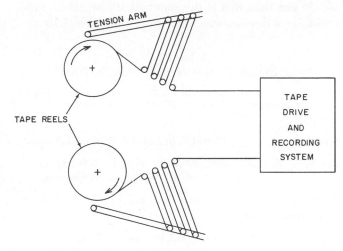

Fig. 7.7l Tension arms for starting and stopping high-speed tape

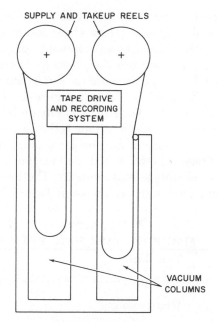

Fig. 7.7m Vacuum columns for starting and stopping high-speed tape

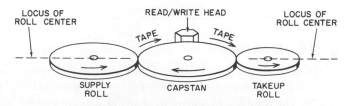

Fig. 7.7n Tape transport based on Newell principle

lengths are usually shorter than those used with reel-to-reel systems, and therefore the average random access times are less. A cartridge system might use endless 100-ft (30 m), $\frac{1}{4}$-in. (6.25 mm) tapes, operate at a speed of 10 in./second (250 mm/s), and have a 7- to 9-bit character transfer rate of about 500 characters/second. Capacities vary, but there are generally one or two orders of magnitude less than that of the standard 10½-in. (262.5 mm) reel-to-reel devices which are shown in Table 7.7o.

Table 7.7o
REEL-TO-REEL MAGNETIC TAPE STORAGE SYSTEMS

Range of transfer rates	2,000 to 150,000 characters/second
Speed range	25 to 125 in./second (625 to 3125 mm/s)
Recording density	200, 400, 556, 800, or 1,600 bits/in. (8, 16, 22, 24, 32, or 64 bits/mm)
Number of recording tracks	Usually 8 or 9; some 10 or 16
Tape width	Usually 0.5 in.; some 0.75 or 1.0 in. (12.5 mm; some 18.75 or 25 mm)

Magnetic tape recording systems are best suited for sequentially organized data. Other forms of processing, where either the data are randomly organized or the processing is done randomly, are inefficiently handled with magnetic tape, because the serial nature of the storage medium produces long average access times (on the order of seconds). For large-capacity on-line storage requirements, the magnetic tape is the lowest-cost storage medium. Because of the smaller average access times, disk storage is better suited than magnetic tape for medium-sized storage requirements. The basic features of magnetic tape units are shown in Table 7.7p.

Table 7.7p
MAGNETIC TAPE UNIT FEATURES

Tape density
Tape speed   Transfer rate
Interrecord gap

Start-stop delays

Rewind speed

Bidirectional read/write

Bubble Memories

The available designs of magnetic bubble memories use the presence or absence of bubbles to represent 1s and 0s. The bubbles are magnetic domains that are moved serially in a film of garnet. The bubbles are propagated along paths determined by patterns of Permalloy deposited on an oxide layer that covers the garnet. When magnetized by a rotating magnetic field, the patterns have alternating magnetic polarities that attract bubbles from one spot to another, as shown in Figure 7.7q.

All commercial devices enclose the bubble-containing material with a pair of coils that generate the rotating magnetic field.[8] The presence or absence of a bubble is

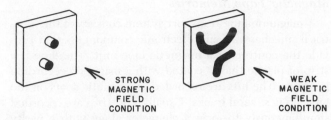

Fig. 7.7q   Magnetic bubble creation

detected by a magnetoresistive element (Figure 7.7r). The passing bubble causes a change in resistance across the detector. The bubbles are created or recorded by current pulses through control loops (Figure 7.7s).

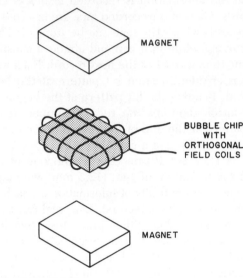

Fig. 7.7r   Bubble memory structure

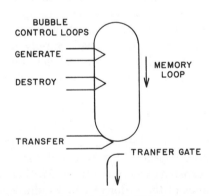

Fig. 7.7s   Single-loop bubble memory

Most bubble memories are organized into major and minor loops, which yield shorter access times than the serial shift-register organization used in earlier designs. Each minor loop is a storage area, while the major loops provide the input and output circuits. A "page" of information is stored by placing one bit in each minor loop. At playback, the bits on the desired page rotate to the tops of the minor loops, where they are transferred—or swapped—to major loops.

Most bubble memories are further organized into block replicates. Each major loop is divided into separate input and output loops, which are placed at opposite ends of the minor loops (Figure 7.7t).

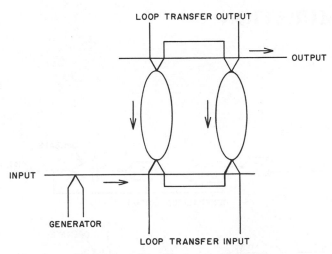

Fig. 7.7t   Multiple-loop bubble memory

The bubble domain technology is destined to play a leading role in the future development of computer hardware. Magnetic bubble memories have the potential of high bit density and large capacity per chip. In addition, they are less expensive to produce since they are a solid state technology and eliminate the mechanical motion associated with today's disk and tape systems. Long-term nonvolatility of the stored information is provided more easily than with most semiconductor devices. Also, logic and display functions are possible. The ability to store, transmit, and process information in the same medium eliminates or reduces the problems of electrical inter-connections, signal conversion, and associated power dissipations.

REFERENCES

1. Hnatek, E.R., *A User's Handbook of Semiconductor Memories*, New York: John Wiley & Sons, 1977.
2. Hordeski, M.F., *Microprocessor Cookbook*, Blue Ridge Summit: Tab, 1979.
3. *Ibid.*
4. Hassberg, C., "Trends in Microcomputer Storage," *Electronic Products*, June, 1979.
5. Hordeski, M.F.
6. *Ibid.*
7. Intel Corporation, "2816 E²PROM Data Sheet," Santa Clara, CA, 1981.
8. Bernhard, R., "Bubbles Take on Disks," *IEEE Spectrum*, May, 1980.

BIBLIOGRAPHY

Abbott, R.A., Regitz, W.M., and Karp, J.A., "A 4K MOS Dynamic Random Access Memory," *IEEE J. Solid State Circuits*, October, 1973.

Amelio, G.E., et al., "Experimental Verification of the Charge Coupled Device Concept," *Bell Systems Technical J.*, Vol. 49, 1970.

Carter, D., "A Case for Bubble Memories," *Instruments & Control Systems*, December, 1981.

Davis, S., "Selection and Application of Semiconductor Memories," *Computer Design*, January, 1974.

Frankenberg, R.J., *Designer's Guide to Semiconductor Memories*, Boston: Cahners, 1975.

Greene, R., and House, D., "Designing with Intel PROMs and ROMs," Intel Application Note AP-6, Intel Corp., Santa Clara, CA, 1975.

Hordeski, M.F., "Fundamentals of Digital Control Loops," *Measurements and Control*, February, 1978.

———, "Using Microprocessors," *Measurements and Control*, June, 1978.

Lenk, J.D., *Handbook of Microprocessors, Microcomputers, and Minicomputers*, Englewood Cliffs, NJ: Prentice-Hall, 1979.

Pro-Log Corporation, *PROM User's Guide*, Monterey, CA, 1979.

Texas Instruments Inc., *The MOS Memory Data Book for Design Engineers*, Dallas, TX, 1980.

7.8 MODELING AND SIMULATION

Process Simulation

The simulation of most processes, reactions, and plants may be carried out by developing a representation or building a model of common elements, where each element is adapted to the particular functions it must simulate. When all the parts have been defined and assembled, the overall effects and constraints may be imposed, such as material and energy balances, operating modes, limitations, instrumentation, and control system characteristics.

A common element in the process and plant simulations is the lag, defined by the transfer function (in Laplace notation)

$$H(s) = \frac{Y(s)}{X(s)} = \frac{K}{1 + \tau s} \qquad 7.8(1)$$

where K is the steady-state gain and τ is the time constant. In general, the gain and time constant will vary with other simulation variables, such as flows, temperatures, and pressures. A generalized circuit as shown in Figure 7.8a may have to be used. Other versions of the lag are also shown with different aspects and advantages, such as independent adjustment of gain and time constant, bipolar output, and simplicity and independent adjustment (but poorly scaled in many cases).

Another commonly used element is the lead-lag, for which the transfer function is

$$H(s) = K \frac{s}{1 + \tau s} \qquad 7.8(2)$$

or

$$H(s) = K \frac{1 + \tau_1 s}{1 + \tau_2 s} \qquad 7.8(3)$$

Figure 7.8b shows a generalized circuit for the type of transfer function in equation 7.8(2), with variable gain K and time constant τ.

The class of lead-lag transfer functions stated in equation 7.8(3) may be simulated with circuits of the type shown in Figure 7.8c. The generalized arrangement used for individual time constants and gain variations by signal

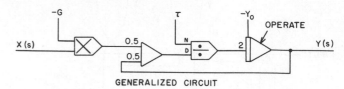

GENERALIZED CIRCUIT

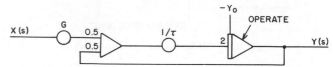

INDEPENDENT GAIN AND TIME CONSTANT ADJUSTMENT

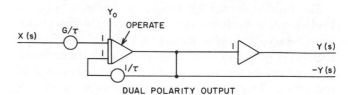

DUAL POLARITY OUTPUT

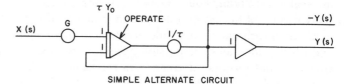

SIMPLE ALTERNATE CIRCUIT

Fig. 7.8a   The lag function

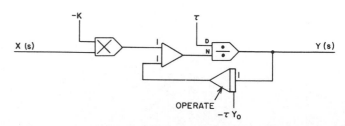

Fig. 7.8b   The lead-lag function for equation 7.8(2)

inputs often results in poor integrator scaling, especially for large τ_2. The common circuit permits the ratio between τ_1 and τ_2 to be adjusted over a wide range.

In many analog computer simulations, a signal delay effect, or simulation of a transport lag, such as the flow of fluids through pipes or the movement of material on a conveyor, is desired. Delays or deadtimes like this often

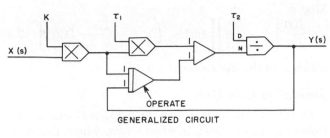

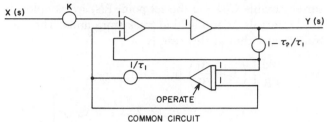

Fig. 7.8c The lead-lag function for equation 7.8(3)

represent important dynamic factors and influence the stability and operation of the plant or process. In general, there is no true analog computer "model" of a transportation delay, and for most applications some form of approximation of the delay function must suffice.

The most commonly known approximations are the Padé functions. These are often unsatisfactory in process and plant simulations, and other empirically determined functions are used. Figure 7.8d shows an example of delay approximation, for which the time domain responses are improved above those of the Padé functions, to give more stable, damped, and consistent performance. The delay approximation is the low-pass type, with the transfer function

$$H_2(s) = \frac{3.45 - 0.345(\tau s)}{3.45 + 3.18(\tau s) + (\tau s)^2} \quad 7.8(4)$$

The transportation delay time τ may be varied by adjusting the coefficient potentiometers in the circuits in Figure 7.8d. If the delay time is a function of a simulation variable, such as the flow rate in a pipe, the potentiometers must be replaced by multipliers. The range of variation is generally limited to less than 100:1 for the type of circuit given, without rescaling or rearrangement of the gain distributions. The shortest delay is determined by the highest input gains available (or gain-integrator speed combinations).

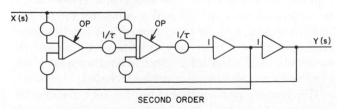

Fig. 7.8d Transport lag approximation

Control System Simulation

The simulation of instrumentation and control systems includes measuring or detecting devices, the controllers, and the actuating or manipulating output elements. Most measuring devices may be simulated by one or more simple lags for dynamic representation, with added nonlinearities for device characteristics, sensitivity, or operating point modeling. For special functions, such as logarithmic input-output relationships (pH measurements), the general purpose function generating units may be used, or the functions generated by implicit techniques. When the measuring device puts out a discontinuous (pulse type) signal, the signal-generating circuits may be used, as discussed later in this section.

A direct, three-mode (PID) controller has the transfer function

$$M(s) =$$
$$= \frac{100}{PB}\left[1 + \frac{1}{T_i s}\right]\left[1 + \frac{T_d s}{1 + T_0 s}\right][R(s) - C(s)] \quad 7.8(5)$$

where

$M(s)$ = controller output in Laplace representation,
$C(s)$ = measured (controlled) input variable,
$R(s)$ = reference (set point),
PB = proportional band (in percent),
T_i = reset (integration) time,
T_d = derivative (rate) time, and
T_0 = stabilizing (filtering) time constant.

The filter time constant is usually made as small as possible and is often a fraction of the derivative time T_d (such as $T_d/16$).

Figure 7.8e shows a simulation circuit for the three-mode controller, with the initial value M_0 corresponding to the controller output in "manual."

The controller transfer function in equation 7.8(5) is of the interacting type, which is the common case for most industrial applications. A mathematically noninteracting controller is expressed by

$$M(s) =$$
$$= \left[\frac{100}{PB} + \frac{1}{T_i s} + \frac{T_d s}{1 + T_0 s}\right][R(s) - C(s)] \quad 7.8(6)$$

in which all the three modes can be adjusted independently.

In general, the proportional band should be an *overall* adjusting effect, which leads to the practical noninteracting controller with the transfer function

$$M(s) =$$
$$= \frac{100}{PB}\left[1 + \frac{1}{T_i s} + \frac{T_d s}{1 + T_0 s}\right][R(s) - C(s)] \quad 7.8(7)$$

with the circuit arrangement shown in Figure 7.8f.

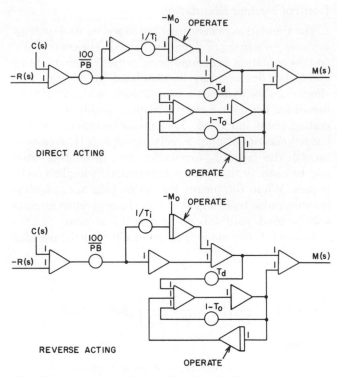

Fig. 7.8e Interacting PID controller described by equation 7.8(5)

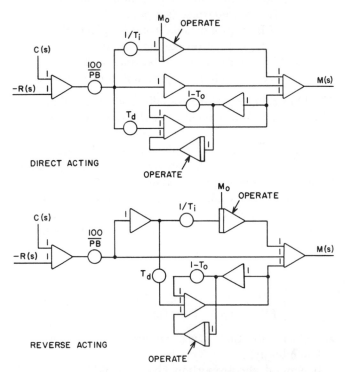

Fig. 7.8f Practical noninteracting controller described by equation 7.8(7)

Modern controllers often limit the derivative action to respond only to measured variable changes, and not to rapid set-point changes, which could upset the process. This is stated in the modified transfer function

$$M(s) =$$
$$= \frac{100}{PB}\left[1 + \frac{1}{T_i s}\right]\left[R(s) - \left(1 + \frac{T_d s}{1 + T_0 s}\right)C(s)\right] \quad 7.8(8)$$

which is of the interacting kind.

Simulation with DDC

Direct digital control may be simulated with a controller circuit as shown in Figure 7.8g, where the measured variable C(s) and the set point R(s) are sampled at fixed intervals Δt and held in track-store units. The controller algorithm in this case is

$$\Delta m = \frac{100}{PB}\left[\left(1 + \frac{\Delta t}{T_i}\right)(r_n - c_n) - r_{n-1} + \\ + c_{n-1} - \frac{T_d}{\Delta t}(c_n - 2c_{n-1} + c_{n-2})\right] \quad 7.8(9)$$

with Δm the change of the output computed in the time interval. This control corresponds to the one given in equation 7.8(8), with derivative action responding to the measured variable only. The track-store operation is controlled by the pulse P occurring once during each time interval Δt to produce an output pulse Δm of amplitude (height) corresponding to the desired change in the control variable.

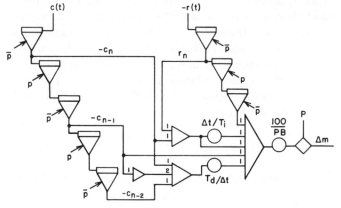

Fig. 7.8g Simulation of DDC based on equation 7.8(9)

Control Valve Simulation

The most important actuating or manipulating output element is the control valve, which may have several distinct functional aspects in an analog computer simulation. The flow characteristics can generally be linear, equal percentage, quick opening, or butterfly type. Except for the linear valve, the flow characteristics must be generated by a special analog computer circuit (by implicit techniques), or programmed in a function-generating unit, for a true representation (see Chapter IV for details). Additional effects such as limiting or backlash in valve stem movements must be included.

The dynamic performance of the control valve may be represented by a time lag of first or second order, with a limited velocity in stem movement. The time lags are simulated by circuits that were described under transfer functions, lags, and lead-lags. Velocity limiting may be expressed by the equation

$$\frac{dm_0}{dt} = \text{LOW}\left(m_L; \frac{dm_i}{dt}\right) \qquad 7.8(10)$$

where m_0 is the stem position, m_L the velocity limit, and m_i the input stem position of an ideal, unconstrained valve. The simulation circuit for velocity limiting is shown in Figure 7.8h.

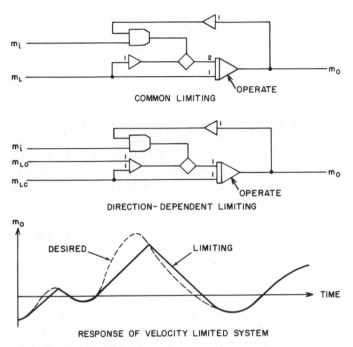

COMMON LIMITING

DIRECTION-DEPENDENT LIMITING

RESPONSE OF VELOCITY LIMITED SYSTEM

Fig. 7.8h   Velocity-limiting circuits and response

Process Reactor Models

The reactor is the heart of the chemical process. The rest of the process consists of the reactant preparation hardware and of the devices required for the separation and purification of the products or materials produced in the reactor. Thus, the plant efficiency hinges on the efficient operation of the reactor. Process reactors are known by different names in different processes: electrolytic cells, reformers, cracking furnaces, and catalytic crackers are some examples.

A mathematical model of the process reactor is the group of equations that describe the operation of the reactor. To write these equations requires a knowledge of thermodynamics, fluid dynamics, heat transmission, molecular diffusion, reaction kinetics, control theory, and economics. However, with the aid of digital and analog computers, the control engineer need only to rely on calculus and a command of differential equations to use these equations in simulation.[1]

If writing the equations that describe the reactor is a science, using these equations to simulate the reactor on a computer is an art. As any art, it must be learned through practice. The detailed mathematical model of a process reactor will often consist of a large number of nonlinear equations, which cannot be practically solved in a computer. The art of simulation consists of picking out those equations that are relevant to the objectives of the simulation and of dropping those terms that contribute little to the answers. In addition, the engineer must determine what variables are to be solved for from what equations.

The type of process reactor determines the type of mathematical model. The models describing stirred tank reactors and plug-flow reactors will be presented in this section after a brief introduction to reaction kinetics. Following this discussion the modifications necessary to represent nonideal flow, heterogeneous reactions, and catalytic reactors will be presented.

Reaction Kinetics

Reaction kinetics deals with the rate of reaction with respect to time. This rate depends on the concentration of the reactants and on temperature. For gas phase reactions the concentrations of the reactants are proportional to the pressure, and therefore the pressure also affects the rate of reaction. Consider the reaction

$$aA + bB \rightleftharpoons qQ + sS \qquad 7.8(11)$$

which means that a moles of A react with b moles of B to produce q moles of Q and s moles of S. A and B are therefore the reactants and Q and S are the products. The arrow pointing to the left indicates that the reaction is reversible, and if the reaction goes on long enough, it will reach equilibrium when the rate of reaction to the right equals the rate of reaction to the left. If the reaction is irreversible, the reaction will stop when either reactant is totally consumed.

For homogeneous reactions the rate of reaction is expressed in terms of the rate of appearance of one of the reactants per unit volume of reacting mixture.

$$r_A = \left(\frac{1}{V}\right)\frac{dN_A}{dt} \qquad 7.8(12)$$

A homogeneous reaction is one that takes place in a single phase, whether liquid, solid, or gas. In equation 7.8(12), r_A is the rate of appearance of reactant A per unit volume, and N_A is the total moles of reactant A present. When A is disappearing, r_A must be a negative quantity. The rates of appearance of the other reactants and products can be expressed in terms of r_A since, from equation 7.8(11), for each mole of A that disappears b/a moles of B will also disappear, and q/a moles of Q will appear, etc.

The equations that express the rate of reaction r_A in terms of the concentrations of the reactants and products are called the *kinetic model* of the reaction. This model is obtained by proposing different models for the reaction and then choosing the one that can best predict the reaction rate data taken in the laboratory or pilot plant. Simulation with the laboratory apparatus is an excellent tool for testing the different models. A model for the reaction represented by equation 7.8(11) could be

$$r_A = -k \left(C_A^\alpha C_B^\beta - \frac{1}{K} C_Q^\gamma C_S^\sigma \right) \qquad 7.8(13)$$

where C_A and C_B are the concentrations of the reactants in moles per unit volume, and C_Q and C_S are the concentrations of the products. The reaction coefficient k is a function of temperature and must be determined experimentally. The equilibrium constant K is also a function of temperature and can be determined from thermodynamic data. The second term inside the parentheses in equation 7.8(13) is the rate of the reverse reaction and is zero for irreversible reactions. The exponents α, β, γ, and σ are part of the kinetic model. When they are equal to a, b, q, and s, respectively, the reaction is said to be elementary. If the exponents differ from the coefficients of equation 7.8(11), the reaction is not elementary, meaning that the actual mechanism of the reaction consists of a series of intermediate elementary reactions that are summarized by equations 7.8(11) and 7.8(13).

The reaction rate coefficient k is usually an exponential function of the absolute temperature expressed by the Arrhenius equation:

$$k = k_0 e^{-(E/RT)} \qquad 7.8(14)$$

where E is called the activation energy of the reaction, R is the ideal gas law constant (1.98 BTU/lb-mole °R or 1.98 cal/gm-mole °K), and T is the absolute temperature of the reacting mixture. The constants E and k_0 must be determined experimentally by conducting the reaction at different temperatures. A kinetic model is essential for the simulation of a process reactor, and without it the simulation is worthless. However, the kinetic model[2] does not have to be as accurate for control simulation as it would have to be for the design of the reactor.

Perfectly Mixed Reactor

A flow diagram for a stirred tank (perfectly mixed reactor) is shown in Figure 7.8i. The reactor shown is a continuous one, since the reactants flow continuously into it and the product stream is continuously withdrawn. The mathematical model is essentially the same for a batch reactor in which the reactants are added at the beginning of the batch, "cooked" for a certain period of time, and then discharged.

As we develop the mathematical model, we shall make repeated use of the accounting equation:

Rate of accumulation = (rate in) − (rate out)   7.8(15)

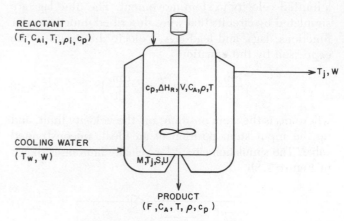

Fig. 7.8i   Perfectly mixed reactor

This equation is applied to any quantity that is conserved, such as total mass, energy, momentum, mass of components that do not participate in the reaction (i.e., solvent), etc. If we include the rate of appearance by chemica reaction as one of the "rate in" terms, we can also write a balance on mass or moles of each of the reactants and reaction products. When applied to mass, equation 7.8(15) is called the *mass, or material balance;* if to energy, then *energy balance,* and so on.

A total mass balance on the reactor in Figure 7.8i is

$$\frac{d}{dt}(V\rho) = F_i \rho_i - F\rho \qquad 7.8(16)$$

where d/dt represents the derivative, or instantaneous rate of change, with respect to time of the mass of reacting mixture, which is the product of its volume V and its density ρ. F_i and ρ_i are the volumetric rate of flow and the density, respectively, of material fed into the reactor, and F is the volumetric rate of flow of material out of the reactor. If the density of the reacting mixture is not changed appreciably by the reaction ($\rho = \rho_i$), then equation 7.8(16) can be further simplified:

$$\frac{dV}{dt} = F_i - F \qquad 7.8(17)$$

A material balance on reactant A is given by

$$\frac{d}{dt}(VC_A) = F_i C_{Ai} + r_A V - FC_A \qquad 7.8(18)$$

where C_{Ai} is the concentration of reactant A in the input stream and C_A is the concentration of A in the reactor and, since it is a perfectly mixed reactor, in the output stream. Similar balances can be made on each of the reactants and reaction products. Differentiation by parts of the left-hand side of equation 7.8(18) gives

$$\frac{d}{dt}(VC_A) = V\frac{dC_A}{dt} + C_A\frac{dV}{dt} \qquad 7.8(19)$$

and from equation 7.8(17),

$$\frac{d}{dt}(VC_A) = V\frac{dC_A}{dt} + C_A(F_i - F) \qquad 7.8(20)$$

Substituting into equation 7.8(18) and rearranging,

$$\frac{dC_A}{dt} = r_A + \frac{F_i}{V}(C_{Ai} - C_A) \qquad 7.8(21)$$

The ratio F_i/V is the reciprocal of the residence time of the reactor, which is also the reactor time constant.

If, as illustrated by the upper portion of Figure 7.8j, there were two inlet streams (F_{iA} and F_{iB}), with corresponding concentrations (C_{iA} and C_{iB}), the flow and concentration equations can be written as

$$dV/dt = F_{iA} + F_{iB} - F \qquad 7.8(22)$$

$$V\frac{dC}{dt} = F_{iA}C_{iA} + F_{iB}C_{iB} - FC \qquad 7.8(23)$$

Then the analog circuit to simulate equations 7.8(22) and 7.8(23) is shown in Figure 7.8j.

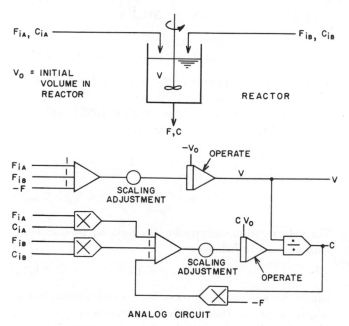

Fig. 7.8j Simulation circuit for stirred-tank reactor

An energy balance on the reactor, illustrated in Figure 7.8i gives

$$\frac{d}{dt}(V\rho C_p T) = F_i\rho C_p T_i + Vr_A(-\Delta H_R) - $$
$$- US(T - T_j) - F\rho C_p T \qquad 7.8(24)$$

where T is the temperature of the reacting mixture, C_p is its specific heat or amount of energy necessary to raise a unit mass by one degree of temperature, T_i is the temperature of the input stream, $-\Delta H_R$ is the heat liberated in the reaction per unit mass of A reacted, T_j is the temperature of the cooling water in the jacket, S is the effective heat transfer surface area of the jacket, and

U is the heat transfer coefficient or rate of heat transfer to the jacket per unit area per degree of temperature difference. U is either experimentally determined or estimated from heat transfer correlations.

Another simple application of differential calculus, with the help of equation 7.8(16), reduces equation 7.8(24) to

$$\frac{dT}{dt} = \frac{F_0}{V}(T_i - T) + \frac{-\Delta H_R}{\rho C_p} r_A - $$
$$- \frac{US}{V\rho C_p}(T - T_j) \qquad 7.8(25)$$

An energy balance on the jacket gives us an equation for T_j:

$$\frac{dT_j}{dt} = \frac{W}{M}(T_W - T_j) + \frac{US}{M}(T - T_j) \qquad 7.8(26)$$

where W is the mass flow rate of cooling water, M is the heat capacity of the water in the jacket and the walls of the reactor, and T_w is the cooling water inlet temperature. It has been assumed that the jacket is also perfectly mixed and that the specific heat of water is unity.

The pressure in the reactor can be calculated, if necessary, from thermodynamic equilibrium relationships, as a function of temperature and concentration. Equations 7.8(17), (21), (25), and (26), coupled with the reaction kinetic equations similar to 7.8(13) and (14), constitute the mathematical model of the system. The volume V, concentrations C_A, C_B, etc., and temperatures T and T_j are the dependent variables, since they depend on time, which is the independent variable. At steady state the dependent variables do not change with respect to time, and therefore the derivatives are equal to zero. In addition to the equations shown, the initial values of each of the dependent variables are necessary to complete the model (Figure 7.8j). The equations listed are ready to be programmed on an analog computer or to be solved digitally by either numerical integration techniques or any of the so-called simulation languages, without any further manipulation.

Plug-Flow Reactor

The flow diagram of a plug-flow reactor is shown in Figure 7.8k. The term "plug flow" arises from the assumption that each element of fluid flows through the reactor as a small "plug," without mixing with the fluid behind or ahead of it. It is also assumed that the concentrations and temperatures are uniform across the cross-sectional area of flow. Cracking furnaces, with very high

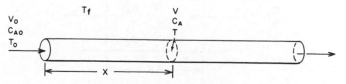

Fig. 7.8k Plug-flow reactor

velocities inside the tubes, approach this condition of plug flow.

Whereas compositions and temperatures are functions of time in the perfectly mixed reactor, in the plug-flow reactor they are functions of time and distance X from the entrance. Thus there will be derivatives with respect to time at each point in the reactor, and derivatives with respect to distance at each instant of time. These are partial derivatives, and the equations they are part of are partial differential equations. To solve a partial differential equation on the analog computer, it must be approximated by a number of ordinary differential equations. It is easier to derive these equations directly by the method that will be shown here.

The reactor is divided into a number of small "pools" that are assumed to be perfectly mixed. As shown in Figure 7.8l, the length of each pool is ΔX and its volume $A\Delta X$, where A is the cross-sectional area of flow. The volumetric rate of flow is equal to the velocity v times the cross-sectional area of A. A magnified sketch of pool number i is shown in Figure 7.8m. Assuming a gas phase reaction, a material balance can be written around each of the pools to give

$$\frac{d}{dt}(A\,\Delta X\,\rho_i) = A(v_{i-1}\rho_{i-1}) - A(v_i\rho_i) \qquad 7.8(27)$$

where ρ_i is the density of the reacting mixture, which can be expressed in terms of the temperature T, pressure P, and average molecular weight M_{av} by the ideal gas law:

$$\rho_i = M_{avi}P_i/RT_i \qquad 7.8(28)$$

or in terms of any other equation of state. R is the ideal gas constant, and M_{avi} can be calculated as a function of the composition of the mixture in the pool. Equation

7.8(27) can be simplified, since A and ΔX are constants, to give

$$\frac{d\rho_i}{dt} = \frac{1}{\Delta X}(v_{i-1}\rho_{i-1} - v_i\rho_i) \qquad 7.8(29)$$

A material balance on reactant A will give us

$$\frac{d}{dt}(A\,\Delta X\,C_{Ai}) = A(v_{i-1}C_{A,i-1}) + \\ + A(\Delta X\,r_A) - A(v_iC_{Ai}) \qquad 7.8(30)$$

where C_{Ai} is the concentration of reactant A in the ith pool. Again we can simplify the equation by dividing out $A\Delta X$:

$$\frac{dC_{Ai}}{dt} = r_A + \frac{1}{\Delta X}(v_{i-1}C_{A,i-1} - v_iC_{Ai}) \qquad 7.8(31)$$

Similar equations can be written for each of the other reactants and reaction products.

An energy balance on the ith pool gives us

$$\frac{d}{dt}(A\,\Delta X\rho_iC_pT_i) = A\,\Delta Xr_A(-\Delta H_R) + \\ + A(v_{i-1}\rho_{i-1}C_pT_{i-1}) + \\ + Up\,\Delta X(T_f - T_i) - A(v_i\rho_iC_pT_i) \qquad 7.8(32)$$

where p is the perimeter of the pipe, so that $p\Delta X$ is the heat transfer area of one pool and T_f is the firing box temperature. With the help of equation 7.8(27) and some calculus, this last equation becomes

$$\frac{dT_i}{dt} = \frac{-\Delta H_R}{\rho_iC_p}r_A + \frac{Up}{A\rho_iC_p}(T_f - T_i) + \\ + \frac{v_{i-1}\rho_{i-1}}{\rho_i\Delta X}(T_{i-1} - T_i) \qquad 7.8(33)$$

The velocity from each pool to the next can be calculated with a pressure balance:

$$P_i - P_{i+1} = k_fv_i^2 \qquad 7.8(34)$$

where the constant k_f is proportional to the friction factor, which can be calculated with fluid dynamics correlations. The pressure can be calculated from equation 7.8(28).

Equations 7.8(28), (29), (31), (33), and (34) must be solved for each one of the pools. Thus an increase in the number of pools requires a greater number of analog computer components or an increase in the time required to simulate the reactor on the digital computer. The greater the number of pools, the closer we approach the partial differential equations that describe the plug-flow reactor.

Nonideal Flow

Process reactors deviate somewhat from the ideal flow conditions of perfectly mixed and plug flow. A stirred tank reactor may contain "pockets" of fluid in which the concentration of reactants is different from that in other

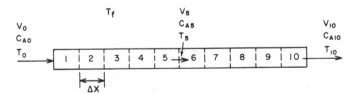

Fig. 7.8l   "Pool" model of plug-flow reactor

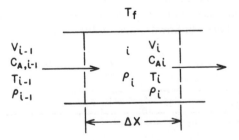

Fig. 7.8m   Pool number i

parts of the reactor. The way to model this type of non-ideality is to divide the reactor into two or more "pools" and apply the material balance to each of the pools. The sum of the volumes of the pools must be equal to the volume of the reactor. If backflow is allowed for from a pool to the one behind it, the rate of recirculation between the pools must be assumed. As this rate is increased, the results of the divided reactor model approach those of the perfectly mixed reactor model. If no backflow is allowed for, as the number of pools is increased, the model approaches that of the plug-flow reactor.

Another type of nonideal flow is created by a tubular reactor in which the velocity, temperature, and concentrations vary across the radius of the tube. If axial symmetry is present, each of the pools representing the reactor can be divided into a number of "rings," and the accounting equations can be written for each of the rings. It is easy to see how the equipment and time requirements for the simulation increase geometrically with the number of pools and rings used. For the transfer of heat and mass by conduction, diffusion and eddy currents between two adjacent rings, appropriate flow models, and heat and mass transfer equations must be used.

Heterogeneous Reactions

When a reaction takes place in more than one phase, it is called heterogeneous. This occurs when the reactants cannot mix completely to form a single phase, that is, when they are immiscible or partially miscible. In heterogeneous reactions, the concentration of reactants and products depends not only on the rate of reaction but also on the rate at which they diffuse from one phase to the other. For fast reactions and slow diffusion rates, the rate of diffusion will have greater influence on the rate at which the reaction occurs than the kinetics of the reaction. In general, material and energy balance equations must be written for each phase and the appropriate diffusion equations used for the transfer of reactants and products between the phases. In addition, equilibrium solubilities and phase equilibrium relationships must be considered.

Catalytic Reactions

A catalytic reaction is one conducted in the presence of a catalyst. A catalyst is a material that speeds up the reaction without being consumed or produced by it. The reaction rate coefficient k is then a function of the temperature, catalyst concentration, and "age." This age is the accumulated time since the catalyst was replaced or regenerated. These functions must be determined experimentally and incorporated into the model. When the catalyst is a solid, the additional complications of heterogeneity that were discussed in the previous paragraph are also present. Sometimes the equations can be sim-

plified considerably by expressing the rate of reaction on the basis of unit mass of catalyst instead of on the basis of unit volume. This is true for fixed catalyst bed reactors.

Conclusions

Presented in this section are methods for developing the equations that describe process reactors. Since the reactor is only a part of the control loop, for the simulation to be complete, it must include the equations that represent the temperature, level, pressure, and concentration sensors, transmission lines, controllers, and control valves. One of the advantages of simulating the reactor on an analog rather than a digital computer is that actual process controllers can be used and tuned in the simulation. This advantage applies to the digital computer if a digital computer is to be used to control the reactor.

A reactor control simulation allows the control engineer to tune the controllers without any loss of production or danger of blowing up the plant. It also provides the perfect tool to train plant operators for smoother and safer startups, it serves as a "live" model with which to try new control ideas that will result in safer and more efficient operation of the reactor, and it gives an understanding of the behavior of the reactor equivalent to several years of reactor operation. This latter advantage derives from the ability to look at variables in the simulation that are impossible or impractical to measure in the process reactor.

REFERENCES

1. Murrill, Pike, and Smith, *Formulation and Utilization of Mathematical Models*, International Textbook Co.
2. O. Levenspiel, *Chemical Reaction Engineering: An Introduction to the Design of Chemical Reactors*, New York: John Wiley & Sons, 1962.

BIBLIOGRAPHY

Adams, J., et al., "Mathematical Modeling of Once-Through Boiler Dynamics," IEEE Winter Power Meeting, Paper 31 TP65-177, 1965.
Chaussard, R., and Grawozel, J., "Large Boilers Dynamic Analysis and Control Systems Adjustment," ISA Proceedings of the Eighth National Power Instrumentation Symposium, Vol. 8, 1965.
deMello, F.P., "Plant Dynamics and Control Analysis," *IEEE Transactions*, Vol. S82, 1963.
Hamell, E., "Modeling and Simulation," *InTech*, November, 1982.
McDonald, J.P., Kwatny, H.G., "Design and Analysis of Boiler-Turbine-Generator Controls Using Optimal Linear Regulatory Theory," *JACC*, 1972.
———, ———, and Spare, J.H., "A Non-Linear Model for Reheat Boiler-Turbine-Generator Systems, Parts I and II," *JACC* 1971.
Mostafa, E.S., "An Operator Training Simulator," *Instrumentation Technology*, January, 1982.
Simpkin, L.J., and Wooden, L.F., "Use of Analog Computer in Investigations of Power Plant Automatic Control Systems," ISA Proceedings of the Ninth National Power Instrumentation Symposium, Vol. 9, 1966.

7.9 PNEUMATIC COMPUTERS

Analog Components

Analog computers solve systems of equations by means of networks in which continuous signals that are "analogous" to problem variables are interconnected in such a way that the network behaves like the real-world event described by the equations.

Network components consist of:

1. Linear computing elements, including amplifiers (inverting and noninverting), adders, subtractors, ratioing devices, integrators, and differentiators.
2. Nonlinear computing elements, including multipliers, dividers, and function generators.
3. Supporting systems: mode control systems, interconnection system, output selection and measurement system, and trouble warning systems.
4. Readout systems: indicators and recorders.

The interconnection of the network devices is the "program." It is always arranged specifically for each problem.

Implementation Technologies

Analog computers can be implemented using electronic or pneumatic components. Their operation may also be simulated on digital computers.

Electronic analog computers are the most common form. The availability since the early 1970s of high-quality, low-cost operational amplifiers has given electronics a substantial advantage in terms of cost, simplicity, and performance.

Since the mid-1970s, the use of digital microprocessors to solve the requisite equations has supplanted the use of analog computers in many applications. The performance of these systems is generally adequate for process control applications. The cost and simplicity of microprocessor systems has afforded them a substantial advantage over all competing technologies, and their share of applications will probably increase.

Pneumatic implementations of analog computers have been commonplace since about 1960, when accurate multipliers and dividers became available. Pneumatic systems are frequently appropriate when the required network is not complex and the conversion from and to pneumatics to enable computation by an electronic analog or a digital computer would result in a more complex total system. While electronic analog and digital computers have taken over applications for which pneumatic analog computers were the best, and possibly only, choice in the past, they have also opened possibilities for hybrid systems.

The most intriguing hybrid approach is one in which digital computers and pneumatic analog computers are used in consort. Pneumatic computers can be used at the process level to solve differential equations or systems of equations that could be time-consuming for a hierarchical process control computer. The superior environmental and interfacing characteristics of the pneumatic components could be used to advantage by locating the pneumatic computer in the field near the signals and actuators, thus saving signal conversion and transmission cost and complexity. The pneumatic analog computer can then exchange information with the digital process control computer to establish operating points or process data indices. Effectively used, it could be possible to significantly reduce the load on a process control computer in some cases. Just as important, it may also be possible to improve the reliability of the overall system by distributing some or all of the mathematics to a lower hierarchical level that, hopefully, will be available so that sophisticated control concepts can be carried out even when the digital computer is unavailable.

Application Qualification

The successful application of pneumatic analog computers depends heavily on qualifying the application carefully. The input and output signals should need to be pneumatic. The equations to be solved must either be simple or be ones that an analog computer would normally handle more advantageously than a digital computer. For example, if solving a differential equation would take too much time on a microcomputer using numerical methods, a pneumatic computer might be an alternative. A second example is in situations where the environmental tolerance of pneumatic instruments would be of value in solving an equation at a particular location, such as near transmitters and valves in a hot, humid, hazardous area. Areas in which strong electromagnetic fields may exist, such as in nuclear equipment, may also be candidates.

Poor applications for analog instruments of either pneumatic or electronic operation are those in which large numbers of function modules must be interconnected, because the difficulty in connections and calibration can quickly become significant. Applications in which solution points are not easily differentiated (those involving small signals or small differences) are also bad applications since the inaccuracies and instabilities of analog components can become significant.

Application Development

Solving a problem on an analog computer begins with a definition of the problem. From the definition, a network diagram is constructed that indicates, by use of the components available, how the equations representing the problem can be solved. The potential application should then be qualified, considering the alternatives, as reasonable for an analog computer in general and a pneumatic analog computer in particular. Considerations should include computational speed, scaling, and accuracy.

Time Scaling

It is necessary to determine if a solution can be developed in sufficient time to be useful. It is frequently necessary to run an analog computation at a rate different than real time (the time at which the process takes place). In most process control applications it is desirable to run the computer at real time. Occasionally, it is advantageous to run the computation cycle at a rate faster than real time as a way of predicting outcome of developing situations. There is seldom a performance-related incentive to run the computation slower than real time, but doing so can be a necessity. The frequency response of pneumatic computing elements is quite low. If signal rates of change required during solution of an equation exceed the bandwidth, amplitude distortion and phase lags are introduced. The speed at which signals can change must be reduced if accuracy is to be maintained.

If time scaling is necessary, it is essential to evaluate whether the solution speed that can be obtained will be of any value in the application.

Magnitude Scaling

All process variables are generally represented by signals varying between 3 and 15 PSIG (0.2 and 1.0 bar), although some equipment is available that will allow signal spans in excess of 100 PSIG (690 kPa). In any case, it is essential that the signals be scaled so the information they contain is spread over the usable range. If they are not, the resolution of the computing elements can become a significant accuracy detractor. For example, if a

pressure used in a computation can vary from 80 to 100 PSIG (552 to 690 kPa) in a particular problem but the transmitter is spanned 0 to 200 PSIG (0 to 1380 kPa), a significant source of error may exist. The range of interest is the 20 PSI (138 kPa) between 80 and 100 (552 to 690 kPa). The range of the transmitter and hence the signal is 0 to 200 PSIG or 200 PSI (0 to 1380 kPa or 1380 kPa). Only 10 percent of the available range is actually used to represent the data of interest.

It is also essential that the signals "make sense" in the context of the problem. For example, assume a certain calculation requires the division of a flow signal by absolute temperature. The signal from the transmitter may have a range of 0 to 100°F (0 to 55°C), however. The divider must be "scaled" to recognize that 0 percent scale is not absolute zero but 460°R (256°K) on a scale of 460–560°R (256–311°K). The denominator of the divider should vary between 460/560 and 1 as the temperature signal moves from 0 to 100 percent of scale or the calcuation will be in error.

Scaling is simplified if calculations are performed on the basis of a 0 to 1.00 span instead of the signal range (3–15 PSIG or 0.2–1.0 bar) or a 0–100 percent span. The problem of carrying the live zero (3 PSIG or 0.2 bar) through scaling calculations simply isn't worth the effort since it has practical value only in the operation of the instruments, not in the problem's mathematics. The value of using 1.00 as span vs. using 100 is seen by considering the operation of a square root extractor (or almost any function module). The square root of span is 1.00 in the case of the 0 to 1.00 span instrument, but it is 10 in the case of the 0 to 100 span instrument. In addition to being confusing, it is easy to complicate scaling problems if a 0–100% span definition is used.

Many applications require compensating process measurement for variations in environmental conditions. An example would be converting a measurement of product density at a variable temperature to specific gravity at a reference temperature. The first step is to model the process from known data, to determine how density varies with temperature, i.e., whether multiplication, division, addition, or subtraction is required. Then the computer must be scaled so that no compensation is applied at reference temperature. In the case of addition or subtraction, the compensating term is 0 at reference conditions, and when multiplying or dividing is required, the compensating term is 1.00 at reference conditions.

Scaling must be checked for intermediate variables as well as inputs and outputs. All intermediate signals should be within the capability of the instruments and large enough so their accuracy is not seriously degraded.

Accuracy Adequacy

Once it is determined that a particular problem can be successfully scaled, an error analysis should be conducted. Accuracy estimation is discussed later in this section. It is important that such an estimate be made and a determination made regarding the usefulness of the accuracy available.

Pneumatic Computing Functions

The devices performing these functions are all described in Section 2.1, but the functions themselves are elaborated on in detail here as a guide to their proper application.

Adding and Subtracting

A device capable of summing is usually also capable of subtraction and averaging, depending on the arrangement of inputs. A general formula for a summing relay would be

$$D = K(A - C) + B \qquad 7.9(1)$$

where A, B, and C are input signals, D is the output, and K is an adjustable gain. Alternatively, one or two of the inputs may be replaced by a spring to apply a constant bias force. To generate a linear model:

$$D = KA + b \qquad 7.9(2)$$

Input C would be replaced with a zero spring and input B replaced with a bias spring b.

Averaging is accomplished by feeding back output D into output C:

$$D = K(A - D) + B \qquad 7.9(3)$$

Solving for D:

$$D = \frac{KA + B}{K + 1} \qquad 7.9(4)$$

Summing relays are often used for adding and subtracting flow signals. For this application, attention must be given to the ranges of inputs and outputs. If adding flows of 0–100 and 0–200 gpm (0–0.38 and 0–0.76 m³/m), for example, should the output scale be 0–200 or 0–300 gpm (0–0.76 or 0–1.13 m³/m)? This depends on whether the sum *can* exceed 200 gpm (0.76 m³/m) in actual practice. If not, greater accuracy will be achieved using the smaller scale, by applying equation 7.9(1) with input C as zero and input B as the 0–200 gpm (0–0.76 m³/m) signal. Otherwise the averaging equation 7.9(4) should be used, whose output scale is the sum of the input scales, or 0–300 gpm (0–1.13 m³/m).

A similar problem may develop when subtracting flow signals. Assume the 0–200 gpm (0–0.76 m³/m) flow is to be subtracted from the 0–300 gpm (0–1.13 m³/m) flow. An output scale of 0–300 gpm (0–1.13 m³/m) is possible using equation 7.9(1) with A as zero. However, if the difference never exceeds 100 gpm (0.38 m³/m), greater accuracy will be achieved by applying the 0–300 gpm (0–1.13 m³/m) signal to both A and B. Then,

$$D = (1 + K)B - KC \qquad 7.9(5)$$

The output scale is then the difference between the input scales. This arrangement was used in the selective control system shown in Figure 1.8e.

Multiplying and Dividing

Pneumatic multipliers typically have only two inputs, but several adjustable coefficients are available to facilitate scaling. The general formula for a multiplier is

$$A = a + f(B - b)[C(1 - c) + c] \qquad 7.9(6)$$

Coefficients a, b, and c are bias or zero adjustments for the three signals, f is the gain of the device with both inputs at 100 percent, and 1 − c is the span of the C input. (For calibrating these devices refer to the last paragraph of this section.) The formula for a divider is found by solving equation 7.9(6) for signal B:

$$B = b + \frac{A - a}{f[C(1 - c) + c]} \qquad 7.9(7)$$

The linear relationship between absolute and Fahrenheit temperatures from the example cited earlier may be solved prior to dividing, by selection of coefficient c as 460/560.

Arbitrary Functions

An arbitrary function is a particular nonlinear relationship between two terms. The most common nonlinear function is the square root. It can be approximated by the motion of an angular mechanical linkage, or by feeding back the output of a divider into the denominator. If B is substituted for C in equation 7.9(7) and a, b, and c are zero and f is 1.0, then

$$B = A/B \qquad 7.9(8)$$

or

$$B = \sqrt{A} \qquad 7.9(9)$$

Similarly, a square function can be made by applying the input to both B and C of a multiplier:

$$A = B^2 \qquad 7.9(10)$$

Curves that do not pass through points (0,0) and (1,1) can also be modeled by multipliers and dividers by using coefficients a, b, c, and f in equations 7.9(6) and 7.9(7). Three types of curves that can be modeled in this way are shown in Figure 7.9a. Curve 1 is a polynomial with positive coefficients, i.e., a positive slope increasing with the input signal. Using equation 7.9(6) for a multiplier and substituting B for C,

$$A = a + f(B - b)[B(1 - c) + c] \qquad 7.9(11)$$

Let b = 0 and expand into the familiar polynomial

$$A = a + fcB + f(1 - c)B^2 \qquad 7.9(12)$$

Curve 2 has a positive but decreasing slope and can be modeled by substituting B for C in divider equation

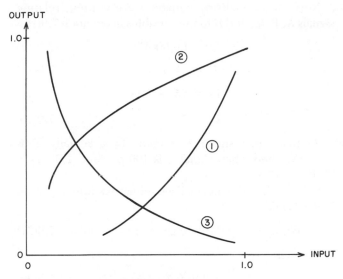

Fig. 7.9a Typical nonlinear functions that can be generated with a multiplier or divider

7.9(7). Its formula is identical to equation 7.9(12) except that A is the input and B is the output. Curve 3 is a hyperbola, described by divider equation 7.9(7), whose input is C, with numerator A fixed.

Cams or four-bar linkages can also be used to generate arbitrary functions, but they lack the flexibility obtainable with multipliers and dividers.

Nonlinear functions modeled by connecting segments of lines can be generated using the selective devices described in Section 2.1. One or more linear functions can be generated with summing devices, and the selectors are then used to switch from one summer to another at the point where these lines intersect. Horizontal segments of the function can be developed by limiters, which are selective relays with one fixed input.

Dynamic Functions

Dynamic functions commonly applied to pneumatic signals include integrating, differentiating, lag, and lead-lag. The lead-lag function was described in detail in Section 1.5, since it is used extensively in feedforward control. The lag function is actually a special case of lead-lag with zero lead time. It, too, is used in feedforward control, and also to filter or dampen fluctuating signals.

Integration is the reset function of a controller. In fact, a pneumatic controller can be used to perform the integrating function by using the pressure in the reset bellows as the output, as shown in Figure 7.9b.

$$D = \frac{100}{(PB)T_i} \int_{t_0} (A - C)dt + B \qquad 7.9(13)$$

Here, input B is the "initial condition" of the integrator, i.e., its output D at time zero. When the block valve admitting signal B is closed, integration begins at a rate determined by the proportional band PB and reset time

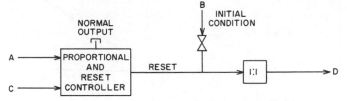

Fig. 7.9b A pneumatic integrator made from a controller

T_i of the controller. The proportional band should be fixed at a high value (500 percent) to avoid saturating the normal output; otherwise inaccuracy will result.

The output of a pneumatic integrator deviates from linearity by about 5 percent due to the *change* in airflow through the reset restrictor with density (output pressure). The 1:1 relay is required at the output because flow cannot be withdrawn from the reset bellows without causing an error. By the same token, the initial-condition valve and associated piping must be free of leaks.

Pneumatic components have also been used for differentiation, as shown in Figure 7.9c. The most serious obstacle to obtaining accurate time differentials is that a differentiator has a very high sensitivity to noise. In fact, differentiators tend to be unstable unless their maximum gain is limited to 20 or less. The effect of this limitation is that a differentiator typically responds with a first-order lag whose time constant is the derivative time constant divided by the maximum gain. This is best expressed using Laplace transforms:

$$G(s) = \frac{\tau s}{1 + \tau_0 s/\alpha} \qquad 7.9(14)$$

The transfer function is G(s), where s is the Laplace operator, τ is the time constant (τ_0 at time t_0) of the resistance-capacity combination (RC), and α is the gain of the unit.

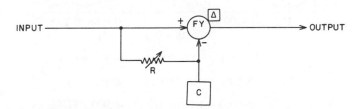

Fig. 7.9c Pneumatic differentiator made from a subtractor, a restrictor, and a capacity tank

The response of a differentiator to a ramp input is shown in Figure 7.9d. The output is superimposed on the input to demonstrate the role of the time constant τ. In actual operation, the output would be zero or some value fixed by a bias spring when the input is steady.

Scaling Procedures

Every analog computer requires scaling to insure compatibility with the input and output signal ranges. Coef-

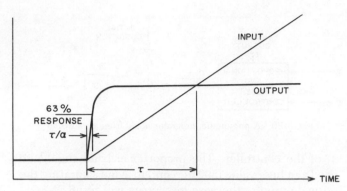

Fig. 7.9d Response of a differentiator to a ramp input

ficients a, b, c, f, K, etc. that appear in the equations of this section must be evaluated according to a carefully organized procedure, or the wrong answers will result. A simple yet highly effective procedure is:

1. Write the equation to be solved, including all conversion factors, with all signals given in the units in which they are to be measured or displayed.
2. Relate each input and output signal (having a range of 0–1.0) to the range of each variable, by a set of "normalizing" equations.
3. Substitute the normalizing equations into the original equation and solve for the output signal.

As an example, follow the application of this procedure to the calculation of the rate of heat transfer, Q, to a liquid cooling medium. The liquid is flowing at rate F, with an inlet temperature T_1, and a higher outlet temperature T_2. The conditions are given in Table 7.9e.

Table 7.9e
HEAT EXCHANGER OPERATING CONDITIONS

Signal	Variable	Range	Normal Value
B	F	0–100 gpm (0–0.38 m³/m)	60 gpm (0.23 m³/m)
A	T_2	25–75°F (−3.9–23.9°C)	50°F (10°C)
C	T_1	0–50°F (−18–10°C)	30°F (−1.1°C)
E	Q	0–60,000 BTU/hr (0–17,400 W)	48,000 BTU/hr (13,920 W)

The equation to be solved is

$$Q = Fk(T_2 - T_1) \qquad 7.9(15)$$

Coefficient k includes liquid density and specific heat, but it can be calculated from the normal operating conditions given in Table 7.9e.

$$k = \frac{48,000}{60(50 - 30)} = 40 \qquad 7.9(16)$$

Next the normalizing equations are written, relating signals A, B, C, and E to the variables in equation 7.9(15):

$$Q = 60,000E \qquad 7.9(17)$$

$$F = 100B \qquad 7.9(18)$$

$$T_2 = 25 + 50A \qquad 7.9(19)$$

$$T_1 = 50C \qquad 7.9(20)$$

Note that when signal A is zero, T_2 is actually 25°F (−3.9°C), and when signal A is 100 percent (1.0), T_2 is 75°F (23.9°C).

Substituting the normalized equation into equation 7.9(15) yields

$$60,000E = (100B)40(25 + 50A - 50C) \quad 7.9(21)$$

Solving for E,

$$E = 3.33B(0.5 + A - C) \qquad 7.9(22)$$

Equation 7.9(22) must be solved by a subtractor and multiplier in combination, as shown in Figure 7.9f. Since it is solved in two operations, equation 7.9(22) must be separated into two pieces. Let the intermediate variable be identified as D:

$$D = 0.5 + A - C \qquad 7.9(23)$$

$$E = 3.33BD \qquad 7.9(24)$$

The entire factor 3.33 in the equation is shown applied to the multiplier. This is not altogether necessary—for example, the subtractor could have a gain of 2.0 and the multiplier a gain of 1.67:

$$D = 1.0 + 2.0(A - C) \qquad 7.9(25)$$

$$E = 1.67BD \qquad 7.9(26)$$

This tends to improve the accuracy of the calculation but increases the danger of saturating the subtractor. Whenever more than one device is used in this way, each operation should be tested for saturation with reasonable combinations of inputs. In this example, a combination of 75°F (24°C) for T_2 and 0°F (−18°C) for T_1 would not be reasonable.

A shortcut method of scaling can be used for compensating computers. Consider the example where a gas flowmeter requires compensation for absolute pressure:

$$W = k\sqrt{hP} \qquad 7.9(27)$$

where W is the mass flow, h is the orifice differential pressure, and P is the absolute pressure. The orifice and

Fig. 7.9f Computer for pneumatic heat transfer calculation

differential range are selected so that 100 percent differential equals 100 percent flow at the normal operating pressure. With this in mind, scaling requires knowing only the normal pressure and the range of the pressure transmitter.

Let the normal pressure be 64.7 PSIA (446.4 kPa), and the transmitter range be 0–75 PSIG (0–517.5 kPa) with a base of 14.7 PSIA (101.4 kPa). Compensation requires that the multiplier exhibit a gain of 1.00 at the normal pressure. The absolute pressure range is 14.7 to 89.7 PSIA (101.4 to 619 kPa); therefore,

$$P = 14.7 + 75C \qquad 7.9(28)$$

Here signal C represents the pressure input to the multiplier. Then the compensating factor is

$$\frac{P}{64.7} = 0.227 + 1.160C \qquad 7.9(29)$$

To conform with equation 7.9(6), the maximum value of the compensating factor is extracted as f = 89.7/64.7 = 1.387

$$\frac{P}{64.7} = 1.387(0.836C + 0.164) \qquad 7.9(30)$$

Then the scaled equation for the multiplier is

$$A = 1.387B(0.836C + 0.164) \qquad 7.9(31)$$

Where A is the multiplier output and B and C are the orifice differential and pressure inputs, respectively. A square-root extractor following the multiplier completes the calculation.

Accuracy Estimation

Three factors must be considered before an estimate of accuracy can be made for a given computer: (1) the sensitivity of the system equation to errors in the input signals, (2) the statistical combination of individual errors, and (3) the contribution of errors in the computing components.

The first consideration requires differentiating the output of the system equation with respect to each input, with all scaling factors included. Equation 7.9(31) can be differentiated as an example:

$$dA = 1.387dB(0.836C + 0.164) + \\ + 1.387B(0.836dC) \qquad 7.9(32)$$

Differential dA is the maximum output error that would result from error dB in signal B and error dC in signal C. Note that the output error depends on signal levels as well as scaling factors.

Equation 7.9(32) ignores probability considerations, however. If both dB and dC are random errors, the resulting error dA with the same probability of occurrence would be related to them by a root-mean-square combination:

$$(dA)^2 = [1.387dB(0.836C + 0.164)]^2 + \\ + [1.387B(0.836dC)]^2 \qquad 7.9(33)$$

The error e contributed by the computing device—a multiplier in the example—is similarly added in rms fashion to equation 7.9(33):

$$(dA)^2 = [1.387dB(0.836C + 0.164)]^2 + \\ + [1.387B(0.836dC)]^2 + e^2 \qquad 7.9(34)$$

Because accuracy changes with the magnitude of the input signals, an estimate of error is valid for only one set of conditions. This is particularly true where a square root is taken:

$$A = \sqrt{B} \qquad 7.9(35)$$

$$dA = \frac{dB}{2\sqrt{B}} = \frac{dB}{2A} \qquad 7.9(36)$$

An error of 0.5 percent in differential pressure B would represent an error in flow A of 0.5 percent at 50 percent flow, 1 percent at 25 percent flow, and 2.5 percent at 10 percent flow.

When flow signals are added, they must, of course, be linear. If one of the streams is shut off, its flow signal will be zero, representing the possibility of a sizeable error. To prevent a large negative error from affecting the summation, high-select relays may be inserted downstream of each square-root extractor, as shown in Figure 7.9g. These selectors prefer a small negative error in differential to a large negative error in flow.

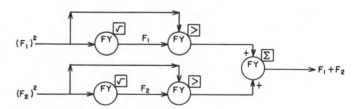

Fig. 7.9g High selectors eliminate large negative errors when flow signals are added

Calibration

Each instrument in a computing system, including indicators, recorders, and transmitters, should be calibrated against a common standard where possible. The most reliable standard to use is a mercury column, accurate to ±0.1 percent of the 3–15 PSIG (0.2–1.0 bar) range.

Calibrating multipliers and dividers is particularly painstaking because the adjustments must be made in a specific order. Referring to equation 7.9(6) for a multiplier:

1. Zero adjustment a must be made with signals B and C at zero.
2. Zero adjustment b must be made with signal B at zero and C at 100 percent.

3. Zero adjustment c must be made with signal B at 100 percent and signal C at zero.
4. Span adjustment 1—c must be made with B at some specified intermediate value and C at 100 percent.

A similar procedure must be followed with dividers.

After all components are calibrated individually, the system must be calibrated as a whole, to offset systematic (nonstatistical) errors. Almost any of the adjustments can be used to calibrate the system at a single operating point, but the wrong choice may cause a greater error at some other point. So the accuracy of the system should be evaluated at several sets of conditions to determine which of the available adjustments would minimize the average error for all sets. Often more than one coefficient may require adjustment.

BIBLIOGRAPHY

Blum, Joseph J., *Introduction to Analog Computation,* New York: Harcourt, Brace & World, 1969.

Carlson, Hannauer, Carey, and Holsberg, *Handbook of Analog Computation,* 2nd ed., Princeton, NJ: Electronic Associates, Inc., 1967.

Hannauer, George, *Basics of Parallel Hybrid Computers*, Princeton, NJ: Electronic Associates, Inc., 1969.

General symbol for recorder or printer. "xx" designates the measured variable and modifier. The letter K is substituted for R to indicate control station.

Distributed control/shared display for an indicator/controller/recorder or alarm point, usually for a video display. Access is limited to the communication link.

An auxiliary operator's interface device for distributed control, normally panel-mounted with an analog faceplate. Normally not mounted on main operator console. Can be a backup controller or manual station.

This is a computer symbol, usually a video display, normally accessible to the operator, which is an indicator/controller/recorder or alarm point in a distributed control scheme.

7.10 TERMINALS AND PRINTERS

Terminals and printers cover several types of devices that are more properly referred to as operator interfaces. Terminal consoles and other devices, such as controller faces, by which the operator communicates with the process are either covered in other sections or require individual systems engineering. Summary information here is provided for printers only.

Types of Printers:	A—Serial
	1—Full character impact
	2—Matrix image impact
	3—Electrographic
	4—Thermal matrix
	5—Ink jet
	B—Line
	1—Matrix image impact
	2—Full character impact
	3—Electrostatic
	4—Xerographic
Speed:	A—Characters per second
	1—10–100
	2—100–900
	3—15–200
	4—15–100
	5—50–50,000
	B—Lines per minute
	1—200–500
	2—300–2,000
	3—300–20,000
	4—4,000–18,000
Cost:	A1—$1,000–$7,000
	A2—$1,000–$4,000
	A3—$500–$3,000
	A4—$500–$5,000
	A5—$5,000–$500,000
	B1—$5,000–$15,000
	B2—$3,000–$75,000
	B3—$5,000–$150,000
	B4—$150,000–$400,000
Partial List of Suppliers:	AccuRay Corp.; Alphacom; Bailey Controls, Div. of Babcok & Wilcox; Burroughs Corp.; Centronics; DataGeneral; DataProducts; Digital Equipment Corp.; Foxboro Co.; Fisher Controls; General Electric; Honeywell; IBM; Lear-Siegler; Measurex; Mead; Printronix; Raytheon, Taylor Instrument Co.; Terminal Data; Westinghouse Electric Corp.; Xerox

The phrase "terminals and printers" describes a class of equipment that allows the operator to communicate with the process. The more correct term for this equipment is "operator interface." The equipment includes keyboards, CRTs, individual controller faces, and traditional printers. This equipment provides the operator with the capability for monitoring the process with either a volatile display or hard copy printout, or modifying certain process parameters by entering changes.

Only the briefest description of analog and digital controllers will be given here since they are discussed in Section 2.3. Recent advances in the field of process control have resulted in the concept of "centralized operations." Process information is concentrated so that it is now feasible for a single operator seated at a compact console to monitor a large process plant. Sections 7.4 and 7.6 discuss control networks but some discussion of how the philosophy of networking affects operator communication with the process will be presented here.

Long control room panels are being replaced by a control console, a video screen, and a printer station for hard copies of CRT displays and data logging. CRTs as display devices are discussed in Section 3.3, but printers and control consoles will be discussed here in detail.

Printers

Printers are categorized according to whether they are impact or nonimpact, shaped character or matrix image, and serial or line. Each category employs its own technology, resulting in certain advantages and limitations, costs, and speeds. Almost every possible combination of categories is available, affording the user a wide choice to suit almost any particular need. Serial printers print one character at a time in either direction, then index. Line printers print essentially an entire line of characters at once.

Impact Printers

Impact printers transfer ink to paper either by having a character strike a ribbon against the paper as does a typewriter, or by having a hammer strike the paper and ribbon against a character.

One of the first popular printers was the cylinder printer. This is an impact, shaped-character, serial printer that has a complete character set embossed on a metal cylinder. When the cylinder is rotated and shifted vertically to position the selected character, a hammer strikes the cylinder (Figure 7.10a). While low in cost ($1,000 to $2,000), these printers give poor print quality, are noisy, and have speeds of only 10 characters per second (cps).

The golf ball printer, which is essentially an updated cylinder printer, uses a metal sphere that rotates in two axes to expose the selected character. The sphere itself then strikes the ribbon and paper. This printer has excellent quality, and large variations in character sets and type fonts are accommodated with interchangeable spheres. Speed is low (15 cps), noise level is high, and cost is $2,000 to $5,000. At least one shape other than a sphere is available; that of a thimble.

A more modern impact, shaped-character, serial printer is referred to as a daisy wheel (Figure 7.10b). Here a number of arms, each with a single character, or petal, at the end, radiate from a central hub. The hub rotates to bring the desired character into position where it is struck by a hammer against the ribbon and paper. It is

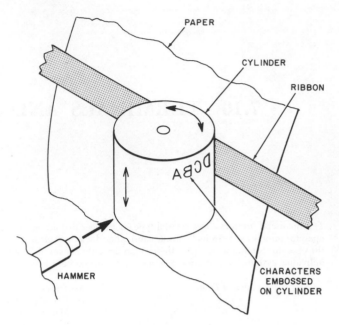

Fig. 7.10a   Cylinder printer

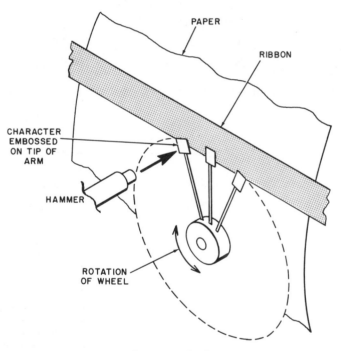

Fig. 7.10b   Daisy wheel printer

more reliable than golf ball printers, and speeds of almost 100 cps are available, with costs $2,000 to $7,000.

The impact, matrix-image, serial printer, popularly referred to as a dot matrix, is a printer that forms characters using a series of small dots. Character fonts are stored in memory. In the case of the serial printer, the print head consists of a vertical column of needle hammers that moves across the page (Figure 7.10c). The hammers are selectively fired against the ribbon and paper. Print heads commonly have nine.needles, print-

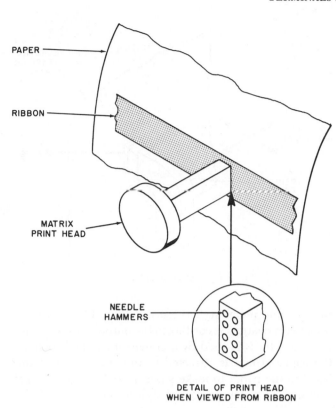

Fig. 7.10c Impact matrix printer

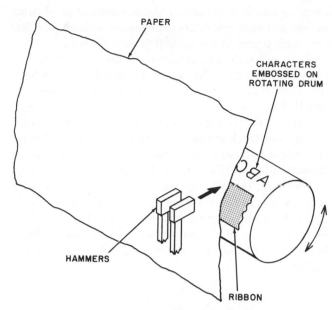

Fig. 7.10d Drum printer

ing six columns horizontally to form each matrix image. The quality of the printed character is a function of the number of rows and columns used to generate the character and the amount of overlapping between dots. Character resolution is improved by either making two or more passes to print each line, or by using high-density (as many as 460 wires) print heads. Also available are multi-mode printers that produce very fast, impact matrix output as well as slower, near-full-character-quality output. Speeds as high as 900 cps are available, with prices in the $1,000 to $4,000 range. Print head life typically averages more than 100 million characters, with resolutions as high as 100 dots per inch (25 mm).

Impact matrix image line printers are also available. Here a set of raised dots is mounted on a wide horizontal bar or comb device. The bar slides horizontally in front of the paper and individual needles strike each selected character dot to produce one row. Characters are formed as the paper is moved to successive rows. Speeds up to 500 lines per minute (lpm) are available, with prices in the $15,000 range.

Full-character impact line printers are available that offer a variety of mechanisms. Drum printers have a rotating cylindrical drum with a complete set of characters embossed around the circumference for each print position. A line of hammers, one for each column, strikes the paper/ribbon/drum sandwich for the required character at each point position as the drum rotates (Figure 7.10d). Reliable operation is achieved at speeds as high

as 2,000 lpm, but with a limited choice of character fonts and some possibility of vertical misregistration of characters. Prices are in the $10,000 to $75,000 range.

Band and belt line printers account for more than half of the impact line printer sales. All band printers have print hammers behind the paper that push the paper against a ribbon that has a continuous steel band behind it. The steel band with etched alphanumeric characters moves horizontally, and the brief hammer strokes imprint characters on the page (Figure 7.10e). Logic circuits in the printer synchronize the firing of the hammers with the moving characters on the band by using magnetic pickups to sense character timing marks also etched on the band. Easily interchangeable type bands with dif-

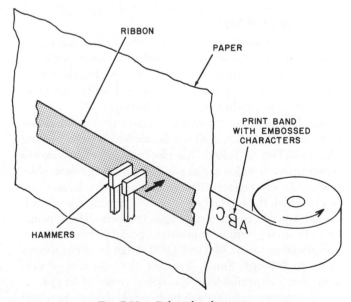

Fig. 7.10e Belt or band printer

ferent character fonts are available, yielding good print quality and high reliability. Speeds are as high as 1,500 lpm, with prices in the $3,000 to $50,000 range.

Chain and train line printers are the last category of impact line printers. The higher-speed train printers have individual character slugs with embossed character images that are pushed along a horizontal track past hammers that strike the paper and inked ribbon (Figure 7.10f). Usually the track contains multiple sets of characters, and only a fraction of a revolution is required to set a character in front of a column. The smaller the character set, the faster the print speed, typically up to 2,000 lpm, with prices in the $50,000 range. Chain printers are quite similar to train printers, but they have character slugs linked together in a chain driven by a sprocket wheel. Speeds of 600 lpm are achievable, with prices in the $25,000 range.

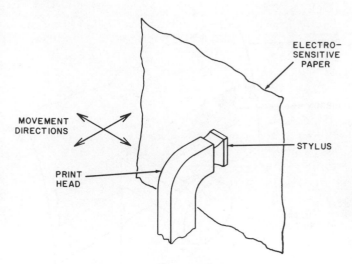

Fig. 7.10g   Electrographic printer

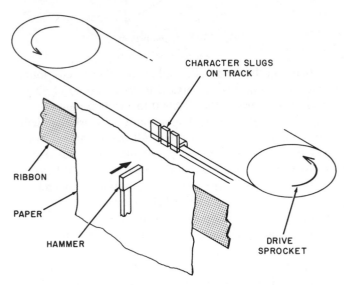

Fig. 7.10f   Chain or train printer

Nonimpact Printers

Nonimpact printers use a variety of techniques to produce copy. These printers offer the user quiet operation at speeds approaching 50,000 lpm, but the fastest ones are expensive, some require special paper, and some are unable to produce multipart forms.

In electrographic printers, a specially coated paper changes color when voltage is applied to the writing element (Figure 7.10g). An electrolytic version uses a wet process in which moist paper is drawn between electrodes. An electro-sensitive version uses dry layered paper in which the top layer burns away to expose a second, darker layer in the desired character form. These printers, of course, require expensive special paper, but they are priced in the $1,000 to $3,000 range for serial speeds of up to 200 cps. Small, low-cost electro-sensitive versions are also available in the speed range of 15 cps.

Electrostatic printers are high-speed devices in which specially coated paper is passed over an array of moving

rows of fine metal styli (Figure 7.10h). Each stylus is selectively charged according to the output required, and the character is formed by a mosaic of charged spots on the paper. Toner is attracted to the charged spots and fused to the paper to make a permanent image. Typical printers use a double row of styli to form a 7 × 9 dot matrix. This printer accommodates plotting as well as printing with a variety of fonts with lower prices than impact models at comparable speeds. Speeds range from 300 to 20,000 lpm, with prices in the $5,000 to $150,000 range.

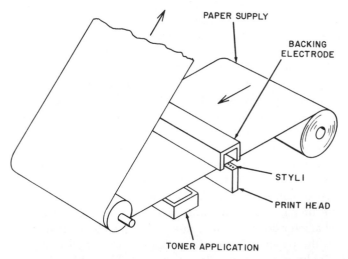

Fig. 7.10h   Electrostatic printer

Thermal-matrix printers typically have a print head containing a 5 × 7 array of dot elements. The print head moves horizontally across specially coated paper, stopping at each column to print a character by heating selected elements in the array. The paper changes color under each element, forming the desired character. These printers are very low cost ($500 to $5,000), quiet, and small, with speeds of 15 to 100 cps.

Ink-jet printers position a spray of electrically charged ink on ordinary paper. One system uses electrostatic deflection plates, forming the characters by manipulating the dot structure (Figure 7.10i), while another uses separate ink injection chambers for each dot. The clarity of the finished copy is a function of the density of the dots, along with the spattering of the ink. As many as 500 dots comprise the ink-jet matrix, resulting in a close resemblance to solid character copy. A variety of character fonts can be stored in ROM, offering high-resolution graphics capabilities, but these printers are relatively expensive. Prices range from $5,000 to $500,000 for a printer with speeds up to 50,000 lpm.

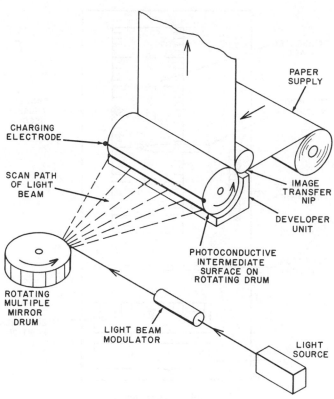

Fig. 7.10j   Xerographic printer

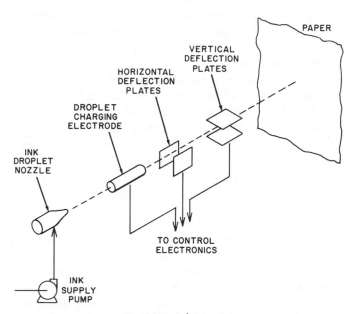

Fig. 7.10i   Ink-jet printer

Xerographic printers use a beam of light to produce a latent image on an intermediate photoconductive surface that is then toned with ink powder. The image is transferred electrostatically to the paper and fused into place. In one instance, a light source and photographic image is used to produce the latent image (Figure 7.10j). In another, a laser beam is used to generate a high-quality dot-matrix character. Speeds range from 4,000 to 18,000 lpm, with high character resolution and quiet operation. Multiple combinations of character sets can be accommodated at the same time, but a high level of maintenance is required for good copy quality, and prices are high ($150,000 to $400,000).

A recent development is the ion deposition printer, with a speed of 3500 lpm. This system replaces the laser, scan mirror, and modulator used in xerographic systems with an ion-projection device that lays charged image dots directly onto the surface of a dielectric-coated drum. The ion-projection device is an X-Y intersection of crossed conductors separated by mica insulators. When a burst of high-frequency alternating current is applied to one

of these intersections, the mica dielectric breaks down and ionizes the air at that point. Once the intersection is ionized, the ions are stripped, focused, and projected through holes in a plate backing the X-Y intersection onto the dielectric surface of the drum. Toner is attracted to the charged ion dots and transferred to the paper by a pressure roller.

Analog Controllers

Analog controllers are still used to a great extent in existing facilities that were built prior to the mid-1970s. They are also frequently used as backup devices for certain control loops in a process plant. A brief description of a typical analog controller faceplate will follow, with reference to Figure 7.10k.

Analog controllers continuously display information about the process variable under control by using a needle and meter arrangement. The needle indicates the value of the process variable, which can be compared to the selected set point indicated on an adjacent scale. The set point is either adjusted manually with the thumbwheel or remotely manipulated. The controller face carries some identification of the loop or process variable and a selector switch for operating the process in manual, automatic, or seal, which is used for bumpless transfer from automatic to manual.

The other meter on the controller face is used to represent the percentage open, or the stroke of the final control element, usually a valve. Frequently a strip-chart

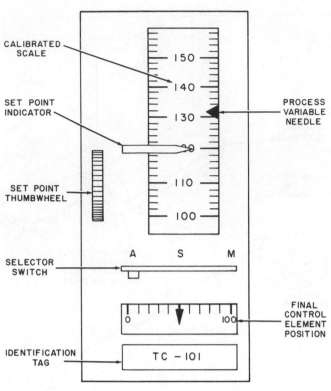

CALIBRATED SCALE

SET POINT INDICATOR

SET POINT THUMBWHEEL

SELECTOR SWITCH

IDENTIFICATION TAG

PROCESS VARIABLE NEEDLE

FINAL CONTROL ELEMENT POSITION

TC – 101

Fig. 7.10k   Analog controller

through, selected, and arranged in the most effective manner. The design of interfaces in the control room must be responsive to a rapidly moving technology base that dictates the increased use of electronic control systems, larger and fewer control rooms with higher instrument densities, and the continued use of computer systems.

The demands on operator skills and know-how are increasing because of more complex and integrated control strategies aimed at operation closer to process and equipment constraints; operating margins with more on-line process improvement investigations; more frequent changes in process design and operating practices; increased multiplexing of equipment to reduce capital investments; greater interaction between units, resulting from increased integration of recovery systems; and stronger governmental and public pressures to reduce industrial pollution.

The operator interface must present information to the operator quickly and in a form permitting rapid decision making and efficient manipulation of needed controls. Continuous monitoring of key operating variables must be provided, with alarming for malfunction of process equipment or control system elements. The alarms must be detected, with prompt access provided to the display and control of pertinent data during upset conditions to ensure proper implementation of emergency procedures. These needs have resulted in various strategies for the design of operator interfaces that are now becoming standardized with regard to ergonomic philosophy. In general, operator input is performed in serial sequences by using a single interface/keyboard/CRT providing all functions at one location within easy reach of a seated operator.

While operator interfaces are used for the precise control of the process, process supervisory control through the use of advanced control and additional computational capabilities with supervisory computers allow for unit process optimization and energy management via a data highway connection. A central plant host computer can advise and coordinate interaction between process supervisory computers to provide plant-wide process optimization. By integrating the control systems of each process with accounting and information-gathering systems, total plant control can lead to better coordination of operations, higher productivity, lower energy consumption, and a more profitable operation.

recorder is incorporated into the controller body for recording the process variable or some other pertinent data. Changes in tuning parameters usually require partial removal of the instrument from the case; such changes are usually neither convenient nor simple.

Digital Controllers

Digital controllers are frequently used as a building block for computer-based process control schemes, and they have developed into quite flexible unit controllers. These devices can control complex systems with multiple loops and controlled outputs, while providing panel instrumentation with annunciation and alarm contact outputs in a compact front panel. Typically microprocessor based, these controllers are designed utilizing a building block concept, permitting easy configuration for a wide range of applications.

As shown in Figure 7.10l, this controller has six automatic/manual stations with four controlled outputs backed up with track and hold logic. Scan rates of four times per second are typical, with trend outputs, a first out alarm annunciator, $4\frac{1}{2}$ digit display, tuning from the panel, and key-lock protection of configuration and tuning.

Operator Interface Philosophy

As technology advances, operators are required to monitor an ever-increasing amount of information. For operators to do their jobs effectively, operator interfaces must be designed so that the needed information is sorted

Information Requirements

It has been found that trend or past history of variables is essential for operator guidance, with graphic forms of display used to convey qualitative information, and digital displays for for quantitative information. Compact, color-coded graphical displays utilizing pattern recognition display techniques are most efficient. Data should

The INCR-DECR pushbuttons under the 4½ digit LED (light emitting diode) display will change the displayed variable. The key switch prevents unauthorized personnel from changing configuration and tuning data. Examples of variables which may be changed are set points and loop outputs in manual. The loop pushbutton cycles through the loop LEDs to identify the loop being displayed. The SERVICE MANUAL LED is lit when the analog tracking circuitry is controlling the outputs.

An eight alarm first out annunciator panel with customized legends and an alarm acknowledge pushbutton.

The DISPLAY pushbutton is used to select the process variable, local setpoint or bias, remote set point, or percent output of the loop identified by the LOOP LEDs.

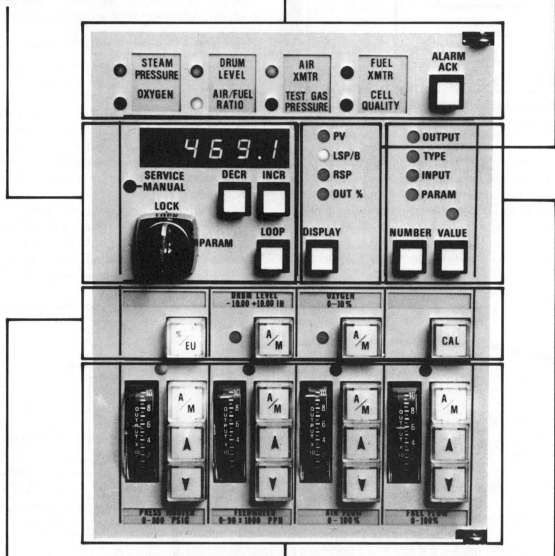

The four auxiliary backlift pushbuttons may be configured to select percent or engineering units, perform automatic oxygen calibration, or serve as remote/local pushbuttons. The two pushbuttons with adjacent loop LEDs may also be configured to be Auto/Manual pushbuttons, and serve as auxiliary loops internal to a control system.

The four controlled outputs are indicated on separate analog meters. Each primary loop has its own backlit Auto/Manual pushbutton and separate increase-decrease pushbuttons. The loop LEDs above the meters are used by the LOOP pushbutton to select the loop being displayed on the digital readout. The customized loop legends are used to specify the process variable and its engineering units.

The NUMBER AND VALUE pushbuttons identify configuration information and are used along with the shared INCR-DECR pushbuttons to enter configuration and tuning data.

Fig. 7.10l   Digital controller (Courtesy Westinghouse Electric Corporation)

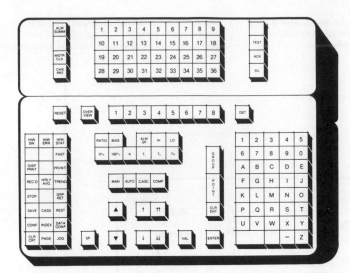

Fig. 7.10m　Typical console keyboard (Courtesy Honeywell, Inc.)

process state. Any individual loop or group of loops can be next selected for examination by entering a short code or depressing one or two keys. All points previously on recorders are available for display using memory stored for preserving the data. In general, random access memory (RAM) is backed up by core memory so that unique application data, such as configuration words, set points, and tuning constants, are always available. Read only memory (ROM) permanently maintains all master instructions necessary for operation, such as control algorithms, diagnostic routines, and executive programs.

Centralized Operator Station Detail

As discussed above, literally hundreds of control loops are available within the grasp of an operator seated at a compact console. The operator can monitor an entire plant, view a plant area more closely, or examine loops in great detail. Modular construction permits reconfiguring of the system to display any new or modified loops, permitting the subsequent addition of optional console bays housing displays, annunciators, or recorder subsystems.

One method for facilitating recognition is to use the familiar format of conventional panel instruments, simulated on color or monochromatic video screens. The operator selects a desired display by pressing appropriate keys on the keyboard, a typical example of which is shown in Figure 7.10m.

For display of process status, the operator has access

be available in groups for monitoring the functioning of small subsystems of the process unit, and alarm systems simplified in hierarchies. In computer control systems, derived variables are used more as a measure of process performance, rather than working with flows, temperature, and pressure.

Systems are now available that enable the operator to examine the deviation of all points under control by simultaneously displaying more than 200 variables on a single CRT monitor for a quick overview of the total

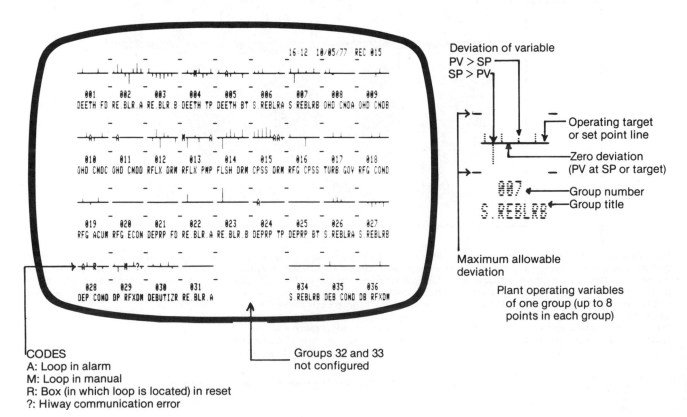

CODES
A: Loop in alarm
M: Loop in manual
R: Box (in which loop is located) in reset
?: Hiway communication error

Groups 32 and 33 not configured

Deviation of variable
PV > SP
SP > PV

Operating target or set point line
Zero deviation (PV at SP or target)
Group number
Group title

Maximum allowable deviation

Plant operating variables of one group (up to 8 points in each group)

Fig. 7.10n　Plant process status display (Courtesy Honeywell, Inc.)

to a presentation of displays in a hierarchy of three or four levels: plant (overview), area, group, and loop. As shown on Figure 7.10n, the plant display is an operation-by-exception presentation of process variables vs. operating targets or set points. Here we have a display with no area level, but with four rows of nine groups with up to eight loops in each group. Each group is identified by a number from 001 to 036 and by the first eight characters of the group title. Limits are scaled to four percentages plus off and are defined during configuration. A detail of group 007 is also shown. Selection of "off" forces a display of zero deviation for that point in the overview.

In another control scheme with four hierarchial levels, a plant alarm summary shows each group by area (Figure 7.10o). Any group with an alarm condition is highlighted, whether acknowledged or not.

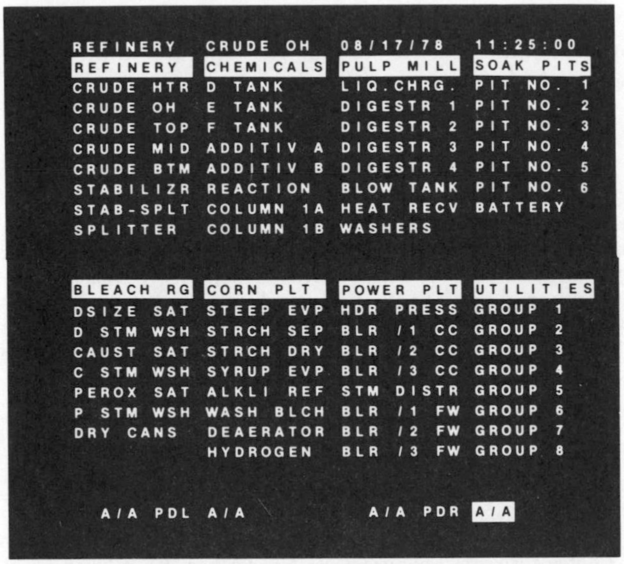

Fig. 7.10o Plant alarm summary (Courtesy The Foxboro Company)

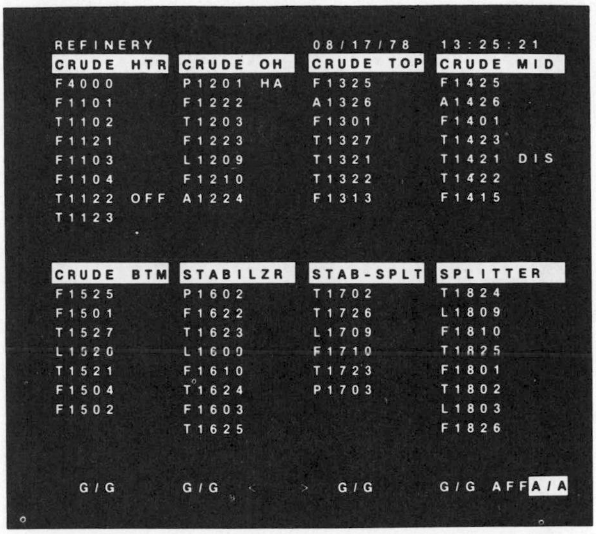

Fig. 7.10p Area alarm summary (Courtesy The Foxboro Company)

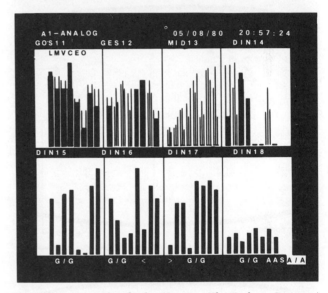

Fig. 7.10q Area value display (Courtesy The Foxboro Company)

In moving from the plant level, each step down provides greater detail on a smaller portion of the plant. At each level, variable-function keys are made available to select another display. At the lowest loop level, keys are also available for manipulating loop parameters. As shown in Figure 7.10p, the area alarm summary lists loop tags by group name, and highlights any loop with an alarm condition.

The area value display in Figure 7.10q provides a graphic indication of the absolute value of each loop set point and measured value. Alarm status or other abnormal conditions are shown above each bar graph.

One level down, the group value display shows the status of each loop in the group. As shown in Figure 7.10r, values associated with a loop are depicted with bar graphs supplemented by numeric values for measurement or input. Contact-type input and outputs are displayed by means of alphanumerics. The loop value display is similar to the group display; however, keys are

now available for operator adjustment of loop parameters and selection of operating mode.

Historical information for process variable data is available to the operator for training, trouble shooting, and optimizing. Typically an hourly average of any variable would be routinely available for the last shift or half day. As shown in Figure 7.10s, a one-hour trend with one-minute averages of a variable is shown. The lower portion of the display permits set point, mode, and output adjustments to be made while the one point trend is displayed. Several time bases are usually available for each point trend, typically 20 minutes, 1, 3, 6, and 12 hours. As shown in the lower half of Figure 7.10s, the operator can also select a related control point from the group and call up a second trend using the same time base. Here the second trend replaces the lower portion of the group display.

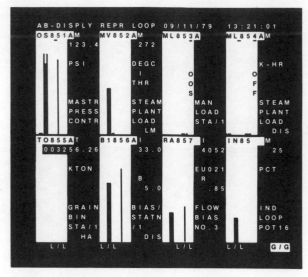

Fig. 7.10r   Group value display (Courtesy The Foxboro Company)

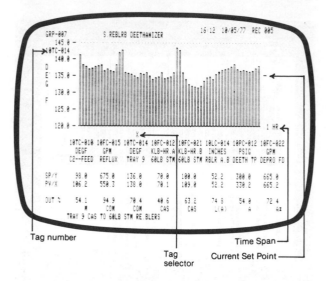

Fig. 7.10s   Trend displays (Courtesy Honeywell, Inc.)

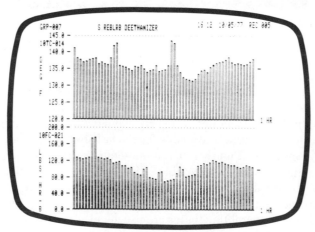

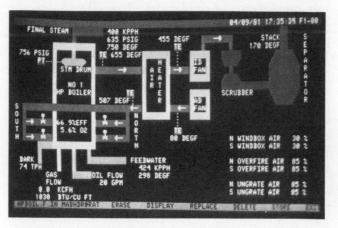

Fig. 7.10t   Graphic display (Courtesy The Foxboro Company)

constructing or displaying previously constructed process units, with the console automatically updating live process values on the screen. The operator can interact with the process in a real-time fashion.

ACKNOWLEDGMENT

By permission of Honeywell, Inc., portions of this section have been prepared from Honeywell Bulletin SY-05-02, *An Evolutionary Look at Centralized Operations/2*, by Henry Marks, 1981. Similarly, material was used from The Westinghouse Electric Corporation Bulletin 106-410 and The Foxboro Company Bulletins N 21-C 15M and PSS 2E-4A2 B.

BIBLIOGRAPHY

Auerbach Computer Technology Reports, Auerbach Publishers, Pennsauken, NJ.

"Byte Printer Directory," *Byte*, March, 1982.

Chilton's Control Equipment Master, Radnor, PA: Chilton Company, 1981.

Computer Review, GML Corporation, Lexington, MA.

Control Engineering, January, 1980, September, 1980, and October, 1982.

DataPro reports on Data Communications, DataPro Research Corporation, Delran, NJ.

Digital Equipment Corporation, *VAX Systems and Options Summary*, Bulletin ED-22960-20, 1982.

Foxboro Company, Bulletins on Fox and Spectrum Process Control Systems.

Harrison, T., ed., *Minicomputers in Industrial Control*, Instrument Society of America, 1978.

Lowe, K., *Practical Computer Applications*, Miller Freeman Publications, 1975.

Marks, H., *An Evolutionary Look at Centralized Operations/2*, Honeywell Bulletin SY-05-02, 1981.

Measurex Corporation, *2002*, Bulletin PA2738-R/10M/1981.

Morrison, D., et al., "Advances in Process Control," *Science*, February 12, 1982.

Powell-Process Systems, Inc., *Micon MDC-200*, Bulletin MDC-200-11S, 1982.

"Printer Survey," *Mini-Micro Systems*, January, 1982.

Smith, A., "Bred for Work," *Business Computer Systems*, December, 1982.

"Terminal Survey," *Mini-Micro Systems*, August, 1981.

Westinghouse Electric Corporation, *Distributed Control System DCS-1700*, Bulletin B-754, 1982.

Wieselman, I., "Trends in Computer Printer Technology," *Computer Design*, January, 1979.

In many instances, multicolor pictorial diagrams of a selected portion of the overall process are available to the operator, as shown in Figure 7.10t. Software in the process control system provides a convenient method for

7.11 WIRING PRACTICES AND SIGNAL CONDITIONING

The ability of a process control system to perform its designed function is directly dependent upon the quality of the signal of the measured variables. This quality is dependant upon the elimination or attenuation of noise that can deteriorate the actual transducer signal. In the worst case, the noise signal can actually be of a higher amplitude than the transducer signal. This type of signal would, of course, be of no value because the useful control information is undetectable. The following wiring practices and signal conditioning techniques are meant to prevent the noise signal from becoming part of the transducer signal, insuring an accurate measured variable signal for process control.

Although there are a multitude of signal noise sources, the four main sources are as follows:

1. Uncontrollable process disturbances that are too rapid to be reduced by control action.
2. Measurement noise resulting from such factors as turbulence around flow sensors, instrument noise, etc.
3. Stray electrical pickup such as from AC power lines, pulses from power switching, RF interference, etc.
4. Analog to digital conversion error.

Noise adversely affects both analog and digital electronic equipment, but in digital systems the noise effects are compounded because of the scanning or snapshot principle of digital systems.

Typical ranges for these noise components are shown in Figure 7.11a. Stray electrical pickup can be minimized by good engineering and installation practices such as proper shielding, screening, grounding, and routing of wires. Even so, noise will still exist at the input, but its effect on controlled and manipulated variables can be significantly reduced through judicious selection of filtering. In most cases, the relatively high-frequency noise (i.e., above approximately 5 Hz) is removed by a conventional analog filter; the remaining noise occupying the range of 0.002 to 5 Hz may require either special analog filtering and/or digital filtering.

It is good practice to follow these three steps to reduce the effects of noise on control quality:

1. Estimate the dynamic characteristics (amplitude and bandwidth) of the noise.
2. Determine its effect on the controlled and manipulated variables.
3. Select the best choice of analog and/or digital filtering.

This section is devoted to a close examination of noise suppression techniques. Noise suppression is a system problem, and discussion of noise therefore should not be limited to the signal transmission medium. The noise suppression methods must be applied consistently from the transducer, through the signal transmission line to the signal conditioner, multiplexer, and/or analog-to-digital converters.

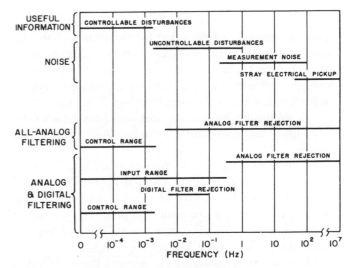

Fig. 7.11a   Frequency range of input signals and filtering functions

Types and Sources of Noise

As shown in Figure 7.11b, signal leads can pick up two types of external noises: normal mode and common mode interferences. Interference that enters the signal path as a differential voltage across the two wires is referred to as normal mode interference. The normal mode noise cannot be distinguished from the signal coming from the transducer. Interference that appears between

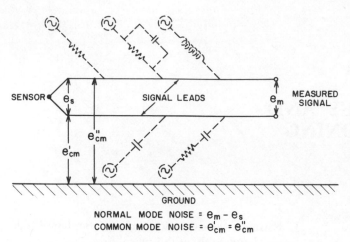

NORMAL MODE NOISE = $e_m - e_s$
COMMON MODE NOISE = $e'_{cm} = e''_{cm}$

Fig. 7.11b Electrical noise interferences on signal leads

ground and both signal leads, as an identical voltage, is referred to as common mode interference.

There are many sources of noise interference. The most common are discussed below.

Inductive pickup from power sources. This includes 60-Hz noise from power lines, 120-Hz noise from fluorescent lighting, as well as high-frequency noise generated by electric arcs and pulse transmitting devices. The worst man-made electrical noise results from the opening or closing of electrical circuits containing inductance. The amplitude of transients generated in an inductive circuit is given by equation 7.11(1):

$$E_i = L(d_i/d_t) \qquad 7.11(1)$$

This amplitude is calculated as the inductance times the rates of change in the switched current. Closing an inductive circuit can cause a transient of twice the input voltage, and opening such a circuit can cause transients as high as ten times the supply voltage. The noise transients are proportional to the instantaneous value of the AC supply when switching inductive loads from an AC supply.

Common impedance coupling "ground loops." Placing more than one ground on a signal circuit produces a ground loop that is a very good antenna, and noise signals induced in the loop are easily coupled into the signal lines. This can generate noise that will completely obscure the useful signal.

Electrostatic coupling to AC signals. The distributed capacity between signal conductors and from signal conductors to ground provides a low-impedance path for cross talk and for signal contamination from external sources.

Ineffective temperature compensation. Ineffective temperature compensation in transducers, lead wire systems, amplifiers and measuring instruments can create changes in system sensitivity and drift of zero line. This is especially important in strain gauge transducers, resistance type temperature sensors, and other balanced-bridge devices.

Loading the signal source. When a transducer or other signal source is connected in parallel with an amplifier or measuring device with a low input impedance, the signal voltage is attenuated because of the shunting effect. This can considerably decrease the sensitivity of the system.

Variable contact resistance. All resistance type transducers, as well as bridge type circuits, are susceptible to changes in contact resistance. The measuring instrument is unable to distinguish between a resistance change in the sensor and a resistance change in the external wiring.

Conducted AC line transients. Large voltage fluctuations or other forms of severe electrical transients in AC power lines (such as those caused by lightning) are frequently conducted into the electronic systems by AC power supply cords.

Conduction pickup. Conduction pickup refers to interfering signals that appear across the receiver input terminals because of leakage paths caused by moisture, poor insulation, etc.

Thermoelectric drift. The junction of two dissimilar metal wires will create thermoelectric drift, which varies with changes in temperature. This is more critical in DC circuits in the microvolt level.

Electrochemically generated corrosion potentials. These potentials can often occur, especially at carelessly soldered junctions.

Failure to distinguish between a ground and a common line. The ground currents that exist in a "daisy-chain" equipment arrangement are primarily noise signals induced into the equipment, or perhaps generated by some equipment, and they find their way back to earth. The resulting voltage drops in the ground wire place each piece of equipment at a different potential with respect to earth. This potential difference is easily coupled into the signal lines between equipment.

Use of the same common line. Using the same common line for both power and signal circuits or between two different signal lines can cause the appearance of transients on the signal and can also cause cross talk between the two signals.

Inadequate common mode rejection. Inadequate common mode rejection due to line unbalance will convert a common mode noise interference to a normal mode noise signal.

Table 7.11c shows the average noise conditions that exist in various industries. Some chemical plants have experienced as high as 60 volts common mode noise interference. Electrical noise interference can be a severe problem in industries such as steel, power, and petroleum, where power consumption is high and complex electrical networks exist.

Reducing Electrical Noise Interference

There are several ways to reduce the effects of noise on control systems, including line filtering, integrating,

Table 7.11c
AVERAGE NOISE CONDITIONS EXPERIENCED IN
VARIOUS INDUSTRIES

Description		Industry		
		Chemical	Steel	Aerospace
Normal Signal Levels (mV)		10–100	10–100	10–1,000
Possible or expected noise level	Normal mode (mV)	1–10	2–7	1–10
	Common mode (Volts)	4–5	4–5	4–5
Desired noise rejection	Normal mode	10^3:1	10^3:1	10^3:1
	Common mode	10^6:1	10^6:1	10^6:1

digital filtering, and improving the signal-to-noise ratio, but none of these will completely solve the problem without proper cable selection and installation.

Careful examination of the various sources of noise shows that some can be eliminated by proper wiring practice. This implies good temperature compensation, perfect contacts and insulation, carefully soldered joints, use of good-quality solder flux, and use of the same wire material. However additional precautions are needed to eliminate pickup noise entirely.

Circuit Arrangement

One of the important practices in eliminating noise in low-level signal systems is the arrangement of external circuits so that noise pickup will appear equally on both sides of the signal pair, remain equal, and appear simultaneously at both input terminals of the differential amplifier (or measuring device). In this case it can be rejected without affecting the frequency response or the accuracy of the system by means of common mode rejection.

If a noise signal is allowed to combine with the useful signal, it can be separated only by selective filtering.

A special discussion of thermocouple signal conditioning is warranted, since it is widely used in various industries. Thermocouple signals are susceptible to serious noise problems because they are often in contact with the device or structure under test. Additional thermo-electric potentials developed in the bimetallic leads should be accounted for when measuring a signal from a thermocouple. The best thermocouple materials are not the best electrical conductors, yet lead wires must be these same materials all the way back to the reference junction. Standard practice to avoid the problem of accounting for each bimetallic junction is to compare the output of the thermocouple with the output of an identical thermocouple in a controlled temperature environment. The balance of the circuit in that case is represented by copper wires creating an identical couple in each lead and producing a zero net effect when both couples are at the same temperature.

The constant-temperature reference junction is the most accurate method, but it is not the most convenient. A simulated constant-temperature reference can be had with devices containing compensating junctions, milli-volt sources, and temperature-sensitive resistors. They permit a transition from thermocouple material leads to copper lead wires.

A more complex conditioning circuitry is shown in Figure 7.11d, which performs the functions of thermocouple signal conditioning, noise discrimination, balancing, ranging, and standardizing. This circuit must be tied to a calibration scheme for the system output. The balance circuit shown accomplishes both variable offset and calibration. The signal-conditioning circuitry makes it possible to convert the transducer output to make use of the optimum range of the measuring system for the best linearity, accuracy, and resolution.

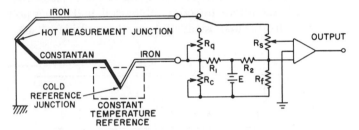

Fig. 7.11d   A typical (complex) thermocouple conditioner

E = A precision isolated power supply
R_1 and R_2 = Precision bridge resistors, low values (5 or 10 ohms)
R_c = Balance control variable resistor, large resistance value
R_f = Reference resistor, large resistance value
R_q = Equivalent line resistor for use during stabilizing and calibrating
R_s = Span adjustment potentiometer

Because the thermocouples are often in electrical contact with the device or structure of which the temperature is being measured and with an intentional ground at the sensor, differential input amplifiers with very good common mode rejection are required to effectively discriminate against noise in the structure or grounding system.[1]

Impedance Level

The common mode rejection of a good low-level signal amplifier is in the order of 10^6 to 1. This value is always decreased when the signal source and input signal leads are connected to the amplifier. Signal source impedance and signal lead impedance have a shunting effect on the signal input to the amplifier. From the point of view of maximum power transfer, equal impedance on both sides of a circuit is an ideal termination. In the transmission of voltage signals, however, current flow has to be reduced to as close to zero as possible. This is achieved by selecting amplifiers with high input impedance and using transducers with low output impedance. Good instrumentation practice dictates that the input impedance of the amplifier be at least ten times the output impedance of the signal source.

However, low signal source impedance has an adverse effect on pickup signals, because these will be shunted by the low impedance of the transducer. An additional step is to ground the transducer whenever possible and to run a balanced signal line to the amplifier input.

There are three basic techniques to condition transducers with high output impedance: high-impedance voltage amplifiers, charge amplifiers, and transducer-integrated unloading amplifiers.

Voltage Amplifiers

Voltage amplifiers must have an amplifier input impedance that is very high compared with the source impedance. In this way, the amplifier's effect on the phase and amplitude characteristics of the system is minimized. A typical transducer with output amplifier is shown in Figure 7.11e. High impedance is practical for an amplifier, but an equally significant portion of the load is in the interconnecting cable itself, and it does not take much cable to substantially lower the available voltage.

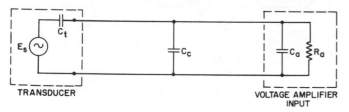

Fig. 7.11e A basic voltage equivalent transducer with output amplifier

E_s = Transducer voltage source
C_t = Transducer capacitance
C_c = Signal cable shunt capacitance
C_a = Amplifier input capacitance
R_a = Amplifier input resistance

In applying a voltage amplifier, low-frequency system response must be considered. The voltage amplifier input resistance (R_a) in combination with the total shunt capacitance forms a high-pass first-order filter with a time constant (T), defined by

$$T = R_a(C_t + C_c + C_a) \qquad 7.11(2)$$

Cutoff frequency can become a problem when it approaches information frequency at low source capacitances (short cable or transducers with very low capacitance) or at lower amplifier-input resistances.

Charge Amplifiers

Charge amplifiers have been widely used in recent years. This approach avoids the cable capacitance effects on system gain and frequency response. The typical charge amplifier shown in Figure 7.11f is essentially an operational amplifier with integrating feedback. A charge amplifier is a device with a complex input impedance that includes a dynamic capacitive component so large that the effect of varying input shunt capacitance is swamped,

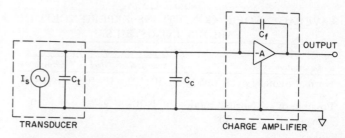

Fig. 7.11f Charge amplifier

I_s = Transducer current source
C_f = Feedback capacitance
C_t = Transducer capacitance
C_c = Signal cable shunt capacitance

and the output is the integral of the input current. Filtering of the resultant signal on both the low and high ends of the information band is desirable at times so that it is possible to get a rather high order of rejection without affecting information by using bandpass filters. The addition of a resistor in parallel with the feedback capacitor in a charge amplifier will decrease the closed-loop gain at low frequencies, resulting in the desired high-pass filter characteristics.

Unloading Amplifiers

Unloading amplifiers integrally mounted in a transducer housing (Figure 7.11g) have become available from transducer manufacturers, because neither voltage amplifiers nor charge amplifiers offer a very satisfactory solution to the conditioning problem for systems with very high input capacitance (usually a result of very long lines). With the voltage amplifier, signal-to-noise ratio suffers, because capacitance loading decreases the available signal. In a charge amplifier, the signal-to-noise ratio suffers because of the increased noise level (input noise is a direct function of input capacitance). Thus, remote signal conditioning appears to offer a satisfactory solution to the accommodation of long data lines. If closely located to the transducer, the voltage-responding or charge-responding amplifiers are equally effective. However, these techniques decrease the dynamic range capability (changing input amplifier gain to accomplish a range change is not possible) and restrict the high-frequency response signal amplitudes due to the limited current capability of the remote amplifier. When these two limitations are

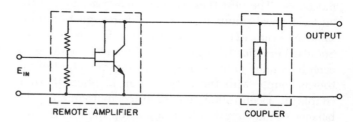

Fig. 7.11g The unloading amplifier, located at the transducer, reduces the input capacitance relative to voltage or charge amplifiers

overcome, almost all of the signal-conditioning equipment will be at the remote location.

In summary the following steps are recommended for low level signals:

1. Select a signal source with low output impedance.
2. Select an amplifier (or measuring device) with high output impedance.
3. Use a balanced line from signal source to amplifier input (maximum allowable unbalance is 100 ohms/1,000 ft or 0.33 ohms/m).
4. Keep signal cables as short as possible.
5. Use remotely located amplifiers when long signal cables are required (except for thermocouple and RTD signals).
6. Select a signal source that can be grounded (thermocouples, center-tapped sensors, etc.).

It is evident that common-mode rejection must be maintained at a high level in order to attain noise-free results from low-level signal sources.

Ground Systems

Good grounding is essential for normal operation of any measurement system. The term "grounding" is generally defined as a low-impedance metallic connection to a properly designed ground grid, located in the earth. In large equipment, it is very difficult to identify where the ground is. On standard all-steel racks, less than 6 ft (1.8 m) in length, differences in potential of up to 15 volts peak-to-peak have been measured. Stable, low-impedance grounding is necessary to attain effective shielding of low-level circuits, to provide a stable reference for making voltage measurements, and to establish a solid base for the rejection of unwanted common-mode signals.

In a relatively small installation, two basic grounding systems should be provided. First, all low-level measurements and recording systems should be provided with a stable system ground. Its primary function is to assure that electronic enclosures and chassis are maintained at zero potential. A satisfactory system ground can usually be established by running one or more heavy copper conductors to properly designed ground grids or directly to earth grounding rods. Signals can be measured with respect to the system reference ground only if the input signals are fully floating with respect to ground. In this case, the stable system ground fulfills the task of providing a base for common-mode noise rejection.

The other important ground is the signal ground. This system is necessary to ensure a low-noise signal reference to ground. This ground should be a low-impedance circuit providing a solid reference to all low-level signal sources and thus minimizing the introduction of interference voltages into the signal circuit. The signal ground should be insulated from other grounding systems, and

it is generally undesirable to connect it to the system ground at any point (Figure 7.11h). In a single-point grounding system, no current flows in the ground reference, and if signal cable is properly selected, noise due to large and hard-to-handle low-frequency magnetic fields will not exist. It should be emphasized that a signal circuit should be grounded at one point and at one point only, preferably at the signal source (Figure 7.11i).

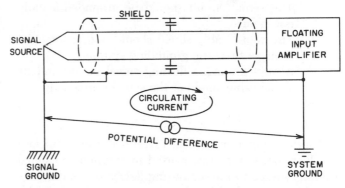

Fig. 7.11h   Incorrect grounding of floating signal circuit. Ground loop is created by multiple grounds in a circuit, by grounding the shield at both ends

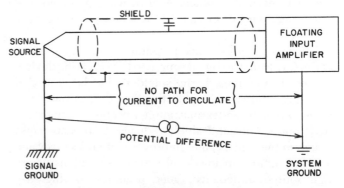

Fig. 7.11i   Correct grounding of floating signal circuit. Ground loop through the signal lead is eliminated by grounding the shield at signal end only

By connecting more than one ground to a single signal circuit, as shown in Figure 7.11h, a ground loop is created. Because two separate grounds are seldom, if ever, at the same potential, this will generate a current flow that is in series with the signal leads. Thus the noise signal is combined with the useful signal. These ground loops are capable of generating noise signals that can be 100 times larger than the typical low-level signal.

In off-ground measurements and recording, the cable shield is not grounded, but it is stabilized with respect to the useful signal through a connection to either the center tap or the low side of the signal source. Since the shield is driven by an off-ground voltage, appropriate insulation is needed between the shield and the outside of the cable.

It is important that electric racks and cabinets be connected to a proper system ground and not allowed to contact any other grounded element in the building. Guidelines on grounding can be summarized as follows:

1. Intentional or accidental ground loops in either the signal circuit or the signal cable shield will produce excessive electrical noise in all low-level circuits and will destroy the useful signal.
2. Every low-level data system should have a stable system ground and a good signal ground.
3. The signal circuit should be grounded at only one point.
4. The signal cable shield should not be attached to more than one grounding system.
5. Always ground a floating signal circuit and its signal cable shield at the signal source only.

Wiring

Another important aspect of reducing noise pickup involves the wiring system used to transmit the signal from its source to the measuring device or computer.

In less demanding low-frequency systems, where the signal bandwidth is virtually steady state and system accuracy requirements are not very high, two-wire signal leads will normally suffice. A third wire, or shield, becomes necessary where any of the above parameters are exceeded.

Where top performance is required, the shield is run all the way from the signal source to the receiving device. As already mentioned, the shield should be grounded the signal source and not at the recorder, because this arrangement provides maximum rejection of common-mode noise. The cable shield reduces electrostatic noise pickup in the signal cable, improves system accuracy and is indispensable in low-level signal applications where high source impedance, good accuracy, or high-frequency response is involved. As the signal frequency approaches that of the noise, which is usually at 60 Hz, filtering can no longer be used to separate noise from the useful signal. Therefore, the only practical solution is to protect the signal lines and prevent their noise pickup in the first place.

Elimination of noise interference due to magnetic fields can also be accomplished by wire-twisting (transpositions). If a signal line consisting of two parallel leads is run along with a third wire carrying an alternating voltage and an alternating current, the magnetic field surrounding the disturbing line will be intercepted by both wires of the signal circuit. Since these two wires are at different distances from the disturbing line, a differential voltage will be developed across them. If the signal wires are twisted (Figure 7.11j), the induced disturbing voltage will have the same magnitude and cancel out.

To prevent noise pickup from electrostatic fields, low-level signal conductors must be surrounded by an effective shield. One type of shield consists of a woven metal

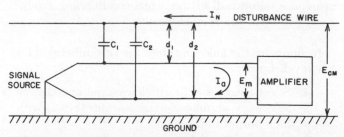

WHERE $d_1 \neq d_2$. THUS, $C_1 \neq C_2$, ∴ DIFFERENCE IN DISTANCE FROM DISTURBANCE WIRE CREATES A DIFFERENTIAL VOLTAGE, AND A NOISE SIGNAL WILL BE INDUCED INTO THE SIGNAL LEADS.

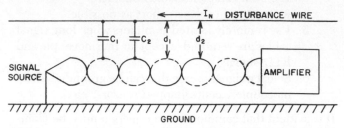

WHERE $d_1 = d_2$. THUS $C_1 = C_2$, ∴ THE INDUCED DISTURBING VOLTAGES WILL HAVE THE SAME MAGNITUDE AND CANCEL OUT WHEN TWISTED WIRES ARE USED.

Fig. 7.11j   Twisted wire eliminates the noise interference due to magnetic field from disturbing wires

braid around the signal pair, which is placed under an outside layer of insulation. This type of shield gives only 85 percent coverage of the signal line and is adequate for some applications if the signal conductors have at least ten twists per foot (33 twists per meter). Its leakage capacity is about 0.1 picofarad per foot (0.3 picofarad per meter). At the microvolt signal levels, this kind of shielding is not satisfactory. Another type of signal cable is shielded with lapped foil (usually aluminum-Mylar tape) shields, plus a low-resistance drain wire. This type of shielding provides 100 percent coverage of the signal line, reducing the leakage capacity to 0.01 picofarad per foot (0.03 picofarads per meter). To be most effective, the wires should have six twists per foot (20 twists per meter). The lapped foil shield is covered with an insulating jacket to prevent accidental grounding of the shield. Care should be exercised to prevent the foil from opening at bends.

This type of low-level signal cable provides the following benefits:

1. An almost perfect shield between the signal leads and ground.
2. Magnetic pickup is very low because the signal leads are twisted.
3. Shield resistance is very low because of the low-resistance drain wire.

Typical costs for the three types of twisted signal wires are shown in Table 7.11k.

The use of ordinary conduit is a questionable means

Table 7.11k
TYPICAL COSTS OF TWISTED SIGNAL WIRES

Source of Noise	Single Pair Cable Type	Cost ($) per Foot (meter)
Magnetic fields	Twisted	.10 (.30)
Magnetic and electrostatic	Twisted with metal braid 85% shield coverage	.23 (.70)
Magnetic and electrostatic	Twisted with aluminum/mylar 100% shield coverage	.15 (.50)

of reducing noise pickup. Its serious limitation is in obtaining or selecting a single ground point. If the conduit could be insulated from its supports, it would provide a far better electrostatic shield.

The following general wiring rules should be observed in installing low-level signal circuits.

1. Never use the signal cable shield as a signal conductor, and never splice a low-level circuit.
2. The signal cable shield must be maintained at a fixed potential with respect to the circuit being protected.
3. The minimum signal interconnection must be a pair of uniform, twisted wires, and all return current paths must be confined to the same signal cable.
4. Low-level signal cables should be terminated with short, untwisted lengths of wire, which expose a minimum area to inductive pickup.
5. Reduce exposed circuit area by connecting all signal pairs to adjacent pins in connector.
6. Cable shields must be carried through the connector on pins adjacent to the signal pairs.
7. Use extra pins in connector as a shield around signal pairs by shorting pins together at both ends and by connecting to signal cable shield.
8. Separate low-level circuits from noisy circuits and power cables by maximum physical distance of up to 3 ft (0.9 m) and definitely not less than 1 ft (0.3 m).
9. Cross low-level circuits and noisy circuits at right angles and at maximum practical distance.
10. Use individual twisted shielded pairs for each transducer. Thermocouple transducers may be used with a common shield when the physical layout allows multiple pair extension leads.
11. Unused shielded conductors in a low-level signal cable should be single-end grounded with the shield grounded at the opposite end.
12. High standards of workmanship must be rigidly enforced at all times.

Filtering

Filtering is required to stabilize the input signal and to remove AC noise components (particularly 60-Hz noise) resulting from direct connection of normal-mode AC signals, from normal-mode noise pickup, or from conversion of common-mode noise to normal-mode noise due to line unbalance. A reliable, low-cost, and effective filter for analog inputs is the balanced resistance-capacitance filter, shown in Figure 7.11l. Its ability to eliminate AC components increases exponentially with the frequency of noise signal components. Common-mode noise rejection of about 40 decibels is possible with this type of filter, with decibel defined as

$$\text{decibel} = 20 \log \frac{\text{inlet noise amplitude}}{\text{outlet noise amplitude}} \qquad 7.11(3)$$

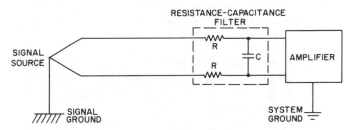

Fig. 7.11l A basic balanced resistance-capacitance filter

Filtering action causes a time delay (time constant, T = RC) between a signal change at the transducer and the time of recognition of this change by the measuring device. As the time delay may be in the order of one second or more, there are situations in which this has to be reduced (systems with high-frequency response). In order not to decrease the filtering efficiency but only the time delay, inductance-capacitance filters may be used (Figure 7.11m). This increases the filter cost.

Since a filter limits the bandwidth of the transmitted signal, it might be desirable to use more complicated and expensive filters when higher-frequency AC transducer signals are involved. Fortunately, this type of situation seldom occurs.

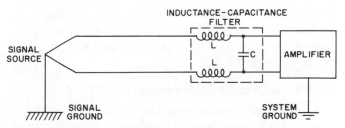

Fig. 7.11m A basic inductance-capacitance filter

Amplifier Guard

In a normal measuring device, a signal line is connected to a differential DC amplifier with floating inputs.

Generally, these amplifiers are provided with an internal floating shield that surrounds the entire input section, as shown in Figure 7.11n. This floating internal shield is called a "guard shield" or simply a "guard."

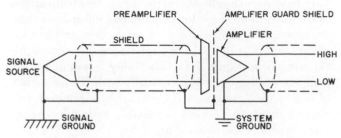

Fig. 7.11n   Isolating amplifier with proper grounding

The guard principle requires that the amplifier guard be driven at the common-mode voltage appearing at the amplifier inputs. The most effective way to do this is as follows:

1. Connect amplifier guard to the signal cable shield and make sure that the cable shield is insulated from chassis ground or from any extension of the system ground.
2. Connect the ground of the signal source to the signal cable shield. In this way, the amplifier guard and signal cable shield are stabilized with respect to the signal from the source.
3. Connect the signal cable shield and its tap to the signal source and also to the signal ground, which should be as close as possible to the signal source. This will limit the maximum common-mode voltage. The signal pair must not be connected to ground at any other point.
4. Connect the amplifier chassis, equipment enclosure, the low side of amplifier output, and output cable shield to the system ground.

Power Transformer Guard

It is important to avoid strong magnetic fields emanating from the power supply transformer. To avoid capacitive coupling, the transformer should be provided with at least two or three shielding systems. The third or final shield should be connected to the power supply output common. The inner shield should be connected to the signal ground.

Low-Level Signal Multiplexing

Conventional noise rejection techniques, such as twisting leads, shielding, and single-point grounding, have been discussed. It was pointed out that input guarding involves isolating the transducer input signal from the common-mode voltage between signal leads and ground. This change from differential to single-ended signal can occur any place in the system after the multiplexer. Early versions of multiplexing systems employed crossbar

switches, relays, and similar electromechanical devices for low-level input signal multiplexing.

A simple passive filter (RC) circuit is shown in Figure 7.11o, which can be designed to reject common-mode noise from about 40 decibels, at the selected frequency. More sophisticated passive networks (such as parallel T or notch filters) improve noise rejection, but it is hard to obtain 60-decibel noise rejection with passive circuits.

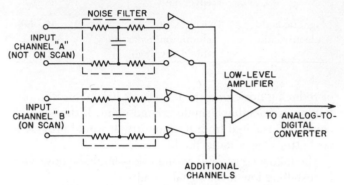

Fig. 7.11o   Passive filter multiplexing circuit

Because of deficiencies in the noise rejection capabilities of the earlier approaches, the limitations in scan rates, and with the ever-increasing use of data acquisition systems in many fields of application, many devices have been developed with extended capability to cover the spectrum of present-day requirements. Each general category contains subsets of devices using the same basic switching element, but offering application-dependent variations. The most important variations are the programmable range (i.e., gas chromatograph signal) switching and input grounding.

There are three commonly used multiplexing techniques: (1) capacitive-transfer-multiplexer, (2) three-wire multiplexer, and (3) solid-state multiplexer.

The capacitive-transfer (flying capacitor) switching arrangement, shown in Figure 7.11p, is simple, economical, and capable of great noise rejection (including random noise spikes). This system is limited to applications where signal bandwidth requirements are narrow (on the

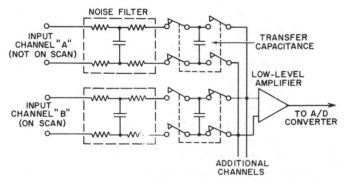

Fig. 7.11p   Capacitive transfer (flying capacitor) coupling multiplexer circuit

order of 0.5 to 1 Hz) due to a necessarily large transfer capacitance to minimize the effect on system resolution of charge losses and of delays during charging, settling, and digitizing.

In the capacitive transfer circuit, between scans, a set of normally closed contacts connect the low-leakage transfer capacitor across the input signal. Common practice is to short the amplifier input during this between-scan period to avoid stray pickup (due to high-impedance open circuit). When the multiplexer selects the input for scan, the amplifier input short is removed, and the contacts switch to connect the transfer capacitor to the amplifier input. This transfer circuit introduces input attenuation and phase lag, which must be considered in circuit design. An additional RC filter is usually necessary to achieve acceptable common mode rejection of 100 to 120 decibels.

The three-wire multiplexing system requires the transducer lead wires to be shielded with the shield terminated at the signal ground, and this guard shield must be carried through the multiplexer, at each point, up where the differential signal is transferred to a ground-referenced signal.

In Figure 7.11q the input amplifier and analog-to-digital converter are enclosed within the guard. The serial digital data are transmitted through a shielded pulse transformer to an output register for presentation in parallel form. Relay coils are matrixed and controlled by an address logic. This system is used when input signal bandwidths of several hertz are required, since filtering is not essential to obtain good common-mode rejection. Common-mode rejection from DC to 60 Hz is about 100 decibels at reduced input bandwidth.

A solid-state multiplexing system is shown in Figure 7.11r. In a typical high-performance solid-state multi-

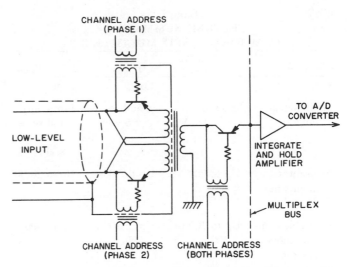

Fig. 7.11r   An integrate and hold amplifier with solid-state multiplexing system for high common-mode rejection

plexer, each input has a matched pair of transistor switches terminating at the primary of shielded input transformers, which are driven through an isolation transformer. One cycle of a square wave of peak amplitude equal to the input signal level is transferred across the transformer by alternately pulsing the switches. This signal is synchronously rectified to preserve original input polarity integrated over the cycle period, and it is amplified and held for digitizing by an integrator and hold amplifier. The cost of the system is relatively high, making the application economically impractical unless the high common-mode tolerance is required. Common-mode rejection from DC to 60 Hz of 12 decibels is easily obtained.

Selection of any multiplexing system should be based on the performance, reliability, and cost of a particular application. Table 7.11s summarizes the features of the systems discussed.

Noise Rejection in A/D Converters

The dominant noise in A/D converters is line frequency noise. One approach toward reducing this noise is to integrate the input signal. The integrating technique relies on A/D converter hardware. The operation of the A/D converter is such that it converts the continuously monitored measurement into a pulse train and totals the number of pulses it receives. If the measurement time is equal to the period of the line frequency, integration yields the true value of signal level, and the line frequency noise effect becomes zero, as shown in Figure 7.11t.

This method is usually applied to slow multiplexers (e.g., 40 points per second scan rate) and is suitable for applications in process monitoring and control where high sampling frequencies are not required. In appropriate situations, it provides noise rejection of about 1,000

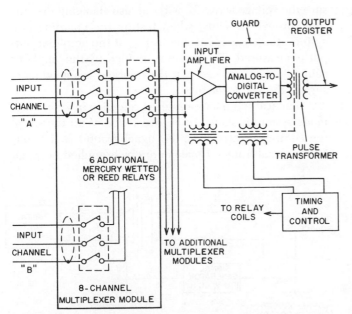

Fig. 7.11q   Circuit diagram of electromechanical, high common-mode voltage rejection, three-wire multiplexing system

Table 7.11s
FEATURE SUMMARY OF
COMMONLY USED MULTIPLEXERS

Features	Capacitive-Transfer	Three-Wire	Solid-State
Cost	Low	Low	High
Input signal ranges	±50 to ±500mV	±5mV to ±5 volts	From 5mV full scale to 100mV full scale
Scan rates (points per second)	Up to 200	Up to 200	Up to 20,000[1]
Operable common-mode environment (volts)	Up to 200 to 300	Up to 200 to 300	Up to 500
Common-mode rejection from DC to 60 Hz (decibels)	100 to 120[2]	100 to 120	120
Accuracy	±0.1% full scale at 1 to 10 samples per second scan rate; ±0.25% full scale at 10 to 50 samples per second	±0.1% full scale for all scan rates	±0.1% full scale[3]

[1] 10-microvolt resolution in high common-mode environments.
[2] With a two-section filter.
[3] Overall accuracy.

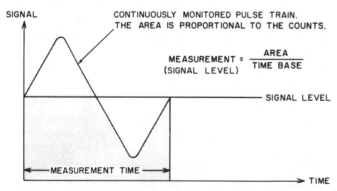

Fig. 7.11t   Line frequency noise reduction by A/D converter integration technique

to 1 (or 60 decibels) at 60 Hz, and offers good rejection of other frequency noises. (See Section 7.3.)

Common-Mode Rejection Measurement

A typical configuration of a guard-shielding measuring system is shown in Figure 7.11u. The basic definition of common-mode rejection is the ratio of the common-mode

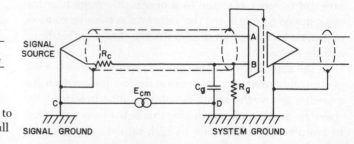

Fig. 7.11u   A typical configuration of a guard-shielded measuring system

R_c = Cable unbalance resistance
C_g = Measuring circuits/system ground capacitance
R_g = Measuring circuits/system ground resistance
E_{cm} = Common-mode voltage

voltage (E_{cm}) between points C and D, to the portion of E_{cm} that appears at the amplifier inputs A and B. Almost all of the remainder of the voltage E_{cm} will appear across C_g and R_g, in which case the common-mode rejection (CMR) is approximately

$$CMR_{(AC)} = \frac{1}{2\, f C_g R_c} \qquad 7.11(4)$$

$$CMR_{DC} = \frac{R_g}{R_c} \qquad 7.11(5)$$

It can be seen from these relations that the common-mode rejection is dependent on the value of R_c, which is the unbalance of the transducer and signal lines. In an ideal case, when R_c is zero, the common-mode rejection will be infinite.

Measuring that portion of E_{cm} that appears across R_c by direct methods is not possible, since it is unlikely that any instrument could be connected to the source to measure the voltage across R_c without also changing the current through it. Thus, when testing the guarding of a system, the component due to E_{cm} at the amplifier output is measured. This value is divided by the amplifier gain, and it is assumed that the quotient is a good measure of the voltage across R_c.

A practical setup for measuring rejection in a common-mode system is shown in Figure 7.11v. With the signal source disconnected from the signal ground, a 100-volt, 60-Hz signal from a signal generator is applied between

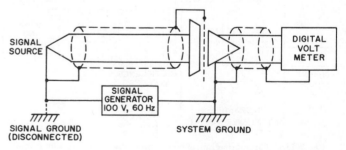

Fig. 7.11v   A practical setup for measuring common-mode rejection

the system ground and signal source. The change in the digital voltmeter (DVM) reading will show the effect of this common-mode signal on the system. To obtain the effective voltage across the amplifier inputs, this change is divided by the amplifier gain. The common-mode rejection is then obtained by dividing the signal generator voltage with the assumed voltage at the amplifier input. For example,

$$\text{Amplifier gain (G)} = 1,000$$
$$\text{Signal generator voltage } (E_{cm}) = 100 \text{ volts}$$
$$\text{DVM reading before applying } E_{cm} \ (e'_{cm}) = 2.0 \text{ volts}$$
$$\text{DVM reading after applying } E_{cm} \ (e''_{cm}) = 2.1 \text{ volts}$$

The increase in voltage at the amplifier output (e_{cm}) due to E_{cm} is

$$e_{cm} = e''_{cm} - e'_{cm} = 2.1 - 2.0 = 0.1 \text{ volts} \quad 7.11(6)$$

The voltage at the amplifier input (e'_{cm}) due to E_{cm} is

$$e'_{cm} = \frac{e_{cm}}{G} = \frac{0.1}{1,000} = 0.0001 \text{ volts} \quad 7.11(7)$$

Common-mode rejection of the system is

$$CMR = \frac{E_{cm}}{e'_{cm}} = \frac{100}{0.0001} = \frac{1,000,000}{1} = 10^6{:}1 \quad 7.11(8)$$

When dealing with AC common-mode signals applied to DC measuring instruments, it is important to consider the effect of inherent noise rejection in the measuring instrument (or amplifier). Many DC instruments have an input filter that allows the undisturbed measurement of DC signals in the presence of AC noise. Such instruments are said to have normal-mode interference rejection. Thus, if common-mode rejection is measured by the indirect method just described, the apparent common-mode rejection will be the product of the actual common-mode rejection and of the filter rejection. In the earlier example, if a filter with 10:1 normal-mode rejection were inserted in front of the amplifier input, the apparent common-mode rejection would increase to $10^7{:}1$.[2]

Conclusions

Electrical noise is not unique to new installations. Even systems that are in satisfactory operation can develop noise as a result of burned or worn electrical contacts, defective suppressors, loose connections, and the like. Any equipment used to identify the noise source must have the frequency response and rise time capability to display noise signals. Minimum requirements would be 10 MHz bandwidth and 35 nanosecond rise time. Once the noise signal is identified, its elimination involves applying the basic rules discussed earlier. The most important ones are listed below:

1. Select signal sources with low output impedance and with grounding capability.
2. Use only top-quality signal cable, in which the signal pair is twisted and protected with a lapped foil shield, plus a low-resistance drain wire.
3. Provide a good low-resistance system ground and a stable signal ground located near the signal source.
4. Ground the signal cable shield and signal cable circuit at the source only, and keep them from contacting ground at any other point.
5. Select appropriate multiplexing and amplifier circuits having input guarding systems of at least $10^6{:}1$ common-mode rejection at 60 Hz.
6. Provide triple-shielded power supplies for measuring devices.
7. Measure the common-mode rejection of the system by an indirect method whenever the rejection capability is in doubt.
8. Pay particular attention to the quality of the workmanship and system wiring practices.
9. Perform the analog-to-digital conversion as close to the signal source as practical.

REFERENCES

1. Lovuola, V.J., "Preventing Noise in Grounded Thermocouple Measurements," *Instruments & Control Systems*, January, 1980.
2. Deavenport, J.E., "EMI Susceptibility Testing of Computer Systems," *Computer Design*, March, 1980.

BIBLIOGRAPHY

Denny, H., "Grounding for the Control of EMI," Virginia: DWCI, 1983.
Kaufman, A.B., "Third Party Junctions in TC's," *Instruments & Control Systems*, December, 1981.
Kells, L., *Differential Equations*, New York: McGraw-Hill, 1968.
Klipec, B., "Reducing Electrical Noise in Instrument Circuits," IEEE Transactions on Industry and General Applications, 1967.
Morrison, R., *Grounding and Shielding Techniques in Instrumentation*, New York: John Wiley & Sons, 1977.
Nalle, D.H., "Signal Conditioning," *Instruments & Control Systems*, February, 1969.
Schmid, H., *Electronic Analog/Digital Conversions*, New York: Van Nostrand, Reinhold, 1970.

Chapter VIII

CONTROL OF UNIT OPERATIONS

R. J. Baker • B. Block

A. M. Calabrese • E. J. Farmer

D. M. Gray • B. P. Gupta

H. I. Hertanu • F. B. Horowitz

D. L. Hoyle • K. J. Jentzen

L. A. Kane • C. H. Kim

B. G. Lipták • T. J. Myron, Jr.

G. A. Pettit • C. J. Santhanam

CONTENTS OF CHAPTER VIII

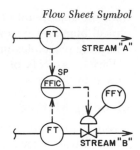

8.1 BLENDING AND RATIO CONTROLS

Type of System:	(a) Analog Mechanical, (b) Analog Pneumatic, (c) Analog Electronic, (d) Digital Blending
	NOTE: In the following feature summary, the letters (a) to (d) refer to the listed designs.
Features Available:	All designs are available with indicating, recording, or remote blend ratio adjustment features. In the case of digital systems, signal conversion is usually required.

Ratio Adjustment Ranges:

	Linear	Square Root
(a)	0.1 to 3.0	0.5 to 1.7
(b)	0 to 3.0	0 to 1.7
(c)	0.3 to 3.0	0.6 to 1.7
(d)	0.001 to 1.999	

Precision:

	Inaccuracy	Repeatability
(a)	±2%	
(b)	±1%	0.25%
(c)	±0.5%	0.25%
(d)	±0.25%	0.1%

Controller Cost Ranges:	(a) $1000–$2000 (b) $1000–$2500 (c) $1500–$3500 (d) $6500–$12,500. In the case of (d), signal-conditioning equipment is included, and the price is given on a per–flow stream basis.
Partial List of Suppliers:	Fischer and Porter Co. (a and d); Foxboro Co. (b, c, and d); Jordan Controls, Inc. (c); Moore Products Co. (b); Neptune Measurement Co. (d); Robertshaw Controls Corp. (c); Taylor Instrument Co. (a and c); Waugh Controls Corp. (d).

Blending systems are applied to a variety of materials in a number of industries: solvents, paints, reactor feeds, foams, fertilizers, soaps, and liquid cleaners in the chemical industry; gasoline, asphalt, lube and fuel oils, and distillates in the petroleum industry; wine, beer, candy, soups, ice cream mix, and cake mixes in the food industry; cement, wire insulation, and asbestos products in the building industry.

These applications provide the processor with economic advantages by controlling the consumption of materials (costly components and additives can be blended more precisely) and by reducing investment in floor space and batching tanks (costly blend tanks are eliminated). Through the use of a continuous system, time lags of batch methods are eliminated, productivity is increased,

manpower needs are reduced, and inventory can be in the form of component base stocks rather than as partially blended or finished products.

Blending systems provide technical advantages by accurately controlling the quality of the product and by providing the flexibility to blend a variety of finished products with a minimum time required to change from one product to another.

These applications have a common denominator, which is the continuous control of the flow of each component with fixed ratios between components, so that when the streams are continuously combined to form the finished blend at a fixed through-put rate, the composition of the finished product is within specifications.

In this section, blending systems will be described

from the standpoint of control techniques. A number of typical blending systems will be described to show the operating principles involved.

Various aspects of analog ratio control are also discussed in Sections 1.5, 1.6, 8.8, and 8.10.

Blending Methods

Automatic, continuous, in-line blending systems provide control of gases, liquids, and solids in predetermined proportions at a desired total blend flow rate. The blending systems consist of flow transmitters (to detect controlled variables), ratio relays (to set proportions), and controllers (to complete the closed-loop control).

A two-stream blending system is illustrated in Figure 8.1a. The component "A" flow controller is set by the total blend flow controller, and the component "B" flow is ratioed to "A." It is also possible to have both blending components ratioed to the total blend flow, as shown in Figure 8.1b. In either case, the blending system maintains the blending ratio as well as the total flow rate. Incorporation of a preset totalizer with automatic system shutdown facilities provides batching capability as well.

Many of the commercially available blending systems provide system options, such as flow alarm indications, system shutdown features, temperature compensation circuitries, scalers for conversion of transmitted signals to easily understood engineering units, pacing controls to slow down or shut down automatically, and manual or automatic adjustment of blend rate and ratio. All of these systems share one type or another of a ratioing mechanism. The methods of blending will be examined here. The working principles of components will be detailed in the latter part of this section.

Rate Blending

The blending system shown in Figure 8.1a is commonly found in the chemical industry for blending gas or liquid flows. A typical example is the manufacture of hydrochloric acid, which is done at fixed concentration and regulated flow rate by the absorption of anhydrous hydrogen chloride gas in water. The flow rate of hydrochloric acid from the absorption tower is measured to set the water flow, thereby maintaining the desired through-put, and the anhydrous hydrogen chloride gas is ratioed to water flow to give constant concentration.

A system in which all the blend components are ratioed to the total blend flow is illustrated in Figure 8.1b. A well-known example of this application is in the continuous or semicontinuous charging of a batch reactor, in which the recipe is given to set only the ratios of each ingredient and the total reactant charging rate. In the semicontinuous batch operation, a preset totalizer is used to terminate the charging operation when all ingredients have been charged. Numerous streams can be blended by incorporation of additional ratio devices and related controls.

Totalizing Blending

When totalized flows are ratioed, the integrated quantity of each component (over a period) is controlled in a direct ratio to the total quantity of the blended product. A schematic diagram of a totalizing blending system is shown in Figure 8.1c. By this system, more precise control over the amount of each component is obtained than is possible with a rate-blending system. In the rate-

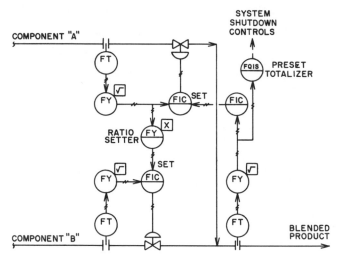

Fig. 8.1a   Analog rate blending of two components

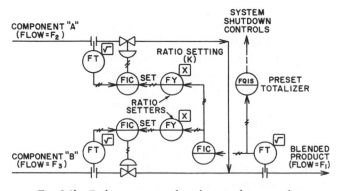

Fig. 8.1b   Both components directly ratioed in an analog
blending system

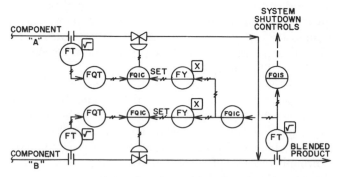

Fig. 8.1c   Analog totalizing blending system

blending system corrections are made to the ratio controller flow only after deviations have occurred (feedback control) and without correction for errors that have already occurred. In other words, the control system has no memory. Totalizing the flows and comparing the totals ensures the precise percentage of each component in the total blend.

Digital techniques have also been applied to digital blending systems, as shown in Figure 8.1d. Turbine meters or other pulse-generating devices can be used to generate digital flow signals. A bi-directional counter is used to integrate the flow measurement and demand signals and to compare them in order to generate a corrective control signal whenever the counts in the bi-directional counter memory are not zero.

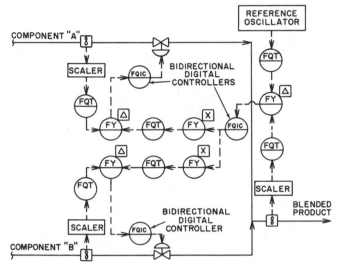

Fig. 8.1d   Digital totalizing blending system

The totalizing-blending system (analog or digital) finds use in continuous in-line blending of petroleum products, such as gasoline and asphalts, in which long runs and batching operations require precise control of ingredients to ensure in-spec blending and uniform end-products. The accuracy of this system allows for on-line blending of many petroleum, chemical, food, and cement products, with the product being sent directly to final shipping and storage containers. Multi-component blending systems can be obtained by the addition of more flow ratio controllers.

Optimizing Blending

Almost all of the commercially available ratio relays and controllers are able to accept remote set points. Therefore, the blend ratios and total blend flow rate can be automatically adjusted by a process variable. Thus, an optimizing blending system has the added capability of automatically manipulating ratio settings or the total rate, or both, based on certain criteria. A schematic of an optimizing blending system is shown in Figure 1.6d,

where the blend analyzer is used to measure composition and to adjust the ratio setting while the level in some downstream tank is used to set the total flow rate.

In the hydrochloric manufacturing process, a densitometer can be used to detect the solution concentration and to correct the ratio settings, if deviations from set point occur. A chromatographic analysis or Reid vapor pressure measurement of gasoline provides automatic adjustment of the blending ratio of components such as butane to give the proper octane number.

Another application is reactor charge rate control of jacketed exthotheric reactors, based on the heat transfer coefficient and plant cooling capacity. For the selection of flow sensors, refer to the Process Measurement volume.

Analog Blending

The heart of the analog blending system is the mechanism for ratio control. This is often a separate component, although it may be housed with the controller. As shown in Figure 8.1a, a blending system can be constructed by ratio controlling the blend components with the total blend flow. Thus, the total blend rate is controlled together with the individual blend ratios. With incorporation of a preset totalizer, system shutdown can be initiated when batch blending is completed.

The ratio control relationship is derived with reference to the system shown in Figure 8.1b. It is assumed that C_1 and C_2 are the flow constants for the flow-measuring orifices. Then,

$$F_1 = C_1 \sqrt{P_1} \text{ or } P_1 = (F_1/C_1)^2 \qquad 8.1(1)$$

and

$$F_2 = C_2 \sqrt{P_2} \text{ or } P_2 = (F_2/C_2)^2 \qquad 8.1(2)$$

where

F_1 = total flow rate,
F_2 = component "A" flow rate,
P_1 = output signal of total flow transmitter, and
P_2 = output signal of component "A" flow transmitter.

If K is the desired ratio setting,

$$P_2 = KP_1 \qquad 8.1(3)$$

By substituting equation 8.1(3) into (2),

$$F_2 = C_2 \sqrt{KP_1} \qquad 8.1(4)$$

and by substituting equation 8.1(1) into (4) and simplifying,

$$F_2 = (C_1/C_2)(\sqrt{K})(F_1) \qquad 8.1(5)$$

Using the same mathematical derivation, a ratioing system with linear input signals can be expressed as

$$F_2 = (C_1/C_2)(K)(F_1) \qquad 8.1(6)$$

A graphical representation of equation 8.1(6) is shown in Figure 8.1e, top, and the inverse-ratio controller characteristic is illustrated on the bottom. From equation 8.1(5) it can be seen that the square root input signals from flowmeters can be used as in Figure 8.1e, with the ratio flow control circuit based on a linear relationship; the only modification is the square root calibration of the ratio setting dial and of the indicating dials. The ratio setting dial is graduated and calibrated as the square root of the linear ratio. Most ratio control mechanisms provide for bias adjustments to change the basic characteristics shown in Figure 8.1e, to meet various process requirements. Figure 8.1f shows some of these biased relationships in which the secondary flow has a preset minimum value or is kept zero until the primary flow reaches some value. Also refer to Figure 8.8d and text for a discussion of the limitations of biased ratio relays.

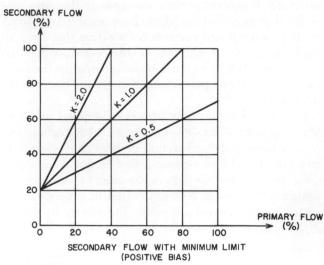

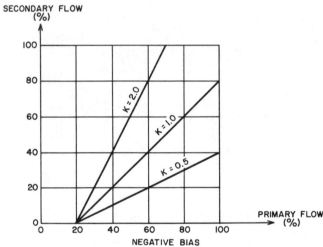

Fig. 8.1f   Biased ratio relationships

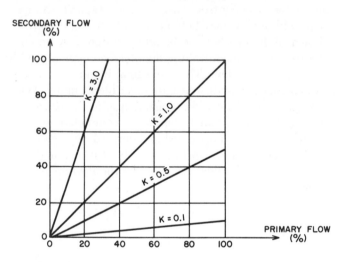

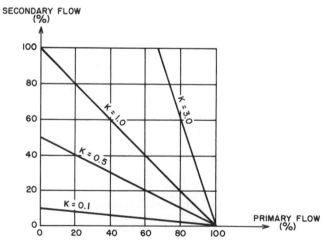

Fig. 8.1e   Direct ratio relationship (*top*) and inverse ratio relationship

Mechanical Ratio Control

The mechanical ratio control system consists of a proportioning mechanism, pneumatic or electronic flow signal receivers, and a case-mounted controller. The adjusting system resets the control point at a preset ratio by means of an adjustable mechanical linkage. The receivers are linked to a pen assembly, and output motion from the pen assembly operates the proportioning mechanism. Subsequent output motion from the proportioning mechanism positions the input lever of the pneumatic controller.

Figure 8.1g illustrates that each pen assembly spindle connects to one of the proportioning mechanism input levers through an adjustable link. Each input lever positions an internal crank arm assembly and raises or lowers one end of the ratio beam. In turn, the link from the ratio beam connects to the input lever of the controller. The overall result is that controller output pressure changes whenever a receiver moves an input lever.

For remote adjustment of ratio set point, the manual set ratio mechanism is replaced by a pneumatic receiver. An external 3 to 15 PSIG (0.2–1.0 bar) set point signal is used to position the receiver in proportion with the desired ratio. The ratio proportioning mechanism is precalibrated at the factory for the specific application, and all that is normally required is to check before use that

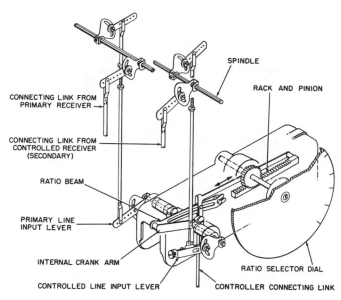

CONNECTING LINK FROM
PRIMARY RECEIVER

CONNECTING LINK FROM
CONTROLLED RECEIVER
(SECONDARY)

RATIO BEAM

PRIMARY LINE
INPUT LEVER

INTERNAL CRANK ARM

CONTROLLED LINE INPUT LEVER

SPINDLE

RACK AND PINION

RATIO SELECTOR DIAL

CONTROLLER CONNECTING LINK

Fig. 8.1g Ratio proportioning mechanism

the match marks are aligned and that the recorder pens and the transmitting meters are synchronized. Lubrication is seldom required, but the mechanism should be periodically inspected, cleaned, and checked so that the proportioning mechanism operates frictionlessly.

The output inaccuracy of ± 2 percent full scale can be obtained. Rangeability of this type of proportioning mechanism is approximately 40 to 1.

Pneumatic Ratio Control

The pneumatic ratio controllers contain no friction-producing mechanical links. The ratio relay modifies the input signal by means of the pneumatic circuit illustrated in Figure 8.1h. The primary variable signal is tubed through a fixed restriction (FO) into an adjustable area restriction. If the variable restriction valve is closed, the signal is not modified. This condition represents 100 percent ratio, because the controlled variable signal (the set point of the secondary flow controller) must equal the

primary variable signal for the control circuit to be satisfied. If the adjustable area restriction is opened, the pressure between the two restrictions will drop until the flow through the fixed area restriction equals the flow through the adjustable area restriction. Thus, the pressure is modified as a function of the opening of the adjustable area restriction. By calibrating the adjustable restriction in terms of percent ratio, one can set the relay for any desired ratio within its limits.

A booster relay should be used to ensure rapid transmission of the modified signal from the ratio relay to the controller. For applications in which the secondary variable set point will always be less than 100 percent of the primary, a 1:1 booster relay is recommended. In other cases a 2:1 or higher booster may be used.

The pneumatic set ratio circuit is identical to the manual set ratio relay, except that the adjustable restriction opening is set by a pneumatic diaphragm motor. This allows for continuous automatic adjustment of the ratio in accordance with a pneumatic signal received from a quality controller or other optimizing device. Incorporation of reset in the quality controller is recommended to eliminate the necessity for vernier adjustments to obtain the exact ratio and to compensate for the linearity limitations of the ratio unit.

The ratio relay should be calibrated at a specific ratio setting under actual operating conditions, even though the ratio setting is to be changed with operating conditions, to obtain maximum system accuracy. The accuracy (secondary flow set point) of ± 1 percent of full scale can be expected. Signal rangeabilities of 50 to 1 can be obtained, but system rangeability is determined by the flowmeters used. Most pneumatic ratio control systems are not designed to operate below 20 percent of full scale flow with square root signals or below 10 percent of full scale flow with linear signals.

Electronic Ratio Control

Electronic ratio control systems operate on the Wheatstone bridge principle, shown in Figure 8.1i. The bridge is said to be in a null, or balanced, condition when the ratios of resistance are such that $R_c/R_1 = R_f/R_2$, and no potential difference exists between points "A" and "B." If the ratio R_c/R_1 changes, then R_f/R_2 must also change by a like amount and in the same direction in order to maintain a null, or balanced, condition.

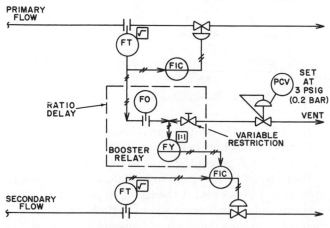

PRIMARY
FLOW

RATIO
DELAY

BOOSTER
RELAY

SET
AT
3 PSIG
(0.2 BAR)

VENT

VARIABLE
RESTRICTION

SECONDARY
FLOW

Fig. 8.1h Pneumatic ratio relay

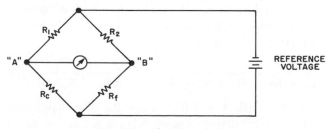

R_1 R_2

"A"   "B"

R_c R_f

REFERENCE
VOLTAGE

Fig. 8.1i Wheatstone bridge

Figure 8.1j illustrates the operating principle of a Wheatstone bridge control system. Here the fixed resistors, or bridge arms, are replaced by potentiometers so that the ratios previously mentioned are easily varied. Assuming an initial balance, an increase or decrease in the setting of the command potentiometer (primary) causes an error signal to appear at the input of a servo amplifier, which supplies power to the driven load. The direction of movement is dependent upon the polarity (or phase) of the error signal. The brush arm of the feedback potentiometer is mechanically geared to the driven load (secondary) and thus rotates until $R_f/R_2 = R_c/R_1$. At this point a null again exists (error signal equals zero), and positioning ceases. The actual position of each pot wiper, expressed as a fraction of its total possible travel, may be written

$$\frac{R_c}{R_c + R_1} \quad \text{and} \quad \frac{R_f}{R_f + R_2} \qquad 8.1(7)$$

At null, $R_c/(R_c + R_1) = R_f/(R_f + R_2)$. This is shown graphically in Figure 8.1k for a 0–100 percent movement of the command.

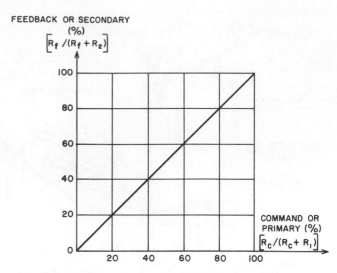

Fig. 8.1k   Command-to-feedback relationship

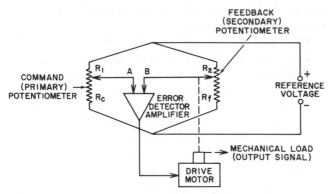

Fig. 8.1j   Wheatstone bridge operation

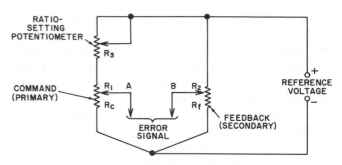

Fig. 8.1l   Electronic ratio circuit

In this example the feedback signal is a measurement of actual movement or displacement, but when used in a flow ratio application, the feedback signal will be related to the secondary set point, and the command becomes the primary (wild) flow variable signal. Introduction of a fixed resistance in series with the command pot causes a change in the slope of the characteristic curve in Figure 8.1k. By making this additional resistance a potentiometer, as shown in Figure 8.1l, one can limit the full range travel of the feedback pot and of the driven load (secondary) to any desired degree for a 0–100 percent movement of the command. In Figure 8.1l the feedback pot position at null will be

$$\frac{R_f}{R_f + R_2} = \frac{R_c}{R_c + R_1 + R_3} = K_1 \left(\frac{R_c}{R_c + R_1} \right) \qquad 8.1(8)$$

where $K_1 = (R_c + R_1)/(R_c + R_1 + R_3)$. The slope of the characteristic curve in Figure 8.1m is K_1, and its value is dependent upon the setting of R_3, which is usually calibrated to represent K_1, the ratio setting.

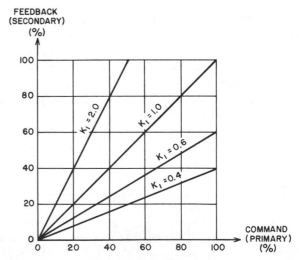

Fig. 8.1m   Command-to-feedback relationships

Electronic ratio control provides fast, accurate, and adjustable ratios between input and output signals. Accuracy of ±0.5 percent of span is attainable.

Care should be exercised to provide a constant supply voltage and frequency. A change of 10 percent from the nominal voltage will cause a zero shift of as much as 0.5 percent of input value, and a change of 10 Hz over the range of 47 to 63 Hz will cause a zero shift of 0.25 percent of input value.

Ratio Dial Setting

The setting of the ratio relay is a function of the ranges of the transmitters. If the transmitters are measuring over the same range and in identical units, the graduations on the ratio dial represent the exact ratio between the primary and secondary flows. However, where maximum capacities and primary meter measurement units differ, the ratio selector dial setting must be calculated for each ratio desired. (For a review of general scaling procedures, refer to Section 7.9.)

Commercially available ratio control units are graduated to handle signals of the same characteristics (either linear or square root) and of the same units.

The following equation is used to calculate ratio dial settings:

$$\text{Ratio dial setting} = \frac{(F_{pm})}{(F_{sm})}(R) \qquad 8.1(9)$$

where

F_p = flow rate through primary,
F_s = flow rate through secondary,
F_{pm} = maximum capacity of primary flow transmitter,
F_{sm} = maximum capacity of secondary flow transmitter, and
R = desired ratio of F_s/F_p.

If F_{pm} is 50 GPM (189 l/m) and F_{sm} is 25 GPM (95 l/m), and it is desired to maintain the secondary at exactly 25 percent of the primary flow, then the ratio dial setting is (50/25)(0.25) = 50%. In selecting the ratio dial settings, one should keep in mind the rangeability limitations of the flow transmitters. (See Chapter II of the Process Measurement volume for details.)

Ratio Controller Tuning

A block diagram of a simple ratio control system is shown in Figure 8.1n. In the ratio control system, the set point of the secondary controller is directly related to the output of the primary flow transmitter. As the primary flow changes, the secondary controller assumes a new set point to maintain the desired ratio. (For a discussion of transfer functions and their use in feedback loops, refer to Sections 1.3 and 1.4.)

If simple characteristics are assumed for the transfer functions in the block diagram, the overall system transfer function for set point disturbances can be expressed as:

$$F_s = RF_p \left(\frac{T_s(s+1)}{T_p(s+1)} \right)$$

$$\times \left(\frac{T_i(s+1)}{(T_iT_s/K_c)s^2 + T_i(1/K_c+1)(s+1)} \right) \qquad 8.1(10)$$

where

T_p = lag of primary flow measuring element,
T_s = lag of secondary flow measuring element,

T_i = integral time of controller, and
K_c = controller gain.

As can be seen from equation 8.1(10), the best regulation of the controlled secondary variable (with primary variable changes) can be obtained when the lags T_p, T_s, and T_i are minimum and the controller gain K_c is maximum, without creating instability. This statement is true not only for ratio loops but also for most feedback loops of all types.

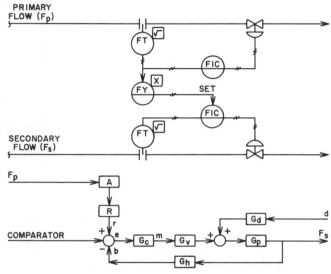

Fig. 8.1n   Ratio control loop.

A = Input element.	G_h = Feedback sensor transfer function.
b = Feedback variable.	
d = Disturbance or load variable.	G_p = Process transfer function.
e = Error (deviation) signal.	G_v = Control valve transfer function.
F_p = Primary flow.	
F_s = Secondary flow.	m = Manipulated variable.
G_c = Controller transfer function.	r = Reference (set-point) input.
G_d = Disturbance or load transfer function.	R = Desired ratio.

Digital Blending System

The application of digital techniques to ratioing and blending may result in the total elimination of control system errors. This system continuously compares the total accumulated flows from each additive line with the total accumulated signal from a master oscillator. If there is a difference between these two values the corresponding control valve is repositioned to correct the deviation.

An overall digital blending system is illustrated in Figure 8.1o. The flow of each component is digitized by a turbine or displacement type flowmeter or by an analog-to-pulse generator, producing a pulse train whose frequency is proportional to flow rate. A standardizer is used to scale the transmitter output frequency to a common reference basis, such as 1000 pulses per gallon (265 pulses per liter). This frequency is compared with a reference frequency produced by a numerically controlled frequency generator, which is commonly referred to as

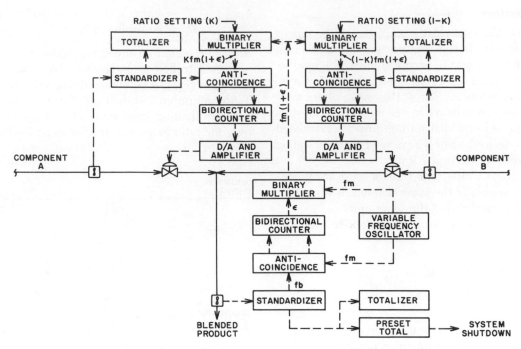

Fig. 8.1o   Schematic diagram for a two-element digital blending system

a *binary multiplier*. The inputs to the multiplier consist of a numerical quantity and a pulse frequency, and the output is a new pulse train whose frequency is the product of the two inputs. The multiplier produces two reference frequencies proportional to the manually set numerical ratio settings of K and $(1 - K)$.

Each digitized flow rate is compared with its corresponding demand signal generated by the ratio set module (binary multiplier). This comparison is performed by a *bi-directional binary counter*. The bi-directional counter counts in the positive direction on pulses from one input and in the negative direction on pulses from the other input. The set point pulses produce "add" pulses, and the measurement pulses produce "subtract" pulses. Hence, if the flow-generated pulses equal the demand pulses, the algebraic sum is zero and no change will occur in the binary memory, and no corrective action is taken. Should the rate from one input exceed that from the other, an error count will accumulate in the memory, causing the valve-control logic to generate a proportional correction. This correction signal, after conversion and amplification, positions the control valve. Thus, the quantitatively controlled flow rates of the blend components are maintained at the prescribed ratio.

For applications requiring precise control of total flow rate as well as of the blend ratios, a further digital control loop is provided, as shown in Figure 8.1o. Here a variable frequency oscillator is manually set so that its frequency is proportional to the desired total blend flow rate. This reference signal (f_m) together with the signal generated by the actual total blended flow rate (f_b) is synchronized by anticoincidence logic and accumulated

in a bi-directional binary counter. The instantaneous counts (accumulation) of this counter are a measure of the difference between the total number of pulses generated by the reference oscillator and by the flowmeter, respectively. Thus,

$$\epsilon = \Sigma f_m - \Sigma f_b \qquad 8.1(11)$$

where

ϵ = instantaneous error accumulation in bi-directional counter,

f_m = reference oscillator frequency, and

f_b = total blend flowmeter-generated frequency.

This instantaneous error (ϵ) serves as the numerical input into a binary multiplier whose input frequency is f_m. Therefore, the multiplier output frequency is ϵf_m. This output is then mixed with f_m in such a manner as to avoid time coincidence, and it thereby yields a pulse train having the average frequency of $f_m (1 + \epsilon)$. Thus, if the blend operation produces a flow rate that is less than the sum of the constituent flow rates, or if the blend output flow rate must be controlled while the blend ratios are kept constant, then the error term (ϵ) provides the necessary augmentation to the total flow rate reference frequency. The frequency input to the ratio setting binary multipliers is $f_m(1 + \epsilon)$, and the resulting ratio demand outputs are

$$Kf_m(1 + \epsilon) \qquad \text{and} \qquad (1 - K)f_m(1 + \epsilon) \qquad 8.1(12)$$

for the two-component blending system.

This principle can also be used automatically to slow

down the total blend flow rate, by substitution of the master demand frequency (f_m) with a component flow frequency as an input to the master bi-directional counter. This feature is useful when one component may fall behind at startup, when a strainer is plugged, or when a pump cannot meet the flow requirements. When this occurs, the component controller takes over the pacing from the master demand unit if a predetermined error has been accumulated and adjusts the total flow rate to a value that the component can maintain. An alarm and automatic shutdown logic circuitry can also be incorporated to signal alarm conditions or automatically to shut down the system if any of the components fall below their preset minimum rates.

The total blended product requirement may be preset on a totalizer to initiate batch shutdown. Analyzers or optimizers can be added to adjust automatically the blend ratios or the total blend flow rate as required.

The inaccuracy of the overall control system can exceed ±0.25 percent, with repeatability of better than 0.1 percent. The blend ratio setting can cover a range from 0.001 to 1.999, using four-digit thumbwheels. In a digital blending system, the dynamic response is limited only by the control valve stroke speeds, since the control system itself has practically no dead time.

Conclusions

Whether the blending system is designed by a user or is a package purchased from a manufacturer, the measuring and transmitting devices should be matched against the selected blending system in accuracy, rangeability, and flexibility. It is inconsistent to install an accurate digital blending system with low-accuracy sensors.

BIBLIOGRAPHY

Muhleisen, E.H., "Digital Flow and Blending," *Instruments and Control Systems*, May 1962.

Myron, T.J., "Ratioing Ingredients into the Blend," *Instrumentation Technology*, June 1973.

Schall, D.C., "Automatic Blending Conserves Refinery Tankage," *Control Engineering*, December 1962.

Walls, W.R., "Digital On-Line Process Control," *Industrial Electronics*, 1967.

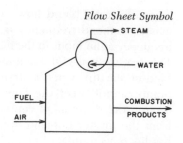

8.2 BOILER CONTROLS AND OPTIMIZATION

Boilers are available in two basic designs: fire-tube and water-tube. Fire-tube boilers are generally limited in size to approximately 25,000 lb/hr (3.15 kg/s) and 250 PSIG, (1.7 MPa), saturated steam. Although they are noted for their ability to respond to changing demands, their size and pressure limitations preclude their use in large industrial facilities. Because of thermodynamic considerations, boilers should produce steam at high pressure and temperature to realize a maximum work efficiency. These conditions are achievable only with water-tube boilers—hence, they will be given prime consideration in this section.

Steam boilers are used industrially both as a power source and in processing. They consist of a furnace, where air and fuel are combined and burned to produce combustion gases, and a water tube system, the contents of which are heated by these gases. The tubes are connected to the steam drum, where the generated water vapor is withdrawn. If superheated steam is to be generated, the steam from the drum is passed through the superheater tubes, which are exposed to the combustion gases. If one defines high pressure as exceeding 100 PSIG

(0.69 MPa), then this section is devoted to the general discussion of high-pressure steam boilers.

Figure 8.2a shows a typical boiler arrangement for gas or oil fuel with measurement points for control indicated.

Steam boilers as referred to in this section are drum-type boilers. Very large, supercritical pressure boilers are the "once-through" type and are found only in the largest electric generating plants.

A steam boiler outlet may be connected to a header in parallel with other boilers or directly to a single steam user. The boiler may be controlled by pressure or flow, depending upon the process requirements. The "load" on a steam boiler refers to the amount of steam demanded by the steam users.

The boiler must follow the needs of the plant in its demand for steam, and its control system must therefore be capable of satisfying rapid changes in load. Load changes can be a result of rapidly changing process requirements or cycling control equipment. Whereas load may be constant and steady over prolonged periods, the utility system must have sufficient "turn-down" to stay in operation at reduced capacities as portions of the plant may be shut

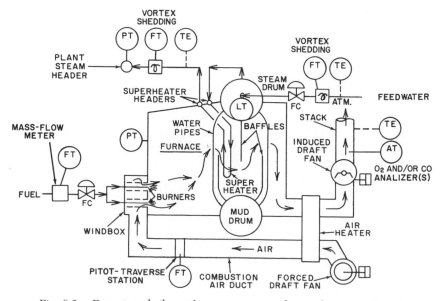

Fig. 8.2a   Drum-type boiler with measurement and control points indicated

down. This consideration usually leads to a greater "turn-down" requirement for the boilers than for any other portion of the plant. In addition, it is desired to maximize boiler efficiency. This is accomplished through optimization, which can result in oil-burning boiler efficiencies in the 85 to 88% range or in gas-burning boiler efficiencies at around 83 percent.

Sensor and Control Performance

The normal "on-line" requirements for steam boilers are to control steam pressure within ±1% of desired pressure; fuel air ratio within ±2% of excess air (±0.4% of excess oxygen), based on a desired "load" versus "excess air" curve; steam drum water level within ±1 inch of desired level; and steam temperature (where provision is made for its control) within ±10°F (5.6°C) of desired temperature. In addition, the efficiency of the boiler should be monitored within ±1%.

In order to reach these performance goals, it is necessary to install accurate sensors and to make sure that the load does not change more than 20% of full scale per minute and that there are no boiler design problems limiting this ability. The various loops tend to interact, so that integration into an overall system is necessary both during design and when the loops are being "field-tuned."

In the area of sensor selection, particular attention must be paid to the flow sensors. Fuel flow should preferably be metered within ±½% error of actual flow over a range of 10:1, using a gyroscopic mass flowmeter. Steam and water flows can be detected by vortex-shedding flow transmitters, having a ±1% error of actual flow over a 10:1 rangeability. Air flow can be detected using the area averaging pitot stations that have nozzle inserts to eliminate wall effects and membrane-type pressure-balancing d/p transmitters.

Safety Interlocks

Operations relating to safety, startup, shutdown, and burner sequencing are basically digital in nature and operate from inputs such as contact closures.

The basic safety interlocks are as follows:

Purge interlock—Prevents fuel from being admitted to an unfired furnace until the furnace has been thoroughly air purged.

Low air flow interlock and/or fan interlock—Fuel is shut off upon loss of air flow and/or combustion air fan or blower.

Low fuel supply interlock—Fuel is shut off upon loss of fuel supply that would otherwise result in unstable flame conditions.

Loss of flame interlock—All fuel is shut off upon loss of flame in the furnace and/or fuel to an individual burner is shut off upon loss of flame to that burner.

Fan interlock—Stop forced draft fan upon loss of induced draft fan.

Low water interlock (optional)—Shut off fuel on low water level in boiler drum.

High combustibles interlock (optional)—Shut off fuel on highly combustible content in the flue gases.

Where fans are operated in parallel, an additional interlock is required to close the shutoff dampers of either fan when it is not in operation. This is necessary to prevent air recirculation around the operating fan.

Automatic startup sequencing for lighting the burners and for sequencing them in and out of operation is common. This is accomplished at present by either relays or solid state hard-wired logic systems. These functions are frequently accomplished by PLC's.

Design and specification of the various safety interlocks are specialized as a result of involvement of insurance company regulations, NFPA (National Fire Protection Association), and state regulations.

Boiler-Pressure Controls

A combustion control system can be broken down for ease of examination into the *fuel control* and *combustion air control* subsystems. The interrelationships between these two subsystems necessitate the use of the *fuel-air ratio controls*. For safety purposes, fuel addition should be limited by the amount of available combustion air, and combustion air may need minimum limiting for flame stability.

The steam drum pressure is an indication of the balance between the inflow and the outflow of heat. Therefore, by controlling the steam supply pressure, one can establish a balance between the demand for steam (process load) and the supply of steam (firing rate). A change in steam pressure will result from a change in firing rate only after a delay of a few seconds to a minute, depending on the boiler and the load level. Therefore, as will be seen in Figure 8.2f, feed forward control can improve pressure control by adjusting fuel as soon as a load change is detected, instead of waiting for pressure to change first.

The various pressure control (or firing rate control) loops will be described in increasing order of sophistication, starting with the most traditional design configurations.

When more than one boiler is operated from the same master controller, a panel station should be provided for *each* boiler, for the purpose of balancing or intentionally unbalancing the loads.

This station provides for hand operation of the entire combustion control system of the associated boiler in addition to the ability to bias the master controller signal up or down when on automatic (Figure 8.2b).

When steam pressure is controlled by other means, steam flow can be the master controller.

If variations in fuel heating value are minor, the master

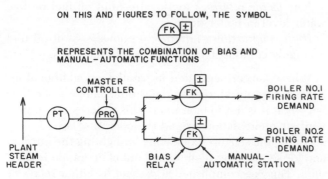

Fig. 8.2b   Load sharing controls among several boilers

flow controller shown might be eliminated and the master load signal generated by a manual loading station.

Situations may arise when it is desirable to have either flow or pressure control. In these cases, a master control arrangement, as shown in Figure 8.2c, can be used. Although it may appear simpler to switch transmitters, it is desirable to transfer the controller outputs so that the controller does not have to be returned each time the measurement is switched.

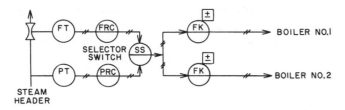

Fig. 8.2c   Boiler control with alternative pressure or flow master

A typical setting for a pressure-controlled "master" is 16 percent for proportional band and 0.25 repeats per minute for reset, whereas for a flow control "master" the comparable settings might be 100 percent proportional band and 3.0 repeats per minute for reset. (See Section 1.11 for details on controller tuning.)

Feedforward Control of Steam Pressure

In single boiler installations in which steam lines are small, steam line velocities are high, and the relative capacity in the steam system is low, it is sometimes necessary for the sake of control stability to correct the pressure "master" in a feedforward manner, as shown in Figures 8.2d and 8.2e.

These systems allow the use of wider proportional bands

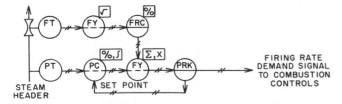

Fig. 8.2d   Boiler control using flow-corrected pressure signals

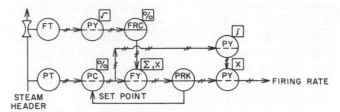

Fig. 8.2e   Flow-corrected pressure control with integral correction in the bypass

and lower reset settings on the pressure controller than would otherwise be possible for desirable plant operation. Of these two, Figure 8.2e is the more sophisticated, because the integral mode is in the by-pass, and therefore it can continuously adjust the fuel and air flow controls to match steam flow, thus providing for quick, stable response to major and rapid load changes.

Neither "master controller" loop in Figures 8.2d and 8.2e can be used with more than one boiler, unless the steam flow measurement is a total from all boilers or a high-flow selector is used, selecting the highest of the steam flows. This is a result of the positive feedback from steam flow to fuel feed rate. In the case of a single boiler, steam flow is only a function of plant demand. In the case of multiple boilers on a header, total steam flow is determined by the plant demand, but individual boiler steam flows are determined primarily by the firing rates.

The actual steam flow at any point is not necessarily a true indication of demand. For example, an increase in steam flow caused by increased firing should not be interpreted as a load increase—this would create a positive feedback loop, capable of destabilizing the boiler. According to Shinskey, the true load on a boiler can be approximated by $\sqrt{h/p}$, where "h" is the differential developed by a flow element and "p" is the steam pressure. Figure 8.2f describes a boiler-pressure control system using this type of feedforward model. Fuel flow is set proportionally to the estimate of load $\sqrt{h/p}$. Dynamic compensation is applied in the form of a lead-lag function to help overcome the heat capacity in the boiler.

The pressure controller adjusts the ratio of firing rate to estimated load, to correct for inaccuracy of the model and variations in heat of combustion of the fuel. The multiplier also changes the gain of the pressure feedback loop proportional to load. This feature is valuable in that boilers seem more difficult to control at lower loads because of lower velocities.

Fuel Controls (Measurable Fuels)

The primary boiler fuels are coal, oil, and gas, but there are a large variety of auxiliary fuels, such as waste gases, waste sludges, and waste wood products (bark, sawdust, hogged fuel, and coffee grounds). In many cases these auxiliary fuels are dumped to the boiler plant on

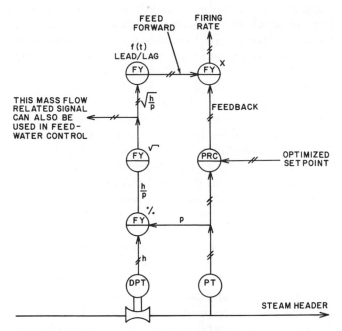

Fig. 8.2f Firing rate determination using feedforward loop with feedback trim

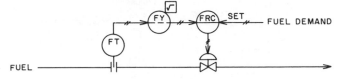

Fig. 8.2h Fuel controller used to keep demand and flow in linear relationship

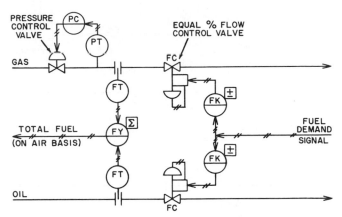

Fig. 8.2i Fuel demand is manually split between the two fuels on an open-loop basis

an uncontrolled basis for immediate burning. There are myriads of these combinations, and only the more common fuel control problems will be covered in the discussion to follow.

Gas and oil involve the simplest controls, since they are easily measured and require little more than a control valve in the fuel line. A valve positioner capable of providing a linear relationship between flow and control signal is desirable (Figure 8.2g). A flow controller is often needed as a means of more precise linearization (Figure 8.2h). In cases in which better than 3:1 turndown and high measurement accuracy is desired, the gyroscopic mass flow sensors should be used instead of orifices. When it is desired to fire the fuels in a predetermined ratio to each other regardless of load, a manually adjustable signal splitter can be used as shown in Figure 8.2i. The most precise and complex method of ratioing fuel (not shown) is to split the demand signal and send that to individual flow control loops. If the fuel is a gas at variable pressure, a pressure control valve is frequently installed upstream to the flow sensor, as shown in Figure 8.2i. Both valves affect both variables (pressure and flow), and therefore they will interact. In order to eliminate the resulting oscillations, one should either leave the pressure unregulated and pressure compensate

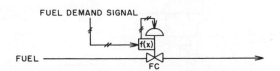

Fig. 8.2g Positioner used to maintain linear relationship between demand and flow

the flow sensor or assign less pressure drop to the pressure control valve than to the flow control valve (thereby using a larger valve for pressure control than for flow control). Because the burner backpressure will increase as flow increases, the available pressure differential for the flow control valve will decrease as the flow rises. In order to obtain an approximately linear relationship between fuel flow and valve position, an equal percentage valve is needed.

In the case of oil fuels, proper atomization at the burner, and therefore complete combustion, will be achieved only if the oil is kept at constant pressure and viscosity. When heavy residual oils are burned, they must be continuously circulated past the burner and back (Figure 8.2j). The difference between the readings of two oscillating gyroscopic mass flowmeters can signal the net flow to the burner. The burner backpressure is controlled by the control valve in the recirculating line, whereas the flow controller set point is adjusted by the firing rate demand signal. The firing rate is controlled by an alteration in the opening of the burner orifice. Atomizing steam is ratioed to the firing rate, and the heating steam is modulated to keep the fuel viscosity constant.

Figure 8.2k illustrates the controls required when the fuel demand is split between two fuels on a closed-loop (automatic) basis.

The instruments shown ("panel stations") provide the means of manual control plus the ability of automatic control, with bias of one fuel with respect to the other.

Since one of the requirements ultimately is to have fuel ratioed to combustion air, any totalization of fuel for

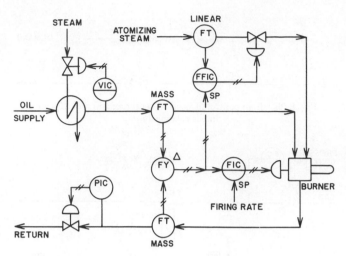

Fig. 8.2j   Oil flow control for recirculating burner provided with steam atomization

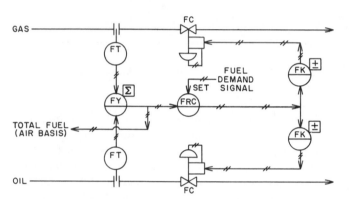

Fig. 8.2k   Fuel demand split between fuels on closed-loop basis

control purposes should be on an "air-required-for-combustion" basis. If totalization is needed on any other basis, such as BTU for other purposes, a separate totalizer should be used.

WASTE OR AUXILIARY FUEL CONTROLS

When auxiliary fuel is burned on an uncontrolled-availability basis, the fuel and air control system needs to be able to accommodate sudden changes in auxiliary flow without upsetting the master controller. The master controller should be designed and used to respond to total load demands only and not to correct for fuel upsets. A typical fuel control system for accommodating variations in auxiliary fuel without upsetting the master is shown in Figure 8.2l.

In the basic system without auxiliary fuel, the signal is relayed directly to the control valve. Addition of auxiliary fuel shifts the primary fuel control valve opening to prevent fuel variations from affecting overall boiler performance. A more precise system is shown in Figure 8.2m. Here the flow controller adjusts the primary fuel control valve to satisfy total fuel demand and prevents auxiliary fuel variations from upsetting the master controller.

Figure 8.2n describes a slightly more advanced system, in which the allowable maximum percentage of waste fuel that can be burned is set on the ratio relay FY-1. This ratio must be set under 100% if the heating value of the waste is so low that it could cause flameout if not enriched by supplemental fuel. For proper operation the subtractor FY-2 must be scaled with the flowmeter ranges taken into consideration, and further scaling is required if the heat flow range of the total heat demand does not match that of the waste fuel flowmeter. When waste fuel gas availability becomes limited, waste fuel gas pressure will drop and PY-3 will select it for control, thereby overriding the waste flow controller. FY-2 will respond to this by increasing the supplemental fuel flow.

Closed-loop control will give greater precision and better linearization, but the performance can be limited by poor hardware selection or poor installation practices. An example of the first case is the problem of transmitter rangeability if orifices are used instead of turbine or mass flowmeters. An example of poor installation practices is the case of long transmission lines being allowed to introduce dead time into the loop because pneumatic leads were installed without boosters.

Whatever type of fuel control is used, the maximum flexibility in design will be present if all flow signals are linear and all control valve characteristics are also linear. In this manner the various flows and signals can be com-

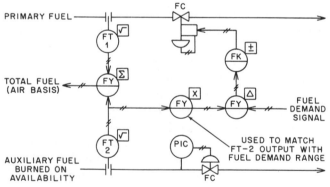

Fig. 8.2l   Auxiliary fuel is burned on uncontrolled-availability basis

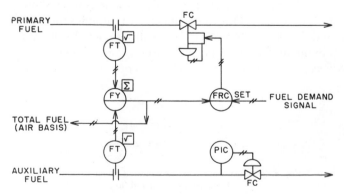

Fig. 8.2m   Closed-loop control of uncontrolled auxiliary fuel

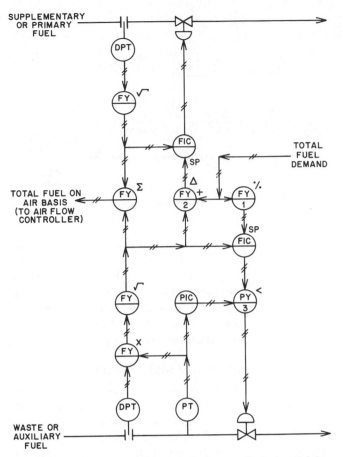

Fig. 8.2n Automatic control system for burning limited-availability waste up to preset maximum percentage in total mixture

In pulverized coal–fired boilers the coal is ground to a fine powder and is carried into the furnace by an air stream. There are normally two or more pulverizers (in parallel) per boiler. Pulverized coal flow is regulated at the pulverizer, and each manufacturer has a different design requiring different controls. One control arrangement is shown in Figure 8.2o. Here the primary air comes from a pressure fan that blows through the pulverizer, picking up the coal and transporting it to the furnace.

In addition to the controls shown, an air temperature control is required. In this loop, cold and hot combustion air is mixed ahead of the primary air fan to control the temperature of the coal air mixture in the pulverizer. This control is necessary to maintain a maximum safe operating temperature in the pulverizer. This is a simple feedback loop, usually involving proportional control only.

A control arrangement for a bowl-type pulverizer is

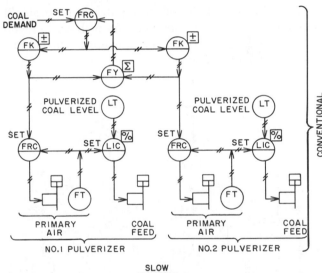

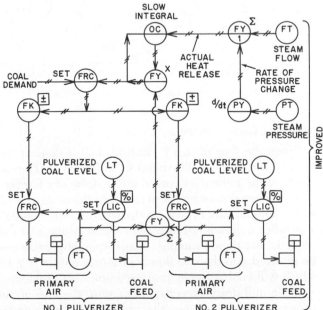

Fig. 8.2o Conventional and advanced control of ball-type pulverizers

bined, subtracted, multiplied, or divided to produce the desired control. One optimal condition is to have the total fuel demand signal linear, with fuel totalized on a basis of required combustion air. The other desired end condition is to have the total fuel control capacity maximum at a value approximately 10 percent greater than that required for maximum boiler-capacity. This excess is necessary for control flexibility at maximum boiler load. Additional excess capacity should not be considered, since it reduces turn-down capability.

Unmeasured Fuels

Coal can be an unmeasured fuel. In such cases coal control systems are open loop, wherein a control signal positions a coal-feeding device directly. This is the case with a spreader stoker or cyclone furnace or indirectly with pulverized coal.

A spreader stoker consists of a coal hopper on the boiler front with air jets or rotating paddles that flip the coal into the furnace, where a portion burns in suspension and the rest drops to a grate. Combustion air is admitted under the grate. There is no way to control fuel to a spreader stoker except in an open-loop manner by positioning a feeder lever that regulates coal to the paddles.

shown in Figure 8.2p. In this type of pulverizer the air
fan sucks air through the pulverizers with the fan (called
an exhauster fan) located between the pulverizer and the
burners. The coal-air temperature loop is similar to the
one described in Figure 8.2o.

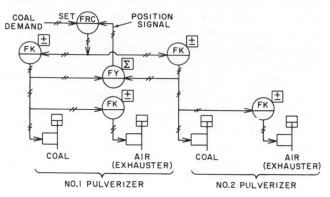

Fig. 8.2p   Coal fuel control using bowl-type dual pulverizers

The control of a ball mill–type pulverizer is again different from a control standpoint. This is shown in Figure
8.2q, including the application of manual compensation
for the number of pulverizers in service.

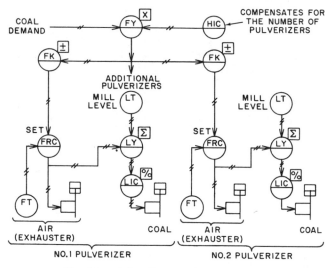

Fig. 8.2q   Ball mill–type pulverizer controls

Metered Coal Controls

When the flow of coal is controlled, it is done at the
inlet to the mill, as illustrated in Figure 8.2r. The capacity of the pulverizer contributes a delay of a few seconds to several minutes, depending on the design of the
mill. Hammer mill delays are the shortest, whereas ball
and roller mill delays can reach several minutes. If the
delay is less than a minute, a corresponding delay can
be inserted into the coal flow transmitter output (FY-1),
which will enable the loop to overcome this problem.

If the mill delay exceeds one minute, the flow of the
primary air that conveys the pulverized coal must be

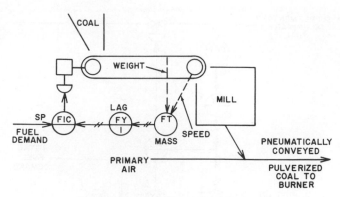

Fig. 8.2r   Compensation for mill capacity

manipulated as a function of coal demand. Because the
coal loading of the air is not uniform, Shinskey recommends the improvements noted in Figure 8.2o. These
include the determination of the actual heat release based
on steam reading (FY-1 in Figure 8.2o) and the slow
integral correction of the total flow of primary air until
the estimated and actual heat release are matched.

Air Flow Measurement and Control

Combustion air for steam boilers may be supplied by
induced draft (suction fan at boiler outlet or stack draft),
forced draft (pressure fan at inlet), or a combination of
forced and induced draft known as balanced draft fans.
With balanced draft boilers, a slight negative pressure
is maintained in the furnace.

In the control of combustion air (if there are both
forced and induced draft fans), one fan should be selected
for basic control of air flow and the other assigned to
maintaining the draft pressure in the furnace. The following discussion is based on a single air flow source (fan)
per boiler. Balanced draft and its effects on air flow control will be covered later in this section.

For successful control of the air-fuel ratio, combustion
air flow measurement is important. In the past it was
impossible to obtain ideal flow detection conditions.
Therefore, the practice was to provide some device in
the flow path of combustion air or combustion gases and
field calibrate it, by running combustion tests on the
boiler.

These field tests, carried out at various boiler loads,
used fuel flow measurements (direct or inferred from
steam flow) and measurements of percent of excess air
by gas analysis, and they also used the combustion equations to determine air flow. Since we are concerned with
a relative measurement with respect to fuel flow, the air
flow measurement is normally calibrated and presented
on a relative basis. Flow versus differential pressure characteristics, compensations for normal variations in temperature, and variations in desired excess air as a function
of load—all are included in the calibration. The desired
result is to have the air flow signal match the steam or

fuel flow signals when combustion conditions are as desired.

The following sources of pressure differential were considered:

> Burner differential
>> Windbox pressure minus furnace pressure
>
> Boiler differential
>> Differential across baffle in combustion gas stream
>
> Air heater differential
>> Gas side differential on air heater
>
> Air heater differential
>> Air side differential on air heater
>
> Venturi section or flow tube
>> Installed in stack
>
> Piezometer ring
>> At forced draft fan inlet
>
> Venturi section
>> Section of forced draft duct
>
> Orifice segment
>> Section of forced draft duct
>
> Air foil segments
>> Section of forced draft duct

Of these, the most desirable are the last four, because they use a primary element designed for the purpose of flow detection and measure flow on the clean air side. Some typical installations are shown in Figure 8.2s.

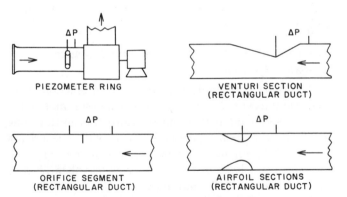

Fig. 8.2s Choices of air flow measurement

The development of the area-averaging pitot stations provided with "hexcel"-type straightening vanes and with membrane-type pressure balancing d/p cells represents a major advance in combustion air flow detection. These units for both circular and rectangular ducts are described in Figures 2.13k and 2.13l in the Process Measurement volume of this handbook.

Regardless of the type of primary element used, the signal obtained will frequently be noisy because of pulsations from the pumping action of the fans or from the combustion process. Provision should be made for dampening the flow signals, since otherwise the controller

cannot be tuned for sensitive response. Normal differential pressure ranges for these measurements are between 1 and 6 inches of water, for conventional sensors and as low as 0.1 inch of water for area-averaging pitot stations.

Damper and Fan Controls

Control devices for boiler air flow control on pneumatic installations are double-acting pistons, but in some cases electric motors are also used. In either case linear relationship is required between control signal and combustion air flow. Characteristics of most constant-speed fans or dampers approximate those given in Figure 5.1b, and the desired relationship is linear.

The relationships between open- and closed-loop control that were noted in connection with fuel control also apply to air flow control. Closed-loop air control is more precise, and self-linearizing, if the integrated system does not compare air flow with fuel flow (or as inferred from steam flow) directly in a ratio or difference controller.

Open-loop air control variations that may be used depending on the arrangement of fans are as shown in Figure 8.2t through Figure 8.2x. Figure 8.2x illustrates a single damper–controlled open loop. Figure 8.2u shows a combination of damper and speed control. A system of this sort is often necessary to increase turn-down (rangeability) where fan speed is variable. Good response of air flow based on fan speed adjustment alone is normally not attainable below approximately $\frac{1}{3}$ maximum speed. Depending on fan design, this may correspond to 50 percent of boiler capacity. Use of a damper in combination with speed adjustment allows further turndown, since fan speed is blocked at approximately $\frac{1}{3}$ speed. Split ranging, as shown in Figure 8.2u, conserves steam or fan power.

As a result of inlet damper leakage that is normally

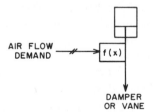

Fig. 8.2t Open-loop air control with single damper

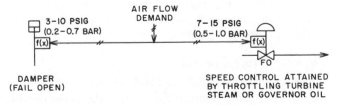

Fig. 8.2u Combination damper and speed control to increase rangeability

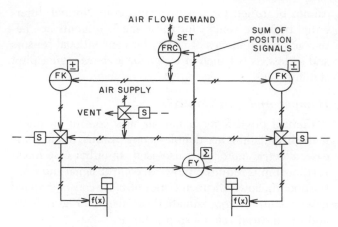

Fig. 8.2v   Parallel fans with automatic air flow–balancing controls

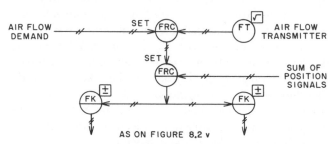

Fig. 8.2w   Closed-loop control of parallel fans with balancing controls

present, it may be necessary for wide-range low-load or startup control to parallel inlet and outlet dampers. To save fan power the inlet damper may be operated over the full 3- to 15-PSIG range (0.2 to 1.0 bar), whereas the discharge damper can be fully open at 3 PSIG (0.2 bar) and closed at 9 PSIG (0.6 bar).

As shown in Figure 8.2v, when two fans are operated in parallel or on single-fan operation, the idle fan should have its damper closed to prevent recirculation from the operating fan.

When one fan of a two-fan system is switched on or off, considerable manual operation is required to prevent serious air flow upsets. The system shown in Figure 8.2v eliminates this problem by automatically compensating the operating fan damper position as the parallel fan is started up or shut down. Separate discharge dampers may be used for shutoff purposes, supplementing the interlocks shown.

Closed-loop versions of the loops illustrated in Figures 8.2t, 8.2u, and 8.2v would consist of flow controllers with air flow feedback superimposed on the components shown. For example, Figure 8.2w shows the closed-loop version of Figure 8.2v.

Furnace Draft Control

Whenever both forced draft and induced draft are used together, at some point in the system the pressure will be the same as that of the atmosphere. Balanced-draft boilers are not normally designed for positive furnace pressure. Therefore, the furnace pressure must be neg-

ative to prevent hot gas leakage. Excessive vacuum in the furnace, however, produces heat losses through air infiltration. The most desirable condition is thus one in which there is a very slight (approximately 0.1 inch of water, or 25 Pa) negative pressure at the top of the furnace.

Pressure taps for measuring furnace pressure may be located some distance below the top. Because of the chimney effect of the hot furnace gases, pressures measured below the top of the furnace will be lower by approximately 0.01 inch of water per foot (0.008 kPa/m) of elevation. Thus, if the pressure tap is 20 feet (6 m) below the top of the furnace, the desirable pressure to maintain is approximately 0.3 in. of water (0.075 kPa) vacuum.

In the case of a balanced draft boiler the maintenance of constant furnace pressure or draft keeps the forced and induced draft in balance. The purpose of this balance is to share properly the duty of providing combustion air and to protect furnaces not designed for positive-pressure operation.

The measurement of furnace draft produces a noisy signal, limiting the loop gain to relatively low values. In order to provide control without undue noise effects, it is desirable to use a full span of approximately 4 to 5 inches of water. This is normally a compound range, such as +1 to −4, or +2 to −3, in. of water (+0.25 to −1 kPa, or +0.5 to −0.75 kPa). Even with this span and with a set point of −0.1 to −0.3 in of water (0.025 to 0.075 kPa), the controller gain can still not exceed 1.0. In some cases it may be necessary to use integral control only to stabilize the loop.

Interaction Between Air Flow and Furnace Pressure Controls

Additionally, stability problems and interactions may occur in the overall system because of measurement lags. It is recommended that the pressure transmitter connections to the boiler furnace be made with pipe at least 1 inch (25 mm) in diameter because of the very low pressures involved. If the distance is less than 25 feet (7.5 m), ¾-inch (18.75 mm) pipe may also be used.

Either the forced or the induced draft fan can be used to control the furnace draft, with the other fan performing the basic air flow control function. Interaction cannot be completely eliminated between these two loops, but it can be minimized by system designs such as those shown in Figures 8.2x and 8.2y.

The common rule is that air flow should be measured and controlled on the same side (air or combustion gas) of the furnace to minimize interaction between the flow and pressure loops.

If the combustion air is preheated, then its temperature will vary substantially and compensation is needed. The mass flow of air is related to $\sqrt{h/T}$, where "h" is a differential across a restriction and "T" is absolute temperature. This loop is shown in Figure 8.2y, together with an excess oxygen trim on the air flow controller.

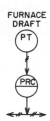

Signal to Forced Draft Fan
If air flow is detected by
  (a) boiler differential,
  (b) gas side ΔP of air heater,
  (c) venturi in stack,
and if air flow is controlled by throttling the induced draft fan.

Signal to Induced Draft Fan
If air flow is detected by
  (a) air duct venturi,
  (b) air duct orifice,
  (c) air side ΔP of air heater,
  (d) piezometer ring in forced draft fan inlet,
  (e) area averaging pitot on air side,
and if air flow is controlled by throttling the forced draft fan.

Fig. 8.2x   Choices of furnace draft control

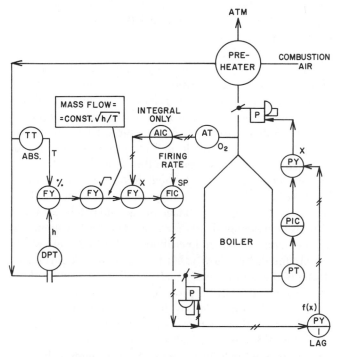

Fig. 8.2y   Parallel control of inlet and outlet dampers reduces interactions

The air flow and furnace pressures interact similarly to the phenomenon in Figure 8.2i. Because in the case of air flow controls the dampers are the same size and therefore their pressure drops are similar, decoupling needs to be applied. One might reduce interaction by connecting the two dampers (or fans) in parallel and using the furnace pressure as a trimming signal. (This can also be used to overcome problems resulting from noisy furnace pressure signals and slow response caused by the series relationship between flow and pressure loops.) This control system is illustrated in Figure 8.2y.

In this arrangement the air flow controls move both dampers equally, and the furnace pressure corrects for any mismatch. The furnace pressure might respond faster to a change in the downstream damper opening, and therefore a dynamic lag (PY-1) is provided.

Fuel-Air Ratio

When the controls for fuel-air ratio are considered, one point is very important. Because of the combustion gas velocity through the boiler, for safety reasons the fuel-air ratio should be maintained on an instant-by-instant rather than a time averaged basis.

As a general rule (except for the case of very slow-changing boiler loads), fuel and air should be controlled in parallel rather than in series. This is necessary because a lag of only one or two seconds in measurement or transmission will seriously upset combustion conditions in a series system. This can result in alternating periods of excess and deficient combustion air. Consequently, the discussion here will be limited to parallel air and fuel control systems.

The simplest control of fuel-air ratio is with a system calibrated in parallel, with provision for the operator to make manual corrections (Figure 8.2z). In this system the operator uses the bias provision of the panel station (FK) to compensate for variations in fuel pressure, temperature, or heating value or for air temperature, humidity, or other factors. A system of this sort should be commissioned with detailed testing at various loads for characterizing and matching fuel and air control devices. In addition, simple systems of this type should be adjusted for higher excess air, since they have no means of automatically compensating for the fuel and air variations.

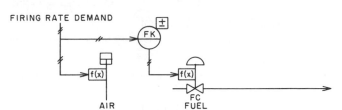

Fig. 8.2z   Simple parallel air-fuel ratio system

The next higher degree of sophistication is a system with simple proportional compensation. This can be done by balancing fuel burner pressure to the differential produced between windbox and furnace pressures.

In the system shown in Figure 8.2aa the burner fuel apertures are used to measure fuel flow, and the burner air throat is used as a primary element to detect air flow. There is no square root extraction (such extraction would not show true flows because of the nature of the primary elements), and therefore the actual controller loop gain changes with load (capacity). Rangeability is limited unless there are multiple burners that can be put into or taken out of service.

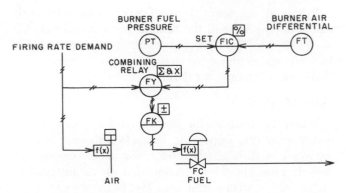

Fig. 8.2aa   Proportional compensation in air-fuel ratio control

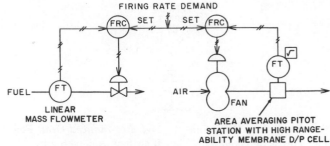

Fig. 8.2cc   Closed-loop air-fuel ratio control

The arrangement in this control system can also be reversed, with firing rate demand directly adjusting the fuel and the correction control being on air flow. This choice will be considered later in this section.

As boilers become larger, the need for precision control becomes greater, together with the potential for savings. The following series of diagrams represents further degrees of system sophistication.

The system in Figure 8.2bb is quite similar to that shown in Figure 8.2aa, except that here flows are measured as accurately as possible and are used in a flow controller to readjust the primary loop through the combining relay (FY).

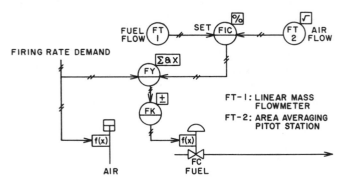

Fig. 8.2bb   Proportional compensation with accurately measured flows

In this system fuel and air flow are open-loop controlled, and only secondary use is made of their measurements. The undesirable consequence is that fuel disturbances need master controller action for correction, and the fuel and air loops can interact with each other. The effect of interaction and disturbances in the fuel and air control loops can be minimized by the use of closed-loop fuel and air control. The system shown in Figure 8.2cc (which uses essentially the same equipment) is more desirable from the noninteracting, self-linearizing standpoint but must be provided with high turndown flow sensors and signal transmission that does not contribute a time lag.

In the system shown in Figures 8.2bb and 8.2cc, fuel flow and air flow signals for proper fuel-air ratio are matched. This is done by matching air to fuel in the field combustion test calibration of the air flow measurement. Field testing is less stringent with the system shown in Figure 8.2cc, since the self-linearizing feature of the closed-loop system reduces the work to characterize the fuel and air control devices.

The systems shown here are for measurable fuels. In burning coal, fuel flow is often inferred from steam flow, and the steam flow–air flow relationship is used as a control index.

In the event that the heat of combustion or flow rate of the fuel varies unpredictably, as frequently happens during feeding of coal, wood, or refuse, air flow should be set in ratio to the heat released by the fuel. For a boiler, heat release is indicated by steam flow combined with the rate of rise of steam pressure. This is the same calculation that was made in Figure 8.2o to recalibrate the coal flow measurement. In the steady state, the heat release, coal flow, and total heat demand signal are identical, so any of the three may be used to set air flow. The faster response to a change in demand will be achieved if total heat demand sets air flow. The best response to an unmeasured change in coal flow or heating value would be obtained if air flow is set by calculated heat release. If the higher of the two signals is selected to set air flow, then excess air will be provided in either event.

A manual adjustment of the fuel-air ratio can be provided with either of the systems illustrated in Figures 8.2bb and 8.2cc. This adjustment can be inserted as shown in Figure 8.2dd to keep from changing the gain of the loop as adjustments are made. If ratio adjustment

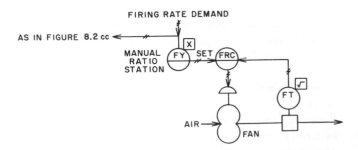

Fig. 8.2dd   Closed loop with manually adjusted ratio station

is manual, flue gas analyzers, such as oxygen, carbon monoxide, or combustibles analyzers, should be provided for the operator's guidance or as feedback control of the ratio desired. A measurement of percent oxygen in the combustion gases is most necessary for operator guidance when multiple fuels are being burned or when fuel properties vary. See Figure 8.2ee for a relationship between flue gas composition and excess air.

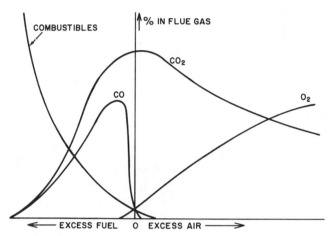

Fig. 8.2ee Major components of flue gas are oxygen, carbon dioxide, carbon monoxide, and unburned hydrocarbons.

Excess Oxygen Trim

Although fuel is measurable and can be totalized (on an air-required basis), variations in fuel BTU content, specific gravity, temperature, and other physical properties of either fuel or air will disrupt the fuel-air ratio. This results in safety and fuel economy problems. To maintain the optimum quantity of excess air, the fuel-air ratio should be automatically adjusted by a controller acting on the measured amount of oxygen in the flue gas. Because the correct air-fuel ratio can be predicted within relatively narrow limits, the range of the adjustment can be restricted. If this correction is required, it should be provided with means of programming the percent oxygen set point as a function of boiler load. In practically all boilers, desired excess air is not constant but decreases as load is increased. An example of this relationship is shown in Figure 8.2ff, together with an illustration of the effect of the type of fuel.

The various types of oxygen analyzers have been discussed in Section 11.21 of the Process Measurement volume of this handbook. Of the various designs, the probe-type "in situ" sensor shown in Figure 8.2gg is the best choice. In order to minimize the effect of duct leakage, the probe should be installed close to the combustion zone. The output signal of these zirconium oxide probes is logarithmic. According to Shinskey, this is not undesirable; because the sensitivity of the signal to oxygen content varies inversely with the oxygen content, the gain of the oxygen control loop could be expected to

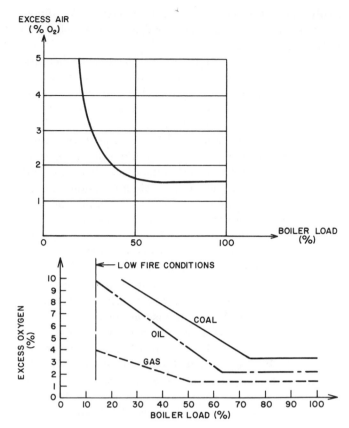

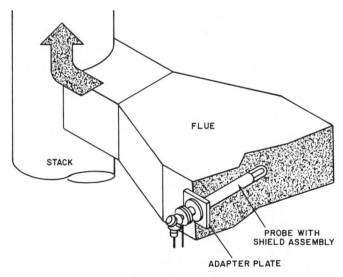

Fig. 8.2ff The ideal amount of excess oxygen provided to a boiler depends on load as well as fuel properties.

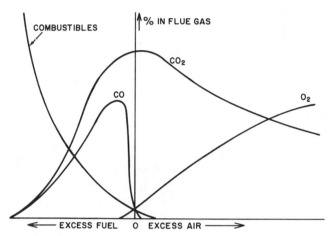

Fig. 8.2gg Installation of probe-type oxygen analyzer

change with load. At minimum load, the required oxygen content may be three times as great as at full load, so the sensitivity of the analyzer is only one third that at full load. However, the delay in response of oxygen content to a change in air flow is longer at low load than at high load. This longer delay increases the gain of the integral controller used on the oxygen signal, which virtually offsets the reduced gain of the logarithmic signal. Consequently, the logarithmic analyzer seems well suited to this application.

As was shown in Figure 8.2ee, there are other flue gas constituents that also reflect the amount of excess air. Gilbert claims that one can achieve superior control of fuel-air ratio by controlling carbon monoxide emission in the 100- to 200-ppm range. His reasoning is that air infiltration or a maladjusted burner could give a satisfactorily high oxygen level while still not providing complete combustion. On the other hand, a carbon monoxide measurement is not affected by air infiltration and is a true indication of the completeness of combustion.

Some of the more advanced microprocessor-based systems operate on a multivariable envelope control principle, simultaneously detecting five or six flue gas variables and keeping all of them within acceptable limits through the use of selective control algorithms. This technique seems to represent the ultimate in control sensitivity and resulting combustion efficiency.

Even the best excess oxygen trim system has some risk of failure, and the boiler controls must be protected from such occurrence. For this reason, limits should be placed on the output of the excess air controller so as to keep the air-fuel ratio in reasonable bounds in the event that the measurement should fail without notice.

Air-Fuel Ratio with Excess Oxygen Trim

In the design of a control system to follow the desired curve (Figure 8.2ff), provision should be made for shifting the curve to the right or the left. Figure 8.2hh shows an example of automatic fuel-air ratio correction based on load and excess air indicated by percent oxygen.

To obtain some of the advantages of the closed fuel loop system, noninteracting oxygen analysis may be used to calibrate continuously the inherently poor fuel flow signal, if it could not otherwise be used with accuracy. An example of how a satisfactory coal flow signal can be

obtained by continuously calibrating a summation signal of pulverizer feeder speeds is shown in Figure 8.2ii.

In both Figure 8.2hh and Figure 8.2ii the set point of the excess oxygen controller is based on the steam flow. This set point must be characterized as a function of steam flow (load) in accordance with Figure 8.2ff.

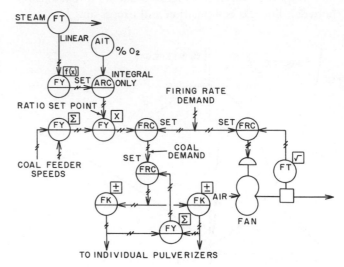

Fig. 8.2ii   Load versus excess air correction on boiler with coal fuel

Fuel and Air Limiting

Maintaining the correct fuel-air ratio also contributes to limiting the fuel rate to available air and limiting air minimum and maximum to fuel flow, air leading fuel on load increase, and air lagging fuel on load decrease. This arrangement also protects against a fan failure or a sticking fuel valve.

Though not normally furnished on smaller boilers, the following limiting actions are desirable for safety purposes:

1. Limiting fuel to available air flow.
2. Minimum limiting of air flow to match minimum fuel flow or to other safe minimum limit.
3. Limiting minimum fuel flow to maintain stable flame.

These limiting features are simple to apply with the basic noninteracting, self-linearizing system in Figure 8.2aa. Figure 8.2jj shows the necessary modifications that can provide these features without upsetting the set point of the fuel-air ratio. The following is accomplished by the illustrated system:

1. If actual air flow decreases below firing rate demand, then the actual air flow signal is selected to become the fuel demand by low selector (FY-1).
2. If fuel flow is at minimum and firing rate demand further decreases, actual fuel flow becomes the air flow demand, because FY-2 will select the fuel signal if it is greater than firing rate demand signal. A manual air flow minimum is also avail-

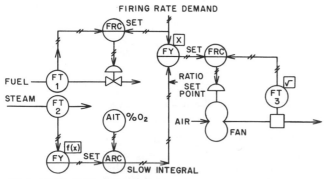

FT-1: LINEAR MASS FLOWMETER
FT-2: LINEAR VORTEX SHEDDING TRANSMITTER
FT-3: AREA AVERAGING PITOT STATION

Fig. 8.2hh   Air-fuel ratio control, with load versus excess
air curve considered

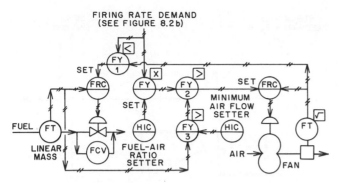

Fig. 8.2jj Parallel, closed-loop control with air and fuel limiting

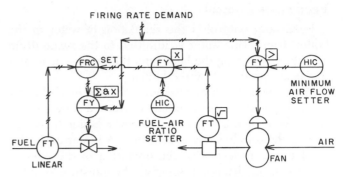

Fig. 8.2ll Firing demand determines air flow while fuel set point is adjusted by air flow

able to come into use through FY-3, such that if fuel flow signal drops below the HIC setting, this manual setting will become the air flow set point.

3. Fuel is minimum limited by separate direct-acting pressure or flow regulator (FCV).

In open-loop systems these limits are more difficult to apply. The application of these limits to an open-loop system is shown in Figure 8.2kk. Here the fuel set point is determined (limited) by actual air flow, and a "fuel cutback" necessitated by reduced firing rate demand is accomplished at the expense of a temporary fuel-air ratio offset.

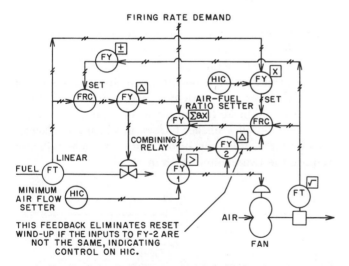

Fig. 8.2kk Parallel, open-loop control with fuel and air limiting applied

Limiting combustion air flow to a minimum or to the rate at which fuel is being burned creates special problems, because when the limit is in force, provision must be made to block the integral action in the air flow controller.

It may seem that a better way to limit fuel would be to have the firing rate demand directly set the air flow, with fuel being controlled through the combining relay (Figure 8.2ll). In this arrangement final fuel-air ratio correction occurs through the integral mode correction of the fuel-air ratio controller. This

system is only partially effective, however, because on a sudden decrease of firing rate demand, the resulting reduction in fuel flow will occur only after the air flow has already been reduced.

A further consideration in setting fuel rather than air directly by the firing rate demand is that the parallel boilers can be more easily kept in balance, because balancing fuel directly balances the heat input without consideration of excess air between boilers.

Figure 8.2mm illustrates a closed-loop control system corrected by oxygen analysis and provided with safety limits to protect against air deficiency. For the dual selector system to function, air and fuel flows must be scaled on the same heat-equivalent basis. The dual selector system forces air flow to lead fuel on an increasing load and to lag on a decreasing load. Then flue-gas oxygen content tends to deviate above the set point on *all* load changes. If the oxygen controller were allowed to react proportionally to these deviations, it would tend to defeat the security provided by the selectors. Consequently, the integral control mode alone is used on the oxygen signal, so that reaction to rapid fluctuations is minimized. The principal function of the controller is to correct for long-term deviations caused by flowmeter errors and variations in fuel quality.

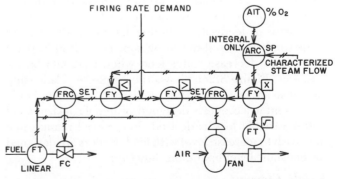

Fig. 8.2mm Feedforward control system that automatically maintains excess air during upsets

Feedwater Control

Feedwater control is the regulation of water to the boiler drum. This water is admitted to the steam drum and after absorbing the heat from the furnace generates the steam produced by the boiler.

Proper boiler operation requires that the level of water in the steam drum be maintained within a certain band. A decrease in this level may uncover boiler tubes, allowing them to become overheated. An increase in this level may interfere with the operation of the internal devices in the drum that separate the moisture from the steam.

The water level in the steam drum is related to, but is not a direct indicator of, the quantity of water in the drum. At each boiler load there is a different volume in the water that is occupied by steam bubbles. Thus, as load is increased there are more steam bubbles, and this causes the water to "swell," or rise, rather than fall, because of the added water usage. Therefore if the drum volume is kept constant, the corresponding mass of water is minimum at high boiler loads and maximum at low boiler loads. The control of feedwater therefore needs to respond to load changes and to maintain water by constantly adjusting the mass of water stored in the system.

Feedwater is always colder than the saturated water in the drum. Some steam is then necessarily condensed when contacted by the feedwater. As a consequence, a sudden increase in feedwater flow tends to collapse some bubbles in the drum and temporarily reduce their formation in the evaporating tubes. Then, although the mass of liquid in the system has increased, the apparent liquid level in the drum falls. Equilibrium is restored within seconds, and the level will begin to rise. Nonetheless, the *initial* reaction to a change in feedwater flow tends to be in the wrong direction. This property, called "inverse response," causes an effective delay in control action, making control more difficult. Liquid level in a vessel lacking these thermal characteristics can typically be controlled with a proportional band of 10 percent or less. By contrast, the drum-level controller needs a proportional band more in the neighborhood of 100 percent to maintain stability. Integral action is then necessary, whereas it can usually be avoided when very narrow proportional band settings can be used.

Control of feedwater addition based on drum level alone tends to be self-defeating, since on a load increase it tends to decrease water feed when it should be increasing. Figure 8.2nn shows the response relationship among steam flow, water flow, and drum level that should be present in a properly designed system if constant level under variable load is desired. For special reasons one may wish to increase level with load. Boilers are designed for constant level operation, however.

Single Element

For small boilers having relatively high storage volumes and slow-changing loads, a simple proportional

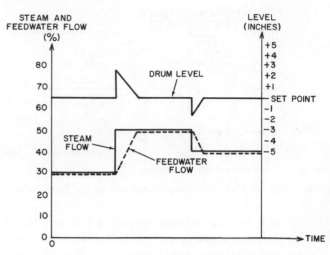

Fig. 8.2nn   Response relationship among steam flow (load), feedwater flow (manipulated variable), and level (controlled variable)

control may suffice, imprecise as it is. Integral action should not be used, because of resulting instability that is a result of integration of the swell on load changes that must later be removed. Control of this type therefore involves the addition of feedwater on straight proportional level control.

Two Elements

For larger boilers, and particularly when there is a consistent relationship between valve position and flow, a two-element system (Figure 8.2oo) can do an adequate job under most operating conditions. Two-element control is primarily used on intermediate-size boilers, in which volumes and capacities of the steam and water system would make the simple level control inadequate because of "swell." Smaller boilers, in which load changes may be rapid, frequent, or of large magnitude, will also require the two-element system.

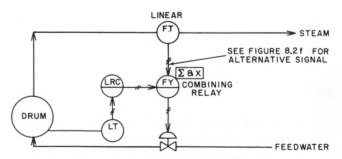

Fig. 8.2oo   Two-element feedwater control

Field testing, characterization, and adjustment of the control valve are required so that the relationship of *control signal to feedwater valve flow matches that of the steam flow to the flow transmitter output*.

Any deviations in this matching will cause a permanent level offset at the particular capacity and less than optimal control (Figure 8.2pp). The level controller gain should

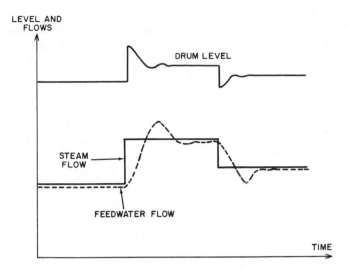

Fig. 8.2pp   The effect of mismatching between steam flow transmitter and valve flow characteristics

be such that, on a load change, the level controller output step will match the change in the steam flow transmitter signal.

Three Elements

As boilers become greater in capacity, economic considerations make it highly desirable to reduce drum sizes and increase velocities in the water and steam systems. Under these conditions the boiler is less able to act as an integrator to absorb the results of incorrect or insufficient control. A three-element system is used on such large boilers.

Three-element control is similar to the two-element system, except that the water flow loop is closed rather than open. There are several ways of connecting a three-element feedwater system, each of which can produce the results shown in Figure 8.2nn. Figure 8.2qq illustrates various ways of connecting this system.

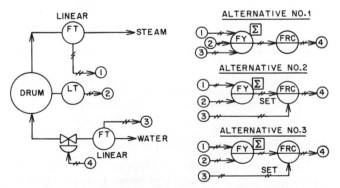

Fig. 8.2qq   Three-element feedwater system. FY is provided with independent gain adjustment capability for level and flow.

In making gain adjustments on a three-element feedwater system, the first step is to determine the relative gains between level and flow loops. By observing a change in boiler load one can note the particular boiler "swell"

characteristics. Maximum system stability results when the negative effect of swell equals the positive effect of flow. For example: If a 20 percent of maximum flow change produces a 2.4-PSI (16.6 kPa) change in flow transmitter output and this flow change also produces a 3-inch (75 mm) swell on a 30-inch (750 mm) range transmitter or a 1.2-PSI (8.3 kPa) transmitter output change, then the gain of the level loop should be double the gain on the flow loop.

A feedforward variation is recommended by Shinskey to maintain a steam-water balance, reducing the influence of shrink-swell and inverse-response phenomena. The system shown in Figure 8.2rr causes feedwater flow to match steam flow in absence of action by the level controller. The two flowmeters have identical ranges, and their signals are subtracted. If the two flow rates are identical, the subtractor sends a 50 percent signal to the flow-difference controller. An increase in steam flow will call for an equal increase in feedwater flow to return the difference signal to 50 percent.

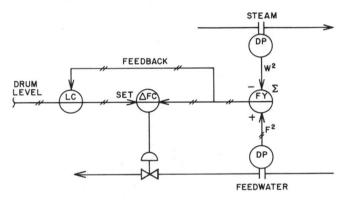

Fig. 8.2rr   Feedforward control of drum level

Errors in the flowmeters, and the withdrawal of perhaps 2.5 percent water as "blowdown" (which is not converted to steam) will prevent the two flow signals from being identical. Any error in the steam-water balance will cause a falling or rising level. Therefore, the level controller must readjust the set point of the flow-difference controller to strike a steady-state balance. The system assumes orifice-type flow sensors and does not use square root extractors, because the period of oscillation and dynamic gain of a two-capacity level process varies directly with flow. The gain of the feedwater control loop without square root extraction seems to compensate correctly for the process gain change.

Figure 8.2rr also shows external feedback from the flow-difference measurement applied to the level controller. This will precondition the level controller during startup or at other times when feedwater is controlled manually or otherwise limited.

Valve Sizing

Linear or linearized flow signals are required for calculation flexibility. To keep the controller gain independ-

ent of load variations, the feedwater valve also should
have linear characteristics. The simplest way to ensure
this is by using a characterizing positioner on the control
valve. It may also be required to dampen the control
valve, to make it less susceptible to noisy control signals.

For control valve sizing a system "head" curve showing
the relationship between system pressures and capacities
should be developed. A typical head curve is shown in
Figure 8.2ss. The head curve demonstrates a basic problem in selecting the flow rate and pressure drop when
the feedwater control valve is sized. Capacity C_2 and
differential X are the most desirable from a control standpoint. Capacity C_3 and differential Y or Z are often used
in an attempt to satisfy a boiler code requirement. The
code requires that sufficient water be furnished to the
drum with safety valves blowing.

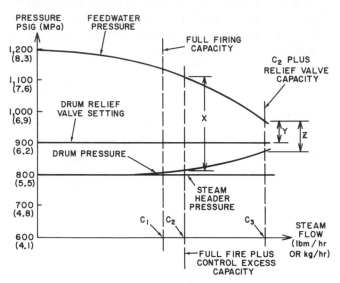

Fig. 8.2ss   Head curve for feedwater valve sizing

This code requirement is interpreted as full capacity
of boiler plus safety valve capacity even though maximum
firing rate could not hold the pressure. Safety valves will
not blow unless the firing rate is higher than capacity.
The code does not require all the capacity to be provided
in the feedwater control valve; therefore there is no need
to degrade the control quality by furnishing an oversized
valve.

It is not uncommon to see a valve that was designed
for more than required capacity, and for a 30-PSI (207
kPa) differential, operating at a 500-PSI (3.45 MPa) differential and at a fraction of its design capacity. The duty
required of a feedwater control valve is quite heavy because of the large energy dissipation as the water passes
through the valve. Very pure water as normally used in
boilers tend to be "metal-hungry." This, combined with
water velocity, produces a corrosion-erosion (or cavitation) effect that calls for a chrome-moly steel valve body
when velocities are in excess of 10 fps (3 m/s).

Pump Speed Control

Control of pump speed to regulate feedwater flow can
be accomplished if the pump is driven by a steam turbine
or is furnished with variable-speed electric drive. This
can be used in place of a control valve to save pump
power on single boiler systems. In systems in which
several boilers are operating in parallel, the speed control
can be used to save pump power by controlling the discharge pressure at a constant differential pressure above
boiler pressure.

When pump speed is being used in place of a feedwater valve on a single boiler system, a large part of the
speed control range is used in developing pump head at
low flow. Characterization of the signal is thus necessary
for good operation and constant gain throughout the operating range.

This is demonstrated in Figure 8.2tt, which shows the
pump characteristics. The reader will note the control
ranges used for 0 load and from 0 to full load. (Further
discussion of variable speed pumping is provided in Section 5.8 of this volume and in Section 9.12 of the Process
Measurement volume of this handbook.)

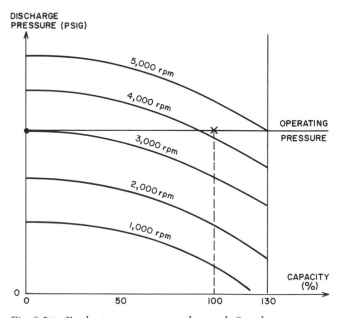

Fig. 8.2tt   Feedwater pump on speed control. Speed range at zero
capacity to reach operating pressure is 0–3000 rpm. Speed range for
zero to 100% capacity is 3000–4200 rpm.

Steam Temperature Control

The purpose of steam temperature control is to improve the thermal efficiency of steam turbines. Its most
common application is for steam turbine electric power
generation. The factors affecting steam temperature in a
convection-type superheater are superheater area, flue
gas flow pattern across the superheater, flue gas mass
flow, temperature of flue gases leaving the furnace, and
steam flow through the superheater. Additionally, fur-

nace temperature affects a radiant superheater. Some superheaters may be designed for a flat curve combining radiant and convection surface, but most superheaters are the convection type. Central-station steam temperatures are limited to approximately 1050°F (566°C), whereas those in industrial units may be considerably less. If these temperatures can be controlled with extreme precision, they can be pushed closer to the allowable limits. Temperature can be controlled by adjustment of the amount of recirculation of the flue gas or by "attemperation," which is an energy-wasting method of superheat removal through feedwater spraying.

The required positions of burners or recirculating and bypass dampers as a function of load are well established for any given boiler design. Therefore, it is common practice to program their positions directly from load. Readjustment to correct for inaccuracies in the program and changes in the characteristics of the boiler is accomplished by feedback control of temperature, using proportional and integral action. Being applied through a multiplier, the feedback loop gain varies directly with load, and according to Shinskey this tends to cancel the inversely varying process gain.

Temperature control by attemperation is more responsive and can be used to supplement flue-gas manipulation. To minimize water usage, however, and to avoid conflict with flue-gas manipulation, proportional-plus-derivative control should be used for attemperation. The controller may be biased to deliver a nominal amount of feedwater at zero temperature deviation. The control system is described in Figure 8.2uu.

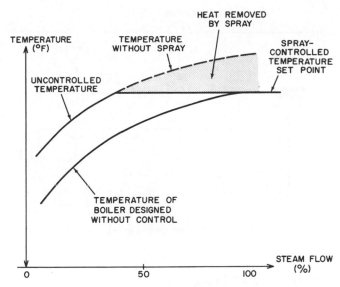

Fig. 8.2vv   Desuperheater characteristics

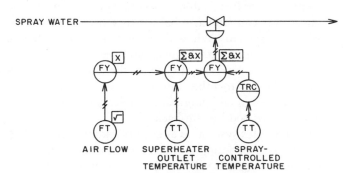

Fig. 8.2ww   Desuperheater spray control system

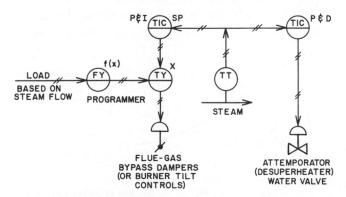

Fig. 8.2uu   Steam temperature control

Desuperheater Spray Controls

To use a desuperheater spray for steam temperature control, the boiler would normally be provided with added superheater area. Figure 8.2vv demonstrates the effect of the water spray (which is usually between a primary and a secondary superheater section) for temperature control.

Provision must be made to prevent reset windup when in the uncontrolled load range. Control of this system is shown in Figure 8.2ww. (Note that in Figure 8.2uu this is a proportional and derivative [P&D] controller.)

Large boilers may have burners that tilt up and down approximately 30 degrees for steam temperature control. This effectively changes the furnace heat transfer area, resulting in temperature changes of flue gases leaving the boiler. Spray is frequently used with these systems as an override control. These systems are used on large power plant boilers and normally require the type of controls illustrated in Figure 8.2uu.

Variable Excess Air

Since mass flow of flue gas affects steam temperature, variation in excess combustion air can be used to regulate steam temperature. This method may be used on a boiler that was not specifically designed for temperature control. Although increasing excess air flow increases boiler stack heat losses, turbine thermal efficiency will also increase, and maintaining steam temperature can provide the greater economic benefit. The control arrangement in Figure 8.2xx implements this method.

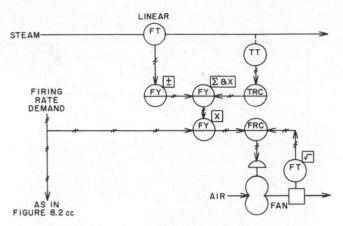

Fig. 8.2xx Steam temperature control by adjustment of excess air

Integration of Loops

By combining the loops and subsystems shown in Figures 8.2b, 8.2jj, 8.2qq, and 8.2ww, one can create an integrated control system. Other combinations of the subsystems can similarly be put together to form a coordinated system. When designing, it is advisable to break the overall system down into these subsystems and examine them individually. Only then should the subsystems be put together into the total system.

Major design and operation problems in complex systems are created by "tie-backs."

In a complex system, as a result of adding, subtracting, multiplying, dividing, and comparing control signals or transmitter signals, it is a major problem to get the system on "automatic" control easily and quickly.

The chief points to remember include:

1. The systems often interact, e.g., air flow affects steam temperature, feedwater flow affects steam pressure, and fuel flow affects drum level and furnace draft. Designs with the minimum of interaction are the most desirable.
2. For flexibility and rangeability linear flow signals are necessary. Control valves and piston operators need linearizing positioners.
3. Fuels should be totalized on an air-required basis.
4. Tie-back arrangements, which simplify the task of getting quickly on automatic control, are very important in complex systems.
5. The flows of fuel and air should be controlled such that the flow rates reaching the burner always represent a safe combination.

Optimization of Boilers

When the yearly boiler fuel cost is in the millions, even a few percentage points of improved efficiency can justify the costs of added instruments and controls. For well-designed large boilers, with blowdown losses and auxiliary pump or fan operating costs disregarded, Con-

solidated Edison* estimates the maximum efficiencies using dry fuels as 89 percent for coal, 87.5 percent for oil, and 82.5 percent for gas-burning units.

In the following paragraphs a number of optimization techniques will be described. The various goals of optimization include:

1. Minimize excess air and flue gas temperature
2. Measure efficiency.
   a. Use the most efficient boilers.
   b. Know when to perform maintenance.

3. Minimize steam pressure.
   a. Turbines thereby open up turbine governors.
   b. Reduce feed pump discharge pressures.
   c. Reduce heat loss through pipe walls.

4. Minimize blowdown.
5. Provide accountability.
   a. Monitor losses.
   b. Recover condensate heat.

Minimization of Stack Losses

As shown in Figure 8.2yy, most heat losses in a boiler occur through the stack. Under air-deficient operations unburned fuel leaves, and when there is an air excess heat is lost as the unused oxygen and its accompanying nitrogen are heated up and then discharged into the atmosphere. The goal of optimization is to keep the total losses at a minimum. This is accomplished by minimizing excess air (see Figure 8.2ff) and by minimizing the stack temperatures (see Figure 8.2zz). The best results can be expected from the multivariable envelope-type boiler optimization systems that simultaneously monitor and control not only the excess oxygen in the flue gas but also the other components (carbon monoxide, carbon

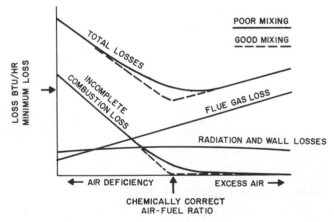

Fig. 8.2yy The minimum point of the total loss curve of a boiler is where optimized operation is maintained

* The help of Mr. L. David Shapiro, Technical Superintendent of the 74th Street Station of Consolidated Edison of New York, is gratefully acknowledged for providing these data.

dioxide, hydrocarbon, water) plus stack temperature and opacity.

Assuming that for a particular boiler design using a particular fuel at normal loading, the optimum flue gas temperature is 400°F with 2 percent oxygen, one can estimate the potential fuel savings through optimization by determining the fuel loss using Figure 8.2zz. The present, unoptimized stack gas conditions are entered.

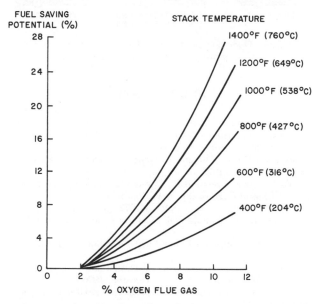

Fig. 8.2zz Fuel-saving potentials in a boiler with optimum conditions corresponding to 400°F stack gas temperature and 2% excess oxygen

Efficiency Monitoring

Boiler efficiency can be monitored indirectly (by measurement of flue gas composition, temperature, combustion temperature, and burner firing rate) or directly (through time-averaged steam and fuel flow monitoring). For the direct efficiency measurement, it is important to select flowmeters with acceptable accuracy and rangeability. In order to arrive at a reliable boiler efficiency reading one must ensure that the error contribution of the flowmeters, *based on actual reading*, does not exceed $\pm\frac{1}{2}$ percent to $\pm\frac{3}{4}$ percent.

The rangeability of the meters must be 10:1, covering the range of steam and fuel flows between 10 percent and 100 percent load. Orifice- or Venturi-type flowmeters cannot meet these performance criteria. For the fuel flow sensor, the oscillating mass flowmeters are the best choice, although turbine or displacement flowmeters can also be considered if they are periodically recalibrated and are compensated for fuel density (temperature) variations. For steam flow metering, the best choice is the vortex-shedding flowmeter with pressure and temperaature compensation. Shunt flowmeter errors are ±2 percent of actual flow over a rangeability of 7:1.

If the boiler efficiency is continuously and reliably monitored, a drop in this reading will signal the need

for maintenance, cleaning, or recalibration. When multiple boilers are used, knowing the actual efficiency to load relationship allows the selection of the most efficient unit(s) to meet a load (Figure 8.2aaa). Given a number of boilers generating steam, the group efficiency can be increased if the most efficient boiler is allocated more of the load and the least efficient is allowed less.

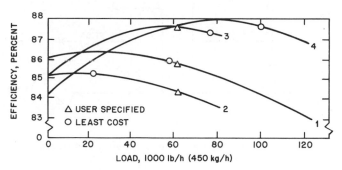

Fig. 8.2aaa In multi-boiler stations, economy is served by selecting the most efficient unit(s) for each load

Steam Pressure Optimization

In co-generating plants, in which the boiler steam is used to generate electricity and the turbine exhaust steam is used as a heat source, one obtains optimization by maximizing the boiler pressure and minimizing the turbine exhaust pressure, so as to maximize the amount of electricity generated.

In plants that do not generate their own electricity, one achieves optimization by minimizing boiler operating pressure. This reduces the pressure drops in turbine governors by opening them up farther, lowers the cost of operating feedwater pumps because their discharge pressure is reduced, and generally lowers radiation and wall losses in the boiler and piping.

Pump power is particularly worth saving in high-pressure boilers, because as much as 3 percent of the gross work produced by a 2400-PSIG (16.56 MPa) boiler is used to pump feedwater. Figure 8.2bbb illustrates the method of finding the optimum minimum steam pressure, which then becomes the set point for the master controller (see Figure 8.2f).

As long as all steam user valves (including all turbine throttle valves) are less than fully open, a lowering in the steam pressure will not restrict steam availability, because the user valves can open further. The high signal selector (TY-1) selects the most open valve and the valve position controller (VPC-2) compares that signal with its set point of, for example, 80 percent. If even the most open valve in the plant is less than 80 percent open, the pressure controller set point is slowly lowered. VPC-2 is an "integral-only" controller; its reset time is at least ten times that of PRC-3. This slow integral action guarantees that only very slow "sliding" of the steam pressure will occur and that noisy valve signals will not upset the

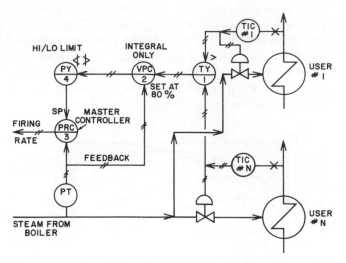

Fig. 8.2bbb   Main steam header pressure optimization

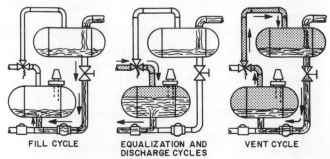

Fig. 8.2ccc   Operation of the pumpless condensate return system (courtesy of the Johnson Corp.)

system because VPC-2 responds only to the integrated area under the error curve. The output signal from VPC-2 is limited by PY-4, so that the steam pressure set point cannot be moved outside some preset limits. This necessitates the external feedback to VPC-2, so that when its output is overriden by a limit, its reset will not wind up.

This load-following steam pressure optimization not only increases boiler efficiency but also:

- prevents any steam valve in the plant from fully opening and thereby losing control;
- opens all steam valves in the plant, thereby moving them away from the unstable (near closed) zone of operation; and
- reduces valve maintenance and increases valve life by lowering pressure drop.

Miscellaneous Optimization Strategies

The amount of blowdown should be controlled automatically using conductivity sensors. Recovering the heat content of the blowdown can also be considered.

The heat content of the returning condensate should not be lost. Pumping water at high temperatures is difficult; therefore, the best choice is to use pumpless condensate return systems. Figure 8.2ccc illustrates the operation of such a system, which uses the steam pressure itself to push back the condensate into the deaeration tank. This approach eliminates not only the maintenance and operating cost of the pump but also the flash and heat losses, resulting in the return of more condensate at a higher temperature.

The metering of steam use throughout the plant not only will contribute to better accountability but also will assist in finding leaking steam traps or isolating plant or header sections with insufficient insulation.

BIBLIOGRAPHY

ASME Research Committee on Industrial and Municipal Wastes, "Combustion Fundamentals for Waste Incineration," American Society of Mechanical Engineers, New York, 1974.

Babcock & Wilcox, "Steam: Its Generation and Use," Babcock and Wilcox, New York, 1972.

Bell, A.W. and Breen, B.P., "Converting Gas Boilers to Oil and Coal," *Chemical Engineering*, April 26, 1976.

"Boiler Controls," *Power*, October 1972, pp. 21–28.

Buntzel, H., "Calibrating Boiler Control Systems," *InTech*, January 1974.

Cho, C.H., "Optimum Boiler Load Allocation," *InTech*, October 1978.

Claypoole, G.T., "The Conceptual Design of a Utility," Paper 76-560, ISA 1976 Annual Conference, Houston, 1976.

Congdon, P., "Control Alternatives for Industrial Boilers," *InTech*, December 1981.

Criswell, R.L., "Control Strategies for Fluidized Bed Combustion," *InTech*, January 1980.

Davis, T.P., "Control Strategies for Fuel Oil Heating," *InTech*, September 1977.

Dukelow, S.G., "Charting Improved Boiler Efficiency," *Factory*, April 1974, pp. 31–34.

———, "Combustion Controls Save Money," *Instruments and Control Systems*, November 1977.

———, "Control of Steam Boilers," *Instrumentation in the Processing Industries*, 1973, Chapter 10.6, pp. 632–667.

Environmental Protection Agency, "Controlled Flare Cuts Smoke, Noise, Saves Steam," *Technology Transfer*, U.S.E.P.A., Washington, D.C.

———, Rep. No. 650/2-74-066, U.S.E.P.A., Washington, D.C.

The Foxboro Company, Differential vapor pressure cell transmitter, Model 13VA, Tech. Inform. Sheet 37–91a, April 15, 1965.

Gilbert, L.F., "Precise Combustion-Control Saves Fuel and Power," *Chemical Engineering*, June 21, 1976.

Guide to Selecting Combustion Stack Analyzers, *InTech*, April 1982.

Hockenbury, W.D., "Improve Steam Boiler Control," *Instruments and Control Systems*, July 1981.

Hurlbert, A.W., "Air Flow Characterization Improves Boiler Efficiency," *InTech*, March 1978.

Johnson, A.J. and Auth, G.H., *Fuels and Combustion Handbook*, McGraw-Hill, New York, 1951, p. 355.

Kaya, A., "Optimum Soot Blowing Can Improve Your Boiler Efficiency," *Instrumentation Technology*, January 1982.

Keller, R.T., "New Boiler Control System," *Instruments and Control Systems*, June 1976.

Kolmer, J.W., "Programmed O_2 Optimizes Combustion Control," *Instruments and Control Systems*, December 1977.

Krigman, A., "Combustion Control," *InTech*, April 1982.

Kuhlman, L.J., "Waste Firing Control in Industrial Boilers," *Instrumentation Technology*, February 1976.

Lipták, B.G., "Save Energy by Optimizing Boilers, Chillers, Pumps," *InTech*, March 1981.

Louis, J.R., "Control for a Once-Through Boiler," *Instruments and Control Systems*, January 1971.

Matthys, W.J., "Retrofitting Industrial Boilers to Meet the Fuels Market," Paper presented at American Power Conference, Chicago, April 21–23, 1975.

May, D.L., "Cutting Boiler Fuel Costs With Combustion Controls," *Chemical Engineering*, December 22, 1975.

"Microprocessors That Run Boilers," *Business Week*, August 13, 1979.

Mockenbury, W.D., "Improve Steam Boiler Control Step by Step," *Instruments and Control Systems*, July 1981.

Nathan, P. and Wasserman, R., "Energy Control Cuts Electric Bills," *Instrumentation Technology*, December 1975.

Neuberger, E.D., "Programmed O_2 Setpoint Optimizes Combustion Control," *Instruments and Control Systems*, December 1977.

Nordmann, B., "Compensate Combustion Controls for BTU Variations," *Instruments and Control Systems*, April 1978.

O'Meara, J.E., "Oxygen Trim for Combustion Control," *InTech*, March 1979.

Perry, D., "Upgrade Boiler Controls With DDC," *Power*, October 1975.

Pocock, R.E., Ross, D.F., and McGandy, E.L., "Oxygen Analyzers for the New Proposed EPA Pollution Stack Monitoring Regulations," 68th Annual Meeting of the Air Pollution Control Association, June 1975.

Pollard, E.V. and Drewry, K.A., "Estimating Performance of Automatic Extraction Turbines," West Lynn, MA, General Electric Co.

Robnett, J.D., "Energy Optimization," *InTech*, June 1983.

Rodman, C.W., "Monitoring Combustion Chamber Pulsations," *InTech*, April 1982.

Roth, J.W., "Development of a Multi-Element Boiler Combustion Control Systems," Southern California Meter Association, October 1961.

Scheib, T.J., "Instrumentation Cuts Boiler Fuel Costs," *Instruments and Control Systems*, November 1981.

Schwartz, J.R., "Carbon Monoxide Monitoring," *InTech*, June 1983.

Shinskey, F.G., "Adaptive Control? Consider the Alternatives," *Instruments and Control Systems*, August 1974.

———, *Energy Conservation Through Control*, Academic Press, New York, 1978.

———, *pH and pIon Control in Process and Waste Streams*, New York, Wiley (Interscience), 1973, pp. 88;–92.

———, *Process Control Systems*, McGraw-Hill, New York, 1967, pp. 188–190.

———, "Process Control Systems with Variable Structure," Control Engineering, August 1974.

———, "Two-Capacity Variable Period Processes," *Instruments and Control Systems*, October 1972.

Tomlinson, L.O. and Snyder, R.W., "Optimization of STAG Combined Cycle Plants," Presented at the American Power Conference, Chicago, April 29–May 1, 1974.

Trimm, T.G., "Perform an Energy Audit On Your Boilers," *InTech*, July 1982.

Walsh, T.J., "Controlling Boiler Efficiency," *Instruments and Control Systems*, January 1981.

Wells, R.D., "Computer Cuts Steam Cost by 1.3 Million," *InTech*, January 1983.

Yarway Steam Conditioning Valve (manual), Blue Bell, PA, Yarway Corp.

Zimmer, H., "Chiller Control Using On-Line Allocation for Energy Conservation," Paper No. 76-522, Advances in Instrumentation, Part 1, ISA/76 International Conference, Houston, 1976.

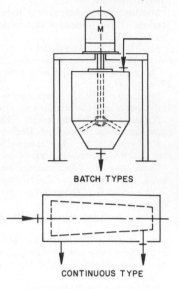

BATCH TYPES

CONTINUOUS TYPE

8.3 CENTRIFUGE CONTROLS

This section provides an overview of current practice in centrifuge controls. Both sedimentation and filtering centrifuges in industrial use are discussed.

Centrifuges are used for liquid-liquid or liquid-solid separation. The machines in operation can be classified into two groups:

Sedimentation centrifuges: These have solid walls, and separation occurs by sedimentation. The operating principle is shown in Figure 8.3a. The feed enters a solid-walled bowl rotating about a vertical axis. The solid and liquid phases are acted upon by centrifugal force and gravity. The former is preponderant, and separation of solid and liquid phases takes place as shown.

Filtering centrifuges: These machines have perforated walls that retain the solids on a permeable surface and through which the liquid can escape. This design is also

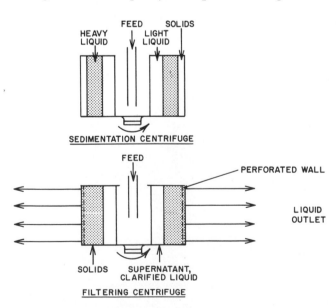

SEDIMENTATION CENTRIFUGE

FILTERING CENTRIFUGE

Fig. 8.3a   Centrifuge types

shown in Figure 8.3a. The action is similar to that of a filter but with a much higher "g force" than possible in gravity or pressure filtration. Nearly all the liquid is removed, leaving behind an almost dry cake.

The centrifugal force obtained in industrial machines is several times the force of gravity. For a particle rotating with an angular velocity w at radius r from the axis of rotation, the centrifugal separating effect is given by

$$G = \frac{wr}{g} \qquad 8.3(1)$$

Filtering centrifuges operate at a g range of 400 to 1800, whereas the g forces for sedimentation units range from 3000 to over 60,000 in laboratory machines (ultracentrifuges).

The critical speed phenomenon must be considered in the design and operation of centrifuges, as with any high-speed machine. At critical speed, frequency of rotation matches natural frequency of the rotating member. At this speed, even the minute vibrations induced by slight imbalances are strongly reinforced. Centrifuges operate well above the critical speed and pass through it during acceleration and deceleration. Except in case of major bowl imbalance, this does not cause problems.

Centrifuge Selection

The selection of a centrifuge involves a balance between several factors: (1) the ability of the machine to process the given feed slurry or emulsion at the desired degree of separation, (2) the reliability of the machine, (3) the operating and maintenance requirements, and (4) the investment.

In the majority of cases, equipment manufacturers have standard machines that are adapted to the applications specified by the customer. Sedimentation machines are usually chosen on the basis of small-scale tests in laboratory centrifuges. Filtering centrifuges are chosen on the basis of tests in batch machines. Based on such tests, the manufacturers will offer specific machines and will outline the anticipated performance.

The instrument engineer should concentrate on two aspects of the process centrifuge installations:

Feed slurry control: Regulation of feed slurry at the correct continuous rate or in the right batch sizes is of utmost importance, since the machine cannot usually tolerate major variations in feed rate or composition. Control of wash liquor feed is equally important.

Sequencing operations: All batch machines are sequentially operated, and the related interlock design is one of the important steps in engineering a system.

The overall approach to centrifuge control is outlined in this section. Also to be considered are instruments on drive and discharge mechanisms, auxiliary devices, and safety interlocks. The following tables and figures provide data on both sedimentation and filtering centrifuges: Table 8.3b provides overall data; Figure 8.3c provides data on application based on particle size; and Figure 8.3d provides data on the applications based on the percentage of solids in feed slurry.

Sedimentation Centrifuges

Sedimentation units are used as clarifiers, desludgers, and liquid-liquid phase separators. Particle size is such that separation usually obeys Stokes law. Sedimentation units are generally of small diameter and run at high speed. These can be classified into the following types: (1) tubular, (2) disk, and (3) solid-bowl. Overall data on present-day machines were presented in Table 8.3b.

The g concept is often used to compare the performance of solid wall centrifuges. This equivalence converts the geometry, size, and speed of the bowl to the area of

Table 8.3b
CLASSIFICATION OF CENTRIFUGES

Rotor Type	Range of Bowl Diameter in inches (mm)	Maximum Centrifugal Force (g)	Method of Solids Discharge	Method of Liquid Discharge	Maximum Capacity gals/hr or ft³/batch short tons/hr	Maximum Capacity (liters/hr in meter batch) (meters tons/hr)
I. Sedimentation Centrifuges						
Tubular	2–6 (50–150)	60,000	Manual (batch)	Continuous	3000 g/hr	(11,250 1/hr)
Disk	9–32 (230–800)	2500–8000	Batch of semi-continuous	Continuous	12,000–24,000	(45 to 90,000 1/hr)
Solid bowl						
Constant speed (horizontal)	14–36 (350–900)	1000–3000	Automatic batch	Continuous overflow	60 ft³/batch	(1.7 m³/batch)
Variable speed (vertical)	12–84 (300–2 m)	Up to 3200	Automatic batch	Continuous	15 ft³/batch	(0.42m³/batch)
Continuous		Up to 3200	Continuous	Continuous	Up to 65 tons/hr solids	(59 tons/hr)
II. Filtering Centrifuges						
Conical screen						
Wide-angle		Up to 1400	Continuous	Continuous	15,000 g/hr	(56,250 1/hr)
Differential scroll		Up to 1800	Continuous	Continuous	70 tons/hr solids	(63.5 tons/hr)
Vibrating screen		Up to 500	Continuous	Continuous	100 tons/hr solids	(90.7 tons/hr)
Pusher		1800	Batch	Continuous	10 tons/hr solids	(9.1 tons/hr)
Cylindrical screen						
Pusher		1500	Continuous	Continuous	40 tons/hr solids	(36.3 tons/hr)
Differential scroll		1500	Continuous	Continuous	40 tons/hr solids	(36.3 tons/hr)
Horizontal		1300	Batch	Intermittent	25 tons/hr solids	(22.7 tons/hr)
Vertical		900	Batch	Intermittent	10 tons/hr solids	(9.1 tons/hr)

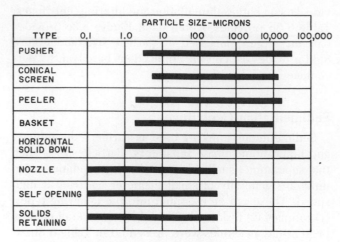

Fig. 8.3c   Application of centrifuge based on size of particles

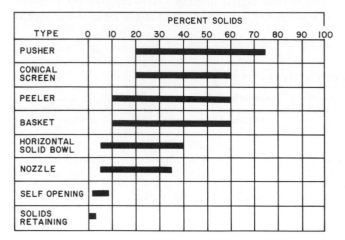

Fig. 8.3d   Type of centrifuge versus percentage of solids

a settling tank theoretically capable of the same separation in unit gravity field.

The control concepts applicable to these designs are best discussed in terms of three categories: batch, semicontinuous, and continuous.

Batch

All laboratory centrifuges and some of the small sedimentation machines in industry are operated as batch units with manual control of feed and manual discharge of solids. In such instances, little automatic control is required, since all sequencing is being done manually.

Semicontinuous

In this category of automatic batch-type centrifuges, one finds the fall disk-type machines, such as the desludging centrifuges, and also some solid-bowl types.

One process control scheme for a desludging, or clarifier-type, centrifuge is shown in Figure 8.3e. The feed to the machine is introduced from a sludge or magma feed tank by gravity. It has been found that if a minimum head of 10 feet (~3 m) is available, gravity flow is ade-

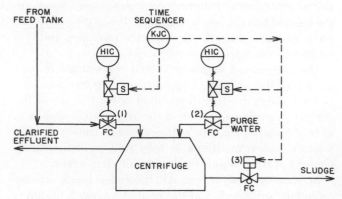

Fig. 8.3e   Desludging, or clarifier, centrifuge. Valve (1) is normally open, except during flushing. Valve (2) is normally closed, except during flushing. Valve (3) is operated intermittently to discharge solids.

quate for control. Feed tank location and size are determined along with pipe size for the specific application.

The feed enters this unit at a predetermined rate and is continuously clarified. The clarified effluent is then discharged. The sludge accumulates in the system and is periodically ejected. The unit may or may not run at full speed while sludge is ejected. The interlock system needs at least one sequence timer and three or more working contacts to control valves on the following lines: feed, sludge, and purge water. Usually added are partial desludging and other refinements, such as water spray in specific parts of the unit.

In these units the choice of control valves is of major importance. Properly chosen ball- or plug-type valves work satisfactorily, as do diaphragm types of flow pressure, on-off service. For throttling valve designs applicable to slurry service, refer to appropriate parts of Chapter IV.

In one modification of this machine, the sludge accumulates in an outer chamber. When the chamber is full of sludge, a sensing device activates the desludging operation. Such a unit can tolerate significant variations in the proportion of sludge in feed liquor.

Continuous

Among the versatile units used in petrochemical processing, polymerization and waste treatment systems are the fully continuous solid-bowl sedimentation centrifuges.

One example of a solid-bowl machine is shown in Figure 8.3f. The slurry is introduced in the revolving bowl of the machine through a stationary feed tube at the center. It is acted upon by centrifugal force, and the solids (denser phase) are thrown against the wall. Inside the rotating bowl is a screw conveyor with a slight speed differential with respect to the bowl rotation; it moves the solids up the beach and out of the liquid layer. Solids and clarified liquid are continuously discharged. Control of these solid-bowl machines requires consideration of the torque developed by the unloading plow and other

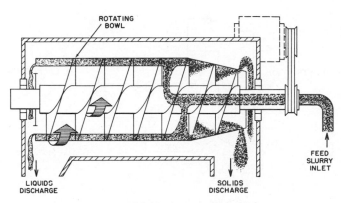

Fig. 8.3f Continuous, solid-bowl centrifuge

internal design considerations. One control scheme shown in Figure 8.3g automatically adjusts the feed rate to maintain the torque. Load cells are frequently applied in such installations as torque detectors.

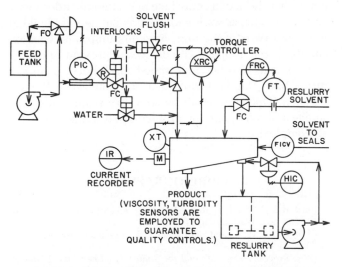

Fig. 8.3g Solid-bowl centrifuge with feed on torque control

With interlock in effect (not shown in Figure 8.3g), feed is closed, flush is opened, and the unit is shut down. This is brought about by (1) low coolant or flush flows to seals, (2) low flow or high temperature in the lubricating oil system, (3) high motor current, and (4) high torque. Other, less severe conditions, such as low reslurry solvent flow, might temporarily stop feed and open flush without shutdown.

It is important that a reasonably uniform slurry be supplied to the machine. For this purpose, a circulating pump with a recycle line is usually installed to keep the slurry in motion and thus to prevent settling out of crystals in the tank or in the pipelines. In case the slurry does not settle quickly, magma extraction can be performed by a siphon.

Another method of controlling a solid-bowl machine is to adjust the relative speed of the conveyor and the bowl to balance the unloading requirements. In such

cases, the process slurry feed to the machine can be on flow control. Balancing of the machine is accomplished by measurement of the torque and adjustment of the differential on the scroll (conveyor).

Filtration Centrifuges

Filtration centrifuges are also of three types: batch, automatic-batch, and fully continuous. In all of them a cake is deposited on a filter medium held in a rotating basket, which is then washed and spun dry. The method of solid discharge distinguishes the type. See Table 8.3b and Figures 8.3c and 8.3d for overall data on some of the more common designs.

Automatic Batch

Two versions of the control scheme around a horizontal cylindrical basket machine (also called a "peeler") on automatic-batch sequence control are shown in Figure 8.3h. In one case the circulating feed pump feeds one or more centrifuges against a head tank, and the overflow from this tank is returned to the crystallizer. In the second scheme the feed is by gravity alone.

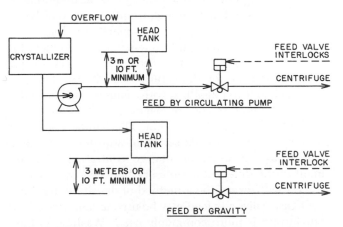

Fig. 8.3h Intermittent feeding of automatic batch centrifuges

A typical automatic batch machine will operate on the following sequence cycle:

1. Screen rinse Residual layer of crystals is rinsed prior to loading.
2. Loading Feed is admitted through the feed valves.
3. Cake rinse After the crystal layer is established, feed is stopped and rinsing is started.
4. Drying The cake (after washing) is spun dry.
5. Unloading The unloading knife is cut in to discharge solids.

In these machines, the feed slurry and wash liquor line pressure are assumed to be reasonably steady, and therefore feed flow controllers are not used. The sequence just discussed is that of a constant-speed machine. A variable-speed machine has a similar sequence, but with periods provided for acceleration and deceleration at appropriate points.

The entire sequence of the batch cycle is governed by one or more sequence timers with contractors to actuate the valves.

Continuous

In a fully continuous machine of the reciprocating pusher type or the differential scroll type, the only control required is that of constant slurry and wash-water feed rate.

When a magma is to be extracted continuously, there is a need to control the quantity of solids. This involves measuring the volumetric feed flow rate, detecting the solid concentration, and throttling a valve to maintain the solid charge rate constant. However, in many cases the expense of installing such systems is not justified; rather, allowing some variation in specifications may be more appropriate.

There are two simple and inexpensive methods by which a crystal magma can be continuously withdrawn at controlled rates: (1) Removal can be accomplished in small batches but at such high frequency as to make the flow virtually continuous. The air-operated pulsing valves with adjustable stoke and frequency may be used in this application. The volumetric feeds of the ball check type are also suitable for these purposes. (2) An elutriation leg attached to a crystallizer with a laundering fluid inflow can also be used to control magma density. One version of this is shown in Figure 8.3i. The laundering fluid is on flow control with its set point adjusted to maintain density as a measure of solid concentration. Measurement of the differential pressure between two points in the crystallizer can be a method of density detection.

Occasionally, raw feed may be used as laundering fluid, but filtrate is more commonly used. Washing in these machines is also fully continuous, and wash fluid flow is on automatic control.

A special application is that of a scraped surface crystallizer, producing the slurry for separation in a continuous centrifuge. In such cases the liquor can be fed by a metering pump, with the centrifuge mounted in series with the crystallizer.

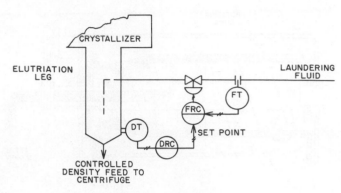

Fig. 8.3i Elutriation leg

Conclusions

Some of the more common methods of controlling a centrifuge were outlined in this section. It should be noted that the actual centrifuge operation includes many auxiliaries, such as wash lines for preventing solid buildup in various parts of the unit, lube oil pumps and coolers, and discharge hoppers with devices to prevent bridging.

As is true of other high-speed machinery, manufacturers have standard units that are adapted to the application. Operating characteristics vary significantly, and sludge and solid handling problems may add to the overall complexity.

BIBLIOGRAPHY

Burd, R.S., "A Study of Sludge Handling and Disposal," Department of the Interior, Washington, D.C., 1968.
Eckenfelder, W.W. and Santhanam, C.J., *Sludge Treatment*, Marcel Dekker, New York, 1982.
EPA, *Process Design Manual for Sludge Treatment and Disposal*, Environmental Protection Agency, Washington, D.C., 1974.
———, *Sludge Treatment and Disposal*, Vols. I and II, Environmental Protection Agency, Washington, D.C., 1978.
Lipták, B., *Environmental Engineers' Handbook*, Chilton Book Co., Radnor, PA, 1974.
McIlvaine Co., *The Sedimentation and Centrifugation Manual*, McIlvaine Co., Chicago, 1981.
Perry, R.H. and Chilton, C.H., *Chemical Engineering Handbook*, McGraw-Hill, New York, 1973.
Schweitzer, P.A., *Handbook of Separation Techniques*, McGraw-Hill, New York, 1979.

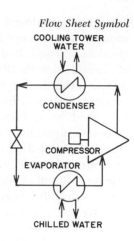

8.4 CHILLER AND HEAT PUMP CONTROLS

In this review of the state of the art of chiller controls, the reader will be first acquainted with the thermodynamic aspects of the refrigeration process. This will be followed by descriptions of conventional chiller controls and refrigeration machine and heat pump optimization control systems.

The "Pumping" of Heat

Just as conventional pumps can lift water from a low to a high elevation, so can refrigeration machines "pump" heat from a low to a high temperature level. Figure 8.4a illustrates this, where the heat pump removes Q1 amount of heat from the chilled water and, at the cost of investing W amount of work, delivers Qh quantity of heat to the warmer cooling tower water.

On the right side of this same illustration, the idealized temperature-entropy cycle is shown for the refrigerator. The cycle consists of two isothermal and two isentropic (adiabatic) processes:

1-2: adiabatic process through expansion valve
2-3: isothermal process through evaporator
3-4: adiabatic process through compressor
4-1: isothermal process through condenser

The isothermal processes in this cycle are also isobaric (constant pressure). The efficiency of a refrigerator is defined as the ratio between the heat removed from the process (Q1) and the work required to achieve this heat removal (W).

$$\beta = \frac{Q1}{W} = \frac{T1}{Th - T1} \qquad 8.4(1)$$

Because the chiller efficiency is much more than 100 percent, it is usually called the "coefficient of performance" (COP).

If a chiller requires 1.0 kWh (3412 BTU/h) to provide a ton of refrigeration (12,000 BTU/h), its coefficient of performance is said to be 3.5. This means that each unit

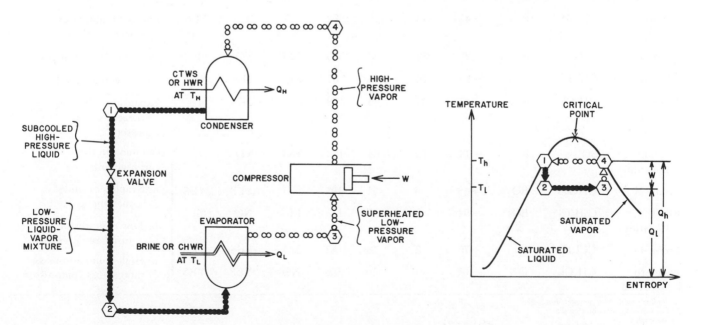

Fig. 8.4a The refrigeration cycle

823

of energy introduced at the compressor will pump 3.5 units of heat energy to the cooling tower water. Conventionally controlled chillers operate with coefficients of performance (COP) in the range of 2.5 to 3.5. Optimization can double the COP by increasing T1, by decreasing Th, and through other methods that will be described later. Figure 8.4a shows the four principal pieces of equipment that make up a refrigeration machine. In path 1-2 (through the expansion valve) the high-pressure subcooled refrigerant liquid becomes a low-pressure, liquid-vapor mixture. In path 2-3 (through the evaporator) this becomes a superheated low-pressure vapor stream, whereas in the compressor (path 3-4) the pressure of the refrigerant vapor is increased. Finally, in path 4-1 this vapor is condensed at constant pressure. The liquid leaving the condenser is usually subcooled, whereas the vapors leaving the evaporator are usually superheated by controlled amounts.

The unit most frequently used in describing refrigeration loads is the ton. Because several tons are referred to in the literature, it is important to distinguish among them:

Standard ton	200 BTU/min (3520W)
British ton	237.6 BTU/min (4182 W)
European ton (Frigorie)	50 BTU/min (880 W)

Table 8.4b
REFRIGERANT CHARACTERISTICS

Refrigerant		Applicable compressor (R = Reciprocating, RO = Rotary, C = Centrifugal)	Boiling point in °F* at atmospheric pressure	Evaporator pressure in PSIA† if operating temperature is 5°F (−15°C)	Condenser pressure in PSIA,† if operating temperature is 86°F (30°C)	Latent heat in BTU/lbm‡ at 18°F (−7.8°C)	Toxic (T), Flammable (F), Irritating (I)	Mixes and/or compatible with the lubricating oil	Chemically inert and noncorrosive	Remarks
Ethane	C_2H_6	R	−127	236	675	148	T&F	NO	YES	For low-temperature service
Carbon Dioxide	CO_2	R	−108	334	1039	116	NO	YES	YES	Low-efficiency refrigerant
Propane	C_3H_8	R	−48	42	155	132	T&F	NO	YES	
Freon-22	$CHClF_2$	R	−41	43	175	92	NO	(1)	YES	For low-temperature service
Ammonia	NH_3	R	−28	34	169	555	T&F	NO	(2)	High-efficiency refrigerant
Freon-12	CCl_2F_2	R	−22	26	108	67	NO	YES	YES	Most recommended
Methyl Chloride	CH_3Cl	R	−11	21	95	178	(3)	YES	(4)	Expansion valve may freeze if water is present
Sulphur Dioxide	SO_2	R	+14	12	66	166	T&I	NO	(4)	Common to these refrigerants:
Freon-21	$CHCl_2F$	RO	+48	5	31	108	NO	YES	YES	a. Evaporator under vacuum
Ethyl Chloride	C_2H_5Cl	RO	+54	5	27	175	F&I	NO	(5)	b. Low compressor discharge pressure
Freon-11	CCl_3F	C	+75	3	18	83	NO	YES	YES	c. High volume-to-mass ratio across compressor
Dichloro Methane	CH_2Cl_2	C	+105	1	10	155	NO	YES	YES	

(1) Oil floats on it at low temperatures.
(2) Corrosive to copper-bearing alloys.
(3) Anesthetic.
(4) Corrosive in the presence of water.
(5) Attacks rubber compounds.

* °C = $\dfrac{°F - 32}{1.8}$
†PSIA = 6.9 kPa
‡BTU/lbm = 232.6 J/kg

Refrigerants

The fluid that carries the heat from a low to a high temperature level is referred to as the refrigerant. Table 8.4b provides a summary of the more frequently used refrigerants.

The data are presented on the assumption that the evaporator will operate at 5°F (−15°C) and the temperature of the cooling water supply for the condenser will allow it to maintain 86°F (30°C). Other temperature levels would have illustrated the relative characteristics of the various refrigerants equally well. It is generally desirable to avoid operating under vacuum in any parts of the cycle because of sealing problems. At the same time, very high condensing pressures are also undesirable because of the resulting structural strength requirements. From this point of view, the refrigerants between propane and methyl chloride in the tabulation display favorable characteristics. An exception to this reasoning is when very low temperatures are required. For such service ethane can be the proper selection in spite of the resulting high system design pressure.

Another consideration is the latent heat of the refrigerant. The higher it is, the more heat can be carried by the same amount of working fluid, and therefore the corresponding equipment size can be reduced. This feature has caused many users in the past to compromise with the undesirable characteristics of ammonia.

One of the most important considerations is safety. In industrial installations, the desirability of nontoxic, nonirritating, nonflammable refrigerants cannot be overemphasized. It is similarly important that the working fluid be compatible with the compressor lubricating oil. The refrigerants that are corrosive are undesirable for the obvious reasons of higher first cost and increased maintenance.

Most working fluids listed are compatible with reciprocating compressors. Only the last four fluids in the tabulation having high volume-to-mass ratios and low compressor discharge pressures can justify the consideration of rotary or centrifugal machines.

When all the advantages and drawbacks of the many refrigerants are considered, overall it is Freon-12 that is suited for the largest number of applications.

Circulated Fluids

In the majority of industrial installations, the refrigerant evaporator is not used directly to cool the process. More frequently, the evaporator cools a circulated fluid, which is then piped to cool the process.

For temperatures below the point at which water can be used for a coolant, brine is frequently used. It is important to remember that weak brines may freeze and that strong brines, if they are not true solutions, may plug the evaporator tubes. For operation around 0°F (−18°C), the sodium brines (NaCl) are recommended,

and for services down to −45°F (−43°C), the calcium brines (CaCl) are best.

Care must be exercised in handling brines because they are corrosive if not at the pH of 7 or if oxygen is present. In addition, brine will initiate galvanic corrosion between dissimilar metals.

Small Industrial Refrigerators

On the left side of Figure 8.4c, the direct expansion-type control is shown. Here a pressure-reducing valve maintains a constant evaporator pressure. The pressure setting is a function of load, and therefore these controls are recommended for constant load installations only. The proper setting is found by adjustment of the pressure-control valve until the frost stops just at the end of the evaporator, indicating the presence of liquid refrigerant up to that point.

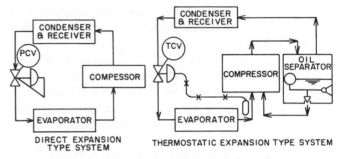

Fig. 8.4c Small industrial refrigerators with throttling control

If the load increases, all the refrigerant will vaporize before the end of the evaporator, causing low efficiency as the unit is "starved." This condition will be relieved only by a change in the pressure setting. When the unit is down, the pressure-control valve closes, isolating the high- and low-pressure sides of the system. This guarantees the desirable high startup torque.

On the right side of the same illustration the thermostatic expansion-type control is shown. This system, instead of maintaining evaporator pressure, controls the superheat of the evaporated vapors. This design is therefore not limited to constant loads, because it guarantees the presence of liquid refrigerant at the end of the evaporator under all load conditions.

Figure 8.4c also shows a typical oil separator.

Expansion Valves

On the left side of Figure 8.4d a fairly standard superheat control valve is shown. It detects the pressure into and the temperature out of the evaporator. If the evaporator pressure drop is low, these measurements (the saturation pressure and the temperature of the refrigerant) are an indication of superheat. The desired superheat is set by the spring in the valve operator, which together with the saturation pressure in the evaporator opposes the opening of the valve. The "superheat

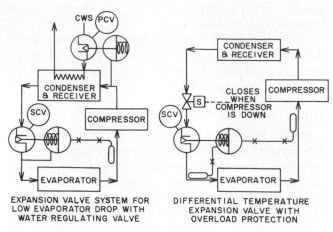

EXPANSION VALVE SYSTEM FOR
LOW EVAPORATOR DROP WITH
WATER REGULATING VALVE

DIFFERENTIAL TEMPERATURE
EXPANSION VALVE WITH
OVERLOAD PROTECTION

Fig. 8.4d   Expansion valve installation

feeler bulb" pressure balances these forces when the unit is in equilibrium, operating at the desired superheat (usually 9°F, or 5°C).

If the process load increases, it causes an increase in the evaporator outlet temperature. An increase in this temperature results in a rise in the "feeler bulb" pressure, which in turn further opens the superheat control valve. This greater flow from the condenser to the evaporator increases the saturation pressure and temperature, and the increased saturation pressure balances against the increased feeler bulb pressure at a new (greater) valve opening in a new equilibrium. To adjust to an increased load condition, the evaporator pressure increases, but the amount of superheat (set by the valve spring) is kept constant.

In this same sketch, the operation of the cooling water regulating valve is also illustrated. This valve maintains the condenser pressure constant and at the same time conserves cooling water. At low condenser pressure, such as when the compressor is down, the water valve closes. It starts to open when the compressor is restarted and its discharge pressure reaches the setting of the valve. The water valve opening follows the load, further opening at higher loads to maintain the condenser pressure constant. This feature too can be incorporated into any one of the refrigeration units.

At very low temperatures, a small change in refrigerant vapor pressure is accompanied by a fairly large change in temperature. For example, Freon-12 at the temperature level of $-100°F$ ($-73°C$) will show a 5°F (2.8°C) temperature change, corresponding to a 0.3 PSIG (2.1 kPa) variation of saturation pressure. Therefore, the use of a thermal bulb to detect indirectly the saturation pressure in the evaporator will result in a more sensitive measurement. A differential temperature expansion valve, taking advantage of this phenomenon, is illustrated on the right side of Figure 8.4d.

On-Off Control of Small Industrial Units

Figure 8.4e shows the controls for a fairly simple and small refrigeration unit. This system includes a conventional superheat control valve, a low-pressure-drop evaporator, a reciprocating on-off compressor, and an air-cooled condenser. The purpose of this refrigeration package is to maintain a refrigerated water supply to the plant within some set limits.

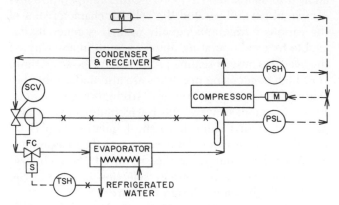

Fig. 8.4e   Small industrial refrigeration unit with electric on-off control

The high-temperature switch (TSH) shown on the illustration is the main control device. Whenever the temperature of the refrigerated water drops below a preset value (e.g., 38°F, or 3.3°C), the refrigeration unit is turned off, and when it rises to some other level (e.g., 42°F, or 5.6°C), it is restarted. This on-off cycling control is accomplished by the temperature switch closing the solenoid valve when the water temperature is low enough. The closing of this valve causes the compressor suction pressure to drop until it reaches the set point of the low-pressure switch, which in turn stops the compressor.

While the unit is running, the expansion valve maintains the refrigerant superheat constant and the safety interlocks protect the equipment. These interlocks include such features as turning off the compressor if the fan motor stops or if the compressor discharge pressure becomes too high for some other reason.

This unit can operate only at full compressor capacity or not at all. This type of machine is referred to as one with two-stage unloading. When varying the cooling capacity of the unit is desired instead of turning it on and off, two possible control techniques are available. One approach involves the multi-step unloading of reciprocating compressors. In a three-step system the available operating loads are 100%, 50%, and 0%, whereas with five-step unloading, 100%, 75%, 50%, 25%, and 0% loads can be handled.

Multi-Stage Refrigeration Units

It is not practical to obtain a compression ratio outside the range of 3:1 to 8:1 with the compressors used in the process industry. This places a limitation on the minimum temperature that a single-stage refrigeration unit can achieve.

For example, in order to maintain the evaporator at $-80°F$ ($-62°C$) and the condenser at $86°F$ ($30°C$)—compatible with standard supplies of cooling water—the required compression ratio would be as follows, if Freon-12 were the refrigerant:

$$\text{Compression ratio} = \frac{\text{Refrigerant pressure at } 86°F}{\text{Refrigerant pressure at } -80°F} = \frac{108}{2.9} = 37$$

$$8.4(2)$$

Such compression ratio is obviously not practical, and therefore a multi-stage system is required.

Large Industrial Refrigerators with High Turn-Down Ratio

The refrigeration unit shown in Figure 8.4f, although far from a "standard" system, does contain some of the features typical to conventional industrial units in the 500-ton (1760 kW) and larger sizes. These features include the capability for continuous load adjustment as contrasted with stepwise unloading, the application of the economizer expansion valve system, and the use of hot gas bypass to increase rangeability.

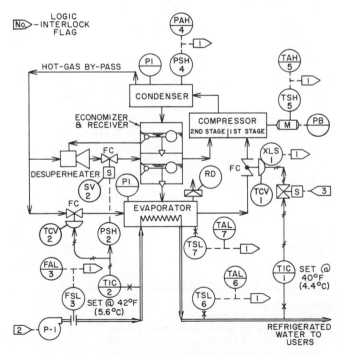

Fig. 8.4f Conventionally controlled industrial refrigeration system with high turn-down ratio

The unit illustrated provides refrigerated water at $40°F$ ($4.4°C$) through the circulating header system of an industrial plant. The flow rate is fairly constant, and there-

fore process load changes are reflected by the temperature of the returning refrigerated water. Under normal load conditions, this return water temperature is $51°F$ ($10.6°C$). As process load decreases, the return water temperature drops correspondingly. With the reduced load on the evaporator, TIC-1 gradually closes the suction vane or the prerotation vane of the compressor. By throttling the suction vane, a 10:1 turndown ratio can be accomplished. If the load drops below this ratio, the hot gas bypass system has to be activated.

The hot gas bypass is automatically controlled by TIC-2. Its purpose is to keep the constant speed compressor out of surge by opening this bypass valve when the load drops to levels sufficiently low to approach surge. If the chilled water flow rate is constant, the difference between chilled water supply and return temperatures is an indication of the load. If full load corresponds to a $15°F$ ($8.3°C$) difference on the chilled water side of the evaporator and the chilled water supply temperature is controlled by TIC-1 at $40°F$ ($4.4°C$), then the return water temperature detected by TIC-2 is also an indication of load.

If surge occurs at 10 percent load, this would correspond to a return water temperature of $41.5°F$ ($5.3°C$). In order to stay safely away from surge, TIC-2 in Figure 8.4f is set at $42°F$ ($5.6°C$), corresponding to approximately 13 percent load. When the temperature drops to $42°F$ ($5.6°C$), this valve starts to open, and its opening can be proportional to the load detected. This means that the valve is fully closed at $42°F$ ($5.6°C$), fully open at $40°F$ ($4.4°C$), and throttled in between. Therefore, it is theoretically possible to achieve a very high turndown ratio by temporarily running the machine on close to zero process load. This can be visualized as a heat pump, transferring heat energy from the refrigerant itself to the cooling water. In the process some of the refrigerant vapors are condensed, resulting in an overall lowering of operating pressures in the system.

The main advantage of a hot gas bypass, therefore, is that it allows the chiller to operate at low loads without going into surge. The price of this operational flexibility is an increased energy cost, because the work introduced by the compressor is wasted as friction drop through the hot gas bypass valve. As will be shown later, optimized control systems eliminate this waste through the use of variable speed compressors, which will respond to a reduction in load by lowering their speed; instead of throwing away the unnecessarily introduced energy in TCV-1 and TCV-2, they do not introduce it in the first place.

The economizer shown in Figure 8.4f can increase the efficiency of operation by 5 to 10 percent. This is achieved by the reduction of space requirements, savings on compressor power consumption, reduction of condenser and evaporator surfaces, and other measures. The economizer shown in Figure 8.4f is a two-stage expansion valve with condensate collection chambers. When the load is above 10 percent, the hot gas bypass system is inactive. Condensate is collected in the upper chamber of the economizer, and it is drained under float level control, driven by the condenser pressure. The pressure in the lower chamber floats off the second stage of the compressor, and it too is drained into the evaporator under float level control, driven by the pressure of the compressor second stage. Economy is achieved as a result of the vaporization in the lower chamber by precooling the liquid that enters the evaporator and at the same time by desuperheating the vapors that are sent to the compressor second stage.

When the load is below 10 percent, the hot gas bypass is in operation, and the solenoid valve SV-2, which is actuated by the high pressure switch PSH-2, opens. Some of the hot gas goes through the evaporator and is cooled by contact with the liquid refrigerant, and some of the hot gas flows through the open solenoid. This second portion is desuperheated by the injection of liquid refrigerant upstream of the solenoid, which protects against overheating the compressor. Operating safety is guaranteed by the following interlocks:

The first interlock system prevents the compressor motor from being started if one or more of the following conditions exist, and it also stops the compressor if any except the first condition listed occurs while the compressor is running.

1. Suction vane is open, detected by limit switch XLS-1.
2. Refrigerated water temperature is dangerously low, approaching freezing as sensed by TSL-6.
3. Refrigerated water flow is low, measured by FSL-3.
4. Evaporator temperature has dropped near the freezing point as detected by TSL-7.
5. Condenser discharge pressure (and, therefore, pressure in the condenser) is high, indicated by PSH-4.
6. Temperature of motor bearing or winding is high, detected by TSH-5.
7. Lubricating oil pressure is low (not shown).

The second interlock system guarantees that the following pieces of equipment are started or are already running upon starting of the compressor:

1. Refrigerated water pump (P-1).
2. Lubricating oil pump (not shown).
3. Water to lubricating oil cooler, if such exists (not shown).

The third interlock usually ensures that the suction vane is completely closed when the compressor is stopped.

Optimization of Refrigeration Machines

As illustrated in Figure 8.4g, the cooling load from the process is carried by the chilled water to the evaporator, where it is transferred to the freon. The freon takes the heat to the condenser, where it is passed on to the cooling tower water, so that it might finally be rejected to the ambient air. This heat pump operation involves four heat transfer substances (chilled water, freon, cooling tower water, air) and four heat exchanger devices (process heat exchanger, evaporator, condenser, cooling tower). The total system operating cost is the sum of the cost of circulating the four heat transfer substances (M1, M2, M3, and M4). In the traditional (unoptimized) control systems, such as the one illustrated in Figure 8.4f, each of these four systems were operated independently in an uncoordinated manner. In addition, conventional control systems were not designed to follow the load but were operated at constant high levels of operating energy with the unnecessarily introduced energy being wasted.

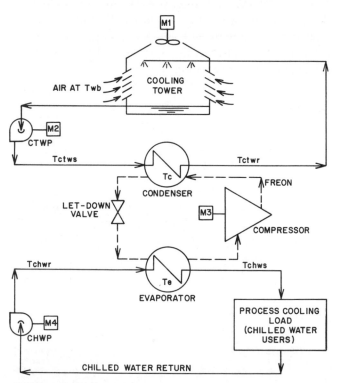

ABBREVIATIONS:

Twb	: WET BULB TEMP.
Tctws	: COOLING TOWER WATER SUPPLY TEMP.
Tctwr	: COOLING TOWER WATER RETURN TEMP.
Tchws	: CHILLED WATER SUPPLY TEMP.
Tchwr	: CHILLED WATER RETURN TEMP.
CTWP	: COOLING TOWER WATER PUMP
CHWP	: CHILLED WATER PUMP
Tc	: FREON TEMPERATURE IN CONDENSER
Te	: FREON TEMPERATURE IN EVAPORATOR

Fig. 8.4g Optimized control of refrigeration machines as an integrated system.

Load following optimization eliminates this waste while operating the aforementioned four systems as a coordinated single process, with the goal of control being to maintain the cost of operation at a minimum. The controlled variables are the supply and return temperatures of chilled and cooling tower waters, and the manipulated variables are the flow rates of chilled water, freon, cooling tower water, and air. By allowing the water temperatures to float in response to load and ambient temperature variations, the waste associated with keeping them at arbitrarily selected fixed values is eliminated and the chiller operating cost is drastically reduced (sometimes cut in half). In the following paragraphs the specific optimization loops and strategies will be described. First those four control loops will be discussed which serve to maintain the four water temperatures (Tchws, Tchwr, Tctws, Tctwr) at their optimum values.

Chilled Water Supply Temperature (Tchws) Optimization

The yearly operating cost of a chiller is reduced by 1.5 percent to 2 percent for each 1°F (0.6°C) reduction in the temperature difference across this heat pump (Tc–Te in Figure 8.4g). In order to minimize this difference, Tc must be minimized and Te maximized. Therefore, the optimum value of Tchws is the maximum temperature that will still satisfy all the loads. Figure 8.4h illustrates the method of continuously finding and maintaining this value.

It should be noted that an energy-efficient refrigeration system cannot be guaranteed by instrumentation alone, because equipment sizing and selection also play an important part in the overall result. For example, the evaporator heat transfer area should be maximized, so that Te is as close as possible to the average chilled water temperature in the evaporator. Similarly, at the compressor the freon flow is to match the load by motor speed control rather than by throttling. TCV-1 and TCV-2 shown in Figure 8.4f represent sources of energy waste, whereas Figure 8.4h shows the energy-efficient technique of motor speed control. The variable speed operation can be achieved by using variable speed drives on electric motors or can be accomplished through the use of steam turbine drives. If several constant-speed motors are used, then all compressors should be driven to their maximum load (TCV-1 and TCV-2 in Figure 8.4f fully open) except one, which is used to match the required load by throttling.

Figure 8.4h shows the proper technique of maximizing the chilled water supply temperature in a load-following, floating manner. The optimization control loop guarantees that all users of refrigeration in the plant will always be satisfied while the chilled water temperature is maximized. This is done by selecting (with TY-1) the most

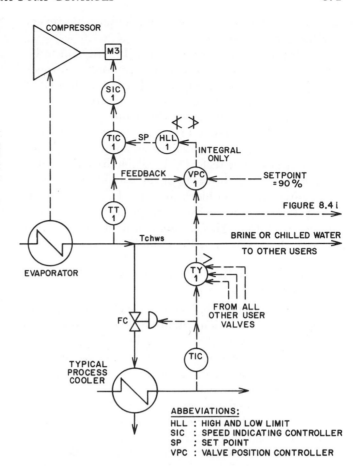

Fig. 8.4h Load following floating control of chilled water supply temperature.

open chilled water valve in the plant and comparing that transmitter signal with the 90 percent set point of the valve position controller, VPC-1. If even the most open valve is less than 90 percent open, the set point of TIC-1 is increased; however, if the valve opening exceeds 90 percent, the TIC-1 set point is decreased. Thereby, a condition is maintained in which all users are able to obtain more cooling (by further opening their supply valves) if needed while the header temperature is continuously maximized. The VPC-1 set point of 90 percent is adjustable. Lowering it gives a wider safety margin, which might be required if some of the cooling processes served are very critical. Increasing the set point maximizes energy conservation at the expense of the safety margin.

An additional benefit of this load following optimization strategy is that because all chilled water valves in the plant are opened up as Tchws is maximized, valve cycling is reduced and pumping costs are lowered. The reduction in pumping costs is a direct result of the opening up of all chilled water valves, which therefore require less pressure drop. Valve cycling is eliminated when the valve opening is moved away from the unstable region near the closed position.

In order for the control system in Figure 8.4h to be stable, it is necessary to use an integral-only controller for VPC-1, with an integral time that is tenfold that of the integral setting of TIC-1. This mode selection is needed to allow the optimization loop to be stable when the valve opening signal selected by TY-1 is either cycling or noisy. The high/low limit settings on the set point signal (HLL-1) to TIC-1 guarantee that VPC-1 will not drive the chilled water temperature to unsafe or undesirable levels. Because these limits can block the VPC-1 output from affecting Tchws, it is necessary to protect against reset windup in VPC-1. This is done through the external feedback signal shown in Figure 8.4h.

Chilled Water Return Temperature (Tchwr) Optimization

The combined cost of operating the chilled water pumps and the chiller itself is a function of the temperature drop across the evaporator (Tchwr − Tchws). Because an increase in this ΔT decreases compressor operating costs while also decreasing pumping costs, the aim of this optimization strategy is to maximize this ΔT.

This ΔT will be the maximum when the chilled water flow rate across the chilled water users is the minimum. If even the most open chilled water valve is not yet fully open, one has a choice of increasing the chilled water supply temperature (set by TIC-1 in Figure 8.4h) or increasing the temperature rise across the users by lowering the ΔP (set by PDIC-1 in Figure 8.4i), or both. Increasing the chilled water supply temperature reduces the yearly compressor operating cost (M3) by approximately 1.5 percent for each °F of temperature increase, whereas lowering the ΔP reduces the yearly pump operating cost (M4) by approximately 50¢/GPM for each PSID (or $0.32 m³/s for each KPa).

The set points of the two valve position controllers (VPC-1s in Figures 8.4h and 8.4i) will determine if these adjustments are to occur in sequence or simultaneously. If both set points are the same, simultaneous action will result, whereas if one adjustment is economically more advantageous than the other, then the set point of the corresponding VPC will be positioned lower than the other.

This will result in sequencing, which means that the more cost-effective correction will be fully exploited before the less effective one is started. In Figures 8.4h and 8.4i, it was assumed that increasing Tchws is the more cost-effective step, and therefore the VPC in Figure 8.4h is shown with a setting of 90 percent, whereas the VPC in Figure 8.4i is set at 95 percent.

This controller is the cascade master of PDIC-1, which guarantees that the pressure difference between the chilled water supply and return is always high enough to motivate water flow through the users but never so high as

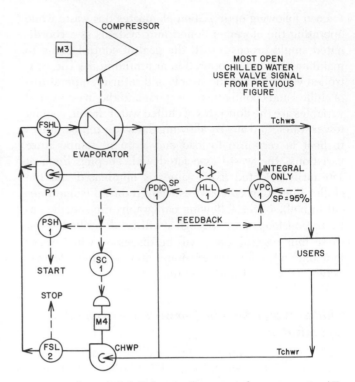

Fig. 8.4i   Floating of chilled water flow rate to keep evaporator ΔT always at optimum values

to exceed their pressure ratings. The high and low limits are set on HLL-1, and VPC-1 is free to float this set point within these limits to keep the operating cost at a minimum.

In order to protect against reset windup when the output of VPC-1 reaches one of these limits, an external feedback is provided from the PDIC-1 output signal to the pump speed controller SC-1. The VPC is an integral-only controller which can be tuned to be more responsive as the VPC in Figure 8.4h.

When the chilled water pump station consists of several pumps, only one of which is variable-speed, additional pump increments are started when PSH-1 signals that the pump speed controller set point is at its maximum. When the load is dropping, the excess pump increments are stopped on the basis of flow, detected by FSL-2. In order to eliminate cycling, the excess pump increment is turned off only when the actual total flow corresponds to less than 90 percent of the capacity of the *remaining* pumps.

This load-following optimization loop will float the total chilled water flow to achieve maximum overall economy. In order to maintain efficient heat transfer and appropriate turbulence within the evaporator, a small local circulating pump (P1) is provided at the evaporator. This pump is started and stopped by FSLH-3, guaranteeing that the water velocity in the evaporator tubes will never drop below the adjustable limit of, for example, 4 ftps (1.2 m/s).

Cooling Tower Supply Temperature (Tctws) Optimization

Minimizing the temperature of the cooling tower water is one of the most effective ways to contribute to chiller optimization. Conventional control systems of the past have been operated with constant cooling tower temperatures of 75°F (23.9°C) or higher. A constant utility condition is the worst enemy of efficiency and therefore of optimization. Each 10°F (5.6°C) reduction in the cooling tower water temperature will reduce the yearly operating cost of the compressor by approximately 15 percent.

Stated another way: If a compressor is operating at 50°F (10°C) condenser water instead of 85°F (29.4°C) it will meet the same load while consuming half as much power. Operation at condenser water temperatures of less than 50°F (10°C) is quite practical during the winter months. Savings exceeding 50 percent have been reported.[1]

As shown in Figure 8.4j, an optimization control loop is required in order to maintain the cooling tower water supply continuously at an economical minimum temperature. This minimum temperature is a function of the wet bulb temperature of the atmospheric air. The cooling tower cannot generate a water temperature that is as low as the ambient wet bulb, but it can approach it. (The temperature difference between Tctws and Twb is called the "approach.")

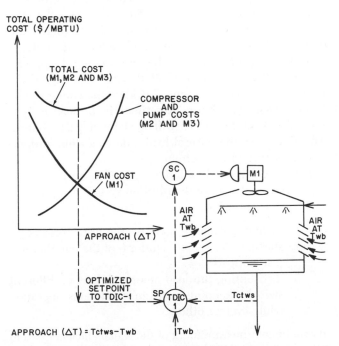

Fig. 8.4j Load following floating control of the cooling tower water temperature

Figure 8.4j illustrates the fact that as the approach increases, the cost of operating the cooling tower fans drops, and the cost of pumping and of compressor operation increases. Therefore, the total operating cost curve

has a minimum point, identifying the optimum approach that will allow operation at an overall minimum cost. This ΔT automatically becomes the set point of TDIC-1. This optimum approach increases if the load on the cooling tower increases or if the ambient wet bulb decreases.

If the cooling tower fans are centrifugal units or if the blade pitch is variable, then the optimum approach is maintained by continuous throttling. If the tower fans are two-speed or single-speed units, then the output of TDIC-1 will incrementally start and stop the fan units in order to maintain the optimum approach. In cases in which a large number of cooling tower cells constitutes the total system, it is also desirable to balance the water flows to the various cells automatically as a function of the operation of the associated fan. In other words, the water flows to all cells whose fans are at high speed should be controlled at equal high rates; cells with fans operating at low speeds should receive water at equal low flows, and cells with their fans off should be supplied with water at equal minimum flow rates.

Cooling Tower Water Return Temperature (Tctwr) Optimization

As depicted in Figure 8.4k, the combined cost of operating the cooling tower pumps and the chiller compressor is a function of the temperature rise across the condenser. Because an increase in this ΔT increases compression costs while decreasing pumping costs, the combined total curve has a minimum cost point. The ΔT corresponding to this minimum automatically becomes the set point of TDIC-1 in the optimized control loop shown in Figure 8.4k. This controller is the cascade master of PDIC-1, which guarantees that the pressure difference between the supply and return cooling tower water flows is always high enough to provide flow through the users but never so high as to cause damage. The high and low limits are set on HLL-1. TDIC-1 freely floats this set point within these ΔP limits, to keep the operating cost at a minimum.

In order to protect against reset windup (when the output of TDIC-1 reaches one of these limits), an external feedback is provided from the PDIC-1 output signal to the pump speed controller SC-1.

When the cooling tower water pump station consists of several pumps, only one of which is variable-speed, additional pump increments are started when PSH-1 signals that the pump speed controller set point is at its maximum. When the load is dropping, the excess pump increments are stopped on the basis of *flow*, detected by FSL-2. In order to eliminate cycling, the excess pump increment is turned off only when the actual total flow corresponds to less than 90 percent of the capacity of the *remaining* pumps.

This load-following optimization loop will float the total cooling tower water flow to achieve maximum overall

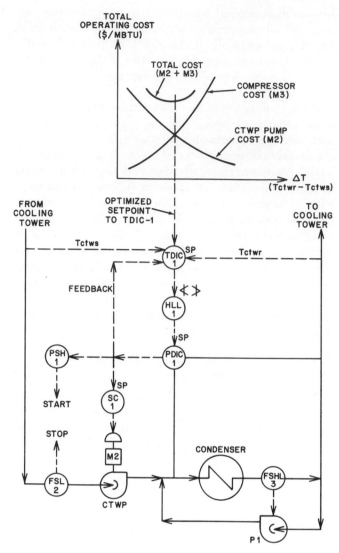

Fig. 8.4k   Floating of cooling tower water flow rate to keep condenser
ΔT always at optimum values

Since coolant can be provided from many sources in a typical plant, another approach to optimization is to reconfigure the system in response to changes in loads, ambient conditions, and utility costs. For example, during some operating and ambient conditions, the cooling tower water may be cold enough to meet the demands of the process directly. Alternatively, if the cooling tower outflow is below the temperature required by the process, the chillers can be operated in a thermosiphon mode. Freon circulation is then driven by the temperature differential rather than by a compressor.

When demand is low, it may be possible to save money by operating the chillers part of the time at peak efficiency rather than steadily at partial loading. Efficiency tends to be low at partial loads because of losses caused by friction drop across suction dampers, prerotation vanes, or steam governors. Cycling is practical if the storage capacity of the distribution headers is enough to avoid frequent stops and starts. When operation is to be intermittent, data such as the heat to be removed and the characteristics of the available chillers are needed to determine the most economical operating strategy.

When chiller cycling is used, one relies upon the thermal capacity of the chilled water distribution system to absorb the load while the chiller is off. For example, if the pipe distribution network has a volume of 100,000 gallons (378,000 l), this represents a thermal capacity of approximately 1 million BTUs for each degree F of temperature rise (1.9×10^9 J/°C). So, if one can allow the chilled water temperature to float 5°F (2.8°C) (e.g., from 40 to 45°F, or from 4.4°C to 7.2°C) before the chiller is restarted, this represents the equivalent of approximately 400 tons (1405 kW) of thermal capacity. If the load happens to be 200 tons (704 kW), the chiller can be turned off for two hours at a time. If the load is 1000 tons (3514 kW), the chiller will be off for only 24 minutes. This illustrates the natural load-following, time-proportioning nature of this scheme.

If the chiller needs a longer period of rest than the thermal capacity of the distribution system can provide, one has the option of:

1. adding tankage to increase the water volume,
2. starting up a second chiller (not the one that was just stopped), or
3. distributing the load among chillers of different sizes by keeping some in continuous operation while cycling others.

By continuous measurement of the actual efficiency ($/ton) of each chiller, all loads can be met through the operation of the most efficient combination of machines for the load.

In plants with multiple refrigerant sources, the cost per ton of cooling can be calculated from direct measurements and used to establish the most efficient combination of units to meet present or anticipated loads.

economy. In order to maintain efficient heat transfer and appropriate turbulence within the condenser, a small local circulating pump (P1) is provided at the condenser. This pump is started and stopped by FSHL-3, guaranteeing that the water velocity in the condenser tubes will never drop below the adjustable limit of, for example, 4 fps. (1.2 m/s).

Other Chiller Optimization Methods

The controls described in Figures 8.4h through 8.4k will automatically minimize the operating cost of the refrigeration machine as a single integrated system. Additional steps toward reducing operating costs will be described in the following paragraphs.

If storage tanks are available, it is cost-effective to generate the daily brine or chilled water needs of the plant at night, when it is the least expensive to do so, because ambient temperatures are low and nighttime electricity is less expensive in some areas.[2]

As with boilers, this accounts for differences among units as well as for the efficiency versus load characteristics of the individual coolant sources, as shown in Figure 8.4l.

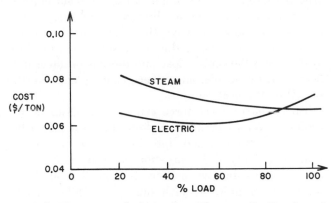

Fig. 8.4l The unit cost of refrigeration is a function of load but depends on the characteristics of each chiller. The indicated cost figures are for representative installed equipment.

The last goal of chiller optimization is applicable only where the chillers are not operated continuously. In such installations, the optimization system "knows" how many BTUs need to be removed before startup and the size and efficiency of the available chillers. Therefore, the length of the "pull-down" period can be minimized and the energy cost of this operation optimized.

Heat Recovery Controls

Figure 8.4m depicts the required optimizing control loop when the heat pumped by the chiller is recovered in the form of hot water.

Similarly to the control system shown for chilled water temperature floating in Figure 8.4h, the hot water temperature can also be continuously optimized in a load-following floating manner. If at a particular load level it is sufficient to operate with 100°F (37.8°C) instead of 120°F (48.9°C) hot water, this technique will allow the same tonnage of refrigeration to be met by the chiller, at 30 percent lower cost. The reason is that the compressor discharge pressure is determined by the hot water temperature in the split condenser.

The optimization control loop in Figure 8.4m guarantees that all hot water users in the plant will always obtain desired results while the water temperature is minimized. TY-1 selects the most open hot water valve in the plant, and VPC-1 compares that transmitter signal with a 90 percent set point. If even the most open valve is less than 90 percent open, the set point of TIC-1 is decreased; if the opening exceeds 90 percent, the set point is increased. Thereby, a condition is maintained in which all users are able to obtain more heat (by further opening their supply valves) if needed while the header temperature is continuously optimized.

Figure 8.4m also shows that an increasing demand for heat will cause the TIC-1 output signal to rise as its

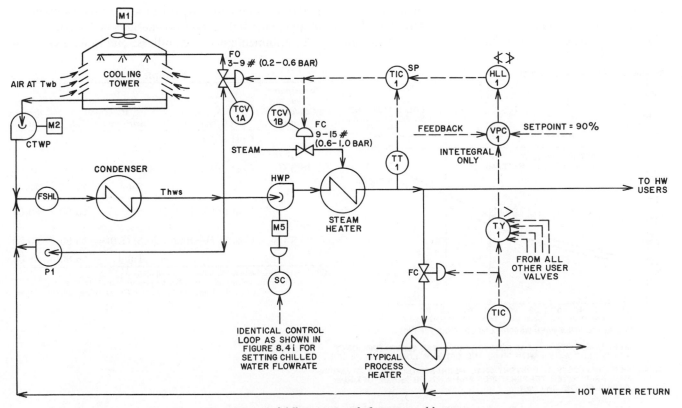

Fig. 8.4m Load following controls for recovered hot water

833

measurement drops below its set point. An increase in heat load will cause a decrease in the heat spill to the cooling tower, since the control valve TCV-1A is closed between 3 and 9 PSIG (0.2 and 0.6 bar). At a 9-PSIG (0.6 bar) output signal, all the available cooling load is being recovered and TCV-1A is fully closed. If the heat load continues to rise (TIC-1 output signal rises over 9 PSIG, or 0.6 bar), this will result in the partial opening of the "pay heat" valve, TCV-1B. In this mode of operation the steam heat is used to supplement the freely available recovered heat to meet the prevailing heat load.

A local circulating pump, P1, is started whenever flow velocity is low. This prevents deposit formation in the tubes. P1 is a small 10- to 15-hp pump operating only when the flow is low. The main cooling tower pump (usually larger than 100 hp) is stopped when TCV-1A is closed.

AUTOMATIC SELECTION OF OPTIMUM OPERATING MODE

The cost-effectiveness of heat recovery is a function of the outdoor temperature, of the unit cost of energy from the alternative heat source, and of the percentage of the cooling load that can be used as recovered heat. According to Figure 8.4n, if steam is available at $7/MMBTU and only half of the cooling load is needed in the form of hot water, it is more cost-effective to operate the chiller on cooling tower water and use steam as the heat source when the outside air is below 65°F (18°C). Conversely, when the outdoor temperature is above 75°F (23.9°C), the penalty for operating the split condenser at hot water temperatures is no longer excessive; therefore, the plant should automatically switch back to recovered heat operation. This cost-benefit analysis is a simple and continually used element of the overall optimization scheme.

In plants in locations such as the southern states, where there is no alternative heat source, another problem can arise because all the heating needs of the plant must be met by recovered heat from the heat pump. It is possible that during cold winter days, there might not be enough recovered heat to meet this load. Whenever the heat load exceeds the cooling load and there is no alternative heat source available, an artificial cooling load must be placed on the heat pump.

This artificial heat source can in some cases be the cooling tower water itself. A direct heat exchanger between cooling tower and chilled water streams is of advantage not only in this situation but also when there is no heat load—but when there is a small cooling load during the winter. At such times, the chiller can be stopped, and the small cooling load can be met by direct cooling from the cooling tower water that is at a winter temperature.

Retrofit Chiller Controls

In plants that are being designed today, it is easy to install variable-speed pumps, to provide the thermal capacity required for chiller cycling, or to locate chillers and cooling towers at the same elevation and near each other so that the cooling tower water pumping costs will be minimized. In existing plants, one has to accept the inherent design limitations and optimize the system without any changes to the equipment. This is quite practical to do and can still produce savings exceeding 50 percent,[1] but certain precautions are needed.

In optimizing existing chillers, one must give careful

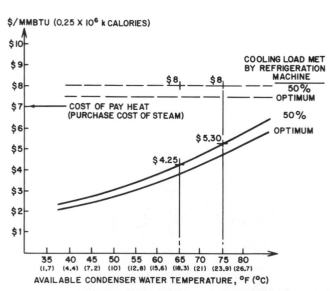

	Example*		
Temperature of Condenser Water	Cost Components	Mechanical Refrigeration Mode	Heat Recovery Mode
65°F (18.3°C)	Cost of Cooling	$4.25	$8.00
	Cost of Heating	(0.5) (7.0) = $3.50	$0.00
	Total	$7.25	$8.00
75°F (23.9°C)	Cost of Cooling	$5.30	$8.00
	Cost of Heating	(0.5) (7.0) = $3.50	$0.00
	Total	$8.80	$8.00

* This example is based on the following assumptions:
    a. The actual cooling load is 50% of chiller capacity. (CL = 0.5 CAP)
    b. The heating load (the demand for hot water) is 50% of cooling load. (HL = 0.5 CL)

— — OPERATING COSTS OF HEAT RECOVERY CHILLERS GENERATING 105 °F (40.6 °C) HOT WATER WITH 50 % AND "OPTIMUM" LOADING OF CHILLER.

—— OPERATING COSTS OF CONVENTIONAL MECHANICAL CHILLER AT VARIOUS CONDENSER WATER TEMPERATURES AND WITH 50 % AND "OPTIMUM" LOADING OF THE CHILLER.

Fig. 8.4n Automatic selection of the most cost-effective mode of refrigeration

consideration to the constraints of low evaporator temperature, economizer flooding, steam governor rangeability, and surge.

SURGE CONTROL

At low loads, not enough freon is circulated. This can initiate a surge condition with its associated violent vibration.

Old chillers usually do not have automatic surge controls and have only vibration sensors for shutdown. If one intends to operate at low loads, it is necessary to add an antisurge control loop. It should be noted that surge protection is always provided at the expense of efficiency. To bring the machine out of surge, the freon flow must be increased if there is no real load on the machine. The only way to provide this increase in flow is to add artificial and wasteful loads (e.g., hot gas or hot water bypasses). Therefore, it is much more economical either to cycle a large chiller or to operate a small one than to meet low load conditions by running the machine near its surge limit.

LOW EVAPORATOR TEMPERATURE

Low temperatures can occur in the evaporator when one optimizes an old chiller that has been designed for operation at 75°F (23.9°C) condenser water and runs it on 45 or 50°F (7.2 or 10°C) condenser water (in the winter). This phenomenon is exactly the opposite of surge because it occurs when freon is being vaporized at an excessively high rate. This occurs because the chiller is able to pump twice the tonnage for which it was designed as a result of the low compressor discharge pressure. In such a situation, the evaporator heat transfer area becomes the limiting factor; furthermore, the only way to increase heat flow is to increase the temperature differential across the evaporator tubes. This shows up as a gradual lowering of freon temperature in the evaporator until it reaches 32°F (0°C), and the machine shuts down to protect against ice formation.

There are two ways to prevent this phenomenon from occurring in existing chillers. The first is to increase the evaporator heat transfer area (a major equipment modification). The second is to prevent the freon temperature in the evaporator from dropping below 33°F (0.6°C) by not allowing the cooling tower water to cool the condenser all the way down to its own temperature. This latter solution requires only the addition of a temperature control loop. This prevents the chiller from taking full advantage of the available cold water from the cooling tower by throttling down on its flow rate, thereby causing its temperature to increase.

ECONOMIZER FLOODING

On existing chillers, the economizer control valves (LCV-1 and LCV-2 in Figure 8.4o) are often sized on the assumption that the freon vapor pressure in the con-

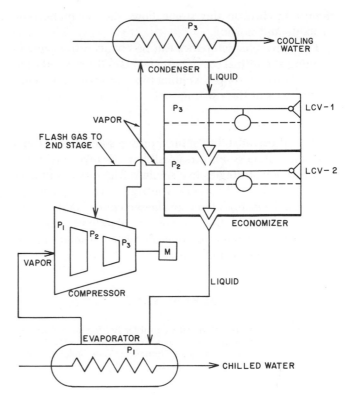

Fig. 8.4o   Economizer flooding can be solved by supplementing LCV 1 and 2 with larger external valves.

denser (P3) is constant and corresponds to a condenser water temperature of 75 or 85°F (23.9 or 29.4°C). Naturally, when such units are operated with 45 or 50°F (7.2 or 10°C) condenser water, P3 is much reduced, as is the pressure differential across LCV-1 and LCV-2. If this occurs—as it easily can—when the freon circulation rate is high, the control valves will be unable to provide the necessary flow rate, and flooding of the economizer will occur (the flow is higher and the ΔP is lower than the basis of valve sizing). The solution is to install larger valves, preferably external and with two-mode control to eliminate offset. (Plain proportional controllers cannot maintain their set points as loads change, whereas the addition of the integral mode eliminates this offset.)

This is important in machines that were not designed originally for optimized, low-temperature condenser water operation, because otherwise the compressor can be damaged by the liquid freon that overflows into it from the flooded flash evaporator.

STEAM GOVERNOR RANGEABILITY

In order for a steam turbine–driven compressor to be optimized, its rotational velocity must be modulated over a reasonably wide range. This is not possible with old, existing machines, because they are provided with quick-opening steam valves. A slight increase in lift from the fully closed position results in a substantial steam flow and therefore a substantial rotational velocity. If one attempts to throttle this steam flow, the valves become unstable and noisy.

The valve characteristics can be changed from quick-opening at minimal cost. One can obtain the desired wide rangeability by simply welding two rings with V-notches to the seats of the existing steam governor valves.

Conclusions

With the availability of inexpensive solid state sensors and microprocessors, it is possible to incorporate all the optimization strategies in a single refrigeration unit controller. Such multivariable controllers can replace the traditional concept of uncoordinated control of flows and temperatures with the new concept of following the load with floating parameters that always correspond to the lowest possible cost of operation.

REFERENCES

1. Romita, E., "A Direct Digital Control for Refrigeration Plant Optimization," *ASHRAE Transactions*, vol. 83, Part 1, 1977.
2. Lipták, B.G., "Save Energy by Optimizing Your Boilers, Chillers and Pumps," *InTech*, March 1981.

BIBLIOGRAPHY

ASHRAE Handbook and Product Directory, 1976 Systems Volume, Chapter 7, ASHRAE, Inc., New York, 1976.

——, 1976 Systems Volume, Chapter 10, ASHRAE Inc., New York, 1976, p. 10.2.

——, 1974 Applications Volume, Chapter 35, ASHRAE Inc., New York, 1974, p. 35.4.

Backus, A.O., "Energy Savings Through Improved Control of Heat Pump Setback," *ASHRAE Journal*, February 1982.

Bower, J.R., "The Economics of Variable Speed Pumping with Speedchanging Drives," *Pumpworld*, Worthington Pump, Inc., Mountainside, NJ.

Carrier System Design Manual, Part 7, Carrier Corporation, Syracuse, NY, 1969.

Cooper, K.W. and Erth, R.A., "Centrifugal Water Chilling Systems: Focus on Off-Design Performance," *Heating/Piping/Air Conditioning*, January 1978, p. 63.

——, "Saving Energy with Refrigeration," *ASHRAE Journal*, December 1978.

——, "Utilizing Water Chillers Efficiently," *International Conference on Energy Use Management, Tucson, AZ*, Pergamon Press, New York, 1977, Vol. 1, p. 233.

Edwards, T.C., "A New Air Conditioning Refrigeration and Heatpump Cycle," *ASHRAE Transactions*, Volume 84, Part 2, 1978.

Freedman, G.M., "Improving ROI from Central Plant Systems," *Specifying Engineer*, January 1980.

Grant, E.L. and Ireson, W.G., *Principles of Engineering Economy*, The Ronald Press Co., New York, 1970.

Hallanger, E.C., "Operating and Controlling Chillers to Save Energy," *ASHRAE Journal*, September 1981.

Lipták, B.G., "Control of Refrigeration Units," *Instrumentation in the Processing Industries*, Chilton, Philadelphia, 1972, Section 10.5, p. 616.

——, "Envelope Control for Coal Gasification," *Instrumentation Technology*, December 1975.

——, "Optimizing Plant Chiller Systems," *InTech*, September 1977.

——, "Reducing the Operating Costs of Buildings," *ASHRAE Transactions*, Volume 83, Part 1, 1977.

——, "Optimizing Controls for Chillers and Heat Pumps," *Chemical Engineering*, October 17, 1983.

Miller, D.K., "The Performance of Water Cooled Lithium Bromide Absorption Units for Solar Energy Applications," *Heating/Piping/Air Conditioning*, January 1976.

"Moore Turns to the Heat Pump," *Power*, November 1973, p. 24.

Newton, A.B., "Optimizing Solar Cooling Systems," *ASHRAE Journal*, November 1976.

Null, H.R., "Heat Pumps in Distillation," *Chemical Engineering Progress*, July 1976.

Polimeros, G., "There Is Energy for Use in Refrigeration Waste Heat," *Specifying Engineer*, January 1980.

Rishel, J.B., "Cut Energy Costs with Improved Controls for Pumping Systems," *Instruments and Control Systems*, August 1976.

Shinskey, F.G., "Energy Conservation," *Instruments and Control Systems*, January 1977.

Stair, W.S., "Air Conditioning System," United States Patent Numbers 2,715,514 and 2,715,515 issued August 16, 1955.

Werden, R.G., "Heat Pump Gets by with Less Energy by Making Ice Summer and Winter," *Architectural Record*, November 1976.

Zimmer, H., "Chiller Control for Energy Conservation," 1976 ISA Conference, Reprint No. 76-522, Instrument Society of America, Research Triangle Park, NC, 1976.

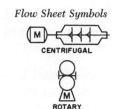

CENTRIFUGAL

ROTARY

RECIPROCATING

COMPRESSOR
(ANY DESIGN)

8.5 COMPRESSOR CONTROLS

Nomenclature:

h = differential head (ft or m)
H = polytropic compressor head (ft or m)
$K_{1,2,3}$ = flow constant
ṁ = mass flow (lbs/hr or kg/h) = $c\sqrt{h\zeta}$
n = polytropic coefficient
P_D = discharge pressure (PSIA or Pa)
P_I = inlet pressure (PSIA or Pa)
Q = volume flow rate (ACFH or m³/h) = $c\sqrt{h/\zeta}$
R = gas constant
T_I = inlet temperature
u = rotor tip speed (ft/sec or m/s)
W = weight flow (lbm/hr or kg/h)
Z = gas compressibility factor
τ = motor torque (ft-lbf or J)
ψ = head coefficient
ω = angular velocity (radians/hr)
ζ = density of gas (lbs/ft³ or kg/m³)

Compressors are gas-handling machines that perform the function of increasing gas pressure by confinement or by kinetic energy conversion. Methods of capacity control for the principal type of compressor are listed in Table 8.5a. The method of control to be used is determined by the process requirements, the type of driver, and the cost.

The primary emphasis should be on the process requirements. Thermodynamic factors affecting compressor performance must be first met through the use of an optimal control system. The selection of the type of driver is usually the next task. Although moderately steady thermodynamic conditions warrant analog control, wide variations of suction pressure, temperature, and molecular weight may require the use of a microprocessor for flexible, reliable, and optimal efficient control. Since the driver constitutes half the compressor installation, careful selection must be made in order to ensure trouble-free performance. When the control involves variable speed, this is easily accomplished by the use of a steam turbine, a gas turbine, a gasoline engine, or a diesel engine. For constant speed control, electric motors are well suited; however, variable speed can be obtained from electric motors by the use of hydraulic coupling or wound rotor induction motors, which are expensive and inefficient. In some cases, the driver may be selected on the basis of overall plant material and energy balance.

The cost of control must be considered. The cost of adding hardware must be weighed against the benefits of energy savings and reliable control. This is especially important in the case of parallel compressor control, in which a slight imbalance between the two compressors can cause one compressor to be fully loaded and another to be recycling, uselessly wasting horsepower. Although startup and shutdown controls are important, proper se-

Table 8.5a
CAPACITY CONTROL METHODS OF COMPRESSORS

Compressor Type	Capacity Control Method
Centrifugal	Suction throttling
	Discharge throttling
	Variable inlet guide vanes
	Speed control
Rotary	Bypassing
	Speed control
Reciprocating	On-off control
	Constant-speed unloading
	Speed control
	Speed control and unloading

lection and installation of valves, transmitters, and lubrication and seal systems cannot be underemphasized to ensure uninterrupted operation for long periods.

Centrifugal Compressors

The centrifugal compressor is a machine that converts the momentum of gas into a pressure head.

$$H = \frac{\tau\omega}{w} = \frac{n}{n-1} ZRT_I, [(P_D/P_I)^{\frac{n}{n-1}} - 1] \quad 8.5(1)$$

This equation is the basis for plotting the compressor curves and for understanding the operation of capacity controls. The nomenclature for the symbols in the equation is furnished at the beginning of this section.

In equation 8.5(1), the pressure ratio (PD/PI) varies inversely with mass flow (W). For a compressor running at constant speed (ω), constant inlet temperature (T_I), constant molecular weight (implicit in R), and constant n, τ, and Z, the discharge pressure may be plotted against weight flow as in Figure 8.5b (curve I). The design point (1) is located in the maximum efficiency range at design flow and pressure.

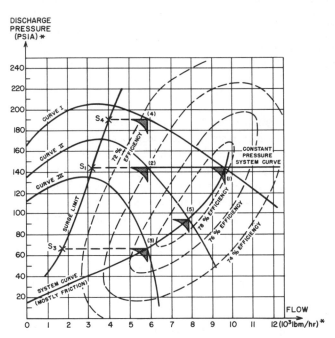

Fig. 8.5b Centrifugal compressor curves. For SI units see Section A.1.

Suction Throttling

One can control the capacity of a centrifugal compressor by placing a control valve in the suction line, thereby altering the inlet pressure (P_I). From equation 8.5(1) it can be seen that the discharge pressure will be altered for a given flow when P_I is changed, and a new compressor curve will be generated. This is illustrated in Figure 8.5b (curves II and III). Consider first that the

compressor is operating at its normal inlet pressure (following curve I) and is intersecting the "constant pressure system" curve at point (1) with a design flow of 9600 lbm/hr (4320 kg/hr) at a discharge pressure of 144 PSIA (1 MPa) and 78 percent efficiency. If it is desired to change the flow to 5900 lbm/hr (2655 kg/h) while maintaining the same discharge pressure, it would be necessary to shift the compressor from curve I to curve II. The new intersection with the "constant pressure system" curve is at the new operating point (2), at 74 percent efficiency. In order to shift from curve I to curve II, one must change the discharge pressure of 190 PSIA (1.3 MPa) at the 5900 lbm/hr (2655 kg/h) flow on curve I, to 144 PSIA (1 MPa) on curve II. If the pressure ratio is ten ($P_D/P_I = 10$), then it would be necessary to throttle the suction by only $\Delta P_I = 46/10 = 4.6$ PSI (32 kPa) to achieve this shift.

It is also important to consider how close the operating point (2) is to the surge line. The surge line represents the low-flow limit for the compressor, below which its operation is unstable as a result of momentary flow reversals. Methods of surge control will be discussed later. At point (2) the flow is 5900 lbm/hr (2655 kg/h), and at the surge limit (S_I) it is 3200 lbm/hr (1440 kg/h). Thus, the compressor is operating at 5900/3200 = 184 percent of surge flow. This may be compared with curve I at point (1), where prior to suction throttling the machine is operating at 9600/3200 = 300 percent of surge flow.

The same method of suction throttling may be applied in a "mostly friction system" also shown in Figure 8.5b. In order to reduce the flow from 9600 lbm/hr (4320 kg/h) to 5900 lbm/hr (2655 kg/h), it is necessary to alter the compressor curve to curve III, so that the intersection with the "mostly friction system curve" is at the new operating point (3), at 77 percent efficiency. In order to do this, one must change the discharge pressure from 190 PSIA (1.3 MPa—on curve I) to 68 PSIA (0.5 MPa—on curve III). Thus, $\Delta P_D = 190 - 68 = 122$ PSI (0.8 MPa), and the amount of inlet pressure throttling for a machine with a compression ratio of ten is $\Delta P_I = 122/10 = 12.2$ PSI (84 kPa). The corresponding surge flow is at 1700 lbm/hr (765 kg/h), which means that the compressor is operating at 5900/1700 = 347 percent of surge flow. Therefore, surge is less likely in a "mostly friction system" than in a "constant-pressure system" under suction throttling control.

Discharge Throttling

A control valve on the discharge of the centrifugal compressor may also be used to control its capacity. In Figure 8.5b, if the flow is to be reduced from 9600 lbm/hr (4320 kg/h) at point (1) to 5900 lbm/hr (2655 kg/h), the compressor must follow curve I and therefore operate at point (4), at 190 PSIA (1.3 MPa) discharge pressure and 72 percent efficiency. However, the "mostly friction system" curve at this capacity requires only 68 PSIA (0.5

MPa) discharge control valve. The surge flow (at S_4) is 4000 lbm/hr (1800 kg/h), and the compressor is therefore operating at 5900/4000 = 148 percent of surge. Thus, surge is more likely to occur in a mostly friction system when discharge throttling is used than when suction throttling is used.

The various parameters involved in suction and discharge throttling of this sample compressor are compared in Table 8.5c.

Table 8.5c
COMPRESSOR PARAMETERS AS A FUNCTION OF THROTTLING METHOD

	Control Valve ΔP(PSI)	Compressor Efficiency	Operation Above Surge By
Suction Throttling "Constant Pressure System"	4.6	74%	184%
Suction Throttling "Mostly Friction System"	12.2	77%	347%
Discharge Throttling "Mostly Friction System"	122	72%	148%

Inlet Guide Vanes

This method of control uses a set of adjustable guide vanes on the inlet to one or more of the compressor stages. By prerotation or counter-rotation of the gas stream relative to the impeller rotation, the stage is unloaded or loaded, thus lowering or raising the discharge head. The effect is similar to suction throttling as illustrated in Figure 8.5b (curves II and III), but less power is wasted because pressure is not throttled directly. Also, the control is two-directional, since it may be used to raise as well as to lower the band. It is more complex and expensive than throttling valves but may save 10 to 15 percent on power and is well suited for use on constant-speed machines in applications involving wide flow variations.

The guide vane effect on flow is more pronounced in constant discharge pressure systems. This can be seen in Figure 8.5b (curve II), in which the intersection with the "constant pressure system" at point (2) represents a flow change from the normal design point (1) of 9600 − 5900 = 3700 lbm/hr (1665 kg/h), whereas the intersection with the "mostly friction system" at point (5) represents a flow change of only 9600 − 7800 = 1800 lbm/hr (810 kg/h).

Variable Speed

The pressure ratio developed by a centrifugal compressor is related to tip speed by the following equation:

$$\psi\, u^2/2g = \frac{ZRT_I}{(n - 1/n)} [(P_D/P_I)^{(n - 1/n)} - 1] \quad 8.5(2)$$

From this relation the variation of discharge pressure with speed may be plotted for various percentages of design speed as shown in Figure 8.5d. The obvious advantage of speed control from a process viewpoint is that both suction and discharge pressures can be specified independently of the flow. The normal flow is shown at point (1) for 9700 lbm/hr (4365 kg/h) at 142 PSIA (0.98 MPa). If the same flow is desired at a discharge pressure of 25 PSIA (173 kPa), the speed is reduced to 70 percent of design, shown at point (2). In order to achieve the same result through suction throttling with a pressure ratio of 10 to 1, the pressure drop across the valve would have to be (142 − 25)/10 = 11.7 PSI (81 kPa), with the attendant waste of power, as a result of throttling. This is in contrast with a power saving accomplished with speed control, since *power input is reduced as the square of the speed*.

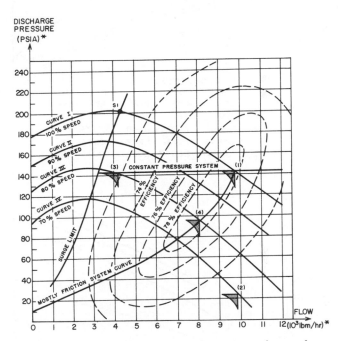

Fig. 8.5d Control of centrifugal compressor capacity by speed variation. For SI units see Section A.1.

One disadvantage of speed control is apparent in constant-pressure systems, in which the change in capacity may be overly sensitive to relatively small speed changes. This is shown at point (3), where a 20 percent speed change gives a flow change of (9600 − 4300)/9600 = 55%. The effect is less pronounced in a "mostly friction system," in which the flow change at point (4) is (9600 − 8100)/9600 = 16%.

Surge Control

The design of compressor control systems is not complete without consideration of surge control, because this affects stability of the machine. Surging begins at the positively sloped section of the compressor curve. In

Figure 8.5d this occurs at S_I on the 100 percent speed curve at 4400 lbm/hr (1980 kg/h). This flow will ensure safe operation for all speeds, but some power will be wasted at speeds below 100 percent because the surge limit decreases at reduced speed. Even for a compressor running at a constant speed, surge point changes as the thermodynamic properties vary at the inlet. This is shown in Figure 8.5e. Although an inaccurate surge control point can put the compressor into deep surge, a conservative surge point results in useless recycling and wasted horsepower. Various schemes to control surge are outlined in the following paragraphs.[1]

Incipient surge can be used with other strategies. ΔP versus h can be used with combinations of temperature, pressure, and molecular weight compensation. A typical example of controlling surge by ΔP versus h is shown in Figure 8.5f. If the value of h is multiplied by k3, this value may be compared with ΔP ($P_D - P_I$) in the anti-surge controller. As long as hk3 is less than $P_D - P_I$, the compressor will be outside the surge region. Since the control is discontinuous, "reset windup" can be a problem. (See section on control modes in Chapter 1 for details.) Depending upon the process conditions under which the compressor may be forced to operate, the most suitable surge control scheme should be chosen. This is done by plotting surge curves for the various schemes outlined earlier and choosing one that approximates a straight line under various process conditions. According to Shinskey, the surge line must be replaced by a surge envelope as the compression ratio is increased.[2]

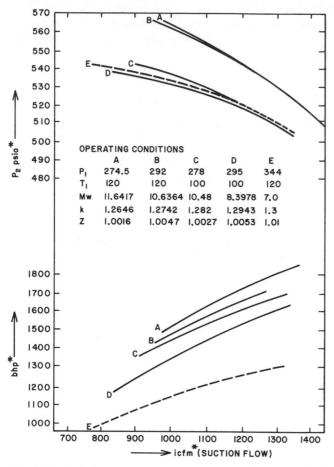

Fig. 8.5e   Typical compressor performance curve for a centrifugal compressor. For SI units see Section A.1.

OPERATING CONDITIONS

	A	B	C	D	E
P_I	274.5	292	278	295	344
T_I	120	120	100	100	120
Mw	11.6417	10.6364	10.48	8.3978	7.0
k	1.2646	1.2742	1.282	1.2943	1.3
Z	1.0016	1.0047	1.0027	1.0053	1.01

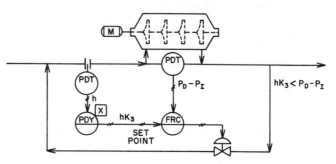

Fig. 8.5f   Centrifugal compressor with anti-surge controller

1. Compressor pressure rise ($\Delta P = P_D - P_I$) versus differential across suction flow meter (h).
2. Pressure ratio (P_D/P_I) versus actual volumetric flow (Q).
3. Break horsepower versus mass flow ($\dot{m}$).
4. Pressure ratio (P_D/P_I) versus Mach number squared.
5. Incipient surge.
6. Minimum flow.
7. Surge spike detection.

Rotary Compressors

Variable Speed

The rotary compressor is essentially a constant-displacement, variable-discharge pressure machine. Common types are the lobe, the sliding vane, and the liquid ring. The characteristic curves for a lobe-type unit are shown in Figure 8.5g. Since the unit is of the positive-displacement type, the inlet flow will vary linearly with the speed. Curves I and II in Figure 8.5g show this. The small decrease in capacity at constant speed with an increase in pressure is a result of slip of the gas at impeller clearances. It is necessary to compensate for this by small speed adjustments as the discharge pressure varies. For example, when the compressor is operating at point (1), it delivers the design volume of 60 ACFM (0.028 m³/s) and 3½ PSIG (24 kPa). In order for the same flow to be maintained when the discharge pressure is 7 PSIG (48 kPa), the speed must be increased from 1420 rpm to 1550 rpm at point (2). This can be accomplished by manipulation of the motor or drive speed by a pneumatic flow controller on the discharge line.

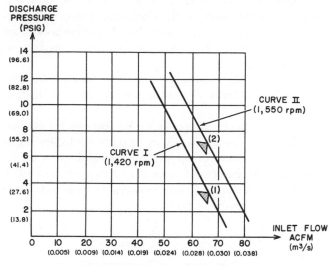

Fig. 8.5g   Rotary compressor curves

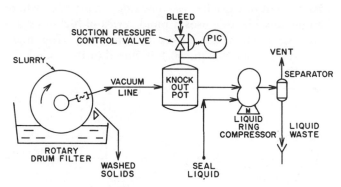

Fig. 8.5i   Liquid ring rotary compressor on suction pressure control

Reciprocating Compressors

The reciprocating compressor is a constant-volume, variable-discharge pressure machine. A typical compressor curve is shown in Figure 8.5j for constant-speed operation. The curve shows no variation in volumetric efficiency in the design pressure range, which may vary by 8 PSIG (53.6 kPa) from unloaded to fully loaded.

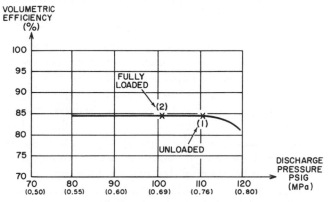

Fig. 8.5j   Reciprocating compressor curve

The volumetric inefficiency is a result of the clearance between piston end and cylinder end on the discharge stroke. The gas that is not discharged re-expands on the suction stroke, thus reducing the intake volume.

The relationship of speed to capacity is a direct ratio, since the compressor is a displacement-type machine. The typical normal turndown with gasoline or diesel engine drivers is 50 percent of maximum speed, in order to maintain the torque within acceptable limits.

Bypass and Suction Control

The discharge flow may be varied to suit process demands by a bypass on pressure control. This is shown in Figure 8.5h. Excess gas is vented to the atmosphere as the temperature control valve closes. The temperature of the outlet gas is controlled to prevent product degradation and to provide the proper product dryness. In systems in which the gas is not vented, it may be returned to the suction of the blower on pressure control.

Flow may also be varied by throttling the suction to the compressor, but this method is limited by the horsepower of the driver and by the temperature rise of the gas.

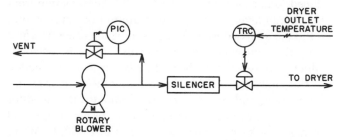

Fig. 8.5h   Rotary blower with bypass capacity control

Suction Pressure Control

An important application of the liquid ring rotary compressor is in vacuum service. The suction pressure is often the independent variable and is controlled by bleeding gas into the suction on pressure control. This is shown in Figure 8.5i, in which suction pressure control is used on a rotary filter to maintain the proper drainage of liquor from the cake on the drum.

On-Off Control

For intermittent demand, where the compressor would waste power if run continuously, the capacity can be controlled by starting and stopping the motor. This can be done manually or by the use of pressure switches. Typical switch settings are on at 140 PSIG (1 MPa), off at 175 PSIG (1.2 MPa). This type of control would suffice for processes in which the continuous usage is less than

50 percent of capacity, as shown in Figure 8.5k, where an air mix blender uses a rapid series of high-pressure air blasts when the mixer becomes full. The high-pressure air for this purpose is stored in the receiver.

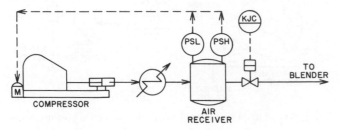

Fig. 8.5k On-off control on reciprocating compressor

Constant-Speed Unloading

In this type of control, the driver operates continuously, at constant speed, and one varies the capacity in discrete steps by holding suction valves open on the discharge stroke or opening clearance pockets in the cylinder. The most common schemes are three- and five-step unloading techniques. The larger number of steps saves horsepower because it more closely matches the compressor output to the demand.

In three-step unloading, the capacity is 100 percent, 50 percent, or 0 percent of maximum flow. This method of unloading is accomplished by the use of valve unloading in the double-acting piston. At 100 percent, both suction valves are closed during the discharge stroke. At 50 percent, one suction valve is open on the discharge stroke, wasting half the capacity of the machine. At 0 percent, both suction valves are held open on the discharge stroke, wasting the total machine capacity.

For five-step unloading, a clearance pocket is used in addition to suction valve control. The capacity can be 100, 75, 50, 25, or 0 percent of maximum flow. This is shown in Figure 8.5l. At 100 percent, both suction valves and the clearance pocket are closed; at 75 percent, only

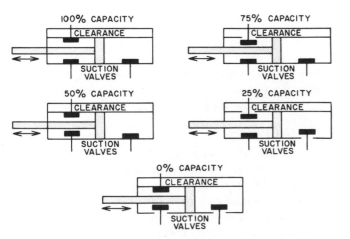

Fig. 8.5l Constant-speed, five-step unloading with valves and clearance pockets

the clearance pocket is open; at 50 percent, only one suction valve is open on the discharge stroke; at 25 percent, one suction valve and the clearance pocket are open; at 0 percent, both suction valves are opened during the discharge stroke.

The use of step unloading is most common when the driver is inherently a constant-speed machine, such as an electric motor.

The flow sheet representation for this control method is shown in Figure 8.5m. The pressure controller signal from the air receiver operates a solenoid valve in the unloader mechanism. The action of the solenoid valve directs the power air to lift the suction valves or to open the clearance port, or both.

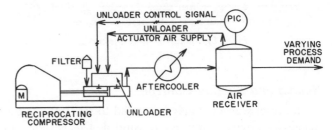

Fig. 8.5m Constant-speed capacity control of reciprocating compressor

For three-step unloading, two pressure switches are needed. The first switch loads the compressor to 50 percent if the pressure falls slightly below its design level, and the second switch loads the compressor to 100 percent if the pressure falls below the setting of the first switch.

For five-step unloading, a pressure controller is usually substituted for the four pressure switches otherwise required, and the range between unloading steps is reduced to not more than 2 PSI (13.8 kPa) deviation from the design level, keeping the minimum pressure within 8 PSI (55 kPa) of design. In cases in which exact pressure conditions must be met, a throttling valve is installed, which bypasses the gas from the discharge to the suction of the compressor. This device smoothes out pressure fluctuations and, in some cases, eliminates the need for a gas receiver in this service. This can prove economical in high-pressure services above 500 PSIG (3.45 MPa), in which vessel costs become significant.

Variable Speed

This method of control is generally used on gas turbines, steam turbines, and gasoline and diesel engines that are easily adaptable to speed control by fuel throttling or steam regulation. In contrast, electric motors are usually constant-speed devices. When large flow changes are anticipated in the process, consideration must be given to loss of torque, which occurs at reduced speed. The gas turbine is less susceptible to losses at reduced speed.

A patented control device known as the "Varsudi Controller" combines unloading control with speed control. This allows practically straight-line capacity control while maintaining the speed within a range at which the motor torque is close to the maximum. By sensing the suction and discharge pressure and relating these to the horsepower curve, the cylinder is unloaded when 3 percent overload occurs.

Lube and Seal Systems

A typical lube oil system is shown in Figure 8.5n. Dual pumps are provided to ensure the uninterrupted flow of oil to the compressor bearing and seals. Panel alarms on low oil level, low oil pressure (or flow), and high oil temperature are provided. A head tank provides oil for coasting down in case of a power failure. The design of these systems is critical, because failure of the oil supply could mean a shutdown of the entire process. In cases in which oil cannot be tolerated in contact with the process gas, an inert gas seal system may be used. This is shown in Figure 8.5o for a centrifugal compressor with balanced seals.

Microprocessor-Based Compressor Control

A dedicated microprocessor is more flexible than an analog system and can be very cost-effective, if wide variations of the suction inlet conditions exist. In a microprocessor, a new control algorithm can be implemented or an existing control scheme improved with only software changes. Fail-safe features can be included in the software, e.g., switching to minimum flow control upon failure of a transmitter that is used to compute the

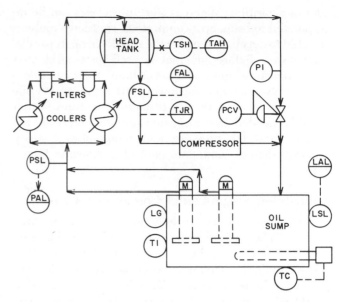

Fig. 8.5n   Compressor lube oil system for bearings and seals

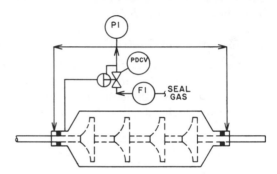

Fig. 8.5o   Compressor seal system

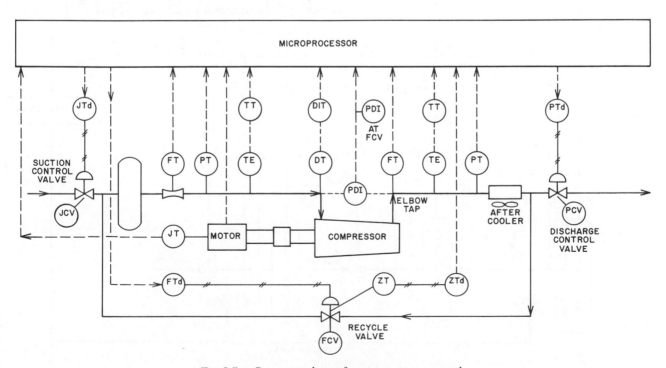

Fig. 8.5p   Compressor layout for microprocessor control

control algorithm. Whereas startup routines can be incorporated to bring up a compressor, shutdown routines are used to decouple the effect of shutdown on its parallel compressor. Before an input is used for control, it is checked against upper and lower limit. If an input does not fall between upper and lower limit, either a safe value is substituted or it reverts to fallback features. Other variables needed for control are computed internally and compensated for pressure and temperature. Microprocessor-based compressor control is shown in Figure 8.5p.

Multi-Inlet, Single-Outlet Compressor Control

These compressors are generally refrigeration compressors. The main problem here is to keep the process temperature within tolerable limits. Feedforward control is based upon the amount of recycle flow, or recycle valve position. The temperature controller trims the feedforward typically to give satisfactory control. A typical control scheme to control a three-inlet and one-outlet compressor is shown in Figure 8.5q.

Parallel Compressor Control

Controlling two or more compressors operating in parallel and having identical characteristics would be relatively simple. It is very difficult, if not impossible, to find two compressors having identical performance char-

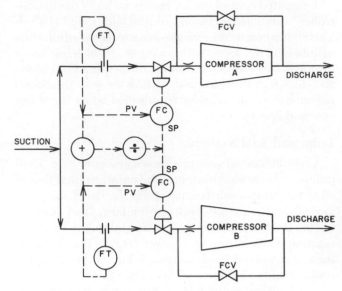

Fig. 8.5q   Multi-inlet compressor control

acteristics. Slight variations in flow cause one compressor to be fully loaded. The parallel machine then has useless recycle. The control scheme shown in Figure 8.5r alleviates that problem. Typically, care is exercised to ensure that the suction valve that receives the lower flow is kept 100 percent open. This prevents both suction valves from going fully closed to balance the flow.

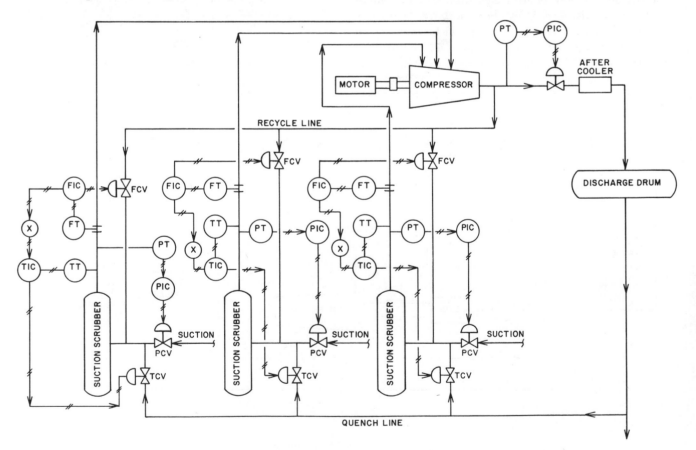

Fig. 8.5r   Parallel compressor control

Installation and General Considerations

A check valve in the discharge line as close to the compressor as possible will protect it from surges. On motor-driven compressors, it is helpful to close the suction valve during starting to prevent overload of the motor. After the unit is operating, it should be brought to stable operating range as soon as possible to prevent overheating. The recycle valve control should be fast opening and slow closing to come out of surge quickly and then stabilize the flow. The transmitter ranges should be such that the recycle valve is able to open in a reasonable amount of time with reasonable proportional and integral control constants of the antisurge controller. The importance of proper controller tuning cannot be underestimated.

REFERENCES

1. Gupta, B.P. and Teague, C.C., "Compressor Surge Control," *1977 Symposium on Instrumentation for the Process Industries*, Texas A & M University, College Station, TX, January, 1977, pp. 7–13.
2. Shinskey, F.G., *Energy Conservation Through Control*, Academic Press, New York, 1978, p. 134.

BIBLIOGRAPHY

Adamski, R.S., "Improving Reliability of Rotating Machinery," *InTech*, February 1982.

Arant, J.B., "Compressor Instrumentation Systems," *Instrumentation Technology*, April 1974.

Austin, W.E., "Packaged Air System Can Give Peak Operation, Cost Savings," *Hydraulics and Pneumatics*, August 1979.

Balje, O.E., "A Study on Design Criteria and Matching of Turbomachines," Part B, Transactions of ASME, *Journal of Engineering Power*, January 1962.

Bently, D.E., "Machinery Protection Systems for Various Types of Rotating Equipment," Bently Nevada Corp., Minden, NV, 1980.

Bogel, G.D. and Rhodes, R.L., "Digital Control Saves Energy in Gas Pipeline Compressors," *InTech*, July 1982, pp. 43–45.

Braccini, G., "Electronic System for the Antisurge Control of Centrifugal Compressors," *Quoderni Pignone* No. 16.

Buzzard, W.S., "Controlling Centrifugal Compressors," *Instrumentation Technology*, November 1973.

———, "Override Strategies for Analog Control," *InTech*, November 1978, pp. 67–69.

Byers, R.H., Snider, M., and Brownstein, B., "Simulator to Test Compressor Research Facility Control System Software," *Military Electronics Defense Expo '80*, Interavia (Geneva, Switzerland), pp. 140–153.

Carrier Corp. *Air Conditioning News*, Carrier Corp., Syracuse, NY, May 12, 1980, pp. 3, 12.

"Centrifugal Compressors," *Bulletin 8282-C*, Ingersoll-Rand Co., Woodcliff Lake, NJ, 1972.

"Centrifugal Compressors for General Refinery Service, 3rd ed.," *API Standard 617*, American Petroleum Institute, Washington, D.C., 1973.

"Compressibility Charts and Their Application to Problems Involving Pressure-Volume-Energy Relations for Real Gases," *Bulletin P-7637*, Worthington-CEI Inc., Mountainside, NJ, 1949.

Compressor Handbook for the Hydrocarbon Processing Industries, Gulf Publishing Co., Houston, 1979.

Cooke, L.A., "Saving Fuel at Pipeline Compressor Stations," *Instrumentation Technology*, November 1977.

Davis, A., Keegan, P.J., and Peltzman, E., "Variable Capacity Compressor Controller," *Report No. DOE/CS/35224-T1*, U.S. Dept. of Energy, 1979.

Dussourd, J.L., Pfannebecker, G., Singhania, S.K., and Tramm, P.C., "Considerations for the Control of Surge in Dynamic Compressors Usisng Close Coupled Resistances," *Centrifugal Compressor and Pump Stability, Stall, and Surge*, 1976, pp. 1–28.

Dwyer, J.J., "Compressor Problems: Causes and Cures," *Hydrocarbon Processing*, January 1973.

"Economical Evaporator Designs by Unitech," *Bulletin UT-110*, Ecodyne Unitech Division, Union, NJ, 1976.

"Elliott Multistage Compressors," *Bulletin P-25*, Elliott Division, Carrier Corp., Jeannette, PA, 1973.

Engineering Data Book, 9th ed., Natural Gas Processors Suppliers Assn., Tulsa, 1972.

Evans, F.L., Jr., *Equipment Design Handbook for Refineries and Chemical Plants*, Vol. 1, Gulf Publishing Co., Houston, 1971, p. 76.

Fallin, H.K. and Belas, J.J., "Controls for an Axial Turboblower," *Instrumentation Technology*, May 1968.

Fehervari, W., "Asymmetric Algorithm Tightens Compressor Surge Control," *Control Engineering*, October 1977.

Filippini, V., "Surge Condition and Antisurge Control of a Centrifugal Compressor," *Quaderni Pignone*, No. 5.

"Gas Properties and Compressor Data," *Form 3519-C*, Ingersoll-Rand Co., Woodcliff Lake, NJ, 1967.

Gaston, J.R., "Centrifugal Compressor Control," ISA Chemical and Petroleum Instrumentation Symposium Proceedings, Research Triangle Park, NC, 1974.

Gupta, B.P. and Jeffrey, M.F., "Compressor Controls Made Easy with Microprocessors," *Instrument Maintenance and Management*, Vol. 13, Research Triangle Park, NC, 1979, pp. 63–67.

Gupta, B.P. and Jeffrey, M.F., "Optimize Centrifugal Compressor Performance," *Hydrocarbon Processing, Computer Optimization*, June 1979.

Hallock, D.C., "Centrifugal Compressors—The Cause of the Curve," *Air and Gas Engineering*, January 1969.

Hassenfuss, F., "Compressor Controls Coordinated for Carbon Dioxide Line," *Oil and Gas Journal*, July 31, 1972, pp. 91–95.

Hiller, C.C. and Glicksman, L.R., *Improving Heat Pump Performance via Compressor Capacity Control—Analysis and Test*, Vol. 1, Massachusetts Institute of Technology, Cambridge Energy Lab., Cambridge, MA, January 1976.

Hume, W.P.F., "Application of Solar Energy to a Heat Pump in a Northern Climate," *Chart. Mech. Eng.*, September 1980, pp. 61–63.

Kolnsberg, A., "Reasons for Centrifugal Compressor Surging and Surge Control," *Journal of Engineering for Power*, ASME, January 1979, pp. 79–86.

Langill, A.W., Jr., "Microprocessor-Based Control of Large Constant-Speed Centrifugal Compressors," *Advances in Test Measurement*, Vol. 19, ISA, Research Triangle Park, NC, 1982.

Lee, W.D. and Fazzolare, R.A., "Refrigerator and New Technologies," *Energy Use Management*, Vol. II, Pergamon Press, New York, 1977, pp. 227–230.

"Lubrication, Shaft-Sealing, and Control-Oil Systems for Special-Purpose Applications," *API Standards 614*, American Petroleum Institute, Washington, D.C., 1973.

Magliozzi, T.L., "Control System Prevents Surging in Centrifugal Flow Compressors," *Chemical Engineering*, May 8, 1967, pp. 139–142.

Moellenkamp, G., Scott, J., and Farmer, F., "Compressor Station Mini-Computer System for Control, Monitoring, and Telemetry," *13th World Gas Conference*, Paper IGU/C, International Gas Union (London, England), 1976.

Neerken, R.F., "Compressors in Chemical Process Industries," *Chemical Engineering*, January 20, 1975.

Nisenfeld, A.E. and Cho, C.H., "Parallel Compressor Control," *Australian Journal of Instrumentation and Control*, December 1977, pp. 108–112.

———, "Parallel Compressor Control . . . What Should Be Considered," *Hydrocarbon Processing*, February 1978, pp. 147–150.

Perry, R.H. and Chilton, C.H., *Chemical Engineers Handbook*, 5th ed., McGraw-Hill, New York, 1973, pp. 6–31.

Rammler, R., "Energy Savings Through Advanced Centrifugal Compressor Performance Control," *Advances in Instrumentation*, Vol. 37, ISA, Research Triangle Park, NC, 1982.

——— and Langill, A.W., Jr., "Centrifugal Compressor Performance Control," *Joint Automatic Control Conference*, AIChE, 1979, pp. 261–265.

"Reciprocating Compressors for General Refinery Service," 2nd ed., *API Standard 618*, American Petroleum Institute, Washington, D.C., 1974.

Roberts, W.B. and Rogers, W., "Turbine Engine Fuel Conservation by Fan and Compressor Profile Control," *Symposium on Commercial Aviation Energy Conservation Strategies*, National Technical Information Service Document PC A16/MF A01, April 2, 1981, pp. 231–256.

Rutshtein, A. and Staroselsky, N., "Some Considerations on Improving the Control Strategy for Dynamic Compressors," *ISA Transactions*, Vol. 16, ISA, Research Triangle Park, NC, 1977, pp. 3–19.

"Scaling Information for Computing Elements," *Application and Installation Data Handbook*, Fisher Controls Co., Marshalltown, IA, January 1976.

Scheel, L.F., *Gas Machinery*, Gulf Publishing Co., Houston, 1972.

Shah, B.M., "Saving Energy with Jet Compressors," *Chemical Engineering*, July 7, 1975.

Shinskey, F.G., "Controlling Unstable Processes, Part I: The Steam Jet," *Instruments and Control Systems*, December 1974.

———, "Effective Control for Automatic Startup and Plant protection," *Canadian Control Instrumentation*, April 1973.

———, "Interaction Between Control Loops, Part II: Negative Coupling," *Instruments and Control Systems*, June 1976.

Smith, G., "Compressor Surge Control Achieved Using Unique Flow Sensor," *Pipeline Gas Journal*, December 1978, pp. 42, 46–47.

Staroselsky, N., "Better Efficiency and Reliability for Dynamic Compressors Operating in Parallel or in Series," *Paper 80-PET-42*, ASME, New York, 1980.

——— and Ladin, L., "Improved Surge Control for Centrifugal Compressors," *Chemical Engineering*, May 21, 1979, pp. 175–184.

Van Ormer, H.P., Jr., "Air Compressor Capacity Controls: A Necessary Evil, Vol. 1. Reciprocating Compressors," *Hydraulics and Pneumatics*, June 1980, pp. 67–70.

Waggoner, R.C., "Process Control for Compressors," *Advances in Instrumentation*, Vol. 31, ISA, Research Triangle Park, NC, 1976.

Warnock, J.D., "Are Your Compressors Wasting Energy?" *Instruments and Control Systems*, March 1977, pp. 41–45.

———, "Typical Compressor Control Configurations," *Advances in Instrumentation*, Vol. 31, ISA, Research Triangle Park, NC, 1976.

White, M.H., "Surge Control for Centrifugal Compressors," *Chemical Engineering*, December 25, 1972.

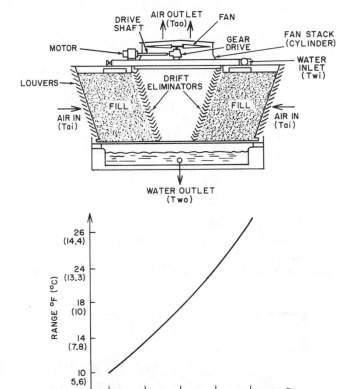

8.6 COOLING TOWER CONTROLS

In order to discuss the control aspects of cooling towers, it is advisable first to define some related terms:

Approach: The difference between the wet bulb temperature of the ambient air and the water temperature leaving the tower. The approach is a function of cooling tower capability; a larger cooling tower will produce a closer approach (colder leaving water) for a given heat load, flow rate, and entering air condition (Figure 8.6a). (Units: °F or °C)

Blowdown: Water discharged to control concentration of impurities in circulated water. (Units: % of circulation rate)

Drift: Water loss due to liquid droplets entrained in exhaust air. Usually under 0.2 percent of circulated water flow rate.

Fan Driver Output: Actual power output (BHP) of driver to shaft.

BHP =

$$= \frac{(Motor\ Effciency)(Amps)(Volts)(Power\ Factor)\ 1.73}{746}$$

Liquid Gas Ratio (L/G): Mass ratio of water to air flow rates through the tower. (Units: lbs/lbs or kg/kg)

Make-up: Water added to replace loss by evaporation, drift, blowdown, and leakage. (Units: % of circulation rate)

Power Factor: The ratio of true power (watts) to the apparent power (amps × volts).

Range (Cooling Range): The difference between the temperatures of water inlet and outlet, as shown in Figure 8.6a. For a system operating in a steady state, the range is the same as the water temperature rise through the load heat exchanger. Accordingly, the range is determined by the heat load and water flow rate, not by the size or capability of the coolinig tower. On the other hand, the range does affect the approach, and as the range increases it will cause a corresponding increase in the approach, if all other factors remain unaltered (see Figure 8.6a).

Speed Reducer: A device for changing the speed of the driver in order to arrive at the desired fan speed.

Water Loading: Water flow divided by effective hor-

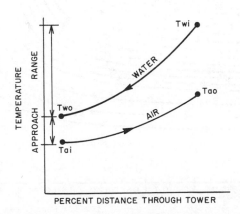

Fig. 8.6a Temperature relationship between water and air in a counterflow cooling tower. An increase in range will cause an increase in approach, if all other conditions remain unaltered. The mechanical cooling tower is a water-to-air heat exchanger.

izontal wetted area of the tower. (Unit: GPM/ft² or m³/[hr] [m²])

The Cooling Process

For operational purposes, tower characteristic curves for various wet bulb temperatures, cooling ranges, and approaches are plotted against the water flow to air flow ratio as shown in the following equation:

$$\frac{KaV}{L} \sim \frac{(L)^n}{(G)} \qquad 8.6(1)$$

where

K = overall unit of conductance, mass transfer between saturated air at mass water temperature and main air stream—lb per hour (ft²) (lb/lb)

a = area of water interface per unit volume of tower—ft² per ft³

V = active tower volume per unit area—ft³ per ft²

L = mass water flow rate—lb per hour

G = air flow rate—lb dry air per hour

n = experimental coefficient—varies from -0.35 to -1.1 and has an average value of -0.55 to -0.65.

From these curves, one can determine tower capability, effect of wet bulb temperature, cooling range, water circulation, and air delivery, as shown in Figure 8.6b. The ratio KaV/L is determined from test data on hot water

and cold water, wet bulb temperature, and the ratio of the water flow to the air flow.

The initial investment required for a cooling tower is a function essentially of the required water flow rate, but this cost is also influenced by the design criteria for approach, range, and wet bulb. As shown in Figure 8.6c, cost tends to increase with range and to decrease with a rise in approach or wet bulb. The initial investment is about $20 per GPM of tower capacity, whereas the energy cost of operation is approximately 0.01 BHP/GPM. This means that the cost of operation reaches the cost of initial investment in 5 to 10 years, depending on energy costs.

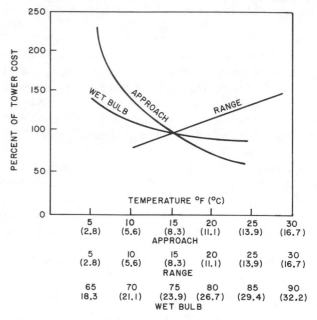

Fig. 8.6c Comparative cooling-tower costs (Courtesy of Foster Wheeler Corp.)

The thermal capability of cooling towers for air conditioning applications is usually stated in terms of nominal tonnage based on heat dissipation of 1.24 kW (15,000 Btu/h) per condenser kW (ton) and a water circulation rate of 0.054 l/s per kW (3 GPM per ton) cooled from 35 to 29.4°C (95 to 85°F) at 25.6°C (78°F) wet bulb temperature. It may be noted that the subject tower would be capable of handling a greater heat load (flow rate) when operating in a lower ambient wet bulb region. For operation at other flow rates, tower manufacturers will usually provide performance curves covering a range of at least 2 to 5 GPM per nominal ton (0.036 to 0.09 lps/kW).

Capacity Controls and Optimization

The cooling tower water temperature controls have already been described in the discussion of chiller optimization, in connection with Figure 8.4j. Similarly, the cooling tower water flow rate controls have been covered in the discussion of Figure 8.4k. These will not be re-

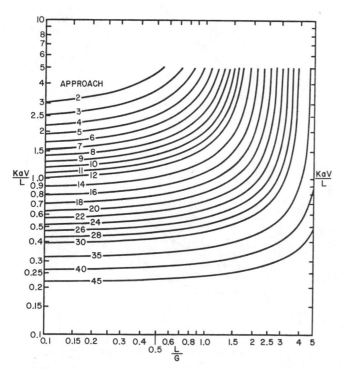

Fig. 8.6b Characteristic curves of cooling towers (based on 64°F [17.8°C] air wet-bulb temperature and 18°F [10°C] cooling range).

peated here, but it is suggested that these pages be reviewed by the reader.

Variations in the cooling load should be matched by a change in the level of operation of the cooling tower fan(s), in accordance with Figure 8.4j. If the fans are single-speed devices, then the only control available is to cycle the fan units on or off as a function of load, as illustrated in Figure 8.6d. This can be done by simple sequencers or by more sophisticated digital systems. In most locations the need to operate all fans at full speed applies for only a few thousand hours every year. Fans running at half speed consume approximately one seventh of design air horsepower but produce over 50 percent design air rate (cooling effect). Since the actual air entering the tower is less than design temperature about 98 percent of the time, two-speed motors are a wise investment for minimizing operating costs.

Freeze Protection

In case of a power failure during subfreezing weather, electrically heated tower basins should be provided with an emergency draining system. Still another method of freeze protection is the bypass circulation method illustrated in Figure 8.6e.

In this system a thermostat detects the outlet temperature of the cooling tower water, and when it drops below 45°F (7.2°C), the bypass circulation is started in a protected indoor area. This circulation is terminated when the water temperature rises to approximately 50°F (10°C). The thermostat is wired first to open the solenoids S1, S2, and S3 and then to start the circulating pump (P2), which usually is a small, ¼- or ⅓-HP (186.5 or 248.7 W) unit. The bypass line containing solenoid S1 is sized to handle the flow capacity of P2, with a pressure drop

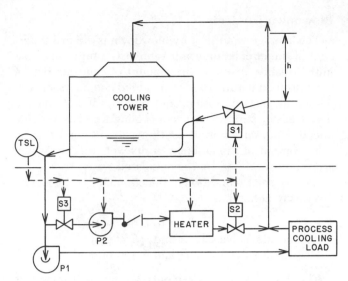

Fig. 8.6e Bypass circulation method of freeze protection for cooling towers

("h") that is less than the height of the riser pipe. This way, when P1 is off, no water will reach the top of the column.

In the northern regions, in addition to the S1 bypass shown in Figure 8.6e, there is a full-sized bypass (not shown in Figure 8.6e). When the main (P1) pump is started in the winter, this full bypass is opened, not allowing water to be sent to the top of the tower until its temperature is approximately 70°F (21°C). At that temperature, the bypass is closed, since the water is warm enough to be sent to the top of the tower without danger of freezing. As the water temperature rises further, fans are cycled on.

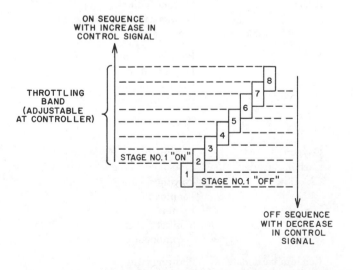

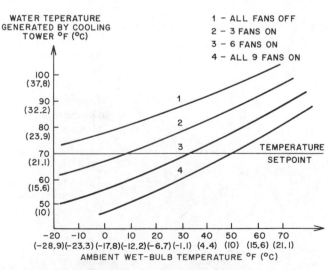

Fig. 8.6d Load following cooling tower control by fan cycling. *Left:* The output of the approach controller (TDIC-1 in Fig. 8.4j) can be sent to a sequencer, which turns fan stages on and off. *Right:* Example showing the effect of fan cycling on a 9-cell tower designed for 160,000 GPM (608,000 l/m) with an 84°F (29°C) inlet and 70°F (21°C) outlet temperature, operating at 50% load. (From Wiston, G., "Cold Weather Operation of Crossflow Cooling Towers," *Plant Engineering*, September 8, 1975.)

Blowdown Controls

The average water lost by blowdown is 0.5 to 3.0 percent of the circulating water rate. It is a function of the initial quality of water, the amount of concentration of the dissolved natural solids and added chemicals for protection againsts corrosion, and buildup of scale on the heat transfer surfaces. Since a cooling tower is a highly effective air scrubber, it continuously accumulates the solid content of the ambient on its wet surfaces, which are then washed off by the circulated water.

The "normal" condition of the circulated water can be arbitrarily defined as:

pH   : 6 to 8
Chloride as Na Cl   : under 750 ppm
Total dissolved solids: under 1500 ppm

The blowdown requirement can be determined as follows:

$$B = \frac{([E + D]/N) - D}{1 - (1/N)} \qquad 8.6(2)$$

where:

B = Blowdown in % (typically, for N = 2 blowdown is 0.9%, for N = 3, 0.4%, for N = 4, 0.24%, etc.)
E = Evaporation in % (typically 1%)
D = Drift in % (typically 0.1%)
N = Number of concentrations relative to initial water quality

In cooling towers, the major causes for concern are delignification (binding agent for the cellulose) by the use of oxidizing biocides, such as chlorine, and excessive bicarbonate alkalinity; biological growth, which can clog the nozzles and foul the heat exchange equipment; corrosion of the metal components (should be less than 3 mils per year without pitting); general fouling by a combination of silt, clay, oil, metal oxides, calcium and magnesium salts, organic compounds, and other chemical products that can cause reduced heat transfer and enhanced corrosion; and scaling by crystallization and precipitation of salts or oxides (mainly calcium carbonate

Table 8.6f
COOLING TOWER PROBLEMS AND PREVENTIVES

Problem	Factors	Causative Agents	Corrective Treatments
Wood Deterioration	Microbiological Chemical	Cellulolytic fungi Chlorine	Fungicides Acid
Biological Growths	Temperature Nutrients pH Inocula	Bacteria Fungi Algae	Chlorine Chlorine donors Organic sulfurs Quatenary ammonia
General Fouling	Suspended solids Water Velocity Temperature Contaminants Metal oxides	Silt Oil	Polyelectrolytes Polyacrylates Lignosulfonates Polyphosphates
Corrosion	Aeration pH Temperature Dissolved solids Galvanic couples	Oxygen Carbon dioxide Chloride	Chromate Zinc Polyphosphate Tannins Lignins Synthetic organic compounds
Scaling	Calcium Alkalinity Temperature pH	Calcium carbonate Calcium sulfate Magnesium silicate Ferric hydroxide	Phosphonates Polyphosphates Acid Polyelectrolytes

and magnesium silicate) on surfaces. The treatment techniques to prevent these conditions from occurring are listed in Table 8.6f. The blowdown must meet the water quality standards for the accepting stream and must not be unreasonably expensive. It is also possible to make the water system a closed-cycle system (no blowdown) by ion exchange treatment.

Miscellaneous Controls

The minimum basin level is maintained by the use of float-type make-up level control valves (see Section 5.10). The maximum basin level is guaranteed by overflow nozzles, sized to handle the total system flow. In critical systems, additional safety is provided by the use of high/low level alarm switches.[2]

The operational safety of the fan and the fan drive can be safeguarded by torque and vibration detectors.[3] Low-temperature alarm switches[4] can also be furnished as warning devices signaling the failure of freeze protection controls.

The need for flow balancing among multiple cells is another reason for using added valving and controls. The warm water is usually returned to the multi-cell system through a single riser, and a manifold is provided at the top of the tower for water distribution to the individual cells. Balancing valves should be installed to guarantee equal flow to each operating cell. If two-speed fans are used, further economy can be gained by lowering the water flow when the fan is switched to low speed. It is also advisable to install equalizer lines between tower sumps to eliminate imbalances caused by variations in flow rates or pipe layouts.

Another reason for considering the use of two-speed fans is to have the ability to lower the associated noise level at night or during periods of low load. Switching to low speed will usually lower the noise level by approximately 15 dB.

REFERENCES

1. Wiston, G., "Cold Weather Operation of Crossflow Cooling Towers," *Plant Engineering*, September 8, 1975.
2. Lipták, B.G., *Instrument Engineers' Handbook*, Process Measurement Volume, revised ed., Chilton Book Co., Radnor, PA, 1982, Chapter 3.
3. ———, *Instrument Engineers' Handbook*, Process Measurement Volume, revised ed., Chilton Book Co., Radnor, PA, 1982, Chapter 9.
4. ———, *Instrument Engineers' Handbook*, Process Measurement Volume, revised ed., Chilton Book Co., Radnor, PA, 1982, Chapter 4.

BIBLIOGRAPHY

Baechler, R.H., Blew, J.O., and Duncan, C.G., "Cause and Prevention of Decay of Wood in Cooling Towers," ASME Petroleum Division Conference, ASME, New York, 1961.
Baker, D.R., "Durability of Wood in Cooling Towers," *Technical Bulletin R-62-W-1*, The Marley Co., Kansas City, MO, 1962.
"Chilled and Dual-Temperature Water Systems," *ASHRAE Handbook*, Systems Volume, ASHRAE, New York, 1976, Chapter 18.
"Cooling Towers," *ASHRAE Handbook & Product Directory*, Equipment Volume, ASHRAE, New York, 1979, Chapter 21.
Donohue, J.M., "Cooling Towers—Chemical Treatment," *Industrial Water Engineering*, May 1970.
Kern, D.Q., *Process Heat Transfer*, McGraw-Hill, New York, 1950, p. 563.
Koch, J., *Untersuchung und Berenchnung und Kuehlwerken*, VDI Forschungsheft No. 404, Berlin, 1940.
Lipták, B.G. (ed.), "Cooling Towers—Their Design and Application," *Environmental Engineers' Handbook*, Vol. 1, Chilton Book Co., Radnor, PA, 1974.
Mathews, R.T., "Some Air Cooling Design Considerations," *Proceedings of the American Power Conference*, 1970.
Nottage, H.B., "Merkel's Cooling Diagram as a Performance Correlation for Air-Water Evaporative Cooling Systems," *ASHRAE Transactions*, Vol. 47, 1941, p. 429.
Nussbaum, O.J., "Dry-type Cooling Towers for Packaged Refrigeration and Air Conditioning Systems," *Refrigeration Science and Technology*, 1972.
Walker, W.H., Lewis, W.K., McAdams, W.H., and Gilliand, E.R., *Principles of Chemical Engineering*, McGraw-Hill, New York, 1937, p. 480.

8.7 CRYSTALLIZER CONTROLS

This section provides a brief overview of control practices for industrial crystalizers. Crystallizations based on production of supersaturation (and product crystals) by evaporation, vacuum, cooling, and reaction are discussed. Salient points concerning the control of multiple-effect crystalizers are also outlined.

Crystallization is the unit operation that is best suited to the recovery of dissolved substances from solutions. The solid product has many desirable properties, such as good flow characteristics, handling ease, suitability for packaging, and pleasant appearance.

Crystallization can be achieved by any one of the following methods: cooling, evaporation, vacuum, or reaction (direct combination). The choice of method depends on the feed liquid composition, product specifications, solubility curve of the product, and other engineering considerations.

Most crystalizers in industrial use operate on a continuous basis, but batch crystalizers are used in certain instances for fine chemicals and pharmaceuticals.

This section is devoted to the subject of crystallizer controls, focusing on measurement and control practices together with a description of the basic crystallization techniques.

Control Basis

Crystallization can occur only if some degree of supersaturation or supercooling has been achieved. Any crystallization operation has three basic steps: (1) achievement of supersaturation, (2) formation of nuclei, and (3) growth of nuclei into crystals.

Supersaturation is defined by

$$S = c/c' \qquad 8.7(1)$$

where

S = degree of supersaturation,
c = actual concentration of the substance in the solution, and
c' = normal equilibrium concentration of the substance in pure solvent.

Nucleation and growth rates are controlled by supersaturation (or supercooling). The ideal process would be a stepwise procedure, but nucleation cannot be eliminated in a growing mass of crystals. The influence of supersaturation on nucleation and growth rates is shown in Figure 8.7a. Whereas growth rate increases linearly with supersaturation, nucleation rate increases exponentially. Figure 8.7a shows the three regions of supersaturation: (1) metastable (nucleation is very low and growth predominates), (2) intermediate (nucleation becomes larger, but growth is still significant), and (3) labile (nucleation predominates).

Since very small crystals are difficult to dewater, dry,

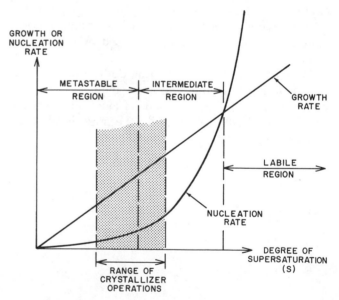

$S = C/C'$, where

S = degree of supersaturation,
C = actual concentration of substance in the solution (parts/100 parts of solvent), and
C' = normal equilibrium concentration of substance in pure solvent.

Fig. 8.7a   Influence of degree of supersaturation on nucleation and growth.

or handle, crystallizer design and control are directed toward reasonably large crystals but minimize nucleation. Usual industrial practice involves sufficiently low supersaturation to minimize nucleation while being adequate for reasonable growth.

Thus, control of supersaturation becomes the basic element of crystallizer system control. Unfortunately, the workable degree of supersaturation is usually 0.5 to 1 percent, and the degree of supercooling associated with it is so small that it cannot be directly measured. Hence, the vast majority of crystallizers use indirect means of control.

The one exception to this rule is sugar, in which degree of supersaturation can be greater and correlations have been developed to measure supersaturation directly. Sugar supersaturation recorders that automatically compile supersaturation are based on these empirical correlations.

To define the maximum number of controllers, let us consider the degrees of freedom for a crystallization system. The variables are:

1. temperature and flow of the process fluid,
2. temperature and flow of coolant,
3. degree of supersaturation, and
4. ratio of mother liquor to crystals. This can be changed by varying recycles of mother liquor to feed stream.

The limitations are:

1. enthalpy balance based on the first law of thermodynamics, and
2. stoichiometric balance based on solubility curves.

Since the system has six variables and two equations, it has four degrees of freedom. Thus, the maximum number of automatic controllers permissible in a crystallizer system (except reaction crystallizers), without overdefining it, is four.

Evaporator Crystallizers

This type of crystallizer produces supersaturation and, hence, crystals by one of three methods: (1) indirect heating, (2) submerged combustion, and (3) spray evaporation. The first two are the dominant types and will be discussed here.

Indirect Heating (Circulating-Magma) Crystallizers

Circulating-magma crystallizers with indirect heating are by far the most important type of crystallizers in use today; both the forced-circulation (FC) and the draft-tube baffle (DTB) belong to this class. A typical FC crystallizer is shown in Figure 8.7b. In this version the feed enters the system on the recirculation loop.

The critical design parameters in FC crystallizers are

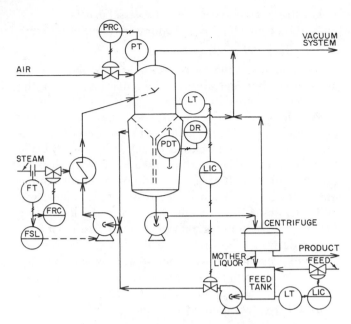

Fig. 8.7b Circulating magma crystallizer with indirect heat

the internal recirculation rate and velocity, the crystallizer hold volume, and the speed of the circulating pump. At internal circulation rates less than the optimum, excessive flashing can occur at the boiling surface, causing excessive supersaturation. If all this supersaturation cannot be relieved by deposition of the solute because of the lack of adequate crystal-surface area in the suspension, excessive nucleation will occur, causing excessive fines and buildup of solids on the crystallizer walls. In general, recirculation rates aim at restricting the flashing of the crystallizer walls. In general, recirculation rates aim restricting the flashing at the boiling surface to approximately 1.7 to 4.5°C (3 to 8°F), whereas a magma-density range of 15 to 25 percent is typical, even though the exact optimum depends on the particular crystal system.

There are differences in design, but their approach to process control is similar. One version of control is shown in Figure 8.7b, where four control loops are involved:

1. Feed liquor is fed on level control to the feed tank. Here it is mixed with mother liquor from the centrifuge.
2. The mixed liquor is fed to the suction side of the crystallizer recirculating pump. This feed is adjusted by a level controller. (For methods of protecting the level sensor from plugging and material buildup, refer to Chapter 3 of the Process Measurement volume of this handbook.
3. Steam flow to the heat exchanger is on flow control. Once the steam rate is fixed, production rate is also fixed, provided that the feed composition does not change.

4. Temperature control in the vessel may be achieved by controlling the evaporator chamber pressure by an air bleed.

Refinements to this basic system are possible, such as an interlock between steam and circulating pump or the addition of a density recorder.

The draft-tube baffle (DTB) crystallizer gives larger crystals than the FC crystallizers under equivalent conditions. It consists of a closed vessel with an inner baffle forming a partitioned settling area, inside which is tapered vertical-draft tube surrounding a top or bottom entering agitator. The agitator is of the axial-flow type and operates at low speeds. An elutriating leg can be fitted to the bottom cone if further classification of crystals is desired. The draft tube is centered by support vanes to prevent body swirl and to minimize turbulence in the circulating magma. Supersaturation may be generated by either evaporation cooling or vacuum control of a DTB-type vacuum crystallizer.

Multiple-Effect Operation

Evaporator crystallizers are often used as multiple-effect systems. Instrumentation on such a system is more involved and depends on the type of process flow. As an example, a simplified control scheme on a triple-effect unit is shown in Figure 8.7c. The unit has these features:

1. Level in each unit is an important process variable, because it determines the residence time. This is usually controlled by throttling the mixed liquor makeup.

2. Steam flow to the first unit is usually on flow control.

3. Feed enters the recirculation vessel on (feed tank) level control. Temperature control in the last unit is obtained by absolute pressure control of the air bleed.

In this system, density recorders can also be used. If boiling point elevation is sufficiently large, these may be used to control the effluent liquor concentration directly.

Submerged-Combustion Crystallizers

Submerged-combustion crystallizers are used for corrosive materials and for salts having inverted solubility. One version of a control scheme is shown in Figure 8.7d, where feed liquor to the system is on the flow control and burner fuel gas may be under flow and pressure control, whereas a bypass controls the combustion flow of air to the burners.

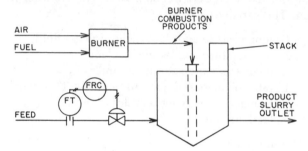

Fig. 8.7d Submerged combustion evaporator crystallizer

Only the basic elements of control are shown here. Safety interlocks for combustion equipment per FIA or other requirements and auxiliaries are also to be considered in developing a control scheme. The reader is therefore referred to Section 8.12 for a review of direct-fired heater controls.

Cooling Crystallizers

Cooling crystallizers operate at substantially atmospheric pressure, and their heat is transferred to air or to a cooling medium by direct or indirect contact.

There are many types of cooling crystallizers, but the majority fall into three categories: (1) controlled-growth magma crystallizers, (2) classifying crystallizers, and (3) direct-contact crystallizers.

Controlled-Growth Magma Crystallizers

The various cradle crystallizers and scraped-surface units belong in this category. The cradle types are used in small applications and involve little instrumental control. The various scraped-surface crystallizers are used in: crystallization from high-viscosity liquors or in open-tank crystallizers, as coolers to induce nucleation. Two versions of controlled-growth magma crystallizers and the associated control instrumentation are shown in Figures 8.7e and 8.7f.

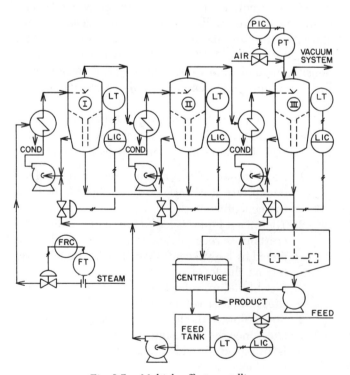

Fig. 8.7c Multiple effect crystallizer

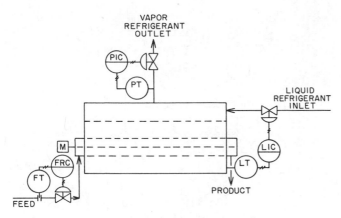

Fig. 8.7e　Control of a cooling crystallizer

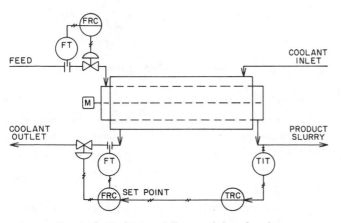

Fig. 8.7f　Cooling crystallizer with liquid coolant

In Figure 8.7f, an evaporating refrigerant in the shell cools and crystallizes the product from the liquor flowing in the tubes. Usually several scraped-surface tubes are used in series.

The control scheme is as follows:

1. Feed liquor is under flow control. (Feedback control from product outlet rate is not practical because of high system lag.)
2. Liquid refrigerant enters under level control and leaves as a vapor under pressure control. In many instances, flow control of the feed liquor is accomplished by a metering pump. Two- and three-mode controllers are used on pressure control.

A more conventional system uses a heat transfer fluid for cooling and can be controlled as shown in Figure 8.7f, where the flow of cooling fluids is controlled by outlet temperature of the process slurry. One refinement of this scheme is to cascade the outlet temperature of process slurry to the flow loop on the coolant.

Classifying Crystallizers

There are various types of classifying crystallizers in which supersaturation is produced entirely by cooling.

One of these is the Oslo, or Krystal-type, cooling crystallizer.

The control scheme for one type of classifying crystallizer is shown in Figure 8.7g. For a given quantity of feed, the ratio of crystals to liquor is fixed by stoichiometry, but frequently a ratio of crystals to mother liquor other than the stoichiometric one is required for good operation. Controls when excess mother liquor is required are shown in Figure 8.7g.

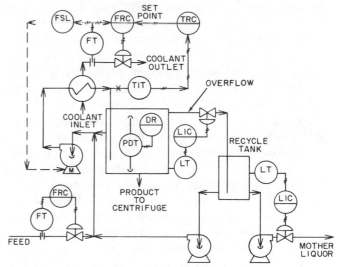

Fig. 8.7g　Classifying crystallizer, cooling-type

- Flow control of feed liquor is provided to maintain constant through-put.
- Flow control on coolant is based on process fluid outlet temperature at the heat exchanger. One refinement is cascade control; another is to interlock the coolant flow with the circulation pump motor.
- Overflow from the crystallizer is on level control. This is a case in which differential pressure–type sensors can be used.
- Mother liquor outflow from the recycle vessel is on level control.

In crystallization operations, one important instrument is the density sensor (differential pressure or other type), which gives an indication of the amount of crystals in suspension. This measurement is based on the fact that the density of the circulating stream is constant. Hence, the measurement of differential pressure between two points in the vessel is a measure of the amount of crystals in suspension, because any change in density is caused by a change in the amount of crystals in suspension.

Direct-Contact Crystallizers

In this unit the coolant is an evaporating refrigerant (or brine) and is in direct contact with the process slurry. Basic control methods are similar to those in controlled-growth crystallizers. These involve constant feed rate and evaporation or refrigerant flow controls, based on process fluid outlet temperature or on tank level.

Vacuum Crystallizers

In vacuum crystallizers, heat input to produce evaporation comes entirely from the sensible heat of the feed liquor and from the heat of crystallization of the product. Thus, supersaturation is produced by a combination of cooling and concentration effects. Two types that are important are forced-circulation (FC) and draft-tube baffle (DTB) crystallizers (discussed earlier).

One control scheme of a vacuum crystallizer is depicted in Figure 8.7h. The system shown achieves pressure control, and hence maintains process temperature by throttling a coolant in the crystallizer jacket. A more conventional and faster method of pressure control is an air bleed. The illustrated technique, however, is probably more suitable for high-vacuum applications.

The control system consists of (1) feed onflow control, (2) process slurry that leaves the system on level control (the level transmitter generally requires protection against plugging), and (3) temperature control accomplished indirectly by manipulation of jacket coolant. Use of suspension density recorder and other refinements can also be considered. The control scheme for classifying crystallizers is roughly analogous to the one illustrated in Figure 8.7h.

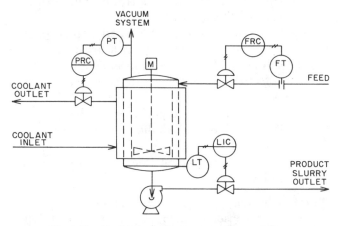

Fig. 8.7h   Draft tube baffle-type vacuum crystallizer

Reaction Crystallizers

In reaction crystallizers, crystallization is associated with a chemical reaction, which frequently provides the heat required for evaporation and crystallization.

One design used in ammonium sulfate processes is shown in Figure 8.7i. With the associated control scheme, the acid-to-feed tank is on flow control, total feed is on level control, and gaseous ammonia is added on pH-cascaded flow control. The pH metering lines should be continuously flushed.

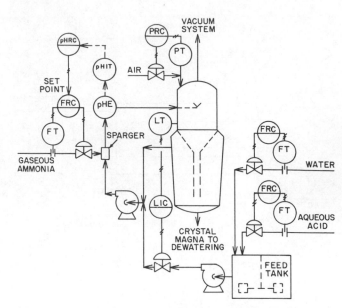

Fig. 8.7i   Reaction crystallizer on ammonium sulfate

Auxiliary Equipment

Control of a crystallizer is affected by the control of the equipment associated with it. There are three systems whose control should be considered in particular:

1. The feed system, including feed liquor and recycle and wash streams.
2. Vacuum control to maintain predetermined pressure in the system as a common means of crystallizer temperature control (discussed earlier).
3. Dewatering system control. This includes filters or centrifuges (the latter considered in Section 8.3).

Conclusions

In this section some of the common methods of crystallizer control have been outlined, but it should be noted that crystallization is still more an art than a science. The system for any particular application should be chosen after a step-by-step analysis of the complete process scheme. Even in well-known systems, manufacturers often recommend pilot plant study because of the tremendous influence of minor constituents in the feed liquor upon crystal growth. The developer of a control system should take these factors into consideration.

BIBLIOGRAPHY

Bamford, A.W., *Industrial Crystallization*, Macmillan, New York, 1966.
McCabe, W.L. and Smith, J.C., *Unit Operations of Chemical Engineers*, McGraw-Hill, New York, 1956.
Mullin, J.W., *Crystallization*, Butterworth & Co., London, 1961.
Perry, R.H. and Chilton, C.H., *Chemical Engineering Handbook*, McGraw-Hill, New York, 1973.
Randolph, A.D., "Crystallization," *Chemical Engineering*, May 4, 1970.
Schweitzer, P.A., *Handbook of Separation Techniques*, McGraw-Hill, New York, 1979.

8.8 DISTILLATION CONTROLS AND OPTIMIZATION

Distillation control and optimization are two of the greatest challenges facing the instrument engineer. Control difficulties arise because of the multitude of potential variable interactions and disturbances that are possible for a single column, as well as the system for which the column is a part. Even seemingly identical columns will exhibit great diversity of operation in the field. Therefore, this section will not attempt to provide control strategies that can be picked from illustrations and applied to columns in a "cookbook" fashion. To do so will result in poor column operation. Instead, discussion will begin with a basic description of the distillation process and equipment, followed by techniques used to derive a mathematical column model, how to implement the model with instrumentation, how to evaluate interactions and alternative control strategies, how to handle disturbance variables, how to improve operation through the use of computers and analyzers, and finally, how to derive and implement optimization equations within the limits and constraints of a particular columns operation.

The goal of this section is to provide the instrument engineer with the tools necessary to design *unique* control strategies that will match the specific requirements of the distillation columns he will encounter.

Distillation separates a mixture on the basis of a difference in the composition of a liquid and that of the vapor formed from the liquid. In the process industry, distillation is widely used to isolate and purify volatile materials. Thus, proper instrumentation of the distillation operation is vital to achieve maximum product of satisfactory purity.

Although one speaks of controlling a distillation tower, many of the instruments actually are associated with equipment other than the tower. For this reason, it might be useful to review the equipment used in distillation.

Distillation Equipment

The tower or column has two purposes: First, it separates a feed into a vapor portion that ascends the column and a liquid that descends the column. Second, it achieves intimate mixing between the two countercurrent flowing phases. The purpose of the mixing is to get an effective transfer of the more volatile components into the as-cending vapor and a corresponding transfer of the less volatile components into the descending liquid.

In continuous distillation, the feed is introduced continuously into the side of the distillation column. If the feed is all liquid, the temperature at which it first starts to boil is called the *bubble point*. If the feed is all vapor, the temperature at which it first starts to condense is called the *dew point*. The column is operated in a temperature range that usually is intermediate to the two extremes of dew point and bubble point. For effective separation of the feed, it is important that both vapor and liquid phases exist throughout the column.

The separation of phases is accomplished by gravity, with the lighter vapor rising to the top of the column and the heavier liquid flowing to the bottom. The intimate mixing is obtained by one or more of several methods. A simple method is to fill the column with lumps of an inert material, or *packing*, that will provide surface for the contacting of vapor and liquid. Another effective way is to use a number of horizontal plates, or trays, which cause the ascending vapor to be bubbled through the descending liquid.

The other equipment associated with the column is shown schematically in Figure 8.8a. The overhead vapor leaving the column is sent to a cooler, or condenser, and is collected as a liquid in a receiver, or accumulator. A part of the accumulated liquid is returned to the column

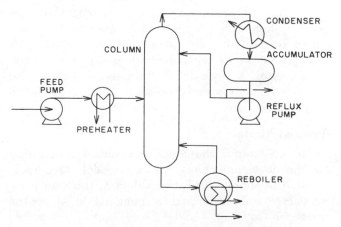

Fig. 8.8a   Distillation equipment

as reflux. The remainder is withdrawn as overhead product or distillate.

The bottom liquid leaving the column is heated in a reboiler. Part of this liquid is vaporized and injected into the column as boil-up. The remaining liquid is withdrawn as a bottom product, or residue.

In some cases, additional vapor or liquid is withdrawn from the column at points above or below the point at which the feed enters. All or a portion of this sidestream can be used as intermediate product. Sometimes, economical column design dictates that the sidestream be cooled and returned to the column to furnish localized reflux. The equipment that does this is called a sidestream cooler. At other times, localized heat is required. Then some of the liquid in the column is removed and passed through a sidestream reboiler before being returned to the column.

Distillation Model and Control Equations

Instrumentation is used to solve equations necessary for the proper control of distillation columns. The primary application of instruments in distillation is to minimize upsets to the unit, caused by a change in process inputs. The instruments calculate the effects of the changes and determine the corrective action needed to counteract them. The control actions are implemented by direct manipulation of the final control elements or by alteration of the set points of the controllers.

The benefits of better control are reduced energy consumption, reduced operating manpower, increased throughput, reduced disturbances to other processing units, increased plant flexibility, and increased product recovery.

The first step in the design of a control system is the derivation of a process model. Knowing these equations, one can select the manipulated variables and develop the operating equations for the control system. The instrumentation is then selected for the correct solution of these equations. The components are scaled, because most instruments act on normalized numbers (0 to 100 percent) rather than on actual process values.

The final control system may contain a single computing relay or may be a complex, multicomponent computer system. Here, a general discussion of the procedures for designing distillation controls is followed by examples of the common applications in distillation column control. A more detailed discussion of alternative strategies and optimized distillation column controls will be presented later.

Process Model

The first step in the design of a control system must be the development of a process model. Frequently omitted in simple distillation columns, this step is essential to minimize field reconfiguration of control strategies.

The model defines the process with equations developed from the material and energy balances of the unit. The model is kept simple by the use of one basic rule: The degrees of freedom for control limit the controlled variables (product compositions) specified in the equations. (See Section 1.1 for a detailed discussion.) Some of the variables that can be manipulated to control a column are shown in Figure 8.8b. For example, for a given feed rate only one degree of freedom exists for material balance control. If overhead product (distillate) is a manipulated variable (controlled directly to maintain composition), then the bottom product cannot be independent but must be manipulated to close the overall material balance according to the following equations:

$$F = D + B \qquad 8.8(1)$$

$$\text{Accumulation} = \text{Inflow} - \text{Outflow} \qquad 8.8(2)$$

$$= F - (D + B) \qquad 8.8(3)$$

If accumulation must be zero, then B is dependent upon F and D, as expressed by equation 8.8(4).

$$B = F - D \qquad 8.8(4)$$

where

  F = feed rate (the inflow)
  D = overhead rate (an outflow)
  B = bottoms rate (an outflow).

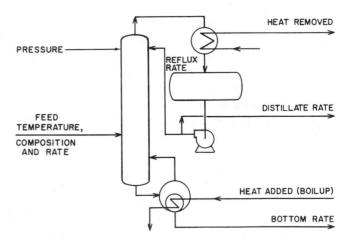

Fig. 8.8b   Variables that fix the distillation operation

Similarly, the criterion for separation is the ratio of reflux (L) to boil-up (V). Manipulating reflux has the same effect (though opposite) as manipulating boil-up. Consequently, only one degree of freedom exists to control separation.

For a column of the type shown in Figure 8.8c, two equations define the process: an equation for separation and an equation for material balance control.

The load and the controlled and manipulated variables are identified first. Load and controlled variables are

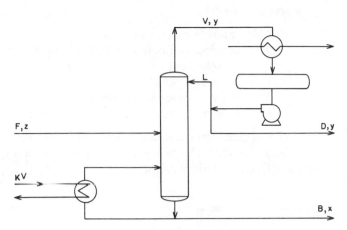

Fig. 8.8c Typical two-product distillation column

usually obvious, but identification of manipulated variables can be difficult. The general guidelines are:

1. Manipulate the stream that has the greatest influence on the associated controlled variable.
2. Manipulate the smaller stream if two streams have the same effect on the controlled variable.
3. Manipulate the stream that has the most nearly linear correlation with the controlled variable.
4. Manipulate the stream that is least sensitive to ambient conditions.
5. Manipulate the stream least likely to cause interaction problems.

The equations are then solved for the manipulated variables in terms of the controlled and load variables. In that form the equations are the mathematical representation of the control systems.

Computing Functions

The form of the control system equations essentially defines the computing functions required. Economies can often be made by combining functions. For example, the solution of the equation

$$C = (A + B)(D) \qquad 8.8(5)$$

for C requires a summing and a multiplying operation. If the computing relays are pneumatic, then two separate instruments must be used. If electronic, only one instrument is required, because an electronic multiplier (divider) algebraically adds several signals and multiplies (divides) the sum by another signal, which can also be a summation. Sections 2.1, 7.1, and 7.9 give a listing of the operations and combinations of operations that can be performed by the available devices.

The prime selection factor, however, is the integrity of the equation, and therefore a *different arithmetic operation cannot be substituted to facilitate the use of less expensive instruments*.

Scaling

Analog, and most digital, instruments operate on normalized values of the process variables, as discussed in Sections 7.8 and 7.9. That is, the measurement signal will vary from 0 to 100 percent as the process variable shifts from zero to its maximum value. Figure 8.8d illustrates the relationship among the various forms of analog signals and some typical process measurements.

The actual value of a process measurement is found by multiplying the analog signal by the calibrated full-scale value (*meter factor*) of the process variable.

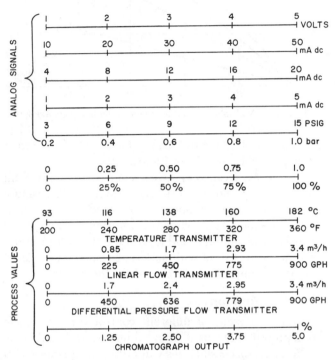

Fig. 8.8d Common analog signals and their relationship to process variables

In the examples of Figure 8.8d the temperature, represented by a 75 percent analog signal, is 320°F (160°C), the linear flow is 775 GPH (2.93 m³/h), the output of the differential pressure transmitter (flow squared) is 779 GPH (2.95 m³/h), and the composition is 3.75 percent.

As an example, let us review a flow ratio system in which the load stream L has the range of 0 to 1000 GPM (0 to 3.79 m³/h), the manipulated stream M has a range of 0 to 700 GPM (0 to 2.65 m³/h), and the ratio range R is 0 to 0.8 (R = M/L).

Let L', R', and M' represent the normalized values of flow L, ratio R, and stream M, respectively. The scaled equation is

$$700M' = (1000L')(0.80R') \qquad 8.8(6)$$

Reducing to the lowest form,

$$M' = 1.143(L')(R') \qquad 8.8(7)$$

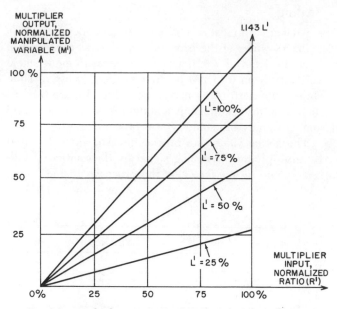

Fig. 8.8e   Multiplier output for the solution of equation 8.8(9)

where

$$L_i = \text{internal reflux rate,}$$
$$L = \text{external reflux rate (GPM, n m}^3/\text{m),}$$
$$C_p = \text{specific heat of external reflux,}$$
$$T_v = \text{vapor temperature,}$$
$$T_l = \text{reflux temperature, and}$$
$$H = \text{heat of vaporization of the external reflux.}$$

Figure 8.8f shows an instrumentation system for solving equation 8.8(8). The temperatures must be subtracted before the multiplication.

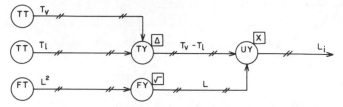

Fig. 8.8f   Control system for solving internal reflux rate

The number 1.143 is the *scaling factor*. M′ is plotted as a function of L′ and R′ in Figure 8.8e.

In applications such as the constant separation system, exact scaling is not critical. The flexible scaling cannot be used (1) when compensation for feed composition is part of the model, (2) when narrow spans must be used for reasons of stability, and (3) when transmitter calibrations are inconsistent with material balance ratios. Exact scaling techniques must be used for these cases.

Control Equations

Literally dozens of different control strategies have been proposed for distillation column control, and many of the more successful ones will be analyzed later in this section. However, for the present discussion, a particular set of control variables is used to illustrate the calculations involved, and the set used, although common, cannot be considered best for all columns.

Internal Reflux Computer

A major heat load in some columns is the energy required to heat the reflux returning to the column up to its boiling point. This heat—and the heat required to vaporize this *external* reflux—is obtained from the rising vapors that condense to become *internal* reflux. The latent heat of vaporization is essentially constant, and the ratio of internal reflux rate to external reflux is a function of the amount of subcooling of the external reflux.

Subcooling can be expressed as the difference between the overhead vapor temperature and the external reflux temperature. The equation solved by an internal reflux[1] system is

$$L_i = L\left[1 + \frac{C_p}{H}(T_v - T_l)\right] \qquad 8.8(8)$$

The subtracter is scaled first. Assuming a ΔT_{max} of 50°F (27.8°C), the span of T_v between 150 and 250°F (65.6 and 121°C), and the span of T_l between 125 and 225°F (51.7 and 107°C), the equation for the subtracter is written

$$\Delta T' = \frac{(150 + 100T'_v) - (125 + 100T'_l)}{50} \qquad 8.8(9)$$

This reduces to the scaled equation

$$\Delta T' = 2(T'_v - T'_l + 0.25) \qquad 8.8(10)$$

If the following assumptions are made,

$$L_{i_{max}} = 15,000 \text{ GPM (0.95 m}^3/\text{s)}$$
$$L_{max} = 10,000 \text{ GPM (0.63 m}^3/\text{s)}$$
$$C_p = 0.65 \qquad H = 250$$

The equation for the multiplier then becomes:

$$L'_i = \frac{10,000L'}{15,000}\left[1 + \left(\frac{0.65}{250}\right)50\Delta T'\right] \qquad 8.8(11)$$

then equation 8.8(11) reduces to

$$L'_i = 0.667L'(1 + 0.13\ \Delta T') \qquad 8.8(12)$$

When $\Delta T'$ is zero, the internal reflux equals 0.667 times the external reflux. The number one (1) within the parentheses therefore sets the minimum internal reflux. When $\Delta T'$ is 100 percent, the ratio of internal reflux to external reflux is at a maximum.

The expression within the parentheses must be normalized. This is done by dividing both terms by the total numerical value, that is, 1.13. To preserve the equality, the coefficient of L is multiplied by 1.13. The scaled equation becomes

$$L'_i = 0.754L'(0.885 + 0.115\ \Delta T') \qquad 8.8(13)$$

Internal reflux systems are designed to compensate for changes outside the column, and it should be understood that a change within the column can introduce a positive feedback. Figure 8.8g shows a typical internal reflux application and its response to an upset *within* the column. The control system reacts in the same way to an increase in vapor temperature and to a decrease in liquid temperature, but the required control actions are in the opposite direction.

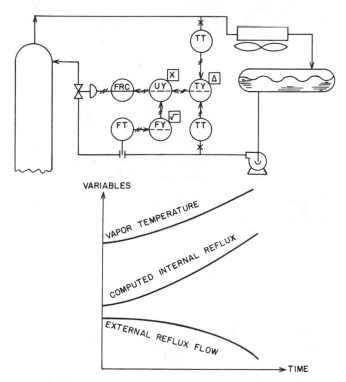

Fig. 8.8g Response of internal reflux control system to an increase in the concentration of heavy components in the overhead vapors

Flow Control of Distillate (Fast Response)

The column interactions that otherwise may necessitate the use of an internal reflux control system can be eliminated in some cases when the flow of distillate product draw-off is controlled and reflux is put under accumulator level control. This is a slower system than one in which flow controls the reflux, and its response is not always adequate. If necessary, the response can be sped up by reduction of the accumulator lag.[2]

The material balance around the accumulator (Figure 8.8h) is expressed by

$$V = L + D \qquad 8.8(14)$$

where V is boil-up (vapor rate), L is reflux rate, and D is distillate rate. To overcome the accumulator lag, L must be manipulated in direct response to a change in D rather than by a level controller. If V is constant (k), equation 8.8(14) is solved for L, which is the manipulated variable in this part of the system:

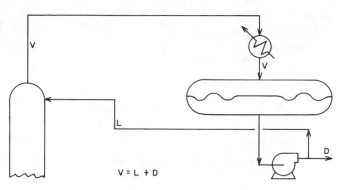

Fig. 8.8h Reflux accumulator material balance

$$L = k - D \qquad 8.8(15)$$

For this equation to be satisfied, L must be decreased one unit for every unit D is increased, and vice versa.

If V is indeed constant and the computations and flow manipulations are perfectly accurate, no level controller is needed. If these conditions cannot be met, a trimming function is introduced. The system equation becomes

$$L = m - KD \qquad 8.8(16)$$

where m is the output of the level controller and K is an adjustable coefficient. The resulting control system is shown in Figure 8.8i.

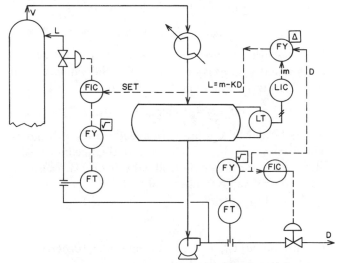

Fig. 8.8i Reflux rate control system for overcoming accumulator lag

The range of coefficient K should be broad enough to allow scaling and adjustment to be done during commissioning. Because the level controller trims the computation, the scaling and the value of K do not alter the steady state value of reflux L, since these factors affect the transient response only. The response of the reflux flow to changes in distillate for several values of K is given in Figure 8.8j. The full-scale values of reflux and distillate flows in this case are 1000 GPM (3.79 m³/m) and 500 GPM (1.89 m³/m), respectively.

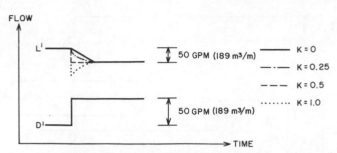

FLOW

50 GPM (189 m³/m) — K = 0
—·— K = 0.25
— — — K = 0.5
········ K = 1.0

50 GPM (189 m³/m)

TIME

Fig. 8.8j Reflux response as function of K

When K = 0, the reflux is adjusted by the level controller. In other cases the reflux flow is immediately altered by some percentage for a change in distillate, and the level controller forces the balance of the change.

If K = 0.5, the reflux flow is changed to the exact new steady state value, because K equals the ratio of D_{max}/L_{max}, and therefore the computation is exact. If K = 1.0, the initial response of reflux is greater than required for the new steady state, and the level controller corrects the flow.

The value of K does not change the steady state flow. It affects the transient response only, and therefore it can be used to adjust the dynamics of the loop. The greater the value of K, the faster the response.

Care must be taken to prevent increasing the response to the point of instability. A rule of thumb is

$$K_{max} = 1.5(D_{max}/L_{max}) \qquad 8.8(17)$$

In some instruments the range of adjustability of K is limited and scaling is necessary. For the values used in the illustration, equation 8.8(16) becomes

$$1000L' = 1000m' - 500KD' \qquad 8.8(18)$$

where L′ and D′ are the normalized values of L and D. The maximum value of m is equal to the maximum value of L, since the level controller by itself can cause the level control valve to open fully.

The scaled equation is

$$L' = m' - 0.5KD' \qquad 8.8(19)$$

K must be adjustable over a range of ±10 percent for satisfactory tuning flexibility.

A similar system can be used on column bottoms where the bottom product is flow-controlled and the [bottoms] level is maintained by manipulation of the heat input or boil-up (V). The equation for that system is

$$V = m - KB \qquad 8.8(20)$$

where V is the boilup and B is the bottom product flow.

Constant Separation (Feedforward)

A distillation column operating under constant separation conditions has one fewer degree of freedom than others, because the energy-to-feed ratio is constant. At a given separation, for each concentration of the key component in the distillate, there is a corresponding concentration in the bottoms.

In other words, holding the concentration of a component constant in one product stream fixes it in the other.[3] Figure 8.8k is an example of a constant separation feedforward system in which distillate is the manipulated variable. If the flow measurements are of the differential pressure type, and since Fz = Dy + Bx = Dy + (F − D)x, the equations are

$$D^2 = F^2\left(\frac{z-x}{y-x}\right)^2 = F^2\left(\frac{D}{F}\right)^2 \qquad 8.8(21)$$

$$Q^2 = F^2[Q/F]^2 \qquad 8.8(22)$$

where

z, y, x = mole fraction of the key light component in feed, overheads, and bottoms, respectively,
D/F = required distillate-to-feed ratio, and
Q/F = required energy-to-feed ratio.

No scaling is required of this equation if a manually (or automatically) adjustable ratio relay is used to allow signal weighting to be done in the field.

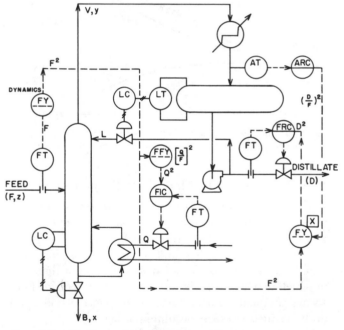

Fig. 8.8k Feedforward distillation control system with constant separation

Normal design practice calls for the output of the trim composition analyzer controller ARC to be at 50 percent when the design or normal distillate-to-feed ratio is required. If the gain of the multiplier is set at 2, the output tracks the load when this normal distillate-to-feed ratio is in force.

In a linear system the gain of the multiplier equals the

scaling factor. In this system, however, the gain of the multiplier equals the square root of the scaling factor. When this rule is applied to the example, the scaled form of equation 8.8(21) is

$$D^{2\prime} = 4.0(F^2)'[(D/F)^2]' \qquad 8.8(23)$$

where $D^{2\prime}, F^{2\prime}$ and $[(D/F)^2]'$ are the normalized values of the respective terms in equation 8.8(21).

The instrument labeled "dynamics" is a special module designed to influence the transient response. In the steady state its output equals its input, so no scaling is needed. See Section 1.5 for a more complete discussion on dynamic compensation elements, such as the one shown in Figure 1.5i.

Maximum Recovery (Feedforward)

In many distillations, one product is worth much more than the other, and the control system is designed to maximize the more valuable stream. The most common equation for this type of system is[3]

$$D = m(KF + K_2F^2) \qquad 8.8(24)$$

where

$$D = \text{distillate rate,}$$
$$F = \text{feed rate,}$$
$$K = \text{adjustable coefficient,}$$
$$K_2 = 1 - K, \text{ and}$$
$$m = \text{feedback trim.}$$

One control diagram for a maximum recovery system is shown in Figure 8.8l. Note that the distillate-to-reflux loop for accelerated response is used also. The summing instrument used to compute $(KF + K_2F^2)$ needs no special scaling. Although the coefficients can be calculated in advance with reasonable accuracy, on-line adjustment is quite easy, and the rigor of the calculations can be avoided. The multiplier may require exact scaling, however.

In a typical system the maximum value of the parenthetical term is equal to the maximum feed flow, e.g., 150 GPM (0.57 m³/m). The values of m are computed from the feed composition. A typical range for m is 0.35 to 0.65. The distillate flow transmitter is calibrated for 90 GPM (0.34 m³/m) maximum flow.

The scaled equation is

$$90D' = (150S')(0.35 + 0.3m') \qquad 8.8(25)$$

where $S' = (KF + K_2F^2)$. Note that in this equation the minimum value of D is $(0.35)(S)$. The number 0.35 is the zero value for m', and 0.30 is the span value.

Reducing 8.8(25) to its normalized form gives

$$D' = \frac{150}{90}(S')(0.35 + 0.30m') \qquad 8.8(26)$$

The zero and span values must be included in the normalized procedure, and the right-hand side of 8.8(26) is

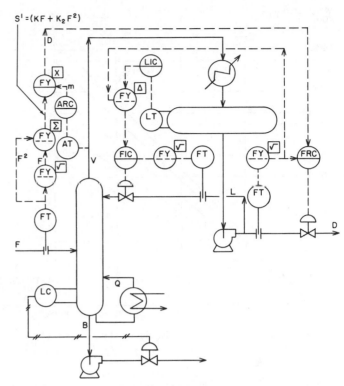

Fig. 8.8l Maximum recovery system: instrumentation solves quadratic equation for distillate rate

therefore multiplied by and divided by the zero plus the span:

$$D' = (0.65)(1.667)(S')\left(\frac{0.35 + 0.30m'}{0.65}\right) \qquad 8.8(27)$$

Reducing the remaining fraction to its simplest form, the scaled equation becomes

$$D' = 1.084(S')(0.538 + 0.462m') \qquad 8.8(28)$$

The response curves of the scaled multiplier are shown in Figure 8.8m.

Composition Control of Two Products

Because many variables that affect product composition are difficult to anticipate or control in many columns (e.g., feed composition), and because composition specifications for both products may be tight, some columns require better control than can be achieved by the previous constant separation example.

One method that can be used on some columns for achieving the required product specifications is to control directly the compositions of both products. Another benefit of dual composition control is minimized energy consumption. However, it is difficult to implement dual composition control on many columns because of severe interaction problems that may exist.

An example of a feedforward dual composition control model follows, after which a helpful method for deter-

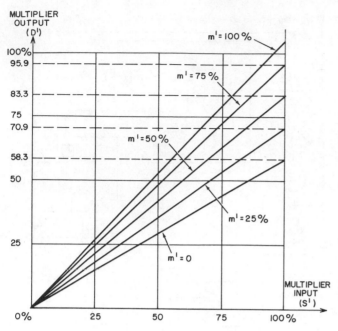

Fig. 8.8m   Multiplier output is solution of equation 8.8(29)

mining the degree of interaction based on actual process data will be discussed.

The control of distillate composition can still be done by manipulating distillate flow as required by

$$D = F\left(\frac{z - x}{y - x}\right) \qquad 8.8(29)$$

However, also to enforce composition control of the bottom product, an additional manipulated variable is needed. Another product stream cannot be independently manipulated without changing the accumulation in the column, and this is not practical. The energy balance must therefore be adjusted to control [bottoms] composition x.

The relationship between x and the energy balance was developed by Shinskey[3] as a function of separation S:

$$S = \frac{y(1 - x)}{x(1 - y)} \qquad 8.8(30)$$

The relationship between separation (S) and the ratio of boil-up to feed (V/F) over a reasonable operating range is

$$V/F = a + bS \qquad 8.8(31)$$

where a and b are functions of the relative volatility, the number of trays, the feed composition, and the minimum V/F. The control system therefore computes V based on the equation

$$V = F\left[a + b\left(\frac{y(1 - x)}{x(1 - y)}\right)\right] \qquad 8.8(32)$$

Since y is held constant, the bottom composition controller adjusts the value of the parenthetical expression if an error should appear in x. Let $V/F = y(1 - x)/x(1 - y)$, and the control equation becomes:

$$V = F(a + b[V/F]) \qquad 8.8(33)$$

where [V/F] = the desired ratio of boil-up to feed.

The system implementing equations 8.8(29) and 8.8(33) is shown in Figure 8.8n. The two instruments respectively labeled "FY − 1" and "FY − 2" must be multipliers. The instrument labeled "dynamics" is a special instrument for dynamic compensation and is discussed in Section 1.5. The scaling of FY − 1 may be done as described in connection with Figures 8.8k or 8.8l. (The illustration is based on the first.) Multiplier FY − 2 must be scaled using the latter technique, outlined in connection with Figure 8.8l.

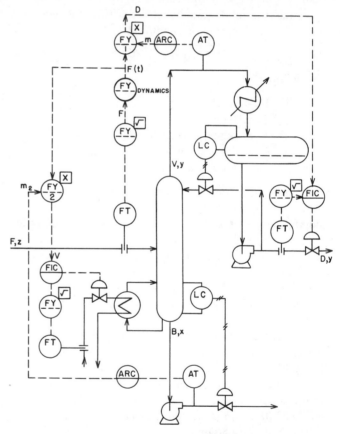

Fig. 8.8n   Feedforward control system provides closed-loop composition control of two product streams

Included in a and b is the relationship between boil-up (vapor rate) and energy flow Q and the minimum ratio of boil-up to feed. Equation 8.8(33) can therefore be written

$$Q = kF([V/F]_{min} + [V/F]) \qquad 8.8(34)$$

where

k represents the proportionality constant.

Let Q = 10,000 lb/hr (4500 kg/hr),
F = 60 GPM (0.23 m³/m),
$[V/F]_{min}$ = 1.0,
$[V/F]_{max}$ = 3.0, and
k = 0.600.

The values of $[V/F]_{min}$ and $[V/F]_{max}$ are used to establish zero and span values for the scaled equation, as explained in connection with equation 8.8(12). The scaled equation is

$$10,000Q' = (0.60)(60F')(1.0 + 2.0[V/F]') \qquad 8.8(35)$$

$$Q' = 0.0036F'(1.0 + 2.0[V/F]') \qquad 8.8(36)$$

$$Q' = 0.01080F'(0.333 + 0.667[V/F]') \qquad 8.8(37)$$

Control of Two Products with Interaction

There is always interaction between the material and energy balances in a distillation column. In some columns this interaction is not severe enough to impede closed-loop composition control of two product streams, but in others it is. The severity is a function of feed composition, product specifications, and manipulated and controlled variable pairing.

The severe interactions frequently occur when the energy balance is manipulated by two independent composition controllers. A column in which reflux flow and steam flow are the manipulated variables is an example of a severely interacting column. The control system equations are

$$L = F([L/F]) \qquad 8.8(38)$$

$$Q = kF([V/F]_{min} + [V/F]) \qquad 8.8(34)$$

where L is the reflux rate and $[L/F]$ is the desired reflux-to-feed ratio.

Note that in the control system described by these two equations, the rate of products leaving the column is dependent on two energy balance terms. Increasing heat input forces the composition controller resetting reflux flow to increase heat withdrawal, and the top and bottom composition controllers therefore "fight" each other. The only way to avoid this "fighting" is by preventing a change at one end of the column from upsetting the other end.

The heat input is changed when the bottom composition controller is upset. If the upset is because of a high concentration of light ends in the bottom product, heat is increased to adjust the separation being performed *and to drive the extra light ends out the top*. The top composition controller does not know how to split the increased vapor load, but it sees a measurement indicating an upset and responds to an increase in heat input by increasing the reflux flow. Theoretically, if the reflux rate is compensated for the change in heat input, the top composition controller upset can be avoided.

One can find the relationship between reflux L and

heat input Q by solving equations 8.8(38) and 8.8(34) for L in terms of Q. The resultant equation is of the form

$$L = k_1Q - k_2F \qquad 8.8(39)$$

The values of k_1 and k_2 are found by deriving equation 8.8(39) using actual process values of $[L/F]$, $[V/F]_{min}$, and $[V/F]$.

The decoupling equation 8.8(39) replaces 8.8(38) in the control model. The resulting system is shown in Figure 8.8o.

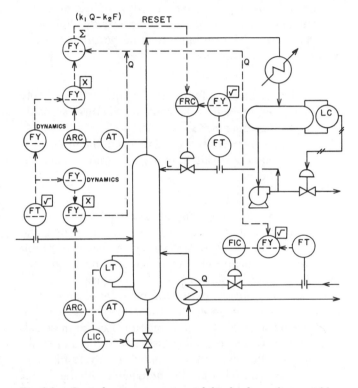

Fig. 8.8o   Control system improves stability by decoupling variables

The system is now half-decoupled: a change in heat input will not upset the top temperature, because the decoupling loop adjusts the reflux independently of the top temperature controller. However, the heat input is still coupled to reflux, because a change in reflux will still cause the bottom temperature controller to adjust steam flow. This degree of decoupling is enough to reduce the interaction approximately 20-fold. The two multipliers are scaled as described previously, and the adder is tuned on line.

Classical decoupling schemes,[4] however, often do not provide a solution to the problem of interaction because of practical problems encountered on real columns.[5,6] In computers, initialization routines may prevent placing classical decoupling schemes on line without upsetting product compositions. Decoupling systems that include overrides can drive to saturation when constraints are encountered. Most seriously, decouplers applied to systems with negative interaction (defined later) may have very little tolerance for errors in decoupler gains. For this group, which always includes the interaction encountered in reflux and boil-up controls, small errors can transform a system that provides complete decoupling into one that provides no control at all. Since the proper decoupler gains depend on the process gains, which inevitably change with variations in feed rate, product specifications, and column characteristics, these systems require constant attention and adjustment beyond the ordinary capability of plant operating personnel.

The difficulties associated with the application of decoupling systems have prompted a re-examination of interaction itself. The problem may be postulated in two ways:

1. For a given column, is the interaction equally strong in each of the control structures available to the designer?
2. For a given control structure, will the interaction be equally strong in every column in which it is applied?

The stumbling block of loop assignment may be in this way converted into a stepping stone by providing the opportunity to use a control structure that will exhibit minimum interaction in any particular application.

Shinskey and Ryskamp have given consistent guidelines for assigning loops to minimize loop interaction. Thus, Shinskey[7] suggests that the controller assigned to the less pure product should manipulate the mass balance, whereas the controller assigned to the more pure product should manipulate separation. Ryskamp[8] suggests that the controller for the component with the shorter residence time should adjust vapor flow, and the controller for the component with the longer residence time should adjust the liquid/vapor ratio. The following analytical approach, which has been refined by L. M. Gordon and taken from reference 6, verifies these guidelines for most columns and identifies those columns in which exception should be taken.

Relative Gain

The concept of relative gain provides a measure of the interaction that may be expected when a particular loop is closed. First proposed by Bristol[9] and developed by Shinskey,[7,10] the concept may be used to analyze possible control scheme configurations for minimum interaction. Given a control scheme by which a set of variables, C, is controlled by manipulating a set of variables, M, the relative gain, λ_{ij}, for the loop that controls c_i by manipulating m_j is defined as:

$$\lambda_{ij} \equiv \frac{\left. \dfrac{\delta c_i}{\delta m_j} \right|_m}{\left. \dfrac{\delta c_i}{\delta m_j} \right|_c} \qquad 8.8(40)$$

The numerator, or open-loop gain with all other loops open, indicates the sensitivity of c_i to m_j when all other manipulated variables are held constant (i.e., when all other controllers are in manual). The denominator, or open-loop gain with all other loops closed, indicates the same sensitivity when all the other controlled variables are held constant (i.e., when all other controllers are in automatic). Since interaction is created by reciprocal disturbances among control loops in automatic, the ratio of these two gains measures the effect of the interaction with other loops on the feasibility of controlling c_i by manipulating m_j. For $\lambda_{ij} = 1.0$, the influence of m_j over c_i is not changed when the other loops are closed, indicating that this loop does not interact with the others. As the value λ_{ij} departs from 1.0, it indicates an increasing degree of interaction. Values greater than 1.0 indicate that the interaction reduces the sensitivity of c_i to m_j, which degrades performance when all control loops are closed. Values of λ_{ij} that approach zero indicate that most of the influence of m_j over c_i is exerted through interaction, by way of the other closed loops, causing poor performance when the other loops are opened. Negative values of λ_{ij} indicate that the interaction reverses the *direction* in which c_i is driven by changes in m_j, causing the stability of this loop to be conditional on the automatic/manual status of other loops.

The relative gains for all possible pairings may be gathered together in a relative gain array (RGA), defined as:

$$\Lambda = \begin{array}{c} c_1 \\ \\ \\ c_n \end{array} \begin{bmatrix} \lambda_{11} & \cdots\cdots\cdots & \lambda_{1n} \\ & & \\ & \lambda_{ij} & \\ & & \\ \lambda_{n1} & \cdots\cdots\cdots & \lambda_{nn} \end{bmatrix} \begin{array}{c} m_1 \cdots\cdots\cdots\cdots m_n \end{array}$$

Bristol has shown that the sum of the numbers in every row and column must equal 1.0. Thus, arrays with values of λ greater than 1.0 must also contain negative values of λ. An array that includes negative values is said to exhibit negative interaction. Conversely, an array in which all relative gains are positive, and therefore less than or equal to 1.0, is said to exhibit positive interaction. In the case of 2×2 arrays with positive interaction, as one value of λ approaches 0.0, the other must approach 1.0, and values of λ that approach 0.5 indicate maximum positive interaction. In the case of matrices with negative interaction, values of λ which approach $\pm \infty$ indicate maximum interaction.

The set of possible controlled and manipulated variables for a distillation column forms a 5×6 matrix,

$$
\Lambda = \begin{array}{c} \\ y \\ x \\ l_a \\ l_b \\ p \end{array}
\begin{array}{|ccccccc}
D & B & S & V & L & Q \\
\hline
\lambda_{yD} & & & & & \\
& & & & & \\
& & & \text{etc.} & & \\
& & & & & \\
& & & & & \\
\end{array}
$$

where

Q = heat removal rate,
l_a = accumulator level,
l_b = bottom level, and
p = column pressure.

However, evaluating the complete set of relative gains suggested by this array gives a misleading representation of the interaction in dual composition control. The definition of relative gain given by equation 8.8(40) requires evaluation of each open-loop gain with *all* other loops open. Since separation actually results from the ratio of internal vapor and liquid rates, if boil-up and reflux are held constant, variations in D or B will have no effect on composition (only on the liquid levels), so that the relative gains for controlling compositions with these variables will be zero, indicating that these assignments will not control. Yet these assignments do control through the action of the level and pressure control loops, which reflect these changes into reflux or boil-up, if distillate or bottoms flow is used to control composition.

To evaluate the relative gains properly for composition control, the overall matrix must be reduced to 2 × 2 subsets that set x and y against possible pairs of manipulated variables, under the assumptions that the remaining variables are used to control both levels and the column pressure and that these loops are closed and holding their measurements at set point. Since these loops are an order of magnitude faster than the composition loops, the assumption that they are able to achieve this is reasonable. The open-loop gains (numerators) for the pairs in these matrices are then evaluated with only the other composition controller in manual. For example, in the subset that assigns D and S to control y and x,

$$
\Lambda_{DS} = \begin{array}{c} \\ y \\ x \end{array}
\begin{array}{|cc}
D & S \\
\hline
\lambda_{yD}(\Lambda_{DS}) & \\
& \\
\end{array}
$$

the relative gain for control of top composition by manipulation of distillate flow is given by:

$$
\lambda_{yD}(\Lambda_{DS}) = \frac{\left.\dfrac{\delta y}{\delta D}\right|_{S,l_a,l_b,p}}{\left.\dfrac{\delta y}{\delta D}\right|_{x,l_a,l_b,p}} \qquad 8.8(41)
$$

In this expression, identification of the subset is essential. The relative gain λ_{yD} appears in more than one subset, and the value of $\delta y/\delta D$ will depend on which other manipulated variable is held constant.

Determining the values of these relative gains is not the formidable task that it seems to be at first. Since the overall mass balance requires equal and opposite changes in D and B,

$$
dD = -dB \qquad 8.8(42)
$$

This is achieved by assigning at least one of these variables for level control to close the overall mass balance. Since this substitution may be made in both the numerator and the denominator of the relative gain, the signs cancel, and the value of the relative gain will not be changed. For example, (dropping the notation for the always-constant levels and pressure),

$$
\lambda_{yD}(\Lambda_{DS}) = \frac{\left.\dfrac{\delta y}{\delta D}\right|_S}{\left.\dfrac{\delta y}{\delta D}\right|_x} = \frac{-\left.\dfrac{\delta y}{\delta B}\right|_S}{-\left.\dfrac{\delta y}{\delta B}\right|_x} = \lambda_{yB}(\Lambda_{BS}) \qquad 8.8(43)
$$

Next, eliminating heat removal, Q, as a candidate for composition control reduces the problem to six unique 2 × 2 composition subsets,

$$
\Lambda_{DS} = \begin{array}{c} \\ y \\ \\ x \end{array}
\begin{array}{|cc}
D & S \\
\hline
& \\
& \\
& \\
\end{array}
\qquad
\Lambda_{SV} = \begin{array}{c} \\ y \\ \\ x \end{array}
\begin{array}{|cc}
S & V \\
\hline
& \\
& \\
& \\
\end{array}
$$

$$
\Lambda_{DL} = \begin{array}{c} \\ y \\ \\ x \end{array}
\begin{array}{|cc}
D & L \\
\hline
& \\
& \\
& \\
\end{array}
\qquad
\Lambda_{SL} = \begin{array}{c} \\ y \\ \\ x \end{array}
\begin{array}{|cc}
S & L \\
\hline
& \\
& \\
& \\
\end{array}
$$

$$
\Lambda_{DV} = \begin{array}{c} \\ y \\ \\ x \end{array}
\begin{array}{|cc}
D & V \\
\hline
& \\
& \\
& \\
\end{array}
\qquad
\Lambda_{LV} = \begin{array}{c} \\ y \\ \\ x \end{array}
\begin{array}{|cc}
L & V \\
\hline
& \\
& \\
& \\
\end{array}
$$

which contain 24 relative gains. However, under the assumption that both levels and pressure are invariable, the requirement that the sum of the elements in each row and column must be 1.0 reduces the problem to determining just six unique values. For example,

$$
\Lambda_{DS} = \quad
\begin{array}{c|cc}
 & D & S \\
\hline
y & \lambda_{yD}(\Lambda_{DS}) & 1 - \lambda_{yD}(\Lambda_{DS}) \\
\\
x & 1 - \lambda_{yD}(\Lambda_{DS}) & \lambda_{yD}(\Lambda_{DS})
\end{array}
$$

Finally, collecting the six unique relative gains together yields the following set of definitions:

$$
\lambda_{yD}(\Lambda_{DS}) = \frac{\left.\dfrac{\delta y}{\delta D}\right|_{S}}{\left.\dfrac{\delta y}{\delta D}\right|_{x}} \qquad
\lambda_{yD}(\Lambda_{DL}) = \frac{\left.\dfrac{\delta y}{\delta D}\right|_{L}}{\left.\dfrac{\delta y}{\delta D}\right|_{x}}
$$

$$
\lambda_{yD}(\Lambda_{DV}) = \frac{\left.\dfrac{\delta y}{\delta D}\right|_{V}}{\left.\dfrac{\delta y}{\delta D}\right|_{x}} \qquad
\lambda_{yS}(\Lambda_{SV}) = \frac{\left.\dfrac{\delta y}{\delta S}\right|_{V}}{\left.\dfrac{\delta y}{\delta S}\right|_{x}}
$$

$$
\lambda_{yS}(\Lambda_{SL}) = \frac{\left.\dfrac{\delta y}{\delta S}\right|_{L}}{\left.\dfrac{\delta y}{\delta S}\right|_{x}} \qquad
\lambda_{yL}(\Lambda_{LV}) = \frac{\left.\dfrac{\delta y}{\delta L}\right|_{V}}{\left.\dfrac{\delta y}{\delta L}\right|_{x}}
$$

which may be derived by differentiating the equations that model the column plus a mass balance across the condenser.

Results

The results of these derivations may be collected into a set of factors that quantify the interaction of each possible dual composition control scheme.

$$
\lambda_{yD}(\Lambda_{DS}) = \frac{1}{1 + \dfrac{(y - z)x(1 - x)}{(z - x)y(1 - y)}} \equiv \lambda \qquad 8.8(44)
$$

$$
\lambda_{yD}(\Lambda_{DL}) = \lambda + (1 - \lambda)\epsilon \qquad 8.8(45)
$$

where

$$
\epsilon \equiv \frac{nEy(1 - y)}{2\left(\dfrac{zL}{D} + 1\right)(y - x)} \qquad 8.8(46)
$$

and

    n = the number of column trays
    E = tray efficiency (assumed constant across the column)

also,

$$
\lambda_{yD}(\Lambda_{DV}) = \lambda + (1 - \lambda)\epsilon\left(1 + \frac{D}{L}\right) \qquad 8.8(47)
$$

$$
\lambda_{yS}(\Lambda_{SV}) = 1 - \lambda + \frac{\lambda}{\epsilon\left(1 + \dfrac{D}{L}\right)} \qquad 8.8(48)
$$

$$
\lambda_{yS}(\Lambda_{SL}) = 1 - \lambda + \frac{\lambda}{\epsilon} \equiv \sigma \qquad 8.8(49)
$$

$$
\lambda_{yL}(\Lambda_{LV}) = 1 - \sigma + \sigma\left(1 + \frac{L}{D}\right)(1 - \epsilon) \qquad 8.8(50)
$$

For example, consider a column having 100 theoretical trays receiving a feed of 30 percent lights and separating it into a top product of 99 percent purity and a bottom product 95 percent pure, using a reflux ratio of 12. Then,

$$
y = 0.99 \qquad\qquad nE = 100
$$

$$
x = 0.05 \qquad\qquad \frac{L}{D} = 12
$$

$$
z = 0.3
$$

From these values, the matrices may be completed as:

$$
\Lambda_{DS} = \quad
\begin{array}{c|cc}
 & D & S \\
\hline
y & 0.07 & 0.93 \\
\\
x & 0.93 & 0.07
\end{array}
$$

$$
\Lambda_{DL} = \quad
\begin{array}{c|cc}
 & D & L \\
\hline
y & 0.176 & 0.824 \\
\\
x & 0.824 & 0.176
\end{array}
$$

$$
\Lambda_{DV} = \quad
\begin{array}{c|cc}
 & D & V \\
\hline
y & 0.186 & 0.814 \\
\\
x & 0.814 & 0.186
\end{array}
$$

$$
\Lambda_{SV} = \quad
\begin{array}{c|cc}
 & S & V \\
\hline
y & 1.496 & -0.496 \\
\\
x & -0.496 & 1.496
\end{array}
$$

$$
\Lambda_{SL} = \quad
\begin{array}{c|cc}
 & S & L \\
\hline
y & 1.543 & -0.543 \\
\\
x & -0.543 & 1.543
\end{array}
$$

$$
\Lambda_{LV} = \quad
\begin{array}{c|cc}
 & L & V \\
\hline
y & 17.221 & -16.221 \\
\\
x & -16.221 & 17.221
\end{array}
$$

For this column, applying traditional reflux and boil-up controls, (Λ_{LV}) will lead to strong negative interaction between the composition loops. Opposite, but equally strong, positive interaction will exist for the material-balance scheme (Λ_{DV}), which assigns distillate to control top composition and boil-up to control bottom composition. (Using B instead of D to control bottom composition suggests an attractive pairing. However, this subset [Λ_{BV}], must be rejected, because no appropriate variable would be available for control of column base level.) For both of these schemes, although the individual loops may work well, performance will be very poor when both loops are closed. The subset Λ_{SL} contains a likely pairing for top composition. However, the assignment of reflux for controlling bottom composition may show poor dynamic performance because of liquid holdup within the column, unless accumulator level or column pressure is controlled by manipulating boil-up. The first two subsets, Λ_{DS} and Λ_{DL}, are also material-balance schemes. Between the two, the subset Λ_{DS} shows excellent pairings after B is assigned in place of D. However, this scheme is subject to a subtle dynamic problem described by Shinskey[10] for the control scheme designed around the Λ_{DV} subset (Figure 8.8p).

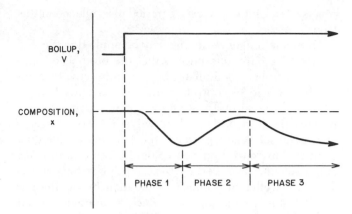

Fig. 8.8q Open-loop response of the bottom composition controller for λ_{xv} (λ_{DV}) = 0.186

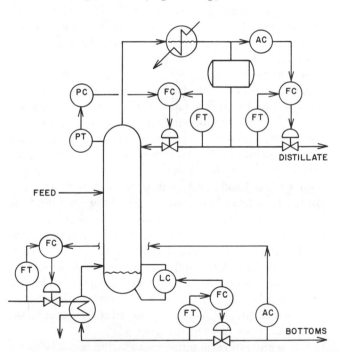

Fig. 8.8p Control scheme for the Λ_{DV} subset (flooded condenser)

The open-loop response of the bottom composition controller in such a scheme is shown in Figure 8.8q and includes three separate phases. In the first, increasing boil-up reduces the impurity in the bottom product, x. At the same time, with the top composition controller in manual, the pressure controller carries the increased heat load by raising reflux to lower the level in the flooded condenser. Later, this increased reflux reaches the bottom of the column and the response enters the second

phase, during which the impurity in the bottom product reverses and begins to increase. Its final value may be predicted from Figure 8.8r. With constant distillate flow, increasing reflux improves separation along a line of constant D/F, making both products more pure. Thus, x will be lower and y will be higher than their original values.

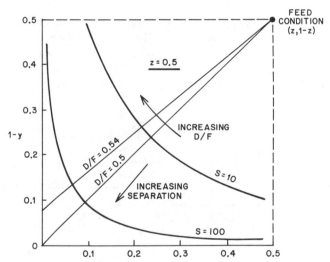

Fig. 8.8r Relation among product purities, separation, and the material balance

If the top composition controller is in automatic, the response then enters the third phase. Since D is over-pure, the top composition controller will increase distillate flow. To hold pressure, the pressure controller simultaneously enforces an equal reduction in reflux, which eventually reaches the bottom of the column and causes x to reverse a second time. Its final value may be determined from the relative gain, which is the ratio of its net change at the end of the second phase to its net change at the end of the third phase. For the example given previously, $\lambda_{xv}(\Lambda_{DV})$ = 0.186. Thus, the final change in x will be 5.4 times its change at the end of the second phase. This double reversal effectively extends the dead time in this response to the beginning of the second reversal, causing an excessive period of oscillation in this loop.

The same line of reasoning may be applied to the Λ_{BS} subset. The only difference is that the bottoms composition controller manipulates boil-up indirectly, through the bottoms level controller. In this case, $\lambda_{xB}(\Lambda_{BS}) = 0.93$, and the second reversal is not so drastic. Nevertheless, the period of this loop will still be excessive.

The subset Λ_{SV} represents the best choice for dual composition control on this column (Figure 8.8s). This is the scheme proposed by Ryskamp.[8] The top composition controller manipulates separation by driving the distillate-to-vapor ratio, D/V, which is uniquely related to the reflux ratio by a mass balance across the accumulator, i.e.,

$$V = L + D \qquad 8.8(14)$$

from which

$$\frac{D}{L} = \frac{D/V}{1 - D/V} \qquad 8.8(51)$$

Bottom composition is controlled by manipulation of boil-up in cascade, which leaves bottoms level to be controlled by manipulation of bottoms flow. Operating with the receiver full of liquid eliminates accumulator level as a controlled variable and allows pressure control by manipulating the total of reflux plus distillate flow.

The dynamics of this scheme will be superior to those of the material-balance subsets for two reasons. First,

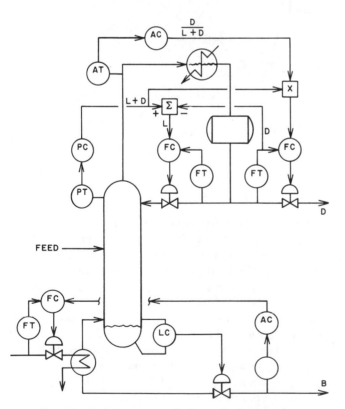

Fig. 8.8s   Reflux ratio control scheme for the Λ_{SV} subset

the feedforward path from D to L avoids any delay from the pressure controller. Second, since $\lambda_{yS}(\Lambda_{SV}) > 1.0$, the second reversal does not occur, and the bottoms composition loop will have a much shorter period of oscillation.

A closer look at equations 8.8(44) through 8.8(50) shows how the design of the control system is influenced by factors that vary from application to application. The fundamental significance of the separation and material balance is embodied to the factor λ, which appears in all other relative gains. Depending only on feed and product specifications, this factor is independent of column design and varies from 0.0 to 1.0. In cases in which bottom product is very pure or the feed is primarily lights, this factor will approach 1.0. In cases in which top product is very pure or the feed is primarily heavies, this factor will approach 0.0.

In the first case, this subset will suggest a control scheme with little interaction. In the second case, substituting B for D will provide attractive pairings. However, since these relative gains are less than 1.0, the double reversal in the bottom composition loops may make the dynamics of this loop unattractive. As the separation becomes more symmetrical, i.e., as top and bottom compositions approach equal purity and the feed approaches an even mix of lights and heavies, λ will approach 0.5, indicating that these material-balance schemes will be highly interactive.

The influence of the column design resides in the value of ϵ, which depends on the number and efficiency of the trays and on the reflux ratio. High purity products and high reflux ratios reduce the value of ϵ. As ϵ approaches 0.0, $\lambda_{yD}(\Lambda_{DL})$ and $\lambda_{yD}(\Lambda_{DV})$ approach λ from above, making the column design less significant in determining the control system design. Conversely, as ϵ approaches 1.0, both $\lambda_{yD}(\Lambda_{DL})$ and $\lambda_{yS}(\Lambda_{SL})$ approach 1.0, with $\lambda_{yD}(\Lambda_{DL})$ approaching from below and $\lambda_{yS}(\Lambda_{SL})$ approaching from above. In this case, both of these subsets offer attractive pairings that require control of bottom composition by manipulating reflux. The dynamics of these arrangements will almost certainly be unacceptable unless accumulator level or column pressure is controlled by manipulating boil-up. Given this assignment, the bottom composition loop of $\lambda_{yD}(\Lambda_{DL})$ will still be subject to the double reversal described earlier.

The reflux ratio also influences the difference between $\lambda_{yD}(\Lambda_{DL})$ and $\lambda_{yD}(\Lambda_{DV})$. At high reflux ratios, these relative gains approach equality, with $\lambda_{yD}(\Lambda_{DL}) < \lambda_{yD}(\Lambda_{DV})$. Similarly, the factors $\lambda_{yS}(\Lambda_{SL})$ and $\lambda_{yS}(\Lambda_{SV})$ approach equality at high reflux ratios, with $\lambda_{yS}(\Lambda_{SV}) < \lambda_{yS}(\Lambda_{SL})$. Finally, large reflux ratios make a strong contribution to $\lambda_{yL}(\Lambda_{LV})$, raising this factor to the large positive numbers, which indicate strong interaction and predict poor performance for traditional controls. This is consistent with the observation that traditional controls have been most successful on columns with low reflux ratios. Decoupling control systems should be considered only when none of these possible subsets predicts reasonable performance.

Dynamics must be considered in applying the results of the relative gain analysis. In general, top composition control by manipulation of a variable at the bottom of the column may be acceptable because of the rapid response of vapor flow to changes in heat input. However, the response of bottom composition to manipulation of variables at the top of the column is usually not feasible unless accumulator level or column pressure is controlled by manipulating boil-up. Given that assignment, the period of the bottom composition loop in systems designed around relative gains slightly greater than 1.0 will be shorter than the period of the same loop in systems designed around relative gains slightly less than 1.0. This is because the double reversal that adds extra dead time into the response is not present for relative gains greater than 1.0.

A final consideration is that the designer should avoid control systems in which the response of either composition loop is inhibited when the other loop is out of service because of analyzer failure or maintenance or because the operator has placed the controller in manual.

Control of Two Products with Sidedraw

The presence of a sidestream product in addition to an overhead and bottom product *adds a degree of freedom* to a control system. This extra degree of freedom can be seen from the overall material balance equation:

$$F = D + C + B \qquad 8.8(52)$$

where C is the sidestream rate. Two of the product streams are available for manipulation, and the material balance can still be closed by the third product stream.

The added degree of freedom makes careful analysis of the process even more essential to avoid mismatching of manipulated and controlled variables. As in previous systems, the analysis involves developing the process model and determining the relationship among the several controlled and manipulated variables. There are, however, several possible combinations of variables that must be explored.

The possible combinations of manipulated variables for the column in which the bottoms composition and the sidestream composition must be controlled are

Distillate and sidestream flows
Distillate and bottom flows
Distillate flow and heat input

Sidestream and bottom flows
Sidestream flow and heat input
Bottom flow and heat input

One frequently successful pairing is to control sidestream composition by manipulation of distillate flow and bottoms composition by manipulation of sidestream flow.

The equations are

$$D = F\left(\frac{z_1 - c_1}{y_1 - c_1}\right) \qquad 8.8(53)$$

$$C = F\left(\frac{z_2 - x_2}{c_2 - x_2}\right) \qquad 8.8(54)$$

The symbols z_1, y_1, and c_1 are the concentrations in the feed, distillate, and sidestream of the component under control in the sidestream. The concentrations of the key component in the bottom are respectively expressed by z_2, x_2, and c_2 for the feed, the bottoms, and the sidestream.

The control system is shown in Figure 8.8t. Note that the ratio of heat input to feed (and, therefore, boil-up to feed) is held constant. Separate dynamic elements are used for the distillate loop and for the heat input and sidestream loops.

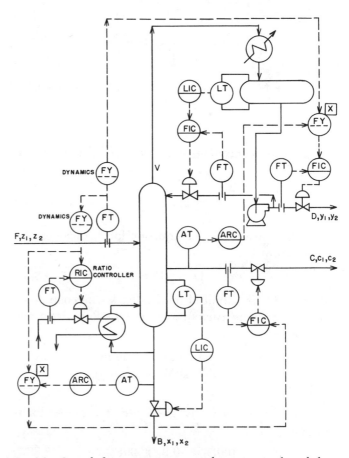

Fig. 8.8t Control of composition in two product streams with a sidedraw

The computing relays may be scaled by either of the methods discussed previously.

Feed Composition Compensation

Occasionally, changes in feed composition occur too fast for feedback control, and compensation for these changes is necessary (Figure 8.8u).

The basic equation 8.8(55) already has a term z rep-

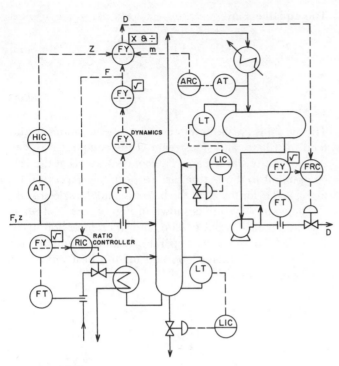

Fig. 8.8u Feed composition measured and used to compute distillate flow

resenting concentration of the key component in the feed:

$$D = F\left(\frac{z - x}{y - x}\right) \qquad 8.8(55)$$

When z is measured, the equation for distillate can be simplified to

$$D = zF/m \qquad 8.8(56)$$

where m is the output of the feedback trim controller.

To scale the instruments, assume the following range of values:

$$D = 0\text{--}100 \text{ GPM } (0\text{--}0.38 \text{ m}^3/\text{m})$$
$$F = 0\text{--}500 \text{ GPM } (0\text{--}1.9 \text{ m}^3/\text{m})$$
$$z = 0\text{--}0.3$$

A value for m will be assigned after the numerator is scaled.

$$100D = (0.3z')(500F') \qquad 8.8(57)$$

$$D = (1.50)(z')(F') \qquad 8.8(58)$$

Under normal conditions the value of the denominator is to be one and the output of the feedback controller will be 50 percent. Based on this, and assuming a reasonable span (20 percent) for the feedback trim, the complete equation is written

$$D' = \frac{(1.50)(z')(F')}{(0.9 + 0.2m')} \qquad 8.8(59)$$

To prevent the denominator from exceeding one, the numerator and the denominator are divided by the maximum value of the denominator (1.1). The result is the scaled equation for the system:

$$D' = \frac{1.363(z')(F')}{(0.82 + 0.18m')} \qquad 8.8(60)$$

The auto manual station (HIC) is used in the event of analyzer failure. Dynamic compensation is placed on the feed signal only.

The Total Model

It is possible to design a system to compensate for all load variables: feed rate, composition, enthalpy, reflux, and bottoms enthalpy. The goal of these systems is to overcome the problems associated with unfavorable interactions and to isolate the column from changes in ambient conditions. These problems can usually be solved by careful system analysis and variable pairing, thus avoiding complicated total energy and material balance control systems.

The complexity of the total material and energy balance systems is made apparent by the list of equations required in the model:

> Feed enthalpy balance
> Bottoms enthalpy balance
> Internal reflux computation
> Reboiler heat balance
> Overall material balance

Batch Distillation

The goal of a batch distillation is to produce a product of specified composition at minimum cost. This means that operating time must be reduced to some minimum while product recovery is maintained.

If product removal is too fast, separation and the quantity of product recovered are reduced. Conversely, if the product is withdrawn so as to maintain separation, its withdrawal rate is reduced and operating time is increased. However, the set point to a composition controller can be programmed so that the average composition of the product is within specifications and thereby withdrawal rate is maximized.[11]

Figure 8.8v shows a system for doing this when a constant vapor rate is maintained from the reactor. The equation is

$$y = mD + y_i \qquad 8.8(61)$$

where y is the fraction of key component in the product, m is the rate of change of y with respect to the distillate (D), and y_i is the initial concentration of the product. The only adjustment required is in setting m. The higher its value, the faster y will change and the smaller will be the quantity of material recovered.

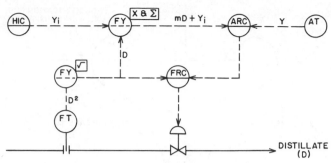

Fig. 8.8v   Control system for batch distillation

Evaluating Alternative Control Strategies

As discussed, many choices confront the design engineer when selecting variables for column control. The first decision involves configuration of the top or bottom control loops, which directly determine product compositions. Once these strategies are tentatively determined, strategies for the remaining variables (e.g., pressure or levels) become much easier to select.

Choices for controlling product compositions include controlling top or bottom composition only (generally suitable for constant separation conditions, where specifications for one product are loose or where effective feedforward/feedback systems can be designed to compensate for load changes) and control of both product compositions (minimizes energy use and provides tight specification top and bottom products for columns in which the problems of interaction are small). These choices can be broken down further into considerations such as manipulation of distillate-boil-up (generally suitable for high reflux columns) or manipulation of reflux-boil-up (generally suitable for low reflux columns), and so forth. Further considerations include the use of decoupling control schemes (can present practical problems, such as insensitive control, operating problems, and high sensitivity to errors) and the use of temperature measurements to infer composition or analyzers to measure composition directly (generally an economic decision based on how well a temperature-sensitive control point can be determined and the costs of analyzer hardware and increased maintenance).

These choices are based on operating objectives of the column, expected disturbance variables, and the degree of control loop interaction.

Column Operating Objectives

Operating objectives primarily involve the composition specifications for the top and bottom product streams. But other information that should be evaluated in determining control system design includes the product composition most important to maintain during disturbances, acceptable variation in product specifications,

and relative values of the product streams and energy used in the separation.[12]

Disturbance Variables

In addition to the disturbance variables already discussed, other common disturbances are steam header pressure, feed enthalpy, environmental conditions (e.g., rain, barometric pressure, and ambient temperature), and coolant temperature.

To handle these disturbances, column controls can be designed so column operation is insensitive to the disturbances, or secondary controls can be designed to eliminate the disturbances (some of these will be discussed later). It is important also to evaluate the expected magnitude and duration of expected disturbances so proper control system scaling and tuning can be achieved.

Control Loop Interaction

Actual selection of which product composition to control (or both, if control of both is possible) as well as which variables will give good control can be aided by calculation of a relative gain array. Gordon's technique was presented previously under relative gains in the section on modeling composition control of two products, and several alternative control strategies were evaluated.[6] Relative gain analysis should be considered the first step in evaluating alternative composition control strategies.

Once interaction of the various variable pairings has been established and the column's operating objectives and disturbance variables are considered, the primary composition control loops of the column can be selected.

Examples of alternative methods for controlling the remaining variables will follow after a discussion of using temperature measurements instead of analyzers to control compositions inferentially.

Inferring Composition from Temperature Measurements

If the higher cost of analyzer hardware and required maintenance precludes their use, or if back-up is desired in case of analyzer failure or maintenance, temperature measurement often can be used to infer composition.

Since distillation separates materials according to their difference in vapor pressures, and since vapor pressure is a temperature-controlled function, temperature measurement often can be used to indicate composition. This presumes that the column pressure remains constant, or that the temperature measurement is compensated for pressure changes. Then any change in composition within a column will be detected as a temperature change.

The best point to locate the temperature sensor cannot be established from generalizations. The important consideration is to measure temperature on a tray that most reflects changes in composition.

When composition of the bottom product is the important consideration, it is desirable to maintain a constant temperature in the lower section. This can be done by letting the temperature measurement set the control point of the reboiler steam supply (Figure 8.8w).

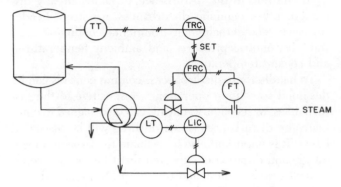

Fig. 8.8w   Temperature cascaded heat addition to the reboiler

When composition of the distillation product is the important consideration it is desirable to maintain a constant temperature in the upper section as in Figure 8.8x. Notice that the point of column pressure control is near the temperature control point. This arrangement helps to fix the relation between temperature and composition at this particular point.

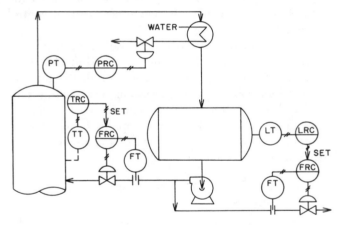

Fig. 8.8x   Temperature cascaded reflux flow for improved overhead composition control

If column temperature profiles for small positive and negative changes in manipulated variables (Figure 8.8y) can be generated, Thurston[13] proposes two criteria helpful in selecting sensor locations from these profiles: (1) A temperature-manipulated variable sensitivity in the range of 0.1 to 0.5°C/% is desired, and (2) equal temperature changes on increasing and decreasing the manipulated variable are preferred.

Distillation temperature is an indication of composition only when column pressure remains constant or if the temperature measurement is pressure-compensated. When separation by distillation is sought between two compounds having vapor pressures close together, tem-

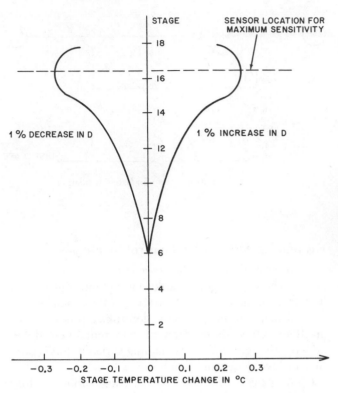

Fig. 8.8y   Changes in column temperature profile for varying distillate flow

perature measurement as an indication of composition is not satisfactory. Fixing two temperatures in a column is equivalent to fixing one temperature and the pressure. Thus, by controlling two temperatures, or a temperature difference, one can nullify the effect of pressure variations. Actually, the assumption used here is that the vapor pressure curves for two components have constant slopes.

Controlling two temperatures is not equivalent to controlling a temperature difference. A plot of temperature difference versus bottom product composition exhibits a maximum. Thus, for some temperature differences below the maximum it is possible to get two different product compositions.

Separation of normal butane and isobutane (in the absence of other components, such as pentanes and heavier substances) can be accomplished very well by using temperature difference control. A schematic for one commercial installation is shown in Figure 8.8z.

Pressure Control

Most distillation columns are operated with constant pressure control. However, several advantages can be achieved through floating-pressure operation in most columns. Resistance to floating-pressure control is largely traditional and is based on using temperature measurements, which are sensitive to pressure changes, to indicate and control compositions. Analyzers are increasingly replacing temperature measurements, thereby

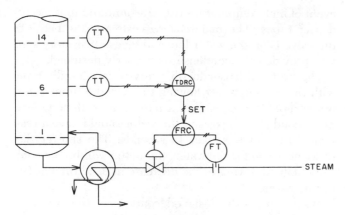

Fig. 8.8z   Heat input controlled by temperature difference

removing constant pressure requirements. However, even when analyzers are not used, temperature measurements can be compensated for pressure variations.

The primary advantage of floating-pressure control is the ability to operate with minimum column pressure within the constraints of the system. Lower pressure reduces the volatility of distillation components, thereby reducing the heat input required to effect a given separation. Other advantages include increased reboiler capacity and reduced reboiler fouling.

After traditional methods of constant pressure control are reviewed, floating-pressure operation control strategies will be discussed. Constant pressure control systems will be described for the following conditions: (1) liquid distillate withdrawn when uncondensables are present, (2) vapor distillate withdrawn when uncondensables are present, and (3) liquid distillate withdrawn, negligible uncondensables.

Liquid Distillate and Inerts

The problem of pressure control is complicated by the presence of large percentages of inert gases. The uncondensables must be removed or they will accumulate and blanket off the condensing surface, thereby causing loss of column pressure control.

The simplest method of handling this problem is to bleed off a fixed amount of gases and vapors to a lower pressure unit, such as to an absorption tower, if such is present in the system. If an absorber is not present, it is possible to install a vent condenser to recover the condensable vapors from this purge stream.

It is recommended that the fixed continuous purge be used wherever economically possible; however, when this is not permitted, it is possible to modulate the purge stream. This might be desirable when the amount of inerts is subject to wide variations over a time.

As the uncondensables build up in the condenser, the pressure controller will tend to open the control valve to maintain the proper rate of condensation. This is done by a change of air-loading pressure on the diaphragm

control valve. The air-loading pressure could also be used to operate a purge control valve, as the pressure passes a certain operating point. This could be done by means of a calibrated valve positioner or a second pressure controller (Figure 8.8aa).

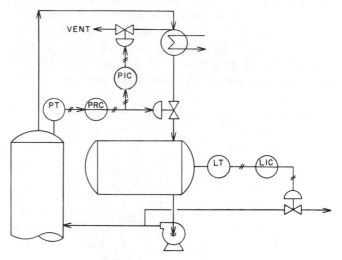

Fig. 8.8aa   Column pressure control with inerts present

Vapor Distillate and Inerts

In this case the overhead product is removed from the system as a vapor and, consequently, the pressure controller can be used to modulate this flow as shown in Figure 8.8bb. The system pressure will quickly respond to changes in this flow.

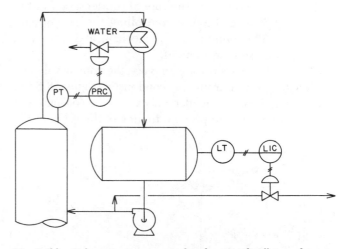

Fig. 8.8bb   Column pressure control with vapor distillate and inerts present

A level controller is installed on the overhead receiver to regulate the cooling water to the condenser. It will condense only enough condensate to provide the column with reflux.

This control system depends upon having a properly designed condenser in order to operate satisfactorily.

The condenser requires a short residence time for the water to minimize the level control time lag.

If the condenser is improperly designed for cooling water control, it is recommended that the cooling water flow be maintained at a constant rate and the level controller regulate a stream of condensate through a small vaporizer and mix it with the vapor from the pressure-control valve. If the cooling water has bad fouling tendencies, it would be preferable to use a control system similar to that in Figure 8.8cc, using the level controller to regulate a vapor bypass around the condenser.

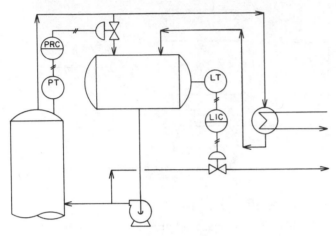

Fig. 8.8cc   Column pressure control using lowered condenser

Liquid Distillate and Negligible Inerts

This situation is the most common and is usually controlled by adjustment of the rate of condensation in the condenser. The method of controlling the rate of condensation will depend upon the mechanical construction of the condensing equipment.

One method of control is to place the control valve on the cooling water from the condenser (Figure 8.8dd). This system is recommended only when the cooling water contains chemicals to prevent fouling of the tubes in the

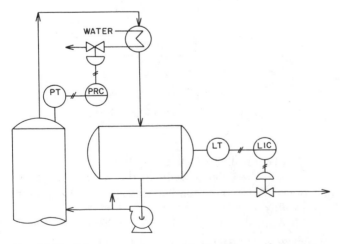

Fig. 8.8dd   Column pressure control by throttling condenser water

event of high temperature rises encountered in the condenser tubes. The maintenance costs are low because the valve is on the water line and gives satisfactory service provided the condenser is properly designed.

The best condenser for this service is a bundle type with the cooling water flowing through the tubes. This water should be flowing at a rate of more than $4\frac{1}{2}$ feet per second (1.35 m/s), and the water should have a residence time of less than 45 seconds. The shorter the residence time for the water, the better will be the control obtained, owing to the decrease in dead time or lag in the system.

With a properly designed condenser, the pressure controller need have only proportional control, because a narrow throttling range is sufficient. However, as the residence time of the water increases, it will increase the time lag of the system, and consequently the controller will require a wider throttling range and will need automatic reset to compensate for the load changes. The control obtained by using a wide proportional band is not satisfactory for precision-distillation columns because of the length of time required for the system to recover from an upset.

It would be impossible to use this control system, for instance, on a condenser box with submerged tube sections, because there would be a large time lag in the system due to the large volume of water in the box. It would take quite a while for a change in water-flow rate to change the temperature of the water in the box and finally affect the condensing rate.

In the presence of such unfavorable time lags, it becomes necessary to use a different type of control system, one that permits the water rate to remain constant and controls the condensing rate by regulating the amount of surface exposed to the vapors. This is done by placing a control valve in the condensate line and modulating the flow of condensate from the condenser. When the pressure is dropping, the valve cuts back on the condensate flow, causing it to flood more tube surface and, consequently, reducing the surface exposed to the vapors. The condensing rate is reduced, and the pressure tends to rise. It is suggested that a vent valve be installed to purge the uncondensables from the top of the condenser, if it is thought that there is a possibility of their building up and blanketing the condensing surface.

Condenser Below Receiver

A third possibility for this type of service is used when the condenser is located below the receiver. This is frequently done to make the condenser available for servicing and to save on steel work. It is the usual practice to elevate the bottom of the accumulator 10 to 15 feet (3 to 4.5 m) above the suction of the pump in order to provide a positive suction head on the pump.

In this type of installation the control valve is placed in a bypass from the vapor line to the accumulator (see

Figure 8.8cc). When this valve is open, it equalizes the pressure between the vapor line and the receiver. This causes the condensing surface to become flooded with condensate because of the 10 to 15 feet of head that exist in the condensate line from the condenser to the receiver. The flooding of the condensing surface causes the pressure to build up because of the decrease in the condensing rate. Under normal operating conditions the subcooling that the condensate receives in the condenser is sufficient to reduce the vapor pressure in the receiver. The difference in pressure permits the condensate to flow up the 10 to 15 feet of pipe that exist between the condenser and the accumulator.

A modification of this latter system controls the pressure in the accumulator by throttling the condenser bypass flow (Figure 8.8ee). The column pressure is maintained by throttling the flow of vapor through the condenser. Controlling the rate of flow through the condener gives faster pressure regulation for the column.

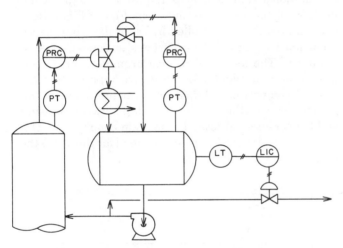

Fig. 8.8ee   High-speed column pressure control

Vacuum Systems

For some liquid mixtures, the temperature required to vaporize the feed would need to be so high that decomposition would result. To avoid this, it is necessary to operate the column at pressures below atmospheric.

The common means for creating a vacuum in distillation is to use steam jet ejectors. These can be used singly or in stages to create a wide range of vacuum conditions. Their wide acceptance is based upon their having no moving parts and requiring very little maintenance.

Most ejectors are designed for a fixed capacity and work best at one steam condition. Increasing the steam pressure above the design point will not usually increase the capacity of the ejector; as a matter of fact, it will sometimes decrease the capacity because of the choking effect of the excess steam in the diffuser throat.

Steam pressure below a critical value for a jet will cause

the ejector operation to be unstable. Therefore, it is recommended that a pressure controller be installed on the steam to keep it at the optimum pressure required by the ejector.

The recommended control system for vacuum distillation is shown in Figure 8.8ff. Air or gas is bled into the vacuum line just ahead of the ejector. This makes the maximum capacity of the ejector available to handle any surges or upsets.

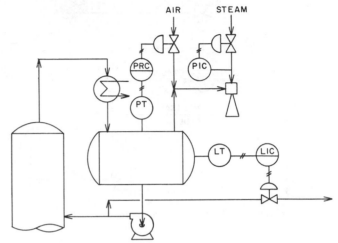

Fig. 8.8ff   Vacuum column pressure control

Because ejectors are fixed capacity, the variable load is met by air bleed into the system. At low loads this represents a substantial waste of steam. Therefore, in processes in which load variations are expected, one can lower the operating costs by installing a larger ejector and a smaller ejector. This makes it possible to switch to the small unit automatically when the load drops off, thereby reducing steam demand.

A control valve regulates the amount of bleed air used to maintain the pressure on the reflux accumulator. Using the pressure of the accumulator for control involves less time lag than if the column pressure were used as the control variable.

Floating-Pressure Control

As mentioned, floating-pressure operation can often offer reduced energy consumption by providing for minimum pressure operation within the constraints of the system. It is possible to operate a total condensing distillation column with no pressure control.[7] Although this provides for optimum operation at steady state, major problems could occur (e.g., flooding the column) during transient upsets. To prevent this, pressure control should be provided.

Minimum pressure operation can be achieved by manual or automatic adjustment of the set point of the pressure controller to keep the condenser fully loaded in the long term. However, to prevent upsets caused by rapid set point changes, Shinskey[7] proposes the valve position control (VPC) scheme shown in Figure 8.8gg. The VPC adjusts the set point of the pressure controller and maximizes cooling by holding the condenser control valve in the fully open or fully closed position, depending on whether the valve bypasses, throttles, or floods the condenser.

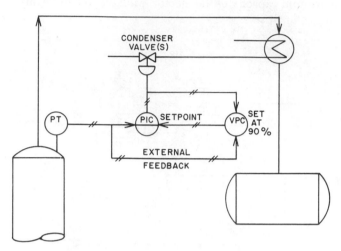

Fig. 8.8gg   VPC provides floating-pressure control

The pressure controller should incorporate proportional plus integral action to provide rapid response to upsets, whereas the VPC should be an integral-only controller, so a rapid change in valve position will not produce a proportional change in the pressure set point. The integral time setting of one VPC should be approximately tenfold that of the overhead composition controller. In addition, it is common practice to limit the range within which the VPC can adjust the PIC set point and to provide the external feedback line shown in Figure 8.8gg to eliminate reset wind-up when the VPC output reaches one of these limits.

Partial condensers require a different approach to floating-pressure control because they are sensitive to cooling and accumulation of noncondensables. Typically, pressure in these systems is controlled by addition or venting of inert gases. Though simple, this practice requires a source of inert gases, does not allow for steady state optimum pressure operation, wastes overhead products that are vented with the noncondensables, and can create problems in downstream units through the addition of noncondensables.

Floating-pressure control for partial condensers with no liquid product and for those with liquid product will both be discussed.

Figure 8.8hh shows a partial condensing system with no liquid product. Here, both level and pressure controllers are used to provide floating-pressure operation.[7] The level controller acts as the VPC in the total condensing system to provide complete flooding in the long term, and the pressure controller handles short-term upsets.

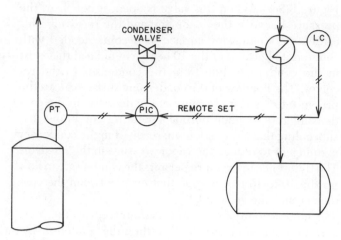

Fig. 8.8hh   Floating-pressure control for a partial condenser with no liquid product

When both liquid and vapor products are withdrawn, an additional control loop is required to control composition of the vapor, as shown in Figure 8.8ii. Here column pressure is controlled by vapor flow, but the set point must be adjusted for changes in condenser temperature.[7] The temperature measurement is characterized to an equivalent vapor pressure representing the desired composition. A bias adjustment should be incorporated to readjust the relationship between pressure and temperature for desired changes in composition. This method completely eliminates the need for throttling the cooling water to the condenser.

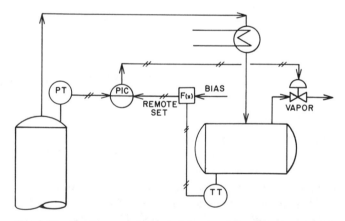

Fig. 8.8ii   Floating-pressure control for a partial condenser with both liquid and vapor products

Feed Control

One of the best means of obtaining stability of operation in almost any continuous-flow unit, including distillation, is by holding the flow rates constant. Therefore, when possible, a controller is used on the feed to maintain a constant rate of flow.

Using a narrow proportional band setting, the controller will cause the control valve to take large corrective action for small changes in flow rate. Therefore, the feed rate will be maintained constant by this high-gain controller.

In some instances, the feed pump of a distillation unit is a steam-driven pump instead of an electrically driven one. In this case the feed flow control valve can be placed on the steam to the driver.

Feed composition has a great influence upon the operation of a distillation unit. Unfortunately, though, feed composition is seldom subject to adjustment. For this reason, it is necessary to make changes elsewhere in the operation of the column in order to compensate for variations in feed composition. The corrective steps that can be taken will be discussed later. For the time being, it is assumed that a constant feed composition exists.

The thermal condition of the feed determines how much additional heat must be added to the column by the reboiler. For efficient separation, it usually is desirable to have the feed at its bubble point when it enters the column. Unless the feed comes directly from some preceding distillation step, an outside source of heat is required.

Steam may be used to heat the feed, and the sensing device may be a thermocouple contained in a thermowell placed inside the feed line. Any change in the temperature of the feed leaving the exchanger will cause a corrective adjustment to the supply of steam into the exchanger. To maintain feed temperature, generally a three-mode controller is used. On startup, the initially large correction provided by rate action also helps to get the unit lined out faster.

As discussed in Section 1.7, the use of a cascade loop (Figure 8.8jj) can provide superior temperature control.

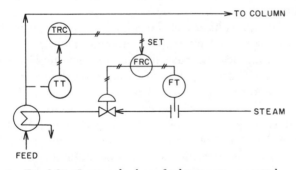

FIG. 8.8jj   Improved column feed temperature control

Constant temperature feed does not necessarily mean constant feed quality. If feed composition varies, its bubble point also varies. It is common practice to set the temperature control at a point that is equivalent to the bubble point of the heaviest feed. As the feed becomes lighter, some of it will evaporate, but this variation can be handled by subsequent controls.

When a furnace is used to add heat to a distillation unit, the temperature controller is used to regulate the amount of fuel to the furnace. For greater stability, a flow controller can be used on the fuel line, and its control setting can be determined by the temperature

controller in a manner analogous to that of the cascade loop in Figure 8.8jj.

Variable Column Feed

Having constant feed conditions simplifies the amount of control required to achieve stable operation. However, suppose the distillate product is to be fed into a second column. Then any inadvertent changes that occur in the first column would be reflected in the quantity and composition of the feed to the second. If the flow of products from the previous column is controlled by liquid level controllers, these controllers can have a wide proportional band so that changes in the reboiler or accumulator level can swing over a wide range without drastically upsetting the flow of products. Nevertheless, a second column will receive a varying flow of feed if it is linked to the first column.

One way to iron out temporary variations caused by liquid level changes is to add flow controllers to the product lines.

With variable feed rates and variable feed compositions, cascade controls are justified. If the feed rate and composition are relatively constant, hand reset of the major control loop is sometimes adequate. In other cases the flow set point is continuously adjusted by the level controller in a cascade arrangement (Figure 8.8kk).

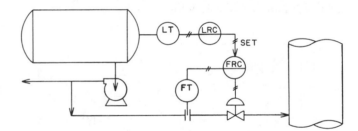

Fig. 8.8kk   Cascade control of feed to second column

If feed rate disturbances must be accepted by the column, a feedforward control system as shown in Figure 8.8ll can be used to minimize these disturbances.[12] The

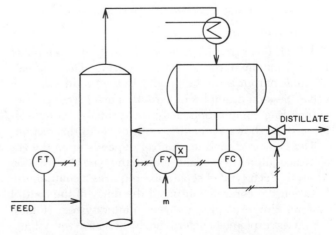

Fig. 8.8ll   Feedforward control minimizes feed rate disturbances

ratio, m, is selected by performance of a simple material balance around the column. Changing a product flow proportional to feed minimizes internal column transients and, thus, the quantity of off-spec material during recovery.

The value of m, however, is accurate only for one feed composition and will have to be readjusted either manually or automatically for different feed compositions.

Analyzers

Analytical or specific composition control is a way to sidestep the problems of temperature control. Although additional investment is needed for the analytical equipment, a savings from improved operation usually results.

Several types of instruments are available for composition analysis. Of these, the chromatograph is the most versatile. (For details, refer to Chapter 10 of the volume on Process Measurement.) Once, the time required for a chromatographic analysis (approximately 15 minutes) was a great barrier to its use for automatic control. Since then the equipment has been modified so that analyses can now be made in less than five minutes. With careful handling, this sampling rate will permit closed-loop distillate control.

Light ends distillation have been satisfactorily controlled by the use of chromatography. The diagram for a superfractionator designed to separate isobutane and normal butane is shown in Figure 8.8mm.

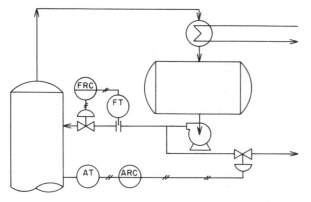

Fig. 8.8mm   Distillate withdrawal controlled by chromatograph

The chromatograph continuously analyzes a sample of vapor from one of the intermediate trays. The chromatographic output is used to modulate the product drawoff valve. Several examples of various control strategies incorporating composition analyzers (primarily chromatographs) are shown in the figures throughout this section.

The choice of analyzer control depends upon the analytical equipment available and on the type of separation desired. Each type of separation requires a compromise between the controllability and the delay of the control system. For example, a three-column system (Figure 8.8nn) was explored to determine the best analyzer system.

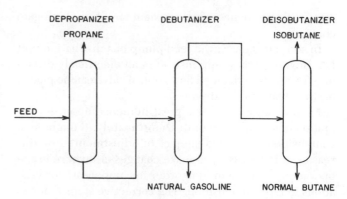

Fig. 8.8nn   Three columns linked together

In the depropanizer (where isobutane was to be measured in the presence of ethane, propane, and normal butane) and in the deisobutanizer (where isobutane was to be measured in the presence of normal butane and isopentane), an infrared analysis was to be preferred.

However, in the debutanizer the aim was to measure the combined isopentane plus normal pentane concentrations in the presence of isobutane and normal butane effectively to control the butane-pentane separation. Here, investigation revealed that measurement of the refractive index of the sample in the gaseous state would give in effect a molecular weight analysis and would not differentiate among the isomers. Uniquely, the nonspecificity of refractive index was found to be desirable.

Boiling point analyzers are reliable enough to be used for on-line control. Normally, cut points between overhead products and sidecuts are maintained by temperature controllers. These controllers generally influence reflux rate to achieve the desired results. Laboratory distillation results are used to adjust the set points to the temperature controllers. This method of control, however, is cyclical because of the time lags involved in temperature control. To avoid exceeding the target cut points and to meet required product specifications, the cut point is set below specification. This results in downgrading of the more valuable product to the product of lesser value part of the time. This downgrading can be minimized through the use of on-line boiling point analyzers. Enough additional profit can be achieved for many columns by minimizing this downgrading to justify a boiling point analyzer, depending upon the value of the products, how much downgrading is occurring, and the ability of the personnel at a particular location to provide the additional maintenance required for analyzer systems.

Viscosity is another property that can be measured continuously to give faster control corrections. In vacuum distillation, the viscosimeter monitors each of the streams for which viscosity is a specification. Any deviation from the desired viscosity is corrected by a change in the set point of the control loop involved. Deviation from desired viscosity and the subsequent downgrading of the product stream occur because of frequent, minor variation in tower operating conditions and feed compositions. In addition to normal operation, using a viscosity analyzer minimizes downgrading during major upsets and large feed composition changes. Low viscosity vacuum bottoms can be detected quickly and diverted to recoverable feed for profitable reprocessing.

Once again, profitability determination requires a thorough analysis of column operation and an assessment of the engineering, operating, and maintenance capabilities at the location where this type of control is to be implemented.

Many other analytical instruments are being moved out of the laboratory and into the processing area. Mobile units containing several different kinds of analyzers can be used to learn the best place to locate on-stream analyzers. In cases in which permanent analyzers cannot be justified, the mobile unit is connected to the process long enough to find the best operating conditions. Then the mobile unit can be moved elsewhere.

Sampling

Proper sampling of material in a column is necessary if in-plant analyzers are to control effectively. The problem of proper sampling has been the subject of several symposia and certainly is worthy of special attention. A poor sampling system often is responsible for the unsatisfactory performance of plant analyzers. For added detail, see Chapter 10 of the Process Measurement volume of this handbook.

The factors favored by sampling at, or very near, the column terminals are (1) freedom from ambiguity in the correlation of sample composition with terminal composition, and (2) improved control loop behavior as a result of reduction of transport lag (dead time) and of the time constants (lags) describing the sampling point's compositional behavior. This assumes that the manipulation for control is applied at the same terminal (steam or reflux) as the manipulated variable.

The factors favored by sampling nearer the feed entry are (1) improved terminal composition behavior as a result of earlier recognition of composition transients as they proceed from the feed entry toward the column terminals, and (2) less stringent analytical requirements as a result of (a) analyzing the control component at a higher concentration and over a wider range and (b) simplifying the multicomponent mixture, since non-key components tend to exhibit constant composition zones in the column. These latter points often make "impossible" analyses possible.

Computer Control

On-line computer control can greatly enhance the profitability of the distillation process. Although much of the gain can be attributed to the ability of the computer to solve complex equations required for optimization quickly

(these equations will be discussed in detail under optimization), operational benefits, data collection improvements, and increased flexibility advantages can often justify computer control even if rigorous on-line optimization is not implemented. It must be emphasized, however, that operating environments vary greatly among installations, and many practical problems can impede successful computer control of the distillation process.

It is useful to review briefly some of the elements required for a successful computer control project before the application of computers specifically to the distillation process is discussed.

Careful analysis of control requirements, potential benefits, operator needs, training requirements, maintenance requirements, computer functions, and project costs should precede any implementation of a computer system.[14] This analysis should involve all personnel who will be involved in the design, specification, operation, and maintenance of the system to ensure their acceptance of the project. In particular, the needs of plant operators should be assessed. Since these individuals will spend the most time with the system after installation, provisions must be made to ensure easy operation and to instill confidence; otherwise, they will bypass the computer system more often than necessary, and much of the benefits of computer control will be lost.

This is also the time to decide how the project will be implemented and by whom. This initial evaluation process is usually the most important phase of a computer project. Many attempts at computerization have failed because the hardware was purchased before the specifics (what needs to be done, why it needs to be done, who will do it, and how much it will cost) were well thought out.

If projected benefits exceed expected costs and potential practical difficulties, detailed engineering design and hardware procurement can begin. In addition to proper computer system design, accurate measurements must be ensured. Provisions for proper training also must be made.

After startup, proper maintenance of the computer system and measurement devices must be provided to ensure continued success of the project.

The main control functions of a computer as applied to distillation include engineering calculations, operating assistance, quality controls, and heat balance controls.[15]

Some of the primary engineering calculations are made from material and heat balances around column sections, as discussed previously under modeling. These include tray loadings, internal vapor flows, internal liquid flows, and heat duties. These calculations are helpful as operating guidelines and as inputs for on-line control. However, they are based on steady state conditions, and input signals must be averaged to make the calculations. Response, although normally fast enough for on-line control, may not be adequate if frequent, short-term disturbances must be handled. However, information gained from these types of calculations can often justify the computer system by providing better operating guidelines, even if it is not used for on-line control.

Controls that make the operator's job easier include balance of heater pass outlet temperatures on multiple-pass heaters and feed rate control by adjustment of pass flows as shown for a crude oil heater in Figure 8.8oo,[15] feedforward adjustments of products and pumparounds from feed rate changes, column bottom level control by adjustments to vacuum heater pass flows, and control of overhead receiver level.

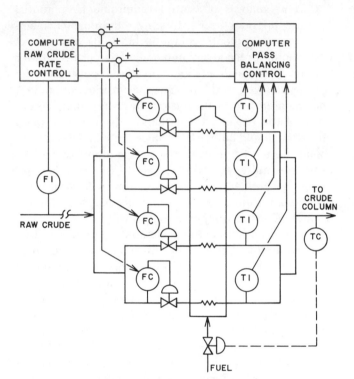

Fig. 8.8oo Computerized heater balancing and charge rate. (From Van Horn, L.D., "Crude Unit Computer Control . . . How Good Is It?" *Hydrocarbon Processing*, April 1980.)

Since many of these situations require changing the set points of several controllers simultaneously, the operator's tasks are reduced.

Product quality controls are enhanced through the use of the computer to adjust local column temperature and side draw flow rates to control distillation properties related to the product specifications. An example is true boiling point (TBP) cut points. TBP cut points approximate the composition of a hydrocarbon and are numerically similar to the American Society for Testing and Materials' (ASTM) 95 percent points. The ASTM laboratory distillate evaluation method is the standard used in the petroleum refining industry for determining the value (composition) of the distillation products.

The computer calculates the TBP cut points from local temperature, pressure, steam flow, and reflux data. Lo-

cal reflux is derived from internal liquid and vapor flows, as discussed previously, and the remaining variables are measured.

Boiling point analyzers can be used to provide the measurement signals. The calculated cut points can be used by themselves if there is no analyzer, or as a fast inner loop with analyzer trim.

Adjustment of pumparound refluxes, as shown in Figure 8.8pp, is an example of a heat balance control. The goal is to maximize heat exchange to feed, subject to certain limits[16] (limits and constraints are detailed under optimization). The problem often simplifies to recovering heat at the highest possible temperature, which means recovering it as low as possible in the column.

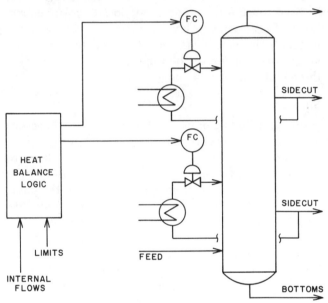

Fig. 8.8pp Computer adjustment of pumparound refluxes

A careful analysis of limits and operating constraints is essential to successful computer control of distillation columns. If the system is not designed to provide limit checks and overrides to handle operating limits, frequent operator intervention will be required during upsets. This lack of confidence in the computer system will cause the operators to remove the column from computer control more often than necessary, thereby reducing the effectiveness of the system.

Many of the primary applications of computers to distillation control discussed here will be detailed in the following section on optimization.

Optimizing Equations

Operating guidelines and recommended control strategies discussed so far have been aimed at providing for optimum distillation operation (e.g., relative gain analysis, dual composition control strategies, and floating-pressure operation).

The following derivations will provide insight into how optimization equations are derived and modeled and how constraints are handled. Again, however, the reader is advised that every distillation column is unique; the examples given are for illustrative purposes only and should not be considered suboptimum or optimum solutions for every column.

Optimization of a single distillation column normally implies a maximum profit operation, but, to achieve maximum profit, the price of the column's products must be known. It is therefore impossible to carry control to this extent for every column in a system, because the price of products for many columns is unknown. Such prices are unknown because the products are feedstreams to other units whose operations would also need to be taken into account to establish the column's product prices.

When product prices are unknown, it is possible to carry optimization only to the stage at which specified products can be produced for the least operating cost. This can be called an optimum with respect to the column involved, but only a suboptimum with respect to the system of which the column is a part.

When column product prices are known, complete economic optimization can be achieved. However, a number of different situations may exist. If there is a limited market for the products, then the control problem will be to establish the separation that results in the maximum profit rate. Such an optimum separation will be a function of all independent inputs to the column involved.

When an unlimited market exists for the products, and sufficient feedstock is available, the optimization problem becomes more difficult. Not only must the optimum separation be established, but also the value of feed must be determined. Optimization for this case will result in operating the column at maximum loading. One of three possible constraints will be involved. Through-put will be limited by the overhead vapor condenser, the reboiler, or the column itself. In some cases the constraint will change, depending upon product prices and other independent variables of the system.

Suboptimization

Application of automatic control systems to single columns should follow three logical steps in the overall hierarchy toward the goal of optimization.

1. Application of basic controls to regulate the most basic functions, such as pressures, temperatures, levels, and flows.
2. Application of controls to regulate the main source of heat inputs, including regulation of internal reflux flow rate, feed enthalpy, and reboiler heat flow rate.
3. Application of controls to regulate the specified separation.

A single column automated through these three stages will have a suboptimized operation. This suboptimum is defined as an operation that will produce close to the specified separation, whether or not that separation is right or wrong with regard to the total system of which the column is a part. If product purities are better than specified, the operation does not approach a suboptimum.

When a single column is automated through the suboptimum operation stage, it will still exhibit up to 5 degrees of freedom. As a basis for proceeding into optimization, Figure 8.8qq is presented as one example of a column automated through the suboptimization stage. Figure 8.8rr is another example of suboptimization. Both figures achieve the goal of a suboptimum; the difference mainly involves the basic controls.

As shown in Figure 8.8qq, the system used to regulate the separation is a *predictive control system*, similar to that described earlier. The function of the predictive control system is to manipulate the energy balance (reflux flow rate) and the material balance (bottom product flow rate) to give the separation specified. The equations derived for these manipulations are called the *operating control equations*.

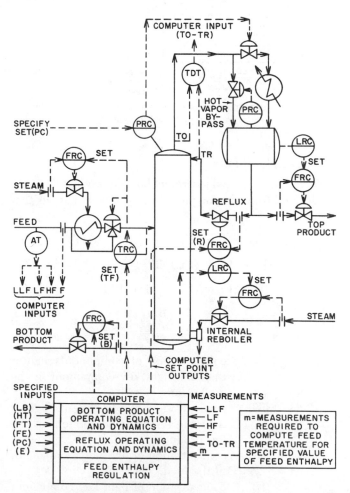

Fig. 8.8qq   Distillation column automated through the suboptimization stage (to produce close to the specified separation)

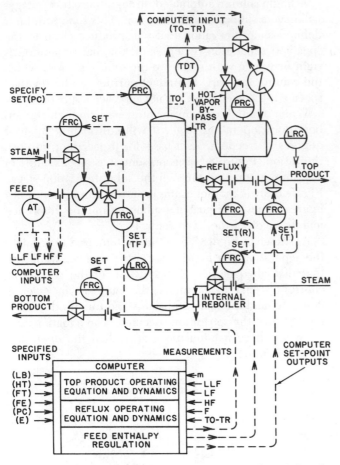

Fig. 8.8rr   Alternative method of column automation through the sub-optimization control stage

Control Equations

The equation used to predict bottom product flow rate is derived from the four material balance equations that follow:

$$1 = (LLT) + (LT) + (HT) \qquad 8.8(62)$$

$$(LLF)(F) = (LLT)(T) \qquad 8.8(63)$$

$$(LF)(F) = (LT)(T) + (\overline{LB})(B) \qquad 8.8(64)$$

$$T = F - B \qquad 8.8(65)$$

where

LLF = concentration of components lighter than the light key component in the feedstream,

LLT = concentration of components lighter than the light key component in the top product,

LF = concentration of light key component in the feed,

LT = concentration of light key component in top product,

F = feed flow rate,

T = top product flow rate,

B = bottom product flow rate,

$(\overline{HT})$ = specified concentration of heavy key component in the top product, and

$(\overline{LB})$ = specified value of light key component in the bottom product.

LLT is eliminated from equation 8.8(62) by equation 8.8(63) and LT is eliminated from equation 8.8(62) by equation 8.8(64). T is then eliminated by equation 8.8(65). The solution for the ratio of bottom product flow rate to feed flow rate will be

$$(B/F) = [1 - (\overline{HT}) - (LLF) -$$
$$- (LF)]/[1 - (HT) - (\overline{LB})] \qquad 8.8(66)$$

Derivation of the internal reflux operating equation is more difficult. Typically this equation is developed in two parts:

$$(RI/F) = (RI/F)t + (RI/F)e \qquad 8.8(67)$$

where

(RI/F) = internal reflux to feed flow rate ratio required to give a specified separation,

$(RI/F)t$ = theoretical part of reflux operating equation, and

$(RI/F)e$ = experimental part of reflux operating equation.

The experimental part of this equation is necessary because the effect of loading on overall separating efficiency (E) is normally unpredictable. Both parts of the reflux operating equation are functions of all independent inputs to the system. However, simplifications are normally considered for the experimental part, as follows:

$$(RI/F)t =$$
$$= f_1[(LLF), (LF), (HF), (\overline{E}), (\overline{FT}), (\overline{FE}) \ldots$$
$$\ldots (\overline{PC}), (\overline{LB}), (\overline{HT})] \qquad 8.8(68)$$

$$(RI/F)e = f_2(RI) \qquad 8.8(69)$$

The theoretical part as given in functional form is normally developed by tray-to-tray runs on an off-line digital computer. A statistically designed set of runs is made, and the information thus obtained is curve-fitted to an assumed equation form. Once the steady-state theoretical equation is developed and placed in service, the experimental part is determined by on-line tests. These tests involve operating the column at different loads to determine the correction required to (RI/F)t for the separation to be equal to that specified. Average overall efficiency $(\overline{E})$ is set to give (RI/F)t required to equal the actual (RI/F) existing. The loading tests are carried out under this condition.

Reflux Equation

One can obtain internal reflux flow rate of a distillation column by making a heat balance around the top tray. The following equation is obtained:

$$(RI) = R[1 + K(TO - TR - dtr)] \quad 8.8(70)$$

Substitute this equation into 8.8(67) to eliminate RI:

$$(R/F) = \frac{[(RI/F)t + (RI/F)e]}{[1 + K(TO - TR - dtr)]} \quad 8.8(71)$$

where

R = external reflux flow rate,
K = ratio of specific heat to heat of vaporization of the reflux,
TO = overhead vapor temperature,
TR = external reflux temperature, and
dtr = difference in temperature between dew point of overhead vapor and boiling point of external reflux.

Dynamics

Equations 8.8(67) and 8.8(66) are steady-state equations. Therefore, if they are applied without alteration, undesirable column response will result, especially for sudden feed flow rate changes. Feed composition changes are less severe than are feed flow rate changes and seldom require dynamic compensation.

One useful criterion for the design of dynamic elements to compensate for feed flow rate changes requires the following: When feed flow rate changes, the column's terminal stream flows should respond *fast*, in the correct direction and without overshoot. The simplest form of dynamics to achieve this criterion involves dead time plus a second-order exponential response. The feed flow rate signal is brought through this dynamic element before being used in the operating equations to obtain bottom product flow rate (B) and reflux flow rate (R) set points. The transfer function for the dynamic element is

$$\frac{(FL)}{(FM)} = \frac{e^{-pt}}{(T1)p + 1} \quad 8.8(72)$$

where

(FL) = feed flow rate lagged,
(FM) = feed flow rate measured,
e = e of log to the base e,
t = dead time,
$T1$ = time constant, and
p = differential operator.

Using equation 8.8(72), F is eliminated from the left side of equations 8.8(66) and 8.8(67) to obtain the complete set of operating equations as used in Figure 8.8qq and as shown following.

$$B = (FM) \left[\frac{e^{-pt}}{(T1)p + 1} \right]$$

$$\left[\frac{1 - (\overline{HT}) - (LLF) - (LF)}{1 - (\overline{HT}) - (\overline{LB})} \right] \quad 8.8(73)$$

$$R = (FM) \left[\frac{e^{-pt}}{(T1)p + 1} \right]$$

$$\left[\frac{(RI/F)t + (RI/F)e}{1 + K(TO - TR - dtr)} \right] \quad 8.8(74)$$

where, in functional form,

$$(RI/F)t + (RI/F)e =$$
$$= f_1[(LLF), (LF), (HF), (\overline{E}), (\overline{FE}), (\overline{PC}) \dots$$
$$\dots (\overline{FT}), (HT), (\overline{LB})] + f_2(RI) \quad 8.8(75)$$

where

$(\overline{E})$ = specified constant average efficiency,
$(\overline{FT})$ = specified value of feed tray location,
$(\overline{FE})$ = specified value of feed enthalpy,
$(\overline{PC})$ = specified value of column pressure, and
LLF, LF, HF = measured values of concentration of feed components.

Application of equations 8.8(73), (74), and (75) will result in a suboptimized operation. This is an operation producing a performance close to the specified one.

Inspection of equations 8.8(73) and (75) shows that the system still has five degrees of freedom. Therefore, feed tray location $(\overline{FT})$, feed enthalpy $(\overline{FE})$, column pressure $(\overline{PC})$, concentration of heavy key component in the top product $(\overline{HT})$, and concentration of light key component in the bottom product $(\overline{LB})$ must all be specified. Although tray efficiency is specified, it remains to be a fixed value, as explained earlier.

Optimization

Product Prices Unknown

Optimization implies maximum profit rate. Unless the terminal products of a distillation column have known prices, however, it is impossible to maximize profit rate for that column without taking into account all other aspects of the process of which the column is a part. Thus, optimization for a single distillation column whose terminal product prices are unknown is defined as an operation producing specification products for the least operating cost. The problem of determining the required separation for the column is a problem in optimizing the overall system of which the column is a part.

Optimization of a single column whose product prices are unknown involves determining the values for $(\overline{FT})$,

$(\overline{FE})$, and $(\overline{PC})$ that result in minimum operating costs for whatever separation is specified. Any applicable mathematical approach can be used to establish values for $(\overline{FT})$, $(\overline{FE})$, and $(\overline{PC})$ that will result in minimum operating costs. Assuming that equations 8.8(73), (74), and (75) are available, it is a relatively simple matter to establish optimum values for these three variables that result in minimum operating costs. Since these variables have specific constraint values, one method involves a search technique. It is usually difficult to justify the search technique for on-line computer control. Therefore, a statistical design study can be made off-line on another computer that allows correlation of the variables with each of the three optimizing variables.

Theoretically, three equations in functional form describe the optimum for $(\overline{FT})$, $(\overline{FE})$, and $(\overline{PC})$:

$$(\overline{FT})_o = f_3(LLF),\ (LF),\ (HF),\ (F),\ (\overline{HT}),\ (\overline{LB}) \qquad 8.8(76)$$

$$(\overline{FE})_o = f_4(LLF),\ (LF),\ (HF),\ (F),\ (\overline{HT}),\ (\overline{LB}) \qquad 8.8(77)$$

$$(\overline{PC})_o = f_5(LLF),\ (LF),\ (HF),\ (F),\ (\overline{HT}),\ (\overline{LB}) \qquad 8.8(78)$$

In a practical sense, the optimum for $(\overline{PC})$, column pressure, will be the minimum value within constraints of the system. Normally $(\overline{PC})$ can be lowered until the condenser capacity is reached, or until entrainment of liquid in the vapor on the trays is initiated.

Table 8.8ss illustrates all the control equations involved to optimize the operation of a column when product prices are unknown. Determination of column pressure is handled by the predictive control equation, 8.8(78).

Table 8.8ss
OPTIMIZING CONTROL FOR SINGLE COLUMN WHEN PRODUCT
PRICES ARE UNKNOWN

(Criterion is to Produce Specification Products for the Least Operating Cost)

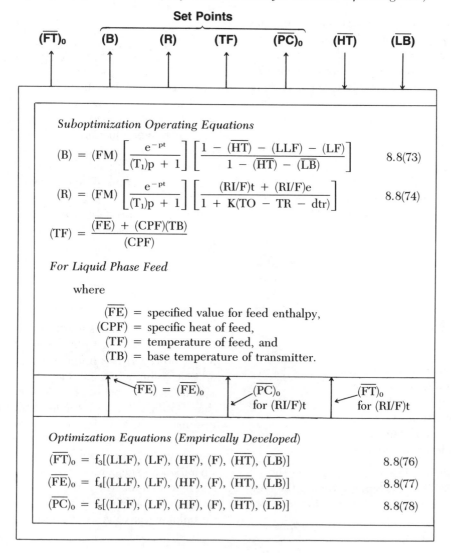

Set Points

| $(\overline{FT})_o$ | (B) | (R) | (TF) | $(\overline{PC})_o$ | $(\overline{HT})$ | $(\overline{LB})$ |

Suboptimization Operating Equations

$$(B) = (FM)\left[\frac{e^{-pt}}{(T_1)p + 1}\right]\left[\frac{1 - (\overline{HT}) - (LLF) - (LF)}{1 - (\overline{HT}) - (\overline{LB})}\right] \qquad 8.8(73)$$

$$(R) = (FM)\left[\frac{e^{-pt}}{(T_1)p + 1}\right]\left[\frac{(RI/F)t + (RI/F)e}{1 + K(TO - TR - dtr)}\right] \qquad 8.8(74)$$

$$(TF) = \frac{(\overline{FE}) + (CPF)(TB)}{(CPF)}$$

For Liquid Phase Feed

where

$(\overline{FE})$ = specified value for feed enthalpy,
(CPF) = specific heat of feed,
(TF) = temperature of feed, and
(TB) = base temperature of transmitter.

$(\overline{FE}) = (\overline{FE})_o$ $(\overline{PC})_o$ for $(RI/F)t$ $(\overline{FT})_o$ for $(RI/F)t$

Optimization Equations (Empirically Developed)

$$(\overline{FT})_o = f_3[(LLF),\ (LF),\ (HF),\ (F),\ (\overline{HT}),\ (\overline{LB})] \qquad 8.8(76)$$

$$(\overline{FE})_o = f_4[(LLF),\ (LF),\ (HF),\ (F),\ (\overline{HT}),\ (\overline{LB})] \qquad 8.8(77)$$

$$(\overline{PC})_o = f_5[(LLF),\ (LF),\ (HF),\ (F),\ (\overline{HT}),\ (\overline{LB})] \qquad 8.8(78)$$

Product Prices Known

When terminal product prices for a single column are known, the design to optimize the operation will start at the level of control just described. In other words, the column is automated to obtain the specified separation for the least operating cost. The problem now is to determine values for the separation that will maximize profit rate. There are, however, a number of different situations that may exist, which will be covered soon. Several general optimizing policies can be stated first. The purpose of these general policies is to reduce the number of variables involved in design of the optimizing control system.

Optimizing Policies

1. The optimum separation to specify for a single distillation column can be determined independently of feed cost.
2. One condition resulting in an optimized operation for a single distillation column is production of that substance with the highest unit price at minimum specified purity.
3. The optimum separation to specify for a single distillation column is not a function of each terminal product price but a function of the price difference between products.
4. When the individual components in the products of a single column have separate assigned prices, the optimum separation is not a function of all component prices but a function of the price difference between the heavy key component in the top and bottom (PHT − PHB) and the price difference between the light key component in the top and bottom products (PLT − PLB).

Proof of these policies can be obtained by evaluation of the partial differential equations that describe the profit rate for a single column with respect to the specified separation, $(\overline{LT})$, $(\overline{HB})$.

Column Operating Constraints

When product prices for a distillation column are known, complete economic optimization almost always requires the operation to be against a constraint. If not against an operating constraint, the optimum will occur when the specified separation $(\overline{LB})$, $(\overline{HT})$ is of such a value that incremental gain in product worth is equal to incremental gain in operating cost. Since the majority of cases will involve operating constraints, it is important to understand the principles involved.

Loading of a distillation column is affected by the specified separation and by the existing feed rate. Loading is increased by specification of a better separation or by an increase in the feed rate at a constant separation. In general, both the feed rate and separation are involved as optimizing variables when an unlimited market exists for the products and when sufficient feedstock is available.

The operating constraints normally involve the respective capacities of (1) the condenser, (2) the reboiler, and (3) the column. As feed rate is increased, or a better and better separation specified, one of these three constraints will be approached.

Condenser Constraint. Capacity of a given condenser at maximum coolant flow rate is a function of the differential temperature between the overhead vapor and the coolant media. One useful approach to operating a column against the condenser constraint requires correlation of maximum vapor flow rate with this temperature difference. Such a correlation can be obtained by column testing. The information obtained by on-line tests is curve-fitted to some general form, such as the one in equation 8.8(79):

$$(V_{oh})_{max} = a_1 + a_2(\Delta T) + a_3(\Delta T)^2 \qquad 8.8(79)$$

where

$(V_{oh})_{max}$ = maximum overhead vapor flow rate that will load the condenser maximally,

ΔT = temperature difference between overhead vapor and coolant to the condenser, and

a_1, a_2, a_3 = coefficients.

Values for feed rate and for the separation can be determined that will result in $(V_{oh})_{max}$ to load the condenser.

Accuracy of the condenser-loading equation can be carried as far as desired. For example, an equation can be developed for temperature of the overhead vapor as a function of all independent variables of the system for use in the ΔT determination. This would result in a completely predictive system for loading the condenser.

If the condenser becomes fouled, new coefficients must be established for equation 8.8(79).

Reboiler Constraint. Capacity of the reboiler at maximum flow rate of the heating media is a function of the temperature difference between the heating media in the reboiler tubes and the liquid being reboiled. Just as with the condenser, tests can be conducted to correlate maximum vapor flow rate out of the reboiler with temperature difference across the reboiler tubes. Also, temperature of the reboiler can be expressed as a function of the column's independent inputs. Column pressure will be one of the major variables that will affect temperature of the reboiler liquid. Also, this will be one of the major variables affecting overhead vapor temperature that are used in the condenser loading equation.

Column Constraint. Capacity of a given column will be a function of liquid and vapor flow rates within the column as well as of the column pressure. In general, capacity is limited by entrainment of liquid by the vapor. At low internal liquid flow rates a higher vapor flow rate can be used. Also, column capacity will be higher at higher pressures. However, if column capacity is limited by the tray downcomers, internal liquid flow rate can be increased by lowering pressures. If capacity is limited by entrainment, then loading can be increased at higher pressures. Therefore, the capacity limiting parameter must be known.

Over a limited range one can assume a linear relationship between column pressure, liquid flow rate (L), and the maximum vapor rate (V_{max}) that will initiate entrainment. Data such as are found in Figure 8.8tt are obtained by column testing, since technology is not sufficiently advanced to predict these effects. An equation can be developed from the test data to cover a limited range of liquid and vapor flow rates, and pressure. The relationship can usually be considered linear. Equation 8.8(80)

$$V_{max} = a_1 + a_2(L) + a_3(PC) \qquad 8.8(80)$$

is useful in predicting the values of feed flow rate and of the separation that will cause maximum vapor flow rate to exist. The use of these loading functions to optimize the operation of a column will be covered soon.

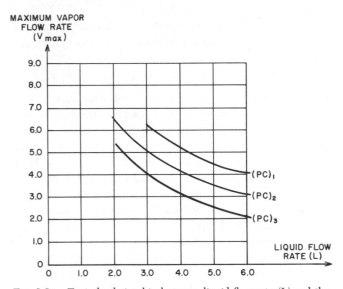

Fig. 8.8tt Typical relationship between liquid flow rate (L) and the vapor flow rate (V_{max}) that will initiate entrainment

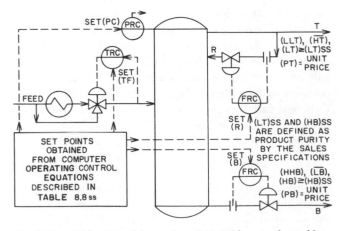

Fig. 8.8uu With prices of the products (PT, PB) known, the problem is to find the separation (HT, LB) that will maximize profit rate

Limited Market and Feedstock

Assuming that the column is already equipped with the operating control functions given in Table 8.8ss, the overall situation is illustrated in Figure 8.8uu. Basic controls are not shown for purpose of clarity, although the basic controls can be assumed to be the same as in Figure 8.8qq.

INPUTS AND OUTPUTS FOR COMPUTER CONTROL

Measured Inputs	Specified Inputs	Constants	Outputs
(LLF)	$(\overline{HT})$	t, $(LT)_{ss}$	$(\overline{FT})_0$
(LF)		T_1, $(HB)_{ss}$	$(PC)_0$
(HF)	$(\overline{LB})$	K	(TF) from $(FT)_0$
(FM)		dtr	(R)
(TO-TR)		CPF	(B)
		TB	
		PF	
		PT	
		PB	

This column has four feed components having concentrations of LLF, LF, HF, and HHF. Three of these components appear in each product. The separation for the column is fixed by specification of the concentration of heavy key component in the top product $(\overline{HT})$ and the concentration of light key component in the bottom product $(\overline{LB})$. The top and bottom products have unit prices of PT and PB, respectively. All the component in the feed that is lighter than the light key component (LLF) will go to the top product, whereas all the component that is heavier than the heavy key component (HHF) will go to the bottom product. Therefore, the operation of the column can do nothing about the distribution of these two components between the top and bottom products. $(\overline{HT})$ will be the impurity component in the top, and $(\overline{LB})$ the impurity component in the bottom. The concentrations of these two components in the products can be fixed by specification of their values as inputs to the predictive controller. (LT) is the purity component in the top product, and (HB) is the purity component in the bottom. Sales specifications for the two products usually state the minimum acceptable purities.

$$(HB) \geq (HB)_{ss} \qquad 8.8(81)$$

$$(LT) \geq (LT)_{ss} \qquad 8.8(82)$$

where $(HB)_{ss}$ and $(LT)_{ss}$ are minimum purity sales specifications.

Control Equations

Top Product More Expensive. As stated earlier, the general optimizing policy is that the product with the highest unit price must be produced at minimum specified purity. The heavy key component in the top product

$(\overline{HT})$ is the control component. Its value can be calculated as a function of feed component concentrations, minimum sales purity specification, $(LT)_{SS}$, and concentration of light key component in the bottom product $(\overline{LB})$.

Optimum Concentration of Heavy Key in Top Product. The following material balance equations are involved in determining the optimum value for the concentration of the heavy key component in the top product, $(\overline{HT})_0$.

$$(\overline{HT})_0 = 1.00 - (LT)_{SS} - (LLT) \qquad 8.8(83)$$

$$(LLF)(F) = (LLT)(T) \qquad 8.8(84)$$

$$T = (F - B) \qquad 8.8(85)$$

$$B = (F) \left[\frac{1 - (\overline{HT})_0 - (LLF + LF)}{1 - (\overline{HT})_0 - (\overline{LB})} \right] \qquad 8.8(86)$$

In equation 8.8(83), (LLT) is eliminated by use of equation 8.8(84). T is eliminated by equation 8.8(85), and B is eliminated by 8.8(86). The following is obtained:

$$(\overline{HT})_0 = \frac{\begin{array}{c}(LF) + (\overline{LB}) \\ [(LLF - 1)] - (LT)_{SS}[(LLF) \\ + (LF) - (\overline{LB})] \end{array}}{(LF) - (\overline{LB})} \qquad 8.8(87)$$

Optimum Concentration of Light Key in Bottom Product. Having obtained the value for optimum concentration of heavy key in the top product, the problem reduces to finding the value for optimum concentration of light key component in the bottom product, $(\overline{LB})_0$.

The maximum value that can be specified for $(\overline{LB})$ to meet the minimum sales specifications $(HB)_{SS}$ will be as follows:

$$(\overline{LB})_{max} = \frac{\begin{array}{c}[1 - (\overline{HT})_0][(LLF) \\ + (LF) + (HF) - (HB)_{SS}] \\ - (LLF + LF)[1 - (HB)_{SS}] \end{array}}{LF - (\overline{HT})_0} \qquad 8.8(88)$$

This equation is derived by the same procedure used for equation 8.8(87). As $(\overline{LB})$ is lowered from its maximum value, the flow rate of bottom product will decrease and the top product flow rate will increase. Since the top product is the highest unit price, profit will increase as $(\overline{LB})$ is lowered. Also, operating costs will increase, since a better and better separation is being specified.

As $(\overline{LB})$ is lowered further, one of two things will occur to establish $(\overline{LB})_0$ (the lowest value for LB). Either the change in operating cost will approach the change in profit rate, or the column will be loaded against an operating constraint. This constraint will be a condenser limit, a reboiler limit, or a column limit. Whichever occurs first will establish $(\overline{LB})_0$.

Assume, for example, that no constraints are involved. Worth of the products will be given by

$$PW = (PT)(T) + (PB)(B) \qquad 8.8(89)$$

where PW is the total product worth rate, dollars/unit time. Eliminate B by

$$B = F - T \qquad 8.8(90)$$

to obtain

$$PW = (PT - PB)(T) + (PB)(F) \qquad 8.8(91)$$

Taking the partial derivative of this equation with respect to $(\overline{LB})$, the following is obtained:

$$\frac{\delta(PW)}{\delta(\overline{LB})} = (PT - PB)\frac{\delta(T)}{\delta(\overline{LB})} \qquad 8.8(92)$$

Operating costs can be approximated closely by

$$(OC) = (CRI)(RI) \qquad 8.8(93)$$

where CRI is the unit cost of operation per unit of internal reflux. Taking the partial derivative of this equation with respect to $(\overline{LB})$, the following is obtained:

$$\frac{\delta(OC)}{\delta(\overline{LB})} = (CRI)\frac{\delta(RI)}{\delta(\overline{LB})} \qquad 8.8(94)$$

When the change in operating cost is equal to the change in product worth, the value for $(\overline{LB})_0$ is determined.

$$(PT - PB)\left[\frac{\delta(T)}{\delta(\overline{LB})}\right] = (CRI)\left[\frac{\delta(RI)}{\delta(\overline{LB})}\right] \qquad 8.8(95)$$

Evaluation of this equation will yield a value for $(\overline{LB})_0$. (T) must be expressed in terms of the specified separation, the feed composition, and flow rate. (RI) must be expressed in terms of the independent inputs to the system as used in the internal reflux control equation. The simultaneous solution of equations 8.8(87) and 8.8(95) will yield optimum values for LB and HT: $(\overline{LB})_0$, $(\overline{HT})_0$. The total solution is indicated in Table 8.8vv.

Bottom Product More Expensive. The general optimizing policy requires that the product with the highest unit price be produced at minimum specified purity.

Optimum Concentration of Light Key in Bottom Product. Since the bottom product must be produced at minimum purity, equation 8.8(88) will give the optimum value to specify for $(\overline{LB})$. The optimum $(\overline{LB})$ will be equal to $(\overline{LB})_{max}$. This in turn will give minimum specified sales purity $(HB)_{SS}$.

Optimum Concentration of Heavy Key in Top Product. The maximum value for heavy key in the top product is given by equation 8.8(87). As $(\overline{HT})$ is reduced from its maximum allowable value, flow rate of the top product will decrease and flow rate of the bottom product will increase. Therefore, total worth of the products will increase together with operating costs. The optimum value for $(\overline{HT})$ will occur when an operating constraint is encountered or when the incremental gain in product worth is equal to the incremental gain in operating cost.

For this case, assume that the column approaches an operating constraint as $(\overline{HT})$ is lowered. Let this constraint be flooding above the feed tray because of excessive entrainment of liquid in the vapor.

First, an equation needs to be developed for vapor

Table 8.8vv
CONTROL HIERARCHY INVOLVED FOR OPTIMIZATION TO MAXIMIZE PROFIT RATE
WHEN NO OPERATING CONSTRAINTS ARE INVOLVED

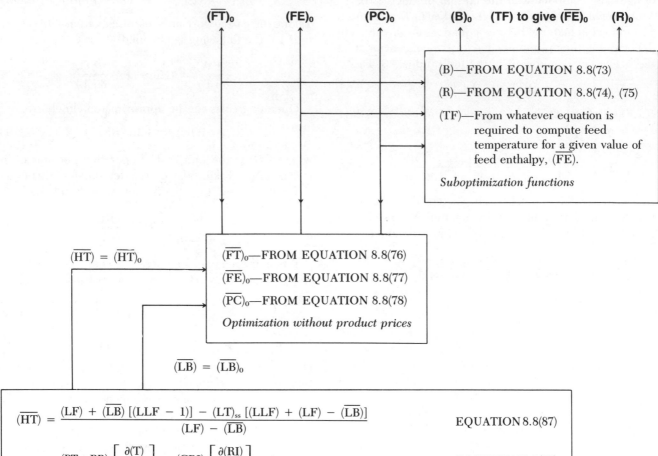

flow rate above the feed tray in terms of variables that are contained in the bottom product and reflux operating equations. The loading equation is developed as follows:

$$RI + T = V_t \qquad 8.8(96)$$

where

     RI = internal reflux flow rate,
     T = top product flow rate (distillate), and
     V_t = vapor flow rate above feed tray.

V_t can have a maximum value as given by equation 8.8(80), which is developed specifically for the column involved. Therefore, equate the right side of equation 8.8(80) with the left side of 8.8(96) to obtain

$$RI + T = a_1 + a_2(L) + a_3(\overline{PC}) \qquad 8.8(97)$$

where

     a_1, a_2, a_3 = coefficients of experimental loading equation,
     PC = column pressure, and
     L = liquid flow rate in column at the point at which flooding occurs.

For this case L will equal (RI). Also, eliminate (T) by substituting $(F - B)$. Substitute

$$RI = \left[\frac{RI}{F}\right] F$$

$$B = \left[\frac{B}{F}\right] F$$

The following is obtained:

$$a_1 + a_3(\overline{PC}) + \left[(a_2 - 1)\frac{RI}{F} + \frac{B}{F} - 1\right]F = 0 \quad 8.8(98)$$

In this equation a_1, a_2, and a_3 are known from the experimental loading equation. (PC) and F are measured, (RI/F) and (B/F) are obtained from the operating control equations for reflux and bottom product flow rate.

It is now possible to find the optimum value for $(\overline{HT})$. As $(\overline{HT})$ is lowered from its maximum allowable value as given by equation 8.8(87), the values for (RI/F) and (B/F) will change. $(\overline{HT})$ can be lowered until loading equation 8.8(98) is equal to zero. Also column pressure $(\overline{PC})$ can be raised to allow a greater loading that will result from a lower $(\overline{HT})$. A point will be found at which maximum profit rate will exist.

The maximum limit for $(\overline{PC})$ will be determined by several factors. As $(\overline{PC})$ is increased, operating costs will increase, because of the resulting smaller differential temperature at the reboiler. Therefore, one possible limit would be a reboiler (or operating cost) limit. Another possibility will be set by the column pressure rating. Another limit for maximum $(\overline{PC})$ may be determined by requirements of upstream processing equipment. Yet another limit for maximum $(\overline{PC})$ could be loading of the downcomers between each tray. This comes about because at higher vapor densities, disengagement of vapor from the liquid becomes more difficult. Therefore, at some maximum pressure, density of the vapor can approach the point at which vapor will not have sufficient time to disengage from the liquid in the downcomers, and a condition known as *downcomer flooding* will occur.

Assume for the purpose of illustration here that the maximum for $(\overline{PC})$ is set by pressure rating of the column. Therefore, $\overline{PC}$ will be set and left at this value.

Figure 8.8ww shows the optimizing control system connected to a typical distillation column. Only part of the basic controls are shown for purpose of clarity. Assume, however, that all basic controls are the same as those shown in Figure 8.8qq.

Unlimited Market and Feedstock

When an unlimited market exists for the products, the feed flow rate and separation resulting in maximum profit rate must be determined. Operation for a column under this condition will always be against an operating constraint. Assume for the purpose of the example to be given that the operating constraint involved will be the overhead vapor condenser capacity.

In general, the overall optimization problem for this case is illustrated in Figure 8.8xx. Values must be determined for the separation, $(\overline{HT})_0$, $(\overline{LB})_0$, for feed flow rate, $(F)_0$, and column pressure, $(\overline{PC})_0$, that will give maximum profit rate. One of the key component spec-

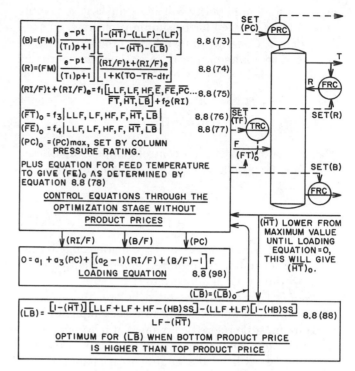

Fig. 8.8ww System for maximizing profit rate when bottom product price is higher than top product price and when entrainment above the feed tray limits loading

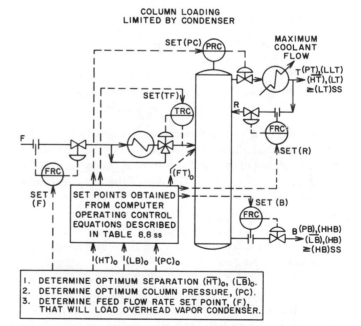

Fig. 8.8xx When an unlimited market exists for the products and the product prices (PT, PB) are known, the problem requires finding the optimum separation, the optimum column pressure, and the value for feed flow rate that will give maximum loading

ifications, $(\overline{HT})$ or $(\overline{LB})$, can be easily determined from the general optimizing policies. In general, the incremental gain in recovery of the most valuable product will not exceed the incremental gain resulting from increasing feed flow rate. This means that both products operated at minimum purity will allow the largest quantity of feed to be charged for maximum profit rate.

Optimum Concentrations. Both products must be produced at minimum purity to achieve the maximum profit operation for this case. Therefore,

$$\overline{LT} = (\overline{LT})_{SS} \qquad 8.8(99)$$

$$\overline{HB} = (\overline{HB})_{SS} \qquad 8.8(100)$$

The concentration of each control component, $\overline{LB}$ and $\overline{HT}$, must be as follows to satisfy 8.8(99) and (100):

$$(\overline{LB})_0 = (\overline{LB})_{max} \qquad 8.8(101)$$

$$(\overline{HT})_0 = (\overline{HT})_{max} \qquad 8.8(102)$$

$(\overline{HT})_{max}$ and $(\overline{LB})_{max}$ are given by equations 8.8(87) and (88). The separation is optimized independently by column pressure $(\overline{PC})$ and feed rate (F).

Loading Constraint. As stated before, the overhead vapor condenser limits loading for this example. Column pressure must therefore be operated at maximum to obtain maximum condensing capacity. (ΔT across the condenser tubes will be the largest at maximum column pressure, and maximum condenser capacity will result.) The maximum pressure that can be used will be determined by one of the following five constraints, assuming operating costs do not become prohibitive before a physical constraint is reached.

1. Column downcomer capacity.
2. Reboiler capacity.
3. Upstream equipment pressure requirements.
4. Pressure rating of the column shell.
5. Fouling of the reboiler or condenser tubes.

As column pressure is increased, capacity of the reboiler and tray downcomers will be approached. Also, pressure rating of the shell and of other processing equipment will be approached. The one of the five constraints that is approached first, as column pressure is raised, will set the maximum pressure. Assume for the purpose of illustration that the shell pressure rating is the limit on column pressure. Pressure can therefore be set constant at the rating value.

Optimum Feed Flow Rate. After the optimum separation $(\overline{LB})_0$ and $(\overline{HT})_0$ has been determined, as well as the optimum column-operating pressure $(\overline{PC})_0$, the feed rate can be increased until the condenser capacity is approached. There are several ways that feed rate can be manipulated to maintain condenser loading. One very useful method involves a predictive control technique.

Overhead vapor flow rate can be expressed in terms of various independent inputs and terms obtained from the operating control equations. Development of the predictive control equation proceeds as follows:

$$V_{oh} = T + R \qquad 8.8(103)$$

where

V_{oh} = vapor flow rate overhead,
T = top product flow rate, and
R = external reflux flow rate.

First, eliminate T by $T = F - B$. Now, the maximum overhead vapor flow rate will be given by equation 8.8(79). Therefore, equations 8.8(79) and (103) can be set equal.

$$a_1 + a_2(\Delta T) + a_3(\Delta T)^2 = R + F - B \qquad 8.8(104)$$

or

$$F = a_1 + a_2(T_{oh} - T_c) + \\ + a_3(T_{oh} - T_c)^2 + B - R \qquad 8.8(105)$$

where

a_1, a_2, a_3 = coefficients for condenser-loading equation that are determined by column tests,
F = feed flow rate,
T_{oh} = overhead vapor temperature,
T_c = temperature of coolant to overhead vapor condenser, and
B and R = bottom product and reflux flow rates from outputs of operating control equations.

Temperature of the overhead vapor will be a function of all independent inputs to the system. However, column pressure is usually the main variable of concern. For this example let

$$T_{oh} = fn(\overline{PC}) \qquad 8.8(106)$$

where $fn(\overline{PC})$ is some function of column pressure.

In many cases $fn(\overline{PC})$ can be considered a linear function such as

$$fn(\overline{PC}) = d_1 + d_2(\overline{PC}) \qquad 8.8(107)$$

This equation can be determined off-line from correlation of data obtained from flash calculations at the average composition of the overhead vapor existing. If changes in composition of the overhead vapor affect temperature of the overhead vapor by a significant amount, then composition would also have to be taken into account. Composition for the overhead vapor can be easily approximated from feed composition analysis. If equation 8.8(105) is carried to this extent, then the feed flow rate can be predicted to be of such a value to keep the condenser against its maximum capacity. For the purpose of illustration here, T_{oh} is assumed to be a function of column pressure only.

Eliminate T_{oh} from 8.8(105) by equations 8.8(106) and (107) to obtain

$$a_1 + a_2[d_1 + d_2(\overline{PC}) - T_c] + a_3[d_1 + \\ + d_2(\overline{PC}) - T_c]^2 + B - R = F_{max} \qquad 8.8(108)$$

$(\overline{PC})$ and (T_c) are measured, B and R are obtained from the operating equations set point calculations. F_{max} will be the feed rate required to load the condenser for the particular values of $(\overline{HT})_0$ and $(\overline{LB})_0$ determined.

Figure 8.8yy shows the overall optimizing control system applied. Only the necessary basic controls are shown. The other controls can be assumed to be as those shown in Figure 8.8qq.

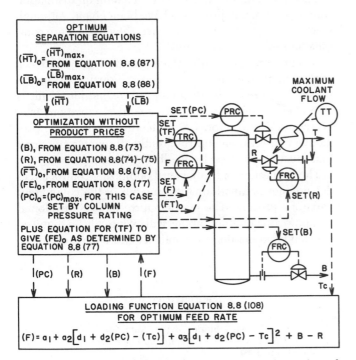

Fig. 8.8yy Optimization for maximum profit rate when unlimited market exists for the products and loading is limited by condenser capacity

Reboiler Limiting. Assume now that loading is limited by the reboiler instead of the condenser, as just covered. Optimum separation remains the same. However, column pressure must be operated at a minimum to gain maximum reboiler capacity. For this example, assume that minimum column pressure is set by pressure requirements of downstream equipment. Therefore, column pressure is set at a constant value and will not be changed unless pressure requirements of downstream equipment are changed.

Having achieved the optimum separation (minimum purity of products) and optimum column pressure, the feed rate can now be increased until the reboiler limits. This then will represent the maximum profit operation.

Again, manipulation of the feed flow rate can be handled by a predictive control technique. Liquid flow rate below the feed tray (LR) is given by

$$(LR) = RI + FI \qquad 8.8(109)$$

where FI is the internal feed flow rate.

$$FI = F[1 + (KF)(T_v - T_f)] \qquad 8.8(110)$$

where

(KF) = a constant equal to the specific heat of the feed divided by the heat of vaporization,

T_v = temperature of vapor above the feed tray, and

T_f = temperature of feed at column entry.

The vapor flow rate out of the reboiler is given by

$$(VB) = (LR) - B \qquad 8.8(111)$$

Now substitute equation 8.8(110) into (109) to eliminate FI. Then substitute 8.8(109) into 8.8(111) to eliminate (LR). The following is obtained:

$$(VB) = RI + F[1 + (KF)(T_v - T_f)] - B \qquad 8.8(112)$$

Next substitute (RI/F)F for RI, and (B/F)F for B. Then solve for F to obtain

$$(F) = \frac{(VB)_{max}}{(RI/F) + 1 + (KF)(T_v - T_f) - (B/F)} \qquad 8.8(113)$$

where

(F) = set point of feed flow controller, and

$(VB)_{max}$ = maximum reboiler heat input rate.

Equation 8.8(113) calculates that feed flow rate which will cause the vapor rate $(VB)_{max}$ to exist for all separations specified. (RI/F) and (B/F) are obtained from the operating control equations used to achieve a suboptimum operation. T_v and T_f are measured. Equation 8.8(113) is used by specifying $(VB)_{max}$ and then evaluating the column operation. After sufficient time for the column to stabilize, the reboiler valve position (output of reboiler heat flow controller) is observed. Say, for example, that the reboiler valve is 85 percent open. $(VB)_{max}$ can then be increased until the reboiler heat control valve is near its maximum opening, say 95 percent open. Enough room must be left to maintain control. $(VB)_{max}$ can be adjusted by the plant operator to maintain the reboiler valve near open or can be handled automatically by logic-type feedback. Once $(VB)_{max}$ is established by experience, few adjustments will be required to maintain the column loaded. Adjustments to $(VB)_{max}$ will be required only as the maximum reboiler duty available varies. The control scheme is illustrated in Figure 8.8zz.

Alternative Method for Maintaining Maximum Feed Rate When a Reboiler or Condenser Constraint Exists. Another approach that can be considered for maintaining maximum feed flow rate involves straight feedback control. For example, the reboiler valve stem position (output of heat flow controller) can be measured and feed flow rate manipulated to maintain the reboiler valve near open.

Likewise, if a condenser limit is involved, stem position of the column pressure control valve can be mea-

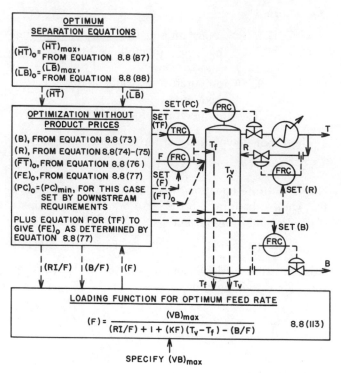

Fig. 8.8zz   Optimization of a distillation column when unlimited market exists for the products, prices of the products are known, and loading is limited by the reboiler

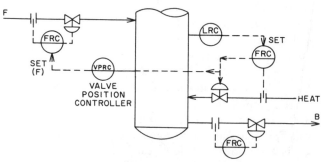

VPC MAINTAINS COLUMN BACK-PRESSURE CONTROL VALVE NEARLY OPEN BY MANIPULATING FEED FLOW RATE TO KEEP CONDENSER LOADED

VPC MAINTAINS REBOILER HEAT VALVE NEARLY OPEN BY MANIPULATING FEED FLOW RATE TO KEEP REBOILER LOADED

Fig. 8.8aaa   Feedback control method for maintaining condenser or reboiler loaded

sured and the feed rate manipulated by a conventional controller to maintain the backpressure valve near open. Also, if a hot vapor bypass around the condenser exists, stem position of the bypass valve will also indicate the state of condensing. These procedures are illustrated in Figure 8.8aaa.

Conclusions

Example solutions to some of the common distillation column optimizing problems have been given. Although many different situations can exist, they usually are combinations of those presented.

It is important to realize that optimization by feedback control methods cannot approach the quality of control possible by predictive (feedforward) techniques. This is true even though the predictive control equations may be required to be updated by feedback. In effect, predictive optimization control greatly attenuates any error that must be handled by feedback (updating).

The application of feedforward optimizing control forces development of mathematical models of the component parts of a process. The mathematical models developed for optimizing unit operations will eventually be required to extend optimization to include an entire plant complex.

REFERENCES

1. Lupfer, D.E. and Johnson, M.L., "Automatic Control of Distillation Columns to Achieve Optimum Operations," *ISA Transactions*, April 1964.
2. Van Kampen, J.A., "Automatic Control by Chromatograph of a Distillation Column," Convention on Advances in Automatic Control, Nottingham, England, April 1961.
3. Shinskey, F.G., *Process Control Systems*, McGraw-Hill Book Co., New York, 1967.
4. Luyben, W.L., "Distillation Decoupling," *AIChE Journal*, Vol. 16, No. 2, March 1970, pp. 198–203.
5. Shinskey, F.G., "The Stability of Interacting Loops With and Without Decoupling," Presented at the IFAC Symposium on Multivariable Control, Fredericton, New Brunswick, July 4–8, 1977.
6. Gordon, L.M., "Practical Evaluation of Relative Gains: The Key to Designing Dual Composition Controls," *Hydrocarbon Processing*, December 1982.
7. Shinskey, F.G., *Distillation Control for Productivity and Energy Conservation*, McGraw-Hill Book Co., New York, 1978, p. 307.
8. Ryskamp, C.J., "New Control Strategy Improves Dual Composition Control," *Hydrocarbon Processing*, June 1980.
9. Bristol, E.H., "On a New Measure of Interaction for Multivariable Process Control," *IEEE Transactions on Automatic Control*, January 1966, pp. 133–134.
10. Shinskey, F.G., "Predicting Distillation Column Response Using Relative Gains," *Hydrocarbon Processing*, May 1981.
11. Converse, A.O. and Gross, G.D., "Optimal Distillate Rate Policy in Batch Distillation," *Industrial Engineering Chemistry*, August 1963.
12. Thurston, C.W., "Computer-Aided Design of Distillation Column Controls: Part 1," *Hydrocarbon Processing*, July 1981.
13. ———, "Computer-Aided Design of Distillation Column Controls: Part 2," *Hydrocarbon Processing*, August 1981.
14. Van Horn, L.D., "Computer Control . . . How to Get Started," *Hydrocarbon Processing*, September 1979.
15. ———, "Crude Unit Computer Control . . . How Good Is It?" *Hydrocarbon Processing*, April 1980.

16. Bannon, R., et al., "Heat Recovery in Hydrocarbon Distillation," *CEP*, July 1978.
17. Black, J.W., "Model Estimates Analyzer Payouts," *Hydrocarbon Processing*, September 1981.
18. Latour, P.R., "Online Computer Optimization: What It Is and Where to Do It," *Hydrocarbon Processing*, June 1979.
19. DiBiano, R.J., "Improve Reliability of Advanced Tower Control," *Hydrocarbon Processing*, October 1980.
20. ———, "Advanced Control: General Purpose Design vs. Working It Out in the Field," *Hydrocarbon Processing*, August 1982.

BIBLIOGRAPHY

Ackley, W.R., "Feedforward Control Strategy for Distillation Towers," *Instrumentation Technology*, February 1974.

Adiutori, E.F., "The New Heat Transfer," Ventuno Press, Cincinnati, 1974, Chapter 7.

Andrews, A.J. and Griffin, D.E., "Performance Audits Evaluate Two Distillation Control Projects," *Oil & Gas Journal*, December 8, 1980.

Black, C., "Distillation Modeling of Ethanol Recovery and Dehydration Process for Ethanol and Gasohol," *Chemical Engineering Progress*, September 1980.

Bojnowski, J.H., Crandall, J.W., and Hoffman, R.M., "Modernized Separation System Saves More Than Energy," *Chemical Engineering Progress*, October 1975.

Boyd, D.M., "Fractionation Column Control," *Chemical Engineering Progress*, June 1975.

Bristol, E.H., "On a New Measure of Interaction for Multivariable Process Control," *IEEE Transactions on Automatic Control*, January 1966.

Buckley, P.S., "Distillation Column Design Using Multivariable Control," *Instrumentation Technology*, September/October 1978.

———, "Controls for Sidestream Drawoff Columns," *Chemical Engineering Progress*, Vol. 65, No. 5, 1969, p. 45.

———, "Material Balance Control in Distillation Columns," Presented at AIChE Workshop on Industrial Process Control, Tampa, FL, November 11–13, 1974.

———, Cox, R.K., and Luyben, W.L., "How to Use a Small Calculator in Distillation Column Design," *Chemical Engineering Progress*, June 1978, pp. 49–55.

———, Cox, R.K., and Rollins, D.L., "Inverse Response in a Distillation Column," *Chemical Engineering Progress*, June 1975.

Chestnut, H. and Mayer, R.W., *Servomechanisms and Regulating System Design*, Vol. 1, John Wiley & Sons, Inc., New York, 1951, pp. 152–153.

Daniels, F., *Outlines of Physical Chemistry*, John Wiley & Sons, Inc., New York, 1948, p. 203.

Denbigh, K., *The Principles of Chemical Equilibrium*, Cambridge University Press, London, 1961, pp. 63 and 116.

Doig, I.D., "Variation of the Operating Pressure to Manipulate Distillation Processes," *Australian Chemical Engineering*, July 1971.

Douglas, J.M., Jafarey, A., and McAvoy, T.J., "Short-Cut Techniques for Distillation Column Design and Control, Part 1: Column Design," *I&EC Process Design and Development*, Vol. 18, April 1979, pp. 197–202.

Doukas, N., "Control of Sidestream Columns Separating Ternary Mixtures," *Instrumentation Technology*, June 1978.

———, and Luyben, W.L., "Economics of Alternative Distillation Configurations for the Separation of Ternary Mixtures," *I&EC Process Design and Development*, Vol. 17, no. 272, 1978.

Ellerbe, R.W., "Steam Distillation Basics," *Chemical Engineering* (NY), March 4, 1974.

Elliott Division, Carrier Corp., "Elliott Multistage Centrifugal Compressors," *Bulletin P–11A*, Jeannette, PA, Elliott Division, Carrier Corp.

Evans, F.L., *Equipment Design Handbook for Refineries and Chemical Plants*, Vol. I, Gulf Publishing Co., Houston, 1973, p. 89.

Evans, L.B., "Impact of the Electronics Revolution on Industrial Process Control," *Science*, Vol. 195, No. 4283, March 18, 1977.

Farwell, W., "More Cooling at Less Cost," *Plant Engineering*, August 1974.

Fauth, C.J. and Shinskey, F.G., "Advanced Control of Distillation Columns," *Chemical Engineering Progress*, June 1975.

Fenske, M.R., "Fractionation of Straight-Run Pennsylvania Gasoline," *Industrial Engineering Chemistry*, May 1932.

Forman, E.R., "Control Systems for Distillation," *Chemical Engineering*, November 8, 1965.

The Foxboro Company, Differential Vapor Pressure Cell Transmitter, Model 13VA, Technical Information Sheet 37–91A, Foxboro, MA, April 1965.

Frank, O. and Prickett, R.D., "Designing Vertical Thermosyphon Reboilers," *Chemical Engineering* (NY), September 3, 1973.

Franzke, A., "Save Energy with Hydraulic Power Recovery Turbines," *Hydrocarbon Process*, March 1975.

Gallier, P.W. and McCune, L.C., "Simple Internal Reflux Control," *Chemical Engineering Progress*, September 1974.

Griffin, D.E., "Tighten Distillation Column Control and Save Energy," *Instruments and Control Systems*, March 1981.

———, et al., "The Use of Process Analyzers for Composition Control of Fractionators," Paper presented at ISA Spring Joint Conference, Houston, TX, May 1978.

Hadley, K., "Control Objectives Analysis," National Petroleum Refiners Association Computer Conference, New Orleans, 1977.

Hengstebeck, R.J., "An Improved Shortcut for Calculating Difficult Multicomponent Distillations," *Chemical Engineering* (NY), January 13, 1969.

Houghton, J. and McLay, J.D., "Turboexpanders Aid Condensate Recovery," *Oil and Gas Journal*, March 5, 1973.

Hughart, C.L., "Designing Distillation Units for Controllability," *Instrumentation Technology*, May 1977.

Instrument Society of America, "Process Instrumentation Terminology," *ISA Standard S51.1*, Instrument Society of America, Pittsburgh, 1976.

Johnson, A.J. and Auth, G.H., "Fuels and Combustion Handbook," McGraw-Hill, New York, 1951, p. 286.

Keenan, J.H., et al., "The Fuel Shortage and Thermodynamics," MIT Industrial Liaison Program, Paper 12-13-73, Massachusetts Institute of Technology, Boston, MA.

Kline, P.E., "Technical Task Force Approach to Energy Conservation," *Chemical Engineering Progress*, February 1974.

Knipp, R.S., "An Advanced System for Process Information Handling," Texas A & M Instrumentation Symposium, January 1972.

Lamb, M.Y., "Computer Control of a Propylene Upgrading Unit," 85th National AIChE Meeting, Philadelphia, 1978.

Latour, P.R., "Composition Control of Distillation Columns," *Instrumentation Technology*, July 1978.

Lupter, D.E., "Distillation Column Control for Utility Economy," Presented at 53rd Annual GPA Convention, Denver, March 25–27, 1974.

Luyben, W.L., "Feedback Control of Distillation Columns by Double Differential Temperature Control," *Industrial and Engineering Fundamentals*, Vol. 8, November 1969.

———, "10 Schemes to Control Distillation Columns with Sidestream Drawoff," *ISA Journal*, Vol. 13, No. 7, 1966, pp. 37–42.

MacFarlane, A.G.J., "A Survey of Some Recent Results in Linear Multivariable Feedback Theory," *Automatica*, Vol. 8, No. 4, 1972, pp. 455–492.

MacMullan, E.C., "Fractionator Control System Using an Analog Computer," U.S. Patent 3,282,799, November 1, 1966.

Matheson Gas Data Book, 4th ed., The Matheson Company, East Rutherford, NJ, 1966.

McCoy, R.D., "Adding Capabilities to Process Chromatography with Microprocessor-based Programmers," Paper presented at ISA Joint Spring Conference, Houston, TX, May, 1978.

McNeill, G.A. and Sacks, J.D., "High Performance Column Control," *Chemical Engineering Progress*, March 1969.

Monroe, E.S., "Vacuum Pumps Can Conserve Energy," *Oil and Gas Journal*, February 3, 1975.

Niagara Blower Company, Niagara Aero Heat Exchanger, Bulletin 169, Niagra Blower Company, Buffalo, NY.

O'Connor, J. and Illing, H., "The Turbine Control Valve . . . A New Approach to High-Loss Applications," *Instrumentation Technology*, December 1973.

Oglesby, M.W. and Hobbs, J.W., "Chromatograph Analyzers for Distillation Control," *Oil and Gas Journal*, January 10, 1966.

Owens, W.R. and Maddox, R.N., "Short-cut Absorber Calculations," *Industrial Engineering Chemistry,* December 1968.

Painter, J.W. and Gonnella, J.L., "Improved Control of a Distillation Column Using a Minicomputer and an On-line Gas Chromatograph," Texas A & M Instrumentation Symposium, January 1978.

Perry, J.H., *Chemical Engineers' Handbook,* 4th ed., McGraw-Hill Book Company, New York, 1963.

Rademaker, O., Rijnsdorp, J.E., and Maarleveld, A., *Dynamics and Controls of Continuous Distillation Units,* Elsevier, Amsterdam, 1975, p. 71.

Rathore, R.N.S., VanWorme, K.A., and Powers, G.J., "Synthesis of Distillation Systems with Energy Integration," *AIChE Journal,* September 1974.

Rijnsdorp, J.E., "Interaction in Two-Variable Control Systems for Distillation Columns," *Automatica,* Vol. 1, Pergamon Press, London, 1965.

Rippin, D.W.T. and Lamb, D.E., "A Theoretical Study of the Dynamics and Control of Binary Distillation," *Bulletin from the University of Delaware,* 1960.

Sayles, J.H., "Computer Control Maximizes Hydrocracker Throughput," *Instrumentation Technology,* May 1973.

Shinskey, F.G., "Avoiding Reset Windup in Cascade Systems," *Instruments and Control Systems,* August 1971.

———, "Controlling Distillation Processes for Fuel-Grade Alcohol," *Instrumentation Technology,* December 1981.

———, "Controlling Surge-Tank Level," *Instruments and Control Systems,* September 1972.

———, "Controlling Unstable Processes, Part II: A Heat Exchanger," *Instruments and Control Systems,* January 1975.

———, "Controlling Variable-Period Once-Through Processes," *Instruments and Control Systems,* November 1972.

———, "Effective Control for Automatic Startup and Plant Protection," *Canadian Controls Instrumentation,* April 1972.

———, "Energy-Conserving Control for Distillation Units," *Chemical Engineering Progress,* May 1976.

———, *pH and pIon Control in Process and Waste Streams,* Wiley-Interscience, New York, 1973, pp. 194–199.

———, "Temperature Control for Gas-Phase Reactors," *Chemical Engineering* (NY), October 5, 1959.

———, "Values of Process Control," *Oil and Gas Journal,* February 18, 1974.

———, "What to Know About Interaction Between Control Loops," *Canadian Controls Instrumentation,* March 1972.

———, "When You Have the Wrong Valve Characteristic," *Instruments and Control Systems,* October 1971.

———, "When to Use Valve Positioners," *Instruments and Control Systems,* September 1971.

Sisson, B., "How to Determine MEA Circulation Rates," *Chemical Engineering* (NY), November 26, 1973.

Smith, D.E., et al., "Distill with Composition Control," *Hydrocarbon Processing,* February 1978.

Smuck, W.W., "Operating Characteristics of a Propylene Fractionating Unit," *Chemical Engineering Progress,* June 1963.

Sternlicht, B., "Low-Level Heat Recovery Takes on Added Meaning as Fuel Costs Justify Investment," *Power,* April 1975.

Technical Data Book—Petroleum Refining, 2d ed., American Petroleum Institute, Washington, DC, 1970, pp. 7–32.

Toijala, K. and Fagervik, K., "A Digital Simulation Study of Two-Point Feedback Control of Distillation Columns," Kem. Teollisuus, January 1972.

Treybal, R.E., *Mass-Transfer Operations,* McGraw-Hill Book Company, New York, 1955, p. 205.

Tyreus, B. and Luyben, W.L., "Control of a Binary Distillation Column with Sidestream Drawoff," *I & E C Process Design and Development,* Vol. 14, No. 4, 1975, p. 391.

———, "Two Towers Cheaper Than One?" *Hydrocarbon Processing,* June 1975.

Union Carbide Corporation, *Conserve Energy by Using High-Flux Tubing in Heat-Pump Applications,* Union Carbide Corporation, Tonawanda, NY.

VanHorn, L.D. and Latour, P.R., "Computer Control of a Crude Still," *Instrumentation Technology,* November 1976, pp. 33–39.

VanWinkle, M., *Distillation,* McGraw-Hill Book Company, New York, 1976.

Wahl, E.F. and Harriott, P., "Understanding and Prediction of the Dynamic Behavior of Distillation Columns," *I & E C Process Design and Development,* Vol. 9, No. 3, 1970, pp. 396–407.

Waller, K.V.T., "Decoupling in Distillation," *AIChE Journal,* May 1974.

White, M.H., "Surge Control for Centrifugal Compressors," *Chemical Engineering* (NY), December 25, 1972.

Wolf, C.W., Weiler, D.W., and Ragi, E.G., "Energy Costs Prompt Improved Distillation," *Oil and Gas Journal,* September 1, 1975.

Wood, C.E., "Method of Selecting Distillation Column Tray Location for Temperature Control," Presented at Joint IMIQ-AIChE Meeting, Mexico City, September 24–27, 1967.

Wright, R.M., "A Better Approach to Distillation Control," *Instruments and Control Systems,* June 1976.

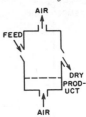

8.9 DRYER CONTROLS

The subject of dryers is so extensive that some limitations must be imposed in order to present a worthwhile discussion. In general, this presentation will be based on these principles:

1. The dryers and drying principles discussed will relate to process requirements, that is, to the removal of a volatile solvent from a solid material. Such processes as the removal of a solvent from an air stream (air drying) or the removal of one solvent from another (drying of an organic solvent) will be considered beyond the scope of this section.
2. The solvent to be removed will be taken as water, although an occasional reference may be made to other solvents.
3. The heating medium used will be steam, except in cases in which a different heat source is customarily used.
4. Safety controls such as flame burnout indicators will not be included in the discussion except when they specifically relate to the process.

Principles

A drying curve of a typical material is shown in Figure 8.9a. It consists of four zones. Section A-B represents the period of product entry into the dryer. Since some heat is necessary to bring the material to the initial drying temperature, evaporation is slow. The next section, B-C, is a period of evaporation of surface moisture or moisture that migrates readily. The temperature during this time is the wet-bulb temperature of the air directly in contact with the product. After the surface moisture has evaporated, the rate of evaporation drops off (section C-D). This phenomenon is due in part to a "case-hardening" of the surface and in part to the long path necessary for the water to migrate to the surface. If the solid is one that has water-of-crystallization, or bound water, the rate will then drop off even more, as shown in section D-E of the curve.

Two other curves are useful for an understanding of dryer control requirements. The first of these, Figure 8.9b, represents the derivative of Figure 8.9a, that is, the variation in rate of drying of the material as a function of time. The other curve, Figure 8.9c, depicts the temperature levels during drying. In each case the identifying letters correspond to those in Figure 8.9a. Section B-C is thus a period of *constant rate*, and the temperature is also constant, as already mentioned. The remainder of the curve is known as the period of *falling rate*. The temperature of the product will rise rather rapidly to some point close to the temperature of the surroundings within the dryer. It will then asymptotically approach a final value.

Each material to be dried will have a unique drying characteristic, depending upon its substance, the sol-

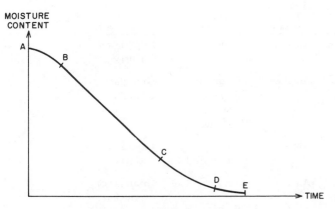

Fig. 8.9a Typical drying curve

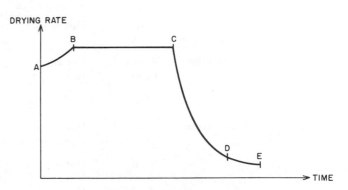

Fig. 8.9b Drying rate curve

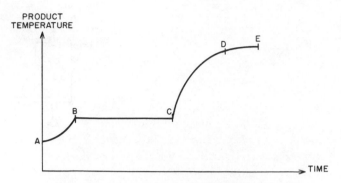

Fig. 8.9c   Variation in product temperature

vent, the affinity of one of these for the other, and the surroundings characteristic of the particular dryer.

The theoretical drying rate depends upon feed rate and driving force. The latter is a combination of temperature difference between the product and the surroundings and the moisture condition prevailing in the dryer atmosphere. Other factors of equal importance are the intimacy of material surface contact with the heating medium and the degree of agitation or surface renewal.

The three essential operating factors necessary to any drying operation are a source of heat, a means of removing the solvent from the environment of the product, and a mechanism to provide agitation or surface renewal. The extensive assortment of dryer types available serves as testimony to the fact that drying is among the most difficult of process problems. A few representative dryer types are presented in Tables 8.9d (batch dryers) and 8.9e (continuous dryers) by way of general illustration. Further elaboration of control schemes for them will be presented in the following paragraphs.

Control

The specific property of interest in dryer operation is moisture content of the final product. This property is difficult to measure directly, particularly for continuous systems.

In batch systems, one sets up an empirical cycle by taking samples for moisture analysis at various periods. Once approximate drying conditions have been established, checking of moisture is required only near the estimated end point.

Table 8.9d
TYPICAL BATCH DRYER CHARACTERISTICS

Type	Examples	Heat Source	Method of Moisture Removal	Method of Agitation	Typical Feed
Atmospheric	Tray	Hot air	Air flow	Manual	Granular or powder
Vacuum	Blender tumbler	Hot surface	Vacuum	Tumbling	Granular or powder
	Tray	Hot surface	Vacuum	None	Solid or liquid
	Rotating blade	Hot surface	Vacuum	Mixing blade	Solid or liquid
Specialty	Fluid bed	Hot air	Air flow	Air	Granular or powder

Table 8.9e
TYPICAL CONTINUOUS DRYER CHARACTERISTICS

Type	Examples	Heat Source	Method of Moisture Removal	Method of Agitation	Typical Feed
Heated cylinder	Double-drum	Hot surface	Air flow	None	Liquid or slurry
Tumbling	Rotary	Hot air	Air flow	Tumbling	Granular
	Turbo	Hot air	Air flow	Wipe to successive shelves	Granular
Air stream	Spray	Hot air or combustion gas	Air flow	Not required	Liquid or slurry
	Flash	Hot air	Air flow	Air	Granular or powder
	Fluid bed	Hot air	Air flow	Air	Granular

A similar procedure is followed for continuous systems. Empirical runs are made to establish dryness of a product with given feed and dryer characteristics so that operating curves can be developed. The control scheme is designed to maintain these conditions with occasional feedback and with corrections from grab samples. There are a few analyzers available for moisture analysis in flow systems that can be adapted for automatic, closed-loop dryer control. The most successful of these units rely on infrared or capacitance detection of material flowing through a sample chamber. Care must be exercised in the selection to allow for changes in bulk density or for void spaces, which will introduce an error. These sensors are sometimes ineffective with relation to bound moisture. The discussion of moisture analyzers in sections 11.14 and 11.15 in the volume on Process Measurement, gives further elaboration.

Although spray and flash dryers have a retention period of less than a second, holdup in the majority of continuous dryers ranges from 30 minutes to an hour. This fact makes automatic feed control so difficult that it is rarely used. It is usually not feasible to measure the variation in moisture content of the feed. The feed rate is metered by a screw conveyor or another similar device. The variation in entering moisture is accounted for by the dryer control system or by periodic analysis of the dried product followed by suitable controller adjustments.

Dryer controllers usually include proportional and reset modes, only because dryer dynamics do not generally warrant the use of the rate response.

Most dryer control systems are still relatively unsophisticated. The drying rate curve levels off in the area where most product specifications lie, so control of final conditions is not critical. Most moisture specifications are somewhat liberal in recognition of the difficulties present in moist and dry material handling. They also allow for a possible change in moisture content as a result of ambient air conditions after drying, a characteristic of most dried materials.

To get a general view of the use of controls as applied to dryers, it is desirable to examine and discuss each of the basic units listed in Tables 8.9d and 8.9e. This is far from a complete listing, but it covers the field thoroughly enough to provide a good background.

Batch Dryers

Although continuous processing implies a more modern approach, batch dryers are still installed in many plants. These are particularly well adapted for drying relatively small quantities, especially in cases in which batch identity might be of value and for processes by which a variety of products are manufactured. The selection and operation of controls for batch units is complicated by the fact that the state of the product changes with time.

Atmospheric

The term *atmospheric dryer* is applied here to designate batch dryers that operate at close to atmospheric pressure and with the heat for drying supplied by the air within the cabinet. The same medium is relied upon to remove the generated moisture. The two most common forms are the tray dryers, in which trays covered with material to be dried are loaded into racks, and the truck dryers, which are similar except that the racks of trays are mounted on trucks.

The available control parameters are the air velocity and distribution, temperature, and humidity. The velocity and distribution of the air are usually set manually by dampers and are not changed. Humidity is also a manual setting by virtue of adjusting the damper in the recirculation line. If recirculation is not used and if product requirements warrant, dehumidification is used on the inlet air. Both of these provisions are shown dotted in Figure 8.9f.

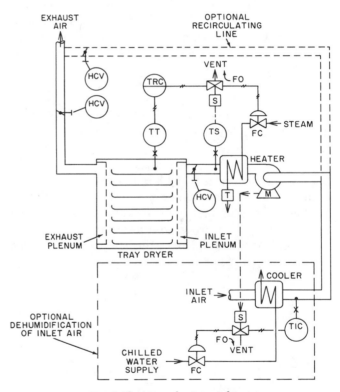

Fig. 8.9f Atmospheric tray dryer

Control of the atmospheric temperature within the dryer is accomplished by regulation of the steam to the heating coil. The thermal bulb can be placed in the dryer or in the outlet air. The sizing of the steam valve is determined by the dryer air flow and temperature requirements as analyzed on psychometric charts (see Figure 11.10a in the volume on Process Measurement) and by the pressure of steam used. It is necessary during the latter portion of the drying cycle to reduce steam flow to a level sufficient only to heat the air to the dryer temperature. The load at this time is limited to the dryer heat loss. A temperature switch is frequently included in the heater discharge to limit this maximum inlet temperature.

It is not uncommon to use a program controller (see Section 2.1 for details) for temperature control of a batch tray dryer, particularly on solvent service. In this way the rate of evaporation is controlled and the concentration of solvent in the air *is held below the explosive range*.

Vacuum

The operation of a vacuum dryer relies on the principle of heat transfer by conduction to the product while it is contained in a vessel under vacuum. In its simplest form this dryer is a tray unit in a vacuum chamber with hollow shelves, through which a heat transfer medium is circulated. Such dryers usually start operation at a very low temperature and are known as *freeze dryers*. Other types are the double-cone *blender-dryer* and the vacuum *rotary dryer*. The blender-dryer tumbles the product in a jacketed vessel. The shell of the rotary dryer is also jacketed, but it is fixed, with material agitation being supplied by a hollow mixing blade (for the passage of heat-transfer fluid).

Parameters that can be controlled in batch vacuum dryers are absolute pressure (vacuum), rate of tumbling, or agitation and temperature of the heat-transfer medium. The general rule is to control the first two manually and concentrate on the transfer fluid temperature for automatic control. A typical scheme for the blender-dryer is shown in Figure 8.9g. The requirements for control of the freeze dryer are somewhat more complex. Refrigeration is required for the heat-transfer fluid in the first part of the cycle to freeze the dryer and its contents. The circulating fluid is then gradually warmed to provide heat for evaporation.

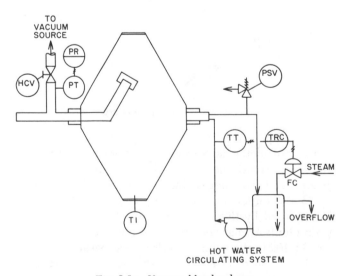

Fig. 8.9g   Vacuum blender-dryer

Special

There are a number of alternative dryers used for batch work that do not fall into the discussed categories. A unit that has become popular in the past few years is the batch fluid-bed dryer.

In a fluid-bed dryer the product is held in a portable cart with a perforated bottom plate. Heated air, blown (pressure) or sucked (vacuum) through the plate, fluidizes the product to cause drying. The air flow is manually adjusted to get proper bed fluidization, and a timer is set for the drying period. Control is restricted to temperature control of the inlet air, as indicated in Figure 8.9h.

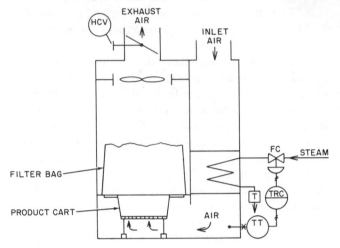

Fig. 8.9h   Batch fluid-bed dryer

A duplex control scheme is sometimes used to take advantage of the high heat transfer properties of the fluid bed. The bed is sufficiently well mixed so that a high inlet temperature can be used at the beginning of the cycle without danger of damaging the product as long as the temperature is reduced during the latter stages of drying. This is accomplished by the use of a controller on the outlet air that pneumatically adjusts the set point on the inlet air controller. As the bed temperature rises during drying, the set point of the inlet air temperature controller is reduced (Figure 8.9i).

Continuous Dryers

Most large-scale industrial processes producing dry solid materials use continuous dryers. The control of a dryer is similar to that of other process equipment, with a few important exceptions. First, as stated earlier, actual dryness (moisture content) of the product is difficult to measure, and therefore control generally depends on secondary variables. The majority of continuous dryers require control of the flow and temperature of an air stream. Temperature detection of an air stream is inherently sluggish in response. On the more positive side, the holdup in the majority of dryers is on the order of

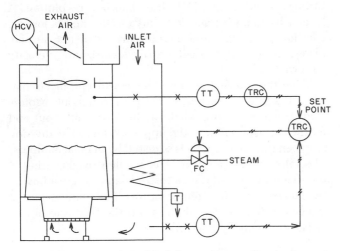

Fig. 8.9i   Batch fluid-bed dryer with cascade control

30 minutes to an hour, effects of input changes are blended, and fast response is therefore not required. Notable exceptions to the last items are flash and spray dryers with very short holdup periods.

Dryer controllers are generally two-mode units with proportional and reset modes. Setting them in the field requires a good deal of patience, since the full effect of a change in a variable may take half an hour or more. Some variables, notably humidity of the inlet air, may defy regulation.

Heated Cylinder

Many years ago, when the concept of continuous drying was first developed, the heated-cylinder dryer was invented. The best known of these units is the double-drum dryer depicted in Figure 8.9j. Liquid is fed into the "valley" between the heated cylinders. The drums, rotating downward at the center, receive a coating of the liquid with the thickness depending upon the spacing between the rolls. The material must be dry by the time it rotates to the doctor knife, where it is cut off the roll.

The variables available for control are the speed of the cylinders, the spacing between them, the liquid level in the valley, and the steam pressure in the cylinders. The

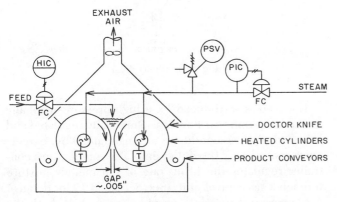

Fig. 8.9j   Double drum dryer

first two (speed and spacing) are usually adjusted manually. The liquid level is maintained by throttling the feed stream. Some attempts have been made to control this level automatically, but they have been thwarted by three basic difficulties: The height of the level is only 6 to 9 inches (150 to 225 mm); the liquid is constantly in a state of extreme agitation, bubbling, and boiling; and the liquid is highly concentrated and tends to plug those level sensors that depend on physical contact for measurement. (See Chapter 3 in the volume on Process Measurement for non–contact-level detectors.) Frequently the control of feed is manual, indicated in the diagram by a manual loading station (HIC). With all of the other controls on manual, only the steam pressure is controlled automatically.

Rotary

The term *tumbling dryers* is used here to designate equipment in which material is tumbled, or mechanically turned over, during the drying process. Two examples in this class are the rotary dryer and the turbo dryer.

The cross section of a rotary dryer is shown in Figure 8.9k. As the shell is rotated, the material is lifted by the flights and then dropped through an air stream. The speed of rotation, the angle of elevation, and the air velocity determine the material holdup time. Variable-speed drives are sometimes incorporated to change the rotation rate, but they are usually manually controlled. Air flow is not varied as a control parameter.

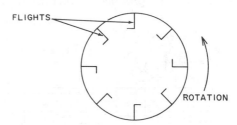

Fig. 8.9k   Cross section of rotary dryer

A typical control scheme for a rotary, counterflow dryer (material direction opposite air direction) is shown in Figure 8.9l. Primary controls maintain the air flow and inlet air temperature. Secondary considerations are pressure control on the outlet air to maintain pressure within the dryer and temperature and level alarms at the dry product outlet. A temperature switch is also provided at the air outlet to prevent dryer over-heating when feed is stopped. Direct setting of the air inlet temperature is based on the supposition that the product approaches this value before it reaches the discharge end. Better control of moisture is obtained if the control is based directly on the temperature of the product, but there are practical difficulties in providing a material holdup for measurement with consistent renewal. The feed rate

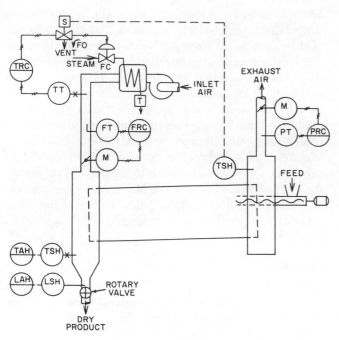

Fig. 8.9l   Rotary dryer

through a slot after a little less than one revolution. A leveler bar then spreads it evenly on the shelf below. In addition to the general countercurrent of hot air, fan blades on a central shaft impart a horizontal velocity pattern.

Since the internal fan provides consistent air circulation, it is feasible to control the air throughout. Motorized dampers are provided for the lower air input and for the combined flow to the upper sections. The division between the upper inlets is adjusted by manual dampers. The inlet air temperature is controlled by throttling of the steam valve, and the motorized dampers are adjusted by the corresponding temperatures. Circulation rates of the center fan and the shelves and the feed rate of material are manually controlled.

Spray

The spray dryer is an exception to the generalization of large sojourn times in dryers, with holdup normally on the order of tenths of a second. A liquid or thin slurry feed is atomized into a chamber in which there is a large flow of hot air (Figure 8.9n). The inlet air temperature must be quite high, and therefore direct-fired heaters are used whenever possible.

is also manually controlled, since the long holdup makes automatic adjustments impractical.

Turbo

In a turbo dryer the material is dried on rotating horizontal shelves (Figure 8.9m). They are arranged in a vertical stack, and the product is wiped from each shelf

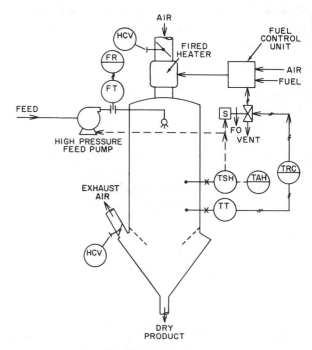

Fig. 8.9n   Spray dryer

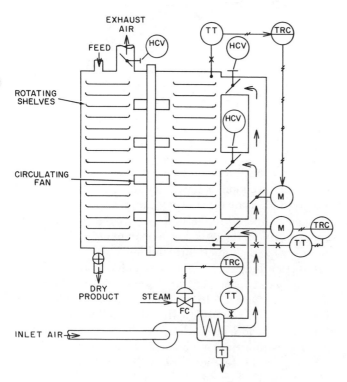

Fig. 8.9m   Turbo dryer

The feed is introduced by a high-pressure manually controlled pump. Air flow is also manually regulated and balanced. Process conditions are maintained by temperature control near the outlet end. The temperature controller regulates the firing rate of the fuel-air mixture through a gas control unit. (See Section 8.12 for details). A temperature switch is provided to shut off both the feed and the fuel in case of fire or other abnormally high temperature conditions.

Fluid Bed

The fluid-bed dryer has a fluidized bed of material maintained by an air flow upward through a perforated plate (Figure 8.9o). Feed is controlled by a variable-speed screw, and discharge is by overflow of the bed through a side arm.

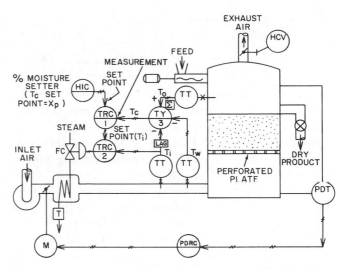

Fig. 8.9p Humidity-corrected feedforward control of fluid-bed dryer

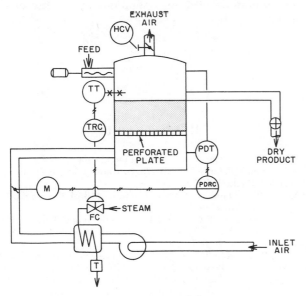

Fig. 8.9o Fluid-bed dryer

The bed is maintained at the desired product moisture level by a temperature controller with its bulb within the fluidized product or in the air space above.

An improved but rather more sophisticated control scheme has been suggested. Since the system of Figure 8.9o is sensitive to the temperature but not to the absolute humidity of the air stream, an increase in humidity of the entering air will cause a reduction in inlet air temperature instead of an increase. The effect of an increased humidity is to reduce the drying rate, which represents less heat loss from the air, and to start to raise the outlet temperature. The controller compensates by lowering the inlet temperature, which further reduces the drying capacity of the air.

Two alternative control schemes are available that do not exhibit this effect. One is an adaptation of the system of Figure 8.9l, in which air flow rate and inlet temperature are directly controlled. The other is the feedforward cascade control scheme illustrated in Figure 8.9p.

Feedforward System

For a condition in which the product moisture content required is in the "falling rate region," ($X_p < X_c$, Figure 8.9q) the product moisture can be expressed as

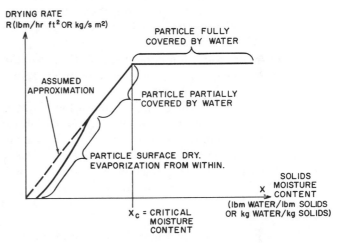

Fig. 8.9q Drying rate versus moisture content

$$X_p = \text{constant } \ln\frac{T_i - T_w}{T_o - T_w} \qquad 8.9(1)$$

where X_p is product wetness and T_i, T_o, T_w respectively are air inlet, outlet, and wet bulb temperatures. Note the absence of the variables: feed rate, air rate, feed moisture, air humidity. Thus, any one of these variables can be changed without affecting product dryness if the relationship of these three temperatures is maintained. Equation 8.9(1) has three temperature variables and therefore gives specific solutions only after incorporation of the dry bulb–wet bulb relationships and of the conservation of energy equation. For a given dryer and a given product moisture content,

$$\frac{T_i - T_w}{T_o - T_w} = K \qquad 8.9(2)$$

is constant. With this definition of K, the outlet air temperature required to compensate for load changes is given by

$$T_c = T_{on} - \frac{\overbrace{a}}{K(T_i - T_{in})} -$$

$$- \frac{\overbrace{b}}{(1 - K)(T_w - T_{wn})} \quad 8.9(3)$$

In this equation the second subscript, n, denotes normal load conditions. As shown in Figure 8.9p, the correcting terms (a and b in equation 8.9(3)) are subtracted from the detected value of T_o in TY-3, and the resulting T_c (corrected) signal becomes the measurement to TRC-1. The set point of TRC-1 is manually set to allow adjustment for new product dryness requirements, and therefore TRC-1 in effect is a moisture controller.

The loop operation is as follows: If feed rate or feed moisture content increases, it reduces T_o, and therefore T_c drops. This causes an increase in TRC-1 output, which raises the set point of TRC-2. If air humidity is increased, this increases T_w, which also reduces T_c and therefore raises the set point of TRC-2. (To stabilize the system, a lag is shown, the function of which is elaborated on in Section 1.5).

To control moisture, temperatures at *both* ends of the dryer *must* change, and the difference between T_i and T_o is a measure of dryer load.

If the product is heat-sensitive, then a high limit is introduced. If the inlet temperature (T_i) is too high for convenient wet bulb (T_w) measurement, this signal can also be obtained by measuring air humidity *upstream* to the heater and by use of a function generator. The wet bulb temperature is then obtained from these data.

Conclusions

The control systems that have been described represent actual current practice. The scheme shown for any particular dryer is not a unique solution. Each application could be handled in several different ways. They are intended not to innovate, but rather to illustrate, combinations taken from existing installations.

Improved analysis of dryer operation is now available through use of computer techniques. The feedforward cascade scheme for fluid-bed dryers presented earlier is one example of the review of basic drying equations that can lead to better control. With the improved methods available, the control engineer will face the new challenge of optimizing dryer operations by the use of automatic controls.

BIBLIOGRAPHY

Hawkins, J.C., "Advances in Mineral Dryer Control Systems," 1970 ISA Conference, Reprint No. 850–70, Instrument Society of America, Research Triangle Park, NC, 1970.

Myron, T.J., "Product Moisture Control," *MBAA Technical Quarterly*, Vol. 12, No. 4, 1975.

———, Shinskey, F.G., and Baker, R., "Inferential Moisture Control of a Spray Dryer," Presented at the ISA Food Industries Symposium, Montreal, June 4–6, 1973.

Perry, R.H. and Chilton, C.H., *Chemical Engineers' Handbook*, 5th ed., McGraw-Hill Book Co., New York, 1973, p. 20–13.

Shinskey, F.G., "Controlling Unstable Processes, Part II: A Heat Exchanger," *Instruments and Control Systems*, January 1975.

———, "Effective Control for Automatic Startup and Plant Protection," *Canadian Control Instrumentation*, April 1973.

———, "How to Control Product Dryness," *Instrumentation Technology*, October 1968.

———, "Process Control Systems with Variable Structure," *Control Engineering*, August 1974.

Weiner, A.L., "Drying Instrument Air," *ISA Journal*, August 1965.

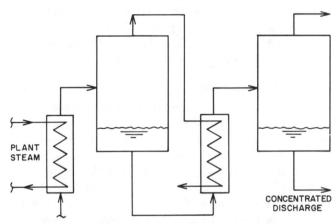

8.10   EVAPORATOR CONTROL

Evaporation is one of the oldest unit operations, dating back to the Middle Ages, when any available energy source was used to concentrate thin brine solutions in open tanks. As archaic as this technique is, it is still widely used during the early spring in northern New England when maple sugar sap is tapped and concentrated in open-fired pans. Solar evaporation and bubbling hot gases through a solution are other examples of concentrating solutions. For our purposes, evaporation will be limited to concentrating aqueous solutions in a closed vessel or group of vessels in which the concentrated solution is the desired product and indirect heating (usually steam) is the energy source. Occasionally, the water vapor generated in the evaporator is the product of interest, such as in desalinization or in the production of boiler feedwater. In other cases neither vapor nor concentrated discharge has any market value, as in nuclear wastes.

Evaporator Terminology

Single-effect evaporation. Single-effect evaporation occurs when a dilute solution is contacted only once with a heat source to produce a concentrated solution and an essentially pure water vapor discharge. The operation is shown schematically in Figure 8.10a.

Multiple-effect evaporation. Multiple-effect evapora-

tions use the vapor generated in one effect as the energy source to an adjacent effect (Figure 8.10b). Double- and triple-effect evaporators are the most common; however, six-effect evaporation can be found in the paper industry where kraft liquor is concentrated, and as many as 20 effects can be found in desalinization plants.

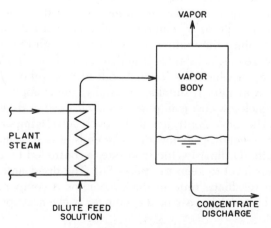

Fig. 8.10b   Multiple-effect evaporator

Boiling point rise. This term expresses the difference (usually in °F) between the boiling point of a constant composition solution and the boiling point of pure water at the same pressure. For example, pure water boils at 212°F (100°C) at 1 atmosphere, and a 35% sodium hydroxide solution boils at about 250°F (121°C) at 1 atmosphere. The boiling point rise is therefore 38°F (21°C). Figure 8.10c illustrates the features of a Dühring plot in which the boiling point of a given composition solution is plotted as a function of the boiling point of pure water.

Economy. This term is a measure of steam use and is expressed in pounds of vapor produced per pound of steam supplied to the evaporator train. For a well designed evaporator system the economy will be about 10% less than the number of effects; thus, for a triple-effect evaporator the economy will be roughly 2.7.

Capacity. The capacity for an evaporator is measured in terms of its evaporating capability, viz., pounds of vapor produced per unit time. The steam requirements

Fig. 8.10a   Single-effect evaporator

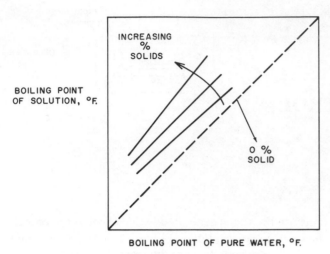

Fig. 8.10c   Dühring plot of boiling point rise

for an evaporating train may be determined by dividing the capacity by the economy.

Co-current operation. The feed and steam follow parallel paths through the evaporator train.

Countercurrent operation. The feed and steam enter the evaporator train at opposite ends.

Types of Evaporators

Six types of evaporators are used for most applications, and for the most part the length and orientation of the heating surfaces determine the name of the evaporator.

Horizontal tube evaporators (Figure 8.10d) were among the earliest types. Today, they are limited to preparation of boiler feed water, and in special construction (at high cost), for small-volume evaporation of severely scaling liquids, such as hard water. In their standard form they

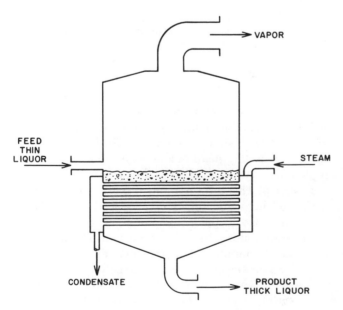

Fig. 8.10d   Horizontal-tube evaporator

are not suited to scaling or salting liquids and are best used in applications requiring low throughputs.

Forced circulation evaporators (Figure 8.10e) have the widest applicability. Circulation of the liquor past the heating surfaces is assured by a pump, and consequently these evaporators are frequently external to the flash chamber so that actual boiling does not occur in the tubes, thus preventing salting and erosion. The external tube bundle also lends itself to easier cleaning and repair than the integral heater shown in the figure. Disadvantages include high cost, high residence time, and high operating costs due to the power requirements of the pump.

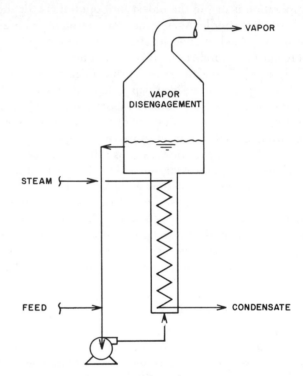

Fig. 8.10e   Forced-circulation evaporator

A short-tube vertical evaporator (Figure 8.10f) is common in the sugar industry for concentrating cane sugar juice. Liquor circulation through the heating element (tube bundle) is by natural circulation (thermal convection). Since the mother liquor flows through the tubes, they are much easier to clean than those shown in Figure 8.10d, in which the liquor is outside the tubes. Thus, this evaporator is suitable for mildly scaling applications in which low cost is important and cleaning or descaling must be conveniently handled. Level control is important—if the level drops below the tube ends, excessive scaling results. Ordinarily, the feed rate is controlled by evaporator level to keep the tubes full. The disadvantage of high residence time in the evaporator is compensated for by the low cost of the unit for a given evaporator load.

A long-tube vertical evaporator, or rising film concen-

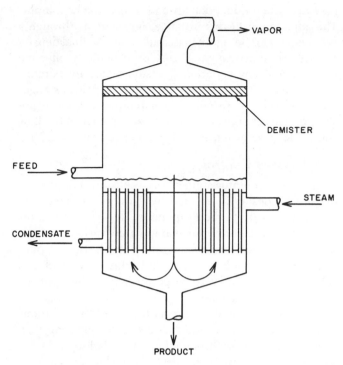

Fig. 8.10f   Short-tube vertical evaporator

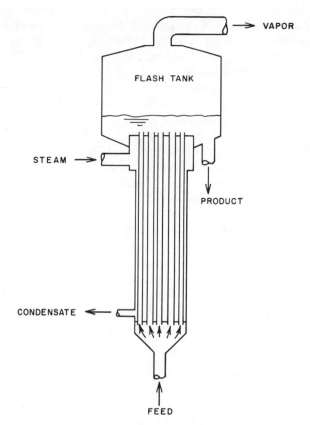

Fig. 8.10g   Long-tube vertical evaporator

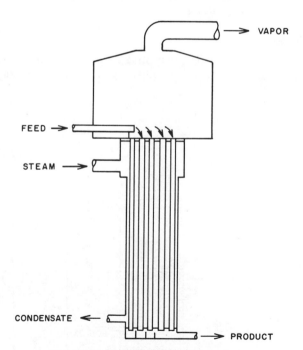

Fig. 8.10h   Falling-film vertical evaporator

trator (RFC), shown in Figure 8.10g is in common use today, because its cost per unit of capacity is low. Typical applications include concentrating black liquor in the pulp and paper industry and corn syrup in the food industry. Most of these evaporators are of the single-pass variety, with little or no internal recirculation. Thus, residence time is minimized. Level control is important in maintaining the liquid seal in the flash tank. The units are sensitive to changes in operating conditions, which is why many of them are difficult to control. They offer low cost per pound of water evaporated and have low holdup times but tend to be tall (20 to 50 feet, or 6 to 15 m), requiring more head room than other types.

A falling film evaporator (Figure 8.10h) is commonly used with heat-sensitive materials. Physically, the evaporator looks like a long-tube vertical evaporator, except that the feed material descends by gravity along the inside of the heated tubes, which have large inside diameters (2 to 10 inches, or 50 to 250 mm).

An agitated film evaporator, like the falling film evaporator, is commonly used for heat-sensitive and highly viscous materials. It consists of a single large-diameter tube with the material to be concentrated falling in a film down the inside, where a mechanical wiper spreads the film over the inside surface of the tube. Thus, large heat transfer coefficients can be obtained, particularly with highly viscous materials.

Control Systems for Evaporators

The control systems to be considered in achieving final product density include (1) feedback, (2) cascade, and (3) feedforward. For ease of illustration, a double effect, co-current flow evaporator will be used. Extension to more or fewer effects will not change the control system configuration.

The choice of system should be based on the needs and characteristics of the process. Evaporators as a process class tend to be capacious (mass and energy storage capability) and have significant dead time (30 seconds or greater). If the major process loads (feed rate and feed density) are reasonably constant and the only corrections required are for variations in heat losses or tube fouling, feedback control will suffice. If steam flow varies because of demands elsewhere in the plant, a cascade configuration will probably be the proper choice. If, however, the major load variables change rapidly and frequently, it is strongly suggested that feedforward in conjunction with feedback be considered.

Feedback Control

A typical feedback control system (Figure 8.10i), consists of measuring the product density with a density sensor and controlling the amount of steam to the first effect by a three-mode controller. The internal material balance is maintained by level control on each effect. (A brief description of the various methods of measuring product density will be found at the end of this section; additional discussion regarding density measurement may be found in Chapter 6 in the volume on Process Measurement). Ritter and associates[1] have modeled the control system configuration shown in Figure 8.10i and have found it to be very stable. They investigated other combinations of controlled and manipulated variables, which provided poor evaporation.

Cascade Control

A typical system is illustrated in Figure 8.10j. This control system, like the feedback loop in Figure 8.10i,

measures the product density and adjusts the heat input. The adjustment in this instance, however, is through a flow loop that is being set in cascade from the final density controller, an arrangement that is particularly effective when steam flow variations (outside of the evaporator) are frequent. It should be noted that with this arrangement the valve positioner is not required. A valve positioner will actually degrade the performance of the flow control loop. (For more information see Section 1.7.)

Feedforward Control

In most evaporator applications the control of product density is constantly affected by variations in feed rate and feed density to the evaporator. In order to counter these load variations, the manipulated variable (steam flow) must attain a new operating level. In the pure feedback or cascade arrangements this new level was achieved by trial and error as performed by the feedback (final density) controller.

A control system able to react to these load variations when they occur (feed rate and feed density) rather than wait for them to pass through the process before initiating a corrective action would be ideal. The technique is termed *feedforward control*. (For more on this technique see Section 1.5.) Figure 8.10k illustrates in block diagram form the features of a feedforward system. There are two types or classes of load variations—measured and unmeasured. The measured load signals are inputs to the feedforward control system, where they compute the set point of the manipulated variable control loop as a function of the measured load variables. The unmeasured load variables pass through the process, undetected by the feedforward system, and cause an upset in the con-

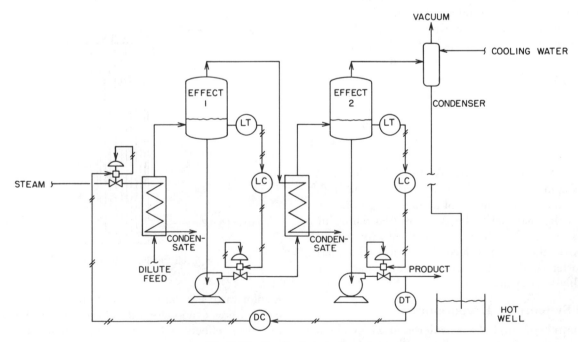

Fig. 8.10i   Feedback control system

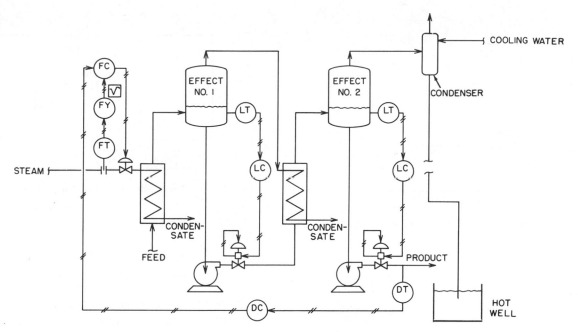

Fig. 8.10j   Cascade control system

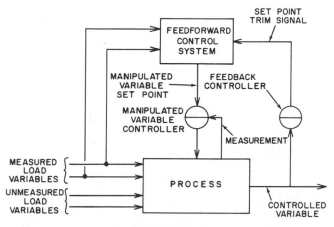

Fig. 8.10k   Feedforward system

trolled variable. The output of the feedback loop then trims the calculated value of set point to the correct operating level. In the limit, feedforward control would be capable of perfect control if all load variables could be defined, measured, and incorporated in the forward set point computation. Practically speaking, the expense of accomplishing this goal is usually not justified.

At a practical level, then, the load variables are classified as either major or minor, and the effort is directed at developing a relationship that incorporates the major load variables, the manipulated variables, and the controlled variable. Such a relationship is termed the steady state model of the process. Minor load variables are usually very slow to materialize and are hard to measure. In terms of evaporators, minor load variations might be heat losses and tube fouling. Load variables such as these are easily handled by a feedback loop. The purpose of

the feedback loop is to trim the forward calculation to compensate for the minor or unmeasured load variations. Without this feature the controlled variable could be off the set point. Thus far we have discussed two of the three ingredients of a feedforward control system, viz., the steady state model and feedback trim. The third ingredient of a feedforward system is dynamic compensation. A change in one of the major loads to the process also modifies the operating level of the manipulated variable. If these two inputs to the process enter at different locations, there usually exists an imbalance or inequality between the effect of the load variable and the effect of the manipulated variable on the controlled variable, i.e.,

$$\frac{\Delta\text{controlled variable}}{\Delta\text{load variable}} \neq \frac{\Delta\text{controlled variable}}{\Delta\text{manipulated variable}}$$

This imbalance manifests itself as a *transient* excursion of the controlled variable from set point. If the forward calculation is accurate, the controlled variable returns to set point once the new steady state operating level is reached. In terms of a co-current flow evaporator, an increase in feed rate will call for an increase in steam flow. Assuming that the level controls on each effect are properly tuned, the increased feed rate will rapidly appear at the other end of the train while the increased steam flow is still overcoming the thermal inertia of the process. This sequence results in a transient decrease of the controlled variable (density), and the load variable passes through the process faster than the manipulated variable. This behavior is shown in Figure 8.10l. The same arrangement is seen in Figure 8.10m, except that the manipulated variable passes through the process faster

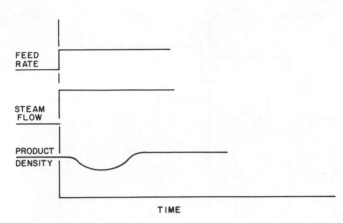

Fig. 8.10l   Load variable faster than manipulated variable

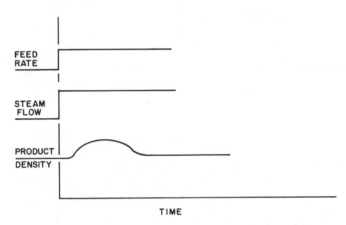

Fig. 8.10m   Manipulated variable faster than load variable

than the load variable. Such behavior may occur in a countercurrent evaporator operation. This dynamic imbalance is normally corrected by inserting a dynamic element (lag, lead-lag, or a combination thereof) in at least one of the load measurements to the feedforward control system. Usually, dynamic compensation of that major load variable, which can change in the severest manner (usually a step change), is all that is required. For evaporators this is usually the feed flow rate to the evaporator. Feed density changes, although frequent, are usually more gradual, and the inclusion of a dynamic element for this variable is not warranted. In summary, the three ingredients of a feedforward system are (1) the steady state model, (2) process dynamics, and (3) addition of feedback trim.

STEADY STATE MODEL

Development of the steady state model for an evaporator involves material and energy balance. A relationship between the feed density and percent solids is also required and is specific for a given process, whereas the material and energy balances are applicable to all evaporator processes. Figure 8.10n illustrates the double-effect evaporator from the standpoint of a material balance.

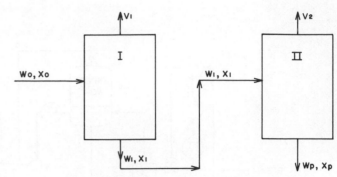

Fig. 8.10n   Double-effect evaporator material balance

Where

W_o = feed rate in lbs (kg) per unit time,

V_1 = vapor flow from Effect I in lbs (kg) per unit time,

X_o = weight fraction solids in feed,

W_1 = liquid flow rate leaving Effect I in lbs (kg) per unit time,

X_1 = weight fraction solids in W_1,

V_2 = vapor flow from Effect II in lbs (kg) per unit time,

W_p = liquid product flow in lbs (kg) per unit time, and

X_p = weight fraction solids in product (the controlled variable).

Overall balance in Effect I:

$$W_o = V_1 + W_1 \qquad 8.10(1)$$

Overall balance in Effect II:

$$W_1 = V_2 + W_p \qquad 8.10(2)$$

Solid balance in Effect I:

$$W_o X_o = W_1 X_1 \qquad 8.10(3)$$

Solid balance in Effect II:

$$W_1 X_1 = W_p X_p \qquad 8.10(4)$$

Substituting 8.10(2) in 8.10(1):

$$W_o = V_1 + V_2 + W_p \qquad 8.10(5)$$

Combining 8.10(3) and 8.10(4):

$$W_o X_o = W_p X_p \qquad 8.10(6)$$

Solving 8.10(6) for W_p and substituting in 8.10(5) gives

$$W_o = V_1 + V_2 + \frac{W_o X_o}{X_p} \qquad 8.10(7)$$

The W_o term of 8.10(7) can be written in terms of volumetric flow (gallons per unit time), the usual method of measuring this variable:

$$W_o = V_o D_o = V_o D_w S_o \qquad 8.10(8)$$

where

V_o = volumetric feed rate in gallons per unit time,

D_o = feed density in lbs per gallon,

D_w = nominal density of water, 8.33 lbs per gallon (1 kg/l), and

S_o = specific gravity of feed.

Substituting for W_o from 8.10(8) in 8.10(7) and combining terms,

$$V_o D_w S_o \left(1 - \frac{X_o}{X_p}\right) = V_1 + V_2 = V_t \qquad 8.10(9)$$

where V_t = total vapor flow in lbs per unit time.

The total vapor flow (V_t) is proportional to the energy supplied to the train (plant steam) and the proportionality constant is the economy (E) of the system, i.e.,

$$V_t = W_s E \qquad 8.10(10)$$

where

W_s = steam flow in lbs steam per unit time, and

E = economy in lbs vapor per lb steam.

Substituting for V_t in 8.10(9)

$$V_o D_w S_o \left(1 - \frac{X_o}{X_p}\right) = W_s E \qquad 8.10(11)$$

Equation 8.10(11) is the steady state model of the process and includes all of the load variables (V_o and X_o), the manipulated variable (W_s), and the controlled variable (X_p). At this point the $W_s E$ portion of equation 8.10(11) may be modified to include heat losses from the system and to include the fact that the feed may be subcooled; this is effectively a heat loss, since in either case a portion of the steam supplied is for purposes other than producing vapor. Typical values of effective heat losses vary from 3 to 5%. If, for example, a 5% heat loss is assumed, equation 8.10(11) becomes

$$V_o D_w S_o \left(1 - \frac{X_o}{X_p}\right) = 0.95 \, W_s E \qquad 8.10(12)$$

For an in-depth discussion relating to methods of computing the economy (E) of a particular evaporator system, see references (2) and (3).

The $S_o (1 - X_o/X_p)$ portion of equation 8.10(11) is a function of the feed density, $f(D_o)$, i.e.,

$$f(D_o) = S_o \left(1 - \frac{X_o}{X_p}\right) \qquad 8.10(13)$$

For each feed material a relationship between the density of the feed material and its solids weight fraction has usually been empirically determined by the plant or is available in the literature. See reference (4), where the

density = percent solids relationship of 70 inorganic compounds is available.

Assume, for example, that a feed material (to be concentrated) has the solids-specific gravity relationship shown in Table 8.10o. If this feed material were to be concentrated so as to produce a product having a weight fraction of 50% (X_p = 0.50), the $f(D_o)$ relationship of 8.10(13) could be generated as shown in Table 8.10p.

Table 8.10o
WEIGHT FRACTION AND
SPECIFIC GRAVITY
RELATIONSHIP

Solids Weight Fraction (X_o)	Specific Gravity (S_o)
0.08	1.0297
0.16	1.0633
0.24	1.0982

Table 8.10p
DENSITY TO WEIGHT FRACTION RELATIONSHIP

Solids Weight Fraction (X_o)	Specific Gravity (S_o)	$\left(1 - \dfrac{X_o}{X_p}\right)$	$S_o\left(1 - \dfrac{X_o}{X_p}\right) = f(D_o)$
0	1.0000	1.0	1.0000
0.08	1.0297	0.840	0.865
0.16	1.0633	0.680	0.723
0.24	1.0982	0.520	0.571

This body of data is plotted in Figure 8.10q. In all the cases investigated, the $f(D_o) = S_o$ relationship is a straight line having an intercept of 1.0, 1.0. The $f(D_o)$ relationship can then be written in terms of the equation of a straight line: $y = mx + b$, i.e.,

$$f(D_o) = 1.0 + m \, (S_o - 1.0) \qquad 8.10(14)$$

where m = slope of line.

Using the data of Table 8.10p the value of m is determined as

$$m = \frac{1.0000 - 0.571}{1.0000 - 1.0982} = -4.37 \qquad 8.10(15)$$

Therefore,

$$f(D_o) = 1.0 - 4.37 \, (S_o - 1.0) \qquad 8.10(16)$$

Substituting 8.10(16) into 8.10(11) and solving for the manipulated variable, steam flow, W_s:

$$W_s = \frac{V_o D_w f(D)}{E} \qquad 8.10(17)$$

SCALING AND NORMALIZING

With the equation for the steady state model defined, it can now be scaled and the analog instrumentation

911

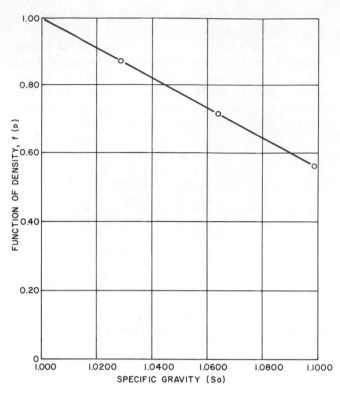

Fig. 8.10q   Density to specific gravity relationship

specified. (For more information see Section 7.1). Scaling of analog computing instruments is necessary to ensure compatibility with input and output signals and is accomplished most effectively by normalizing, i.e., assigning values from 0 to 1.0 to all inputs and outputs. The procedure involves (1) writing the engineering equation to be solved, (2) writing a normalized equation for each term variable of the engineering equation, and (3) substituting the normalized equivalent of each term in (2) into (1).

The first step of the procedure has already been done—equation 8.10(17) is written. To illustrate, let V_o = 0 to 600 GPH (0 to 2.3 m^3/h), W_s = 0 to 2500 lbs per hr (0 to 1125 kg/h), S_o = 1.000 to 1.1000, E = 1.8 lbs vapor per lb of steam (1.8 kg vapor per kg of steam), and D_w = 8.33 lbs per gallon (1 kg/l). The scaled equations for each input are

$$V_o = 600 \ V_o' \qquad\qquad 8.10(18)$$

$$W_s = 2500 \ W_s' \qquad\qquad 8.10(19)$$

$$S_o = 1.0000 + 0.10000 \ S_o' \qquad 8.10(20)$$

where

V_o' = volumetric flow transmitter output, 0 to 1.0 or 0 to 100%,

W_s' = steam flow transmitter output, 0 to 1.0 or 0 to 100%,

S_o' = specific gravity transmitter output, 0 to 1.0 or 0 to 100%,

The values of D_w and E need not be scaled, since they are constants. Since $f(D_o)$ is already on a 0 to 1.0 basis, the $f(D_o)$ term for the sake of completeness can be written

$$f(D_o) = 1.0 \ f(D_o)' \qquad\qquad 8.10(21)$$

Operating on the $f(D_o)$ equation first, equations 8.10(21) and 8.10(20) are substituted into equation 8.10(16):

$$(D_o)' = 1.0 - 4.37 \times$$

$$\times \ (1.0000 + 0.1000 \ S_o' - 1.0) \quad 8.10(22)$$

$$(D_o)' = 1.0 - 0.437 \ S_o' \qquad\qquad 8.10(22)$$

Substituting 8.10(18), 8.10(19), and 8.10(22) into 8.10(17) as well as the values of E and D_w,

$$2500 \ W_s' = \frac{600(V_o') \ 8.33 \ (1.0 - 0.437 \ S_o')}{1.8} \quad 8.10(23)$$

$$W_s' = 1.11 \ V_o' \ (1.0 - 0.437 \ S_o') =$$

$$= 1.11 \ V_o' \ f(D_o)' \qquad\qquad 8.10(23)$$

PROCESS DYNAMICS

The dynamics of the co-current evaporator, in which steam is the manipulated variable, require that a lead-lag dynamic element be incorporated in the system to compensate for the dynamic imbalance between feed rate and steam flow. In the example, it was arbitrarily assumed that steam flow is the manipulated variable resulting in equation 8.10(17). In some applications evaporators are run on waste steam, in which case the feed rate is proportionally adjusted to the available steam, which makes feed the manipulated variable and steam the load variable. Solving 8.10(17) for feed rate,

$$V_o' = \frac{W_s E}{D_w(D)} \qquad\qquad 8.10(24)$$

In this arrangement the dynamics do not change, but the manipulated variable advances through the process faster than the load variable, which requires a dynamic element having first-order lag characteristics. The instrument arrangement for each case is shown in Figures 8.10r and 8.10s.

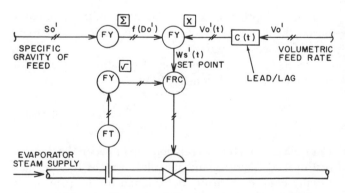

Fig. 8.10r   Feedforward with dynamics (Equation 8.10 [23])

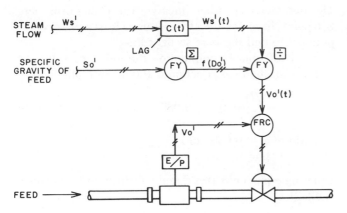

Fig. 8.10s   Feedforward with dynamics (Equation 8.10 [24])

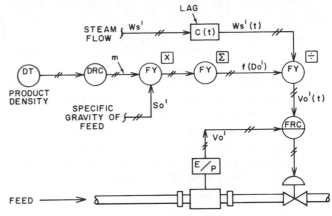

Fig. 8.10u   Feedforward-feedback with dynamics (Equation 8.10 [24])

FEEDBACK TRIM

The final consideration is feedback trim. As a general rule, feedback trim is incorporated into the control system *at the point at which the set point of the controlled variable appears*. For the evaporator the set point is the slope of the $f(D_o)$ relationship (Figure 8.10q). If the value of X_p changes, the slope of the line changes too. To this point the value of m was assumed to be a constant (0.437), which is incorporated into the summing relay or amplifier (Figures 8.10q and 8.10r). The instrumentation was scaled to make one grade of product, e.g., 50% solids. If a more or less concentrated product were desired, the gain term would have to be changed manually. In order to increase the flexibility of the control system, a multiplier and a final product density control loop are added. The controller output is now variable not only to permit changing the concentration of the product (slope adjust) but also to adjust the steam flow set point to compensate for the minor load variations, which up to this point were not considered. Each arrangement (comparable with Figures 8.10r and 8.10s) is shown in Figures 8.10t and 8.10u.

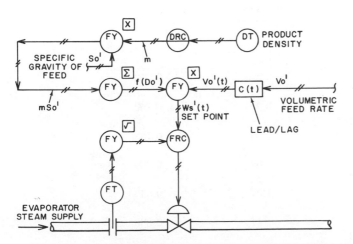

Fig. 8.10t   Feedforward-feedback with dynamics (Equation 8.10 [23])

Other Control Loops

STEAM ENTHALPY

So far we have discussed only load changes related to feed density and feed flow rate and have specified the instruments to accomplish the control objective. One additional load variable that, if allowed to pass through the process, would upset the controlled variable, i.e., product density, is steam enthalpy. In some applications the steam supply may be carefully controlled so that its energy content is uniform; in other applications substantial variations in steam enthalpy may be experienced. The objective is to consider the factors that influence the energy level of the steam supplied and to design a system that will protect the process from these load variables.

For saturated steam the energy content per unit weight is a function of the absolute pressure of the steam. If the flow of steam to the process is measured with an orifice meter, the mass flow of steam is

$$W_s = K_1 \sqrt{\frac{h}{v}} \qquad 8.10(25)$$

where

    W_s = steam flow in lbs per hour (kg/hr),
    h = differential head measurement in feet (m),
    v = specific volume in cu ft per lbm (m³/kg), and
    K_1 = orifice coefficient dependent on the physical characteristics of the orifice.

The total energy to the system is

$$Q = W_s H_s \qquad 8.10(26)$$

where

    Q = energy to system per BTU per hour (J/h),
    W_s = steam flow in lbs per hour (kg/h), and
    H_s = heat of condensation in BTU per lb, or J/kg (enthalpy of saturated vapor minus enthalpy of saturated liquid).

Substituting Equation 8.10(25) into Equation 8.10(26) gives

$$Q = K_1 \sqrt{\frac{(H_s)^2}{v}h} \qquad 8.10(27)$$

For any particular application the steam pressure will vary within limits around a normal operating pressure. To demonstrate the design of a control system to compensate for variations in energy input to the process, assume that the steam pressure varies between 18 and 22 PSIA (124 and 152 kPa), with a normal operating value of 19.7 PSIA (135.9 kPa). The pressure transmitter has a range of 0 to 25 PSIA (0 to 172.5 kPa). These values of pressure variation and operating pressure are typical for evaporator operation. The value of the $(H_s)^2/v$ term appearing in equation 8.10(27) will vary, depending on the steam pressure. Over a reasonably narrow range of pressures, the value of $(H_s)^2/v$ can be approximated by a straight line with the general form of

$$(H_s)^2/v = bP + a \qquad 8.10(28)$$

where

P = absolute pressure in PSIA (Pa), and
a and b = constants.

Table 8.10v shows the typical values of H_s, v, $((H_s)^2/v)'$ and p and p' selected from the specified range of pressures and from where the steam is condensed at 1.5 PSIG (10.4 kPa).

$$(H_s^2/v) \text{ normal} = \frac{(970.7)^2}{20.4} = 46,208$$

Rewriting equation 8.10(28) in scaled form:

$$\left(\frac{(H_s)^2}{v}\right)' = bp' + a \qquad 8.10(29)$$

The designer can either linearize the data—using any two points from Table 8.10u—or use a least squares computation to find the best straight line.

In a least square computation the following values of a and b constants of equation 8.10(29) were obtained:

$$\left(\frac{(H_s)^2}{v}\right)' = 0.085 + 1.161 p' \qquad 8.10(30)$$

where a = 0.085, and
b = 1.161.

Squaring equation 8.10(27) and substituting in equation 8.10(30):

$$(Q^2)' = (1.161 p' + 0.085) hk_1 \qquad 8.10(31)$$

The parenthetical portion of equation 8.10(31) can be rewritten so as to make the sum of the two coefficients equal 1.0, which simplifies its implementation using conventional analog hardware. This is done by multiplying and dividing each term in the parenthesis by the sum of the two coefficients.

$$(Q^2)' = 1.246 \left(\frac{1.161 p'}{1.246} + \frac{0.085}{1.246}\right) hk_1 \qquad 8.10(32)$$

$$(Q^2)' = 1.246 (0.932 p' + 0.068) hk_1 \qquad 8.10(32)$$

The instrumentation to implement equation 8.10(32) is shown in Figure 8.10w.

INTERNAL MATERIAL BALANCE

The feedforward system described earlier imposes an external material balance as well as an internal material balance on the process. The internal balance is maintained by liquid level control on the discharge of each effect.

Analysis of a level loop indicates that a narrow pro-

Table 8.10v
SPECIFIC VOLUME-ENTHALPY DATA

(p) Steam Supply Pressure PSIA (kPa)	(H_s) Heat of Condensation BTU/lbm (MJ/ka)	(v) Specific Volume ft³/lbm (m³/kg)	$((H_s)^2/v) = \dfrac{(H_s)^2/v}{((H_s)^2/v)_{normal}}$	$p' = \dfrac{p}{p_{max}} = \dfrac{p}{25}$
18 (124.2)	969.1 (225.4)	22.2 (1.38)	0.916	0.720
19 (131.1)	970.2 (225.7)	22.1 (1.37)	0.965	0.760
19.7* (135.9)	970.7 (225.8)	20.4 (1.26)	1.00	0.788
20 (138)	971.2 (225.9)	20.1 (1.25)	1.02	0.800
21 (144.9)	972.1 (226.1)	19.2 (1.19)	1.07	0.840
22 (151.8)	973.0 (226.3)	18.4 (1.14)	1.11	0.880

* Normal operation

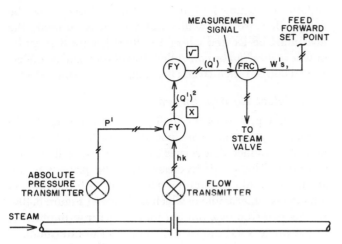

Fig. 8.10w Enthalpy compensation system

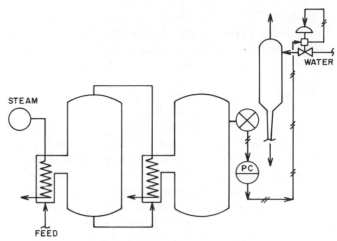

Fig. 8.10x Absolute pressure control

portional band (<10%) can achieve stable control. However, because of the resonant nature of the level loop that causes the process to oscillate at its natural frequency, a much lower controller gain must be used (proportional bands 50 to 100%).[5] With the wider proportional bands, reset is required to help maintain set point. A valve positioner is also recommended to overcome the usual limit cycle characteristics of an integrating process and the nonlinear nature of valve hysteresis. (See Section 1.7 for more information.)

ABSOLUTE PRESSURE

The heat to evaporate water from the feed material is directly related to the boiling pressure of the material. In most multiple-effect evaporations each effect is held at a pressure less than atmospheric in order to keep boiling points below 212°F (100°C). The lowest pressure is in the effect closest to the condenser, with pressures increasing slightly in each effect away from the condenser. Three possible methods of controlling the absolute pressure are (1) controlling the flow of water to the condenser, (2) bleeding air into the system with the water valve wide open, and (3) locating the control valve in the vapor drawoff line, manually setting the water flow rate and air bleeding as necessary.

Method 3 requires an extraordinarily large valve, since the vapor line may be 24 or 30 inches (600 or 750 mm) in diameter. Method 2 is uneconomical, because the expense of pumping the water offsets the savings realized by using a smaller valve on the air line. Method 1 represents the best compromise between cost and controllability and is preferred (Figure 8.10x). (For more on condenser pressure control see Section 8.13.)

Product Density Measurement

Perhaps one of the most controversial issues in any evaporator control scheme is the method to measure the product density. Common methods include (1) temper-

ature difference, boiling point rise; (2) conductivity; (3) differential pressure; (4) gamma gauge; (5) U-tube densitometer; (6) buoyancy float; and (7) refractive index.

Each method has its strengths and weaknesses. In all cases, however, care must be taken to select a representative measurement location to eliminate entrained air bubbles or excessive vibration, and the instrument must be mounted in an accessible location for cleaning and calibration. The location of the product density transmitter with respect to the final effect should also be considered. Long runs of process piping for transporting the product from the last effect to the density transmitter increase dead time, which in turn reduces the effectiveness of the control loop.

BOILING POINT RISE

Perhaps the most difficult and controversial method of product density measurement is by temperature difference or boiling point rise. Dühring's rule states that a linear relationship exists between the boiling point of a solution and the boiling point of pure water at the same pressure. Thus, the temperature difference between the boiling point of the solution in an evaporator and the boiling point of water at the same pressure is a direct measurement of the concentration of the solution. Two problems in making this measurement are location of the temperature bulbs and control of absolute pressure.

The temperature bulbs must be located so that the measured values are truly representative of the actual conditions. Ideally, the bulb measuring liquor temperature should be just at the surface of the boiling liquid. This location can change, unfortunately, if the operator decides to use more or less liquor in a particular effect. Many operators install the liquor bulb near the bottom of the pan where it will always be covered, thus creating an *error due to head effects* which must be compensated for in the calibration.

The vapor temperature bulb is installed in a condensing chamber in the vapor line. Hot condensate flashes over the bulb at an equilibrium temperature dictated by the pressure in the system. This temperature minus the liquid boiling temperature (compensated for head effects) is the temperature difference reflecting product concentration.

Changes in absolute pressure of the system alter not only the boiling point of the liquor but also the flashing temperature of the condensate in the condensing chamber. Unfortunately, the latter effect occurs much more rapidly than the former, resulting in transient errors in the system that may take a long time to resolve. Therefore, it is imperative that absolute pressure be controlled closely if temperature difference is to be a successful measure of product density. These systems were more effective installations when control of water rate to the condenser rather than an air-bleed system was used.

CONDUCTIVITY

Electrolytic conductivity is a convenient measurement to use in relationships between specific conductance and product quality (concentration), such as in a caustic evaporator. Problem areas include location of the conductivity cell so that product is not stagnant but is flowing past the electrodes; temperature limitations on the cell; cell plugging; and temperature compensation for variations in product temperature.

DIFFERENTIAL PRESSURE

Measuring density by differential pressure is a frequently used technique. The flanged differential pressure transmitter is preferred for direct connection to the process; otherwise, lead lines to the transmitter could become plugged by process material solidifying in the lines. Differential pressure transmitters are more frequently used on feed density than on final density measurements.

GAMMA GAUGE

This measurement is popular in the food industry because the measuring and sensing elements are not in contact with the process. It is very sensitive and not subject to plugging. Periodic calibration may be required because of the half-life of source materials. Occasionally, air is entrained especially in extremely viscous solutions. Therefore, the best sensor location is a flooded low point in the process piping.

U-TUBE DENSITOMETER

The U-tube densitometer, a beam-balance device, is also a final product density sensor. Solids can settle out in the measuring tube, causing calibration shifts or plugging.

BUOYANCY FLOAT

Primarily used for feed density detection, the buoyancy float can also be applied to product density if a

suitable mounting location near the evaporator can be found. Because flow will affect the measurement, the float must be located where the fluid is almost stagnant or where flow can be controlled (by recycle) and its effects zeroed out. A Teflon-coated float helps reduce drag effects.

Auto-Select Control System

In many processes the final product is the result of a two-step operation. The first step produces an intermediate product that serves as the feed to a final concentrator. The aim is to ensure that the process is run at the maximum throughput consistent with the process limitations, an example of which is shown in Figure 8.10y. In this two-step evaporation there are three limitations to the process: (1) the steam supply to the intermediate concentrator can be reduced as a result of demands in other parts of the plant; (2) the steam supply to the final concentrator can be reduced as a result of demands in other parts of the plant; and (3) the final concentrator can accept feed only at or below a certain rate, and it is desired to run this part of the process at its limit.

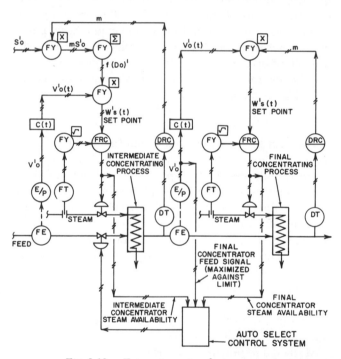

Fig. 8.10y Evaporator auto-select system

Each part of the process has its own feedforward-feedback system. The intermediate concentrating process has the feedforward system shown in Figure 8.10t. The final concentrating process does not require feed density compensation, so that it is necessary only to establish the ratio of steam to feed and to adjust the ratio by feedback. Ordinarily, the system is paced by the final concentrator feed signal, which is fed to the auto-select system. The final concentrator controller output manipulates the feed to the intermediate process so as to maintain the final concentrator feed at the set point limit.

Should either steam supply become deficient, its output would be selected to adjust the feed to the intermediate process. Thus, the process is always run at the maximum permissible rate consistent with the process limitation(s). (For more on auto-select control systems, see Section 1.8.)

REFERENCES

1. Ritter, R.A. and Andre, H., "Evaporator Control System Design," *Canadian Journal of Chemical Engineering*, 48: 696, December 1970.
2. McCabe, W.L. and Smith, J.C., *Unit Operations of Chemical Engineering*, McGraw-Hill, New York, 1956.
3. Brown, G.G., et al., *Unit Operations*, John Wiley & Sons, New York, 1950.
4. Perry, J.H., et al., *Chemical Engineers' Handbook, 4th ed.*, McGraw-Hill, New York, pp. 3–72 to 3–80.
5. Shinskey, F.G., *Process Control System*, 2nd ed., McGraw-Hill, New York, 1979, p. 74.

BIBLIOGRAPHY

Chen, C.S., "Computer Controlled Evaporation Process," *Instruments and Control Systems*, June 1982.

Economical Evaporator Designs by Unitech, Bulletin UT–110, Ecodyne United Division, Union, NJ, 1976.

Farin, W.G., "Control of Triple-Effect Evaporators," *Instrumentation Technology*, April 1976.

Hoyle, D.L. and Nisenfeld, A.E., "Dynamic Feedforward Control of Multi-effect Evaporators," *Instrumentation Technology*, February 1980.

Spencer, G.L. and Meade, G.P., *Cane Sugar Handbook*, 8th ed., New York, 1948, p. 699.

Stickney, W.W. and Fosbery, T.M., "Treating Chemical Wastes by Evaporation," *Chemical Engineering Progress*, April 1976.

8.11 EXTRUDER CONTROLS

Extruder Types

Single-screw extruders commonly convert granular resin feeds into sheets, films, and shapes such as pipe, and are described by screw diameter in inches and L/D ratio—L being the screw length and D the screw diameter. They are supplied in sizes of 1, $1\frac{1}{2}$, $2\frac{1}{2}$, $3\frac{1}{2}$, $4\frac{1}{2}$, 6, 8, and 12 inches (25, 37.5, 62.5, 87.5, 112.5, 150, 200, and 300 mm). L/D ratios from 20 to 30 are common. The single-screw extruder is by far the type most frequently manufactured.

Machines using twin screws are generally large-volume production units for pelletizing resins in petrochemical plants and are equipped with various combinations of intermeshing and nonmeshing screws that corotate or contrarotate. Screw design features allow compounding, devolatilizing, melting, blending, and other processing in a single machine. Twin screw machines are often melt-fed directly from polymerization reactors and perform multiple functions on the polymer prior to pelletizing and packaging it as a finished product.

Extruder Dies

The shape and ultimate use of the output are defined by the die shape. Dies are broadly classified as (1) sheet dies extruding flat sheets as much as 120 inches (3 m) wide and $\frac{1}{2}$ inch (12.5 mm) thick; (2) shape dies for making pipe, gasketing, tubular products, and many other designs; (3) blown film dies, using an annular orifice to form a thin-walled envelope. The diameter of the envelope is expanded with low pressure air to roughly three times the annular orifice diameter and forms a thin film. The process is for films 5 mils thick (0.005 inch, or 0.125 mm) at the upper limit; (4) spinnerette dies for extrusion of single or multiple strands of polymer for textile products, rope, tire cord, or webbing; (5) pelletizing dies for granular products in resin production, synthetic rubbers, and scrap reclaiming. These dies from multiple strands roughly $\frac{1}{8}$ inch (3.125 mm) in diameter. Rotating knives continuously cut the strands in short lengths, after which the pellets drop into water for cooling; and (6) cross-head dies for wire coating in which the bare wire or cable enters the die and emerges coated with semimolten polymer. The wire enters and leaves the die at an angle of 90 degrees to the extruder axis—hence the term cross-head.

Many special arrangements of extruder dies are used for composite films. Two polymers enter the die from two extruders and exit as a sheet; the top and bottom polymers are of different chemical composition so as to obtain a film or sheet of two colors or so as to have desirable characteristics of both materials. Dies for rigid foam production are similar to blown film dies; special dies with moving parts extrude netting continuously. Most dies require a short connecting pipelike piece, commonly called the adapter, to connect to the extruder head.

Polymer Types and Characteristics

Any list of commercial polymers is incomplete owing to the rapid progress of polymer science and production. The common resins, such as polyethylene and styrene, extrude rather simply and are relatively stable and reasonably predictable. Polymers in the vinyl chloride group are not stable at extrusion temperatures. If not promptly removed from the process they decompose, emitting hydrogen chloride gas, which is very corrosive to metal. Consequently, dies for vinyl chloride are commonly chrome-plated and designed so as to eliminate pockets or crevices where the material may be entrained and decompose.

An important characteristic of synthetic polymers that complicates the extrusion is sensitivity to shear rate, a phenomenon known as non-Newtonian behavior (see Section 7.1 in the Process Measurement volume of this handbook). The relationship between apparent viscosity and shear rate is also dependent on temperature and must be considered when control equipment is designed for extruders.[1] Certain polymers, such as polyvinyl chloride (PVC) and acrylonitrile-butadiene-styrene (ABS), are much more sensitive to shear rate than are more crystalline materials, such as nylon and acrylics.

The basic formula for die output is

$$Q = k\Delta P/\mu \qquad 8.11(1)$$

where

Q = output capacity in cu in. per second (cm³/s),

k = die constant in cu in. (cm³),

ΔP = pressure difference between inlet and outlet (lb per sq in., or kg/cm²), and

μ = viscosity in lb sec per sq in. (kg s/cm²).

Die pressure is frequently assumed to be a linear function of screw speed. As previously stated, the apparent viscosity of a given material is a function of temperature as well as of shear rate, and melt temperature measured at the adapter is also an important operational guide to extruder operation.

Extruder Barrel Temperature Control

The barrel is usually divided into temperature control zones roughly 15 to 18 inches (375 to 450 mm) long (Figure 8.11a). A 4½-inch (112.5 mm) extruder, for example, may have 4 to 6 barrel zones. Conventional control systems use a separate temperature control loop in-

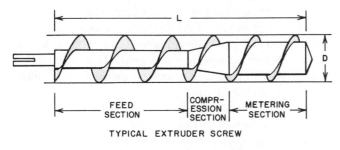

TYPICAL EXTRUDER SCREW

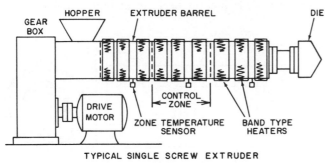

TYPICAL SINGLE SCREW EXTRUDER

Fig. 8.11a   Typical single-screw extruder

cluding power control device, temperature sensor, and heater for each zone. Band-type resistance heaters are of two-piece construction to facilitate removal.

Extruders require large heater ratings to decrease heating time. After extrusion has begun, heat is internally generated from the friction and shear of the rotating screw. This heat is a function of the screw speed squared. At high production speeds the internal heat equals the external losses, resulting in an adiabatic operation requiring no conducted heat. At high speeds the melt temperature rises above the accepted maximum, and barrel cooling is required. The amount of internal heat generated is also influenced by screw design, head pressure, and resin viscosity.

Controllers for the extrusion industry have developed very specialized forms. Rarely are recorder-controllers used. Instead the instruments are almost always of the electronic type, of small size and with combination outputs for extruder barrel heating, cooling, alarms and start-up interlocks.

The available choices of control instrumentation features are summarized in Table 8.11b.

Electric heating directly at the barrel can also be accomplished by induction, using a coil of copper wire wound about the barrel to induce an electromagnetic field. The barrel steel is heated by internal circulating currents, a much more responsive technique than resistance-type heating. The coil is energized with 60-cycle current and is usually controlled by contactors.

Fan Cooling

Many extruders use aluminum heating shells with cast-in heaters of encased nichrome coiled throughout the aluminum jacket. The shells are in two segments clamped together to surround the extruder barrel. Shells for fan cooling have fins cast at the outside surface to increase surface area. Motor-driven fans or blowers are positioned directly below the zone and when switched on (manually or automatically) remove barrel heat by forced convection. A zone fan on a 2½-inch (62.5 mm) extruder can typically remove the equivalent of 5 kW per hour (Figure 8.11c).

Table 8.11b
CONTROLLER FEATURE SELECTION

Sensor Type	Temperature Display	Set Point Display	Control Mode	Power Device
Thermistor	Hi-lo lamps	Arbitrary scale	On-off	Contactors
Thermo-couple	Deviation meter	Calibrated dial	Time proportioning	SCR
Resistance	Calibrated scale	Calibrated scale	Two mode	Saturable reactor
	Digital display	Digital display	Three mode	Valves

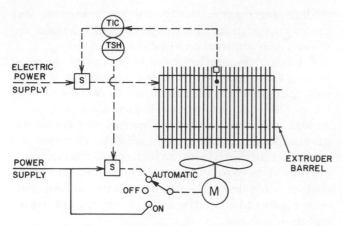

Fig. 8.11c   Heater with fins for electric heating and fan cooling

The control panel is usually equipped with a three-position selector switch. In the automatic position the temperature controller turns the fan on at approximately 25% heater power level (in the proportioning band); the fan then stays on throughout the balance of the range to zero heat. The ON position allows continuous fan operation for shutdown and is operated by an auxiliary circuit in the control instrument.

Water Cooling

Extruders using water cooling (Figure 8.11d) have the aluminum shells with cast-in tubing as well as heaters. Treated water is continuously circulated through the coils by a common pump and heat exchanger. Solenoid valves, regulated by the zone temperature controller, allow circulation to each zone in order to maintain the proper barrel zone temperature. The valves are operated from an auxiliary switch in the zone temperature control instrument in a time proportioning manner with extra slow cycle rate and capability for very short pulses.

Water cooling is frequently too effective compared with fan cooling. Cooling water is usually flashed into steam at most barrel temperatures used for processing thermoplastic materials. Some machine builders use compressed air to clear the water from the passages and eliminate trapped fluid and erratic cooling. By this method, a very short pulse of cooling water charging is followed by immediate removal to obtain less severe cooling. Running the heat exchanger sump at higher temperatures also reduces the severity of water cooling. Some extruder barrels have grooves machined in the outside surface, with cooling coils imbedded in the grooves and band heaters clamped directly over the barrel and tubing.

Hot Oil Systems

Extruders 6 inches (150 mm) and larger, both single and twin screw designs, frequently use circulated oil heating around the barrel through an outside jacket. The upper temperature limit is roughly 400°F (204°C). Such design is primarily for rubber, PVC, styrenes, polyethylene, and acrylics. The extruder is zoned and pumped continuously from a common sump that contains both electric heating elements and cooling coils. Sump temperature is maintained by simple, self-contained controllers. The extruder barrel is heated or cooled by the oil, depending on internal heat generation. Circulation is at constant flow rate, and zone temperature controllers are not used (Figure 8.11e).

Water is sometimes substituted for hot oil when the operation, although primarily cooling, requires initial heating. Water is capable of considerably greater heat transfer than is hot oil because of high specific heat,

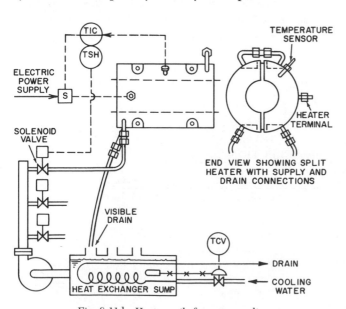

Fig. 8.11d   Heater coils for water cooling

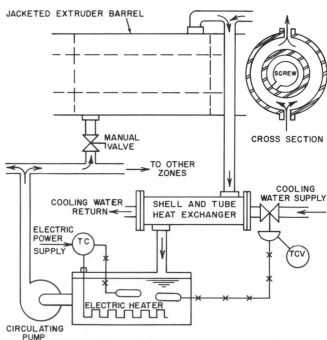

Fig. 8.11e   Oil-heated extruder zone

efficient thermal conductivity, and low viscosity (which promotes turbulent flow). In these water circulating systems, the extruder zones are encased, each zone with its own pump and sump tank. The control system circulates water at constant temperature, which is returned through a shell and tube cooler to the sump.

Melt Temperature Cascade System[2]

The output melt temperature is a function of internal shear energy (converted to heat energy) plus conducted heat or minus the barrel cooling, depending on the operation. The temperature of the melt is as important as the output rate for quality extrusion.

The cascade control to reset the zone controller set points achieves continuous temperature control of the melt and approaches the ultimate in one-knob operation. Figure 8.11f shows the arrangement for cascade feedback. One special feature is that the system allows only depression of the zone temperature controller set points— a safety consideration because of heat degradation and pressure buildup in polymer systems. A safety interlock with the extruder screw drive is usually incorporated in order to prevent both polymer freezing during shutdowns and drive damage at startup.

Zone controllers provide both heating and cooling, and their set points are regulated by the melt controller. Each zone can be adjusted individually to follow a certain percentage of the feedback cascade signal so that a preset program of zone depression will follow a definite barrel temperature zone profile curve. The extruder metering zones are capable of the greatest heat transfer and therefore receive the greatest percentage of feedback and most set point depression. Feeding and compression zones are most effective at relatively constant temperature to provide optimum wall friction for most efficient feedings; therefore, they usually receive a small percentage of the feedback signal.

The melt controller must have proportional and integral control; according to Meissner,[3] rate action is also beneficial. The zone controllers should be proportional for simplicity only, because reset action at the primary melt controller achieves the desired melt temperature without droop. Some cascade systems incorporated a tachometer feedforward signal to supply set point depression proportional to screw speed, a function that reduces temperature departures from the control point caused by screw speed drift or intentional changes.

Extruder Pressure Measurement

Most extruders run at pressures from 500 to 10,000 PSIG (3.5 to 69 MPa), and the die output flow is directly proportional to extruder output pressure, which in turn depends on screw speed, die restriction, and resin viscosity. Knowledge of extruder pressure is very important for efficient extrusion. Some common extruder pressure sensors are:

Grease Sealed Gauges. These are direct reading bourdon tube gauges with a tube tip capillary for complete filling with high temperature grease, usually silicone. The grease remains viscous at high temperatures and prevents the molten polymer from entering and solidifying in the gauge or piping. Disadvantages include the need for periodic greasing and occasional contamination of the product. (For other designs of volumetric pressure seal elements see Figure 5.2f in the volume on Process Measurement volume of this Handbook. These may require temperature compensation.)

Pneumatic Force Balance Transmitters. These provide linear output to low pressure indicators, recorders, and controllers. Their force balance principle makes them less sensitive to process temperature variations (see Section 2.7 for more details).

Strain Gauge Transducers. These are high-quality pressure sensors, widely used in extruder applications as discussed in Section 5.7 in the volume on Process Measurement volume of this handbook.

Pressure Control of Extruders

Synthetic fiber processes frequently use an extruder to melt and transport nylon and polyesters to a bank of gear pumps feeding individual spinning die heads. The

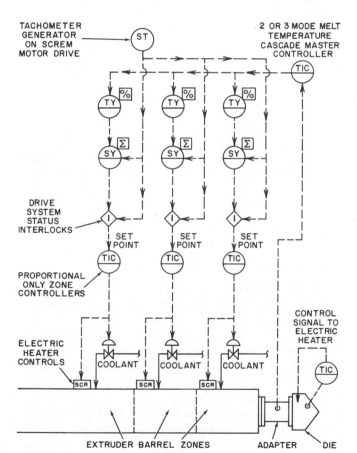

Fig. 8.11f   Melt temperature cascade system

shear characteristics of these polymers are near-Newtonian and thus display relatively constant viscosity at wide variataions in shear rate. This characteristic is responsible for pressure nearly proportional to screw speed, which is necessary for pressure control by regulation of the screw speed.

In contrast, PVC can be pumped through a restricted orifice at increasing rates without an increase in back pressure because its apparent viscosity drops with increased pumping shear rate. The pressure of such a process is nearly impossible to control by manipulating screw speed.

Figure 8.11g illustrates the control scheme in fiber processes. Because some of the fiber spinning pumps frequently fail or are intentionally stopped, the pressure controller must immediately slow the extruder screw to a new output rate determined by the constant volume pumps remaining operable.

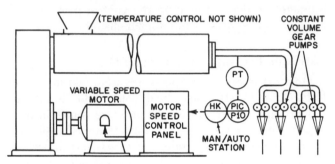

Fig. 8.11g   Extruder pressure control in synthetic fiber production

The screw drive motors are generally DC drives or eddy current clutches (Section 5.8) capable of accepting 4 to 20 madc control signals. For stable control the response of the entire control loop must be virtually instantaneous. Hysteresis in mechanical drive systems can be a source of control problems. Pressure control of extruders feeding extrusion dies directly has been used experimentally with variable results, depending on the shear to viscosity relationships of the polymers.

Use of control valves between extruder and die heads is popular with extrusion theorists but has not been commercialized because of the light demand and high cost of rapidly responding, high pressure and high viscosity control valves.

Produce Dimension Control

Blown film lines producing quality tubular and flat film to 5 mils (0.125 mm) thickness commonly use thickness measurement and control. Thickness can be adjusted by the speed of the takeup rolls if the thickness variations (in machine direction) are consistent across the film. Increasing the takeup speed reduces the film gauge; decreasing the speed increases its thickness (Figure 8.11h).

Thickness variations in die direction (across the width)

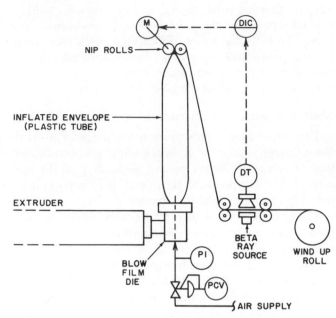

Fig. 8.11h   Film thickness control through nip roll speed manipulation

can be measured, but most attempts to automate sheet die lip adjustments have been unsuccessful. Sheet dies extrude materials up to $\frac{1}{2}$ inch (12.5 mm) thick and 120 inches (3 m) wide.

Measurement of film thickness to 100 mils (2.5 mm) is made with radiation instruments, using beta rays (Section 9.1 in the Process Measurement volume of this handbook.) Among design variations are scanning heads that measure and record thickness over the entire width of materials. Infrared and mechanical devices have also been used. The viscosity of the polymer in the extruder and die head constitutes the most difficult aspect of extruder design, simulation, and control. The art is steadily advancing, and certain viscosity analyzers are available for extruder applications. (For more information see Section 7.11 in the volume on Process Measurement). In most applications the viscosity measurement controls the polymerization, not the rate or dimensional quality of the pelletizing extruder output.

Computer Control of Extrusion

The digital computer with supervisory control has been successfully applied to composite film systems[4] and assumes functions such as speed, temperature, gauge, profile, and web tension. The role of the computer will be important in manipulating extruder pressure and temperature according to the correct mathematical model and will correct itself or adapt to variations in the process material. Product gauging that is not readily applicable to shapes, such as synthetic fiber and blow molding applications, appears to be a much more difficult process to model and therefore is less likely to be controlled by computers.

REFERENCES

1. Bernhart, E.C., *Processing of Thermoplastic,* Van Nostrand Reinhold, New York, 1959.
2. Pettit and Ahlers, *SPE Journal,* Vol. 24, November 1968.
3. Meissner, S.C. and Meneges, E.P., Technical Papers, Volume XVII, Society Plastic Engineering.
4. *Plastic World,* June 1971, p. 84.

BIBLIOGRAPHY

Hopkins, B., "Temperature Control of Polymerization Reactors," *Instrumentation Technology,* May 1973.

Martin, T.L., "Temperature Control in Plastics Processing," *Plastics,* April 1967.

Mober, C., "Control Schemes for Plastic Molding Machines Conserve Energy," *Instrument and Control Systems,* April 1982.

8.12 FURNACE AND REFORMER CONTROLS

PROCESS MEDIUM

FUEL

The discussion that follows covers a wide range of process furnaces, including heaters, cracking furnaces, and reformers. Each type of furnace is described from a mechanical and process standpoint, in enough detail to make the recommended control strategies understandable to the reader. Enough examples are presented so that easy extrapolation can be made to services not specifically discussed here. The recommended control strategies are broken into process controls, fuel firing controls, and safety controls. Applications of advanced analyzer and digital controls are outlined. The safe operation of the furnace is concluded to be a prime objective; every effort should be made in the control system design to achieve this goal.

General Considerations

Furnaces and heaters are devices in which heat energy is transferred to a charge or feed in a controlled manner. The typical furnace, or heater (the terms will be used interchangeably here), usually takes the form of a metal housing lined with a heat-conserving refractory. The charge can enter as a solid, liquid, or gas and may or may not be transformed to a different state by the energy supplied. The charge can be carried through the furnace or heater continuously or through metal tubes or troughs or may be batch-heated by remaining stationary after entering. The functions of furnaces can be broken down into three categories:

1. To heat or vaporize the charge.
2. To provide heat of reaction to reacting feeds.
3. To provide an elevated and controlled temperature for the physical change of charge materials (e.g., change of grain structure).

An example of a heating and vaporizing furnace is a refinery crude oil heater, in which crude oil is heated and partially vaporized in preparation for distillation. Typical of a furnace in which reaction heat is supplied is an ammonia reformer furnace, in which hydrocarbon vapors and steam are reacted to form hydrogen and carbon monoxide. An annealing furnace is an example of a furnace in which physical change takes place. A metal charge is conveyed through or placed in the furnace for a fixed time at various temperatures, resulting in the desired final grain structure. In many instances a combination of these functions is carried out in the same furnace. For example, a reformer furnace first brings the charge up to reacting temperature and then supplies the reaction heat.

Control System Functions

The primary functions of furnace control systems are:

1. To maintain the desired rate of energy transfer to the charge.
2. To maintain a controlled and efficient combustion of fuel.
3. To maintain safe conditions in all phases of furnace operation.

The control system must ensure that the charge receives the heat energy at the proper rate. Most commonly, the temperature of the charge is the index used as the measure of the heat transferred. When vaporization occurs along with the transfer of sensible heat, the charge temperature may not be a good index. In this case the total final heat content of the charge is important, but this is a more difficult condition to measure.

Proper combustion of fuel, the second purpose of the control system, involves many factors, such as the regulation of combustion air, the preferential burning of one fuel over another, and the control of atomizing steam when fuel oil is burned. Such factors will be discussed more fully under the description of furnace types.

Safety is always an important consideration in any process, but it takes on added importance in furnaces in which combustion takes place. There always exists the possibility of an explosive mixture being formed in the furnace, which can have disastrous effects. When a combustible charge is being handled, the danger is even greater. Although such a condition can develop during normal operations, its likelihood is greatest during furnace startup and shutdown. To minimize these hazards, industry and local government agencies have developed various codes and practices. These tend to set minimum standards in control equipment and in acceptable startup and operating practices. Strong reliance is placed on instrumentation to warn or to shut down the unit if a dangerous condition arises. Local codes and industry practices, usually set by the insurers of the plant, should be consulted before firing controls are designed and installed.

One danger that exists in any control system using air-operated control valves is the loss of motivating instrument air. If this occurs, the valves will usually go completely open or completely closed. The failure positions are shown on the control system sketches next to each valve symbol. FC indicates that the valve fails closed, and FO indicates it fails open. The air failure actions are chosen by the instrument engineer to give as safe an emergency operation as possible. The choice is dependent on the valve function in the process stream.

The majority of furnaces in operation today burn some type of fossil fuel (such as fuel gas, fuel oil, and coal), and the selection of the particular fuel or combination of fuels is generally a matter of overall plant economics. One of the factors to be considered in selecting a fuel is the lower thermal efficiency of natural gas versus the inefficiency in burning fuel oil as a result of poor atomization and need for more excess air.

Many installations burn gas and oil simultaneously or with one or the other as a standby. The more common approach is to burn fuel gas preferentially, because of the difficulty in storing it, and to use fuel oil as the standby. For these reasons, the systems described in this section will cover arrangements for both fuel gas and fuel oil firing as well as combination firing (i.e., firing both fuels simultaneously). Coal firing is rarely used in process furnaces but is commonly used in electric power generation stations and metallurgical furnaces.

Combustion Air Requirements

To burn the combustible portion of any given fuel completely, there is an ideal quantity of oxygen, and therefore an ideal quantity of air, required. Under real conditions, the inefficiencies in combustion require that some additional air be added to ensure complete combustion. The quantity over and above the ideal amount of air is called the "excess air," and it is usual to talk about "percent excess air," which is the ratio of excess air to ideal combustion air times 100.

Each fuel has a practical minimum limit on percent excess air, and probably the most important function of the firing control system is to maintain the excess air percentage above that minimum. The excess air requirements are usually satisfied by maintenance of a suitable furnace draft (e.g., a suitable low pressure at the stack) to ensure that sufficient air is drawn into the combustion chamber. For a more accurate calculation of excess air, the measurement of the oxygen content of the flue gases is required (usually by a continuous automatic analyzer), since oxygen content of the flue gas is directly related to percent excess air.

General Control Systems and Instrumentation

In discussing controls for the more common types of process furnaces, this section is subdivided by type of furnace rather than by the type of control system. Each type of furnace will be described from a process standpoint first, since the process function of the heater naturally determines the furnace operation and therefore the control requirements. Next, the important process variables will be noted and the methods used to measure and control them outlined. The various types of firing controls applicable to the particular type of furnace will then be described. Finally, specific safety considerations and instrumentation will be presented.

The control systems to be outlined have been empirically developed, and most of them are tried and reliable methods. The question of control dynamics is a necessary consideration in most industrial processes, but for the most part, these problems are solved in practice by the use of versatile instrument hardware.

The analog process controller has a wide range of gain adjustment and at least two modes of dynamic compensation (integral and derivative), which also are adjustable over a wide range. Transmitters are available with suppressed ranges and with noise-damping and range-change adjustments that are either built in or provided by component substitution. Final control elements, usually control valves or dampers, may be selected with a wide variety of characteristics, and in many cases these characteristics are field modified by the appropriate cams on the valve operator. These components are fully described in Chapter 4.

Normally, the instrument engineer can select the proper components to cover the range of system dynamics, and final "tuning" of the control loop is done in the field under actual operating conditions. For the great majority of the cases, this approach, although not the most systematic, is most efficient in that little time is expended in lengthy analytical calculations based on questionable assumptions.

For the special cases in which it is recognized that long dead times or highly interacting process gains will cause instabilities, the use of mathematical analysis and process simulation becomes a necessity.

The intent of this section is to emphasize the process considerations and related control concepts in furnace operation. The instrumentation is shown in symbolic form in the various illustrations. References to actual instrument component types are avoided unless a particular type of instrument is vital to the operation of the system. The various types of instruments for measuring and controlling flow, pressure, and temperature used in the systems to be outlined are described in detail in the Process Measurement volume of this handbook.

Startup Heaters

Startup heaters are treated first because of their simplicity. These units are usually required, as their name implies, at the startup of a process unit, and their span of use is usually from a few hours to, at most, a few weeks. They are usually vertical, cylindrical units with vertical process tubes along the inner walls and with a single burner centered in the floor. Draft is by natural convection induced by a stack mounted on the top of the heater. Some process systems that require startup heaters are catalytic cracking units to heat up fluidized catalyst beds, ammonia units to heat up the ammonia converter catalysts, and fixed-bed gas-drying units to regenerate the dryer beds.

Process Considerations

The startup heater usually heats an intermediate stream, such as air or natural gas, which in turn heats another fluid or solid, such as a reactor catalyst. As an example, an ammonia unit startup heater will be described. This unit is used to heat "synthesis gas" (primarily a mixture of hydrogen and nitrogen), which in turn is used to heat the ammonia converter catalyst bed to a temperature of 700°F (371°C) by burning natural gas. The synthesis gas is recirculated through the catalyst bed, gradually bringing it up to operating temperature. After normal operation has begun, the exothermic nature of the reaction keeps the bed at temperature and the startup heater can be shut down.

Process Controls

The important variables are "synthesis gas" flow and temperature. Figure 8.12a shows the necessary controls for the operation of the unit. The flow of synthesis gas is measured by the flow transmitter (FT-1) and is indicated on a flow indicator (FI-1A) mounted on a panel board. The desired gas flow is manually adjusted in the

field of operating the hand valve (HV-1) and observing the local flow indicator (FI-1B). The effluent synthesis gas temperature is maintained by the temperature recorder controller (TRC-1), which measures the gas temperature by means of a thermocouple and controls the quantity of fuel gas burned by modulating the control valve (TV-1).

Firing Controls

The fuel gas firing is set by the process temperature controller (TRC-1). The heater draft (i.e., negative pressure in the fire box) is produced by the stack; the operator sets it by observing the draft gauge (PI-1), usually an inclined manometer, and by manually adjusting the position of the stack damper. Once initially set, the damper is rarely adjusted again, unless furnace conditions or loads change drastically.

Safety Controls

The major hazards involved with these heaters are (1) the interruption of process gas flow and the resulting possible tube rupture and (2) flame failure and the generation of dangerous fuel-air mixtures in the fire box.

The loss of process gas flow is detected by the low-flow alarm switch (FSL-1), which trips the solenoid (FY-1) to vent the diaphragm on control valve (TV-1). This immediately cuts off fuel and thus terminates firing. Two other devices may cut off the fuel supply in the same manner: the flame detector switch (BS-1), which activates on loss of flame, and the low-pressure switch (PSL-1), which activates on low fuel gas pressure. The loss of flame or the loss of fuel pressure requires that the fuel gas be cut off to prevent the formation of a dangerous air-fuel gas mixture in the fire box, should the fuel gas pressure be restored. The solenoid valve (FY-1) should be the manually reset type (requiring manual opening after trip) to ensure that the control valve stays shut until the operator desires to reopen it. The control valve (TV-1) should be a single-seated valve for tight shutoff, to prevent leakage of fuel into the fire box when it is closed. Failure of instrument air will cause the control valve (TV-1) to fail closed (FC), as indicated next to the valve symbol. This is the safest possible action, since it terminates the furnace firing.

Fired Reboilers

The fired reboiler provides heat input to a distillation tower by heating the tower bottoms and vaporizing a portion of it. Normally a tower reboiler uses steam or another hot fluid as a source of heat, but where heat duties are great or where tower bottom temperatures are high, the fired reboiler is used. In construction, depending on its size, the reboiler may be of the vertical, cylindrical type or the larger, conventional horizontal-type furnace.

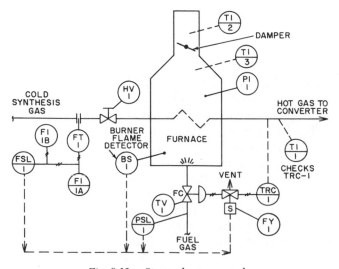

Fig. 8.12a   Startup heater controls

Process Considerations

The fired reboiler heats and vaporizes the tower bottoms as this liquid circulates by natural convection through the heater tubes. The coils are generously sized to ensure adequate circulation of the bottoms liquid. Temperature of the reboiler return fluid is generally used as the means of controlling the heat input to the tower. Overheating the process fluid is a contingency which must be guarded against, since most tower bottoms will coke or polymerize if under excessive temperatures for some length of time.

Process Controls

A common control scheme is shown in Figure 8.12b; this depicts the tower bottom along with the fired reboiler. It is not usually practical to measure the flow of tower bottoms to the reboiler, first, because the liquid is near equilibrium (near the flash point), and second, because it is usually of a fouling nature, tending to plug most flow elements. Proper circulation of the fluid is provided for in the careful hydraulic design of the interconnecting piping. The other important variable is the reboiler return temperature, which is controlled by TRC-1 throttling the fuel gas control valve (TV-1). The high-temperature alarm (TAH-1) is provided to warn the operator that the process fluid has suddenly reached an excessive temperature, indicating that manual adjustments are needed to cut back on the firing.

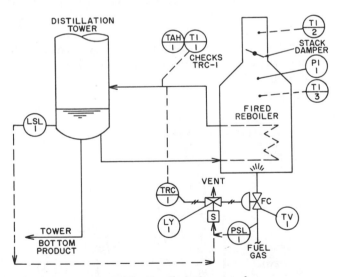

Fig. 8.12b Fired reboiler controls

Firing Controls

The firing controls are relatively simple and are similar to those of the startup heater described earlier. The process temperature controller (TRC-1) actuates the fuel valve (TV-1) to satisfy process requirements. The fuel in this case is gas, but it could just as well be fuel oil. The furnace draft is set by means of the stack damper observing the draft gauge (PI-1). The stack temperature

(TI-2) and the firebox temperature (TI-3) are detected as a check for excessive temperatures that may develop during periods of heavy firing.

Safety Controls

The major dangers in this furnace type are the interruption of process fluid flow or the stoppage of fuel. The loss of process fluid can occur if the liquid level in the tower bottom is lost. If this happens, flow will stop in the reboiler tubes, and a dangerous overheating of tubes may result. To protect against this, the low-level switch (LSL-1) is wired to trip the solenoid valve (LY-1), which vents the diaphragm of the control valve (TV-1). This causes the fuel valve to close.

A momentary loss of fuel can be dangerous because the flames can be extinguished, and on resumption of fuel flow a dangerous air-fuel mixture will develop in the fire box. To prevent this occurrence, the low-pressure switch (PSL-1) trips the solenoid (LY-1) on low fuel pressure, thereby closing the fuel valve (TV-1). The solenoid valve (LY-1) should be the manual reset type, which remains vented and thereby causes the valve (TV-1) to remain closed even on return of fuel gas pressure.

The air failure action of the control valve (TV-1) is closed (FC) on loss of motivating instrument air. This action on fuel valves is the safe mode, since firing is discontinued during the emergency.

Process Heaters and Vaporizers

The unit feed heater of a crude oil refinery is representative of this class of furnaces. Crude oil, prior to distillation in the "crude tower" into the various petroleum fractions (gasoline, naphtha, gas oil, heavy fuel oil, and so forth) must be heated and partially vaporized. The heating and vaporization is done in the crude heater furnace, which consists of a fire box with preheat coils and vaporizing coils. The heating is usually done in coils in the convection section of the furnace, which is the portion that does not see the flame but is exposed to the hot flue gases on their way to the stack. The vaporizing takes place at the end of each pass in the radiant section of the furnace (where the coils see the flame and the luminous walls of the fire box). The partially vaporized effluent then enters the crude tower, where it flashes and is distilled into the desired "cuts." Other process heaters that fall into this category are refinery vacuum tower preheaters, reformer heaters, hydrocracker heaters, and dewaxing unit furnaces. The control systems presented will also be applicable to these types of process heaters.

Process Controls

The prime variables with regard to the process are

1. flow control of feed to the unit,
2. proper splitting of flow in the parallel paths through

the furnace (to prevent overheating of any one stream and its resultant coking), and

3. correct amount of heat supplied to the crude tower.

Figure 8.12c shows the typical process controls for this type of furnace. The crude feed rate to the unit is set by the flow controller (FRC-2). The flow is split through the parallel paths of the furnace by remote manual adjustment of the control valves (HV-1 through HV-4) via the manual stations (HIC-1 through HIC-4). The proper settings are determined by equalization of the flow indications on FI-1 through FI-4. The temperature indicators TI-1 through TI-12 are periodically observed to determine if there are any rising or falling trends in any one of the passes. If such trend develops, the flow through that pass is altered slightly to drive the temperature back toward the norm.

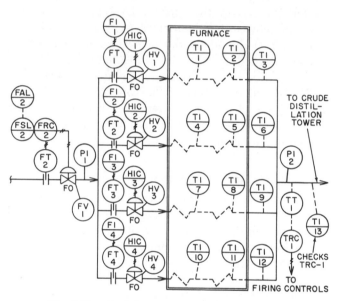

Fig. 8.12c   Crude heater-vaporizer process controls

The desired heat input into the feed stream is more difficult to control, because the effluent of the furnace is partially vaporized and the feed stock varies in composition depending on its source. If the feed were only heated, with no vaporization taking place, the control would require only that an effluent temperature be maintained. If complete vaporization and superheating occurred, this too could be well handled by straight temperature control. But in the case of partial vaporization with a variable feed composition, effluent temperature control alone is not a reliable approach. The composition and, hence, the boiling point curve of the feed is not constant with time, and the required control temperature itself varies. Additional information is thus required, and it is obtained from the distillation downstream of the furnace. By observation of the product distribution from the fractionation, the need for a change in heat input can

be determined. This approach is slow and not precise, but to do a more accurate job would require a great deal of sophisticated instrumentation, involving special analyzers to measure feed composition and a computer to optimize the mathematical model and thereby determine the required heat input to the feed. Current practice is to achieve approximate control with a temperature controller (TRC-1), whose set point the unit operator periodically changes to account for feed variations. He depends on his experience and on the results in the fractionator to determine the proper temperature setting.

Fuel Gas Firing Controls

Figure 8.12d shows the firing controls for the furnace using fuel gas only. The local pressure controller (PIC-1) maintains the burner header pressure. The fuel gas headers serve many burners spaced equally along the floor of the fire box. The burners, being essentially fixed-diameter orifices, will pass more or less fuel depending on the header pressure. In order to change the rate of firing (furnace heat input), the burner header pressure must be altered. This is accomplished by the effluent temperature controller (TRC-1), which resets the set point on PIC-1 as it attempts to maintain a constant furnace effluent temperature. Cascading the TRC-1 output to PIC-1 is beneficial in that the pressure controller compensates for fast local disturbances (e.g., changes in fuel supply pressure) without allowing them to upset the effluent temperature, whereas the TRC-1 provides the slower correction required to correct for process and ambient disturbances.

The flow recorder (FR-1) is necessary to make efficiency checks on the furnace and also to determine the heat input.

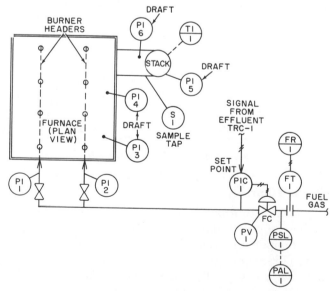

Fig. 8.12d   Crude heater firing controls using fuel gas

Fuel Oil Firing Controls

Figure 8.12e shows the firing controls for fuel oil. The process effluent temperature control, cascaded to the burner pressure controller (PIC-1), operates as described under fuel gas firing. The major difference with oil firing is the introduction of atomizing steam, which is required to disperse the fuel oil sufficiently to produce efficient burning. The pressure differential controller (PDIC-1) maintains the atomizing steam header at a fixed pressure above the fuel oil header to ensure that it will always have enough atomizing capability. The differential pressure controller (PDIC-1) is necessary because the pressure in the fuel oil header varies because of the automatic control by TRC-1. PI-1 and PI-2 are required to set PDIC-1 initially and to check its accuracy during operation.

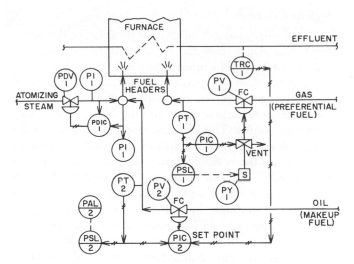

Fig. 8.12f Crude heater firing controls using both oil and gas

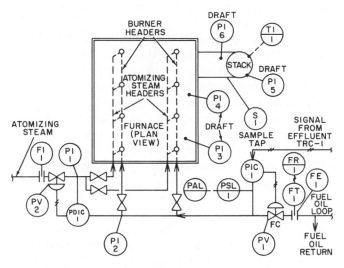

Fig. 8.12e Crude heater firing controls using fuel oil

Multiple Fuel Firing Controls

Figure 8.12f shows the more common arrangement for mixed fuel firing, with the fuel oil the make-up fuel and fuel gas burned preferentially. This situation is common when variable but insufficient amounts of fuel gas are available from a process unit such as a petroleum refinery. The fuel gas pressure is reduced, and the burner header pressure is maintained constant by PIC-1. In effect, the amount of fuel gas burned is held constant as long as the fuel gas availability is constant. The set point of PIC-1 is adjusted to change the rate of gas firing. If the pressure in the gas header falls below a predetermined minimum limit for stable burner operation, the fuel gas supply is shut off by the low-pressure switch (PSL-1), tripping the solenoid valve (PY-1) and causing the control valve (PV-1) to close. When this occurs, the full load is taken up by the fuel oil system.

Although the fuel oil is the trim medium, it must be available in sufficient quantity to take the whole load in the event of a fuel gas interruption. The temperature controller (TRC-1) varies the set point of the fuel oil header pressure controller (PIC-2) to satisfy process requirements. On any burner there is a "turn-down" limitation, which is the ratio of minimum to maximum fuel burning capacity (a common ratio is 3:1). Low pressure in the fuel oil header is indicative of the approach of a minimum firing condition; therefore, the low-pressure alarm (PAL-2) is used to warn the unit operator when this situation arises. The operator then has the option of going out to the furnace and manually turning off the individual fuel oil burners or reducing the fuel gas firing rate by lowering the set point of PIC-1. Either change will increase the fuel oil demand and thus bring the system into a stable operating zone.

Furnace draft (see Figure 8.12e) is normally maintained by free convection in this type of furnace (i.e., the stack produces a negative pressure in the fire box), and air is drawn in through louvers along the sides of the furnace. Furnace draft is set at crucial points in the furnace by means of the draft points PI-3 through PI-6 and by manual setting of the louvers on the furnace. These louver adjustments are made initially at startup and are changed only occasionally when firing is being optimized. PI-3 through PI-6 are switchable points on an inclined manometer mounted on the outside wall of the furnace.

The sample tap (S-1) is used to take samples of flue gas to determine its oxygen and combustible content. This serves to indicate the efficiency of the furnace and to warn of the development of dangerous conditions. The samples may be taken periodically and analyzed in the laboratory, or they may be taken continuously and analyzed by an on-stream process analyzer. The analysis of the gas usually is the easiest part of the operation. The delivery of a sample for analysis, properly conditioned (i.e., free of soot and water), is the difficult job.

Safety Controls

The primary hazards in process heaters are the interruption of charge flow and the interruption of fuel flow and loss of flame. The interruption of crude charge below a minimum flow will result in overheating and possible rupturing of the tubes. Resumption of flow may cause hydrocarbon charge leakage into the fire box, with catastrophic results. The low flow alarm (FAL-2 in Figure 8.12c) alerts the operator to this impending condition; he then has the option of correcting the fault or terminating firing. The control valves (FV-1 and HV-1 through HV-4) on instrument air failure will fail open, maintaining the flow through the furnace coils.

Interruption of fuel can cause burner flameout, and the resumption of fuel flow can cause an explosive mixture in the fire box. Flame failure can be detected by sensors, and the shutdown of fuel firing can be automated. This approach is not generally followed in process heaters because of the expense (there can be scores of burners to scan) and because operation is relatively steady (startups and shutdowns occur a year or two apart). Usually a low-pressure alarm (PAL-1, Figure 8.12d) on the fuel header is relied upon to warn the operator of fuel loss, leaving the shutdown decision to him.

On instrument air failure, the fuel firing valve PV-1 closes, shutting down the furnace firing.

Reformer Furnaces

The purpose of a typical reforming furnace, the primary reformer in an ammonia plant, is to produce hydrogen for combination with nitrogen in the synthesis of ammonia (NH_3). Hydrogen is produced by reforming hydrocarbon feed (usually methane or naphtha) using high-pressure steam, as shown in the reaction

$$CH_4 \quad + \quad H_2O \quad = \quad 3H_2 \quad + \quad CO \qquad 8.12(1)$$
(Methane) (Steam) (Hydrogen) (Carbon monoxide)

$$\text{General: } C_nH_{(2n+2)} + nH_2O = (2n+1)H_2 + nCO$$
$$8.12(2)$$

The carbon monoxide is removed by further reaction later in the process. The reaction is endothermic (absorbs heat) and takes place at pressures of approximately 450 PSIG (31.5 MPa) and at temperatures of around 1500°F (865°C). The reaction takes place as the feed gas and steam pass through tubes filled with a nickel catalyst, which are heated in the radiant section of the reformer furnace. Steam must be provided in excess of the reaction requirements to prevent the side reaction of coke formation on the catalyst. The coking of the catalyst deactivates it, resulting in expensive replacements. To minimize the coking, steam is usually supplied in a ratio of 3.5:1 by weight, relative to feed gas. Special precaution must be taken to maintain the excess steam at all times, since even a few seconds' interruption while feed gas continues can completely ruin the whole catalyst charge.

Process Controls

As indicated, the major process variables to be held are the feed gas flow, the reforming steam flow, and the effluent temperature and composition. As illustrated in Figure 8.12g, the feed gas flow is maintained by means of the flow controller (FRC-1), which is pressure-compensated via PT-1 and FY-1. Pressure compensation of flow corrects the measurement for fluctuations in feed gas pressure. The steam rate is maintained by means of FRC-2, and the ratio of steam to feed gas flow is continually monitored through FY-2. The ratio computer FY-2 accepts a gas flow signal from FT-1 and a steam flow signal from FT-2, and its output is scaled in the ratio of steam to carbon. If this ratio falls below the limit of approximately 3:1, the low flow ratio alarm FAL-3 is sounded in the control house. If the ratio continues to fall below approximately 2.7:1, the feed gas is shut off via FSL-4, which trips the solenoid FY-3 and causes the air diaphragm operator on the feed control valve (FV-1) to be vented, thereby closing the valve. The valve (FV-1) is a quick-closing valve (4 to 5 seconds for full closure) so that the flow of gas can be almost instantly stopped on reforming steam failure, thus protecting the reformer catalyst. It must also be a single-seated tight shutoff valve, to prevent leakage during the time it takes to close the electric-operated shut-off valve, MV-1. The operation of MV-1 is relatively slow because of the electric motor and speed reduction gearing used. An alternative approach to this ratio control arrangement is shown in Figure 8.12h. Here the feed gas flow is measured by flow transmitter (FT-1), as is the steam by FT-2, and a constant ratio of steam to gas is maintained. The steam is on straight flow control, and the gas flow tracks the steam flow. In principle this appears to be an improvement over the previous control system, but in practice the dynamics of

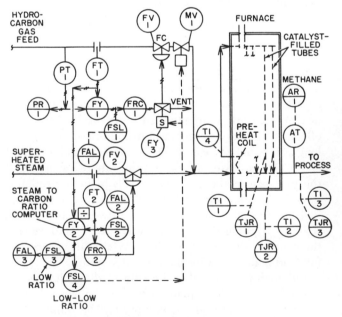

Fig. 8.12g   Reforming furnace process controls

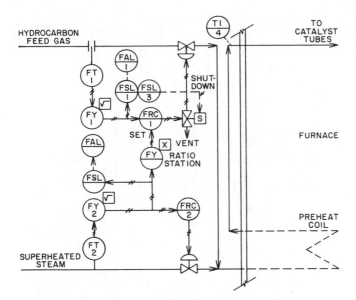

Fig. 8.12h Reformer furnace control of steam-gas flow ratio

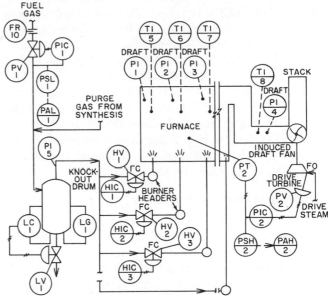

Fig. 8.12i Reformer furnace firing controls. (PV-2 is the speed governor operator.)

measurement and the process make it less stable. The steam flow measurement noise and lags may result in cycling the gas flow signal, such that the integrated error of gas flow in the latter system is considerably greater than in the former.

The effluent analyzer AT-1 (see Figure 8.12g) is used to determine the extent of reaction completion by measuring the methane (CH_4) content of the stream. An optimum furnace temperature profile can be arrived at by manipulating furnace temperatures to achieve the desired degree of conversion. This analyzer can be either an infrared or a chromatographic analyzer. The analysis is relatively simple, involving the measurement of 0 to 10 percent methane in a background of hydrogen and carbon monoxide. The only difficulty is the high water content in the stream caused by excess steam used in the reaction; therefore, water separating components are required in the analyzer sampling system.

Firing Controls

Firing of a reformer furnace (Figure 8.12i) because of its massive design and great heat inertia, is essentially manual. The furnace has approximately a dozen fuel headers with about 20 individual burners per header. Process temperature indicators TI-1 and TI-2 (see Figure 8.12g) are provided at the exit of the reaction tubes. These are constantly monitored, and periodic adjustment of firing is made by manipulation of the fuel header control valves (HV-1, HV-2, HV-3) via the remote manual stations (HIC-1, HIC-2, HIC-3). The pressure upstream of the header control valves is controlled, so that once the valve stroke is set, fuel flow remains constant.

The fuel generally used is natural gas, with a small amount of purge gas from the NH_3 synthesis loop. The fuel gas is pressure-controlled by PIC-1 before it enters the knockout drum. The knockout drum removes the

small trace of water that is carried by the fuel gas. The level controller (LC-1) is a snap-acting on-off controller that periodically expels the water accumulated through its valve (LV-1). The flow recorder (FR-10) measures the fuel consumption and is used for efficiency checks and energy balances. Draft is usually maintained in these furnaces by an induced draft fan, which is usually steam-driven, and the speed of the fan sets the furnace draft (negative pressure) in the fire box. The pressure indicator controller (PIC-2) measures the pressure in the furnace (via PT-2) and controls the speed of the fan by adjusting the operator of the fan turbine governor to hold the desired furnace pressure. The draft points PI-1 through PI-4 are used to set the air inlet louvers manually, on the side of the furnace, to balance the drafts at various points in the furnace.

Safety Controls

The main safety considerations are loss of furnace draft and interruption of fuel supply. The loss of furnace draft may result from a malfunction of the induced draft fan or from loss of steam to the fan drive turbine. The effect is drastically to restrict combustion air, causing hazardous fuel air ratios; in addition, the fire-box pressure will most likely rise above atmospheric, forcing flames out into the area around the furnace housing. The high-pressure alarm (PAH-2 in Figure 8.12i) signals the operator that fire-box pressure is rising. The operator at this point begins to cut firing to minimum levels by means of the hand control stations (HIC-1, HIC-2, HIC-3).

Interruption of fuel supply can result in the possibility of a burner flameout, and the resumption of fuel flow may cause a dangerous air-fuel mixture. Flame detection on this type of furnace is impractical, mainly because of the number required (there are in most cases in the neighborhood of 100 burners). Therefore, fuel pressure below a minimum (as indicated by PAL-1) implies a flame-out, and firing should be cut. The manual procedure, using HIC-1, HIC-2, and HIC-3 to cut back on fuel to the burner headers, is normally used.

On instrument air failure (see Figures 8.12g and 8.12i) the gas feed valve (FV-1) closes to prevent coking the reformer catalyst. The fuel valves (PV-1, HV-1, HV-2, and HV-3) close to stop firing during emergency. The steam valve (FV-2) fails open to maintain the flow of cooling steam through the furnace coils.

Cracking (Pyrolysis) Furnaces

An example of a cracking furnace is an ethylene pyrolysis unit. Feed stock, which can vary from heavy gas oil to ethane, is preheated and vaporized in a preheat coil in the furnace; then it is mixed with steam and cracked. Steam is added to the feed in a fixed ratio to hydrocarbon to reduce the partial pressure of the hydrocarbon feed. This tends to maximize the amount of olefins produced and to minimize the coke buildup in the coils. The feed is heated to 1500°F (865°C) in the pyrolysis coil, which causes the cracking of the long chain hydrocarbons into shorter chain molecules and initiates the forming of unsaturated (olefin) molecules. The severity of cracking is dependent on the temperature achieved and on the residence time in the pyrolysis coils. Therefore the distribution of furnace products is dependent on the degree of firing and on the temperature profile in the furnace. Effluent from the furnace is quickly quenched to prevent recombination of products into undesirable polymers.

Process Controls

The important variables to be controlled are hydrocarbon feed flow, steam flow, coil temperatures and fire box temperature, which results in the desired effluent product distribution. The process control system (Figure 8.12j) is typical of cracking furnace instrumentation. The charge to the unit is fixed by the flow controllers (FRC-1 through FRC-3) on the individual passes, and the total flow is recorded on FR-4. It is important to keep the flow through the coils constant, since coking gradually builds up the pressure drop in the coils. If distribution were left purely to hydraulic splitting, the coking would start in one coil and reduce the flow through that coil, which in turn would over-heat, producing more coke. Eventually the flow would reduce to such a low rate that overheat of the coil will cause its rupture. The individual flow control valves introduce the variable pressure drop that allows all coils to coke at an almost equal rate.

The steam flow controllers (FIC-5 through FIC-7) maintain the appropriate amount of steam to match the flow of hydrocarbon feed. The temperature controller (TRC-1) sets the total heat input to the process by bringing the effluent temperature to the desired value. The

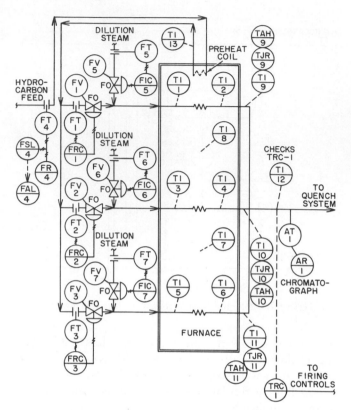

Fig. 8.12j   Pyrolysis furnace process controls

temperature indicator points (TI-1 through TI-6 as well as TI-7 through TI-11) are monitored by the process operator to maintain a certain furnace temperature profile, and hence a certain product distribution in the furnace effluent. The temperature relationships are accomplished by manually trimming burners at the required place in the fire box.

In order to determine whether the desired product specifications are being achieved, an analyzer is commonly used at the exit of the quench system (AT-1). This analyzer is usually a chromatograph, measuring most of the components in the stream that are lighter than butane. This analysis is a difficult one, primarily because of the sample handling requirements. The sample has a very high water content which must be condensed and removed prior to entering the analyzer. It also has a large amount of entrained coke and tars which likewise must be scrubbed out.

Firing Controls

Figure 8.12k shows the firing controls with gas as fuel. The temperature controller (TRC-1) maintains the process gas effluent temperature at the desired value by resetting the set points of the local pressure controllers (PIC-1 and PIC-2). These pressure controllers maintain the desired burner header pressure to satisfy the heat liberation required to sustain the degree of cracking being achieved. The details of the controls are the same as described under Firing Controls for Process Heaters and Vaporizers, if the fuel is gas only. Fuel oil and multiple fuel cases are also applicable (see Figures 8.12d, e, and i).

Safety Controls

The major hazards in operating cracking furnaces are interruption of feed flow, interruption of fuel flow, coking of individual coils, and instrument air supply failure. Interruption of feed flow can result in a dangerous situation if firing is maintained at the normal rate, because the tubes are not designed for the excessive temperatures that result if charge is stopped or is drastically reduced, and the danger of tube rupture is pronounced. The low-flow alarm (FAL-4, Figure 8.12k) is provided in the feed stream to warn the operator of impending danger. Once he verifies that the danger is real he can trip the vent solenoid valve (HY-1) by operating the pushbutton (HS-1), which vents the diaphragm on the emergency valve (HV-1), causing the complete stoppage of fuel gas. This process can be automated, so that on continued fall of feed flow a low-flow switch (FSL-4) automatically actuates the trip solenoid (HY-1), which cuts off the fuel flow.

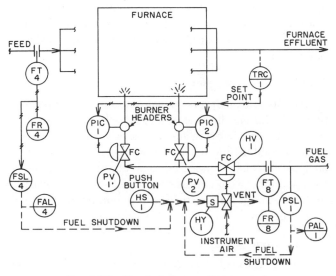

Fig. 8.12k Pyrolysis furnace firing controls

Interruption of fuel flow will cause burner flameout and on resumption of fuel flow may result in a dangerous fuel-air mixture. To prevent this occurrence, the low-pressure switch (PSL-1) is installed (see Figure 8.12k), which trips the emergency shutoff valve (HV-1) via the solenoid valve (HY-1). The solenoid valve is the manual reset type; therefore, once tripped it must be manually reopened.

Excessive coking of an individual furnace coil can occur as a result of the restriction of flow through it. This is a self-worsening effect and tends to cause dangerous overheats. The prevention of this situation is of prime con-

cern to the operator. Should such a condition occur, it will be detected by the high-temperature alarms (TAH-9 through TAH-11, Figure 8.12j), which warn the operator. The operator can increase the flow setting on the appropriate feed flow controller, thus forcing more fluid through the hot tube hoping to bring the temperature down. If this does not alleviate the condition, he has no alternative but to shut down the firing via HS-1 in Figure 8.12k.

Failure of instrument air will result in the control valve actions shown under the individual valves in Figures 8.12j and 8.12k. In Figure 8.12j the feed valves and the steam valves open on air failure to continue flow through the furnace coils and prevent overheating and possible rupture of the coils. In Figure 8.12k the fuel control valves close, extinguishing the fire in the furnace, which results in the safest condition in an emergency.

Combustion Air Preheat

The dramatic increase in fuel costs in recent years has prompted plant operators to search for more fuel-efficient equipment. With regard to furnaces, the overall efficiency is the ratio of the amount of heat absorbed by the process to the heat liberated by the fuel. One of the major heat losses in a furnace is the heat going up the stack in the form of hot flue gases. Therefore, much attention has been devoted to reducing the flue gas exit temperature as much as is practical (approximately 300°F, or 150°C). A common method of achieving this is to add a combustion air preheater to the furnace. This device is a heat exchanger that transfers heat from the hot flue gases going to the stack to the cooler combustion air going to the burners.

In addition to the combustion air preheater, it will usually be necessary to add a forced draft fan to push the combustion air through the preheater (exchanger). The instrumentation necessary to control this equipment is relatively simple.

Figure 8.12l shows a typical combustion air preheat system. The forced draft fan is driven by a steam turbine drive equipped with variable speed governor. A pressure controller PIC-1, sets the discharge pressure (and consequently the air flow) by setting the appropriate fan speed. A low-pressure alarm (PAL-1) is included to alert the operator of possible fan malfunction. A trip solenoid, FY-1, is provided to trip the forced draft fan to minimum speed in the event of fanning of the induced draft fan. This precaution is necessary to prevent positive pressure on the still hot combustion chamber, which can cause hot gas to blow out burner openings and other leaks in the furnace.

Temperature indicators, TI-1, TI-2, TI-3, and TI-4, are provided on both streams in and out of the combustion air preheater to check the operation of the units. The pressure measurements PI-1 and PI-2 are also in-

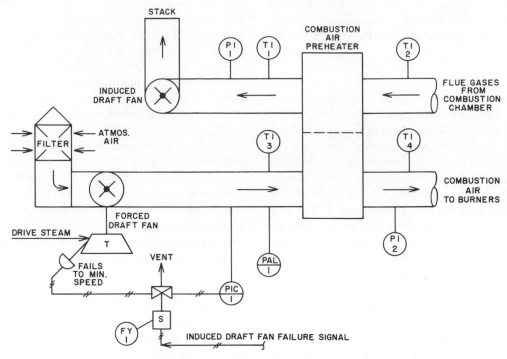

Fig. 8.121   Combustion air preheat controls

cluded to check for excessive pressure drop across the unit.

Analytical Instruments

There are two important analyzer measurements used in furnace controls, oxygen analysis and combustibles detection. A third, less commonly used, measurement is a calorific analysis of the fuel gas being burned. Here the intent is briefly to indicate how and when these analyzers are tied into combustion control systems.

Oxygen Analyzers

The oxygen analyzer is a common instrument for furnaces consuming large amounts of fuel. Its main function is to measure the oxygen content of the flue gas from which the actual excess air being delivered to the furnace may be computed. Keeping the excess air down to the desired minimum results in the conservation of heat, since the sensible heat of excess air is an outright loss. By trimming down on the air intake into the furnace with louvers or dampers while observing the oxygen content of flue gas, one can optimize the efficiency of combustion with respect to air consumption. The analyzer has a secondary function in that it can be tied to a low oxygen alarm and warn the operator of an impending hazardous furnace atmosphere. Whether to acquire an oxygen analyzer is generally decided on economic grounds. The cost of the analyzer and its daily upkeep must be saved, as a minimum, for the installation to be justified. For large fuel consumers, the time required to pay for the analyzer installation can be a matter of months. Where furnace

operation is such that there are frequent changes in firing loads and/or frequent changes in types of fuel, the analyzer could be considered a necessity from a safety standpoint alone.

Combustible Analyzers

Combustible analyzers, which measure the unburned fuel in the furnace flue gas, are essentially safety devices. If rapid changes in firing and frequently varying fuels are used, a combustible analyzer may be a necessity. The analyzer is usually connected to an alarm device to warn the operator of a dangerous situation. It may also be used to shut down the furnace firing if the combustibles exceed some safe limit.

Calorimetric Analyzers

Calorimetric analyzers are used to measure the heating value of the fuel being fed to the furnace. When gas streams of varying heating values are combined, the resulting mixture will have a varying heating value. For stable furnace firing, the heating value of the fuel should be kept as constant as possible. The usual method of accomplishing this is to have a supplemental fuel with high and constant heating value as a control stream. If the heating value of the total stream drops, more of the supplemental fuel is blended in. Like the other analyzers mentioned earlier, this instrument too can serve as a safety device by giving an alarm when the heating value of the fuel reaches such a low point that it will only marginally support combustion. Both fuel blending and emergency shutdown of firing caused by low heating values can be automated (Figure 8.12m). Liquid propane is the supplemental fuel under flow control by FRC-1, being reset by the BTU analyzer controller ARC-1. The level controller LC-1 on the vaporizer adds just the right amount of steam required to vaporize the supplemental fuel. The knockout drum is required to keep any unvaporized fuel from reaching the burners, which are designed for gas firing only. If the analyzer detects a dangerously low heating value, the fuel supply is shut down by ASL-1, actuating the shutoff valve (AV-1).

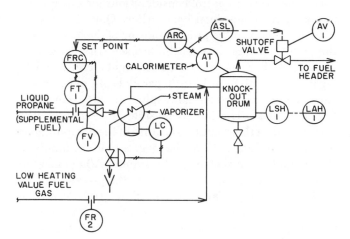

Fig. 8.12m   Calorimeter analyzer application

Advanced Controls

Typical advanced control systems discussed here include feedforward control, analog computer control, and digital computer control as applied to furnaces. An example of each type of control will be presented to give the reader an idea of the possibilities of this approach. The divisions made are arbitrary in that the control systems can be designed to include two or all three techniques in the overall system. Thus, a digital computer control system could incorporate the feedforward technique with an analog computer or a dedicated microprocessor to supply input data.

Feedforward Control

In feedforward control, known relationships between variables are used to adjust for disturbances before they enter the process. This is contrary to feedback control, which must first detect an error in the process variable before corrective action is initiated.

A simple example of feedforward control is given in Figure 8.12n. The normal control is a feedback, with the temperature controller (TRC-1) providing the set point for the fuel pressure controller (PIC-1). On change of feed flow to the furnace, which is an independent variable in this system, there is a definite need to change the rate of fuel firing. As shown, the flow transmitter (FT-1) detects the change in feed flow and sends an altered signal to the multiplying relay (FY-1). The mul-

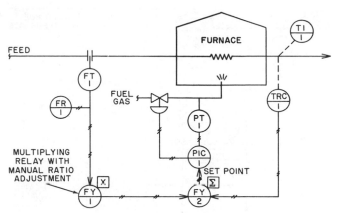

Fig. 8.12n   Feedforward control

tiplying relay sets the relationship between a change in feed flow and the required change in fuel header pressure. This is an empirically determined value and can be field-adjusted. The modified signal then enters the summing relay (FY-2), which adds the signal to the temperature controller output, thereby setting a new fuel gas header pressure via PIC-1. In effect, advance information is fed forward to the firing controller (PIC-1), indicating that a change in process load is taking place and that firing conditions should begin to change. Without the feedforward leg of the loop (FY-1 and FY-2), the required change in firing conditions would take place much later, after the temperature controller (TRC-1) detected a change in the controlled variable (the effluent temperature of the furnace). If a constant ratio existed between feed flow and header pressure, the temperature controller would not be necessary, but in practice, with changing ambient and process conditions, this ratio changes with time. The TRC-1 thus acts as a slow trim-controller, keeping the controlled variable at the desired value.

Analog Computer or Dedicated Microprocessor Controls

There are certain variables that are impractical to measure yet are important parameters in a control system. These may be obtained through computation, using one or more easily measured variables. Relatively inexpensive computing components are available to do such computations. They are available in both pneumatic and electronic versions, and some instrument suppliers offer these as assembled systems.

An example of the use of an analog computer is in the calculation of the heating value of a fuel gas mixture, blended from varying quantities of two known fuel gases. Figure 8.12o depicts this computation. The heating values of fuel gases #1 and #2 are known, but the percentage of each component in the final mixture is variable, resulting in a varying heating value of the total mixture. The computation shown is quite straightforward, using two multiplying relays, FY-1 and FY-2, two

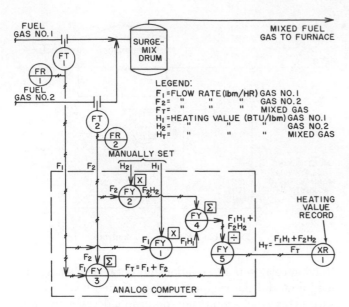

Fig. 8.12o Analog computer determines heating value of mixed gas

Digital Computer Control

Digital computers have been used successfully in process control in a wide range of applications. Their potential in future installations is enormous. The degree of complexity of hardware and software can vary widely depending on the functions of digital computers. Here the attempt is to give an example of digital computer control for the purpose of indicating the potential of this tool in furnace operations. Figure 8.12q represents a computer control system for an ethylene cracking furnace. The process control considerations for this type of furnace have been described earlier. One of the main considerations was the proper splitting of the feed among the parallel passes through the furnace. In a large ethylene plant there may be from six to eight identical cracking furnaces with from 10 to 20 parallel coils per furnace, and maintaining proper flow splits in these coils is a formidable task. This job is well suited to computer control because of the repetitious nature of the computations involved. In Figure 8.12q the signals that are inputs to the computer are shown by the symbol A/D (analog to digital), and the computer outputs to the process by D/A (digital to analog). The total feed flow (A/D-1) and the individual coil flows (A/D-2, -3, -4, etc.) are taken in by the computer, from which the proper valve settings are calculated. The proper valve adjustments are sent out (through signals D/A-2, -3, -4, etc.), which causes the valves to assume their required position to divide the total feed flow. At the start, the total flow will be divided equally among the passes, but as time progresses, coke will gradually build up in the individual coils, causing outlet temperatures to vary from coil to coil. The com-

summing relays, FY-3 and FY-4, and a dividing relay, FY-5. The results are continually recorded on the BTU recorder XR-1. This system may be combined into the fuel-firing system shown in Figure 8.12p. The fuel gas header pressure for stable burner operation, as controlled by PIC-1, is dependent on the heating values of the fuels consumed. This is so because the two fuels differ greatly in heating value, and further, the amounts of each gas consumed can vary considerably with time. The combined heating value is therefore computed using the same analog computer described earlier, and its signal provides the set point of the fuel gas pressure controller PIC-1. The signal-limiting relays FY-6 and FY-7 are installed to maintain the header pressure within safe operating limits even if the computer calls for extreme pressure settings.

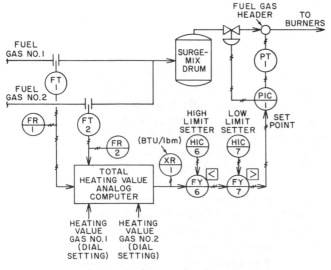

Fig. 8.12p Application of analog computer

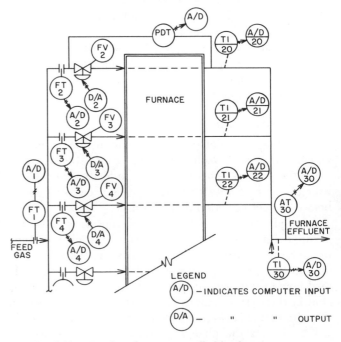

Fig. 8.12q Pyrolysis furnace controlled by digital computer

puter is programmed to keep the effluent temperatures the same (within some tolerance) by altering the flow of the feeds through each coil. It does this by taking in the effluent coil temperature information (A/D-20, -21, -22, etc.), comparing the temperatures, and modifying the "feed-splitting" computation. The results of the computation are again sent out to reposition the feed valves (FV-2, FV-3, FV-4, etc.). The analyzer, measuring the effluent stream composition (A/D-30), also sends its signal to the computer, where it is compared with the desired values. Based on the differences, the degree of fuel firing at various zones in the furnace is computed. These may be printed out as instructions to the operator or sent out as outputs to the fuel valves, changing the heat flow pattern throughout the furnace and thus optimizing the cracking operation.

Many additional functions can be performed using much of the same input data. Some of these are "off-normal" alarms on feed flow, "off-normal" alarms on effluent temperatures, alarms to signal excessive pressure drop across any coil, high coil-metal temperatures and other scanning tasks. Both the control and scanning functions are done on a periodic basis, with the cycle period determined by the nature of the measurement. Flows are usually handled at much higher frequencies than temperature or pressure inputs. Other computations, such as optimization and furnace efficiency calculations, can be performed at a much slower rate.

The question of when to use a computer is a difficult one to answer, but when a high-quantity throughput or an expensive product is involved, there is a good possibility that justification exists.

Summation

The examples given in this section are representative of a wide spectrum of furnaces throughout the process industry. Important aspects of furnace control were covered, with stress being placed on the process requirements.

Safety of operation is a requirement that cannot be compromised. Although many risks are involved in furnace operation, many beyond the control of the instrument engineer, those that can be controlled should obviously be. This does not always mean the use of additional instrument hardware, but rather the effective use of minimum amounts of components to do the practical job.

The use of excessive interlocks has, in many instances, made a unit so cumbersome to operate that all interlocks were deactivated, leaving the unit virtually unprotected. There is no substitute for careful planning in the design of furnace safety controls and their mode of operation.

Finally, with the expansion of industrial facilities all over the world and with the use of larger furnace units, thought should be given to the use of the more advanced forms of controls. Proper use of feedforward, analog, and digital computer controls can offer a furnace operation that is more stable, more efficient, and therefore more profitable.

BIBLIOGRAPHY

Bailey, E.G., *How to Save Coal*, Bailey Meter Company, Bulletin No. 41, April 1918.

Baxley, R.A., "Computer Control of Severity in Ethylene Cracking Furnaces," *Instrumentation Technology*, November 1971.

Boyle, W.E. and Hoyt, P.R., "Experimental Application of Combustion Controls to a Process Heater," Annual ISA Conference, September 13–17, 1948.

Calabrese, A.M. and Krejci, L.D., "Safety Instrumentation for Ammonia Plants," *Chemical Engineering Progress*, February 1974.

Clelland, P.J., "Furnace Implosions: Causes and Cures," *Instrumentation Technology*, December 1977.

DeLorenzi, O., *Combustion Engineering Reference Book on Fuel Burning and Steam Generation*, Combustion Engineering, New York, 1955.

Dukelow, S.G., "Energy Conservation Through Instrumentation," University of Houston Energy Conservation Seminar, February 7,

———, "Fired Process Heaters," presented at ISA/CHEMPID, April 1980.

———, "How Control Improvements Save Process Heater Fuel," 1979 Conference on Industrial Energy Conservation Technology.

Harrison, T.J., *Microcomputer in Industrial Control: An Introduction*. Instrument Society of America, North Carolina, 1978.

Johnson, R.K., "Modern Approach to Process Heater Control Systems," Parkersburg-Marietta ISA Section, Parkersburg, West Virginia, February 8, 1977.

Knapp, Russell V., inventor, U.S. Patent 1,940,355, Patented December 19, 1933.

Kostiw, W.B., "How to Instrument Fuel Oil Systems," *Instrumentation Technology*, October 1974.

Luetuge, J.A., "Measurement and Control of Temperatures in Rotary Kilns," *Instrumentation Technology*, March 1968.

Marchant, G.R., "Digital Controls for Continuous Copper Smelting," *Instrumentation Technology*, June 1978.

Shinskey, F.G., "Feedforward Controls Applied," *ISA Journal*, November 1963.

———, "Analog Computing Control," *Control Engineering*, November 1962.

8.13 HEAT EXCHANGER CONTROLS

Practically all chemical plants include some form of heat transfer equipment. The control of these devices will be covered in this section. Both the nature of heat exchange and the quality of control desired play a part in determining the proper instrumentation for the unit. As far as the nature of heat transfer is concerned, the following paragraphs will discuss liquid-to-liquid exchangers, steam heaters, condensers, and reboilers as the basic forms of heat exchange. Next, the more sophisticated control systems—such as cascade and feedforward—are covered, and the section ends with a brief evaluation of multi-purpose heat transfer systems.

Variables and Degrees of Freedom

All control loops function on the basis of controlling one variable by manipulating the same or some other process variable(s). It is important to be able to determine the maximum number of independently acting automatic controllers that can be placed on a process. By definition, this is called the number of degrees of freedom. (See Section 1.1 for more details.)

Figure 8.13a shows a steam heater with its variables and defining parameters. The temperatures and flows are variables, and the specific and latent heats are parameters. One arrives at the available degrees of freedom by subtracting the number of system-defining equations from the number of variables.

Degrees of freedom =

$$= \text{(number of variables)} - \qquad 8.13(1)$$
$$- \text{(number of equations)}$$

In this case, there are four variables and one equation, which is obtained from the first law of thermodynamics, stating the conservation of energy:

$$\lambda W_s = C_p W (T_2 - T_1) \qquad 8.13(2)$$

Therefore, this system has three degrees of freedom, which at the same time is the maximum number of automatic controllers that can be used.

In a liquid-to-liquid heat exchanger there are four temperature and two flow variables with still only one (conservation of energy) defining equation, (8.13(3)) resulting in five degrees of freedom.

$$C_p F (T_2 - T_1) = C_{pc} F_c (T_{1c} - T_{2c}) \qquad 8.13(3)$$

In a steam-heated reboiler or in a condenser cooled by a vaporizing refrigerant (assuming no superheating or supercooling) there are only two flow variables and one defining equation (8.13(4)) allowing for only a single degree of freedom or one automatic controller.

$$\lambda W_s = \lambda_1 W_1 \qquad 8.13(4)$$

In the majority of installations fewer controllers are used than there are degrees of freedom available, but every once in a while problems associated with over-defining a system are also likely to arise.

Liquid-to-Liquid Heat Exchangers

In Figure 8.13b the pneumatic transmitters are of the indicating, filled-bulb type, and the indicating controller is panel-mounted. These features have been uniformly used on all illustrations in this section for reasons of consistency only.

The purpose of this section is to evaluate control techniques and not to recommend hardware selection. Therefore, it should be clearly understood that it is not the writer's intent to express a preference for filled thermal bulbs over thermocouples or for pneumatic indicating instruments over electronic recording ones. In fact,

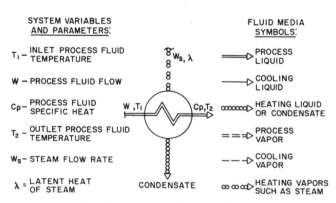

SYSTEM VARIABLES
AND PARAMETERS:

T_1 – INLET PROCESS FLUID TEMPERATURE

W – PROCESS FLUID FLOW

C_p – PROCESS FLUID SPECIFIC HEAT

T_2 – OUTLET PROCESS FLUID TEMPERATURE

W_s – STEAM FLOW RATE

λ – LATENT HEAT OF STEAM

W_s, λ

W, T_1 C_p, T_2

CONDENSATE

FLUID MEDIA
SYMBOLS:

⟹ PROCESS LIQUID

→ COOLING LIQUID

ꝏꝏꝏ▷ HEATING LIQUID OR CONDENSATE

==⇒ PROCESS VAPOR

- - -▷ COOLING VAPOR

∞ ∞ ∞▷ HEATING VAPORS SUCH AS STEAM

Fig. 8.13a   Heat exchanger variables

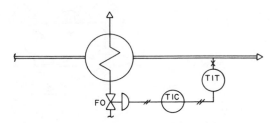

Fig. 8.13b   Feedback control by throttling coolant inlet

the repeatability, sensitivity, and speed of response of electronic systems with thermocouple elements are considered to be superior. For more details on temperature detectors, see Chapter 4 in the Process Measurement volume of this Handbook.

Figures 8.13b and 8.13c illustrate cooler and heater installations with the control valve mounted on the exchanger inlet and outlet, respectively. From a control quality point of view, it makes little difference whether the control valve is upstream or downstream to the heater. The location is normally based on the desirability of operating the heat-transfer-medium side of the exchanger under supply or return header pressures.

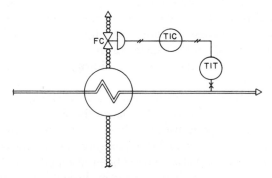

Fig. 8.13c   Feedback control by throttling heating media outlet

Component Selection

It is generally recommended to provide positioners for these valves to minimize the valve friction effects. The use of equal percentage valve trims (Section 4.6) is also recommended, because it usually contributes to maintaining the control system gain constant, under changing throughput conditions. This is a result of the equal percent trim, which keeps the relationship between valve opening and temperature change (reflecting load variations) constant.

In the majority of installations, a three-mode controller would be used for heat exchanger service. The derivative or rate action becomes essential in long time-lag systems or when sudden changes in heat exchanger throughput are expected. Because of the relatively slow nature of these control loops, the proportional band setting must be wide to maintain stability. This means that the valve

will be fully stroked only as a result of a substantial deviation from desired temperature set point. (For controller tuning, see Section 1.11.) The reset or integral control mode is required to correct for temperature offsets caused by process load changes. Besides the changes in process fluid flow rate, other variables can give the appearance of load changes, such as inlet temperature or header pressure changes of the heat transfer medium.

The selection and location of the thermal element are also important. It must be placed in a representative location, without increasing measurement time lag. In reference to Figure 8.13b, this would mean that the bulb should be located far enough from the exchanger for adequate mixing of the process fluid but close enough so that the introduced delay will not be substantial. If the process fluid velocity is 3 ft/second (0.9 m/s), then a one-second distance-velocity lag is introduced for each 3 ft (0.9 m) of pipe between the exchanger and the bulb. This lag can be one of the factors that will limit the dynamic performance of the system, but it should also be clearly understood that all thermal elements are dynamically imperfect to start with. This inherent limitation results from the fact that in order to change the temperature of a sensor, heat must be introduced, and it has to enter the bulb through a fixed area.

Let us calculate the dynamic lag of a typical filled bulb having the area of 0.02 sq ft (0.0018 m²) and the heat capacity of 0.005 BTU/°F (9.49 J/°C). If this bulb is immersed in a fluid with a heat transfer coefficient (based on flow velocity) of 60 BTU/h/°F/ft² (1.23 MJ/h/°C/m²) and then the process temperature is changed at a rate of 25°F/minute (13.9°C/min), the dynamic lag can be calculated. First the amount of heat flowing into the element under these conditions will be determined:

$$q = \text{(rate of temperature change) (bulb heat capacity)} =$$
$$= (25)(60)(0.005) =$$
$$= 7.5 \text{ BTU/h (7.9 kJ/h)} \qquad 8.13(5)$$

The dynamic measurement error is calculated by determining the temperature differential across the fluid film surrounding the bulb that is required to produce a heat flow of 7.5 BTU/h (7.9 kJ/h).

$$q = Ah\,\Delta T \qquad 8.13(6)$$

and therefore,

$$\Delta T = q/Ah = 7.5/(0.02)(60) =$$
$$= 6.25°F \ (3.47°C) \qquad 8.13(7)$$

If the rate of process temperature change is 25°F/minute (14°C/m) and the dynamic error based on that rate is 6.25°F (3.47°C) then the dynamic time lag is

$$t_0 = 6.25/25 = 0.25 \text{ minutes} =$$
$$= 15 \text{ seconds} \qquad 8.13(8)$$

This lag can also be calculated as

$$t_0 = \frac{\text{(bulb heat capacity)}}{\text{(bulb area)(heat transfer coefficient)}} =$$
$$= \frac{60 \times 0.005}{0.02 \times 60} = 0.25 \text{ minutes} \qquad 8.13(9)$$

Bulb time lags vary from a few seconds to minutes, depending on the nature and velocity of the process fluid being detected. Measurement of gases at low velocity involves the longest time lags, and water (or dilute solutions) at high velocity the shortest. From equation 8.13(9), it is clear that one method of reducing time lag is by miniaturizing the sensing element.

These numbers were based on a bare bulb diameter of $\frac{3}{8}$ inch (9.375 mm).

The addition of a thermowell will further increase the lag time, but in most industrial installations, thermowells are necessary for reasons of safety and maintenance. When they are used, it is important to eliminate any air gaps between the bulb and the socket.

The measurement lag is only part of the total time lag of the control loop. For example, an air heater might have a total lag of 15 minutes, of which 14 minutes is the *process lag*, 50 seconds is the bulb lag, and 10 seconds is the time lag in the control valve.

Three-Way Valves

The limits within which process temperature can be controlled are a function of the nature of load changes expected and of the speed of response for the whole unit. In many installations, the process time lag in the heat exchanger is too great to allow for effective control during load changes. In such cases, it is possible to circumvent the dynamic characteristics of the exchanger by partially bypassing it and blending the warm process liquid with the cooled process fluid, as shown in Figure 8.13d. The resulting increased system speed of response together with some cost savings are the main motivations for considering three-way valves in such services. The "bulb time lag" discussed in the previous paragraph has an increased importance in these systems, because it represents a much greater *percentage* of total loop lag time than in the previously discussed installations.

As illustrated in Figures 8.13d and 8.13e either a diverter or a mixing valve can be used for this purpose.

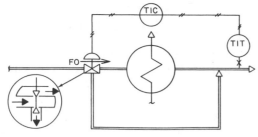

Fig. 8.13d   Diverter valve used to control cooler

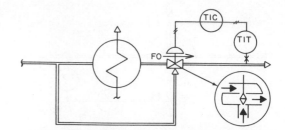

Fig. 8.13e   Mixing valve used to control cooler

Stable operation of these valves is achieved by *flow* tending to *open* the plugs in both cases. If a mixing valve is used for diverting service or a diverting one for mixing, the operation becomes unstable because of the "bathtub effect" resulting from the flow to close path across the valve. Therefore, it is not good enough just to install a three-way valve; it has to be either the mixing or the diverting design to match the particular service.

Three-way valves are unbalanced designs and are normally provided with linear ports. The unbalanced nature places a limitation on allowable shutoff pressure difference across the valve, and the availability with linear ports only prevents the relationship between valve movement and temperature change from being constant. (Equal percentage characteristics are required for that, because a change in detected temperature usually requires an exponential change for the flow rate of the heat transfer fluid to reach a new equilibrium at the changed load level.)

Misalignment or distortion in a control valve installation can cause binding, leakage at the seats, high deadband, and packing friction. Such conditions commonly arise as a result of high-temperature service on three-way valves. The valve, having been installed at ambient conditions and rigidly connected at three flanges, cannot accommodate pipe line expansion because of high process temperature, and therefore distortion results. Similarly, on mixing applications, when the temperature difference is substantial between the two ports, the resulting differential expansion can also cause distortion. For these reasons, the use of three-way valves at temperatures above 500°F (260°C) or at differential temperatures exceeding 300°F (167°C) is not recommended.

The choice of three-way valve location relative to the exchanger (Figures 8.13d and 8.13e) is normally based on pressure and temperature considerations, with the upstream location (Figure 8.13d) being usually favored for reasons of uniformity of valve temperature. When the overriding consideration is to operate the exchanger at a high pressure, the downstream location might be selected.

Cooling Water Conservation

Figure 8.13f modifies the previous sketch to include the additional feature of cooling water conservation. This

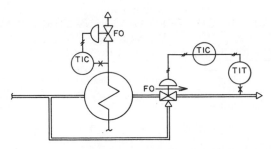

Fig. 8.13f Conservation of cooling fluid

system tends to maximize the outlet cooling water temperature, thereby minimizing the rate of water usage. The application engineer, when using this concept, should be careful in evaluating the temperature levels involved and to make sure that the water contains chemicals to prevent tube fouling, if high temperature operation is planned. If the cooling water supply is sufficient this system not only will conserve water but also will protect against excessively high outlet water temperatures.

Balancing the Three-Way Valve

It is recommended that a manual balancing valve be installed in the exchanger bypass, as shown in Figure 8.13g. This valve is so adjusted that its resistance to flow equals that of the exchanger.

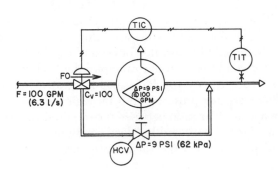

Fig. 8.13g Use of balancing valve in the exchanger bypass

When sizing the pump it should be kept in mind that the *resistance* to flow in such installation will be *maximum when one of the paths is closed* and the other is fully open, whereas minimum resistance will be experienced when the valve equally divides the flow between the two paths. This is not necessarily self-evident, and therefore it will be illustrated in an example.

It is assumed that the flow rate of process fluid is 100 GPM (6.3 l/s), the pressure drop at full flow through either the exchanger or the balancing valve is 9 PSI (6.21 kPa), and the diverting valve has a valve coefficient of $C_v = 100$. The equivalent coefficient for the exchanger (or balancing valve) is calculated as:

$$C_e = \frac{\text{flow}}{\sqrt{\text{pressure drop}}} = \frac{100}{\sqrt{9}} =$$
$$= 33.3 \qquad\qquad 8.13(10)$$

Therefore, in either extreme position (closed or full bypass), the total system resistance expressed in valve coefficient units is

$$\frac{1}{(C_t)^2} = \frac{1}{(C_v)^2} + \frac{1}{(C_e)^2} =$$
$$= \frac{1}{(100)^2} + \frac{1}{(33.3)^2} \therefore C_t = 31.7 \quad 8.13(11)$$

When the valve divides the flow equally between the two paths, because of the linear characteristics of all three-way valves, its coefficient at each port will be $C_v = 50$. The equivalent coefficient, $C_e = 33.3$ of the exchanger and balancing valve being unaffected, the total system resistance in valve coefficient units, is $2C_t$, where

$$\frac{1}{(C_t)^2} = \frac{1}{(C_v)^2} + \frac{1}{(C_e)^2} =$$
$$= \frac{1}{(50)^2} + \frac{1}{(33.3)^2} \therefore 2C_t = 55.6 \quad 8.13(12)$$

If we calculate the total pressure drop through the system when the valve is in its extreme and when it is in its middle position, handling the same 100-GPM (6.3 l/s) flow,

$$\Delta P \text{ extreme} = \left(\frac{\text{flow}}{C_t}\right)^2 = \left(\frac{100}{31.7}\right)^2 =$$
$$= 10 \text{ PSI (69 kPa)} \qquad 8.13(13)$$

$$\Delta P \text{ middle} = \left(\frac{100}{55.6}\right)^2 = 3.25 \text{ PSI (22.4 kPa)}$$
$$8.13(14)$$

the results indicate that the system drop in one of the extreme positions is more than three times that of the middle position.

Two Two-Way Valves

When for reasons of temperature or because of other considerations, three-way valves cannot be used, but it is desired to improve the system response speed by the

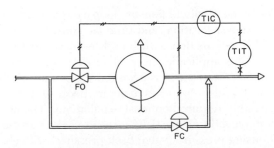

Fig. 8.13h Exchanger bypass control using two-way valves

use of exchanger bypass control, the installation of two two-way valves is the logical solution.

As illustrated in Figure 8.13h, the two valves should have opposite failure positions. Therefore, when one is open the other is closed, and at a 9 PSI (0.6 bar) signal both are halfway open. In order for these valves to give the same control as a three-way valve would, it is necessary to provide them with linear plugs.

It is logical for the reader to ask, Why not always use this two two-way valve system; why consider three-way valves at all? The answer is basically cost. The price of a three-way valve is about 65 percent that of two two-ways, the installation cost is similarly lower, and when positioners are required (the majority of cases), only one needs to be purchased, instead of two. There are some other considerations, which might tend to modify this impression of substantial cost savings. One such consideration is that the capacity of a three-way valve is the same as the capacity of a single-ported two-way valve, which is only 70 percent of the capacity of a double-ported one. This could mean that instead of a 10-inch (250 mm) three-way unit, two 8-inch (200 mm) double ported two-way valves can be considered, and therefore the cost advantage of a three-way installation is not as substantial as implied earlier. This is true if the comparison is based on double-ported two-way valves, but the inherent leakage (approximately $\frac{1}{2}$ percent of full capacity) of these units makes them unsuitable for some installations. Therefore, where tight shutoff is required, only the three-way or the single-ported two-way valves can be considered, and their capacity is about the same.

As touched upon earlier, three-way valves are not recommended for high temperature or high-pressure differential service. In addition to the reasons noted earlier, the hollow plug design of three-way valves also contributes to this limitation because it is more subject to heat expansion and is more difficult to harden than the solid plugs.

To summarize, bypass control is applied to circumvent the dynamic characteristics of heat exchangers and thereby improve their controllability. Bypass control can be achieved by the use of either one three-way valve or two two-way valves, and Table 8.13i summarizes the merits and drawbacks of using one or the other.

Steam Heaters

The general discussion on loop components, accessories, sensor location, and time lag considerations presented in connection with liquid-to-liquid heat exchangers is also applicable here. The desirability of equal percentage valve trims is even more pronounced than it was on liquid-to-liquid exchangers because of the high rangeability required on most installations. The need for high rangeability is partially a result of the variations in condensing pressure (valve back-pressure) with changes in process load. This can be best visualized by an example.

Table 8.13i
MERITS OF TWO-WAY VERSUS THREE-WAY
VALVES IN
EXCHANGER BYPASS INSTALLATIONS

	One Three-Way Valve	Two Single-Seated Two-Way Valves	Two Double-Ported Two-Way Valves
Most economical	Yes		
Provides tight shutoff	Yes	Yes	
Applicable to service above 500°F (260°C)		Yes	Yes
Applicable to differential temperature service above 300°F (167°C)		Yes	Yes
Applicable to operation at high pressure and pressure differentials		Yes	Yes
Highest capacity for same valve size			Yes

In Figure 8.13j both the high and the low load conditions are shown. When the steam flow demand is the greatest, the back-pressure is also the highest, leaving the lowest driving force (pressure drop) for the control valve. High flow and low pressure drop result in a large valve that might be throttled beyond its capability under low load conditions. In our example, the back-pressure at low loads is only 100 PSIA (690 kPa), allowing some 16 times greater pressure drop through the valve than at high loads. The ratio between the required valve coefficients for the high and low load conditions represents the rangeability that the valve has to furnish.

$$\text{Rangeability} = S \frac{F_{s,max}}{F_{s,min}} \sqrt{\frac{[(P_1 - P_2)(P_1 + P_2)]_{min}}{[(P_1 - P_2)(P_1 + P_2)]_{max}}} =$$
$$= 1.5 \times 5 \sqrt{\frac{100 \times 300}{7 \times 393}} =$$
$$= 25.5 \qquad\qquad 8.13(15)$$

The letter "S" with the numerical value of 1.5 in the above equation represents the safety factor that is applied in selecting the control valve. A rangeability requirement of this magnitude can create some control problems, which can best be solved by the installation of a smaller control valve in parallel with the large one. If this is not done, the control quality will suffer for two reasons:

1. At low loads the valve will operate near to its clearance flow point where the flow-versus-lift

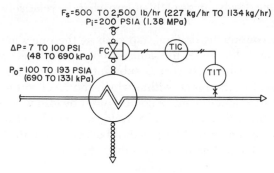

Fig. 8.13j Feedback control of steam-heated exchanger

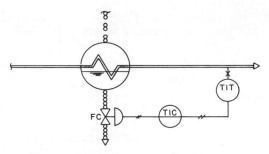

Fig. 8.13k Controlling the rate of condensate removal

curve changes abruptly, contributing to unstable or possibly on-off cycling valve operation.

2. For good control the system gain should not vary with changes in load, which an equal percentage control valve can guarantee only if control valve ΔP is not a function of load. This being the case, the only way to guarantee constant system gain is to install two valves for the two load conditions, both sized to maintain the same gain.

For a more detailed discussion on control valve rangeability, refer to Section 4.13.

Minimum Condensing Pressure

As it has been shown in connection with Figure 8.13j, the condensing pressure is a function of load when the temperature is controlled by throttling the steam inlet. This at low loads and low operating temperatures can result in below atmospheric condensing pressures. If this occurs, then the condensing pressure will not be sufficient to discharge the condensate through the steam trap, which therefore accumulates inside the heater. As condensate accumulation progresses, more and more of the heat transfer area will be covered up, resulting in a corresponding increase in condensing pressure. When this pressure rises sufficiently to discharge the trap, the condensate is suddenly blown out and the effective heat transfer surface of the exchanger increases several fold, instantaneously. Such upsets of course make the control of temperature impossible, and methods of improving this situation have to be considered. In the following paragraphs, the merits of some of these techniques will be discussed.

Control Valve in the Condensate Line

Mounting the control valve in the condensate line, as shown in Figure 8.13k, is sometimes proposed as a solution to minimum condensate pressure problems. An additional motivation to consider this technique is the cost advantage of purchasing a small condensate valve instead of a larger one for steam service.

On the surface this appears to be a very convenient solution, since the throttling of the valve causes varia-

tions only in the condensate level inside the partially flooded heater and has no effect on the steam pressure, which stays constant. Therefore, there are no problems in condensate removal.

What is important to keep in mind is that the criterion for a successful heater installation is not whether the accumulated condensate is removable but whether it allows for precise temperature regulation. The answer is negative; the valve in the condensate line does not allow for accurate control of temperature. The reason for this is the dynamics of the system.

When the load is decreasing, the valve is likely to close completely, before the condensate builds up to a high enough level to match the new lower load with a reduced heat transfer area. In this direction, the process is slow, because steam has to condense before level can be affected. When the load increases, the process is fast, because just a small change in control valve opening is sufficient to drain off enough condensate to expose an increased heat transfer surface. Having these "non-symmetrical" dynamics, control is bound to be poor. If the controller is tuned for the fast response speeds of the increasing load direction, substantial overshoot can result when load is decreasing, whereas if it is tuned for the slow part of the cycle, the overshoot and possible cycling occur in the opposite direction.

For these reasons, it is recommended that control valves not be mounted in the condensate lines on services in which temperature control is important.

Pumping Traps

It is possible to prevent condensate accumulation in heaters operating at low condensing pressures by the use of lifting traps. This device is illustrated in Figure 8.13l and depends on an external pressure source for its energy.

The unit is shown in its filling position, in which the liquid head in the heater has opened the inlet check valve of the trap. Filling progresses until the condensate overflows into the bucket, which then sinks, closing the equalizer and opening the pressure source valve. As pressure builds in the trap, the inlet check valve is closed, and the outlet one opens when the pressure exceeds that of the condensate header. The discharge cycle follows; during which the bucket is emptied. When near empty,

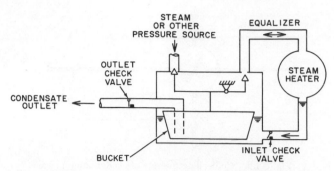

Fig. 8.13l Lifting or pumping trap shown in its "filling" position

the buoyant force raises the bucket, which then closes off the steam valve and opens the equalizer. Once the pressure in the trap is lowered the condensate outlet check valve is closed by the header back pressure, and the inlet check is opened by the liquid head in the heater, which then is the beginning of another fill cycle.

The aforementioned pumping trap guarantees condensate removal regardless of the minimum condensing pressure in the heater. If such a trap is placed on the exchanger illustrated in Figure 8.13j, it will make temperature control possible, even when the heater is under vacuum. This of course does not relieve the rangeability problems discussed earlier, and the use of two valves in parallel might still be necessary.

Level Controllers

Because the low condensing pressure situation is a result of the combination of low load and high heat transfer surface area, it is possible to prevent the vacuum from developing by reducing the heat transfer area. One method of achieving this is shown in Figure 8.13m, in which the steam trap has been replaced by a level control loop. With this instrumentation provided, it is possible to "adjust the size" of the heater by changing the level set point to match the process load. This technique gives good temperature control if the level setting is correctly made and if there are no sudden load variations in the system. The response to load variations is non-symmetrical, just as it was outlined in connection with mounting the temperature control valve in the condensate line.

One disadvantage of this approach is the relatively high cost.

A continuous drainer trap such as the one shown in Figure 8.13n serves the same purpose as the level control just described. Its cost is substantially lower, but it is limited in the range within which the level setting can be varied and its control point is offset by load variations. For this reason, it is unlikely to be considered for installation on vertical heaters or reboilers, where the range of level adjustment can be substantial.

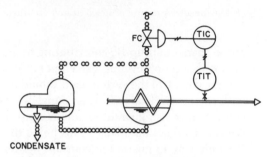

Fig. 8.13n Continuous drainer trap

Bypass Control

Table 8.13i in the preceding text summarized some of the features of three-way valves when installed to circumvent the transient characteristics of the cooler by bypassing it. Figure 8.13o shows the same concept applied to a steam heater. The advantages and limitations of this system are the same as discussed in connection with liquid-liquid exchangers, but there is one additional advantage. This has to do with the fact that the bypass created an additional degree of freedom, and therefore steam can now be throttled as a function of some other property. The logical decision is to adjust the steam feed so that it maintains the condensing pressure constant. This then eliminates problems associated with condensate removal. One should also realize that in case of full bypass operation, the stagnant exchanger contents will be exposed to steam heat, and therefore, unless protected, it is possible to boil this liquid.

Table 8.13p summarizes some of the features of the various techniques discussed, which can be applied to combat problems created by low condensing pressures.

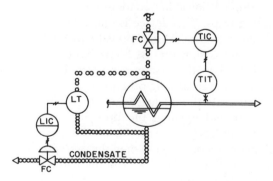

Fig. 8.13m Steam trap replaced by level control

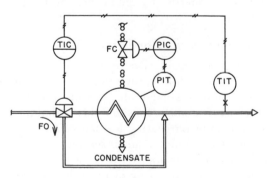

Fig. 8.13o Bypass control on steam-heated exchanger

Table 8.13p
FEATURES OF HEATER CONTROL SYSTEMS

	System Cost	Condition for Giving Precision Control		
		Small or Slow Load Variations	Fast or Large Load Variations	Low (Vacuum) Condensing Pressures
Valve in condensate line throttled by temperature	Low	Questionable	No	No
Valve in steam line throttled by temperature	Medium	Yes	No	No
Two valves in steam line throttled by temperature	Medium	Yes	Yes	No
Valve in steam line throttled by temperature and condensate removed by drainer trap	Medium	Questionable	No	Yes
Valve in steam line throttled by temperature and condensate removed by pumping trap	Medium	Questionable	No	Yes
Valve in steam line throttled by temperature and condensate removed by level controller	High	Yes	No	Yes
Three-way valve controls temperature by throttling bypass, and steam inlet is controlled to maintain condensing pressure	High	Yes	Yes	Yes

Condensers

The control of condensers and reboilers as part of larger systems is covered in Section 8.8. For this reason, only a brief discussion is presented here.

Depending on whether the control of condensate temperature or of condensing pressure is of interest, the systems shown in Figures 8.13q and 8.13r can be considered. Both of these throttle the cooling water flow through the condenser, causing a potential for high temperature rise that is acceptable only when the water is chemically treated against fouling. For good, sensitive control the water velocity through the condenser should be such that its residence time does not exceed one minute.

When it is not desirable to throttle the cooling water,

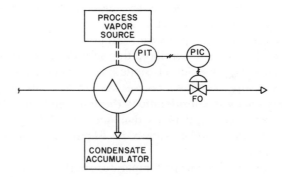

Fig. 8.13r Condenser on pressure control

the system illustrated in Figure 8.13s can be considered. Here the exposed condenser surface is varied to control the rate of condensation. Where noncondensables are present, a constant purge may serve to remove the inerts. One drawback of this system is the same "non-symmetricity" that has been discussed in connection with Figure 8.13k.

To reduce the problems associated with non-symmetrical process dynamics, the "hot gas bypass" systems shown in Figures 8.13t and 8.13u can be considered. In case of Figure 8.13t, the opening of the bypass valve results in pressure equalization between condenser and accumulator, causing partial flooding of the condenser, as a result of the relative elevations. When the condensing pressure is to be reduced, the valve closes, re-

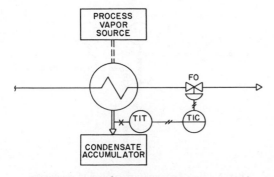

Fig. 8.13q Condenser on temperature control

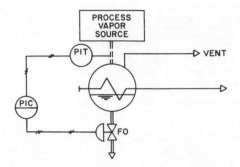

Fig. 8.13s Condenser control by changing the wetted surface area

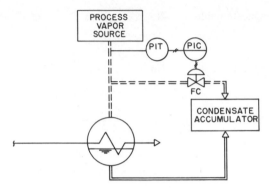

Fig. 8.13t Hot gas bypass control

sulting in an increase of exposed condenser surface area. In order to expose more area, the condensate is transferred into the accumulator, which can occur only if the accumulator vapor pressure has been lowered because of condensation. Therefore, the system speed in this direction is a function of condensate super-cooling. Increased super-cooling increases system response speed. Based on these considerations the unit might or might not be symmetrical in its dynamics. If high speed is desired, the controls shown in Figure 8.13u can be considered.

When the condensing temperature of the process fluid is low, water is no longer an acceptable cooling medium. One standard technique of controlling a refrigerated con-

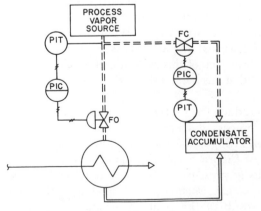

Fig. 8.13u High-speed hot gas bypass control

denser is illustrated in Figure 8.13v. Here the heat transfer area is set by the level control loop and the operating temperature is maintained by the pressure controller. When process load changes, it affects the rate of refrigerant vaporization, which is compensated for by level-controlled makeup. Usually the pressure and level settings are manually made, although there is no reason why these set points could not be automatically adjusted as a function of load, if required. (See Section 8.4 for refrigeration units.)

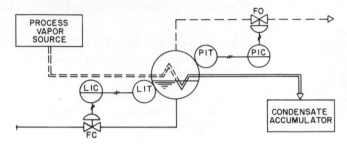

Fig. 8.13v Condenser controls with refrigerant coolant

Reboilers and Vaporizers

As noted at the beginning of this section, in case of a steam-heated reboiler there is only one degree of freedom available, and therefore only one controller can be installed without overdefining the system. This one controller is usually applied to adjust the rate of steam addition. As to the minimum condensing pressure considerations, the same applies here as has been discussed earlier in connection with liquid heaters. Figures 8.13w and 8.13x show the two basic alternatives for controlling the reboiler either to generate vapors at a controlled superheat temperature or to generate saturated vapors at a constant rate, set by the rate of heat input.

A number of more sophisticated alternates are discussed in Section 8.8, where the possibilities of control by temperature difference, by composition, or by various cascade and computer methods are covered.

Cascade Control

The reader is referred for a general discussion of cascade control to Section 1.7. Probably the most frequent use of cascade loops is in connection with heat transfer

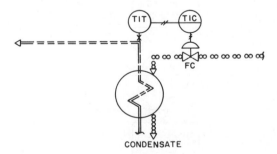

Fig. 8.13w Temperature control of reboilers

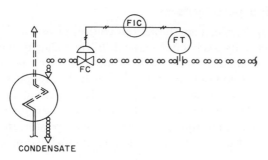

Fig. 8.13x   BTU control of reboilers

units. Cascade systems by definition consist of two controllers in series. The master in case of heat exchangers detects the process temperature, and the slave is installed on a variable that may cause fluctuations in the process temperature. The master adjusts the slave set point, and the slave throttles the valve to maintain that set point. It cannot be overemphasized that a cascade loop controls *a single* temperature, and the slave controller is there only to assist in achieving this. In other words, the cascade loop does not have two independent set points.

Cascade loops are invariably installed to prevent outside disturbances from entering the process. An example of such would be the header pressure variations of a steam heater. The conventional single controller system (see Figure 8.13j) cannot respond to a change in steam pressure until its effect is felt by the process temperature sensor. In other words, an error in the detected temperature has to develop before corrective action can be taken. The cascade loop, in contrast, responds immediately, correcting for the effect of pressure change before it can influence the process temperature (Figure 8.13y).

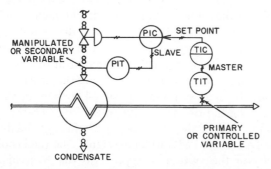

Fig. 8.13y   Temperature-pressure cascade loop on steam heater

The improvement in control quality due to cascading is a function of relative speeds and time lags. A slow primary (master) variable and a secondary (slave) variable that responds quickly to disturbances represent a desirable combination for this type of control. If the slave can quickly respond to fast disturbances, then these will not be allowed to enter the process and therefore will not

upset the control of the primary (master) variable. Some typical cascade applications for heat exchangers are illustrated in the following sketches.

In Figures 8.13y and 8.13z, the controlled variable is temperature, whereas the manipulated variable is the pressure or flow of steam. The primary variable (temperature) is slow, and the secondary (manipulated) variable is capable of quickly responding to disturbances. Therefore, if disturbances occur (sudden change in plant steam demand), upsetting the manipulated variable (steam pressure), it will be immediately sensed and corrective action will be taken by the secondary controller so that the primary variable (process temperature) will not be affected. As to the nature of the possible disturbances, they usually have to do with the properties of the heating or cooling media supply. These have been covered in connection with the discussion on system degrees of freedom at the beginning of this section, and it can be said that the use of cascade control on heat transfer equipment contributes to fast recovery from load changes or from other disturbances.

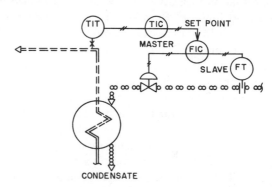

Fig. 8.13z   Temperature-flow cascade loop on steam reboiler

Feedforward Control

As discussed in Section 1.5, feedback control involves the detection of the controlled variable (temperature) and the counteracting of changes in its value relative to a set point, by the adjustment of a manipulated variable (coolant flow). This mode of control necessitates that the disturbance variable first affect the controlled variable itself before correction can take place. Hence, the term "feedback" can imply a correction "back" in terms of time, a correction that should have taken place earlier, when the disturbance occurred.

In this manner of terminology, feedforward is a mode of control that responds to a disturbance such that it instantaneously compensates for an error that the disturbance would have caused otherwise in the controlled variable, later in time.

Figure 8.13aa illustrates a steam heater under feedforward control. All variables that can affect the heat balance relationship are measured, and the manipulated variable (steam flow) is adjusted when upsets occur. The

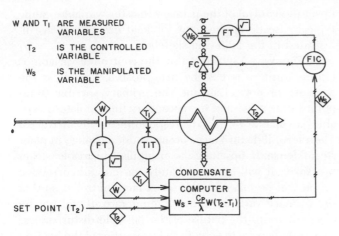

W AND T_1 ARE MEASURED
      VARIABLES

T_2 IS THE CONTROLLED
      VARIABLE

W_s IS THE MANIPULATED
      VARIABLE

SET POINT (T_2) ———

COMPUTER
$$W_s = \frac{C_P}{\lambda} W(T_2 - T_1)$$

Fig. 8.13aa   Feedforward control of heat exchangers

Multipurpose Systems

In the earlier paragraphs the control of isolated heat transfer units has been discussed. In the majority of critical installations, the purpose of such systems is not limited to the addition or removal of heat, but to make use of both heating and cooling in order to maintain the process temperature constant. Such task necessitates the application of multipurpose systems, incorporating many of the features that have been individually discussed earlier.

Figure 8.13bb, for example, depicts a design that uses hot oil as its heat source and water as the means of cooling, arranged in a recirculating system. The points made earlier in connection with three-way valves, cascade systems, and so forth also apply here, but there are a few additional considerations worth noting.

conservation of energy equation (equation 8.13(2)) describes the relationship among the four variables. A few computing modules allow this equation to be solved for W_s, such that any variations in W and T_1 (the load) are compensated for, and therefore T_2 is maintained constant. Figure 1.5i illustrates an improved version of this system incorporating feedback trimming and dynamic lag-balancing features.

The advantages of feedforward control are similar to those of cascade, because the load upset or supply disturbance is corrected for before its effect is felt by the controlled variable. As can be expected, feedforward control contributes to stable, dampened response to load changes and to fast recovery from upsets.

In view of the remarkable improvements in the response to load and set point changes that can be attributed to the feedforward method of control, the reader is justified in questioning why this technique should not be standardized. This writer would answer that in two parts:

First, the cost of such installation is not always justified. Secondly, the implications of high precision control must be consistently accepted. This means that in order to keep T_2 within, say, $\pm\frac{1}{2}\%$ of full scale, the total accumulated error in flow measurements, temperature sensing, transmission, conversion, and computation should not exceed this amount. Therefore, in a computerized chemical plant, where the sensing elements have already been selected for precision consistent with the computer and where the inclusion of equation 8.13(2) in the computer program represents only a small added calculation, feedforward exchanger control could be standardized. In conventional plants, the use of this technique is likely to be reserved for only the most critical heat transfer units, where the expense of additional detector and computing hardware can be justified by the resulting stable and accurate control.

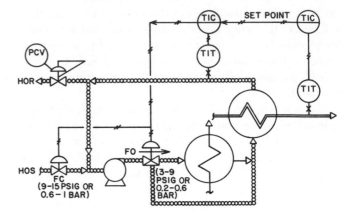

Fig. 8.13bb   Recirculating multipurpose heat transfer system

Probably the most important single feature of this design is that it operates on a "split-range signal." This means that when the process temperature is above the desired set point, the valves will receive a dropping signal. While the value of this signal is between 9 and 15 PSIG (0.6 and 1 bar), the three-way valve is fully open to the exchanger bypass and the two-way hot oil supply valve is partially open. If the reduction in the two-way valve opening is not sufficient to bring the process temperature down to set point, the signal further decreases, fully closing the two-way valve at 9 PSIG (0.6 bar) and beginning to open the cooler flow path through the three-way valve. At a 3 PSIG (0.2 bar) signal, the total cooling capacity of the system is applied to the recirculating stream, which in that case flows through the cooler without bypass.

The implications of such split-range operations should be fully realized:

1. At a signal level near 9 PSIG (0.6 bar), the system can be unstable and cycling, because this is the point at which the three-way valve is just beginning to open to the cooler and the system might

receive alternating slugs of cooling and heating due to the limited rangeability of the valves. (Zero flow is not enough; minimum flow is too much for the particular load condition.)

2. While the signal is in the 9 to 15 PSIG (0.6 to 1 bar) range, the cooler shell side becomes a reservoir of cold oil. This upsets the controls twice: once when the three-way valve just opens to the cooler and once when the cold oil has been completely displaced and the oil outlet temperature from the cooler suddenly changes from that of the cooling water to some much higher value.

3. Most of these systems are "non-symmetrical" in that the process dynamics (lags and responses) are different for the cooling and heating phases.

To remedy the aforementioned problems, several steps can be and should be considered:

1. As shown in the illustration, a cascade loop should be used, so that upsets and disturbances in the circulating oil loop are prevented from entering the process and thereby from upsetting its temperature.

2. A slight overlapping of the two valve positioners is desirable, which offsets the beginning of cooling and the termination of heating phases, so that they will not both occur at 9 PSIG (0.6 bar). The resulting sacrifice of heat energy is well justified by the improved control obtained.

3. In order to protect against the development of an extremely cold oil reservoir in the cooler, a set minimum continuous flow through this unit can be maintained.

In connection with the recirculating design shown in Figure 8.13bb, it is also important to realize that this is a flooded system, and therefore when hot oil enters it, a corresponding volume of oil must be allowed to be removed. The pressure control valve shown serves this function. The same purpose can be achieved by elevational head on the return header, the important consideration being that whatever means are used, the path of least resistance for the oil must be back to the pump suction to keep it always flooded and thereby to prevent cavitation.

Most multipurpose systems represent a compromise of various degrees, which is perfectly acceptable if the application engineer realizes the features that are lost and gained as the result of the compromise. Figure 8.13cc, for example, illustrates a design in which low cost and high response speed to load changes were the main considerations. This was achieved by the use of minimum hardware and by circumventing the transient characteristics of the exchangers. The price paid in this compromise involves the full use of utilities at all times, the

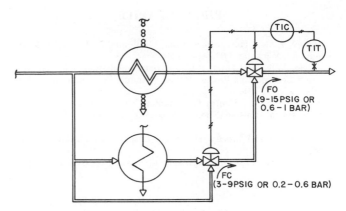

Fig. 8.13cc Multipurpose temperature control system using the blending of process streams at differing temperatures

development of hot and cold reservoirs, and the necessity for supply disturbances to affect the process temperature before corrective action can be initiated. The reader will find it to be a valuable and educational exercise to select arbitrarily control system features that are desired and others that are of lesser consequence and then to design a system that will satisfy those requirements.

Conclusions

The greatest danger in connection with heat exchanger control system design is oversimplification. Once it is realized that there are many factors entering the design problem, the good quality of control is half achieved. It is hoped that no reader of this section will put the book down and say that "you control a cooler by throttling the water and control a heater by manipulating the system to it." Once a complex design challenge is recognized for what it is, a step-by-step analysis of desirable features will yield the desired results.

Some of the considerations that always pay to look into are:

The effect of supply disturbances on system performance.

The response speed of the system.

Rangeability considerations.

The quality of cooling water available.

Potential problems due to non-symmetrical dynamics and to low minimum condensing pressures.

Few generalizations are true in all cases, but there are some that apply to the great majority of heat exchanger controls. In connection with the control valve used, for example, it can be said that the use of a positioner and of equal percentage ports usually improves the performance. Similarly, it is true almost without exception that cascade loops will give better control than conventional ones, and feedforward control will represent yet a further improvement.

BIBLIOGRAPHY

Bailey, E.G., "How to Save Coal," Bulletin No. 41, Bailey Meter Company, Wickliffe, OH, April 1918.

Boyle, W.E. and Hoyt, P.R., "Experimental Application of Combustion Controls to a Process Heater," Annual ISA Conference, September 13–17, 1948.

Dukelow, S.G., "Energy Conservation Through Instrumentation," University of Houston Energy Conservation Seminar, February 7, 1975.

———, "Fired Process Heaters," presented at ISA/CHEMPID, April 1980.

———, "How Control Improvements Save Process Heater Fuel," 1979 Conference on Industrial Energy Conservation Technology, Houston, Texas, April 23, 1979.

Forman, E.R., "Control Dynamics in Heat Transfer," *Chemical Engineering*, January 3, 1966.

Johnson, R.K., "Modern Approach to Process Heater Control Systems," Parkersburg-Marietta ISA Section, Parkersburg, West Virginia, February 8, 1977.

Lipták, B.G., "Control of Heat Exchangers," *British Chemical Engineering*, July 1972.

Mathur, J., "Performance of Steam Heat-Exchangers," *Chemical Engineering*, September 3, 1973.

Shinskey, F.G., "Feedforward Control Applied," *ISA Journal*, November 1963.

8.14 HVAC CONTROLS*

The instrumentation used in HVAC applications used to be called commercial-quality. This was the code word for the inexpensive and inaccurate devices that this industry could afford. Process control–quality instruments were limited to a few critical loops, if any were used at all, because of cost considerations.

During the last decade the cost gap between commercial- and process control–quality systems has disappeared. This development removed the last obstacle that was preventing the HVAC industry from taking full advantage of the state of the art in instrumentation and control. Advanced and optimized controls can cut the cost of building operations in half. This paper gives some examples of the control strategies to be used toward that end.

The purpose of HVAC controls is to provide comfort in offices and manufacturing spaces. Supply air is the means of providing comfort in the conditioned zone. The air supplied to each zone must provide heating or cooling, raise or lower humidity, and provide air refreshment. To satisfy these requirements, it is necessary to control the following properties of the supply air: temperature, humidity, and fresh air ratio.

Figure 8.14a illustrates the main components of an airhandler. The term "airhandler" refers to the total system, including fans, coils, dampers, ducts, and instruments. The system operates as follows:

Outside air is admitted by OAD-05 and is then mixed with the return air from RAD-04. The resulting mixed air is filtered (F), heated (HC) or cooled (CC), and humidified (H) or dehumidified (CC) as required. The resulting supply air is then transported to the conditioned zones by the variable-volume supply fan station. In each zone, the variable air volume damper (VAV-23) determines the amount of air required, and the reheat coil (RHC) adjusts the air temperature as needed. The return air from the zones is transported by the variable-volume return-air fan station. If the amount of available return

air exceeds the demand for it, the excess air is exhausted by EAD-03.

As can be seen, the HVAC process is rather simple. Its process material is clean air, its utility is water or steam, and its overall system behavior is slow, stable, and forgiving. For precisely these reasons, it is possible to obtain acceptable HVAC performance using inferior-quality instruments, which are configured into poorly designed loops. Yet there is an advantage in applying the state of the art of process control to the HVAC process because it can provide a drastic reduction in operating costs, attributable to increased efficiency of operation. Some of these more efficient control concepts will be described in this paper.

Fan Controls

The VAV box openings in the various zones determine the total demand for supply air. The pressure in the SA distribution header is controlled by PIC-19, which modulates the supply-air fan station to match demand (Figure 8.14b). When the PIC-19 output has increased the fan capacity to its maximum, PSH-19 actuates and starts an additional fan. Inversely, as the demand for supply air is dropping, FSL-15 will stop one fan unit, whenever the load can be met by fewer fans than the number in operation. The important point to remember is that in cycling fan stations, fan units are started on pressure and are stopped on flow control. The operating cost of such a fan station is 20 percent to 40 percent lower than if constant-volume fans with conventional controls are used.

Because the conditioned zones are pressurized slightly, some of the conditioned air will leak into the atmosphere, creating pressurization loss. Being able to control the pressurization loss is one of the advantages of the control system described in Figure 8.14b. The flow ratio controller FFIC-14 is set at 90 percent, meaning that the return-air fan station is modulated to return 90 percent of the air supplied to the zones. Therefore, pressurization loss is controlled at 10 percent, which corresponds to the minimum fresh-air make-up requirement, resulting in a minimum-cost operation.

Because the conditioned zones represent a fairly large

* Reprinted by permission of the American Society of Heating, Refrigerating, and Air-Conditioning Engineers, Inc. Paper published in *ASHRAE Transactions*, Volume 89, Part 2a, January 1984.

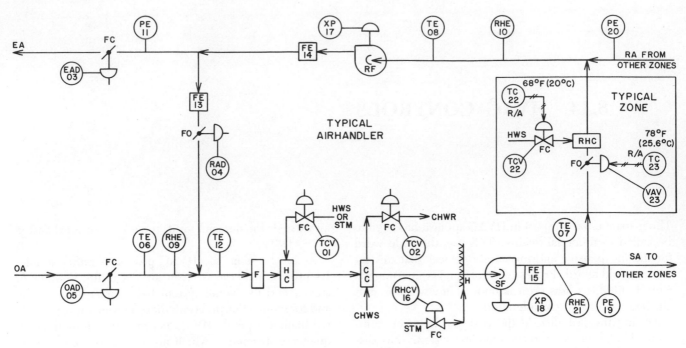

Fig. 8.14a   Components of a typical major airhandler, including main controls.

Abbreviations:
CC = cooling coil;
CHWR = chilled water return;
EA = exhaust air;
EAD = exhaust air damper;
F = filter;
FE = flow element;
FC = fail closed;
FO = fail open;
H = humidifier;
HC = heating coil;
HWS = hot water supply;
OA = outside air;
OAD = outside air damper;
PE = pressure element;

RA = return air;
RAD = return air damper;
RF = return fan;
RHC = reheat coil;
RHCV = relative humidity control valve;
RHE = relative humidity element;
SA = supply air;
SF = supply fan;
STM = steam;
TCV = temperature control valve;
TE = temperature element;
VAV = variable air volume damper;
XP = positioner for fan volume control, such as a
blade pitch positioner.

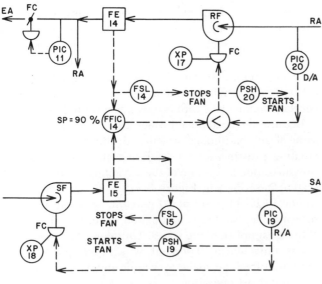

Fig. 8.14b   Variable-volume fan controls

capacity, a change in supply air flow will not immediately result in a need for a corresponding change in the return air flow. Thus, PIC-20 (see Figure 8.14b) is included in the system to prevent the flow-ratio controller from increasing the return air flow rate faster than required. This dynamic balancing eliminates cycling and protects against collapse of the ductwork under excessive vacuum. Closure of the exhaust-air damper by PIC-11 indicates that the control system is properly tuned and balanced and is operating at maximum efficiency. Under such conditions, the outside air admitted into the airhandler matches exactly the pressurization loss, and no return air is exhausted.

To maximize the benefits of such an efficient configuration, the dampers must be of tight shutoff design. With a pressure difference of 4 in H_2O (996 Pa), a closed, conventional damper will leak at a rate of approximately 50 cfm/ft^2 (15.2(m^3/min)/m^2). In the HVAC industry, a 5 cfm/ft^2 (1.52 (m^3/min)m^2) leakage rate is considered to represent a tight shutoff design. Actually, it is cost-effective to install tight shutoff dampers with leakage rate of less than 0.5 cfm/ft² (0.15(m³/min)m²), because the resulting savings over the life of the buildings will be much greater than the increase in initial investment for better dampers.

Temperature Controls

A substantial source of inefficiency in conventional HVAC control systems is the uncoordinated arrangement of temperature controllers. Two or three separate temperature control loops in series are not uncommon. For example, one of these uncoordinated controllers may be used to control the mixed air temperature, another to maintain supply-air (SA) temperature, and a third to control the zone-reheat coil. Such practice can result in simultaneous heating and cooling and therefore in unnecessary waste. Using a fully coordinated split-range temperature control system, such as that shown in Figure 8.14c, can reduce yearly operating costs by more than 10 percent.

output signal is low, 3 to 6 PSIG (20.7 to 41.3 kPa), heating is done by TCV-01. As the output signal reaches 6 PSIG (41.3 kPa), heating is terminated; if free cooling is available, it is initiated at 7 PSIG (48.2 kPa). When the output signal reaches 11 PSIG (75.8 kPa—the point at which OAD-05 is fully open), the cooling potential represented by free cooling is exhausted and, at 12 PSIG (82.7 kPa), "pay cooling" is started by opening TCV-02. In such split-range systems, the possibility of simultaneous heating and cooling is eliminated. Also eliminated are interactions and cycling.

Figure 8.14c also shows some important overrides. TIC-12, for example, limits the allowable opening of OAD-05, so that the mixed-air temperature will never be allowed to drop to the freezing point and permit freeze-up of the water coils.

The minimum outdoor air requirement signal (from Figure 8.14e) limits the amount of outdoor air to be admitted.

The economizer signal (from Figure 8.14e) allows the out signal of TIC-07 to open OAD-05 only when "free cooling" is available. (Free cooling exists when the enthalpy of the outdoor air is below that of the return air.)

Last, the humidity controls (in Figure 8.14d) will override the TIC-07 signal to TCV-02 when the need for

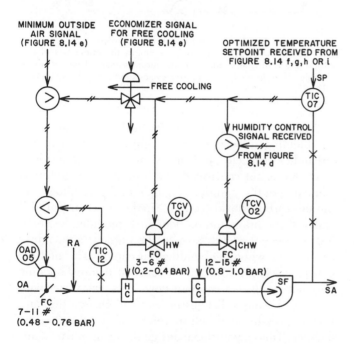

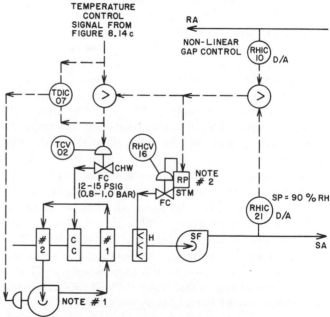

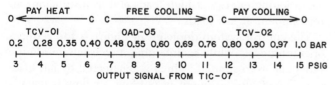

Fig. 8.14c Split-range temperature control system

Fig. 8.14d Humidity controls.
Note #1: When the need for dehumidification (in the summer) overcools the supply air and therefore increases the need for reheat at the zones, this pumparound economizer loop is started. TDIC-07 will control the pump to "pump around" only as much heat as is needed.
Note #2: This reversing positioner functions as follows:

Input from RHIC-10	Output to RHCV-16
3	15
9	3

In this control system, the SA temperature set point (TIC-07) is continuously modulated to follow the load, as in Figures 8.14f through 8.14i. The loop automatically controls all heating or cooling modes. When the TIC-07 dehumidification requires that the supply-air temperature be lowered *below* the set point of TIC-07.

Humidity Controls

Humidity in the zones is controlled according to the moisture content of the combined return air (see Figure 8.14d). The process controlled by RHIC-10 is slow and contains large dead time and transport-lag elements. In other words, a change in the SA humidity will not be detected by RHIC-10 until some minutes later. During this period (in the winter), it is possible for RHIC-10 to demand more and more humidification. To prevent possible saturation of the supply air, the RHIC-10 output signal is limited by RHIC-21. In this way, the moisture content of the supply air is never allowed to exceed 90 percent RH.

For best operating efficiency, a nonlinear controller with a neutral band is used at RHIC-10. This neutral band can be set, say, between 30 percent RH and 50 percent RH and, if the RA is within these limits, the output of RHIC-10 is at 9 PSIG (62 kPa) and neither humidification nor dehumidification is demanded. This arrangement can lower the cost of humidity control during the spring and fall by approximately 20 percent.

The same controller (RHIC-10) controls both humidification (RHCV-16) and dehumidification (TCV-02) on a split-range basis. As the output signal increases, the humidifier valve closes, between 3 and 9 PSIG (20.7 and 62 kPa). At 9 PSIG (62 kPa), RHCV-16 closes and remains so as the output signal increases to 12 PSIG (82.7 kPa). At this condition, TCV-02 starts to open. Dehumidification is accomplished by cooling through TCV-02. This chilled-water valve is controlled by humidity (RHIC-10) or temperature (TIC-07). The controller that requires more cooling will be the one allowed to throttle TCV-02.

Subcooling the air to remove moisture can substantially increase operating costs if this energy is not recovered. The dual penalty incurred for overcooling for dehumidification purposes is the high chilled-water cost and the possible need for reheat at the zone level. The savings from a pumparound economizer can eliminate 80 percent of this waste. In this loop, whenever TDIC-07 detects that the chilled water valve (TCV-02) is open more than would be necessary to satisfy TIC-07, the pumparound economizer is started. This loop in coil 1 prevents overcooling of the supply air, using this cooling capacity to precool the air in coil 2 before it enters the main cooling coil.

In this way, the chilled-water demand is reduced in the cooling coil (TCV-02), and the need for reheating at the zones is eliminated. Although Figure 8.14d shows a modulating controller setting the speed of a circulating pump, it is also possible to use a constant-speed pump operated by a gap switch.

Outdoor Air Controls

Outdoor air is admitted to satisfy requirements for fresh air or to provide free cooling. Both control loops are shown in Figure 8.14e.

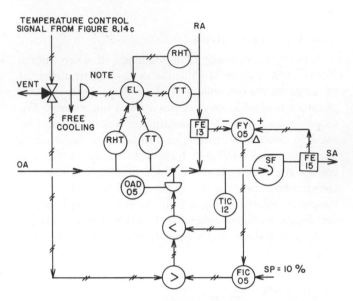

Fig. 8.14e   Outside air control loops.
Note: The enthalpy logic unit (EL) compares the enthalpies of the outside and return airs and vents its output signal if free cooling is available. Therefore, the economizer cycle is initiated whenever Hoa < Hra.

The minimum requirement for fresh outdoor air while the building is occupied is usually 10 percent of the airhandler's capacity. In more advanced control systems, this value is not controlled as a fixed percentage but as a function of the number of people in the building or of the air's carbon dioxide content. In the most conventional systems, the minimum outdoor air is provided by keeping 10 percent *of the area* of the outdoor air damper always open when the building is occupied. This method is very inaccurate, because a constant damper opening does not result in a constant flow. This flow varies with fan load, because changes in load will change the fan's suction pressure and will therefore affect ΔP across the damper. This conventional design results in waste of air-conditioning energy at high loads and in insufficient air refreshment at low loads.

The control system depicted in Figure 8.14e reduces operating costs while maintaining a constant minimum rate of air refreshment, which is unaffected by fan loading. Direct measurement of outdoor air flow is usually not possible because of space limitations. For this reason, Figure 8.14e shows the outdoor air flow as being determined as the difference between FE-15 and FE-13. FIC-05 controls the required minimum outdoor air flow by throttling OAD-05.

The full use of free cooling can reduce the yearly air-conditioning load by more than 10 percent. The enthalpy logic unit (EL) in Figure 8.14e will allow the temperature-controller signal (TIC-07 in Figure 8.14c) to operate the outdoor air damper whenever free cooling is available. This economizer cycle is therefore activated whenever the enthalpy of the outdoor air is below that of the return air.

Free cooling can also be used to advantage while the building is unoccupied. Purging the building with cool outdoor air during the early morning results in cooling capacity being stored in the building structure, reducing the daytime cooling load.

Temperature Optimization

In the traditional HVAC control systems, the set point of TIC-07 (in Figure 8.14c) is manually set and is held as a constant. This practice is undesirable, because for each particular load distribution the optimum SA temperature is different. If the manual setting is less than that temperature, some zones will become uncontrollable. If it is more than optimum, operating energy will be wasted.

Winter Mode

In the winter, the goal of optimization is to distribute the heat load between the main heating coil (HC in Figure 8.14c) and the zone reheat coils (RHC in Figure 8.14f) most efficiently. The highest efficiency is obtained if the SA temperature is close to being high enough to meet the load of the zone with the minimum load. In this way, the zone reheat coils are used only to provide the difference between the loads of the various zones, whereas the base load is continuously followed by TIC-07. In Figure 8.14f, the set point of TIC-07 is adjusted to keep the least-open TCV-22 only about 10 percent open. First, the least open TCV-22 is identified and its

opening in the valve position controller (VPC-07) is compared with the desired goal of 10 percent. If the valve opening is more than 10 percent, the TIC-07 set point is increased; if less, the set point is lowered. Stable operation is obtained by making VPC-07 an integral-only controller with an external feedback to prevent reset windup.

In Figure 8.14f, it is assumed that the hot-water supply (HWS) temperature is independently adjustable and will be modulated by VPC-22 so that the *most*-open TCV-22 will always be 90 percent open. The advantages of optimizing the HWS temperature include:

1. minimizing heat-pump operating costs by minimizing HWS temperature,
2. reducing pumping costs by opening all TCV-22 valves in the system, ánd
3. eliminating unstable (cycling) valve operation by opening all TCV-22 valves.

In Figure 8.14f, a valve position alarm (VPA-07) is also provided to alert the operator if this "heating" control system is incapable of keeping the openings of all TCV-22 valves between the limits of 10 percent and 90 percent. Such alarm will occur if the VPCs can no longer change the TIC set point(s), because their maximum (or minimum) limits have been reached. This condition will occur only if the load diistribution was not correctly estimated during the design phase of the project or if the mechanical equipment was not correctly sized.

If the HWS temperature cannot be modulated to keep the most open TCV-22 from opening to more than 90 percent, then the control loop depicted in Figure 8.14g should be used. Here, as long as the most-open valve is less than 90 percent open, the SA temperature is set to keep the least-open TCV-22 at 10 percent opening

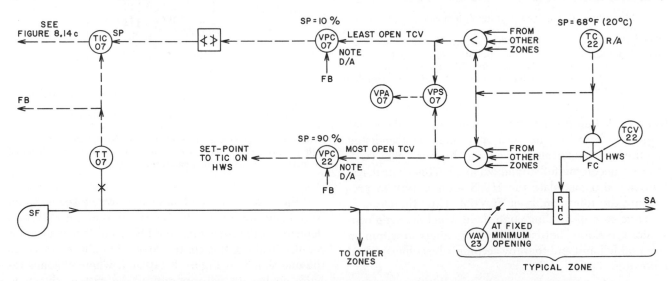

Fig. 8.14f Optimization of air and water temperatures during heating (winter) mode of airhandler operation.
Note: Valve position controllers (VPCs) are provided with integral action only for stable floating control. The integral time is set to be ten times the integral time of the associated TIC. External feedback is provided to eliminate reset wind-up when VPC output is overruled by set point limit on TIC.

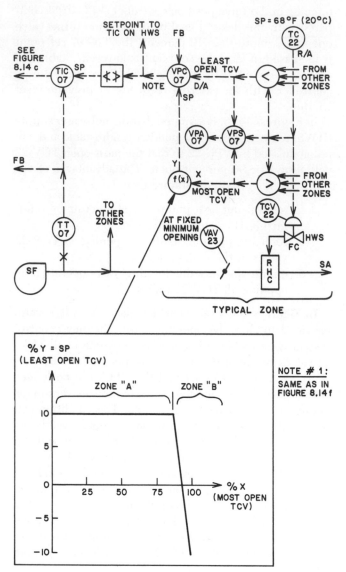

Fig. 8.14g   Alternative method of air supply temperature optimization in winter.
Note: Valve position controllers (VPCs) are provided with integral action only for stable floating control. The integral time is set to be ten times the integral time of the associated TIC. External feedback is provided to eliminate reset wind-up when VPC output is overruled by set point limit on TIC.

(zone A). When the most-open valve reaches 90 percent opening, control of the least-open valve is abandoned and the loop is dedicated to keeping the most-open TCV-22 from becoming fully open (zone B). This control system can also modulate the HWS temperature in preventing the full opening of TCV-22. This, therefore, is a classic case of herding control, in which a single constraint envelope "herds" all TCV openings to within an acceptable band and thereby accomplishes efficient load following.

Summer Mode

In the cooling mode during the summer, the SA temperature is modulated to keep the most-open variable

volume box (VAV-23) from fully opening. Once a control element is fully open, it can no longer control; therefore, the occurrence of such a state must be prevented. On the other hand, it is generally desirable to open throttling devices such as VAV boxes to:

1. reduce the total friction drop in the system,
2. eliminate cycling and unstable operation (which is more likely to occur when the VAV box is nearly closed), and
3. allow the airhandler to meet the load at the highest possible supply-air temperature.

This statement does not apply if air transportation costs exceed cooling costs (e.g., undersized ducts, inefficient fans). In this case, the goal of optimization is to transport the minimum *quantity* of air. The amount of air required to meet a cooling load will be minimized if the cooling capacity of each unit of air is maximized. Therefore, if fan operating cost is the optimization criterion, the SA temperature is to be kept at its achievable minimum, instead of being controlled as in Figure 8.14h.

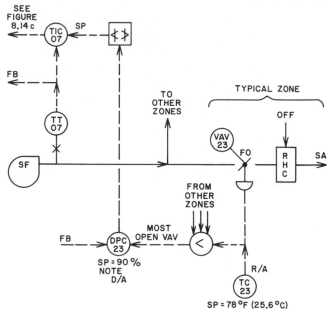

Fig. 8.14h   Air supply temperature optimization in summer (cooling mode).
Note: The damper position controller (DPC) has integral action only, with its setting being ten times the integral time of TIC-07. External feedback is provided to eliminate reset wind-up.

If the added feature of automatic switchover between winter and summer modes is desired, the control system depicted in Figure 8.14i should be used. When all zones require heating, this control loop will behave exactly as the one shown in Figure 8.14g and, when all zones require cooling, it will operate as the system shown in Figure 8.14h. In addition, this control system will operate automatically with maximum energy efficiency during the transitional periods of fall and spring. This high

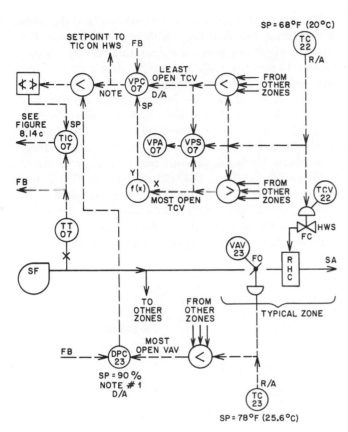

Fig. 8.14i   Control system optimizes the water and air temperatures
in both summer and winter.

Note: See Figures 8.14g and 8.14h.

efficiency is a result of the exploitation of the self-heating effect. If some zones require heating (perimeter offices) and others require cooling (interior spaces), the airhandler will automatically transfer this free heat from the interior to the perimeter zones by intermixing the return air from the various zones and moving it through the 10-degree zero energy band (ZEB) between the settings of TC-22 and TC-23. When the zone temperatures are within this comfort gap of 68°F (20°C) to 78°F (26°C), no pay energy is used and the airhandler is in its self-heating or free-heating mode. This is a very effective means of reducing operating costs in buildings. The savings can amount to more than 30 percent during the transitional seasons.

When the temperature in one of the zones reaches 78°F (26°C), the air supply temperature set point will be lowered by DPC-23 and the air-side controls will be automatically switched to cooling (as depicted in Figure 8.14h). If at the same time some other zone temperatures are below 68°F (20°C), requiring heating, their heat demand will have to be met by the heat input of the zone-reheat coils only. This mode of operation is highly inefficient because of the simultaneous cooling and reheating of the air. Fortunately, this combination of conditions is highly unlikely because under proper design

practices, the zones served by the same airhandler should display similar load characteristics. The advantage of the control loop in Figure 8.14i is that it can automatically handle any load or load combination, including this unlikely, extreme case.

Self-Heating and Zero Energy Band

Figure 8.14j illustrates the concept of comfort envelopes. The combination of temperature and humidity conditions within such envelopes is considered to be comfortable. Therefore, as long as the space conditions fall within this envelope, there is no need to spend money or energy to change those conditions. This comfort gap is also referred to as zero energy band (ZEB), meaning that if the space is within this band, no pay energy of any type will be used. This concept is very cost-effective. When a zone of the air-handler in Figure 8.14a is within the comfort gap, its reheat coil is turned off and its VAV box is closed to the minimum flow required for air refreshment. When all the zones are inside the ZEB, the HW, CHW, and STM supplies to the airhandler are all closed and the fan is operated at minimum flow. When all other airhandlers are also within the ZEB, the pumping stations, chillers, cooling towers, and HW generators are also turned off.

With larger buildings having interior spaces that are heat-generating even in the winter, ZEB control can

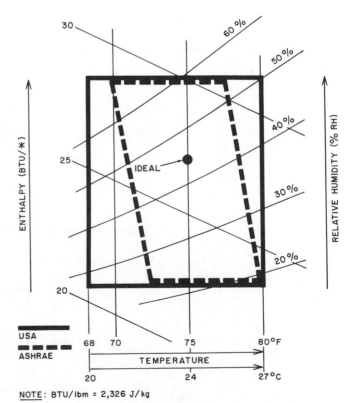

Fig. 8.14j   Definitions of "comfort zones" in terms of temperature and humidity.

make the building self-heating. Optimized control systems in operation today are transferring the interior heat to the perimeter without requiring *any pay heat* until the outside temperature drops below 10°F to 20°F (−12.3°C to −6.8°C). In regions in which winter temperature does not drop below 10°F (−12.3°C), ZEB control can eliminate the need for pay heat altogether. In regions farther north, ZEB control can lower the yearly heating fuel bill by 30 percent to 50 percent.

Elimination of Chimney Effects

In high-rise buildings, the natural draft resulting from the chimney effect tends to pull in ambient air at near ground elevation and tends to discharge it at the top of the building. During cold days, this added outdoor air

will substantially increase the heating load and the costs of operating the building. Although eliminating the chimney effect can lower the operating cost by approximately 10 percent using a relatively simple control system, few such systems are yet in operation.

Figure 8.14k shows the required pressure controls. The key element of this control system is the reference riser, which allows all pressure controllers in the building to be referenced to the barometric pressure of the outside atmosphere. Using this pressure reference allows all zones to be operated at 0.1 in H_2O (25 Pa) pressure (PC-7) and permits maintaining of this constant pressure at both ends of all elevator shafts (PC-8 and 9). If the space pressure is the same on the various floors of a high-rise building, there will be no pressure gradient to motivate

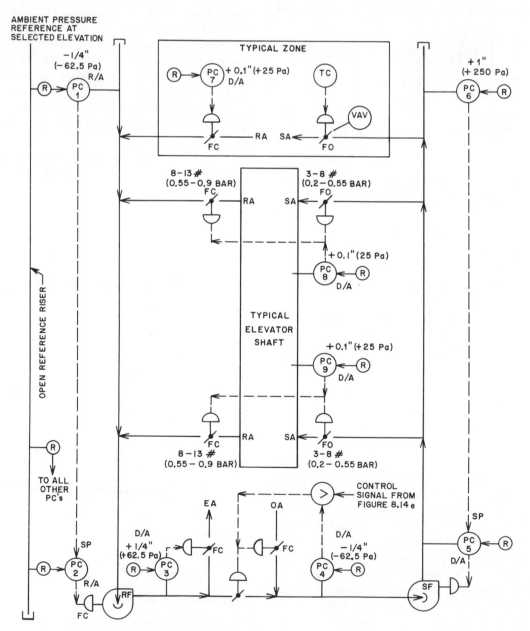

Fig. 8.14k   Elimination of chimney effects in high-rise buildings

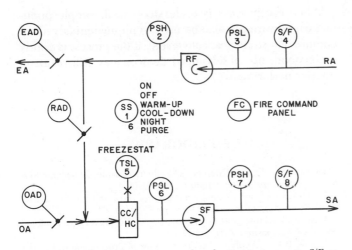

Fig. 8.14l Safety and operating mode selection instruments. *S/F* = Smoke and fire detector(s). *FC* = The Fire Command Panel provides the Fire Chief with access to all fans and dampers in the building. This panel is used during fire fighting and building evacuation.

and PC-6) remains constant. This control approach results in the most efficient operation of variable-air-volume fans.

Interlocks

Figure 8.14l shows the major interlock instruments on a typical airhandler. The airhandler can be in one of seven states, or operation modes: on, off, warm-up, cool-down, night, purge, and emergency.

The emergency mode can be signaled by any of the following categories of safety instruments: smoke/fire detectors, freezestats, and abnormal pressure sensors.

When a smoke or fire condition is detected (S/F-4 and 8), the fans stop, the OAD and RAD dampers close, the EAD damper opens, and an alarm is actuated. The operator can switch the airhandler into its purge mode, so that the fans are started, OAD and EAD are opened, and RAD is closed. If the smoke/fire emergency requires the presence of fire fighters, the fire command panel is used. From this panel the fire chief can operate all fans and dampers as needed for safe and orderly evacuation and protection of the building.

In another emergency condition a freezestat switch on one of the water coils is actuated. These switches are usually set at approximately 35°F (1.5°C) and serve to protect from coil damage resulting from freeze-ups. Multistage freezestat units might operate as follows:

At 38°F (3°C)—close OAD
At 36°F (2°C)—fully open water valve
At 35°F (1.5°C)—stop fan

If single-stage freezestats are used, they will stop the fan, close the OAD, and activate an alarm.

Yet another type of emergency is signaled by excessive

the vertical movement of the air and, as a consequence, the chimney effect will have been eliminated. A side benefit of this control strategy is the elimination of all drafts or air movements between zones, also minimizing the dust content of the air. Another benefit is the capability of adjusting the "pressurization loss" of the building by varying the settings of PC-7, 8, and 9.

Besides reducing operating costs, the use of pressure-controlled elevator shafts increases comfort because drafts and the associated noise are eliminated.

Figure 8.14k also shows the use of cascaded fan controls. The set points of the cascade slaves (PC-2 and PC-5) are so programmed that the air pressure at the fan is adjusted as the square of flow, and the pressure at the end of the distribution headers (cascade masters PC-1

Table 8.14m
AIRHANDLER INTERLOCK TABLE
Actuated Devices

Operating Mode or Emergency Condition	Supply Fan	Return Fan	Outside Air Damper	Exhaust Air Damper	Return Air Damper	Coil Control Valves	Alarm
Off	–	–	C	C	O	C	–
On	On	On	Modulating				–
Warm-up	On	On	C	C	O	O(HC)	–
Cool-down	On	On	C	C	O	O(CC)	–
Night	Cycled to maintain required nighttime temperature						
Purge	On	On	O	O	C	Modulating	–
PSH-2	–	Off	–	C	–	–	Yes
PSL-3	–	Off	–	C	–	–	Yes
S/F-4	Off	Off	C	O	C	C	Yes
TSL-5	Off	–	C	–	O	C	Yes
PSL-6	Off	–	C	–	O	–	Yes
PSH-7	Off	–	C	–	O	–	Yes
S/F-8	Off	Off	C	O	C	C	Yes

pressures in the ductwork, on the suction or discharge sides of the fans, resulting from operation against closed dampers or from other equipment failures. When this happens, the associated fan is stopped and an alarm is actuated.

Table 8.14m summarizes the status of the fans, valves, and dampers under the various emergency and operating mode conditions. In the warm-up or cool-down mode, the airhandler is operated with 100 percent return air as it is bringing the zones to comfort levels. The purge mode can be used not only to remove smoke or chemical fumes but also to accomplish nighttime cool-down by using free cooling. When the system is in the night mode during the winter, the zone thermostat set points are lowered substantially and the airhandler is periodically started to prevent the zone temperatures from dropping below these limits.

The airhandler operating mode selection can be done manually by the operators' setting the six-position switch SS-1 or automatically by a central computer.

Conclusions

The process of air conditioning is similar to all other industrial processes. Fully exploiting the instrumentation and control state of the art results in dramatic improvements. There are few processes in which the use of instrumentation know-how alone can halve the operating cost of a process. Building conditioning is one of those few processes.

The control and optimization strategies described in this paper can be implemented by pneumatic or electronic instruments and can be controlled by analog or digital systems. The type of hardware used in optimization is less important than the control concepts implemented.

When the process is understood well, simple pneumatic hardware can also be used to implement advanced optimization strategies; conversely, if the process is poorly understood, use of advanced devices may still result in inferior performance.

BIBLIOGRAPHY

Daryani, S., "Design Engineer's Guide to Variable Air-Volume Systems," *Actual Specifying Engineer*, July 1974.

DHO/Atlanta Corp., Conserving Energy, Powered Induction Unit (Data Sheet #1), Atlanta, DHO/Atlanta Corp., 1974.

Kusuda, T., "Intermittent Ventilation for Energy Conservation," ASHRAE Symposium #20, Paper #4.

Lipták, B.G., "Applying the Techniques of Process Control to the HVAC Process," *ASHRAE Transactions*, Paper No. 2778, Volume 89, Part 2 A and B, 1983.

——— (ed.), "Carbon Dioxide Sensors," *Environmental Engineer's Handbook*, Volume II, Section 4.6, Chilton Book Co., Radnor, PA, 1975.

———, "Control of Compressors," *Instrument Engineer's Handbook*, Volume II, Section 10.13, Chilton Book Co., Radnor, PA, 1970.

———, "Reducing the Operating Costs of Buildings by the Use of Computers," *ASHRAE Transactions* 83:1, 1977.

———, "Save Energy by Optimizing Your Boilers, Chillers and Pumps," *Instrumentation Technology*, March 1981.

———, "Savings Through CO₂ Based Ventilation," *ASHRAE Journal*, July 1979.

Luciano, J.R., "Energy Conservation Techniques for Hospital Operating Rooms," *ASHRAE Journal*, May 1983.

Nordeen, H., "Control of Ventilation Air in Energy Efficient Systems," ASHRAE Symposium #20, Paper #3.

Shih, J.Y., "Energy Conservation and Building Automation," ASHRAE Paper #2354.

Spielvogel, L.G., "Exploding Some Myths About Building Energy Use," *Architectural Record*, February 1976.

Stillman, R.B., "Systems Stimulation," (Engineering Report prepared for IBM-RECD), 1971.

Turk, A., "Gaseous Air Cleaning," *ASHRAE Journal*, May 1983.

Woods, J.E., "Impact of ASHRAE Ventilation Standards 62-73 on Energy Use," ASHRAE Symposium #20, Paper #1.

8.15 ORP CONTROLS

A description is given of oxidation-reduction (redox) reactions and their control by oxidation-reduction potential (ORP) measurement. A comparison of ORP and pH instrumentation is made. Example titration curves provide an understanding of ORP response. The two major industrial applications for ORP control are described: cyanide oxidation and chrome reduction, explained for both batch and continuous waste treatment systems.

An oxidation-reduction reaction involves the transfer of electrons from one material (oxidation) to another (reduction). In industrial applications an oxidizing or reducing agent is used to promote the desired reaction. An example is the reduction of plating waste hexavalent chromium ion by ferrous ion given in reaction 8.15(1).

$$Cr^{6+} + 3Fe^{2+} \longrightarrow Cr^{3+} + 3Fe^{3+} \qquad 8.15(1)$$

The reducing agent, ferrous ion, donates electrons to the chromium. The chrome is reduced while the iron is oxidized.

ORP Measurement

During this type of reaction, an inert metal electrode in contact with the solution will detect the solution's ability to accept or donate electrons. The resulting oxidation-reduction potential, also known as redox potential, is directly related to the progress of the reaction. A reducing ion (ferrous) provides electrons and tends to make the electrode more negative. An oxidizing ion (Cr^{6+}) accepts electrons and tends to make the electrode more positive. The resulting net electrode potential is related to the ratio of concentrations of oxidizing and reducing ions in solution.

ORP is extremely sensitive in measuring the degree of treatment of the reaction. However, it cannot be related to a definite concentration (only the ratio) and therefore cannot be used as a monitor of final effluent concentration.

The exact potential to ensure complete treatment can be theoretically calculated but in practice is subject to variations in reference electrode potential, pH, other waste stream contaminents, temperature, purity of reagents, and so forth.[1] It is usually determined empirically by testing the treated wastewater for trace levels of the material to be eliminated. The optimum control point is the ORP existing when just enough reagent has been added to complete the reaction. Suggested control points given later in this section are approximate and should be verified on-line by testing of actual samples.

Instrumentation

ORP is sensed by a platinum- or gold-measuring electrode. The measuring circuit is completed by a reference electrode identical to the type used for pH measurement. ORP measurement is very similar to pH determination except that readout is in millivolts and temperature compensation is generally not used.

Polarity of ORP measurements can be a source of confusion. Electrodes that measure pH produce a negative potential with upscale pH readings. Thus, the reference electrode connection of a pH instrument is the positive input. When ORP equipment is based on a pH instrument design, it is usually necessary to connect the ORP electrode to the reference input and the reference electrode to the measuring terminal. These connections provide on-scale readings of positive ORP for the common applications described in this section.

ORP electrodes must be maintained with a very clean metal surface. Routine cleaning of electrodes with a soft cloth, dilute acids, or cleaning agents is often needed to promote fast response.

ORP instruments are calibrated like voltmeters, measuring absolute millivolts, although a standardized (zero) adjustment is often available on instruments designed also for pH measurement. Instruments are zeroed with measuring and reference electrode inputs shorted.

To verify operation of electrodes, it is useful to have a known ORP solution as a reference. Manufacturers' electrode instructions often give reference solution compositions using quinhydrone and pH buffer solutions for this purpose. These must be made up fresh to prevent air oxidation and deterioration.

A more stable ORP reference solution has been developed, consisting of 0.1 M ferrous ammonium sulfate, 0.1 M ferric ammonium sulfate, and 1.0 M sulfuric acid.

Its ORP is $+476$ mV when measured with a silver-silver chloride, saturated potassium chloride reference electrode.[2]

Control

As with pH, reliable ORP control requires vigorous mixing to ensure uniform composition throughout the reaction tank. For continuous control the tank should provide adequate retention time (process flow rate divided by filled tank volume), typically 10 minutes or more.[3]

Complete treatment requires a slight excess of reagent and a control point slightly beyond the steep area of the titration curve. Control in this plateau area, where process gain is relatively low, is often handled by simple on-off control. Reagent feeders are typically metering pumps or solenoid valves. A needle valve in series with a solenoid valve can be used to set the reagent flow more accurately and to improve on-off control.

Chrome Waste Treatment

Chromates are used as corrosion inhibitors in cooling towers and in various metal finishing operations, including bright dip, conversion coating, and chrome plating. The resulting wastewater from rinse tanks, dumps, or cooling tower blowdown contains toxic soluble chromium ion, Cr^{6+}, which must be removed before discharge.

The most frequently used technique for chrome removal is a two-stage chemical treatment process. The first stage lowers the pH and adds reducing agent to convert the chrome from soluble Cr^{6+} to Cr^{3+}. The second stage neutralizes the wastewater, forming insoluble chromium hydroxide, which can be removed.

First Stage

Sulfuric acid is used to lower the pH to approximately 2.5 to speed up the reduction reaction and ensure complete treatment. The reducing agent may be sulfur dioxide, any of various sulfites, or ferrous compounds. Reaction 8.15(2) describes the reduction with chrome expressed as chromic acid, CrO_3, which has a $6+$ charge on the chromium. The reducing agent is expressed as sulfurous acid, H_2SO_3, generated by sulfites at low pH. The result is chromium sulfate, $Cr_2(SO_4)_3$, which has a $3+$ charge on the chromium.

$$2CrO_3 + 3H_2SO_3 \longrightarrow Cr_2(SO_4)_3 + 3H_2O \qquad 8.15(2)$$

As shown in Figure 8.15a, the first-stage reaction is monitored and controlled by independent control loops: acid addition by pH control and reducing agent addition by ORP control. Additional acid is called for in pH control whenever the pH rises above 2.5. Additional reducing agent is called for in ORP control whenever the ORP rises above approximately $+250$ mV.

The ORP titration curve, Figure 8.15b, shows the entire millivolt range covered if Cr^{6+} chrome were treated as a batch. With continuous treatment, operation is maintained in the completely reduced portion of the curve near the nominal $+250$ mV control point. The exact set point for a given installation should be at a potential where all the Cr^{6+} has been reduced but without excess sulfite consumption, which is accompanied by the odor of sulfur dioxide.

Chrome reduction is slow enough that 10 to 15 minutes may be required for complete reaction. This time increases if pH is controlled at higher levels. There is also

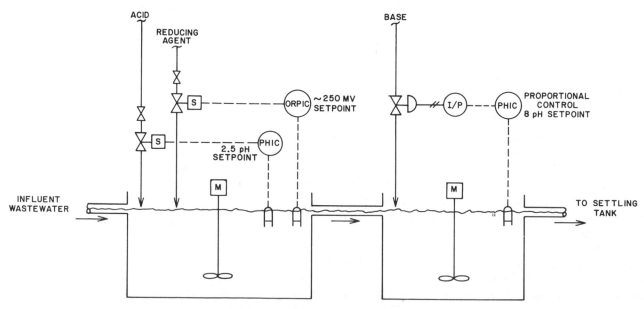

Fig. 8.15a Continuous chrome treatment

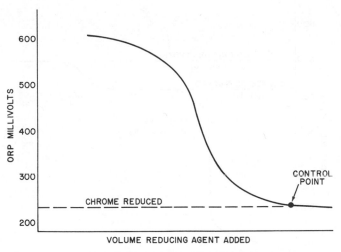

Fig. 8.15b   Chrome reduction titration curve

$$Cr_2(SO_4)_3 + 6NaOH \longrightarrow 3Na_2SO_4 + 2Cr(OH)_3$$
$$8.15(3)$$

In the second stage, pH control is more difficult than in the first, since the control point is closer to the sensitive neutral area. Although this reaction is fast, retention time of at least 10 minutes is usually needed for continuous treatment to achieve stability. Proportional pH control is often used in this stage.

A subsequent settling tank or filter removes the suspended chromium hydroxide. Flocculating agents have been found helpful in this separation.

Batch Chrome Treatment

Figure 8.15c shows the arrangement for batch treatment in which all steps are accomplished in a single tank with one pH and one ORP controller. The steps of the treatment are sequenced, changing the pH set point as needed. Acid is added to lower pH to 2.5, then reducing agent is added to lower ORP to approximately +250 mV. After a few minutes to ensure complete reaction and possibly a grab sample test for Cr^{6+}, base reagent is added to raise pH to 8. A settling period then follows, or the batch is pumped to a separate settling tank or pond.

Cyanide Waste Treatment

Cyanide solutions are used in plating baths for brass, cadmium, copper, silver, gold, and zinc. The toxic rinse

some effect of pH variation on measured ORP. Thus, pH must be stable to achieve consistent ORP control.

Second Stage

The wastewater is neutralized to precipitate Cr^{3+} as insoluble chromium hydroxide, $Cr(OH)_3$, as well as to meet discharge pH limits. Sodium hydroxide or lime is used to raise the pH to 7.5 to 8.5, given by reaction 8.15(3).

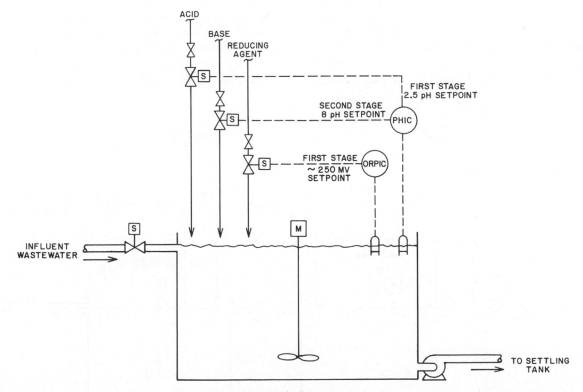

Fig. 8.15c   Batch chrome treatment

waters and dumps from these operations require destruction of the cyanide before discharge.

The most frequently used technique for cyanide destruction is a one- or two-stage chemical treatment process. The first stage raises the pH and oxidizes cyanide to less toxic cyanate. When required, the second stage neutralizes and further oxidizes cyanate to harmless bicarbonate and nitrogen. The neutralization also allows the metals to be precipitated and separated from the effluent.

First Stage

Sodium hydroxide is generally used to raise the pH to approximately 11 to promote the oxidation reaction and ensure complete treatment. The oxidizing agent is generally chlorine or sodium hypochlorite, NaOCl. The overall reaction for the first stage is given here, with cyanide expressed in ionic form, CN^-, and the result as sodium cyanate, NaCNO, and chloride ion, Cl^-.

$$NaOCl + CN^- \longrightarrow NaCNO + Cl^- \qquad 8.15(4)$$

As shown in Figure 8.15d, the first-stage reaction is monitored and controlled by independent control loops: base addition by pH control and oxidizing agent addition by ORP control. The pH controller calls for additional base whenever the pH falls below 11. The ORP controller calls for additional oxidizing agent whenever the ORP falls below approximately $+450$ mV.

The ORP titration curve, Figure 8.15e, shows the entire millivolt range covered if cyanide is treated as a batch. With continuous treatment, operation is maintained in the oxidized, positive region of the curve near the $+450$ mV set point. The exact set point should be determined empirically as the potential when all the cyanide has been oxidized but without excess reagent

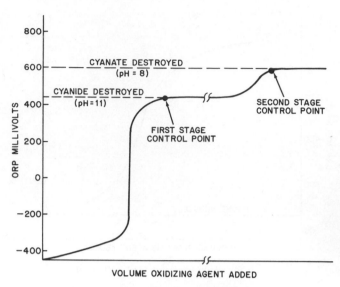

Fig. 8.15e Cyanide oxidation titration curve

feed. This point can be verified with a sensitive colorimetric test.

In this reaction, pH has a strong inverse effect on the ORP. Thus, pH must be closely controlled to achieve consistent ORP control, especially if hypochlorite is used as the oxidizing agent. Hypochlorite addition raises pH, which—unchecked—lowers the ORP, calling for additional hypochlorite and causing a runaway situation. Careful pH control above the level influenced by hypochlorite and separation of the ORP electrodes from the hypochlorite addition point can prevent this difficulty.

Gold ORP electrodes have been found to give more reliable measurement than platinum for this application.[4] Platinum may catalyze some additional reactions at its surface and is more subject to coating than gold. The

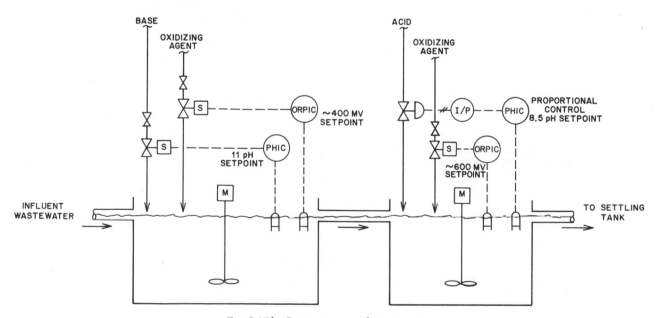

Fig. 8.15d Continuous cyanide treatment

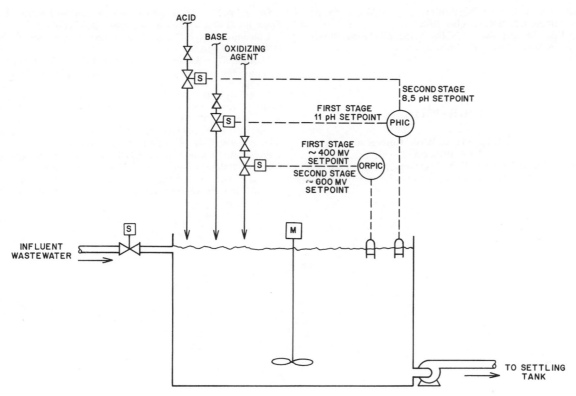

Fig. 8.15f   Batch cyanide treatment

solubility of gold in cyanide solutions does not present a problem, since it is in contact primarily with cyanate. Any slight loss of gold actually serves to keep the electrode clean.

Second Stage

The wastewater is neutralized to promote additional oxidation as well as to meet discharge pH limits. Sulfuric acid is typically used to lower the pH to approximately 8.5 where the second oxidation occurs more rapidly. Acid addition must have fail-safe design, since pH levels below 7 can generate highly toxic hydrogen cyanide if the first-stage oxidation was not complete.

Hypochlorite is added either in proportion to that added in the first stage or by separate ORP control to complete the oxidation to sodium bicarbonate, $NaHCO_3$, in the following reaction:

$$2NaCNO + 3NaOCl + H_2O \longrightarrow$$
$$2NaHCO_3 + N_2 + 3NaCl \quad 8.15(5)$$

ORP control in the second stage is very similar to that in the first, except that the control point is near $+600$ mV. In the second stage, pH control is more difficult than in the first, since the control point is closer to the sensitive neutral area. Proportional pH control is often used here.

A subsequent settling tank or filter can remove sus-

pended metal hydroxides, although further treatment may be required.

Batch Cyanide Treatment

Figure 8.15f shows the arrangement for batch treatment with all steps accomplished in a single tank with one pH and one ORP controller. The steps are sequenced, changing the pH and ORP set points to obtain the required treatment, with the added assurance that treatment is complete before going on to the next step. Caustic is added to raise the pH to 11. Hypochlorite is added to raise the ORP to approximately $+450$ mV, simultaneously adding more caustic, as required, to maintain 11 pH. An interlock must be provided to prevent acid addition before positive oxidation of all cyanide to cyanate. Then acid can be added to neutralize the batch, and further hypochlorite oxidation completes the cyanate-to-bicarbonate conversion. A settling period can then remove solids, or the batch can be pumped to another settling tank or pond.

REFERENCES

1. Latimer, W.M., *Oxidation Potentials*, Prentice-Hall, New York, 1952, Chapter 1.
2. Light, T.S., "Standard Solution for Redox Potential Measurements," *Analytical Chemistry*, 44:6, May 1972, pp. 1038–1039.

3. *pH Controllability*, Leeds & Northrup Co., North Wales, PA, Application Bulletin C2.0001, October 1975.
4. Shinskey, F.G., *pH and pIon Control in Process and Waste Streams*, John Wiley & Sons, New York, 1973, p. 120.

BIBLIOGRAPHY

Cali, G.V. and Galetti, B.J., "Plating Waste Control Instrumentation," *Pollution Engineering*, March 1976, pp. 48–50.

Chamberlin, N.S. and Snyder, H.B., "Technology of Treating Plating Wastes," Tenth Industrial Waste Conference, Purdue University, May 1955.

Greer, W.N., "Measurement and Automatic Control of Etching Strengths of Ferric Chloride," *Plating*, October 1961.

Lanouette, K.H., "Heavy Metals Removal," *Chemical Engineering*, October 1977, pp. 73–80.

Mattock, G., "Automatic Control in Effluent Treatment," *Transactions of the Society of Instrument Technology*, December 1964, pp. 173–189.

Record, R.G.H., "The Use of Redox Potentials in Chemical Process Control," *Instrument Engineer*, October 1965, pp. 65–75, continued April 1966, pp. 95–102.

Shinskey, F.G., *pH and pIon Control in Process and Waste Streams*, John Wiley & Sons, New York, 1973, Chapter 5.

Weast, R.C., *Handbook of Chemistry and Physics*, 50th ed., The Chemical Rubber Company, Cleveland, Ohio, 1969.

Yorgey, W.B., "Updating a Wastewater Treatment Plant with Automatic Controls," *Plant Engineering*, February 1973, pp. 178–182.

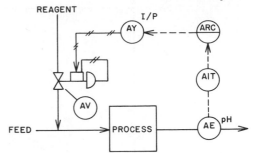

8.16 pH CONTROL

Flow Rangeability of Reagent Delivery Devices:	Metering pumps (20:1 to 200:1), linear and equal percentage valves (50:1 to 500:1), two sequenced valves (over 1000:1), three valves in cascade residence control (near 100,000:1).
pH Swing That Valves Can Handle:	Can be determined as the base 10 logarithm of the rangeability for unbuffered titrations; if rangeability is 1000:1, the controllable pH swing is 3, and for 100,000:1 it is 5.
Feedforward Control Model:	If process flow rate variations do not exceed a 3:1 range, flow need not be part of the model.
Design Considerations:	Inline mixer or a stirred tank? Number and size of tanks; agitation, baffling, reagent addition and sensor location.
Reaction Tanks:	Liquid depth should equal diameter, retention time should not be less than 3 minutes (5 to 30 minutes for lime), and dead time should be 1/20th of the retention time. For strong acid–strong base neutralizations, one, two, or three tanks are recommended for influent pH limits of 4 to 10, 2 to 12, and 0 to 14, respectively. Influent should enter at top, and effluent should leave at the bottom.
pH Sensor Choices:	Submersible sensors are preferred over flow-through designs.
Agitator Choices:	Propeller (under 1000 gallon [3780 l] tanks) or axial-flow impellers (over 1000 gallons [3780 l]) are preferred. Flat-bladed radial flow impellers should be avoided. Acceptable impeller-to-tank diameter ratio is from 0.25 to 0.4. Peripheral speeds of 12 fps (3.6 m/s) for large tanks and 25 fps (7.5 m/s) for tanks with volumes less than 1000 gallons (3780 l) are acceptable.

In the entire spectrum of industrial process control, the pH control problem is the most demanding. Process design considerations for continuous treatment can be of prime importance and are fully covered later in this section. In order to cope with the chemistry of the process, the designer must undertake an exhaustive investigation of the characteristics of the material to be treated to determine the measurable ion load on the process. It is important to know how much the pH of the treatable material varies, the frequency with which the load changes, and the representative titration curves at various ion loadings (see equation 8.16(4)). The titration data obtained will reflect the difficulty of the control problem and will show whether the control problem becomes more or less severe with ion loading. Ion loading is high when the solution pH is far from neutrality.

Figure 8.16a illustrates the titration characteristics of a strong acid–strong base system with and without buffering. A strong acid or base is a compound that is completely dissociated in aqueous solutions. A buffered aqueous solution is one that resists a change in its hydrogen ion concentration when either an acid or an alkali is added to the solution. The titration curves of Figure 8.16a were obtained experimentally, using the reagents

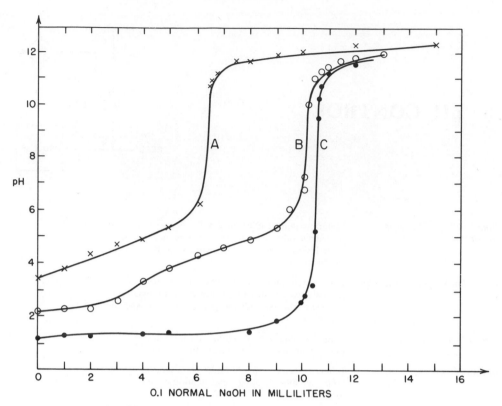

Fig. 8.16a   Typical acid-base titration curves. *Key:* A = 9.9 ml HCl + 50 ml of 0.1N KHC₈H₄O₄ per 100 ml solution; B = 6.7 ml of 0.1N HCl + 50 ml of 0.1N KHC₈H₄O₄ per 100 ml solution; C = 100 ml of 0.1N HCl.

indicated. The buffered solutions were prepared[1] with hydrochloric acid and potassium acid phthalate ($KHC_8H_4O_4$).

The problem of controlling pH is aptly demonstrated by these titrations. The control point is usually 7.0. In this range the slopes of the titration curves are steep, i.e., there is a large change in pH for a small change in reagent addition. The slope is especially steep for curve C, representing a strong acid–strong base system without buffering, and becomes less severe with buffering. This characteristic results in a high process gain (the process being sensitive) about the set point (pH of 7), which must be countered in an automatic control system with a low controller gain (high proportional band in the control instrument) if control loop stability is to be maintained. However, if a load upset drives the pH measurement below 4 or above 10 (for curve C), the process gain is very low (it is no longer sensitive), and small changes in pH result from large changes in reagent addition. In an automatic control system this requires a controller with high gain if the system is to be responsive. This is one of the problems that must be resolved in connection with most pH installations.

In Figure 8.16b a strong acid has been added to water to achieve a pH of 6 and a pH of 2 (curves A and B). The reagent flow (10 percent NaOH) requirement for each solution is plotted on separate scales (lower and

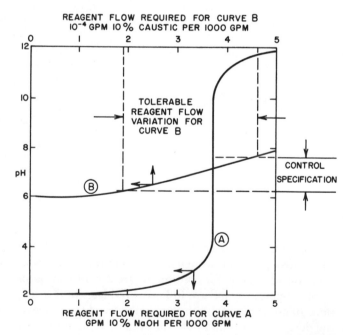

Fig. 8.16b   Relationship of accuracy and rangeability to ion loading

upper X axes). Assuming a control specification of pH 7.0 ± 0.5, a reagent flow of ± 28 percent variation can be tolerated when the pH of the inlet material is 6.0. When the pH is 2.0, the tolerable reagent flow variation is ± 0.0028 percent, and the problem is 10,000 times more difficult. This example also points up the other problem associated with pH control; reagent flow rangeability. If faced with a pH treatment problem as shown by this example, the reagent system must be capable of a 10,000 to 1 flow turndown.

Reagent Addition Requirements

Reagent addition requirements can be handled in diverse ways, depending on the process loads (flow of material to be neutralized) into the neutralization facility and the variation of the hydrogen or hydroxyl ion concentration, or both, in that flow. It should be recognized at the outset that because of the logarithmic nature of the pH measurement, a pH change of one unit can cause a tenfold change in load, whereas a 100 to 300 GPM (378 to 1134 l/m) change in flow (assuming no change in pH) is only a threefold change. Thus, the consequences of flow variations in waste streams can be relatively minor in comparison with ion concentration variations.

The equipment to deliver reagents to the process under automatic control includes a metering device or a control valve. Metering devices as a choice for reagent delivery are very accurate; however, delivery rangeability capability is limited to approximately 20 to 1 if speed is manipulated. Both speed and stroke can be manipulated to yield 200 to 1 rangeability, but the resulting relationship is squared and may require characterization. This means that where speed alone is manipulated, pH variations for a strong acid–strong base reaction greater than ± 0.65 will result in cyclic or inadequate control. (A pH change of 1.3 means a 20-fold change in reagent requirement.) When pH load variations are minor (0.4 or less) and flow variations are less than 4 to 1, the choice of a metering pump with speed control is sufficient.

Control valves, like the metering pump, have limited rangeability. In this category two types of internal plug forms are usually considered for throttling service. They are the linear and the equal percentage* throttling characteristics.

Both valves are available with minimum turndown of 50 to 1, and some have recently been developed mainly in the smaller sizes with reported rangeabilities as high as 500 to 1.

Digital valves that can furnish rangeabilities of 2000 to 1 are also available. Cost, size, complexity and materials of construction limit the application of these devices.

Valve Sequencing

A technique for increasing valve rangeability[3] involves the sequencing of a pair of equal percentage valves (Fig-

---

* The term *equal percentage* means that the valve will produce a change in flow rate corresponding to a unit change in lift (valve plug movement), which is a fixed percentage of the flow rate at that point.

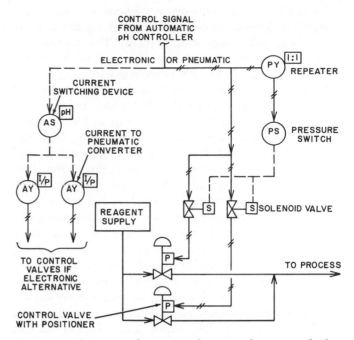

Fig. 8.16c Electronic and pneumatic alternatives for sequenced valve arrangement

ure 8.16c) so as to achieve an overall rangeability approaching the product of the individual valves rangeabilities, e.g., $50 \times 50 = 2500$. The loss of rangeability is mainly caused by the amount of overlap between valves. A plot of the performance characteristics of this pair of valves is shown in Figure 8.16d. The valve positioner of

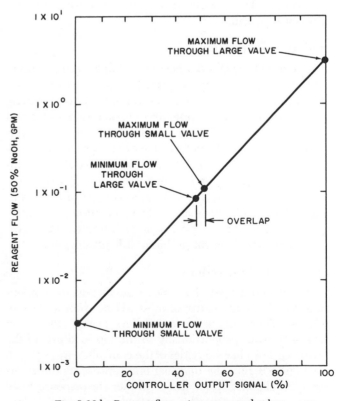

Fig. 8.16d Reagent flow using sequenced valves

the smaller sequenced valve is calibrated for full stroke over 0 to 52 percent controller output signal (closed at 0 percent; fully open at 52 percent). The positioner of the larger sequenced valve is calibrated for full stroke over the range of 48 to 100 percent of controller output. Transfer between the valves (as the controller output changes) can be implemented with either pneumatic or electronic control elements, as shown in Figure 8.16c. Since only one valve at a time is operating while the other valve is closed, the characteristic of the pair is equal percentage, as the semilog plot of Figure 8.16d illustrates. If the smaller valve, for example, were permitted to remain open when the larger valve came into service, the valve characteristic curve would have a discontinuity at the transfer point that could result in an unstable control system. There is a small flow transient at the transfer point, but the characteristic curve is maintained.

Sequencing of linear valves to provide an overall linear characteristic with wide rangeability is generally not satisfactory, because the transfer point occurs at 10 percent or less of the controller output signal. This means that the larger valve is essentially doing all the work. A high gain relay is also required to expand the controller output signal to operate the smaller valve.

Linear valves with rangeabilities of 200 to 1 or greater can be sequenced successfully using convential split-range techniques. The wide rangeability minimizes the step change in reagent delivery at the transfer point where the small valve remains open as the large valve begins to open. To counteract the large gain change at the transfer point, a characterizer must be used to ensure constant control loop gain.

Valve Linearization

The drawback of sequencing is a high gain contribution to the overall loop gain, especially at high reagent flows, a condition that is typical of the variable-gain (sensitivity is a function of valve opening) characteristics of equal percentage valves. One approach to countering this variable gain is by a characterizer with an input-output opposite that of the equal percentage valve(s). This approach is illustrated in Figure 8.16e. The resultant valve characteristic using this technique is approximately linear, which is highly desirable from an automatic control point of view, since the variable gain nature of the process makes the control problem difficult enough.

Nonlinear Controller

Special controllers have been developed to compensate for the nonlinearity of most pH neutralization processes. These nonlinear controllers change their gain characteristics proportionally to the ion load(pH) of the process. The characteristics of the controller are as shown in Figure 8.16f. The diagonal line represents the error-output relationship for the controller (in response to an error, a corrective signal is generated—the output—which

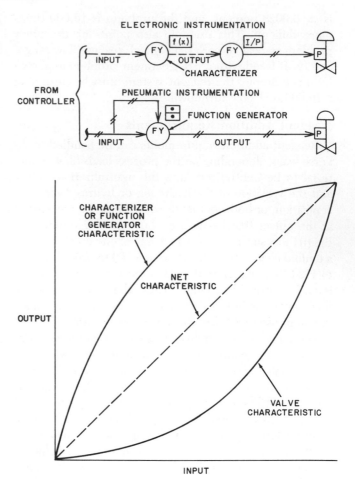

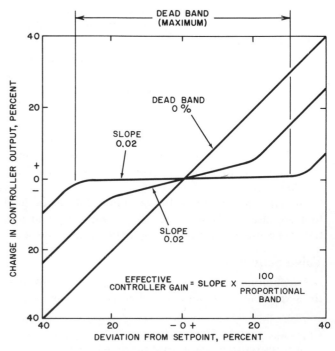

Fig. 8.16e   Linearization of equal percentage valves

Fig. 8.16f   Nonlinear controller characteristics

eliminates the deviation from set point) with a 100 percent proportional band (gain = 1.0) without the nonlinear adjustments available with this controller. The first available adjustment is a slope adjustment that allows the proportional band to be increased (gain reduced) about the zero deviation point by a factor of 50. This means that when the gain setting of the controller is 1.0 (100 percent proportional band), the effective proportional band is 5000 percent, or a gain of 0.02 (insensitive controller) at the zero deviation point. The slope can be adjusted manually or by an external signal. The second adjustment is the error deviation range, over which the slope adjustment is operative; this is referred to as "dead-band."

The dead-band is adjustable from 0 to ± 30 percent error (deviation from set point). This latter feature allows the gain of the control loop to be adapted proportionally to the ion load. If the process to be controlled resembles Figure 8.16a, a reagent flow rate or valve position signal can automatically adjust the dead-band. At high ion loadings (curve A) the controller gain will be low, a desirable condition when the process valve gains are high. At lower ion loadings (curve B) the dead-band can be reduced, thereby increasing the gain of the controller, a condition that is desirable when the process and valve gains are low. The effectiveness of this type of controller and the benefits achieved by adapting the control loop characteristics to those of the process have been demonstrated[3,4] on operating installations.

Control Systems

The choice of which type of control system should be used is dictated in large measure not only by the process loads, particularly ion loading, but also by the rate and the frequency with which the loads are changing. Again, knowledge of the characteristics of the process stream, particularly the type of information supplied by titration curves, is essential. The reagent delivery requirements deserve extensive consideration, as do the characteristics of the device used to bring the reagents into the process as discussed later in this section.

Batch pH Process

When the flow rate of material to be treated is reasonably small (perhaps less than 100 GPM, or 378 l/m), batch treatment may be a cost-effective pH control approach. As the flow rate increases, the tankage required rapidly shifts the economics in favor of a continuous pH control arrangement. Two unique characteristics of the pH batch process are:[3]

1. The measurement (actual pH) and the set point (desired pH) are away from each other most of the time.
2. When the measurement and set point are equal

(end point), the load on the process (reagent requirement) and, hence, the controller output are zero.

The controller characteristics for the batch pH control application should be proportional plus derivative.[8] Reset must not be used, since reset windup[8] will result in overshoot of the controlled variable. In a proportional controller, the corrective action generated is proportional to the size of the error; in a reset controller, to the area under the error curve; and in a rate controller, to the rate at which the error is changing. Once the measurement goes past the set point there is no way for the control system to bring it back to the set point, unless, of course, two controllers and two reagent supplies are used. In the absence of the reset control mode (proportional-only), a controller is usually supplied with a 50 percent bias so that the controller output is 50 percent when the measurement and the set point are equal. For the batch application with a proportional plus derivative controller, the bias must be 0 percent so that when measurement and set points are equal, the controller output is zero percent.

The effect of secondary lags in the valve, process vessel, and measurement are compensated for by the derivative action of the controller. If, for example, reagent is added but its effect has not yet been seen by the pH electrode, when measurement and set are equal, then too much reagent will have been added. With the derivative-time setting properly adjusted, the controller will shut off the reagent valve while the measurement is still away from set point, thereby allowing the process to come gradually to equilibrium.

Too much derivative time in the controller is preferable to too little. When there is too much, the valve will close prematurely but will open again when the measurement does not reach set point. Too little derivative allows the valve to remain open too long, resulting in overshooting the desired pH target.

The variable gain characteristic of the equal percentage valve is an asset to this type of control system. When the measurement is far away from set point, the valve will be wide open, permitting essentially unrestricted reagent flow to the process. As the measurement approaches set point and the valve closes, the decreasing gain of the valve counters the increasing gain of the process. Figure 8.16g illustrates the measurement-valve behavior of the batch process.

Although the installation and process design considerations for the batch process are not as severe or demanding as the continuous operation, care should be taken to ensure that (1) adequate mixing is provided, (2) tank geometry precludes the existence of stagnant areas, (3) reagent delivery piping between valve and process is as short as possible, and (4) electrodes are placed in responsive locations.

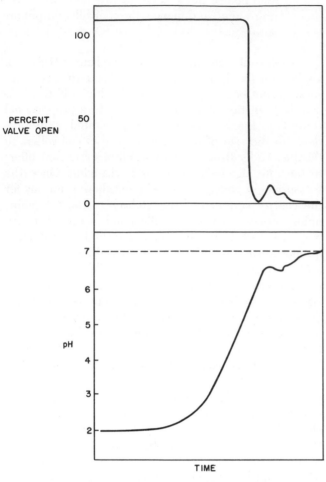

Fig. 8.16g Measurement-valve behavior for a batch process

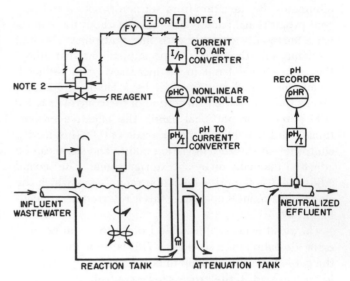

Fig. 8.16h Feedback control of pH.
Note 1: For linearization of equal percentage characteristic, commercially available divider or function generator may be used here.
Note 2: Characteristics linear or equal percentage, depending on reagent delivery requirements. Positioner recommended for either choice.

Feedback Control Systems

Feedback control can be used very effectively in wastewater neutralization, provided the process is not subjected to dramatic or frequent load variations, or both. Maintained step changes in either load or set point can be handled effectively. Figure 8.16h illustrates a feedback control system in which the reagent flow rangeability requirements are not severe and can be handled by a single valve having linear characteristics. This system can accommodate (for a strong acid–strong base) inlet pH variations of approximately ± 0.9 units around some normal value. If a linear valve is unavailable because of material or size limitations, an equal percentage valve can be used but should be characterized to provide linear reagent delivery as shown in Figure 8.16h. A valve positioner is required to eliminate valve hysteresis (difference in opening and closing characteristics) and to provide responsive valve movement.

The feedback controller in Figure 8.16h is a nonlinear controller with the characteristics shown in Figure 8.16f. The overall loop stability depends on the characteristics of the treatable process material. For a process with a titration curve like that shown in part A of Figure 8.16b,

there is no question that the nonlinear characteristic will be helpful in achieving loop stability.

As the normal value of inlet pH increases toward neutrality (for an acid material), the titration curve approaches that shown in part B of Figure 8.16b. For this case, the value of the nonlinear characteristic diminishes, and therefore the nonlinearity should be dialed out, or a standard controller should be used. If the buffering characteristic of the material is variable, there is no choice other than to adjust the nonlinearity of the controller for the severest case—which is usually the case of little or no buffering. The point is that the availability of the nonlinear feature markedly increases the flexibility of the control system at a moderate cost.

In Figure 8.16h two vessels or tanks are shown for illustration. A single divided vessel would also suffice. The objective is to provide a reaction section and an attenuation section. The former should be as small as possible but should provide efficient backmixing with the minimum agitation cost, thus permitting a tight control loop for the reaction portion of the process. If the accuracy capability of the valves is less than required, a noisy measurement will result, and the attenuation portion provides a smoothing effect. A more detailed description of process equipment design requirements is included later in this section.

For those control conditions in which the set point is low or high, say 2 or 12, the process gain is very low, i.e., it takes a large change in reagent flow to cause a small change in measured pH, and a linear controller with a high gain (high sensitivity, narrow proportional band) suffices. In fact, on-off control (reagent valve is either fully open or closed) may be adequate. Low values of pH set point are used for the destruction of hexavalent chromium. The destruction proceeds rapidly when the pH is controlled at about a value of 2.0. Higher values of pH lengthen the process. See Section 8.15 for a more detailed discussion of this type of process treatment.

Sequenced Valves

A wider reagent delivery capability can be obtained by using the sequenced valve approach (Figure 8.16i). The arrangement is virtually the same as that shown in Figure 8.16h, except that the controller output can be switched to either valve by a pressure switch (PS) or its electronic equivalent, the function of which was described in Figure 8.16c. Valve positioners must be used, since each valve must be calibrated to stroke over only a portion of the controller output signal range. Figure 8.16j illustrates various combinations of different pairs of sequenced valves. Table 8.16k lists the various flow rangeabilities for some valve pairs, assuming a constant pressure drop across the valves (equivalent to 9.5 feet, or 2.85 m, of 66° Be sulfuric acid) and assuming an individual valve rangeability of 35 to 1. The valve size coefficients (CVs) are 1.13, 0.14, 0.08, and 0.04, respectively, for CV-1, 2, 3, and 4.

The overlap between each valve pair becomes smaller as the rangeability increases. The pressure switch to transfer the valves can be set anywhere in the overlap region, because in this region the process loads can be satisfied by either valve.

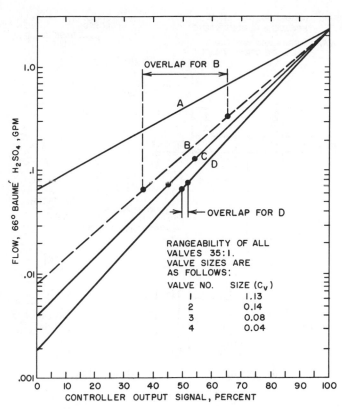

Fig. 8.16j   Delivery capability for various valve pairs.
Key: A = CV-1 alone; B = CV-1 + CV-2; C = CV-1 + CV-3; D = CV-1 + CV-4.

Table 8.16k
REAGENT DELIVERY TURNDOWN
(RANGEABILITY) FOR SEQUENCED PAIRS
OF EQUAL PERCENTAGE VALVES

Valve Pair	Line on Figure 8.16j	Turndown	Log Turndown*	Valve Positioner Calibration(s) (%)
CV-1 (alone)	A	35 to 1	1.54	0 to 100
CV-1 + CV-2	B	275 to 1	2.44	0–63; 37–110
CV-1 + CV-3	C	570 to 1	2.76	0–58; 44–100
CV-1 + CV-4	D	1150 to 1	3.06	0–51; 50–100

* Signifies the approximate pH swing that valves will accommodate.

Two Reagent Systems

Situations may arise wherein the influent may enter the system on either side of neutrality. Figure 8.16l illustrates the two-sided feedback control system. Although only one valve for each side is shown, it would be possible to have a sequenced pair for one side of neutrality and a single valve for the other, or a sequenced pair for both sides. Since this is a feedback control system, load changes cannot be frequent or severe in order for this system to give acceptable performance. For those applications in which load changes are frequent and severe, a combination feedforward-feedback should be considered.[5] If sequencing is used, the reagent delivery system will have a high gain characteristic, since the

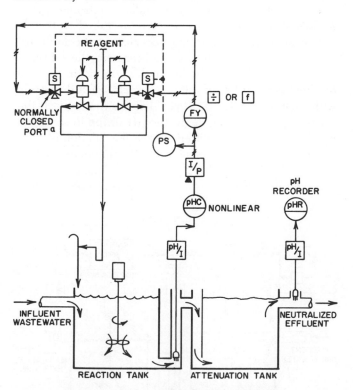

Fig. 8.16i   Wide-range feedback control of pH.
^aThis port is closed when the coil is de-energized because the pressure switch (PS) did not close the electric circuit to supply current to the solenoid coil.

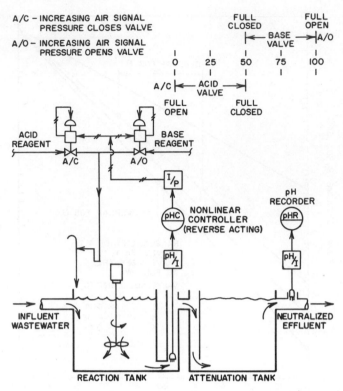

Fig. 8.16l   Two-sided feedback control of pH

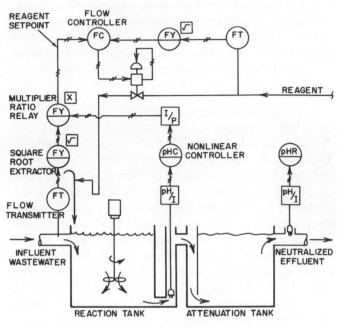

Fig. 8.16m   Ratio control of pH

stroking of the pair (moving from closed to open) is accomplished with only half the controller output signal, thereby doubling the gain (making it twice as sensitive). The valve gain will vary with the turndown, and a characterizer will be required for each set of sequenced valves to provide uniform loop gain.

Ratio Control Systems

Ratio control of pH can be extremely effective when the process flow rate is the major load variable, and the objective is to meet increased flow with a corresponding increase in reagent. Since flow measurements may be in error and reagent concentration may vary, a means for on-line ratio adjustment must be provided. Figure 8.16m illustrates a ratio control system in which the reagent set point is changed proportionally to changes in process flow. A feedback signal supplied by the feedback controller (pHC) also adjusts the reagent flow set point proportionally to a nonlinear function of the deviation between desired and actual effluent pH.

Note that the rangeability of the ratio system is limited by that of the flow meters, typically 4 to 1 for orifice meters to 30 to 1 for some turbine meters.

Cascade Control Systems

Cascade control (the output of one controller—the master or primary—is the set point of another) as applied to pH control systems can take two forms. In addition to the usual condition in which the output of one con-

troller serves as the set point to another controller, it is also possible to have two vessels arranged in series, each with its own control system. The latter arrangement is referred to as cascaded residences.[6]

The conventional cascade control system[8] is shown in Figure 8.16n wherein the output of controller pHC-1 is the set point of the slave, or secondary controller, pHC-2. This arrangement is particularly useful when lime is the reagent. In this instance, because of the finite reaction time between the acid and reagent, the set point of pHC-2 may have to be lower than the desired pH of the final effluent because the materials are still reacting with each other after they have left the first tank. If the set point pHC-2 is too high, the pH of the final stream will be greater than desired. When flocculation is to be carried out downstream of the pH treatment facility, stable pH values can be extremely important.

A delicate balance must be struck in this type of system

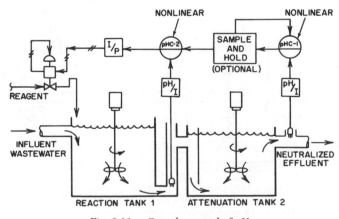

Fig. 8.16n   Cascade control of pH

with respect to the size of the first vessel. A long residence time in the first tank ensures long contact time between reagents, thereby producing an effluent pH which is close to the desired value, but at the same time it may result in a sluggish control loop around this vessel. For efficient cascade control, response of the inner loop (control loop around the first tank) must be fast. The other control loop (pHC-2), sometimes referred to as the master, or primary, control loop, is usually tuned (control mode adjustments such as proportional band are set) so as to be less responsive than the inner loop. The tuning of pHC-1 will be a result of the dead time (a delay between a change in reagent flow and the time when its effect is first felt), capacity, and process characteristics.

When this part of the process is dominated by dead time, the technique of sample data control may be useful in stabilizing the control system by a sample and hold device (Figure 8.16n). This device may be a timer that automatically switches the controller[7] between automatic and manual modes of operation. This can allow the controller to be in automatic for a fraction (x) of the cycle time (t) and then can switch it to a fixed-output, manual condition for the rest of the cycle (1 − x)t.

In the second form of cascade (cascaded residences) (Figure 8.16o), each vessel has its own feedback loop. This approach is recommended when the incoming material is very strongly acidic or basic (pH values less than 1 or greater than 13).[6] The first stage controls the effluent at a pH of approximately 4 or 10, and the second stage brings the effluent to its final value, near 7. The choice of pH set point for the first stage depends on the characteristics of the material. For example, material having a titration characteristic like that of A in Figure 8.16b may have a set point of approximately 3.5. The purpose is to make the control problem as simple as possible by staying on as linear a portion of the titration curve as possible.

This approach is logical when one considers the process gain characteristic as well as the accuracy limitation of a reagent delivery system. The remainder of the neutralization control problem is then similar to that illustrated in part B of Figure 8.16b. In this manner, the control system does not have to cope with the entire nonlinear characteristic of the process all at once. A sequenced pair of valves is shown in conjunction with the first stage in order to handle the pH load variations. A single valve would probably suffice for the second stage, because its influent is pH-controlled. Depending on the valve sizes and on the individual valve rangeabilities, this three-valve arrangement has a maximum possible reagent flow turndown of approximately 125,000 to 1 (50 × 50 × 50).

Feedforward Control Systems

A feedforward control (the consequences of upsets are anticipated and counteracted before they can influence the process) system is dedicated to bringing about control by correcting for changes in process load as they occur. The corrective action is implemented using a control system that is essentially a mathematical model of the process. Ordinarily, the inclusion in the model of each and every load to which the process is subjected is neither possible nor economically justifiable. This means that a feedback control loop (usually containing the nonlinear controller for pH applications) is required in conjunction with the feedforward system (see Section 1.5). The function of the feedback controller is to trim and correct for minor inaccuracies in the feedforward model.

Feedforward Model

The model for the case in which a base reagent is added to an acid wastewater will be developed here. This can be extended to the case in which an acid reagent is added to a basic waste.

A waste acid flowing at rate F and normality N_a (concentration) is neutralized with a basic reagent having a normality N_b, at a flow rate B:

$$BN_b = FN_a \qquad 8.16(1)$$

When computing hardware is used to perform arithmetic operations, the technique of normalizing or scaling inputs and outputs (converting into consistent engineering units) must be considered. The starting point for instrument scaling[8] is the engineering equation 8.16(1). The next step is to write a set of normalized equations for each signal associated with the engineering equation. Normalization refers to expressing all inputs and outputs as a number beween zero and one or, equivalently, between zero and 100 percent of instrument or transmitter input or output. The reagent and waste flows written in normalized form are

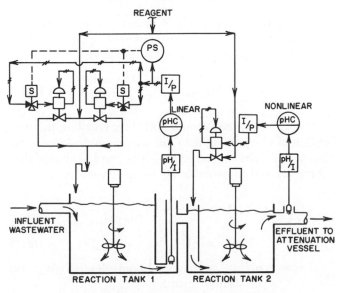

Fig. 8.16o Cascade residence control of pH

$$B = B'B_{max} \qquad 8.16(2)$$

where

> B = actual reagent flow, in engineering units;
> B' = fractional output of measuring device; and
> B_{max} = maximum value of measurable reagent flow, in engineering units.

$$F = F'F_{max} \qquad 8.16(3)$$

where

> F = actual waste acid flow, in engineering units;
> F' = fractional output of measuring device; and
> F_{max} = maximum value of measurable acidic wastewater flow, in engineering units.

Substituting equations 8.16(2) and 8.16(3) into 8.16(1):

$$B' = CF'N_a \qquad 8.16(4)$$

where $C = \dfrac{F_{max}}{B_{max}N_b}$

Assuming a strong acid is being neutralized, its normality is approximately equal to its measurable hydrogen ion concentration:

$$N_a = [H+] = 10^{-pHi} \qquad 8.16(5)$$

where

> $[H+]$ = hydrogen ion concentration (mol per liter), and the subscript
> i = inlet conditions.

Substituting equation 8.16(5) into 8.16(4) and writing the resulting relationship in logarithmic form:

$$\log B' = \log C + \log F' - pH \qquad 8.16(6)$$
$$B' = R^{X_B - T}$$

where

> R = rangeability of the valve(s): 35:1, 50:1, and so on.
> X_B = fractional control signal to valve(s), 0 to 1.0.

Writing equation 8.16(7) in logarithmic form, substituting into equation 8.16(6) for log B′, and solving for X_B, the fractional control signal is:

$$X_B = 1.0 + \frac{\log F'}{\log R} + \frac{1}{\log R}(\log C - pH_i) \qquad 8.16(8)$$

Letting:

$$1.0 + \frac{\log F'}{\log R} = f(F') \qquad 8.16(8a)$$

where $f(F') + \dfrac{1}{\log R}(\log C - pH_i) \qquad 8.16(9)$

Equation 8.16(9) is a mathematical representation of the process in which inlet flow and pH are the loads. In those instances in which flow variations are less than approximately 3 to 1, the inclusion of flow as a load variable is not warranted, since a flow variation of approximately 10 to 1 is equivalent to a pH change of one unit. If in this instance F′ is assumed to be equal to 1.0, equation 8.16(9) becomes:

$$X_B = 1.0 + \frac{1}{\log R}(\log C - pH_i) \qquad 8.16(10)$$

Expressing pH_i on a scaled basis:

$$pH_i = S\, pH'_i \qquad 8.16(11)$$

where S = span of the inlet pH transmitter, and
pH_i = fractional transmitter output.

Substituting equation 8.16(11) into 8.16(10) and factoring:

$$X_B = 1.0 + \frac{S}{\log R}\frac{\log C}{S} - pH_i' \qquad 8.16(12)$$

The parenthetical portion of equation 8.16(12) including the gain term (S/log R) is the form of a proportional controller (described by equation 8.16(13)):

$$m_c = \frac{100}{PB}(r - c) \qquad 8.16(13)$$

where

> $\dfrac{100}{PB}$ = controller gain = $\dfrac{S}{\log R}$,
> PB = proportional band,
> r = set point = $\dfrac{\log C}{S}$,
> c = measurement = pH_i, and
> m_c = controller output signal.

By returning to equation 8.16(1) and solving for the condition in which the maximum amount of ions are entering the process, equation 8.16(14) is obtained:

$$N_{A(max)} = \frac{B_{max}N_B}{F_{max}} \qquad 8.16(14)$$

One develops equation 8.16(15) by recalling the definition of C, applying it to the maximum acid flow condition that is being considered here, and assuming that normality and measurable [H+] are equivalent:

$$C = \frac{1}{N_{A(max)}} = \frac{1}{10\text{-}pH_{min}}$$

$$\log C = pH_{min} = S\, pH'_{min} \qquad 8.16(15)$$

where pH'_{min} = fractional equivalent of pH_{min} on a pH scale of span S.

Substituting for log C from 8.16(15) into 8.16(12):

$$X_B = 1.0 + \frac{S}{\log R}(pH'_{min} - pH'_i) \quad 8.16(16)$$

An examination of equation 8.16(16) shows that when the inlet pH is at its minimum value ($pH'_{min} = pH_i$), equation 8.16(16) equals 1.0, which indicates that the fractional control signal to the reagent delivery system is 100 percent, corresponding to full reagent flow.

Equation 8.16(16) must now be modified to include feedback trim. Ideally, the feedback controller, which provides the trim signal, would not be required if the feedforward computation as developed were perfect. But inlet flow rate is not constant: Reagent concentration may change, instrument measuring errors are present, and the strong acid assumption will not be completely true 100 percent of the time. Since these loads are not likely to change suddenly or at a high frequency, feedback trim to adjust for their effect should be adequate. Ideally (when the feedforward model is perfect), the output of the feedback controller should be 50 percent, or 0.5 on a scaled, or normalized basis. This gives the feedback controller maximum flexibility in each direction. If the signal to the valves is to be 1.0 or 100 percent when the influent wastewater pH is at its minimum value and the output of the feedback controller is 0.5, then equation 8.16(16) becomes:

$$X_B = 2.0 + \frac{S}{\log R}(pH'_{min} - pH'_i) - 2m_b \quad 8.16(17)$$

where m_b = output of feedback controller, which ideally has a value of 0.5.

Equation 8.16(17) is difficult to implement with conventional analog hardware because of the 200 percent bias required, i.e., the 2.0 term in equation 8.16(17). Instead of working with the minimum value of pH, one can work with the maximum value of influent pH corresponding to minimum reagent demand. The minimum controllable amount of reagent corresponds to a flow rate that does not yet cause the smaller sequenced valve to cycle on and off. Where basic reagent is being added to an acid waste, the maximum value of pH that can be controlled by the system is estimated from the valve rangeability and from a knowledge of the minimum expected influent pH. For example, if a pair of sequenced valves has an installed rangeability of 1500 to 1 and the minimum pH expected is 3.0, the maximum controllable influent pH value is 2.0 + log 1500 ≈ 5.2 = pH_{max}.

Equation 8.16(17) in terms of pH'_{max} is:

$$X_B = 1.0 + \frac{S}{\log R}(pH'_{max} - pH'_i) - 2m_b \quad 8.16(18)$$

The gain term ($S/\log R$) is adjusted to give X_B a value of 1.0 when pH'_i is at the minimum expected value. For example, let the span (S) of the pH transmitter be 0 to 9.0, pH_i = 2.0 and pH_{max} = 5.2. Then pH'_i = 2.0/9.0 = 0.222; pH'_{max} = 5.2/9.0 = 0.577. Then:

$$1.0 = 1.0 + \frac{S}{\log R}(0.577 - 0.222) - 2(0.5)$$

$$S/\log R = 2.81$$

Since $S/\log R = 100/PB$ (equation 8.16(13)), the proportional band setting required for this feedback controller is:

$$PB = \frac{100}{2.81} = 35.5 \text{ percent}$$

In those instances in which flow compensation is required, the 1.0 term of equation 8.16(18) is replaced by the $f(F')$ expression as defined in equation 8.16(8a). The $f(F')$ signal is easily generated by characterization of the wastewater flow signal from a primary flow device. When the inlet pH is greater than pH_{max}, the sequenced valves are completely closed and a separate linear feedback controller can be used to operate a third valve, referred to as the trim valve. Since the reagent delivery accuracy requirement is dramatically reduced as the inlet pH approaches neutrality (Figure 8.16b), this type of control

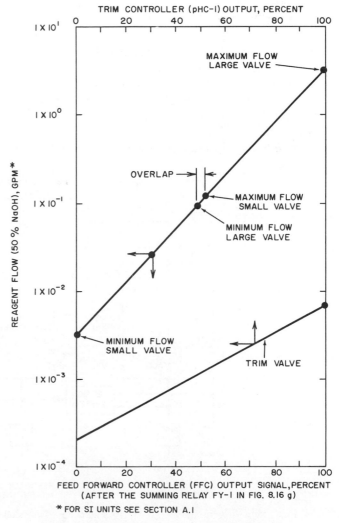

Fig. 8.16p Feedforward pH control using three valves.

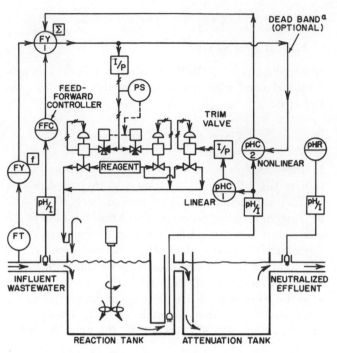

Fig. 8.16q   Three-valve feedforward pH control system.
^aIf pHC-2 is provided with a dead band, it will be inactive when the
trimming controller (pHC-1) alone is operating.

arrangement in which the flow characterization of the
influent wastewater flow signal and the use of the dead-
band adjustment feature previously discussed are shown.
A system essentially as outlined in Figure 8.16p has demonstrated the need for the combination of feedforward-
feedback control, because each type of control was tested
individually and was found to be unsatisfactory.[4]

Figure 8.16r illustrates the feedforward-feedback control system arrangement for a two-sided waste neutralization task. The pH can be on either side of neutrality
with severe measurement (load) variations. A logic system is also required in order to establish whether each
(or neither) one of the two systems is needed to be in
operation. A combination of feedforward-feedback on one
side of neutrality and conventional feedback on the other
is also possible. The nature and characteristics of the
problem to be solved will indicate the nature of the
solution.

Process Equipment Considerations

Composition processes, whether pH or "pIon" (such
as pCl and pAg), should be recognized as having two
distinct aspects: one chemical and the other physical.
This was noted in 1950 by Chapman[9] and still holds true
today.

is sufficient. The trim valve is sized to deliver a maximum
reagent flow, which is slightly in excess of the minimum
reagent capability of the smaller sequenced valve. Figure
8.16p illustrates the reagent flow–controller output relationship for two sequenced valves in connection with
a trim valve.

Figure 8.16q illustrates a feedforward control system

Many pH control applications have become
standardized to the point of catalog items while
others introduce special features characteristic
of their individuality. Problems of one application may be wholly physical while others may
have chemical limitations. Only by rigid separation and classification of all factors involved

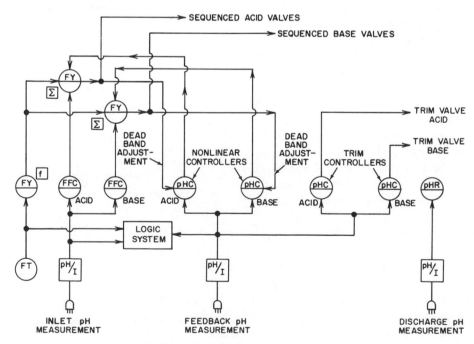

Fig. 8.16r   Two-sided feedforward control of pH

in a pH process, together with an understanding of their characteristics, can the process be most effectively instrumented.

Several physical or process design considerations are associated with composition (pH or pIon) control applications. The most important is the primary device used to mix the reagent with the process stream. This can be as simple as reagent addition upstream of a pump or inline mixer with a downstream measurement point or as complex as two mixed reaction vessels followed by an attenuation vessel.

Where reaction vessels are required, design decisions must be made to determine (1) size and number, (2) baffling, (3) agitation (how much and what type), (4) measurement probe location(s), and (5) reagent addition point location.

The ease or difficulty of most industrial control applications is closely related to a property of the process referred to as "dead time." Analogous terms, such as "transport time," "pure delay," and "distance velocity lag," describe the same effect.[7] Dead time is defined as the interval between the introduction of an input disturbance to a process and when a measuring device *first* sees the effect of that disturbance. Qualitatively, the relationship between dead time and controllability is simple: The more dead time, the more difficult the problem of control. The presence of dead time in pH or pIon processes is extremely detrimental to controllability. The major reason is the severe sensitivity of the measurement of interest at the control point. The pH process as described earlier in this section illustrates this point. One

of the major goals of system design is to eliminate the dead time or to reduce it to an absolute minimum.

The Inline Mixer

Since agitated vessels are expensive, simple devices such as inline mixers are often considered for composition control systems. Properly applied, these devices are effective, but careful attention to the following design criteria is required: reagent delivery hysteresis, loop gain, and neutralization stage interaction.

As a general rule, an inline mixer will afford control between pH 6 and 8 for unbuffered influent streams between pH 4 and 10. The response will always be oscillatory following a disturbance. Buffering extends the controllable pH limits.

Reagent Delivery Hysteresis

Consider a reagent delivery device such as a control valve, a metering pump, or a dry feeder. The smallest incremental change that these devices can make is approximately 1 percent. Converted to the logarithmic pH scale and using an influent pH of 14 and a set point of 7, 1 percent excess acid produces a pH of 2, and 1 percent too little yields pH 12. These values were derived from Figure 8.16s, where 1×10^6 reagent units are required to neutralize pH 14 to 7. One percent of this total is 10,000 reagent units, which correspond to pH 2 and 12.

In a similar fashion, the effect of the same error can be estimated for any other set point. Using a setpoint of 12, for example, $1 \times 10^6 - 10,000$, or 990,000, reagent units are required for neutralization from a 14 pH in-

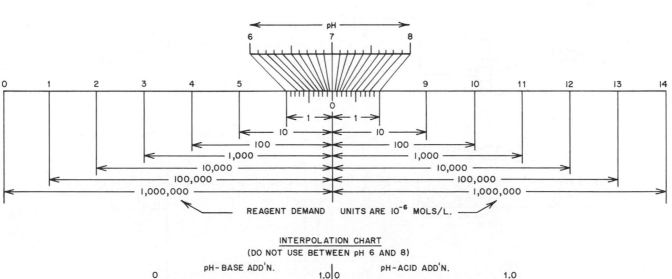

Fig. 8.16s   pH vs. reagent demand: strong acid–strong base

fluent. One percent excess acid (9900 reagent units) corresponds to 10,000 − 9900, or 100 reagent units, which is pH 10. One percent too little acid corresponds to 19,900 reagent units, or approximately 12.3 pH using the interpolation chart on Figure 8.16s.

The same procedure is used to illustrate the sensitivity of the process to hysteresis. Increasing the error to 1.5% for a set point of 12 gives a low pH of 3.13 and a high of 12.39. (10,000 − 14,850 = 4850 on acid side and 10,000 + 14,850 on caustic side.)

Methods of reducing valve hysteresis, such as pulse interval control and the uses of digital valves,[10] have been proposed. Although these techniques add cost and complexity to the control system, they should be investigated as alternatives to the installation of stirred tanks.

If hysteresis cannot be eliminated, it can profoundly influence pH loop performance unless some other element can be introduced to smooth out this incremental response or, in effect, to reduce the gain of the control loop. A typical control system using an inline mixer is shown in Figure 8.16t.

Loop Gain

The gain of a control loop (in this case, measurement, controller, reagent delivery device, and process) is the product of the gains of the individual elements. The gain of an inline mixer will be compared with that of a mixed tank, since these are the only elements of the control loops that differ in the proposed neutralization systems.

The gain of an inline mixer is unity, since it is a pure dead time device. A step change in pH at the inlet results in the same step change at the outlet one dead time later. Because the inline mixer has unity gain, the hysteresis errors pass through unattenuated and the effluent cycles at the limits of the hysteresis band of the reagent delivery device. Typically, for an influent pH of 14 and 1 percent hystersis, a pH cycle develops between the limits of 10 and 12.3 for a set point of 12. Controller tuning can change the period of oscillation but cannot reduce the amplitude below these limits.

The gain of a mixed tank is a function of the dead time and capacity. If the dead time is about one-twentieth of the retention time, its gain is approximately 0.033, or 30 times less than that of the inline mixer. (Derivation is given at the end of this section.)

This gain reduction diminishes the effects of hysteresis by a proportionate amount. For an influent pH of 14, then, a mixed tank as described previously with closed loop control and a set point of 7 will cycle between the limits of 3.5 and 10.5. For a set point of 2 or 12, one-quarter amplitude damping can usually be achieved.

Note that this comparison of process devices (inline mixer versus mixed tank) is based on an unbuffered pH system. Buffering as a result of weak acids and bases moderates the pH curve, which reduces the process gain term in the control loop. A 30-fold reduction in process gain has the same effect as substituting a mixed tank for an inline mixer.

Neutralization Stage Interaction

As discussed earlier, a single-stage neutralization within effluent limits of 6 to 9 pH is not possible with current technology when the influent is strong acid or base at a pH of 0 or 14. Since several stages are required, they

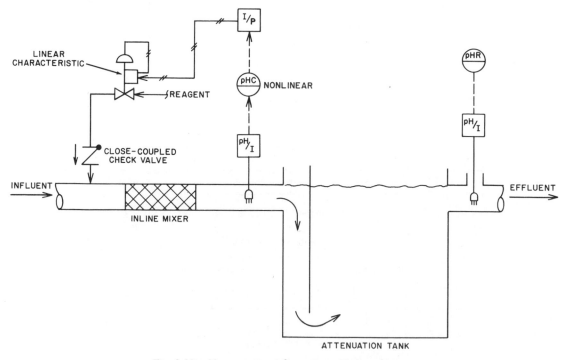

Fig. 8.16t   Narrow range inline mixer pH control system

must be arranged to minimize both interaction and complexity.

It is estimated that 4 or 5 inline mixers would be needed if loop gain were the only consideration. Interaction, however, is a much more serious problem, because the inline mixers would have identical characteristics. An upset in an upstream mixer control loop would set all the other downstream loops into oscillation. Early field experience with trying to control similar processes was completely unsuccessful, and this concept of pH control was abandoned.[3]

Mixed tanks can be selected to minimize interaction both by size and by degree of agitation. The best arrangement is to follow a small tank with one five times larger. The larger volume attenuates upsets from the first tank and minimizes interaction. The second tank is also stirred less vigorously to lengthen its natural period of oscillation. Field experience has proved that pH 0 to 14, strong acid or base influent can be successfully neutralized with effluent limits of 6 to 9 pH using three tanks. Two of these would be mixed, and one would be used as an unmixed attenuation vessel. The remainder of this section is devoted to the design criteria essential for successful pH control using this approach.

Tank Size and Number

In this section the discussion will be directed at vessel(s) or tank(s), and the terms "treatment vessel(s)," "reaction vessel(s)," and "attenuation vessel(s)" will be used. "Treatment vessels" refers to all of the vessels constituting the facility. "Reaction vessels" defines the vessel in which the reaction between process stream and reagent takes place. "Attenuation vessel" refers to any vessel after the reaction vessel the sole purpose of which is to provide or add capacity to the treatment process.

The reaction vessel should be of cubic dimensions. If a cylindrical tank is used, the depth of liquid should equal the tank diameter. The size of the vessel depends on the rate of reaction to be carried out. There is, however, a minimum size limitation. For neutralization, experience indicates that the reaction vessel time constant (retention time) should not be less than three minutes ($\tau_1 = 3$ min.), and the dead time (τ_d) to time constant ratio should be approximaely 0.05 ($\tau_d/\tau_1 \simeq 0.05$). For a vessel with a three-minute constant, then, the observed dead time should be on the order of nine seconds.

Tank size should be increased if the reaction time between influent and reagent is extended, an example of which is the use of lime for neutralization. With high-calcium lime, a five-minute time constant is required. If, however, a dolomitic lime were used, a 20- or 30-minute time constant would be required because of the low solubility and reaction rate characteristics of this reagent. A larger tank should be used if an insoluble precipitate, such as iron hydroxide, is formed. The precipitate traps unreacted reagent or influent material, or

both, and causes an extended reaction time, since the trapped material must diffuse from the precipitate before reaction can occur.

The number of treatment vessels required is related to the difficulty of the control problem. The maximum or minimum pH or pIon values, or both, coupled with the degree of buffering or complexation at the control point, determine the degree of difficulty. For example, with strong acid–strong base neutralizations and with a dead time to retention time ratio of 0.05, one stirred tank will be sufficient if the influent pH is between 4 and 10. Two vessels are required, one stirred and one unstirred, if the influent pH is as low as 2 or as high as 12. If the influent pH is less than 2 or greater than 12, experience shows that three treatment vessels (two reaction and one attenuation) are required. The reaction vessels should be stirred, and the third vessel should be unstirred. The unstirred vessel serves to damp the cyclic upsets that can occur in the effluent pH from the second stirred vessel. Although this description pertains to separate vessels, partitioning of an existing vessel will serve the same purpose. Figure 8.16u illustrates the attenuation effect.

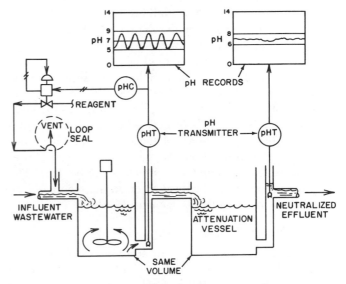

Fig. 8.16u   Effect of attenuation vessel

Tank Connection Locations

The inlet and outlet in the treatment vessel should be located at opposite sites—one high and one low—with respect to the bottom of the tank. Generally, it is most convenient to introduce the influent stream on the surface of the tank and to locate the outlet at the bottom of the vessel.

Variations in the location of the inlet and outlet can considerably change the dead time. Reversing the flow through the tank so that the inlet is on the bottom and the outlet is at the surface, for example, causes the dead time to increase by a factor of 2 or 3. Examination of the flow patterns in the tanks (Figure 8.16v) shows that the path from inlet to outlet can be doubled by this change. The additional dead time is attributable to the swirl effect of the agitator, which is minimized, but not eliminated, by baffling.

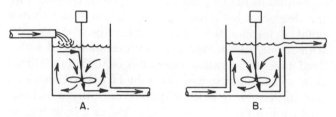

Fig. 8.16v   Flow patterns in stirred tanks. A. Recommended flowpath. B. Undesirable flowpath.

Sensor Locations

The location of the measuring electrodes also deserves serious consideration. The general guidelines are that the locations should be responsive and the information supplied by them should be timely. Submersible-type electrode assemblies are preferred when the measurement is used as an input to a control system. This preference is not always possible because of physical constraints. If flowthrough assemblies have to be used, the sampling time, i.e., the time required physically to transport the sample from the process to the electrodes (which is essentially dead time), should be kept to a minimum. Figure 8.16u shows a submersible-type assembly on the reaction vessel located as close as possible to the vessel exit. Location within the tank proper increases the measurement noise, principally because of concentration gradients. The requirements of the monitoring electrodes shown on the attenuation vessel are not as severe. Either flowthrough or submersible detectors can be used. The information supplied by these electrodes provides a clean record for any regulatory agencies involved.

Equalization Tanks

Upstream of a stirred neutralization vessel, a lagoon or a holding tank can be very useful because it serves to smooth out upsets in influent pH and flow, thus allowing the use of a simple feedback system rather than a more costly feedforward control system. A lagoon can also be used to store the material that is bypassed around the neutralization process in case of failure, a very important consideration if off-specification effluent causes a plant shutdown. The one thing that a lagoon cannot do is replace a mixed vessel as part of a control system. Any attempt to control the pH of a lagoon by closed-loop feedback control can only result in an effluent pH value on the opposite side of neutrality. The period of oscillation of such pH swings will depend on the dead time of the lagoon, but typically it will be on the order of hours.

Mixing and Agitation

The two types of mixing that are important to the control system are intermixing and backmixing. The reagent must be intermixed with the process stream to furnish complete elimination of the areas of unreacted reagent or untreated influent. Adequate intermixing between influent and reagent can be readily achieved by adding the reagent at a point of small cross-sectional area where there is some turbulence.[9] Figure 8.16u illustrates the reagent being added in the pipeline before the influent enters the treatment facility. This is a desirable practice because it eliminates poor intermixing, which can cause a noisy signal to be observed in the effluent pH.[4] A loop seal arrangement, particularly when long reagent transfer lines are required, allows the reagent line to remain full up to the point of introduction to the process and thus eliminates a potential source of process dead time.

Backmixing is more important than intermixing to close pH control. The treated stream must be held in a vessel sufficiently long for the reagent to react and be backmixed. In general, the degree of backmixing can be defined in terms of the pumping capacity of an agitator with respect to the flow and volume of the neutralization vessel. In practice, however, this definition has limited usefulness because of variables such as agitator construction and blade pitch, baffling of the neutralization vessel, and placement of inlet and outlet measuring electrodes. Experience shows that the best way to define backmixing for control purposes is by the ratio of the system dead time to retention time of the neutralization vessel. The retention time is the volume of the vessel divided by the flow through the vessel. A ratio of dead time to retention time equal to 0.05 is adequate for good control.

Suitable baffles or agitator positioning should be used in mixed neutralization vessels to avoid a whirlpool effect. The power supplied by the impeller must be used to turn the contents of the vessel over, not to whirl them about. With these effects in mind, a propeller or an axial-flow impeller should be selected to direct the flow of the vessel contents toward the bottom of the tank. The flat bladed radial-flow impeller should be avoided, since it generally tends to divide the vessel into two sections and increases system dead time.

Figure 8.16w is a plot of tank size against agitator pumping capacity per unit tank volume on logarithmic coordinates. The family of curves shown for various dead times was developed from empirical data in tanks with capacities of 200, 1000, 10,000, and 18,000 gallons (756, 3780, 37,800, and 68,040 l). They apply to baffled tanks of cubic dimensions with the inlet at the surface and the outlet at the bottom on the opposite side of the tank. The ratio of impeller diameter to tank diameter varies from 0.25 to 0.4. Square pitch propellers at an average peripheral speed of 25 fps were used in up to 1000-gallon capacity tanks. Axial-flow turbine impellers at an average

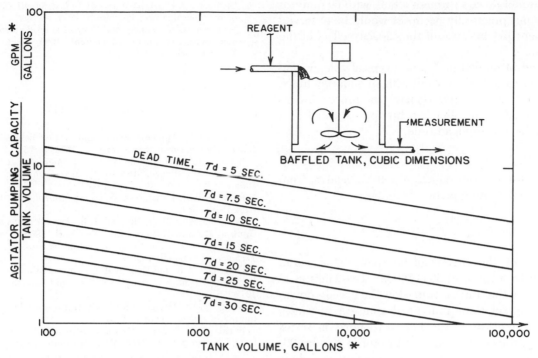

Fig. 8.16w   Dead time (τ_d) as a function of mixing intensity
*For SI units see Section A.1.

peripheral speed of 12 fps (3.6 m/s) were used in the larger tanks.

Control Dynamics

The performance of a stirred tank to periodic disturbances can be evaluated by consideration of the dead time and time constant properties of the tank.

For example, if the total system dead time is τ_{dt}, it can be defined as:

$$\tau_{dt} = \tau_{d1} + \tau_{d2} \qquad 8.16(19)$$

where

τ_{d1} = tank dead time, inlet to outlet, and
τ_{d2} = remaining loop dead time (sampling system and control valve motor).

Given $\tau_{dt} = 0.05 \ V/F \qquad 8.16(20)$

where

V = vessel volume, and
F = flow through vessel.

The time constant (τ_1) for an agitated vessel with dead time (τ_{d1}) can be expressed as:[7]

$$\tau_1 = V/F - \tau_{d1} \qquad 8.16(21)$$

Assuming that the stirred tank has the minimum 3.0-minute time constant previously mentioned and that the total dead time is divided 80 percent to (τ_{d1}) and 20

percent to τ_{d2}, equation 8.16(21) can be restated:

$$\tau_1 = 0.96 \ V/F$$

Expressing τ_1 in terms of dead time by combining equations 8.16(22) and 8.16(20):

$$\tau_1 = 19.2 \ \tau_{dt} \qquad 8.16(23)$$

The dynamic gain of a stirred tank to periodic disturbances[7] is given by equation 8.16(24):

$$G_d = \frac{\tau_0}{2\pi\tau_1} \qquad 8.16(24)$$

where

G_d = dynamic gain of the stirred tank =
$= \dfrac{\text{percent change in output}}{\text{percent change in input}}$,
τ_0 = period of oscillation of the disturbance, and
τ_1 = first-order time constant of the tank; approximately equal to (tank volume/flow through the tank − system dead time).

To visualize the effect of dynamic gain, consider a flowing stream whose pH falls from 7 to 4 and returns to 7 in one minute. If the stream flowed through a tank with one minute retention time (volume/flow), the spike in pH would pass through virtually unchanged, and the effluent pH would closely track the influent pH. If, however, the stream flowed through a tank with 60 minutes retention time, practically no upset would be observed in the effluent pH because of the capacity effect of the large volume.

The period of oscillation, τ_0 of a typical composition process under closed loop control with an optimally tuned (controller settings adjusted to match the process it controls) three-mode controller can be approximated as a function of the system dead time.[3]

$$\tau_0 \simeq 4\,\tau_{dt} \qquad\qquad 8.16(25)$$

Substituting for τ_1 from equation 8.16(23) and τ_0 from equation 8.16(25) into equation 8.16(24):

$$G_d = \frac{4\tau_{dt}}{2\pi(19.2\tau_{dt})} = 0.033$$

In this example the stirred tank has reduced the overall process gain by a factor of 30 (1/0.033). Two tanks used in series reduce the process gain (slow the process down) by the product of their individual gains. Assuming a second tank identical to the first, two tanks in series would reduce the process gain by a factor of 30^2, or 900. With the stirred tank, therefore, it is possible to reduce the process gain to a controllable level. An added benefit of an increased tank capacity is to smooth out high-frequency errors in reagent delivery caused by measurement noise.

This example is readily related to Figure 8.16u, in which the output of the reaction vessel is the input disturbance in the attenuation vessel. If the frequency or period (τ_0) of the input disturbance can be kept short (on the order of seconds) by virtue of a *tight* control loop around the reaction vessel, then the dynamic gain number of the attenuation vessel will be very low (0.033 for the example), thereby increasing its attenuation capability. This results in a stable effluent pH that averages the input disturbance.

REFERENCES

1. Lange, N.A., *Lange's Handbook of Chemistry*, 10th ed., McGraw-Hill, New York, 1967.
2. Shinskey, F.G., "How Difficult is pH Control?" Publication 230A, The Foxboro Co., Foxboro, MA., 1971.
3. ——, *pH and pIon Control in Process and Waste Streams*, John Wiley & Sons, New York, 1973.
4. —— and Myron, T.J., "Adaptive Feedback Applied to Feedforward pH Control," *Water and Waste Engineering*, February 1972.
5. Myron, T.J., "Guidelines for Effective pH Control System Design," *Instrumentation Technology*, January 1972.
6. Hoyle, D.L., "Designing for pH Control," *Chemical Engineering*, November 8, 1976.
7. Shinskey, F.G., *Process Control Systems*, 2nd ed., McGraw-Hill, New York, 1979.
8. Lipták, B.G., ed., *Instrument Engineers' Handbook*, Process Measurement volume, revised edition, Chilton Book Co., Radnor, PA, 1982.
9. Chapman, A.L., *Applications of Industrial pH Control*, Instrument Publishing Co., Pittsburgh, 1950.
10. McMillan, G.K., *Tuning and Control Loop Performance*, Instrument Society of America Monograph, 1983, Chapter 7. Research Triangle Park, North Carolina.

BIBLIOGRAPHY

Arant, J.B., "Applying Ratio Control to Chemical Processing," *Chemical Engineering*, September 18, 1972, p. 155.

Babcock, R.H., "Industrial Waste Instrumentation, a Contemporary Engineering Challenge," ISA Conference, Sarina, Ontario, Canada, 1968.

Buckley, P.S., *Techniques of Process Control*, John Wiley & Sons, New York, 1964.

Coughanowr, D.R. and Koppel, L.B., *Process Systems Analysis and Control*, McGraw-Hill Book Co., New York, 1965.

Davalloo, P., "Neutralizing Industrial Wastes," *Instrumentation Technology*, September 1979.

Dunnigan, A.R. and Dennis, R.A., Control System for a Very Wide Range of pH Effluent Stream, The Foxboro Company, Foxboro, MA, 1972.

Field, W.B., "Design of a pH Control System by Analog Simulation," *ISA Journal*, January 1959.

Fraade, D.J., "Using a Microprocessor to Solve pH Control Problems," *Instrumentation Technology*, November 1978.

Gates, L.E., et al., "Liquid Agitation—C.E. Refresher," *Chemical Engineering*, December 1975 to December 1976.

Harriot, P., *Process Control*, McGraw-Hill Book Co., New York, 1964.

Hoffmann, F., "How to Select a pH Control System for Neutralizing Waste Acids," *Chemical Engineering*.

Hoyle, D.L., "The Effect of Process Design on pH and pIon Control," 18th ISA-AID Symposium, San Francisco, CA, 1972.

Iverson, A.A., "Controlling Effluent pH," *Instrumentation Technology*, May 1977.

—— and Orchiai, S., "A Scheme for Effective Control of Effluent pH," ISA Conference, Houston, TX, 1976.

Jungck, P.R. and Woytowicz, E.T., "Understanding pH Control," *Instrumentation Technology*, November 1971, pp. 43–46.

Lopez, A.M., "Optimization of System Response," Ph.D. Thesis, Louisiana State University, Baton Rouge, LA, 1968.

Mellichamp, D.A., Coughanowr, D.R., and Koppel, L.B., "Characterization and Gain Identification of Time Varying Flow Processes," *AIChE Journal*, January 1966, pp. 75–82.

——, "Identification and Adaptation in Control Loops with Time Varying Gain," *AIChE Journal*, January 1966, pp. 83–89.

Meyer, J.R., Whitehouse, G.D., Smith, C.L., and Murrill, P.W., "Simplifying Process Response Approximation," *Instruments and Control Systems*, December 1967, p. 76.

McAvoy, T.J., "Time Optimal and Ziegler-Nichols Control, Experimental and Theoretical Results," *Industrial and Engineering Chemistry Process Design and Development*, Vol. 11, No. 1, 1972.

Nowroozi, M., "Process Control and Dynamic Behavior of Industrial Acidic Sumps," M.S. Thesis, Pahlavi University, Shiraz, Iran, 1975.

Prasad, C.C., "A Dual-loop Predictor for pH Control," *Instrumentation Technology*, July 1975.

Rowton, E.E., "Sampled-Data Control of pH," *Instrumentation Technology*, June 1968.

Shinskey, F.G., "Adaptive pH Controller Monitors, Nonlinear Processes," *Control Engineering*, February 1974.

——, "A Self-Adjusting System for Effluent pH Control," The Foxboro Company, Foxboro, MA, 1972.

——, "Feedforward Control of pH," *Instrumentation Technology*, June 1968.

Stansbury, J.E., "A Good, Tight pH Control," *Instruments and Control Systems*, August 1980.

Wilson, H.S. and Wylupek, W.J., "Design of pH Control Systems," *ISA Journal*, July 1965, pp. 41–46.

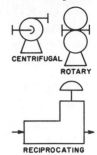

8.17 PUMP CONTROLS

The first part of this section is devoted to the topic of conventional pump controls, and the second part deals with pumping system optimization. Other related topics are discussed in the following sections:

A general discussion of pumps, pumping stations, metering pumps, and variable-speed drives can be found in Section 5.8 in this volume.

In the Process Measurement volume of this handbook, there is an even more detailed treatment of the subjects of metering pumps and variable-speed drives in Sections 2.11 and 9.12, respectively.

Control of Pumps

Capacity control of pumps involves the incompressibility of liquids, because changes in the volume rate of flow throughout the system occur simultaneously and equally, and density is constant at constant temperature, regardless of pressure. Pump capacity may be affected by (1) a control valve in the discharge of a pump, (2) on-off switching, (3) variation in the speed of the pump, or (4) stroke adjustment. Flow control by on-off switching provides only zero flow or full flow, whereas the other control methods provide continuously adjustable flows in the system. The applicability of these four methods of capacity control is related to the pump type, such as centrifugal, rotary, or reciprocating. The possible types of capacity controls for the various pumps are summarized in Table 8.17a.

Centrifugal Pumps

The centrifugal pump is the most common type of process pump, but its application is limited to liquids with viscosities up to 3000 centistokes (0.003 m^2/s). The capacity-head curve is the operating line for the pump at constant speed and impeller diameter. Three types of curves are shown in Figure 8.17b, illustrating various relationships of capacity to discharge pressure. The capacity varies widely with changes in discharge pressure for all curves, but the shape of the curve determines the type of control that may be applied.

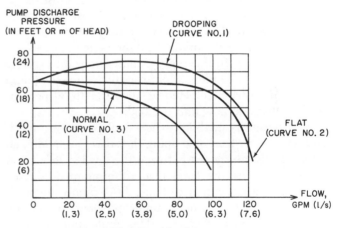

Fig. 8.17b Centrifugal pump curves

For on-off switching control, curves #1 and #2 are satisfactory as long as the flow is above 100 GPM (6.3 l/s). Below this flow rate, curve #1 allows for two flows to correspond to the same head, and curve #2 may drop to zero flow to obtain a small head increase. Both are therefore unstable in this region. Curve #3 is stable for all flows and is best suited for throttling service in cases in which a wide range of flows is desired.

The following process loops illustrate typical control applications of centrifugal pumps for both on-off and throttling control.

Table 8.17a
PUMP CONTROL METHODS

Method of Control	Possible Types of Control	
	On-Off	*Throttling*
On-Off Switch	Centrifugal, rotary, or reciprocating	
Throttling Control Valve		Centrifugal or rotary
Speed Control		Centrifugal, rotary, or reciprocating
Stroke Adjustment		Reciprocating

On-Off Level Control

Figure 8.17c illustrates the use of the level switch for on-off pump control. In order to prevent overheating of the motor, the number of pump starts should not exceed 15 per hour.

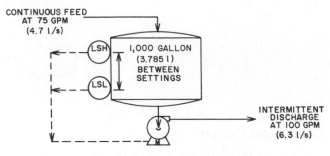

Fig. 8.17c   On-off level control

In connection with Figure 8.17c this can be checked by calculation of the filling and emptying times. At a feed rate of 75 GPM (4.7 l/s), it takes 13 minutes to fill the 1000-gallon (3785 l) volume between LSL and LSH, when the discharge pump is off. The pump capacity being 100 GPM (6.3 l/s), it takes 40 minutes to discharge this same volume while the feed is on. Therefore the pump will start approximately once an hour and will run 75 percent of the time.

Therefore the 1000-gallon (3785 l) tank is satisfactory. The switch prevents tank overflow and dry pumping, which would damage the seals. Since the pump starts once an hour, it will not overload the motor by excessive starts and stops. In this type of application, the capacity of the tank is the independent variable, since the pump capacity is usually selected at about 25 to 30 percent greater than the flow into the tank.

This design is modified when dual pumps are used. In this case, instead of keeping the spare pump idle, an alternator is interposed between the level switches and the motors. The alternator places the pumps in service in an alternating sequence. Thus, if two pumps are used, each will start half as many times per hour, permitting the tank size to be reduced. This application is shown in Figure 8.17d, where a cooling water return sump is illustrated. The minimum sump size is to be found allowing 10 starts per pump per hour. In this system, each pump is designed to handle the total feed flow into the

sump by itself. The pumps do not operate together except when abnormally high flow rates are charged into the sump.

The design is for 10 starts per hour for each pump. Because there are two pumps, the sump capacity can allow 20 starts per hour, or 3 minutes per empty and refill cycle. Let x = sump capacity. The minimum sump capacity is calculated by allowing the sum of feed and discharge times to equal 3 minutes.

$$\frac{x}{4000} + \frac{x}{1000} = 3 \text{ minutes}$$

$$x = 2400 \text{ gallons } (9.09 \text{ m}^3)$$

In addition to the reduction of sump volume, a further advantage is that a spare pump is available for emergency service. The purpose of a LSHH level switch is to put both pumps on line at once should the continuous flow exceed 5000 GPM (0.32 m³/s), resulting in the rise of the level to this point. The combined pump flow of 10,000 GPM (0.63 m³/s) will then prevent overflow of the sump. When the level drops to LSL, both pumps will be off, and the normal alternating cycle is resumed.

On-Off Flow Control

Illustrated in Figure 8.17e is a tandem pump arrangement that responds to varying flow demands on pump outlet. Pump I is normally operating at point (1) (at 80 GPM and 36 ft, or 5 l/s and 10.8 m). When flow demand increases to 120 GPM (7.6 l/s), the head drops to 22 ft at point (2), and FSH starts pump II. the combined characteristic gives 120 GPM at 40 ft (7.6 l/s at 12 m) at point (3). In this control scheme, a wide range of flows is possible at high pump efficiency without serious loss in pressure head.

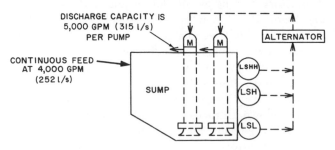

Fig. 8.17d   On-off level control of dual pump station

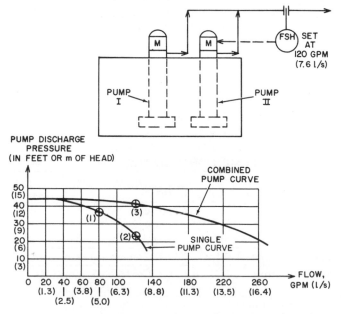

Fig. 8.17e   On-off flow control of pumps

On-Off Pressure Control

A pressure switch may be used to start a spare pump in order to maintain pressure in a critical service when the operating pump fails. In this case, a low pressure switch would be used to actuate the spare pump, piped in parallel with the first pump. A second function, as illustrated in Figure 8.17f, is to boost pressure. Pump I is normally operating at point (1). When the discharge pressure rises to 50 ft (15 m) at point (2), the flow is reduced from 53 to 20 GPM (3.3 to 1.3 l/s). At this point the pressure switch will start pump II and close the bypass valve. The system will now operate at point (3) on the combined characteristic, delivering 60 GPM at 50 ft (3.8 l/s at 15 m) of head.

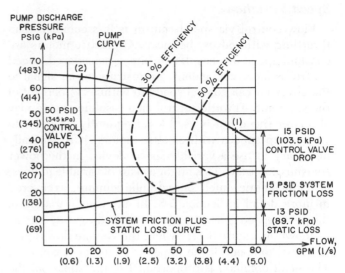

Fig. 8.17g Throttling control of centrifugal pump

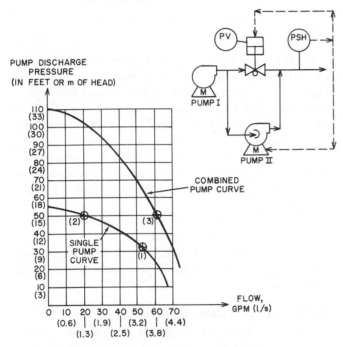

Fig. 8.17f On-off pressure control of pumps

Throttling Control

Throttling control may be achieved by use of control valves. In Figure 8.17g, design point (1) is near the maximum efficiency of the pump. Therefore, when throttling to point (2), the efficiency will drop, but the pump will still be stable.

For good controllability, the control valve is usually sized to pass the design flow with a pressure drop equal to the system dynamic friction losses excluding the control valve but not less than 10 PSID (69 kPa) minimum. (For more details on assigning sizing pressure drops to control valves, refer to Section 4.13). Pump flow is controlled by varying the pressure drop across the valve. This relationship is shown at point (1). When the flow is throttled to 15 GPM (0.95 l/s) at point (2), the control valve must burn up a differential of 50 PSID (345 kPa). However, the pump must not be run at zero flow or

overheating will occur and the fluid will vaporize, causing the pump to cavitate. To avoid this, a bypass line can be provided with a back-pressure regulator. It is set at a pressure that will guarantee minimum flow as the pump is throttled toward zero flow.

A typical process loop on flow control is shown in Figure 8.17h. The rangeability of the control valve is assumed to be 25:1 (see Section 4.13 for details). Thus, if the maximum flow required is 70 GPM (4.4 l/s) through the flow control valve (Figure 8.17h), then about 3 GPM (0.2 l/s) would be the minimum controllable flow. If lower flows are anticipated, then a second flow control valve should be installed in parallel with the first.

One can calculate the minimum flow needed through the pump to prevent vaporization from a heat balance on the pump by assuming that the motor horsepower is converted to heat. If this flow were calculated to be 20 GPM (1.3 l/s), then the PCV in the bypass line would be sized to pass 20 GPM with a corresponding set pressure of 63 PSIG (435 kPa). These same principles apply to process loops, in which pump capacity is varied to maintain the level, pressure, or temperature of other pieces of equipment.

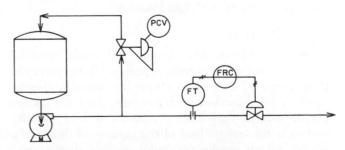

Fig. 8.17h Throttling control with pressure kickback

Speed Variation

Flow control via speed control is less common than throttling with valves, because AC electric motors are constant-speed devices. If a turbine is considered, speed control is more convenient. An instrument air signal to the governor can control speed to within $\pm\frac{1}{2}$ percent of the set point. Throttle control on a gasoline engine may also alter speed but is used less frequently. In order to vary pump speeds with electric motors, it is generally necessary to use a variable-speed device in the power transmission train. This might consist of variable pulleys, gears, magnetic clutch, or hydraulic coupling, as covered in more detail in Section 5.8. In any case, variation of the pump speed generates a family of head-capacity curves, as shown in Figure 8.17i, where the volume flow is proportional to speed if the impeller diameter is constant. The head obtained is proportional to the square of the speed. The intersection of the system curve with the head curve determines the flow rate at point (1), (2), or (3).

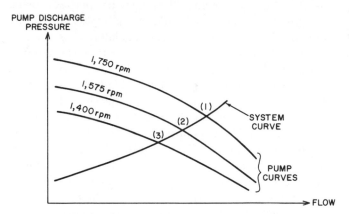

Fig. 8.17i   Centrifugal pump with speed variation

Rotary Pumps

The typical pump characteristics for a rotary pump, such as the gear, lobe, screw, or vane type, show a fairly constant capacity at constant speed with large changes in discharge pressure. This is shown in Figure 8.17j. These pumps cover the viscosity range from less than 1 centipoise up to 500,000 centipoises (0.001 Pa·s up to 500 Pa·s). The usual application of this type of pump is for the highly viscous liquids and slurries that are beyond the capabilities of centrifugal pumps.

On-Off Control

The operation of rotary pumps with on-off control is similar to that with centrifugal pumps. The criterion for the maximum number of starts per hour must be checked carefully for motors on rotary pumps, since the starting torque may be very large as a result of fluid viscosity, and so is the inertia load of the column of fluid in the piping, which accelerates under positive displacement each time the pump starts.

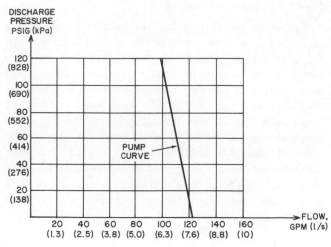

Fig. 8.17j   Rotary pump curve

In slurry service, on-off control creates problems caused by settling of solids, and it is therefore not recommended. Instead, a circulating loop is used with a pressure-controlled bypass back to the feed tank. For example, intermittent flow to feed a centrifuge is obtained by opening an on-off valve via a signal from the cycle timer. Such a loop is shown in Figure 8.3g. The pressure-controlled bypass allows the normal pump flow to be maintained in the loop when the centrifuge feed valve is closed.

The on-off control in this case is applied to the fluid rather than to the pump motor.

Manual on-off control is often applied to rotary pumps in bulk storage batch transfer services with local level indication.

Safety and Throttling Controls

A safety relief valve is always provided on a rotary pump to protect the system and pump casing from excessive pressure should the discharge line be blocked while the pump is running. The relief valve may discharge to pump suction or to the feed tank. In cases when slurries and viscous materials may not be able to pass through the relief valve, a rupture disk is placed on the discharge line.

The output of a rotary pump may be continuously varied to suit process demands by use of a pressure-controlled bypass in combination with a flow control valve. The bypass is necessary to accommodate changes in flow to the process, since the total flow through the bypass plus to the process is constant at constant speed.

The capacity of a rotary pump is proportional to its speed, neglecting the small losses due to slippage. The flow can therefore be controlled by speed-modulating devices, which are fully described in Section 5.8.

In this case, no bypass is needed. The rangeability is limited by the speed-control device, which for pulleys and magnetic drives is approximately 4 to 1. The response is slightly slower than with a control valve because of the inertia of the system; however, this type of control would be favored when it is desired to avoid the use of control valves on slurry or gummy services. Such installation is shown in Figure 8.17k, where a screw pump feeds latex slurry to a spray dryer.

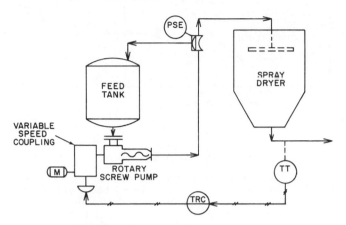

Fig. 8.17k   Speed control of rotary pumps

Reciprocating Pumps

Reciprocating pumps, such as the piston and diaphragm types, deliver a fixed volume of fluid per stroke. The control of these pumps is based on changing the stroke length, changing the stroke speed, or varying the interval between strokes. In all cases, the discharge from these pumps is a pulsed flow, and for this reason it is not suited to control by throttling valves. In practice, the volume delivered per stroke is less than the full stroke displacement of the piston or diaphragm.

This hysteresis is a result of high discharge pressures or high viscosity of the fluid pumped. Under these conditions, the check valves do not seat instantaneously. A calibration chart must therefore be drawn for the pump under actual operating conditions. A weigh tank or level-calibrated tank is usually the reference standard. Since the discharge is a pulsed flow, it must be totalized and divided by the time interval to get average flow rate for a particular speed and stroke setting. Metering inaccuracy is approximately ±1 percent of the actual flow with manual adjustment and ±1.5 percent with automatic positioning. Methods of stroke and speed adjustment are covered in detail in Section 5.8, and other features of metering pumps are discussed in Section 2.11 in the Process Measurement volume of this handbook.

On-Off Control

Once the pump has been calibrated, it can be programmed via a timer or counter to deliver a known volume of fluid to a process. The pump may deliver one full stroke and then stop until the next electrical signal is received from the timer or may continuously charge a specified number of pump strokes and then be shut

down by a counter. The flow may be smoothed by a pulsation dampener and the system and pump protected from over-pressure by a relief valve.

Throttling Control

Continuously variable flow control at constant speed may be accomplished by automatic adjustment of the stroke length. The range of flow control by stroke adjustment is zero to 100 percent. However, in order to maintain accuracy, the practical range is 10 to 100 percent of design flow. The flow is related to stroke length through system calibration.

In some applications the reciprocating pump combines the measuring and control functions, receiving no independent feedback to represent flow. In other cases it is used as a final control element, except that flow detection is performed by an independent sensor. Figure 8.17l illustrates an installation in which the pump is both the measuring and the control device for ratio control and is provided with automatic calibration capability.

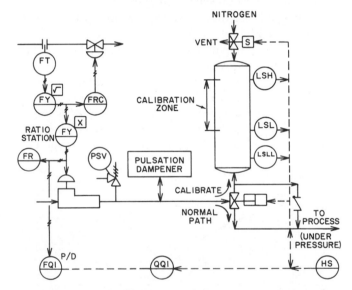

Fig. 8.17l   Ratio and calibration controls for reciprocating pump. When level has reached LSH, the three-way valve returns to the "normal" path and nitrogen enters the tank to initiate discharge. When level drops to LSLL, discharge is terminated by venting off the nitrogen. Counter QQI is running while rising level is between LSL and LSH. Total count, when compared with known calibration volume, gives total error. Hand switch HS initiates calibration cycle by diverting the three-way valve to the "calibrate" path.

Variable speed control is usually applied in multiple-head pumps in which all the pumps are coupled mechanically. A control signal may adjust all flows in the same proportion simultaneously. The rangeability and accuracy of flow control depend on the method of speed variation chosen.

Control Valves Versus Variable-Speed Pumps

When transporting or distributing fluids in control conventional systems, constant-speed pumps were used with kickback control valves serving to bypass that excess water, which should not have been pumped in the first place. Variable-speed pumping eliminates this waste, because only as much water is being pumped as is demanded by the load.

The main source of savings through such optimization is from transporting only as much water as is needed and thereby eliminating all energy waste from bypass or three-way valve throttling.

The power consumption of a pump is related to the product of flow and head, divided by efficiency. The size of the pump motor can be approximated as:

$$HP = \frac{(GPH)(PSI)}{100,000}$$

Pump selection is affected by the nature of the load. All system head curves are parabolas ($H \sim Q^2$), but they differ in the steepness of these curves and in the ratio of static head to friction drop. As shown in Figure 8.17m, the value of variable-speed pumping increases as the system head curve becomes steeper.

Studies indicate that on *mostly friction* systems, such as zone 4 in Figure 8.17m, the savings represented by variable-speed pumping will increase with reduced pump loading.[1] If on the yearly average the pumping system operates at *not more than 80 percent* of design capacity, the installation of variable-speed pumps will result in a payback period of approximately three years. As pump loading drops, the benefits of variable-speed pumping will increase as shown by Figure 8.17n. The energy sav-

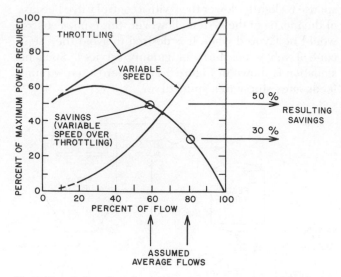

Fig. 8.17n   Savings through variable-volume water transportation. (From Liu, T., "Controlling Pipeline Pumps for Energy Efficiency," *InTech*, June 1979.)

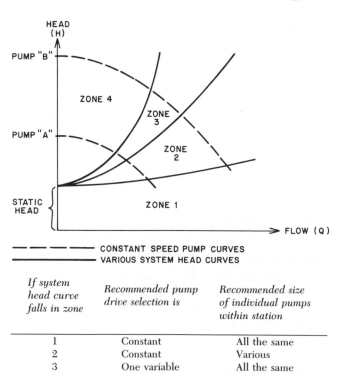

If system head curve falls in zone	Recommended pump drive selection is	Recommended size of individual pumps within station
1	Constant	All the same
2	Constant	Various
3	One variable	All the same
4	All variable	Various

Fig. 8.17m   Pump selection

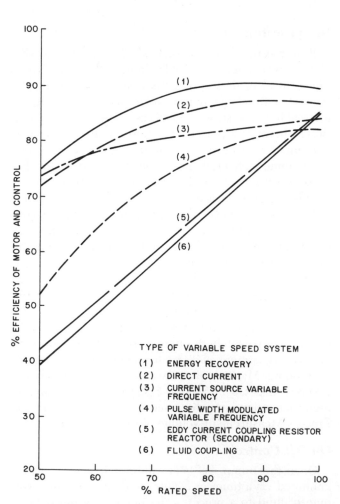

Fig. 8.17o   Energy efficiency of pump and fan motor and control combinations, for representative units 150 hp and larger. (From Schroeder, E.G., "Choose Variable Speed Drives for Pump and Fan Efficiency," *InTech*, September 1980.)

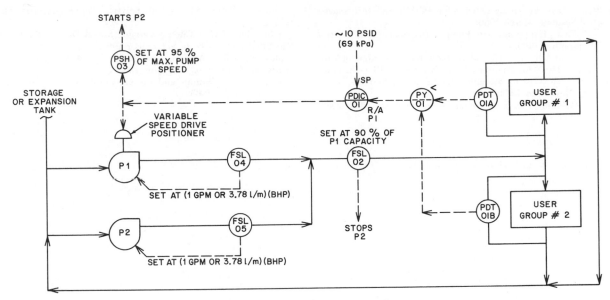

Fig. 8.17p   Optimized pump controls

Optimized Control Systems

Figure 8.17p illustrates the instrumentation require-ments of an optimized, variable-speed pumping system. PDIC-01 maintains a minimum of 10 PSID, or 69 kPa, (adjustable) pressure difference between supply and re-turn from each user. Therefore, no user will be denied more fluid if its control valve opens up further. On the other hand, if demand drops, PDIC-01 will slow down the variable-speed pump to keep the differential at the users from rising beyond 10 PSID (69 kPa).

When the variable-speed pump approaches its maxi-mum speed, PSH-03 will automatically start pump #2. When the load drops down to the set point of FSL-02, the second pump is stopped. One important feature to remember in connection with multiple-pump station controls is that extra increments of pumps are started by *pressure*, but they are stopped by *flow* controls.

FSL-04 and 05 are safety devices that will protect the pumps from overheating or cavitation, which might occur if the pump capacity is very low. As long as the pump flow rate is greater than 1 GPM (3.78 l/m) for each pump break horsepower, the operation is safe. If the flow rate drops below this limit, the pumps are stopped by the low flow switches.

Variable-speed pumping will never completely elim-inate the use of control valves. However, their use will increase, because they do represent a more energy-ef-ficient method of fluid transportation and distribution than did the constant-speed pumps used in combination with control valves. Therefore, if the fluid is being dis-tributed in a mostly friction system and the yearly av-erage loading is under 80 percent of pump capacity, the

installation of variable-speed pumps will reduce oper-ating costs.

REFERENCES

1. Langfeldt, M.K., "Economic Considerations of Variable Speed Drives," ASME Paper 80-PET-81, 1980.
2. Liu, T., "Controlling Pipeline Pumps for Energy Efficiency," *InTech*, June 1979.
3. Schroeder, E.G., "Choose Variable Speed Drives for Pump and Fan Efficiency," *InTech*, September 1980.

BIBLIOGRAPHY

Baumann, H.D., "A Case for Butterfly Valves in Throttling Appli-cations," *Instruments and Control Systems*, May 1979.
——, "Control Valves vs. Speed Controlled Pump," Texas A&M Symposium, 1981.
——, "How to Assign Pressure Drop Across Liquid Control Valves," Proceedings at 29th Annual Symposium on Instrument Engineering for the Process Industry, Texas A&M University, January 1974.
Bower, J.R., "The Economics of Variable-Speed Pumping with Speed-Changing Devices," *Pump World*.
Fischer, A.K., "Using Pumps for Flow Control," *Instruments and Control Systems*, March 1983.
Gottliebson, M., "Explore the Use of Variable Speed Water Booster Pumps," *Water and Wastes Engineering*, May 1978.
Hall, J., "Motor Drives Keep Pace with Changing Technology," *Instruments and Control Systems*, September 1982.
——, "Pump Primer," *Instruments and Control Systems*, April 1977.
Holland, F.A., "Centrifugal Pumps," *Chemical Engineering*, July 4, 1966.
Janki, C., "What's New in Motors and Motor Controls?" *Instruments and Control Systems*, November 1979.
Lipták, B.G., "Costs and Benefits of Boiler, Chiller, and Pump Optimization," *Instrumentation in the Chemical and Petroleum Industries*, vol. 16, Instrument Society of America, Research Triangle Park, NC, 1980.
——, "Save Energy by Optimizing Your Boilers, Chillers, Pumps," *InTech*, March 1981.

Merritt, R., "Energy Saving Devices for AC Motors," *Instruments and Control Systems*, March 1980.

———, "What's Happening with Pumps," *Instruments and Control Systems*, September 1980.

Papez, J.S. and Allis, L., "Consideration in the Application of Variable Frequency Drive for Pipelines," ASME Paper 80-PET-78, 1980.

Rishel, J.B., "The Case for Variable-Speed Pumping Systems," *Plant Engineering*, November 1974.

———, "Matching Pumps to System Requirements," *Plant Engineering*, October 1975.

———, "Water System Head Analysis," *Plant Engineering*, October 1977.

Schroeder, E.G., "Choose Variable Speed Devices for Pump and Fan Efficiency," *InTech*, September 1980.

Stewart, R.F., "Applying Adjustable Speed AC Drives," *Instruments and Control Systems*, July 1981.

Thurlow, C., "Pumps and the Chemical Plant," *Chemical Engineering*, May 24, 1965.

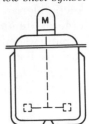

8.18 REACTOR CONTROL AND OPTIMIZATION

Two types of reactors are used in chemical plants: continuous reactors and batch reactors. Continuous reactors are designed to operate with constant feed rate, withdrawal of product, and removal or supply of heat. If properly controlled, the composition and temperature can be constant with respect to time and space. Figure 8.18a illustrates the flow control of multi-phase reactants based on reactor inventory. In batch reactors, measured quantities of reactants are charged in discrete quantities and allowed to react for a given time, under predetermined controlled conditions.

Temperature Control

Reaction temperature is frequently selected as the controlled variable in reactor control. It may be necessary to control reaction rate, side reactions, distribution of side products, or polymer molecular weight and molecular weight distribution. All of these are sensitive to temperature. It is frequently necessary to control reaction temperature to within $\frac{1}{2}°F$ (0.28°C). Many reactions are exothermic. In order to control reaction temperature, the released heat must be removed from the system as it is liberated by the reactants. A simple temperature control scheme is depicted in Figure 8.18b. The reaction temperature is sensed, and flow of heat transfer medium to the reactor jacket is manipulated.

In case of a large number of installations this scheme is considered to be unsatisfactory because, under throttled conditions, the flow of heat transfer medium can be inadequate to maintain a good heat transfer coefficient, and because the temperature gradient in the heat trans-

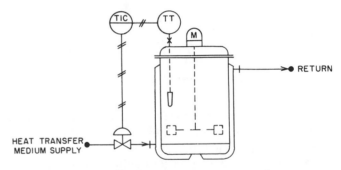

Fig. 8.18b   Reactor temperature control

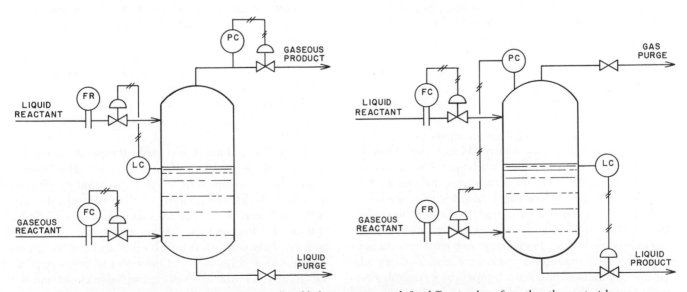

Fig. 8.18a   One reactant can be automatically added as it is consumed if it differs in phase from the other materials.
Note: In Figures 8.18a through h, all control loops are shown as conventional pneumatic loops. The intent is to represent reactor control concepts in the simplest form.

fer medium across the jacket may be large enough to keep different areas of the jacket heat transfer surface at different temperatures. This can result in localized temperature differences within the reactor (hot and cold spots), that are both uncontrollable and undesirable.

A more desirable arrangement is shown in Figure 8.18c. In this scheme, the liquid heat transfer medium is recirculated at a high rate through the jacket by way of an external pumping loop. The fluid velocity in the reactor jacket is maintained high enough to produce satisfactory film coefficients for heat transfer. In addition, a sufficient volume of liquid is circulated to keep the temperature gradient in the heat transfer medium as it passes through the jacket at a low enough level to maintain the jacket wall temperatures uniform throughout the reactor.

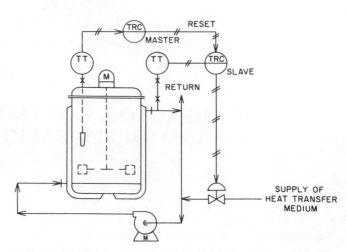

Fig. 8.18d   Cascade temperature control of reactor with recirculation

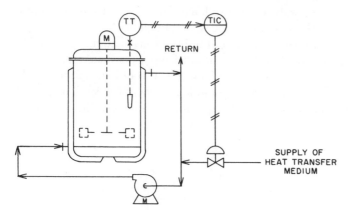

Fig. 8.18c   Reactor temperature control with recirculation

Both of these temperature control systems suffer from deficiencies that relate to time lags inherent in most reactor systems. First, there is a time lag in the response of the loop of the heat transfer medium in adjusting the temperature of the cooling (or heating) medium. Second, there is a time lag resulting from the physical mass of the reactor itself and from the heat load imposed on the cooling system to readjust the reactor temperature. Third, a very significant time lag is caused by the reactant mass and by the relatively large quantity of heat that must be removed to bring about a small change in reactant temperature level. Because of these process lags, a simple temperature control system tends to overcompensate for system disturbances. Each time an upset occurs, there can be cycling and poor control before the controller compensates for the system disturbance. Usually the period of this oscillation is several times the reactor heat transfer time lag. When the controller is properly adjusted, there still may be three cycles of oscillation before the product temperature returns to its control point. Periods of cycling up to one hour can exist, resulting in poor reactor temperature control for extended periods. A superior method of reactor temperature control, a cascade loop, is depicted in Figure 8.18d. Here the controlled process variable (reactor batch temperature), whose response is slow to changes in the heat transfer medium

flow (manipulated variable), is allowed to adjust the set point of a secondary loop, whose response to changes is rapid. In this case, the reactor batch temperature controller varies the set point of the jacket temperature control loop.

An essential feature of a successful cascade control system installation is that the secondary control loop should be able to correct for disturbances in the heat transfer medium source without allowing its effects to be felt by the master controller. For example, a change in cooling water supply temperature is corrected for in the slave loop and is not allowed to upset the master controller. As pointed out in the detailed discussion of cascade systems in Section 1.7, the process lags should be distributed between master and slave loops in such a way that the time constant of the slave is one-tenth that of the master. Cascade loops will not function properly if the master is faster than the slave.

In most processes, a certain temperature has to be reached in order to initiate the reaction. In the case of a steam-cooling water system, the steam may be directly injected into the cooling water circulating loop by way of a ring heater or a steam-water eductor. Where other heat transfer fluids are involved, indirect means of heating the circulating heat transfer medium are used. A heat exchanger in the circulating loop is used to heat the fluid indirectly.

Figure 8.18e depicts a cascade temperature control system with provisions for batch heat-up. The heating and cooling medium control valves are split-range controlled, such that the heating medium control valve operates between the air signal values of 9 and 15 PSIG (0.6 and 1.0 bar) and the cooling medium control valve operates between 3 and 9 PSIG (0.2 and 0.6 bar). This is a fail-safe arrangement. In the event of instrument air failure, the heating medium control valve closes and the cooling medium control valve opens to provide emergency cooling for the batch.

Figure 8.18e also shows an arrangement whereby an

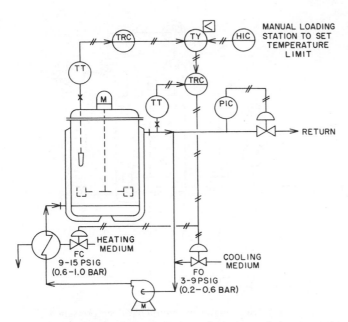

Fig. 8.18e Cascade temperature control system with heating and cooling capability

upper temperature limit is set on the recirculating heat transfer medium stream. This is an important consideration if the product is temperature-sensitive or if the reaction is adversely affected by high reactor wall temperature. In this particular case, the set point to the slave controller is prevented from exceeding a preset high temperature limit. Another feature shown is a back-pressure control loop in the heat transfer medium return line. This may be needed to impose an artificial system back pressure, so that during the heat-up cycle no water leaves the recirculation loop and therefore the pump does not experience cavitation problems.

The complexity and details of control for all types of heat exchangers are discussed in Section 8.13.

If the temperature control loops are tuned to give optimal control during the reaction phase (cooling cycle), then the control loop will not be sufficiently damped to prevent overshoot of set point during the heat-up period. If the loop is damped to minimize overshoot during heat-up, then control during reaction will suffer. Frequently, operators will manually control the approach to the temperature set point in order to prevent overshoot, which otherwise can result in undesirable product properties or in uncontrollable reaction rates. Instrumentation is available to permit rapid automatic heat-up without temperature overshoot, in conjunction with optimal reaction control.

One such system operates as an on-off controller during the heat-up cycle. When the set point has been approached within some small margin, the control is first reversed to temporary cooling to remove the thermal inertia from the system. After a brief period of full cooling, the loop is switched from on-off to PID control, where the three modes of the controller have already

been tuned for the dynamics of the cooling cycle. This method of temperature overshoot prevention is referred to as *dual mode control*. Other techniques for the prevention of reset windup are discussed in detail in Sections 2.3 and 2.6.

Sometimes it is necessary to control the reaction at several temperatures or to control the rate of temperature rise during a reaction. For this purpose, clock-actuated cams or other types of function generators are used to regulate the controller set point as a function of time. For a more detailed discussion of this hardware refer to Section 2.1.

Occasionally the design engineer must use his imagination in developing indirect temperature control systems. As an example of this, high-pressure polyethylene processes operate at pressures sufficiently high (e.g., 20,000 to 50,000 PSIG, or 138 to 345 MPa) that the resultant wall thickness of the tubular reactor is too thick to permit good temperature control because of the poor heat transfer through the reactor wall. In such a case, temperature control can be obtained by having reaction temperature control the catalyst flow to various points in the reactor, which in turn controls reaction rate.

In a process in which the reactor pressure is a function of temperature (e.g., the reactor pressure is essentially the vapor pressure of one of the major components in the reaction), this pressure may be sensed and used to control temperature.

Pressure Control

Certain types of chemical reactions require, in addition to temperature control, some form of pressure control. Typical of these reactions are oxidation and hydrogenation reactions, in which the concentration of oxygen and hydrogen in the liquid reactants and the consequent reaction rate are a function of pressure. The reaction rate is also a function of pressure in gas phase reactions. In high-pressure polyethylene polymerization, both the reaction rate and the resultant polymer properties are sensitive to reaction pressure.

Figure 8.18f depicts a batch reaction in which the process gas is wholly absorbed in the course of the reaction. Here, the concentration of process gas in the reactants is related to the partial pressure of the process gas over the reactants. Pressure may be sensed and controlled, thereby controlling the concentration of process gas in the reactants and the resultant reaction rate. In this mode of control, system pressure responds quickly to changes in controller output, and therefore a fairly narrow proportional band may be used. Derivative response is not required. In addition to pressure control, the reactor will require one of the previously discussed temperature control systems. Figure 8.18f is simplified in this regard.

Certain reactions not only absorb the process gas feed but also generate by-product gases. Such a process might involve the formation of carbon dioxide in an oxidation

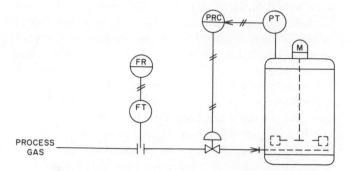

Fig. 8.18f   Reactor pressure control by modulating gas makeup

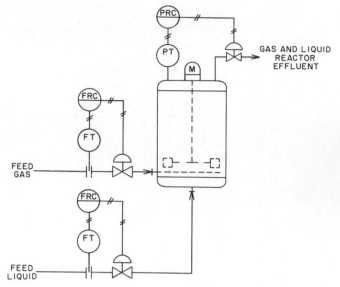

Fig. 8.18h   Continuous control of reactor pressure

reaction. Figure 8.18g illustrates the corresponding pressure control system. Here, the process gas feed to the reactor is on flow control, and reactor pressure is maintained by throttling a gas vent line. This particular illustration also shows a vent condenser, which is used to minimize the loss of reactor products through the vent.

In the case of a continuous reactor, a system such as is shown in Figure 8.18h is often used. Here the reactor is full of liquid, and both the reactor liquid and any unreacted or resultant by-product gases are relieved through the same outlet line. Reactor pressure is sensed, and the overflow from the reactor is throttled to maintain the desired operating pressure. Process gas feed and process liquid feed streams are on flow control.

All of these illustrations are simplified. For instance, it may be desirable to place one of the flow controllers on ratio in order to maintain a constant relationship between feed streams. If the reaction is hazardous and there is the possibility of an explosion in the reactor (e.g., oxidation of hydrocarbons), it may be desirable also to add safety devices, such as a high-pressure switch, to stop the feed to the reactor automatically.

Microprocessor-Based Controls

A batch process varies with time. The starting of a pump, the opening of a valve, the completion of charging of a particular material into the reactor, and the reaching

of the end point of a reaction are all events in time, requiring different control actions. In the design of a batch control system one must deal with time-based process conditions and transition phenomena.

The control system will frequently serve the creation of multiple products, accommodate idle times between batches or, in some instances, include automatic startup and shutdown procedures.

Sequential control encompasses all control functions (discrete and continuous/regulatory) triggered by events (time-based or process-based), whether normal or abnormal.

There are various ways of diagramming sequential control, such as ladder diagrams (change-oriented), matrix diagrams (status-oriented), and flow charts (decision flow–oriented).

An adequate compromise representation of sequential control is the time-sequence diagram, which is shown in Figure 8.18i. For a better understanding of the time-sequence diagram representation, a simple sequence is described in both ladder diagram form and time-sequence diagram form in Figure 8.18j.

The time-sequence diagram clearly shows the sequential changes in control actions and their time relationships. The sequence of events can be followed along dotted lines for vertical time coincidence and along the solid lines in a horizontal direction for sequence control.

The diamond symbol is used to indicate a trigger event, and a vertical dotted line indicates the time coincidence of trigger events. When two or more diamonds occur in the same relative time line an "AND" logic condition is assumed.

The format of the time-sequence diagram proceeds from left to right in discrete time steps with relative time being the horizontal coordinate.

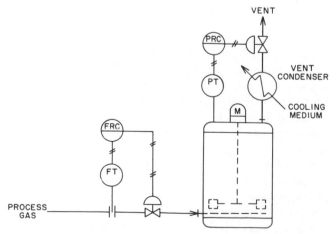

Fig. 8.18g   Reactor pressure control by throttling flow of vent gas

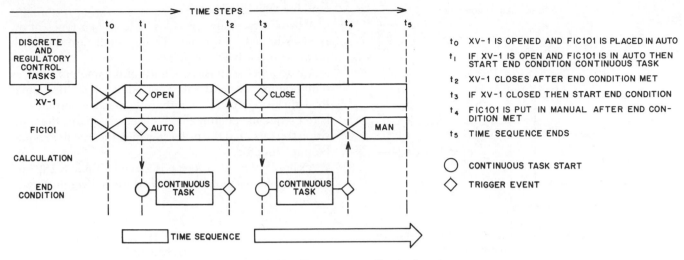

t_0 XV-1 IS OPENED AND FIC101 IS PLACED IN AUTO

t_1 IF XV-1 IS OPEN AND FIC101 IS IN AUTO THEN START END CONDITION CONTINUOUS TASK

t_2 XV-1 CLOSES AFTER END CONDITION MET

t_3 IF XV-1 CLOSED THEN START END CONDITION

t_4 FIC101 IS PUT IN MANUAL AFTER END CONDITION MET

t_5 TIME SEQUENCE ENDS

○ CONTINUOUS TASK START

◇ TRIGGER EVENT

Fig. 8.18i Time sequence diagram format

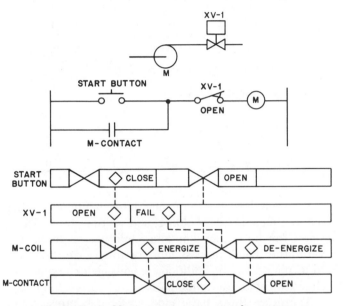

Fig. 8.18j Ladder versus time-sequence diagramming

Basic Discrete Control Functions

Discrete functions are developed, using hardware or software, to control discrete devices such as on/off valves, pumps, or agitators, based on status (ON/OFF) of equipment or values of process variables. The following types of discrete control are distinguished:

DIRECT ON/OFF CONTROL

This is the simplest form of discrete control that is used when direct human intervention is required to activate a device based on process information acquired from instruments and human judgment.

INTERLOCK CONTROL

This type of control allows automatic actuation of a particular device only if certain process conditions sensed by various instruments are met. The two categories of interlocks are safety and permissive interlocks.

Safety interlocks are designed to ensure the safety of operating personnel and to protect plant equipment. These types of interlocks are associated with equipment malfunction or shutdown.

In some situations locking devices can prevent personnel access into certain process areas if ambient conditions are hazardous. In other installations, equipment can be protected from mechanical damage by interlocks that will, for example, stop a pump when its discharge valve is closed. This way, the pump is protected from mechanical damage.

Permissive interlocks establish orderly startup and shutdown of equipment. For example, a conveyor feeding an elevator should not be started before the elevator is running, because accumulation of material may occur if the conveyor is operating before the elevator starts. A permissive interlock is provided to fulfill this requirement.

SEQUENTIAL CONTROL

A discrete device can have only two states, such as ON/OFF, OPEN/CLOSE, ENABLED/DISABLED. The triggers can be time-related or process-related (temperature, pressure, level, and so forth). Sequencing requires an end condition (trigger) to be reached before the system can proceed to the next step. In sequential control one trigger or a combination of triggers may determine a step transition.

Continuous/Regulatory Functions Related to Batch Processes

Changing Control Algorithms

An example of changing control algorithms while a particular batch is in progress is shown in Figure 8.18k. In this example, one of three reactor coolant fluids can be selected, depending on the state that the batch process is in at a particular time.

For instance, during a reactor cleaning it is not nec-

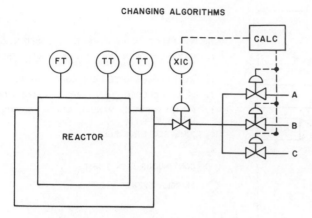

CHANGING ALGORITHMS

Fig. 8.18k   One of three coolant fluids is selected, depending on the reactor temperature. The valve for the coolant fluid is controlled by an algorithm based on the coolant selected.

essary to control temperature, and therefore the algorithm must be disabled. Whenever temperature control becomes necessary the algorithm is automatically reactivated.

Anti-Reset Wind-Up Function

Occasionally, in idle or disabled reactors the PID algorithm can accumulate the error until the controller output is saturated. This can be corrected by:

1. Setting limits that may turn the integral function on or off, according to the status of the batch process.
2. Switching the PID algorithm to manual, causing the set point to track the process variable. Whenever the PID algorithm is switched back to automatic the algorithm will start with a zero error and the integral function will start operating on real process conditions.
3. Ramping the set point up or down. This method is similar to the tracking method.

Unit Control

A process unit consists of a group of mechanical equipment; each piece performs, quasi-independently, a portion of the chemical process. Examples are filtration units, separation units, reactors, and distillation columns.

In batch process units the batch sequence is subdivided into *process states*. Each state is given a unique name, such as CHARGE, REACT, HEAT, COOL, HOLD, DISCHARGE, WASH, and EMPTY. Within each state, discrete control functions, continuous-regulatory control functions, and safety and permissive interlocks are performed.

It is important for one to consider the impact of common resources when designing a multi–batch process control system. For example, a solid material conveying system may be shared among several batch reactors or steam demand must be scheduled to satisfy a variable load consisting of several simultaneous batch processes.

Batch Procedure

A batch procedure provides at least two types of directives to make a product. These usually are the batch recipe and the batch sequence.

A batch recipe consists of a list of parameters, such as temperature set points, flow set points, operation time, total quantities of various ingredients, and profiles. A batch sequence defines all the necessary states, the transition triggers, and the order in which those states should be implemented.

A sample of a batch recipe is given in Table 8.18l, and a sample of a batch process procedure is given in Figure 8.18m.

Table 8.18l
BATCH RECIPE
Product XYZ

Parameter Name		Value	Units
Ingredient A	Grade	–	lb
Ingredient B	Generic	85	lb
Ingredient C	Generic	15	lb
React Temp	Grade	–	°F

	Grade		
Grade Data	Standard	Extra	Super
Ingredient A	10	15	30
React Temp	200	180	190

Between the active-sequence states mentioned earlier one can include idle, holding, and waiting states. These states can be required to allow information exchanges with other batch process units or to receive directives from operators. In addition, a fail-safe emergency or shutdown state can be defined.

There are three basic modes of batch process unit operation: manual, semi-automatic, and automatic.

In the manual mode, stepping through the batch sequence is done by the plant operator via specific commands. In the semi-automatic mode, the batch sequence is initiated by the operator, but then it proceeds automatically through all the states of the sequence until the operation ends. In the automatic mode, once a sequence is initiated, it can be repeated a predetermined number of times without any operator intervention.

When many batch process units are operating simultaneously in a plant it is necessary to track the status of each batch procedure and to schedule the common resources adequately to service all batch process requirements. A supervisory/scheduling program must be developed to accomplish this task. This program will require adjustment based on operating data collected by the plant control system.

The control engineer should design a reporting system that will generate adequate batch historical information

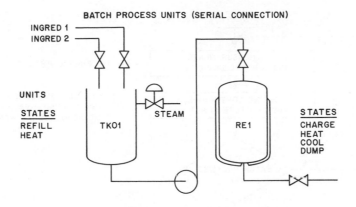

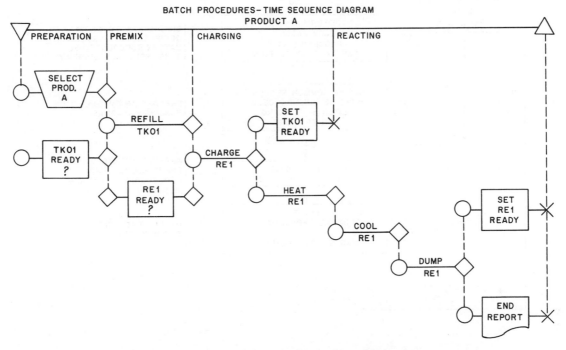

Fig. 8.18m   Batch procedure time sequence for a multi-product/single-stream process

Engineering a Batch Process Control Strategy

to permit future optimization of the entire process operation in order to produce each product at minimum possible cost and maximum through-put.

A step-by-step approach for engineering a batch control strategy is shown in Figure 8.18n and the associated Table 8.18o.

One starts with the basic engineering document, namely the *Process and Instrumentation-Control Diagram* (P&IC-D). On the P&IC-D the control engineer must identify the discrete and analog measurements and control devices (step 1).

In the next step (step 2), one must define all individual continuous (regulatory) and discrete functions using adequate symbology or developing control documents such as logic diagrams (ladder diagrams), boolean equations, and time sequence diagram(s), which must be referenced on the P&IC-D. All necessary calculations, such as ma-

terial and heat balances and recipe corrections based on quality control data, should also be defined.

In step 3, a batch procedure is determined for each batch process unit.

In step 4, the common resources are defined and the scheduling and tracking of the multiple batch operations are determined.

In step 5, an overall failure analysis must be done, with the plant considered in its entirety. Various adjustments of the previously designed strategies may be required.

In step 6, all the historical data trending and reporting necessary for adequate process management are designed and formatted.

Various microprocessor- and computer-based systems are specifically designed to perform the wide variety of control tasks involved in batch processes.

In the area of continuous (regulatory) control a fill-in-the-blank type language supported by CRT-based op-

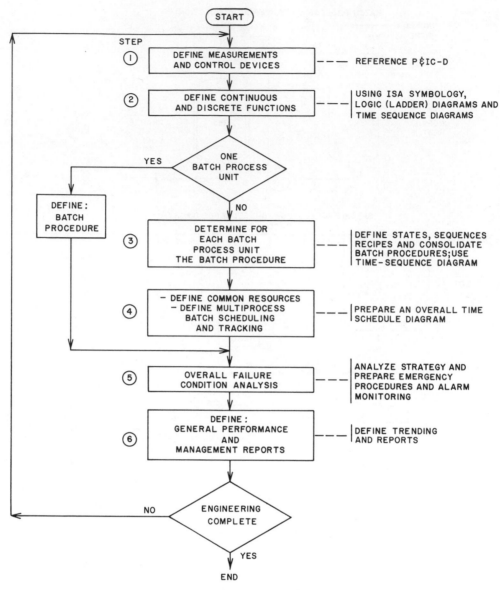

Fig. 8.18n   Batch process control step by step

erator consoles has become an accepted standard. The fill-in-the-blank language is used to configure the control system data base, to create the necessary control algorithms, and to generate the necessary displays (plant overviews, group and loop displays) for plant/operator interface.

The discrete functions (direct ON/OFF control, interlock control, sequence control) require *programming*.

Many programming languages available for generating control software are of the "high-level" type, meaning that the instructions are expressed in English. Various structures and procedures contained in the programming languages offered for control applications include:

Processing discrete inputs and outputs.
Timing and counting routines.
Special routines applied to startup electric motors

sequentially, in order to limit the inrush current imposed on the power supply.
Integration of analog variables.
Message triggering.
Transfer routines for synchronization of dual processors (whenever 100 percent redundancy is required).
Dynamic links of discrete variables to P&IC-D–type graphic displays for monitoring batch processes.
Interface routines for linking recipe data files with the main control program.

Process management functions are programmed using high-level languages that are well suited for data manipulations, such as BASIC and FORTRAN.

The application software development for a certain project is considered finished when all the programs are

Table 8.18o
TYPICAL SEQUENCE: CONTROL
TASKS AND EVENTS

1. Sequence start prerequisites and permissives (e.g., unit available, previous batch complete, interlocks satisfied)

2. Discrete element and devices (e.g., outlet valve, inlet valve, manifold, agitator)

3. Regulatory control loop set point changes, mode changes, output changes, profiles, etc. (A/M-PID algorithm, SP profile, output)

4. End conditions

5. Timers

6. Calculations (e.g., heat or mass balance)

7. Operator action request/response (e.g., clean, take laboratory analysis, check batch end)

8. Process failure conditions (e.g., level too high, pressure too low)

9. Failure actions (branch to alternative sequence)

debugged and are running within the constraints of the operating system and the hardware configuration.

Simulation

All the control programs must be tested under simulated plant conditions before the control system is connected to the process. A step-by-step simulation and check-out must take place in order to ensure the validity of all control strategies designed for a particular application.

This is done by connection of a hardwired simulator provided with analog input signals (DC source and individual potentiometers), analog output loads (DC meters), discrete input signals (AC/DC source and individual switches), and discrete output signals (AC/DC lights). The first test includes all continuous (regulatory) loops under open and closed conditions while the discrete controls are deactivated. Usually, open-loop testing (without process dynamics) is adequate for inherently stable fast processes. Whenever dynamic process conditions must be simulated, a program is required to generate it. Depending on the hardware configuration of the system, the control strategy and the model software can be running in the same processor or, if this is not possible, then an additional processor is required for the process model. This processor must communicate with the control system processor while the application software strategy and the process model program are running. The process variables of the models are monitored to determine the response and stability of algorithms to set point changes and simulated process load changes.

The next step is the simulation and checkout of the discrete functions, namely direct on/off control, interlock control, and sequential control. Using the simulator one manually activates the switches to imitate normal or abnormal conditions and checks the system response by watching the lights connected to the discrete outputs. A color CRT with P&IC-D graphics, if available, may replace the set of lights connected to discrete outputs.

If a very large number of discrete I/O is required in order to check a very complicated sequential control strategy, a simulator program can be generated, and the test can be conducted in the same manner as a dynamic test for continuous (regulatory) control described earlier. Simulation should be conducted on a module-by-module basis and then repeated while all application software programs are running.

In addition to the previously described simulation, when the system is fully installed in the field a test run is recommended prior to actual startup.

Reactor Models

The reactor is the heart of the chemical process. The rest of the process consists of the reactant preparation hardware and of the devices required for the separation and purification of the products or materials produced in the reactor. Thus, the plant efficiency hinges on the efficient operation of the reactor. Process reactors are known by different names in different processes: electrolytic cells, reformers, cracking furnaces, and catalytic crackers are some examples.

A mathematical model of the process reactor is the name we give to the group of equations that describe the operation of the reactor. To write these equations the control engineer must use his knowledge of thermodynamics, fluid dynamics, heat transmission, molecular diffusion, reaction kinetics, control theory, and economics. With the aid of modern digital computers he will need only to refresh his calculus and command of differential equations to use these equations in simulation.[1]

If writing the equations that describe the reactor is a science, using these equations to simulate the reactor on a computer is an art. As any art, it must be learned through practice. The detailed mathematical model of a process reactor will often consist of a large number of nonlinear equations. The art of simulation consists of picking out those equations that are relevant to the objectives of the simulation and of dropping those terms that contribute little to the answers. In addition, the engineer must determine what variables are to be solved for from what equations.

The type of process reactor determines the type of mathematical model. The models describing stirred tank reactors and plug-flow reactors will be presented in this section after a brief introduction to reaction kinetics. Following this discussion the modifications necessary to represent non-ideal flow, heterogeneous reactions, and catalytic reactors will be presented.

Reaction Kinetics

Reaction kinetics deals with the rate of reaction with respect to time. This rate depends on the concentration of the reactants and on temperature. For gas phase reactions the concentrations of the reactants are proportional to the pressure, and therefore the pressure also affects the rate of reaction. Consider the reaction

$$aA + bB \rightleftharpoons qQ + sS \qquad 8.18(1)$$

which means that a moles of A react with b moles of B to produce q moles of Q and s moles of S. A and B are therefore the reactants and Q and S are the products. The arrow pointing to the left indicates that the reaction is reversible, and if left reacting long enough, it will reach equilibrium when the rate of reaction to the right equals the rate of reaction to the left. If the reaction is irreversible, the reaction will stop when either reactant is totally consumed.

For homogeneous reactions the rate of reaction is expressed in terms of the rate of appearance of one of the reactants per unit volume of reacting mixture.

$$r_A = \left(\frac{1}{V}\right)\frac{dN_A}{dt} \qquad 8.18(2)$$

A homogeneous reaction is one that takes place in a single phase, whether liquid, solid, or gas. In equation 8.18(2), r_A is the rate of appearance of reactant A per unit volume, and N_A is the total moles of reactant A present. When A is disappearing, r_A must be a negative quantity. The rates of appearance of the other reactants and products can be expressed in terms of r_A since, from equation 8.18(1), for each mole of A that disappears b/a moles of B will also disappear, and q/a moles of Q will appear, etc.

The equations that express the rate of reaction r_A in terms of the concentrations of the reactants and products are called the *kinetic model* of the reaction. This model is obtained by proposing different models for the reaction and then choosing the one that can best predict the reaction rate data taken in the laboratory or pilot plant. Simulation with the laboratory apparatus is an excellent tool for testing the different models. A model for the reaction represented by equation 8.18(1) could be

$$r_A = -k\left(C_A^\alpha C_B^\beta - \frac{1}{K}C_Q^\gamma C_S^\sigma\right) \qquad 8.18(3)$$

where C_A and C_B are the concentrations of the reactants in moles per unit volume, and C_Q and C_S are the concentrations of the products. The reaction coefficient k is a function of temperature and must be determined experimentally. The equilibrium constant K is also a function of temperature and can be determined from thermodynamic data. The second term inside the parentheses in equation 8.18(3) is the rate of the reverse reaction and is zero for irreversible reactions. The exponents α, β, γ,

and σ are part of the kinetic model. When they are equal to a, b, q, and s, respectively, the reaction is said to be elementary. If the exponents differ from the coefficients of equation 8.18(1), the reaction is not elementary, meaning that the actual mechanism of the reaction consists of a series of intermediate elementary reactions that are summarized by equations 8.18(1) and 8.18(3).

The reaction rate coefficient k is usually an exponential function of the absolute temperature called the Arrhenius equation:

$$k = k_0 e^{-(E/RT)} \qquad 8.18(4)$$

where E is called the activation energy of the reaction, R is the ideal gas law constant (1.98 BTU/lb-mole °R) and T is the absolute temperature of the reacting mixture. The constants E and k_0 must be determined experimentally by conduction of the reaction at different temperatures. A kinetic model is essential for the simulation of a process reactor, and without it the simulation is worthless. However, the kinetic model[2] does not have to be as accurate for control simulation as it would have to be for the design of the reactor.

Perfectly Mixed Reactor

A flow diagram for a stirred tank (perfectly mixed reactor) is shown in Figure 8.18p. The reactor shown is a continuous one, since the reactants flow continuously into it and the product stream is continuously withdrawn. The mathematical model is essentially the same for a batch reactor in which the reactants are added at the beginning of the batch, "cooked" for a certain period, and then discharged.

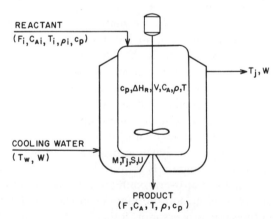

REACTANT
$(F_i, C_{Ai}, T_i, \rho_i, c_p)$

T_j, W

$c_p, \Delta H_R, V, C_A, \rho, T$

COOLING WATER
(T_w, W)

M, T_j, S, U

PRODUCT
(F, C_A, T, ρ, c_p)

Fig. 8.18p   Perfectly mixed reactor

As we develop the mathematical model, we shall make repeated use of the accounting equation:

Rate of accumulation = (rate in) − (rate out)   8.18(5)

This equation is applied to any quantity that is conserved, such as total mass, energy, momentum, and mass of components that do not participate in the reaction (i.e., solvent). If we include the rate of appearance by chemical reaction as one of the "rate in" terms, we can also write a balance on mass or moles of each of the reactants and reaction products. When applied to mass, equation 8.18(5) is called the *mass*, or *material, balance*; if to energy, then *energy balance*, and so on.

A total mass balance on the reactor in Figure 8.18p is

$$\frac{d}{dt}(V\rho) = F_i\rho_i - F\rho \qquad 8.18(6)$$

where d/dt represents the derivative, or instantaneous rate of change, with respect to time of the mass of reacting mixture, which is the product of its volume V and its density ρ. F_i and ρ_i are the volumetric rate of flow and the density, respectively, of material fed into the reactor, and F is the volumetric rate of flow of material out of the reactor. If the density of the reacting mixture is not changed appreciably by the reaction ($\rho = \rho_i$), then equation 8.18(6) can be further simplified:

$$\frac{dV}{dt} = F_i - F \qquad 8.18(7)$$

A material balance on reactant A is given by

$$\frac{d}{dt}(VC_A) = F_iC_{Ai} + r_AV - FC_A \qquad 8.18(8)$$

where C_{Ai} is the concentration of reactant A in the input stream and C_A is the concentration of A in the reactor and, since it is a perfectly mixed reactor, in the output stream. Similar balances can be made on each of the reactants and reaction products. Differentiation by parts of the left-hand side of equation 8.18(8) gives

$$\frac{d}{dt}(VC_A) = V\frac{dC_A}{dt} + C_A\frac{dV}{dt} \qquad 8.18(9)$$

and from equation 8.18(7),

$$\frac{d}{dt}(VC_A) = V\frac{dC_A}{dt} + C_A(F_i - F) \qquad 8.18(10)$$

Substituting into equation 8.18(8) and rearranging,

$$\frac{dC_A}{dt} = r_A + \frac{F_i}{V}(C_{Ai} - C_A) \qquad 8.18(11)$$

The ratio F_i/V is the reciprocal of the residence time of the reactor, which is also the reactor time constant.

If, as illustrated by Figure 8.18q, there were two inlet streams (F_{iA} and F_{iB}), with corresponding concentrations (C_{iA} and C_{iB}), the flow and concentration equations can be written as

$$dV/dt = F_{iA} + F_{iB} - F \qquad 8.18(12)$$

$$V\frac{dC}{dt} = F_{iA}C_{iA} + F_{iB}C_{iB} - FC \qquad 8.18(13)$$

An energy balance on the reactor, illustrated in Figure 8.18p gives

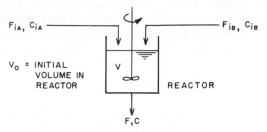

Fig. 8.18q Parameters involved in reactor–material balance calculations

$$\frac{d}{dt}(V\rho C_pT) = F_i\rho C_pT_i + Vr_A(-\Delta H_R) - \\ - US(T - T_j) - F\rho C_pT \qquad 8.18(14)$$

where T is the temperature of the reacting mixture, C_p is its specific heat or the amount of energy necessary to raise a unit mass by one degree of temperature, T_i is the temperature of the input stream, $-\Delta H_R$ is the heat liberated in the reaction per unit mass of A reacted, T_j is the temperature of the cooling water in the jacket, S is the effective heat transfer surface area of the jacket, and U is the heat transfer coefficient or rate of heat transfer to the jacket per unit area per degree of temperature difference. U is either experimentally determined or estimated from heat transfer correlations. Another simple application of differential calculus, with the help of equation 8.18(6), reduces equation 8.18(14) to

$$\frac{dT}{dt} = \frac{F_0}{V}(T_i - T) + \frac{-\Delta H_R}{\rho C_p}r_A - \\ - \frac{US}{V\rho C_p}(T - T_j) \qquad 8.18(15)$$

An energy balance on the jacket gives us an equation for T_j:

$$\frac{dT_j}{dt} = \frac{W}{M}(T_W - T_j) + \frac{US}{M}(T - T_j) \qquad 8.18(16)$$

where W is the mass flow rate of cooling water, M is the heat capacity of the water in the jacket and the walls of the reactor, and T_w is the cooling water inlet temperature. It has been assumed that the jacket is also perfectly mixed and that the specific heat of water is unity.

The pressure in the reactor can be calculated, if necessary, from thermodynamic equilibrium relationships, as a function of temperature and concentration. Equations 8.18(7), (11), (15), and (16), coupled with the reaction kinetic equations similar to 8.18(3) and (4), constitute the mathematical model of the system. The volume V, concentrations C_A, C_B, etc., and temperatures T and T_j are the dependent variables, since they depend on time, which is the independent variable. At steady state the dependent variables do not change with respect to time, and therefore the derivatives are equal to zero. In addition to the equations shown, the initial values of each of the dependent variables are necessary to complete the model (Figure 8.18q). The equations listed are to be programmed, on a computer, to be solved digitally by either numerical integration techniques or any of the so-called simulation languages, without any further manipulation.

Plug-Flow Reactor

The flow diagram of a plug-flow reactor is shown in Figure 8.18r. The term plug-flow arises from the assumption that each element of fluid flows through the reactor as a small "plug," without mixing with the fluid behind or ahead of it. It is also assumed that the concentrations and temperatures are uniform across the cross-sectional area of flow. Cracking furnaces, with very high velocities inside the tubes, approach this condition of plug-flow.

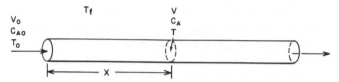

Fig. 8.18r   Plug-flow reactor

Whereas compositions and temperatures are functions of time in the perfectly mixed reactor, in the plug-flow reactor they are functions of time and distance X from the entrance. Thus, there will be derivatives with respect to time at each point in the reactor and derivatives with respect to distance at each instant of time. These are partial derivatives, and the equations they are part of are partial differential equations. To solve a partial differential equation it must be approximated by a number of ordinary differential equations. It is easier to derive these equations directly by the method that will be shown here.

The reactor is divided into a number of small "pools" that are assumed to be perfectly mixed. As shown in Figure 8.18s, the length of each pool is ΔX and its volume A ΔX, where A is the cross-sectional area of flow. The volumetric rate of flow is equal to the velocity v times the cross-sectional area of A. A magnified sketch of pool i is shown in Figure 8.18t. Assuming a gas phase reaction, a material balance can be written around each of the pools to give

$$\frac{d}{dt}(A \Delta X \rho_i) = A(v_{i-1}\rho_{i-1}) - A(v_i\rho_i) \quad 8.18(17)$$

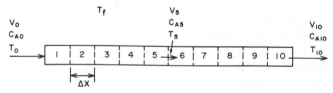

Fig. 8.18s   "Pool" model of plug-flow reactor

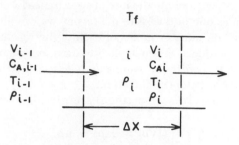

Fig. 8.18t   Pool number i

where ρ_i is the density of the reacting mixture, which can be expressed in terms of the temperature T, pressure P, and average molecular weight M_{av} by the ideal gas law:

$$\rho_i = M_{avi}P_i/RT_i \quad 8.18(18)$$

or in terms of any other equation of state. R is the ideal gas constant, and M_{avi} can be calculated as a function of the composition of the mixture in the pool. Equation 8.18(17) can be simplified, since A and ΔX are constants, to give

$$\frac{d\rho_i}{dt} = \frac{1}{\Delta X}(v_{i-1}\rho_{i-1} - v_i\rho_i) \quad 8.18(19)$$

A material balance on reactant A will give us

$$\frac{d}{dt}(A \Delta X C_{Ai}) =$$
$$= A(v_{i-1}C_{A,i-1}) + A(\Delta X r_A) - A(v_iC_{Ai}) \quad 8.18(20)$$

where C_{Ai} is the concentration of reactant A in the ith pool. Again we can simplify the equation by dividing A ΔX out:

$$\frac{dC_{Ai}}{dt} = r_A + \frac{1}{\Delta X}(v_{i-1}C_{A,i-1} - v_iC_{Ai}) \quad 8.18(21)$$

Similar equations can be written for each of the other reactants and reaction products.

An energy balance on the ith pool gives us

$$\frac{d}{dt}(A \Delta X \rho_i C_p T_i) = A \Delta X r_A(-\Delta H_R) +$$
$$+ A(v_{i-1}\rho_{i-1}C_pT_{i-1}) +$$
$$+ Up \Delta X(T_f - T_i) - A(v_i\rho_iC_pT_i) \quad 8.18(22)$$

where p is the perimeter of the pipe, so that p ΔX is the heat transfer area of one pool and T_f is the firing box temperature. With the help of equation 8.18(17) and some calculus, this last equation becomes

$$\frac{dT_i}{dt} = \frac{-\Delta H_R}{\rho_iC_p}r_A + \frac{Up}{A\rho_iC_p}(T_f - T_i) +$$
$$+ \frac{v_{i-1}\rho_{i-1}}{\rho_i\Delta X}(T_{i-1} - T_i) \quad 8.18(23)$$

The velocity from each pool to the next can be calculated with a pressure balance:

$$P_i - P_{i+1} = k_f v_i^2 \qquad 8.18(24)$$

where the constant k_f is proportional to the friction factor that can be calculated with fluid dynamics correlations. The pressure can be calculated from equation 8.18(18).

Equations 8.18(18), (19), (21), (23), and (24) must be solved for each one of the pools. Thus, an increase in the number of pools requires an increase in the time required to simulate the reactor on the digital computer. The greater the number of pools, the closer we approach the partial differential equations that describe the plug-flow reactor.

Non-Ideal Flow

Process reactors deviate somewhat from the ideal flow conditions of perfectly mixed and plug-flow. A stirred tank reactor may contain "pockets" of fluid in which the concentration of reactants is different from that in other parts of the reactor. The way to model this type of non-ideality is to divide the reactor into two or more "pools" and to apply the material balance to each of the pools. The sum of the volumes of the pools must be equal to the volume of the reactor. If backflow is allowed for from a pool to the one behind it, the rate of recirculation between the pools must be assumed; as this rate is increased, the results of the divided reactor model approach those of the perfectly mixed reactor model. If no backflow is allowed for, as the number of pools is increased, the model approaches that of the plug-flow reactor.

Another type of non-ideal flow is created by a tubular reactor in which the velocity, temperature, and concentrations vary across the radius of the tube. If axial symmetry is present, each of the pools representing the reactor can be divided into a number of "rings" and the accounting equations can be written for each of the rings. It is easy to see how the equipment and time requirements for the simulation increase geometrically with the number of pools and rings used. For the transfer of heat and mass by conduction, diffusion and eddy currents between two adjacent rings, appropriate flow models and heat and mass transfer equations must be used.

Heterogeneous Reactions

When a reaction takes place in more than one phase, it is called heterogeneous. This occurs when the reactants cannot mix completely to form a single phase, that is, when they are immiscible or partially miscible. In heterogeneous reactions, the concentration of reactants and products depends not only on the rate of reaction but also on the rate at which they diffuse from one phase to the other. For fast reactions and slow diffusion rates, the rate of diffusion will have greater influence on the rate at which the reaction occurs than the kinetics of the reaction. In general, material and energy balance equations must be written for each phase and the appropriate diffusion equations used for the transfer of reactants and products between the phases. In addition, equilibrium solubilities and phase equilibrium relationships must be considered.

Catalytic Reactions

A catalytic reaction is one conducted in the presence of a catalyst. A catalyst is a material that speeds up the reaction without being consumed or produced by it. The reaction rate coefficient k is then a function of the temperature, catalyst concentration, and "age." This age is the accumulated time since the catalyst was replaced or regenerated. These functions must be determined experimentally and incorporated into the model. When the catalyst is a solid, the additional complications of heterogeneity that were discussed in the previous paragraph are also present. Sometimes the equations can be simplified considerably by expressing the rate of reaction on the basis of unit mass of catalyst instead of on the basis of unit volume. This is true for fixed catalyst bed reactors.

Conclusions on Modeling

Presented in this section are methods for developing the equations that describe process reactors. Since the reactor is only a part of the control loop, for the simulation to be complete, it must include the equations that represent the temperature, level, pressure, and concentration sensors, transmission lines, controllers, and control valves.

A reactor control simulation allows the control engineer to tune the controllers without any loss of production or danger of blowing up the plant. It also provides him with the perfect tool to train plant operators for smoother and safer startups, it serves as a "live" model with which to try new control ideas that will result in safer and more efficient operation of the reactor, and it gives him an insight into the behavior of the reactor equivalent to several years of reactor operation. This last advantage derives from the ability to look at variables in the simulation that are impossible or impractical to measure in the process reactor.

When modeling on digital computers, one must recognize the following:

1. The digital computer is not a continuous device; therefore, all the calculations required to implement a particular algorithm must be broken down into finite steps.
2. Numerical integration involves approximating continuous differential equations with discrete finite-difference equations. The accuracy and stability of these approximating equations must be constantly considered.
3. If the model algorithm contains nonlinear equa-

tions or functions, analytical solutions cannot be used and an iterative trial-and-error procedure must be devised. A procedure of trial and error approximations is repeated until the iteration converges to the right value.

Optimization

It is sometimes beneficial to optimize the performance of the reactor. For example, it may be desirable to maintain the reactor throughput at as high a rate as the heat removal capabilities of the system will permit. This is a broad topic, and, within the scope of this discussion, it is difficult to do more than give a few specific examples.

One form of optimization is frequently applied in the batch polymerization of copolymers. In copolymer polymerization, reactivity ratios are generally such that one of the monomers is depleted at a faster rate than the other. Product requirements sometimes demand that the ratio of monomer to co-monomer in the final product be uniform. That is, as the polymerization progresses, the ratio of unreacted monomer and comonomer must be kept constant. In such cases, the more reactive monomer or a mixture of monomers is fed continuously to the reactor to maintain the desired ratio of monomers in the reactor. In order to do this properly it is necessary to know how far polymerization has advanced. This can be determined by measuring the total heat released by the system and relating this heat release to the degree of polymerization. If the flow of coolant makeup is multiplied by its temperature rise, this will give the instantaneous rate of heat removal from the system.

This rate of heat removal can be used to reset a flow controller maintaining the monomer feed rate to the reactor. This rate of heat evolution need only be integrated by time to give the total heat released by the batch reaction at any instant in time. This total is related to the total conversion at a given time, and therefore its measurement can be useful for the batch addition of certain ingredients. For example, modifiers must be added at certain percent monomer conversions to polymer during the reaction. This approach is applicable to many rubber polymerization processes.

Another type of reactor optimization may involve maintaining the reaction at as high a rate as the system will tolerate. Figure 8.18u depicts such a system. Here a monomer is polymerized essentially at the same rate at which it is fed to the reactor. The reactor is controlled to maintain a feed rate high enough that polymerization will occur as rapidly as the cooling system can remove the reaction heat. The cooling system capacity will, of course, vary with ambient temperature conditions as they affect the corresponding cooling water temperature. There may also be a variation in the heat transfer coefficient as the concentration of product solids and its viscosity increases.

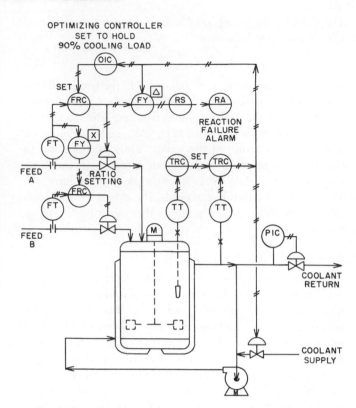

Fig. 8.18u   Continuous reactor with optimized control system

The system shown in Figure 8.18u uses a conventional cascade temperature control loop. The two feed streams are on ratio control. Reaction optimization is achieved by charging feed A at a high enough rate that the coolant makeup to the circulating cooling loop is maintained at near its maximum capacity (90 percent open valve). This is achieved by sensing the air signal to the coolant control valve and using this information to reset the feed flow set point on the master feed control loop (in this case, feed A). By maintaining the coolant makeup at as high a rate as is practical, the jacket temperature is maintained at as low a temperature as the coolant source permits, thereby assuring maximum permissible heat removal at any point in the reaction. Coolant supply temperature levels may fluctuate. This is the case with tower water, which depends on ambient humidity and temperature conditions and on cooling tower heat load to establish the cooling water supply temperature. The performance of the illustrated control system is unaffected by these fluctuations. In order to ensure that charging is interrupted if for any reason the reaction ceases, the air signals to the control valve on feed A and to the coolant makeup are compared. When the reaction is proceeding normally, there is a predictable relationship between these air signals. In the event that the reaction ceases, the generated reaction heat drops, and the control signal to the monomer feed control valve will change to admit more and more feed, such that the valve would approach a fully open condition. The difference between the two air signals can be used to actuate an audible alarm or to initiate a safe and automatic shutdown procedure.

The performance of the reactor can also be optimized by reduction of the total batch cycle time. Techniques to achieve this include the speeding up of metering and weighing by increasing the flow rates of additives and by increasing the rates of heat-up or cool-down, or both.

Whenever common resources are servicing several reactors one must also optimize the distribution of these resources so as to reduce the waiting time for a particular reactor before it can obtain a particular resource during its batch sequence.

REFERENCES

1. Murrill, Pike, and Smith, *Formulation and Utilization of Mathematical Models*, International Textbook Co.
2. Levenspiel, O., *Chemical Reaction Engineering: An Introduction to the Design of Chemical Reactors*, John Wiley & Sons, Inc., New York, 1962.

BIBLIOGRAPHY

Bennett, C.E., "Evaluating Controls for Batch Processing," *Instrumentation Technology*, July 1979.

Buckley, P.S., "Designing Override and Feedforward Controls," *Control Engineering*, Part I, August 1971; Part II, October 1971.

———, "Protective Controls for a Chemical Reactor," *Chemical Engineering*, April 20, 1970.

Carlo-Stella, G., "Distributed Control for Batch Systems," *InTech*, March 1982.

Corrigan, T.E. and Young, E.F., "General Considerations in Reactor Design," *Chemical Engineering*, October 1955.

Fihn, S.L., "Specifying Batch Process Control Strategies," *InTech*, October 1982.

Fisher Controls, "Batch & Sequence Control Concept," Bulletin 4:006, August 1982.

Forman, E.R., "Control Systems for Process Reactors," *Chemical Engineering*, December 6, 1965.

Hall, C.J., "Don't Live with Reset Windup," *Instrumentation Technology*, July 1974.

Harriott, P., *Process Control*, McGraw-Hill Book Co., New York, 1964, p. 311.

Hopkins, B. and Alford, G.H., "Temperature Control of Polymerization Reactors," *Instrumentation Technology*, May 1973, pp. 39–43.

Krigman, A., "Guide to Selecting Weighing, Batching and Blending Systems," *InTech*, October 1982.

Luyben, W.L., "Batch Reactor Control," *Instrumentation Technology*, August 1975.

Luyben, W.L., *Process Modelling, Simulation and Control for Chemical Engineers*, McGraw-Hill Book Co., New York, 1973.

Lynch, E.P., *Applied Symbolic Logic*, John Wiley & Sons, New York, 1980.

Marroquin, G. and Luyben, W.L., "Practical Control Studies of Batch Reactors Using Realistic Mathematical Models," *Chemical Engineering Science*, vol. 28, 1973, pp. 993–1003.

Mayer, F.X. and Spencer, E.H., "Computer Simulation of Reactor Control," *ISA Journal*, July 1961, pp. 58–64.

Millman, M.C. and Katz, S., "Linear Temperature Control in Batch Reactors," *Industrial & Engineering Chemistry Process Design and Development*, vol. 6, October 1967, pp. 447–451.

Schreiner, F.H., "Sequence Control for Polymerization," *Control Engineering*, September 1968.

Schrock, L.J., "Minimum Time Batch Processing," *ISA Journal*, October 1965, pp. 75–82.

Scott, D.H., "A Systems Approach to Batch Control," *Instrumentation Technology*, August 1970.

Shaw, W.T., "Computer Control of Batch Processes," *EMC Controls*, 1982.

Shinskey, F.G., *Process-Control Systems*, McGraw-Hill Book Co., New York, 1979.

——— and Weinstein, J.L., "Dual Mode Control System for a Batch Exothermic Reactor," Paper No. 6-4 1-65, *Advances in Instrumentation*, Proceedings of the 20th Annual ISA Conference and Exhibit, Los Angeles, vol. 20, 1965.

Siebenthal, C.D. and Aris, R., "Application of Pontryagin's Method to the Control of Batch and Tubular Reactors," *Chemical Engineering Science*, vol. 19, 1964, pp. 747–761.

Ward, J.C., "Digital Batch Process Control," *InTech*, October 1982.

Wensley, J.H., "Fault-Tolerant Control for Batch Processes," *InTech*, October 1982.

8.19 SEMICONDUCTOR PLANT OR CLEAN-ROOM CONTROLS

The process, or production, area of a typical semiconductor manufacturing laboratory is between 100,000 and 200,000 ft² (9290 and 18,580 m²). The market value of the daily production from this relatively small area is over 1 million dollars. Plant productivity is increased if the following conditions are maintained in the process area:

- No drafts ($+0.02''$ H₂O $\pm$ 0.005" or 5 Pa $\pm$ 1.3 Pa)
- No temperature gradients ($72°F \pm 1°F$ or $22°C \pm 0.6°C$)
- No humidity gradients (35% RH $\pm$ 3%)
- No air flow variations (60 airchanges/hr $\pm$ 5%)

Therefore, the goal of optimization is to maximize plant productivity continuously through the accurate control of these parameters. The secondary goal of optimization is to conserve energy. The main control elements and control loop configurations required for arriving at a high productivity semiconductor manufacturing plant are described here.

Clean-Room Control

Pressure

In order to prevent contamination through air infiltration from the surrounding spaces, the clean-room pressure must be higher than that of the rest of the building. As shown in Figure 8.19a, the clean-room is surrounded by a perimeter corridor. The pressure in this corridor is the reference for the clean-room pressure controller DPS-1 in Figure 8.19b. This controller is set to maintain a few hundredths of an inch of positive pressure relative to the corridor. The better the quality of building construction, the higher this set point can be, but even with the lowest quality buildings, a setting of approximately 0.02" H₂O (5 Pa) can be easily maintained. Because at such near-atmospheric pressures, air behaves as if it were incompressible, the pressure control loop shown in Figure 8.19b is both fast and stable. When the loop is energized, DPS-1 quickly rotates the return air control damper (RAR-1) to the required control position.

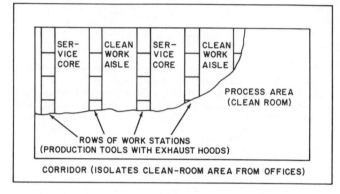

Fig. 8.19a Floor plan of semiconductor manufacturing clean room. *Note:* Each row of work stations is also called a "zone." Each work station is also called a "subzone."

This position will remain unaltered as long as the air balance in the area remains the same:

Return Air Flow = Supply Air Flow −
− (Exhaust Air Flow + Pressurization Loss)

When this air flow balance is altered (for example, as a result of a change in exhaust air flow), it will cause a change in the space pressure, and DPS-1 will respond by modifying the opening of RAR-1.

Plant productivity is maximized if drafts are eliminated in the clean work aisle. Drafts would stir up the dust in this area, which in turn would settle on the product and cause production losses. There are approximately 200 work stations in a typical semiconductor producing plant. When a DPS-1 unit is provided to control the pressure at each work station, the result is a uniform pressure profile throughout the clean room. Both pressure differences and the cause of drafts are eliminated, and therefore the result is a draft-free, high-productivity plant. Although the process or clean work aisle is where DPS-1 controls the space pressure, it is important to make sure that *all* points, including the service aisle, be under positive pressure. Because the local circulating fan within the work station draws the air in from the service core and discharges it into the clean work aisle, the pressure in the service core will always be lower than on the process side. Therefore, it is possible for localized vac-

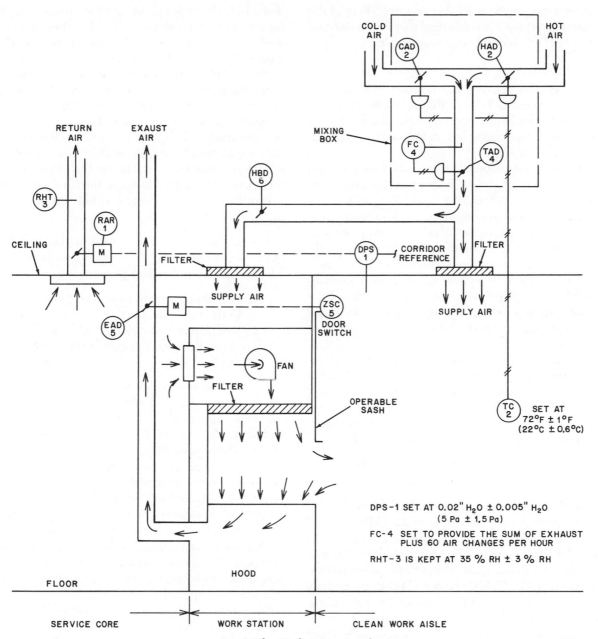

Fig. 8.19b Work station controls

uum zones to evolve in the service core, which would cause contamination by allowing air infiltration. To prevent this from happening, several solutions have been proposed.

One possibility is to have DPS-1 to control the service core pressure. This is not recommended, because a draft-free process area can be guaranteed only if there are no pressure gradients on the work aisle side; this can be achieved only by locating the DPS-1 units on the work aisle side of the work stations.

Another possibility is to leave the pressure controls on the work aisle side but raise the set point of DPS-1 until all the service cores in the plant are also at a positive pressure. This solution cannot be universally recom-

mended either, because the quality of building construction might not be high enough to allow operation at elevated space pressures. This is because the pressurization loss in badly sealed buildings can prevent reaching the elevated space pressures. Yet another solution is to install a second DPS controller, which would maintain the service core pressure by throttling the damper HBD-6. This solution will give satisfactory performance but will also increase the cost of the control system by the addition of a few hundred control loops (one per work station).

A more economical solution is shown in Figure 8.19b. Here a hand-operated bypass dampepr (HBD-6) is manually set during initial balancing. This solution is reasonable for most applications, because the effect of the work station fan is *not* a variable, and therefore a constant setting of HBD-6 should compensate for it.

Temperature

The temperature at each work station is controlled by a separate thermostat, TC-2 in Figure 8.19b. This temperature controller adjusts the ratio of cold air to hot air within the supply air mixing box, as required, to maintain the space temperature.

Unfortunately, conventional thermostats cannot be used if the temperature gradients within the clean-room are to be kept within $\pm 1°F$ of $72°F$ ($\pm 0.6°C$ of $22°C$). The conventional thermostats cannot meet this requirement either from the measurement accuracy point of view or from the control quality point of view. Even after individual calibration, one should not expect less than a ± 2 or $3°F$ (± 1 or $1.7°C$) error in overall loop performance, if HVAC-quality thermostats are used. Part of the reason for this is the fact that the "offset" cannot be eliminated in plain proportional controllers, such as in thermostats.

This means that in order for the thermostat to move its output from the midscale value (50%–50% mixing of cold and hot air) an error in room temperature must first develop. This error is the permanent offset. One can determine the size of this offset error for TC-2 in Figure 8.19b when it requires maximum cooling as follows:

$$\text{Offset Error} = \frac{\text{Spring Range of CAD}}{2 \text{ (Thermostat Gain)}}$$

Assuming that CAD-2 has an 8 to 13 PSIG (55 to 90 kPa) spring (or a spring range of 5 PSI, or 34.5 kPa) and TC-2 is provided with a maximum gain of 2.5 PSI/°F (31 kPa/°C), the offset error is 1°F (0.6°C). Under these conditions, therefore, the space temperature must permanently rise to 73°F (22.8°C) before CAD-2 can be fully opened. The offset error will increase as the spring range increases or as the thermostat gain is reduced. Sensor and set point dial errors are always additional to the offset error.

Therefore, in order to control the clean-room temperature within $\pm 1°F$ ($\pm 0.6°C$), it is necessary to use an RTD-type or a semiconductor transistor–type temperature sensor and a proportional-plus-integral controller, which will eliminate the offset error. This can be most economically accomplished through the use of microprocessor-based shared controllers, which communicate with the sensors over a pair of telephone wires, serving as a data highway.

Humidity

The relative humidity sensors are located in the return air stream (RHT-3). In order to keep the relative humidity in the clean-room within 35% RH $\pm$ 3% RH, it is important to select a sensor with a lower error than $\pm$ 3% RH. The repeatability of most human hair element sensors is approximately $\pm$ 1% RH. These units can be used for clean-room control applications if they are *individually calibrated* for operation at or around 35% RH. Without such individual calibration, they will not perform satisfactorily, because their off-the-shelf inaccuracy, or error, is approximately $\pm$ 5% RH. The controller associated with RHT-3 is not shown in Figure 8.19b, because relative humidity is not controlled at the work station (sub-zone) level but at the "zone" level. A zone is a row of work stations. The control action is based on the relative humidity reading in the combined return air stream from all work stations within that zone.

Flow

The proper selection of the mixing box serving each work station (sub-zone) is of critical importance. Each mixing box serves the dual purposes of:

1. providing accurate control of the total air supply to the sub-zone, and
2. modulating the ratio of "cold" and "hot" air to satisfy the requirements of the space thermostat TC-2.

The total air supply flow to the sub-zone should equal 60 air changes per hour *plus* the exhaust rate from that sub-zone. This total air supply rate must be controlled within $\pm$ 5% of actual flow by FC-4, over a flow range of 3:1. The rangeability of 3:1 is required because as processes change, their associated exhaust requirements will also change substantially. FC-4 in Figure 8.19b can be set manually, but this setting must change every time a new tool is added to or removed from the sub-zone. The setting of FC-4 must be done by individual in-place calibration against a portable hot wire anemometer reference. Settings based on the adjustments of the mixing box alone (without an anemometer reference) will not provide the required accuracy.

Some of the mixing box designs available on the market are not acceptable for this application. The design features that are unacceptable for this application include:

1. "Pressure-dependent" designs. This means that the total flow will change if air supply pressures vary. Only "pressure-independent" designs can be considered, because both the cold and the hot air supply pressures to the mixing box will vary over some controlled minimum.
2. Low rangeability designs. A 3:1 rangeability with an accuracy of $\pm$ 5% of *actual* flow is required.
3. Selector or override designs cannot be used. It is unacceptable to use such designs in cases in which the same damper can be controlled by either flow or temperature on a selective basis. Such override designs will periodically disregard the requirements of TC-2 and thereby induce upsets or cycling, or both.

If the mixing box is selected to meet the aforementioned criteria, both air flow and space temperature can be accurately controlled.

Control and Optimization at the Zone Level

Each row of work stations shown in Figure 8.19a is called a zone, and each zone is served by a cold deck (CD), a hot deck (HD), and a return air (RA) subheader. These subheaders are frequently refered to as "fingers." The control devices serving the individual work stations (sub-zones) will be able to perform their assigned control tasks if the "zone finger" conditions make it possible for them to do so.

For example, RAR-1 will be able to control the sub-zone pressure as long as the ΔP across the damper is

high enough to remove all the return air without requiring the damper to open fully. As long as the dampers are throttling (neither fully open nor completely closed), DPS-1 in Figure 8.19b is in control.

PIC-7 in Figure 8.19c is provided to control the vacuum in the RA finger and thereby to maintain the required ΔP across RAR-1. PIC-7 is a nonlinear controller with a fairly wide neutral band. This protects from changing the CD finger temperature (TIC-6 setpoint) until a sustained and substantial change takes place in the detected RA pressure.

Similarly, the mixing box in Figure 8.19b will be able to control sub-zone supply flow and sub-zone temperature as long as its dampers are not forced to take up extreme positions. Once a damper is fully open, the

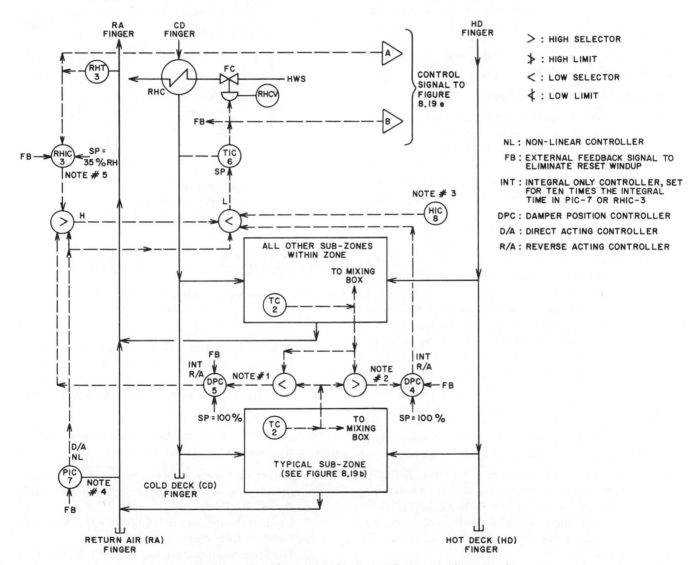

Fig. 8.19c   Typical zone control system implements envelope control.

Note #1: This is the opening of the most open hot air damper. HAD-2 is fully open when the TC-2 output is 3 PSIG (21 kPa).

Note #2: This is the opening of the most open cold air damper. CAD-2 is fully open when the TC-2 output is 15 PSIG (104 kPa).

Note #3: HIC-8 sets the maximum limit for the CD temperature at approximately 70°F (21°C).

Note #4: As vacuum increases (and pressure decreases) the output of PIC-7 also increases.

Note #5: D/A in summer, R/A in winter.

associated control loop is out of control. Therefore, the purpose of the damper position controllers (DPCs) in Figure 8.19c is to prevent their corresponding dampers from having to open fully.

Lastly, the relative humidity in the return air must also be controlled within acceptable limits.

Therefore at the "zone level" there are five controlled, or limit variables, but only one manipulated variable.

Limit or Controlled Variables	Manipulated Variables
RA pressure	TIC-6 set point
RA relative humidity	
Max. CAD opening	
Max. HAD opening	
Max. TIC-6 set point	

Figure 8.19c shows the required control loop configuration to accomplish the aforementioned goals. It should be noted that dynamic lead/lag elements are not shown, and it should also be noted that this loop can be implemented in either hardware or software.

Constraint or Envelope Control

Whenever the number of control variables exceeds the number of available manipulated variables, it is necessary to apply multivariable envelope control. This means that the available manipulated variable (TIC-6 set point) is not assigned to serve a single task but is selectively controlled to keep many variables within acceptable limits, within the "control envelope" shown in Figure 8.19d.

By adjusting the set point of TIC-6, one has a means of changing the cooling capacity represented by each unit of CD air. Because the same cooling can be accomplished by using less air at lower temperatures or by using more air at higher temperatures, one has a means of *affecting the overall material balance by manipulating the set point of the heat balance controls*. Return air humidity can similarly be affected by modulating TIC-6 set point, because when this set point is increased, more CD air will be needed to accomplish the required cooling. By increasing the ratio of humidity-controlled CD air in the zone supply (HD humidity is uncontrolled), one is also bringing the zone closer to the desired 35% RH set point.

In this envelope control system the following conditions will cause an increase in the TIC-6 set point:

1. Return air shortage detected by a drop in the RA finger pressure (increase in the detected vacuum) measured by PIC-6. An increase in TIC-6 set point will increase the CD demand, which in turn lowers the HD and, therefore, the RA demand.
2. Hot air damper in mixing box (HAD-2 in Figure 8.19b) is near to being fully open. This too necessitates the aforementioned action to reduce HD demand.
3. RA humidity does not match RHIC-3 set point. This condition also necessitates an increase in the

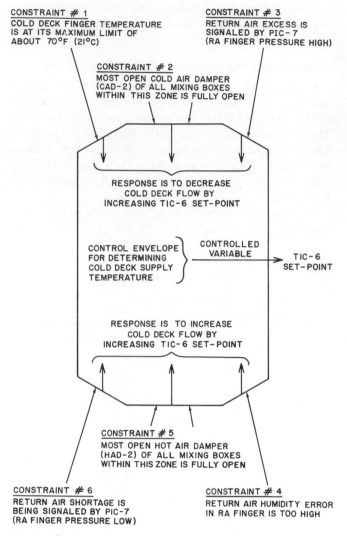

CONSTRAINT # 1
COLD DECK FINGER TEMPERATURE IS AT ITS MAXIMUM LIMIT OF ABOUT 70°F (21°C)

CONSTRAINT # 3
RETURN AIR EXCESS IS SIGNALED BY PIC-7 (RA FINGER PRESSURE HIGH)

CONSTRAINT # 2
MOST OPEN COLD AIR DAMPER (CAD-2) OF ALL MIXING BOXES WITHIN THIS ZONE IS FULLY OPEN

RESPONSE IS TO DECREASE COLD DECK FLOW BY INCREASING TIC-6 SET-POINT

CONTROL ENVELOPE FOR DETERMINING COLD DECK SUPPLY TEMPERATURE

CONTROLLED VARIABLE → TIC-6 SET-POINT

RESPONSE IS TO INCREASE COLD DECK FLOW BY INCREASING TIC-6 SET-POINT

CONSTRAINT # 5
MOST OPEN HOT AIR DAMPER (HAD-2) OF ALL MIXING BOXES WITHIN THIS ZONE IS FULLY OPEN

CONSTRAINT # 6
RETURN AIR SHORTAGE IS BEING SIGNALED BY PIC-7 (RA FINGER PRESSURE LOW)

CONSTRAINT # 4
RETURN AIR HUMIDITY ERROR IN RA FINGER IS TOO HIGH

Fig. 8.19d   Constraint control envelope for cold deck temperature optimization

proportion of the CD air in the zone supply. Because the moisture content of the CD air is controlled, an increase in its proportion in the total supply will help in controlling the RH in the zone.

On the other hand, the following conditions will require a decrease in the TIC-6 set point:

1. Return air excess detected by a rise in RA finger pressure (drop in the detected vacuum), measured by PIC-6.
2. Cold air damper in mixing box (CAD-2 in Figure 8.19b) is fully open.
3. CD finger temperature exceeds 70°F (21°C). This limit is needed to keep the CD always cooler than the HD.

Plant-Wide Control and Optimization

In conventionally controlled semiconductor plants both the temperature and the humidity of the main CD supply air are fixed. This can severely restrict the performance of such systems, because as soon as a damper or a valve is fully open (or closed), the conventional system is out of control. The optimized control system described here does not suffer from such limitations, because it automatically adjusts both the main CD supply temperature and humidity, so as to follow the load smoothly and continuously and never allow the valves or dampers to lose control by fully opening or closing. The net result is not only increased productivity but also reduced operating costs.

A semiconductor production plant might consist of two dozen zones. Each of the zones can be controlled as shown in Figure 8.19c. The total load represented by all the zones is followed by the controls shown in Figure 8.19e. Consequently, the overall control system is hierarchical in its structure: The sub-zone controls in Figure 8.19b are assisted by the zone controls in Figure 8.19c, and the plantwide controls in Figure 8.19e guarantee that the zone controllers can perform their tasks. The interconnection among the various levels of the hierarchy is established through the various valve and damper position controllers (VPCs and DPCs). These guarantee that no throttling device, such as mixing boxes or RHC valves anywhere in this plant, will ever be allowed to reach an extreme position and thereby lose control. Whenever a control valve or a damper is approaching the point of losing control (nearing full opening), the load-following control system at the next higher level modifies the air or water supply conditions so that it will not need to open fully. This hierarchical control scheme is depicted in Figure 8.19f.

The overall plant-wide control system shown in Figure 8.19e can be viewed as a flexible combination of material

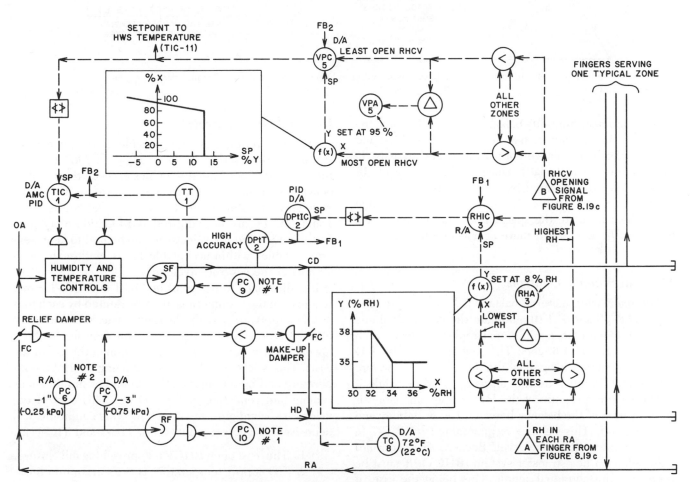

Fig. 8.19e   Plant-wide optimization control system.

Symbols: DPtIC = Dew point controller. PID = Controller with proportional, integral, and derivative control modes. AMC = Controller with positions for automatic-manual-cascade modes of operation. VPC = Valve position controller. INT = Controller has integral control action only. Its setting is to be ten times the integral time setting of the associated process controller(s).

Note #1: This pressure controller keeps the lowest of all finger pressures at a value above some minimum. It modulates the fan volume or speed and starts extra fan units to meet the load. When the load drops, the unnecessary fans are stopped by flow (not pressure) control.

Note #2: If the pressure rises above $-1''$ H_2O $(-0.25$ kPa), the relief damper starts to open, and if the pressure drops below $-3''$ H_2O $(-0.75$ kPa), the make-up damper opens. This opening is limited by TC-8, which prevents the HD temperature from dropping below 72°F (22°C). Between the settings of PC-6 and PC-7, both dampers are tightly closed and the suction pressure floats.

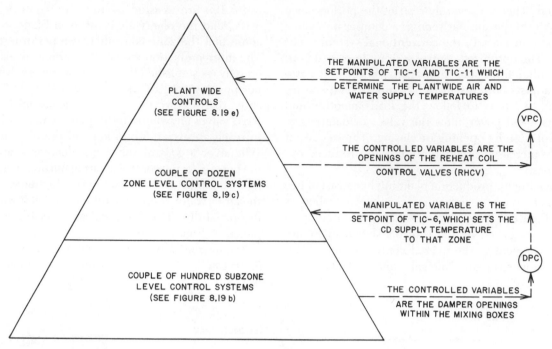

Fig. 8.19f   Illustration of the hierarchical control architecture in a semiconductor manufacturing plant

balance and heat/humidity balance controls. Through the load-following optimization of the set points of TIC-1 and TIC-11, the heat balance controls are used to assist in maintaining the material balance around the plant. If, for example, the material balance requires an increase in air flow while the heat balance requires a reduction in the heat input to the space, both requirements will be met by the admitting of more air at a lower temperature.

Material Balance Controls

The plant-wide material balance is based on pressure control. PC-9 and PC-10 modulate the variable volume fans so as to maintain a minimum supply pressure in each of the CD and HD fingers. The suction side of the CD supply fan station is open to the outside, and therefore it will draw as much outside air as the load demands.

The suction pressure of the HD supply fan station is an indication of the balance between RA availability and HD demand. This balance is maintained by PIC-7 in Figure 8.19c at the zone level. Because this controller manipulates a heat transfer system (RHCV), it must be tuned for slow, gradual action. This being the case, it will not be capable of responding to sudden upsets or to emergency conditions, such as the need to purge smoke or chemicals. Such sudden upsets in material balance are corrected by PC-6 and PC-7. PC-6 will open a relief damper if the suction pressure is high, and PC-7 will open a makeup damper if it is low. In between these limits, both dampers will remain closed and the suction pressure will be allowed to float.

Heat Balance Controls

This also requires a multivariable envelope control system (similar to that shown in Figure 8.19d), because the number of controlled variables exceeds the number of available manipulated variables. Therefore, the plant-wide air and water supply temperature set points (TIC-1 and TIC-11) are not adjusted as a function of a single consideration but are selectively modulated to keep several variables within acceptable limits, within the constraints of the control envelope.

By adjusting the set point of TIC-1 and TIC-11, one is adjusting the cooling capacity represented by each unit of CD supply air and the heating capacity of each unit of hot water. This then provides not only for load-following control but also for minimization of the operating costs. This is accomplished by minimizing the HWS temperature and by minimizing the amount of simultaneous cooling and reheating of the CD air. This control envelope is so configured that the following conditions will decrease the TIC set points (both TIC-1 and TIC-11):

1. The least open RHCV is approaching full closure.
2. CD supply temperature is at its maximum limit.

On the other hand, the following conditions will require an increase in the TIC set points:

1. The most open RHCV is approaching full opening.
2. The CD supply temperature is at its minimum limit.

The set point of VPC-5 is produced by a function generator, f(x). The dual purposes of this loop are:

1. Prevention of any of the reheat coil control valves (RHCVs) from fully opening and thereby losing control.
2. Forcing the least open RHCV toward a minimum opening (to minimize wasteful overlap between cooling and reheat) but keeping it from full closure as long as possible.

This is accomplished by f(x), which keeps the least open valve at approximately 12 percent opening as long as the most open valve is less than 90 percent open. If this 90 percent opening is exceeded, f(x) prevents the most open valve from fully opening. This is accomplished by lowering of the set point of VPC-5, which in turn will increase the set points of the TICs. Increasing the set point of TIC-1 results in reducing the load on the most open RHCV, whereas the increase in the TIC-11 set point increases the heating capacity at the reheat coil. Through this method of load following and through the modulation of main supply air and water temperatures, all zones in the plant will be kept under stable control if the loads are similar in each zone.

All control systems—including this one—will lose control when the design limits of the associated mechanical equipment are reached or if the dissimilarity in load distribution is greater than was anticipated by the mechanical design. Therefore, if a condition would ever arise in which one zone requires large amounts of cooling (associated RHCV closed) while some other zone at the same time requires its finger reheat coil valve (RHCV) to be fully open, then the control system must decide which condition it is to correct, because it cannot correct both. The control system in Figure 8.19e is so configured that it will give first priority to preventing the RHCV valve from fully opening. Therefore, if as a result of mechanical design errors or misoperation of the plant there is no CD supply temperature that can keep all valves from either fully closing or fully opening, then the control system will allow some RHCV valves to close while keeping all of them less than 100 percent open. Whenever such condition is approaching (and, therefore, the difference between the openings of the most open and the least open valves reaches 95 percent), a valve position alarm (VPA-5) is actuated. This will allow the operator to check the causes of such excessively dissimilar loads between zones and will allow him to take corrective action by revising processes, relocating tools, or modifying air supply ducts and adding or removing mixing boxes.

By keeping all RHCV valves away from being nearly closed, this control system simultaneously accomplishes the following goals:

1. Eliminates unstable (cycling) valve operation by not allowing them to operate nearly closed.
2. Minimizes pumping costs by minimizing pressure losses through throttling valves.

3. Minimizes heat pump operating costs by minimizing the required hot water temperature.
4. Provides a means of detecting and thereby smoothly following the variations in the plant-wide load.

Humidity Controls

At the zone level the RA humidity is controlled by RHIC-3 in Figure 8.19c. In addition, the dew point of the main CD air supply must be modulated, because the main CD supply temperature is being modulated to follow the load. This is accomplished by measurement of the relative humidity in all RA fingers (RHT-3 in Figure 8.19c) and selection of the one finger that is farthest away from the control target of 35 percent RH.

The control system is similar to the TIC set point optimization loop discussed earlier. The first elements in the loop are the selectors. They pick out the return air fingers with the highest and the lowest relative humidities. The purpose of the loop is to "herd" all RH transmitter readings in such a direction that *both* the highest and the lowest readings will fall within the acceptable control gap limits of 35% ± 3% RH. This is accomplished by sending the highest reading to RHIC-3 in Figure 8.19e as its measurement and also sending the lowest reading to this RHIC-3 as its modified set point.

Using this technique, a humidity change in either direction is recognized and is corrected through this herding technique. The set point of RHIC-3 is produced by the function generator, f(x). Its dual purposes are:

1. To prevent the most humid RA finger from exceeding 38 percent RH.
2. To keep the dryest RA finger humidity from dropping below 32 percent RH, as long as this can be accomplished without violating the goals of number 1 above.

This is done by keeping the set point of RHC-3 at 35 percent as long as the dryest finger reads 34 percent RH or more.

If it drops below that value, the set point is raised to the limit of 38 percent RH in order to overcome this low humidity condition, but without allowing excessive humidity in the return air of some other zone. This control strategy and the RHIC-3 controller in Figure 8.19c will automatically respond to seasonal changes and will give good RH control as long as the loads in the various zones are similar. If the humidity loads are substantially dissimilar, this control system is also subject to mechanical equipment limitations.

In other words, the mechanical equipment is so configured that addition or removal of moisture is possible only at the main air supply to the CD. Consequently, if some zones are moisture-generating and others require humidification, this control system can respond to only one of these needs. Therefore, if the lowest RH reading is below 32 percent while the highest is already at 38 percent, the low humidity condition will be left uncontrolled, and the high humidity zone will be controlled to prevent it from exceeding the 38 percent RH upper limit. Allowing the minimum limit to be temporarily violated while maintaining control on the upper limit is a logical and safe response to humidity load dissimilarities between zones. It is safe because the intermixing of the return airs will allow self-control of the building, by transferring moisture from excess humidity zones to ones with humidity deficiency.

Whenever the difference between minimum and maximum humidity readings reaches 8 percent RH, the RHA-3 alarm is actuated. This will alert the operator that substantial moisture load dissimilarities exist and will cause him to find and eliminate the causes.

Exhaust Air Controls

The exhaust air controls at the subzone level are shown in Figure 8.19b. At this level the control element is a two-position damper, EAD-5. When the work station is functioning, its operable sash is open. This condition is detected by the door position limit switch, ZSC-5, which in that case fully opens the two-position damper. When the work station is out of service, its operable sash is closed, and therefore the door switch closes EAD-5 to its minimum position. This minimum damper position still provides sufficient air exhaust flow from the work

station to guarantee the face velocity needed for operator safety. In other words, when the sash is closed, the air will enter the work station over a smaller area, and therefore less exhaust air flow is required to provide the air inflow velocity that is needed to keep the chemical fumes from leaking out. By this technique the safety of operation is unaffected and operating costs are lowered, because less outside air needs to be conditioned if the exhaust air flow is lowered.

In order for EAD-5 to maintain the required exhaust air flows accurately, it is necessary to keep the vacuum in the exhaust air collection ductwork at a constant value. This pressure control loop is depicted in Figure 8.19g. In each zone exhaust finger, the PC-1 shown in Figure 8.19g keeps the vacuum constant by throttling the EAD-1 damper as required. In order for these dampers to stay in control while the exhaust air flow varies, it is important to keep them from fully opening. This is accomplished first by identification of the most open finger damper (low selector) and then by comparison of its opening with the set point of the damper position controller, DPC-2. The task of this controller is to limit the opening of the most open EAD-1 to 80 percent and to increase the vacuum in the main EA header if this opening would otherwise exceed 80 percent.

Therefore, if the measurement of DPC-2 exceeds 80 percent (drops below 9 PSIG, or 62.5 kPa), the vacuum set point of PC-3 is increased (pressure setting lowered).

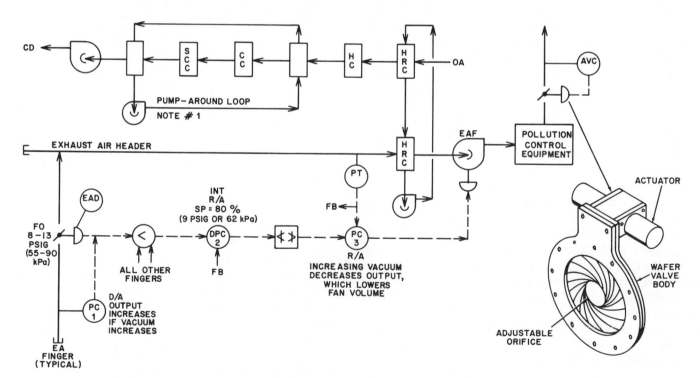

Fig. 8.19g   Exhaust air optimization controls and heat recovery.
Note #1: When the need for dehumidification in the summer results in excessive overcooling here and therefore in increased demand for reheat at the CD fingers, the pumparound economizer loop is started to lower operating cost and increase efficiency by reducing the degree of overcooling and thereby lowering the need for reheat.

This in turn will increase the operating level of the exhaust fan station (EAF).

The limits on the set point of DPC-2 will prevent mechanical damage, such as collapsing of the ducts as a result of excessive vacuum or manual misoperation. The reasons for integral action and external feedback are the same as in the case of all other damper and valve position controllers discussed earlier.

Figure 8.19g also shows a glycol-circulating heat recovery loop, which can be used as shown to preheat the entering outside air or can be used as a heat source to a heat pump in the winter (not shown). In either case, the plant operating cost is lowered by recovering the heat content of the air before it is exhausted in the winter.

The discharging of chemical vapors into the atmosphere is regulated by pollution consideration. The usual approach is to remove most of the chemicals by adsorption and scrubbing prior to exhaustion of the air. An added measure of safety is provided by exhausting the air at high velocity, so as to obtain good dispersion in the atmosphere. Because the volume of air being exhausted varies, an air velocity controller (AVC in Figure 8.19g) is used to maintain the velocity of discharge constant. This is accomplished by modulation of a variable orifice iris damper, also illustrated in Figure 8.19g.

Conclusions

The productivity of semiconductor manufacturing plants can be greatly increased and the operating costs can be lowered through the instrumentation and control methods described previously and by such added control system features as:

- Use of low leakage dampers (0.5 CFM/ft^2 at 4" H$_2$O ΔP, or 2.5 l/s/m^2 at 1 kPa ΔP)
- Accurate air flow metering for material balance and pressurization loss control
- Pumparound economizers (see Figure 8.19g)

The initial cost of the previously described control system is not higher than the cost of conventional systems, because the added expense of more accurate sensors is balanced by the much-reduced installation cost of distributed shared controls. Therefore, the benefit of increased productivity is a result not of higher initial investment but of better control system design.

It should be emphasized that the described load-following optimization envelope control strategy is far from being typical of today's practices in the semiconductor manufacturing industry. In fact, we are not aware of a single plant in which it is fully implemented, and we know of many plants that are operated under conventional HVAC controls. Therefore, our conclusion that yields will increase and operating costs will drop when such improved control systems are installed is merely an assumption. This appeares to be supported by the results of partial system implementations and by the experiences in other industries but will remain to be an assumption until the first semiconductor manufacturing plant with such modern controls is started up.

BIBLIOGRAPHY

Daryanani, S., "Design Engineer's Guide to Variable Air-Volume Systems," *Actual Specifying Engineer*, July 1974.

DHO/Atlanta Corp., "Conserving Energy," *Powered Induction Unit, Data Sheet #1*, May 1, 1974.

Kusuda, T., "Intermittent Ventilation for Energy Conservation," ASHRAE Symposium #20, Paper #4.

Lipták, B.G. (ed.), "Carbon Dioxide Sensors," *Environmental Engineer's Handbook*, Volume II, Section 4.6, Chilton Book Co., Radnor, PA, 1975.

———, "Reducing the Operating Costs of Buildings by the Use of Computers," *ASHRAE Transactions*, Volume 83, Part 1, 1977.

———, "Savings Through CO$_2$ Based Ventilation," *ASHRAE Journal*, July 1979.

———, "Save Energy by Optimizing Your Boilers, Chillers, and Pumps," *Instrumentation Technology*, March 1981.

Nordeen, H., "Control of Ventilation Air in Energy Efficient Systems," ASHRAE Symposium #20, Paper #3.

Shih, J.Y., "Energy Conservation and Building Automation," ASHRAE Paper #2354.

Spielvogel, L.G., "Exploding Some Myths About Building Energy Use," *Architectural Record*, February 1976.

Stillman, R.B., *Systems Simulation Engineering Report*, Prepared for IBM-RECD, December 10, 1971.

Woods, J.E., "Impact of ASHRAE Ventilation Standards 62-73 on Energy Use," ASHRAE Symposium #20, Paper #1.

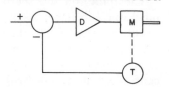

8.20 SPEED CONTROL OF MULTIPLE-DRIVE PROCESS LINES

Types of Drives:	Eddy-current clutches, DC static converters, AC variable frequency
Control Techniques:	Analog, digital, programmable
Partial List of Suppliers:	Allen-Bradley Co., Avtek Systems, Eaton Corp., General Electric, Louis Allis Co., Parametrics Div. of Zero-Max Industries, Inc., Reliance Electric Co., Wer Industrial

Multiple-drive systems for process lines are available using both analog and digital techniques and are continually being advanced through the use of programmable digital hardware. These systems can use all available types of electrical drives and can be applied wherever speed is used for control of process variables.

Applications Requiring Multiple Drives

The simplest application of two drives in a speed-controlled system is a conveyor line in which more than one set of rollers must handle a common strip of material. This strip can be a belt or a web of product, such as paper, plastic film, or foil. Roll and motor speeds must be maintained either at identical levels or at a constant ratio to allow for gearing differences.

Many process lines require the ability to vary and control speed relationships between machine sections. This ability is a key part of the process, regulating such things as stretch ratios of calendaring systems, proportions of various ingredients in a product, and printing synchronization. Regulation of some process variables is accomplished through coordination of drive speed as well as of torque.

Multiple-drive systems are also used in advanced process control, in which speed is varied in response to changes in product properties, such as temperature or thickness. The refinement of process control computers has increased the potential applications of speed-controlled systems as effective control loops are developed for manufacturing lines.

Basic Electrical Drive Systems

Several techniques can be used for actual motor speed control. Those most commonly used in multiple systems are the eddy-current clutch, the DC static drive, and the AC variable-frequency drive.

An eddy-current clutch drive applies variable voltage to a field coil, regulating the slip of an output rotor with respect to a constant-speed input rotor. The result is a smoothly operating unit with wide torque capability and simple, compact electronic controls. Its low efficiency, however, has limited its use to a few specialized applications.

The DC static converter is probably the most widely used of the speed control devices. Since DC motor speed is proportional to voltage and motor torque is proportional to current, the output of the system is easily regulated. The available horsepower in these drives ranges from fractional to in the thousands; models currently on the market can give full output torque over the entire speed range with proper motor cooling. High efficiencies are obtained from the use of solid-state components, and speed regulation to 0.1% is available using conventional tachometers. The DC drive also lends itself to tension control because of its simple current-torque characteristic. It can also provide full negative, or braking, torque to a load when equipped with regenerative capability. One disadvantage of this type of drive is its poor power factor at low speeds. Installation of correction capacitors is sometimes required to avoid penalties from the utility. Direct current motors are also typically more expensive than corresponding induction motors.

Alternating current variable-frequency drives are the beneficiary of many recent electronic advances. In these drives, the speed of the AC motor is proportional to the output frequency of the controller, with voltage varied in proportion. The development of power transistors and logic devices has enabled these drives to produce variable-frequency power compatible with virtually all AC motors up to 40 hp (30 kW). In larger power ranges, SCR choppers are used in place of transistors; efficiency deteriorates, however, since the power waveform is degraded. The AC transistorized drive has a high power factor (typically over 90 percent), and the ability to produce full motor torque over the entire speed range with proper motor cooling. Variable-frequency drives are inherently self-regulating when used with synchronous motors. Speed regulation with induction motors depends on slip and is typically 2 percent. Closer regulation requires tachometer feedback.

Analog Techniques in Multiple Systems

A conventional analog drive is speed-regulated through the comparison of a desired speed, represented by a reference voltage, with an actual speed, represented by a tachometer-generated voltage. The resulting error is then used to raise or lower the actual drive power output to the proper level. Speed coordination between two drives is easily implemented using the wide variety of amplifiers, multipliers, and meters developed for use with analog voltages. The techniques described here can be used with any of the basic drive types listed previously, and systems are widely available.

A typical application of speed control, shown in Figure 8.20a, is the regulation of one set of draw rolls at a differential but constant ratio to the speed of another, prior set of rolls. The voltage used as a reference for the first, or lead, drive is fed into a potentiometer. The reference for the second drive is taken from the wiper of the pot, giving a reference that will vary with both the reference of the first drive and the wiper position. The ratio will be constant between the two drives for a given position. An amplifier is often used between the two drives to give the second drive the ability to go faster than the first as well as to prevent interaction between drives. Additional amplifiers and control potentiometers enable the coordination of any number of drives; the system illustrated will maintain the desired speed ratios of the three sets of draw rolls as the process speed is changed. A variation often seen is a speed-follower system, in which the actual speed of a lead motor becomes the reference for the next drive or set of drives. This actual speed is generated by the lead motor tachometer.

Motor-operated reference potentiometers, acceleration control amplifiers, and upper and lower limit settings are usually incorporated into multiple-drive systems to give total control of reference voltages. The components used in these systems typically give accur-

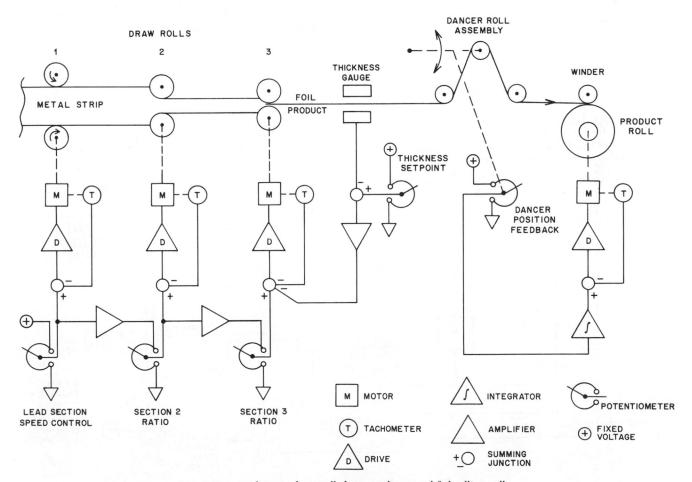

Fig. 8.20a   Analog speed-controlled process line metal foil rolling mill

acies of 5 percent to 0.1 percent, depending on quality and complexity.

Winding and unwinding of a web of material such as paper is a common application of multiple drives. These systems often use movable rolls, called dancers, which accumulate a certain length of material and hold a preset tension on it. The dancer assembly is located between a constant-speed output roll and a roll of material being wound, as in Figure 8.20a. As the diameter of the roll being wound increases, its rotational speed must decrease in order to maintain a constant surface speed. Gradual surface speed increases on the winder pull the dancer down, since the input speed to the assembly from the draw rolls has not changed. The dancer motion then mechanically drives a feedback potentiometer, producing a voltage that is integrated and used to correct winder speed to the proper level. Dancer systems can also be used in the same way between two machine rolls to maintain speed synchronization in a system with otherwise low accuracy. Slight changes in the speed of one section will raise or lower the dancer, which will produce a voltage correcting the speed error.

Analog systems can also be adapted to more complex process control. The thickness control loop shown in Figure 8.20a is an example of the control of two process variables in an interrelated system. The operation of its digital counterpart will be discussed in the following paragraphs.

Digital Control Techniques

The advent of sophisticated and cost-effective digital hardware has greatly expanded the capability of process systems. Digital signals are superior in many applications to analog signals, because of their precision and high noise immunity. They may also be processed easily by high-speed computers.

A digitally regulated drive and motor will typically use a pulse generator as a feedback device. The pulse generator is driven by the motor and produces a frequency proportional to motor speed. The speed reference for such a drive will be either a precision frequency or a numerical value. Systems using a frequency reference often include a phase-locked loop for comparison of the reference and actual speeds. These systems can also use up-down counters to detect any buildup in speed error. Systems using a numerical reference compare the desired speed with the number of pulses generated by the feedback device in a given period; this allows numerical treatment of any speed error and great flexibility in processing the correction. Digital systems typically operate with zero average speed error. No speed difference is allowed to accumulate.

Two or more motor drives are easily coordinated with digital references. Analog components and their accuracy problems are replaced by thumbwheel switches and microprocessor-based computation, as in Figure 8.20b. Nu-

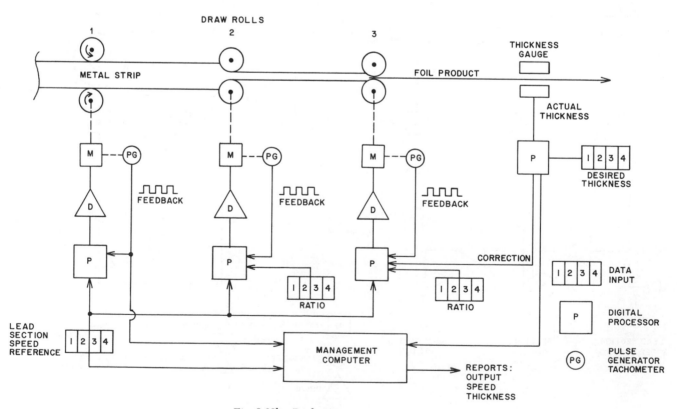

Fig. 8.20b   Production management system

merical speed ratios can be set and used precisely, and synchronization is held exactly, since no pulses are lost. The history of the speed of the machine is always available for subsequent correction.

The computation capability of a digital system allows complex functions to be easily implemented. An example is the control of the thickness of a strip of metal being drawn to a given gauge between sets of rollers, shown in Figure 8.20b. A thickness measuring device outputs a number representing the actual gauge; a comparison is made with the desired gauge, and a correction signal is fed to the speed control of the draw rolls. An increase in the speed of the last roll will reduce the thickness of the product foil, since the input rate has not changed; a decrease in the speed of the last roll will increase thickness. Limits are set upon the maximum and minimum speeds and their rates of change. Tuning of the control system is facilitated, since parameters are entered numerically, giving complete understanding of the control scheme. Operator access to the system is also enhanced, since changes can be made directly through a keyboard or a video display using easily learned commands.

Digital techniques in speed control also allow the monitoring of all functions of a production process. As shown in Figure 8.20b, the information gathered regarding speed and thickness can be stored and used to generate reports and inform equipment operators of the quantity and quality of the foil being produced. The precise control of motor speed has thus led to accurate information, which is vital to production management.

BIBLIOGRAPHY

Dorf, R.C., *Modern Control Systems*, Addison-Wesley, Reading, MA, 1974.

Editors of Instrumentation Technology, *Instrumentation and Control Systems Engineering Handbook*, Instrument Society of America, New York, 1978.

Fink, D.G. (ed.), *Standard Handbook for Electrical Engineers*, McGraw-Hill, New York, 1978.

Rembold, U., et al., *Computers in Manufacturing*, Marcel Dekker, New York, 1977.

Weiss, H.L., *Coating and Laminating Machines*, Converting Technology Corp., New York, 1977.

8.21 STEAM TURBINE CONTROL

Steam turbines are energy conversion machines. They extract energy from steam and convert it to shaft work. Sizes range from shaft output energies of a few kilowatts to well over 1000 megawatts, and there is no reason still larger machines could not be built. No other prime mover can achieve the shaft output capability easily attained by large steam turbines.

Rotational speeds vary from approximately 1800 to 12,000 rpm and, when the turbine is so designed, can be easily varied over a substantial range. This is frequently useful in driving pumps and compressors. With appropriate control equipment these devices are also capable of excellent speed stability, which is desirable in applications such as prime mover service for electric generators.

Industry experience indicates that steam turbines are very reliable compared with other prime movers. Availability factors are high, and maintenance costs are quite low. This is largely a result of the inherently balanced design, which is completely free of reciprocating or rubbing parts (except, of course, bearings).

The amount of energy that is extracted from the steam depends on the enthalpy drop across the machine. Since the enthalpy of steam is a function of temperature and pressure, and since operating conditions are generally known in terms of inlet and outlet temperature and pressure, energy considerations are facilitated by use of a graphical aid that relates these parameters. A Mollier diagram is such a tool (Figure 8.21a).

Inlet temperature and pressure may be plotted as a point and labeled P1. Exhaust conditions may be similarly plotted and labeled P2. A line connecting the two points is called the "expansion line" and represents the operation of the turbine in extracting energy from the steam. In actuality, the line is not straight. Its shape depends on the internal operation of the turbine.

An ideal turbine would expand steam isentropically—that is, at constant entropy. If a straight line were drawn vertically down the diagram to the exhaust pressure line, a point, P3, could be located that would represent the exhaust conditions of an ideal machine with inlet conditions represented by P1.

The change in enthalpy from P1 to P2 may be referred to as Δh_{1-2}. The change in enthalpy between P1 and P3 may be referred to as Δh_{1-3}. Once this is established, a number of useful calculations can be easily performed.

The efficiency of the turbine, neglecting mechanical losses, may be found by:

$$\eta = \frac{\Delta h_{1-2}}{\Delta h_{1-3}} \qquad 8.21(1)$$

The steam rate that an ideal machine would require to operate as assumed is called the theoretical steam rate, Q_T, and may be found as:

$$Q_T = \frac{2545}{\Delta h_{1-3}} \frac{\text{lbm}}{\text{HP-hr}} \left(\frac{\text{kg}}{\text{W-hr}}\right) \qquad 8.21(2)$$

The actual steam rate, Q_A, will be larger by an amount that depends on the efficiency, hence:

$$Q_A = Q_T/\eta \frac{\text{lbm}}{\text{HP-hr}} \left(\frac{\text{kg}}{\text{W-hr}}\right) \qquad 8.21(3)$$

This somewhat simplified approach will provide the control engineer with a useful tool for analyzing steam turbine applications.

It is frequently possible to design steam turbines into steam systems in ways that make energy costs attractive compared with alternatives such as electric motors and gas turbines. Use of these systems can sometimes save an energy conversion and the inherent losses. For example, less energy will usually be required for a steam turbine prime mover to drive a load than would be necessary if the size of a turbine driving an electrical generator were increased so an electric motor could be used to drive the same load.

Even though the initial cost of steam turbines is frequently considerably higher than alternative prime movers, there are some benefits that can mitigate at least some of the cost difference. Especially in larger sizes, these devices are physically smaller than most other prime movers and, consequently, require less floor space. This can decrease the cost of the building in which they are to be installed. Since the inherently balanced design

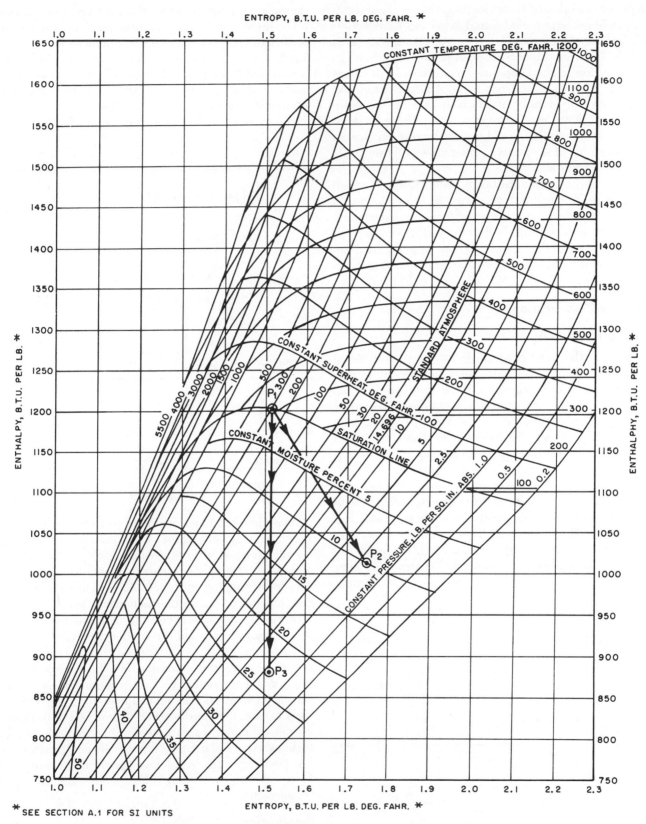

Fig. 8.21a   Mollier diagram showing performance of a steam turbine.

produces considerably less vibration than do reciprocating machines, equipment foundations can be considerably lighter. Since they are less likely to initiate fires than other prime movers, they also have an advantage in applications around flammable materials. In situations in which load characteristics make it necessary for the prime mover to endure substantial overloads, the steam turbine can have an advantage. It can tolerate, without damage, overloads that would eventually severely shorten the service life of alternative prime movers if they could tolerate them at all.

Besides initial cost, the major disadvantage of steam turbines is their low tolerance for wet or contaminated steam. Wet steam can cause rapid erosion, and contaminants can cause fouling. Both reduce efficiency and can shorten the life of the equipment. Steam monitoring can be important to the reliability and operating cost of steam turbines.

Turbine Classifications

Process Conditions

Probably the most descriptive way to discuss steam turbines for control and instrument engineering work is by referring to the context in which they are designed to be used. This involves making a statement regarding input and output conditions. The input may be from a single source or may be "mixed" (in which the steam comes from more than one source, each of which may be at a different pressure). The output may be "condensing," in which the exhaust pressure is sub-atmospheric, or "noncondensing" (sometimes called "backpressure"), in which the exhaust pressure is greater than atmospheric (Figure 8.21b). There may be more than one output. A second, and sometimes a third, outlet may be provided to allow the "extraction" of steam for other

uses at a pressure between the inlet and exhaust pressures (Figure 8.21c). Extraction turbines can be condensing or noncondensing. A rare combination of a "mixed input" and an "extraction" turbine exists, in which a port is provided through which steam can be removed when load conditions permit, or steam can be inducted when additional energy is necessary. This unit is usually referred to as an "induction-extraction" turbine.

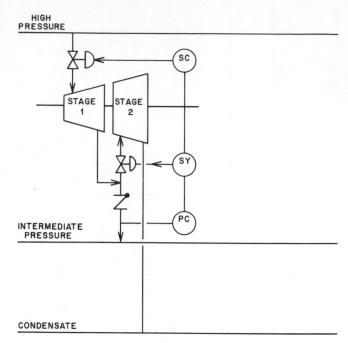

Fig. 8.21c　Typical installation of extraction turbine

Sometimes turbines are referred to by their location in the plant steam system. A backpressure turbine with its inlet connected to the plant's high pressure header and its outlet supplying steam to an intermediate header is called a "topping" turbine, since it is using steam from the "top" of the plant's thermodynamic cycle. Similarly, a turbine installed between the noncondensed exhaust of another machine and the condensate system is called a "bottoming" turbine. A topping turbine could also be described as a "noncondensing," or "backpressure," turbine, and a bottoming turbine is merely a condensing turbine. The terms *topping* and *bottoming* generally are used to describe a turbine that is designed to work over an unusual range of pressures. For example, a bottoming turbine is unusual in that it is generally designed for very low inlet pressures, and hence the term "bottoming" is more descriptive than the more generic term "condensing."

Application

The next most descriptive way to discuss a steam turbine for control and instrumentation work is in terms of its application. The broad classifications are "generator drive" and "mechanical drive."

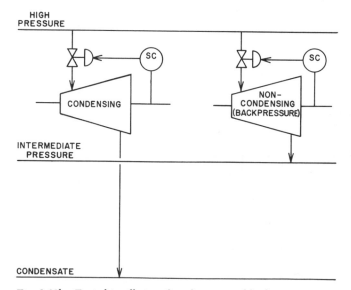

Fig. 8.21b　Typical instillation of condensing- and backpressure-type steam turbines

Internal Construction

The most basic internal feature is the operating principle, which may be either "impulse" or "reaction." A related, but not especially useful, descriptor is the direction of steam flow. In the United States, almost all turbines are "axial flow," and a small number are "tangential flow." In Europe, a significant number of turbines are "radial flow." Mechanical features are generally useful in describing turbines. It is useful to know if a unit is "single-stage" or "multi-stage," and, if multi-stage, how many stages are involved. The number of parallel exhaust stages can be used to describe turbines as "single-flow," "double-flow," and so forth. The casing and shaft arrangement is also an important way of categorizing turbines. The common categories are "single casing," in which there is one casing and one shaft, "tandem," in which there are two or more casings connected end to end by a shaft, and "cross-compound," in which there are two or more casings connected by multiple shafts. Obviously, any turbine can be described using all three methods of classification.

Control System Functions

There are two broad classifications of steam turbine control systems. They are "safety systems" and "process systems." Safety systems are intended to eliminate or, at least, to minimize the possibility of damage to the machine or hazard to personnel in the event of an unusual operating condition. Process systems are intended to control the operation of the machine in accordance with the requirements of the application (Figure 8.21d).

Safety Systems

The master safety element is the steam supply valve. This valve may be a block valve or the function may be provided by the throttling valve used for speed control. Taking the turbine off-line is accomplished by closing this valve; consequently, the control system should be designed to require all interlocks to be satisfied in order for it to open.

The safety interlocks generally consist of lube oil failure, bearing temperature, overspeed, and vibration. Lube oil is generally monitored by a pressure switch in the case of pressure lubrication systems or by a level switch in nonpressure systems. This "switch" may be a pneumatic device or an electrical one, although electrical devices are more common. When the switch is not satisfied, it causes the turbine shutoff valve to close.

Sudden loss of load will cause the turbine and its driven load to overspeed. This can happen in mechanical drive

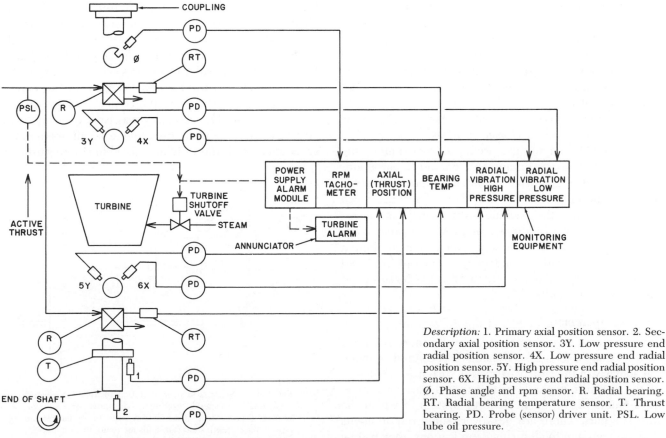

Description: 1. Primary axial position sensor. 2. Secondary axial position sensor. 3Y. Low pressure end radial position sensor. 4X. Low pressure end radial position sensor. 5Y. High pressure end radial position sensor. 6X. High pressure end radial position sensor. Ø. Phase angle and rpm sensor. R. Radial bearing. RT. Radial bearing temperature sensor. T. Thrust bearing. PD. Probe (sensor) driver unit. PSL. Low lube oil pressure.

Fig. 8.21d Turbine "Health-Monitoring" System. Adapted from Figure C-1 of American Petroleum Institute Standard 670, Noncontacting Vibration and Axial Position Monitoring System.

applications, but it is commonplace in electrical generator drives. Abnormal electrical conditions in a distribution system can cause protective devices to separate the generator from the system. Suddenly, a generator that may have been supplying megawatts of power and, consequently, magawatts of load to its prime mover can dissipate only a small amount of energy in the friction of its bearings and in windage on its moving parts. If the control system is not fast enough to reduce steam to the turbine it will overspeed. This condition can be detected by a variety of devices, such as centrifugal switches, electronic tachometers, or strain detecting devices installed on or near components of the machine that are affected by the overspeed condition.

Bearing temperature is an additional indication that lubrication is occurring. It is also a way of detecting deterioration in bearings before complete mechanical failure. Mechanical wear accelerates as a result of improper lubrication or because of mechanical stresses that cause deformation of the bearing's component parts. In either case, the bearing begins to dissipate abnormally large amounts of energy, which in turn results in heating. Consequently, a sudden rise in bearing temperature is generally an indication of incipient failure. It is important to discontinue operation of the machine quickly when a potential bearing failure is detected. Some turbine designs maintain very small clearances between stationary and rotating parts. If the bearing deforms it may mean the total destruction of the machine.

Especially on larger machines, a stationary vibration monitoring system is usually installed. Such a system generally consists of accelerometers or proximity sensors located radially in each bearing and axially on the end of the shaft or the thrust collar. Two sensors positioned at right angles are typically used in bearings. Electronic monitoriing equipment is used to measure the acceleration or displacement that occurs at each monitoring point. The monitoring equipment generally incorporates an "alarm" setting, which is intended to warn an operator of an impending problem, and a "danger" setting, which is intended to shut the machine down. In many instances, a vibration-initiated shutdown can prevent major damage in situations in which, without prompt action, equipment could be lost. Vibration monitoring equipment is covered extensively in Section 9.13 in the Process Measurement volume of this handbook.

Process Systems

Since steam turbines are so versatile, there is quite a variety of application-related methods of controlling them. Particularly in the case of extraction turbines, the design of the controls depends in substantial part on the requirements of the overall energy system of which the unit is a part. There are, however, some requirements that are common to all turbine applications; they will be discussed first.

One valve is common to all turbine applications: the valve between the inlet and the supply. This valve is the primary means of controlling the unit. When its position is changed, the energy introduced to the unit is matched with the energy required from it. This match is generally defined as the turbine operating at some intended speed. If the supply valve is too far open, the turbine will run at a speed above that desired. If the valve is too far closed, the turbine will slow down. In essence, the valve is controlling the flow of steam, generally measured in pounds per hour, into the turbine with the assumption that inlet and outlet conditions are constant. Under these conditions, there is a functional relationship between steam flow and shaft horsepower.

In extraction turbines, a second "valve" is required. It controls the steam extracted from the turbine and may be a control valve or an assembly within the turbine. Pressure is usually the measured variable for this function, although it can be shaft speed, or a combination of the two. If the turbine incorporates the controls as a built-in feature, the turbine is referred to as an "automatic-extraction" type. Such turbines are generally designed to deliver 100 percent shaft power and to provide extraction steam only if the load requirements permit. This is the most common type of extraction machine.

Basic speed control is almost always provided by a "governor." Governors come in a variety of types, including mechanical, hydraulic, and electrical. They all include some means of operating a pilot valve, which controls the turbine inlet valve in response to shaft speed.

Mechanical governors have developed from James Watt's original flyball governor shown in Figure 8.21e. The assembly consisted of two weights on the end of short arms with a hinge in the center to allow vertical motion of each. The assembly was caused to rotate by gearing of the weights to the machine's shaft. As the shaft speed increased, the weights tended to rise toward hor-

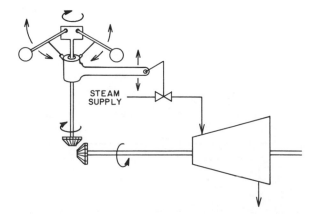

Fig. 8.21e Schematic representation of flyball governor. As shaft changes speed, rotating fly balls move up and down. Linkage then controls steam supply valve to regulate steam rate. If shaft speed is too fast, balls move up, raising linkage, which then closes down inlet valve. If shaft speed is too slow, balls drop, lowering collar and opening inlet valve.

izontal. A linkage was then used to control the throttling valve to admit less steam as the weights rose and more as they fell. This system was the beginning of machine control and is used almost unchanged in modern mechanical governors.

Hydraulic governors are more correctly referred to as "mechanical-hydraulic" devices. The shaft speed is generally detected by the flyball technique. Instead of a direct mechanical linkage between the position of the flyballs and the control valve, a hydraulic system is used to amplify the effect of small changes in flyball position and to condition the signal to produce specific control effects. The amplification feature allows very small changes in flyball position, corresponding to small changes in shaft speed, to produce effective control actions. This is essential for precise speed control. Signal conditioning allows control problems to be dealt with as well as the implementation of special characteristics.

There are two basic governor characteristics: (1) "isochronous," in which the objective is to maintain a certain speed regardless of load, and (2) "droop," in which the speed of the machine is deliberately decreased as load increases (Figure 8.21f). Droop governors can be essential for system stability in some applications. For example, if steam turbines are used to drive two electric generators that are electrically connected in parallel, a droop characteristic will allow the generators to share load, whereas an isochronous governor would not. If both generators were to run at exactly the same speed, the division of load between them would depend only on the electrical characteristics of the generators. If they were not precisely identical, the division of load would be unequal. If the speeds were not perfectly matched, one might carry all of the load. A droop governor, however, would cause the one with the heavier load to tend to slow down. When the load carried by the first generator matched that of the other unit and began to surpass it, the first unit would slow down. An equilibrium would be quickly established with each unit carrying a share of the load.

The problems requiring compensation generally are instabilities that occur during speed changes. Obviously, a high-gain control system can be susceptible to instability. A technique called "buffering" is used to eliminate such problems. In effect, the governor introduces droop on all speed changes and controls the rate at which the temporary droop characteristic is removed. In this way, speed transitions can be made smoothly. The droop due to buffering can be built into a governor whether the device is a droop type or not.

Electronic governors perform the same functions as their mechanical-hydraulic counterparts, but in a somewhat different way. The flyballs are replaced by an electronic tachometer input that is usually generated by a magnetic sensor. The sensor is triggered when the teeth of a gear connected to the machine's shaft pass by it. The varying reluctance of the magnetic circuit is used to generate a periodic function with a frequency proportional to the rotational speed of the shaft. Actuation is achieved by a control signal that operates a valve actuator, which may be pneumatic, electric or, most commonly, hydraulic. The characteristics of the controller itself depend on the transfer function built by the manufacturer into the unit's circuits. Generally, the governor is designed to perform the same function as older governor designs. The main difference is that a wider range of features can be built into a single unit, and one design can be easily adapted to a variety of applications. For example, one such unit, in addition to speed control, can maintain either inlet or exhaust pressure or can respond to some

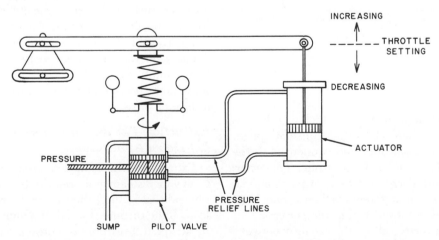

Fig. 8.21f Basic droop governor. On an increase in speed the fly balls move out, which raises the stem on the pilot valve. This movement is opposed by the spring. Pressure ports from the pilot valve to the top of the actuator, the bottom of the actuator drains to the sump through the pilot valve. This decreases the throttle setting. As the linkage moves down, the spring force increases and the force provided by the fly balls is exactly opposed. This moves the pilot valve back to the null position, which maintains the new lower speed.

other process condition. Further, it can automatically parallel generator sets, provide overspeed protection, and monitor other machine safety devices so a shutdown can be effected in the event of an unsafe condition. Electronic governors provide versatility rather than improved performance.

Steam turbine performance and speed control are addressed in API Standard 611, *General Purpose Steam Turbines for Refinery Services*, API Standard 612, *Special Purpose Steam Turbines for Refinery Services;* NEMA Standard SM21, *Multistage Steam Turbines for Mechanical Drive Service* (Table 8.21g); and NEMA Standard SM22, *Single Stage Turbines for Mechanical Drive Systems*. Performance objectives of governors are covered in NEMA Standard SM22-3.13. Governor systems are classified as A, B, C, or D, depending on performance objectives. Table 8.21g summarizes the basis for these classifications.

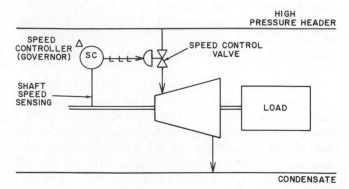

Fig. 8.21h Simple mechanical drive

Table 8.21g
GOVERNOR CLASSIFICATION AND
PERFORMANCE PER NEMA SM21

Class	Speed Range, %*	Maximum Speed Regulation, %†	Maximum Speed Variation, ±%‡	Maximum Speed Rise, %§	Trip Speed Setting, %‖
A	10, 20, 30, 50, 65	10	0.75	13	115
B	10, 20, 30, 50, 65, 80	6	0.50	7	110
C	10, 20, 30, 50, 65, 80	4	0.25	7	110
D	10, 30, 50, 65, 80, 85, 90	0.5	0.25	7	110

* Governor may be adjusted to produce any speed within this percentage of rated speed.
† Maximum speed variation from no load to full load.
‡ Maximum speed variation when operation is at constant load.
§ Maximum overspeed that can occur under any operating conditions.
‖ Proper overspeed trip setting to coordinate with governor maximum speed rise.

Applications

The simplest application is one in which a turbine is used to operate a mechanical load at constant speed. Steam is supplied from a header and condensed in the turbine (Figure 8.21h). A speed controller (governor) senses shaft speed and corrects it to a set point by manipulating the valve supplying steam to the turbine. Variations in load, caused by either shaft loading or variations in supply header pressure, affect the balance between the energy supplied to the turbine from the steam system and the work removed from the turbine's shaft. If the net change is positive (that is, more energy is available than is being used), the shaft will speed up. The governor will detect the increase in speed and act to eliminate it.

Its only way of doing so is to reduce the energy supplied to the turbine by closing the supply valve. If the net change were negative, the shaft would slow down. The governor would respond by opening the supply valve.

Pressure Let-Down Application

A noncondensing turbine is generally less expensive to buy and to operate than a condensing one, since the energy is extracted from steam with a higher enthalpy and, hence, smaller volume per unit of energy. This has the desirable effect of reducing the size of the turbine and, frequently, increasing the efficiency. Something must be done with the low-pressure exhaust steam, however. Since plants usually have a requirement for low-pressure steam for various loads, such as heating, noncondensing turbines can be used to advantage.

If the low-pressure header were supplied by throttling high-pressure steam to the lower pressure, considerable energy would be lost. If instead the header is supplied by a noncondensing turbine, a minimum of energy is lost. Figure 8.21i shows such an application.

The low-pressure header is supplied preferentially by the steam turbine, but since the objective of the turbine's speed controller is constant shaft speed, the amount of steam available from the turbine depends on its load. If demand on the low-pressure header exceeds the turbine's ability to supply it, additional steam is supplied through the pressure let-down valve. This valve responds to header pressure.

Extraction Turbine in Pressure Let-Down Application

Extraction turbines may be visualized as two-stage units from which steam at a pressure between that of the supply and that of the exhaust may be removed. When the load on the turbine is small, the high-pressure stage may be adequate to supply it; consequently, a large amount of extraction steam may be available to supply the low-pressure header. As load increases, the second stage becomes necessary and begins to compete for the steam previously being extracted. The control system must allow for this.

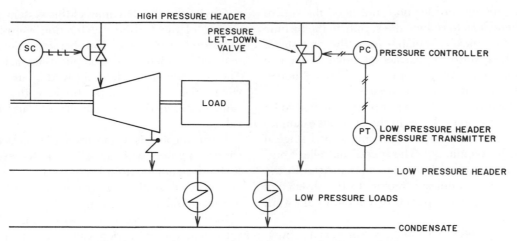

Fig. 8.21i Backpressure turbine in a pressure let-down application

Actually, there are even more constraints than are immediately obvious. Generally, at least a minimal amount of steam must be maintained through the second stage to prevent overheating. This requirement may necessitate limiting extraction, but it can also mean maintaining a specific second-stage discharge pressure. These requirements are identified by study of the manufacturer's operating specifications.

Figure 8.21j shows an extraction turbine in a pressure let-down application. In this example, the exhaust of the first stage is used to supply a low-pressure header. The second stage is assumed to be condensing. The speed controller (governor) is arranged by hydraulic or mechanical linkage to close (or nearly close) the extraction valve if speed control can be maintained by means of the first stage and the steam inlet valve. If speed cannot be

maintained, the extraction valve is opened, admitting more steam to the second (condensing) stage and, consequently, starving the low-pressure header. As in the previous example, a pressure-operated valve is used to maintain pressure on the low-pressure header when the available extraction steam is not sufficient.

Pressure-Regulating Extraction Application

So far it has been assumed that the low-pressure header will take whatever amount of steam is available from the turbine. If the supply is in excess of the low-pressure header's requirements, it would have to be condensed or vented to avoid overpressuring of the header. An alternative to both supplying an otherwise useless condenser and losing treated water by venting would be to operate the extraction turbine slightly differently (Figure 8.21k).

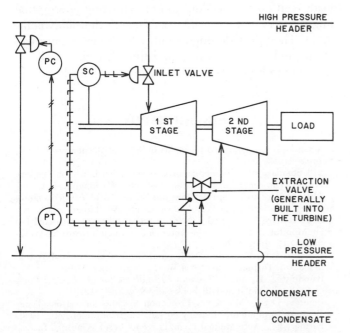

Fig. 8.21j Extraction turbine in pressure let-down application

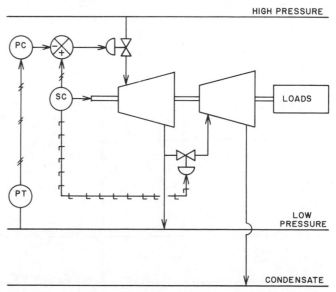

Fig. 8.21k Pressure-regulating extraction turbine application

If the low-pressure header does not need the steam, it might be beneficial to reduce the supply to the turbine and, hence, the energy available from the first stage. This would then cause the extraction valve to open so the condensing stage could supply the additional horsepower necessary to maintain shaft speed.

The speed controller in this example is assumed to be linked to the extraction valve as in the previous example, so that extraction steam is used in the second stage if speed cannot be maintained. The speed controller's output to the supply valve, however, goes instead to the positive input of an adder-subtractor. In the absence of a signal on the negative input, this scheme would work as in the previous example.

A pressure controller with low-pressure header pressure as its measured variable is used as the negative input to the adder-subtractor in Figure 8.21l. If the pressure controller is direct-acting, its output will increase on high header pressure. This will be subtracted from the speed controller's output and will force the inlet valve to close slightly. This will cause the turbine to begin to slow down. The speed controller will attempt to open the inlet valve, but the change will again be resisted by the pressure controller. As speed continues to fall off, the speed controller will begin opening the extraction valve. The pressure will then decrease, so the pressure controller will reduce its output, closing the extraction valve somewhat. Eventually, an equilibrium will be achieved.

Obviously, the dynamics of this system are dependent on use of the turbine's extraction control system as a pressure control valve. The characteristics of this "valve" may not be suitable for all applications (see Figure 8.21j). They can sometimes be improved by use of a characterizing device, such as a positioner with a characterizing cam, between the speed controller and the extraction valve.

If the turbine is a nonautomatic extraction type, the control engineer will have access to the extraction valve. If the designer is careful to observe the manufacturer's operating constraints, a system that works as described in Figure 8.21k can be constructed.

In this configuration, low speed opens both the inlet valve and the extraction valve, and high speed closes them. High pressure closes the inlet valve and opens the extraction valve. Low pressure closes the extraction valve and opens the inlet valve. Since the response to changing pressure requirements always produces complementary control operations on the inlet and extraction valves, speed control is not adversely affected by responses to pressure disturbances.

An intriguing aspect of this configuration is the possibility of eliminating the need to throttle steam at all if the turbine's operating capacities match low-pressure header load sufficiently. It uses the turbine's extraction capability to supply a range of low-pressure header loads. Obviously, the size and characteristics of the turbine have everything to do with the viability of such an approach.

Conclusion

Steam turbines are versatile energy conversion devices that, in addition to powering a variety of mechanical loads, can do an excellent job of extracting energy that might otherwise be wasted from a plant's thermodynamic cycle. They provide opportunities for process improvements and energy savings, and therefore their application should be carefully and insightfully considered.

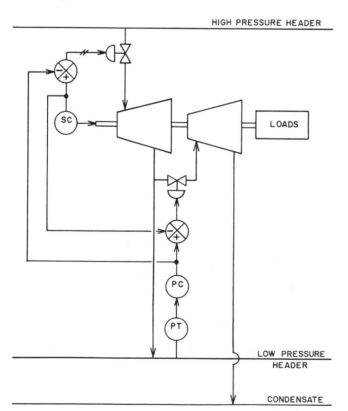

Fig. 8.21l   Condensing pressure let-down turbine

BIBLIOGRAPHY

Adamski, R.S., "Improved Reliability of Rotating Machinery," *InTech*, February 1983.

API Standard 611, "General Purpose Steam Turbines for Refinery Services," The American Petroleum Institute, Washington, D.C.

API Standard 612, "Special Purpose Steam Turbines for Refinery Services," The American Petroleum Institute, Washington, D.C.

API Standard 670, 1st ed., "Noncontacting Vibration and Axial Position Monitoring System," The American Petroleum Institute, Washington, D.C., June 1976.

Batsonis, P.B., Conrad, J.D., Jr., and Giras, T.C., "Digital System Will Control Turbine," *Electrical World*, November 1970.

Baumeister, T., *Marks' Standard Handbook for Mechanical Engineers*, 8th ed., McGraw-Hill Book Co., New York, 1978.

Bentley, D.E., "Machinery Protection Systems for Various Types of Rotating Equipment," Bentley Nevada Corp., Minden, NV, 1980.

Farmer, E., "Auto/Manual Logic Prevents Reset Windup," *Instruments and Control Systems*, October 1979, pp. 134–135.

Feuell, J., "Single Stage Steam Turbine–Generator Set Replaces Pressure Reducing Station to Reduce Plant Energy Costs," *Turbomachinery International*, January–February 1980.

———, "Steam Turbine Induction Generator Set," *Turbomachinery International*, May–June 1982.

Gires, T.C. and Birnbaum, M., "Digital Control for Large Steam Turbines," Proceedings of the American Power Conference, Chicago, April 1968.

NEMA Standard No. SM21, "Multistate Steam Turbines for Mechanical Drive Service."

NEMA Standard No. SM22, "Single Stage Turbines for Mechanical Drive Service."

Oetkin, A., "Steam Turbine Applications of 2301 Control Systems," The Woodward Governor Co., Ft. Collins, CO.

Osborne, R.L., "Controlling Central Station Steam Turbine Generators," *Instruments and Control Systems*, November 1975.

Perry, R.H. and Chilton, C.H., *Chemical Engineer's Handbook*, 5th ed., McGraw-Hill Book Co., New York, 1973, pp. 24–16 to 24–30.

Podolsky, L.B., "A Feedforward System for Digital Electrohydraulic Turbine Control," 1973 ISA Conference, Reprint No. 73–661, 1973.

———, Osborne, R.L., and Heiser, R.S., "Digital Electro-Hydraulic Control for Large Steam Turbines," Proceedings of the 14th International ISA Power Instrumentation Symposium, New York, May 1971.

Power Magazine, vol. 126, no. 4, April 1982, pp. 344–349.

Shapiro, L.J., "Condensing Turbine Can Improve Economics of Co-Generation," *Power Magazine*, vol. 126, no. 8, August 1982, pp. 73–74.

Staniar, W., *Plant Engineering Handbook*, 2nd ed., McGraw-Hill Book Co., New York, 1959, pp. 12–154 to 12–189 and 12–222 to 12–241.

The Woodward Governor Co., "Analytic Representation of Mechanical-Hydraulic and Electro-Hydraulic Governors," *Bulletin 25067*, The Woodward Governor Co., Ft. Collins, CO, 1978.

———, "43027 Electric Governing Extraction Control," *Bulletin 82802*, The Woodward Governor Co., Ft. Collins, CO, 1976.

———, "43027 Electric Control for Steam Turbine Applications," The Woodward Governor Co., Ft. Collins, CO, 1978.

8.22 WATER TREATMENT CONTROLS

Not many processes used in the treatment of water and wastewater are well suited to automatic process control. This is because of a notable absence of continuous analyzers reliable enough to measure all the necessary parameters of water quality. Chief among these are the lack of means to control coagulation and flocculation of water or the biological treatment of wastewater. Both of these processes are, for the most part, an art, not a science, still requiring some human judgment to determine chemical application rates and process control parameters. There are only four control measurements commonly in use: pH, oxidation-reduction potential (ORP), residual chlorine, and flow rate.

Other properties that are measured to assist the operator in controlling the system are conductivity, alkalinity, temperature, suspended solids, dissolved oxygen, and color.

Water and wastewater treatment consists of unit operations that may be classed as mechanical, chemical, biological, and any combinations of these. The mechanical operations most often used are screening, filtration, and separation by gravity.

The majority of treatment processes involve chemicals in continuous rather than batch systems. Emphasis will be placed on this aspect under these general headings: (1) chemical oxidation, (2) chemical reduction, (3) neutralization, (4) precipitation, and (5) biological control.

Critical design factors for all chemical treatment involve time, pH, concentration of contaminant(s), and chemical dosage rate.

Temperature is an effect in all chemical reactions, although most treatment processes currently in use are not substantially influenced by temperature variations. There are specific exceptions, but generally temperature control is not practiced.

Chemical Oxidation

Water and wastewater are treated by chemical oxidation in specific instances when the contaminant can be destroyed, its chemical properties altered, or its physical form changed. Examples of chemicals that can be destroyed are cyanides and phenol. Sulfides can be ox-idized to sulfates, thus changing their characteristics completely. Iron and manganese can be oxidized from the soluble ferrous or manganous state to the insoluble ferric or manganic state, respectively, permitting their removal by sedimentation. Strong oxidants, such as chlorine, chlorine dioxide, ozone, and potassium permanganate are used. Chlorine is preferred when it can be used because it is the least expensive and is readily available.

All chemical reactions of this type are pH-dependent in relation to the time required for the reaction to proceed to completion and for the desired end products. Residual oxidant or ORP measurement is used to control the process.

Oxidation-reduction implies a reversible reaction. Since these reactions are carried to completion and are not reversible, the term is misleading. In practice, control is by what may be called "electrode potential readings." An illustration is the oxidation of cyanide into cyanate with chlorine, according to the following reaction:

$$
\begin{array}{ll}
2Cl_2 & \text{Chlorine} \\
+ & \\
4NaOH & \text{Sodium hydroxide} \\
+ & \\
2NaCN & \text{Sodium cyanide} \\
\downarrow & \\
2NaCNO & \text{Sodium cyanate} \\
+ & \\
4NaCl & \text{Sodium chloride} \\
+ & \\
2H_2O & \text{Water} \qquad 8.22(1)
\end{array}
$$

The electrode potential of the cyanide waste solution will be on the order of -200 to -400 millivolts. After sufficient chlorine has been applied to complete the reaction according to equation 8.22(1), the electrode potential will be on the order of $+200$ to $+400$ mv. The potential value will not increase until *all* cyanide has been oxidized. Control of pH is essential, with the minimum being 8.5. The reaction rate is faster at higher values.

Complete oxidation (destruction of cyanide) is a two-step reaction. The first step is oxidation to the cyanate level described in equation 8.22(1). The end point of the reaction is again detected by electrode potential readings and will be on the order of +700 to +800 mv. The overall reaction is:

$$
\begin{array}{ll}
5Cl_2 & \text{Chlorine} \\
+ & \\
10NaOH & \text{Sodium hydroxide} \\
+ & \\
2NaCN & \text{Sodium cyanide} \\
\downarrow & \\
2NaHCO_3 & \text{Sodium bicarbonate} \\
+ & \\
N_2 & \text{Nitrogen} \\
+ & \\
4H_2O & \text{Water} \qquad\qquad 8.22(2)
\end{array}
$$

Although the process and process control are generally similar in all chemical oxidation reactions, the oxidation of cyanide is used as an example because it alone involves a two-step process, whereas the other applications cited involve only one. Because of this added complexity, the toxicity of cyanide, and the rigid requirements on waste discharge, a batch-type treatment is recommended. This affords the assurance of complete treatment before discharge. In Figure 8.22a, chlorine is charged at a constant rate, with automatic pH control of caustic addition maintaining the batch pH at a value of 9.5.

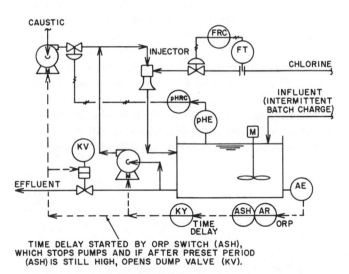

TIME DELAY STARTED BY ORP SWITCH (ASH), WHICH STOPS PUMPS AND IF AFTER PRESET PERIOD (ASH) IS STILL HIGH, OPENS DUMP VALVE (KV).

Fig. 8.22a   Batch oxidation of cyanide waste with chlorine

When the ORP set point of +750 mv is reached on the ORP controller, a delay timer is actuated and the chemical feed systems are shut down. After a 30-minute delay, the tank contents are discharged if the potential reading is at or above the set point. If not, it would indicate that further reaction has been taking place in the delay period, resulting in a subsequent drop in potential value. In this event, the system is reactivated and the cycle is repeated. Additional, usually duplicate, tanks

are required to receive the incoming waste while others are being used for treatment.

Continuous flow-through systems offer the advantage of reduced space requirements, but this is often offset in capital costs by the additional process equipment that is required. In the system shown in Figure 8.22b, the two reaction steps previously mentioned are separated. In the first step, the ORP controller set point is approximately +300 mv. It controls the addition of chlorine to oxidize the cyanide into cyanate. The pH is maintained at approximately 10. Reaction time is on the order of five minutes. Since the second step (that of oxidation of the cyanates) requires an additional amount of chlorine to be charged at nearly the same rate as in the first step, one obtains the chlorine flow rate signal by measuring the chlorine feed rate to the first step and multiplying it by a constant to control the chlorine feed rate in the second step. Caustic requirement is dependent solely on the chlorine rate (pH control is not necessary), and therefore the same signal can be used to adjust caustic feed. The ORP instrument that is sampling the final effluent can signal process failure if the potential level drops below approximately +750 mv.

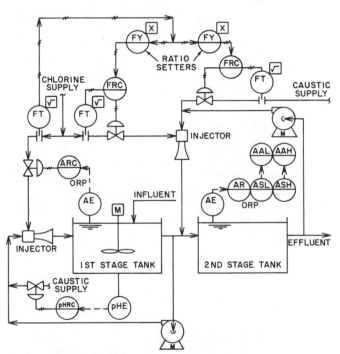

Fig. 8.22b   Continuous oxidation of cyanide waste with chlorine. Influent here has continuous constant flow rate and variable quality.

It would appear that a feedback loop from the second ORP analyzer to the second chlorinator is desirable. Practice has not shown this to be necessary, because the ratioing accuracy available between first-stage chlorine rate and secondary addition (approximately 1:1) is sufficiently high. At worst, the system as shown will apply a little more chlorine than is actually required. Table

Table 8.22c
SET POINTS AND PARAMETERS

Parameters	Process Steps	
	Cyanide to Cyanate	Destruction of Cyanate
pH	10–12	8.5–9.5
Reaction Time (Minutes)	5	45
ORP (mv) Set Point	+300	+750
Maximum Concentration of Cyanide (Cyanate) That Can Be Treated	1000 milligrams per liter	1000 milligrams per liter

8.22c lists the set points and variables applicable to this oxidation process. Fixed flow rate systems are preferred to provide constant reaction times.

The use of residual chlorine analyzers is not applicable to this process, since the metal ions usually present in the waste and the intermediate products interfere with accurate determinations. They are used in processes in which the presence of excess residual chlorine indicates a completed reaction. The set point is usually 1 milligram per liter or less.

Figure 8.22d is a schematic typical of systems with variable quality and variable flow rate. The chlorinator has two operators: one controlled by feedforward, the other by a feedback loop. Most reactions are completed within five minutes, and except for cyanide treatment, most all other chemical oxidation operations are carried out simultaneously with other unit operations, such as

coagulation and precipitation, which govern the pH value. Thus, whereas the pH value affects the rate of reaction, it is seldom controlled solely for the oxidation process.

Chemical Reduction

Chemical reduction is quite similar to chemical oxidation, except that reducing reactions are involved. Commonly used reductants are sulfur dioxide and its sodium salts, such as sulfite, bisulfite, and metabisulfite. Ferrous iron salts are infrequently used. Typical examples are reduction of hexavalent chromium, dechlorination, and deoxygenation.

Figure 8.22e is a schematic of a typical system for the reduction of highly toxic hexavalent chromium to the innocuous trivalent form according to the following reaction:

$$
\begin{array}{ll}
3SO_2 & \text{Sulfur dioxide} \\
+ & \\
2H_2CrO_4 & \text{Chromic acid} \\
\downarrow & \\
Cr_2(SO_4)_3 & \text{Chromic sulfate} \\
+ & \\
2H_2O & \text{Water} \qquad\qquad 8.22(3)
\end{array}
$$

Most hexavalent chrome wastes are acid, but the rate of reaction is much faster at very low pH values. For this reason pH control is essential. Sulfuric acid is preferred because it is cheaper than other mineral acids. The set point of the pH controller is approximately 2. As in the treatment of cyanide, the chemical reaction is not reversible, and the control of sulfur dioxide addition is by electrode potential level, using ORP instrumentation. The potential level of hexavalent chromium is +700 to +1000 mv, whereas that of the reduced trivalent chrome is +200 to +400 mv. The set point on the ORP controller is approximately +300 mv.

The control system consists of feedback loops for both

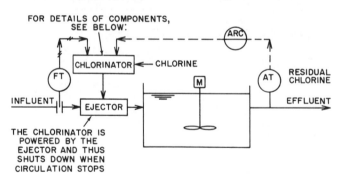

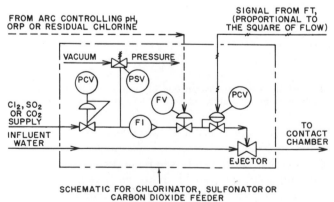

Fig. 8.22d Variable quality and flow rate of waste oxidized by chlorine

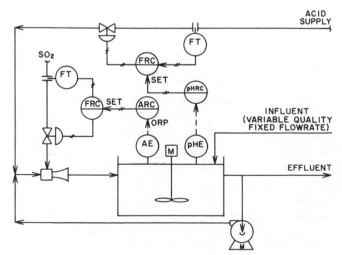

Fig. 8.22e Reduction of chromium waste with sulfur dioxide

pH and ORP. The common and preferred design is for fixed flow systems.

The trivalent chromic sulfate is removed from solution by subsequent raising of the pH to 8.5, at which point it will precipitate as chromic hydroxide. The control system for this step is identical with the one used in Figure 8.22k. Critical process control factors are summarized as follows:

	Set Points and Parameters
Variable	*Value*
pII	2.0
ORP (mv) set point	+300
Reaction time	10 at pH 2.0
(minutes)	5 at pH 1.5

In the other examples cited, dechlorination and deoxygenation, control consists of adding the reducing agent in proportion to the oxidant concentration but maintaining a slight excess. In most cases, a slight excess of reducing agent is not detrimental. The pH value is not critical and is determined by other factors. Corrosion control is the most common of these factors.

Dechlorination to a fixed residual value is controlled, as illustrated in Figure 8.22d, except that sulfur dioxide is used instead of chlorine.

Neutralization

Strong alkalis react quickly and efficiently to neutralize strong acids. The simplicity of this fact is misleading in its consequences. Acid neutralization is a common requirement in wastewater treatment, but few operations can be as complex. Essential information required for proper design includes: (1) flow rate and range of flow variations; (2) titratable acid content and variations in acid concentration; (3) rate of reaction; and (4) discharge requirements for suspended solids, dissolved solids, and pH range.

(The phenomenon of pH and its detection is covered in Section 10.14 in the Process Measurement volume of this Handbook and in Section 8.16 of this volume.)

When the purpose of a control system is automatically to neutralize plant wastes, an understanding of the neutralization phenomenon is necessary. Figure 8.22f shows the pH values corresponding to the changing mixtures of a strong acid and a strong base. The slope of this pH curve is so great near neutrality (pH = 7) that there is no likelihood of controlling such a system. Fortunately, plant effluents usually contain weak acids or bases that are neutralized by strong reagents. (The dotted pH curve shows that these are much easier processes to control.) The slope of the pH curve is affected by the ionization constants of the acid and base involved and by buffering. Buffering compounds are those that contain no hydrogen or hydroxyl ions but are capable of suppressing the re-

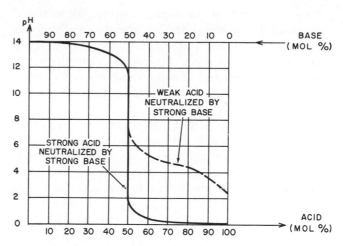

Fig. 8.22f   pH curve of a strong acid and base combination

lease of these ions from other solutes and thereby affect the solution acidity or alkalinity.

Neutralization control of plant wastes is difficult because of likely variation in:

acid or base contents by several decades.
the type of acid (or base), thereby changing the applicable pH curve.
amount of buffering.
effluent from acidic to basic, which would require two reagents.
flow rates of effluent.

As a consequence, a reagent addition rangeability of several hundred to one may be required. This is accomplished by the use of two or more control valves in parallel. As shown in Figure 8.22g, the smaller valve has equal percentage characteristics and is throttled by a proportional controller. This is desirable to match the pH characteristics near neutrality with that of the valve. In order to maintain the relationship between pH and valve opening, the reset mode had to be eliminated. If

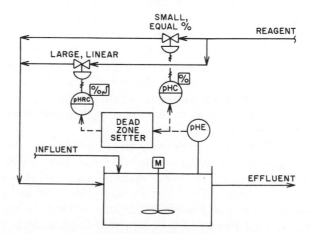

Fig. 8.22g   Addition of reagent with high rangeability for precise neutralization. (From F. G. Shinskey, *Process Control Systems*, McGraw-Hill Book Co., New York, 1967.)

pH measurement moves outside a preset "dead zone," this causes the second controller to make an adjustment in the opening of the large linear valve, thereby compensating for load changes. This second controller is provided with two control modes, with the integral action serving to bring the system back to set point after a load change.

It should be emphasized that pH is a measure of hydrogen ion activity and not acid concentration. A weak sulfuric acid solution will have a low pH value because of the high degree of hydrogen ion activity (disassociation), whereas some strong organic acids may show a pH value as high as 3 or 4.

An equalizing basin should be installed ahead of the neutralizing system whenever possible. This will tend to level out fluctuations in influent flow and concentration. This point cannot be overemphasized, because the lack of such a basin has been the cause of many failures. Pumping from an equalizing basin at a constant rate eliminates the need for high rangeability flow rate instrumentation. This, in combination with reduced variations in base or acid content, reduces the reagent feed range requirements. The obvious disadvantage is in the capital cost for large basins. Most systems are designed with as large an equalization tank as possible, consistent with available space. Any equalization that can be installed will pay dividends.

Using acid wastes as an example, the maximum acid concentration and maximum flow rate will determine the capacity of the alkali feed system. The ratio between this and the requirement at minimum flow and minimum acid concentration will determine the range requirements of the reagent feed system. When an equalization tank is used, a more dependable operation with lower investments in high rangeability equipment can be expected.

It is essential that reaction rates be determined so that suitable reaction tank sizes (residence time) can be calculated. One plots these rates by determining the total amount of alkali, in case of an acid waste, that is required to neutralize a sample of the waste, making sure the reaction has gone to completion. This amount is then added to a second sample as a single dose, and the pH rise versus time is plotted. Typical curves that can be expected are shown in Figure 8.22h. Sizing to provide at least 50 percent more holding time than that shown by the curve is recommended.

The ideal aim is sufficient equalization to provide a homogeneous waste at a constant flow rate. To neutralize such a system, a lime feeder operating at a preset constant rate would suffice. Unfortunately, this seldom occurs, and provision must be made for reagent throttling. Figure 8.22i illustrates a system that can handle both flow and acid concentration changes.

The three treatment tanks and the final control elements for lime slurry feeding have identical capacities.

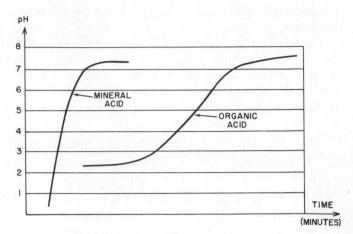

Fig. 8.22h   Acid neutralization reaction rates

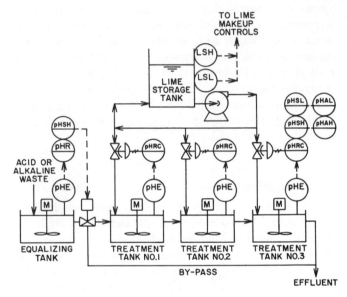

Fig. 8.22i   Acid waste neutralization system

Assuming a 10:1 range for each lime slurry control element, the range of the system is 30:1. Thus, it can handle any combination of flow and acid concentration within that range.

The set point on the three pH controllers is at the final pH value desired. At periods of low flow, or when the acid content is low, treatment tank number 1 can handle the entire requirement. Under other conditions, all three tanks may be required, with the first one or two satisfying a major portion of the reagent requirement and the third serving a "polishing," or final trim, function.

One must keep two factors in mind in designing an acid neutralization control system: (1) pH is a logarithmic function, and (2) pH expresses hydrogen ion activity, not acid concentration.

Each tank in this system should be sized for a minimum of 50 percent of the total retention time determined at maximum flow rate. Where mixtures of acids are involved, the maximum time (not average) must be used.

Figure 8.22i has provision for the occasional case in which incoming streams may be self-neutralizing. If this occurs, the treatment system is bypassed.

There are obviously many variations of this schematic that may be practical after a careful study of the waste characteristics. Tank sizes and number may be varied, the final control elements for lime feed can be of different sizes, and provision for recycle can be made if necessary.

Before design is started, regulations regarding discharge must be known. A common allowable range for pH values is between 6 and 8. There may also be limits on suspended or dissolved solids. The latter may be expressed as conductivity limits. A solution to these restrictions may be found in the choice of neutralizing chemicals. For instance, the resulting product of neutralizing sulfuric acid with lime (the cheapest alkali available) is calcium sulfate, a relatively insoluble product. This can be removed by sedimentation to reduce the quantity of suspended solids. Caustic soda reacts with sulfuric acid to form soluble sodium sulfate, requiring no subsequent sedimentation, but the effluent will contain substantial quantities of dissolved solids.

High maintenance costs of pH electrodes have been reported when lime is used as the reagent, caused by the formation of calcium sulfate coatings on the electrodes. Daily maintenance may be needed. (See Section 10.4 in the Process Measurement volume of this handbook for details on devices for cleaning electrodes.)

In those systems in which equalizing basins or other averaging techniques cannot be applied and accurate pH control is required, the concept of feedback control is insufficient, and feedforward schemes should be considered. Assuming that the reagent control valve is equal percentage in its characteristics, this can be used to advantage in a feedforward system to generate the forward loop pH function. Because the relationship between influent pH and reagent required to neutralize it is variable, feedback trimming of any feedforward loop is essential. A generalized feedforward-feedback model for the equal percentage valve position (not flow through it, but the signal it receives), represented by X, can be written as

$$X = (K_c)(\text{set point} - \text{measurement}) + \\ + \log (Fa) \quad 8.22(4)$$

where K_c is the proportional gain of the feedforward controller, F is the influent flow rate, and a is the output signal from the feedback controller.

The nature of a neutralization system is such that the process gain (slope of the pH curve) is likely to be the highest at set point (around pH = 7) and decreases as deviation from set point increases. To compensate for this inverse relationship between process gain and deviation, the feedback controller is to be a nonlinear one, with its gain directly varying (increasing) with deviation.

With this controller, as shown in Figure 8.22j, the greater the deviation from set point, the larger the controller correction, but this does not create an unstable condition, because at high deviations the process gain is low.

If the reagent flow rangeability exceeds the capability of a single valve, the technique illustrated in Figure 8.22g can be applied.

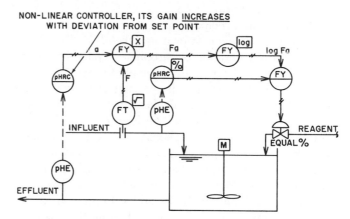

Fig. 8.22j Feedforward-feedback neutralization control

Precipitation

Precipitation is the creation of insoluble materials by chemical reactions that provides treatment through subsequent liquid-solids separation. Typical of these operations is the removal of sulfates, removal of trivalent chromium, and softening of water with lime. Iron and manganese are removed by a variation of this process following the treatment discussed earlier, in connection with the chemical oxidation process. Lime softening is a common process and will be used as an example.

The reaction involved is

$$Ca(HCO_3)_2 \quad \text{Calcium bicarbonate} \\ + \\ Ca(OH)_2 \quad \text{Calcium hydroxide} \\ \downarrow \\ 2CaCO_3 \quad \text{Calcium carbonate} \\ + \\ 2H_2O \quad \text{Water} \quad 8.22(5)$$

The calcium carbonate formed by this reaction is relatively insoluble and can be removed by graviity separation (settling). Typical settling time is 30 minutes or less, but most systems are designed for continuous operation with typical detention times of one hour.

Water treatment using this process is called *excess lime softening* derived from the application of lime in excess of that required for the reaction described in equation 8.22(5). Control consists of adding sufficient calcium hydroxide to maintain an excess hydroxide alkalinity of 10 to 50 milligrams per liter, as shown by equation 8.22(6):

$$2P = MO + 10 \text{ to } 50 \quad 8.22(6)$$

where P is the phenolphthalein alkalinity and MO is the methyl orange alkalinity. This results in a pH value of 10 to 11, but pH control is not satisfactory for economical operations. This is an example in which suitable analytical instrumentation is not available for continuous system control. If the quality of the untreated water is variable, operator control of lime dosage is essential. Pacing of manual dosage by feedforward control from flow rate is most frequently practiced.

A factor in the precipitation of calcium carbonate is a chemical phenomenon known as *crystal seeding*. This involves the acceleration of carbonate crystal formation by the presence of previously precipitated crystals. This is accomplished in practice by passing the water being treated through a "sludge blanket" in an upflow treatment unit shown schematically in Figure 8.22k. The resulting crystals of calcium carbonate are hard, dense, and discrete, and they separate readily. When colloidal suspended material is also to be removed, which would be the case where surface waters are softened, a coagulant of aluminum or iron salts is also added to precipitate the colloids. Dosage is variable, depending on the quantity of suspended material. Application of both coagulant and calcium hydroxide is controlled by flow ratio modulation.

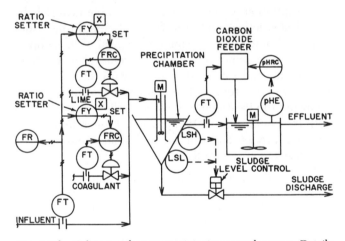

Fig. 8.22k   Calcium carbonate precipitation control system. Details of carbon dioxide feeder are shown in Figure 8.22d.

The resulting sludge, consisting of calcium carbonate, aluminum, or iron hydroxides, and the precipitated colloidal material are discharged to waste continuously. As previously noted, the presence of some precipitated carbonate is beneficial in order to remove all the sludge. An automatic sludge level control system is used in Figure 8.22k to control the sludge level at an optimum.

Water softened by the excess lime treatment is saturated with calcium carbonate and therefore is unstable. Stability is achieved by addition of carbon dioxide to convert a portion of the carbonates into bicarbonate, according to the following equation:

$$CO_2 \quad \text{Carbon dioxide}$$
$$+$$
$$CaCO_3 \quad \text{Calcium carbonate}$$
$$+$$
$$H_2O \quad \text{Water}$$
$$\downarrow$$
$$Ca(HCO_3)_2 \quad \text{Calcium bicarbonate} \qquad 8.22(7)$$

In contrast with the softening reaction, this process is suited to automatic pH control. Figure 8.22k shows this control system. The carbon dioxide feeder has two operators, one controlled by feedforward on influent flow and the other by feedback on effluent pH. The set point is on the order of 9.5 pH.

Fouling of the electrodes is likely to occur as a result of precipitation of crystallized calcium carbonate. Daily maintenance may be expected, unless automated cleaners are used. The farther downstream (from the point of carbon dioxide application) the electrodes can be placed, consistent with acceptable loop time delays, the less will be the maintenance requirement.

Biological Control

Nearly every process for water or wastewater treatment uses chlorination. It may be used either to reduce the possibility of pathogenic bacterial contamination of the receiving water or to prevent interference of biological growth with other processes. Disinfection of domestic waste and of potable water supplies is almost universally practiced. Control of biological slimes that interfere with heat exchange in cooling water systems is readily accomplished with chlorine. The pulp and paper industry and the food industry typically use chlorination to prevent product deterioration.

Chlorine is a strong oxidant with many other uses, as was noted earlier in this section. It is a very effective bactericidal agent readily available at reasonable cost. Its effectiveness has been proved by over 60 years of use.

Several factors add to the complexity of chlorination systems for biological control. Materials in the water that can be chemically oxidized cause an immediate reduction of the available active chlorine into an ineffective chloride form. Sufficient chlorine must be added to the water to account for this reaction and, in addition, to provide a residual amount, reserving sufficient chlorine for subsequent reactions. The amount of chlorine involved in the initial reduction is called *chlorine demand*. A dosage greater than this amount provides excess chlorine, which is called *residual chlorine*.

When nitrogenous material, particularly ammonia, is present in the water, the residual chlorine will be altered. Two kinds of residual chlorine are recognized: (1) Free residual is that remaining after the destruction with chlorine of ammonia or of certain organic nitrogen compounds. (2) Combined residual is produced by the reaction of chlorine with natural or added ammonia or with certain organic nitrogen compounds. This subject is dealt with adequately in "Water Treatment Plant Design," American Water Works Association Inc., 1969.

Since these two compounds (free residual chlorine and combined residual chlorine) are completely different in their ability to control bacterial organisms, it is important to differentiate which of the two forms of chlorine are involved in a process. Laboratory methods are available for both the measurement and the differentiation of these two. Continuous analyzers suitable for control are also available to measure either the level of free residual chlorine or the level of total residual chlorine, meaning the combination of the two when both are present. The presence of residual chlorine in water does not ensure either disinfection or biological control. A bacteriological analysis requires several hours, or even days. Through years of experience, it has been determined that measurement of residual chlorine can be a suitable inferential indicator of the effectiveness of biological control. For this reason, such processes are suitably controlled by analysis of residual chlorine.

An example of a typical control system used in the disinfection of wastewater is shown schematically in Figure 8.22l. The chemical feed system for applying chlorine must be of sufficient capacity both to satisfy the demand of the waste and to provide sufficient residual after the contact time required for the disinfection action. Typical of this for wastewater treatment would be a chlorine demand of 5 milligrams per liter.

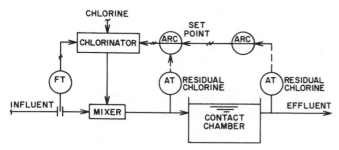

Fig. 8.22l Chlorine disinfection of waste water with variable flow rate and variable quality. Details of chlorinator are shown in Figure 8.22d.

Another important factor is that the amount of residual chlorine will decline with time. There is no assurance that the initial residual concentration will persist for the length of time required for disinfection to be accomplished. It is important that residual chlorine be present during this entire period. Typical design is for a contact chamber sized to permit retention for 30 minutes at maximum flow. Local regulations usually require that there be a minimum of 1 milligram per liter of available chlorine at the end of this contact time. It is not uncommon for flows to vary over a 6:1 range. To account for these variables and for the continuing chemical reactions, the control system automatically controls the residual after a short (approximately 5 minutes) fixed contact time. This

is accomplished by building the 5-minute retention time into the sampling system. Feedback control, in conjunction with flow-proportioning feedforward control, establishes a constant residual value at the inlet of the contact chamber. Because of the variable rate at which residual decay occurs, assurance that the residual at the exit of the tank is maintained at a minimum value requires a second analyzer to change the set point on the controller as required.

The amount of residual chlorine decay varies widely as a result of variations in temperature, quality of wastewater, and detention times. It may vary from a minimum of 0.5 to 5 milligrams per liter. To maintain an effluent residual of 1 milligram per liter, the set point on the controller for the chlorinator may be anywhere from 1.5 to 6 milligrams per liter.

Chlorination of wastewater for disinfection is unique in that it is usually the final process unit prior to discharge. For this reason detention time is provided as a part of the process. For most other biological control applications, other subsequent unit operations provide sufficient contact time, and the residual decay can be reasonably well predicted. A typical system is shown in Figure 8.22d.

Conclusions

Most water and wastewater treatment systems are designed for continuous operation, using several process units. Many of the processes consist of chemical treatment. Relatively few of these are suited to automatic

Table 8.22m
CONTROL INSTRUMENTATION APPLICABLE TO VARIOUS EFFLUENT AND WATER TREATMENT SYSTEMS

Treatment Process	Instrumentation			
	Flow	pH	ORP	Residual Chlorine
Chemical Oxidation of				
Cyanide		✓	✓	
Iron	✓			✓
Manganese	✓			✓
Hydrogen Sulfide	✓			✓
Chemical Reduction of				
Chromium		✓	✓	
Residual Chlorine	✓			✓
Precipitation of				
Chromium	✓	✓		
Iron	✓	✓		
Manganese	✓	✓		
Hardness	✓			
Neutralization of				
Acid and Alkali	✓	✓		
Alkalinity (Recarbonation)	✓	✓		
Biological Control	✓			✓

control because of the lack of reliable sensing devices. The development of *selective ion electrodes* suitable for continual analysis will, it is hoped, increase the range of automatic control.

Some of the features of automated effluent and water treatment systems are summarized in Table 8.22m.

The natural laws governing chemical reactions dictate the design considerations. The most critical of these involve (1) pH, (2) reaction rates, (3) ratios (chemical dosage), (4) concentration, and (5) temperature.

A discussion of coagulation and flocculation of water for suspended solids removal is conspicuously absent. It is because these have not yet been reduced from an art to a science. The same is true for biological waste treatment, although oxygen measurement (see Section 11.21 in the Process Measurement volume of this handbook) is coming into wider use for guidance to operator control.

BIBLIOGRAPHY

Bethlehem Regulator Selection Handbook, The Bethlehem Corporation, Bethlehem, Pennsylvania.

Butler, J.N., *Ionic Equilibrium: A Mathematical Approach*, Addison-Wesley, Reading, Massachusetts, 1964.

Corrigan, T.E. and Young, E.F., "General Considerations in Reactor Design—II," *Chemical Engineering*, October 1955.

Docherty, P.J., Jr., Automatic pH Control: Neutralization of Acid Wastes by Addition of Lime Slurry, Masters Thesis, Dartmouth College, Hanover, New Hampshire, June 1972.

Guisti, A.L. and Hougen, J.O., "Dynamic of pH Electrodes," *Control Engineering*, April 1961.

Hoyle, D.L., "The Effect of Process Design on pH and pIon Control," Presented at the 18th National Symposium of the Analytical Instrumentation Division of the Instrument Society of America, San Francisco, May 3–5, 1972.

"Ion-Selective Measuring Systems," *Technical Information Sheet* 43-31a, The Foxboro Company, Foxboro, Massachusetts.

Light, T.S., "A Standard Solution for Oxidation-Reduction Potential (ORP) Measurements," *Analytical Chemistry*, May 1972.

Moore, F.E., "Taking Errors out of pH Measurement by Grounding and Shielding," *Instrument Society of America Journal*, February 1966.

Negus, L.E. and Light, T.S., "Temperature Coefficients and Their Compensation in Ion-Selective Measuring Systems," Presented at the 18th National Symposium of Analytical Instrumentation Division of the Instrument Society of America, San Francisco, May 3–5, 1972.

Parsons, W.A., *Chemical Treatment of Sewage and Industrial Wastes*, National Lime Association, Washington, D.C., 1965.

Shinskey, F.G. and Myron, T.J., "Adaptive Feedback Applied to Feedforward pH Control," Instrument Society of America Paper No. 565-70.

APPENDIX

The contents of this appendix are intended to complement those of the Process Measurement volume of this HANDBOOK. It is assumed that the owner of this volume already possesses the volume on Process Measurement and therefore has ready access to its appendix containing information on conversion factors, chemical resistances of materials, compositions of metals, steam and water tables, pipe friction, tank volumes, and abbreviations.

CONTENTS OF APPENDIX

A.1 INTERNATIONAL SYSTEM OF UNITS

The decimal system of units was conceived in the 16th century when there was a great confusion and jumble of units of weights and measures. It was not until 1790, however, that the French National Assembly requested the French Academy of Sciences to work out a system of units suitable for adoption by the entire world. This system, based on the metre as a unit of length and the gram as a unit of mass, was adopted as a practical measure to benefit industry and commerce. Physicists soon re-alized its advantages and it was adopted also in scientific and technical circles. The importance of the regulation of weights and measures was recognized in Article 1, Section 8, when the United States Constitution was written in 1787, but the metric system was not legalized in this country until 1866. In 1893, the international metre and kilogram became the fundamental standards of length and mass in the United States, both for metric and customary weights and measures.[1]

Table A.1a
INTERNATIONAL SYSTEM OF UNITS

Quantity	Unit	SI Symbol	Formula	Quantity	Unit	SI Symbol	Formula
BASE UNITS				energy	joule	J	$N \cdot m$
length	metre	m	—	entropy	joule per kelvin	—	J/K
mass	kilogram	kg	—	force	newton	N	$kg \cdot m/s^2$
time	second	s	—	frequency	hertz	Hz	(cycle)/s
electric current	ampere	A	—	illuminance	lux	lx	lm/m^2
thermodynamic temperature	kelvin	K	—	luminance	candela per square metre	—	cd/m^2
amount of substance	mole	mol	—	luminous flux	lumen	lm	$cd \cdot sr$
luminous intensity	candela	cd	—	magnetic field strength	ampere per metre	—	A/m
SUPPLEMENTARY UNITS:				magnetic flux	weber	Wb	$V \cdot s$
plane angle	radian	rad	—	magnetic flux density	tesla	T	Wb/m^2
solid angle	steradian	sr	—	magnetomotive force	ampere	A	—
DERIVED UNITS:				**DERIVED UNITS:**			
acceleration	metre per second squared	—	m/s^2	power	watt	W	J/s
activity (of a radio-active source)	disintegration per second	—	(disintegration)/s	pressure	pascal	Pa	N/m^2
				quantity of electricity	coulomb	C	$A \cdot s$
angular acceleration	radian per second squared	—	rad/s^2	quantity of heat	joule	J	$N \cdot m$
				radiant intensity	watt per steradian	—	W/sr
angular velocity	radian per second	—	rad/s	specific heat	joule per kilogram-kelvin	—	$J/kg \cdot K$
area	square metre	—	m^2				
density	kilogram per cubic metre	—	kg/m^3	stress	pascal	Pa	N/m^2
				thermal conductivity	watt per metre-kelvin	—	$W/m \cdot K$
electric capacitance	farad	F	$A \cdot s/V$	velocity	metre per second	—	m/s
electrical conductance	siemens	S	A/V	viscosity, dynamic	pascal-second	—	$Pa \cdot s$
electric field strength	volt per metre	—	V/m	viscosity, kinematic	square metre per second	—	m^2/s
electric inductance	henry	H	$V \cdot s/A$				
electric potential difference	volt	V	W/A	voltage	volt	V	W/A
				volume	cubic metre	—	m^3
electric resistance	ohm	Ω	V/A	wavenumber	reciprocal metre	—	(wave)/m
electromotive force	volt	V	W/A	work	joule	J	$N \cdot m$

The following tables of conversion factors are intended to serve two purposes:

1. To express the definitions of miscellaneous units of measure as exact numerical multiples of coherent "metric" units. Relationships that are exact in terms of the base unit are followed by an asterisk. Relationships that are not followed by an asterisk are either the results of physical measurements or are only approximate.

2. To provide multiplying factors for converting expressions of measurements given by numbers and miscellaneous units to corresponding new numbers and metric units.

Conversion factors are presented for ready adaptation to computer readout and electronic data transmission. The factors are written as a number equal to or greater than one and less than ten with six or less decimal places. This number is followed by the letter E (for exponent), a plus or minus symbol, and two digits which indicate the power of 10 by which the number must be multiplied to obtain the correct value. For example:

$$3.523\ 907\ E-02 \text{ is } 3.523\ 907 \times 10^{-2}$$

or

$$0.035\ 239\ 07$$

Similarly:

$$3.386\ 389\ E+03 \text{ is } 3.386\ 389 \times 10^3$$

or

$$3\ 386.389$$

An asterisk (*) after the sixth decimal place indicates that the conversion factor is exact and that all subsequent digits are zero.

When a figure is to be rounded to fewer digits than the total number available, the procedure should be as follows:

1. When the first digit discarded is less than 5, the last digit retained should not be changed. For example, 3.463 25, if rounded to four digits, would be 3.463; if rounded to three digits, 3.46.

2. When the first digit discarded is greater than 5, or if it is a 5 followed by at least one digit other than 0, the last figure retained should be increased by one unit. For example 8.376 52, if rounded to four digits, would be 8.377; if rounded to three digits, 8.38.

3. When the first digit discarded is exactly 5, followed only by zeros, the last digit retained should be rounded upward if it is an odd number, but no adjustment made if it is an even number. For example, 4.365, when rounded to three digits, becomes 4.36. The number 4.355 would also round to the same value, 4.36, if rounded to three digits.

Where less than six decimal places are shown, more precision is not warranted.

Table A.1b
ALPHABETICAL LIST OF UNITS
(Symbols of SI units given in parentheses)

To convert from	To	Multiply by	To convert from	To	Multiply by
A			British thermal unit (59°F)	joule (J)	1.054 80 E+03
abampere	ampere (A)	1.000 000*E+01	British thermal unit (60°F)	joule (J)	1.054 68 E+03
abcoulomb	coulomb (C)	1.000 000*E+01	Btu (International Table)·ft/h·ft²·°F (k, thermal conductivity)	watt per metre kelvin (W/m·K)	1.730 735 E+00
abfarad	farad (F)	1.000 000*E+09	Btu (thermochemical)·ft/h·ft²·°F (k, thermal conductivity)	watt per meter kelvin (W/m·K)	1.729 577 E+00
abhenry	henry (H)	1.000 000*E−09	Btu (International Table)·in/h·ft²·°F (k, thermal conductivity)	watt per meter kelvin (W/m·K)	1.442 279 E−01
abmho	siemens (S)	1.000 000*E+09	Btu (thermochemical)·in/h·ft²·°F (k, thermal conductivity)	watt per metre kelvin (W/m·K)	1.441 314 E−01
abohm	ohm (Ω)	1.000 000*E−09	Btu (International Table)·in/s·ft²·°F (k, thermal conductivity)	watt per metre kelvin (W/m·K)	5.192 204 E+02
abvolt	volt (V)	1.000 000*E−08	Btu (thermochemical)·in/s·ft²·°F (k, thermal conductivity)	watt per metre kelvin (W/m·K)	5.188 732 E+02
acre foot (U.S. survey)[1]	metre³ (m³)	1.233 489 E+03	Btu (International Table)/h	watt (W)	2.930 711 E−01
acre (U.S. survey)[1]	metre² (m²)	4.046 873 E+03	Btu (International Table)/s	watt (W)	1.055 056 E+03
ampere hour	coulomb (C)	3.600 000*E+03	Btu (thermochemical)/h	watt (W)	2.928 751 E−01
are	metre² (m²)	1.000 000*E+02	Btu (thermochemical)/min	watt (W)	1.757 250 E+01
angstrom	metre (m)	1.000 000*E−10	Btu (thermochemical)/s	watt (W)	1.054 350 E+03
astronomical unit	metre (m)	1.495 979 E+11	Btu (International Table)/ft²	joule per metre² (J/m²)	1.135 653 E+04
atmosphere (standard)	pascal (Pa)	1.013 250*E+05	Btu (thermochemical)/ft²	joule per metre² (J/m²)	1.134 893 E+04
atmosphere (technical = 1 kgf/cm²)	pascal (Pa)	9.806 650*E+04	Btu (thermochemical)/ft²·h	watt per metre² (W/m²)	3.152 481 E+00
B			Btu (thermochemical)/ft²·min	watt per metre² (W/m²)	1.891 489 E+02
bar	pascal (Pa)	1.000 000*E+05	Btu (thermochemical)/ft²·s	watt per metre² (W/m²)	1.134 893 E+04
barn	metre² (m²)	1.000 000*E−28	Btu (International Table)/in²·s	watt per metre² (W/m²)	1.634 246 E+06
barrel (for petroleum, 42 gal)	metre³ (m³)	1.589 873 E−01	Btu (International Table)/h·ft²·°F (C, thermal conductance)	watt per metre² kelvin (W/m²·K)	5.678 263 E+00
board foot	metre³ (m³)	2.359 737 E−03	Btu (thermochemical)/h·ft²·°F (C, thermal conductance)	watt per metre² kelvin (W/m²·K)	5.674 466 E+00
British thermal unit (International Table)[2]	joule (J)	1.055 056 E+03			
British thermal unit (mean)	joule (J)	1.055 87 E+03			
British thermal unit (thermochemical)	joule (J)	1.054 350 E+03			
British thermal unit (39°F)	joule (J)	1.059 67 E+03			

1. Since 1893 the U.S. basis of length measurement has been derived from metric standards. In 1959 a small refinement was made in the definition of the yard to resolve discrepancies both in this country and abroad, which changed its length from 3600/3937 m to 0.9144 m exactly. This resulted in the new value being shorter by two parts in a million.

At the same time it was decided that any data in feet drived from and published as a result of geodetic surveys within the U.S. would remain with the old standard (1 ft. = 1200/3937 m) until further decision. This foot is named the U.S. survey foot.

As a result all U.S. land measurements in U.S. customary units will relate to the metre by the old standard. All the conversion factors in these tables for units referenced to this footnote are based on the U.S. survey foot, rather than the international foot.

Conversion factors for the land measures given below may be determined from the following relationships:

1 league = 3 miles (exactly)
1 rod = 16½ feet (exactly)
1 section = 1 square mile (exactly)
1 township = 36 square miles (exactly)
1 chain = 66 feet (exactly)

2. This value was adopted in 1956. Some of the older International Tables use the value 1.055 04 E+03. The exact conversion factor is 1.055 055 852 62*E+03.

3. The SI unit of thermodynamic temperature is the kelvin (K), and this unit is properly used for expressing thermodynamic temperature and temperature intervals. Wide use is also made of the degree Celsius (°C), which is the SI unit for expressing Celsius temperature and temperature intervals. The Celsius scale (formerly called centigrade) is related directly to thermodynamic temperature (kelvins) as follows:
   1. The temperature interval one degree Celsius equals one kelvin exactly.
   2. Celsius temperature (t) is related to thermodynamic temperature (T) by the equation $t = T - T_0$,
      where $T_0 = 273.15$ K by definition.

4. This is sometimes called the moment of inertia of a plane section about a specified axis.

Table A.1b

ALPHABETICAL LIST OF UNITS continued

(Symbols of SI units given in parentheses)

To convert from	To	Multiply by	To convert from	To	Multiply by
Btu (International Table)/ s·ft²·°F	watt per metre² kelvin (W/m²·K)	2.044 175 E+04	clo	kelvin metre² per watt (K·m²/W)	2.003 712 E−01
Btu (thermochemical)/s·ft²·°F	watt per metre² kelvin (W/m²·K)	2.042 808 E+04	cup	metre³ (m³)	2.365 882 E−04
Btu (International Table)/lb	joule per kilogram (J/kg)	2.326 000*E+03	curie	becquerel (Bq)	3.700 000*E+10
Btu (thermochemical)/lb	joule per kilogram (J/kg)	2.324 444 E+03	**D**		
Btu (International Table)/ lb·°F (c, heat capacity)	joule per kilogram kelvin (J/kg·K)	4.186 800*E+03	day (mean solar)	second (s)	8.640 000 E+04
Btu (thermochemical)/ lb·°F (c, heat capacity)	joule per kilogram kelvin (J/kg·K)	4.184 000 E+03	day (sidereal)	second (s)	8.616 409 E+04
bushel (U.S.)	metre³ (m³)	3.523 907 E−02	degree (angle)	radian (rad)	1.745 329 E−02
C			degree Celsius	Kelvin (K)	$t_K = t_C + 273.15$
caliber (inch)	metre (m)	2.540 000*E−02	degree centigrade	[see footnote 3]	$t_C = (t_F - 32)/1.8$
calorie (International Table)	joule (J)	4.186 800*E+00	degree Fahrenheit	degree Celsius	
calorie (mean)	joule (J)	4.190 02 E+00	degree Fahrenheit	kelvin (K)	$t_K = (t_F + 459.67)/1.8$
calorie (thermochemical)	joule (J)	4.184 000*E+00	degree Rankine	kelvin (K)	$t_K = t_R/1.8$
calorie (15°C)	joule (J)	4.185 80 E+00	°F·h·ft²/Btu (International Table) (R, thermal resistance)	kelvin metre² per watt (K·m²/W)	1.761 102 E−01
calorie (20°C)	joule (J)	4.181 90 E+00	°F·h·ft²/Btu (thermochemical) (R, thermal resistance)	kelvin metre² per watt (K·m²/W)	1.762 280 E−01
calorie (kilogram, International Table)	joule (J)	4.186 800*E+03	denier	kilogram per metre (kg/m)	1.111 111 E−07
calorie (kilogram, mean)	joule (J)	4.190 02 E+03	dyne	newton (N)	1.000 000*E−05
calorie (kilogram, thermochemical)	joule (J)	4.184 000*E+03	dyne·cm	newton metre (N·m)	1.000 000*E−07
cal (thermochemical)/cm²	joule per metre² (J/m²)	4.184 000*E+04	dyne/cm²	pascal (Pa)	1.000 000*E−01
cal (International Table)/g	joule per kilogram (J/kg)	4.186 800*E+03	**E**		
cal (thermochemical)/g	joule per kilogram (J/kg)	4.184 000*E+03	electronvolt	joule (J)	1.602 19 E−19
cal (International Table)/g·°C	joule per kilogram kelvin (J/kg·K)	4.186 800*E+03	EMU of capacitance	farad (F)	1.000 000*E+09
cal (thermochemical)/g·°C	joule per kilogram kelvin (J/kg·K)	4.184 000*E+03	EMU of current	ampere (A)	1.000 000*E+01
cal (thermochemical)/min	watt (W)	6.973 333 E−02	EMU of electric potential	volt (V)	1.000 000*E−08
cal (thermochemical)/s	watt (W)	4.184 000*E+00	EMU of inductance	henry (H)	1.000 000*E−09
cal (thermochemical)/cm²·min	watt per metre² (W/m²)	6.973 333 E+02	EMU of resistance	ohm (Ω)	1.000 000*E−09
cal (thermochemical)/cm²·s	watt per metre² (W/m²)	4.184 000*E+04	ESU of capacitance	farad (F)	1.112 650 E−12
cal (thermochemical)/cm·s·°C	watt per metre kelvin (W/m·K)	4.184 000*E+02	ESU of current	ampere (A)	3.335 6 E−10
carat (metric)	kilogram (kg)	2.000 000*E−04	ESU of electric potential	volt (V)	2.997 9 E+02
centimetre of mercury (0°C)	pascal (Pa)	1.333 22 E+03	ESU of inductance	henry (H)	8.987 554 E+11
centimetre of water (4°C)	pascal (Pa)	9.806 38 E+01	ESU of resistance	ohm (Ω)	8.987 554 E+11
centipoise	pascal second (Pa·s)	1.000 000*E−03	erg	joule (J)	1.000 000*E−07
centistokes	metre² per second (m²/s)	1.000 000*E−06	erg/(cm²·s)	watt per metre² (W/m²)	1.000 000*E−03
circular mil	metre² (m²)	5.067 075 E−10	erg/s	watt (W)	1.000 000*E−07
			F		
			faraday (based on carbon-12)	coulomb (C)	9.648 70 E+04
			faraday (chemical)	coulomb (C)	9.649 57 E+04
			faraday (physical)	coulomb (C)	9.652 19 E+04

Table A.1b

ALPHABETICAL LIST OF UNITS *continued*
(Symbols of SI units given in parentheses)

To convert from	To	Multiply by
fathom	metre (m)	1.828 8 E+00
fermi (femtometer)	metre (m)	1.000 000*E−15
fluid ounce (U.S.)	metre³ (m³)	2.957 353 E−05
foot	metre (m)	3.048 000*E−01
foot (U.S. survey)[1]	metre (m)	3.048 006 E−01
foot of water (39.2°F)	pascal (Pa)	2.988 98 E+03
ft²	metre² (m²)	9.290 304*E−02
ft²/h (thermal diffusivity)	metre² per second (m²/s)	2.580 640*E−05
ft²/s	metre² per second (m²/s)	9.290 304*E−02
ft³ (volume; section modulus)	metre³ (m³)	2.831 685 E−02
ft³/min	metre³ per second (m³/s)	4.719 474 E−04
ft³/s	metre³ per second (m³/s)	2.831 685 E−02
ft⁴ (moment of section)[4]	metre⁴ (m⁴)	8.630 975 E−03
ft/h	metre per second (m/s)	8.466 667 E−05
ft/min	metre per second (m/s)	5.080 000*E−03
ft/s	metre per second (m/s)	3.048 000*E−01
ft/s²	metre per second² (m/s²)	3.048 000*E−01
footcandle	lux (lx)	1.076 391 E+01
footlambert	candela per metre² (cd/m²)	3.426 259 E+00
ft·lbf	joule (J)	1.355 818 E+00
ft·lbf/h	watt (W)	3.766 161 E−04
ft·lbf/min	watt (W)	2.259 697 E−02
ft·lbf/s	watt (W)	1.355 818 E+00
ft·poundal	joule (J)	4.214 011 E−02
free fall, standard (*g*)	metre per second² (m/s²)	9.806 650*E+00
G		
gal	metre per second² (m/s²)	1.000 000*E−02
gallon (Canadian liquid)	metre³ (m³)	4.546 090 E−03
gallon (U.K. liquid)	metre³ (m³)	4.546 092 E−03
gallon (U.S. dry)	metre³ (m³)	4.404 884 E−03
gallon (U.S. liquid)	metre³ (m³)	3.785 412 E−03
gallon (U.S. liquid) per day	metre³ per second (m³/s)	4.381 264 E−08
gallon (U.S. liquid) per minute	metre³ per second (m³/s)	6.309 020 E−05
gallon (U.S. liquid) per hp·h (SFC, specific fuel consumption)	metre³ per joule (m³/J)	1.410 089 E−09
gamma	tesla (T)	1.000 000*E−09
gauss	tesla (T)	1.000 000*E−04
gilbert	ampere (A)	7.957 747 E−01
gill (U.K.)	metre³ (m³)	1.420 654 E−04
gill (U.S.)	metre³ (m³)	1.182 941 E−04
grad	degree (angular)	9.000 000*E−01
grad	radian (rad)	1.570 796 E−02

To convert from	To	Multiply by
grain (1/7000 lb avoirdupois)	kilogram (kg)	6.479 891*E−05
grain (lb avoirdupois/7000)/ gal (U.S. liquid)	kilogram per metre³ (kg/m³)	1.711 806 E−02
gram	kilogram (kg)	1.000 000*E−03
g/cm³	kilogram per metre³ (kg/m³)	1.000 000*E+03
gram-force/cm²	pascal (Pa)	9.290 304*E−02...
gram-force/cm²	pascal (Pa)	9.806 650*E+01
H		
hectare	metre² (m²)	1.000 000*E+04
horsepower (550 ft·lbf/s)	watt (W)	7.456 999 E+02
horsepower (boiler)	watt (W)	9.809 50 E+03
horsepower (electric)	watt (W)	7.460 000*E+02
horsepower (metric)	watt (W)	7.354 99 E+02
horsepower (water)	watt (W)	7.460 43 E+02
horsepower (U.K.)	watt (W)	7.457 0 E+02
hour (mean solar)	second (s)	3.600 000 E+03
hour (sidereal)	second (s)	3.590 170 E+03
hundredweight (long)	kilogram (kg)	5.080 235 E+01
hundredweight (short)	kilogram (kg)	4.535 924 E+01
I		
inch	metre (m)	2.540 000*E−02
inch of mercury (32°F)	pascal (Pa)	3.386 38 E+03
inch of mercury (60°F)	pascal (Pa)	3.376 85 E+03
inch of water (39.2°F)	pascal (Pa)	2.490 82 E+02
inch of water (60°F)	pascal (Pa)	2.488 4 E+02
in²	metre² (m²)	6.451 600*E−04
in³ (volume; section modulus)[5]	metre³ (m³)	1.638 706 E−05
in³/min	metre³ per second (m³/s)	2.731 177 E−07
in⁴ (moment of section)[4]	metre⁴ (m⁴)	4.162 314 E−07
in/s	metre per second (m/s)	2.540 000*E−02
in/s²	metre per second² (m/s²)	2.540 000*E−02
K		
kayser	1 per metre (1/m)	1.000 000*E+02
kelvin	degree Celsius	$t_C = t_K - 273.15$
kilocalorie (International Table)	joule (J)	4.186 800*E+03
kilocalorie (mean)	joule (J)	4.190 02 E+03
kilocalorie (thermochemical)	joule (J)	4.184 000*E+03
kilocalorie (thermochemical)/min	watt (W)	6.973 333 E+01

5. The exact conversion factor is 1.638 706 4*E−05.

Table A.1b

ALPHABETICAL LIST OF UNITS *continued*

(Symbols of SI units given in parentheses)

To convert from	To	Multiply by
kilocalorie (thermochemical)/s	watt (W)	4.184 000*E+03
kilogram-force (kgf)	newton (N)	9.806 650*E+00
kgf·m	newton metre (N·m)	9.806 650*E+00
kgf·s²/m (mass)	kilogram (kg)	9.806 650*E+00
kgf/cm²	pascal (Pa)	9.806 650*E+04
kgf/m²	pascal (Pa)	9.806 650*E+00
kgf/mm²	pascal (Pa)	9.806 650*E+06
km/h	metre per second (m/s)	2.777 778 E−01
kilopond	newton (N)	9.806 650*E+00
kW·h	joule (J)	3.600 000*E+06
kip (1000 lbf)	newton (N)	4.448 222 E+03
kip/in² (ksi)	pascal (Pa)	6.894 757 E+06
knot (international)	metre per second (m/s)	5.144 444 E−01
L		
lambert	candela per metre² (cd/m²)	1/π *E+04
lambert	candela per metre² (cd/m²)	3.183 099 E+03
langley	joule per metre² (J/m²)	4.184 000*E+04
league	metre (m)	[see footnote 1]
light year	metre (m)	9.460 55 E+15
liter[6]	metre³ (m³)	1.000 000*E−03
M		
maxwell	weber (Wb)	1.000 000*E−08
mho	siemens (S)	1.000 000*E+00
microinch	metre (m)	2.540 000*E−08
micron	metre (m)	1.000 000*E−06
mil	metre (m)	2.540 000*E−05
mile (international)	metre (m)	1.609 344*E+03
mile (statute)	metre (m)	1.609 3 E+03
mile (U.S. survey)[1]	metre (m)	1.609 347 E+03
mile (international nautical)	metre (m)	1.852 000*E+03
mile (U.K. nautical)	metre (m)	1.853 184*E+03
mile (U.S. nautical)	metre (m)	1.852 000*E+03
mi² (international)	metre² (m²)	2.589 988 E+06
mi² (U.S. survey)[1]	metre² (m²)	2.589 998 E+06
mi/h (international)	metre per second (m/s)	4.470 400*E−01
mi/h (international)	kilometre per hour (km/h)	1.609 344*E+00
mi/min (international)	metre per second (m/s)	2.682 240*E+01
mi/s (international)	metre per second (m/s)	1.609 344*E+03
millibar	pascal (Pa)	1.000 000*E+02
millimetre of mercury (0°C)	pascal (Pa)	1.333 22 E+02

To convert from	To	Multiply by
minute (angle)	radian (rad)	2.908 882 E−04
minute (mean solar)	second (s)	6.000 000 E+01
minute (sidereal)	second (s)	5.983 617 E+01
month (mean calendar)	second (s)	2.628 000 E+06
O		
oersted	ampere per metre (A/m)	7.957 747 E+01
ohm centimetre	ohm metre (Ω·m)	1.000 000*E−02
ohm circular-mil per foot	ohm millimetre² per metre (Ω·mm²/m)	1.662 426 E−03
ounce (avoirdupois)	kilogram (kg)	2.834 952 E−02
ounce (troy or apothecary)	kilogram (kg)	3.110 348 E−02
ounce (U.K. fluid)	metre³ (m³)	2.841 307 E−05
ounce (U.S. fluid)	metre³ (m³)	2.957 353 E−05
ounce-force	newton (N)	2.780 139 E−01
ozf·in	newton metre (N·m)	7.061 552 E−03
oz (avoirdupois)/gal (U.K. liquid)	kilogram per metre³ (kg/m³)	6.236 021 E+00
oz (avoirdupois)/gal (U.S. liquid)	kilogram per metre³ (kg/m³)	7.489 152 E+00
oz (avoirdupois)/in³	kilogram per metre³ (kg/m³)	1.729 994 E+03
oz (avoirdupois)/ft²	kilogram per metre² (kg/m²)	3.051 517 E−01
oz (avoirdupois)/yd²	kilogram per metre² (kg/m²)	3.390 575 E−02
P		
parsec	metre (m)	3.085 678 E+16
peck (U.S.)	metre³ (m³)	8.809 768 E−03
pennyweight	kilogram (kg)	1.555 174 E−03
perm (0°C)	kilogram per pascal second metre² (kg/Pa·s·m²)	5.721 35 E−11
perm (23°C)	kilogram per pascal second metre² (kg/Pa·s·m²)	5.745 25 E−11
perm·in (0°C)	kilogram per pascal second metre (kg/Pa·s·m)	1.453 22 E−12
perm·in (23°C)	kilogram per pascal second metre (kg/Pa·s·m)	1.459 29 E−12
phot	lumen per metre² (lm/m²)	1.000 000*E+04
pica (printer's)	metre (m)	4.217 518 E−03
pint (U.S. dry)	metre³ (m³)	5.506 105 E−04
pint (U.S. liquid)	metre³ (m³)	4.731 765 E−04
point (printer's)	metre (m)	3.514 598*E−04
poise (absolute viscosity)	pascal second (Pa·s)	1.000 000*E−01
pound (lb avoirdupois)[7]	kilogram (kg)	4.535 924 E−01

6. In 1964 the General Conference on Weights and Measures adopted the name litre as a special name for decimetre. Prior to this decision the litre differed slightly (previous value, 1.000028 dm³) and in expression of precision volume measurement this fact must be kept in mind.

7. The exact conversion factor is 4.535 923 7*E−01.

Table A.1b
ALPHABETICAL LIST OF UNITS continued
(Symbols of SI units given in parentheses)

To convert from	To	Multiply by
pound (troy or apothecary)	kilogram (kg)	3.732 417 E−01
lb·ft² (moment of inertia)	kilogram metre² (kg·m²)	4.214 011 E−02
lb·in² (moment of inertia)	kilogram metre² (kg·m²)	2.926 397 E−04
lb/ft·h	pascal second (Pa·s)	4.133 789 E−04
lb/ft·s	pascal second (Pa·s)	1.488 164 E+00
lb/ft²	kilogram per metre² (kg/m²)	4.882 428 E+00
lb/ft³	kilogram per metre³ (kg/m³)	1.601 846 E+01
lb/gal (U.K. liquid)	kilogram per metre³ (kg/m³)	9.977 633 E+01
lb/gal (U.S. liquid)	kilogram per metre³ (kg/m³)	1.198 264 E+02
lb/h	kilogram per second (kg/s)	1.259 979 E−04
lb/hp·h (SFC, specific fuel consumption)	kilogram per joule (kg/J)	1.689 659 E−07
lb/in³	kilogram per metre³ (kg/m³)	2.767 990 E+04
lb/min	kilogram per second (kg/s)	7.559 873 E−03
lb/s	kilogram per second (kg/s)	4.535 924 E−01
lb/yd³	kilogram per metre³ (kg/m³)	5.932 764 E−01
poundal	newton (N)	1.382 550 E−01
poundal/ft²	pascal (Pa)	1.488 164 E+00
poundal·s/ft²	pascal second (Pa·s)	1.488 164 E+00
pound-force (lbf)[8]	newton (N)	4.448 222 E+00
lbf·ft	newton metre (N·m)	1.355 818 E+00
lbf·ft/in	newton metre per metre (N·m/m)	5.337 866 E+01
lbf·in	newton metre (N·m)	1.129 848 E−01
lbf·in/in	newton metre per metre (N·m/m)	4.448 222 E+00
lbf·s/ft²	pascal second (Pa·s)	4.788 026 E+01
lbf·s/in²	pascal second (Pa·s)	6.894 757 E+03
lbf/ft	newton per metre (N/m)	1.459 390 E+01
lbf/ft²	pascal (Pa)	4.788 026 E+01
lbf/in	newton per metre (N/m)	1.751 268 E+02
lbf/in² (psi)	pascal (Pa)	6.894 757 E+03
lbf/lb (thrust/weight [mass] ratio)	newton per kilogram (N/kg)	9.806 650 E+00
Q		
quart (U.S. dry)	metre³ (m³)	1.101 221 E−03
quart (U.S. liquid)	metre³ (m³)	9.463 529 E−04
R		
rad (radiation dose absorbed)	gray (Gy)	1.000 000*E−02
rhe	1 per pascal second (1/Pa·s)	1.000 000*E+01
rod	metre (m)	[see footnote 1]

To convert from	To	Multiply by
roentgen	coulomb per kilogram (C/kg)	2.58 E−04
S		
second (angle)	radian (rad)	4.848 137 E−06
second (sidereal)	second (s)	9.972 696 E−01
section	metre² (m²)	[see footnote 1]
shake	second (s)	1.000 000*E−08
slug	kilogram (kg)	1.459 390 E+01
slug/ft·s	pascal second (Pa·s)	4.788 026 E+01
slug/ft³	kilogram per metre³ (kg/m³)	5.153 788 E+02
statampere	ampere (A)	3.335 640 E−10
statcoulomb	coulomb (C)	3.335 640 E−10
statfarad	farad (F)	1.112 650 E−12
stathenry	henry (H)	8.987 554 E+11
statmho	siemens (S)	1.112 650 E−12
statohm	ohm (Ω)	8.987 665 E+11
statvolt	volt (V)	2.997 925 E+02
stere	metre³ (m³)	1.000 000*E+00
stilb	candela per metre² (cd/m²)	1.000 000*E+04
stokes (kinematic viscosity)	metre² per second (m²/s)	1.000 000*E−04
T		
tablespoon	metre³ (m³)	1.478 676 E−05
teaspoon	metre³ (m³)	4.928 922 E−06
tex	kilogram per metre (kg/m)	1.000 000*E−06
therm	joule (J)	1.055 056 E+08
ton (assay)	kilogram (kg)	2.916 667 E−02
ton (long, 2240 lb)	kilogram (kg)	1.016 047 E+03
ton (metric)	kilogram (kg)	1.000 000*E+03
ton (nuclear equivalent of TNT)	joule (J)	4.184 E+09[9]
ton (refrigeration)	watt (W)	3.516 800 E+03
ton (register)	metre³ (m³)	2.831 685 E+00
ton (short, 2000 lb)	kilogram (kg)	9.071 847 E+02
ton (long)/yd³	kilogram per metre³ (kg/m³)	1.328 939 E+03
ton (short)/yd³	kilogram per metre³ (kg/m³)	1.186 553 E+03
ton (short)/h	kilogram per second (kg/s)	2.519 958 E−01
ton-force (2000 lbf)	newton (N)	8.896 444 E+03
tonne	kilogram (kg)	1.000 000*E+03
torr (mm Hg, 0°C)	pascal (Pa)	1.333 22 E+02
township	metre² (m²)	[see footnote 1]

8. The exact conversion factor is 4.448 221 615 260 5*E+00.

9. Defined (not measured) value.

Table A.1b

ALPHABETICAL LIST OF UNITS *continued*

(Symbols of SI units given in parentheses)

To convert from	To	Multiply by	To convert from	To	Multiply by
U			**Y**		
unit pole	weber (Wb)	1.256 637 E−07	yard	metre (m)	9.144 000*E−01
			yd²	metre² (m²)	8.361 274 E−01
W			yd³	metre³ (m³)	7.645 549 E−01
W·h	joule (J)	3.600 000*E+03	yd³/min	metre³ per second (m³/s)	1.274 258 E−02
W·s	joule (J)	1.000 000*E+00	year (365 days)	second (s)	3.153 600 E+07
W/cm²	watt per metre² (W/m²)	1.000 000*E+04	year (sidereal)	second (s)	3.155 815 E+07
W/in²	watt per metre² (W/m²)	1.550 003 E+03	year (tropical)	second (s)	3.155 693 E+07

A.2 GRAPHIC SYMBOLS FOR DISTRIBUTED CONTROL/SHARED DISPLAY INSTRUMENTATION, LOGIC, AND COMPUTER SYSTEMS

Purpose

The purpose of this Standard is to establish documentation for that class of instrumentation consisting of computers, programmable controllers, minicomputers and microprocessor-based systems that have shared control, shared display, or other interface features. Symbols are provided for interfacing field instrumentation, control room instrumentation, and other hardware to the above. Terminology is defined in the broadest generic form to describe the various categories of these devices.

It is not the intent of this Standard to mandate the use of each type symbol for each occurrence of a generic device within the overall control system. Such usage could result in undue complexity in the case of a P&ID drawing. If, for example, a computer component is an integral part of a distributed control system, the use of the computer symbol would normally be an undesirable redundancy. If, however, a separate general purpose computer is interfaced with the system, the inclusion of the computer symbol may provide the degree of clarity needed for control system understanding.

This Standard attempts to provide the users with defined symbolism and rules for usage, which may be applied as needed to provide sufficient clarity of intent. The extent to which these symbols are applied to various types of drawings remains with the users. The symbols may be as simple or complex as needed to define the process.

Scope

This Standard satisfies the requirements for symbolically representing the functions of distributed control/shared display instrumentation, logic, and computer systems. The instrumentation is generally composed of field hardware communication networks and control room operator devices. This Standard is applicable to all industries using process control and instrumentation systems.

No effort will be made on the flow diagram to explain the internal construction, configuration, or method of operation of this type of instrumentation, logic and computer systems. Personnel needing to understand flow diagrams must have a basic understanding of the total system in order to correctly interpret the diagram. The type of computation or the use of the process variable within a program is not indicated except in those cases where the process variable is an integral part of the control strategy. In applications where all instrument system data base information is available to the computer via the communication link, the depiction of the computer interconnections is optional in order to conserve space on flow diagrams.

Application to Work Activities

This Standard is intended for use whenever any reference to an instrument is required. Such references may be required for the following uses as well as others:

Flow diagrams, process and mechanical
Instrumentation system diagrams
Specifications, purchase orders, manifests, and other lists
Construction drawings
Technical papers, literature, and discussions
Tagging of instruments
Installation, operation, and maintenance instructions, drawings, and records

Relationship to Other ISA Standards

This Standard complements ISA Standard 5.1, "Instrumentation Symbols and Identification," for symbols and formats representing functional identification codes. For clarification of examples, a limited amount of ISA-S5.1 symbology has been included in this document.

Relationship to Other Standards

Where applicable, definitions not included below are in accordance with ANSI X3/TR-1-77, "American National Dictionary for Information Processing," and/or ISA-S5.1.

Definitions and Abbreviations

Accessible—A system feature that is viewable by and interactive with the operator, and allows the operator to perform user permissible control actions, e.g. set point changes, auto-manual transfers, or on-off actions.

Assignable—A system feature that permits an operator to channel (or direct) a signal from one device to another, without the need for changes in wiring, either by means of switches or via keyboard commands to the system.

Communication Link—The physical hardware required to interconnect devices for the purpose of transmitting and/or receiving data.

Computer Control System—A system in which all control action takes place within the control computer. Single or redundant computers may be used.

Configurable—A system feature that permits selection through entry of keyboard commands of the basic structure and characteristics of a device or system, such as control algorithms, display formats, or input/output terminations.

C.R.T.—Cathode Ray Tube

Distributed Control System—That class of instrumentation (input/output devices, control devices, and operator interface devices) which in addition to executing the stated control functions also permits transmission of control, measurement, and operating information to and from a single or a plurality of user specifiable locations, connected by a communication link.

I/O—Input/Output

Shared Controller—A control device that contains a plurality of pre-programmed algorithms which are user retrievable, configurable, and connectable, and allows user defined control strategies or functions to be implemented. Control of multiple process variables can be implemented by sharing the capabilities of a single device of this kind.

Shared Display—The operator interface device used to display signals and/or data on a time shared basis. The signals and/or data, i.e., alphanumeric and/or graphic, reside in a data base from where selective accessibility for display is at the command of a user.

Software—Digital programs, procedures, rules, and associated documentation required for the operation and/or maintenance of a digital system.

Software Link—The interconnection of system components or functions via software or keyboard instruction.

Supervisory Set Point Control System—The generation of set point and/or other control information by a computer control system for use by shared control, shared display or other regulatory control devices.

Symbols

General

Standard instrumentation symbols as shown in ISA-S5.1 are retained as much as possible for flow diagram use, but are supplemented as necessary by the new symbols 1 through 10. Symbol size should be consistent with ISA-S5.1, Section 3. The symbol descriptions listed to the right of each symbol are intended as guidelines for applications and are not intended to be all inclusive. The symbol may be used if one or more of the descriptions apply. Shared signal lines can be expressed by the symbol for a system link (see symbol 10).

Distributed Control/Shared Display Symbols

Advances in control systems brought about by microprocessor-based instrumentation permit shared functions such as display, control, and signal lines. Therefore, the symbology defined here should be "Shared Instruments," which means shared display and/or shared control. The square portion of this symbol, as shown in symbols 1 through 3, has the meaning of shared type instrument.

1. Normally Accessible to Operator—Indicator/Controller/Recorder or Alarm Point. Usually used to indicate video display.

   (1) Shared display.
   (2) Shared display and shared control.
   (3) Access limited to communication link.
   (4) Operator interface on communication link.

2. Auxiliary Operator's Interface Device

   (1) Panel mounted; normally having an analog faceplate; not normally mounted on main operator console.
   (2) Can be a backup controller or manual station.
   (3) Access may be limited to communication link.
   (4) Operator interface via the communication link.

3. Not Normally Accessible to Operator

   (1) Shared blind controller.
   (2) Shared display installed in field.
   (3) Computation, signal conditioning in shared controller.

(4) May be on communication link.

(5) Normally blind operation.

(6) May be altered by configuration.

Computer Symbols

The following symbols should be used where systems include components identified as computers, as distinct from an integral processor, which drive the various functions of a distributed control system. The computer component may be integrated with the system via the data link, or it may be a stand-alone computer.

4. Normally Accessible to Operator—Indicator/ Controller/Recorder or Alarm Point. Usually used to indicate video display.

5. Not Normally Accessible to Operator

   (1) Input/output interface.

   (2) Computation/signal conditioning within a computer.

   (3) May be used as a blind controller or software calculation module.

Logic and Sequential Control Symbols

6. General Symbol—For undefined complex interconnecting logic or sequence control. (Also see ISA-S5.1.)

7. Distributed control interconnecting logic controller with binary or sequential logic functions.

   (1) Packaged programmable logic controller, or digital logic controls integral to the distributed control equipment.

   (2) Not normally accessible to the operator.

8. Distributed control interconnecting logic controller with binary or sequential logic functions.

   (1) Packaged programmable logic controller, or digital logic controls integral to the distributed control equipment.

   (2) Normally accessible to the operator.

Internal System Function Symbols

9. Computation/Signal Conditioning

   (1) For block identification refer to ISA-S5.1, Table 2, "Function Designations for Relays."

   (2) For extensive computational requirements, use designation "C." Explain on supplementary documentation.

   (3) Used in conjunction with function relay bubbles per ISA-S5.1.

Common Symbols

10. System Link

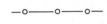

   (1) Used to indicate either a software link or manufacturer's system supplied connections between functions.

   (2) Alternatively, link can be implicitly shown by contiguous symbols.

   (3) May be used to indicate a communication link at the user's option.

Recorders and Other Historical Data Retention

Conventional hard-wired recording devices such as strip chart recorders shall be shown in accordance with ISA-S5.1. (See Figure A.2g.)

For assignable recording devices use Symbol 1.

Long term/mass storage of a process variable by digital memory means such as tape, disc, etc., shall be depicted in accordance with Distributed Control/Shared Display Symbols or Computer Symbols of this Standard, depending on the location of the device.

Identification

For purposes of this Standard, identification codes shall be consistent with ISA-S5.1, with the following additions.

Software Alarms

Software alarms may be identified by placing ISA-S5.1 Table 1, letter designators on the input or output signal lines of the controls, or other specific integral system component. See Alarms below.

Contiguity of Symbols

Two or more symbols can adjoin to express the following meanings in addition to those shown in ISA-S5.1:

1. Communication among the associated instruments, e.g., hard wiring, internal system link, backup.

2. Instrument integrated with multiple functions, e.g., multipoint recorder, control valve with integrally mounted controller.

The application of contiguous symbols is a user option.

If the intent is not absolutely clear, contiguous symbols should *not* be used.

Alarms

General

All hard-wired standard devices and alarms, as distinct from those devices and alarms specifically covered in this standard, shall be shown in accordance with ISA-S5.1, Table 1.

Figures A.2a and b below illustrate principles of the methods of symbolization and identification. Additional applications that adhere to these principles may be devised as required. The location of the alarm identifiers is left to the discretion and convenience of the user.

Instrument System Alarms

Multiple alarm capability is provided in most systems. Alarms covered by this standard should be identified as shown by the examples in Figures A.2a and b.

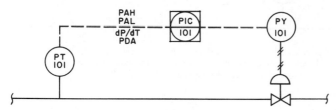

Fig. A.2a   Alarms on measured variables shall include the variable identifiers, i.e.:

Pressure: PAH   (high)
           PAL   (low)
           dP/dt (rate of change)
           PDA (deviation from set point)

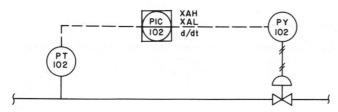

Fig. A.2b   Alarms on controller output shall use the undefined variable identifier X, i.e.:

        XAH (high)
        XAL (low)
        d/dt (rate of change)

Examples of Use

The following Figures A.2c through A.2o, illustrate some of the various combinations of symbols presented in this Standard and ISA-S5.1. These symbols may be combined as necessary to fulfill the needs of the user.

Controllers located in the diagram main information line are to be considered the primary controllers. All devices outside the main line provide a backup or secondary function.

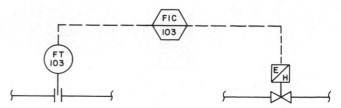

Fig. A.2c   Computer control: No backup, shared display

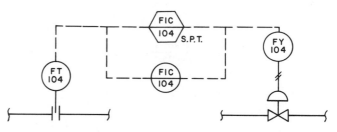

Fig. A.2d   Computer control with analog backup

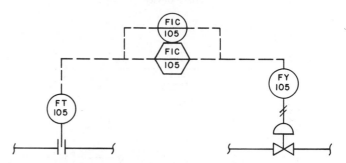

Fig. A.2e   Computer control: Full analog backup through set point tracking (SPT)

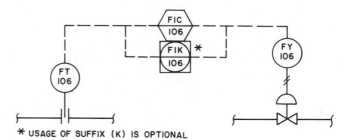

✱ USAGE OF SUFFIX (K) IS OPTIONAL

Fig. A.2f   Computer control: Full backup from distributed control instrumentation. Computer uses instrument system communication link.
*Use of suffix K is optional.

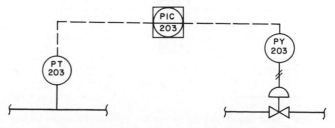

Fig. A.2g   Shared display/shared control, no backup

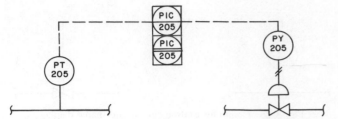

Fig. A.2h   Shared display/shared control with auxiliary operator's interface device

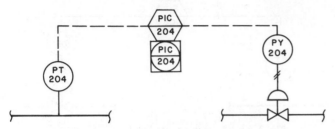

Fig. A.2i   Analog control interfaced with shared display/shared control backup

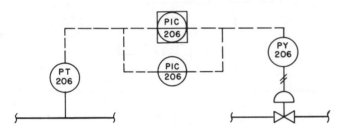

Fig. A.2j   Shared display/shared control with analog controller backup

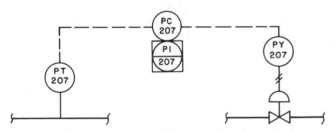

Fig. A.2k   Analog control: Blind controller, shared display

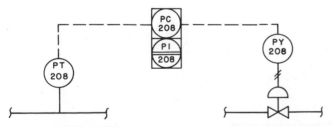

Fig. A.2l   Blind shared control with auxiliary operator's interface backup

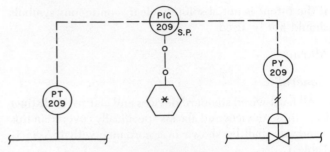

Fig. A.2m   Supervisory set point control: Analog controller with conventional faceplate. Computer supervisory set point via communication link.
*User identification is optional. Use of suffix K is optional.

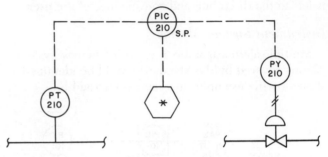

Fig. A.2n   Supervisory set point control: Analog controller complete with conventional faceplate. Computer supervisor set point hard-wired.
*User identification is optional. Use of suffix K is optional.

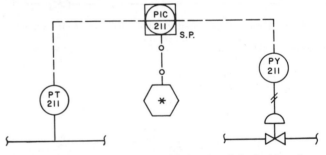

Fig. A.2o   Supervisory set point control: Shared display/shared control with full computer access via the communication link
*User identification is optional. Use of suffix K is optional.

Typical Flow Diagrams

Figure A.2p combines the basic symbols of this Standard in a simplified drawing. It is intended to provide a hypothetical example and to stimulate the user's imagination in the application of symbolism to this equipment. Figure A.2p is arranged in the following manner:

- Volumetric fuel and air flows provide inputs for combustion system firing rate and fuel air ratio via distributed control instrumentation. Set points for both rate and ratio can be computer generated.
- Combustion air and gas pressures are monitored by pressure switches which control the gas safety

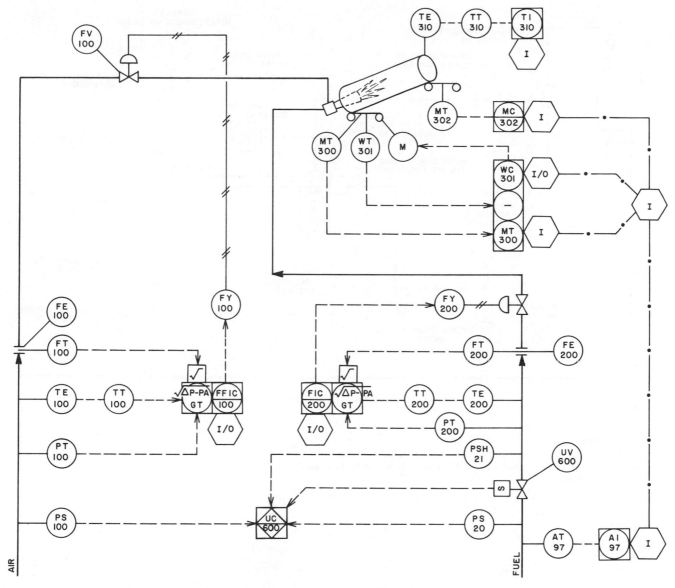

Fig. A.2p Example of simplified drawing

shutoff valve via UC-600 "distributed control interconnecting logic."

• Material moisture content is measured, dry weight of the input material is calculated, and feed rate is controlled by MT-300 and WC-301. Discharged material moisture content is read by MT-302. At this point, firing rate and/or feed rate could be controlled by the Distributed Control System (DCS) instrumentation or by the computer taking other process variables into consideration.

• British thermal unit or Joules (Btu or J) analysis (AT-97) is input to the computer system to generate feed forward control adjusting firing rate, in Btu/hr (J/h). The set point is calculated by the computer, based on feed rate, weight, and moisture content.

• Internal system links are shown for selected computer input/output, while the firing rate and ratio set points are implied. Shown in the same manner, the links between the calculation modules and the controllers are implied by contiguous symbols, while the wild flow to the ratio control is shown in the system link symbol.

Figure A.2q combines the symbols to depict a cascade loop with alarms. Notes are added on the diagram itself for clarification purposes only.

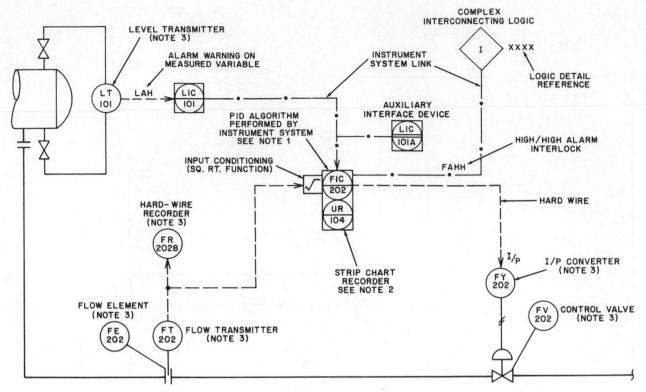

Fig. A.2q   Typical flow diagram cascade control loop

Notes: Shared Display
1. Display/adjustments on console. Communication via data link.
2. Located in console. Signal selected from instrument system data base.
3. Field mounted.

A.3 CONTROL VALVE SIZING EQUATIONS

Scope

This standard establishes equations for flow through control valves throttling either compressible or incompressible fluids. The equations are not, however, intended for use with mixed phases. Use of these equations for mixed phases, wet or dry slurries, or for non-Newtonian liquids may result in inaccuracies.

This standard does not apply to fluid power components as defined in the National Fluid Power Association (NFPA) Standards.

General

The equations of this standard are based on the use of sizing factors obtained by the physical testing of control valve specimens in a laboratory using the procedures of ISA S39.02, "Control Valve Capacity Test Procedures." The sizing factors for throttling valves refer to the fully open or "rated" condition, unless they are otherwise identified.

Liquid Sizing Equations

The equation for the flow of a liquid through a valve takes either of the following forms:

$$q = N_1 F_P F_Y F_R C_V \sqrt{\frac{p_1 - p_2}{G_f}} \qquad \text{A.3(1)}$$

$$w = N_{10} F_P F_Y F_R C_V \sqrt{(p_1 - p_2)G_f} \qquad \text{A.3(2)}$$

The pressure drop used for calculating the flow is limited to Δp_T (refer to equation 12 or 13) because of choking.

Gas or Vapor Sizing Equations

The flow of gas or vapor through a valve may be calculated using any of the following forms of the basic equations:

$$w = N_6 F_P C_V Y \sqrt{x p_1 \gamma_1} \qquad \text{A.3(3)}$$

$$q = N_7 F_P C_V p_1 Y \sqrt{\frac{x}{G_g T_1 Z}} \qquad \text{A.3(4)}$$

$$w = N_8 F_P C_V p_1 Y \sqrt{\frac{xM}{T_1 Z}} \qquad \text{A.3(5)}$$

$$q = N_9 F_P C_V p_1 Y \sqrt{\frac{x}{M T_1 Z}} \qquad \text{A.3(6)}$$

$$\text{where } Y = 1 - \frac{x}{3 F_k x_T} \qquad \text{A.3(7)}$$

$$\text{and } F_k = \frac{k}{1.40} \qquad \text{A.3(8)}$$

In all the above gas formulas x may not exceed $F_k x_T$, even though the actual pressure drop ratio is greater, because of the effect of flow choking.

Numerical Constant, N

The numerical constant N is a function of the measurement units used. Values for N are listed in Table A.3a.

Piping Geometry Factor, F_p

The Piping Geometry Factor F_p takes into account fittings such as reducers, expanders, or other devices attached to the valve body that are not in accord with the test manifold as specified in ISA S39.02. The value is the ratio of the valve C_v installed with reducers to the rated C_v, of the valve installed in a standard manifold. Both C_v's are to be tested under otherwise identical service conditions.

To meet a designated tolerance of ± 5 percent in the calculated flow rate, the factor F_p must be determined by test. When estimated values are permissible, the following equation may be used:

$$F_P = \left[1 + \frac{\Sigma K}{N_2} \left(\frac{N_3 C_v}{d^2} \right)^2 \right]^{-\frac{1}{4}} \qquad \text{A.3(9)}$$

The factor ΣK is the algebraic sum of the effective velocity head coefficients of all devices attached to, but not including, the control valve.

TABLE A.3a
NUMERICAL CONSTANTS

	N	w	$q*$	$p***$	γ	ν	T	d,D
N_1	0.0865	–	m³/h	kPa	–	–	–	–
	0.865	–	m³/h	bar	–	–	–	–
	1.00	–	gpm	psia	–	–	–	–
N_2	0.00214	–	–	–	–	–	–	mm
	890	–	–	–	–	–	–	in
N_3	645	–	–	–	–	–	–	mm
	1.00	–	–	–	–	–	–	in
N_4	76,000	–	m³/h	–	–	**Centistokes	–	mm
	17,300	–	gpm	–	–	**Centistokes	–	in
N_5	0.00241	–	–	–	–	–	–	mm
	1000	–	–	–	–	–	–	in
N_6	2.73	kg/h	–	kPa	kg/m³	–	–	–
	27.3	kg/h	–	bar	kg/m³	–	–	–
	63.3	lb/h	–	psia	lb/ft³	–	–	–
N_7	4.17	–	m³/h	kPa	–	–	deg. K	–
	417	–	m³/h	bar	–	–	deg. K	–
	1360	–	scfh	psia	–	–	deg. R	–
N_8	0.948	kg/h	–	kPa	–	–	deg. K	–
	94.8	kg/h	–	bar	–	–	deg. K	–
	19.3	lb/h	–	psia	–	–	deg. R	–
N_9	22.4	–	m³/h	kPa	–	–	deg. K	–
	2240	–	m³/h	bar	–	–	deg. K	–
	7320	–	scfh	psia	–	–	deg. R	–
N_{10}	86.5	kg/h	–	kPa	–	–	–	–
	865	kg/h	–	bar	–	–	–	–
	500	lb/h	–	psia	–	–	–	–

* The standard cubic foot is taken at 14.73 psia and 60°F, and the standard cubic meter at 101.3 kilopascals and 15°C.

** Centistokes $= 10^6 \left(\dfrac{m^2}{s} \right)$

*** All pressures are absolute

$$\Sigma K = K_1 + K_2 + K_{B_1} - K_{B_2} \qquad A.3(10)$$

where

K_1 = Upstream resistance coefficient
K_2 = Downstream resistance coefficient
K_{B_1} = Inlet Bernoulli coefficient
K_{B_2} = Outlet Bernoulli coefficient
C_V = Rated C_V of the valve selected

Equations 23, 24 and 25 give the approximate values which may be used where experimental values of the K's are not available.

When inlet and outlet fittings are identical $K_{B_1} = K_{B_2}$ and hence drop out of the expression. In those cases in which the piping diameters approaching and leaving the valve are different, the values of K_{B_1} and K_{B_2} are calculated by use of:

$$K_{B_1} \text{ or } K_{B_2} = 1 - \left(\frac{d}{D} \right)^4 \qquad A.3(11)$$

Liquid Sizing Factors

Liquid Critical Pressure Ratio Factor, F_F

The Liquid Critical Pressure Ratio Factor is the ratio of the apparent vena contracta pressure at choked flow conditions to the vapor pressure of the liquid at inlet temperature. Although F_F varies with different fluids, an approximate evaluation can be made using the curve of Figure A.3k or Equation 26. Experimental values should be used when possible to maintain maximum sizing accuracy.

Cavitation and Flashing

The procedure used to size control valves for liquid service should consider the possibility of cavitation or flashing since these phenomena can limit valve capacity as well as produce physical damage to the valve. With increasing pressure drop across the valve, vapor cavities form when the vena contracta pressure drops below the

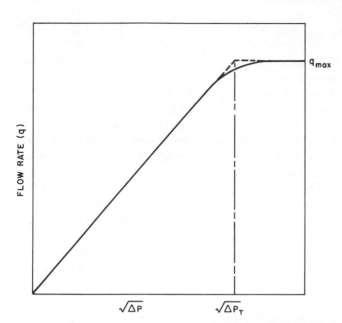

Fig. A.3b Typical flow rate vs. $\sqrt{\Delta p}$ curve (constant upstream pressure and vapor pressure)

liquid's vapor pressure. Without vapor cavity formation, the flow rate is proportional to the square root of the pressure drop. The proportionality deteriorates with the vapor formation and the flow becomes completely choked when sufficient vapor has formed so that there will be no increase in flow as the pressure drop is increased. This relationship between flow rate and pressure drop for a typical valve is shown in Figure A.3b.

The following equation can be used to determine the maximum allowable, or terminal pressure drop that is effective in producing flow.

$$\Delta p_T = \left(\frac{F_{LP}}{F_P}\right)^2 (p_1 - F_F p_v) \qquad \text{A.3(12)}$$

Or when the valve is used without reducers, this reduces to

$$\Delta p_T = F_L^2 (p_1 - F_F p_v) \qquad \text{A.3(13)}$$

It should be noted, however, that this limitation on the sizing pressure drop, Δp_T, does not imply that this is the maximum drop that may be handled by the valve.

Equations 12 and 13 may be incorporated into the general sizing equations giving two equations. The first equation is for a Newtonian liquid under non-cavitating or non-flashing conditions. The second equation gives the maximum flow rate at choked conditions.

The equations are:

$$q = N_1 F_P F_R C_V \sqrt{\frac{p_1 - p_2}{G_f}} \qquad \text{A.3(14)}$$

$$q_{max} = N_1 F_R F_{LP} C_V \sqrt{\frac{p_1 - F_F p_v}{G_f}} \qquad \text{A.3(15)}$$

Equation 14 yields a q which must be compared with the q_{max} determined from Equation 15. The smaller value will predict the actual flow rate.

Liquid Pressure Recovery Factor, F_L

The Liquid Pressure Recovery Factor, F_L, is a measure of the ability of the valve to convert the kinetic energy of the flow stream at the vena contracta back into pressure. The factor is a function of the valve internal geometry. Under nonvaporizing flow conditions it is defined by the equation

$$F_L = \sqrt{\frac{p_1 - p_2}{p_1 - p_{vc}}} \qquad \text{A.3(16)}$$

F_L is measured under choked flow conditions when it is given by the expression

$$F_L = \sqrt{\frac{\Delta p_T}{p_1 - F_F p_v}} \qquad \text{A.3(17)}$$

The factor F_L is determined experimentally as specified in ISA S39.02 using cold water as the test fluid.

When the valve is installed with reducers the F_L factor and F_p factor are treated together as F_{LP}. The equation then for choked flow in a valve installed in other than line size is

$$q_{max} = N_1 F_{LP} F_R C_v \sqrt{\frac{p_1 - F_F p_v}{G_f}} \qquad \text{A.3(18)}$$

To meet the desired maximum deviation of ±5 percent established for this standard, F_L or F_{LP} must be determined by testing. When estimated values are permissible, the following equation may be used for F_{LP}:

$$F_{LP} = \left[\frac{K_i}{N_2}\left(\frac{N_3 C_V}{d^2}\right)^2 + \frac{1}{F_L^2}\right]^{-\frac{1}{2}} \qquad \text{A.3(19)}$$

In this equation K_i is the head loss coefficient $(K_1 + K_{B_1})$ of the reducer or other device between the upstream pressure tap and inlet face of the valve only.

Liquid Choked Flow Factor, F_y

The Liquid Choked Flow Factor accounts for the effect of the different valve geometries and fluid properties on choked flow. F_y is the ratio of actual choked flow through the valve to the calculated flow through the valve assuming nonvaporizing, incompressible flow. It may be calculated by using the equation

$$F_y = \frac{F_{LP}}{F_P} \sqrt{\frac{p_1 - F_F p_v}{\Delta p}} \quad \lim F_y = 1 \qquad \text{A.3(20)}$$

or read from Figure A.3c. Calculated values of F_y greater than 1.0 indicate that choked flow conditions do not exist. F_y for use in the general equations must be limited to values equal to or less than 1.0.

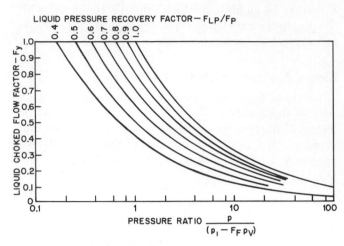

Fig. A.3c　Liquid Choked Flow Factor, Fy

$$Re_V = \frac{N_4\,F_d q}{\nu F_{LP}^{\frac{1}{4}} C_V^{\frac{1}{2}}} \left[\left(\frac{N_3 F_{LP} C_V}{N_2 D^2} \right)^2 + 1 \right]^{\frac{1}{4}} \quad \text{A.3(21)}$$

The factor F_d allows a single curve to correlate the data from a number of valve styles. F_d values of 1.0 can be used for V-notch ball valves and single ported globe valves. An F_d of 0.7 may be used for valves with two parallel flow paths such as double-ported globe valves and butterfly valves. Caution must be used in applying the universal curve for other valve styles when the F_d has not been established.

The bracketed quantity in Equation 21 accounts for the "velocity of approach." Except for wide-open ball or butterfly valves, this refinement has only a slight effect on the Re_V calculation and can generally be omitted.

Most flows in process plants have Reynolds numbers in excess of 10^5, hence the Reynolds number factor may be taken as 1.0.

It should be noted that when valve is operated with a pressure drop close to Δp_T, a rate of flow less than predicted by Equations 1 or 2 may result because of some vaporization within the valve.

Reynolds Number Factor, F_R

Reynolds number effects may be significant when the viscosity of the fluid is high or the flow velocities become small. When these conditions exist there is no longer turbulent flow through the valve and actual flow rates are considerably below those for turbulent flow. The F_R flow factor is the ratio of the actual flow rate in nonturbulent flow to the flow rate calculated for turbulent conditions in the same manifold.

Test results have shown the F_R can be found using a Reynolds number and universal curve shown in Figure A.3d.

The valve Reynolds number is given by the expression:

Gas Sizing Factors

Expansion Factor, Y

The factor Y accounts for the change in fluid density as it passes from the valve inlet to the vena contracta and also for the change in area of the vena contracta as the pressure drop is varied (contraction coefficient). Theoretically, Y is affected by all of the following:

    a. Ratio of port area to body inlet area
    b. Shape of the flow path
    c. Pressure drop ratio, x
    d. Reynolds number
    e. Ratio of specific heats, k

The influence of items a, b, and c is defined by the factor x_T, which may be established by air test. (Refer to ISA S39.02.) Alternatively x_T may be determined inferentially from F_L. (See Appendix D.) Test data indicates that Y may be taken as being a linear function of x as shown by Equation 7 with the restriction on size of x to x_T.

Reynolds number effects may be disregarded for all practical purposes in the case of compressible flow.

The effect of the ratio of specific heats of the fluid is considered in the ratio of specific heats factor F_k.

Pressure Drop Ratio Factor, x_T

If the inlet pressure (p_1) is held constant and the outlet pressure (p_2) is progressively lowered, the mass flow rate through a valve will increase to a maximum limit. A further reduction in p_2 will produce no further increase in flow. This limit is reached when x reaches a value of $F_k x_T$. The value of x used in any of the equations 3 through 7 must be held to this limit even though the actual pressure drop is greater. This means that numerically the value of Y ranges between 1.0 and 0.667, and the slope of the Y versus x curve is established by the

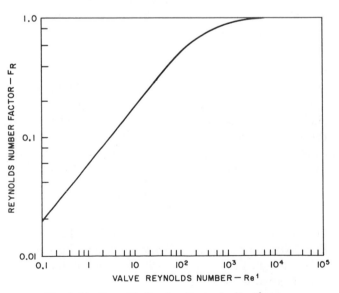

Fig. A.3d　Reynolds Number Factor for liquid sizing

product $F_k x_T$. If a valve is installed with reducers, expanders, or other devices attached to it, the value of x_T is affected and it should be replaced in the equations by x_{TP}.

To meet the specified tolerance limitation, the valve and attached fittings must be tested as a unit. When calculated values are acceptable, the following equation may be used:

$$x_{TP} = \frac{x_T}{F_p{}^2} \left[1 + \frac{x_T K_i}{N_5} \left(\frac{N_3 C_v}{d^2}\right)^2 \right]^{-1} \quad \text{A.3(22)}$$

In this equation x_{TP} and x_T are the installed value and rated value respectively of the pressure drop ratio factor, and K_i is the inlet velocity head coefficients ($K_1 + K_{B_1}$) of the reducer or other device located between the upstream pressure tap and the inlet face of the valve.

Ratio of Specific Heats Factor, F_k

The ratio of specific heats of a compressible fluid affects the flow rate through a valve or orifice and the factor F_k accounts for this effect. It has a value of 1.0 for air, which has a ratio of specific heats (k) of 1.40. Both theoretical and experimental evidence indicate that a linear relationship to k is adequate for valve sizing calculations, and that:

$$F_k \simeq \frac{k}{1.40}$$

Compressibility Factor, Z

Equations 4, 5, and 6 do not contain a term for the actual specific weight of the fluid at upstream conditions. Instead, this term is derived from the laws of perfect gases, based on the inlet temperature and pressure. Under certain conditions, deviations from these laws can result in gross error. The compressibility factor is included to correct for real gas effects. The value of Z can be determined by curves found in many reference sources if the critical data of the fluid are known.

Tolerances

When the equations of this standard are used with the coefficients evaluated according to ISA S39.02, the expected tolerance for any test specimen is ± 5 percent, except where noted otherwise. The C_v determined for liquids may, in certain instances, differ from that determined for gases or vapors. As the resistance of the valve decreases to where it approaches the resistance of a pipe of equal length and diameter, the tolerance of C_v measurement increases. For example, the ± 5 percent holds for $N_3 C_v/d^2 \leq 40$, but reaches ± 7 percent at $N_3 C_v/d^2 = 80$.

Nomenclature

A	Pipe flow area
C_v	Valve sizing coefficient
d	Nominal valve size
D	Internal diameter of the piping
F_d	Valve Style Modifier, dimensionless
F_F	Liquid Critical Pressure Ratio Factor, dimensionless
F_k	Ratio of Specific Heats Factor, dimensionless
F_L	Rated Liquid Pressure Recovery Factor, dimensionless
F_{LP}	Combined Liquid Pressure Recovery Factor and Geometry Factor of valve with reducers (when there are no reducers, F_{LP} equals F_L), dimensionless
F_P	Piping Geometry Factor, dimensionless
F_R	Reynolds Number Factor, dimensionless
F_y	Liquid Choked Flow Factor, dimensionless
g	Local acceleration of gravity
G_f	Specific gravity (ratio of densities) of the liquid at flowing temperature compared to water at base conditions, dimensionless
G_g	Specific gravity of gas relative to air, with both at standard conditions, dimensionless
h	Fluid head or liquid column
k	Ratio of specific heats, dimensionless
K	Head loss coefficient of a device, dimensionless
M	Molecular weight, dimensionless
N	Numerical constant
p_1	Upstream absolute static pressure
p_2	Downstream absolute static pressure
p_v	Vapor pressure absolute of liquid at inlet temperature
p_{vc}	Vena contracta absolute pressure
Δp	Pressure drop ($p_1 - p_2$) across the valve
Δp_T	Terminal pressure drop (see Figure A.3b)
q	Volume rate of flow
q^{max}	Maximum flow rate (choked flow conditions) at a given upstream pressure
R	Gas constant
Re_v	Valve Reynolds number, dimensionless
T	Absolute temperature
T_1	Absolute upstream temperature (deg. K or deg. R)
V	Velocity at valve inlet
w	Weight rate of flow
x	Ratio of pressure drop to absolute inlet static pressure ($\Delta p/p_1$), dimensionless
x_T	Rated Pressure Drop Ratio Factor, dimensionless
x_{TP}	Value of x_T for valve/fitting assembly, dimensionless
Y	Expansion Factor, ratio of flow coefficient for a gas to that for a liquid at the same Reynolds number, dimensionless
Z	Compressibility Factor, dimensionless
γ	Specific weight
ν	Kinematic viscosity, centistokes

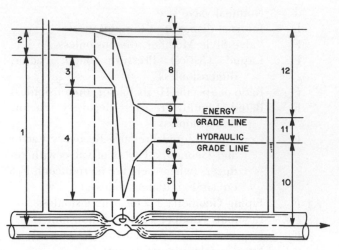

Fig. A.3e Head changes in a control valve-piping system. See Table A.3f for definitions.

Approximate Values of Loss Coefficient for Abrupt Contraction or Expansion

An understanding of the various loss mechanisms involved in a control valve-piping system can be obtained by looking at the energy grade lines and the hydraulic grade lines for a system containing abrupt contractions and expansions. These are shown schematically in Figure A.3e. Each of the pressure drops associated with this figure is given in Table A.3f. Some of the pressure drops are nonrecoverable and some are recoverable as shown in the hydraulic grade line. Table A.3f also defines the various coefficients associated with the system. The Bernoulli coefficients K_{B1} and K_{B2} account for the two-dimensionality of the flow, hence, relate the total kinetic energy to that calculated using the valve inlet velocity.

Where experimental values for K are not available the following expressions may be used to calculate the ap-

proximate values for abrupt contraction and expansion when the reducer and expander are the same size:

$$K_1 + K_2 = 1.5 \left(1 - \frac{d^2}{D^2} \right)^2 \qquad \text{A.3(23)}$$

Reducer only:

$$K_1 = 0.5 \left(1 - \frac{d^2}{D^2} \right)^2 \qquad \text{A.3(24)}$$

Expander only:

$$K_2 = 1.0 \left(1 - \frac{d^2}{D^2} \right)^2 \qquad \text{A.3(25)}$$

Table A.3g gives some computed values for various valves.

Liquid Critical Pressure Ratio Factor

An approximate value for the liquid critical pressure ratio factor which may be used when experimental data are not available is given in Figure A.3j. Alternatively the following equation may be used:

$$F_F = 0.96 - 0.28 \sqrt{\frac{p_v}{p_c}} \qquad \text{A.3(26)}$$

Other Sizing Equations

A number of formulas have been published for sizing valves and many are in wide usage. These formulas are repeated here for reference and comparison. It is recommended, however, that the equations contained in this standard be used.

The "Recommended Voluntary Standard Formulas for Sizing Control Valves" contains the only prior U.S. standard control valve sizing formulas for compressible fluids. (Fluid Controls Institute FCI-62-1.) In U.S. customary units the basic FCI equation is:

Table A.3f
DEFINITIONS OF HEAD CHANGE TERMS

Reference Number*	Head Changes Defined	Reference Number*	Head Changes Defined
1	Inlet pressure head $\dfrac{p_1}{\gamma}$	7	Reducer loss $= K_1 \dfrac{V^2}{2g}$
2	Inlet velocity head $\dfrac{V_1^2}{2g}$	8	Valve loss $h_v = N_2 \left(\dfrac{d^2}{N_3 C_v} \right)^2 \dfrac{V^2}{2g}$
3	Reducer drop $= (K_1 + K_{B1}) \dfrac{V^2}{2g}$	9	Expander loss $= K_2 \dfrac{V^2}{2g}$
4	Δh to vena contracta $= \dfrac{h_v}{F_L^2}$	10	Outlet pressure head $\dfrac{p_2}{\gamma}$
5	Pressure recovery (4) − (8)	11	Outlet velocity head $\dfrac{V_2^2}{2g}$
6	Expander recovery $= (K_2 - K_{B2}) \dfrac{V^2}{2g}$	12	Total loss h_L

*See Fig. A.3e

Table A.3g
COMPUTED F_P FACTORS USING EQUATION 9
(For Reducer and Expander of the Same Size)

d/D	$\frac{N_3C_v}{d^2}$					
	10	15	20	25	30	80
0.67	0.98	0.95	0.91	0.87	0.83	.48
0.50	0.96	0.91	0.85	0.79	0.73	.38
0.25	0.93	0.87	0.79	0.72	0.65	.31

Table A.3h
REPRESENTATIVE x_T FACTORS AT FULL RATED OPENING FOR FULL SIZE TRIM

Valve Type	Trim Type	Flow Direction	x_T
Globe			
Single-ported	Ported plug	Either	0.75
	Contoured plug	Flow-to-open	0.72
		Flow-to-close	0.55
	Characterized cage	Flow-to-open	0.75
		Flow-to-close	0.70
Double-ported	Ported plug	Either	0.75
	Contoured plug	Either	0.70
Angle	Contoured plug	Flow-to-open	0.72
	Characterized cage	Flow-to-open	0.65
		Flow-to-close	0.60
	Venturi	Flow-to-close	0.20
Ball	Characterized	–	0.25
	Regular port (dia. 0.8d)	–	0.15
Butterfly	60 deg opening	–	0.38
	90 deg opening	–	0.20

The above values are typical only for the types shown. Variations may occur due to other types of trim, full or reduced ports, and among manufacturers.

Table A.3i
REPRESENTATIVE F_L FACTORS AT FULL RATED TRAVEL FOR FULL SIZE TRIM

Valve type	Flow Direction	Trim Type	F_L
Globe			
Single-port or double-port	Either	Ported plug	0.85-0.95
Double-port	Either	Contoured plug	0.85-0.90
Single-port	Flow-to-open	Contoured plug	0.85-0.90
Single-port	Flow-to-close	Contoured plug	0.75-0.85
Angle	Flow-to-open	Contoured plug	0.85-0.90
Angle	Flow-to-close	Venturi	0.45-0.55
Ball		Characterized	0.60-0.65
Ball		Regular port	0.55-0.60
Butterfly		60 degree	0.55-0.65

The above values are typical only for the types shown. Variations may occur due to other types of trim, full or reduced ports, and among manufacturers.

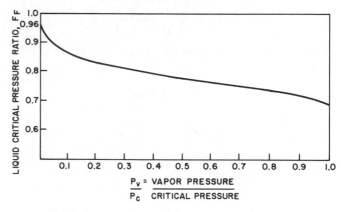

Fig. A.3j   Liquid Critical Pressure Ratio Factor, F_F

$$q = 963 \, C_v \sqrt{\frac{(p_1 - p_2)(p_1 + p_2)}{G_f T}} \qquad \text{A.3(27)}$$

where $p_2 \le 0.5 \, p_1$

Flow predicted by this equation is almost always more than the true flow, the size of the error depending upon the x_T value of the valve under consideration. If the error is to be limited to 10 percent, the pressure-drop ratio must not exceed the limits given in Table A.3k.

Manufacturers of control valves publish other sizing equations, one of which is in U.S. customary units:

$$q = C_v C_1 p_1 \sqrt{\frac{520}{G T_1}} \sin \left[\frac{3417}{C_1} \sqrt{\frac{\Delta p}{P_1}} \right]_{\text{deg}} \qquad \text{A.3(28)}$$

The bracketed expression is limited to a maximum value of 90 degrees, representing terminal flow conditions. The factor C_1 is derived from physical testing with air and water, and, when equated to the ISA equations, is equal to $40\sqrt{x_T}$ (within testing tolerance). The results of this equation agree with Equation 2 within the tolerance of measurement if the compressibility factor Z is included in both.

Another manufacturer's formula which is widely used is, in U.S. customary units:

$$q = 834 \, \frac{C_v C_f p_1}{\sqrt{G T_1}} (y - 0.148 y^3) \qquad \text{A.3(29)}$$

where

$$y = \frac{1.63}{C_f} \sqrt{\frac{\Delta p}{p_1}} \le 1.50$$

Table A.3k
PRESSURE DROP LIMITS FOR EQUATION 27

Rated Valve x_T	0.80	0.75	0.70	0.60	0.50	0.40	0.30	0.20
Maximum x	1.0	0.44	0.38	0.28	0.21	0.15	0.10	0.065

$$C_f = \sqrt{\frac{(p_1 - p_2)}{(p_1 - p_{vc})}}$$

This equation is based on the proposition that

$$C_f = \sqrt{\frac{x_T}{0.84}}$$

hence

$$x_T = 0.84 C_f^2 \qquad\qquad A.3(30)$$

The factor C_f is determined by testing with water or air under choked flow conditions and is taken to be equivalent to the factor F_L.

A.4 CONTROL VALVE CAPACITY TEST PROCEDURE

1 Scope

This test standard utilizes the mathematical equations outlined in ANSI/ISA S75.01, "Control Valve Sizing Equations" [see Section A.3], in providing a test procedure for obtaining the following:

1. Valve Sizing Coefficient, C_v
2. Liquid Pressure Recovery Factor, F_L
3. Reynolds Number Factor, F_R
4. Liquid Critical Pressure Ratio Factor, F_F
5. Piping Geometry Factor, F_p
6. Pressure Drop Ratio Factors, x_T and x_{TP}

This standard is intended for control valves used in flow control of process fluids and is not intended to apply to fluid power components as defined in the National Fluid Power Association Standard NFPA T.3.5.28-1977.

2 Test System

2.1 General Description

A basic flow test system as shown in Figure A.4a includes:

1. Test specimen
2. Test section
3. Throttling valves
4. Flow measuring device
5. Pressure taps
6. Temperature sensor

2.2 Test Specimen

The test specimen is any valve or combination of valve, pipe reducer and expander or other devices attached to the valve body for which test data are required. Modeling of valves to a smaller scale is an acceptable practice in this standard, although testing of full-size valves or models is preferable. Good practice in modeling requires attention to significant relationships such as Reynolds Number, the Mach Number where compressibility is important, and geometric similarity.

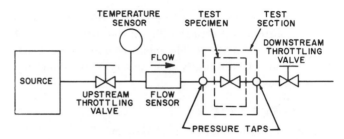

Fig. A.4a Basic flow test system

2.3 Test Section

The upstream and downstream piping adjacent to the test specimen shall conform to the nominal size of the test specimen connection and to the length requirements of Table A.4b.

The piping on both sides of the test specimen shall be Schedule 40 pipe for valves through 250 mm (10 in.) size having a pressure rating up to and including ANSI Class 600. Pipe having 10 mm (0.375 in.) wall may be used for 300 mm (12 in.) through 600 mm (24 in.) sizes. An effort should be made to match the inside diameter at the inlet and outlet of the test specimen with the inside diameter of the adjacent piping for valves outside the above limits.

The inside surface shall be reasonably free of flaking rust or mill scale and without irregularities which could cause excessive fluid frictional losses.

2.4 Throttling Valves

The upstream and downstream throttling valves are used to control the pressure differential across the test section pressure taps and to maintain a specific downstream pressure. There are no restrictions as to style of these valves. However, the downstream valve should be of sufficient capacity to ensure that choked flow can be achieved at the test specimen for both compressible and incompressible flow. Vaporization at the upstream valve must be avoided when testing with liquids.

2.5 Flow Measurement

The flow measuring instrument may be any device which meets specified accuracy. This instrument will be used to determine the true time average flow rate within

Table A.4b***
PIPING REQUIREMENTS STANDARD TEST SECTION

A*.**	B	C	D	Standard Test Section Configuration
At least 18 nominal pipe diameters	2 nominal pipe diameters	6 nominal pipe diameters	At least 1 nominal pipe diameter	

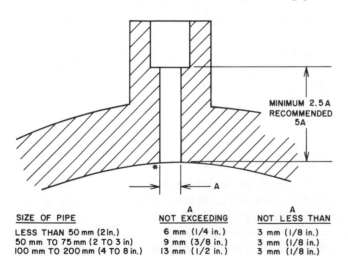

*Dimension "A" may be reduced to 8 nominal diameters if straightening vanes are used.
Information concerning the design of straightening vanes can be found in "ASME Performance Test Code PTC 19.5–1972, Applications, Part II of Fluid Meters, Interim Supplement on Instruments and Apparatus."
**If an upstream flow disturbance consists of two ells in series and they are in different planes Dimension "A" must exceed 18 nominal pipe diameters unless straightening vanes are used.
***See Section 2.2 for definition of the test specimen.

an error not exceeding ±2 percent of the actual value. The resolution and repeatability of the instrument shall be within ±0.5 percent. The measuring instrument shall be calibrated as frequently as necessary to maintain specified accuracy.

2.6 Pressure Taps

Pressure taps shall be provided on the test section piping in accordance with the requirements listed in Table A.4b. These pressure taps shall conform to the construction illustrated in Figure A.4c.

Orientation:

> Incompressible fluids—tap center lines shall be located horizontally to reduce the possibility of air entrapment of dirt collection in the pressure taps.
>
> Compressible fluids—tap center lines shall be oriented horizontally or vertically above pipe to

reduce the possibility of dirt or condensate entrapment.

2.7 Pressure Measurement

All pressure and pressure differential measurements shall be made to an error not exceeding ±2 percent of actual value. Pressure measuring devices shall be calibrated as frequently as necessary to maintain specified accuracy.

Pressure differential instruments are required in the measurement of the pressure differential across the test specimen to avoid additional inaccuracies resulting from taking the difference of two measurements. Exceptions to this are the procedures in Section 4.2 and 6.2 for determining maximum flow rates for incompressible and compressible flow respectively.

2.8 Temperature Measurement

The fluid temperature shall be measured to an error not exceeding ±1°C (±2°F) of actual value.

The inlet fluid temperature shall remain constant within ±3°C (±5°F) during the test run for any one specific test point.

2.9 Installation of Test Specimen

The alignment between the center line of the test section piping and the center line of the inlet and outlet of the test specimen shall be as follows:

Pipe Size	Allowable Misalignment
15 mm thru 25 mm ($\frac{1}{2}$ in. thru 1 in.)	0.8 mm ($\frac{1}{32}$ in.)
32 mm thru 150 mm (1-$\frac{1}{4}$ in. thru 6 in.)	1.6 mm ($\frac{1}{16}$ in.)
200 mm and larger (8 in. and larger)	1 percent of the diameter

When rotary valves are being tested, the valve shafts shall be aligned with test section pressure taps.

Each gasket shall be positioned so that it does not protrude into the flow stream.

SIZE OF PIPE	A NOT EXCEEDING	A NOT LESS THAN
LESS THAN 50 mm (2 in.)	6 mm (1/4 in.)	3 mm (1/8 in.)
50 mm TO 75 mm (2 TO 3 in)	9 mm (3/8 in.)	3 mm (1/8 in.)
100 mm TO 200 mm (4 TO 8 in.)	13 mm (1/2 in.)	3 mm (1/8 in.)
250 mm AND GREATER (10 In. AND GREATER)	19 mm (3/4 in.)	3 mm (1/8 in.)

Fig. A.4c   Recommended pressure connection
*Edge of hole must be clean and sharp or slightly rounded, free from burrs, wire edges or other irregularities. In no case shall any fitting protrude inside the pipe. Any suitable method of making the physical connection is acceptable if above recommendations are adhered to. Reference: "ASME Performance Test Code PTC 19.5-1972, Applications, Part II of Fluid Meters, Interim Supplement on Instruments and Apparatus."

2.10 Accuracy of Test

A tolerance of ±5 percent is expected for the flow coefficient C_v of the test specimen for valves having a $N_3 C_v/d^2$ ratio up to 30.

3 Test Fluids

3.1 Incompressible Fluids

Fresh water within a temperature range of 20°C ± 14°C (68°F ± 25°F) shall be the basic fluid used in this test procedure. Inhibitors may be used to prevent or retard corrosion and to prevent the growth of organic matter. The effect of additives on density or viscosity shall be evaluated by computation using the equations in this standard. The sizing coefficient shall not be affected by more than 0.1 percent. Other test fluids may be required for obtaining F_R and F_F.

3.2 Compressible Fluids

Air or some other compressible fluid shall be used as the basic fluid in this test procedure. Vapors which may approach their condensation points at the vena contracta of the specimen are not acceptable as test fluids. Care shall be taken to avoid internal icing during the test.

4 Test Procedure—Incompressible Fluids

The following instructions are given for the performance of various tests using incompressible fluids.

The procedures for data evaluation of these tests follow in Section 5.

4.1 C_v Test Procedure

The following test procedure is required to obtain test data for the calculation of the flow coefficient C_v. The data evaluation procedure is provided in Section 5.1.

4.1.1 Install the specimen without reducers or other attached devices in accordance with piping requirements in Table A.4b.

4.1.2 Flow tests will include flow measurements at three pressure differentials in increments of at least 15 kPa (0.15 bar) (2.0 psi). These pressure differentials, measured across the test section pressure taps, shall be between 35 and 100 kPa (0.35 and 1.0 bar) (5.0 and 15.0 psi) with the valve at the rated travel.

In the case of very small valve capacities, nonturbulent flow may occur at the recommended pressure differentials. When this occurs larger pressure differentials must be used to assure turbulent flow. A minimum valve Reynolds number, Re_v, of 10^5 is recommended (See Equation 4.).

For large valves where flow source limitations are reached, lower pressure differentials may be used optionally as long as turbulent flow is maintained. Deviations from standard requirements shall be recorded.

4.1.3 In order to keep the downstream portion of the test section liquid filled and to prevent vaporization of the liquid, the upstream pressure must be maintained equal to or greater than the minimum values in Table A.4d. This minimum upstream pressure is dependent on the Liquid Pressure Recovery Factor, F_L, of the test

Table A.4d
MINIMUM UPSTREAM TEST PRESSURE

Δp								
	kPa	35	40	55	70	80	95	100
	bar	0.35	0.40	0.55	0.70	0.80	0.95	1.0
	psi	5.0	6.0	8.0	10.0	12.0	14.0	15.0
F_L		ABSOLUTE UPSTREAM PRESSURE* p_1						
	kPa	280	320	440	560	640	760	800
0.5	bar	2.8	3.2	4.4	5.6	6.4	7.6	8.0
	psia	40	45	60	80	95	110	120
	kPa	190	220	300	380	440	520	550
0.6	bar	1.9	2.2	3.0	3.8	4.4	5.2	5.5
	psia	25	30	40	55	65	75	80
	kPa	150	160	220	280	320	380	400
0.7	bar	1.5	1.6	2.2	2.8	3.2	3.8	4.0
	psia	22	25	30	40	45	55	60
	kPa	150	160	170	210	250	290	310
0.8	bar	1.5	1.6	1.7	2.1	2.5	2.9	3.1
	psia	22	23	25	30	35	40	45
	kPa	150	160	170	190	200	230	240
0.9	bar	1.5	1.6	1.7	1.9	2.0	2.3	2.4
	psia	22	23	25	27	29	31	35

*a. Minimum upstream pressures have been calculated to provide a downstream gauge pressure of at least 14 kPa (0.14 bar) (2 psig) above atmospheric pressure.

b. Upstream pressures were calculated using $p_1 \, min = \dfrac{2 \, \Delta p}{F_L^2}$

c. Upstream pressures were rounded down to the next 10 kPa (0.10 bar).

d. Upstream pressures were rounded down to the next 5 psia (35kPa) while still maintaining a minimum pressure as specified in note (a).

specimen. If F_L is unknown, a conservative estimate for the minimum inlet pressure should be made.

4.1.4 The valve flow tests shall be performed at the following valve travels:

1. At 100 percent rated valve travel
2. Optionally—at each 10 percent of rated valve travel
3. Optionally—at points less than 10 percent of rated travel to more fully determine the inherent characteristic of the specimen, if applicable.

4.1.5 The following data shall be recorded:

1. Valve travel (measurement error not exceeding ± 0.5 percent of rated travel)
2. Upstream pressure (p_1) (measurement error not exceeding ±2 percent of actual value)
3. Pressure differential (Δp) across test section pressure taps (measurement error not exceeding ± 2 percent of actual value)
4. Volumetric flow rate (q) (measurement error not exceeding ± 2 percent of actual value)
5. Fluid inlet temperature (T) (measurement error not exceeding ± 1°C(± 2°F))
6. Barometric pressure (measurement error not exceeding ± 2 percent of actual value)
7. Physical description of test specimen (i.e., type of valve, flow direction, etc.)

4.2 F_L Test Procedure

The maximum flow rate, q_{max}, is required in the calculation of the Liquid Pressure Recovery Factor, F_L. For a given upstream pressure, the quantity q_{max} is defined as that flow rate at which a decrease in downstream pressure will not result in an increase in the flow rate. The test procedure required to determine q_{max} is included in this section. The data evaluation procedure including the calculation of F_L is contained in Section 5.2. The test for F_L and corresponding C_v must be conducted at identical valve travels. Hence, the tests for both these factors at any valve travel shall be made while the valve is locked in a fixed position.

4.2.1 The standard test section as shown in Table A.4b shall be used with the valve under test at 100 percent of rated travel.

4.2.2 The downstream throttling valve shall be in the fully open position. Then, with a preselected upstream pressure, the flow rate will be measured and the downstream pressure recorded. This test establishes a "maximum" pressure differential for the test specimen in this test system. A second test run shall be made with the pressure differential maintained at 90 percent of the pressure differential determined in the first test with the same upstream pressure. If the flow rate in the second test is within 2 percent of the flow rate in the first test,

the "maximum" or choked flow rate has been established. If not, the test procedure must be repeated at a higher upstream pressure. If choked flow cannot be obtained, the published value of F_L must be based on the maximum measurement attainable, with an accompanying notation that the actual value exceeds the published value, e.g., $F_L > 0.87$. Note that values of upstream pressure and pressure differential used in this procedure are those values measured at the pressure taps.

4.2.3 Record the following data:

1. Valve travel (measurement error not exceeding ± 0.5 percent of rated travel)
2. Upstream pressure (p_1) and downstream pressure (p_2) (measurement error not exceeding ± 2 percent of actual value)
3. Volumetric flow rate (q) (measurement error not exceeding ± 2 percent of actual value)
4. Fluid temperature (measurement error not exceeding ± 1°C (± 2°F))
5. Barometric pressure (measurement error not exceeding ± 2 percent of actual value)

4.3 F_p Test Procedure

The Piping Geometry Factor, F_p, modifies the valve sizing coefficient for reducers or other devices attached to the valve body that are not in accord with the test section. It is the ratio of the installed C_v with these reducers or other devices attached to the valve body to the rated C_v of the valve installed in a standard test section and tested under identical service conditions. This factor is obtained by replacing the valve with the desired combination of valve, reducers, and/or other devices and then conducting the flow test outlined in Section 4.1 treating the combination of the valve and reducers as the test specimen for the purpose of determining test section line size. For example, a 100 mm (4 in.) valve between reducers in a 150 mm (6 in.) line would use pressure tap locations based on 150 mm (6 in.) nominal diameter. The data evaluation procedure is provided in Section 5.3.

4.4 F_R Test Procedure

To produce values of the Reynolds Number Factor, F_R, nonturbulent flow conditions must be established through the test valve. Such conditions will require low pressure differentials, high viscosity fluids, small values of C_v or some combination of these. With the exception of valves with very small values of C_v, turbulent flow will always exist when flowing tests are performed in accordance with the procedure outlined in Section 4.1, and F_R under these conditions will have a value of 1.0.

Determine values of F_R by performing flowing tests with the valve installed in the standard test section without reducers or other devices attached. These tests should follow the procedure for C_v determination except that:

1. Test pressure differentials may be any appropriate values provided that no vaporization of the test fluid occurs within the test valve.
2. Minimum upstream test pressure values shown in Table A.4d may not apply if the test fluid is not fresh water at 20°C ± 14°C (68°F ± 25°F).
3. The test fluid should be a Newtonian fluid having a viscosity considerably greater than water unless instrumentation is available for accurately measuring very low pressure differentials.

Perform a sufficient number of these tests at each selected valve travel by varying the pressure differential across the test valve so that the entire range of conditions, from turbulent to laminar flow, is spanned. The data evaluation procedure is provided in Section 5.4.

4.5 F_F Test Procedure

The Liquid Critical Pressure Ratio Factor, F_F, is ideally a property of the fluid and its temperature. It is the ratio of the apparent vena-contracta pressure, at choked flow conditions, to the vapor pressure of the liquid at inlet temperature.

The quantity F_F may be determined experimentally by using a test specimen for which F_L and C_v are known. The standard test section without reducers or other devices attached will be used with the test specimen installed. The test procedure outlined in Section 4.2 for obtaining q_{max} will be used with the fluid of interest as the test fluid. The data evaluation procedure is in Section 5.5.

5 Data Evaluation Procedure—Incompressible Fluids

The following procedures are to be used for data evaluation of the test procedures described in Section 4.

5.1 C_v Calculation

5.1.1 Using the data obtained in Section 4.1, calculate C_v for each test point at a given valve travel using the equation:

$$C_v = \frac{q}{N_1} \sqrt{\frac{G_f}{\Delta p}} \tag{1}$$

Round off the calculated value to no more than three significant digits.

5.1.2 The rated C_v of the valve is the arithmetic average of the calculated values for 100 percent of rated travel obtained from the test data in Section 4.1.5. A critical examination of the individual values calculated should reveal equal values of C_v within the tolerance given in Section 2.10.

5.2 F_L Calculation

$$F_L = \frac{q_{max}}{N_1 C_v \sqrt{(p_1 - 0.96 p_v)/G_f}} \tag{2}$$

where p_1 is the pressure at the upstream pressure tap for the q_{max} determination (Section 4.2).

5.3 F_p Calculation

Calculate F_p as follows at each test point:

$$F_p = \frac{q}{N_1 \sqrt{\Delta p / G_f}\,(C_v \text{ rated})} \tag{3}$$

5.4 F_R Calculation

Use test data, obtained as described under Section 4.4, in Equation (1), Section 5.1 to obtain values of an apparent C_v. This apparent C_v is equivalent to $F_R C_v$. Therefore, F_R is obtained by dividing the apparent C_v by the experimental value of C_v determined for the test valve under standard conditions at the same valve travel. Although the data may be correlated in any manner suitable to the experimenter, a method which has proven to provide satisfactory correlations involves the use of the valve Reynolds Number which may be calculated from:

$$Re_v = \frac{N_4 F_d q}{\nu F_{LP}^{\frac{1}{4}} C_v^{\frac{1}{4}}} \left[\frac{F_{LP}^2 C_v^2}{N_2 D^4} + 1 \right]^{\frac{1}{4}} \tag{4}$$

where:

F_d = Valve Style Modifier and accounts for the effect of geometry on Reynolds Number. F_d has been found to be proportional to $1/\sqrt{\eta}$

η = the number of similar flow paths (i.e., $\eta = 1$ for single ported valves, $\eta = 2$ for double ported, etc.)

ν = Kinematic viscosity in centistokes

Plotting values of F_R versus Re_v will result in the curve which appears as Figure 3 in ANSI/ISA S75.01, "Control Valve Sizing Equations (1977)". [See Figure A.3d.]

5.5 F_F Calculation

Calculate F_F as follows:

$$F_F = \frac{1}{p_v} \left[p_1 - G_f \left(\frac{q_{max}}{N_1 F_L C_v} \right)^2 \right] \tag{5}$$

where:

p_v is the fluid vapor pressure at the inlet temperature

$F_L C_v$ is determined for the test specimen by the standard method (Section 4.2 – F_L Test Procedure).

6 Test Procedure—Compressible Fluids

The following instructions are given for the performance of various tests using compressible fluids.

The procedures for data evaluation of these tests follow in Section 7.

6.1 C_v Test Procedure

The determination of the flow coefficient, C_v, requires flow tests using the following procedure to obtain the necessary test data. The data evaluation procedure is in Section 7.1. An alternative procedure for calculating C_v is provided in Section 6.3.

6.1.1 Install the test specimen without reducers or other devices in accordance with the piping requirements in Table A.4b.

6.1.2 Flow tests will include flow measurements at three pressure differentials. In order to approach flowing conditions which can be assumed to be incompressible, the pressure drop ratio ($x = \Delta p/p_1$) shall be ≤ 0.02.

6.1.3 The valve flow tests will be performed at the following valve travels:

1. At 100 percent rated valve travel
2. Optionally—at each 10 percent of rated valve travel
3. Optionally—at points less than 10 percent of rated travel to more fully determine the inherent flow characteristic of the specimen, if applicable

6.1.4 The following data shall be recorded:

1. Valve travel (measurement error not exceeding $\pm$ 0.5 percent of rated valve travel)
2. Upstream pressure (p_1) (measurement error not exceeding $\pm$ 2 percent of actual value)
3. Pressure differential (Δp) across test section pressure taps (measurement error not exceeding $\pm$ 2 percent of actual value)
4. Volumetric flow rate (q) (measurement error not exceeding $\pm$ 2 percent of actual valuve)
5. Fluid temperature (T_1) upstream of valve (measurement error not exceeding $\pm$ 1 °C ($\pm$ 2°F))
6. Barometric pressure (measurement error not exceeding $\pm$ 2 percent of actual value)
7. Physical description of test specimen (i.e., type of valve, flow direction, etc.)

6.2 x_T Test Procedure

The maximum flow rate, q_{max}, (referred to as choked flow) is required in the calculation of x_T the Pressure Drop Ratio Factor. This factor is the terminal ratio of the differential pressure to absolute upstream pressure, ($\Delta p/p_1$), for a given test specimen installed without reducers or other devices. The maximum flow rate is defined as that flow rate at which, for a given upstream pressure, a decrease in downstream pressure will not produce an increase in flow rate. The test procedure required to obtain q_{max} is contained in this section with the data evaluation procedure in Section 7.2. An alternative procedure for determining x_T is provided in Section 6.3.

6.2.1 The standard test section shall be used with the test valve at 100 percent of rated travel.

6.2.2 Any upstream supply pressure sufficient to produce choked flow is acceptable as is any resulting pressure differential across the valve provided the criteria for determination of choked flow specified in Section 6.2.3 are met.

6.2.3 The downstream throttling valve will be in the wide-open position. Then, with a preselected upstream pressure the flow rate will be measured and the downstream pressure recorded. This test establishes the maximum pressure differential for the test specimen in this test system. A second test shall be conducted using the downstream throttling valve to reduce the pressure differential by 10 percent of the pressure differential determined in the first test (with the same upstream pressure). If the flow rate of this second test is within 0.5 percent of the flow rate for the first test, then the maximum flow rate has been established.

Although the absolute value of the flow rate must be measured to an error not exceeding $\pm$ 2 percent, the repeatability of the tests for x_T must be better than $\pm$ 0.5 percent in order to attain the prescribed accuracy. This series of tests must be made consecutively, using the same instruments, and without alteration to the test setup.

6.2.4 Record the following data:

1. Valve travel (measurement error not exceeding $\pm$ 0.5 percent of rated travel)
2. Upstream pressure (p_1) (measurement error not exceeding $\pm$ 2 percent of actual value)
3. Downstream pressure (p_2) (measurement error not exceeding $\pm$ 2 percent of actual value)
4. Volumetric flow rate (q) (measurement error not exceeding $\pm$ 2 percent of actual value)
5. Fluid temperature upstream (T_1) of valve (measurement error not exceeding $\pm$ 1°C ($\pm$ 2°F))
6. Barometric pressure (measurement error not exceeding $\pm$ 2 percent of actual value)

6.3 Alternative Test Procedure for C_v and x_T

6.3.1 The standard test section will be used with the test valve at 100 percent of rated travel.

6.3.2 With a preselected upstream pressure, p_1, measurements shall be made of flow rate, q, upstream fluid temperature, T_1, downstream pressure, p_2, for a minimum of five well-spaced values of x (the ratio of pressure differential to absolute upstream pressure).

6.3.3 From these data points calculate values of the product YC_v using the equation:

$$YC_v = \frac{q}{N_7 p_1} \sqrt{\frac{G_g T_1}{x}} \qquad (6)$$

where Y is the expansion factor defined by:

$$Y = 1 - \frac{x}{3 F_k x_T}$$

where:

$$F_k = \frac{k}{1.40}$$

6.3.4 The test points shall be plotted on linear coordinates as (YC_v) vs x, and a linear curve fitted to the data. If any point deviates by more than 5 percent from the curve, additional test data shall be taken to ascertain if the specimen truly exhibits anomalous behavior.

6.3.5 At least one test piont, $(YC_v)_1$, must fulfill the requirement that:

$$(YC_v)_1 \geqslant 0.97(YC_v)_0$$

where $(YC_v)_0$ corresponds to $x \approx 0$.

6.3.6 At least one test point, $(YC_v)_n$, must fulfill the requirement that:

$$(YC_v)_n \leqslant 0.83(YC_v)_0$$

6.3.7 The value of C_v for the specimen shall be taken from the curve at $x = 0$, $Y = 1$. The value of x_T for the specimen shall be taken from the curve at $YC_v = 0.667C_v$.

6.4 F_p Test Procedure

The Piping Geometry Factor, F_p, modifies the valve sizing coefficient for reducers or other devices attached to the valve body that are not in accord with the test section. The factor F_p is the ratio of the installed C_v with the reducers or other devices attached to the valve body to the rated C_v of the valve installed in a standard test section and tested under identical service conditions. This factor is obtained by replacing the valve with the desired combination of valve, reducers and/or other devices and then conducting the flow test outlined in Section 6.1 treating the combination of valve and reducers as the test specimen for the purpose of determining test section line size. For example, a 100 mm (4 inch) valve between reducers in a 150 mm (6 inch) line would use pressure tap locations based on a 150 mm (6 inch) nominal diameter. The data evaluation procedure is provided in Section 7.3.

6.5 x_{TP} Test Procedure

Perform the tests outlined for x_T in Section 6.2 replacing the valve with the desired combination of valve and pipe reducers or other devices, treating the combination of valve and reducers as the test specimen. The data evaluation procedure is provided in Section 7.4.

7 Data Evaluation Procedure— Compressible Fluids

The following procedures are to be used for data evaluation of the test procedures described in Section 6.

7.1 C_v Calculation

Using the data obtained in Section 6.1, assuming the expansion factor $Y = 1.0$, calculate the flow coefficient, C_v, for each test point using:

$$C_v = \frac{q}{N_7 p_1} \sqrt{\frac{T_1 G_g}{x}} \qquad (7)$$

Calculate the arithmetic average of the three test values obtained at rated travel to obtain the rated C_v.

7.2 x_T Calculation

Calculate x_T as follows:

From Equation (6):

$$q = \frac{N_7 Y C_v p_1 \sqrt{x}}{\sqrt{G_g T_1}}$$

When $x = F_k x_T$, then $q = q_{max}$

$$q_{max} = N_7 Y C_v p_1 \sqrt{\frac{F_k x_T}{G_g T_1}}$$

and:

$$x_T = \left[\frac{q_{max}}{N_7 Y C_v p_1} \right]^2 \frac{G_g T_1}{F_k} \qquad (8)$$

Assuming air as test fluid and substituting $Y = 0.667$, $G_g = 1.0$ and $F_k = 1.0$:

$$x_T = \left[\frac{q_{max}}{0.667 N_7 C_v p_1} \right]^2 T_1 \qquad (9)$$

7.3 F_P Calculation

Calculate F_P as follows using average values:

$$F_P = \frac{q}{N_7 p_1 (C_v \text{ rated})} \sqrt{\frac{x}{T_1}} \qquad (10)$$

7.4 x_{TP} Calculation

Calculate x_{TP} as follows:

From Equation (6):

$$q = N_7 F_P Y C_v p_1 \sqrt{\frac{x}{G_g T_1}}$$

with F_P added to account for reducers and other devices.
When $x = x_{TP}$, $q = q_{max}$

$$q_{max} = N_7 F_P Y C_v p_1 \sqrt{\frac{x_{TP}}{G_g T_1}}$$

Assuming air as the test fluid:

$$Y = 0.667$$
$$G_q = 1.0$$
$$F_k = 1.0$$

$$x_{TP} = \left[\frac{q_{max}}{0.667 N_7 F_P C_v p_1} \right]^2 T_1 \qquad (11)$$

Table A.4e
NUMERICAL CONSTANTS

	N	q^*	p	ν	T	d,D
N_1	0.0865	m³/h	kPa	—	—	—
	0.865	m³/h	bar	—	—	—
	1.00	gpm	psia	—	—	—
N_2	0.00214	—	—	—	—	mm
	890	—	—	—	—	in.
N_3	645	—	—	—	—	mm
	1.00	—	—	—	—	in.
N_4	76,000	m³/h	—	Centistoke**	—	mm
	17,300	gpm	—	Centistoke**	—	in.
N_7	4.17	m³/h	kPa	—	K	—
	417	m³/h	bar	—	K	—
	1,360	scfh	psia	—	°R	—

* The standard cubic foot is taken at 14.73 psia and 60°F, and the standard cubic meter at 101.3 kiloPascals and 15.6°C

** Centistoke = 10^{-6} m²/sec

All pressures are absolute.

8 Numerical Constants

The numerical constants, (N), depend on the measurement units used in the general sizing equations. Values for N are listed in Table A.4e.

A.5 BIBLIOGRAPHY FOR INSTRUMENT ENGINEERS

Adams, L.F., *Engineering Measurements and Instrumentation*, The English Universities Press, Ltd., London, 1975.

Adiutori, E.F., *The New Heat Transfer*, Ventuno, Cincinnati, Ohio, 1974.

Andrew, W.G. and Williams, B., *Applied Instrumentation in the Process Industries*, vol. I. 2nd ed., Gulf Publ., 1979.

Athans, M., and Falb, P., *Optimal Control: An Introduction to the Theory and Its Applications*, McGraw-Hill Book Co., New York, 1966.

Bates, R.G., *Determination of pH—Theory and Practice*, John Wiley & Sons, Inc., New York, 1964.

Beveridge, G.S.G., and Schechter, R.S., *Optimization: Theory and Practice*, McGraw-Hill Series in Chemical Engineering, McGraw-Hill Book Co., New York, 1970.

Buckley, P.S., *Techniques of Process Control*, John Wiley & Sons, Inc., New York, 1964.

Butler, J.N., *Ionic Equilibrium: A Mathematical Approach*, Addison-Wesley Publishing Co., Inc., Reading, Mass., 1964.

Caldwell, W.I., Coon, G.A., and Zoss, L.M., *Frequency Response for Process Control*, McGraw-Hill Book Co., New York, 1959.

Cheremisinoff, P.N., and Perlis, H.J., eds., *Analytical Measurements and Instrumentation for Process and Pollution Control*, Ann Arbor Science, 1981.

Coakley, W.A., *Handbook of Automated Analysis, Continuous Flow Techniques*, Dekker, 1981.

Combs, C.F., ed., *Basic Electronic Instrument Handbook*, McGraw-Hill Book Co., New York, 1972.

Considine, D.M., *Process Instruments and Controls Handbook*, McGraw-Hill Book Co., New York, 1957.

Daniels, F., *Outlines of Physical Chemistry*, John Wiley & Sons, Inc., New York, 1948.

Diefenderfer, A.J., *Principles of Electronic Instrumentation*, W.B. Saunders Co., Philadelphia, 1972.

Doebelin, E.O., *Measurement Systems—Application and Design*, rev. ed., McGraw-Hill Book Co., New York, 1975.

Eckman, D.P., *Automatic Process Control*, John Wiley & Sons, Inc., New York, 1958.

E.E.U.A., *Installation of Instrumentation and Process Control Systems*, Handbook No. 34, Constable and Co., Ltd., London, 1973.

Evans, F.L., Jr., *Equipment Design Handbook for Refineries and Chemical Plants*, Gulf Publishing Co., Houston, Texas, 1971.

Gibson, J.E., *Nonlinear Automatic Control*, McGraw-Hill Book Co., New York, 1963.

Gregory, B.A., *An Introduction to Electrical Instrumentation*, Macmillan, Inc., New York, 1973.

Harriott, P., *Process Control*, McGraw-Hill Book Co., New York, 1964.

Herrick, C.N., *Instrumentation and Measurement for Electronics*, McGraw-Hill Book Co., New York, 1972.

Hutchinson, J.W., *ISA Handbook of Control Valves*, Instrument Society of America, Pittsburgh, Pa., 1971.

Johnson, A.J., and Auth, G.H., *Fuels and Combustion Handbook*, McGraw-Hill Book Co., New York, 1951.

Johnson, C.D., *Process Control Instrumentation Technology*, 2nd ed., Wiley & Sons, New York, 1982.

Jones, B.E., *Instrumentation, Measurement and Feedback*, McGraw-Hill Book Co., New York, 1977.

Jones, E.B., *Instrument Technology*, Butterworth, Inc., Woburn, Mass., vol. 1 (revised), 1965, vol. 2, 1956, vol. 3, 1974.

Kaller, H., *Handbook of Instrumentation and Controls*, McGraw-Hill Book Co., New York, 1961.

Kwakernaak, H., and Swan, R., *Linear Optimal Control Systems*, John Wiley & Sons, Inc., New York, 1972.

Levenspiel, O., *Chemical Reaction Engineering*, John Wiley & Sons, Inc., New York, 1962.

Lipták, *Instrument Engineers' Handbook*, Chilton Book Co., Radnor, Pa., vol. I, 1969, vol. II, 1970, Supplement, 1972.

———, *Instrument Engineers' Handbook: Process Measurement*, Chilton Book Co., Radnor, Pa., 1982.

———, *Instrumentation and Automation Experiences in Wastewater Treatment Facilities*, EPA-600/2-76-198, October, 1976.

———, *Instrumentation in the Processing Industries*, Chilton Book Co., Radnor, Pa., 1973.

———, *Overview of Coal Conversion Process Instrumentation*, ANL-FE-49628-TM01, DOE, May, 1980.

Lupfer, D.E., and Johnson, M.L., "Automatic Control of Distillation Columns to Achieve Optimum Operation," *ISA Trans.*, Instrument Society of America, Pittsburgh, Pa., April, 1974.

Meditch, J.S., *Stochastic Optimal Linear Estimation and Control*, McGraw-Hill Book Co., New York, 1969.

Murrill, P.W., *Automatic Control of Processes*, International Textbook Co., Scranton, Pa., 1967.

Oliver, B.M., and Cage, J.M., *Electronic Measurements and Instrumentation*, McGraw-Hill Book Co., New York, 1971.

Parsons, W.A., *Chemical Treatment of Sewage and Industrial Wastes*, National Lime Association, Washington, D.C., 1965.

Perry, J.H., *Chemical Engineers' Handbook*, 4th ed., McGraw-Hill Book Co., New York, 1963.

Shell Flow Meter Engineering Handbook, Waltman Publishing Co., 1968.

Shinskey, F.G., "Controlling Multivariable Processes," Instrument Society of America, Research Triangle, N.C. 1981.

———, *Distillation Control*, 2nd ed., McGraw-Hill Book Co., New York, 1984.

———, *Energy Conservation Through Control*, Academic Press, Inc., New York, 1978.

———, *pH and pLon Control*, John Wiley, New York, 1973.

———, *Process Control Systems*, McGraw-Hill Book Co., New York, 1979.

Smith, C.L., *Digital Computer Process Control*, International Textbook Co., Scranton, Pa., 1972.

Spink, L.K., *Principles and Practices of Flowmeter Engineering*, 9th ed., The Foxboro Company, Foxboro, Mass., 1967.

Spitzer, F., and Howarth, B., *Principles of Modern Instrumentation*, Holt, Rinehart and Winston, New York, 1972.

Tagg, C.F., *Electrical Indicating Instruments*, Butterworth, Inc., Woburn, Mass., 1974.

Technical Data Book—Petroleum Refining, American Petroleum Institute, Washington, D.C., 1970.

Treybal, R.E., *Mass-Transfer Operations*, McGraw-Hill Book Co., New York, 1955.

Van Winkle, M., *Distillation*, McGraw-Hill Book Co., New York, 1967.

Wightman, E.J., *Instrumentation in Process Control*, Butterworth, Inc., Woburn, Mass., 1972.

Wipke, T.W., et al., *Computer Representation and Manipulation of Chemical Information*, Krieger, 1980.

Wolf, S., *Guide to Electronic Measurements and Laboratory Practice*, Prentice-Hall, Inc., Englewood Cliffs, N.J., 1973.

INDEX